FOOD CHEMISTRY AND NUTRITIONAL BIOCHEMISTRY

FOOD CHEMISTRY AND NUTRITIONAL BIOCHEMISTRY

Charles Zapsalis   **R. Anderle Beck**

Framingham State College, Framingham, Massachusetts

JOHN WILEY & SONS

New York   Chichester   Brisbane   Toronto   Singapore

Library of Congress Cataloging in Publication Data

Zapsalis, Charles.
  Food chemistry and nutritional biochemistry.

  Includes bibliographies and index.
  1. Food—Composition. 2. Nutrition. 3. Metabolism.
I. Beck, R. Anderle.   II. Title.
TX545.Z38   1986   641   85-6249
ISBN 0-471-86129-4

Printed in the United States of America

10 9 8 7 6 5 4 3 2 1

Notice
All application data, drug dosages, dietary quantities, food ingredient listings,
engineering principles, and similar data discussed herein are believed to be accurate. There is no guarantee of accuracy, however, and other reference sources,
clinical test values, and specific citations of manufacturer's data for product
listings and applications should be consulted. Information supplied by specified
corporations does not imply any use or product warranty whatsoever.

CONTRIBUTORS

JOHN J. BRINK
Clark University
Chapter 21 Food, Drug, and Immune System Interactions,
pp. 1069–1084

RICHARD M. MILASZEWSKI
Framingham State College
Chapter 6 Carbohydrates: Chemistry, Occurrence, and Food
Applications, pp. 318–342

CAROL L. RUSSELL
Framingham State College
Chapter 11 Nutritional Energetics, pp. 621–637

P R E F A C E

Food Chemistry and Nutritional Biochemistry is intended for students of food science and nutrition. It may also appeal to professional food scientists, practicing dietitians, and other medical professionals who desire a comprehensive knowledge of all foods from chemical, biochemical, and nutritional perspectives.

Food chemistry, biochemistry, and nutrition have been traditionally taught at the undergraduate and graduate levels as distinctly different courses, and the student has typically been left to integrate the significance of these three areas. As teaching scientists in the field of food science and nutritional biochemistry, we recognized the need for a book that integrates these areas. According to this objective, an effort has been made to provide a salient and balanced presentation of key principles underlying food chemistry, biochemistry, and nutrition.

The textbook consists of three parts. Part 1 details the basic chemistry of food constituents including water, proteins, enzymes, vitamins, carbohydrates, lipids, natural colorants and flavors, plus fermentation products in foods. Analytical methods that underlie the proximate analysis of food composition (e.g., fat, carbohydrate, protein, moisture analysis), flavor analysis, enzymatic methods of food analysis, nutritional assessments of protein quality, and vitamin assays are provided. Moreover, included are detailed discussions of nutritional energetics based on the laws of thermodynamics, photosynthesis as it pertains to the production of primary food resources, and colloidal food systems as they relate to food industrial applications.

Part 2 outlines the integrated metabolism of all food constituents including the possible roles for trace elements. Food toxicants and some basic concepts of food toxicology are also detailed. The metabolism section also surveys some important nutritional and etiological factors that may be linked to atherosclerosis, hypertension, certain cancers, and diabetes. Effects of hormonal control mechanisms on the overall scheme of nutritional biochemical processes are cited along with some common food–drug interactions.

Part 3 offers some of the basic information on molecular genetics that serves as a basis for the field of genetic engineering as it pertains to the development of new food resources.

Any textbook of this scope is liable to a variety of compositional pitfalls, overt errors, and oversights, for which we are most apologetic. We welcome comments and suggestions.

Acknowledgments

We are most grateful for the attention, guidance, and support afforded to the production of this textbook by Katie Vignery, who served as a superb editor. Thanks are also in order for all the other fine professionals at John Wiley & Sons, including Ray O'Connell, Page Mead, and Butch Cooper as well as the whole production staff who made this book possible. The patient oversight and serious attention to production details offered by Maryellen Costa and Susan Winick were especially appreciated.

Appreciation is extended to both Dr. Pericles Markakis, Department of Food Science and Human Nutrition, Michigan State University, and Dr. Michael Quinn, associated with the New York State Institute for Basic Research in Developmental Disabilities and the Nutrition Department at Rutgers University. Their critique of the whole manuscript and helpful suggestions were very instrumental in the development of this book.

We wish to thank many of our colleagues at Framingham State College for their valuable opinions and helpful discussions during the early drafts of the present textbook. Special thanks are extended to Dr. Nancy Jane Bowden for advice concerning nutritional information; Dr. Constance B. Jordan for supplying certain nutritional references; Dr. Thomas Koshy for his mathematical assistance dealing with orders of reaction; Dr. Paul D. Peterson for his advice concerning X-ray diffraction; Dr. Dana N. Jost for his helpful critique of the photosynthesis chapter; and Dr. Paul F. Cotter for his review of genetic modifications in agricultural resources.

We owe considerable debt to many others. Included among these are Barbara W. LeDuc for the acquisition of all literature dealing with minerals; as well as the assistance afforded by Linnane Hallberg; and Dr. John Kapsalis of the U.S. Army Natick Laboratories, who supplied valuable information regarding water in foods.

Many individuals contributed directly or indirectly in obtaining illustrative reference material used in this text, including Dr. Dorothy C. Hodgkin for her personal photographic material related to vitamin B_{12}; Brian D. Holden; Dr. Arthur G. Rand and Dr. Kenneth L. Simpson (University of Rhode Island); Dr. Fred A. Liberatore (DuPont (New England Nuclear)); and all individuals responsible for giving reprint permission in numerous tables and figures cited throughout the text.

Also, we wish to thank Mary A. Zapsalis for her secretarial help.

CHARLES ZAPSALIS
R. ANDERLE BECK
Framingham Centre
Framingham, Massachusetts

O V E R V I E W

C O N T E N T S

PART · 1

Food Constituents, Food Sources, and Energetics

C H A P T E R · 1

Water in Foods: Physical and Chemical Phenomena

INTRODUCTION

The basic constituents of living systems include proteins, carbohydrates, lipids, nucleic acids, vitamins, minerals, and water. Although these substances are all fundamental to the existence of living organisms, water is the most ubiquitous constituent and often comprises 50 to 90% of plant and animal weight as well as foods (Table 1.1). Water is of basic importance to all known forms of life because it provides a physical environment for essential biochemical reactions; it serves as a transport medium for nutrients and waste products of metabolism; and it facilitates the transport of the respiratory gases, oxygen and carbon dioxide.

While water is the single most important factor for the maintenance of living organisms, an understanding of its occurrence in foods is fundamental to the technological principles of food production and preservation. The moisture content of natural, processed, or manufactured foods may determine the food's legal standard of identity, texture, palatability, consumer acceptability, quality control evaluation, and duration of preservation for the marketed product.

The occurrence of water in foods depends on the physical and chemical composition of the natural food tissue or processed food product. Water can be widely distributed throughout a food as a solvent system for sugars, salts, organic acids, phenols, and hydrophilic macromolecular carbohydrates, gums, or proteins to form colloidal systems.

Forces of hydrogen bonding and capillary action also characterize the occurrence of water in foods. These two factors are responsible for the establishment of mono- and polymolecular layers of water on both the internal and external surfaces of foods.

Water molecules display uniquely different physical properties from all other compounds formed during the geochemical evolution of the earth. These properties become apparent when water is compared with the hydrides of other Group VI elements in the periodic table such as H_2S, H_2Se, and H_2Te. An inspection of physical constants in Table 1.2 shows that water should have a boiling point of $-100°C$ instead of $+100°C$. This temperature is not realized since the electronegative oxygen in the water molecule withdraws electrons from its covalently bonded hydrogens. The shifting of these electrons produces partially negative charges designated as $2\delta-$, leaving the protons with

Table 1.1. Moisture Contents of Major Food Groups and Some Specific Foods

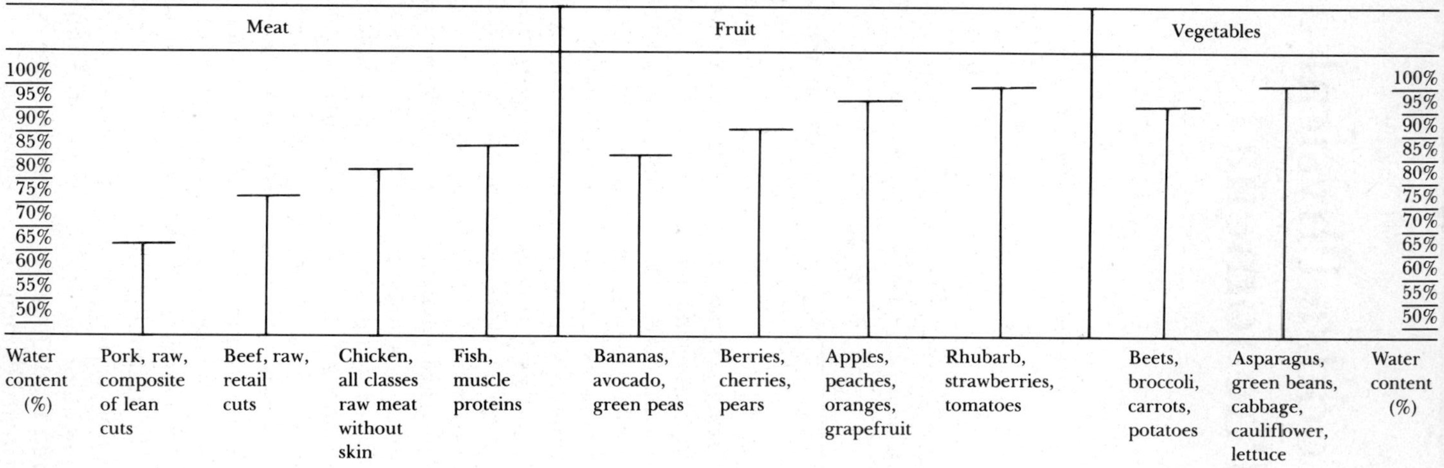
Meat
Fruit
Vegetables
100%
95%
90%
85%
80%
75%
70%
65%
60%
55%
50%
Water content (%)
Pork, raw, composite of lean cuts
Beef, raw, retail cuts
Chicken, all classes raw meat without skin
Fish, muscle proteins
Bananas, avocado, green peas
Berries, cherries, pears
Apples, peaches, oranges, grapefruit
Rhubarb, strawberries, tomatoes
Beets, broccoli, carrots, potatoes
Asparagus, green beans, cabbage, cauliflower, lettuce
100%
95%
90%
85%
80%
75%
70%
65%
60%
55%
50%
Water content (%)

Food	Percentage water
Dairy products	
Butter	15
Cheeses (wide range of moisture content depending on type)	
Cheddar cheese	40
Cottage cheese	75
Cream	60–70
Dry milk powders	4
Fluid dairy products (whole milk, nonfat milk, butter milk)	87–91
Ice cream and sherbet	65
High lipid foods	
Margarine	15
Mayonnaise	15
Pure oils and fats	0
Salad dressings	40
Fruits and vegetables	
Avocado	65
Beans (green, lima)	67
Berries	81–90
Citrus fruits	86–89
Cucumbers	96
Dried fruits	up to 25
Fresh fruits (edible portion)	90
Fresh fruits, juices, and nectars	85–93
Guavas	81
Legumes (dry)	10–12
Melons	92–94
Ripe olives	72–75
Potatoes	
White potatoes	78
Sweet potatoes	69

Food	Percentage water
Tap roots	
Radishes (highest)	93
Parsnips (lowest)	79
Cereals and cereal products	
Breakfast cereals	<4
Macaroni	9
Milled grain products: flour, grits, semolina germ	10–13
Whole grained cereals	10–12
Baked cereal products	
Bread	35–45
Crackers and pretzels	5–8
Pies	43–59
Rolls	28
Nut meats	
Ripe raw nuts	3–5
Fresh chestnuts	53
Meat, seafood, and poultry products	
Animal flesh and seafoods (wide range depending on fat content and age of specimen)	50–85
Eggs	
Fresh eggs	74
Dried eggs	5
Poultry meats	
Geese	50
Chicken	75
Sugar and sugar-based products	
Honey and other syrups	20–40
Fruit jellies, jams, and marmalades	≤35
White sugar (cane or beet), hard candy, plain chocolate	≤1

Table 1.2. Important Physical Properties for Water and Other Substances

Chemical substance	Melting point (°C)	Boiling point (°C)	Heat of vaporization (cal g^{-1})	Specific heat capacity (cal g^{-1})	Heat of fusion (cal g^{-1})	Dielectric constant at 25°C
Water	0.0	100.0	540	1.000	80.0	78.5
Ethanol	− 114.0	78.0	204	0.581	24.9	24.3
Methanol	− 98.0	65.0	263	0.600	22.0	32.6
Acetone	− 95.0	56.0	125	0.528	23.0	20.7
Ethyl acetate	− 84.0	77.0	102	0.459	—	6.0
Chloroform	− 63.0	61.0	59	0.226	—	—
Ammonia	− 78.0	− 33.0	327	1.120	84.0	16.9
Hydrogen sulfide	− 83.0	− 60.0	132	—	16.7	—
Hydrogen fluoride	− 92.0	19.0	360	—	54.7	—

partially positive charges indicated as $2\delta+$ (Figure 1.1B). This intramolecular electron delocalization contributes to the maintenance of a dipolar molecule with an average interatomic hydrogen–oxygen distance of 0.0965 nm and a calculated bond angle of 104.5° (Figure 1.1B). The observed bond angle for water represents the combined effects of intramolecular repulsion between the two positively charged hydrogen atoms and forces of repulsion attributable to the lone pair electron orbitals of oxygen. The interplay of these forces also gives the water molecule an overall pyramidal symmetry shown in Figure 1.1C.

INTERACTION OF WATER WITH CRYSTALLINE SUBSTANCES

The dipolarity of the water molecule allows it to approach positively charged cation or negatively charged anion species. For those instances in which cations and anions comprise crystal lattices, the dipolar water molecule can effectively dissolve ions electrostatically held in crystalline structures. The tendency of these ions to interact with water supersedes their electrostatic forces of attraction responsible for the maintenance of the crystal lattice.

The cations and anions liberated from crystalline substances are subsequently stabilized in the aqueous solvent by counter-charged portions of dipolar water molecules. This principle is conventionally referred to as *hydration*.

The effective solvent properties of water are mirrored in its high *dielectric constant* (*D*) (Table 1.2). This parameter is a measure of a solvent's ability to counteract the attractive forces between two countercharged ions. The calculation of the dielectric constant is based on the relationship

$$F = \frac{e_1 e_2}{D r^2}$$

Essential factors required for this calculation include the forces of attraction between two oppositely charged ions (*F*); the magnitude of the electrical charge exhibited by ions (e_1 and e_2); and the distance between the ionic species (*r*)

HYDROGEN BONDING AND OTHER PROPERTIES OF WATER

The dipolar nature of water molecules enables them to participate in *hydrogen bonding*. Hydrogen bonding refers to the interaction of a hydrogen atom that is covalently bonded to one electronegative atom with a second electronegative atom. In such circumstances,

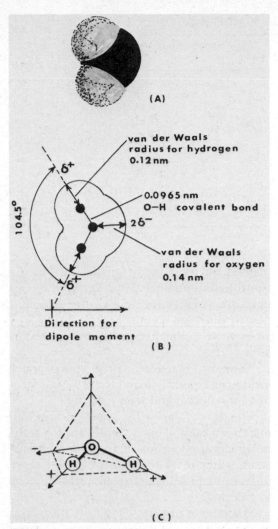

FIGURE 1.1. Molecular structures for water: (A) molecular shape, (B) bond angles and atomic radii, and (C) pyramidal symmetry.

the hydrogen atom tends to associate with the second electronegative atom by sharing the nonbonded electron pair of that atom. Hydrogen bonding interactions of water molecules may strictly involve water–water interactions, but water molecules may also associate with functional groups of organic molecules that contain nitrogen, oxygen, fluorine, or chlorine.

From an energetic perspective, hydrogen bonds are considerably weaker than covalent bonds (Table 1.3). The weak electron pair

sharing responsible for this type of bonding demonstrates a bond energy of ~4.5 kcal/mol, which is in marked contrast to 100 kcal/mol required to sever the covalent O—H bonds in water.

Although hydrogen bonding is characteristic of water's molecular interactions, the actual bond length may be somewhat variable depending on the physical state of the water and the presence of other chemical substances that undergo hydrogen bonding interactions with water. For example, the most common noncrystalline form of ice, known as "ice I," displays a molecular lattice structure in which one water molecule is hydrogen bonded to four neighboring molecules. These four water molecules indicate oxygen–oxygen distances equivalent to 0.276 nm and hydrogen bond lengths of 0.177 nm (Figures 1.2 and 1.3).

The hydrogen bonds established in solid, liquid, and vapor phases of water are very temporal phenomena. The typical hydrogen bond has an estimated half-life ($t\frac{1}{2}$) of ~10^{-11} s. Although short-lived, the incessant formation and severing of hydrogen bonds among neighboring water molecules accounts for the high *internal cohesion* and *surface tension* of water and the *structural rigidity* of ice.

Hydrogen bonding also contributes to the notable *density* (ρ) of water. As illustrated by Figure 1.4, water has a maximum density of 3.98°C and becomes significantly less dense at temperatures other than 3.98°C. This density relationship reflects the buoyant behavior of solid ice on liquid water, since the highly ordered molecular structure of water molecules in ice creates a volumetric expansion of water accompanied by a density decrease.

When ice melts there is a volume contraction of about 9%. The tetrahedral arrangement of ice partially breaks down in water, permitting a closer molecular packing, although water is far from being a close-packed structure. The number of nearest neighbors for individual water molecules (coordination number) increases. That

Table 1.3. A Summary of Important Bond Energies

	Approximate bond energy	
Type of bond	kcal mol^{-1}	kJ mol^{-1}
Noncovalent bond types		
Van der Waals bond	1	4.18
Hydrophobic bond	1–3	4.18–12.55
Hydrogen bond	4.5	18.83
Ionic bond	5	20.92
Covalent bond types		
H_3C—CH_3	88	368.19
H—H	104	435.14
H_3C—H	104	435.14
H_2C=CH_2	163	681.99
N≡N	226	945.58

is, ice has a coordination number of 4, while water has corresponding values of 4.4 at 3.98°C and 4.9 at 83°C. As the temperature rises from 0 to 100°C, the distance between nearest neighbors increases because of thermal expansion. The two factors, coordination number and the distance between nearest neighbors, contribute their influence to the density changes of water (solid and liquid). Specifically, as the temperature rises from 0 to 3.98°C, the density increase is explained on the basis of an increase in the coordination number. However, as the temperature rises above 3.98°C, the observed density decrease is explained on the basis of an increase in the distance of nearest neighbors.

Additional noteworthy properties of water include its especially high heat of fusion, heat of vaporization, and heat capacity.

The *heat of fusion* for water is approximately 80 cal/g. This is significant for technological applications of food freezing because water must relinquish 80 cal/g more in freezing at 0°C than from 1 to 0°C prior to freezing.

Specific heat capacity is typically measured as the calories (or joules) required to raise the temperature of 1.0 g of some substance by 1°C. Increases in the specific heat values for substances normally parallels their increasing resistance to temperature changes

● Hydrogen	○---●	0.177 nm
○ Oxygen	○---○	0.276 nm
— Covalent bond	●○	0.0965 nm
---- Hydrogen bond	↵	109° angle

FIGURE 1.2. Molecular bonding interactions for water in the form of ice.

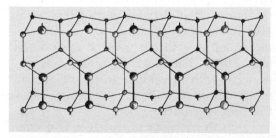

FIGURE 1.3. Crystal lattice structure for ice.

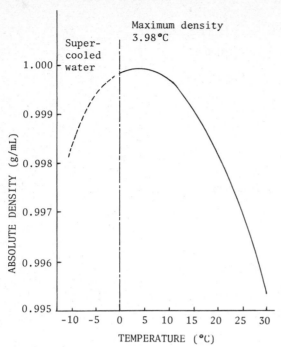

FIGURE 1.4. Density characteristics for water at various temperatures.

when a given amount of heat is absorbed. For pure water, the specific heat value is especially high (Table 1.2) with respect to most common solvents and approximates 1.0 cal/g.

The removal of substantial amounts of heat from foods during dehydration is related to the high specific *heat of vaporization* of water. Water has a 540 cal/g heat of vaporization at 100°C and displays higher values at lower temperatures.

INTERACTION OF WATER WITH ORGANIC MOLECULES

Solvent characteristics of water are not merely restricted to its solvation of crystalline substances at the ionic level, as previously discussed. The dipolar properties of water allow it to participate in hydrogen bond interactions with almost any polar molecule or functional group. This enables water to dissolve nonionic organic molecules, including sugars and alcohols containing hydroxyl groups, and aldehydes and ketones, which contain carbonyl groups.

Water also acts as a dispersing medium for *amphipathic molecules* such as salts of fatty acids and a wide variety of substances broadly classified as polar lipids, proteolipids, glycolipids, and nucleic acids. Amphipathic molecules are characterized by the presence of both hydrophilic and hydrophobic groups within the same molecule (Figures 1.5A–D). The apparent "solubilization" of amphipathic molecules results from the association of water with hydrophilic moieties such as carboxyl, phosphate, hydroxyl, carbonyl, or certain nitrogen-containing groups.

Unlike crystalloids, which are readily solvated by water at the individual molecular or ionic level, amphipathic molecules form a macromolecular aggregate in the aqueous environment known as a *micelle*. Individual molecules participating in a micelle number from several hundred to many thousands (Figure 1.5E).

Micelle formation is significantly influenced by entropy considerations. Since the nonpolar portion of amphipathic molecules is directed toward an intramicellar position, polar portions of the same molecules become oriented toward the aqueous surroundings. Although the inherent forces of hydrogen bonding and internal cohesion of water do contribute stability to the micelle structure, maximization of entropy for water molecules in the aqueous environment abets the structural maintenance of the micelle. The entropy factor is significant since introduction of hydrophobic portions of amphipathic molecules into the aqueous environment would necessarily require an energy expenditure and contribute to a more "ordered" aqueous molecular environment, accompanied by an unfavorable concomitant decrease in entropy.

Other than entropy considerations, nonstoichiometric van der Waals forces exist

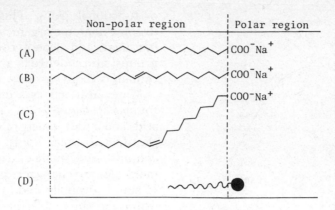

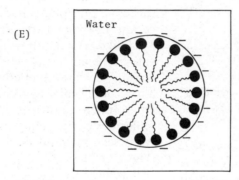

FIGURE 1.5. Various types of structures for typical amphipathic fatty acid salts (A–C); (D) generalized structure for an amphipathic molecule; and (E) micelle structure for generalized amphipathic molecules in water.

within the micelle structure. These attractive forces among the hydrophobic portions of amphipathic molecules collectively contribute added stability to micelle architecture (Figure 1.5E).

WATER ACTIVITY IN FOODS

The interaction of water with electrolyte or nonelectrolyte solutes generally results in an *entropy decrease* for water. This decrease in entropy is accompanied by diminished water vapor pressure and an alteration in the colligative properties of water. Solute influence on water behavior has been widely investigated and can be predicted to a limited extent based on the behavior of an ideal solute. For example, 1.0 g molecular weight of a nonassociable, nondissociable solute depresses the freezing point of water by 1.86°C and elevates the boiling point by 0.52°C. The relationship of ideal solutes to water vapor pressure has been customarily explained by *Raoult's Law:*

$$\frac{P_0 - P}{P_0} = \frac{n_1}{n_1 + n_2} \qquad (1.1)$$

where

P = vapor pressure of the solution
P_0 = vapor pressure of the solvent
n_1 = moles of solute
n_2 = moles of solvent

Since a 1.0 m concentration of an ideal solute in a kilogram of water results in a system having a vapor pressure equivalent to 98.23% of pure water, the effective water vapor pressure decrease is calculated to be 1.77% (equation 1.2).

Multivalent cations and anions typically exert more of an effect on water vapor pressure than similar molal concentrations of substances yielding only monovalent ions upon dissociation.

Although mathematical concepts outlined for the behavior of water provide fundamental principles for understanding A_w in foods, it should be recognized that these are merely idealized relationships. The potential interaction of water contained in foods is as varied as the chemical and physical com-

$$\frac{P_0 - P}{P_0} = \frac{(1.0 \text{ mole of ideal solute})}{(1.0 \text{ mole of ideal solute}) + (55.51 \text{ moles kg}^{-1} \text{ of water})} \times 100$$

$$\frac{P_0 - P}{P_0} = 1.77\%$$

$$(1.2)$$

Accordingly, the solution to solvent vapor pressure ratio may be expressed as *water activity* (A_w):

$$A_w = \frac{P}{P_0} \qquad (1.3)$$

where

P = vapor pressure of the solution
P_0 = vapor pressure of pure water at a specific temperature

Apart from the ideal mathematical prediction of A_w based on Raoult's Law, the solute complexity of foods produces a range of perfectly unpredictable effects on observed water vapor pressure. Nonelectrolyte solutes at ≤ 1.0 m concentrations create a minimal decrease in A_w, but strong electrolytes decidedly decrease A_w. The depression of A_w by specific solutes increases in accord with the principles of ionic strength (μ):

$$\frac{\Gamma}{2} = \mu = \frac{1}{2} \sum_i C_i Z_i^2 \qquad (1.4)$$

where

C_i = molar concentration of a specific ion
Z_i = electrical charge of each ion in solution

plexity of foods. Estimation of food water content based on proximate methods of analysis does not accurately relate to the concentration of water available for participation in solvent activities. Some portion of the water content acts in the solvation of solutes, but significant amounts of water remain bound to functional groups of insoluble food constituents or reside deep in the microphysical environment of the food structure. The combination of these phenomena contribute major analytical difficulties to the reliable assessment of A_w in foods.

SORPTION ISOTHERMS FOR WATER

Water activity of a food system is inherently related to the relative humidity of the physical space surrounding the food. Relative humidity refers to the amount of water vapor contained in a specific volume of air compared to the maximum amount of water vapor the air could hold at a specific temperature.

A graphic plot of water adsorbed to a food

substance versus relative humidity of the surroundings at a given temperature is described as an *adsorption isotherm*.

Adsorption isotherms can be established by determining the increased weight for a specific volume of food under conditions of increasing relative humidity. Conversely, the progressive decrease in food weight resulting with vaporization of water from a wet food under conditions of steadily decreasing relative humidity provides the basis for a graphic plot known as a *desorption isotherm* (Figure 1.6). Sorption isotherms for foods are normally divided into three regions, I, II, and III, which respectively represent sorption interactions of water monolayers, multilayers of water, and water condensation in food pores and capillaries.

The shape of sorption isotherms varies significantly among foods depending on chemical composition, concentration of spe-cific constituents, and the macro- and microphysical architecture of the food. Moreover, sorption isotherms within a specific food sample may not demonstrate uniformity throughout due to variabilities in the distribution of proteins, fats, carbohydrates, and minerals.

Theories postulated in an effort to explain water sorption phenomena abound. Foremost among theories are those of Brunauer *et al.* (1938), Freundlich (1926), Harkins and Jura (1944), Henderson (1952), Langmuir (1918), Sharp (1962), and Zsigmondy (1911).

One traditionally popular explanation of water sorption by foods involves the concept of the *Brunauer–Emmett–Teller* (BET) *isotherm* (Brunauer *et al.*, 1938). Although the BET isotherm archetypically describes surface interactions of nonpolar gases with catalytic agents, it has been applied to sorption

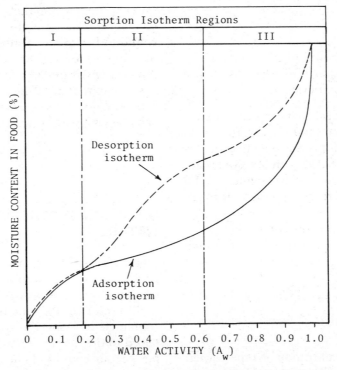

FIGURE 1.6. Comparison of adsorption and desorption isotherms for a food substance.

interactions between water and foods with the following considerations (Labuza, 1968):

1. Sorption phenomena involve only specific sites on a physical surface.

2. The heat of sorption (Q_s) for a water monolayer on a surface is constant and equivalent to the sum of the heat of vaporization (ΔH_v) and the heat of site interactions.

3. The heat of sorption (Q_s) for sequential layers of water above the first water monolayer is equal to the heat of vaporization (ΔH_v).

A general expression of the BET isotherm, accounting for kinetic, thermodynamic, and statistical mechanics assumptions, has been detailed by Adamson (1963) and Labuza (1968):

$$\frac{A_w}{(1-A_w)V} = \frac{1}{V_mC} + \left[\frac{A_w(C-1)}{V_mC}\right] \quad (1.5)$$

where

A_w = water activity
V = volume of water adsorbed per gram of solid
V_m = monolayer volume of water adsorbed per gram of solid
C = $K \exp(Qs/RT)$, constant related to the heat of adsorption
K = accommodation coefficient/frequency factor $\simeq 1.0$
R = gas constant
T = absolute temperature

Plotting $A_w/(1-A_w)V$ versus A_w results in a straight line whose slope and ordinate intercept can be used to calculate coverage of the water monolayer (Figure 1.7).

Although sorption isotherms as depicted in Figure 1.6 often show sigmoidal relationships to some extent, a variety of possible graphic manifestations exist for sorption isotherms involving water and various foods.

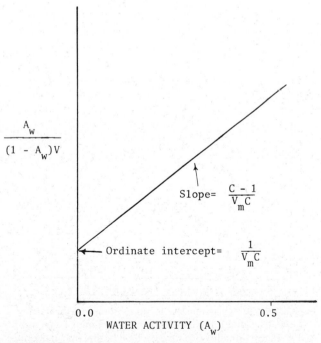

FIGURE 1.7. Graphic representation of $A_w/(1 - A_w)V$ versus A_w; the slope of the resulting line can be used to determine coverage of a water monolayer.

This variability is primarily dependent on the hygroscopic nature and properties of the particular food substance. As illustrated in Figure 1.8, exemplary isotherms such as curve *A* reflect the behavior of a food substance having *significant hygroscopic properties* with only minor increases in relative humidity; curve *B* represents *moderate sensitivity to increasing relative humidity;* while curve *C* demonstrates a food system characterized by *negligible water absorption* until relative humidity reaches a point where a significant hygroscopic state develops.

In some cases adsorption isotherms such as *A, B,* and *C* (Figure 1.8) may rarely coincide with water desorption isotherms for food systems, but generally, water adsorption isotherms do not realistically predict desorption phenomena. The disparity between desorption and adsorption isotherms is referred to as *hysteresis*. Figure 1.9 displays an accentuated hysteresis loop for a nondescript food. Hysteresis loops for foods vary widely and identical foods often display different patterns at different temperatures (Figure 1.10).

PHYSICAL EXPLANATIONS FOR HYSTERESIS

Obvious variations in the graphic representation of hysteresis loops result from concentration differences of individual food constituents and especially variations in the capillary porosity factors that characterize foods.

In addition to water solute interactions, a wide spectrum of theories has been devel-

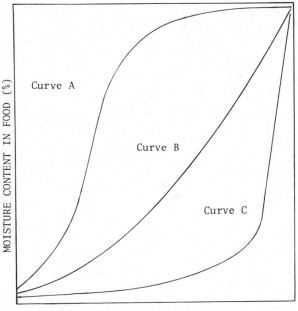

FIGURE 1.8. Characteristic sorption isotherms for a hygroscopic food that is highly sensitive to only minor increases in the relative humidity (Curve A); a food having only moderate sensitivity to increasing levels of relative humidity (Curve B); and a food system characterized by negligible water adsorption until relative humidity reaches a point where a hygroscopic state develops (Curve C).

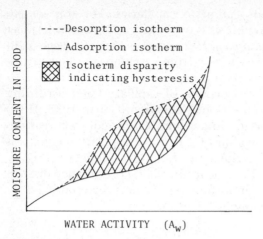

FIGURE 1.9. An accentuated hysteresis loop for a nondescript food substance.

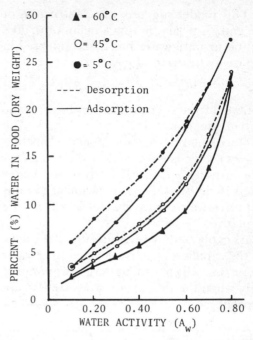

FIGURE 1.10. Adsorption and desorption isotherms for a single food substance at three different temperatures. Note that the area indicating hysteresis is different for each temperature level.

oped to explain hysteresis from a physical perspective. All of these theories are related to principles of capillary condensation for water, mathematically based on the Kelvin equation as reviewed by Kapsalis (1979):

$$RT \ln P/P_0 = -2\sigma V \cos \theta/r_\mathrm{m} \qquad (1.6)$$

where

P = water vapor pressure above curved meniscus

P_0 = saturation vapor pressure for water at a prescribed temperature (T)

σ = surface tension

θ = *contact angle* for water on a solid surface; for complete wetting $\theta = 0$ and $\cos \theta = 1.0$ (Figure 1.11)

V = molar volume for water

r_m = mean radius of curvature for meniscus, mathematically described as $2/r_\mathrm{m} = 1/r_1 + 1/r_2$; both r_1 and r_2 represent the radii of curvature at the liquid–vapor interface

T = absolute temperature

R = gas constant

Two of the most popular theories explaining hysteresis include the "ink bottle" and "open pore" theories.

Ink Bottle Theory

This theory assumes a structural analogy between an ink bottle and the pores of various foods. The neck of the ink bottle corresponds to a small radius (r_1) entry way, which leads to a blind cavity having a larger radius (r_2) (Figure 1.12).

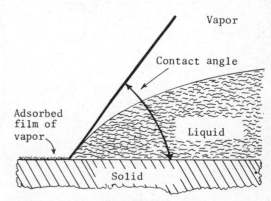

FIGURE 1.11. Physical interaction of water on a solid surface, which demonstrates the origin of the contact angle (θ). (Based on Kapsalis, 1979.)

This model suggests that obstruction of the entry capillary, having a radius of r_1, does not occur until water has first condensed in the cavity described by radius r_2 at

$$P/P_0 = \exp(-2\sigma\,V/r_2\,RT) \qquad (1.7)$$

obtained from substitution of r_2 for r_m in the Kelvin equation.

Since water desorption from the "ink bottle" structure is governed by the small capillary (radius r_1), which leads to the larger blind cavity (radius r_2), the capillary radius (r_1) effectively controls water loss from the structure.

As explicated by Kapsalis (1979), the neck of the capillary pore is occluded by a water meniscus, which can undergo evaporation only when the desorption pressure (P_d) approaches

$$P_d = \exp(-2\sigma\,V/r_1\,RT) \qquad (1.8)$$

at which the pore immediately empties.

Open Pore Theory

The open pore theory involves not only the fundamental considerations of the ink bottle theory but also accounts for multilayer water adsorption to capillaries. For this model, hysteresis is explained by *differences* in *water adsorption* (P_a) and *desorption* (P_d) *vapor pressures* arising from variations in water meniscus geometry.

Adsorption of capillary water according to the Cohan equation (Cohan 1938, 1944) occurs in a cylindrical fashion, whereas desorption of a concave capillary meniscus is described by the Kelvin equation (Figure 1.13). The mathematical presentation of these relationships has been clearly outlined by Kapsalis (1979), where

$$P_a = P_0 e^{-\sigma V/rRT} = P_0 e^{-\sigma V/(r_c - D)RT} \qquad (1.9)$$

where

r_c = capillary pore radius
D = thickness of water film adsorbed

Capillary condensation of water, followed by eventual occlusion and meniscus formation in the pore, requires a desorption pressure described by the Kelvin equation:

$$P_d = P_0 e^{-2\sigma V/rRT} \qquad (1.10)$$

Note: Equation 1.10 assumes complete wetting and $\cos\theta = 1.0$.

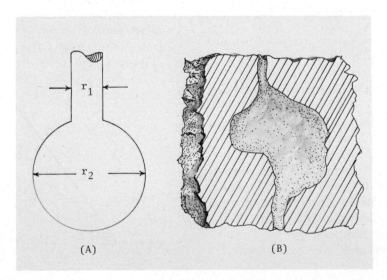

FIGURE 1.12. Idealized structure of an "ink bottle" showing the entry pore radius of r_1, and blind cavity having a radius of r_2 (A); in comparison with a more typical "ink-bottle-type" cavity found in foods (B).

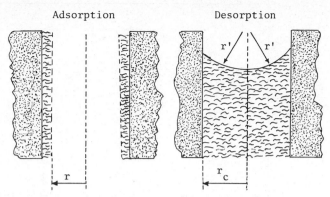

FIGURE 1.13. Principal schematic diagrams illustrating adsorption and desorption of capillary water in foods as explained in terms of the Cohan equation.

For those instances where capillary pore radius (r_c) is greater than two times the thickness of adsorbed film (D), the adsorption vapor pressure (P_a) exceeds desorption vapor pressure (P_d) and hysteresis occurs. In those cases where the capillary radius (r_c) is equivalent to two times the thickness of the adsorbed water film (D), hysteresis will not occur.

HEAT OF ADSORPTION AND FOOD MOISTURE

Isotherms for foods subjected to a variety of experimental temperature conditions can be mathematically defined in terms of the Clausius–Clapeyron relationship:

$$\frac{d(\ln A_w)}{d(1/T)} = \frac{Q_s}{R} \qquad (1.11)$$

where

A_w = water activity
T = absolute temperature
R = gas constant
Q_s = heat of adsorption

Application of the above equation to a graphic plot of the logarithm of A_w versus the reciprocal of the absolute temperature at a constant moisture level results in a straight line with a slope of $-Q_s/R$ (Figure 1.14A). Foods that have less than complete monolayer coverage by water often display a strong binding interaction with water. Typical Q_s value calculations for these systems based on slope analysis range from 2.0 to 10 kcal/mol.

The theory of the BET isotherm indicates that the heat of sorption (Q_s) will be constant for foods with less than monolayer sorption of water. The theoretical graphic plot based on this consideration has been illustrated in Figure 1.14B. It must be noted, however, that experimentally determined Q_s values, plotted versus percent moisture content for a food specimen displaying less than uniform water monolayer coverage, will show Q_s values significantly higher than theoretical values (Figure 1.14B).

Labuza (1968) has suggested that this apparent departure from the BET isotherm at low moisture levels actually results from the combined influence of both Q_s and the heat of vaporization (ΔH_v). That is, the latent heat of vaporization, which approximates 10.4 kcal/mol, must be added to the calculated heat of sorption in order to determine the "total" heat of sorption for a food system with less than monolayer water sorption. Those food systems having more than monolayer conditions exhibit a total heat requirement for water removal equivalent to ΔH_v.

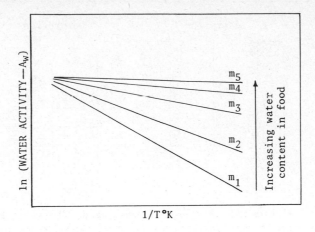

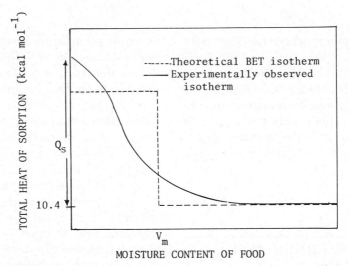

FIGURE 1.14. (Top) Graphic representation of the Clausius-Clapeyron relationship where $\ln A_w$ is plotted against $1/T°K$ and the calculated slope yields $-Q_s/R$; and (Bottom) graphic comparison of a theoretical BET isotherm with a corresponding experimentally observed isotherm for a food substance.

PHYSICAL BEHAVIOR OF WATER AND PRINCIPLES OF FOOD PRESERVATION

The preservation of foods for future consumption requires the control of food moisture. Foods may deteriorate because of the growth of saprophytic and heterotrophic organisms that are indigenous to a food or happen to inoculate the food during proc-

essing. Food deterioration can also result from enzymatic and nonenzymatic reactions owing to the chemical complexity of most foods.

Whether food deterioration is the sole or a combined result of microorganisms or biochemical or chemical reactions, the impact of these deteriorative factors can be minimized by controlling the concentration levels of food moisture.

The occurrence of water in foods depends largely on the physical and chemical characteristics of the natural food substance or processed food product. Water is normally distributed as a fluid system solvating or dispersing food constituents. In other cases, water may simply be adsorbed as mono- or polymolecular layers to internal and external food surfaces as a result of hydrogen bonding or capillary action. Outermost polymolecular water layers, and any water that does not tenaciously interact with food constituents as water of hydration, promote food deterioration. The deteriorative involvement of this water can be minimized by freezing the water or removing it from the food system by dehydration.

Freezing of Water in Foods

The overall favorable impact of food preservation by freezing is actually a composite effect resulting from three important considerations.

First, the freezing of foods decreases the availability of water necessary to support microbial growth and minimizes the involvement of water as a solvent medium for enzymatic and biochemical deterioration of foods.

Second, the lower temperatures required for effective freezing of a food inhibit the reproduction of most deteriorative and pathogenic organisms. Although low temperatures of $-29°C$ $(-20°F)$ can exert a lethal effect on many microorganisms and food parasites, a significantly longer period is required to kill organisms at low temperatures than during exposure to heat. One typical example is the nematode *Trichinella spiralis,* responsible for trichinosis. This organism can be killed in infected pork by elevating its internal temperature to $58.3°C$ ($137°F$); however, a temperature of $-17.8°C$ ($0°F$) must be sustained for at least 20 days to ward off survival of the nematode.

It should be noted that many psychroduric organisms, capable of surviving the freezing process, will actively resume growth and reproduction during the thawing of frozen foods. These organisms may be strict psychrophiles or mesophilic organisms that display limited psychrophilic growth characteristics.

Third, freezing of water in food systems results in increased food solute concentrations. This is significant from food preservation perspectives because decreased availability of food water coincides with increased water crystallization. The increased concentration of solutes in a diminishing fluid aqueous environment subjects microorganisms to severe osmotic, pH, and vapor-pressure stresses that can be lethal. Despite a wide range of microorganisms for which these stresses are lethal, the osmophilic nature of many bacteria, yeasts, and molds allows them to survive freezing. Notable osmophiles include yeasts and molds belonging to the genera *Botrytis, Fusarium, Mucor, Penicillium, Rhizopus,* and *Stemphylicum.*

The freezing curve for water and its relationship to foods: Since water is a major constituent of foods, the observed freezing curves for foods are reminiscent of the freezing curve for water. A freezing curve for water is determined from temperature–time readings for liquid water until ice has been deposited in an experimental system.

The freezing point (T_{fp}) for pure water can be determined from a cooling curve by extrapolating the horizontal *YZ* portion of the curve until it intersects with line *WX* (Figure 1.15). The depression of the cooling curve below the apparent freezing point (T_{fp}) for water demonstrates the occurrence of supercooling. This brief period precedes nucleation of water and ensuing ice crystal formation, graphically depicted as a temperature increase due to the latent heat of fusion of water. A freezing curve for an aqueous solution generally does not demonstrate a horizontal region after liberation of the heat of fusion because the solution becomes increasingly more concentrated as water freezes out. The progressive concentration of solute in the aqueous system resulting from water crystallization continuously contributes to a decrease in the T_{fp} for the aqueous system. This is indicated as a declining slope on the cooling curve between points Y and Z (Figure 1.15).

The extrapolation of the line YZ to an intersection with line WX (Figure 1.15) indicates the temperature at which the aqueous solution would have frozen in the absence of supercooling. Although supercooling and the liberation of the heat of fusion occur during food freezing, the actual freezing curves for water in foods rarely follow the idealized cooling curve for water presented in Figure 1.15. The lack of identical correspondence between freezing curves for pure water and water contained in foods parallels obvious disparities in the freezing points for water in these two systems. Pure water technically freezes at 0°C (32°F), but the freezing point for most foods is less than 0°C because of the presence of various solutes.

For most foods, the temperature range of 0 to −5°C corresponds to the extraction period for the latent heat of fusion. This temperature range is also referred to as the *thermal-arrest region* (Figure 1.16). Water actively crystallizes to form ice during this period, but the longer a food is maintained in the thermal-arrest temperature range, the worse is the effect on its quality. An extended thermal-arrest time is destructive because of intra- and extracellular migration of water prior to its actual crystallization. Patterns of water movement during crystallization are dictated by distribution of solute concentration levels in the food, diffusion forces, and architectural attributes of the food that favor directional fluid channeling. Ice crystals formed under these conditions tend to be extremely large and invariably result in intra- and extracellular ice crystals that pierce individual cells and create excessive drip-loss when the food substance is defrosted. The inconsistent formation and distribution of large ice crystals in the marginally adequate freezing range of 0 to −5°C characterizes the phenomenon of *discontinuous freezing*. Discontinuous freezing not only leads to the loss of textural integrity in the food caused by increased cellular flaccidity and separation of contiguous cells, but also encourages abnormal patterns of solute distribution and the release of deteriorative enzymes in the drip during defrosting.

Contrary to discontinuous freezing, rapid heat removal from a food substance and minimization of the thermal-arrest time favor the uniform distribution of small ice

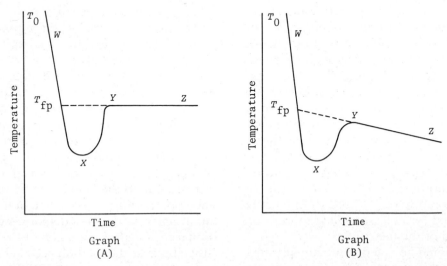

FIGURE 1.15. Freezing curves for pure water (A) and water plus at least one solute (B).

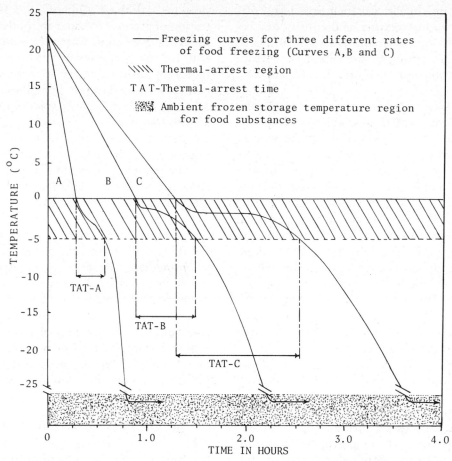

FIGURE 1.16. Freezing curves for three food specimens that depict thermal-arrest times (TAT) of different time durations, where TAT-A is the shortest and TAT-C is the longest.

crystals throughout a food. This is known as *continuous freezing*. Application of current food-freezing technologies favors continuous freezing since it avoids the severe problems associated with discontinuous freezing and maintains the desired esthetic, organoleptic, and palatability characteristics of the original food prior to freezing.

A comparison of discontinuous versus continuous freezing phenomena indicates that the undesirable features of discontinuous freezing parallel the advantageous destruction of microorganisms that inhabit or contaminate foods. Microorganisms subject to a discontinuous freezing environment experience cell rupture and other related

physical stresses that are antagonistic to their vegetative cell structure. Unfortunately, the advantages of discontinuous freezing are more counterproductive to food quality than beneficial, so conventional implementation of continuous freezing technologies inadvertently contribute to higher survival rates for microorganisms.

Prefreezing treatments for foods: Application of current food-freezing technologies causes some dehydration or chemical and structural damage to most foods. Many of these adverse effects are evident only when the food is defrosted as (1) cellular dehydration and loss of nutrients in the drip from the ruptured cells; (2) food browning, off-flavors, loss of texture and consistency resulting from enzymatic reactions and air–food contact; or (3) rancidity arising from the deterioration of food fats and oils.

Glazes, polyphosphates, alcohols, and saccharides. Some problems associated with the dehydration of defrosted foods can be averted by providing frozen foods with protective coatings or glazes. Glazes are applied by dipping previously frozen foods in water or some other solution such as alginate gel (e.g., carageenan). The superficial coating provided by the glaze is especially effective for decreasing food tissue dehydration in foods exposed to air-blast freezing methods. Dehydration of foods by air-blast freezing may result in "freezer burn," which is evident as external toughening of meat tissues, while high carbohydrate foods undergo hardening, caking, or segmentation.

In addition to glazes, cell rupture and dehydration of food tissues can be minimized by prefreezing treatments using polyphosphates, polyhydroxy alcohols (e.g., glycerol), or mono- and disaccharides.

The permeation of polyphosphates into the myomeric tissues of fish prior to freezing characterizes the ability of this additive to obviate protein denaturation and excessive drip loss from defrosted tissues. Normal freezing of foods causes localized increases in the concentrations of salts, organic acids, and other cell constituents that induce isoelectric precipitation or ionic strength denaturation of food proteins. Polyphosphates not only minimize protein denaturation and associated precipitation, but facilitate water resorption by defrosted tissues. The effects of polyphosphate addition have been widely observed; however, the favorable chemical and histological mechanisms responsible for the maintenance of food quality are not entirely understood.

Polyhydroxy alcohols such as glycerol and some mono- and disaccharides also minimize food protein denaturation by maintaining the hydration layer that surrounds proteins. Administration of a $1:3\ M$ concentration of glycerol to water in foods prevents the freezing of some water. This is especially true for many species of fish. The maintenance of residual water in a fluid phase, contrary to an entirely crystalline state in a frozen food, provides a solvent system for solutes that would otherwise become highly concentrated and cause ionic strength or salt-induced precipitation of proteins. The advantageous preservation roles of saccharides are similarly related to the action of glycerol and other alcohols.

Blanching and sugar syrups. In addition to the fact that drip loss (from defrosted foods) leads to immediate flaccidity and dehydration, the aqueous drip disperses a wide range of intracellular solutes and enzymes. The ensuing extracellular reactions invariably contribute undesirable organoleptic and esthetic problems to the food. Among the most deteriorative enzymes are the polyphenoloxidases (polyphenolases, phenolases), peroxidases (catalases), pectin esterases, and proteolytic enzymes.

The deteriorative activity of these enzymes in the aqueous drip is controlled by exposing foods to a brief prefreezing treatment of steam or boiling water. This treatment, known as *blanching*, denatures enzyme proteins, thereby minimizing or terminating their deteriorative activity.

Blanching is widely applied to many fruits and vegetables as a prefreezing treatment, but its implementation is highly discretionary since blanching can adversely affect the quality and texture of many heat-sensitive foods (e.g., berries, melons). As an alternative to blanching for heat-sensitive foods, the activity of deteriorative oxidative enzymes can be moderated by decreasing the air–food contact. This is achieved by mixing the food with sugar or sugar syrups before freezing.

Food Dehydration

Removal of food moisture by dehydration represents one of the most traditional food preservation methods. Earliest applications of food dehydration were subject to the whims of sun exposure, wind, and other environmental factors. Duration and termination of the drying period in conjunction with food quality were empirically evaluated.

Contemporary food dehydration techniques employ a scrupulously controlled environment of dried air, inert gas, superheated steam, or vacuum (reduced pressure). Additional dehydration techniques may also supply heat to foods by dielectric, infrared, or microwave techniques.

Among the currently available dehydration methods, air drying is still the most widely used. Industrial predilection for air drying is based on proven economic and food-quality considerations. Air-drying methods not only provide minimal operational expense but permit a wide flexibility suitable for a variety of foods.

Food dehydration by air methods requires:

1. Control of the surface-to-volume ratio and the initial temperature of the food before exposure to drying conditions.

2. Temperature control of the food during the drying period.

3. Control of velocity, volume, and relative humidity of the air used for drying purposes, thereby obtaining optimum dehydration rates with minimal impairment to food quality.

The latter consideration is of great importance to food quality because severe drying rates foster excessive surface (or outer layer) dehydration of foods. This situation encourages *case hardening* of the food because the superficial dehydration rate exceeds the actual rate of internal water migration and diffusion toward the outer regions of the food. Case hardening in foods is evidenced as protein denaturation, gelatinization of carbohydrates, or increased concentrations of free sugars that form a physical boundary wholly or partially impervious to continued moisture diffusion. Formation of this boundary effectively retards continued internal dehydration of foods under a prescribed set of drying conditions.

Food dehydration using air: Water evaporation from foods occurs according to predictable physical properties of air under various temperature and relative humidity conditions. A specific volume of air normally shows a doubling of its moisture-carrying capacity for every rise of 15°C (59°F) and a 1/273 increase in volume for every 1.0°C increase in temperature at standard pressure. Although the heat of vaporization varies with temperature, ~4400 kilogram calories (kgc) are required to vaporize 454 g of water in conventional dehydration operations (*note:* 1 kgc = 4 Btu). Water evaporation rates from food depend on surface-to-volume relationships; physical or chemical interactions of food water; food porosity; relative humidity; and velocity of air used for drying. Assuming physical and chemical interactions of water to be negligible, the water evaporation rate from surfaces can be approximated using the equation

$$W = 0.093 \left(1 + \frac{V}{230}\right)(P_T - P_a) \quad (1.12)$$

where

W = weight (g) of water evaporated m^{-2} h^{-1}

V = linear air velocity on a surface in m min^{-1}

P_T = water vapor pressure at a specific evaporation temperature, "T"

P_a = water vapor pressure in atmosphere

Application of this relationship demonstrates that a linear air velocity of 70 m min^{-1} (230 ft min^{-1}) effects an evaporation rate twice that for stagnant air, while a 140 m min^{-1} (460 ft min^{-1}) velocity evaporates water three times faster.

Freeze-drying (lyophilization) of foods: Freeze-drying or lyophilization is an applied form of vacuum distillation whereby the food substance to be dried is solidly frozen before being subjected to a very low absolute pressure (high vacuum) and a controlled heat input.

Under pressure conditions of ≤4.0 mm Hg and a temperature of <0.0°C, water locked in an ice matrix undergoes a transformation into a vapor by passing the intermediate formation of a liquid phase. This type of transformation, known as sublimation, coincides with the triple-point, three-phase plot of water illustrated in Figure 1.17.

The essential prerequisites for freeze-drying foods include the following conditions:

1. The product must be solidly frozen below its lowest possible freezing point (*eutectic point*).

2. A condensing surface for water vapor must be provided that has a temperature of ≈ −40°C.

3. The freeze-drying system must provide evacuation to an absolute pressure of between 5 and 25 μm Hg.

4. A source of controlled heat input into the product over a temperature range of −40 to +65°C must be available to drive water from the solid to the vapor state (heat of sublimation).

The operation of a freeze-drying system fulfilling these prerequisites relies on a constantly changing state of unbalance between

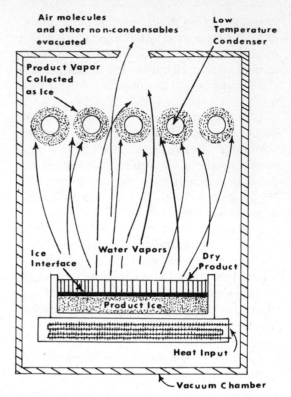

FIGURE 1.18. Schematic diagram for a lyophilization system, showing the sublimation of water from the food product ice followed by its collection on low-temperature condensers. (With permission of the Virtis Co., Inc., Gardiner, New York.)

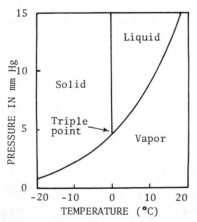

FIGURE 1.17. Triple-point curve for the three phases of water.

the food product ice and the pressure/temperature conditions of the drying system.

The sublimation of water originating from the product ice continues as long as a state of unbalance exists and the product ice is at a higher energy level than the rest of the system (Figure 1.18). Therefore, freeze-drying operations are designed to provide an isolated set of controlled conditions effecting and maintaining optimum pressure/temperature differences for a given product, which effectively dry the product in the least amount of time. The limit of unbalance is determined by the maximum amount of heat that can be applied to the product without causing a change from the solid ice matrix to the liquid state, that is, "melt back." This can occur even though the chamber pressure is low since the product dries from the surface closest to the area of lowest pressure. This surface is known as the "ice interface." The arrangement of dry particles above this interface offers resistance to the vapors released from below, raising the product pressure/temperature. To avoid "melt back," heat energy applied to the product must not exceed the rate at which heat energy directed into the ice matrix (and carried away by the migrating vapors) is removed by the condenser refrigeration system. The maintenance of an extremely low condenser temperature traps sublimed water as ice, effectively removing it from the system. Air and other noncondensable molecules within the chamber, as well as mechanical restrictions located between the product ice and the condenser, often contribute additional resistance to the movement of migrating vapors toward the condenser.

Interpretation of freeze-drying curves. A representative freeze temperature curve for milk has been illustrated to show a typical sublimation drying cycle (Figure 1.19), where vacuum, shelf temperature, product temperature, and condenser temperature are plotted as they reflect equipment performance. The preliminary portion of the drying curve (0–2.5 h) indicates product and shelf temperature as the volume of milk was frozen on the refrigerated shelf of the drying chamber. Actual freeze-drying was not initiated until the product temperature reached −40°C.

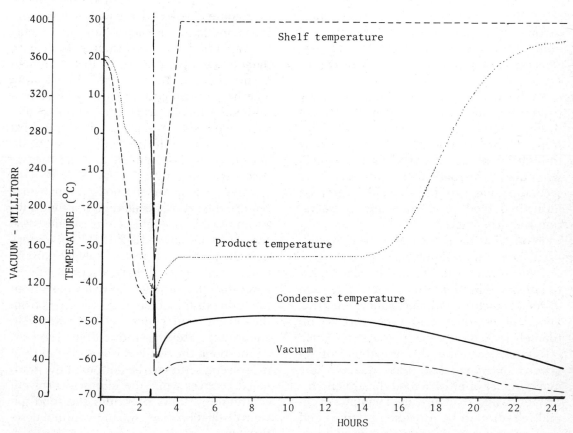

FIGURE 1.19. A representative freeze-drying curve for milk showing the interrelationship of shelf, product, and condenser temperatures, as well as vacuum conditions over the course of a freeze-drying cycle measured in hours. (Courtesy of the Virtis Co., Inc., Gardiner, New York, and adapted for presentation.)

After beginning the drying cycle, the shelf heating system was set in this example to +30°C and the refrigerated condenser stabilized at a temperature where it could absorb the thermal loading of the condensing water vapor. Correspondingly, the system vacuum also rose slightly and stabilized. The linear nature of product, condenser, and vacuum parameters (4–14 h) suggests the presence of a product with relatively unrestricted vapor passage above the ice interface as the interface recedes.

Completion of ice sublimation (14–24 h) from the food product is evidenced by a sensible heat gain due to continued shelf heat input. At this stage, the condenser is condensing a decreasing amount of water vapor, which indicates a drop in temperature, seeking to institute a new thermal balance. As the refrigerated vapor condenser is exposed to lower pressure, the system vacuum follows. Primary drying of the product was indicated as the product temperature paralleled shelf temperature (23–24 h) (Figure 1.19).

At the conclusion of this drying stage, many products would initiate a secondary freeze-drying stage lasting another 2 to 24 h. This step removes additional water intrinsically imbedded in the physical and chemical structure of the food. The requirements for secondary-stage drying depend largely on individual food production and preservation requirements.

Foods that have been effectively lyophilized show a number of quality advantages. Foremost among the favorable characteristics of lyophilization are excellent retention of food structure and biochemical integrity. Food solutes, solvents, and structure are ultimately locked into one solid matrix, which effectively decreases the probability of reactions detrimental to food quality. The maintenance of normal food constituent distributions and structure also facilitates eventual rehydration by providing textures and flavors similar to the prelyophilized food. An added advantage of freeze-drying rests on the fact that the main requirement for long-term food storage is the ability of the storage container to prevent moisture reentry into the product. In many other cases, however, long-term food storage can be ensured only by the added elimination of oxygen as well as water.

PROXIMATE ANALYSIS OF FOOD WATER

A wide variety of analytical procedures has evolved for moisture analysis including indirect distillation, direct distillation, and chemical and instrumental techniques.

Indirect distillation methods are actually gravimetric analytical methods since moisture content is based on weight loss from foods due to water evaporation. Water may be evaporated from food samples by heating them to a temperature of 70 to 130°C in an oven containing air, inert gas, or a vacuum. The heat transfer to foods may be accomplished by forced draft, convection, or dielectric methods. Normally, samples are dried until 2.0 to 5.0 g of dried residual solids do not differ by more than 1 to 3 mg for two successive weighings.

Direct distillation methods for moisture require the reflux distillation of a food with an immiscible solvent such as xylenes (B.P. 138–144°C), toluene (B.P. ~111°C), or heptane (B.P. 99°C). The reflux solvent must be water immiscible and have a specific gravity lower than that of water. Solvent vapors from the reflux distillation and water vapors from the food are both condensed and trapped in a graduated Sterling–Bidwell-type receiver. At this point the distilled water, denser than the solvent, settles to the bottom of the graduated receiver while the immiscible solvent continuously flows back into the food-solvent distillation flask during continued refluxing.

Aside from distillation procedures, the Fischer titration and calcium carbide methods represent chemical approaches for estimating food moisture.

The Fischer method relies on titration of a methanolic solution of iodine, pyridine, and sulfur dioxide until a free iodine color is evident. Normally, iodine is not visible during the titration as long as water is present in the food system. A two-step chemical reaction is involved in this assay:

STEP 1:

$$H_2O + (I_2 + SO_2 + 3 \; C_5H_5N) \longrightarrow$$
$$2 \; (C_5H_5N \cdot HI) + C_5H_5NOSO_2$$

STEP 2:

$$C_5H_5NOSO_2 + CH_3OH \longrightarrow C_5H_5N \; (HCH_3SO_4)$$

The Fischer titration, along with a variety of modifications, is widely employed in the food industry for moisture assessment in fats, waxes, alcohols, esters, flour, starch, vegetable purées, grains, confectionaries, honey, dehydrated foods, and a host of processed foods.

The reaction between food water and calcium carbide provides a basis for another assay technique. Acetylene gas generated from a mixture of calcium carbide and water in a closed system indicates calculable increases in gas pressure, gas volume, and corresponding sample weight loss solely attributable to the formation of acetylene from the carbide reaction with food water.

Instrumental analysis of food water is rapid, yet principally complex. These methods include nuclear magnetic resonance spectrometry and gas–liquid chromatography, in addition to instrumental methods that assess the proportional relationships between food moisture and electrical frequency, resistance, and dielectric properties.

Detailed analytical considerations and other important assay methods are discussed by Joslyn (1970), Pearson (1971), Winn (1965), The Milk Industry Foundation (1959), Cleland and Fetzer (1942), Sair and Fetzer (1942), and McCombs and Wright (1954).

Each analytical method described above has application limitations and methodological idiosyncracies. For example, the application of direct or indirect distillation assays to heat labile foods may lead to decomposition and volatilization of food components in addition to water evaporation.

Actually an absolute quantitative determination of water in foods is difficult since water may be operationally defined as "free" or "bound." So-called free water can be easily assessed by gravimetric methods of analysis relying on evaporation or direct distillation. Free water in a food substance represents the concentration of water in excess of the critical amount necessary for maintaining the physical structure and integrity of the food. Loss of water below a critical level leads to protein denaturation, salt and sugar crystallization, plus wholesale changes in the food–solute equilibria. Free water is usually frozen at about 0°C or readily evaporated from foods under atmospheric pressures at 103°F.

Bound water, however, represents the integral water required for maintenance of stable monosaccharide, disaccharide, and salt hydrates as well as protein and polysaccharide gels. Water operationally defined as "bound" is not frozen in the immediate temperature range below 0°C. In fact, studies have shown that water bound to biocolloids may remain unfrozen at −230°C. In the case of evaporation, bound water may not vaporize at temperatures far in excess of 103°C unless a strong vacuum is also applied. Even if evaporation occurs under these conditions, it is not unlikely for food structure to undergo some decomposition and volatilization.

WATER ACTIVITY AND FOOD PRESERVATION

Water activity has a profound affect on the liability of a food to undergo microbial, enzymatic or nonenzymatic spoilage. The levels of water activity that contribute to food spoilage vary depending on the type of food, food solute concentrations, pH, temperature, the presence of humectants, and many other factors. It has generally been observed that dried foods with 5 to 15% moisture correspond to the lower regions of sorption isotherms depicting monolayer and multilayers of water sorption. Foods in this water activity range usually require minimal processing or preservation in order to achieve good storage stability.

Intermediate moisture foods, having 20 to 40% moisture ($A_w \geq 0.5$), correspond to the upper regions of sorption isotherms for which multilayered water and capillary water predominate. These foods are almost invariably subject to rapid spoilage unless the raw or processed food is quickly preserved by physical or chemical preservation methods (e.g., canning, dehydration, freezing, blanching, antibiotics).

Water activities between 1.0 and 0.65 are ideal for the growth of many microorganisms, but water activities of 0.75 to 0.65 largely restrict the growth of most organisms except halophiles and osmophiles (Figure 1.20).

Enzymatic food spoilage reactions can occur over a wide gamut of water activity levels, but this type of spoilage is most prevalent at $A_w = 0.3$. Fruits and vegetables are typically subject to the action of amylases, polyphenol oxidases, and peroxidases, while enzymatic spoilage of lipid-containing foods is mediated by lipoxygenase, lipases, or phospholipases. In most cases, the highly de-

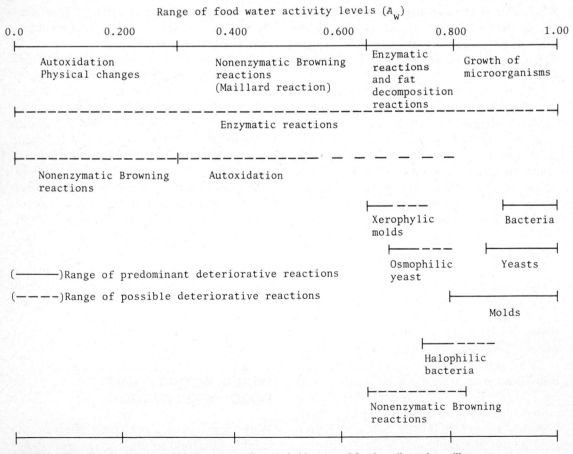

FIGURE 1.20. Water activity and the corresponding probable type of food spoilage that will occur.

structive nature of these enzymes can be effectively minimized by prepreservation treatments such as blanching.

Nonenzymatic deterioration of foods occurs from water activities of 1.0 to 0. Foremost among these reactions are nonenzymatic browning caused by the Maillard reaction, and the autoxidation of unsaturated lipid constituents in foods to form destructive free radicals.

The reactions of amino group substituents contained in amino acids with carbohydrates lead to a series of browning reactions over a water activity range of 0.8 to 0.4. Water activity levels of less than 0.4 generally lead to decreases in nonenzymatic browning.

Contrary to many food preservation techniques that rely on lowering water activity, low water activity levels may foster lipid oxidation. The reaction of molecular oxygen with unsaturated lipids to form free radicals can be moderated by increasing water activities in foods above 0. It should be noted, however, that the advantages of increased water activity do not progressively increase beyond water activities greater than or equal to water monolayer coverage of the food product.

In addition to the deteriorative reactions discussed above, water activity should also be recognized as a contributing factor to protopectin hydrolysis; reduction of pectin esters; pectin demethylation; autocatalytic hydrolysis of lipids; and degradation of chlorophylls and benzopyran pigments (e.g., anthocyanins, flavonoids).

REFERENCES

Adamson, A. W. 1963. *Physical Chemistry of Surfaces.* Interscience, New York.

Brunauer, S., P. H. Emmett, and E. Teller. 1938. Adsorption of gases in multimolecular layers. *J. Amer. Chem. Soc.* **60:**309.

Cleland, J. E., and W. R. Fetzer. 1942. Historical view of distillation methods of moisture. *Ind. Eng. Chem. Anal. Ed.* **14:**242.

Cohan, L. H. 1938. Sorption hysteresis and vapor pressure of concave surfaces. *J. Amer. Chem. Soc.* **60:**433.

Cohan, L. H. 1944. Hysteresis and capillary theory of adsorption of vapors. *J. Amer. Chem. Soc.* **66:**98.

Duckworth, R. B. (ed.). 1975. *Water Relations of Foods.* Academic Press, New York.

Freundlich, H. 1926. *Colloid and Capillary Chemistry.* Methuen, London.

Harkins, W. D., and G. Jura. 1944. A vapor adsorption method for determination of the area of a solid without assumption of molecular area. *J. Amer. Chem. Soc.* **66:**1366.

Henderson, S. M. 1952. A basic conception of equilibrium moisture. *Agr. Eng.* **33:**24.

Joslyn, M. A. 1970. *Methods of Food Analysis.* Academic Press, New York.

Kapsalis, J. G. 1979. Moisture sorption hysteresis. In *Second International Symposium on Properties of Water in Relation to Food Quality and Stability, 10–16 September 1978, Osaka, Japan.*

Labuza, T. P. 1968. Sorption phenomena in foods. *Food Technol.* **22:**263.

Langmuir, I. 1918. Adsorption of gases on plane surfaces of glass, mica and platinum. *J. Amer. Chem. Soc.* **40:**1361.

McCombs, E. A., and H. M. Wright. 1954. Formamide as a Karl Fischer solvent. *Food Technol.* **8:**73.

Pearson, D. 1971. *The Chemical Analysis of Food.* Chemical Publishing Company, New York.

Rockland, L. B., and G. F. Stewart (eds.). 1981. *Water Activity: Influences on Food Quality.* Academic Press, New York.

Sair, L., and W. R. Fetzer. 1942. Summary of moisture methods used in the wet milling industry. *Cereal Chem.* **19:**633.

Sharp, J. G. 1962. Nonezymatic browing deterioration in dehydrated meat. In *Recent Advances in Food Science* (J. Hawthorne and M. Leitch, eds.), Vol. 2. Butterworths, London.

The Milk Industry Foundation. 1959. *Laboratory Manual—Methods of Analysis of Milk and Milk Products,* 3rd ed. Milk Industry Foundation, Washington, D.C.

Winn, P. N. 1965. *Principles and Methods of Measuring Moisture in Liquids and Solids,* Vol. 4. Reinhold, New York.

Zigmondy, R. 1911. Über die Struktur des Gels der Kieselsäure. Theorie der Entwasserung. *Z. Anorg. Chem.* **71:**356.

CHAPTER · 2

Proteins: Chemistry, Structure, and Analysis

INTRODUCTION

Proteins occupy a very prominent position in the study of biochemistry. Their versatile functions are unequaled among the organic compounds that are synthesized by all living entities. For this reason, they were named "proteins" based on the Greek word *protos,* which means "first."

The analysis of proteins at the elemental level reveals the presence of carbon, hydrogen, oxygen, and nitrogen in all proteins along with the variable presence of sulfur, phosphorous, and metals (e.g., zinc, copper, iron, manganese, and other trace metals).

Proteins are generally categorized as biopolymeric substances where monomeric α-amino acids are covalently linked according to the dictates of current genetic theory. Infinite possibilities for the covalent linkage of 20 different amino acids commonly found in proteins account for an unlimited variety of possible protein structures. The diversity of protein structures is paralleled only by their wide range of different biochemical functions in the living cells of microorganisms, plants, and animals. Some of these functions are indicated in Figure 2.1. For example, enzymes mediate the biosynthesis

and degradation of biochemical compounds; antibodies offer protection to an animal from undesirable or lethal effects of certain antigens; hemoglobin facilitates respiratory gas transport *to* and *from* cells that are actively oxidizing food constituents for energy purposes; some proteinaceous-based agents such as ferritin store iron in the spleen; insulin and glucagon jointly participate in the regulation of carbohydrate metabolism; certain proteinaceous substances such as venoms from snakes and other organisms serve as survival or protective mechanisms; structural proteins such as collagen are partially responsible for the tensile strength of individual bones, the smooth articulation between bones at joints, and a variety of other physical chondral features; while muscle proteins facilitate movement of the whole organism and its various organs. The biochemical and functional versatility of the proteins is undeniable and, therefore, the name bestowed upon them is most appropriate.

STEREOISOMERISM

All amino acids, with the exception of glycine, exhibit optical activity, and it is gen-

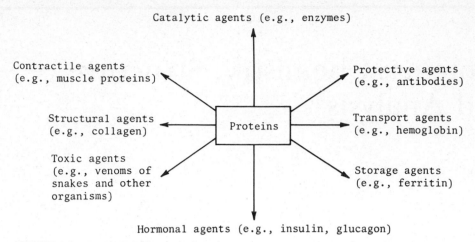

FIGURE 2.1. Some typical protein functions.

erally recognized that L-amino acids are solely important for the normal nutrition of man and most higher organisms. The absolute configuration for the "D" and "L" forms of amino acids are related to the respective "D" and "L" isomeric forms of glyceraldehyde as indicated below.

$$
\begin{array}{cc}
\text{CHO} & \text{CHO} \\
| & | \\
\text{HO}-\text{C}-\text{H} & \text{H}-\text{C}-\text{H} \\
| & | \\
\text{CH}_2\text{OH} & \text{CH}_2\text{OH} \\
\textbf{L-Glyceraldehyde} & \textbf{D-Glyceraldehyde}
\end{array}
$$

Asymmetric α-carbons present in glyceraldehyde

$$
\begin{array}{cc}
\text{COOH} & \text{COOH} \\
| & | \\
\text{H}_2\text{N}-\text{C}-\text{H} & \text{H}-\text{C}-\text{NH}_2 \\
| & | \\
\text{CH}_3 & \text{CH}_3 \\
\textbf{L-Alanine} & \textbf{D-Alanine}
\end{array}
$$

Asymmetric α-carbons present in alanine

The "L" designation for an amino acid indicates that its α-amino group is positioned to the left only when the carboxyl group on the Fischer projection formula for the amino acid appears in the *upward* direction. The "D" designation correspondingly indicates that the amino group is oriented to the right of the α-carbon as seen in the above diagram for alanine.

Although alanine is representative of an amino acid that has only *one* asymmetric carbon and *two* possible stereoisomeric forms—

since only one of its carbon atoms has four different substituents attached to it—*most* other amino acids have *at least two* asymmetric carbons. Examples of these amino acids include threonine, isoleucine, and hydroxyproline. In accordance with the general rule that the number of stereoisomers for a substance is equivalent to 2^n, where n is the number of asymmetric carbons, it is clear that threonine, isoleucine, and hydroxyproline can display four possible stereoisomeric structures (Figure 2.2).

The amino acid cystine, which is the oxidized form of cysteine, has two asymmetric carbons and theoretically it should also display four stereoisomeric structures, but in reality, it has only three stereoisomers as a result of the physical phenomenon known as *internal compensation*. That is, both of the "one-half sides" of the structure of *meso*-cystine (Figure 2.3) rotate light the same number of degrees, but in opposite directions. Consequently, one manifestation of optical activity cancels the other, with the net result being an absence of overall optical activity.

The presence of one or more asymmetric carbons in organic compounds parallels their property for rotating a plane of polarized light. This phenomenon is shown by amino acids, peptides, or proteins, and the exact amount of light rotation can be measured

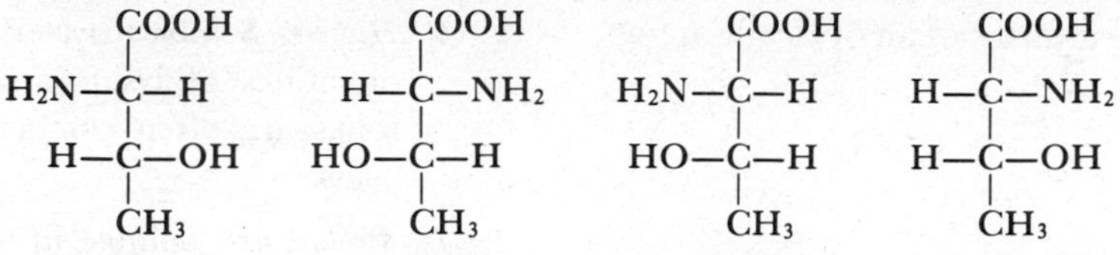

$$
\begin{array}{cccc}
\text{COOH} & \text{COOH} & \text{COOH} & \text{COOH} \\
| & | & | & | \\
\text{H}_2\text{N—C—H} & \text{H—C—NH}_2 & \text{H}_2\text{N—C—H} & \text{H—C—NH}_2 \\
| & | & | & | \\
\text{H—C—OH} & \text{HO—C—H} & \text{HO—C—H} & \text{H—C—OH} \\
| & | & | & | \\
\text{CH}_3 & \text{CH}_3 & \text{CH}_3 & \text{CH}_3 \\
\textbf{L-Thr} & \textbf{D-Thr} & \textbf{L-\textit{allo}-Thr} & \textbf{D-\textit{allo}-Thr}
\end{array}
$$

FIGURE 2.2. Four stereoisomers for threonine (Thr) are possible since it contains two asymmetric carbons. The *allo-* designation refers to those stereoisomers of the respective D and L forms, which are also mirror images (enantiomers). For example, L-threonine and L-*allo*-threonine show the same configuration around the α-carbon, but the configurations around the second asymmetric carbons are the exact opposite.

by using a polarimeter. The optical rotation of polarized light is measured in degrees from a specific visual reference point. Light rotation depends on a number of factors including the nature of the asymmetric compound, its concentration in solution, the pH of the solution, the path length of the polarized light through the optically active solution, and the particular wavelength chosen for the polarized light. All of these factors can be arranged into a mathematical relationship that is equated to a value known as the *specific rotation* ([α]). The specific rotation is a constant for certain solutes measured under prescribed polarimetric conditions of temperature and wavelength of polarized light:

$$[\alpha]_D^t = \frac{a \times 100}{l \times c}$$

where

$[\alpha]_D^t$ = *specific rotation* constant evaluated using the *D*-line of a sodium lamp as a source of polarized light, while *t* refers to the sample temperature in °C

l = *solute pathlength* that must be traversed by polarized light as measured in dm

c = *solute concentration* in g/100 mL

a = *observed degrees of polarized light rotation* caused by fixed concentration levels of an optically active solute

The knowledge of the *l*-value for pathlength along with any two of the three remaining variables permits calculation of any remaining unknown value in the relationship, in-

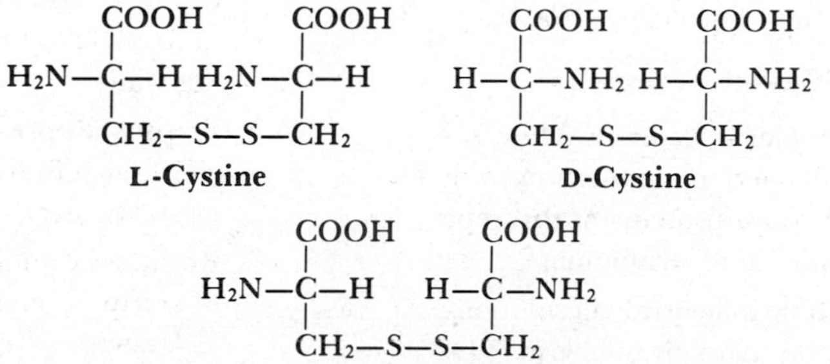

$$
\begin{array}{cc}
\text{COOH} \quad \text{COOH} & \text{COOH} \quad \text{COOH} \\
| \qquad\quad | & | \qquad\quad | \\
\text{H}_2\text{N—C—H} \;\; \text{H}_2\text{N—C—H} & \text{H—C—NH}_2 \;\; \text{H—C—NH}_2 \\
| \qquad\quad | & | \qquad\quad | \\
\text{CH}_2\text{—S—S—CH}_2 & \text{CH}_2\text{—S—S—CH}_2 \\
\textbf{L-Cystine} & \textbf{D-Cystine}
\end{array}
$$

$$
\begin{array}{c}
\text{COOH} \quad\quad \text{COOH} \\
| \qquad\qquad | \\
\text{H}_2\text{N—C—H} \quad \text{H—C—NH}_2 \\
| \qquad\qquad | \\
\text{CH}_2\text{—S—S—CH}_2 \\
\textit{meso}\textbf{-Cystine} \\
\textbf{(no optical activity)}
\end{array}
$$

FIGURE 2.3. L-Cystine and D-cystine exhibit optical activity, whereas *meso*-cystine exhibits none due to *internal compensation*.

cluding the *concentration* of an optically active solute.

CLASSIFICATION OF PROTEINS

Proteins have traditionally been classified according to the chemical constituents responsible for their overall structure, their relative solubilities in both organic and inorganic solvents, and their relative shapes. Some of the noteworthy criteria underlying these classification schemes have been itemized below.

I. **Categories of proteins based on individual chemical constituents**

   A. *Simple proteins:* Hydrolysis yields only amino acids.

   B. *Conjugated proteins:* Hydrolysis yields amino acids plus other organic and inorganic compounds.

      1. Nucleoproteins: Nucleic acids and proteins.

      2. Phosphoproteins: Proteins + phosphoric acid.

      3. Lipoproteins: Proteins + lipids.

      4. Glycoproteins: Proteins + carbohydrates.

      5. Chromoproteins: Colored prosthetic group (e.g., heme) + protein.

   C. *Derived proteins:* These biopolymers are *derived* from the previous two classes.

II. **Solubility classification of proteins**

   A. *Albumins:* Soluble in water in dilute solution; precipitated by saturation with $(NH_4)_2SO_4$. Typical proteins of this type are egg albumin and lactalbumin.

   B. *Globulins:* Soluble in neutral salt solutions such as NaCl and sparingly soluble in H_2O. Exemplary proteins include serum globulins, β-lactoglobulin in milk, actin and myosin in meats, and globulins in plant seeds.

   C. *Histones:* Soluble in water and contain large quantities of lysine and arginine. Histones are often combined with nucleic acids.

   D. *Prolamines:* Soluble in 50–90% ethanol. Examples include zein from corn, gliadin obtained from wheat, and hordein present in barley.

   E. *Protamines:* These proteins are commonly associated with nucleic acids. Notable concentrations of protamines occur in the ripe sperms of various fishes.

   F. *Scleroproteins:* Insoluble in water and neutral solutions. Scleroproteins are fibrous proteins that serve important structural and binding purposes in tissues.

III. **Shape classification of proteins**

   A. *Fibrous proteins:* These proteins tend to display tightly linked and elongated polypeptide structures that are largely insoluble in most common solvents, and they are especially resistant to the hydrolytic actions of proteolytic enzymes. Examples of these proteins include silk, wool, hair, and other portions of the integumentary systems such as horn, hoofs, nails, and skin. The fibrous proteins are divided into three groups, namely, *collagens, elastins* and *keratins.*

      1. Collagen proteins: Constitute the major portion of connective tissues and are not water soluble although they can be converted to gelatin by the actions of dilute acids, alkalies, or boiling water.

      2. Elastins: Represent the major protein constituent found in tendons and other elastic tissues.

      3. Keratins: Contain a large percentage of sulfur-containing amino acids such as cysteine. Keratins are the major structural contributors to hair, nails, and wool.

   B. *Globular proteins:* Represent a class of protein structures that are soluble in water or aqueous media containing acids, bases, ethanol, or various salts. The shape of the proteins is spherical or ellipsoidal (e.g., enzymes and hemoglobin).

AMINO ACIDS—THE STRUCTURAL UNITS OF PROTEINS

More than 100 amino acids occur naturally, but only about 20 of these are consistently used for the majority of common protein biosyntheses. All amino acids share some common structural features including a free carboxyl group, a free amino group, a hydrogen atom, and an "R group," all located on the α-carbon. The "R group" is important since it imparts key structural differences to individual amino acids that account for their distinctly different chemical and physical properties (Figure 2.4).

Amino acids that commonly occur in proteins have been structurally detailed in Table 2.1 according to the substituents on the α-carbon of the respective amino acids. Although proline and hydroxyproline are classified with the α-amino acids, they are actually *imino acids* owing to the location of their nitrogen atom.

Acid–Base Properties of Amino Acids

Amino acids are *amphoteric* substances because they can either donate or accept protons according to the definition of Brönsted acids or bases. The amphoteric nature of the amino acids including the simplest amino acid,

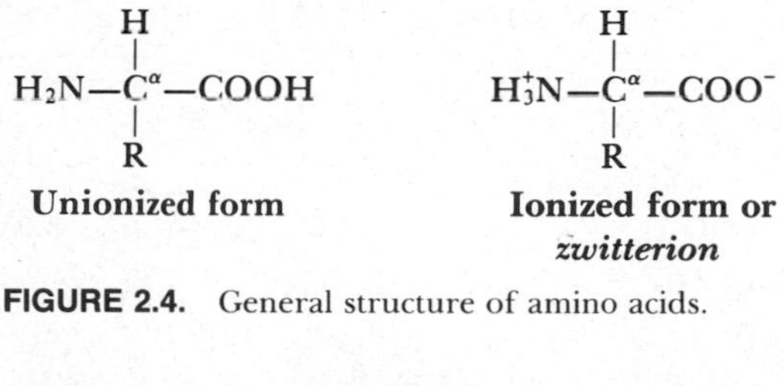

glycine, depends on their ability to exist as a dipolar-ionic species called a *zwitterion,* which simultaneously exhibits positive and negative charges of equal magnitude.

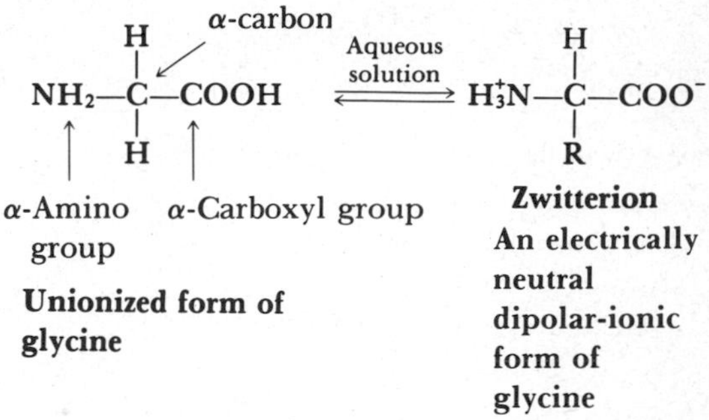

$$NH_2-\overset{\overset{H}{|}}{\underset{\underset{H}{|}}{C}}-COOH \underset{}{\overset{\text{Aqueous solution}}{\rightleftharpoons}} H_3\overset{+}{N}-\overset{\overset{H}{|}}{\underset{\underset{R}{|}}{C}}-COO^-$$

α-Amino group   α-Carboxyl group

Unionized form of glycine

Zwitterion
An electrically neutral dipolar-ionic form of glycine

Charge acquisition by the α-amino group is attributable to the ability of the nitrogen atom to share its electrons with any available proton (H^+), while the negatively charged α-carboxyl group results from its loss of a proton (H^+) to the surrounding aqueous medium. Although glycine displays two hydrogen atoms on the α-carbon, all other amino acids have the general structural formula where one hydrogen atom is replaced

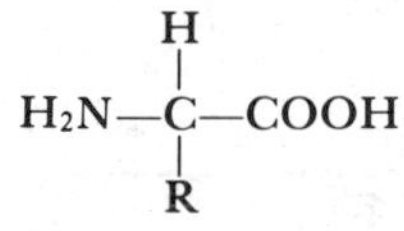

$$H_2N-\overset{\overset{H}{|}}{\underset{\underset{R}{|}}{C}}-COOH$$

by any one of a number of structural substituents usually referred to as an *R group*. These R groups account for the structural diversity of the amino acids illustrated in Table 2.1 and contribute characteristic neutral, acidic, or basic properties to all of the structures of the amino acids. Since ionization phenomena, according to the Brönsted acid–base theory, necessarily involves the presence of varying conjugate acid–base pair ratios over the pH range from 0 to 14, it is important to recognize that each amino acid will naturally display a characteristic set of pK_a values depending on its ionizable groups. The specific pK_a values for these ionizable groups will generally reflect their acidic or basic properties.

$$H_2N-\overset{\overset{H}{|}}{\underset{\underset{R}{|}}{C^\alpha}}-COOH \qquad H_3\overset{+}{N}-\overset{\overset{H}{|}}{\underset{\underset{R}{|}}{C^\alpha}}-COO^-$$

Unionized form

Ionized form or zwitterion

FIGURE 2.4.   General structure of amino acids.

Table 2.1. Names and Structures of Amino Acids in the form of Zwitterions

Name	Abbreviation	Structure	α-COOH pK_a	α-NH$_3^+$ pK_a	R group pK_a	pI
1. Monoamino and monocarboxylic amino acids						
Glycine	L-Gly(G)	H$_3^+$NCH$_2$COO$^-$	2.34	9.60	—	5.97
L-Alanine	L-Ala(A)	H$_3^+$NCHCOO$^-$	2.35	9.69	—	6.02
L-Valine	L-Val(V)	H$_3^+$NCHCOO$^-$	2.32	9.62	—	5.97
L-Leucine	L-Leu(L)	H$_3^+$NCHCOO$^-$	2.36	9.60	—	5.98
L-Isoleucine	L-Ile(I)	H$_3^+$NCHCOO$^-$	2.36	9.68	—	6.02
2. Monoamino-dicarboxylic amino acids and their amides						
L-Aspartate	L-Asp(D)	H$_3^+$NCHCOO$^-$	1.88	9.60	3.65	2.77
L-Glutamate	L-Glu(E)	H$_3^+$NCHCOO$^-$	2.19	9.67	4.25	3.22
L-Asparagine	L-Asn(B)	H$_3^+$NCHCOO$^-$	2.02	8.80	—	5.41
L-Glutamine	L-Gln(Q)	H$_3^+$NCHCOO$^-$	2.17	9.13	—	5.65
3. Hydroxy-amino acids						
L-Serine	L-Ser(S)	H$_3^+$NCHCOO$^-$	2.21	9.15	—	5.68
L-Threonine	L-Thr(T)	H$_3^+$NCHCOO$^-$	2.09	9.10	—	5.00
4. Thio(sulfur)-containing amino acids						
L-Cysteine	CysH(C)	H$_3^+$NCHCOO$^-$	1.96	10.28	8.18	5.07
L-Cystine	Cys-sCy		1.65	7.85	—	5.06

Table 2.1. (*Continued*)

Name	Abbreviation	Structure	α-COOH pK_a	α-NH$_3^+$ pK_a	R group pK_a	pI
L-Methionine	L-Met(M)	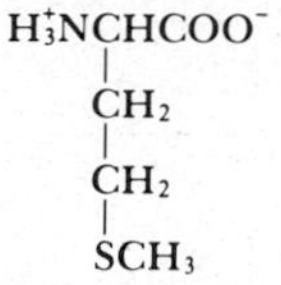 H$_3^+$NCHCOO$^-$ CH$_2$ CH$_2$ SCH$_3$	2.28	9.21	—	5.74

5. *Diamino-monocarboxylic amino acids*

Name	Abbreviation	Structure	α-COOH pK_a	α-NH$_3^+$ pK_a	R group pK_a	pI
L-Lysine	L-Lys-(K)	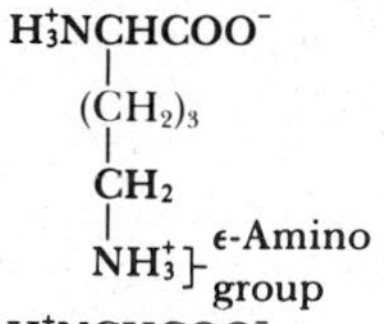 H$_3^+$NCHCOO$^-$ (CH$_2$)$_3$ CH$_2$ NH$_3^+$ } ε-Amino group	2.18	8.95	10.53	9.74
L-Arginine	L-Arg(R)	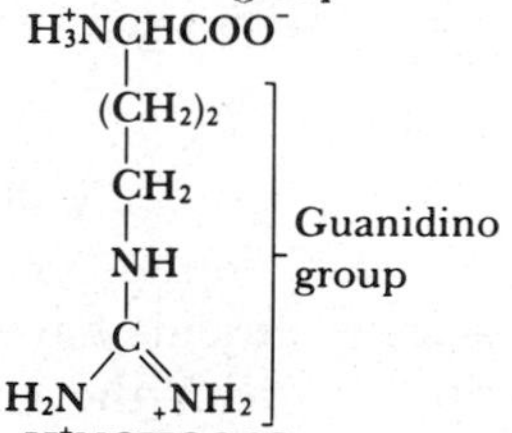 H$_3^+$NCHCOO$^-$ (CH$_2$)$_2$ CH$_2$ NH C H$_2$N $_+$NH$_2$ — Guanidino group	2.17	9.04	12.48	10.76
L-Histidine	L-His(H)	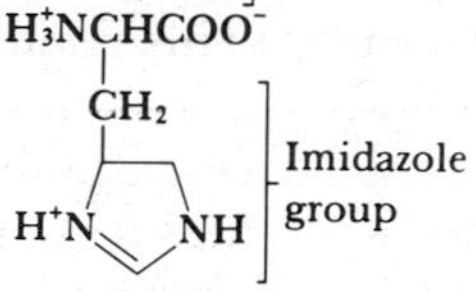 H$_3^+$NCHCOO$^-$ CH$_2$ H$^+$N NH — Imidazole group	1.82	9.17	6.0	7.59

6. *Aromatic acids*

Name	Abbreviation	Structure	α-COOH pK_a	α-NH$_3^+$ pK_a	R group pK_a	pI
L-Phenylalanine	L-Phe(F)	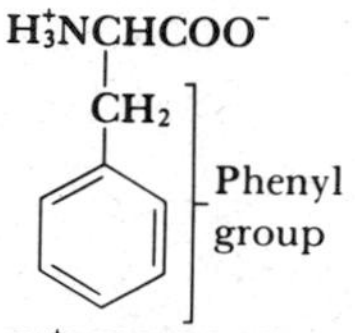 H$_3^+$NCHCOO$^-$ CH$_2$ } Phenyl group	1.83	9.13	—	5.48
L-Tyrosine	L-Tyr(Y)	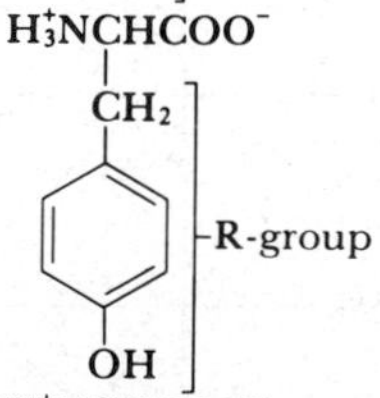 H$_3^+$NCHCOO$^-$ CH$_2$ —R-group OH	2.20	9.11	10.07	5.66
L-Tryptophan	L-Trp(W)	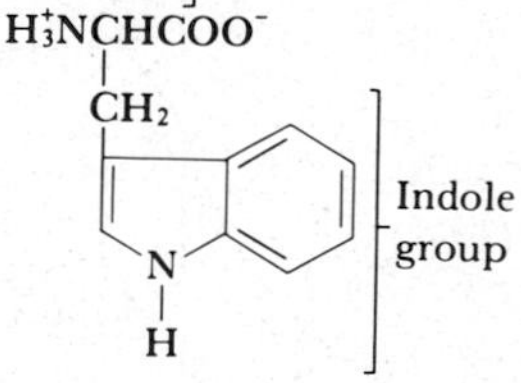 H$_3^+$NCHCOO$^-$ CH$_2$ N H — Indole group	2.83	9.39	—	5.89

Table 2.1. *(Continued)*

Name	Abbreviation	Structure	α-COOH pK_a	α-NH$_3^+$ pK_a	R group pK_a	pI
7. *Imino acids*						
L-Proline	L-Pro(P)		1.99	10.60	—	6.30
L-Hydroxyproline	L-Hyp		1.92	9.73	—	5.82

Acid–Base Titration of Amino Acids

The amphoteric behaviors of amino acids are clearly revealed in their acid–base titration plots where the observed pH for an aqueous solution containing a fixed amount of amino acid is plotted versus equivalents of acid or base added. An example of a titration plot for glycine appears in Figure 2.5. As base (OH$^-$) is added to a 5.97 pH aqueous solution of glycine, *its zwitterionic form is progressively deprotonated* at the α-amino position (α-NH$_3^+$) until all of the α-NH$_3^+$ substituents have been stripped of a proton (H$^+$) with the addition of *one equivalent of base*. This results in the conversion of all amino acid molecules to their anionic form. The corresponding addition of *one equivalent of acid* (H$^+$) to the zwitterion form of glycine eventually *converts all of the amino acid into its cationic form*. The ratio of the concentrations for the conjugate acid–base pair of an amino

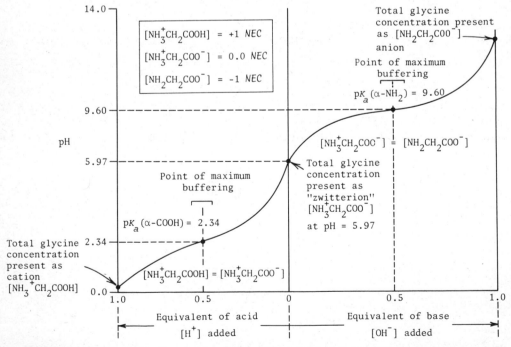

FIGURE 2.5. Acid-base titration plot for glycine and *net electrical charge(s)* (*NEC*) for its various ionization species.

Table 2.2. Characteristic pK_a Values for Some Important Functional R Groups Present on Amino Acids

Functional R group	pK_a value	Occurrence in specific amino acids
β-Carboxyl	3.6	Aspartic acid
γ-Carboxyl	4.3	Glutamic acid
Imidazole	6.0	Histidine
Sulfhydryl	8.3	Cysteine
Phenolic	10.1	Tyrosine
ε-Amino	10.3	Lysine
Guanidinium	12.5	Arginine

acid is equal to 1.0 when 0.5 equivalent of acid or base has been added. Therefore, the pH that corresponds to that point where equivalent amounts of a conjugate acid–base pair exist is defined as the pK_a value for that particular ionization step, and it also represents the pH where that amino acid species will display maximum buffering capacity. Amino acids more complex than glycine, such as those containing an R group with marked acidic or basic properties, provide similar, albeit more complex, titration curves with pK_a values corresponding to all additional ionizable groups (Table 2.2). Titration curves for two totally protonated amino acids containing three ionizable groups (i.e., two trifunctional amino acids) are shown in Figure 2.6 along with their corresponding ionization steps during titration. The titration plots for the amino acids illustrated in Figure 2.6 graphically indicate the predictability of the pK_a values for all ionizable species on an amino acid when a selected species has been quantitatively titrated. Other pH points plotted during the course of an amino acid titration can also be predictably calculated using the Henderson-Hasselbalch equation

$$pH = pK_a + \log \frac{[base]}{[acid]}$$

Isoelectric Point for an Amino Acid

A simple titration plot for glycine (Figure 2.5) indicates that the totally protonated form of glycine displays a *net electrical charge* (*NEC*) of +1, while its totally deprotonated form has a −1 *NEC*. The dipolar-ionic zwitterion of the amino acid represents the only electrically neutral species, and it predominates at a pH of 5.97. The pH where an amino acid assumes a *NEC* of 0.0 is called its *isoelectric point* (p*I*). Normally the p*I* can be calculated on the basis of the pK_a values on either side of its electrically neutral species:

$$p\!I \text{ glycine} = \frac{pK_{a_1} + pK_{a_2}}{2}$$

$$= \frac{2.34 + 9.60}{2} = 5.97$$

The isoelectric points for amino acids having more than two pK_a values, such as aspartic acid and lysine, *must be calculated only from the* pK_a *values on either side of their electrically neutral zwitterionic species* (see Figure 2.6 for ionization species):

p*I* for aspartic acid (an acidic amino acid)

$$p\!I = \frac{2.09 + 3.86}{2} = 2.98$$

p*I* for lysine (an alkaline amino acid)

$$p\!I = \frac{8.95 + 10.53}{2} = 9.74$$

Acidic amino acids generally have p*I* values <7.0, while *basic amino acids* have a p*I* > 7.0.

Since any pH conditions that favor a *NEC* of 0.0 correspond to the amino acid's zwitterionic species, amino acids in this form will not display any movement toward an anode or cathode if they are exposed to direct current.

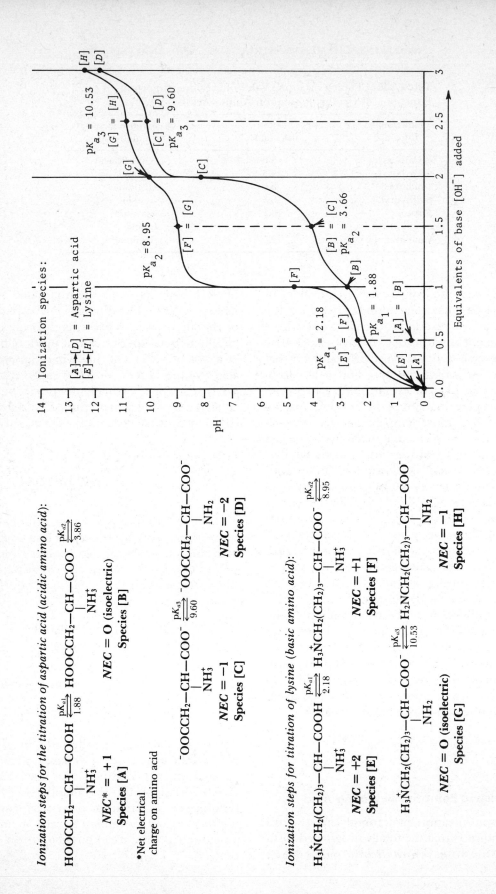

FIGURE 2.6. Stepwise ionization for acidic and basic amino acids along with the corresponding locations of their ionization species on a titration curve.

PEPTIDE STRUCTURES

Two or more amino acids can be sequentially linked by amide linkages known as *peptide bonds:*

$$
H_3^+N-\underset{\underset{R_1}{|}}{\overset{\overset{H}{|}}{C}}-COO^- + H_3^+N-\underset{\underset{R_2}{|}}{\overset{\overset{H}{|}}{C}}-COO^- \xrightarrow{\ H_2O\ }
$$

$$
H_3^+N-\underset{\underset{R_1}{|}}{\overset{\overset{H}{|}}{C}}\underset{\underset{H}{\underset{|}{}}}{\overset{\overset{O}{\overset{||}{}}}{C}}-N-\underset{\underset{R_2}{|}}{\overset{\overset{H}{|}}{C}}-COO^-
$$

Peptide bond

These peptide bonds can be repeatedly established for a multitude of similar or different amino acids. The sequence and number of amino acids linked in peptide bonds is not a random event within living systems. Instead, it is scrupulously controlled by genetic directives of protein biosynthesis.

Peptide structures are conventionally written so that the *N terminus* of one amino acid residue is at the head of the structure, while the last peptide-linked amino acid residue constituting the *C terminus* appears at the tail of the structure. Peptide-linked amino acids are customarily called *residues* because an established peptide bond results from the elimination of water between the carboxyl (—COOH) and amino (—NH₂) groups present on two different amino acids.

A peptide that contains between two and ten amino acid residues is known as an *oligopeptide,* but a peptide having more than ten amino acid residues is called a *polypeptide.* Peptides with only two amino acid residues are called *dipeptides;* those with three residues are *tripeptides;* four residues constitute *tetrapeptides,* and so on.

Peptides are named with reference to their N-terminal and C-terminal ends, and all amino acids involved in peptide bonds are designated by the suffix *-yl.* For example, the tripeptide having a residue sequence of aspartic acid, glycine, and lysine—where aspartic acid and lysine respectively account for N- and C-terminal acids—would be named as

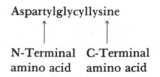

Regardless of the number of amino acids linked in a specific peptide sequence, all peptide bonds have *very characteristic bond rotational properties and spatial orientations* centered around the α-carbon (C$_\alpha$) of the amino acid residues (Figure 2.7).

The angles of bond rotation surrounding the *alpha*-carbon are respectively designated ψ or φ depending on their location as seen in Figure 2.8. The ψ bond represents the rotational angle about the C$_\alpha$—C bond and the φ designation refers to bond rotation

$$
H_3^+N-\underset{\underset{R_1}{|}}{\overset{\overset{H}{|}}{C}}-\overset{\overset{O}{||}}{C}-\underset{\underset{H}{|}}{N}-\underset{\underset{R_2}{|}}{\overset{\overset{H}{|}}{C}}-\overset{\overset{O}{||}}{C}-\left[\underset{\underset{H}{|}}{N}-\underset{\underset{R}{|}}{\overset{\overset{H}{|}}{C}}-\overset{\overset{O}{||}}{C}\right]_n-\underset{\underset{H}{|}}{N}-\underset{\underset{R}{|}}{\overset{\overset{H}{|}}{C}}-COO^-
$$

| **N-Terminal amino acid residue** | **Second amino acid residue** | **Unspecified number (*n*) of similar or different amino acids linked in peptide bonds** | **C-Terminal amino acid residue** |

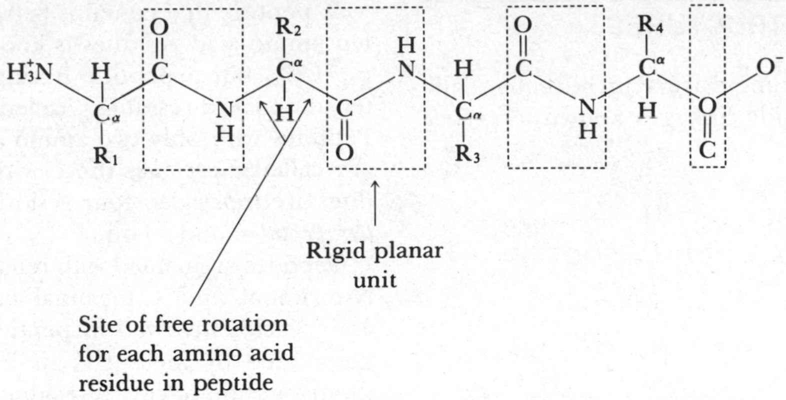

FIGURE 2.7. Schematic diagram indicating bonds from the α-carbon of amino acid residues in a tetrapeptide that are able to rotate with respect to the whole peptide structure. The area designated by the broken line represents that portion of each amino acid residue that is *not* subject to bond rotation, and necessarily exists as a rigid planar unit.

about the C_α—N bond. Rotations about the C_α—C and C_α—N bonds are in marked contrast to the limited degree of rotation around the C—N bond axis. This lack of rotational freedom is caused by a partial double bond (*pi* bond) character originating from tautomeric or resonance structures between the carbon and nitrogen atoms engaged in the peptide bond. This *pi* bond character not only limits free rotation around the bond axis but conveys a planar spatial orientation to every peptide bond (Figure 2.8).

The unique rotational properties of peptide-linked amino acids have proven to be fundamentally important to the geometric properties of the simplest and most complex peptide-linked structures. These structures include the *right-handed helix,* the *left-handed helix, parallel-pleated sheets,* and *antiparallel-pleated sheets,* which are discussed in the following sections.

Examples of Peptide Structures

The physical and chemical properties observed for a peptide represent the composite behavior of all its constituent amino acids. Therefore, depending on the number of amino acids, the various types of R groups and the characteristic interactions among all R groups comprising a peptide, different peptide structures will show characteristically (1) different titration curves; (2) different electrical charges at prescribed environmental pH conditions; (3) different ranges of buffering activity; and (4) different isoelectric points. The characteristic isoelectric points for peptides are significant because they not only represent that pH where the peptide shows a lack of movement in a direct electrical current but also the pI that corresponds to the pH where the peptide shows *maximum insolubility* in an aqueous system.

The amino acids comprising various peptide structures certainly have potential nutritional value, but simple peptide structures found in *nature* have little, if any, overall importance to the nutritional value of foods. Instead, many of the simplest peptides including glucagon, insulin, gastrin, secretin, calcitonin, prolactin, adrenocorticotropic hormone (ACTH), thyroid stimulating hormone (TSH), oxytocin, bradykinin, vasopressin, angiotensin, and still others exert profound effects on the metabolism of humans and other animals. Many peptides including actinomycin D, gramicidin S, bacitracin A, and numerous related cyclic peptides display unusual antibiotic effects while still other simple peptide structures such as as-

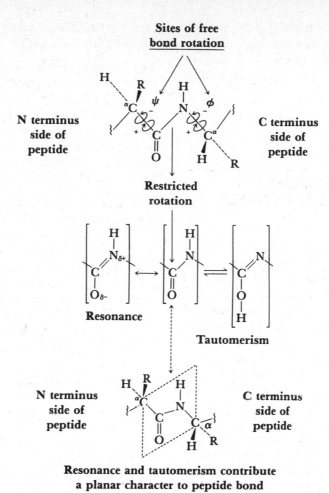

FIGURE 2.8. Schematic diagram showing sites of bond rotation in a peptide structure in addition to peptide resonance and tautomer structures responsible for maintaining the planar spatial properties of a peptide bond.

partame (the methyl ester of L-aspartyl-L-phenylalanine) produce characteristically strong sweetening powers.

Some peptide structures such as glutathione (GSH) can also participate in readily reversible oxidation–reduction reactions. The reduced form of this tripeptide, also recognized as γ-glutamylcysteinylglycine (GSH), can be oxidized to yield a disulfide linkage between two glutathione molecules as shown below:

$$2 \text{ GSH} \underset{+2 \text{ H}}{\overset{-2 \text{ H}}{\rightleftharpoons}} \text{GS—SG}$$

Reduced glutathione   **Oxidized glutathione**

Reduced glutathione is especially important for maintaining the structural integrity of bioactive proteins that have been genetically programmed to contain sulfhydryl groups at specific sites. The reduced glutathione serves as an agent to ward off the effects of oxidizing agents that could destroy critical conformational features of proteins that depend on sulfhydryl group presence. Thus, reduced glutathione preferentially yields its hydrogens as indicated above. This action spares key proteins from sulfhydryl oxidation, which could ultimately produce detrimental or lethal effects within a cell. The oxidized glutathione (GSSG) is again converted to two molecules of reduced glutathione (GSH) by the reductive actions of NADPH along with the enzyme *glutathione reductase.*

ANALYTICAL BIOCHEMISTRY FOR AMINO ACIDS, PEPTIDES, AND PROTEINS

The fundamental contribution of amino acids, peptides, and proteins to life processes at all levels of organization has prompted the development of many analytical methods designed to (1) elucidate both amino acid composition and sequences held in peptide structures and (2) establish the molecular shape of biochemically functional peptide structures. Some analytical rationales for achieving these goals are simple while others are quite complex. Moreover, the analytical biochemistry developed for qualitative and quantitative assessments of amino acids and protein structures are essential for clinical diagnostic work related to the human condition in states of health and disease, the assessment of food protein adequacy for nutritional purposes, and further technological development of old and new food protein resources.

The analytical methods for peptides and proteins can be operationally subdivided into five categories:

1. Nondiscriminative analyses of peptide structures.

2. Quantitative analysis and identification of individual amino acids.

3. Determination of amino acid sequence in peptides and proteins.

4. Conformational structures of proteins and their analysis.

5. Separation techniques for complex protein mixtures.

Nondiscriminative Analyses of Peptide Structures

The nondiscriminative assay of peptide structures ranging from simple peptides to complex proteins has commonly relied on *absorption principles of electromagnetic radiation,* or the assay of a *specific element* or *functional group* that has a nearly ubiquitous, yet predictable, occurrence in peptide-linked structures.

Absorption of electromagnetic radiation: The physical interaction of electromagnetic radiation (light) with molecules under carefully controlled conditions has served as an effective tool for quantitating many inorganic, organic, and biochemical substances (e.g., nucleic acids, carbohydrates, lipids, vitamins). So too is the case for amino acids, peptides, and proteins, which can be quantitatively assessed by their absorptive interactions with light.

When a discrete "packet" of light energy having a given frequency known as a *photon* collides with a molecule, light is absorbed. Thus the *probability of light absorption* or the *amount of light absorbed* is clearly proportional to the number of molecules that traverse a light path. In the case of solutions, the number of light-absorbing solute molecules is proportional to the product of their concentration c and the light path length b that traverses the solution. These factors are considered in the expression

$$\log \frac{I_0}{I} = a\,b\,c$$

where I_0 is the intensity of the *incident light,* I is the intensity of the *transmitted light,* and a is an *absorptivity coefficient,* which is characteristic for a light-absorbing species. If solute concentration levels (c) are expressed in moles per liter, a is denoted as the *molar absorption* or *extinction coefficient* indicated as ϵ. Since the $\log I_0/I$ is commonly expressed as light *absorbance* (A), the light absorption law can be expressed as

$$A = \epsilon\,b\,c$$

Use of this mathematical expression assumes that absorbance (A) is a linear function for a light-absorbing solute concentration at any *given* wavelength. The validity of the expression also depends on incident light (I_0), which is a monochromatic, parallel light interaction with an isotropic medium (randomly distributed solute molecules). Note too that ϵ varies considerably with respect to various wavelengths of incident light and the solvent medium.

The absorbance for biochemicals having an *unknown molecular weight* can also be expressed as an $E_{1\,cm}^{1\%}$ absorption unit. This useful unit corresponds to the absorptivity of a 1.0-cm length of a 1.0% (w/v) solution of a light-absorbing species and

$$A = E_{1\,cm}^{1\%} \cdot c$$

where c is expressed in percentage concentration (e.g., $10^{-3}\% = 10^{-3}$ g/100 mL).

When the values for extinction coefficients (ϵ) or the $E_{1\,cm}^{1\%}$ values are known for specific proteins, their unknown concentration levels can be calculated on the basis of their light absorbance at a specific wavelength. Some typical molar absorptivities appear in Table 2.3. Since solvent conditions as well as pH widely affect the validity of these values, calculations dealing with protein concentration should strictly adhere to the information tabulated in any reference source outlining absorptivity constants.

Although amino acids absorb electromagnetic radiation in some portion of the ultra-violet region, the quantitation of amino acids and proteins by means of direct spectrophotometric analysis can generate substantial analytical errors. This is especially true when proteins are present in conjunction with strongly adsorbing ultraviolet species such as nucleic acids (which show maximum absorption at 260 nm) or other substances that have a multitude of double bonds. Assuming that highly purified proteins can be isolated from the ultraviolet coabsorbing effects of other compounds, a λ_{max} of 280 nm can be used as an index for quantitating a *pure protein*. This absorbance maximum reflects the presence of aromatic amino acids in the protein structure such as tyrosine, tryptophan, and phenylalanine (Figure 2.9). The direct quantitation of proteins based on 280 nm absorbance must be carefully interpreted since different proteins display varying ratios of phenylalanine, tyrosine, and tryptophan with respect to their overall amino acid constitution. In fact, the comparative differences in light absorption for the same concentration level of different proteins may vary by a factor of five to six times at a wavelength of 280 nm.

As a result of the analytical errors inherent in the direct photometric quantitation of proteins, along with poor sensitivity at low protein concentration levels, a number of *chromogenic* (color-producing) reactions have been developed that detect proteins. These methods employ chemical reagents and conditions that develop a measurable colored reaction product (chromophore) whose con-

Table 2.3. Typical Molar Absorptivities for Some Proteins at Specific Wavelengths

Protein molecule	$\epsilon_M \times 10^{-4}$	Analytical λ
1. Albumin (from human serum, M.W. = 69,000)	4.0	280
2. Bovine caseins		
α_s	2.73	280
β	1.15	280
κ	2.44	280
3. Gliadin (α form from wheat)	2.9	276
4. Ovalbumin (from eggs)	2.85	290

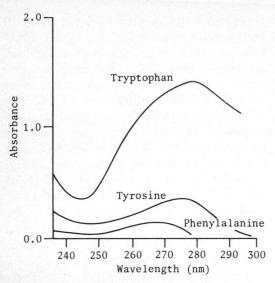

FIGURE 2.9. Graphic relationship for the relative absorbance spectra of the amino acids that notably absorb light over the range of 260–285 nm as a result of aromatic substituents in their respective structures.

centration is directly dependent on the presence of (1) peptide linkages or (2) specific amino acid residues contained within protein structures. Some typical traditional methods of protein assay include:

1. The *biuret reaction,* which depends on the quantitative reaction of alkaline copper with peptide linkages.

2. The *Folin–Ciocalteau reaction,* which similarly depends on the interactions of alkaline copper and a phenol reagent to react with aromatic amino acids and other groups.

3. The *dye-binding protocols* that exploit the ability to quantitatively bind certain colored organic dyes such as amido black 10B, acid orange 12, cochineal red A, buffalo black, or the disulfonic anionic dye known as orange G.

The historical interest in protein substances has led to the development of many other related assay techniques especially for the evaluation of food proteins. Many of the most important methods are detailed by Joslyn (1970), Pomeranz and Meloan (1978),

and the Association of Official Analytical Chemists.

Elementary analysis of peptides and proteins: The quantitative evaluation of proteins can be achieved by determining the presence of a specific element, or functional group, provided that the one or the other occurs with a certain degree of quantitative predictability among protein or peptide substances. Furthermore, these methods require an assay specimen that is uncontaminated by other extraneous peptides, proteins, or other substances quite unrelated to the specimen that may interfere with its reliable evaluation. Purified iron-containing proteins such as hemoglobin may be evaluated on the basis of iron content; iodine may be used to determine thyroglobulin; and still other proteins may be evaluated on the basis of their natural phosphorous, sulfur, or nitrogen content.

The quantitative assay of total protein, single purified proteins, and peptides has been widely employed since 1883, when the Danish investigator Kjeldahl developed a method for evaluating protein changes in grains used for brewing purposes. Many adaptations of the method have evolved since the 1880s, but the basic assay procedure depends on the quantitative conversion of organic nitrogen to ammonium sulfate. The method requires the use of concentrated sulfuric acid (sp. gr. 1.84), which digests the protein until all its carbon and hydrogen are oxidized, while the nitrogen is reduced and retained within a reaction vessel as ammonium sulfate. Many metals have been used in order to catalyze this wet oxidation of the protein including mercury and selenium. The protein digest is then admixed with sodium hydroxide and heated in order to liberate ammonia from the ammonium sulfate. The ammonia is trapped in a fixed, known volume of a standard acid solution (typically, 5% boric acid). The unreacted acid is then assayed by means of titration to an indicator endpoint and the percentage of nitrogen that emanated from the original sample is calculated.

Since most pure proteins contain 16% nitrogen, the protein content of a particular sample may be obtained by multiplying the percent nitrogen value by the factor 6.25 (or 100/16). This general factor (6.25) has been widely implemented for the proximate analysis of proteins in foods, but certain other factors must be used if proteins are known to contain more or less nitrogen per unit weight of protein. Some alternative factors for determining the percent protein content of certain foods are indicated in Table 2.4.

Aside from the Kjeldahl procedure, another classical nitrogen assay technique known as the Dumas method depends on the pyrolytic liberation of nitrogen from protein and peptide substances. Although traditional applications dating as early as 1831 relied on volumetric measurement of freed elemental nitrogen, more current methods depend on gas chromatographic analysis of nitrogen in gases liberated from nitrogen-containing samples.

Further details on the current application of these nitrogen and protein assays to foods and other specimens are surveyed by many current sources including Joslyn (1978), Pomeranz and Meloan (1970), as well as the Association of Official Analytical Chemists.

Quantitative Analysis and Identification of Individual Amino Acids

All amino acids are subject to a variety of analytical chemical reactions (Figure 2.10).

Effective applications of these reactions often require that the amino acids exist in their free form, or as the N-terminal amino acid residue on a peptide-linked structure. For those cases where the applications of the reactions are designed to identify or quantitate amino acids occurring between the N-terminal and C-terminal residues of the peptide-linked structure, a preliminary step must be undertaken to hydrolyze the structure, thereby freeing all amino acids from the peptide linkages that bind them. Hydrolysis is usually achieved by placing the peptide linked structure in an evacuated glass tube along with $6.0\ N$ HCl for a period of 24 h. The temperature of the acidified peptide is maintained at 110°C by means of a hot-air oven or an oil bath surrounding the tube. All of the *free amino acids* obtained from the peptide or protein during this incubation step are referred to as the *hydrolyzate* mixture.

The unidentified free amino acids contained in the hydrolyzate may be identified by any one of a variety of chromatographic methods coupled with the use of a specific chromogenic reagent such as one of those listed in Figure 2.10.

Paper and thin-layer chromatography: Paper and thin-layer chromatographies have been widely used to detect the presence of free or chromogenically derivatized amino acids that are present in complex mixtures. Both techniques depend on the differential *ascending* or *descending migration* of individual amino acids on a stationary phase. The dif-

Table 2.4. Typical Factors Used for Converting Percentage Nitrogen in Foods to Percentage Protein

Protein sources	Factor ($\times$ Kjeldahl % N = % protein)
Grains (e.g., barley, oats, wheat, rye, millet)	5.38
Milk and milk products	6.38
Nuts and seeds	5.30
Gelatin	5.55

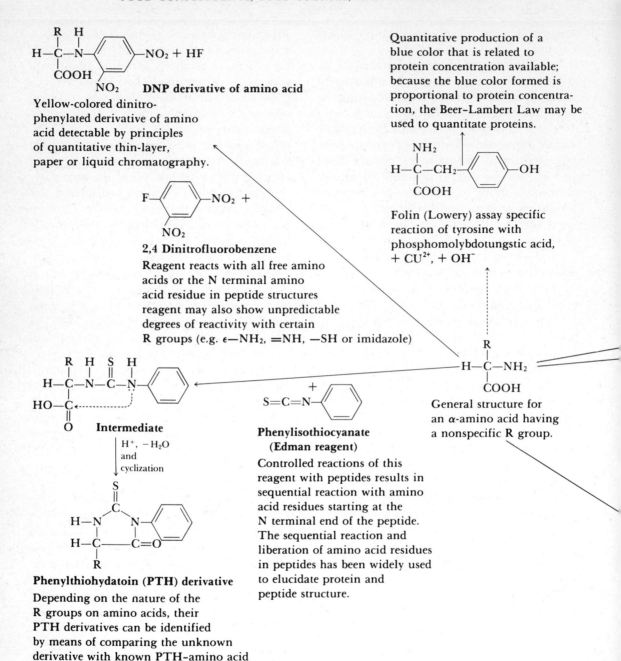

DNP derivative of amino acid

Yellow-colored dinitro-
phenylated derivative of amino
acid detectable by principles
of quantitative thin-layer,
paper or liquid chromatography.

Quantitative production of a
blue color that is related to
protein concentration available;
because the blue color formed is
proportional to protein concentration, the Beer–Lambert Law may be
used to quantitate proteins.

2,4 Dinitrofluorobenzene

Reagent reacts with all free amino
acids or the N terminal amino
acid residue in peptide structures
reagent may also show unpredictable
degrees of reactivity with certain
R groups (e.g. ϵ—NH_2, =NH, —SH or imidazole)

Folin (Lowery) assay specific
reaction of tyrosine with
phosphomolybdotungstic acid,
+ CU^{2+}, + OH^-

Intermediate

H^+, $-H_2O$
and
cyclization

General structure for
an α-amino acid having
a nonspecific R group.

**Phenylisothiocyanate
(Edman reagent)**

Controlled reactions of this
reagent with peptides results in
sequential reaction with amino
acid residues starting at the
N terminal end of the peptide.
The sequential reaction and
liberation of amino acid residues
in peptides has been widely used
to elucidate protein and
peptide structure.

Phenylthiohydatoin (PTH) derivative

Depending on the nature of the
R groups on amino acids, their
PTH derivatives can be identified
by means of comparing the unknown
derivative with known PTH–amino acid
derivatives on a thin-layer or
paper chromatogram. An alternative
method of analysis reproducibly
employs high-pressure liquid
chromatography.

FIGURE 2.10. Some important chemical reactions for free amino acids and N-terminal amino acids of peptides.

"Ruheman's blue" (λ_{max} = 570 nm) for all amino acids except proline, which produces a yellow color with a λ_{max} = 440 nm. Amino acid reactions with ninhydrin may be quantitatively evaluated by using the principles of the Beer–Lambert Law.

$$2 \left[\text{Ninhydrin} \right] +$$

Ninhydrin

$$\text{H}-\overset{\overset{\displaystyle R}{|}}{\underset{\underset{\displaystyle COOH}{|}}{C}}-\overset{\overset{\displaystyle H}{|}}{N}-\overset{\overset{\displaystyle O}{||}}{C}CF_3 + HCl$$

Trifluoroacetylated derivatives of amino acids permit vaporization of the amino acids thereby providing a form of the various amino acids that can be assayed by means of gas–liquid chromatography.

$$+$$

$$\overset{\overset{\displaystyle O}{||}}{CF_3C}Cl$$

Trifluoroacetyl chloride reagent

$$+ \quad ClSO_2-\text{[naphthalene]}-N(CH_3)_2$$

5-Dimethylamino naphthalenesulfonyl chloride (*dansyl* chloride)

Dansyl reagent reacts with individual amino acids and the N terminal amino acid residues of peptides to yield a highly fluorescent dansyl amino acid, which can be identified against standard dansylated amino acids using thin-layer chromatography or a high-pressure liquid chromatograph equipped with a fluorescence detector.

$$\text{H}-\overset{\overset{\displaystyle R}{|}}{\underset{\underset{\displaystyle COOH}{|}}{C}}-\overset{\overset{\displaystyle }{|}}{\underset{\underset{\displaystyle H}{|}}{N}}-SO_2-\text{[naphthalene]}-N(CH_3)_2$$

Dansylated derivative of amino acid

ferential migration is promoted by the movement of carefully formulated solvents such as acetic acid, pyridine, butanol and water, over processed cellulosic sheets of paper, or a glass plate superficially coated with a dried slurry composed of an inorganic binding agent plus alumina, silica gel, or some other component. In either paper or thin-layer techniques, reliable ascending or descending migration and separation of amino acids requires enclosure of the chromatographic system in a chamber saturated with solvent vapors. The relative movement of "unknown" amino acids partitioned on the chromatography medium is compared with other "known" amino acids on the basis of their respective R_f values. The R_f values used for comparative estimates of sample migration are calculated according to the formula

$$R_f = \frac{\text{distance traveled by "unknown" or "standard" amino acid}}{\text{distance traveled by solvent from origin to solvent front}}$$

The origins of the numerator and the denominator for this equation are schematically shown in Figure 2.11. Substances that display similar calculated R_f values are assumed to be identical provided that their chromatography conditions are consistently the same. As indicated, free colorless amino acids may be partitioned and identified by these chromatographic procedures only when they have been colorimetrically derivatized before or after partitioning. Although postchromatographically partitioned amino acids are commonly detected by ninhydrin, many prechromatographic detection methods have been employed. These procedures typically include reactions of amino acids with dinitrofluorobenzene, phenylisothiocyanate, or dansyl chloride.

Liquid chromatographic analysis of phenylthiohydantoin (PTH)-amino acids: The advent of high-pressure liquid chromatography (HPLC) has expanded the analytical use of chromogenic reactions for amino acids such as the Edman reagent to new vistas.

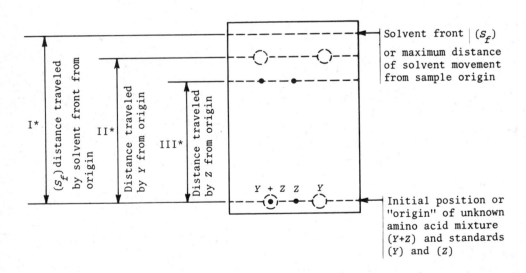

$$R_f \text{ for } Y = II^*/I^* \quad R_f \text{ for } Z = III^*/I^*$$

FIGURE 2.11. Schematic diagram illustrating origin of R_f values for a mixture of amino acids that has been chromatographically separated into its Y and Z components. (Migration distances are conventionally expressed in cm or mm.)

This chromatographic method relies on the differential migration of amino acids and many other nonprotein-related molecules over a compact, stationary phase situated in a stainless-steel or glass column. The differential migration and characteristic separation of molecules on the stationary phase is promoted by a solvent phase whose polarity is carefully controlled with respect to the stationary phase. As the name of the technique implies, the permeation of a solvent phase through the stationary phase contained in the column requires substantial pressure (75–5000 psi). This high pressure is required since the stationary phase contained in the column is composed of tightly packed, small-diam-

eter (~8–50 μm) particles that offer considerable resistance to the solvent's fluid flow. Since different substances such as derivatized amino acids demonstrate different affinities for the mobile solvent phase as opposed to the stationary phase, they elute from the column at different times. The elution of different molecules is usually detected by their light absorption at specific wavelengths in the ultraviolet or visible region of the spectrum, but more recent methods employ fluorescent or electrochemical detection. A schematic illustration of a typical instrument appears in Figure 2.12. As further illustrated in Figure 2.13, the *phenylthiohydantoin* (PTH) derivatives of amino acids produced

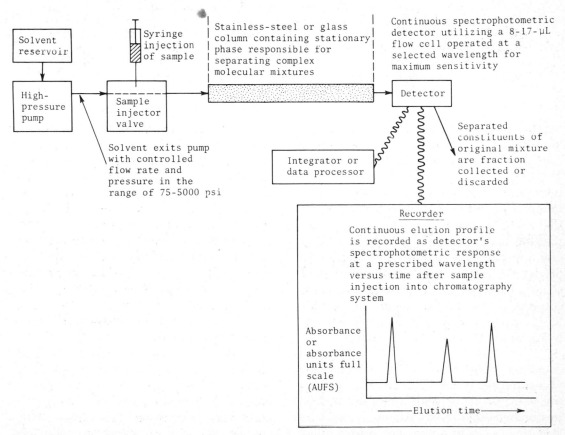

FIGURE 2.12. Schematic diagram for the operation of a high-pressure liquid chromatograph (HPLC). The system illustrated utilizes only one type of solvent that has a fixed ionic strength, polarity, and composition and is characteristic of "isocratic" HPLC procedures, whereas time controlled regulation of solvent phase ionic strength, polarity and composition typifies "gradient elution" HPLC operations.

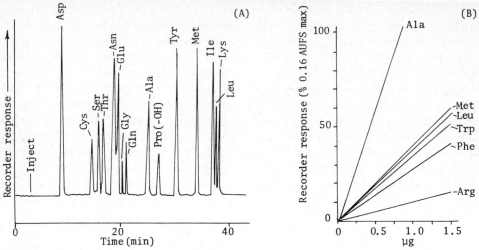

FIGURE 2.13. Hydrolysis of peptides into amino acids followed by formation of PTH derivatives of amino acids serves as a basis for determining amino acid composition in peptides. (A) PTH–amino acid derivatives can be easily separated using a C_{18}-reverse phase column (4 mm × 60 cm); a 254-nm UV detector (0.16 AUFS); and a solvent program $10/90 \rightarrow 90/10$ acetonitrile/0.01 M NaHOAc, pH 4.0. (B) The detector response for HPLC analysis demonstrates a linear concentration-dependent response specific for each type of PTH derivative. Thus, integration of chromatographic peaks for unknown concentrations of an identified amino acid can be compared to standard peaks of similar identity in order to permit amino acid quantitation.

by the Edman reagent can be separated, identified, and quantitated by reverse-phase HPLC. The key to the effective separation of PTH–amino acid derivatives depends on the column packing material, which has a monomolecular layer of octadecyltrichlorosilane chemically bonded to an inorganic packing material whose particle size diameter approximates 10.0 μm. The interaction of the C_{18} (octadecyl) moiety superficially coating each particle of packing material is responsible for the partitioning of complex PTH–amino acid mixtures during an analytical time frame of 15 to 30 min.

Amino acid analysis by ion-exchange chromatography: Acid hydrolysis of peptides and proteins yields a mixture of free protonated amino acids. Although all amino acids may exist in a protonated form, they invariably display net differences in their relative positive charges. These fundamental differences in electrical charge permit their

separation, identification, and quantitation according to highly automated methods of ion-exchange chromatography. This type of chromatographic separation relies on the ability of certain insoluble, synthetic-polymeric resins (R_p) to exchange their electrostatically bound ions for similarly charged ions present in a sample. Polymeric ion-exchange resins may have acid groups (e.g., R_p—COOH or R_p—SO_3H) or basic groups (e.g., R_p—NH_2 or R_p—N^+ (CH_3)$_3$). Those resins with acidic groups are called *cationic resins* while resins with basic groups are termed *anionic resins*.

In the case of either an anionic or cationic exchange resin, it is often necessary to activate the resin before it can be used for purposes of analytical separation. This may require initial electrostatic association of the anionic exchange resin with an anion such as Cl^-, while a cationic exchange resin may require Na^+. These preexisting states of the respective ion-exchange resins then allow them to exchange the bound ion for more negative or positive ions present in a sample:

Since amino acids loaded onto a resin can be separated in a quantitative fashion under

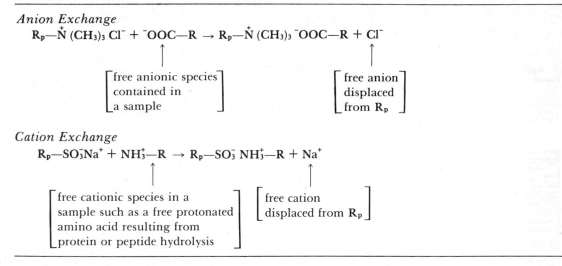

Anion Exchange

$$R_p{-}\overset{+}{N}(CH_3)_3\ Cl^- + {}^-OOC{-}R \rightarrow R_p{-}\overset{+}{N}(CH_3)_3\ {}^-OOC{-}R + Cl^-$$

$$\begin{bmatrix} \text{free anionic species} \\ \text{contained in} \\ \text{a sample} \end{bmatrix} \qquad \begin{bmatrix} \text{free anion} \\ \text{displaced} \\ \text{from } R_p \end{bmatrix}$$

Cation Exchange

$$R_p{-}SO_3^-Na^+ + NH_3^+{-}R \rightarrow R_p{-}SO_3^-\ NH_3^+{-}R + Na^+$$

$$\begin{bmatrix} \text{free cationic species in a} \\ \text{sample such as a free protonated} \\ \text{amino acid resulting from} \\ \text{protein or peptide hydrolysis} \end{bmatrix} \begin{bmatrix} \text{free cation} \\ \text{displaced from } R_p \end{bmatrix}$$

Anionic exchange resins are certainly important to many forms of analytical chemistry, but the cationic exchange resins are critical for amino acid analyses.

Because of inherent differences in the net positive charges among amino acids in a hydrolyzate, all of the cationic amino acid species will displace some of the sodium ions from the $R_p{-}SO_3^-\ Na^+$ form of the resin as indicated in the equation shown above. In general, the most alkaline amino acids are most tightly bound to the resin by electrostatic forces, while less alkaline (more acidic) varieties interact only loosely with the cationic exchange resin. Owing to these differences in electrostatic interactions of amino acids with a resin, the pH of an aqueous system surrounding a resin may be adjusted over the range of 3.2 to 6.8 in order to selectively neutralize resin-bound amino acids. As individual amino acids become more neutral, their electrostatic affinities for the resin diminishes, and they are progressively released from the resin in a quantitative, orderly fashion. Acidic amino acids such as glutamic and aspartic acids are readily displaced from the resin only to be followed by the neutral and alkaline amino acids—in that order.

scrupulously controlled conditions using an automated amino acid analyzer, the amino acids may be spectrophotometrically quantitated using a ninhydrin reaction as they are discharged from the resin. Ninhydrin reacts with all amino acids to give a purple color that is detectable at 540 to 570 nm, while proline and hydroxyproline yield a yellow color that can be monitored at 440 nm. In accordance with the principles of quantitative spectrophotometry (the Beer–Lambert Law), the absorbances at these wavelengths are detected and automatically recorded on a strip chart. The ninhydrin response for each amino acid is indicated as the change in absorbance at 540 to 570 nm or 440 nm with respect to their increasing elution time from the resin (Figure 2.14).

Amino acid analyzers employed for nutritional, clinical, or food analyses must be able to evaluate hydrolyzed amino acids resulting from peptides or proteins, as well as naturally occurring, free amino acids that happen to exist in biochemically complex samples (e.g., raw foods, cooked foods, blood, urine, or other body fluids). Hydrolyzate assays for amino acids are quite simple compared to the more protracted assay of naturally occurring amino acids where sample

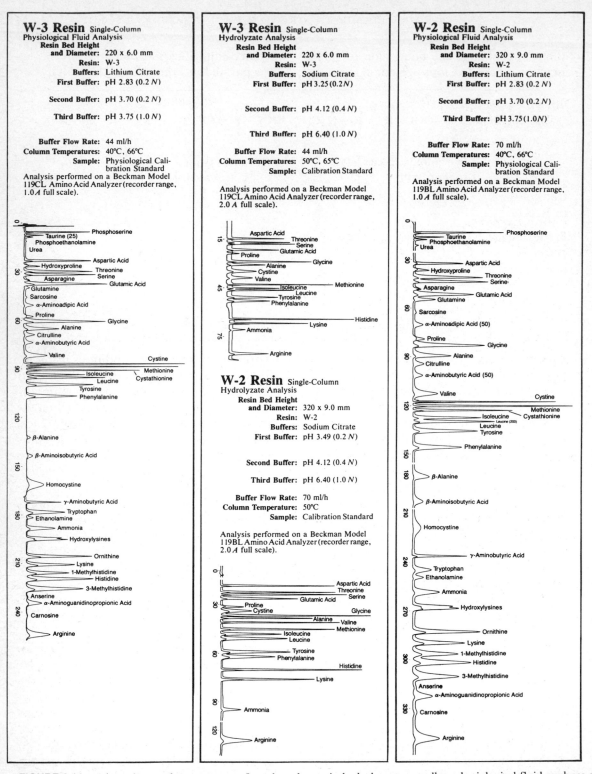

FIGURE 2.14. Ion exchange chromatograms for selected protein hydrolyzates as well as physiological fluid analyses of acidic and basic amino acids using three types of resins developed for automated amino acid analysis. (Methods developed and chromatograms furnished by Robert Slocum and Patrick Lee, Beckman Instrument Co., Applications Research Department, Palo Alto, California.)

preparation involves only preliminary deproteinization and "no" peptide hydrolysis. This latter analytical assay of amino acids is referred to as a *physiological mode of amino acid analysis* and unlike hydrolyzate analyses of amino acids, which identify the 20+ common amino acids, the physiological assay can evaluate 250 different amino acids and related compounds (Figure 2.14).

Determination of Amino Acid Sequence in Peptides and Proteins

Frederick Sanger was the first biochemist to elucidate the sequence of amino acids in a biologically active protein. Although Sanger's determination of the amino acid sequence in insulin served as a benchmark in historical analytical biochemistry, it also proved to be an outstanding accomplishment because his work demonstrated that *specific functional proteins* have an amino acid sequence that is *similar* or *identical* for all members of the same species. At present, the analytical methods employed by Sanger have become more technologically advanced and automated, but the fundamental analytical rationale for determining protein structures has remained similar since 1955.

The biochemical and structural idiosyncracies of naturally occurring proteins are nearly infinite, but there are a set of general steps that can be followed in order to elucidate the amino acid sequence of proteins.

Assuming that a polypeptide or protein structure exists as a series of amino acids linked by peptide bonds, one of the most fundamental structural determinations involves identification of its N-terminal and C-terminal amino acid residues.

N-Terminal amino acids can be identified by their selective reactions with any one of a variety of reagents including the Sanger reagent (2,4-dinitrofluorobenzene), dansyl chloride, or the Edman reagent (phenylisothiocyanate) (Figure 2.10). In all cases, the N-terminal amino acid residue on the peptide reacts with the selected reagent to yield an N-terminal amino acid derivative. The derivatized amino acid is eventually removed from its parent peptide linkage and then identified. Since all three reagents mentioned produce N-terminal amino acid derivatives that are easily detected by visible or fluorescent spectrophotometry, the hydrolyzed N-terminal amino acid derivatives may be identified by comparison with similarly derivatized amino acid standards. These standards are based on R_f values obtained from thin-layer or paper chromatography, or HPLC in the case of the Edman reagent.

The *determination of the C-terminal amino acid* residue of a peptide can be achieved by using a *carboxypeptidase enzyme* or *hydrazine reagent*. The carboxypeptidase ideally liberates C-terminal amino acids in a sequential fashion from the peptide structure, while the hydrazine reagent forms a hydrazine derivative from all the amino acid residues involved in peptide linkages except the C-terminal amino acid.

Free hydrazine derivatives of all amino acids *except* the C-terminal residue of the peptide.

The C-terminal amino acids produced by carboxypeptidase or hydrazinolysis can be separated from the remainder of the reaction products and identified by common methods of amino acid analysis.

After identification of the N-terminal and C-terminal amino acids, a new peptide or protein sample is used for structural study. This new sample is selectively cleaved into smaller fragments through the action of other enzymes or reagents that attack the carboxyl side of certain amino acids engaged in a peptide linkage. *Trypsin* cleaves the carboxyl side of peptide linkages involving lysine or arginine; *chymotrypsin* and *pepsin* cleave the carboxyl side of bulky nonpolar amino acid residues such as phenylalanine, tryptophan, tyrosine, or several others, while *cyanogen bromide* (CNBr) cleaves those peptide bonds where the carboxyl group belongs to methionine. In the last case, methionine's presence as an internal amino acid residue of the peptide is detectable as the conversion product of methionine to a homoserine lactone.

Methionine residue contained within a peptide structure

Homoserine lactone produced as C terminus of peptide after CNBr reaction

Amino acyl peptide
+
CH_3SCN
Methyl thiocyanate
+
Br^-

In general it is worth noting that selective fragmentation of peptides "with a single reagent" or "enzyme" cited above will usually yield *one more fragment than the actual number of specific amino acid peptide linkages that are broken*. For example, a peptide containing five methionine residues will yield six peptide fragments after reacting with CNBr.

Overall peptide structure A–L containing five methionine residues

Peptide cleavage at site of five methionine residues yields six peptide fragments.

	A–B	C–D	E–F	G–H	I–J	K–L
Fragments	#1	#2	#3	#4	#5	#6

Many other chemical reagents and enzymes including *aminopeptidase, thermolyase,* and others have been widely implemented for the selective fragmentation of protein structures, but those methods cited above have received the most notable application in the historical investigation of proteins and peptides.

Determination of internal amino acid sequence: Small peptides produced by chemical or enzymatic peptidolysis must be individually segregated using any one of the quantitative analytical methods, discussed elsewhere in this chapter, before further study of the peptide-linked structure can be undertaken. Typical isolation procedures for these peptides include *ion-exchange, electrophoresis, ultracentrifugation, HPLC, molecular exclusion chromatography, isoelectric focusing,* or some other combination of methods.

Assuming that an isolated peptide fragment contains approximately 12 to 20 amino acid residues, these amino acids can be sequenced by using the Edman reagent (phenylisothiocyanate). Although this reagent is also used for the singular determination of the N-terminal amino acid residue on a peptide, its controlled reaction with an isolated peptide leads to a stepwise degradation of the amino acids comprising the peptide, starting at the N-terminal amino acid. The sequential production of PTH–amino acid derivatives according to their occurrence in the peptide, coupled with the identification of the PTH–amino acid derivatives using standards, serves as a basis for establishing the amino acid sequence in the overall peptide structure. One round in the application of the Edman reagent to amino acid sequence determination for a generalized peptide structure appears in Figure 2.15.

After each peptide fragment obtained from a specific fragmentation enzyme or reagent has had its amino acid sequence determined, another sample of the original peptide or protein is subjected to hydrolysis at selected points other than those in the previous effort. These peptide fragments are then sequenced using the Edman reagent along with any other analytical approaches necessary for deciphering C- or N-terminal amino acid residues.

The repetitive fragmentation of a polypeptide structure and amino acid sequence determination within each of the resulting fragments permits the tabulation of numerous sequenced peptide fragments. Once the tabulated amino acid sequences of peptide fragments have been matched with each other, regions of overlapping amino acid sequences serve as a basis for elucidating the primary structure of the original peptide

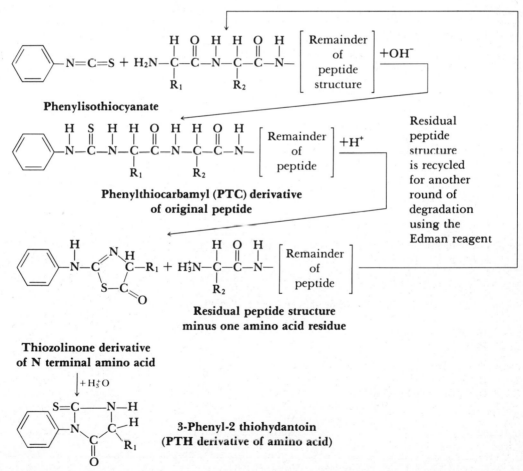

FIGURE 2.15. Application of the Edman reagent to a peptide structure results in its reaction with the N-terminal amino acid of the peptide and produces a PTH derivative of the amino acid, which can be identified.

structure. This traditional approach for peptide sequencing has been illustrated in Table 2.5 for a decapeptide. Figure 2.15 also details the key steps necessary for the Edman reaction, which is used in step 3 of the decapeptide sequence determination in Table 2.5.

Conformational Structures of Proteins and Their Analysis

Conformational structures for proteins: The structures for proteins are infinitely variable and complex, but despite this com-plexity, most proteins can be superficially categorized according to their individual *primary, secondary tertiary,* or *quaternary* structures.

Primary structure: Every protein is ultimately constructed according to the dictates of DNA-directed protein biosynthesis. This genetic control is maintained in proteins at their lowest level of molecular organization, namely, the sequence and number of amino acids in their structures. The sequence of peptide-bonded amino acid residues is generally referred to as the protein's *primary structure*.

Table 2.5. Elucidation of the Amino Acid Sequence for a Decapeptide

Structure for original decapeptide
H_3^+N–Gly–Ala–Lys–Val–Met–Phe–Lys–Met–Val–Ala–COOH
Analytical logic for establishing the decapeptide's structure

1. Hydrolysis and total amino acid analysis of the decapeptide using $6N$ HCl yields
   (2 Ala, Gly, 2 Val, 2 Lys, 2 Met, Phe)[a]

2. N-Terminal (N-T) and C-terminal (C-T) amino acid residues using the Sanger and the hydrazine reagents, respectively, produce the following information:
   N-T: H_3^+N–Gly–[b](2 Ala, 2 Lys, 2 Val, 2 Met, Phe)
   C-T: (H_3^+N–Gly, Ala, 2 Lys, 2 Val, 2 Met, Phe)–[b]Ala–COOH

3. Trypsin fragmentation of the decapeptide along with N-T and C-T determinations as in step 2 yield
   T_1: H_3^+N–Gly–(Ala)–Lys-COOH
   T_2: H_3^+N–Val–(Met, Phe)–Lys-COOH
   T_3: H_3^+N–Met–(Val)–Ala-COOH

4. Edman degradations of three trypsin fragments formed in step 3 indicate the following amino acid sequences:
   T_1: H_3^+N-Gly–Ala–Lys-COOH
   T_2: H_3^+N-Val–Met–Phe–Lys-COOH
   T_3: H_3^+N-Met–Val–Ala-COOH

5. Second fragmentation with CNBr for overlapping and N-T and C-T as in step 2
   CB_1: H_3^+N-Gly–Ala–Lys–Val–Met[c]
   CB_2: H_3^+N-Phe–Lys–Met[c]
   CB_3: H_3^+N-Val–Ala-COOH

6. Assemble data from all steps and deduce sequence
   T_1: H_3^+N-Gly–Ala–Lys
   CB_1: H_3^+N-Gly–Ala–Lys–Val–Met
   T_2:   Val–Met–Phe–Lys
   CB_2:   Phe–Lys–Met
   T_3:   Met–Val–Ala-COOH
   CB_3:   Val–Ala-COOH

H_3^+N–Gly–Ala–Lys–Val–Met–Phe–Lys–Met–Val–Ala-COOH
Deduced amino acid sequence of the decapeptide

[a]Parentheses and commas denote that the sequence of amino acids is not known.
[b]When dashes are used, it means the amino acid sequence is known.
[c]Detected as homoserine lactone at C terminus of fragment.

Using methods of amino acid residue analysis for polypeptides such as Edman degradation and others, it is clear that functionally similar proteins within a species demonstrate similar primary protein structures. Minor variations in amino acid sequences may occur in parts of molecules that are not critical to protein function, but the sequence of virtually every amino acid in the primary structure is governed one way or another by DNA-directed processes of protein synthesis.

Although the primary protein structure has now been established for many proteins, a feasible experimental rationale for elucidating the complete amino acid sequence of proteins was not introduced until the 1950s by Frederick Sanger. Sanger's work outlined the primary structure of bovine insulin, an accomplishment that has since served as a historical benchmark for other protein sequencing efforts.

As shown in an earlier section, the α-carbons on any two consecutive amino acid residues involved in a peptide bond necessarily lie in a single plane, but the plane of the peptide bond can undergo bond rotation around the C_α—N (ϕ) and C_α—C (ψ) axes. Regardless of axial rotation possibilities, the rigidity of the peptide bond is uniformly maintained, and this ensures that a distance of 3.6 Å will be consistently retained between two consecutive α-carbons in the primary structure.

Secondary structure: The primary structure of a protein may assume *a variety of different conformational forms* depending on the types of R groups represented among its constituent amino acid residues. These conformational forms of primary structures account for the *secondary structures* of proteins. The secondary structures can be further defined according to whether or not they display *helical coiling* around a central axis or a *pleated sheet* structure.

The α-helix. The helical coiled forms of primary protein structures are often described

in terms of an α-helix. The discovery of this structure is a result of the X-ray diffraction studies of proteins conducted during the late 1940s by L. Pauling and R. Corey. The coiled structure of the α-helix typically displays 3.6 amino acid residues for every turn in its coiled structure (Figure 2.16). This form of protein structure is also defined as a *right-handed helix* since it can be visualized by clenching one's right hand into a fist while having the thumb extended. The projection of the thumb, pointing to the left, corresponds to the direction of the end of the coiled primary structure. The other clenched fingers in the fist correspond to the *counterclockwise* yet upward coiling of the helical structure.

The α-helix also shows a predictable 5.4 Å rise for each turn of 3.6 amino acid residues along the vertical axis of the coiled structure. These size dimensions coupled with the planar nature of the peptide bond result in a coiled secondary structure. This structure is stabilized by spontaneous formation of intrachain hydrogen bonds between the carbonyl oxygen of each peptide bond and the amide hydrogen of the peptide bond situated four residues away. Hydrogen bonding of this type stabilizes the α-helix provided that other amino acid residues within the secondary coiled structure do not generate forces of attraction or repulsion that exceed the attractive forces of hydrogen bonding. Certain amino acid residues such as proline or hydroxyproline in particular contain an α-N that is part of a rigid ring. Peptide bonding of these amino acids invariably causes sharp bends in the α-helix and detracts from its idealized coiled structure. The coiled conformation of the helix can also be destabilized or prohibited from forming if certain other amino acids consecutively occur two or more times within a polypeptide sequence. Consecutive aspartyl and/or glutamyl residues serve to *destabilize* the α-helix because the close proximities of their negatively charged carboxyl groups generate forces of repulsion. Forces of repulsion can also occur if successive amino acid residues happen to be arginine and/or

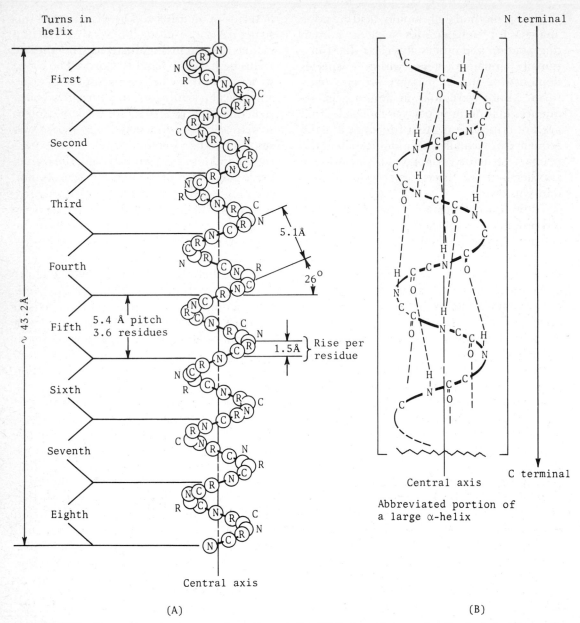

FIGURE 2.16. Two schematic representations of an α-helix. (A) Each residue contained in the helix shows a rise of 1.5 Å and one turn on the structure is completed for every 3.6 residues, which produces a pitch of 5.4 Å. (B) The α-helix is stabilized by hydrogen bonding between —C=O and an —NH group located four residues below.

lysine. Destabilization of the helix can also be encouraged by the close proximity of amino acid residues that display bulky R groups such as isoleucine, threonine, or valine that are characterized by structural branching on the β-carbon. Serine, on the other hand, destabilizes the α-helix for other reasons. Destabilization in this case reflects the ability of a seryl hydroxy group to readily form hydrogen bonds with water molecules that may surround the secondary structure.

Although bulky R groups notably destabilize the α-helix, the presence of amino acid residues having small R groups do not invariably convey added stability to the coiled structure. Glycine residues in particular have only a hydrogen atom in the place of the R group on other amino acids, yet the small size of the hydrogen atom allows for a very wide range of structural rotation about the α-carbon. This wide range of structural rotation ultimately serves to destabilize overall helical structure.

β-*Pleated sheet structures.* In addition to the α-helix structure described by Pauling and Corey, further studies indicated the existence of other secondary protein structures known as β-pleated sheets. These protein sheets are constructed of parallel or antiparallel polypeptide chains that are held together by hydrogen bonding (Figure 2.17). The β-pleated secondary structures are favored especially when the primary amino acid sequence of a protein contains repeating sequences of amino acids having small R groups (e.g., alanine, glycine).

Tertiary structure: Tertiary levels of protein structure are a result of *primary or secondary structures that have become intramolecularly folded.* The architecture of the tertiary structure is thermodynamically favored and quite stable owing to forces of attraction and repulsion between various R groups on amino acid residues. These forces arise from the close proximity of R groups that are engaged in electrostatic interactions, hydrogen bonding, hydrophobic interactions of nonpolar side chains, dipole–dipole interations, and disulfide linkages (Figure 2.18).

The conformations of tertiary protein structures may appear quite random, but their complex three-dimensional forms reflect the DNA-directed assembly of individual amino acids held in a primary amino acid sequence. The genetically prescribed locations of amino acid residues necessarily increase the probability that R groups on certain amino acid residues will interact to produce forces of attraction or repulsion. These forces then interact to stabilize precise conformational shapes of biochemically active proteins. Thus, the precise tertiary conformational shapes of proteins are inherently tied to the primary amino acid sequence of the protein.

Another interesting feature of compact tertiary structures centers around the containment of hydrophobic R groups within the central region of the protein molecule while polar R groups are directed toward the protein's surface. This structural behavior is important because the polar R groups have a tendency to increase the water solubility of the protein while ensuring that water will be excluded from the internal reaches of the molecule by internal hydrophobic R groups. The exclusion of water from the inside of the protein structure does not have great importance by itself, but it should be recognized that the biochemical activity for many of the most important proteins occurs within a very limited portion of the whole tertiary molecular structure. This particularly reactive portion of the protein is generally referred to as its *active site.* The active site may be responsible for promoting biocatalytic activities displayed by enzymes; respiratory gas transport exhibited by proteins such as myoglobin; biochemical transport mechanisms of metabolic intermediates, or other functions that simply require the maintenance of a hydrophobic environment.

Allosterism and tertiary structure. Allosterism describes the ability of some proteins to display graded degrees of biochemical activity caused by changes in their tertiary conformational structures. Conformational changes in a tertiary structure can be incited by one or more small *effector molecules* that bind to some site on the protein quite apart from that responsible for its biochemical activity. Once bound to the protein, the effector concomitantly *causes* alterations within the existing tertiary protein structure. These alterations may reflect the *formation or*

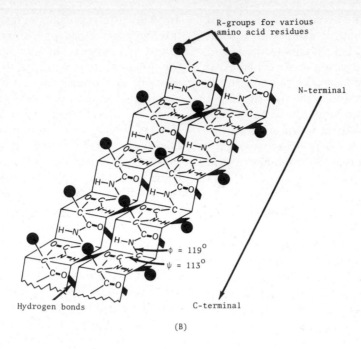

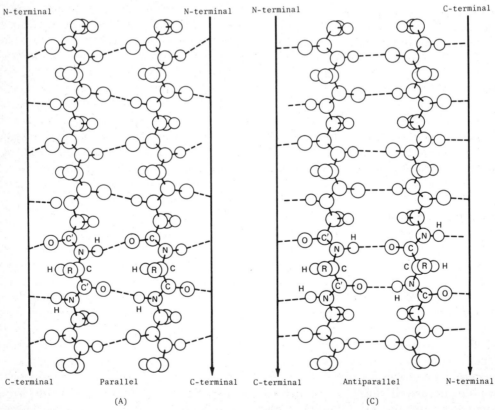

FIGURE 2.17. Molecular structures for two types of β-pleated sheet structures found in proteins: (A) and (B) show two views of a parallel pleated sheet; and (C) shows the secondary structure for an antiparallel sheet.

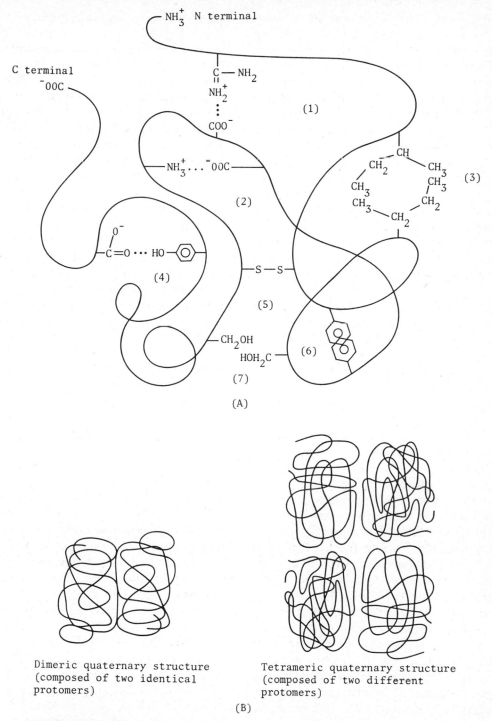

FIGURE 2.18. (A) Tertiary structure for a protein stabilized by (1,2) electrostatic interactions (salt links); (3,6) hydrophobic interactions; (4) hydrogen bonding; (5) covalent disulfide bonds formed from the oxidation of two cysteine residues; and (7) dipole–dipole interactions. (B) Quaternary structures for proteins composed of two and four protomers.

interruption of electrostatic interactions between key amino acid residues held in a primary amino acid structure, as well as an *increase or decrease in certain bond strains* within any existing conformation. Depending on the nature and extent of these interactions, the protein will assume a new thermodynamically preferred conformation that either enhances or diminishes its biochemical activity. Substances that enhance the biochemical activity of allosteric proteins are known as *positive effectors,* and those that decrease biochemical activity are *negative effectors.* The concept of protein allosterism in tertiary conformations has provided a useful basis for explaining the graded degrees of kinetic activity displayed by regulatory enzymes.

Quaternary structure: The noncovalent association of two or more tertiary protein structures (*protomers*) forms the descriptive basis for quaternary levels of protein structure. If the protomers are identical molecules, the quaternary structure is said to be *homogeneous,* while dissimilar protomers form a *heterogeneous* quaternary structure.

The cohesive forces responsible for maintaining quaternary structures are principally the same as those that stabilize tertiary conformations. These forces typically originate from hydrophobic and electrostatic interactions between the various protomers.

The complexity of quaternary protein structures parallels their biochemical functions and fundamental importance in many of the most critical metabolic pathways. Among all of the levels of protein structure, none are more important than the allosteric proteins that display a quaternary structure.

Unlike tertiary levels of protein structure where a single molecule displays intramolecular allosteric behavior, allosteric changes on one protomer of an allosteric quaternary protein are translated into conformational changes in other protomers. This conformational translation is *communicated* to other protomers in the quaternary structure as a result of *alterations in hydrophobic or electro-static interactions* existing between protomers. These conformational changes are not the result of random interactions among various R groups in the quaternary structure. Instead, studies of hemoglobin have shown that conformational changes in its four protomers involve the formation or disruption of eight predictably located electrostatic interactions (*salt links*) that exist among the four protomers—α_1, α_2, β_1, and β_2 (Figure 2.19).

As a result of structural interactions such as these, it is now clear that spatially distinct events involving one protomer of a quaternary structure can affect the conformations of its associated protomers. The consequences of this intermolecular "communication system" are extremely important since it provides a mechanism whereby the presence or absence of metabolic intermediates can control the performance of essential regulatory (allosteric) proteins and enzymes.

Fibrous and globular protein structures: The primary to quaternary hierarchy of protein classification is useful for describing the molecular conformations of proteins, but their physical forms can also be classified as *fibrous* or *globular.*

Fibrous proteins are noted especially for their tensile strength and water insolubility. The physical strength of these proteins is derived from extensive lateral cross-linking of primary and/or secondary levels of protein structure. Several types of bonding interactions may be possible, but hydrogen bonding and disulfide linkages are most important.

Classic examples of fibrous proteins include α-keratin present in hair, fur, hoofs, nails, feathers, and other integumentary structures. The study of X-ray diffraction data has revealed that the main structural unit of α-keratin consists of three right-handed α-helices that are jointly "twisted" into a left-handed coil to form a *protofibril.* The protofibril structure is stabilized by disulfide linkages, but the progressive parallel-type aggregation of protofibrils into larger

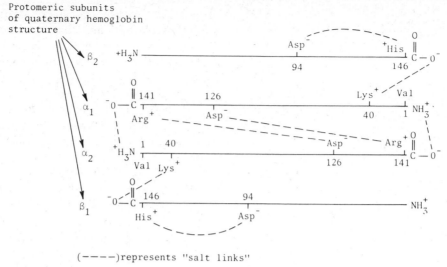

FIGURE 2.19. Schematic illustration of the eight electrostatic interactions that unite the four protomers in a tetrameric (four-subunit) hemoglobin molecule. Based on this type of a linkage system, an allosteric change in the tertiary conformation of a single protomer can be relayed to the other protomers, thereby altering their conformational states.

macrofibrils present in wool and other fibrous proteins is stabilized by hydrogen bonding phenomena.

The application of moist heat to α-keratin can disrupt its structure and promote the formation of β-keratin. The forces of hydrogen bonding that formerly stabilized the α-keratin protofibrils are broken and the keratin assumes the secondary conformation of a parallel β-pleated sheet.

Silk, on the other hand, represents another type of fibrous protein structure which has an *antiparallel β-pleated sheet structure*. As in the case of many β-pleated sheet structures, the presence of amino acid residues with small R groups prove to be especially critical to the overall structure. Unlike keratin, hydrogen bonding interactions among repeating structural units of (Gly–Ser–Gly–Ala–Gly–Ala)$_n$ act to stabilize the fibrous antiparallel structure of silk since cysteinyl residues are not present.

The structure of *tropocollagen* provides still another insight into the structural architecture of fibrous proteins. As mentioned in an earlier section, proline, hydroxyproline, and glycine do not favor the formation or stabilization of the α-helix. Nevertheless, the regular repetitive occurrence of these three amino acid residues in the polypeptide structure of tropocollagen account for 50% of its total amino acid residues (25% glycine and 25% combined proline and hydroxyproline). The critical placement of these amino acid residues encourages the unique helical structure of tropocollagen, which displays three intertwined left-handed helices stabilized by hydrogen bonding. This structure for tropocollagen naturally undergoes further intertwining with similar structures in the form of a *right-handed super-helix* that characterizes collagen. The structural organization demonstrated in collagen is typical of animal connective tissues, cartilage, and portions of bone structure. Moreover, the fibrous level of protein structure is also responsible for the toughness of dietary meats and associated myofibrillar proteins. These protein sources often show graded degrees of increased toughness as an animal ages because of increased amounts of cross-linking among collagen molecules (Figure 2.20).

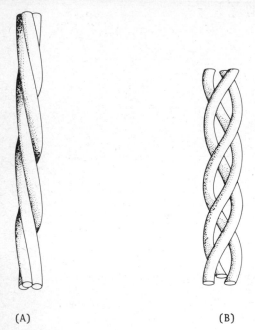

(A)   (B)

FIGURE 2.20. Portions of typical structures for the tertiary conformations of (A) keratin, which has three α-helical coils, and (B) triple-stranded helix found in tropocollagen.

The functions of fibrous proteins are limited primarily to the maintenance of tissue structure in animals apart from the fact that they can also provide a physical barrier between the animal and its environment. Globular proteins, which represent the other major division of protein classification, have relatively few structural functions. Their importance rests on their ability to carry out a wide range of very specific biochemical reactions. The reaction specificity and function of each globular protein is determined by its unique integration of primary, secondary, and tertiary levels of protein structure.

Globular proteins may include those proteins that are responsible for electron transport mechanisms (e.g., cytochrome c); serum albumins; glycoproteins; immunoglobulins (e.g., antibodies); globular protein hormones (e.g., adrenocorticotropic hormone, glucagon, insulin); enzymes; respiratory gas transport proteins (e.g., hemoglobin, myoglobin); and many other proteins having basic

importance to life processes. Many of the globular proteins are characterized by the presence of nonpolar R group amino acid residues directed toward the inside of their compact tertiary conformation, while the polar R group residues are superficially oriented in order to permit hydration of the molecule. The functions of globular proteins studied to date seem very obvious once their individual structures and conformations have been elucidated, but it is absolutely impossible to project the function of a globular protein based on theorized combinations of primary, secondary, or tertiary conformational interactions.

Analysis of protein conformational structures: The conformational structures of proteins and their helicity properties are not readily obvious from routine methods of chemical analysis. Instead, the elucidation of protein structure requires the use of sophisticated analytical methods including *X-ray diffraction, optical rotatory dispersion,* and *circular dichroism.*

X-Ray diffraction analysis. X-ray diffraction analysis has provided detailed knowledge about the three-dimensional structure of proteins. A monochromatic X-ray beam is aimed at a pure crystalline molecule of a protein, which can be accurately adjusted in any of three spatial dimensions (precessed). Interactions of X rays (λ = 1.54 Å) with the array of atoms in the crystalline planes scatter the X rays. Photographic intensity records of X-ray diffraction spots (caused by scattering centers) are then analyzed on the basis of the Bragg equation where $n\lambda = 2d \sin \theta$ (Figure 2.21).

As seen in this figure, this equation links crystal layer spacing (d) to the angle θ of the incident and reflected X ray; λ is the wavelength of the incident beam, while the order of reflection (n) assumes integral values.

According to key diffraction angles (θ) over the crystalline planes, a basis is established for generating an *electron density map* or a *Fourier synthesis.* Fourier synthesis provides a

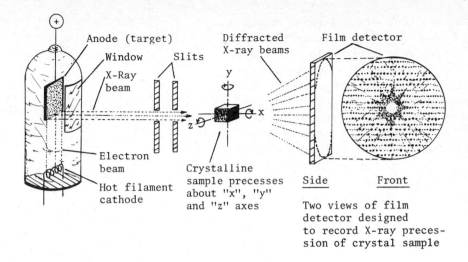

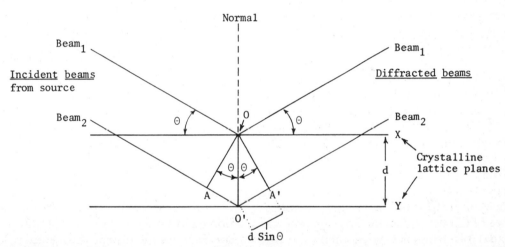

FIGURE 2.21. (Top) Illustration showing the important features of an apparatus designed to record the diffraction of X rays by crystals of pure proteins or other cyrystallizable substances. The intereaction of monochromatic X ray beams 1 and 2 are detailed in their interaction with the equidistant parallel planes of a crystal structure (bottom). Note that the X-ray beam impinging on the lattice plane at angle Θ is reflected from the plane at an equal angle. The distance $AO'A'$ is the extra distance the X-ray beam 2 must travel over that of the X-ray beam diffracted by the first crystal plane (X). The angles of AOO' and $O'OA$ are both equal to Θ. If the distance $AO' + O'A' = n\lambda$, where n is an integer, then the scattered radiation will be in phase at OA' and the crystal will appear to reflect the X radiation. Since $AO' = d \sin \Theta$ and $O'A' = d \sin \Theta$, where d is the interplanar distance of the crystal, $n\lambda = d \sin \Theta + d \sin \Theta$ or $n\lambda = 2d \sin \Theta$ (Bragg equation).

three-dimensional representation of electron distribution about the individual atoms that make up molecules in the crystal structure.

Apart from the highly resolved X-ray diffraction patterns provided by 1.54-Å X rays, the complexity of thcsc patterns produced by many molecules precludes any meaningful structural insights or interpretation. Thus early investigators recognized that complex molecules such as proteins could be prepared with inclusions of heavy metals (lead or uranium). Since these so-called isomorphous replacements in proteins failed to distort the protein molecular shape, the heavy metal also served as a key benchmark in the orientation of any diffraction pattern that was demonstrated. Spatial perspectives permitted by isomorphic replacements have allowed not only the accurate determination of individual amino acid residues in complex structures, but also key spatial understanding of complex secondary through quaternary features of molecular structures.

Optical rotatory dispersion and circular dichroism: Both *optical rotatory dispersion* (ORD) and *circular dichroism* (CD) represent useful methods for elucidating the secondary conformational structures of proteins. Both methods rely on the interaction of plane polarized light with chiral molecules.

Plane polarized light can be resolved into *two* chirally polarized light beams that display helical paths of opposite chiralities. *One* light beam is polarized in a right-handed sense (R) and the *second* in a left-handed direction (L). When a plane polarized light passes through a medium containing chiral molecules, the left- and right-handed chirally polarized beams exhibit different molar extinction coefficients ($\epsilon_R \neq \epsilon_L$) and different indices of refraction ($n_R \neq n_L$). Measurements of CD depend on the first case, whereas circular birefringence phenomena are tied to differences in the indices of refraction. ORD phenomena reflect the dependence of circular birefringence on wavelength.

CD spectra can be plotted as the difference ($\Delta\epsilon$) between molar absorptivities for left (ϵ_L)- and right (ϵ_R)-handed polarized light as a function of wavelength (λ) or wavenumber. However, current literature practices often report CD spectra in terms of molar ellipticity (θ) (Figure 2.22).

ORD is closely related to CD but ORD spectra depend on differences in the refractive index (Δn) between left (n_L)- and right (n_R)-handed polarized light. Data are usually expressed in terms of specific rotation calculated with respect to solute concentration

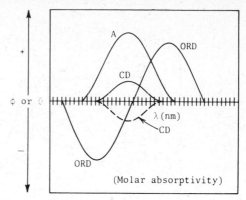

FIGURE 2.22. Comparison of absorption spectrum (A) along with circular dichroism (CD) and optical rotatory dispersion (ORD) spectra. The abscissa corresponds to wavelength (λ) measured in nanometers (nm), while the ordinate can be expressed in terms of *molar ellipticity* (θ) or *molar rotation* (ϕ) according to the following relationships:

$$\theta = 3299\ \Delta\epsilon \qquad (\Delta\epsilon = \text{change in the observed molar absorptivity coefficient})$$

$$\phi = 180\ l\ (n_L - n_R)/\lambda$$

n_L = index of refraction for left-handed circularly polarized light

n_R = index of refraction for right-handed circularly polarized light

l = optical path of sample (often reported in dm)

in moles per liter or molar rotation (Figure 2.22).

With respect to an optically active absorption band for some substance, ORD and CD spectra have the unique theoretical relationships shown in Figure 2.22. The unequal absorption (CD) and unequal velocity of transmission (optical rotation) phenomena illustrated have been recognized as evidence of the Cotton effect.

Cotton effects displayed by simple proteins are attributable to polypeptide absorption bands in the 190 to 250 nm range in addition to the 278 to 295 nm range, which reflects the presence of aromatic side chains of tyrosine and tryptophan. As seen in Figure 2.23, the relative shapes and magnitudes of CD versus ORD spectra are often useful for detecting minor differences in protein conformation or the chemical environments of chromophoric side chains.

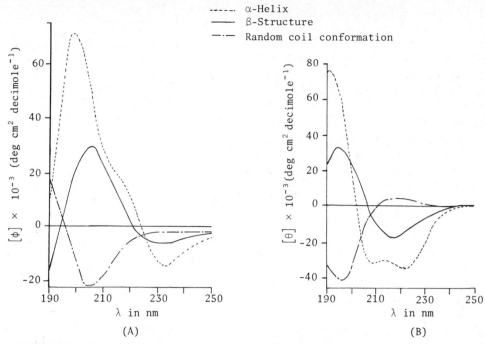

FIGURE 2.23. Comparison of three conformational forms of poly-L-lysine using ORD and CD techniques. Both ORD (A) and CD (B) spectra were obtained under the following conditions: α-helix and β-pleated structure were subjected to pH 11.3, heated at 48°C for 15 min, and then cooled to 22°C; the random coiled conformation reflects the influence of pH 5.5. For precise ORD and CD spectra dealing with poly-L-lysine under more refined conditions, consult Greenfield *et al.*, 1967, *Biochemistry* **6**:1630; and Greenfield and Fasman, 1969, *Biochemistry*, **8**:4108.

Protein Denaturation: The natural conformations of proteins undergo structural changes when treated with acids, bases, reducing agents, oxidizing agents, guanidine hydrochloride or urea, detergents, and/or heat.

Acids and bases affect peptide bonds and alter the ionization state of carboxyl groups on peptide-linked amino acid residues. The main effects of treating proteins with reducing agents involves changes in naturally occurring disulfide linkages. Urea and guanidine hydrochloride may break intramolecular hydrogen bonds since both reagents are very dipolar and readily interrupt the natural patterns of bonding responsible for normal protein conformations. Detergents have a significant effect on protein conformations since they interfere with normal intramolecular hydrophobic interactions, whereas the application of heat to proteins can sever noncovalent and covalent bonds.

The loss of protein conformation due to any of these treatments can ultimately promote the unraveling of the natural protein conformation into a randomly coiled primary protein structure. The progressive destruction of a *native protein*—a naturally occurring protein—into a randomly coiled, biologically inactive protein is referred to as *denaturation*.

The denatured protein is often damaged so severely that it can never regain its original conformation or biochemical function even if the denaturing influence is eliminated. A permanent change in protein conformation is described as *irreversible denaturation*. As a rule, most proteins can be irreversibly denatured quite easily, but the denaturation process can be reversible for some proteins under certain circumstances.

In 1972, Anfinsen showed that the pancreatic enzyme ribonuclease A could be denatured by reducing its four disulfide linkages with β-mercaptoethanol in the presence of guanidine hydrochloride and urea. The denatured enzyme formed a random coiled structure as a result of this treatment (Figure 2.24) but removal of the denaturing agents led to the reestablishment of the original enzyme structure and a significant part of its original enzymatic activity. This example of *protein renaturation* has been cited as firm evidence that the reestablishment of protein conformation and function is a *thermodynamically preferred event* that is determined by the protein's fundamental primary amino acid sequence.

Assuming that the energy differences between the native and denatured states of proteins are not too great, the predominant occurrence of one form over the other can be theoretically projected using the following equation:

$$\Delta G = \Delta G^{0\prime} + RT \ln K_{eq}$$

At equilibrium,

ΔG = 0
$\Delta G^{0\prime}$ = $- RT \ln K_{eq}$
R = universal gas constant
T = absolute temperature
K_{eq} = $\dfrac{[D]}{[N]}$ in the case where
$[N] \underset{K_{eq}}{\overset{}{\rightleftharpoons}} [D]$ and
$[N]$ = native state of protein (genetically specified bioactive conformation)
$[D]$ = denatured state of protein (random opened coil conformation)

The values for $[N]$ and $[D]$ in the equation are not always easy to determine, but an approximate ratio of one protein form to the other can be assessed on the basis of decreased biochemical activity of the protein (e.g., enzyme activity), changes in light absorption at a specific wavelength, viscosity, protein sedimentation constants, crystallization abilities, or other measurable param-

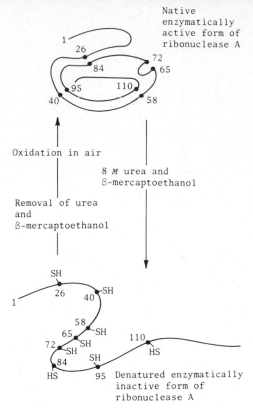

FIGURE 2.24. Denaturation and renaturation of ribonuclease A.

eters indicative of physical damage to protein structure.

Denaturation of proteins by heat. The denaturation of proteins by heat serves many practical purposes in the preservation and processing of foods. The controlled application of heat to raw foods eliminates deteriorative, pathological, or toxigenic organisms, while blanching of vegetables and fruits denatures naturally occurring deteriorative enzymes. In addition, the application of heat to animal and fish proteins destroys the natural enzyme activities of various tissues and improves the palatability and digestibility of these protein sources.

Regardless of the uses for heat in food processing, all heat applications denature structural and enzymatic proteins to some extent. The denaturation of skeletal proteins in livestock and fish markedly improves their food quality and acceptability by eliminating structural connections among native proteins. The denaturation of enzymes has two distinct advantages. First, enzyme denaturation eliminates enzyme-mediated destruction of food colors, textures, and flavors during storage. The second desirable effect of enzymatic denaturation deals with the elimination of food microorganisms. The denaturation of enzymes that are critical for the maintenance of microbial life ensures the annihilation of organisms in fresh or preserved foods. This effect prohibits the possibilities of toxic or pathological consequences due solely to microorganisms when the food is eventually consumed.

The heat used to denature proteins may be in the form of dry heat or moist heat (e.g., $121 + °C$ steam at 15 lb psi). Dry heat is usually less effective for denaturing proteins because it relies on the oxidation of native proteins, contrary to moist heat, which rapidly coagulates proteins. The denaturing effects of moist heat have been widely employed as an important component of traditional food preservation and canning methods, which involve retorting.

Since protein denaturation undermines the ability of all microorganisms to exist, almost any heating method can serve as a weapon controlling food microorganisms. The heat resistance of microorganisms varies widely according to the food's characteristic pH, ionic strength, and viscosity, the presence of lipids, and many other factors. Based on these considerations, a reference microbial strain is often chosen for study to demonstrate microbial survival patterns in heat-treated foods. Reference organisms are usually selected on the basis of their similarity to organisms that may occur naturally in foods and present a realistic threat to food quality or human health if consumed. Foods inoculated with a specific reference strain are ordinarily subjected to a series of heating time/temperature studies in order to assess microbial heat resistance and these data are then expressed as a *thermal-death-time* (TDT)

curve (Figure 2.25). The TDT curve for the reference organism serves as a useful index for assessing microbial survival rates in the food following a specified amount of heat treatment.

Most canning and retorting operations for food preservation select a time–temperature relationship that achieves a 10^{12} log cycle reduction of spore population numbers. Spores of toxigenic bacteria such as *Clostridium botulinum* that have wide technological significance throughout the food industry are not uncommonly used for safety and viability guidelines. Both c and z values obtained from TDT curves serve as useful indexes of a microorganism's heat resistance (Figure 2.25). The c values represent the number of minutes necessary to destroy the reference microorganism at 121°C, while the z value corresponds to the number of degrees Celsius that are required to produce a 1.0 log cycle reduction in the population number of microorganisms. Although vegetative organisms can be used for these studies, the high thermal resistance of spores makes them more desirable for TDT curve indexes.

Problems associated with protein denaturation in foods. The obvious advantages of protein denaturation, including the elimination of enzymatic activity and undesirable microorganisms, can be accompanied by a variety of objectionable side effects that produce unnatural food colors, odors, and textures. The application of heat to proteins can encourage browning reactions of the food, the production of odorous amines, the degradation of disulfide linkages, and the production of volatile sulfur-containing compounds from amino acid residues such as cysteine. Actually, the variety of chemically related food-quality changes are limited only by the severity of the heat treatment in relation to the total protein composition. Although the denaturation of many proteins will occur over the range of 55 to 75°C, many of the most unacceptable food quality problems develop at temperatures in the range of 100 to $121 + °C$.

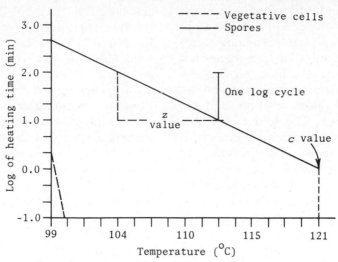

FIGURE 2.25. Thermal-death-time curves depicting the heat resistance of spores and vegetative organisms.

Protein denaturation can also be accomplished by

1. Exposure to *ionizing radiation,* which produces many random free radical reactions involving proteins and their chemical environment.

2. *Freezing,* which crystallizes water of hydration surrounding a protein and causes ultimate protein denaturation by dehydration mechanisms and increased *ionic strength* of the surrounding solvent phase.

3. Mechanically produced heat associated with *food processing* or solar sources that dehydrate the natural water content of proteins.

4. Many other industrial processes including *microwave heating, severe pH fluctuations,* and the application of *extreme pressures.*

Separation Techniques for Complex Protein Mixtures

Industrial uses for food proteins may require their isolation from blended, homogenized, or cellularly disrupted natural sources before they can be used to attain certain nutritional objectives. In other cases, the biochemical characterization of molecular function(s), structure(s), amino acid profiles, and other features of protein structure may require preliminary isolation and purification of a protein before any meaningful data can be obtained.

Regardless of the practical objectives for isolating specific proteins, several properties of proteins act as a basis for many of the most useful protein-isolation methods in the analytical laboratory as well as the industrial preparation of some food protein isolates.

Preliminary Isolation of Proteins from Crude Cellular Extracts: Individual solubility differences for native proteins can be used as a preliminary step in the selective isolation of proteins from biochemically complex aqueous solutions. Solubility differences among proteins are caused by the collective physicochemical behavior of each protein's precise and characteristic amino acid composition. Differences in protein solubility can be readily demonstrated by varying the pH, ionic strength, dielectric constant, and temperature properties of aqueous solvents.

Solvent pH, ionic strength, and protein solubility. If the pH of an aqueous solvent approximates the pH value of a protein that exists as its isoelectric species ($NEC = 0.0$), the protein will precipitate from solution.

This behavior stems from diminished electrostatic repulsions between similar adjacent protein molecules as they assume electrical neutrality, which in turn permits them to coalesce and precipitate from solution. This is in contrast to electrical charges on a protein *above or below its characteristic isoelectric point* where resultant negative or positive electrical charges normally stabilize protein molecules in solution.

The careful selection of minimal protein solubility conditions based merely on the manipulation of solvent pH provides the underlying analytical rationale for all methods of isoelectric precipitation (Figure 2.26A).

The ionic strength of an aqueous solvent also exerts an influence on the solubility of proteins (Figure 2.26B). Low molar concentrations for certain salts can actually improve the solubility of some proteins according to the principle of *salting in*. Salts typically used for this purpose include sodium chloride, potassium chloride, ammonium chloride, magnesium chloride, ammonium sulfate, and some divalent zinc salts. The improved solubility of proteins such as globular proteins is believed to be caused by an increased tendency of protein R groups to ionize and interact with aqueous solvents. The effects of increasing salt concentrations on protein solubility do not increase indefinitely. Instead, at some specific ionic strength for each protein, progressive increases in ionic strength will be met with decreases in protein solubility and protein precipitation. This precipitation of the protein is referred to as a *salting-out* effect. Although many physical factors are at work during salting-out phenomena, the main effect of the high salt concentration is believed to involve removal of the water of hydration from proteins, thereby decreasing their solubility. In most cases it has been recognized that salts of divalent neutral salts are more effective than monovalent salts for many salting-out procedures.

Dielectric constant and temperature effects on protein solubility. Dielectric constant and temperature properties can also influence the solubility of specific proteins in aqueous solutions. Since the protein solubility features of water are tied to its high dielectric constant, the addition of certain water-miscible organic solvents such as alcohols or ketones (e.g., ethanol, methanol, butanol, or acetone) can be used at low temperatures (2.0–5.0°C) to augment decreases in the dielectric constant of aqueous solvents. For protein solutions, a decrease in the dielectric constant inevitably (1) decreases the ability of the solvent to counteract attractive forces between electrically charged solutes; (2) lessens the degree of R-group ionizations on the protein; and (3) increases the probability of protein aggregation at the molecular level, which may be followed by precipitation of protein molecular aggregates.

Solvent temperature has a notable influence on protein solubilities, but temperature regulation alone is not well suited as a sole protein fractionation method. The most effective use of temperature for protein purification lies in conjunction with joint manipulations of temperature, dielectric constant, pH, and/or ionic strength of solvent media that lead to the selective precipitation of proteins.

In general terms, protein solubilities usually increase over 0 to 40°C but temperatures of 40 to 50 + °C usually cause increased destabilization and denaturation of proteins—especially in the neutral pH region (Figure 2.26C).

Depending on the type of proteins being fractionated, all isolation methods for proteins can inadvertently lead to some denaturation and losses in the native protein conformation; however, temperatures of 2 to 3°C often minimize many of these isolation side effects.

Analytical Characterization of Individual Proteins: Aside from the preliminary methods for isolating proteins cited above, more rigorous and reliable techniques are required for the analytical purification and characterization of many proteins. These methods include column chromatography

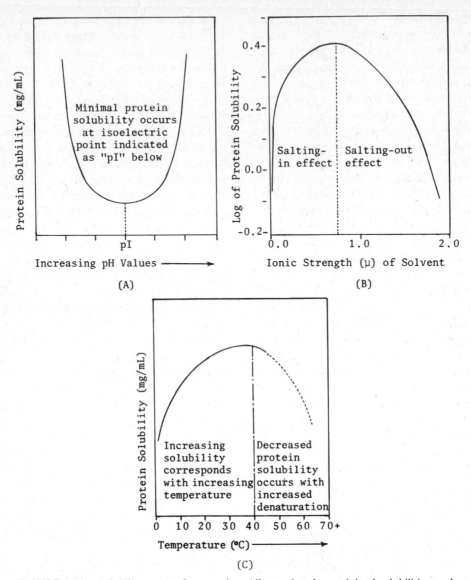

FIGURE 2.26. Solubility curves for proteins. All proteins show minimal solubilities under specific (A) pH, (B) ionic strength, and (C) temperature conditions of the aqueous solvent medium. These conditions are usually different for dissimilar proteins and, therefore, careful selection of minimal protein solubility conditions can be used to aid the preliminary isolation of proteins from biochemically complex mixtures. The solubility responses of proteins increases with increases in the ionic strength of the aqueous solvent (salting in) up to a point where their solubility markedly decreases (salting-out).

on modified celluloses such as diethylaminoethyl (DEAE)-cellulose (an anion exchanger) and carboxymethyl cellulose (CMC) (a weak cation exchanger); specific adsorption of proteins from biochemically complex mixtures to barium sufate, various processed aluminum oxides, calcium phosphate (hydroxyapatite), or charcoal; and a host of more sophisticated analytical purification and identification methods afforded by electrophoresis, ultracentrifugation, molecular exclusion, and affinity chromatographies.

The stepwise and progressive isolation of a protein corresponds to some measure of its increased purity or biochemical activity. This assumes that protein denaturation has been held to a minimum. A measure of a single protein's biochemical activity per unit weight in a protein isolate (measured according to total weight) is generally reported as the protein's *specific activity*.* The specific activity for a protein may be reported in terms of enzyme activity, enzyme inhibition, or some other analytically significant feature of the protein when it is studied under specified *in vitro* or *in vivo* assay conditions.

Electrophoresis of amino acids, peptides, and proteins: Amino acids, peptides, and proteins display positive, negative, or neutral electrical charges depending on the pH of their chemical environment. Negatively charged species will migrate toward a positively charged anode; positively charged species migrate in the direction of the negative cathode; and electrically neutral structures will remain stationary if they are placed in a direct current. The mobility of charged molecules in an electrical field represents the operational principle for many electroanalytical procedures generically referred to as *electrophoresis*.

Exclusive of many physical variables, the electrophoretic mobility (μ) for an ideal spherically charged particle can be described by

$$\mu = \frac{Q}{6\,\pi\eta r}$$

where

μ = electrophoretic mobility
Q = *NEC* on the particle
r = particle radius in cm
η = viscosity of the liquid medium where electrophoresis occurs

*The specific activity term is not relegated strictly to biochemical descriptions of protein activities and can be used to describe purity levels for certain vitamins, pharmaceutical substances, and many other bioactive compounds.

Assuming that the electrically charged particle does not become engaged in any strong electrostatic interactions with other charged species during its migration in an electrical field, the particle movement will be determined on the basis of its charge-to-size ratio (Q/r). The greater the Q/r ratio for a particle, the greater its rate of electrophoretic mobility. If two or more particles happen to have the same sizes and dimensions, then electrophoretic mobility is governed only by the individual electrical charges of each particle.

Methods of electrophoresis generally fall into two broad categories called *moving boundary* and *zonal* methods.

Moving-boundary electrophoresis, described by Tiselius in the 1920s, represents the simplest form of an electrophoretic method, and its operational principles serve as the underlying basis for more recent methods. Using moving-boundary electrophoresis, Tiselius demonstrated that simple mixtures of charged particles (i.e., charged solutes) contained in an electrolyte or a buffer solution could be separated in the liquid medium. Solute separations using the method were far from perfect; however, the movement of charged particles was clearly evident as a *moving boundary* (Figure 2.27). A pure electrically charged solute contained in an electrolyte solution showed only *one* "boundary" between the solvent and the charged solute species migrating in the direct current. If a mixture of two different charged solute species was subjected to direct current in the solvent, then *two* distinct "boundaries" would be shown. Moving-boundary methods have little application to current practices of electrophoresis since complete partitioning of proteins is rarely possible. If complete partitioning is achieved, there are many practical difficulties encountered in recording the electrophoretic migration patterns of the charged solutes. Furthermore, it is even more difficult to physically isolate one electrophoretically pure substance from another without completely disturbing the whole electrophoresis cell.

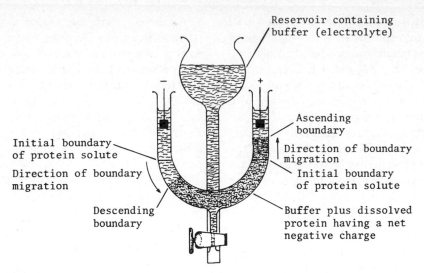

FIGURE 2.27. Moving-boundary electrophoresis of a negatively charged protein showing the direction of its migration when placed in a direct electrical current. The development of early electrophoresis techniques by Tiselius and others employed this fundamental apparatus for moving boundary electrophoresis and served as a basis for the future development of zonal methods (*Diagram note:* Apparatus consists of a U tube connected to a buffer reservoir through a stopcock path that is not shown.)

Methods of *zonal electrophoresis* occur on a fixed solid support that is saturated with the aqueous electrolyte or buffer solution as opposed to the electrolyte solution used for moving-boundary methods. The foremost advantage of the solid support rests on its ability to prohibit mechanical disturbances and convection currents within the electrolyte solution. These phenomena easily occur in moving-boundary electrophoresis cells and inevitably lead to poor separations of solute mixtures. The solid support is also important because it can be manufactured to have specific molecular adsorption or molecular sieving properties. The electrically induced migration of charged solutes, coupled to their individual interactions with different solid supports, have been extensively used for selectively enhancing the separation of certain charged solutes while retarding the separation of others. Not only does this improve the resolution power of electrophoresis methods for separating complex mixtures of charged molecules, but separations become more reproducible and highly quantitative.

All practical forms of zone electrophoresis used for amino acid, peptide, and protein separations can be schematically represented by Figure 2.28. Buffer saturation of the solid support is maintained since each of its ends is immersed in a buffer reservoir. One reservoir remains in contact with the anode while the other stays in contact with the cathode. The solid support may be situated in a horizontal or vertical position depending on the specific zonal electrophoretic method chosen.

Solid support media for zonal electrophoresis. Typical solid supports include refined and processed cellulose papers, cellulose acetate, and a wide variety of processed carbohydrate (e.g., starch, agarose) or polyacrylamide gels.

Although electrophoresis carried out using paper saturated with buffer characterized the earliest methods of zonal electrophoresis, paper supports have a tendency to inadvertently adsorb many proteins and peptides. This problem led to the develop-

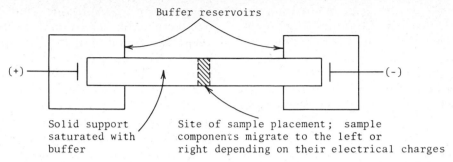

FIGURE 2.28. Schematic diagram representing the essential components of a zonal electrophoresis apparatus.

ment of cellulose acetate supports in which the hydroxyl groups of cellulose responsible for adsorbing electrically charged molecules have been converted to nonadsorbing acetate groups. Cellulose-based electrophoresis supports depend largely on the differential migration rates of charged molecules in an electrical current. The development of starch gels by O. Smithies and polyacrylamide gels by L. Orenstein and B. J. Davis during the 1950s were especially significant because they led to the evolution of highly refined electrophoresis methods. These methods then minimized the limited and traditional dependence on cellulose supports. These gel media also became widely accepted because they permitted the *differential migration* of *charged molecules* in a current *in addition to* partitioning charged molecules on the basis of their *individual molecular sizes*. The porous three-dimensional structure of these gels is responsible for their molecular sieving actions, and this permits the reversible diffusion of the charged molecules "into" or "out of" the gel matrix depending on their size.

Polyacrylamide gels, in particular, have assumed a commanding role in nutritional, clinical, and investigative studies of peptide structures since the porosity of the gels can be carefully regulated by photopolymerized cross-linking of polyacrylamide with *N,N'*-methylene-*bis*-acrylamide. This permits a choice of gels each having different molecular sieving properties. The cross-linked polyacrylamide gels are usually prepared in columns (discs) or slabs.

The ability of polyacrylamide gels to regulate the electrophoretic migration of peptides and proteins based on molecular size is notably shown by methods of sodium dodecyl sulfate (SDS)–polyacrylamide gel electrophoresis. Under pH conditions of 7.0, a 1.0% concentration level of the detergent SDS will bind to most proteins. If the proteins are multichained structures, the SDS leads to their eventual dissociation. Further treatment of the dissociated proteins with 0.1 M mercaptoethanol severs any disulfide linkages contained in the protein or peptide structures. Moreover, this reagent destroys any organized coiled structure (secondary structure) of the bioactive protein. The randomly coiled protein resulting from these steps now displays an electrical charge caused by protein-bound SDS instead of the proteins constituent amino acid residues. Since SDS bound to a unit weight of protein is relatively constant ($\sim$1.4 g SDS/g protein), the electrophoretic mobility of proteins will depend only on their size and ability to migrate through the manufactured polyacrylamide gel. A typical calibration curve for an SDS–gel separation of proteins over a molecular weight range of 13,500 to 68,000 appears in Figure 2.29.

After being separated on any type of solid electrophoresis support, proteins, peptides, or amino acids can be detected by a variety of methods. Common methods include surveying the absorbance of the solid support at particular wavelengths or staining the separated substances with specific dyes or color-producing reactants (e.g., ninhydrin). Radioactively labeled substances may be located on electrophoretic separation supports by radioautography or, alternatively, the support medium may be carefully segmented and each fragment may be counted for its radioactivity using liquid scintillation methods.

Prediction of electrophoresis separation patterns. The direction of migration for amino acids, peptides, and proteins can be predicted with limited degrees of accuracy if the pI value for each structure is known, or if the numbers, types, and pK_a's for all ionizable groups on the structures are known. The net electrical charge (*NEC*) for any amino acid or peptide-related structure is intrinsically related to its immediate environment, and any effort to predict its electrophoretic behavior *must* also account for this variable.

For example, the pattern of electrophoretic migration of simple amino acids can be predicted at a given pH assuming that all factors other than a direct current through an electrolyte solvent have been excluded. If four amino acids including aspartic acid, glycine, histidine, and lysine are placed in the middle of an electrophoresis migration

field (Figure 2.30A) having a buffer pH of 6.0, these amino acids would eventually develop the separation pattern shown in Figure 2.30B. This pattern develops because glycine is nearly electrically neutral at a pH of 6.0, while the *NEC* on aspartic acid is $\simeq$ -1.0; the *NEC* for lysine is $\simeq$ $+1.0$; and histidine displays a *NEC* of $\simeq$ $+0.5$. The direction of virtually any amino acid can be predicted in this way provided that its pI is known along with the pH of the electrophoresis environment. The relative migration for peptides and proteins is more difficult to project since their overall migration is a reflection of all ionizable residues contained in their structures.

Isoelectric focusing. The ability of amino acids, peptides, and proteins to display certain electrical charges based on the pH of the surrounding environment necessarily requires them to be classified as *ampholytes*. Although ampholytes are positively charged at the low end of the pH scale and negatively charged at the high end of the scale, every ampholyte will assume a *NEC* of $\simeq 0.0$ at some pH value that corresponds to its isoelectric point.

Using a method reminiscent of density gradient ultracentrifugation except for the fact that a pH gradient is established between two electrodes instead of a density gradient formed by centrifugal force, ampholytes can be *forced to migrate in a current* until the pH of a pH gradient happens to match the pI of the ampholyte.

This behavior of ampholytes has been analytically exploited for the separation of proteins and peptides using the method known as *isoelectric focusing*. It should be noted here that the electrolyte medium employed between the direct current electrodes in this method is very different from the electrolyte media employed for moving-boundary electrophoresis. For isoelectric focusing methods, a stable pH gradient serves as an electrolytic medium. This medium is prepared from carefully manufactured polymers of aliphatic amino and carboxylic acids as op-

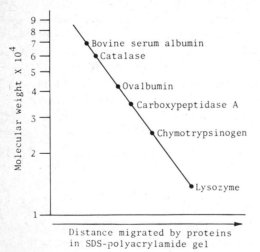

FIGURE 2.29. Molecular weight calibration plot for selected protein standards separated using SDS–polyacrylamide gel electrophoresis.

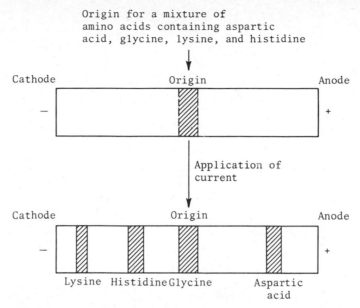

FIGURE 2.30. Anticipated electrophoresis migration pattern for four amino acids subjected to an electrical current in a buffer of pH 6.0.

posed to simple electrolyte salts. These polyampholytes usually display molecular weights over a range of 300 to 650 and have the ability to maintain a pH gradient throughout the pH range of 3.0 to 11.0 (Figure 2.31). One typical polyampholyte used for maintaining pH gradients in isoelectric focusing methods is available as "Ampholine" (LKB Industries, Inc.).

Complex protein mixtures administered to a pH gradient, maintained by polyampholytes such as Ampholine, electrophoretically migrate in the pH gradient until the pH of the medium matches their respective pI values. The equivalency of the protein pI values with that of the polyampholyte pH condition causes the proteins to individually "focus" into narrow bands. Proteins differing in pI values of only 0.01 pH units can be independently resolved using this technique.

Applications for electrophoresis methods. Among the gamut of electrophoresis techniques available for studies of food proteins, clinical nutritional specimens, or food product and quality control efforts, isoelectric focusing is

used far less on a routine basis than other electrophoretic methods. Furthermore, since none of the electrophoretic methods have been developed on a large enough scale to provide purified bulk proteins from crude protein sources, the application of electrophoresis has necessarily been limited to analytical efforts.

The list of past uses for electrophoresis is long, and future uses of the method are nearly unlimited. Past applications have involved its use in cellulose acetate methods for detecting the nature of various thickening or gelling agents; electrophoresis on paper has been used for investigating changes in albumin proteins during egg storage; and both disc and slab polyacrylamide gel techniques have been widely applied to the routine identification and screening of different cheeses (e.g., cow milk versus sheep milk) and fish or animal species. The protein separation reliability and reproducibility of gel methods in these assays has made them especially useful for detecting the illicit adulteration of expensive protein foods with less expensive protein adjuncts. Typical practices of this type would normally include the intermingling of

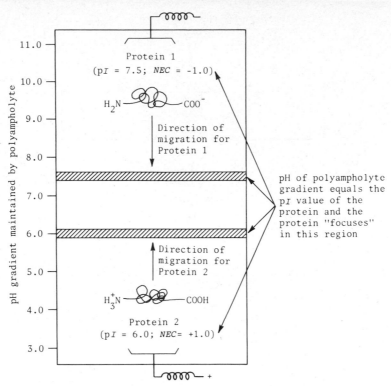

FIGURE 2.31. Examples of two different proteins isoelectrically focused in distinct regions of a polyampholyte pH gradient that corresponds to the pI value of each protein.

langostino with lobster meat, cod meat with processed crab meat, processed skate meat with scallops, or various other bulk fish species with more economically valued species (Coduri and Rand, 1972; Coduri *et al.*, 1979). Figure 2.32 shows a schematic diagram for disc gel separations of various fish proteins according to species as well as the electrophoretic separation rationale for suspecting the adulteration of one processed fish species with another species.

Although the context of electrophoretic methods has been presented here in relation to the analysis of proteins, it should be clearly noted that its uses are not entirely restricted to protein analysis. Other applications of electrophoresis extend to the separation of water-soluble coal tar dyes and certain vitamin and vitamin-like substances, along with many other compounds.

Ultracentrifugation: The application of high centrifugal forces to complex protein mixtures eventually results in their mutual separation based on differences in molecular size, density, and shape. Current applications of this analytical principle are an outgrowth of Svedberg's development of the ultracentrifuge during the 1920s. This instrument has the ability to routinely spin samples at centrifugal speeds approaching $250,000 \times G$. When separations or large protein volumes are necessary as a prelude to more elaborate analytical investigations (up to 50.0 mL), a *preparative ultracentrifuge* may be used. Centrifugation of proteins for other purposes such as determining molecular weights, purity, or density may require the use of smaller sample volumes (1.0–3.0 mL) in an *analytical centrifuge*.

The earliest centrifugation methods achieved separation of substances having different densities using a *uniform density medium* throughout the length of the centrifugal solvent medium. Later developments in centrifuge methods indicated that more ef-

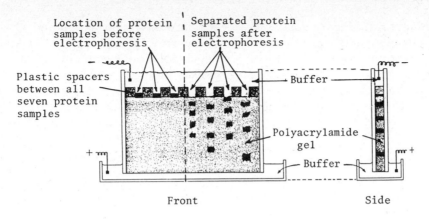

(A) Slab gel electrophoresis

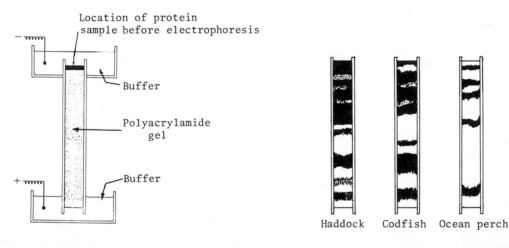

(B) Disc gel electrophoresis

(C) Comparative electrophoretic separation patterns for three different fish species.

FIGURE 2.32. (A) Slab gel electrophoresis: Two views of a slab-type polyacrylamide gel electrophoresis unit depicting the positions of protein sample mixtures before and after electrophoresis. (B) Disc gel electrophoresis: Schematic design for a disc gel polyacrylamide electrophoresis unit. The protein sample mixture is placed on the surface of the gel and all of its individual protein components separate according to their electrophoretic mobilities and sizes. (C) Examples of disc gel electrophoretic separations for fish proteins obtained from different species. The uniform application of slab or disc gel methods to fish and other proteins can serve as a basis for identifying certain animal species, studying food protein adulteration, or assessing changes in food quality.

fective and rapid separations of compounds could be achieved when centrifugation media increased in density with increasing distance from the axis of centrifugal rotation. This discovery spurred the development of *density gradient methods*, which are widely used in current applications of centrifugation.

Density gradients for centrifugation may be routinely prepared by careful layering of different density solutions containing a solute such as sucrose into a centrifuge tube. Alternatively, a density gradient can be formed by centrifugal force applied to a homogeneous solution of a dense inorganic salt such as cesium chloride (Figure 2.33). The sucrose gradients are often used when different protein substances are evaluated on the basis of their respective sedimentation velocities, while the cesium chloride media are used widely for sedimentation equilibrium studies (Figure 2.33).

Sedimentation-velocity (SV) method. According to this method, a preformed sucrose gradient is prepared so that the density of the sucrose is greatest at the bottom of the centrifuge tube. In order to determine the molecular weight of an unknown protein, the unknown specimen is placed on the top of the preformed gradient and is jointly centrifuged with selected known-molecular-weight substances. The known compounds act as future benchmarks for determining the molecular weight of the unknown substance. Prolonged centrifugation of the known standards and unknown protein causes the heaviest compounds to migrate rapidly and reside at the greatest distances from the axis of centrifugal rotation. After centrifugation has ended, the centrifuge tube is removed from the rotor and the stratified compounds are extricated from the tube. The orderly removal of the individually partitioned substances can be carried out by puncturing the bottom of the tube and monitoring the 280-nm absorbance of sucrose gradient fractions as the tube contents drain. This step will provide a molecular weight profile of proteins and standards beginning with the highest molecular substances in the first fractions that drain from the tube. Figure 2.34 shows a typical molecular weight profile developed during the centrifugation of an unknown protein along with reference standards.

Density-gradient sedimentation equilibrium (DGSE) or isopycnic equilibrium. Unlike the previous method, a preformed centrifugation medium is not used with this method. Instead, a dense inorganic salt such as cesium chloride is admixed with substances of standard molecular weight and a protein of unknown molecular weight. This homogeneous solution eventually develops a salt density gradient after a period of centrifugation. Naturally, the greatest salt density appears at the bottom of the centrifuge tube and decreases toward the top. The density gradient established during centrifugation then causes the various unknown proteins and standards to separate from each other depending on their equivalency with different density regions present in the salt gradient. Once the different substances have reached different density equilibrium positions, they can be recovered in the order of their separation as detailed in the previous case.

Molecular weight determination using an analytical ultracentrifuge. Molecular weights for unknown protein substances can be estimated using an analytical ultracentrifuge using either the sedimentation-velocity or the density-gradient equilibrium methods.

The value of the analytical ultracentrifuge lies in its ability to optically follow the sedimentation of substances when they are subjected to carefully controlled centrifugal forces. If two or more substances are significantly different in size, shape, and density, an analytical ultracentrifuge will detect differences in light absorption over the length of the centrifuge cell as different substances become partitioned. The "boundaries" separating substances are usually detected by ultraviolet light absorption over the length of the centrifuge cell, while the centrifuge rotor actually spins at speeds of 25,000 to 40,000+ rpm. The light absorbance can be recorded over the whole length of the centrifugation cell, but it is often more useful to detect the presence of solute separations (i.e., protein separations) on the basis of a derivative plot. This graphic representation is obtained by calculating the changes in light absorption ($d A$) with respect to unit increases in centrifuge cell length ($d x$) for the entire length of the whole centrifuge cell (x) (Figure 2.35).

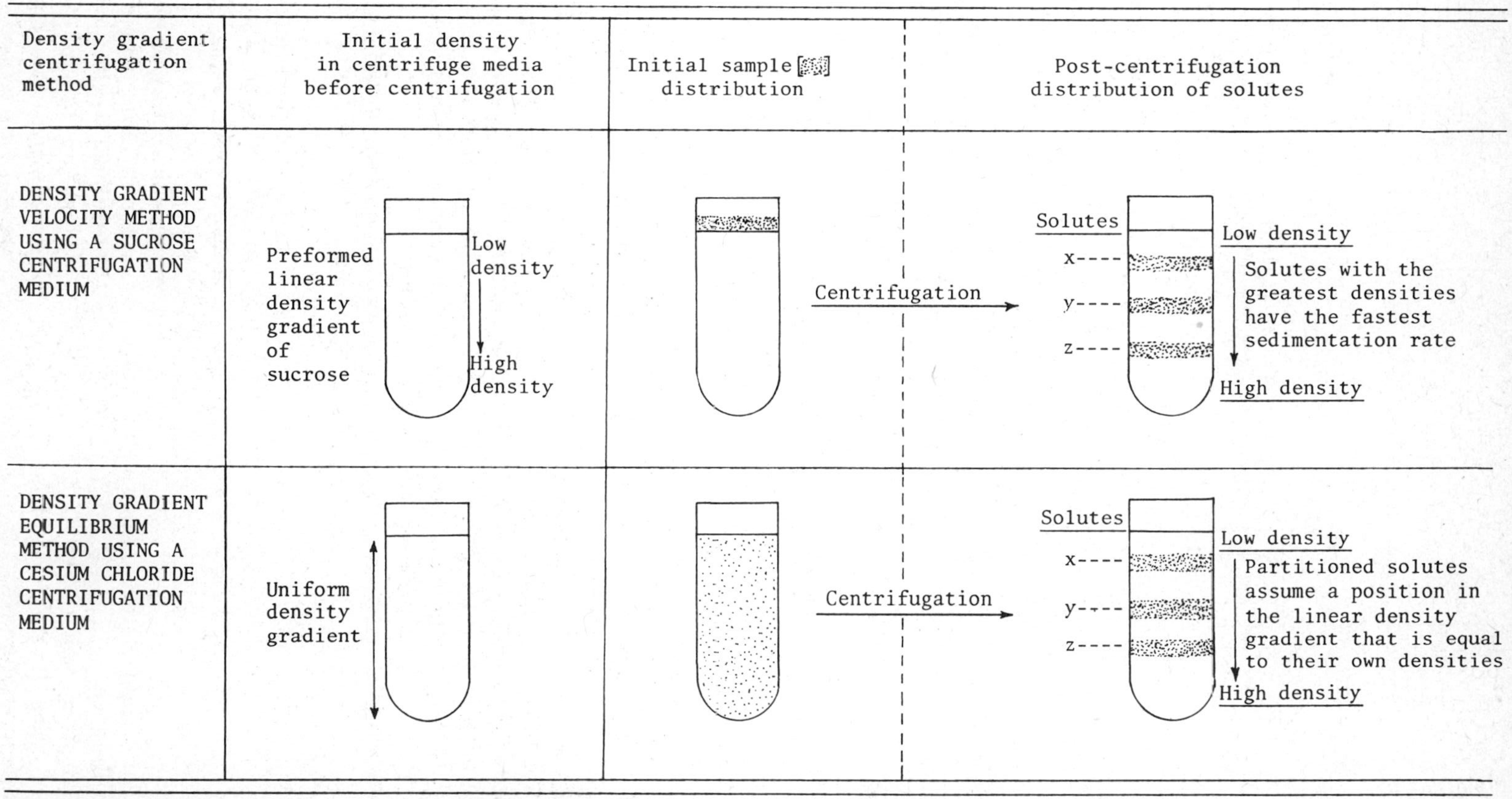

FIGURE 2.33. Comparative sample preparation and postcentrifugation separation patterns for density-gradient velocity and density-gradient equilibrium methods employed in ultra-centrifugation.

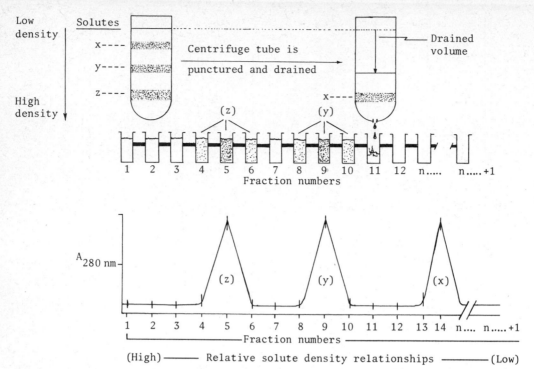

FIGURE 2.34. Elution of ultracentrifugally separated solutes into partitioned volumes by puncturing and draining the centrifuge tube. The joint centrifugation of substances of known and unknown molecular weight followed by careful establishment of an elution profile forms the basis for the comparative assessment of substances of unknown molecular weight.

Derivative plots are especially useful because each peak expressed on the graph corresponds to a boundary separation for a sedimenting solute (i.e., protein) contained in the centrifugation medium. The measurment of the distances moved by solutes over a certain period of time provides sedimentation velocity data that can be used in the following formula for calculation of unknown molecular weights:

$$M = \frac{RTs}{D\,(1 - \bar{V}\rho)}$$

where

M = molecular weight of the unknown solute substance

R = universal gas constant (8.314×10^7 ergs mol^{-1} deg^{-1})

T = absolute temperature (°K)

D = diffusion constant of the solute (e.g., 10^{-7} cm^2 s^{-1} for most proteins)

$\bar{V}$ = partial specific volume for solute expressed as mL g^{-1}; this value is based on the volume that results from adding 1.0 g of solute to a volume of solvent; the partial specific volume for most proteins is ~0.70–0.75 mL g^{-1}

ρ = density of solvent medium in g mL^{-1}

s = sedimentation coefficient defined as $(d\,x/d\,t)/\omega^2\,x$, where

x = distance of the solute boundary from the center of centrifugal rotation in cm

t = time in seconds

ω = angular velocity of centrifuge rotor in rad/s obtained from instrument specifications; these values are usually calculated according to rotor type operating at a prescribed speed

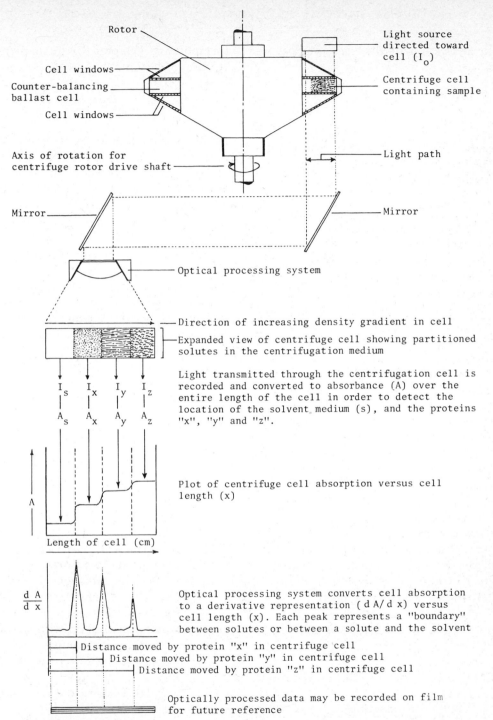

FIGURE 2.35. Density gradient separation for a solute mixture (e.g., three proteins x, y, and z) using an ultracentrifuge. After solutes have been definitively separated, the centrifuge tube is punctured and a prescribed volume of the contents is permitted to drain into each of a series of fraction collection tubes. As long as the system is stabilized against convection by maintenance of a solvent density gradient, each of the drops will represent successive layers in the centrifuge tube. This method can also serve as the basis for estimating the molecular weight of an unknown substance provided that the unknown is centrifuged jointly with a series of substances of known molecular weight. The relative position of the unknown and known substances can provide an approximate molecular weight for the unknown.

The derivation of the mathematical relation

$$M = \frac{RTs}{D(1 - \bar{V}\rho)}$$

is based on the kinetic behavior of a circularly sedimenting particle in a liquid medium. Rotation of particles suspended in a liquid medium at a uniform speed around a circular path generates a centrifugal force upon the particles that is equal to $M\omega^2 x$. M is the molecular weight of the particles, and $\omega^2 x$ is the acceleration, where ω is the velocity of rotation (angular velocity) and x is the radius from the axis of rotation. The particles displace the solution into which they sediment, resulting in a buoyant force that is equal to $M\omega^2 x \bar{V}\rho$, where $\bar{V}$ is the partial specific volume of the particles, ρ is the density of the liquid medium, and the terms M, ω, and x are as indicated in the centrifugal force expression. The suspended particles generate a frictional force as they migrate in the liquid medium that is equal to $f(dx/dt)$, where f is the frictional coefficient and dx/dt (change in x with time) is the rate of sedimentation. The frictional force, generated by the sedimenting particles in the liquid medium placed in a centrifuge operating at a uniform speed, maximizes with time as the rate of sedimentation reaches constancy; therefore, the centrifugal force is equal to frictional force plus the buoyant force. Mathematically, the expression is

$$M\omega^2 x = f\frac{dx}{dt} + M\omega^2 x \bar{V}\rho$$

Solving this equation for M, the expression

$$M = \frac{f}{(1 - \bar{V}\rho)} \cdot \frac{dx/dt}{\omega^2 x}$$

is derived. Expressing $dx/dt/\omega^2 x$ as s, the *sedimentation coefficient* M becomes $fs/(1 - \bar{V}\rho)$. Since f is related to the diffusion constant D by the expression $D = RT/f$ or $f = RT/D$

(presented in Chapter 8), the mathematical relation M becomes

$$\frac{RT/D \; s}{(1 - \bar{V}\rho)} \quad \text{or} \quad M = \frac{RTs}{D(1 - \bar{V}\rho)}$$

The sedimentation coefficients for various proteins are usually expressed in units of 10^{-13} s, but these values are conventionally multiplied by 10^{+13} s and then reported as a *Svedberg constant* (S). Therefore, an s value of 5.5×10^{-13} s would be reported as an S value of $5.5 \; S$.

The reliable use of the sedimentation velocity method for determining unknown molecular weights requires preliminary knowledge of the protein diffusion constant (D). Since an accurate estimation of this term is not always available, an alternative mathematical approach has been developed for the determination of unknown molecular weights by using the following equation in conjunction with the sedimentation equilibrium method of ultracentrifugation.

$$M = \frac{2 \, RT \, (\ln C_2 - \ln C_1)}{\omega^2 \, (1 - \bar{V}\rho) \, (x_2^2 - x_1^2)}$$

where

M = molecular weight of unknown solute
R = universal gas constant
T = absolute temperature
C_1 = solute concentration at distance x_1 from the axis of centrifugal rotation
C_2 = solute concentration at distance x_2 from the axis of centrifugal rotation
$\bar{V}$ = partial specific volume
ρ = density of the solvent medium in g mL^{-1}
ω = angular velocity of the centrifuge rotor in rad/s

The equilibrium distribution of the solute concentration in the centrifuge cell is determined by optical methods, as previously discussed. The optical data then provide a basis for establishing solute concentrations C_1 and

C_2 at two respective points x_1 and x_2 within the centrifuge cell. These measurements alone provide the basis for calculating molecular weight for the solute without regard to the protein diffusion constant.

Gel permeation chromatography (or molecular exclusion chromatography). Complex mixtures of proteins in addition to many other high-molecular-weight substances can be separated and purified by using gel permeation chromatography. This principle of molecular separation depends on the differential migration of various sample solutes through a porous stationary chromatography matrix held in a column. The movement of the sample solutes through the porous matrix is promoted by concurrent downward flow of an aqueous solvent along with the Brownian motion of the various solute molecules in and out of the matrix.

One of the most typical porous stationary media used for fractionating materials of different molecular weight is Sephadex. This extremely hydrophilic bead-formed gel is commercially manufactured by cross-linking bacterially produced dextrans with reagents such as epichlorohydrin. Differing degrees of dextran cross-linking result in different degrees of gel swelling in the presence of water and electrolyte solutions. This, in turn, produces a stationary carbohydrate matrix that has predictable abilities to fractionate substances of different molecular weight. Very highly cross-linked dense gels are useful for separating substances with molecular weights as low as 700, while other less dense gels can efficiently separate substances over the range of 5000 to 600,000.

Regardless of the amount of dextran cross-linking, the principle of molecular fractionation for gels depends on the limited abilities of large molecules to enter swollen gel particles, while the smaller molecules penetrate more deeply into the three-dimensional network of the gel matrix. Molecular exclusion of large molecules from the internal volume of the swollen gel beads encourages their rapid elution from the column prior to smaller molecules that are able to penetrate the gel structure. The eventual exit of all fractionated specimens from the column can be monitored spectrophotometrically at 270 to 280 nm and recorded as an elution profile (Figure 2.36).

Since the volume of solvent required to elute protein substances of known molecular weight from a certain type of gel is a reproducible event under carefully controlled conditions, the elution profiles for proteins of unknown molecular weight can be compared with similar profiles obtained for known proteins. Although cursory comparisons of elution profiles for standard and unknown proteins provide an approximate idea of unknown molecular weights for proteins, a more accurate estimate is usually obtained by plotting the V_e/V_0 ratios for standard and unknown proteins versus the logarithm of the molecular weights for protein standards (Figure 2.37). According to these methods, V_e represents the *elution volume* required to elute individual known and unknown proteins from the gel column, while V_0 represents the *void volume* of the gel column. The void volume is significant because it defines the elution volume necessary for discharging molecules from the gel column whose molecular size exceeds the largest pore size of the gel. Since the effective partitioning capacity of most gels rarely exceeds molecular weights of 2×10^6, a colored dextran (e.g., blue dextran, M.W. = 2×10^6) is often used for determining the V_0 value.

Depending on protein separation requirements, the partitioning capabilities of gel permeation chromatography media can be enhanced by incorporating cationic or anionic groups into the gel structure. This modification couples principles of ion exchange chromatography with the molecular sieving action of the gel.

Exclusive of gel permeation chromatography media, other related methods have been developed for molecular weight partitioning of proteins using HPLC. These

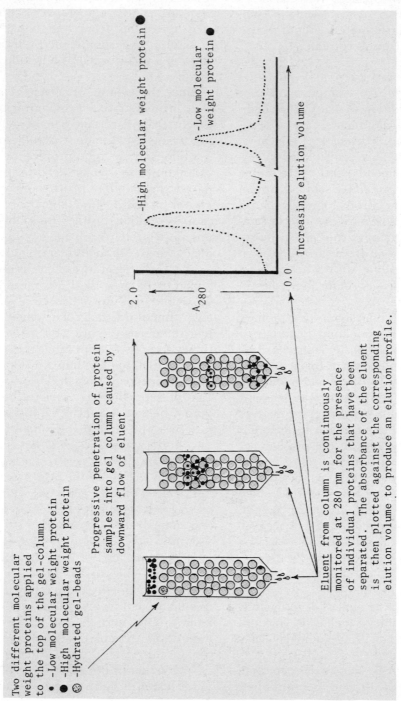

FIGURE 2.36. Schematic steps in the progressive penetration of low- and high-molecular-weight substances through a gel permeation chromatography column along with the development of an elution profile.

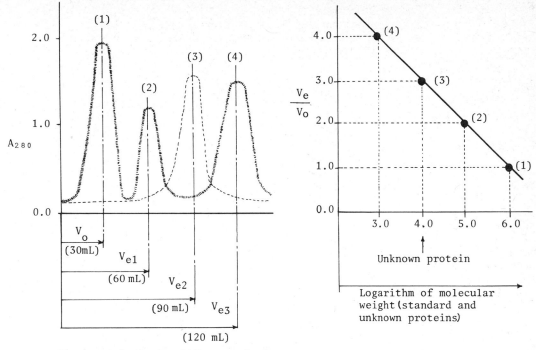

Elution volumes in mL for standards
( ) and an unknown protein ( )
after their application to a gel
permeation chromatography column

FIGURE 2.37. Determination of molecular weight for an unknown protein based on its elution from a gel permeation chromatography column as compared with standards of known molecular weight protein (e.g., molecular weights for standards: (1) = $\sim10^6$; (2) = 10^5; and (4) = 10^3). The unknown protein is eluted between standards (2) and (4), which produces a V_e/V_0 (90 mL/30 mL) equal to 3.0, which corresponds to a log molecular weight value of 4.0 or antilog value of 10^4.

methods are based on principles of molecular exclusion in addition to concurrent electrical charge interactions between proteins and the stationary chromatography phase. The stationary phase is manufactured from processed silica covered with a covalently bonded hydrophilic moiety. Residual acidic silanol groups on the surface of the stationary phase provide a site for electrical charge interactions with proteins since they tend to be negatively charged under most operational pH conditions.

Protein separation by size will occur for proteins that have acidic, neutral, or mildly basic (to 8.5) pI's. Small proteins are able to penetrate the pores of the bonded stationary phase while large proteins are excluded from the pores. However, in those cases in which proteins are more basic, the principles of molecular exclusion phenomena become less important than the strong electrostatic retention of the basic protein to the stationary phase. Figure 2.38 illustrates the range of possible protein separation using a silica-based column in a typical HPLC assay.

Affinity chromatography: Affinity chromatography relies upon the specific binding affinity between a substance that is to be isolated and a molecule that it can specifically bind (a ligand).

The ligand may be covalently bonded to a chromatographic matrix such as a cross-linked agarose gel or some other chemically

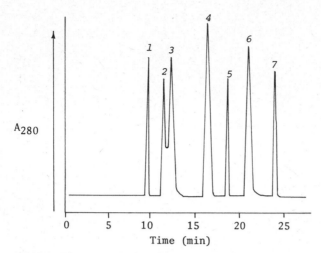

Chromatographic key:

(1) Ferritin (540,000)   *(4)* Myoglobin (17,000)
(2) Bovine serum albumin (68,000) *(5)* Ribonuclease A (13,700)
(3) Egg albumin (45,000)   *(6)* Cytochrome *c* (12,500)
                              (7) Guanosine (283)

FIGURE 2.38. Rapid separation of a complex protein mixture using HPLC and a silica-based column. This analytical method combines the resolution of gel electrophoresis with large-scale preparation capability of gel filtration at a rapid speed.

modified polysaccharide structure held in a chromatography column. If a solution containing a diversity of biochemical compounds is passed over the ligand, only those substances that have biospecific adsorption affinities will selectively adsorb to the ligand. All other substances that have negligible adsorption affinities for the ligand pass in an unrestricted fashion from the chromatography column.

This discriminative adsorption process can be used for the effective concentration of important proteins from dilute or complex mixtures of proteins and nonproteins. Mutual binding interactions between a protein and a certain ligand may be tenacious or weak, depending upon the participating substances and the chemical conditions present during the period of binding interaction (e.g., pH, ionic strength, temperature). If interactions are very strong, it may be necessary to modify the chemical conditions that

foster protein–ligand interactions. As a result of this modification, protein–ligand interactions may become less favorable and actually facilitate the eventual recovery of the protein for some use. In other cases when protein interactions are weak, the adsorbed protein may simply exit the chromatography column after all other non-ligand-interacting substances have been eluted from the column. Concentrated protein eluted from the column in this latter situation can also be directed toward some use.

The biochemical specificity of ligand–protein interactions can be simplistically visualized as complementary binding of topologically distinct sites on both substances (Figure 2.39). The actual binding interactions are also encouraged and stabilized by steric, electrostatic, and hydrophobic interactions in combination with other forces that are not unlike those binding interactions between an enzyme and its substrate.

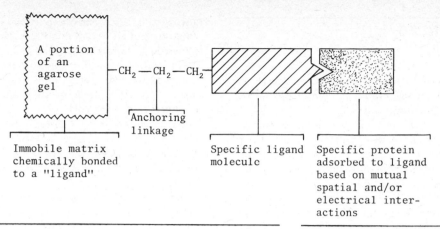

Examples of Specific Ligands Anchored to Matrices | Substance Adsorbed to Ligand

I. Metal chelating affinity supports:

 a. Zn^{2+} chelate of iminodiacetic acid,
  gel matrix- Sepharose ®

          I. a. Ceruloplasmin, interferon,
            α_2-macroglobulin, α-anti-
            trypsin, acid glycoprotein

 b. Cu^{2+} chelate of iminodiacetic acid,
  gel matrix- Sepharose ®

          I. b. Lactoferrin, prealbumin,
            transferrin, β-lipoprotein,
            γ-globulins

II. p-Chloromercurobenzoate, gel matrix-
  agarose, anchor linkage-
  n-alkylcarbamate

          II. Papaya peptidases A and B,
            wheat germ porphobilinogen
            deaminase, streptolysin O

III. Spermine, gel matrix-agarose, anchor
  linkage- n-alkylcarbamate

         III. Separation of plant specific
           tRNA's, aminoacyl-tRNA
           synthetases

IV. 1,6-Diaminohexyl-glycoholic acid,
  gel matrix- agarose, anchor
  linkage- n-alkylcarbamate

         IV. Isolation of membrane
           proteins and glycoproteins,
           isolation of receptor
           proteins used for entero-
           hepatic bile acid transport

V. 1,6-Diaminohexyloxamate, gel matrix-
  agarose

         V. Separates lactic acid
           dehydrogenase isoenzymes,
           and purifies various
           microbial neuramidases

FIGURE 2.39. Principle of affinity chromatography and examples of some ligands and their uses. Although affinity chromatography has many uses for isolating proteins, it must be noted that many other nonprotein substances can be isolated using the principle, provided that a suitable ligand can be isolated and immobilized on a chromatographic or gel matrix.

The use of affinity chromatography for the isolation, analysis, and industrial preparation of proteins and, particularly, enzymes important to the food industry are limited only by the availability of biochemically specific ligands. It should also be noted at this point the method is not strictly limited to proteins and that other versions of affinity chromatography can be used for the isolation of nonprotein substances including certain vitamin substances, nucleotides, and nucleosides.

REFERENCES

Association of Official Analytical Chemists (Current and revised editions). *Official Methods of Analysis*. Washington, D.C.

Coduri, R. J., K. Bonatti, and K. L. Simpson. 1979. Application of vertical plate gel electrophoresis to the separation of pigmented and non-pigmented trout and salmon species. *J. Ass. Offic. Agr. Chem.* **62:**269.

Coduri, R. J., and A. G. Rand. 1972. Vertical plate gel electrophoresis for the differentiation of fish and shellfish species. *J. Ass. Offic. Agr. Chem.* **55:**464.

Dickerson, R. E., and I. Geis. 1969. *The Structure and Action of Proteins*. Harper & Row, New York.

Freifelder, D. 1976. *Physical Biochemistry Applications to Biochemistry and Molecular Biology*. W. H. Freeman, San Francisco.

Glusker, J. P., and K. N. Truebold. 1972. *Crystal Structure Analysis*. Oxford University Press, New York.

Haurowitz, F. 1962. *The Chemistry and Function of Proteins*, 2nd ed. Academic Press, New York.

Holmes, K. C., and D. M. Blow. 1965. *The Use of X-Ray Diffraction in the Study of Protein and Nucleic Acid Structure*. Wiley, New York.

Joslyn, M. A. 1970. *Methods in Food Analysis*. Academic Press, New York.

Mahler, H. R., and E. H. Cordes. 1971. *Biological Chemistry*, 2nd ed. Harper & Row, New York.

Marshall, A. G. 1978. *Biophysical Chemistry Principles, Techniques, and Applications*. Wiley, New York.

Metzler, D. E. 1977. *Biochemistry. The Chemical Reactions of Living Cells*. Academic Press, New York.

Neurath, H., and R. Hill (Eds.). 1975. *The Proteins*, 3rd ed. Academic Press, New York.

Pomeranz, Y., and C. E. Meloan. 1978. *Food Analysis Theory and Practice*. Avi Publishing Co., Westport, Conn.

Schroeder, W. A. 1968. *The Primary Structures of Proteins*. Harper & Row, New York.

Wood, W. B., J. H. Wilson, R. M. Benbow, and L. E. Hood. 1974. *Biochemistry—A Problem Approach*. Benjamin, Menlo Park, Calif.

C H A P T E R · 3

Proteins: Sources and Nutritional Evaluation

NUTRITIONAL EVALUATION OF PROTEIN QUALITY

Proteins are indispensable biochemical constituents of all cells. They are found in the nucleus, cytosol, cell membranes, and all extracellular fluids of living organisms. Proteins consumed by humans as food are predominantly digested in the gastric cavity and the small intestine. Amino acids produced as a result of protein digestion are absorbed directly into the lymph system, while other intestinally absorbed amino acids are directed to the liver for anabolic and metabolic purposes. Although most proteins undergo preliminary digestion before absorption, some peptides and small proteins may pass directly through the intestinal mucosal cells and into the blood stream. This type of absorption is significant because it can represent the initial step in food allergy responses.

ESSENTIAL VERSUS NONESSENTIAL AMINO ACIDS

The biosynthesis of proteins occurs in the ribosomes of all cells. Human cells as well as the cells of all other organisms synthesize proteins from a combination of nonessential and essential amino acids.

Nonessential amino acids can be produced in required amounts for protein biosynthesis by the organism from amino acid or non-amino acid precursors that happen to be produced during the normal course of metabolism. Essential amino acids required for protein biosynthesis must be obtained from an exogenous source such as the diet of an animal. Essential amino acids are termed *essential* because the organism displays an absolute inability to synthesize certain amino acids or, *if the organism can manufacture these substances,* the biochemically required amounts outstrip the amino acids' normal rate of production by the organism. Moreover, it should be recognized that *essential* amino acids for one organism are *not necessarily essential* for the normal growth, reproduction, and maintenance of life processes for all other organisms. Therefore, essential amino acid requirements are often *species specific.* Regardless of the types and numbers of amino acids that are essential for different organisms, the net result of their deficiency is inevitably demonstrated as malnutrition involving decreased growth rate, wasting of the biophysical structure, increased susceptibility to disease, and biochemical dysfunctions along with ultimate death.

A consensus of current nutritional opinion indicates that ten amino acids—*arginine, histidine, isoleucine, leucine, lysine, methionine, phenylalanine, threonine, tryptophan,* and *valine*—are considered to be essential amino acids in the overall nutrition of all mammals and, naturally, man. The essentiality of arginine and histidine in human nutrition may be variable depending on the stage of the human life cycle studied (Holt and Snyderman, 1965; Heird *et al.,* 1972; Rose and Wixom, 1955; Weller *et al.,* 1971; Kopple and Swendseid, 1975).

PROTEIN QUALITY BIOASSAYS

Because the amount of protein required by organisms for maintenance of their normal life processes, growth, and reproduction increases as the nutritive quality of dietary protein declines, any study related to animal nutrition must be jointly concerned with *quantity* and *quality* of dietary protein. In order to assess the quantitative requirements for a dietary protein, a nutritionist must determine its quality. Amino acid analysis of the protein can provide useful protein quality information, but a verification of protein quality can be established only on the basis of bioassays using animal test subjects. The test subjects are typically evaluated according to their growth, fertility, repair and maintenance of tissues, and an array of related biological criteria when they are nurtured on a particular protein.

It has been recognized by many nutritionists and food scientists (Hopkins and Steinke, 1978) that animal-based nutrition studies designed to evaluate the quality of human food proteins may be subject to great inaccuracies. Due to moral, ethical, and practical problems regarding in-depth nutritional studies with human subjects, the quality of proteins has historically been based largely on rat nutrition studies. Rats provide a useful biological tool for nutritional studies but the extrapolation of their nutritional requirements to the formulation of prescribed nutritional indices for all stages of the human life cycle can be inaccurate for a variety of reasons:

1. Rats grow at a faster rate than humans— a process that necessitates high dietary concentrations of amino acids.

2. Rats have more hair than humans and necessarily require higher dietary quantities of sulfur-containing amino acids such as cystine and methionine.

3. Rats recycle fewer amino acids than humans in the overall scheme of metabolism and, therefore, their total essential amino acid requirement *relative to protein* is higher than human requirements.

In spite of these obvious experimental difficulties, the Food and Agricultural Organization of the United Nations/World Health Organization (FAO/WHO) as well as the Food and Nutritional Board of the National Academy of Sciences (FNB/NAS) have compiled useful information for protein quality measurements in terms of rat studies. Table 3.1 outlines typical comparative amino acid patterns for humans and rats as percent of protein. It is worth noting that the FAO/WHO and FNB/NAS recommendations for methionine and cystine in humans as well as phenylalanine and tyrosine are dissimilar, but other values are nearly equivalent. Rats, on the other hand, seem to require higher concentration levels of methionine and cystine than humans along with arginine, which they are unable to synthesize.

The historical evaluation of protein quality has taken many routes since the 1900s. Methods such as those that determine the *biological value* (BV) of proteins have provided the assay standard for many years, but recent investigations of these procedures have indicated that they represent the most un-

Table 3.1. Amino Acid Requirements for Humans and Rats

Amino acids	For humans, average requirements based on combined FNB of NAS[a] and FAO/WHO[b] estimates (%)	For rats, based on NRC-NAS[c] (%)
Arginine[d]	—	5.0
Histidine[d]	1.7	2.5
Isoleucine	4.1 ± 0.10	4.6
Leucine	7.0	6.2
Lysine	5.3 ± 0.20	7.5
Methionine + cystine[e]	3.05 ± 0.45	5.0
Phenylalanine + tyrosine[f]	6.65 + 0.65	6.7
Threonine	3.75 ± 0.25	4.2
Tryptophan	1.05 + 0.05	1.2
Valine	4.9 ± 0.10	5.0
Total essential amino acids (EAA)	36.65 ± 0.65	47.9
Total nonessential amino acids $100 - EAA =$	63.35 ± 0.65	52.1

[a]Williams *et al.* (Food and Nutrition Board of National Academy of Sciences) (1974). Used by nutritionists concerned with the U.S. dietary goals.

[b]FAO/WHO (Food and Agricultural Organization/World Health Organization) (1973). Used by nutritionists concerned with international nutritional goals.

[c]NRC-NAS (National Research Council of National Academy of Sciences) (1972).

[d]Required only for infants.

[e]The sulfur-containing amino acids, methionine and cystine, are grouped together since cystine can replace part of the requirements for methionine.

[f]The aromatic amino acids, phenylalanine and tyrosine, are grouped together since tyrosine can replace part of the requirements for phenylalanine.

satisfactory method for estimating protein quality. The pitfalls of BV assays have been averted to some extent by the adoption of more specific protein assays in recent years. Quite apart from analytical disputes surrounding any of the older or newer methods, some notable assay principles have been outlined below.

Biological Value (BV)

The study of protein requirements for humans or test animals is actually a study of the animal body's amino acid requirements and overall nitrogen metabolism.

Proteins, amino acids, nitrogenous metabolic intermediates, and waste nitrogen substances are normally in a dynamic state of interconversion from one form to another. These interconversion processes for nitrogen are not in a perpetual state of balanced equilibrium. Rather, the dietary nitrogen intake of an animal in the form of amino acids and proteins must be equivalent to the metabolic load of nitrogen excreted in urine (e.g., urea, creatinine, uric acid) and lost in feces, sweat, body secretions, or sloughed portions of the integumentary system (e.g., hair, nails, skin). Amino acid intake that exceeds nitrogen and protein biosynthetic requirements of animal tissues results in nitrogen excretion as urea; alternatively, organic acid metabolites of amino acids may be deaminated and used for energy or assimilated into processes of fat and carbohydrate metabolism.

The BV index for a protein is based on studies of nitrogen balance in the animal body. Such studies mathematically compare the amount of food protein-nitrogen *ingestion* with the amounts of fecal and urinary nitrogen *excretion*.

The BV for a specific food protein is expressed as a percentage of the dietary protein nitrogen retained by the test subject. The BV index for a protein can be calculated according to the equation

$$BV = \frac{\text{food N} - (\text{fecal N} + \text{urinary N})}{\text{food N} - \text{fecal N}} \times 100$$

Amino Acid Score

The quality of a protein can be designated by an amino acid score that reflects the comparison of the amino acid composition of a test protein to that of an amino acid reference pattern established by the FAO/WHO (1973).

Amino acid score =
$$\frac{\text{mg amino acid/g test protein (100)}}{\text{mg amino acid/g reference pattern}}$$

The reference pattern of amino acids established as a standard for amino acid comparison consists of the amino acid concentration levels outlined in Table 3.2. The indexing of individual dietary protein qualities is based on the amino acid concentration, which acts as a limiting factor with respect to its prescribed concentration level in the reference protein.

Protein Efficiency Ratio (PER)

The PER method has been widely employed by the Association of Official Analytical Chemists (AOAC) and has been recognized as a credible guideline for protein quality evaluation by the U.S. Food and Drug Administration (USFDA), the U.S. Department of Agriculture (USDA), and various counterparts of the Canadian government. According to this bioassay method for assessing protein quality, rats are fed a 10% protein for four weeks and their individual weight gains are recorded. The protein ration usually fed to the test subjects is in the form of casein because it is inexpensive and quite pure. The PER value for the protein

Table 3.2. Amino Acid Scoring Reference Pattern[a]

Amino acid	mg of amino acid/g of nitrogen
Isoleucine	250
Leucine	440
Lysine	340
Methionine + cystine	220
Phenylalanine + tyrosine	380
Threonine	250
Tryptophan	60
Valine	310

[a]FAO/WHO Energy Requirements and Protein Requirements, *WHO Tech. Ref. Ser.* No. 522, 1973, p. 63.

is then calculated from experimental data as indicated below:

$$PER = \frac{\text{weight gain (g) of animal}}{\text{protein (g) intake by animal}}$$

The PER values for all types of dietary protein sources are calculated in a similar fashion and assigned values relative to the arbitrary value of 2.5 chosen for casein. Hegsted (1977) claims that this method can distinguish good dietary proteins from poor proteins, but it does not provide a quantitative measure of differences in protein quality since PER values are not proportional to nutritive quality.

Net Protein Ratio (NPR)

The NPR method for estimating relative protein qualities is similar to the PER method; however, an additional group of rats is fed a protein-free diet. The control protein for these studies is typically an albumin (e.g., lactalbumin, egg albumin) or in some cases casein may be substituted for albumin. The NPR value is calculated according to the following equation:

$$NPR = \frac{\text{weight gain test group} + \text{weight loss group fed nonprotein diet}}{\text{dietary protein consumed}}$$

The NPR values of the tested proteins are expressed as percentages, with the control protein assigned a value of 100%. A comparison of PER and NPR values (Hopkins and Steinke, 1978) has revealed that protein sources such as lean beef have the same PER and NPR values (88%), whereas white flour has a PER value of 28 and NPR value of 51. The difference in NPR and PER values may be related to the fact that the NPR method considers protein not only for growth but also for maintenance of normal life processes. Hence, the NPR method is considered more accurate than the PER method.

Net Protein Utilization (NPU)

The NPU method requires the study of two groups of test animals. One group is fed a test protein while the other group receives a protein-free diet. After completion of a prescribed feeding period, the nitrogen content for both groups of test animals is determined by the use of chemical methods. The NPU for the test protein originates from the relationship

$$NPU = \frac{body\ N\ test\ group\ -\ body\ N\ group\ fed\ nonprotein\ diet}{N\ consumed}$$

The equation for calculating the NPU is similar to the NPR, but nitrogen content is used as a criterion for evaluating various proteins. Most nutritionists believe that the NPU and NPR methods provide protein quality data of equivalent significance. For additional remarks concerning the NPU method consult Miller and Bender (1955).

Relative Nutritive Value (RNV) and Relative Protein Value (RPV)

Both of these protein quality bioassays involve feeding rats graded amounts of test or control proteins for comparative purposes. Depending on the type of protein being tested, weight gains in test animals are plotted against corresponding amounts of dietary protein intake. Least-squares analysis of these growth data, followed by slope analysis of the linearly plotted data, serves as a basis for comparison of test and control proteins.

The ratio of the individual linear slopes for each type of protein ration, with respect to the control data, yields a series of ratios that characterize the nutritional value of test proteins:

$$slope\ ratio = \frac{slope\ of\ response\ curve\ for\ animals\ fed\ test\ protein}{slope\ of\ response\ curve\ for\ animals\ fed\ control\ protein}$$

Nutritional studies using this basic method of slope ratio analysis for relative protein quality evaluations can provide either a series of animal response lines that show a common axis intercept in the case of RNV measurements, or response lines that do not display any common interception point (RPV).

According to a number of authorities (Hegsted, 1977; Samonds and Hegsted, 1977), the slope-ratio methods provide more reliable estimates of protein quality because they are *multiple-point assays* as opposed to *two-point assays* employed in the PER, NPR, and NPU methods. Contrary to the advantages of RPV and RNV methods, nutritionists remain skeptical of their significance since inherent shortcomings of statistical analysis may cause credibility problems in the final slope-ratio values that indicate protein quality.

Although bioassays may be unrivaled in terms of their legitimacy for assessing food protein qualities, food scientists often prefer *in vitro* assays over *in vivo* studies, which are protracted and expensive. Satterlee *et al.* (1977) and others have outlined *in vitro* protein quality evaluation methods that are practical, economical, and rapid compared to most *in vivo* assays. Many *in vitro* assays are based on concentration levels of the essential amino acid profile found in a test protein. This profile is then transformed into computed PER (C-PER) values that approximate rat-PER bioassays (Table 3.3).

SOURCES OF DIETARY PROTEINS

Dietary proteins are obtained from many different species of animals and plants. Species that have historically provided sources of dietary protein are regarded as "conventional" protein sources. The use of plant and animal species other than those considered to be "conventional" protein sources are alternatively called "unconventional." Unconventional protein sources may include the use of new or improved genetic hybrids of existing plant or animal species; newly discovered high-protein-producing species; or the production of single-cell protein (SCP) sources from bacteria, fungi, or algae.

A survey of worldwide food resources reveals that some countries are in dire need of food whereas other technologically developed countries seem to display excessive per capita food consumption. The food consumption pattern of the United States is normally cited as excessive since the 99.0 g/day per capita intake of dietary protein is nearly twice the 56.0 g/day amount for a 70.0-kg adult male suggested by the Recommended Dietary Allowances (1980).

The predominant sources of protein in the United States are supplied by red meats, poultry, and fishes, while dried beans, soy flour, nuts, peas, corn grits, potatoes (sweet

and white), vegetables, and fruits augment dietary animal proteins (USDA, 1975). Those countries other than the United States that maintain similar levels of protein nutritional status often rely more heavily on plant protein sources and less on animal sources.

The prodigious consumption of meat in the United States has been indicted as a wasteful practice according to the opinion of economically poor nations and other critics of American agricultural practice. The gist of these arguments centers about the fact that one acre of cereal grain to be consumed by people can produce up to five times more protein than an acre of cereal grain devoted to red meat production. The magnitude of cereal grain production destined for raising meat-producing animals is reflected by their consumption of nearly 90% of the 27 million metric tons of cereal grain, legumes, and protein produced. This animal feed ration produces only 5.3 million metric tons of protein suitable for human diets. In terms of other numbers, only ~1.0 kg of animal is produced for ~5.0 kg of plant protein fed to livestock (Pimentel and Pimentel, 1977). This 1:5 feed conversion ratio is a rather wasteful process, but it is often justified by the superior organoleptic and palatability features of animal proteins over plant proteins. Furthermore, *many* of the feedstocks supplied to cattle and livestock cannot be consumed by humans.

The abundance of high-quality protein sources in the nutritional pattern of the developed countries is not easily matched by

Table 3.3. Comparison of Computed Protein Efficiency Ratio (C-PER) Values Obtained on the Basis of *In Vitro* Assays and Rat Bioassays (rat-PER)[a]

Sample	C-PER	Rat-PER	Differences in PER estimates
Cornmeal	0.76	0.70	0.06
Whole corn	1.20	1.36	0.16
Nonfat dry milk	2.76	2.69	0.07
Regular pizza ingredients	2.29	2.02	0.26
Beef and noodle dinner	2.16	1.80	0.36

[a]Based on data reported by Satterlee *et al.* (1977).

poorer countries. Nevertheless, economically disadvantaged countries have successfully used carefully concocted mixtures of plant proteins to treat protein-calorie malnutrition or kwashiorkor. Incaparina, for example, has been prepared by the Institute of Nutrition of Central America and Panama from a mixture of 29.0 g of whole ground sorghum grain in conjunction with 38% cotton seed flour, 3% Torula yeast, 1% calcium carbonate, and 4500 I.U. of vitamin A per 100 g. A proximate analysis of the protein content for this formulated product shows a level of ~27.5% protein (Burton, 1976) and a food protein quality index similar to milk.

Meat Proteins

"Meat" is a generic term that refers to the edible muscle tissues of an animal body. On the basis of dietary and anatomical considerations, the musculature of animals can be divided into *three* principal *types* that include *smooth muscle* of the intestinal tract; *cardiac muscle* of the heart; and *skeletal muscle* that contains lean meat. Some of the noteworthy organizational features of muscle structure have been depicted in Figure 3.1.

Muscle structure: The basic unit of muscle structure is the *fiber,* which is a long cylindrical cell (~10 to 100-μ diameter × ~2–500 mm). The cell normally exists in a state of syncytium or multinucleation since each cell can contain ~100 to 200 nuclei.

Groups of fibers are arranged into bundles that constitute the overall muscle structure. Each fiber is also enclosed by an electrically excitable membrane called the *sarcolemma,* which is involved in the transmission of electrical impulses in a manner reminiscent to membranes that surround nerves.

The bundles of individual muscle fibers or *myofibrils* are also immersed in a liquid matrix called the *sarcoplasm.* The sarcoplasm has a complex mixture of meat juices and contains glycogen, phosphocreatine, glycolytic enzymes, myoglobin, cytochromes, amino

acids, peptides, and magnesium salts of ATP. The colors of the myofibrils are red to colorless *depending on the abundance and distribution of mitochondria.* Among common dietary protein sources, few proteins display more typical color differences in musculature than the fowl. Birds that engage in sustained flight, for example, have red muscles owing to the presence of abundant mitochondria in regular patterns, myoglobin, and a high cytochrome concentration. This is in marked contrast to gallinaceous birds (e.g., turkeys, chickens, grouse), which have white breast muscles resulting from low numbers of mitochondria and a lack of their patterned distribution.

Myofibrils have a contractile unit of structure called the *sarcomere* that repeatedly occurs along their structure at specific length intervals. A longitudinal survey of myofibrillar structure reveals an alternating sequence of dark and light bands. The dark bands are called *anisotropic* or A bands whereas the light bands are designated as *isotropic* or I bands. The A bands are optically anisotropic since they display double refraction or birefringence as opposed to the I bands, which do not exhibit birefringence.

Within the I bands, two lines occur that extend across the width of the myofibril. These two lines represent the point of myofibrillar joining and maintain the A and I bands in a state of functional coordination. Contained in the middle of the A band lies the H zone, which is bisected by the M line that contains a protein called *M protein* (Figure 3.1).

The thick filaments of muscle proteins contain *myosin,* and thin filaments are predominantly composed of *actin* along with small amounts of the proteins known as *tropomyosin* and *troponin* (Figure 3.1).

Myosin, whose molecular weight is ~470,000 daltons, consists of two identical polypeptides each of which has 1800 amino acid residues and a molecular weight of ~200,000 daltons. Each polypeptide forms a head that extends into a rod. The heads of the respective rods form a double-stranded

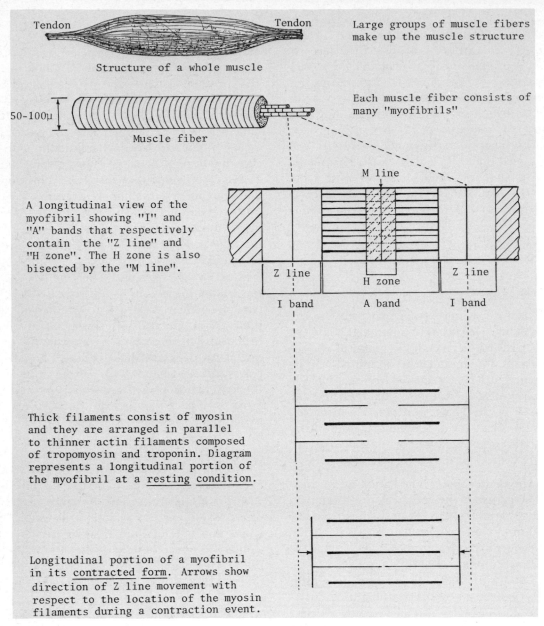

FIGURE 3.1. Hierarchy of structural levels in a muscle and schematic representation of the contraction process involving the interaction of myosin and actin filaments.

α-helical structure. In addition, the two heads can be further subdivided into four light polypeptides chains—two of which are identical and have a molecular weight of ~18,000 daltons. The two remaining light polypeptides have a molecular weight of 16,000 to 21,000 daltons (Figure 3.2).

The chief protein found in thin filaments is actin, which occurs as a monomer, G-actin

(globular), and a polymer, F-actin (fibrous), which are interconvertible forms. The function of actin is complex, but it is generally recognized to enhance ATPase activity of myosin subject to the modulating influences of light polypeptide chains attached to the heads (Taylor, 1972).

The association and dissociation of myosin, actin, ATP, and its hydrolytic products (ADP,

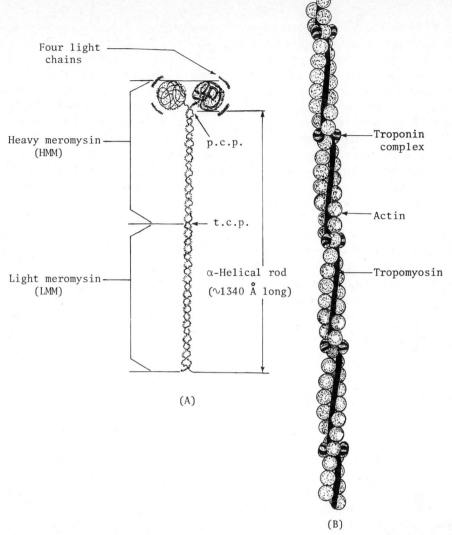

Four light
chains

Heavy meromysin
(HMM)

p.c.p.

t.c.p.

α-Helical rod
(∼1340 Å long)

Light meromysin
(LMM)

(A)

Troponin
complex

Actin

Tropomyosin

(B)

FIGURE 3.2. Schematic molecular diagrams for the structure of myosin (A) and a thin filament (B) that jointly participate in muscle contraction. (A) Myosin can be cleaved into two fragments by trypsin at a specific point (t.c.p.), which produces *light meromysin* (LMM) and *heavy meromysin* (HMM). About 90% of the LMM (∼850 Å) exists as a two-stranded α-helical structure according to optical rotatory methods of analysis, while the HMM is a double-stranded helical rod that terminates in two *globular heads*. The HMM can be additionally cleaved by papain at a specific point (p.c.p.) to produce two free globular subfragments along with an α-helical protein rod. Both of the round globular heads contain ATPase active sites as well as an actin binding site. According to most contemporary studies the four light chains appear to modulate the ATPase activities of myosin. (B) Actin, tropomyosin, and a troponin complex represent the major subdivisions of the thin filament. The troponin complex actually consists of three polypeptide chains designated as TpC, TpI, and TpT, which respectively display molecular weights of 18,000, 24,000 and 38,000 daltons. The presence of Ca^{2+} initially regulates muscle contraction when it binds to the TpC polypeptide chain in the troponin complex. This action produces conformational changes that are allosterically transmitted to the tropomyosin *and then* actin—in that order. This sequence of allosteric effects facilitates the interaction of actin with myosin. Furthermore, this interaction culminates in the *hydrolysis of ATP* (see Figure 3.4) and muscle contraction events are maintained until Ca^{2+} is removed as an allosteric stimulus in the overall process described by

$Ca^{2+} \longrightarrow$ troponin $\longrightarrow$ tropomyosin $\longrightarrow$ ⎧actin (low myosin interaction form)
⎩actin (readily interacts with myosin to pro-
     duce a muscle contraction event)

P_i) produce the force required for contracting individual sarcomeres. Although the precise mechanism for muscle contraction remains to be unassailably described using current information, some of the most significant steps responsible for muscle contraction are known with certainty.

One of the initial steps in muscle contraction begins when a nerve impulse reaches a neuromuscular junction. The acetylcholine released into the junction acts to increase the sodium ion permeability of the adjoining plasma membrane. The altered permeability of the membrane results in a rapid influx of sodium ions along the membrane along with an associated and concurrent sweep of depolarization that moves along the membrane of the fiber. This sweep of depolarization travels not only on the surface of the membrane but also by way of the so-called *T tubules* within the fiber and eventually reaches a large bulbous cistern-like structure in the sarcoplasmic reticulum (parallel to the myofibril and perpendicular to T tubules). The depolarization of cisternae results in a release of calcium ions that are attached to a calciquesterin protein found in the sarcoplasmic reticulum. The calcium ions then act as physiological activators because they are intermediates between the initial nerve impulse and the actuation of a muscle contraction.

Ebashi *et al.* (1969) have reported that the effect of the calcium ion on the interaction of myosin and actin is mediated by troponin and tropomyosin located in the grooves of the F-actin. Calcium ions first attach themselves to troponin, which subsequently interacts with myosin. The myosin–actin interaction complex undergoes repeated association and dissociation in a cyclic reaction controlled by ATP (Figure 3.3).

The hydrolysis of ATP yields free energy that is utilized in the muscle for its contractile movement and ability to do mechanical work. Electron micrographic analysis and X-ray diffraction studies have revealed that crossbridge associations formed by myosin heads with actin protein produce actomyosin. When ATP is added, crossbridges are released and thin filaments slide by the thicker filaments, thereby producing muscle contraction.

The energy unleashed from ATP for muscle contractions is not the result of spontaneous ATP hydrolysis in the ultrastructure of the muscle. Instead, energy produced from ATP is controlled by the action of the enzyme called *adenosine triphosphatase* (ATPase), which resides in that portion of the myosin known as heavy meromysin.

$$ATP + H_2O \xrightarrow[\substack{\Delta G^{0'} = -7300 \text{ cal} \\ (\text{pH } 7.0)}]{\text{ATPase}} ADP + P_i + H^+$$

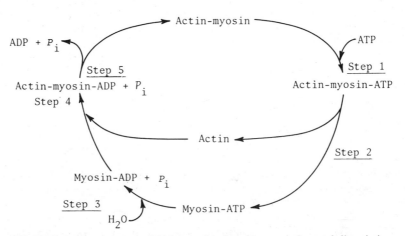

FIGURE 3.3. A sequence of reactions showing the association and dissociation of actin and myosin controlled by ATP hydrolysis. (Based on Taylor, 1972.)

Depleted stores of ATP in many muscle tissues of vertebrates can be replenished by a guanidinium phosphate such as *phosphocreatine,* which participates in the conversion of ADP to ATP in the presence of a *creatine kinase* enzyme. Invertebrates, on the other hand, utilize *phosphoarginine* in a similar reaction.

$$\text{Phosphocreatine + ADP} \xrightarrow[\substack{\Delta G0' = -3.0 \text{ kcal} \\ (\text{pH } 7.0)}]{\text{creatine kinase}}$$

$$\text{creatine + ATP}$$

Exclusive of the roles of phosphagens such as phosphocreatine and phosphoarginine, the ultimate sources of ATP for muscle contraction arise from glycolysis, the tricarboxylic acid cycle, and oxidative phosphorylation.

A proposed mechanism for muscle contraction: Contraction phenomena displayed by muscle tissues requires the sliding of the thin filament portion of the muscle structure by the thick filaments. The *rowboat* model of this contraction process has been postulated by H. E. Huxley. According to this operational concept, a portion of the myosin rod extending 70 to 100 nm from its tail swings and makes contact with an actin filament. When ATP is added, it binds to the head of the myosin protein and leads to the detachment of the head from actin since the ATP-myosin complex has a low affinity for interacting with actin.

The bound ATP hydrolyzes at this point and yields ADP, P_i, and free energy. The split ATP products bound to the myosin active site eventually associate with actin.

The crossbridges between myosin and actin undergo a change in angle from 90° to 45° (Figure 3.4) because of conformational changes occurring in the head brought about by the energy liberated during ATP hydrolysis. It is believed by many investigators that the conformational changes occur in the "hinge" portion of the flexible myosin rod and not in the head of the myosin protein. The "hinge" also represents a part of the

myosin rod that is notably susceptible to trypsin fragmentation.

The observed change in angles is sufficient to move the actin filament by ~10 nm or the length of two actin subunits. Since the hydrolysis products of ATP—namely, ADP + P_i—have departed from the myosin head at this point, another ATP can bind to the

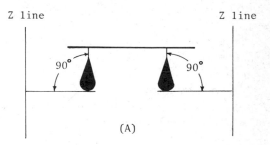

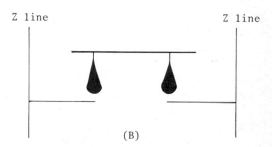

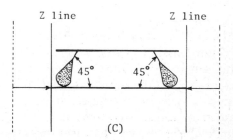

FIGURE 3.4. Schematic representation of actin and myosin interactions during muscle contraction. (A) Attachment of myosin heads (solid black region) to actin forming right-angled crossbridges *in the absence of Mg^{2+} activated ATP.* (B) Detachment of myosin crossbridges from actin *in the presence of ATP.* (C) Attachment of myosin heads as crossbridges to actin resulting during muscle fiber contraction. The myosin heads here (dotted region) include the hydrolytic products of ATP, that is ADP and P_i. The energy produced by *ATP hydrolysis brings about conformational changes in the angle of the myosin heads from 90° to 45°* and a relative contractile movement between actin and myosin (direction of movement indicated by arrows).

myosin head during a concurrent dissociation of the crossbridged actin–myosin complex. This step allows a relative sliding of individual actin and myosin components, but the existence of many other crossbridges in the muscle ultrastructure keep the thin filaments from slipping back into their original positions.

When the depolarization stimulus to the muscle structure ceases, the membranes return to their original polarized state. The outer portion of the membrane becomes more positive than the inside; calcium ions are transported back to the cisternae of the sarcoplasmic reticulum by an ATP-dependent "pump"; and the contractile unit again assumes its resting state.

Postmortem changes in slaughtered animals and meat tenderness:

Postmortem changes in slaughtered animals and meat tenderness: Many biochemical and physicochemical changes occur after the death of an animal. In general, animals are usually bled after slaughtering in order to avoid bacterial spoilage of the edible flesh. As the blood volume of the animal decreases during bleeding, the amount of oxygen transported by hemoglobin to the cells diminishes and aerobic biochemical reactions responsible for the conversion of glycogen to carbon dioxide and water cease. As a consequence of decreased aerobic respiration in the animal tissues, glycogen is converted to lactic acid with a limited synthesis of ATP. Lack of ATP in the muscles inevitably leads to rigor mortis in the muscle structure because actin and myosin interact to form inflexible *actomyosin,* which causes the meat to become tough.

If ATP happens to be present in large quantities, it becomes interspaced between actin and myosin and contributes to the softness and pliability of meat. The usual time frame for the onset of rigor mortis is 8 to 12 h in most animals; however, fish may require 1 to 10 h.

In the event that an animal has physically struggled prior to slaughtering, the onset of rigor mortis is expedited since glycogen stores are depleted along with ATP. This possibility makes it highly desirable to have all animals well fed and rested before slaughtering. Aside from the preslaughter resting period, the actual slaughtering process must be conducted by the most efficacious methods available so as to minimize the animal's anxiety, stress, and physical struggling, all of which deplete glycogen reserves.

Anaerobic reactions associated with lactic acid formation generally cause a lowering of the animal's physiological pH of 7.2–7.4 to 5.1–5.3. After the onset of rigor mortis, meat usually undergoes a variety of changes, including the liberation of *cathepsin D* and other enzymes. The roles and activities of cathepsins have been widely studied with respect to their potential for tenderizing edible meats that may be tough. Although mechanisms for cathepsins remain to be firmly established, Robbins *et al.* (1979) have reported that bovine spleen extracts of cathepsin D alter the Z line structure of muscle subunits under postmortem pH conditions (pH 5.1–5.3) and reduce myosin–actin interactions. Further, it has also been suggested that the cathepsin activity may produce both heavy and light chains from myosin in addition to altering the normal troponin–myosin complex.

Apart from the postmortem events that affect meat tenderness, tenderness is also related to the presence of connective tissues such as collagen that exists among myofibrils in addition to the individual lengths of sarcomeres.

The degree of collagen cross-linking among myofibrils parallels the age of an animal. The older the animal, the greater the degree of collagen cross-linking and the greater the meat toughness. Sarcomeres, on the other hand, which are short, display a compact, contracted structure and contribute to a tough meat, contrary to long, relaxed sarcomeres that produce tender meat. The correlation between sarcomere length and meat tenderness makes sarcomere length a reliable index of meat tenderness and has been successfully used to predict suitable aging periods for meat. There are different techniques available for measuring sarcomere length, but laser diffraction yields fast, accurate results.

Toughness of meats may also be attributed to other factors such as *gap filaments,* which affect the tensile strength of raw or heat-denatured myofibrils (Locker *et al.,* 1977). A *calcium-activated factor* may be involved in eliminating the integrity of gap filaments during the aging process of meats, but this requires further study.

In general, many meats show improved tenderness during aging. Unfortunately, the benefits of aging may be countered by increased fat rancidity in pork and other high-fat meats.

Classification of muscle proteins: Muscle proteins can be classified as sarcoplasmic, stroma, or myofibrillar proteins depending on their characteristic solubilities in selected aqueous solvents.

Sarcoplasmic proteins. Sarcoplasmic proteins display notable solubilities in pH neutral solutions having ionic strengths of ≤0.1. These proteins occur throughout the cytoplasm of the muscle cell and generically refer to ~100 to 200 different types of proteins designated as *myogen.* Sarcoplasmic proteins account for 20 to 34% of skeletal muscle proteins depending upon the animal species studied. Included among those proteins is myoglobin, which happens to be responsible for oxygen transport to the respiring muscle tissues as well as the color of red meats.

Stroma proteins. Proteins that are insoluble in pH neutral aqueous solvents are recognized as stroma proteins. These proteins account for 10 to 15% of the total protein in the skeletal muscle and normally include the lipoproteins, mucoproteins, and connective tissues such as collagen and elastin.

Myofibrillar proteins. Myofibrils contain a variety of water-extractable proteins that are quite stable under conditions of high ionic strength and may comprise 52 to 56% of the skeletal muscle. The tenderness of meats is partially attributable to actin, myosin, and other myofibrillar proteins that readily interact with water as a result of electrical charges on their constituent amino acid residues. Myofibrillar proteins are also important to meats because they contribute to their emulsifying capacities and contain a high proportion of the essential amino acids in the meat tissue.

Slaughtering operations produce a variety of edible animal parts that are a composite mixture of sarcoplasmic, stroma, and myofibrillar proteins. The percentage of proteins in different parts of the carcasses of various animal species is quite variable, as indicated in Table 3.4.

Fish and Oceanic Food Proteins

The recent realization of inadequate high-quality protein resources throughout the world has inevitably directed food science and technology toward fish and marine-food resources. The potential protein productivity of these resources may come to fruition at some time in the future, but technological success in producing the constant, vast, yet economical amounts of protein required for the human population has been meager.

Fish, shellfish, and crustaceans all represent potentially good sources of dietary proteins since they display 15 to 20% protein (Table 3.5 and 3.6) that is easily digestible and rich in essential amino acids. Moreover, their dietary acceptability as protein sources is buttressed by their high concentrations of minerals, A, D, and B series vitamins, and a high concentration of unsaturated fatty acids. The fat content of aquatic (freshwater) or marine (saltwater) organisms relative to protein content varies depending on their natural habitats. *Pelagic* organisms that inhabit the open ocean (e.g., herring, mackerel, salmon, tuna, sardines, anchovies) may contain up to 20% fat while the *demersal* organisms (e.g., cod, haddock, whiting, flat fishes such as halibut or ocean perch) have ~5.0% fat.

Table 3.4. Percentage of Protein Contained in 100-g Edible Portions of Various Animal Tissues[a]

Name	Percentage	Name	Percentage
Beef		Varieties of meat and mixtures	
Chuck, cooked	26.0	Brains	10.4
Hamburger, cooked	22.0	Chili con carne	10.3
Porterhouse, cooked	23.0	Heart, beef, raw	15.0
Rib roast, cooked	24.0	Kidney, beef, raw	19.7
Round, cooked	27.0	Liver, beef, fried	23.6
Corned beef, end	13.7	Liver, calf, raw	19.0
Dried or chopped beef	34.3	Liver, pork, raw	19.6
Roast beef, end	25.0	Sausage, bologna	14.8
Lamb		Sausage, frankfurters, cooked	14.0
Leg roast, raw	18.0	Sausage, liverwurst	16.7
Leg roast, cooked	24.0	Sausage, pork, raw	10.8
Rib chop, raw	14.9	Sweetbreads, cooked	22.7
Rib chop, cooked	24.0	Tongue, beef	16.4
Pork			
Bacon, fried	25.0		
Ham, fresh raw	15.2		
Ham, cured cooked	23.0		
Veal			
Veal cutlet, cooked	28.0		
Stew meat, cooked	25.0		

[a]Based on data reported in USDA Handbook Number 8 (1963, revised edition) and Burton (1976).

The protein content of fish and other marine organisms consists of ~20 to 30% sarcoplasmic proteins, 70 to 80% structural proteins, and only 2 to 3% connective tissue. The myofibrillar structures of most organisms including the fish are similar to that found in mammals. The myofibrils are striated and contain myosin, actin, actomyosin, and tropomyosin. However, unlike mammalian proteins, the muscles of these organisms may demonstrate a rapid loss of ATPase activity, along with a rapid aggregation and dehydration of the constituent structural proteins.

Fish proteins can be divided into "red" and "white" muscle protein classifications. The proteins comprising the white muscles of fishes have traditionally been regarded as the most desirable dietary form of fish proteins. This preference no doubt stems from their mild organoleptic properties, flaky texture, and suitability for market distribution as raw or precooked fillets, breaded frozen fish sticks, or a canned product. Although white muscle proteins of fishes are of great-est dietary and nutritional significance, the red muscles are *no less important* than white muscles in the overall biochemical and physiological realm of fishes' existence. In contrast to white muscles, however, red muscles tend to have higher concentrations of unsaturated lipids, fatty acids, myoglobin, respiratory enzymes, and potentially deteriorative enzymes such as *lecithinases* or *succinic dehydrogenase*. This biochemical complexity of red muscles encourages many serious technological problems associated with their processing, storage, and marketability as a protein source. In fact, the high fatty acid content, high enzyme concentrations, and high rate of lipid oxidative rancidity in these

Table 3.5. Typical Composition for Fish and Related Food Resources

Moisture	78–83%
Protein	15–20%
Fat	0.2–20.0%
Mineral matter	1.0–1.8%
Total solids	18.0–35.0%

Table 3.6. Percentage Protein Composition for a Variety of Fish and Seafood Resources[a]

Name	Percentage	Name	Percentage
Bluefish	27.8	Lobster, raw	16.2
Clams, raw	12.8	Oysters, raw	9.8
Cod, raw	16.5	Oysters, stewed	5.3
Cod, dried	81.8	Salmon, raw	17.4
Flounder, raw	16.4	Sardines, canned	25.7
Haddock	18.7	Scallops, raw	14.8
Halibut, raw	18.6	Swordfish, raw	27.4
Halibut, cooked	22.6	Tuna, canned	29.0
Herring, raw	18.3	Shrimp	22.5

[a]Tabulated values based on data reported in USDA Handbook Number 8 (1963) and Burton (1976).

protein-rich tissues cause such serious organoleptic problems that most red meats are suitable only for pet and agricultural feeds.

Red fish meats normally comprise ~1 to 2% of demersals and ~10.0% of the pelagic fishes. The cartilaginous fishes or elasmobranchs, including the sharks, show muscular features quite unlike other finfishes. It is not only difficult to ascribe distinct red and white meat designations to their muscle structure, but they also lack the flaky texture of finfishes and contain ≥10.0% collagen (consult Figure 3.5 for comparative differences in transverse muscle structures of fishes).

The postmortem changes in the texture, odor, and dietary acceptability of fish and marine food proteins are enormously complex and varied. Fish proteins are readily subject to the proteolytic attack of psychrophilic and mesophilic bacteria as well as oxidative rancidity reactions mediated by enzymes. These reactions culminate in the production of objectionable odors, off-flavors, and textural degradation of the muscle accompanied by dehydration and increased muscle toughness.

It is interesting to note in passing that the proteins of fatty fishes tend to be somewhat more resistant to denaturation than low-fat species if denaturation is expressed as salt-soluble proteins. The reasons for these observations are unclear, but it is believed that certain native lipids, or possibly lipoproteins, may physically associate with proteins in high-fat species to retard denaturation processes, or competitive binding among muscle proteins and neutral lipids for naturally available free fatty acids may quantitatively control desolubilization of proteins.

The textural changes in fish musculature and its food acceptability are also linked to the indirect effects and presence of substances such as *trimethylamine oxide* (TMAO). For example, enzymatic or nonenzymatic conversion of TMAO to *dimethylamine* (DMA) may produce formaldehyde that will attack the normal texture of structural and sarcoplasmic proteins. Although reactions such as these are recognized, the actual significance of formaldehyde in altering fish textures requires additional study.

Fish protein concentrate, modified fish proteins, and nonfish protein sources: Fish meals and flours represent a facet of fish processing that produces a high-protein product called *fish protein concentrate* (FPC). This product is typically produced from abundant types of *underutilized fish species* that are not suitable for filleting or other types of economical marketing.

To produce FPC, 70 to 90% of fish flesh is mechanically removed from bone structure and the flesh is coarsely ground, dehydrated, defatted, and then deodorized by the use of carefully selected extraction solvents. Many FPC products have historically

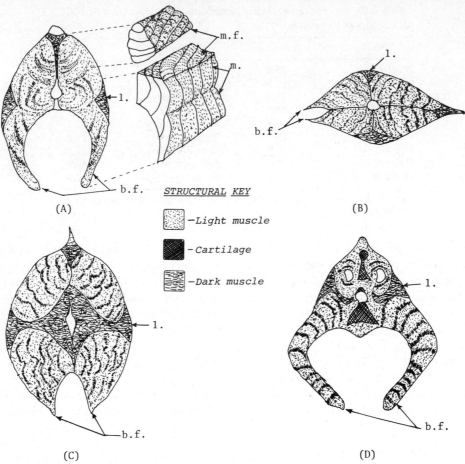

FIGURE 3.5. Transverse sections of selected fish species illustrating the locations of key flesh structures and other anatomical features (l., lateral line; m.f., muscle fibers; m.-mycommata; b.f., belly flaps). Fish types detailed include (A) demersal fishes (cod, haddock, pollock, hake); (B) demersal flat-fishes (flounder); (C) pelagic fishes (tuna, swordfish); and (D) elasmobranchs (cartilage fishes such as sharks).

suffered from high bacterial counts and residual fishy odors that make them barely palatable for human consumption. However, recent improvements in fish processing, sterilization, and solvent stripping methods have minimized most organoleptic problems associated with early FPC products. The coarse deodorized and defatted flour is further milled into a bland-tasting flour before it is labeled as a true fish protein concentrate. Although the product can be consumed directly as a protein source, its future uses probably lie in its supplemental admixture to milled cereals and low-quality protein grains.

Aside from FPC, the functional properties, nutritional benefits, and applications of fish muscle proteins can also be improved by *acylating* fish proteins. Protein acylation is achieved by reacting acetic or succinic anhydrides (electrophiles) with amino groups (nucleophiles) to produce amide products. Acylated proteins formed by amide production not only increase the temperature stability of fish proteins over the range of 20 to 40°C but improve the possibilities for fish protein usage in baked products, whipped desserts, confections, and other processed foods.

The future of drum-dried fish muscle, fish muscle emulsions for fish sausages, fish jellies (*kamaboko*), fermented strips of skipjack (*katsuwo-bushi*), and a host of other processed and modified fish proteins have yet to be technologically and economically exploited in most Western and European countries.

The protein quality of edible portions of the king crab, Pacific (Dungeness) crab, Atlantic blue crab, lobster, shrimp, scallop, oyster, and clam rivals or occasionally exceeds that of the finfishes (Table 3.6). Unfortunately, the delectability, high market price, and comparatively low productivity of these species versus fishes ensures that they will never provide a significant yield of protein to counteract world food shortages.

One of the possible marine resources that may offer a rich source of protein are the small crustaceans called *krill*. Each organism ranges from 40 to 60 mm in length and weighs 0.6 to 1.5 g. One kilogram of krill is composed of 1300 to 7500 individual organisms, which have a protein content in the range of 11 to 16%. Unlike many other marine protein resources, krill offers high protein content coupled with high levels of natural productivity in the open ocean. It is estimated, for example, that the annual krill catch can easily exceed 100–400 million tons, and there is little doubt that food-technology innovations will utilize this resource in some way.

Poultry Proteins

Poultry meat: The protein composition of mammalian red meats is similar to those concentrations found in poultry meats. Chicken and turkey usually contain a higher percentage of protein than geese and ducks as a result of the rigorous genetic cross-breeding and hybridization efforts of poultry science. Aside from the high protein content of poultry meats (Table 3.7), they are preferred in many diets as a result of their favorable digestibility, their high B vitamin and mineral contents, plus their low concentration levels of saturated fatty acids and cholesterol.

Poultry meats are less expensive than most other animal meats to produce since poultry show a very good feed-to-protein conversion ratio of ~2.3 : 1.0; that is, ~2.3 lb of feed are required to produce 1.0 lb of bird.

The tenderness of poultry is evaluated on the basis of the birds live weight and age, which corresponds to a scale of "smaller and younger" to "larger and older." The graded tenderness of most poultry such as chickens is mirrored by their designation as a broiler or fryer, roaster, capon, stag, stewing chicken, or old rooster, where the meat tenderness is maximum for broilers and fryers.

Egg proteins: The shell, white, and the yolk of eggs (Figure 3.6), respectively, contain ~4.0, ~10.5, and ~17.4% protein, which totally account for an ~11.8% protein content in the entire egg. The egg yolk proteins consist of a variety of glycoproteins, lipoproteins, phosphoglycoproteins, and phosphoglycolipoproteins. In contrast to these proteins, the egg white consists of at least 40 different types of globular glycoproteins. In spite of the significant dietary consumption of egg proteins, relatively little is known about

Table 3.7. Percentage of Protein in 100 g of an Edible Portion of Poultry Meat[a]

Poultry type or meat portion	Protein (%)
Chicken, raw fryers	20.5
Chicken, raw roasters	20.2
Chicken, roasted dark meat	28.0
Chicken, roasted white meat	32.0
Chicken liver	22.1
Duck	16.1
Goose	15.9
Turkey	20.1

[a]Based on data reported in USDA Handbook Number 8 (1963) and Burton (1976).

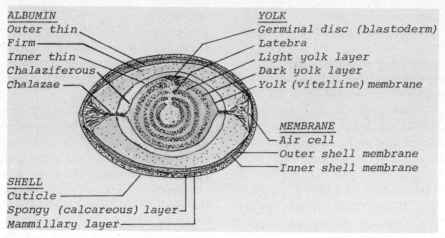

FIGURE 3.6. Important structural features of a hen's egg.

their individual amino acid sequences or native biochemical functions in the fertile egg. Some of the notable egg proteins are listed in Table 3.8.

Blood plasma proteins: The blood of slaughtered animals including poultry may provide an excellent source of dietary protein; however, the dietary acceptability of blood-derived protein fractions in foods offers a major marketing challenge. The notable ease of digestion for chicken blood plasma proteins (~90%) is coupled with a

PER value of ~2.8 as opposed to 2.5 for casein (Del Rio de Reys *et al.*, 1980). The future of plasma proteins probably rests in supplementing animal feeds or as an adjunct to low-protein foods. For example, dialyzed plasma fortification of wheat flour at 2.5 to 5.0% concentration levels can raise the PER value from ~0.87 to ~2.0.

Milk Proteins

Milk is a lacteal fluid secreted by the mammary glands of mammals (Figure 3.7). Al-

Table 3.8. Egg White Proteins Tabulated According to Their Percentage Occurrence in an Egg, Their Isoelectric Points, and Their Notable Properties[a]

Protein	Percentage	pI	Notable properties
Ovalbumin	54.0	4.5	Contains sulfhydryl groups that easily denature
Ovotransferrin	12.0	6.5	Antimicrobial substance; complexes with metals
Ovomucoid	11.0	4.1	Trypsin inhibitor
Ovoinhibitor	1.5	5.1	Inhibits serine proteinases
Ficin inhibitor	0.05[b]	~5.1	Inhibits thiol proteinases
Ovomucin	3.5	4.5–5.0	Involved in egg deterioration
Lysozyme	3.4	10.7	Cleaves polysaccharides
Ovoflavoprotein	0.8	4.0	Binds riboflavin
Ovomacroglobulin	0.5	4.5	Displays strong antigenic properties
Avidin	0.5	10.0	Binds biotin

[a]Adapted from Osuga and Feeney (1977).
[b]From Fossum and Whitaker (1968) and Sen and Whitaker (1973).

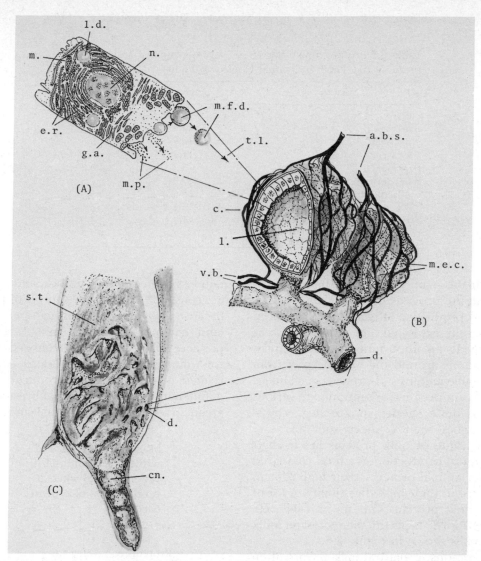

FIGURE 3.7.   Lacteal secretions in the cow mammary gland involve the simultaneous production of milk proteins (m.p.) and milk-fat droplets (m.f.d.) by many lactating cells (A). Liberated fat droplets and milk proteins proceed toward the lumen (t.l.) of an alveolus (B). The lumenal accumulations of milk secretions from billions of alveoli are introduced into a duct (d.) network, which ultimately directs the milk into a large common duct and reservoir recognized as the cistern (cn.) (C). Other notable features of the figure include: a.b.s., arterial blood supply; c., capillary network; e.r., endoplasmic reticulum; g.a., golgi apparatus; l.d., lipid droplet; m., mitochondria; m.e.c., myoepithelial cellular network; n., nucleus; s.t., secretory tissue; and v.b., venous blood. Milk protein production is intracellularly fabricated by lactating cells according to recognized processes of RNA transcription and translation prior to their release within the lumen. The m.p. production originates from glucose, amino acids, and proteins supplied by the hematogenous circulation. Acetate and β-hydroxybutyrate supplied by the blood to the mammary tissues are converted to C_4–C_{16} fatty acids. All of these lipids clearly contribute to the overall fat content of the milk. Lactose contained in the lacteal secretions originates from glucose transformation and subsequent assembly of the dissacharide. Some milk proteins pass relatively unchanged from the blood into milk via specialized transport systems. These proteins may include serum albumins and immunoglobulins. The initiation of milk production or lactogenesis is partially tied to the suppression of progesterone. The constancy of milk production is also contingent on the removal of milk from the mammary gland according to a regular time frame; otherwise lactation diminishes and eventually ceases.

Table 3.9. Approximate Percentage Composition of Major Milk Proteins and Their Individual Isoelectric Points (p*I*)

Name	Percentage	p*I*
Caseins	78	4.6
α,1 casein	42.9	5.1
β-casein	19.5	5.3
κ-casein	11.7	3.7–4.2
γ-casein	3.9	5.8
Whey (or milk serum) proteins	17.0	
β-lactoglobulin	8.5	5.3
α-lactalbumin	5.1	5.1
Immunoglobulins	1.7	4.0–6.0
Serum albumin	1.7	4.7

though the main purpose of the milk is to nourish the progeny of a particular species, the milk production of many species has been successfully exploited and maximized so as to provide an important food for human populations around the world. Apart from the notable amounts of proteins in all milks, the percent protein concentration in various milks is quite species specific (e.g., cow—3.35%; goat—3.70%; sheep—6.30%).

The value of milk proteins lies in their economical production as a food compared to other animal proteins, their high protein quality, and their high digestibility. A list of major milk proteins appears in Table 3.9. Furthermore, a partitioning of skim milk proteins is shown in Figure 3.8.

Alterations in natural milk flavors often result from protein denaturation that occurs during heat processing. Milk pasteurization alone causes little significant change in protein structure or flavors, but prolonged exposure to more severe heat can cause notable structural changes and off-flavors in whey proteins; exposure of β-lactoglobulin sulfhydryl groups; alterations of caseins; and an overall increased tendency for proteins to coagulate.

Plant Protein Resources

The leaves, seeds, nuts, and root systems of plants contain 2 to 25% protein. Plants con-

tain less protein than livestock, and yet they are considered to be good sources of protein because they yield more protein per acre of land than meat-producing animals. The low per acre "plant-to-protein" conversion ratio for animals reflects tremendous expenses related to livestock food production, land maintenance, and wintering of livestock that are incurred before a marketable meat prod-

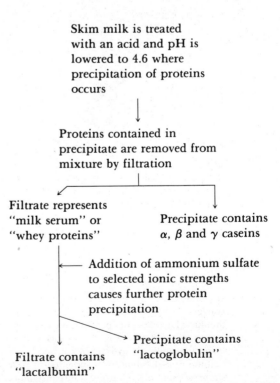

FIGURE 3.8. Schematic operations required for the partitioning of skim milk proteins.

uct is produced (e.g., beef). In fact, plant protein yields may be five to nine times richer than animal yields per acre. Unfortunately, however, the low concentrations of certain amino acids in plants requires that large quantities of the material be consumed by humans to meet their dietary protein requirements.

The practical inefficacies of consuming only plant foods to satisfy normal dietary protein requirements has prompted technological efforts to extract, purify, and concentrate plant proteins before they are used as a food. Many of these efforts are presently unfeasible and surpass the technological capabilities of nations that already have the greatest deficits in protein and fiscal resources for developing new food technology. The following discussion details some of the most important plant protein sources.

Cereals: The main cereal crops of the world include wheat, corn, rice, rye, oats, barley, sorghum, millet, and triticale. The protein content and quality of cereals is generally poor (~8–16 + % protein) (Table 3.10), but many of these deficiencies can be ameliorated by adding amino acids such as lysine or methionine that otherwise limit protein quality; or cereals may be nutritionally supplemented by the proper admixture of other plants.

Wheat. The protein content and functional versatility of wheat makes it suitable for the production of breads, rolls, cookies, crackers, macaroni, noodles, doughnuts, pancakes, and many other products. The protein content of wheat is species dependent and subject to variability depending on environmental or climatic conditions where it is grown (Table 3.11). *Common* (hard) *wheats* contain 12 to 13% protein; *club* (soft) *wheats* contain 7.5 to 10% protein; and *durum* (very hard) *wheats* have 13.5 to 15% protein. Most wheats can provide necessary levels of dietary protein at a reasonable caloric intake value, but their inherent lysine deficiency requires the admixture of lysine to wheat products in order to minimize this deficiency and improve protein quality.

Nearly 82% of the protein content found in the starchy wheat endosperm, known as *gluten,* is composed of *gliadin* and *glutenin,* which are particularly rich in glutamine and proline. The high proline content of these proteins clearly prohibits their formation of a helical secondary structure. The hydrated forms of both proteins contribute to their functional uses in many foods. Glutenin (M.W. 20,000–100,000 daltons) consists of a subunit structure linked by disulfide linkages that impart elastic-cohesive properties to doughs, while gliadin (M.W. 16,000–50,000 daltons) displays viscous properties and low elasticity and contributes to the volume of wheat doughs.

Corn. The 9 to 10% protein of corn resides in both the germ and the endosperm of the corn kernel. The predominant species of corn are *deficient* in lysine and tryptophan, but newer mutant varieties such as "Opaque-2" have been able to yield higher lysine concentrations.

According to the principles of protein solubility, corn proteins can be fractionated into four groups, namely, (1) *albumins,* which are water soluble; (2) *globulins,* which are soluble in salt solutions; (3) *prolamines,* which are soluble in 70% ethanol; and (4) *glutelin* proteins, which dissolve in 0.1 *M* sodium hydroxide.

The endosperm proteins of corn contain ~40 to 50% zein, which is classed as a prolamine, and 30 to 40% glutelins. Recent innovations in corn germ processing have led to production of corn germ flours, which offer protein concentrations of ~25%.

Oats. Oats offer a protein content of 15 to 22% accompanied by a good balance of essential amino acids including lysine and sulfur-containing amino acids. Since Captain Bartholomew Gosnold first planted oats on Cuttyhunk Island, Massachusetts, in 1602, oat harvests measured in bushels have as-

Table 3.10. Percentage Protein Content of Selected Cereals, Grains, and Their Related Food Products[a]

Grain, cereal, or product	Protein (%)	Grain, cereal, or product	Protein (%)
Corn		**Wheat**	
Whole-ground cornmeal		Bulgar, canned	5.9(56)
unbolted, dry	9.0(12)	Farina, quick-cooking,	
		enriched, cooked	1.2(89)
Degermed cornmeal,			
cooked		Whole-wheat flour	13.3(12)
Enriched	1.3(88)		
Unenriched	1.3(88)	All-purpose flour,	
		enriched	10.0(12)
Corn (hominy) grits,			
degermed, cooked		Cake or pastry	
Enriched	1.2(87)	(general)	7.3(12)
Unenriched	1.2(87)	Angel food	8.4
		Sponge	7.9
Corn flakes, added			
nutrients		Doughnuts	6.6
Plain	8.0(4)		
Sugar-covered	5.0(2)	Puffed wheat, added	
		nutrients	13.3(3)
Puffed corn,			
presweetened, added		Shredded wheat	8.0(7)
nutrients	3.3(2)		
		Wheat flakes, added	
Oats		nutrients	10.0(4)
Oatmeal or rolled		Macaroni, cooked tender	
oats, cooked	2.1(87)	Enriched	3.6(72)
		Unenriched	3.6(72)
Puffed oats, added			
nutrients	12.0(3)	Noodles (egg), cooked	
		Enriched	4.4(70)
Rice		Unenriched	4.4(70)
White rice, dry	7.5		
Brown rice, dry	7.6	Pretzels	8.8
Enriched, cooked	1.9(73)		
Unenriched, cooked	1.9(73)	**Rye**	
Puffed rice, added		Rye flour, dark	12.0
nutrients	6.7(4)		
Cassava beans		**Soy beans**	
Tapioca	0.6	Soy bean flour defatted	44.7

[a]Percentage protein is expressed with respect to the percentage water content of a given sample in the parentheses where necessary.

sumed a fourth-place rank in the United States following soybeans, wheat, and corn. Of the average 160 bushels per acre yield of oats produced in the United States, approximately 90% of the yield has traditionally been used to feed livestock, but recent advances in protein extraction from oats by centrifugation to produce protein concentrates having up to 80% protein will inevitably lead to increased use of oat proteins in beverages, meat extenders, and baked goods.

Rice, barley, rye, sorghum, and millet. These sources of plant protein represent minor

Table 3.11. Comparison of Protein Content in Spring and Winter Wheats

Protein (%)	Glutenin	Gliadin	Globulin	Leucosin
Spring wheat	4.7	3.96	0.62	0.40
Winter wheat	4.20	3.90	0.64	0.63

contributors to the total dietary protein supplies of the United States and most Western European countries, although rice happens to be the world's most important food crop in developing countries and Asia.

The protein content of rice varies from approximately 7.5 to 14% depending on the variety; rye contains 11 to 12% protein; and barley has 11 to 13%. Unlike rice, which serves as a nutritional staple in many diets, rye and barley serve principally as animal feeds and have rather limited applications in foods such as barley's role in beer production and rye's use in selected types of crackers and breads.

Sorghum offers a relatively poor source of proteins compared to other cereals because of significant imbalances in native amino acid content and its poor digestibility. With the exception of special high-lysine sorghum mutants, lysine and sulfur-containing amino acids are low, although cystine concentrations may be adequate.

The role of millet as a key protein source is minor for most sectors of the human population.

Soybeans. Soybeans have clearly served as a food since 2838 B.C., when King Chan Nonag of China mentioned soybeans in his medical treatise. Since 1904, when soybeans were brought to the United States from Europe, their agricultural production has steadily increased, principally because of their high protein and oil contents.

Unlike the average maximum of ~16% protein found in most plants, seeds, and nuts, soybeans naturally contain up to 38% protein, which rivals or exceeds the protein content and food values of meats, milk, fish, and eggs. Soy proteins contain nearly optimal proportions of all the essential amino acids for animals and humans including lysine and tryptophan, but the protein is poor in cystine, methionine, and valine.

There are three general types of soy proteins that can be classified according to their protein content.

Soy flours and grits are products that contain 40 to 60% protein. Grits are coarse products if ground larger than 100 mesh size, while soy flours are classified as 100 mesh size or finer. The protein content of flours and grits may be accompanied by natural soy oils, although moisture content may be variable.

Soy protein isolate (SPI) represents the major proteinaceous fraction of flaked soybeans that have had most nonprotein substances removed. The SPI is typically spray dried to a protein content of 90% on a dry weight basis.

Soy protein concentrate (SPC) is prepared by removing most water-soluble nonprotein constituents from defatted soy flakes or flours including soluble sugars such as sucrose, stachyose, and raffinose. The aqueous extraction of these sugars plus other nonprotein substances using selected alcohols leave an insoluble protein product (SPC) that can have approximately 75% protein measured as a moisture-free product.

The textural and functional properties of soy proteins permit them to be modified into fibers if they are subjected to prescribed heat and moisture treatments. The proteins can be alternatively acid hydrolyzed to yield hydrolyzed vegetable protein.

Many studies have been carried out in order to assess the nutritional value of soy proteins. The value of soy-protein-containing products for human consumption has been widely recognized and supported by many nutritional investigations. Typical among soy nutritional studies are those of Young *et al.* (1977) and INCAP's Center for Advanced Studies in Nutrition and Food Science (University of San Carlos, Guatemala) where children (2–4 years) were fed a soy product (Supra 620, Ralston Purina & Co.) whose digestibility was high and equivalent to whole milk and superior to whole egg.

The nutritional suitability of soy protein in infant nutrition is also demonstrated by its use in milk-free infant formulas manufactured for infants who are intolerant to either milk protein or lactose. For these applications, soy protein properly supplemented with methionine has been shown to provide growth patterns and nitrogen retention equivalent to cow's and breast milks (Graham *et al.*, 1970; Fomon *et al.*, 1973). An ingredient survey of essential amino acids in Prosobee (Mead Johnson & Co.) typifies the

breadth of amino acid balance afforded by the use of soy protein in a formulated milk substitute compared with the essential amino acid spectra of human and bovine milks (Table 3.12). Moreover, the use of processed soy proteins in certain infant formulas is desirable because they can be made hypoallergenic for infants who are otherwise sensitive to milk or food proteins such as gluten.

Aside from positive nutritional features, soybeans may contain various undesirable substances including trypsin inhibitors, hemagglutenins, saponins, goitrogens, estrogenic compounds, and possibly aflatoxins (produced as a result of soy bean storage conditions) that can have antinutritional consequences (Wolf, 1975). Fortunately, however, many of these factors can be inactivated or eliminated by careful heat treatment in order to maximize the potential dietary utilization of soy proteins and eliminate any potential health hazards.

STRUCTURES OF SOY PROTEINS. Soybeans naturally contain a mixture of pro-

Table 3.12. Comparison of Essential Amino Acid Contents in a Soy-Protein-Substituted Infant Formula (Prosobee, Mead Johnson & Co.) as Compared with Infant Requirements for Essential Amino Acids, Human Milk, and Bovine Milk[a]

Essential amino acids (mg/100 kcal)	Infant requirement	Soy protein milk substitute (Prosobee)	Human milk	Bovine milk
Histidine	25	78	29	144
Isoleucine	62	159	91	144
Leucine	144	273	133	530
Lysine	92	203	97	420
Methionine	—	70	33	133
Cystine	—	33	29	48
Methionine + cystine	52	103	62	181
Phenylalanine	—	173	64	261
Tyrosine	—	141	81	271
Phenylalanine + tyrosine	112	314	145	532
Threonine	78	125	67	248
Tryptophan	15	44	24	74
Valine	83	159	93	371

[a]Information adapted from *Handbook—Infant Formulas* (1977), Mead Johnson and Co., Evansville, Ind., and used with permission.

teins ranging in molecular weight from 8000 to 600,000 daltons. These proteins are soluble in aqueous solvents and can be ultracentrifugally partitioned into four fractions:

found to contain 12 N-terminal residues that consisted of 6 glycines on the basic subunits, while 2 phenylalanines, 2 isoleucines, and 2 leucines occur on the acidic subunits. Elec-

Water Soluble Protein Fractions (WSPF) of Soybeans
Obtained by Ultracentrifugation

~30% WSPF		~70% WSPF	
2 S	15 S	7 S	11 S
Contain cytochrome, trypsin inhibitor, and a variety of globulins		Contains hemagglutinins, lipoxygenase, α-amylase, and more than one-half of the 7 S fraction is globulin	Globulins make up the majority of the 11 S fraction

Both the 7 S and 11 S fractions exhibit very rapid, reversible, association–dissociation abilities under certain conditions of ionic strength and pH.

Although the dissociation phenomena and structural-functional biochemical relationships for soy protein are not clearly understood, Koshiyama (1968) has reported isolation of a monomeric 7 S globulin (M.W. 180,000 to 210,000 daltons) at a pH of 7.6 and an ionic strength of 0.5, but alteration of the ionic strength to 0.1 causes the monomeric globulins to undergo dimerization.

The 7 S fraction of soy globulin displays nine N-terminal amino acid residues and may be dissociated using 8 M urea into a series of subunits each having a molecular weight of 22,000 to 24,000 daltons (Koshiyama, 1971).

The 11 S protein fraction shows a similar pattern of association and dissociation as the 7 S fraction at ionic strengths of ≈ 0.1. Catsimpoulas (1969) in particular has reported that the 11 S fraction of *glycinin* (a globulin) contains three acidic (A) subunits and three basic (B) subunits, each of which has a different isoelectric point. Three pairs of these different subunits were found to occur twice in the macromolecular assembly of the 11 S globulins. Badely *et al.* (1975) further confirmed this observation when the protein was

tron microscopy has further revealed that the 11 S protein fraction has a two-layered structure, with each layer consisting of alternating acid and basic subunits in a hexagonal structure (Figure 3.9).

Other Plant Protein Sources

Lentils, beans, peas, chickpeas, and related leguminous plants are consumed for their protein content throughout many parts of the world. In addition to acceptable NPU values (Table 3.13), many of the legumes such as peas can be dehulled and pin-milled to produce a protein-rich flour (32% protein, dry weight) that has many diverse applications. Peanuts offer other possibilities as a protein source but they are inferior to soybeans due to their lysine and methionine deficiencies. In spite of these inadequacies, peanuts can be imaginatively formulated into FPC or skim milk (whole or dried) to enhance their protein character.

Promising sources of plant protein are also present in hybrid cereal grains and leaves of plants.

Of the newer cereal grains, *triticale,* offers a higher protein content and better amino acid balance than many traditional species of wheat. Triticale is a hybrid cereal produced by cross-breeding rye and durum

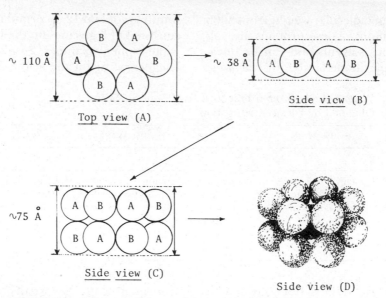

FIGURE 3.9. Schematic diagram of the protein glycinin (a globulin) present in the 11 S fraction obtained from soybeans. The subunits of the glycinin are labeled as "A" or "B," respectively, depending on whether they are acidic (pI = 4.75, 5.15, 5.40) or basic (pI = 8.00, 8.25, 8.50). Four views of the molecule are shown: *Top view* (A): hexagonal shape of alternating A and B units; *side view* (B): side view of alternating A and B units that form *monomeric hexagonal* subunit of glycinin's dimeric quaternary structure; *side view* (C): side view of *dimeric quaternary structure*; and *side view* (D): three-dimensional view of quaternary protein structure.

wheat. The hard qualities of durum wheat that limit its use to non-bread-making applications such as alimentary pastes are moderated by the genetic traits of rye. Rye on the other hand, offers a flour that contains the protein gliadin but does not form dough films sufficiently strong to support an ideal bread structure. The genetic traits of durum wheat counteract this problem. It is clear then that the functional properties of triticale over rye or durum wheat make it suitable for breads, noodles, pancake mixes, cereal flakes, and a host of related applications. Further, the positive features of rye and durum wheat that make each resistant to certain environmental stresses (arid soil, extreme climatic conditions, etc.) are mirrored in the triticale hybrid.

Plant leaves from alfalfa, grasses, and other leafy vegetables naturally contain large amounts of protein and potentially abundant amounts of food proteins. The possi-

bilities for incorporating leaf proteins into conventional dietary patterns is limited by two factors, namely, (1) economic feasibility and (2) health and organoleptic factors. New technologies must be developed at high cost to produce a commercially marketable protein product that will have uncertain consumer acceptance and marketability. Moreover, without regard to the well-balanced amino acid composition of plant leaves such as alfalfa and others, many plant protein isolates are often plagued by the presence of undesirable toxic compounds, peculiar colors, poor textures, and unacceptable flavors.

"Unconventional" and Single-Cell Protein (SCP) Sources

Difficulties associated with the development of new plant protein sources are shared by the so-called unconventional and SCP sources. These technological approaches for devel-

Table 3.13. Approximate Net Protein Utilization (NPU) Values of Some Foods

Net protein utilization[a] (%)	Type of food	Essential amino acids	
		Adequate	Poor
95	Eggs	Trp, Lys, Met, Cys	
85	Cow's milk	Trp, Lys	
82	Fish	Lys	
73	Cottage cheese	Lys	
70	Turkey	Lys	
68	Pork, beef	Lys	
67	Wheat germ	Lys	Trp
64	Chicken, lamb	Lys	
64	Oatmeal		Lys
63	Corn		Trp, Lys
63	Asparagus		Met, Cys
57	Potato	Trp	Met, Cys
61	Soybeans	Lys, Trp	Met, Cys, Val
50	Lima beans	Trp, Lys	Met, Cys
41	Peanuts		Lys, Met, Cys, Thr
36	Kidney beans	Lys	Trp, Met, Cys

[a]For more complete information concerning percent NPU values, consult Scrimshaw and Young (1976).

oping protein-rich food substances rely heavily on controlled mass culture techniques for yeasts, bacteria, molds, and algae. The desirability of protein production by these organisms lies in their rapid growth and ability to transform basic nutrients in a defined culture medium into a nutritionally significant protein.

Mass cultures of bacteria and yeasts have been prominently cited for their potential uses in unconventional protein production since their culture requirements and biochemical compositions are well known (Figure 3.10).

Bacterial populations tend to show higher rates of protein production along with higher concentrations of sulfur-containing amino acids than yeasts, but these advantages are often overshadowed by bacterial proteins' lack of functional versatility in conventional foodstuffs. Yeast proteins, however, may be produced at a slower rate than bacterial proteins but the protein-rich culture product is adaptable to many conventional food products. In addition, the biochemical by-products of the yeast culture do not share the psychological stigma of bacterial proteins.

Fungi and algae have received comparatively little attention as possible protein sources in spite of their successful growth in continuous mass cultures using inexpensive waste carbohydrates and mixtures of inorganic salts. Both classes of organisms display slower growth and protein production rates than bacteria or yeasts, which jades their possible economic values. Moreover, algae may require additional energy expenditures for culture aeration, high-intensity illumination, and other *in vitro* culture necessities. Added to these considerations, both groups of organisms can produce undesirable flavors, toxic compounds ranging from aflatoxins in fungi to geosins (cyclic peptides) and other toxic substances in algae, and unsavory food textures, as well as diarrhea, nausea, and other problems when consumed in significant dietary amounts.

IMPROVEMENT OF DIETARY PROTEIN QUALITY

Protein calorie malnutrition is the most common nutritional deficiency in the world; it

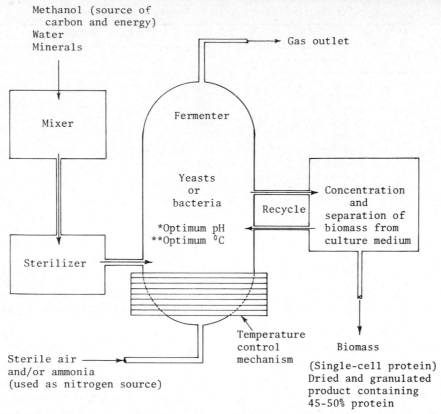

FIGURE 3.10. Synthetic protein from methanol. Protein-rich bacteria and yeasts can be cultured in a growth medium composed of selected petrochemical derivatives such as natural gas hydrocarbons, petroleum, specially processed coal derivatives, and methanol in order to produce proteins. Methanol is especially desirable because it is water soluble and readily available to microorganisms in the culture (*optimum pH, 3.0–5.0 for yeasts; 6.0–7.5 for bacteria; and **optimum °C, 27–30 for yeasts; 35–43 for bacteria).

is predominantly associated with developing countries. When protein calorie malnutrition strikes at an early age in the form of kwashiorkor or marasmus, the consequences may be evidenced as stunted growth, physical impairment, loss of learning abilities, increased incidence of mental retardation, or possibly death.

Since the protein supplies of many developing countries hinge on the availability of cereals and legumes that are naturally deficient in one or more essential amino acids, this problem has encouraged nutritionists and food scientists to upgrade low-quality protein sources by means of amino acid *fortification,* protein *supplementation,* or protein *complementation.*

Amino Acid Fortification

Amino acid fortification deals with improving the quality of food proteins by adding prescribed amounts of *at least* one deficient amino acid that naturally limits the food protein's quality. For most human diets the nutritionally limiting amino acids are the essential amino acids that particularly include *lysine, methionine, tyrosine,* and *tryptophan.*

The method for adding nutritionally limiting amino acids to foods depends on the type of food. Dry powdered food products (e.g., cereal flours, dried milk) or liquid foods can, respectively, be fortified by adding the desired amino acids as a powder or dissolving them in the liquid. In cases of whole grain wheat or rice fortification, however, it is sometimes necessary to superficially abrade the surfaces of dried grains and then steep them in a solution such as 35% lysine for periods up to three hours at 71°C so they absorb the amino acid.

The well-intended effects of food protein fortification using added amino acids produces mixed nutritional and chemical results. Although methionine, lysine, cystine, and other essential amino acids may limit the quality of proteins, this is not a license for the nutritionist or food scientist to over-zealously add unbounded amounts of deficient amino acids to foods. The intrinsic biochemical importance of essential amino acids is not synonymous with the dietary rationale that "more is better." Since all essential amino acids are critical to metabolic pathways, they can be among the most toxic amino acids and their megadoses in foods may cause deleterious effects in animals ranging from growth inhibition to eye, skin, and liver lesions. Furthermore, the excessive addition of certain fortifying amino acids can lead to the production of an *unbalanced* protein. In order to avert this problem, agencies such as the U.S. Food and Drug Administration (USFDA) have suggested that the safety criteria for adding amino acids should be determined by taking into account the differences in the quantities of deficient amino acids in a food with respect to the amounts of the same amino acids contained in an idealized reference protein source such as whole egg.

The scrupulous addition of deficient amino acids to poor-quality protein foods can have decided nutritional advantages for animals as well as humans. For example, weanling rats fed white bread plus lysine, and white bread with lysine and threonine, showed improved weight gains, PER, and NPU values over unfortified diets at all levels of energy intake (Jansen and Verburg, 1977). Studies related to methionine have also attested to the advantages of amino acid fortification in humans. Infants fed soy protein isolate reportedly fared in an inferior way to other infants that received breast milk, milk-based formula, or soy protein isolate containing methionine during the first six weeks of life (Fomon, 1979).

Regardless of the positive nutritional advantages accrued from food fortification, the addition of chemically labile or reactive amino acids to foods can alter the normal food flavor profile as well as its palatability. This is especially true for methionine. In some amino acid fortification efforts, less reactive derivatives of certain amino acids may be added to foods that offer the same nutritional benefits of the pure amino acid without sacrificing food quality. One notable example of a derivatized amino acid is N-acetyl-DL-methionine, which can be substituted for pure methionine.

Some chemical consequences of adding free amino acids to foods: The addition of free amino acids to foods may promote nonenzymatic browning reactions such as the Maillard and Strecker degradation reactions.

The Maillard reaction is characterized by a multistep reaction sequence that is initiated by the formation of an addition product involving glucose (reducing sugar) and a free amino acid. The resulting glycosylamine undergoes an Amadori rearrangement and eventually yields *hydroxymethylfurfural* (HMF). HMF serves as a precursor to *melanoidin* formation when additional amines are present (Figure 3.11).

The imminent production of melanoidins and spectrophotometric evidence of nonenzymatic browning is readily obvious at 277 to 285 nm, which seems to be indicative of furfural formation.

The undesirable effects of nonenzymatic browning can be limited in many practical situations where amino acids are added to foods by blocking the functional carbonyl groups on reactive sugars. Successful blocking reagents for sugars include a variety of sulfite salts that inhibit Schiff's base formation in the overall reaction and thereby prohibit the eventual formation of melanoidins.

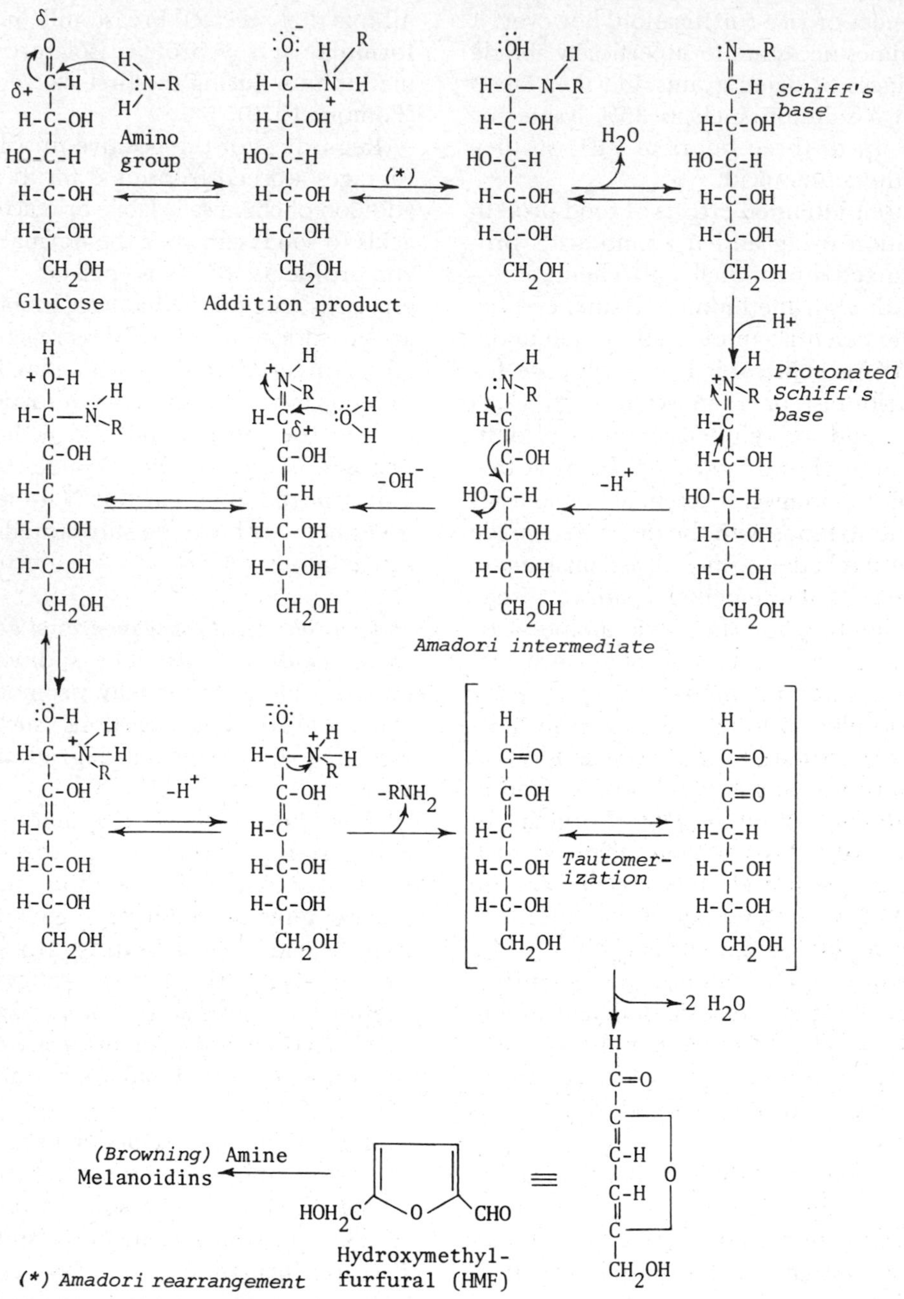

FIGURE 3.11. Some important reaction steps in the Maillard reaction including the initial reaction of a free amino acid with a reducing sugar (glucose) to produce glycosylamine, and then an Amadori intermediate that feeds hydroxymethylfurfural (HMF) formation. The HMF subsequently contributes to the formation of melanoidins in the presence of additional amines.

Proposed blocking effects of sulfites are attributable to the reaction outlined below:

radation aldehydes formed in bread crust from various amino acids.

$$
\begin{array}{c}
\underset{\text{C}}{\overset{\text{H}}{\underset{|}{\text{C}}}}\!\!\overset{\text{O}}{\diagup}\\
\text{H—C—OH} + \text{NaHSO}_3 \rightleftharpoons \\
\underset{\perp}{}
\end{array}
\quad
\begin{array}{c}
\text{OH}\\
|\\
\text{H—C—SO}_3^-\text{Na}^+\\
|\\
\text{H—C—OH}\\
\underset{\perp}{}
\end{array}
\quad + \quad
\begin{array}{c}
\text{H}\\
|\\
\text{R—C—COOH}\\
|\\
\text{NH}_2
\end{array}
$$

End of   **Sodium**
reducing sugar   **bisulfite**

Further reaction between amino acid and reducing sugar is inhibited

In contrast to the Maillard reaction, the Strecker degradation involves the reaction of free α-amino acids with di- or tricarbonyl-containing compounds at high temperatures. The products of this reaction typically include the corresponding aldehyde of the reacting amino acid with one less carbon, carbon dioxide, and a pyrazine substance (Figure 3.12). According to proposed mechanisms for the Strecker degradation, alanine can react with a dicarbonyl compound to produce a Schiff's base, which undergoes decarboxylation. The decarboxylated intermediate species is then protonated and hydrolyzed to yield acetaldehyde and tautomeric substances. The tautomers self-condense to form a *proaromatic* compound through the elimination of water. The proaromatic compound then serves as a precursor to the formation of a *pyrazine derivative* in the presence of oxygen.

Melanoidins produced as a result of amino acid reactions with sugars may be esthetically objectionable and distasteful. Furthermore, they may decrease the effective concentration levels of amino acids that are designed to fortify a food. Apart from the negative effects of the Strecker degradation on nutritionally important amino acids in foods, it should be noted that the Strecker reaction has been commercially used for producing favorable flavors in chocolate, honey, maple syrup, cooked foods, and baked products (Hodge, 1953; Hodge *et al.*, 1972). Table 3.14 outlines some characteristic Strecker deg-

Protein Supplementation

The practice of protein supplementation involves the addition of a "small" amount of a protein "rich" in a particular amino acid to another protein that lacks the same amino acid(s). Supplementation procedures usually increase the efficiency of protein utilization, but it should be recognized that the supplemental protein also adds to the total protein content of the food and may inadvertently decrease protein utilization efficiency in some cases. In other situations, the admixture of two or more different proteins can result in a synergistic nutritional improvement of the proteins; that is, the total nutritional value of the mixed proteins supersedes their individual nutritional values when separately consumed.

The nutritional benefits of protein supplementation are clearly illustrated by the admixture of sesame protein to navy bean protein. The intermingling of these proteins reportedly produces a PER value for the mixture that exceeds the PER values of the individual proteins when fed to rats. The nutritional advantages of this supplementation prove to be beneficial because navy beans are deficient in methionine and cystine yet rich in lysine, while sesame protein has significant concentrations of sulfur-containing amino acids and a poor amount of lysine (Tables 3.15 and 3.16) (Boloorforooshan and Markakis, 1979). Similar improvements in the protein quality of rice, wheat,

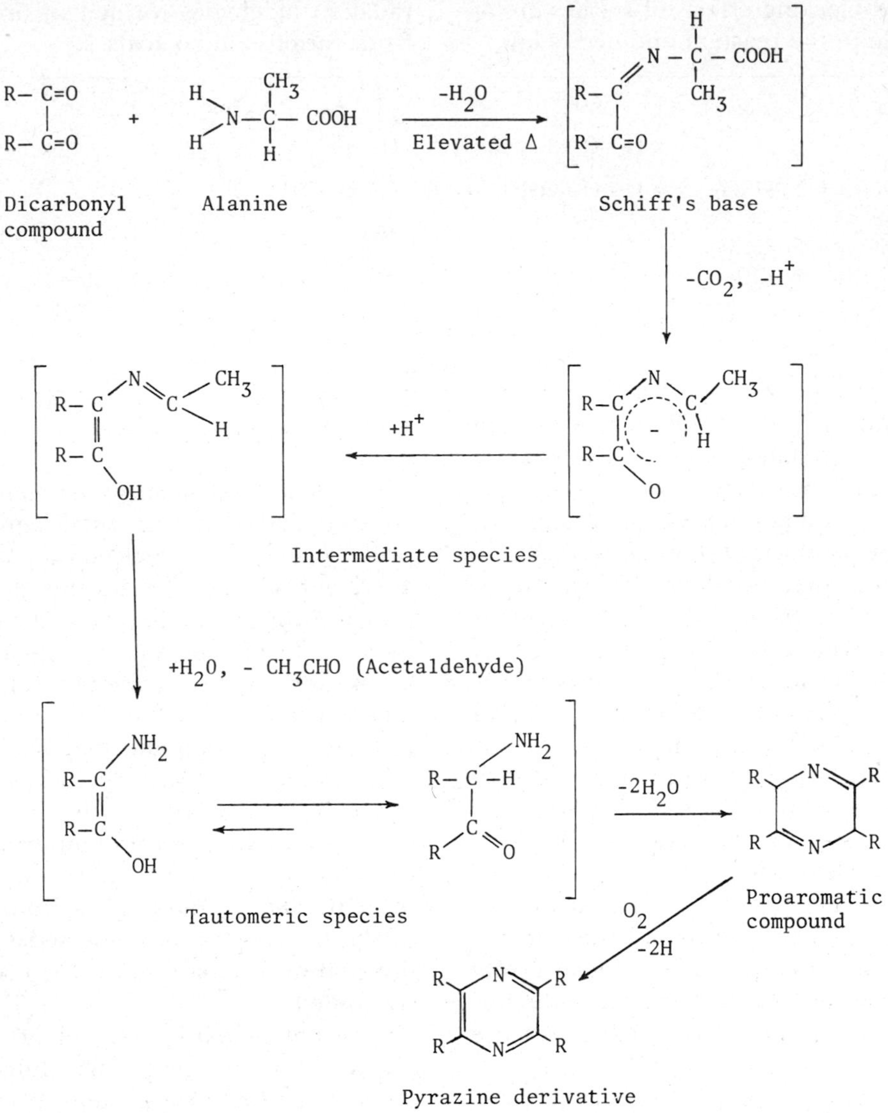

FIGURE 3.12. The role of a dicarbonyl compound in the Strecker degradation sequence involving the amino acid alanine. (Adapted from Rizzi, 1969.)

maize, sorghum, cassava beans, and other vegetative proteins have been achieved by supplementing these poor plant protein staples of the rural Kenyan diet with high-quality proteins and amino acids (Kilonzo, 1978).

The beneficial effects of supplementing notoriously low-quality protein diets throughout the world do not always meet with unmitigated success. Actually, some human populations that have historically and chronically consumed low-quality protein sources respond in an uncertain fashion to protein supplementation. The explanation for such lackluster performances has been explained as *population adaptation*. In these cases, adaptation of the population to a low-grade protein is so great that any favorable improvement in food protein fails to produce better health, or apparent improvement of nutritional status.

Studies with Nigerian men have also shown that the NPU of their traditional rice,

Table 3.14. Strecker Degradation Aldehydes Formed in Bread Crust from Various Amino Acids[a]

Aldehyde	Amino acid
Acetaldehyde	Alanine
Formaldehyde	Glycine
2-Methylbutanol	Isoleucine
Isovaleraldehyde	Leucine
Methional	Methionine
Phenylaldehyde	Phenylalanine
2-Hydroxypropanal	Threonine
Glyoxal	Serine

[a]Adapted from Johnson *et al.* (1966).

sorghum, and cassava root diets did not yield significantly different effects than a minimally adequate native protein diet supplemented with whole egg (Nicol and Philips, 1978). Further evidence of possible *adaptation* effects within a population are mirrored in studies conducted among children (3–6 years) in two Indian villages. A survey of the study population indicated dietary deficiencies of lysine, tryptophan, tyrosine, and sulfur-containing amino acids without any obvious manifestation of clinical deficiency effects. Other investigations by Baker *et al.* (1978), related to kwashiorkor, have also provided circumspect evidence regarding the benefits of supplementing low-quality protein foods with high-quality proteins to eliminate kwashiorkor symptoms in children.

Apart from the lack of irrefutable evidence supporting the merits of protein supplementation, it is still uncertain whether supplementation produces undetectable improvements in brain development, struc-

ture, or function. Progressive research into brain nutrition, growth, and development indicates that amino acid availability is critical to the maintenance of normal cell structure, ion balance, proper neurotransmission and neuroexcitatory functions, behavior patterns (Scrimshaw, 1967), and other central nervous system functions that remain to be established. Unfortunately the answers to many of these questions are sluggish in their evolution owing to the necessary dependence on studies of protein calorie malnutrition in animal models that must be extrapolated to seemingly similar conditions in humans (Kaladhar and Narasinga Rao, 1977).

Protein Complementation

Protein complementation is a term that refers to the admixture of two or more protein sources, based on their amino acid excesses and deficiencies, so that the nutritional weaknesses of each protein will counterbalance the other proteins. The nutritional advantages and disadvantages of protein complementation are not unlike protein supplementation, although complementary formulation may entail the mixing of larger relative proportions of different protein-containing foods.

Food protein quality can occasionally be improved by methods other than fortification, supplementation, or complementation. Sweet potatoes contain 4 to 6% protein but they also contain up to 3.5 times the recommended daily allowance for vitamin A.

Table 3.15. Comparison of PER Values for Navy Bean Flour, Sesame Flour, and Supplemented Combination Mixtures of These Flours[a]

Dietary substance	PER value
1. Navy bean flour	1.56
2. Sesame flour	1.19
3. 87.5% navy bean flour + 12.5% sesame flour	1.79
4. 75% navy bean flour + 25% sesame flour	2.26
5. 50% navy bean flour + 50% sesame flour	2.30

[a]Adapted from data reported by M. Boloorforooshan and P. Markakis, 1979, *Journal of Food Science* **44**:391. Copyright © by Institute of Food Technologists.

Table 3.16. Amino Acid Composition for Navy Bean Flour and Defatted Sesame Flour Expressed in g/16 N[a]

Amino acid	Navy bean protein	Defatted sesame flour
Alanine	3.3	3.5
Arginine	5.1	11.6
Aspartic acid	10.6	6.7
Cysteine	0.7	1.6
Glutamic acid	12.8	17.1
Glycine	3.1	3.6
Histidine	2.4	2.2
Isoleucine	3.7	3.0
Leucine	6.7	5.4
Lysine	5.7	2.3
Methionine	1.0	3.1
Phenylalanine	5.3	3.9
Proline	3.3	3.0
Serine	5.1	3.6
Threonine	4.1	3.0
Tryptophan	1.2	1.5
Tyrosine	3.1	2.9
Valine	4.4	3.9

[a]Reprinted from M. Boloorforooshan and P. Markakis, 1979, *Journal of Food Science* **44:**391. Copyright © by Institute of Food Technologists.

Therefore, sweet potatoes can be fortified with soy flour, cotton seed flour, and gluten to make high-quality protein products that have ~17 to 24% protein (dry weight) jointly accompanied by a significant concentration of vitamin A (Walter and Purcell, 1978). The addition of legume flours to wheat and corn products yield similar nutritional advantages since the legumes elevate ascorbic acid and riboflavin concentration levels in the cereal products. The protein quality of foods can also be improved or at least affected by dietary mineral content. In some studies of rat nutrition, it has been speculated that dietary minerals may become the "most limiting factors" in conjunction with available dietary protein (Ebihara *et al.,* 1979).

Conclusions Regarding Protein Quality Improvement

The nutritional benefits of amino acid fortification, protein supplementation, and protein complementation have been widely studied. In spite of these studies, protein quality ratings for many foods do not produce steadfast improvements in the nutritional status of individuals who are *clearly lacking* proper levels of protein intake and proper dietary amino acid breadth.

This would suggest that current methods designed for assessing food protein qualities are based on

1. Species specific results of test animal populations that have little relationship to actual dietary requirements of humans.

2. Methods for protein quality assessment that may be unduly superficial owing to important dietary protein interrelationships with minerals, vitamins, and other nutrients.

These potential problems, coupled with a lack of knowledge regarding dietary protein intake with respect to ideal levels of calorie consumption, have prompted many nutritional queries about the efficacy of enriching foods with specific amino acids and proteins (Hegsted, 1978; McLaren, 1974).

REFERENCES

Badely, R. A., D. Atkinson, H. Hauser, D. Oldani, J. P. Green, and J. M. Stubbs. 1975. The structure, physical and chemical properties of the soybean protein glycinin. *Biochim. Biophys. Acta.* **412**:214.

Baker, R. D., S. S. Baker, G. M. Margo, and H. H. Reuter. 1978. Successful use of soyamaize mixture in the treatment of kwashiorkor. *South Afr. Med. J.* **53**:674.

Bender, A. E., and B. H. Doell. 1975. Note on the determination of net protein utilization by carcass analysis. *Brit. J. Nutr.* **11**:138.

Boloorforooshan, M., and P. Markakis. 1979. Protein supplementation of navy beans with sesame. *J. Food Sci.* **44**:390.

Brunner, J. R. 1977. Milk proteins. In *Food Proteins* (J. R. Whitaker and S. R. Tannenbaum, eds.), pp. 175–208. Avi Publishing Co., Westport, Conn.

Burton, B. T. 1976. *Human Nutrition,* 3rd ed. H. J. Heinz Co. McGraw-Hill. A Blakiston Publication, New York.

Catsimpoulas, N. 1969. Isolation of glycinin subunits by isoelectric focusing in urea-mercaptoethanol. *FEBS Lett.* **4**:259.

Del Rio de Reys, M. T. E., S. M. Constanides, V. C. Sgarbieri, and A. A. El-Dash. 1980. Chicken blood plasma proteins: physiological, nutritional and functional properties. *J. Food Sci.* **45**:17.

Ebashi, S., M. Endo, and I. Ohtsuki. 1969. Control of muscle contraction. *Quart. Rev. Biophys.* **2**:164.

Ebihara, K., Y. Imamura, and S. Kiriyama. 1979. Effect of dietary composition on nutritional equivalency of amino acid mixtures and casein in rats. *J. Nutr.* **109**:2106.

Fomon, S. D. 1979. Methionine fortification of a soy protein formula fed to infants. *Amer. J. Clin. Nutr.* **32**:2460.

Fomon, S. J., L. N. Thomas, L. J. Filer, T. A. Anderson, and K. E. Bergman. 1973. Requirements for protein and essential amino acids in early infancy. Studies with a soy-isolate formula. *Acta Paediatr. Scand.* **62**:33.

Fossum, K., and J. R. Whitaker. 1968. Ficin and papain inhibitor from chicken egg white. *Arch. Biochem. Biophys.* **125**:367.

Graham, G. G., R. P. Placko, E. Morales, G. Acevedo, and A. Cordano. 1970. Dietary protein quality in infants and children. *Amer. J. Dis. Child.* **120**:419.

Hegsted, D. M. 1977. Protein quality and its determination. In *Food Proteins* (J. R. Whitaker and S. R. Tannenbaum, eds.), pp. 347–362. Avi Publishing Co., Westport, Conn.

Hegsted, D. M. 1978. Protein calorie malnutrition. *Amer. Sci.* **66**:61.

Heird, W. C., J. F. Nicholson, J. M. Driscoll, Jr., J. N. Schullinger, and R. W. Winters. 1972. Hyperammonemia resulting from intravenous alimentation using a mixture of synthetic L-amino acids: a preliminary report. *J. Pediatr.* **81**:162.

Hodge, J. E. 1953. Chemistry of browning reactions in model systems. *Agr. Food Chem.* **1**:928 (1).

Hodge, J. E., F. D. Mills, and B. E. Fisher. 1972. Compounds of browned flavor derived from sugar amine reactions. *Cereal Sci. Today* **17**:34.

Holt, L. E., Jr., and S. E. Snyderman. 1965. Protein and amino acid requirements of infants and children. *Nutr. Abstr. Rev.* **35**:1.

Hopkins, D. T., and F. H. Steinke. 1978. Updating protein quality measurement techniques. *Cereal Foods World* **23**(9): 539.

Jansen, G. R., and D. T. Verburg. 1977. Amino acid fortification of wheat diets fed at varying levels of energy intake to rats. *J. Nutr.* **107**:289.

Johnson, J. A., L. Rooney, and A. Salem. 1966. Chemistry of bread flavor. In *Flavor Chemistry* (I. Hornstein, ed.). ACS Advances in Chemistry Series 56, Washington, D.C.

Kaladhar, M., and B. S. Narasinga Rao. 1977. Experimental protein and energy deficiencies: effects on brain free amino acid composition in rats. *Brit. J. Nutr.* **38**:141.

Kilonzo, J. M. 1978. The improvement of the biological value of food plants from Kenya by combinations with vegetable proteins, animal proteins, and limiting amino acids. *Nutr. Rep. Int.* **815**:599.

Kopple, J. D., and M. E. Swendseid. 1975. Evidence that histidine is an essential amino acid in normal and chronically uremic man. *J. Clin. Invest.* **55**:881.

Koshiyama, I. 1968. Factors influencing conformation changes in a 7S protein of soybean globulins by ultracentrifugal investigations. *Agr. Biol. Chem.* (Tokyo) **32**:879.

Koshiyama, I. 1971. Chemical and physical properties of a 7S protein in soybean globulins. *Agr. Biol. Chem.* (Tokyo) **35**:385.

Locker, R. H., G. T. Daines, W. A. Carse, and N. G. Leet. 1977. *Meat Sci.* **1**:87.

McLaren, D. S. 1974. The great protein fiasco. *Lancet* **2**:93.

Miller, D. S. and A. E. Bender. 1955. The determination of net utilization of proteins by a shortened method. *Brit. J. Nutr.* **9**:382.

Nicol, B. M., and P. G. Philips. 1978. The utilization of proteins and amino acids in diets based on casava, rice or sorghum by young Nigerian men of low income. *Brit. J. Nutr.* **39**:271.

Osuga, D. T., and R. E. Feeney. 1977. Egg proteins. In *Food Proteins* (J. R. Whitaker and S. R. Tannenbaum, eds.), pp. 209–266. Avi Publishing Co., Westport, Conn.

Pimentel, D., and M. Pimentel. 1977. Food protein production: land, energy and economics. In *Food Proteins* (J. R. Whitaker and S. R. Tannenbaum, eds.), pp. 542–553. Avi Publishing Co., Westport, Conn.

Rizzi, G. P. 1969. The formation of tetramethylpyrazine and 2-isopropyl-4,5-dimethyl-3-oxazoline in the Strecker degradation of DL-valine with 2,3-butanedione. *J. Org. Chem.* **34**:2002.

Robbins, F. M., J. E. Walker, S. H. Cohen, and S. Chatterjee. 1979. Action of proteolytic enzymes on bovine myofibrils. *J. Food Sci.* **44**:1672.

Rose, W. C., and R. L. Wixom. 1955. The amino acid requirements of man. XVI. The role of the nitrogen intake. *J. Biol. Chem.* **217**:997.

Samonds, K. W., and D. M. Hegsted. 1977. Animal bioassays: a critical evaluation with specific reference to assessing nutritive value for the human. In *Evaluation of Proteins for Humans* (C. E. Bodwell, ed.), pp. 68–80. Avi Publishing Co., Westport, Conn.

Satterlee, L. D., J. G. Kendrick, and G. A. Miller. 1977. Rapid *in vitro* methods of measuring protein quality. *Food Technol.* **31**:77.

Scrimshaw, N. S. 1967. Malnutrition, learning and behavior. *Amer. J. Clin. Nutr.* **20**:493.

Scrimshaw, N. S., and V. R. Young. 1976. The requirements of human nutrition. In *Food and Agriculture*, pp. 27–40. Freeman, San Francisco.

Sen, L. C., and J. R. Whitaker. 1973. Some properties of a ficin-papain inhibitor from avian egg white. *Arch. Biochem. Biophys.* **1958**:623.

Taylor, E. W. 1972. Chemistry of muscle contraction. *Annu. Rev. Biochem.* **41**:577.

USDA. 1975. National food situation. *Econ. Res. Serv.* NFS-154.

Walter, W. M., and A. E. Purcell. 1978. Preparation and storage of sweet potato flakes fortified with plant protein concentrate and isolates. *J. Food Sci.* **43**:407.

Weller, L. A., D. H. Calloway, and S. Margen. 1971. Nitrogen balance of men fed amino acid mixtures based on Rose's requirements, egg white protein and serum free amino acid patterns. *J. Nutr.* **101**:1499.

Wolf, W. J. 1975. Effects of refining operations on the composition of foods. 2. Effects of refining operations on legumes. In *Nutritional Evaluation of Food Processing* (R. S. Harris and E. Karmas, eds.), 2nd ed. Avi Publishing Co., Westport, Conn.

Young, V. R., W. M. Rand, and N. S. Scrimshaw. 1977. Measuring protein quality in humans: a review and proposed method. *Cereal Chem.* **54**:929.

C H A P T E R · 4

Enzymes: Kinetics, Properties, and Applications

INTRODUCTION

An enzyme is a biocatalyst that has the ability to increase the rate of a thermodynamically feasible reaction required for the normal maintenance of living cells. The "substance" transformed into "product" during a thermodynamically feasible reaction involving an enzyme is known as the *substrate*. Many enzymes have a simple protein structure such as the hydrolytic enzymes: lysozyme, pepsin, trypsin, and ribonuclease. In most cases, however, enzymes exist as a conjugated protein structure. These enzymes display biocatalytic activity only when the protein component of the enzyme, known as the *apoenzyme,* is bound to some nonprotein moiety. If the nonprotein moiety is firmly attached to the apoenzyme (e.g., nondialyzable), it is regarded as a *prosthetic group* and the whole active conjugated enzyme protein is called a *holoenzyme*.

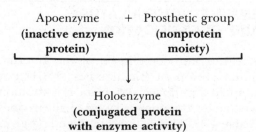

The nonprotein moiety is usually an organic molecule that facilitates the transfer of functional groups or biochemical intermediates fundamental to the overall enzyme reaction. Although many of these nonprotein substituents called *coenzymes* can exist as prosthetic groups on an apoenzyme, many coenzymes are only tenuously associated with the protein and therefore cannot be classified as a prosthetic group. The normal activity of some enzymes may also require the presence of *metal cofactors* prior to enzyme activation. Typical metal cofactors participating in enzyme activation include Cu^{2+}, Fe^{2+}, Mg^{2+}, Mo^{6+}, Zn^{2+}, and many other ions for which the biochemical role has not yet been determined.

The specific biocatalytic activity of enzymes is governed by the tertiary and quaternary structure of the simple enzyme protein, or the apoenzyme protein structure in conjunction with any necessary coenzymes and cofactors. Since activity of an enzyme depends on the three-dimensional configuration and structure of the polypeptides forming the enzyme, it must be recognized that enzymes are under stringent genetic controls. The idiosyncratic catalytic activity of enzymes is actually the manifestation of genetic programming where the *primary amino acid sequence contributes to the formation of a three-dimensional active site* on the protein responsible for biocatalytic activity. The *active site* is usually a relatively small portion of the enzyme structure and typically provides a crevice that *complements* the molecular geometry of a particular substrate prior to catalysis. This concept of complementary geometric binding at the active site provides enzymes with *specificity characteristics;* that is, the molecular geometry of only one substrate may be suited to the rigid catalytic active site of one enzyme. Emil Fischer first introduced this theory, referred to as the *lock and key concept,* but other theories followed in order to explain the behavior of many enzymes that did not adhere to this principle. These theories fostered enzyme–substrate interaction concepts such as the *induced-fit theory* of enzyme specificity, which suggests that the active site of all enzymes is not invariably rigid, but rather, undergoes dynamic three-dimensional alterations in the presence of certain substrates. This flexibility of the active site seems to account for the broad substrate specificity displayed by some enzyme molecules.

Genetic control of protein synthesis not only regulates the characteristic specificity of an enzyme active site but also governs the actual presence of enzymes in a cell at all times. These genetic directives control the two broad classes of all enzyme proteins, which are known as *constitutive* and *induced enzymes. Constitutive* enzymes are always present within a cell regardless of substrate availability. However, *induced enzymes* are present only when an enzyme's substrate activates genetic control mechanisms leading to the intracellular presence of the enzyme.

In addition to these considerations, all classes of enzymes share other unique features.

1. The biocatalytic activity of enzymes results from their ability to *lower the activation energy required for a reaction* to proceed: enzymes *do not alter the equilibrium constant* for any potential biochemical reaction.

2. Enzyme-catalyzed reactions mediate the conversion of a substance to a product in a "clean-reaction" that *results in a minimal variety of undesirable side products.*

3. Some *enzymes mediate* only *unidirectional reactions* where substance A is converted to product B (i.e., $A \rightarrow B$); *but many other enzymes are capable of encouraging both forward and reverse reactions* (i.e., $A \rightleftarrows B$).

4. Enzymes are *not consumed* while mediating a reaction.

5. *Enzymes are crystallizable proteins,* as demonstrated by Sumner in 1926, which can catalyze cell-free reactions provided that adequate *in vitro* conditions have been provided (i.e., pH, ionic strength, temperature, etc.).

Life as known on this planet is governed by fundamental principles of genetic theory involving polynucleotide governance by deoxyribonucleic acid (DNA) and ribonucleic acid (RNA). The species-specific biochemical details stored in these polynucleotides can only be expressed as a life form only by complex enzyme–substrate interactions.

For man, essential substrates for enzyme activity are ultimately obtained through the diet. Deficits of particular substrates or chemical constituents contributing to the activity of enzymes can cause simple reversible pathologic disorders or irreversible disorders and eventual death. Moreover, the biochemical dysfunction of checks and balances governing normal enzyme activity account for many inborn errors in human metabolism, marginal, or severe latent pathologies, organic manifestations of viral activity, aberrations in normal biochemical pathways leading to tumor growth or malignancies, as well as many other problems yet to be discovered.

ENZYME NOMENCLATURE AND CLASSIFICATION

The Enzyme Commission of the International Union of Biochemists (1961) has adopted a uniform, systematic classification scheme for the nomenclature of all enzymes (Table 4.1). This classification method gives

Table 4.1. Some Important Class and Subclass Designations for the International Enzyme Classification Scheme

1. *Oxidoreductases*

   Enzymes involved in oxidation–reduction reactions, where the substrate may contain the following functional groups. The reduced forms of respective substrates have been cited below except for 1.11.

1.1 —$\overset{\mid}{C}$HOH	1.5 —$\overset{\mid}{\underset{\mid}{C}}$—NH—	1.9 Heme groups
1.2 Aldehyde or keto-	1.6 NADH or NADPH	1.10 Diphenols and related structures
1.3 $\overset{\diagdown}{\underset{\diagup}{C}}$H—C$\overset{\diagup}{\underset{\diagdown}{}}$H	1.7 Other nitrogenous compounds	1.11 Act on hydrogen peroxide as an acceptor
1.4 $\overset{\diagdown}{}$CH—NH$_2$	1.8 Sulfur groups	1.12 Hydrogen acts as a donor

2. *Transferases*

   Functional group transfers include

2.1 Single-carbon groups	2.5 Alkyl or related groups
2.2 Aldehyde or ketone groups	2.6 Nitrogen groups
2.3 Acyl groups	2.7 Phosphate groups
2.4 Glycosyl groups	2.8 Sulfur-containing groups

3. *Hydrolases*

3.1 Esters	3.4 Peptide bonds	3.7 C—C bonds
3.2 Glycosidic bonds	3.5 C—N bonds other than peptide bonds	3.8 Halide bonds
3.3 Thioether bonds	3.6 Acid anhydrides	3.9 P—N bonds

4. *Lyases*

   Addition reactions to substrates containing the following types of double bonds

   4.1 —$\overset{\mid}{C}$=$\overset{\mid}{C}$— 4.2 —$\overset{\mid}{C}$=O 4.3 —$\overset{\mid}{C}$=N—   4.5 Carbon-halides

5. *Isomerases*

   Catalyze geometric changes within a substrate molecule

5.1 Racemases and epimerases	5.4 Intramolecular transferases
5.2 *Cis–trans* isomerases	5.5 Intramolecular lyases
5.3 Intramolecular oxidoreductases	5.99 Other isomerases

6. *Ligases*

   Formation of bonds between two molecules accompanied by ATP cleavage

   6.1 C—O   6.2 C—S   6.3 C—N   6.4 C—C

four descriptive numbers to each enzyme. For example, the number 1.1.1.1 assigns a respective class, subclass, sub-subclass, and serial number to an enzyme. A cursory class, subclass survey (e.g., 1.1, 1.2, etc.) characterizing the enzyme activities for six major classes has been outlined in Table 4.1, whereas Table 4.2 details some enzymes that occur in each enzyme class. Before the development of this classification scheme, enzyme nomenclature was inconsistent. Enzymes were merely named in terms of their preferred substrate with an *-ase* suffix to designate enzyme activity. Accordingly, "sucrose" would be a substrate for the "sucrase" enzyme. Since the systematic nomenclature for enzymes can

be awkward, the older enzyme nomenclature based on an *-ase* suffix is still widely used in trivial nomenclature. The trivial names for some broad classes of enzymes are detailed in Table 4.3.

CHEMICAL KINETICS

Chemical thermodynamics and kinetics are useful for studying changes that occur during the course of a chemical reaction. Specifically, thermodynamics can be used to predict the feasibility of a chemical reaction, whereas chemical kinetic principles are used

Table 4.2. Enzyme Classification and Trivial Nomenclature for Some Enzymes Based on the Reactions That They Catalyze

Major class and subclasses	Trivial enzyme nomenclature
1. *Oxidoreductases* (enzymes of biological oxidation and reduction)	
1.1 Acting on the $\diagdown$CH—OH group of donors	
1.1.1 With NAD^+ or $NADP^+$ as acceptor	Alcohol and lactate dehydrogenases
1.1.3 With O_2 as acceptor	Glucose oxidase
1.2 Acting on the aldehyde or keto group of donors	
1.2.1 With NAD^+ or $NADP^+$ as acceptor	Glyceraldehyde 3-phosphate dehydrogenase
1.2.3 With O_2 acceptor	Xanthine oxidase
1.3 Acting on the $\diagdown$CH—CH group of donors	
1.3.1 With NAD^+ or $NADP^+$ as acceptor	Dihydrouracil dehydrogenase
1.3.2 With cytochrome acceptor	Acyl CoA dehydrogenases
1.4 Acting on the $\diagdown$CH—NH₂ group of donors	
1.4.3 With O_2 as acceptor	Amino acid oxidases
2. *Transferases* (group transferring enzymes)	
2.1 Transferring C_1-groups	
2.1.1 Methyltransferases	Guanidinoacetate methyltransferase
2.1.2 Hydroxymethyl and formyltransferases	Serine hydroxymethyltransferase
2.1.3 Carboxyl- and carbamyltransferases	Ornithine carbamyltransferase
2.3 Acyl transferases	Choline acetyltransferase
2.4 Glycosyl transferases	
2.6 Transferring N-containing groups	
2.6.1 Aminotransferases, etc.	Transaminases
3. *Hydrolases* (enzymes catalyzing hydrolytic cleavage)	
3.1 Cleaving ester linkages	
3.1.1 Carboxylic ester hydrolases	Esterases, lipases
3.1.3 Phosphoric monoester hydrolases	Phosphatases
3.1.4 Phosphoric diester hydrolases	
3.2 Cleaving glycosides	
3.2.1 *O*-Glycosides	Amylase, β-glycosidase
3.2.2 *N*-Glycosides	Nucleosidases
3.4 Cleaving peptide linkages	
3.4.11 α-Aminoacyl-peptide hydrolases	Aminopeptidase
3.4.12 Peptidyl amino acid hydrolases	Carboxypeptidases
3.4.14–15 Peptidylpeptide hydrolases, etc.	
4. Lyases (enzymes cleaving C—C, C—O, C—N bonds)	
4.1 C—C lyases	
4.1.1 Carboxy-lyases	Pyruvate decarboxylase
4.1.2 Aldehyde-lyases	Aldolases
4.2 C—O lyases	
4.2.1 Hydro-lyases	Fumarate hydratase
4.3 C—N lyases	
4.3.1 Ammonia lyase	Histidine ammonia-lyase(histidase)
5. Isomerases (enzymes catalyzing geometric changes within one molecule)	
5.1 Racemases and epimerases	Ribose 5-phosphate epimerase
5.2 *Cis–trans* isomerases	Maleylacetoacetate isomerase
5.3 Intramolecular oxidoreductases	
5.3.1 Interconverting aldoses and ketoses	Glucose-phosphate isomerase
5.4 Intramolecular transferases	Methylmalonyl CoA mutase

Table 4.2. (*Continued*)

Major class and subclasses	Trivial enzyme nomenclature
6. Ligases (enzymes that link two molecules, also called synthetases)	
6.1 Forming C—O bonds	
6.1.1 Amino acid–RNA ligases	Amino acid activating enzyme
6.3 Forming C—N bonds	
6.3.1 Acid–ammonia ligases	Glutamine synthetase
6.4 Forming C—C bonds	
6.4.1 Carboxylases	Acetyl CoA carboxylase

to determine the time requirements for conversion of a reactant to a product at various concentration levels.

Kinetic principles have basic importance in the overall activities of enzymes as well as their roles in all phases of metabolism.

Empirical Rate Laws

The rate of a chemical reaction may be expressed as the change of concentration of reactants and products with time. It is influenced by temperature, concentrations of molecules (atoms or ions), pressure, and catalysts.

Consider a simple single-step reaction where $A + 2B \rightarrow 2C$. The molar concentrations of A, B, and C are represented as $[A]$, $[B]$, and $[C]$, respectively. The concentration of B changes at twice the rate of change in the concentration of A, while the concentration of C increases at twice the rate of decrease in A. The relationship of the time derivatives for molecules (atoms or ions) participating in the chemical reaction is

$$- \frac{d[A]}{dt} = - \frac{1}{2} \frac{d[B]}{dt} = + \frac{1}{2} \frac{d[C]}{dt}$$

The concentration of the reactants and products is reflected by Figure 4.1, which is consistent with the statement of the differentials; that is, reactant concentrations decrease and product concentrations increase with reaction time.

Any one of the derivatives may be chosen to express the rate of a reaction, but in order to avoid ambiguity, it must be stated which time derivative has been chosen. In the re-

action of $A + 2B \rightarrow 2C$, the decrease in the concentration of reactant A can be used to define the rate law expressed mathematically as

$$- \frac{d[A]}{dt} = k \, [A] \, [B]^2$$

where

k = rate constant at a specified temperature

$[A]$ = concentration of reactant A

$[B]^2$ = concentration of reactant B raised to the power of 2, which denotes the order of the reaction for B

Hence, the rate of the reaction is proportional to the concentration of the reactants that are raised to some power. The mathematical equation of the rate law is governed by the order of reaction, which can provide an insight into the mechanism controlling the chemical reaction. The order of a reaction cannot be determined merely by studying the stoichiometry of the chemical equation but must be determined experimentally.

The rate laws provide basic information related to the time that it takes for reactants to be converted to products. Similarly, rate laws provide information regarding product quantities that result from a given reaction.

Orders of Reaction

The kinetic properties governing almost any reaction can be described according to the *order of reaction*. The observed rate of a re-

Table 4.3. Trivial, Common Designations Often Used for Various Types of Enzymes and Their Functions

Type of enzyme	Biochemical or catalytic function
Aldolase	$-\overset{\mid}{C}-\overset{\mid}{C}-$ bond cleavage to form an aldehyde
Carboxylase	CO_2 or HCO_3^- addition to substrate yielding a carboxyl group
Decarboxylase	Carboxyl group cleavage from substrate resulting in liberation of CO_2
Dehydrogenase	Removal of hydrogen atoms from a substrate
Esterase	Joint formation of an acid and an alcohol resulting from ester hydrolysis
Hydratase	Water addition to $-\overset{\mid}{C}=\overset{\mid}{C}-$ without breaking the bond, *or* removal of water from $-\overset{\mid}{\underset{H}{C}}-\overset{\mid}{\underset{OH}{C}}-$ to form a double bond
Hydroxylase	Formation of a hydroxyl group using an oxygen atom derived from O_2
Isomerase	Converts certain substrates between their *cis* $\rightleftarrows$ *trans*, D $\rightleftarrows$ L, or aldose $\rightleftarrows$ ketose forms
Kinase	Phosphate group transfer from a high-energy phosphate compound such as ATP to the enzyme's substrate
Ligase	Two molecules are bonded together using energy released from pyrophosphate hydrolysis of a high-energy substance such as ATP
Lyase	Addition or removal of functional groups from a substrate where the mechanism does not involve hydrolysis
Mutase	Intramolecular shifting of a group, e.g., methyl group transfer = $-\overset{\mid}{\underset{CH_3}{C}}-\overset{\mid}{\underset{H}{C}}- \longrightarrow -\overset{\mid}{\underset{H}{C}}-\overset{\mid}{\underset{CH_3}{C}}-$
Oxidase	Addition of oxygen (O_2) to hydrogen atoms (obtained from oxidation of a substrate) thereby forming H_2O, hydrogen peroxide (H_2O_2), or a superoxide anion (O_2^-).
Oxygenase	Incorporation of molecular oxygen (O_2) into a substrate
Peptidase	Peptide bond hydrolysis to form free amino acids and/or peptides
Phosphatase	Liberation of inorganic phosphate (P_i) from hydrolysis of phosphoric acid esters
Phosphorylase	Splits bond through introduction of inorganic phosphate (phosphorolysis)
Reductase	Adds hydrogen atoms to substrate
Sulfatase	Sulfate liberation from hydrolysis of sulfuric acid esters
Synthase	Two molecules are bonded together, but a pyrophosphate bond is not hydrolyzed. This type of enzyme–substrate reaction is the opposite of ligase or synthetase reactions
Synthetase	Same as ligase
Transaminase	Amino group transfer to a keto acid
Transferase	Group transfer reactions from one molecule to another. Group transfers may include phosphates, methyls, or other moieties, but not hydrogen atoms

action under defined conditions, where only the concentrations of reactants are allowed to vary, serves as a basis for assessing the order of a chemical reaction. Reaction orders are specified as *zero, first, second,* and *third*. Determination of reaction order is the most basic aspect of a kinetic investigation, and although reaction orders may entail both fractional and integral numbers, the integral approach has been presented in this discussion.

Zero-order reaction: The rate in this order of reaction is independent of reactant concentration $[A]$, which is eventually converted to product $[P]$. Consider the reaction where

$$[A] \rightarrow [P]$$

$$\text{Rate} = -\frac{d[A]}{dt} = k$$

$$-d[A] = k\, dt$$

$$d[A] = -k\, dt$$

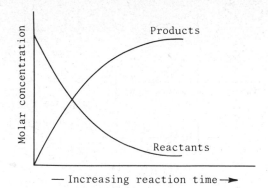

FIGURE 4.1. Relationship of reactants to products plotted versus reaction time.

Integration of the differential equation between the limits of $t = 0$ and $t = t$ yields

$$\int_{[A]_0}^{[A]_t} d[A] = -k \int_0^t dt$$

$$[A]_t - [A]_0 = -kt$$

$$[A]_t = [A]_0 - kt \quad or \quad [A]_0 - [A]_t = kt$$

The value of the rate constant, k, for this zero-order reaction can be graphically determined from the slope of the line when $[A]_0 - [A]_t$ is plotted versus time (Figure 4.2).

Provided that the concentration of reactant $[A]$ is not a limiting factor in the overall conversion of $[A]$ to product $[P]$, zero-order reactions are governed *only* by the availability of an inorganic or biochemical catalyst such as an enzyme. This is contrary to

first-, second-, and third-order reaction discussions that follow.

First-order reaction: First-order reactions represent those reactions in which the reaction *rate is directly proportional* to the concentration of a single reactant. Consider the reaction

$$[A] \rightarrow [P]$$

The rate of this reaction depends on the decreasing concentration of $[A]$:

$$\text{Rate} = -\frac{d[A]}{dt} = k[A]$$

$$= -\frac{d[A]}{[A]} = k\, dt$$

Integration of the differential equation between limits $t = 0$ and $t = t$ yields

$$\int_{[A]_0}^{[A]_t} \frac{d[A]}{[A]} = -k \int_0^t dt$$

$$\ln \frac{[A]_t}{[A]_0} = -kt \quad or \quad \ln \frac{[A]_0}{[A]_t} = kt$$

$$\frac{[A]_t}{[A]_0} = e^{-kt} \quad or \quad \frac{[A]_0}{[A]_t} = e^{kt}$$

$$[A]_t = [A]_0\, e^{-kt} \quad or$$

$$[A]_0 = [A]_t\, e^{kt} \quad or \quad \log \frac{[A]_0}{[A]_t} = \frac{kt}{2.303}$$

The graphic plot of $\ln([A]_0/[A]_t)$ versus time in Figure 4.3 indicates the determination of the rate constant.

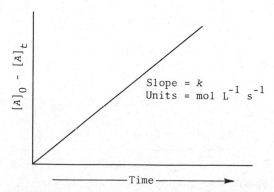

FIGURE 4.2. Graphic determination of the rate constant, k, for a zero-order reaction.

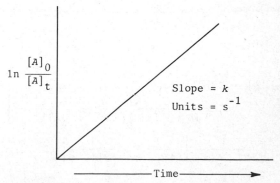

FIGURE 4.3. Determination of the rate constant for a first-order reaction based on slope measurement.

Second-order reaction: Second-order reactions are different from first-order reactions since the observed *rate for a reaction is proportional to the product of the concentrations of two reactants*, or, alternatively, the second power of a single reactant concentration. In this case, consider the reaction in which the concentration of $[A]$ is at least equal to the concentration of $[B]$, and $[A] + [B] \rightarrow [P]$. Let $[A]_0$ and $[B]_0$ be the initial concentration of A and B, respectively. Also, let $([A]_0 - x)$ and $([B]_0 - x)$ be the concentrations of $[A]_t$ and $[B]_t$, respectively.

Note: "x" is the quantity of A and B lost at time t.

Rate of reaction

$$= -\frac{d[A]}{dt} = k[A][B]$$

$$-\frac{d[A]}{dt} = k([A]_0 - x)([B]_0 - x)$$

But

$$\frac{d[A]}{dt} = \frac{d([A]_0 - x)}{dt}$$

$$\frac{d[A]}{dt} = -\frac{dx}{dt}$$

since $[A]_0$ is a constant.
Therefore,

$$\frac{dx}{dt} = k([A]_0 - x)([B]_0 - x)$$

Separating the variables, we have

$$\frac{dx}{([A]_0 - x)([B]_0 - x)} = kdt$$

But

$$\frac{1}{([A]_0 - x)([B]_0 - x)}$$

$$= \frac{1}{[B]_0 - [A]_0} \left[\frac{1}{[A]_0 - x} - \frac{1}{[B]_0 - x} \right]$$

$$\int_0^t \frac{1}{[B]_0 - [A]_0} \left[\frac{1}{[A]_0 - x} - \frac{1}{[B]_0 - x} \right] dx$$

$$= k \int_0^t dt$$

$$\frac{1}{[B]_0 - [A]_0} \left[-\ln([A]_0 - x) + \ln([B]_0 - x) \right]_0^t$$
$$= kt$$

$$\frac{1}{[B]_0 - [A]_0} \left[\ln \frac{[B]_0 - x}{[A]_0 - x} \right] \Big|_0^t = kt$$

Note: $[x(0) = 0]$

$$\frac{1}{[B]_0 - [A]_0} \left[\ln \frac{[B]_0 - x}{[A]_0 - x} - \ln \frac{[B]_0}{[A]_0} \right] = kt$$

$$\frac{1}{[B]_0 - [A]_0} \ln \frac{[A]_0([B]_0 - x)}{[B]_0([A]_0 - x)} = kt$$

Since $[B]_0 - x = [B]_t$ and $[A]_0 - x = [A]_t$, we have

$$\frac{1}{[B]_0 - [A]_0} \ln \frac{[A]_0[B]_t}{[B]_0[A]_t} = kt$$

A graphic plot of

$$\ln \frac{[A]_0[B]_t}{[B]_0[A]_t}$$

versus time illustrated in Figure 4.4 permits calculation of the rate constant from the slope of the line.

The calculation of the rate constant for a second-order reaction can be alternatively calculated if reactant concentrations $[A]$ and $[B]$ are absolutely equivalent. If $[A] = [B]$, and $2[A]$ represents the sum of the concentrations for both reactants. This simplification suggests that

$$2[A] \rightarrow [P]$$

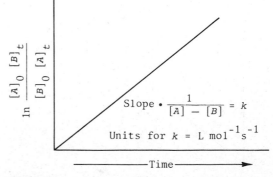

FIGURE 4.4. Calculation of the rate constant for a second-order reaction.

and the second-order reaction rate is

$$\text{Rate} = -\frac{d[A]}{dt} = k[A]^2$$

Subsequent integration of this equation yields

$$\int_{[A]_0}^{[A]_t} \frac{-1}{[A]^2} \cdot d[A] = k \int_0^t dt$$

and

$$\frac{1}{[A]_t} - \frac{1}{[A]_0} = kt$$

Therefore

$$[A]_t = \frac{[A]_0}{1 + kt[A]_0}$$

Solving this equation for t and the rate constant k, respectively, results in the expressions

$$t = \frac{1}{k} \left(\frac{1}{[A]_t} - \frac{1}{[A]_0} \right)$$

$$k = \frac{1}{t} \left(\frac{1}{[A]_t} - \frac{1}{[A]_0} \right)$$

Calculation of the rate constant in this treatment does not require the use of separate variables as in the previous calculation for the rate constant k; and k can also be evaluated graphically by plotting $1/[A]_t$ versus time. The slope of the resulting line can be used to determine the rate constant, k.

It should be mentioned that some second-order reactions may deceptively appear to be first-order reactions in certain instances. For the reaction of

$$[A] + [B] \rightarrow [P]$$

extremely high concentrations of $[A]$ or $[B]$ may cause a rate of reaction that could appear to be dependent on the concentration of only one reactant. These reactions are referred to as *apparent first-order reactions* or *pseudo-first-order reactions*.

Third-order reaction: Third-order reactions are extremely complex reactions in which the overall rate of reaction depends on the product of three participants in a reaction. For example,

$$[A] + [B] + [C] \rightarrow [P]$$

These reactions are comparatively rare with respect to zero-, first-, and second-order reactions and they are beyond the scope of this text.

A summary of important rate equations and their respective graphic forms has been outlined in Table 4.4.

Half-life for first- and second-order reactions: Half-life of a reaction, denoted by $t_{1/2}$, is the time that it takes for the concentration of a reactant to reach *one-half of its initial value*.

For a first-order reaction, the time for the concentration of A to fall from $[A]_0$ ($[A]_0$ represents the initial concentration of A) to $\frac{1}{2}[A]_0$ is calculated from the equation

$$t_{1/2} = \frac{\ln 2}{k} \quad or \quad \frac{0.693}{k}$$

The $t_{1/2}$ value for a first-order reaction is independent of concentration, but the $t_{1/2}$ for a second-order reaction is calculated from the expression $1/[A]_0 k$. In this latter case, $t_{1/2}$ clearly depends on the concentration of a reactant.

Mechanisms, Molecularity, and Orders of Reaction

The conversion of reactants to products in a chemical reaction follows a pathway referred to as the *mechanism of the reaction*, which describes the detailed information for the transformations that occur. Many chemical reactions proceed in a number of steps called *elementary reactions* (single-step reactions). The term used to describe each elementary reaction is known as *molecularity*, which refers to the number of molecules, atoms, or ions that react with each other to form products. When one, two, or three molecules, atoms, or ions react to form the products of the elementary reaction, they are called *mon-*

Table 4.4. Summary of Important Rate Equations for Reaction Orders

Order	Differential form	Integrated form[a]	Graphical investigation of the overall order	Half-life $(t_{1/2})$
Zero	$-\dfrac{d[A]}{dt} = k$	$[A]_0 - [A]_t = kt$	Plot of $[A]_0 - [A]_t$ vs. Time; Slope $= k$, Units $=$ mol L^{-1}s^{-1}	$t_{1/2} = \dfrac{[A]_0}{2k}$
First	$-\dfrac{d[A]}{dt} = k[A]$	$[A]_0 = [A]_t\, e^{kt}$	Plot of $\ln\dfrac{[A]_0}{[A]_t}$ vs. Time; Slope $= k$, Units $=$ s^{-1}	$t_{1/2} = \dfrac{\ln 2}{k} = \dfrac{0.693}{k}$
Second Type 1	$-\dfrac{d[A]}{dt} = k[A]^2$	$\dfrac{1}{[A]_t} - \dfrac{1}{[A]_0} = kt$		$t_{1/2} = \dfrac{1}{[A]_0 k}$
Second Type 2	$-\dfrac{d[A]}{dt} = k[A][B]$	$\dfrac{1}{[A]_0 - [B]_0}\ln\dfrac{[A]_t[B]_0}{[A]_0[B]_t} = kt$	Plot of $\ln\dfrac{[B]_0[A]_t}{[A]_0[B]_t}$ vs. Time; $k = \dfrac{1}{[A]-[B]}\cdot$ slope, Units $=$ L mol^{-1}s^{-1}	

[a] $[A]_0$ and $[B]_0$ represent initial concentration levels while the designations $[A]_t$ and $[B]_t$ represent concentration levels at a time other than "zero"—namely, time t.

omolecular, bimolecular, and termolecular re-actions, respectively. A reaction of some specified order may be accounted for in terms of a sequence of several unimolecular, bi-molecular, or termolecular steps. Hence, molecularity has a different meaning than the order of reaction. The order of reaction is an empirical quantity, obtained from the rate law, which is deduced from the mech-anisms that have rate-limiting steps.

Molecularity and order of reaction will be illustrated by considering the reaction

$$A + B \rightarrow P + Q$$

A two-step reaction is postulated for this re-action. One elementary reaction occurs slowly while the other proceeds rapidly.

Step 1: $\quad\quad A + A \xrightarrow{k_1} A' + P$ (slow)

Step 2: $\quad\quad A' + B \xrightarrow{k_2} A + Q$ (fast)

Overall reaction: $\quad A + B \longrightarrow P + Q$

Both elementary reactions in the postulated mechanism are bimolecular since two mol-ecules undergo reaction. The rate law of the overall reaction is based on the first step, which is the rate-determining step. There-fore,

$$\text{rate} = k_1[A][B]$$

The first step for the overall reaction is slow and it accordingly controls the rate of the total reaction. The reaction is a second-order reaction, which is consistent with the pos-tulated two-step mechanism.

The order of reaction in this example co-incides with molecularity in the selected re-action, but in most cases this is not true. Note that the interpretation of a rate law is full of pitfalls, one reason being that a second-or-der reaction can also form a reaction scheme more complex than a simple bimolecular col-lision.

Rate of Reaction and Temperature

The rate constant for chemical reactions de-pends on a number of factors including the frequency of molecular collisions, kinetic energy of the colliding molecules, and steric factors linked to the spatial orientation of colliding mole-cules. Temperature influences all of these factors and, therefore, exerts a significant effect on the rate of a chemical reaction. Within specified limits for a given reaction, increases in temperature are usually associ-ated with increases in the rate of the reac-tion.

The Arrhenius equation has convention-ally been used to describe the influence of temperature on the rate constant for a re-action as follows:

$$k = Ae^{-E_a/RT}$$

where

k = rate constant for a reaction

A = collision frequency factor for mol-ecules involved in a reaction

E_a = energy of activation—the minimum energy that must be available in a molecular collision between reac-tants to initiate product formation

The energy of activation term (E_a) rep-resents the minimum energy that must be available in a molecular collision between two reactants to initiate product formation. The energy involved in molecular collisions is im-portant since chemical bonds must be bro-ken and new ones formed if a reaction is to occur.

Collisions of reactants having high kinetic energies lead to the formation of a reactant intermediate or transition state complex that dis-plays a large but temporal potential energy state. The potential energy of this species is eventually dissipated as kinetic energy with the concomitant formation of product (Fig-ure 4.5). In those instances in which the kinetic energy of molecules is normally inadequate to form reaction intermediates, additional heat energy must be applied in order to increase kinetic energy of the reac-tants so that they achieve the necessary en-ergy of activation (E_a). For further details regarding the principles of chemical and biochemical energetics as well as activation energy, consult Chapter 11.

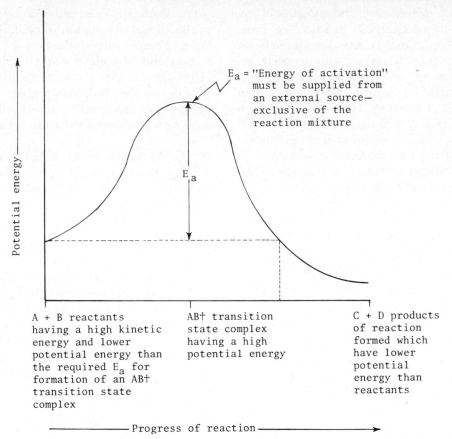

E_a = "Energy of activation" must be supplied from an external source—exclusive of the reaction mixture

E_a

Potential energy →

A + B reactants having a high kinetic energy and lower potential energy than the required E_a for formation of an AB† transition state complex

AB† transition state complex having a high potential energy

C + D products of reaction formed which have lower potential energy than reactants

—————— Progress of reaction ——————→

FIGURE 4.5. Energy activation and its relationship to potential energy during a reaction sequence where reactants *A* and *B* interact to produce an intermeidate transition state complex (*AB†*) prior to product formation in the forms of *C* and *D*.

Although the Arrhenius equation presented above may be adequate for describing rates of reaction for simple molecules and atoms, the equation must be slightly modified for application to reactions dealing with complex molecules. Modification of the equation requires the introduction of *P* and *Z* factors that respectively account for *probability* or *steric factors* (*P*) and the *number of collisions* (*Z*). The equation now becomes

$$k = PZe^{-E_a/RT}$$

where *P* and *Z* have effectively replaced the collision frequency factor in the original Arrhenius equation.

The original Arrhenius equation may be converted to a more useful logarithmic form by multiplying both sides of the equation by $\log_{10}$ to give

$$\log k = -\frac{E_a}{2.303\ R}\frac{1}{(T)} + \log A$$

A plot of the log *k* versus $1/(T)$ terms in this equation yields a straight line having a slope equivalent to $-E_a/2.303\ R$. This type of plot has been illustrated in Figure 4.6.

If the experimental data are minimal and rate constants k_1 and k_2 exist only for respective temperatures T_1 and T_2, the energy of activation for the reaction can be calculated from the logarithmic form of the Ar-

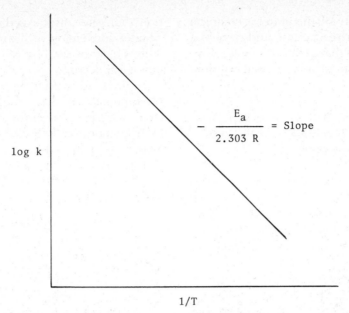

FIGURE 4.6. Graphic determination of the energy of activation (E_a) required to initiate a chemical reaction.

rhenius equation. The derivation of the equation follows:

(1) $\log k_1 = \log A - \dfrac{E_a}{2.303\, RT_1}$

(2) $\log k_2 = \log A - \dfrac{E_a}{2.303\, RT_2}$

Subtracting equation 1 from equation 2 yields

$\log k_2 - \log k_1 = \log A - \log A$
$\qquad - \dfrac{E_a}{2.303\, RT_2} - \left(- \dfrac{E_a}{2.303\, RT_1} \right)$

$\log \dfrac{k_2}{k_1} = \dfrac{E_a}{2.303\, R} \left(\dfrac{1}{T_1} - \dfrac{1}{T_2} \right)$

$\log \dfrac{k_2}{k_1} = \dfrac{E_a}{2.303\, R} \left(\dfrac{T_2 - T_1}{T_1 T_2} \right)$

Solving for E_a results in

$E_a = 2.303\, R \left(\dfrac{T_1 T_2}{T_2 - T_1} \right) \log \left(\dfrac{k_2}{k_1} \right)$

The calculation of the energy of activation (E_a) is important because it provides information concerning the energy changes that occur during an effective molecular collision.

ENZYME KINETICS

Biochemical and nonbiochemical reactions are governed by the same set of thermodynamic and kinetic principles. Aside from the academic interest offered by these principles, they serve as an operational basis for enzyme actions in metabolism, normal health, disease, industry, and biotechnology.

Enzyme Catalysts

As discussed in the previous section, chemical reactants are converted to products provided that

1. The reaction is thermodynamically feasible.

2. Sufficient energy exists for the reactants to attain an adequate energy of activation.

Enzymes are proteins that display catalytic properties in biochemical reactions. These catalytic agents promote reactions by *decreasing the energy of activation* and encouraging product formation.

A cursory study of the general chemical reaction in which reactants A and B yield a product(s) reveals that a transition state complex (AB†) is formed before final product formation:

$$A + B \longrightarrow \underset{\substack{\textbf{Transition} \\ \textbf{state complex}}}{AB\dagger} \longrightarrow \underset{\textbf{Products}}{C + D}$$

$$\underset{\textbf{Reactants}}{A + B}$$

Catalyzed and uncatalyzed energy relationships for this reaction have been depicted as energy profiles in Figure 4.7. The tops of the energy peaks for both reactions signify the presence of an unstable transition state complex (AB†). This complex tends to assume a lower energy level by undergoing product formation. The final-product energy is also lower than the energy of the original reactants. The catalyzed reaction shows significantly lower energy of activation than the uncatalyzed reaction and facilitates conversion of A and B to products. This is true for catalysts of biological or nonbiological origin.

In order to further demonstrate catalytic phenomena, consider the following three cases, which deal with the decomposition of hydrogen peroxide (H_2O_2) without a catalyst (case 1), with an iodide catalyst (case 2), and with the enzyme known as *catalase* (case 3).

Case 1: *Without a catalyst*

$$2H_2O_2 \longrightarrow 2H_2O + O_2$$

The decomposition of H_2O_2 is very slow. The value of E_a is 76 kJ mol^{-1} at room temperature.

Case 2: *With iodide (I^-) as the catalyst*

$$2H_2O_2 \xrightarrow{\ I^-\ } 2H_2O + O_2$$

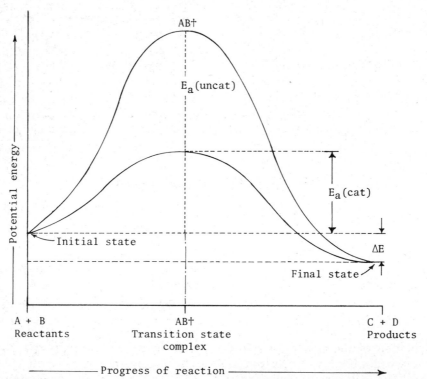

FIGURE 4.7. Transition state of reactants leading to product formation from the perspective of potential energy: overall reaction—$A + B \rightarrow AB\dagger \rightarrow C + D$. Note the significantly higher energy of activation required for an uncatalyzed reaction (E_a(uncat)) as opposed to a corresponding catalyzed reaction (E_a(cat)).

The reaction rate

$$= -\frac{d[H_2O_2]}{dt} = k[H_2O_2][I^-]$$

The reaction occurs in two steps:

(1) $H_2O_2 + I^- \longrightarrow H_2O + IO^-$

(2) $H_2O_2 + IO^- \longrightarrow H_2O + O_2 + I^-$

The *first reaction* is the rate-determining step, which *controls the overall reaction;* but in the second step, iodide is regenerated for the support of another catalytic reaction. The E_a value for this catalyzed reaction is 57 or 19 kJ mol^{-1} less than the uncatalyzed reaction.

Case 3: *With catalase as a catalyst*

$$2\ H_2O_2 \xrightarrow{\text{catalase}} 2\ H_2O + O_2$$

The E_a value for *catalase*-mediated H_2O_2 destruction is only 8 kJ mol^{-1}. The efficiency of catalyzed H_2O_2 decomposition in cases 2 and 3 versus the uncatalyzed reaction can be determined by E_a analysis.

Case 2, with iodide (I^-) as a catalyst, gives

$$\frac{k\ (\text{catalyst})}{k\ (\text{no catalyst})}$$

$$\sim \frac{\exp\ (-57\ \text{kJ mol}^{-1}/2.5^*\ \text{kJ mol}^{-1})}{\exp\ (-76\ \text{kJ mol}^{-1}/2.5\ \text{kJ mol}^{-1})}$$

$$\simeq \frac{e^{-22.8}}{e^{-30.4}} \simeq \frac{1.25 \times 10^{-10}}{6.27 \times 10^{-14}} \simeq 1998$$

In this instance, the iodide catalyst increases the rate of H_2O_2 decomposition by a factor of ~1998.

Case 3, with catalase as a catalyst, gives

$$\frac{k\ (\text{catalyst})}{k\ (\text{no catalyst})}$$

$$\simeq \frac{\exp\ (-8\ \text{kJ mol}^{-1}/2.5\ \text{kJ mol}^{-1})}{\exp\ (-76\ \text{kJ mol}^{-1}/2.5\ \text{kJ mol}^{-1})}$$

$$\simeq \frac{e^{-3.2}}{e^{-30.4}} \simeq \frac{40.7 \times 10^{-3}}{6.27 \times 10^{-14}} \simeq 6.5 \times 10^{11}$$

Thus, catalase increases the rate of H_2O_2 decomposition by a factor of ~3.25×10^8 beyond that where I^- is used as a catalytic agent.

*When RT is ~2.5 kJ mol^{-1}.

As a rule, enzyme-catalyzed reactions proceed at rates that are 10^8 to 10^{20} times faster than the corresponding uncatalyzed reaction.

The enzyme-negotiated acceleration of reactants into product formation is desirable and essential for the existence of all known metabolic pathways. However, enzyme mediated reactions cannot be too fast so that they incite a total upset in normal metabolic schemes. For example, oxidative reactions responsible for yielding energy from carbohydrates in an organism must occur within the reins of energy requirements at any given instant. Thus, excessive enzyme-negotiated oxidation of carbohydrates may be antagonistic to the biochemical operation of the organism if only moderate oxidative reactions are necessary.

From other perspectives, some reactions cannot occur too rapidly. For example, the superoxide anion that is produced during purine metabolism exhibits highly reactive properties that are probably unfavorable for the normal operation of certain biochemical pathways. Some studies have suggested that the high reactivity and errant disposition of these anions may spur on the development of certain arthritic ailments in humans. Recognition of this possibility plus evidence of other potentially destructive reactions suggest that these anions cannot be removed too swiftly from normal biochemical systems. Thus, the disposition of superoxide anions is ensured by a very efficient enzyme mechanism (Figure 4.8). In this case superoxide anion is *processed* in a reaction with hydrogen ions, which results in hydrogen peroxide formation. This product is also very toxic, but it can be readily converted to water and oxygen by catalase.

In this situation catalase may process up to 5.6×10^6 hydrogen peroxide molecules each second. An enzyme reaction rate less than this may prove to be lethal for a living, actively metabolizing cell.

Other notable examples of catalytic activity by enzymes abound. One of the more rapid reactions mediated by enzymes in-

1. Purine metabolism leads to xanthine formation

2. Xanthine $+ H_2O + O_2 \rightarrow$ Uric acid $+ \cdot O_2^-$ (Superoxide anion)

$$\downarrow$$

Urate (excreted in the urine)

3. $2 \cdot O_2^- + 2\,H^+ \longrightarrow H_2O_2 + O_2$
(hydrogen peroxide)

4. $H_2O_2 \xrightarrow{\text{catalase}} H_2O + \frac{1}{2}\,O_2$

FIGURE 4.8 Schematic reaction sequence for role of catalase in the elimination of the superoxide anion obtained from purine metabolism.

volves catalase activity in the bombardier beetle. The posterior abdominal tip of this organism contains an "activator gland" with 25% (w/v) hydrogen peroxide. Predatory advances on the organism cause a glandular conversion of the hydrogen peroxide to nearly boiling water in a second by catalase. Oxygen jointly formed during the reaction directs a "pressurized jet" of nearly boiling water toward the predator as a deterrant. The catalytic activity of enzymes cannot occur too fast for the survival of this beetle!

Determination of Enzyme Reaction Rates

Analysis of enzyme reaction rates and kinetic principles can be approached by graphic analysis of *initial velocity* (v_0) for a reaction, when plotted against reaction time for a number of *fixed but different substrate concentrations*. The enzyme-catalyzed reaction may be monitored by spectrophotometry or other instrumental methods to determine the rate of substrate disappearance ($-d[S]$) or product formation ($+d[P]$).

The initial velocities for the enzyme-catalyzed reactions are graphically determined for different concentrations of substrate at a specified time, usually the one minute mark of the reaction. Thus, the velocities for S_1, S_2, S_3, and S_4, respectively, correspond to initial velocities v_{01}, v_{02}, v_{03}, and v_{04} (Figure 4.9). Estimated initial velocities may then be plotted against corresponding substrate concentrations [S] as illustrated in Figure 4.10.

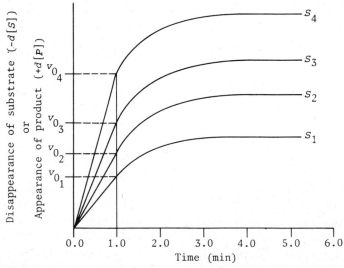

FIGURE 4.9. Determination of initial velocity (v_0) for a fixed concentration of enzyme in the presence of four different but steadfast concentrations of substrate represented as S_1, S_2, S_3 and S_4.

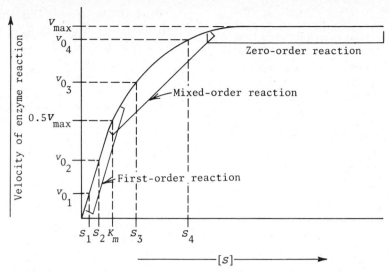

FIGURE 4.10. A plot of initial velocities for an enzyme reaction versus substrate concentrations in moles $[S]$.

Experimental studies of most enzyme-mediated reactions result in a *hyperbolic reaction velocity* when plotted versus $[S]$, with the *maximum velocity* for the reaction (V_{max}) asymptotically approached by the hyperbola. Generally, the v_0 values increase in proportion to $[S]$ up to a certain point *obeying a first-order reaction* since the rate of the reaction depends on $[S]$. However, when $[S]$ becomes *very large, the observed velocity reaches a maximum.* From this point on, the evident flattening of the reaction curve is indicative of a *zero-order reaction.* Obviously, first-order reactions are co-mingled during the transition from a first- to zero-order reaction. Under zero-order reaction conditions, the *catalytic activity and concentration of the enzyme* limit the maximum velocity (V_{max}).

The Michaelis–Menten Equation

The hyperbolic nature of a normal enzyme plot, asymptotically approaching V_{max}, can be mathematically defined according to the Michaelis–Menten equation. The Michaelis–Menten explanation of enzyme behavior was eventually modified by Briggs and Haldane for the reaction of a substrate with an enzyme. The expression

$$v_0 = \frac{V_{max}\,[S]}{K_m + [S]}$$

where

v_0 = initial velocity of the enzyme reaction

V_{max} = maximum velocity

K_m = Michaelis constant

$[S]$ = molar concentration of substrate

has been developed to explain this enzyme kinetic behavior. The derivation of this mathematical concept is based on a number of assumptions such as the *steady-state concept* for the formation of the $[ES]$ complex. For example, in the typical schematic enzyme–substrate reaction

$$[E] + [S] \underset{k_{-1}}{\overset{k_1}{\rightleftharpoons}} [ES] \underset{k_{-2}}{\overset{k_2}{\rightleftharpoons}} [P] + [E] \quad (4.1)$$

The *rate of formation* for the $[ES]$ intermediate leading to eventual product formation is equal to the *rate of $[ES]$ decomposition*, that is,

rate of $[ES]$ formation = rate of $[ES]$ decomposition

or

$$+ \frac{d[ES]}{dt} = - \frac{d[ES]}{dt}$$

Note:

$$\frac{d[ES]}{dt} = \text{change in the concentration of the } [ES] \text{ complex per unit time}$$

sign (+ or −) = denotes the appearance or disappearance, respectively

Considering equation 4.1, expression 4.2 can be written and used ultimately to derive the Michaelis–Menten equation by basing it on certain assumptions and definitions:

$$k_1 [E][S] + k_{-2} [E][P]$$
$$= k_{-1} [ES] + k_2 [ES] \quad (4.2)$$

$$[E] (k_1 [S] + k_{-2} [P]) = [ES](k_{-1} + k_2) \quad (4.3)$$

$$\frac{[ES]}{[E]} = \frac{k_1 [S] + k_{-2} [P]}{k_{-1} + k_2} \quad (4.4)$$

$$\frac{[ES]}{[E]} = \frac{k_1 [S]}{k_{-1} + k_2} + \frac{k_{-2} [P]}{k_{-1} + k_2} \quad (4.5)$$

Assuming that $[P] = 0$, since the rate of the reaction is measured before product formation, the last term of the equation can be neglected and

$$\frac{[ES]}{[E]} = \frac{k_1 [S]}{k_{-1} + k_2} \quad (4.6)$$

Grouping the rate constants (k_1, k_{-1}, and k_2) in equation 4.6 according to Briggs and Haldane as

$$K_m = \frac{k_{-1} + k_2}{k_1}$$

$$\frac{[E]}{[ES]} = \frac{K_m}{[S]} \quad (4.7)$$

In order to account for the *total enzyme concentration* ($[E]_t$) distributed throughout the reacting system, it is assumed that

$$[E]_t = [E] + [ES]$$

where

[E] = free enzyme exclusive of that in the [ES] complex

[ES] = concentration of enzyme bound to substrate

Since $[E] = [E]_t − [ES]$, this term can be appropriately inserted into the left-hand terms of equation 4.7:

$$\frac{[E]}{[ES]} = \frac{[E]_t − [ES]}{[ES]}$$

$$= \frac{[E]_t}{[ES]} − \frac{[ES]}{[ES]} = \frac{[E]_t}{[ES]} − 1 \quad (4.8)$$

$$\frac{[E]_t}{[ES]} − 1 = \frac{K_m}{[S]} \quad \text{or}$$

$$\frac{[E]_t}{[ES]} = \frac{K_m}{[S]} + 1 \quad (4.9)$$

The rate determining step for the enzyme reaction (slowest step) is the decomposition of the [ES] complex and, by definition

$$v_0 \propto [ES] \quad \text{or} \quad v_0 = k_2 [ES]$$

The V_{max} for the enzymatic reaction is attained only when the [ES] complex reaches a maximum concentration and essentially all of the enzyme ($[E]_t$) is in the form of [ES]. Accordingly, $V_{max} = k_2 [E]_t$. Based on the previous expression of v_0 and V_{max}, respectively,

$$[ES] = \frac{v_0}{k_2} \quad \text{and} \quad [E]_t = \frac{V_{max}}{k_2}$$

$$\frac{[E]_t}{[ES]} = \frac{\dfrac{V_{max}}{k_2}}{\dfrac{v_0}{k_2}} \quad \text{or} \quad \frac{[E]_t}{[ES]} = \frac{V_{max}}{v_0} \quad (4.10)$$

$$\frac{V_{max}}{v_0} = \frac{K_m}{[S]} + 1 \quad (4.11)$$

$$\frac{V_{max}}{v_0} = \frac{K_m}{[S]} + \frac{[S]}{[S]} \quad (4.12)$$

$$\frac{V_{max}}{v_0} = \frac{K_m + [S]}{[S]} \quad (4.13)$$

$$V_{max} [S] = v_0 (K_m + [S]) \quad (4.14)$$

Dividing both sides of equation 4.14 by ($K_m + [S]$) results in the Michaelis–Menten equation:

$$v_0 = \frac{V_{max} [S]}{K_m + [S]} \quad (4.15)$$

The K_m value previously defined as ($k_{-1} + k_2)/k_1$ can be calculated from equation 4.15 at the point of $\frac{1}{2}V_{max}$, where the K_m is expressed in units of the substrate concentration. This can be substituted since

$$\frac{V_{max}}{2} = \frac{V_{max}[S]}{K_m + [S]} \qquad (4.16)$$

Simplification of equation 4.16 shows that

$$K_m = [S] \qquad (4.17)$$

The K_m value is a highly significant characteristic of an enzyme reaction when it is investigated under well-defined experimental conditions. The K_m value provides

1. An index of the substrate concentration necessary to *saturate or occupy one-half of the available enzyme active sites* present.

2. An *indication of substrate binding affinity to the active site of the enzyme*—the smaller the K_m value for a given enzyme reaction, the larger the enzyme-substrate binding affinity. For example, a K_m value of $1.0 \times 10^{-6} M$ shows an enzyme substrate binding affinity 1000 times greater than an enzyme with a $K_m = 1.0 \times 10^{-3} M$.

The K_m values for enzymes are widely variable and often relate to enzyme reactions only under very specific experimental conditions. Lack of strict adherence to experimental conditions during enzyme studies can lead to an almost infinite variety of different K_m values. For this reason, many *in vitro* enzyme assays do not actually represent the *in vivo* kinetic performance of an enzyme.

Table 4.5 outlines representative K_m values for some important enzymes.

Graphic Plots for Enzyme Kinetics

The Michaelis–Menten equation certainly defines the observed kinetic behavior of enzymes in terms of observed velocity and $[S]$. Due to practical and analytical considerations, it is often useful to discuss enzyme kinetic behavior in terms of a double reciprocal plot based on the Michaelis–Menten equation 4.15. The reciprocal of this equation (i.e., 4.15) gives the so-called Lineweaver–Burk formula where

$$\frac{1}{v_0} = \frac{K_m}{V_{max}}\left(\frac{1}{[S]}\right) + \frac{1}{V_{max}} \qquad (4.18)$$

This equation is equivalent to the elementary algebraic straight-line relationship expressed by $y = mx + b$ (m = slope; b = ordinate intercept) and facilitates the determination of V_{max} and K_m for an enzyme reaction when $1/v_0$ (ordinate) is plotted against $1/[S]$. Furthermore, this graphic method obviates the necessity of measuring reaction velocities for especially high substrate concentrations that approach V_{max}, and it avoids some of the quantitative uncertainties associated with standard Michaelis–Menten plot of v_0 versus $[S]$. A typical Lineweaver–Burk plot with its important features appears in Figure 4.11A.

The practical utility of the Lineweaver–Burk plot is equaled only by the Woolf and Eadie–Hofstee plots. A plot of $[S]/v_0$ (ordi-

Table 4.5. K_m Values for Some Notable Enzymes

Enzyme	Substrate(s)	K_m value (mM)
Catalase	Hydrogen peroxide (H_2O_2)	25.0
Carbonic anhydrase	Bicarbonate (HCO_3^-)	9.0
Chymotrypsin	Acetyl-L-tryosinamide	32.0
Chymotrypsin	Glycyltyrosinamide	122.0
β-Galactosidase	Lactose	4.0
Hexokinase	Glucose	0.15
Hexokinase	Fructose	1.50
Pyruvate carboxylase	Pyruvate	0.40

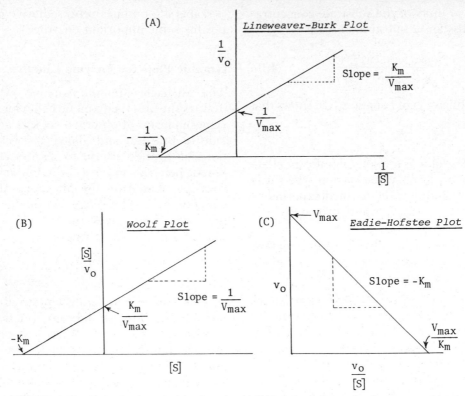

FIGURE 4.11. Comparative graphic plots that provide a basis for evaluating enzyme kinetic data and determining the K_m value and V_{max} for an enzyme under various reaction conditions.

nate) versus $[S]$ serves as the basis for the Woolf plot (Figure 4.11B), which adheres to the formula

$$\frac{[S]}{v_0} = \frac{K_m}{V_{max}} + \frac{1}{V_{max}}[S] \qquad (4.19)$$

This method of plotting aids enzyme studies because the more reliable experimental kinetic measurements occurring at low substrate concentrations assume the greatest importance when the graphic plot is extrapolated to the left of the ordinate.

The Eadie–Hofstee plot, on the other hand, is based on a plot of v_0 (ordinate) versus $v_0/[S]$, which is based on the straight-line equation

$$v_0 = -K_m \frac{v_0}{[S]} + V_{max} \qquad (4.20)$$

This type of a graph not only facilitates the determination of K_m and V_{max} but also

helps detect deviations in the linearity of experimental data that would otherwise go undetected in other methods (Figure 4.11C).

Enzyme Reactions Involving Two Substrates

The study of enzymes to this point has considered enzyme reactions with only one substrate. However, other reactions do exist where an enzyme will consecutively or simultaneously interact with two different substrates. Two substrate reactions may display *random, compulsory-order,* or *ping-pong mechanisms.* A comparison of these three mechanisms can be schematically represented with reference to the reaction

$$S_1 + S_2 \overset{E}{\rightleftharpoons} P + Q$$

Substrates   **Products**

Note: S_1 and S_2 are two different substrates.

The occurrence of this reaction can adhere to any of the three following reaction sequences:

Case 1: *Random mechanism*

$$[E] + [S_1] \rightleftharpoons [ES_1] \qquad\qquad [EP] \rightleftharpoons [E] + [P]$$

$$[ES_1 S_2] \rightleftharpoons [PEQ]$$

$$[E] + [S_2] \rightleftharpoons [ES_2] \qquad\qquad [EQ] \rightleftharpoons [E] + [Q]$$

The enzyme combines at random with $[S_1]$ and $[S_2]$ to form a common $[ES_1S_2]$ complex before the products $[P]$ and $[Q]$ are produced.

Case 2: *Compulsory-order mechanism*

$$[S_1] \qquad\qquad [S_2]$$
$$\Updownarrow \qquad\qquad \Updownarrow$$
$$[E] \rightleftharpoons [ES_1] \rightleftharpoons [ES_1 S_2] \rightleftharpoons$$

$$[P] \qquad\qquad [Q]$$
$$\Updownarrow \qquad\qquad \Updownarrow$$
$$[EPQ] \rightleftharpoons [EQ] \rightleftharpoons [E]$$

The enzyme combines with $[S_1]$ first and then $[S_2]$ before the products $[P]$ and $[Q]$ are produced.

Case 3: *Ping-pong mechanism*

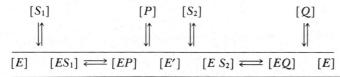

The enzyme combines with $[S_1]$ and forms $[P]$, then it combines with $[S_2]$ and forms $[Q]$. Note that $[E']$ may represent a modified version of the enzyme formed by phosphorylation or some similar activation reaction.

Although kinetic idiosyncracies exist among the three reaction mechanisms outlined above, it is difficult to distinguish one type of a two-substrate reaction from another, especially random versus compulsory-

order reactions. Any distinction of one reaction mechanism from another requires determination of the rate equations responsible for the *steady-state occurrence* of an $[ES]$ complex.

Establishment of the rate equation for the compulsory-order reaction that follows outlines the necessary considerations involved in developing a hypothetical rate equation for multisubstrate reactions.

Graphic Distinctions between Compulsory Order and Ping-Pong Mechanisms

Consider the compulsory-order mechanism detailed below, along with its individual rate constants.

$$[S_1] \qquad\qquad [S_2]$$
$$\Updownarrow \qquad\qquad \Updownarrow$$
$$[E] \underset{k_2}{\overset{k_1}{\rightleftharpoons}} [ES_1] \underset{k_4}{\overset{k_3}{\rightleftharpoons}} [ES_1S_2] \underset{k_6}{\overset{k_5}{\rightleftharpoons}}$$

$$[P] \qquad\qquad [Q]$$
$$\Updownarrow \qquad\qquad \Updownarrow$$
$$[EPQ] \underset{k_8}{\overset{k_7}{\rightleftharpoons}} [EQ] \underset{k_{10}}{\overset{k_9}{\rightleftharpoons}} [E]$$

Establishment of a rate equation here assumes that the reaction rate is measured before product formation and prior to any possible reverse reactions. Therefore, the rate equation is calculated where $[P]$ and $[Q]$ equal zero. This assumption eliminates the possibility of reverse reactions for $[E]$ to $[EQ]$ and $[EQ]$ to $[EPQ]$ along with their respective k_{10} and k_8 constants. Using a steady-state perspective, the overall reaction rate is governed by enzyme distribution among its four possible complexed forms, namely $[ES_1]$, $[ES_1S_2]$, $[EPQ]$, and $[EQ]$. A detailed mathematical analysis of these four complexes based on a steady-state treatment has been presented by Engel. The overall rate equation for the compulsory-order reaction becomes

$$v_0 = \frac{V_{max}[S_1][S_2]}{K_{S_1S_2} + K_{S_1}[S_2] + K_{S_2}[S_1] + [S_1][S_2]}$$

Note:

$K_{S_1} = K_m$ for

$S_1 = \dfrac{k_5k_7k_9}{k_1(k_5k_7 + k_5k_9 + k_6k_9 + k_7k_9)}$

$K_{S_2} = K_m$ for

$S_2 = \dfrac{k_9(k_4k_6 + k_4k_7 + k_5k_7)}{k_3(k_5k_7 + k_5k_9 + k_6k_9 + k_7k_9)}$

$K_{S_1S_2} = $ constant

$= \dfrac{k_2k_9(k_4k_6 + k_4k_7 + k_5k_7)}{k_1k_3(k_5k_7 + k_5k_9 + k_6k_9 + k_7k_9)}$

The corresponding rate equation for a ping-pong mechanism is

$$v_0 = \frac{V_{max}[S_1][S_2]}{K_{S_1}[S_2] + K_{S_2}[S_1] + [S_1][S_2]}$$

The kinetic equations for both the compulsory-order and ping-pong mechanisms may be transformed into a Lineweaver–Burk relationship, which conveniently distinguishes one mechanism from another (Figure 4.12).

Factors Controlling the Rate of Enzyme Reactions

The rates of enzyme-mediated reactions are determined by a number of factors, including *enzyme* and *substrate concentrations; temperature, pH,* and *ionic strength;* and *concentrations of specific inhibitors* and *metal activators.*

Enzyme and substrate concentrations: Increasing amounts of enzyme in the presence of a fixed substrate concentration cause an overall increase in reaction activity (Figure 4.13). Increases in substrate concentrations in the presence of a steadfast enzyme concentration also create an apparent increase in reaction velocity up to a point when the reaction assumes zero-order kinetics (Figure 4.10).

Temperature, pH, and ionic strength effects: The velocity of an enzyme reaction plotted against variations in temperature, pH, or ionic strength properties of the reaction medium usually produces a *bell-shaped curve*

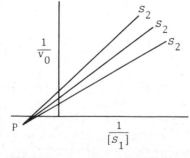 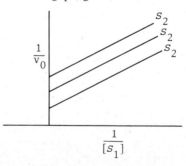

FIGURE 4.12. Exemplary Lineweaver–Burk plots for compulsory-order and ping-pong mechanism. Steadfastly fixed, yet different concentration levels of S_2 for compulsory-order mechanisms yields a series of straight lines radiating from a central convergence point (P). Ping-pong mechanisms, however, are indicated as a series of parallel lines intersecting the ordinate at different points.

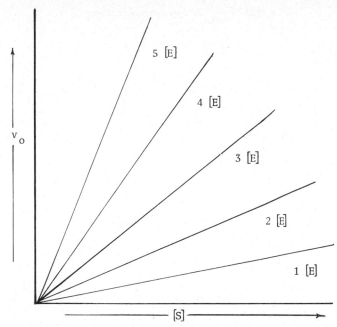

FIGURE 4.13. Plots of enzyme-mediated reaction velocities versus [S] concentrations for five different enzyme concentrations [E].

(Figure 4.14). The peak of the curve represents that particular characteristic of the medium that promotes optimum enzyme reaction velocity. Furthermore, the optimum enzyme reaction conditions usually reflect the *in vivo* requirements of the native enzyme. The optimum conditions for some enzymes are restricted to very limited ranges, while other enzymes may display a uniform V_{max} over a wide range of temperature, pH, or ionic strength conditions. For example, the activity of papain used in meat tenderization is not merely restricted to a small pH range but extends over a wide range from pH 4.0 to 8.0.

Extremes beyond optimum temperature, pH, and ionic strength conditions may significantly alter the electrical charges of R groups on amino acids in the enzyme structure. Since enzyme catalysis is intrinsically related to tertiary and quaternary protein structure, modification of electrical charges on R groups can impair or eliminate enzyme activity altogether by denaturing the protein.

The application of extreme temperatures to enzymes serves as an effective method for controlling enzymatic reactions in foods and food tissues. Although the application of heat is usually more destructive to enzymes than very low temperatures, both can ultimately contribute to denaturation of enzyme proteins. Temperatures of 70 to 80°C for 2 to 5 min destroy most enzymes. Accordingly, temperatures ≥80°C are safely used to denature enzymes in fruits and vegetables by blanching, while food pasteurization and sterilization temperatures (50–120°C) are geared to denature proteins, including enzymes, which are critical for the sporulation and/or growth of microorganisms.

Denaturation of enzymes using very high temperatures is usually an *irreversible* form of denaturation. Enzymes lowered to temperatures less than their optimum, but above 0°C, can also lead to enzyme denaturation, but the loss of enzyme activity is not always permanent. Low-temperature deactivation of enzymes may be followed by resumption of enzyme activity when higher tempera-

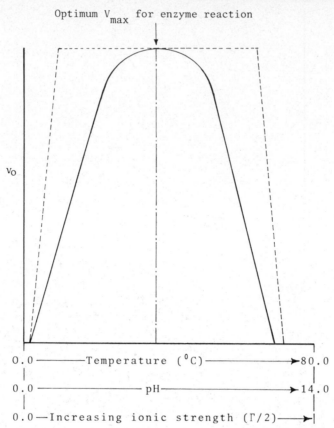

FIGURE 4.14. "Bell-shaped curve" demonstrating the dependence of an enzyme-mediated reaction on conditions of temperature, pH, and ionic strength. The solid line (——) traces the reaction velocity where the highest velocity (V_{max}) is limited to small ranges of temperature, pH, or ionic strength conditions, whereas the dotted line (----) indicates the highest velocity over a wide range of corresponding conditions.

tures (e.g., 0–37°C) again prevail. Actually, most suboptimum temperature deactivations of enzymes occurring at $\geq 0°C$ are the result of two related factors:

1. Dehydration of enzyme proteins caused by *water crystallization* during freezing of water.

2. Permanent or temporary denaturation of enzyme proteins caused by *high solute concentrations* (ionic strength) in any remaining unfrozen water contained in cells and tissues.

ENZYME INHIBITORS AND METAL ACTIVATORS

The activity of enzymes can be inhibited by many different chemical substances. Some of these inhibitory compounds are produced by normal cellular biochemical pathways where they regulate key enzymes in metabolism. Many other inhibitors have nonbiological origins. Inhibitors can decrease enzyme activity by physically binding to the *active site* of an enzyme *or* by binding at a location *other than the active site.* In either case, inhibitors interfere with the normal chemical or conformational requirements of the enzyme's functional catalytic site. In some cases inhibitor binding to an enzyme may be a permanent affair or only temporary.

Although enzyme inhibitors have never been intentionally added to natural or processed foods on a production basis, their biochemical action is the underlying basis for some forms of food intoxications, the effects of antibiotics on animals and bacteria, and direct or indirect target effects of certain agricultural pesticides and environmental pollutants. Inhibitors are also of interest be-

cause they have been extensively used to investigate the properties of enzyme active sites, elucidate biochemical pathways, and chemotherapeutically treat various pathologies (Figure 4.15).

Enzyme inhibitors may be classified as competitive (CI), noncompetitive (NCI), and uncompetitive (UCI).

Competitive Inhibition

This form of inhibition is characterized by the ability of an enzyme active site to bind with substrate *or* inhibitor molecules; however, the binding of substrate *or* inhibitor

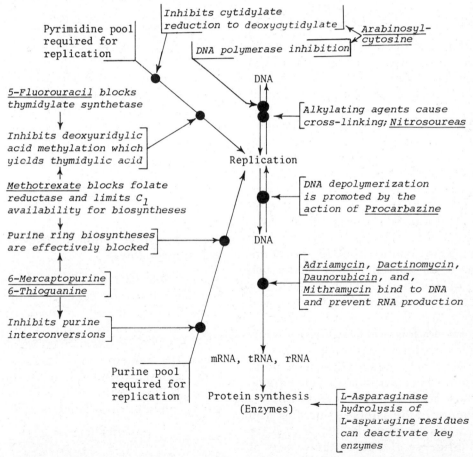

FIGURE 4.15. The strategic interruption of cellular replication mechanisms at the level of nucleic acid activity serves as one possible rationale for controlling tumor growth. A variety of classical antitumor agents along with their schematic roles as *blocking or inhibitory agents* in normal nucleic acid activities are illustrated.

necessarily excludes binding of the other substance at the same active site. Competitive inhibitors are permitted to bind at the enzyme active site because they are wholly or partially reminiscent of the substrate's molecular structure.

The effect of a CI is clearly demonstrated by the enzyme activity of *succinic acid dehydrogenase* (SAD), which is responsible for oxidizing the two methylene carbons of succinic acid to form fumaric acid:

$$
\begin{array}{c}
\text{COO}^- \\
| \\
\text{CH}_2 \\
| \\
\text{CH}_2 \\
| \\
\text{COO}^-
\end{array}
\quad + \text{FAD} \underset{\text{enzyme}}{\overset{\text{SAD}}{\rightleftharpoons}}
\begin{array}{c}
\text{COO}^- \\
| \\
\text{CH} \\
\| \\
\text{HC} \\
| \\
\text{COO}^-
\end{array}
\quad + \text{FADH}_2
$$

Succinate   **Fumarate**

Although succinic acid is the primary substrate for SAD, the structural similarity of malonic acid ($^-\text{OOC}-\text{CH}_2-\text{COO}^-$) permits it to competitively bind with succinic acid at the active site of the enzyme. Regardless of the absolute molar concentration of malonic acid to the succinic acid substrate, a 1:50 ratio (malonic acid:succinic acid) causes a 50% reduction in reaction velocity. The effects of a CI such as malonic acid can be overcome by substantially increasing the concentration of the substrate over that of the inhibitor.

Competitive inhibitors are of special importance to the operation of antibiotics such as sulfa drugs. The sulfanilamide moiety of sulfa drugs resembles *p*-aminobenzoic acid (PABA), which is an essential biochemical prerequisite for folic acid biosynthesis in many pathogenic bacteria. Accordingly, the competitive binding of the sulfa drug moiety and the *p*-aminobenzoic acid at a single enzyme step in folic acid biosynthesis eventually leads to the demise of the pathogen since folic acid requirements can never be met by PABA.

The influence of a CI on the velocity of an enzyme reaction can be mathematically predicted, based on the formation of an *EI* complex; that is,

$$[E] + [I] \underset{k_{-1}}{\overset{k_1}{\rightleftharpoons}} [EI]$$

Since the $[EI]$ complex is unable to form a product, the term K_i is introduced to describe the dissociation constant of the $[EI]$ complex where

$$K_i = \frac{k_{-1}}{k_1}$$

and the dissociation constant for the $[EI]$ complex becomes

$$K_i = \frac{[E][I]}{[EI]}$$

The velocity of an enzyme reaction in the presence of a CI can be mathematically expressed in terms of an equation that is similar to the Michaelis–Menten relationship:

$$v_0 = \frac{V_{\max}[S]}{K_m\left(1 + \dfrac{[I]}{K_i}\right) + [S]}$$

The reciprocal of this expression gives the Lineweaver–Burk plot equation given below, and the corresponding plot of $1/v_0$ (ordinate) versus $1/[S]$ for a series of increasing inhibitor concentrations is shown in Figure 4.16.

$$\frac{1}{v_0} = \frac{K_m}{V_{\max}}\left(1 + \frac{[I]}{K_i}\right) \cdot \frac{1}{[S]} + \frac{1}{V_{\max}}$$

Noncompetitive Inhibition (Reversible and Irreversible)

A reversible NCI inhibits an enzyme reaction but it does not "compete" with the substrate for the active site of an enzyme. Binding of the NCI to an enzyme occurs at a site *other than the active site* and may involve binding to the free enzyme molecule or the enzyme–substrate complex. Both situations have been outlined below along with the appropriate inhibitor dissociation constant:

$$(1)\ [E] + [I] \rightleftharpoons [EI]$$

$$K_i^{EI} = \frac{[E][I]}{[EI]}$$

$$(2)\ [ES] + [I] \longrightarrow [ESI]$$

$$K_i^{ESI} = \frac{[ES][I]}{[ESI]}$$

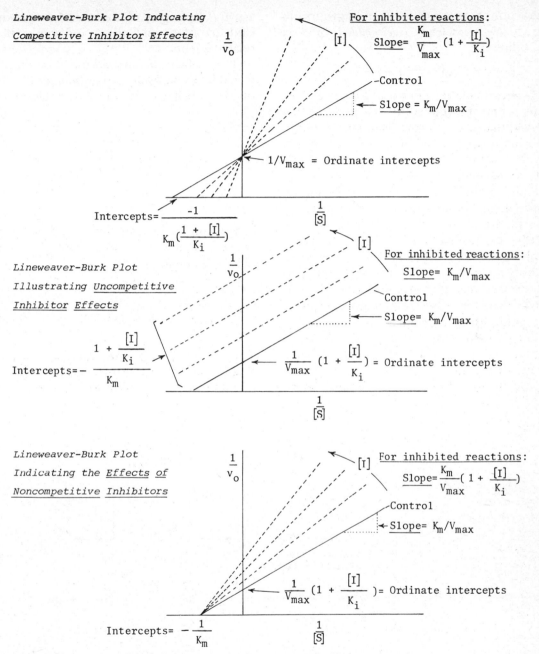

FIGURE 4.16. Comparative graphic analyses for Lineweaver–Burk plots depicting the graphic effects of *competitive*, *uncompetitive*, and *noncompetitive* inhibitors.

Unlike a CI, the effects of an NCI cannot be minimized by substantially increasing substrate concentrations over that of the inhibitor. In fact, NCI effects are independent of substrate concentration.

Some of the most common examples of these inhibitors involve the actions of heavy metal ions such as Cu^{2+}, Hg^{2+}, and Ag^+ on the sulfhydryl groups (—SH) of cysteine residues held in enzyme structures. These interactions between the sulfhydryl group and the heavy metal lead to *mercaptide* formation as follows:

$$E—SH + Ag^+ \rightleftarrows E—S—Ag + H^+$$

The rapid equilibrium existing between free metal ions and the sulfhydryl group is believed to cause unfavorable modifications in the conformation of the enzyme active site.

The action of heavy metals on enzyme reactions may be reversible; however, organophosphorous compounds such as diisopropyl phosphofluoridate (DIPF) are representative of irreversible inhibitors that *react* at the active site of an enzyme. Most organophosphorous compounds are believed to covalently bind with the hydroxyl groups of serine residues that are crucial for the catalytic activities of enzymes (Figure 4.17).

Although many similar cases of irreversible NCI exist, the organophosphorous reactions are of special interest because they mirror the action of organophosphorous insecticides that decrease acetylcholinesterase activity. Examples of these insecticides include parathion, malathion, and many other related agents.

The velocity of an enzyme reaction affected by an NCI may be mathematically defined in Michaelis–Menten terminology as

$$v_0 = \frac{\dfrac{V_{max}}{\left(1 + \dfrac{[I]}{K_i}\right)}}{K_m + [S]} \cdot [S]$$

Expression of this equation as a double-reciprocal (Lineweaver–Burk) plot gives the following equation:

$$\frac{1}{v_0} = \frac{K_m}{V_{max}} \left(1 + \frac{[I]}{K_i}\right) \cdot \frac{1}{[S]} + \frac{1}{V_{max}} \left(1 + \frac{[I]}{K_i}\right)$$

The graphic plot of this equation for $1/v_0$ versus $1/[S]$ for increasing concentrations of inhibitors is illustrated in Figure 4.16.

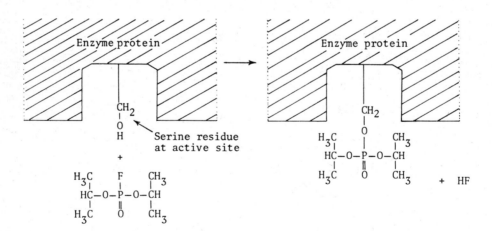

Diisopropyl phosphofluoridate (DIPF) plus enzyme (DIPF serves as an irreversible noncompetitive inhibitor)

Diisopropyl phosphofluoridated enzyme (inactivated by DIPF at the active site)

FIGURE 4.17. Example of diisopropyl phosphofluoridate (DIPF) as an irreversible noncompetitive inhibitor (NCI) at the active site of an enzyme. The enzyme active site illustrated is typical of that found in acetylcholinesterase, which is critical for neurotransmission processes. Note that the DIPF reacts with a serine residue contained within the active site of the enzyme.

Uncompetitive Inhibition

An ideal UCI is a substance that reversibly binds to an enzyme–substrate (ES) complex to yield an ESI complex:

$$[ES] + [I] \rightleftharpoons [ESI]$$

The inhibitor binds only to the ES complex as opposed to free enzyme and it effectively blocks substrate conversion to product. The K_i constant for an UCI is defined as

$$K_i = \frac{[ES][I]}{[ESI]}$$

and the associated velocity equation for an enzyme reaction with an UCI becomes

$$v_0 = \frac{\dfrac{V_{max}[S]}{\left(1 + \dfrac{[I]}{K_i}\right)}}{[S] + \dfrac{K_m}{1 + \dfrac{[I]}{K_i}}}$$

A reciprocal transformation of this velocity expression gives its corresponding Lineweaver–Burk equation.

$$\frac{1}{v_0} = \frac{K_m}{V_{max}} \frac{1}{[S]} + \frac{1}{V_{max}}\left(1 + \frac{[I]}{K_i}\right)$$

The graph of $1/v_0$ (ordinate) versus $1/[S]$ for a series of increasing inhibitor concentrations directs the positioning of a series of parallel lines (Figure 4.16).

A tabular survey of the characteristic mathematical features that define the position of Lineweaver–Burk plots for various inhibitor effects appear in Table 4.6.

Metal Activators and Enzymes

The *in vivo* catalytic activities of many enzymes show a *dependent* and *graduated* response to the availability of certain metal ions. The mechanism of metal ion involvement in enzyme reactions is not always clear and may be extremely complex depending on the nature of the biochemical reaction. From the perspectives of food technology, enzyme activation by metal ions warrants special attention since these ions may promote the rate of deteriorative enzymatic reactions in mildly processed or unprocessed foods. Metal ions can originate from natural environmental sources, metal food-preparation surfaces, storage containers, metal piping, or water used in food processing.

Some of the most thoroughly understood enzyme reactions depend upon the availability of Mg^{2+} and Mn^{2+}. These ions are in-

Table 4.6. Characteristic Ordinate and Abscissa Intercepts as Well as Slopes for Lineweaver–Burk Plots under Normal Reaction Conditions in Addition to Selected Types of Inhibitors

Type of inhibitor	Linear slope	Ordinate intercept	Abscissa intercept
No inhibitor	$\dfrac{K_m}{V_{max}}$	$\dfrac{1}{V_{max}}$	$-\dfrac{1}{K_m}$
Competitive	$\dfrac{K_m}{V_{max}}\left(1 + \dfrac{[I]}{K_i}\right)$	$\dfrac{1}{V_{max}}$	$\dfrac{-1}{K_m\left(\dfrac{1 + [I]}{K_i}\right)}$
Uncompetitive	$\dfrac{K_m}{V_{max}}$	$\dfrac{1}{V_{max}}\left(1 + \dfrac{[I]}{K_i}\right)$	$-\dfrac{1 + \dfrac{[I]}{K_i}}{K_m}$
Noncompetitive (both K_i values are identical)	$\dfrac{K_m}{V_{max}}\left(1 + \dfrac{[I]}{K_i}\right)$	$\dfrac{1}{V_{max}}\left(1 + \dfrac{[I]}{K_i}\right)$	$-\dfrac{1}{K_m}$

trinsically related to enzymatic phosphorylation reactions, which also involve ATP:

A plot of this equation in terms of v_0 (ordinate) versus increasing metal concentra-

$$
\text{Adenosine}-\text{O}-\overset{\overset{\text{O}}{\|}}{\underset{\underset{\text{O}^-}{|}}{\text{P}}}-\text{O}-\overset{\overset{\text{O}}{\|}}{\underset{\underset{\text{O}^-}{|}}{\text{P}}}-\text{O}-\overset{\overset{\text{O}}{\|}}{\underset{\underset{\text{O}^-}{|}}{\text{P}}}-\text{O}^-
$$

$$
\text{Mg}^{2+}
$$

ATP (Adenosine triphosphate)

$$
\text{Adenosine}-\text{O}-\overset{\overset{\text{O}}{\|}}{\underset{\underset{\text{O}^-}{|}}{\text{P}}}-\text{O}-\overset{\overset{\text{O}}{\|}}{\underset{\underset{\text{O}^-}{|}}{\text{P}}}-\text{O}^- + \text{HO}-\overset{\overset{\text{O}}{\|}}{\underset{\underset{\text{O}^-}{|}}{\text{P}}}-\text{O}^- + \text{Mg}^{2+}
$$

ADP (Adenosine diphosphate)

Two exemplary phosphorylation reactions that require the presence of Mg^{2+} or Mn^{2+} have been outlined below for *hexokinase* (1) and *pyruvate kinase* (2):

(1) Glucose + ATP + Mg^{2+} $\underset{}{\overset{\text{hexokinase}}{\rightleftharpoons}}$
   glucose 6-phosphate + ADP + Mg^{2+}

(2) Pyruvate + ATP + Mn^{2+} $\underset{}{\overset{\text{pyruvate kinase}}{\rightleftharpoons}}$
   phosphoenolpyruvate + ADP + Mn^{2+}

A schematic enzyme reaction sequence for an enzyme reaction that requires a metal ion can be described as

tions for a series of fixed but increasing substrate concentrations is evident as a series of hyperbolic curves with corresponding increases in the values of V_{max} (Figure 4.18).

MODULATION OF ENZYME ACTIVITY

The previous discussion of Michaelis–Menten enzyme kinetics has shown that pH, ionic

$$
[E] + [\text{Met}] \underset{k_{-\text{met}}}{\overset{k_{\text{met}}}{\rightleftharpoons}} [E\text{Met}] \overset{S}{\underset{S}{\overset{k_s}{\underset{k_{-s}}{\rightleftharpoons}}}} [E\text{Met } S] \overset{k}{\longrightarrow} [E] + [P] + [\text{Met}]
$$

where
   Met = metal
   k_{met} = rate constant for metal involvement
   k_s = rate constant for substrate involvement
   k = rate constant for breakdown of EMet$S$ complex to product

Mathematical analysis of this reaction according to steady-state concepts reveals a velocity equation:

$$
v_0 = \frac{V_{max}\,[S][\text{Met}]}{[S][\text{Met}] + K_s\,[\text{Met}] + K_s + K_m}
$$

Note:

$$
K_m = \frac{k_{-\text{met}}}{k_{\text{met}}} = \text{dissociation constant of the } [E\text{Met}] \text{ complex}
$$

$$
K_s = (k_{-s} + k/k_s) = \text{apparent Michaelis–Menten constant for substrate binding}
$$

strength, coenzymes, cofactors, and many other factors can alter enzyme reaction rates. Furthermore, fixed concentrations of enzymes display rates of overall product formation governed by a variety of metabolic inhibitors and cellular substrate concentrations.

It should be clearly recognized that the activity of an enzyme below prescribed limits of maximum velocity are not necessarily incompatible with its participation in normal metabolic activities. Actually, the interplay and restrictions of complex metabolic checks, balances, and counterbalances ensure the consistent operation of most *in vivo* enzyme systems at *kinetic rates of less than maximum velocity*.

Exclusive of the obvious kinetic effects outlined above for enzymes obeying Michaelis–Menten kinetics, most enzymes responsible for regulating metabolic pathways require more sophisticated control mecha-

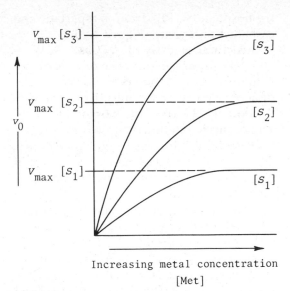

FIGURE 4.18. Plot of v_0 versus metal ion concentration. The $[S]$ values are fixed for any curves but $[S_3] > [S_2] > [S_1]$.

nisms for their normal operation. These enzymes, known as *regulatory enzymes,* display evidence of kinetic activity controls involving principles of (1) *noncovalent modulation,* (2) *covalent modulation,* (3) *intermediate structural modulation,* and (4) *zymogen activation.*

Noncovalent Modulation and Allosteric Enzymes

Enzymes that demonstrate Michaelis–Menten kinetics are generally theorized as having a single catalytic site that is responsible for converting substrate to product. Although the catalytic actions of these enzymes can be influenced by inhibitors, they fail to show *graduated* rates of kinetic activity due to alterations in their three-dimensional molecular structures.

This pattern of behavior is in marked contrast with specialized *regulatory enzymes known as allosteric enzymes.* These enzymes normally experience spatial changes in their molecular structure that lead to corresponding changes in their kinetic rate for converting

substrate to product. The relationship between molecular structure and kinetic rates for allosteric enzymes attests to the cooperative behavior of *at least one catalytic site on two or more discreet, yet joined,* protein subunits composing the entire enzyme structure.

Alterations in the spatial structures of allosteric enzymes, followed by changes in their kinetic activities, result from the noncovalent binding of specific effector (modulator) molecules to sites other than the enzyme's catalytic site. Although effector molecules are highly specific for a given allosteric enzyme, the *effector may exert a positive or negative kinetic effect. Positive modulators enhance the kinetic activity of the allosteric enzyme, while negative modulators decrease activity.*

The regulatory nature of allosteric enzymes may include sensitivity to one or more modulators. Those enzymes influenced by a single specific modulator are referred to as *monovalent,* while *polyvalent modulation* refers to the mutually exclusive noncovalent bonding potential of two or more different types of modulators.

The chemical nature of modulators is highly variable. Depending on the intrinsic metabolic role of an enzyme, the modulator may be protein, lipid, carbohydrate, nucleotide, or almost any other possible metabolic intermediate.

Since most biochemical pathways require two or more enzymes, at least one enzyme in the early stages of the pathway has regulatory significance. Therefore, allosteric modulators can actively participate in *feedback* inhibition (end-product inhibition) for specific biochemical pathways. In these cases, especially high levels of product, originating from a series of enzymatic reactions, or key intermediates toward final product formation, allosterically interact with the regulatory enzyme as a negative modulator to decrease the rate of product formation (Figure 4.19).

Allosteric enzymes modulated by substances other than their substrate are known as *heterotropic enzymes.* If the modulator affecting the enzyme is the same as the en-

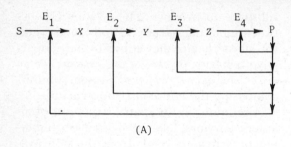

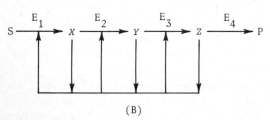

(A)

(B)

FIGURE 4.19. Any one of a series of enzymes participating in product formation represents a possible site for feedback inhibition by the final product (P) (A) or any intermediate (X, Y, or Z) (B) depending on the location of the regulatory enzyme.

zyme's substrate, the enzyme is called a *homotropic allosteric enzyme*. Homotropic enzymes contain a minimum of two substrate binding sites since kinetic rate control in these enzymes depends on the number of occupied substrate binding sites. Some allosteric enzymes are purely homotropic or heterotropic, but the majority of allosteric enzymes are heterotropic/homotropic hybrids in which a variety of metabolites and substrates act as modulating agents.

It should be recognized that a structural similarity between a modulator and the substrate for an allosteric enzyme is not a prerequisite for modulator activity. The model of aspartate transcarbamylase (ATCase) clearly demonstrates the extreme dissimilarity that can exist between an allosteric modulator and the substrate involved in an allosteric enzyme reaction.

Aspartate transcarbamylase is responsible for one of the preliminary biosynthetic steps leading the nucleotide structure of cytidine

triphosphate (CTP) from L-aspartate and carbamyl phosphate (Figure 4.20).

Structural analysis of ATCase (M.W. ~ 305,000) shows the presence of at least two trimeric catalytic subunits and three smaller dimeric regulatory subunits. Excessive production of CTP from the biosynthetic pathway results in binding of approximately two CTP molecules for each regulatory dimer. The CTP binding then initiates a change in regulatory subunit conformation that is ultimately transmitted as a spatial change in the conformation of the catalytic subunits. The net effect of these spatial interactions is the decreased production of N-carbamyl-L-aspartate. Excessive concentrations of adenosine triphosphate (ATP) have also been found to exert a modulation effect related to overall CTP production. In this case, ATP binding to the regulatory subunits of ATCase initiates conformational changes in the catalytic subunits, which enhances N-carbamyl-L-aspartate production leading to CTP. Therefore, ATP acts as a positive modulator.

In both instances, modulation of ATCase results from substances (CTP, ATP) that do not resemble the structure of any fundamental substrates required for ATCase activity, and yet these substances control the kinetic behavior of the allosteric enzyme. This type of enzyme activity also represents a notable departure from structural similarities, which are often prerequisites for competitive inhibition of enzymes strictly obeying Michaelis–Menten kinetics.

Allosteric Enzyme Kinetics

The kinetic behavior of allosteric enzymes becomes evident when their kinetic activities are compared with nonregulatory enzymes that adhere to Michaelis–Menten kinetics. A study of both types of kinetics over a velocity range of 0.1 V_{max} and 0.75 V_{max} (i.e., $[S]_{0.1}$ to $[S]_{0.75}$) graphically displays the differences between these types of enzymes. For example, an ideal Michaelis–Menten hyperbola for a velocity versus $[S]$ plot (Figure

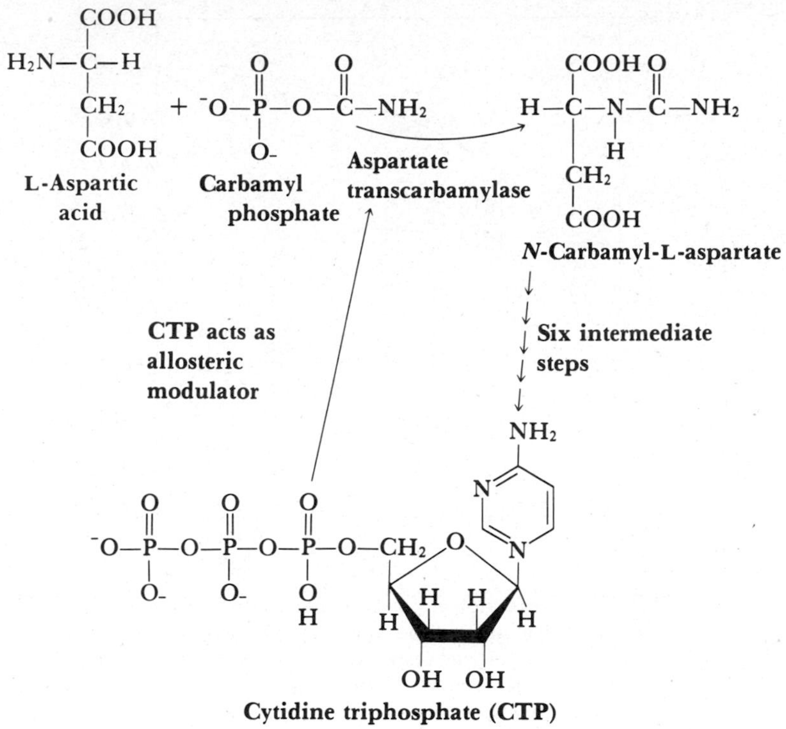

COOH
|
H₂N—C—H
|
CH₂   O   O
| + ⁻O—P—O—C—NH₂
COOH |
O⁻

L-Aspartic acid   **Carbamyl phosphate**

Aspartate transcarbamylase

COOH O
| ||
H—C—N—C—NH₂
| H
CH₂
|
COOH

N-Carbamyl-L-aspartate

CTP acts as allosteric modulator

Six intermediate steps

NH₂
|
N
|
O   O   O
|| || ||
⁻O—P—O—P—O—P—O—CH₂   O
| | |
O⁻   O⁻   O
H   H   H   H

OH OH

Cytidine triphosphate (CTP)

FIGURE 4.20. Allosteric modulator activity of cytidine triphosphate on aspartate transcarbamylase (ATCase).

4.21) requires a 27 $[S]$ increase in order to go from $0.1\,V_{max}$ to $0.75\,V_{max}$. This is contrary to allosteric enzymes that traverse the equivalent velocity range with only a 2.3 $[S]$ increase. In other words, allosteric enzymes that display sigmoidal kinetics show lower rates of velocity at lower substrate concentrations than nonregulatory enzymes obeying Michaelis–Menten kinetics. The apparent kinetic disadvantage for the enzyme obeying sigmoid kinetics eventually becomes insignificant once the enzyme is exposed to some critical substrate concentration, at which time only minor substrate increases lead to rapid increases in velocity.

The sigmoid kinetic response for allosteric enzymes permits sensitive regulation of metabolic pathways based on enzyme velocity and provides a graduated enzyme response to available substrate concentrations. Evidence of sigmoid kinetics also supports the concept that allosteric regulatory enzymes function on the basis of cooperative substrate binding. This indicates that substrate binding at one site on an allosteric enzyme sequentially facilitates binding of substrate molecules at other sites on the same enzyme. Substrate binding at one site that enhances substrate binding at other sites is known as *positive cooperativity*. Although many regulatory enzymes fall into this category, some display negative cooperative effects, when binding of a substrate molecule at one location diminishes the substrate binding at other sites on the enzyme. These regulatory enzymes may require at least a 6000 $[S]$ increase in order to traverse the velocity range between $0.1\,V_{max}$ and $0.9\,V_{max}$ (Figure 4.22).

Kinetic Considerations for Allosteric Modulation

The kinetic behavior of an allosteric enzyme is complicated further by the influence of

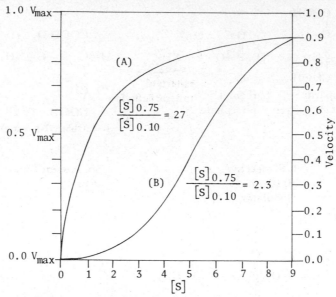

FIGURE 4.21. Comparison of velocity curves for two different types of enzymes that demonstrate the same velocities at $[S] = 9.0$. (A) Indicates a typical hyperbolic response while (B) shows a sigmoidal response. It is also clear from the examples given that the $0.75\ V_{max}/0.10\ V_{max}$ ratio for the hyperbolic response is 27 as opposed to a 2.3 ratio for the sigmoidal plot of enzyme action.

positive and negative modulators on the *apparent* K_m and maximum velocity of enzyme activity.

Positive modulators appear to enhance the activity of allosteric enzymes by causing a decrease in the apparent K_m, while negative modulators cause an increase in the apparent K_m and a corresponding decrease in enzyme–substrate affinities. Allosteric enzymes that respond to these modulator effects, but retain the same V_{max} in all cases, are known as *K enzymes* (Figure 4.23).

This type of modulator response is in contrast to *M enzymes,* which exhibit variations in V_{max} and maintain a constant apparent K_m value. In this case, *negative modulators decrease* the level of V_{max} and *positive modulators increase* V_{max} (Figure 4.23).

It must be clearly recognized that the apparent K_m for an allosteric enzyme that displays sigmoidal kinetics is not the same as the K_m value used for enzymes obeying Michaelis–Menten kinetic relationships. The apparent K_m similarly refers to that substrate concentration responsible for one-half maximum velocity ($[S]_{0.5}$); but the $[S]_{0.5}$ value for the apparent K_m cannot be substituted into the Michaelis–Menten rate equation to calculate initial velocity of an allosteric enzyme. The v versus $[S]$ kinetic relationship for Michaelis–Menten enzymes is typically hyperbolic and the same graphic relationship for allosteric enzymes is sigmoidal.

Since the sigmoidal kinetic plots for allosteric enzymes reflect the cumulative effects of modulators plus the multiple interaction of cooperative substrate binding sites within a single enzyme structure, the apparent K_m is more aptly called K' (*intrinsic binding constant*). The K' value is a measure of allosteric protein–ligand (allosteric enzyme–substrate) binding based on substrate concentration. Alterations in the concentrations of positive or negative modulators for *K* enzymes directly translate into corresponding changes in K' (Figure 4.23).

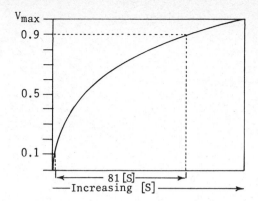

*Nonregulatory enzymes
obeying Michaelis-
Menten rate equation*

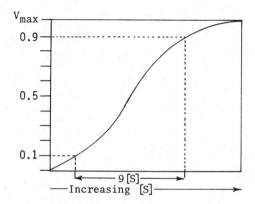

*Regulatory enzymes
displaying positive
cooperativity*

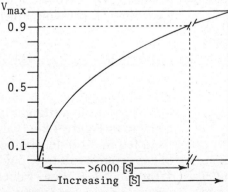

*Regulatory enzymes
exhibiting negative
cooperativity*

FIGURE 4.22. Comparative graphic illustrations for those enzymes displaying different types of kinetic plots. The relative substrate concentration increases ($[S]_{0.1} \rightarrow [S]_{0.9}$) required to elevate each enzyme from 10 to 90% V_{max} ($0.1\,V_{max} \rightarrow 0.9\,V_{max}$) have been indicated.

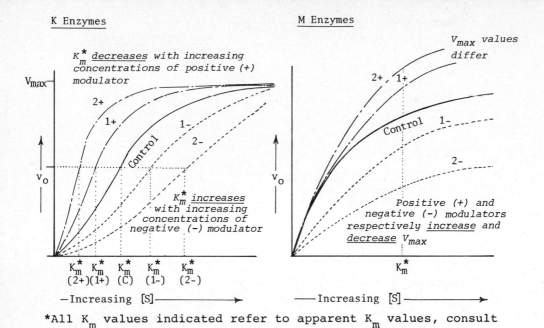

K Enzymes

K_m^* _decreases_ with increasing concentrations of positive (+) modulator

K_m^* _increases_ with increasing concentrations of negative (-) modulator

K_m^* K_m^* K_m^* K_m^* K_m^*
(2+)(1+) (C) (1-) (2-)

—Increasing [S]——→

M Enzymes

V_{max} values differ

Positive (+) and negative (-) modulators respectively _increase_ and _decrease_ V_{max}

K_m^*

——Increasing [S]——→

*All K_m values indicated refer to apparent K_m values, consult legend and text for details

FIGURE 4.23. Graphic comparisons of modulator effects on substrate–velocity curves for two types of heterotropic regulatory enzymes. For those enzymes classified as K enzymes, the presence of a positive (+) or negative (−) modulator, respectively, increases or decreases enzyme–substrate affinity without influencing the V_{max}. In the case of M enzymes, effectors alter the V_{max} but the *apparent K_m* remains relatively unchanged from control enzyme behavior (i.e., enzyme–substrate affinities remain unaffected). *It is very important to recognize that allosteric enzymes do not display actual K_m values and those K_m's indicated in the figure actually represent apparent K_m values for illustrative purposes only.* Since allosteric sigmoidal kinetic relationships do not follow a hyperbolic Michaelis-Menten relationship, the concept of an *apparent K_m* is usually expressed by the symbol $[S]_{0.5}$ or $K_{0.5}$. These designations symbolize that substrate concentration that gives half maximum velocity for an allosteric enzyme.

Allosteric Cooperativity and the Hill Plot

A variation of the equation outlined by A. V. Hill in 1910 provides the basis for dealing with the sigmoidal kinetic relationships exhibited by allosteric enzymes. The so-called Hill equation is

$$\frac{v}{V_{max}} = \frac{[S]^n}{[S]^n + K'}$$

This equation is reminiscent of the fundamental Michaelis–Menten equation, but it accounts for the intrinsic binding constant (K') and the number of substrate binding sites found on a single enzyme molecule as the Hill coefficient (n). The Hill equation

can be rearranged and transformed into a logarithmic expression, which is very useful:

$$\log \frac{v}{V_{max} - v} = n \log [S] - \log K'$$

A plot of $\log v/(V_{max} - v)$ as a function of $\log [S]$ results in a straight line with a slope equal to n (Figure 4.24). Values for n range from 0.0 to $\geq$1.0 depending on the degree of cooperative substrate binding. If n is 1.0, <1.0, or >1.0, these values indicate *noncooperative, negative cooperative,* or *positive cooperative* substrate binding, respectively.

The Hill plot can also be used to determine the substrate concentration that will

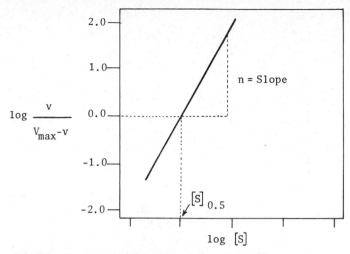

FIGURE 4.24. Generalized plot for a Hill equation (i.e., $\log[S]$ versus $v/(V_{max}-v)$). The $v/(V_{max}-v)$ reflects the amount of the ES complex formed compared to that condition in which the enzyme may be completely saturated with substrate. The slope of the linear plot (n) reflects the effect of one molecule of substrate *or* effector binding to the enzyme on the binding of the next substrate molecule. When $n = 1.0$, the linear plot becomes indicative of a nonregulatory enzyme obeying the classical Michaelis–Menten equation. Although $\frac{1}{2}V_{max} = K_m$ for non-regulatory enzymes, this is not true for regulatory enzymes. The dynamic behavior of regulatory enzymes must be considered with respect to K', where $K' = [S]_{0.5}^{n}$. The value of K' is altered as different concentrations of positive and negative modulators become available. Positive modulators reduce $[S]_{0.5}$ values and cooperativity to a point where the enzyme exhibits classical Michaelis–Menten kinetics.

give $0.5\ V_{max}$ (i.e., $[S]_{0.5}$) for an enzyme. For example, when the ordinate value for $\log v/(V_{max} - v) = 0.0$, the corresponding substrate concentration on the abscissa defines the $\log [S]_{0.5}$ for the enzyme. This value is related to the calculation of K' by the expression

$$K' = \sqrt[n]{[S]_{0.5}}$$

The foregoing explanation of sigmoid kinetic relationships has been based on the sequential interaction model for enzyme–substrate binding. Further details regarding the precepts of this model along with alternative mathematical treatments for allosteric enzyme kinetics can be found in Segel (1976), Piszkiewicz (1977), or other similar references.

Covalent Modulation of Regulatory Enzymes

The structural interconversion between active and inactive forms of a regulatory enzyme by a completely different enzyme serves as a basis for *covalent modulation*. Although many regulatory enzymes are probably controlled by covalent modulation, few are thoroughly understood. The archetype of covalent modulation involves interconversion of inactive phosphorylase-*b* to active phosphorylase-*a* (Figure 4.25). The function of this enzyme entails the mobilization of glucose 1-phosphate from glycogen during the early stages of carbohydrate metabolism.

Two inactive dimers of phosphorylase-*b* are converted to an active phosphorylase-*a* tetramer by *phosphorylase kinase*, which phos-

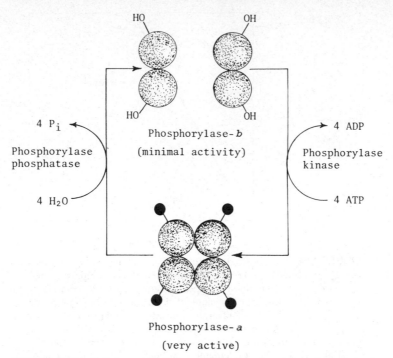

FIGURE 4.25. Covalent modulation of phosphorylase-*b* to -*a*. The two minimally active phosphorylase-*b* dimers each display two —OH groups that represent a serine residue capable of undergoing phosphorylation at the expense of ATP. The two phosphorylated dimers combined as a tetramer form phosphorylase-*a*, and each solid black circle represents a phosphorylated serine residue (see diagram).

phorylates the serine residue on each unit of the tetramer at the expense of ATP. This phosphorylation gives phosphorylase-*a* the ability to activate glucose 1-phosphate formation from glycogen. If excessive amounts of glucose 1-phosphate result, the interplay of metabolic pathways may promote the activity of yet another enzyme known as phosphorylase phosphatase. This enzyme mediates the hydrolytic dephosphorylation of phosphorylase-*a* and yields phosphorylase-*b*, which has little ability to cleave glycogen.

To further complicate the principles of covalent enzyme modulation, it is now eminently clear that the activity of many enzymes is sequentially regulated by other enzymes. For example, phosphorylase kinase activity is *promoted* by another enzyme known as *adenyl cyclase,* which is activated by epinephrine in the muscles and glucagon in the liver. The activation of one enzyme that con-

secutively modulates one or more enzymes in a series is recognized as an *amplification cascade*. In the example cited, the action of nonenzymes such as epinephrine or glucagon on specific receptor sites of cell membranes precedes adenyl cyclase and protein kinase activity. The amplification cascade effect has great biochemical significance because a one-enzyme, one-substrate reaction can lead to a multiple log-cycle production of different enzyme molecules, each of which may negotiate the conversion of thousands of substrate molecules into product.

Intermediate Structural Modulation of Regulatory Enzymes

Multiple structural forms of a single type of enzyme among different cells or tissues characterizes those enzymes known as *isozymes* or *isoenzymes*. The oligomeric structures of these enzymes (more than one protein subunit) permits variations in their quaternary structures and distinct differences in their abilities to convert substrate into product.

The two or more oligomers constituting one in a series of isoenzymes may all be of one type (*homogeneous*) or different (*heterogeneous*). One group of isoenzymes, displaying both homogeneous and heterogeneous quaternary structures, are the lactate dehydrogenases (LDH enzymes). This category of enzymes displays five distinct quaternary forms, all of which catalyze the following reaction in varying degrees:

$$\text{Lactate} + \text{NAD}^+ \rightleftharpoons \text{pyruvate} + \text{NADH}(\text{H}^+)$$

Genetic studies have revealed that different genes are responsible for LDH isoenzyme synthesis, and five possible quaternary structures may be formed from only two different protein subunits designated as "M" or "H" (Figure 4.26).

A homogeneous tetramer of the M subunit (M_4) is most prevalent in skeletal and embryonic tissues, where it rapidly reduces pyruvate to lactate. The corresponding homologous tetramer constructed from H subunits (H_4) is found predominantly in cardiac tissue, where it displays a low rate of pyruvate reduction to lactate. The higher concentration of one isoenzyme form over another reflects the normal metabolic demands of certain cells or tissues. In the case of cardiac muscle, LDH (M_4) isoenzyme activity is inhibited by high concentrations of lactate, and any tendency toward continued production of lactate from pyruvate is decreased. These factors, in turn, cause the cardiac tissue to rely on an aerobic metabolic pathway that oxidizes pyruvate to carbon dioxide without formation of a lactate intermediate. Muscle tissue, on the other hand, contains the isoenzyme (M_4), which is not significantly inhibited by high concentrations of lactate. This permits continued reduction of pyruvate to yield lactate, thereby supplying energy via anaerobic metabolism (glycolysis).

A comparative study of LDH isoenzymes in cardiac and muscle tissues suggests that the LDH in muscle tissues allows a considerable depletion of carbohydrate energy reserves plus a concomitant production of high lactate levels. This is in marked contrast to the isoenzyme constitution of cardiac muscle, which does not favor accumulation of lactate and energy production from anaerobic glycolytic pathways.

Since the isoenzyme structures for any general class of an enzyme are dictated by different genes, the amino acid composition and isoelectric properties of isoenzymes vary. These fundamental variations allow quantitative analytical determination of specific isoenzymes on a routine basis using electrophoretic and selected chromatographic methods. Isoenzyme separation patterns obtained from these methods can have clinical diagnostic significance for blood serum and plasma samples. In the case of LDH isoenzymes, high concentrations of M_4 released from cardiac tissue parallel the occurrence of a severe myocardial infarction, along with other isoenzymes representing the creatine phosphokinase group (CPK isoenzymes). Other than cardiac problems, elevated levels of M_4 and M_3H forms of LDH in the blood can also substantiate the presence of diseases such as infective hepatitis (jaundice) and related liver damage.

A given set of isoenzymes may display a

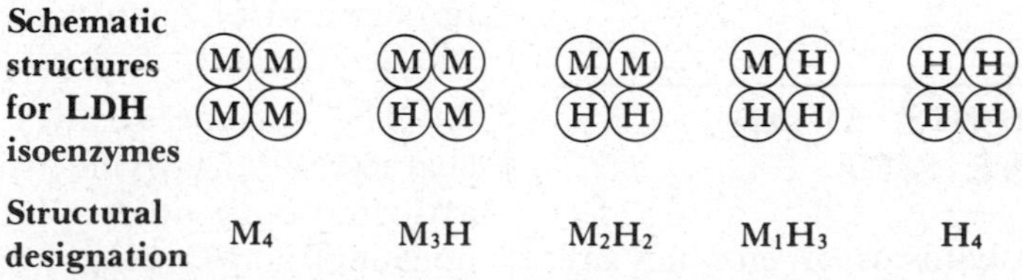

Schematic structures for LDH isoenzymes	$M M$ / $M M$	$M M$ / $H M$	$M M$ / $H H$	$M H$ / $H H$	$H H$ / $H H$
Structural designation	M_4	M_3H	M_2H_2	M_1H_3	H_4

FIGURE 4.26. Possible subunit combinations for LDH isoenzymes.

wide range of V_{max} and apparent K_m values under the same experimental (assay) conditions. Furthermore, since some isoenzymes display allosteric behavior, ideal Michaelis–Menten kinetic principles may not prevail.

Zymogen Activation

A number of important gastric and pancreatic proteases are synthesized as inactive enzyme precursors known as *zymogens* prior to their activation by other enzymes. Typical zymogens include *chymotrypsinogen, trypsinogen,* and *pepsinogen.* The enzymes responsible for the *regulatory conversion* of these *inactive* zymogens to *active* digestive enzymes often require fairly specific conditions such as a low pH. Zymogen activation is often accompanied by the formation of one or more peptides liberated from the N terminal end of the original zymogen protein. Typical examples of zymogen activation include

detailed mechanisms of enzyme action responsible for decreasing the energy of activation in biochemical reactions are still largely obscure. In fact, possible mechanisms for enzyme actions on substrates are limited only by the imagination of the theorist. According to most evidence, however, a number of facts concerning enzyme catalysis are clear:

1. There is *no single enzyme mechanism* that explains how enzymes cause a decrease in energy of activation for biochemical reactions.

2. The enzyme protein is a large structure that has a relatively small active site where catalysis of an organic reaction occurs. The conformational and electrochemical properties of the enzyme active site are thought to contribute stress on any substrate molecule present in the *ES* complex. The substrate then undergoes effects of bond strain at catalytically strategic points that incite a chemical reaction.

Type of zymogen	Enzyme activating zymogen	Enzyme activated from zymogen		Residues from activated zymogen
Pepsinogen	$\xrightarrow[\text{pH 2.0–3.0}]{\text{Pepsin}}$	Pepsin	+	Peptides
Trypsinogen	$\xrightarrow{\text{Enterokinase}}$	Trypsin	+	Hexapeptide
Chymotrypsinogen	$\xrightarrow{\text{Trypsin}}$	Chymotrypsin	+	Two dipeptides

The storage of these enzymes as inactive enzyme precursors is significant because it *minimizes* the proteolytic digestion of the cells responsible for their synthesis. When released into the digestive tract, these inactive enzymes undergo irreversible activation and then demonstrate their characteristic proteolytic activities.

CATALYTIC EVENTS AT AN ENZYME ACTIVE SITE

The catalytic mechanisms of enzymes are rooted in the basic laws of thermodynamics as well as kinetic principles. Unfortunately,

3. The characteristic catalytic activities of an enzyme active site are the result of very specific functional groups. The spatial positions of these functional groups are a consequence of the genetically dictated primary structure of the enzymatically active protein. Furthermore, the catalytic activity of these groups is promoted by their relative spatial orientations to each other as directed by the tertiary and/or quaternary structure of the whole enzyme.

4. The specific catalytic properties that characterize an enzyme-active site are closely affiliated with nucleophilic (electron-pair donating) and/or electrophilic (electron-pair acceptor) interactions between the enzyme and its bound substrate. Typical electrophilic groups present in enzyme structures are di- and trivalent metal cations (Mg^{2+}, Mn^{2+}, and Fe^{2+}) or $—NH_3^+$ groups of amino acids. Notable nucleophilic groups in enzyme structure may include amino acids with hydroxyl, sulfhydryl (thiol), and/or imidazole groups.

Although some amino acid residues seem to be more important and conspicuously present at the active sites of enzymes than others, there is no defined amino acid or peptide sequence ubiquitously present in all enzymes. Some common amino acid residues that seem to be involved in the catalytic activities of enzyme active sites include *serine residues* in phosphoglucomutase and acetylcholinesterase; the *thiol-containing groups* of cysteine in papain; *glutamic* and *aspartic acids* in lysozyme; and the *imidazole groups of histidine* in ribonuclease and chymotrypsin. Clearly one of the most important amino acids found at enzyme active sites seems to be histidine. The imidazole group of this amino acid displays highly nucleophilic properties that are only enhanced by physiological and cellular pH conditions in the vicinity of neutrality. The importance of this residue has been widely supported by chemically altering the histidine residues that are strategically located in the active site of many enzymes. The outcome of these studies almost invariably leads to deactivation of the enzyme.

5. The reaction mechanism of many enzymes *results in the formation of an unstable covalent bond between a portion of the enzyme active site and the substrate.* The covalent enzyme–substrate intermediate has a temporal existence and is readily converted to product and free enzyme in most cases.

Chymotrypsin—A Model for Enzyme Mechanisms

α-Chymotrypsinogen is representative of a zymogen-type enzyme produced by the pancreas before it assumes an active enzyme form in the gastrointestinal tract known as α-*chymotrypsin*. The enzymatically active α-chymotrypsin is a digestive protease that is re-

sponsible for cleaving peptide bonds on the carboxyl side of amino acids with bulky hydrophobic (e.g., methionine) or aromatic (phenylalanine, tryptophan, and tyrosine) **R** groups.

The enzymatically inactive α-chymotrypsinogen, containing 245 amino acid residues, acts as a substrate for trypsin. This enzyme attacks the Arg–Ile (15–16) peptide bond and causes the formation of π-chymotrypsinogen. The π-chymotrypsinogen subsequently attacks other similar polypeptides at amino acid residues 13–14, 146–147, and 148–149. This causes the release of two dipeptides, namely, Ser–Arg (14–15) and Thr–Asn (147–148) (Figure 4.27).

The α-chymotrypsin formed from this activation sequence has a catalytic center with three amino acids (Asp 102, His 57, and Ser 195) that participate in peptide cleavage. Each of these amino acids has a specific function, but the critical feature of the enzyme mechanism resides in the nucleophilic nature of the Ser 195 hydroxyl group. Its nucleophilic property is enhanced by a chain of hydrogen bonds that link it through the imidazole group of His 57. The initiator for this catalytic event is the negative charge of Asp 102, which delocalizes the electrons in the imidazole ring of His 57. This encourages the formation of a charge relay system as illustrated in Figure 4.28.

Although there is no catalytic mechanism that is common to all enzymes as previously discussed, the theorized mechanism for α-chymotrypsin provides a model of complexity and insight into the chemistry surrounding the behavior of one very basic and important enzyme. For further details regarding the chemical mechanisms of other enzymes, the reader is encouraged to consult any one of the references cited at the end of the chapter.

MULTIENZYME SYSTEMS

The kinetic principles and discussion of enzymes here have dealt largely with the con-

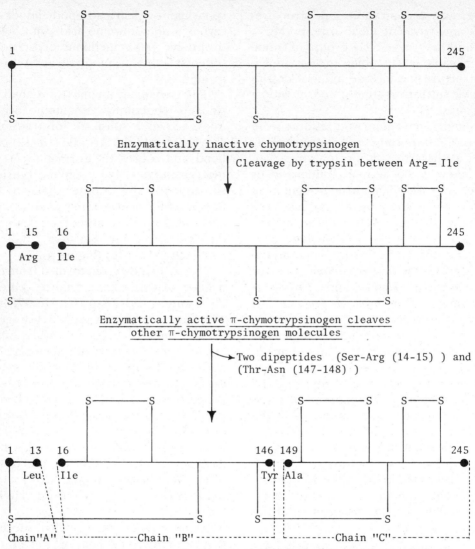

FIGURE 4.27. Activation of α-chymotrypsin from enzymatically inactive α-chymotrypsinogen. A series of three peptide chains, designated *A*, *B*, and *C*, are formed from π-chymotrypsinogen to give α-chymotrypsin. The α-chymotrypsin is the *active form* of this enzyme in the digestive system; note that the *A*, *B*, and *C* chains remain linked in the active form of the enzyme by disulfide bonds (—S—S—).

version of substrate to product by enzymes obeying Michaelis–Menten or allosteric kinetic principles. There are many other enzymatic reactions that superficially appear to convert substrate to product. However, the enzyme apparently responsible for product formation is, not a single enzyme, but a complex aggregate of several enzymes that operate in a joint fashion to yield product from substrate. Such multienzyme complexes rarely

have predictable kinetic behavior unless studied in the realm of their native biochemical milieu, which is practically impossible.

Multienzyme systems demonstrate sequential cooperative behaviors involving the interaction of one enzyme with others. Such interactions facilitate and maximize the potential transfer of key metabolic intermediates among enzymes.

A number of *multienzyme-complex* models

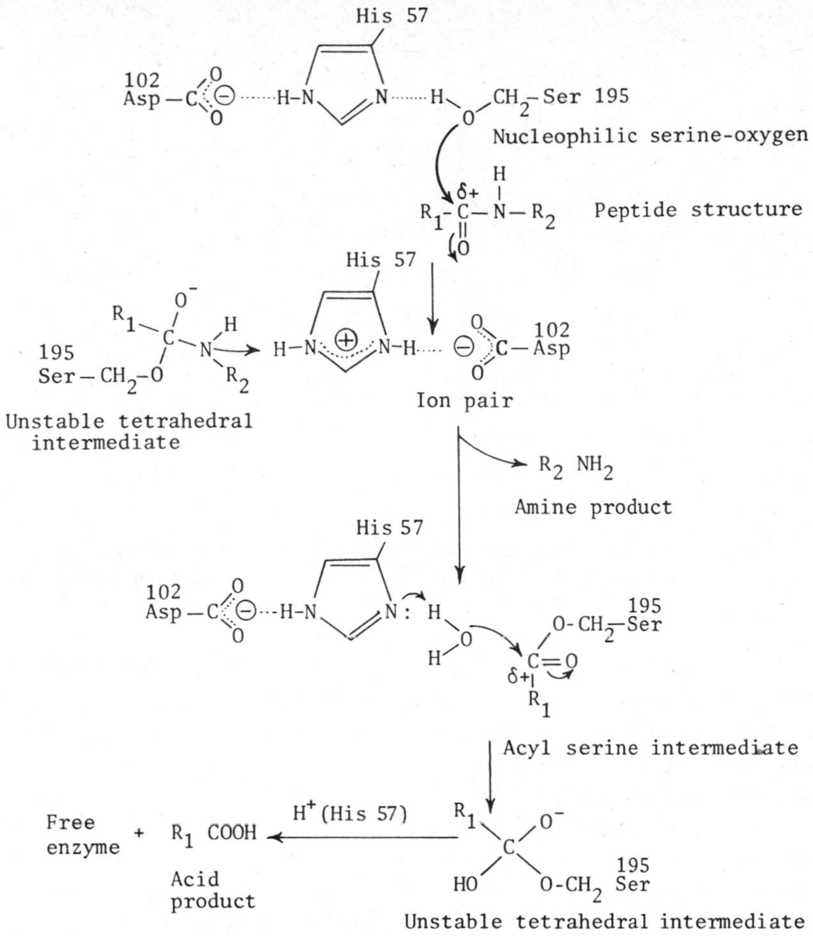

FIGURE 4.28. Catalytic mechanism for the action of α-chymotrypsin on a peptide. The Asp 102 forms an ion pair with the imidazole ion of His 57, inducing Ser 195 to perform a nucleophilic attack on the carbonyl carbon of the peptide. This attack yields an unstable tetrahedral intermediate before covalent catalysis commences. Covalent catalysis subsequently leads to formation of an acyl-Ser intermediate and an amine product. The carbonyl carbon of the acyl-Ser intermediate is attacked by water whose nucleophilicity is enhanced by a charge relay system, and a second unstable tetrahedral intermediate is formed. The unstable tetrahedral intermediate receives a H^+ from the charge relay system (His 57) and is converted to an acid product, R_1COOH. The chymotrypsin, with its Ser 195 nucleophilic site, is free to act again on another peptide molecule.

have been elucidated including the *fatty acid synthetase complex* discussed in Chapter 17; however, one of the best models for the *modus operandi* of a multienzyme complex involves the oxidative decarboxylation of α-keto acids such as pyruvate. In *Escherichia coli* the *pyruvic acid dehydrogenase* (PAD) *multienzyme complex* brings about the oxidation of pyruvate to carbon dioxide and acetyl-CoA. The PAD complex (M.W. $= 4.0 \times 10^6$)

is composed of three physically associated enzyme components denoted as PAD_{E1}, PAD_{E2}, and PAD_{E3}. These three components represent pyruvate dehydrogenase, dihydrolipoyl transacetylase, and dihydrolipoyl dehydrogenase (Figure 4.29), respectively. In this multienzyme complex, the PAD_{E2} enzyme component serves as a stoichiometric association core for PAD_{E1} and PAD_{E3}.

The actual physical association of sepa-

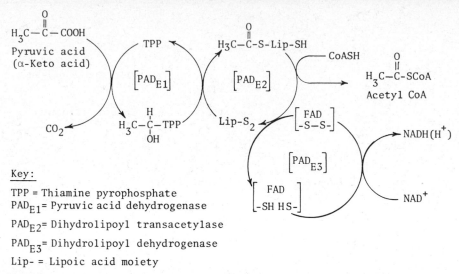

Key:

TPP = Thiamine pyrophosphate
PAD_{E1} = Pyruvic acid dehydrogenase
PAD_{E2} = Dihydrolipoyl transacetylase
PAD_{E3} = Dihydrolipoyl dehydrogenase
Lip- = Lipoic acid moiety

FIGURE 4.29. Schematic interactions involved in the pyruvic acid dehydrogenase (PAD) multienzyme complex. Pyruvate oxidation ultimately yields acetyl CoA and carbon dioxide.

rate enzymes around a core structure is often an essential prerequisite for the activity of individual enzymes comprising a multienzyme structure. Still other enzyme systems exist in which individual kinetic activities of enzymes are not *totally* dependent on the association of enzymes. In these cases, the contact of one enzyme with another serves to accelerate the rate of individual enzyme reactions contributing to overall final product formation. Tryptophan synthetase (M.W. = 135,000) serves as a representative example of one such multienzyme system where two α-polypeptide chains and two β-polypeptide chains, which exist as a dimer (β_2), individually catalyze the reactions:

(1) Indole 3-glycerophosphate $\xrightarrow{2 \text{ α-polypeptides}}$ indole + glyceraldehyde 3-phosphate

(2) Indole + serine $\xrightarrow{\text{β2-polypeptide (dimer)}}$ tryptophan + H_2O

The admixture of the α and β chains to form an $\alpha_2\beta_2$ enzyme complex gives the following reaction:

(3) Indole 3-glycerophosphate + serine $\xrightarrow{\alpha_2\beta_2 \text{ complex}}$ tryptophan + glyceraldehyde 3-phosphate + H_2O

This reaction (3) progresses at a rate 25 to 100 times faster than the rate of final product formation by separate enzyme reactions.

ENZYME ASSAY AND PURIFICATION

The quantitative evaluation of a particular enzyme is based on measurements of its catalytic activity. An enzyme assay is normally conducted by placing either a liquid sample or a cell extract, from a quantitatively defined sample, in a carefully formulated solution that contains the enzyme's substrate(s). *Enzymatic conversion of the substrate to product may be monitored by measuring the disappearance of substrate or the appearance of product.* The monitoring step may use almost any analytical technique capable of detecting differences in the concentration levels of substrate *or* product. Normal monitoring methods may include electrometric methods such as polarographic or potentiometric measurements; liquid or gas chromatography; densimetric assay of electrophoretic or

Table 4.7. Three Representative Methods for Measuring Enzyme Activity Based on Product Formation

Substrate for enzyme	Enzyme responsible for mediating product formation	Product of enzyme action on substrate	
		Primary product +	Secondary product
(A) p-Nitrophenyl phosphate	Alkaline phosphatase	p-Nitrophenol (410 nm)[b]	+ Phosphate[a]
(B) Ethanol + NAD$^+$	Alcohol dehydrogenase (enzyme requires NAD$^+$ coenzyme)	Acetaldehyde[a]	+ NADH(H$^+$) (340 nm)[b]
(C) Lecithin + H$_2$O	Lecithinase or phospholipase D	Phosphatidic acid[a] +	Choline[a] (choline formation followed by its extraction and conversion to a red-violet reineckate (520 nm)[b]

[a]This product species does not permit reliable assay or estimation of enzyme activity.
[b]Optimum wavelength for spectrophotometric assay of product formation.

thin-layer chromatographic separations; or spectrophotometric measurements in the visible or ultraviolet regions (250–750 nm). The most widely used enzyme assays often rely on spectrophotometric methods since at least one product or substrate in a reaction will absorb light at a discriminative wavelength (λ). Differences in light absorption (at a specific λ) over a range of 0.0–2.0 absorbance units usually provide the basis for evaluating the activity of an enzyme (Table 4.7, A and B). In those cases in which the substrate or product does not permit direct spectrophotometric assays, a key product of an enzyme reaction may be used to produce a quantitatively significant spectrophotometric species (Table 4.7 C). Assays of this type are called *coupled assays*.

Regardless of the analytical technique employed, the disappearance of substrate ($-d[S]$) or the appearance of product ($+d[P]$) plotted versus time in minutes may appear as Figure 4.30.

As discussed previously, the $-d[S]$ or $+d[P]$ rates for an enzyme reaction will significantly depend on pH, ionic strength buffering capacity, and other properties of the enzyme reaction medium. For these reasons, it is very important to conduct enzyme assays *only* under the *most scrupulously con-* *trolled conditions possible*. These standardized assay conditions ensure that a specific amount of enzyme will consistently provide the same amount of catalytic activity from one assay to the next. This consideration is especially important since the calculated quantity of any enzyme is proportional to the conversion rate of substrate to product, as indicated in Figure 4.31.

Many enzymes can be directly and reliably assayed in quantitative extracts obtained from plant or animal cells, but many other assays require preliminary purification of the enzyme prior to assay. Since enzymes are structurally complex proteins, almost any prin-

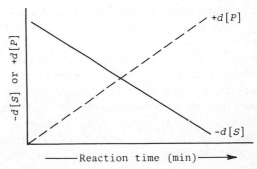

FIGURE 4.30. The appearance of product (P) or disappearance of substrate (S) can be used as an effective index of enzyme activity and a basis for enzyme assay.

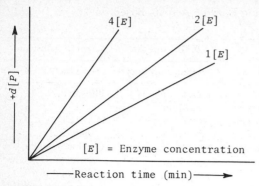

FIGURE 4.31. The rate of product formation from a specific substrate used in an enzyme assay is dependent on enzyme concentration ($[E]$).

ciple suited for protein isolation discussed in Chapter 2 has a potential application for enzyme purification.

Once the enzyme has been quantitatively isolated in a pure form, its *catalytic activity may be determined in relation to the actual weight of the enzyme protein in milligrams.* In many analytical situations, it may be impractical or technically impossible to completely purify an enzyme prior to assay. Actually, rigorous enzyme purification efforts can be counterproductive to reliable enzyme assays since important coenzymes and cofactors may be lost during the course of purification. In those instances in which complete enzyme purification is impossible, a "less than completely purified" enzyme protein is evaluated on the basis of its per mg ability to catalyze a reaction under standard assay conditions.

Various *units* describing enzyme activity have been used, but the most popular designations are based on the *international unit* (I.U.). *One I.U.* defines the quantity of an enzyme required for the transformation of 1.0 μM of a substrate to product under standard assay conditions. *Specific activity* is another term that describes enzyme activity from a perspective of purity. The greater the number of I.U.'s per milligram of protein isolated, the higher the degree of enzyme purity.

Another measure of enzyme activity involves the terms *molecular activity* or some-

times *turnover number.* Both of these terms refer to the number of substrate molecules transformed per unit time by a single enzyme (or by a single active site) when the enzyme is the only rate-limiting factor. Turnover numbers for enzyme activity can vary widely from a high value of 36×10^6 substrate molecules transformed per minute for carbonic anhydrase, to a lower value of 1240 for phosphoglucomutase.

APPLICATIONS FOR ENZYME ASSAYS

Enzyme assays have many important uses in the food industry. From an analytical perspective, highly specific enzymes can be used for quantitating a certain food ingredient where the ingredient happens to be the preferred substrate for the enzyme. Measurement of enzyme activity according to the representative assays in Tables 4.7 and 4.8 then allow precise determination of the original concentration of the food ingredient.

The discriminative advantages of enzyme assays cannot be overemphasized where a particular food ingredient may be present in the midst of many other complex biochemicals. For example, the enzyme known as *invertase* may be used to quantitatively determine sucrose in the absence of raffinose, while a mixture of *melibiase* and *invertase* can be employed for determining sucrose where raffinose is present.

The assay of enzyme activity in foods, subsequent to their heat or irradiation exposure required for preservation, serves as a quality control index for food pasteurization, sterilization, and deactivation of deteriorative enzymes. Some important enzymes monitored in foods include *phosphatases* in milk; *lipases* in fats and oils; *pectinases, phenolases,* and *peroxidases* contained in fruits, vegetables, and their juices; *proteases* contained in high-protein seeds, fish, red meats, fowl; and a host of other enzymes beyond the scope of this discussion.

Table 4.8. Highly Discriminative and Reliable Enzymatic Methods of Analysis Available for the Quantitative Analysis of Food Constituents*

Acetic acid

Principle

In the presence of the enzyme acetyl-CoA synthetase (ACS), acetic acid (acetate) is converted with adenosine 5'-triphosphate (ATP) and coenzyme A (CoA) to acetyl-CoA (1).

Acetyl-CoA reacts with oxaloacetate to citrate in the presence of citrate synthase (CS) (2).

(1) Acetate + ATP + CoA $\xrightarrow{ACS}$
$$\text{acetyl-CoA} + \text{AMP} + \text{pyrophosphate}$$

(2) Acetyl-CoA + oxaloacetate + H_2O $\xrightarrow{CS}$
$$\text{citrate} + \text{CoA}$$

The oxaloacetate required for reaction (2) is formed from malate and nicotinamide adenine dinucleotide (NAD) in the presence of malate dehydrogenase (MDH) (3). In this reaction NAD is reduced to NADH.

(3) Malate + NAD^+ $\xrightleftharpoons{MDH}$
$$\text{oxaloacetate} + \text{NADH} + H^+$$

The determination is based on the formation of NADH measured by the increase in absorbance at 340, 334, or 365 nm. Since a preceding indicator reaction is used, the amount of NADH formed is not linearly proportional to the acetic acid concentration (for calculations, see below).

Ascorbic acid (Vitamin C)

Principle

L-Ascorbic acid (ascorbate) and some more reducing substances (x–H_2) reduce the tetrazolium salt MTT [3-(4,5-dimethylthiazolyl-2)-2,5-diphenyltetrazolium bromide] in the presence of the electron carrier PMS (5-methylphenazinium methyl sulphate) at pH 3.5 to a formazan (1).

(1) Ascorbate (x–H_2) + MTT $\xrightarrow{PMS}$
dehydroascorbate (x) + MTT-formazen$^-$ + H^+

In a blank determination, only the ascorbate fraction as part of all reducing substances present in the sample is oxidatively removed by ascorbic acid oxidase (AAO) in the presence of oxygen (2). The dehydroascorbate formed does not react with MTT/PMS.

(2) Ascorbate + $\frac{1}{2}O_2$ $\xrightarrow{AAO}$ dehydroascorbate + H_2O

The absorbance difference of the sample minus the absorbance difference of the sample blank is equivalent to the quantity of ascorbate in the sample. The MTT-formazan is the measuring parameter and is determined by means of its absorbance in the visible range at 578 nm.

Cholesterol

Principle

Cholesterol is oxidized by cholesterol oxidase to cholestenone (1). In the presence of catalase, the hydrogen peroxide produced in this reaction oxidizes methanol to formaldehyde (2). The latter reacts with acetylacetone forming a yellow lutidine dye in the presence of NH_4^+ ions (3).

(1) Cholesterol + O_2 $\xrightarrow{\text{chol. oxidase}}$
$$\Delta^4\text{-cholestenone} + H_2O_2$$

(2) Methanol + H_2O_2 $\xrightarrow{\text{catalase}}$ formaldehyde + 2 H_2O

(3) Formaldehyde + NH_4^+ + 2 acetylacetone $\longrightarrow$
$$\text{lutidine} + 3 H_2O$$

The concentration of the lutidine dye formed is stoichiometric with the amount of cholesterol and is measured by the increase of absorbance in the visible range at 405 nm.

Citric acid

Principle

In the reaction catalyzed by the enzyme citrate lyase (CL), citric acid (citrate) is converted to oxaloacetate and acetate (1).

(1) Citrate $\xrightarrow{CL}$ oxaloacetate + acetate

In the presence of the enzymes malate dehydrogenase (MDH) and lactate dehydrogenase (LDH), oxaloacetate and its decarboxylation product pyruvate are reduced to L-malate and L-lactate, respectively, by reduced nicotinamide adenine dinucleotide (NADH) (2) (3).

(2) Oxaloacetate + NADH + H^+ $\xrightarrow{MDH}$ L-malate + NAD^+

(3) Pyruvate + NADH + H^+ $\xrightarrow{LDH}$ L-lactate + NAD^+

The amounts of NADH oxidized in reactions (2) and (3) are stoichiometric with the amount of citrate. It is NADH that is measured at a wavelength of 334, 340, or 365 nm.

Ethanol

Principle

Ethanol is oxidized in the presence of the enzyme alcohol dehydrogenase (ADH) by nicotinamide adenine dinucleotide (NAD) to acetaldehyde (1).

(1) Ethanol + NAD^+ $\xrightleftharpoons{ADH}$
$$\text{acetaldehyde} + \text{NADH} + H^+$$

The equilibrium of this reaction lies on the side of ethanol and NAD. It can, however, be completely displaced to

---

*NOTE: This table details the analytical principles for many of these food assay methods. Methods detailed in here have been developed and made available by Boehringer Mannheim. (Reprinted with permission from *Methods of Enzymatic Food Analysis*, 1983 (copyright), Boehringer-Mannheim GmbH, Biochemica, Mannheim.)

Table 4.8. (*Continued*)

the right at alkaline conditions and by trapping of the acetaldehyde formed. Acetaldehyde is oxidized in the presence of aldehyde dehydrogenase (Al-DH) quantitatively to acetic acid (2).

(2) Acetaldehyde $+$ NAD$^+$ $+$ H$_2$O $\xrightarrow{\text{Al-DH}}$ acetic acid $+$ NADH $+$ H$^+$

The amount of NADH formed in the reactions (1) and (2) is stoichiometric with half the amount of ethanol. NADH is determined by means of its absorbance at 334, 340, or 365 nm.

D-Gluconic acid

Principle

In the presence of the enzyme gluconate kinase, D-gluconic acid (D-gluconate) is phosphorylated by adenosine 5'-triphosphate (ATP) to gluconate 6-phosphate (1).

(1) D-Gluconate $+$ ATP $\xrightarrow{\text{gluconate kinase}}$ gluconate-6-P $+$ ADP

In the reaction catalyzed by 6-phosphogluconate dehydrogenase (6-PGDH), gluconate 6-phosphate is oxidized by nicotinamide adenine dinucleotide phosphate (NADP) to ribulose 5-phosphate with the formation of reduced nicotinamide adenine dinucleotide phosphate (NADPH) (2).

(2) Gluconate-6-P $+$ NADP$^+$ $\xrightarrow{\text{6-PGDH}}$ ribulose-5-P $+$ NADPH $+$ CO$_2$ $+$ H$^+$

The amount of NADPH formed in the above reaction is stoichiometric with the amount of gluconate. The increase in NADPH is determined by means of its absorbancy at 340, 334 or 365 nm.

D-Glucono-δ-lactone (GdL) is determined by the same principle, but only after alkaline hydrolysis (3).

(3) D-Glucono-δ-lactone $+$ H$_2$O $\longrightarrow$ D-gluconate

Glucose/Fructose

Principle

Glucose and fructose are phosphorylated to glucose 6-phosphate (G-6-P) and fructose 6-phosphate (F-6-P) by the enzyme hexokinase (HK) and adenosine 5'-triphosphate (ATP) (1) (2).

(1) Glucose $+$ ATP $\xrightarrow{\text{HK}}$ G-6-P $+$ ADP

(2) Fructose $+$ ATP $\xrightarrow{\text{HK}}$ F-6-P $+$ ADP

In the presence of the enzyme glucose 6-phosphate dehydrogenase (G6P-DH), G-6-P is oxidized by nicotinamide adenine dinucleotide phosphate (NADP) to gluconate 6-phosphate with the formation of reduced nicotinamide adenine dinucleotide phosphate (NADPH) (3).

(3) G-6-P $+$ NADP$^+$ $\xrightarrow{\text{G6P-DH}}$ gluconate 6-phosphate $+$ NADPH $+$ H$^+$

The amount of NADPH formed in this reaction is stoichiometric with the amount of glucose. It is NADPH that is measured by the increase in absorbance at 334, 340, or 365 nm.

On completion of reaction (3), F-6-P is converted to G-6-P by phosphoglucose isomerase (PGI) (4).

(4) F-6-P $\xrightleftharpoons{\text{PGI}}$ G-6-P

G-6-P reacts in turn with NADP forming gluconate 6-P and NADPH. The amount of NADPH obtained in this reaction is stoichiometric with the amount of fructose. The increase in NADPH is measured by means of its absorbance at 334, 340, or 365 nm.

Glutamic acid

Principle

In the presence of the enzyme glutamate dehydrogenase (GlDH), L-glutamic acid (L-glutamate) is deaminated oxidatively by nicotinamide adenosine dinucleotide (NAD) to α-ketoglutarate (1). In the reaction catalyzed by diaphorase the NADH formed converts iodonitro tetrazolium chloride (INT) to a formazan, which is measured in the visible range at 492 nm (2).

(1) L-Glutamate $+$ NAD$^+$ $+$ H$_2$O $\xrightleftharpoons{\text{GlDH}}$ α-ketoglutarate $+$ NADH $+$ NH$_4^+$

(2) NADH $+$ INT $+$ H$^+$ $\xrightarrow{\text{diaphorase}}$ NAD$^+$ $+$ formazen

The equilibrium of reaction (1) lies far on the side of glutamate. By trapping the NADH formed with INT (2), the equilibrium is displaced in favor of α-ketoglutarate.

Glycerol

Principle

In the reaction catalyzed by glycerokinase (GK), glycerol is phosphorylated by adenosine 5'-triphosphate (ATP) to glycerol 3-phosphate (1).

(1) Glycerol $+$ ATP $\xrightarrow{\text{GK}}$ glycerol 3-phosphate $+$ ADP

The adenosine 5'-diphosphate (ADP) formed in the above reaction is reconverted by phosphoenolpyruvate (PEP) into ATP with the formation of pyruvate (2).

(2) ADP $+$ PEP $\xrightarrow{\text{PK}}$ ATP $+$ pyruvate

In the presence of the enzyme lactate dehydrogenase (LDH) pyruvate is reduced to lactate by reduced nicotinamide adenine dinucleotide (NADH) with the oxidation of NADH to NAD (3).

(3) Pyruvate $+$ NADH $+$ H$^+$ $\xrightarrow{\text{LDH}}$ lactate $+$ NAD$^+$

The amount of NADH consumed in the above reaction is stoichiometric with the amount of glycerol. NADH is determined by means of its absorption at 334, 340, or 365 nm.

Table 4.8. (*Continued*)

Isocitric acid

Principle

D-Isocitric acid (D-isocitrate) is oxidative decarboxylized by nicotinamide adenine dinucleotide phosphate (NADP) in the presence of the enzyme isocitrate dehydrogenase (ICDH) (1).

(1) D-Isocitrate + NADP$^+$ $\xrightarrow{\text{ICDH}}$

$\qquad$ 2-ketoglutarate + CO_2 + NADPH + H$^+$

The amount of NADPH formed in reaction (1) is stoichiometric with the amount of D-isocitrate. NADPH is determined by means of its absorbance at 334, 340, or 365 nm.

Lactate

Principle

In the presence of L-lactate dehydrogenase (L-LDH), L-lactic acid (L-lactate) is oxidized by nicotinamide adenine dinucleotide (NAD) to pyruvate (1).

(1) L-Lactate + NAD$^+$ $\underset{\text{}}{\overset{\text{L-LDH}}{\rightleftharpoons}}$

$\qquad$ pyruvate + NADH + H$^+$

The equilibrium of this reaction lies almost completely on the side of lactate. However, by trapping the pyruvate in a subsequent reaction catalyzed by the enzyme glutamate-pyruvate transaminase (GPT) in the presence of L-glutamate, the equilibrium can be displaced in favor of pyruvate and NADH (2).

(2) Pyruvate + L-glutamate $\underset{\text{}}{\overset{\text{GPT}}{\rightleftharpoons}}$

$\qquad$ L-alanine + α-ketoglutarate

The amount of NADH formed in the above reaction is stoichiometric with the concentration of L-lactic acid. The increase in NADH is determined by means of its absorbance at 334, 340, or 365 nm.

Malic acid

Principle

In the presence of L-malate dehydrogenase (L-MDH), L-malic acid (L-malate) is oxidized by nicotinamide adenine dinucleotide (NAD) to oxaloacetate (1). The equilibrium of this reaction lies far on the side of malate. Removal of oxaloacetate from the reaction system causes displacement of the equilibrium in favor of oxaloacetate. In the reaction catalyzed by the enzyme glutamate-oxaloacetate transaminase (GOT), oxaloacetate is converted to L-aspartate in the presence of L-glutamate (2).

(1) L-Malate + NAD$^+$ $\underset{\text{}}{\overset{\text{L-MDH}}{\rightleftharpoons}}$

$\qquad$ oxaloacetate + NADH + H$^+$

(2) Oxaloacetate + L-glutamate $\underset{\text{}}{\overset{\text{GOT}}{\rightleftharpoons}}$

$\qquad$ L-aspartate + α-ketoglutarate

The amount of NADH formed is stoichiometric with the concentration of L-malate. It is NADH that is measured at a wavelength of 334, 340, or 360 nm.

Raffinose

Principle

In the presence of the enzyme α-galactosidase, raffinose is hydrolyzed at pH 4.5 to galactose and sucrose (1).

(1) Raffinose + H_2O $\xrightarrow{\text{α-galactosidase}}$

$\qquad$ galactose + sucrose

Galactose is oxidized by nicotinamide adenine dinucleotide (NAD) to galactonolactone in the presence of the enzyme galactose dehydrogenase (Gal-DH) (2).

(2) Galactose + NAD$^+$ $\xrightarrow{\text{Gal-DH}}$

$\qquad$ galactonolactone + NADH + H$^+$

The amount of NADH formed in the above reaction is stoichiometric with the amount of raffinose. The increase in NADH is determined by means of its absorption at 334, 340, or 365 nm.

This assay allows the determination of raffinose besides sucrose and glucose.

Starch

Principle

In the presence of the enzyme amyloglucosidase (AGS) starch is hydrolyzed to glucose at pH 4.6 (1).

(1) Starch + $(n - 1)$ H_2O $\xrightarrow{\text{AGS}}$ n glucose

The glucose formed is determined with hexokinase (HK) and glucose 6-phosphate dehydrogenase (G6P-DH) at pH 7.6. Glucose is phosphorylated to glucose 6-phosphate (G-6-P) by adenosine 5′-triphosphate (ATP) (2) in the presence of hexokinase.

(2) Glucose + ATP $\xrightarrow{\text{HK}}$ G-6-P + ADP

In the presence of G6P-DH glucose 6-phosphate is oxidized by nicotinamide adenine dinucleotide phosphate (NADP) to gluconate 6-phosphate under formation of NADPH (3).

(3) G-6-P + NADP$^+$ $\xrightarrow{\text{G6P-DH}}$

$\qquad$ gluconate 6-phosphate + NADPH + H$^+$

The amount of NADPH formed in the above reaction is stoichiometric with the amount of glucose. NADPH is determined by means of its absorbance at 334, 340, or 365 nm.

Sorbitol

Principle

In the enzymatic reaction catalyzed by sorbitol dehydrogenase (SDH), D-sorbitol is oxidized with nicotinamide adenine dinucleotide (NAD) to fructose (1) with the formation of reduced nicotinamide adenine dinucleotide (NADH).

(1) D-Sorbitol + NAD$^+$ $\xrightarrow{\text{SDH}}$

$\qquad$ fructose + NADH + H$^+$

Table 4.8. (*Continued*)

Under the conditions of the assay the equilibrium of the reaction is completely in favor of fructose and NADH. The amount of NADH formed in the above reaction is stoichiometric with the amount of sorbitol. The increase in NADH is measured by means of its absorbance at 334, 340, or 365 nm.

Sucrose/Glucose

Principle

The glucose concentration is determined before and after enzymatic hydrolysis.

Determination of glucose before inversion:

At pH 7.6, the enzyme hexokinase (HK) catalyzes the phosphorylation of glucose by adenosine triphosphate (ATP) (1). In the presence of glucose 6-phosphate dehydrogenase (G6P-DH) the glucose-phosphate (G6P) produced is specifically oxidized by nicotinamide adenine dinucleotide phosphate (NADP) to gluconate-phosphate with the formation of reduced nicotinamide adenine dinucleotide phosphate (2).

(1) Glucose + ATP $\xrightarrow{HK}$ G6P + ADP

(2) G6P + NADP$^+$ $\xrightarrow{G6P-DH}$

$$\text{gluconate-phosphate} + NADPH + H^+$$

The NADPH formed in this reaction is stoichiometric with the amount of glucose and is measured by the increase in absorbance at 334, 340, or 365 nm.

Enzymatic Inversion

At pH 4.6, sucrose is hydrolyzed by the enzyme β-fructosidase (invertase) to glucose and fructose (3).

(3) Sucrose + H$_2$O $\xrightarrow{\beta\text{-fructosidase}}$ glucose + fructose

The determination of glucose after inversion (total glucose) is carried out simultaneously according to the principle outlined above.

The sucrose content is calculated from the difference of the glucose concentrations before and after enzymatic inversion.

Urea/Ammonia

Principle

In the presence of the enzyme urease urea is hydrolyzed to ammonia and carbon dioxide (1).

In the presence of glutamate dehydrogenase (GlDH) and reduced nicotinamide adenine dinucleotide (NADH) ammonia reacts with α-ketoglutarate to L-glutamate, whereby NADH is needed (2).

(1) Urea + H$_2$O $\xrightarrow{urease}$ 2 NH$_3$ + CO$_2$

(2) α-Ketoglutarate + NADH + NH$_4^+$ $\xrightarrow{GlDH}$

$$\text{L-glutamate} + NAD^+ + H_2O$$

The amount of NADH formed in the above reaction is stoichiometric with the amount of ammonia or with half the amount of urea, respectively. NADH is determined by means of its absorbance at 334, 340, or 365 nm.

Enzyme assays are also of basic importance in assessing the quantity of an enzyme necessary to implement specific production steps in selected food processing, manufacturing, or fermentation industries.

INDUSTRIAL ENZYME TECHNOLOGY

The overall effects of enzymes are responsible for processes of fermentation, which have been known for centuries. Although traditional fermentations related to lactic acid fermentations of milks, cheeses, and pickling, along with alcoholic fermentations used for beer, wines, and other beverages, rely on the presence of living yeast cells, the Büchners (1897) determined that cell-free extracts of yeast could convert glucose into ethanol. This discovery was the foundation for many current applications of enzymes to food technology. Clearly, many industries still rely on viable, living organisms to facilitate food processing and production. However, growing numbers of industries depend on the *in vitro* performance of enzymes under carefully controlled conditions to make partial or complete food products.

Enzymes extracted from living tissues or large populations of certain microorganisms (e.g., bacteria, yeasts, fungi) may be applied to whole or unit food industrial processes in two ways:

1. Enzyme extracts that contain the purified enzyme protein may be admixed to a *batch* of a specific substrate under prescribed reaction conditions in order to form a product.

2. The extracted enzyme protein may first be *immobilized* on a supporting substance by physical forces such as *adsorption phenomena* or *chemical bonding*. Chemical bonding methods for immobilizing enzymes are limited only by the tactical methods of biochemical engineering and the negative effects on enzyme reactivity that result after an enzyme is "fixed" to a stationary phase or solid support. The most successful methods of enzyme immobilization require the convenient yet gentle bonding of the enzyme protein to the solid support while minimizing overt changes in the demonstrated *in vivo* kinetic behavior of the enzyme. Although many bonding reactions are available, the attachment of enzymes to aldehydic matrices on a solid support by reductive alkylation is exemplary of typical immobilization techniques (Royer *et al.*, 1977). This particular approach to enzyme immobilization, along with alternative methods, is illustrated in Figures 4.32 and 4.33.

Another variation on the theme of enzyme immobilization can involve the *physical enclosure* or *encapsulation* of the enzyme in a

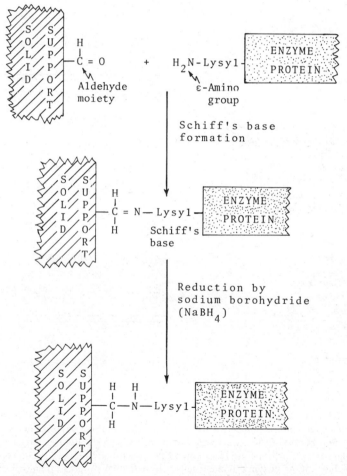

FIGURE 4.32. One type of an enzyme immobilization method in which the ε-amino group of an amino acid residue contained within an enzyme is reacted with an aldehyde present on a solid support. The Schiff base formed during this linkage reaction is eventually reduced, thereby serving as the final step in the overall enzyme immobilization process.

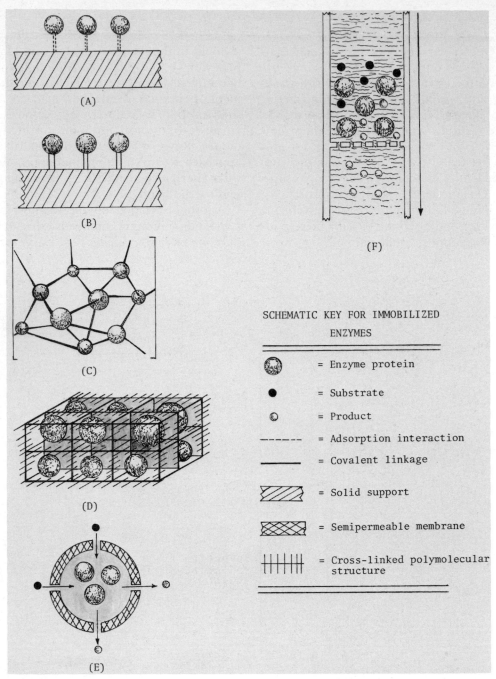

FIGURE 4.33. Schematic diagrams of methods used to immobilize enzymes. (A) *Adsorption interaction*. Physical adsorption interaction between solid support and enzyme acts as a linkage mechanism; the solid support may consist of surface active alumina, carbon, clay (bentonite), ion exchange resins, cellulose(s), or glass. (B) *Covalent attachment*. Linking of enzyme to solid support is achieved by reacting nonessential residues on the enzyme to chemically activated solid supports. The activated support may consist of a potentially reactive azide, isocyanate, diol, carbodiimide, and so on. The solid support may consist of a synthetic polymer, processed cellulose, or glass. (C) *Enzyme cross-linking*. Intermolecular covalent bonding between enzyme molecules is facilitated by alkylating agents, diisothiocyanate, aldehydes, and so on. (D) *Structural entrapment of enzymes*. Enzymes are structurally entrapped by cross-linked polyacrylamides, starches, silica gels, or other substances. The influx of substrate and the efflux of product from the entrapped enzyme occurs freely without loss of the enzyme. (E) *Spherical or cylindrical microencapsulation*. Enzymes are entrapped in a spherical or cylindrical semipermeable membrane that permits the free influx of substrate and efflux of product through very small pores, while averting the loss of the enzyme protein. (F) *Semipermeable membrane ultrafiltration*. The immobilized enzyme is "retained above" or "entrapped on" a semipermeable membrane that permits the passage of the enzyme–substrate reaction product, P.

physical structure. This *allows both the entrance of substrate and the exit of product* while retaining the enzymatic protein inside a cage structure.

The enzymes employed for batch processing of substrates or immobilization techniques may originate from plant, animal, or microbiological sources. In the cases of plants and animals, *endoenzymes* or enzymes contained *within* cells are generally harvested by means of cell disruption and subsequent removal of cellular debris. Microbiological sources often secrete enzymes into their growth media that are recognized as *exoenzymes*. Removal of the cells from their growth media, followed by isolation of the enzyme from the growth liquor, can yield an important variety of enzymes depending on the microbiological species. Varieties of enzymes produced by microorganisms are very great since they also act as sources of endoenzymes that can be harvested in the same fashion as enzymes from plant and animal cells. Apart from crude extraction steps, it is sometimes necessary to purify enzymes further by means of centrifugation, filtration, ion-exchange, gel or affinity chromatography, isoelectric precipitation, salting out, or any other available methods in the arsenal of protein purification. It should be recognized, however, that the industrial enzyme purification requirements are rigorous, but this is *not meant to imply that only 99% pure* or *crystallized enzymes* are a prerequisite for successful industrial enzyme technology. In general, enzymes destined for industrial use are often considered to be *crude extracts* from the perspectives of analytical biochemistry.

A survey of recent and past food industrial enzyme technology shows that the bulk of utilitarian enzymes are degradative extracellular hydrolases (e.g., amylases, proteases) that do not require the presence of cofactors. There is little doubt that this limited range of industrial enzyme availability will expand as biochemical engineering, enzyme harvesting, and culturing techniques along with economic pressures force innovation in this currently "juvenile" technology.

Batch Application of Enzymes

The concept of *batch application* has been widely employed in the food industry for many years. The use of enzymes in this form does not readily permit their recovery after substrate is converted to product, but since many enzymes used in these processes are relatively inexpensive and widely available, enzyme recovery is not particularly important.

The minimal requirements for a successful batch-type enzyme reaction often necessitate control of pH, temperature, and continuous mixing or agitation of an enzyme–substrate reaction liquor. Crude enzyme sources such as barley malt may be admixed with the substrate-containing liquor, or a crude extract of endo- or exoenzymes may be added to the reaction liquor.

Typical enzyme extracts used for batch-processing operations include *rennin* used for milk coagulation, *amylases* derived from barley malt for preparation of fermentation liquors in brewing, and *invertase* used in the confectionary industry, to name only a few (Table 4.9).

Although many notable examples of batch enzyme technology exist, one of the more interesting and potentially significant applications involves the batch enzymatic conversion of cellulose to free glucose and other sugars.

It is well known that the β-1,4 linkages between glucose molecules incorporated in the structure of cellulose are resistant to the attack of all α-1,4 and α-1,6-glycosidases that attack starches. Although cellulose digestive enzymes are rarely produced naturally in bulk quantities, selected mutant strains of the fungus *Trichoderma reesei* have been successfully grown en mass and shown to produce four times the amount of cellulase present in an original wild fungal strain. This mutant fungus produces a complex mixture of "exo-" and "endo-" β-1,4-glycosidases that

Table 4.9. Survey of Industrial Enzyme Applications

Type of enzyme	Application method[a]	Enzyme function or use objective
α-Amylase	BP, I	Glucose production from starch
β-Amylase	BP, I	High maltose syrup production
Bromelain	SA	Meat tenderizer
Catalase	BP	Decomposition of hydrogen peroxide in cold pasteurization of milk used for cheese production
Cellulase	BP	Glucose production from cellulose
Glucose isomerase	I	Conversion of glucose to fructose in syrup production
Glucose oxidase	BP	Removal of glucose and oxygen from foods such as eggs to prevent browning
Invertase	BP, I	Conversion of sucrose to glucose and fructose; the conversion product can be used to increase sweetness in foods (syrup)
Lactase	BP, I	Used for degradation of lactose which inhibits undesirable lactose crystal formation in ice cream (sandy ice cream)
Lipase	BP	Flavor production in certain cheeses; retards certain types of bread spoilage
Naringinase	BP, I	Debittering of citrus products
Papain	BP, SA	Meat tenderizer; stabilizes and serves as a chill-proofing agent in beer production
Pectinase	BP	Prevents cloudiness in wines and fruit juices which may be caused by pectins
Pepsin	BP	Rennet extender
Polyphenol oxidase	BP	Contributes to production of brown colors required in the processing of tea and coffee
Rennin	BP	Cheese manufacturing where milk coagulation is required at a predictable time
Ribonucleases	BP	Production of nucleotides which may be suitable as flavor enhancers

[a]Preferred or possible method of enzyme application: BP, batch processing; I, immobilization; SA, superficial application to food product.

are able to convert *amorphous cellulose* or *soluble cellulose* derivatives *into glucose and cellobiose,* while other related enzymes hydrolyze insoluble crystalline cellulose. Cellulase enzymes cultivated and isolated from the fungal culture are then admixed to a batch of cellulose substrate under stringently controlled reaction conditions that promote cellulose digestion. The typical reaction sequence for cellulose conversion to glucose is shown in Figure 4.34, while Figure 4.35 demonstrates the likely composition of the cellulose digestion reaction liquor during the period of batch enzymatic hydrolysis. As expected, cellulase activity causes a depletion of the cellulose substrate along with

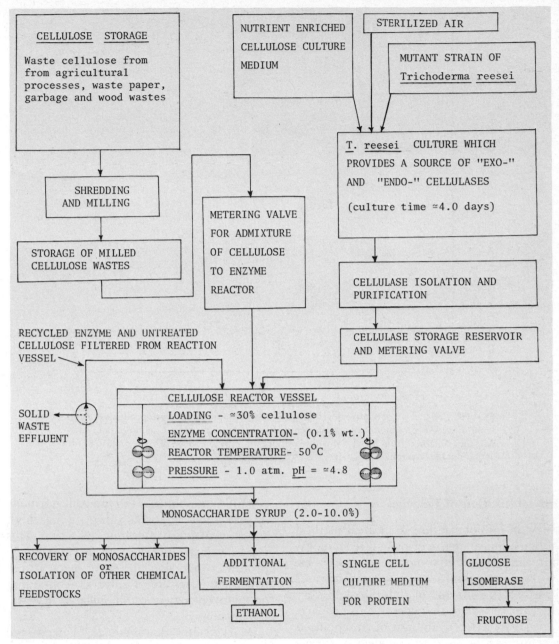

FIGURE 4.34. Waste cellulose conversion to glucose, other monosaccharides, and disaccharides by means of cultivated enzymes.

a concomitant increase in cellobiose, glucose, and xylose (Andren *et al.*, 1975; Nystrom *et al.*, 1978) (Figure 4.35).

Glucose generated from cellulose offers great promise for providing substrate to ethanolic fermentations, the cultivation of single-cell proteins, or other microorganisms that yield important natural products. Furthermore, enzymatic cellulose degradation provides a feasible method for the productive elimination of bountiful agricultural cellulosic wastes.

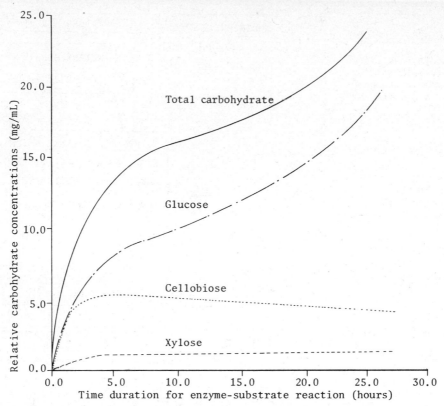

FIGURE 4.35. Typical relative development of fermentation breakdown products derived from a nonspecific amount of a waste cellulose substrate through the action of cultivated enzymes obtained from fungi. Consult Andren *et al.* (1975) and Nystrom *et al.* (1978) for actual details regarding similar fermentative product profiles.

Immobilization of Enzymes

Enzyme immobilization is an ideal method of biochemical engineering and technology, but its use is limited to only certain enzymes. Enzyme immobilization is favored in cases in which the enzyme of choice may be very difficult to isolate or expensive to prepare. Moreover, a fixed amount of immobilized enzyme can be used to *continuously* process substrate to product, thereby extending the activity and life span of an enzyme to a larger amount of substrate than possible in batch enzyme treatments.

Enzymes that are suitable for immobilization must be able to survive the immobilization process. This requirement singularly eliminates immobilization possibilities for many of the most delicate enzymes. A suc-

cessfully immobilized enzyme will normally display a low K_m for its substrate (i.e., high substrate affinity) after immobilization. This not only facilitates the rate of substrate to product conversion but encourages a high hydraulic flow rate of substrate-containing liquor over enzymes immobilized on a stationary matrix.

Marked alterations in the K_m of an enzyme almost always follow its immobilization. Although the K_m value may increase or decrease, immobilization usually causes an apparent reduction in the enzyme–substrate affinity by 25 to 60% or more. The causes for this lessened biocatalytic activity are varied, but they may be related to

1. Enzyme *protein denaturation* that occurs during the immobilization process.

2. *Alterations in the conformational structure* of the immobilized enzyme that inhibits effective substrate binding at the active site, or the exit of product from the active site region.

3. Close *proximity of immobilized enzymes* on the solid support *that mutually and sterically inhibit* the *biocatalytic activity* of neighboring enzymes.

4. *Inadequate maintenance of optimum conditions* for enzyme activity (e.g., pH, ionic strength, temperature, metal cofactor availability).

Food industrial applications of immobilized enzymes may be implemented in two ways. *First, enzymes immobilized on a solid support may be admixed to a substrate-containing liquor* provided with a prescribed amount of substrate contact time, *and then removed from the reaction liquor for further use.* The removal of the immobilized enzyme from the reaction liquor is achieved by gravitational settling, centrifugation, or filtration. The supernatant liquor resulting from these respective enzyme recovery methods usually contains both product and residual unreacted substrate. The product may be segregated from the remaining substrate, which is then reacted further with the immobilized enzyme, or if the concentration of the substrate is too low, substrate-containing liquor may simply be discarded.

A *second method* for utilizing immobilized enzymes involves the principles of *continuous substrate processing.* In this situation, substrate-containing liquor is passed over a fixed amount of immobilized enzyme that is entrained against a porous membrane or screen at the bottom of a reaction containment vessel.

Substrate can be processed by a multiple series of consecutive exposures to immobilized enzymes (Figure 4.36A) or a series of singular but tandem exposures (Figure 4.36B). The downward pressure caused by fluid flow of substrate over the immobilized enzyme may cause eventual compaction of the enzyme and its stationary matrix. This,

in turn, diminishes the rate of enzyme–substrate contact along with any debris that may become entrained over the immobilized enzyme. The downward compaction of the immobilized enzyme can be minimized by using fluidized bed techniques. Here, substrate is introduced into a bed of immobilized enzyme contained in a reaction vessel. Substrate then percolates *upward* through the enzyme and avoids compaction of the immobilized enzyme plus restricted flow rates (Figure 4.36C).

One typical commercial application of immobilized enzyme technology is demonstrated by the production of high-fructose corn syrups. In this process D-glucose obtained from wet corn-milling operations is converted into its corresponding keto-sugar recognized as D-fructose. This conversion is accomplished by the action of glucose isomerase, which may be immobilized on DEAE-cellulose. This is an economically important reaction since the resulting glucose–fructose mixture yields sweetening properties equivalent or superior to sucrose obtained from beets or cane. High-fructose corn syrups prepared in this manner usually consist of ~71% solids, of which 42% is D-fructose, 50% remains as D-glucose, and the remaining 8% is accounted for as miscellaneous saccharides. Intermediate concentrations of glucose and fructose up to these concentration levels may be alternatively formulated in order to produce the characteristic sweetness or requisite sweetening demands for any of a variety of food products such as fruit flavored beverages, canned fruits, and carbonated beverages. Moreover, high-fructose corn sweeteners prepared in this way act as desirable substitutes for food formulations that require invert sugar (composition: 40% fructose, 40% glucose, and 20% sucrose).

A general serial reaction sequence for industrial implementation of immobilized enzymes such as glucose isomerase and other types of enzymes is presented in Figure 4.37.

Food processing can avail itself of many additional immobilized enzyme applications; however, these applications are largely

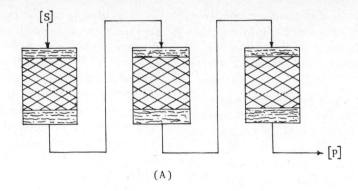

(A)

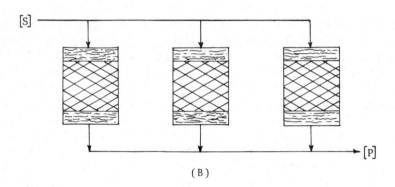

(B)

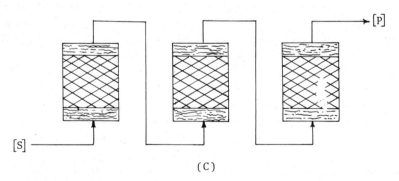

(C)

FIGURE 4.36. Schematic diagrams for possible flow schemes where a liquid containing a specific substrate is passed over immobilized enzymes. (A) Represents the conversion of S → P through a multiple and consecutive series of enzyme/substrate exposures; (B) represents S → P conversion using singular but tandem exposures of enzyme and substrate; and (C) shows the conversion of S → P by means of method outlined in "A" above, but the contact process between enzyme and substrate involves a fluidized bed. The cross-hatched area in all cases represents the effective possible volume for immobilized enzyme.

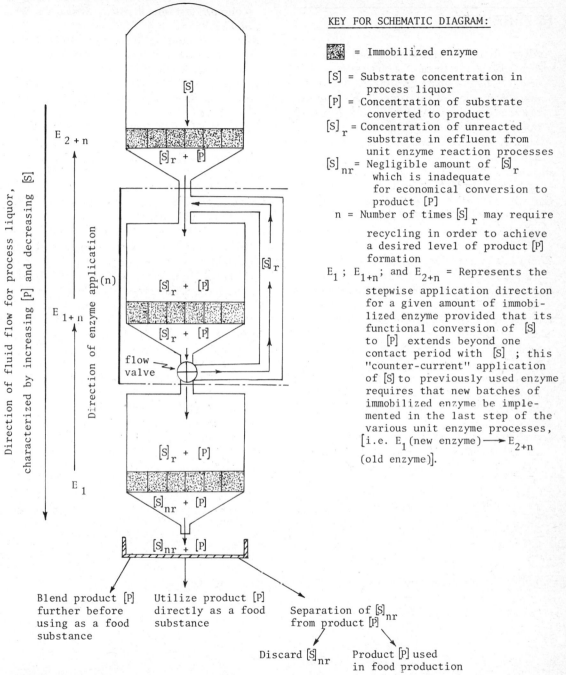

FIGURE 4.37. Ideal serial reaction sequence for the industrial application of immobilized enzymes.

in their infancy and have little current use. Enzymes that may have significant immobilization applications, especially with regard to food processing, are *amylases, aminoacylases, cellulases, α- and β-galactosidases, lactases,* and *invertases.* Surveys of enzyme immobilization applications techniques, and technology have been presented by Olson and Richardson (1974), Stanley and Olson (1974), Weetall (1975), and Hultin (1983).

REFERENCES

Andren, R. K., M. H. Mandels, and J. E. Medeiros. 1975. Production of sugars from waste cellulose by enzymatic hydrolysis. Part II. Primary evaluation of substrates. U.S. Army Natick Laboratories (MA). Presented at 170th National ACS Meeting, Chicago, IL. Aug. 24–29.

Atkins, P. W. 1978. *Physical Chemistry*. Freeman, San Francisco.

Barman, T. E. 1966. *Enzyme Handbook* Vol. I. Springer-Verlag, New York.

Carasik, W., and J. O. Carroll. 1983. Development of immobilized enzymes for production of high fructose corn syrup. *Food Technol.* **37**(10):85–91.

Conn, E. C., and P. K. Stumpf. 1976. *Outlines of Biochemistry*, 4th ed. Wiley, New York.

Engel, P. C. 1981. *Enzyme Kinetics*. Chapman & Hall, London.

Hultin, H. O. 1983. Current and potential uses of immobilized enzymes. *Food Technol.* **37**(10):68–78, 80–82, 176. (*Excellent status report dealing with enzyme immobilization technology.*)

Mahler, H. R., and E. H. Cordes. 1971. *Biological Chemistry*, 2nd ed. Harper & Row, New York.

Marshall, A. G. 1978. *Biophysical Chemistry*. Wiley, New York.

Nystrom, J. M., R. K. Andren, and A. L. Allen. 1978. Enzymatic hydrolysis of cellulosic waste: The status of the process technology and an economic assessment. *AICE Symposium Series* **74:**172, 82–88.

Olson, N. F., and T. Richardson. 1974. Immobilized enzymes in food processing and analysis. *J. Food Sci.* **39:**653–659.

Piszkiewicz, D. 1977. *Kinetics of Chemical and Enzyme Catalyzed Reactions*. Oxford University Press, New York.

Pitcher, W. H. 1980. *Immobilized Enzymes for Food Processing*. CRC Press, Boca Raton.

Pomeranz, Y., and C. E. Meloan. 1978. *Food Analysis Theory and Practice*. Avi Publishing Co., Westport, Conn.

Pulley, J. E. 1969. Enzymes simplify processing. *Food Eng.* **41**(2):68.

Royer, G. P., W. E. Schwartz, and F. A. Liberatore. 1977. Complete hydrolysis of polypeptide with insolublized enzymes. In *Methods of Enzymology* (C. H. W. Hirs, and S. N. Timasheff, eds.), pp. 40–45. Academic Press, New York.

Segel, I. H. 1976. *Biochemical Calculations*. Wiley, New York.

Sprössler, B., and H. Plainer. 1983. Immobilized lactose for processing whey. *Food Technol.* **37**(10):93–95.

Stanley, W. L., and A. C. Olson. 1974. The chemistry of immobilizing enzymes. *J. Food Sci.* **39:**660.

Weetall, H. H. 1975. Immobilized enzymes and their application in the food and beverage industry. *Process Biochem.* **10:**3–6.

Whitaker, J. R. 1972. *Principles of Enzymology for the Food Sciences*. Dekker, New York.

CHAPTER · 5

Vitamins and Vitamin-like Substances

PART I
The Water Soluble Vitamins

INTRODUCTION

The term *vitamine* was introduced by Funk in 1911 to describe essential accessory chemical factors for life processes. According to Funk's belief, the antiberiberi factor and other factors were amines, but later studies did not substantiate this belief and the terminal "*e*" was eventually removed, leaving the word *vitamin*.

The progressive meaning and significance of the term became more confused in the ensuing years when it was recognized that no single set of distinguishing features defined the "vitamin" substances. Actually, the only features shared by vitamins was their ability to maintain and promote both growth and normal health in animals, including man, at dietary concentrations of 0.005 to 0.00002%, depending on the vitamin. Later studies, however, did reveal that many of the so-called vitamin substances were intrinsic structural components of coenzymes involved in carbohydrate, fat, and protein metabolism and the functioning of the nervous system. Accordingly, for this discussion, the fundamental relationships between vitamins and coenzymes will be emphasized.

VITAMIN REQUIREMENTS

Vitamins are considered to be essential biochemical substances that certain organisms, including man, are *unable to synthesize at concentration levels required for growth and maintenance of normal health*. Many substances labeled vitamins are vitamins in the true sense of the word only for certain organisms. Vitamin C (ascorbic acid) is a required exogenous nutrient for man because the body *lacks* the enzyme known as L-gulonoxidase which converts L-gulonolactone to L-ascorbic acid. Since most other higher plants and animals can manufacture ascorbic acid in required amounts, ascorbic acid cannot be considered a vitamin for these organisms.

For man and other animals, dietary sources provide requisite concentrations of most vitamins, while intestinal-tract flora complement the dietary availability of many B vitamins. Some vitamins such as niacin can be synthesized within the animal body through conversion of tryptophan to nicotinic acid, but the maintenance of vitamin requirements in this fashion is usually of little significance for other types of vitamins.

The dietary requirements for various vitamins are variable since humans and other animal subjects may display different requirements depending on their *health, stage of growth,* and *living habits.* A complete deficiency of a particular vitamin (*avitaminosis*) may be indicated as a *gross* or *subclinical* deficiency. From diagnostic perspectives, gross vitamin deficiencies such as advanced scurvy caused by lack of vitamin C, can be easily determined; but a marginal, subclinical deficiency for the same vitamin may be indicated only as a slow healing wound or ulcer in a subject who otherwise displays good health. Marginal deficiencies in other vitamins may become evident as headache, loss of appetite, insomnia, nervousness, and/or a host of many seemingly unrelated problems—all of which indicate *no significant* cause–effect relationship between the problem and the symptomatic effect. In spite of detailed knowledge concerning gross vitamin deficiencies, the clinical and biochemical record is still disjointed and incomplete regarding the effects and diagnosis of these subclinical situations.

Vitamin requirements for humans also become obfuscated because of a wide spectrum of life-styles and living habits. For instance, excessive alcohol consumption can interfere with the utilization of vitamins B_1, B_2, and folic acid; implementation of an oral contraceptive regimen by women may require added amounts of vitamins B_1, B_2, B_{12}, C, and folic acid; heavy smoking depresses vitamin C serum levels by 30 to 40%; while weight loss via calorie restriction can lead to marginal or gross vitamin deficiencies.

Vitamin deficiencies resulting from insufficient dietary intake of a vitamin may cause marginal or gross vitamin deficiencies; however, both eventualities are recognized as *primary* or *dietary* deficiencies. This is in contrast with *secondary* or *conditioned* vitamin deficiencies whereby absorption or biochemical utilization of the vitamins is hampered as a result of predisposing pathological or stress conditions. Typical examples of secondary deficiencies include malabsorption of vitamin B_{12} because of inadequate levels of *intrinsic factor* or broad spectrum vitamin deficiencies caused by chronic alcoholism, continuous pregnancies, lactation, and so on.

Even in the face of all recognized variabilities for vitamin requirements in humans, estimates of vitamin requirements for all normal states of the human condition have been formulated. The *minimum daily requirement* (MDR) represents the minimum intake of a vitamin necessary for preventing the initial symptoms of a particular vitamin deficiency. These values have been promulgated by selected governmental agencies to be used as a guide for *assessing* the nutritional value of food commodities, food-labeling requirements, ranking the potency of vitamin preparations, and other consumer guideline indices.

The *recommended daily allowance* (RDA) for a vitamin is an estimate of vitamin intake, which reportedly provides for the maintenance of normal health in the majority of healthy persons throughout the United States. Unlike MDR values, RDA values "guarantee" an adequate vitamin supply for most individuals and provide a margin of nutritional safety if vitamin availability sporadically falls below the specified RDA value.

VITAMIN NOMENCLATURE

Osborne and Mendel (1915) and McCollum and Davis (1915) demonstrated the presence of fat-soluble and water-soluble vitamin factors. The fat-soluble factor A cured eye disease in rats, while the water-soluble B factor was recognized as an *antiberiberi* substance.

J. C. Drummond (1920) suggested that vitamin A should be reserved for the eye factor; vitamin B should refer to the antiberiberi factor; and vitamin C ought to designate the *antiscurvy factor*. Subsequent letters of the alphabet were assigned to other vitamin substances as they were discovered. Vitamin nomenclature was eventually made more complex in following years for two reasons. First, numerical designations of vitamins within an alphabetical division developed (vitamin B_1, B_2, etc.) when the initial vitamin B substance was found to represent *more than one vitamin-active substance*. The second reason for complex vitamin nomenclature stems from the early years of vitamin investigation when a myriad of clinical symptoms were unable to be recognized as being caused by deficiency of the same vitamin-active substance. Because of these problems, a complex vitamin nomenclature was a logical result until structural properties and chemistry of the vitamins were clearly defined. A detailed listing of noteworthy vitamins and vitamin-like substances reported since 1920 have been outlined in Table 5.1.

In addition to currently recognized vitamins, a group of substances called *provitamins* further complicate perspectives of vitamin biochemistry and nutrition. Provitamins are organic substances that undergo a chemical transformation to yield a vitamin-active substance. The only recognized provitamins are provitamins A and D, which are converted into vitamins A and D, respectively. Carotenoids in plants, such as β-carotene, can be enzymatically altered to form vitamin A in animals, while provitamin D, located in epidermal tissues, undergoes an ultraviolet photoconversion to vitamin D in the presence of sunlight.

Still other vitamin-like substances exist that display biochemical activities related to vitamins but cannot be truly classified as vitamins since man is able to synthesize them in adequate quantities. Owing to food distribution, biochemical functions, and analytical considerations, these vitamin-like substances are often studied in conjunction with the B vitamins. Foremost among these substances are choline, myo (*meso*)- inositol, *para*-aminobenzoic acid (PABA), carnitine, lipoic acid, and bioflavonoids.

For simplistic reasons, vitamins can also be classified as water-soluble or fat-soluble compounds. Water-soluble vitamins, including vitamins B_1, B_2, B_6, and B_{12}, lipoic acid, biotin, pantothenic acid, nicotinic acid, and folic acid, are constantly required by most higher animals and man since prolonged "warehousing" of these vitamins for future nutritional requirements is impossible. Concentrations of water-soluble vitamins in excess of nutritional biochemical requirements often results in their prompt urinary or fecal excretion. On the other hand, fat-soluble vitamins A, D, E, and K are not subject to urinary excretion and can be stockpiled in moderate quantities. This fact obviates the necessity for a daily ration of these vitamins in the diet of most animals. The storage of fat-soluble vitamins is not unlimited, and excessive dietary intake can result in serious health complications such as infantile hypercalcemia due to vitamin D or chronic vitamin A overdoses, which can cause bone fragility in children and abnormal fetal development.

VITAMIN ASSAYS

Vitamin assays have fundamental importance for establishing the vitamin content in foods; determination of crucial vitamin concentration levels in test animals that coincide with the symptomatic effects of vitamin deficiency; diagnosis of certain clinical ailments; and pharmaceutical evaluations of vitamin potency.

A myriad of vitamin assay techniques exist that are based on chemical, microbiological, and biological principles, each of which has its set of idiosyncratic advantages and disadvantages.

Chemical Assays

Because most vitamin substances contain a functional group or chemical structure that interacts with ultraviolet, visible, and/or infrared radiation at a specific wavelength, these interactions provide a quantitative basis for many older methods of vitamin assay. If the vitamin structure does not lend itself to these requirements, it may be chemically derivatized to some other form that permits a more

Table 5.1. A Historical Survey of Vitamins and Their Designations

Vitamin group	Characterization	Common name
Vitamin A group Vitamin A_1	Antixerophthalmic factor, necessary for prevention of night blindness, for proper reproduction, and egg production	Vitamin A[a]
Vitamin A_2 Neovitamin A etc.		Dehydrovitamin A *cis* isomers of vitamin A
α-Carotene β-Carotene γ-Carotene Cryptoxanthin etc.	The carotenes and cryptoxanthin are converted to vitamin A by enzymes found, for example, in the intestinal mucosa	Carotene[a] (mixt.)
Vitamin B	Original name given to the crude preparations that relieved beriberi; later found to contain many factors	
Vitamin B complex	Factors isolated from yeast, liver, and other sources of the original "water-soluble B"; includes all the vitamins listed from this point up to vitamin C	The B vitamins
Vitamin B_1	Antiberiberi factor; a component of cocarboxylase	Thiamine[a,b]
Vitamin B_2	A component of Warburg's "yellow enzyme"; essential for cellular oxidation and reduction	Riboflavin,[a] riboflavine[b]
(Vitamin B_3)[c]	Necessary for the growth of pigeons; possibly the same as pantothenic acid	
(Vitamin B_4)	Prevents muscular weakness in rats and chicks; probably a mixture of riboflavin and pyridoxine, but also described as a mixture of arginine and glycine	
(Vitamin B_5)	Necessary for growth of pigeons; probably nicotinic acid	
Vitamin B_6 group Pyridoxine[a]	Prevents a specific dermatitis in young rats and in man, convulsions in infants; component of amino acid decarboxylation and transamination coenzymes	Vitamin B_5 Pyridoxine[a]
(Vitamin B_7)	A factor in rice polishings that prevents digestive disturbances in pigeons	
(Vitamin B_8)	Adenylic acid, a participant in phosphate transfer in many types of organisms; not usually classified as a vitamin;	
(Vitamin B_9)	Number unused because nine B vitamins were recognized when vitamins B_{10} and B_{11} were announced	
(Vitamin B_{10})	Chick feathering factor ⎱ probably a mixture of folic acid	
(Vitamin B_{11})	Chick growth factor ⎰ and vitamin B_{12}	
Vitamin B_{12} group Vitamin B_{12} Vitamin B_{12a} Vitamin B_{12b} Vitamin B_{12c} (Vitamin B_{12d})	Antipernicious anemia principle Vitamin B_{12a}; vitamin B_{12b}, and vitamin B_{12d} reported to be identical	Vitamin B_{12} Cyanocobalamin[a,b] Hydroxocobalamia[b] Nitrocobalamin; nitritocobalamin[b]
(Vitamin B_{13})	An uncharacterized factor, discovered in distillers' dried solubles, that promotes growth of rats, pigs, chicks, etc.	
(Vitamin B_{14})	Originally thought to be a metabolite of xanthopterin (related to folic acid); data not confirmed	

Table 5.1. (*Continued*)

Vitamin group	Characterization	Common name
(Vitamin B_{15})	Reported to facilitate oxygen uptake in the presence of anoxic or histiotoxic anoxia in rabbits; not confirmed	Pangamic acid
(Vitamin B_c)	Chick factor; identical with folic acid	
(Vitamin B_p)	Antiperosis factor for chicks, considered replaceable by manganese and choline	
(Vitamin B_t)	Growth factor for meal worm (*Tenebrio*); identified as carnitine	
(Vitamin B_w)	Name applied to biotin	
(Vitamin B_x)	Name applied to either pantothenic or *p*-aminobenzoic acid	
Nicotinamide[a,b]	Prevents pellagra; a component of hydrogen-transferring coenzymes I and II	Niacinamide[a]
Nicotinic acid[a,b]	Converted to nicotinamide in the body	Niacin[a]
Pantothenic acid[a,b]	Prevents dermatitis in chicks and gray hair in rats; a component of acetyl-transferring coenzyme A	Pantothenic acid[a,b]
Biotin[b]	Cures alopecia (baldness) and "spectacle eye" in rats fed a biotin-deficient diet supplemented by raw egg whites; prevents perosis in chicks	Biotin[b]
Biocytin	Source of biotin for certain microorganisms	
Choline[a,b]	Required for transmethylation	Choline[a,b]
Inositol[a]	Prevents baldness and "spectable eye" in mice and other animals	Inositol[a], mesoinositol[b]
p-Aminobenzoic acid[a,b]	Prevents gray hair in black rats and promotes growth in chicks; component of folacin	*para*-Aminobenzoic acid[a]
Folic acid group	Involved in transfer and metabolism of "single-carbon" fragment; prevents macrocytic (megaloblastic) anemia	Folic acids[b]
Pteroylglutamic acid[a]		Folic acid[a]
Pteroyltriglutamic acid		Teropterin
Pteroylheptaglutamic acid		
Pteroic acid[b]		Pteroic acid[b]
10-Formylpteroic acid		Rhizopterin
10-Formylpteroylglutamic acid		
Folinic acids	Growth factor for *Leuconostoc citrovorum*	
Leucovorin, etc.		Folinic acid-SF
Lipoic acid group	Pyruvate and α-ketoglutarate oxidation factor; acetate formation factor	
α-Lipoic acid		α-Lipoic acid
β-Lipoic acid		β-Lipoic acid
γ-Lipoic acid		γ-Lipoic acid
Vitamin C	Antisorbutic factor	Ascorbic acid[a,b]
(Vitamin C_2)	Vitamin J	
Vitamin D group	Antirachitic vitamin	Vitamin D
Vitamin D_2	Plant-type vitamin D, active in man and four-footed animals	Calciferol[a]
Vitamin D_3	Animal-type vitamin D, active in man and all species of animals, including chicks	Activated (or irradiated) 7-dehydro-cholesterol[a]
Vitamin E group	Prevents sterility, blood and vascular system disorders, central nervous system degeneration, and muscular dystrophy in experimental animals	Vitamin E

Table 5.1. (*Continued*)

Vitamin group	Characterization	Common name
α-Tocopherol[a,b]		2,5,7,8-Tetramethyl-2-(4,8,12-trimethyltridecyl)-6-chromanol; 5,7,8-trimethyltocol
β-Tocopherol[a,b]		5,8-Dimethyltocol
γ-Tocopherol[a,b]		7,8-Dimethyltocol
δ-Tocopherol[a]		8-Methyltocol
(Vitamin F)	Obsolete designation for essential fatty acids; also an obsolete name for vitamin B_1	
(Vitamin G)	Obsolete name for riboflavin	
(Vitamin H)	Obsolete name for biotin	
(Vitamin I)	Vitamin B_7	
(Vitamin J)	Postulated as antipneumonia principle	Vitamin C_2
Vitamin K group	Blood-clotting factor	Vitamin K
Vitamin K_1		Phylloquinone[b]
Vitamin K_2		Farnoquinone[b]
Menadione[a]		Menadione[b]
(Vitamins L_1 and L_2)	Postulated as necessary for maturation of lactation tissue	
(Vitamin M)	Obsolete name for folic acid; factor for prevention of anemia and leucopenia in monkeys	
(Vitamin P group)	A group of factors that decrease capillary fragility; vitamin character uncertain	Citrin; permeability factors including the bioflavonoids, as certain flavanone glycosides, e.g., hesperidin, rutin[a]
(Vitamin R)	Promotes bacterial growth; probably one of the folic acid group	
(Vitamin S)	Promotes bacterial growth; probably biotin; also applied to streptogenin, a chick growth factor	
(Vitamin T)	Reported to improve protein assimilation in rats and produce gigantism in insects	Termite factors
(Vitamin U)	Promotes bacterial growth; probably one of the folic acid group; term "vitamin U" also used for anti-ulcer factor in cabbage juice	
(Vitamin V)	Promotes bacterial growth; probably diphosphopyridine nucleotide, the nicotinamide coenzyme	
(Vitamin W)	Promotes bacterial growth; probably biotin	
(Vitamin X)	Promotes bacterial growth; probably biotin	
(Vitamin Y)	Probably identical with pyridoxine	

[a]Official names of U.S.P. XV, N.F. X, and N.N.R.

[b]Names approved by I.U.P.A.C. (see *J. Chem. Soc.*, 1951, 3526; *Bull. Soc. Chim. Biol.*, **33**, 1663 (1951); *Chem. Eng. News*, **30**, 104 (1952); **33**, 2433 (1955).

[c]Names no longer used as in parentheses.

SOURCE: Reprinted with permission from R. E. Kirk, and D. F. Othmer (eds.). 1955. *Encyclopedia of Chemical Technology*, Vol. 14, Table I, pp. 778–783. Wiley, New York.

sensitive and specific method of instrumental analysis.

Chemical methods are usually quite accurate, but the reliability of all techniques *depends on the efficacy of vitamin extraction and preparation techniques prior to* the actual *assay,* as well as the unavoidable presence of *biologically inactive vitamin derivatives* that chemically *interfere with the assay.* The latter consideration is especially true in the case of food analyses, in which assays are unable to decipher differences between chemically *or* physically bound forms of the vitamin versus those forms that are biologically available.

Many advances in direct and indirect chemical assays of vitamins are still necessary along with strict requirements for procedural uniformity before chemical assays of vitamins can serve as sole vitamin evaluation techniques.

Microbiological Assays

Since many microorganisms have a species-specific requirement for certain vitamins, it is possible to use these organisms for measuring vitamin concentrations in body fluids, tissue, or food extracts. These assays require the establishment of a standard growth curve whereby some measurement of the microorganisms' growth response (e.g., cell numbers, generation time) is plotted against measured increases in the concentration of pure vitamin in the culture medium.

Microbiological assays have served as the analytical basis for the evaluation of most vitamins in the clinical and nutritional domains for many years and most RDA and MDR as well as tabulated values for vitamin concentrations in foods are based on bioassays.

Unfortunately, bioassays are subject to many pitfalls. A credible assay assumes that the sample specimen has been properly prepared prior to assay. Inadequate extraction of the vitamin from its native biochemical environment will lead to quantitation on the low side, while rigorous extraction procedures may destroy the vitamin substance altogether. In some cases, *analogues* of the actual vitamin, or chemical substances completely divorced from the vitamin, may cause a positive microbiological response *completely unrelated* to the vitamin's availability. In still other situations, antibiotics contained in a sample or chemical remnants of reagents required for preliminary extraction of the vitamin may suppress the growth of microorganisms in spite of a high sample vitamin concentration. In recent years, requirements for more expeditious rates of vitamin assay, lack of test reproducibilities, and lack of protocol uniformity among laboratories have led to queries about the validity of these vitamin assay methods.

Animal Assays

Vitamin assays using animals are enormously complex and time consuming. These assays have great value since animals respond to the biologically active forms of the vitamin as opposed to inactive vitamin analogues or extraneous biochemical compounds that are unrelated to the vitamin. Despite the value of animal assays, the stringent experimental requirements for these assays do not encourage their use in routine nutritional or clinical vitamin investigations.

Four principal types of animal experiments provide the basis for quantitating vitamin requirements as well as determining actual vitamin concentrations. These methods include *growth assays, reaction time assays, graded response assays,* and *all-or-none assay responses* to a particular vitamin. All four of these animal assay methods require the use of a *basal diet* for the test subjects that includes the necessary ingredients for a completely balanced diet with the exception of the experimental variable (i.e., the vitamin of interest).

Chicks, dogs, guinea pigs, and rats are the primary test subjects for most growth assays. Usually, the response of the test subject is measured in increased body weight as an increasing dose of the vitamin is added to the basal diet. These growth assays may be classified as *curative* or *prophylactic*. Curative assays require measurement of weight increases in animals that have been vitamin deprived prior to administering dietary dosages of the vitamin in the basal diet. The prophylactic regimen entails measurement of weight increases in young test animals that have been given various dietary allowances of the specific vitamin.

Reaction time assays are based on depriving the test subject of a particular vitamin until a significant clinical or pathological symptom is shown. The test subject is then administered a metered food specimen containing the vitamin and time is provided (1–3 days) until evidence of the deficiency symptom disappears. Elimination of the deficiency symptom is followed by a *second* period of vitamin deprivation until the deficiency symptom reappears. A measure of the original vitamin concentration given to the test subject is then based on the time duration before a reappearance of the vitamin deficiency occurs.

Graded response assays rely on quantitating a definable characteristic of a vitamin deficiency in test animals when they are subjected to graded concentration levels of the vitamin. In this case, measurement of animal response will mirror changes in nutritional vitamin availability. Graded response assays have been classically applied to the estimation of vitamin D in chicks and rats. Since vitamin D availability regulates calcium deposition in bones, vitamin D may be assayed according to ash content of the animal's leg bone(s) because high levels of vitamin D correlate with high bone ash content.

All-or-none vitamin assays have limited utility, since somewhat arbitrary qualitative or quantitative estimates of test animal response to vitamin deficiencies must be defined before a response can be tabulated. The percent of test animals that react in a positive fashion to unknown concentrations of a vitamin provides the basis for animal response measurements in this assay. For instance, because vitamin E affects rat fertility, vitamin E may be estimated from the fertility percentage of mated female rats.

Following discussions of individual vitamins will reveal the most noteworthy assays used for certain vitamins. It should also be clearly noted that sole reliance on a single vitamin assay technique is *risky at best* and the most credible estimates of vitamin concentration are the product of two or more different analytical methods.

The units for measurement of vitamin activity and potency are variable. Although all calculations for vitamin quantitation can be expressed in whole or fractional milligram units (i.e., mg, μg, ng, or pg units), vitamin units for A, D, and E are still reported as an *International Unit* (I.U.). An International Unit refers to the observed biological activity attributable to a definite weight of a highly purified form of a vitamin. Since biological effects provide the underlying criterion for International Units, it is clear that 1.0 I.U. for one vitamin is not equivalent to 1.0 I.U. for another vitamin (Table 5.2).

Table 5.2. Vitamins A, D, and E Weight Equivalents for Corresponding International Units

Vitamin	International units	Weight equivalents
Vitamin A	1 I.U. of all-*trans*-retinol	0.300 μg
	1 I.U. of all-*trans*-retinyl acetate	0.344 μg
	1 I.U. of β-carotene	0.600 μg
Vitamin D	1 I.U. of pure crystalline vitamin D_3	0.025 μg
Vitamin E	1 I.U. of synthetic *dl*-α-tocopherol acetate	1 mg
	1.1 I.U. of *dl*-α-tocopherol	1 mg
	1.36 I.U. of the natural form *d*-α-tocopherol acetate	1 mg

VITAMIN ANTAGONISTS AND ANTIVITAMINS

The biochemical activity of vitamins can be inhibited by a number of compounds that are known as *antivitamins* or *vitamin antagonists*. Antivitamins were recognized by D. D. Woods in 1920 when the bacteriostatic effects of sulfanilamide were counteracted by *p*-aminobenzoic acid (PABA). The activity of an antivitamin substance may be measured in terms of an *inhibition index* (II). This index is based on the weight ratio of the antivitamin substance required to eliminate the effectiveness of a unit weight of a particular vitamin. The II for antivitamins or vitamin antagonists is a constant value provided that competitive antagonism occurs. For example, a PABA/sulfanilamide antagonism that indicates an II value of 4000 suggests that 4000 times more sulfanilamide than PABA is required to diminish bacterial growth.

Antivitamin activity is attributable to many different types of substances for many different reasons. Antivitamins may actually *bind* the vitamin, as in the case of biotin binding by *avidin,* which is found in egg whites, or antivitamin substances may simply *destory* the vitamin by enzymatically digesting it into a number of smaller vitamin-inactive products. The latter problem is clearly shown in Chastek paralysis, in which the enzyme, *thiaminase,* destroys thiamine. This disease principally affects captive foxes and minks that are fed thiaminase-containing entrails of stock animals or raw fish. In still other cases, antivitamins may inhibit the coenzyme functions of a specific vitamin.

The historical study of antivitamin activities suggests that microorganisms may be affected more by antivitamin substances than man or higher animals. Actually, higher animals and man are not immune to the same array of antivitamin substances that affect microorganisms, but the chronic low-level dietary exposure to these antivitamin substances may be inadequate for the full-blown development of clinical evidence.

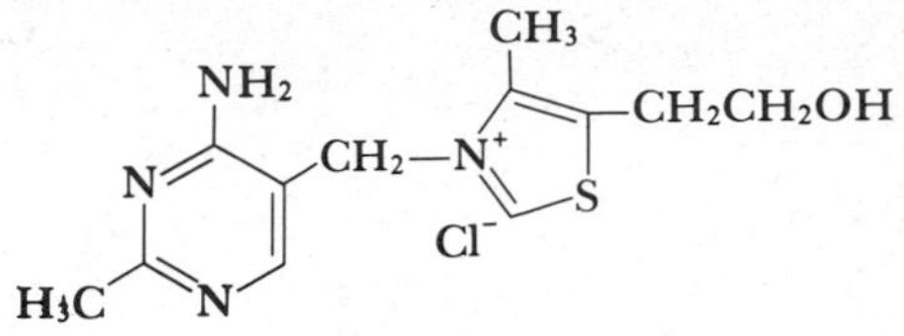

FIGURE 5.1.   Vitamin B_1.

Aside from unfavorable dietary effects, antivitamins such as the folic acid antagonist aminopterin may provide a logistical basis for chemotherapeutic agents designed to combat leukemia, encourage tumor regression, or control similar pathological conditions.

VITAMIN B_1

Vitamin B_1 or thiamine can be isolated in a free form (Figure 5.1) or a hydrochloride derivative (Figure 5.2) when prepared by industrial pharmaceutical processes.

A variety of alternative designations exist for vitamin B_1 hydrochloride including thiamine hydrochloride; thiamine chloride hydrochloride; betabion hydrochloride; aneurin hydrochloride; thiaminium chloride hydrochloride; and 3-(4-amino-2-methyl-pyrimidyl-5-methyl)-4-methyl-5-(β-hydroxyethyl) thiazolium chloride hydrochloride (Table 5.3).

In nature, thiamine commonly exists as its coenzyme, thiamine pyrophosphate (TPP) (Figure 5.3).

Sources

Significant vitamin B_1 concentrations are found in many species of plants and animals.

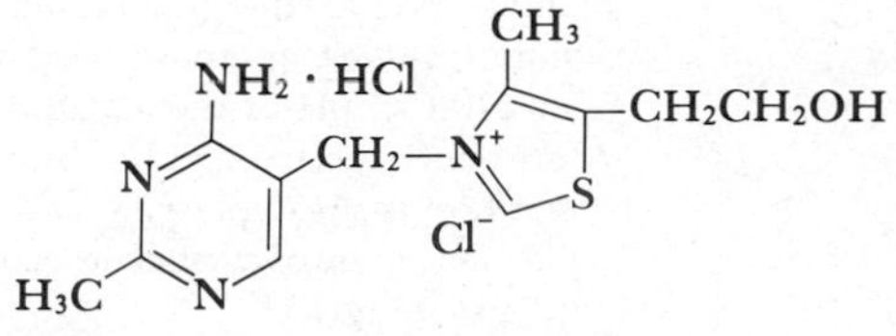

FIGURE 5.2.   Vitamin B_1 hydrochloride.

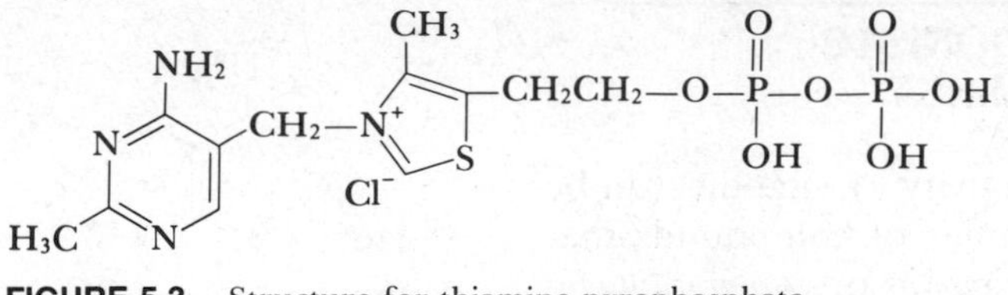

FIGURE 5.3.   Structure for thiamine pyrophosphate.

Some of the richest sources are whole grains and enriched cereals, legume seeds, pork, green vegetables, yeast, liver, eggs, milk, tubers, and roots. Because thiamine is located in both the germ and outer seed coats of cereal grains, a large amount of the vitamin can be lost during milling and refining.

Human Requirements and Deficiency Effects

The RDA for thiamine is 0.4 mg/1000 kcal for all ages, but an added 0.2 mg daily is encouraged during the second and third trimester of pregnancy and 0.5 mg during lactation. One milligram of the vitamin is recommended for all age groups including geriatric patients when caloric intake is less than 2000 kcal. Vitamin B_1 deficiencies are reportedly controlled by prophylactic oral doses of 1.0 to 2.0 mg/day, whereas therapeutic oral or intramuscular injections may range from 10 to 50 mg/day.

Moderate deficiency effects of thiamine in humans can cause peripheral neuropathy due to upset carbohydrate metabolism in neurons. This problem is recognized as *dry beriberi.*

Wet beriberi, on the other hand, results from severe thiamine deficiencies, which ultimately contribute to high-output congestive heart failure due to impaired vascular and cardiac carbohydrate metabolism.

Thiamine deficiencies are most common among rice-eating populations in Asia. In the United States, alcoholics are prone to develop Wernicke–Korsakoff syndrome accompanied by ataxia, sudden dementia, and ophthalmoplegia. The lack of expeditious treatment for this syndrome may result in irreversible consequences.

Table 5.3.   Chemical Characterization and Properties for Thiamine Hydrochloride

Empirical formula	$C_{12}H_{18}ON_4SCl_2$
Molecular weight	337.26
Melting point	246–250°C (decomposes)
pH of 1% solution in water	3.13
pH of 0.1% solution in water	3.58
Solubility (g/100 mL of solvent)	
Water	100 g
Ethanol (95%)	1.0 g
Ethanol (100%)	0.3 g
Glycerol	5.0 g

Insoluble in acetone, benzene, chloroform, ethyl ether, hexane

Stability: Very stable (desiccated); not subject to atmospheric oxidation; vitamin is stable at atmospheric and autoclaving pressures to a temperature of 110°C if pH of the aqueous environment is <5.5.

Absorbance spectrum: The absorbance spectrum for thiamine in an aqueous solution is pH dependent and shown in Figure 5.4.

Units: One gram of crystalline vitamin B_1 hydrochloride corresponds to 333,000 I.U. or 3 μg/I.U.

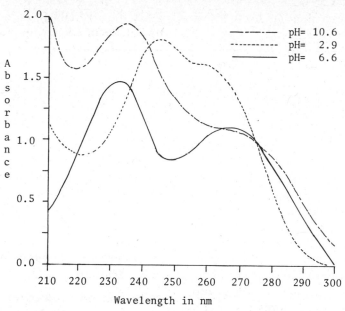

FIGURE 5.4. Characteristic absorption spectra for thiamine in a variety of aqueous solutions having different pH conditions.

Biochemistry

Thiamine pyrophosphate acts as a coenzyme for a wide variety of enzymes including α-ketoacid decarboxylases, α-ketoacid oxidases, phosphoketolases, and transketolases.

The biochemical activity of thiamine is attributable to the No. 2 carbon contained in the thiazole ring since it acts as a temporal carrier of a covalently bound *active aldehyde* group. The coenzyme activity of TPP also requires Mg^{2+} as a cofactor.

The No. 2 carbon in the thiazole ring is significant because its hydrogen dissociates as a proton, leaving the remaining thiazole moiety as a carbanion (Figure 5.5). Carbanion formation is a key step for TPP involvement in α-ketoacid decarboxylation exemplified by the formation of acetaldehyde from pyruvic acid. This reaction typifies the essential role that TPP plays along with a number of other coenzymes, vitamins, and cofactors that integrate the glycolytic reactions of the Embden–Meyerhoff pathway with the tricarboxylic acid (TCA) cycle. The entire process of pyruvic acid decarboxylation and acetaldehyde formation is the result, not of a single enzyme reaction, but of three enzymes that comprise a multienzyme complex (M.W. $= 4.8 \times 10^6$). A five-step reaction

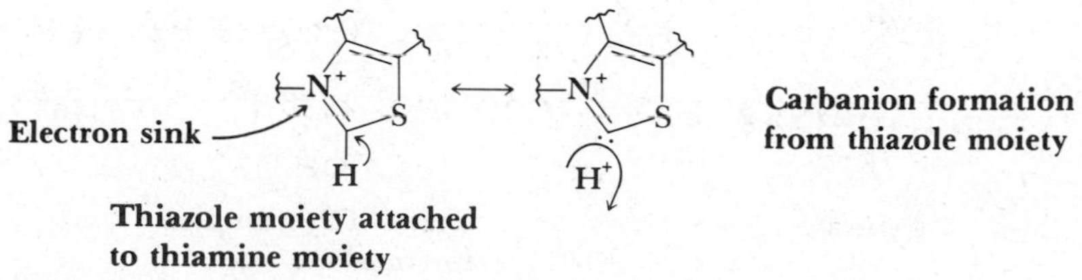

Electron sink

Carbanion formation from thiazole moiety

Thiazole moiety attached to thiamine moiety

FIGURE 5.5. Thiazole ring transformation into a reactive carbanion.

sequence for this reaction has been shown below.

$$H_3C-\underset{\underset{O}{\|}}{C}-COOH + CoA-SH \xrightarrow[\text{Mg}^{2+},\text{ FAD}]{\substack{\text{Lipoic acid,}\\ \text{Thiamine}\\ \text{pyrophosphate}}} H_3C-\underset{\underset{O}{\|}}{C}-S-CoA + NADH(H^+) + CO_2$$

Pyruvate   **Acetyl-CoA**

$$\Delta G^{0'} = -8.0 \text{ kcal/mol}$$

The overall pyruvic acid decarboxylation reaction is initiated by the *pyruvic acid dehydrogenase* enzyme. Thiamine pyrophosphate is a prosthetic group on this enzyme that facilitates the formation of an intermediate *acetol–TPP* complex along with carbon dioxide derived from the carboxyl group of pyruvic acid (Figure 5.6). The acetol group, containing two carbon atoms, is subsequently conveyed to oxidized lipoic acid, which is covalently linked to the second of the three enzymes, *dihydrolipoyl transacetylase* (Figure 5.7). This second step results in the formation of a reduced lipoic acid, thioester, with the two-carbon unit obtained from pyruvic acid.

The acetyl-lipoic acid thus formed participates in a transfer of the acetyl group to coenzyme A, thereby forming acetyl-CoA, which is discharged as an end product of the multienzyme reaction sequence (Figure 5.8).

The flavoenzyme *dihydrolipoyl dehydrogenase,* which contains flavin adenine dinucleotide (FAD), acts as the third enzyme in the multienzyme sequence. This enzyme reconstitutes the formation of oxidized lipoic acid (cyclic lipoic acid form) and yields reduced FAD (i.e., $FADH_2$) (Figure 5.9). The reduced flavin coenzyme is then oxidized by nicotinamide adenine dinucleotide (NAD^+) to yield $NADH(H^+)$.

$$FADH_2 + NAD^+ \rightleftharpoons FAD + NADH(H^+)$$

The initial decarboxylation of pyruvic acid is not reversible, and the overall $\Delta G^{0'}$ is about -8.0 kcal/mol.

Miscellaneous Considerations

Thiamine is one of many vitamins synthesized by intestinal flora; however, the quantities produced are insufficient as a sole source of the vitamin for humans and most other animals.

Although vitamin B_1 is stable within pH bounds prescribed above, it readily undergoes destruction in the presence of alkalies and alkaline drugs such as phenobarbital (sodium), and the vitamin is precipitated by tannins found in wine and other foods.

Thiamine is also subject to enzymatic attack and destruction by the enzyme known as *thiaminase*. This enzyme, found in the raw entrails of stock animals, raw fresh- or salt-water fishes, and certain plants, is presumed to sever the thiazole ring from the overall vitamin B_1 structure.

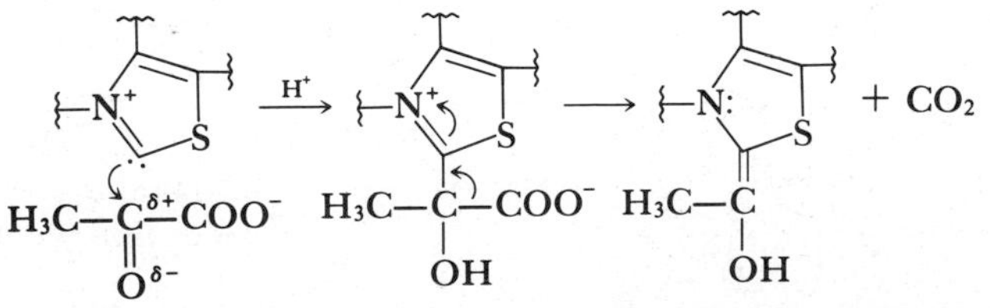

FIGURE 5.6.   Acetol–TPP complex formation.

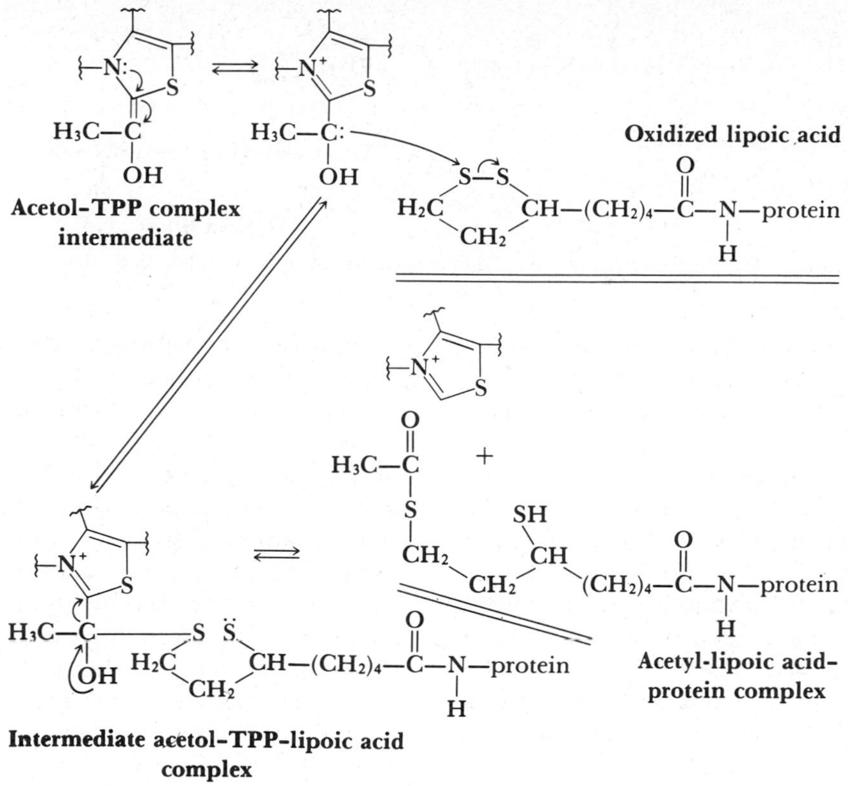

FIGURE 5.7. Acetol group reactivity and its role in dihydrolipoyl transacetylase.

A number of experimental antagonists have been produced in order to study short- and long-term effects of thiamine deficiencies. *Oxythiamine* and *pyrithiamine* represent two of the most widely used antagonists. The pyrimidine moiety of oxythiamine has a hydroxyl group substituted for the normally occurring amino group, whereas pyrithiamine contains a pyrimidine ring in place of the thiazole ring of the vitamin.

The antivitamin activity for the two substances differs. Oxythiamine easily forms a pyrophosphate derivative similar to the structure of TPP. The pyrophosphate derivative then behaves as a competitor in enzyme systems requiring TPP. Pyrithiamine, how-

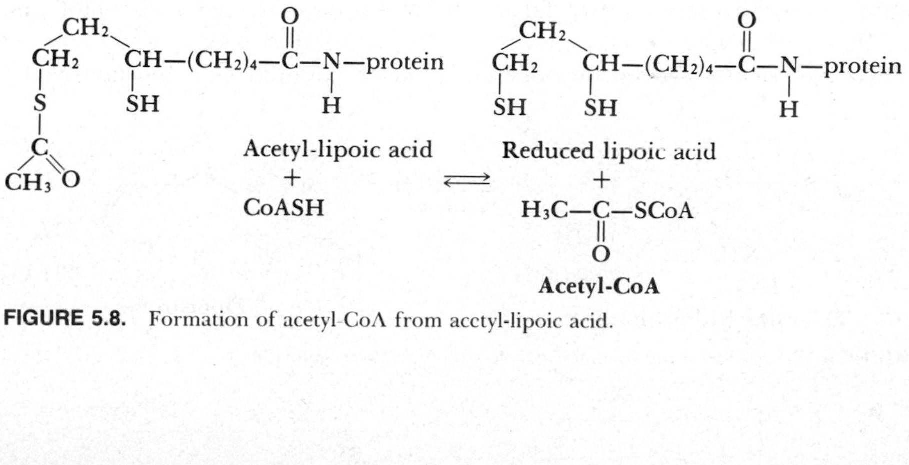

FIGURE 5.8. Formation of acetyl-CoA from acetyl-lipoic acid.

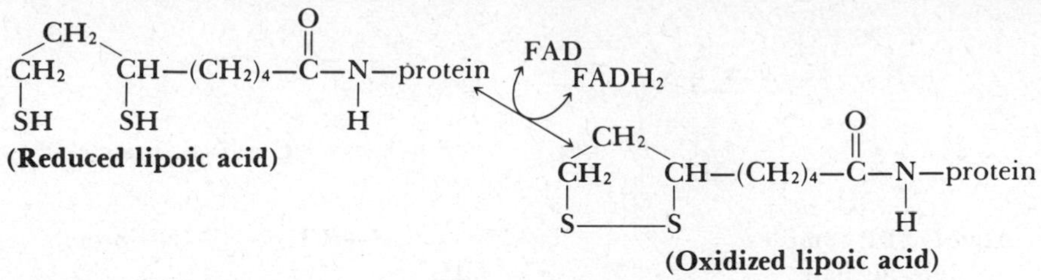

FIGURE 5.9. Regeneration of oxidized lipoic acid by dihydrolipoyl dehydrogenase.

ever, competitively binds with thiamine at the substrate binding site on thiamine kinase, which is responsible for TPP formation.

An additional antivitamin agent known as *amprolium,* the 2-*n*-propyl pyrimidine analogue of thiamine, is used as an effective coccidiostat.

Exclusive of antivitamin agents, considerable amounts of dietary thiamine can be lost from cooked vegetables when pot liquors are drained from food. For meats, leaching of thiamine from tissues is not as significant a problem as the *in situ* destruction of thiamine during roasting. Frying and boiling methods for cooking meats are far less vitamin destructive. From a standpoint of food storage, thiamine loss is minimized by storage temperatures $\leq 0°F$ ($-18°C$).

The lability of thiamine under alkaline conditions serves as a basis for one of the most widely used methods of vitamin B_1 assay. Oxidation of thiamine in an alkaline solution forms a thiochrome derivative that displays an intense blue fluorescence that is directly related to vitamin concentration (Figure 5.10). This assay method is suitable for evaluating many different foodstuffs, organs, body fluids, pharmaceuticals, and natural products having low concentrations of

vitamin B_1 along with high amounts of unrelated vitamin substances that interfere with other assay methods. Reproducibility for the method is $\pm 5-10\%$ depending on the nature of the sample. Other physical and chemical assays based on paper, thin-layer, or ion-exchange chromatography have been widely used in the past, but the underlying analytical rationales for these methods have been adapted for modern high-pressure liquid chromatographic procedures. Bioassays employing *Ochromonas danica* (Pringsheim 933-7), *Lactobacillus fermentii* 36 (ATCC 9338), *Kloeckera brevis* B768 (NC1B), and many other organisms have also been widely used. However, these methods are lengthy (3 h to 3 days) and often give a positive response to breakdown products of thiamine as well as to many nonthiamine substances.

VITAMIN B_2

Vitamin B_2 or riboflavin is a slightly water-soluble pigment that has a minute but ubiquitous occurrence in all plant and animal cells. Riboflavin is synonymous with *lactoflavine,* vitamin G, 7,8-dimethyl-10-(D-ribo-

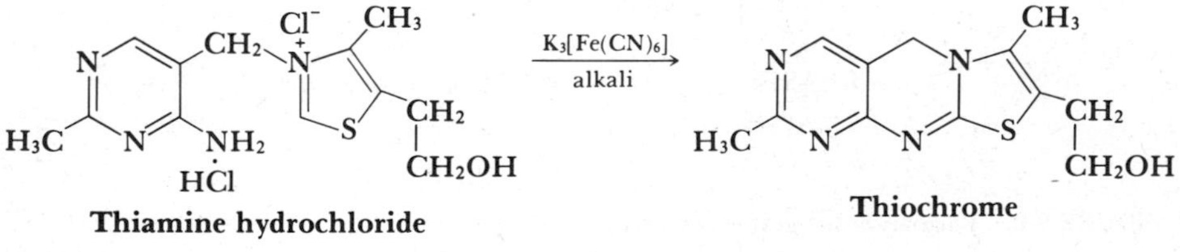

Thiamine hydrochloride

Thiochrome

FIGURE 5.10. Thiochrome formation from thiamine hydrochloride.

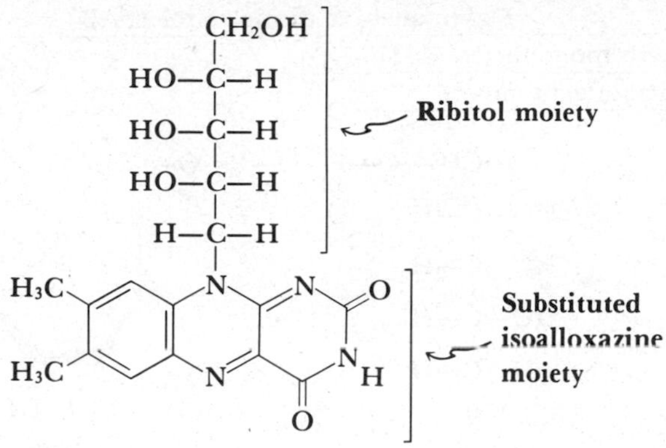

FIGURE 5.11. Structure of riboflavin involves ribitol and an isoalloxazine ring system.

2,3,4,5-tetrahydroxypentyl) isoalloxazine, 7,8-dimethyl-10-ribitylisoalloxazine, Flavaxin, Ribipca, and Beflavin.

The riboflavin structure is a combination of a substituted *isoalloxazine* ring and the sugar alcohol of ribose, known as *ribitol* (Figure 5.11 and Table 5.4). Riboflavin occurs in a wide variety of tissues, but it exists almost exclusively as riboflavin-phosphoric acid (*flavin mononucleotide* or FMN) or riboflavin-

Table 5.4. Chemical Characterization and Properties for Riboflavin

Empirical formula	$C_{17}H_{20}N_4O_6$
Molecular weight	376.36
Melting point	274–282°C (decomposes)
Solubility (g/100 mL of solvent)	
Water (25°C)	0.012 g
Ethanol (27.5°C) (100%)	0.0045 g
Insoluble in acetone, benzene, chloroform ethyl ether, hexane	
Optical rotation: $[\alpha]_D^{20}$	− 114° for 0.1 N NaOH

Stability: The vitamin is oxidized by chromic acid, destroyed by $KMnO_4$ in 0.1 N acetic acid in ~10 min at 25°C, but undergoes only 10% destruction in the presence of $KMnO_4$ at pH 4.5. Riboflavin is also stable in the presence of many other oxidizing agents such as H_2O_2, Br_2H_2O, and HNO_3 (conc.); and it is stable in the presence of strong mineral acids. A combination of H_2O_2 and ferrous iron effectively decomposes riboflavin. Riboflavin is also characterized by its reversible reduction to its leuco-base by a variety of reagents that include alkaline sulfides, active hydrogen, sodium hydrosulfite, or stannous or titanous chlorides. Both visible and ultraviolet wavelengths are destructive to the vitamin, and this consideration necessitates its controlled exposure to light during preservation and assay. The vitamin displays significant yellow-green fluorescence when irradiated with ultraviolet light, but the intensity of fluorescence is dependent on the concentration of riboflavin and solvent pH. Maximum fluorescence is usually observed at pH values of 6.0–7.0.

Absorbance spectrum: The absorbance spectrum for riboflavin shows notable absorption peaks at 266, 371, 444, and 475 nm, as illustrated by Figure 5.13. Maximum absorption occurs at 220–225 nm, but this is not suitable for routine assay of the vitamin.

Units: Riboflavin is conventionally measured in terms of milligram quantities of the chemically pure substance.

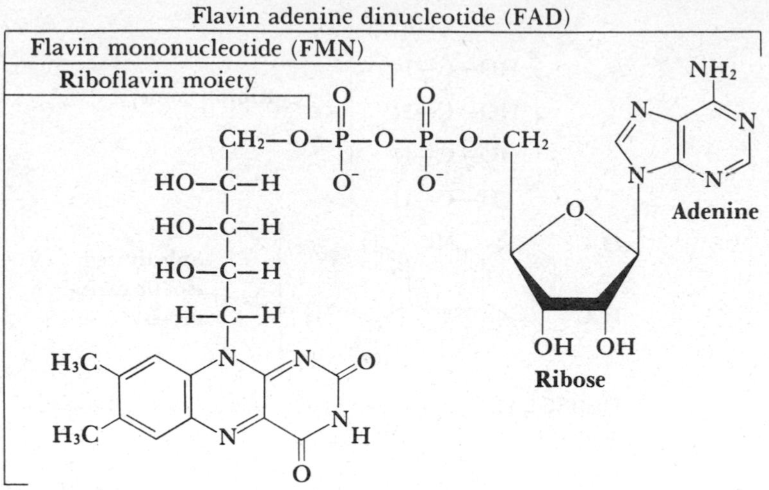

Flavin adenine dinucleotide (FAD)

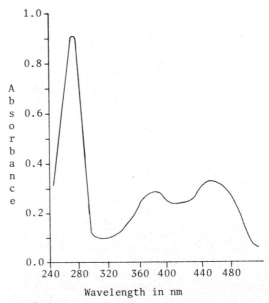

FIGURE 5.12. Structural relationships for the flavin nucleotides.

phosphoric acid-adenine dinucleotide (*flavin adenine dinucleotide* or FAD) (Figure 5.12).

Sources

High concentrations of riboflavin occur in plant and animal foods. Noteworthy sources include anaerobic fermentation bacteria, heart, kidney, liver, eggs, malted barley, yeast, and green vegetables.

Human Requirements and Deficiency Effects

The United States RDA for riboflavin is 1.7 mg for adult males and 1.3 mg for adult females. The second and third trimesters of pregnancy require an added allotment of ~0.3 mg, while the lactation period requires an additional 0.5 mg. Riboflavin requirements for children relatively supersede those of adults; therefore, 0.6 mg is the RDA for infants graduating upward to ~1.5 mg for adolescents. Clinically significant evidence of riboflavin deficiency in adults probably does not exist unless chronic intake is less than 0.3 mg/1000 kcal.

Unlike some vitamins, *ariboflavinosis* does not lead to lethal consequences for humans, but it does cause a wide variety of lesion effects on the body. Angular stomatitis evi-

dent as fissures at the angles of the mouth represents one of the most common deficiency symptoms. These lesions may eventually encroach onto the skin or the mucous membrane in the mouth. Riboflavin deficiencies have also been linked to the denudation of tongue papillae (*bald tongue*); magenta tongue color; seborrheic dermatitis of the nose, nasolabial folds, and canthi of the

FIGURE 5.13. Absorbance spectrum for riboflavin.

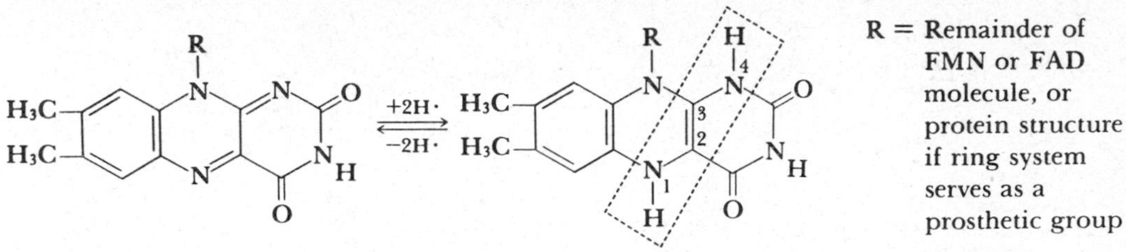

FIGURE 5.14. Structures for 1,4-oxidized/reduced forms of the isoalloxazine ring system.

eyes; and scrotal or vulval dermatitis, which leads to itching and eventual desquamation. Advanced deficiencies may also show proliferation of blood vessels from the limbic plexus of the eye into corneal regions. This vascular proliferation into the cornea ceases and blood vessels actually regress upon therapeutic administration of riboflavin.

The prophylaxis and treatment of ariboflavinosis may require 2.0 to 15.0-mg dosages of the vitamin. Since excessive amounts of riboflavin are discharged in the urine, toxic effects of riboflavin overdoses have not been rigorously detailed in the literature.

Biochemistry

The substituted isoalloxazine ring serves as a key component of flavin coenzymes such as FMN and FAD, as well as the flavin-linked enzymes. Biochemical activity of the isoalloxazine ring system depends on its reversible 1,4-oxidation/reduction in both coenzyme and prosthetic forms as shown in Figure 5.14.

Oxidative biochemical reactions are required in every cell of living organisms, especially oxidations that require the removal of hydrogen atoms on two adjacent carbon atoms as shown in Figure 5.15. In this case,

an integral enzyme of the TCA cycle known as *succinic dehydrogenase*, containing FAD as a prosthetic group, mediates the oxidation of succinate to fumarate. Many other biochemical oxidations are carried out in a similar vein, albeit with different enzymes. Actually some enzymes that have a vitamin B_2 requirement also require the added participation of metals like iron or molybdenum.

Contrary to early theories that suggested simultaneous 1,4-reduction of the positions in the isoalloxazine ring, contemporary theories of riboflavin activity favor a two-step reduction scheme wherein the first reduction step yields a *semiquinone* formed by the addition of a singular hydrogen atom. The semiquinone formed by this step has an unpaired electron and marginal stability due to its resonance forms. Although the actual roles of metals such as iron or molybdenum are far from lucid, it is plausible that the unpaired electron of these metals may be shared with the semiquinone to give added stability (Figure 5.16).

The stabilization effects of metals on the semiquinone structure should not be confused with another group of flavoproteins known as *metalloflavoproteins*. Metalloflavoproteins can be purified as a single structure without the loss of flavin and metal com-

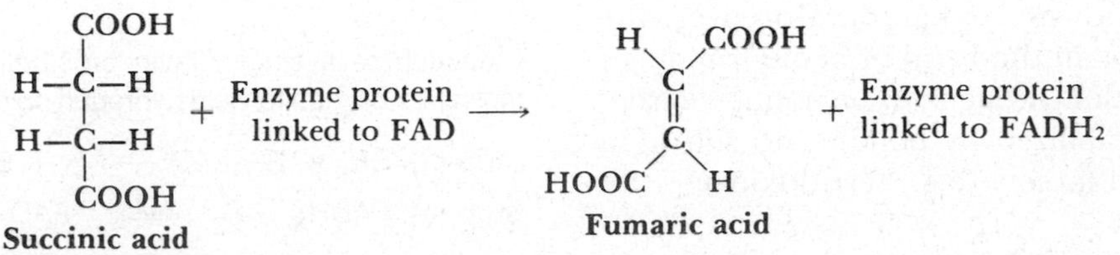

FIGURE 5.15. Succinic acid is oxidized by a FAD-linked enzyme to yield fumaric acid and reduced flavin nucleotide.

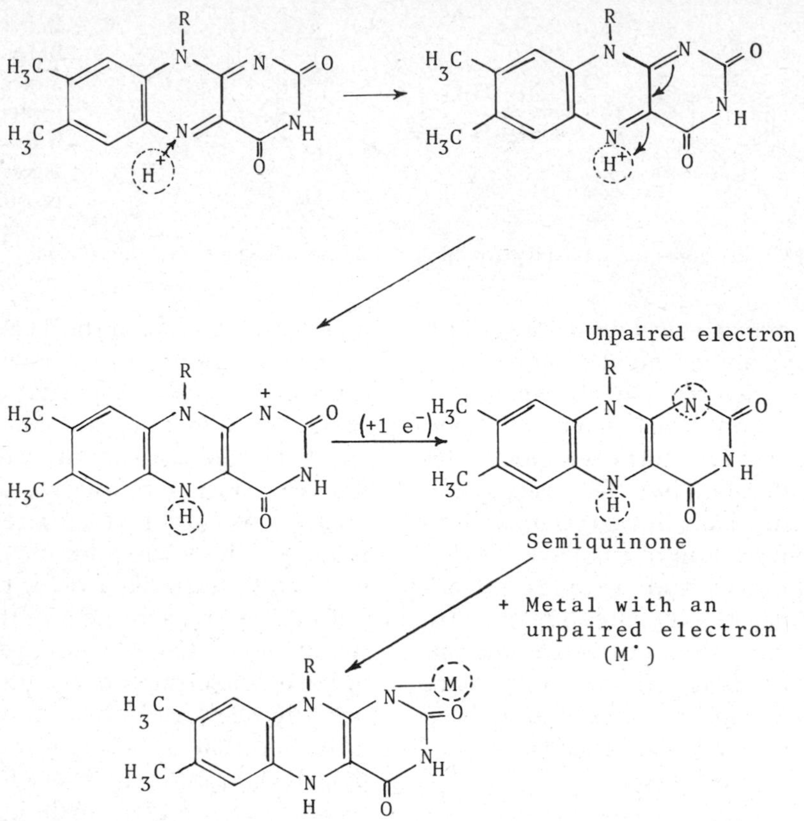

FIGURE 5.16. Stabilization of the semiquinone structure of riboflavin by metals having an unpaired electron.

ponents, unlike simple metal–semiquinone interactions, which tend to separate. Metalloflavoproteins also demonstrate important biochemical functions since they facilitate electron transfers from certain substrates to a variety of oxidants including ferricytochrome-*c*, nicotinamide nucleotides (NAD^+), nitrites (NO_2^-), and nitrates (NO_3^-).

Although molybdenum is a constituent of many metalloflavoproteins, iron is also found either singularly in the structure or in combination with molybdenum. Moreover, the iron may be in the form of heme-iron (e.g., lactic acid dehydrogenase) or nonheme-iron, which is stabilized by bonding to sulfur in the protein moiety (e.g., ferridoxins).

Flavoprotein interactions with oxygen: The flavin prosthetic group of certain flavopro-

teins is able to undergo a 1,4-type reduction with hydrogens derived from a substrate only to be rapidly reoxidized by molecular oxygen. This behavior is notably dissimilar from other vitamins and coenzymes.

The reaction sequence for molecular oxygen with flavoproteins varies depending on whether the protein contains *one* or *two* flavin prosthetic groups as demonstrated below:

Case 1:

Oxidation of a single flavin prosthetic group by molecular oxygen to yield hydrogen peroxide (H_2O_2):

Step (a): $SH_2 + E\text{—}FAD \longrightarrow S + E\text{—}FADH_2$

Step (b): $FADH_2 + O_2 \longrightarrow E\text{—}FAD + H_2O_2$

Here, the flavin moiety is first reduced and then reoxidized to form H_2O_2.

Case 2:

Water is formed instead of H_2O_2 as a result of the action of flavin-containing monooxygenases on a substrate during a single-step mechanism:

$$SH + E{-}FADH_2 + O_2 \longrightarrow S{-}OH + FAD + H_2O$$

In this situation, a single atom of molecular oxygen is instituted as a hydroxyl group on the substrate, while the other oxygen is released in a water molecule.

Miscellaneous Considerations

Dietary antagonists to riboflavin have not been widely reported in the literature, although a natural substance in the ackee nut of Jamaica may be antagonistic to normal riboflavin metabolism. Other antagonists have been developed or produced to achieve extended periods of riboflavin deficiency in animals. These substances include diethylriboflavin, 7-chloro-6-methyl-riboflavin, and galactoflavin. The latter substance is formed merely by substituting galactose for the ribitol moiety of the normal riboflavin.

Vitamin B_2 is chemically stable during most food-processing efforts, but similar to vitamin B_1, it becomes prone to destruction under alkaline conditions. Furthermore, vitamin sensitivity to light is enhanced by the presence of vitamin C. Normal photosensitivity of riboflavin should encourage the use of amber glass containers, metal cans, or opaque food-packaging methods for foods high in vitamin B_2 wherever possible.

Vitamin B_2 assays can be performed by polarographic, spectrophotometric, fluoro-metric, and bioassay methods. Polarographic assays are of particular importance to pharmaceutical evaluations of the vitamin and not food substances.

Fluorometric assays for riboflavin are usually more sensitive than direct photometric measurements, but both methods have applications to selected natural products, foods, or pharmaceuticals. Maximum fluorescence of the vitamin occurs at pH 6 to 7, but the most reliable fluorometric assays are conducted in the pH range of 3 to 5, where fluorometric responses of the vitamin are not greatly dependent on minor pH variations. An alternative to direct fluorometric measurement of vitamin B_2 relies on the irradiation of chloroform-insoluble riboflavin contained in an alkaline medium. This reaction is selective for riboflavin since extraneous flavin compounds are unable to produce the irradiation reaction product known as *lumiflavin* (Figure 5.17). This irradiation product of riboflavin may be photometrically evaluated at 450 nm or fluorometrically assayed at 513 nm.

Limits of riboflavin detection using microbiological assays tout a sensitivity of 0.5 ng, but these assays can erroneously respond to the unpredictable presence of starches, glycogen, free fatty acids, and proteins that commonly occur with riboflavin in foods. *Lactobacillus helveticus* has served as one of the most common bioassay organisms.

Modern developments in high-pressure liquid chromatography using advanced fluorometric, electrochemical, and spectrophotometric detectors will undoubtedly challenge the historical use of bioassays for riboflavin.

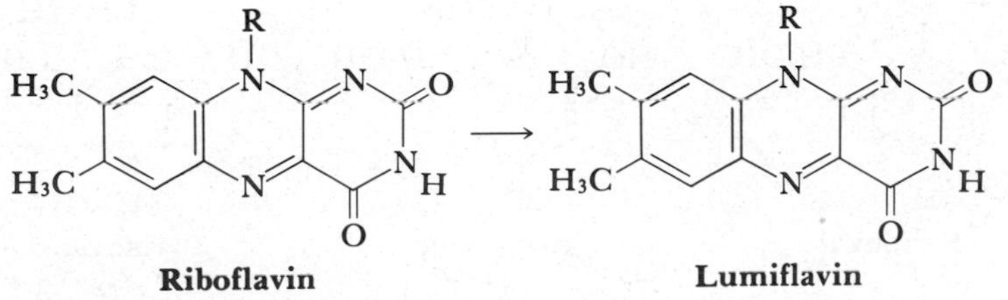

FIGURE 5.17. Lumiflavin formation from riboflavin.

VITAMIN B₆

This vitamin is a widely distributed, colorless member of the B complex that was first recognized as the *anti-acrodynia rat factor* or the *rat antidermatitis factor*. References to the structure of vitamin B₆ can pertain to any one of three possible pyridine derivatives that occur in foods (Figure 5.18).

The *pyridoxine form* was synonymous with vitamin B₆ activity until it was recognized that (1) many foods deficient in pyridoxine could serve as very effective sources of vitamin B₆ and (2) classical assays for vitamin B₆ could not differentiate among the three forms of the vitamin substance. Although evidence suggests that pyridoxine is not the ultimate biochemically active form of vitamin B₆, its hydrochloride derivative is effectively administered as an adjunct in the prophylaxis and treatment of multiple vitamin B complex deficiencies.

The notable properties for pyridoxine hydrochloride are outlined in Table 5.5.

Literary references to pyridoxine hydrochloride include pyridoxium chloride, pyridoxol hydrochloride, hexabione hydrochloride, aderman-hydrochloride, 5-hydroxy-6-methyl-3,4-pyridine dimethanol hydrochloride, Hexabetalin, Hexavibex, Pyridipca, Benedon, Hexermin, Campoviton 6, and Hexobion.

Sources

Vitamin B₆ has a widespread occurrence in plants and animals. Especially good sources include whole grain cereals, milk, yeast, liver, and kidney.

Human Requirements and Deficiency Effects

An induction of vitamin B₆ deficiency can lead to a lack of appetite, weight loss, lassitude, cheilosis, peripheral neuritis, pellagra-like skin changes, and other vitamin B-type deficiency symptoms.

Acute juvenile deficiencies can cause anemia, hyperirritability, and convulsions. In spite of evidence for juvenile vitamin B₆ deficiency symptoms, marginal to moderate adult deficiencies often lack clearly identifiable syndromes.

The availability of vitamin B₆ levels in the body may be assessed by evaluating the urinary excretion of tryptophan metabolites such as *xanthurenic acid*. Since the vitamin is required for the activity of the kynureninase enzyme, which normally converts tryptophan to nicotinic acid, a deficiency of vitamin B₆ can lead to an upset of tryptophan metabolism, decreased levels of metabolically derived nicotinic acid, and abnormal diversion of tryptophan to xanthurenic acid.

The RDA for this vitamin is set at ~2.1 mg/day for adults with an additional 0.6 mg/day for periods of pregnancy and lactation.

Additional vitamin intake may be dictated during tuberculosis treatment with isonicotinic and acid hydrazides. Painful and disabling polyneuritis caused by this therapeutic drug treatment is reportedly controlled by 10 to 20 mg/day doses of pyridoxine.

Because toxicity of the vitamin may not occur in humans until a dosage level of almost 300 mg/day is reached, the administration of vitamin B₆ above recommended levels does not pose an imminent health hazard. Excessive concentrations of the vitamin

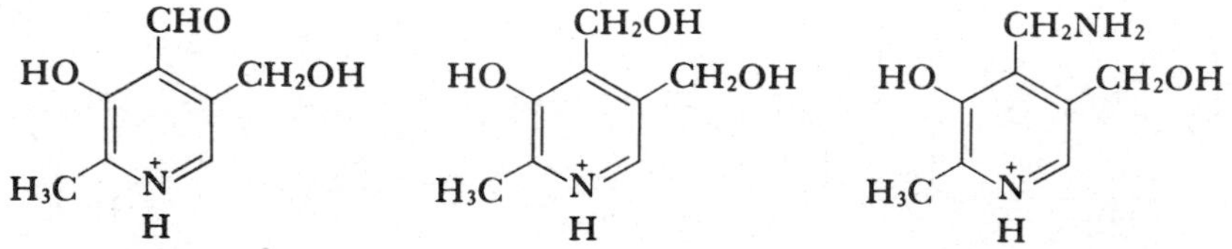

FIGURE 5.18. Some typical forms of vitamin B₆ structures.

Table 5.5. Characterization and Properties of Pyridoxine Hydrochloride

Empirical formula	$C_8H_{11}NO_3 \cdot HCl$
Molecular weight	205.64
Melting point	204–206°C (decomposition)
Melting point (free base)	160°C
pH of 10% (w/v) solution in water	3.2
Solubility (g/100 mL of solvent)	
Water	22.2 g
Ethanol	1.1 g
Slightly soluble in acetone	
Insoluble in ether and chloroform	
Optical rotation	inactive

Stability: Acidic solutions of pyridoxine hydrochloride in water are stable and capable of standing heat for 30 min at 120°C.

Absorption spectrum: Absorption maxima are 291 nm at pH 2.1 and 255 and 326 nm at pH 7.0. Absorption spectra for vitamin B_6 substances are highly variable depending on the pH of the solvent and type of solvent.

Units: The activity of pyridoxine is reported as milligram units of the pure synthesized substance.

seem to be destroyed by the body, and a paucity of information is available concerning the corporal storage of the vitamin.

Biochemistry

The actual biochemical activity of the B_6-group vitamin substances (i.e., pyridoxal, pyridoxine, and pyridoxamine) depends on their conversion to a coenzyme form by a phosphorylation step that yields pyridoxal phosphate and/or pyridoxamine phosphate (Figure 5.19).

The initial conversion of dietary pyridoxine to pyridoxol phosphate occurs in the liver and serves as a prerequisite to the formation of both coenzyme forms. It should also be noted that pyridoxal formation precedes the excretion product of vitamin B_6, which is called 4-pyridoxic acid. Over 85% of pyri-

doxine administered to humans undergoes a rapid conversion to this derivative (Figure 5.20).

Pyridoxal phosphate is an especially important coenzyme derivative of the vitamin for enzymes that mediate transamination, decarboxylation, racemization, desulfhydration, amine oxidation, and deamination reactions:

1. Transamination

$$RCHNH_2COOH + R'COCOOH \rightleftharpoons$$
$$RCOCOOH + R'CHNH_2COOH$$

2. Decarboxylation

$$RCHNH_2COOH \rightleftharpoons RCH_2NH_2 + CO_2$$

3. Racemization of a single amino acid

D-amino acid $\rightleftharpoons$ L-amino acid

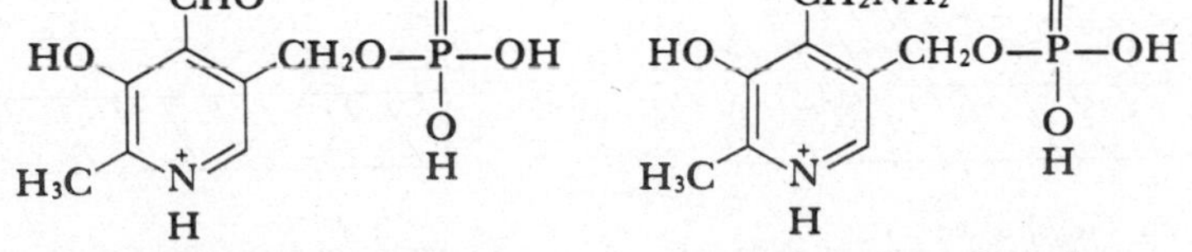

Pyridoxal phosphate **Pyridoxamine phosphate**

FIGURE 5.19. Phosphorylation of B_6 vitamin substances yields coenzyme forms.

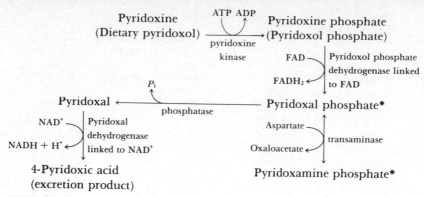

FIGURE 5.20. Conversion sequence for dietary pyridoxine (pyridoxol) to coenzyme active derivatives as indicated by an asterisk.

4. Desulfhydration

$$RCHSCHNH_2COOH \longrightarrow$$
$$RCH_2COCOOH + NH_3 + H_2S$$

5. Amine oxidation

$$RCH_2NH_2 + O_2 + H_2O \longrightarrow$$
$$RCHO + NH_3 + H_2O_2$$

6. Deamination

$$RCHOHCHNH_2COOH \dashrightarrow$$
$$RCH_2COCOOH + NH_3$$

The coenzyme is covalently bonded to the ε-amino group of a lysyl-residue located *at* or *near* the active site of an apoenzyme, which is solely responsible for mediating any specific reaction detailed above. Regardless of the apoenzyme, this type of covalent interaction is called a *Schiff's base*, and it represents an underlying prerequisite for enzyme activity (Figure 5.21).

Since Schiff's base formation seems to be essential for many enzyme reactions that require pyridoxal phosphate, the various differences in product formation from these enzymes are ascribed to characteristic electronic rearrangements of the *apoenzyme/Schiff's base/substrate complex*. The exemplary behavior of the Schiff's base in a characteristic

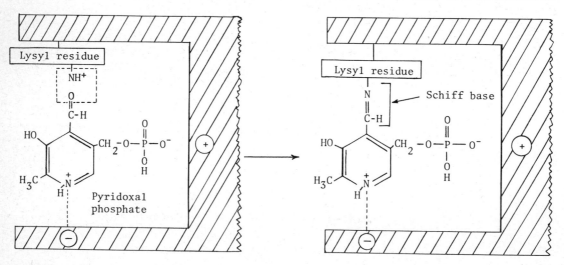

Active site on an enzyme

FIGURE 5.21. Involvement of pyridoxal phosphate coenzyme in Schiff's base formation at the active site of an enzyme.

transamination reaction has been outlined in Figure 5.22.

Miscellaneous Considerations

Many chemical substances appear to be vitamin B_6 antagonists. Foremost among these antagonists in experimental studies on man has been *4-deoxypyridoxine* (Figure 5.23).

The increased requirement for vitamin B_6 during isoniazid treatment for tuberculosis, discussed previously, represents the antivitamin activity of *hydrazines*, which react with pyridoxal to form hydrazones. This sequence of reactions obviates the activity of pyridoxine kinase, which is necessary for coenzyme formation. Other antituberculosis drugs such as cycloserine upset normal pyridoxine biochemistry and ultimately lead to urinary excretion of the vitamin. Amphetamines, birth control pills (Mason *et al.*, 1969; Brown *et al.*, 1969; Brown, 1972), chlorpromazine hydrochloride, estrogen, and reserpine all seem to influence vitamin B_6-dependent enzyme activity as well as its distribution among various body tissues.

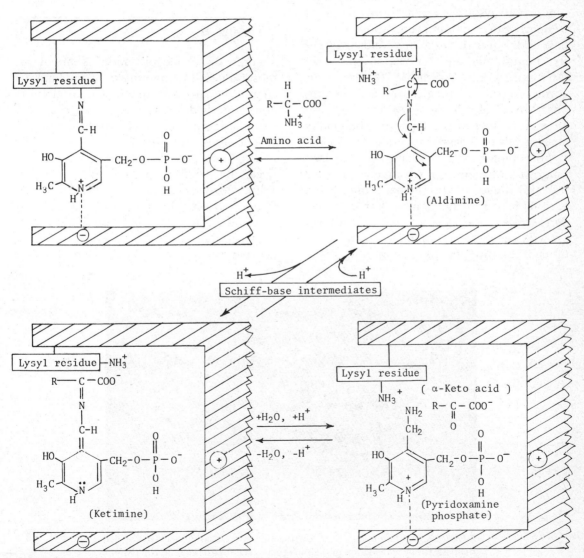

FIGURE 5.22. Proposed mechanism for a transamination reaction involving a vitamin B_6 coenzyme.

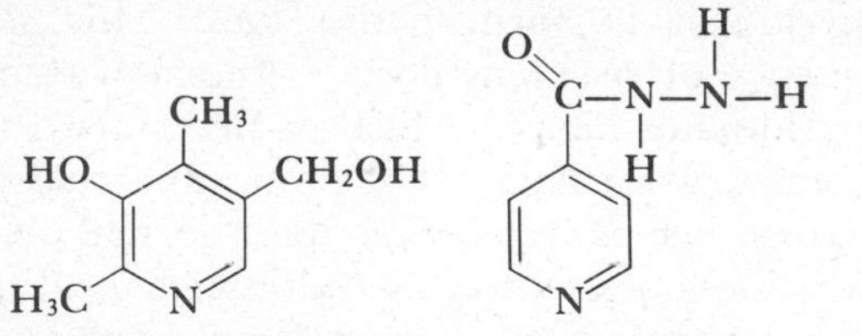

4-Deoxypyridoxine   **Isoniazid**

FIGURE 5.23. Some structures that exhibit vitamin B_6 antagonism.

Numerous assays for vitamin B_6 have been reported, but most methods have limited applications to certain samples. Chemical and physical assays are acceptable only for pure or nearly pure vitamin preparations, so they have little value for analyzing foodstuffs. Instrumental methods for vitamin B_6 assay (McCormick and Wright, 1970), thin-layer chromatography, and electrophoretic methods (Ahrens and Korytnyk, 1969) have also been described in the literature, but many analysts rely on bioassays for evaluating foods.

Microbiological assays have been favored as a result of their high sensitivities (0.5–1.0 ng) and ability to discriminate among the various forms of the vitamin. For example, *Saccharomyces carlsbergensis* has a uniform growth response to equimolar concentrations of the three forms of vitamin B_6; *Strep-*

tococcus faecalis is insensitive to pyridoxine, while pyridoxal displays 36% of pyridoxamine's growth effect; and, *Lactobacillus helveticus* responds only to pyridoxal (Strohecker and Henning, 1965).

Although acid aqueous solutions of vitamin B_6 can withstand 120°C heat for up to 30 minutes, the cooking or processing of foods can destroy initial vitamin concentrations by 50% or more. Freezing of foods at 0°F (-18°C) or less does not harm the vitamin and aids in its retention.

VITAMIN B_{12}

Vitamin B_{12} (cyanocobalamin) is one of the most remarkable vitamins known to man. Its chemical structures and biochemical functions underly the mechanisms responsible for maintaining normal health and nutrition.

The disease of *pernicious anemia* was recognized and described in historical medical circles by Addison in England as early as 1849. However, it was not until 1926 that Minot and Murphy demonstrated that half a pound a day of raw liver served as an effective tool in combating severe megaloblastic anemias and the accompanying central

FIGURE 5.24. Exemplary X-ray diffraction pattern that ultimately provided the basis for the elucidation of the molecular structure for vitamin B_{12}. (Kindly provided by Dr. Dorothy C. Hodgkin.)

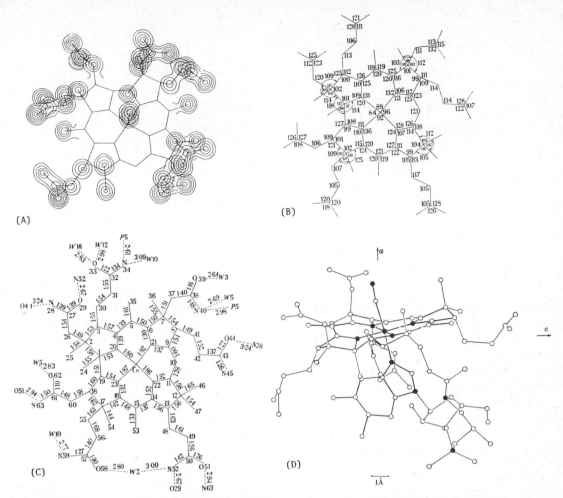

FIGURE 5.25. (A) Typical electron density map calculated from X-ray diffraction data (5.24) that indicates the molecular structure of the vitamin B_{12} corrin ring along with its major substituents. These density maps along with other analytical parameters allow the calculation of (B) key bond angles exhibited by the molecule as well as (C) bond lengths. The ultimate molecular skeletal and spatial model for the vitamin appears in illustration (D). (Kindly supplied by Dr. Dorothy C. Hodgkin, used with permission of The Royal Society (London) and adapted from *Proc. Roy. Soc. A*, Vol. 278, pp. 1–26, 1964; and *Proc. Roy. Inst.*, Vol. 42, No. 199, pp. 377–396, 1968.)

nervous system demyelination of the disease. A classic series of studies was designed to isolate the *anti-pernicious anemia factor* (anti-PA factor) in both the United States and England (Rickes *et al.*, 1948; Smith and Parker, 1948). The chemical structure of vitamin B_{12} was elucidated by Dorothy Hodgkin and co-workers through a series of complex X-ray diffraction analyses of wet and dry crystals (Figure 5.24). These diffraction patterns provided the mathematical basis for establishing electron density maps of the vi-

tamin (Figure 5.25A) and its molecular structure (Figures 5.25B–D and 5.26).

Today, the structure of the vitamin is still recognized as being unique among biochemical substances since it has the highest molecular weight of any nonprotein, nonpolymeric substance having natural origins.

Vitamin B_{12} has been ascribed no less than 50 names since its discovery, including all chemical and conventional vitamin trade names. Nearly all name variations refer to its most common cyanocobalamin form, which

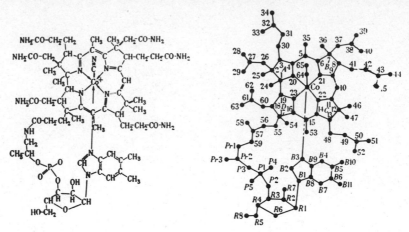

FIGURE 5.26. Planar structure and numbering scheme for the structure of vitamin B_{12} (cyanocobalamin). (Structures kindly provided by Dr. Dorothy C. Hodgkin.)

is widely used in therapeutic vitamin preparations and as a supplement to many foods. Important alternative names for vitamin B_{12} (cyanocobalamin) include the *Lactobacillus lactis Dorner factor* (LLD factor), *extrinsic factor,* and the *anti-pernicious anemia factor.* The cyanocobalamin form of the vitamin is the most commonly prepared pharmaceutical form, but this form occurs only rarely in nature under circumstances in which man or other organisms may be exposed to unusually high cyanide concentrations. In the case of pharmaceutical preparations, the cyanocobalamin form is a chemically stable artifact of the commercial isolation process (Table 5.6).

The *cobalamin* term refers to the entire structure of the cyanocobalamin molecule (Figure 5.26) with the exception of the cyanide group. However, the term *corrin ring* (Figure 5.27) refers solely to the tetrapyrrolic structure of the vitamin, with the exception of the central cobalt atom and all the associated side chains. Corrin rings differ from the tetrapyrrolic rings found in hemoglobin and chlorophyll because there are *only three methene bridges* connecting three pyrroles. The first and fourth pyrroles in the corrin ring are directly linked without a methene bridge to form a closed *tetrapyrrolic* structure.

The vitamin B_{12} molecule is also unique since it is the only known biologically important compound that requires cobalt. The cobalt may assume an oxidation state of $1+$, $2+$, or $3+$ depending on the environment of the corrin ring or its biochemical function. The cobalt atom is coordinated to all four pyrrolic nitrogens in the corrin ring in addition to a ribonucleotide that contains *5,6-dimethylbenzimidazole* (DMB). This ribonucleotide is of particular interest since DMB is not especially common among naturally occurring nucleotide bases. Furthermore, the DMB moiety is bound to ribose in a novel α-*N*-glycosyl linkage. This linkage is contrary to β-type linkages that typify most nucleotide structures.

The coordination bonds from cobalt to four pyrrolic nitrogens *plus* the DMB-containing ribonucleotide account for five of six possible coordination bonds. Depending on the chemical environment, the vitamin B_{12} structure may have a number of other groups *substituted for cyanide* that normally occupies the sixth coordination position. Alternative groups bound at this position include an —OH in *hydroxocobalamin;* —NO_2 in *nitrocobalamin;* and —HSO_3 in the case of *sulfitocobalamin.* Corresponding ammonia or histidine substitutions at the same position respectively yield *ammonia cobalichrome* and

Table 5.6. Chemical Characterization and Properties of Vitamin B_{12} (Cyanocobalamin)

Empirical formula	$C_{63}H_{88}CoN_{14}O_{14}P$
Molecular weight	1355.42
Melting point	Darkens at 210–220°C but not melted at 300°C
Hydroscopicity	Moderate and may absorb 12–13% water
Solubility	1.0 g dissolves in 80 mL of water; insoluble in acetone, chloroform, and ether
pH properties	Aqueous solutions are pH neutral
Optical rotation	$[\alpha]_{656}^{23} = 59 \pm 9°$ for dilute aqueous solutions

Stability: In the pH range of 4.5–5.0, cyanocobalamin can be autoclaved for up to 20 min at 120°C with minimal loss. Aqueous solutions of the vitamin can be greatly stabilized by the addition of ammonium sulfate. Destabilization and deterioration of the vitamin are encouraged by food constituents such as aldehydes, acacia, ascorbic acid, ferrous gluconate, ferrous sulfate, and vanillin.

Absorption spectrum: Absorption maxima in water appear at 361, 278, and 550 nm. Excessive amounts of cyanide added to cyanocobalamin under alkaline conditions converts the red colored vitamin to a purple color with a corresponding absorption spectrum shift to the right. Cyanide occupies the fifth and sixth coordination positions in the dicyanide complex and the visible spectrum absorption maximum becomes 582 nm. Consult Figure 5.28 for an illustration of this relationship.

Units: The vitamin is assayed and expressed in milligram, microgram, or nanogram units of the pure crystalline substance.

histidine cobalichrome. The structural variations outlined above are conventionally called *corrinoids* because they are related to vitamin B_{12}; they contain a corrin nucleus, but they are *not identical* to vitamin B_{12}, which is solely recognized as *cyanocobalamin*.

More complex organic substitutions for the cyanide ligand of vitamin B_{12} may include substituents such as 5'-deoxyadenosine. This form of the vitamin is known as *coenzyme* B_{12} (CoE-B_{12}). The CoE-B_{12} readily

decomposes in the presence of cyanide or light to respectively form cyanocobalamin or hydroxocobalamin.

Another complex variation in the structure of the vitamin B_{12} corrinoid includes *pseudovitamin* B_{12} and its coenzyme. The structure of pseudovitamin B_{12} was uncov-

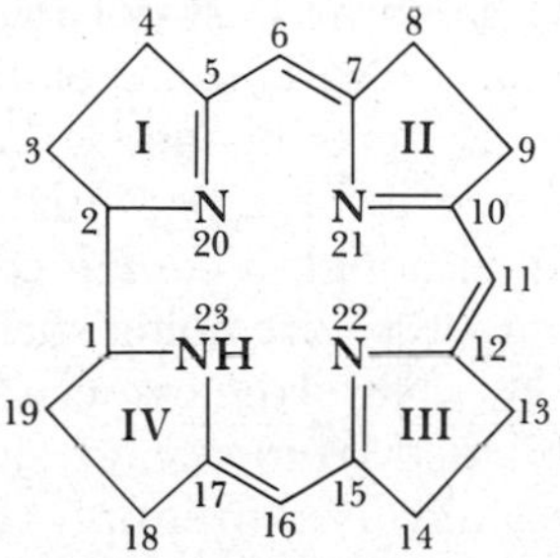

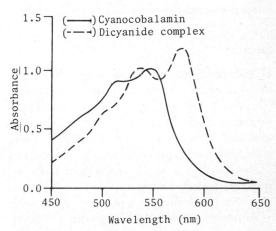

FIGURE 5.27. The structure of a *corrin ring* is characterized by the absence of a methene bridge between the first (I) and fourth (IV) pyrrolic rings, whereas all other rings are linked by methene bridges.

FIGURE 5.28. Spectrophotometric absorption curves for vitamin B_{12} (cyanocobalamin) and its dicyanide complex.

ered in the 1950s when H. A. Barker observed that *Clostridium tetanomorphum*, an anaerobic mud bacterium, contained adenine in place of DMB. The coenzyme form of this pseudovitamin B_{12} form also has a 5'-deoxyadenosyl group bound as a ligand to the cobalt atom.

These coenzyme forms of vitamin *and* pseudovitamin B_{12} are particularly important for methyl-transfer reactions because methyl groups can occupy the sixth coordination position of the cobalt atom as opposed to the 5'-deoxyadenosyl moiety. These forms of the vitamin are known as *methylcobalamins*.

Other variations on the theme of these corrinoids occur widely in nature and can include additional nitrogenous bases such as 2-methyladenine, guanine, uracil, and others.

Aside from the cobalamin analogues described, which vary in their respective heterocyclic bases, some naturally occurring vitamin B_{12}-related corrinoids lack both the cobalt atom and the heterocyclic bases. The biochemical function of these substances is not clear, and they have been isolated primarily from bacteria such as *Chromatium* (Toohey, 1965). Other novel forms of the vitamin have also been isolated from sewage sludge (Friedrich, 1979; Neujahr, 1955; Smith, 1965) and pharmaceutical cyanocobalamin preparations (Katada *et al.*, 1979).

Sources

Highly productive sources of vitamin B_{12} are limited almost entirely to microorganisms contained in soils, muds, and the rumen or intestinal flora of animals. Most plants contain only traces of the vitamin, and many reported concentrations for the vitamin in plants probably result from inaccurate assays. Although traces of vitamin B_{12} have been found in the roots of some plants such as tomatoes (0.002–0.01 ng/g) these concentrations are probably a reflection of vitamin B_{12} absorption from surrounding soil. Seaweed sources may indicate the presence of the vitamin, but its trace vitamin B_{12} amounts are largely the result of the indigenous or symbiotic superficial bacteria on the plant. Notable exceptions of this rule may include some *Cyanophycean, Phaeophycean,* or *Rhodophycean* algae, but even their synthetic abilities for vitamin B_{12} are uncertain.

Vitamin B_{12} is found in nearly every animal tissue and invariably originates from ingested dietary sources or commensal digestive-tract flora. Nonruminant, noncarnivorous animals satisfy all their vitamin B_{12} requirements by absorption from dietary sources or vitamin B_{12} synthesized by gut flora. Humans seem to be entirely dependent on dietary intake, since vitamin synthesis by intestinal flora occurs beyond the point of physiological vitamin absorption. Moreover, there does not seem to be any defined biochemical mechanism responsible for effecting the release of vitamin B_{12} from its synthesizing organisms. Burrowing or grubbing animals presumably obtain adequate amounts of the vitamin from the soil, while nonruminant herbivors either absorb the vitamin from the intestine or some ingested food source or engage in deliberate or incidental coprophagy.

Although liver, particularly ox liver, contains the most abundant stores of the vitamin in animals, kidney and brain tissues also have notable concentration levels. Vitamin B_{12} is also present in egg yolk, milk, and colostrum, which has appreciably higher concentrations than milk. Many seafoods contain significant stores of the vitamin, especially clams and related mollusks (Table 5.7).

During early efforts to develop commercial vitamin B_{12} production, sewage sludge was considered to be a potential source of the vitamin, although it contained high levels of vitamin B_{12} corrinoids such as pseudovitamin B_{12}. Nevertheless, crude vitamin extracts obtained from sewage sludge have been successfully used to supply the vitamin requirement of growing piglets at a 2% dietary concentration level. The presence of vitamin B_{12} and related cobalamins in sewage sludge appears to be inversely related to

Table 5.7. Mean Cyanocobalamin Concentrations for Selected Seafoods Based on Radioassay Methodology That Detects Total Cyanocobalamin

Seafood species	Fresh weight	Dry weight
Mya arenaria (soft-shelled clam)	186.60	1018.50
Gadus morhua (cod)	1.28	6.08
Melanogrammus aeglefinus (haddock)	0.64	4.67
Clupea harengus harengus (herring, salted)	13.15	26.79
Homarus americanus (lobster)	0.97	3.90
Pectin (sp.) (scallops)	1.86	10.18
Penaeus (sp.) (shrimp)	1.71	7.46

NOTE: Cyanocobalamin concentrations are expressed as micrograms per 100 g of edible seafood.

oxygen availability during the growth and reproduction of sewage sludge bacteria. The relatively high concentrations of the vitamin in silage are governed by similar factors.

Numerous foods have been tabulated according to their vitamin B_{12} content. It should be clearly recognized, however, that many reported concentration levels for this vitamin are actually a composite of procedurally diverse assay methods. These methods often lack steadfast analytical reproducibility or an ability to discriminate for the nutritionally significant form(s) of the vitamin.

Human Requirements and Deficiency Effects

Approximately 3.0 μg/day of dietary vitamin B_{12} is adequate for satisfying normal adult vitamin requirements. Because of geographic and socioeconomic distribution of high-grade fish and animal protein foods, vitamin consumption often ranges from 3 to 31 μg/day. Depending on the biochemical demands of the body, vitamin absorption can range from 20 to 70% of the initial vitamin intake.

Severe cases of pernicious anemia caused by a strict B_{12} deficiency will show significant improvement with B_{12} injections at concentration levels of 1.0 μg, but complete symptomatic remission may require parenteral injections of 0.5 to 2.0 μg/day. Liver stores of the vitamin may be reestablished with minimum doses of 2.0 to 4.0 μg/day for variable periods of time depending on the severity of the clinical symptoms. After the vitamin has saturated the body tissues, an intake of ~1.5 μg/day is sufficient for supplying most physiological and biochemical demands. Although the RDA for normal adults is 3.0 μg/day, pregnancy conditions warrant vitamin intake in the range of 4.0 μg/day, whereas the elderly require at least 4.0 μg/day.

The onset of pernicious anemia in humans is not directly related to chronic dietary B_{12} deficiencies but, rather, to malabsorption of the vitamin caused by a lack of the glycoprotein called *intrinsic factor* (IF) (M.W. = 50,000). This substance, secreted by cells located in the fundus of the stomach, eventually combines with vitamin B_{12} liberated from dietary sources to form a B_{12}–IF complex. The B_{12}–IF complex subsequently facilitates the absorption of the B_{12} in the distal ileum.

Vitamin B_{12} absorption is diagnostically assessed by using the conventional *Schilling test*. This clinical procedure requires oral administration of 0.5 to 1.0 μCi of radioactive ^{57}Co- or ^{58}Co-labeled cyanocobalamin along with a joint intramuscular injection of up to 1.0 mg of nonradioactive vitamin to saturate the B_{12}-binding capacity of the plasma and liver, in addition to assuring adequate conditions for vitamin excretion. After 24 h the liver is externally scanned in order to assess its accumulation of vitamin B_{12} or else the patient's serum, feces, and/or urine may be studied for levels of radioactivity. The normal urinary excretion for healthy individuals is more than 7.5% of the administered dose.

Clinical symptoms of vitamin B_{12} deficiency also include leukopenia and thrombocytopenia, which are reminiscent of folate deficiencies, but contrary to folate deficiencies, B_{12} deficiencies impair the maintenance and formation of myelin through biochemical reactions that have not been entirely elucidated. In severe cases, these problems are accompanied by lesions on the posterior and lateral spinal cord along with degeneration of the peripheral nerves. It is also interesting to note that vitamin B_{12} activities may involve disorders of the cranial nerves, as evidenced by *tic douloureux,* since excruciating neuralgic pain along the distribution of the fifth cranial nerve can be minimized, or eliminated in some cases, by increased availability of the vitamin. Serious deficiencies of the vitamin may initiate the onset of cardiovascular problems such as palpitations, but this is caused most often by anemia rather than nervous system degeneration. Other characteristic symptoms of both early and advanced deficiencies include tongue inflammation, papillary atrophy, and oral lesions on both the lips and mucous membranes.

Deficiency symptoms are uncommon among vegetarians who consume dairy products and eggs (*lacto-ovo vegetarians*), but this is not true for strict vegetarians who refrain from these sources of the vitamin.

Aside from strict vegetarians, B_{12} deficiency symptoms including megaloblastic anemia may be caused by the fish tapeworm known as *Diphyllobothrium latum*. This organism has an innate ability to absorb the vitamin from dietary substances transiting the intestine prior to host absorption.

Contrary to the well-documented range of deficiency symptoms that have appeared in biomedical literature over the years, vitamin B_{12} is definitely involved in many other organic disorders in which the mechanisms of vitamin activity have not been recognized.

Biochemistry

A prerequisite to biochemical activity of B_{12} in humans and many other higher animals involves its absorption from the digestive tract by the glycoprotein known as *intrinsic factor* (IF). This substance is secreted by the fundic cells of the stomach and has the novel ability to discriminatively bind vitamin B_{12} and promote its absorption by the intestine.

Since the molar ratio between B_{12} binding to IF is 2:1, it has been speculated that IF is dimeric or else it contains two binding sites for the vitamin on each molecule. The binding of vitamin B_{12} to IF seems to diminish the pH lability of the glycoprotein and makes it immune to digestive proteolytic enzyme activity.

The formation of a B_{12}–IF complex then promotes B_{12} absorption into the body by a physical absorption event between the B_{12}–IF complex and the microvilli located on the ileal mucosal cells. Binding of the B_{12}–IF complex to the membrane of these cells is probably caused by a mutual binding interaction between the glycoprotein moiety of IF and the mucopolysaccharide portion of the mucosal cells.

The binding phenomenon between the IF–B_{12} complex and mucosal cells seems to be augmented by the presence of calcium ions. The molar concentration balance between both IF and calcium ion concentration levels may determine successful absorption of B_{12} (Herbert *et al.,* 1964). The mechanism by which the vitamin transgresses the ileal membrane is still shrouded in mystery, although many conjectures have been postulated. These suppositions include: (1) the possible operation of a *releasing factor* that liberates B_{12} from the IF–B_{12} complex after its binding to the mucosal membrane; (2) *pinocytic absorption* of the B_{12}–IF complex or B_{12} alone; and (3) the biochemical *dismantling* of the extramucosal B_{12}–IF complex of B_{12}, followed by B_{12}'s subsequent intramucosal cell reassembly.

Regardless of the transport mechanism across the ileal membrane, it is certain that B_{12} is complexed with B_{12}-binding proteins known as *transcobalamins* (TC) before the vitamin can be distributed throughout the body. Three types of TC substances designated as TC I, TC II, and TC III have been identified, but the integration of their biochemical roles is not known. Both TC I and TC III contain sialic acid and, therefore, they are neuraminidase labile, whereas TC II is unaffected by this enzyme. TC II seems to be necessary for the transport of B_{12} to cells, and although definite roles for TC I and TC III remain to be established, relative concentration fluctuations in these three B_{12} binders seem to have some clinical implications. For example, TC I is synthesized and released by granulocytes at all stages of differentiation, but especially high levels are observed during chronic myeloid leukemia and acute promyelocytic leukemia. Elevated TC II levels, on the other hand, are believed to be generated along with the stimulation and proliferation of the mononuclear phagocyte system during autoimmune- or inflammatory diseases (e.g., lupus erythematosus, rheumatoid arthritis, colitus ulcerosa), chronic monocytic leukemia, and Gaucher's disease (Rachmilewitz *et al.*, 1979). High concentrations of TC III originating from mature granulocytes also seem to correspond with pathological problems such as polycythemia vera (PV) and nonleukemic leukocytosis. A high ratio of TC III : TC I clinically suggests PV and nonleukemic leukocytosis while discounting the probability of chronic myeloid leukemia (Rachmilewitz *et al.*, 1979).

Although TC II is involved in the intestinal absorption of B_{12}, the transport of 25% of plasma cobalamins after equilibrium, and the entry of the vitamin into metabolizing cells, biochemical roles for TC I are poorly understood. In fact, an absolute deficiency of TC I appears to be benign, although it carries up to 75% of the normal human plasma cobalamin.

Apart from TC I, II, and III, additional plasma proteins have been identified that bind cobalamins. These compounds include a variety of B_{12}-binding agents called *cobalophilins* or R-type binders; two separate and different immunoglobulins (Ig), which are respectively complexed with TC I and TC II; plus a vitamin B_{12}-binder that has α_2-globulin properties known as TC 0 (Carmel, 1979).

Important biochemical reactions: Vitamin B_{12} (cyanocobalamin) is the most common pharmaceutical form of the vitamin, but it is not the most biochemically active or important form. The biochemically active *coenzyme form of the vitamin* B_{12} (CoE-B_{12}) must be manufactured from cyanocobalamin. This conversion requires the two-step participation of an NADII–flavoprotein–disulfide (R–S–S–R) protein reducing system that ultimately yields a dithiol that reduces the structural intermediates leading to CoE-B_{12} formation (Figure 5.29).

After the formation of B_{12r} (Co^{2+}) and finally B_{12s} (Co^{1+}), the B_{12s} serves the substrate for an alkylation reaction with ATP. This reaction is directed by B_{12s} (Co^{1+}) adenosyltransferase, which catalyzes a C-5′ nucleophilic displacement on ATP and yields coenzyme B_{12} (CoE-B_{12}) plus an inorganic tripolyphosphate (Figure 5.29).

The importance of CoE-B_{12} is clearly recognized but many of its most important biochemical mechanisms remain to be completely established. Coenzyme B_{12} is a requirement for four categorical reactions that include (1) *intramolecular shifts* of a hydrogen atom and a substituent group on adjacent carbon atoms, (2) *carbon–oxygen bond cleavages,* (3) *carbon–carbon bond cleavages,* (4)

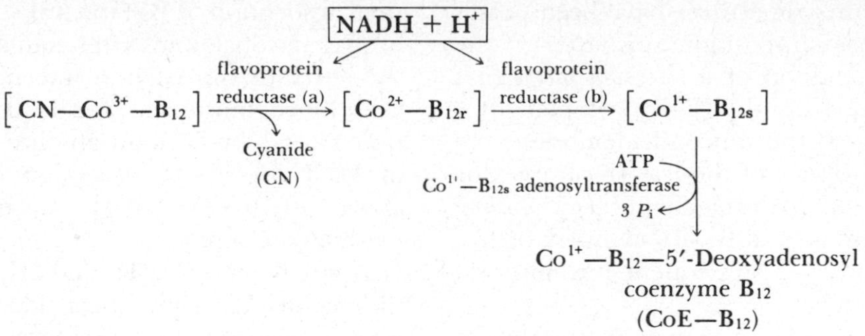

$$\boxed{\text{NADH} + \text{H}^+}$$

$$\left[\text{CN—Co}^{3+}\text{—B}_{12} \right] \xrightarrow[\text{Cyanide} \atop (\text{CN})]{\text{flavoprotein} \atop \text{reductase (a)}} \left[\text{Co}^{2+}\text{—B}_{12r} \right] \xrightarrow{\text{flavoprotein} \atop \text{reductase (b)}} \left[\text{Co}^{1+}\text{—B}_{12s} \right]$$

$$\text{Co}^{1'}\text{—B}_{12s} \text{ adenosyltransferase} \left. \right) \begin{array}{c} \text{ATP} \\ 3\,P_i \end{array}$$

$$\text{Co}^{1+}\text{—B}_{12}\text{—5}'\text{-Deoxyadenosyl}$$
$$\text{coenzyme B}_{12}$$
$$(\text{CoE—B}_{12})$$

FIGURE 5.29. Two-step reduction of cyanocobalamin to Co^{1+}-B_{12s} followed by the formation of coenzyme B_{12} mediated by Co^{1+}-B_{12s} adenosyltransferase.

methyl group transfer reactions, and a host miscellaneous reactions.

1. Coenzyme B_{12} is required for the *1,2-shift* of a hydrogen atom on one carbon atom to an adjacent intramolecular carbon atom, along with a corresponding *2,1-shift* of the substituent group on the other carbon (Figure 5.30).

These reactions probably require a preliminary removal of a hydride ion from the substrate by a specific enzyme. This event then promotes dislocation of the methylene carbon linking the cobalt atom of CoE-B_{12} with its 5′-deoxyadenosyl moiety. The hydride ion is then received by C-2 along with a concomitant transfer of the C-2 substituent group to C-1 (Figure 5.30).

2. The cleavage of carbon–oxygen bonds as demonstrated by *dioldehydrase* reactions also requires the presence of CoE-B_{12}. The postulated mechanism for these reactions has been widely based on the CoE-B_{12}–dependent conversion of 1,2-propanediol to propylaldehyde (Figure 5.31).

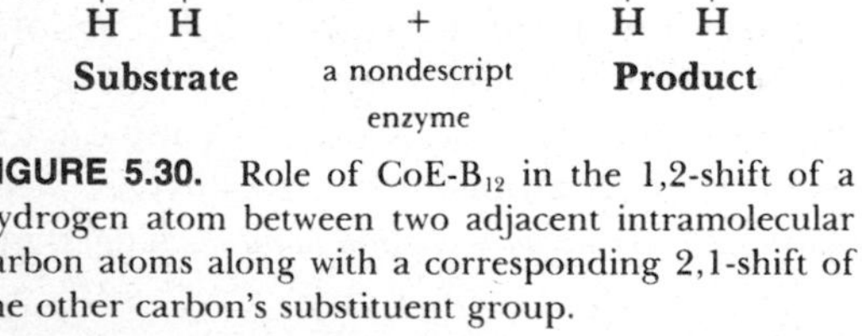

Theorized mechanisms for such reactions depend on the release of 5′-deoxyadenosine from the CoE-B_{12} structure, but the mechanics of this release mechanism and the relinking of the 5′-deoxyadenosyl moiety to the corrinoid structure are open to wide speculation. Most evidence, however, points to the retention of the 5′-deoxyadenosyl group on the enzyme protein by means of hydrogen bonding (Metzler, 1977).

3. Carbon–carbon bond cleavages, such as those required for the enzymatic interconversion of methylmalonyl-CoA to succinyl-CoA by *methylmalonyl-CoA mutase*, require the presence of CoE-B_{12}. As in the previous discussion, this mechanism is far from lucid, but a proposed mechanism has been contrived on the basis of current evidence (Figure 5.32).

4. Methyl group transfer reactions are crucial for the operation of many biochemical pathways. Metabolic schemes in many species can effectively transfer methyl groups as the methyltetrahydrofolate (N^5-methyl THF) derivative of folic acid. For other organisms, methyl transfer reactions necessarily involve the joint interaction of both folic acid and CoE-B_{12}. This possibility is best exemplified by the activity of N^5-*methyl THF-homocysteine transmethylase,* which singularly effects methonine formation from homocysteine in higher plants, fungi, and some bacteria. In man and some species of bacteria, the same reaction will not occur unless

X = Substituent hydroxyl, amino, alkyl or carbonyl group

$$\underset{\text{Substrate}}{\overset{\displaystyle \text{H} \;\boxed{\text{X}}}{\underset{\displaystyle \text{H} \;\; \text{H}}{\text{R—C}_1\text{—C}_2\text{—R}'}}} \xrightarrow[\substack{+ \\ \text{a nondescript} \\ \text{enzyme}}]{\text{coenzyme B}_{12}} \underset{\text{Product}}{\overset{\displaystyle \boxed{\text{X}}\;\text{H}}{\underset{\displaystyle \text{H}\;\;\text{H}}{\text{R—C}_1\text{—C}_2\text{—R}'}}}$$

FIGURE 5.30. Role of CoE-B_{12} in the 1,2-shift of a hydrogen atom between two adjacent intramolecular carbon atoms along with a corresponding 2,1-shift of the other carbon's substituent group.

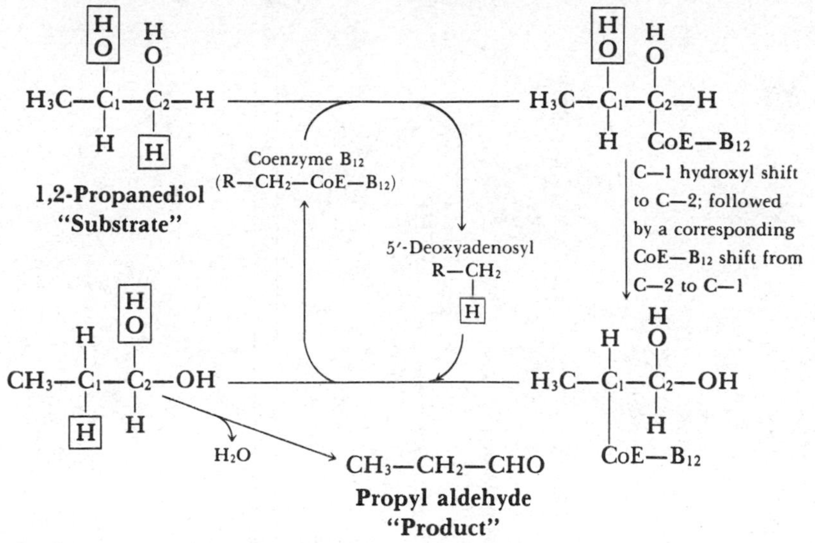

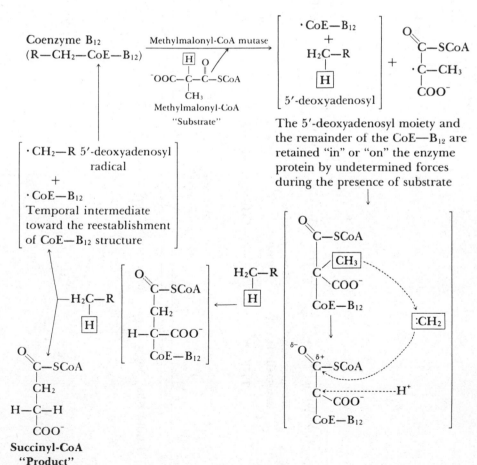

FIGURE 5.31. Role of CoE-B_{12} in the conversion of a diol into its corresponding aldehyde.

FIGURE 5.32. Proposed mechanism for the role of CoE-B_{12} during enzymatic cleavage of a carbon–carbon bond.

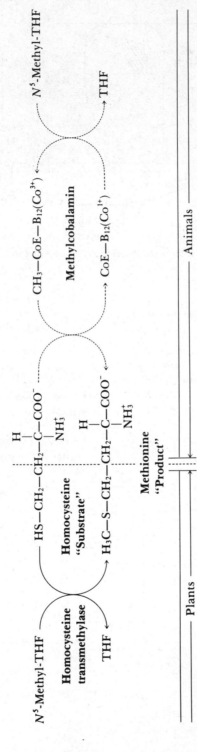

FIGURE 5.33. Methyl group transfer reactions in plants and animals.

methyl cobalamin (CH_3–CoE–B_{12}) is present (Figure 5.33).

Many other types of methyl transfer reactions are found in nature but they are often limited to select groups of microorganisms while still other types of methyl transfer reactions occur only under certain conditions in man and other species.

Methane production. Methane can be produced in significant concentrations by *methanogenic* bacteria. These organisms transform methyl groups contained in acetate, methanol, or N^5-methyl-THF into methane, or they produce the gas by reducing carbon dioxide, formaldehyde, or formate. In all cases, methane production in these organisms requires the presence of a vitamin B_{12}-related corrinoid such as *pseudovitamin* B_{12} (ψ-vitamin B_{12}) or *Factor A*. These vitamin B_{12} analogues respectively contain adenine and 2-methyladenine in place of the DMB moiety in the nucleotide portion of vitamin B_{12}.

According to mechanisms that are not clear, these vitamin B_{12} analogues form methylcobalamin intermediates prior to methyl group reduction and methane pro-

duction. Methane formation in this step does not occur as a simple reduction. It requires the intervention of a low-molecular-weight, heat-stable dialyzable compound called coenzyme M (CoE-M) to produce a methyl-CoE-M derivative before methane formation. Studies indicate that CoE-M is present in all methanogenic organisms and there is little doubt that it is required by all organisms that produce methane as a primary metabolic end product (McBride and Wolf, 1971). The oxidized form of CoE-M is believed to undergo reduction, as indicated in Figure 5.34, after which methylcobalamin donates its methyl group to CoE-M. The methyl-CoE-M may then be reduced by hydrogen in a reaction that requires ATP plus Mg^{2+} or additional oxidoreductase reactions.

Acetate formation. Acetate formation from carbon dioxide, along with the cooperation of a reducing system, earmarks an interesting energy-producing mechanism found in some anaerobic bacteria such as *Clostridium aceticum*. As is the case in many B_{12}-related reactions, the mechanism has not been deciphered, but it seems clear that a carboxy-

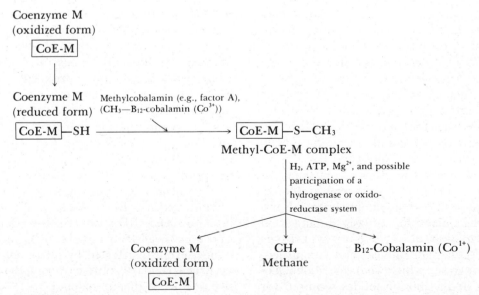

FIGURE 5.34. Involvement of coenzyme M (CoE-M) with methylcobalamin during methane production.

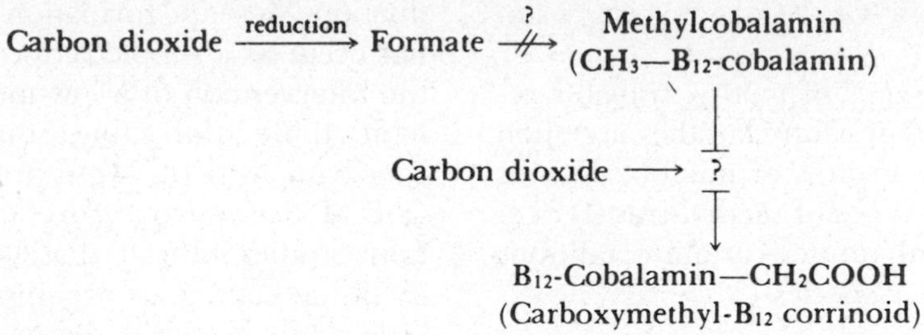

$$\text{Carbon dioxide} \xrightarrow{\text{reduction}} \text{Formate} \xrightarrow{\;\;?\;\;} {\LARGE/\!\!/} \begin{array}{c} \text{Methylcobalamin} \\ (CH_3\text{—}B_{12}\text{-cobalamin}) \end{array}$$

$$\text{Carbon dioxide} \longrightarrow \;?$$

$$\begin{array}{c} B_{12}\text{-Cobalamin—}CH_2COOH \\ (\text{Carboxymethyl-}B_{12} \text{ corrinoid}) \end{array}$$

FIGURE 5.35. Acetate formation in *Cl. aceticum* from carbon dioxide requires the formation of a carboxymethyl-B_{12} cobalamin intermediate.

methyl-B_{12} cobalamin intermediate may be formed prior to acetate formation (Figure 5.35).

Miscellaneous biological methylation reactions. In addition to methylation reactions required for the synthesis of biochemical compounds, methylcobalamins also contribute methyl groups to inorganic substances such as arsenic, mercury, selenium, and tellurium. Methylation reactions of mercury and arsenic are of special importance since they are common environmental pollutants originating from industrial processes or agricultural insecticides for which they are used as stomach poisons. Living organisms biochemically cope with these toxic substances by methylating them into mono- or dimethylated derivatives of mercury or arsenic. Unfortunately, this conversion often enhances the bioaccumulation of organometallic metabolites in many organisms to a point where species-specific lethal effects may result. A schematic survey of methylation reactions for mercury and arsenic have been detailed in Figures 5.36(A) and (B).

In addition to the previous reactions it is clear that vitamin B_{12} coenzymes are necessary adjuncts to ethanolamine-lyase, L-β-lysine mutase, ornithine mutase, glutamic acid transformation of threo-β-methyl-L-aspartate by glutamate mutase, and ribonucleotide reductase, which reduces ribonucleotides to deoxyribonucleotides required for DNA synthesis.

Miscellaneous Considerations

Vitamin B_{12} provides an example of the classic difficulties in correlating a vitamin's clinical deficiency symptoms, such as pernicious anemia, with its universally important biochemical role within individual cells. In spite of 30 years of study, it should be realized that paucities of data exist in matters as simple as this vitamin's tissue distribution in animals during states of normal health and disease. Myasishchiva *et al.* (1979) have indicated that proliferative characteristics of some tumors may be linked to methyl and 5'-deoxyadenosyl cobalamin biosynthesis in malignant cells as well as alterations in the activity of cobalamin-dependent methionine synthetase. However, it is difficult to assess the significance of these findings since there is a lack of hard-core data concerning nominal concentration levels for 5'-deoxyadenosyl-, methyl-, and hydroxocobalamins in healthy cells. Furthermore, there is no doubt that currently recognized mechanisms for this vitamin hardly represent the tip of an iceberg with respect to its overall involvement in the maintenance of good health and the production of selected food resources.

Vitamin B_{12} does not appear to cause any notable toxicity effects when its nutritional intake exceeds RDA values. Moreover, the nutritional availability of vitamin B_{12} in foods is not seriously affected by other naturally occurring food constituents during the practice of normal dietary habits.

Herbert and others (1974, 1979) have

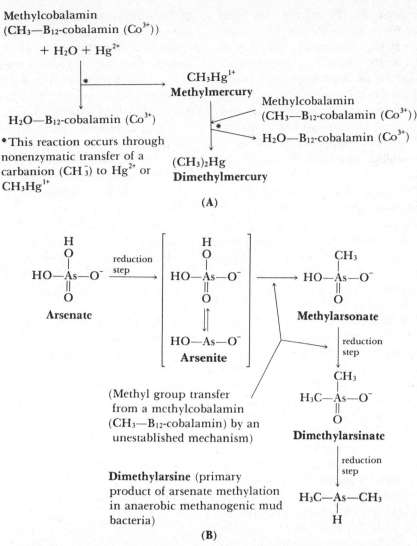

FIGURE 5.36 A and B. Schematic methylation sequences for mercury and arsenic-containing substances.

claimed that 95% of all dietary B_{12} may be destroyed in a meal that contains 500 mg of ascorbic acid, but these studies have met with experimental counterclaims by Newmark *et al.* (1976) and Thenen (1979). Although iron and some other metals may act as deterrents to the destruction of B_{12} by vitamin C, the possibility remains that the multigram ingestion of vitamin C along with vitamin B_{12} may provide the basis for eventual vitamin B_{12} deficiency. It is also interesting to note that

combinations of thiamine and nicotinic acid may also destroy B_{12} in solutions, but their destructive influence on B_{12} as separate entities is not significant (Smith, 1965). Nearly all of the conceived notions about the nutritional incompatabilities of vitamin B_{12} are still widely disputed and inconclusive.

The reliable assay of vitamin B_{12} at picogram per milliliter levels has been a major stumbling block for clinical, nutritional, and biochemical investigations. Direct spectrophotometric assays for the vitamin have been devised, but these methods are limited almost entirely to the pharmaceutical assay of B_{12} and its quality control during production. The advent of microbiological assays that employ *Lactobacillus lactis* Dorner (LLD), *Lactobacillus leichmanii, Euglena gracilis,* and other organisms have provided requisite quantitative growth responses at the picogram per milliliter concentration levels of vitamin B_{12}. Unfortunately, however, these bioassays also respond to thymidine, deoxyribonucleosides, and the unpredictable presence of vitamin B_{12} analogues. This difficulty has invariably led to inaccurate assessments of the vitamin in many biochemically complex foodstuffs where bioassays have been employed.

Routine clinical B_{12} assays for tissues, body fluids, and nutritional evaluations of patients, as well as food resources, have generally relied upon microbiological assays in the past, but these techniques have yielded in more recent years to routine radioassays for the vitamin. Radioassays for vitamin B_{12} are based on the competitive binding of radioactive $[^{57}Co]B_{12}$ and nonradioactive B_{12} to purified IF. For these assays, a measured amount of liquid sample containing B_{12} is mixed with a fixed amount of $[^{57}Co]$vitamin B_{12} (cyanocobalamin) and the mixture is boiled at an acid pH (4.0) in the presence of cyanide ions. This step releases the vitamin from its *in vivo* binding proteins and converts all the vitamin to its cyanocobalamin form. The nonradioactive cyanocobalamin in the liquid sample is then incubated with a fixed amount of radioactive $[^{57}Co]$ cyanocobalamin while in the presence of IF. This incubation permits competitive but nondiscriminative binding of both cyan-

ocobalamin types to the IF. After incubation, the radioactive and nonradioactive forms of cyanocobalamin bound to IF are then separated from the unbound forms by adsorption of the unbound forms onto charcoal or some other agent. Radioactive $[^{57}Co]$ cyanocobalamin bound to IF is counted by using a gamma spectrometer. Finally a standard curve is also prepared that uses standard nonradioactive cyanocobalamin concentrations in place of the sample. For increases in nonradioactive B_{12} standard concentrations (or samples), an inverse relationship will be noted for radioactive B_{12} bound to IF. A schematic representation of the binding among these assay ingredients has been outlined in Figure 5.37, whereas Figure 5.38 details a typical set of graphic data.

Although vitamin B_{12} radioassays are simple in principle, they are highly sensitive and subject to many analytical pitfalls and inherent errors. This is true especially where serum samples and food specimens serve as assay subjects (Kolhouse *et al.,* 1978; Beck, 1979). High-pressure liquid chromatographic methods of analysis hold future promise for the assay of vitamin B_{12} and its cobalamin analogues, but most current methods lack the necessary picogram sensitivity of radioassays.

NIACIN

Niacin, also known as nicotinic acid, is one of the most important vitamins in human and animal nutrition. Although an understanding of the biochemical function for niacin has been a product of relatively recent

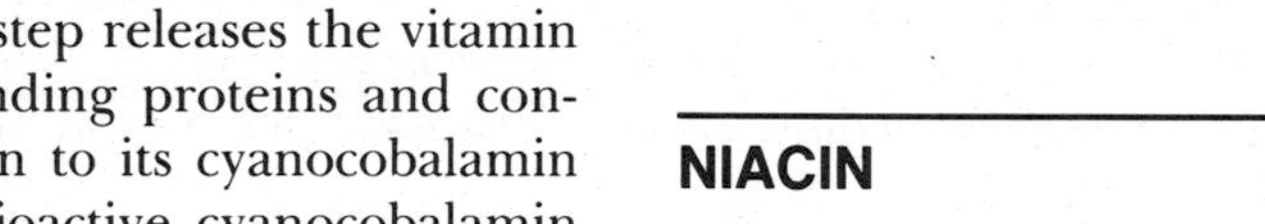

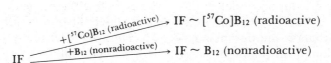

FIGURE 5.37. Purified intrinsic factor (IF) nondiscriminatively binds both radioactive $[^{57}Co]$cyanocobalamin and nonradioactive cyanocobalamin.

*Counts per minute (cpm) for a fixed amount of radioactive B_{12} that has competitively bound to a fixed amount of intrinsic factor (IF) in the presence of various standard concentrations of nonradioactive B_{12}.

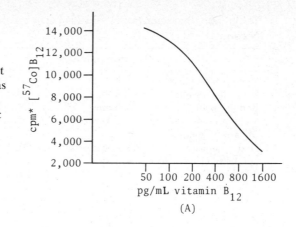

(A)

Linear regression analysis showing the correlation of B_{12} concentrations obtained from radioassay with intrinsic factor (IF) and a microbiological assay.

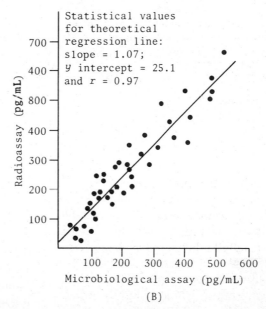

(B)

FIGURE 5.38 A and B. (A) Typical standard curve used as a basis for evaluating unknown cyanocobalamin (vitamin B_{12}) concentration levels. (B) Comparability of radioassays and microbiological assays for the evaluation of standard cyanocobalamin concentrations; the concentration range of B_{12} on the ordinate and abscissa is in pg/mL.

research, the chemical and physical properties of the substance have been defined for many years.

Alternative literature designations for niacin are Akotin, Nicacid, Niacangin, Nicoanacid, Nicotinipca, Nicyl, vitamin PP—pellagra preventive factor, antipellagra vitamin, and pyridine-β-carboxylic acid. The term *niacin* is the official name for the nicotinic acid form of the vitamin, but the amide derivative of niacin represents the most im-

portant biochemical form (vitamin B_3) (Table 5.8 and Figure 5.39).

This vitamin exists in cells as *nicotinamide nucleotide coenzymes*. Two coenzymes are of interest, *nicotinamide adenine dinucleotide* (coenzyme I, or diphosphopyridine nucleotide, DPN) and *nicotinamide adenine dinucleotide phosphate* (coenzyme II, or triphosphopyridine nucleotide, TPN). Exclusive of the synonyms for the coenzyme nucleotides in parentheses above, these coenzymes are re-

Table 5.8. Characterization and Properties of Nicotinamide and Nicotinic Acid

	Nicotinic acid	Niacinamide
Empirical formula	$C_6H_5O_2N$	$C_6H_6ON_2$
Molecular weight	123.11	122.12
Boiling point	Sublimes	150–160°C at 5×10^{-4} mm
Melting point	235.5–236.5°C	128–131°C
pH at 1 % solution	3.0	6.0
Hydroscopicity	Nonhygroscopic	Slightly hydroscopic
Solubility (g/100 mL of solvent)		
Water (25°C)	1.67 g	100 g
Ethyl alcohol	0.73 g	66.6 g
Glycerin		10 g
Ethyl ether	Insoluble	Slightly soluble
Stability: Both substances display moderate resistance to alkaline and acid conditions as well as heat.		
Absorption maximum	385 nm	212 nm
Units: Vitamin is measured in terms of milligrams of the pure synthesized substance.		

spectively recognized as NAD$^+$ and NADP$^+$ in current literature (Figure 5.40).

Sources

Niacin is found widely among vegetable foods, whereas nicotinamide occurs in many foods derived from animal sources. Appreciable amounts of the vitamin exist in liver, yeast, white meats, alfalfa, leguminous plants, corn, and whole cereals. Cereal-based staples can often act as a dietary source of the vitamin if the cereal grain is whole or lightly milled. Whole wheat flour, for example, contains about 60 μg/g niacin while the same unenriched patent flour may contain only 16 μg/g. Adequate dietary availability of niacin is ensured in many manufactured products by admixing the vitamin to the final milled grain.

Tuberous roots, fruits, common vegeta-

Nicotinic acid   **Nicotinamide**

FIGURE 5.39. Structures for nicotinic acid and nicotinamide.

bles, and milk are relatively poor sources of the vitamin.

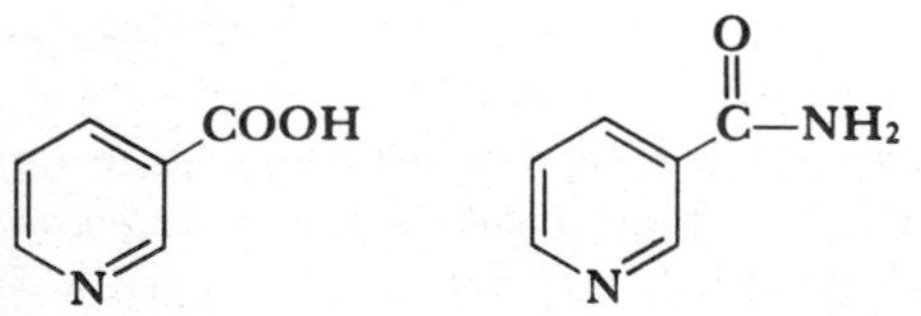

Human Requirements and Deficiency Effects

The human body has a limited capacity for synthesis of niacin from tryptophan unlike some other vitamin substances. Experiments show that ~60 mg of dietary tryptophan yields about 1.0 mg of niacin. Based on a United States dietary average of 600 mg/day for tryptophan, this conversion augments the dietary niacin requirement by up to 10.0 mg. Therefore, human populations with substandard dietary protein intake and correspondingly low dietary tryptophan levels of less than 150 mg are apt to develop pellagra.

Pellagra is often recognized as the *Four-D disease* since it is associated with *dermatitis, diarrhea, dementia,* and *death.* The initial evidence of dermatitis and lesions materializes on portions of the body exposed to solar radiation or heat. The skin lesions progress from an initial state of erythema or hyperpigmentation to a scaled and cracked condition that undergoes desquamation. Evidence of lesions on the upper chest to the lower neck are pathognomonic of advanced

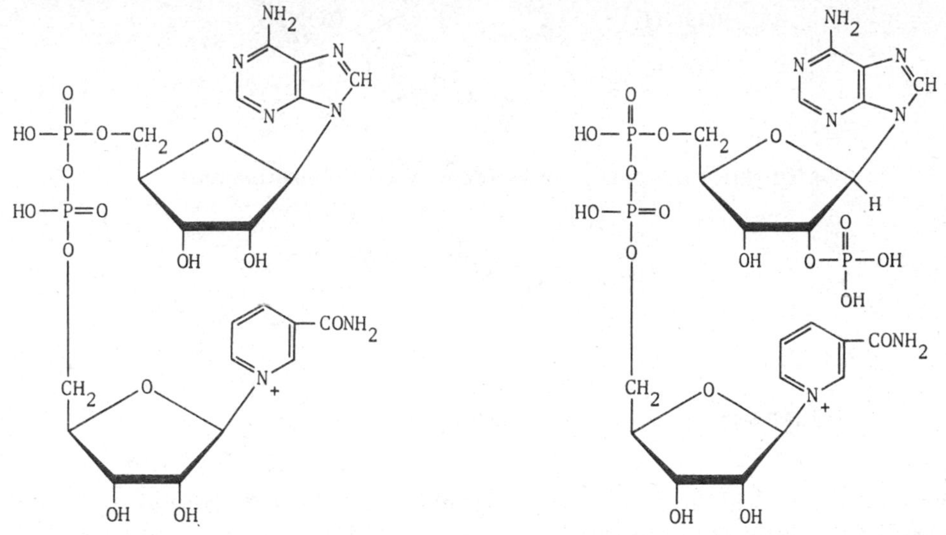

Nicotinamide adenine dinucleotide (NAD+)   Nicotinamide adenine dinucleotide phosphate (NADP+)
   Diphosphopyridine nucleotide (DPN+)   Triphosphopyridine nucleotide (TPN+)
           or Coenzyme I   or Coenzyme II

FIGURE 5.40.   Structures for the nicotinamide coenzymes, NAD$^+$ and NADP$^+$.

pellagra. This phenomenon is known as *Casal's Necklace*. Although pellagra symptoms such as stomatitis, cheilosis, and raw tongue parallel those of riboflavin, abdominal pain and diarrhea also accompany pellagra. Furthermore, the loss of memory, hyperirritability, and insomnia all attest to nervous system involvement in severe niacin deficiencies.

Advanced cases of pellagra may be successfully treated by administering 50 to 100 mg of nicotinamide for three or four days, three times a day by an intramuscular injection route. Nicotinamide is then given orally in similar quantities until recovery is evident along with a minimum of 100 g/day of high-quality protein.

The recommended allowances for tryptophan are expressed as *niacin equivalents* (NE) since tryptophan is convertible *into* niacin. Therefore, 1.0 NE corresponds to at least 60.0 mg of available tryptophan, which is convertible into 1.0 mg of niacin. According to conventional dietary allowances, males and females, respectively, require 18.0 and 13.0 mg NE. The female allowance is increased by 2.0 mg during the second and third

trimesters of pregnancy, while an additional 5.0 mg is encouraged during the lactation period. As in the case of riboflavin, the niacin requirement for children relatively supersedes adult requirements. A range of 5.0 to 9.0 mg gradually increasing to 18.0 mg is respectively encouraged for infants and adolescent boys.

From caloric perspectives, 6.6 mg of niacin are required for 1000 kcal and not less than 13.0 mg based on a 1000 to 2000-kcal diet. The minimum dietary allotment for prevention of pellagra symptoms is about 4.4 mg/1000 kcal/day, which corresponds to 8.8 mg for a 2000-kcal diet.

The toxicity of nicotinic acid and nicotinamide is marginal for nearly all animals. Rats for example, display an LD$_{50}$ of 7.0 g/kg. In nearly all cases, excessive amounts of nicotinamide are discharged into the urine as methylated derivatives of nicotinamide. These include *N*-methylnicotinamide, 2- or 6-pyridone derivatives of *N*-methylnicotinamide, nicotinuric acid, and a wide spectrum of related metabolites (Figure 5.41).

Aside from toxic effects at high concen-

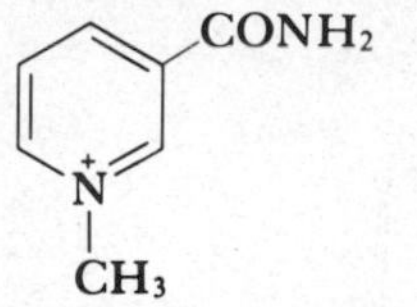

N-Methylnicotinamide 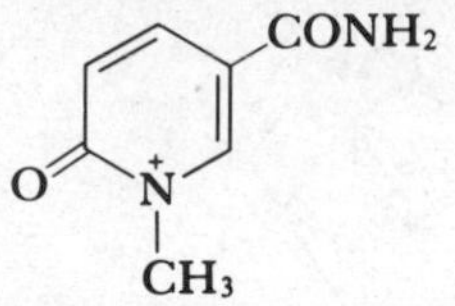   6-Pyridone-N-methylnicotinamide

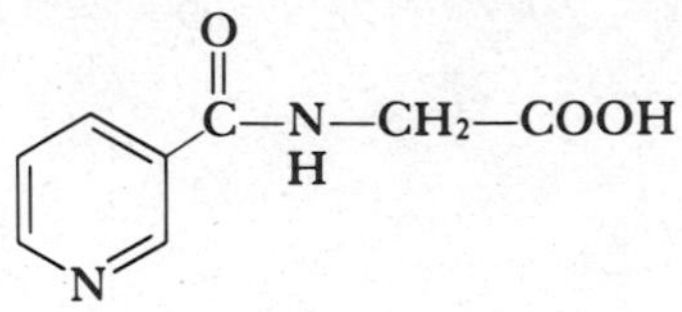

Nicotinuric acid

FIGURE 5.41. Some notable metabolites of nicotinic acid.

tration levels, nicotinic acid (not nicotinamide) has been used as an effective vasodilator. Although it does have side effects such as dizziness, nausea, and flushing of the skin, these symptoms are recognized as nonharmful, temporary phenomena that are unrelated to the vitamin's toxicity. Nicotinamide (not nicotinic acid), on the other hand, can lower blood cholesterol and triacylglycerols when administered in high oral concentrations.

Biochemistry

Nicotinamide nucleotides serve as coenzymes for dehydrogenase enzymes, which mediate a wide variety of oxidation-reduction reactions.

The importance of NAD^+ is demonstrated by its participation with alcohol dehydrogenase, which catalyzes ethanol oxidation to acetaldehyde.

$$CH_3CH_2OH + NAD^+ \xrightleftharpoons{\text{alcohol dehydrogenase}}$$
$$CH_3CHO + NADH(H^+)$$

Based on the law of mass action, the theoretical equilibrium constant for this reaction can be expressed as

$$K_{eq} = \frac{[CH_3CHO]\,[NADH]}{[CH_3CH_2OH]\,[NAD^+]}$$

Since calculation of the K_{eq} at pH 7.0 and 9.0 reveals respective values of 10^{-4} and 10^{-2}, it is clear that the K_{eq} for the reaction is de-

pendent on H^+ concentrations. This is not unexpected because H^+ is a product of ethanol oxidation. According to the law of mass action, a low hydrogen concentration or a high pH can only encourage a left-to-right reaction that yields more acetaldehyde and $NADH(H^+)$. Alternatively, high concentrations of hydrogen ions or low pH would shift the reaction from right to left, thereby yielding ethanol and NAD^+.

The presence of the hydrogen ion during this reaction stems from the fact that ethanol conversion to acetaldehyde requires a loss of two hydrogen atoms. This oxidation of ethanol may produce *two hydrogen atoms with their full complement of electrons:* two protons and two electrons, or a single proton (H^+) accompanied by a hydride ion (H:).

A study of the nicotinamide moiety of NAD^+ or $NADP^+$ shows that it can be reduced by one proton and two electrons (Figure 5.42). This reduction could occur by a single-step addition reaction of H: to the pyridine ring (*para* to the nitrogen) while the H^+ counterion resulting from hydride formation simply goes into solution (Figure 5.43).

Another possible mechanism proposed by Pullman in 1954 suggests that nicotinamide reduction is caused by its behavior as a quaternary base. As one example of this mechanism, NAD^+ is reduced when one of the two hydrogens released from ethanol enters the pyrimidine ring of the nicotinamide moiety (*para* to the nitrogen) (Figure 5.44).

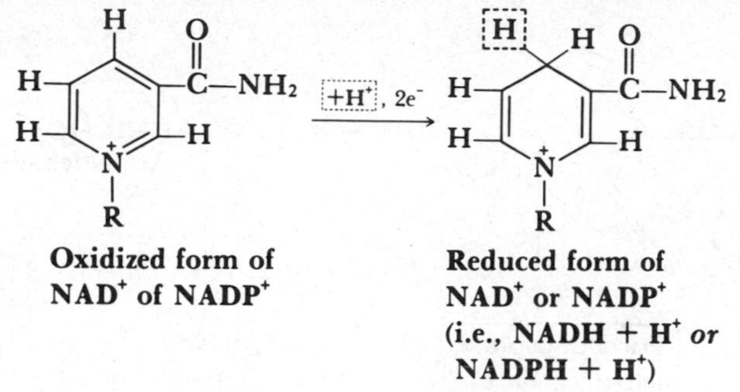

FIGURE 5.42. Oxidized and reduced forms of the pyridine moiety present in both NAD^+ and $NADP^+$.

The second hydrogen from ethanol is theorized as splitting into a proton (H^+) and an electron. This electron then pairs off with the positively charged nitrogen, and the proton becomes associated with a residue of phosphoric acid on the reduced NAD^+ structure.

The actual incorporation of hydrogen into the pyrimidine ring of NAD^+ or $NADP^+$ is spatially specific *depending on the type of dehydrogenase enzyme*. That is, hydrogen atoms oriented toward the front of the nicotinamide moiety, as drawn above, characterize *A-type dehydrogenases* such as alcohol dehydrogenase of yeasts and the cardiac isoenzyme of LDH. The so-called *B-type dehydrogenases* such as hepatic glucose dehydrogenase and yeast glucose 6-phosphate dehydrogenase exhibit a spatial orientation of the reducing hydrogen atom that is contrary to the *A-type enzymes*.

Regardless of the presence of NAD^+ or $NADP^+$, dehydrogenase enzymes linked to these nicotinamide nucleotides mediate reactions that may be reversible or irreversible. Reaction reversibility depends largely on mass action affects as well as the equilibrium constant for a given reaction.

Reduced nicotinamide nucleotides (i.e., NADH and NADPH) do not remain in a

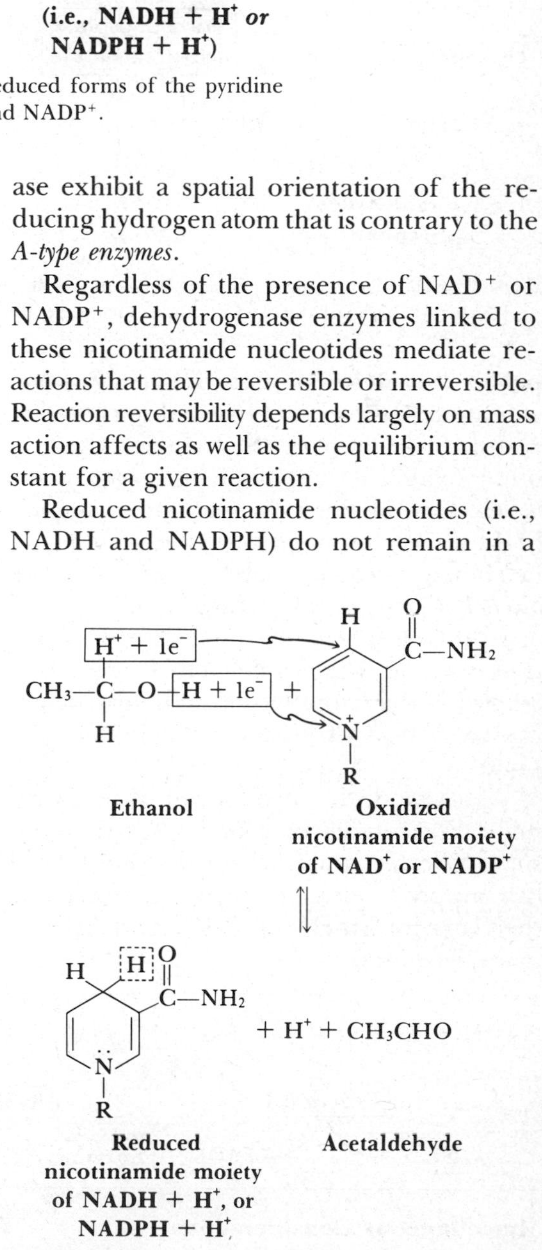

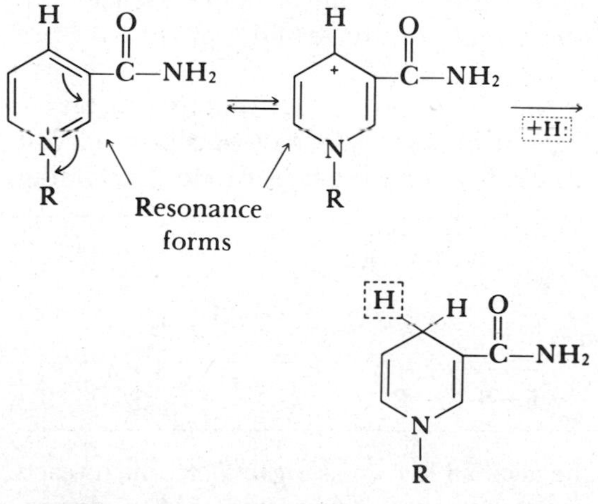

FIGURE 5.43. Reduction of the nicotinamide moiety by a hydride species.

FIGURE 5.44. Oxidation of ethanol to acetaldehyde along with the reduction of NAD^+ ($NADP^+$).

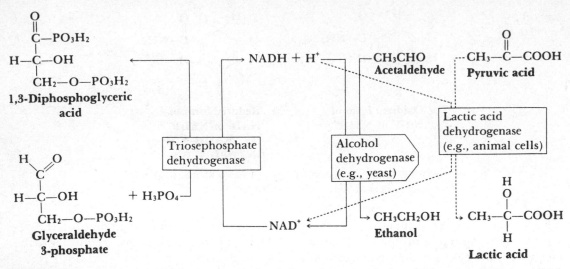

FIGURE 5.45. Examples of coupling between biochemical reactions occurring in animals and yeast where oxidized and reduced forms of reduced pyridine nucleotides are critical.

permanent reduced state after a substrate oxidation. Instead, the reduced nucleotides readily yield their hydrogens (electrons) to other oxidized compounds in a *coupled reaction*. One of the most common examples of these interactions involves the coupling of biochemical reactions by *triosephosphate* and *alcohol dehydrogenases* (Figure 5.45).

In this instance, the acetaldehyde reduction to ethanol by alcohol dehydrogenase is *coupled* to the triosephosphate dehydrogenase oxidation of glyceraldehyde 3-phosphate.

Apart from the involvement of nicotinamide nucleotides in coupled reactions, it should be recognized that these substances can supply a source of reducing power for the flavin prosthetic groups (e.g., FAD) on many enzymes:

nicotinamide, pyridine 3-sulfonamide, 7-aminonicotinamide, and 4-acetylpyridine.

Diets high in maize reportedly increase daily requirements for niacin. The reasons for this apparent increase in demand are clouded, but they have been attributed to a naturally occurring niacin-binding substance or an antiniacin agent. These compounds have not been definitely isolated.

Efforts to improve niacin content in maize have been based on treating the maize with lime. This treatment is supposed to decrease effects of niacin-binding substances or/and antiniacin agents, but actual evidence supporting the advantage of this practice is quite conflicting.

Nicotinic acid concentrations can be increased in some foods during post-harvest processing. This is true for foods containing

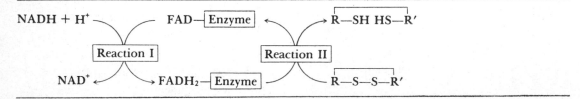

Miscellaneous Considerations

A variety of *niacin antagonists* have been identified including 3-acetylpyridine, 6-amino-

the alkaloid known as *trigonelline*, which reacts in the presence of heat and acid to give nicotinic acid (Figure 5.46). Occurrence of this reaction during coffee bean roasting can

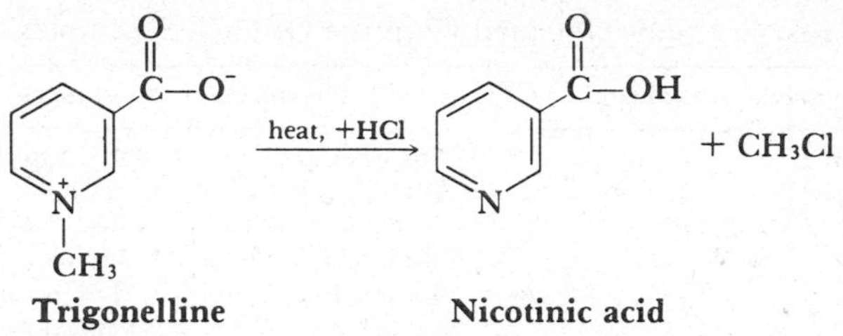

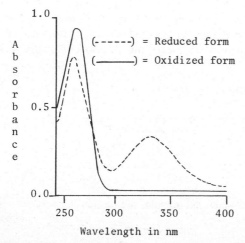

Trigonelline   **Nicotinic acid**

FIGURE 5.46. Formation of nicotinic acid from trigonelline.

produce a 30-fold increase in nicotinic acid content (i.e., 1.0 to 30.0 mg/100 g), making it one of the best sources of the vitamin. Many cereal grains, peas, and other plant sources contain this alkaloid.

Traditional estimations of pure nicotinic acid and nicotinamide in solutions have relied heavily on spectrophotometric analysis at 262 nm. The isolated absorption spectra for these substances are characteristic (Figure 5.47) yet pH variable and, therefore, they have little application to the analysis of biochemically complex samples that contain the vitamin (e.g., foods, skeletal and organ meats, natural products).

Biochemically, complex samples may be more accurately assayed by using a total nicotinic acid assay. This method requires the hydrolysis of coenzyme-bound nicotinamide and nicotinic acid with sulfuric acid. The pyrimidine ring of these substances is then cleaved by cyanogen bromide to produce a fission product that reacts with sulfanilic acid. This reaction generates a yellow-colored polymethine dye that can be assayed at 436 nm.

Bioassays for nicotinamide and nicotinic acid are reported to be accurate for concentrations as low as 0.5 ng. *Lactobacillus arabinosus* 17-5 (ATCC 8014) responds to both nicotinamide and nicotinic acid, whereas *Leuconostoc mesenteroides* (ATCC 9135) responds only to nicotinic acid. Arecoline, quinolinic acid, nicotinic acid diethylamide (coramine) along with both isomers of nicotinic acid—picolinic and isonicotinic acids—do not interfere with the reliability of these assays.

Thin-layer chromatography, polarography, fluorometry, and gas chromatography have also been successfully used for assay of this vitamin, but future assays for the vitamin will rely more heavily on advanced high-pressure liquid chromatographic techniques.

Detailed methods for niacin assay are reported by Baker and Frank (1968), McCormick and Wright (1971), Snell and Wright (1941), and Strohecker and Henning (1965).

FIGURE 5.47. Absorption spectra for oxidized and reduced forms of pyridine nucleotides found in NAD⁺ (NADH) and NADP⁺ (NADPH). The adenine moiety contained in the nucleotides absorbs strongly at 340 nm while the observed 360 ± 20-nm absorption corresponds to the peak absorption range for the reduced pyridine ring.

PANTOTHENIC ACID

Recognition of this vitamin was uncertain before the 1930s because vitamin B activity was attributed to a single substance. The so-

Table 5.9. Fractions Obtained from the Original B complex

Thermolabile substances	Thermostable substances	
Vitamin B_1	Adsorbed to Fuller's earth	Filtrate from Fuller's earth
	Vitamin B_1	Biotin
	Vitamin B_6	Myo-inositol
	Niacin	Pantothenic acid

called vitamin B substance was eventually partitioned on Fuller's earth into a number of additional vitamin and vitamin-like substances, currently recognized as vitamins B_1, B_2, B_6, niacin, biotin, pantothenic acid, and myo-inositol (Table 5.9).

The technical name of pantothenic acid is D(+)-*N*-(2,4-dihydroxy-3,3 dimethylbutyryl)-β-alanine, but *obsolete* designations include the *chick antidermatitis factor,* the *chick antipellagra factor,* or the *filtrate factor,* owing to its nonadsorption to Fuller's earth at an acid pH.

The important chemical and physical properties for pantothenic acid and its calcium salt, which is used for therapeutic applications, have been itemized in Table 5.10. The natural dextrorotatory form of the vitamin is the only form that has vitamin activity (Figure 5.48).

Other vitamin active forms for animals include the acetate, benzoate, diphosphate, and ethyl esters of pantothenic acid, but these same derivatives are often inactive for certain bacteria, such as lactic acid bacteria. Substitution of a hydroxy group for one of the two methyl groups of the vitamin decreases activity somewhat, but the pantothenol form exhibits almost 86% of pantothenic acid activity in chick growth assays. For humans the activity of pantothenol exceeds that of an equal amount of pantothenic acid by 7 to 17% (Figure 5.49).

The actual role of pantothenic acid became apparent in 1947, when Lipmann recognized that a coenzyme substance composed of >10% pantothenic acid was necessary for the *acetylation of aromatic amines* in liver tissue. Added study also demonstrated that this was probably the identical coenzyme that was responsible for acetylating choline in brain tissue. Pantothenic acid was then recognized as an important constituent of coenzyme A (CoASH)—its most

Table 5.10. Characterization of Pantothenic Acid and Calcium Pantothenate

	Pantothenic acid	Calcium pantothenate
Physical appearance	Colorless, viscous oil	White, microcrystalline
Empirical formula	$C_9H_{17}O_5N$	$(C_9H_{16}O_5N)_2Ca$
Molecular weight	219.2	476.5
Solubility: Both forms are readily soluble in water and insoluble in benzene and chloroform.		
Optical rotation	$[\alpha]_D^{25} = +37.5°$	$[\alpha]_D^{25} = +28.2°$
Sterilization method	Filtration is preferred	Filtration is necessary
Unstable in	Acid, alkali, heat	Acid, alkali, heat
Units: Pantothenic acid activity is usually measured in mg of the pure, synthesized substance. For chick assays, 1 g of pantothenic acid equals 70,000–75,000 chick units and 1.0 chick unit corresponds to 14 µg of pantothenic acid.		

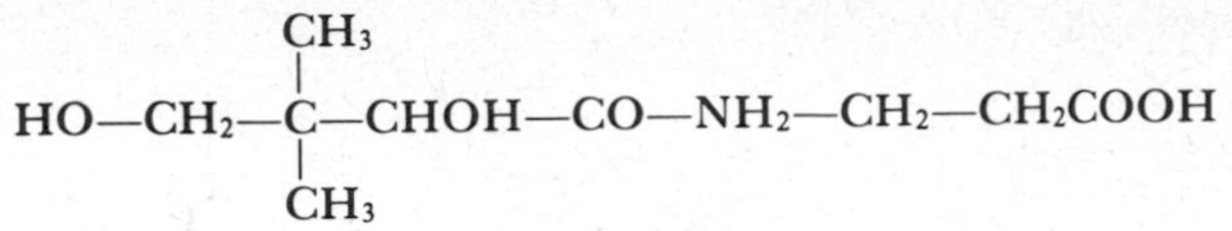

$$CH_3$$
$$|$$
$$HO—CH_2—C—CHOH—CO—NH_2—CH_2—CH_2COOH$$
$$|$$
$$CH_3$$

Pantothenic acid

FIGURE 5.48. Pantothenic acid.

common form of occurrence in nature (Figure 5.50).

Sources

Although pantothenic acid occurs widely in plants and animals, most pantothenic acid exists in a chemically bound form. The richest common sources of the vitamin are liver and kidney tissues; however, the jelly of the queen bee contains at least six times the amount found in liver. Muscle tissue generally contains much lower concentrations than kidney or liver tissues. Selected portions of plants such as the bran fraction of cereal grains and molasses provide good sources of the vitamin. Because molds and microorganisms have an innate ability to synthesize pantothenic acid, dense populations of these organisms in food substances can cause elevated food pantothenic acid content.

Representative pantothenic acid concentrations for some foods are dry brewer's yeast 200 μg/g; beef liver, 76 μg/g; egg yolk, 63 μg/g; kidney, 35μg/g; wheat bran, 30 μg/g; buckwheat, 26 μg/g; fresh spinach, 26 μg/g; roasted peanuts, 25 μg/g; whole milk powder, 24 μg/g; and white bread, 5 μg/g.

Human Requirements and Deficiency Effects

The wide distribution of pantothenic acid makes deficiency symptoms of this vitamin in man a rarity, although gross vitamin B-complex deficiencies in man are accompanied by pantothenic acid levels that may be 25% below normal. Apart from obvious dietary sources, some pantothenic acid may be obtained from intestinal flora since this route is known to occur in the rumen of sheep and cattle.

Human requirements for pantothenic acid are probably in the range of 10 mg/day, but less than nutritionally balanced diets may contain $\geq$15 mg/day. Toxicity effects of this vitamin are not known to occur in man.

Experimental deficiencies of pantothenic acid in humans have been demonstrated by using a pantothenic acid antagonist. The antagonist, called ω-*methyl-pantothenic acid*, reportedly causes tiredness, cramps, abdominal pain, nausea, flatulence, and parathesia of the feet and hands. The severe parathesia and tenderness of the feet experienced by World War II prisoners of war were probably a result of diets low in pantothenic acid.

Biochemistry

The importance of pantothenic acid is centered about its key structural contribution to CoASH. Coenzyme A has many important biochemical functions owing to its reactive terminal thiol (—SH). This functional group can readily undergo reversible thioester formation with the carboxyl portion of acetate and similar acid intermediates in metabolism. Thioester formation yields a product

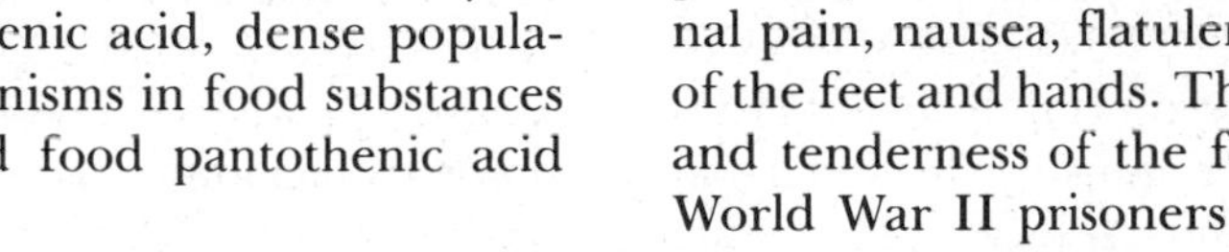

$$CH_3$$
$$|$$
$$HO—CH_2—C—CHOH—CO—NH—CH_2—CH_3—CH_2OH$$
$$|$$
$$CH_3$$

Pantothenol

FIGURE 5.49. Pantothenol.

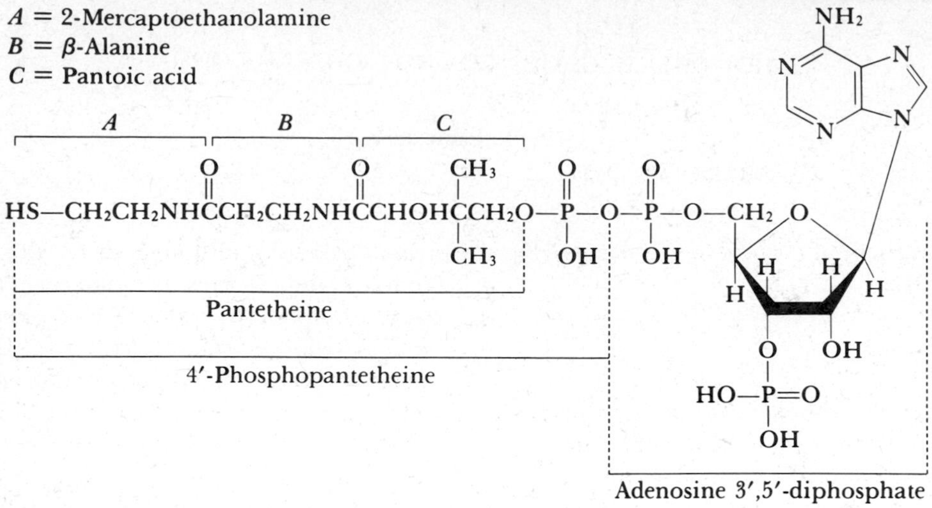

A = 2-Mercaptoethanolamine
B = β-Alanine
C = Pantoic acid

FIGURE 5.50. Structure of coenzyme A (CoASH).

that displays limited resonance possibilities, unlike corresponding oxygen esters (Figure 5.51).

Limited thioester resonance incites the development of a carbonyl character where the carboxyl carbon becomes partially positive (δ^+), and the oxygen assumes a partially negative (δ^-) charge. The fractional positive charge on the carboxyl carbon encourages dissociation of hydrogen on the α-carbon atom as a proton, thereby leaving the α-carbon with a fractional negative charge (Figure 5.52). Since thioesters exhibit less resonance than corresponding oxygen esters, they liberate a higher free energy when they undergo hydrolysis.

The resonance behavior and properties of coenzyme A thioesters make this species susceptible to *nucleophilic or electrophilic attack,* as shown in Figure 5.53.

Thioester formation between acetates and CoASH results in activated acetates that are responsible for condensing two-carbon fragments with oxaloacetate in the TCA cycle. Since this cycle has central importance in the cellular oxidation of nutrients, the biochemical influence of CoASH is felt in the far reaches of overall carbohydrate, lipid, and protein metabolism. Moreover, the coenzyme serves as a key factor in fatty acid oxidation as well as the biosynthesis of isoprenoids including steroids, carotenoids, and terpenes.

Pantothenic acid also serves as a structural feature in the 4'-phosphopantetheine moiety of *acyl carrier protein* (ACP). ACP is a coordinating protein for six enzymes in the multienzyme complex called *fatty acid synthetase* (FAS), which is responsible for fatty acid biosynthesis. Apart from the coordinating activity of ACP, its 4'-phosphopantetheine moiety offers an anchor site for stabilizing fatty acyl thioester intermediates as they undergo FAS-mediated fabrication into in-

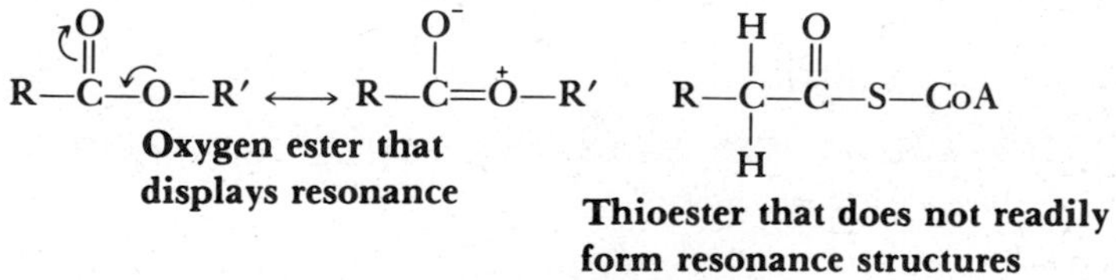

FIGURE 5.51. Examples of oxygen and thioesters.

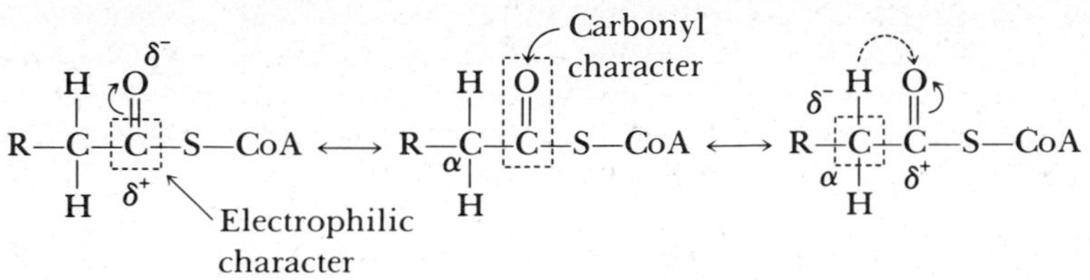

FIGURE 5.52. Fractional charge distribution within a thioester structure.

dividual fatty acids. Studies of *Escherichia coli* have revealed that the 4′-phosphopantetheine structure is linked to ACP residue 36 (serine) (Figure 5.54).

Miscellaneous Considerations

A variety of methods have been developed for the assay of pantothenic acid in foodstuffs, but the uniformity of assay protocols from one analyst to another, the wide occurrence of the vitamin, and its tightly bound biochemical nature in CoASH all contribute to analytical difficulties.

Successful assays require preliminary liberation of the vitamin from CoASH through the use of enzymes. For chemical assays, the liberated pantothenic acid or pantothenyl alcohol is then subjected to acid or basic hydrolysis. This treatment gives free pantoic acid plus β-alanine (from pantothenic acid) or 3-amino-1-propanol (from pantothenyl alcohol). The latter two substances are colorimetrically evaluated following a reaction with 1,2-naphthaquinone-4-sulfonate. Since this method is not always specific for the vitamin, its results should be checked against reliable but cumbersome chick assays or microbiological assays. *Saccharomyces carlsbergensis* (ATCC 9080) and *Lactobacillus arabinosus* (*plantarum*) (ATCC 8014) can be used for the microbiological assay of pantothenic acid and its salts in a concentration range of 10 ng, whereas *Acetobacter suboxydans* (ATCC 621 H) is sensitive only to ≥50 ng concentrations of pantothenyl alcohol (Strohecker and Henning, 1965; Freed, 1966).

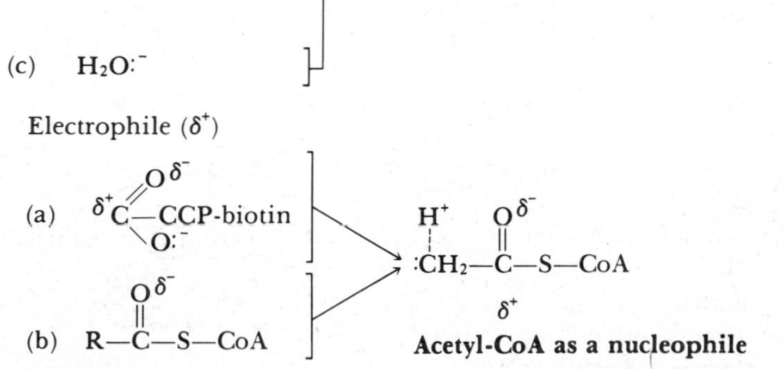

FIGURE 5.53. Thioester (acetyl-CoA) attack by *electrophiles* or *nucleophiles*.

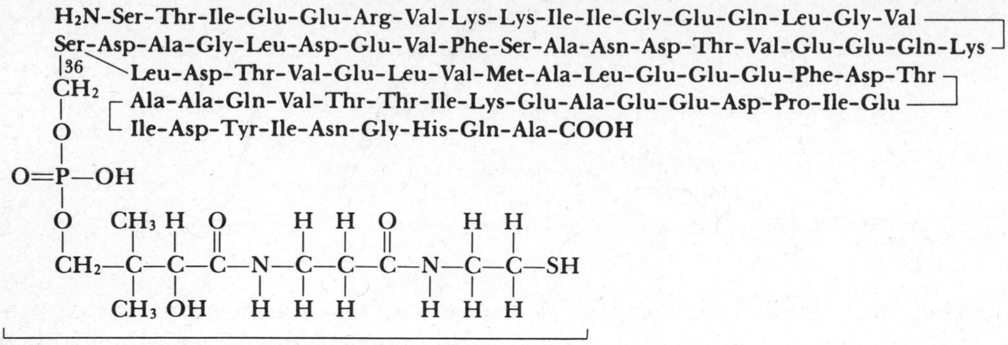

H₂N–Ser–Thr–Ile–Glu–Glu–Arg–Val–Lys–Lys–Ile–Ile–Gly–Glu–Gln–Leu–Gly–Val ⌐
Ser–Asp–Ala–Gly–Leu–Asp–Glu–Val–Phe–Ser–Ala–Asn–Asp–Thr–Val–Glu–Glu–Gln–Lys ⌐
|36
CH₂ ⌐ Ala–Ala–Gln–Val–Thr–Thr–Ile–Lys–Glu–Ala–Glu–Glu–Asp–Pro–Ile–Glu ⌐
        Leu–Asp–Thr–Val–Glu–Leu–Val–Met–Ala–Leu–Glu–Glu–Glu–Phe–Asp–Thr ⌐
O   ⌐ Ile–Asp–Tyr–Ile–Asn–Gly–His–Gln–Ala–COOH

$$O=P-OH$$

structure of the 4′-phosphopantetheine moiety

4′-Phosphopantetheine moiety

FIGURE 5.54. Speculated structure for acyl carrier protein (ACP) isolated from *Escherichia coli*. Note that the 4′-phosphopantetheine moiety is linked to the Ser (36) position of the overall structure.

BIOTIN

The essential biochemical activity of biotin was first recognized with the discovery that it protected certain experimental animals from *egg-white injury*. Rats, for example, which had been fed copious amounts of raw egg whites developed a characteristic *spectacle-eye syndrome* or *alopecia* around the eyes, hind leg paralysis, and eczema-type dermatitis. Further research by György (1939) and Parsons *et al.* (1937) ascertained that yeast and liver contained an *anti-egg-white injury factor*. This substance was soon designated as *vitamin H*, but this name was short-lived because it was found to be identical to *bios* and *coenzyme R*, which had been previously identified as growth factors for microorgannisms (Kögl and Tönnis, 1936). Purification of the *egg-white injury factor* led to the eventual discovery of the heat labile glycoprotein, known as *avidin*, which has a specific binding

affinity for the *anti-egg-white injury factor* now recognized as *biotin*.

The molecular structure of biotin has been defined as 2′-keto-3,4-imidazolido-2-tetrahydrothiophene-*n*-valeric acid. This structure occurs rarely in food and instead, it is usually found tightly yet flexibly bound to the ε-amino group of a lysine residue contained in the protein structure of certain biotin-dependent enzymes. As a result of the covalent bonding between biotin and the ε-amino group of lysine (a lipoamide linkage), *in vivo* biotin is largely nondialyzable. Accordingly, it is customarily liberated from hydrolyzed biotin-containing proteins as *biocytin*, which incorporates the *in vivo* lipoamide between the valeric acid portion of biotin and the lysine residue from the original biotin-containing protein (Lane and Lynen, 1963) (Figure 5.55).

Additional systematic investigations of biotin have resulted in the identification of

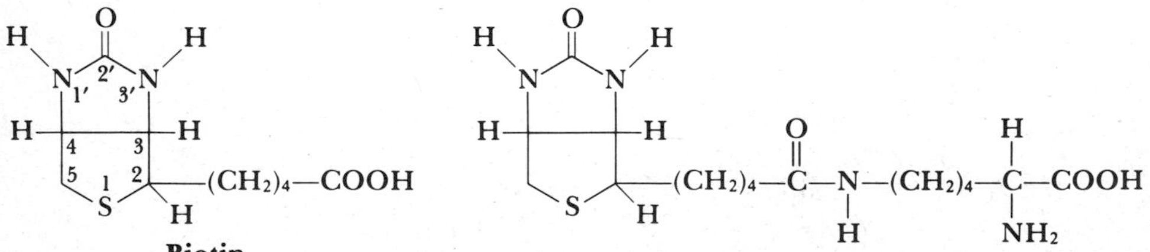

Biotin
(2′-Keto-3,4-imidazolido-2-
tetrahydrothiophene-*n*-valeric acid)

Biocytin

FIGURE 5.55. Structures for biotin and biocytin.

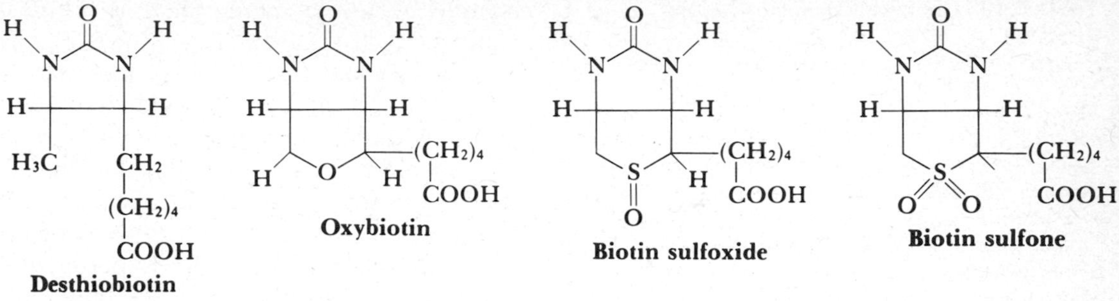

FIGURE 5.56. Selected structures for derivatives of biotin.

many biotin derivatives including *desthiobiotin, oxybiotin, sulfoxide,* and *sulfone biotin derivatives* (Figure 5.56). These biotin derivatives have selected abilities for promoting the growth of biotin-requiring yeast, lactobacilli, and animal species as indicated in Table 5.11. In the cases of those organisms that positively respond to the growth-stimulating effects of biotin derivatives, it is interesting to note that they do so without prior conversion of the derivative to biotin. Furthermore, current evidence shows that dextrorotatory (+)-biotin is the only one of eight possible stereoisomers that has biological activity. (*Note:* biotin has three chiral carbons.) The sulfone and sulfoxide derivatives of biotin have little significance among the nutritional demands of nearly all organisms, but it is important to note that the cooking of foods can inevitably convert a significant portion of food biotin content to these oxidized derivatives. Table 5.12 outlines the notable characteristics of biotin exclusive of its numerous derivatives.

Sources

A nutritionally balanced diet provides 100 to 300 µg of biotin per day, which satisfies the nutritional demands of the body. The actual necessity for dietary biotin is open to speculation since up to *six times* the daily biotin intake is discharged in the urine. This suggests that biotin produced by intestinal flora, nurtured on dietary constituents, may supersede the dietary value attributed to biotin *contained* in foods—assuming that the human subject is in a normal healthy state. A representative tabulation of biotin levels in foods has been outlined in Table 5.13.

Human Requirements and Deficiency Effects

Overt symptoms of biotin deficiencies in humans are rare, especially in cases where a well-balanced diet prevails. However, infants are most susceptible to *biotin avitaminosis* because of the excessive or poor discretionary use of sulfa drugs. These antibiotics

Table 5.11. Biotin Derivatives and Their Biological Activity on Selected Organisms[a]

Biotin derivatives	Saccharomyces cerevisiae	Lactobacillus casei	Animals	Lactobacillus arabinosus	Streptococcus faecalis R.
Biotin sulfoxide	(+)	−	−	N.S.	N.S.
Biotin sulfone	(+)	−	−	N.S.	N.S.
Desthiobiotin	+	−	−	−	N.S.
Oxybiotin	+	+	−	+	−

[a]Growth responses: "+" = positive growth effect; "−" = no growth effect; (+) = partial positive growth effect; and N.S. = no significant effect of biotin derivative based on available data.

Table 5.12. Chemical Characterization of Biotin

Empirical formula	$C_{10}H_{16}O_3N_2S$
Molecular weight	244
Melting point	230–232°C (decomposition)
Optical rotation $[\alpha]_D^{22}$	+92° (in 0.1 N NaOH)

Solubility: Soluble in water and ethanol but insoluble in chloroform, ethyl ether, and petroleum ethers.

Stability: Biotin shows stability toward strong acids but it is unstable to alkali.

Units: The activity for biotin is reported as micrograms of the chemically pure substance.

and others inadvertently diminish the resident biotin-producing flora of the intestine. Biotin deficiencies in infants may also be linked to various forms of seborrheic dermatitis and Leiner's disease. The most distinctive evidence of these maladies is seborrheic dermatitis of the scalp in infants and children less than 4 years old. The progressive dermatitis of the scalp can lead to local *balding* patterns known as *cradle cap* in the child, along with scaling of the cheeks, neck, umbilicus, and groin areas. In contrast with widespread yet localized patches of dermatitis, Leiner's disease involves the whole skin surface. These problems can be largely remediated by intramuscular administration of 2 to 5 mg of biotin for 1 to 2 weeks or similar amounts of biotin via an oral route for a period of 2 to 3 weeks. Seborrheic dermatitis observed in older children and adults rarely if ever responds to biotin treatment and, therefore, probably originates from a completely different factor.

For adults, induced biotin deficiencies are clinically mild and nonspecific but generally include the development of nonpruritic dermatitis, which appears after approximately 11 days. By the fifth week of biotin deficiency, typical problems include muscular pains, lassitude, and hyperanesthesia. These symptoms are then upstaged by the appearance of anorexia, anemia, hypercholesterolemia, and nausea by the tenth week of induced biotin deficiency. Experimental evidence suggests that even the most advanced deficiency symptoms stemming from a lack of biotin can be effectively regressed in approximately 5 days using ~150 μg/day doses of the vitamin.

Biochemistry

Biotin is bound through a lipoamide linkage to specific proteins that are largely responsible for executing carboxylation reactions. These reactions generally involve a two-step mechanism that requires the participation of three proteins including a *carboxyl carrier protein* (CCP) that *contains biotin* (CCP–biotin) and serves as a carrier for activated molecules of carbon dioxide ($CO_2 \sim$ CCP–biotin), which are ultimately destined for transfer. The remaining two proteins involved in carboxylation reactions are *biotin carboxylase* and *transcarboxylase* enzymes. The activated carbon dioxide contained on its carrier protein ($CO_2 \sim$ CCP–biotin) may be transferred to an α-keto acid in the early steps of gluconeogenesis at which point pyruvic acid is converted into oxaloacetate or the activated carbon dioxide may be added to the molecular structure of certain acyl-CoA derivatives. Exemplary reactions involving biotin

Table 5.13. Concentration Levels of Biotin in Selected Foods (in μg/100 g)

Apples	~ 0.9	Mushrooms	~16.0
Beans	~ 3.0	Oranges	~ 2.0
Beef (muscle)	~ 2.6	Peanuts	~30.0
Beef (liver)	~96.0	Potatoes	~ 0.6
Cheese	~ 1.0–8.0	Spinach	~ 7.0
Lettuce	~ 3.0	Tomatoes	~ 1.0
Milk	~ 1.0–4.0	Wheat	~ 5.2

Pyruvate carboxylase reaction

Step 1

$$ATP + HCO_3^- + CCP\text{-biotin} \xrightleftharpoons[\text{acetyl-CoA}]{Mg^{2+}} CO_2 \sim CCP\text{-biotin} + ADP + P_i$$

Bicarbonate   **Biotin**
                                    carboxylase

Step 2

$$CO_2 \sim CCP\text{-biotin} + Pryuvate \xrightleftharpoons{Mn^{2+}} CCP\text{-biotin} + Oxaloacetate$$

(α-keto-acid)

Pyruvate:oxaloacetate
transcarboxylase

Acetyl-CoA carboxylase reaction

Step 1

$$ATP + HCO_3^- + CCP\text{-biotin} \xrightleftharpoons{Mg^{2+}} CO_2 \sim CCP\text{-biotin} + ADP + P_i$$

Biotin
carboxylase

Step 2

$$CO_2 \sim CCP\text{-biotin} + Acetyl\text{-CoA} \rightleftharpoons CCP\text{-biotin} + Malonyl\text{-CoA}$$

Acetyl-CoA:malonyl-CoA
transcarboxylase

FIGURE 5.57.   Role of biotin in the formation of oxaloacetate and malonyl-CoA.

participation have been detailed in Figure 5.57 for the production of oxaloacetate and malonyl-CoA.

Using the two preceding reactions as examples of biotin behavior in carboxylations, the reaction scheme outlined in Figure 5.58 has been established. The possible mechanisms involved in the two-step reaction sequence have been further detailed in Figure 5.59.

Both the biotin-linked CCP and biotin carboxylase are dimers respectively constructed from subunits with molecular weights of 2.2×10^4 and 5.1×10^4. Although the precise linkages responsible for constructing these dimers are not entirely certain, at least two lysyl bridges seem to be responsible for the subunit linkages in CCP. Transcarboxylase, on the other hand, has a molecular weight of 1.3×10^5 and is divisible into four subunits, which have molecular weights of $3.0–3.5 \times 10^4$.

The activity observed for pyruvate carboxylase and acetyl-CoA carboxylase not only depends on a triad–protein interaction of the structures cited above, but their actual biochemical activity is also contingent upon the association of the three proteins into multitriad aggregates before any significant carboxylation reaction can occur. For example, in the case of pyruvate carboxylase, the increased availability of cellular *citrate* levels acts as an *allosteric activator* to initiate the aggregation of 10 to 20 pyruvate carboxylase protein triads. This active carboxylase aggregate displays a molecular weight in the range of $3.5–8.0 \times 10^6$.

Moreover, the activity of carboxylases such as pyruvate carboxylase may be subject to the influence of additional allosteric activators such as acetyl-CoA or some of its related acyl-CoA derivatives. In the absence of these activators, the prerequisite carboxylation of biotin cannot occur and the overall conversion of pyruvate to oxaloacetate will remain at a standstill. Such temporal fluctuations of acyl-CoA derivatives, citrate, and other activators allosterically regulate the oxaloace-

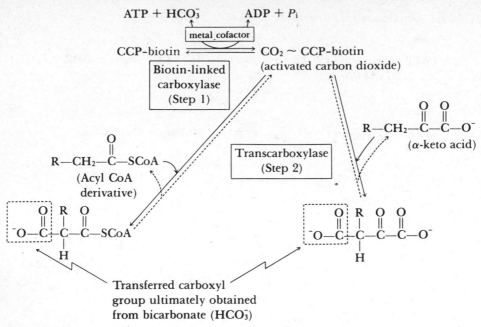

FIGURE 5.58. Two-step reaction sequence underlying the participation of biotin in carboxylation reactions.

tate concentration levels within cells. This type of interplay is crucial because oxaloac-etate acts as both an intermediate in the TCA cycle and a stoichiometric intermediate in gluconeogenesis.

Acetyl-CoA carboxylases (M.W. = 8.0 × 10^5) found in chicken and rat livers, along with propionyl transcarboxylase (M.W. = 8.0 × 10^5) also mirror the aggregative behavior of smaller subunits (18–20 subunits) in a fashion reminiscent of pyruvate carboxylase. These enzymes also require the presence of metals such as Zn^{2+} or Co^{2+}.

Apart from the previously discussed car-boxylation reactions, other biotin-depend-ent enzymes have been identified. These in-clude *carbamyl phosphate synthetase* (Table 5.14A) in *Escherichia coli*, which produces carbamyl phosphate from glutamine and ATP, as well as *urea carboxylase* (Table 5.14B) found in yeast (*Saccharomyces cerevisiae*), which carboxylates urea to *N*-carboxyurea (allo-phanate). The latter reaction has special im-portance to this yeast because its sole sources of nitrogen, obtained from urea, allantoic

acid, allantoin, and some other substances, require a preliminary formation of carbox-ylated urea before hydrolysis to ammonia and carbon dioxide. The essential presence of biotin-containing enzymes also extends to the oxidation and utilization of propionic acid in animals where propionyl-CoA is con-verted to succinyl-CoA. Succinyl-CoA sub-sequently serves as an important factor in the TCA cycle and a key intermediate in many biosynthetic reactions (Table 5.14C).

Nearly all carboxylase enzymes that re-quire biotin also require carbon dioxide and an energy-rich compound such as ATP to activate the carboxylation sequence, but this is not true in all cases. For example, *methyl-malonyl:oxalacetate transcarboxylase (methylma-lonyl-CoA pyruvate carboxyl transferase)* repre-sents at least one common exception to such observations (Table 5.14D). This enzyme mediates a direct transfer of the carboxyl portion of methylmalonyl-CoA to pyruvate, thereby yielding oxaloacetate. This reaction is no doubt significant in certain mammalian cells, but it is probably subordinate to those

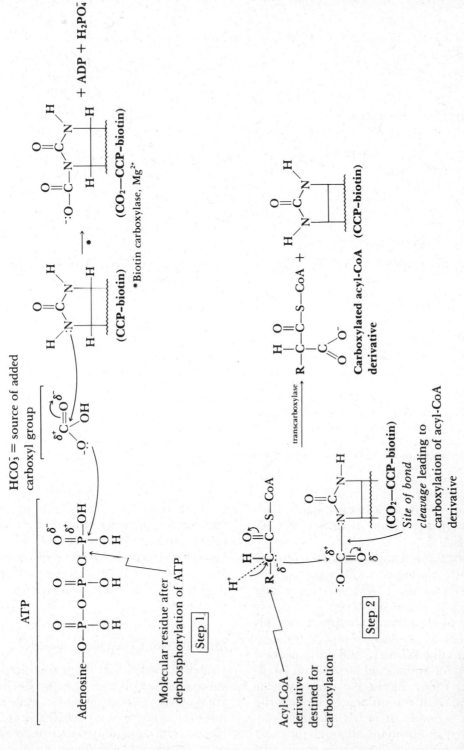

FIGURE 5.59. Speculated role of biotin in the carboxylation of an acyl-CoA derivative.

Table 5.14. Biotin-Dependent Enzyme Reactions

(A) *Carbamyl phosphate synthetase:*

$$\text{Glutamine} + 2\text{ ATP} + \text{HCO}_3^- \xrightarrow{a} \text{glutamate} + 2\text{ADP} + \text{H}_2\text{N}-\overset{\overset{\displaystyle O}{\|}}{\text{C}}-\text{O}-\overset{\overset{\displaystyle O}{\|}}{\underset{\underset{\displaystyle OH}{|}}{\text{P}}}-\text{O}^-$$

Carbamyl
phosphate

(B) *Urea carboxylase:*

$$\underset{\underset{\displaystyle \text{Urea}}{}}{\underset{\text{H}_2\text{N}\quad\text{NH}_2}{\overset{\overset{\displaystyle O}{\|}}{\text{C}}}} + \text{HCO}_3^- \xrightarrow[a]{\text{ATP}\quad\text{ADP}+P_i} \underset{\underset{\displaystyle \substack{N\text{-Carboxyurea} \\ \text{(allophanate)}}}{}}{\underset{\text{H}_2\text{N}\quad\underset{\displaystyle \text{H}}{\text{N}}}{\overset{\overset{\displaystyle O\quad\ O}{\|\quad\ \|}}{\text{C}\quad\text{C}}}-\text{O}^-} \xrightarrow[\substack{\text{allophanate} \\ \text{amidolyase}}]{+\text{H}_2\text{O}} 2\text{HCO}_3^- + 2\text{NH}_4^+$$

(C) *Propionyl-CoA carboxylase:*

$$\text{Propionyl-CoA} + \text{HCO}_3^- + \text{ATP} \xrightarrow[a]{\text{Mg}^{2+}} \text{D-methylmalonyl-CoA} + \text{ADP} + P_i$$

$$\updownarrow \substack{\text{methylmalonyl-CoA} \\ \text{racemase}}$$

L-methylmalonyl-CoA

$$\updownarrow \substack{\text{methylmalonyl-CoA mutase} \\ +\text{ coenzyme B}_{12}}$$

succinyl-CoA
(many fates in intermediate
metabolism)

(D) *Methylmalonyl:oxalacetic transcarboxylase:*

$$\text{Methylmalonyl-CoA} + \text{pyruvate} \underset{a}{\rightleftharpoons} \text{propionyl-CoA} + \text{oxaloacetate}$$

[a]Site of biotin-dependent enzyme activity.

reactions that involve carbon dioxide fixation through the action of propionyl-CoA carboxylase.

The biochemical implications of biotin activity are clearly linked to all facets of intermediary metabolism. Reactions involving carbamyl phosphate synthetase may simply appear to yield carbamyl phosphate, but this substance is an *essential prerequisite* for normal pyrimidine synthesis and an important constituent of amino acid metabolism (Wellner *et al.*, 1968). Biotin also exerts similar direct and indirect influences among the metabolic schemes of nearly all lipids, carbohydrates, and proteins, although the exact mechanism of its involvement remains to be established in many cases. Among current interests are the roles of biotin deficiency in relation to hypercholesterolemia and hyperglycemia; biotin–lipid metabolism interactions controlled by genetic factors; and the high levels of biotin associated with cancerous tumors.

Miscellaneous Considerations

Under normal dietary regimens, biotin is not subject to the influence of many biochemical antagonists. Certain bacteria and insects, however, seem to be especially susceptible to biochemical antagonists such as desthiobiotin, desthioisobiotin, biotin sulfone, 4-imidazolidone-2-caproic acid, and ureylenecyclohexylvaleric acid (Kodicek, 1966; Langer and György, 1968).

Foremost among all naturally occurring biotin antagonists is the tetrameric glycoprotein known as avidin (M.W. = 6.8×10^4). As indicated earlier, this agent was partially responsible for the discovery of biotin since it was found to counteract the biochemical availability of the vitamin in animal digestive processes. Avidin is produced in the hen oviduct, where it is introduced into egg white. Its production is under strict control of endocrine hormones such as progesterone and estrogen, but the evolutionary or biochemical reasons for its presence in egg white are uncertain. The binding affinity of avidin for biotin probably involves the interaction of the imidazolidone of the biotin structure with any one of *four* possible sites on the avidin structure. When fully saturated, avidin can bind approximately 4 biotin molecules per tetrameric molecular structure.

The toxicity effects of biotin lack hard-core documentation, but most evidence points to rapid urinary excretion of any excessive biotin concentration levels in the body. The physiological excretion of biotin in mammals is under strict control of dietary biotin availability as well as the degree of biotin adsorption from indigenous intestinal flora. Biotin is excreted largely in an unmetabolized form, but occasional *traces* of *sulfoxide* derivatives formed in the liver, along with *bisnorbiotin* and *tetranorbiotin* produced in other tissues, may be present along with biotin. The latter two metabolites are generally recognized by their respective loss of two or four carbon atoms off the biotin side chain prior to excretion.

The most reliable quantitative analyses of biotin have traditionally relied on bioassays that employ yeasts, molds, and bacteria. Bioassay organisms typically include *Saccharomyces cerevisiae* (Snell *et al.,* 1940), *Lactobacillus casei* (Shull *et al.,* 1942; Shull and Peterson, 1943; Landy and Dicken, 1942), *Clostridium butyricum* (Lampen *et al.,* 1942), and *Lactobacillus arabinosus* (Wright and Skeggs, 1944). Of all bioassay organisms available, *L. arabinosus* is often preferred because it produces very reproducible growth curves in well-defined basal growth media where biotin is the only variable nutrient. Furthermore, *L. arabinosus* bioassays are desirable because *L. casei* requires certain undefined growth factors that are likely to vary from sample to sample, thereby yielding data of querulous significance. The singular use of *S. cerevisiae* also has its analytical pitfalls and credibility problems because it responds to biotin, biotin sulfone, desthiobiotin, and other biotin derivatives (Genghof *et al.,* 1948). A more recent adjunct to traditional bioassay organisms used in assessing blood urine and animal tissues is *Ochromonas danica* (Baker *et al.,* 1962).

Instrumental and chemical methods of analysis are less discriminative and sensitive than bioassays, but their advantage is primarily one of convenience, although recent advances in high-pressure liquid chromatography should provide new analytical perspectives for the vitamin. The application of spectrophotometric techniques for evaluating avidin and biotin concentrations have been detailed by Green (1970).

L-ASCORBIC ACID (VITAMIN C)

Vitamin C (L-ascorbic acid) has special historical significance among the vitamins. Since the 1700s it was clear that scurvy could be eliminated or prevented in humans through the dietary consumption of fresh fruits and vegetables. The dietary factor attributable to their *antiscorbutic effect* was first designated as the *antiscorbutic principle* and, later, *vitamin C*. Firm scientific investigations into the nature of this *principle* were not encouraged until a serendipitous discovery by Holst and Frölich (1907, 1912), which illustrated that guinea pigs were subject to induced vitamin C deficiencies. This work led to the use of the guinea pig as the first bioassay tool for evaluating the concentrations of a vitamin

Table 5.15. Ascorbic Acid Concentrations in Selected Plants in mg/100 g Edible Portion as Purchased

Broccoli	113	Kale	500
Black current	200	Parsley	190
Cabbage	47	Potatoes	73
Citrus fruits	220	Spinach	220
Guava	300	Squashes	90 (average)
Green peppers	120	Tomatoes	100

substance, and it was a most fortunate discovery since present-day evidence has shown that vitamin C is a dietary requirement only for man, guinea pigs, monkeys, the Indian fruit bat, certain Passiform birds (Chaudhuri and Chatterjee, 1969; Chatterjee, 1970), and some varieties of catfishes (Wilson, 1973).

Despite the recognition of an *antiscorbutic principle* throughout the early 1900s, the actual chemical characterization and isolation of the vitamin did not materialize until 1932 when Waugh and King isolated *hexuronic acid* (L-ascorbic acid) from lemon juice and demonstrated its antiscorbutic activity in guinea pigs. This work, viewed in perspective with studies reported by Szent-Györgyi in 1928, which identified hexuronic acid as a component of oxidation-reduction activities in the adrenal cortex, provided the basis for many future concepts explaining the biochemical roles of ascorbic acid.

Unlike most other vitamin substances, L-ascorbic acid displays relatively high concentration levels in most animal and plant tissues. Although 20 mg/day of L-ascorbic acid will counteract the onset of scurvy in man, normal blood plasma contains about 0.6 mg/100 mL. Plant tissues, on the other hand, contain varied concentrations ranging up to 500 mg/100 g (Table 5.15).

The structure of L-ascorbic acid is reminiscent of hexose derivatives with two asymmetric carbons that account for *four* stereoisomeric configurations. The most notable derivatives of L-ascorbic acid include *D-isoascorbic acid*, which displays only 5 to 20% of L-ascorbic acid vitamin activity depending on the assay organism employed, and *D-erythro-3-ketohexuronic acid*, which lacks vitamin potency in nearly all organisms studied to date (Figure 5.60). An additional 2-sulfate ester of L-ascorbic acid exists within cells, but this does not eliminate the biological activity of the vitamin substance.

The biochemical importance of L-ascorbic acid is primarily related to its oxidation–reduction reactions in plants and animals. Oxidation of the vitamin substance inevitably leads to the loss of its enolic hydrogens and the concomitant formation of *dehydroascorbic acid* (Figure 5.61).

Literary and pharmaceutical references to vitamin C or L-ascorbic acid include over 35 different names, with the most common

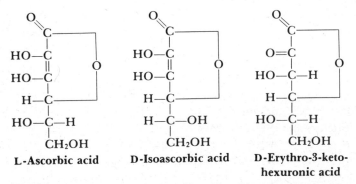

FIGURE 5.60. Structures for ascorbic acid and its related derivatives.

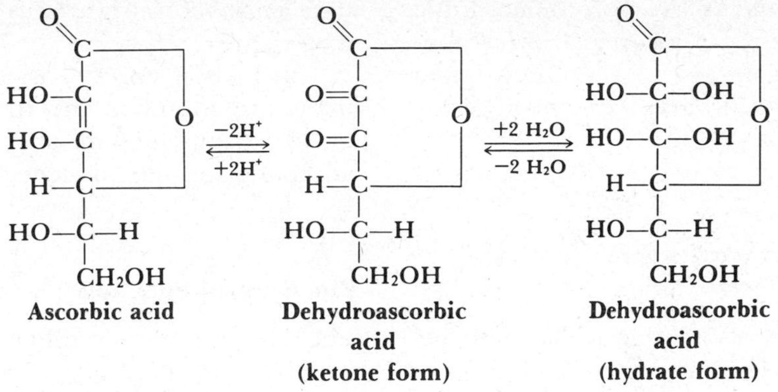

Ascorbic acid   Dehydroascorbic acid (ketone form)   Dehydroascorbic acid (hydrate form)

FIGURE 5.61. Important reactions for ascorbic acid.

designations being L-xyloascorbic acid, 3-oxo-L-gulofuranolactone (enol form), L-3-keto-threohexuronic acid lactone, and ceritamic acid. Some important properties for ascorbic acid or vitamin C are outlined in Table 5.16.

Sources

Meager dietary concentrations of ascorbic acid are found in fish, meat, pasteurized milk, and eggs; and it is entirely absent in dried cereal grains as well as cereal grains unless they have begun to sprout. Some of the most significant sources of ascorbic acid have been outlined in Table 5.15.

Rosehips, paprika, brussels sprouts, and pine needles also have high but variable concentrations of ascorbic acid, although they are not major dietary sources. Regardless of tabulated ascorbic acid values for food

Table 5.16. Chemical Characterization and Properties of L-Ascorbic Acid

Empirical formula	$C_6H_8O_6$
Molecular weight	176.0
Melting point	192°C
pK values	pK_1 = 4.2; pK_2 = 11.6
pH	2.0 at a concentration of 50 mg/mL
Solubility (g/100 mL of solvent)	
Water	33.3
Absolute alcohol	2.0
Glycerol	1.0
Insoluble in ethers, benzene, chloroform, and fat solvents.	
Specific rotation	$[\alpha]_D^{25}$ = 21.0° in water
Absorption maximum	245 nm in acid solution and 265 nm in neutral solution

Stability: L-Ascorbic acid crystals are stable to dry air environments, room temperatures, and sunlight; aqueous solutions ≤7.6 pH do not cause its oxidation when exposed to air, provided that copper and other substances are not available to catalyze a reaction; pH values ≤4.0 contribute added stability to dehydroascorbic acid while pH values >4.0 can facilitate the irreversible conversion to a biologically inactive substance.

Units: L-Ascorbic acid[a] is usually measured in mg, g, or international units (I.U.), where 1 I.U. is equivalent to the activity of 0.05 mg of ascorbic acid (i.e., 1.0 g = 20,000 I.U.)

[a]Hereafter, L-ascorbic acid will be designated as ascorbic acid.

sources, it should be clearly recognized that the concentration of ascorbic acid in foods is subject to plant-growth conditions, plant maturity, and post-harvest treatment of the plant materials.

Human Requirements and Deficiency Effects

The deficiency of ascorbic acid in humans causes scurvy, which inevitably leads to muscular weakness, listlessness, dry skin, arthritic joints, aching bones, and oral hemorrhagic symptoms that typically involve the gums. The hemorrhaging gums eventually become spongy, swollen, and increasingly susceptible to infection (Bourne, 1949) and gangrene.

Clinical and subclinical scurvy is avoided by a dietary ascorbic acid intake of 60 mg for adult males and females. Lactation and pregnancy require up to an additional 20 to 40 mg/day. Dietary levels of ascorbic acid for infants should equal or exceed 35 mg, whereas adolescents require at least 55 mg. These recommended values are subject to some flexibility since some authorities have provided for an allowance of 70 to 75 mg/day for adults (King, 1975). Although this amount may exceed by 50% the amount of ascorbic acid required for the maintenance of normal plasma ascorbate concentrations (0.6 mg/100 mL) and a normal body pool of ascorbate (~1500 mg), there is no firm toxicological evidence to support the contention that chronic levels of ascorbic acid above the RDA values may be harmful. Moreover, any individual's physiological demand for ascorbic acid may be markedly influenced by his exposure to heavy metals (e.g., mercury, copper, lead), environmental temperatures, consistent exposure to drug treatments (e.g., steroid contraceptives, aminopyrines, phenobarbital), and many other factors. Exclusive of these stress influences that increase the demand for ascorbic acid, it is speculated that an adult human body, initially saturated with ascorbic acid, requires at least 90 days of absolute ascorbate

deficiency before signs of scurvy are apt to materialize.

The highest stores of ascorbic acid in humans are found in the thymus, pituitary gland, corpus luteum, and, especially, the adrenal glands and aqueous humor.

Biochemical Function

The absence of a synthetic pathway for ascorbic acid in primates and some other animals stems from the lack of an enzyme known as L-*gulonoxidase* that converts L-gulonolactone to 3-keto-L-gulonolactone (Figure 5.62). This inadequacy necessitates dietary availability of the vitamin in these organisms if normal healthy life processes are to be maintained. Assuming adequate dietary availability, ascorbic acid is absorbed into the body during the transit of foods through the small intestine until the ascorbic acid holding capacity of the body has been saturated. Concentrations of ascorbic acid beyond requisite metabolic demands are promptly metabolized or excreted in the urine. Unlike some other vitamin substances, high concentrations of ascorbic acid cannot be warehoused for future use.

Apart from the obvious effects of ascorbic acid in eliminating the symptoms of scurvy, its biochemical activities are both *defined* and yet *obscure*. The biochemical activity of ascorbic acid is *defined* because it displays notable reducing properties that permit its oxidation to *dehydroascorbic acid*, but the mechanism underlying the actual *integration* of this behavior into biochemical pathways is still *obscure* in many cases and open to wide speculation.

The oxidation–reduction activity of ascorbic acid is very important for ensuring the integrity of substances such as glutathione (GSH). This tripeptide serves as (1) a reducing agent in certain prostaglandin syntheses (e.g., PGE_2), (2) an enzyme activator, and (3) a participant in amino acid transport mechanisms. Since oxidized glutathione (GSSG) is required at some point in a wide range of biochemical reactions, the

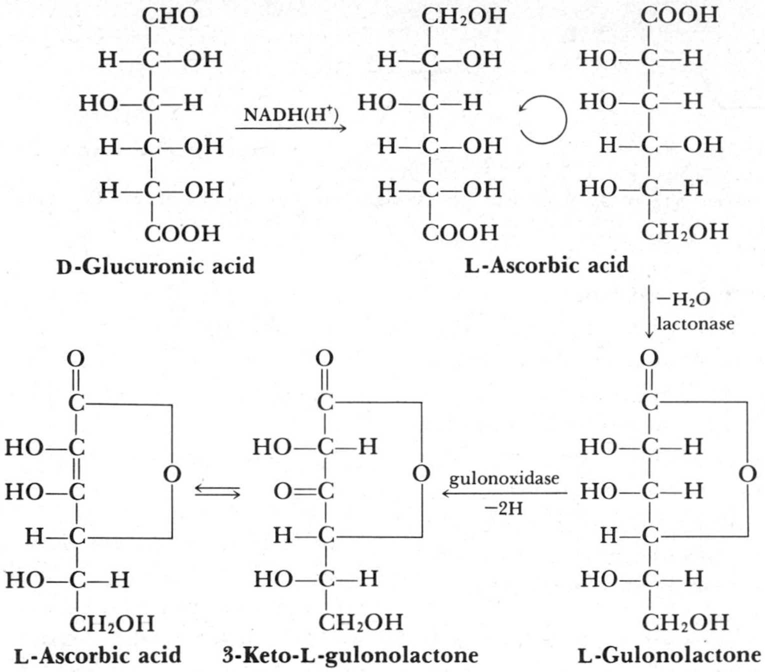

FIGURE 5.62. Biosynthetic pathway for L-ascorbic acid from D-glucuronic acid. Those species that require a dietary source of L-ascorbic acid do not have the L-gulonoxidase enzyme responsible for converting L-gulonolactone to 3-keto-L-gulonolactone.

interconversion of reduced and oxidized ascorbate forms can provide the biochemical means that establishes GSSG (Figure 5.63).

Ascorbic acid is also fundamentally important for collagen biosynthesis, wherein it acts as an external reductant in the conversion of proline to hydroxyproline. Hydroxyproline is critical for collagen because 25% of its structure consists of proline/hydroxyproline residues.

The activities of hydroxylase enzymes can also depend on the availability of ascorbic acid. In some cases, the ascorbic acid maintains necessary metal cofactors in a reduced state (e.g., Fe^{2+}), while other enzyme reactions such as those mediated by *dopamine β-hydroxylase*, found in nerve, brain, and adrenal tissues, uses ascorbic acid as a cosubstrate in a reaction that yields norepinephrine (noradrenaline) from dopamine (Figure 5.64).

It is also clear that the oxidative degradations of phenylalanine and tyrosine both rely on ascorbic acid as an external reductant

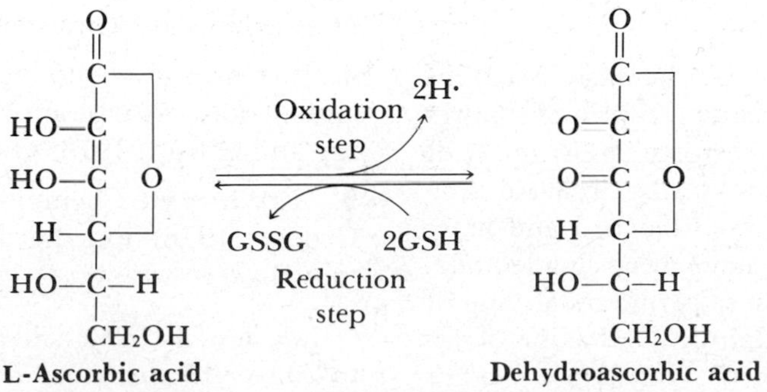

FIGURE 5.63. Interaction of ascorbate with glutathione.

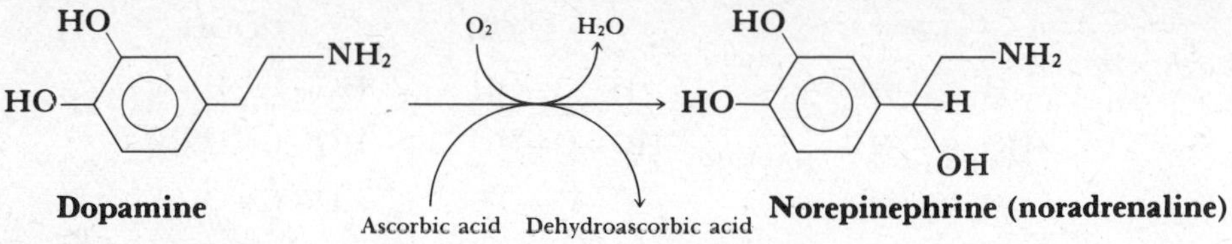

FIGURE 5.64. Site of ascorbic acid involvement during norepinephrine formation from dopamine.

in connection with the enzyme actions of 4-hydroxyphenylpyruvic acid dioxygenase, as well as homogentisic acid 1,2-dioxygenase, which also requires GSH (Figure 5.65). For guinea pigs, any *in vivo* biochemical deficit in this ascorbic acid oxidation sequence leads to increased urinary excretion of 4-hydroxyphenylpyruvic and homogentisic acids.

More general observations on the biochemical roles of ascorbic acid relate to

1. The biosynthesis of mucopolysaccharides for which ascorbate 2-sulfate facilitates sulfate transfer to connective tissues (Bond, 1975).

2. Deficits of ascorbic acid that encourage diabetogenic conditions caused by elevated concentrations of dialuric acid or alloxan; while concentrations of ascorbic acid far in excess of metabolic and physiological requirements may lead to corresponding excesses in production of peroxide, hydroxyl, or superoxide radical concentrations (Heikkila and Cohen, 1975).

3. Redox reactions of ascorbic acid that link it to electron transport (Weis, 1975).

4. The role of ascorbic acid in *intercellular* and *extra-* to *intracellular* solute transfers over *in vivo* membrane barriers (Mann and Newton, 1975).

There are many biochemical reactions other than those already recognized that require ascorbic acid. Unfortunately, many of these remain a mystery since reliable analytical methods for ascorbic acid and its *in vivo* redox species have been inadequate. Furthermore, an essential understanding of the *in vivo* electrochemical behavior of ascorbic acid is jaded by the coexistence of other cellular constituents that behave in a chem-

ically similar fashion. In view of these problems, plus the highly reactive nature of ascorbic acid, credible nutritional studies of its roles are extremely challenging.

Miscellaneous Considerations

Ascorbic acid intake beyond known metabolic requirements causes few if any severely toxic effects. Claims by Pauling (1970) encourage multigram (2.3–9.0 g/day) intake of ascorbic acid to avoid the onset and severity of the common cold, but these claims have been widely contested by medical authorities (Anderson *et al.*, 1972; Anderson, 1975; Coulehan *et al.*, 1975).

Some other substantiated effects of multigram ascorbic acid ingestion suggest that there is a potential likelihood for

1. Increased urinary oxalate excretion and possible oxalate stone formation in the renal system (Briggs *et al.*, 1973; Takenouci *et al.*, 1966).

2. Gastrointestinal tract disturbances resulting from large-scale ascorbic acid intake prior to and after meals (Coulehan *et al.*, 1975; Korner and Weber, 1972; Greer, 1955).

3. Increased thrombocytosis and decreased coagulation time following high doses of ascorbic acid (Schrauzer and Rhead, 1973).

4. Counteractivity to the effects of heparin and dicoumarol type anticoagulants (Schrauzer and Rhead, 1973; Owen *et al.*, 1970).

5. Erythrocyte hemolysis (Mengel and Greene, 1976) and the destruction of vitamin B_{12} (see vitamin B_{12} section).

Levels of ascorbic acid in foods are subject to 20 to 80% decreases depending on food-handling and -processing techniques. Causes

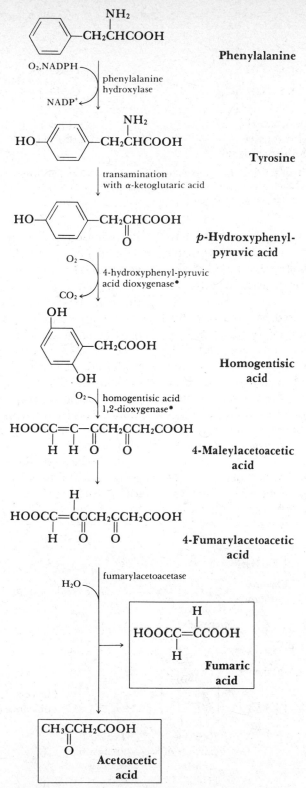

FIGURE 5.65. Steps in the oxidative degradation of phenylalanine and tyrosine that require the presence of ascorbic acid along with oxygenase enzymes. The specific ascorbic acid-requiring steps are denoted by an asterisk.

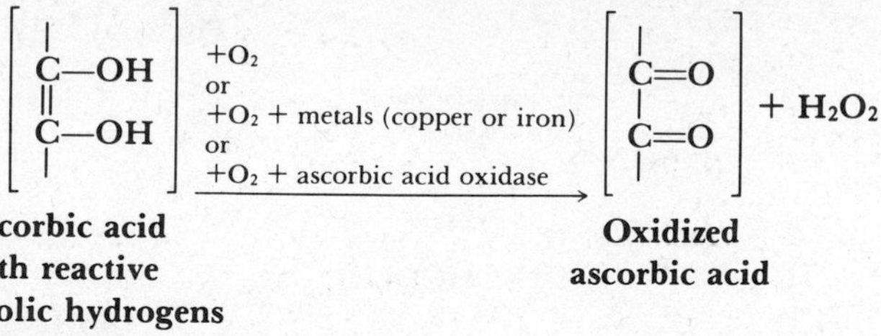

$$\begin{bmatrix} | \\ C-OH \\ || \\ C-OH \\ | \end{bmatrix} \xrightarrow[\substack{+O_2 \\ \text{or} \\ +O_2 + \text{metals (copper or iron)} \\ \text{or} \\ +O_2 + \text{ascorbic acid oxidase}}]{} \begin{bmatrix} | \\ C=O \\ | \\ C=O \\ | \end{bmatrix} + H_2O_2$$

**Ascorbic acid
with reactive
enolic hydrogens**

**Oxidized
ascorbic acid**

FIGURE 5.66. Direct oxidation of ascorbic acid to dehydroascorbic acid by oxygen; oxygen and metals (Cu^{2+} or Fe^{2+}); or ascorbic acid oxidase and oxygen.

for these losses may be as simple as the leaching of ascorbic acid from food tissues during blanching or cooking steps, while other significant losses in ascorbic acid are attributable to oxidation.

The oxidation of ascorbic acid to dehydroascorbic acid is singularly promoted by the presence of molecular oxygen, although the availability of copper or iron affects the rate of oxidation. Oxidations of ascorbic acid to dehydroascorbic acid are *directly* executed by enzymes such as *ascorbic acid oxidase* (Figure 5.66) or *indirectly* by *copper-containing peroxidase, phenolase,* or *cytochrome oxidase enzymes.*

Peroxidase enzymes, which occur widely in plants, indirectly promote the deterioration of ascorbic acid levels in plant tissue.

Peroxidase enzymes inevitably generate hydrogen peroxide, which oxidizes plant flavones to flavone oxides. Flavone oxides then oxidize ascorbic acid, which, in turn, reforms the original flavones and dehydroascorbic acid (Figure 5.67). Although the preceding oxidations of ascorbic acid require the presence of oxygen, this latter route proceeds under anaerobic conditions. Phenolases and cytochrome oxidases operate in similarly indirect fashions, respectively, by oxidizing mono- or dihydroxyphenols to quinones, or oxidizing cytochrome, which then oxidizes ascorbic acid.

Regardless of the oxidation path for ascorbic acid to dehydroascorbic acid, this oxidized form (dehydroascorbic acid) is liable to undergo added degradation in neutral

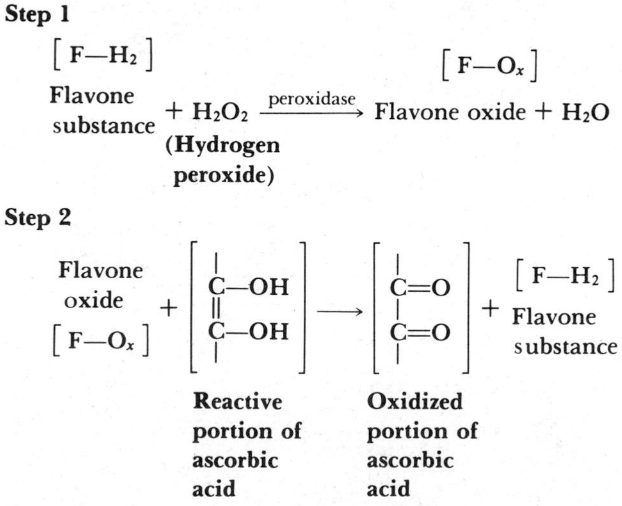

Step 1

$$\begin{bmatrix} F-H_2 \end{bmatrix}$$

**Flavone
substance** $+ H_2O_2 \xrightarrow{\text{peroxidase}}$ $\begin{bmatrix} F-O_x \end{bmatrix}$ Flavone oxide $+ H_2O$

**(Hydrogen
peroxide)**

Step 2

**Flavone
oxide**

$$\begin{bmatrix} F-O_x \end{bmatrix} + \begin{bmatrix} | \\ C-OH \\ || \\ C-OH \\ | \end{bmatrix} \longrightarrow \begin{bmatrix} | \\ C=O \\ | \\ C=O \\ | \end{bmatrix} + \begin{bmatrix} F-H_2 \end{bmatrix}$$

**Flavone
substance**

**Reactive
portion of
ascorbic
acid**

**Oxidized
portion of
ascorbic
acid**

FIGURE 5.67. Indirect oxidation of ascorbic acid by the action of peroxidase on hydrogen peroxide in conjunction with a flavone substance.

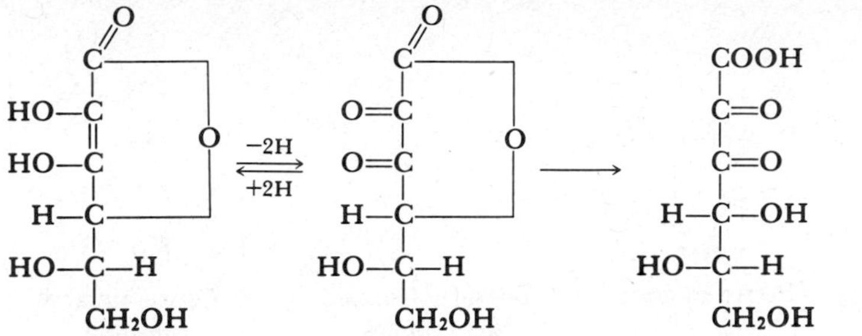

Ascorbic acid **Dehydroascorbic acid** **Diketogulonic acid**

FIGURE 5.68. Formation of a 2,3-diketohexuronic acid from ascorbic acid.

media to yield 2,3-diketohexuronic acid (Figure 5.68). Prerequisites for this conversion are not stringent since only mild heat (~60°C) for about 10 min, under anaerobic conditions can completely destroy the lactone ring of the dehydroascorbic acid. The diketo- derivative of hexuronic acid can then undergo an additional step to yield furfurals, which polymerize to form dark colors or combine with available amino acids, thereby producing a browning reaction (Figure 5.69).

In view of the foregoing degradative pathways, it is clear that ascorbic acid losses in foods can generally be minimized by effective food blanching, which eliminates enzyme activities, by minimizing metal ion availability in food-processing water, and by minimizing air–food contact.

The assay of ascorbic acid in foods has been based largely on its reactions with 2,4-dinitrophenylhydrazine (Roe *et al.*, 1948; Roe and Kuether, 1943) and 2,6-dichlorobenzenone indophenol, but a variety of other methods have been reported and surveyed by Roe (1961), Olliver (1971), Strohecker and Henning (1965), and Freed (1966).

Among the methods reported, the 2,6-dichlorobenzenone indophenol method has been most widely adapted to titrimetric and optical methods of analysis for ascorbic acid in foods. This assay is based on the reducing properties of ascorbic acid, which allow it to reduce the blue-colored indophenol dye to a corresponding, colorless (*leuco-*) form of the dye. A credible, quantitative reduction step here requires the prior extraction of ascorbic acid from any *in vivo* protein binders and stabilization of the ascorbic acid in 6% metaphosphoric acid. The quantitative formation of the *leuco* form of the indophenol dye results in a decreased absorption at 515 nm.

FOLIC ACID

This vitamin has been the key element of many investigations dealing with a variety of microorganisms and animal species. Along with these diverse investigations came a confusing roster of names that synonymously and unwittingly described the same vitamin substance common to many different species of organisms. These designations included vitamin M, a protective agent against cyto-

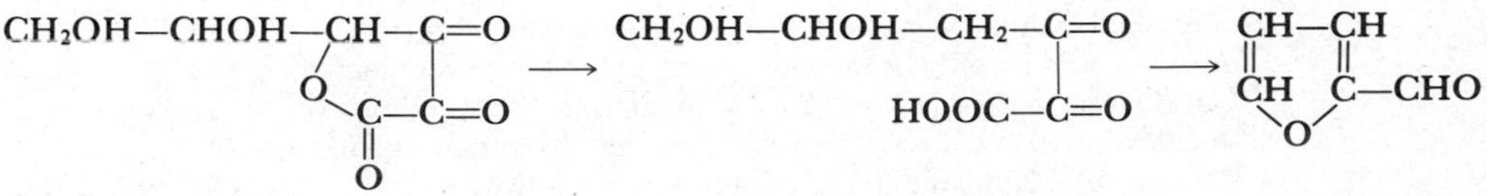

Dehydroascorbic acid **2-Keto-3-ketohexuronic** **Furfural**

FIGURE 5.69. Furfural formation from dehydroascorbic acid.

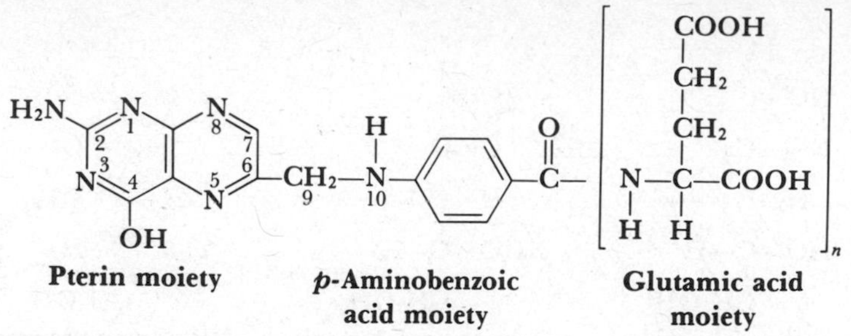

FIGURE 5.70. Pteroylglutamic acid (PGA) when $n = 1$; conjugates of PGA exist when $n = 3-7$ residues.

penia in monkeys; vitamin B_c, an *anti-anemia factor* for chicks; factor U, a growth substance required by chicks; the *Lactobacillus casei* (*L. casei*) factor, a factor required for the growth of this bacterium; and the "citrovorum factor," a required growth factor for *Leuconostoc citrovorum*.

Studies in the 1930s by Dr. Lucy Wills demonstrated a dominant occurrence of megaloblastic anemia in pregnant women who consumed large amounts of white rice and bread; however, the clinical anemia signs diminished when patients were administered autolyzed yeast supplements. The mysterious anti-anemia agent was called the *Wills factor* until studies in 1941 by Mitchell *et al.* (1941) isolated a substance known as *folic acid* from spinach leaves (Latin, *folium*), which paralleled the behavior of the *Wills factor*.

The ensuing years of research indicated that folic acid was one of a number of chemically related, nutritionally active substances derived from *pteroylglutamic acid* (PGA) or *N-{p-*{[(2-amino-4-hydroxy-6-pteridinyl)methyl]-amino}benzoyl}glutamic acid. Alternative common names often observed in the older literature are Cytofol, Foliamin, Folipac, Folsan, Folsäure, Foluite, and Inca Folic (Figure 5.70 and Table 5.17).

In addition to PGA, two to six additional glutamic acid molecules may be linked by peptide bonds to the initial glutamic acid residue. These *tri-* or *hepta*-glutamyl peptides are known as *conjugates*. *Conjugase* enzymes located in the chick pancreas, kidney, and other animal tissues permit animals to use the conjugates as a source of PGA, but the growth of microorganisms is rarely prompted

Table 5.17. Characterization and Properties of Folic Acid

Although many derivatives of folic acid exist, the properties for the most common form (PGA) have been detailed below:

Empirical formula	$C_{19}H_{19}N_7O_6$
Molecular weight	441.40
Melting point	None, but chars at 250°C
Optical rotation	$[\alpha]_D^{25} = +23°$

Solubility (g/100 mL of solvent)
    Water (25°C)   0.16 g
    Water (100°C)   1.00 g
  Insoluble in acetone, benzene, ether, and chloroform
Stability: Stable in 0.1 N sodium hydroxide solutions but less stable in acid solutions; stability of PGA increases as the pH of the solvent increases, and both the acid and salt forms of PGA are extremely photosensitive.
Absorption spectrum: Characteristic ultraviolet absorption maxima are observed at 257, 282, and 365 nm in 0.1 N sodium hydroxide.
Units: Folic acid activity is usually reported as µg or milligrams of the pure crystalline substance.

by these substances. Instead, microorganisms require only the *p*-aminobenzoic acid (PABA) moiety as a vitamin since they are able to synthesize the rest of the larger molecule. The PABA moiety occurs widely in plants and animals but its vitamin-like activity seems relegated entirely to microorganisms and not man or higher animals. Exceptions to this general rule are PABA-deprived rats that reportedly develop gray hair and chicks that seem to require PABA as a growth factor.

Parallels in structural chemistry between PABA and sulfa drugs, such as sulfanilamide, allow the sulfa drugs to exert antivitamin activity in pathogenic bacteria since they compete with PABA in the fundamental biochemical pathways leading to folate synthesis (Figure 5.71).

Sources

Yeast, mushrooms (*Basidiomycetes*), kidney, liver, and especially vegetables having deep-green leaves provide the richest vitamin sources for humans. Lesser amounts of folic acid are found in meats, cereals, fruits, and certain roots. Poultry, eggs, and milk are not good sources of the vitamin. Food storage and cooking markedly decrease concentration levels of folic acid.

Human Requirements and Deficiency Effects

Few vitamins have promoted as much confusion and scientific debate as folic acid. In spite of detailed research, it is not certain whether most current folic acid deficiencies epidemiologically stem from extremely poor dietary behaviors, lack of normal folic acid absorption from the intestine, or dietary adjuncts that interfere with normal absorption and utilization of the vitamin. The scientific record is far from clear.

Folic acid deficiency in humans is often exhibited as macrocytic anemia, accompanied by alimentary tract lesions, stomatitis, glossitis, diarrhea, and alterations in intestinal function. The onset of megaloblastic anemia and vomiting during pregnancy can stem from elevated folic acid requirements along with metabolic aberrations in folic acid coenzyme synthesis. Folic acid treatment of patients is often accompanied by significant reticulocyte responses and higher red blood cell counts in some cases of macrocytic anemia including sprue (Darby and Jones, 1945; Sheehym and Baggs, 1962) and pernicious anemia (Moore *et al.*, 1945; Waters and Mollin, 1961). Folic acid therapy *does not* have an impact on the neurological lesions that accompany pernicious anemia.

Notably poor diets and diets jointly deficient in vitamin C and folic acid, such as 100% milk diets for infants, encourage the development of megaloblastic anemia. Moreover, folic acid deficiencies are not uncommon in alcoholics since they have poor dietary habits and a retarded ability to *deconjugate polyglutamates* that occur in foods.

Experimental folate deficiencies and anemias are difficult to produce in mammals unless dietary deficiencies of the vitamin are supported with a folic acid antagonist, an ascorbate deficiency, excessive dietary concentrations of methionine, and a flora-free intestinal tract.

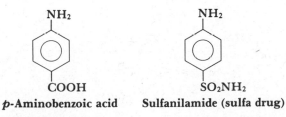

FIGURE 5.71. Comparative structures for *p*-aminobenzoic acid (PABA) and sulfa drugs.

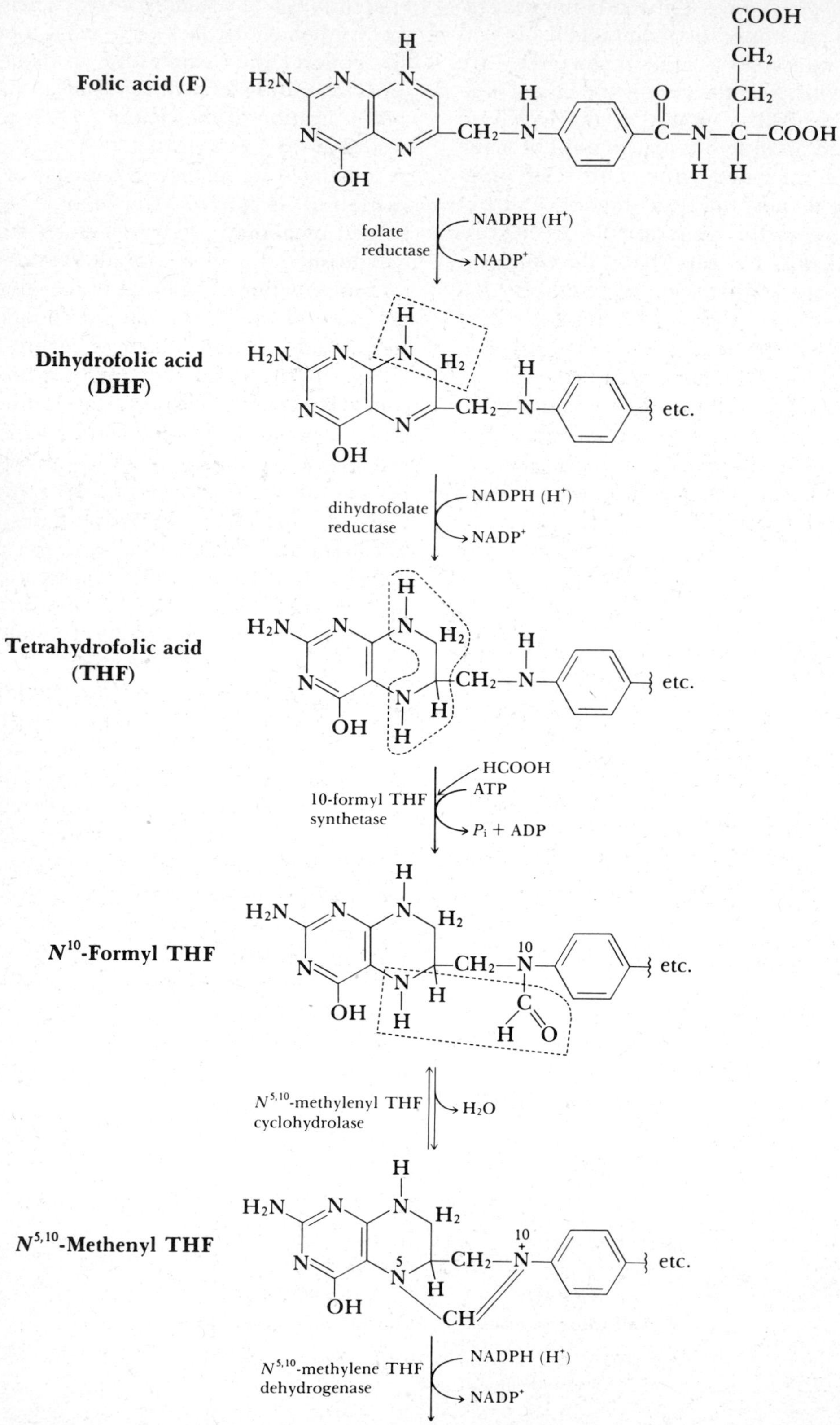
Folic acid (F)
folate reductase
NADPH (H⁺)
NADP⁺
Dihydrofolic acid (DHF)
dihydrofolate reductase
NADPH (H⁺)
NADP⁺
Tetrahydrofolic acid (THF)
10-formyl THF synthetase
HCOOH
ATP
P_i + ADP
N^{10}-Formyl THF
$N^{5,10}$-methylenyl THF cyclohydrolase
H_2O
$N^{5,10}$-Methenyl THF
$N^{5,10}$-methylene THF dehydrogenase
NADPH (H⁺)
NADP⁺
etc.

$N^{5,10}$-**Methylene THF**

N^5-**Methyl THF**

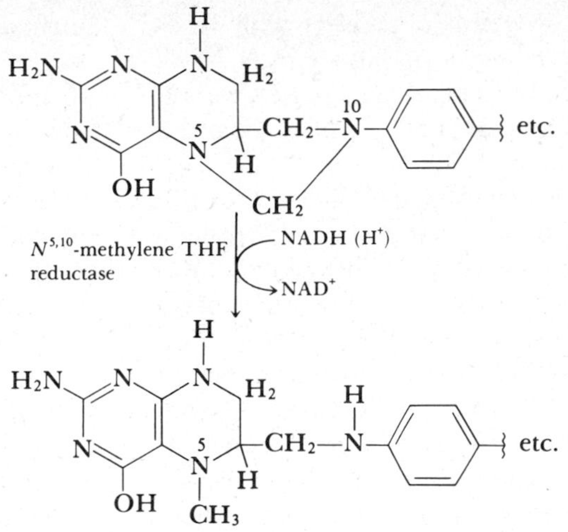

FIGURE 5.72. Significant biochemical events outlining the formation of N^5-methyl THF from folate.

Although folic acid is primarily stored in the liver, the serum, blood, urine and other body tissues contain an extremely diverse variety of folic acid forms and derivatives that can methodologically hamper biochemical investigations of the vitamin.

The recommended dietary allowances for folic acid are 0.4 mg/day for adults, along with an additional 0.1 mg during lactation and 0.4 mg during pregnancy. A regimen of oral contraceptive therapy reportedly stimulates folic acid requirements as a result of alterations in estrogen and/or progesterone levels.

According to most available evidence, folic acid is probably not toxic at levels in excess of the daily nutritional demands, but high doses may be counterproductive in some

is normally reduced to *dihydrofolic acid* (DHF) by the enzyme known as *L-folate reductase*. The DHF can be further reduced by *dihydrofolic reductase* to form a product known as *tetrahydrofolic acid* (THF). The NADPH coenzyme is required as a reducing agent for both reactions (Figure 5.72). The primary role of THF revolves around its activity in mobilizing one-carbon units such as *methyl* (—CH_3), *methylene* (—CH_2) *formyl* (—CHO), and *formimino* (—CH=NH) groups. The addition or removal of carboxyl groups (—COOH), respectively, depends on biotin, and thiamine pyrophosphate or the vitamin B_6-group vitamins. Folic acid is not involved in these reactions.

The role of THF as a transfer agent for single-carbon units is exemplified by the ac-

$$\text{THF} + \text{HCOOH} \xrightarrow[\text{Formic acid}]{\text{ATP} \quad \text{ADP} + P_i} \text{Formyl } N^{10}\text{—THF}$$
$$\text{Formic acid} \qquad\qquad\qquad\qquad \text{THF-formate complex}$$

conditions such as pernicious anemia where they can mask coexisting neurological problems.

Biochemistry

The reduction products of folic acid constitute its various coenzyme forms. Folic acid

tivation of formic acid, which gives a THF–formate structure. The formate, bound to the 10 position of folic acid, serves as a substrate for $N^{5,10}$-*methylenyl THF cyclohydrolase*, which achieves ring closure between the number 5 and 10 positions on the folic acid structure. The product of the ring closure is $N^{5,10}$-methenyl THF. The reducing action of NADPH, working in concert with $N^{5,10}$-*methylene THF dehydrogenase* yields $N^{5,10}$-methylene THF, which can be further reduced to N^5-methyl THF by NADH and a reductase (Figure 5.72).

The single carbon transfer activity of folate is fundamental to the *de novo* synthesis of purines (Warren *et al.*, 1959), pyrimidines (Luzzati and Guthrie, 1955), many amino acids (McIntosh *et al.*, 1957; Davis, 1951; Kaufman, 1958), and pantothenic acid (Greenberg and Humphreys, 1958). A schematic diagram for folic acid involvement in these reactions and others appears in Figure 5.73. Furthermore, the vitamin is involved in many novel reactions such as the prodigious development of carbon monoxide from N^{10}-formyl THF in the gas gland of the Portugese man-of-war, and biosphere production of carbon monoxide and methane by bacteria. The complex roles of folic acid also extend into the realm of vitamin B_{12} coenzyme activities during the synthesis of methionine and related mechanisms.

Miscellaneous Considerations

According to Burchenal (1955), *antifolic acid* activities are attributable to five general classes of chemical compounds. These classes are: (I) 4-amino pteroylglutamic acids and related derivatives; (II) 9-methyl or 10-methyl glutamic acid derivatives; (III) 2,4-diamino pteridines; (IV) dihydrotriazines, and (V) diamino-dichlorophenyl pyrimidines. The mode of antifolic acid activities for these substances are not uniform or indiscriminate of species. Classes II and V are effective antagonists for both bacteria and animals, class III compounds are effective against bacteria, and class IV substances act as noncompetitive inhibitors for folic acid in bacteria such as *Streptococcus faecalis* and *Lactobacillus citrovorum*. Class I substances, including *aminopterin*, have been widely employed to augment experimental folic acid deficiencies, whereas amethopterin (Methotrexate) has been used as a chemotherapeutic agent for certain cancers. Both substances act as *competitive inhibitors of dihydrofolate reductase*, which

normally produces THF. The resulting decrease in stores of THF can impede the proliferation of certain malignant cells in the bone marrow, dermal epithelium, intestinal and buccal mucosa, and urinary bladder since THF is required for the very active rate of nucleic acid synthesis and cell replication accompanying malignant cells.

Folate antagonists usually have a relatively short biological half-life. Although they may be partially metabolized by the body, 55 to 80% of 15 to 30 mg/day administrations of the antagonist is excreted in the urine within 24 hours.

Assay techniques for folic acid are diverse and debatable from many analytical perspectives. Most problems center around the chemical instability of naturally occurring folates; reliability and uniformity of biological and chemical assays for folates; quantitative extraction of folates from tissues and biological fluids prior to assay; and analytical detectability of polyglutamate folate derivatives.

Chick bioassays are reliable but protracted methods of analysis that have been largely bypassed by microbiological assays whenever routine assessments of folic acid are necessary. Typical microbiological assays employ *S. faecalis* or *L. casei*. The sensitivities of these methods fall in the minimum range of 0.1 ng (Strohecker and Henning, 1965; Freed, 1966).

Chemical methods such as polarography, column chromatography, and enzymatic assays have been designed primarily for pharmaceutical assays of folates (Strohecker and Henning, 1965; McCormick and Wright, 1971), but they have marginal or negligible applications to most food systems. The most recent advances in folate assay rely on radioanalytical principles. These methods are based on sequential competitive binding (Rothenburg *et al.*, 1972) of a radioactive ^{125}I-folic acid derivative and nonradioactive folic acid (and its derivatives) to a *folate-binding protein* (FBP) (β-lactoglobulin(s)) obtained from milk. The binding relationship for these components has been detailed in Figure 5.74.

Following a brief incubation period of the

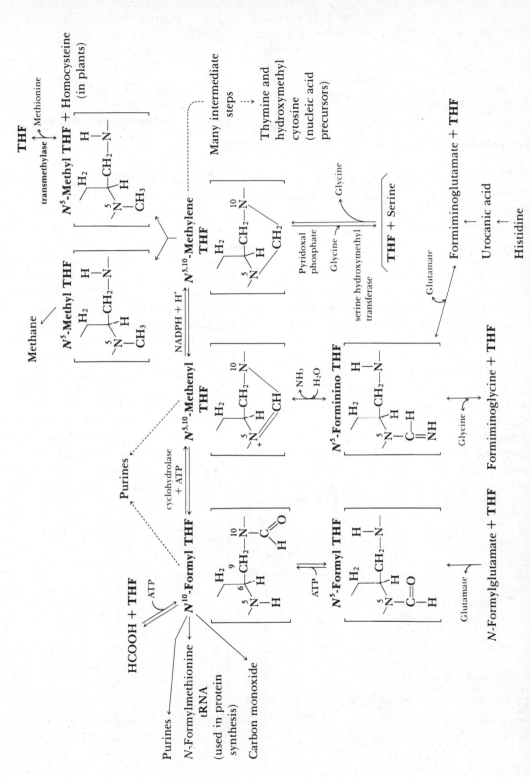

FIGURE 5.73. Possible biochemical roles for folate and its various coenzyme forms ([ ] = partial structure).

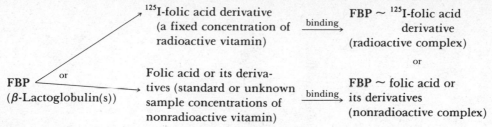

FIGURE 5.74. Schematic basis for radiometric folate assays.

FBP with a fixed concentration of radioactive ^{125}I-folic acid and a series of standard but different nonradioactive folate concentrations, the FBP-bound folate forms (radioactive *and* nonradioactive) are *separated from unbound* folates by adsorption to a binding agent. The radioactivity of the resulting supernatant liquor is counted for all standard but different folic acid concentrations over a typical range of 0.0 to 25.0 ng/mL. This radioactive counting step then serves as a basis for the establishment of a standard curve (Figure 5.75). The similar treatment of unknown specimens and comparison of their radioactivity levels with the standard curve reveals their folic acid concentration.

Although these radioassays have met with qualified acceptance in the clinical domain for evaluating serum and other body fluids, radioassays are considered to have uncertain analytical credibility for the analysis of food-stuffs (Ruddick *et al.*, 1978). Most problems indicate a lack of uniformity and reproducibility in folate assays caused by inconsistencies in the type of folate present in unknown samples, as well as the particular folate derivative employed for the establishment of the standard curve with which the unknown sample is to be compared. Unfortunately, the derivatives of folic acid display idiosyncratic behaviors during their quantitative radioassay. Moreover, additional confusion stems from what form of folic acid seems to be the most appropriate assay standard, since the terms *folic acid* or *folate* refer, not only to PGA, but also the generic classification of all types of PGA involved in nutrition. The latter generic classification includes those forms of PGA having varying degrees of glutamate polymerization, various N^5 and N^{10} derivatives of folic acids, and various degrees of pyrazine moiety reduction.

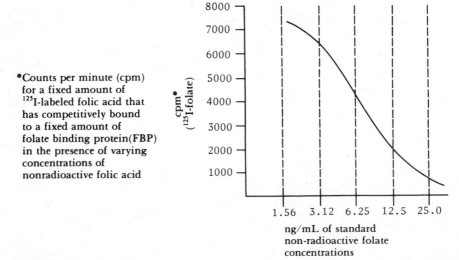

*Counts per minute (cpm) for a fixed amount of ^{125}I-labeled folic acid that has competitively bound to a fixed amount of folate binding protein(FBP) in the presence of varying concentrations of nonradioactive folic acid

FIGURE 5.75. Typical standard curve used as a basis for evaluating unknown folate concentration levels.

PART II
Fat Soluble Vitamins and Miscellaneous Vitamin-Like Substances

INTRODUCTION

The maintenance of normal biochemical activities in man and many other organisms requires certain fat-soluble vitamins such as A, D, E, and K as well as the *vitamin-like* substances known as *lipoic acid, choline,* the *ubiquinones, coenzyme* Q_n, and *myo-inositol*, which are discussed in the following pages.

All of these substances effectively mediate life-supporting reactions that would otherwise be biochemically impossible or enormously complex in their absence. Contrary to *true* vitamins, however, the availability of *vitamin-like* substances does not depend solely on exogenous dietary acquisition. For humans, it is clear that available *in vivo* biosynthetic pathways *plus* dietary sources provide more than adequate concentrations of these substances for all life processes. In fact, since few if any deficiencies have ever been clinically recorded, the importance of dietary availabilities for vitamin-like substances is very questionable.

VITAMIN A

McCollum and Davis (1913, 1915) and Osborne and Mendel (1913, 1915) observed that an accessory growth factor, contained in certain fats, was required as a nutritional supplement for the growth and normal development of rats. This growth factor was designated as the *fat-soluble "A" factor* to distinguish it from *water-soluble "B" growth factors* required by animals. The name *fat-soluble "A" factor* was modified to *vitamin A* after the structure of the growth-active substance was elucidated by Karrer and others (Karrer *et al.,* 1931, 1933; Heilbron *et al.,* 1932).

The vitamin A substance was eventually found to occur as vitamin A_1, or *retinol,* in mammals and the oils of marine fishes; and vitamin A_2, or *3-dehydroretinol,* in the oils of certain freshwater fishes (e.g., pike). Both structures of the vitamin are similar in that they contain a β-*ionone ring* and a polyisoprenoid (polyprenyl) side chain. Vitamin A_2, however, contains an additional double bond in the 3,4 position of the β ring and demonstrates only 40% of the bioassay activity shown by A_1. Nevertheless, both substances are conventionally referred to as vitamin A.

Vitamin A is required by all vertebrates; it can be extracted from the unsaponifiable fraction of animal fats. Plants do not contain vitamin A, but they have a wide range of naturally occurring carotenoids, *some of which may be converted to vitamin A.* Carotenoid precursors of vitamin A, or one of the *provitamins A* such as β-carotene (Figure 5.76), are conventionally modified to active vitamin A (retinol) in the animal body through a series of very specific enzyme reactions further linked to reduction of the reaction products (Figure 5.77).

The symmetrical 40 carbon structure of β-carotene can theoretically produce *two units of vitamin A activity per molecule.* This 1:2 ratio between β-carotene and retinol production is hardly realized *in vivo* since (1) whole β-carotene may be absorbed into the lymphatic system prior to retinol conversion; (2) less than 100% of the provitamin may actually be absorbed into the body from potential dietary supplies; or (3) whatever β-carotene is cleaved to yield retinol can undergo further oxidation to retinoic acid, which is excreted. Moreover, since the presence of a β-ionone ring is critical to retinol activity, any provitamin substances having *less than* the two β-ionone rings present in β-carotene are automatically inferior provitamin A sources. Accordingly, vitamin A activity of most fruits and vegetables is contingent upon their β-carotene content, while additional α- and γ-

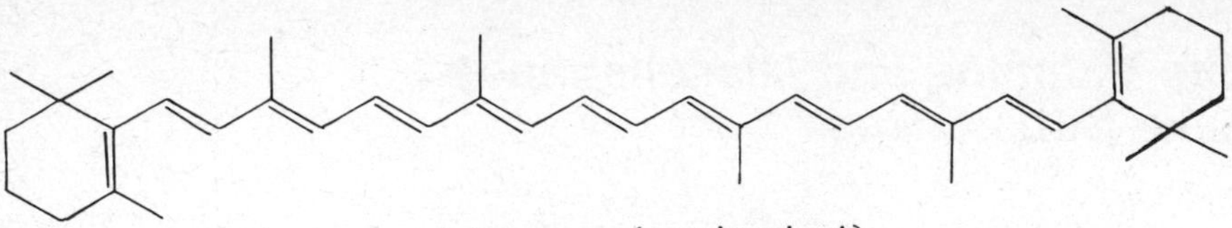

β-Carotene (provitamin A)

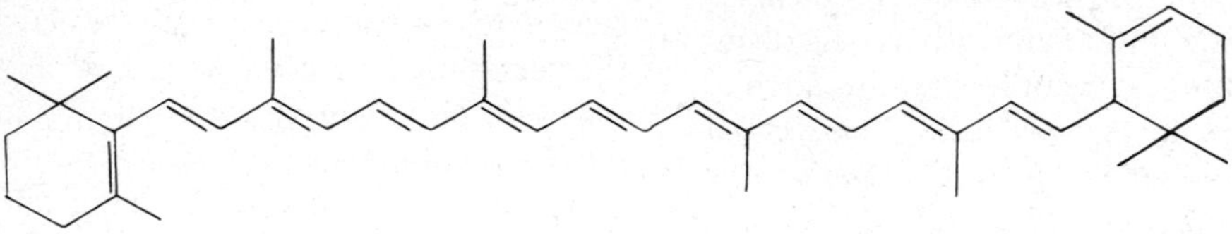

α-Carotene

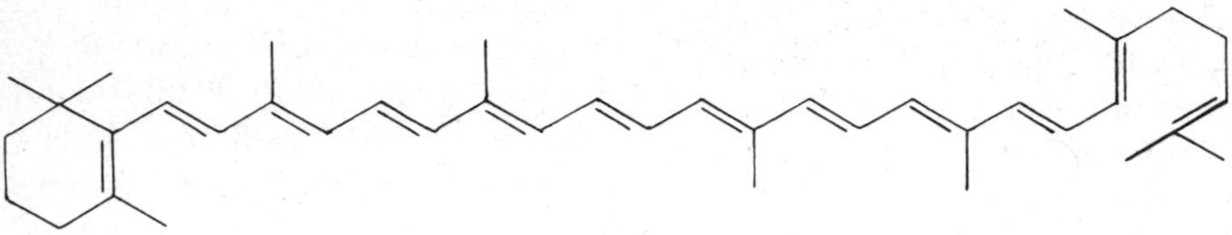

γ-Carotene

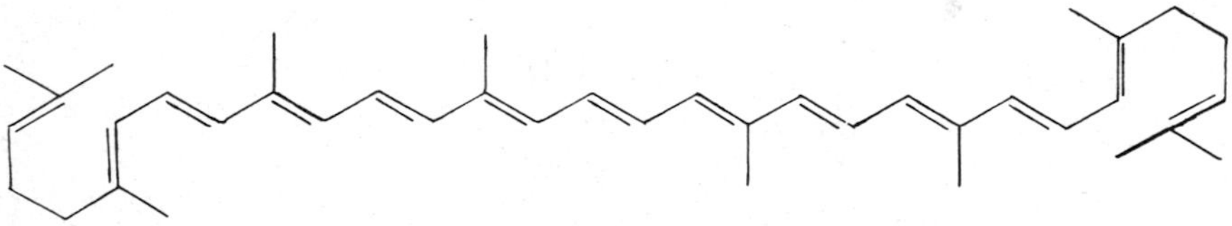

Lycopene

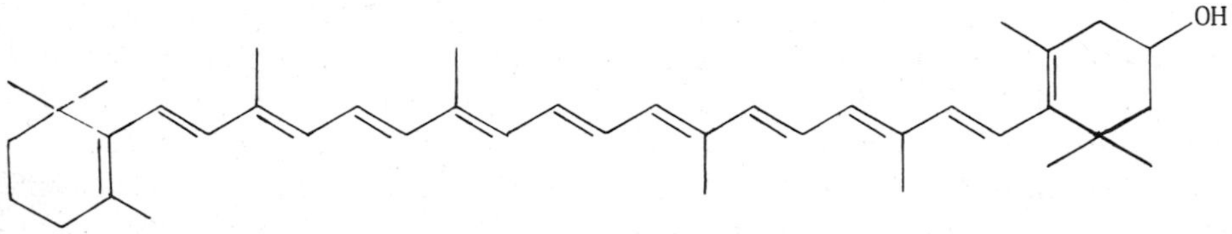

Cryptoxanthin

FIGURE 5.76. Some selected carotenoid pigments that occur in plants.

carotenoids, cryptoxanthins, and other carotenes (e.g., neo-β-isomers) contribute only one-half the vitamin A equivalent of β-carotene under the best conditions of physiological conversion. *Lycopenes* found widely in foods and most notably in foods such as tomatoes, do not exhibit any provitamin A activity whatsoever since they have no β-ionone rings. It should be noted that an accurate assessment of vitamin A activity for all provitamins A is impossible to establish since biological responses of test animals depend

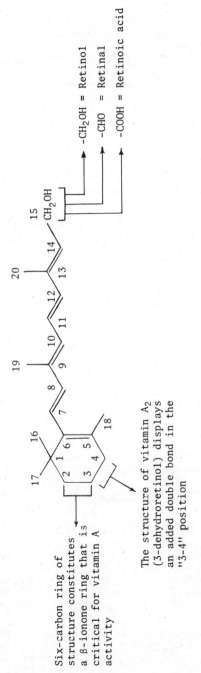

Structure for vitamin A₁ (retinol)
or
all-trans-retinol

FIGURE 5.77. Structure for retinol and some other important reduction products of provitamin A precursors.

Table 5.18. Relative Biological Potencies of Carotenoid Stereoisomers

Carotenoid stereoisomer	Relative biological potency (%)	Carotenoid stereoisomer	Relative biological potency (%)
all-*trans*-β-Carotene	100	all-*trans*-γ-Carotene	27–42
all-*trans*-Cryptoxanthin	57	neo-β-Carotene U	38
neo-β-Carotene B	53	neo-Cryptoxanthin U	27
all-*trans*-α-Carotene	53	neo-γ-Carotene P	19
neo-Cryptoxanthin A	42	neo-α-Carotene U	13

on animal species, nutritional status of the test animal at the time of provitamin assay, and the ability of test animals to utilize certain vegetable sources that contain one or more provitamins. As indicated by Freed (1966), these difficulties have necessitated the United States Pharmacopeia (U.S.P.) to arbitrarily equate 0.6 μg of β-carotene as the equivalent of 1.0 U.S.P. unit of vitamin A. An exemplary range of biological activities for various carotenoid stereoisomers has been tabulated with respect to all-*trans*-β-carotene, which exhibits 100% of potential biological potency (Zechmeister, 1949) (Table 5.18).

Chemical Characterization of Provitamins A and Vitamin A

The importance of provitamins A and vitamin A to the nutritional value of foods requires attention to the chemical properties of both classes of substances. In general, provitamins A are all soluble in fats and fat solvents although reticent solubility occurs in alcohol; they are liable to autoxidation, enzymatic oxidation (e.g., lipoxygenase), and photodestruction. Provitamins A are heat stable in a deoxygenated environment with the main exception being stereoisomeric changes, and they all display closely allied

Table 5.19. Characterization for Some Provitamin A Compounds

Substance	Provitamin A			Absorption maxima for various solvents (nm)	
	Empirical formula	Molecular weight	Melting point (°C) evacuated tube	Wavelength(s) (nm)	Solvent
α-Carotene	$C_{40}H_{56}$	536.85	187.5	485, 454	(Chloroform)
β-Carotene	$C_{40}H_{56}$	536.85	183	497, 466	(Chloroform)
γ-Carotene	$C_{40}H_{56}$	536.85	152–153.5	437, 462, 494	(Petroleum ether)
Cryptoxanthin (hydroxy-β-carotene)	$C_{40}H_{56}O$	552.85	158–159	452, 480	(Chloroform)
Lycopene	$C_{40}H_{56}$	536.35	172–173	All-*trans* form 446, 472, 505 15,15′ *cis* form 361, 444, 470, 502 (carbon disulfide)	

absorption spectra that vary only slightly depending on the type of solvent employed. Some of the most notable properties for the provitamins A as well as vitamins A_1 and A_2 have been detailed in Tables 5.19 and 5.20.

Sources

The occurrence of vitamin A or retinol is limited to products such as milk, butter, eggs, and the livers of animals or certain marine fishes. The majority of vitamin A in humans probably originates from the conversion of provitamins A, although only one-sixth of dietary carotene undergoes retinol conversion and absorption into the body.

Highly pigmented fruits and vegetables including papayas, pumpkins, sweet potatoes, carrots, red palm oil produced in the tropics, and the dark green foliage of vegetables contain notable provitamin A concentrations. It is interesting to note that the level of carotenoids present in some foods is lower than might otherwise be expected based on visual inspection alone. Tomatoes, for example, have a $9:1$ ratio between lycopene having no vitamin A activity and provitamin A-active β-carotene.

A generalized prospectus of retinol content measured in terms of International Units per 100 g of plant material indicates that most yellow and green plants have approximately 1500 to 2000 I.U.; butter fat, 100 to 4500 I.U.; and animal livers, 12,000 to 17,000 I.U.

Table 5.20. Notable Properties for Vitamins A_1 and A_2

	Vitamin A_1	Vitamin A_2
Physical form	Pale yellow needles	Golden yellow oil
Empirical formula	$C_{20}H_{30}O$	$C_{20}H_{30}O$
Molecular weight	286.00	284.42
Boiling point	120–125°C at 5×10^{-3} mm	—
Melting point	62–64°C	17–19°C (all *trans*) 73–74°C (13-*cis*) 77–79°C (4-*cis*) < -30°C (9,13-di-*cis*)
Solubility properties	Insoluble in water and glycerol, but soluble in fats and most organic solvents (e.g., ethers and chloroform)	Soluble in oils, fats, and nonpolar solvents
Optical activity	Absent	Absent
Absorption maxima	324–325 nm	351, 287 nm
Fluorescence	Characteristic green fluorescence in ultraviolet light	Absent
Stability	Vitamin A is destroyed by ultraviolet light and shows extreme sensitivity to air by being oxidized except when dissolved in oils, in which it is quite stable; heat stable only in inert atmospheres or in alkaline solutions.	Readily destroyed by oxygen or ultraviolet light
U.S.P. units of vitamin A activity/g	3.33×10^6 I.U./g (by definition)	1.33×10^6 I.U./g

Human Requirements and Deficiency Effects

Clinical deficiency effects of vitamin A (or retinol) in humans have been recognized for many years; however, its exact biochemical roles in physiological events remain obscure and enigmatic.

Deficiencies of the vitamin result in a variety of symptoms such as dessication of the hair, skin, and conjunctiva of the eye; growth retardation; and depreciated resistance to infection (Bernard and Halpern, 1968; Bieri *et al.*, 1969). Foremost among classic vitamin A deficiency effects are *night-blindness,* or *xerophthalmia,* and *keratinization* of the epithelial cells lining the respiratory passages and alimentary tract. These internally keratinized tissues become very reminiscent of keratinized squamous cells located on the epidermis.

The roles of vitamin A in reproductive processes as well as epidermal biochemistry are indisputable, but most evidence supporting its actual roles are circumstantial or speculative. It has also been suggested by many investigators that the clinical diversity of vitamin A deficiency effects may actually be reflected by enzymes that directly or indirectly depend on the vitamin in some way. Anomalies in the availability of steroid enzymes involved in the synthesis and metabolism of glucocorticoids, mineral corticoids, androgens, and estrogens are specifically thought to contribute to growth and reproductive maladies (Pohanka *et al.*, 1973; Juneja *et al.*, 1969; Rogers and Bieri, 1968); while enzymes required for synthesis of mucopolysaccharides in epithelial tissues may also be impaired (Rogers, 1969; Olsen, 1969; Deluca *et al.*, 1972).

Other postulated roles for vitamin A include its regulatory controls on the transcription of RNA into proteins, thereby affecting the differentiation of epithelial cells, biosynthetic availability of steroids as well as other processes (Wasserman and Corradino, 1971).

The action of vitamin A as an electron acceptor has been proposed. This possibility has yet to be firmly established, but the theory is supported somewhat by one of the best understood light-sensitive membrane proteins—*bacteriorhodopsin*—found in *Halobacter halobium.* This bacterium utilizes 11-*cis*-retinal to gather light energy, which is consequently used to pump ions for energy processes within the organism. Research by Lucy and Lichti (1969) and others have also demonstrated the *in vitro* electron transport capabilities of vitamin A.

The fat-soluble property of vitamin A allows the body to stockpile excessive amounts of the vitamin in adipose tissues and various organs. It is believed that the body can store enough vitamin A to satisfy biochemical demands for 1 to 2 years. The efficient storage mechanism for retinol is coupled with toxic side effects in cases where chronic intake is ≥50,000 I.U./day.

Hypervitaminosis effects include excessive sweating and hyperpigmentation on the soles of the feet, palms of the hands, and around nasolabial folds. Advanced symptoms are evidenced as increased intracranial pressure, edema, impaired liver function, and a variety of maladies too numerous to list. Well over 200 references and studies have been presented in the literature that deal with retinol toxicity, but some of the more useful references are those by Moore (1957, 1967), Anonymous (1959), Stimson (1961), Krause (1965), Rubin *et al.* (1970), Muenter *et al.* (1971), Hayes and Hegsted (1973), and Nieman and Klein Obbink (1954). As in cases of retinol deficiency, the biochemical explanations for vitamin A hypervitaminosis symptoms require additional study.

Biochemical Function

The only well documented biochemical functions for retinol involve its roles in the vision process of man and the eyes of most other animals. On the basis of present studies, the participation of retinol in vision processes resides strictly in the retina of the eye, which contains two types of light receptor cells known as *rods* and *cones*. The rods are responsible for *scotopic vision*, or that portion of visual perception that allows distinction of gray shades, while cone cells provide spectral sensitivity to the eye and the basis for color vision. The role of retinol in these processes is unquestionable, but known bio-

chemical processes provide a meager explanation of the overall vision process.

The participation of retinol in photo-optical stimulation of the brain depends initially upon its dietary availability in the intestine as provitamin A or retinyl esters (Figure 5.78). These substances are con-

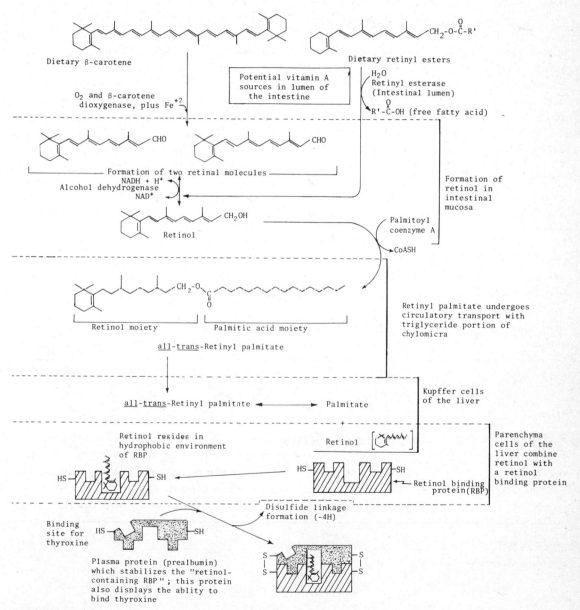

FIGURE 5.78. Biochemistry and serum transport of retinol and its subsequent utilization in photomechanisms of the eye.

verted to retinol in the intestinal mucosa and then converted to all-*trans*-retinyl palmitate. Although small amounts of free retinol may exist in the body, the palmitoyl ester of retinol is most common.

The all-*trans*-retinyl palmitate is eventually introduced into the circulation along with the triacylglycerol portion of the chylomicra and then stored in the Küpffer cells of the liver. Here the all-*trans*-retinyl palmitate undergoes repetitive conversion to palmitate plus retinol and reesterification to form all-*trans*-retinyl palmitate. In cases of physiological demand, retinol is *packaged* into a transport protein that also binds and transports thyroxine. These substances are bound at different sites on the transport protein, called a *retinol-thyroxine binding protein* (R-TBP). The blood plasma typically contains 35 to 43 mg/L R-TBP.

The appearance of retinol-containing transport protein at the retina is followed by a dynamic and rapid set of events that involve

1. The release of retinol as a retinyl-ester.

2. Lipolytic hydrolysis of the retinyl-ester to yield retinol.

3. Oxidation of retinol by a specific dehydrogenase along with NAD$^+$ to form all-*trans*-retinal.

5. Combination of the 11-*cis*-retinal with specific retinyl-binding proteins, called *opsins* that are present in both rods and cones. The opsins plus 11-*cis*-retinal are linked, probably by means of a Schiff base that involves a lysine residue on the opsin. This linkage forms the basis for an extremely photosensitive visual pigment known as *rhodopsin* (visual purple).

The rhodopsin is contained within a series of disc-like membranous structures stacked end to end in order to form columnar structures located at the interior end of the rods. Light energy that intercepts any given rod on the retina is then absorbed as it passes along the length of the rod. The energy dissipated along the path of photon absorption initiates a conformational shift in the rhodopsin structure such that the opsin-linked 11-*cis*-retinal is transformed into all-*trans*-retinal. The retinal interactions are outlined below for the visual cycle (Figure 5.79).

The strictly photochemical transition of 11-*cis*-retinal to all-*trans*-retinal, sometimes referred to as *rhodopsin bleaching*, culminates in three profound consequences:

1. Conversion of 11-*cis*-retinal to all-*trans*-retinal seems to alter the ionization charac-

4. Conversion of all-*trans*-retinal to 11-*cis*-retinal by the action of a suspected retinal isomerase.

teristics found on certain portions of the rhodopsin protein structure.

2. The altered ionization properties stim-

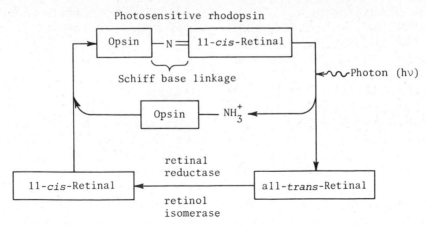

FIGURE 5.79. Interconversions for retinal during photoabsorption in the rod cells.

ulate the uptake of protons by the protein structure.

3. The all-*trans*-retinal then develops a very tenuous association with the formerly photoactive rhodopsin and eventually separates from the complex to yield the *free opsin protein* and all-*trans*-retinal.

Through mechanisms that are far from lucid, these events sequentially *stimulate* further *cytoplasmic movement of calcium ions* from the rod vesicles and *alterations in the flux of sodium ions* across the plasma membrane of the rods; *an action potential is developed* and propagated along the optic nerve toward the optic center of the brain. For a more detailed prospectus on the vision process exclusive of vitamin A activities, consult Metzler (1977).

The preceding discussion has dealt with the role of vitamin A in scotopic vision. It should be noted, however, that the role of vitamin A in color or *photobic* vision is just as complex. The cone cells also contain 11-*cis*-retinal linked and positioned in a biochemical milieu that permits three different types of cone cells to respectively absorb photostimuli having wavelength maxima of ~440, ~535, and 575 nm. Photostimulation of these cells results in spectral discrimination and blending of the primary colors at the vision center of the brain and serves as a primary impetus for color vision. Although free retinol displays an absorption maximum in the ultraviolet range, protonation of the Schiff base linking 11-*cis*-retinal with key lysine residues, contained in cone proteins, causes a shift in the maximum absorbance of retinal to ~440 nm. Studies by Arnaboldi *et al.* (1979) and Honig *et al.* (1979) have further demonstrated that the ability of the Schiff base-linked retinal chromophore to detect primary colors may also depend on the localization of external charges around the visual pigments that superficially protrude from surrounding functional groups on cone proteins. Photoactivation of the cone cells then presents the vision center of the brain with a rainbow of colors as opposed to scotopic sensations produced by the rods. The visual refinement of photobic and scotopic images by the brain is not understood.

Miscellaneous Considerations

The reliable assay of vitamin A can be a protracted affair due to its photosensitivity and susceptibility to undergo rapid oxidation. The Carr–Price reaction has served as the historical basis for assessments of vitamin A since it characteristically reacts with antimony trichloride in chloroform to yield a blue chromophore with characteristic absorption at 620 nm (Carr and Price, 1926). Successful vitamin A assays have also been accomplished by fluorometry (Hansen and Warwick, 1969); thin-layer chromatography

(Targen *et al.*, 1969; Deutsch *et al.*, 1964; Friedman, 1960; McRoberts, 1962; Wilkie and Jones, 1954), and magnesia (Friedman, 1960). Additional details on vitamin A assay have been detailed by Drujan (1971), Ames and Lehman (1960), Strohecker and Henning (1965), and the most current edition of the *Official Methods of Analysis*.

Provitamin A content as β-carotene can also be useful for estimating the potential vitamin A content of foods and many raw feedstocks, but care must be exercised to eliminate inactive provitamin A–like carotenes such as lycopene before any assay of provitamin A–active forms. Such assays are effectively conducted through the use of reverse-phase high-pressure liquid chromatographic analysis (Zakaria *et al.*, 1979). The analytical evaluation of provitamins A in this fashion, owing to high resolution of lycopenes and α- and β-carotenes, certainly has the potential for replacing historical methods of provitamin A analyses (Freed, 1966; Strohecker and Henning, 1965).

Notable provitamin A losses in foods often parallel the free radical destruction and rancidity reactions of fats and oils. This action commonly occurs during the extended storage of margarines, butter, and related products. Deterioration of provitamins A by direct or indirect actions of lipoxidases also contributes to decreases in vitamin A activity. Therefore, it is often desirable to preserve provitamin A carotenes in fruits and vegetables by processes of deaeration or steam blanching prior to canning or dehydration. Furthermore, preservation of both vitamin A and provitamins A are also encouraged by minimizing the light transparency of food-packaging materials.

Antagonists or *antivitamin A* substances are not widely reported, although it has been suspected that certain essential oils, citral, and bromobenzene may influence vitamin A biochemical activity in rats, while errant halogenated naphthalenes in cattle feed may initiate vitamin A deficiencies along with bovine hyperkeratosis. From perspectives of human health, it is uncertain, yet doubtful,

that antagonistic links in rats similarly affect humans (Meunier *et al.*, 1949).

VITAMIN D

The earliest accounts of rickets seem to originate with Whistler in the 1640s, but the complexities of this affliction in conjunction with the biochemical roles of vitamin D have been detailed with certainty only since 1969.

Research by Mellanby (1919) first reported scientific data supporting the antirachitic effects of fish oils and certain other foods on children, but persistent beliefs by others suggested that the antirachitic effects of fish oils were really attributable to their vitamin A content.

McCollum *et al.* (1922), however, succeeded in oxidizing vitamin A in fish oils only to find that they still retained their characteristic antirachitic properties. In spite of mounting evidence to substantiate the existence of an antirachitic factor, the nature of the "factor" became more enigmatic in view of additional research (Steenbock, 1924) that showed that photo-irradiation of foods and rachitic children eliminated the symptoms of severe rickets. Now, long after these studies, it is clearly recognized that the active *antirachitic factor*, or *vitamin D substance*, evolves only *after* provitamin D sterols are sequentially absorbed from the intestine, photo-irradiated, and hydroxylated.

Ultraviolet irradiation (275–300 nm) of $\Delta^{5,7}$-unsaturated provitamin D sterols such as *ergosterol* or *7-dehydrocholesterol* respectively yield *ergocalciferol* (vitamin D_2) or *cholecalciferol* (vitamin D_3). The natural occurrence of 7-dehydrocholesterol is limited to animal sources (e.g., fats, oils) while ergosterol is found in yeast, fungi, and many vegetables. Although cholecalciferol-type sterols are effective in ameliorating rachitic symptoms, their rate of effectiveness is inferior to *25-hydroxycholecalciferol* (25-HCC) and *1,25-dihydroxycholecalciferol* (1,25-DHCC), both of which are progressive metabolic derivatives of *cholecalciferol* (CC). The conversion of CC to 25-HCC occurs in the liver, and the structural transition of 25-HCC to 1,25-DHCC takes place in the kidney. Since the 1,25-DHCC shows the most biological activity of all provitamins D, there is little doubt that this structure represents the culmination of provitamin D modifications and the actual vitamin D substance.

The vitamin D content of foods is usually assessed on the basis of its D_2 and D_3 forms, which are present. The most significant properties of these two forms are outlined in Table 5.21.

The occurrence of vitamin D in foods is somewhat limited when compared to other vitamins. Exemplary distributions of the vitamin D_3 form, measured as International Units per 100 g include butter, 40 I.U.; cheese, 10 I.U.; eggs, 50 to 100 I.U.; margarine, 300 I.U.; and fatty marine fish oils 10–55,000 I.U. Actual amounts of vitamin D in processed foods may be somewhat variable as a result of enrichment and fortification efforts. For example, natural milk has very little potential as a natural vitamin D source, but ~93% of milk available in the marketplace contains 400 I.U./quart.

Human Requirements and Deficiency Effects

A steadfast requirement for vitamin D in humans is largely speculative. However, it is believed that required amounts of vitamin D can be satisfied by usual dietary habits and occasional exposure to sunlight. It is generally estimated that 400 I.U. (10 μg of cholecalciferol) will provide an adequate margin of safety for the vitamin D requirements of the elderly, children, and pregnant or lactating women. Premature infants have a more probable predilection for developing rickets than full-term infants, but infant vitamin D levels and demands ought to be carefully monitored in all cases.

The biochemical role of vitamin D is relegated to the normal intestinal absorption and assimilation of calcium into the skeletal bone structure. Therefore, severe deficiencies of the vitamin in children invariably lead to soft, pliable bones and *rickets* while similar deficiencies in adults cause *osteomalacia*. The onset of early juvenile rickets is evident as *craniotabes,* or softening of the skull in the areas of the occipital and parietal bones at the site of the lambdoidal sutures; enlargement of chostrochondral junctions; irregular development of primary dentition; deformed growth of the long bones; development of bowed legs; as well as striated muscle and postural deformation. Adult osteomalacia, on the other hand, is characterized by increasing rarefaction of the pelvic, thoracic, and long bones, along with an increased probability for spontaneous fractures.

The biochemical reactions for vitamin D

Table 5.21. Noteworthy Properties of Vitamins D_2 and D_3

	Vitamin D_2	Vitamin D_3
Empirical formula	$C_{28}H_{44}O$	$C_{27}H_{44}O$
Molecular weight	396.6	384.6
Melting point	115–118°C	84–85°C
Optical rotation $[\alpha]_D^{20}$	82.6 in acetone	83.3 in acetone
Absorption maximum	264.5 nm	264.5 nm

Stability: Both forms are insoluble in water, slightly soluble in vegetable oils, and very soluble in usual organic solvents (e.g., chloroform, petroleum ether); commercial carrier solvents may employ sesame oil or propylene glycol.

Units: One I.U. of vitamin D is based on the biological activity of 0.025 μg of activated 7-dehydrocholesterol (World Health Organization, 1949)

| Potency | 40×10^6 I.U./g | 40×10^6 I.U./g |

beyond its preliminary conversions to 1,25-DHCC are not certain, but research has linked its behavior with *hormonal activities* rather than classical vitamin/coenzyme actions. This concept seems justified since the 1,25-DHCC produced in the kidneys is translocated by the blood to effector targets that control Ca^{2+} absorption along the small intestine, as well as Ca^{2+} residence in bones (Figure 5.80). This chemical "messenger" effect is quite differ-

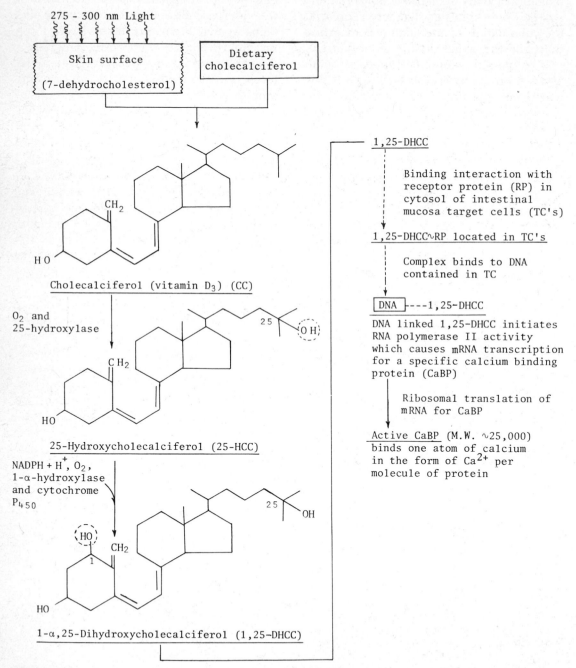

FIGURE 5.80. Schematic diagram outlining the role of vitamin D and its various intermediates in eliciting the production of a specific calcium-binding protein that is responsible for the intestinal absorption of Ca^{2+}.

ent from other vitamins that either contribute to coenzyme structures or serve as prosthetic groups on key enzymes. The biochemical role of 1,25-DHCC is also integrated with the availability of *parathyroid hormone* (PTH). That is, low levels of blood Ca^{2+} activate the parathyroid gland production of PTH. This polypeptide hormone jointly accelerates renal phosphate excretion and increases the production of 1,25-DHCC from 25-HCC. Since PTH seems implicated with the availability of hormone-like 1,25-DHCC, the action of PTH may be defined as a *tropic hormone* for 1,25-DHCC.

The physiological interrelations and theories on vitamin D activity are complex, and many aspects remain to be unraveled. For further detailed discussions on vitamin D, consult Wasserman and Taylor (1966), Fraser and Kodicek (1970), DeLuca (1971), Norman *et al.* (1969), Harrison and Harrison (1970a,b), and especially Melancon *et al.* (1970) or Harrison and Harrison (1974).

Miscellaneous Considerations

The lipophilic nature of vitamin D encourages its accumulation in the body. Levels of vitamin D acquired through excessive voluntary intake or food product enrichment can yield toxic results in infants and adults especially where chronic intake is ≥ 2000 I.U./kg body weight/day. Characteristic effects of hypervitaminosis D include irritability, vomiting, brittle bones, hypertension, renal insufficiency, and development of a systolic heart murmur. Blood serum levels for Ca^{2+} and phosphorous often become elevated in severe cases of hypervitaminosis, only to be followed by progressive calcification of the vascular system and other soft tissues.

On the basis of all current evidence, it is clear that intake of four to five times the current RDA for vitamin D (400 I.U./day) is not harmful, but an intake of >400 I.U./day does not produce any noteworthy health benefits. Foremost among many recent reports on vitamin D's range of toxic effects and biochemistry are those by DeLuca (1974),

Omdahl and DeLuca (1973), Taussig (1966), Fraser *et al.* (1966, 1973), Kenny *et al.* (1963), Hayes and Hegsted (1973), and the National Research Council, Food and Nutrition Board (1975).

Quantitative methods of analysis for vitamin D include a gamut of biological, chemical, and instrumental approaches. Owing to the direct effects of vitamin D on Ca^{2+} deposition in the skeletal structure, some historical bioassay methods have been based on quantitating the ash content from fat-free, dry tibiae obtained from 21-day-old chicks fed graded amounts of standard and unknown vitamin D levels in feed. Other similar methods have been outlined by the Estimation of Vitamins (1947), Campbell *et al.* (1945), Bliss (1945), Baker and Wright (1940), and DeLuca and Blunt (1971).

Chemical methods for vitamin D assay have relied on spectrophotometric detection of its reaction products produced in conjunction with antimony trichloride (Ewing *et al.*, 1943; Nield *et al.*, 1940; Milas *et al.*, 1941; Zimmerli *et al.*, 1943; DeWitt and Sullivan, 1946), glycerol dichlorohydrin (Sobel *et al.*, 1945; Campbell, 1948; Lyness and Quackenbush, 1955), and other reagents. However, the reliability of these methods often rests on the preliminary elimination of vitamin A content in foods. Instrumental methods for vitamin D analysis have employed gas chromatography (Sheppard and Hubbard, 1971) and, more recently, high-pressure liquid chromatography. The latter method is particularly useful since it can distinctly separate and quantitate vitamin D while in the presence of vitamin A.

THE QUINONES: UBIQUINONES, VITAMINS E AND K

Many different quinone structures having a nearly ubiquitous occurrence in living cells have been isolated since 1955. These quinones, now commonly called *ubiquinones*, are characterized by a central benzoquinone

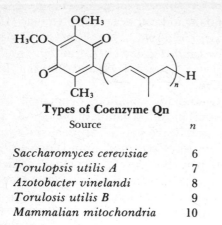

Types of Coenzyme Qn

Source	n
Saccharomyces cerevisiae	6
Torulopsis utilis A	7
Azotobacter vinelandi	8
Torulosis utilis B	9
Mammalian mitochondria	10

FIGURE 5.81. Ubiquinones (coenzyme Q) structures.

structure accompanied by an unsaturated isoprenoid side chain of variable length (n) depending on the biological source. In more recent terminology the ubiquinones have been conventionally referred to as *coenzyme* Q_n, where the subscript "n" is used to designate the number of isoprene units linked to the structure as a single aliphatic side chain (e.g., coenzyme Q_6, Q_7, . . . , Q_{10}) (Figure 5.81).

Aside from microorganisms and animal mitochondria, plants exhibit a similar series of related quinones located within the chloroplasts which are known as *plastoquinones*. These structures differ from the aforementioned quinones only by the substitution of two methyl (—CH_3) groups for the two methoxy (—OCH_3) groups, and n consists of *nine* isoprene units.

The complexity of quinone structures is not limited solely to methyl group substitutions or variations in n, but, rather, the quinones may undergo further cyclization involving the hydroxy (—OH) group of a

reduced ubiquinone, with the double bond located in the primary isoprene substituent linked to the benzoquinone ring. This cyclization provides the structural foundation for the *ubichromanols* (6-chromanols) while more oxidized naturally occurring ubichromanol derivatives are recognized as *ubichromenols* (Figure 5.82).

The recognition of chromanols has proved to be especially significant since they have a structural similarity to the tocopherol family, which includes vitamin E. The occurrence of tocopherols is biosynthetically limited to plants, plant products, and oils. Eight important structural varieties of these chromanols exist (Figure 5.83) although only α-, β-, γ-, and δ-tocopherols have been intensively studied. Of these four tocopherols, α-tocopherol seems to be the most abundant and has been of special interest in animal and human nutrition.

Additional unsaturation of the isoprene side chains on α-, β-, γ-, and δ-tocopherol structures results in a series of corresponding α-, β-, γ-, and δ-*trienol derivatives* (Figure 5.84).

Other structural variations of the tocopherols occur naturally in the chloroplasts as *tocopherolquinones* (Figure 5.85). These may undergo additional reduction to yield *tocopherol hydroquinones*.

The fundamental complexities of quinone structure extend still further into the realm of other vitamin substances, namely, the *vitamins K*. The vitamins K are structurally reminiscent of the quinones described earlier, and they principally occur in two separate families depending on the saturation status of their isoprenoid substituents. The *vitamin K₁ series* (*phylloquinones*, n = 4) has a single double bond only in the isoprene

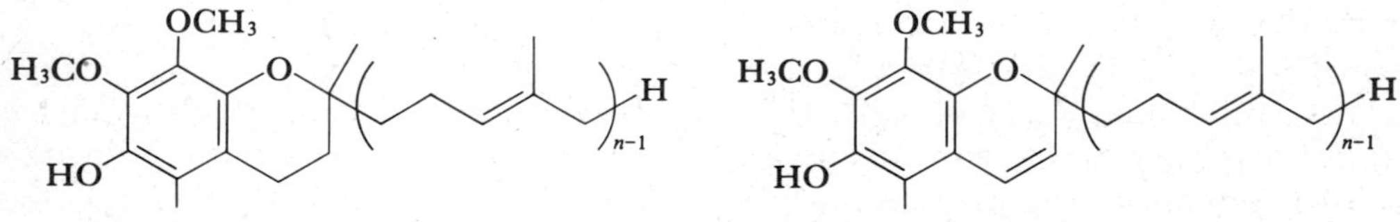

Ubichromanol   **Ubichromenol**

FIGURE 5.82.   Structures for ubichromanol and ubichromenol.

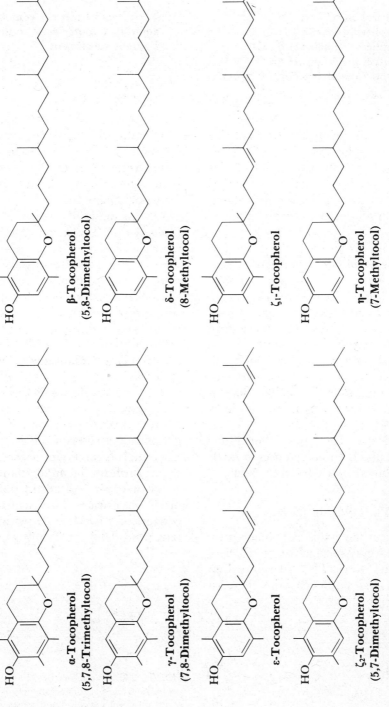

FIGURE 5.83. Eight naturally occurring structures for the tocopherols.

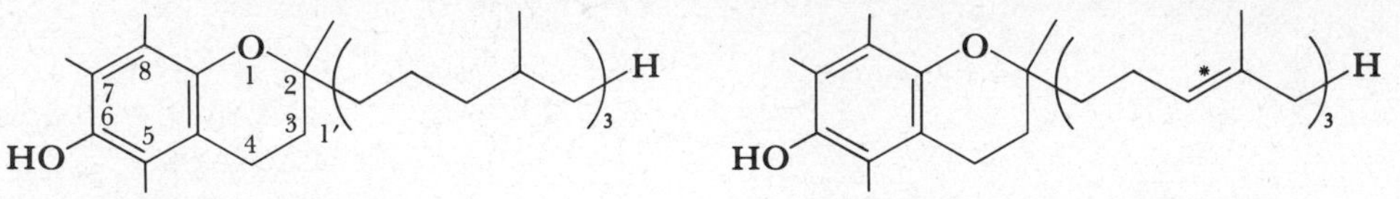

Tocopherol structures:
α-Tocopherol (vitamin E)
β-Tocopherol contains H on C_7
γ-Tocopherol contains H on C_5
δ-Tocopherol contains H on C_5 and C_7

Tocotrienol structures contain an additional double bond in the isoprene substituent (*).

FIGURE 5.84. Comparative structures for tocopherols and tocotrienols.

moiety bound to the ring structure (similar to chromanols), while the *vitamin K_2 series* (*menaquinones*, n = 5) contain double bonds in all substituent isoprenes. Phylloquinones predominantly occur in plants and menaquinones are limited to bacteria and animals (Figure 5.86).

Both vitamins K_1 and K_2, which can also be classified as *napthoquinones,* are the result of natural biochemical pathways, but synthetic vitamin K derivatives known as *vitamin K_3* and *phthiocol,* have been formulated. Unlike the vitamins K, these structures (Figure 5.87) lack the polyisoprenoid chain found in K_1 and K_2, but both structures exhibit full vitamin K activities. These structures are presumably convertible into active vitamin K forms in the animal body by the addition of isoprenoids.

Details regarding the types of tocopherols, vitamin K, and their occurrences in foods have been detailed in Tables 5.22–5.24.

Biochemical Functions of Quinones

The biochemical mechanisms promoted by quinones are largely speculative and elusive, but their abilities to readily undergo reversible oxidation–reduction reactions seem to be very important (Figure 5.88).

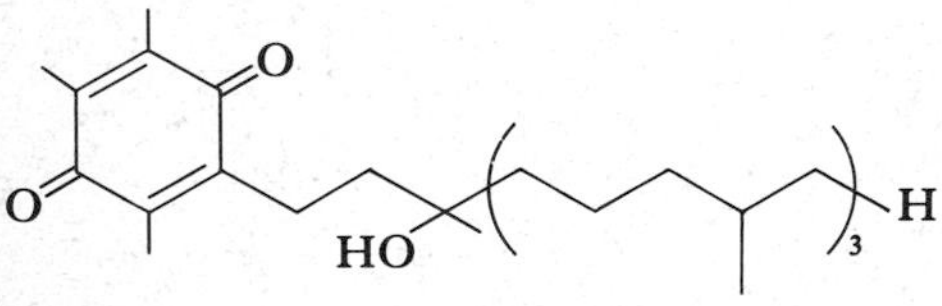

Ubiquinone, for example, acts as a fundamental electron carrier in aerobic respiration where it is situated between the flavoproteins and cytochromes in the electron transport chain.

On the basis of present evidence, tocopherols are also believed to participate in *reversible* oxidation–reduction reactions that mediate intracellular antioxidant mechanisms. Antioxidants are crucial to the "housekeeping activities" of cells because they suppress or terminate the propagation of free radical reactions involving unsaturated fatty acids contained in membranes. Moreover, the same antioxidant mechanisms displayed in *in vivo* tocopherols account for their advantageous use in foods to prevent the autoxidation and incipient rancidity of unsaturated fats. A proposed mechanism for autoxidation has been detailed in Figure 7.13. Tocopherols can limit these reactions by acting as preferential antioxidants. That is, instead of continued abstraction of ·H from other unsaturated fatty acids, or continued conversion of RH → R·, with the concomitant production of ·H to propagate the au-

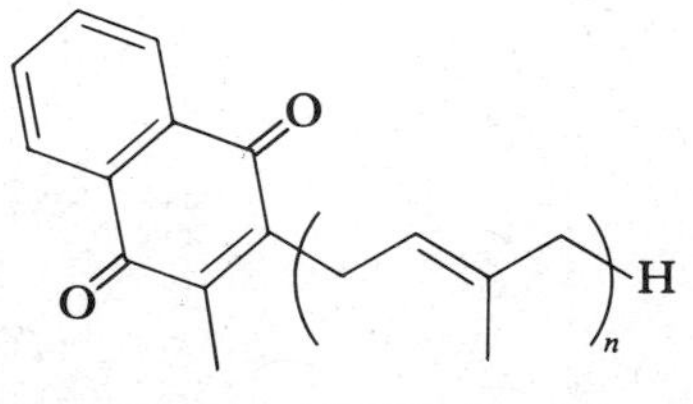

FIGURE 5.85. Structure of a tocopherolquinone.

FIGURE 5.86. Vitamins K: The *n* value = 4,5 isoprene substituents; for the phylloquinone (vitamin K_1) series only one double bond exists in the innermost isoprene unit.

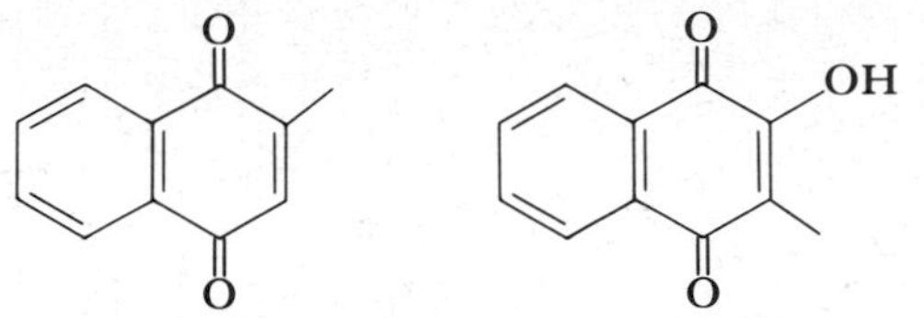

Menadione (vitamin K₃)   **Phthiocol**

FIGURE 5.87.   Synthetic vitamin K derivatives.

tocatalytic process, tocopherols and certain chromanols yield ·H to *limit* continued autocatalysis. The organic free radicals formed in these cases are relatively stable and do not promote autoxidation. The proposed interconversions of tocopherol structures that limit these reactions have been outlined in Figure 5.89.

Although steadfast mechanisms have not been dictated for the biochemical roles of less complex hydroquinones, it is reasonably certain that they are also able to form stable free radicals (Figure 5.90).

Oxidation–reduction roles for vitamins K are quite certain, but in view of established biochemical reactions, specific biochemical roles remain to be established. Evidence supporting the roles for vitamins K in oxidation–reduction reactions are circumstantial and related to the ability of *dicoumarol* (an antivitamin K substance) to serve as a competitive inhibitor against some K vitamin forms. This action is believed to uncouple conventional electron transport mechanisms in the mammalian mitochondria. Other supporting evidence for the oxidation–reduction reactions of K vitamins stem from the fact that some forms may be reversibly reduced to a *quinol*. Since the vitamins K are omnipresent in microorganisms, plants, and animal tissues, albeit at occasionally low concentration levels, the ability to form quinols may be extremely important.

Ubiquinones and Coenzyme Q₁₀

The long isoprenoid side chain of these structures encourages both coenzyme Q_{10} and the ubiquinone group to behave as neutral lipids. Therefore, their dietary occurrence is most prominent in plant oils (e.g., corn oil, wheat germ oil), Drackett protein, and whole soybeans.

It is risky at best to make any assessment of dietary requirements that will ensure an adequate supply of vitamin K substances since many animal tissues seem capable of supplying adequate amounts. Furthermore, it is

Table 5.22. Chemical Properties and Characterization of the Tocopherols

	Types of Tocopherols			
	α	β	γ	δ
Empirical formula	$C_{29}H_{50}O_2$	$C_{28}H_{48}O_2$	$C_{28}H_{48}O_2$	$C_{27}H_{46}O_2$
Molecular weight	430	416	416	402
Absorption maximum in alcohol (nm) *and*	292	297	298	298
$E_{1\,cm}^{1\%}$	70–74	86.5	91.5	91.2
Activity of tocopherols compared with *dl*-α-tocopherol acetate = 1.0	*dl*-α = 1.1 *d*-α = 1.49	*dl*-β = 0.3 *d*-β = 0.4	*dl*-γ = 0.15 *d*-γ = 0.20	*dl*-δ = 0.012 *d*-δ = 0.016

Chemical characteristics: Tocopherols are generally pale-yellow, viscous oils, soluble only in organic solvents suitable for fats.

Stability: Tocopherols are stable to marginal heat, alkali, and visible light in the absence of oxidizing agents, but they are readily destroyed by ultraviolet light, easily oxidized by chemical agents such as nitrate, auric, ceric, and ferric ions. Food stability of tocopherols is good, but frying is detrimental (50–75% loss); a 1-min boiling period for food causes only minor losses providing there is an absence of other oxidizing agents, and baking decreases levels by 5.0–20.0%.

Table 5.23. Characterization of Vitamin K Active Compounds

	Absorption maxima and corresponding $E_{1\,cm}^{1\%}$	Density	Stability	Solubility	Melting point	Molecular weight	Empirical formula	Alternative names
Vitamin K₁	425, 428, 424, 424, and 350 at respective wavelengths of 243, 248, 268, 269, and 325 nm in isooctane	$d_{25}^{25} = 0.967$	Stable in air and moisture, but decomposes in sunlight; destroyed by alkaline hydroxides and reducing agents; long-term stability absolutely requires controlled exposure to sunlight	Insoluble in water, and "sparingly" soluble in polar organics; soluble in almost all non-polar organic (fat) solvents	$-20°C$ (yellow, viscous oil at room temperature)	450.68	$C_{31}H_{46}O_2$	Antihemor-rhagic vita-min; Mono-Kay; Mephy-ton; Kona Kion; Phyto-menadione; "Koagula-tions vitamin"
Vitamin K₂ (properties variable depending on "n")	278, 295, 266, 267, and 48 at respective wavelengths of 243, 248, 261, 270, and 325–328 nm in petroleum ether	—	Similar to vita-min K₁	Soluble in non-polar solvents such as petroleum ether	50–54°C (crystals are light yellow micro-plates)	658.00	$C_{46}H_{64}O_2$	Antihemor-rhagic vita-mins, $n = 7$ although n does naturally vary
Vitamin K₃ (Menadi-one-synthetic vita-min K-like sub-stance)	—	—	Destroyed by al-kalies, reduc-ing agents, and sunlight	One gram dis-solves in 60 mL alcohol, 10 mL benzene, and 50 mL vegetable oil.	105–107°C (bright yellow crystals)	172.17	$C_{11}H_8O_2$	Kaylot; Kativ-G; Kipca; Oil Solubles; Kappaxin; He-modal; Ka-reon; Kolklot Synkay; Thy-loquinone

Table 5.24. Summary of Tocopherol and Vitamin K Concentrations in Selected Food Substances

Food source	Total tocopherol content assayed as α-tocopherol (mg/100 g of food)	Food source	Vitamin K reported as menadione units/100 g of edible food
Butter	~ 1.5–4.3	Cabbage	~ 55
Cabbage	~ 2.0–3.0	Cauliflower	~ 24
Carrots	~ 0.15	Carrots	~ 5–7
Cheese	~ 0.5–1.0	Honey	~ 25
Lard	~ 0.2–2.5	Liver	
Liver (beef)	~ 1.0–1.7	Chicken	~ 15
Margarine	~ 30–100	Pork	~110
Milk bovine	~ 0.05–0.15	Milk (bovine)	~ 8–10
Oils		Peas	~ 50
Coconut	~ 4.0–8.5	Potatoes	~ 10–12
Corn	~ 40.0–60.0	Spinach	~160
Corn germ	~ 90.0–250.0	Tomatoes (ripe)	~ 25
Olive	~ 5.0–35.0	Wheat	
Palm	~ 5.0–55.0	Milled	~ 18
Peanut	~ 15.0–60.0	Bran	~ 36
Rapeseed	~ 45.0–50.0	Germ	~ 18
Soybean	~100.0–270.0		
Sunflower	~ 70.0		
Peas	~ 4.0–6.0		
Potatoes	~ 0.13		
Spinach	~ 0.5–6.0		
Tomatoes (ripe)	~ 0.6–1.0		

difficult to determine ubiquinone and coenzyme Q_{10} demands in the animal body because the biochemical behavior supposedly reserved for α-tocopherol seems to overlap with these quinones. Coenzyme Q_{10}, for example, is unable to prevent fetal resorption in female rats, but the 6-chromanol derivative of hexahydrocoenzyme Q_4 prevents fetal resorption and allows development of the fetus to term along with the birth of live progeny. The 6-chromanol of hexahydrocoenzyme Q_4 also ameliorates induced anemic and dystrophic conditions in α-tocopherol-deficient monkeys.

From perspectives on human health, it has been reported that gum tissues afflicted with peridontal disease have low concentrations of ubiquinone, plus the fact that urine specimens of some patients suffering from diabetes or atherosclerotic heart disease may have high concentrations of coenzyme Q_{10}. Clearly, all claims and counterclaims regarding the requirements for ubiquinone in the animal body, its clinical significance, and the biochemical rationale for its apparent functional overlap with α-tocopherol must be more clearly established. Apart from these debatable functions, the electrochemical role of

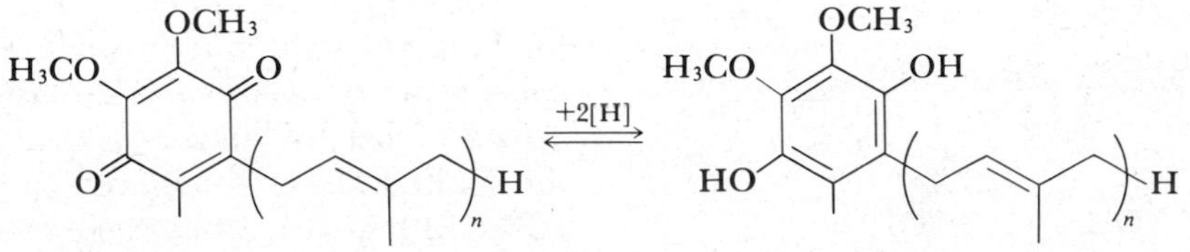

FIGURE 5.88. Reversible oxidation–reduction reactions for ubiquinones (coenzyme Q).

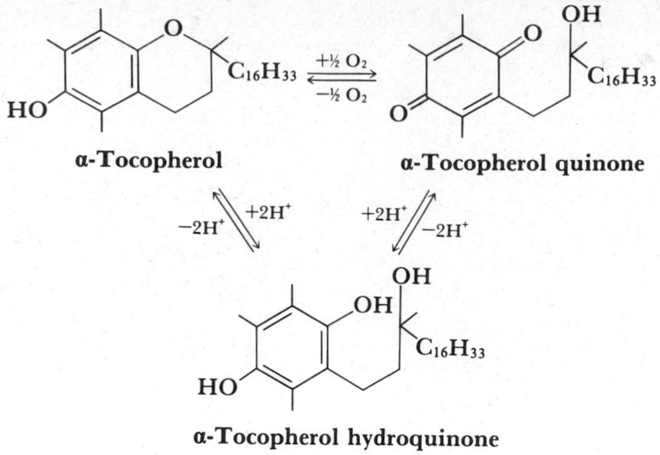

FIGURE 5.89. Possible interconversions of tocopherol structures.

coenzyme Q_{10} has been well established with reference to the electron transport chain (see oxidative phosphorylation).

The human body is believed to contain 0.5–1.5 g of total cellular coenzyme Q_{10}. The highest concentration levels are found in organs with high levels of aerobic respiration including the heart, adrenal glands, kidney, liver, and pancreas. Organ meats and meat by-products derived from livestock offer similarly high dietary sources of the substance.

Vitamin E and the Tocopherols

The human requirements for vitamin E are difficult to establish because there are no universally defined clinical symptoms to mark the onset of a deficiency. Moreover, in view of the antioxidant activities of tocopherols, it is difficult to quantitatively assess the human body requirement for such antioxidant substances. In most cases, vitamin E deficiency is considered nonexistent if 0.5 to 1.2 mg/100 mL of *total* plasma tocopherols are present, although α-tocopherol levels may be 10 to 15% lower (Bieri and Prival, 1965). A dynamic response between increasing vitamin E intake and apparent plasma levels is very limited according to animal studies. Test subjects administered varying tocopherol concentrations over a range of 200 to 800 I.U./day have shown plasma concentrations about double those levels in animals having no tocopherol supplement (Farrell and Bieri, 1975).

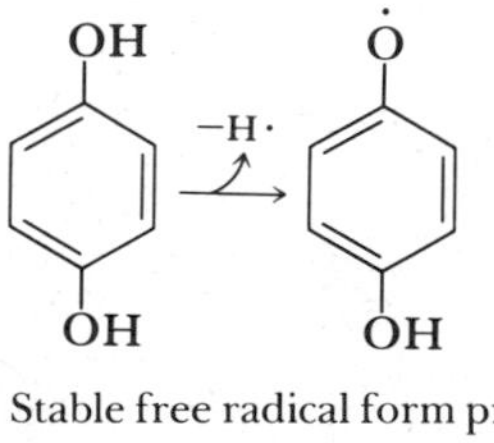

The absolute minimum tocopherol requirement for humans is about 5.0 mg/day, along with an added requirement of 0.6 mg for each gram of dietary polyunsaturated fatty acid. The latter restriction assumes special importance if dietary unsaturated fatty acids have been processed so as to diminish their naturally high levels of tocopherol (Table 5.24). Assuming an absence of tocopherol stripping from natural vegetable oils, high intake of vitamin E usually parallels the consumption of dietary unsaturated fatty acids. The daily vitamin E consumption in the United States is believed to range from 10 to 30 mg with an average amount being 14 mg.

Vitamin E activity has traditionally been

FIGURE 5.90. Stable free radical form produced from a hydroquinone.

described according to its antisterility effects since 1922 when Evans and Bishop reported its ability to avert fetal resorption in female rats as well as degeneration of the germinal testicular epithelium in male rats. Corroboration of these functions in humans remain to be substantiated.

Other animal studies report that tocopherol deficiencies promote (1) nutritional muscular dystrophy, (2) development of anomalous membranes in the endoplasmic reticulum and lysosomes, and (3) obvious physical degeneration of the animal that can lead to death. Actually, the range of vitamin E deficiency effects claimed for animals is almost as varied as the species studied. Conservative researchers, however, attribute tocopherol effects to two types of activities—namely, (1) those biochemical processes in which it acts as a definite antioxidant and (2) activities in which its antioxidant actions are not observed, but scientific observations support unknown, yet active involvement of the tocopherols in a biochemical process. A useful survey of vitamin E-related maladies and activities has been outlined by Wasserman and Taylor (1972).

In the case of humans, administration of α-tocopherol has been observed to decrease creatinuria associated with cystic fibrosis and xanthomatous biliary cirrhosis, whereas prolonged investigational deficiencies seem to lower plasma tocopherol levels and increase the *in vitro* hemolytic susceptibility of red blood cells when they are exposed to hydrogen peroxide. The *in vitro* behavior of red blood cells may mirror the lessened ability of the red blood cell membrane to resist the action of certain oxidizing agents (Rose and György, 1952; Century and Horwitt, 1965). This hemolytic behavior has also been used as the basis for tocopherol bioassays discussed later.

More definite human requirements for tocopherols are shown by infants who are fed diets high in polyunsaturated fatty acids, also accompanied by high levels of iron. In these situations, hemolytic anemia is a likely eventuality.

Speculated *in vivo* antioxidant activities of tocopherols are supported by the fact that tocopherol deficiency symptoms can be remarkably diminished by nontocopherol antioxidants such as *N,N*-dimethyl-*p*-phenylenediamine. There are also reports that up to three generations of rats can be reared on this *substitute* for tocopherol. Moreover, *in vitro* and *in vivo* studies affirm the ability of tocopherols to detoxify lipid peroxides and hinder the autocatalytic destruction of unsaturated fats, as shown in Figure 7.13.

Aside from tocopherol involvement in antioxidant mechanisms per se, it is interesting to note the intrinsic functional link between the tetrameric selenoprotein, *glutathione peroxidase* (M.W. = 88,000, containing 4 atoms of selenium) and vitamin E. The functional relationship between these two substances became evident when dietary selenium levels as low as 0.05 ppm/day were found to partially allay the symptomatic responses of certain vitamin E deficiencies. The reasons for this activity remain to be entirely explained. Nevertheless, it is clear that selenium-glutathione peroxidase acts as a stopgap agent for inhibiting intracellular propagation of lipid peroxides and minimizing hydrogen peroxide accumulation when vitamin E levels are inadequate or oxidative stress is high (Tappel, 1974; Noguchi et al., 1973) (Figure 5.91).

In summary then, increased availability of dietary selenium may encourage the activity of selenium-glutathione peroxidase, thereby functionally supporting antioxidant activities related to tocopherol. It should be clearly recognized that wholesale doses of *selenium cannot substitute for tocopherol activity*, and *tocopherols cannot substitute for dietary selenium requirements*. However, there is a distinct complementary overlapping of some fundamental protective biochemical reactions (Noguchi et al., 1973). Additional biochemical mechanisms for antioxidant actions of selenium have been detailed by Metzler (1977).

Human toxicity effects of vitamin E overdoses are not well documented in the lit-

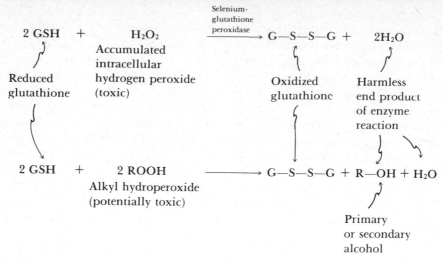

FIGURE 5.91. Role of selenium-glutathione peroxidase.

erature, but it has been reported (Farrell and Bieri, 1975) that 100 to 1000 I.U. of vitamin E/day produces no ill effects or impairment of blood coagulation, blood glucose levels, or obvious clinical upsets in liver, kidney, thyroid gland, muscle, leukocyte, or erythrocyte functions. Correspondingly, it should be noted that there is no distinct scientific evidence to justify therapeutic vitamin E intake beyond levels of 150 mg (Witting, 1972) or 150 to 204 I.U. depending on the form of vitamin E (consult Table 5.2 for milligram and I.U. equivalencies).

Attempts to assay vitamin E and tocopherols have employed chemical, biological, and instrumental techniques. Unfortunately, the most expedient methods are chemical and instrumental methods, which cannot accurately assess the biological potency of selected tocopherols. Time-honored bioassays, on the other hand, are meticulous, exhaustive efforts unsuitable for routine analyses of patient specimens or food substances. These assays customarily involve correlation studies between available vitamin E concentrations and testicular degeneration in male rats or fetal resorption in females. Another quasi-biological assay technique, somewhat more simple than sterility studies, involves the measurement of toco-

pherol concentrations that inhibit erythrocyte hemolysis. For this test, erythrocytes are obtained from tocopherol-deprived rats and the cells are incubated with hydrogen peroxide or dialuric acid (Rose and György, 1952; Century and Horwitt, 1965). Gas chromatography (Slover et al., 1968), ceric titration, thin-layer or column chromatography, and many other methods have also been outlined since the 1930s (Strohecker and Henning, 1965; Freed, 1966), but the future of most reliable instrumental methods of analysis will probably depend on high-pressure liquid chromatography. High-pressure liquid chromatography offers many advantages, including rapid quantitative and analytical identification of tocopherols, coupled with minimal heat destruction via oxidative routes that destroy the vitamin during gas chromatographic analyses.

Vitamin K

The human requirements for vitamin K are usually satisfied by a combination of a well-balanced nutritional regimen plus absorption of vitamin K from natural intestinal flora. Dietary deficiencies of the vitamin are rare in the American population, but postpartum infants can provide an exception to the general rule. This group is of special concern if maternal vitamin K deficiencies precede the nonexistent to low levels of vitamin K available from the infant's intestinal flora.

Deficiency effects in infants reportedly encourage hemorrhages that cause the infant to suffer from anoxia, but this problem can often be ameliorated by a 10 to 25-mg dose of vitamin K_1 (phylloquinone) to the pregnant woman prior to delivery or a 5.0-mg dose to the postpartum infant.

The most significant block to normal vitamin K absorption in the general population stems from intestinal malabsorption of fats. This can be caused by a variety of ailments including biliary obstruction(s), idiopathic steatorrhea, celiac disease, sprue, pancreatic fibrosis, ileitis, or ulcerative colitis. Aside from these maladies, adult vitamin K deficiencies may be produced or encouraged by the chronic use of antibiotics that typically decrease the population numbers of intestinal flora.

The only well-documented biochemical deficiency of vitamin K deals with its role in the biosynthesis of the hepatic enzyme known as *proconvertin*. Proconvertin is important because it mediates a single chemical event in the overall conversion of the highly soluble blood protein *fibrinogen* to an insoluble component of blood clots called *fibrin*.

The activity of *proconvertin* itself relies on several other preliminary interactions among calcium ions, plasma globulins, prothrombin kinase, and thrombokinase to form a *convertin* intermediate. This convertin intermediate *plus* prothrombin (a plasma zymogen) yield the active proteolytic enzyme known as *thrombin*.

A deficiency of vitamin K causes decreased levels of prothrombin in the blood and adversely influences the availability of proconvertin produced by the liver. The ensuing steps of the blood-clotting mechanism seem to occur independent of vitamin K availability. A summary of the clotting process has been outlined in Figure 5.92.

The exact mechanism of vitamin K participation in the process of blood coagulation is still hazy, but it is clear that glutamic acid residues 7, 8, 15, 17, 20, 21, 26, 27, 30, and 33 present in normal prothrombin happen to include γ-carboxyglutamate (Figure 5.93). According to most evidence, vitamin K appears to facilitate posttranscriptional addition of the γ-carboxyl groups to glutamate residues in the preformed prothrombin structure (Metzler, 1977). The γ-carboxylate group added in this fashion improves the ability of key protein clotting factors to bind calcium ions, thereby contributing to the eventual development of *protein cross-linking and a "hard" clot.*

Although vitamin K occurs in many leafy vegetable tissues and other foods, the detrimental effects of its dietary overdoses are rarely documented. In fact, most literature surveys on vitamin K overdoses deal with the therapeutic administration of vitamin K_3 (menadione) to infants. The most notable problems include interactions of this highly water soluble substance (menadione) with the sulfhydryl substituents in neonatal tissues, as well as associated brain damage (kernicterus) and hemolytic anemia at dose concentrations in excess of 5.0 mg (Owen, 1971; Hayes and Hegsted, 1973; Anonymous, 1961; Campbell and Link, 1941; AAP, 1971).

The normal activity of vitamin K in the animal body may be upset by certain antivitamin substances such as coumarin derivatives, but these do not pose a nutritional threat to the normal dietary availability of vitamin K. As indicated by Owen (1971), the daily vitamin K requirement for adults is ≤ 1.0 mg and possibly as low as 100 μg. This amount contrasts with an ordinary dietary vitamin K availability of ≥ 500 μg/day. Therefore, with vitamin K consumption far exceeding requirements, a significant chronic dietary intake of any natural vitamin K antagonist would be necessary before the onset of clinical deficiency symptoms in man.

The implications of antivitamin K substances are far different for cattle feeds contaminated with spoiled sweet clover, which can contain *dicoumarol*. The potent antivitamin activity of this substance in certain an-

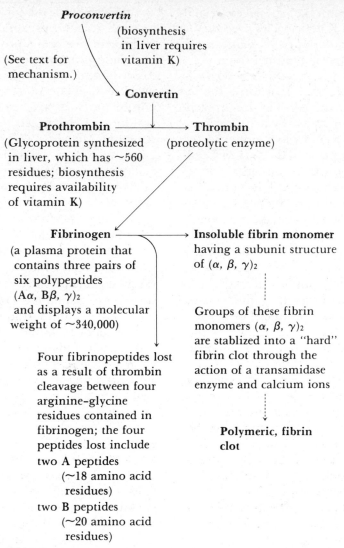

FIGURE 5.92. Reaction cascade for the formation of a polymeric fibrin clot showing the participation of vitamin K-dependent proconvertin and prothrombin.

imals is detrimental to the *in vivo* oxidation of vitamin K to phylloquinone-2,3-epoxide and the epoxide's subsequent reconversion to vitamin K. Dicoumarol is thought to competitively inhibit epoxide reconversion to vitamin K in the liver, but the complex reasons for this inhibition and the eventual development of fatal hemorrhagic diseases in cattle will require added study.

Apart from the fatal hemorrhagic effects of dicoumarol on cattle, rodents, and other animals, synthetic dicoumarol derivatives serve as the basis for therapeutic anticoagulants in humans in addition to the rodent poison known as warfarin. It is believed that both of these dicoumarol derivatives similarly inhibit the γ-carboxylation of glutamate residues required for the cross-linking of crucial blood clotting proteins.

Recognition of these activities for vitamin K does not signal an end to its possible biochemical or nutritional functions. These

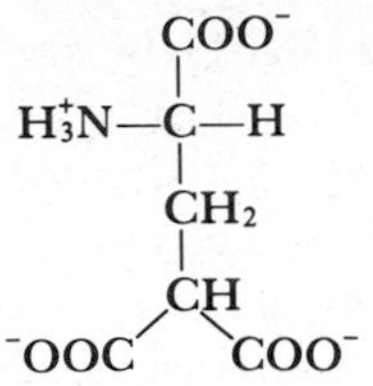

FIGURE 5.93. γ-Carboxyglutamate.

functions represent only those aspects of vitamin K behavior that are best understood.

Reliable quantitative assays for K vitamins are challenging owing to their variety of forms and the overlapping of their similar biochemical activities. The vitamin K_2 series alone can have a minimum of nine different forms (Table 5.25).

Foods are rarely assayed for vitamin K content on a routine basis, with the main exceptions being poultry or specialty animal feeds. The principles of older chemical assays outlined by Almquist (1941) along with modern methods of instrumental analysis provide the analytical basis for some current assays (Sommer and Kofler, 1966; Strohecker and Henning, 1965; Sheppard and Hubbard, 1971). Rat bioassays have also provided the basis for many vitamin K assays, but these efforts require the scrupulous maintenance of a low vitamin K rodent diet and the elimination of coprophagy (Metta *et al.*, 1961; Mameesh and Johnson, 1959). Successful adherence to these conditions permits the establishment of a straight-line relationship between prothrombin clotting time and vitamin K availability (Mameesh and Johnson, 1959). Chick bioassays similarly involve blood clotting time measurements, but the analyst must be cognizant of misconstruing low vitamin K_1 levels with the lower activity levels of less biologically active forms of vitamin K.

Table 5.25. Structures and Activities for Vitamins K_1 and K_2

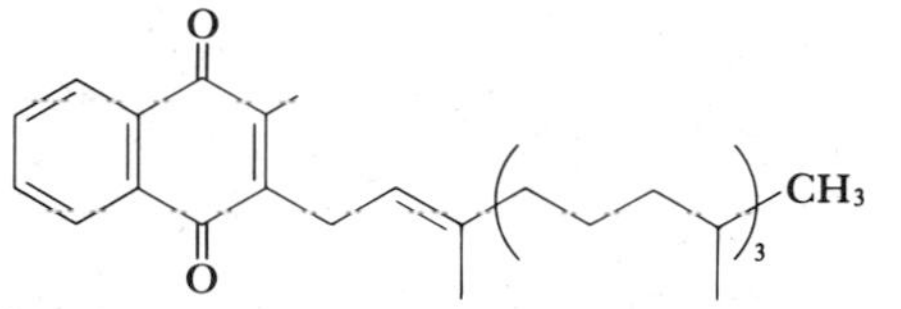

Vitamin K_1 (2-methyl-3-phytyl-1,4-naphthoquinone)

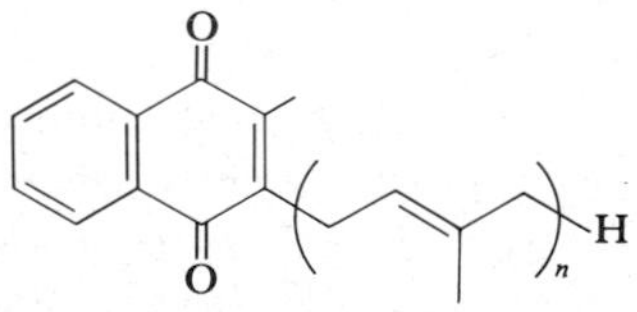

Vitamin K_2 ($n = 1$–9 repeating units depending on source; the relative activities of vitamin K_2 series for which n varies from 1 to 9 are compared to the biological activity of vitamin K_1, which is designated as 100% of any possible vitamin K activity)

Vitamin K_2 series where $n =$	Relative activity to chick bioassay where $K_1 = 100\%$	Possible sources for certain vitamin K_2 substances
0	<5%	
1	~15	
2	~40	
3	~100	Animal and human tissues
4	~120	
5	~100	Decomposed fish meal
6	~70	Fish meal and certain bacteria
7	~68	
8	~60	
9	~25	*Mycobacteria* sp.

LIPOIC ACID

In 1949 it was recognized that culture of the ciliate *Tetrahymenia geleii* required an accessory growth factor in addition to the ingredients of a casein growth medium. Further research indicated that the accelerated growth of *Lactobacillus casei* due to sodium acetate could be paralleled by a "nonacetate containing extract" of yeast; and still other studies pointed to the conclusion that *Streptococcus faecalis* could metabolize pyruvate only in the presence of a *pyruvate oxidation factor*. By 1949, these circuitous investigations eventually centered on the central recognition that the same growth factor was involved in all three organisms.

This undefined growth-active substance progressively assumed many names including Factor II, Factor IIA, the *acetate-replacing factor,* protogen, thioctic acid, 6-thioctic acid, and α-lipoic acid. Thioctic acid and lipoic acid are currently the most common designations.

Lipoic acid can occur in two biochemically active forms, which include an oxidized disulfide form (—S—S—) and a reduced sulfhydryl form (—SH HS—). Lipoic acid is interconvertible between these two forms, and this serves as a chemical basis for its biochemical reactivity (Figure 5.94).

The important coenzyme activities of lipoic acid often occur only so long as it is linked to an ε-*N*-lipoyl-lysine residue in an enzyme structure. This is reminiscent of ε-*N*-biotinyl-lysine occurrence as biocytin in biotin-containing proteins (Figure 5.95A).

The reactive nature of oxidized lipoic acid reflects its high intrinsic ring strain over the disulfide-containing ring. This oxidized structure readily reacts with thiol groups and cyanide ions (Figures 5.95B and C).

Dietary sources of lipoic acid are diverse, but the compound is present only in small amounts, as demonstrated by the work of Lester Reed in the late 1940s, in which only 30 mg of lipoic acid was obtained from 10 tons of a water-soluble liver residue. Traditional studies have indicated that humans probably synthesize required amounts of lipoic acid, although dietary sources from yeast and liver may augment any prevailing physiological requirement.

Notable properties of lipoic acid are outlined in Table 5.26.

The widespread occurrence of lipoic acid in biochemical systems is unchallenged. Nonetheless, its detailed roles in many important biochemical reactions remain unknown or very superficially understood at best.

Most contemporary mammalian models for lipoic acid activities are based on biochemical studies of *Escherichia coli*. This bacterium requires a lipoic acid moiety linked to its dehydrogenase enzymes, which are responsible for the oxidative decarboxylation of α-keto acids such as pyruvate, α-ketoglu-

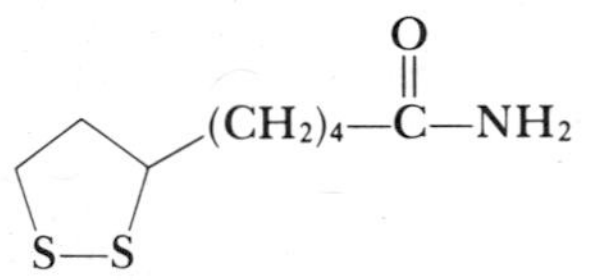

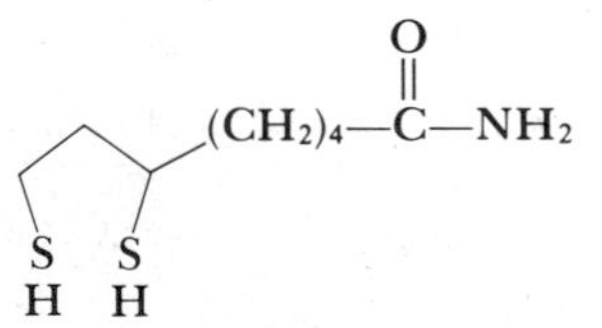

FIGURE 5.94. Oxidized and reduced forms of lipoic acid.

Table 5.26. Noteworthy Properties for Lipoic Acid

Empirical formula	$C_8H_{14}O_2S_2$
Molecular weight	206.32
Natural isomer	*d*-lipoic acid
Melting point	46–48°C
Specific rotation	$[\alpha]_D^{23} = +104$

pK_a: 5.4 (*Note:* lipoic acid's sodium salt displays a pH of 7.4 in an aqueous solution)

Solubility: Soluble in fat solvents, and sparingly soluble to insoluble in water

Absorbance maximum: 333 nm ($\epsilon = 150$ for a solvent of methanol)

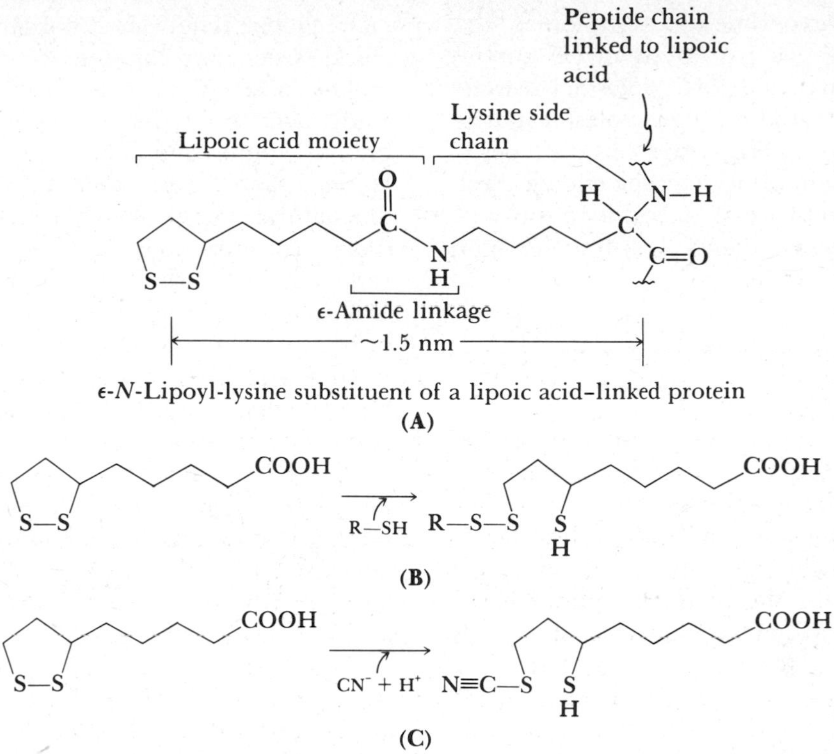

FIGURE 5.95. (A) structural linkage of lipoic acid to a lipoic acid-requiring protein; reactions of oxidized lipoic acid with thiol moieties (B) and cyanide ions (C).

tarate, and α-keto acids with branched side chains metabolically derived from valine, leucine, and isoleucine. Although these decarboxylation reactions are principally similar for the acids mentioned, specific α-keto acid dehydrogenases are required for decarboxylation of any specific substrate. Cleavage of α-keto acids typically yields CO_2 and an acyl-CoA derivative of the α-keto acid that has one less carbon than the original substrate:

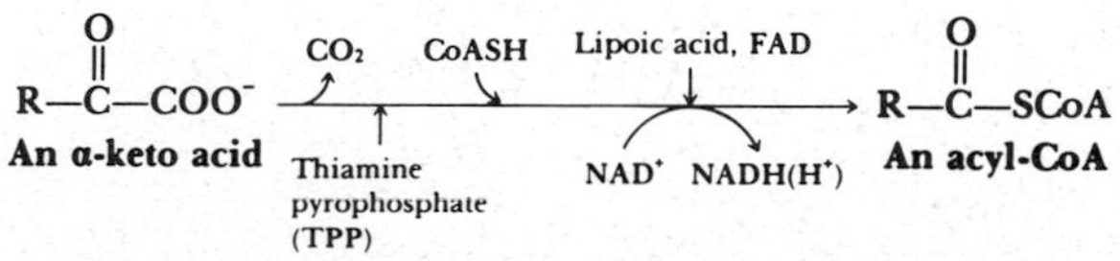

mine diphosphate, coenzyme A, FAD, and especially NAD^+, which behaves as an oxidant in the overall reaction. A more detailed survey of such a reaction scheme has been shown for the decarboxylation of pyruvate in the previous section, dealing with thiamine biochemistry as well as multienzyme systems.

Certain microorganisms intractably require lipoic acid in their growth media, but rats, chicks, and humans fed clinically con-

Decarboxylation reactions are not simple affairs. They often require the close interplay of multienzyme systems—two enzymes of which are linked to lipoic acid—and a prerequisite availability of lipoic acid, thia-

trolled diets fail to respond to the presence or absence of lipoic acid. Furthermore, uncertainty surrounding the extent of lipoic acid participation in many biochemical reactions only spurs on added confusion regarding experimental testimonies that promote its positive effects in (1) limiting liver-, plasma-, and aortic-lipid increases for animals nurtured on high-cholesterol diets; (2) its ability to encourage certain types of tumor growth; (3) its neurobiochemical role in the vision process; and (4) its positive effects in helping patients overcome the rigors of hepatic coma.

CHOLINE

Choline is another vitamin-like substance universally found in plant and animal cells where it can exist as free choline or a component of phospholipids such as lecithin. In animals, choline also serves as a structural component of the neurotransmitter *acetylcholine* (Figure 5.96). Choline is chemically designated as β-hydroxyethyltrimethyl-ammonium hydroxide; however, more common names include amanatine, bursine, fagine, gossypine, luridine, and vidine. Noteworthy chemical and physical properties for choline include a molecular weight of 121.18, a solubility that is nearly limited to water and alcohols (insoluble in ethers), and an ability to readily absorb CO_2 from air.

The lipotropic activities associated with choline draw special attention to this substance as a dietary factor. Moreover, it is of interest to humans because deficiencies have never been steadfastly documented and most suspected ramifications of choline deficiencies in humans are based on established di-

etary choline requirements observed in rats, chicks, swine, and canines.

The most notable sources of choline include bile, brain tissue, egg yolks, whole grains, legumes, muscular tissues, hops, belladonna, and *Strophanthus*. The probability of a natural choline overdose resulting from dietary constituents is slim, based on studies with rabbits that show the subcutaneous lethal dose to be in the range of ~500 mg/kg of body weight.

Unlike most vitamin substances, choline is required in relatively large concentrations. According to some generalized estimates, choline is required at 300 times the prevailing species requirement for riboflavin. For humans, the daily choline requirement is believed to be in the range of 0.5 to 1.0 g/day and somewhere between 0.3 and 0.6 g is obtained directly from the diet.

This relatively high demand for choline undoubtedly reflects its important roles in animal metabolism. Although choline deficiency effects may be largely species specific, choline deficiencies are generally believed to cause a fatty liver, hemorrhagic kidney degeneration, and cardiovascular lesions. Specific deficiencies of choline in poultry contribute to slipped tendon or perosis problems that can be overcome only by dietary choline; swine exhibit poor locomotor coordination and inferior reproductive potential; while a deficiency in canines leads to a fatty, cirrhotic liver, weight loss, anemia, and dermal and peptic ulcers.

The essential nutritional and biochemical activity of choline lies partially but promi-

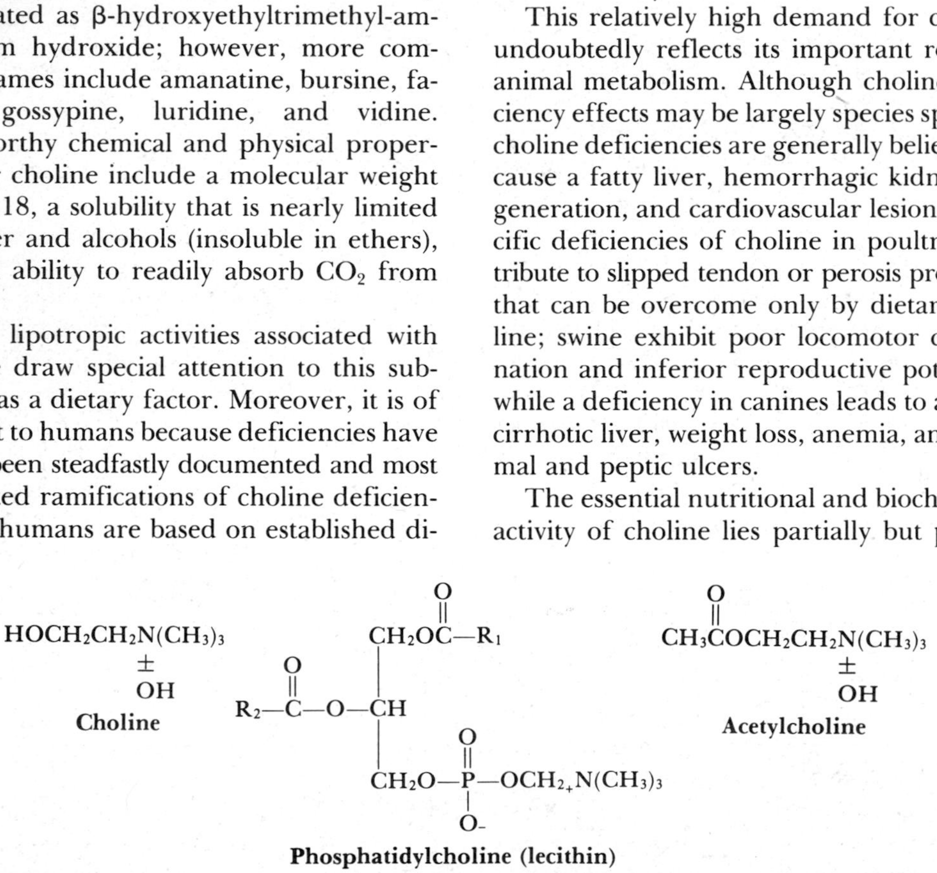

$$HOCH_2CH_2N(CH_3)_3$$
$$\pm$$
$$OH$$
Choline

$$R_2-\overset{\displaystyle \overset{O}{\|}}{C}-O-CH$$

$$
\begin{array}{c}
\overset{\displaystyle \overset{O}{\|}}{CH_2OC-R_1} \\
| \\
\\
| \\
\overset{\displaystyle \overset{O}{\|}}{CH_2O-P-OCH_{2+}N(CH_3)_3} \\
| \\
O-
\end{array}
$$

$$CH_3\overset{\displaystyle \overset{O}{\|}}{C}OCH_2CH_2N(CH_3)_3$$
$$\pm$$
$$OH$$
Acetylcholine

Phosphatidylcholine (lecithin)

FIGURE 5.96. Structure of choline, acetylcholine, and lecithin, which incorporates the choline structure.

nently in its ability to warehouse a readily available supply of methyl groups (e.g., —CH$_3$ groups). Scientific evidence for this concept was first supported by choline deficiency studies in rats conducted by Best and Huntsman (1932) and Griffith and Wade (1939), which showed progressive, hemorrhagic kidney degeneration as well as the onset of a fatty liver. Symptomatic regression of these problems were later shown to result when methionine or betaine (an oxidation product of choline) had been supplied to the test subjects. Folic acid and vitamin B$_{12}$, both associated with methyl transfer reactions, were also found to avert choline deficiency effects, but this was caused by their cooperative *de novo* participation in synthesizing methionine from homocysteine. Methionine formed from this reaction can provide an alternative supply of necessary methyl groups for most reactions (Figure 5.97).

Choline is biochemically employed in at least four important ways. First, choline may be oxidized (dehydrogenated) to betaine, a quaternary nitrogen compound that is capable of donating methyl groups to other substances such as homocysteine, thereby forming methionine. A second utilization pathway for choline leads to its phosphorylation by ATP to yield phosphorylcholine. This substance can then be transformed into cytidine diphosphate choline, which is subsequently esterified to an appropriate acyl glyceride to produce a phospholipid such as phosphatidylcholine. Both of the foregoing pathways have been outlined in Figure 5.98. It should be noted that the choline pathway leading to betaine, along with its associated methyl transfer reactions, is favored primarily when choline is in great supply.

Aside from these two pathways, a third realm of choline activity involves its integral structural contribution to *sphingomyelin*. This substance is the only phospholipid constituent of membranes that is derived from an alcohol other than glycerol. Instead of glycerol, sphingomyelin employs the 18-carbon unsaturated amino alcohol known as sphingosine for the esterification backbone of phosphorylcholine plus any one of many possible fatty acids (Figure 5.99).

A fourth possible role for choline involves

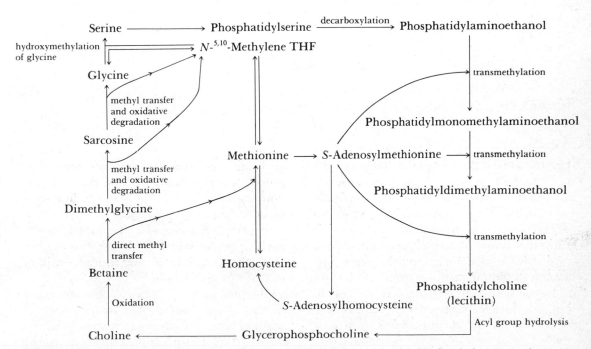

FIGURE 5.97. Summary of choline metabolism and interrelationships with other biochemical compounds.

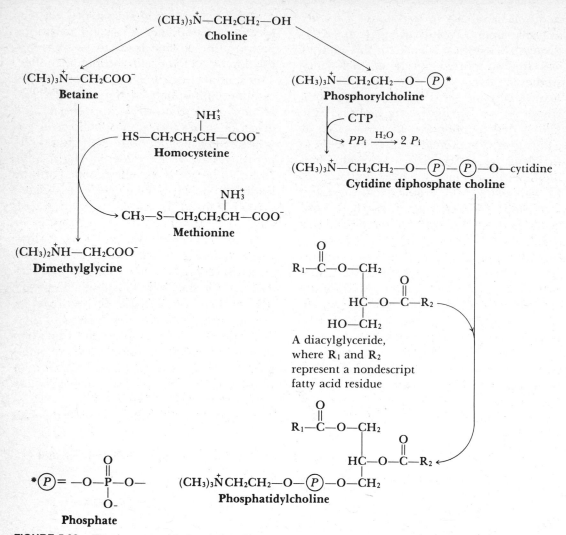

FIGURE 5.98. Two important biochemical pathways for choline utilization.

its strategic activity in processes of neuro-transmission. It is well known that nerve cell transmissions rely on temporal but progressive ion depolarization along the electrically excitable membrane surrounding individual neurons. Electrical depolarization of this type is unidirectional and solely accountable for conduction of neuronal impulses. The transmission of neural impulses from one neuron to another as well as from one neuron to striated muscle, respectively, require the presence of a spatial junction known as a *synapse* or a *neuromuscular junction*. Communication of a neural depolarization event between parasympathetic nerves or from a

parasympathetic nerve to an effector muscle are both based on the neutrotransmission action of acetylcholine. Nerve communication across a synapse via acetylcholine constitutes a *cholinergic junction*. This situation is in contrast to those synaptic junctions occurring in the sympathetic nervous system that utilize norepinephrine (noradrenaline) in place of acetylcholine.

When a wave of electrical depolarization sweeps along a nerve fiber (axon), it eventually reaches the presynaptic membrane of the neuron (Figure 5.100). This event trigers the rupture of synaptic vesicles that contain acetylcholine. The neurotransmitter is then

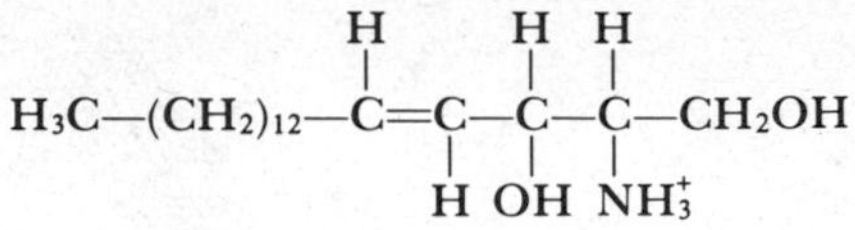

$$H_3C-(CH_2)_{12}-\overset{\overset{\displaystyle H}{|}}{C}=\overset{\overset{\displaystyle H}{|}}{\underset{\underset{\displaystyle H}{|}}{C}}-\overset{\overset{\displaystyle H}{|}}{\underset{\underset{\displaystyle OH}{|}}{C}}-\overset{\overset{\displaystyle H}{|}}{\underset{\underset{\displaystyle NH_3^+}{|}}{C}}-CH_2OH$$

Sphingosine

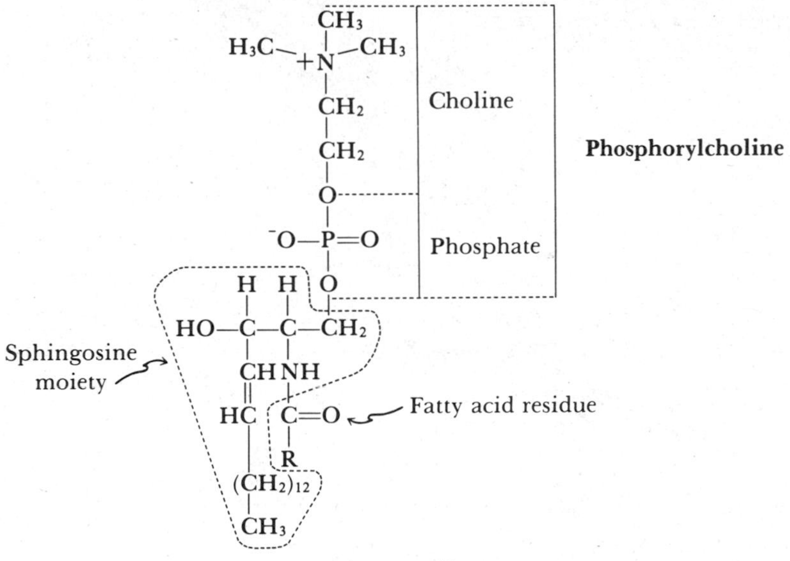

FIGURE 5.99. Comparative structures for sphingosine and sphingomyelin.

discharged into the *synaptic cleft* between neurons. This space represents a distance of ~500 Å.

Acetylcholine is synthesized from acetyl-CoA and choline at the presynaptic end of the axon as shown below:

postsynaptic membrane. The discharge of acetylcholine (~8000–12,000 molecules/300 to 400-Å-diameter synaptic vesicle) is followed by a pinocytic-type resealing of the synaptic vesicle by a very small portion of the presynaptic membrane.

$$\underset{\textbf{Acetyl-CoA}}{H_3C-\overset{\overset{\displaystyle O}{\|}}{C}-SCoA} + \underset{\textbf{Choline}}{HO-CH_2CH_2-\overset{+}{N}(CH_3)_3} \underset{\text{acetyltransferase}}{\overset{\text{choline}}{\rightleftharpoons}} \underset{\textbf{Acetylcholine}}{H_3C-\overset{\overset{\displaystyle O}{\|}}{C}-O-CH_2CH_2\overset{+}{N}(CH_3)_3}$$

All synthesized acetylcholine is not packaged into synaptic vesicles; therefore, significant unvesicularized quantities remain in the cytosol of the nerve cell for future use.

The diffusionary migration of acetylcholine across the synaptic cleft leads to the eventual attachment of this positively charged, quaternary nitrogen compound to an anionic (*negatively charged*) receptor site on the

Within 0.1 ms of binding at the anionic receptor site, there is a marked alteration in cation permeability over the neuron. This change in permeability incites sodium ion migration into the cell and an efflux of potassium ions from the cell. The dynamic ion flux inevitably alters the preexisting electrical potential of the nerve cell, the outside of the nerve cell being ~60 to 80 mV more

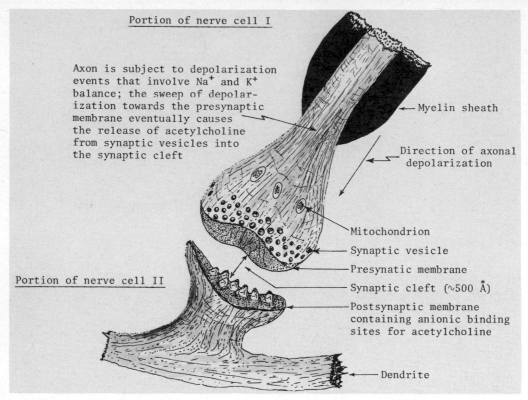

Portion of nerve cell I

Axon is subject to depolarization events that involve Na^+ and K^+ balance; the sweep of depolarization towards the presynaptic membrane eventually causes the release of acetylcholine from synaptic vesicles into the synaptic cleft

Myelin sheath

Direction of axonal depolarization

Mitochondrion

Synaptic vesicle

Presynatic membrane

Synaptic cleft ($\sim$500 Å)

Portion of nerve cell II

Postsynaptic membrane containing anionic binding sites for acetylcholine

Dendrite

FIGURE 5.100. Schematic diagram for a sectioned portion of two nerve cells and synapse as viewed from a possible electron micrograph. The direction of nerve impulse conduction is from nerve cell I toward nerve cell II.

positive than the inside. This depolarization event then sweeps the second nerve until it encounters the next presynaptic membrane and the presynaptic cleft, at which point acetylcholine activity is reenacted.

Acetylcholine bound to an anionic receptor site of the postsynaptic membrane cannot remain interminably bound. If it did, this would prohibit future nerve impulse conductions. Thus, bound acetylcholine acts as a substrate for an enzyme called *acetylcholinesterase*, which hydrolyzes acetylcholine to acetate and choline (Figure 5.101). This step permits restoration of the original cation membrane permeabilities on the postsynaptic membrane and promotes eventual repolarization of the whole nerve cell. The acetylcholine is believed to react with acetylcholinesterase at a serine-active site on the enzyme to yield free choline and an acetylated enzyme. The free choline returns to a choline pool, and the acetylated enzyme

eventually undergoes hydrolysis to yield the free enzyme and acetate. The turnover number for acetylcholinesterase is in the range of $\sim$20–27,000 s^{-1}, or about one molecule of acetylcholine cleared in 40 to 60 μs. This rapid turnover number ensures that the synaptic cleft and nerve can become electrically repolarized enough to mediate up to 1000 neural impulses s^{-1}.

The wide spectrum of biochemical functions and relatively high concentrations of choline in biochemical systems should seemingly simplify its quantitative analysis, but most assays are made complex because choline may be present in a variety of biochemical forms. Free choline is easily extracted from food substances using 70 to 80% alcohol or cold water, but effective determination of total choline requires vigorous extraction. Once extracted, choline may then be assayed by chemical methods that require further purification and precipitation of

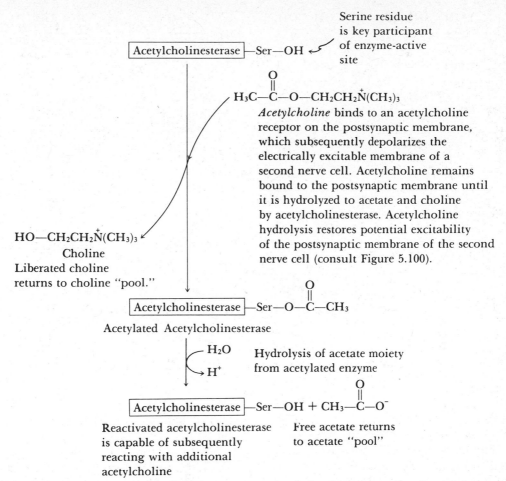

FIGURE 5.101. Enzymatic hydrolysis of acetylcholine by acetylcholinesterase leading to dislocation of acetylcholine from the anionic receptor site of a postsynaptic membrane (*see Figure 5.100*).

choline as phosphotungstate or reineckate salts, or choline may be quantitated by microbiological assays (Lim and Schall, 1964; Levine and Chargaff, 1951; Engel *et al.*, 1954; Hodson, 1945; Horowitz and Beadle, 1943).

MYO-INOSITOL

Inositols constitute a group of vitamin-like compounds that are chemically characterized as polyhydroxy cyclic alcohols (*cyclohexitols*), which are structurally reminiscent of monosaccharides. Inositols can occur in *nine* possible stereoisomeric forms (Figure 5.102), but the myo-inositol form, also called

meso-inositol or *i*-inositol, represents the form of most interest from the perspectives of vitamin activity. Other common names for this stereoisomer include phaseomannite, dombose, nucite, bios I, *rat antispectacled eye factor*, and mouse anti-alopecia factor; because of myo-inositol's sweet taste, it has also been called *meat sugar*.

Myo-inositol is recognized as a key biochemical substance for most microorganisms, plant cells, and animal cells. This undoubtedly accounts for its wide distribution in foodstuffs obtained from animal organs, yeasts, seeds, and whole grains. Myo-inositol is rarely available in foods as a free molecular entity, but, rather, it is bound to phospholipids (*phosphoinositides*). For plants, this

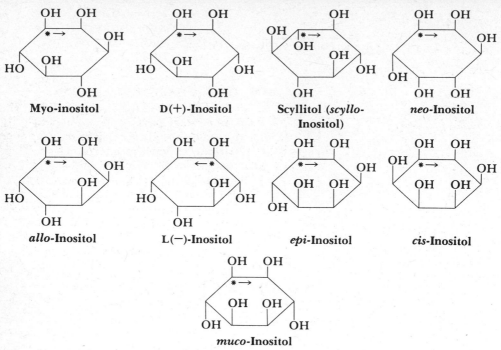

FIGURE 5.102. Nine stereoisomeric forms of cyclohexitols (inositols). Asterisk (*) and arrow (→) indicate the C_1 position and the preferred direction for numbering consecutive carbons in the ring structure.

inositol may contribute to cell wall polysaccharides as a structural precursor to galactinol. The myo-inositol structure can also be isolated from plants as a hexaphosphoric acid ester known as *phytic acid*. Calcium or magnesium salts of phytic acid commonly occur in plants as *phytins*. The hydrolysis of phytic acid followed by crystallization of myo-inositol from water serves as a commercial route for inositol production. In addition to commercial efforts, highly specific *phytase* enzymes are able to liberate myo-inositol from phytic acid *in vivo*. Other notable properties of myo-inositol appear in Table 5.27.

Although clear-cut clinical effects of myo-inositol deficiencies in humans require additional study, its wide distribution throughout the body in phospholipids attests to its essential biochemical importance (Abels *et al.,* 1943). The basic import of myo-inositol seems to be more clearly demonstrated in hamster, rat, and mouse studies (Woolley, 1940, 1944). For example, a joint deficiency of *p*-aminobenzoic acid and myo-inositol can lead to the

death of hamsters, while rats and mice exhibit poor growth, characteristic bilateral alopecia around the eyes, and general hair loss with only strict myo-inositol deprivation. Actually, the hair loss demonstrated by rodents, presumably caused by myo-inositol deficiency, is reminiscent of biotin deficiencies. Accordingly, it has been tacitly proposed in some studies that myo-inositol elevates intestinal biotin synthesis, but this is not certain.

Aside from dietary sources, synthesis of

Table 5.27. Noteworthy Properties for Myo-inositol

Empirical formula	$C_6H_{12}O_6$
Molecular weight	180.16
Melting point	225–227°C

Solubility: Myo-inositol is soluble in 5.7 parts of water (24°C), slightly soluble in alcohol, and insoluble in ether or organic solvents; crystallizes from water >50°C as anhydrous crystals

pH: Approximately neutral to litmus

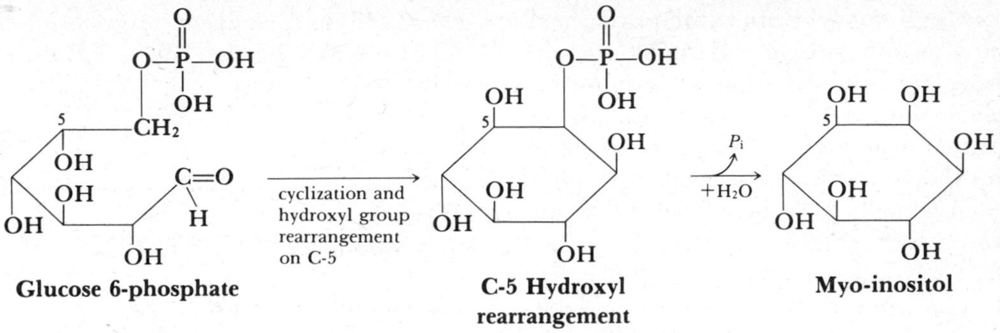

FIGURE 5.103. Myo-inositol formation from glucose 6-phosphate.

myo-inositol in animal cells is believed to occur through the cyclization of glucose 6-phosphate; rearrangement of the C-5 hydroxyl group on cyclized derivatives; and subsequent phosphate hydrolysis from the ring (Figure 5.103).

The metabolism of myo-inositol in animals is highly varied and may involve its conversion into phospholipids, glucose, or oxidation into carbon dioxide. The linkage of myo-inositol to cytidine diphosphate diacylglycerol yields *phosphatidylinositol* (Figure 5.104). This phospholipid may then undergo phosphorylation at one or two of the hydroxyl groups on inositol to produce phosphatidylinositol mono- or diphosphates. The functions of these phosphorylated deriva-

tives are not completely known, but especially high concentrations are located in the brain and other central nervous system tissues. Moreover, it is presumed that the lipotropic properties associated with myo-inositol are significantly related to the biochemical pathway detailed in Figure 5.104 (Milhorat, 1971; Gavin and McHenry, 1941).

The involvement of myo-inositol in gluconeogenesis requires its preliminary conversion into D-glucuronic acid, followed by the formation of L-gulonic acid, 3-oxogulonic, xylitol, D-xylulose, and eventually glucose via the pentose phosphate pathway (Alam, 1971).

There is little evidence from present studies that myo-inositol is toxic to animals, al-

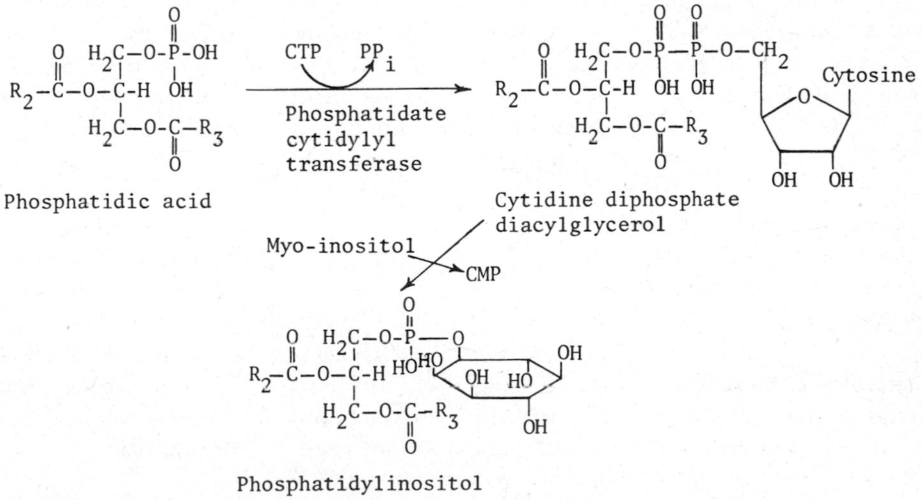

FIGURE 5.104. Synthesis pathway for phosphatidylinositol from myo-inositol and phosphatidic acid.

though high concentrations of phytic acid (inositol hexaphosphate) can inhibit iron absorption from the intestine by forming insoluble iron–phytate complexes. Aside from this difficulty, it has been suggested by some studies that phosphates derived from phytates can actually have beneficial cariostatic effects on teeth (McClure, 1964).

The analytical evaluation of inositol is difficult and time consuming since reliable methods require up to 6 h of acid refluxing merely to liberate combined forms of inositol from phytic acid, soybean cephalin, and similar substances prior to analytical quantitation (Snell, 1950). Liberated inositol may then be assayed by conventional bioassay methods using test organisms such as *Saccharomyces carlsbergensis* (ATCC 9084); (Atkin

et al., 1943) or *Neurospora crassa* "inositolless" (ATCC 9683) (Beadle, 1944; Tatum *et al.*, 1946).

OTHER VITAMIN-LIKE SUBSTANCES

It is incumbent to point out here that numerous *vitamin-like* substances exist in addition to the ones previously discussed. This presentation has been intentionally limited since the chemical nature, biochemical roles, and virtual necessity of these substances in the scheme of human nutrition are not recognized. Some of these substances have been outlined below.

Name	Source	Speculated Function
Vitamin B_{15} (Pangamic acid)	Ubiquitously found in seeds and may be isolated from rice bran or apricot kernels; substance is also found in ox and horse livers, brewers yeast, and often in association with other members of the B-complex vitamins	It has been suggested that B_{15} is useful in treatment of rheumatic and cardiovascular diseases
Vitamin B_{17} (Generic designations include *nitrilosides* or β-cyanogenic glucosides; composed of a sugar, acetone, or benzaldehyde moiety along with hydrogen cyanide)	Common nitriloside sources include *amygdalin* obtained from fruit seeds and bitter almonds; *linamarin* (phaseolunatin) from cassava and lima beans; and *dhurrin* found in fodder grasses, millet, and sorghum grains	Vitamin B_{17}-type substances have been implicated with the occurrence of sickle cell anemia, hypertension, certain circulatory disorders, and the uncertain treatment of some neoplastic diseases
Vitamin L L_1: *o*-aminobenzoic acid L_2: 7-tetrahydro-3,4-dihydroxy-5-(methylmercaptomethyl)-2-furyladenine	—	*Vitamin-like* substance that has been claimed to augment development of lactation tissue
Vitamin P (Rutin, rutoside—a bioflavonoid substance chemically defined as the 3-rhamnoglucoside (3-rutinoside) of 3,5, 7,3′,4′-pentahydroxyflavone	Found in many plants but especially high concentrations (3%, w/w, dry basis) can be obtained from initial bloom stages of the buckwheat plant; for reliable assay consult Strohecker and Henning, (1965)	This vitamin-like substance has been pharmaceutically produced to decrease the fragility of capillaries

Name	Source	Speculated Function
Vitamin U (Cabigen)	Vitamin U can be isolated from cabbage leaves, leaves of other green vegetables, and cabbage leaf juice	This substance has been claimed as useful for the therapy of peptic ulcer (anti-ulcer vitamin)

Another vitamin-like substance, formerly referred to as "vitamin B_t", now recognized as *carnitine,* has been regarded as an essential growth factor for certain organisms such as the mealworm, *Tenebrio molitor.* Although carnitine is nearly omnipresent in animal tissues, it is not strictly classified as a vitamin substance for humans. This lack of vitamin status does not intimate that its role is unimportant in humans, but rather, more than adequate concentration levels of carnitine are obtained from the diet. Therefore, carnitine rarely serves as a limiting factor to the maintenance of normal health in humans. The critical role of carnitine as a transport vehicle for acetyl-CoA-activated long-chain fatty acids into the mitochondrial matrix, where their oxidation occurs, has been detailed in Chapter 17.

USE OF VITAMINS IN THE FOOD INDUSTRY

Concentrations of vitamins in foods differ widely for many reasons. Some raw natural foods display uniform amounts of vitamin throughout all members of the same plant or animal species, while other food sources display random variations. These fluctuations in the vitamin contents of foods may be caused by many factors including environmental growth conditions, seasonal species variations, maturity of the raw food, and disease factors. Added to these conditions are variations in food vitamin concentrations originating from effects of storage, cooking, and postharvest or postmortem processing conditions.

Improved unit food handling, processing, and engineering technologies provide only partial answers for minimizing vitamin losses in foods. In many cases, the rate and magnitude of vitamin losses may be so great that it is economically more feasible to admix industrially prepared vitamins to processed foods than attempt to minimize processing losses.

Conventional addition of vitamins to foods throughout the food industry can usually be described in any one of the following four ways:

1. *Enrichment,* which involves addition of vitamins to "pre" or "post" processed foods at a concentration level well in excess of the amounts naturally present.

2. *Vitamin fortification,* which describes those cases in which foods are used as vehicles for carrying vitamins otherwise absent in the natural food.

3. *Revitaminization,* which requires postprocessing addition of vitamins to foods in order to reestablish their preprocessing vitamin content.

4. *Standardization,* which specifies those additions of vitamins to foods so as to compensate for natural vitamin fluctuations.

It is clear that the objectives of revitaminization and standardization are designed to maintain predictable vitamin distributions and availability among various foodstuffs. Enrichment, on the other hand, is often employed to ensure adequate availability of vitamins in the habitual or preferential dietary regimes of populations that are recognized to be uniformly lacking in one or more vitamin substances. The practice of vitamin enrichment in foods may involve regional or continental proportions, but the practice is usually controlled by legal restrictions of governmental agencies or recognized health organizations in the world theater. Vitamins typically admixed to foods in order to ameliorate or minimize vitamin deficiencies in population segments of certain countries have been detailed in Table 5.28.

Fortification of foods with vitamins is also subject to legal restrictions, but it has assumed great importance in at least two areas of food technology and production. These include the formulation of complete or therapeutic dietary infant formulas designed as substitutes for breast milk and that portion of the food industry that thrives on the production of "table-ready" foods and snack foods.

The infant formulas typically consist of a carefully formulated protein–fat–carbohydrate base inherently low in native vitamin content. This base is then strategically fortified with vitamins comparable to human

Table 5.28. Exported or Fortified Foods in Other Countries

Food	Country	Nutrients
Superamine (wheat–chickpeas–lentil–milk)	Algeria	A, B_2, D, Ca
Golden elbow macaroni (corn–soy–wheat)	Brazil	B_1, B_2, B_6, B_{12}, D, niacin
Cerealina weaning mix (soy–corn starch–milk)	Brazil	A, B_1, B_2, B_6, C, D, niacin, minerals
Mandioca flour (mandioca–soy)	Brazil	Vitamins, amino acids, minerals
Saci (soy beverage)	Brazil	A, B_1, B_2, B_6, B_{12}, D, niacin
Duryea (corn–soy–milk)	Colombia	A, B_2, C, niacin, Fe, Ca, P
Protone (corn–milk–yeast)	Congo	Vitamins, minerals
Frescavida beverage powder (sesame–wheat germ–sugar)	El Salvador	A, B_1, B_2, C, D, niacin, amino acids, minerals
Faffa (peas–milk–sugar–tef)	Ethiopia	A, B_1, B_2, B_6, B_{12}, D, E, niacin, folacin, calpan
INCA Parina (corn–cottonseed–soy–yeast)	Guatemala (also Colombia, Nicaragua, El Salvador, Honduras, Brazil, Mexico)	A, B_2, niacin, lysine
Puma (soy beverage)	Guyana	A, B_1, B_2, B_6, B_{12}, niacin
Vitasoy (soy beverage)	Hong Kong	A, B_1, B_2, B_6
Bal Ahar (wheat–peanut–milk)	India	A, B_1, B_2, Fe, Ca
Bal-Amul weaning mix (milk–vegetable–protein–cereal)	India	Vitamins, minerals
Modern bread (wheat flour)	India	A, B_1, B_2, niacin, lysine, minerals
Atta (wheat flour–peanut flour)	India	A, B_1, B_2, niacin, Fe, Ca
Nutro biscuits (peanut–wheat flour)	India	Vitamins, lysine, minerals
Fortified table salt (trial basis)	India	A, B_1, B_2, B_6, niacin, lysine, Fe
Lac-tone (milk–peanut–sugar)	India	Vitamins, minerals

Table 5.28. (*Continued*)

Food	Country	Nutrients
Banana-Soy Powder (infant beverage feeding)	Israel	Vitamins, minerals
Bread (school lunch)	Japan	Vitamins, lysine
Rice (school lunch)	Japan	B_1, B_2, lysine, minerals
Noodles	Japan	B_1, lysine
Flour	Japan	A, B_1, B_2, Ca, P
Supro (barley–yeast–milk)	Kenya	A, B_{12}, Fe
Laubina weaning mix (wheat–peas–milk)	Lebanon (Middle East)	A, D, Ca, P
Arlac (peanut–milk)	Nigeria	B_1, B_2, B_{12}, D, minerals
Bulgur	Nigeria	Vitamins, lysine, minerals
Wheat flour (protein concentrate)	Pakistan	A, Ca
Peruvita (cottonseed-quinua–milk)	Peru	A, B_1, B_2, Ca
Aliment de Sevrage (millet–peanut–milk–sugar)	Senegal	A, D, Ca
Pro Nutro (soy–corn–peanut–milk–wheat germ)	South Africa	A, B_1, B_2, niacin, minerals
Kupagani biscuit (wheat–soy–milk-sugar–fat)	South Africa	Vitamins, minerals
Vitabean (soy beverage)	Southeast Asia (Singapore)	A, B_1, B_2, B_6, C, D_3, niacin, calpan
Samson (soy beverage)	Surinam	Vitamins
Rice	Thailand	Vitamins, amino acids, minerals
Kaset textured proteins (mung–soy)	Thailand	Vitamins, methionine, tryptophan
Wheat	Tunisia	Vitamins, lysine, minerals
MCH (wheat–peas–lentil)	Tunisia	A, B_1, B_2, B_6, niacin, methionine, Ca
Fortified noodles	Tunisia	Vitamins, minerals
PL beverage (milk–sugar)	Venezuela	Vitamins, minerals
Ceplapro (corn–wheat–soy–milk)	Vietnam	Vitamins, minerals
Milk biscuit (milk–wheat–soy–sugar)	Zambia	A, B_1, B_2, D, niacin, minerals
WSB (wheat–soy blend)	(Over 19 overseas countries)	A, D, E, B_1, B_2, B_6, B_{12}, niacin, calpan, folacin, C
CSM (corn–soy–milk)	(Over 90 overseas countries)	A, B_1, B_2, B_6, B_{12}, C, D, E, folacin, calpan, antioxidants, minerals
Protein fortified flour (flour, 70-wheat protein, 30)	Overseas (many countries)	A, Ca
Non-fat dry milk	Overseas (many countries)	A, D

NOTE: The foods along with their characteristic formulations and/or distribution practices are a matter of past historical record as well as a present practice. Nonetheless, these practices are subject to change at any time.

SOURCE: Reprinted with permission from *Vitamin Fortification and Nutrified Foods*, Hoffmann-LaRoche Inc., Nutley, N.J. Additional information can be found in Agren *et al.* (1969), Bacigalupo (1969), Bressani and Elias (1966), Brooke (1968), Cowan and Pellett (1969), Hoover and Senti (1969), Kapsiotis (1969), Mitsuda (1969), and Senti (1969).

milk in nutritional breadth and balance, or the formula is manufactured so as to contend with certain nutritional or therapeutic conditions present in an infant (Tables 5.29A, B, C, D).

The influx of "table-ready" precooked and processed foods, along with snack foods, into the contemporary diet has contributed to diminished vitamin intake among individuals who rely on these foods to satisfy their daily nutritional needs. Technological efforts to allay the low concentration levels of vitamins in these foods and improve their nutritional image often requires some type of vitamin fortification. Selected foods that can be fortified with vitamins to enhance their nutritional breadth have been outlined in Table 5.30.

MENSURATION CRITERIA FOR ADDING VITAMINS TO FOODS

The amount of vitamin used to fortify foods can be gaged according to a *caloric density* approach or the principle of *metabolic self-sufficiency,* as indicated by Borenstein (1971).

The caloric density method recommends that a daily intake of 2500 to 2800 kcal should have 100% MDR for all vitamins. For example, one pound of cookies that have 2750 kcal should contain 100% of the MDR for all vitamins. Consumption of several cookies each day will not provide an adequate supply of vitamins, but the increased intake of cookies will at least correspond to a caloric intake that is proportional with respect to the MDR for vitamins. According to this rationale, consumers could conceivably comingle foods in any ratio to account for their daily calorie requirement and still realize 100% of the MDR for vitamins.

The metabolic self-sufficiency method dictates that vitamins be added to foods only in amounts that are directly necessary to metabolize the macronutrient constituents with which they are mixed. If 0.10 mg of vitamin B_1 is required for the carbohydrate in 50.0 g of sucrose, 2.0 mg of vitamin B_1 should be distributed throughout every kilogram of sucrose. Although vitamin fortifications based on this approach are self-regulating from economic and safety vantage points, the justification for adding vitamins to foods according to vitamin–calorie ratios is often speculative. Speculation arises here simply because the analytical pitfalls inherent in established MDR values may overestimate or underestimate actual vitamin demands for a certain population segment, or the ability of the body to absorb vitamins from selected fortified foods may not meet theoretical projections. In addition, since recommended food storage conditions designed to maximize vitamin retention in foods cannot be legally enforced, the practice of food vitamin fortification may produce uncertain nutritional effectiveness.

VITAMIN OVERAGE

The amount of excess vitamin necessarily added to foods is often based on the extrapolated destruction rates for known amounts of vitamins with reference to prescribed food-processing treatments. Based on this rationale, it has been possible to project how much excess vitamin should be added to a food in order to compensate for normal processing losses and still ensure adequate levels of vitamin dictated by labeling requirements, standards of identity, or nutritional guidelines. This excess amount of vitamin necessarily added to foods that compensates for processing losses is called *overage*.

The vitamin overage added to preprocessed foods has often been established on the basis of controlled laboratory experiments that assess the destruction of individual pure vitamins or vitamin mixtures admixed to foods. Countless studies of this type have appeared in past and recent literature. Unfortunately, the biochemically simplistic

Table 5.29. Exemplary Vitamin Fortification Levels Contained in Infant Formula Diets That Serve as Substitutes for Human Breast Milk or Bovine Milk or Serve as a Therapeutic Source of Nutrition for Certain Infants

A. Enfamil. Bovine milk protein (1.5 g/100 mL); lactose (7.0 g/100 mL); and fat (3.7 g/100 mL), which consists of 80% soy oil and 20% coconut oil. Since the dietary requirement for vitamin E is related to the intake of polyunsaturated fatty acids (PUFA), the Enfamil vitamin E level of 12.0 I.U./quart of formula gives a vitamin E (I.U.) to PUFA (g) ratio of 0.7:1.0, which is in line with levels recommended (Fomon, 1974; Committee on Nutrition/ American Academy of Pediatrics, 1976). Formulation is used as a supplement to breast feeding, or as sole source of nutrition for full-term normal infants (calorie distribution parallels that of human milk).

	Percentage of Calories from		
	Protein	Fat	CHO
Enfamil formula	9	50	41
Human milk	6	56	38

Vitamin and mineral content per quart[a]	Human milk	Enfamil[b]
Vitamin A, I.U.	1797	1600
Vitamin D, I.U.	21	400
Vitamin E, I.U.	6	12
Vitamin C (ascorbic acid), mg	41	52
Folic acid (folacin), µg	49	100
Thiamine (vitamin B_1), mg	0.15	0.5
Riboflavin (vitamin B_2), mg	0.3	0.6
Niacin, mg	1.4	8
Vitamin B_6, mg	0.1	0.4
Vitamin B_{12}, µg	Trace	2
Pantothenic acid, mg	1.7	3
Choline, mg	—	45
Calcium, mg	322	520
Phosphorus, mg	133	440
Iodine, µg	28	65
Iron, mg	0.5	1.4[c]
Sodium, mg	152	265
Potassium, mg	507	660
Magnesium, mg	38	45
Copper, mg	0.4	0.6
Zinc, mg	3.5	4
Manganese, mg	Trace	1

[a]Normal feeding dilution, 20 kcal per fluid ounce.

[b]Biotin, vitamin K, and certain minerals present in breast milk are supplied by the nonfat milk in Enfamil.

[c]Enfamil With Iron contains 12 mg iron per quart.

NOTE: The osmolar load of Enfamil is in the range of that of human milk, and curd tension is reduced to a nonmeasurable level ("zero") comparable to that of human milk. Fat is a blend of vegetable oils for good tolerance, good absorption, avoidance of sour regurgitation odor.

Table 5.29. (*Continued*)

B. ProSobee. Infant formula for routine feeding of infants intolerant to either milk protein or lactose. Formulation contains soy protein isolate (2.7 g/100 mL); sugar (3.9 g/100 mL); corn syrup solids (2.4 g/100 mL); and soy oil (3.2 g/100 mL).

Each quart of ProSobee formula (20 kcal/fl oz) supplies 640 kcal and the following vitamins and minerals:	Per 100 kcal (5 fl oz)	Per quart
Vitamin A, I.U.	250	1600
Vitamin D, I.U.	62.5	400
Vitamin E, I.U.	2.2	14
Vitamin C (ascorbic acid), mg	8	52
Folic acid (folacin), μg	16	100
Thiamine (vitamin B_1), mg	0.08	0.5
Riboflavin (vitamin B_2), mg	0.09	0.6
Niacin, mg	1.25	8
Vitamin B_6, mg	0.06	0.4
Vitamin B_{12}, μg	0.31	2
Biotin, mg	0.008	0.05
Pantothenic acid, mg	0.47	3
Vitamin K_1, μg	15.6	100
Choline, mg	13	85
Calcium, mg	117	750
Phosphorus, mg	78	500
Iodine, μg	7	45
Iron, mg	1.9	12
Magnesium, mg	10.9	70
Copper, mg	0.09	0.6
Zinc, mg	0.78	5
Manganese, mg	0.16	1
Chloride, mg	62.5	400
Potassium, mg	109	700
Sodium, mg	54.7	350

C. Pregestimil. A nutritionally complete formula providing all major nutrients in simple easily absorbable forms for infants and children with disaccharidase deficiency and malabsorption problems. Formulation contains charcoal-treated, enzymatically hydrolyzed casein to reduce allergenicity (2.2 g/100 mL); fractionated triacylglycerols from coconut oil along with corn oil (2.8 g/100 mL); and carbohydrate (8.8 g/100 mL) from dextrose, modified tapioca starch, plus corn syrup solids.

Each quart in normal dilution (20 kcal/ fl oz, or 0.67 kcal/mL) supplies the following vitamins and minerals:	Per 100 kcal (5 fl oz)	Per quart
Vitamin A, I.U.	250	1600
Vitamin D, I.U.	62.5	400
Vitamin E, I.U.	1.6	10
Vitamin C (ascorbic acid), mg	8	52
Folic acid (folacin), μg	16	100
Thiamine (vitamin B_1), mg	0.08	0.5
Riboflavin (vitamin B_2), mg	0.09	0.6
Niacin, mg	1.2	8
Vitamin B_6, mg	0.06	0.4
Vitamin B_{12}, μg	0.3	2
Biotin, mg	0.008	0.05
Pantothenic acid, mg	0.5	3

Table 5.29 (*Continued*)

Each quart in normal dilution (20 kcal/ fl oz, or 0.67 kcal/mL) supplies the following vitamins and minerals:	Per 100 kcal (5 fl oz)	Per quart
Vitamin K_1, µg	16	100
Choline, mg	13	85
Inositol, mg	4.7	30
Calcium, mg	94	600
Phosphorus, mg	70	450
Iodine, µg	7	45
Iron, mg	1.9	12
Magnesium, mg	11	70
Copper, mg	0.09	0.6
Zinc, mg	0.6	4
Manganese, mg	0.16	1
Chloride, mg	70	450
Potassium, mg	101	650
Sodium, mg	47	300

D. Enfamil Premature Formula. Infant formula designed for feeding rapidly growing low-birth-weight infants. Formula contains protein (2.2 g/100 mL) from skim milk; fat (4.1 g/100 mL) from corn oil and fractionated coconut oil; and carbohydrate (9.2 g/100 mL) from sucrose and lactose.

Each quart of Enfamil Premature Formula (24 kcal/fl oz) supplies 768 kcal and the following vitamins and minerals:	Per 100 kcal (4.2 fl oz)	Per quart
Vitamin A, I.U.	250	1920
Vitamin D, I.U.	62.5	480
Vitamin E, I.U.	1.9	14.4
Vitamin C (ascorbic acid), mg	8	62
Folic acid (folacin), µg	15.6	120
Thiamine (vitamin B_1), mg	0.08	0.6
Riboflavin (vitamin B_2), mg	0.09	0.7
Niacin, mg	1.25	9.6
Vitamin B_6, mg	0.06	0.5
Vitamin B_{12}, µg	0.3	2.4
Pantothenic acid, mg	0.5	3.6
Vitamin K_1, µg	9.4	72
Choline, mg	7	54
Calcium, mg	156	1200
Phosphorus, mg	78	600
Iodine, µg	8	61
Iron, mg	0.22	1.7
Magnesium, mg	10.4	80
Copper, mg	0.09	0.7
Zinc, mg	0.6	4.8
Manganese, mg	0.16	1.2
Sodium, mg	39.1	300

NOTE: The nutrient levels meet Food and Drug Administration requirements for infant formula requiring supplemental iron. The premature infant should also receive supplemental vitamin E.

NOTE: Enfamil, ProSobee, Pregestimil, and Enfamil Premature Formula are all registered trademarks of Mead Johnson and Company, Evansville, Indiana.

SOURCE: Reprinted with permission from *Handbook, Infant Formulas*, 1977, Mead Johnson & Co., Evansville, Indiana 47721.

Table 5.30. Some Consideration Factors Regarding the Addition of Vitamins to Foods[a]

Food	Vitamins technically capable of being added	General remarks
Breakfast cereal	Thiamine, riboflavin, and niacin	As an enrichment factor
	Folacin	Good carrier
	Cyanocobalamin	Good carrier
	Pyridoxine	Loss of natural in processing
	Vitamin E	Loss of natural in processing
	Vitamin A	Good carrier
Cheese products		
Primary	β-Carotene	Added yellow color and increases vitamin A value
Processed	β-apo-8'-Carotenal and/or β-Carotene or colorless vitamin A	Added orange or yellow color and vitamin A value
Corn meal products	Thiamine, riboflavin, and niacin	As an enrichment factor
	Pyridoxine	Loss of natural in processing
	Vitamin E	Loss of natural in processing
	Vitamin A	Good carrier
Flour, all types (at the mill)[b]	Niacin, thiamine, and riboflavin	As an enrichment factor
	Pyridoxine	Loss of natural in milling
	Vitamin E	Loss of natural in milling and bleaching processes
	Vitamin A	Good carrier, no color
	β-Carotene	Good carrier, colors yellow
Fruit gelatin powder	Ascorbic acid	Fruit associated, good carrier
Margarine	Vitamin A, β-carotene, and vitamin D	Added as fortification factors
	Vitamin E	High unsaturated fat
Milk products		
Fluid skim milk	Vitamin A	Milk fat removed
Low-fat fluid milk	Vitamin D	Calcium product
Nonfat dry milk	Vitamin E	Low in nonfat dry milk
Skim-milk yogurt	Vitamin A	Milk fat removed
Mellorine (vegetable fat ice cream)	Vitamin A	Milk fat removed
Filled milks (vegetable fat fluid milk)	Vitamin A	Milk fat removed
Evaporated milk (canned)	Vitamin E, ascorbic acid	Loss of natural in processing
Winter butter	β-Carotene, vitamin A	Loss of natural color and vitamin A in winter
Noncitrus juices, fruit beverages[c]	Ascorbic acid	No natural or variable level, good carrier
	β-Carotene or β-apo-8'-carotenal or vitamin A	Provides added color and vitamin A value
Pasta products	Thiamine, riboflavin, and niacin	As an enrichment factor
	Pyridoxine	Loss of natural in processing
Peanut butter	Thiamine	Loss of natural in nut roasting
	Ascorbic acid	Good carrier
	Vitamin A	Spread for breads
Potato products		
Chips	Thiamine, riboflavin, and niacin	As enrichment factors
	Pyridoxine	Low natural level
	Vitamin E	High fat (unsaturated) level
Instant	Ascorbic acid	Loss of natural in processing
	Vitamin A	Good carrier
Puddings and dessert mixes	Thiamine, riboflavin, and niacin	Calorie/food nutrient concept

Table 5.30. (*Continued*)

Food	Vitamins technically capable of being added	General remarks
Rice products	Thiamine, riboflavin, and niacin	As enrichment factors
	Pyridoxine, vitamin E	Loss of natural in processing
Snacks, cereal type	Thiamine, riboflavin, and niacin	As enrichment factors
	Pyridoxine	Low natural level
Soups, chowders (dried, frozen, canned)	Water-soluble and fat-soluble vitamins	Vitamin addition to meet the breadth of a normal balanced meal
Tomato juice	Ascorbic acid	Natural content varies widely
Vegetable salad, cooking oil	Vitamin E	High unsaturated fat

[a]Foods containing little or no vitamin addition are biscuits, biscuit mixes, cakes, cake mixes, candy, carbonated beverages, cheese spreads, processed chicken products, coffee cakes, confectionery, cookies, crackers, dessert toppings, processed fish products, foreign food dishes, processed fruits, ice cream, jams, jellies, processed meat products, mini-sandwich snacks, muffins, muffin mixes, pancake mixes, pastries, pastry snacks, preserves, potato chip snacks, pretzels, processed cheese, puff snacks, stews, syrups, sweet bakery products, TV dinners, wafers, etc. The use of convenience foods is increasing in the American diet (Bivens, 1967).

[b]Mill addition of vitamins to all types of flour would be assurance that they were present in all bread, rolls, sweet goods, dry premixes, frozen dough products, cookies, crackers, snacks, etc.

[c]Dried beverage powders, citrus semblance drinks, and fruit carbonated drinks are included.

NOTE: Greater concern is being shown for folacin and cyanocobalamin (B_{12}) adequacy of diets. Fewer data exist on their respective additions to foods. Folacin is stable to heat in acid medium but is destroyed more easily in alkaline or neutral conditions. Significant amounts of natural folacin in foods is destroyed during heat processing. Cyanocobalamin has been associated with animal foods and as more foods of plant origin become the major portion of the diet, cyanocobalamin levels will become more critical. The meal replacers such as infant meals, adolescent meals, geriatric meals, allergenic meals, pregnancy and lactation meals, weight-watcher meals, muscle-builder meals, etc., should receive attention for any special vitamin requirements. For example, infant foods of soy protein origin require consideration for zinc, cyanocobalamin, and vitamin K_1 addition (Filer, 1968, 1969; Hollenbeck et al., 1955).

SOURCE: Reprinted with permission and adapted from *Vitamin Fortification and Nutrified Foods,* Hoffmann-LaRoche Inc., Nutley, N. J.

and often esoteric approaches that accompany these studies fail to account for (1) synergistic chemical effects that cause the demise of vitamins in foods; (2) unique biochemical compositions and physical states of foods at their time of processing; and (3) other idiosyncracies of specific food engineering processes that happen to be employed. In many cases, there are no credible substitutes for setting required levels of vitamin overage in foods short of pre- and postprocessing vitamin assays based on actual food-processing operations.

Once calculated, the required concentration level of vitamin overage for a food substance can be achieved by adding commercially prepared crude vitamin extracts or purified vitamin in the form of salts; fatty acid esters of certain vitamins (e.g., acetate or palmitate derivatives of vitamin A); stabilized aqueous solutions of pure vitamin;

provitamin forms; or concentrated oil solutions of fat-soluble vitamins. The most appropriate form of a vitamin for admixture to foods is a discretionary decision that must take into account the possible development of off-flavors originating from the vitamin or its carrier; the ease of vitamin disperison throughout a specific food formulation; economic factors; and commercially available bulk forms of vitamin available to the food industry.

The concentration levels of vitamins added to foods for purposes of overage, fortification, enrichment, revitaminization, or standardization should require the scrupulous attention of the food processor. In many cases, vitamin assays for foods may be a very simple matter, but vitamin monitoring assays can also be problematical for a variety of reasons. Limited numbers of food samples can obviously lead to random sampling errors in evaluating the vitamin content of a food, especially if the vitamin admixed to food is inadequately dispersed. Moreover, "tried and true" vitamin assays for some foods are not necessarily the most appropriate methods for all foods. For example, assay reliability may be upset by the natural presence of unimportant vitamin analogues or nonvitamin substances present in certain foods, as well as the failure of a prescribed method of vitamin assay to effectively extract vitamins from a food prior to assay. All of these factors in conjunction with the idiosyncracies of large-scale industrial food production can easily lead to a ±10 to 35 + % range of assay variability from projected vitamin concentrations anticipated in batch-processed foods.

CONCLUSION

Although a considerable amount of evidence has been accumulated regarding the distribution, biochemistry, and nutritional requirements of vitamins, wide investigational frontiers still exist. Our current lack of knowledge is not an indictment of past research efforts, but a summons to progressively improve vitamin assay methods as well as conceptual approaches toward the biochemical mechanics of vitamins. Unfortunately the inability of researchers to steadfastly define the roles of some vitamin substances has been siezed by pseudoscientists and food faddists as a license to promulgate their own self-proclaimed mechanisms and functions for vitamins. In recent years these efforts have only served to confuse an already complex topic further; they have heightened the scope of mysticism surrounding the actual roles of vitamins; they have promoted the widespread acceptance of megavitamin dietary intake; and they have promoted the concept that vitamins can act as a panacea for the maladies of mankind. There is no doubt that vitamins participate in wondrous biochemical reactions in many life processes and there is no doubt that fresh, creative scientific approaches to their further understanding are crucial, but this noble effort must not be confused with unfounded claims and counterclaims.

REFERENCES

AAP (American Academy of Pediatrics) Committee on Nutrition. 1971. Vitamin K supplementation for infants receiving milk substitute infant formulas and for those with fat malabsorption. *Pediatrics* **48**:483.

Abels, J. C., C. W. Kupel, G. T. Pack, and C. P. Rhoads. 1943. Metabolic studies of patients with cancer of the gastrointestinal tract. XV. Lipotropic properties of inositol. *Proc. Soc. Exp. Biol. Med.* **54**:157.

Agren, G., Y. Hofvander, R. Selenius, and B. Valquist. 1969. Faffa: a supplementary cereal-based weaning food in Ethiopia. In *Protein-Enriched Cereal Foods for World Needs*, pp. 278–287. American Association of Cereal Chemists, St. Paul, Minn.

Aherns, H., and W. Korytnyk. 1969. Pyridoxine chemistry. XXI. Thin-layer chromatography and thin-layer electrophoresis of compounds in the vitamin B_6 group. *Anal. Biochem.* **30**:413.

Alam, S. Q. 1971. Inositols IX. Biochemical systems. In *The Vitamins* (H. S. Sebrell and R. S. Harris, eds.), Vol. III, pp. 380–394. Academic Press, New York.

Almquist, H. J. 1941. Method of assay for K supplements. *J. Ass. Offic. Agr. Chem.* **24**:405.

Ames, S. R., and R. W. Lehman. 1960. Estimation of the biological potency of vitamin A sources from their maleic values. *J. Ass. Offic. Agr. Chem.* **45**:21.

Anderson, T. W., D. B. W. Reid, and G. H. Beaton. 1972. Vitamin C and the common cold: a double blind trial. *Can. Med. Ass. J.* **107**:503.

Anderson, T. W. 1975. Large scale trials of vitamin C. *Ann. N. Y. Acad. Sci.* **258**:498.

Anonymous. 1959. Congenital hypervitaminosis A. *Nutrition* **17**:276.

Anonymous. 1961. Toxicity of vitamin K substitutes in premature infants. *Nutr. Rev.* **19**:75.

Anonymous. 1968. Nutritional folate deficiency. *Brit. Med. J.* **2**:377.

Arnaboldi, M., M. G. Motto, K. Tsujimoto, V. Balogh-Nair, and K. Nakanishi, 1979. Hydroretinals and hydrorhodopsins. *J. Amer. Chem. Soc.* **101**:7082.

Atkin, L., A. S. Shultz, W. L. Williams, and C. N. Frey. 1943. Yeast microbiological methods for the determination of vitamins. Pyridoxine. *Ind. Eng. Chem. Anal. Ed.* **15**:141.

Bacigalupo, A. 1969. Protein-rich cereal foods in Peru. In *Protein-enriched Cereal Foods for World Needs*, American Association of Cereal Chemists, St. Paul, Minn.

Baker, A. Z., and M. D. Wright. 1940. Biological assays of vitamin D₃. *Analyst* **65**:326.

Baker, H., O. Frank, V. B. Matovich, I. Pascher, S. Aronson, S. H. Hutner, and H. Sobotka. 1962. A new assay method for biotin in blood, serum, urine and tissues. *Anal. Biochem.* **3**:31.

Baker, H., and O. Frank. 1968. *Clinical Vitaminology.* Wiley-Interscience, New York.

Barker, H. A., H. Weissbach, and R. D. Smith. 1958. A coenzyme containing pseudovitamin B₁₂. *Proc. Natl. Acad. Sci.* **44**:1093.

Beadle, G. W. 1944. An inositolless mutant strain of *Neurospora* and its use in bioassays. *J. Biol. Chem.* **156**:683.

Beck, R. A. 1979. Comparison of two radioassay methods for cyanocobalamin in seafoods. *J. Food Sci.* **44**:1078.

Beck, R. A. 1979. Essential prerequisites for the analysis of cyanocobalamin in biochemically complex samples using radiometric competitive binding assays. In *Vitamin B₁₂. Proceedings of the Third European Symposium on Vitamin B₁₂ and Intrinsic Factor, March 5–8, 1979* (B. Zagalak and W. Friedrich, eds.), pp. 675–679. Walter de Gruyter, Berlin.

Bernard, R. A., and B. P. Halpern. 1968. Taste changes in vitamin A deficiency. *J. Gen. Physiol.* **52**:444.

Bessey, O. A., and C. G. King. 1933. The distribution of vitamin C in plant and animal tissues and its distribution. *J. Biol. Chem.* **103**:687.

Best, C. H., and M. E. Huntsman. 1932. The effects of the components of lecithin upon deposition of fat in liver. *J. Physiol. London* **75**:405.

Bieri, J. G., and E. L. Prival. 1965. Serum vitamin E determined by thin-layer chromatography. *Proc. Soc. Exp. Biol. Med.* **120**:554.

Bieri, J. G., E. G. McDaniel, and W. E. Rogers. 1969. Survival of germfree rats without vitamin A. *Science* **163**:574.

Bivens, G. E. 1967. Household Use of 32 Convenience Foods, 1955 and 1965. USDA Agr. Res. Serv. Consumer and Food Econ. Res. Div. Presentation at 45th Agr. Outlook Conference. Washington, D. C., November 14.

Bliss, C. I. 1945. The combined slope in comparative tests of tibia and toe ash in chick assay for vitamin D. *Poultry Sci.* **24**:534.

Bond, A. D. 1975. Ascorbic-2-sulfate metabolism by human fibroblasts. *Ann. N. Y. Acad. Sci.* **258**:307.

Borenstein, B. 1971. Rationale and technology of food fortification with vitamins, minerals, and amino acids. In *CRC Critical Reviews in Food Technology*, Vol. 2, Issue 2. Chemical Rubber Company, Boca Raton, Fla.

Bourne, G. H. 1949. Vitamin C and immunity. *Brit. J. Nutr.* **2**:341.

Bressani, R., and L. G. Elias. 1966. All-vegetable protein mixture for human feeding. The development of INCAP vegetable mixture 14 based on soybean flour. *J. Food Sci.* **31**:626.

Briggs, M. H., P. Garcia-Webb, and P. Davies. 1973. Urinary oxalate and vitamin C supplements. *Lancet* **2**:201.

Brooke, C. L. 1968. Fortification of food products with vitamin A. *Proceedings of West. Nutr. Conf.*, San Juan, Puerto Rico, August 1968, p. 137.

Brown, R. R., D. B. Rose, J. M. Price, and H. Wolf. 1969. Tryptophan metabolism as affected by anovulatory agents. *Ann. N. Y. Acad. Sci.* **166**:44.

Brown, R. R., 1972. Normal pathological conditions which may alter the human requirement for vitamin B₆. *Agr. Food Chem.* **20**:498.

Burchenal J. H. 1955. Folic acid antagonists. *Amer. J. Clin. Nutr.* **3**:311.

Campbell, H. A., and K. P. Link. 1941. Studies on the sweet clover disease. IV. The isolation and crystallization of the hemorrhagic agent. *J. Biol. Chem.* **138**:21.

Campbell, J. A., B. B. Migicovsky, and A. R. C. Emslie. 1945. Chick assay for vitamin D. *Poultry Sci.* **24**:3,72.

Campbell, J. A. 1948. Modified glycerol dichlorohydrin reaction for vitamin D₃. *Anal. Chem.* **20**:766.

Carmel, R. 1979. Large vitamin B₁₂-binding proteins and complexes in human serum. In *Proceedings of the Third European Symposium on Vitamin B₁₂ and Intrinsic Factor, March 5–8, 1979* (B. Zagalak and W. Friedrich, eds.), pp. 776–790. Walter de Gruyter, Berlin.

Carr, F. H., and E. A. Price. 1926. Color reactions attributed to vitamin A. *Biochem. J.* **20**:497.

Century, B., and M. K. Horwitt. 1965. Biological availability of various forms of vitamin E with respect to different indices of deficiency. *Fed. Proc.* **24**:960.

Chatterjee, I. B. 1970. Biosynthesis of L-ascorbate in animals. In *Methods of Enzymology* (D. B. McCormick and L. D. Wright, eds.), Vol. 18, Part A, pp. 28–34. Academic Press, New York.

Chaudhuri, C. R., and I. B. Chatterjee. 1969. L-Ascorbic acid synthesis in birds; phylogenetic trend. *Science* **164**:435.

Committee on Nutrition/American Academy of Pediatrics. 1976. Commentary on breast feeding and infant formulas, including proposed standards for formulas. *Pediatrics* **57**:278.

Coulehan, J. L., L. Kapner, S. Eberhard, F. H. Taylor, and K. D. Rogers. 1975. Vitamin C and upper respiratory illness in Navajo children: Preliminary observations (1974). *Ann. N. Y. Acad. Sci.* **258**:513.

Cowan, J. W., and P. L. Pellett. 1969. The Development of Laubina, infant food mixture for the Middle East. In *Protein-Enriched Cereal Foods for World Needs*, pp. 305–314. American Association of Cereal Chemists—St. Paul.

Darby, W. J., and E. Jones. 1945. Treatment of sprue with synthetic *L. casei* factor (folic acid, vitamin M). *Proc. Soc. Exp. Biol. Med.* **60**:259.

Davis, B. D. 1951. Aromatic biosynthesis. III. Role of *p*-aminobenzoic acid in the formation of vitamin B_{12}. *J. Bact.* **62**:221.

DeLuca, H. F. 1971. The role of vitamin D and its relationship to parathyroid hormone and calcitonin. *Recent Progr. Hormone Res.* **27**:479.

DeLuca, H. F. 1974. Vitamin D: the vitamin and the hormone. *Fed. Proc.* **33**:2211.

DeLuca, H. F., and J. W. Blunt. 1971. Vitamin D. In *Methods in Enzymology* (D. B. McCormick and L. D. Wright, eds.), Vol. XVIII, Part C, pp. 709–733. Academic Press, New York.

Deluca, L., N. Maestri, F. Bonanni, and D. Nelson. 1972. Maintenance of epithelial cell differentiation; the mode of action of vitamin A. *Cancer* **30**:1326.

Deutsch, M. J., S. W. Jones, J. B. Wilkie, D. Duffy, and H. W. Loy. 1964. Assay of oil-soluble vitamins. Part I. Vitamin A and carotene: suggestions for improving assay. *J. Ass. Offic. Agr. Chem.* **47**:756.

DeWitt, J. B., and M. X. Sullivan. 1946. Quantitative estimation of vitamin D. *Ind. Eng. Chem., Anal. Ed.* **18**:117.

Drujan, B. D. 1971. Determination of vitamin A. In *Methods in Enzymology* (D. B. McCormick and

L. D. Wright, eds.), pp. 565–573. Academic Press, New York.

Drummond, J. C. 1920. The nomenclature of the so-called accessory food factors. *Biochem. J.* **14**:660.

Engel, R. W., W. D. Solamon, and C. J. Ackerman. 1954. Chemical estimation of choline. In *Methods of Biochemical Analysis* (D. Glick, ed.), Vol. I, pp. 265–286. Wiley-Interscience, New York.

Estimation of the Vitamins. 1947. Biological Symposia XII, p. 452. Jaques Cattell Press, Lancaster, PA.

Ewing, E. T., G. V. Kingsley, R. A. Brown, and A. D. Emmett. 1943. Determination of vitamins D in fish liver oil. *Ind. Eng. Chem., Anal. Ed.* **15**:301.

Farrell, P. M., and J. G. Bieri. 1975. Megavitamin E supplementation in man. *Amer. J. Clin. Nutr.* **28**:1381.

Filer, L. J. 1969. Fat-soluble vitamins. Presentation at symposium, *Total Synthesis of Food*, AAAS Meeting, Boston, December 27, 1969.

Filer, L. J. 1968. Enrichment of special dietary food products. *Agr. Food Chem.* **16**:148.

Fomon, S. J. 1974. *Infant Nutrition*, 2nd ed. Saunders, Philadelphia.

Fraser, D., B. S. L. Kidd, S. W. Kook, and L. Paunier. 1966. A new look at infantile hypercalcemia. *Pediatr. Clin. North Amer.* **13**:503.

Fraser, D. R., and E. Kodicek. 1970. Unique biosynthesis by kidney of a biologically active vitamin D metabolite. *Nature* **228**:764.

Fraser, D., S. W. Kook, H. P. Kind, M. F. Holick, Y. Tanaka, and H. F. DeLuca. 1973. Pathogenesis of hereditary vitamin D-dependent rickets. *N. Engl. J. Med.* **289**:817.

Freed, M. (ed.). 1966. *Methods of Vitamin Assay*, 3rd ed. The Association of Vitamin Chemists Inc. Wiley-Interscience, New York.

Friedman, L. 1960. Vitamin A in oleomargarine. *J. Ass. Offic. Agr. Chem.* **43**:6.

Friedrich, W. 1979. Concerning a new corrinoid from municipal sludge. In *Vitamin B_{12}. Proceedings of the Third European Symposium on Vitamin B_{12} and Intrinsic Factor, March 5–8, 1979* (B. Zagalak and W. Friedrich, eds.), pp. 163–168. Walter de Gruyter, Berlin.

Gavin, G., and E. W. McHenry. 1941. Inositol: A lipotropic factor. *J. Biol. Chem.* **139**:485.

Genghof, D. S., C. W. H. Partridge, and F. H. Carpenter. 1948. An agar plate assay for biotin. *Arch. Biochem.* **17**:413.

Green, N. M. 1970. Spectrophotometric determination of avidin and biotin. In *Methods in Enzymology* (D. B. McCormick and L. D. Wright, eds.), Vol. XVIII, Part A, pp. 418–424. Academic Press, New York.

Greenberg, D. M., and G. K. Humphreys. 1958. Biosynthesis of the thymine methyl group. *Fed. Proc.* **17:**234.

Greer, E. 1955. Vitamin C in acute polio myelitis. *Med. Times.* **83:**1160.

Griffith, W. H., and N. J. Wade. 1939. Choline metabolism. I. The occurrence and prevention of hemorrhagic degeneration in young rats on low choline diet. *J. Biol. Chem.* **131:**567.

György, P. 1939. The curative factor (vitamin H) for egg white injury, with particular reference to its presence in different foodstuffs and in yeast. *J. Biol. Chem.* **131:**733.

Harrison, H. E., and H. C. Harrison. 1970a. Dibutyryl cyclic AMP, vitamin D and intestinal permeability to calcium. *Endocrinology* **86:**756.

Harrison, H. E., and H. C. Harrison. 1970b. Role of vitamin D, parathyroid hormone and cortisol in the intestinal transport of calcium and phosphate. In *The Fat Soluble Vitamins* (H. F. DeLuca and J. W. Suttie, eds.), pp. 39–54. The University of Wisconsin Press, Madison.

Harrison, H. E., and H. C. Harrison. 1974. Calcium. In *Intestinal Absorption, Biomembranes.* (D. H. Smyth, ed.), Vol. 4B, pp. 793–846. Plenum Press, New York/London.

Hansen, L. G., and N. J. Warwick. 1969. A fluorometric micromethod for serum vitamins A and E. *Clin. Chem.* **39:**538.

Hayes, K. C., and D. M. Hegsted. 1973. Toxicity of the vitamins. In *Committee on Food Protection. Toxicants Occurring Naturally in Foods*, 2nd ed., pp. 235–253. National Academy of Science, Washington, D. C.

Heibron, I. M., R. A. Morton, and E. T. Webster. 1932. The structure of vitamin A. *Biochem. J.* **26:**1194.

Heikkila, R. E., and G. Cohen. 1975. Cytotoxic aspects of the interaction of ascorbic acid with alloxan and 6-hydroxydopamine. *Ann. N. Y. Acad. Sci.* **258:**221.

Herbert, V., R. R. Streiff, and L. W. Sullivan. 1964. Notes on vitamin B_{12} absorption: autoimmunity and childhood pernicious anemia; relation of intrinsic factor to blood group substance. *Medicine* **43:**679.

Herbert, V., and E. Jacob. 1974. Destruction of vitamin B_{12} by ascorbic acid. *J.A.M.A.* **230:**241.

Herbert, V., L. Landau, R. Bash, S. Grosberg, and N. Colman. 1979. Ability of megadoses of vitamin C to destroy vitamin B_{12} and cobinamide and to reduce absorption of vitamin B_{12} (with a note on vitamin B_{12} radioassays). In *Vitamin B_{12}. Proceedings of the Third European Symposium on Vitamin B_{12} and Intrinsic Factor, March 5–8, 1979* (B. Zagalak and W. Friedrich, eds.), pp. 1069–1077. Walter de Gruyter, Berlin.

Hodson, A. Z. 1945. The use of *Neurospora* for the determination of choline and biotin in milk products. *J. Biol. Chem.* **157:**383.

Hollenbeck, C. M., R. Monahan, W. L. Benson, and J. F. Mahoney. 1955. Some preliminary considerations of vitamin B_{12} fortifications of foods. *Amer. Ass. Cereal Chem. Trans.* **13:**233.

Holst, A., and T. Frölich. 1907. Experimental studies relating to ship beri-beri and scurvy. II. *J. Hyg.* **7:**634.

Holst, A. and T. Frölich. 1912. Experimental scurvy. *Z. Hyg. Infektionskrankh.* **72:**1.

Honig, B., U. Dinur, K. Nakanishi, V. Balogh-Nair, M. A. Gawinowicz, M. Arnaboldi, and M. G. Motto. 1979. An external point-charge model for wavelength regulation in visual pigments. *J. Amer. Chem. Soc.* **101:**7084.

Hoover, S. R., and F. R. Senti. 1969. Enrichment and fortification of U. S. A. donated foods. Presentation 8th International Congress of Nutrition, Prague. August 28–September 5.

Horowitz, N. H., and G. W. Beadle. 1943. A microbiological method for the determination of choline by use of a mutant of *Neurospora. J. Biol. Chem.* **150:**325.

Juneja, H. S., N. R. Moudgal, and J. Ganguly. 1969. Studies on vitamin A. The effect of hormones and gestation in retinoate-fed female rats. *Biochem. J.* **111:**97.

Kapsiotis, G. D. 1969. History and Status of Specific Protein-Rich Foods. FAO/WHO/UNICEF Protein Food Program and Products. In *Protein-Enriched Cereal Foods for World Needs*, pp. 255–265. American Association of Cereal Chemists, St. Paul, Minn.

Karrer, P., R. Morf, and K. Schöpp. 1931. *Zur Kenntnis des Vitamins-A aus Fischtranen. Helv. Chim. Acta* **14:**1036.

Karrer, P., R. Morf, and K. Schöpp. 1933. *Synthesis des Perhydrovitamins-A. Helv. Chim. Acta* **16:**625.

Katada, M., S. Tyagi, A. Nath, R. L. Petersen, and R. K. Gupta. 1979. A novel form of vitamin B_{12} and its derivatives. *Biochim. Biophys. Acta* **584**:149.

Kaufman, S. 1958. The participation of tetrahydrofolic acid in the enzyme conversion of phenylalanine to tyrosine. *Biochem. Biophys. Acta* **27**:428.

Kenny, F. M., T. Aceto, Jr., M. Purisch, H. E. Harrison, and B. M. Blizzard. 1963. Metabolic studies in a patient with idiopathic hypercalcemia of infancy. *J. Pediatr.* **62**:531.

King, G. C. 1975. Current status of vitamin C and future horizons. *Ann. N. Y. Acad. Sci.* **258**:491.

King, C. G. 1975. Current status of vitamin C and future horizons. *Ann. N. Y. Acad. Sci.* **258**:540.

Kodicek, E. 1966. Antivitamins of nicotinic acid and of biotin. In *Antivitamins*, Bibliotheca Nutritio et Dieta (J. C. Somogyi, ed.), p. 8. Karger, Basel.

Kögl, F., and B. Tönnis. 1936. *Über das Bios-Problem. Darstellung von Krystallisiertem Biotin aus Eigelb. Z. Physiol. Chem.* **242**:43.

Kolhouse, J. F., H. Kondo, N. C. Allen, E. Podell, and R. Allen. 1978. Cobalamin analogues are present in human plasma and can mask cobalamin deficiency because current radioisotope dilution assays are not specific for true cobalamin. *N. Engl. J. Med.* **299**:785.

Korner, W. F., and F. Weber. 1972. Tolerance of high ascorbic acid doses. *Int. J. Vitamin Nutr. Res.* **42**:528.

Krause, R. F. 1965. Liver lipids in a case of hypervitaminosis. *Amer. J. Clin. Nutr.* **16**:455.

Lampen, J. O., G. P. Bahler, and W. H. Peterson. 1942. The occurrence of free and bound biotin. *J. Nutr.* **23**:11.

Landy, M., and D. M. Dicken. 1942. A microbiological assay method for the six B vitamins using *Lactobacillus casei* and a medium of essentially known composition. *J. Lab. Clin. Med.* **27**:1086.

Lane, M. D., and F. Lynen. 1963. The biochemical function of biotin. VI. Chemical structure of the carboxylated active site of propionyl carboxylase. *Proc. Natl. Acad. Sci.* **49**:379.

Langer, B. W., and P. Györgi. 1971. Biotin VIII. Active compounds and antagonists. In *The Vitamins* (W. H. Sebrell and R. S. Harris, eds.), Vol. II, pp. 294–322. Academic Press, New York.

Levine, C., and E. Chargaff. 1951. Procedures for the microestimation of nitrogenous phosphatide constituents. *J. Biol. Chem.* **192**:465.

Lim, F., and E. D. Schall. 1964. Determination of choline in feeds. *J. Ass. Offic. Agr. Chem.* **47**:501.

Lucy, J. A., and F. U. Lichti. 1969. Reactions of vitamin A with acceptors of electrons. Interactions with iodine and the formation of iodine. *Biochem. J.* **112**:231.

Luzzati, D., and R. Guthrie. 1955. Studies of a purine- or histidine-requiring mutant of *Escherichia coli. J. Biol. Chem.* **216**:1.

Lyness, W. I., and F. W. Quackenbush. 1955. New color reaction for vitamin D. *Anal. Chem.* **27**:1978.

Mameesh, M. S., and B. C. Johnson. 1959. Dietary vitamin K requirement of the rat. *Proc. Soc. Exp. Biol. Med.* **101**:467.

Mann, G. V., and P. Newton. 1975. The membrane transport of ascorbic acid. *Ann. N. Y. Acad. Sci.* **258**:243.

Mason, M., J. Ford, and H. L. C. Wu. 1969. Effects of steroid and non-steroid metabolites on enzyme conformation and pyridoxal phosphate binding. *Ann. N. Y. Acad. Sci.* **166**:170.

McBride, B. C., and R. S. Wolf. 1971. A new coenzyme of methyl transfer, coenzyme M. *Biochemistry* **10**:2317.

McClure, F. J. 1964. Cariostatic effect of phosphates. *Science* **144**:1337.

McCollum, E. V., and M. Davis. 1913. The necessity of certain lipids in the diet during growth. *J. Biol. Chem.* **15**:167.

McCollum, E. V., and M. Davis. 1915. The essential factors in the diet during growth. *J. Biol. Chem.* **23**:231.

McCollum, E. V., J. E. Simmonds, J. E. Becker, and P. G. Shipley. 1922. Studies on experimental rickets. XXIII. The production of rickets in the rat by diets consisting of essentially purified food substances. *J. Biol. Chem.* **54**:249.

McCormick, D. B., and L. D. Wright (eds.). 1970. *Methods in Enzymology*, Vol. XVIII, Part A. Academic Press, New York.

McCormick, D. B., and L. D. Wright (eds.). 1971. *Methods in Enzymology*, Vol. XVIII, Part B. Academic Press, New York.

McIntosh, E. N., M. Purko, and W. A. Wood. 1957. Ketopantoate formation by a hydroxymethylation enzyme from *Escherichia coli. J. Biol. Chem.* **228**:499.

McRoberts, L. H. 1962. Determinations of vitamin A and carotene in oleomargarine. *J. Ass. Offic. Agr. Chem.* **45**:442.

Melancon, M. J. Jr., H. Morii, and H. F. DeLuca. 1970. Physiologic effects of vitamin D, parathyroid hormone and calcitonin. In *The Fat Soluble Vitamins* (H. F. DeLuca and J. W. Suttie, eds.), pp. 111–123. The University of Wisconsin Press, Madison.

Mellanby, E. A. 1919. An experimental investigation on rickets. *Lancet* **1**:407.

Mengel, C. E., and H. L. Greene. 1976. Ascorbic acid effects on erythrocyte. *Ann. Intern. Med.* **84**(4):490.

Metta, V. C., L. Nash, and B. C. Johnson. 1961. A tubular coprophagy-preventing cage for the rat. *J. Nutr.* **74**:473.

Metzler, D. E. 1977. *Biochemistry: The Chemical Reactions of Living Cells.* Academic Press, New York.

Meunier, P. R., J. Ferrando, and G. Thomas. 1949. Influence de la vitamine A sur la detoxication du benzoate de sodium par l'organisme du rat. *Compt. Rend.* **228**:1254.

Milas, N. A., R. Heggie, and J. A. Raynolds. 1941. Quantitative estimation of vitamin D. *Ind. Eng. Chem. Anal. Ed.* **13**:227.

Milhorat, A. T. 1971. Inositol. XI. Deficiency in human beings. In *The Vitamins* (H. S. Sebrell and R. S. Harris, eds.), Vol. III, pp. 398–405. Academic Press, New York.

Minot, G. R., and W. P. Murphy. 1926. Treatment of pernicious anemia by special diet. *J.A.M.A.* **87**:470.

Mitchell, H. K., E. E. Snell, and R. J. Williams. 1941. The concentration of 'folic acid.' *J. Amer. Chem. Soc.* **63**:2284.

Mitsuda, H. 1969. New Approaches to Amino Acid and Vitamin Enrichment in Japan. In *Protein-Enriched Cereal Foods for World Needs.* American Association of Cereal Chemists—St. Paul. pp. 208–219.

Moore, C. V., O. S. Bierbaum, A. D. Welch, and L. D. Wright. 1945. The activity of *Lactobacillus casei* factor (folic acid) as an antipernicious anemia substance. I. Observations on four patients: Two with Addisonian pernicious anemia, one with non-tropical sprue and one with pernicious anemia of pregnancy. *J. Lab. Clin. Med.* **30**:1056.

Moore, T. 1957. *Vitamin A.* Elsevier, New York.

Moore, T. 1967. Pharmacology and toxicity of vitamin A. p. 294. In *The Vitamins* (W. H. Sebrell and R. S. Harris, eds.), 2nd ed. Academic Press, New York.

Muenter, M. D., H. O. Perry, and J. Ludwig. 1971. Chronic vitamin A intoxication in adults. *Amer. J. Med.* **30**:129.

Myasishcheva, N. V., E. V. Quadros, Y. V. Vares, and J. C. Linnell. 1979. Interference with cobalamin metabolism and tumour growth by an analogue of methylcobalamin. In *Proceedings of the Third European Symposium on Vitamin B$_{12}$ and Intrinsic Factor. March 5–8, 1979* (B. Zagalak and W. Friedrich, eds.), pp. 1125. Walter de Gruyter, Berlin.

National Research Council, Food and Nutrition Board. 1975. Hazards of overuse of vitamin D. *Nutr. Rev.* **33**:61.

Neujahr, H. Y. 1955. On vitamins in sewage sludge. *Acta Chem. Scand.* **9**:622.

Newmark, H. L., J. Scheiner, M. Marcus, and M. Prabhudesai. 1976. Stability of vitamin B$_{12}$ in the presence of ascorbic acid. *Amer. J. Clin. Nutr.* **29**:645.

Nieman, C., and H. B. Klein Obbink. 1954. The biochemistry and pathology of hypervitaminosis A. *Vitamins and Hormones* **12**:69.

Nield, C. H., W. C. Russell, and A. Zimmerli. 1940. A spectrophotometric determination of vitamins D$_2$ and D$_3$. *J. Biol. Chem.* **136**:73.

Noguchi, T., A. H. Cantor, and M. L. Scott. 1973. Mode of action of selenium and vitamin E in prevention of exudative diathesis in chicks. *J. Nutr.* **103**:1502.

Norman, A. W., M. R. Haussler, T. H. Adams, J. F. Myrtle, P. Roberts, and K. A. Hibberd. 1969. Basic studies on the mechanism of action of vitamin D. *Amer. J. Clin. Nutr.* **22**:396.

Osborne, T. B., and L. B. Mendel. 1913. The relation of growth to the chemical constituents of the diet. *J. Biol. Chem.* **15**:311.

Osborne, T. B., and L. B. Mendel. 1915. Further observations on the influence of natural fats upon growth. *J. Biol. Chem.* **20**:379.

Olliver, M. 1971. Ascorbic Acid. IV. Estimation. In *The Vitamins* (W. H. Sebrell and R. S. Harris, eds.), Vol. I, pp. 338–385. Academic Press, New York.

Olsen, J. A. 1969. Metabolism and function of vitamin A. *Fed. Proc.* **28**:1670.

Omdahl, J. L., and H. F. DeLuca. 1973. Section B, Vitamin D. In *Modern Nutrition in Health and Disease, Dietotherapy* (R. S. Goodhart and M. E. Shils, eds.), 5th ed. Lea and Febinger, Philadelphia.

Owen, C. A., G. M. Tyce, and E. V. Flock. 1970. Heparin-ascorbic acid antagonism. *Mayo Clin. Proc.* **45**:140.

Owen, C. A., Jr. 1971. Vitamin K group. XIII. Requirements of human beings. In *The Vitamins* (W. H. Sebrell and R. S. Harris, eds.), 2nd ed.,

Vol. III, pp. 521–522. Academic Press, New York.

Parpia, H. A. B. 1969. Protein foods of India based on cereal legumes and oil seeds. In *Protein-Enriched Cereal Foods for World Needs*, pp. 129–139. American Association of Cereal Chemists, St. Paul, Minn.

Parsons, H. T., J. G. Lease, and E. Kelley. 1937. Interrelationship between dietary egg white and the requirement for protective factor in cure of nutritional disorder due to egg white. *Biochem. J.* **31**:424.

Pauling, L. 1970. *Vitamin C and the Common Cold.* Freeman, San Francisco.

Pohanka, D. G., H. Smiciklas-Wright, and R. L. Pike. 1973. Δ^5-3-β-Hydroxysteroid dehydrogenase activity in the adrenal cortex of the sodium-restricted rat. *Proc. Soc. Biol. Med.* **142**:1092.

Rachmilewitz, B., M. Schlesinger, R. Rabinowitz, and M. Rachmilewitz. 1979. The origin and clinical implications of vitamin B_{12}-binders—the transcobalamins. In *Proceedings of the Third European Symposium on Vitamin B_{12} and Intrinsic Factor, March 5–8, 1979* (B. Zagalak and W. Friedrich, eds.), pp. 765–776. Walter de Gruyter, Berlin.

Rickes, E. L., N. G. Brink, F. R. Koniuszy, T. R. Wood, and K. R. Folkers. 1948. Crystalline vitamin B_{12}. *Science* **107**:396.

Roe, J. H., and C. A. Kuether. 1943. The determination of ascorbic acid in whole blood and urine through the 2,4-dinitrophenylhydrazine derivation of dehydroascorbic acid. *J. Biol. Chem.* **147**:399.

Roe, J. H., M. B. Mills, M. J. Oesterling, and C. M. Damron. 1948. The determination of diketo-L-gulonic acid, dehydro-L-ascorbic acid and L-ascorbic acid in the same tissue extract by the 2,4-dinitrophenylhydrazine method. *J. Biol. Chem.* **174**:201.

Roe, J. H. 1961. Appraisal of methods for the determination of ascorbic acid. *Ann. N. Y. Acad. Sci.* **92**:277.

Rogers, W. E., Jr., and J. G. Bieri. 1968. Adrenal Δ^5-3-β-hydroxysteroid dehydrogenase related to vitamin A. *J. Biol. Chem.* **243**:3404.

Rogers, W. E., Jr. 1969. Re-examination of enzyme activities thought to show evidence of a coenzyme role for vitamin A. *Amer. J. Clin. Nutr.* **22**:1003.

Rose, C. S., and P. György. 1952. Specificity of hemolytic reaction in vitamin E deficient erythrocytes. *Amer. J. Physiol.* **168**:414.

Rothenburg, S. P., M. DaCosta, and Z. Rosenberg. 1972. A radioassay for serum folate. Use of a two phase sequential incubation, ligand binding system. *N. Engl. J. Med.* **286**:1335.

Rubin, E., A. L. Florman, T. Degnan, and J. Diaz. 1970. Hepatic injury in chronic hypervitaminosis A. *Amer. J. Dis. Child.* **119**:132.

Ruddick, J. E., J. Vanderstoep, and J. F. Richards. 1978. Folate levels in food—a comparison of microbiological assay and radioassay methods for measuring folate. *J. Food Sci.* **43**:1238.

Schrauzer, G. N., and W. H. Rhead. 1973. Ascorbic acid abuse: effects on long term ingestion of excessive amounts on blood levels and urinary excretion. *Int. J. Vitamin Nutr. Res.* **43**:201.

Senti, F. R. 1969. Formulated cereal foods in the USA Food for Peace Program. In *Protein-Enriched Cereal Foods for World Needs*, pp. 246–254. American Association of Cereal Chemists, St. Paul, Minn.

Sheehym, T. W., B. Baggs, E. Perez-Santiago, and M. H. Flock. 1962. Prognosis of tropical sprue. A study on the effect of folic acid on the intestinal aspects of acute chronic sprue. *Ann. Int. Med.* **57**:892.

Sheppard, A. J., and W. D. Hubbard. 1971. Gas chromatography of vitamins D_2 and D_3. In *Methods in Enzymology* (D. B. McCormick and L. D. Wright, eds.), Vol. XVIII, Part C, pp. 733–746. Academic Press, New York.

Sheppard, A. J., A. R. Prosser, and W. D. Hubbard. 1971. Gas chromatography of vitamin E. In *Methods in Enzymology* (D. B. McCormick and L. D. Wright, eds.), Vol. XVIII, Part C, pp. 356–365. Academic Press, New York.

Shull, G. M., B. L. Hutchings, and W. H. Peterson. 1942. A microbiological assay for biotin. *J. Biol. Chem.* **142**:913.

Shull, G. M., and W. H. Peterson. 1943. Improvements in the *Lactobacillus casei* assay for biotin. *J. Biol. Chem.* **151**:201.

Slover, H. T., J. Lehman, and R. J. Valis. 1968. Vitamin E in foods: determination of tocols and tocotrienols. *Amer. Oil Chem. Soc.* **46**:417.

Smith, E. L., and L. F. J. Parker. 1948. Purification of anti-pernicious anemia factor. *Biochem. J.* **43**:vii–ix.

Smith, E. L. 1965. *Vitamin B_{12}*, 3rd ed. Methuen, London.

Snell, E. E., R. E. Eakin, and R. J. Williams. 1940. A quantitative test for biotin, its occurrence and properties. *J. Amer. Chem. Soc.* **62**:175.

Snell, E. E., and L. D. Wright. 1941. A microbiological method for the determination of nicotinic acid. *J. Biol. Chem.* **139:**675.

Snell, E. E. 1950. *Vitamin Methods I.* Academic Press, New York.

Sobel, A. E., A. M. Mayer, and B. Kramer. 1945. Colorimetric reaction of vitamins D_2 and D_3 and their provitamins. *Ind. Eng. Chem. Anal. Ed.* **17:**160.

Sommer, P., and M. Kofler. 1966. Physiocochemical properties and methods of analysis for phylloquinones, menaquinones, ubiquinones, plastoquinones, menadione and related compounds. *Vitamins Hormones* **24:**349.

Steenbock, H. 1924. The induction of growth promoting and calcifying properties in a ration by exposure to sun-light. *Science* **60:**224.

Stimson, W. H. 1961. Vitamin A intoxication in adults: report case with a summary of the literature. *N. Engl. J. Med.* **265:**36.

Strohecker, R., and H. M. Henning. 1965. *Vitamin Assay—Tested Methods.* The Chemical Rubber Co., Cleveland.

Szent-Györgyi, A. 1928. Observations on the function of peroxidase systems and the chemistry of the adrenal cortex. *Biochem. J.* **22:**1387.

Takenouchi, K., K. Aso, K. Kawase, H. Ichikawa, and T. Shiomi. 1966. On the metabolites of ascorbic acid, especially oxalic acid, eliminated in urine, following the administration of large amounts of ascorbic acid. *J. Vitaminol. Japan* **12:**49.

Tappel, A. L. 1974. Selenium-glutathione peroxidase and vitamin E. *Amer. J. Clin. Nutr.* **27:**960.

Targan, S. R., S. Merrill, and A. D. Schwabbe. 1969. Fractionation and quantification of β-carotene and vitamin A derivatives in human scrum. *Clin. Chem.* **15:**479.

Tatum, E. L., M. G. Ritchie, E. V. Cowdry, and L. F. Wicks. 1946. Vitamin content of mouse epidermis during methylcholanthrene carcinogenesis. I. Biotin, choline, inositol *p*-aminobenzoic acid and pyridoxine. *J. Biol. Chem.* **163:**675.

Taussig, H. B. 1966. Possible injury to the cardiovascular system from vitamin D. *Ann. Intern. Med.* **65:**1195.

Thenen, S. W. 1979. High ascorbic acid intake and vitamin B_{12} status in the rat. In *Vitamin B_{12}. Proceedings of the Third European Symposium on Vitamin B_{12} and Intrinsic Factor. March 5–8, 1979.* (B. Zagalak and W. Friedrich, eds.), pp. 1065–1068. Walter de Gruyter, Berlin.

Toohey, J. I. 1965. A vitamin B_{12} compound containing no cobalt. *Proc. Natl. Acad. Sci.* **54:**934.

Wagner, A. E., and K. Folkers. 1966. *Vitamins and Coenzymes.* Wiley-Interscience, New York.

Warren, L., J. G. Flaks, and J. M. Buchanan. 1959. Biosynthesis of purines. XX. Integration of enzymatic transformylation reactions. *J. Biol. Chem.* **229:**627.

Waters, A. H., and D. L. Mollin. 1961. The folic acid activity of serum in normal subjects and patients with megaloblastic anemia. *Proc. Congr. European Soc. Hematol.* **8:** No. 332.

Wasserman, R. H., and A. N. Taylor. 1966. Vitamin D_3-induced calcium-binding protein in chick intestinal mucosa. *Science* **152:**791.

Wasserman, R. H., and R. A. Corradino. 1971. Metabolic role of vitamins A and D. *Annu. Rev. Biochem.* **40:**501.

Wasserman, R. H., and A. N. Taylor. 1972. Metabolic roles of fat soluble vitamins D, E, and K. *Annu. Rev. Biochem.* **41:**179.

Weis, W. 1975. Ascorbic acid and electron transport. *Ann. N. Y. Acad. Sci.* **258:**190.

Wellner, V. P., J. I. Santos, and A. Meister. 1968. Carbamyl phosphate synthetase. A biotin enzyme. *Biochemistry* **7:**2848.

Wilkie, J. B., and S. B. Jones. 1954. Standardization of alumina adsorbents for vitamin A chromatography. *J. Ass. Offic. Agr. Chem.* **37:**880.

Wilson, R. P. 1973. Absence of ascorbic acid synthesis in channel catfish, *Ictalurus punctatus* and blue catfish, *Ictalurus frucatus. Comp. Biochem. Physiol.* **46B:**635.

Witting, L. A. 1972. Recommended dietary allowance for vitamin E. *Amer. J. Clin. Nutr.* **27:**960.

Woolley, D. W. 1940. Nature of anti-alopecia factor. *Science* **92:**384.

Woolley, D. W. 1944. The nutritional significance of inositol. *J. Nutr.* **28:**305.

Wright, L. D., and H. R. Skeggs, 1944. Determination of biotin with *Lactobacillus arabinosus. Proc. Soc. Exp. Biol. Med.* **56:**95.

Zacharia, M., K. Simpson, P. R. Brown, and A. Krstulovic. 1979. Use of reversed-phase high-performance liquid chromatographic analysis for the determination of provitamin A carotenes in tomatoes. *J. Chrom.* **176:**109.

Zechmeister, L. 1949. Stereoisomeric provitamins A. *Vitamins Hormones* **7:**57.

Zimmerli, A., C. H. Nield, and W. C. Russell. 1943. Antimony trichloride reagent for the determination of certain sterols and vitamins D_2 and D_3. *J. Biol. Chem.* **148:**245.

CHAPTER · 6

Carbohydrates: Chemistry, Occurrence, and Food Applications

INTRODUCTION

Carbohydrates represent a generic class of substances that display wide biochemical diversity and unparalleled abundance throughout the biosphere. Carbohydrates contribute to the fundamental structures of plants, animals, and microorganisms in addition to supplying a source of biochemical energy that supports life processes and reproduction. The human population relies on carbohydrates alone for more than 50% of its entire caloric requirements, mainly in the form of cereal grains. Complex polymeric carbohydrates such as cereal starches provide at least 75% of these calories, while the remaining 25% of energy requirements are obtained from sucrose and simple sugars such as glucose, fructose, and others.

The noncaloric properties of carbohydrates are no less important than their abilities to supply energy. Animals and humans both rely on nondigestible carbohydrates as a source of roughage. This ensures that digestible food nutrients consumed along with roughage will transit the digestive system in concert with the normal pattern of intestinal peristaltic activity. Furthermore, the role of nondigestible carbohydrates is critical in providing a physical growth matrix for digestive tract flora that augment the availability of many dietary nutrients that are present in foods.

The noncaloric chemical properties of carbohydrates are also closely knit within efforts of modern food science and technology to develop and improve existing food systems. Here, the properties of individual carbohydrates are exploited for their abilities to suspend, emulsify, or produce emollient effects in foods while at the same time improving the organoleptic and esthetic acceptance of foods through careful control of mouth feel, gelation, and clarity. The subjective evaluation of organoleptic and esthetic food properties can be assessed by controlled taste panels, and experience has shown that one or more physicochemical effects of food carbohydrates seem to be linked to the overall acceptance of many foods. These acceptability judgments stem largely from fundamental, yet enormously complex interactions of polar carbohydrate molecules with inorganic ions, lipids, proteins, other carbohydrates, and various decomposition products of organic substances present in foods. These interactions are further complicated by the characteristic osmotic, solubility, diffusion, hygroscopic, viscosity, and rheologic properties of food carbohydrates. The acceptability of the most delectable french pastries, fruit, jams, gravies, chocolate confections, honeys, and countless other foods rest solely on sensory evaluation of physicochemical properties of carbohydrates.

The term *carbohydrate* evolved during early studies suggesting that these compounds were hydrates of carbon [*hydrate de carbone* (French) or *Kohlenhydrate* (German)], which descriptively adhered to the empirical formula of $C_n(H_2O)_n$. Although many carbohydrates including lactose, sucrose, starch, and cellulose mirrored this formula, many noncarbohydrate substances such as formaldehyde (CH_2O), acetic acid ($C_2(H_2O)_2$), and lactic acid ($C_3(H_2O)_3$) also demonstrated the same empirical relationship. Conversely, there are some carbohydrates such as rhamnose ($C_6H_{12}O_5$) whose empirical formula defies the "hydrate of carbon" concept.

The inadequacy of empirical descriptions for carbohydrates led to more comprehensive definitions. One popular definition distinguishes carbohydrates from other organic substances because they contain carbon, hydrogen, and oxygen along with the *saccharose* group (Figure 6.1). This group, or its first reaction product, often occurs in carbohydrate molecules that contain a hydrogen to oxygen ratio found in water.

More contemporary perspectives of carbohydrates define them as *polyhydroxyaldehydes or ketones* and their derivatives (e.g., deoxy- and amino sugars). This classification encompasses a wide variety of substances such as D-glucose, D-fructose, sucrose, starch and many other substances that are further classified according to their molecular sizes.

Monosaccharides (e.g., D-glucose, D-fructose) display the fundamental physical and chemical properties of the carbohydrates, but unlike more complex structural forms of the carbohydrates, they cannot be hydrolyzed further into smaller carbohydrate units. Many common monosaccharides including D-glucose and D-fructose act as specific examples of simple sugars and occur freely in nature, where they are principally found in plant tissues, saps, fruit pericarps, and honeys.

Disaccharides represent the next organization level of molecular complexity beyond the monosaccharides. Sucrose is a classic example of a disaccharide since its hydrolysis products include the two monosaccharides, D-glucose and D-fructose. Carbohydrates such as raffinose, on the other hand, are classified as *trisaccharides* because they produce three monosaccharide units on hydrolysis, namely, D-glucose, D-fructose, and D-galactose. Trisaccharides and more complex carbohydrates that yield three to ten monosaccharide units are generally called *oligosaccharides*.

Polysaccharides offer the ultimate complexity in the molecular hierarchy of carbohydrate structure. These carbohydrates are complex polymers that may contain several thousand monosaccharidic monomers per molecule.

As a consequence of the long historical development of carbohydrate chemistry, many of the most common carbohydrates have assumed common or trivial names. The majority of these trivial names have a nominal prefix designating the source of the carbohydrate followed by the suffix *-ose*. For example, fructose, maltose, lactose, and xylose respectively denote *fruit sugar, malt sugar, milk sugar,* and *wood sugar*. These familiar names indicate little about the chemical

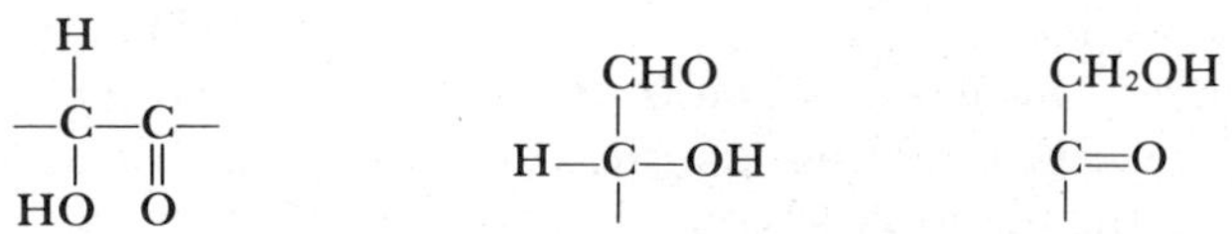

Saccharose group   **"Ald-"**   **"Ket-"**

Prefixes and structures used
for monosaccharide nomenclature

FIGURE 6.1. Some key structural features of monosaccharides.

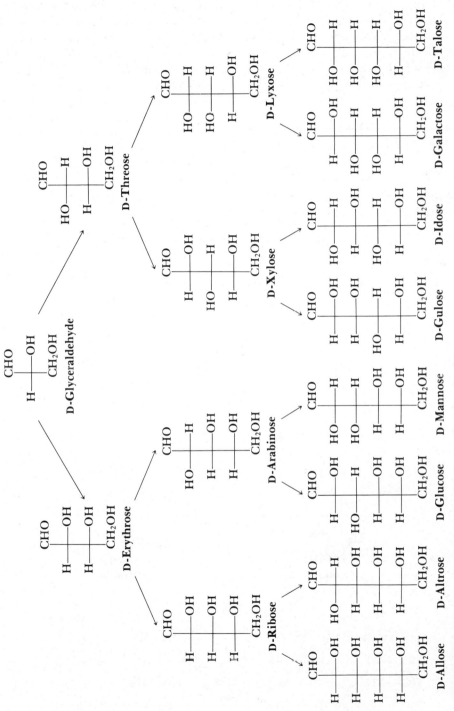

FIGURE 6.2. Fischer projection formulas for the acyclic series of D-aldoses.

structure of a certain carbohydrate, yet the names are so frequently used that no attempt will be made here to uniformly apply systematic organic nomenclature to carbohydrates. Detailed information concerning the systematic nomenclature can be found in the references cited at the end of the chapter.

MONOSACCHARIDES

Classification

Monosaccharides that are polyhydroxy aldehydes are referred to as *aldoses*. The prefix

ald- denotes the aldehyde carbonyl function and the suffix *-ose* signifies a carbohydrate. Monosaccharides that are polyhydroxy ketones are *ketoses*, the prefix *ket-* referring to the ketone carbonyl function (Figure 6.1). Each of these classes of compounds can be further divided according to the number of carbon atoms in the monosaccharide, namely: *tri-* (3), *tet-* (4), *pent-* (5), *hex-* (6), and *hept-* (7). When these terms are combined, a more descriptive definition of a carbohydrate results. For example, D-glucose is an *aldohexose* or a six-carbon aldose (Figure 6.2). Ketoses are given the suffix *-ulose*. Therefore, D-fructose is an example of a *hexulose* (Figure 6.3).

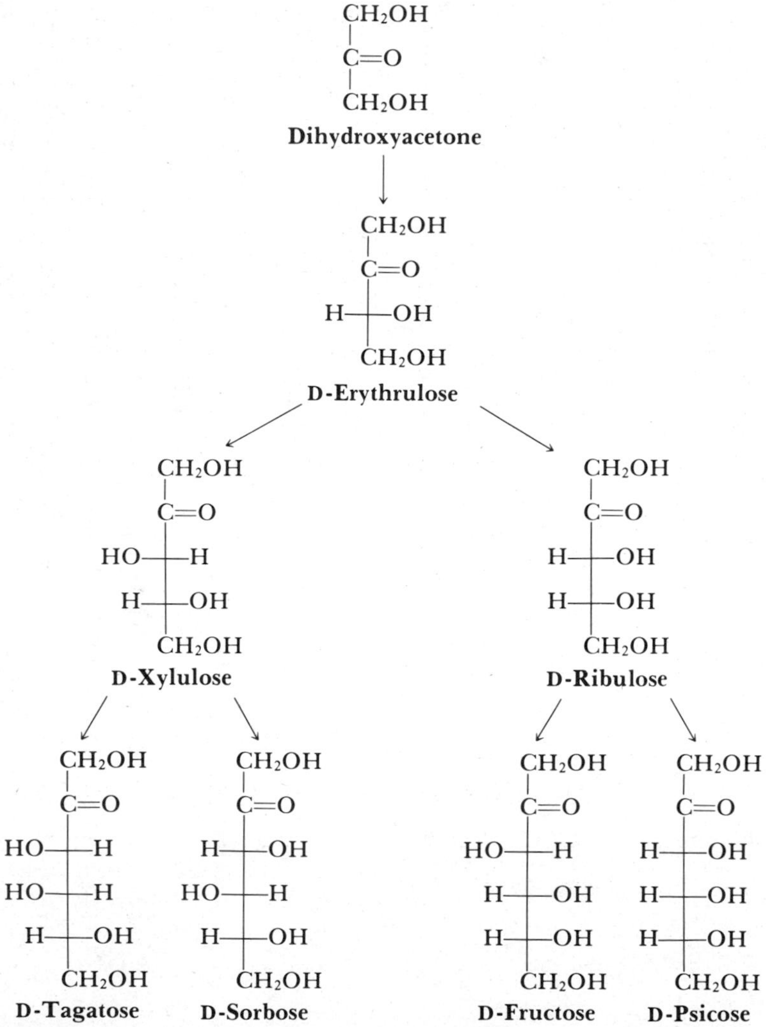

FIGURE 6.3. Fischer projection formulas for the acyclic series of D-ketoses.

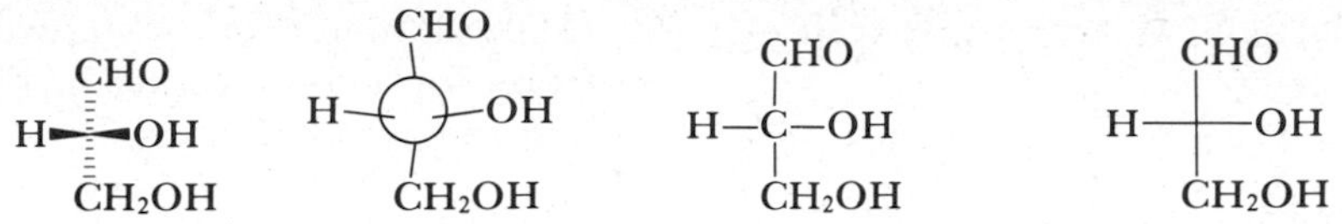

CHO

H——OH

CH$_2$OH

**Wedge-bond
formula**

CHO

H—◯—OH

CH$_2$OH

**Ball-and-stick
formula**

CHO

H—C—OH

CH$_2$OH

**Fischer projection
formula**

CHO

H——OH

CH$_2$OH

**Abbreviated Fischer
projection formula**

FIGURE 6.4. Various structural representations for (+)-glyceraldehyde.

Monosaccharides that have both an aldose and ketose carbonyl group are named as derivatives of aldoses. The final *e* of the name of the parent aldose is replaced by the suffix *-ulose*. D-Arabinosone is classified as a pentosulose. Monosaccharides having two ketone carbonyl groups are named by adding the suffix *-diulose* to the appropriate prefix such as *pento-* (5), *hexo-* (6), and so on.

Structure and Stereochemistry

Acyclic structures of monosaccharides—Fischer projection formulas: With the exception of dihydroxyacetone, all of the monosaccharides contain one or more chiral carbon atoms. Molecules that possess chiral carbons have isomeric forms that differ in the spatial arrangement of groups about the chiral centers; they are called *stereoisomers*. Stereoisomers that are nonsuperimposable mirror images of each other are called *enantiomers*. Glyceraldehyde exists as an enantiomeric pair, (+)-glyceraldehyde and (−)-glyceraldehyde. Each enantiomer can be represented by a wedge-bond formula or a ball-and-stick formula.

Consider (+)-glyceraldehyde. According to the wedge-bond formula, the wedges rep-

in the plane of the paper. The hydroxyl group and the hydrogen point toward the viewer and the formyl and hydroxymethyl groups point away from the viewer (Figure 6.4).

With carbohydrates, it is conventional to use Fischer projection formulas, which represent a projection of either the wedge-bond or ball-and-stick formulas onto the plane of the paper. By convention, the *carbonyl group* is placed at the *top* of the Fischer projection. The *horizontal bonds* are projected *out* of the paper toward the viewer and the *vertical bonds* are projected *behind* the paper away from the viewer. When using the abbreviated version of the Fischer projection, the chiral carbon atoms are outlined.

A Fischer projection formula cannot be lifted from the plane of the paper (projection plane), but it may be rotated 180° as illustrated for (+)-glyceraldehyde.

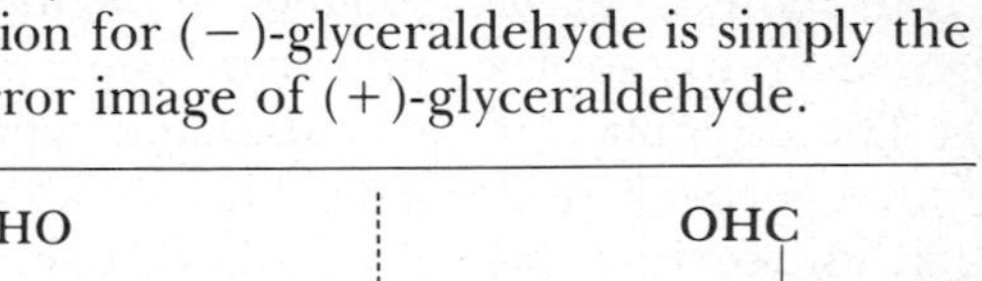

CHO

H——OH

CH$_2$OH

(+)-Glyceraldehyde

$\xrightarrow{180° \text{ rotation}}$

CH$_2$OH

HO——H

CHO

(+)-Glyceraldehyde

Since (+)-glyceraldehyde and (−)-glyceraldehyde are enantiomers, the Fischer projection for (−)-glyceraldehyde is simply the mirror image of (+)-glyceraldehyde.

CHO

H——OH

CH$_2$OH

(+)-Glyceraldehyde

**Mirror
plane**

OHC

HO——H

HOH$_2$C

(−)-Glyceraldehyde

resent bonds that protrude toward the viewer, while the dashes represent bonds that point away from the viewer. In the ball-and-stick formula, the circle represents a chiral carbon

Configuration: D and L convention. In carbohydrate chemistry there is a configurational system that is based on labeling (+)-glyceraldehyde as D-(+)-glyceraldehyde and

(−)-glyceraldehyde as L-(−)-glyceralde-
hyde.

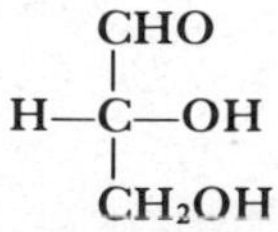

```
        CHO                    CHO
         |                      |
   H—C—OH              HO—C—H
         |                      |
       CH₂OH                 CH₂OH
  (+)-Glyceraldehyde    (−)-Glyceraldehyde
  D-(+)-Glyceraldehyde  L-(−)-Glyceraldehyde
```

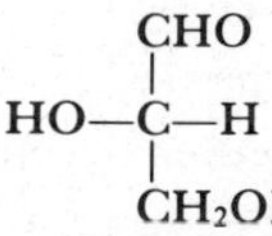

On the basis of this system, all monosaccharides that have the same stereochemistry
as D-(+)-glyceraldehyde at the highest numbered chiral carbon belong to the *D series*. All
those that have the same stereochemistry
as L-(−)-glyceraldehyde at the highest
numbered chiral carbon belong to the *L
series*. Examples of some D- and L-sugars are
given below. The highest numbered chiral
carbon is signified by an asterisk.

Establishment of *R,S* configurational
notation for a chiral center involves the following steps:

1. Assign priority to each of the substituents attached to the chiral center according to atomic number; the higher the atomic
number, the higher the priority.

2. Orient the molecule so that the group
of lowest priority points away from the viewer
(e.g., hydrogen in the following two diagrams).

3. Trace a path from the highest to the
lowest priority; if the path proceeds in
a clockwise rotation, the chiral carbon is
designated as *R*, while a corresponding
counterclockwise rotation indicates a chiral
carbon where the configuration is *S*.

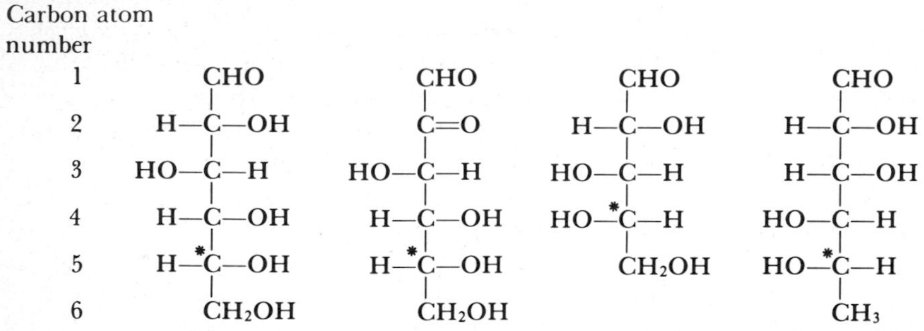

Carbon atom number				
1	CHO	CHO	CHO	CHO
2	H—C—OH	C=O	H—C—OH	H—C—OH
3	HO—C—H	HO—C—H	HO—C—H	H—C—OH
4	H—C—OH	H—C—OH	HO—C̲*—H	HO—C—H
5	H—C̲*—OH	H—C̲*—OH	CH₂OH	HO—C̲*—H
6	CH₂OH	CH₂OH		CH₃
	D-Glucose	**D-Fructose**	**L-Arabinose**	**L-Rhamnose**

Most of the sugars found in nature belong
to the D series; however, there are also some
important sugars in the L series, such as L-
arabinose and L-rhamnose. Both L-rhamnose and L-arabinose are monomers derived
from polysaccharides found in legumes,
cereals, nuts, apples, and sugar beets.

The D families of the aldoses and ketoses
are respectively presented in Figures 6.3 and
6.4. The corresponding L families are simply
constructed by recalling their enantiomeric
relationship with the D families.

Configuration: R,S convention. The configuration of any chiral center can be
determined according to the *R,S* system
of configurational notation introduced
by R. S. Cahn, C. Ingold, and V. Prelog.

As an example for the application of these
rules, consider D-(+)-glyceraldehyde

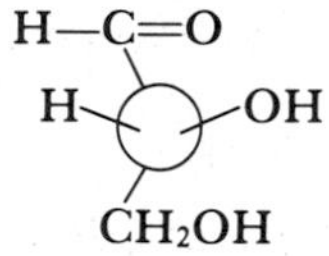

```
        H—C=O
       H——OH
        CH₂OH
```

The priority sequence from high to low is
OH > CHO > CH₂OH > H. Orienting the
molecule so that the group of lowest priority
is away from the viewer gives the following
structure:

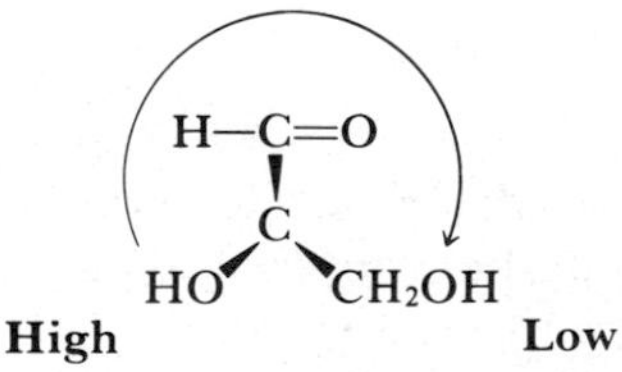

```
        H—C=O
           C
       HO     CH₂OH
   High            Low
```

Tracing a path from the highest to the lowest priority substituents results in a clockwise rotation and the D-(+)-glyceraldehyde is the *R* isomer or *R*-(+)-glyceraldehyde. The corresponding *S*-(+)-glyceraldehyde is the mirror image of *R*-(+)-glyceraldehyde.

Applying the same *R,S* configuration system to (+)-glucose permits each chiral carbon to be defined according to its individual *R,S* configuration.

Carbon atom number	Fischer projection	Configuration
1	H‑C=O	—
2	H——OH	R
3	HO——H	S
4	H——OH	R
5	H——OH	R
6	CH₂OH	—
	(+)-Glucose	

Optical activity. Plane polarized light is chiral in nature and therefore interacts with chiral molecules. A chiral molecule that rotates the plane of polarization in a clockwise manner is called *dextrorotatory* and it commonly given the prefix (+). A chiral molecule rotating the plane of polarization in a counterclockwise manner is called *levorotatory* and is given the prefix (−).

Optical rotations are measured with a *polarimeter.* The observed angle of rotation is proportional to the concentration of the optically active compound, the length of the sample cell where optical rotation is measured and the wavelength of light used for optical activity measurements. The rotation of sugars is commonly expressed as *specific rotation* $[\alpha]_D^t$:

$$[\alpha]_D^t = \frac{a}{1 \cdot c}$$

where

a = observed angular rotation (degrees)
1 = path length of the cell (dm)
c = concentration of solute (g/mL)
D = light of the sodium D line
t = temperature (°C)

Successful analytical estimates of solute concentrations using measures of optical rotation rely on precise rotational data. Only minor errors in determining rotational degrees can lead to major errors in calculated solute concentrations.

Cyclic structures for monosaccharides

Fischer formulas. As illustrated in previous sections, the simple sugars such as glucose, fructose, and others may be represented by acyclic structures. However, the chemistry of sugars requires that the acyclic form must exist in equilibrium with the cyclic form of the same sugar.

Aldehydes and ketones react with alcohols to form *hemiacetals* and *hemiketals* as indicated in the following reaction:

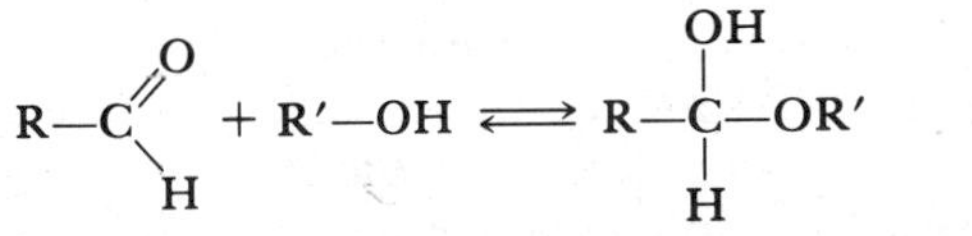

$$R-\underset{H}{\overset{O}{C}} + R'-OH \rightleftharpoons R-\underset{H}{\overset{OH}{\underset{|}{C}}}-OR'$$

Hemiacetal structure

D-(+)-Glucose, which possesses both a carbonyl group and a hydroxyl group, is capable of reacting in an intramolecular manner to yield a cyclic hemiacetal. First consider the reaction between carbon atom C-1 and C-5 of D-(+)-glucose. Intramolecular ring closure would produce a six-membered cyclic hemiacetal. Close examination reveals that a new chiral carbon is produced at C-1; thus two cyclic forms are possible for D-(+)-glucose. In one case, the hydroxyl group is represented to the right on the Fischer projection, and in the other structure, it is written to the left. This clearly results in the formation of a diastereomer pair. When dealing with carbohydrates, these types of diastereomers are referred to as *anomers*

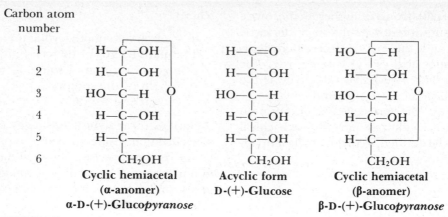

FIGURE 6.5. Hemiacetal formation within D-(+)-glucose to form a six-membered ring called a pyranose.

and C-1 is the anomeric carbon. The two anomeric forms of the sugar are designated as α or β.

For the D series of monosaccharides, writing the hydroxyl group at C-1 to the right in a Fischer projection indicates the α-anomer and writing the hydroxyl group to the left produces the β-anomer.

The formation of a hemiacetal within a monosaccharide to produce a six-membered ring is called a *pyranose.* The open chain form can be distinguished from its cyclic counterpart by including the ring designation in the name. Accordingly, the cyclic forms of (+)-glucose are α-D-(+)-glucopyranose and α-D-(+)-glucopyranose (Figure 6.5).

D-(+)-Glucose can also produce a five-membered cyclic hemiacetal by the reaction of the hydroxyl group at C-4 and the car-

bonyl group. A five-membered ring is called a *furanose;* therefore, D-(+)-glucose may display an equilibrium involving α-D-(+)-glucofuranose and β-D-(+)-glucofuranose (Figure 6.6). Despite the potential for forming these five-membered rings, thermodynamic and stability factors typical of the glucose structure usually favor the six-membered pyranose structures.

Haworth perspective formulas. The Fischer projection adequately illustrates the configuration at the chiral centers of a sugar, but it fails to provide a good picture of either the pyranose or furanose forms of a sugar.

Haworth developed perspective formulas that offered a better representation of the cyclic molecular structures for sugars. In the case of a pyranose ring, Haworth depicted

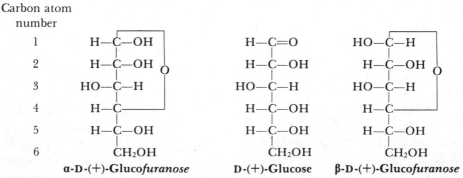

FIGURE 6.6. Potential hemiacetal formation within the D-(+)-glucose molecule to form a furanose (although furanose rings are possible for glucose molecules, stability factors usually favor the pyranose form).

a hexagonal ring perpendicular to the plane of the paper, whereas the furanose ring was indicated as a pentagon. The ring numbering proceeds from the anomeric carbon, noted below by an asterisk, in a clockwise fashion. The correct placement of the individual functional groups on the pyranose or furanose rings can be determined from the corresponding Fischer formula.

Pyranose ring   **Furanose ring**

As an example consider the conversion of α-D-(+)-glucopyranose. In the Fischer projection, all the ring atoms are not in the same plane and, therefore, the Fischer projection must be modified. Placement of the ring atoms into the same plane can be accomplished by rotating the whole structure about C-5 until the oxygen and the ring carbon atoms are in the same plane.

Conformational pyranose rings. Although the Haworth perspective formulas are far more useful for conceptual purposes than the corresponding Fischer projections, the Haworth formulas do not accurately depict the geometry of the pyranoid ring. The pyranose ring is structurally reminiscent of the cyclic structure of cyclohexane, and it follows that the pyranose ring may adopt a *chair, half-boat, boat,* and *twist* conformations. Like cyclohexane, the chair conformation is favored for the pyranoid ring.

Two chair conformations are possible, including the $^4C_1$ and the $^1C_4$ conformations. This symbolism refers to the atoms in the conformation that lie above and below a reference plane. The reference plane is defined by carbon atoms 2, 3, and 5 and the oxygen atom. By convention, the ring atom that lies above the reference plane is written as a superscript and precedes the letter (C), while the ring atom that lies below the plane is written as a subscript (Figure 6.8). The $^4C_1$ conformer is the same as the C1 designation according to the conformational

Carbon atom number				
1	H—C—OH		H—C—OH	CH₂OH
2	H—C—OH		H—C—OH	
3	HO—C—H O	rotate about C-5	HO—C—H O ≡	
4	H—C—OH		H—C—OH	
5	H—C		HOH₂C—C	
6	CH₂OH		H	

"Conventional" Fischer projection   **"Modified" Fischer projection**   **Equivalent Haworth formula**

Since there are six carbons in the ring, a pyranose ring is drawn. All the groups that are to the left of the modified Fischer projection are placed above the plane of the ring, and those to the right are placed below the plane. Haworth formulas for the pyranose and furanose forms of selected α-D-ketose and α-D-aldoses are detailed in Figure 6.7.

nomenclature of Reeves, and the $^1C_4$ is equivalent to the 1C conformer. Both of these systems are often used.

In the case of α-D-glucopyranose, there are two possible chair conformations. In one of the conformations, the hydroxyls at C-2 to C-4 and the hydroxymethyl group at C-5 are in the equatorial position while the hydroxyl at C-1 is in the axial position. This

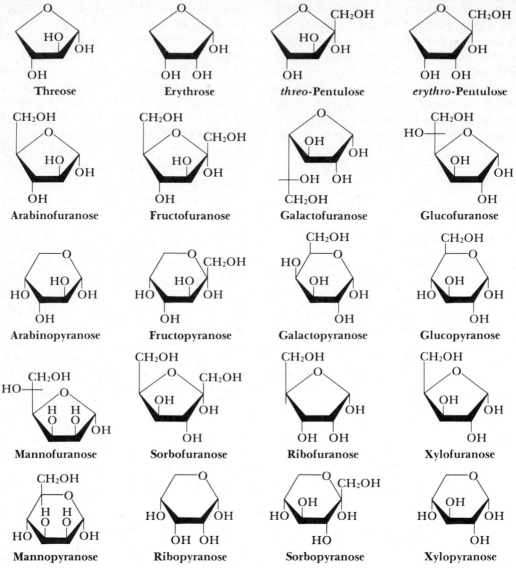

FIGURE 6.7. Haworth formulas for a selected group of furanose and pyranose rings.

conformation is often referred to as the $^4C_1$ or C1 conformer. The other chair conformation is designated as the $^1C_4$ or 1C conformer. It has the hydroxyls at C-2 to C-4, the hydroxymethyl group at C-5 in the axial position, and the C-1 hydroxyl group in the equatorial position (Figure 6.9).

In both the α- and β-D-glucopyranose structures, the $^1C_4$ conformation is not favored. The presence of the hydroxyl groups and the hydroxymethyl group in the axial positions creates unfavorable nonbonded 1,3-diaxial interactions. These interactions are relieved in the $^4C_1$ conformer when the same groups occupy the equatorial positions.

The relationship between D- and L-pyr-

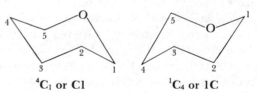

$^4C_1$ or C1   **$^1C_4$ or 1C**

FIGURE 6.8. Conformational designations for the pyranose ring.

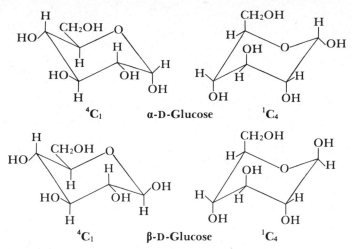

FIGURE 6.9. $^4C_1$ and $^1C_4$ conformations for α- and β-D-glucopyranose.

anoid sugars is illustrated in Figure 6.10. It should be noted that the anomeric designation for the hydroxyl group in the L series is opposite to that of the D series.

Haworth formulas can be readily converted to conformational formulas. For example, consider the conversion of α-D-(+)-glucopyranose. The so-called α-configuration requires that the hydroxyl at C-1 be placed in the axial position (Figure 6.11).

The stereochemistry of the hydroxyl groups at C-1 and C-2 is *cis*. Since the hydroxyl group at C-1 is axial, the hydroxyl group at C-2 becomes equatorial (Figure 6.12).

The relationship of the hydroxyl groups at C-2 to C-3, C-3 to C-4, and C-4 hydroxyl to the C-5 hydroxymethyl group is *trans* and, therefore, equatorial to equatorial (Figure 6.13).

Anomeric effect. The anomeric effect arises from the interaction of an electronegative group at the anomeric carbon with that of the ring oxygen atom. It is described in terms of an electrostatic repulsion between the nonbonded electrons of the ring oxygen and that of the group at the anomeric carbon atom. The angle between the dipole moments is small when the group is placed at the equatorial position and the repulsion is high. When the group is in the axial position, the electrostatic repulsion is less (Figure 6.14).

The anomeric effect and nonbonded 1,3-diaxial interactions play an important role in determining the stability of the conformers. For example, α-D-(+)-glucopyranose would be expected to be a much less favorable structure than β-D-(+)-glucopyranose as a result of the unfavorable nonbonded

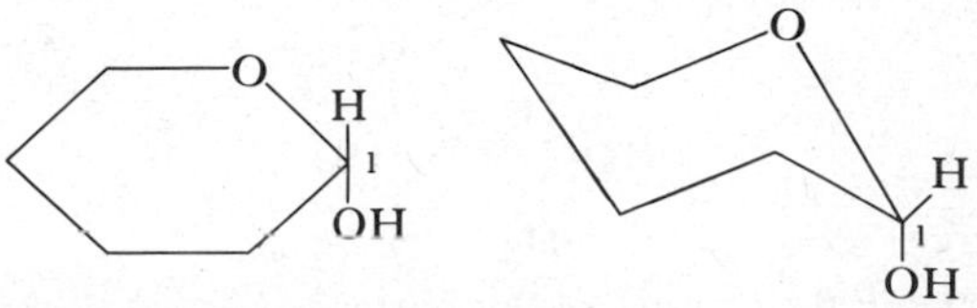

D-Family

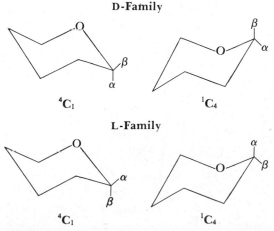

L-Family

FIGURE 6.10. Conformation of the pyranoid ring in the D- and L-sugars.

FIGURE 6.11. α relationship at the anomeric carbon.

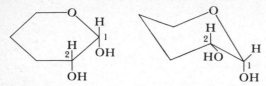

FIGURE 6.12. *cis* relationship of C-1 and C-2 hydroxyl groups.

1,3-diaxial interactions between the C-1 hydroxyl and the axial protons at C-3 and C-5. The fact that the α-anomeric form represents 36% of the pyranose under equilibrium conditions serves as positive evidence for the anomeric effect that encourages the presence of α-D-(+)-glucopyranose.

The anomeric effect becomes even more pronounced as the electronegativity of the C-1 group is increased, and the observed phenomenon is also affected by changes in solvent polarity.

Although the anomeric effect accounts for the notable occurrence of α-D-(+)-glucopyranose (~36%), the 1,3-diaxial interactions with the C-1 hydroxyl group of the α-conformer exceed the forces of electrostatic repulsion between the C-1 hydroxyl group and the oxygen in the ring. Therefore, the net effect of these combined intramolecular interactions leads to the predominant presence of β-D-(+)-glucopyranose (~64%) under equilibrium conditions.

Furanose ring conformations. Conformational studies on the structure of cyclopentane,

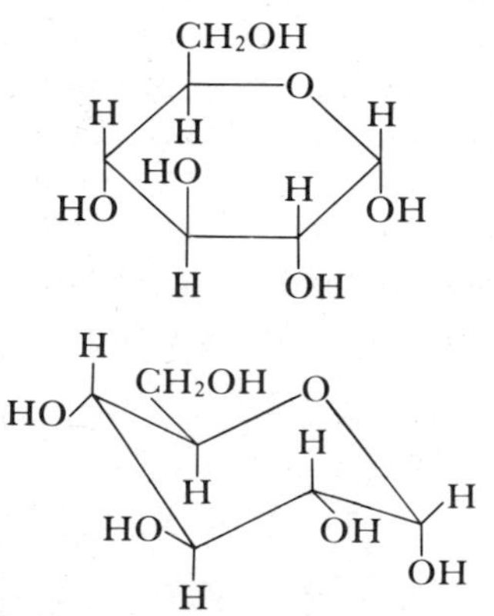

High electrostatic repulsion

Low electrostatic repulsion

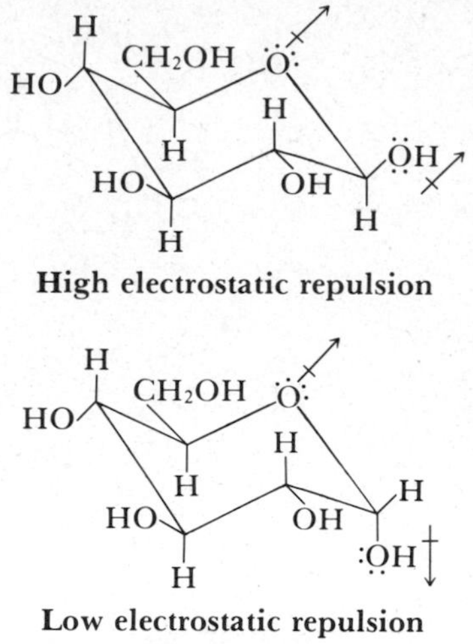

FIGURE 6.14. Anomeric effect illustrated for D-(+)-glucopyranose.

which serves as a model of furanose ring structures for monosaccharides, have shown that the molecule is not planar. It can exist either in the form of *envelope* or *twist conformations*, which are both energetically similar. Since the comparable energy differences between these conformational forms is small, both conformers are probably present in solution.

Aside from the recognition of these two fundamental conformers, detailed investigation of the furanose ring will reveal that there are actually ten envelope and ten twist

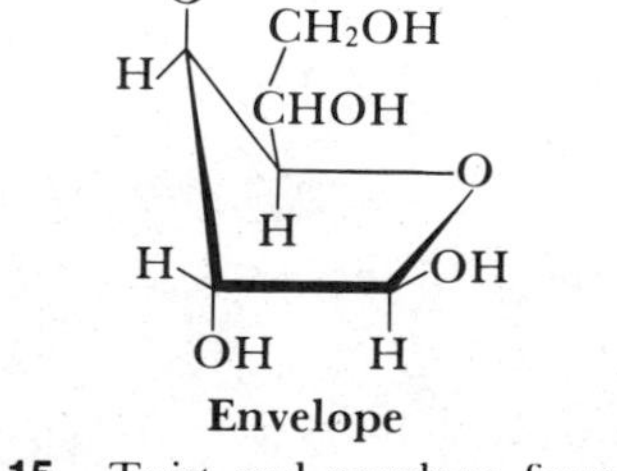

FIGURE 6.13. α-D-(+)-Glucopyranose.

Twist

Envelope

FIGURE 6.15. Twist and envelope forms of D-(+)-glucofuranose.

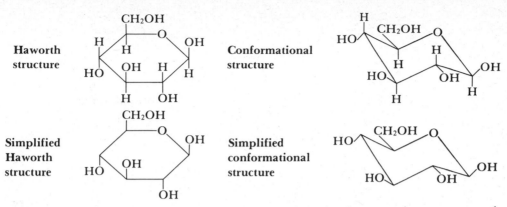

FIGURE 6.16. Corresponding simplified Haworth and conformational pyranose ring structures used for indicating α-D-(+)-glucopyranose as well as other types of monosaccharides in the text.

forms that are interconvertible by means of pseudorotation (Stoddart, 1971).

Examples of a twist and envelope conformation for D-(+)-glucofuranose are given in Figure 6.15.

A Note on Carbohydrate Structures for This Text

In an effort to simplify structural expressions of monosaccharides, oligosaccharides, and polysaccharides, the Haworth and conformational structures that follow in the text will omit the presence of bonded hydrogen atoms. These hydrogen atoms will be part of the molecular structure only in those instances where they are critical for the understanding of important reactions. Figure 6.16 includes simplified Haworth and conformational structures.

Mutarotation Involves Interconversion of α- and β-Anomeric Forms

An aqueous solution of α-D-glucose has an initial specific rotation of +112°. The specific rotation of the solution decreases slowly until it reaches an equilibrium value of +52.7°. An aqueous solution of β-D-glucose has an initial specific rotation of +19°, which increases slowly until it reaches an equilibrium value of +52.7°. These individual behaviors for sugars reflect the phenomenon known as *mutarotation*, which occurs in all solutions of reducing sugars.

Mutarotation is caused by a slow interconversion of α and β anomeric forms for pyranose and furanose sugars, through the open chain form of the sugar, until a thermodynamic equilibrium among the forms is established (Figure 6.17). For glucose, a typ-

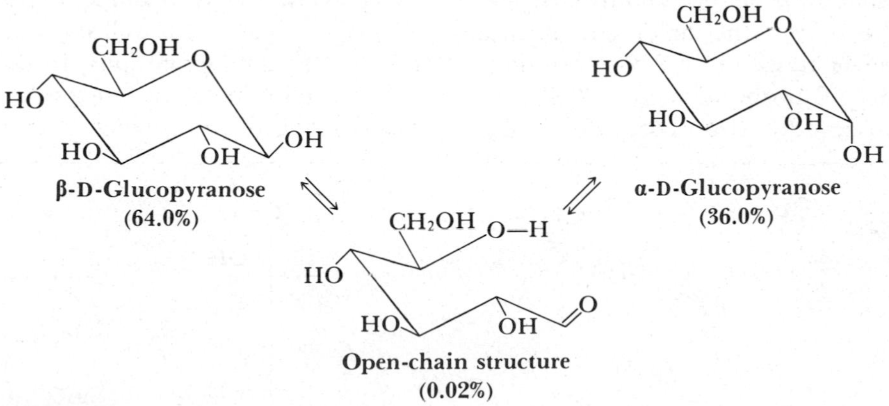

β-D-Glucopyranose
(64.0%)

α-D-Glucopyranose
(36.0%)

Open-chain structure
(0.02%)

FIGURE 6.17. Mutarotation.

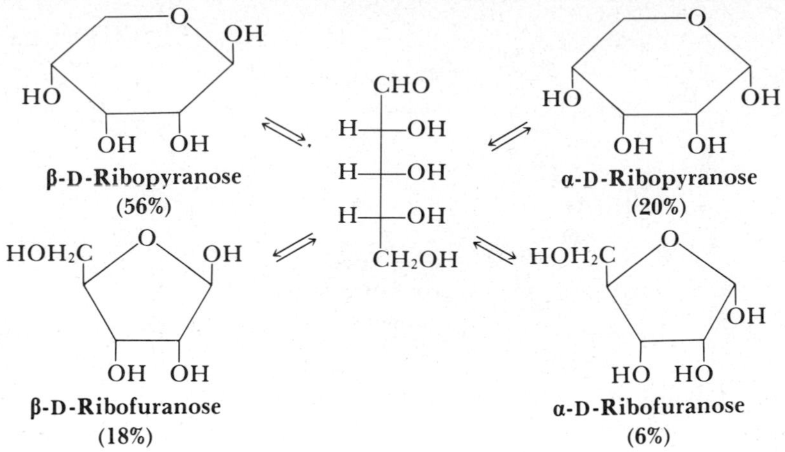

FIGURE 6.18. Possible structures and distribution of ribopyranoses and ribo-furanoses produced during mutarotation.

ical equilibrium distribution corresponds to about 64% of β-D-glucopyranose, 36% of α-D-glucopyranose and a trace amount (~0.02%) of the open-chain form of glucose.

If mutarotation begins with the α-gluco-pyranose form, a hydrogen atom is eliminated from the anomeric carbon (C-1) and a proton other than that eliminated from the C-1 position is translocated to the oxygen in the pyranose ring. This activity results in the cleavage of the C—O bond and the formation of a semicyclic free aldehyde form of the hexose. An analogous but reverse process is required for the conversion of β-D-glucopyranose to the aldehyde.

The predominant form of glucose in aqueous solutions is the pyranose form; however, there are a number of monosaccharides that have large amounts of the furanose form present under equilibrium conditions. As an example, ribose solutions display 56% β-D-ribopyranose, 20% α-D-ribopyranose, 18% β-D-ribofuranose, 6%

α-D-ribofuranose, and a trace amount of the open-chain form (Figure 6.18).

Mutarotation of sugars is catalyzed by the presence of protons or hydroxyl ions (i.e., acids or bases) and certain enzymes generically categorized as *mutarotases* that occur in some molds, kidney, and liver tissues.

Some Important Reactions and Derivatives of Monosaccharides

Oxidation: The carbonyl group of most sugars can be readily oxidized, whereas the oxidation of primary hydroxyl groups (—CH_2OH) and secondary hydroxyl groups require more rigorous conditions.

Sugars that have an aldehyde group are easily oxidized by alkaline cupric citrate (Benedict's reagent) as well as alkaline cupric tartrate (Fehling's solution). In the process Cu^{2+} is reduced to Cu^+ in the form of cuprous oxide (Cu_2O), which is evident as a

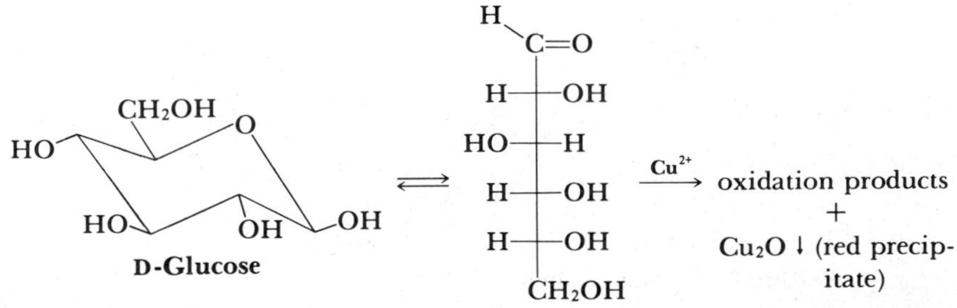

Table 6.1. Structures for Some Reducing and
Nonreducing Sugars

Sugar	Classification
β-D-Glucose	Reducing sugar
β-D-Fructose	Reducing sugar
Maltose	Reducing sugar
Vanillin-β-D-glucoside	Nonreducing sugar
Sucrose	Nonreducing sugar

NOTE: Note that the hemiacetal or hemiketal function is indicated within the circled area of structures.

distinct red precipitate. Those sugars that react with Fehling's or Benedict's reagents are classified as reducing sugars. Therefore, sugars can be crudely classified as reducing and nonreducing sugars. Reducing sugars are generally recognized by the presence of a hemiacetal group. The structures of some reducing and nonreducing sugars are illustrated in Table 6.1. Glucose is readily recognizable as a reducing sugar.

Fructose isomerizes in alkaline solution to a mixture of glucose and mannose, which also react with alkaline copper (Cu^{2+}) to produce Cu_2O, typical of reducing sugars (Figure 6.19). Glycosidic bonds involving the C-1 hydroxyl group do not display the hemiacetal function and cannot display reducing sugar activity.

Aldoses readily react with bromine in a buffered solution at pH 5 to 6 to yield aldonic acids.* In acidic solution, aldonic acids

*Aldonic acids are named by substituting -*onic* acid as a suffix in place of -*ose* of the parent sugar.

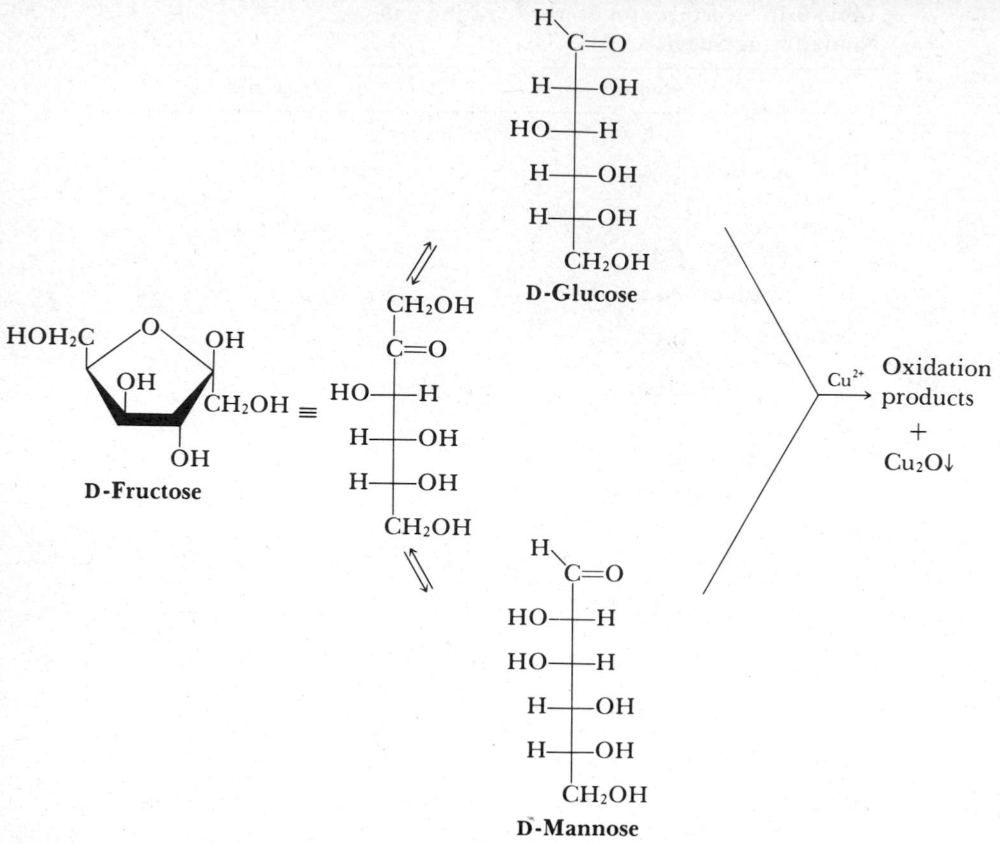

FIGURE 6.19. Fructose isomerizes in alkaline solution resulting in the reduction of available Cu^{2+} to produce Cu^{1+} in the form of a cuprous oxide precipitate.

cyclize to form lactones. The reaction of D-glucose with bromine water yields an equilibrium mixture of D-glucuronic acid and D-gluconic acid-γ-lactone.

Strong oxidizing agents will attack both the carbonyl group and the primary hydroxyl group of a sugar to yield polyhydroxy dicarboxylic acids called *aldaric acids*.* For

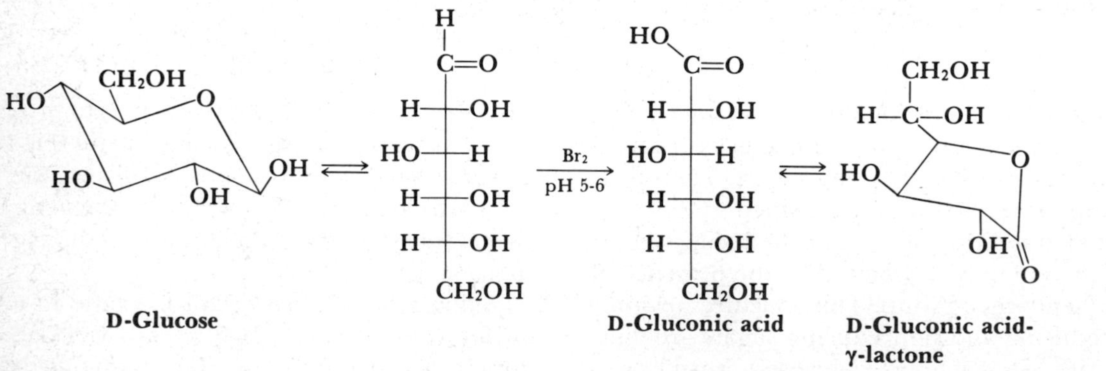

*Aldaric acids are named by substituting -*aric* acid as a suffix in place of -*ose* of the parent sugar.

example, nitric acid reacts with D-glucose to give glucaric acid and with D-erythrose to yield *meso*-tartaric acid

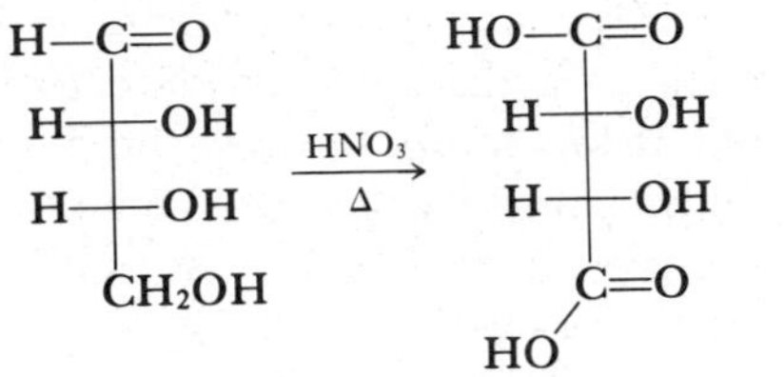

D-Erythrose   **meso-Tartaric acid**

When the hemiacetal function is protected, as in the cases where glycoside formation has occurred, the primary alcohol group can be oxidized to produce *uronic acids*.* Methyl-α-D-glucopyranoside is readily converted to methyl-α-D-glucopyranosiduronic acid below to illustrate uronic acid formation.

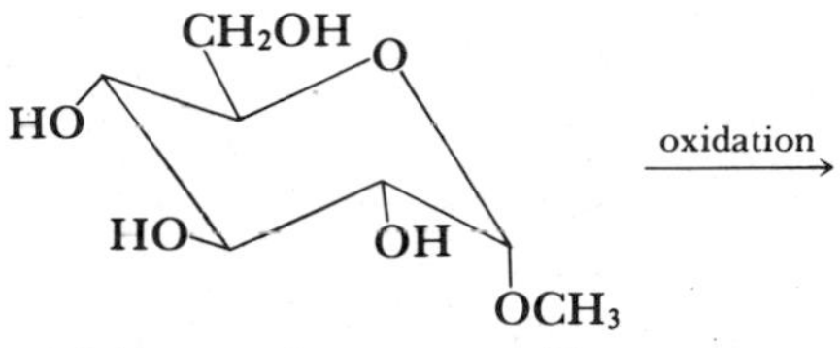

Methyl-α-D-glucopyranoside

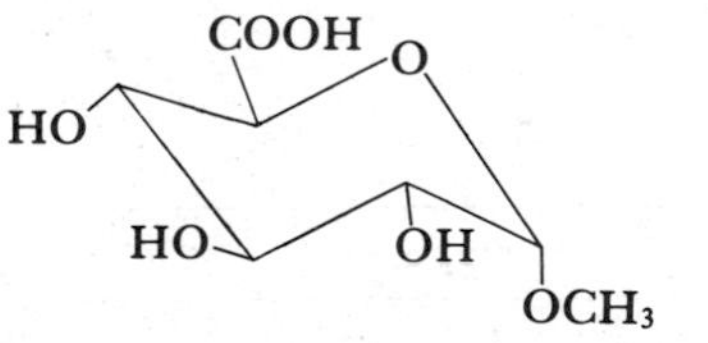

Methyl-α-D-glucopyranosiduronic acid

Protection of the amino group present in amino sugars allows for the oxidation of amino sugars into their uronic acid derivatives. Hexuronic acid units are found in

polysaccharides such as hemicelluloses, pectins, gums, and mucopolysaccharides.

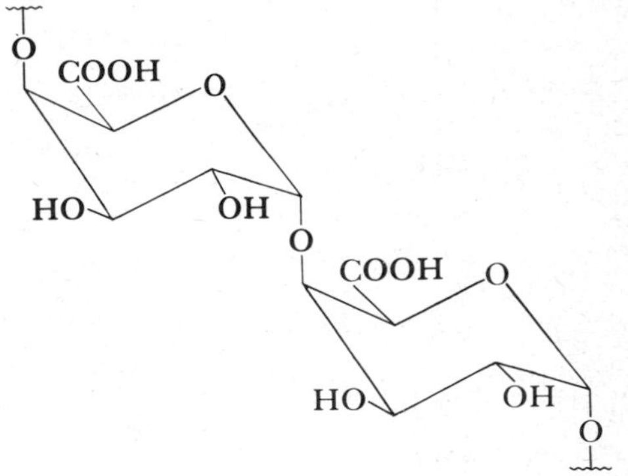

Galacturonic acid units contribute to the structure of pectin

Reduction: The reduction of monosaccharides by catalytic hydrogenation, LiAlH₄ or NaBH₄, yields *polyhydric alditols*.** Reduction of D-glucose yields D-glucitol (sorbitol).

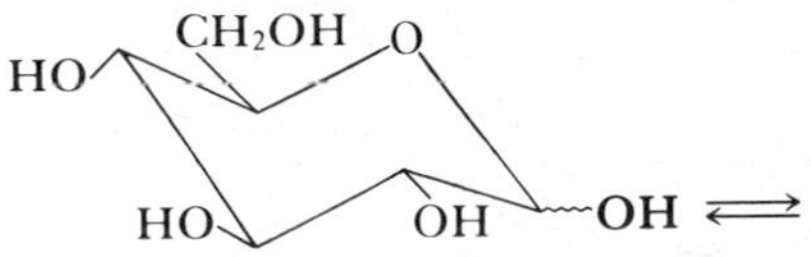

D-Glucose

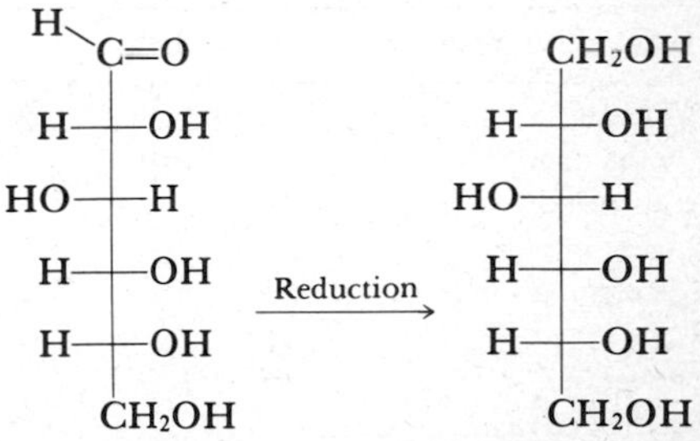

D-Glucitol (sorbitol)

---

*Uronic acids are named by substituting -*uronic* acid as a suffix in place of -*ose* of the parent sugar.

**Alditols are named by substituting -*itol* as a suffix in place of -*ose* of the parent sugar.

Reduction of fructose generates a new chiral carbon center and therefore produces two alditols, namely, D-glucitol and D-mannitol.

The alditols are found in many fruits and vegetables (Table 6.2) as naturally occurring compounds, and they readily participate in hydrogen bonding interactions with water (owing to the presence of numerous hydroxyl groups). For this reason, alditols such as sorbitol may display very useful properties as humectants in foods. Many alditols also produce a sweet taste sensation. Sorbitol

and mannitol are added to many sugarless chewing gums and dietetic foods in which sucrose or glucose may be undesirable.

Esters: Hydroxyl groups of sugars react with acetic anhydride in pyridine to form acetates. At low temperatures, the configuration at the anomeric carbon is maintained. For example, the reaction of β-D-glucopyranose at temperatures below 0.0°C gives β-D-glucopyranose pentaacetate. At elevated temperatures, the selectivity is lost and a mixture of α- and β-acetates form.

Table 6.2. Occurrence of Alditols in Some Foods

Food substances[a]	Arabitol	Xylitol	Mannitol	Sorbitol	Galactitol
Bananas	0.020				
Cherry (red)	0.040			1.125	
Chestnuts (edible)		0.015	0.020	0.010	
Egg plant		0.180	0.270		
Peaches				0.950	
Pears			4.600		
Pumpkin		0.097	0.200		
Strawberries		0.360			
White mushroom	0.350	0.130	0.475		0.050
Wine (apple)		0.120		0.225	
Yogurt					0.900

[a]Expressed as milligrams per 100 grams of dry matter or, in the case of liquids, milligrams per 100 grams of liquid.

These acetate derivatives usually form crystalline derivatives that can be easily isolated from a reaction mixture. This reaction is also important because the acetates may be converted back to the parent form of the sugar by treatment with catalytic amounts of sodium methoxide in methanol. Therefore, this route offers a convenient method for derivatizing, isolating, and purifying most carbohydrates.

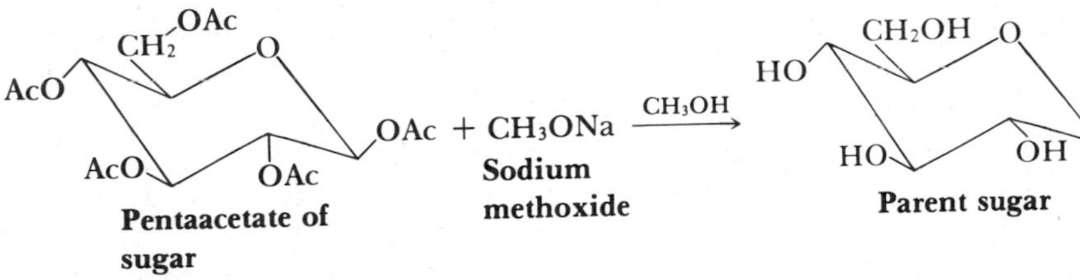

mediates in plant, animal, and microbial metabolism.

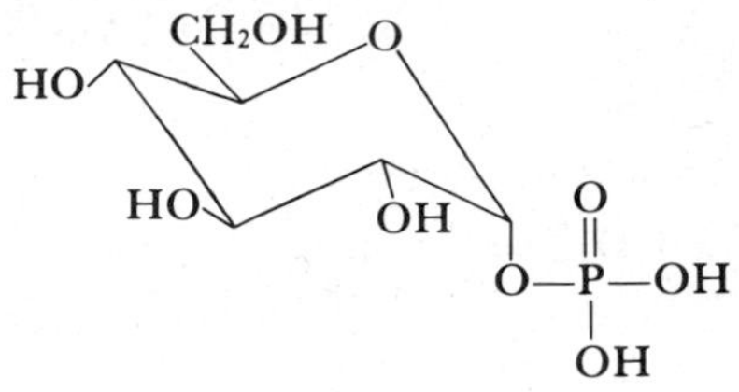

α-D-Glucopyranosyl phosphate

Pentaacetate of sugar + CH$_3$ONa **Sodium methoxide** $\xrightarrow{\text{CH}_3\text{OH}}$ **Parent sugar**

Other esters such as benzoate and *p*-nitrobenzoate esters are respectively formed by treating sugars with benzoyl chloride or *p*-nitrobenzoyl chloride. Sulfonate esters of sugars can also be prepared by reacting a sugar with sulfonyl chlorides of *p*-toluenesulfonate (*tosylate*), methanesulfonate, or trifluoromethanesulfonate in the presence of pyridine.

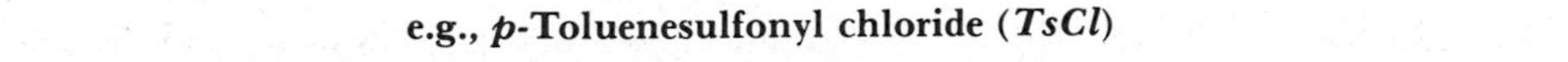

Ethers: Hydroxyl groups can be readily converted to methyl ethers by reaction with aqueous sodium hydroxide and dimethylsulfate or with silver oxide and methyliodide. With sugars, the anomeric hydroxyl group must be protected during etherification since it is sensitive to base. Protection is accomplished by conversion of the sugar to a methyl glycoside, which can be

Tosylated sugar

e.g., **p-Toluenesulfonyl chloride (*TsCl*)**

Sugar phosphates may also be formed from monosaccharides, and many of these phosphorylated compounds serve as key inter-

removed at a later time under mild acidic conditions to give the etherified parent sugar.

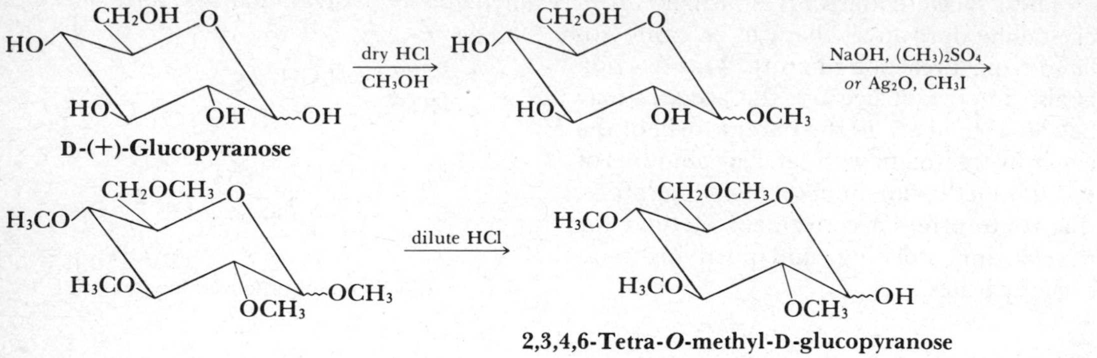

D-(+)-Glucopyranose

2,3,4,6-Tetra-O-methyl-D-glucopyranose

Reactions with base: The rearrangements that occur when monosaccharides are treated with dilute aqueous base were described by Lobry de Bruyn and Alberda van Ekenstein. Based on theories offered by these researchers, aldoses and ketoses can undergo transformations including isomerization and epimerization. For example, treating D-(+)-glucose with dilute base yields an equilibrium mixture of D-glucose, D-mannose, and D-fructose. Under these conditions, the base abstracts the acidic α-hydrogen to produce an enolate that gives rise to products. In aqueous base, yields of D-fructose are ap-

proximately 30%. The yield can be increased to approximately 45% by using glucose isomerizate.

scribed by Lobry de Bruyn and Alberda van Ekenstein. Based on theories offered by these researchers, aldoses and ketoses can undergo transformations including isomerization and epimerization. For example, treating D-(+)-glucose with dilute base yields an equilibrium mixture of D-glucose, D-mannose, and D-fructose. Under these conditions, the base abstracts the acidic α-hydrogen to produce an enolate that gives rise to products. In aqueous base, yields of D-fructose are approximately 30%. The yield can be increased to approximately 45% by using glucose isomerizate.

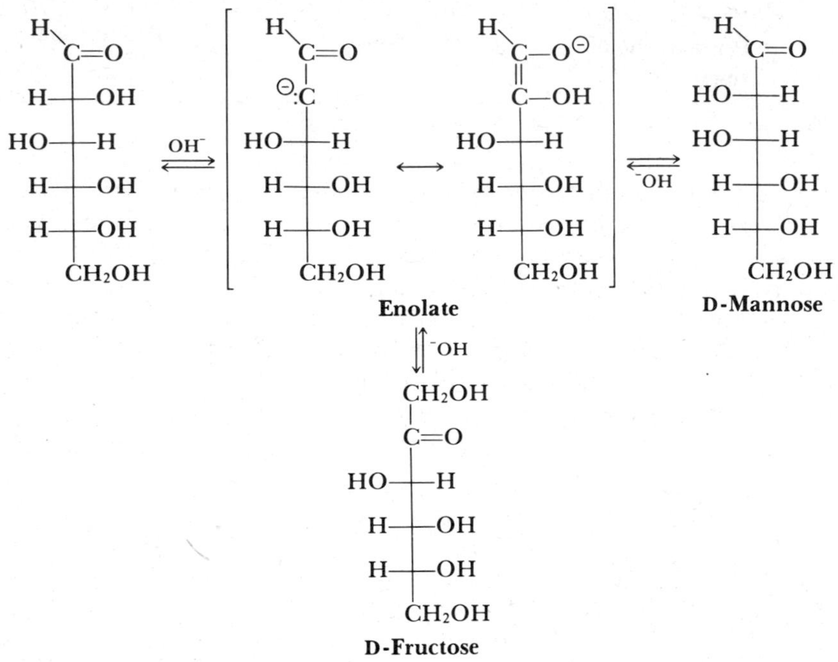

Under the influence of increasing temperature and stronger alkaline solutions, the sugars can undergo further reactions ultimately leading to saccharinic acids, fragmentation products such as lactic acid, formic acid, glycolic acid, acetoin, and various furan derivatives including furfuryl alcohol (Figure 6.20).

Reactions with acids: Under the influence of acid and heat, hexoses are degraded

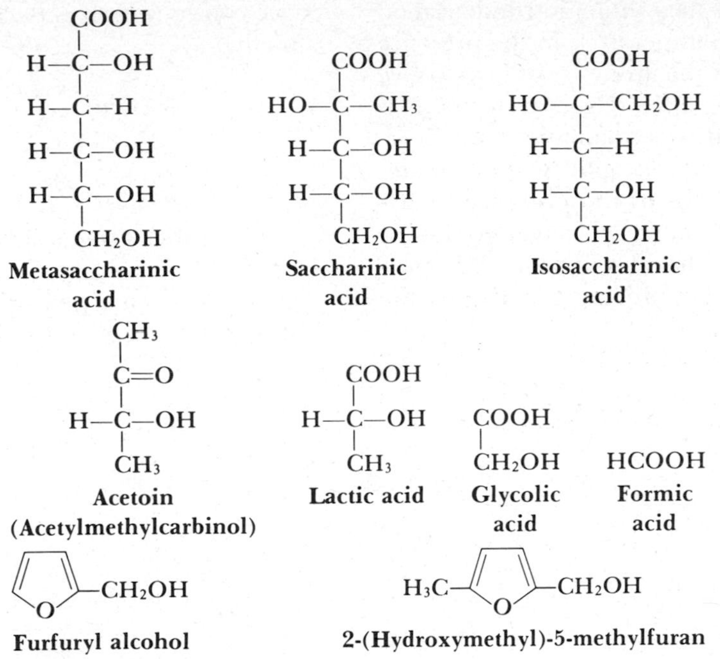

COOH
H—C—OH
H—C—H
H—C—OH
H—C—OH
CH₂OH
**Metasaccharinic
acid**

COOH
HO—C—CH₃
H—C—OH
H—C—OH
CH₂OH
**Saccharinic
acid**

COOH
HO—C—CH₂OH
H—C—H
H—C—OH
CH₂OH
**Isosaccharinic
acid**

CH₃
C=O
H—C—OH
CH₃
**Acetoin
(Acetylmethylcarbinol)**

COOH
H—C—OH
CH₃
Lactic acid

COOH
CH₂OH
**Glycolic
acid**

HCOOH
**Formic
acid**

—CH₂OH
Furfuryl alcohol

H₃C— —CH₂OH
2-(Hydroxymethyl)-5-methylfuran

FIGURE 6.20. Some typical products formed during the treatment of sugars with strong basic conditions.

to 2-furaldehydes. For example, in the presence of acid, glucose and fructose are converted to a 1,2-enediol, which then undergoes dehydration and cyclization to form 5-(hydroxymethyl)-2-furaldehyde. Upon further reaction with acid, 5-(hydroxymethyl)-2-furaldehyde can be converted to levulinic and formic acids.

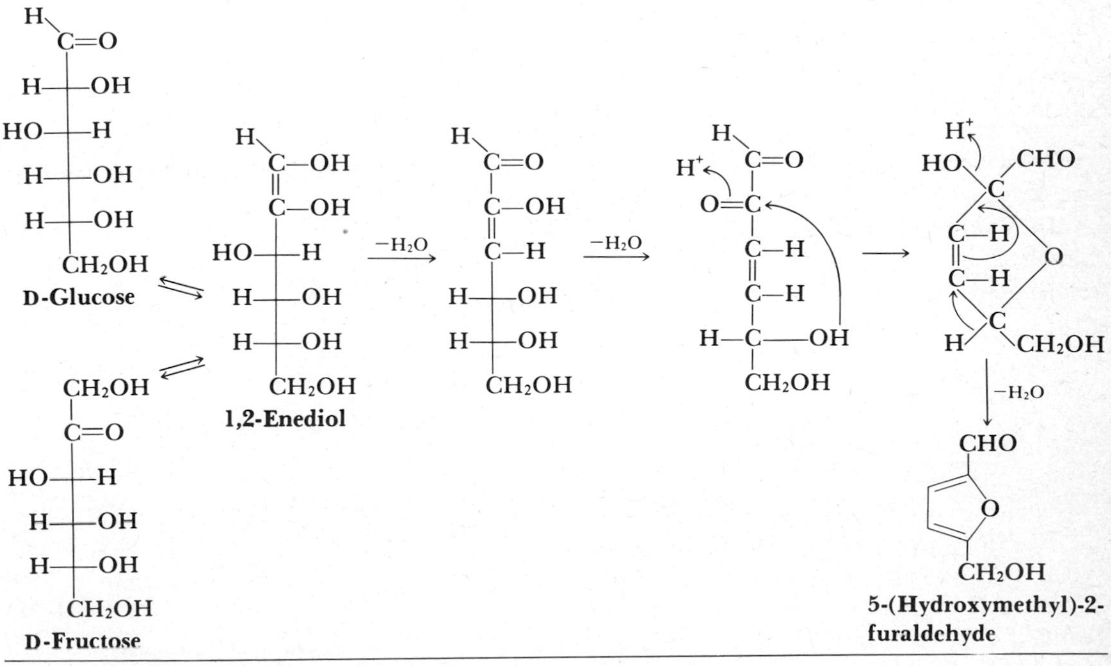

H
C=O
H——OH
HO——H
H——OH
H——OH
CH₂OH
D-Glucose

CH₂OH
C=O
HO——H
H——OH
H——OH
CH₂OH
D-Fructose

H
C—OH
C—OH
HO——H
H——OH
H——OH
CH₂OH
1,2-Enediol

−H₂O→

H
C=O
C—OH
C—H
H——OH
H——OH
CH₂OH

−H₂O→

H
C=O
O=C
C—H
C—H
H—C—OH
CH₂OH

H⁺

→

H⁺
HO   CHO
C
C—H
O
C—H
C
H   CH₂OH

−H₂O

CHO
O
CH₂OH
**5-(Hydroxymethyl)-2-
furaldehyde**

Sugar syrups may undergo condensation either at room temperature in the presence of acid or under the influence of heat to give disaccharides and other oligosaccharides. This type of reaction is commonly known as *reversion*. As an example, glucose, held at ambient temperatures in the presence of hydrochloric acid, may be converted into a mixture of disaccharides of which β-isomaltose (4.2%) and β-gentiobiose (3.4%) are the principal products.

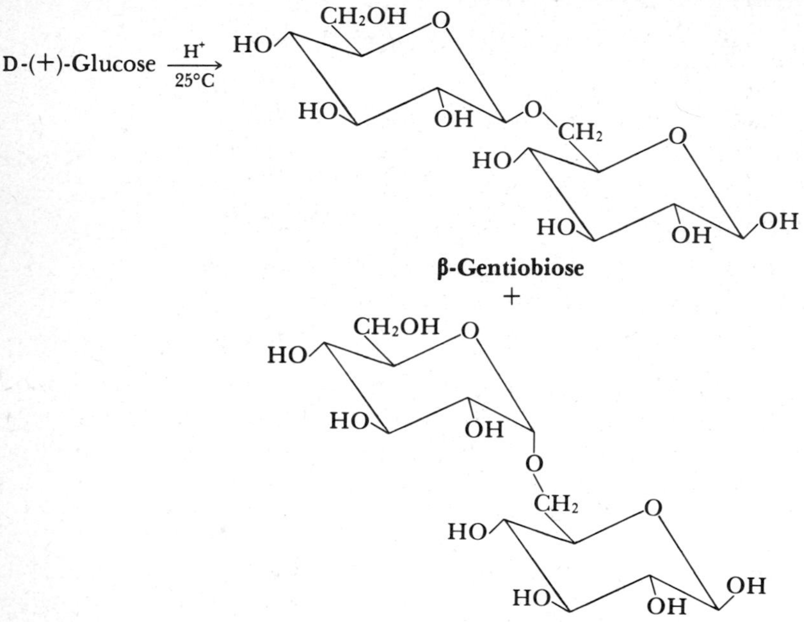

meric configuration is β-, and the R group is methyl-.

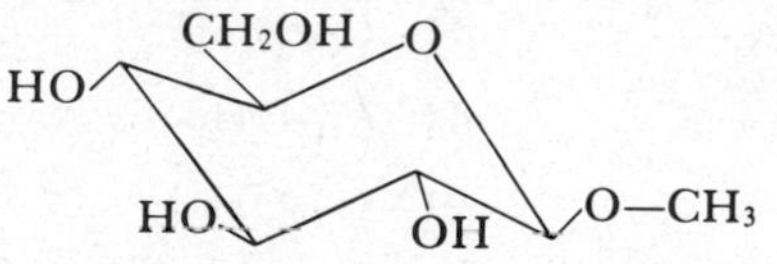

Methyl-β-D-glucopyranoside

O-Glycosides are generally stable in basic solutions, but in aqueous acid, glycosides hy-

β-Gentiobiose

+

β-Isomaltose

Glycosides: Sugar acetals formed at the anomeric carbon are called *glycosides*. The R group is referred to as the *aglycone* and the carbohydrate moiety is denoted as a *glycosyl group*.

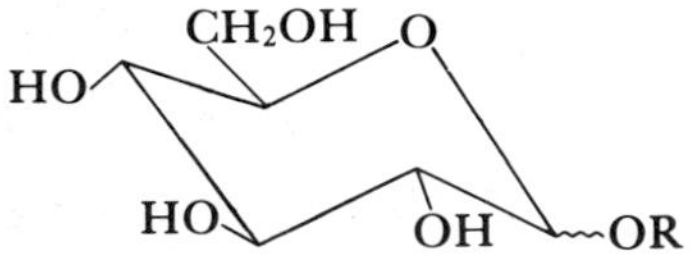

Glycosides are named by replacing the final *-e* of the parent sugar with *-ide* preceded by the name of the R group. For example, in methyl-β-D-glucopyranoside, the glycosyl group is that of D-glucopyranose, the ano-

drolyze readily to produce the parent sugar and an aglycone.

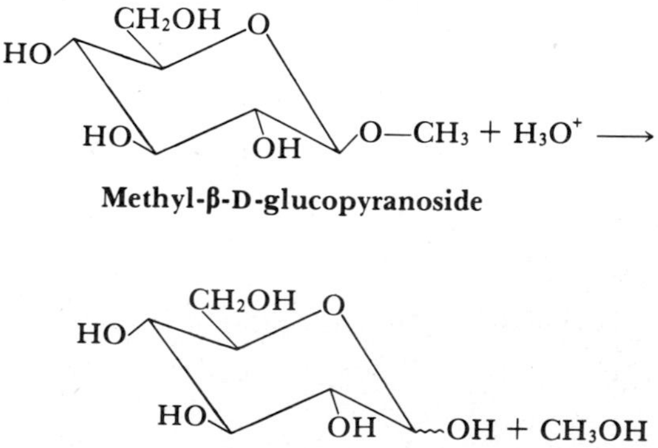

Methyl-β-D-glucopyranoside

α- and β-D-Glucose

The hydrolysis of glycosides can be accomplished by the use of enzymes. The advantage of enzyme hydrolysis lies in the high degree of hydrolytic specificity that various enzymes display; α-glucosidases catalyze only the hydrolysis of α-glucosidic linkages, and β-glucosidases hydrolyze only β-glucosidic linkages.

Glycosides are widespread in nature. Anthocyanidin glycosides contribute to the coloration of many fruits and flowers. Cyanogenic glycosides such as amygdalin are found in seed pits and other vegetative plant structures. These substances have assumed their name on the basis that hydrolysis of the glycoside liberates hydrocyanic acid.

The saponins represent another important group of steroid glycosides that are very widespread among higher plants. The steroid aglycone moiety contributes to the notable detergent activities of these compounds in aqueous solutions. Another group of interesting glycosides, classified as *thioglycosides*, occur in mustard seeds, horseradish roots, and many other members of these species. Hydrolysis of thioglycosides releases *isothiocyanates*, which are volatile and generally pungent and produce lachrymatous sensations. The structures for some glycosides are illustrated in Figure 6.21.

Sugar acetals can also react intramolecularly to produce a glycosidic form recog-

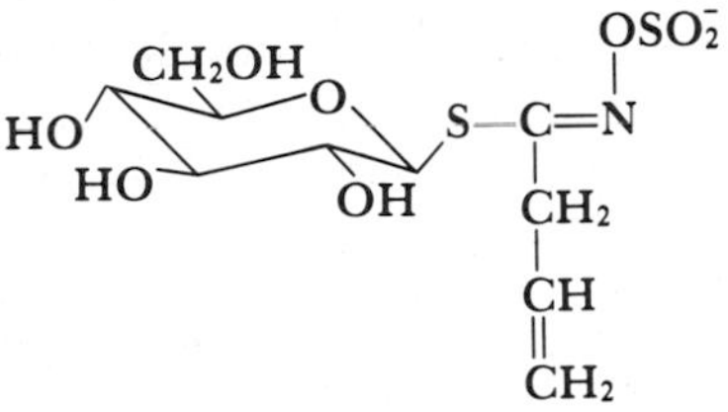

Sinigrin naturally occurs in mustard seed

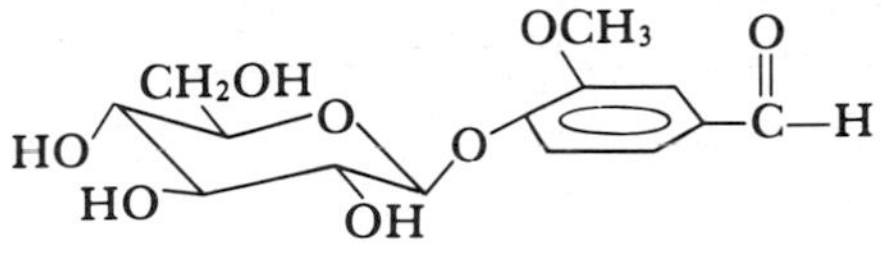

Vanillin-β-D-glucopyranoside occurs naturally in vanilla beans

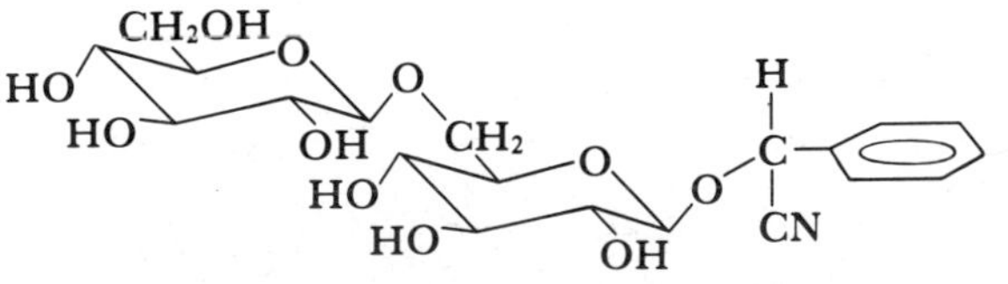

Amygdalin—present in apricot, peach, and plum pits as well as apple seeds

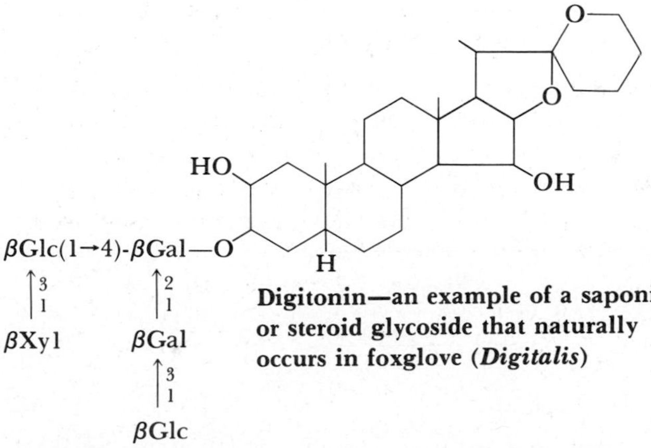

βGlc(1→4)-βGal—O

Digitonin—an example of a saponin or steroid glycoside that naturally occurs in foxglove (*Digitalis*)

FIGURE 6.21. Examples of some common glycosides.

nized as a *glycosan*. For example, the elimination of a water molecule from C-6 of the glucose and the anomeric carbon produces a 1,6-anhydro-D-glucopyranose, commonly called *levoglucosan*.

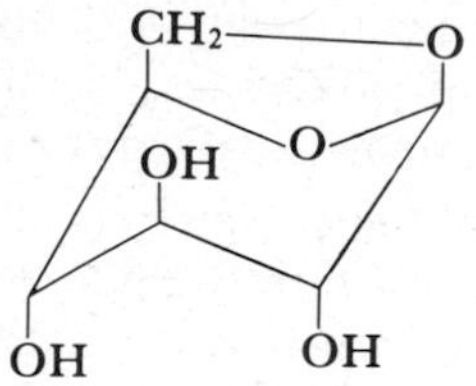

Levoglucosan

Acetals and ketals: The hydroxyl groups of sugars react with aldehydes and ketones to respectively form acetals and ketals.

Acetone preferentially reacts with vicinal hydroxyls having a *cis* relationship to form isopropylidine ketals. With α-D-galactopyranose, the C-1 and C-2 hydroxyl groups along with those residing at C-3 and C-4 have *cis* geometries, and they react with acetone to form 1,2:3,4-di-*O*-isopropylidine-α-D-galactopyranose.

The stability of sugar acetals and ketals under neutral and basic conditions allows for the chemical transformation of the remaining hydroxyl groups. After structural modification involving other hydroxyl groups has been completed, treatment with acid removes the acetal or ketal protecting group. The preparation of ascorbic acid (vitamin C) from α-L-sorbofuranose using the principle of ketal protecting groups has been illustrated in Figure 6.22.

OLIGOSACCHARIDES

Sucrose, which is obtained from sugar beets, sugar cane, and other plant sources, as well as *lactose,* which occurs in the lacteal secretions of mammals, represent the most common disaccharides found in nature. *Maltose,* on the other hand, is usually found as a component of syrups and hydrolyzed starch polysaccharides. The trisaccharide *raffinose* and

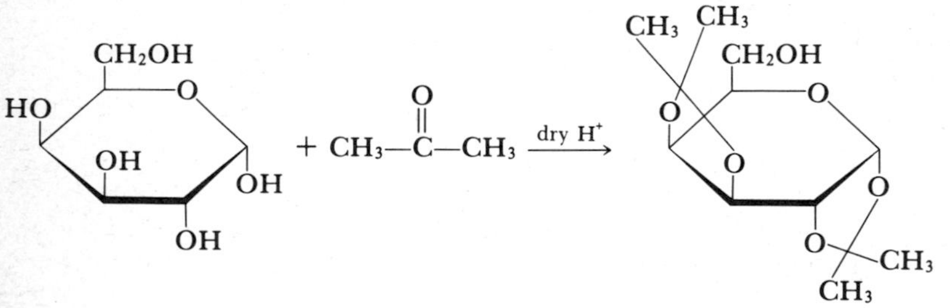

1,2:3,4-D-*O*-isopropylidine-α-D-galactopyranose

α-D-Glucopyranose reacts with benzaldehyde to give 4,6-*O*-benzylidine-α-D-glucopyranose. The hydroxyl groups at C-4 and C-6 can adopt a geometry that is favorable for the formation of a six-membered cyclic acetal.

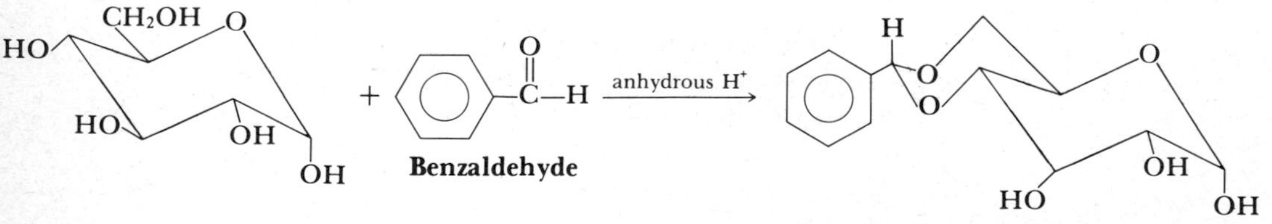

the tetrasaccharide *stachyose* are principally found in leguminous plants. Maltotriose is another derivative of starch hydrolysis, which may commonly occur as a result of enzymatic starch hydrolysis. A variety of other oligosaccharides may be produced when glucose

Benzaldehyde

4,6-*O*-Benzylidine-α-D-glucopyranose

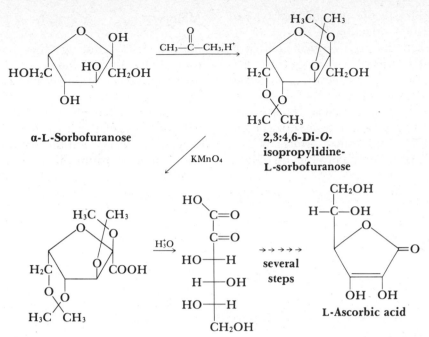

α-L-Sorbofuranose

CH₃—C—CH₃,H⁺

2,3:4,6-Di-*O*-isopropylidine-L-sorbofuranose

KMnO₄

H₂O

several steps

L-Ascorbic acid

FIGURE 6.22. Use of protecting groups in the chemical synthesis of ascorbic acid (vitamin C).

and sucrose syrups are heated in the presence of acid.

Lactose

Lactose is a disaccharide composed of galactose and glucose in which the C-4 hydroxyl group of glucose is joined to the anomeric carbon of galactose by a β-glycosidic bond. This type of glycosidic linkage is also conventionally indicated by the notation β(1→4). Two distinct forms of the lactose are possible, namely, α- and β-lactose.

Milk at body temperature contains an equilibrium mixture composed of approximately 40% α-lactose and 60% β-lactose. Lactose is commercially produced from whey concentrates, and the α-anomer may be selectively crystallized from the whey concentrate at temperatures below 95°C.

Since the glucopyranose component of the disaccharide may undergo mutarotation, the lactose disaccharide is considered to be a reducing disaccharide.

According to the IUPAC system of nomenclature, a reducing disaccharide such as lactose is named as an *O*-glycosyl glucose; thus α-lactose is named as 4-*O*-β-D-galactopyranosyl-α-D-glucopyranose. The prefixes

4-*O*-β-D-Galactopyranosyl-α-D-glucopyranose
α-Lactose

4-*O*-β-D-Galactopyranosyl-β-D-glucopyranose
β-Lactose

4- and β- refer to the position and type of linkage at C-4 of glucose.

Lactose can undergo controlled hydrolysis to form syrups of varying glucose, galactose, and lactose concentration. In recent years, it has been proposed that these lactose syrups may serve as a suitable replacement for sucrose in ice cream, as a replacement for invert sugar in canned fruits, or as a fermentable syrup in the production of beer and wine.

Lactose can also be reduced to *lactitol* according to the following reaction:

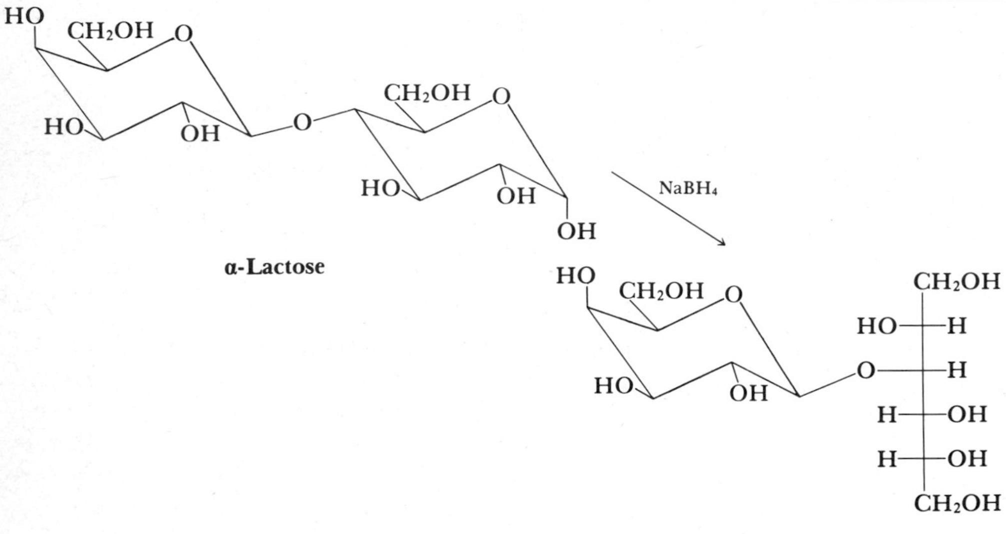

Maltose

Maltose is a reducing disaccharide composed of two glucose molecules. The C-4 hydroxyl group of one glucose molecule is joined to the anomeric carbon of the other glucose molecule by an α glycosidic bond. The β-maltose form of the disaccharide is readily available, and in solution it mutarotates to form an equilibrium mixture of α- and β-maltose.

Maltose is not present in appreciable quantities in most native biochemical sys-

α-Lactose

Lactitol

Several potential applications for lactitol in foods have been proposed based on its ability to retard crystallization of key food components, decrease water activity while improving water retention in foods, and demonstrate excellent temperature stability.

tems, although it is found in small amounts in grains such as barley and fruits including grapes, peaches, and plums. Maltose arises in these sources largely from the action of some amylases on starches.

Maltose is also produced as a by-product

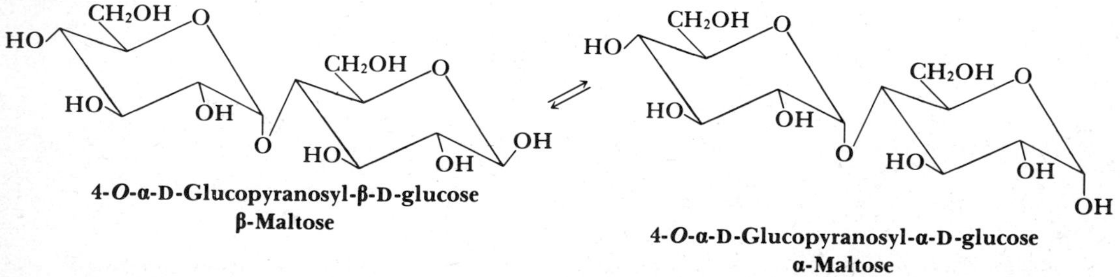

4-*O*-α-**D**-Glucopyranosyl-β-**D**-glucose
β-Maltose

4-*O*-α-**D**-Glucopyranosyl-α-**D**-glucose
α-Maltose

of baked doughs and breads. In fresh un-sweetened breads, for example, maltose may be present in varying concentrations from 1.7 to 4.3%.

Maltose is commercially prepared by the hydrolysis of starches to produce syrups containing varying percentages of maltose, maltotriose, and glucose. Maltose may be obtained in a crystalline form from these syrups. The high-maltose syrups are also used to maintain the proper moisture content of candy. Maltose syrup exerts this effect by preventing excessive hardening due to the loss of water and also prevents excessive softness as a result of the absorption of large amounts of water.

Many of the specially prepared maltose syrups offer functional properties suited to the formulation of stable icings and desert preparations.

Sucrose

Unlike maltose, sucrose is very widespread in nature and represents the major disaccharide (or oligosaccharide) component of many fruits and vegetables.

Since mutarotation is impossible, owing to the glycosidic linkage of the monosaccharide components at anomeric carbons, sucrose is classified as a *nonreducing disaccharide*. The glycosidic bond at glucopyranose is α and that of the fructofuranose is β. Nonreducing disaccharides are named as *glycosyl glycosides*. Using this system, there are two acceptable names for sucrose including β-D-fructofuranosyl-α-D-glucopyranoside or α-D-glucopyranosyl-β-D-fructofuranoside.

Sucrose is commercially produced from sugar beets or sugar cane according to steps outlined in following sections. This disaccharide has major importance in the food industry since it is used in large quantities for baking, soft drinks, and confectionery products.

The objective for adding sucrose to foods depends on the nature of the food product. It may be used as a sweetener and filler in the preparation of chocolates, a texture modifier in cakes, a preservative agent in jams and canned fruits, or a gelling agent in conjunction with acid for pectin gels; it may also be introduced into microbial fermentations for the production of ethyl alcohol or selected dextrans.

The hydrolysis of sucrose by acid or *invertase*, which is an enzyme specific for hydrolyzing the β-fructofuranoside bond, yields equimolar mixtures of glucose and fructose. The hydrolyzed sucrose product is commonly called invert sugar and is the major recognized sugar component of honeys.

Sucrose $\xrightarrow{\text{acid or invertase}}$
$[\alpha]_D^{20} = 66.5°$

$$\underbrace{\begin{array}{cc} \text{D-glucose} & + \quad \text{D-fructose} \\ [\alpha]_D^{20} = +52.7° & [\alpha]_D^{20} = -92.4° \end{array}}_{\text{Invert sugar}}$$

Net specific rotation following inversion $= [\alpha]_D^{20} = -19.9°$

The term *invert sugar* arises from the fact that an equimolar mixture of D-glucose and D-fructose has a specific rotation of $-19.9°$ as opposed to the parent sucrose structure, which displays a specific rotation of $+66.5°$.

α-D-Glucopyranosyl-β-D-fructofuranoside or
β-D-Fructofuranosyl-α-D-glucopyranoside
Sucrose

Thus, an inversion in the sign of specific rotation occurs when sucrose is hydrolyzed.

The presence of fructose in invert sugars accounts for a notable increase in the sweetness of the inverted mixture over that sweetness displayed by the parent sucrose before its inversion. Fructose also increases the water retention of the mixture and minimizes the potential for sugar crystallization. These properties alone have encouraged the wide use of invert sugar in foods such as candies, jams, jellies, and bakery goods.

The fundamental sucrose disaccharide also cosidically linked D-galactopyranose, D-glucopyranose, and D-fructofuranose.

The tri- and higher oligosaccharides are named by beginning with the first nonreducing component. The configuration of the glycosyl group is noted, and this is followed by two numbers that indicate the position of the glycosidic bond. The numbers (in parentheses) are separated by an arrow pointing *from* the glycosyl carbon atom *to* the number of the hydroxy group on the next ring to which it is bonded. Accordingly, raffinose is named as O-α-D-galactopyranosyl-(1→6)-

O-α-D-Galactopyranosyl-(1→6)-O-α-D-glucopyranosyl-(1→2)-β-D-fructofuranoside
Raffinose

occurs in trisaccharides such as raffinose. This nonreducing trisaccharide consists of gly-

O-α-D-glucopyranosyl-(1 → 2)-O-β-D-fructofuranoside. Maltotriose, which is a reducing

O-α-D-Glucopyranosyl-(1→6)-O-α-D-glucopyranosyl-(1→4)-α-D-glucopyranose
α-Maltotriose

oligosaccharide, is named in a similar manner, and the last portion of the name identifies the reducing unit.

POLYSACCHARIDES

Plant and animal cells contain high-molecular-weight carbohydrates that are generally called polysaccharides. Distinct differences exist among individual polysaccharides that include

1. The *distribution and types of monosaccharides* present in the structure.

2. The *site*(s) *of glycosidic linkage* between monosaccharide constituents.

3. The *residue sequence of monosaccharides* in polysaccharide structures.

Those polysaccharides containing identical saccharidic monomers are called *homopolysaccharides* (or *homoglycans*), whereas *heteropolysaccharides* (or *heteroglycans*) have more than one type of monomer. These monomers may be common simple sugars such as hexoses, discussed in earlier sections, or possibly *amino-, N-acetyl-, carboxyl-, sulfated, methyl esterified*, and/or *peptidyl derivatives* of these simple sugars and their corresponding acids.

In spite of the variety of saccharidic monomers, the majority of polysaccharides are composed of only one to three monomer types in a predominantly repetitive sequence. This is quite unlike polynucleotides or polypeptides where a fundamental alphabet of monomers is used to dictate specific genetic programs or protein structures. Polysaccharides occur widely throughout plant and animal tissues as pure carbohydrate structures or covalently bonded to lipids (*glycolipids*), peptides (*glycopeptides*), or protein (*glycoprotein*) moieties.

Regardless of the saccharidic subunits that can contribute to polysaccharide structures, the glycosidic linkages common to polysaccharides generally limit the conformational shapes of these macromolecules. This feature reflects the limited angular possibilities for torsion angles ϕ and ψ associated with the glycosidic linkages between two monosaccharides (Figure 6.23). In the case of poly-β-D-glucose or cellulose, conformational possibilities limited by the ϕ and ψ torsion angles *encourage* an extended polymeric structure. The limited flexibility of this fundamental structure, in addition to the equatorial positions of hydroxyl groups on the D-glucose monomers, promote hydrogen bonding interactions among neighboring chains, which, in turn, accounts for the strength of cellulose fibers.

For other polyglucose structures including starch and glycogen, which rely on $\alpha(1\rightarrow4)$glycosidic linkages, the strong extended linear conformation of cellulose is not possible. Instead the polysaccharide tends to form a helical coil as in the case of amylose. This structure has a 14-nm-diameter, left-handed helical structure, which has a calculated pitch of 0.85 nm. Although long helical coils may characterize the amylose molecule during its biosynthesis or in certain *in vitro* aqueous systems, more tightly coiled parallel or antiparallel double helical chains, stabilized by hydrogen bonding, probably occur in the native crystalline arrays of amylose in starch granules.

As indicated earlier, the potential varia-

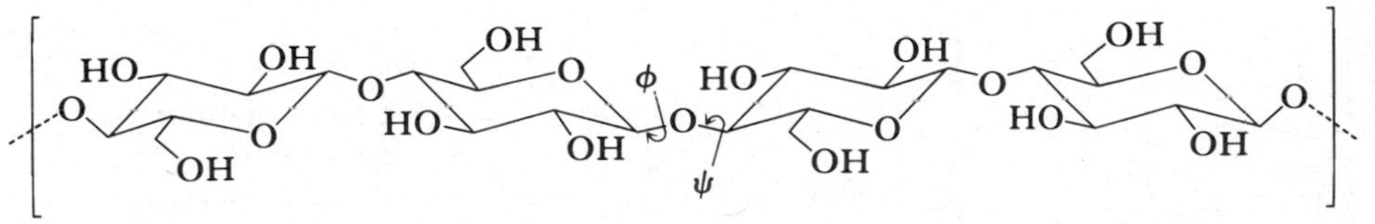

FIGURE 6.23. A portion of the conformational structure for cellulose illustrating the locations of the ϕ and ψ torsion angles.

bility among polysaccharides is nearly limitless, but the underlying patterns of their physicochemical behavior are relatively limited to their restricted conformational forms, their characteristic hydrogen bonding interactions, and the largely repetitive occurrence of similar subunits in their unique structures.

Polysaccharide Nomenclature

The nomenclature for polysaccharides can be complex according to the specific systematic methods but, as in the cases of monosaccharides, the common trivial names for many polysaccharides still persist. Many names end in the suffix *-in;* typical examples include chit*in*, dextr*in*, inul*in*, and pect*in*.

Many other algal, plant gum, and bacterial polysaccharides, however, terminate their names with *-an,* such as laminar*an,* carrageen*an,* guar*an,* xanth*an,* and dextr*an.* Added to this nomenclature is the generic prefix *glyc-,* which refers to sugars and sugar derivatives. Sugars (*glyc-*) that have undergone successive polymerization (*-an*) are called *glycans. Homoglycans* are polysaccharides composed of identical sugar residues, while *heteroglycans* are composed of two or more different sugars. In many cases the *glyc*-term is eliminated in preference of the abbreviation for the particular polymerized sugar residues (generic nomenclature):

Polysaccharides can be conveniently classified as storage or structural carbohydrates. Starch, glycogen, inulin, floridian starch, and many other polysaccharides serve as readily available energy stores for certain cells. Other structural polysaccharides such as chitin and cellulose, respectively, act as exoskeletons for arthropods or offer structural rigidity to plants. Polymers of hyaluronic acid along with a host of mucopolysaccharides serve as *intercellular cements* and provide a protective function in the interstices of animal cells, while pectins serve a similar function in plant and fruit tissues. The following discussion is designed to outline the notable dietary, nutritional, and industrial relevance of these storage and structural polysaccharides. Furthermore, the important properties of polysaccharide plant gums, saps, exudates, and hydrocolloid substances that occur in natural or processed foods will be surveyed.

Storage Polysaccharides of Plants

Starches represent the chief energy reserve for green plants. Starches occur throughout the leaves, stems, roots, and other plant organs in varying amounts. The highest concentrations of starch usually reside in cereal grains and underground tubers, although legumes and nuts also contain large amounts of polysaccharides. Plants often hold their most significant reservoirs of starch in the form of *granules* that have species-specific,

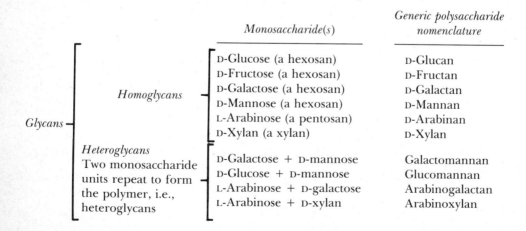

		Monosaccharide(s)	*Generic polysaccharide nomenclature*
Glycans	*Homoglycans*	D-Glucose (a hexosan)	D-Glucan
		D-Fructose (a hexosan)	D-Fructan
		D-Galactose (a hexosan)	D-Galactan
		D-Mannose (a hexosan)	D-Mannan
		L-Arabinose (a pentosan)	D-Arabinan
		D-Xylan (a xylan)	D-Xylan
	Heteroglycans Two monosaccharide units repeat to form the polymer, i.e., heteroglycans	D-Galactose + D-mannose	Galactomannan
		D-Glucose + D-mannose	Glucomannan
		L-Arabinose + D-galactose	Arabinogalactan
		L-Arabinose + D-xylan	Arabinoxylan

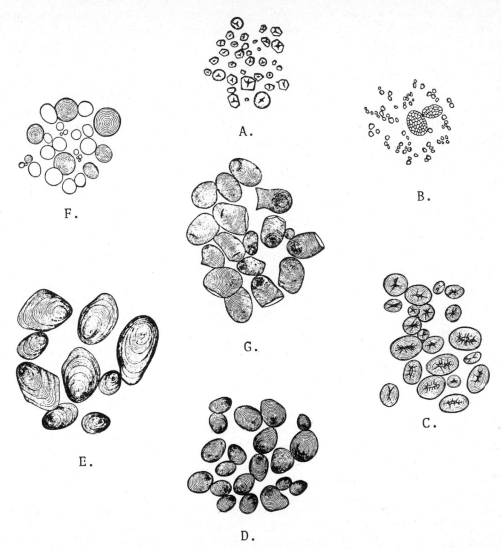

FIGURE 6.24. A selected group of starch grains including (A) wheat, (B) rice, (C) sage, (D) potato, (E) West Indian arrowroot, (F) corn, and (G) bean.

genetically controlled shapes, sizes, and optical characteristics (Figure 6.24).

For those plants in which starches are predominantly stored, starch granules first appear as minute points that rapidly grow to eventually fill the cell. Microscopic inspections of granules show that stratified layers of starch form around a nucleus called a hilum. The progressive deposition of starch layers around the hilum leads to the formation of a granule that is characteristic of the plant species. Diameters for granules vary widely (e.g., tapioca, 5–35 μm; potato starch, 15–100 μm; corn, 5–25 μm; rice, 3–8 μm) and assume geometric shapes that are reminiscent of polygons, ellipsoids, perfect spheres, and disks.

The optical properties of granules are especially characteristic when they are observed between two polarizing filters mutually positioned at right angles to a light beam. Under these conditions, the granules are optically *anisotropic* and display very characteristic *birefringence** patterns similar to the shape of a maltese cross.

Most native plant starches are heterogeneous mixtures of the two homopolysaccharides, namely, *amylose* and *amylopectin*. Although the ratios of these two starch types in nutritionally important plants and cereal grains are influenced by plant growth conditions, the potential ratio of amylose to amylopectin in plants is ultimately governed by genotypic factors. Corn species, for example, offer great genotypic versatility for producing different relative percentages of amylose and amylopectin within the corn kernels on a plant. *Normal* or *dent-maize* species have an amylose content of ~28%; *waxy-maize* (named with reference to the waxy luster of its kernels) has ~1% amylose and the 90+% remainder of starch is amylopectin; while *amylo-maize* produces >50% amylose. Other starch-producing plants show similar but smaller dynamic ranges with respect to the two starch mixtures; for example, waxy and normal species of sorghum have <1 to ~28% amylose, wheat has ~28% amylose, and so on.

The molecular structure of amylose is linear and its D-glucose monomers are repetitively linked at α-1,4 positions. Molecular weights for amylose range from a few thousand to ~150,000. Amylopectin also has a linear structure identical to amylose, but in addition to the α-1,4 glycosidic linkages, periodic D-glucose residues show α-1,6 glycosidic bonding to linear amylo-type (α-1,4) chains. This type of bonding encourages the development of a dendritic (treelike) macromolecular structure. The molecular weights for amylopectin vary widely and often exceed 500,000.

A *single* amylose molecule is characterized by the presence of *one reducing end* and *one nonreducing end*. An amylopectin molecule, on the other hand, has one reducing end and $n + 1$ reducing ends, where n is the number of randomly branched linear D-glucose chains bonded by α-1,6 linkages to the principal amylose structure (approximately one branch in 25–30 D-glucose units). Unlike the coiled linear conformation of amylose, however, amylopectin does not have a steadfastly preferred conformation.

It has been classically observed that amylose will assume a characteristic blue-black color when iodine is added to an aqueous suspension of the starch. This blue color is attributable to the physical positioning of iodide ions within the central cores of coiled amylose molecules. The absence of helicity in amylopectin markedly diminishes polyiodide alignment throughout a central molecular core; therefore, instead of blue-black color, amylopectin produces a purple-red color with iodine. The phenomenon of iodine binding can be monitored by using spectrophotometric or potentiometric titration methods that permit the establishment of *starch–iodine binding curves* (Figure 6.25). The value of these curves rests on their ability to depict the amylose/amylopectin homogeneity of uncharacterized starches. Polyiodide binding for pure amylose experimentally follows a hyperbolic relationship, whereas amylopectin molecules superficially bind little iodine and indicate a straight line.

Iodine binding curves for any starch mixture having less than 100% amylose or amylopectin content will display intermediate amounts of iodine binding between the two standard starches. The actual percentage of amylose in a starch mixture can be estimated by

$$\text{percentage amylose} = \frac{\text{iodine binding capacity (unknown)}}{\text{iodine binding capacity for pure amylose}}$$

Note that substitution values for the expression are obtained by linear extrapolation of the experimental data for amylose, amylopectin, and the uncharacterized starches *back to zero* (Figure 6.25).

The amylose fraction of many common starches including corn, rice, and potato have

---

Birefringence refers to a difference in refractive index for incident light plane polarized into perpendicular directions.

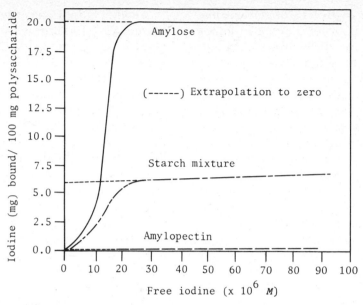

FIGURE 6.25. Typical iodine-binding curves for amylose, amylopectin, and an unknown mixture of these two starches. Polyiodide binding (assayed by potentiometric iodine titration) is about 20 mg/0.10 g for amylose, while the corresponding value for amylopectin approaches 0.0 mg/0.10 g. The significant data regarding the iodine binding of starches are obtained by extrapolating the respective plots back to "zero."

amylose as a minor component of the total starch (~17–30%), while certain genetically selected pea and corn species may have up to 75% amylose. Still other cereal species, especially waxy corn, show the predominant presence of amylopectin. Although amylose and amylopectin composition of starch granules for a plant species are quite characteristic, relative fluctuations in starch components may occur. These variations may depend on the genotype of the plant, plant maturity, and its growth conditions.

It has often been observed that barley, potato, wheat, maize, and pea plant maturation are associated with increases in amylose content, starch branching, and the number-average *degree of polymerization* ($\overline{DP}$) evidenced by starch molecules. The $\overline{DP}$ value for starches has importance because it indicates the average number of monomers present in a given polysaccharide, and this value serves as an index for evaluating the probable functional properties and appli-

cations of a particular starch to a food system. The $\overline{DP}$ value may be estimated for specific isolated fractions of starch according to light scattering or osmotic pressure methods, which respectively give a *weight-average* molecular weight or a *number-average* value.

Starch Hydrolysis

Starch molecules may be fragmented by a variety of methods, but the foremost methods include acid hydrolysis conducted in the range of pH 2.0 or enzymatic hydrolysis.

Acid hydrolysis of starches has special importance in industrial carbohydrate processing wherein the primary objective is to randomly fragment starch molecules, thereby improving their desired functional properties for specific food applications. The process can be *controlled* within limits to produce significant amounts of D-glucose, disaccharides, oligosaccharides, or mixtures of these carbohydrate forms.

The random hydrolysis of starches mediated by acids is overshadowed in importance only by enzymatic hydrolysis methods. These methods have industrial importance and nutritional significance, and they offer a more discriminative method of polysaccharide fragmentation. In fact, the fairly predictable hydrolytic behavior of enzymes is responsible for a significant amount of data regarding the molecular structure of starches and their components.

The α-*amylases*, which are universally activated by the presence of a chloride ion, are especially notable because they are widely distributed among plants and animals (e.g., saliva, pancreatic secretions). These amylases are classified as *endoglucosidases* because they randomly attack amylose at α-1,4 linkages to produce simpler sugars such as maltose and glucose, as well as short polysaccharide chains called *dextrins*.

Other enzymes known as β-*amylases* are present in plants (e.g., barley malt, sweet potatoes, wheat, soybeans), but these enzymes exhibit *exoglycosidic* activity. That is, they attack the nonreducing end of amylose to produce successive units of maltose.

Amylopectins are subject to enzymatic hydrolysis by α- and β-amylases, but the α-1,6 glycosidic bonds responsible for molecular branching cannot be hydrolyzed by these enzymes. Moreover, the singular use of β-amylase on amylopectin results in hydrolysis of the linear molecular portions (from the reducing end) to within two or three glucose residues of the α-1,6 linkage. Since dextrins of this sort impede further enzymatic hydrolysis, these branching structures are called *limit dextrins*.

Limit dextrins can be enzymatically hydrolyzed at the α-1,6-position, however, through the action of α-*1,6-glucosidase,* and this will permit further polysaccharide hydrolysis by β-amylase. Enzymes that sever these α-1,6 glycosidic linkages are generally called *debranching enzymes*. In perspective then, the complete hydrolysis of amylopectin to a mixture of glucose and maltose can be effectively carried out by the combined actions of α-amylase plus α-1,6-glucosidase.

Another important enzyme called *glucoamylase* (α-1,4-glucan glucohydrolase) displays the joint glucosidic hydrolysis properties of α- and β-amylases. This enzyme consecutively hydrolyzes α-1,4 linkages found in starches and also hydrolyzes and liberates glucose residues held in α-1,6 linkages at a somewhat lower rate. The enzyme is classified as an exoenzyme and initiates its hydrolysis activity at the nonreducing end of starch structures. Although glucoamylases are important because of their association with the food degradative activities of bacteria and fungi, these enzymes also have great industrial significance during the initial production stages of corn starch conversion to glucose, corn syrup, and high-fructose corn syrups.

Starches are also subject to the *in vivo* attack of *phosphorylase* enzymes that are present in both animals and plants. Although the actions of phosphorylase are discussed elsewhere in more detail, phosphorylase enzymes differ from amylases since they are responsible for mobilizing phosphorylated glucose from starch molecules prior to the entry of glucose into main-line biochemical pathways. Amylases cleave starch molecules by introducing a water molecule into a glycosidic linkage, whereas phosphorylases introduce a molecule of phosphoric acid at a glycosidic linkage to produce glucose 1-phosphate.

Gelatinization and Retrogradation

Although isolated plant starches appear to be soft, dry, and white to cream-colored powders, the processed starch particles largely retain the physical integrity of the parent starch granule. In spite of the physical strength of starch granules in resisting the rigors of abrasion, attrition, and pulverizing during the initial isolation stages from plant sources, the characteristic properties of starch granules can be easily and irreversibly deranged by exposing them to water, along with gentle agitation and heat. Under these conditions granules imbibe water (Figure 6.26), and their constituent polysaccharides

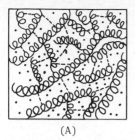 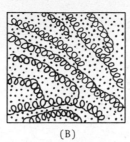

| (A) | (B) |

FIGURE 6.26. Schematic interaction of coiled polysaccharide molecules in (A) native starch granules and (B) swollen hydrated starch granules. *Note:* Coils represent $\alpha(1\rightarrow4)$-linked glucose molecules having an occasional $\alpha(1\rightarrow6)$-branched linear chain of glucose molecules; dashed lines represent hydrogen bonding; and dots represent water molecules distributed among the starch molecules.

become *hydrated*. The principal water–polysaccharide interaction involves mutual hydrogen bonding between water molecules and the hydroxyl groups on the polysaccharide.

During this hydration process, the crystallites within the starch granules lose their birefringent properties, swell, and eventually lose their absolute structural identity. Moreover, a typical aqueous suspension of starch granules increases in viscosity, the opacity of granules is lost, and a paste is ultimately formed. This overall process is called *gelatinization*.

Provided that pH, ionic strength, rate of heating, and concentrations of aqueous starch suspensions are carefully controlled, the gelatinization temperature can be used as a criterion that reflects, not only the type of starch,

but also the strength of intermolecular forces that exist within a specified type of starch granule. Since gelatinization temperature is determined from that temperature at which the granule loses its birefringent properties, the birefringent endpoint temperature (B.E.P.T.) offers a basis for comparing starch gelatinization properties (Table 6.3). The B.E.P.T. values for different starches also offers a practical index for predicting the behavior of starches according to specific dictates of industrial or food formulation applications.

The gelatinization temperature for a starch also marks the point at which up to 50% of small *amylose polymers* will *begin to* preferentially *leach out of the granule*. This physical process is joined with the eventual rupture or collapse of the starch granules since part of their hydrating capacity has been lost in the form of leached polymers. Collapse of the granules is also encouraged because the swollen fragile granules are fragmented because of mutual kinetic collisions and associated shear stress.

The initial process of granule hydration and subsequent swelling causes an inevitable increase in the viscosity of the aqueous starch suspension. The hydration of the granule may go unchecked provided that water for hydration is unlimited. When granules finally collapse, the viscosity or body of the system decreases, but it finally achieves an equilibrium value on continued cooking. The equilibrium viscosity ultimately depends on the pH, on ionic strength of the system and, more importantly, on the type of raw fractionated starch.

As the decrease in viscosity occurs, the starch dispersion undergoes a transition from a pastey salve to a more fluid, elastic sol. The properties of cooked pastes are often useful for the categorization of starches, as shown in Table 6.4.

Table 6.3. B.E.P.T. Values for Selected Starches

Type of starch	Estimated amylose content[a]	B.E.P.T.
Barley		
High amylose	45	63–67
Normal	29	53–60
Waxy	<2	55–60
Maize		
High amylose	56	>73
Normal	30	60–65
Waxy	<2	62–65

[a]Amylose content of starch based on iodine binding assays. Note that high amylose content correlates with high gelatinization temperature ranges.

Retrogradation of Starch

Aqueous solutions of amylose leached from native starch granules during heating may readily undergo reaggregation as the solu-

Table 6.4. Cooked Pasting Behavior for Different Starches

Category of starch paste	Examples[a]	Gel properties of cooked paste
Viscous short-bodied	Cereal starches—corn, sorghum, rice, wheat	Opaque gel
Viscous long-bodied	Tuber and root starches—potato, cassava (tapioca)	Usually weak, clear gels
Heavy bodied/stringy	Waxy starches—waxy corn, sorghum, rice	Clear gel, poor ability to form gels

[a]Examples cited are obviously subject to natural or selected variations that may have higher or lower amylose/amylopectin ratios that affect expected behavior.

tion is cooled. Rapid cooling results in the establishment of intermolecular associations among two or more amylose molecules wherever compatible hydrogen bonding possibilities exist and large amounts of solvent are entrapped within the gelled molecular network. Gradual cooling of amylose solutions, on the other hand, leads to a more exacting formation of hydrogen bonds among amylose molecules. This results in the formation of solid, sphero-crystalline molecular associations that effectively exclude solvent and may ultimately lead to the formation of a two-phase system (solid—starch/liquid—solvent). The aggregation of molecules according to these events represents *retrogradation*.

Linear polysaccharides having about 2000 glucose units (e.g., potato starch) and dextrins of less than 30 units usually display poor retrogradation, while starch molecules in the range of 300 to 450 glucose units (e.g., corn starch) show significant binding interactions and retrogradation. Concentrated solutions of amylopectin will undergo macromolecular aggregation, especially if they experience repeated freeze–thaw transitions, but the typical α-1,6 branching of these starch molecules diminishes their molecular aggregation when compared with amylose (Figure 6.27).

Freezing significantly affects starch retrogradation since the formation of ice crystals concomitantly withdraws the water of hydration from polysaccharide molecules. Once water molecules have been effectively withdrawn from intermolecular spaces and im-

mobilized in water crystal matrices, the polysaccharide molecules are naturally encouraged to undergo mutual hydrogen bonding interactions where possible and display retrogradation. Frozen aqueous food systems in which retrogradation has occurred typically lose a significant amount of water when thawed and still more water if the thawed, concentrated starch paste is subjected to pressure.

Leavened baked products such as breads probably represent the most common example of retrogradation when they undergo *staling*. Although natural linear starches are retrograded to some extent during the baking of doughs, the linear region of branched starches show time–temperature-dependent retrogradation after the initial baking process. This is detectable as a hardening of normally elastic baked doughs and crumbs. Retrogradation occurs at an increasing rate as storage temperature for bread approaches 0°C; however, the staling process is nearly eliminated at ≤0°C.

Modified Starches Present in Foods

The molecular structures of raw starches can be transformed into a wide variety of starch by-products that are extensively used by food processors and found widely in Western diets. The rationale for modifying starches rests on food technological and economic requirements of the food industry plus the product demands dictated by consumers, especially in the case of convenience foods. *Modified starches* can be prepared from nearly

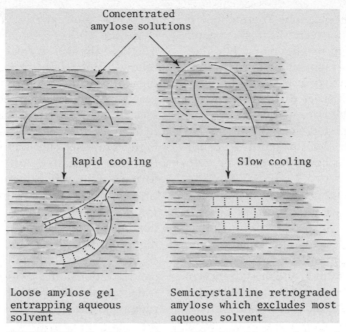

FIGURE 6.27. Comparison of a loose amylose gel network formed by fast cooling of a concentrated amylose suspension versus a semicrystalline retrograded amylose structure formed by slow cooling.

any type of starch, but economic factors related to raw starch productivity from cereal crops largely governs most commercial starch modification efforts. The large production of corn throughout the United States, along with modern processing equipment suitable for handling corn, has spurred the development and production of many modified starches that are now fundamental to food development, processing, and consumer acceptability of foods.

Dextrinization and hydrolysis of starch: The natural state of raw starches including

both amylose and amylopectin display molecular characteristics that make them largely insoluble in water. A variety of chemical, biochemical, and heat treatments can be applied to these starches in order to modify their solubility properties. In most cases, this is achieved by fragmenting the glucan structure. The progressive fragmentation of starch molecules leads to starch-based products that display ever *increasing solubility, sweetness,* and *hydroscopicity* while their *cohesiveness* and *viscosity decrease.* In other words, these treatments affect the graded destruction of starches according to the reaction scheme:

amylose and/or $\longrightarrow$ fragmented poly- $\longrightarrow$ oligosaccharides
amylopectin   saccharides smaller   (dextrins)
(native polysaccharides)   than the native
                 starch molecules $\searrow$
                                               disaccharides
                                                (maltose)
                                                $\searrow$
                                              monosaccharides
                                              (D-glucose)

Natural starches can be fragmented to molecular sizes between the native starch polysaccharide and its oligomers. These intermediate starch fragmentation products are called *dextrins*. Dextrins are conventionally produced by industrial processes in which aqueous starch slurries undergo acid-mediated hydrolytic bond cleavage, or starches may be heated under low-moisture conditions to form so-called *pyrodextrins* or *torrefaction* dextrins.

The acid treatment of starches involves the admixture of acid (hydrochloric or nitric acid) to an aqueous suspension of starch held at a temperature less than the gelatinization point (~50–55°C) for the specific starch. Neutralization, washing, and dehydration of the suspension follow the controlled exposure of starch to acid. Since the fragmentation of starch polysaccharides can lead to the leaching of the dextrins from starch granules, moisture content of the acidic extragranular solution must be carefully controlled during these steps. Controlled acidic treatment of starches produces *thin-boiling* or *acid-fluidity starches* due to hydrolytic debranching of amylopectin and, to a lesser degree, the hydrolysis of linear amylose molecules. Starches processed in this way demonstrate a low viscosity during cooking and yet they undergo rapid retrogradation on cooling to form firm gels.

The viscosities of acid-modified starches are often indexed according to a *fluidity number*. This value is calculated from the number of milliliters of a standard alkaline paste of a starch, flowing through a special calibrated funnel, in the time required for 100 mL of water to flow through the identical funnel.

The controlled heating of dry starches (5–7% moisture), as opposed to aqueous starch slurries, causes a series of irreversible modifications in the molecular structures of starch molecules. Pyrodextrins prepared in this way have variable physical properties depending upon the acidity of the reaction mixture during the torrefaction process, but most starches do display reduced viscosity, increased cold-water solubility, and a reduced tendency to gel. Three typical pyrodextrin categories include the *white dextrins, yellow (canary) dextrins,* and the *British gums*. Starch hydrolysis followed by additional polysaccharide polymerization and some transglycosidation reactions may occur especially where polysaccharides or oligosaccharides formed on hydrolysis interact to produce new, more highly branched starch molecules than those found in the native starch. Simple dextrins formed in this way generally do not produce firm gels when dispersed in aqueous or sugar solutions as do the fluidity starches. White dextrins form soft pastes; yellow dextrins are highly branched and exhibit very low viscosities even in solutions having high solids concentrations; and British gums provide a wide range of dextrin dispersion possibilities ranging from viscosities similar to premodified starches to the viscosities shown by yellow dextrin.

Although the application of commercial dextrins to new food systems is expected to expand such as in the case of tapioca dextrins, many of these products are marginally used in foods because of their distinctive off-flavors and caramelized odors, poor clarity, and undesirable amber to brown colors.

Aside from the production of acid fluidity starches and pyrodextrins, starches can also be effectively degraded into marginally small oligosaccharides and di- and monosaccharides by the concerted action of acids (pH ~2.0), enzymatic treatments (α- and β-amylases, glucoamylase), or combined acid–enzymatic treatments.

The ratio of each hydrolysis product to the others can be controlled by prescribed reaction conditions dependent upon acid concentration, reaction time, and types of hydrolytic enzymes employed. Less than complete degradation of starches by these methods produces smaller, less complex dextrins or oligomers (~20 glucose units) than the dextrinization methods described above. Under the most effective conditions, the glycosidic linkages of starches can be entirely severed to produce only free D-glucose.

Polysaccharide hydrolysis and the dextrose equivalent value: The degree of polysaccharide hydrolysis is often expressed in terms of its *dextrose equivalent* (*DE*). This reference term originates from the expression

$$DE = \frac{\text{weight of reducing sugar expressed as dextrose}}{\text{total weight of dry solids expressed as dextrose}} \times 100$$

The *DE* terminology requires the assumption that the *sugar weight represents dextrose,* in addition to the fact that the weight of all dry matter associated with the dextrose can potentially be transformed *into* dextrose.

A *DE* value of 0.0 corresponds to unmodified starch in which no degradation has occurred, and a *DE* of 100 represents complete transformation of raw starch to dextrose (i.e., D-glucose). Those starch hydrolysis products having an approximate *DE* of ≥20 are classified as glucose syrups when posthydrolysis products of starch (nutritive saccharides) are purified and concentrated. This product can be further concentrated by vacuum, drum, or spray drying to yield a product known as *dried glucose syrup*. Crystalline glucose monohydrate and anhydrous glucose can also be prepared from concentrated and refined glucose syrups having *DE* values of 93 to 98. A concentrated aqueous solution of nutritive saccharides obtained from starch, or a dehydrated (e.g., spray dried) product obtained from such a solution, is called *maltodextrin* provided that it has a *DE* value <20.

The occurrence and use of these starch hydrolysis products in contemporary foods is largely at the discretion of the food processor or manufacturer; and the selection of any one hydrolysis product over another must consider organoleptic appeal, palatability, consumer acceptability, and economic factors associated with the final food product. Moreover, it should be noted that many difficulties would be encountered in food preparation or preservation if hydrolyzed starches were not available.

Low-*DE* (10–35) hydrolyzed starches are widely used as vehicles for carrying flavors because they have little taste, while offering increased cohesiveness and thickness to foodstuffs in which any degree of sweet flavor is partially or wholly unacceptable. Intermediate-*DE* (40–60) glucose syrups can be used to modify food consistency too, but they also provide the sweetening potential of sucrose when necessary. Furthermore, these products can also be advantageously used in jellies and boiled confections as an impurity to minimize undesirable opacity and textural problems accompanying sucrose crystallization. The high concentration of fermentable sugars also makes syrups with *DE* of 40 to 60 ideal substrates to augment brew fermentations. The glucose syrups that have *DE* values of >60 are used extensively in baking, preservation, and canning operations. The high concentrations of glucose and maltose promote yeast leavening of doughs; the hygroscopic nature of these sugars encourages a moist texture in baked doughs; and they promote nonenzymatic browning reactions on the surfaces of baked doughs. Jams and fruit preserves can also use high-*DE* syrups to minimize sucrose crystallization, and more importantly, the high sugar concentration of *DE* >60 syrups increases the osmotic stress on potentially destructive microbial contaminants and controls the growth of all but the most osmophilic organisms. Canning operations for fruits employ high-*DE* syrups as an effective sweetening agent that does not eliminate the characteristic yet delicate flavor sensations associated with fruits. This feature is coupled with a minimization of volatile flavor loss and decreased food–air contact that can be responsible for ruining or at least diminishing food qualities.

Although a litany of specific applications and advantages could be presented that tout the merits of starch-derived sugars, disaccharides, and oligosaccharides, Table 6.5 and Figures 6.28 and 6.29 summarize the most desirable functional properties, compositions, and respective *DE* values for hydrolyzed starches. Note, too, that those *DE* val-

Table 6.5. Compositional Comparison of Maltodextrins with Corn Syrups Based on Their Relative Dextrose Equivalent (DE) Values

| | Percentage composition | | | | | | |
Dextrose equivalent (DE)	Dextrose (glucose)	Disaccharides	Trisaccharides	Tetrasaccharides and higher polymers	Density (kg/ft^3)	Average pH	Solution properties
Maltodextrins[a]							
6	—	—	—	—	17.3	4.5	Partially opaque
10	1.0	4.0	6.0	89.0	14.5	4.4	30–40% solids, clear
15	1.0	3.0	8.0	88.0	14.5	4.4	60% solids, clear
Corn syrup solids[a]							
20	1.0	6.0	8.0	85.0	15.4	4.4	70% solids, clear
25	6.0	4.0	4.0	86.0	20.0	4.7	70% solids, clear
36	7.0	32.0	13.0	48.0	20.9	4.9	70% solids, clear

[a]Maximum percentage moisture: 6.0%.

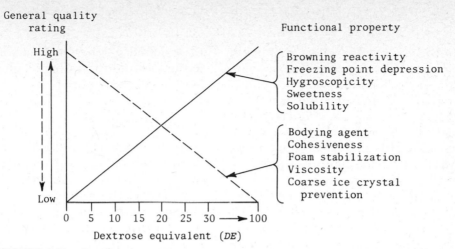

General quality rating

High

Low

Functional property

{ Browning reactivity
Freezing point depression
Hygroscopicity
Sweetness
Solubility

{ Bodying agent
Cohesiveness
Foam stabilization
Viscosity
Coarse ice crystal
  prevention

0   5   10 15 20 25 30 ——→ 100

Dextrose equivalent (DE)

FIGURE 6.28. Functional properties of starch, maltodextrins, and sugars based on their apparent dextrose equivalent (DE) value. (*Note:* that the graph shows a direct relationship for schematic purposes only; actual relationships between DE values and functional properties may not be linear for all products.)

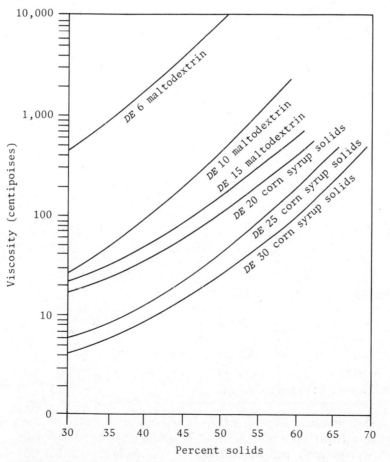

DE 6 maltodextrin

DE 10 maltodextrin

DE 15 maltodextrin

DE 20 corn syrup solids

DE 25 corn syrup solids

DE 30 corn syrup solids

Viscosity (centipoises)

Percent solids

FIGURE 6.29. Viscosity of various maltodextrins and corn syrup solids ranging from 40 to 60+ % solids for DE values of 6 to 36 measured at 28.0°C.

ues approaching zero reflect maximum bodying agent, cohesion, foam stabilization, viscosity, and coarse ice crystal prevention properties for polysaccharidic starch, whereas *DE* values approaching "zero" also represent "minimal" levels of solubility, sweetness, hygroscopicity, freezing point depression, and possibilities for the support of browning reactions.

Oxidized starches: Starches can be oxidized by a variety of reagents including hydrogen peroxide, periodate, or ozone; however, alkaline hypochlorite is widely used. The oxidation of starches usually involves the oxidation of hydroxyl groups on glucose residues to yield carbonyl and then carboxyl groups. The C-2 and C-3 positions are most vulnerable to attack. The oxidized starch can eventually undergo a β-elimination reaction, which depolymerizes and therefore severs the polysaccharide structure.

the oxidized starches have only minor uses in the food industry, where high solids and low viscosity are desired.

Starch Derivatives

In contrast with the modified forms of starch discussed so far, derivativization of starches may or may not reduce the viscosity of native starches and, most often, derivativization is designed solely to impart different properties to native starches. Although a myriad of starch derivativization methods are described in the scientific and patent literature, relatively few of these methods have importance in food commerce and industry.

Unlike those starch modifications involving hydrolysis and dextrinization, chemical derivativization of starch is accomplished by a very slight alteration of the polysaccharide structure. These alterations are *usually* pro-

Primary reaction sites

NaOCl (sodium hypochlorite) oxidation

Intervening steps in excess sodium hydroxide promote β-elimination

Polysaccharide chain is broken

*Reacting sugar residue

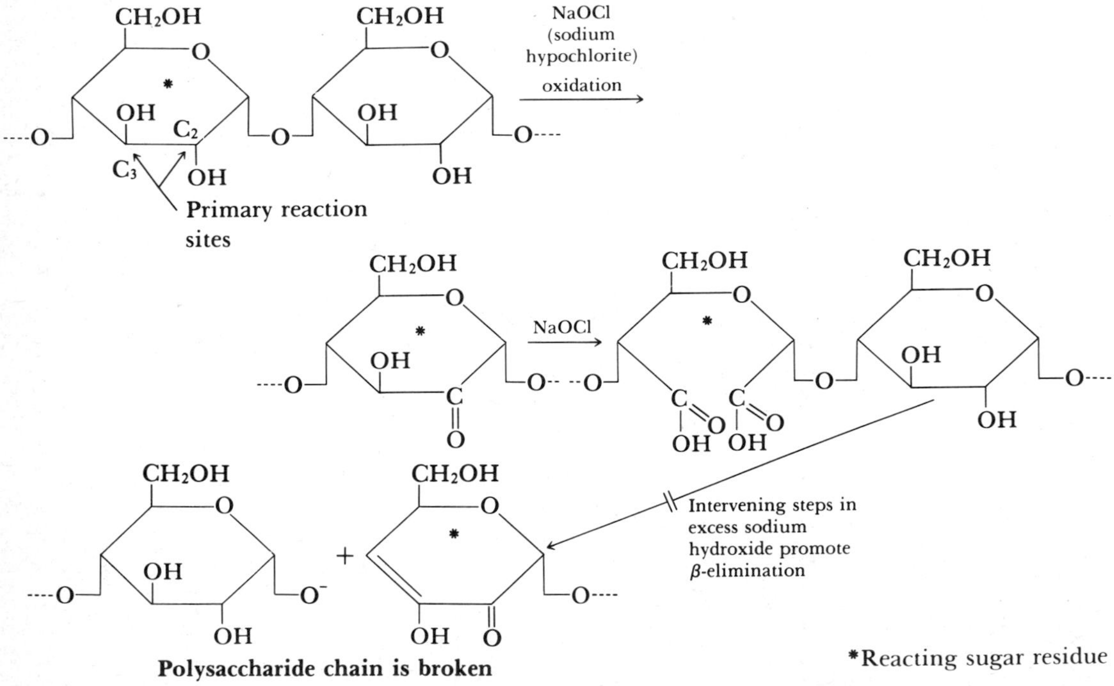

Random oxidation using alkaline hypochlorite leads to an increased starch solubility, reduces its viscosity, and minimizes molecular tendencies to associate and retrograde into gels. Compared to the majority of modified starches that are manufactured,

duced at the sites of the primary and secondary hydroxyl groups.

The conditions for the chemical modification of starches are usually so mild that aqueous suspensions of starch granules often retain their integrity after derivativization.

Starches are derivatized by many variations on the general theme where the pH of an aqueous starch slurry is controlled with alkali; reagent and/or a catalyst is added; the ungelatinized starch temperature is slightly elevated; and after a prescribed reaction time, the starch derivatives are isolated by filtration or centrifugation, washed with water, dried, and packaged.

Derivativization can involve two different strategic approaches including (1) the formation of *cross-linked starches* or (2) the formation of *stabilization derivatives* of starches.

Cross-linked starches are produced by exposing starch granules to difunctional reagents that ideally react with the hydroxyl substituents on two different starch polysaccharide molecules within the granule. The collective effect of cross-linking different polysaccharide molecules together minimizes the susceptibility of starch sols to shear stresses that may be encountered during food processing operations. Moreover, minor amounts of cross-linking reagents such as sodium trimetaphosphate or phosphorous oxychloride can exert a marked effect on the rate and extent of starch swelling during cooking. Cross-linking of starches is also recognized to decrease the sensitivity of starch sols to temperature, agitation, and acids. The total effect of these modifications enhances resistance of the starch to losses in viscosity.

Stabilization derivatives of starches differ from cross-linked starches because they are the product of monofunctional reagents. That is, there is a one-to-one ratio between the monofunctional derivatizing reagent and the hydroxyl group on the starch molecule where it reacts.

The introduction and bonding of these functional reagents into the architecture of starch polysaccharides interferes with their intermolecular associations. Other reagents not only interfere with intermolecular associations but may impart additional hydrophilic properties and viscosity, or in the case of cationic starches, tertiary or quaternary amines can be employed to produce quaternary ammonium or amino alkyl starches.

Some notable starch derivatives

Hydroxyalkyl starches. A variety of hydroxyalkyl starches can be industrially prepared, but many of these starches have limited food applications. Hydroxypropyl starch is somewhat an exception, however, since it has been widely used as a food thickener when clarity and paste stability are required.

Alkene oxides (e.g., ethylene oxide, propylene oxide) along with basic catalysts promote the establishment of *hydroxyalkyl derivatives* at the sites of the three functional hydroxyl groups on the glucose monomers of a starch molecule.

Native starch

$+ CH_2 - CH_2$
Ethylene

*R—OH = Potentially reactive hydroxyl group

Hydroxyalkylated starch

$a,b,c = 0,1,2,$ etc.

Although hydroxyalkyl substituents can undergo progressive reactions to produce poly(alkene oxide) derivatives, hydroxy-ethyl- and -propyl starches represent the most commercially available forms. Hydroxy-alkylation reactions readily occur at the C-2 position on anhydroglucose monomers of starches, followed later by other reactive hydroxyl groups on C-3 and C-6.

The formation of alkyl halides followed by titrimetric analysis of these products from hydroxyalkyl starches serves as an index for reporting the *degree of alkyl substitution*. This index, called the *molar substitution (MS) value*, specifically refers to the calculated mole fraction of a particular substituent group present on each anhydroglucose residue in the polysaccharide structure.

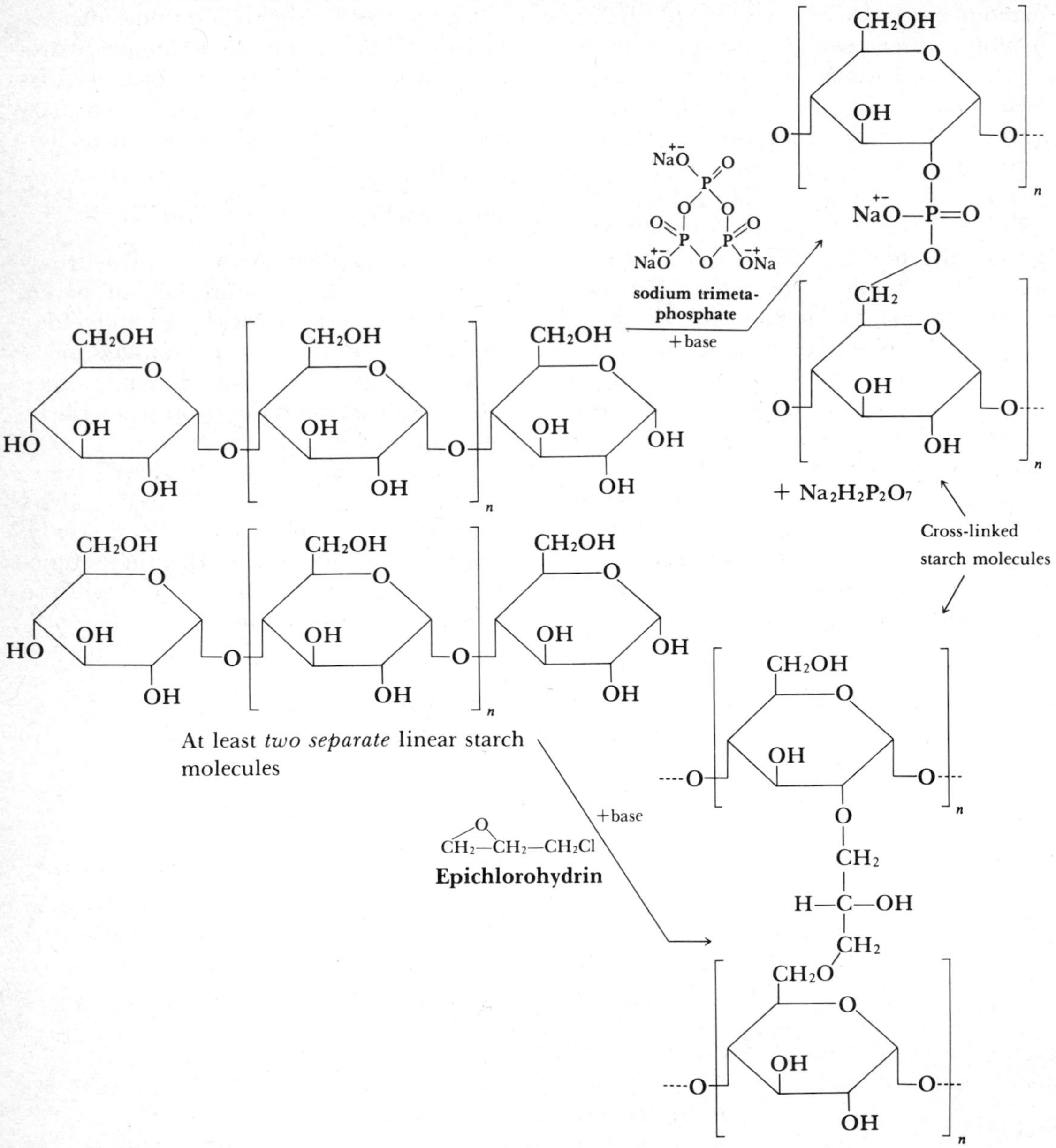

sodium trimeta-
phosphate
+ base

$+ Na_2H_2P_2O_7$

Cross-linked
starch molecules

At least *two separate* linear starch molecules

$CH_2-CH_2-CH_2Cl$ (O)
Epichlorohydrin

+base

Cross-linked starches. Many starches can undergo intermolecular cross-linking involving esterification. Esterification reactions may rely on the use of trimetaphosphate, or etherification may be conducted through the use of epichlorohydrin. Although many cross-linking reagents exist, relatively few are used for food starch production, and the choice of a cross-linking agent is governed by toxicity and carcinogenicity factors as well as industrial practicality.

Cross-linking reactions are typically conducted under alkaline conditions (pH ~10.0–11.0) where the starch molecules are held at subgelatinization temperatures along with the selected cross-linking agent.

Cross-linked starches are used in food systems (e.g., salad dressings) in which starch paste stability must be maintained under conditions of high-temperature cooking, retorting, acidity, high shear stress, and other conditions in which stable swollen starch granules are desired.

Starch acetates and succinates. A variety of starches, and especially corn starch, can be acetylated by acetic anhydride, vinyl acetate, or succinic anhydride. Under prescribed chemical engineering conditions involving pH, temperature, and reaction time, the *MS* values for these respective substituents on the starch molecule can be controlled. Starch acetates are very useful for minimizing the retrogradation of starch pastes, and acetylated derivatives of phosphorous oxychloride cross-linked starches offer excellent food thickening, texturizing, and stabilizing properties. Succinylated derivatives of starch are used to thicken selected foods, and the 1-octenyl succinic ester has also been successfully used for the improvement of starch affinity to nonpolar substances including fats and oils.

Starch phosphates. Starches can be esterified with sodium ortho-, pyro-, and tripolyphosphates to produce starch gels that are significantly more stable than those produced from native starches. Potato and corn starches are commonly subjected to these phosphate esterification procedures. The most common reactions typically involve the following steps:

*R—OH = potential starch reactive sites

Since the reaction conditions associated with the use of tripolyphosphates are less severe than other methods that may also promote uncontrollable hydrolysis of starch, this reagent is usually preferred. The merits of this derivativization process rest on improving the functional properties of starches. The phosphate moieties increase the stability of viscous starch pastes since similarly charged phosphate groups may mutually repel each other and prohibit the association of one polysaccharide molecule with another.

These viscosity properties along with the high clarity properties of starch phosphates make them ideal candidates for use in frozen foods, in which resistance to polysaccharide aggregation is critical to food quality.

Pregelatinized starches. Although native starches show marginal to negligible cold-water solubility, suspensions of most starches can be gelatinized prior to their actual use, dried, and later used as a pregelatinized starch that does exhibit cold-water solubility. When rehydrated, these pregelatinized starches also form homogeneous viscous pastes that are comparable to cooking raw starch, and they permit the formulation of many unique food products without the presence of heat (i.e., instant puddings).

Clearly the most undesirable property of pregelatinized starches is revealed when they are subjected to rapid rehydration. This produces a pastey mass with a large number of lumps. This undesirable situation can be minimized in many cases by using wetting agents that increase the swelling rate of the pregelatinized starch.

Bleached starches. Exposure of starches to small amounts of hydrogen peroxide, peracetic acid, sodium chlorite or sodium hypochlorite, and other agents will bleach starches to a stark white appearance. The bleached starches function in a manner similar to the parent starch, but a concerted effort must be made to remove all traces of the bleaching agents and inorganic salts formed during bleaching before these starches can be used.

Important Factors That Influence Starch Behavior in Foods

Gelation, retrogradation, viscosity, and film formation are important characteristic properties of all raw and modified starches. These properties are quite specific for pure starches until starches interact with other food ingredients. Many of these food ingredients typically include sugars, acids, salts, fats, oils, and other factors that affect the shear behavior of starches.

Effects of sugar solids: Many food products contain complex mixtures of starches that act as thickeners or stabilizers while present with variable amounts of sugar solids. Sugar-solid concentrations may run as high as 60% depending upon the food. Since sugar molecules are hydrophilic, increasing concentrations of sugars can be expected to compete for available water that would ordinarily interact with coexisting starches. Dissolved sugars also have a tendency to decrease water activity in concert with their increasing concentration levels. This action exerts an inhibitory effect on the rate of starch granule swelling, in the case of pregelatinized starches as well as in raw starches exposed to cooking. Figure 6.30 graphically displays the Brabender cooking and cooling curves for a marginally cross-linked waxy starch. Portions of the water system have been replaced by sugar in samples B, C, D, E, and F; note that sample A contains no sugar and sample AA shows the behavior of a 55% sugar system.

Inspection of the plotted data reveals that initial stages of starch gelatinization (indicated by the *initial rise* (*IR*)) are achieved with an increasing time/temperature relationship as sugar concentration (%) increases. The final viscosity for a starch–sugar system is a function of the percentage starch based on available water up to 50% sugar concentrations (Figure 6.30, sample E); beyond this

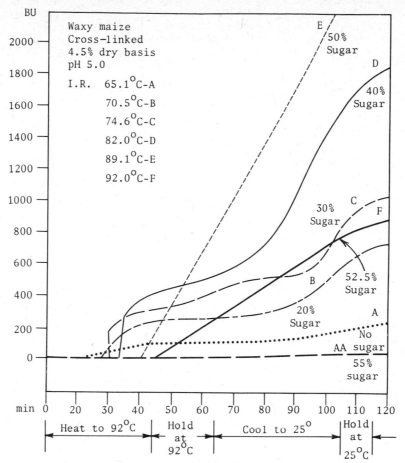

FIGURE 6.30.   Effects of different sugar concentrations on the cooking and cooling curves of a marginally cross-linked waxy maize starch expressed in terms of Brabender units. (Courtesy of the American Maize-Products Company, Corn Processing Division, Hammond Indiana.)

point, swelling properties are inhibited for most practical purposes. All sugars show similar qualitative effects on gel strength to differing extents, but disaccharides have a greater effect than monosaccharides at identical percentage weight/concentration levels.

Effects of pH and acidity: Foods having a pH of <4.0 encourage marked losses in starch viscosity. Viscosity losses result from acid hydrolysis of glucosidic linkages, and the extent of hydrolysis is related to the pH of the system and the type of starch.

Granule swelling rates and starch viscosities are not significantly affected in the pH range of 4.0 to 7.0, but acidic foods having a pH of ≤4.0, such as salad dressings, tomato-based soups and sauces, and fruit pie fillings, often upset the predictable behaviors of food starches. Figure 6.31 exhibits the relationship of pH on the gelatinization and degradation of starch. Note that a pH increase from 4.0 to 7.0 increases the ability of the starch to hydrate and gelatinize. Gelatinization of starch at pH 2.5 appears to be similar to the pH 7.0 curve, but the acidic conditions promote glucosidic cleavage of the starch molecules, granule swelling, granule fragmentation, and a decrease in potential starch viscosity, and the starch granule undergoes prompt degradation on standing. It is instructive to note, too, that high alka-

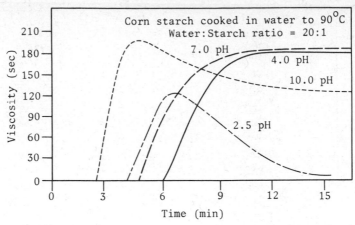

FIGURE 6.31. Effects of pH on the viscosity of corn starch. (Reprinted by permission of the Corn Refiners Association, Washington, D.C.)

linity promotes the overall hydration rate and gelatinization of starch molecules, but the high pH also contributes to the rupture of swollen starch granules and a corresponding loss in viscosity. Aside from their physico-chemical behavior, alkaline starch systems have limited relevance to most food systems.

The detrimental effects of low pH food systems on starch behavior can be combated by using chemically modified or cross-linked starches. Only moderate amounts of inter-molecular cross-linking are necessary to en-sure a starch behavior that readily gelatinizes and counteracts many acid effects.

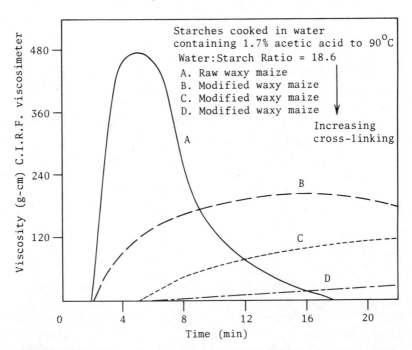

FIGURE 6.32. Effects of starch cross-linking on its viscosity properties. (Cour-tesy of the American Maize-Products Company, Corn Processing Division, Ham-mond, Indiana.)

According to Figure 6.32, starch cross-linking for a waxy maize starch progressively increases for curves B, C, and D. Moderately cross-linked, raw waxy maize shows prompt gelatinization while resisting acidic hydrolysis during four to 6 min of normal cooking.

Additional cross-linking necessitates higher cooking temperatures and/or times to ensure the maximum development of starch viscosity. In some cases such as curves C and D in Figure 6.32, the highly cross-linked starch may not display its maximum viscosity in an acidic food system until the product is pressure cooked or retorted.

Effects of salts, proteins, and lipids: With the exception of potato starches and some other specialty starches that contain phosphate substituents, most starches are relatively unaffected by salt concentrations that normally exist in foods. The physical properties of starches are not subject to major alterations when admixed to protein solutions and milk-based liquids unless the protein content or proteinaceous milk solids display concentrations far above the levels found in the natural food. In the cases of those foods that are completely formulated, starch behavior cannot be predicted with certainty and may reflect the synergistic influences of the collective food ingredients.

Amphipathic and lipid components influence the starch behavior in foods in different ways. Monoacylglycerols that obviously contain a hydrophobic fatty acid esterified to a polar glycerol moiety interact with the coiled structure of amylose in a fashion similar to that of iodine—but not with amylopectin. This property of monoacylglycerols is believed to account for minimizing retrogradation and staling of baked doughs. In addition to prohibiting gel formation of starches in high-starch canned goods, monoacylglycerols can also decrease the physical associations among cooked strands of alimentary pastes and cooked rice grains. The use of amphipathic substances, especially derivatives of 16+ carbon aliphatic fatty acids, have certain processing trade-offs when used with starches. These agents tend to increase the minimum temperature at which starch granule hydration begins; they increase the minimum temperature at which maximum starch viscosity will occur; and unfortunately these agents decrease the starch gel strength.

The influence of lipids on starch behavior is less dramatic than for amphipathic compounds. Generally, increasing concentrations of lipids, such as 16+ carbon triacylglycerols, are inversely related to the temperature at which maximum viscosity for a specific starch occurs. Lipids are also used to modify the physical powdering properties of starch granules. Starch granules modified by an aerosol exposure to food-grade oils improves the packing properties of granules and permits the starch to be used as a molding powder for certain foods (e.g., candies). Starch prepared in this way not only serves as a mold but eventually removes moisture from the final confectionery product and hastens the setting process.

Factors that influence shear properties: Starches vary in their abilities to withstand shearing and mixing since they differ in their fundamental natural or modified molecular structures. The ability of starches to resist shear stresses brought about by mechanical forces is generally proportional to the degree of intermolecular cross-linking. As illustrated in Figure 6.33, unmodified waxy starch is considerably less stable than regular starch, and the stability of regular corn starch is less than that of cross-linked starch. Clearly these relationships are generally true, but since the interactions of complex food ingredients vary greatly among different foods, as well as their individual exposures to mechanical processing stresses (e.g., pumping, cooking, cooling, high-speed mixing, and homogenization), similar concentrations of identically cross-linked starches usually have food system-specific responses to shear stresses.

Freeze–thaw effects: Repetitive freeze–thaw cycles can also influence the behavior of starches contained in food systems. At room

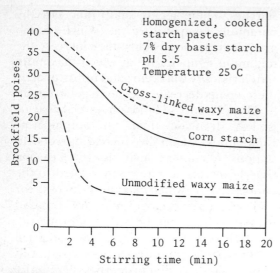

FIGURE 6.33. Shear properties for cross-linked starches plotted with reference to Brookfield poises and stirring time in minutes. (Courtesy of the American Maize Products Company, Corn Processing Division, Hammond, Indiana.)

temperature, water molecules actively engage in hydrogen bonding with polysaccharides. The water molecules establish an orderly immobilized layer around the polysaccharide, which gives way to an increasingly random distribution of the water molecules as the distance from the polysaccharide increases. These hydrated polysaccharides naturally participate in random inter- and intramolecular folding, coiling, and entwining conformations. The linear components of starches readily interact with each other to form molecular aggregates (micelles) that are stabilized by hydrogen bonding.

The progressive establishment of micelle structures leads to the eventual exclusion of immobilized water layers around starch molecules, along with the demise of the immobilized layer of water that originally stabilized the polysaccharides in solution. The other important activity of water–polysaccharide systems involves multiple intermolecular hydrogen bonding interactions among polysaccharides that form three-dimensional gel matrices that effectively trap large amounts of water.

The progressive increases in hydrogen bonding among polysaccharides during sol to gel transformations, coupled with the exclusion of water from the gel, often results in water drainage from the gel or *syneresis*. Pastry and pie fillings, sauces, gravies, and other foods that contain polysaccharides in the form of starch thickeners are potential candidates for micelle formation and the syneresis phenomena described.

Aside from the undesirable consequences of freeze–thaw transitions outlined above, the desirable gel properties of many native starches may be similarly upset by holding starch-containing foods at or near freezing conditions (1.7–4.4°C). Most starches show an inherent tendency toward retrogradation and syneresis under these conditions with the net result being a loss of desired food viscosity accompanied by a separation or curdling of food starches from other food ingredients.

Repeated freeze–thaw cycles cause solute (salt, sugar, etc.) partitioning or freeze-out as water crystallization progresses throughout the system. Thawing of these disrupted systems results in considerable syneresis and drip-loss from the food. A variety of proprietary starches have been developed to combat these undesirable freeze–thaw effects, and that starch that indicates the lowest percent liquid separation from starch gels for a series of successive freeze–thaw cycles is considered to be the starch of choice where freeze–thaw stability is required (Figure 6.34).

GLYCOGEN

Glycogen is the major energy storage polysaccharide of animal cells and the structural and functional analogue of starches in plant cells. The glycogen molecule is a high-molecular-weight compound composed of $\alpha(1 \rightarrow 4)$ glycosidically bonded D-glucose molecules. The typical molecular weight of glycogen may range into the millions. Since each 8 to 12 glucose residues are glycosidi-

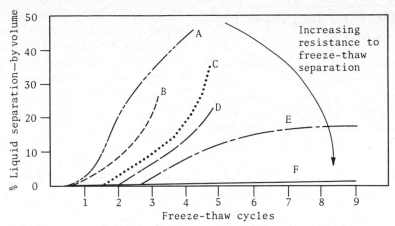

FIGURE 6.34. Graphic profiles for a typical range of different food starches (denoted as A→F) to resist liquid separation in an aqueous–starch food system. It is obvious that the most successful starches for resisting the rigors of freeze–thaw separations will closely parallel the abscissa of the plot. (Courtesy of the American Maize-Products Company, Corn Processing Division, Hammond, Indiana.)

cally linked by an α(1→6) linkage to another α(1→4)-linked glucan chain, glycogen molecules are usually more highly branched yet structurally more compact than amylopectins.

The large numbers of nonreducing ends on each glycogen molecule are fundamentally important to the biochemical roles of glycogen in energy storage and glucose mobilization into biochemical pathways for oxidation. For example, *glycogen synthetase* and *glycogen phosphorylase* activities both require the presence of a nonreducing end on highly branched glycogen molecules. The synthetase implements added glucose polymerization to glycogen molecules, and the phosphorylase mediates the phosphorylation of glucose residues stored in glycogen to meet cellular energy demands.

Depending on the nutritional status of animals and humans, glycogen may account for up to 10% of the wet weight of the liver and 1 to 2% of the skeletal muscle. The sheer bulk of muscle tissues in most animals, however, ensures that the majority of glycogen reserves will occur in muscle tissues.

Food scientists are interested in the glycogen content of muscles for a variety of reasons, but mainly because the antemortem

glycogen levels of muscles can have an important bearing on the keeping quality and texture of butchered and dressed meats. In general terms, both the minimization of antemortem animal stress and adequate resting of animals will encourage the retention of muscle glycogen. The postmortem glycogen is then subject to anaerobic metabolic processes that produce lactic acid. The mild acidity of lactic acid then provides a mild preservative action that enhances the keeping quality of meats during cold storage. Similar conservation of glycogen reserves would be desirable prior to fish processing, but the natural struggle of fishes during a catch results in glycogen utilization and a miniscule preservative action attributable to lactic acid. The livers of all animals also rapidly hydrolyze glycogen to D-glucose after slaughter.

Aside from postmortem hydrolysis of glycogen by *in vivo* enzymes, the molecule is also enzymatically attacked by α- and β-amylases that respectively produce glucose and maltose. The activity of β-amylase also leads to the production of a limit dextrin. The response of glycogen to iodine is similar to that of amylopectin in that it shows a red-violet color.

STRUCTURAL POLYSACCHARIDES FOUND IN PLANTS

The earliest stages of plant growth are accompanied by cell-wall formation. This activity involves the entwining of complex polysaccharides into a matrix that is itself largely composed of polysaccharides. The primary cell wall synthesized by green plants is principally composed of cellulose microfibrils that are $\beta(1\rightarrow4)$-linked polyglycans ($\overline{DP} = 8000-13,000$), although certain algae are also rich in xylan and mannan fibrils.

Continued development of the plant matrix coincides with the appearance of polygalacturonic acid derivatives generically referred to as *pectic substances*. These substances later coexist with developing xylans and neutral polysaccharides called *hemicelluloses*.

Dicotyledonous or woody plants eventually display still other polysaccharide structures. Mature primary cell walls of these plants contain linear glucan chains having branches composed of xylose, fucose, and galactose units. These so-called *xyloglucans* are *covalently linked* to other coexisting arabinogalactans, rhamnogalactans (polygalacturonic acid chain with periodic rings of rhamnoses), and cellulose microfibrils.

In addition to this menagerie of polysaccharides, plant cell walls often incorporate a minor amount of *extensin*. This is a 4-hydroxyproline-rich glycoprotein that has mixed arabinose and galactose oligomer (~8–10) units attached to the protein hydroxyl groups. Mature woody plants also have large amounts of aromatic ringed polymers called *lignins* that contribute to the strength of cell walls.

Aside from plant structural roles, many of these polysaccharides have dietary importance and significance in foods as native or chemically modified forms.

Cellulose

Cellulose is a high-molecular-weight polymer ($\overline{DP} \geq 10,000$) of anhydro-D-glucopyranose units linked by $\beta(1\rightarrow4)$ glycosidic bonds (Figure 6.35). The chemical resistance and tough native structure of cellulose stems from its molecular ability to establish partially crystalline microfibrils stabilized by mutual hydrogen bonding interactions. Noncrystalline forms of cellulose called *amorphous cellulose* display little or no fibrous structure as in the case of vegetable pulps. The tenacious hydrogen bonding interactions found in partially crystalline celluloses have a major influence on the behavior and physical properties of cellulose including (1) its characteristic density; (2) its failure to markedly swell in water; and (3) its apparent inert reactivity toward most hydrolytic enzymes and other chemical reagents. Conversely, the absence of comparable hydrogen bonding interactions in amorphous cellulose promotes its swelling in water and decreases the potential tensile strength of its collective microfibrils. This feature of amorphous cellulose structure accounts for its increased elasticity when compared with crystalline forms of cellulose.

Cellulose is subject to hydrolysis reactions that fragment the structure into shorter ol-

The repeating subunit in cellulose is β-cellobiose. The $\beta(1\rightarrow4)$ linkage within the disaccharide is displayed between all subunit bonding sites.

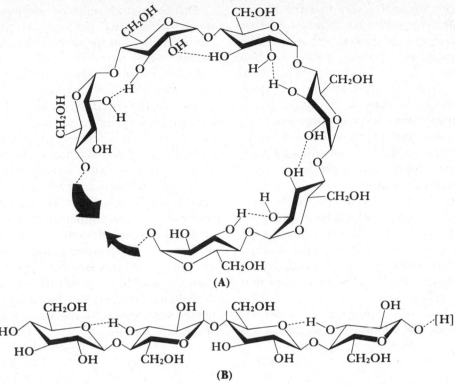

FIGURE 6.35. Partial conformational structures for (A) the α-maltose repeating subunits of starch and (B) the β-cellobiose subunits of cellulose. Note the profound influence of α(1→4) linkages on starch as opposed to β(1→4) linkages in cellulose. The coiled molecular conformation, stabilized by hydrogen bonds, is favored in starch, while cellulose exists as a linear structure.

igomers, disaccharides (cellobiose), or, ultimately, its full complement of D-glucose units. Hydrolysis of cellulose can be effectively conducted in industry by strong mineral acids (e.g., 72% sulfuric acid, 40% hydrochloric acid, 85% phosphoric acid), but these conditions are far too severe for promoting the hydrolysis of cellulose during the *in vivo* digestive process of living organisms. Instead, those organisms that can digest cellulose require specific enzymes, namely β(1→4)-glucosidases, to liberate D-glucose from the cellulose polymer. Man and carnivorous animals do not have these enzymes, and dietary celluloses largely survive the digestive process. This is in contrast to *cellulase*-producing fungi, other selected microorganisms, and some bacteria that reside in the digestive tracts of ruminants. These organisms produce β(1→4)-glucosidases that promote the liberation of D-glucose from

cellulose as a nutritional staple. The simple presence of β(1→4) linkages in cellulose as opposed to α(1→4) linkages of starches accounts for a major distinction in the nutritional availability of *empirically identical* and *calorically similar* polymers of the same monomer, D-glucose.

Modified Celluloses

Native celluloses are of interest to food scientists and nutritionists because they can affect food texture, toughness, digestibility, mouth feel, and dietary bulk. Furthermore, celluloses can also be modified and introduced into contemporary foods for a variety of reasons.

These modified celluloses commonly assume two possible forms, (1) *hydrolyzed celluloses* and (2) *derivatized celluloses*.

Controlled hydrolysis of celluloses can be

carried out in such a way that the microcrystalline regions are *separated from* noncrystalline amorphous cellulose. The microcrystalline cellulose isolate is then dried and added to foodstuffs to improve noncaloric bulk; produce free-flowing powders of sugar syrups (e.g., invert liquid sugars, honey) or lipid-containing foods (e.g., cheese solids and extracts); as well as to augment the stabilization of colloidal food dispersions with the aid of emulsifiers (e.g., creamy salad dressings).

Modified forms of cellulose can be produced by reacting the potentially reactive C-2, C-3, and C-6 hydroxyl groups of anhydroglucose residues to form methyl, ethyl, hydroxymethyl, hydroxyethyl, hydroxypropyl, and carboxymethyl substituted cellulose derivatives.

If all three hydroxyl groups are derivatized, the *degree of hydroxyl substitution* is 3.0

Other related cellulose derivatives require the reactive presence of propylene oxide in addition to methylchloride, which permits hydroxypropyl substitution on anhydroglucose residues in the cellulose molecule. The hydroxypropyl substituent ($-OCH_2CH(OH)CH_3$) contains a secondary hydroxyl group and, accordingly, this product of cellulose is often regarded as a propylene glycol ether of cellulose. *Varying ratios* of propylene glycol ether substitution to methoxyl substitution account for the characteristic differences in the solubility and thermal-gel point temperature displayed by aqueous systems of these various cellulose derivatives. The general structure for these modified celluloses is shown below, but to reiterate, a wide range of different cellulose derivatives are possible depending on the presence of methoxyl and propylene glycol ether substitution.

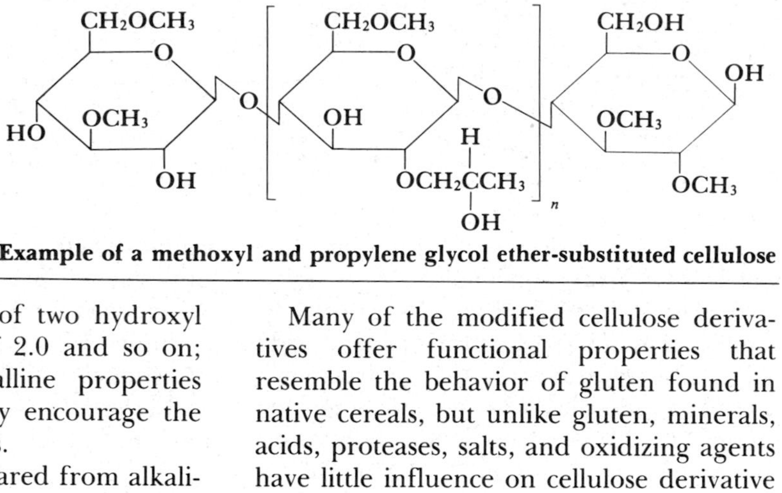

Example of a methoxyl and propylene glycol ether-substituted cellulose

(DS = 3.0); substitution of two hydroxyl groups produces a DS of 2.0 and so on; however, the microcrystalline properties of cellulose do not readily encourage the formation of 3.0 DS values.

Methyl cellulose is prepared from alkali-treated cellulose that results in an alkali cellulose product. This is then treated with methylchloride to form a methyl ether of cellulose.

Many of the modified cellulose derivatives offer functional properties that resemble the behavior of gluten found in native cereals, but unlike gluten, minerals, acids, proteases, salts, and oxidizing agents have little influence on cellulose derivative behavior.

The notable gelling point of modified celluloses only at baking temperatures encourages their use for gas retention during the

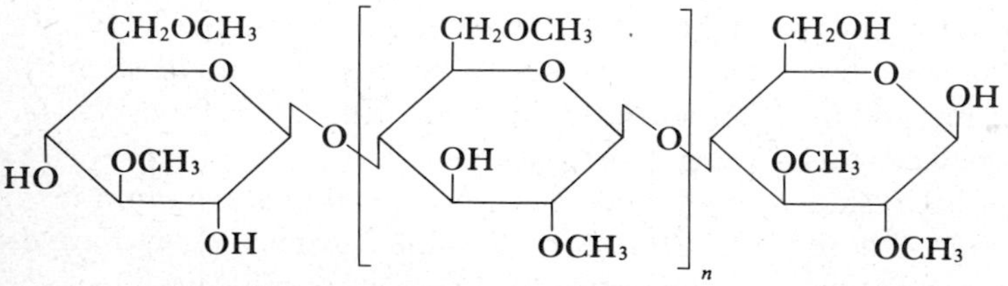

Example of a methyl ether derivative of cellulose

baking of doughs, especially when increased toughness of the final baked product must be avoided.

Additional properties of these derivatives include their abilities to act as surfactants, promote moisture retention, and exhibit negligible fat and oil absorption. The surfactant properties appear to retard starch retrogradation, thereby minimizing the staling of baked doughs. Many cellulose derivatives can imbibe nearly 40 times their weight in water. This property improves mixing tolerances during food formulation processes and certainly promotes moisture retention in food products. The poor interaction properties of cellulose derivatives with fats and oils, on the other hand, can be exploited to great advantage because their presence in doughnuts and fried cakes minimizes lipid absorption.

Aside from these modified celluloses, other cellulose derivatives, such as the sodium salt of carboxymethylcellulose (CMC), have been widely applied to food formulations. This compound is often admixed to microcrystalline cellulose in order to improve the functional properties of microcrystalline cellulose. Together these modified celluloses can form fragile gel systems sufficient to prohibit the stratification of suspended food solids in liquids. They also have the joint effect of suppressing the coalescence of emulsified droplets in two-phase systems (e.g., salad dressings). Ice cream and other frozen foods may also use significant amounts of CMC at low concentration levels in order to improve the texture, body, and melting-point characteristics of the product. More importantly, however, CMC decreases water migration during freezing processes and retards the formation of large ice crystals in foods.

Pectins and Pectic Substances

The cellular interstices or middle lamellae of land plants contain colloidal heteropolysaccharides known as *pectins*. High concentrations of pectins between neighboring cells ensure the existence of water-permeable channels throughout plant tissues, and this is thought to permit the translocation of water throughout plant tissues at rates exceeding simple intercellular osmosis. The apparent diversity of pectins, based on their recognized chemical structures and behaviors, has resulted in a series of definitions that account for various forms of pectin, more properly termed *pectic substances,* in addition to their derivatives.

Pectic substances are complex colloidal carbohydrate derivatives that are prepared from or occur in plants and have a large percentage of total anhydrogalacturonic acid residues in a chain-like polymeric structure. Carboxyl groups on these polygalacturonic acid structures may exhibit partial esterification to methyl groups or they may be partly or wholly neutralized by one or more bases.

Protopectin represents the water-insoluble, parent-pectic substance that occurs in plants. Restricted hydrolysis of protopectin produces pectinic acids and pectin.

Pectinic acids are those polygalacturonic acids that contain a significant number of esterified methyl groups. These polymers will form gels if acid and sugar concentrations are carefully controlled; or alternatively, low-methoxyl polymers may undergo gel formation with the help of certain metallic ions. The salts of pectinic acids are defined as either *normal* or *acid pectinates*.

Pectin (or pectins) define those water-soluble pectinic acids that are capable of gel formation under suitable sugar and acid conditions but may display varying degrees of methyl ester content and degrees of neutralization.

Pectic acid is a term that pertains to those pectic substances composed principally of colloidal polygalacturonic acids that are free from methyl ester groups. Salts of pectic acids are normal or acid pectates.

Aside from the roles of pectic substances in plant structure and physiology, pectic substances are also responsible for imparting desirable and undesirable textures to fruits and vegetables. Moreover, pectic substances are widely used as gelling agents for keeping jams, jellies, preserves, confectioneries, beverage solids, cold setting flan jellies, and other foods in a controlled form.

Pectic substances exist in close association with cellulosic components of plant cells in the form of insoluble protopectin. Exposure of protopectin to heat and a low pH aqueous medium causes protopectin hydrolysis to eventually form pectin. Although citrus peel provides a major commercial source of pectin, all fruits and vegetables have varying amounts (Figure 6.36).

The hydrolysis of protopectin by acids is paralleled during normal ripening processes of fruits and vegetables by native plant enzymes known as *protopectinases*. As protopectin levels decrease during ripening, the concentration of soluble pectins increases and the firmness of the fruit and vegetable tissues is lost. Still further production of soluble pectins leads to very mealy tissues whose only remaining texture is attributable

to the resident turgor pressure, native lignin, and cellulose content of the plant tissue.

Molecular structure and functional properties of pectic substances: The principal structure of pectic substances depends on α-D-(1→4) linkages between anhydrogalacturonic acid residues. The linear arrangement of these residues encourages the maintenance of a threefold screw-type molecular conformation. Some of the carboxyl groups exist as free acids, others are esterified to methyl groups, some carboxyl groups are neutralized, and still others have secondary acetyl groups. The last possibility accounts for the presence of so-called *pectin acetyls* in certain plants.

Pectin acetyls range in concentration from 0.15% in citrus and apple to as much as 4.0% for pear and peach pectins. Pectin acetyls hold special importance because a 1:8 ratio of acetyl groups to galacturonic acid residues prohibits sugar–acid–water gel formation of pectins discussed in earlier sections.

The majority of plants contain combinations of pectic substances comprised of pure galacturonans having neutral sugars such as arabinose, galactose, rhamnose, xylose, and others as part of their structures. Significant differences in the composition of pectic substances in different plants is not always species specific, but rather reflects plant cell growth and cultivation.

The chemical properties of pectic substances are dissimilar from other polysaccharides owing to the presence of carboxyl groups. The esterification of carboxyl groups promotes the fragmentation of the pectin under alkaline conditions since the glycosidic bonds become very labile (carboxylic acid ester groups are electron withdrawing). Alkali-mediated reactions of this type progress by means of a β-elimination reaction that fragments glycosidic bonds and eventually produces an unsaturated structure having a C-4 to C-5 double bond (Neukom and Deuel, 1958, 1960) (Figure 6.37).

In addition to the alkaline reactivity of pectic substances, they can also be oxida-

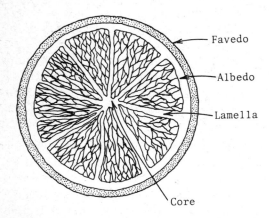

Favedo

Albedo

Lamella

Core

FIGURE 6.36. Parts of citrus fruits used in the manufacture of pectin. Citrus pectins are derived from the peel of lemon and lime, and to a lesser extent grapefruit and orange. Citrus peel is a by-product of juice and oil pressing and it contains a high concentration of pectin, which has very desirable functional properties. Apple pomace, which is the residue from pressing apples for apple juice, serves as the raw material for commercial apple pectins. These pectins are normally darker in color (brownish) than citrus pectins, but the functional properties are nearly identical.

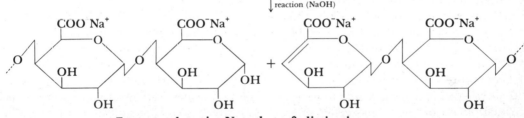

FIGURE 6.37. Some common native forms of pectic substances and their derivatives.

tively degraded by peroxides, dichromate, halogens, periodate, ascorbic acid, and some other agents.

The bulk of commercially supplied pectins find uses in the manufacturing of jellies, jams, preserves, and related food systems. The desirable gelling and viscosity properties of pectins stem from limited rotation of the galacturonic acid residues about glycosidic bonds throughout the pectin molecule, as well as the ability of the pectins to establish

fairly predictable gel strengths under prescribed sugar (soluble solids):acid (pH 3.05–3.5):water concentration ratios. The sugar:acid:water ratio required for pectin gelation serves as a basis for the commercial tabulation of *pectin grades,* that is, the number of parts of sugar that must be added to one part of pectin to produce a gel of standard strength and firmness set by industry criteria.

The gelling properties of pectins are further complicated by their total molecular weight, their degree of subunit polymerization, and the *degree of methylation (DM)* exhibited by carboxyl substituents.

Those pectins having 50% of available carboxyl groups methylated have a *DM* 50 index, a 75% methylated pectin has a *DM* 75 index, and so on. These *DM* values have important significance since progressively higher *DM* values for pectins parallel higher temperatures where gel formation occurs.

The characteristic properties of pectin molecules coupled with their individual requirements for establishing and maintaining a gel have resulted in their somewhat arbitrary classification as *rapid-set, slow-set,* or *low-methoxyl pectins.*

Rapid set pectins have a *DM* value of ≥70, and at an optimum acid pH of 3.0 to 3.4 plus sugar, they produce a strong gel in a short time. Since molecular weight is proportional to gel strength, the high molecular weight of these pectins ensures a strong gel, but it should be recognized that the degree of methoxylation alone does not influence gel strength. Rapid set pectins are especially useful for maintaining suspended solids, fruit pulps, and homogenized fruit and vegetable tissues in a controlled form before they have an opportunity to partition from other essential constituents of a food system.

The physical chemistry involved in pectin gel formation is complex, but certain fundamental controlling factors are recognized. As indicated earlier, sugar, acid, and water are required for the formation of any pectin gel. For example, a typical jelly may have 65 to 70% sugar (soluble solids) and a pH value in the range of 2.9 to 3.2. Since the pK values of pectins range from 3.4 to 4.4 depending on the *DM* value, increases in food acid content tend to shift the ionization equilibrium of pectic carboxyl groups toward an unionized state. This condition shortens the length of individual pectin molecules by decreasing their mutual electrostatic interactions, pectin–water interactions are physically decreased, and a well-hydrated gel is no longer favored.

Gel viscosity lost by acid effects can be countered by the addition of sugar or similar hydrophilic solutes. These agents reduce pectin solvation, which favors the association of pectin molecules into a gel network by fostering hydrogen bonding between unionized carboxyl and hydroxyl groups.

The stabilizing forces for the gel are believed to originate mainly from hydrogen bonding forces between unionized carboxyl and hydroxyl groups on adjoining pectin molecules. Also, because of the importance of ionizable carboxyl groups in the gelling properties of pectin, it should be noted that fully esterified pectins exhibit viscosity properties that are largely pH independent, whereas the viscosity of low-ester pectins is pH dependent.

Slow-set pectins have *DM* values in the range of 50 to 70, and they establish a gel structure with aqueous sugar media in the optimum pH range of 2.8 to 3.1. In accord with the lower *DM* value, a gel is formed at a lower temperature than for rapid-set pectins. These pectins are suitable for many uses, but *DM* 60 slow-set pectins are useful for jellies.

Low-methoxyl pectins, which have ≤50% methylation, do not form anything other than a loose, flaccid, and irregular gel at best when added to nearly all sugar–acid–water media. Instead, gel formation requires the availability of divalent cations such as calcium, which effectively cross-links intra- and intermolecular carboxyl groups of pectin. Here gel strength depends on the degree of methoxylation but not on the molecular weight of the pectin molecules. Low-methoxyl pectins have many specialized food applications as gelling or thickening agents for jams, jellies, bakery glazings, topping preparations, variegated syrup production, yogurt, fruit preparations, confectionery jellies, and many other products. Low-methoxyl pectins are especially useful for implementing desired thickness and texture properties in foods where sugar must be eliminated in favor of flavor or dietary restrictions.

Pectic substances may be attacked by enzymes: Many bacteria, fungi, fruits, and vegetables have varying concentrations of enzymes that can deesterify or depolymerize pectins. *Pectinesterases* (PE's), also known as *pectases, pectin methoxylases, pectin demethoxylases, pectylhydrolases,* and still other names, progressively deesterify methyl (but not acetyl) groups from pectin molecules. The major reaction products are low-methoxyl pectins and pectic acids. Those enzymes that hydrolyze the $\alpha(1\rightarrow4)$ glycosidic bonds of pectin and pectic acids are called *polygalacturonases* (PG's). These enzymes have also been known by many other names including pectases, pectolases, pectin glycosidases, pectin depolymerases, and other terms.

Fully methylated pectins are not substrates for PG activity, but a combination of PE and PG enzymes can split all glycosidic bonds thereby producing free D-galacturonic acid. The $\alpha(1\rightarrow4)$ cleavage of pectins can also proceed by means of a β-elimination reaction mediated by *transeliminase enzymes*.

The PG and PE enzymes have respective optimum reaction pH's of ~3.5 to 4.0 and ~7.5. Depending on the fruit, vegetable, or food system undergoing processing, these enzymes may or may not attack the pectin components of foods. Given ideal reaction conditions, however, it is not unusual for products to lose their desired pectin gels. Some food industries actually permit pectin-digesting enzymes to react with food pectins before the enzymes are destroyed by retorting or other deactivation methods. When the desired destruction of pectin and modification of food texture has been achieved during the reaction period, often called a "cold break," pectin-digesting enzymes are then deactivated. Contrary to this approach, many fruits and vegetable products are subjected to heated deactivation of pectin-digesting enzymes in early processing steps, known as a "hot break," thereby ensuring the retention of a firm pectin-stabilized food structure. Hot breaks, if properly implemented, will contribute to the retention of a firm, pectin-containing food, but care must be taken to avoid excessively high or prolonged heat, which will destroy its gelling power.

Hemicelluloses and Pentosans

Hemicelluloses and pentosans accompany other constituents of plant cell wall structure including cellulose, pectic substances, some glycoproteins, and varying amounts of lignin depending on the plant species. Although the name *hemicellulose* implies that these compounds may be associated with cellulose molecules in some way, they are not synthetic precursors or derivatives of cellulose. The *pentosan* term stems from the fact that these compounds readily yield five carbon sugars on hydrolysis. Hemicelluloses and pentosans are both complex, nonstarchy heteropolysaccharides that often differ in their water solubility. Hemicelluloses are water insoluble but they do display solubility in aqueous alkaline solutions, whereas pentosans are usually water insoluble.

The structural diversity of the hemicelluloses is very great since they offer many variations in molecular branching and are composed of many different sugars and sugar derivatives. Hemicelluloses typically have molecular weights of ~600,000, which reflect 150 to 200 unit mixed polymers of hexoses, pentoses, D-galacturonic acid, L-arabinose, 4-O-methyl-D-glucuronic acid, and minor concentrations of L-rhamnose, L-fucose, and various O-methylated neutral sugars. The glucuronic acid residues in particular have an important influence on the solubility of these polysaccharides.

The xylans represent the most common group of these plant polysaccharides. They are polymers of ~75 $(1\rightarrow4)$-linked β-D-xylopyranosyl units, which have a $(1\rightarrow3)$ linkage to a polysaccharide branch on some nonspecific sugar residue in a chain. Notable xylan concentrations are found in grain hulls, corn cobs, plant stems, and annual plants at concentrations in the range of 20 to 35%; hardwood plants may contain 20 to 25%; and soft woods have only about 10%.

Aside from xylans, mannans represent another important hemicellulose that is found in plant seeds and vegetable ivory (i.e., tagua palm seed). The mannans are linearly linked $(1\rightarrow4)$-β-D-mannose polymers with little or no branching, which is quite unlike the branched mannans found in yeasts.

Clearly, the structure and presence of hemicelluloses is more than an inconsequential vestige of plant evolution, but the principal function of hemicelluloses remains to be fully discovered.

The water-soluble pentosans definitely affect the mouth feel of many fruit and vegetable tissues. Since they are often branched, they form viscous gels and impart mucilagenous properties to these vegetable tissues. Furthermore, the availability of hemicelluloses and pentosans in yeast-raised doughs are believed to contribute to the structure of the baked dough as opposed to water-insoluble pentosans, which impart poor baking properties to flour.

PLANT HYDROCOLLOIDS, GUMS, AND MICROBIAL HYDROCOLLOIDS

Gums have traditionally denoted a variety of compounds including proteins, terpenes, polysaccharides, and synthetic polymers. The contemporary connotation for gums refers to polysaccharides and their derivatives that *form viscous solutions or dispersions in cold or hot water*. These hydrophilic polysaccharides or hydrocolloids may be divided into two groups, (1) *natural gums* and (2) *modified gums*.

Natural gums include seaweed extracts, plant exudates and saps, seed and seed root gums, and gums obtained from microbial fermentations. Modified gums include any chemically altered form of a natural gum such as propylene glycol alginate, triethanolamine alginate, carboxymethyl locust bean gum, or carboxymethyl guar gum.

Gum classification can be enormously complex depending on the classification criteria. Gums may be classed according to their natural sources as indicated above; they can be classed according to the dominant sugar constituent in the gum or its characteristic functional group (e.g., sulfate ester); or the

ionic properties of the polysaccharide may serve as a basis for classification (Table 6.6).

Plant hydrocolloids are generally polyuronides containing more than one monosaccharide unit while a single monosaccharide serves as the structural unit for seaweed hydrocolloids. Table 6.6 offers a chemical, structural, and ionic breakdown of some important natural hydrocolloids.

The hydrocolloids share similar characteristics that are responsible for their ability to form viscous aqueous solutions and dispersions that possess suspending and stabilizing properties. Some hydrocolloids perform better than others, however, de-

Table 6.6. Chemical Classification of Natural Plant Hydrocolloids Used in Foods

Anionic Seaweed Polysaccharides

Agar Linear polygalactose sulfuric acid ester.
Alginates Linear polymer of mannuronic and guluronic acids.
Carrageenan Complex of a high anhydrogalactose to galactose ratio molecule containing sulfate esters, generally thought to be a combination of κ and λ fractions of carrageenan.

Anionic Exudate Polysaccharides

Arabic Complex of arabic acid, a highly branched polymer composed of galactose, rhamnose, arabinose, and glucuronic acid.
Ghatti Complex of ghatti acid, which consists of hexuronic acids and pentoses.
Karaya Complex polymer composed of galactose, rhamnose, and glucuronic acid, which may be partially acetylated.
Tragacanth A polysaccharide mixture, tragacanthin and bassorin polymers of fucose arabinose, xylose, and glucuronic acid.

Nonionic Seed Polysaccharides

Guar Straight-chain polymer of mannan residues having relatively consistent branching on every second mannose residue by a single galactose residue.
Locust bean Straight-chain polymer of mannan residues that display relatively consistent branching on every fourth mannose residue by single galactose molecules.
Tamarind Polysaccharide composed of galactose, xylose, and glucose.

NOTE: For further details consult text.

pending on the desired result. Some specific objectives include their ability to act as emulsifiers, adhesives, binders, flocculants, film formers, and friction reducers.

Occasionally the dispersing and suspending properties of hydrocolloids can be enhanced by *synergistic effects* arising from the mixture of two different hydrocolloids. In these cases the performance of the hydrocolloid mixture supersedes that of the individual hydrocolloids alone. For example, synergistic hydrocolloid mixtures may improve a gel or pudding to make it more firm or strong, modify its texture, and/or alter its gelling temperature. Furthermore, the stability of emulsions may be improved by making them heavier bodied and more inversion resistant; the heavy suspending properties of sols may be enhanced; food foams having lighter, fluffier properties plus added stability may be formulated; and frozen foods can be produced that offer greater structural stability in the face of freeze–thaw shock stresses and offer added resistance to food ingredient crystallization.

Some hydrocolloids are used more than others in food systems. These include tragacanth, arabic, karaya, locust bean, guar, and ghatti gums along with algal hydrocolloids such as carrageenan and other alginates. Aside from all of these so-called universal gums, gum arabic is the only gum that is readily soluble in water to 50% concentration levels. The notable hydrophilic properties of the remaining hydrocolloids cause them to swell in water and form colloidal solutions (sols).

Those hydrocolloids that display swelling at 2 to 5% concentration levels produce viscous jellies (gels), and the alginates in particular will form a rigid gel structure at 1% concentrations in aqueous media.

Because food uses for hydrocolloids revolve around their ability to maintain stable dispersions and viscous aqueous systems, viscosity and gel strength are commonly used parameters in grading these hydrocolloids. Tragacanth, karaya, locust bean, and guar gums swell in aqueous media to produce high-viscosity sols. These display nearly Newtonian behavior since their viscosities vary in a linear fashion as concentration levels of the gums traverses a 0.3 to 0.5% concentration range. Beyond these concentrations, the viscosity of the gums undergoes a geometric increase and they assume the typical behavior of non-Newtonian liquids. This non-Newtonian behavior is attributed to complex surface interactions and attractions at these higher gum concentrations (Figure 6.38).

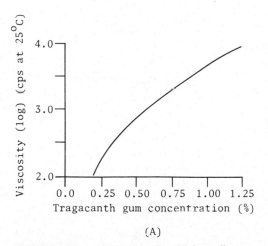

(A)

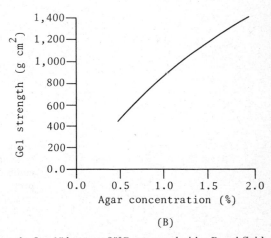

(B)

FIGURE 6.38. (A) Viscosity for a high-quality gum tragacanth after 15 hours at 25°C measured with a Brookfield viscometer (Model LVF) using a No. 3 spindle at 30 rpm. (B) Gel strength for a highly purified sample of agar at 25°C, measured according to the force required to penetrate the gel using a 1.0 cm² plunger.

Table 6.7. Some Typical Viscosities Measured in cps Units for High-Quality Hydrocolloids Held at Various Concentrations for 20 h at 25°C

Hydrocolloid	Viscosity	Hydrocolloid	Viscosity
Quince seed (1%)	50	Carrageenan (1%)	1400
Arabic (10%)	100	Alginate, sodium	
Stach paste (12%)	100	salt (1%)	1750
Chondrus crispus		Karaya (1%)	3000
(Irish moss) (1%)	200	Guar (1%)	3200
Tamarind (2%)	400	Locust bean	
Ghatti (5%)	650	(carob) (1%)	3200
Methylated		Tragacanth (1%)	3400
cellulose (1%)	1100		

NOTE: Values in parentheses indicate hydrocolloid concentration.

Table 6.7 outlines some of the comparative viscosities for high-grade hydrocolloids, and Table 6.8 summarizes many of the most prevalent uses for hydrocolloid in foods.

A more detailed prospectus of hydrocolloids and their properties follows. These hydrocolloids include seaweed hydrocolloids; exudate gums, seed gums, and extracted gums; and microbial hydrocolloids.

Seaweed Hydrocolloids

The *Rhodophyta,* or red algae, and *Phaeophyta,* or brown algae, are the major marine species responsible for the production of algal hydrocolloids used in foods. The Phaeophycean genera including *Ascophyllum, Laminaria,* and *Macrocystis* produce algin and alginate polysaccharides, and six genera of

Table 6.8. Some Hydrocolloids and Their Uses in Foods

Type of hydrocolloid	Applications	Important properties
Agar	Icings, glazes	Forms firm gels at 1% concentrations
Alginates	Desserts, ice cream, sherbet ices, beverage emulsions	Low-viscosity emulsifier, increasing amounts of calcium ions increase viscosity to produce short flow properties
Arabic	Bulking agent, beverage emulsion, protective colloid, crystallization inhibitor	One of several gums that require high concentration to produce increased viscosity, up to 50% concentration; forms coacervates with gelatin and other proteins

Table 6.8. (*Continued*)

Type of hydrocolloid	Applications	Important properties
Carrageenan	Contributes viscosity and gelling properties to aqueous food systems and demonstrates similar milk reactivity; used in ice creams, yogurts, toppings, syrups, cocoa particle suspension in chocolate milk	κ-Carrageenan forms rigid thermally reversible high-strength gels ι-Carrageenan-thixotropic dispersions; elastic thermally reversible gels with high salt tolerance λ-Carrageenan is a cold soluble, nongelling, anionic dispersant stabilizer
Furcelleran	Desserts, flans	Forms stable gels without refrigeration
Ghatti	Imparts stability to foods through its binding and emulsifying properties	Similar to gum arabic in many ways but has a tan color
Guar	Ice cream, sauces, processed cheeses; versatile thickener, viscosity modifier, free water binder, suspending agent and stabilizer	Rapid hydration at pH 8.0 and displays notable stability and buffering action over a pH range 4.0–10.5
Karaya	Ice cream and fruit ices	Demonstrates good adhesive properties and acid resistance
Locust bean gum	Ice cream, sauces, bakery products, processed cheeses, and ground or emulsified meat products where it offers improved stabilization and water binding	Viscosities of these aqueous gums increase upon heating to about 83°C
Tamarind	Confectioneries, jujubes, ice cream stabilizer	Wide pH range stability, may act as a fruit pectin substitute and endures high sugar concentrations
Tragacanth	Sauces, beverage emulsions, pour-dispensable dressings	Acid resistant, emulsifier

Rhodophycean algae produce the hydrocolloids such as agar or agarose (*Gracilaria, Gelidium*); carrageenan (*Chondrus, Eucheuma, Gigatina*); and furcellaran (*Furcellaria*).

Agar: The structural elucidation of agar in 1967 revealed that an "agarose" fraction exhibited superior gelling properties as compared with other "agaropectin" fractions. The neutral copolymeric structure of agarose consists of four (1→4)-linked 3,6-anhydro-α-L-galactopyranose units and three (1→3)-linked β-D-galactopyranose units. Variations on this structural theme are exhibited by agaropectin, but it has varying inclusions of 4,6-*O*-(1-carboxyethylidene)-D-galactopyranose as well as esters of sulfated galactans, methylated sugar residues, and minor amounts of pyruvic acid.

The concentration ratio of agarose to agaropectin is not fixed in seaweeds and is subject to wide variation. Molecular weights are similarly unpredictable, and depending on the analytical method employed, molecular weights may range from 5000 to 110,000.

Among all the hydrocolloids, agar has a notable ability to form detectable gels at 0.04% aqueous concentration levels, plus establishing a very strong gel at a well defined transition temperature. An agar sol held at 95°C will not undergo gelation until it is cooled to 36 to 40°C. This gel resists dissolution due to heat or hot water up to temperatures of 95°C. Although hard gels are formed at 1.0% agar concentrations in aqueous media, food applications often use 0.5 to 2.0% concentrations depending on requirements. Agar gel strengths are progressively increased by the admixture of sugars and peptones. The admixture of certain gums such as locust bean gum to agar sols prior to cooling also enhances agar gel resiliency to physical deformation and markedly improves the breaking strength of agar gels.

The food stabilization properties of mixed agarose–agaropectin gels are desirable in foods because they offer better stability than either of the polysaccharides used separately. The desirable food uses of agars also stem from their ability to form resilient gels during heating and cooling cycles, and although agar gels contract and exhibit syneresis with age, syneresis can be decreased by increasing agar concentrations.

Algin: Algin is a generic term that refers to the salts of alginic acids. Algin is found in all members of the *Phaeophycean* algae (brown seaweeds) where it acts as a structural component of cell walls. Here it is not free but exists as an insoluble salt of alginic acid involving calcium, magnesium, sodium, or potassium cations. The alginic acid component consists of D-mannuronic and L-guluronic acids, but it has three distinct types of polymeric segments that contain (1) D-mannuronic acid, (2) L-guluronic acid, and (3) a mixture of alternating D-mannuronic and L-guluronic acid residues. The occurrence of these polymeric segments in some commercially important alginic acid sources have been outlined in addition to the mannuronic acid and guluronic acid composition of alginic acid in Table 6.9.

The similarities in polymannuronan and polyguluronan structures for *Macrocystis pyrifera* and *Ascophyllum nodosum,* as opposed to the structure of *Laminaria hyperborea*, account for the functional differences in the behavior of "brown seaweed" alginates.

The edible gels formed by alginates require the presence of calcium ions. Calcium may be supplied as calcium carbonate, phosphate, tartrate, chloride, or some other salt form. High availability of calcium ions in foods promotes rapid formation and precipitation of calcium alginate. Low levels of calcium ions or calcium ions supplied from dicalcium phosphate, which liberates calcium only at 94 to 106°C, retards the rate of algin gel formation. Both methods are used for supplying calcium ions to food–alginate systems depending on the desired effect (i.e., fast versus slow gel formation). Since total substitution of sodium alginate by calcium has been stoichiometrically pegged at ~7.2% of the sodium alginate weight, it may be necessary to adjust concentrations of calcium

Table 6.9. Mannuronic and Guluronic Acid Content of Alginic Acids (A) and Polymeric Proportions of These Acids and Their Alternating Segments in Alginic Acids (B)

A. Species source	Mannuronic acid (% by weight) (M)	Guluronic acid (% by weight) (G)	M:G ratio	Possible M:G ratio range
Ascophyllum nodosum	65	35	1.85	1.45–1.95
Ecklonia cava and				
Eissenia bicyclis	62	38	1.60	—
Laminaria digitata	59	41	1.45	1.45–1.60
Laminaria hyperborea (stipes)	31	69	0.45	0.45–1.00
Macrocystis pyrifera	61	39	1.56	—

B. Species source	Polymannuronan	Polyguluronan	Alternating
Ascophyllum nodosum	38.4	20.7	41.0
Laminaria hyperborea	12.7	60.5	26.8
Macrocystis pyrifera	41.0	17.0	42.0

depending on food formulation requirements. The moderation of excessive calcium levels to achieve a desired level and gel structure can be implemented through the use of *sequesterants*. These cation binding agents chelate unwanted divalent calcium and thereby produce the desired calcium balance in the algin system. Typical sequesterants include sodium polyphosphates such as sodium tripolyphosphate or sodium hexametaphosphate.

Except for very high temperature conditions, alginate viscosity decreases with increasing temperature, and these viscosity decreases are reversible on cooling. The pH affects on the viscosity of alginates is negligible over the range of pH 5.0 to 10.0, but pH values below 4.5 increase the viscosity. At pH 3.0, increasing viscosity gives way to alginic acid precipitation. This is in contrast to strongly alkaline solutions that produce sodium alginate gels. Ionic strengths associated with algin gels also influence their behavior. As in the cases of many polymeric solutions, ionic strength increases are followed by contraction of the polymer.

Carrageenan: Complex mixtures of galactans having different proportions of half-ester sulfate groups, linked to one or more hydroxyl groups of galactose units, defines the general structural features of these hydrocolloids.

Carrageenans have high molecular weights and linear macromolecular structures, and they display alternating $\alpha(1\rightarrow3)$ and $\beta(1\rightarrow4)$ linkages between galactose residues.

It has been widely recognized that two heterogeneous groups of carrageenans can be distinguished on the basis of their solubility in potassium chloride solutions. The most soluble fraction is nongelling and is referred to as λ-carrageenan. The other fraction that undergoes gelling upon the addition of potassium chloride is called κ-carrageenan.

Each of these fractions has a heterogeneous composition of carrageenan structures, and recognition of these molecular forms serves as a basis for conventional nomenclature of carrageenans as opposed to simple solubility differences.

At least seven different types of carrageenans are recognized; they are designated as μ (mu), κ (kappa), ν (nu), ι (iota), λ (lambda), θ (theta), and ξ (xi), but only the κ-, λ-, and ι-carrageenans hold commercial importance.

The composition of carrageenans depends on the type of seaweed from which the hydrocolloid has been extracted as well as its geographic growth location. The galactan structures of carrageenans follow the galactose residue linkage pattern outlined below.

to their gelation properties, even in the cases where κ- and ι-carrageenans, respectively, form gels with potassium and calcium ions

A	B	A	B	A	B
---Gal	β(1→4) Gal	α(1→3) Gal	β(1→4) Gal	α(1→3) Gal	β(1→4) Gal (1→ ---

Galactose residue		Type of carrageenans
A Gal	B Gal	
D-Galactose-2-sulfate	D-Galactose-2-sulfate	ξ
	D-Galactose-2,6-disulfate	λ
	3,6-Anhydro-D-galactose-2-sulfate	θ
	3,6-Anhydro-D-galactose-2-sulfate	ι
D-Galactose-4-sulfate	3,6-Anhydro-D-galactose	κ
	D-Galactose-2,6-disulfate	ν
	D-Galactose-6-sulfate	μ

The κ-carrageenan structure contains D-galactose-4-sulfate, which has 1→3 intermolecular linkages, and this galactose residue is involved in a 1→4 linkage with 3,6-anhydro-D-galactose. The biochemical predecessor of κ-carrageenans is the μ-form, which differs from κ-carrageenan since it participates in a 1→4 linkage with D-galactose-6-sulfate. Transformation of μ- to κ-carrageenan is mediated in seaweeds by the enzyme known as *dekinkase*. In a similar vein, ν-carrageenan is thought to be the biochemical predecessor of ι forms. The ν- and ι-carrageenans differ from their respective μ and κ-counterparts since they have an additional sulfate group situated at the C-2 position of the 1→4-linked unit.

In the case of λ-carrageenans, 70% of the 1→3-linked galactose residues are sulfated at the C-2 position and the 1→4-linked residue is D-galactose-2,6-disulfate. When subjected to alkaline conditions, the 6-sulfate moiety of λ-carrageenan can be eliminated to form 3,6-anhydrogalactose-2-sulfate. This alkali-treated λ-carrageenan is called θ-carrageenan. Figure 6.39 and 6.40 illustrate the structural relationships that exist among carrageenans.

The molecular conformation and shape of macromolecular carrageenans is related to their gelation properties, even in the cases where κ- and ι-carrageenans, respectively, form gels with potassium and calcium ions (λ-carrageenans do not form gels). The contribution of these ions to carrageenan gels has been conceptualized as an electrostatic linking mechanism between alternating sulfate residues located on aligned galacturonan polymers. It has been speculated that monovalent ions such as sodium may not be as effective as potassium and ammonium because of its relatively large ionic diameter. Characteristic sol–gel transformations of ι- and κ-carrageenans encourage the formation of double helices in aqueous media, contrary to λ-carrageenan, which prohibits intermolecular gelling phenomena.

The thickening and gelling properties of natural heterogeneous carrageenans are predictable within certain bounds, but the natural heterogeneity of carrageenans imparts different physical properties to carrageenan solutions or gels than the ultrapure molecular types. With respect to viscosity of aqueous carrageenan systems, observed viscosity is a function of hydrocolloid concentration, temperature, predominant type of carrageenan, and the type of metallic ions associated with sulfate esters (i.e., Ca^{2+}, K^+, or Na^+). From general perspectives, viscosities for aqueous carrageenan systems will decrease by 10% for every 10°F rise in temperature. Water and milk systems containing

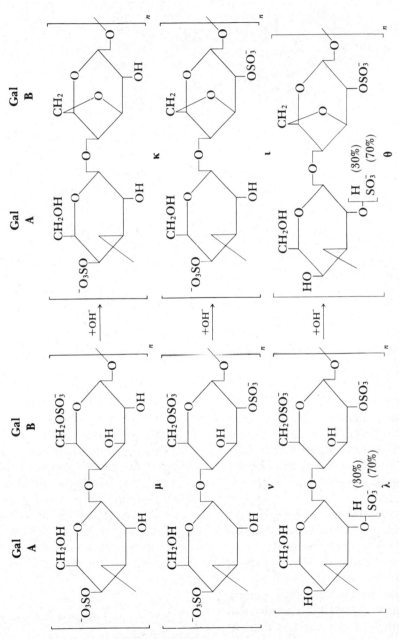

FIGURE 6.39. Haworth structures for the various forms of carrageenans.

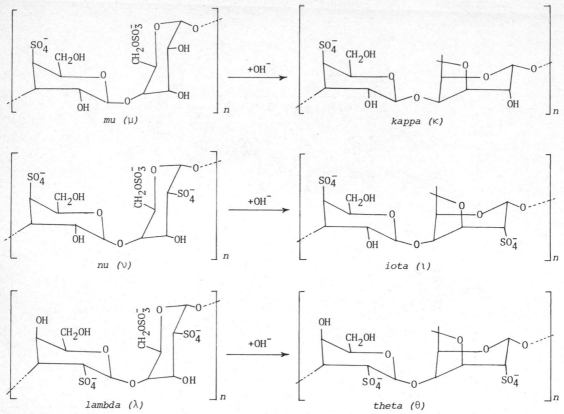

FIGURE 6.40. Conformational structures for carrageenans that correspond to the Haworth structures detailed in Figure 6.39.

carrageenan display similar rheological properties, but the rheology of milk systems may be slightly different owing to the interactions of the hydrocolloids with milk proteins. That is, water solutions are slightly thixotropic, but milk solutions can be extremely thixotropic.

Carrageenans are usually stable throughout the range of pH 5.0 to 10.0. Colloidal properties of these carrageenans are often retained even after heating if pH is held above 6.0. Values less than pH 4.0 cause an expressed decrease in viscosity and gelling properties. At a pH of ≥10.0, viscosity will also decrease but at a rate far less than in the acidic range.

The application range of carrageenans to aqueous milk- and ice cream-based food systems is nearly limitless. Carrageenan is used as a stabilizer in ice creams (0.01–0.03%), sherbets (0.01–0.03%), chocolate milks (0.03–0.04%), and whipped creams (0.05–0.15%). It is also used in a variety of syrups (0.3–0.5%) and toppings (0.1–0.3%). In the processing of beer, carrageenans may be added to boiling wort in order to produce a "hot break"; furthermore, a "cold break" occurs that will precipitate residual proteins responsible for chill haze. The body and mouth feel of many creamed soups and chowders are improved by the addition of carrageenan.

Exudate Gums, Seed Gums, and Extracted Gums

Many land plants naturally produce polysaccharide gums that are generically classified as exudate gums, seed gums, or extracted gums. The properties of these gums are widely different among members of the same plant species depending on the harvest time, growth climate, preliminary extraction, and gum purification methods. The overlying differences among gums obtained from different species accounts for their characteristic viscosities as well as their food uses.

Exudate gums: Many species of shrubs or low-growing trees, especially native to the Middle East and North Africa, offer a ready source of gum exudates. These exuded gums often display species characteristic globular (gum arabic), flake, or thread-like ribbon (gum tragacanth) shapes. The quality and aging of gums may be associated with darkened colors, off-odors, and peculiar tastes, especially in those cases in which residual tannins exist in gums. Of all known botanical exudates—and there are many—ghatti, guar, karaya, and tragacanth have the greatest food uses.

Gum arabic. Dried exudates from *Acacia senegal* and *A. seyal* trees, native to the Sudan and the Senegal regions of Africa, provide the major world resources of this gum. The *A. senegal* gums are the most desirable for food applications. Bark stripping, sap exudation over the wounded surfaces of the tree, followed by bacterial growth over the exudate-covered surfaces are typical of the production scenario for gum arabic from cultivated *Acacia* trees. Conditions for gum arabic production are optimal under conditions of severe moisture deficits, hot temperatures, and poor plant nutrition.

A molecular prospectus of gum arabic polysaccharide indicates that it is a slightly acidic or neutral salt of a complex polysaccharide containing calcium, magnesium, and potassium. The arabic acid of the gum polysaccharide contains L-arabinose, L-rhamnose, D-glucuronic acid, and D-galactose. The coiled molecular structure is studded with numerous side chains, and the whole molecular weight is 240,000 to 300,000.

Gum arabic is one of the most widely used plant hydrocolloids because it is compatible with proteins, starches, carbohydrates, and other hydrocolloids with the exception of gum tragacanth. Mixtures of these two gums results in lower viscosities than that displayed by either one of the gums at the same concentration level (80:20 mixtures of tragacanth gums:arabic gums have minimal viscosity). Pure arabic gum is soluble in aqueous systems up to a 50% concentration, and no significant viscosity is observed at ≤30% gum concentrations at maximum viscosity pH values in the vicinity of pH 7.0. Viscosities attributable to gum arabic are stable over a wide pH range, but viscosity is lowered by electrolytes and increased temperatures.

Food uses of gum arabic exploit its action as a protective colloid and stabilizer as well as the adhesiveness of its water solutions to improve the body, viscosity, and texture of foods. Confectionery uses are largely centered around the ability of the gum to thicken candies, jellies, glazes, and chewing gums plus inhibit sugar crystallization. The emulsification properties of the gum are used in liquid flavor emulsions, citrus-fruit-flavoring emulsions, and solid confections containing fat.

As a result of low cost, high solubility, and superior performance over other compounds, gum arabic acts as a fixative for spray-dried flavors by forming superficial yet impenetrable films around flavor particles, thereby protecting them from evaporation, oxidation, and efflorescent phenomena.

Gum karaya (or sterculia gum). The dried exudate of the *Sterculia urens* tree cultivated in India provides the natural source of this gum. The gum exuded after tapping the trees during April to June contains a partially acetylated, high-molecular-weight heteropolysaccharide composed of L-rhamnose, D-galacturonic acid, and D-galactose. Along with a hexuronic acid content of about 36%, as well as calcium and magnesium cations, the karaya polysaccharide has labile *O*-acetyl groups that liberate free acetic acid in the presence of atmospheric moisture and some trimethylamine when hydrolyzed. This gum is among the least soluble exudates. It does not dissolve in water to form a true solution, but rather it swells in cold water at 4% concentration levels to give viscous sols although its solubility may be elevated to 25% by heating. The notable insolubility of the gum has been attributed to the presence of the *O*-acetyl groups.

The viscosity of gum karaya decreases with increasing ionic strength and temperatures. Gum sols are also sensitive to alkali and offer their maximum viscosity at pH 8.5, above which the sols become *stringy*.

Karaya gum finds many food uses as a thickening and suspending agent, and as a stabilizer for salad dressings, natural and artificial whipped creams, sherbets, ice cream, and other frozen desserts (0.2–0.4% concentrations), in which it minimizes ice crystal formation. The gum also acts as an emulsifier and processing aid for cheese spreads at 0.5 to 0.8% concentration levels and imparts both smooth cohesive properties and binding action to sausage emulsions at 0.25 to 1.0% concentrations. The gum can also be used in conjunction with alginates or carrageenans to retard staling of bread, doughnuts, and other bakery goods and improve the tolerance of doughs to overmixing. The typical karaya gum/alginate or carrageenan concentration ratio is 0.1–0.9%:0.02–0.1%.

Gum tragacanth. Dry mountainous regions of Iran, Syria, and Turkey offer several tree species, of the genus *Astralagus*, which are among the best commercial sources of this gum exudate. The most desirable commercial grades of the gum have been produced in Iran.

Gum extrusions from incisions on the plant are permitted to dry and are then collected as high-quality ribbons (*Maftuli*) or as flakes (*Kharmoni*).

Tragacanth gums contain a soluble *tragacanthin* fraction and an insoluble *bassorin* fraction that accounts for 60 to 70% of the total gum. As a whole, gum tragacanth is composed of a neutral arabinogalactan, which is structurally associated with complex heteroglycans of D-galacturonic acid, D-xylose, L-fucose, and D-galactose. The polymethoxylated state of bassorin gum is believed to produce tragacanthin after its demethoxylation. The tragacanthin yields a colloidal hydrosol with water while the bassorin fractions simply swell into a gel. The cellulose, starch, and protein substances that accompany commercial tragacanth gums are not intrinsically related to their molecular structure.

Although tragacanth gum has been widely used in foods for its thickening, water-binding, water-suspending, and emulsification properties, it has been largely replaced by xanthan gum and propylene glycol alginate derivatives. The viscosities of tragacanth gums are probably the highest of all the natural colloids. The gum is compatible with other hydrocolloids, but it does not display any relationship between pH and gelling over the pH range of 1.9 to 8.5.

The long shelf life and resistance to hydrolysis exhibited by this gum have encouraged its use in salad dressings, fruit fillings, tomato and barbecue sauces, and citrus beverages, in which it improves mouth feel. The gum also improves fruit/cream-filling textures, transparency, sheen, and flavor distribution.

Viscosities of tragacanth sols are diminished by the presence of di- or trivalent cations, although trivalent anions are most effective in controlling gelling. Gum tragacanth sols do not display thixotropy.

Gum ghatti. Exudates from *Anogeissus latifolia,* a tree native to India and Sri Lanka, provide the natural source of this gum. Gum exuded from artificial incisions on the tree is dried and ground into a fine powder.

This particular gum is a complex calcium, magnesium polysaccharide salt composed of L-arabinose, D-galactose, D-mannose, D-xylose, D-glucuronic acid, and a 6-deoxyhexose in a molar ratio of 10:6:2:1:2:trace (<1%). The molecular weight is approximately 12,000.

Partial hydrolysis of ghatti gum produces two aldobiuronic acids, namely, 6-*O*-(β-D-glucopyranosyluronic acid)-D-galactose and 2-*O*-(β-D-glucopyranosyluronic acid)-D-mannose. These are accompanied by 50% pentose and 12% galactose or galacturonic acid. The major backbone of the native gum exhibits 1→6-linked β-D-galactopyranosyl units with acid-labile side chains linked to the main structure by L-arabinofuranose residues.

When dispersed in water, gum ghatti forms viscous solutions intermediate between arabic and karaya gums; however, the emulsification and dispersion properties of ghatti are equal or superior to those of gum arabic.

Ghatti gum has wide hydrocolloid compatability, but its viscosity is pH dependent. Maximum viscosity occurs at pH 8.0 and markedly diminishes on either side of this optimum in the range of pH 2.0 to 12.0. Organic acids and minerals also diminish sol viscosity.

Ghatti gum can be formulated into many foods for many different reasons analogous to those outlined for gum arabic, but the tan color of its solutions must always be considered before using it in food systems.

Seed gums and extracted gums: The exudate gums from trees and bushes are different from natural seed gums mainly as a result of their diverse monosaccharidic constituents and acidity. Although seeds contain starch as their main energy reserve, their polysaccharide gums have been used for many years in the food industry. In fact, guar gum has been consumed by humans more than all the other gums combined.

Guar gum. The pod-bearing nitrogen-fixing legume known as *Cyamopsis tetragonolobus* produces seeds containing guar gum. The gum itself resides in the endosperm of the seed, and it is necessary to isolate and pulverize this endosperm before its carbohydrate portion can be used as guar gum. Commercial gums contain ~2.5% crude fiber, 10 to 15% moisture, 5 to 6% protein, and 0.5 to 0.8% ash.

Unlike more complex exudate gums, guar gum is a straight D-galacto-D-mannoglycan chain (Figure 6.41A) studded with many individual galactose branches. Units of D-mannopyranose are joined by β(1→4) linkages and single D-galactopyranose units are bonded by α(1→6) linkages. Contrary to locust bean gum, which has single side branches of galactose on every fourth mannose unit, guar has branching on every second mannose unit. The greater branching of guar gums over locust bean gums accounts for its superior cold-water hydration plus its superior hydrogen-bonding activity.

Due to the absence of many polyuronic acid residues, the pH of guar sols is ~5.0 to 7.0 for a 1.0% solution. Guar sols are stable over a pH range of 4.0 to 5.0, but the fastest hydration rate occurs at pH 8.0. Rates of hydration can be greatly affected, too, by the availability of salts, water-miscible solvents, or low-molecular-weight sugars.

Guar gum is compatible with nearly all natural gums, starches, water-soluble proteins, pectins, celluloses, and their derivatives. Furthermore, guar gum exerts synergistic viscosity increases when admixed to xanthan gum.

The food applications for this gum are far more diverse than all other gums. The gum serves as a thickener, a viscosity modifier, a free-water binder, a suspending agent, and a stabilizer. Combinations of guar gum and carrageenan are especially effective as suspending agents for cocoa beverages. The use of the pure gum also retards the accumulation of moisture by icings and baked goods packaged in transparent wrappings; and it is useful as a bodying agent for sugarless dietetic foods and beverages.

Locust bean gum. Seeds of the leguminous evergreen plant *Ceratonia siliqua,* also known as the carob tree, can be milled to produce locust bean gum. The gum is present in the endosperm of the seeds, which are contained in a pod. Eight to ten seeds reside in the 20 to 25-cm-long pods. The pod or "kibble" alone has little value as a gum source

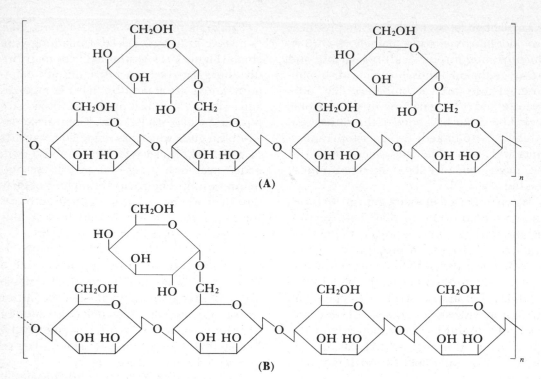

FIGURE 6.41. Repeating structural subunits for (A) guar gum and (B) locust bean gum.

but it is rich in sugar content and may be used as a flavor in many foods. The locust bean is cultivated throughout the Mediterranean countries, but the tree grows most extensively in Spain.

Compared to other gums, the molecular architecture of locust bean gum is relatively simple. It is a galactomannan composed of a main mannose chain (Figure 6.41B) with single-unit galactose residues on almost every fourth mannose residue. The molecular weight is ~310,000. The gum does not contain any uronic acids and little or no pentose.

Locust bean gum is not readily soluble in cold water. Instead, stable solutions having high viscosities attributable to this gum at low concentrations must be heated to 82°C and then cooled. The gum is compatible with other plant hydrocolloids, carbohydrates, and proteins. Increasing ionic strengths may affect sol viscosity, and high electrolyte concentrations, tannic acids, and polyvalent ions may precipitate the gum. By itself, locust bean gum has very little or no gelling ability in aqueous media, but it can enhance the elasticity and gel-breaking strengths of other gels such as those formed by agar, carrageenan, and xanthan gums. Contrary to the behavior of many pure gels, gel mixtures including locust bean gum show reduced tendencies to undergo syneresis.

Locust bean gum has been widely used as a stabilization aid for ice creams where smooth melt-down is desirable. For bakery goods such as biscuits, cakes, and breads, the water-binding properties are used to ensure soft consistent textures, especially when gluten contents of bread doughs are subject to variation.

The gum obtained from locust beans also improves the extrusion properties of meat emulsions as a result of its water-binding and stabilization effects. It also acts as an emulsifier and stabilizer for dressings and sauces, and the gum retards the shrinkage and fracture of frozen pie fillings. In the most general of terms, locust bean gum has properties similar to guar gums.

Other extractable gums. In addition to guar and locust bean gums, gums from tamarind, psyllium, and flax seeds have been used for many specialized food applications, but due to expense and unpredictable sources, these gums have been largely replaced in foods by other hydrocolloids.

Tamarind seed polysaccharides are characterized by their gel formation over a wide pH range while in the presence of high sugar concentrations (>65%, w/w), and their gel strength is nearly twice that of fruit pectins on an equal weight basis when sucrose concentrations are 65 to 72%.

Psyllium seed gum is obtained from plants of the genus *Plantago;* its most current production stems from Indian grown *Plantago ovata.* Boiling an aqueous extraction of the seed coat yields a mixture of neutral and acidic polysaccharides that are species specific. The gum undergoes slow hydration at 1% concentration levels to produce viscous solutions and gelatinous masses at higher concentrations.

Quince seed gum is extracted from the fruit seeds of the quince trees *Cydonia vulgaris* or *C. oblonga.* The gum is soluble in cold water, but hot water gives highly viscous solutions at 1.5% gum concentration levels. Higher concentrations of the gum produce viscous systems that defy dispersal.

Larch gum. Aqueous extraction of the western larch tree, *Larix occidentalis,* yields 5 to 35% larch gum from heart wood based on a dry weight basis. The gum is a complex arabinogalactan with a large amount of branching. The arabinose and galactose occur in a ratio of 1:6. Two fractions of the arabinogalactan are recognized, one having a molecular weight of 16,000 and the other one, a weight of 100,000. Larch gum produces an amber-colored solution in aqueous media along with a pH 4.0 to 4.5. The gum also has interesting viscosity properties since solutions of >40% solids can be readily prepared, but the viscosity of these solutions is far lower than that of other gums. Furthermore, these larch gum solutions retain Newtonian flow properties.

One of the most useful properties of larch gum stems from its ability to diminish surface tension of aqueous solutions as well as interfacial tension of water–oil mixtures. These desirable surfactant properties, coupled with pH stability and electrolyte tolerance, have prompted many food uses for this gum, especially in areas of flavor technology. Here, the bodying properties of the gum plus its compatibility for binding essential oils, flavor bases, and nonnutritive sweeteners, have encouraged its use in dressings, pudding mixes, and other foods for which minimum ingredient bulk is required.

Microbial Hydrocolloids

The fermentative growths of certain microorganisms produce high-molecular-weight polysaccharides as a consequence of their growth. These polysaccharides are detectable as exocellular slimes generated during microbial action on selected growth media and many of these compounds act as *in vivo* cell wall coatings for microorganisms. The potential uses for these exocellular microbial polysaccharides in foods is nearly limitless in view of the wide variety of natural and genetically manipulated polysaccharides, yet relatively few of these substances have been used in foods. The main exceptions to the rule, however, include the dextrans and xanthan gum.

Dextrans are polysaccharides composed of α-D-glucose units. Most contemporary dextrans are produced by a specific strain of *Leuconostoc mesenteroides,* which is grown in a modified sucrose growth medium. The hydrocolloid produced under these conditions consists of 95% $\alpha(1\rightarrow6)$-linked D-glucose residues while the remaining D-glucose residues are bonded by $\alpha(1\rightarrow3)$ linkages, thereby producing a branched polymer. Exclusive of this form, other dextrans may display $(1\rightarrow4)$ and $(1\rightarrow2)$ glycosidic bonds at branching sites.

The desirable properties of these hydrocolloids can be controlled to some extent so as to yield polysaccharides over the range of 50,000 to 100 million daltons, although

100,000-dalton varieties are preferred for food uses.

Xanthan gum is also an exocellular bio-polysaccharide, but it is synthesized by selected strains of *Xanthomonas campestris*. The growth of this organism is well-aerated nutrient media containing glucose, trace elements, an appropriate nitrogen source, di-potassium phosphate, and other compounds results in a constant production of the hy-drocolloid situated in the cell wall coatings of the microorganisms. Molecular weight of the polysaccharide may range from 2 million to 50 million, but the smaller weight is prob-ably the more accurate value owing to the presence of extraneous polymer associations at high weights.

The polysaccharide (Figure 6.42) may ex-ist as the potassium, sodium, and calcium salt of D-glucuronic, D-mannose-branched struc-tures where the fundamental repeating structure has a total of five sugars—that is, two glucose units, two mannose units, and one glucuronic acid.

The principal backbone of the gum is built from β-D-glucose units glycosidically linked by a β(1→4) linkage reminiscent of cellulose. The glucuronic acid unit and the two man-nose residues are part of the side chain. Fur-thermore, the terminal β-D-mannose unit is glycosidically bonded to the C-4 position of β-D-glucuronic acid, which in turn is glyco-sidically bonded to the C-2 position of α-D-mannose. This noteworthy side chain is present at the C-3 position on every other glucose residue in the main chain. The struc-ture is further complicated by the presence of ketalic-linked pyruvic acid residues at the C-4 and C-6 positions on terminal D-man-nose residues. The actual quantitative distribution of these pyruvate groups is not known. Nonterminal D-mannose units in side chains display C-6 acetyl groups.

The two important features of xanthan gum structure involve its branching and the β(1→4) linkages present in the main glucan chain. This characteristic branching is thought to minimize the susceptibility of the gum to enzymatic attack and the β(1→4) linkages are believed to account for the polymer's ri-gidity. Also, as opposed to the behaviors of most other anionic hydrocolloids, salt addi-

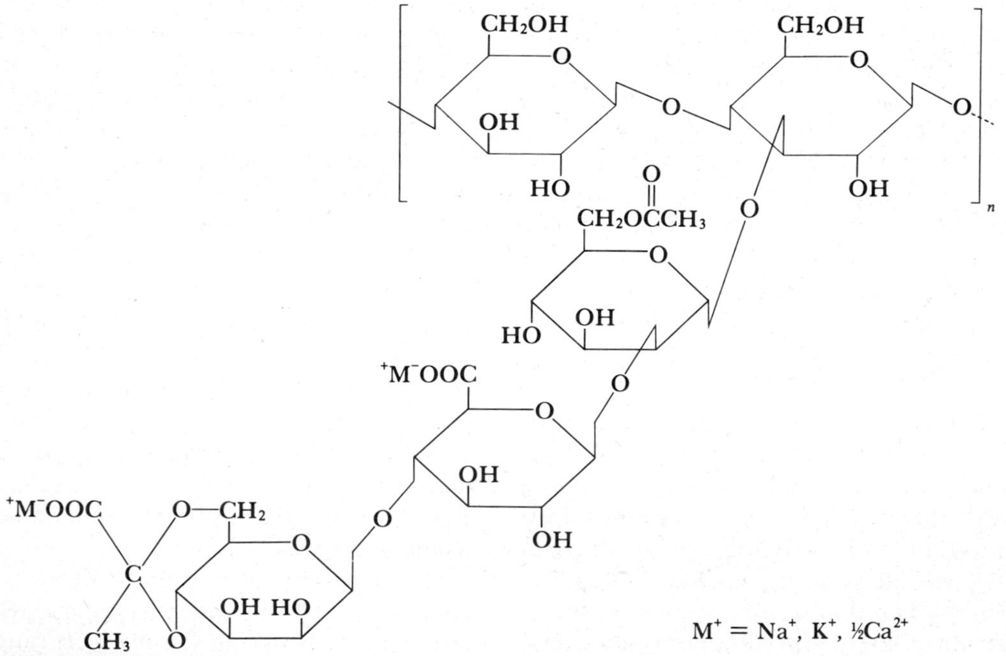

$M^+ = Na^+, K^+, \frac{1}{2}Ca^{2+}$

FIGURE 6.42. Average repeating unit for xanthan gum structure.

tion to xanthan solutions results in an increased viscosity at levels >0.15% gum concentrations. The application of heat to xanthan gum solutions causes its initial viscosity to decrease, but it soon increases markedly due to the unwinding of ordered conformational helices followed by their subsequent formation of random coils. These conformational changes increase the hydrodynamic volume for any gum–water system and consequently increase its viscosity. The observed optical rotation for xanthan gums usually decreases as unwinding of the random coil proceeds.

Xanthan gum is compatible with locust bean gum, with which it augments gel formation, but not with guar gum and other conventional hydrocolloids. The favorable interaction of locust bean and xanthan gums results from strong noncovalent interactions between helical regions on the xanthan gum with unbranched portions of the locust bean gum. Other hydrocolloids do not encourage these gel-supporting interactions.

"Old" and "New" Generation Hydrocolloids

The functional diversity of natural gums and hydrocolloids has been successfully demonstrated for many years. There is no doubt, however, that conventional uses for these substances will expand to include new food applications. This is especially true for new hydrocolloid derivatives as well as synthetic polysaccharides such as *polydextrose*.

Thermal polymerization of D-glucose under acid conditions yields randomly crosslinked glucose polymers with nearly every type of glycosidic bonding. Although 1→3, 1→4, and 1→6 bonds do exist, the 1→6 variety often predominates. The structure of the polymer may also contain some sorbitol end groups as well as monoester bonds involving citric acid. This unique structure of the compound accounts for many of its special properties.

Unlike most other carbohydrates, the structure of polydextrose is stubborn toward intestinal and microbial digestion processes, and only about 25% of the carbohydrate can be utilized. This is in marked contrast to other carbohydrates, which can provide up to 4 kcal/g or 100% of their potential dietary carbohydrate content. Moreover, polydextrose lacks sweetness, its high water solubility (≤80% in water) lends useful bodying properties to the polymer, it resists crystallization phenomena, and it may be used to effectively reduce food fats ordinarily required to complement sugar, flour, and starch components of foods. Therefore, the potential uses for this hydrocolloid in calorie-restricted dietetic foods, diabetic diets, and low-fat foods as well as other new variations on the theme of hydrocolloid chemistry have yet to be fully exploited. For a useful summary regarding polydextrose properties and their food applications plus other polysaccharides consult Torres and Thomas (1981) and Sanderson (1981).

Preservation of Hydrocolloids

Nearly all of the hydrocolloids discussed in this section are subject to some bacterial attack that may ruin their desired thickening, gelling, stabilizing, or emulsification properties in foods. Almost all hydrocolloids can be preserved by methyl- and propyl-*o*- or propyl-*p*-hydroxybenzoates. These preservatives are respectively used at concentration levels of 0.17 and 0.03%. Other conventional preservatives may include 0.1% sodium benzoate (arabic, ghatti, karaya, locust bean gums), 0.1% sodium pentachlorophenate (guar and locust bean gums), sorbic acid, glycerin, or propylene glycol, depending on the circumstances and the food product.

MISCELLANEOUS STRUCTURAL AND CONJUGATED POLYSACCHARIDES

In contrast with cellulose and other carbohydrates that are directly or indirectly responsible for plant cell wall construction, the exoskeletons of many invertebrates and the connective tissues of animals all display a variety of interesting polysaccharides.

Chitin, a homoglycan of N-acetyl-D-glucosamine, is closely related to cellulose since it offers a firm architecture to arthropod and crustacea exoskeletons as well as lower forms of plant life, especially the fungi. The total biomass of linear β-linked N-acetyl-D-glucosamine residues is second only to cellulose productivity. Note, too, that the molecular structure of the substance differs from cellulose only in that the C-2 hydroxyl function has been replaced by an acetamido function.

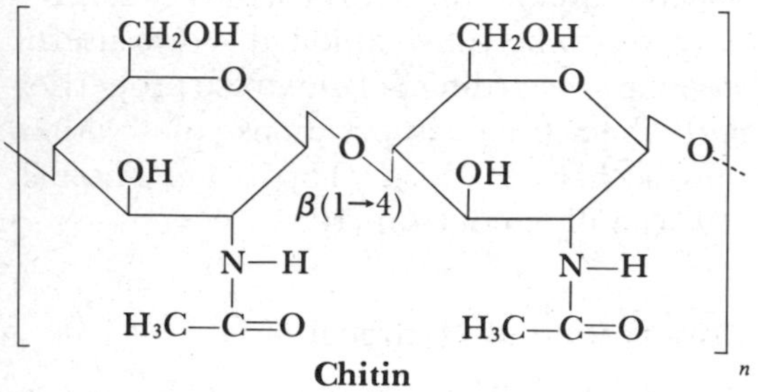

Chitin

Acid Mucopolysaccharides

Structural polysaccharides such as the acid mucopolysaccharides are found throughout all levels of human intercellular and tissue organization.

Hyaluronic acid, one of the most common heteroglycans, is composed of alternating D-glucuronic acid and N-acetyl-D-glucosamine units. The fundamental repeating disaccharide structure of this heteroglycan displays a β(1→3) glycosidic linkage while

the disaccharide subunits are linked through β(1→4) positions.

Hyaluronic acids are noted for water-binding, jelly-like properties that favor their intercellular cementing properties. These acids also impart lubricating properties to synovial fluids and contribute to the impact-absorbing properties of articulated joints. The highly viscous properties of the acid mucopolysaccharides are attributed to interchain hydrogen bonding and a high degree of hydration, which is fostered by the anionic character of these molecules at physiological pH values. The hyaluronic acid polymer is subject to enzymatic attack by *hyaluronidase*, which catalyzes the hydrolysis of β(1→4) linkages. The action of this enzyme can eventually cause a decrease in the native viscosity of the hyaluronic acid.

The hyaluronic heteroglycan is also found in the vitreous humor of the eye plus those structures intimately associated with reproductive processes such as the umbilical cord, ova, and so on.

In addition to hyaluronic acid, chondroitin and dermatan sulfates offer further examples of acid mucopolysaccharides that have intercellular cementing actions in animal tissues. These substances occur widely in connective tissues, and although the two compounds are structurally similar, chondroitin sulfate does have two possible forms denoted as A or C.

The chondroitin sulfates consist of equimolar amounts of D-glucuronic acid, N-acetyl-D-glucosamine and sulfate—with the position of the sulfate ester being the most discriminative feature. Dermatan sulfate,

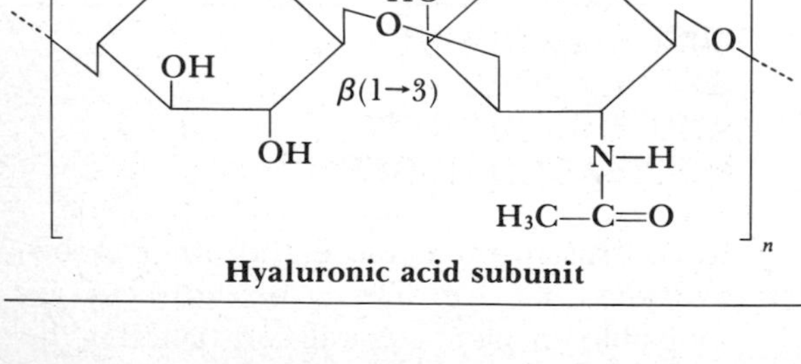

Hyaluronic acid subunit

however, has L-iduronic acid in place of D-glucuronic acids that are present in type A and C chondroitin sulfates. The individual repeating subunits are linked by $\beta(1{\rightarrow}4)$ glycosidic bonds.

mucopolysaccharide has distinctive anticoagulant properties with respect to whole blood, and it is secreted into the circulatory system by the mast cells of the lungs, liver, and other tissues.

Chondroitin sulfates

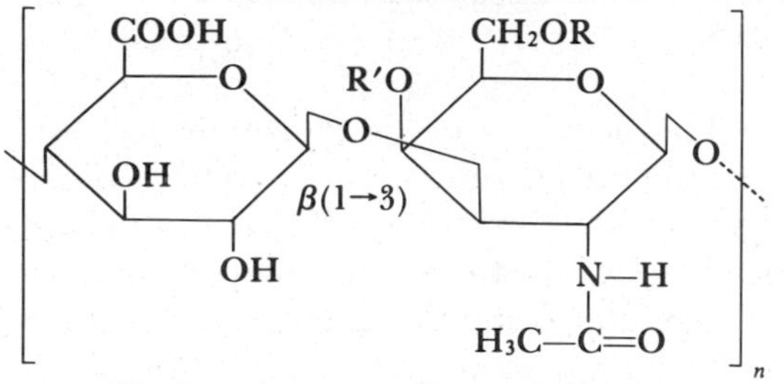

D-Glucuronic acid   N-Acetyl-D-galactosamine

Type A: R = −H, R′ = −SO₃H
Type C: R = −SO₃H, R′ = −H

Dermatan sulfate

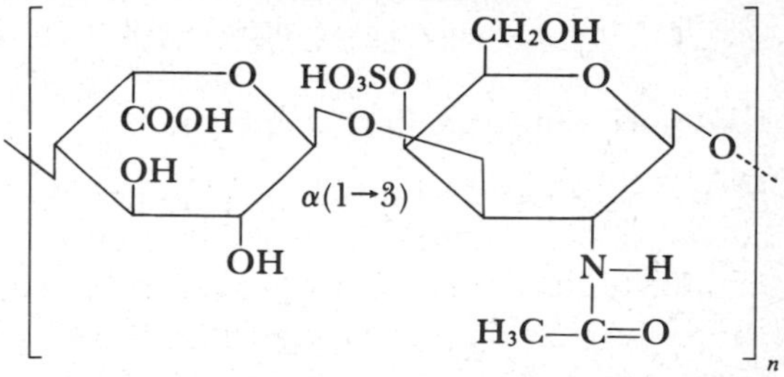

L-Iduronic acid   N-Acetyl-D-galactosamine

(or Type B chondroitin sulfate)

The chondroitin sulfates are structurally related to keratosulfate, which is present in both integumentary and connective tissues. The fundamental repeating subunit here involves an alteration of D-galactose with 2-acetamido-2-deoxy-D-glucose-6-sulfate residues, which are united by jointly alternating $\beta(1{\rightarrow}4)$ and $\beta(1{\rightarrow}3)$ glycosidic bonds.

Some structural properties of acid mucopolysaccharides may also involve their covalent linkage to silicon. Concentrations of this element exist at levels of 0.02 to 0.04% in chondroitin 4-sulfate and dermatan sulfate. The presence of one silicon atom for nearly every 200 monomeric units of a polysaccharide suggests that it may have a

Keratosulfate

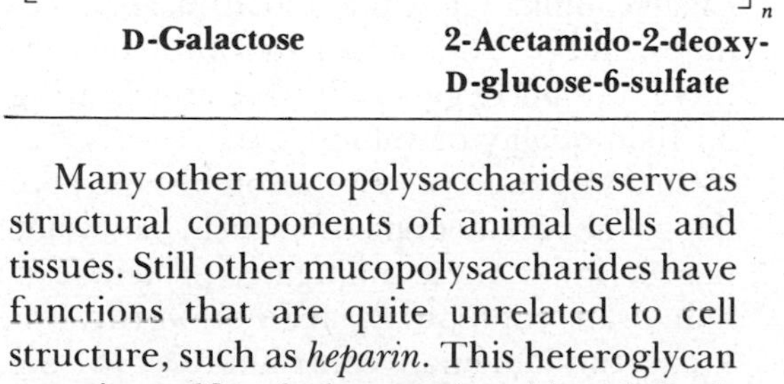

D-Galactose   2-Acetamido-2-deoxy-D-glucose-6-sulfate

Many other mucopolysaccharides serve as structural components of animal cells and tissues. Still other mucopolysaccharides have functions that are quite unrelated to cell structure, such as *heparin*. This heteroglycan contains sulfated glucuronic acid and disulfated glucosamine residues that are consistently linked through $\alpha(1{\rightarrow}4)$ bonds. The

critical involvement in interchain polysaccharide linkages. Based on the behavior of orthosilicic acid and the plausible covalent linkage between polysaccharide chains via silicon ethers (Figure 6.43), it is speculated that silicon could have important trace nutrient significance in animals as well as humans.

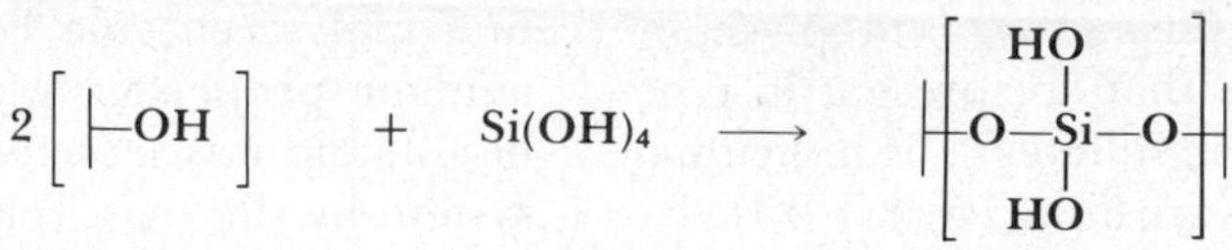

$$2\left[-\text{OH}\right] + \text{Si(OH)}_4 \longrightarrow \left[-\text{O}-\overset{\overset{\displaystyle \text{HO}}{|}}{\underset{\underset{\displaystyle \text{HO}}{|}}{\text{Si}}}-\text{O}-\right]$$

| Partial structure for a polysaccharide chain | Orthosilicic acid | Ether linkage between polysaccharide chains |

FIGURE 6.43. Typical orientation of silicon in the formation of a silicon ether linkage between polysaccharide chains.

Glycoproteins and Peptidoglycans

Many of the heteropolysaccharides discussed above such as chondroitin and dermatan sulfates are actually covalently linked to some protein structure. This covalent linkage often occurs through an *N*- or *O*-acylglycosylamine linkage and may commonly involve a serine hydroxyl group on the protein. These conjugated protein–carbohydrate structures are called glycoproteins or proteoglycans. Actual protein content normally ranges from 1 to 80 + % for these substances depending on the compound. Glycoproteins are not strictly structural proteins, but many are intrinsically associated with the protective function of cell coats or they augment the functional properties of membranes. Other glycoproteins have structural significance at the biochemical level as opposed to their physical contribution to macromolecular structures and tissues. Glycoproteins can be conveniently subdivided into three groups:

1. The *mucopolysaccharides*, which contain a short polypeptide chain that is covalently linked to a long linear polysaccharide (e.g., widely distributed in cartilage, eyes, tendons, and skin).

2. *Mucin glycoproteins* (mucoproteins), which consist of acid mucopolysaccharides covalently linked to specific proteins, thereby forming sticky, slippery substances (e.g., salivary and mucous membrane secretions).

3. *Plasma glycoproteins* that occur in blood sera.

Examples of these typical glycoprotein structures are presented in Figure 6.44.

The activities of glycoproteins are well documented throughout the archives of biochemistry where some of their most significant roles involve intercellular contact and recognition responses. For example, a heterocellular *in vitro* mixture of dispersed kidney and liver cells will normally aggregate with similar cells prior to their tissue culture. After these recognition responses, the cells may then undergo cellular multiplication. The exact biochemical mechanism of glycoprotein recognition among cells is still largely speculative but critical to the overall process. Singular cells grow unchecked until some specific surface area of the individual cells contacts or impinges upon neighboring cells. At this juncture, overriding cellular growth ceases. Such normal states of cell recognition and contact responses do not readily occur among cancerous cells.

Coupled with these activities, which demand cellular "self-recognition" at the biochemical level, glycoproteins are similarly fundamental to the antigen–antibody responses observed among higher animals. These complex involvements of glycoproteins in animals are not paralleled by their importance to food systems at this time. Clearly, milk, egg white, blood plasma, legumes, and cereals contain glycoproteins, but these substances generally have little bearing on food quality or value.

In addition to the glycoproteins, the peptidoglycans are not noted for their food value per se; rather, these conjugated protein–carbohydrate compounds are noteworthy for their fundamental contributions to the structural matrices of bacterial cell walls. Here, complex polymers of polysaccharides are intertwined and covalently bonded to similarly

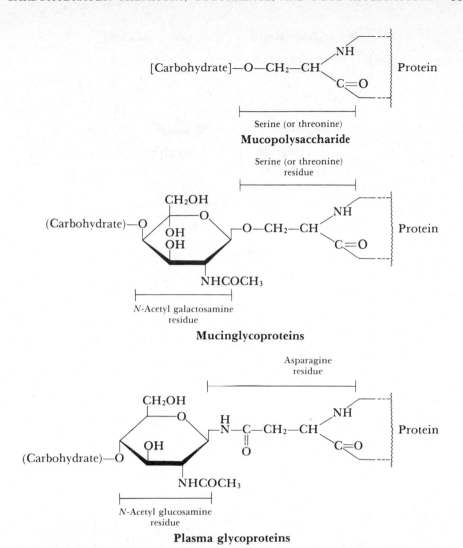

FIGURE 6.44. Some typical examples of basic glycoprotein structures.

complex peptide chains. The major carbohydrate constituents of the peptidoglycans are *N*-acetyl-D-glucosamine and *N*-acetylmuramic acid, which are jointly linked through a β(1→4) glycosidic bond to produce a structurally repetitious disaccharide. The *N*-acetylmuramic acid residue is a product of *N*-acetyl-D-glucosamine that has an ether linkage between its C-3 hydroxyl group and the α-hydroxyl group of lactic acid. The other peptidyl amino acid residues that are quite common to these structures include amide bonded L-alanine, L-lysine (or *meso*-

diaminopimelic acid), γ-D-glutamic acid, and D-alanine. Note the joint presence of D- and L-amino acid residues in the tetrapeptide structure shown in Figure 6.45.

Polymerization of the structure illustrated in Figure 6.46 is complicated by cross-linking between the terminal D-alanine carboxyl group and an adjacent peptide chain. Although these linkages can be direct as in the case of *Escherichia coli*, interchain linkages are often connected through an additional peptide chain such as the pentaglycine structure present in *Staphylococcus aureus* (Figure

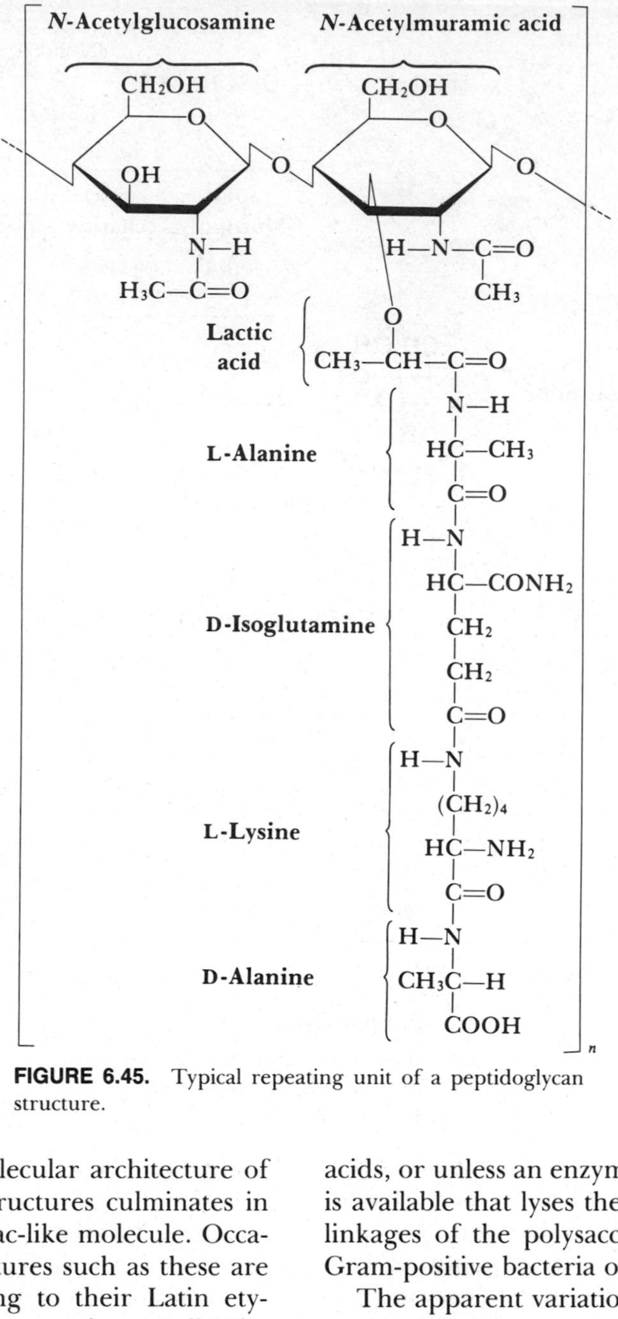

N-Acetylglucosamine   **N-Acetylmuramic acid**

CH₂OH   CH₂OH
O   O

OH   O   O

N—H   H—N—C=O
H₃C—C=O   CH₃

Lactic acid   { O
CH₃—CH—C=O

L-Alanine   { N—H
HC—CH₃
C=O

D-Isoglutamine   { H—N
HC—CONH₂
CH₂
CH₂
C=O

L-Lysine   { H—N
(CH₂)₄
HC—NH₂
C=O

D-Alanine   { H—N
CH₃C—H
COOH $_n$

FIGURE 6.45.   Typical repeating unit of a peptidoglycan structure.

6.46). The macromolecular architecture of these complicated structures culminates in the formation of a sac-like molecule. Occasionally, whole structures such as these are called a *murein* owing to their Latin etymology, that is, *murus,* meaning a wall. The three-dimensional murein gridwork is essentially the same for all bacteria, yet variations on the theme do exist. It should also be recognized that these cell wall constructions are largely immune to enzymatic attack unless an enzyme can hydrolyze peptide linkages that involve D- as opposed to L-amino acids, or unless an enzyme such as lysozyme is available that lyses the $\beta(1\rightarrow4)$ glycosidic linkages of the polysaccharide chain (e.g., Gram-positive bacteria only).

The apparent variation of peptidoglycan structures seems even more complex and confounding because $\geq50\%$ of the cell wall is composed of still other polymers including teichoic acids, polysaccharides, proteins, and polypeptides. In fact, teichoic acids alone may account for $\geq35\%$ of the dry cell weight of Gram-positive bacteria. As illustrated in Figure 6.47, teichoic acid polymers have ribitol

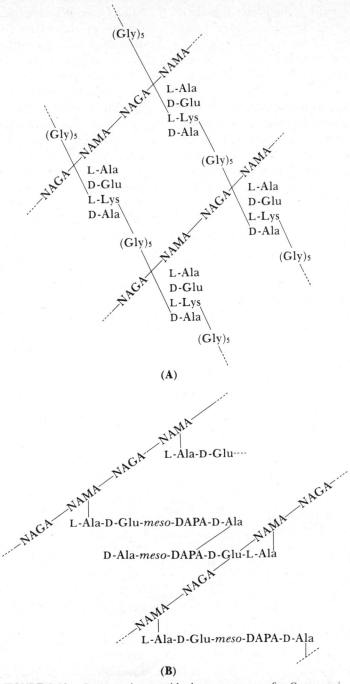

FIGURE 6.46. Comparative peptidoglycan structures for Gram-positive (A) bacterial cell wall and a Gram-negative (B) bacterial wall. Note that the tetrapeptide chains of *S. aureus* (A) are linked by pentaglycine bridges, whereas *E. coli* (B) displays direct peptide bonds between the amino function of DAPA and the carboxyl group of the terminal carboxyl group of D-alanine (DAPA, diaminopimelic acid; NAMA, *N*-acetylmuramic acid (Figure 6.45); NAGA, *N*-acetylglucosamine; (Gly)₅, glycine pentapeptide).

or glycerol molecules bonded together through intervening phosphodiester linkages, whereas the free functional hydroxyl group of the glycerol phosphate may be alternatively bonded to N-acetyl-D-glucosamine or D-glucose (Figure 6.47).

Other polysaccharide components of cell walls may also contain glucose, rhamnose, galactose, or mannose along with their amide sugar derivatives to form cell capsular structures that are elastic, flexible, or highly rigid.

To this point, the structural generalities detailed have basically involved the Gram-positive bacteria. This group of bacteria is generically distinct from Gram-negative bacteria since the former group binds Gram's iodine and stains purple, whereas iodine decolorization and a resulting pink stain imparted by safranin characterizes the Gram-negative bacteria. These obvious visual distinctions among bacterial species superficially reflect major structural differences in their cell walls.

The Gram-negative bacteria offer far more complexity than Gram-positive types because they contain additional lipopolysaccharides, polypeptides, and lipoproteins. The lipopolysaccharide known as "lipid A" typically contains 3-deoxy-D-mannooctulosonic acid, glucosamine, fatty acids, and phosphate. This lipid A component is joined to a two-part core. One part of the core contains a repetitious trisaccharide linked by phosphate groups to similar groups. Each trisaccharide unit has an ethanolamine moiety, two heptose sugars in the form of D-mannoheptose, and a single eight-carbon sugar–acid known as octulosonic acid. The second part of the structure displays a glucose, galactose, and galactosamine oligosaccharide. Polysaccharides attached to the two-part core contain the serologically complex yet distinct determinants for O antigen that underly the classification scheme for the *Enterobacteriaceae*. These antigenic determinants appear similar for all O antigens, and

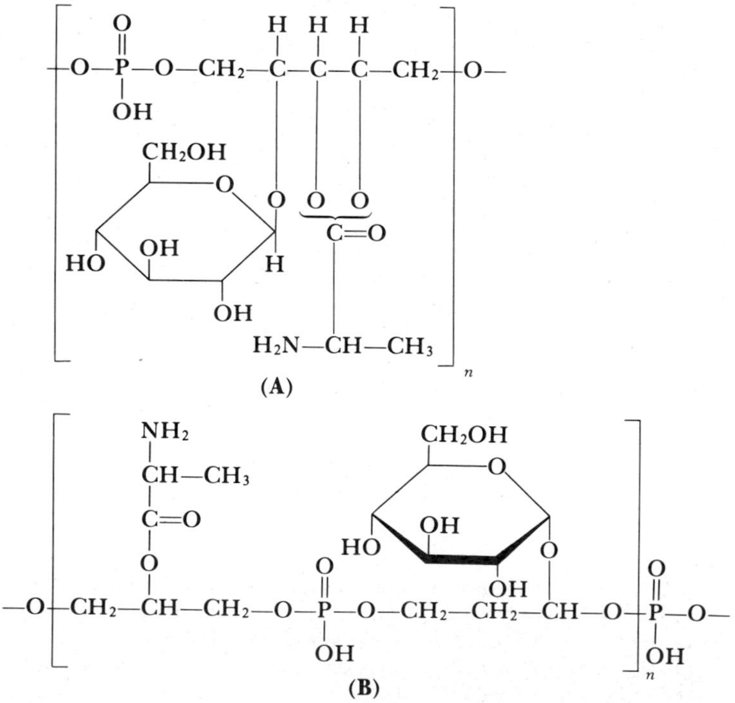

FIGURE 6.47. Typical repeating subunits in teichoic acid structures based on structures of the (A) five-carbon sugar, ribitol, and (B) glycerol; some organisms may have N-acetyl-D-glucosamine in place of the D-glucose shown.

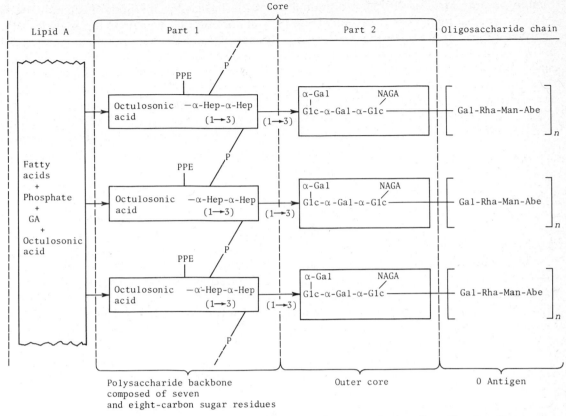

FIGURE 6.48. Generalized structure for the lipopolysaccharide structure of *Salmonella typhimurium*. Hep, D-mannoheptose; PPE, moiety containing (phosphatidyl)ethanolamine; GA, glucosamine; Man, mannose; Rha, rhamnose; Gal, galactose; Glc, glucose; NAGA, *N* acetylglucosamine; Abe, abequose (3,6-dideoxy-D-xylohexose).

in the case of *Salmonella typhimurium*, the tetrasaccharide structure of the O antigen chain is composed of galactose, rhamnose, mannose, and abequose (Figure 6.48).

PROXIMATE ANALYSIS OF FOOD CARBOHYDRATES AND CRUDE FIBER

The quantitative methods used for the proximate analysis of food carbohydrates are historically rooted in wet chemical procedures. These methods were first reported during the 1840s, when Trommer demonstrated that monosaccharides could bring about the reduction of cupric ions (Cu^{2+}) under alkaline conditions with the concomitant production of copper oxide (Cu^{1+}). Fundamentally similar reactions are still employed for detecting reducing sugars in foods, but the quantitative reliabilities have been improved by better analytical reagents and instrumentation.

Contemporary methods for sugar analyses are largely based on specific color reactions of some reagents with reducing mono-, di-, and polysaccharides, or furfural-type sugar degradation products contained in strong acid (e.g., 96% sulfuric acid). Numerous methods for sugar analysis including polarimetry, polarography, ultraviolet/visible and infrared spectrophotometry, nuclear magnetic resonance spectroscopy, and many others have become widely available. Many of the classical and modern approaches to sugar and carbohydrate analysis have been surveyed by Joslyn (1970) and

Pomeranz and Meloan (1971). Additional details concerning classical carbohydrate analyses appear in the following references, and in spite of their early publication dates, many still hold timely analytical information: Bates *et al.* (1942), Bell (1955), Browne and Zerban (1941), Dische (1955, 1962), Hodge and Hofreiter (1962), International Commission for Uniform Methods of Sugar Analysis (1966), Montreuil and Spik (1963), Staněk *et al.* (1963), Hodge and Davis (1952), Dubois *et al.* (1956), and Whistler and Wolfrom (1962).

Unless established methodological precedents already exist, a food analyst can become mesmerized in selecting the best possible analytical method for a specific carbohydrate assay. Fortunately, though, most current methods rely on oxidation of sugars in the presence of alkaline copper solutions, phenols, organic acids, or some other compound. The reduction of alkaline copper (Cu^{2+}) solutions to form reduced copper oxides (Cu^{1+}) serves as the basis for traditional Fehling, Barfoed, and unified Munson–Walker methods. The precipitated copper oxide can be gravimetrically determined after a sugar reaction, or the color of the postreaction mixture can be spectrophotometrically determined in order to measure the reducing sugar content of food samples. Aside from these methods, the colorimetrically measured reduction of 3,5-dinitrosalicylic acid by reducing sugars, as well as sugaracidified phenol reactions, can offer sensitive methods of analysis. The salicylic acid method has sensitivity in the range of 0.45 to 2.5 mg for reducing sugars, whereas the phenol method detects quantities as low as 10.0 μg with ±2% accuracy. Other methods such as the reaction of acidified anthrone with mono-, di-, and polysaccharides can be used for the colorimetric estimation of total monosaccharide content in a sample. This reaction can also be used to evaluate the monosaccharidic constituents of glucosides, dextrans, dextrins, gums, and plant polysaccharides including cellulosic substances. The sensitivity of this method parallels that of the acidified phenol method.

All the classical methods of sugar analysis will undoubtedly remain in the analytical arsenal of food analysts for many years, but as a result of increasing awareness concerning the roles of carbohydrates as human food, it is becoming ever more important to rapidly identify and quantitate the presence of individual sugars. It is not sufficient merely to quantitate presence of sugars as reducing or nonreducing on the basis of classical wet chemical methods.

Too often in the past, analysts and nutritionists have succumbed to the assumption that hexose and pentose contents of foods mirrored total reducing sugars, while oligosaccharide content of foods merely reflected nonreducing sugar components expressed as sucrose. This nebulous generalization is not accurate because the nonreducing fraction of foods may contain, as the major sugar present, higher-molecular-weight and more complex homologues of sucrose such as verbascose, stachyose, and raffinose.

The early developments of paper, thin-layer, silicic-acid, ion-exchange, and other chromatographic methods for separating and identifying oligo- and monosaccharides have been detailed by Joslyn (1970) and many others including Bell (1955), Binkley and Wolfrom (1948), Hough (1954), Cassidy (1951, 1957), Lederer and Lederer (1957), Block *et al.* (1958), Knapman *et al.* (1958–1965), Heftman (1961), Whistler and Wolfrom (1962), Bishop (1964), Mangold *et al.* (1964), Montreuil *et al.* (1967), Montreuil and Spik (1968), Kirchner (1967), Shaw *et al.* (1980), Damon and Pettitt (1980), Wilson *et al.* (1981), Hurst and Martin (1980), and Macrae (1983). Most chromatographic techniques offer excellent analytical potential, but until the burgeoning development of high-pressure liquid chromatography (HPLC), many of these methods were protracted and unsuitable for routine laboratory uses.

The adoption of these fundamental separation principles by modern HPLC techniques permits the rapid resolution and quantitation of individual sugars, oligosaccharides, polysaccharides, starches, and some hydrocolloid gums. Not only does the advent of these methods allow rapid quantitative saccharide assays for food research, development, labeling, or quality-control purposes (Figures 6.49 and 6.50), but electronic acquisition and processing of chromatographic data can be used to characterize complex carbohydrate polymers (Figure 6.51) according to their average molecular weight distributions.

Aside from assessing the digestible carbohydrate contents of foods, food scientists and nutritionists have drawn increased attention to the nutritional consequences of indigestible dietary carbohydrates generally classified as *crude fiber*. Crude fiber substances are chemically diverse and *may* include cellulose, hemicelluloses, certain hydrocolloids, pectin, and lignin. Lignin is not a carbohydrate but, rather, it is an enormously complex statistical polymer composed of oxyphenylpropane units that occur in close association with the structural car-

bohydrate and hydrocolloid components of many plant cells. Some of the phenyl constituents of lignin along with a representative structure for lignin are illustrated in Figure 6.52. By virtue of this close physical association between lignin and indigestible cell wall constituents, it is not uncommon to see this compound included in discussions dealing with crude fiber.

Nondigestible components of foods contribute bulk to the diet and promote intestinal peristalsis and elimination in both humans and other animals. The presence of dietary fiber in human foods may have epidemiological significance since low levels of fiber are believed to parallel increased incidences of diverticulosis, cardiovascular disease, colonic cancer, and diabetes. In spite of the available hard-core data, however, more supporting scientific data must be acquired in order to affirm the fundamental dietary and health benefits of crude fiber.

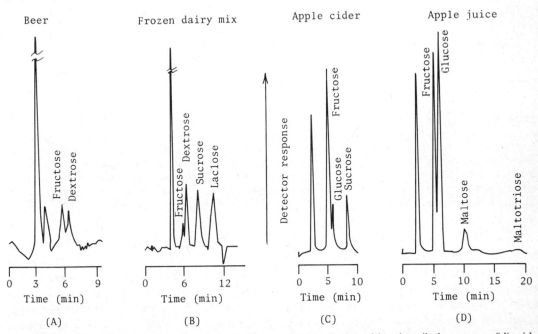

FIGURE 6.49. Resolution and quantitation of monosaccharides can be achieved easily by means of liquid chromatography (Shaw et al., 1980; Damon and Pettitt, 1980; Wilson et al., 1981; Hurst and Martin, 1980) as indicated for samples of (A) beer, (B) a frozen dairy mix product, (C) apple cider, and (D) apple juice. [Assay conditions: Column-Waters Associates Carbohydrate Column (3.9 mm ID × 30 cm); solvent—water/acetonitrile (range 15/85 → 20/80 → 35/65 depending on sample complexity); detector-refractive index; flow—1.7–4.0 mL/min; sample size injection ≈20 μL.] (Figures courtesy of Waters Associates, Division of Millipore Corporation.)

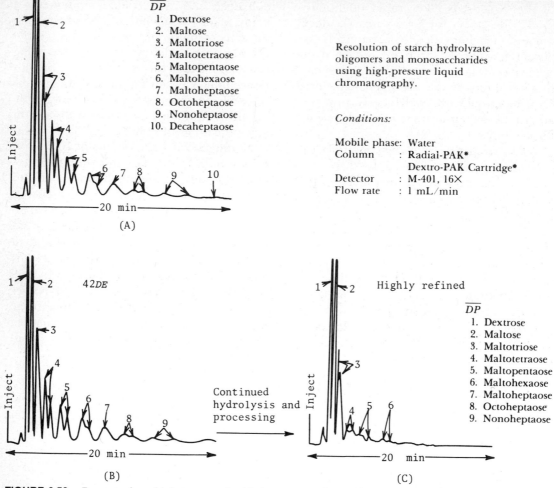

FIGURE 6.50. Reverse-phase-high-pressure liquid chromatography can be used to separate carbohydrate oligomers up to *DP* 10 as well as their enantiomers by using water as a mobile phase (chromatogram A). This has important significance for the analysis of complex carbohydrates such as starch hydrolyzates, or in those studies designed to monitor hydrolyzate processing in which oligomer composition is in a constant state of flux (chromatograms B and C). Chromatogram B shows a starch processed to 42 *DE* and continued processing of the same starch eventually reveals the formation of a highly refined product that has a predominant degree of polymerization (*DP*) of 1 and 2. (Figures courtesy of Waters Associates, Division of Millipore Corporation.)

Crude fiber measurements of foods have also been used as a convenient index of nutrient values for poultry and livestock feeds since the low digestibility of high-fiber feeds provides animals with nothing short of a poor nutritional potential. Ruminants are an exception to the general rule because they utilize cellulose, whereas other animals do not; however, the lignin component of crude fiber cannot be used by any agricultural animals or humans.

The crude fiber contents of grain and stockfeeds, cereal grains, and their products (wheat, macaroni, noodles, breads, baked products), nuts, nut products, spices, cocoa, and many other food substances are estimated from the amount of dilute acid- and alkali-soluble matter that resides in foods. The remaining insoluble food residue after acid and alkali extraction represents crude fiber. This fraction is mainly composed of cellulose and lignin (~97%) along with traces of mineral matter. The detectable amounts of cellulose and lignin often vary consider-

Rapid characterization
of natural starches

Various starches exhibit different molecular weight
averages and *MWD* values depending on their origins

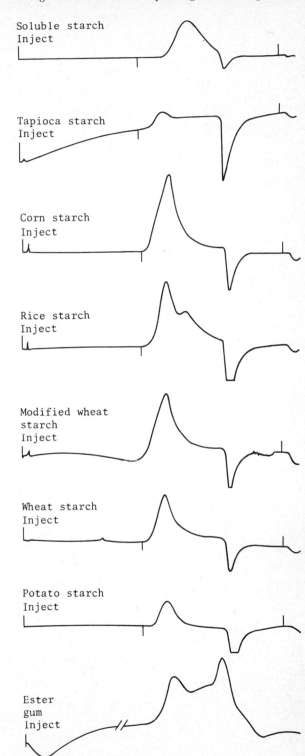

Soluble starch

$\overline{M}_n$	1,618
$\overline{M}_w$	8,727
$\overline{M}_z$	25,873
$\overline{M}_v$	8,727
MWD	5.40

Tapioca starch

$\overline{M}_n$	2,397
$\overline{M}_w$	116,153
$\overline{M}_z$	371,292
$\overline{M}_v$	116,123
MWD	48.46

Corn starch

$\overline{M}_n$	5,736
$\overline{M}_w$	193,137
$\overline{M}_z$	481,543
$\overline{M}_v$	193,137
MWD	33.67

Rice starch

$\overline{M}_n$	3,992
$\overline{M}_w$	107,142
$\overline{M}_z$	336,941
$\overline{M}_v$	107,122
MWD	26.84

Modified wheat starch

$\overline{M}_n$	4.411
$\overline{M}_w$	205,783
$\overline{M}_z$	565,364
$\overline{M}_v$	205,738
MWD	46.65

Wheat starch

$\overline{M}_n$	3,567
$\overline{M}_w$	164,879
$\overline{M}_z$	438,482
$\overline{M}_v$	164,846
MWD	46.22

Potato starch

$\overline{M}_n$	75,721
$\overline{M}_w$	205,493
$\overline{M}_z$	362,754
$\overline{M}_v$	205,449
MWD	33.67

Ester gum

$\overline{M}_n$	682
$\overline{M}_w$	2,705
$\overline{M}_z$	15,342
$\overline{M}_v$	2,705
MWD	4.0

FIGURE 6.51. High-temperature gel-permeation chromatography coupled with a refractive-index detection system on a high-pressure liquid chromatograph offers an effective means for the rapid prediction of polysaccharide molecular weight averages ($\overline{M}_v, \overline{M}_n, \overline{M}_z, \overline{M}_w$) as well as the complete molecular weight distribution (MWD). (Figures courtesy of Waters Associates, Division of Millipore Corporation.)

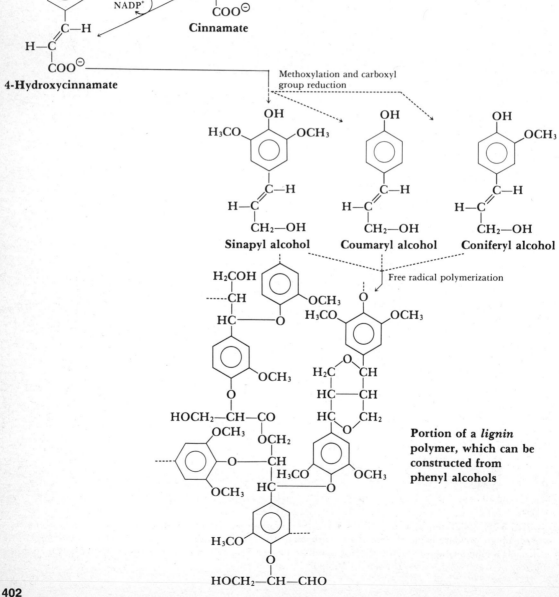

FIGURE 6.52. Synthesis of lignin involves a complex series of reactions that probably begin with the deamination of tyrosine or phenylalanine. Deamination of phenylalanine requires the activity of cinnamate-4-hydroxylate to produce 4-hydroxycinnamate, but the deamination step involving tyrosine does not require the enzyme to produce the same product. The 4-hydroxycinnamate does not serve as synthesis stock for the formation of sinapyl, coniferyl, and coumaryl alcohols. These and possibly other related alcohols are produced by hydroxylation of the phenyl rings; methylation of the hydroxyl groups to produce methoxylated phenyl ring derivatives (—OCH₃) at the expense of S-adenosylmethionine; and the final reduction of carboxyl groups to produce their corresponding alcohol. Polymerization of these phenyl alcohols is probably encouraged by free radical reactions that culminate in a physically rigid covalently bonded structure known as lignin.

402

ably from the amounts present. Only 55 to 80% of cellulose or 3 to 68% of the lignin may be recovered. Care must be taken to realize that crude fiber estimations do not measure absolute amounts of structural constituents in foods, and the idiosyncratic behavior of different foods toward established crude fiber assays may result in querulous data. Crude fiber assays must be conducted with strict methodological uniformity, and even in these cases, the variable responses of foods to alkali and acid treatments may cause inconstant degrees of oxidative hydrolysis for cellulose and unpredictable lignin degradation. There are no absolute species-specific values for crude fiber in natural foodstuffs; even the most widely reported crude fiber values are highly empirical, and there is no recommended level of dietary fiber intake that will ensure proper nutrition.

MOLECULAR HETEROGENEITY MEASUREMENTS FOR OLIGO- AND POLYSACCHARIDES

Most biopolymers such as starches are non-homogeneous in their molecular weights and are mixtures of macromolecules that have varying degrees of polymerization. Numerous analytical and preparative procedures permit the separation of these macromolecules and act as basic tools for characterizing the probable physical and behavioral properties of biopolymers. Information indicating molecular heterogeneity of a biopolymer sample can be expressed in terms of the calculated statistical modes or averages of molecular weight distribution within the sample. Such calculations take into account the following factors:

1. The molecular weight (M) of the polymeric molecule containing i-repeating units is indicated as M_i (i is also recognized as the degree of polymerization).

2. The number of molecules of weight M_i present in the polymer sample is designated as N_i.

The first mode of number-average molecular weight, $\overline{M}_n$, is defined by

$$\overline{M}_n = \frac{\Sigma N_i M_i}{\Sigma N_i}$$

Experimental determination of $\overline{M}_n$ can be obtained from analytical methods that count individual macromolecules in the sample solution. These methods would include end-group analysis, measurement of osmotic pressure, and colligative property measurements related to boiling-point elevation or melting-point depression.

The second mode or weight-average molecular weight, $\overline{M}_w$, may be defined as

$$\overline{M}_w = \frac{\Sigma N_i M_i^2}{\Sigma N_i M_i}$$

The $\overline{M}_w$ value can be experimentally determined from turbidity measurements of a macromolecular solution or by joint diffusion or sedimentation methods.

A third mode or z average molecular weight, $\overline{M}_z$, can be estimated from sedimentation equilibrium studies using ultracentrifugation or calculated from

$$\overline{M}_z = \frac{N_i M_i^3}{N_i M_i^2}$$

As a practical simplified example for the use of these statistics, consider the situation in which four molecules having a molecular weight of 200 and four molecules having a molecular weight of 400 are present in a sample. The critical molecular weight averages based on statistical calculations become

$$\overline{M}_n = \frac{(4)(200) + (4)(400)}{4 + 4} =$$

$$\frac{2.4 \times 10^3}{8} = 300.00$$

$$\overline{M}_w = \frac{[(4)(200)^2] + [(4)(400)^2]}{[(4)(200)] + [(4)(400)]} =$$

$$\frac{8.0 \times 10^5}{2.4 \times 10^3} = 333.33$$

$$\overline{M}_z = \frac{[(4)(200)^3] + [(4)(400)^3]}{[(4)(200)^2] + [(4)(400)^2]} =$$

$$\frac{2.88 \times 10^8}{8.0 \times 10^5} = 360.00$$

According to these calculations, it is clear that the z-average ($\overline{M}_z$) is the largest and the n-average ($\overline{M}_n$) is the smallest. For those situations in which polymers are completely homogeneous, the three average molecular weights will be identical, but increasing heterogeneity or nonuniformity among the polymers causes the $\overline{M}_z:\overline{M}_w:\overline{M}_n$ ratios to increase. When molecular weight distribution is attributable to complete randomness as a result of synthesis (e.g., polysaccharide formation) or degradation (e.g., natural or industrial starch hydrolysis) of linear molecules, the "normal" ratios for $\overline{M}_z:\overline{M}_w:\overline{M}_n = 3:2:1$. Therefore, the estimation of these average molecular weight values affords a very useful insight as to whether or not structural distribution of similar polymer mixtures is narrow, broad, or "normal."

A fourth statistical feature of polymers involves calculating the viscosity–average molecular weight, $\overline{M}_v$. Although this value can be estimated by exacting viscosity techniques, it can be calculated from

$$\overline{M}_v = \alpha\sqrt{\frac{\Sigma N_i M_i^{1+\alpha}}{\Sigma N_i M_i}}$$

The derivation of this statistical relationship is complex and closely allied with the calculation of intrinsic viscosity $[\eta]$, presented in Chapter 8, where $[\eta] = KM^\alpha$. This expression for the intrinsic viscosity, often called the Mark-Houwink equation, involves parameters K and α, which are mathematical constants for a given polymer solvent system at a given temperature. Specifics regarding the calculation are reviewed elsewhere (Blackley, 1968; Seymour, 1971). Since values for the α terms can range from 0.5 to 2.0, reflecting tightly coiled or rod-shaped molecules in solutions, respectively, there are as many $\overline{M}_v$ values for a given sample as there are α values for the polymer. It is fortunate, however, that $\overline{M}_v$ is not very sensitive to values of α over the range of practical interest. When $\alpha = 1.0$, the $\overline{M}_v$ statistic is identical to $\overline{M}_w$. For the unlikely possibility that $\alpha = -1.0$, $\overline{M}_v$ equals $\overline{M}_n$. If $\alpha = 0.0$, as

in the case of dilute solutions of perfect spherical molecules, then the so-called intrinsic viscosity is independent of molecular size and little if any information can be obtained about molecular size. In general, α values usually lie between 0.6 and 0.8, and, therefore, $\overline{M}_v$ values fall between $\overline{M}_n$ and $\overline{M}_w$, being closer to the $\overline{M}_w$ value than to $\overline{M}_n$.

A typical "normal" distribution profile for molecular weight averages of a nonhomogeneous polydisperse system is shown in Figure 6.53, in which the values for $M_z > M_w > M_v > M_n$.

The current interest in average molecular weight distributions is more than academic. The advent of manufactured and processed foods requires the availability of statistical perspectives that define functional and structural attributes of food ingredients such as natural starches and gums. The practical value of these estimations is often the prerogative of experienced food formulation experts and engineers, but the physical characteristics and functional properties of polymers reflect molecular weight averages in the following ways:

$\overline{M}_z$ = flexibility and stiffness
$\overline{M}_w$ = tensile strength and hardness
$\overline{M}_v$ = extrudability and molding properties
$\overline{M}_n$ = brittleness and flow properties

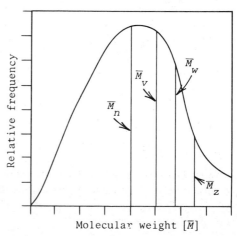

FIGURE 6.53. Typical molecular weight distribution for a natural biopolymer.

NATURALLY OCCURRING MONO-, DI-, AND POLYSACCHARIDES

From the perspective of the food and nutritional analyst, it is clear that the hexoses, D-glucose, and D-fructose comprise the bulk of free monosaccharides in natural foodstuffs.

Notable sources of free glucose include ripe fruits, honey, flower nectars, leaves, saps, root tissues, and blood. D-Fructose, on the other hand, occurs in many plant juices and in honey as a hydrolysis product of sucrose. The pentose sugars, namely D-xylose and L-arabinose, occur in minor but nevertheless noteworthy amounts in plants. Free xylose is present in many varieties of apples, asparagus, barley malts, brewhouse worts, corn, grapes, apricots, bamboo shoots, maple syrup, pears, juniper berries, strawberries, prunes, tomatoes, alfalfa, and both wines and beers in which it is not fermented by yeasts. Arabinose is similarly present in onions, grapes, strawberries, commercial beer, corn, and alfalfa.

A variety of other sugars having little nutritional value also occur freely in plant tissues such as L-sorbose (keto sugar) found in passion fruit; L-rhamnose (deoxyketo sugar) in poison ivy; and heptuloses including the keto sugars known as D-sedoheptulose and D-mannoheptulose that are, respectively, present as a trace intermediate in nearly all plants and avocado pears. With the exception of the aforementioned simple sugars and miniscule amounts of many other free monosaccharides, a diverse variety of monosaccharides participate in one or more glycosidic linkages with similar and/or different sugars. Whereas the sugars mentioned above may be present only in trace amounts as free sugars, the same sugars are often quantitatively important constituents of di-, oligo-, and polysaccharides (Tables 6.10–6.12).

The sugar and carbohydrate contents of edible fruits and vegetables often reflect plant species characteristics, plant growth conditions, and the maturity of the plant tissue(s) at the time of harvest. Four distinct fruit and vegetable classifications can be established on the basis of the last consideration, and the discriminative application of these classes accounts for the wide spectrum of apparent carbohydrates present in dietary regimens:

1. Those fruits and vegetables harvested *prior* to *in vivo* conversion of their free sugars into more complex, texturally undesirable oligo- and polysaccharides (e.g., green beans, green peas, sweet corn).

2. Those legumes that are harvested after conversion of their starches into di- and oligosaccharides such as raffinose, stachyose, or verbascose (e.g., soybeans).

3. Those fruits and vegetables that show marginal changes in free sugars after natural ripening and harvest (e.g., citrus fruits).

4. Those edible plant tissues that are harvested before peak free sugar availability is produced at the expense of storage polysaccharides during postharvest storage, ripening (e.g., apples, bananas, pear, potatoes), or malting in the case of cereal grains.

Not only do these basic differences lead to distinct differences in the free sugar–carbohydrate profiles of foods, but recognition of the differences in sugar availabilities augment the proper grading (e.g., color), processing, and handling of fruits and vegetables.

SUGARS OF COMMERCE AND INDUSTRY

The direct use of sucrose as a food industrial sugar is certainly unrivaled in developed countries, and its bulk use is generally followed only by D-glucose, D-fructose, maltose, and lactose.

Cane Sugar

Sugar-bearing juices are expressed from sugar cane by crushing the cane between

Table 6.10. Types, Composition, and Sources of Some Important Food Carbohydrates

Carbohydrate type	Class[a]	Monosaccharide components										Some common dietary sources
		L-Arabinose	D-Fructose	L-Fucose	D-Glucose	D-Glucuronic acid	D-Galactose	D-Galacturonic acid	D-Mannose	L-Rhamnose	D-Xylose	
Lactose	D				X		X					Milk, cheese, and other dairy products
Maltose, isomaltose	D				X							Hydrolyzed starch syrups, malt, and honey
Sucrose	D		X		X							Sugar beets, sugar cane, fruits, vegetables, and processed sweetened foods
α,α-Trehalose	D				X							Fungi
Fructosylsucroses	O		X		X							Cereals, leeks, and onions
Maltooligosaccharides	O				X							Amylodextrins, malt, and starch syrups

Carbohydrate	Type[a]	Source
Raffinose, stachyose	O	Cereals, legume seeds, tubers, and molasses
Cellulose	P	Plant cell walls and fibers
Glycogen	P	Animal tissues and liver
Hemicelluloses	P	Plant fiber and cell walls, bran, cereals, legumes, nuts, and flours
Inulin (a fructan)	P	Tubers of jerusalem artichoke
Mannans	P	Tagua palm seed (commercial source)
Mannogalactans	P	Alfalfa, clover, carob seeds, coffee seeds, and guar gum
Pectic substances	P	Fruits such as apple, quince, citrus, etc., sugar beets, and vegetables
Pentosans	P	Occurs with hemicelluloses and pectic substances
Starch, dextrins	P	Cereals, legumes, roots, and tubers

[a] D, disaccharide; O, oligosaccharide; P, polysaccharide.

Table 6.11. Approximate Monosaccharide and Reducing Sugar Content in Some Edible Fruits

Fruit	Percentage			
	D-Fructose	D-Glucose	Sucrose	Reducing sugars
Apple	5.0	1.7	3.1	8.3
Cherries	7.2	4.7	0.1	12.5
Grapes (Concord)	4.3	4.8	0.2	9.5
Melon (cantaloupe)	0.9	1.2	4.4	2.3
Oranges (composite)	1.8	2.5	4.6	5.0
Peaches	1.6	1.5	6.6	3.1
Pear (Bartlett)	5.0	2.5	1.5	8.0
Pineapple (ripe)	1.4	2.3	7.9	4.2
Plum (sweet)	2.9	4.5	4.4	7.4
Raspberries	2.4	2.3	1.0	5.0
Tomatoes	1.2	1.6	—	3.4

SOURCE: Adapted from Hardinge *et al.* (1965).

rollers. Although a wide variety of cultivated hybrids have been developed, most sugar canes are hybrids of *Saccharum officinarum,* the "noble cane," along with hardier species. The opaque juice contains significant amounts of acids, proteins, extraneous solids, and gums. Lime neutralization of the organic acids, heated coagulation of proteins, and integrated filtration and centrifugation steps are used to produce a raw sugar extract that contains up to 97% sucrose. This raw sucrose liquor requires further refining through

Table 6.12. Disaccharide, Polysaccharide, and Fiber Contents in Edible Portions of Some Vegetables, Dry Legumes, and Cereals

Foodstuff	Percentage						
	Sucrose	Cellulose	Natural dextrins	Hemicellulose	Pectin	Pentosans	Starch
Beans (snap)	0.5	0.5	0.3	1.0	0.5	1.2	2.0
Beets (sugar)	12.9	0.9	—	0.8	—	—	—
Cabbage (raw)	0.3	0.8	—	1.0	—	—	—
Carrots (raw)	1.7	1.0	—	1.7	0.9	—	—
Corn (fresh)	0.3	0.6	0.1	0.9	—	1.3	14.5
Lettuce	0.2	0.4	—	0.6	—	—	—
Peas (green)	5.5	1.1	—	2.2	—	—	4.1
Potatoes (white)	0.1	0.4	—	0.3	—	—	17.0
Sweet potatoes	4.1	0.6	—	1.4	2.2	—	16.5
Soybeans	7.2	2.6	1.4	6.6	—	4.0	2.0
Navy beans	—	3.1	3.7	6.4	—	8.2	35.2
Peanuts	4.5	2.4	2.5	3.8	—	—	4.0
Barley grain (hulled)	—	2.6	—	6.0	—	8.5	62.0
Oats (hulled)	—	—	—	—	—	6.4	56.4
Rice, polished (raw)	0.4	0.3	0.9	—	—	1.8	72.9
Rye (grain)	—	3.8	—	5.6	—	6.8	57.0
Sorghum (grain)	—	—	—	—	—	2.5	74.0
Tapioca (cassava)	—	—	—	—	—	1.0	27.5
Wheat (grain)	1.5	2.0	2.5	5.8	—	6.6	71.5

SOURCE: Adapted partially from Hardinge *et al.* (1965); for added details on oligosaccharides in legumes see Solsulski *et al.*, *J. Food Sci.* **47**:498, 1982.

washing, dissolution, adsorptive processes, filtration, concentration, crystallization, and centrifugation to yield a white product that is 99 + % sucrose.

The progressive purification of refined crystalline sucrose results from recycling mother liquors to yield some useful amounts of minor sugars before being discarded. Moreover, when impurities of extraction liquors become too high for the production of quality sucrose, molasses is produced, and specifically *blackstrap molasses* may be produced from final mother liquors.

Refiners syrup, sugar-house molasses, and *treacle* are also produced by refiners for a variety of purposes, including the commercial fermentative production of alcohols, organic acids (e.g., citric acid), and animal feeds. Those additional sugars, called *brown sugars*, merely represent the intermediate isolation of small sucrose crystals tainted with molasses during the overall production of 99 + % crystalline sucrose.

Beet Sugar

The commercial production of beet sugar has occasionally rivaled or even exceeded that of cane sugar at various times, and it still offers a potentially important source of industrial sugar. Cultivated varieties of the beet species, *Beta vulgaris*, are sliced into cossettes that are subjected to a heated aqueous countercurrent extraction. The raw extract is purified by liming, filtration, concentration, adsorptive processes, crystallization, centrifugation, and drying to produce the final refined sucrose product.

Assuming the adequate refining of raw beet sugar, the composition and appearance of the final product should be identical to raw cane sugar. Molasses can also be produced from beets, but it contains less invert sugar than sugar cane-based molasses.

Liquid Sugar

Washing the molasses coating from crude sucrose crystals by means of a saturated sugar liquor acts as a first step in the production of liquid sugar. The washed crude sucrose is subsequently clarified by liming; it may be passed over bone char or granular vegetable carbon and then concentrated by evaporation to about 67% solids. So-called *premium liquid sucrose* can be produced by ion-exchange techniques, which offer ≤0.005% ash.

Fructose

The production of commercial D-fructose relies principally on the isomerization of D-glucose with an isomerase enzyme. The D-glucose stock for this reaction can be prepared from starches by any one or a combination of the hydrolysis methods discussed earlier.

This enzymatic conversion ultimately provides a 54:46% mixture of D-glucose: D-fructose, whose sweetness compares favorably with inverted sucrose prepared by hydrolysis or an invertase enzyme. The 54:46% D-glucose:D-fructose mixture is often called *high-fructose syrup* (HFS), and because of recent advances in food technology and biochemical engineering, this product challenges many of the traditional uses of sucrose in foods.

The most significant commercial sources of high-fructose syrups are produced from steam jet-pasted corn starch. The viscosity of the cooked paste may be decreased with diastase and be passed over a stationary bed of glucoamylase to yield a high D-glucose syrup. Passage of this product over an immobilized bed of isomerase enzyme produces an equilibrium mixture of D-fructose and D-glucose. The equilibrium mixture may be marketed as a high-fructose syrup or a liquid sugar, or through the use of ion-exchange and other advanced purification methods such as *reverse osmosis*, the D-fructose can be concentrated to still higher levels. Extensive concentration of D-fructose is obviously a prerequisite for its eventual crystallization. The bulk of HFS used in the food industry results from corn starch processing, and this product is known as *high-fructose corn syrup* (HFCS).

Lactose

The lacteal secretions of mammals contain 2 to 8% lactose. The disaccharide can be produced in crystalline form from acid-treated wheys from which proteins have been effectively eliminated. Filtration of these wheys is followed by carbon decolorization and manipulation of the solvent dielectric properties and volume to produce crystalline lactose. Commercial lactose represents a monohydrated α-form of the disaccharide, but the β-form will be produced if crystallization occurs at ≥93.3°C.

Maltose

Malt sugars are largely the product of industrial or preliminary fermentation processes, and they occur rarely in natural food products. Significant amounts of malt sugars are present in sprouted grains such as barley, which is introduced into the malting stage of brewing processes. Exclusive of natural sources, high-maltose corn syrups can be manufactured through β-amylase hydrolysis of starch. These high-maltose corn syrups contain a maltose:D-glucose:oligosaccharide ratio of about 56:10:44. These syrups are widely employed for the manufacture of preserves.

Invert Sugars

Disaccharides such as sucrose can be inverted for the preparation of invert syrups by using acid hydrolysis methods. A prescribed amount of acid, typically tartaric or hydrochloric acid, is added to a granulated sugar solution held at 100°C. After a holding period, acid neutralization is achieved by addition of sodium bicarbonate. The inverted liquid syrup has a final invert sugar concentration of ~15 to 90%. The favorability of this product rests on its wide potential application range to food products (e.g., bread baking, canning, carbonated beverages, glacé fruits) coupled with minimal problems of sugar recrystallization prior to use and its commercial availability as a product having higher densities than sucrose.

Corn Syrups and Corn Sweeteners

Corn syrup results from the partial hydrolysis of corn starch according to acid or acid–enzyme mechanisms discussed in earlier sections (Figure 6.54). Complete hydrolysis of corn starch obviously produces pure glucose. Neutralization of the starch hydrolysis process at different stages of completion yields a combination of glucose, maltose, and higher saccharides. The controlled, graded hydrolysis of corn starches produces a range of corn syrups, each characterized by its own *DE* value, reducing-sugar content, viscosity (Figure 6.29), and density. As expected, high *DE* values for syrups parallel increasing availability of reducing sugars (i.e., low-molecular-weight sugars) and lower syrup viscosities (Table 6.13). Added to *DE* categorizations, corn syrups can be specified according to their Baumé degree (Be°) index, solids concentration, or sweetness.

Baumé values depend on the weight per unit volume or specific gravity of samples according to the formula

$$\text{specific gravity*} = 145/(145 - Be°)$$

The Baumé degree range is from 0.0 to 70°Be but the normal range for corn syrups is between 42 and 45°Be.

The solids concentrations of corn syrups can be quantitated by using vacuum oven or azeotropic distillation methods, and the total solids are expressed as a percent of the total syrup.

From the comparative perspective of Baumé degrees and *DE* values for corn syrups, it is useful to note that at a specific Baumé value, two syrups having different *DE* values will also display marked differences in total solids. Increases in the total-solids level of corn syrups coincides with increases in *DE* values. This effect originates from more water

*Formula given is for samples heavier than water, not to be confused with specific gravity = 140/(130 + Be°), which is used for fluids lighter than water.

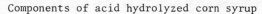

Components of acid hydrolyzed corn syrup

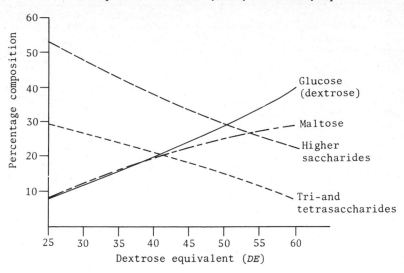

Components of acid-enzyme hydrolyzed corn syrup

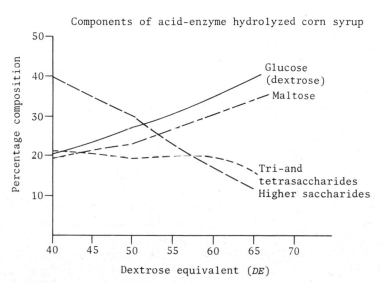

FIGURE 6.54. Comparative percent compositions of corn syrups produced by acid hydrolysis and combined acid–enzyme hydrolysis of corn starch with respect to the apparent dextrose equivalent (*DE*). (Reprinted by permission of the Corn Refiners Association, Washington, D.C.)

molecules chemically interacting with the corn syrup solids. Furthermore, since water density is far less than that of corn syrup solids, more solids are required to produce a specified density of Be°.

Increases of glucose and maltose contained in corn syrups also coincide with increased *DE* values and syrup sweetness. Although the sweetness of corn syrups and sugars varies with concentration, taste perception of individuals, and chemical impurities present in samples during testing, 43°Be corn syrups are commonly ranked in sweetness with respect to invert sugar, sucrose, and dextrose. Ranking figures for 43°Be syrups have invert sugar at the "high sweetness"

Table 6.13. Typical Compositional Properties for Corn Syrups

Sample	Dextrose equivalent	Saccharides, carbohydrate basis (%)						
		DP_1^a	DP_2	DP_3	DP_4	DP_5	DP_6	DP_{7+}
Maltodextrin	12	1	3	4	3	3	6	80
Corn syrup AC[b]	27	9	9	8	7	7	6	54
Corn syrup AC	36	14	12	10	9	8	7	40
Corn syrup AC	42	20	14	12	9	8	7	30
Corn syrup AC	55	31	18	12	10	7	5	17
Corn syrup HM,[b] DC[b]	43	8	40	15	7	2	2	26
Corn syrup HM, DC	49	9	52	15	1	2	2	19
Corn syrup DC	65	39	31	7	5	4	3	11
Corn syrup DC	70	47	27	5	5	4	3	9
Corn syrup DC	95	92	4	1	1	Sum of $DP_{5,6,7+}$		2
High-fructose corn syrup	—	94[c]	4	—	—	Sum of $DP_{3,4,5,6,7+}$		2

[a] DP, degree of polymerization.
[b] AC, acid conversion; DC, dual conversion (acid and enzyme treatment of starch), HM, high maltose.
[c] Dextrose plus fructose.

SOURCE: Reprinted by permission of the Corn Refiners Association, Washington, D.C.

end of the scale (120°), whereas regular *DE* corn syrups represent the low end of the scale (40°):

invert sugar > sucrose > glucose >
  (120)   (100)   (75)

   high *DE* > intermediate *DE* > regular *DE*
    (60)   (50)   (40)

Synergistic sweetening effects have also been documented for sucrose–corn syrup mixtures. In these cases the perceived sweetness is higher than projected calculations based on total solids.

CONCLUSION

Although proteins, fats, minerals, vitamins, and water are essential components of a bal-
anced human diet, none of these broad categories has more basic importance to foods than the carbohydrates. Carbohydrates represent a common energy link between the biochemical energy reserves of plants and the basic source of nutritional energy for animals.

The majority of the world's human population owes its continued existence to carbohydrates, whether in the form of mono-, di-, or polysaccharides. Of all these structural forms, however, the polysaccharides demand special attention because they represent the major caloric reservoir of most human diets; they are responsible for the textural and rheological properties of many foods; and they perform essential thickening, stabilizing, gel-forming, and emulsification actions in foods.

The dietary implications of excessive or inadequate carbohydrate consumption by animals and humans are well documented, but the ability of humans to convert amino acids and glycerol moieties of fats into glucose confounds serious efforts to establish specific dietary requirements for carbohydrates. The practical establishment of a minimum dietary requirement for carbohydrate intake is also clouded by ketotic events, the degradation of tissue proteins, involuntary dehydration, and an overall upheaval of the total cation pool in the bodies of selected test subjects.

The long classical history of carbohydrate chemistry is woven throughout the rich developmental history of organic chemistry, physical chemistry, biochemistry, and nutrition. More significantly, however, the basic nutritional importance and availability of carbohydrates may ultimately control the survival of the human species throughout many regions of the world.

REFERENCES

General references

Advances in Carbohydrate Chemistry and Biochemistry. 1945–Present. Academic Press, New York.

Lineback, D. R., and G. E. Inglett. 1982. *Food Carbohydrates*. Avi Publishing Co., Westport, Conn.

Pancoast, H. M., and W. R. Junk. 1980. *Handbook of Sugars*. Avi Publishing Co., Westport, Conn.

Pigman, W., and D. Horton (eds.). 1970, 1976, 1979, 1980. *The Carbohydrates*, 2nd ed., Vols. IA, IIA, IB, IIB. Academic Press, New York.

Rules of carbohydrate nomenclature. 1963. *J. Org. Chem.* **28:**281.

Schallenberger, R. S. 1982. *Advanced Sugar Chemistry*. Avi Publishing Co., Westport, Conn.

Staněk, J., M. Cherný, and J. Pacák. 1965. *Oligosaccharides*. Academic Press, New York.

Stoddart, J. F. 1971. *Stereochemistry of Carbohydrates*. Wiley, New York.

Some classical and useful analytical references for sugars

Bates, F. J., and Associates. 1942. *Polarimetry Saccharimetry and the Sugars*. U.S. National Bureau of Standards Circular 440.

Bell, D. J. 1955. Mono- and oligosaccharides and acidic monosaccharide derivatives. In *Moderne Methoden der Pflanzenanalyse* (K. Paech and M. V. Tracey eds.), Vol. 2, pp. 1–54. Springer, Berlin.

Binkley, W. W., and M. L. Wolfrom. 1948. *Chromatography of Sugars and Related Substances*. Sugar Research Foundation, New York.

Bishop, C. T. 1964. Gas–liquid chromatography of carbohydrate derivatives. *Adv. Carbohydrate Chem.* **19:**95.

Block, R. J., E. L. Durrum, and G. Zweig. 1958. *A Manual of Paper Chromatography and Paper Electrophoresis*, 2nd ed., pp. 170–214. Academic Press, New York.

Browne, C. A., and F. W. Zerban. 1941. *Physical and Chemical Methods of Sugar Analysis*. Wiley, New York.

Cassidy, H. G. 1957. Fundamentals of chromatography. In *Technique of Organic Chemistry* (A. Weissberger, ed.), Vol. 10. Wiley, New York.

Cassidy, H. G. 1951. Adsorption and chromatography. In *Technique of Organic Chemistry* (A. Weissberger, ed.), Vol. 5. Wiley, New York.

Damon, C. E., and B. C. Pettitt. 1980. HPLC determination of fructose, glucose and sucrose in molasses. *J. Ass. Offic. Anal. Chem.* **63:**476.

Dische, Z. 1962. General color reactions. *Methods Carbohydrate Chem.* **1:**477.

Dische, Z. 1955. New color reactions for determination of sugars in polysaccharides. *Methods Biochem. Anal.* **2:**313.

Dubois, M., K. Gilles, J. K. Hamilton, P. A. Rebers, and F. Smith. 1956. Colorimetric method for determination of sugars and related substances. *Anal. Chem.* **28:**350.

Hardinge, M. G., J. B. Swarner, and H. Crooks. 1965. Carbohydrates in foods. *J. Amer. Dietet. Ass.* **46:**197.

Heftmann, E. (ed.). 1961. *Chromatography*. Reinhold, New York.

Hodge, J. E., and B. T. Hofreiter. 1962. Determination of reducing sugars and carbohydrates. *Methods Carbohydrate Chem.* **1:**380.

Hodge, J. E., and H. A. Davis. 1952. Selected methods for determining reducing sugars. *U.S. Dept. Agr. Bur. Agr. Ind. Chem.* **AIC 333.**

Hough, L. 1954. Analysis of mixtures of sugars by paper and cellulose column chromatography. *Methods Biochem. Anal.* **1:**205.

Hurst, W. J., and R. A. Martin. 1980. HPLC determinations of carbohydrates in chocolate: A collaborative study. *J. Ass. Offic. Anal. Chem.* **63**:595.

International Commission for Uniform Methods of Sugar Analysis. 1966. Report of the Proceedings of the 14th Session, Washington, D. C.

Kirchner, J. G. 1967. Thin layer chromatography. In *Technique of Organic Chemistry* (A. Weissberger, ed.), Vol. 12. Wiley, New York.

Knapman, C. E., *et al.* (eds.) 1958–1965. *Gas Chromatography Abstracts.* Butterworth, London.

Lederer, M. (ed.). 1958–1963. *Chromatographic Reviews*, Vols. 1–5. Elsevier, Amsterdam.

Lederer, E., and M. Lederer. 1957. *Chromatography*, 2nd ed. Elsevier, Amsterdam.

Macrae, R. 1983. *HPLC in Food Analysis.* Academic Press, New York.

Mangold, H. K., H. H. O. Schmidt, and E. Stahl. 1964. Thin-layer chromatography (TLC). *Methods Biochem. Anal.* **12**:393.

Montreuil J., and G. Spik. 1968. Methodes chromatographiques et electrophoretiques de dosage des glucides constituant les glycoproteines: Oses "neutres," acides sialques, acides uroniques, osamines. In *Microdosage de Glucides*, Vol. 2. Publ. Fac. Sci. Lille, Lille, France.

Montreuil, J. G. Spik, and A. Konarsaka. 1967. Methodes chromatographiques de dosage des oses "neutres." In *Microdosage des Glucides*, Vol. 3. Publ. Fac. Sci. Lille, Lille, France.

Montreuil, J., and G. Spik. 1963. Methodes colorimetriques de dosage de glucides totaux. In *Microdosage de Glucides*, Vol. 1. Publ. Fac. Sci. Lille, Lille, France.

Shaw, P. E., C. W. Wilson, and R. J. Knight. 1980. High performance liquid chromatographic analysis of D-mannoheptulose, perseitol, glucose and fructose in avacado cultivers. *J. Agr. Food Chem.* **28**:379.

Staněk, J., M. Cerný, J. Kocourek, and J. Pacák. 1963. *The Monosaccharides.* Academic Press, New York.

Trommer, H. 1841. *Unterscheidung von Gummi, Dextrin, Traubenzucker und Rohrzucker. Ann. Chem.* **39**:360.

Whistler, R. L., and M. L. Wolfrom (eds.). 1963. *Methods in Carbohydrate Chemistry*, Vol. 2. Academic Press, New York.

Whistler, R. L., and M. L. Wolfrom (eds.). 1962. *Methods in Carbohydrate Chemistry*, Vol. 1. Academic Press, New York.

Wilson, A. M., T. M. Work, and R. J. Bushway. 1981. HPLC determination of fructose, glucose and sucrose in potatoes. *J. Food Sci.* **46**:300.

Selected references for polymeric carbohydrates and gums

Blackley, D. C. 1968. Molecular weight determination. In *Addition Polymers: Formation and Characterization* (D. A. Smith, ed.), pp. 131–219. Butterworths, London.

Chemistry of Plant Gums and Mucilages. 1959. ACS Monograph 141. Reinhold, New York.

Code of Federal Regulations, Title 21, Parts 121.104 and 121.105, Vol. 39, No. 185, pp. 34181–34183, 34201–34211. September 23, 1974.

Davidson, R. L. (ed.). 1976. *Handbook of Water-Soluble Gums and Resins.* McGraw-Hill, New York.

Glicksman, M. 1969. *Gum Technology in the Food Industry.* Academic Press, New York.

Lawrence, A. A. 1973. *Edible Gums and Related Substances. Food Technology Review.* No. 9. Noyes Data Corp., Park Ridge.

Mantel, C. L. 1947. *The Water Soluble Gums.* Reinhold, New York.

Meer, G., W. A. Meer, and J. Tinker. 1975. Water soluble gums—their past, present and future. *Food Technol.* **29**(11):22.

Physical Functions of Hydrocolloids. 1960. ACS Monograph 25. Reinhold, New York.

Neukom, H., and H. Deuel. 1960. Über den abbau von Pektinstoffen bei Alkalischer Reaction. *Beih. Z. Schweiz. Forstver.* **30**:233.

Neukom, H., and H. Deuel. 1958. Alkaline degradation of pectin. *Chem. Ind.* (London), p. 683.

Sanderson, G. R. 1981. Polysaccharides in foods. *Food Technol.* **35**(7):50.

Seymour, R. B. 1971. *Introduction to Polymer Chemistry.* McGraw-Hill, New York.

Torres, A., and R. D. Thomas. 1981. Polydextrose . . . and its applications in foods. *Food Technol.* **35**(7):44.

Whistler, R. L. (ed.). 1973. *Industrial Gums—Polysaccharides and Their Derivatives*, 2nd ed. Academic Press, New York.

Traditional surveys of food carbohydrate analyses (textbooks)

Joslyn, M. A. 1970. *Methods in Food Analysis.* Academic Press, New York.

Pomeranz, Y., and C. E. Meloan. 1971. *Food Analysis: Theory and Practice.* Avi Publishing Co., Westport, Conn.

Triebold, H. O., and L. W. Aurand. 1963. *Food Composition and Analysis.* Van Nostrand Reinhold, New York.

CHAPTER · 7

Lipids: Chemistry, Structure, and Analysis

INTRODUCTION

Living systems contain a diverse group of substances that are commonly recognized as lipids. Foremost among the biochemical functions of lipids are their abilities to act as

1. Storage and transport vehicles for metabolic energy.

2. Components of cellular membrane structures.

3. Protective coatings on organisms exposed to environmental forces.

4. Essential components in the overall mechanism of cell-surface recognition phenomena.

Other than the uniform solubility of lipids in ethyl and petroleum ethers, carbon disulfide, and carbon tetrachloride, they do not share any common chemical or structural denominator. From a quantitative perspective, fats and fatty oils comprise the bulk of the lipid content in vegetables, land animals, and marine animals.

The reference to fats and fatty oils is entirely generic since both categories represent water-insoluble hydrophobic substances; however, fats are *solids* at ambient temperatures, whereas oils are *liquids*. Natural fats and fatty oils are composed of esters of glycerol with fatty acids that are commonly referred to as *glycerides* or *acylglycerols*. The melting points for individual acylglycerols determines whether they will exist as a fat or a fatty oil; and the melting point for any acylglycerol reflects the carbon-chain length of its esterified fatty acids (up to three) as well as their individual degrees of unsaturation.

Native fats and fatty oils coexist with a variety of nonglyceride lipids. Some of these differ from acylglycerols only in the backbone structure to which individual fatty acids are esterified. For example, in place of a glycerol backbone, phosphoglycerides have glycerol 3-phosphate, sphingolipids contain the fatty alcohol known as sphingosine (4-sphingenine), and waxes incorporate a long-chained alcohol.

Since the acylglycerols along with the phosphoglycerides, sphingolipids, and waxes contain esterified fatty acids, they are subject to the effects of alkaline hydrolysis or *saponification*, which yields individual salts of fatty acids known as *soaps*. This reactive property alone has served as a basis for the classification of all saponifiable lipids as *complex lipids*.

Along with the nonglyceride lipids cited above, native fats and fatty oils also contain unsaponifiable steroid alcohols, isoprenoid-based structures known as *terpenes,* quinone derivatives, and derivatives of unsaturated fatty acids such as the *prostaglandins.* These substances are only minor constituents in the total composition of lipids, but they are noteworthy because many of them are implicated with dietary and health considerations, vitamin activities, fat and oil colors, flavor and aroma of foods, and other recognized biochemical, nutritional, and organoleptic properties of foods.

Many other lipid substances are found in living cells linked by covalent bonds or weak bonding interactions to carbohydrates or proteins. These associations lead to the respective formation of *glycolipids* and *lipoproteins,* which display very specialized biological functions.

From the viewpoints of food scientists and nutritionists, dietary lipids represent both *processed* and *unprocessed fats* and *fatty oils.*

Processed lipids are described as those forms that have been physically or chemically isolated from some natural source, refined, fractionated, hydrogenated, deodorized, bleached, or purified in other ways to yield a final lipid product that adheres to some prescribed set of specifications. These processed lipids have specialized uses in food preparation, manufacturing, and processing. Furthermore, since the acylglycerol contents of fats and fatty oils are basic to the functional properties of processed lipids, minor concentrations of nonacylglycerol lip-

ids are often removed from processed lipids to eliminate the potential development of undesirable food flavors, odors, colors, and textures.

Unprocessed lipids differ from processed varieties because they are subject to little if any refinement before their introduction into a dietary regimen. These lipids may reside in the natural cellular or biochemical milieu as in the case of fats and fatty oils in peanuts, soybeans, whole milk and its marginally processed products, fresh fishes, or meats. In addition, unprocessed lipids also include those forms that have undergone only minor processing to the extent that they have been crudely extracted and collected from their natural sources before being introduced into foods or the diet.

THE GLYCERIDES OR NEUTRAL ACYLGLYCEROLS OF FATS AND FATTY OILS

According to the simplest classification scheme, an acylglycerol or glyceride is the esterification reaction product of glycerol with one to three fatty acid molecules. Since up to three fatty acids may be esterified to the glycerol structure, *mono*acyl-, *di*acyl-, and *tri*acylglycerols may be formed. Note that the number of water molecules produced as a by-product of the esterification reaction corresponds to the number of fatty acids esterified.

$$
\begin{array}{ccc}
& \overset{O}{\underset{\|}{HO-C-R'}} & \\
\underset{\substack{| \\ HO-C-H \\ | \\ CH_2OH}}{CH_2OH} + \underset{\substack{O \\ \| \\ HO-C-R'' \\ O \\ \| \\ HO-C-R'''}}{} \longrightarrow \underset{\substack{O \\ \| \\ R''-C-O-C-H \quad O \\ | \quad \| \\ CH_2O-C-R'''}}{\overset{CH_2O-C-R'}{}} + 3HOH
\end{array}
$$

L-Glycerol   **Three fatty acids: R groups**   **Triacylglycerol**
             may be the same but they are   **(or triglyceride)**
             often different; note too that
             if R′ and R‴ are equivalent,
             the resulting triacylglycerol
             will not be asymmetric

The lack of any important ionizable groups in the fatty acid chains accounts for the classification of acylglycerols as *neutral lipids*.

A careful survey of many acylglycerol structures indicates that L-acylglycerols predominate in the biological world. This fact is related to the stereospecific enzymatic activity of glycerol kinase, which is responsible for the phosphorylation of L-glycerol to L-glycerol 1-phosphate, an essential *in vivo* precursor to acylglycerol formation.

Since the aliphatic side groups of esterified fatty acids on the acylglycerol structure are often different, it is necessary to positionally discriminate among each of the carbon positions within the glyceryl (C_3H_5) moiety. The presence of three different fatty acid esters on a glyceryl structure produces a *mixed* triacylglycerol as opposed to a *simple* triacylglycerol in which all esters involve identical fatty acid residues. Depending on the fatty acid that occupies the middle carbon of the glyceryl moiety (β or 2) as well as the outside positions (α or 1 and α' or 3), mixed triacylglycerols having two different fatty acids can display up to *six* different molecular structures. Consider the possible esterification of palmitic (P) and oleic (O) acids to the glycerol structure shown below:

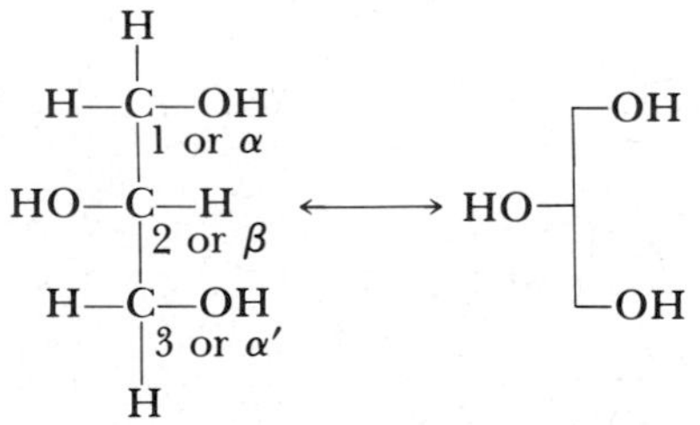

Fischer projection formula for L-glycerol   **Abbreviated stick structure**

Since three hydroxyl sites are available for esterification to individual fatty acids, the possible triacylglycerol structures can be envisioned as

Three different fatty acids available for synthesis of a mixed triacylglycerol offer 18 different molecular structures and so on.

Depending on the fatty acid residues at the 1, 2, or 3 positions of the glyceryl moiety, trivial or common names can be assigned to mixed triacylglycerols, effectively indicating their respective structures.

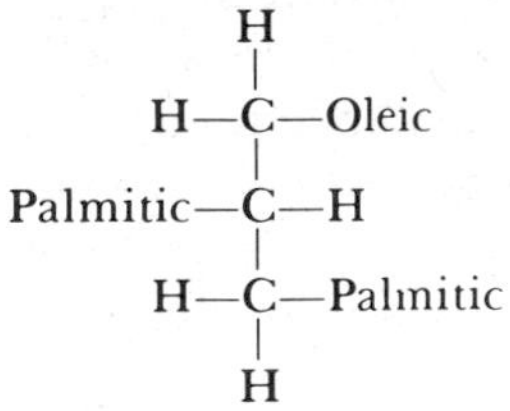

1 (or α)-Oleodipalmitin 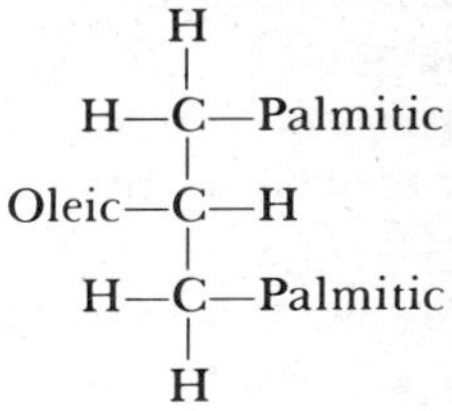   **2 (or β)-Oleodipalmitin**

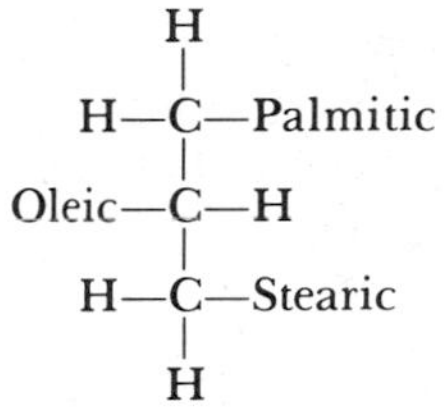

2 (or β)-Oleopalmitostearin

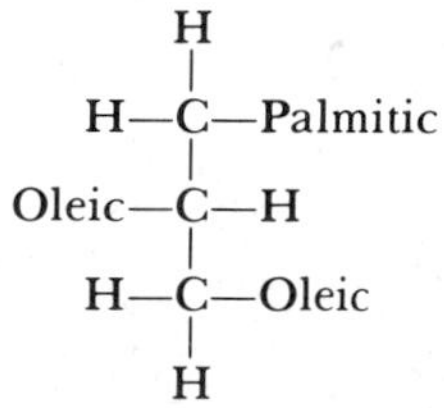

1 (or α)-Palmitodiolein 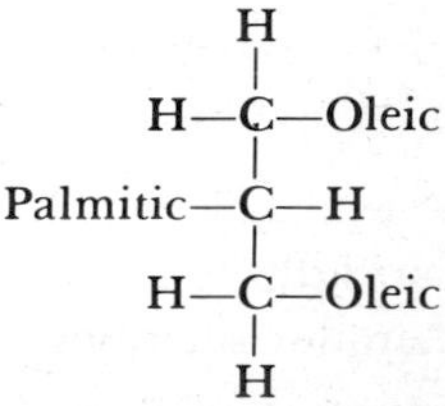   **2 (or β)-Palmitodiolein**

Because native fats and oils principally contain triacylglycerols, mono- and diacylglycerols are usually present only as partial hydrolysis breakdown products of triacylglycerol structures owing to inadequate or long-term storage. Depending on food moisture content, fatty acid hydrolysis from triacylglycerols may approach a level of 50%. The mono- and diacylglycerols of food commerce and industry are commonly present

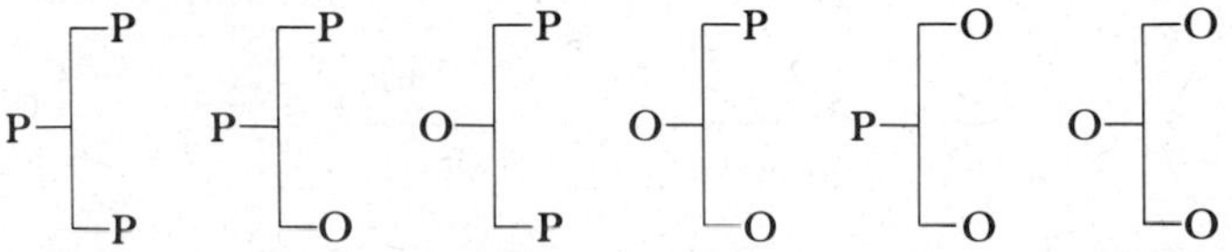

PPP   **PPO**   **POP**   **POO**   **OPO**   **OOO**

in processed foods, but these are largely prepared by industrial molar ratio-dependent esterification reactions using one or more selected fatty acids plus free glycerol. Many of these acylglycerol forms have important applications as emulsifiers for ice cream, salad dressing, peanut butter, and innumerable other food uses. Another important route for monoacylglycerol production in particular involves the glycerolysis of fats and oils through a *transesterification* reaction. The products of this reaction, namely monoacylglycerols, depend on the fat-to-glycerol content used during the reaction. Reaction products are commonly described as "40% monos," "60% monos," and so on. Preparation of "90% monos" can be achieved by means of molecular distillation. Although 1 (or α)- or 2 (or β)-monoacylglycerols may exist, 1 (or α)-monoacylglycerols are notably more stable than the 2 (or β) isomers. In fact, the 2 (or β) isomers often undergo rapid conversion or isomerization into 1 (or α) forms.

Triacylglycerol Composition and Variation

Generalizations concerning the fatty acid and triacylglycerol compositions of fats and oils in similar members of plant and animal species can be made, but these compositional profiles may vary widely for many reasons. Singular or collective effects of genetic factors, environmental conditions, and health and nutrition states of organisms are all partially responsible for obvious variations.

A superficial survey of normal fatty acid concentrations in many organisms reveals an immense mathematical potential for different triacylglycerol structures. As indicated by Litchfield, many seed fats and oils have the potential to contain 125 to 1000 different triacylglycerol structures; animal fats offer the potential for 1000 to 64,000 triacylglycerol species; and on the basis of 142 different fatty acids, butterfat could theoretically contain 2,863,288 different triacylglycerol structures. According to established data, however, hard-core compositional information for native acylglycerol structures is available for only ~300 seed fats and ~60 animal fats. Moreover, according to detailed surveys of fatty acid distribution alone, about 6000 lipid substances have been studied, 1500 of which are animal and marine oils with the remaining 4500 representing plant and seed oils. As impressive as the scope of these surveys appears, they are really miniscule since at least 250,000 species alone make up the botanical kingdom.

Table 7.1 details the distribution of sat-

Table 7.1. Occurrence of some Simple (GS$_3$, GU$_3$) and Mixed (GS$_2$U, GSU$_2$) Triacylglycerols in Animal and Vegetable Fats and Oils[a]

Type of fat or oil	Mole %				
	G$_3$	GS$_2$U	GSU$_2$	GU$_3$	S
Safflower oil	—	—	18	82	6
Olive oil	—	—	45	55	15
Peanut oil	—	1	56	43	19
Herring oil	—	4	61	35	23
Lard	2	26	54	18	37
Shea butter	4	36	55	5	46
Beef tallow	15	46	37	2	58
Cocoa butter	2	77	21	0	61

[a]S and U, respectively, represent saturated and unsaturated fatty acid esters and G refers to the glyceryl moiety.

SOURCE: Adapted from Sonntag (1979).

urated and unsaturated fatty acids among some triacylglycerols that have relevance to food systems.

The observed distribution of fatty acids in acylglycerol structures of a single species or organism is not a random event guided solely by fatty acid availability at the time of their esterification to the glycerol structure. Many theories have been developed in order to deal with fatty acid distribution among acylglycerol structures, but few of these theories have survived rigorous examination. One notable exception that may have some relevance is the *"even" distribution theory* proposed by Hilditch and his co-workers. This theory indicates that component fatty acids of triacylglycerols are distributed as broadly as possible among all acylglycerol molecules in a cell or organism.

Using numerical perspectives for the theory, this means that (1) when fatty acid A accounts for ~35 mole % or more of the total fatty acids (A + X) in a fat or fatty oil, *that* acid will occur once (indicated as GAX_2) in nearly all acylglycerol molecules; (2) if fatty acid A constitutes ~35 to 65 mole % of the total fatty acid content, it may occur twice (GA_2X) in any given acylglycerol molecule or more as its relative concentration to other fatty acids increases; and (3) when a fatty acid accounts for ≥70 mole % of the total fatty acids available for triacylglycerol formation, the occurrence of simple triacylglycerols is favored (GA_3) and remaining fatty acids suitable for esterification will not occur more than once in any acylglycerol structure.

Structural studies of many triacylglycerols in animal body fats, fruit coat fats, and other samples, however, do not invariably support the "even" distribution theory. For example, many notable stearic acid-rich saturated triacylglycerols are found in fats, although the native saturated fatty acid pool does not exceed 60%.

The advent of improved analytical methods for studying acylglycerol structures, including enzyme-specific hydrolysis activities of pancreatic lipase, have also affirmed that biosynthesis of triacylglycerols may be different for plants and animals. Lipase-specific hydrolysis of fatty acids esterified to the 1 and 3 or α and α' positions prior to the 2 or β positions has shown that 95% of the 2-acylglyceryl positions in oil seeds contain oleic and/or linoleic acids; and excess amounts of oleic, linoleic, unsaturated, and saturated fatty acids occupy the 1 and 3 positions of the glycerol moiety. Acylglyceryl esterification of fatty acids in animals is less predictable. Superficial studies of animal fats have indicated that 20 to 80% of the 2-acylglyceryl positions are occupied by a saturated acid such as palmitic.

The inefficacies of the "even" distribution theory have led to other theories such as the *"restricted random" distribution* theory detailed by Kartha and the *"1,3 random" theory* of Vander Wal.

Aside from a number of exceptions and corollaries, the "restricted random" theory assumes that (1) fully saturated triacylglycerols (GS_3) can exist in each species of plant and animal fat until their *in vivo* solubility is prohibited; (2) random distribution of saturated and unsaturated fatty acids occurs among acylglycerol structures as long as GS_3 has *in vivo* solubility; and (3) when GS_3 solubility reaches a critical level, saturated fatty acid esterification at the 2-glyceryl position is limited, to ensure fat solubility, and any saturated fatty acids are randomly esterified to 1 and 3 positions.

The "1,3 random" distribution theory, on the other hand, claims that fatty acids esterified at the 2-acylglyceryl positions are the product of random distribution. Furthermore, the 1-,3-acylglyceryl positions are identical and fatty acid distribution at these sites is random. Favorable acceptability of this theory stems from experimental data, yet many exceptions to the theory do exist.

Stereospecific Analysis of Triacylglycerols

Principles of stereospecific analysis have suggested that fatty acids having short carbon chains and unsaturation are favorably esterified at the 2-acylglyceryl position with some very specific exceptions. Porcine fats, for example, deviate from this general rule since 70 to 80% of the 2-acylglyceryl positions may be occupied by palmitic acid. For marine animals, long-chain fatty acids are directed to the 1 and 3 positions, whereas polyunsaturated and short-chained fatty acids are esterified at the 2 position. This is contrary to avian fats, whose triacylglycerol synthesis seems to be directed by the preferential esterification of unsaturated fatty acids in the 2-acylglyceryl position.

Experimental observations such as these dealing with the location of fatty acids on the glycerol backbone can be achieved by specific enzymatic techniques, as indicated in Figure 7.1 (Brockerhoff and Yurkowski, 1966).

Pancreatic lipase–mediated hydrolysis of a triacylglycerol produces the free fatty acids R' and R''' plus a mixture of 1,2- and 2,3-diacylglycerols, and 2-acylglycerol. Analytical separation of these hydrolysis derivatives can be easily accomplished, and, if necessary, the free fatty acids may be identified.

Owing to the reactivity of the two diacylglycerols with phenyldichloro-phosphate, corresponding phospholipid-like derivatives may be synthesized. The 1,2-diacylglycerols yield an L form derivative and the 2,3-diacylglycerol produces a D form. The application of phospholipase A can now be used to selectively hydrolyze fatty acids residing at the 2-acylglyceryl position of L forms while leaving the D forms untouched. The so-called lyso structure resulting from fatty acid hydrolysis may be effectively isolated from the unreacted phospholipid remnants and the freed fatty acids. These steps in stereospecific analysis are then followed by GLC quantitation and identification of fatty acids esterified to (1) all glyceryl positions ($C_{1,2,3}$); (2) the 1,3 positions ($C_{1,3}$); and (3) the 2 positions (C_2). Careful analysis of the lyso derivative will reveal the identity and concentration of the 1-acylglyceryl fatty acid (C_1). The identification of the fatty acid in the

3 position can be determined by one of the following routes where

$$C_3 = (C_{1,3} - C_1) = (C_{1,2,3} - C_{1,2}) = (C_{2,3} - C_2)$$

Further projections concerning the theoretical number of triacylglycerols that may occur, without precise reference to stereospecific esterification directives during their biosynthesis, may be calculated as a function of the number of fatty acid structures available for esterification (Table 7.2).

Saturated and Unsaturated Fatty Acids

As discussed in earlier sections, fatty acids are esterified to the glyceryl portion (C_3H_5) of a triacylglycerol molecule and rarely exist as naturally free carboxylic acids at large concentrations within living cells.

The term *fatty acid* is used to define a whole series of carboxylic acids commencing with formic (methanoic) acid (C_1). The successive members of this structural series include acetic (ethanoic) acid (C_2), propionic (propanoic) acid (C_3), butyric (butanoic) acid (C_4), and other homologous acids that have carbon numbers up to C_{18} or higher.

Although the C_1 to C_4 fatty acids are hydrophilic, this miscibility property yields to increasing hydrophobic characteristics and complete water insolubility as the homologous series proceeds from C_5 to C_{18+}.

A study of fatty acid occurrence indicates

Table 7.2. Theoretical Number of Triacylglycerols Based on the Number of Fatty Acids

Types of different fatty acids available (n)	Number of possible triacylglycerols (N)
1	1
2	4
3	9
4	20
7	84
10	220
$\downarrow$	$\downarrow$
n	N^a

[a]$N = n + n(n - 1) + n(n - 1)(n - 2)/6.$

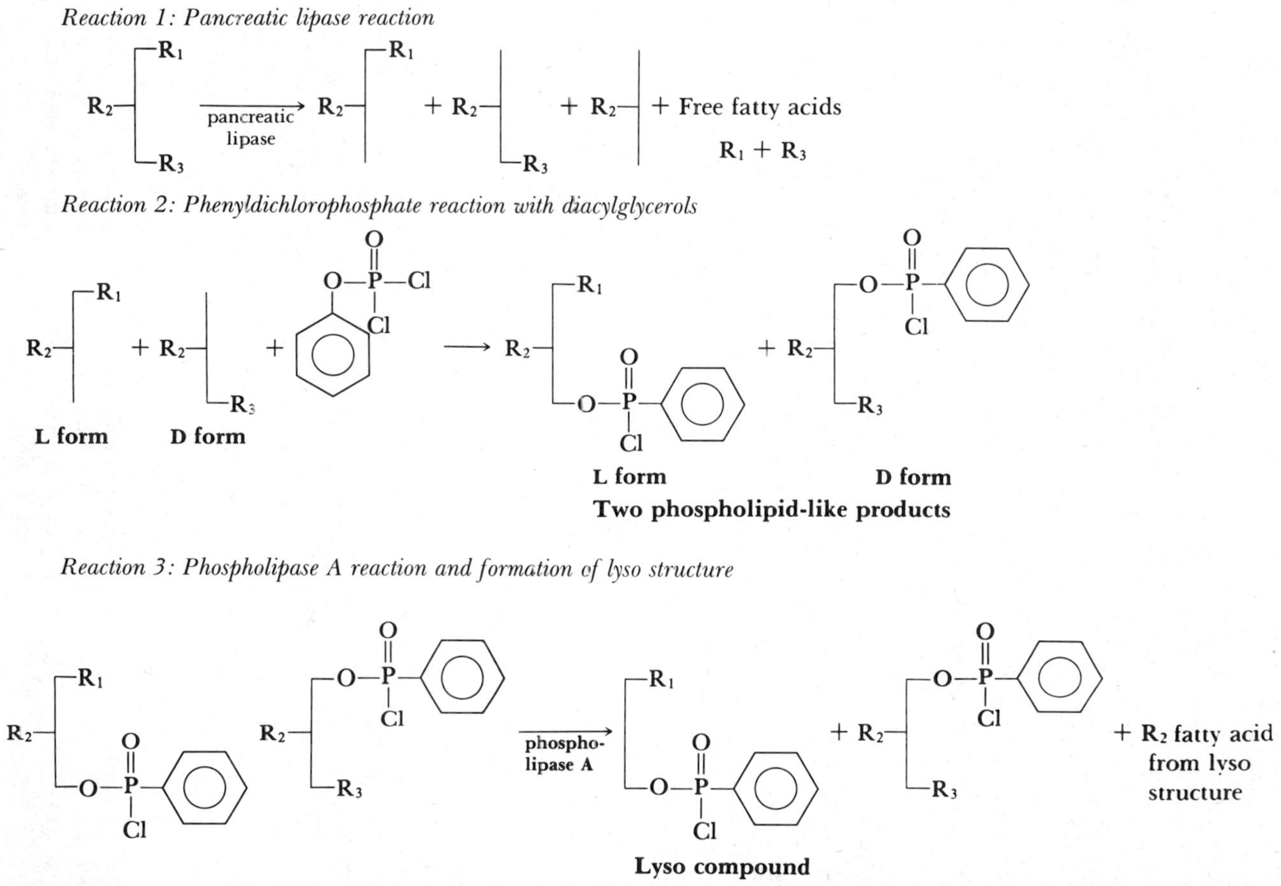

FIGURE 7.1. Key steps in the stereospecific analysis of triacylglycerols. Note that R_1, R_2, and R_3 represent fatty acids that many be identical or different, and reactions are not balanced since *possible* reaction products of each step are of primary interest.

that they are largely straight-chained and display an even number of carbon atoms. Recognition of branched chain *iso* fatty acids and odd-numbered fatty acids did not become a widely accepted fact until about 1954.

Fatty acids are described as *saturated* if they do not contain a singular double bond, whereas any fatty acid that contains one or more double bonds is respectively called *mono*unsaturated (monenoic) or *poly*unsaturated (polyenoic). The *degree of unsaturation* within a fatty acid is often cited with reference to the average number of double bonds in the fatty acid.

The common occurrence of both saturated, mono-, and polyunsaturated fatty acids necessitates uniform methods of nomenclature and short-hand designations that indicate the individual structures. Moreover, since unsaturated fatty acids can display *cis* or *trans* isomerization about a double bond, notations often indicate molecular configuration. Since relatively few unsaturated fatty acids are naturally *trans,* it should be gen-

erally understood for this text that configurations of unsaturated fatty acids are *cis* unless noted otherwise (Figure 7.2).

The systematic nomenclature for successively longer numbers of carbon atoms in straight-chained fatty acids follows the Geneva system. As an example, consider simple C_{12}, C_{14}, C_{16}, and C_{18} fatty acids that assume the corresponding prefixes such as *dodec-, tetradec-, hexadec-,* and *octadec-.* The suffix for saturated fatty acid nomenclature is *-anoic.* Therefore, stearic acid, which is a C_{18} fatty acid, would be named as *octadecanoic acid,* and other related acids would be named in a similar fashion.

Unsaturated fatty acid nomenclature follows analogous guidelines, but the suffix indicating the presence of one or more double bonds is *-enoic,* which is preceded by a *di-, tetra-, penta-,* and so on, prefix in order to indicate the number of double bonds in the carbon chain. A numerical indication of the double bond location is established by counting from the carboxyl group as *number 1,*

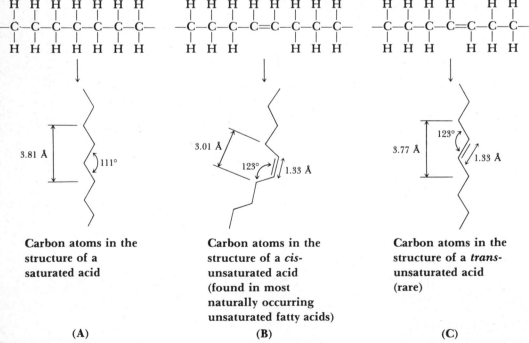

Carbon atoms in the structure of a saturated acid

(A)

Carbon atoms in the structure of a *cis*-unsaturated acid (found in most naturally occurring unsaturated fatty acids)

(B)

Carbon atoms in the structure of a *trans*-unsaturated acid (rare)

(C)

FIGURE 7.2. Structural configurations of (A) saturated, (B) *cis*-unsaturated, and (C) *trans*-unsaturated fatty acids.

and the rest of the carbons are numbered consecutively. Based on these guidelines, then, the structure shown below would be named as 9,12-octadecadienoic acid.

Those polyunsaturated fatty acids having a methylenic carbon ($-CH_2-$) located between two double bonds have higher potential reactivity than single or widely separated

$$H-\overset{\overset{\displaystyle H}{|}}{\underset{\underset{\displaystyle H}{|}}{C}}-\overset{\overset{\displaystyle H}{|}}{\underset{\underset{\displaystyle H}{|}}{C}}-\overset{\overset{\displaystyle H}{|}}{\underset{\underset{\displaystyle H}{|}}{C}}-\overset{\overset{\displaystyle H}{|}}{\underset{\underset{\displaystyle H}{|}}{C}}-\overset{\overset{\displaystyle H}{|}}{\underset{\underset{\displaystyle H}{|}}{C}}-\overset{\overset{\displaystyle H}{|}}{\underset{\underset{13}{}}{C}}=\overset{\overset{\displaystyle H}{|}}{\underset{12}{C}}-\overset{\overset{\displaystyle H}{|}}{\underset{\underset{\displaystyle H}{|}}{C}}-\overset{\overset{\displaystyle H}{|}}{\underset{10}{C}}=\overset{\overset{\displaystyle H}{|}}{\underset{9}{C}}-\overset{\overset{\displaystyle H}{|}}{\underset{\underset{\displaystyle H}{|}}{C}}-\overset{\overset{\displaystyle H}{|}}{\underset{\underset{\displaystyle H}{|}}{C}}-\overset{\overset{\displaystyle H}{|}}{\underset{\underset{\displaystyle H}{|}}{C}}-\overset{\overset{\displaystyle H}{|}}{\underset{\underset{\displaystyle H}{|}}{C}}-\overset{\overset{\displaystyle H}{|}}{\underset{\underset{\displaystyle H}{|}}{C}}-\overset{\overset{\displaystyle H}{|}}{\underset{\underset{\displaystyle H}{|}}{C}}-COOH$$

9,12 — Octadecadienoic acid

Double bonds C_{18} **Two double bonds**
follow the
C_9 **and** C_{12}
positions

The shorthand version for indicating fatty acid structures of both saturated and unsaturated fatty acids is based on the number of carbon atoms in the chain plus the unsaturated bonds between carbons. The structure of stearic acid, a saturated C_{18} fatty acid, is symbolized as 18:0; oleic acid, which is also a C_{18} fatty acid, contains one double bond (*cis*) between carbons 9 and 10 and is symbolized $18:1\Delta^9$; linoleic acid, a diunsaturated C_{18} fatty acid with points of unsaturation between C_9–C_{10} and C_{12}–C_{13}, is indicated as $18:2\Delta^9,12$ and so on, depending on the number of double bonds and the length of the carbon chain.

The polyunsaturated fatty acids may occur as *conjugated* or *nonconjugated* forms. Conjugated forms demonstrate alternating double–single–double bonds (1,3-unsaturation), whereas nonconjugated structures have more than one single bond that intervenes between successively occurring double bonds (e.g., 1,4-unsaturation). Among the known unsaturated fatty acids, the substituted allene positioning of double bonds or 1,2-unsaturation ($-CH_2=C=CH_2-$) does not exist.

double bonds. It should also be recognized that the reactivities of unsaturated fatty acids are influenced by the *cis* or *trans* configurations on either side of a double bond. Since *trans* forms encourage minimal irregularity in the straight hydrocarbon chain, these forms offer less reactivity than corresponding *cis* forms. Furthermore, *trans* forms usually display a higher melting point than *cis* forms.

The average degree of unsaturation for a fatty acid or a mixture of fatty acids can be expressed as an *iodine value* (*IV*), while the average molecular weight for acylglycerols can be indicated by a saponification value. Detailed methods for these indices are specified by the American Oil Chemists' Society (AOCS).

By definition, the *IV* is expressed as the number of centigrams of iodine that are required to iodinate 1 g of unsaturated fat (% iodine absorbed). Since iodination of alkenes follows the generalized reaction which re-

$$-\overset{|}{C}=\overset{|}{C}- + X_2 \longrightarrow -\overset{|}{\underset{X}{C}}-\overset{|}{\underset{X}{C}}-$$

$X_2 =$ a halogen; $I_2,\ Br_2,$ or Cl_2

$$-\overset{\overset{\displaystyle H}{|}}{\underset{\underset{\displaystyle H}{|}}{C}}-\overset{\overset{\displaystyle H}{|}}{C}=\overset{\overset{\displaystyle H}{|}}{C}-\overset{\overset{\displaystyle H}{|}}{\underset{\underset{\displaystyle H}{|}}{C}}-\overset{\overset{\displaystyle H}{|}}{C}=\overset{\overset{\displaystyle H}{|}}{C}-\overset{\overset{\displaystyle H}{|}}{\underset{\underset{\displaystyle H}{|}}{C}}-$$

Nonconjugated fatty acid chain

$$-\overset{\overset{\displaystyle H}{|}}{\underset{\underset{\displaystyle H}{|}}{C}}-\overset{\overset{\displaystyle H}{|}}{C}=\overset{\overset{\displaystyle H}{|}}{C}-\overset{\overset{\displaystyle H}{|}}{C}=\overset{\overset{\displaystyle H}{|}}{C}-\overset{\overset{\displaystyle H}{|}}{\underset{\underset{\displaystyle H}{|}}{C}}-$$

Conjugated fatty acid chain

sults in a vicinal dihalide, the total consumption of iodine by a given amount of fat or oil reflects its degree of unsaturation. Although classical methods cite *IV* indices, pure iodine is rarely used since it fails to react in a purely quantitative fashion with all unsaturated sites. Therefore, many historically *wet* methods of analysis for *IV* estimations such as the Wijs (Wys) and Hanus methods, respectively, use ICl or IBr for halogenation. The results of halogen consumption by samples using these reagents is then expressed as an *IV*. The higher the *IV* for a lipid, the greater will be its sites of unsaturation.

Saponification values are also based on classical *wet* chemical methods of analysis in which a weighed sample of fat (usually 1.0 g) is treated with a measured volume of alcoholic potassium hydroxide and saponified by heating. Back-titration of the excess potassium hydroxide using a standard acid (e.g., 0.5 *N* acid) permits calculation of the weight (mg) of potassium hydroxide required to bring about alkaline hydrolysis of the fat, and calculation of a saponification value. This measurement provides an average index of the molecular weight of glycerides that comprise a lipid sample. Low-molecular-weight (short-chained fatty acid) acylglycerols require larger amounts of potassium hydroxide for the saponification of a gram of fat than those acylglycerols that contain higher-molecular-weight or long-chained fatty acids. As a rule, the saponification value for glyceryl moieties esterified to fatty acids is inversely proportional to the mean molecular weight of the glycerides that comprise the fat or oil.

The *IV* and saponification values for various lipids can serve as useful criteria for the industrial and dietary classifications of lipids, but present methods for establishing these indices are often calculated from chromatographic and nuclear magnetic resonance data.

Natural Occurrence of Fatty Acids

Saturated fatty acids: Formic (C_1), acetic (C_2), and propionic (C_3) acids represent the lower numbered homologues of all the important saturated fatty acids, but they are not commonly found in fats and oils. Butyric (C_4) acid and higher saturated homologues of these acids do occur widely among food lipids (Table 7.3). For example, butyric acid may account for 2 to 4% of the total fatty acid content of milk fats obtained from var-

Table 7.3. Boiling Points of Saturated Fatty Acids and Melting Points of Acids and Corresponding Simple Triglycerides

Number of carbon atoms	Acid	Acid B.P. (°C at 16 mm)	Acid M.P. (°C)	Triglyceride M.P. (°C)[a]
4	Butyric	163 (at 760 mm)	−8	—
6	Caproic	107	−3.4	—
8	Caprylic	135	16.7	—
10	Capric	159	31.6	31.5
12	Lauric	182	44.2	46.4
14	Myristic	202	54.4	57.0
16	Palmitic	222	62.9	65.5
18	Stearic	240	69.6	73.1
20	Arachidic	—	75.4	—
22	Behenic	—	80.0	—
24	Lignoceric	—	84.2	—

[a]Melting point of the highest melting, most stable crystal modification.

SOURCE: From *Bailey's Industrial Oil and Fat Products* (4th ed., Vol. 1, D. Swern (ed.). Copyright © 1979, John Wiley & Sons. Reprinted with permission.

ious mammals; caproic (octanoic) acid comprises 1 to 4% of milk fats, <10% of coconut and palm kernel oils, and up to 70% of the seed fat in *Cuphea hookeriana;* and capric (decanoic) acid occurs at 2 to 4% levels in milk fats and 48% levels in Palmae seed oils. Pelargonic (nonanoic) acid, however, occurs at nearly undetectable concentration levels in most seed oils and can only be produced in commercial quantities by the ozonolysis of oleic acid.

Lauric (dodecanoic) acid complements palmitic (hexadecanoic) and stearic (octadecanoic) acids as the most widely distributed saturated fatty acids in living organisms. Although the seed fats of the *Lauraceae* (laurel) family contain up to 90% lauric acid, palm kernel oil, coconut oil, and babassu butters that have 40 to 50% lauric acid are among the most common plant sources. Palmitic and stearic acids are more widely distributed at higher concentrations than lauric acid. These two fatty acids often coexist at varying ratios in seed oils of single plant species, mainly as a result of geographic influences, soil conditions, and other growth influences. Genetic effects are also of central importance in controlling the distribution of saturated fatty acids in oil seeds.

Many of the traditionally recognized levels of fatty acids in plant sources can be genetically altered as demonstrated in the case of safflower seeds. Seeds normally exhibiting a palmitic (C_{16}):stearic (C_{18}) acid ratio of 6%:3% have been modified to produce a 6%:≥10% ratio. For animals, too, the C_{16}:C_{18} ratio may be variable owing to different anatomical residence sites for fats. Based on piglet studies in which palmitic acid content averages 22.5% of all fatty acids, typical C_{16}:C_{18} ratios may reflect a 24:1 ratio for the outer shoulder; ~2:1 for the inner shoulder, inner and outer loins; and, 1.8:1 for kidney tissue.

Although the ratios of various saturated fatty acids to stearic acid have been widely surveyed in many foods, the fatty acid ratios present in commercial "stearic acids" have been largely ignored. These commercial stearic acids are often derived from extensive processing of hydrogenated soybean oils, tallows, or other fats and oils.

It should be noted at this point that heptadecanoic (margaric) acid had been regarded as a rare fatty acid whose occurrence was limited to a few species. However, about 20 years ago this acid was recognized as a component of many animal fats. In fact, the presence of this fatty acid in high-purity vegetable oils is so minor that its presence has been used as an analytical index for detecting the adulteration of pure vegetable oils with animal fat.

Arachidic (C_{20}), behenic (C_{22}), and lignoceric (C_{24}) saturated fatty acids appear as minor components of common fats and oils including soybean, olive, corn, cotton seed, peanut, and safflower oils.

For further information detailing the saturated fatty acid distribution profiles of some important food fats and oils, consult Table 7.4.

Mono- and polyunsaturated fatty acids: Unsaturated fatty acids occur widely throughout the plant and animal world. Many historical problems in understanding their chemistry and biochemistry stem from analytical difficulties related to their isolation and purification. Furthermore, many of the most interesting unsaturated fatty acids are notably less stable than their saturated counterparts and readily display geometric or positional isomerism involving their double bonds. As indicated earlier, the *cis* position is preferred around a double bond, and the most common site of unsaturation is between the C_9 and C_{10} positions. Most of the important unsaturated fatty acids are even numbered (e.g., C_{18}) but odd-numbered fatty acids do exist. Butterfat is probably the best example of a common dietary fat that contains some unsaturated, odd-numbered carbon fatty acids while certain fishes such as mullet may contain over 20%. Also, contrary to the general rule that *trans*-unsaturated fatty acids are generally uncommon, the catalytic hydrogenation of unsaturated fatty acids by

Table 7.4. Typical Compositions and Chemical Constants of Common Edible Fats and Oils

Column headers (Carbon atoms: double bonds — common name): 4:0 Butyric, 6:0 Caproic, 8:0 Caprylic, 10:0 Capric, 11:0 Undecanoic, 12:0 Lauric, 13:0 Tridecanoic, 14:0 Myristic, 14:1 Myristoleic, 15:0 Pentadecanoic, 15:1 Pentadecenoic, 16:0 Palmitic, 16:1 Palmitoleic, 17:0 Margaric, 17:1 Margaroleic, 18:0 Stearic, 18:1 Oleic, 18:2 Linoleic, 18:3 Linolenic, 19:0 Nonadecanoic, 20:0 Arachidic, 20:1 Gadoleic, 20:2 Eicosadienoic, 20:4 Arachidonic, 22:0 Behenic, 22:1 Erucic, 22:2 Docosadienoic, 24:0 Lignoceric.

Fat/Oil	4:0	6:0	8:0	10:0	11:0	12:0	13:0	14:0	14:1	15:0	15:1	16:0	16:1	17:0	17:1	18:0	18:1	18:2	18:3	19:0	20:0	20:1	20:2	20:4	22:0	22:1	22:2	24:0	Iodine value range	Saponification value range
Babassu		0.4	5.3	5.9		44.2		15.8				8.6				2.9	15.1	1.7			0.1								13–18	247–254
Butterfat	3.8	2.3	1.1	2.0	0.1	3.1	0.1	11.7	0.8	1.6		26.2	1.9	0.7	0.2	12.5	28.2	2.9	0.5			0.2							25–42	210–240
Chicken fat								1.3	0.2			23.2	6.5	0.3	0.1	6.4	41.6	18.9	1.3					0.1					76–80	194–204
Citrus seed oil						0.2		0.5				28.4	0.2			3.5	23.0	37.8	5.7		0.8								99–106	192–197
Cocoa butter						0.1						25.8	0.3			34.5	35.3	2.9			1.1								32–40	190–200
Coconut oil		0.5	8.0	6.4		48.5		17.6				8.4				2.5	6.5	1.5			0.1								7–13	248–264
Cohune oil		0.3	8.7	7.2	0.1	47.3		16.2				7.7				3.2	8.3	1.0											8–14	250–260
Corn oil						0.1						12.2	0.1			2.2	27.5	57.0	0.9		0.1								110–128	186–196
Cottonseed oil								0.9				24.7	0.7	0.1		2.3	17.6	53.3	0.3		0.1								99–121	189–199
Lard								1.5		0.2		24.8	3.1	0.5	0.3	12.3	45.1	9.9	0.1		0.2	1.3	0.1	0.4					53–68	192–203
Murumurú tallow		0.1	1.3	1.5		46.2		32.4				5.6	0.1			2.2	8.9	1.5			0.2								8–13	237–247
Oat oil								0.2				17.1	0.5			1.4	33.4	44.8			0.2	2.4							105–110	180–198
Olive oil												13.7	1.2			2.5	71.1	10.0	0.6		0.9								76–90	188–196
Palm oil						0.3		1.1				45.1	0.1			4.7	38.8	9.4	0.3		0.2								45–56	195–205
Palm kernel oil		0.3	3.9	4.0		49.6		16.0				8.0				2.4	13.7	2.0			0.1								14–24	243–255
Peanut oil								0.1				11.6	0.2	0.1		3.1	46.5	31.4			1.5	1.4	0.1		3.0			1.0	84–102	188–196
Rapeseed oil								0.1				2.8	0.2			1.3	23.8	14.6	7.3		0.7	12.1	0.6		0.4	34.8	0.3	1.0	97–110	168–183
Rapeseed oil (low erucic)												3.9	0.2			1.9	64.1	18.7	9.2		0.6	1.0			0.2			0.2	110–115	—
Rice bran oil				0.1		0.4		0.5				16.4	0.3			2.1	43.8	34.0	1.1		0.5	0.4			0.2			0.1	92–109	181–195
Safflower oil						0.1						6.5				2.4	13.1	77.7			0.2								138–151	186–198
Safflower oil (high oleic)								0.1				5.5	0.1			2.2	79.7	12.0	0.2		0.2								85–93	185–195
Sesame seed oil								0.1				9.9	0.3			5.2	41.2	43.2	0.2		0.3								104–118	187–196
Soybean oil								0.1				11.0	0.1			4.0	23.4	53.2	7.8		0.3				0.1				125–138	188–195
Sunflower seed oil				0.1		0.5		0.2				6.8	0.1			4.7	18.6	68.2	0.5		0.4								122–139	186–196
Tallow (beef)						0.1		3.3	0.2	1.3	0.2	25.5	3.4	1.5	0.7	21.6	38.7	2.2	0.6	0.1	0.1			0.4					33–50	190–202
Tallow (mutton)				0.2		0.3		5.2	0.3	0.8	0.3	23.6	2.5	2.0	0.5	24.5	33.3	4.0	1.3	0.8				0.4					35–46	192–198
Tucum oil		0.2	2.9	2.3		51.8	0.3	22.0				6.8				2.3	9.3	2.4											10–14	240–250
Ucuhuba tallow		0.3	0.3	0.8		16.3		70.8	1.3			4.3	0.5			0.7	3.7	0.6	0.3		0.1								6–17	215–232

[a]Fatty acid compositions were determined by gas–liquid chromatography and are expressed as mean average weight percent compositions on a fatty acid basis. Trace acids (less than 0.1%) are excluded.

NOTE: Data supplied through the courtesy of Durkee Industrial Foods, SCM Corporation, Cleveland, Ohio.

industrial processes, as well as the partial hydrogenation of polyunsaturated fatty acids by rumen bacteria, do result in *trans* forms. The latter reaction, mediated by rumen bacteria, accounts for the presence of up to 9% *trans*-unsaturated fatty acids in milk fat that are reported as elaidic acid.

$$CH_3(CH_2)_7 \quad (CH_2)_7COOH$$
$$C=C$$
$$H \qquad H$$

Oleic (*cis*-9-octadecenoic) acid

$$CH_3(CH_2)_7 \quad H$$
$$C=C$$
$$H \qquad (CH_2)_7COOH$$

Elaidic (*trans*-9-octadecenoic) acid

Mono- and diunsaturated fatty acids are not uncommon in the C_{18} unsaturated fatty acid pool of land plants and animals, but the occurrence of four or more double bonds seems entirely limited to C_{20} to C_{24} fatty acids that have marine origins.

Depending on the number of carbons in a fatty acid structure, it will contain two to ten fewer hydrogen atoms than its corresponding saturated fatty acid. This can be empirically calculated for various unsaturated fatty acids using the following formulas:

Type of fatty acid	Formula
Monounsaturated	$C_nH_{2n-2}O_2$
Diunsaturated	$C_nH_{2n-4}O_2$
Triunsaturated	$C_nH_{2n-6}O_2$
Tetraunsaturated	$C_nH_{2n-8}O_2$
Pentaunsaturated	$C_nH_{2n-10}O_2$

It should be noted that naturally occurring unsaturated fatty acids have at least ten carbons with the exception of crotonic acid ($CH_2CH=CH—COOH$). Of all the naturally occurring monounsaturated fatty acids outlined in Table 7.4, palmitoleic and oleic acids account for the bulk of unsaturated fatty acids. Many of those fatty acids that do not have double bonds in the 9 position (between C_9 and C_{10}) are common, but they exist at low concentrations in native fats and oils. Important exceptions to this rule are petroselinic and erucic acids.

Unsaturated fatty acids that have two (di-) and three (tri-) unsaturated carbon–carbon bonds commonly occur. The distribution of these polyunsaturated fatty acids in terrestrial plants and animals is limited to fatty acids with a minimum of 18 carbons. This is not true for marine animal oils, however, in which polyunsaturated C_{14}–C_{16} fatty acids are known. Of the common unsaturated fatty acids currently recognized, linoleic (*cis,cis*-9,12-octadecadienoic) and linolenic (*cis,cis,cis*-9,12,15-octadecatrienoic) acids are the most important and ubiquitous in their occurrence. In many cases, the linoleic acid content of edible vegetable oils has been improved relative to its coexisting native fatty acids. Most of these improvements reflect the effects of plant selection and other genetic modifications within plant species. This is especially true for safflower oils in which linoleic acid content has been improved to the point where it accounts for up to 78% of oil seed composition. The important food-industrial roles for sunflower oil plus its functional attributes have also encouraged the genetic development of high-linoleic-acid seeds. Exclusive of the fact that linoleic acid content in a sunflower head shows increased concentration from the periphery of the composite flower toward its center, an average linoleic content of 70% is not uncommon for many plants grown in the northern United States. Southern-grown sunflower seeds, on the other hand, may have up to 40% less simply as a result of geographic and climatic variations.

Linolenic (*cis,cis,cis*-9,12,15-octadecatrienoic) acid, which is a triunsaturated fatty acid, also occurs widely as a significant component of many vegetable oils usually noted for their high degree of unsaturation. This fatty acid may account for 35% of hempseed, ≥45% of linseed, and up to 65% of perilla oils, but these sources are more noteworthy as a source of drying oils than food oils.

Of all the depot fats in animals, linolenic acid makes up <1% of the fatty acids, although horse fat may contain concentration levels of ~15%.

Tetraunsaturated fatty acids do occur in plant and fish tissues, but they are quite rare when compared to concentrations of linoleic and linolenic acids. Aside from these generalizations, the tetraunsaturated fatty acid known as parinaric (9,11,13,15-octadecatetraenoic) acid can contribute up to 50% of the fatty acid content in *Parinarium laurinum*, and the bulk of herring oil is composed of the C_{18} polyene fatty acid denoted as 6,9,12,15-octadecatetraenoic acid. Arachidonic (5,8,11,14-eicosatetraenoic) acid is especially important as a tetraunsaturated fatty acid that is present in animal fats and may have essential nutritional importance. Arachidonic acid was first recognized to occur in mosses and ferns around 1965, but it is now regarded to be a wide-ranging component of many plant seeds. In addition to the tetraunsaturated oils cited here, many others are found in fish oils.

Pentaunsaturated fatty acid occurrence is also relegated mainly to marine animal oils. A typical representative of these fatty acids is clupanodonic (4,8,12,15,19-docosapentaenoic) acid, which occurs in nearly all marine oils. These pentaunsaturated oils may also exist with hexaunsaturated fatty acids such as 4,7,10,13,16,19-docosahexaenoic acids, which exist in sardine, cod liver, herring, and coho salmon oils.

For additional information dealing with the hierarchy of mono- and polyunsaturated double bonds, consult Tables 7.5 and 7.6.

Table 7.5. Fatty Acids with One Double Bond

Formula	Systematic name	Common name	M.P. (°C)	Principal source
$C_{10}H_{18}O_2$	4-Decenoic	Obtusilic	—	*Lindera obtusiloba* fat
$C_{10}H_{18}O_2$	9-Decenoic	Caproleic	—	Animal milk fats
$C_{12}H_{22}O_2$	4-Dodecenoic	Linderic	1.3	*Lindera obtusiloba* fat
$C_{12}H_{22}O_2$	9-Dodecenoic	Lauroleic	—	Animal milk fats
$C_{14}H_{26}O_2$	4-Tetradecenoic	Tsuzuic	18.5	*Litsea glauca* fat
$C_{14}H_{26}O_2$	5-Tetradecenoic	Physteric	—	Sperm whale oil (14%)
$C_{14}H_{26}O_2$	9-Tetradecenoic	Myristoleic	—	Animal milk fats, sperm whale oil, *Pycnanthus kombo* fat (23%)
$C_{16}H_{30}O_2$	9-Hexadecenoic	Palmitoleic	—	Animal milk fats, marine oils (60–70%), seed fats (*Tricospidaria lanceolata*), *Doxantha unguiscati*, mink and beef fat
$C_{17}H_{32}O_2$	9-Heptadecenoic	—	—	Tallow, Canadian musk ox
$C_{18}H_{34}O_2$	6-Octadecenoic	Petroselinic	30	Umbelliferae, especially parsleyseed oil (75%)
$C_{18}H_{34}O_2$	9-Octadecenoic	Oleic	14, 16	Olive oil (80%), pecan oil (85%), and generally in all vegetable and animal fats
$C_{18}H_{34}O_2$	*trans*-9-Octadecenoic	Elaidic	44	Beef fats, many animal fats
$C_{18}H_{34}O_2$	*trans*-11-Octadecenoic	Vaccenic	44	Butter, tallow
$C_{20}H_{38}O_2$	*cis*-5-Eicosenoic	—	—	*Limnanthes* genus
$C_{20}H_{38}O_2$	9-Eicosenoic	Gadoleic	—	Marine oils
$C_{20}H_{38}O_2$	11-Eicosenoic	—	—	Jojoba wax
$C_{22}H_{42}O_2$	11-Docosenoic	Cetoleic	—	Marine oils
$C_{22}H_{42}O_2$	13-Docosenoic	Erucic	33.5	Cruciferae (mustard family) (over 40%)
$C_{24}H_{46}O_2$	15-Tetracosenoic	Selacholeic	—	Marine oils
$C_{26}H_{50}O_2$	17-Hexacosenoic	Ximenic	—	*Ximenia americana* fat
$C_{30}H_{58}O_2$	21-Triacontenoic	Lumequeic	—	*Ximenia americana* fat

SOURCE: From *Bailey's Industrial Oil and Fat Products* (4th ed., Vol. 1), D. Swern (ed.). Copyright © 1979, John Wiley & Sons. Reprinted with permission.

Table 7.6. Fatty Acids with More Than One Double Bond

Formula	Systematic name	Common name	M.P. (°C)	Sources
$C_{18}H_{32}O_2$	*cis,cis*-9,12-Octadecadienoic	Linoleic	−5	Safflower oil (75%), sunflower oil (60%), cottonseed oil (45%), and many seed fats
$C_{18}H_{30}O_2$	*cis,cis,cis*-9,12,15-Octadecatrienoic	Linolenic	−11	Linseed oil (45–50%), perilla oil (65%), and other drying oils and seed fats
$C_{18}H_{30}O_2$	6,9,12-Octadecatrienoic	—	—	Boraginaceae family (0–27%)
$C_{18}H_{30}O_2$	*cis,trans,trans*-9,11,13-Octadecatrienoic	α-Eleostearic	49	Tung oil (85%), poyoak, and a few other oils
$C_{18}H_{30}O_2$	*trans,trans,trans,*-9,11,13-Octadecatrienoic	β-Eleostearic	71	Isomerization of α-isomer
$C_{18}H_{28}O_2$	9,11,13,15-Octadecatetraenoic	Parinaric	86(α) 96(β)	*Parinarium laurinum*
$C_{20}H_{32}O_2$	5,8,11,14-Eicosatetraenoic	Arachidonic	ca. −50	Animal depot fats and phosphatides of liver and brain
$C_{22}H_{34}O_2$	4,8,12,15,19-Docosapentaenoic	Clupanodonic	—	Marine oils

SOURCE: From *Bailey's Industrial Oil and Fat Products* (4th ed., Vol. 1, D. Swern (ed.). Copyright © 1979, John Wiley & Sons. Reprinted with permission.

Some novel fatty acids found in oils and fats: Prior to 1929 it was generally recognized that fats and oils were composed of even-numbered, straight carbon chains. In later years it became evident that butterfat, ox fat, mutton, marine oils, and other animal fats also contained odd-numbered fatty acid chains plus some methyl branched fatty acids.

Still other variations on the theme of fatty acid structure are shown by their hydroxy and dihydroxy derivatives. Most of these fatty acids have low concentrations in foods. For example, the free and bound forms of long-chained hydroxy acids account for only 0.12% of the fatty acid content of butterfat. Ricinoleic (12-hydroxy-9-octadecenoic) acid is somewhat of an exception in its occurrence because it comprises up to 90% of the mixed fatty acids in castor oil.

Vicinal (1,2-system) dihydroxy acids also exist as glycols in selected plants and may be saturated or unsaturated. Fatty acids of this type are probably more widespread than presently recognized owing to their difficult isolation and structural analysis. Notable amounts of vicinal dihydroxy acids including 9,10-dihydroxydocosanoate and 15,16-dihydroxytetracosanoate are reported to exist at respective concentrations of 16.5 and 6.0% in vegetable oils of *Cardimine impatiens*. Even higher concentrations of a novel triunsaturated hydroxy fatty acid known as kamolenic (18-hydroxy-9,11,13-octadecatrienoic) acid can be obtained from oils of the *Kamala* species and seed fats of *Mallotus*. These sources may contain up to 70% kamlolenic acid.

Many other types of interesting fatty acid structures are present in living organisms, but their nutritional and biochemical importance is either small or simply unknown. Some of these acids display acetylenic (triple bond), cyclopropenoid, epoxy, keto-, cyclopentenoid and furanoid features as indicated in Figure 7.3.

Analytical Detection and Quantitation of Fatty Acids

Acylglycerols may be separated and identified by means of specialized thin-layer chromatographic (TLC) methods. However, the complexity of most oils and fats diminishes the usefulness of TLC as a singular method for the routine distributional analysis and quantitation of fatty acids. With the excep-

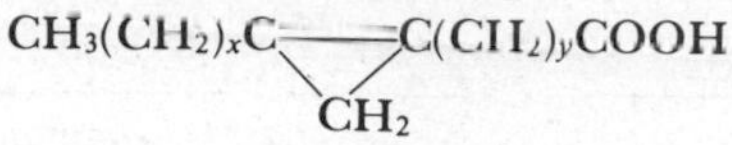

$$CH_3(CH_2)_xC \underset{CH_2}{=\!=\!=} C(CH_2)_yCOOH$$

$$x + y = 13$$

Malvalic acid: A cyclopropenoid acid obtained from seed oils of **Malva verticilla** and **M. parviflora**

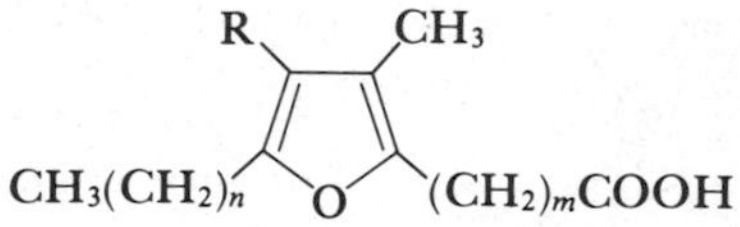

$$CH_3(CH_2)_n \quad O \quad (CH_2)_mCOOH$$

(with R and CH₃ substituents)

m	n	R
6	2	CH_3
8	4	H
10	4	CH_3

Some furanoid fatty acids typically occur in the liver oils of cod fish, dog fish, and capelin

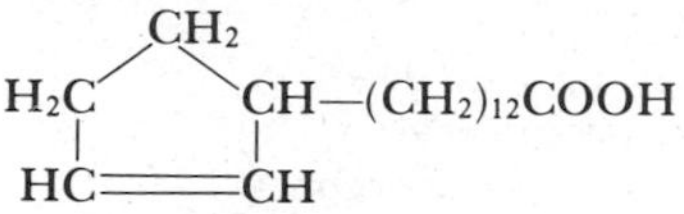

$$H_2C \quad CH-(CH_2)_{12}COOH$$
$$HC =\!=\!= CH$$

Chaulmoogric acid: A cyclopentenoid fatty acid obtained from the **Flacourtaceae**

FIGURE 7.3. Some unusual fatty acid structures.

tion of preparative methods, which effectively use classical methods of TLC, this approach to lipid analysis began to wane in 1952 when James and Martin reported a gas–liquid chromatographic (GLC) method for the separation of saturated fatty acids having up to 12 carbons. The low volatility of fatty acids greater than C_{12} limited early applications of this method. Cropper and Heywood (1953) then found that GLC separations of fatty acids up to C_{34} were possible after individual fatty acids were converted to their corresponding methylester derivatives ($R-CO-O-CH_3$). In more recent years the use of capillary chromatographic columns and the development of new GLC technology have permitted the quantitative analysis of virtually any fatty acid mixture.

Aside from determining fatty acid profiles for sample oils and fats, the characteristic vapor phase separations of GLC also permit the microgram introduction of fatty acids into a mass spectrometer. Such GLC–mass spectrometer interfaced instruments currently provide a rapid and powerful tool for characterizing complex mixtures of fatty acids present in native fats and oils; detecting contaminants in so-called pure fatty acids; as well as detecting levels of hydrolytic, oxidative, or thermal fatty acid decomposition (Leemans and McCloskey, 1967).

In addition to these instrumental methods, the advent of reverse-phase, high-pressure liquid chromatography (HPLC) has encouraged the development of routine screening methods for natural or manufactured fatty acid mixtures. The desirability of this method lies in its nondestructive nature; volatilization of individual fatty acid and lipid components of samples is not necessary; and the method offers high resolution of individual components in fatty acid mixtures. Derivatization of fatty acids is not a prerequisite to fatty acid analyses using HPLC, but the detection of free fatty acids at nanomole concentrations can be accomplished by producing their corresponding *p*-bromophenacyl derivatives (Plattner *et al.*, 1977). Figure 7.4 illustrates the typical fatty acid resolution range for this method and some specific applications to dairy food products.

PHYSICAL PROPERTIES OF FATS AND FATTY OILS

Viscosity and Oiliness

The relatively high viscosities demonstrated by fatty oils result from the intermolecular attractions of long fatty acid chains esterified

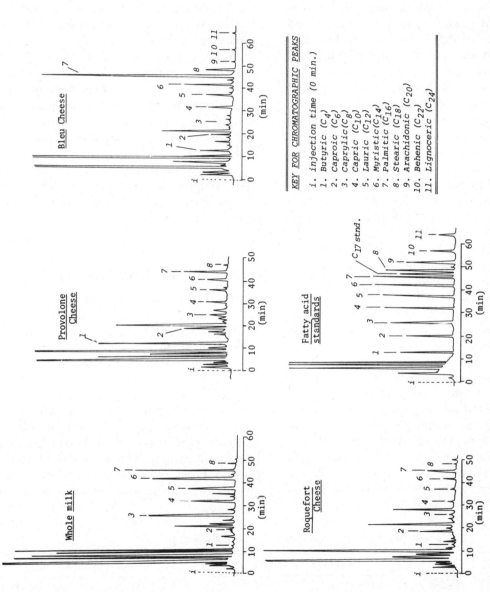

FIGURE 7.4. Fatty acid profiles for selected dairy products based on high-pressure liquid chromatographic analysis. Conditions: column, C_{18}-reverse phase; detector, 254 nm, 1.0 AUFS; solvent gradient, water/acetonitrile; sample, p-bromophenylacyl esters of fatty acids. Since fatty acids impart characteristic flavors and aromas to many dairy products, fatty acid profiles such as those illustrated may be useful as a quality control or analytical index for food production.

KEY FOR CHROMATOGRAPHIC PEAKS

i. injection time (0 min.)
1. *Butyric (C_4)*
2. *Caproic (C_6)*
3. *Caprylic(C_8)*
4. *Capric (C_{10})*
5. *Lauric (C_{12})*
6. *Myristic(C_{14})*
7. *Palmitic (C_{16})*
8. *Stearic (C_{18})*
9. *Arachidonic (C_{20})*
10. *Behenic (C_{22})*
11. *Lignoceric (C_{24})*

to acylglycerol structures. Oils that contain low-molecular-weight fatty acids are slightly less viscous than those oils that have a similar degree of unsaturation but higher-molecular-weight acids. The viscosity of an oil also shows increases when unsaturated sites of the fatty acid structures are hydrogenated so that saturation of molecules is increased.

The coefficient of viscosity (η) for oils is defined in terms of the force per unit area required to maintain a unit difference of velocity between two parallel (laminar) layers that are separated by a unit distance. If the force per unit area is expressed in dynes per square centimeter, velocity is reported in centimeters per second and distance is in centimeters; then the value for (η) may be expressed in poise units.

For many classical instrumental methods of oil viscosity measurement such as efflux, falling ball, or rising bubble viscometers, the weight of the liquid provides the necessary force to cause flow. The use of these instruments also results in the measurement of *kinematic viscosity,* in which *stokes* are measured. Since stokes are equal to the dynamic viscosity in poises divided by the fluid density, centipoises (cP) and centistokes (cSt) are used in preference to poises or stokes. Each of these respective units (i.e., cP and cSt) represent 1/100th of the standard units. These units of measurement are contrary to the Saybolt viscometer measurements where a fixed volume of oil effluxes from the instrument over a period of time measured in seconds. Therefore viscosities for oils are indicated in seconds. Some typical viscosities

for important fats and fatty oils at different temperatures (100 and 200°F) based on kinematic and Saybolt viscosity values are outlined in Table 7.7.

It should be noted that fatty acids, their esters, and oils demonstrate a near linear relationship between the log of the viscosity and the reciprocal of the absolute temperature (Figure 7.5).

All samples of fatty oils show Newtonian behavior, but many oils demonstrate thixotropic plastic properties at high rates of shear. The reasons for this behavior are somewhat speculative, but high shear stresses are thought to cause orientation of micelles and molecular aggregates.

In the most practical terms, viscosity measurements are useful for evaluating oxidative or thermal polymerization changes in fats and fatty oils.

Surface and Interfacial Tension

Surface and interfacial tension data for many organic compounds including fats and oils fall in the range of 20 to 40 dynes/cm. Increases of surface tension parallel corresponding increases in carbon chain length and decrease with increasing temperature.

Other than the general absence of significant surface activity, this property of neutral triacylglycerols is not important. This is quite contrary, however, to the surface and interfacial tension effects caused by monoacylglycerols. When added to neutral acylglycerols, monoacylglycerols markedly re-

Table 7.7. Some Typical Viscosities for Important Fats and Fatty Oils

Food oil	Specific gravity	Kinematic viscosity (cSt)		Saybolt viscosity (s)	
		100°F (37.8°C)	210°F (98.9°C)	100°F (37.8°C)	210°F (98.9°C)
Cotton seed	0.9187	35.88	8.39	181	52.7
Soybean	0.9228	28.49	7.60	134	50.1
Sunflower	0.9207	33.31	7.68	156	50.3
Coconut	0.9226	29.79	6.06	140	45.2
Palm kernel	0.9190	30.92	6.50	145	46.5
Lard	0.9138	44.41	8.81	206	54.2
Cod liver	0.9138	32.79	7.80	153	50.7

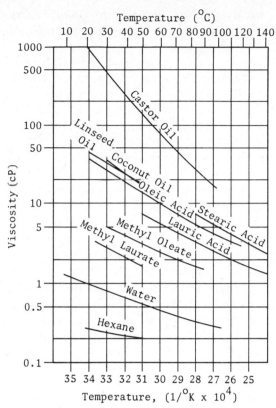

FIGURE 7.5. Viscosities of fatty oils, fatty acids, and methyl esters compared with water and hexane. (From *Bailey's Industrial Oil and Fat Products* (4th ed., Vol. 1). D. Swern (ed.). Copyright © 1979, John Wiley & Sons, and reprinted with permission.)

duce their interfacial tension. This principle accounts for the functional improvement of shortenings, to which monoacylglycerols have been added since the 1930s. The emulsifying activities of monoacylglycerols is somewhat limited compared to the actions of phospholipids, which effectively decrease the interfacial tension between fat and water phases in food products such as margarine, mayonnaise, chocolates, creams, and other foods.

Specific Gravity

Specific gravities for common fats and oils are mainly limited to the range of 0.914 to 0.964 at 15°C. Many studies have been undertaken to accumulate and tabulate specific gravity data for different fats and oils. These values correlate closely with the specific gravity values calculated by the formula

$$\text{specific gravity} = 0.8475$$
$$+ 0.00030 \text{ (saponification value)}$$
$$+ 0.00014 \ (IV)$$

According to normal processing temperature ranges of 65 to 250°C, density displays a linear decrease of 0.64 g/L/°C. Solid-state densities for fats are much higher than those of liquid fats and range from 1.00 to 1.06 kg/L.

Melting Point and Crystal Structure

Commercial fats, oils, and pure fatty acids can be crudely characterized and occasionally identified on the basis of their melting and freezing points. Melting point determinations also have technical importance for assessing food fat behavior in processed foods. As a general rule, the melting points for fatty acids increase with increasing chain length and decrease as acids become more unsaturated. Melting points for even- and odd-numbered carbon chain lengths show an interesting alternation between any two successive numbers of a homologous series, as illustrated in Figure 7.6.

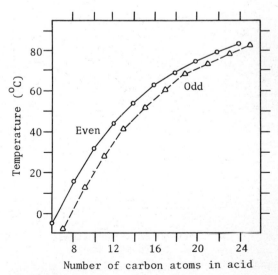

FIGURE 7.6. Melting point alteration of straight-chain saturated fats. (From *Bailey's Industrial Oil and Fat Products* (4th ed., Vol. 1). D. Swern (ed.). Copyright © 1979, John Wiley & Sons, and reprinted with permission.)

Pure acylglycerols of single fatty acids and well-characterized mixed triacylglycerols usually demonstrate reproducible melting and freezing points that reflect the collective effects of their composite fatty acid esters (Figure 7.7). The significance of these data, however, is often difficult to assess since many commercial fats show *polymorphic behavior*. *Polymorphism* refers to the different and sometimes unpredictable patterns of molecular packing that are displayed by fat crystals when their constituent triacylglycerol molecules are chilled to the point of solidification at different chilling rates. Because the thermal history of solidified triacylglycerols effects the crystalline packing of molecules, variations in chilling rates can produce different density relationships and melting points in chemically identical triacylglycerol samples.

Three polymorphic forms are generally recognized; they are identified as α, β′, and β forms. As an example, consider the crystalline forms of tristearin for which the α form represents the least dense and lowest melting form; the β′ form exhibits inter-

mediate density and melting point; and the β form has the greatest density and the highest melting point.

For many single-acid triacylglycerols, increasing crystalline organization for the α→β′→β forms also corresponds to increasing stability; but in the cases of mixed triacylglycerols, the β form may actually have less stability (lower melting point) than the β′ form. Still other cases exist where multiple β and β′ forms exist depending on the triacylglycerol structures present. This variety of possible polymorphs for simple triacylglycerols results in a correspondingly wide range of melting points (e.g., −44.6 to 73.1°C or greater) depending on the fatty acids and polymorphs present.

Regardless of the individual melting point properties that characterize polymorphs, X-ray diffraction studies of polymorphic cross sections for long-chained compounds suggest that random-oriented chain axes are associated with the α form (hexagonal crystals); alternating axial/planar rows occur in the β′ form (orthorhombic crystals); and axially pitched but parallel/planar rows exist in the β form (triclinic crystals). These forms are schematically depicted in Figure 7.8. Although most natural fats are completely solid only at very low temperatures, enough of a solid crystalline matrix containing the liquid portion of a fat usually exists so that characteristic crystals can be observed. The α forms contain plate-like crystals of ~5 μm in size; the β′ forms consist of needle-shaped crystals in the size range of ~1 μm; and β forms are the largest, averaging 20 to 45 μm in size.

The ratio of the solid crystal matrix for a fat relative to the liquid portion is responsible for the characteristic density and structural and textural properties of fats. The crystal matrix itself is held by weak but attractive van der Waals forces that can be disrupted by physical shear stresses.

Once disrupted, many but not all van der Waals forces can be restored upon standing. Fat products that display this behavior are described as *thixotropic*.

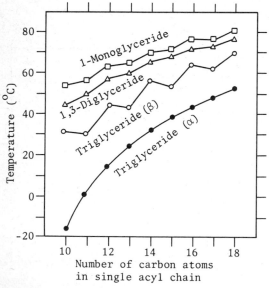

FIGURE 7.7. Melting point alteration of glycerides. (From *Bailey's Industrial Oil and Fat Products* (4th ed., Vol. 1). D. Swern (ed.). Copyright © 1979, John Wiley & Sons, and reprinted with permission.)

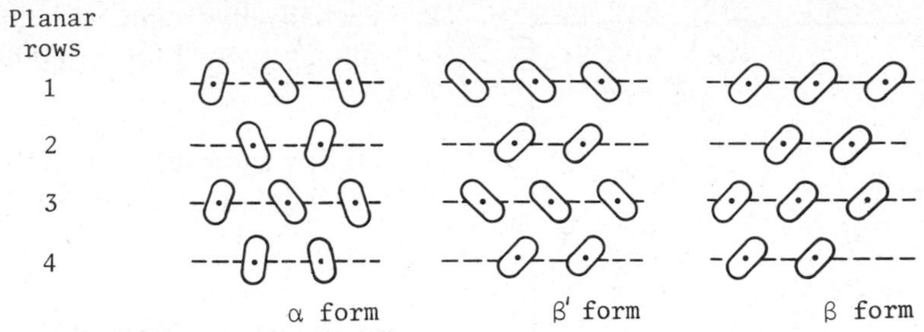

FIGURE 7.8. Schematic positions of long-chain molecules in the cross-sectional portions of α-, β′-, and β-polymorphs based on X-ray diffraction studies.

The crystal structures of margarines, shortenings, and specialty fats have special food-quality significance. Although β′ forms of fat crystals are preferred for many food fats, it should be recognized that many commercial fats have the propensity to assume the β form. This is regarded as an undesirable transition because the food fat develops a grainy, coarse texture. Margarines and shortenings are especially susceptible to these textural changes.

Natural fats and various foods containing them usually consist of mixed acylglycerols. Thus, density and melting point properties are far more complex than similar values for triacylglycerols having only one type of fatty acid. On this basis alone it is also clear why melting points have inexact analytical significance.

Freezing points for various food fats are not generally useful for any detailed analytical measurements.

Solubility and Miscibility

Fats and oils are readily miscible with most organic solvents (but not alcohols) provided that temperatures are above their respective melting points. Detailed physical studies indicate that refined liquid oils may dissolve 0.07% of their own weight of water at −1°C and up to 0.14% at 32°C. In terms of gas solubility, liquid oils dissolve 62% of their own volume of carbon dioxide at 140°C and 92% at 64°C. Other gas solubilities (e.g., carbon monoxide, hydrogen, nitrogen, and oxygen) jointly increase along with increases in oil temperature. Air, however, dissolves in liquid oils to a greater extent at higher temperatures than at lower temperatures (e.g., 8 vol% at 30°C and 13 vol% at 150°C).

Smoke, Fire, and Flash Points

The thermal stability of fatty material when heated in the presence of air can be measured according to smoke, fire, and flash points. These values vary widely and depend on the solvent, fatty acid, and mono- and diacylglycerol impurities in fats. These parameters also act as important technical criteria for the purchase and applications of commercial oils.

The *smoke point* refers to that specific temperature under standardized test conditions at which a thin but continuous smoke stream is observed; the *flash point* is the temperature at which standard test conditions cause a flash to appear when a test flame is applied to a lipid specimen; and the *fire point* describes the temperature at which volatile fatty materials will support continued combustion.

Since fatty acids are more volatile than acylglycerols, fatty acid concentrations often determine the smoke, fire, and flash points of most food oils. A free fatty acid concentration of 0.01% in corn oil, cotton seed oil, or peanut oil may cause the smoke point to vary from 450°F to only 200°F when free fatty acid content approaches 100%. Flash points have a similarly wide temperature range of 625 to 380°F while fire points range from 685 to 430°F. The relative significance of these terms is summarized in Figure 7.9.

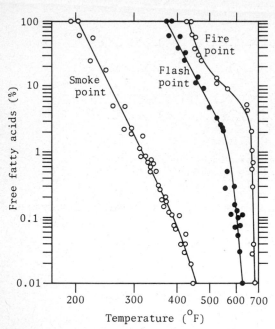

FIGURE 7.9. Smoke, fire, and flash points of miscellaneous crude and refined fats and oils, as functions of the content of free fatty acids. (From *Bailey's Industrial Oil and Fat Products* (4th ed., Vol. 1). D. Swern (ed.). Copyright © 1979, John Wiley & Sons, and reprinted with permission.)

The unsaturation of most common food oils imparts little effect on smoke, fire, and flash points.

Specific Heat of Fusion

Specific heat and heat of fusion are closely linked to the size and shape of molecules. Usually the increasing specific heat of liquid fats parallels increases in the length of fatty acid chains. This relationship is reversed, however, as unsaturation decreases. Heats of fusion for liquid fats also increase with chain length and decrease with unsaturation. Specific heat (C_p) can be calculated with reasonable accuracy according to these formulas:

1. Liquid fats: $C_p = 1.93 + 0.0025t$ ($t = 15$ to 60°C).

2. Partially hydrogenated fats (e.g., palm oil, tallow): $C_p = 1.99 + 0.0023t$ ($t = 40$ to 70°C).

3. Highly saturated fats ($IV < 10$): $C_p = 1.92 + 0.003t$ ($t = 60$ to 80°C).

Vapor Pressure, Thermal Conductivity, and Heats of Combustion

The vapor pressures and thermal conductivities for fats and oils are very low. In the case of pure fatty acids, the vapor pressure is inversely related to increasing chain lengths; and for triacylglycerols, a linear relationship exists between the logarithm of the vapor pressure and the reciprocal of the absolute temperature over a specified temperature and pressure range studied. The difficulties inherent in accurately determining thermal conductivities do not permit any wide-ranging generalizations concerning their significance.

The heats of combustion for saturated fatty acids show parallel increases in their heats of combustion as carbon chains become longer (e.g., butyric acid (5.9 kcal/g) < lauric acid (8.9 kcal/g) < stearic acid (9.6 kcal/g). The heats of combustion for triacylglycerols generally reflect the collective presence of their esterified fatty acids.

The influence of unsaturated fatty acids on the heat of combustion from a triacylglycerol is minor since values for unsaturated fatty acids are only marginally lower than the corresponding values for saturated fatty acid chains (e.g., linoleic acid (9.35 kcal/g) < oleic acid (9.45 kcal/g)).

Estimations for the heats of combustion for fatty oils adhere closely to the equation

$$\text{heat of combustion} = 11,380 - (IV)$$
$$- 9.15(\text{saponification value})$$

Based on a constant volume at 15°C, calculated values range from 9020 for coconut oil to 9680 for rapeseed oil.

Refractive Index

The refractive index (n) for fats and fatty oils shows a corresponding increase with either the increase in molecular weight or the degree of unsaturation in neutral fats. Since fatty acids, mono- and diacylglycerols, conjugated double bonds, and fatty acid oxidation products cause detectable positive deviations in refractive index measurements for fats, n values can be used as an expeditious quality-control method for monitoring the hydrogenation, isomerization, or purity of fats and oils. Studies show that refractive indices for neutral oils are typically 1.4468, 1.4568, and 1.4687 for respective iodine values of 0, 100, and 200.

NOTABLE CHEMICAL REACTIONS OF FATS AND FATTY OILS

The chemical reactions of fats and fatty oils mirror the classic reactions of esters, olefins, and hydrocarbons. Triacylglycerols may undergo basic or acid hydrolysis to yield free fatty acids or soaps (fatty acid salts) plus glycerol. *Interesterification* or *alcoholysis* represents another common reaction of triacylglycerol structures whereby individual positions of esterified fatty acids can be interchanged on the glyceryl moiety. The major reactions of

Many of these characteristic reactions were first recognized as important because of their fundamental chemical significance, but current interest in these reactions stems from their important contributions to fat and oil processing, lipid deterioration, and the analysis of fats and fatty oils.

Interesterification

A specific melting point range, solid fat index, and crystalline habit for a fat often determines its suitability as a food fat or food ingredient. Commercial and food industrial demands for triacylglycerols with specific properties are not always readily met by natural fats, and it may be necessary to prepare them by means of interesterification reactions.

These reactions are initiated by the interaction of native triacylglycerols with glycerol (at ~80°C) and catalytic agents such as sodium methoxide or a sodium-potassium alloy. The ensuing *glycerolysis* reaction produces an assortment of mono- and diacylglycerols. When little or no glycerol is available, fatty acids formerly esterified to native triacylglycerols undergo migration and reshuffling to other acylglycerol structures (*interesterification*), or the rearrangement may occur within the native triacylglycerol itself (*intraesterification*).

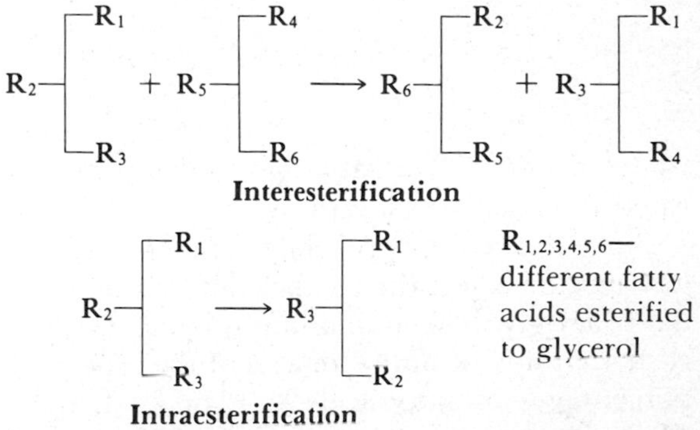

Interesterification

Intraesterification

$R_{1,2,3,4,5,6}$— different fatty acids esterified to glycerol

triacylglycerols containing unsaturated fatty acids involve oxidation, reduction, and *cis-trans* isomerizations of the esterified acids.

Based on the recognition of these specific types of reactions, the full gamut of possible reshuffling for fatty acyl groups among triacylglycerols can become very complex. As an example consider the formation of an interesterification equilibrium mixture for a triacylglycerol of β-oleostearolinolein.

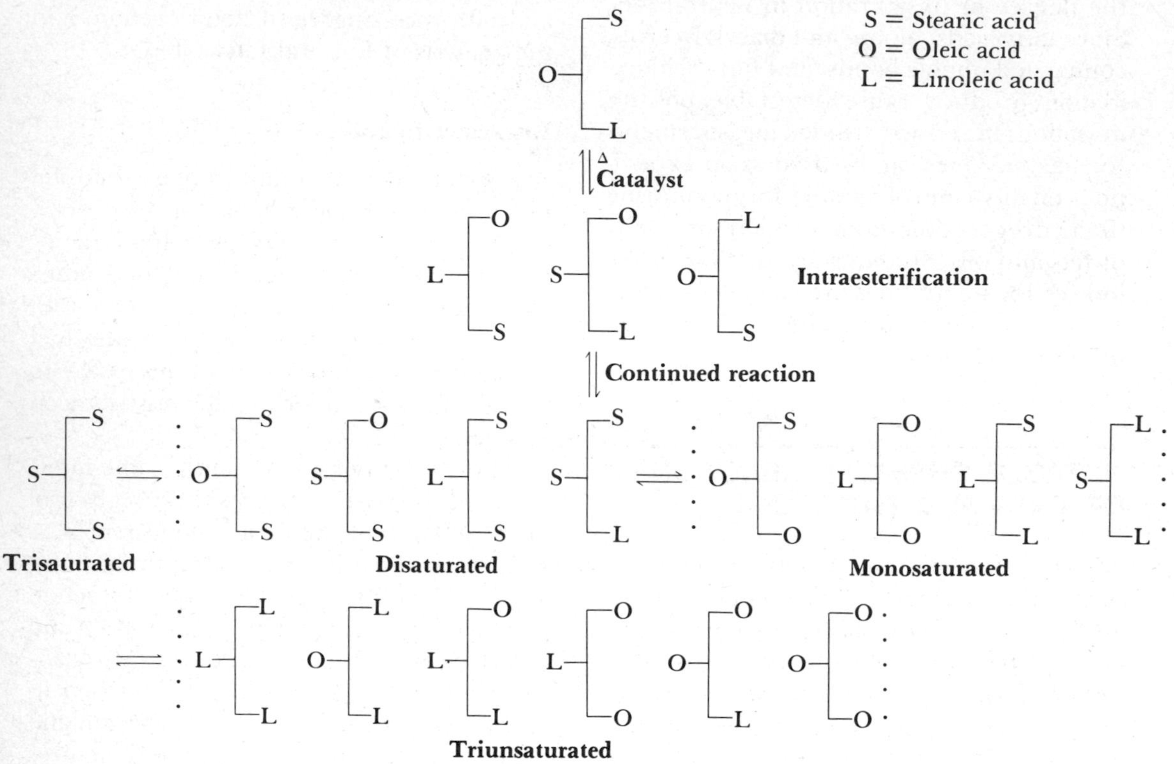

the corresponding molar percentage for those triacylglycerols having two acids will be

$$\% \, XXY = 3X^2Z : 10{,}000$$

S = Stearic acid
O = Oleic acid
L = Linoleic acid

Intraesterification

Continued reaction

Trisaturated   **Disaturated**   **Monosaturated**

Triunsaturated
Interesterification equilibrium mixture

As illustrated, the shorthanded notation for the overall reaction sequence is

$$SSS \rightleftharpoons (SUS \rightleftharpoons SSU) \rightleftharpoons (SUU \rightleftharpoons USU)$$
$$\rightleftharpoons UUU$$

where S and U are saturated and unsaturated fatty acids, respectively.

Since the laws of probability and random distribution affect the composition of the final triacylglycerol equilibrium mixture, it is clear that if X, Y, and Z represent the molar percentages of fatty acids X, Y, or Z, then the molar percentage of triacylglycerols containing only one acid will be

$$\% \, XXX = X^3 : 10{,}000$$

and the molar percentage of triacylglycerols containing three acids is

$$\% \, XYZ = 6XYZ : 10{,}000$$

Using this mathematical basis for studying the randomized equilibrium interesterification of equimolar percentages of stearic, oleic, and linoleic acids, the probable composition of the mixture would approximate the percentages detailed in Table 7.8. Aside from intraesterification reactions, interesterification has the most important commercial significance because it can yield (1) new triacylglycerol structures having different esterified fatty acid distributions than the native glyceryl structures and (2) variable amounts of mono- and diacylglycerols. An

Table 7.8. Triacylglycerol Structures Resulting from the Randomized Interesterification of β-Oleostearolinolein to a Point of a Complete Equilibrium Mixture

Stearic–stearic–stearic $= (33.3 \times 33.3 \times 33.3):10,000 = 3.7\%$
Oleic–oleic–oleic $= (33.3 \times 33.3 \times 33.3):10,000 = 3.7\%$
Linoleic–linoleic–linoleic $= (33.3 \times 33.3 \times 33.3):10,000 = 3.7\%$
Stearic–stearic–oleic $= (33.3 \times 33.3 \times 33.3)3:10,000 = 11.1\%$
Stearic–stearic–linoleic $= (33.3 \times 33.3 \times 33.3)3:10,000 = 11.1\%$
Stearic–oleic–oleic $= (33.3 \times 33.3 \times 33.3)3:10,000 = 11.1\%$
Stearic–linoleic–linoleic $= (33.3 \times 33.3 \times 33.3)3:10,000 = 11.1\%$
Oleic–oleic–linoleic $= (33.3 \times 33.3 \times 33.3)3:10,000 = 11.1\%$
Oleic–linoleic–linoleic $= (33.3 \times 33.3 \times 33.3)3:10,000 = 11.1\%$
Stearic–oleic–linoleic $= (33.3 \times 33.3 \times 33.3)6:10,000 = 22.2\%$

increasing ratio of free glycerol to native triacylglycerols yields interesterification products with increasing proportions of monoacylglycerols until an equilibrium maximum limits their formation beyond a ~70% limit.

The basic importance of monoacylglycerols in foods as emulsifiers requires the production of technical monoacylglycerols having a 40 to 70% concentration level in reaction mixtures. Higher concentrations than these for 1-monoacylglycerols, approaching 90%, can be achieved only by molecular distillation.

Interesterification reactions can also be controlled by conducting the reaction above or below the melting point of a fat. Based on the condition where the temperature is held above the melting point for a fat, triacylglycerol structures can have their native esterified fatty acids reshuffled and randomized among all available hydroxyl positions on glycerol molecules; or two or more fats may be interesterified to produce new acylglycerol products with the interesterification reaction conducted at temperatures less than the melting point of the fat. These two interesterification methods describe *randomized interesterification* (reaction occurs above fat melting point) and *directed interesterification* (reaction occurs below fat melting point), respectively. In the latter case, the interesterification reaction will involve only the liq-

uid portion of a fat, and the specific reaction temperatures can be used to selectively crystallize higher-melting-point acylglycerols from others. This practice prevents occurrence of more randomized reactions. Directed interesterification of this type results in an equilibrium displacement during successive interesterification reactions so that a variety of different triacylglycerol structures, each with varying structural properties, will be produced.

It ought to be noted that the use of glycerol is not an absolute prerequisite for interesterification, and, in fact, the use of catalysts alone (with heat) may be adequate.

Although lard has been extensively processed in the United States by interesterification methods, many European countries have employed the method for the production of custom-made fats and margarine oils.

Hydrogenation

Hydrogenation of fats and oils represents the largest single chemical reaction conducted by the fatty oil processing industry. In the most general way, the process involves the progressive conversion of unsaturated double bonds within fatty acid chains to more saturated species. Complete hydrogenation results in a fully saturated fatty acid. These industrial reduction reactions are executed in the presence of hydrogen gas and a metal catalyst (nickel, platinum, palladium) or an

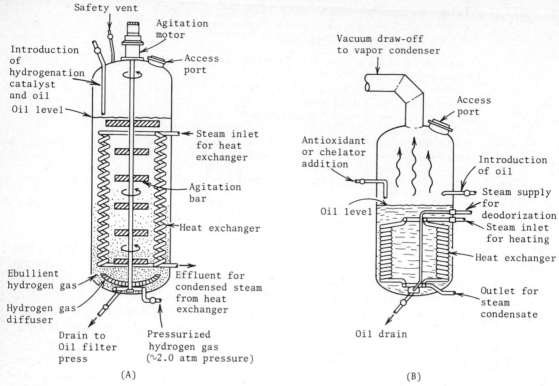

FIGURE 7.10. Schematic representations for an industrial hydrogenation reactor (A) and an oil deodorizer (B).

organometallic catalyst, depending on the process.

The solid (catalyst), gas (hydrogen), and liquid (oil) phases are stirred in a heated reactor vessel where gas diffusion into the liquid phase is encouraged by pressurization of the reactor headspace (Figure 7.10A).

At least one of the reactants must be chemisorbed on the catalyst. Evidence also suggests that unsaturated sites in the liquid phase (oil) react with hydrogen by a route of surface organometallic intermediates. The progress of the idealized hydrogenation reaction can be conceived as

$$\begin{array}{c} \text{H} \ \text{H} \\ | \ \ | \\ \text{—C=C—} + \text{H}_2 \longrightarrow \end{array} \quad \begin{array}{c} \text{H} \ \text{H} \\ | \ \ | \\ \text{—C—C—} \\ | \ \ | \\ \text{H} \ \text{H} \end{array}$$

This reaction is fostered by the fact that the two σ bonds (C—H) formed during the reaction are stronger than the σ bond (H—H in hydrogen gas) and the π bonds that are being eliminated.

As a result of kinetic and contact idiosyncracies that occur between unsaturated fatty acid sites and the catalytic surface, the reduction of unsaturated sites in fatty acids may be complete or only partial. This leads to the combined formation of saturated bonds as well as new geometric forms of fatty acids and their isomers.

In those cases where mono-, di-, and polyene fatty acid mixtures are comingled during hydrogenation, competitive reduction among the species occurs at the catalytic surface. That is, dienes may be preferentially adsorbed to the catalyst surface, where they are hydrogenated to monoenes and/or isomerized before their desorption and diffusion back into the oil. When di- and polyene concentrations are low, then monoene species undergo surface adsorption interactions at the catalytic site where they then react.

Since hydrogenation events proceed from polyenes to dienes to monoenes, the hydrogenation process has been described as a *selective process.* "Selectivity" here should not be confused with the *selectivity* classifications of catalysts used in hydrogenation reactions. According to this usage, selectivity describes the ability of a catalyst to produce an oil of a softer consistency or lower melting point at a given iodine value (*IV*).

As a practical example of hydrogenation effects on triacylglycerols, consider the following. If trioleates were present in an oil undergoing hydrogenation, the process would ideally yield triacylglycerols containing stearic acid; but it should be recognized that limited *cis–trans* isomerization can occur if certain metal catalysts such as nickel are used. Both of these eventualities resulting from hydrogenation result in a higher melting point for the triacylglycerols, but the melting point is not as high for the hydrogenated product when *cis–trans* isomers are present.

For those cases where linoleic, oleic, and saturated acid triacylglycerols are present, competitive hydrogenation reactions of mono- and diene species occur in a selective fashion (e.g., cottonseed oil). The conversion of linoleic to oleic (and its *trans* isomer, elaidic) acid preempts the conversion of oleic acid to stearic acid. Depending on application requirements, selectivity ratios exceeding 50:1 may be desired for the hydrogenation rate of linoleic acid to oleic acid. Corresponding relationships exist during the hydrogenation sequence of linolenic acid—containing oils such as soybean oil, although the probability of isomerization is markedly enhanced when triunsaturated fatty acids are present.

The hydrogenation of vegetable oils is a desirable reaction because it permits the technical production of many different edible fats (e.g., confectionary fats, shortenings, margarine) from a wide spectrum of partially solid or liquid fats. This process also permits the controlled hydrogenation of many different types of unsaturated fatty acids so that they can be substituted for other fats depending on their practical availability and economic cost. Furthermore, as a consequence of well controlled hydrogenation reactions, less-saturated fats and oils can be substituted for more-saturated natural fats without compromising the final quality of the food in which they are used.

In addition to the obvious advantages for converting edible liquid fats and fatty oils into edible plastic forms, hydrogenation is also important because it minimizes the polyunsaturation of fatty acids. This is desirable in one way because it minimizes the susceptibility of unsaturated fatty acids to autoxidative processes that cause the spoilage of food fats and oils, and create off-flavors and odors. On the other hand, the hydrogenation process has detrimental nutritional ramifications because it leads to the reduction of linoleic and linolenic acids.

Interesterification, Hydrogenation, and Solid Content Index (*SCI*) Values

The effects of interesterification are reflected by the *solid content index* (*SCI*), or solid fat content of fat mixtures as they are affected by temperature. Since the solid portion of a fat often serves as a criterion for its use as an edible fat, this index is useful for estimating the mouth feel and functional properties of fats processed by ester–ester interchange reactions (e.g., interesterified shortening, margarine, confectionery fats).

The native and interesterification forms of cocoa butter display major differences in their *SCI* values. Natural cocoa butter, for example, is noted for its sharp melting characteristics. This form of cocoa normally displays a steep *SCI* curve (Figure 7.11). Interesterification of the same cocoa butter results in a wider melting point range and a flattened *SCI* curve. According to Figure 7.11, these effects are notable for the interesterified butter, whose solids occur up to 50°C. Natural cocoa butter that normally melts at mouth temperature can be interesterified to

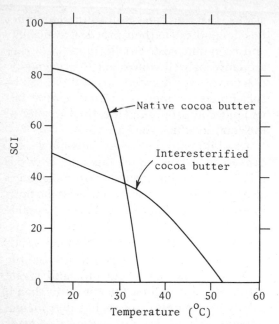

FIGURE 7.11. Solid content index (*SCI*) curves for native and interesterified cocoa butters. Note that the interesterified form exhibits solids up to 50°C.

resist melting by the formation of higher-melting triacylglycerols.

The *SCI* curve for hydrogenated fats also acts as a useful criterion for projecting their functional behavior in foods (Figure 7.12). A steep *SCI* curve is especially desirable if it is included in a margarine formulation or a confectionery fat in which well-defined mouth-melting is necessary. Those hydrogenated fats having broad *SCI* curves, on the other hand, may be most useful for use in pastry and bakery goods.

Oxidation and Rancidity Reactions of Fats and Fatty Oils

In vivo fats and oils are protected by many antioxidants and biochemical pathways that ensure their chemical and biochemical integrity. From the instant that a plant or animal is harvested or slaughtered for food purposes, fat and oil contents of living tissues begin to deteriorate.

Since the deterioration of lipid content acts as a major criterion for food accepta-

bility, it is clear that the destruction of food fats and oils must be minimized. The battle to ensure fat stability centers about minimizing (1) *bacterial activities*, (2) *enzyme-catalyzed hydrolysis and oxidation reactions*, and (3) the direct chemical attack of atmospheric oxygen on fats and oils by *autoxidative* routes.

Major technical and mechanical advances involving food refrigeration, freezing, packaging, efficient transportation, and sterilization have circumvented many difficulties caused by (1) and (2) above, but (3) is another matter. Autoxidation mechanisms require low activation energies to initiate the oxidative deterioration of lipids and since total exclusion of oxygen from stored or processed foods is not always possible, the autoxidation of fats may proceed regardless of low temperature food storage or sterilization efforts. Based on current perspectives of fat and oil degradation, their deterioration can be categorized as purely *hydrolytic* or *oxidative*.

Hydrolytic reactions: Under proper conditions of fat–water miscibility and high temperature (225–280°C), all fats and oils may undergo chemical hydrolysis to yield free fatty acids (FFA) and a variety of isomeric di-

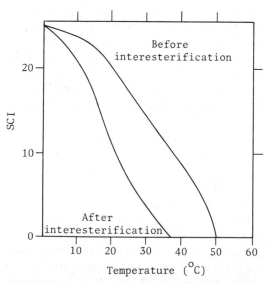

FIGURE 7.12. Influence of interesterification on a 80:20 mixture of lightly hardened oil and palm oil based on *SCI* values.

acylglycerols, α- and β-monoacylglycerols and free glycerol depending on the extent of hydrolysis. A complete hydrolysis reaction would be represented as

$$
C_3H_5(OOCR)_3 + 3HOH \longrightarrow \begin{array}{c} C_3H_5(OH)_3 \\ \textbf{Glycerol} \\ + \\ 3RCOOH \\ \textbf{Free fatty} \\ \textbf{acids (FFA)} \end{array}
$$

Triacylglycerol

Similar hydrolytic results are achieved by less rigorous physical conditions during the digestive hydrolysis of fatty acids from triacylglycerols in animals and other organisms. In these cases the hydrolysis is conducted by *lipolytic enzymes* or *lipases*.

Carbon chains of $\leq C_{12}$ are more readily hydrolyzed than those with 12 to 18 carbons, but these generalizations depend on reaction conditions. The free fatty acids that exist in natural or hydrolyzed fats or oils can be measured according to the *acid* or *neutralization number*. This *number* specifically refers to the number of milligrams of potassium hydroxide required to neutralize FFAs in 1 g of fat. High acid numbers coincide with high FFA concentrations.

Free fatty acids are more or less acceptable in foods, food fats, or oils depending on the food and the ultimate uses of the fats or oils. For butter, margarine, palm kernel, and coconut oils, only 0.1 to 1.0 mg/100 g levels of FFA content can cause very undesirable flavors. Off-flavors produced by this route indicate evidence of *hydrolytic rancidity*. This is contrary to foods such as beef that may have a FFA content of 12 to 15% and still retain its palatability. Olive oils too may display up to 1.5 to 2.0% FFA content without major destruction of oil quality.

Given proper ambient temperature conditions and adequate water availability, most foods have sufficient chemical and biochemical complexity to support microbial growth, sporulation, and/or reproduction. Therefore, lipid-containing foods ranging from whole grains to butchered meats and fat products such as butter, margarines, milk, and mayonnaise are all liable to undergo microbial attack and the subsequent effects of hydrolytic rancidity (plus autoxidation processes that follow). The microbial production of hydrolytic enzymes by *Penicillium, Aspergillus, Serratia, Pseudomonads,* and countless other organisms typically cause food lipids to become "soapy." This action reflects hydrolysis of acylglycerols containing medium-chain-length ($\leq C_{10}$) fatty acids. Souring properties are also attributed to the presence of hydrolyzed fatty acids since short-chain varieties are not only volatile but they exhibit pK_a values in the range of 4.76 to 5.00.

Atmospheric oxidation (autoxidation) and oxidative rancidity

The primary products of autoxidation. Aside from the basic structural deterioration of acylglycerols by hydrolysis along with the attendant consequences of rancidity, autoxidation of fats and oils culminates in a variety of new and modified products that arise from unsaturated fatty acids. Autoxidation reaction products not only cause obvious deterioration of native unsaturated fatty acids, but the reaction spurs the destruction of nutritionally important unsaturated fatty acids, not to mention the joint development of objectionable lipid tastes and odors.

The autoxidation of individual unsaturated fatty acids is affected by differences in their respective molecular structures including the length of the carbon chain and the presence of both geometric (*cis–trans*) and positional isomers. Reaction rates for autoxidative processes are also influenced by the presence of *prooxidants* such as trace metals or antioxidants, either added or indigenous; available oxygen levels; and temperature conditions.

Detailed studies of autoxidation reactions have revealed that the process proceeds more rapidly as fatty acid unsaturation increases. For many of the most common edible oils, this means that their main polyunsaturated component, linoleic acid, will act as a prime candidate for autoxidation. Higher orders of polyunsaturation among fatty acids (four to six double bonds) in fish and marine animals suffer still higher autoxidation rates. For example, classical studies of a progressively unsaturated series of methylated C_{18}-fatty acids, including methyl oleate, methyl linoleate, and methyl linolenate, show relative autoxidation rates of $1:12:25$. The increasing reactivity here is attributable to the activation of methylenic carbons by surrounding double bonds as unsaturation of fatty acids increases.

Autoxidation of unsaturated fatty acids proceeds by a free radical chain mechanism that results in hydroperoxide formation. This step, in turn, culminates in a decomposition reaction that produces a variety of secondary oxidative products. Figure 7.13 illustrates a possible autoxidation sequence for linoleate. Note that the structure can yield three major resonance forms of the peroxyl radical and, therefore, three peroxides may be formed. As a result of thermodynamic stability within the conjugated system, C-9 and C-13 peroxides account for 95 to 98% of the hydroperoxide yield and C-11 hydroperoxide yields are 3 to 5%.

The autoxidation of unsaturated fatty acids by the route outlined is typical of many common reactions for unsaturated fatty acids, but it is not all-inclusive and other oxidative pathways are possible. For example, it is recognized that photooxidation mechanisms can yield hydroperoxides without preliminary free radical reactions. Many of these photosensitized oxidation routes are believed to proceed with the participation of singlet oxygen.

Molecular oxygen may exist in a singlet state, $^1O_2$ (antiparallel electron spins), or a triplet state, $^3O_2$ (parallel electron spins). In the excited state ($^1O_2^*$), one electron leaves its orbital and moves into the orbital of its paired electron as indicated below:

Molecular oxygen state	Symbol	Electron spin and orbital	
Singlet (ground)	$^1O_2$	↑↓ ↑↓	
Singlet (excited)	$^1O_2^*$	↑↓ —	(reactive empty orbital)
Triplet (ground)	$^3O_2$	↑ ↑	

The vacant orbital of excited singlet oxygen is readily filled by an electron from a double bond (where electron density is high). Studies reveal that oxygen in the excited state reacts almost 1500 times faster with methyl linoleate than oxygen in its ground state. Energy absorption by oxygen in the ground state is facilitated by photosensitizers (S). These agents include natural dyes and tetrapyrrolic pigments such as chlorophyll, pheophytin, and myoglobin, along with many other compounds can undergo photoactivation. The activated sensitizer (S*) can incite lipid peroxidation as shown in the following sequence (Rawls and Van Santen, 1970):

$$^1S + h\nu \longrightarrow {}^1S^* \rightsquigarrow {}^3S^*$$

$$^1S^* + {}^3O_2 \longrightarrow {}^1O_2^* + {}^1S$$

$$^1O_2^* + RH \longrightarrow ROOH$$

Note: RH = unsaturated fatty acid

Secondary products resulting from autoxidation. Assuming that a hydroperoxide [R′—CH(OOH)—R″] has been formed from an unsaturated fatty acid such as linoleic, the hydroperoxide may participate in a number of degradative reactions that cleave the fatty acid, propagate the autoxidative sequence, or terminate the autoxidative process by forming aldehydes, alcohols, and/or ketones. The genesis of these substances, known as *secondary autoxidative products,* have been cursorily outlined in Figure 7.14.

These idealized pathways for alkoxy radical reactions are greatly simplified and do not reflect the total random complexity of product formation or the range of potential reactants involved. For example, it is clear that derivatives of the free radicals produced by autoxidation can also yield polymers, ethers, peroxides, lactones, aromatics, and various acids.

Although hydroperoxides alone do not account for objectional tastes and odors in foods, all of the reaction products stemming from hydroperoxide presence, cited above, can cause organoleptic problems. Problems of this type are generically classified as *oxidative rancidity.*

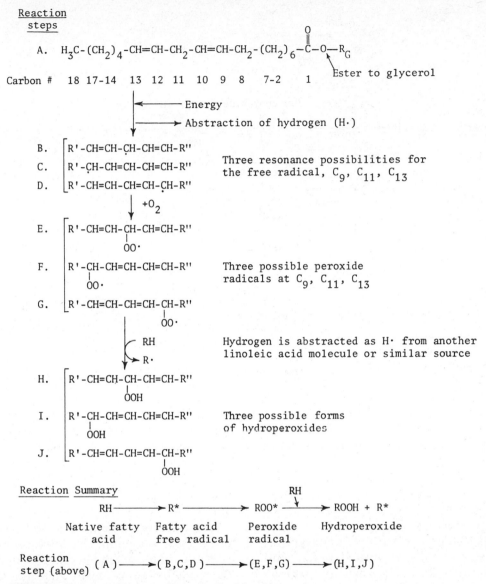

FIGURE 7.13. Autoxidation pathway for linoleic acid.

Many of these objectionable compounds display part per million (ppm) and part per billion (ppb) taste threshold values (*TV*). Table 7.9 reports some typical values for water, milk, vegetable, and paraffin oils.

It should be noted, too, that peroxides alone are not responsible for imparting flavor changes to foods. Flavor changes occur with the appearance of secondary oxidative reaction products such as aldehydes, ketones, and so on.

Oxygen uptake during autoxidation. Recognition of the important role for oxygen in autoxidation mechanisms has resulted in many studies designed to measure oxygen absorption by fats, fatty oils, and food systems. Figure 7.15A illustrates two classical examples of what can occur. In nearly all cases oxygen uptake is slow but variable depending on the physical state of the food and/or unsaturated fatty acid(s). This first stage of oxygen absorption represents a hydroperoxide *in-*

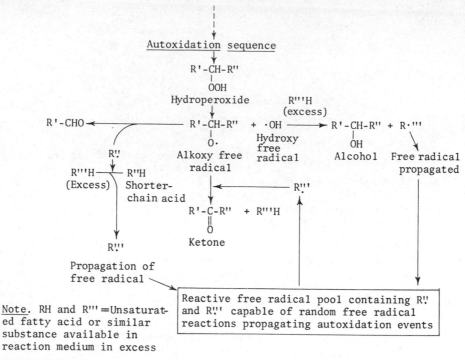

FIGURE 7.14. Some reaction products resulting from hydroperoxide degradation.

duction period. According to Curve I (Figure 7.15A) a specimen shows miniscule uptake of oxygen during the induction period as opposed to Curve II, which clearly absorbs more measurable oxygen during the induction period. After a certain induction period, hydroperoxide content shows a very rapid increase.

Based on carefully prepared oxygen uptake curves and coordinated organoleptic studies of foods, it is possible to establish a correlation between an oxygen uptake value by a food and its corresponding organoleptic score. This relationship has practical value for projecting the incipient occurrence of flavor deterioration, but it must be recognized that a low *TV* for many secondary autoxidation products will appear before detectable increases in oxygen uptake are apparent.

Peroxide formation in foods reflects oxygen uptake to a point, and it has been used as a status index of limited value for studying the oxidation processes of unsaturated lipids.

Peroxides generally liberate iodine from

Table 7.9. Some Typical Flavor Threshold Values (*TV*'s) for Aliphatic Aldehydes Expressed in ppm

	Water	Milk	Vegetable oil	Paraffin oil
Hexanal	0.04	0.05	0.31	0.65
Propionaldehyde	0.18	0.42	0.20	1.00
Lauraldehyde	0.001	—	0.85	0.40
trans-2-Nonenal	0.001	0.004	0.09	0.42
trans,cis-2,6-Nonadienal	0.00009	—	—	0.0012
trans-6-Nonenal	—	—	—	0.0004
Valeraldehyde	0.08	0.15	—	0.16

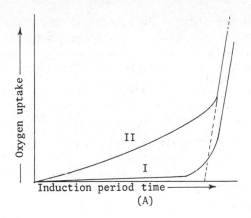

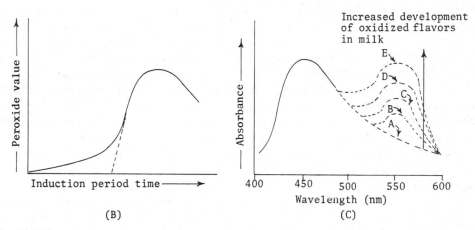

FIGURE 7.15. (A) Relationship between oxygen uptake and lipid oxidation. Curve I indicates no significant oxygen uptake during the induction period, whereas Curve II shows significant oxygen uptake over the induction period. (B) Peroxide occurrence during lipid oxidation. (C) Spectral relationships for thiobarbituric acid (TBA) test of milk (A) as it undergoes progressive development of oxidized flavors (B ⟶ E).

potassium iodide in a glacial acetic acid solution. This reaction serves as the basis for the calculation of a *peroxide number*. Strictly defined, the peroxide number is a measure of reactive oxygen in terms of millimoles of peroxide or milliequivalents of oxygen per 1000 g of fat (1.0 mmol = 2.0 mEq).

Peroxide development in lipid systems adheres to the curvulinear relationship shown in Figure 7.15B for oxygen uptake, but it eventually decreases as oxygen absorption decreases and the formation sites of reactive peroxide intermediates diminish.

Aside from monitoring autoxidative reactions in foods with peroxide values, it may be better to monitor the appearance of secondary autoxidative products that cause off-odors. Again, a problem arises since many of the compounds responsible for off-odors and flavors are far below the limits of analytical detection. Nevertheless, one useful test for determining the presence of secondary reaction products is the *thiobarbituric acid* (TBA) *test*. Oxidation products of unsaturated fatty acids, principally linoleic acid, seem to be responsible for a color reaction with TBA that is spectrophotometrically detected at 535 nm. Exclusive of academic debates concerning the natural occurrence of malonaldehyde, it is generally conceded that this substance reflects evidence of peroxide decomposition.

opposed to identical hydrogenated oils. Some studies claim that unsaponifiable and/or sa-

$$R-CH_2-CH=CH-CHO \xrightarrow[+O_2, +H\cdot]{} R-CH-CH=CH-CHO$$

Aldehyde product resulting from autoxidative degradation of unsaturated fatty acid

$$\underset{\text{Hydroperoxide}}{\overset{|}{OOH}}$$

peroxide decomposition

$$R-CHO + OHC-CH_2-CHO$$
Malonaldehyde
(or malonic dialdehyde)

The presence of the malonaldehyde condensation product with TBA can often be detected against a spectral backdrop for the unoxidized food substance, as in the case of milk (Figure 7.15C).

Other analytical methods for assaying the occurrence of autoxidation reactions include deviations in refractive indices; instrumental analysis by gas chromatography and mass spectrometry; color development of autoxidized fats with concentrated base by means of α-dicarbonyl condensations; and monitoring ultraviolet absorption as conjugation changes develop during fat oxidation.

Flavor reversion: Objectionable flavors and odors may be produced in oils that are indicative of *flavor reversion*. This problem is different from autoxidative processes, however, because the phenomenon occurs under oxidative conditions far inferior to those that incite autoxidation. Therefore, flavor reversion in oils may occur anywhere in the peroxide value range of 1 to 100.

Descriptive reports of flavor reversion often cite flavors as being "beany," "grassy," "fishy," or "painty."

Flavor reversion is generally attributed to the linoleic or linolenic acid components of soybean oils, rapeseed oils, and all marine animal oils. The mechanisms for flavor reversion are highly speculative since less than 1.0% of the oxygen required for autoxidation will cause flavor reversion. Apart from the difficulties involved in studying nearly undetectable and complex reactions, different reversion mechanisms are believed to occur in natural unhydrogenated oils as

ponifiable substances are involved in reversion. Other studies suggest that fishy flavors in unhydrogenated vegetable oils result from oxidative interactions between nitrogen compounds and oil constituents.

As though speculated mechanisms for flavor reversion were not complicated enough, "ordinary" flavor reversion must be discerned from "heat reversion" phenomena. *Heat reversion* seems to be associated with glycerol fractions of oils.

Owing to the importance of soybean oils, reversion flavor studies have suggested the production of isolinoleic acid (a 9,15-diene). Other substances associated with reversion include 3-hexenal; *trans*-2-hexenal; *trans*-2-nonenal; *trans,cis*-2,6-nonadienal; 2-heptanal; *n*-butyraldehyde; ethylformate; and many other odorous substances.

Regardless of conventional deodorization procedures designed to ensure oil quality, such treatments do not prohibit reversion. Moreover, the elimination of oxygen from unsaturated and saturated fats or oils will not prevent the formation of ketone derivatives as a result of high temperatures and photo-induced reactions.

Oxidative-enzymatic breakdown of unsaturated fatty acids: Enzyme-mediated oxidations of unsaturated fatty acids proceed by the action of *lipoxygenases* (lipoxidases) that occur in the herbaceous tissues, oil seeds, and grains of higher plants. These enzymes have not been recognized in animal tissues.

Lipoxygenases promote the introduction of oxygen into unsaturated fatty acids such as linoleic, linolenic, and arachidonic, which typically contain a *cis,cis*-1,4-pentadiene system

$$(\text{—CH}\underset{1}{=}\text{CH}\underset{2}{—}\text{CH}_2\underset{3}{—}\text{CH}\underset{4}{=}\text{CH}\underset{5}{—})$$

Peroxide formation mediated by lipoxygenase activity probably begins at the site of activated methylene sites ($\text{—CH}_2\text{—}$) and then oxidative reactions proceed by free radical formation. Although the mechanisms for lipoxygenase-induced oxidation of fatty acids are not identical to autoxidative mechanisms, hydroperoxides and conjugated *cis,trans*-hydroperoxides can ultimately be formed. The active nature of these substances often promotes other nonenzymatic reactions such as oxidations of carotenoids and other unsaturated fatty acids.

Additional enzyme-mediated oxidative routes for lipids are possible. These routes are demonstrated by many saprophytic organisms such as molds, which oxidize fatty acids having fewer than 14 carbons after their hydrolysis from acylglycerols. The oxidative mechanism here resembles β-oxidation of fatty acids in normal metabolism, but the complete oxidation of fatty acids is preempted by the formation of odorous methyl ketones. Concentrations of C_5, C_7, and C_9 methyl ketones impart flavorful properties to many cheeses ripened by molds, but concentrations of ≥40.0 μg/g are objectionable. Aside from cheeses, pork lard, butterfat, and palm and coconut oils may suffer rancidity as a result of these oxidative products.

Prooxidants and antioxidants: The biochemical maintenance of unsaturated fatty acid structures is largely contingent upon minimizing free radical reactions.

Those fats and oils that (1) have a high degree of unsaturation or (2) exist in a comingled fashion with reactive substances usually undergo rapid decomposition. In the latter case (2), agents that promote oxidative processes such as heavy metal cations and unstable organics are called *prooxidants,* whereas agents that minimize the activity of prooxidants are known as *antioxidants* or *oxidation inhibitors.*

Naturally occurring fats and oils contain indigenous antioxidants that protect the unsaturated lipids from free radical destruction in their native vegetative (seeds, fruits, etc.) and animal sources. Although animal and vegetable oils both contain antioxidants, the highest concentrations of antioxidants are present in vegetable oils.

The most commonly occurring natural antioxidants in plants are the tocopherols (vitamin E). These 6-chromanol derivatives are not synthesized by animals, including humans, but the dietary consumption of plant lipids ensures the maintenance of tocopherol levels in most cases.

As illustrated earlier, these substances inhibit the random reaction behavior exhibited by fatty acid free radicals (Figure 7.13) as well as peroxy free radicals. A model role for the action of an antioxidant during the oxidation of a hydroquinone has been indicated in Figure 7.16. The intermediate free radical forms of antioxidants are not destructive to unsaturated fatty acids and they are eventually converted to harmless quinones. Similar natural antioxidant activities (and limited antimicrobial properties) are attributed to natural phenolics that occur in nearly all spices including ginger, vanilla pods, woodruff, marjoram, and cloves to name only a few (Figure 7.17).

The natural occurrence of tocopherols in fats and oils reflects the concentration at which they exhibit maximum antioxidant potentials. Above these natural levels their antioxidant activities may be reversed and they can actually function as prooxidants. Therefore, it is clear that increasing concentration levels of antioxidants do not impart added margins of antioxidant safety to fats and oils.

Antioxidant effects for tocopherol-related substances can be enhanced by a variety of *synergists.* Synergists include substances such as citric, ascorbic, and phosphoric acids as well as ethylenediaminetetraacetic acid (EDTA). Whereas tocopherols inhibit oxidative free radical chain mechanisms, these agents act as metal *chelators* or *deactivators.* This action suppresses the oxi-

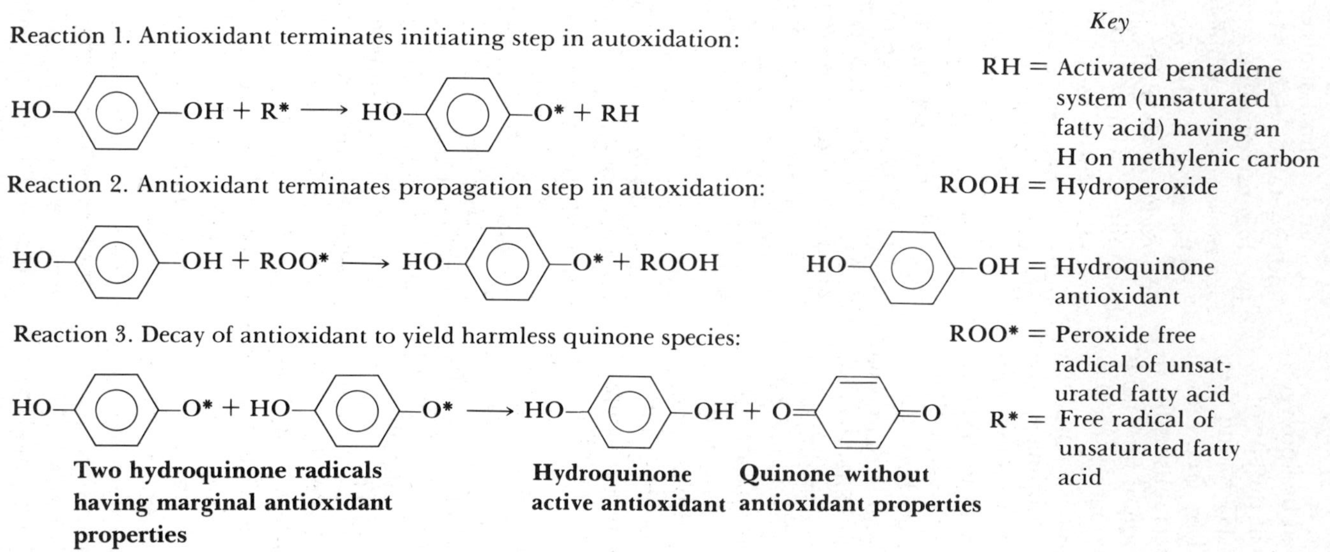

Key

RH = Activated pentadiene system (unsaturated fatty acid) having an H on methylenic carbon

ROOH = Hydroperoxide

OH = Hydroquinone antioxidant

ROO* = Peroxide free radical of unsaturated fatty acid

R* = Free radical of unsaturated fatty acid

Reaction 1. Antioxidant terminates initiating step in autoxidation:

HO—⬡—OH + R* ⟶ HO—⬡—O* + RH

Reaction 2. Antioxidant terminates propagation step in autoxidation:

HO—⬡—OH + ROO* ⟶ HO—⬡—O* + ROOH

Reaction 3. Decay of antioxidant to yield harmless quinone species:

HO—⬡—O* + HO—⬡—O* ⟶ HO—⬡—OH + O=⬡=O

Two hydroquinone radicals having marginal antioxidant properties

Hydroquinone active antioxidant **Quinone without antioxidant properties**

FIGURE 7.16. A possible antioxidant mechanism.

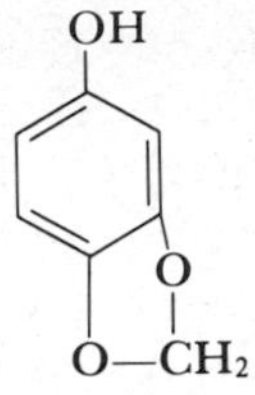

OH

Sesamol
(A)

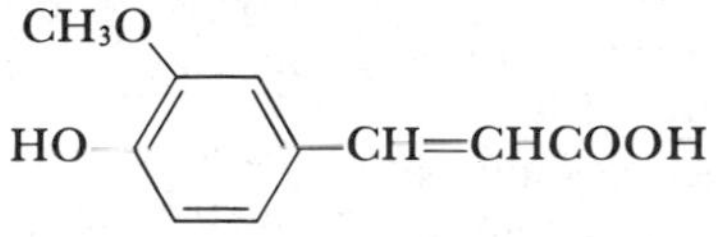

CH₃O

HO —⟨ ⟩— CH=CHCOOH

Ferulic acid
(B)

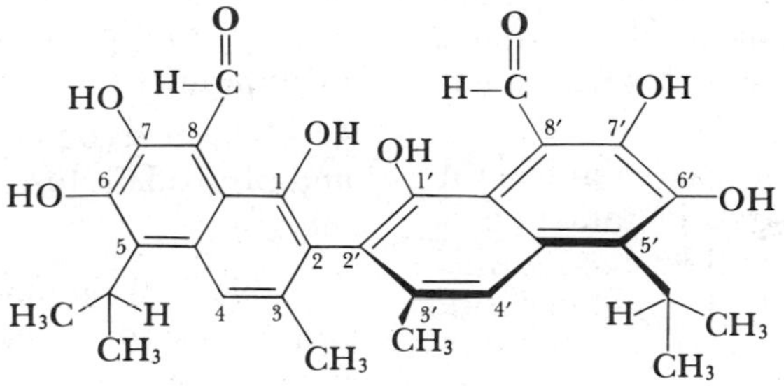

Gossypol
(C)

FIGURE 7.17. Structures for some natural phenolic antioxidants: (A) sesame oil contains 0.3–0.5% sesamoline, a glucoside of the phenolic compound sesamol; (B) rice bran oil, corn, wheat, and oats contain various forms of ferulic acid esterified to triterpene alcohols or steroids such as sitosterols depending on the source; and (C) gossypol is a common phenolic substance present in crude but not refined cotton seed oil.

dation and reduction reactions between organic peroxides and metals that are spurred on by electrochemical differences in redox potentials.

and ≤ 0.5 ppm is necessary to lower by one-half the oxidation rate of lard, which is notably deficient in antioxidants. Aside from the notable prooxidant behavior of copper,

$$Met^{1+} + ROOH \xrightarrow{\quad /\!/ \quad} Met^{2+} + R-O:^- + \cdot OH$$

Chelators block reaction

$$Met = \text{Metal cation}$$

$$Met^{2+} + ROO:^- \xrightarrow{\quad /\!/ \quad} ROO\cdot + Met^{1+}$$

Of all the metal ions, copper is one of the most important prooxidants. A concentration of ≤ 0.1 ppm will promote oxidation and flavor problems in margaine oils; <0.02 ppm is necessary to ensure vegetable oil quality;

other metal cations including iron, manganese, chromium, nickel, vanadium, zinc, and aluminum all display varying and unpredictable prooxidant activities.

The limited availability of bulk natural

antioxidants for the food industry has resulted in the purification of many natural antioxidant substances and the chemical synthesis of others that effectively suppress free radical mechanisms. Some of the substances include gum guaiac, propylgallate, nordihydroguaiaretic acid (NDGA), butylated hydroxyanisole (BHA), and butylated hydroxytoluene (BHT). The chemical archives list hundreds of potential antioxidants, but few meet the prerequisites for food uses such as toxicity standards, long-term heat stability (as is needed for frying oils), lipid solubility, nonvolatility, and other factors. Individual mechanisms for these antioxidants are found elsewhere in the food and chemical literature.

Antioxidant activity in foods can be evaluated according to the stability ratio of a fat containing antioxidant to the same fat without antioxidant (control). The stability ratio is designated as the *protection factor* (*PF*) (Figure 7.18).

$$PF = \frac{\text{stability of fat with antioxidant}}{\text{stability of fat without antioxidant (control)}}$$

For lard and other fats or oils, peroxide values studied over the course of oxidation show that the necessary induction period for lipid rancidity can be extended by antioxidants, thereby improving their keeping quality.

The chemical events leading to hydrolytic and oxidative rancidity are enormously complex and any understanding of the phenomenon must consider

1. The dynamic character of the reactions involved in lipid oxidation.

2. The miniscule levels of peroxide decomposition necessary to produce off-flavors and destroy nutritionally important unsaturated fatty acids.

3. A realistic correlation of objective chemical tests for lipid oxidation with organoleptic indicators.

4. A keen recognition of the deteriorative and prooxidant forces that undermine lipid integrity.

Table 7.10 summarizes the major factors responsible for accelerating lipid deterioration and oxidation.

Polymerization

The unsaturated fatty acid components of fats and oils can undergo polymerization reactions. These reactions may occur with or without the presence of oxygen under high temperatures typical of commercial frying

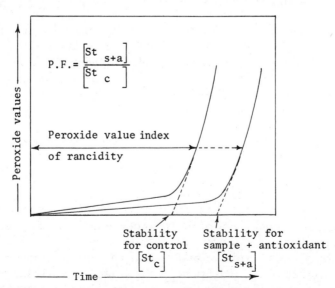

FIGURE 7.18. Relationships between fat stability and antioxidant effects using the peroxide value as a rancidity index.

Table 7.10. Corresponding Accelerating and Inhibiting Factors Involved in the Deterioration of Lipids

Accelerating factors	Inhibitors
High temperatures	Refrigeration and freezing
Polyunsaturated fatty acids	Hydrogenation of unsaturated fatty acids
Peroxides (oxidized fats)	Antioxidants
Heavy metals (cations)	Chelating agents
Enzymatic hydrolysis	Blanching
Oxygen availability	Decrease partial pressure of oxygen (physically or by increasing inert gas presence, e.g., nitrogen gas)
Photo-induced oxidation	Minimize lipid exposure to ultraviolet (blue) light

operations or other heated processing steps. Not only does intense heat promote polymerization reactions of unsaturated fatty acids, but concentrations of free fatty acids may also increase to a level of 1%, fatty acid unsaturation appears to decrease in concert with the *IV,* and the color of the fat or oil becomes progressively darker.

Thermal polymerization reactions for unsaturated fatty acids remain a matter of spec-ulation, but 1,4- and 1,3-dienes probably undergo a Diels–Alder-type reaction. By this route, new carbon-to-carbon bonds unite two fatty acids to form a dimeric structure, presumably involving fatty radical formation in at least one conjugated double-bonded structure (Figure 7.19A). In those cases in which nonconjugated polyunsaturated fatty acids undergo dimerization, most evidence indicates an isomerization into a reactive

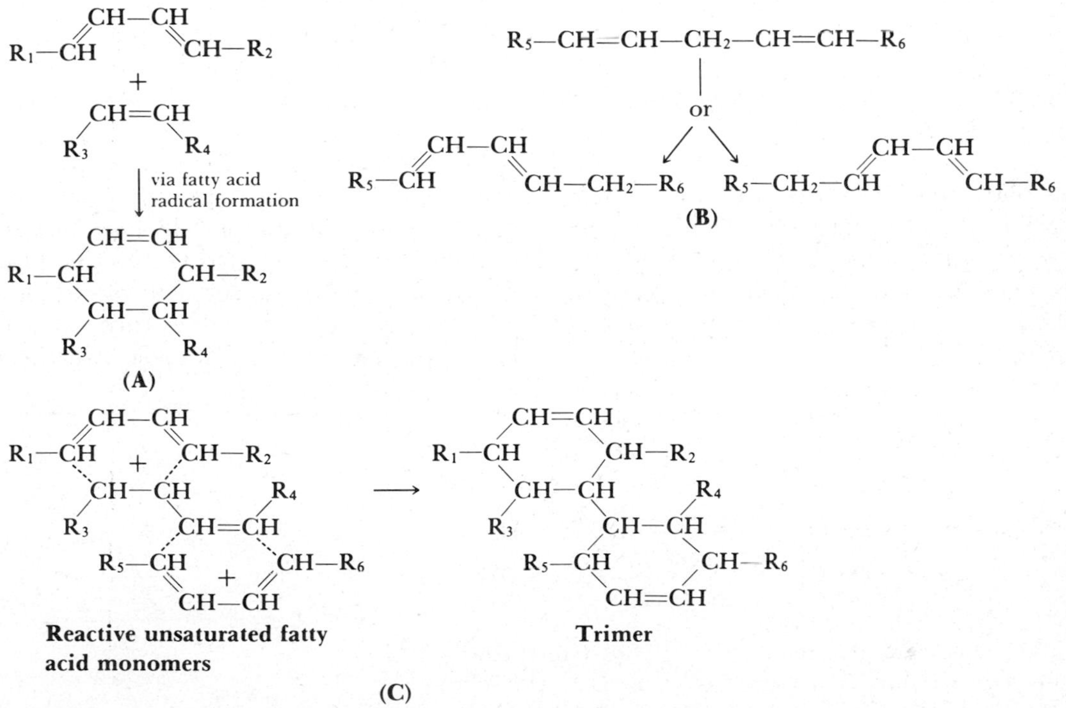

FIGURE 7.19. Some key reactions necessary for the thermal polymerization of unsaturated fatty acids. (A) Thermal dimerization of unsaturated fatty acids. (B) Formation of 2 conjugated unsaturated fatty acid from a nonconjugated structure. (C) Polymer formation among unsaturated monomers.

conjugated species before thermal polymerization reactions can proceed (Figure 7.19B).

For model systems where monoester polymerization of polyunsaturated acids occurs, polymerization may yield polycyclic structures. Multiple Diels–Alder reactions (two) may result in trimer formation (Figure 7.19C), but wherever triacylglycerols as opposed to monoesters are involved, the ultimate reaction products become very complicated. Furthermore, increasing concentrations of any polymeric product in oils will usually accompany their increased foaming and viscosity. This is especially true for frying oils.

Oxidative polymerization of polyunsaturated fatty acids, such as linoleate or more unsaturated fatty acids, also produce complex fatty acid polymers. Many of these mechanisms are still open to wide speculation. Although hydroperoxide intermediates of unsaturated fatty acids seem crucial, actual fatty acid polymerization probably involves the formation of carbon–carbon and carbon–oxygen linkages.

OTHER GLYCEROL-CONTAINING SUBSTANCES

Triacylglycerols have three hydroxyl groups, each esterified to a fatty acid. In addition to these saponifiable lipids, glycerol may provide an esterification "backbone" for *alkyl ether acylglycerols*, *glycosylacylglycerols*, and *phosphoglycerides*.

Alkyl Ether Acylglycerols and Glycosylacylglycerols

Alkyl ether acylglycerols are characterized by the presence of two fatty acid esters and the third hydroxyl position on the glyceryl moiety is joined by an ether (R—O—R') linkage to a long alkyl or alkenyl chain. Alkaline or enzymatic hydrolysis of alkyl ether acylglycerols yields *glyceryl ethers*.

The *glycosyl diacylglycerols* occur less widely than the alkyl ethers, but they are found in plants and the neural tissues of vertebrates. This group typically contains at least one sugar held in a glycosidic linkage with one hydroxyl group on a diacylglycerol (Figure 7.20).

Phosphoglycerides (or Glycerol Phosphatides)

The fats and oils of many animal and plant tissues as well as their cell membranes may contain large amounts of phosphorous-containing lipids. Since many of these lipids often (but not always) contain the polyhydric alcohol glycerol, the group is often designated as *glycerol phosphatides* or *phosphoglycerides*. Phospholipids in this group are also generically called *phosphatides* for convenience.

In contrast to the triacylglycerols, the glycerol phosphatides typically have two esterified fatty acids on glycerol, but its third hydroxyl group is esterified to phosphoric acid. The phosphoric acid, in turn, is combined with one of many possible structures

An alkyl ether diacylglycerol
$R_1 = $ —$CH_2(CH_2)_{14}CH_3$ (hexadecyl—) --------> Chimyl alcohol
$R_2 = $ —$CH_2(CH_2)_{16}CH_3$ (octadecyl—) --------> Batyl alcohol
$R_{3,4} = $ Fatty acids

Glyceryl ether

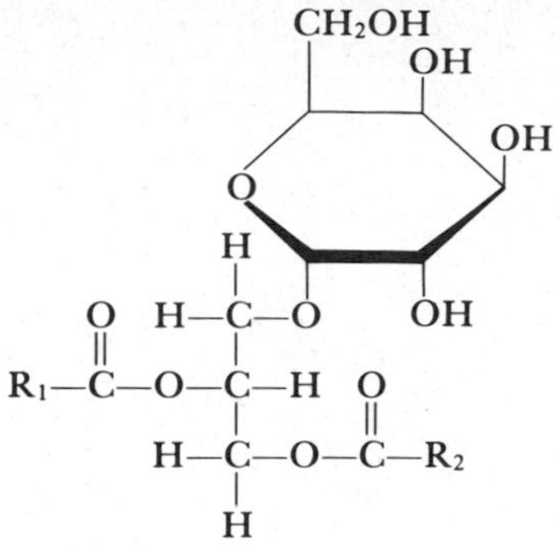

CH_2OH
OH
OH
O
H
O $H-C-O$ OH
$R_1-C-O-C-H$ O
$H-C-O-C-R_2$
H

R_1 and R_2 = Fatty acids

FIGURE 7.20. A galactosyl diacylglyceride.

tional approach instituted in 1967 by the IUPAC-IUB Commission on Biochemical Nomenclature, numbers 1 and 3 may not be interchanged for the same primary hydroxyl group of a glycerol structure. Furthermore, the second hydroxyl group must be directed to the left of the C-2 position in the Fischer projection structure for glycerol. The carbon atoms above and below the C-2 position are designated as C-1 and C-3, respectively. Stereospecific numbering according to these guidelines is indicated by the prefix *sn* in the stem name of the compound. Using stereospecific numbering and nomenclature for

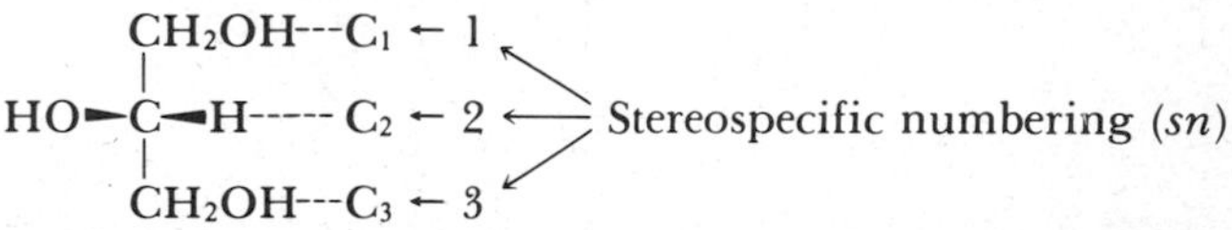

$CH_2OH\text{---}C_1 \leftarrow 1$
$HO-C-H\text{-----}C_2 \leftarrow 2 \quad$ Stereospecific numbering (*sn*)
$CH_2OH\text{---}C_3 \leftarrow 3$

such as choline, serine, or ethanolamine (Table 7.11).

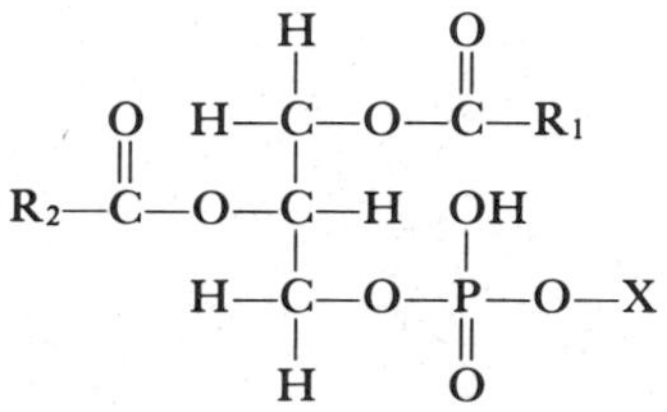

H O
O $H-C-O-C-R_1$
$R_2-C-O-C-H$ OH
$H-C-O-P-O-X$
H O

Since the phosphoric acid ester of glycerol serves as the principal parent structure of phosphatides, the presence of an asymmetric carbon atom (C-2) permits the formation of two equivalent stereochemical forms, namely, D-glycerol 1-phosphate and L-glycerol 3-phosphate:

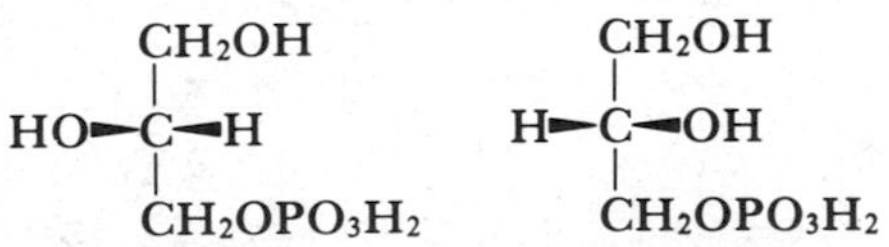

CH_2OH CH_2OH
$HO-C-H$ $H-C-OH$
$CH_2OPO_3H_2$ $CH_2OPO_3H_2$
L form   **D form**
Glycerol 3-phosphoric acids

In an effort to avoid confusion between different stereochemical forms of glycerol phosphate esters, glycerol derivatives are named according to *stereospecific numbering* (*sn*) of the carbon atoms. Using a conventional

glycerol the corresponding *sn* nomenclature for the two optical isomers of L-glycerophosphate below would be denoted as *sn*-glycerol 3-phosphoric acid (A) and *sn*-glycerol 1-phosphoric acid (B).

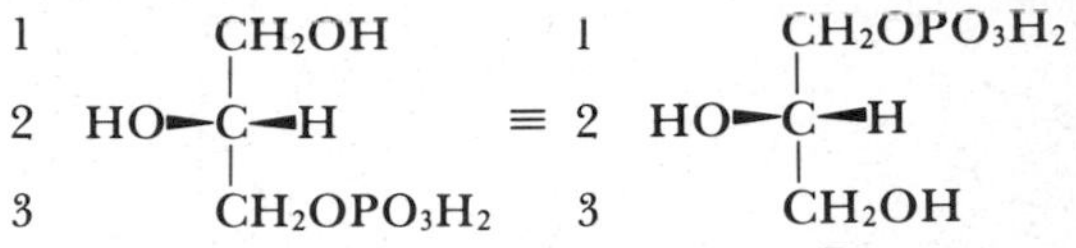

1   CH_2OH 1 $CH_2OPO_3H_2$
2   $HO-C-H$ $\equiv$ 2 $HO-C-H$
3    $CH_2OPO_3H_2$ 3 CH_2OH
     (A)   **(B)**

Application of the stereospecific nomenclature to other phospholipids appears in Table 7.11, but notice that many of these structures are named for simplicity with reference to phosphatidic acid, which has the structure

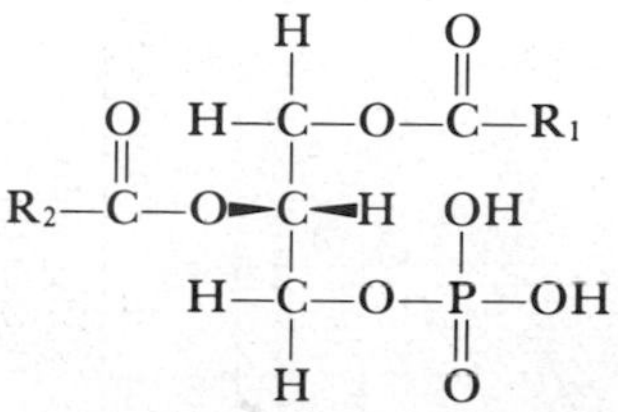

H O
O $H-C-O-C-R_1$
$R_2-C-O-C-H$ OH
$H-C-O-P-OH$
H O

Phosphatidic acid structure
(3-*sn*-phosphatidic acid)

Phosphoglycerides display *important* amphipathic properties owing to the presence

Table 7.11. Some Important Glycerol Phospholipids Found in Foods

Phospholipid	Usual fatty acid (nonpolar component)	Base (polar component)	Common name

3-*sn*-Phosphatidylcholine
CH_2OCOR_1
|
R_2COOCH O
| ‖
$CH_2-O-P-OCH_2CH_2N^+(CH_3)_3$
       |
       O^-
	Stearic or palmitic (R_1) polyunsaturated (R_2)	Choline	Lecithin

3-*sn*-Phosphatidylaminoethanol
CH_2OCOR_1
|
R_2COOCH O
| ‖
$CH_2-O-P-OCH_2CH_2\overset{+}{N}H_3$
       |
       O^-
	Stearic or palmitic (R_1) polyunsaturated (R_2)	Aminoethanol	Cephalin

3-*sn*-Phosphatidylserine
CH_2OCOR_1
|
R_2COOCH O $\overset{+}{N}H_3$
| ‖ |
$CH_2-O-P-OCH_2CH_2NH_3$
       | COO^-
       OH
	Stearic or palmitic (R_1) polyunsaturated (R_2)	Serine	Cephalin

3-*sn*-Phosphitalaminoethanol
α $CH_2OCH=CHR_1$
|
R_2COOCH β O
| ‖
$CH_2-O-P-OCH_2CH_2\overset{+}{N}H_3$
       |
       O^-
	Unsaturated ether (α) Linoleic (β)	Aminoethanol	Plasmalogen

Inositol phospholipid

Myoinositol replaces base

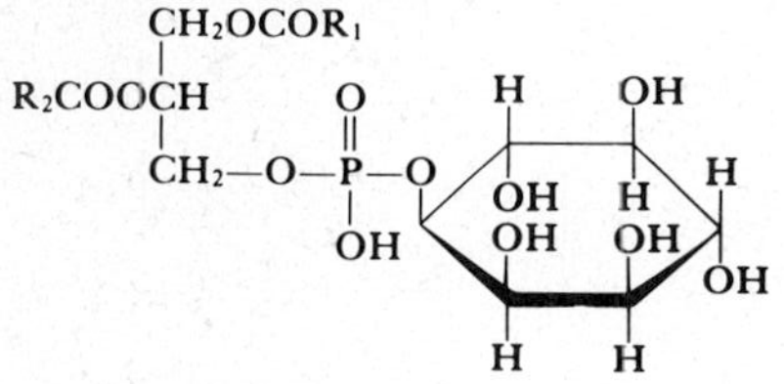

Palmitic (R_1)
Arachidonic (R_2)

³-*sn*-Phosphatidylinositol

—

Glycerol replaces base

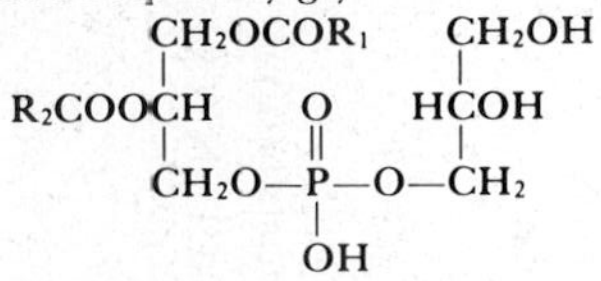

Polyunsaturated fatty acid (R_1, R_2)

³-*sn*-Phosphatidylglycerol

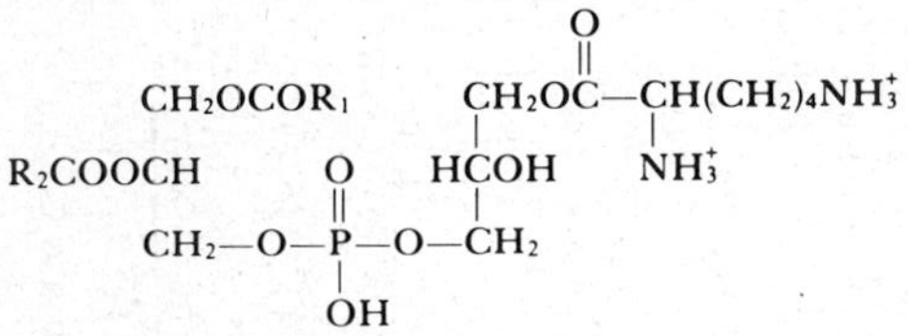

Aminoacyl phosphatidyl glycerol

(3-*sn*-Phosphatidyl)-3-*O*-L-lysyl glycerol

of *both* nonpolar hydrocarbon tails on ester-ified fatty acids and polar structures com-bined to the phosphate moiety. This feature also contributes important emulsification properties to the phosphoglycerides and en-hances their contribution to membrane structure and function.

Stereospecific analysis similar to that con-ducted on triacylglycerols, but using more specific hydrolytic enzymes (Figure 7.21), reveals that each type of phosphoglyceride may occur in many different forms depend-ing on its fatty acid substituents. Most data suggest that mixtures of saturated and un-saturated fatty acids are present, with the unsaturated acid residing at the 2-acylgly-ceryl position.

Aside from the phosphoglycerides al-ready discussed, many other forms do occur including *cardiolipins,* which are integral components of the inner mitochondrial membrane in cardiac muscle; *plasmalogens,* which correspond to phosphoglyceride an-alogs of alkyl ether acylglycerols found in the membranes of nerve and muscle cells; and *lysophosphoglycerides,* which result from the singular hydrolysis of a fatty acid from phosphoglycerides. Although common as metabolic intermediates, these substances can exert membrane toxicity at very high con-centrations. *Phosphatidyl sugars* also exist and have a very polar carbohydrate instead of an alcohol; these reside widely among plants and microorganisms.

SPHINGOLIPIDS

Another group of phosphatides known as sphingolipids are closely associated with plant and animal membrane components. In an-imals, high concentrations of these lipids oc-cur in the brain and neural tissues while only trace amounts exist in depot fats. The fatty alcohol called *4-sphingenine* serves a central structural function similar to that of glycerol in the preceding groups. The synthesis of 4-sphingenine stems from a complex biosyn-thesis requiring palmityl-CoA and serine. Functional sphingolipids are subsequently produced from the esterification of one fatty acid to the 4-sphingenine as well as the link-age of a very polar group.

In addition to 4-sphingenine, mammals may also employ *sphinganine (dihydrosphin-gosine)* as an esterification base; plants and yeasts employ *4-hydroxysphinganine (phytos-phingosine)*; and marine invertebrates exhibit diunsaturated bases such as 4,8-sphinga-diene.

Those sphingosine bases that have an amide link to a long saturated or monoun-saturated fatty acid form a *ceramide.* Struc-tures of this type exhibit two nonpolar hydrocarbon chains and serve as the fundamental basis for all sphingolipids. Var-iation among sphingolipids does occur, however, since different polar substituents may be linked to the 1-hydroxyl position of the sphingosine base (Figure 7.22).

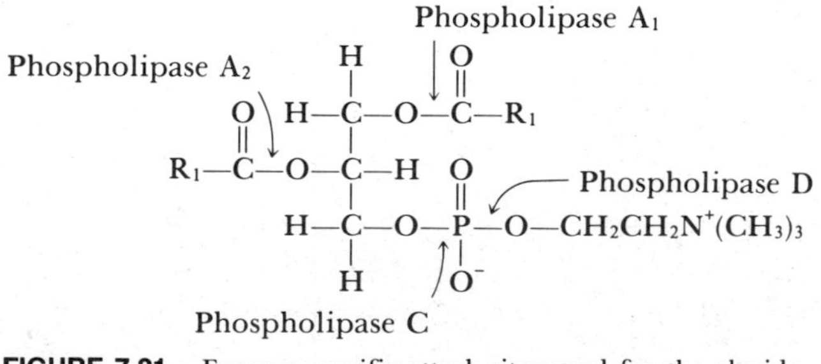

FIGURE 7.21. Enzyme-specific attack sites used for the elucida-tion of phosphatidic acid structures.

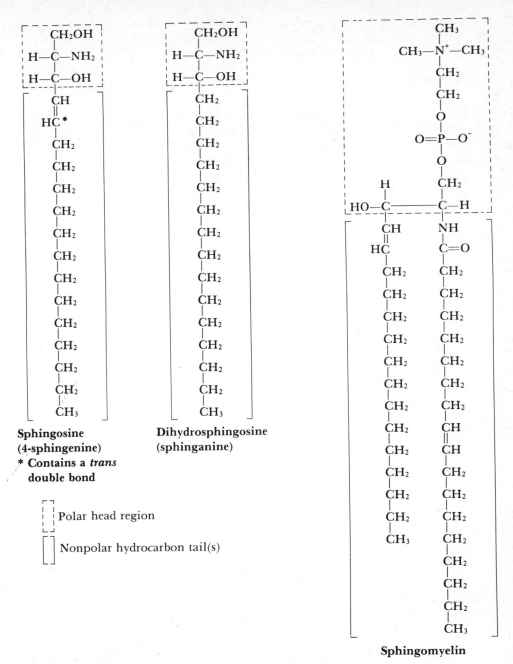

FIGURE 7.22. Structures for sphingosine, sphinganine, and sphingomyelin.

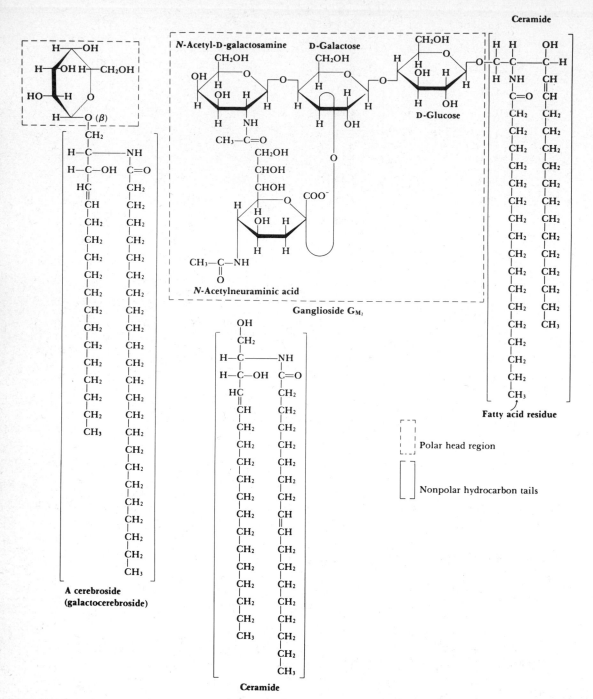

FIGURE 7.23. Typical structures for ceramide, a cerebroside, and ganglioside.

Sphingomyelin, the most common sphingolipid in animals, has either phosphorylcholine or phosphorylethanolamine as the polar 1-hydroxy substituent on *ceramide* (Figure 7.23).

positions and number of sialic acid residues and hexose components. Furthermore, a β-glycosidic linkage to ceramide and an N-acetyl-D-galactosamine are common to almost all gangliosides.

GLYCOLIPIDS

Those lipids that contain a polar carbohydrate head such as D-glucose or D-galactose and no phosphate group are classified as *glycolipids*. The glycolipids may contain a backbone of glycerol (glycosyldiacylglycerols) or sphingosine to form *cerebrosides*.

Simple glycolipids such as the cerebrosides are synthesized from a ceramide (sphingosine plus a fatty acid) by the addition of a sugar (Figure 7.23). Cerebrosides that reside in the brain and myelin sheaths of nervous tissues contain D-galactose (galactocerebrosides) as opposed to the nonneural tissues, which contain D-glucose (glucocerebrosides).

Cerebrosides may contain a variety of long-chained fatty acids (C_{22}–C_{26}), although lignoceric acid (C_{24}) seems to be one of the most prevalent acids. Both glycerol and sphingosine-based glycolipids have distinct dipolarity owing to the presence of two nonpolar hydrocarbon chains and a polar, but *neutral,* carbohydrate moiety.

In contrast to the cerebrosides, gangliosides represent a far more complicated group of sphingosine-based glycolipids that have a *polar* oligosaccharide substituent. The oligosaccharide may contain one or more residues of sialic acid, which imparts a negative charge to the overall structure at a neutral pH. For human gangliosides, N-acetylneuraminic acids serve as the terminal residues in oligosaccharide side chains. Minute concentrations of gangliosides are present in nonneural tissues, but concentrations of >5% occur in the gray matter of the brain and outer surfaces of neurocellular membranes. At least 25 types of gangliosides exist owing to fundamental differences in the relative

NONSAPONIFIABLE LIPIDS AND MINOR CONSTITUENTS OF FATS AND FATTY OILS

Polyprenyl (Isoprenoid) Lipids

Many types of nonsaponifiable lipids that fall into the classification of polyprenyl compounds are formed from a simple repeating unit described as an *isoprenoid unit* (Figure 7.24).

Isoprenoid units that are condensed to form complex *polyprenyl structures* originate from condensation of two acetyl-CoA units to produce acetoacetyl-CoA (Figure 7.25). An enzyme-mediated ester condensation of a third acetyl-CoA unit subsequently results in 3-hydroxy-3-methyl-glutaryl-CoA. Reduction of the thioester by very common biochemical mechanisms yields a key intermediate in all isoprenoid structures known as *mevalonic acid.* Following three phosphoryl transfers from ATP to mevalonic acid along with a decarboxylative step, *isopentenyl pyrophosphate* or *prenyl pyrophosphate* is formed. At this point, initiation of polyprenyl biosynthesis depends on the isomerization of one isopentenyl pyrophosphate to dimethylallyl pyrophosphate. The dimethylallyl pyrophosphate then serves as an initiator for subsequent prenyl unit additions as pyrophosphates are cleaved (Figure 7.25). Prenyl units polymerized by this route can result in the production of many different high-molecular-weight structures including terpenes, steroids, and plant pigments such as carotenoids. Further extensive polymerization of prenyl units may also lead to high polymeric forms of natural rubber. Figure 7.24 illustrates the location of these repeating isoprenoid structures in some common polyprenyl compounds.

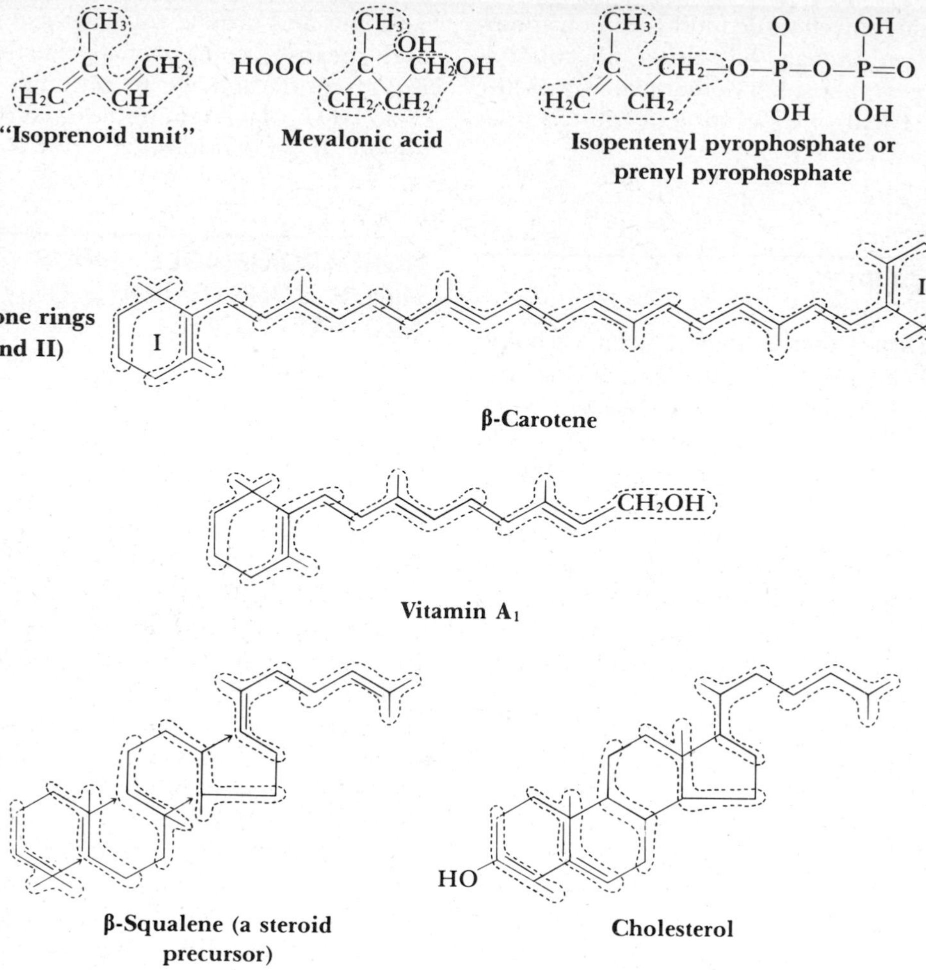

FIGURE 7.24. The isoprenoid structure and its occurrence in some important compounds.

The terpenes: Isopentenyl pyrophosphate-based compounds are produced by many plants, animals, and bacteria. A survey of these substances, generically referred to as terpenes, reveals that they are multiples of the isoprenoid unit. The simplest structures, known as *monoterpenes,* contain two isoprene units, while *sesquiterpenes, diterpenes, triterpenes,* and *tetraterpenes,* respectively, contain three, four, six, and eight isoprenoid units. Linear as well as cyclized molecular forms are common among the terpenes (Figure 7.26).

Plants in particular have a wide variety of monoterpenes and sesquiterpenes. Furthermore, the characteristic flavors and fragrances of plant foods and spices are often attributable to these terpenes and their immediate derivatives. Due to the oily properties of many aromatic terpenes, they are often referred to as *essential oils.*

Although plants may contain an abundance of a single terpene such as α-pinene, which constitutes up to 65% of terpentine oil, most plant and spice oils are very complex mixtures of *d*-limonene, camphor, α- and β-pinenes, menthone, phellandrene, sylvestrene, terpenines, and other terpenes.

The mono-, di-, and higher polyterpenoids are present in flowers, roots, stems, and leaves of plants, but their fruits and superficial tissues are noted for essential oils. Cop-

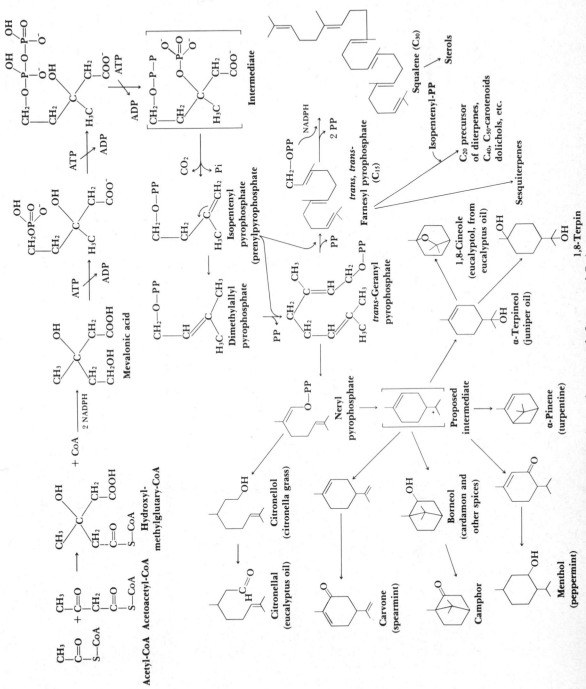

FIGURE 7.25. Pathways for the biosynthesis of polyprenyl compounds from acetyl-CoA.

Common terpene structures

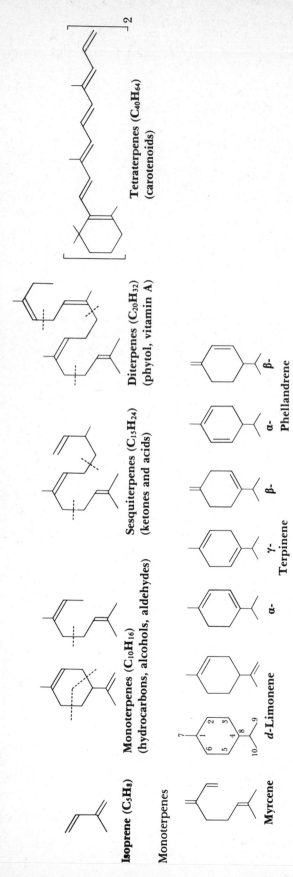

Isoprene (C_5H_8)

Monoterpenes

Monoterpenes ($C_{10}H_{16}$)
(hydrocarbons, alcohols, aldehydes)

Sesquiterpenes ($C_{15}H_{24}$)
(ketones and acids)

Diterpenes ($C_{20}H_{32}$)
(phytol, vitamin A)

Tetraterpenes ($C_{40}H_{64}$)
(carotenoids)

Myrcene

d-Limonene

α-　**γ-**　**β-**　**β-**
　　Terpinene　　**Phellandrene**

Ocimene　**α-Pinene**　**β-Pinene**　Camphene　Sylvestrene　*d*-Δ³-Carene

Oxygenated monoterpenes

Menthol　Menthone　Carvone　Cineole-I,4　Cineole-I,8　*d*-Camphor　Pulegone

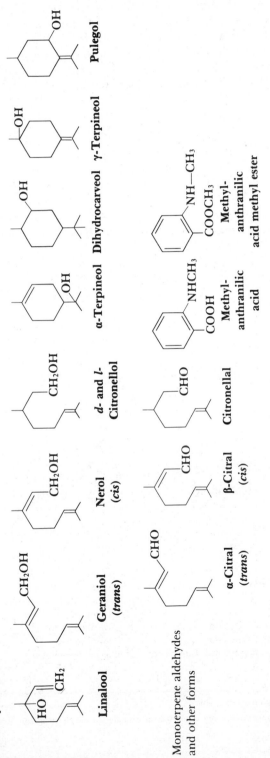

FIGURE 7.26. Some monoterpenes, sesquiterpenes, and their derivatives, which occur in essential oils.

465

ious amounts of these oils often reside in resinous duct tissues and very specialized oil glands.

Biochemical functions for these plant terpenes are not fully understood, but it has been proposed and roundly debated that they may counteract insect invaders, promote plant pollination, act as waste products of plant metabolism, or serve some other plant survival mechanism. Suspicions concerning biochemical functions of plant terpenes are also paralleled with doubts concerning their actual step-by-step biosynthesis from isopentenyl pyrophosphate.

Many oxygenated derivatives of the monoterpenes and sesquiterpenes also exist, and these play a far greater role in the actual flavors and fragrances associated with essen-

tial oils than the parent unoxygenated terpenes. Most of these oxygenated terpene derivatives assume the forms of aliphatic alcohols, aldehydes, and their esters and/or ethers. Some of the most common oxygenated terpenes are illustrated in Figure 7.26, whereas Table 7.12 outlines many of the oxygenated and unoxygenated terpenes found in numerous essential oil sources.

Structures directly or indirectly synthesized from isoprenoid units are also responsible for the molecular structure of nonsaponifiable constituents of fats, fatty oils, biological membranes, and substances involved in photosynthetic processes. This diverse group includes the key sesquiterpene intermediate *farnesol;* the diterpene alcohol *phytol,* which serves as a structural compo-

Table 7.12. Some Important Essential Oil Sources Indicating the Major Terpene Constituent as a Percentage of Total Essential Oils in Source[a]

Essential oil source	Major terpene constituent(s)	Essential oil source	Major terpene constituent
Basil, sweet	Methyl ester of chavicol (estragole) (50–60%)	Ginger	Zingiberene (a sesquiterpene)
Bay	Eugenol, chavicol (55–65%)	Lavandin	Linaloöl and linalyl acetate (66%), esterified terpenes (20–25%)
Birch, sweet	Methyl salicylate glycosides (98%)	Lavender	Linalyl acetate ester (50%)
Caraway	Carvone (50–60%), *d*-limonene (50–40%)	Lemon	*d*- and *dl*-limonene (90%), citral (2.5–6%)
Cardamon	Cineole, terpineol, terpineol acetate	Lemon grass	Citral (75–85%)
Celery seed	*d*-Limonene (50+%) and sedanolide, sedanonic anhydride	Mint, Japanese	*l*-Menthol (70–90%)
Cinnamon bark	Cinnamaldehyde (60–75%), eugenol (15%)	Neroli (orange flower)	Methyl anthranilate, indole, and many others
Cinnamon leaves	Eugenol (80–96%)	Nutmeg	α-Pinene and other oxygenated terpenes
Clove	Eugenol (free and acetate forms) (95%)	Orange, sweet	*d*-limonene (90%), capraldehyde, and other oxygenated terpenes
Coriander	Linaloöl (60–70%)	Origanum	Carvacol (60%)
Dill seed and weed	Carvone (20–60%)	Palmerosa	Geraniol (84–94%)
Tarragon (estragon)	Methyl ether of chavicol (estragole) (65%)	Pennyroyal, European	Pulegone (90–95%)
Fennel	Anethole and fenchone		

[a]For those sources with no expressed percentage, concentrations are variable.

NOTE: Note that the major terpene constituent of these essential oils is not necessarily attributable to the characteristic flavors and fragrances associated with the oil.

nent of chlorophyll; and the triterpene *squalene* (first isolated from elasmobranch fishes), which serves as an important precursor to cholesterol and steroid structures. Tetraterpene structures are exhibited by *carotenoids* (C_{40}) and their oxygenated derivatives called *xanthophylls*. The central cleavage of those carotenoids containing β-ionone rings may serve as precursors of vitamin A formation, while still other structures lacking β-ionone rings, but otherwise reminiscent of the carotenoids, form *lycopene*-type pigments (e.g., tomatoes).

Other vitamin-active terpenes considered to be fat-soluble vitamins, namely vitamins E and K, as well as terpenoids displaying coenzyme functions in mitochondrial oxidations such as *ubiquinone* or *coenzyme Q*, have been discussed elsewhere in the text. Counterparts of coenzyme Q that exhibit similar biochemical functions in chloroplasts are called *plastoquinones*.

Added to this diverse collection of terpenes are the polyisoprenoids called *polyprenols*. These are long-chained linear structures that contain a terminal primary alcohol and 55 to 95 carbon atoms contributed by 11 to 19 isoprenoid units. *Bactoprenol* or *undecaprenyl* alcohol (C_{55}) in prokaryotes and *dolichol* (C_{95}) present in animal tissues along with their corresponding phosphate esters participate in important coenzyme activities. Biochemical functions for these substances seem to be linked to enzyme-mediated transfers of cytoplasmic monosaccharides across cell membranes to sites of peptidoglycan, lipopolysaccharide, and teichoic acid synthesis in cell walls and their surfaces.

There are many other fungi, plant, and insect hormone substances that develop from isoprenoid-based farnesyl pyrophosphate (C_{15}). Included among these are *gibberellins* and *abscisic acid*, which have plant growth-hormone activities. *Juvenile hormones* of insects that regulate their egg-hatching and larval stages of development also show polyprenyl origins. Evidence indicates too that plants synthesize some isoprenoid derivatives, such as *juvebione* in the balsam fir, which can have profound effects on the developmental physiology of insects. Therefore, it is clear that at least some plant species may ultimately discourage potential insect enemies by synthesizing biochemical substances that are functionally reminscent of insect hormones.

Provided that many species of insects are susceptible to the hormone-like effects of isoprenoid derivatives, the production of these substances by natural or synthetic routes may lead to a new dimension of insect control other than the use of hard-core toxicants.

Aside from the insect-active effects of juvenile hormones and some plant isoprenoids, chemical communication among insects is mediated by long-chained alcohols and ketones known as *pheromones*. These compounds elicit specific flight, fight, food recognition, reproductive, and mating responses among insects of a similar species.

The steroids: The steroids represent a large class of compounds that contain a characteristic four-ringed nucleus composed of three fused six-membered rings and a single five-membered ring. For purposes of classification, the steroids may be considered as derivatives of a fused, reduced ring system known as *perhydrocyclopentanophenanthrene*. Note that rings A, B, and C are fused cyclohexane rings that display the classical nonlinear arrangement of *phenanthrene*, while the D ring is a terminal cyclopentane structure.

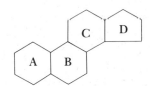

Perhydrocyclopentanophenanthrene

Phenanthrene

Using *cholestanol* (*dihydrocholesterol*) as an example of a typical steroid alcohol or *sterol*, the conventional numbering system follows the succession indicated in the formula

and *progesterone*. The naturally occurring steroids *androgen* and *estrogen* do not contain any aliphatic C-17 substituents.

As a result of *trans* fusions to rings A and

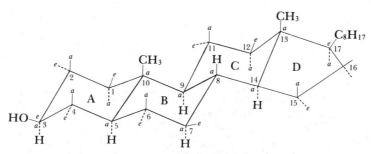

Cholestanol

A cursory inspection of most naturally occurring steroids reflects the common occurrence of an oxygen atom at the C-3 position. For the sterols, the oxygen is present in the form of a hydroxyl group, whereas other steroids may display a carbonyl group at the same site.

Many steroids also contain two axially oriented methyl groups at the C-18 and C-19 positions. Conventional molecular projection formulas assume the extension of these groups out of the plane of the page and toward the viewer. The C-3 hydroxyl group is viewed in the same way since it is equatorial to the overall structure. The majority of sterols also have an aliphatic substituent extending from the C-17 position. This substituent, too, projects from the plane of the page toward the viewer when inspecting the projection formula. The C-17 bonded structure may have eight to ten carbons in the case of *sterols,* five carbons in the *bile acids,* and two carbons in the *adrenal cortical steroids*

D, rings B and C are firmly held in characteristic chair conformations. This ensures that the whole molecule is more or less rigid, with the exception of ring A, which can form a boat or chair conformation depending on substituent group interactions at C-10 and C-3 positions.

Molecular asymmetry is also an important feature of steroid structures. For cholestanol, nine asymmetric centers (C-3, 5, 8, 9, 10, 13, 14, 17, and 20) exist. Since methyl groups at C-18 and -19 along with the C-3 hydroxy group and aliphatic chain are on the same side of the steroid projection model, these groups display a β orientation (to the front). Any groups that project from the opposite side (to the back) are said to have α orientation. It has become a convention to use dotted lines for those substituents that are α oriented and solid lines for β-oriented substituents on the steroid structure. With reference to Figure 7.27, it is clear that the C-10 and C-13 methyl groups, the C-3 hy-

FIGURE 7.27. Conformational structure of cholestanol.

droxyl group, the C-8 hydrogen, and the C-17 aliphatic side chain are all β oriented.

The C-5 and C-10 positions of molecular asymmetry have special importance since the C-5 hydrogen atom and the C-10 methyl group may be on similar or different sides of the molecular plane. When rings A and B are *cis* to each other, the C-5 hydrogen and the C-10 methyl group are on the same side of the molecule. This is recognized as *normal* molecular configuration. The opposite condition, when the A and B rings are *trans* to each other, results in a corresponding *allo* configuration (Figure 7.28). Note too that the C-3 hydroxyl group may assume an α or β position (Figure 7.29). This C-3 hydroxyl position has important analytical significance since the glycosidic steroid derivative known as *digitonin* precipitates 3-β-hydroxy steroids. This characteristic reaction can be used as a preparative approach to quantitative steroid analysis as well as the configurational study of steroids.

The sterols. Sterols are crystalline, neutral, nonsaponifiable alcohols that display high melting points and properties that rcscmble those exhibited by *cholesterol*. In the case of fats and fatty oils, sterols account for most of the nonsaponifiable lipid fraction. For example, wheat germ oil can contain 1.5 to 3.0% of nonsaponifiable lipid of which 70 to 85% may be classified as sterols.

The predominant sterol in animal fats and oils is cholesterol, but the most common sterols in vegetable oils are generically called *phytosterols*.

In addition to a hydrocarbon chain at the C-17 position and a β-hydroxyl group at C-3, cholesterol exhibits a double bond at the 5,6 position.

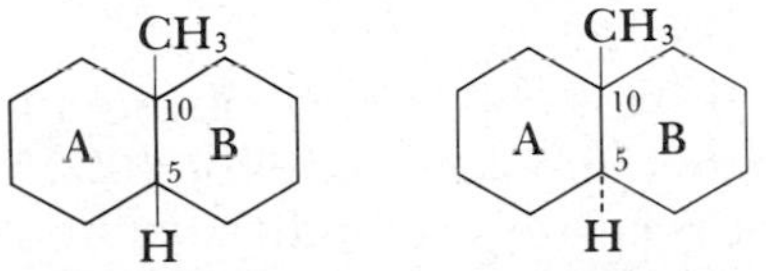

FIGURE 7.29. Conformational structures for the C-3 hydroxyl group on steroids.

Cholesterol pervades the animal body at different concentration levels depending on the tissue and its lipid content. Bile and blood plasma also contain cholesterol, but in the latter case, about 66% is esterified to an unsaturated fatty acid such as linoleic acid while the rest remains as the free sterol.

The reduction of the 5,6 double bond in cholesterol results in two commonly occurring derivatives of cholesterol, namely, β-*cholestanol* (β, *allo*) and *coprosterol* (β, *normal*). Small amounts of β-cholestanol occur as minor constituents of blood plasma and tissue sterols, but coprosterol is the predominant fecal sterol.

As indicated in earlier sections, *7-dehydrocholesterol* (vitamin D_1) results from the oxidation of cholesterol, and this structure is characterized by a conjugated pair of double bonds. Significant amounts of this sterol derivative are located in the skin.

The conjugated double bonds shown by 7-dehydrocholesterol are also reflected in the yeast sterol known as *ergosterol.* Both ergosterol and 7-dehydrocholesterol serve as precursors of vitamin D-active compounds since the B ring, containing the conjugated double bonds, can rupture to yield vitamin D_2 and eventually vitamin D_3. Additional types of sterols widely occur; the characteristic R groups at C-17 along with the double bond locations within the rigid steroid nucleus are cited in Table 7.13.

Among the phytosterols, *stigmasterol* and β-*sitosterol* are the most common. Stigmasterol is the major sterol in soybean oil and β-sitosterol is the chief sterol in cottonseed oil. Depending on the type of oil, oils may contain 50 to 500 mg of sterol per 100.0 g of oil.

FIGURE 7.28. Conformational structures for C-5 and C-10 substituents of steroids.

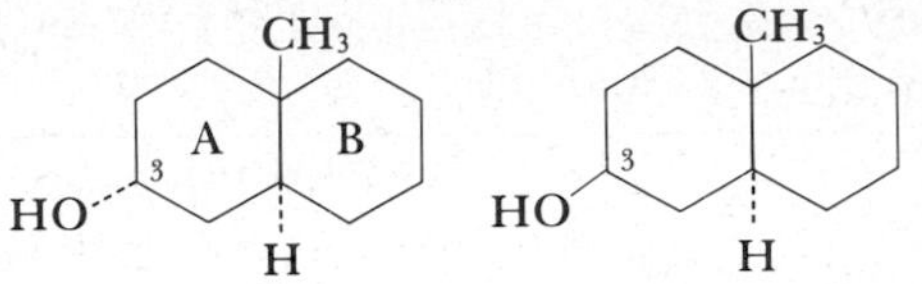

Table 7.13. Nomenclature and C-17 Aliphatic Side Chain Present in Various Sterols That Are Present in Fats and Oils

R = $-CH(CH_3)(CH_2)_3CH(CH_3)_2$	Cholesterol
R = $-CH(CH_3)CH_2CH_2CHCH(CH_3)_2$ $\quad\quad\quad\quad\quad\quad\|$ $\quad\quad\quad\quad\quad C_2H_5$	β-Sitosterol
R = $-CH(CH_3)CH{=}CHCHCH(CH_3)_2$ $\quad\quad\quad\quad\quad\quad\quad\|$ $\quad\quad\quad\quad\quad\quad C_2H_5$	Stigmasterol
R = $-CH(CH_3)CH{=}CHCHCH(CH_3)_2$ $\quad\quad\quad\quad\quad\quad\quad\|$ $\quad\quad\quad\quad\quad\quad CH_3$	Campesterol (no double bond at 5,6)
R = $-CH(CH_3)CH{=}CHCH_2CH(CH_3)_2$	Brassicasterol
R = $-CH(CH_3)CH_2CH_2CCH(CH_3)_2$ $\quad\quad\quad\quad\quad\quad\quad\|\|$ $\quad\quad\quad\quad\quad\quad CH$ $\quad\quad\quad\quad\quad\quad\|$ $\quad\quad\quad\quad\quad\quad CH_3$	Avenasterol (double bond at 5,6 or 7,8 only)
R = $-CH(CH_3)CH{=}CHCH_2CCH(CH_3)_2$ $\quad\quad\quad\quad\quad\quad\quad\quad\|$ $\quad\quad\quad\quad\quad\quad\quad C_2H_5$	α-Spinasterol (double bond at 7,8)
R = $-CH(CH_3)CH{=}CHCHCH(CH_3)_2$ $\quad\quad\quad\quad\quad\quad\quad\|$ $\quad\quad\quad\quad\quad\quad CH_3$	Ergosterol (double bonds at 5,6 and 7,8)
R = $-CH(CH_3)CH_2CH_2CH{=}C(CH_3)_2$	Lanosterol (double bond at 8,9)

Isomers of β-sitosterol do exist such as γ-*sitosterol,* which is a minor sterol constituent of wheat and rye germ oils. Other variations in β-sitosterol structure may involve the C-17 hydrocarbon as demonstrated by *campesterol.* This sterol occurs in soybean, wheat germ, and rapeseed oils.

In terms of widespread occurrence, ergosterol is also one of the preeminent sterols of plants apart from the fact that it occurs in animals and animal products such as egg yolk. Stigmasterol, also named dihydro-β-sitosterol, exists as a minor sterol constituent of sitosterols in cottonseed, wheat germ, and corn.

The advent of highly sensitive gas chromatographic analysis for rare oils has uncovered many new sterols. A survey of rapeseed oils indicates that sterols may comprise about 55 + % of the 0.75% nonsaponifiable lipid fraction. An analytical profile of sterols in many plant oils indicates the presence of β-sitosterol (55%), campesterol (25%), *brassicasterol* (15%), and cholesterol (few percent). Although cholesterol is not regarded as a major plant sterol, it may occur in trace amounts depending on the plant species.

Other minor plant sterols include *avenasterol, α-spinasterol, stigmastenol,* and *24-methylcholest-7-enol.* Spinach seed, alfalfa, tea, and shea butter have significant concentrations of these sterols and only minor concentrations of the more typical and common sterols. Avenasterol, having a C-5 double bond, is common in sesame seed, wheat germ oils, rice bran, and coffee (>90 mg/100 g of oil), while alfalfa, pumpkin, and spinach seeds have another form of avenasterol with a C-7 double bond at concentrations of >40 mg/100 g. The α-spinasterol concentrations of these same seed oils as well as shea butters may exceed 45 mg/100 g of oil. Similar concentrations are recognized for stigmasterol in these sources.

In the cases of refined oils, sterol reduction may approach a level of 40%, but the relative occurrence of individual sterols with respect to each other remains quite constant. In the case of margarines, the sterol content of individual oil ingredients is reflected in the final product, but corn oil–based margarines seem to contain higher phytosterol concentration levels (400–550 mg/100 g) than soybean oil margarines (120–400 mg/100 g). All steps of oil processing and especially hydrogenation probably contribute in one way or the other to decreases in the concentrations of native oil sterols. For example, it is recognized that the oils of well-hydrogenated stick margarine contain a lower phytosterol concentration than the less hydrogenated versions of the same oil formulation.

Although many phytosterols are present in foods as free sterols, they frequently exist as esterified or glycosidic steroid derivatives.

In addition to cholesterol and the phytosteroids, there are many other sterols that commonly occur. Lanosterol, first isolated from the lipid coating of wool, serves as an important intermediate of cholesterol biosynthesis in animals. The nonvascular plants such as the algae and fungi synthesize a wide spectrum of *phycosterols* and *mycosterols*, respectively. Many of these are structurally reminiscent of ergosterol, but most of the minor sterols in these sources remain to be characterized.

The C_{24} steroids. Cholesterol serves as an immediate precursor of bile acids synthesized in the liver. These steroid derivatives have extraordinary amphipathic or surface-active properties that promote emulsification of dietary triacylglycerols and fat-soluble vitamins. Emulsification of these dietary constituents not only promotes the enzymatic hydrolysis of triacylglycerols but promotes their physiological absorption as well. The bile acids also act as the functional vehicle for cholesterol excretion in humans since its oxidation to carbon dioxide and water is impossible. Therefore, bile acids generated in the liver are routinely discharged in the bile by way of the gall bladder into the intestine.

At least four bile acids have been isolated in humans; these include *cholic, deoxycholic, chenodeoxycholic,* and *lithocholic acids.* The A and B rings of these steroids may be *cis* or *normal* in their configuration, but all of the hydroxyl groups display an α configuration. Although species variability is common, the most common human bile acid is cholic acid (Figure 7.30).

Note that the C-17 hydrocarbon chain has five carbons and a terminal carboxyl group, which is critical for the biosynthesis of amide-linked amino acids. The linkage of sterols to glycine and taurine at the carboxyl group yields *glycocholic* and *taurocholic acids* whose salts are responsible for the noted emulsi-

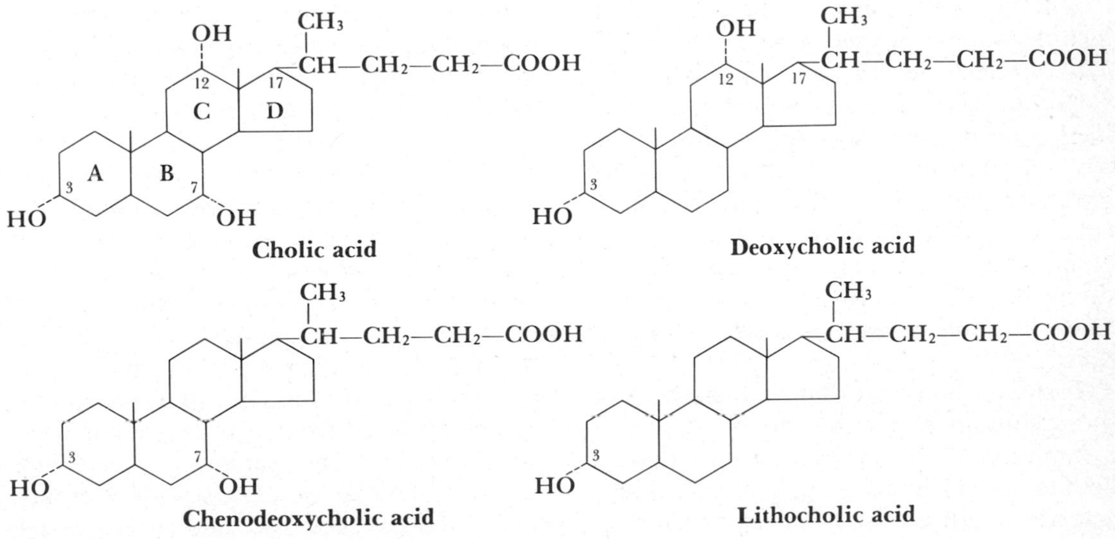

FIGURE 7.30. Structures for four bile acids.

$$\overset{\displaystyle O}{\underset{\displaystyle \|}{C_{23}H_{26}(OH)_3C}}\overset{\displaystyle H}{\underset{}{-N}}-CH_2-COOH$$

Glycocholic acid (cholylglycine)

$$\overset{\displaystyle O}{\underset{\displaystyle \|}{C_{23}H_{26}(OH)_3C}}\overset{\displaystyle H}{\underset{}{-N}}-CH_2-CH_2-SO_3H$$

Taurocholic acid (cholyltaurine)

FIGURE 7.31. Empirical formulas for glycocholic and taurocholic acids indicating characteristic differences in the C-17-linked groups.

fication and detergent activities of the bile acids (Figure 7.31).

The C_{21} steroids. *Progesterone* and *adrenal cortical steroids*, which have endocrinological importance, are derived from the cholesterol nucleus; however, the C-17 side chain has only two carbon atoms.

Among these steroids are progesterone, which is produced by the corpus luteum, and the mineral and glucocorticoids, which are produced by the adrenal cortex. Under basal conditions, human subjects produce 15 to 30 mg of adrenal corticosteroids per day with peak production between 4 and 8 A.M. At least 40 different forms have been recognized, but only about seven are considered to be active forms. Among the active mineral corticoids are *aldosterone, desoxycorticosterone,* and *17-hydroxy-11-desoxycorticosterone,* which influence sodium and chloride ion resorption plus potassium ion excretion in the renal tubules.

The glucocorticoids, on the other hand, include *cortisone, cortisol, corticosterone,* and *dehydrocorticosterone.* Those corticoids are involved in increasing glucose formation and increasing protein catabolism to foster glucose synthesis in the liver. Excessive activity of these agents may cause a dystrophic muscular condition. However, the therapeutic administrations of some corticoids (e.g., cortisone, cortisol) are effective for *reducing* collagen in joints as well as tissue inflammations.

It should be noted that characteristics of these substances include double bonds at C-3, and C-4,5 to form an α,β-unsaturated ketone; a C-11 hydroxyl group; and a characteristic oxidized C-20, -21 side chain (Figure 7.32).

The C_{18} and C_{19} steroids. The male sex hormones known as *androgens* display a characteristic C_{19} steroid structure that does not have a C_{17} side chain. Testosterone produced by the testis is a typical steroid in this class. The corresponding female sex hormones synthesized by the ovary are different from male hormones since the A ring of the steroid is aromatic and the C-3 hydroxyl group assumes phenolic properties. Because of this structural difference, estrogens display the behavior of weak acids that are extractable from benzene solutions using weak bases. One estrogen typically produced by the ovary is *estradiol-17β.*

Aside from these estrogens, *estrogenically active substances* do occur in some plants. These substances are known as *phytoestrogens,* but they have molecular structures that are quite unlike the steroids and therefore should not be confused with them. Their association with the estrogens stems from their physiological effects on animals, which are similar to those of estrogens. These effects include stimulation of uterine hypertrophy in immature rodents, stimulation of protein synthesis in ovariectomized rats, and other mammary, cervical, and vulvular effects normally associated with estrogens.

Many other miscellaneous types of steroids exist. Among these are the stimulating cardioactive glycosides of squill, foxglove, and others that produce a steroid aglycone upon hydrolysis of their sugar moieties. The γ-lactone ring containing steroids such as *digitoxigenin* are typical of this group.

Another interesting group of plant compounds derived from steroids are known as the *saponins.* The saponins, which are glycone derivatives of *sapogenin* steroids, are noted for their lytic and surface-active properties.

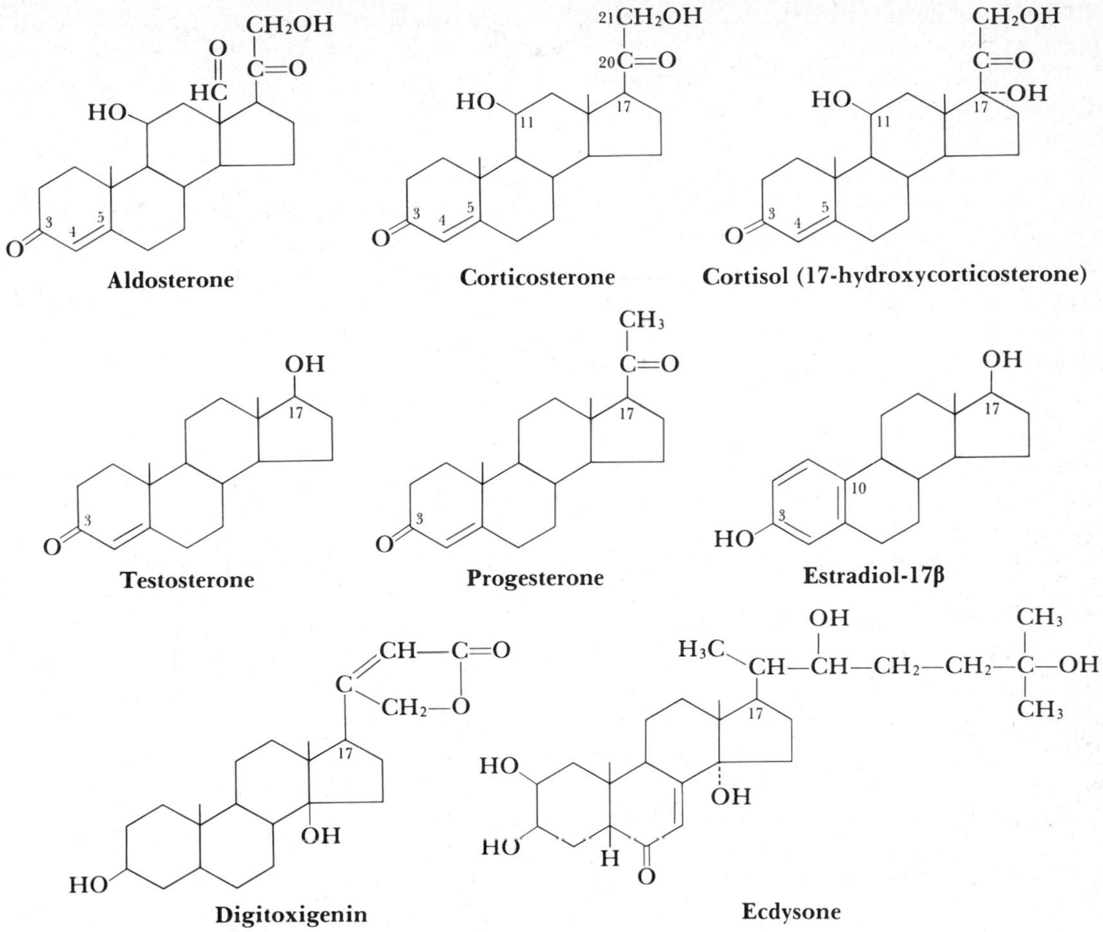

FIGURE 7.32. Structures for some important steroids.

Other organisms, including insects, are not without their full complement of steroid hormone-active substances such as ecdysone that affect molting and metamorphosis.

Additional varieties of steroid-based compounds may serve as growth-stimulating agents for livestock and poultry production, and many related forms are recognized as carcinogenic.

PROSTAGLANDINS, THEIR DERIVATIVES, AND LEUKOTRIENES

The Swedish Nobel laureate von Euler is credited with the isolation of hormonal sub-stances known as prostaglandins, which are derived from arachidonic acid. Although knowledge of the normal functions of many prostaglandins is speculative, it is clear that they occur in miniscule concentrations throughout animal tissues and are involved in the contraction of smooth muscle, the ac-tivites of enzymes involved in lipid metab-olism, normal functioning of the central nervous system, blood pressure and pulse rate regulation, steroid hormone function, mobilization of adipose tissue-stored fats, uterine muscle contraction, and other com-plex physiological mechanisms. Unlike many other hormonal substances, prostaglandins are synthesized and discharged from the majority of mammalian cells and tissues, but they are not stored in large bulk quantities prior to their release.

The major classes of primary prostaglandins (PG) are the A, E, and F series, which correspond to PGA, PGE, and PGF, respectively, and they are all structurally similar to prostanoic acid (Figure 7.33). Prostaglandin structures are relatively complex as a result of their many different functional groups. Functional groups for the PGA series reflect the presence of α,β-unsaturated ketones, the PGE series displays a β-hydroxyketone, and the PGF series exhibits 1,3-diols. Conventional prostaglandin nomenclature also employs numerical subscripts of 1, 2, and 3 to indicate the specific number of double bonds in side chains. Since all double bonds are *trans,* there is no need to specifically indicate the presence of *cis* or *trans* isomers except for unusual structural forms or when clarity is questionable. An α subscript is commonly used, however, to indicate the configuration of the C-9 hydroxyl group when it is directed down from the plane of the prostaglandin ring.

Linoleic acid serves as the major dietary predecessor of the prostaglandins. For humans, linoleic acid cannot be synthesized *de novo,* and so it is considered to be an *essential fatty acid.* Of the average 10 g/day adult dietary intake of linoleic acid, only a small portion undergoes hepatic conversion into arachidonic acid and only about 1.0 mg/day of prostaglandins are metabolized. It is clear that the quantitative production of prostaglandins from polyunsaturated fatty acids has relatively little importance compared to the constant dietary requirement for linoleic acid, which serves as a prostaglandin precursor.

The elongation and desaturation of lin-

oleic acid-based precursors to arachidonic acid and prostaglandins proceeds by way of tri-, tetra-, and penta-, unsaturated intermediates. These precursors undergo cyclization and incorporate oxygen into the unsaturated molecule. Based on currently recognized biosynthetic pathways, PGE_3 and $PGE_{3\alpha}$ develop from eicosapentaenoic acid (C_{20}-Δ^5,8,11,14,17); PGE_2 and $PGF_{2\alpha}$ originate from arachidonic acid (C_{20}-Δ^5,8,11,14); and PGE_1 and $PGF_{1\alpha}$ evolve from dihomo-γ-linoleic acid (C_{20}-Δ^8,11,14). In the case of mammals, prostaglandins derived from arachidonic acid or other unsaturated acids having 18 carbons and a double bond six carbons from the methyl terminus (Δ^6) of the acid serve as a substrate for the principal prostaglandin synthesis pathways. As an example, consider the structure of linoleic acid and its conversion to the prostaglandin compounds in Figure 7.34.

Depending on the prostaglandin synthesis route, crucial unsaturated fatty acid synthesis stock, such as arachidonic acid, is enzymatically hydrolyzed from cellular phospholipids by phospholipase A. The liberated fatty acid subsequently undergoes a transformation and reaction that yield a cyclic endoperoxide. This step is mediated by a microsomal endoperoxide synthetase or cyclooxygenase that incorporates molecular oxygen at C-11 through a free radical mechanism.

As a sequel to this step, a cyclic endoperoxide–15-hydroperoxide known as PGG_2 is formed. The PGG_2 is then converted to prostaglandin H_2 (PGH_2) by a reduced glutathion-dependent peroxidase. Specific biochemical pathways remain to be established for the synthesis of all known prostaglandins, but tissue and cell-specific enzymes are responsible for the appearance of D-, E-, and F-type prostaglandins as well as the conversion of these prostaglandins into prostacyclin (PGI_2) or *thromboxanes.*

Specific cellular tissue differences in prostaglandin syntheses obviously account for notably high concentrations of PGE_2 and $PGE_{2\alpha}$ in the spleen and kidney; PGI_2 dis-

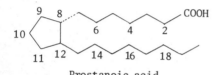

Prostanoic acid

FIGURE 7.33. Structure of prostanoic acid and numerical sequence for counting its various carbon positions.

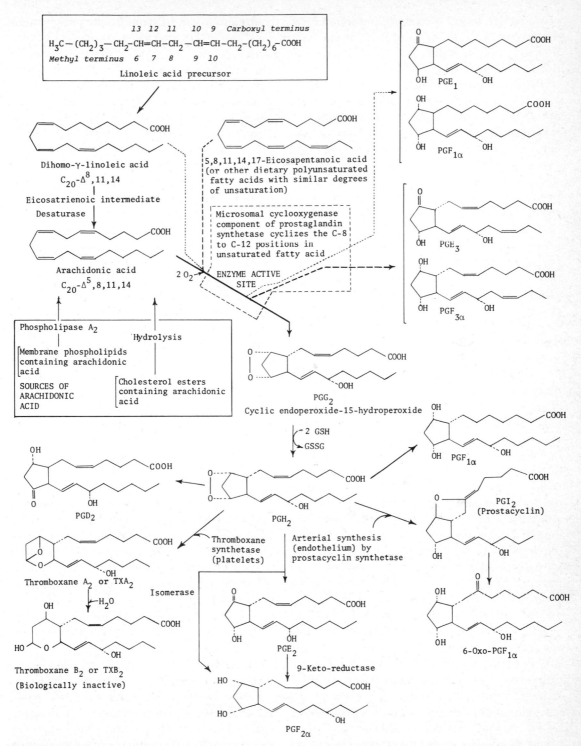

FIGURE 7.34. Possible biochemical pathways for the synthesis of prostaglandins. The key biosynthetic pathways involve arachidonic acid precursors but elongation and desaturation of linoleic acid can serve similar prostaglandin precursor functions.

charged by blood vessels; equimolar formation of PGE_2, $PGF_{2\alpha}$, and PGI_2 by cardiac tissues; and thromboxane A_2 (TXA_2) synthesized by platelets.

Prostaglandin synthesis is significantly affected by both steroid and nonsteroid anti-inflammatory agents. Furthermore, arachidonic acid availability and cyclooxygenase activity are both prerequisites to prostaglandin synthesis. As a consequence of these key requirements, hydroxycortisone, prednisone, and other therapeutic steroids inhibit phospholipase A_2, which liberates arachidonic acid from its esterified phospholipid reservoirs. On the other hand, nonsteroidal agents including indomethacin, acetylsalicylic acid (aspirin), and other compounds can interfere with cyclooxygenase activity. In the case of aspirin, this may be caused by the acetylation of the enzyme-active site.

Unlike hormones, prostaglandins have an immediate hormone-like effect on the sites that surround the cells that produce and discharge the prostaglandins. The biochemical half-life for prostaglandins is short but variable depending on the type. Estimated half-lives for TXA_2, PGI_2, and PGH_2 are estimated to be 0.5, 1.0, and 5.0 min, respectively, although some of these values may be debatable on either the high or low side. The short-lived existence yet profound physiological effects of prostaglandins and their biochemical relatives seem to involve a series of cascade-like reactions that require the participation of cyclic adenosine monophosphate (cAMP) (Samuelsson and Paoletti, 1978). The overall effects of these cascade reactions are different for different target cells, thus accounting for a full gamut of different physiological responses for cardiopulmonary/vascular functions as well as lung, kidney, reproductive, and gastrointestinal tract activities. Vasoconstriction is promoted by PGE_2; PGE_1 induces vasodilation; PGE-based structures show abilities to moderate gastric acid secretion in gastric mucosal cells by possibly decreasing cAMP levels; TXA_2 encourages arteriole constriction, platelet

aggregation, and thrombus-forming potentials; and PGI_2 serves as an antagonist to TXA_2 activities. In spite of rapid advances in the understanding of prostaglandins, some noteworthy references for prostaglandin actions include Wolfe (1981), Kinsella (1981), Vane and Bergström (1979), Samuelsson and Paoletti (1976–1980), Kharash and Fried (1977), and Bergström *et al.* (1968).

As indicated above, PGI_2 or prostacyclin serves as an antagonist to TXA_2 activities. Clearly, then, the prostacyclins can suppress platelet aggregation in the blood, and it is generally conceded that infusion rates of only 16 ng/kg body weight/min will completely but transiently suppress platelet aggregation in humans. Since platelet aggregation on blood vessel walls at the site of an injury serves as an initial stage in the formation of permanent fibrin clots, the therapeutic or dietary control over prostacyclin activity can have major consequences in eliminating certain blood-clotting mechanisms. In spite of a very short-half-life, prostacyclins are unlike prostaglandins since they persist in the blood after its circulation through the lungs. Prostaglandins, on the other hand, are normally cleared in a very rapid fashion during their circulation through the lungs, liver, and kidneys.

In addition to all the foregoing prostaglandin-related compounds, another group of short-lived (~30 s half-life) unsaturated fatty acid derivatives have been recognized. These compounds, known as *leukotrienes,* develop directly from arachidonic acid instead of the biosynthetic pathways responsible for the appearance of prostaglandins.

Leukotrienes exhibit a series of three double bonds (trienes) that produce the characteristic ultraviolet absorption properties of these molecules. The generic name for these compounds stems from their early recognition as components of leukocytes (white blood cells). Leukotriene structures display a structural similarity reminiscent of prostaglandins but some forms such as leukotriene C have a covalently bonded molecule of cysteine (Figure 7.35).

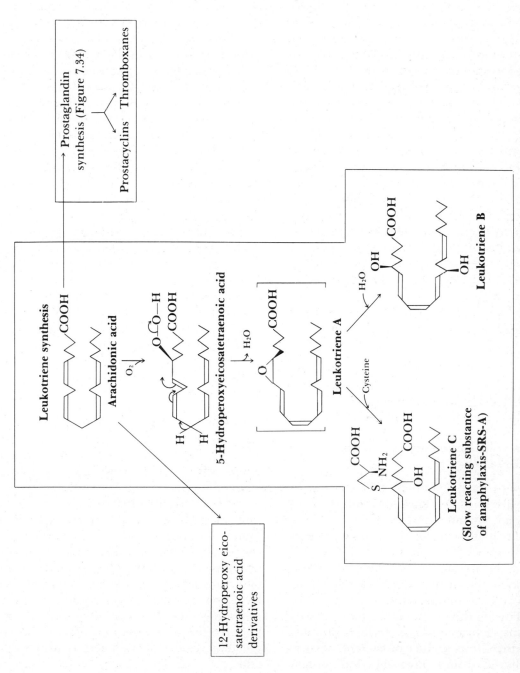

FIGURE 7.35. Leukotriene synthesis from arachidonic acid and the relative relationship to prostaglandin-derived products.

Biochemical mechanisms for the physiological modes of leukotriene action are enigmatic and complex since minute concentrations and very short half-lives make experimentation difficult. Nevertheless, most evidence suggests that at least one of the leukotrienes is certainly involved in human allergic responses as well as the bronchiole constriction of air passages during asthma attacks. It has also been discovered that the so-called slow reacting substance of anaphylaxis (SRS-A), which has been recognized for some time, is also identical to leukotriene C.

Some common theories for leukotriene action on the bronchi suggest that leukotrienes are produced when arachidonic acid is released in the mast cells. Stimuli for this action may depend on the interaction of specific agents such as certain drugs, pollens, or food components with antibodies on the mast cell surface. This action culminates in a burst of leukotrienes that have a potent physiological constrictive effect on the bronchioles.

LIPOPROTEIN OCCURRENCE, STRUCTURES, AND FUNCTIONS

Lipoprotein Transport Vehicles of Blood Plasma

Although small quantities of free fatty acids and glycolipids and small concentrations of nonpolar vitamins and hormones may be present in plasma, nearly all plasma lipids are transported by specific proteins synthesized in the liver and intestinal mucosa. These plasma-lipid transport vehicles are generally categorized as lipoproteins.

Following the ingestion of a fatty meal by humans or primates, lymph fluids sampled from the thoracic duct or mesenteric sites are altered from a normally clear, watery fluid to a milky, opalescent fluid. The opacity of the lymph depends on the amount of dietary fats that enter the duodenum of the stomach and subsequently cause the en-gorgement of the lymphatic channels with triacylglycerols. Moreover, since the immediate products of lipid digestion, known as *chyle*, exist as tiny lipid globules or *chylomicra*, the effective discharge of chyle into the venous blood circulation results in a temporal period of *absorptive lipemia*. Under these conditions blood lipid concentrations are higher than in the prefat absorptive state.

The heterogeneous nature of dietary lipids is reflected in the classes of lipoproteins that are responsible for plasma lipid transport. The protein component of these lipoprotein complexes have a relatively high proportion of nonpolar amino acid residues that permit their effective interaction and binding with blood lipids.

The characteristic lipid-binding properties of these plasma proteins, called *apoproteins*, are thought to depend in part on London–van der Waals dispersion forces and, especially, hydrophobic interactions between apoproteins and lipids. Persistent analytical studies of plasma lipoprotein complexes have revealed that they can be individually categorized by their electrophoretic mobility, hydrated density, or apolipoprotein composition.

Lipoprotein classification: Electrophoretic mobility has served as the most common classification method for plasma lipoproteins. This method employs electrophoretic support media such as paper or agarose, and plasma proteins are separated according to the corresponding migration of serum proteins. Some partitioned lipoprotein fractions may remain at the origin of the electrophoretic medium, but those lipoproteins that migrate into the α_2-globulin region are called pre-β-lipoproteins; those migrating into the β-globulin zone are β-lipoproteins; and those moving into the α_1-globulin region are denoted as α-lipoproteins.

In contrast to this practical approach to lipoprotein classification, the relative buoyant densities of lipoproteins can be determined by ultracentrifugation. Five broad density classes are recognized and these include (1) chylomicrons, (2) very-low-density lipoproteins (α_2 VLDL), (3) low-density lipoproteins (LDL), high-density lipoproteins (HDL), and very-high-density lipoproteins (VHDL). The relative densities for these lipoproteins are outlined in Table 7.14, along with their corresponding classification according to electrophoretic subdivisions.

The LDL fraction can be fractionated still further into LDL_1 or intermediate density lipoprotein (IDL) (density 1.006–1.019 g/mL), and LDL_2 (density 1.019–1.063 g/mL). The HDL fraction has also been separated into HDL_2 (density 1.063–1.125 g/mL) and HDL_3 (density 1.125–1.21 g/mL).

Plasma lipoproteins display wide heterogeneity in size and composition for each density fraction studied. Not only does this heterogeneity stem from the lipid fractions carried by various plasma apoproteins, but the individual apoproteins themselves offer wide immunological, structural, and molecular weight diversities.

Aside from electrophoretic and buoyant density properties, lipoproteins have been classified according to their apoprotein composition. The classification criteria in this case considers that lipoprotein families exist, each of which represents a polydisperse mixture of lipid–apolipoprotein associations that have a distinct apolipoprotein or distinctive constitutive polypeptide(s). The five fundamental *lipoprotein families* include LpA, LpB, LpC, LpD, and LpE. The A-I and A-II apolipoproteins occur in the LpA family; β-apolipoprotein(s) occur in the LpB family; C-I, C-II, and C-III apolipoproteins are present in the LpC family; the LpD family has a single D-type apolipoprotein; while an arginine-rich apolipoprotein is found in the LpE family.

Macromolecular complexes of lipoproteins, comprised of more than a single lipoprotein family, are often found at a hydrated density of <1.030 g/mL. Lipoprotein complexes such as these are regarded as *associated families*.

The majority of lipoprotein complexes having a density of <1.006 g/mL have B-, C-, and E-type lipoproteins, whereas the isolation of lipoproteins from higher-density fractions (HDL) indicates the presence of several lipoprotein families including LpA, LpC, LpD, and LpE.

Aside from the contributions of apolipoproteins to the transport of plasma proteins, knowledge of their functions remains highly speculative. The C-I and A-I forms are claimed to activate *lecithin cholesterol acetyltransferase* (LCAT), an enzyme that mediates cholesteryl esterification. The C-I and C-II forms are thought to act as cofactors for the lipolytic enzyme lipoprotein lipase. The other functions for native apolipoproteins remain to be firmly established.

Macromolecular structure of plasma lipoproteins: It is clear from detailed sequencing and conformational studies of plasma lipoproteins that the strategic placement of nonpolar amino acid residues, held in α-helical protein structures, probably promote phospholipid interactions and the development of effective macromolecular amphipathic structures. Although the specific orientation of the apolipoprotein helix with respect to fatty acid chains on key phospholipid components is speculative, many investigators favor a fluid mosaic model for plasma lipoprotein structures. Based on this theory, which is reminiscent of cell membrane structural theories in later sections of this text, protein structures coexist with lipids (amphipathic lipids) as an integrated matrix. In the purest of structural forms, an amphipathic plasma lipoprotein may strictly display secondary α-helical protein constituents as illustrated in Figure 7.36 (left); but most evidence points to the presence of tertiary and quaternary protein structures in the mosaic model in addition to the classic helical protein component of plasma lipoproteins (Figure 7.36, right).

VLDL, LDL, and HDL functions and interactions: In addition to the confusion and uncertainties surrounding plasma lipoprotein functions, they are responsible for the

Table 7.14. Classification of Plasma Lipids

Electrophoretic classification	Origin	Pre-β	β	α	α
Hydrated density classification	Chylomicrons	Very low density (VLDL)	Low density (LDL)	High density (HDL)	Very high density (VHDL)
Diameter (angstroms)	5000–800	800–300	200	100	—
Density (g/mL)	<0.95	0.95–1.006	1.006–1.063	1.063–1.210	>1.21
Composition					
Protein (%)	0.5–2.0	12.0	25.0	50.0	55.0
Lipid (total) (%)	98.0–99.5	88.0	75.0	50.0	45.0
Cholesterol (free + esters)	3–13% of total lipid	13–45% of total lipid	50% of total lipid	20% of total lipid	
Major lipid	Triacylglycerol	Triacylglycerol, phosphatidyl choline, sphingomyelin	High cholesteryl linoleate	High cholesteryl linoleate and phosphatidyl choline	Cholesteryl esters and phosphoglycerides
Speculated lipoprotein origins	Intestine	Liver and intestine	Metabolism product of VLDL	Liver and intestine	
Apolipoprotein	B, C-I, C-II, C-III	B, C-I, C-II, C-III	B	A-I, A-II, C-II	A-I, A-II
Concentration (mg/mL)	100–250	130–200	210–400	50–130	290–400
Function	Exogenous triacylglycerol transport	Endogenous triacylglycerol transport; high dietary carbohydrate increases this fraction	Cholesterol and phospholipid transport to peripheral cells	May transport cholesterol from peripheral cells to liver	—

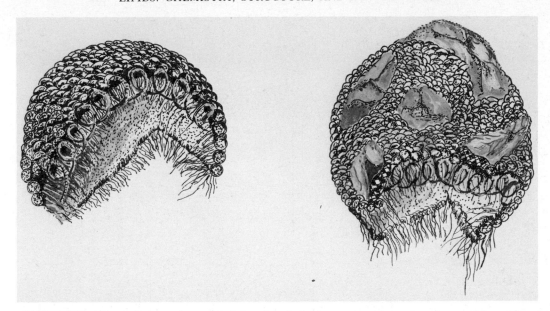

FIGURE 7.36. Plasma lipoprotein models. Polar phospholipid components are oriented toward the surface of the lipoprotein while the hydrocarbon chains participate in countless hydrophobic interactions. According to the most general models (left), polar lipids are intertwined by several proteins that principally display secondary protein (helical) structures. According to other advanced models (right), fluid mosaic lipoprotein structures are suspected to include key proteins conventionally associated with plasma lipoproteins, and these proteins display additional tertiary and quaternary levels of protein structure.

transport of lipids to sites of metabolism. Exogenous (dietary) lipids are transported as chylomicrons and endogenous lipids are transported in LDL fractions. Conceptual views of lipoprotein metabolism based on experimental evidence indicate that triacylglycerol-rich particles are discharged by the hepatic tissues in union with β-apolipoproteins (density <1.006 g/mL). These β-apolipoprotein–triacylglycerol complexes subsequently associate with C- and E-type lipoproteins obtained from HDL. Resulting VLDL forms undergo *gradual* lipolysis to yield free glycerol and fatty acids; and as the lipid component is lost, VLDL density increases (Figure 7.37). Evidence also indicates that C proteins lost from VLDL appear in the HDL fractions. As illustrated in Figure 7.37, VLDL shows progressive density increments during its catabolism to form IDL and LDL.

The triacylglycerol fraction of lipoproteins is eliminated by a lipoprotein lipase that is thought to reside on the endothelial surface of adipose and muscle tissues. The C-II apolipoprotein is able to activate this lipase. These overall events may involve the

presence of a C-I apolipoprotein that activates another lipoprotein lipase; a hepatic lipase may also be involved.

The catabolism of LDL may occur in hepatic tissues as well as peripheral cells (smooth muscle and fibroblast cells). One popular LDL catabolic concept suggests that LDL becomes affixed to a specific cell-membrane binding site, followed by its endocytosis into the cell. The characteristic protein of the LDL is either biochemically deactivated or degraded by lysosomal enzymes and lipases in such a way that cholesteryl esters are hydrolyzed and contributed to the existing intracellular pool of cholesterol. The increased availability of intracellular cholesterol (1) minimizes 3-hydroxy-3-methylglutaryl-CoA reductase activity, which serves as a key rate-governing enzyme in cholesterol biosynthesis; (2) promotes fatty acyl-CoA/cholesteryl acyltransferase activity, which produces new intracellular cholesteryl esters; and (3) reduces the activity of LDL receptor sites so that cell membranes refrain from further LDL interactions.

The precise role of HDL, exclusive of the

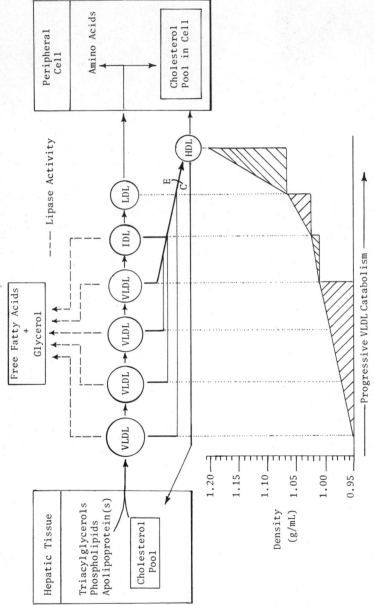

FIGURE 7.37. Plasma lipoproteins with β-apolipoproteins released from hepatic cells undergo progressive lipase-mediated hydrolysis and protein degradation to yield higher density lipoproteins. A variety of LDL species may materialize before the appearance of IDL, LDL, and HDL. Each lipoprotein species increases in its relative density to other forms as catabolism of VLDL proceeds; HDL displays the highest density. It is speculated that HDL may be responsible for the transport of cholesterol between peripheral cells and hepatic tissues, where it is metabolized and discharged from the body.

preceding events, in which it acts as a source of C and E lipoproteins, must be firmly established.

A variety of abnormalities in cholesterol metabolism are recognized to involve the normal functioning of LDL receptor protein. For example, one form of familial hypercholesterolemia reflects a mutation in the LDL receptor protein that forbids the normal binding interaction and LDL-mediated uptake of cholesteryl esters into cells. In still other cholesteryl ester storage diseases, lysosomal lipase is absent. There is little doubt that these and other similarly complex biochemical problems are contributing factors to some forms of atherosclerosis.

In normal individuals the VLDL fraction may fluctuate toward higher concentration levels owing to the dietary presence of high-carbohydrate foods, ethyl alcohol, or triacylglycerol consumption. The LDL fraction contains about 75% of the serum cholesterol, and both the cholesterol and phospholipid contents of the LDL fraction are freely exchangeable with the HDL fraction.

Lipoprotein Membrane Systems

Biological membranes are necessary for the containment of unit biochemical processes and organelles that exist within cells. Membranes also serve as a boundary that separates the aqueous contents of plant and animal cells from their external environments. Membrane functions are not limited to combating the aqueous solvent or environmental stresses that surround cells; membranes also regulate the influx and efflux of cellular nutrients, ions, and metabolic intermediates. Many of these activities are affected by membrane–hormone interactions, enzyme-catalyzed reactions, electrical depolarization phenomena, antibody interactions, and the consumption or production of high-energy phosphates.

The apparent functions of cell membranes are understood much better than their actual biochemical operation or the structural integration of their chief constituents.

Crude fractionations for many types of membranes, from different species and cellular organelles, show that they consist of ~55.0% protein, ~40% amphipathic lipid, and ~5% carbohydrates, although wide variations occur. The carbohydrate fraction is present as a mixture of glycoproteins and glycolipids as opposed to pure polysaccharidic structures.

Electron microscopy, X-ray diffraction, and other advanced analytical methods support the contention that membrane structures are actually fluid phenomena in which a solution of globular proteins is dispersed throughout a fluid lipid matrix. The key membrane components are not statically bound in any type of rigid structure, and it is theorized that both lipid and protein components are laterally mobile throughout the structure. This feature probably has great significance regarding the functional properties that characterize membranes.

The protein constituents of membranes appear either loosely or firmly integrated into the matrix of amphipathic lipid, and both types of protein occurrence are common. These proteins can be experimentally categorized as *peripheral* or *integral* depending on the amount of associated lipid with the proteins when they are extricated from membranes by the use of nonionic surfactants. Peripheral proteins can be isolated without significant amounts of associated lipid while the integral proteins appear to be inextricably associated with membrane lipids.

Lipid components of membranes are oriented in such a way that their hydrophobic moieties (e.g., fatty acid residues) interact to form a lipid bilayer. This concept has been promulgated by S. J. Singer, D. Wallach, and others since the 1960s as a *fluid mosaic model* of membrane structure. In spite of wide scrutiny since its introduction, this model of membrane structure is still widely recognized as plausible.

The positions of hydrophobic moieties on lipid molecules according to this model are significant because they are directed toward the inner reaches of the membrane while hydrophilic portions of the lipids, such as carboxylate groups and phosphodiesters, remain exposed to the aqueous medium. An opposite arrangement is unlikely since interjection of hydrophobic groups into the aqueous medium would not contribute to maximum entropy according to the laws of thermodynamics, and biochemical energy would necessarily be expended in order to maintain this type of structure.

The hydrophilic portions of lipid components are not only better suited for aqueous solvent interactions, but these polar moieties promote lipid bilayer interactions with charged protein molecules that border membranes. This relationship is exhibited in Figure 7.38. In those obvious cases in which proteins penetrate the fluid bilayer structure of membranes, it is not unexpected that hydrophobic portions of the proteins will associate with other hydrophobic groups inside the lipid bilayer.

Distributions of key enzyme and receptor proteins over the width and internal area of the lipid bilayer are not noted for their symmetrical distributions. In fact, experimental evidence indicates that protein locations may be linked to their characteristic biochemical activities by virtue of their fluid movement and exposure to extra- or intracellular aqueous media.

Normal membrane functions are intrinsically tied to the translocation of small hydrophilic organic molecules, water, ions, and

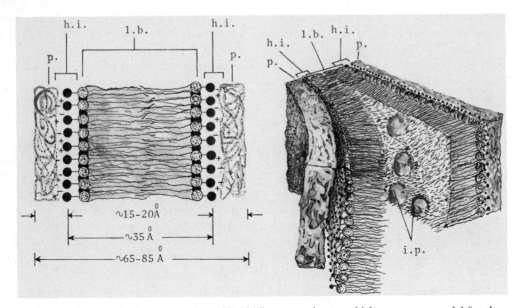

FIGURE 7.38. (Left) Transverse view of lipid bilayer membrane, which serves as a model for the fluid mosaic concept of membrane structure. (Right) Perspective view of membrane structure split down the median of physical lipid bilayer interaction. p., associated membrane protein(s); l.b., lipid bilayer with hydrophobic chains (e.g., fatty acid residues) on outer and inner lipid layers that interact; h.i., hydrophilic interactions between polar moieties of the lipid bilayer and protein; i.p., incorporated proteins that reside in the lipid matrix and contribute to enzymatic, transport, and cell-receptor functions that affect cell physiology and biochemistry. The polar lipid components of membranes include phosphoglycerides (e.g., phosphatidylethanolamine, phosphatidylcholine) and sphingomyelin. Although endoplasmic reticulum and other organelle membrane structures contain miniscule amounts of cholesterol or triacylglycerols, many plasma membranes of higher animal cells display considerable amounts of cholesterol. The fatty acid constituents of the lipid bilayer largely reflect nutritional considerations as do the other polar lipid components of cell membranes. These nutritional factors plus species- and organelle-specific differences serve to produce many complicated membrane structures with highly specific functions.

radicals into or out of cell structures; but the actual mechanisms responsible for these observed phenomena largely remain an enigma. Conventional concepts for solute translocations across membranes are explained in terms of *carriers* or *channeling*. Carriers are simplistically explained as mobile proteins whose superficial exposure on the inner or outer surfaces of a membrane will bind specific transportable solutes. The protein-bound solute complex is then permitted to migrate through the membrane to the opposite side, where the solute is released. Alternative translocation theories such as channeling suggest that solute transport is mediated by the formation of channels or pores that transiently appear across the membrane structure. The energy required to initiate and accomplish such "channeled translocations" of solutes, as well as the mechanism, remain to be established.

FATS AND FATTY OILS OF FOOD COMMERCE AND INDUSTRY

Natural and processed lipid constituents of the human diet represent a complex variety of nonpolar and polar lipids. These dietary lipids are obtained from sources such as oil seeds, fruit pulps, animal tissues, and fishes, all of which display enormous intra- and interspecies differences in their relative lipid components (Table 7.15). Depending on food industrial requirements, those plant and animal tissues richest in lipids are processed for the manufacture of specific edible lipids or lipid fractions.

Fat and Fatty Oil Extraction from Oil Seeds and Fruit Pulps

Oil seed and fruit pulp extraction involves the mechanical and magnetic removal of foreign material from raw oil seeds, fruit pulp, and other oil stock prior to the actual oil extraction process. In most sources such as oil seeds, moisture content should not exceed 7 to 13%, and in the case of high-fat oil seeds, decortication (hull removal) may also be necessary prior to the extraction of oils.

Mechanical and screw pressing methods are used to express oils from vegetable matter. This can be conducted at ambient temperatures (*cold pressing*), but higher oil yields are achieved after the oil source is warmed to ~70°C (*hot pressing*) with steam. Unfortunately, the latter method can lead to the coextraction of gums, undesirable colors, flavors, and free fatty acids with the oil.

The mechanical expression of the oil from a natural source results in a press-cake that still remains rich in oil (5–15%); therefore, it is often flaked and subjected to countercurrent (lipophilic) solvent extraction(s). These solvent-extraction steps may employ hexanes, heptanes, or similar light petroleum fractions. Repeated mechanical and solvent extractions yield processed flakes that have about 1.0% oil content.

The residues of oil-extracted flakes are stripped of solvent, and since they contain high protein concentration, they can be recycled into animal feeds. The *miscella* or oil–solvent mixture is often centrifuged, filtered or otherwise purified, and then subjected to a vacuum stripper that removes most traces

Table 7.15. Common Sources of Extractable Fats and Fatty Oils Expressed as Percentage Composition of the Following Sources

Source	%	Source	%
Soybeans	19–20%	Palm kernel	42–52%
Sunflower	45–63%	Corn germ	47–50%
Peanut	45–50%	Safflower	47–55%
Cottonseed	32–38%	Palm fruit pulp	30–55%
Coconut	64–68%	Olive pulp	39–59%
Rapeseed	25–47%	Animal fat tissues	60–90%
		Whole fish	10–20%

of organic solvent. Since edible oil quality deteriorates with repetitive pressing, oils from second- and third-stage pressings are often relegated to non-food uses.

Extraction of oils from fruit pulps such as palm fruits is similar to oil seed extraction but more expeditious processing is required to minimize the unfavorable actions of native pulp lipases. To further ensure the high quality of the extracted oil, these enzymes are often deactivated by steam sterilization, which promotes the collection of high-quality fats having less than 2.0% free fatty acid content.

Apart from the extraction of oil seeds through the use of organic solvents, more advanced oil extraction methods have been developed. Under supercritical conditions of high pressures and temperatures above 31°C, carbon dioxide exhibits diffusion properties of a gas while its density is similar to that of a liquid. Moreover, this physical state of carbon dioxide has the ability to extract oil from oil seeds in a fashion similar to a hexane solvent. For example, flaked soybeans extracted from 50°C supercritical carbon dioxide at 5000 psi will absorb ~1.3% of its weight in oil, whereas 8000 psi will result in a 2.7% yield.

In spite of obvious advantages of this method for minimizing phosphorous gum extraction and the production of a more desirable light-colored oil than degummed hexane oil extracts, capital expenditures and retooling costs will ultimately determine the expansion of this technology throughout the oil seed industry.

Extraction of Fats and Fatty Oils from Animal Tissues

Rendering is a process designed to extract fat from fat-rich animal tissues or other substances that have a high fat-to-tissue ratio. The highest animal fat contents occur in the leaf fat of hogs (92–95%); internal fats of cattle (60–80% fat) that are ultimately used

for the production of oleostock; and "cutting fats" or other back-fats from hogs (80–85% fat). Low-fat stock having 10 to 15% fat such as bone stock and other sources can also be rendered, but economic feasibility for rendering these low-fat sources depends on their availability in high volumes.

Although whole animal carcasses are rarely if ever used as a source of edible fats for humans, whole small pelagic fishes such as herring, sardines, menhaden, and other species are rendered. Unfortunately, the high rendered fatty oil yields from whales (up to ~30,000 lb/animal) and seals has threatened the very existence of these species. Many whale species may contain up to ~70.0% fat in blubber tissues.

Packing houses and rendering plants may employ *dry rendering* or *wet rendering* methods. Dry rendering involves the expulsion of fat from fatty tissues by the application of heat. In fact, many of the industrial processes for dry rendering are reminiscent of frying bacon in the domestic setting (direct heat liquifaction of fats), but the fatty tissue feedstock may approximate 5000 to 10,000 lb and heat is applied to the tissues in a large steam-jacketed tank. Heated and liquified fatty oils are drained from the rendered feedstock. The rendered feedstock is subsequently compressed to express all traces of residual liquid fat, and all freed fatty liquors are combined and purified for use.

Residual dry rendered fatty tissues are dehydrated by the end of the rendering process, and these residues are often suitable for later use as an animal or poultry protein feed supplement. Although many conventional dry-rendering operations employ atmospheric pressures, improved fat and fatty oil yields are often obtained under vacuum conditions.

Contrary to dry-rendering processes that are principally geared to the production of inedible products, wet rendering is used where flavor, color, keeping quality, and edibility of rendered fats and oils are important. Wet-rendering processes use direct (fresh) steam or hot water in open or closed (40–70 psi pressure) steam boilers (digesters) to melt the fat out of animal tissues. The heated fats and oils undergo flotation in the digester, leaving tankage (a layer of solids) and "stick water" at the bottom.

Rendered fats are drawn from the surface of the rendered liquor, which may be filtered or centrifuged to remove water and particulates.

Up to 99.5% of fats can be recovered from animal tissues by wet-rendering methods. Steam rendering is not as energy efficient in terms of heat consumption as dry rendering, but this negative feature is overshadowed by equipment simplicity and the high effective yields of rendered fat from almost all types of rendering stock.

As expected, some triacylglycerol hydrolysis occurs during wet rendering, but assuming careful storage of prerendered fatty animal tissues, these problems can be largely controlled. In the case of steam-produced lards, free fatty acids are rarely below a concentration level of 0.35%, yet the processing and handling of lard after rendering affects its stability, quality, and oxidative deterioration far more than the sole increases in free fatty acids developed during wet rendering. In recent years, oxidative stability of wet-rendered fats has also been improved by addition of antioxidants such as BHA, BHT, propylgallate, citric acid, and their various combinations.

Many mechanical and process developments for rendering have been developed such as the Kingman Process, the Titan Expulsion System, the Delaval Centrifugation Process, and others. Most of these operations are similar to the processes outlined above in one way or another, but modifications are employed to improve rendered fat yields and the protein content of post-rendered tissues. Other rendering processes may proceed with the use of chemicals or enzymes that promote the hydrolysis and dissolution of connective tissues, which hinder the expulsion of fats from tissues.

Refining Treatments for Extracted and Rendered Fats and Fatty Oils

Aside from the removal of suspended solids from extracted or rendered fats and oils by filtration and centrifugation, it is often necessary to further refine edible oils. Many freshly pressed olive oils, seed oils, and milk fat products do not require refining, but rendered lipids may contain undesirable mono- and diacylglycerols as well as free fatty acids derived from triacylglycerols. Carotenoids and other lipid-soluble pigment derivatives, gums, lipoproteins, lipopolysaccharides, sterols, phosphoglycerides (e.g., lecithins, phosphoinositides), foul-smelling ketones, and aldehydes all have the potential for contributing added problems to edible fats and oils. Other undesirable products originating from lipid autoxidation, polymerization, residues of steroid animal growth hormone, residues of lipid-soluble hydrocarbon pesticides, and mycotoxins produced by fungal invasions of the rendering stock may also be liberated into fat and oil extracts.

Modern fat- and oil-refining methods permit the effective elimination of these substances in most edible oils. However, this elimination is not without some losses of nutritionally important unsaturated fatty acids, tocopherols, and phosphatides depending on the intensity of refining steps. Some of the most common refining steps designed to purify and improve oil and fat qualities include *neutralization, degumming, bleaching, deodorizing,* and *winterization*.

Neutralization and degumming: Proteins, protein fragments, phosphatides, mucilaginous substances, and free fatty acids are almost always present in crude oil extracts. These contaminants may detract from the potential high quality of an oil or contribute to its inferior keeping qualities.

Degumming processes remove mucilaginous substances, proteins, and especially phosphatides from crude fats. This may be accomplished by the addition of dilute phosphoric acid, brine, or alkaline phosphate solutions as well as by steaming or hydrating oils. In some cases, degumming processes are applied prior to advanced refining steps for oils in order to isolate commercial lecithins (crude lecithins) that serve as common edible emulsifiers in margarines, pasta, and other fat-containing foods.

Following degumming steps, oils are subjected to high-speed centrifugation that effectively separates degummed oil from lecithins, and both the lecithin and the degummed oils are then vacuum dried.

The improved keeping quality of degummed oils stems from the removal of carbohydrate-based gums that act as growth nutrients for microorganisms. Moreover, the industrial functional properties of oils are usually improved by degumming since hydrogenation catalysts are not fouled by the presence of impurities that normally occur in crude oil extracts.

Along with degumming, *neutralization* of edible oils may also be carried out to remove free fatty acids. Neutralization involves the addition of alkali (13–20°Be) to crude oil extracts and the subsequent formation of oil-insoluble soaps from the free fatty acids that may be present. These insoluble neutralization products, called "foots," are commercially classified as "soap stock" when they occur in bulk amounts. Acidic substances other than free fatty acids also react in a similar way, and there is a tendency for the added removal of these and other additional impurities by adsorption onto insoluble soaps formed during neutralization. Miscella refining may also involve the addition of alkali, but in many cases alkali and gum conditioners are added to the miscella before oil is stripped of its solvent fraction.

Neutralization of free fatty acids can also be achieved by high-temperature steam distillation of oils under reduced pressures. This process removes fatty acids from acylglycerols since fatty acids are far more volatile than acylglycerols under the same conditions. Liquid–liquid extraction of free fatty acids using propane and other liquid agents has been reported, but these methods are not economically feasible for most edible oils.

Bleaching, deodorizing, and winterization: The elimination of undesirable colors in oils can be achieved by bleaching and decolorizing procedures. Conventional bleaching processes involve treating edible oils with bleaching earths (e.g., bentonite, activated clay, or natural clay) or activated charcoal. These agents not only adsorb undesirable colors in oils but they can be effective in eliminating residues of soaps, gums, heavy-metal contaminants arising from the oil source or catalytic hydrogenation procedures, phosphatides, and water. Since many obnoxious chlorophyllous derivatives and carotenoids are not entirely heat stable, their decoloration or bleaching may be incidental to hydrogenation and other heat-related processes.

Bleaching can also be accomplished through the use of specialized oxidative or thermal treatments of oils. Unfortunately, many of these processes lead to the formation of odorous by-products and polymerization structures that are unacceptable in edible oils.

The *deodorizing* of oils is carried out to remove flavor and odorous compounds that either occur naturally in oils or exist as secondary oxidative or hydrolytic reaction products of oils. Few oils have reaped the economic and improved-use benefits afforded by deodorizing more than coconut and palm kernel oils. These oils have large concentrations of medium-chain-length fatty acids; they are liable to the development of aromatic methylketones; and oxidation of their unsaturated fatty acids leads to peroxide decomposition and the concomitant production of unsaturated aldehydes, keto acids, hydroxy acids, alcohols, and low-molecular-weight fatty acids. The majority of these compounds produce objectionable odors and tastes that limit food and food industrial uses for these oils unless they are removed.

Nearly all edible fats are deodorized before their dietary consumption. Olive oil and some prime animal fats are clear exceptions to the rule, however, since they happen to offer desirable odors and flavors.

In principle, deodorizing can be crudely classified as a specialized form of steam distillation. The process is often conducted at temperatures of 210 to 280°C and low pressures (1–6 mm Hg) (Figure 7.10B). As in the case of similar methods for neutralization, previously discussed, the more extensive application of vacuum distillation in deodorizing processes removes volatile fatty acids to a level of 0.02 to 0.05% in addition to some sterols and tocopherols. Lower-molecular-weight hydrocarbons, peroxides, aldehydes, ketones, and fatty acids derived from oxidative deterioration are usually removed to less than detectable organoleptic levels. Carotenoids are also decomposed by the high temperatures.

Depending on the actual industrial operation, deodorizing can be conducted as a batch or continuous method, but in general, heated fat is allowed to descend through a countercurrent flow of steam. Additional hydrolysis of fat triacylglycerols is minimal during this process, and assuming that an oxygen-deficient atmosphere is maintained, degradation of the oil will be minimized. Added protection of the oil can also be ensured by the addition of citric acid or similar agents, which chelate metal prooxidant activities in oils during processing.

Winterization of fats is a process whereby fats are cooled in order to precipitate fractions that would ordinarily cause clouding of the liquid fat when it is stored at refrigeration temperatures. Warm solid-free oils are slowly cooled in this process. This careful cooling coupled with slight mechanical mixing promotes the gradual growth of large fat crystals. Centrifugal or filtered removal of the crystals follows, and the remaining liquid-fat fraction is evaluated on the basis of a cold test. The results of cold testing are rated according to the length of time required for the winterized fat to show cloudiness at 0.0°C.

The term "winterization" has historical roots in the early cottonseed oil industry where winter storage of the oil in holding tanks resulted in the deposition of solids or stearine (12–25%). Liquid portions of the oil called "winter oil" were then removed and used for bottling or the manufacture of mayonnaise. Although early applications for winterization were geared to cottonseed oil production, the process has become conventional in many oil industries. For example, a significant portion of hydrogenated and winterized soybean oil (110 *IV*) is used as salad oil, which is expected to remain clear at normal refrigeration temperatures. Other oils including olive oil, corn oil, and sunflower seed oils are not usually winterized unless dictated by very specific uses.

Some Food Industrial Classifications of Fats and Oils

Milk fats: These fats have a low degree of unsaturation and a wide variety of short-chained fatty acids. Those fats generically described as butterfats are the most important and are used mainly for edible purposes. The polyunsaturated fat content of milk may be improved for special applications by feeding cows encapsulated safflower oils or similar polyunsaturated fats that prohibit the microbial reduction of fat in the rumen. This action is reflected in the production of increased polyunsaturates in milk and meat tissues.

Lauric acid oils: Seeds of cultivated or noncultivated palms yield copious amounts of these oils, which are low in unsaturation and high in concentrations of lauric acid plus other short-chained fatty acids. The low melting point of these oils is attributed to their high lauric acid content. Compared to many other oils, these oils are light colored and low in nonacylglycerol components. The most important edible representatives of the group are palm kernel and coconut oils.

Vegetable butters: This category of vegetable fats is noted for its low unsaturation and high C_{14}, C_{16}, and C_{18} fatty acid constituents. Many vegetable butters exhibit distinct melting points at relatively low temperatures owing to the uniform distribution of both unsaturated and saturated fatty acids. It should be noted that the melting points for vegetable butters are not the sole reflection of low-molecular-weight fatty acids as in the case of lauric acid oils. Most vegetable butters originate from the seeds of tropical trees, and cocoa butter has traditionally been one of the most important ingredients in the confectionery industry. Shea butters and other less expensive but highly functional vegetable butters have become progressively more important for contemporary foods.

Oleic–linoleic acid oils: Certain fruit pulps produced by perennial plants (e.g., olive, palm) and seeds of selected annual plant (e.g., corn, safflower, sunflower, sesame, peanut, and cotton) offer key sources of oleic–linoleic acid oils. The crude oils from these sources normally exist in a liquid form, and the total complement of oil constituents shows variable but medium degrees of unsaturation. As examples of this variability, palm, corn, and safflower oils usually exhibit respective iodine values of ≈ 50, ≈ 120, and ≈ 145. None of the fatty acids are more unsaturated than linoleic acid, which has two double bonds. The edibility and food uses of these oils are encouraged by their low to medium cost as well as their suitability for industrial hydrogenation. As discussed earlier, this process converts the C_{18} components of the oil into plastic fats. Extensive food uses for these oils have also been prompted by breeding and genetic developments among plant species that yield high oleic acid oils. Some safflower oils, for example, yield oils having an *IV* of 90. Additional plant varieties producing relatively low-*IV* oils are selected for growth in warmer climates. This has been especially useful for certain sunflower and soybean varieties.

Erucic acid oils: The name for this category of oils reflects the presence of a predominant unsaturated (C_{22}) fatty acid known as erucic acid. From a perspective of composition, these oils are otherwise similar to oleic–linoleic acid oils and they do contain a small amount of linolenic acid (6–10%). Rapeseed oil—also known as rape oil, colza oil, and more recently canbra oil—is obtained from the seed of *Brassica napus* (rape) and *B. campestris* (turnip rape). The wide climatic survival and vigor of several rape types allows them to grow in the widely different climatic zones of Sweden, China, Canada, and South America. Three types of rapeseed oils are recognized: (1) high erucic acid oils (20–55%) from *B. campestris* "brown sarson" variety; (2) low erucic acid (0–5%) or canbra oils having less than 3.5% erucic acid; and (3) zero erucic acid oils from the "Oro" and "Zephyr" varieties of *B. napus*. Traditional rapeseed oils, currently called high eurcic acid oils, display pale yellow colors similar to other oils. Rapeseed oils are used in the United States for industrial sources of behenic and erucic acids (lubricant additives and other industrial uses), but many Asian countries use erucic acid oils for edible purposes. Although these oils have been used as salad oils as well as in the manufacture of shortenings and margarines, food uses in the United States are minimal for several reasons. High erucic acid rapeseed oils have pungent mustard-like odors and sulfur compounds that can be removed by deodorizing, but the deodorized oils have a tendency to revert in flavor and taste to the undeodorized oil. Apart from more complex refining requirements, compared to other major food oils in the United States, erucic acid oils may be physiologically harmful to animals. Dietary studies on some test subjects attest to increased fatty infiltration of cardiac tissues and the development of long-term cardiac lesions. Concerns about these effects have prompted the development and culture of zero erucic acid rapeseed varieties in Canadian and European sectors, but these oils are still not consumed in the United States except as a fully hydrogenated product.

Linolenic acid oils: Linolenic acid oils are extracted from cultivated annual plants that otherwise exhibit a fatty acid profile similar to the oleic–linoleic acid oils. Soybean oil is probably the most important linolenic acid-rich edible oil. Special care must be taken during the extraction and refining of linolenic acid-rich oils because they are plagued with flavor instability problems and flavor reversions after deodorizing.

Land animal fats: Native and rendered animal fats display very low degrees of unsaturation, with most fatty acids of the C_{16} and C_{18} varieties. The unsaturated fatty acids found in land animals are usually oleic or linoleic acids. These fatty acids are uniformly distributed among triacylglycerols, all of which have melting points that are relatively high when compared with plant and marine triacylglycerols. Most native fats have low proportions of nonacylglycerol constituents and appear light-colored. In a contrary vein, poorly handled fats are often dark and have a high proportion of free fatty acids accompanied by a comparably wide range of mono- and diacylglycerol derivatives. *Lard* is the major meat-packing by-product of hog body fat, and *tallow* is the corresponding product produced from cattle and sheep. These fats, widely employed for edible purposes, serve as a commercial-industrial source of fatty acids.

Marine oils: These oils display broad variability in terms of their fatty acid chain lengths and their unsaturation. According to most studies it is clear that fatty acid chains above and below C_{18} are common, and polyunsaturated fatty acids having four to five double bonds are not uncommon. Saturated fatty acids may comprise up to 25% of the total acids in marine oils. Marine oils are generally edible, but the wide diversity of their fatty acid components, high degrees of fatty acid unsaturation, and proclivity for undergoing rancidity reactions do not encourage their use where food fats are required. Furthermore, the liability of most marine tissues to undergo rapid protein degradation yielding objectionable amines discourages the use of marine oils in human foods. Oils obtained from the hepatic tissues of fishes such as cod have served as important dietary supplements of vitamin A, and in recent years, consumption of these oils has increased owing to the speculative importance of polyunsaturated fats in the diet. Demersal fishes are usually used as a source of liver oils as opposed to pelagic fish species since the latter group has a relatively smaller hepatic oil content.

Selected polymorphic food fats: Depending on the specific use for an edible fat, it may be expected to (1) entrap and hold air; (2) demonstrate predictable patterns of consistency and plasticity under predetermined conditions; and (3) display a predictable solid-to-liquid ratio over a specified temperature range. Very few natural fats and oils offer optimal functional properties for common food uses unless key polymorphic forms are selected by industrial processing.

Since physical air (gas)–fat interactions are responsible for creaming properties of many fats, as well as the texture, tenderness, and volume characteristics of baked batters, only those polymorphic forms of fats that have superior abilities to entrap gas are used. The β′ crystals of shortenings, for example, are noted for their ability to hold large volumes of tiny gas bubbles (air or nitrogen), whereas β crystalline forms hold a relatively small volume of individually large gas bubbles. Accordingly, for many shortening applications in cakes, icings, and related products for which even consistencies and textures are desirable, only β′ forms of shortening are used.

The suitability of many fats or oils can be significantly improved to meet specific shortening prerequisites through processes of industrial hydrogenation. This procedure is successful only so long as partial hydrogenation yields a fat product having the molecular heterogeneity that actually favors β′ forms of fats. Hydrogenation is especially important for fats such as cottonseed and soybean, which demonstrate good air-entrapment abilities but whose crystalline structures at room temperatures are so inadequate that they show poor functional performance.

Tallows and lards offer wide possibilities as shortening ingredients in foods, but their plastic and functional properties are often unsuited to the final product. These application problems for tallows can be minimized by *destearinization* and their subsequent controlled hydrogenation to produce functionally improved, flavor-stable shortening products. In the case of lards, application improvements can also be produced by interesterification of the oleopalmitostearin-rich fat, which has notable aplastic properties. Following interesterification, the rearranged fat may be lightly hydrogenated to give the desired plastic properties.

Modifications of shortenings, which encourage β′ crystal formation and creaming properties, can also be produced by adding flaked hydrogenated fats. The admixture of tallow or cottonseed oil flakes to shortening fats causes the post-mixed shortening to display properties more reminiscent of the flakes than the initial fat to which they were added. The effectiveness of this treatment is not shared by lard, but flake addition to interesterified lard, cottonseed oil, and tallow retards the characteristic conversion of lard into its β crystalline structure.

Margarines are produced by the formation of a water-in-oil emulsion. Carefully calculated amounts of melted fat, skim milk, and other ingredients (e.g., emulsifiers and/or skim milk substitutes for water volume) are blended into a homogeneous mass and then partially solidified on a cooling surface before fat crystallization occurs. Controlled crystallization of the fat at this point governs the final plastic properties demonstrated by the margarine. Hydrogenated cottonseed and soybean oils have been widely used in margarine manufacture—hydrogenated cottonseed oil because it favors the β′ polymorphic fat structure and soybean because it has predictable and economically favorable availability.

Recent public demands for polyunsaturated dietary fats have also led to increased use of safflower and sunflower oils in margarines. In all cases of margarine production, hydrogenated oils of several hardnesses must be available to ensure a product having an *SFI* suitable for food uses.

Fat–water emulsions often serve as the basis for many cream fillings and icings. The type of fat and the water content of these emulsions is variable. It is not unusual, however, for the water content to approximate a concentration level of ~20%, and the fat used may be selected on the basis of its predominant crystal structure.

Few newly processed fats meet the demands of confectionery product ingredients, and it is often necessary to select the polymorph that best serves food formulation specifications. Selection of a specific fat polymorph over other forms of the same fat can be achieved by *tempering*. In this process, heat of transformation, liberated during fat crystallization, is removed at different temperatures to promote the formation of mixed fat crystals that show a wide range of melting points. The number of potential polymorphic transformations for fats parallels the structural diversity among their triacylglycerol structures. Therefore, the higher the number of triacylglycerol varieties, the greater the number of possible polymorphic structures. Low structural diversity among triacylglycerols in fats leads to correspondingly limited numbers of polymorphs and promotes the rapid conversion of fats into their β forms (e.g., soybean oil flakes, lard flakes, and cocoa butter). Whereas these fats readily undergo transformation into the β polymorph, those fats having wide structural diversity in their triacylglycerols undergo very gradual transformation from one polymorph to another.

Proper tempering of *enrobing fats* such as cocoa butter, cocoa butter substitutes, and extenders is critical to their functional confectionery properties (e.g., melting point). Enrobing fats must exhibit a distinct hardness and brittleness at room temperature and a definite absence of liquid fat, and yet undergo rapid melting at mouth temperatures.

ANALYTICAL AND QUALITY-CONTROL TESTS FOR FATS AND FATTY OILS

Most of the analytical methods currently used for testing fats and fatty oils in the United States are published by the Association of Official Analytical Chemists (AOAC), the American Oil Chemists' Society (AOCS); the International Union for Pure and Applied Chemistry (IUPAC); Codex Alimentarius Commission (CAC); the International Olive Oil Council (IOOC); and other similar organizations. Some of the key analytical criteria for fats and oils are based on the following tests with reference to the *Tentative and Official Methods of the American Oil Chemists Society*, third edition (AOCS).

Gas–liquid chromatography (GLC) (AOCS method Ce 2-66): This method is used for analysis of fatty acid composition of triacylglycerols. Methyl esters of component fatty acids are produced from several milligrams of triacylglycerol sample; the fatty acid methyl esters can be identified on the basis of their column retention time and quantitated by integrating the area under the chromatographic peak with reference to standard concentration levels of specific fatty acid esters. Determination of fat type and estimates of iodine values for fats can also be made using GLC (AOCS method Ce 1-62).

Iodine value (IV) (AOCS method Cd 1-25): The number of centigrams of iodine adsorbed by a gram of fat (percentage iodine absorbed) serves as an empirical estimate of fat unsaturation.

Saponification value (AOCS method Cd 3-25): This gives the number of milligrams of potassium hydroxide necessary for the hydrolysis or saponification of 1.0 g of fat.

Flash point (AOCS method Cc 9b-55): The temperature at which a fat is likely to flash when exposed to a test flame under standard conditions is used as an index of *solvent residues* in crude oils. Flash points serve as an industrial safety index, especially for those manufacturing processes that may not be designed for use with volatile organic solvents.

Moisture analysis: Depending on the type of fat or oil, its moisture content may be determined by (1) Karl-Fischer titration (AOCS method Ca 2e-55); (2) gravimetric methods (AOCS methods Ca 2b-38, Ca 2c-25, Ca 2d-25); or toluene distillation (AOCS method Ca 2a-45). The insoluble impurities in fats and oils can be determined according to gravimetric methods such as AOCS method Ca 3-46.

Unsaponifiable content of fats: Controlled hydrolysis of fats in the presence of ethanolic alkali followed by extraction of the unsaponifiable components into petroleum or ethyl ether serves as a basis for this test. Evaporation of the ether from the unsaponifiables followed by gravimetric analysis of the dried residue serves as an index of unsaponifiable matter. High concentrations of unsaponifiables in marine fats can be determined by AOCS method Ca 6b-53 while other fats having nominal unsaponifiable content can be assayed by AOCS method Ca 6a-40.

Free fatty acid content: The free fatty acid content of many fats can be estimated by their direct titration with standard sodium hydroxide solution (AOCS method Ca 5a-40). Apparent reactive free fatty acids are often expressed in terms of *percent oleic acid,* but in the cases of coconut and palm kernel fats, free fatty acids are calculated as lauric acid, and palmitic acid is used as a reference for palm oils.

Refining loss and bleaching tests: The treatment of crude fats with alkali leads to decreases in their free fatty acid content and other base reactive impurities that can be estimated according to AOCS method Ca 9a-52. Following the refining loss assay, bleaching tests on the fat can be carried out using AOCS methods Cc 8a-52 and 8b-52, which pertain to cottonseed and soybean fats, respectively. Standard AOCS specified bleaching earths are employed and filtered fats are chromerically assayed by AOCS methods Cc 13b-45 or Cc 13c-50.

Peroxide value (PV) (AOCS method Cd 8-53): This assay value refers to the milliequivalents of peroxide per kilogram of a fat sample that oxidizes potassium iodide under standard conditions.

Melting point, congeal point, and cold tests: Melting points for cold tests may be accurately determined by conventional capillary tube melting point techniques (AOCS Cc 1-25) or more empirical methods. The solidification temperature for a fat is measured as its congeal point (AOCS method Cc 14-59). A fat sample is heated until it becomes clear and is placed in a 15 or 20°C cooling bath depending on whether congeal points are respectively <35°C or ≥35°C. Cooling involves two steps: (1) fat crystallization is initiated at the temperature of the cold water bath, and when the first indications of crystallization are evident, (2) the sample is transferred to a 20°C air bath. Sample temperatures are then recorded at one-minute intervals. The maximum temperature recorded during this period, resulting from the heat of crystallization represents the congeal point that is a characteristic of most fats. So-called cold tests of liquid fats may also be conducted (AOCS Cc 11-53) to determine the clarity of a fat sample when held at 0.0°C. Assays such as this serve as an index of liquid fat resistance to crystallization.

Solid content index (SCI) [or solid fat index (SFI)] (AOCS method Cd 10-57): Empirical measure of solid fat content can be calculated from specific volumes of fats at several temperatures by using precision dilatometric methods. The index has important commercial significance since SCI values can serve as a trading criterion of fats. Typical SCI measurements for margarine and shortening are evaluated at 10.0, 21.1, 26.7, 33.3, and 37.8°C. The *SCI* values for fats are also useful for evaluating the progress of ester–ester interchanges in lard, edible oils in margarines, cocoa butter, shortening, and hard butter.

Fat stability and the active oxygen method (AOM): According to this assay (AOCS method Cd 12-57), a predetermined *peroxide value (PV)* is attained in a liquid fat held at 97.8 ± 0.2°C by the ebullition of clean air through the liquid fat. The time (hours) necessary to reach the prescribed *PV* (20–100 mEq of peroxide/kg) is used as an index of oxidative fat stability for a sample, but since there are no universally accepted values, the specified *PV* often reflects the educated discretion of the analyst. As a reference, however, the initial stages of distinct rancidity for lard, hydrogenated lard shortenings, oleo oil, and hydrogenated vegetable oils from soybean and cottonseed oils are respectively 20, 40, 60, and 80. For many other hydrogenated oils such as olive and peanut oils, the significant *PV* may be in the range of 50 to 60. Some other oils including corn, sunflower, and soybeans may show rancidity as high as 125 to 150. Many oxygen-uptake methods are reported in the literature, and additional details on these methods including the Kreis test, the oxygen bomb method (OBM), the thiobarbituric acid (TBA) test, the oven stability or Schaal test may be found in authoritative references such as Swern (1982).

Absorption spectra for fats and oils: Visible or near ultraviolet (UV) light absorption by triacylglycerols is minimal, but nonconjugated, unsaturated fats absorb in the far

UV. The near UV wavelengths, however, can be used to detect the extent of conjugated polyunsaturation that occurs in some oils or that unsaturation induced by catalytic isomerization reactions. Conjugated dienes show notable absorption at 232 nm, conjugated trienes at 260 to 280 nm (triplets), and tetraenes at 290 to 320 (triplets), but the locations of these peaks also depend on the geometric isomers present. Although infrared (IR) absorption is not especially characteristic of individual fats, the carbonyl stretching band can be used as a basis for the estimation of triacylglycerol concentrations assuming that all other ester bonds are absent. Absorption in the IR region serves a more important criterion, however, for determining the isolated occurrence of *trans* unsaturation at 965 to 975 cm^{-1}. The loss of conjugation among unsaturated fatty acids in foods caused by lipoxygenases and other oxidations may also be detected by the relative UV absorption of the lipids at key wavelengths over a prescribed time-test interval, but the accuracy of these tests is subject to the absence of coabsorbing substances at critical assay wavelengths.

Proximate Analysis of Crude Fats in Foods

Continuous extraction of dry comminuted or ground foods with a nonpolar solvent such as petroleum spirits (60°C boiling point) or ethyl ether serves as a practical basis for estimating the crude lipid content of foods. Soxhlet extractors, commonly used for these methods, operate according to the process detailed in Figure 7.39. Other more sophisticated types of lipid extractors have been developed, but most forms operate on principles similar to the soxhlet extraction pro-

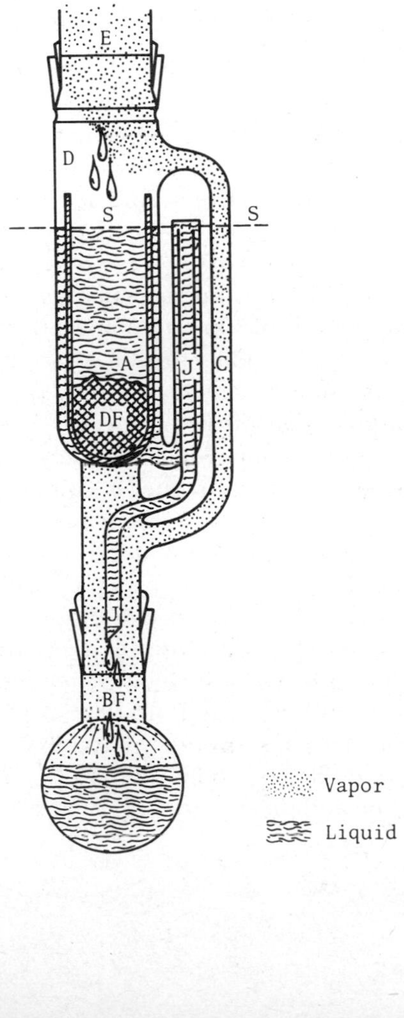

FIGURE 7.39. Apparatus and principles of soxhlet extraction for crude lipids from food.

1. The dried food product (DF) is placed in a porous cellulose or glass fiber thimble (A) that only permits passage of solvent(s) such as ethers and lipid fractions originally contained in the food.

2. Ether solvent contained in the boiling flask (BF) is heated. The solvent vapor passes through tube (C) into the extraction chamber containing the thimble and food specimen. The solvent vapor rises into the condenser (E), where it is cooled to a liquid state.

3. The condensed solvent drips into the thimble (A), which may already contain condensed solvent, and the condensate gradually fills the thimble holder (D). This solvent bathing-action on the sample extracts the lipid content of the food.

4. When the volume of the condensed liquid in (D) and (A) reach the top of a siphon tube (J) at the height of S–S, the liquid solvent and its extracted lipid solutes are siphoned through (J) into the boiling flask.

5. Intermittent siphoning action during steps 1 to 4 above can continue for 4 to 48+ hs depending on the sample and the objective.

6. At the completion of the extraction period, lipid formerly in (DF) will reside in the liquid solvent volume of the boiling flask (BF). The solvent may be stripped from the tared flask (BF) and the amount of extracted fat calculated; or other analytical steps to quantitate extracted lipid may be undertaken.

cedure (e.g., Bailey–Walker or Bolton-type extractors).

For the proximate assay of lipids in liquid foods, it may be possible to extract lipids by shaking the sample with a nonpolar solvent in a separator, but multiple extractions are usually required. Once the lipid–solvent phase has partitioned from the aqueous fractions of a liquid food sample, the evaporation of the solvent yields crude lipid that may be gravimetrically assayed.

The direct extraction of lipids from foods may be complicated by proteins, but preliminary treatment of foods with acids (Werner–Schmid method) or alkali (Rose-Gottlieb and Majonnier methods) may minimize these analytical problems.

Volumetric methods for the estimation of crude lipid content and fat in foods have been widely used for milks, cheeses, and meats (e.g., ground meats, sausages). For these methods, a prescribed weight or volume of food is digested with a strong acid (1.84 sp. gr. sulfuric acid or 60% perchloric acid), and the fat component of the food is permitted to rise into a calibrated tube. The calibrated tube often has units expressed in percentage that corresponds directly to the weight or volume of the food sample digested by the acid. Gerber and Babcock methods for milk employ concentrated sulfuric acid, but the Paley bottle methods for ground meats use perchloric acid. Other acid digestion mixtures for ice creams such as 50:50 acetic acid—perchloric acid mixtures may be used in conjunction with Paley bottle assays.

Apart from these classical proximate methods of analysis for food lipids, the characteristic refractive indices, specific gravities, capacitance, and other physical properties of lipids may be analytically exploited. Detailed surveys of the proximate methods of analysis for foods are detailed in Triebold and Aurand (1963), Pearson (1970), and Joslyn (1970). Very specific methodologies for lipid assays in foods are also outlined in official methods of analysis by the AOCS and the AOAC referenced at the end of the chapter.

POSSIBLE ROLES FOR FATS IN NUTRITION AND HEALTH

Almost 90% of the global production of fats and oils is used for foods. Their uses and functions are widely variable since they provide:

1. Sources of dietary energy.

2. Supplies of important dietary unsaturated fatty acids.

3. Leavening effects to baked batters by air entrapment.

4. Functional and lubricating actions that characterize dressings, mayonnaises, and the like.

5. Carrier vehicles for fat-soluble vitamins.

6. Textural, mouth feel, palatability, satiation, and flavor properties to many foods.

Caloric versus Noncaloric Functions for Fats and Oils

Caloric functions: Excluding erythrocytes and those cells of the central nervous system, free dietary fatty acids and those freed by the digestive hydrolysis of triacylglycerols can be used as a source of dietary energy. Carbohydrates normally serve as an energy source for the central nervous system, but during periods of starvation and nutritional duress, ketone bodies produced from fatty acids and amino acids may supply the brain with a potential source of energy.

Fat consumption, which is widely variable on the international scene, is composed of both *visible* and *invisible fat constituents.* Invisible fats are those contained naturally in meat, fish, eggs, and other dietary products, whereas visible fats represent those fats that have been extracted from their natural origin and processed such as vegetable oils, lard, and butter. Based on a per capita annual dietary consumption of 125 lb of fat in the United States, about 72 lb (58%) are invisible fat and the remainder visible fat. Only in the Netherlands has visible fat consumption superseded per capita U.S. consumption, and only in France and the United Kingdom has butter consumption generally surpassed that of margarine. On a per diem basis, the per capita U.S. consumption of fats is ~160 g, which supplies ~1440 kcal and about 45% of the total caloric intake.

Fats are regarded as the richest source of food energy because they average ~9.0 kcal/g as opposed to proteins and carbohydrates, which each have ~4.0 kcal/g. The high nutritional caloric values for fat reflect a combination of factors including the highly reduced chemical structure of fatty acids, efficient biochemical oxidation routes for fatty acids in cells, and the high caloric density of many food fats owing to their relatively low water content.

Many studies on caloric consumption in the United States suggest that fats will soon provide more than 50% of the per capita dietary energy requirements if current trends do not reverse. Sharp dietary decreases in grain carbohydrate consumption compounded with relative increases in food fats have been blamed for apparent increases in fat-based calorie consumption.

Noncaloric functions: Aside from the caloric functions of food fats, nearly all animals including humans depend on dietary sources of certain polyunsaturated fatty acids. The nutritional demand for these fatty acids is recognized far more clearly than their actual biochemical functions are understood.

The main essential fatty acid (EFA) for humans is linoleic acid, which has both plant and animal origins. The assimilation of linoleic acid into biochemical pathways of the body results in its conversion to longer-chained fatty acids having three, four, and five double bonds.

Exclusion of linoleic and other polyunsaturated fatty acids from the diet results in a spectrum of deficiency effects, but dietary insufficiency is rarely observed with the notable exception of infant patients subjected to intravenous feeding. Recognized symptoms of *essential fatty acid deficiency* (EFAD) include scaly skin (eczema), poor growth and weight gains, erythrocyte hemolysis, increased potential for internal hemorrhaging, poor energy yields from mitochondrial oxidative processes, and other nondescript degenerative phenomena. The EFA availability is also an important factor in pregnancy, lactation, and capillary vitality, but biochemical explanations for EFA activities are elusive. Rodent studies have also indicated that proper levels of EFA promote growth, early sexual maturity, improved nitrogen-sparing effects, increased resistance to internal and external stresses, normal water balance, and minimization of persistent kidney lesions incurred by prolonged periods of EFA deficiency.

Among the unsaturated fatty acids, linoleic and arachidonic acids seem to be the most important EFAs. The EFA activity of linolenic acid is relatively poor compared to linoleic and arachidonic acids, and polyunsaturated fatty acids may show more or less EFA activity. Saturated fatty acids and oleic acid do not demonstrate important EFA activities.

The EFA content of foods can be indexed according to the concentration of *cis,cis*-9,12-linoleic acid, but it should be recognized that conventional methods of partial hydrogenation yield geometric and positional isomers of linoleic acid. These fatty acids do not show significant EFA activity, although they do contribute to the C_{18}-diene content of oils detectable by many conventional analytical methods.

Since the body is unable to synthesize *cis,cis*-9,12-linoleic acid, requisite nutritional levels must come from dietary fats or nutritional supplements. The exact amount of linoleic acid necessary to prohibit clinical and biochemical deficiency signs is uncertain, yet some reputable studies dealing with human and animal subjects indicate that 1 to 2% of the dietary calories is sufficient. The wide availability of vegetable oils in the American diet ensures that EFA deficiencies are unlikely. In fact, the USDA estimates that nearly 23.0 g of linoleic acid per capita occurs in most food supplies or ~6.0% of total dietary energy. Even when EFA content of food drops below desirable dietary levels, at least some limited human studies have indicated that EFA deficiencies of ≥6 months may be necessary before clinical effects become evident.

Although limited amounts of polyunsaturated fatty acids are stored in adipose tissues, pools of polyunsaturated fatty acids may be mobilized from structural complex lipids during periods of EFA deficiency.

Nominal versus High Dietary Levels of Unsaturated Fatty Acids

As discussed in earlier sections, unsaturated fatty acids and their EFA components are essential for normal health including the biosynthesis of prostaglandins. Most evidence, however, does not support the contention that the increased consumption of unsaturated fatty acids parallels increased health benefits. In fact, the opposite may be true.

Apart from growing data that seem to support the beneficial aspects of linoleic acid and other unsaturated fatty acids for minimizing thrombosis and coronary heart disease (CHD), these acids may be linked to a variety of unfavorable clinical and subclinical conditions. These conditions may include abnormalities in cellular metabolism; ceroid body accumulation; enhancement of biliary calculi formation; acceleration of the aging process; and encouragement of carcinogenesis instigated by certain compounds. Nonspecific metabolic disruptions stemming from *in vivo* peroxidation of unsaturated fatty acids may also cause a range of potential problems.

Although much of the evidence to date is speculative, *in vitro* studies show that at least three competitive *in vivo* reactions may occur after peroxy radicals are formed from polyunsaturated fatty acids. These include (1)

the abstraction of hydrogen atoms that yield various hydroperoxides; (2) the production of carbon radicals and isomerized carbon radicals by β-scission of a carbon–oxygen bond; as well as (3) the cyclization of hydroperoxide radicals to produce a cyclic peroxy radical (Figure 7.40). Depending on the antioxidant availability of tocopherols and other reducing agents, spontaneous nonbiological oxidation reactions may be minimized in favor of fatty acid hydroperoxides. In addition to the obvious production of hydroperoxides, these substances probably serve as precursors to leukotriene synthesis (Figure 7.35).

Whether or not *in vivo* formation of lipid peroxides and free radicals represent strictly random reactions, enzymatic reactions, or a combination of both routes remains a matter of conjecture. Nonetheless, these radicals and peroxides are not uncommonly associated with incidences of stroke, arteriosclerosis, inflammation, and emphysema.

Circumstantial evidence for the existence of transient endoperoxides in humans has been indicated by TBA assays of blood sera. Fluctuations in the results of routine assays within and among test subjects have been interpreted as a reflection of tissue variations in the concentrations of prostaglandin precursors. Further, it has been proposed that at least some diseases, the aging process, and exposure to pollutants may be responsible for these apparent fluctuations.

The biochemical transformation of unsaturated fatty acids into prostaglandins as well as the functions of prostaglandins are naturally complex, but these events become still more complex in the face of xenobiotic–prostaglandin interactions. *Xenobiotic substances* are potentially lethal substances that undergo transformation into toxic or deleterious compounds as a result of chemical or enzymatic reactions in a cell or organism. In cases such as the oxidation of benzapyrene into an active carcinogen, prostaglandins show concurrent increases in their concentration levels that are seemingly linked to enzymatic activities. In still other cases, xenobiotic concentrations of nitrogen

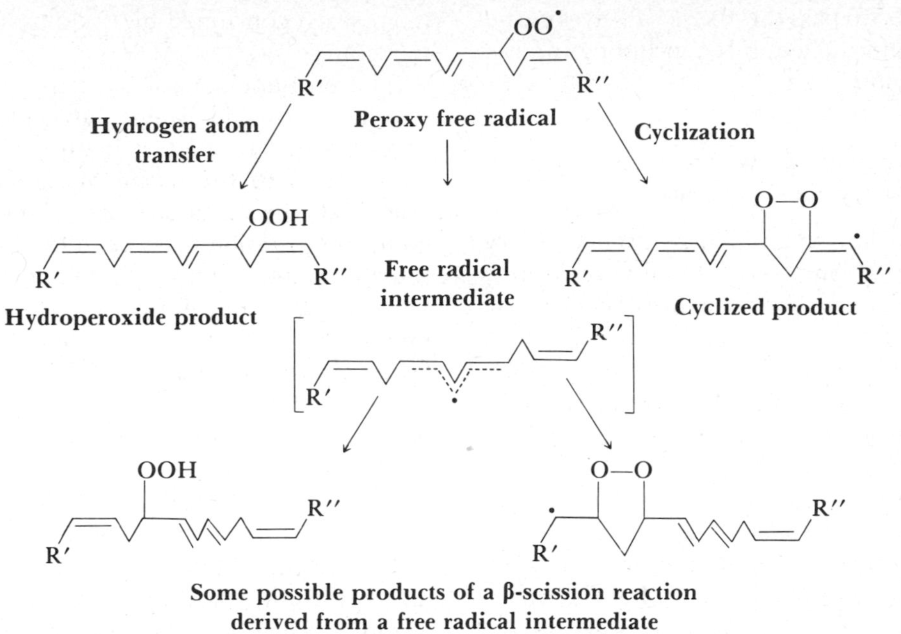

Some possible products of a β-scission reaction
derived from a free radical intermediate

FIGURE 7.40. Three possible paths for the autoxidation of polyunsaturated fatty acids.

dioxide free radicals, which occur as air pollutants, may instigate the oxidation of critical cellular enzymes and enzyme systems or upset prostaglandin synthesis by promoting an autoxidation sequence among polyunsaturated fatty acids.

The degree of unsaturation is dietary fats, the origins of fats, and the gross amount of the fat consumed in the diet may affect the potency of at least some carcinogens and promote tumorigenesis as well as vascular lesions.

A cursory survey for 30 years of studies shows considerable interest in the roles of certain vegetable oils related to tumorigenesis. When supplied to rodent test subjects, dietary availability of selected unsaturated and saturated fatty acids in conjunction with recognized carcinogens and xenobiotics have been ranked according to evidence of neoplasia and other tumorigenic consequences. Exemplary studies include investigations of potent chemical hepatocarcinogens such as 2-acetylaminofluorene (AcAF) and its enzymatically produced (*in vivo*) mutagenic metabolite *N*-hydroxyacetylaminofluorene. According to these investigations, hepatic

necrosis and neoplasms seem to be most severe for animals fed 20% corn oil as opposed to 5% dietary rations of tallow or corn oil, or 20% concentration levels of tallow.

Apart from tumorigenesis, the development of vascular lesions and atherosclerotic tendencies in some primates may be linked to unsaturation of dietary fats and oils. For example, some monkeys fed a high-cholesterol and high-fat diet, with fat coming from highly unsaturated peanut oil, corn oil, butterfat, or butterfat and coconut oil, show most impressive lesion development in peanut oil diets. Little evidence for atherosclerotic damage is evidenced in corn oil-fed animals but moderate numbers of vascular lesions may appear in butterfat-fed animals.

Free radical reactions linked to unsaturated fatty acids available from the diet must be considered among the aberrant routes of nutritional biochemistry not only because of their potentially damaging health consequences but also because of their exceptionally high reactivity. Regardless of the fact that fatty acid free radicals demonstrate a half-life of less than the 10^{-9} s of a hydroxyl free radical, their effective nucleophilic properties surpass the 10^{-5} s half-lives of the most potent nucleophiles, including the hydroxyl anion.

Polyunsaturated Fatty Acids, Autacoids, and Coronary Heart Disease

Popular beliefs that polyunsaturated dietary fatty acids promote the fluidity, transport, and liquid crystalline structure of cholesteryl esters of fatty acids thereby reducing atherosclerotic tendencies may not be so simple. Apart from these possibilities, unsaturated fatty acids may exert their anti-atherosclerotic effects as a basic "biosynthesis stock" for *autacoids*. Autacoids are synthesized in the body from unsaturated fatty acids; typical examples of these compounds include prostaglandins, thromboxanes, and prostacyclin. Since these fatty acid derivatives greatly affect the circulatory system, it has been proposed that the relative abundance of different dietary unsaturated fatty acids may determine the relative autacoid balance within an individual.

Some thromboxanes encourage the aggregation of platelets, whereas prostacyclin and some other prostaglandins counteract platelet aggregation. Assuming that dietary changes cause dynamic shifts in the ratio of thromboxanes to prostaglandins, these dietary changes could encourage or decrease possibilities for developing a thrombosis and its related coronary heart disease or vascular consequences.

Evidence supporting such interactions has been uncovered in blood studies of Eskimos in Danish Greenland, where ischemic heart disease is very low, heart disease accounts for ≤5.0% of the mortality rate for 60 + -year-old portions of the population, and most individuals exhibit low plasma cholesterol and triacylglyceride levels. As opposed to most Western countries, which have high consumption of arachidonic acid, Eskimo diets are rich in eicosapentaenoic acid. As discussed earlier, this polyunsaturated fatty acid has great abundance in the marine food chain including the fish and fish-eating mammals

(e.g., seals) consumed by Eskimos as a protein source.

The eicosapentaenoic acid is a pentaunsaturated fatty acid, whereas arachidonic acid is a tetraunsaturated structure. Counting from the methyl terminus of eicosapentaenoic acid, the extra site of carbon–carbon bond unsaturation is between the third and fourth position. This apparent minor difference exerts a profound impact on the ultimate biochemical products that arise from either of the polyunsaturated fatty acids. For example, the prostaglandin-3 and thromboxane-3 groups originate from eicosapentaenoic acid, and the -2 families stem from an arachidonic acid precursor (Figure 7.41).

Armed with the current information, it is increasingly clear that unsaturated fatty acids have profound biochemical effects on autacoid production. The relative balance of these autacoids ultimately affects thrombosis formation and possibly the severity of incipient coronary heart diseases. The recognition of low coronary heart disease in Eskimos coupled with the historical observation by explorers that these populations have a tendency to bleed, offers still more support for the antiaggregative properties of autacoids derived from polyunsaturated fatty acids. It is also clear from the studies of Hauge and Nicolaysen (1959) that EFA content of diets alone is not the key factor for depressing serum cholesterol levels. According to their studies, rodents having a depleted EFA pool and high serum cholesterol showed greater hypocholesterolemic responses to either cod-liver oil or linolenate than corresponding results dealing with linoleate.

The early speculation and recognition of the hypocholesterolemic activities of marine oils has been surveyed by Peifer (1967) and Stansby (1969).

An Overview of Fats and Coronary Heart Disease

As early as 1838 Lecanu demonstrated the presence of cholesterol in the blood, and in 1843 Vogel observed that large amounts of

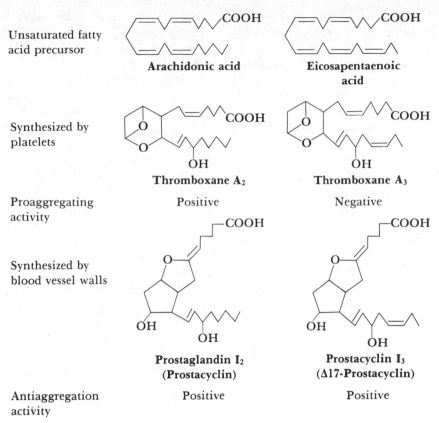

FIGURE 7.41. Dietary fatty acids and their corresponding autacoids.

cholesterol were present in the arteries of atherosclerotic victims. By 1856 Virchow suggested that atherosclerosis occurred when cholesterol precipitated out in the vascular system as a solid layer. Ignatowski then demonstrated in 1908 that atherosclerosis could be readily induced in rabbits by diets rich in meat and dairy products and it was assumed that *in vivo* cholesterol resulted from ingested foods. Rittenberg and Schonheimer (1937) along with Bloch and Rittenberg (1942) suggested the metabolic production of cholesterol within the body.

After Widal *et al.* (1913), who speculated on the relationship between heart disease, cholesterol, and high dietary fat, the 1940s saw marked changes in the epidemiological occurrence of atherosclerosis (Malmros, 1950). These changes were so significant that major studies were undertaken to relate diet

and heart disease. With this objective in mind, Kinsell *et al.* (1952, 1953, 1954) were among the pioneers who made key observations regarding serum cholesterol levels and degrees of saturation displayed by dietary fats.

Over a century of recorded scientific and medical research has produced a collage of conclusions that undeniably link cholesterol with atherosclerosis. Contrary to these voluminous and sometimes controversial research reports, however, chemical and biochemical mechanisms that explain the role of cholesterol in promoting atherosclerosis remain to be firmly established.

A simple perspective of the problem suggests that atherosclerosis is a metabolic disturbance characterized by the deposition of cholesterol and its derivatives on arterial walls. Events such as this ultimately cause a restricted blood flow throughout portions of the body. Typical clinical signs of this problem may include angina pectoris, heart infarct (coronary arteries), cerebral thrombosis (occlusion of a blood vessel to or in the brain), claudicatio intermittens (blockage in a leg), elevated blood pressure (blockage of a kidney artery), cataract, and/or the development of xanthomatosis (yellow plaques under the skin).

Based on the most superficial concept of atherosclerosis, it is recognized that cholesterol occurs in the serum principally as a fatty acid ester. Very little cholesterol exists as a free alcohol. Since many common dietary animal fats have 40 to 70% saturated fatty acids, such as lard, beef tallow, and butter, it is not unreasonable to expect that the resulting cholesteryl esters will reflect higher melting points than those esters of cholesterol involving polyunsaturated fatty acids. It follows too that more saturated cholesteryl esters of fatty acids will have progressively higher melting points than their polyunsaturated counterparts; and both serum emulsification and transport will become more challenging throughout the vascular system. The composite effects of these events on the arteries revolves around the initial deposition of cholesterol on arterial walls and its subsequent site promoter effect for the growth of a plaque.

The major potential impact of dietary unsaturated versus saturated fatty acids has been recognized by many health interest groups and recommendations have been established as a basis for human nutrition. Typical among these groups have been the Intersociety Commission for Heart Disease Resources (1970) and the American Medical Association (1972).

Many of the time-honored experimental principles supposedly linking cause–effect relationships between fatty acid saturation and atherosclerosis have been challenged by Reiser (1973), Kaunitz (1975) and many others including Howard *et al.* (1965), Walker *et al.* (1960), Yudkin (1957), Friend (1967), and Gortner (1975). It has also been noted

that increases in coronary heart disease have shown historic parallels with the consumption of animal protein relative to other dietary components.

A baseline value for proteins of 100 g/day since around 1900 has remained quite steadfast, but animal to vegetable protein ratios reportedly doubled from about 1.1 in 1910 to nearly 2.4 in 1972. Moreover, the animal to vegetable fat ratio decreased from ~4.9 to ~1.6 over the same period while overall fat consumption increased from 125 to 158 g/day. Carbohydrate and fiber consumption respectively decreased by ~21 and ~30% over this same time span. Considering these complex nutritional variables plus the increased dietary availability of refined sugars, minerals, life-style changes, and other unknown predisposing factors, it is nearly impossible to establish clear epidemiological and etiological foundations that decidedly link dietary factors to atherosclerosis and coronary heart disease. Many other perspectives regarding the roles of fats on vascular diseases have been succinctly summarized by Formo (1979).

Hydrogenation and Oxidation May Affect the Dietary Value and Safety of Some Fats

Positional and geometric isomerizations of unsaturated fatty acids resulting from hydrogenation processes have been recognized for many years. One of the most common results of this process involves the conversion of natural *cis* double bonds to unnatural *trans* forms. Since nearly 35 + % of many hydrogenated shortenings and margarines occur as *trans*-unsaturated fatty acids, much concern has been expressed for the biochemical and health consequences linked to these fats. Most evidence indicates, however, that the oxidation of *trans*-unsaturated fatty acids to carbon dioxide parallels that of the corresponding *cis* forms; but *trans* isomers may be assimilated into adipose tissues and effect cell membranes in ways that are not consistent with their *cis*-unsaturated counterparts (Johnston *et al.*, 1958; Kummerow, 1974). Furthermore, *cis,trans* isomers of linoleic acid may eventually be converted into an isomer of arachidonic acid, yet the *trans,trans* isomer of the acid fails to exhibit essential fatty acid activity.

Studies of oxidative and thermal fatty acid polymers on rodents in particular have demonstrated a growth retarding effect and it is recognized that very high consumption of these polymers may be fatal. Because oxidatively produced polymers have obvious rancid odors and flavors, the dietary and health consequences resulting from their consumption by humans may be minimal. Thermal polymers on the other hand are not always indicated by organoleptic measures and may offer a veiled threat to normal human health. The toxic potential of these thermal polymers may rest on chemical mechanisms that lead to the formation of dimeric polymers, lactones, epoxides and cyclized fat derivatives.

Since 1953, the toxic perils surrounding the consumption of heated fats by animals have been vigorously debated (Crampton *et al.*, 1953, 1956). Other studies repeatedly contradict this early work and suggest that heated fats used in the preparation of fried human foods are the nutritional equivalent of unheated fats (Nolen *et al.*, 1967). Nevertheless, health and biochemical ramifications of low-level chronic exposures to highly reactive fat derivatives and polymers produced by heat are not firmly established and many suspicions regarding their importance still surface in the literature. Queries regarding the importance of these substances in foods have resulted in numerous analytical methods for the detection of polymers in oils (Waltking *et al.*, 1975).

REFERENCES

American Medical Association (Council Statement). 1972. Diet and coronary heart disease. *J.A.M.A.* **222:**1647.

Bergström, S., L. A. Carlson, and J. R. Weeks. 1968. The prostaglandins: A family of biologically active lipids. *Pharmacol. Rev.* **20:**1.

Bloch, K., and D. Rittenberg, 1942. The biological formation of cholesterol from acetic acid. *J. Biol. Chem.* **143:**207.

Brockerhoff, H., and M. Yurkowski. 1966. Stereospecific analysis of several vegetable fats. *J. Lipid. Res.* **7:**62.

Crampton, E. W., R. H. Common, F. A. Farmer, A. F. Wells, and D. Crawford. 1953. Studies to determine the nature of the damage to the nutritive value of some vegetable oils from heat treatment. *J. Nutr.* **49:**333.

Crampton, E. W., R. H. Common, E. T. Pritchard, and F. A. Farmer. 1956. Studies to determine the nature of the damage to the nutritive value of some vegetable oils from heat treatment. IV. Ethyl esters of heat polymerized linseed, soybean, and sunflower seed oils. *J. Nutr.* **60:**13.

Cropper, F. R., and A. Heywood. 1953. Analytical separation of the methyl esters of the C_{12}-C_{22} fatty acids by vapor-phase chromatography. *Nature* **172:**1101.

Erickson, D. R., and R. H. Powers. 1974. Objective determination of fat stability in prepared foods. In *Objective Methods for Food Evaluation, Proceedings of a Symposium. November 7–8, 1974*, sponsored by U.S. Army Natick Research and Development Command. NAS, Washington, D.C.

Formo, M. W. 1979. Fats in the diet. In *Bailey's Industrial Oil and Fat Products* (D. Swern, ed.), 4th ed., Vol. I, pp. 233–270. Wiley, New York.

Friend, B. 1967. Nutrients in the United States food supply. A review of trends, 1909–1913 to 1965. *Amer. J. Clin. Nutr.* **20:**907.

Gortner, W. A. 1975. Nutrition in the United States, 1900–1974. *Cancer Res.* **35:**3246.

Hauge, J. G., and R. Nicolaysen. 1959. The serum cholesterol depressive effect of linoleate, linolenic acids, and of cod liver oil in experimental hypercholesterolemic rats. *Acta Physiol. Scand.* **45:**26.

Howard, A. N., G. A. Gresham, and D. Jones. 1965. The prevention of rabbit atherosclerosis by soya bean meal. *J. Atheroscler. Res.* **5:**330.

Intersociety Commission for Heart Disease Resources. 1970. Primary prevention of atherosclerotic diseases. *Circulation* **42:**A55.

James, A. T., and A. J. P. Martin. 1952. Gas–liquid partition chromatography; the separation and microestimation of volatile fatty acids from formic acid to dodecanoic acid. *Biochem. J.* **50**:679.

Johnston, P. V., O. C. Johnson, and F. A. Kummerow. 1958. Deposition in tissues and fecal excretion of *trans* fatty acids in the rat. *J. Nutr.* **65**:13.

Joslyn, M. A. 1970. *Methods in Food Analysis*, 2nd ed. Academic Press, New York.

Kaunitz, H. 1975. Dietary lipids and arteriosclerosis. *J. Amer. Oil Chem. Soc.* **52**:293.

Kharash, N., and J. Fried (eds.). 1977. *Biochemical Aspects of Prostaglandins and Thromboxanes.* Academic Press, New York.

Kinsell, L. W., G. D. Michaels, G. C. Cochrane, J. W. Partridge, J. P. Jahn, and H. E. Balch. 1954. Effect of vegetable fat on hypercholesterolemic and hyperphospholipidemia. *Diabetes* **3**:113.

Kinsell, L. W., G. D. Michaels, J. W. Partridge, L. A. Boling, H. E. Balch, and G. C. Cochrane. 1953. Dietary modifications of plasma cholesterol and phospholipid levels in diabetic patients. *J. Clin. Nutr.* **1**:295.

Kinsell, L. W., J. Partridge, L. Boling, S. Morgen, and G. D. Michaels. 1952. Dietary modification of serum cholesterol and phospholipid levels. *J. Clin. Endocrinol.* **12**:909.

Kinsella, J. E. 1981. Dietary fat and prostaglandins: Possible beneficial relationships between food processing and public health. *Food Technol.* **35**(5):89.

Kritchevsky, D. 1976. Diet and atherosclerosis. *Amer. J. Pathol.* **84**:615.

Kummerow, F. A. 1974. Current studies on relation of fat to health. *J. Amer. Oil Chem. Soc.* **51**:255.

Lecanu, L. R. 1838. Chemical studies of human blood. *Ann. Chim. Phys.* **67**:54.

Leemans, F. A., and J. M. McCloskey. 1967. Combination gas chromatography–mass spectroscopy. *J. Amer. Oil Chem. Soc.* **44**:11.

Malmros, H. 1950. A statistical study of the effect of war-time on arteriosclerosis, cardiosclerosis, tuberculosis and diabetes. *Acta Scand. Suppl.* **246**:137.

Nolen, G. A., J. C. Alexander, and N. R. Artman. 1967. Long-term rat feeding study with used frying fats. *J. Nutr.* **93**:337.

Pearson, D. 1970. *The Chemical Analysis of Foods* 6th ed. Chemical Publishing Co., New York.

Peifer, J. J. 1968. Disproportionately higher levels of myocardial docosahexaenoate and ele-

vated levels of plasma and liver arachidonate in hyperthyroid rats. *J. Lipid Res.* **9**:193.

Peifer, J. J. 1967. Hypocholesterolemic effects of marine oils. In *Fish Oils* (M. E. Stansby, ed.). Avi Publishing Co., Westport, Conn.

Plattner, R. D., G. F. Spencer, and R. Kleiman. 1977. Triglyceride separation by reverse phase high performance liquid chromatography. *J. Amer. Oil Chem. Soc.* **54**:511.

Pomeranz, Y., and C. E. Meloan. 1971. *Food Analysis: Theory and Practice.* Avi Publishing Co., Westport, Conn.

Rawls, R., and P. J. Van Santen. 1970. The role of singlet oxygen in the initiation of fatty acid autoxidation. *J. Amer. Oil Chem. Soc.* **47**:121.

Reiser, R. 1973. Saturated fats in the diet and serum cholesterol concentration: a critical examination of the literature. *Amer. J. Clin. Nutr.* **26**:524.

Rittenberg, D., and R. Schonheimer. 1937. Deuterium as an indicator in the study of intermediary metabolism. XI. Further studies on the biological uptake of deuterium into organic substances with special reference to fat and cholesterol formation. *J. Biol. Chem.* **121**:235.

Samuelsson, B., and R. Paoletti (eds.). 1976–1980. *Advances in Prostaglandin and Thromboxane Research*, Vols 1–8. Raven Press, New York.

Singer, S. J., and G. L. Nicolson. 1972. The fluid mosaic model of the structure of cell membranes. *Science* **175**:720.

Sonntag, N. O. V. 1979. Structure and composition of fats and oils. In *Bailey's Industrial Oil and Fat Products*, (D. Swern, ed.), Vol. I, pp. 1–98. Wiley, New York.

Stansby, M. E. 1969. Nutritional properties of fish oils. In *World Review of Nutrition and Dietetics*, Vol. 11, pp. 46–105. Karger, Basel/New York.

Swern, D. (ed.). 1979. *Bailey's Industrial Oil and Fat Products*, Vols. I and II. Wiley, New York.

Triebold, H. O., and L. W. Aurand. 1963. *Food Composition and Analysis*. Van Nostrand Reinhold, New York.

Vane, J. R., and S. Bergström (eds.). 1979. *Prostacyclins*. Raven Press, New York.

Virchow, R. 1856. *Gasammelte Abhandl. zur Wissenschaftlichen Medizin*. Meidinger, Frankfurt a. M.

Vogel, J. 1843. *Erläuterungstafen zur Pathologischen Histologie. In Icones; Histologiae Pathologie*, p. 52. Voss, Leipzig.

Walker, G. R., E. H. Morse, and V. A. Oversley.

1960. The effect of animal protein diets having the same fat content on the serum lipid levels in young women. *J. Nutr.* **72:**317.

Waltking, A. E., W. E. Seery, and G. W. Bleffert. 1975. Chemical analysis of polymerization products in abused fats and oils. *J. Amer. Oil Chem. Soc.* **52:**96.

Widal, F., A. Weill, and M. Laudat. 1913. Comparative study of the contents of free choles-

terol and its esters in blood serum. *C. R. Soc. Biol.* **74:**882.

Wolfe, L. S. 1981. Prostaglandins and thromboxanes in the nervous system. Chapter 16. In *Basic Neurochemistry* (G. J. Siegel, R. W. Albers, B. W. Agranoff, and R. Katzman, eds.). Little, Brown, Boston.

Yudkin, J. 1957. Diet and coronary thrombosis. Hypothesis and fact. *Lancet* **273:**155.

C H A P T E R · 8

Food Colloids

INTRODUCTION

In 1861 Thomas Graham published his findings dealing with the diffusion of substances in solution through membranes. He reported that dissolved crystalline substances could easily diffuse through membranes while the rates of diffusion for noncrystalline substances were significantly retarded. As a consequence of these findings, Graham classified these substances respectively as *crystalloids* or *colloids*. The "colloid" term originates from the Greek word, *kolla,* which refers to "glue-like" substances. It is now recognized that the slow rates of diffusion observed for these colloidal substances are attributable to the physical sizes of various substances in the colloid that range from 1 to 1000 nm.

For many years colloidal substances were thought to deviate from known principles of chemical and physical behavior, and they were considered to be outside the realm of well-behaved chemical systems. Contemporary understanding of colloidal substances, however, has clearly established that they actually represent a subdivision of the normal physical state of matter that can be attained by almost any material under suitable, controlled conditions.

Particles displaying colloidal dimensions can be categorized as lyophilic, lyophobic, or association colloids. Not all colloids fit into these three groups, and still others have characteristics common with more than one of these groups.

SURFACE PHENOMENA

The behavior of colloidal particles is intrinsically linked to the study of surface chemistry and physical surface phenomena.

A surface is defined as the boundary between two phases involving liquids, solids, or a gas. Boundaries such as those between two different solids, liquids, or a liquid and a solid are known as *interfaces*. Figure 8.1 illustrates a gas–liquid system having a surface and interactions among molecules within the liquid. Within the body of the liquid, *each molecule* is surrounded by other similar molecules. The forces exerted on individual molecules are the same in all directions or omnidirectional. For molecules located at the liquid surface, however, this situation is quite different. The molecules in the liquid phase are not surrounded entirely by identical

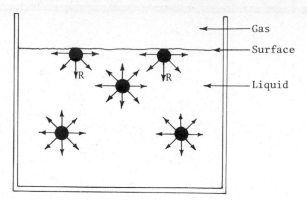

FIGURE 8.1. Forces acting on molecules of a liquid.

molecules because the surface is in contact with a gas. The molecules comprising the gas are relatively far apart compared to the liquid and, consequently, the gas does not exert any significant force of attraction on molecules of the liquid's surface. The net resulting force, R, among molecules at the liquid surface is directed toward the interior body of the liquid. This inward attraction of the surface molecules tends to reduce the liquid surface to a minimum. The imbalance in intermolecular forces of the liquid surface gives rise to the phenomenon of *surface tension* (γ) (Figure 8.1).

Surface tension can be defined as a measure of the amount of free energy required to create 1.0 cm² of new surface. The total surface energy (E_s) required to create 1.0 cm² of new surface for a pure liquid is

$$E_s = \gamma - T \frac{d\gamma}{dT}$$

where

γ = surface tension
T = absolute temperature (K)
$T \frac{d\gamma}{dT}$ = amount of heat energy required to maintain the created surface at a constant temperature

SURFACE ACTIVE SUBSTANCES

The presence of a solute in a liquid such as water influences the degree of surface ten-

sion. The migration of solute molecules either toward or away from the surface of a liquid depends on the chemical nature of the solute. As a rule, highly hydrated ionic species migrate away from the surface of the solution and slightly hydrated substances tend to migrate toward the surface. The latter substances are known as *surface active* materials. The relationship between the excess of solute that migrates to the liquid's surface (Γ) and the change in the surface tension as the concentration of the solute changes ($d\gamma/dc$) can be expressed by the *Gibbs adsorption equation:*

$$-\Gamma = \frac{d\gamma}{dc} \frac{c}{RT}$$

where

c = concentration of solute
R = universal gas constant
T = absolute temperature (K)

When $d\gamma/dc$ is a negative value, Γ is positive and there is a greater concentration of the solute in the surface region than in the body of the liquid. The opposite is true for solutions where $d\gamma/dc$ is positive.

As indicated by Figure 8.2, the surface tension of a liquid such as water can be changed whenever solutes are added. Curve I shows that the surface tension of water increases along with increasing concentrations of inorganic salts. According to Curve II, the surface tension of water decreases

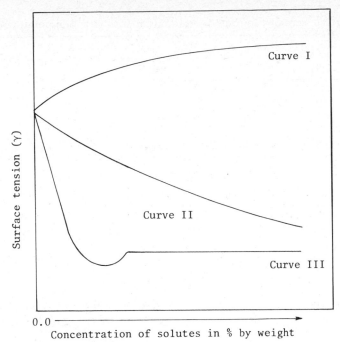

FIGURE 8.2. Influence of increasing solute concentration on the surface tension of water. Consult text for explanation regarding the significance of the individual curves.

with increasing molar concentrations of alcohols, carboxylic acids, and other substances having a large number of hydroxyl groups such as sugars (e.g., sucrose). For Curve III, the surface tension of the liquid progressively decreases until it reaches a "dip" due to impurities in a surfactant solute. The surface tension then increases slightly before leveling off with the further addition of surfactant. Beyond the "dip" on Curve III, the surfactant molecules begin to aggregate into micelle structures. These micelles principally remain in the bulk of the solution since they are hydrophilic and do not cause an infinite decrease in surface tension.

MEASUREMENT OF SURFACE TENSION

Surface tension can be measured by employing the DeNöuy tensiometer, which consists of a platinum ring having a radius r attached to a torsion balance. This ring is immersed in the solution whose surface tension is to be measured. The force required to remove the ring from the liquid's surface is measured directly on the tensiometer. The surface tension acts over the entire circumference of the platinum ring and thereby results in an applied force (F) equal to $4 \pi r \gamma$. For precise measurements of surface tension, a correction factor is customarily invoked because of certain volume of liquid that is adsorbed to the rising ring. Hence,

$$\gamma = \frac{f_c F}{4 \pi r}$$

where

γ = surface tension
F = applied force
f_c = correction factor (ranges from about 0.75 to 1.45)

Surface tension for various substances differs according to temperature (Table 8.1). One mathematical expression designed to

Table 8.1. Some Surface Tensions for Selected Liquids at Different Temperatures

Liquid	Temperature (°C)					
	0	20	40	60	80	100
Water	72.6	72.8	69.6	66.2	62.6	58.9
Ethyl alcohol	24.1	22.3	20.6	19.0	—	—
Cotton seed oil	—	35.4	—	—	28.4	—
Coconut oil						—
Olive oil	—	33.4	—	—	31.3	—

estimate the variation of surface tension with temperature is that of Ramsey and Shields:

$$\gamma \left(\frac{M}{\rho}\right)^{2/3} = k' \, (T_c - T - 6)$$

where

γ = surface tension
M = molecular weight of liquid
$\dfrac{M}{\rho}$ = molecular volume
k' = a constant, independent of T
T_c = critical temperature of the liquid
T = temperature of the liquid phase
ρ = density of liquid

Since M/ρ is the molecular volume of the liquid, $(M/\rho)^{2/3}$ represents a quantity proportional to the molecular free surface of the liquid; therefore, $(M/\rho)^{2/3}$ is often called the *molecular free surface energy*.

ASSOCIATION COLLOIDS

Approximately 50 to 100 molecules or ions can aggregate to form micelles of colloidal size. These multimolecular aggregates, called *association colloids*, generally display spherical or lamellar shapes. The actual number of molecules or ions contained in the micelle depends on the nature and size of the participating species, their concentration, and the temperature of the solution.

The critical concentration required for establishing a micelle structure corresponds to the beginning of the horizontal portion on the type III curve (Figure 8.3) discussed previously. The number of molecules or ions involved in micelle structures can be estimated by the law of mass action as shown by the equation

$$mU \longrightarrow A$$

where

m = number of molecules
U = unaggregated molecules
A = aggregated molecules forming micelle

This equation indicates that m number of molecules are involved in the formation of a micelle from a solution having a definite solute concentration (C). Letting X represent the interacting molecules in the micellar form, the concentration of aggregated molecules is $(X/m)C$ and the concentration of the unaggregated molecules is $(1 - X)C$. Using these terms, the equilibrium constant (K) for micelle formation is

$$K = \frac{\left(\dfrac{X}{m}\right)C}{[(1 - X)C]^m}$$

If $K = 1.0$, the effect of concentration on the degree of aggregation at various values of m can be computed. When m is ≤ 2.0, the formation of micelles is gradual (Figure 8.3), but when m is ≥ 50, micelle formation is very rapid.

COLLOIDAL DISPERSIONS

Colloidal dispersions are considered to be two-phase systems composed of one dispersed *solvent medium* and another phase

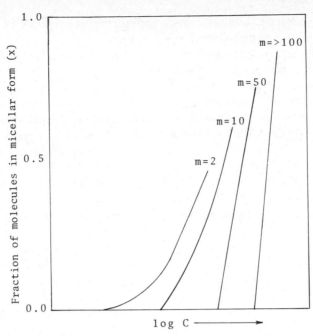

FIGURE 8.3. The effect of concentration on the degree of micellar formation for molecules capable of aggregating into micelle structures.

present as *dispersed particles*. There are nine types of colloidal dispersions, but only five of these have practical relevance to food systems (Table 8.2).

CLASSIFICATION OF COLLOIDAL DISPERSIONS

Sols

Dispersions of solids in liquids are generally defined as lyophobic or lyophilic sols. *Lyophobic sols* (meaning "solvent fearing") are represented by those dispersions that exhibit a minimal attraction between the dispersed phase and the dispersing medium. This is in marked contrast to *lyophilic sols* (meaning "solvent loving"), which refers to systems showing a *definite affinity* between the dispersed phase and the dispersing medium, such as aqueous solutions of proteins, polysaccharides, and nucleic acids. If the lyophobic dispersed phase happens to be in water, the term *hydrophobic sol* is used; but when the lyophilic dispersed phase is in water, the sol is described as *hydrophilic*.

Preparation of sols: Lyophilic sols as well as association colloids can form sponta-

Table 8.2. Typical Colloidal Dispersions Found in Some Foods

Dispersed phase	Dispersed medium	Name of dispersion	Example
Solid	Liquid	Sol	Skim milk
Solid	Gas	Solid aerosol	Smoke
Liquid	Liquid	Emulsion	French dressing
Gas	Liquid	Foam	Whipped cream
Gas	Solid	Solid foam	Foam candy

neously to produce relatively stable systems. Lyophilic colloids represent true solutions of macromolecules including proteins, polysaccharides, nucleic acids, and certain synthetic polymers that naturally exhibit colloidal dimensions. Lyophilic sols also include solutions of molecules or ions that mutually associate to form lyophilic micelles having colloidal dimensions.

Lyophilic sols are prepared by placing a finely pulverized solid into a solvent that demonstrates a strong attraction for the solid. Lyophobic sols are rather difficult to prepare because of their characteristic instability as a physical system. Attempts to prepare lyophobic sols necessarily rely on techniques of dispersion and condensation.

Dispersion is a process whereby substances are pulverized to colloidal dimensions or, alternatively, coarse suspensions are disintegrated by high-speed mixing. In either case, the main objective is to decrease the surface-to-volume ratio for the substance dispersed in the solvent medium, thus increasing the sol's stability. Condensation, on the other hand, is designed to condense smaller particles until they reach colloidal dimensions.

The establishment and preparation of sols may require the removal of noncolloidal substances and impurities in the sol system. The nature of the noncolloidal substances may be ionic or nonionic species that are inorganic and/or organic. Since these species inevitably abet the destabilization of the sol, their removal from a sol is imperative and can be achieved by dialysis, electrodialysis, or ultrafiltration (Figure 8.4).

Colloidal dispersions and the electrical double-layer theory: Colloidal dispersions possess electrical properties that are intimately associated with their ability to adsorb ions and molecules, singularly or in combination, from the solvent medium. Lyophilic sols primarily adsorb the ions of electrolyte solutions with which they are prepared. This behavior of lyophilic sols permits the formation of an electrical double layer around the individual colloidal particles. Figure 8.5

depicts this principle of an electrical double layer.

The stationary zone of electrically charged species that is immediately adjacent to the colloidal solid represents the *Stern layer*. As noted in Figure 8.5, these electrical species are countercharged with respect to the overall charge of the particle. Aside from the electrical species forming the Stern layer, the remainder of similarly charged species are distributed as a mobile phase throughout the electroneutral bulk of the solution in decreasing occurrence as distance increases from the Stern layer.

The boundary of the Stern layer, which gives way to a mobile zone of charged species (*Gouy–Chapman layer*), constitutes the so-called *shear boundary*. The electrical potential present between the shear boundary and the Stern layer is measured as the *zeta* (ζ) *potential*. The magnitude of the ζ potential is a function of an electrolytic species' valence properties as well as the adsorbability of its ions. If the repulsive force generated by the charged double layer effectively disperses two or more charged particles, particle aggregation is prevented since a high ζ potential produces a stable sol. Conversely, when the value of the ζ potential is lowered by progressively adding more electrolytes, the charged double layer surrounding a colloidal particle collapses and precipitation occurs.

In addition to affecting the behavior of colloidal systems, the electrical double layer also acts as a significant influence for electrokinetic phenomena including *electroosmosis, streaming potential, migration potential,* and *electrophoresis* (Table 8.3). Of these four electrokinetic phenomena, electrophoresis is especially important because it provides the usual method for establishing ζ potential values since it is related to the *electrophoretic velocity* (v) of particles by

$$v = \frac{DE\zeta}{4\pi\eta}$$

where

E = electrical field of strength (volts)
D = dielectric constant of the medium
η = viscosity of the medium

Schematic key for membrane purification methods

- Non-filterable macromolecule or particle constituent of sol
- Impurities which are filterable through a semipermeable membrane
- ⊕ Electrically charged (positive) impurities that are filterable through a semipermeable membrane
- Solvent medium (eg. water)
- Semipermeable membrane which permits passage of small particles

A. Dialysis:

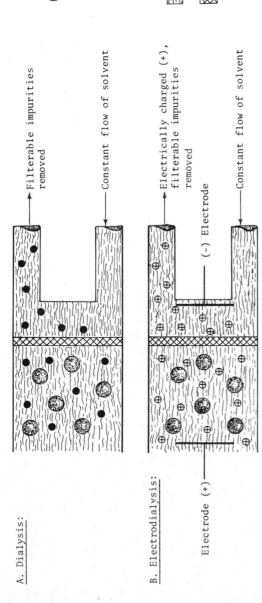

B. Electrodialysis:

C. Ultrafiltration:

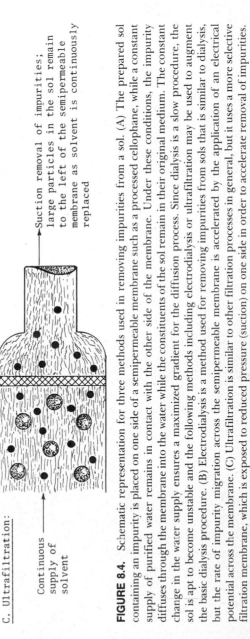

FIGURE 8.4. Schematic representation for three methods used in removing impurities from a sol. (A) The prepared sol containing an impurity is placed on one side of a semipermeable membrane such as a processed cellophane, while a constant supply of purified water remains in contact with the other side of the membrane. Under these conditions, the impurity diffuses through the membrane into the water while the constituents of the sol remain in their original medium. The constant change in the water supply ensures a maximized gradient for the diffusion process. Since dialysis is a slow procedure, the sol is apt to become unstable and the following methods including electrodialysis or ultrafiltration may be used to augment the basic dialysis procedure. (B) Electrodialysis is a method used for removing impurities from sols that is similar to dialysis, but the rate of impurity migration across the semipermeable membrane is accelerated by the application of an electrical potential across the membrane. (C) Ultrafiltration is similar to other filtration processes in general, but it uses a more selective filtration membrane, which is exposed to reduced pressure (suction) on one side in order to accelerate removal of impurities.

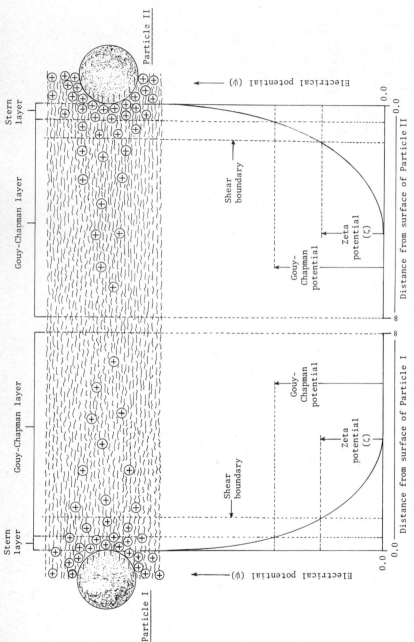

FIGURE 8.5. Diffuse double-layer concept for charged particles illustrating the decay potential within the Stern layers and Gouy–Chapman layers. According to this theory, anything less than an infinite distance between similarly charged particles increases their possible electrical interactions, which may lead to their stabilization or destabilization. Exclusive of the fixed layer of countercharged ions surrounding a particle known as the Stern layer, the remaining counterion population surrounding particles is known as the Gouy–Chapman layer. The potential of the Gouy–Chapman layer displays a zero value only at an infinite distance from the solid particle. For practical reasons, the distance between the Stern layer and the Gouy–Chapman layers can be designated as ψ_d, while the corresponding potential at the site of the shearing boundary is called the *zeta potential* (ζ). This illustration shows two electrically charged particles (negatively charged) that are separated by an infinite distance of an intervening solution containing counterions. When the intervening distance between these two particles is less than an infinite distance, there is an increasing likelihood of an electrical interaction between or among charged particles.

Table 8.3. Electrokinetic Phenomena for which the Presence of a Charged Double Layer Influences Colloidal Behavior

Electroosmosis	A liquid containing charged particles passes through a porous plug under the influence of an electric field.
Streaming potential	A potential difference that arises from forcing a liquid through a porous plug or capillary.
Migration potential	Charged particles suspended in a liquid, when permitted to sediment, create a different electrical potential at each level of the sedimenting system and between the sediment and the liquid. The electric field produced in this way is called the "Dorn effect."
Electrophoresis	Charged colloidal particles migrate under the influence of a direct current.

Stability of sols: The stability of a hydrophobic sol is attributable to the surface charges on particles because those having the same charge, positive or negative, will show mutual repulsion. This contrasts with hydrophilic sols, which owe their stability not only to surface charges but also to the solvation or hydration of particles if the dispersing medium is water. Since hydrophilic sols involve the stabilizing influences of solvation, they can be added to hydrophobic sols in order to protect their stability properties.

The degree of repulsion or attraction among colloidal particles depends on factors such as their chemical nature and size; the distance between individual particles; and the ionic strength of the dispersing medium where the particles reside. As long as colloidal particles remain dispersed, their mutual forces of repulsion exceed their forces of attraction. These interaction forces are typically of the dipole–dipole type called London–van der Waals forces. The dipole–dipole interaction may fall into one of three possible categories: (1) attraction between permanent dipoles; (2) attraction between permanent and induced dipoles; and (3) attractions between induced dipoles, which are commonly known as London dispersion forces.

The energy of interaction for London–van der Waals forces is proportional to the sixth power of the distance separating particles. The forces of attraction are expressed by the equation

$$V_{attr} = -\frac{A}{6}\left[\frac{2a^2}{X^2 - 4a^2} + \frac{2a^2}{X^2} + \ln\left(1 - \frac{4a^2}{X^2}\right)\right]$$

where

V_{attr} = potential energy of attraction between particles

a = radius of the spherical particle

X = distance between the centers of individual particles

A = Hamaker constant

The force of repulsion between two identical particles having a radius a at various distances of separation denoted as h is expressed by the relationship

$$V_{rep} = \frac{Ba}{Z^2}\left[\frac{\exp(Ze\psi_d/2kT) - 1}{\exp(Ze\psi_d/2kT) + 1}\right]^2 \exp(-Kh)$$

where

B = calculated constant (depends on the temperature and dielectric constant of the medium)

Z = valence on the identically charged particle

e = electronic charge

ψ_d = electrical potential at the boundary of the fixed layer of ions (Stern layer) surrounding the particle and the mobile ion layer surrounding the particle

k = Boltzmann's constant

K = $2n_0Z^2 e^2 (\varepsilon kT)^{1/2}$

where

n_0 = bulk concentration of each species

ε = permitivity of the medium

T = absolute temperature

The combined calculations for the van der Waals forces of attraction and repulsion lead to the three distinctive curves shown in Figure 8.6. These curves were independently calculated in the 1940s by Verwey and Overbeek in Holland as well as Deryaguin and Landau in Russia based on earlier studies by Hamaker and Freundlich.

The y axes for the graphs in Figure 8.6 indicate the interaction energy of colloidal particles, while the x axes represent the distance between the particle surfaces. The total potential energies for attraction (V_{attr}) and repulsion (V_{rep}) are indicated on graph I (Figure 8.6) as dotted lines. This total potential energy can also be expressed mathematically as the summation of the attraction and repulsion potential energies (V):

$$V = V_{attr} + V_{rep}$$

Since the total potential energy is a composite of repulsive and attractive forces, these factors are indicated as a solid line on all three graphs (I, II, and III). Each of these

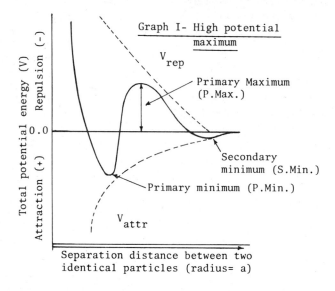

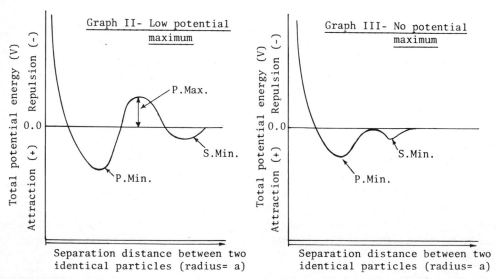

FIGURE 8.6. Total potential energy profiles for three different systems in which two sol particles, each having a radius equal to a, mutually interact at various separation distances (h).

curves displays a characteristic maximum and two minima, yet each curve does retain some semblance of similarity with the others.

The graphic height of the potential energy maxima (P.Max) is especially significant because it indicates the coagulation tendency of a sol, that is,

I. A high potential maximum indicates high repulsion among particles and a sol having a notable longevity.

II. A low potential maximum indicates less repulsion and a longevity less than the previous case (above).

III. An absence of a potential maximum is indicative of the sol's tendency to rapidly coagulate.

The primary minimum (P.Min.) depicted on the potential energy curve occurs when the separation distance between particles is small and particles coagulate, thereby destabilizing the suspension.

The secondary minimum (S.Min.) is represented on the potential energy curve as a shallow depression when particle diameters of 200 nm or larger are separated by large distances. The stability of the system is affected in such a manner that a porous yet compact floc readily forms. These flocculated systems can be redispersed (peptized) by vigorous agitation. It should be noted that the terms *coagulation* and *flocculation* are often used interchangeably, but they represent different physical phenomena.

Destabilization of colloidal sols: The precipitation of sols can be achieved by any conditions that lead to *instability* within the system. Lyophobic sols, for example, are destabilized by removal of both surface charges and the adherence of the solvent layer surrounding particles. Lyophobic sols can lose their surface charges if salt concentrations are added. This is in contrast to lyophilic sols for which coagulation can be accomplished by organic solutes such as certain ketones (acetone) and alcohols (ethanol) that have a high affinity for water, followed by the salt solution. The addition of salts for sol destabilization purposes is called *salting-out*. The salting out powers of various ions are ranked according to the Hofmeister series or *lyotropic series* in which ions are arranged in order of decreasing precipitation effectiveness.

Lyotropic series for

Anions

$$\text{Citrate}^{3-} > \text{Tartrate}^{3-} > \text{SO}_4^{2-} >$$
$$\text{Acetate}^{1-} > \text{NO}_3^{1-} > \text{I}^{1-}$$

Cations

$$\text{Mg}^{2+} > \text{Ca}^{2+} > \text{Sr}^{2+} > \text{Ba}^{2+} >$$
$$\text{Li}^{1+} > \text{Na}^{1+} > \text{K}^{1+} > \text{NH}_4^{1+}$$

The order of these ions may be altered depending on the nature of the colloid undergoing precipitation.

Ions in the lyotropic series react with the colloid surface to decrease surface charges or may reduce the diffuse layer of ions around the colloid, thereby decreasing the ζ potential. Ionizable residues on very-long-chained polymers can also bring about the removal of particles through adsorption and interparticle bridging. These so-called *polyelectrolytes* can be divided into two categories, namely, synthetic or natural. Polyelectrolytes of natural origin derived from starch products, cellulose derivatives, and alginates have special significance to food systems. Synthetic polyelectrolytes, on the other hand, consist of simple monomers that are polymerized into high-molecular-weight substances. Depending on whether the charge of polyelectrolytes happens to be negative, positive, or neutral when placed in water, they may be classed as anionic, cationic, or nonionic, respectively.

Gels

Gels found in foods represent two-phase systems composed of a continuous phase such as water and a dispersed phase (gelling agent) whose chemical composition varies widely depending on the gel. The water molecules are typically held within the three-dimensional network of the gel by forces of hydrogen bonding in addition to physical entrapment in interstitial spaces of the gel structure by capillary action.

Food gels are semisolid systems that show various degrees of elasticity, brittleness, and rigidity, depending on the gelling substance employed. Some of the substances commonly found in food gels under controlled conditions are plant gums, pectins, starches, flour proteins such as gluten and gelatin, and proteins from eggs or milk.

Factors influencing gel structure: Temperature, mechanical agitation, pH variation, salt concentration, and sugars influence the behavior of gels.

Temperature changes influence the bonds responsible for maintaining a gel structure. As temperature rises, bonds become increasingly labile until the semisolid state of the gel assumes viscous liquid properties. This conversion can be reversed upon subsequent cooling in many cases. Gels of this type are called *thermoreversible.*

As a rule, the melting points for a reformed gel, such as gelatin systems, are higher than their original gelatinization points.

Aside from thermoreversible gels, there are other gel types that do not reform on cooling, which are called *thermoirreversible.* Thermoirreversible gels do not reform on cooling because heat actually destroys certain covalent bonds (e.g., disulfide bonds) responsible for the gel structure. This is quite different from thermoreversible gels in which a gel structure is maintained principally by hydrogen bonding.

Mechanical agitation causes gel liquefaction, but the liquid gel can often reform if left undisturbed. This reversible property of some gels is called thixotropy.

The adjustment of pH and introduction of sugars into food systems significantly influences the solvation of gelling substances such as proteins and pectin. A pH change away from the isoelectric points of proteins is particularly responsible for electrical charge retention on the gel particles; the charged proteins display mutual repulsion, and this in turn leads to the physical immobilization of large water volumes. The addition of mono- or multivalent ionic species to a potential gel system containing proteins also alters the charges on molecules, which encourages their crossbridging and ultimate formation of a three-dimensional structure.

The addition of sugars to gels formulated from pectin solutions removes water from the dispersed pectin particles and thereby favors the production of a gel.

Although gels can be formulated so as to retain their original volume for a long period of time, many gels experience some shrinkage of volume upon standing. The net result of the shrinkage is evident as a spontaneous exudation of liquid. This liquid exudation event is called *syneresis,* and the progressive dessication of the gel leads to the formation of a *xerogel* (dry gel).

Examples of gelling substances

Plant hydrocolloids. Plant hydrocolloids are natural polysaccharides that are derived from the saps and exudates of (1) various bushes or trees (acacia, ghatti, karaya, and tragacanth); (2) seed extracts (guar, locust bean, quince, and tamarind); and (3) various marine algae (agar, carrageenan, and furcelleran).

The biochemical composition of plant hydrocolloids is largely polyuronides, which contain more than one monosaccharide unit, whereas the algal gums are "primarily" composed of a single monosaccharide unit.

Of all the gum types mentioned above, gum arabic generally represents the singular water-soluble gum, which is soluble to concentration levels of 50% (w/v). Other varieties of hydrocolloids swell in water to form sols or colloidal solutions. Those hydrocolloids that swell at 2 to 5% concentration levels in water produce viscous jellies or gels. The marine algae such as *Chondrus crispus* or Irish moss, on the other hand, produce hydrocolloid extracts that form firm gels at 1 to 2% concentrations in water.

The most important properties of gums and hydrocolloids (e.g., agar) are respectively their viscosities and gel strengths. Guar, karaya, locust bean gums, and tragacanth swell in water to produce high-viscosity sols whose viscosities vary directly with corresponding increases in gum concentrations over the range of 0.3 to 0.5%. Gum concentrations above the 0.5% level show geometric increases in viscosity because the gums then act as non-Newtonian liquids (see rheology section). This behavior is largely a manifestation of complex surface interactions that occur at high gum concentrations.

A model of hydrocolloid behavior—carrageenan. Exclusive of individual dissimilarities in the hydrocolloid structures of various gums, the chemical and physical interactions responsible for their overall colloidal behaviors are principally similar. Carrageenans, which are widely used in food systems, mirror some of the most common properties of important gums.

Carrageenans are polysaccharides containing sulfated galactans linearly linked in alternating α-1,3 and β-1,4 units.

Aqueous carrageenan solutions gel at room temperature and demonstrate thermoreversible gelling properties. Carrageenan gels do not melt at room temperature, and unlike some gel types, they do not toughen on standing. The strengths of these gels are especially enhanced by cations that interact with the sulfated galactan residues.

Aqueous carrageenan solutions are stable within the pH range of 5 to 10 and their colloidal properties are retained even after heating provided that the pH is held at >6.0.

At temperatures above 50°C, aqueous solutions of carrageenan exist as random coils. As the temperature of aqueous carrageenan systems falls below 50°C, the polysaccharide molecules lose their random coiled structure and assume the form of double helical structures. These structures associate with each other to form a three-dimensional gel network as shown in Figure 8.7.

Carrageenans can interact with proteins that are present in many food systems, es-

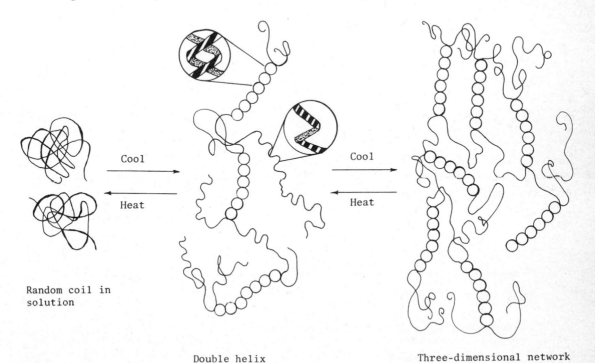

Random coil in solution

Double helix

Three-dimensional network

FIGURE 8.7. Carrageenan molecules exist as random coils in aqueous solutions above 50°C and show an increasing degree of helical structure upon cooling.

pecially milk-based products. These interactions depend on carrageenan concentration, protein type, the isoelectric point of the protein, temperature, pH, as well as the position and number of sulfate groups occurring on the hydrocolloid structure.

As illustrated in Figure 8.8, pH values above the isoelectric point of a protein permit cations such as calcium (Ca^{2+}) to bridge the charged sulfate groups of carrageenan with the carboxylate ions of the protein, but below the isoelectric point of the protein, sulfate groups of carrageenan interact with protonated amino groups of the protein.

Starch gels. The admixture of starch to cold water is followed by gradual penetration of water into the granule. Water imbibition may reach a level of 25 to 30% in some starch granules and yet the integrity of its individ-

ual micelles remains. As the temperature of the starch–water mixture is raised, some of the intermolecular hydrogen bonds contained in amorphous regions of the granule disintegrate and the granule begins noticeable swelling. During this swelling, the natural radial orientation of micelles is lost along with their birefringence properties. Additional swelling of the granules accompanied by high temperature and/or agitation can lead to their eventual disintegration. The temperature at which starch granules begin to swell and lose birefringence is called the *gelatinization temperature*. This temperature has special importance because it can be used as a criterion for identifying starches.

A starch gel is usually formed when a proper starch concentration particularly rich in amylose is heated, stirred, and allowed to cool while undisturbed.

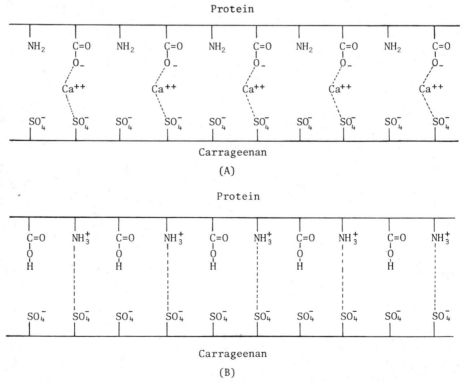

FIGURE 8.8. Typical interactions of carrageenan with a food protein: (A) when the pH value of the protein's aqueous environment is above its isoelectric point, divalent calcium cations bridge the sulfate groups of the carrageenan with the protein; and (B) when the pH value of the environment is below the protein's isoelectric point, an interaction occurs between the sulfate groups of the carrageenan and protonated amino groups.

The amylose molecules of the starch form a three-dimensional network resulting from mutual hydrogen bonding interactions as well as amylose–water interactions in interstitial spaces of the starch structure.

Exclusive of gelling properties, starch molecules do demonstrate an ability to aggregate and form crystals in a process known as *retrogradation*. Retrogradation occurs most readily at ~0°C; therefore, both processed and unprocessed frozen foods may be subject to this phenomenon. The staling of breads and many other baked products also involves retrogradation processes.

Retrogradation does not necessarily destroy the intrinsic food or caloric values of starches, but crystallized starches may be less susceptible to enzyme hydrolysis, which is necessary for normal digestion processes.

The strength of starch gels varies depending on the type of starch used (e.g., dent corn starches; waxy amylose starches; tapioca or potato starches). Waxy-maize starch is primarily amylopectin, but corn starch contains both amylopectin and 25–65% amylose, depending on the variety. When waxy-maize starches are cooked, they form clear, stable, fluid, nongelling cohesive pastes while regular corn starch pastes are cloudy and show a notable tendency to retrograde or *gel*. The relative gel strengths of these two starches alone exemplify the wide range of possible textures and bodying characteristics that can be created in foods merely by using a singular purified starch or comingling varying concentrations of different starches (Figure 8.9).

Pectin gels. Pectin gels are examples of lyophilic colloidal systems in which pectin acts as a gelling agent when stabilized by sugar, water, and acid. Pectins are essentially long chains of polymerized galacturonic acid residues that are partially esterified to methyl alcohol.

Pectin concentrations of 0.2 to 1.75% that have 50 to 70% methoxylation typically require 65 to 70% sugar and a pH of 2.8 to 3.2 as a prerequisite for gelling. Pectin gels forming from hot solutions, having the ingredients listed, may show signs of firmness or gelling at approximately 85 or 55°C depending on whether the pectin is a "rapid set" or "slow set" pectin.

The setting time, amount of sugar, and the pH required for pectin solutions to gel corresponds closely with the degree of pectin methoxylation. The higher the degree of pectin methoxylation, the higher the gelation temperature; for example, according to the gel formation criteria listed in Table 8.4, case 3 would normally show a higher gelation temperature than case 4.

The mechanism of pectin gel formation is complex, but it generally involves the attraction of protons from the acid to the carboxylate groups of the polygalacturonic acid residues. This activity produces a net reduction in the negative charges on the pectin molecules and correspondingly decreases the ability of the pectin to remain dispersed throughout the pregelled solution.

The addition of sugar to protonated pectin molecules results in the solvation of the sugar with an accompanying decrease in the pectin hydration and the establishment of a gel on cooling.

The strengths and stabilities of pectin gels are pH dependent over a range of ~3.5 to 2.6 depending on the type of pectin, but especially low pH values usually encourage liquid exudation (syneresis) of the gel on standing.

Pectin gels are of great value to foods because they have the ability to be carefully formulated according to their projected use in a certain food substance. For example, various types of tomato sauce and catsup consistencies can be carefully regulated by controlling calcium pectate concentrations. Other types of pectin formulations can be used for jellied preserves and jams according to gel setting times in order to maintain even distributions of fruit chunks or pulps in jars after their initial preparation.

Gelatin gels. The bodies of animals are very rich in the protein known as collagen, which

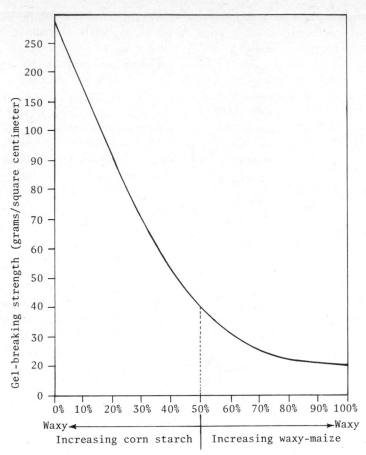

FIGURE 8.9. Gel-breaking strength for waxy maize starch, normal corn starch, and varying mixtures of these two constituents (5.5% starch, dry basis at 25°C). Combinations of these two starches can be used to produce desired texture and body characteristics for many different types of foods.

has an especially high concentration of the amino acids proline and hydroxyproline. Collagen can be extracted from integumentary tissues of animals (e.g., skin) as well as skeletal ligaments and bones when they are heated in water to temperatures of 62 to 63°C. Heating causes the denaturation and partial hydrolysis of the collagen structure and promotes its eventual solubilization.

Many proteins typically demonstrate hydration values of one part water to five parts protein, but gelatin gels can produce a gel structure having water concentration levels equivalent to ten times the weight of the hydrated protein.

Gelatin gels are formed only during cooling of gelatin-containing solutions as collagen molecules become partially renatured to form a three-dimensional gel network. Stabilization of the gel involves hydrogen bonding among amino acids other than proline and hydroxyproline. This process, called *collagen folding*, occurs over a pH range of 5 to 10.

Although hydroxyproline does not directly contribute to hydrogen bonding involved in the gelatin structure, its hydroxy group does augment gel stability by forming hydrogen bonds with water or other substituents in the folded collagen structure.

It should also be noted that gelatins formed as a result of normal meat cooking processes

Table 8.4. Gelation Requirements for Pectins

Case	Percentage methoxylation for pectin	Gelling requirements
1	100	High methoxylated pectins form a gel with sugar alone
2	50	Low methoxylated pectins form gels with the addition of cations such as Ca^{2+}
3	70	"Rapid set" pectins form a gel with the addition of sugar and acid, which will produce pH in the range of 3.0–3.4
4	50–70	Pectin forms a gel in the presence of sugar and a pH range of 2.6–3.2

immobilize water, yet the water entrapped in the three-dimensional gel structure displays solubility, vapor pressure, and colligative properties reminiscent of normal water.

Emulsions

An emulsion is an intimate mixture of two immiscible liquids with one of the liquids present as very fine droplets dispersed in another liquid. The dispersed droplets constitute the *internal phase* of the emulsion, whereas the liquid surrounding the droplets represents the *external phase*.

Due to the immiscible nature of internal and external phases, there is a physical tendency for the dispersed droplets to coalesce into larger droplets and either sediment ("fall") or cream ("rise"). The sedimentation or creaming behavior of dispersed droplets is determined by the relative density of the continuous external phase.

Emulsions can be stabilized for long periods of time with the aid of a colloid mill that mechanically shears the internal phase into very fine droplets, or by the use of a third ingredient called an emulsifier. Both of these methods are intended to achieve the same objectives, that is, minimize creaming or sedimentation of an internal phase.

Mechanical methods "shear" or reduce the droplet size of the internal phase so that its surface-to-volume ratio is maximized within the range of practical limitations. This reduces the tendency of individual droplets to coalesce and eventually partition from the external phase.

Emulsifier substances, on the other hand, are usually long-chained amphipathic organic compounds. These substances are called *amphipathic* because portions of the molecules show distinctly lipophilic (oil soluble, nonpolar) and hydrophilic (water soluble, polar) properties. When dissimilar solubility differences such as these exist on the same molecule, the hydrophilic portion of the molecule is free to interact with an aqueous phase while the lipophilic moiety can interact with a nonpolar oil phase. The long intervening molecular structure between lipophilic and hydrophilic portions of the emulsifier then acts as a link between two immiscible phases and effectively increases their mutual interactions along with the permanence of the emulsified system.

Emulsifiers can also act as surfactants that reduce the surface tension of an aqueous phase. This action makes it possible for water-insoluble substances to be more easily dispersed in an aqueous phase, thereby forming a stable emulsion.

The requirements for establishing stable emulsions are complex and variable depending on their ingredients, yet a number of factors are generally responsible for influencing emulsion stability. These include:

1. *Droplet size of the dispersed internal phase.* The smaller the size, the greater the stability of the emulsion.

2. *Viscosity of the external phase.* High-viscosity external phases are most desirable for stability.

3. *Charge on dispersed phase.* Emulsion stability is favored when charges are similar.

4. *Interfacial tension between phases.* Low surface tensions between an emulsion's external and internal phases favor stability.

5. *Density differences between phases.* Emulsion stability is favored when the density differences between external and internal phases are minimal.

6. *Cohesion of the internal phase.* Emulsion stability is encouraged when cohesiveness of the internal phase is minimal.

7. *Percentage of solids in the emulsion.* A low percentage of the internal phase promotes emulsion stability.

8. *Temperature.* Extremely high or low temperatures are detrimental to emulsion stability.

The individual droplets of the internal phase contained in an emulsion often show diameters of 100 to 10,000 nm; for those instances when droplets have a diameter of less than 100 nm, the dispersion of the internal phase is referred to as a *microemulsion*.

Shorthand designations are conveniently used for describing emulsified systems; these are

1. Oil-in-water, symbolized as O/W.

2. Water-in-oil, symbolized as W/O.

The general properties of emulsions are primarily determined by the individual nature of their external phases. An O/W emulsion functions as a water system that can be diluted with water rather than oil and can be thickened with water-soluble hydrocolloids. The opposite is true for W/O emulsions.

Physical behavior and detection of emulsions: The observed consistency of an emulsion depends on the viscosity of the external phase, the concentration of the dispersed internal phase, and the attractive forces among the droplets. In most cases, emulsion viscosity can be increased by increasing the concentration of individual droplets, decreasing droplet size, lowering temperature, or adding thickeners (e.g., agar, glycerol, sodium carboxymethyl cellulose, or other processed polysaccharides).

The appearance of emulsions depends on the refractive indices and optical dispersive properties of the two comingled phases in addition to droplet size. In the case of coarse emulsions in which droplet sizes are ≥ 1000 nm, the observed color is generally milky white, but a progressive decrease in droplet size produces a blue-white, gray-semitransparent, and finally transparent appearance when dispersed droplets are ≤ 100 nm. Regardless of droplet sizes, many emulsions will appear translucent and colorless if the internal and external phases have similar optical properties.

Many emulsions have an inherent tendency to break down when they are exposed to heat, electrolytes, centrifugal forces, or mechanical agitation. Physical indication of emulsion breakdown is noticeable as an inversion, that is, the internal phase becomes external and vice versa.

Ideally, when droplets of the internal phase of an emulsion have identical sizes, their "close packing" effect accounts for a maximum volume of the internal phase equivalent to ~74% while the external phase approximates ~26% of the volume. When the internal phase of the emulsion accounts for $\geq 74\%$ of the volume, inversion or breakdown of the emulsion is greatly enhanced.

Aside from empirical observations that indicate the presence of an emulsion, most emulsified systems can be objectively detected using one or more of the following tests:

1. *Conductivity test.* Most oils are poor conductors of electricity; therefore, W/O emulsions demonstrate low electrical conductivity properties.

2. *Dilution test.* A drop of oil placed on a microscopic slide blends well with an emulsion of the W/O type.

3. *Dye test.* When an oil-soluble dye is added to an emulsion and the color spreads rapidly, the emulsion is likely to be the W/O type.

4. *Fluorescence test.* A W/O emulsion is indicated as uniform fluorescent field when exposed to ultraviolet light; O/W emulsions are indicated as a nonuniform fluorescent field under similar conditions.

Emulsifiers: Many natural and synthetic chemical substances display abilities to disperse and stabilize emulsified food systems. All living organisms of food importance contain emulsifier-active substances, although their *in vivo* biochemical functions are not always readily associated with the functions they perform in natural or formulated food substances. Natural phospholipids, in particular, demonstrate notable emulsifier properties that have been widely exploited for uses in food processing. These uses for phospholipids are quite unlike their *in vivo* functions in animals or other organisms where they often contribute to lipid mono- and bilayer membrane structures along with proteins. Such membrane systems are responsible for the selective permeability of intra- and extracellular nutrient, waste, and water transport. The functions of lipoproteins in plants are not always well defined, but plants do provide commercially important sources of natural emulsifiers including soya lecithin. The commercial preparation of soya lecithins actually represents a mixture of lecithin and other phospholipids, which can be produced in large quantities by degumming soya oil.

Many other emulsifier-active substances can be chemically produced from naturally occurring glycerides. Of all possible emulsifier-active glycerides, 1-monoacylglycerols are probably most applicable to food systems. The 1-monoacylglycerols do not occur at commercially abundant concentration levels in natural products, but they can be effectively produced in bulk for food applications by glycerolysis of fats followed by the vacuum-catalytic reesterification of selected fatty acids to glycerol.

Exclusive of natural emulsifiers, many additional emulsifiers have been synthetically developed or formulated from organic compounds. These emulsifiers often show superior adaptability to a wide spectrum of different food systems. Since they are relatively pure compounds, they chemically behave in a quasi-predictable fashion depending on their concentration and the type of food system in which they are used.

Classification of emulsifiers and their selection: Emulsifiers are conveniently classified on the basis of their *hydrophile–lipophile balance* (*HLB*). The *HLB* value for an emulsifier mirrors its solubility properties. That is, a *low HLB* value indicates that the emulsifier tends to be oil soluble while a *high HLB* value suggests water solubility.

Numerical *HLB* values range from 0.0 to 20.0, which respectively refer to 100% lipophilic (zero) and 100% hydrophilic behavior (20.0) for an emulsifier.

For applications to food systems, those emulsifiers having *HLB* values of 4.0 to 6.0 are used for producing W/O emulsions, and those in the range of 8.0 to 18.0 are applied to O/W emulsions. Emulsion requirements for certain food ingredients correspond to their specific *HLB* values, shown in Table 8.5.

The criteria for defining a good emulsion of any food ingredients may include factors related to emulsion stability, clarity, viscosity, ease of preparation, palatability, and many other considerations. Using stability as a single criterion, a comparison of the required *HLB* value for a food ingredient can be made with respect to emulsifiers of the same chemical class. Figure 8.10A illustrates a typical situation in which the required *HLB* value for a food ingredient is 12.0 and two emulsifiers having *HLB* values of 12.0 and 11.0 may be available for use. In this case, the emulsifier having the *HLB* value of 12.0 would be preferred for use since it forms a more stable emulsion at a 4% concentration level than the emulsifier with an *HLB* equal to 11.0. The latter emulsifier requires an 8.0% concentration to produce emulsion stability that is still significantly less than the other emulsifier at a 4% concentration.

Added to *HLB* compatability of emulsion ingredients and the emulsifier, emulsion sta-

Table 8.5. Required HLB for O/W Emulsions for a Variety of Emulsion Ingredients (± ~1.0)

Ingredient	HBL value	Ingredient	HBL value
Acid, isostearic	15–16	Corn oil	10
Acid, lauric	16	Cottonseed oil	5–6
Acid, linoleic	16	Glycerol monostearate	13
Acid, oleic	17	Hydrogenated peanut oil	6–7
Alcohol, lauryl	14	Lard	5
Alcohol, oleyl	13	Menhaden oil	12
Alcohol, stearyl	15–16	Palm oil	16
Bees wax	9	Rapeseed oil	6
Cocoa butter	6		

SOURCE: HLB values adapted from *The HLB System, a Time-Saving Guide to Emulsifier Selection*, ICI Americas Inc., Wilmington, Del.

bility is affected by the chemical nature of the emulsifier. Assume for Figure 8.10B that an *HLB* value of 12.0 will be required to form an emulsion. If three chemical types of emulsifier blends A, B, and C are available for use at identical concentrations, that emulsifier blend of a chemical family that produces highest stability in the desired *HLB* range (A) is chosen in preference to other choices (B and C).

Calculation of HLB values: *HLB* values for most nonionic emulsifiers (e.g., polyol fatty acid esters) can be estimated from their compositional and analytical data, which includes the *saponification* number* for the ester (*S*) and the *acid number*** of the recovered acid (*A*) according to the formula:

$$HLB = 20 \left(1 - \frac{S}{A} \right)$$

**Saponification number:* Acts as a numerical estimate of the milligrams of potassium hydroxide required to "neutralize" the fatty acids contained in 1.0 g of a saponified substance. This value is a measure of free fatty acids as well as those esterified to glycerol, and it indicates the molecular size of a saponifiable substance since a high saponification number corresponds to a low molecular weight and vice versa.
***Acid number:* Provides a numerical index regarding the concentration of free fatty acid. The value is based on the number of milligrams of potassium hydroxide required to neutralize the free fatty acids in 1.0 g of sample.

EXAMPLE: Calculate the *HLB* value for TWEEN 20 polyoxyethylene sorbitan monolaurate where $S = 45.5$ and $A = 276$.

$$HLB = 20 \left(1 - \frac{45.5}{276} \right) = 16.7$$

For those cases in which the hydrophilic portion of a manufactured emulsifier consists entirely of ethylene oxide (e.g., polyoxyethylene stearates), the *HLB* formula becomes

$$HLB = \frac{E}{5}$$

where

E = weight percentage of oxyethylene content

$\frac{1}{5}$ = arbitrary value for handling smaller numbers

Calculating emulsifier ratios for a desired HLB value: The preparation of an O/W or W/O emulsion often requires the selection of an emulsifier that will ensure the production and maintenance of a stable emulsion. A single emulsifier can be selected for this purpose, but it is usually preferable to blend two compatible surfactants. One surfactant will typically have a low *HLB* value and the other will have a high *HLB* value. These two substances can be mixed to calculated proportions that provide interme-

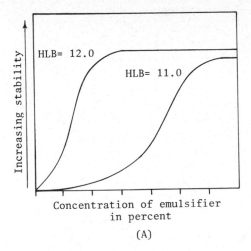

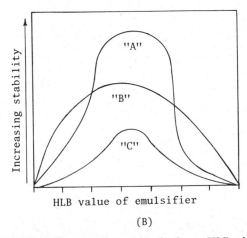

FIGURE 8.10. (A) That emulsifier having an HLB value that contributes to maximum emulsion stability at the lowest concentration level is often preferred for use as a stabilizing ingredient in a food system (consult text for details). (B) That chemical class of an emulsifier that provides the maximum degree of emulsion stability over a desired yet limited range is also preferred for use in stabilizing an emulsified food system.

diate *HLB* values suitable for application to a specific emulsified system.

An O/W emulsion involving corn oil (*HLB* = 10), for example, can be produced by using one or a combination of two emulsifiers depending on which approach will provide the optimum emulsion conditions for a specified objective. Assuming that

emulsifiers SPAN 20* ($HLB_{(B)}$ = 8.3) and TWEEN 20* ($HLB_{(A)}$ = 16.5) are available, it is often necessary to determine how much of each emulsifier ought to be mixed to achieve a necessary *HLB* value (*HLB* = 10) that will promote emulsion stability. This proportion of individual emulsifiers to each other can be determined using equations similar to following ones:

$$\% \text{ TWEEN 20} = \% \text{ (A)} = \frac{100\,(X - \text{HLB}_{(B)})}{(\text{HLB}_{(A)} - \text{HLB}_{(B)})}$$

$$\% \text{ SPAN 20} = \% \text{ (B)} = 100\% - \% \text{ (A)}$$

Substituting the appropriate numbers into these equations yields an emulsifier concentration of 20.7% for TWEEN 20 and 79.3% for SPAN 20:

$$\% \text{ TWEEN 20} = \frac{100(10 - 8.3)}{(16.5 - 8.3)} = 20.7\%$$

$$\% \text{ SPAN 20} = 100\% - 20.7\% = 79.3\%$$

This mathematical approach for calculating the percentage composition of a single emulsifier formulated by admixing two different emulsifiers can be markedly simplified by using an *HLB Computagraph* (ICI Americas, Inc., Wilmington, Delaware) (Figure 8.11). Surfactant *HLB* numbers 0.0 to 9.0 are represented on the left of the Computagraph; *HLB* numbers 9.0 to 20.0+ appear on the right; and the percentage composition (0.0–100.0%) of the surfactant mixture in terms of the "high *HLB* surfactant" (right side of Figure 8.11, TWEENS 60 to 20) is indicated across the bottom of the graph. Again, as an example, suppose that SPAN 20 and TWEEN 20 must be used to produce an *HLB* value of 10.0. Using the computagraph, draw a straight line connecting the *HLB* values for SPAN 20 and TWEEN 20. Next, draw a horizontal line across the computagraph for the desired *HLB* value (*HLB* = 10.0), and then draw a perpendicular line through the intersection of

*SPAN and TWEEN are registered trademarks of ICI Americas Inc., Wilmington, Delaware.

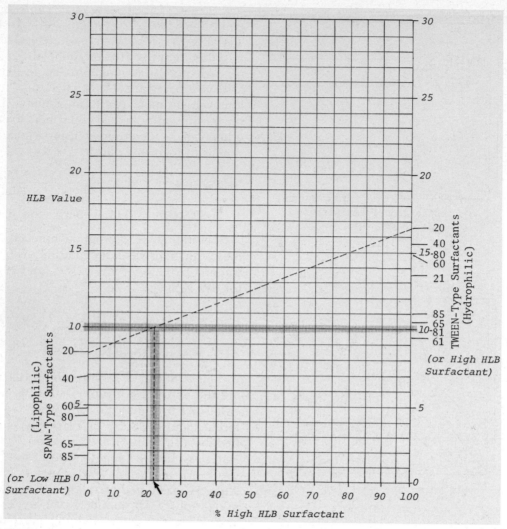

FIGURE 8.11. HLB computagraph used for calculating the ratio of emulsifiers to reach any desired HLB value. As illustrated, the graph demonstrates the calculation of an HLB = 10 surfactant mixture using the surfactants SPAN 20 and TWEEN 20. The surfactants detailed on the figure refer to specific SPAN and TWEEN surfactants manufactured by ICI Americas Inc. (registered trademarks), which have wide uses in solving emulsification problems. See text for further details. (Computagraphic grid principle, copyright, ICI Americas, 1976, and reprinted with permission from ICI Americas Inc., Wilmington, Delaware, but adapted for specific application to this textbook.)

the previous two lines. The value at the bottom of the graph indicates the percentage of the high *HLB* surfactant required—in this case, $\simeq 21.7\%$. This value should approximate the value obtained by more accurate calculations outlined above.

Emulsifier applications in foods: Emulsifiers are customarily added to foods for many technical and nutritional objectives. Technical reasons for their use often center around improving the esthetic, textural, and organoleptic attributes of foods besides ensuring a uniform distribution of certain food ingredients such as lipids, vitamins (e.g., A, D, and E), gases, soluble proteins, and so on. Figure 8.12 and Table 8.6 illustrate some common structures and uses for natural and synthesized emulsifiers in foods.

In addition to these emulsifiers, specially

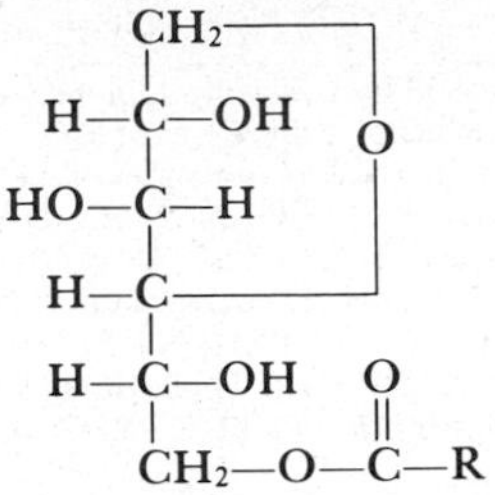

$$CH_2$$

Sorbitan ester
  Sorbitan monostearate
  RCO— = Stearic acid

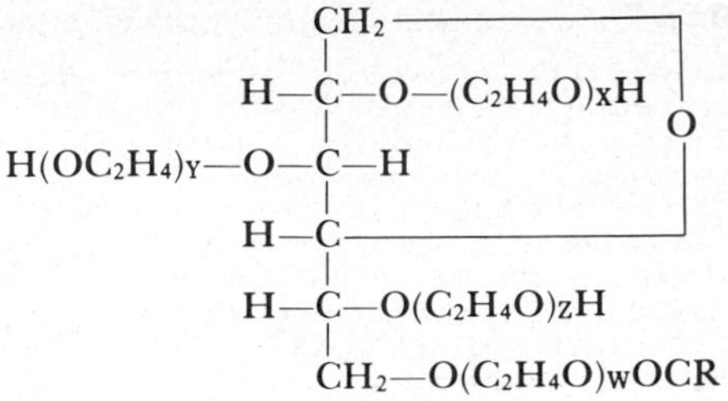

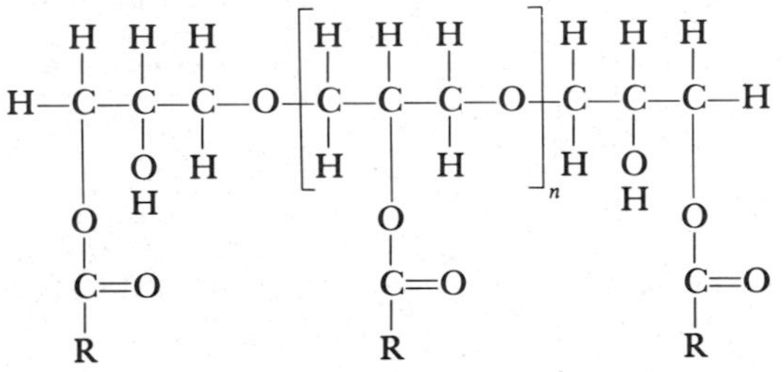

Polyglycerol esters of fatty acids
  RCO— = Fatty acid moiety
  n = number of glycerol units

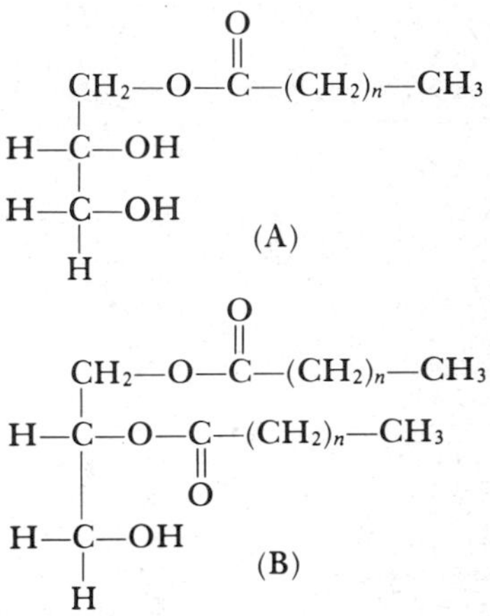

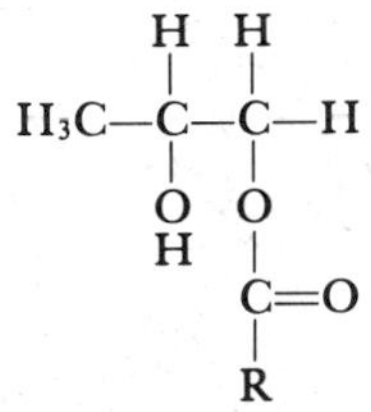

Propylene glycol esters
  RCO— = Stearic acid

Polyoxyethylene sorbitan esters
  Polyoxyethylene sorbitan monoester
  Moles of ethylene oxide = X + Y + Z =
    ~20 moles/mole of sorbitol and
    its mono-, di-, and trianhydrides
  RCO— = Stearic acid (Polysorbate 60)
  RCO— = Oleic acid (Polysorbate 80)

(A)

(B)

Mono- and diglycerides
  (A) α-(or 1-) monoglyceride
  (B) α-,β-(or 1,2-) diglyceride—when
    n = 16 the fatty acid ester is
    stearic acid

FIGURE 8.12.   Molecular structures for some natural and synthetic emulsifiers commonly used in foods.

processed malto-dextrins can provide an effective avenue for emulsifying certain food nutrients and ingredients (Amino Products, Philadelphia, Penn.). These dextrins are specially processed and spray dried to produce a totally water-soluble matrix that has a high surface area, due to large numbers of microporous channels. Many of these processed malto-dextrins can absorb or enrobe ≥300% of their weight in terms of nonaqueous liquids including natural flavoring and oleoresin spice oils, vegetable oils, melted shortening, fat-soluble vitamins, and so on. The characteristic water solubility of these dextrins, coupled with their excellent absorption and enrobing properties for nonpolar substances, permits them to act as excellent vehicles for achieving emulsified food systems that are otherwise difficult or expensive to produce.

Aside from the inherent chemical toxicity of some emulsifier-active substances, years

Table 8.6. Food Applications for Some Common Emulsifiers

At least 23 types of emulsifiers can be used for the preparation and formulation of food products. The following table is based on the use of DREWSORB (SE), DREWPONE (PSE), DREWPOL (PGE), and DREWMULSE (G), which are all *registered trademarks* of PVO International, New York, New York 10005. The respective chemical equivalents of these emulsifiers are indexed below. All application data are used with permission of PVO International as reported in *Food Emulsifiers and Specialty Products*. ALL DATA DISCUSSED HEREIN ARE BELIEVED TO BE CORRECT. HOWEVER, THIS SHOULD NOT BE ACCEPTED AS A GUARANTEE OF THEIR ACCURACY, AND CONFIRMING TESTS SHOULD BE CONDUCTED IN THE LABORATORY FOR ANY EMULSIFIER BEFORE ADOPTING ITS USE IN ANY FOOD PRODUCT OR SYSTEM. NO STATEMENT SHOULD BE CONSTRUED AS A RECOMMENDATION FOR ANY USE THAT WOULD VIOLATE ANY PATENT RIGHTS. NOTHING CONTAINED HEREIN SHALL CONSTITUTE A GUARANTEE OR WARRANTY WITH RESPECT TO THE PRODUCTS DESCRIBED OR THEIR USE.

Table Index No.	Emulsifier description	Table Index No.	Emulsifier description
SE-1	Sorbitan monostearate	PGE-9	Decaglycerol tetraoleate
PSE-2	Polysorbate 60 (polyoxyethylene (20) sorbitan monostearate)	PGE-11	Decaglycerol decastearate
		PGE-12	Decaglycerol decaoleate
		G-14	Glycerol monofat
PSE-3	Polysorbate 65 (polyoxyethylene (20) sorbitan tristearate)	G-16	Glycerol monoshortening
		G-17	Glycerol monooleate
		G-18	Glycerol monostearate
PSE-4	Polysorbate 80 (polyoxyethylene (20) sorbitan monooleate)	G-19	80% Glycerol monostearate, 20% Polysorbate 65
		G-21	80% Glycerol monostearate, 20% Polysorbate 80
PGE-7	Triglycerol monoleate		
PGE-8	Hexaglycerol distearate	G-22	Glycerol monostearate

Food product	Emulsifier type	Recommended use level (%)	Function
Cake, bakers	SE-1 and SE-2	0.30–0.35 based on weight of sponge cake batter	Produces a sponge cake with good volume, texture, and shelf-life
Beverages	PSE-2	0.03–1.0	Foaming agent
Ice cream, ice milk, and soft serve desserts	PSE-3	0.05–0.10	Provides controlled fat destabilization, dryness, and overrun
Chicken and turkey processing	PSE-4	0.2–0.5 (USDA regulation) added to scald tank water	Aids in removal of feathers
Pickle products	PSE-4	0.05	Color and flavor dispersant
Cheese spreads	G-16	0.5–1.0	Emulsion stabilizer, resists syneresis
Chocolate syrup with vegetable fat	G-18	0.1–0.6	Improves flow properties
Confectionery coatings	PGE-7	0.5–1.0	Provides gloss to coating, inhibits fat recrystallization to retain gloss
Coffee whiteners, liquid	PGE-8	0.2–0.5	Produces the proper oil-in-water emulsion and improves dispersibility
Defoaming, process foods	G-17, PGE-9, or PGE-12	0.02–0.05	Prevents foaming and reduces existing foam in sugar–protein syrups
Flavors and colors	PGE-11	0.10–1.0	Improves solubility
Ice cream, ice milk, soft serve desserts, and frozen desserts	G-19	(1) Bulk ice cream, 0.1–0.3 (2) Novelties, Mellorine, and ice milk, 0.15–0.40	Provides dryness and controlled overrun and therefore may be blended with stabilizers, sugar, etc.
Margarine	G-14	0.5	Improves emulsion stability, shelf-life, and antispattering properties
Macaroni and related products	G-22	1.0–2.5 based on the weight of the flour	Complexes with starch to prevent sticking and clumping
Sherbet	G-21	0.10–0.20	Provides aeration, ice crystal control
Salad dressing	PSE-2	0.3	Minimizes oil–water separation
Sweet doughs	G-16	0.25–0.50 based on flour weight	Retards crumb firming, improves texture and eating qualities

of consistent use indicate that most food emulsifiers are very safe for human and animal consumption at conventional usage levels. It has actually been observed in some cases that a "cocktail" of surfactants given to piglets may promote their growth rate well in excess of 10%. Reasons for this observation are debatable, but it is believed by some investigators that emulsifiers may promote enzymatic digestion of lipids in a manner similar to the bile acids. Some other surfactants not necessarily designed for food use and whose actual toxicity remains to be firmly established have been cited for their ability to chelate toxic heavy-metal ions. These substances typically include nitrilotriacetic acid (NTA), which by way of industrial or domestic pollution sources could conceivably transport heavy metals into the dietary regimen or food process water of humans and animals. This possibility has special significance where this surfactant or others may be present in renovated wastewater resources.

Foams

A dispersion of gas bubbles in a liquid or semisolid phase is called a *foam*. The individual gas bubbles are separated by liquid films having a thickness of several centimeters to ≤1.0 nm. Foams are responsible for the textural properties of nearly all baked products, pastries, ice cream, whipped creams, marshmallows, meringues, and the foam "heads" on alcoholic or nonalcoholic beverages.

Foam formation: A gas can be dispersed in a liquid by using suitable mechanical beaters or by bubbling the gas through a large number of small orifices into the liquid. In either case, the actual formation of the foam occurs when the interjected gas volume expands the volume and surface area of the liquid. Wherever a foam is produced, the amount of work expended corresponds to the product of the foam area and the surface tension. Depending on the concentration of the foam, the gas bubbles may assume spherical or polyhedral shapes. For dilute foams, spherical shapes predominate whereas concentrated foams show polyhedral shapes with the individual gas bubbles being separated by very thin liquid films that are easily deformed by mutual pressures from adjoining bubbles (Figure 8.13).

Foam stability: Foams are characterized by their elastic properties as well as their rigidity. Some foams are very delicate and break down under infinitesimal physical stresses, whereas other foams must be cut like solids.

In general, foams are thermodynamically unstable since their collapse is accompanied by a decrease in total free energy. The collapse of most foams often involves drainage of the liquid phase as a consequence of the film-like walls separating bubbles. This behavior of a foam is encouraged by many factors including gravitational forces, a suction effect at the periphery of the film due to the high curvature of individual gas bubbles, evaporation of the liquid phase by mechanical and/or thermal agitation, and collision forces occurring within the gas bubbles. The thinning of the liquid film component of gas bubbles is checked briefly when the film reaches a thickness of ∼1000 Å. At this point electrostatic repulsion of the film charges often becomes an effective force preventing further thinning and prevents drainage of the liquid phase.

The stability of a foam can be enhanced by the addition of a foaming agent (foamer). These agents are termed *amphiphilic* because they are substances (e.g., proteins) capable of adsorbing to the wall of a gas bubble and increase the viscosity of the liquid phase. This action reduces liquid drainage from the foam in addition to reducing gas permeability from bubbles.

Foam stability can also be enhanced by increasing the elasticity of gas bubble walls. The stability of a film is determined by elasticity of the film described in terms of the equation

$$E = 2A \frac{(d\gamma)}{(dA)}$$

where

E = elasticity
A = surface area
γ = surface tension

For this equation, the elasticity term is related to changes in surface tension caused in response to the deformation of film walls.

Stretching of a film containing a foaming agent leads to a decrease in the concentration of the foamer over the surface of the

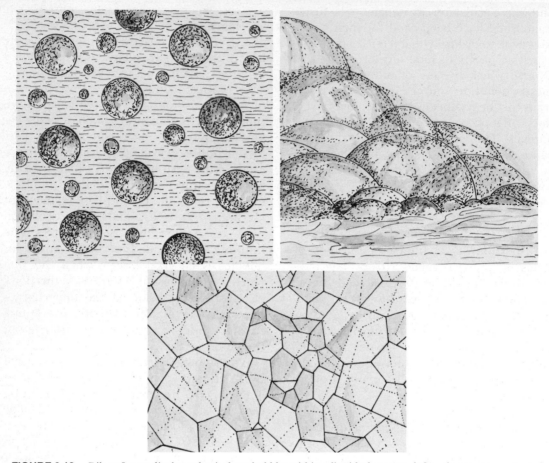

FIGURE 8.13. Dilute foams display spherical gas bubbles within a liquid phase (top left), whereas concentrated surface foams display hemispherical shapes on the top of the bubble (top right), and the internal reaches of concentrated foams display polyhedral shapes because of multiple spatial interactions with neighboring bubbles (bottom middle).

film. In many instances this effect causes increased surface tension and a counteraction to external forces, and further extension of the film is inhibited. In a similar vein, film contraction corresponds to a decrease in surface tension, an increase in potential surface volume of the liquid phase, and a tendency of the foam to oppose further contraction.

From a historical perspective, many agents, such as cobalt compounds in the brewing industry, have been used to stabilize the foaming properties of foods. The inherent toxicity of many early foam stabilizers inevitably led to the development and use of similarly effective, nontoxic substances, gums,

sugars, and proteins. When used at concentration levels of ~0.5 to 2.0%, these agents increase the viscosity of the foam and prevent drainage of the liquid phase. Other substances such as finely pulverized spices, cocoa, primary alcohols, glycerol, and glyceryl ethers along with a host of other compounds are now conventionally used to improve foam life by increasing the surface viscosity.

The longevity of foams can also be improved when proteins involved in air–water interfaces (e.g., egg white-based meringues) become denatured. Mechanical denaturation used for establishing these foams causes an uncoiling of the protein structure and a concomitant orientation of the denatured protein's hydrophobic groups toward the air while the hydrophilic moieties interact with the aqueous phase of the foam. Denatured proteins oriented in this fashion then undergo aggregation and effectively increase the viscosity of the foam. Proteins at their appropriate p*I* values can similarly promote the stability of foams.

Many food substances are characterized by their foam characteristics, but the presence of foams can be detrimental to the expeditious formulation and processing of certain foods including maple and corn syrups, fermentation products, fruit juice concentrates, vegetable oils, tea or coffee extracts, and many other products. These obnoxious foams can be largely eliminated by *antifoaming agents,* which cause a rapid collapse of the foam and inhibit future recurrence of the foam. The effects of these agents stem from their tendency to undergo monolayer spreading, which destabilizes the foam film. Furthermore, these agents often decrease the surface tension of the foam's liquid phase to a point where bubble walls are so thin that they burst.

Water-insoluble silicone oils (e.g., dimethylpolysiloxanes) have been widely used for their antifoaming properties at concentration levels of 10 to 100 ppm, but long-chained primary alcohols, amides, phosphate esters, fatty acids, or fatty acid esters may be used as defoamers depending on the nature of the foaming agent and application problem.

OPTICAL AND KINETIC PROPERTIES OF COLLOIDS

Colloidal particles are in a state of ceaseless random motion. This movement is caused by the unending bombardment of the colloidal particles by molecules of the solvent medium, called *Brownian motion.* In addition to this feature, many of the most important physical properties of colloidal systems involve phenomena of light scattering, osmosis, diffusion, rheological behaviors, and directional migrations of particles in gravitational or electrical fields.

Light Scattering

When a beam of light passes through a colloidal dispersion (e.g., sol, aerosol, emulsion), its path is illuminated in a darkened environment because the particles scatter the light in an omnidirectional pattern. This behavior, usually recognized as the *Tyndall effect,* is often evident as the emergence of the transmitted light beam at a 90° angle to the path of the incident beam.

As early as 1871 Rayleigh made a mathematical effort to define the light-scattering phenomenon of isotropic particles whose dimensions were small compared to the wavelength (λ) of an incident light beam (I_0). During these studies it was noted that the intensity of the light (I_θ) scattered by a particle through an angle of θ depends on the intensity of the incident light, the light path distance (γ_s) through the light scattering volume, and the polarizibility (α) of the particle. For unpolarized light the equation is

$$R_\theta \left(1 + \cos^2\theta\right) = \frac{I_\theta \gamma_s^2}{I_0} = \frac{8\pi^4}{\lambda^4}\left(1 + \cos^2\theta\right)$$

The $R_\theta \left(1 + \cos^2\theta\right)$ term is called the *Rayleigh ratio.* This equation mathematically substantiates the observation that light scattering is influenced by wavelength of the incident light beam. Indeed, light scattering in the visible spectrum ($\sim$400–700 nm) is nearly ten times more intense for shorter wavelengths (violet, blue) than longer ones (yellow, red). This phenomenon alone causes a dispersed particulate system irradiated with white light to appear blue by scattered light and redder by transmitted light. This wavelength dependence of scattering intensity is called *opalescence.* Dispersions of very small molecules show little opalescence since I_θ depends on the square of the particulate size, whereas coarse particles fail to show opalescence because light is scattered by reflection.

Opalescence has a significant visual effect for certain foods, especially high-protein liquids such as diluted milk products. The turbid nature of dilute proteinic lacteal fluids causes diffusion of short light wavelengths, whereas preferential transmission of longer wavelengths makes the diluted milk appear pink.

The total omnidirectional scattering of incident light by a system of dispersed particles can be measured by its *turbidity* (τ). Turbidity can be defined in mathematical terms similar to Beer's law where

$$I_\tau = I_0 e^{-\tau L}$$

where

I_τ = intensity of transmitted light
I_0 = intensity of incident light
L = length of the light path
τ = turbidity

The turbidity term is related to the Rayleigh ratio for unpolarized light by the expression

$$\tau = 2\pi \int_0^\pi R_\theta \sin\theta \, d\theta = (16\pi/3)\, R_{90}$$

where

R_{90} = value of Rayleigh ratio at $\theta = 90°$.

The application of the Rayleigh theory was extended to macromolecular solutions by Peter Debeye in 1947. For "very dilute" solutions of particles having sizes smaller than the wavelength of incident light, the expression would be

$$R_\theta = \frac{2\pi^2 \, \eta_0^2 \, [(\eta - \eta_0)/C]^2}{L \, \lambda^4} \, CM = K_\theta CM$$

where

η = refractive index of the solution
η_0 = refractive index of the solvent
C = concentration of solute in terms of mass per unit volume
M = molecular weight
L = Avogadro's number
$K_\theta = \dfrac{2 \, \pi^2 \eta_0^2 \, [(\eta - \eta_0)/C]^2}{L \, \lambda^4}$

In those instances when concentrated polydispersed systems exceed a size of $\lambda/20$,

the foregoing equation can be modified to account for differences in molecular sizes and shapes:

$$K_\theta \frac{C}{R_\theta} = \frac{1}{\overline{M}_w P_\theta} + 2\, BC$$

where

$\overline{M}_w$ = average mass of molecular weight
P_θ = particle-scattering factor that corrects for scattering from different parts of the same molecule
B = an empirical "second virial coefficient" (Popiel, 1978)
C = concentration of solute in terms of mass per unit volume

These mathematical relationships serve as keystones for graphic determinations of the mean molar mass ($\overline{M}_w$) based on light scattering at a specified θ through solutions of known concentrations (C). Data recorded for such observations are plotted as a *Zimm plot* (Figure 8.14). These various data points are extrapolated to zero concentration and an angle of zero degrees. The ordinate inter-

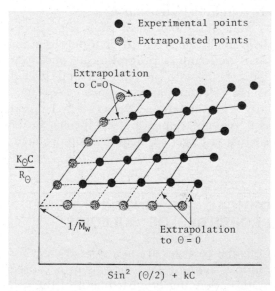

FIGURE 8.14. A typical Zimm plot showing the double extrapolation for $C = 0$ and $\Theta = 0$; the k term represents an arbitrary constant used for spreading out data points in order to facilitate graphing. A graph of this type permits calculation of average molecular weight (mean molar mass) for an unknown substance based on light-scattering measurements.

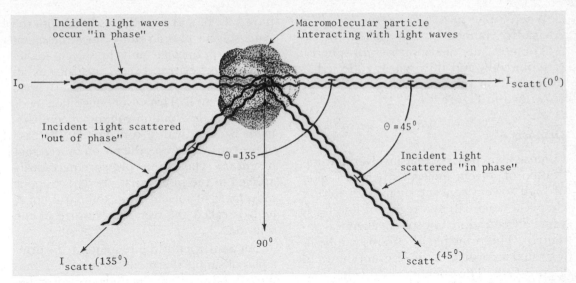

FIGURE 8.15. Phase relationship between incident (I_0) and scattered light (I_{scatt}) waves. The incident waves may occur in coherent phase after interacting with a macromolecule $I_{scatt}(0°)$ or $I_{scatt}(45°)$; or the light may display "out of phase" scattering at $I_{scatt}(135°)$.

cept ($1/\overline{M}_w$) is then used to accurately compute the value of $\overline{M}_w$.

Light scattering and particle shapes: A beam of light undergoes scattering if it is directed toward a solution of very large molecules whose sizes approach the wavelength of the incident light. Some parts of the molecule scatter light according to a pattern of destructive interference; that is, the scattered wavelengths of light cancel each other. In other parts of the molecule, however, light is scattered in a pattern of constructive interference, where the wavelengths undergo

mutual reinforcement. Figure 8.15 shows some typical light-scattering events.

Light-scattering measurements can be a valuable asset for experimentally determining the shapes of molecules. These shape determinations principally depend on calculating a dissymmetry ratio between the light scattered at some forward angle θ and that light scattered at its supplementary angle $180 - θ$ ($I_θ/I_{180-θ}$) (Figure 8.15). Of all the possible light-scattering angles that may be surveyed, the angles of 45° and 135° often serve as common reference points (Figure 8.16).

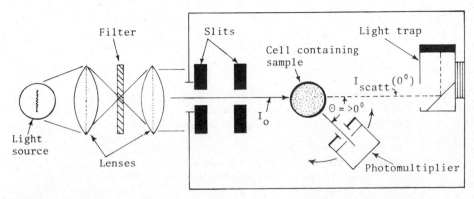

FIGURE 8.16. Schematic diagram for a basic instrument used to measure light-scattering phenomena by molecules.

When a plot of $I_\theta/I_{180-\theta}$ versus L/λ ($L =$ Avogadro's number) is constructed, the extent of molecular dissymmetries for spheres, random coils, and thin rods is reflected as differences in plotted curves such as those illustrated in Figure 8.17.

Osmosis

Osmosis is a process that involves the migration of solvent molecules from a region of high solvent concentration (high solvent chemical potential) to a region of lower solvent concentration (low solvent chemical potential). The principle of osmosis can be illustrated according to the diagram shown in Figure 8.18. The container is divided into two compartments by a semipermeable membrane and each compartment (A and B) has a vertical tube, one of which contains an inverted piston. Compartment A contains pure solvent and the other is filled with the same solvent plus a solute. Since the membrane is permeable only to solvent molecules, the relative excess of solvent in compartment A spontaneously enters compartment B. This solvent migration tends to increase the solution volume of compartment B along with a concomitant increase of pressure that raises the piston by a certain height (h). The counterpressure necessarily applied to the piston that prevents osmotic movement of solvent from compartment A to B is called the osmotic pressure of the solution.

For a sufficiently dilute solution, the principles of simple thermodynamics can be employed for describing osmotic pressure phenomena:

$$\pi = \frac{c\,RT}{M}$$

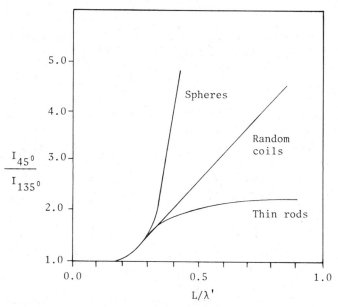

FIGURE 8.17. Relationship between light-scattering data and molecular shapes, where L corresponds to the length of a thin rod, the diameter of a sphere, and the root mean square between the ends of a randomly coiled molecule, and λ' is the wavelength of incident light on the particle divided by the refractive index of the medium. Molecular dissymmetry will be most pronounced where light interacts with the greater pairs of light-scattering points on a molecule, while the light scattering for the smallest approaches unity (i.e., Rayleigh scatter is symmetrical).

where

π = osmotic pressure
T = absolute temperature
M = molar mass
c = mass of solute per unit volume (g/L)
R = universal gas constant

This mathematical expression for osmotic pressure can then be rearranged for practical applications so that the molecular weight (M) of solutes can be estimated:

$$M = \frac{c\,RT}{\pi} \quad \text{or} \quad \frac{\pi}{c\,RT} = \frac{1}{M}$$

Measurement of osmotic pressures for a series of solute concentrations at a given temperature can act as a basis for a plot of ($\pi/c\,RT$) versus c (Figure 8.18). Extrapola-

tion of the linear plot to a concentration of zero results in an ordinate intercept of ($1/M$) that permits calculation of the molecular weight for the macromolecular solute.

Apart from the use of osmotic pressure for estimating molecular weights of certain solute molecules, the principle of osmotic pressure has great biological significance for living cells. Since very small concentrations of solutes contained on one side of an elastic semipermeable biological membrane can generate large pressure differences between the outside and the inside of a cell, osmotic pressure notably accounts for the cell turgor of plant cells. The elastic membranes of a living plant protoplast are semipermeable to water, but they are quite selective in terms of solute (sugars, organic acids, etc.) transfer across the membrane. This property con-

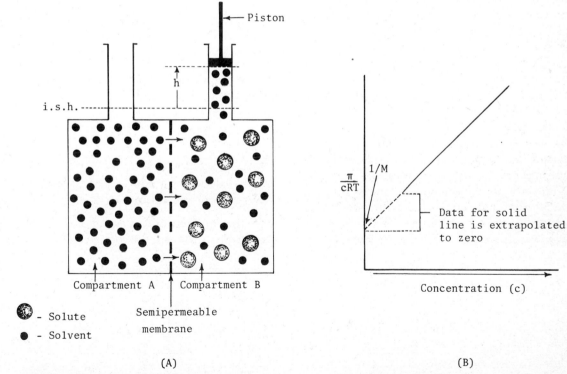

(A)

(B)

FIGURE 8.18. (A) Solvent migration across a semipermeable membrane during osmosis. The increased migration of solvent from compartment A to compartment B drives the piston upward a distance of h above the initial solvent height (i.s.h.) of the tube leading from compartment A. (B) Graphic illustration of osmometry data for π/cRT plotted versus solute concentration (c). Extrapolation of experimentally acquired data (solid line) toward an infinite dilution permits calculation of $1/M$, where M is the molecular weight for a particular substance.

tributes to a sustained turgid state within plant cells for as long as they live. So long as osmotic pressure exists within individual cells, plant cells can collectively exert enough pressure to push germinating seedlings through hard-pack soil; leaves on established plants remain fully spread; and fruits or vegetables remain plump, succulent, and crisp.

Osmometry and the Gibbs–Donnan effect:

The use of osmometry for determining the molecular weights of colloidal electrolytes such as proteins may be subject to certain errors caused by the Gibbs–Donnan effect. This effect describes an equilibrium condition that leads to an unequal distribution of ions across a semipermeable membrane (Figure 8.19).

When a container partitioned by a semipermeable membrane contains a colloidal electrolyte in compartment A, such as sodium proteinate (Na^+P^-) at a concentration of C_1, while compartment B contains sodium chloride (Na^+Cl^-) at an equimolar concentration of C_2, the protein anion (P^-) will be trapped in compartment A due to its size and inability to pass through the membrane. The sodium and chloride ions, on the other hand, freely migrate across the membrane. An initial amount (X) of chloride ions passes across the membrane into compartment A as directed by the concentration gradient (Figure 8.19). This action encourages a corresponding migration of sodium ions (Na^+), which attempt to maintain electrical neutrality in the system. Finally, the newly cre-

ated sodium ion concentration in compartment A favors gradient migration of sodium ions back toward compartment B. This exchange and interplay of ions across the membrane is regarded as the Gibbs–Donnan effect.

Under idealized conditions of constant temperature and pressure, a change in free energy (ΔG) occurs in the system when n moles of sodium chloride have migrated across the semipermeable membrane:

$$\Delta G = d\,n\,RT \left(\ln \frac{[Na^+]_A}{[Na^+]_B} + \ln \frac{[Cl^-]_A}{[Cl^-]_B} \right)$$

At equilibrium conditions when $\Delta G = 0$, this relationship can be simplified for monovalent ions into

$$[Na^+]_A\,[Cl^-]_A = [Na^+]_B\,[Cl^-]_B$$

or

$$x\,(C_1 + x) = (C_2 - x)^2$$

Solving for x, the equation can be transformed into

$$x = C_2^2/(C_1 + 2C_2)$$

Where polyvalent ions are involved, the relationship is described by

$$\left(\frac{[anion]_A}{[anion]_B} \right)^{1/z_a} = \left(\frac{[cation]_B}{[cation]_A} \right)^{1/z_c}$$

where

z_a = valence of anion
z_c = valence of cation

Since the equilibrium concentrations for ionic species are different on both sides of the

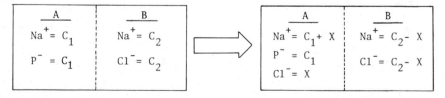

A	B
$Na^+ = C_1$	$Na^+ = C_2$
$P^- = C_1$	$Cl^- = C_2$

Initial solute distribution across a semipermeable membrane (------)

A	B
$Na^+ = C_1 + X$	$Na^+ = C_2 - X$
$P^- = C_1$	$Cl^- = C_2 - X$
$Cl^- = X$	

Solute distribution across a semipermeable membrane after a period of equilibration

FIGURE 8.19. Illustration of the Gibbs–Donnan effect.

membrane, a membrane potential is formed that can be measured by a potentiometer.

The principle of the Gibbs–Donnan equilibrium can contribute to large errors in determining the molecular weight of proteins due to the osmotic influence of ions that have migrated across the semipermeable membrane. In many cases the influence of this error can be minimized if the concentration of the colloidal electrolyte (C_1) is low relative to salt concentration present on both sides of the membrane.

Diffusion

Colloidal systems can display a pattern of behavior in which particles or molecules flow from a region of high concentration to a region of lower concentration until an equilibrium is achieved. This kinetic behavior is generally called *diffusion*.

The diffusion rate for a substance is characterized by its diffusion coefficient (D) according to Fick's Law. According to this law, the amount of solute (dS) diffusing across an area (A) during a prescribed period of time (dt) is proportional to the concentration gradient dc/dx at that point:

$$\frac{dS}{dt} = -DA\left(\frac{dc}{dx}\right)$$

Note: dc = change in concentration over a distance x.

The *minus sign* denotes that diffusion is in the direction of a lower concentration while the coefficient of diffusion is a function of the frictional resistance brought about by the solvent medium's kinetic activity in addition to the size and shape of the diffusing molecule. A high frictional coefficient (f) is associated with low rates of diffusion because the value of the diffusion coefficient is related to the frictional coefficient by

$$D = \frac{kT}{f}$$

where

k = Boltzmann's constant
T = absolute temperature

A combination of the foregoing equation with the expression for Stoke's Law where $f = 6\pi\eta r$ yields a combined equation:

$$D = \frac{kT}{6\pi\eta r}$$

where

r = radius of a spherical particle
η = viscosity of the dispersed medium

Solving the foregoing equation for the radius of a spherical particle leads to the equation

$$r = \frac{kT}{6\pi\eta D}$$

Since the volume of a spherical particle is $4/3\ \pi r^3$, equal to $M\bar{v}/L$, where M is the molar mass, $\bar{v}$ is the partial specific volume ($1/\rho$) of the particle, and L is Avagadro's number, the radius of a spherical particle is expressed as

$$r = \left(\frac{3M\bar{v}}{4\pi L}\right)^{1/3}$$

Equating the preceding equation with $kT/6\pi\eta r$ (above) and solving for M, it is possible to calculate the molar mass for a particle if the diffusion coefficient is known:

$$M = \left(\frac{4\pi L}{3\bar{v}}\right)\left(\frac{kT}{6\pi\eta D}\right)^3$$

Aside from calculating the value for M, the shapes of particles or molecules can be postulated if the value for D is known. Using the equation $D = kT/f$ allows calculation of the frictional coefficient (f), and assuming that a particle is spherical in shape, a value of f_0 can be obtained from Stokes relation where $f_0 = 6\pi\eta r$. When the ratio of f/f_0 is unity, the particle shape is likely to be spherical; for other shapes the f/f_0 ratio will be correspondingly different.

Rheology

When certain forms of matter are subjected to an applied pressure at a specified temperature, they may exhibit spatial deformation and flow patterns over a period of time. Studies dealing with the deformation and flow of matter fall into the realm of rheology. Rheological properties of matter involve many complex concepts; however, the present discussion will deal only with principles of viscosity. Viscosity has special importance for colloidal food systems because it is responsible for characteristic textural and behavioral properties of many natural (honey, molasses, cane, or maple syrups; butter; chocolate), formulated (mayonnaise, salad dressings), or processed (marshmallow cream) dietary constituents.

Viscosity: Forces of internal friction that cause certain substances to resist flow cause viscosity. Consider a layer of liquid between two plates, P_1 and P_2, each having an area (A) of 1 cm^2 and a 1.0-cm distance between them. A force of some type is required to maintain plate P_1 at a constant velocity (V) with respect to P_2 (Figure 8.20).

For conceptual purposes it is presupposed that each plate has an adherent layer of liquid whose velocities are each "zero" with regard to their respective plates (P_1 and P_2). The next layer of liquid moves slowly with respect to the previous fixed layer of liquid, the third layer of liquid moves slowly over the second, and so on. The relative variation in speeds of one layer with respect to another produces a continuous deformation of the liquid so long as some type of force is applied to the system. If the thickness (S) of the liquid between surfaces P_1 and P_2 is increased, application of any equivalent force produces a greater speed, and the velocity (V) will be proportional to S (i.e., $V \propto S$). If the area of one plate is increased over the other, there is a corresponding decrease in speed and the velocity (V) will be inversely proportional to A [i.e., $V \propto (1/A)$]. In terms of force (F), an increase in F produces a proportional increase in velocity (i.e., $F \propto V$). Accordingly, the observed velocity is proportional to F, S, and $1/A$ for a system having a prescribed viscosity (η):

$$(1) \ V = \frac{FS}{\eta A} \qquad (2) \ \eta = \frac{FS}{AV} \qquad (3) \ F = \eta \frac{AV}{S}$$

The second equation here expresses the principle that fluid friction is measured in terms of its coefficient of viscosity (η) (the Greek letter *eta*), which can be defined as the ratio of the tangential force per unit area of surface (F/A) to the velocity gradient (V/S) between two planes of fluid in a laminar flow (Newtonian flow).

Since the liquid between plates P_1 and P_2 ideally consists of a series of discrete individual layers moving relative to each other with different speeds when a shear force, F, is applied to A, incremental expressions for V and S at each individual layer can be designated as dv_1/ds_1, dv_2/ds_2, dv_3/ds_3, and so on ... *ad infinitum.* For this presentation only, assume that the total effect of n individual layers can be indicated as an overall velocity gradient dv_n/ds_n, which is designated as the "rate of shear." Consequently, the expression of the *shear force* is

$$F = \eta \, A \, \frac{dv_n}{ds_n}$$

Measurement of viscosity. The viscosities of liquid substances can be measured by using an Ostwald viscosimeter. This instrument consists of two bulbs connected by a U tube where one side of the U tube has a very narrow bore. A standard quantity of liquid to be studied is placed in the viscosimeter, which is held in a thermostatically controlled bath, and it is subsequently drawn up to the upper bulb (Figure 8.21).

The time (t) is then recorded for the liquid meniscus to fall between the two graduation marks above and below the upper bulb as a result of its own weight. The same type of measurement is then conducted with an equal volume of a reference liquid such as water whose viscosity (η) and density (ρ)

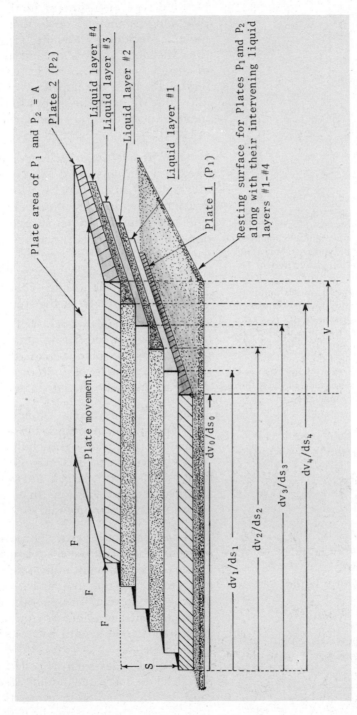

FIGURE 8.20. Two plates designated P_1 and P_2 having identical surface areas (A) of 1 cm² are separated by an intervening layer of liquid (S) having a thickness of 1 cm. The application of a force (F) to P_2 displaces it, along with its underlying liquid layers 1–4 from an initial resting position indicated at dv_0/ds_0. The respective individual rates of velocity for each subdivision of the intervening liquid layer are each designated dv_1/ds_1, dv_2/ds_2 . . . , and so on, depending on the number of idealized layers, and so long as a force (F) is applied to P_2. The overall net movement of P_2 during the period of applied force is indicated as V.

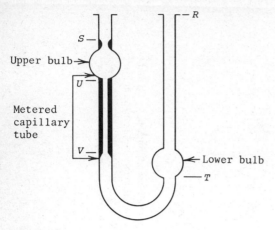

FIGURE 8.21. Ostwald viscosimeter. Many classical viscosity measurements involve the use of an Ostwald viscosimeter, which acts as an instrument for determining the time required for a specific volume of liquid to flow through a metered length of capillary tubing. The sample liquid to be measured is introduced into the U-shaped viscosimeter at point R, and after it has been equilibrated to a specific temperature by a surrounding bath, the sample is sucked up into the left side of the instrument so that it fills the upper bulb. At this stage, the meniscus is at point S while the lower meniscus is at point T. The time is then carefully recorded for the flow of the liquid sample through the capillary tube over the length of U to V. Increasing liquid viscosity corresponds with increasing transit time through the measured length of the capillary tube.

are known. These data then permit calculation of the viscosity for the first liquid by

$$\eta = \frac{\rho t}{\rho_0 t_0} \eta_0$$

where

ρ_0 = density of water
ρ = density of unknown liquid
η_0 = viscosity of water*
η^* = viscosity of unknown liquid
t_0 = time for water
t = time for unknown liquid

The Ostwald viscosimeter is usually defined as a *kinematic viscosimeter* because the

*The unit of viscosity measurement according to cgs system is the *poise*—the term originates from an early French investigator named Poiseville who developed methods for viscosity measurement. The eta term (η) defines the ratio of the shearing stress per square centimeter to the velocity gradient produced by it.

viscosities of substances are defined as viscosity divided by density. Exclusive of the Ostwald-type viscosimetry method, many different commercial or technical viscosimeters, including torsion viscosimeters (MacMichael, Brookfield) and others not discussed here (Stormer, electroviscosimeters), have been developed over the years. The MacMichael and Brookfield instruments indicate unknown viscosities of liquids in two different but related ways.

1. *MacMichael viscosimeter.* A liquid is rotated at a constant speed in a cup in which a cylinder is suspended by a torsion wire. Friction generated by the rotating liquid against the immersed yet suspended cylinder causes twisting of the attached torsion wire. A record of the number of degrees turned by the twisted torsion wire then serves as a measure of the torque of the wire and viscosity of the sample being studied.

2. *Brookfield viscosimeter.* Torque developed by a calibrated spring on a spindle is measured as the spindle turns at a constant speed in a sample of unknown viscosity. Readings are made directly in centipoise.

Torsion viscosimeters similar to those described above are especially useful for evaluating food products including catsup, creamstyle corn, custards, mayonnaise, mustards, salad dressings, pea slurries, starch, and hydrocolloid solutions that cannot be accurately studied by Ostwald-type methods.

Viscosity of food systems: The viscosity observed in many food systems is largely caused by the concentration of individual sugars or macromolecules that interact with other solutes or the solvent. In general, very high sugar concentrations increase the viscosity of an aqueous food system because they are hydrophilic.

Macromolecules such as coiled proteins may lead to increased liquid viscosity because of their ionization properties in the aqueous environment in which they exist. These protein–solvent interactions lead to an eventual uncoiling of the polypeptide and an increasingly high concentration of linear macromolecules. Linear protein structures such as these become physically entwined or cohesively bound by electrical interactions. Similar behaviors also govern the viscosity behavior of starches and pectins.

The influence of temperature critically affects the viscosities of colloidal systems that include protein, carbohydrate, or polysaccharide solutes. In general, low concentrations of colloidal substances and high temperatures correspond with low viscosities. Tapioca starch and alginates such as carrageenan (Figure 8.22) demonstrate the typical dependence of viscosity on concentration or temperature.

Characterization of macromolecules using viscosity measurements: The viscosity of macromolecular solutions largely mirrors the individual molecular size and shape of a solute or particle. As early as 1906, Einstein derived an expression relating viscosity to molecular shapes and sizes, but certain assumptions and hydrodynamic considerations were necessarily established. These assumptions required that the formula be applicable only to dilute suspensions of rigid, uncharged, wettable particles, which are notably larger than the molecules of the medium suspending the particles:

$$\eta = \eta_0 \, (1 + K \, \Phi)$$

where

η = viscosity of the solution
η_0 = viscosity of the pure solvent
Φ = fraction of the total volume occupied by suspended particles
K = 2.5

According to Einstein's investigations dealing with spherical particles and dilute lyophobic sols having spherical dimensions, the K value for the mathematical relation was found to be $\simeq 2.5$. A K value of >2.5, on the other hand, has been interpreted as an indication of elongated particles or highly solvated lyophilic sols that have larger sizes and different shapes than the idealized Einstein particle due to solvation or other phenomena.

About 32 years after the presentation of Einstein's equation, Shima further modified the mathematical relationship so as to effectively deal with ellipsoidal particles. The application of these viscosity relationships to molecular shapes and sizes are notably responsible for many current concepts of protein molecules (Figure 8.23).

For practical considerations, molecular weights of $\geq 30,000$ have customarily been determined for macromolecules from viscosity data based on semiempirical formulas such as

$$[\eta] = KM^{\alpha}$$

where

$[\eta]$ = limiting viscosity number derived from the equation developed by Shima for ellipsoidal particles:
$$\lim_{C \to 0} \left[\frac{(\eta/\eta_0 - 1)}{C} \right]$$
C = particulate concentration (g/mL)
M = molecular weight
K and α = empirical constants

According to this expression, the logarithm of $[\eta]$ should respond as a linear function of molecular weight for a particle when graphed. In the case of coiled macromolecules, the α term typically falls between 0.5 and 1.0, whereas rod-shaped molecules having a fixed diameter display an α value of $\simeq 2.0$. Globular molecules such as proteins typically display an α value equal to zero.

Newtonian versus non-Newtonian systems: When a fluid system is in motion, it demonstrates a viscosity or resistance to shear that can be studied in terms of internal friction. These viscous properties for many liquids have been described in terms of the equation $F = \eta A \, (dv/dx)$, which was first developed by Isaac Newton. Fluids that obey this equation are described as systems that display *Newtonian flow*. Many concentrated

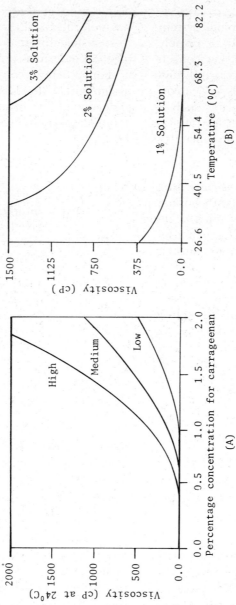

FIGURE 8.22. (A) Representative curves for high-, medium-, and low-viscosity carrageenans with respect to various concentration levels. (B) Viscosity of 1, 2 and 3% aqueous concentration levels of carrageenan at different temperatures.

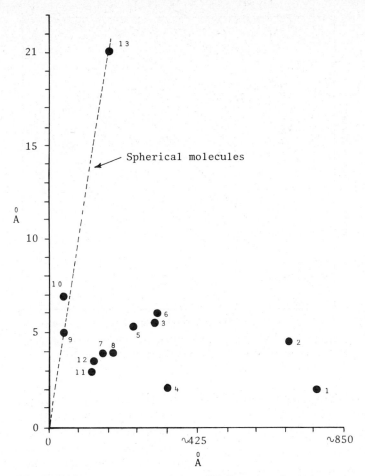

FIGURE 8.23. Approximate relative axial dimensions for various proteins where the width of the molecule as seen in projection is represented on the ordinate and the molecular length is represented on the abscissa (Å = angstrom units). When both molecular axes are equivalent, spherical molecules exist as shown by the dotted line (–––), whereas all other plotted points are indicative of three-dimensional ellipsoidal molecules. Graphic key (protein name/molecular weight): (1) gelatin/350,000; (2) fibrinogen/400,000; (3) edestin/310,000; (4) zein/50,000; (5) γ-globulin/156,000; (6) α$_1$-lipoprotein/200,000; (7) albumin/69,000; (8) β$_1$-globulin/90,000; (9) insulin/36,000; (10) hemoglobin/68,000; (11) egg albumin/42,000; (12) β-lactoglobulin/40,000; and (13) β$_1$-lipoprotein/1,300,000.

colloidal systems such as food gels, pastes, and creams that contain asymmetric particles, however, fail to obey these Newtonian principles of fluid flow and are described as *non-Newtonian systems.*

Non-Newtonian fluids can be subdivided into five discrete patterns of non-Newtonian flow based on shear rate and shear stress behavior, which include Bingham plastic, pseudoplastic, dilatant, thixotropic, and rheoplastic flows. As illustrated in Figure 8.24, Newtonian fluids have a straight-line relationship between shear rate and shear stress that is wholly independent of time. The in-

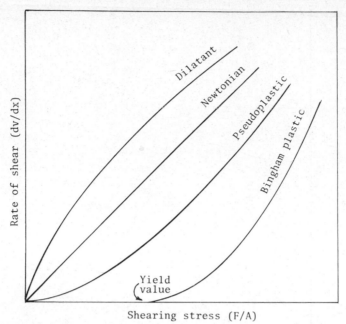

FIGURE 8.24. Notable graphic differences for the rates of shear and shearing stress among systems displaying Newtonian and non-Newtonian behaviors.

dividual non-Newtonian systems are defined by the following considerations:

Dilatant. The shear rate increases more rapidly than in the case of Newtonian systems and the viscosity tends to increase with an increase in shear stress.

Pseudoplastic. The observed viscosity of these non-Newtonian systems decreases with an increase in shear stress.

Bingham plastic. Initial resistance to flow, called *yield stress* must be overcome before flow starts; after flow begins, it follows the same pattern of behavior as Newtonian systems.

Thixotropic. An undefined period of time is required for viscosity of a colloidal system to decrease to a constant value for a certain shear stress. After succumbing to the shear stress, a reversible gel—sol transformation occurs and the system reestablishes its former viscosity on standing.

Rheopexic. Shows converse behavior of thixotropic systems in that a time-dependent increase in viscosity occurs during gentle agitation.

In the most general terms for food systems, very dilute hydrophobic and hydrophilic sols exhibit viscosities similar to water and show Newtonian behavior. Non-Newtonian systems have few unifying features, although they do occur widely throughout many common food systems (Table 8.7). The behavior of non-Newtonian systems largely

Table 8.7. Occurrence of Newtonian and Non-Newtonian Systems in Foods

Type of System	Occurrence
Newtonian	Maple syrup, corn syrup, broth and bouillon soups, homogenized and skim milks, carbonated beverages
Non-Newtonian	
Dilatant	Honey containing dextran impurities
Pseudoplastic	Gelled desserts and puddings
Bingham plastic	Chocolate, butter, cheese, icings, spreads
Thixotropic	Mayonnaise, catsup
Rheopexic	Beaten egg white, whipped cream

depends upon particle size, shape, topology, and solvation by the dispersing medium.

Concluding remarks on viscosity: Viscosity can be operationally defined as the internal friction of a liquid substance or, alternatively, as the resistance experienced by molecules as they move around the interior of a substance owing to the presence of intermolecular forces. Viscosity properties of foods have special significance as a matter of academic interest, but more importantly, measurement of food viscosities may be used as:

1. A quality control or engineering criterion for monitoring specific stages of food processing for certain foods.

2. An index of jellying ability, gel strength, emulsifying ability, and other important physical characteristics tied to the behavior of pectins, proteins, gums, and so on.

3. An evaluation method for moisture content in syrupy foods (e.g., honey), or as a method for reporting the depolymerization activity of certain enzymes that attack colloidal particles (e.g., proteins may be attacked by endoproteases; starches, by amyloclastic enzymes; and pectinic acids may be digested by polygalacturonases, etc.).

4. Viscosity measurements can serve as an indicator of average molecular weight and molecular shape for large natural or synthetic macromolecules.

Aside from light scattering, osmosis, diffusion, and rheological phenomena, colloidal systems display other notable properties. For example, dispersed particles are normally subject to the influence of gravitational forces, which can cause their precipitation from a dispersing medium, whereas Brownian motion encourages maintenance of particulate suspension.

Under the influence of induced gravitational force by centrifugation, colloidal systems can be partitioned according to the densities or molecular weights of the colloidal constituents.

Electrokinetic phenomena including electrophoresis can also be employed for studying and characterizing colloidal systems. The operational principles for both of these concepts have been dealt with in the earlier chapter on proteins, although both concepts equally pertain to nonprotein particles and systems.

REFERENCES

Becher, P. 1965. *Emulsions: Theory and Practices*, 2nd ed. Reinhold, New York.

Davies, E. A. 1967. *Quantitative Problems in Biochemistry*, 4th ed. Williams & Wilkins, Baltimore.

Glicksman, M. 1969. *Gum Technology in the Food Industry*. Academic Press, New York.

Griffin, W. C., and M. J. Lynch. 1972. *Handbook of Food Additives* (C. T. E. Furia, ed.), pp. 397–429. CRC Press, Cleveland.

Marshall, A. G. 1978. *Biophysical Chemistry Principles, Techniques and Applications*. Wiley, New York.

Meyer, R. J., and L. Cohen. 1959. The rheology of natural and synthetic hydrophilic polymer solutions as related to suspending ability. *J. Soc. Cosmetic Chemists* **10:**143.

Moore, W. J. 1972. *Physical Chemistry*. Prentice-Hall, Englewood Cliffs, N.J.

Osipow, L. I. 1962. *Surface Chemistry. Theory and Industrial Applications*. Reinhold, New York.

Popiel, W. J. 1978. *Introduction to Colloid Science*. Exposition Press, Hicksville, NY.

Veis, A. 1964. *The Macromolecular Chemistry of Gelatin*. Academic Press, New York.

Vold, M. J., and R. D. Vold. 1964. *Colloid Chemistry*. Reinhold, New York.

C H A P T E R · 9

Pigments and Natural Colorants

INTRODUCTION

The native colors of plant and animal tissues depend largely on the species-specific presence of key porphyrin-, phenolic-, and isoprenoid-based compounds. The uniquely different patterns of electronic configurations, numbers of conjugated double bonds, and organometallic interactions exhibited by these molecular substructures account for the presence of distinctly different chromophores. Depending on the chromophoric structure, many types of visible light (350–750 nm) interactions are possible such that some spectral wavelengths are absorbed while others are reflected.

Human perception of the chromophoric constituents in foods relies on the visual sensations caused by reflected wavelengths of light as opposed to those that are absorbed. Green, blue, yellow, and red compounds generally reflect spectral wavelengths that correspond to these observed colors, whereas other light wavelengths are absorbed.

Photobiochemical interactions of chromophores *do not* share common functions in sustaining the existence of living organisms. The photo interactions of chlorophyllous plant pigments within green plant cells are tied to their acquisition of energy for photosynthetic operations, but few other compounds exhibit such a direct chromophoric role in cellular energetics.

Other colored biochemical compounds have distinct ecological importance since they may serve as an "innate releasing mechanism" for the orchestration of ethological (behavioral) activities in animals. These behavioral activities may involve insects, birds, marine life, or mammals. For example, the brilliant red, blue, and yellow colors of flowers may attract specific nectar-feeding and pollinating actions of insects, seasonal plumage colors of birds can spur on mating behavior, and so on.

Other colored biochemical substances that appear in foods seem to have a quiescent existence. The pink-, red-, and yellow-colored compounds respectively found in watermelons, tomatoes, and peaches are key visual characteristics of these fruits, but they fail to display any notable biochemical functions within native plant tissues.

Still other colored constituents of foods display colors that are incidental to their biochemical functions. Typical among these examples would be the oxygenated myoglobin structure of interstitial red muscle fluids. This protein is responsible for the *in vivo* oxygenation of red muscle, and it happens to impart a pink color to freshly cut meat tissues. The color and presence of oxygenated myoglobin is also recognized by the food consumer as an index of meat freshness and meat acceptability as a raw cut.

Color is a critical factor in the human sensory evaluation and acceptance of foods, quite apart from food flavor. The following sections cursorily survey the most prevalent colored biochemical constituents of foods, their normal occurrence, and some reactions underlying their destruction in food tissues.

PORPHYRIN-BASED CHROMOPHORES

Porphyrins are derivatives of porphin (a tetrapyrrole), which is comprised of four individual *pyrrolic subunits* joined by methine bridges (Figure 9.1). The heterocyclic nature of the pyrroles, coupled with their unique assembly, encourages the presence of a conjugated double-bond pattern. The chromophoric properties of porphyrins are complicated by their unique abilities to incorporate metals in the center of the complex ring structure.

Chlorophyllous plant pigments contain a coordinated magnesium atom in the center of the porphyrin, whereas the oxygen-carrying properties of myoglobin and hemoglobin depend on the corresponding presence of a coordinated iron atom.

In addition to conjugated bond structures and the coordination of key metallic atoms, chromophoric properties of porphyrin compounds are complicated by their association with proteinaceous molecular components.

Chlorophylls

The mechanism for the photoactivation of chlorophyll, its structure, and its roles in photosynthetic processes have been detailed in Chapter 12.

The green color of chlorophyll is a ubiquitous property of most plant leaves and many fruits. In the case of immature or unripe fruits, the green color impact of chlorophyll may decline as ripeness progresses. During this process other chromophoric effects of carotenoids and anthocyanins supersede the color impact of chlorophyll to produce red, yellow, or purple colorations. These color changes result from the *in vivo* degradation of chlorophyll, which unmasks coexisting pigments; or chlorophyll production ceases in favor of other ancillary plant pigments. In many fruits, including bananas as an example, the disappearance of chlorophyll corresponds closely to the liberation of free sugars from starch, increasing sweetness, and better consumer acceptability.

Since chlorophylls coexist in plant tissues with many other light-absorbing compounds, including carotentoids and xanthophylls, the intensity of green coloration in any plant tissue is a function of chlorophyll concentration ratios to other colored species. Moreover, the relative proportion of chlorophyll-*a* to -*b* can markedly influence the apparent green color of native plant material.

Chlorophyll destruction can proceed as a result of acid/base or enzymatic factors. During normal food processing, cooking and storage of green vegetables, a distinct color transition is often noticed. The natural green color is transformed to a bright green color, followed by a dull green or olive color. Although the native structure of chlorophyll is linked to lipoproteins and shielded from the effects of native phyto-acids, cooking effects expedite chlorophyll susceptibility to native acid effects. Weak acids readily liberate the magnesium atom from chlorophylls to form brown *pheophytin*. This process is also abetted by the coagulation effect of heat on protective lipoproteins.

Chlorophyll is also liable to *chlorophyllase* activity, which frees phytol alcohol from its native chlorophyll ester to produce methyl

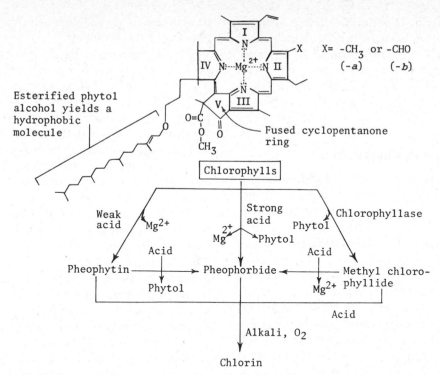

FIGURE 9.1. Chlorophylls-*a* and -*b* can undergo degradation by a variety of pathways. Some of the notable products include *methyl chlorophyllide*, loss of phytol due to *chlorophyllase* action; *pheophorbide*, loss of magnesium and phytol alcohol moiety due to strong acid action on chlorophyll; *pheophytin*, loss of magnesium caused by the action of weak acid on chlorophyll; and *chlorin* reduction of ring IV of chlorophyll promoted by the actions of alkali and oxygen on pheophytin or pheophorbide or the combined actions of these agents and acid on methyl chlorophyllide. All steps contribute to the successive decoloration of native chlorophyll. Note that the phytol alcohol contributes hydrophobic properties to the structure of chlorophyll; the alcohol is probably responsible for fixing the photoactive pigment to the hydrophobic structure of the lamellar membrane in the chloroplast.

chlorophyllide. Since the phytol alcohol is largely responsible for the lipophilic properties of intact chlorophylls, loss of the phytol moiety increases the aqueous solubility of the remaining porphyrin structure.

Strong acid effects on chlorophylls strip both magnesium and phytol alcohol from the structure to produce *pheophorbide*. This process can also occur as a sequel to pheophytin formation. Although chlorophylls exhibit some alkaline stability, the phytol alcohol may also be removed by strong saponification reactions. Pheophytin, pheophorbide, and methyl chlorophyllide derivatives of chlorophyll, however, are all susceptible to chlorin formation by the combined

actions of alkali and/or oxygen effects (Figure 9.1).

Using model systems such as spinach purées, kinetic degradation of chlorophylls seem to reflect first-order reaction kinetics over a temperature range of 260 to 300°F. For more complicated food systems containing chlorophylls, unpredictable factors may modify this degradative kinetic order.

Stable organometallic compounds can be formed from acid-modified chlorophyll structures using zinc or copper ions. The intensely green metallochlorophyll complexes formed from this treatment can serve as an effective masking method for hiding evidence of chlorophyll degradation in foods.

This treatment is generally illegal owing to the toxicity of both metals. Metallic substitution may involve replacement of the magnesium ion, but copper or zinc interactions with the porphyrin may involve the carbonyl moiety linking phytol and/or methyl esters.

Myoglobin and Hemoglobin

Mammals, poultry, and many fishes contain oxygen-carrying proteins that have distinct pink or red colors. For mammals and poultry, the principal coloring agent of edible muscle is *myoglobin*. Residual traces of hemoglobin can exist in these tissues, but effective abattoirial bleeding of animal carcasses minimizes its appearance in most commercial meats. The pink-red color transitions displayed by myoglobin and hemoglobin correspond to the dynamic electronic and molecular changes that occur from their respective oxygenations and deoxygenations.

Structure–function features of myoglobin and hemoglobin: Myoglobin is a bipartite molecule that exhibits an iron-containing porphyrin denoted as a *heme structure* along with a 153-residue polypeptide. The polypeptide displays eight α-helical segments (labeled A → H), which are linked by intervening nonhelical peptides. The porphyrin structure is not unlike that found in chlorophyll, but its pyrrolic rings are studded with four methyl, two vinyl, and two propionate side chains.

Since the iron atom in the heme has a coordination number of six, it can participate in six electron-pair interactions and form six bonds. Four bonds are responsible for orienting the iron atom in the porphyrin to form the heme structure. The remaining fifth and sixth coordinations are oriented above and below the heme plane. The fifth position is linked to the so-called F-8 imidazole residue of the globular protein associated with the polypeptide. The *sixth coordination position* remains available for the *reversible oxygenation* of the molecule.

The occupation of the sixth coordination position of myoglobin by molecular oxygen is a complex and somewhat paradoxical proposition. This problem stems from the fact that the topological and conformational structure of deoxygenated myoglobin is not wide enough to permit free oxygen access to the iron in the heme plane. Thus, the essential step of myoglobin oxygenation is augmented by a unique process.

The iron contained in the heme plane of deoxymyoglobin exists in the 2+ oxidation state and displays a high-spin configuration. The electronic radius of the high-spin ion is too large to fit into the region surrounded by the pyrrolic nitrogens. Therefore, the ion resides nearly 0.6 Å over the heme plane toward the imidazole nitrogen.

The approach of molecular oxygen toward the deoxygenated position of heme iron reverts the coordinated iron atom to a low-spin configuration. This in turn shrinks its electronic radius and allows the iron to move toward the heme plane. The angle between the internuclear axis of oxygen and the iron–oxygen bond is about 120°.

This motion of the iron atom during oxygenation spurs a series of spatial–molecular interactions that ensure (1) oxygen accessibility to the oxygen-carrying site of heme; and (2) an oxy-/deoxymyoglobin equilibrium, which is contingent on oxygen supplied by oxygenated hemoglobin and existing partial pressures of oxygen.

The functional oxygenation of myoglobin requires the maintenance of an Fe^{2+} state within the myoglobin structure. Oxidation of Fe^{2+}-myoglobin to its Fe^{3+} (*ferrimyoglobin*) form by water or other agents obviates any efficient reversibility in the oxygenation–deoxygenation process. Therefore, biochemical evolution has ensured the maintenance of a hydrophobic microclimate about the heme-iron imparted by the inner portion of the surrounding myoglobin polypeptide.

Whereas myoglobin represents a monomeric heme-containing protein structure, hemoglobin is a tetrameric protein that displays allosteric oxygenation–deoxygenation behavior. In addition to oxygen transport, hemoglobin molecules also carry carbon dioxide and hydrogen ions. The basic structure of hemoglobin is fundamentally similar to four mutually interacting myoglobin molecules united by salt links.

Meat color and the role of myoglobin: The heme protein of myoglobin (Mb) is responsible for the purple color of freshly cut meat. Exposure of meat to air induces myoglobin oxygenation to form *oxymyoglobin* (MbO_2), which has a bright red-pink color. Following the deoxygenation of MbO_2, this red-pink color of meat reverts back to its former purple color. This reversible oxygenation can be expressed as

$$\underset{\text{Purple-red}}{Mb} + O_2 \underset{\text{deoxygenation}}{\overset{\text{oxygenation}}{\rightleftarrows}} \underset{\text{Red-pink}}{MbO_2}$$

In the event that the Fe^{2+} oxidation state of heme becomes oxidized to an Fe^{3+} state, *ferrimyoglobin* or *metmyoglobin* (MetMb) results. The MetMb displays a very noticeable brown-gray color:

$$\underset{\text{Purple-red}}{Mb(Fe^{2+})} \underset{\text{reduction}}{\overset{\text{oxidation}}{\rightleftarrows}} \underset{\text{Brown-gray}}{MetMb(Fe^{3+})}$$

If MetMb comes in contact with a reducing substance in meat tissues, it is possible to temporally transform some of the MetMb back to Mb.

The spectral properties of chromophoric species displayed by MbO_2, Mb, and MetMb are interconvertible. Furthermore, all forms of these pigments contribute to raw meat colors depending on their individual occurrences. A major absorbance peak at ~555 nm characterizes Mb, dual absorption maxima at ~540 and ~575 nm are displayed by MbO_2, and MetMb has its absorbance maximum shifted to ~505 nm with a minor peak at 627 nm (Figure 9.2).

Nitrates (NO_3^-) or nitrites (NO_2^-) have historically been added to meats as a compo-

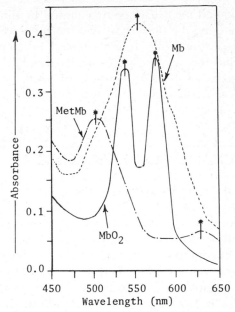

FIGURE 9.2. Relative absorption maxima for myoglobin (Mb), oxymyoglobin (MbO_2), and metmyoglobin (MetMb) (sample concentrations are not equimolar).

nent of curing ingredients. A combination of pH conditions and bacterial actions combine to produce nitrous oxide as shown below:

$$2\ NaNO_3 \xrightarrow{\text{bacteria}} 2\ NaNO_2 + O_2$$

$$2\ NaNO_2 + H_2O \xrightarrow{\text{pH } 5.4–6.0} HNO_2 + NaOH$$

$$3\ HNO_2 \longrightarrow 2\ NO + H_2O + HNO_3$$

The ultimate interaction of nitrous oxide (NO) with myoglobin results in the formation of *nitrosomyoglobin* (NOMb), which displays an unstable reddish color typical of uncooked but cured meats:

$$NO + Mb \longrightarrow NOMb$$

Upon heating at temperatures of 70°C, a red, heat-stable *nitrosohemochrome* derivative is produced from NOMb. Aside from the inherent heat stability of the nitrosohemochrome color, it does undergo photo-induced color fading.

Exclusive of curing processes, the favorable color impact of MbO_2 in commercially packaged meats can be maintained by film-packaging methods that allow marginal oxygen permeability. It is estimated by some studies that oxygen permeability should approach 5–6 L/m²/day/atm in order to prevent the formation of brown MetMb.

PHENOLIC PLANT PIGMENTS

Phenolic constituents of plants impart a diverse range of red to blue color variations commonly associated with leaves, flowers, and fruits. Foremost among those phenols having chromophoric effects are the *flavonoids.* These compounds display wide variations in the number and location of methoxyl (—OCH₃) and hydroxyl (—OH) substituents throughout their structural series of *anthocyanidins, flavonols, flavones,* and *catechins.* Apart from the flavonoids, hydrolyzable and condensed *tannins* plus other minor phenolic compounds contribute to the colors of plant tissues.

Biosynthesis of Flavonoids

The botanical synthesis of flavonoids is largely dependent on the conversion of phenylalanine to *trans*-cinnamic acid. A subsequent sequence of hydroxylation and methylation reactions yield a variety of substituted acids. Conversion of these acids to their coenzyme A esters serves as a prelude to the appearance of numerous flavonoid structures as well as *lignins, stilbene* precursors to *pinosylvin* (found in most pines), benzoic acids, and many other compounds. One popularized route of flavonoid biosynthesis suggests that 4-coumaryl-CoA, produced from phenylalanine, undergoes a chalcone synthase-mediated reaction with malonyl-CoA's to yield *chalcone* (Figure 9.3).

Chalcone is a critical precursor for the biosynthesis of flavanones, flavones, flavonols, and the anthocyanidin structures of native anthocyan*in* plant pigments. Although some biosynthetic steps resulting in flavo-

noids are still speculative, radiotracer, enzyme, metabolic inhibition, and mutant studies of plants have affirmed the biosynthesis scheme outlined above.

The botanical appearance of flavonoids in different plant tissues is certainly controlled by mechanisms of gene expression. The biochemical mechanisms underlying the tissue-specific appearances of flavonoids remain to be charted. For many plants, carbohydrate status in cells may affect the biosynthesis of flavonoids. In other cases such as strawberries, the side of the fruit exposed to the sun reddens faster than the shaded side. Still other plant types, such as red radishes, display uniform superficial red colors over their tap roots regardless of sun exposure.

Uncertainties also surround the subcellular organelle responsible for anthocyanin biosynthesis. Anthocyanogenic subcellular structures called *anthocyanoplasts* may be more or less critical to the formation of certain flavonoids in plants since they have been detected in mono- and dicotyledonous species.

Anthocyanidins and Anthocyanins

Classical studies conducted by Willstäter, Robinson, and Karrer have characterized the basic physical and structural features of natural anthocyanins.

A cursory survey indicates that anthocyanins are glycosylated polyhydroxy and polymethoxy derivatives of 2-phenylbenzopyrylium (flavylium) salts (Figure 9.4).

In more simplistic terms, anthocyanins are *glycosides* composed of the "*anthocyanidin aglycone*" plus one or more *glycosidically bonded mono- or oligosaccharidic units.* The saccharidic moieties of anthocyanins may be structurally complicated by the presence of *p*-coumaric, ferulic, caffeic, or sinapic acids, and occasionally minor amounts of malonic, acetic, or *p*-hydroxybenzoic acids (Figure 9.4). The most common monosaccharides include arabinose, galactose, glucose, rhamnose, and xylose. Carbohydrate attachment to the aglycone often involves the C-3 position, but

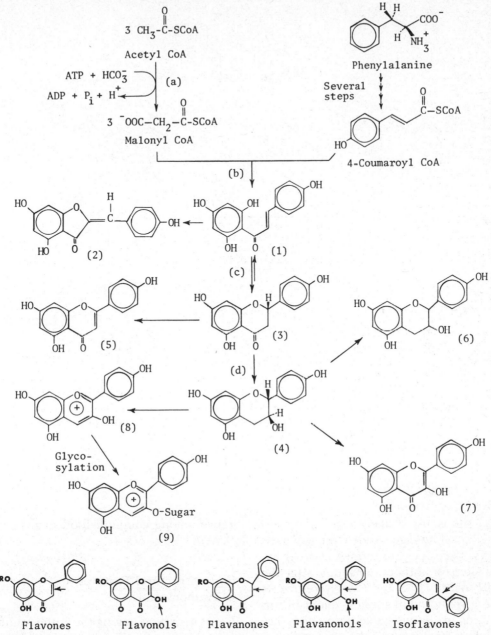

FIGURE 9.3. General biochemical derivations of flavonoids. Note that the structural key for the group of compounds is indicated at the bottom of the figure: *flavones*, position 2,3 is unsaturated; *flavonols*, C-3 hydroxyl is present; *flavanones*, 2,3-site is saturated; *flavanonols*, C-3 hydroxyl is present and site 2,3 is saturated; and *isoflavones*, C-3 phenyl ring substituent is displayed. Pathway Key: (a) acetyl-CoA carboxylase; (b) chalcone synthase; (c) chalcone isomerase; (d) flavanone 3-hydroxylase; (1) chalcone; (2) aurone; (3) flavanone; (4) dihydroflavonol; (5) flavone; (6) catechin; (7) flavonol; (8) anthocyanidin; (9) anthocyanin. Isoflavone in the interim formation of chalcones (1) to flavanones (2).

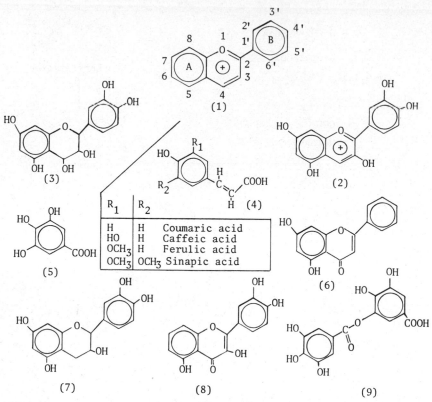

FIGURE 9.4. Notable structures of some flavonoids, their key constituents and related phenolics. (1) 2-Phenylbenzopyrylium (flavylium) salts serve as the basis for many flavonoid structures especially the anthocyanidins; the structure is recognized to exist as flavylium salts as well as a flavylium cation. (2) Cyanidin (an anthocyanidin); (3) leukoanthocyanidin (a flavan-3,4-diol); (4) some acid components of anthocyanin structures; (5) gallic acid; (6) 5,7-dihydroxyflavone; (7) catechin (a flavanol); (8) quercetin (a flavonol); (9) *m*-digallic acid (a depside).

the C-5 site is also a likely location for another glycosyl linkage. Aside from some exceptions, glycosides are formed at other hydroxyl sites on anthocyanins only after the establishment of C-3 and C-5 glycosides.

The glycosidic linkages exhibited by anthocyanins are quite stable unless they are exposed to strong mineral acid conditions (e.g., ~2.0 *M*) to produce anthocyanidins.

Generic forms of nomenclature have been largely retained for many of the most common anthocyanidins owing to their isolative predominance in certain flowers. Six of the most typical anthocyanidins are *pelargonidin, peonidin, malvidin, cyanidin, delphinidin,* and *petunidin*. Although many other anthocyanidins exist, they have a rather limited occur-

rence among common food plants (Figure 9.5A).

Some Structural Features of Anthocyanins

Color variations among the anthocyanins reflect their respective structural differences in the number of hydroxyl groups, the presence or absence of methylation and glycosylation, as well as the distribution of positive charge about the aryl-substituted chroman ring system (Figure 9.5B). As indicated in the figure, the flavylium cation is electron deficient and clearly liable to nucleophilic attack.

Anthocyanins undergo dynamic struc-

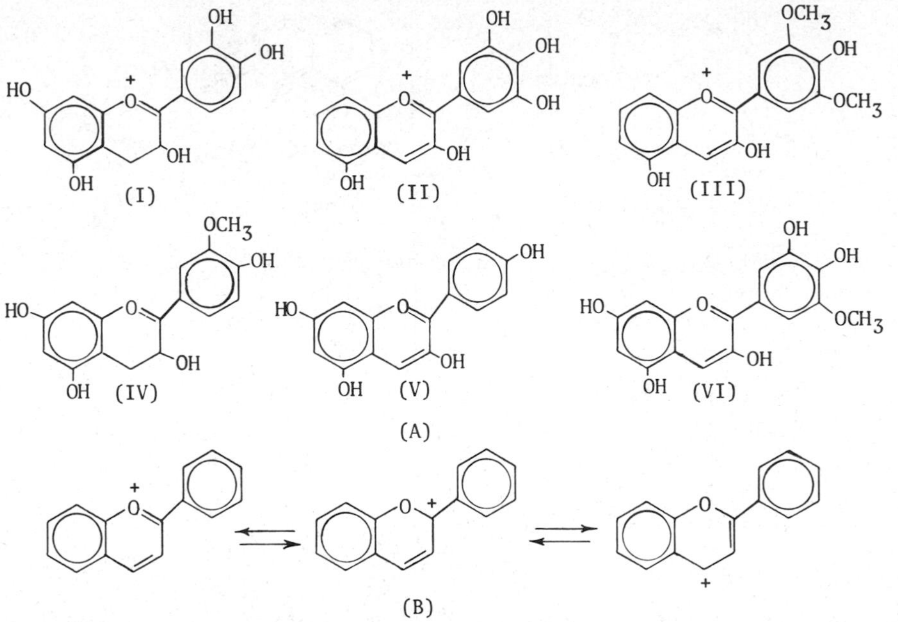

FIGURE 9.5. (A) Selected structures for six of the most common anthocyanidins. All anthocyanidins have hydroxyl groups in the 3,5, and 7 positions although each structure may have its own characteristic hydroxyl or methoxyl groups on the so-called B ring: (I) Cyanidin, (II) delphinidin, (III) malvidin, (IV) peonidin, (V) pelargonidin, and (VI) petunidin. A representative occurrence of these compounds is varied throughout species of fruits and vegetables: apple (I); blackcurrent (I, II); blueberry (I, II, III, IV, VI); cherry (I, IV); grape (I–VI); orange (I, II); peach (I); plum (I, IV); radish (V); raspberry (I); red cabbage (I); strawberry (V-major, I). (B) Possible positive charge distribution over a chroman ring system found in the structure of anthocyanidins.

tural alterations depending on the pH and ionic strengths of their aqueous environments. These structural changes generally involve modifications of the *benzopyrylium moiety*. Four possible structural forms are recognized, with the most prevalent forms being dictated by their thermodynamic stability at a given pH. Recognized structural forms of anthocyanins include (1) *quinoid* (basic), (2) *flavylium* (cationic), (3) *carbinol* (basic), and (4) *chalcone species* (Figure 9.6).

The individual occurrences of anthocyanin species can be justified on the basis of pH-induced acid–base, hydration, or tautomerization phenomena involving the benzopyrylium structure. Interconvertibility of one species to another readily occurs according to the scheme outlined in Figure 9.6. Notice that the mutual existence of any two interconvertible species will be dictated by

the respective equilibrium constant for the acid–base transition (K_{ab}), hydration (K_h), and tautomerization (K_t). These three equilibrium constants display considerable variation from one anthocyanin to another, but as indicated by Brouillard and others, structural behaviors of anthocyanins are mathematically and thermodynamically predictable if these values are known.

Based on pH studies over the range of 0.0 to 6.0 for a natural anthocyanin such as malvidin 3-glucoside, the flavylium cation (FS$^+$) generally predominates at low pH values (e.g., 100% at pH 0.5). A rise in pH produces a sigmoidal concentration decrease in FS$^+$ until it approaches a nearly nonexistent concentration at pH 6.0. The decrease in FS$^+$ results in a concomitant increase in the carbinolic species (CAR), the chalcone (CAL), and the quinonoid base (QS).

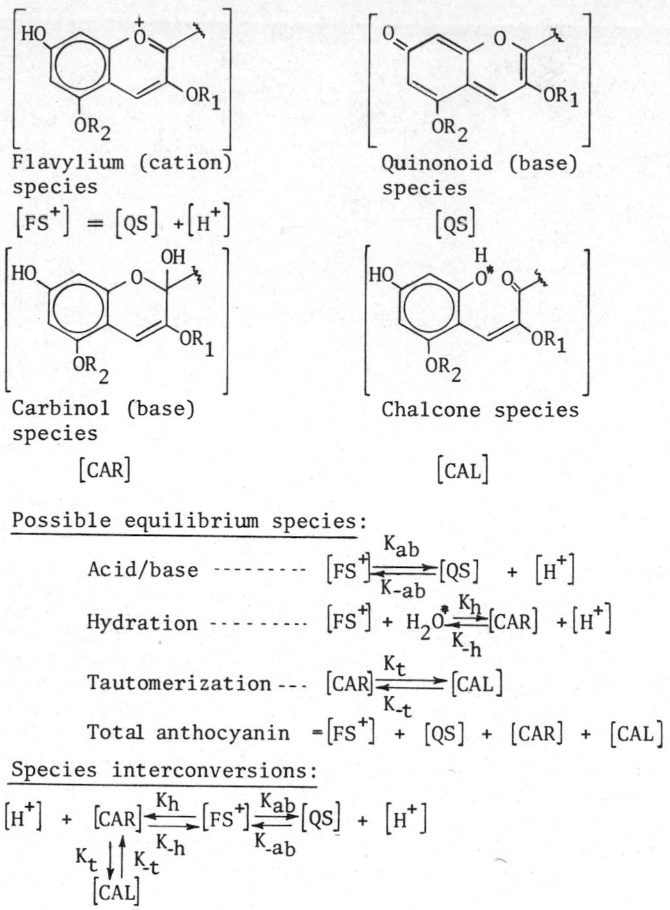

FIGURE 9.6. Acid/base, hydrated, and tautomeric species of a general anthocyanin structure (2-phenyl ring structure is omitted); R_1 = glycosyl group, R_2 = glycosyl group *or* hydrogen (—H). Each species interconversion has its own equilibrium constant as indicated and the interconversion of one species to another is directed by the sequence shown at the bottom of the illustration. Note that the total anthocyanin concentration is the sum of all individual species concentrations.

The CAL and QS species are minor contributors to the equilibrium distribution of the pigment at pH 6.0 where the total concentration ratio of CAR:CAL:QS is ~86.6%:11.0%:2.4%. Moreover, the CAL and QS species may only begin to appear in the pH ranges of 1.0–2.5 and 2.0–3.0, respectively.

Studies of model anthocyanin systems also show that their notable colorations rest mainly on the presence of the flavylium and quinonoid species. Since these are the principal species responsible for visible light absorption, the impact of anthocyanic coloration can be assessed by determining the concentrations of these species at a specific pH value.

Aside from the pH dependency of anthocyanin coloration, it should also be noted that progressive transformations for QS → FS$^+$, FS$^+$ → CAR, and CAR → CAL are endothermic. Therefore, as the thermal environment of anthocyanin(s) increases above 25°C to ≤100°C, the chalcone species will be favored.

Anthocyanins Are Chemically Unstable

Anthocyanidins (aglycones) are insoluble in water, photochemically unstable, and alkali labile. With the exception of transiently occurring anthocyanidins encountered during normal plant metabolism, few if any anthocyanidins commonly exist in a free state. For *in vivo* botanical systems, the glycosidic derivatives of anthocyanidins known as "anthocyanins" are favored. The anthocyanin structure displays good stability within the acid milieu of cellular saps and good water solubility.

Regardless of their acid stability, anthocyanins are subject to the assault of numerous chemical agents that diminish or annihilate their characteristic colors. This reactivity has plagued their potential use as food-coloring agents in the food industry and prompted inventive methods for their preservation in canned (retorted) foods and jellies, as well as fresh fruits and vegetables.

The enzymatic attack of anthocyanins may involve peroxidases, phenolases (phenoloxidases, polyphenol oxidases), or possibly glycosidases.

Since a variety of phenols coexisting with anthocyanins seem to accelerate pigment decolorization, it has been speculated that *phenolase*-mediated destruction of anthocyanins may be an indirect process. That is, phenolase may oxidize a 1,2-benzenediol-type structure to the corresponding *o*-benzoquinone. The quinone subsequently oxidizes the anthocyanin and thereby leads to its destruction. The accessory effects of chlorogenic acid and catechin have specifically been implicated in the phenolase-mediated destruction of eggplant and cherry anthocyanins.

Glycosidase effects may also have an indirect impact on anthocyanin destruction. Here the enzyme first yields an anthocyanidin (an aglycone), which is promptly destroyed.

Regardless of either enzyme mechanism, blanching processes minimize many losses attributed to phenolase activities.

Light, oxygen, metallic ions, traces of sulfur dioxide, and the existence of electrocouples all enhance the discoloration and destruction of anthocyanins. Ascorbic acid may serve as a degradative culprit for anthocyanins, especially if copper ions are present. Oxidation of ascorbic acid in the presence of copper and oxygen yields hydrogen peroxide (H_2O_2). The H_2O_2 undoubtedly induces decoloration of the pigment.

In addition to all of the foregoing factors, lactose, sorbose, and arabinose have also been cited for their production of furfural and 5-hydroxymethyl furfural under heated acidic conditions. These and other carbonyl compounds, including acetaldehyde, all contribute to the potential loss of anthocyanin colors.

Anthocyanin Complexation and Copigmentation

One or more glycosidic forms of the anthocyanidins may be responsible for the primary colors of many flowers, fruits, and vegetables (Figure 9.5). Unfortunately, from analytical perspectives, the occurrence of anthocyanins is complicated by the multiple coexistence of different aglycones and multiple variations in the 3-, 3,5-, or 3,7-mono- or oligosaccharidic groups.

Many types of anthocyanins self-associate with their own structural types or undergo condensation with other organic compounds. The latter case is recognized as a copigmentation phenomenon that usually increases pigment absorption and results in a bathochromic shift.

Associative behaviors of anthocyanins often result in colored complexes that are more stable than their occurrence as free species, especially when exposed to some forms of food processing. Self-association and condensation phenomena are also important factors in the coloration of wines. It is theorized that anthocyanins and coexisting phenolics in red grape juices undergo complexation. These *antho–phenolic* interactions are stabilized by intermolecular hydrogen bonding, until ethanol from fermentation reaches some critical concentration level. The high ethanol concentration during the last phases of vinification ultimately disrupts hydrogen-bonding interactions. The effects of this action are mirrored as a loss in red wine color but this is a change that may be somewhat reversible with the extraction of some alcohol.

In addition to the reactions outlined above, the aging of red wines may provide an added opportunity for pigment polymerization and condensations. The products of the reactions not only exhibit a persistent reddish color but demonstrate *improved* pH stability and resistance to the bleaching effects of sulfur dioxide.

Natural floral, fruit, and vegetable colorations are complicated still further by the photoabsorptive mechanisms of metalloanthocyanin complexes, acetylated anthocyanin structures, as well as copigmentation interactions. The authoritative work by Osawa (1982) cited in the references serves as a guide to these anthocyanin forms.

Leuko- and Proanthocyanidins

Proanthocyanidins have been defined as flavanoids that liberate anthocyanidins upon treatment with acid. Although the colorless anthocyanidin moiety of these flavanoids was anticipated to involve the *idealized structure* of 3,5,7,3′,4′-pentahydroxy-2-flaven or possibly -3-flaven ($C_{15}H_{12}O_{2+n}$; $n = 2-5$), these structures have doubtful natural occurrences. Further research revealed that many "colorless" anthocyanidins, or *leukoanthocyanidins*, freed by acid actions on flavanoids, exhibit some type of a polyhydroxy-3,4-flavandiol having the "leuko-" properties of the *nonexistent* flavens. Since the latter structures (i.e., the -3,4-flavandiols) contain more elements of water than the idealized flaven structures of leukoanthocyanidins, native leukoanthocyanidin structures have generally been denoted as leukoanthocyanidin hydrates ($C_{15}H_{14}O_{3+n}$; $n = 2-5$) (Figure 9.7).

Proanthocyanidin structures are also complicated by the occurrence of (I) dimeric proanthocyanidins (also generally called dimeric leukoanthocyanidins but preferably proanthocyanidinocatechins) and (II) dehydrodicatechins.

The first group (I), initially isolated from cacao beans, was found to consist of a C_{30}-acid labile structure that yields cyanidin and an (−)-epicatechin. Other related dimers of anthocyanidin and catechins have been recognized in many fruits since their first isolation from cacao. Group II, or dehydrocatechins, on the other hand, are dimeric structures too, but acid effects fail to produce anthocyanidins. This group (II) results from the oxidative enzymatic linkage of catechins (polyhydroxy-3-flavanols) (Figure 9.7).

Another group of leukoanthocyanidins are commonly recognized as "unusual leukoanthocyanidins." These structures do not display the features of the 3,4-flavandiol but, instead, they have additional substituents on the A and B rings (Figure 9.7). Many of these leukoanthocyanidins reside in the heartwoods of various tree species.

Leukoanthocyanidins are key monomolecular contributors to the polymeric structures of tannins. Condensed tannins (*flavolans*) notably display leukoanthocyanidic linkages between the 4 position of one residue and the 6 or 8 position of another subunit. These tannins are chemically stubborn to physiological digestive conditions, and upon severe treatment, they yield less soluble polymeric "phlobaphanes" or, possibly, monomeric flavanoids or catechins. The condensed tannin structures should not be confused with tannic acid, which is a representative of the hydrolyzable tannins.

Flavones, Flavonols, and Minor Flavonoids

The flavones, flavonols, and minor flavonoids contribute to the yellow colors of numerous flowers and certain leaves. The molecular structures of these pigments are reminiscent of anthocyanidins, but the flavones and flavonols respectively differ in the absence and presence of a C-3 hydroxyl group, whereas both structures display a 4-keto substituent.

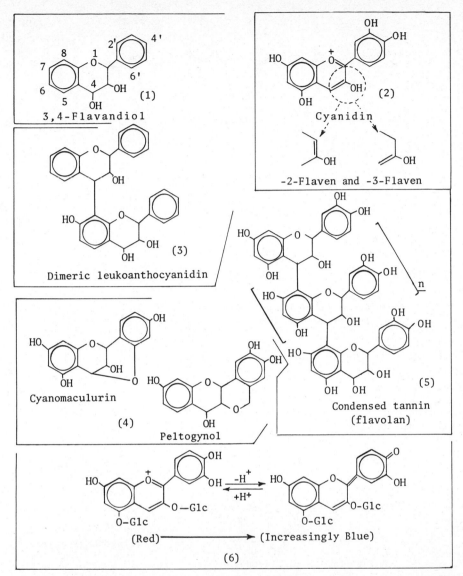

FIGURE 9.7. (1) 3,4-Flavandiol structure that is common to many leukoanthocyanidin structures. (2) The structure of cyanidin and the relative positions of the double bond in the 2,3 and 3,4 positions contribute to the formation of an idealized leukocyanidin (*or* leukoanthocyanidin). (3) Structure for a dimeric leukoanthocyanidin, or according to revamped nomenclature, a "proanthocyanidinocatechin" (structure may be subject to hydroxylation). (4) Two characteristic structures for "unusual leukoanthocyanidins." (5) Multiple (*n*) leukoanthocyanidins contribute to condensed tannin structures. (6) A diglucosyl-cyanidin can undergo progressive color changes from red to blue to near black as pH conditions become increasingly basic.

The flavones and flavonols are also similar to anthocyanins in that they consist of an aglycone, a sugar moiety, and may contain acetylated groups. Apigenin, luteolin, and tricetin are typical representatives of the flavones; kaempferol, quercetin, and my-recetin are exemplary of the flavonols. A variety of minor flavonoids also include chalcones, aurones, isoflavones, and biflavonyls (Figure 9.8).

Based on the recognized biosynthetic schemes for flavonoids, flavones probably

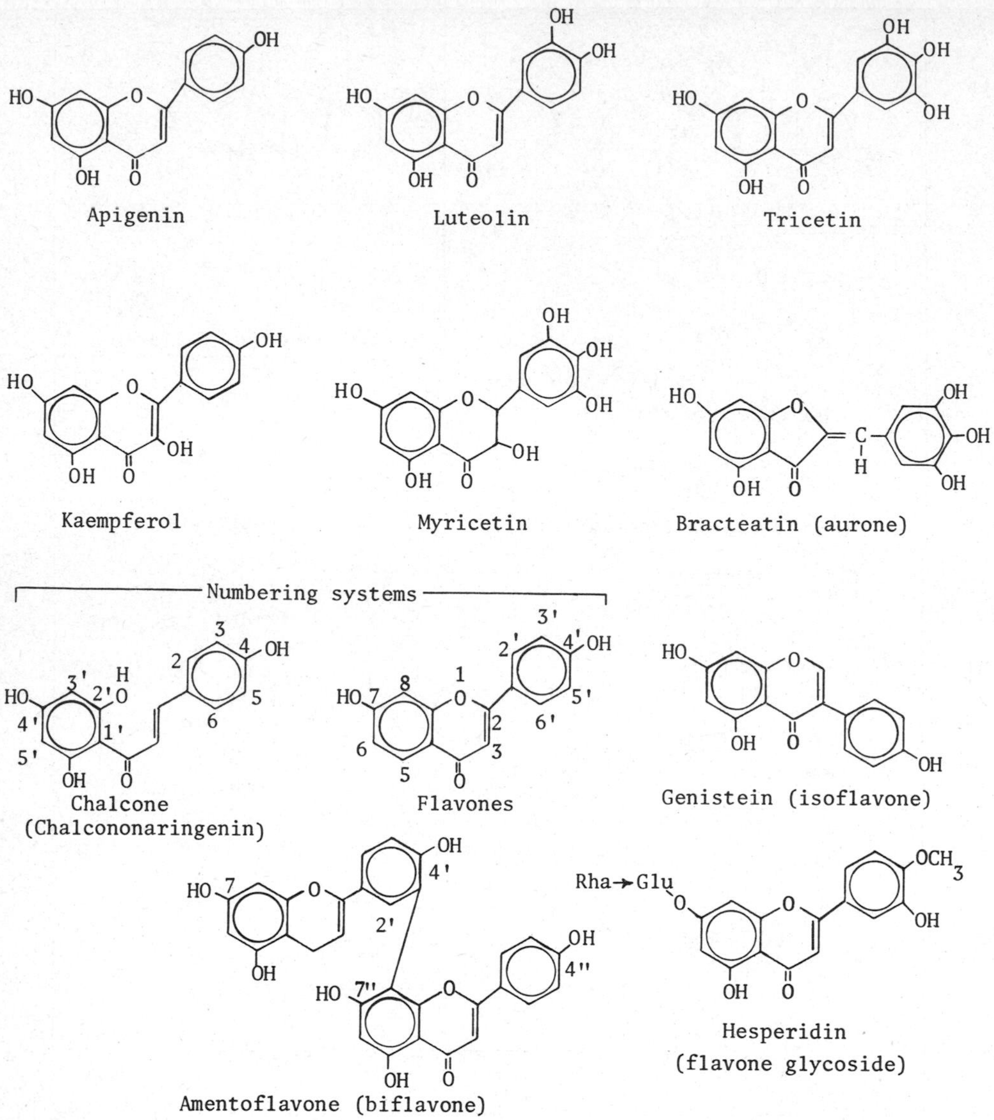

FIGURE 9.8.   Some representative structures of flavones, flavonols, chalcones, aurones, isoflavones, and biflavonyl. The numbering systems for chalcones versus flavones have been indicated for reference purposes.

materialize from a flavanone 3-hydroxylase-mediated conversion of flavanone, while aurones and isoflavones develop from chalcone precursors.

Flavonoid Contributions to Foods

Since the visual appearance of foods is critical to their acceptance, the flavonoids definitely contribute to the apparent visual qualities of foods. So long as the key flavonoids of fruits and vegetables remain in a chemical state indicative of ripeness or edible suitability, this appearance obviously favors their acceptance as foods. Green strawberries, Delicious apples, concord grapes, and similar foods are judged as undesirable until they assume their respective red, dark-red, or mauve hues. In a contrary vein, the degradation of flavonoids and anthocyanins in particular serve as an indictment of vegetable or fruit qualities. For example, losses of food acidity may prompt the dissociation of the 4'-hydroxyl group of the cyanin (diglucosyl cyanidin) detailed in Figure 9.7. Progressively higher pH conditions spur dissociation of additional hydroxyls until a blue color is very prominent. Both native and dissociated anthocyanins provide numerous resonance structures, and assuming that severe pH conditions do not ruin the pigments, color changes may be quite reversible.

Among the yellow flavonols, the 3-rhamnosyl-D-glycosyl derivative of quercetin is most common in flowering plant tissues and some pericarpal tissues of ripe fruits. This derivative is also known as rutin.

The phenolic properties of many flavonolic plant constituents also contribute to astringency sensations in the mouth. Here, the culprits are often leukoanthocyanidins and tannins. Characteristic mouth-puckering sensations caused by these agents may be caused by cross-linking actions of polymeric phenols with protein. Leukoanthocyanidins arc comon constituents of quince, fruits, wine, cider, some legumes, teas, cocoa, and other foods.

Some flavonoids are bitter tasting, whereas others are quite sweet. *Naringin*, which is a L-rhamnose-D-glucosyl derivative of *naringenin* common to grapefruit and some other citrus fruits, produces a taste nearly as bitter as quinine. Neohesperidino-dihydrochalcone, on the other hand, which is a naringin derivative, produces a very sweet taste sensation. Sweetness of the latter compound is dictated by the AH-B theory of sweetness as in the case of any sweet-tasting substance.

Flavonoid effects on animals and humans are uncertain from health perspectives, although many of these polyphenolics exert pharmacological effects. Some flavonoids including the isoflavones, genistein, and diadzein elicit weak estrogenic effects and may induce infertility in sheep and goats fed isoflavone-rich clover. In a contrary vein, the same compounds may actuate higher rates of body growth and milk production in cows grazing over spring pastures.

The flavone glycoside known as hesperidin occurs at 7 to 9% concentrations in dry orange peel. Still unsubstantiated studies claim that this citrus bioflavonoid, also called vitamin P, has a beneficial effect on human health.

Other flavones, including the dihydrochalcone called phlorhizin, are known to occur widely in the root barks of apples and pears, and trace amounts may occur in other plant tissues of the *Rosaceae*. Phlorhizin counteracts glucose resorption in the renal tubules and incites glucosuria.

The established record of flavonoid-related health effects is far from certain, and many claims regarding their potential carcinogenicity remain to be studied.

ISOPRENOID-BASED CHROMOPHORES

The isoprene group (CH_2=$C(CH_3)CH$=CH_2) provides the fundamental structural component for the carotenoid pigments. The name of these pigments (i.e., carotenoids) reflects their classic association with the carrot root (*Daucus carota*) from which they were first isolated. Carotenoids are responsible for many yellow, orange and red colors displayed throughout plant and animal species. The carotenoids are mainly lipid-soluble substances and over 350 carotenoid pigment structures have been recognized.

Structure and Nomenclature of Carotenoids

Carotenoid pigments can be conveniently subdivided into two groups known as (1) *carotenes,* which are hydrocarbons, and (2) *xanthophylls,* which are oxygenated derivatives of carotenes. The oxygen contained in xanthophylls (also called *oxycarotenoids*) may occur as a carboxyl, furanoxy, hydroxyl, methoxyl, keto, or epoxy groups as well as some other more uncommon groups. Carotenoid structures can also be surveyed according to their acyclic, monocyclic, and bicyclic features respectively displayed by lycopene, γ-carotene, and β-carotene.

As indicated above, the carotenoids are aliphatic or aliphatic alicyclic structures comprised of combined C_5-isoprene units. As a classical example of the carotenoids, β-carotene is comprised of eight C_5-isoprene units to produce a single C_{40} structure; however,

C_{30} to C_{50} carotenoids have also been recognized.

During the process of *in vivo* or *in vitro* carotenoid synthesis, isoprenoid units are joined in such a way that their arrangements are reversed in the middle of the molecules (Figure 9.9).

According to the 1971 dictates of the Commission on Biochemical Nomenclature, guidelines were promulgated for naming carotenoids based on the stem "carotene."

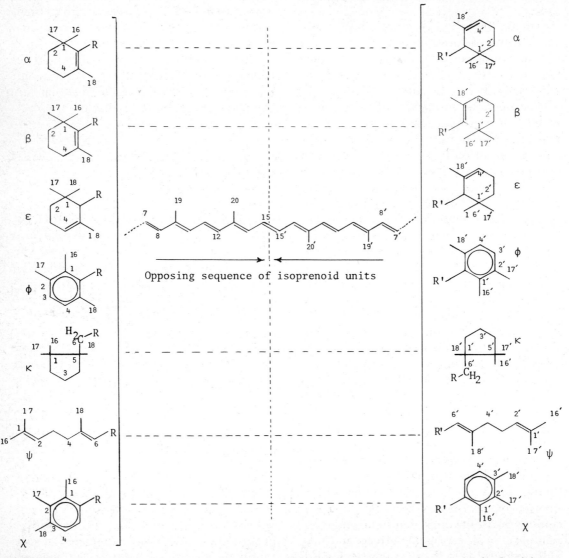

FIGURE 9.9. Carotenoid structures are designated according to their respective end groups by the Greek letters detailed in the figure. Note that the 15,15′ midpoint of the molecule links the identical but reversed structural sequences of isoprenoid units within each respective molecule.

The specific designations for carotene end groups were given by Greek letters (α, β, ε, κ, φ, χ, and ψ) as indicated in Figure 9.10. Since carotenes can exist as carboxylic acids; esters of carotenoid acids, aldehydes, ketones, alcohols; or esters of carotenols, a suitable prefix or suffix is often used for accurate nomenclature. Oxygen bridges are denoted by the "epoxy" prefix applied to the appropriate hydrogenated stem name.

Geometrical isomerism is also considered in modern carotenoid nomenclature. For example, if the structure is *cis*, it is denoted as such; otherwise the molecule is assumed to have an all-*trans* configuration without any designation.

The carotene prefixes such as *neo-*, *pro-*, and *apo-* are used to define, respectively,

1. Carotenoid stereoisomers having at least one *cis* configuration in the characteristic conjugated double-bond sequence (*neo-*).

2. Some poly-*cis* double-bond configurations (*pro-*).

3. Carotenoids derived from another carotenoid that has lost a structural component by some degradation process (*apo-*).

The number of conjugated double bonds determines the characteristic absorption spectra for carotenoids. Those carotenoids having the greatest number of conjugated

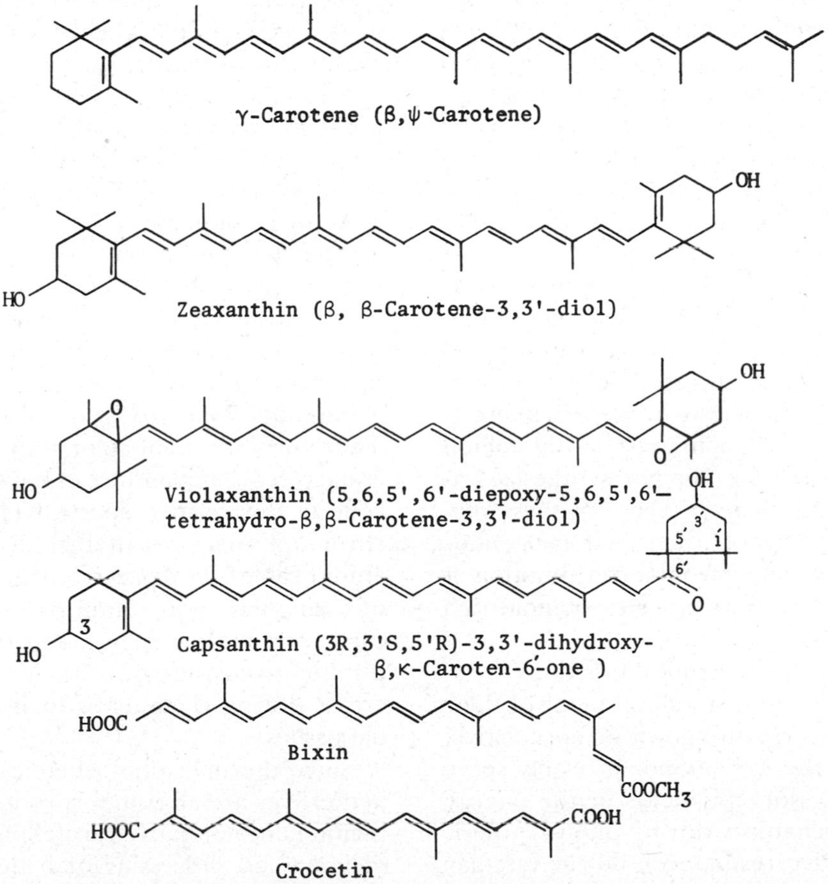

FIGURE 9.10. Some carotenoids named according to their representative structural components (in Greek letters) along with two smaller representatives of carotenoid structure.

double bonds display absorption maxima shifted toward longer wavelengths. Although the absorption of each carotenoid displays its own spectral nuances, the typical absorbance range is usually 375 to 500 nm with a maximum at ~450 nm. On the basis of human visual perception, increases in conjugated numbers of double bonds, as well as *trans* configurations about these bonds, both foster increasingly red hues. Therefore, any reductive actions inflicted upon carotenes, or acid, thermal, or photoeffects that produce *trans* to *cis* conversions, all cause distinct color losses.

Biosynthesis, Bioformation, and Occurrence of Carotenoids

The biosynthesis of carotenoids proceeds according to the synthetic route of all polyprenyl compounds as outlined in the lipid chapter. Isopentenyl pyrophosphate(s) formed from acetyl-CoA ultimately yield geranylgeranyl pyrophosphates (C_{20}), which dimerize to a phytoene (C_{40}) structure. The phytoene then serves as a basis for carotenoid structures.

Vascular and certain nonvascular plants (including the algae), and selected photosynthetic bacteria, can synthesize a plethora of carotenoid structures. The per annum carotenoid biosynthesis in the biosphere is prolific and may reach levels of 100 million tons. Since animals do not synthesize carotenoids, the tissue presence of these pigments is contingent upon their exogenous acquisition by heterotrophic consumption of plants, bacteria, and other carotenoid-rich animal species.

Some *in vivo* biochemical functions of carotenoids are understood, yet many of their functions remain unknown or speculative. For plants, the carotenoids probably serve as photoaccessory pigments in the energy transfer mechanism during photosynthesis (e.g., inductive resonance); but in vascular and nonvascular plants (including fungi) carotenoids may also play key roles in both photoresponses (e.g., phototropisms) and the

origin of plant-growth-regulating substances (e.g., abscisic acids).

Further speculation suggests that carotenoids may be critical factors in the oxygen cycle, in which singlet oxygen is quenched. Carotenoid contributions to the biochemistry of vitamin A (specifically β-carotene) are widely recognized, and these actions in the visual cycle rely on the presence of the β-ionone rings.

The teleological impact of carotenoids in flowering plants is also touted as a stimulus for the pollinating actions of insects.

Regardless of all the possible carotenoid functions cited, hard-core evidence regarding *in vivo* biochemical mechanisms is largely inadequate. Furthermore, in view of recent claims regarding the protective anticarcinogenic actions of some carotenoids in animals, mechanistic studies of carotenoid biochemistry are clearly overdue.

Carotenoids are responsible for the appearances of many edible fruits, vegetables, poultry tissues, animal fats, and the yellow to reddish colors of fish and marine animals.

Many carotenoids in botanical tissues are masked by overbearing color-reflective effects of chlorophylls. In the case of ripening fruits, however, the senescence of chlorophyll is often matched by the unmasking and/or enhanced biosynthesis of α- and γ-carotenoids plus other oxygenated carotenoids. Therefore, the total color impact of carotenoids on the appearance of botanical tissues reflects the relative spectral effects of all chromophores present including anthocyanins (Table 9.1). Apart from the occurrence of singular or multiple forms of carotenes, it should be noted that xanthophyllic-glycoside concentrations also increase during ripening of fruit and vegetable tissues.

Since thermal, photochemical, and oxygen effects are all counterproductive to the maintenance of native carotenoid colors, judicious food processing of fruits and vegetables may dictate the addition of antioxidants that prevent carotenoid bleaching. Carotenoid integrity during food storage may

Table 9.1. Some Coexisting Carotenoids in Foods

Apricot	β- and γ-carotene, lycopene
Carrots	α- and β-carotenes, lycopene, xanthophylls
Corn	β-Carotene and other carotenes (25–30%), xanthophylls including lutein, cryptoxanthan, and zeaxanthin (75–70%)
Vegetable oils	0.03–0.25% carotenoids (these carotenoids are subject to catalytic hydrogenation in processed oils; most colored species will be annihilated)
Egg yolk	Lutein, zeaxanthin, and cryptoxanthin
Peaches	Violaxanthin, cryptoxanthin, β-carotene, and over 20 additional carotenoid structures

also be enhanced by (1) maintenance of a nitrogen-saturated atmosphere surrounding carotene-rich foods (especially dehydrated foods); (2) blanching foods to eliminate lipoxygenase-type reactions that bleach carotenoids; or (3) effective freezing of food tissues.

Demonstrated consumer preference for well-pigmented, yellow poultry products, including egg yolks, dictates the economic advantages of producing heavily pigmented poultry cuts and products. Similar consumer acceptance and economic quality criteria are levied against animal products including butterfat, some milks, and cheeses. These desirable yellow colors in food substances are ensured by *pigmenters*. Pigmenters are generally defined as carotenoids, which when added to animal feeds, result in the yellow coloration of fat, skin, eggs, butter, or cheeses.

Carotenoid pigmenters are species specific for poultry versus ruminants. For example, poultry pigmentation can be achieved using vitamin A precursors such as cryptoxanthin; β-apo-8′-carotenal and β-apo-8′-carotenoic acid ethyl ester; or nonvitamin A precursors including lutein, zeaxanthin, and neoxanthin. All chickens (roasters, broilers, capons) rapidly convert β-carotene into vi-

tamin A, which does not serve as a pigmenter. The structure of β-carotene provides pigmenter effects to cattle, however, in addition to its provitamin A activity. This bipartite effect of β-carotene is a concurrent process. Whatever β-carotene is not metabolized to vitamin A will be deposited in adipose tissues and introduced into lacteal secretions.

Carotenoids also contribute to the characteristic orange, yellow, and red colors of shells, exoskeletons, and skins of crustaceans, mollusks, and fishes. Astaxanthins and numerous other carotenoids are probably produced by animal metabolism of ingested bicyclic carotenoids including β-carotene and lutein. However, heterotrophic nutrition of marine and freshwater species on lower life forms may result in the absorption and deposition of canthaxanthin and other intermediate carotenoid structures plus astaxanthins (Figure 9.11). The direct accumulation of these carotenoids in the predator animal species is possible, although the carotenoids may be subsequently altered.

Certain marine and freshwater animal species also exhibit carotenoid structures that are esterified to long- or short-chain fatty acids, and yet other invertebrates contain *carotenoproteins*. Carotenoproteins are key chromophoric factors in the blue and purple colors of numerous crustaceans. The native blue colors of certain carotenoproteins readily undergo heat destruction and free the characteristic astaxanthin or red canthaxanthin from their protein complexes. This action accounts for the familiar red color produced in cooked shrimp, lobster, and other related species.

Metabolism of Carotenoids

The mammalian metabolism of carotenoids largely revolves around their role in the visual cycle, but carotenoids may have other critical nutritional functions that have not been discovered.

It is generally believed that vitamin A biosynthesis from β-carotene requires the en-

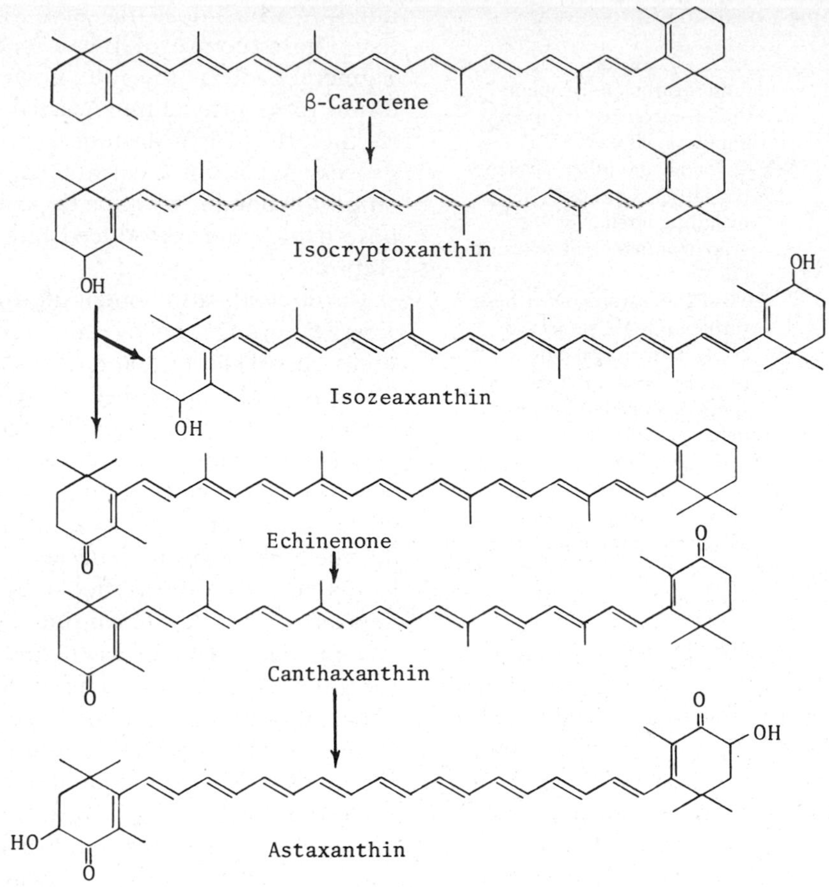

FIGURE 9.11. Structural transformations involving β-carotene and the biogenesis of other carotenoids in freshwater and marine animal species.

zyme-mediated scission of the C_{40} structure by a 15,15'-dioxygenase. The resulting C_{20}-retinals are reduced to retinols and subsequently esterified to palmitic (predominant), stearic, oleic, and linoleic acids to form retinyl esters.

β-Carotene conversions seem to occur exclusively in the intestinal mucosa. Furthermore, retinyl ester depositions in the body proceed by their assimilation into lymph chylomicrons. Lymphatic retinyl esters are then translocated to the liver following their vascular appearance, where they are contributed to the dynamic turnover of warehoused esters.

Aside from the intestinal formation of retinyl esters from β-carotene, humans have a limited ability to absorb β-carotene. Once absorbed, albeit in small amounts, β-carotene undergoes association with the LDL fraction of the blood lipoproteins. Chronic disease effects on vitamin A and carotenoid metabolism are still poorly understood when compared with many other dietary constituents. Alternative oxidative routes for β-carotenes may lead to their terminal oxidation, as indicated in Figure 9.12.

QUINONES

Quinone pigments are widely distributed among the fungi, algae, lichens, and bacte-

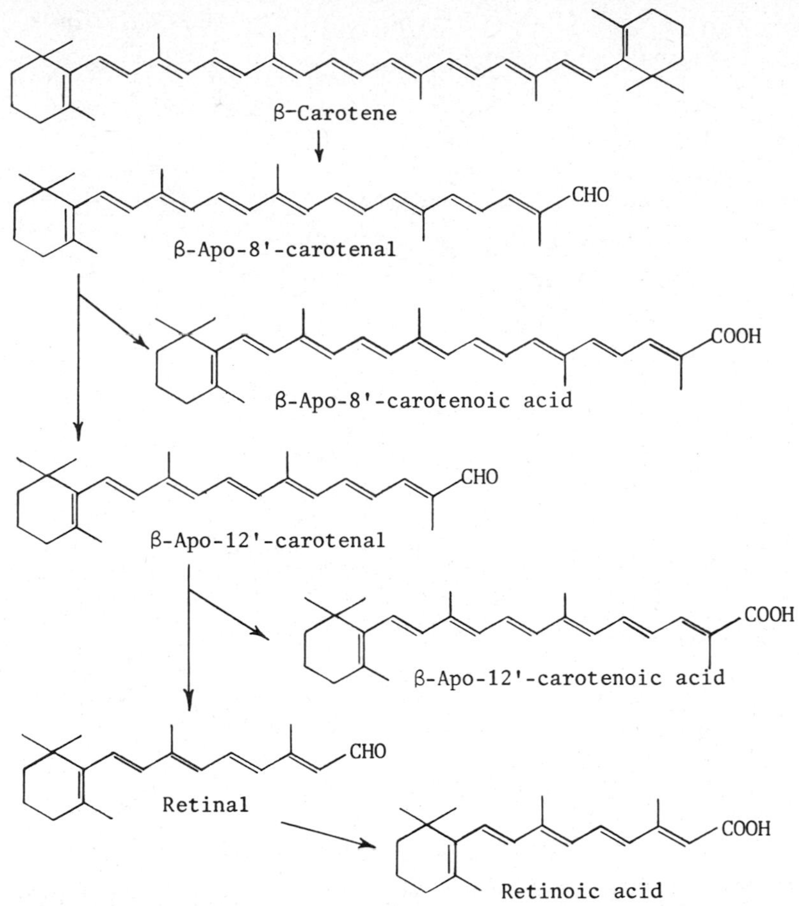

FIGURE 9.12. Some carotenoid degradation products that result from the directed terminal oxidation of β-carotene.

ria. These pigments also reside in roots, bark, and to a limited extent within leaves and flowers. Pigment colors displayed by the group ranges from pale yellows to nearly black.

Quinones may be generically subdivided into three groups designated as *benzoquinones, naphthaquinones,* and *anthraquinones.* Representatives of these three groups appear in Figure 9.13.

The biochemistry of native quinones is complex, and despite their wide occurrences, they are poorly understood. Biosynthesis may follow the acetate–malonate pathway or in the case of higher plants, a cinnamate–flavonoid pathway may be involved.

A variety of anthraquinones, including rhein, are recognized for their potent purgative actions, but apart from trace quantities in some foods, these agents probably have little importance other than their color effects.

BETALAINS

Betalains are classified as a structural group of *chromoalkaloids* that display red and yellow colors. Those structures having red hues are called *betacyanins* and the yellow varieties are *betaxanthins* (Figure 9.13). Note that the inversion of the —H,—COOH configuration

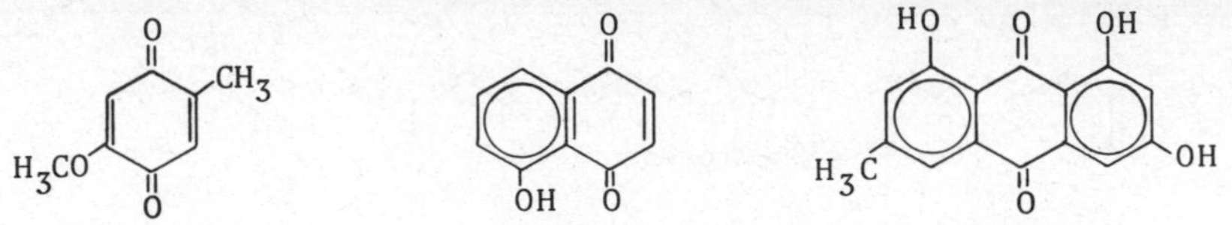

Copronin
(Benzoquinone)

Juglone
(Naphthaquinone)

Emodin
(Anthraquinone)

Some common quinone structures

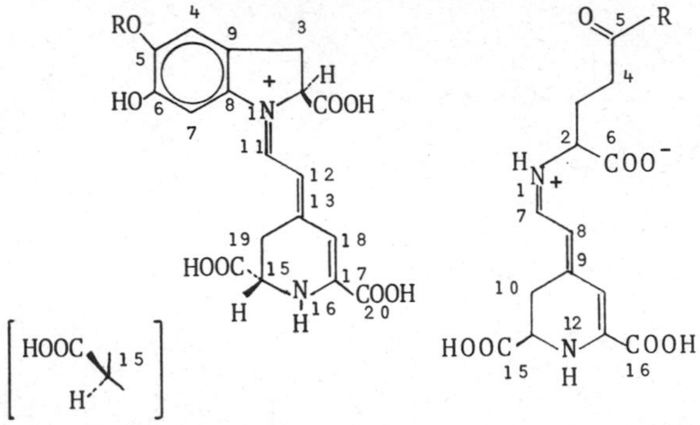

"iso-" Structures

Betanidin R=-H
Betanin R=glucose (β(1→0-))

Isobetanidin R=-H
Isobetanin R=glucose (β(1→0-))

Vulgaxanthin I   R=-NH$_2$
Vulgaxanthin II R=-OH

Betalain structures

FIGURE 9.13. Structures for some notable quinones and betalains. Betalain structures include the aglycones of betanidin and isobetanidin as well as the glycosidic epimers betanin and isobetanin. Inversion of the C-15 configuration gives rise to two isomeric molecules (i.e., betanin and isobetanin).

at C-15 determines the structural difference between the iso- (epimeric) forms of these pigments. Two yellow betaxanthins are also recognized: *vulgaxanthins I* and *II*.

In addition to these pigments, the mono-sulfate esters of *betanin* and *isobetanin* lead to the corresponding formation of prebetanin and isoprebetanin. Betalain structures may also exist as acylated structures with acids including caffeic, citric, ferulic, 3-hydroxy-3-methyl glutaric, malonic, *p*-coumaric, or sinapic acids.

Red beets are the most familiar source of betalain pigments. Nearly 80 to 95% of total beet pigmentation is attributable to betanin, followed by the contributions of isobetanin, prebetanin, and isoprebetanin—in that order.

The biochemical function and synthesis of the betalains remains to be firmly established. Moreover, their biological functions are unclear and usually parallel the "insect- and bird-attractive" teleologies that promote plant pollination and seed dispersal. However true, these explanations fail to firmly demonstrate the reason for high betalain concentrations in the corms, tubers, tap-roots, and stems of many plants. Other common sources of betalains include poke-berries, cactus fruits, amaranthus, and bougainvillaea.

Special interest in betalains has surfaced owing to the toxic properties of many synthetic red dyes historically used in foods. The nontoxic properties of natural chromoal-kaloids offer tempting possibilities for their introduction into foods as an alternative to synthetic dyes. In view of this objective, betanin can be prepared from beet roots as a liquid extract and spray dried into a powdered form. This powdered extract can then be introduced into foods according to government-dictated guidelines. Unlike the anthocyanins, betalains (e.g., betanin) are largely pH stable, but they are subject to deteriorative effects of ultraviolet light, humidity, oxygen, and temperature.

NATURAL COLORANTS USED IN FOODS

Natural food colorings include extracts from plant materials, vegetable extractives, juices from selected fruits, spices, oleoresins, and several *synthetic* carotenoids. Based on congressional Food, Drug and Cosmetic (FD&C) legislation of 1960, these "natural" colorings are also designated as "uncertified color additives." This permanent status permits their acceptable use as coloring agents for animal feeds and human foods. Although synthetic carotenoids are clearly "unnatural," they are "nature identical" and thus considered to be safe and "natural."

Natural Yellow to Orange Hues

Alkaline or oil extractions of seed coats produced by the tree *Bixa orellana,* provide the principal source of annatto. This extract is red in concentrated form, or when diluted, it produces a peach to butter-yellow color. Annatto extracts serve as the basis for two commercial forms of annatto coloring. One lipophilic fraction contains the C_{24}-carotenoid *bixin,* whereas a potassium hydroxide saponification extract yields hydrophilic *norbixin.*

Annatto has historical preeminence as an early food-coloring agent owing to its wide use in standardizing butter colors, producing yellow colors in dairy products (e.g., cheeses, ice creams), and coloring specialty products (e.g., batters, imitation milks, cof-

fee whiteners, whipped toppings, confections).

The effective coloring potential of annatto is dictated by the percentage of norbixin present in the commercially prepared coloring agent (e.g., color intensity), admixed pigment carriers, and dispersants associated with the key coloring ingredient.

So long as a $\leq 1{:}400$ ratio between annatto and food product is maintained, flavor contributed by annatto will be minimal. In cases of saponified annatto, the residual potassium hydroxide may impart some minor flavor effects to water-soluble annatto.

Vegetable-oil carriers for annatto extracts ensure the homogeneous coloring of high-fat and oil-based foods. Vegetable-oil carriers can also serve as a vehicle for coloring aqueous food systems, provided that the coloring is dispersed over salt or sugar before its admixture to the food system.

Bixin-type carotenoids are quite stable toward potential oxidative reactions; however, like all carotenoids, they must be protected from temperature and photo-induced effects that cause bleaching and fading.

Rhizomal extraction of *curcumin* from the turmeric plant (*Curcuma longa*) serves as the basis for another natural yellow coloring agent. Turmeric colorants are insoluble in distinctly polar or nonpolar food systems, but they are soluble in acetone and ethanol. In the case of good-quality dried turmeric powders, the curcumin content should approach concentration levels of 90 to 95%.

Dispersable and liquid forms of turmeric are commonly employed in food processing. Dispersable turmeric may be carried in a vehicle of propylene glycol or polysorbate 80, or alternatively spray-dried on one of numerous hydrophilic polymers such as dextrins or gum arabic. Liquid turmerics are usually dark brown owing to the high concentration of the yellow pigment. Contrary to the recognized spicing effects of turmeric, its use as a coloring agent does not modify food taste in most cases. Turmeric extracts are susceptible to heat-, photo-, oxygen-, and humidity-induced decomposition effects. In those cases in which intermediate bixin-turmeric colors are necessary in foods, blends of annatto and turmeric in oils or propylene glycol containing a suitable emulsifier may be employed.

Orange to Red Hues

Extraction of the dried fruits of *Capsicum annum,* followed by dispersion of the extract in water, polar solvents, or vegetable oils, yield a bright red coloring agent recognized as paprika. This coloring has many food applications after it is dessicated on a carrier vehicle and/or mixed with an appropriate food emulsifier in a dry form.

The paprika not only imparts color to foods, but it also contributes a sweet slightly pungent spice flavor. Meat products, salad dressings, complex spice mixtures, condiments, cheeses, and ethnic foods commonly rely on the coloring effects of paprika.

The synthetic carotenoids including β-carotene, β-apo-8′-carotenal, and canthaxanthin all produce yellow-orange-reddish hues in foods depending on pigment concentration levels. The effective use of these colorants is contingent upon *minimizing* their

reactivity and degradation in foods as well as their effective dispersal (Figure 9.14).

Carotenoid dispersion throughout foods can be ensured by their dissolution in vegetable-oil solvents or the manufacture of carotenoid-containing beadlets. Dilute suspensions of the desired carotenoid colorant then serves as the principal means for coloring foods. Aside from colorant properties, β-carotene and apocarotenal both demonstrate vitamin A activity.

As indicated in an earlier section dealing with betalains, both betanin and vulgaxanthin offer food-coloring potentials. Their application to foods has been plagued by color-uniformity problems and pigment-stability problems within food systems. Nevertheless, beet juices have commercial availability as either a food colorant stabilized with food acid or in dried solid forms on powdered hydrophilic macromolecules (e.g., gum arabic, dextrins).

Additional colored fruit juices and botanical extracts have potential uses in foods. Unfortunately, long-term stability, narrow pH tolerances, and lack of visual clarity jades their more common use. Saffron has been used as an effective colorant, but its high cost

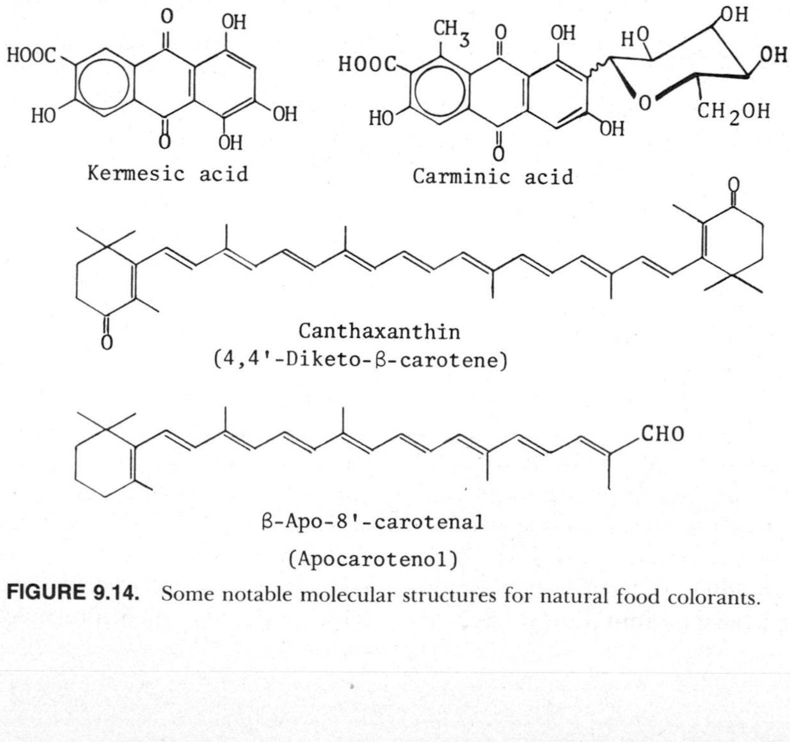

FIGURE 9.14. Some notable molecular structures for natural food colorants.

precludes common use throughout the food industry.

The demise of many synthetic red food dyes has rekindled interest in the use of *cochineal dyes*. This dark red colorant is expressed from the bodies of the cochineal insect (*Dactylopius coccus*), which is most common to Peru and the Canary Islands at this time (Figure 9.14).

The red chromophores of scale insects are largely attributable to the native structures of anthroquinones such as *kermesic* and *carminic acids*. The polyketidic structure of the red colorant is stable in water, alcohols, ethers, acid, and alkaline solutions, but insoluble in nonpolar solvents such as petroleum ether or chloroform. Carminic acid structures exhibit pH-dependent colorations and high lability at low temperatures. The colorant is yellow at pH 4.8, bluish-red above 6.2, and a brilliant carmine between these two extremes. At 135°C, the anthroquinone-based pigment decomposes. Carminic acid accounts for about 10% of dry insect weight; in insects it may serve as a fundamental biochemical energy source in adults as well as eggs.

Food colorant applications for carminic acid have been successfully stabilized by the formation of an aluminum lake that has water-soluble properties. The pink to red hues of carminic acid-based colorants are stable in many proteinaceous food systems that must be subjected to retorting, but the current cost of the pigment counters its widespread use.

Miscellaneous Food Colorants

Controlled heating, sintering, or browning reactions of carbohydrates such as sucrose, dextrose, lactose, invert sugar, molasses, or starch hydrolyzates (*DE* 90–95) yield various forms of caramel coloring. Liquified or dried caramel colorings are commonly used depending on application requirements. Liquid forms are used to produce dark rye breads and chocolate colors in baked goods and fillings, whereas powdered caramels are used to produce dark colors in dried food mixes. Defatted and toasted cottonseed flours also serve as a quasi-caramelized coloring agent for dried mixes.

Extraction of the mauve coloring from grape skins provides *enocianina*. Economic and availability problems counter the popular use of this colorant in products other than nonalcoholic beverages and beverage mixes.

The commercial and food uses of these and other natural colorants will probably continue to grow as existing synthetic colorants continue to be indicted for health and safety reasons.

OBJECTIVE QUANTITATION OF COLOR APPEARANCE

Human perception of color depends on the characteristic distributions of light energies that interact with retinal cones and rods. Only visible wavelengths of light between the violet and red ends of the spectrum can be detected.

It is generally recognized that all food solids and liquids owe their respective colors to the selective absorption of different amounts of light energies intervening between the range of 400 and 700 nm. Those wavelengths that are not absorbed undergo reflection or transmission from various food substances and may be visible to an observer. Superficial study shows that red substances characteristically absorb green light, yellow substances absorb blue, and so on.

Food appearances are a composite effect of their total morphology, but apart from this, food appearance depends on both *chromatic* and *geometric distributions* of absorbed or reflected light.

Chromatic attributes of food appearance can be expressed and measured by spectrophotometric curves where 400 to 700-nm wavelength is plotted against the fractions of incident light (transmitted or reflected). Some ideal spectrophotometric curves for colored substances have been detailed in Figure

9.15A. The uniform reflection (no absorption) of specific wavelengths over the entire visible spectrum produces a near horizontal plot for white light, whereas "uniform partial" or "complete" light absorption respectively correspond to gray or black perceptions. Remaining plots in Figure 9.15A show that chromatic reflectance is greatest over the wavelengths corresponding to a given color and distinctly lower at other wavelengths.

The spatial distribution of reflected light over a food substance markedly affects food appearance. Four types of spatial or geometric light distributions are possible depending on the solid or liquid structure of a food substance. These classifications include specular (I) or diffuse (II) *transmissions* of light and diffuse (III) or specular (IV) *reflections* of light (Figure 9.16A).

Specular transmission (I) is characterized by the unaltered passage of a light beam through a clear solid or liquid medium. Reflection and dispersion of the transmitted light is minimal, although a colored medium may alter the wavelength of transmitted light. As an example of specular transmission, consider the incidence of white light that passes

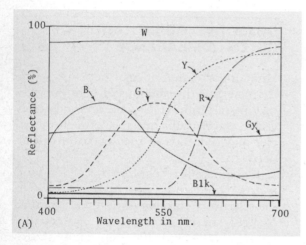

(A)

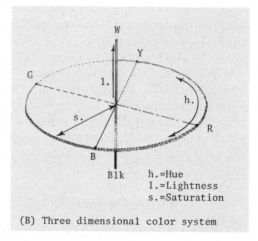

(B) Three dimensional color system

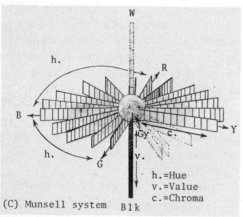

(C) Munsell system

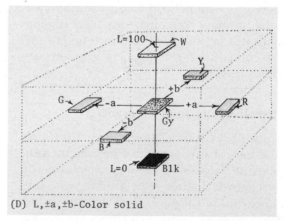

(D) L,±a,±b-Color solid

FIGURE 9.15. (A) Typical light reflectance curves for blue (B), green (G), yellow (Y), red (R), gray (Gy), white (W), and black (Blk) objects from 400 to 700 nm. (B) Three-dimensional color system illustrating the three important attributes of color. (C) The Munsell color solid showing the relative correspondence to B. (D) The opponent color scale concept using the L,a,b color solid. (Note that $L = 10.0\sqrt{Y}$; $a = 17.5(1.02X - Y)/\sqrt{Y}$; $b = 7.0(Y - 0.847 Z)/\sqrt{Y}$.)

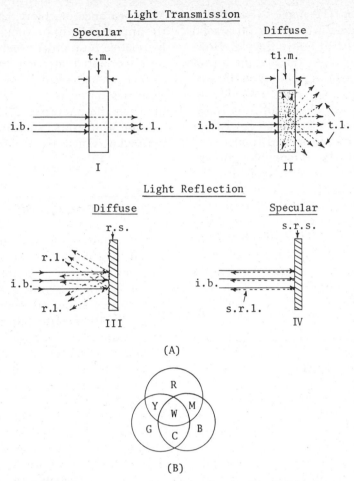

FIGURE 9.16. (A) Types of transmitted and reflected light. See text for details. Key: (I) i.b., incident beam of white light; t.m., light-transmitting medium, which is totally transparent or colored (solid or liquid); *note*: for colored solutions, the apparent color of solution is always the complement of the "color" absorbed (e.g., a yellow solution absorbs in the blue region (465–485 nm); t.l., transmitted light. (II) tl.m., translucent solid or liquid medium, which causes diffusion of the transmitted light (t.l.). (III) incident beam of light is redirected as diffuse reflected light (r.l.) from a reflective surface (r.s.) (c.g., most solid foods). (IV) Incident beam of light is totally reflected (s.r.l.) from a specularly reflective surface (s.r.s.) back over the path of the incident light beam (mirror-like reflection). (B) Additive mixing of light, R + B + G yields white light (W), whereas the mixing of R + B produces magenta (M), B + G produces cyan (C), and R + G gives yellow light.

through cranberry juice to produce red transmitted light.

Diffuse transmission (II) of incident light occurs when a beam of white light traverses a translucent medium. Food liquids such as citrus fruit juices having high suspended solids exhibit this behavior. The incident light beam is not only transmitted as the apparent color of the translucent medium, but the translucent medium produces an omnidirectional dispersion of the incidental transmitted light beam.

Diffuse reflection (III) implies that the interaction of incident white light on a solid food will be absorbed and reflected. However, the reflected light can assume an infinite range of angles depending on the surface structure and texture of the solid. This behavior is typical of almost all food solids including meats, vegetables, baked goods, and countless varieties of processed foods.

Specular reflection (IV) is rare among foods, but it indicates that the incident light is directionally reflected from a liquid or solid medium. The highly directional reflection may permit a mirror-like perception of the incident light. This aspect of reflection is responsible for the glossy appearance of some foods including certain icings, glazes, lipid or sugar bastings, and so on.

ATTRIBUTES OF COLOR APPEARANCE

Spectrophotometric curves for reflected and transmitted lights offer an accurate method for evaluating its chromatic characteristics, but they are an inadequate method for reporting the human perception of color. The human eye can detect three distinct primary light colors namely red (R), green (G), and blue (B), and according to psycho- and neurochemical processes that remain to be established, colors are perceived. Exclusive of R, G, and B perceptions, the extent of additive mixing among the light primaries can be detected as yellow (Y), cyan (C), and magenta (M). The collective additive mixing effect of R + G + B lights leads to the perception of white light (Figure 9.16B).

The collective psycho- and photoneurochemical responses to perceived light on the retina of an observer permits the description of three color attributes: (1) hue, (2) saturation, and (3) value or brightness. These color-appearance indices can be visualized according to a three-dimensional color system (Figure 9.15B).

Hue corresponds to the R, Y, G, C, B, and M appearance of color. Saturation defines the apparent lightness of a hue according to its relative position between the gray (lightness) axis and a saturated hue. A pure color (hue) has a high saturation, but a pastel color (hue) has low saturation. Value or brightness indicates the intensity of light reflection or transmission from a substance. Value is also defined as a measure of luminous intensity.

COLOR APPEARANCE SCALES

Since surface color specifications as perceived by a human observer cannot be accurately reported by simple spectrophotometric methods, other types of color appearance scales have been developed. These scales include the classical Munsell system, the CIE (*tristimulus*) system, and the opponent-colors (L, $\pm a$, $\pm b$) color scales.

The Munsell system of color notation is based on a tridimensional system (Figure 9.15C) that considers hue, chroma (saturation), and value. The hue scale is based on five color primaries (R, Y, G, B, and purple (P)) plus their respective intermediate hues (YR, GY, BG, PB, and RP). Similarly, the coordinates of value and chroma are also denoted by numbers. Value ranges from black (0) to perfect white (10), and the chroma scale ranges from achromatic (0) to some number that indicates maximum saturation.

These fractional perceptive attributes of color have been published in the *Munsell Book of Color* as colored "chips." The hues are arranged in this book according to rows that correspond to *chroma* scales and in columns according to *value* scales. Using the color attributes of the Munsell system, surface colors of foods and agricultural products (raw or refined) can be compared to the charted colors. A complete color notation is typically given as hue-value/chroma (e.g., 9BG 6/10). Modifications of the Munsell system including colored chip mounting on rotating disks have also been devised for specifying color perception.

This historical development of color science and measurement has extensively relied on the tristimulus system and the recommendations of the International Commission on Illumination (Commission Internationale de L'Eclairage or CIE). In 1931, the CIE established a so-called *tristimulus color system*. This system permitted the graphic coordinate plotting of color appearance and defined the color-matching response functions of a standard human observer—all of which were designed to specify spectrophotometric color data.

Using trichromatic systems, any color can be matched by a suitable mixture of three primary colors: X (red), Y (green), and Z (blue), using Y as an index of luminosity. Accordingly, the relative amounts of X, Y, and Z necessary to match a color are reported as color tristimulus values (*TV*'s). Based on the CIE system the *amount* of any primary (X,Y,Z) at a specific wavelength can be given as $\bar{x}$, $\bar{y}$, and $\bar{z}$. The $\bar{x}$, $\bar{y}$, and $\bar{z}$ values, also known as *distribution coefficients* (*DC*'s), are "unitless" values that can be obtained from CIE tables or graphic plots. These unitless values serve as multiplication values that specify the amounts of radiant energy at every contributing wavelength of a color dictated by the interaction of its X, Y, and Z primaries (Hardy, 1936; Judd and Wyszecki, 1963; Wright 1958; Committee on Colorimetry, 1953). Using this rationale, specific *TV*'s can be calculated for reflected or transmitted colors over the range of 400 to 700 nm.

$$X = \sum \bar{x}_\lambda E_\lambda R_\lambda \Delta_\lambda \quad Y = \sum \bar{y}_\lambda E_\lambda R_\lambda \Delta_\lambda$$

$$Z = \sum \bar{z}_\lambda E_\lambda R_\lambda \Delta_\lambda$$

The respective *DC* values are given as $\bar{x}$, $\bar{y}$, or $\bar{z}$; E_λ is the spectral energy distribution of the illuminant; R_λ corresponds to the spectral reflectance of the sample; and Δ_λ is the wavelength interval. Inspection of the previous mathematical expressions indicates that X, Y, and Z are integrals of $\bar{x}$, $\bar{y}$, and $\bar{z}$ each multiplied by the spectral distribution of the color stimulus.

Plots of $\bar{x}_\lambda E_\lambda R_\lambda$, $\bar{y}_\lambda E_\lambda R_\lambda$ . . . , and so on, versus wavelength followed by integrating the curve areas permits *TV* calculations. Using the respective *TV*'s for X, Y, and Z, the relative proportions of each primary can be calculated from the ratios:

$$x = \frac{X}{X + Y + Z} \quad y = \frac{Y}{X + Y + Z}$$

$$z = \frac{Z}{X + Y + Z}$$

The x, y, and z values are specifically recognized as *chromaticity coordinates* or *trichromatic coefficients*. Since the summation of chromaticity coordinates (x, y, and z) totals unity (1.0), chromaticity can be defined using only x and y. It should also be recognized that (1) color specifications are often made with respect to Y, y, and x and (2) *TV*'s may be expressed as the percentage of X, Y, and Z (X%, Y%, Z%) of a standard white reference surface.

UNIFORM COLOR SCALES

Apart from the absolute definition of color by CIE indices, these scales do not provide a routinely useful method for defining perceived colors. This inefficacy has resulted in the development of specialized "uniform color scales" based on the CIE system. These scales are useful because they serve as "appearance scales," not as "color-matching scales" of the CIE system.

One of the most popular uniform color scales is based on the *opponent-color theory* or color vision. According to this theory, retinal perception of red versus green, and blue versus yellow are mutually exclusive. Neural mechanisms operating between the retinal cones and the optic nerve somehow serve to produce red-to-green and yellow-to-blue color dimensions. Although the neuronal switching mechanisms responsible for these color dimensions have not been isolated, the empirical presence of these color dimensions has been operationally recognized since the work of Hering in 1878. Moreover, neuronal impulse monitoring over the optic nerves of primates support the presence of an opponent-color vision theory as opposed to X, Y, Z signal (neuronal) correlations.

One of the most popular and routinely useful opponent-color scales was introduced by Hunter in 1942. According to this system, optical sensors produce voltages proportional to CIE $x, y,$ and z values for color, which are then processed into X, Y, Z and $L, \pm a, \pm b$ scales. The $L, \pm a, \pm b$ scale is a tricoordinate scale that corresponds to black–white (L), red–green ($\pm a$), and yellow–blue ($\pm b$).

Extreme values of the $\pm a$ and $\pm b$ scales correspond to saturated red ($+a$), green ($-a$), yellow ($+b$), and blue ($-b$), respectively (Figure 9.15D). The lightness term is a nonlinear function such as the square or cube root of Y (i.e., Y is the percentage reflectance or transmittance). A variety of other color systems and instruments are available for the evaluation of foods, but their treatment is beyond the scope of this book. Moreover, the $L, \pm a, \pm b$ scale has broad applications to many food systems as opposed to some other methods, which have restricted applications.

CONCLUSION

Color is a critical component in the overall sensory evaluation of foods. For this reason, objective instrumental methods of analysis, geared to quantitating food appearances, have gained wide use in the food industry. Instrumental evaluations of color appearance can serve as a basis for objectively determining food color changes caused by food freezing, repeated freeze–thaw cycles, acid–base fluctuations, reconstitution of freeze-dried foods, radiation preservation, momentary or prolonged air contact, mechanical handling, and many other unit-processing steps involved in food handling. The object of such data acquisition is usually designed to detect color abnormalities in foods or somehow *improve* or *reestablish* the optimal color of a foodstuff. Objective quantitation of food appearances also underlies industry- and government-wide standards of identity for certain cheeses, dairy products, citrus juices, syrups, sausages, and cured meats as well as other foods.

Aside from determining reflected or transmitted light by tristimulus or uniform color scales, spectrophotometric analyses are most critical for the analytical quantitation and characterization of key food pigments. The peculiar infrared, ultraviolet, and visible absorption spectra of isolated or mixed pigments not only serve as a basis of their quantitation and identification but also provide evidence of pigment reactivity and deterioration in foods.

A detailed quantitative account of pigments in plant materials requires painstaking and refined isolative procedures. These methods typically require the chromatographic isolation of key pigments followed by their subsequent spectrophotometric and/or spectroscopic analyses. For a detailed prospectus of analytical pigment chemistry and identification, the following authoritative references should be consulted: Bauernfeind (1981, carotenoids), Markakis (1982, anthocyanins, anthocyanidins, leukoanthocyanidins, etc.), von Elbe and Maing (1973) and von Elbe *et al.* (1974, betalains).

The advent and wide applications of high-pressure liquid chromatography have also expedited the isolation, identification, and quantitative resolution of plant pigments. For studies involving crop quality control or plant breeding objectives, this can be a very useful analytical method.

Genetic plant breeding studies in particular benefit from such a direct analytical method since multiple ratio changes between chlorophyll-a content and accessory pigments (neoxanthin, violaxanthin, lutein, β-carotene, and chlorophyll-b) mirror greening rates and the photochemical dynamics of chloroplast structures.

REFERENCES

Adams, J. B. 1973. Thermal degradation of anthocyanins with particular reference to the 3-glycosides of cyanidin: In acidified aqueous solutions at 100°C. *J. Sci. Food Agr.* **24**:747.

Baranyovits, F. L. C. 1978. Cochineal carmine: An ancient dye with a modern role. *Endeavor* (new series) **2**(2):85.

Bauernfeind, J. C. 1981. *Carotenoids as Colorants and Vitamin A Precursors.* Academic Press, New York.

Brouillard, R. 1982. Chemical structure of anthocyanidins. In *Anthocyanins as Food Colors* (P. Markakis, ed.), pp. 1–40. Academic Press, New York.

Brown, Jun., K. S. 1975. The chemistry of aphids and scale insects. *Chem. Soc. Rev.* **4**(2).

Committee on Colorimetry, Optical Society of America. 1953. *The Science of Color.* Crowell, New York.

Fox, J. B. 1966. The chemistry of meat pigments. *J. Agr. Food Chem.* **14**:207.

Francis, F. J. 1982. Analysis of anthocyanins. In *Anthocyanins as Food Colors* (P. Markakis, ed.), pp. 182–207. Academic Press, New York.

Francis, F. J. 1969. Pigments content and color in fruits and vegetables. *Food Technol.* **23**:32.

Goodwin, T. W. 1976. *Chemistry and Biochemistry of Plant Pigments.* Academic Press, New York.

Hardy, A. C. 1936. *Handbook of Colorimetry.* MIT Press, Cambridge.

Harborne, J. B. 1967. *Comparative Biochemistry of the Flavonoids.* Academic Press, New York.

Hunter, R. S. 1975. *The Measurement of Appearance.* Wiley–Interscience, New York.

Inoue, N., and T. Motohiro. 1971. A cause and mechanism of blue discoloration of canned crab meat. VII. Further study on the causative substance of blue meat. VIII. Some observation of clotting of crab haemolymph. *Bull. Jap. Soc. Sci. Fisheries* **37**(10):1007–1014.

Judd, D. B., and G. Wyszecki. 1963. *Color in Business, Science and Industry,* 2nd ed. Wiley, New York.

Landrock, A. H., and G. A. Wallace. 1955. Discoloration of fresh red meat and its relationship to film oxygen permeability. *Food Technol.* **9**:194.

Markakis, P. (ed.). 1982. *Anthocyanins as Food Colors.* Academic Press, New York.

National Academy of Sciences. 1976. Appearance. In *Objective Methods for Food Evaluation, Proceedings of a Symposium. November 7–8, 1974,* pp. 215–243. NAS, Washington, D.C.

Osawa, Y. 1982. Copigmentation of anthocyanins. In *Anthocyanins as Food Colors* (P. Markakis, ed.), pp. 41–68. Academic Press, New York.

Simpson, K. L., T. Katayama, and C. O. Chichester. 1981. Carotenoids in fish feeds. In *Carotenoids as Colorants and Vitamin A Precursors,* pp. 463–538. Academic Press, New York.

Simpson, K. L., and T. Kamata. 1979. Use of carotenoids in fish feeds. *Proceedings of the World Symposium on Fin Fish Nutrition and Fish Feed Technology, Hamburg, 20–23 June, 1978,* Vol. II.

Thomson, R. H. 1971. *Naturally Occurring Quinones,* 2nd ed., pp. 418–472. Academic Press, New York.

von Elbe, J. H., I.-Y. Maing, and C. H. Amundson. 1974. Color stability of betanin. *J. Food Sci.* **39**:334.

von Elbe, J. H., and I.-Y. Maing. 1973. Betalains as possible food colorants of meat substitutes. *Cereal Sci. Today* **18**:263–264, 316–317.

Waldman, G. 1983. *Introduction to Light, the Physics of Light, Vision and Color.* Prentice-Hall, Englewood Cliffs, N.J.

Wright, W. D. 1958. *The Measurement of Colour.* Macmillan, New York.

Zapsalis, C., and F. J. Francis. 1965. Cranberry anthocyanins. *J. Food. Sci.* **30**:396–399.

CHAPTER · 10

Food Flavors and Flavor Perception

INTRODUCTION

Flavor is the composite sensation of taste on the tongue and odor in the olfactory center of the nose when food is placed in the mouth. In addition, the total sensation of food flavor depends on the mouth feel of the food. Mouth feel for a food substance is a psychophysical sensation or interpretation of smoothness, stickiness, brittleness, viscosity, or any other surface features that are identified with the texture and physical characteristics of a food.

Mouth feel represents a *kinesthetic* property of foods that becomes apparent only during food mastication. During this action, muscles of the oral cavity including those that regulate tongue, cheek, and swallowing actions plus those senses responsible for *haptic* (touch or tactile) impressions all contribute to a mouth feel impression of a food. Kinesthetic and haptic sensations are clearly important in flavor evaluation and the appreciation of food. For example, the heavy-bodied but smooth texture of peanut butter is a recognized property of that food, whereas a very smooth hydrated puree of fresh apples, having lost their natural texture and succulence, may offer little appeal in spite of retaining the natural fruit flavor.

Vision and hearing also have roles in fla- vor sensation. The taste perception of green-colored mashed potatoes does not parallel the taste of the same mashed potatoes in their natural uncolored state. Furthermore, the auditory "crunch" caused by biting into an apple adds flavor appeal to the fruit.

The complexities in defining the nature and character of food flavors have led to a variety of classic definitions. One typical definition claims that

> *Flavor is the sum of those characteristics of any material taken in the mouth, perceived principally by the senses of taste and smell and also the general pain and tactile receptors in the mouth, as received and interpreted by the brain.*

This definition, attributed to Hall, assigns most importance to the sensation and stimulus aspects responsible for flavor, but the Society of Flavor Chemists defines flavor from a more universal perspective:

> *A flavor is a substance which may be a single chemical entity, or a blend of chemicals of natural or synthetic origin whose primary purpose is to provide all or part of the particular flavor effect to any food or other product taken in the mouth.*

Both of the definitions cited are accurate depending on the food or food system being considered.

Chemical compounds recognized to have tastes and odors are numbered in the thousands, and this number will undoubtedly continue to grow as flavor technology progresses. The importance of a nearly limitless spectrum of flavor compounds may be of only academic importance, however, since many distinctly different types of compounds exhibit similar flavor qualities. Outside of being identical in their apparent flavors, the relative *intensities* of similarly flavored compounds may account for their most significant differences. Moreover, it should be recognized that there are far more compounds with flavor than detectable types of flavor sensations.

The complexity of flavor compounds expands to enormous dimensions when foods are cooked, processed, prepared, or stored. Characteristic flavor compounds in a raw foodstuff may be partially retained, but complex reactions usually produce a host of chemical by-products. Some of these by-products elicit delightful flavor sensations and others are nothing short of nauseating. These reactions can be mediated by enzymes, heat, light, retorting, radiation, dehydration, and fermentation reactions.

Apart from occasional references to the origin of flavors during cooking as well as the kinesthetic, visual, and cultural factors that may affect food flavors, this chapter is devoted to the principal taste and flavor factors present in raw food materials.

TASTE PERCEPTION

Four true taste sensations are recognized: *salty, sour, bitter,* and *sweet.* These sensations originate from chemoreceptors located in the taste buds. Most of the taste buds reside on the tongue surface, but others are located on the hard and soft palates, the epiglottal mucosa, the bottom of the tongue, and the floor of the mouth depending on the individual. Many of the taste buds are embedded in a variety of papillary structures that cover the dorsum (top) of the tongue. Four types of papillary structures are recognized (e.g., vallate (the largest), foliate, fungiform, and filiform), but these different papillae show little individual discrimination for specific tastes.

Vallate papillae are endowed with the greatest number of buds (~100–175), whereas most other forms have fewer numbers. Taste buds are speculated to undergo replacement every 200+ h and decline in overall numbers from infancy toward old age. Infants in particular have taste buds that cover a large portion of the mouth and may account for their otherwise unreasonable and obtuse acceptance of many foods.

Some type of a parallel between the four basic taste sensations and the four distinct types of chemoreceptors have been widely suspected for years, but this remains to be convincingly demonstrated. Nevertheless, a regional survey of tongue sensitivity reveals discrete locations for maximum taste sensitivities to sour, sweet, salty, and bitter agents (Figure 10.1).

Since distinctly different chemoreceptors have not been experimentally isolated, many neurophysiologists claim that a single receptor in taste buds responds to three or more tastes. Depending on the apparent sensitivity of any taste bud then, its chemoreceptor activity may simply reflect a greater sensitivity to one taste than others. The taste discrimination of a single receptor is probably linked to its own specific neural code and its ability to initiate a specific neural firing pattern originating from taste-specific receptors. Fundamental differences in firing patterns may cause perception differences in sweetness, saltiness, and so on. Thus, specialized taste maps that depict tongue taste regions probably reflect specialized clusters of certain receptors.

The overall taste sensation depends on the (1) spatial and temporal patterns of nerve activity as well as (2) genetically dictated chemoreceptor functions over the surface of the tongue.

In the first case, individual taste receptors

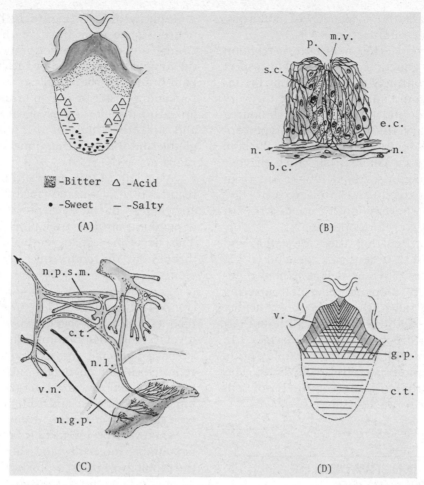

FIGURE 10.1. Key sensory structures in tongue taste sensations. (A) Taste bud sensitivity regions; (B) cellular anatomy of a taste bud; (C) neural innervation of peripheral taste receptors; and (D) regions of the tongue innervated by specific nerves. Illustration key: (B) b.c., basement cells of the taste bud; m.v., microvillus structure; n., nerve to chemoreceptors within microvilli; p., taste bud pore, which permits molecular interaction with receptors(s); s.c., sensory cells, which are surrounded by epithelial cells (e.c.). (C) Primary nerves innervating the tongue: c.t., chorda tympani; n.g.p., glossopharyngeal nerve; n.l., lingual nerve; n.p.s.m., superficial petrosal nerve; v.n., vagus nerve. (D) Gray area of tongue (v.) is innervated by the vagus nerve (v.n.), the cross-hatched region (g.p.) is controlled by the glossopharyngeal nerve (n.g.p.), and the horizontal-lined region relies on the chorda tympani (c.t.).

that have qualitatively different taste responses must eventually converge their stimuli into primary neurons. The collective neural response of these primary neurons as interpreted by the brain probably determines the quality of taste. The temporal pattern of taste perception is caused by the physical location of taste buds. As an example, consider the fact that taste receptors responding to sweetness are clustered on the tip of the tongue and bitter receptors are located toward the rear. If a food contains bitter and sweet substances, taste perception will proceed from sweet to bitter as the food progresses to the base of the tongue. This sequence of perception has distinct effects on the qualitative evaluation of multicomponent taste mixtures.

The second effector of taste perception involves a genetic component. This aspect has been routinely demonstrated by the inability of some subjects to detect the very bitter compound known as phenylthiourea and other aryl thiocarbamates. This perception failure occurs without upsetting the taste thresholds for sweet, salt, sour, and even other bitter compounds. About 40% of American Caucasians demonstrate this recessive genetic trait, whereas only 3% nontasters exist among the Korean population.

Other factors that affect the extent and character of taste sensations depend on pH variations in saliva as well as its native enzyme activity. Depending on pH buffering capacity of saliva, the perception of acidity can be assessed differently by different test subjects. Moreover, salivary enzymes can create rapid changes in certain peptide structures and especially carbohydrates. This action clearly has impact on the flavor sensations of various test subjects.

NEURAL INNERVATION OF TASTE RECEPTORS

An individual taste bud contains 20 to 35 gustatory cells (Figure 10.1). These cells reside between the basal membrane of the tongue epithelium and slightly below its upper layer.

Each bud has an opening called a gustatory pore that contains microvilli ($0.1–0.2$ μm). These microvilli are actually villiferous extensions of gustatory cells and may serve as the actual site of taste reception.

Individual taste buds may contain 4 to 20 sensory cells. Innervating nerve fibers extend through the basement membrane of the receptor and terminate as an arborized network over the surface of individual taste cells. Nerve fibers that penetrate through the basement membrane are actually dendritic branches of neurons innervating a

multiplicity of cells (Figure 10.1). The key neurons arise from the seventh and ninth cranial nerves. The anterior two-thirds of the tongue is innervated by the chorda tympani branch of the facial nerve (VII). The remaining posterior region of the tongue is linked to the glossopharyngeal nerve (IX), and still other buds of the pharynx, epiglottis, and soft palate are innervated by the vagus nerve (X).

Apart from these nerves, the remaining features of taste sensation including thermal and haptic (tactile) receptors are largely dependent on the trigeminal nerve (V). This nerve has great sensory significance throughout the entire buccal mucosal region.

Microvillar perception of a specific taste compound requires a 25 to 30 ms response time. It is theorized that taste perception may involve molecular binding interactions between a molecule and its characteristic receptor locus on microvilli. Such "lock-and-key" interactions are energetically weak phenomena at best and cannot be considered irreversible binding events.

Whatever the interaction between taste-perceptible molecules and their receptors, the topology of the receptor surface and/or its physical chemistry are somehow altered. This action conceivably spurs a depolarization event over the receptor membrane, which is then transmitted as a neural impulse and translated as a taste sensation.

Although enzymes such as acetylcholinesterase are involved in gustatory neural responses, and inhibitors of this enzyme enhance acid and salt sensitivities, enzyme roles in the neurochemistry of taste are not known for certain.

THE FOUR TASTE PRIMARIES

Key receptor sites in taste buds have a propensity or ability to detect a specific taste quality such as sour, sweet, salty, or bitter. These four qualities are often referred to as primary tastes or simply "primaries." Therefore, specific receptor sites can detect a primary taste factor. These primary receptor sites can be independently excited, yet depending on the chemical nature of a compound, more than one primary site can be affected. Such multiple perception phenomena permit the simultaneous sensory detection of several taste qualities.

With regard to the four primaries of salt, sour, bitter, and sweet, four corresponding compounds are recognized as eliciting these specific responses including sodium chloride, citric acid, quinine, and sucrose. Sucrose produces a singularly sweet sensation, quinine is bitter, citric acid is sour, and sodium chloride is salty. Figure 10.2 outlines the structural differences among these and other taste-active compounds.

The respective taste primaries are associated with *limited* parallels to ions or molecular structures. For example, sourness reflects the presence of hydrogen ions, but any effect on a receptor that these may have is also influenced by coexisting anions. Equivalent molecular concentrations and pH conditions for citric acid and acetic acid result in the perception that acetic acid is sourer than citric acid. Therefore, sourness is not synonymous with pH conditions or acidity, and sour sensations reflect a combination of pH and molecular concentrations of acid molecules.

With respect to primary perceptions of salt, the greatest contribution to taste stems from the anionic components. Salts of alkali metals (sodium chloride) are most commonly associated with this perception. However, not all salts taste salty. Some salts such as cesium chloride and potassium iodide are bitter, potassium bromide produces a mixed sensation of salt and bitter, and still others can exhibit sweetness. As in the case of sourness, salty sensations are affected by the presence of different anions and the number of cations present in a food system.

Sweetness and bitterness are generally associated with organic molecules, although some organics remain tasteless as a result of poor water solubility. Moreover, any study of these primaries can be confounded by the myriad of compounds that exhibit similar qualities and still defy structural or molecular similarities. As an example, consider the case of quinine and β-D-mannose; both perceived as being bitter, yet a study of β-D-mannose and α-D-mannose reveals that the α form is sweet and the β structure is bitter. Subtle changes in the substituted molecular structures of saccharin provide another important example of structural diversity among molecules and their relative abilities to produce a scale of sweetness perceptions (Figure 10.3). Substituted 1-propoxybenzene derivatives (Figure 10.4) and many other compounds show similar consequences involving minor molecular differences and major taste differences.

The prediction of a sweet taste sensation elicited by a molecule cannot be unerringly determined from its structure. However, one theory known as the AH-B (acid hydrogen-base) theory suggests that *distinct* spatial interactions between a "sweet" compound and its "receptor" must exist. That is, a sweet-tasting molecule must have a proton attached to an electronegative atom (or region) on a receptor site, at a distance of 0.3 nm from another complementary electronegative interaction (Figure 10.5). The AH-B theory probably spells out at least one prerequisite for many sweet-tasting molecules, but it does not explain all sweet sensations attributed to known molecular structures. For this reason, some versions of the theory have been broadened to include a hydrophobic site in the vicinity of the AH-B system. Figure 10.5 details the common AH-B unit of some diverse chemical compounds that clearly contribute to their sweetness.

The molecular assessment of sweetness becomes even more complicated by surveys of D- and L-amino acid isomers. Amino acids of the L-type (Asp, Glu, His, Ile, Leu, Met, Phe, Ser, and Tyr) have an assortment of tastes including every sensation from marginally bitter or sweet, to insipid or tasteless.

Name	Structure	Taste quality
Quinine	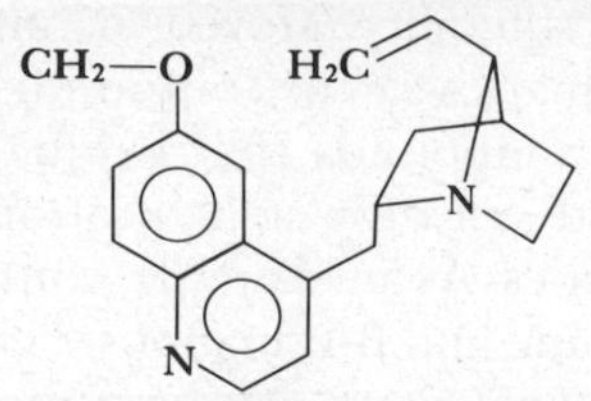 CH₂—O H₂C (structure) N ... N	Bitter
β-D Mannose	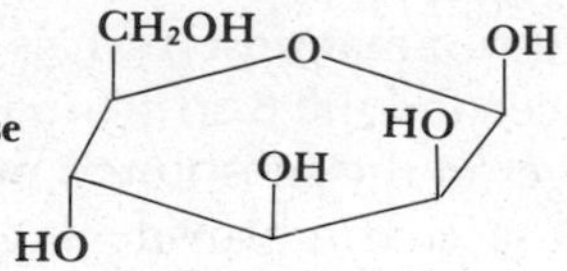 CH₂OH O OH HO OH HO	Bitter
α-D-Mannose	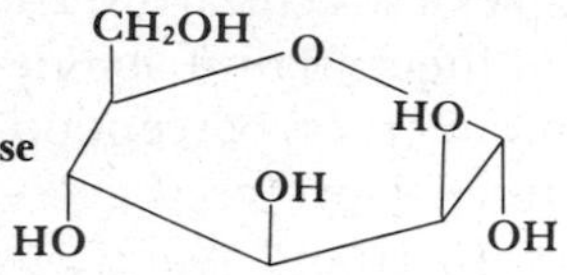 CH₂OH O HO OH HO OH	Sweet
β-D-Glucose	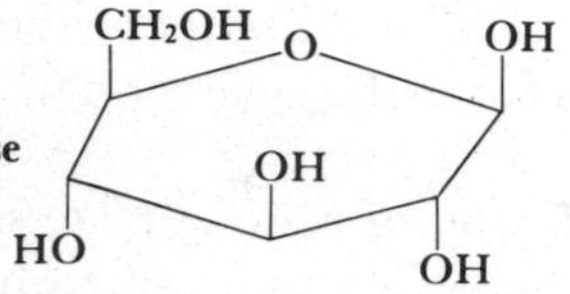 CH₂OH O OH OH HO OH	Sweet
Neohesperidine dihydrochalcone	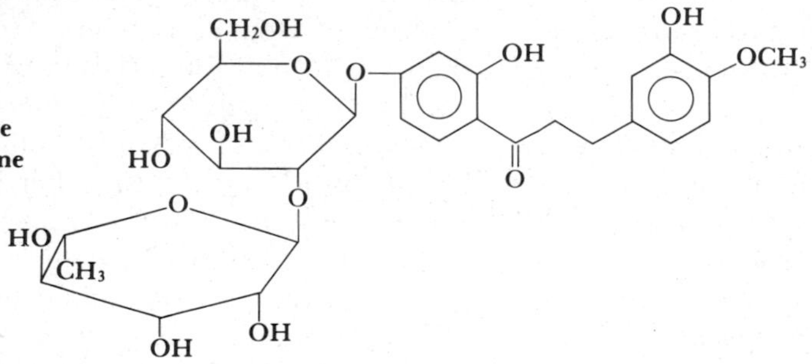 CH₂OH O O OH OH OCH₃ HO OH HO CH₃ OH OH (structure)	Sweet: 1500 times sweeter than sucrose and 50 times sweeter than saccharin
Acesulfame-K Nonmetabolizable 1,2,3-oxathiazine ring	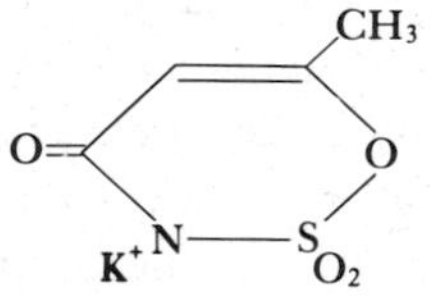 CH₃ O O K⁺ N S O₂	Sweet: 4% aqueous solution is 130 times sweeter than a comparable amount of sucrose
Aspartame Methyl ester of L-aspartyl-L-phenylalanine	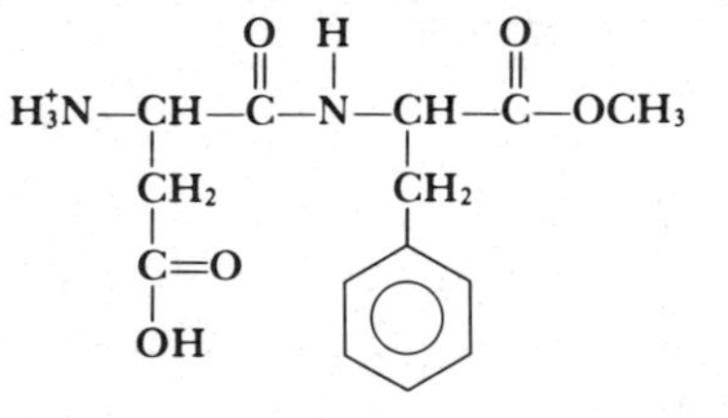 H₃N⁺—CH—C—N—CH—C—OCH₃ / CH₂ CH₂ / C=O / OH (phenyl)	Sweet: 4% aqueous solution is ~185 times sweeter than a comparable amount of sucrose
Thaumatin or Talin	207 amino acid residue polypeptide	Sweet: 4% aqueous solution is ~5000 times sweeter than a comparable amount of sucrose
Hydrogen ion(s)	H⁺	Sour
Sodium ion	Na⁺	Salty
Potassium ion	K⁺	Salty
Potassium iodide	KI	Bitter

FIGURE 10.2. Some notable compounds having taste. Note the diversity of molecular structures that produce a sweet taste sensation.

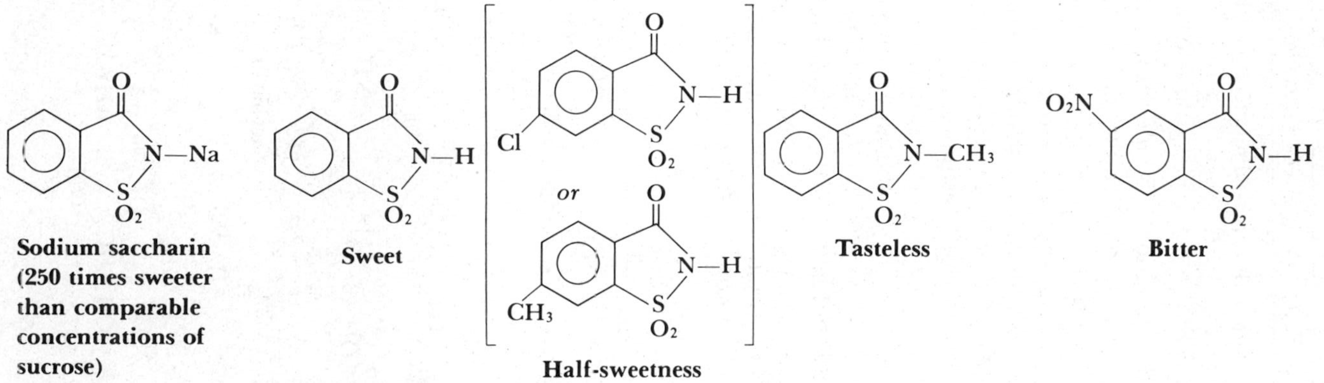

FIGURE 10.3. Substituted saccharin structures and sweetness.

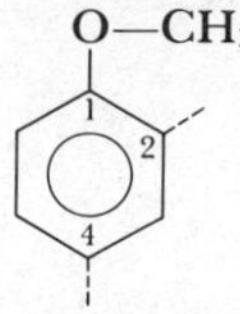

O—CH₂—CH₂—CH₃	2,4 Substitutions		
2 Position	—NH₂	—NO₂	—NO₂
4 Position	—NO₂	—NH₂	—NO₂
Taste	Sweet	Tasteless	Bitter

FIGURE 10.4. Propoxybenzene substitutions and their taste effects.

The corresponding D isomers of these amino acids, however, exhibit notably sweet characteristics. Far more complicated molecular structures exert sweet-tasting sensations such as thaumatin. This compound is a 207-amino acid residue polypeptide (M.W. 22,000) that naturally occurs in ripe fruit of the West African plant known as *Thaumatococcus danielli*.

Aside from the four recognized primary tastes, three others (alkaline, meaty, and metallic) have stirred attention, but specific receptors have not been isolated. This is particularly true for metallic taste perceptions.

By mechanisms that are not precisely understood, metal salts containing iron, copper, tin, and other metals can produce a neuronal sensation akin to imminent pain, or at least a metallic presence. Similar but unprecise locales are thought to be responsible for stimulating the pain fibers instigated by pungency and cryosensory (cool) perceptions in the mouth. Principal structures for some of the latter compounds are detailed in Figure 10.6, including piperine in black pepper, capsaicin from red peppers, zingerone from ginger, and the cooling principle of mint oils, which is menthol.

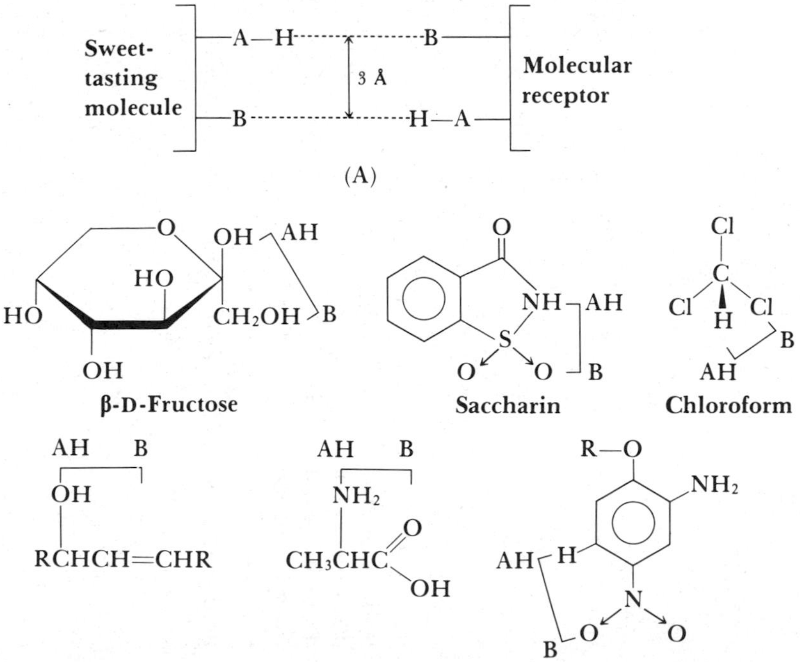

FIGURE 10.5. (A) AH-B model for sweetness perception. (B) AH-B unit of some diverse molecular structures that can be detected as a sweet taste sensation.

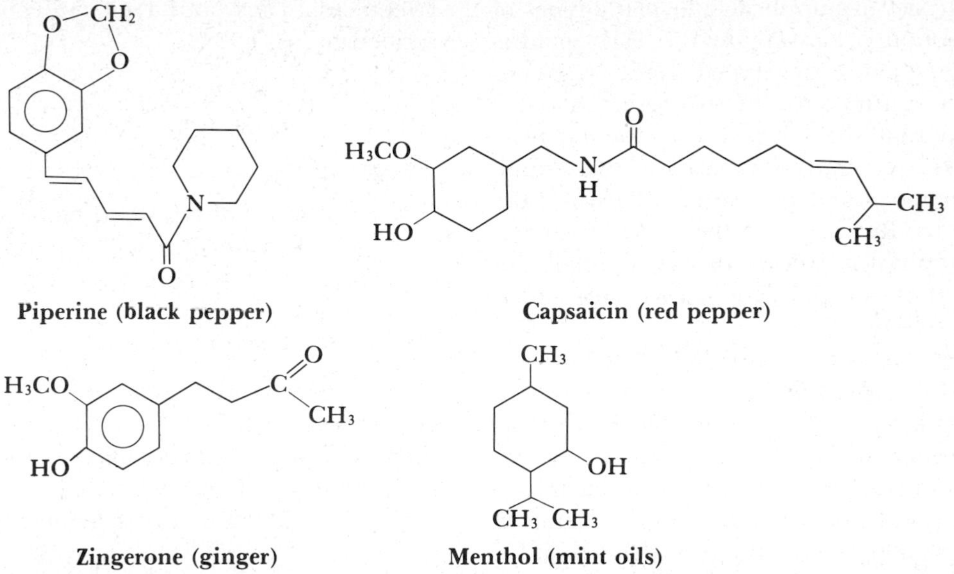

Piperine (black pepper)

Capsaicin (red pepper)

Zingerone (ginger)

Menthol (mint oils)

FIGURE 10.6. Principal structures of pungent spices and the cooling sensation of mint oils. The heat sensations instigated by nonvolatile amides, especially capsaicin, can be expressed according to "heat units" called *Scoville units*. Heat sensation thresholds in the throats of test subjects (using carefully diluted samples) serve as the basis for assigning Scoville units and pungency strengths to selected spices and their extracts.

Depending on water quality and other factors, the minimum concentrations for detecting the major primary tastes have been widely reported and cited as *taste threshold values* (*TTVs*). Some typical *TTVs* have been surveyed in Table 10.1.

Until the physiology of taste is completely unraveled, *TTVs* will remain enigmatic, and assessing temperature sensation and pungency effects as well as understanding sweetness ranking among sugars will be difficult. For example, cold and warm sensations affect the tongue according to two descriptive subdivisions, namely, sour, cold, and salty; and bitter, warm, and sweet. In the case of sugar sweetness ranking for mono- and disaccharides, it still remains to be established why single fibers in the chorda tympani nerve bundle seem to respond differently to different sugars. For 0.5 M sugar concentrations, the most impressive stimuli are produced by D-fructose followed by sucrose, D-glucose, maltose, D-galactose, and lactose.

Clearly, no taste sensation involves only a perception of primary taste unless labora-

tory conditions exist. In fact, nearly all tastes represent a composite effect of neural stimulation that reflects simultaneous blending, interaction, and modification of primary taste stimuli. Depending on the circumstances, some of these stimuli take preference over others and the collective sensory perception is recognized as "taste."

Psychophysics of Taste Perception

The chemical stimuli that incite certain taste perceptions are studied as psychophysical phenomena. The precepts underlying psy-

Table 10.1. Approximate Taste Threshold Values (*TTVs*) for Basic Taste Sensations (Aqueous Solvent)

Taste	Percent by weight in water	Test standard
Bitter	0.0002	Quinine sulfate
Sour	0.004	Citric acid
Salt	0.04	Sodium chloride
Sweet	0.34	Sucrose

chophysics are applicable to many types of perception according to the early mathematical models developed by E. H. Weber over a century ago. Weber's research determined that the *minimal detectable difference* (*MDD*) between the intensity of two stimuli was proportional to the total intensity of the stimulus. Reference to the *MDD* in current psychophysical studies of flavor usually involve the concept of the *just noticeable difference* (*JND*).

For taste assessments, *JND* values are obtained by discrimination tests in which human subjects attempt to perceive a discrete difference in the same stimulus at two different intensities. Thus, the *JND* value largely corresponds to the *MDD*. Refinements of Weber's model have also served as a basis for Steven's law, which accounts for the impact of psychological variables on the perceived intensity of sensations. Although many *JND* values exist for any particular taste primary (or odor), there is not an infinite range of *JND* values. Studies as early as 1908 in Vienna demonstrated only 25 *JNDs* exist for sucrose solutions from a threshold concentration range to a strong intensity. More recent studies with odorous chemical compounds such as anethole also support these early studies.

Notwithstanding the complexities of neural responses initiated by specific taste stimuli, a mathematical model has been developed by L. H. Beidler that relates stimulus concentration to the magnitude of taste response. This model has many technical idiosyncracies but its use is not limited to only certain taste-evoking substances. Historical studies show that increases in stimulus concentration coincide with increasing response at a decreasing rate. Finally, a critical point is reached at which further increases in a stimulus fail to produce any additional increase in response. According to the Beidler theory, a taste-active substance is adsorbed by specific receptors on an excitable membrane, or it is somehow distributed between oral-liquid and key primary taste receptors. This action produces an equilibrium be-

tween adsorbed and free molecules described by

$$\frac{n}{S - n} = K \cdot c$$

where

n = number of adsorbed molecules
S = maximum number of particles that can occupy taste receptors
c = moles per liter
K = equilibrium constant

Furthermore, in the extreme case where receptor responses (*N*) is at a maximum (*Nm*), there will be proportionality to the maximum number of particles that can be adsorbed (*S*) and the Beidler equation becomes

$$\frac{c}{N} = \frac{1}{Nm} \cdot c + \frac{1}{KNm}$$

Aside from similarity to the Langmuir adsorption isotherm equation, this equation is noteworthy because it considers the impact of the equilibrium constant (*K*) and the maximum response (*Nm*, measured in terms of electrical impulse output from a specific receptor) produced by a specific substance. Since *K* and *Nm* values vary among substances, these terms markedly affect *TTVs* as well as *JND* values.

In most cases the detectable taste of a substance exhibits its own time–intensity curve. That is, most taste sensations rapidly increase to a peak and then rapidly decline. If graphically plotted as intensity (*y* axis) versus time (*x* axis), a taste-sensation curve can be depicted as an inverted arch. This taste sensation may span several seconds and yet abide in principle with Beidler's theories. In cases where taste sensations leave an aftertaste, Beidler's concepts usually have little application. Aftertastes are not "miniature" expressions of a whole taste or flavor, but instead, they reflect delayed effects (e.g., dryness and astringency) or long-lasting factors (e.g., bitterness).

Principles of Weber's and Steven's laws as well as the mathematical underpinnings of taste actions outlined by Beidler and others are important in the psychophysical study of tastes and flavors. In industrial practice, however, the psychophysical assessment of whole food tastes and individual components rely heavily on the Arthur D. Little *flavor profile* (FP) and the *quantitative descriptive analysis* (QDA) methods. These topics are beyond the scope of this text. For a review of these principles consult Cairncross and Sjöström (1950) and Stone *et al.* (1974).

OLFACTORY SENSATION

The olfactory epithelium of the nasal cavity is responsible for detecting volatile stimuli that occur in the air or emanate from foods in the mouth. The ciliary projections of olfactory epithelial cells are thought to contain chemoreceptors responsible for interacting with and subsequently detecting odorous molecules. When chemical stimuli present as a vapor phase pass over these receptors, neural impulses are fed to the olfactory bulb of the central nervous system (CNS).

Olfactory nerve fibers are dispersed in the nasal epithelium along with other nonreceptor epithelial cells (Figure 10.7). Each fiber is associated with 6 to 12 olfactory cilia that protrude from the epithelium in a sepal-like arrangement. These cilia are consistently bathed in fluids secreted by Bowman's glands. Some studies suggest that cilia alone are not

essential for olfactory cell activity but the aqueous fluids that surround the cilia are critical for maintaining functional epithelial tissues. The importance of this secretion and its attendant mucal factors may be linked to the adsorption, solvation, and dissolution of odor-active volatiles that activate receptor fibers.

The receptors of the olfactory epithelium are located high in the nasal cavity where they occupy ~ 250 mm^2 within each nostril as well as the two olfactory clefts partitioned by the nasal septum.

Neuronal activation by olfactory receptors obviously requires some type of molecular interaction between an odorous compound and a receptor site. Odorous compounds reach receptors by combined effects of diffusion throughout the nasal cavity and eddy currents that reach high into the cavity during inspiration. Some odor–receptor interaction also results during expiration or when food is being chewed.

All odor reception is based on a backdrop of air, which ordinarily lacks any distinct odor sensation. So long as only a single type of an odorous component exists in the air medium, the neural stimulation of olfactory receptors is probably a simple affair. The appearance of the odorous compound in the vicinity of any odor receptor disturbs the temporal equilibrium. The odor molecule may elicit biologic or metabolic effects through

1. Positive or negative catalytic actions that evoke electrochemical changes in membrane potentials.

2. Alterations of receptor-site enzyme activities.

3. Or other conversion of structural–molecular features of a volatile molecule into neural stimulation and odor perception.

Any one of the foregoing olfactory mechanisms is complex, using a model system having one odorous compound, apart from multiple component odor systems in which

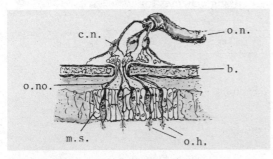

FIGURE 10.7. Cross section of olfactory epithelial tissues and region illustrating: o.n., olfactory nerve; o.no., olfactory neuron; c.n.—connecting neuron; b., bone; m.s., mucus-secreting and supporting cells; o.h., olfactory hairs (cilia).

individual components exhibit *different* vapor pressures, molar concentrations, mucal solubilities, and adsorptive phenomena.

Odor detection along the olfactory epithelium displays (1) spatial and (2) temporal patterns of behavior similar to taste responses already discussed. According to these principles, (1) stimulation of the olfactory epithelium corresponds to its progressive exposure to odorous compounds, and (2) receptor stimulation depends on receptor sensitivity at any time.

Odor detection can be upset, however, by *adaptation phenomena*. This problem develops with prolonged exposure to odors and commonly results in a complete inability to detect an odor. Adaptation is a selective process since it does not necessarily elevate the detection threshold for other compounds. The selective property of adaptation is taken as evidence that olfactory receptors (or their neural pathways) must be qualitatively selective in function. Whatever the adaptive mechanism reflects in neurochemical terms, it is clearly unrelated to a generalized fatigue of the olfactory system.

The dynamics of odor detection are also complicated by the fact that the perceived strength of many odors may be unrelated to their threshold value. Vanillin, for example, demonstrates a very low threshold yet fails to produce an overbearing olfactory sensation at high concentrations. Ethyl mercaptan behaves in a counter fashion in that it has a significantly higher threshold than vanillin, but its olfactory perception becomes progressively overpowering with higher concentrations. Early olfactory studies by Kruger and others have suggested that olfactory strength for a compound may decrease as molecular weight reaches some critical value. For studies on aliphatic compounds, heavier molecules generally show greater odor intensities once their vapor pressures are uniformly equilibrated.

Compared to taste perceptions, which involve four receptor types as well as haptic sensations, olfaction can be considered a keener chemical sense. This sensory mechanism operates regardless of food textural quality estimations, visual perception, or other predisposing psychological variables.

Although there are at least a million possible types of subjective odor sensations, there have been numerous efforts to classify odors. These classifications are quite analogous to the four primary tastes, but they are referred to as the *primary odor characteristics* (POCs). The earliest effort to detail POCs was that of Linnaeus in the eighteenth century. Others including Zwaardemaker, Henning, Crocker, and Amoore have progressively altered the POC designations to suit the historical era—some characteristics being retained and others eliminated (Table 10.2).

Table 10.2. Some Historical Olfactory Ranking Schemes for Primary Odor Characteristics (POCs)

Zwaardemaker	Henning	Crocker	Amoore
Ethereal	Fruity		Ethereal
Aromatic	Spicy		Camphor
Fragrant	Flowery	Fragrant	Floral
Ambrosiac			Musky
Alliaceous			
Empyreumatic	Burnt	Burnt	
Hircine		Goaty	
Foul	Foul		Pungent
Nauseous			
	Resinous		Putrid
		Acid	
			Pepperminty

MOLECULAR BASIS OF OLFACTORY RESPONSE

The psychological aspects of odor detection are linked to characteristics of molecular structure that make one volatile molecule distinctly different from others. Although the exact mechanism behind olfaction is unknown, there are no shortages of innovative theories. From chemical perspectives it is known that very few of the 30 most common chemical elements have any odor. Moreover, most of the molecules containing only hydrogen and oxygen have few or no characteristic odors. This condition exists until nitrogen and/or sulfur are introduced into an organic compound. Oxygen, nitrogen, and sulfur atoms are particularly important because they serve as the basis for key functional groups that have *odor-bearing* properties. Such odor-inciting functional groups can be described as *osmophores,* and they are common features of alcohols, aldehydes, amines, esters, ketones (carbonyl), thiol, carboxyl, and other specialized structures.

Apart from the recognition that osmophoric groups are a necessary prerequisite for olfactory phenomena, their existence alone does not explain all demonstrated cause–effect relationships in olfactory sensations. Indeed, there are many dissimilar molecules that have similar odors and the opposite is also true.

The olfactory character of even the most elementary odorous stimuli are difficult to predict from molecular perspectives when their structures are known. As an example consider the dominant perception of odor transitions for a structural series of ethyl, propyl, butyl, and pentyl acetates. The prevalent fruity (banana/pineapple) odor of pentyl (or amyl) acetate decreases in the direction of ethyl acetate (Figure 10.8), which has little fruity odor. An acetic odor characterizes ethyl acetate, but this is tempered and modified toward a more fruity sensation in the direction of pentyl acetate.

Although it is difficult to anticipate the odor of any molecule based solely on its structure, stylized models of olfactory receptor–odorant interactions can be useful for instructive purposes. For example, consider a hypothetical occurrence of nine olfactory epithelial receptors. Each of these receptors responds to odorant molecules that roughly adhere to the crude structure silhouettes detailed in Figure 10.9. Note that the structure–odor correspondence adheres to the following scheme:

A. Sweet-aromatic
B. Naphthalene—camphor-like
C. Musk
D. Camphor
E. Jasminic—fungus-like
F. Aniseed-like
G. Fatty
H. Floral—woody
K. Woody

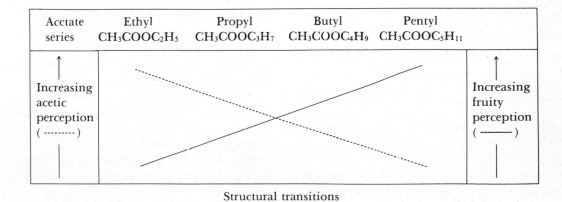

Acetate series	Ethyl $CH_3COOC_2H_5$	Propyl $CH_3COOC_3H_7$	Butyl $CH_3COOC_4H_9$	Pentyl $CH_3COOC_5H_{11}$

Increasing acetic perception (--------)

Increasing fruity perception (———)

Structural transitions

FIGURE 10.8. Relative odor transitions in an acetate ester series.

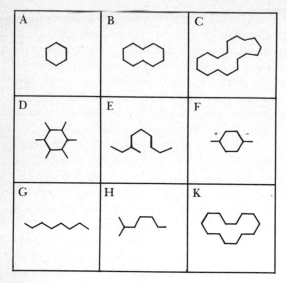

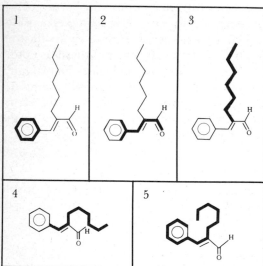

FIGURE 10.9. Idealized molecular structures that will activate key olfactory receptors on the olfactory epithelium are indicated as A to K. For simplistic reasons, only generalized molecular forms are shown. Structures 1 to 5 detail five specific forms of α-hexyl cinnamaldehyde, where the intense dark portions of the structure are emphasized for matching purposes with the corresponding receptor site(s) in A to K above. See text for details. (Structural models kindly provided by Haarmann & Reimer GmbH, Holzminden, and reprinted with permission.)

Using these specific receptor models, the odorant notes of various α-hexyl cinnamaldehyde structures can be projected. The six-membered phenyl ring of cinnamaldehyde (Formula 1), indicated by the more *intense*

black lines, incites an olfactory response similar to that which stimulates receptor A. The perceived odor corresponds to "sweet–aromatic."

From the perspective of Formula 2, the enhanced part of its structure resembles receptor H of the model system and initiates a floral–woody odor. The aliphatic chain of the structure in Formula 3 corresponds to the G sensitive receptor and the odor produced is a fatty base note. The structure for Formula 4 has a corresponding similarity to receptor E sensitivity and generates a jasminic–fungus-like note; whereas a woody note is triggered by the structure in Formula 5 at receptor K.

Since α-hexyl cinnamaldehyde is unlikely to demonstrate any structural conformations that activate receptors C, F, and D, their corresponding odor notes will not be detected.

The correlation between odor and structure offered here may be more specific than the actual state of affairs, but the model is realistic. Moreover, it is clear that a single organic molecule may display a wide range of structural mutability. In any given molecular population, the *statistical probability* of respective *molecular conformations* will affect the olfactory perception of a specific compound. In general, the greater the structural mutability of a compound, the more complex the number of odor characteristics and nuances that it will display. This recognition supports observations that conformationally rigid structures such as camphor will display relatively few odor facets compared to the more flexible molecules.

Once a correlation between known molecular structures and perceived odors is established, it is possible to rationalize structure–odor relationships among an assortment of compounds as previously observed. Unfortunately, however, it is not readily obvious what odor characters will be realized from an a priori knowledge of molecular structures.

There is no steadfast explanation for all neurochemical perceptions of organic molecules, but some of the following molecular properties do seem to influence olfactory recognition:

1. *Molecular weight.* Most odorous molecules are small, low-molecular-weight compounds that include carboxylic acids, phenols, thiols, aldehydes, and so on.

2. *cis/trans and other stereomolecular isomers.* Only minor structural variations influence odor characteristics of molecules (e.g., *cis*-3-hexenol has a fresh green odor versus the *trans* isomer, which is chrysanthemumlike).

3. *Length of unsaturated hydrocarbon chains.* Short-chain fatty acids have stronger odor perception than long-chain structures.

4. *Molecular polarity and aromaticity.*

5. *Molecular distribution of electrical charges* (and/or possible charge–transfer reactions involving a neuroreceptor).

6. *Discriminative electrochemical responses* of olfactory receptors to different molecular structures.

7. *Conformational flexibility of molecular structures.* Odor variation in flexible molecular structures is greater than structurally rigid molecules such as camphor.

A stereochemical basis for olfactory perception has been proposed by Amoore, which depends on principles popularized in lock-and-key concepts of enzyme–substrate interactions. This theory may be compatible with many of the points outlined above, but it is far from perfect. Still other theories have been offered that rely on nonchemical reactions such as olfactory membrane puncturing phenomena by odorous molecules, molecular vibration, or infrared vibrational properties of molecules. These theories are largely questionable and remain to be proven.

MOLECULAR COMPONENTS OF NATURAL FLAVORS

A component analysis of almost any natural flavor by the use of instrumental methods (e.g., gas–liquid chromatography (GLC), high-pressure liquid chromatography (HPLC), and/or mass spectrometry (MS)) reveals the presence of at least several primary flavor contributors. These primary flavor contributors, often called *characteristic impact compounds* (CICs), are usually embedded in a diverse milieu of many other coexisting compounds (Table 10.3). The sensory effect of these complex organic mixtures as they pass over the nasal olfactory epithelium results in our perception of characteristic food flavors. Although taste is important in almost every aspect of flavor chemistry, the volatile organics have critical importance. These compounds include:

1. *Aliphatic compounds* containing different functional groups such as hydroxyl(s), carbonyl(s), esters(s) and/or diol(s); and unsaturated molecular counterparts.

2. *Aromatic compounds* exhibiting the same structural alternatives noted above.

3. *Aliphatic and aromatic compounds containing nitrogen and sulfur.*

4. *Isoprenoids* (mono-, di-, and higher terpenes plus modified terpene derivatives such as β-ionines).

5. *Lactones* (internal molecular esters of hydroxy acids).

Aliphatic Compounds

In spite of an almost infinite molecular series of possibilities, fatty acids, alcohols, and aldehydes are some of the most important flavor contributing compounds. Fatty acids demonstrate odor and taste qualities that reflect their characteristic hydrocarbon chain lengths. Lengthening of the fatty acid chain in a structural series beginning with formic acid results in decreasing sourness, a reduced irritating aroma, and decreased rancidity odors (using butyric acid as a rancidity benchmark).

The ethyl and some higher ester derivatives of fatty acids (from C_1 to at least C_7 types) all share somewhat fruity odors. This is not to say that they are all desirable odors,

Table 10.3. Selected Character Impact Compounds (CICs) for a Variety of Foods

Flavor character	Flavoring compound	Flavor character	Flavoring compound
Almond	5-Methylthiophen-2-carboxaldehyde	Maple	2-Hydroxy-3-Methyl-2-cyclopenten-1-one
Anise	Anethole	Meat	Methyl-5-(β-hydroxyethyl)thiazole
	Methyl chavicol (estragol)	Melon	2-Methyl-3-p-tolypropionaldehyde
Apple	Isoamyl acetate		Hydroxycitronellal dimethyl acetal
	Ethyl 2-methylbutyrate		2,6-Dimethyl-5-heptenal
Bergamot	Linalyl acetate		2-Phenylpropionaldehyde
Blueberry	Isobutyl-2-butenoate		2-Methyl-3-(4-isopropylphenyl)propionaldehyde
Blue cheese	2-Heptanone	Mushroom	1-Octen-3-ol
Butter	Diacetyl	Mustard	Allyl isothiocyanate
Caramel	2,5-Dimethyl-4-hydroxy-3-dihydrofuranone	Peach	γ-Undecalactone
Caraway	d-Carvone		6-Amyl-α-pyrone
Cassia	p-Mentha-8-thiol-3-one	Peanut	2,5-Dimethylpyrazine
Celery	3-Propylidene-1-trihydroisobenzofuranone		2-Methoxy-5-methylpyrazine
Cherry	Benzaldehyde	Pear	cis,trans-Ethyldecane-4,2-dienoate
	Tolyl aldehyde	Peppermint	Menthol
Chocolate	5-Methyl-2-phenyl-2-hexenal	Pineapple	Allyl caproate
	Isoamyl butyrate		Methyl β-methylthiopropionate
	Vanillin		Ethyl butyrate
	Ethyl vanillin	Popcorn	Methyl 2-pyridyl ketone
	2-Methoxy-5-methylpyrazine	Potato	Methional
Coconut	γ-Nonalactone		2,3-Dimethylpyrazine
	α-Octalactone	Prune	Benzyl-4-heptanone
Coffee	Furfuryl mercaptan		Dimethylbenzylcarbinyl isobutyrate
	Furfuryl thiopropionate	Raspberry	6-Methyl-α-ionone
Cognac	Ethyl oenanthate		trans-α-Ionone
Clove	Eugenol		p-Hydroxyphenyl-2-butanone
Coriander	Linalool	Red currant	trans-2-Hexenol
Cucumber	trans,cis-Nona-2,6-dienal	Seafood	Pyridine
Garlic	Diallyl sulfide		Piperidine
Grape (concord)	Methyl anthranilate		Trimethylamine
Grapefruit	Nootkatone	Smoke	Guaicol
Green bell pepper	2-Methoxy-3-isobutylpyrazine		2,6-Dimethoxyphenol
Green leafy	cis-3-Hexanol	Spearmint	1-Carvone
Hazelnut	(Methylthio)pyrazine	Tomato	Isobutylthiazole
Horseradish	1-Penten-3-one		cis-4-Heptenal
Jasmin	Benzyl acetate	Vinegar	Ethyl acetate
			Acetic acid

NOTE: The CICs cited represent a single factor or the chief factor(s) when combined with other key compounds that will produce a taste and/or an odor that is reminiscent of the flavor indicated.

SOURCE: From *Flavors and Spices* by J. A. Rogers and F. Fischetti in *Kirk-Othmer Encyclopedia of Chemical Technology*, 3rd ed., Vol. 10. Copyright 1980, John Wiley & Sons, New York. Reprinted with permission.

but certainly less obnoxious than the free acids.

Corresponding aldehyde derivatives of fatty acids also lose their pungency with increasing hydrocarbon length. The lower alcohols including methanol and ethanol generally exhibit spirit-like odors, but as in the case of higher-molecular-weight aldehydes and free fatty acids, the tendency of higher alcohols to exist as oily liquids decreases their contribution to flavor. This phenomenon is partly tied to the depression of molecular vapor pressure as molecular size increases.

Ketones impart significant flavor qualities to many fruits and plant materials plus dairy products from which they develop as fermentation by-products. The aromas of ketones run the gamut from a bleu cheese type aroma for methyl amyl ketone to the distinctive diacetyl sensation of buttermilk, soured cream, cottage cheese, and related products (Figure 10.10).

Ionones and their derivatives also display important roles in the delicate flavor sensations associated with berries and fruits. Some other ketones serve as important flavor contributors, but they are considered in following sections (see Figure 10.14).

Aromatic Structures

Aromatic ring compounds display a mystifying variety of odors. These odors range from sweet/aromatic for a phenyl ring, to moth-ball-like and woody or musky odors for increasingly complex substituted multi-ringed systems.

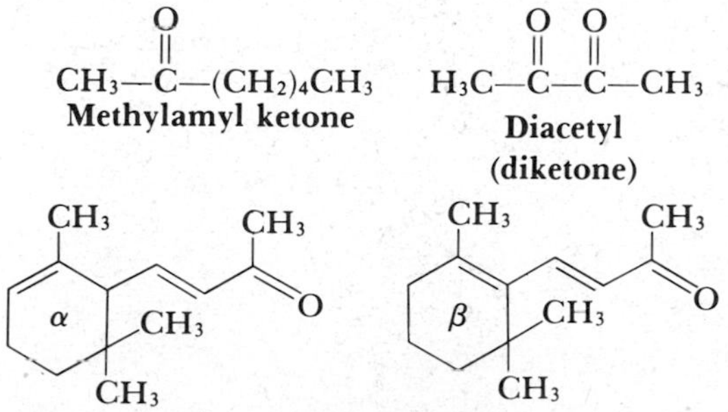

Aldehyde, alcohol (e.g., phenol) and ester derivatives of the ringed aromatic compounds are important contributors to food flavors, plant extracts, and the perfume industry in general. Those aromatic-ringed structures containing carbon, hydrogen, and oxygen substituents, which permit some polarity and a reasonable vapor pressure at room temperature, seem to serve as important stimuli. Carboxyl substituents do not preclude favorable odors from aromatic compounds, but the odor stimulation of salicylic acid (no odor) versus its methyl ester (wintergreen odor) is typical of many carboxylic group effects on odor. Figure 10.11 outlines some aromatic-ring structures and their diversity of seemingly unrelated odor sensations.

Nitrogen- and Sulfur-Containing Molecules

Aliphatic and aromatic compounds containing sulfur and nitrogen are critical elements of flavor chemistry.

Sulfur compounds deserve special recognition for their high volatility, reactivity, and strong and characteristic odors coupled with their very low threshold values. Only 0.2 to 0.5 µg/L (in water) of many sulfur compounds are quite adequate to achieve a very noticeable threshold. The popular recognition of onion, horseradish, and garlic flavors mirrors the potential flavor strength of sulfur compounds. The chemistry of sulfur compounds in food flavors is complicated by their chemical lability and wide range of structural possibilities including

Disulfides	Thiazoles	Thiophenes
Isothiocyanates	Thiepanes	Thiols
Methionals	Thiocyanates	Thiosulfinates
Monosulfides	Thiolanes	Thiosulfonates
Thianes		Trisulfides

The *Cruciferous* vegetables, which include the *Brassicas,* offer pungent flavor proper-

$$CH_3-\overset{\overset{\displaystyle O}{\|}}{C}-(CH_2)_4CH_3$$
Methylamyl ketone

$$H_3C-\overset{\overset{\displaystyle O}{\|}}{C}-\overset{\overset{\displaystyle O}{\|}}{C}-CH_3$$
Diacetyl (diketone)

α- and β-Ionone (odor of violets)

FIGURE 10.10. Some important odorous ketone structures.

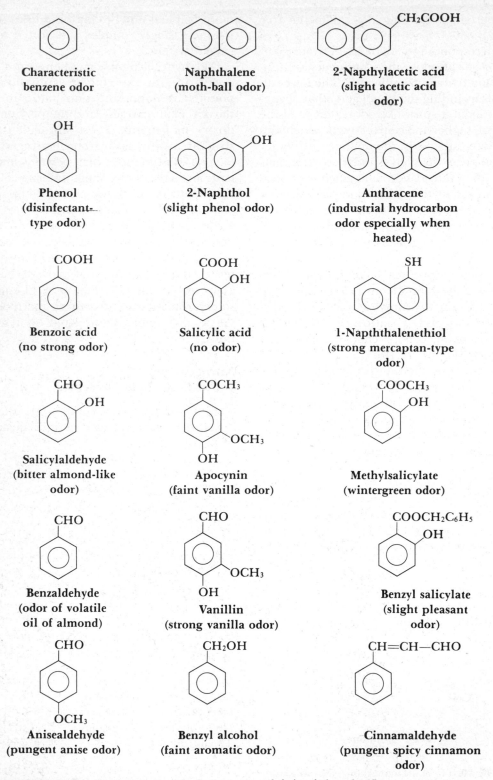

FIGURE 10.11. Some aromatic-ringed structures and their relative odor/flavor contributions.

ties linked to allyl isothiocyanate produced from sinigrin. Related isothiocyanates are responsible for the characteristic flavor of radish, horseradish, and watercress. A more tempered flavor sensation is attributable to the flavor of *Brassicas* including cabbage, cauliflower, rutabaga, and brussels sprouts especially after cooking. Vegetables of the *Allium* family including onions, leeks, and garlic have tremendously strong and persistent flavors that are attributable to disulfides, trisulfides, and other compounds. Allyl disulfide and allyl propyl disulfide are typical components of garlic oil while allyl propyl disulfide occurs in onions.

$$CH_2{=}CH{-}CH_2{-}N{=}C{=}S$$
Allyl isothiocyanate

$$(CH_2{=}CH{-}CH_2)_2S_2$$
Allyl disulfide

Aside from these plant groups and volatile sulfur compounds produced from microbiological or autolytic digestion of animal proteins, most fruits, dairy products, and alcoholic beverages contain few volatile sulfur constituents. Many of the volatile sulfur compounds that appear in these foods, however, arise from heat treatments used during preparation or storage. Other organic sulfur compounds materialize as a result of accidental organic sulfur contaminants that end up in food products.

Dimethylsulfide imparts "cowy" perceptions to milk, which may be caused by improper processing, whereas still other off-flavors may be attributable to benzylmethyl sulfide and/or benzyl mercaptan. The latter compounds arise from grazing land weeds such as landcress consumed by cows along with other herbage. The active metabolizable sulfur compounds in these plants are benzyl glucosinolates. Trace sulfur compounds can occur in wines, beers, or whiskeys for a variety of reasons. One major source of sulfur in all these fermented products may stem from yeast activities on methionine when it is present in fermentation liquors. In wine, however, elemental sulfur and bisulfite are routinely used to check wild yeast activities, and the inadvertent conversion of these sulfur compounds into organic forms is possible. Sulfurated ketones are recognized features of certain fruits and berries. These compounds include two distinct diastereomeric thiols commonly found as 8-mercapto-*p*-menthan-3-ones in buchu oil. Of the (+) and (−) isomers of the menthanones, the (+)-form is most characteristic of buchu oil (black current oil).

Many types of aromatic amines, amino acid derivatives, and heterocyclic nitrogen compounds have considerable impact on foods. The heterocyclic compounds known as *pyrazines* have garnered special attention in recent years because of their desirable food flavor properties. Nearly all of the most important pyrazine-based flavors and aromas develop from heat-treating foods. This is true for bakery products, roasted grains, cocoa products, coffee, and some other foods such as dairy products, dried mushrooms, popcorn, potato products (especially pyrazines developed in the skins on baking, frying, or roasting), rum, Scotch whisky and other whiskeys.

Vegetables in their native states contain few pyrazines, although green bell peppers and volatile chili pepper oils do contain at least one important pyrazine (2-isobutyl-3-methoxypyrazine). Legumes and tomatoes may contain several alkoxypyrazines as a result of natural biosynthetic pathways. Examples include 2,6-dimethylpyrazine in tomatoes and 2-butyl-, 2-isobutyl- and 2-isopropyl-3-methoxypyrazines (Figure 10.12).

For a detailed perspectus of organic sulfur and nitrogen compounds, the reader should consult *Fenaroli's Handbook of Flavor Ingredients* or a similar reference dedicated to nitrogen and sulfur flavor factors.

Isoprenoids, Lactones, and Other Flavor Factors

Isoprenoid derivatives are natural constituents of plant metabolism that develop from

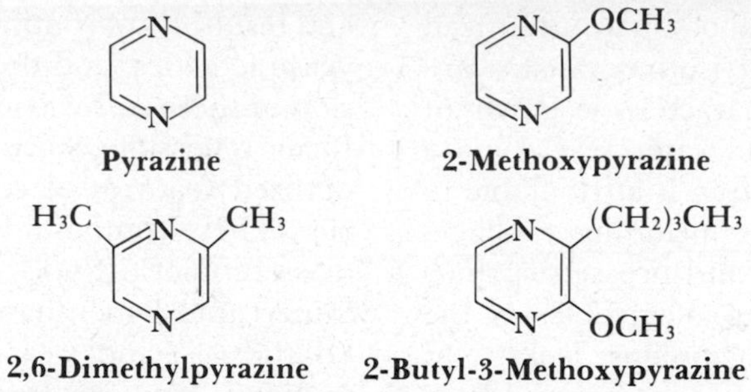

Pyrazine   **2-Methoxypyrazine**

2,6-Dimethylpyrazine   **2-Butyl-3-Methoxypyrazine**

FIGURE 10.12. Examples of some notable pyrazines.

acetyl-CoA-consuming biosynthetic reactions. These compounds exist in complex mixtures in spices and virtually all essential oil mixtures. The biosynthetic origins and range of molecular structures have been presented in the chapter dealing with lipid properties.

Since hydroxy acids contain both —OH and —COOH groups, some of these structures can undergo cyclic ester formation to yield lactones, Lactone formation occurs spontaneously or especially under the influence of heat. Lactone products may be denoted as α, β, γ, or δ depending on the number of members in the lactone ring system. Examples of typical γ- and δ-lactone formations are illustrated in Figure 10.13. Structural justification for lactone formation rests on the stability of the five- and six-membered rings that result. Odor characters range from peachy (R = —C_6H_{13}) to that of coconut (R = —C_5H_{11}) in the case of γ-lactones. Other enolactone derivatives can also impart bouillon-type odors to protein hydrolyzates.

Furanones, pyrones, furyl ketones, and phthalides also impart characteristic odors to foods at low concentrations. With the exception of certain phthalides that contribute to the volatile oils of some *Umbelliferae* (including celery), most of these substances occur when foods are heated one way or another. The burnt or roasted sensory perception of furanones is a component of many cooked meat tissues and their heated drippings. Pyrones and furyl ketones occur in the respective forms of maltol (in roasted malt) and isomaltol (a trace constituent of bread). Formations of the last two compounds are encouraged by caramelization or heat-induced transformations of saccharides (Figure 10.14).

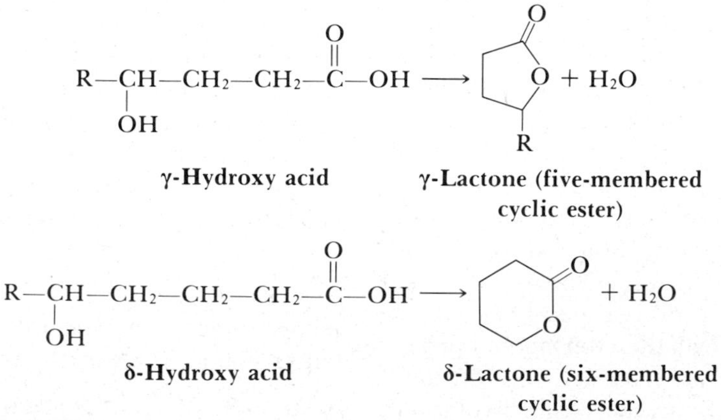

γ-**Hydroxy acid** γ-**Lactone (five-membered cyclic ester)**

δ-**Hydroxy acid** δ-**Lactone (six-membered cyclic ester)**

FIGURE 10.13. Lactone formation.

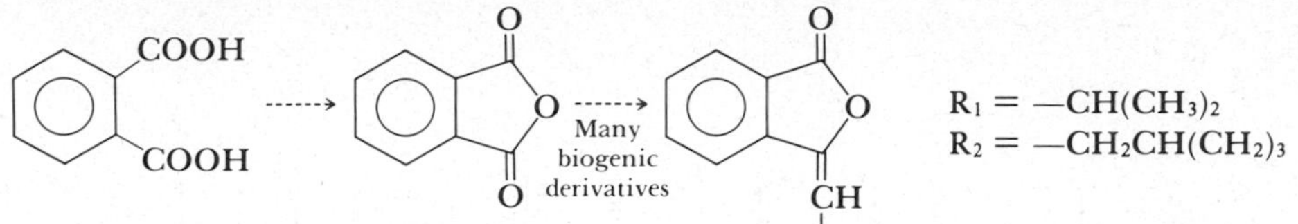

Phthalic
acid

Many
biogenic
derivatives

Phthalic
anhydride

$R_1 = -CH(CH_3)_2$
$R_2 = -CH_2CH(CH_2)_3$

R_1 or R_2

Two examples of phthalides:
if $R_1 = $ 3-isobutylidene phthalide
$R_2 = $ 3-isovalidene phthalide

(A)

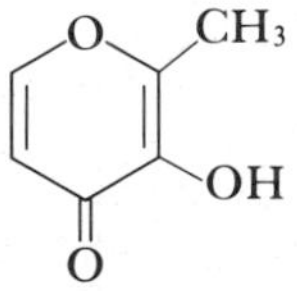

3-Hydroxy-2-methyl-
4-pyrone
(maltol)
(B)

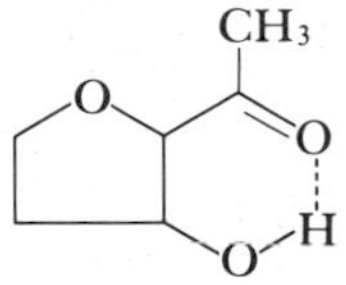

3-Hydroxy-2-furyl-
methyl ketone
(isomaltol)
(C)

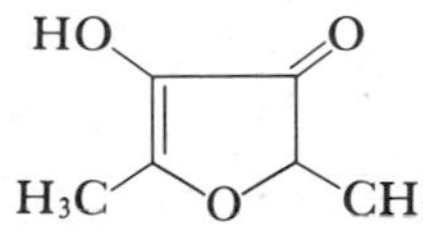

4-Hydroxy-2,5-dimethyl-
3-dihydrofuranone
(D)

FIGURE 10.14. Some typical representative structures of (A) phthalides, which reflect origins from phthalic acid and phthalic anhydride; (B) a pyrone; (C) a furyl ketone; and (D) one of at least 80 to 100 furanone derivatives.

FLAVOR COMPONENTS OF SOME IMPORTANT FOOD GROUPS

For many spices, herbs, fruits, berries, and vegetables, only one or several key compounds can exert a monumental influence on a perceived flavor. Of course such *characteristic impact compounds* (CICs) must coexist in a genetically dictated backdrop of minor flavor contributors to round off the authenticity of any CICs. Recognition of CICs in many natural food substances has served as a basis for controlling food flavor sensations as well as the technological creation of flavors (Table 10.3). Aside from the respective CICs detailed in Table 10.3, most foods such as milk, meat, fish, poultry, baked doughs, and cooked foods reflect flavors having great chemical complexities. The precursors of such flavors are genetically determined by the native biochemical components of animal and plant species, and final flavor formation from these precursors depends on food cooking, curing, and preservation methods (Table 10.4).

NOTABLE FLAVOR CONSTITUENTS OF FRUITS AND VEGETABLES

Apart from biosynthetically produced isoprenoid derivatives that make up essential oils, fruits and vegetables produce many other characteristic volatile aldehydes, alcohols, esters, and ketones. Some of these volatile compounds develop from normal plant metabolism by biogenesis and still other volatile compounds are released only when plant tissues are cut, blended, or chewed.

As examples of the last category, consider onion and garlic. Cut or exposed onion tissues contain S-alkyl-L-cysteine sulfoxides (e.g., S-methyl- and S-propyl-compounds), which can be degraded by alliinase. This results in the development of propanethial-S-oxide ($CH_3CH_2CH{=}S{=}O$), which exhibits the lachrymatory activity of fresh chopped onions. This unstable molecular structure can be further degraded to compounds such as dipropyl disulfide, which has a characteristic onion aroma. Garlic, on the other hand, contains alliin, that is converted into allicin

Table 10.4. Selected Food Groups and the Chemical Precursors (a, b, c . . .) That Ultimately Lead to the Production of Important Flavor-Contributing Compounds

	Bread	Coffee, cocoa	Dairy products		Fish and seafood		Fruit	Meat		Vegetables, herbs, and spices		Wine, beer, and spirits		Yeast	Nuts, roasted
Formation:	H	H	E, H	F	E, H	B	B	H	B	B or E	H/O	F[a]	H	B[a]	H
Flavor compounds															
Acids				a			c	a c						c	
Acetals												c		c	
Aldehydes		a c	b		b		b	a c	b	a c				c	a c
Amines				a	a							a			
Cyanides											a				
Esters							c b					c			
Furans							c	a c							
Heterocycles		a c													
Isoprenoids									c						
Ketones		a c	b		b		b	a c						c	a c
Lactones		a c	b				b		b			a			
Phenols								a c	c						
Pyrans							c								
Pyrazines	a c							a c		a c			a c		a c
Pyridines	a c														
Pyrrolles	a c														
Terpenes							c								
Terpenoids							c								
Sulfur compounds (general):		a c			a			a c		a c	a				a c
Disulfide											o r				

Hydrogen sulfide		d	a c d q		m
Isothiocyanates				h	
Methionals	k		k		k
Monosulfides	k	k e	k	k l f	k
Thianes				m	
Thiazoles		d	d q		d
Thiepanes				m	
Thiocyanates				j j	
Thiolanes			m		
Thiophenes		d q	d q g		f
Thiols		k			
Thiosulfinates				n p	
Thiosulfonates		k q		r	
Trisulfides				f	

[a]Many volatile flavor-contributing compounds occur as fermentation products including

acetal	butyl acetate	ethyl isobutyrate	isobutanol	methyl acetate	pentanol
acetaldehyde	butyric acid	ethyl propionate	isobutyl acetate	2-methyl butan-2-ol	2-pentanol
acetic acid	diacetyl	furfural	isobutyric acid	3-methyl butan-2-ol	3-pentanol
acetone	ethanol	hexanol	isopentanal	2-methylpentan-1-ol	propanol
butanal	ethyl acetate	isoamyl acetate	isopropanol	3-methylpentan-1-ol	propioninc acid
butanol	ethyl butyrate	isobutanal	methanol	pentanal	propyl acetate

NOTE: Many variations in fermentation biochemistry and fermentative organisms produce a wide range of potential products. Some of the more common volatile constituents that may influence flavor perception are indicated in footnote *a*. Also, note that most of the important flavor compounds in meat are generated by the action of cooking and heat (H).

Chart Key

Formation

B Bioformation
E Enzymatic
F Fermentation
H Heat
O Oxidation

a General amino acids
b Fats, fatty acids
c Sugars
d Cysteine
e Dimethyl-β-propiothetin

Flavor Precursor Compounds

f Disulfide	m Peptides
g Furanones	n S-Alkylcysteine
h Glucosinolate(s)	o Sulfide
i Hydrogen sulfide	p Sulfoxide
j Isothiocyanate	q Thiamine
k Methionine	r Thiosulfinates
l S-Methylmethione	

$(CH_2=CHCH_2—SO—S—CH_2CH=CH_2)$ and then diallyl disulfide—the typical garlic odor.

Almost any physical action imposed on plant tissues including tissue damage brought on by bruising can cause the commingling of otherwise isolated chemical compounds in plant tissues. This action may also lead to the uncontrolled evolution of numerous volatile compounds. Such events are not uncommon among tomatoes, melons, and cucumbers. Any additional heat treatments ordinarily result in tempering the intensity of characteristic plant flavors owing to heat-induced protein, carbohydrate, and/or lipid reactions.

Many of the most important aliphatic plant volatiles have their origins in the oxidative degradation of unsaturated fatty acids such as linoleic and linolenic acids. Speculation suggests that the genesis of many plant volatiles may rely on the activity of lipoxygenase. Peroxides produced on unsaturated fatty acids by this enzyme act as a prelude to a complex series of peroxide isomerizations and additional enzyme-mediated reactions. Although unsaturated fatty acids are only minor components of many fruits and vegetables, oleic, linoleic, and linolenic acids serve as potential "feed chemicals" for volatile flavor-producing reactions. These acids are rarely free and usually exist in triacylgly-

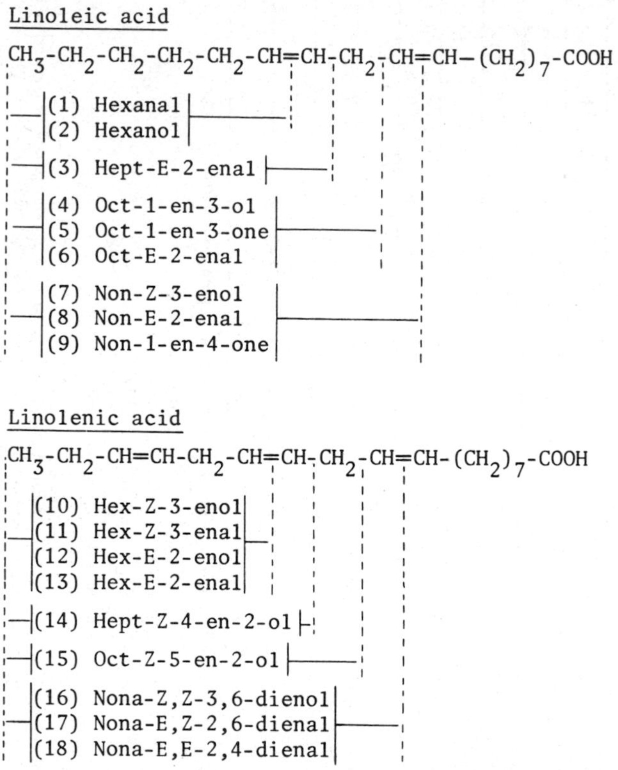

Linoleic acid

$CH_3-CH_2-CH_2-CH_2-CH_2-CH=CH-CH_2-CH=CH-(CH_2)_7-COOH$

(1) Hexanal
(2) Hexanol
(3) Hept-E-2-enal
(4) Oct-1-en-3-ol
(5) Oct-1-en-3-one
(6) Oct-E-2-enal
(7) Non-Z-3-enol
(8) Non-E-2-enal
(9) Non-1-en-4-one

Linolenic acid

$CH_3-CH_2-CH=CH-CH_2-CH=CH-CH_2-CH=CH-(CH_2)_7-COOH$

(10) Hex-Z-3-enol
(11) Hex-Z-3-enal
(12) Hex-E-2-enol
(13) Hex-E-2-enal
(14) Hept-Z-4-en-2-ol
(15) Oct-Z-5-en-2-ol
(16) Nona-Z,Z-3,6-dienol
(17) Nona-E,Z-2,6-dienal
(18) Nona-E,E-2,4-dienal

FIGURE 10.15. Some possible origins for volatile components of fruit- and vegetable-flavor components. The following numbered compounds are important contributors to the fruits and vegetables cited: (3, 4) many fruits and vegetables; (4) potato oil volatiles, green beans, dry beans, etc.; (4, 5) mushrooms; (7) peas and potatoes; (8) carrots (cooked) and many other vegetables; (10, 11) tomatoes; (12, 13) blueberries; (13) apples; (14) bananas and corn; (15) dry beans; (16) melons; and (17) cucumbers.

cerol, phospholipid, or lipoprotein structures. Selective cleavage of unsaturated fatty acid derivatives, followed by their conversion to aldehydes, alcohols, *cis/trans* isomeric structures, volatile alkyl esters, and other structures are *idiosyncracies* of *plant genera* and *species*. The pH of plant saps and fluids, ambient environmental conditions, and contingent metabolic processes of plant tissues all influence the formation of volatile compounds. Figure 10.15 details some possible derivations of important volatile plant constituents. Note, too, that only minor differences in the geometric isomerism about unsaturated sites can have a critical effect on the flavor properties of volatile plant constituents. These isomeric structures are often denoted according to the *Z/E* system detailed below for an exemplary hexadiene:

Many other volatile compounds arise from amino acids, glucosinolates, lignins, and sugars as a result of enzyme- and/or heat-mediated reactions.

VOLATILE FLAVOR CONSTITUENTS OF RED MEATS AND POULTRY

Few foods are more important than meat in the dietary habits of the American people. Despite this fact, relatively little is firmly understood about the chemical constituents that account for characteristic meat flavors. In general, fresh meats have little aroma, and much of their apparent taste reflects dissociated physiological salts, soluble carbohydrates, and/or iron-containing proteins exuded from muscle tissues. The complex-

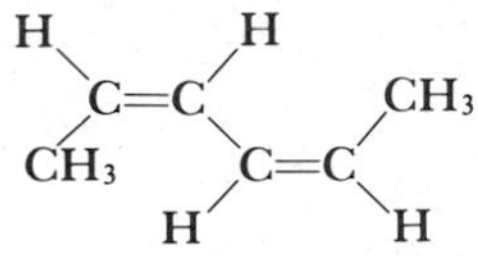

cis, cis-2,4-Hexadiene
or
Hex-Z,Z-2,4-diene

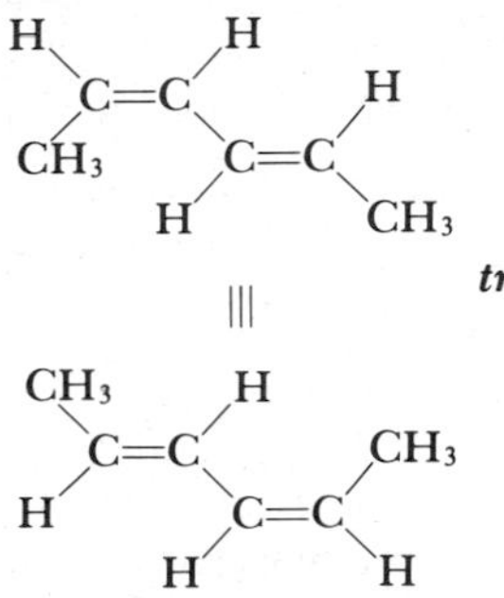

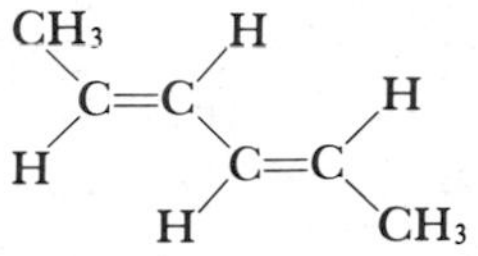

trans, trans-2,4-Hexadiene
or
Hex-E,E-2,4-diene

cis, trans-2,4-Hexadiene
or
Hex-E,Z-2,4-diene

Many aliphatic esters whose origins may be linked to oxidative reactions of unsaturated fatty acids include ethyl acetate, ethyl butyrate, ethyl hexanoate, butyl hexanoate, and a variety of branched esters.

The C_6 derivatives of unsaturated fatty acids, especially hexan*al* (*-ols*), are widespread among all members of the plant kingdom, whereas nonanal is strikingly absent from nearly all plants except olive oil, in which it is a major component. Oleic acid is the most probable source of this C_9-aldehyde.

The decanals are widely distributed among water-soluble aroma factors of citrus fruits, as are the heptanals—albeit at lower concentration levels.

ities of meat flavor chemistry develop during the cooking of meats when a myriad of reactions occur. These reactions culminate in products typical of browning reactions, protein hydrolyzates (peptides, free amino acids, etc.), aliphatic, aromatic, and heterocyclic compounds that perceptibly blend into a meat flavor sensation when eaten. No single compound seems to be responsible for cooked meat flavors, although thiane, thiazine, thiolane, thiol, pyrazine, and furanone factors have been commonly observed. The production of these compounds is beyond the scope of this text, but examples for some of these structures are detailed in Figure 10.16.

Beyond these considerations, flavors are

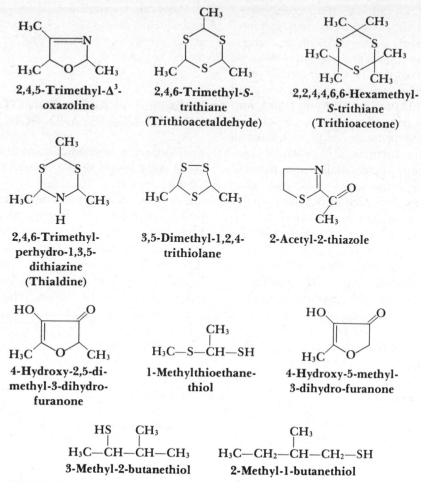

FIGURE 10.16. Examples of important chemical compounds that contribute to the flavor of cooked meats, especially cooked red meats.

imparted to meats by unsaturated fatty acid oxidations as well as hydrogen sulfide-dependent reactions. In the case of most meats, simple cooking of their defatted lean muscle extracts produces a similar meaty flavor irrespective of the animal source (lamb, veal, pork, or beef). Different species-specific flavor sensations often arise only when fatty tissues and their water-soluble fractions coexist with specific meat proteins during cooking. Water-soluble factors include no less than seven broad categories of compounds including (1) amino acids, (2) amines, (3) mono- and disaccharides, (4) sugar phosphates and amines, (5) lactic, succinic, and other organic acids, (6) nucleosides, and (7) nucleotide sugars, sugar amines, and acetylated sugar amines.

Some of the most important flavor contributions of unsaturated fatty acids may rest on their oxidative degradation to yield volatile aldehydes and ketones. Typical among the carbonyl-containing compounds are hexanal and 2,4-decadienal produced from heated linoleic acid, but hundreds of related possible structures exist. Since chicken fat has 14 to 17% of its fatty acids in the form of linoleic acid, it is clear that this acid plays an important role in the production of flavor volatiles in cooked fowl.

Hydrogen sulfide-dependent reactions, such as those between a furanone (e.g., 4-hydroxyl-2,5-dimethyl-3-dihydrofuranone) and hydrogen sulfide, seem to produce a variety of meat-like flavor products. The furan and thiophene products of such reactions have questionable ubiquities among cooked meats, but experiments show that these reactions serve as useful models of flavor-genic meat chemistry.

The contributions of hydrogen sulfide are basic to the flavor of cooked chicken, in which it develops from cystine and cysteine residues of muscle proteins. Without hydrogen sulfide and the flavor-genic reactions that it supports, the already bland flavor of chicken would be reduced even further. Volatile sulfur compounds also augment those flavor contributions caused by an array of volatile carbonyl compounds plus alcohols, ammonia, amines, and sulfides. The major carbonyl-containing volatiles of poultry often result from the Strecker degradation of amino acids or degradative oxidations of linoleic and other unsaturated fatty acids.

Off-flavors in red meats and poultry develop from similar reactions. The production of off-flavors in red meats usually reflects composite effects of lipolytic enzyme activities on triacylglycerols, oxidative rancidity of unsaturated fatty acids, plus the accumulation of protein-degradation products in adipose tissues. Simple enzymatic hydrolysis of saturated fatty acids from triacylglycerols has little effect on meat flavors unless the process is complicated by the joint proteolytic and lipolytic activities of microorganisms. As might be expected, the liberation and autoxidation of unsaturated fatty acids also serves as the root of many off-flavors in meats. This is very true for chickens, turkeys, and other fowl where off-flavor perceptions correlate closely with the dietary assimilation of unsaturated fatty acids (e.g., high fish oil-containing poultry feeds). High levels of tocopherols and/or cooking poultry under conditions that are devoid of oxygen tends to minimize the development of off-flavors in fowl carcasses.

FISH AND SEAFOOD FLAVORS

The postmortem biochemistry of fish is essentially the same as that of warm-blooded animals, but major differences exist in the rate of fish-flesh perishability. Neutral to basic pH values of the flesh promote rampant autolysis of the flesh after a bout of rigor mortis, although this can be controlled to some extent by icing fresh fish as soon as possible.

Apart from this action, both enzymatic and microbiological attack of fish tissues is prompted by gutting and filleting operations. Proteases, lipases, and oxidative enzymes seep over fish tissues and actively promote autolysis. Psychrophilic bacterial contamination from the fish epidermis or careless gutting procedures also serve to inoculate exposed fish tissues. Primary bacterial culprits include *Pseudomonas putrafaciens*, *P. fluorescens*, and *Achromobacter*. Digestive actions of these and other bacteria are difficult to control, and rapid freezing or icing of fish augmented by antibiotics (e.g., chlortetracycline) may offer the only alternatives for checking their slow but progressive growths.

Digestion of fish tissues via proteases and oxidases yield at least 40 different amines, alcohols, aldehydes, carboxylic acids, and sulfur compounds. Foremost among the foul-smelling compounds is *trimethylamine* (TMA). Bacterial proteases produce TMA by way of trimethylamine oxide and TMA-oxidase activity. Since TMA mirrors to some extent the perishability level of fish tissues. TMA assay could serve as a basis for *fish quality indexing* (*FQI*). Unfortunately, most *FQI* values based on TMA assessments have met with practical failure and no uniform maximum TMA level has been routinely used to commercially indict fish quality.

Many other volatile constituents of fish flesh are recognized, but only a few of the commonly recognized contributors are summarized in Table 10.5.

Aside from the volatile constituents of fish flavors, nonvolatile flavors are imparted by alanine, glutamic acid, methionine, valine, and 5'-ribonucleotides. Of all the amino acids,

Table 10.5. Some Important Contributing Flavor Compounds Found in Fish and Seafoods

Carbonyl compounds—at least 30 compounds are commonly recognized:
(a) C_1–C_{12} aldehydes and ketones (generally) Deca-2,4,7-Z,E,E-trienal 3,6,9-dodecatrienal (b) Combined volatile compounds give Good flavor 2,4-Pentadienal and 2,4,5-trimethyl-Δ^3-oxazoline Off-flavors (from frozen storage) Hept-E-4-enal Hept-Z-4-enal Hept-Z,E-2,4-dienal

Alcohols—generally favorable flavor factors:	*Carboxylic acids*—off-flavor impact:	*Sulfur compounds*—some of the numerous types:
2-Phenylethanol 3-Methyl-1-butanol 1-Penten-3-ol Hex-Z-3-enol Hept-Z-4-enol	Acetate[a] Butyrate[a] Pentanoate[a]	Methyl mercaptan Ethyl mercaptan Methyl sulfide Methyl disulfide Dimethyl disulfide Dimethyl trisulfide

[a]Trimethylamine may augment the obnoxious odors of these acids.

histidine has a key flavor effect on pelagic fish tissues including herring, tuna, and mackeral.

The flavor chemistry of crustaceans (lobster, shrimp, crab) and mollusks (oyster, squid, octopus, scallop, clams, etc.) are enigmatic owing to the natural sweetness and mild flavor nuances that characterize one animal versus another. The high protein and glycogen levels of these animals support autolytic and bacterial spoilage at a very high rate. Many of these bacteria have their obvious origins in benthic sediments and muds. The major deteriorative enzymes include proteases, deaminases, and arginases. All three of the enzymes have a potential for producing foul odorous compounds, and arginase produces ammonia from arginine along with ornithine. The source of sweetness perceptions common to many seafoods is not certain, but it is speculated that 5'-ribonucleotides and organic acid derivatives may play important roles.

The high unsaturated fatty acid content of all seafoods promotes off-flavor formation and rancidity as a result of autoxidative processes. These consequences may be spurred on by enzyme actions, ultraviolet light, and/or ozone formed from germicidal lamps.

DAIRY PRODUCTS

Flavor profiles for dairy products represent complex perceptual effects attributed to volatile acids, aldehydes, mono- and diketones, and mouth feel sensations caused by aqueous fat emulsions. Volatile and nonvolatile flavor compounds arise from the (1) heated interaction of milk constituents during sterilization or pasteurization, (2) microbial fermentation, and (3) unchecked enzymatic activity.

Heat-treated milk develops an assortment of flavorful compounds. Depending on the extent of heating, these may range from initial stages of caramelized sugars to the induced formation of alkylpyrazines. Most of the reported pyrazines exist as ethyl or vinyl derivatives of mono- or dimethylated pyrazines. Spray drying and related industrial processes notably encourage these products. Microbial contaminants of milk or those organisms surviving pasteurization processes can produce a host of volatile acids, alcohols, and ketones. These fermentatively produced compounds have been addressed in the chapter dealing with fermentation.

Hydrolytic and oxidative degradations of lipids also produce many characteristic milk flavors. The hydrolytic production of short-chained fatty acids from triacylglycerols can produce butyric acid, which typifies rancid butter perceptions. Some of these lipases originate from the serum fraction of milk, whereas others are irreversibly adsorbed to the interface of fat globules.

In those instances in which long-chained fatty acids happen to be liberated by enzymatic hydrolysis, they may serve as precursors for oxidative reactions and for ketone, alcohol, aldehyde, ester, lactone, and other enzyme-mediated product formations.

Off-flavors in milk are also linked to seasonal fluctuations in grazing herbage or silage quality. Some grassland weeds including landcress can impart definite off-flavors to milk. These flavors are attributed to benzylmethyl sulfide and benzyl mercaptans that develop from benzyl glucosinolate constituents of plants. Other dairy management difficulties causing off-flavors in milk may result from (1) inadequate cow grazing over green pastures, (2) too much dry feed, and (3) the late stages of lactation in milk-producing cows.

In spite of the most concerted protective measures, fresh-drawn milk can develop oxidized flavors shortly after cooling. This may reflect phospholipid oxidations, which are dependent on the presence of critical concentration ratios of ascorbic acid to dehydroascorbic acid, concentrations of dissolved oxygen, the presence of copper-containing protein complexes, or other undefined factors. So-called sun-struck or sunlight off-flavors in milk are often caused by photo-induced riboflavin-dependent conversion of methionine to methanal. Still other undesirable flavors evolve from heating milk, with the concurrent destruction of thiamine to yield hydrogen sulfide, benzothiazoles, and thiophenes.

SOME FUNDAMENTALS OF FLAVOR TECHNOLOGY

The advent of food technology, detailed specifications for food formulations, toxicological considerations, and bulk food ingredient requirements have all spurred on the industrial production of food flavors. Historically, natural sources of food flavors have been subject to climatic, geopolitical, plant-pathological, sanitation, economic, and compositional fluctuations, which are all counterproductive to food manufacturing processes.

Apart from these considerations, modern agriculture could scarcely meet the demands of the most sought-after flavor ingredients if modern flavor technologies were nonexistent. American food demands for strawberry flavors alone would necessitate a 10% dedication of all farmland to this crop.

Industrially produced flavors may consist of (1) a *single synthetic* flavor ingredient, (2) compounded mixtures of *synthetic* and/or *naturally* occurring flavors, and (3) specially *prepared natural flavors* or flavor extracts.

Synthetic flavor production for any food relies on the chemical recognition of key character impact compounds (CICs) that may exist. This process is a complicated effort that involves the chemical analysis of a natural flavor using modern instrumental methods of analysis (e.g., GLC, GLC-MS).

Once the relative concentrations of different-molecular-weight constituents are identified in a "target" flavor, efforts can be undertaken to *create* or *compound* the same flavor sensation in the laboratory. Natural and/or synthetically produced chemicals may be used in this effort so long as toxicological and other health considerations of potential consumers are not compromised.

The perceptual acceptability of any concocted laboratory flavor must be evaluated from the perspective of human test subjects. Instrumental analysis may be an important key to flavor characterization, but the sensory and organoleptic assessment of flavor properties by the human element is an overriding factor in deciding flavor formulation success. Regardless of how well a formulated flavor matches the chemical specification of a target flavor, instrumentation cannot assess off-flavors, rancidity indications, and obnoxious flavor notes caused by impurities that jade flavor authenticity.

Some flavor perceptions are relatively simple to produce since only one or two CICs are necessary. This is *not* the case with most natural flavors. Strawberries, for example, contain about 250 recognized volatile compounds, whereas coffee has over 500. Recognition of flavor complexities necessitates expert analytical chemistry coupled with human flavor evaluations in order to reproduce any flavor. The creation of a desired flavor involves the qualitative and quantitative admixture of substances from any or all of the listed categories to create a "target" flavor: (1) natural plant extracts, (2) fruit juices, (3) naturally occurring essential oils, or (4) synthetic compounds.

Poor natural supplies of popular flavors, their high cost, and lack of chemical uniformity make synthetic flavor constituents a boon for modern food technology. In these instances, modern synthetic organic chemistry can convert abundant raw chemical feedstocks into bulk quantities of flavors. As only one example, consider the industrial synthesis of terpene flavor compounds from crude sulfate terpentine. Copious quantities of this feedstock are a by-product of Kraft paper processing.

Although the most desirable synthetic precursors in terpentine feedstock may not be predictable in all cases (~62.5% α-pinene; ~22.5% β-pinene; 5–10% diterpenes; and 5–10% low volatile compounds), fractional distillation can ensure a reliable industrial supply of good quality and optically pure flavor precursors.

For many flavor-producing terpenes, the β-pinene fraction is critical for a host of synthesis reactions. Pyrolysis of the β-pinene fraction can yield myrcene, but an alternative acid-catalyzed isomerization produces *l*-limonene. Hydrochlorination of myrcene can go on to supply feedstock for acyclic terpenes including geranyl, neryl, and linalyl chlorides. Corresponding terpene alcohols and their acetate esters can be respectively produced by hydrolysis or reactions with sodium acetate. Directed hydrogenation, on the other hand, can yield still other high-purity compounds including citronellol and dimethyl octanols; or oxidative processes can convert acyclic alcohols (e.g., geraniol, nerol, and linalool) to their aldehydes. The principal structures in these reactions are largely illustrated in the lipid classification chapter.

This elementary survey of terpene syntheses from a natural chemical feedstock serves as only one example of many synthetic routes for producing flavor-active compounds. Although rational schemes of organic synthesis and chemical engineering serve as the underpinnings of flavor compounds, these methods are largely a matter of proprietary information within corporate archives. Moreover, the most detailed information is routinely found in the patent literature.

PRACTICAL CONSIDERATIONS OF FLAVOR COMPOUNDING

Flavor creation and compounding involve a variety of important considerations. Foremost among these are the use of only high-purity (99.0 + %) flavor-active ingredients for compounding. These ingredients must assure low sulfur content and exhibit poor post-compounding reactivities that result in off-flavors.

Compounding may employ natural or synthetic constituents to produce some semblance of a target flavor. The initial compounded flavor or "base" flavor should demonstrate a sensory perception similar to the "target" flavor. As previously indicated, the criterion of sensory perception overrides flavor specifications based on instrumental analysis of a target flavor. The base flavor can then be modified by adding or deleting key compounds to achieve the flavor target.

Table 10.6 outlines five creative approaches for producing desired target flavors for a variety of food-product uses. In the case of the chewing gum flavor, "burst" and "trail-out" features are important in producing a desirable aftertaste. The apricot and lemon formulations are designed to augment the natural flavor strengths, whereas the fortified pineapple flavor enforces the natural flavor strength of pineapple to meet the flavor demands of almost any type of product.

Flavor development and compounding are as much an art as a science. Moreover, it should be recognized that admixtures of two

Table 10.6. Flavor Compounding Is a Skilled Art and Science Designed to Produce a Flavor Perception That Seems to be *Authentic* or at Least *Derived* from a Natural Source

Artificial Fruit Flavor for Chewing Gum		Lemon with Other Natural Flavors	
(a) Heliotropin	0.05	Lemon oil	98.80
Ethyl vanillin	1.00	Citral (distillate fraction	
Vanillin	2.00	from lemongrass oil)	1.00
C_{16}-Aldehyde	0.55	Orange aldehydes (from	
(b) Clove oil	1.00	rectified orange oil)	0.20
Peppermint oil	2.00	Total	100.00
(c) Orange oil	37.00		
Lemon oil	5.00	**Apricot with Other Natural Flavors**	
Tangerine oil	5.00	Apricot juice (concentrate—	
(d) Ethyl acetate (plasticizer		65° Brix)	99.60
for chewing gum base)	16.40	Absolute carob	0.10
Ethyl butyrate	15.00	Vanilla extract, tenfold	0.10
Isoamyl acetate	10.00	Orange oil, tenfold	0.10
Isoamyl butyrate	5.00	Petitgrain oil	0.05
Total	100.00	Oil of bitter almond	0.05
		Total	100.00
Fortified Artificial Pineapple Flavor			
Pineapple juice concentrate	60.0	**A Simplified Grape Flavor**	
Apple juice concentrate	23.0		
Tincture St. John's bread	0.5	Methyl anthranilate	92.60
Malt syrup concentrate	0.2	Ethyl caproate	5.00
Vanilla extract, tenfold	0.1	Benzyl propionate	2.20
Citric acid	2.0	Benzilidene acetone	0.20
Ethyl alcohol (95%)	14.0	Total	100.00
Pineapple fortifier	0.2		
Total	100.0		

NOTE: So long as compounded and formulated flavors are not used for deceptive economic purposes, they can provide an important contribution to many forms of food manufacturing processes. All flavors used in the United States must be classified by governmental regulations under the G(enerally) R(ecognized) As S(afe) designation. Depending on advances in research, this classification frequently changes. A flavor ingredient's potential for causing cancer (e.g., cinnamyl anthranilate), mutagenicity, and other toxigenic effects must all be considered as well as the health consequences of impurities that may coexist with key flavor compounds. The table outlines some typical compounded flavors designed for very specific uses. See text for details. Composition expressed as percentage by weight.

SOURCE: From *Flavors and Spices* by J. A. Rogers and F. Fischetti in *Kirk-Othmer Encyclopedia of Chemical Technology*, 3rd ed., Vol. 10. Copyright 1980, John Wiley & Sons, New York. Reprinted with permission.

or more highly purified flavor ingredients can interact in some interesting ways:

1. The individual flavors of two compounds may *fuse into a new unified flavor sensation,* which has detectable features of either or both ingredients.

2. *A flavor blend may develop* with the admixture of two flavor factors.

3. Flavor mixtures may cause an *ascendence of one flavor* over the other.

4. One flavor component of a pair may *annihilate* the perception of the other component.

Polemic disputes have erupted over the years regarding the possibility that flavor compounds may mutually interact so as to neutralize both of their flavor effects. This phenomenon is virtually unknown and certainly does not occur on a routine basis.

FLAVOR CATEGORIES

Flavors can be introduced into foods by a variety of ingredients, which may include (1) condiments, (2) concentrated fruits and juices, (3) spices, (4) process flavors, (5) essential oils, and (6) oleoresins or solid extracts (Table 10.7).

The first three categories have served as food flavor sources for many years, while the remaining classes simply represent advanced industrial methods for introducing concentrated flavor constituents into foods. Process flavors are concentrated flavor-active by-products of conventional food roasting, cooking, or fermentation operations. Essential oils may be obtained from the distillation or physical expression of spices or oil-rich plant material. Oleoresins, on the other hand, are potent flavor-active ingredients prepared from spices or herbs by extraction with volatile organic solvents. The solvent stripping of the plant extract leaves a residue of essential oils, organically soluble resins, nonvolatile fatty acids, and any other solvent-extractable plant constituents. Solvent residues of methylene chloride, ethylene dichloride, trichloroethylene, acetone, isopropanol, methanol, or hexane typically range from 25 to 50 ppm in the final oleoresin product. Storage, product uniformity, and application factors have made oleoresins very important food ingredients. Fatty oils and resins of oleoresin extracts also serve as natural fixatives for volatile flavor constituents. This feature promotes flavor dispersions throughout foods in which oleoresins are used, and also decreases flavor losses during high-temperature food-processing operations.

Table 10.7. Some Important Categories of Flavor Ingredients in Foods and Their Respective Contributions to Food Flavors

Food flavor ingredient (category)	Percentage flavor contribution to food	Approximate concentration in food (ppm)
Synthetic and/or compounded flavors	$40 \longrightarrow 100$	$10^{-2} \longrightarrow 50$
Condiments	$10 \longrightarrow 40$	$10^4 \longrightarrow 8 \times 10^4$
Concentrated juices and processed fruits	$35 \longrightarrow 70$	$10^4 \longrightarrow 10^5$
Essential oils	$80 \longrightarrow 100$	$5 \longrightarrow 50$
Oleoresins (e.g., paprika, turmeric, capsicum)	$40 \longrightarrow 90$	$10 \longrightarrow 8 \times 10^2$
Process flavors	$25 \longrightarrow 100$	$10^2 \longrightarrow 5 \times 10^4$
Spices	$35 \longrightarrow 75$	$5 \times 10^2 \longrightarrow 5 \times 10^3$

PREPARATION OF NATURAL FLAVORS FOR FOOD USES

Flavor-active food ingredients may be introduced directly into certain foods or admixed with the aid of a diluent or "carrier." For example, diluents such as fatty oils and propylene glycol are used for the uniform distribution of oleoresins. Contrary to oleoresins, essential oils (and similar formulated flavor mixtures) may be admixed to foods as beaded or powdered flavors. These flavoring forms can be produced by a number of industrial methods including spray drying, spray chilling, or "plating" procedures. Spray drying involves the atomization of a 125 to 150°F slurry containing the flavor oil through a hot gas stream (350–450°F). A flavor-rich powder is collected at the end of the drying process. Spray chilling can also be used to encapsulate flavors. According to this process, 150 to 180°F fluids are injected into a closed chamber. An air vector component of the chamber atmosphere (32–40°F) is then responsible for associating a vegetable fat or emulsifier component with the principal flavor compound(s). Both of these flavor-encapsulation processes can promote the dispersion of flavors in food systems by dissolving actions, melting, or mechanical shearing actions and/or protect important flavor constituents from chemically reactive environments. Plating processes are less complex than encapsulation since flavors such as those from whole spices are blended at very high speed in a special carrier. The homogeneously distributed flavor–carrier mixture can then be used for admixture to foods.

FLAVOR ENHANCERS AND MODIFIERS

The perception of food flavors involves "quality" and "additive" factors. The taste of sodium chloride is salty, whereas citric acid is sour. These quality differences are substitutive phenomena. However, when the concentrations of salt *or* acid increase respectively, their individual taste perceptions increase in intensity. This occurs in an additive fashion until taste perception is maximized.

Efforts to expand the intensity of limited amounts of desirable food flavors can be achieved by *flavor enhancers*. Since 1909, monosodium glutamate (MSG) has been used for this purpose, but in more recent years, 5'-ribonucleotides were recognized to achieve the same effect.

Among the various 5'-ribonucleotides, disodium 5'-inosinates (5'-IMP) and 5'-guanylates (5'-GMP) offer the most impressive flavor enhancing effects. Although the largest industrial supplies of these compounds are obtained from the enzymatic hydrolysis of yeast RNA, many other plant and animal tissues also contain high potential sources of ribonucleotides. Mushrooms, sardines, bonito, salmon, pork, chicken, and several other organisms contain notable amounts of nucleotides, but *compared to microbial sources,* these are not commercially suited to high-volume production of nucleotides.

Since 5'-GMP has nearly three times the enhancing potential of 5'-IMP, practical economic conditions favor a 50:50 ratio of these two compounds in most flavor enhancers using nucleotides.

Apart from glutamates and nucleotides, other similar agents are recognized including (1) tricholomic and ibotenic acids produced by selected fungi and (2) maltols produced from the decomposition of heated di- and trisaccharides plus isomaltose and protein–lactose systems.

The flavor enhancement properties of these compounds can display synergistic effects. That is, carefully concocted mixtures of different enhancers may produce a superior flavoring effect over an equivalent mixture of a single type of enhancer.

Ribonucleotides are essentially tasteless in most circumstances, and depending on individual taste sensitivities, L-monosodium glutamate can have a salty/sour yet characteristic flavor (D-MSG isomers fail to demonstrate enhancer effects). Taste thresholds in water are in the range of 0.2 to 0.5 g/L for MSG and 0.04 to 0.12 g/L for nucleotides (5′-GMP → 5′-IMP). Although the action mechanisms for all enhancers are not known, most of their effects are contingent to some extent on the pH of the food system as well as buffering effects.

Exclusive of the desirable flavor effects of MSG as an enhancer in meat, poultry, soups, seafoods, and vegetables, its exhorbitant application to foods may have health implications. These problems range from transient headaches to lethargy. More severe problems including hypothalamic and retinal lesions have been detected in rodents and some primates, but the present toxicological data do not yet warrant a wholesale indictment of MSG in foods.

Certain compounds in addition to those outlined above may modify specific taste perceptions. Natural D-glucuronides of acetylated gymnemagnins are known to eliminate the perception of all sweet tastes including the intensity of bitter tastes. Another natural plant product obtained from the berries of *Synsepalum dulcificum* (miracle fruit) is also known to modify taste perceptions of sourness. This glycoprotein is speculated to exert pH-dependent interactions on membranes at the site of sweet-taste receptors. As a result, the sourness is markedly reduced to a sweeter sensation for acid foods. This taste-modifying protein should not be confused with thaumatin, which truly produces a sweet taste sensation.

BRIEF OVERVIEW OF FLAVOR ANALYSIS

Only 10^{-8} to $<10^{-14}\%$ of food composition is responsible for flavor and aroma compounds. Such small amounts of material serve as major obstacles to the analytical concentration and molecular analysis of flavor-active food ingredients. Many different analytical hurdles have been eliminated by the advent of sophisticated gas chromatography (GC) and mass spectrometry (MS). Although MS is crucial for structural identification of flavor compounds, GC is critical for (1) routine partitioning analysis of complex flavors and (2) prepartitioning of complex flavor compounds for MS analysis.

For relatively crude or routine assessments of volatile-flavor compounds, packed GC columns containing selected nonvolatile liquid phases mounted on a stationary packing (GLC) permit adequate identification of many flavor constituents. However, sophisticated instrumental analysis of complex volatile flavors can be achieved only by the high resolving power of capillary GC columns. The internal diameter of these columns (also called open-tubular GC columns) is coated with a suitable stationary phase that permits a high vapor phase resolving ability among flavor components introduced into the column. Theoretical plate potentials for these columns are 10 to 20 times greater than conventional packed GC columns and may demonstrate plate values in the range of 300,000+. Unlike the 1/8- to 1/16-in. by 6-ft dimensions of many conventional GC columns, capillary columns achieve separation over a column diameter of 0.01- to 0.03-in. and 50- to 500-ft lengths depending on application requirements (Figure 10.17).

For a detailed prospectus of analytical flavor chemistry, the reader is encouraged to consult the authoritative work of Teranishi, Merritt, and others cited in the reference section.

In those analytical circumstances where flavor quality control must be carefully monitored, GC or high-pressure liquid chromatographic (HPLC) methods may be employed. The use of either method presumes that an ideal flavor property has been chromatographically "fingerprinted" and then correlated with a taste panel acceptance value (i.e., flavor or "nasal" appraisals). Detectable chromatographic variations of sample flavors from the "ideal" chromatographic fingerprint serve as a basis for accepting or re-

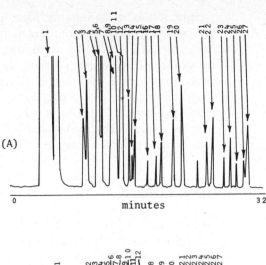

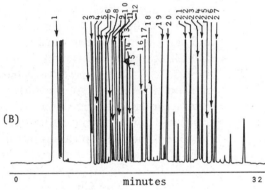

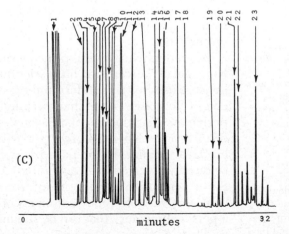

Figure A and B

1. solvent-methylene chloride
2. α-thujene
3. α-pinene
4. camphene
5. sabinene
6. β-pinene
7. myrcene
8. -------
9. p-cymene
10. limonene
11. -------
12. γ-terpinene
13. terpinolene
14. nonanal
15. linalool
16. -------
17. terpinene-4-ol
18. α-terpineol
19. neral
20. geranial
21. neryl acetate
22. geranyl acetate
23. β-caryophyllene
24. α-bergamotene
25. -------
26. -------
27. β-bisabolene

Figure C

1. solvent-methylene chloride
2. α-pinene
3. camphene
4. β-pinene
5. myrcene
6. α-phellandrene
7. 1,4-cineole
8. α-terpinene
9. p-cymene
10. d-limonene
11. γ-terpinene
12. terpinolene
13. linalool
14. borneol
15. terpinene-4-ol
16. α-terpineol
17. neral
18. geranial
19. neryl acetate
20. geranyl acetate
21. β-caryophyllene
22. α-bergamotene
23. β-bisabolene

FIGURE 10.17. Examples of gas chromatographic separations for flavor components of two citrus oil extracts: (A) and (B) lemon oil extract; (C) lime oil extract. The chromatogram illustrated in A shows fairly good resolution of oil constituents using a packed column (⅛ in. × 6 ft) but the higher efficiency of a capillary column (0.25 mm × 25 m) produces superior resolution of oil constituents. Chromatogram C represents a capillary column separation of lime oil components.

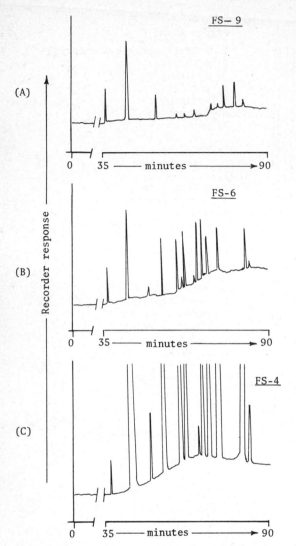

FIGURE 10.18. Gas chromatographic analysis of an *ideal* quality vegetable oil (A) provides a chromatographic fingerprint having very few volatile constituents and a high flavor score (*FS*) of 9.0 on a possible scale range of 1 to 10. Lower *FS* values as determined by the subjective evaluation of taste panel participants correspond to higher amounts of undesirable volatile constituents in oils. The decrease in *FS* values mirrors the appearance of more chromatographically identifiable impurities in oils that have less refinement and/or those obtained from a poor source (B–C). Microprocessor analysis of chromatographic data can be used for the routine monitoring of sample data to produce a calculated *FS* value without the use of a panel. Samples such as those having data (chromatogram C) with high volatile content are automatically recognized as being inferior and are assigned a low *FS* value in lieu of actual taste-panel assessment. This form of quality control can

jecting a flavor sample or food product. Subjective evaluations of chromatography data can be executed by an analyst or preprogrammed judgments based on microprocessor logic. As indicated in Figure 10.18, panel acceptance values usually have an inverse relationship (lower scores) with the number of detectable chromatographic peaks that occur in a given sample.

CONCLUSION

Appearance, mouth feel, flavor, and nutritive values of a food collectively influence its dietary appeal. From sensory perspectives, nutritive value is the least important factor and sight is the most important—assuming that all other factors remain equal. Nevertheless, food flavor serves as a basic factor of food acceptance. At the instant a food enters the mouth, flavor constituents are translated into hedonic responses. That is, the food is judged as pleasant or unpleasant.

For practical purposes, food flavors are studied from the aspect of *stimulus* and *response* phenomena. Chemoreceptors in the mouth and the olfactory epithelium are activated by molecules (stimuli) that elicit taste and flavor responses. Our *responses* take the form of a judgment call that is favorable or unfavorable.

Beyond psychophysical principles underlying flavor perception, little scientific evidence exists for an understanding of this enigmatic sense. Other than occasional experiments that document neuronal stimulation as a result of chemoreceptor activation by flavor stimuli, flavor perception cannot be directly observed in any organism. At best, flavor perception can only be inferred from behavior, verbally reported, or encoded as a personal memory that can be subliminally recalled for future reference.

be employed only when the composition of a food substance is highly correlated with fingerprinted data, and it requires a great deal of analytical expertise and confidence in sophisticated instrumentation.

REFERENCES

Acree, T. E. 1980. Flavor characterization. In *Kirk-Othmer Encyclopedia of Chemical Technology*, 3rd ed., pp. 444–455. Wiley, New York.

Adrian, E. D. 1963. *Olfaction and Taste: Proceedings-First International Symposium Werner-Gren Center* (Y. Zotterman, ed.). Macmillan, New York.

Amoore, J. E. 1952. The stereochemical specificity of olfactory receptors. *Perfum. Essent. Oil. Rec. Yb.* **43**:321–322, 330.

Amoore, J. E. 1963. Stereochemical theory of olfaction. *Nature* **198**:271.

Amoore, J. E., J. W. Johnston, and M. Rubin. 1964. The stereochemical theory of odor. *Sci. Amer.* **210**(Suppl. 2):42.

Amoore, J. E. 1965. Psychophysics of odor. *Cold Spring Harbor Symp. Quant. Biol.* **30**:623.

Amoore, J. E. 1967. Stereochemical theory of olfaction. In *Symposium on Foods: The Chemistry and Physiology of Flavors* (H. W. Schultz, E. A. Day, and L. M. Libbey, eds.). Avi Publishing, Westport, Conn.

Andersen, H. T., M. Funakoshi, and Y. Zotterman. 1963. Electrophysiological responses to sugars and their depression by salt. In *Olfaction and Taste* (Y. Zotterman, ed.). Macmillan, New York.

Beck, L. H., and W. R. Miles. 1947. Some theoretical and experimental relationships between infrared absorption and olfaction. *Science* **106**:511.

Beidler, L. M. 1954. A theory of taste stimulation. *J. Gen. Physiol.* **38**:133.

Beidler, L. M. 1957. Facts and theory on the mechanism of taste and odor perception. In *Chemistry of Natural Food Flavors*. United States Quatermaster Food and Container Institute for the Armed Forces—Chicago.

Beidler, L. M., M. S. Nejad, R. L. Smallman, and H. Tateda. 1960. Rat taste cell proliferation. *Fed. Proc.* **19**:302.

Beidler, L. M. 1966. Chemical excitation of taste and odor receptors. In *Flavor Chemistry* (I. Hornstein, ed.), ACS Advances in Chemistry Series 56, Washington, D.C.

Cairncross, S. E., and L. B. Sjöström. 1950. Flavor profiles—a new approach to flavor problems. *Food Technol.* **4**:308.

Charalambous, G. (ed.). 1980. *The Analysis and Control of Less Desirable Flavors in Foods and Beverages*. Academic Press, New York.

Charalambous, G., and G. E. Inglett (eds.). 1978.

Flavor of Foods and Beverages, Chemistry and Technology. Academic Press, New York.

Cohen, M. J., S. Hagiwara, and Y. Zotterman. 1955. The response spectrum of taste fibres in the cat: A single fibre analysis. *Acta Physiol. Scand.* **33**:316.

de Lorenzo, A. J. 1963. Studies on the ultrastructure and histophysiology of cell membranes, nerve fibers and synaptic junctions in chemoreceptors. In *Olfaction and Taste* (Y. Zotterman, ed.), pp. 5–17. Macmillan, New York.

Dravnieks, A. 1967. Theories of olfaction. In *Symposium on Foods: The Chemistry and Physiology of Flavors* (H. Schultz, E. A. Day, and L. M. Libbey, eds.). Avi Publishing, Westport, Conn.

Edmund, W. J., and D. A. Lillard. 1977. Sensory comparison of aroma precursors in marine and terrestrial animals. *J. Food Sci.* **42**:843.

Fishman, I. Y. 1957. Single fiber gustatory impulses in rat and hamster. *J. Cell Physiol.* **49**:319.

Formácek, V., and K.-H. Kubeczka. 1982. *Essential Oils Analysis by Capillary Gas Chromatography and Carbon-13 NMR Spectroscopy*. Wiley, New York.

Furia, T. E., and N. Bellanca (eds.). 1975. *Fenaroli's Handbook of Flavor Ingredients*, Vols. 1 and 2. Chemical Rubber Company Press, Boca Raton, Fl.

Geldard, F. A. 1953. *The Human Senses*. Wiley, New York.

Gold, H. J., and C. W. Wilson. 1963. The volatile flavor substances of celery. *J. Food Sci.* **28**:484.

Hall, R. L. 1968. Flavor and flavoring, seeking a consensus of definition. *Food Technol.* **22**:1496.

Hayashi, T., K. Yamaguchi, and S. Konosu. 1981. Sensory analysis of taste-active components in the extract of boiled snow crab meat. *J. Food Sci.* **46**:479–483, 493.

Henning, H. 1916. Die Qualitatenreike des Geshmacks. *Z. Psychol.* **74**:203.

Henning, H. 1924. *Smell; a Handbook for the Spheres of Psychology, Physiology, Zoology, Botany, Chemistry, Physics, Neurology, Ethnology, Language, Literature, Aesthetics and History*, pp. 453–458. J. A. Barth, Leipzig.

Hornstein, I., and P. F. Crowe. 1960. Flavor studies on beef and pork. *J. Agric. Food Chem.* **8**:494.

Johnson, J. A., L. Rooney, and A. Salem. 1966. Chemistry of bread flavor. In *Flavor Chemistry* (I. Hornstein, ed.). ACS Advances in Chemistry Series 56, Washington, D.C.

Johnston, J. W. 1953. Infrared loss theory of olfaction untenable. *Phys. Zool.* **23**:943.

Kaiser, R., D. Lamparsky, and P. Schudel. 1975. Analysis of buchu leaf oil. *J. Agric. Food Chem.* **23:**943.

Kimura, K., and L. M. Beidler. 1961. Microelectrode study of taste receptors of rat and hamster. *J. Cell. Physiol.* **58:**131.

Kistiakowski, G. B. 1950. On the theory of odours. *Science* **112:**154.

Kruger, L., A. L. Feldzamen, and W. R. Miles. 1955. A scale for measuring supra-threshold olfactory intensity. *Amer. J. Psychol.* **68:**117–123.

Kruger, L., A. L. Feldzamen, and W. R. Miles. 1955. Comparative olfactory intensities of the aliphatic alcohols in man. *Amer. J. Physiol.* **68:**386.

Kurihara, K., and L. M. Beidler. 1968. Taste-modifying protein from miracle fruit. *Science* **161:**1241.

Langmuir, I. 1918. The adsorption of gases on plane surfaces of glass, mica and platinum. *J.A.C.S.* **40:**1361.

Lund, E. D., and W. L. Bryan. 1977. Commercial orange essence: comparison of composition and methods of analysis. *J. Food Sci.* **42:**385.

Miller, A., III, R. A. Scanlan, J. S. Lee, and L. M. Libbey. 1973. Volatile compounds produced in sterile fish muscle (*Sebastes melanops*) by *Pseudomonas putrafaciens, Pseudomonas fluorescens* and an *Achromobacter* species. *Appl. Microbiol.* **26:**18.

Moncrieff, R. W. 1951. *The Chemical Senses.* Wiley, New York.

Ochi, H. 1980. Production and applications of natural seafood extracts. *Food Technol.* **34:**51–53, 68.

Parliment, T. H. 1982. Capillary gas chromatographic analysis of N and S compounds in flavors and fragrances. *Amer. Lab.* **14**(5):35.

Pepper, F. A., and A. M. Pearson. 1971. Possible role of adipose tissue in meat flavor—the nondialyzable aqueous extract. *J. Agric. Food Chem.* **19:**964.

Pfaffmann, C. 1941. Gustatory afferent impulses. *J. Cell Physiol.* **17:**243.

Pippen, E. L., A. A. Campbell, and I. V. Streeter. 1954. Origin of chicken flavor. *J. Agric. Food Chem.* **2:**364.

Pippen, E. L. 1967. Poultry flavor. In *Symposium on Foods—Chemistry and Physiology of Flavors* (E. Schultz, A. Day, and L. M. Libbey, eds.). Avi Publishing, Westport, Conn.

Potts, A. M., K. W. Modrell, and C. Kingsbury. 1960. Permanent fractionation of the electroretinogram by sodium glutamate. *Amer. J. Ophthalmol.* **50:**900.

Reynolds, W. A., N. Lemkey-Johnston, L. J. Filer, and R. M. Pitkin. Monosodium glutamate: Absence of hypothalmic lesions after ingestion by newborn primates. *Science* **172:**1342.

Rogers, J. A., and F. Fischetti. 1980. Flavors and spices. In *Kirk-Othmer Encyclopedia of Chemical Technology*, 3rd ed., Vol 10, pp. 456–488. Wiley, New York.

Schallenberger, R. S., and M. S. Lindley. 1977. A lipophilic-hydrophobic attribute and component in the stereochemistry of sweetness. *Food Chem.* **2:**145.

Schallenberger, R. S. 1980. Predicting sweetness from chemical structure and knowledge of chemoreception. *Food Technol.* **34:**65.

Schreier, P. (ed.). 1981. *Flavour '81.* 3rd Weurman Symposium, Proceedings of the International Conference. Munich, April 28–30. Walter de Gruyter, New York/Berlin.

Sheldon, R. M., L. A. Walters, and W. E. Druell. 1972. Synthesis of flavor chemicals and oils from domestic raw materials. *164th National Meeting ACS Symposium on Industrially Produced Food Ingredients, August 29, New York.*

Sjöström, L. B. 1972. *The Flavor Profile.* Arthur D. Little, Cambridge, Mass.

Society of Flavor Chemists. 1969. Flavor chemists define flavor two ways. *Food Technol.* **23:**1360.

Stansby, M. E. 1971. Flavors and odors of fish oils. *J. Amer. Oil. Chem. Soc.* **48:**820.

Stöcklin, W., E. Weiss, and T. Reichstein. 1967. Gymnemic acid, the antisaccharic principle of *Gymnema sylvestre* R. Br. Isolation and identification. *Helv. Chim. Acta* **50:**474.

Stevens, S. S. 1958. Measurement of man. *Science* **127:**383.

Stone, H., J. Sidel, S. Oliver, A. Woolsey and R. Singleton. 1974. Sensory evaluation by quantitative descriptive analysis. *Food Technol.* **28:**24.

Teranishi, R. 1958. Odor and molecular structure. In *Gustation and Olfaction* (G. Ohloff, and A. F. Thomas, eds.). Academic Press, New York.

Thomas, C. P., P. S. Dimick, and J. H. MacNeil. 1971. Sources of flavor in poultry skin. *Food Technol.* **25:**407.

Torrey, S. (ed.). 1980. *Fragrances and Flavors, Recent Developments.* Noyes Data Corp., Park Ridge, N.J.

Van der Heijden, A., L. B. P. Brussel, and H. G. Peer. 1978. Chemo-reception of sweet-tasting dipeptide esters: A third binding site. *Food Chem.* **3:**207.

Vernin, G. (ed.). 1982. *Chemistry of Heterocyclic Compounds in Flavors and Aromas.* Wiley, New

York. (Excellent reference on the formation, occurrence, and use of heterocyclic flavoring and aroma compounds.)

von Békésy, G. 1966. Taste theories and the chemical stimulation of single papillae. *J. Appl. Physiol.* **21**(Suppl. 1):1.

Wasserman, A. F., and A. M. Spinelli. 1972. Effect of some water-soluble components on aroma of heated adipose tissue. *J. Agric. Food Chem.* **20**:171.

Wilson, R. A., and I. Katz. 1972. Review of the literature on chicken flavor and report of isolation of several new chicken flavor components from aqueous cooked chicken broth. *J. Agric. Food Chem.* **20**:741.

Wright, R. H. 1957. Odor and molecular vibration. In *Molecular Structure and Organoleptic Quality,* Soc. Chem. Ind. (London), Monograph 1.

Wright, R. H. 1964. Odour and molecular vibration: The infrared spectra of some perfume chemicals. *Ann. N. Y. Acad. Sci.* **116,** Art. 2, 552.

Wright, R. H., J. R. Hughes, and D. E. Hendrix. 1967. Olfactory coding. *Nature* **216**:404.

Yeh, C. S., R. Nickelson II, and G. Finne. 1978. Ammonia-producing enzymes in white shrimp tails. *J. Food Sci.* **43**:1400–1401, 1404.

Zwaardemaker, H. 1925. *L' odorat.* Doin, Paris.

Some notable references dealing with analytical flavor chemistry

Buttery, R. G., R. M. Seifert, D. G. Guadagni, D. R. Black, and L. C. Ling. 1968. Characterization of some volatile constituents of carrots. *J. Agric. Food Chem.* **16**:1009.

Cronin, D. A. 1982. Techniques of analysis of flavours—Chemical methods including sample preparation. In *Food Flavours.* Part A. *Introduction* (I. D. Morton and A. J. Macleod, eds.), pp. 15–78. Elsevier, Amsterdam/New York.

Dimick, K. P., and J. Corse. 1956. Gas chromatography—A new method for the separation and identification of volatile materials in foods. *Food Technol.* **10**:360.

Gohlke, R. S. 1959. Time-of-flight mass spectrometry and gas–liquid partition chromatography. *Anal. Chem.* **31**:535.

Guadagni, D. G., S. Okano, R. G. Buttery, and H. K. Burr. 1966. Correlation of sensory and gas–liquid chromatographic measurements of apple volatiles. *Food Technol.* **20**:518.

McFadden, W. H., and R. Teranashi. 1963. Fast-scan mass spectrometry with capillary gas–liquid chromatography in investigation of fruit volatiles. *Nature* **200**:329.

Merritt Jr., C., and D. H. Robertson. 1982. Techniques of analysis of flavors. In *Food Flavours.* Part A. *Introduction* (I. D. Morton, and A. J. Macleod, eds.), pp. 49–79. Elsevier, Amsterdam/New York.

Morton, I. D., and A. J. Macleod (eds.). 1982. *Food Flavours.* Part A. *Introduction.* Elscvier, Amsterdam/New York.

Teranishi, R., and R. A. Flath. 1981. *Flavor Research—Recent Advances.* Dekker, New York.

Teranishi, R., C. C. Nimmo, and J. Corse. 1960. Gas–liquid chromatography—programmed temperature control of the capillary column. *Anal. Chem.* **32**:1384.

C H A P T E R · 11

Nutritional Energetics

INTRODUCTION

All living organisms require energy to maintain their life-support mechanisms. *Autotrophic* organisms obtain their energy requirements from sunlight, which interacts with photoactive pigments to cleave water and reduce carbon dioxide to organic carbon compounds, later to be used in their metabolic pathways. All *heterotrophic* organisms, including man, obtain their energy from autotrophs or, alternatively, from other heterotrophs. The ingested compounds are involved in a series of enzyme-catalyzed reactions, referred to as *metabolism*, yielding energy to carry on life-support functions within the organism and to perform work. Excess energy may also be stored within the organism for later use. The energy within the cell undergoes transformations that are based on the laws of thermodynamics.

THE LAWS OF THERMODYNAMICS

Since living organisms are governed by the laws of thermodynamics, a familiarization with these laws becomes a necessity for the reader. The laws of thermodynamics will be presented in this book in a concise manner, but for a more rigorous and complete discussion, the reader is encouraged to consult other references. Following the introductory material, the application of thermodynamic principles to biochemical systems will be presented. The basic laws of thermodynamics will be extrapolated to deal with the more complex biochemical systems.

The First Law of Thermodynamics

The first law of thermodynamics is, in essence, a conservation of energy statement. The first law of thermodynamics may be presented mathematically as follows:

$$\Delta E = q - w$$

In this expression, ΔE is the change in the internal energy of a system, q is the heat absorbed by that system, and w is the work done by the system. Note that one could discuss changes in internal energy by considering the heat absorbed and the work done by that system in any process without knowledge of the actual internal energy of the system. Thus, it is not necessary to know the actual energy or even the types of energy terms involved in the total energy in order to evaluate energy changes. Considering the first law, q is positive for a system that absorbs heat such that the contribution to ΔE will be positive and the internal energy will be increased by the process. The work term, w, is positive when the system does work such that the change in the internal energy will be negative as a result of the work done. The system may use its internal energy to do work, resulting in a lower energy. The total change in the internal energy of a system in any process will depend on both the heat absorbed and the work done.

The energy of a system depends on the state of that system and does not depend on how the system achieved that state. It is possible to go from some initial state to a final state by a number of different paths. Heat and work are not properties of a system, but rather properties that depend on the manner in which a change in the state of the system is accomplished. The first law may be written in differential form as

$$dE = đq - đw$$

The đ notation indicates an inexact differential. Since q and w are path-dependent properties, the manner in which a system changes its state must be specified exactly before q or w can be evaluated. Even though q and w have inexact differentials, $đq - đw$ is equal to dE, an exact differential. The state of a system is characterized by the observable properties of that system, such as pressure, temperature, and volume. If a system undergoes a change in its state, and if its initial and final states are known, then it is possible to calculate the change in the internal energy. However, it is not possible to compute the total heat absorbed by the system or the total work done by the system in that change of state unless specific path information is available.

Work related to volume changes: Specific consideration will now be given to the work term. From mechanics, work may be defined as the product of a force and a displacement in the direction of that force. Since many types of forces exist, such as forces due to electric or magnetic fields, the force due to gravity, and frictional forces, different types of work could be considered. Consider the work done by a system when it expands against an opposing pressure. Note that a pressure is a force per unit area and that a volume may be described in terms of an area times a length. A differential expression for this expansion work may be written:

$$đw = P_{opp} \, dV$$

In order to understand why it is the opposing pressure and not the system pressure that is involved in the differential expression for expansion work, consider the following example. An ideal gas is confined to a cylinder topped with a weightless, frictionless piston. First consider the cylinder placed in a vacuum with stops holding the piston in place. The external pressure in this case is zero. If the stops were removed, the piston would be pushed rapidly upward by the expanding gas, but no work would be done in the expansion since the pressure opposing the expansion is zero:

$$w = \int đw = \int_{V_1}^{V_2} P_{opp} \, dV = 0$$

Next consider this same cylinder containing the ideal gas placed in a chamber that allows adjustment of the external pressure. If the gas is to do work on the surroundings, it must expand against an opposing pressure. Assume that the initial pressure of the gas is 10 atm and that the external pressure is maintained constant at 2 atm. The gas is then allowed to expand until the final pressure of the gas is 2 atm. Certainly, the gas must do more work to move the piston against this constant opposing pressure of 2 atm than it did when there was zero opposing pressure. Since the volume change here occurs against a constant pressure, the path is specified and the work done may be computed.

$$w = \int_{V_1}^{V_2} P_{opp} \, dV = P_{opp} \int_{V_1}^{V_2} dV$$

$$w = P_{opp}(V_2 - V_1) = P_{opp} \, \Delta V$$

In the above example, the external pressure was a constant and, accordingly, it was removed from the integral. As the pressure opposing expansion of the gas was increased, the work done in the expansion process was also increased. Note that volume is a system property, not a path-dependent property. Therefore, the integral of the exact differential dV is the finite difference ΔV for the particular state change. In contrast, note that the integral of the inexact differential $đw$ is the total quantity, w.

Consider a final example involving the same gas cylinder discussed above. One can compute the maximum amount of work that can be done on the surroundings by this expanding gas. In order to do the maximum amount of work, the motion of the piston must be opposed with the maximum pressure that will still allow the gas to expand. This requires that the external pressure must be varied so that it is always differentially less than the gas pressure throughout the entire expansion. In this case of maximum expansion work for the system, the opposing pressure is not constant, but varies, as the system pressure varies. Specifically, $P_{opp} = P - dP$, and the opposing pressure is differentially less than the system pressure at all times. In a sum or a difference, a differential may be neglected, since it is so much smaller than the value of that property of which it is the differential. For the case of maximum work, also termed reversible work, P_{opp} may be set equal to P, the system pressure. The work done for this example may be calculated:

$$w = \int_{V_1}^{V_2} P_{opp}\, dV = \int_{V_1}^{V_2} P\, dV = \int_{V_1}^{V_2} \frac{nRT\, dV}{V}$$

$$w = nRT \ln\left(\frac{V_2}{V_1}\right)$$

In the above calculation, the pressure is not constant, and in order to integrate the expression, the pressure was written in terms of the variable V. The pressure and system volume are related by the equation of state for an ideal gas, $PV = nRT$. The temperature of the system is assumed to be maintained constant for this example.

To this point the work term has been discussed along with the maximum amount of work possible for a specific expansion process. This maximum work, as previously mentioned, is also termed the reversible work. It was noted that the system must expand against a pressure just differentially less than the system pressure in order to do the maximum amount of work in that expansion process. Thus, the external pressure and the system pressure are the same to within a differential amount. In a reversible process such as this, the system is differentially close to an equilibrium state at all times. No real system actually changes its state reversibly, but many processes closely approximate reversibility.

Enthalpy: Given information on the specific path by which the state of a system changes, the work done or the heat absorbed by the system in that state change may be calculated. The calculation of w was considered for various paths. It was shown that no path information is necessary to determine the change in the value of a state property. Only knowledge of the initial and final states is required.

Enthalpy is a property that depends only on the state of the system and is defined as follows:

$$H = E + PV$$

In this definition, H is the enthalpy, E is the internal energy, and P and V are the system pressure and volume, respectively. The general definition for H given above is applicable to any type of process that occurs. However, it is found that H is a particularly useful state function for those processes that occur under conditions of constant pressure. The definitional equation for enthalpy may also be written in differential form.

$$dH = dE + P\, dV + V\, dP$$

Substitution for dE from the first law yields the following equation:

$$dH = đq - đw + P\, dV + V\, dP$$

At this time, only that work associated with expansion or compression of the system will be considered.

$$dH = đq - P_{opp}dV + P\,dV + V\,dP$$

In general, the đ w term is not the same as the $P\,dV$ term, which relates to system properties. Under conditions of constant pressure, however, both the system pressure and the external pressure must be the same and, thus, $P_{opp} = P$. Also, if the pressure is a constant, then $dP = 0$; there is no change in pressure for a constant pressure process. It follows that under constant pressure conditions

$$dH = đq \qquad \text{(constant pressure)}$$

Alternatively, consider $H = E + PV$ for the constant pressure case:

$$\Delta H = \Delta E + P\,\Delta V$$
$$\Delta E = q - w$$

Again, for the constant pressure case, $w = P\,\Delta V$. Substituting into the above equations, the following relationship results:

$$\Delta H = q - P\,\Delta V + P\,\Delta V$$
$$\Delta H = q \qquad \text{(constant pressure)}$$

Enthalpy changes for chemical reactions: From the foregoing discussion, it is seen that the heat absorbed during a constant-pressure process is equal to the change in enthalpy for that process. This fact allows the heat absorbed during a chemical reaction to be related to the enthalpy change for that reaction. When calculating the change in enthalpy for a chemical reaction, ΔH_{rexn}, only the enthalpy change associated with the transformation of reactants to products is of interest. Enthalpy changes associated with temperature or pressure changes in the system are not included in this ΔH_{rexn}. Furthermore, the ΔH_{rexn} refers to the heat absorbed during the transformation of a set of reactants at a specified temperature and pressure to a set of products at the same temperature and pressure.

Consider the ΔH_{rexn} for the generalized chemical reaction given below, where a moles of species A react with b moles of species B to form products:

$$a\,A + b\,B \rightleftharpoons c\,C + d\,D$$

If the enthalpies of A, B, C, and D are known, then the change in enthalpy for the reaction would be given by adding the enthalpies of the species present in the final state and subtracting from that the enthalpy of the initial state:

$$\Delta H_{rexn} = c\,\overline{H}_C + d\,\overline{H}_D - a\,\overline{H}_A - b\,\overline{H}_B$$

In this expression, for example, the symbol $\overline{H}_C$ is used to indicate the molar enthalpy of species C. The enthalpy of species C is then given by the number of moles of C present times the molar enthalpy of C. Although the above equation does indicate that the change in enthalpy for a chemical reaction may be computed from the enthalpy of the final state and the enthalpy of the initial state, the value of the actual enthalpy for a species is not known. Since $H = E + PV$ and since the internal energy of a system is not known, the actual value of the enthalpy for a system is also unknown. When discussing state changes or chemical reactions in particular, it is only the change in enthalpy that is of importance.

The evaluation of the ΔH_{rexn} for any chemical reaction could be attempted experimentally. However, many reactions are not suitable for calorimetric study. It was previously noted that H is a state function such that the change in enthalpy for any process depends only on the initial and final states and is independent of the path by which the state change is accomplished. This permits ΔH_{rexn} values to be obtained for reactions that are not suitable for calorimetric study. Chemical reactions may be added and subtracted algebraically to obtain the desired initial and final states from a set of reactions that are experimentally suitable. Another convenience provided by the fact that H is dependent only on the state of the system is the ability to obtain and tabulate relative enthalpy values for compounds. In order to do this, some conventions must be established on which the relative enthalpy values will be based. By convention, the enthalpy of every element in its most stable form and physical state at 25°C and one atmosphere pressure is assigned the value of zero. In theory, the formation of compounds from elements at 1 atm pressure and 25°C could be investigated experimentally and the resulting ΔH values for the formation reactions could be associated with the relative enthalpies of the compounds formed. Consider the following formation reaction:

$$C \text{ (graphite)} + O_2 \text{ (g)} \rightleftharpoons CO_2 \text{ (g)}$$
$$\Delta H_{rexn} = -94.05 \text{ kcal/mol}$$

Based on the conventional assignment of zero enthalpy values for the elements under the standard conditions stated previously, this formation reaction may be considered as providing the relative enthalpy for the compound CO_2.

$$\Delta H_{rexn} = 1 \overline{H}_{CO_2} - 1 \overline{H}_{O_2} - 1 \overline{H}_C$$

$$\Delta H_{rexn} = 1 \overline{H}_{CO_2} - O$$

The relative enthalpies of compounds based on the conventional assignment discussed above are frequently termed the enthalpies of formation and are given the symbol ΔH_f. Tabulations of these enthalpies of formation are available and are given as molar enthalpies of formation at 25°C and 1 atm, ΔH_f^0. These tabulations allow calculation of the enthalpy changes for many chemical reactions at 25°C and 1 atm. An example is given below.

$$C_3H_6 \text{ (g)} + H_2 \text{ (g)} \rightleftharpoons C_3H_8 \text{ (g)}$$

$$\Delta H_f^0 (C_3H_6(g)) = 4.879 \text{ kcal/mol}$$

$$\Delta H_f^0 (H_2(g)) = 0$$

$$\Delta H_f^0 (C_3H_8(g)) = -24.820 \text{ kcal/mol}$$

$$\Delta H_{rexn} = 1 \text{ mol} (-24.820 \text{ kcal/mol})$$
$$- 1 \text{ mol} (4.879 \text{ kcal/mol})$$

$$\Delta H_{rexn} = -29.699 \text{ kcal}$$

The change in enthalpy for the above reaction was calculated from tabulated enthalpy of formation data. The ΔH_{rexn} for the

given reaction is negative. Since under conditions of constant pressure $\Delta H = q$, heat is actually evolved in this reaction. If the ΔH for a chemical reaction is negative, that reaction is termed an exothermic reaction; correspondingly, if the ΔH for a chemical reaction is positive, heat is absorbed during that reaction and the reaction is termed endothermic.

It is possible to determine the ΔH for many chemical reactions from tabulated data* and to relate the change in enthalpy for a reaction to the heat absorbed during that reaction. It would be very useful to be able to discuss the spontaneity of chemical reactions with knowledge of the reactant and product systems only. Thermodynamics can provide that information. In order to consider the spontaneity of chemical reactions, additional thermodynamic laws and state functions must first be introduced.

The Second Law of Thermodynamics

The second law of thermodynamics, also based on observation, may be stated in a number of ways. Experience shows that heat flows from a hotter to a colder body in macroscopic systems. This fact provides a basis for a second law statement. Experience also indicates that all natural or spontaneous changes that occur do so in an irreversible manner. The second law of thermodynamics does provide information on the spontaneity of processes.

The second law may be stated in various equivalent ways. The following statement, which refers to cyclic processes, is due to the work of Clausius. *It is impossible to devise an engine that, working in a cycle, produces no effect other than the transfer of heat from a colder to a hotter body.* A cyclic process is a process in which the initial and final states of the system are identical. For example, an engine operates in a cycle such that the processes by which the engine does work may be repeated

*From R. S. Weast (ed.), *Handbook of Chemistry and Physics*, 63rd ed., The Chemical Rubber Co., Cleveland, 1982–1983.

over and over again. Note that for a cyclic process the total change in any state function over the full cycle will be zero. This is true since the initial and final states in the cycle are identical and, thus, the value of the state property is the same at the beginning and at the end of the cycle. This fact may be stated mathematically for some state function X:

$$\Delta X_{\text{cycle}} = 0$$

If dX is taken as the differential of the state function, then the resulting integral around the full cycle will be zero since the initial and final states are the same.

$$\oint dX = 0$$

In fact, it is a general property of any exact differential that the value of its cyclic integral will be zero.

Entropy: It may be demonstrated that for any reversible cyclic process,

$$\oint \frac{\text{d} q}{T} = 0 \qquad \text{(reversible process)}$$

This expression is true for cyclic processes in which all steps are reversible. Thus the term $\text{d} q_{\text{rev}}/T$ may be considered as the differential of some function that depends only on the state of the system. The differential of the thermodynamic function entropy may be defined as follows:

$$dS = \frac{\text{d} q_{\text{rev}}}{T}$$

If a system undergoes a change of state from some initial state, I, to some final state, II, the change in the system entropy may be computed:

$$\Delta S = \int_{\text{I}}^{\text{II}} \frac{\text{d} q_{\text{rev}}}{T}$$

Entropy is a state function; thus, calculation of changes in entropy are independent of the actual path. No matter how an actual process is accomplished, the entropy change is computed in the same way from infor-

mation about the initial and final states of the system only. The change in entropy associated with a change in the state of the system from an initial state, I, to a final state, II, is computed from

$$\int_{\text{I}}^{\text{II}} \frac{\text{d} q_{\text{rev}}}{T}$$

regardless of whether the actual path is reversible or irreversible. Consider the first law of thermodynamics and consider only expansion or compression work:

$$dE = \text{d} q - \text{d} w$$
$$dE = \text{d} q - P_{\text{opp}} dV$$

Consider a system that changes its state reversibly; then

$$\text{d} q_{\text{rev}} = dE + P \, dV$$

since, as previously mentioned,

$$P_{\text{opp}} \cong P$$

for a reversible process. Thus, regardless of the actual path by which a state change is accomplished, entropy changes may be computed with knowledge of only the initial and final states of the system

$$dS = \frac{dE + P \, dV}{T}$$

Entropy and spontaneity: It was mentioned that the second law provides information about the reversibility or spontaneity of a process. In order to see how the second law could be used for this type of information, consider the Clausius inequality. It has been stated that

$$\oint \frac{\text{d} q_{\text{rev}}}{T} = 0$$

for all reversible, cyclic processes. If any part of a cyclic process is irreversible, then it may be shown that

$$\oint \frac{\text{d} q}{T} < 0$$

A consideration of the above two cyclic integrals and the definition for dS leads to the following:

$$dS = \frac{đq_{rev}}{T}$$

$$dS > \frac{đq}{T}$$

In the second equation, consideration is given to the actual path by which an irreversible process occurs. These relationships may be combined:

$$dS \geq \frac{đq}{T}$$

This inequality is referred to as the Clausius inequality. The equal sign relates to a process that is reversible, whereas the greater than sign refers to an irreversible process. If a process actually occurs irreversibly, then the computed entropy change for that process will be greater than the $\int đq/T$ evaluated for the process

$$\Delta S = \int \frac{đq_{rev}}{T} > \int \frac{đq}{T} \quad \text{(for a specific process)}$$

In view of these considerations, the second law of thermodynamics, formulated in terms of cyclic processes, may be used to distinguish between processes that occur naturally and processes that are reversible. It was shown that the entropy change that is associated with a process may be used to give us an indication of whether or not that process will occur naturally:

$$\Delta S \geq \int \frac{đq}{T} \quad \text{(for a given transformation)}$$

A problem is encountered with the use of ΔS as a criterion for spontaneity. If ΔS is to be used as a criterion for spontaneity, specific path information for the process must be known. For example, in chemical reactions, detailed information on exactly how the set of reactants was transformed to products would be needed before evaluation of the actual $\int đq/T$ for the entire process could be attempted. It would be possible to calculate ΔS for the overall chemical reaction with only a knowledge of the initial and final states of the system. In order to use the calculated ΔS to relate to the spontaneity of the reaction, however, a comparison of the calculated ΔS with the value of the integral of the actual $đq/T$ for the process must be made. This comparison requires specific path information. Thus, the entropy changes associated with various processes do not serve as useful indications of whether or not the processes are naturally occurring.

For the special case of an isolated system, evaluation of the entropy changes involved does provide information about the spontaneity of the processes that occur. An isolated system is one that does not exchange matter or energy with the surroundings. In an isolated system no heat will flow into or out of the system and no work will be done on or by the system; therefore,

$$dS \geq \frac{đq}{T}$$

becomes

$$dS \geq 0$$

if the system is isolated. For finite changes in an isolated system,

$$\Delta S \geq 0$$

In an isolated system, even though $\Delta E = 0$, any naturally occurring process results in $\Delta S > 0$ for that process. The entropy of an isolated system will increase until a maximum entropy is reached and no further spontaneous changes occur. When there is no further change of state in the system, $\Delta S = 0$ and the system is at equilibrium.

Free energy as a criterion for spontaneity: Most processes of interest in the study of food chemistry or biochemistry do not occur in isolated systems. Therefore, an attempt must be made to develop more useful criteria for spontaneity. One important application of thermodynamics to biochemical systems is in the treatment of chemical reactions. The energetics of chemical reactions are frequently studied under conditions of constant temperature and constant pressure. These physical constraints will be applied to the general statement of spontaneity or equilibrium.

$$dS \geq \frac{\text{đ} q}{T}$$

$$T \, dS \geq \text{đ} q$$

$$\text{đ} q - T \, dS \leq 0$$

From the first law,

$$dE = \text{đ} q - \text{đ} w$$

If for now only that work associated with the expansion or compression of systems is considered, the following expressions may be written:

$$dE + P_{\text{opp}} \, dV = \text{đ} q$$

$$dE + P_{\text{opp}} \, dV - T \, dS \leq 0$$

If a constant pressure process is considered, then both P_{opp} and P are constant and at the same equilibrium value. Thus, $P_{\text{opp}} = P$ and since P is constant, the term $P_{\text{opp}} \, dV$ may be written as $P \, dV$ or as $d(PV)$. If this system is restricted to constant temperature, then $T \, dS$ may be written as $d(TS)$. With these restrictions on the physical conditions, the relation for equilibrium or spontaneity may be written as follows:

$$dE + d(PV) - d(TS) \leq 0$$

This is mathematically equivalent to

$$d(E + PV - TS) \leq 0$$

Noting that $H = E + PV$, the equation may again be written as

$$d(H - TS) \leq 0$$

A new state function will now be introduced that is merely a combination of other functions dependent on the state of the system only.

$$G = H - TS$$

In this definitional equation, G is the Gibbs free energy, which is more frequently termed the free energy. The free energy is dependent on both the enthalpy and the entropy of a system. Using the newly defined state function, an equation may be written to determine whether a system is under conditions of equilibrium or whether a process will occur naturally.

$$dG \leq 0$$

For a finite change that occurs under conditions of constant temperature and constant pressure

$$\Delta G \leq 0$$

For chemical reactions or other processes that occur under constant temperature and constant pressure conditions, changes in the free energy of the system will serve as a useful criterion for spontaneity. Those changes of state that occur naturally or spontaneously will be accompanied by a decrease in the free energy of the system. Natural changes will continue to occur until the free energy of the system is at a minimum. When the minimum free energy is attained, no further change in state will take place. Then, ΔG will equal 0, and the system will be at equilibrium.

For example, if the free energy change for a chemical reaction is computed and determined to be negative, then the reaction in question is a spontaneous reaction. If the ΔG computed for a reaction is equal to zero, an equilibrium situation exists. Finally, if the computed ΔG for a reaction is a positive number, the reaction would proceed spontaneously, but in the reverse direction. Thus, computation of the free energy change associated with a chemical reaction—a constant temperature and constant pressure process—could be used to determine whether or not the reaction is spontaneous. The position of equilibrium and the equilibrium constant for a chemical reaction could also be determined from free energy calculations.

Free energy changes for chemical reactions: It is possible to compute free energy changes for reactions from tabulated data. If values for the free energies of all products and reactants were available, then the free energy change associated with a chemical reaction could be determined by subtracting the total free energy of the reactants from the free energy of the product system. By definition, $G = H - TS$. Since the free energy is dependent on the enthalpy, no actual or absolute free energy values are possible. It is possible, however, to obtain and tabulate relative free energy values for various compounds.

The free energy change associated with a chemical reaction, ΔG_{rexn}, could be computed from relative free energy values of products and reactants. The method is analogous to that used in the evaluation of the enthalpy change for a chemical reaction. The relative free energy of a chemical substance is termed the free energy of formation. The standard free energy of formation of a compound is defined as the free energy change for the reaction in which that compound is formed from the elements in their standard states. By convention, every element in its most stable form and physical state at 1 atm, and 25°C is assigned a free energy value of zero. Thus, if the free energy change is known for the reaction in which a compound is formed from the elements that compose it, this ΔG_{rexn} is the relative free energy of the compound, ΔG_f. Hence, free energies of formation may be obtained if free energy changes are determined for formation reactions under standard conditions. Tabulations of free energies of formation are available and are frequently listed as molar values for a temperature of 25°C and a pressure of 1 atm*. Use may then be made of these tabulated values to determine the free energy changes associated with various chemical reactions. An example calculation is now given.

C_3H_6 (g) $+$ H_2 (g) $\rightleftharpoons$ C_3H_8 (g)

ΔG_f (C_3H_6 (g)) $= 14.990$ kcal/mol

ΔG_f (H_2 (g)) $= 0$

ΔG_f (C_3H_8 (g)) $= -5.614$ kcal/mol

$\Delta G_{rexn} = 1$ mol (-5.614 kcal/mol)
$\qquad\qquad - 1$ mol (14.990 kcal/mol)

$\Delta G_{rexn} = -20.604$ kcal

*From R. S. Weast (ed.), *Handbook of Chemistry and Physics*, 63rd ed., The Chemical Rubber Co., Cleveland, 1982–1983.

Based on the tabulated ΔG_f values, the calculated free energy change for the given reaction is negative, and thus, the reaction is spontaneous.

Tabulated values of free energies of formation, ΔG_f values, are most often given for one mole of the material at 25°C, 1 atm, and for the chemical species in their standard states, termed ΔG_f^0 values. Several examples of standard states are as follows. For an ideal gas, the standard state is 1 atm pressure at the temperature specified. A real gas is in its standard state at a given temperature when it is in a state in which it has unit fugacity and exhibits ideal behavior. Pure liquids and solids at 1 atm pressure and at the temperature in question are in their standard states, frequently referred to as states of unit activity. For a solution, the standard state is a hypothetical state in which the solution behaves ideally and has unit concentration. For very dilute solutions that approach ideal behavior, the activity is approximately equal to the concentration. Note that in practice, fugacities for real gases and activities for solutions are frequently approximated by pressures in atmospheres and concentrations in moles per liter, respectively.

The dependence of the free energy on pressure for ideal gases: As mentioned, tables of molar ΔG_f^0 values for 1 atm pressure, 25°C and for chemical species in their standard states are available. It is possible to calculate ΔG_f values for other states. Consider the free energy change associated with a change in pressure for an ideal gas maintained at a constant temperature.

$$G = H - T S$$

This definitional equation may be written in differential form.

$$dG = dH - T\, dS - S\, dT$$

Remember that $H = E + PV$ and, therefore, $dH = dE + P\, dV + V\, dP$.

$$dG = dE + P\, dV + V\, dP - T\, dS - S\, dT$$

From the definition of dS, a substitution can be made into the above equation.

$$dS = \frac{\delta q_{rev}}{T} = \frac{dE + P\,dV}{T}$$

$$T\,dS = dE + P\,dV$$

Substituting this expression for $T\,dS$ into the expression for dG, the following equation is obtained:

$$dG = dE + P\,dV + V\,dP$$
$$- (dE + P\,dV) - S\,dT$$
$$dG = V\,dP - S\,dT$$

For a constant temperature process, dT is equal to zero, that is, there is no change in temperature. Under conditions of constant temperature

$$dG = V\,dP$$

For an ideal gas, $PV = nRT$. Thus, the free energy change associated with an isothermal change in pressure for an ideal gas may easily be computed.

$$\int dG = nRT \int_{P_1}^{P_2} \frac{dP}{P}$$

$$\Delta G = nRT \ln\left(\frac{P_2}{P_1}\right)$$

In this expression, ΔG is the difference in free energy of the gas between that state associated with pressure P_2 and that state associated with pressure P_1, $G_2 - G_1$. No actual free energy values are available, but one may relate the above free energy difference to relative free energy values. If the temperature is maintained at 25°C and P_1 is 1 atm, then G_1 for this ideal gas is the tabulated free energy of formation for that ideal gas, based on convention. Since the standard state for an ideal gas is a pressure of 1 atm, the free energy of formation can be written as ΔG_f^0, where the superscript zero refers to the material in its standard state. Then, the free energy associated with the gas at a pressure P_2 may be designated as $\Delta G_f^{P_2}$, the free energy of formation at pressure P_2. Using these designations, the above equation may be written as follows:

$$\Delta G_f^{P_2} - \Delta G_f^0 = nRT \ln\left(\frac{P_2}{1}\right)$$

$$\Delta G_f^{P_2} = \Delta G_f^0 + nRT \ln (P_2)$$

The above treatment may be extended to gas phase chemical reactions. Consider the following reaction in which all reactants and products are assumed to be ideal gases:

$$a\,A + b\,B \rightleftharpoons c\,C + d\,D$$

The relative free energies of each ideal gas, A, B, C, and D at pressures P_A, P_B, P_C, and P_D, respectively, may be considered with reference to the preceding derivation. These free energy values may be written as follows, noting that $iRT \ln (P_i) = RT \ln (P_i)^i$.

$$\Delta G_{f_A}^{P_A} = \Delta G_{f_A}^0 + RT \ln(P_A)^a$$

$$\Delta G_{f_B}^{P_B} = \Delta G_{f_B}^0 + RT \ln(P_B)^b$$

$$\Delta G_{f_C}^{P_C} = \Delta G_{f_C}^0 + RT \ln(P_C)^c$$

$$\Delta G_{f_D}^{P_D} = \Delta G_{f_D}^0 + RT \ln(P_D)^d$$

For a chemical reaction,

$$\Delta G_{rexn} = \underset{products}{\Sigma \Delta G_f} - \underset{reactants}{\Sigma \Delta G_f}$$

Thus, for the reaction of ideal gases A and B to form products C and D, the ΔG_{rexn} may be expressed as follows:

$$\Delta G_{rexn} = \Delta G_{f_C}^0 + \Delta G_{f_D}^0 - \Delta G_{f_A}^0$$
$$- \Delta G_{f_B}^0 + RT \ln\left(\frac{(P_D)^d (P_C)^c}{(P_A)^a (P_B)^b}\right)$$

$$\Delta G_{rexn} = \Delta G_{rexn}^0 + RT \ln\left(\frac{(P_D)^d (P_C)^c}{(P_A)^a (P_B)^b}\right)$$

In this final expression, the free energy change associated with the chemical reaction is related both to the free energy change for that reaction with all species present at the standard pressure of 1 atm, and to the actual pressures of the gases involved in the reaction.

Free energy changes and the equilibrium constant: It is possible to extend the above treatment, which relates to the variation in the pressure of an ideal gas under conditions of constant temperature, to systems involving solids, liquids, real gases, and solutions. For these systems, a standard state is defined for each species present in a chemical reaction. The ΔG^0_{rexn} may then be computed from tabulated ΔG^0_f data. The relationship between ΔG_{rexn} and ΔG^0_{rexn} will be given in a form similar to that developed for reactions involving only ideal gases. For the more general chemical reaction,

$$a\,A + b\,B \rightleftharpoons c\,C + d\,D$$

where A, B, C, and D may be solids, liquids, gases, or solutions,

$$\Delta G_{rexn} = \Delta G^0_{rexn} + RT \ln \left(\frac{(a_D)^d\,(a_C)^c}{(a_A)^a\,(a_B)^b} \right)$$

$$\Delta G_{rexn} = \Delta G^0_{rexn} + RT \ln Q$$

In this expression, ΔG_{rexn} is the free energy change associated with the chemical reaction when the species present are at the specified activities. The ΔG^0_{rexn} term may be computed for the reaction when all species are present at unit activity or in their standard states. The activity quotient, Q, is introduced into the above equation. For any chemical reaction, Q is defined in terms of a ratio of the product of the activities of reaction products to that of reactants, as indicated below.

$$Q = \frac{\prod_i (a^{n_i}_{products_i})}{\prod_j (a^{n_j}_{reactants_j})}$$

In this equation, the n_i values relate to the stoichiometric coefficients of the reaction products and the n_j values to the stoichiometric coefficients of the reactants. Note that if a real gas is involved in the reaction, the activity designation would be replaced by the fugacity of that gas. Likewise, if an ideal gas is involved in the reaction, the activity designation in the general expression given previously would be replaced by the pressure of that gas.

It has been mentioned that the computation of free energy changes associated with a chemical reaction—in a constant temperature and constant pressure process—could be used to determine whether or not the reaction in question is a spontaneous reaction. It has also been mentioned that the free energy change for a chemical reaction could be used to determine the position of equilibrium and the equilibrium constant for that reaction.

If the free energy change for a chemical reaction is equal to zero, a state of equilibrium exists. If $\Delta G_{rexn} = 0$, then

$$\Delta G^0_{rexn} = -RT \ln \left(\frac{(a_D)^d\,(a_C)^c}{(a_A)^a\,(a_B)^b} \right)_{equilibrium}$$

Under equilibrium conditions, the activity quotient, Q, which is defined for any arbitrary activities of products and reactants, becomes a constant. This equilibrium value of Q is termed K_{eq}, the equilibrium constant. The activities that appear in the expression for K_{eq} are the equilibrium values, and not arbitrary activities. Thus,

$$\Delta G^0_{rexn} = -RT \ln K_{eq}$$

The free energy change for a chemical reaction in which all species are present in their standard states, ΔG^0_{rexn}, may be computed from tabulated data. Thus, evaluation of ΔG^0_{rexn} from tabulated ΔG^0_f data allows calculation of the equilibrium constant for that reaction.

The determination of free energy changes from electrochemical cell data: It is possible to experimentally determine ΔG_{rexn}, ΔG^0_{rexn}, or K_{eq} for chemical reactions using data obtained with electrochemical cells. If the change in free energy associated with a particular chemical reaction is negative, that reaction is spontaneous. In theory, this chemical reaction could be used to do work other than work due to expansion or compression. The free energy change for a reaction is related to the amount of useful work, work other than expansion or compression work, that could be obtained from that reaction. In view of this, the work done using an electrochemical cell may be related to free energy changes accompanying the chemical reaction occurring in that cell.

In order to relate electrochemical work to free energy changes for chemical reactions, it is convenient to extend the general treatment used in the development of criteria for spontaneity to systems that involve useful work. Again, the general statement for equilibrium or spontaneity may be written and the first law of thermodynamics incorporated into that general statement.

$$dS \geq \frac{\partial q}{T}$$

$$T \, dS \geq \partial q$$

$$T \, dS \geq dE + \partial w$$

$$dE + \partial w - T \, dS \leq 0$$

Since useful work is to be considered, ∂w may be defined as the sum of the work done by the system due to expansion and the useful work done by the system.

$$\partial w = P_{opp} \, dV + \partial w_u$$

$$dE + P_{opp} \, dV + \partial w_u - T \, dS \leq 0$$

A relationship is desired between the useful work done and the free energy change associated with a chemical reaction. The term ΔG_{rexn} relates to the free energy change for the transformation of reactants to products at a specified temperature and pressure. The term ΔG_{rexn} does not relate to any temperature or pressure changes occurring in the system. The conditions of constant temperature and constant pressure will now be applied to the general equilibrium or spontaneity statement for systems involving useful work:

$$dE + P_{opp} \, dV - T \, ds \leq -\partial w_u$$

As previously mentioned, for a constant pressure process, $P_{opp} = P$ and $P \, dV = d(PV)$. Also, for a constant temperature process, $T \, dS$ may be written as $d(TS)$.

$$dE + d(PV) - d(TS) \leq -\partial w_u$$

Noting that $H = E + PV$ and $G = H - TS$, the above equation may be written in a more convenient form.

$$d(H - TS) \leq -\partial w_u$$

$$dG \leq -\partial w_u$$

This equation may now be written for finite changes that occur under conditions of constant temperature and constant pressure.

$$\Delta G \leq -w_u$$

If the constant temperature and constant pressure process under consideration is a chemical reaction, then the free energy change for that reaction is related to the useful work done by the system. If the work done is electrochemical work, then the ΔG_{rexn} is less than or equal to the negative of the electrochemical work done. If an electrochemical cell is operated reversibly, then the free energy change for the chemical reaction is equal to the negative of the useful work done. In the case of reversible operation of the cell, the work done by the system is the maximum amount of work for that particular cell. Thus, the following equality may be written:

$$\Delta G_{rexn} = -w_{u_{max}}$$

If the cell is operated irreversibly, then the free energy change for the reaction is less than the negative of the work done in that irreversible process. It is thus possible to relate the maximum useful work done in an electrochemical cell to the free energy change for the cell reaction. In order to equate ΔG_{rexn} with $-w_{u_{max}}$, the electrochemical cell must be operated reversibly.

The cell reactions and emf's of electrochemical cells: Specific consideration will now be given to electrochemical work and electrochemical cells. This will allow a more thorough investigation of the relationship of ΔG_{rexn} to electrochemical work.

First consider an electrochemical cell, with all species present at unit activity (Figure 11.1). The chemical reaction $Zn + Cu^{2+} \rightleftharpoons Zn^{2+} + Cu$ will occur if the leads are connected. In this reaction, electrons are transferred. If all the species involved in the reaction are mixed together, the reaction will still occur, but no useful work will be done by the reaction. If the solutions and metal electrodes are separated, as in the diagram, so that electron transfer will occur only

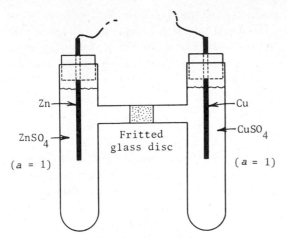

FIGURE 11.1. Electrochemical cell.

through an electrical conductor, electrons will flow through the wire and could do work in the process. The fritted disc pictured does not completely prevent the movement of ions between the solutions, but does retard mixing of the solutions. The given cell could be diagrammed in the following manner:

$$\text{Zn}|\text{ZnSO}_4(a = 1)\|\text{CuSO}_4(a = 1)|\text{Cu}$$

In this representation, the solid lines indicate phase boundaries in each half-cell and the boundary between the two half-cells.

As mentioned, if an electrical connection is made between the two half-cells pictured, a current will flow. Thus, there is a difference in electrical potential between the $\text{Zn}|\text{ZnSO}_4$ half-cell and the $\text{Cu}|\text{CuSO}_4$ half-cell. This potential difference exists whether or not an electrical connection is made. An electrical potential difference exists between the Zn and Zn^{2+} and between the Cu and Cu^{2+} in the individual half-cells as well, but it cannot be experimentally determined by direct measurement. Since there exists a phase difference between the Zn^{2+} in solution and the metallic Zn electrode, a measurement of the electrical potential difference between these species involves not only the potential difference between the Zn^{2+} and the Zn, but also the differences between the different media. However, it is possible to measure the electrical potential difference between the two wire leads, made of the same ma-

terial, which are used to connect the two half-cells. The potential difference of the electrochemical cell is equal to the difference in the potentials of the wire leads. The cell voltage could be measured by putting a voltmeter across the wire leads connecting the half-cells. The measured voltage or potential difference would vary depending on the current drawn by the voltmeter. A real voltmeter does have a finite resistance and, thus, a small amount of current will flow through the meter.

It has been mentioned that the current flowing through the connecting leads could be used to do useful work. If the useful work done by the chemical reaction occurring in the cell is to be related to ΔG_{rexn}, then the reaction should be carried out reversibly. It has been noted that no useful work is done if the half-cell materials for both half-cells are mixed in a single container. If the half-cells are connected by external wire leads, then the reaction proceeds with electron transfer through the leads. If the flow of electrons is impeded by an opposing voltage, the amount of work done in moving the electrons increases as the opposing voltage is increased. The maximum amount of work done by the electrochemical cell is that work done against the maximum opposing voltage, or that voltage that is just differentially less than the cell voltage. In this case, the current flow would be infinitesimal and the cell voltage would be at its maximum or limiting value. The limiting value of the potential difference is termed the cell emf. It is possible to measure the cell emf, the limiting value of the potential difference, with a potentiometer.

The Nernst equation. It is possible to measure the emf of a variety of electrochemical cells under reversible conditions. This reversible emf, E, is related to the maximum useful work done by that electrochemical cell. The electrical work done by a cell when it drives a charge of q coulombs through a potential difference of V volts is qV. If the cell is operated reversibly, the cell voltage, or emf under these limiting conditions, is equal to the opposing voltage to at least within a differential amount. Further, if the cell emf is E, the work done under reversible conditions is qE. The charge of 1 mole of electrons is the Faraday, $\mathfrak{F}$, and is equal to 96,490 coulombs (C). The electrochemical work done by a cell operating reversibly may thus be calculated from the following relation.

$$w_{u_{max}} = n\mathfrak{F}E$$

In this expression, n is the number of moles of electrons transferred during the reaction. It may be noted that the maximum useful work done is related to the free energy change accompanying the chemical reaction.

$$\Delta G_{rexn} = -w_{u_{max}}$$

$$\Delta G_{rexn} = -n\mathfrak{F}E$$

If all species involved in the chemical reaction are present at unit activity, then the ΔG_{rexn} is the standard value, ΔG^0_{rexn}, and the cell emf measured is the standard cell emf, E^0.

$$\Delta G^0_{rexn} = -n\mathfrak{F}E^0$$

These expressions relating the free energy change for a chemical reaction to the useful work done by the reversible operation of an electrochemical cell utilizing that reaction could be combined. In a previous section, the following relationship was developed:

$$\Delta G_{rexn} = \Delta G^0_{rexn} + RT \ln\left(\frac{(a_D)^d(a_C)^c}{(a_A)^a(a_B)^b}\right)$$

$$\Delta G_{rexn} = \Delta G^0_{rexn} + RT \ln Q$$

Substituting for ΔG_{rexn} and ΔG^0_{rexn}, the following equations may be written:

$$-n\mathfrak{F}E = -n\mathfrak{F}E^0 + RT \ln\left(\frac{(a_D)^d(a_C)^c}{(a_A)^a(a_B)^b}\right)$$

$$E = E^0 - \frac{RT}{n\mathfrak{F}} \ln Q$$

This equation is known as the Nernst equation and may be used to calculate cell emf values for systems with a known E^0 value and with known activities for the species involved in the reaction. The Nernst equation is most frequently used for systems maintained at a temperature of 25°C, the temperature at which most standard electrode potentials are tabulated. At this temperature, the above equation may be written in the following form, noting that the natural logarithm and the base 10 logarithm are related by $\ln = 2.303$ log.

$$E = E^0 - \frac{0.0591 \text{ V}}{n} \log Q$$

As previously noted, the free energy change for a chemical reaction is equal to zero when the system is at equilibrium. Under equilibrium conditions, $\Delta G_{rexn} = 0$ and thus, $E = 0$. When a system is at equilibrium, the following equations may be written:

$$E^0 = \frac{RT}{n\mathfrak{F}} \ln\left(\frac{(a_D)^d(a_C)^c}{(a_A)^a(a_B)^b}\right)_{equilibrium} = \frac{RT}{n\mathfrak{F}} \ln K_{eq}$$

For a temperature of 25°C and converting to a base 10 logarithm, this equation becomes

$$E^0 = \frac{0.0591 \text{ V}}{n} \log K_{eq}$$

The equilibrium constant for a chemical reaction may be determined from the E^0 value for that reaction.

E^0 values may be experimentally determined by connecting various half-cells and measuring the emf of the resulting cell when all species are present at unit activity. Measurements of the emf under standard conditions are most frequently made at a temperature of 25°C.

Standard electrode potentials. A large number of electrochemical cells may be constructed from various half-cell combinations. Each half-cell contains one electrode. It is not possible to measure the potential of a single electrode, although it would be more convenient to tabulate the half-cell potentials rather than the emf values for electrochemical cells made up of two electrodes. If half-cell potentials could be obtained, then it would be possible to calculate an emf for a cell constructed of any two half-cells. Since the cell emf is the difference in electrical potential between the two half-cells when the cell is operated reversibly, only differences in electrical potential and not absolute values for electrical potential are important. Thus, a reference electrode may be used to obtain relative values for half-cell potentials. The reference electrode is normally considered to be the standard hydrogen electrode. In this electrode, hydrogen gas is bubbled around a platinum electrode in contact with a solution containing hydrogen ions. The platinum electrode is coated with platinum black. The emf of this electrode at 25°C is assigned the value of zero, when the hydrogen gas is at a pressure of 1 atm, assuming ideal behavior, and when the activity of H^+ in solution is equal to one. This standard electrode may be diagrammed as follows:

$$Pt|H_2 \ (1 \text{ atm})|H^+ \ (a = 1)$$

With the conventional assignment of $E^0 = 0$ for this electrode, the relative emf of other single half-cells may be obtained. This is done by measurement of the emf of the electrochemical cell formed when the standard hydrogen electrode is combined with another half-cell with all species in both half-cells present at unit activity. It is obvious that the standard hydrogen electrode need not be actually combined with every other half-cell to obtain the relative emf of that half-cell. Once the relative emf value for any half-cell has been determined, with a cell emf measurement using the standard hydrogen electrode as the reference electrode, this half-cell may then be combined with other half-cells and emf values may be determined for the cell. Since the cell emf is the difference in electrical potential between the half-cells employed, the knowledge of the cell emf and one relative half-cell potential allows calculation of the other half-cell emf.

Calculation of cell emf's Tables of emf values are available for a large number of electrodes at 25°C and with all species present at unit activity. These half-cell E^0 values may then be used to determine E^0 values for electrochemical cells formed by combination of two half-cells. Consider the following example, using the electrochemical cell previously described:

$$Zn|ZnSO_4 \ (a = 1)\|CuCO_4 \ (a = 1)|Cu$$

Tabulations of the half-cell reactions and E^0 values based on the conventional $E^0 = 0$ assignment for the standard hydrogen electrode are available.

Electrode	Electrode Reaction	E^0 (V)*
$Zn^{2+}|Zn$	$Zn^{2+} + 2e \rightleftarrows Zn$	-0.7628
$Cu^{2+}|Cu$	$Cu^{2+} + 2e \rightleftarrows Cu$	$+0.3402$

From this tabulated data and noting that, by convention, the electrode at which oxidation takes place is diagrammed on the left, we may write the overall cell reaction and compute the E^0 value.

$$E^0 = E^0_{\text{right}} - E^0_{\text{left}}$$
$$E^0 = 0.3402 \text{ V} - (-0.7628 \text{ V})$$
$$E^0 = 1.1030 \text{ V}$$

An alternate procedure used for emf calculation is to simply write the half-cell reactions, writing the reaction at the left electrode as an oxidation and the reaction at the right electrode as a reduction. Then reverse the algebraic sign of E^0 if the reaction is reversed and add the resulting E^0 values.

Left Electrode:
$$Zn \rightleftarrows Zn^{2+} + 2e \qquad\qquad +0.7628 \text{ V}$$

Right Electrode:
$$Cu^{2+} + 2e \rightleftarrows Cu \qquad\qquad +0.3402 \text{ V}$$

Cell Reaction:
$$Zn + Cu^{2+} \rightleftarrows Zn^{2+} + Cu \qquad +1.1030 \text{ V}$$

Since $\Delta G^0_{\text{rexn}} = -n\mathscr{F}E^0$, the reaction is spontaneous if the E^0 computed for the cell reaction is positive. If the E^0 value is negative for the reaction as written, the cell reaction would actually proceed spontaneously in the reverse direction. At this time it may be noted that tables of emf data for conditions other than standard conditions are also available.

*From R. S. Weast (ed.), *Handbook of Chemistry and Physics*, 63rd ed., The Chemical Rubber Co., Cleveland, 1982–1983.

In biochemical applications, for example, most solutions considered have a pH equal to 7.00. For these applications, tables of $E^{0'}$ values are available. These $E^{0'}$ values are half-cell emf's for systems with $[H^+] = 10^{-7} M$ designated as the standard state, rather than for systems in which the standard state activity of H^+ is one. $E^{0'}$ values will be further discussed in this chapter.

Since the E^0 values are available for large numbers of electrodes, E^0 values for electrochemical cells may be readily obtained. If the system considered contains chemical species that are not at unit activity, the E^0 value for the system together with the Nernst equation may be used to determine the E for that cell.

Consider the previous example, again at a temperature of 25°C, but with the activities for $ZnSO_4$ and $CuSO_4$ unspecified.

$$Zn|ZnSO_4\ (a)\|CuSO_4\ (a)|Cu$$

The overall cell reaction as written involves the transfer of 2 moles of electrons per mole of Zn^{2+} formed. The activities of pure solids are equal to one. Noting these facts, the Nernst equation may be written:

$$Zn + Cu^{2+} \rightleftarrows Zn^{2+} + Cu$$

$$E = E^0 - \frac{RT}{2\mathcal{F}} \ln \left(\frac{(1)\ (a_{Zn^{2+}})}{(1)\ (a_{Cu^{2+}})} \right)$$

$$E = 1.1030\ V - \frac{0.0591\ V}{2} \log \left(\frac{(a_{Zn^{2+}})}{(a_{Cu^{2+}})} \right)$$

Once the activities for the Zn^{2+} and the Cu^{2+} are specified, E may be determined.

The values for E^0 and E may be determined for electrochemical cells, and in turn, these values may be used for calculating ΔG^0_{rexn}, ΔG_{rexn}, and K_{eq}. Electrochemical cells provide data that permits the calculation of properties related to both the position of equilibrium for chemical reactions and the spontaneity of chemical reactions.

The Third Law of Thermodynamics

Free energy changes have been discussed for various processes and for chemical reac-tions. Changes in free energy have been related to the spontaneity of processes or to the equilibrium state of systems. These free energy changes are related to changes in enthalpy and changes in entropy for the processes considered. The development of the relationship between changes in free energy and spontaneity for various processes involves a consideration of both the first and second laws of thermodynamics. For a constant temperature process

$$\Delta G = \Delta H - T\ \Delta S$$

The change in free energy is related to an energy factor, ΔH, and an entropy factor, ΔS. If the process considered was also a constant-pressure process, for example a chemical reaction, the ΔG calculated would relate to the spontaneity of that process. Thus, a constant temperature, constant pressure process that proceeds with a negative ΔH value and a positive value for ΔS is a spontaneous process.

For a chemical species in its standard state, we may write

$$\Delta G^0_f = \Delta H^0_f - T\ S^0$$

The ΔG^0_f and ΔH^0_f values for compounds are based on the conventional assignment of zero for the free energy and enthalpy of the elements in their standard states at 25°C and 1 atm pressure.

A consideration of the third law of thermodynamics will illustrate the basis for the tabulated S^0 values for elements and compounds and will give additional insight into the state property, entropy. A statement of the third law proposed by Lewis and Randall follows:

> If the entropy of each element in some crystalline state be taken as zero at the absolute zero of temperature, every substance has a finite positive entropy; but at the absolute zero of temperature the entropy may become zero, and does so become in the case of perfect crystalline substances.*

*G. N. Lewis and M. Randall, *Thermodynamics and the Free Energy of Chemical Substances*, 1st ed. McGraw-Hill, New York, 1923.

Frequently, tabulated entropy values for elements and compounds are based on the third law.

In order to establish conventional entropy values, S_0 is set equal to zero for the elements. The entropies of many compounds will then equal zero at $0\ K$. It is then possible to calculate S_T, the entropy at some temperature T. Knowledge of the heat capacity of a chemical species in each of the phases present from $0\ K$ to T as well as ΔH values for any phase transitions in this temperature range are needed for the computation of S_T. Calculation of S_T would involve the summation of a number of positive terms to S_0. Entropy values computed in this manner are termed third-law entropies. Under standard conditions, the third-law entropy is also the standard entropy, S_T^0, at the temperature T. The entropy at $0\ K$, S_0, is only actually equal to zero for perfect crystalline substances. If any disorder exists at the absolute zero of temperature, S_0 will not be equal to zero, but will be greater than zero. In a perfect crystalline substance, there is only one arrangement of the atoms, ions, or molecules present. If there is any randomness or disorder in the system, S_0 will not equal zero.

Entropy may be related to the disorder in a system. The more randomness or disorder present, the greater the entropy. The entropy of a substance in a gas phase is greater than the entropy of that substance in the liquid phase, which in turn is greater than the entropy of the corresponding solid. The phase change from solid to liquid, in general, may be considered as a change from a more ordered to a less ordered state. Likewise, the change from the liquid to the gas phase is also a change to a less ordered state.

In general, as the temperature of a system is increased, the system will go to a state of higher entropy, a state of greater disorder. As the temperature is increased, more randomness is possible, even if no phase change occurs. At higher temperatures, more energy states are significantly populated and a greater number of distributions among the various energy states is possible.

Processes that result in an increase in the randomness or disorder of a system will be accompanied by an increase in the entropy of that system. Thus, from a consideration of free energy changes related to the spontaneity of processes, it can be seen that, in general, systems tend toward states of lower energy and greater disorder.

Relationship of Thermodynamic Principles to Biochemical Systems

The cellular activities of living organisms are examples of open systems. Open systems exchange both mass and energy with the surroundings as opposed to so-called closed systems in which only energy is exclusively exchanged with the surroundings—but not mass; or alternatively, the isolated system may exchange neither mass nor energy with the surroundings.

In view of the previous distinctions, it is only logical to question whether or not classical thermodynamic laws are applicable to living systems. The laws of thermodynamics frequently deal with systems that display equilibrium conditions, while living systems can never attain a true state of equilibrium. Living systems appear to exist in equilibrium with their surroundings, but in fact, they are systems that operate according to a dynamic state. Chemical composition and cellular activities of living systems are held at a seemingly static state by means of a balanced intake and exit of both matter and energy. When living systems do happen to achieve an equilibrium state, all living activities cease.

The cells of living systems have the inherent ability to detect and counteract changes in steady state concentrations of chemical substances owing to the presence of complex feedback mechanisms that control the utilization and flow of metabolites. Biochemical components of living systems may be considered on an individual basis in order to predict the direction of reactant to product conversion for a specific reaction. The feasibility or probability that a certain biochemical reaction will proceed in one direction or another—or remain in a tempered state of equilibrium—can be determined by calculating the free energy change or ΔG for the reaction. As presented in previous sections, it is clear that the change in free energy (ΔG) reflects:

1. The molar concentrations of products/reactants during the course of the reaction and at equilibrium the K_{eq} for the specific reaction.

2. The oxidation–reduction potentials that exist between the biochemical participants in a reaction.

Equilibrium constant for biochemical reactions and free energy changes: The pertinent mathematical relationships that permit the calculation of free energy and thermodynamic relationships in or among biochemical reactions are summarized in Table 11.1. These relationships have been developed in

the previous sections. Foremost among these mathematical expressions is

$$\Delta G_{rexn} = \Delta G_{rexn}^0 + RT \ln Q$$

$$\Delta G_{rexn} = \Delta G_{rexn}^0 + 2.303 \, RT \log Q$$

$\Delta G_{rexn}^0 =$ standard change in free energy for a chemical reaction

$R = 1.987$ cal mol^{-1} K^{-1}, universal gas constant

$T =$ absolute temperature

$Q =$ activity quotient

As is the case with other mathematical expressions that deal with Q or K_{eq} values, calculations of the change in free energy for a reaction are correctly calculated according to the actual molar activities of reactants and products. It should be recognized, however, that since activity coefficients for biochemical reactions are rarely known, molarities are normally used to approximate activities in biochemical applications. Based on the equation for ΔG, it is clear that the free energy change is a function of reactant and product concentrations as well as the standard free

Table 11.1. Some Important Thermodynamic Equations That Are Useful for Studying Biochemical Reactions and Systems

$\Delta G_{rexn} = \Delta G_{rexn}^0 + RT \ln Q = \Delta G_{rexn}^0 + 2.303 \, RT \log Q$

ΔG_{rexn}^0 (or $\Delta G_{rexn}^{0'}$) $= -RT \ln K_{eq} = -2.303 \, RT \log K_{eq}$

$\Delta G_{rexn}^0 = -n\mathfrak{F}E^0$ or $\Delta G_{rexn}^{0'} = -n\mathfrak{F}E^{0'}$

$$E \text{ (or } E') = E^0 \text{ (or } E^{0'}) - \frac{0.0591 \text{ V}}{n} \log \frac{\prod_i [\text{products}]^{n_i} (\gamma)^{n_i}}{\prod_j [\text{reactants}]^{n_j} (\gamma)^{n_j}}$$

where

$\gamma =$ activity coefficient

n_i and $n_j =$ indicate the number of moles of products and reactants, respectively, in the chemical reaction

Note: In biochemical reactions occurring in dilute solutions, the activity may be considered equal to concentration.

$E_{cell}^0 = E_{oxidation}^0 + E_{reduction}^0$ or $E_{cell}^{0'} = E_{oxidation}^{0'} + E_{reduction}^{0'}$

$\Delta G_{rexn}^0 = \Sigma \Delta G_{f \text{ (products)}}^0 - \Sigma \Delta G_{f \text{ (reactants)}}^0$

$\Delta H_{rexn}^0 = \Sigma \Delta H_{f \text{ (products)}}^0 - \Sigma \Delta H_{f \text{ (reactants)}}^0$

NOTE: For the general presentation of equations that deal with chemical reactions, the subscript "rexn" has been included. However, throughout most of the chapter, only chemical reactions and not other processes will be discussed and the subscript will, therefore, be dropped.

energy change, ΔG^0. Therefore, given the reaction

$$A \rightleftarrows B$$

it is possible to calculate ΔG^0 if the system is at equilibrium or if $\Delta G_{rexn} = 0$.

Because there is no net conversion of A to B under equilibrium conditions, the resulting change in free energy (ΔG) is zero. Furthermore, the concentration ratio of $[B]$ to $[A]$ is the ratio at equilibrium or K_{eq}. Using this assumption ($\Delta G = 0$), and for $T = 298$ K,

$$0 = \Delta G^0 + 2.303 \, RT \log K_{eq}$$

$$\Delta G^0 = -2.303 \, RT \log K_{eq}$$

$$\Delta G^0 = -(2.303)\left(1.987 \, \frac{cal}{K}\right)(298 \text{ K}) \log K_{eq}$$

$$\Delta G^0 = -1364 \text{ cal} \log K_{eq}$$

As long as the K_{eq} value can be determined for a reaction according to conventional analytical methods, the standard change in free energy for a reaction can be calculated. To illustrate this principle, the ΔG^0 has been determined for a series of K_{eq} values ranging from 10^{-3} to 10^3 and presented in Table 11.2.

If $\Delta G = 0$, equilibrium has been attained and thus ΔG^0 equals $-RT \ln K_{eq}$. If the K_{eq} is large, it indicates that the reaction achieved equilibrium with high product concentrations, or that the reaction proceeded far to the right before attaining equilibrium. Con-

versely, a small value for K_{eq} indicates that equilibrium is attained before a significant amount of product is formed. Thus, a reaction with a small K_{eq} does not produce significant amounts of products before equilibrium is attained.

As illustrated and detailed in the development of the thermodynamic laws in earlier pages, standard state conditions occur only when reactants and products are present at unit activity. The standard state for solutes in solution is approximated by unit molarity; for gases, the standard state is approximated by 1.0 atm; and for pure solvents or solid components, the activity is assumed to be unity. Note, if hydrogen ions are produced or used in a biochemical reaction, the standard concentration is 1 M or pH $= 0$. Clearly, these assumptions for standard state conditions are unrealistic for studies dealing with the thermodynamic relationships within living systems. The pH of living systems is never zero and often approaches the neutral pH range (pH 7.0). Since many biochemical reactions occur at a pH of ≈ 7, it is convenient to define a new standard state for H^+. For these biochemical systems, rather than employing the common standard state, the chosen standard state sets $[H^+] = 10^{-7} \, M$. This choice of standard state, therefore, gives rise to a new set of ΔG^0 values that will be termed $\Delta G^{0\prime}$. Thus, $\Delta G^{0\prime}$ values are based on this biochemically appropriate standard state.

As an example for calculating the $\Delta G^{0\prime}$ value for a biochemical reaction, suppose that

Table 11.2. Relationship of Equilibrium Constant (K_{eq}) to ΔG^0_{rexn}

Position of equilibrium	K_{eq}	$\log K_{eq}$	ΔG^0_{rexn} (cal/mol)
	0.001	-3	$+4092$
Products < Reactants	0.01	-2	$+2728$
	0.1	-1	$+1364$
Products = Reactants	1.0	0	0
	10.0	1	-1364
Products > Reactants	100.0	2	-2728
	1000.0	3	-4092

0.015 M of glucose 1-phosphate is converted to glucose 6-phosphate through the enzymatic action of phosphoglucomutase:

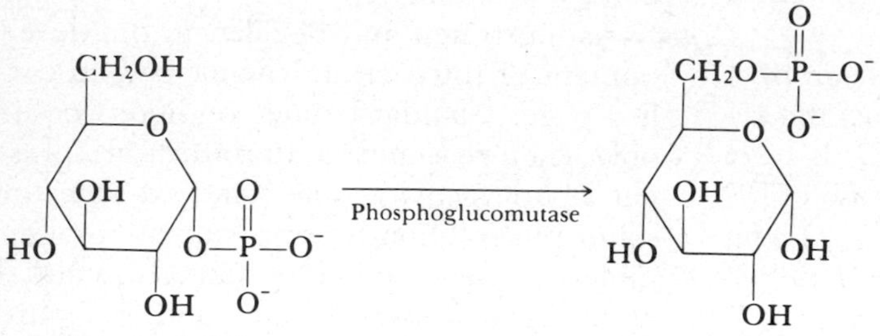

(i.e., human erythrocyte) are respectively $61.2 \times 10^{-6}\,M$ and $4.3 \times 10^{-6}\,M$. The projected change in free energy for this reaction

Glucose-1-PO₄ → (Phosphoglucomutase) → **Glucose-6-PO₄**

Subsequent to the enzyme-mediated reaction at 25°C and pH 7, the final reaction mixture contains 0.013 M of glucose 6-phosphate and 0.002 M glucose 1-phosphate. The $\Delta G^{0\prime}$ for the reaction can be calculated from the K_{eq} resulting from the [glucose 6-phosphate]/[glucose 1-phosphate] concentration ratio, that is, 0.013/0.002 or 6.5.

$$\Delta G^{0\prime} = -2.303\ RT \log K_{eq}$$
$$= -(1364.0\ \text{cal})\,(\log 6.5)$$
$$= -(1364.0\ \text{cal})\,(0.8129)$$
$$= -1109\ \text{cal}$$

Hence, the $\Delta G^{0\prime}$ of the biochemical reaction, with the hypothetical concentration of reactants and products, yielded a value of -1109 cal.

Since the criterion of a spontaneous reaction is that ΔG and not $\Delta G^{0\prime}$ is negative, an example of a reaction is given for the calculation of ΔG.

3-Phospho-glycerate ⇌ (Phosphoglycero-mutase) 2-Phospho-glycerate

This enzyme-mediated reaction has a K_{eq} value of 0.167 at 25°C and a pH of 7.0. The actual concentrations of the 3-phosphoglycerate and 2-phosphoglycerate in the cell

can be estimated according to the solution below:

$$\Delta G = \Delta G^{0\prime} + RT \ln Q = \Delta G^{0\prime} + 2.303\ RT \log Q$$

First, determine $\Delta G^{0\prime}$:

$$\Delta G^{0\prime} = -RT \ln K_{eq} = -2.303\ RT \log K_{eq}$$
$$\Delta G^{0\prime} = -2.303 \left(1.987\ \frac{\text{cal}}{\text{K}}\right)(298\ \text{K})\log 0.167$$
$$\Delta G^{0\prime} = +1060\ \text{cal}$$

Since the actual concentration for 3-phosphoglycerate is $61.2 \times 10^{-6}\,M$ and for 2-phosphoglycerate is $4.3 \times 10^{-6}\,M$,

$$\Delta G = \Delta G^{0\prime} + 2.303\ RT\ \frac{[\text{2-phosphoglycerate}]}{[\text{3-phosphoglycerate}]}$$
$$\Delta G = +1060\ \text{cal} + 2.303 \left(1.987\ \frac{\text{cal}}{\text{K}}\right)$$
$$\times (298\ \text{K})\log \frac{[4.3 \times 10^{-6}]}{[61.2 \times 10^{-6}]}$$
$$\Delta G = -513.0\ \text{cal}$$

Since ΔG has a negative value, the reaction is said to be exergonic and it is thermodynamically feasible. A small value of ΔG indicates that a small change in energy in the system may influence the flow of metabolites in either direction.

It is instructive to note one last example regarding the relationship between the K_{eq} value for a given reaction and $\Delta G^{0\prime}$. In this case if the $\Delta G^{0\prime}$ for a reaction is known, the K_{eq} can be calculated as shown in the following example.

$$\text{Pyruvate} + \text{NADH(H}^+) \rightleftharpoons \text{lactate} + \text{NAD}^+$$

$$\Delta G^{0\prime} = -6000 \text{ cal at pH} = 7.0, T = 38°C$$

The K_{eq} for the reaction can be calculated.

$$\Delta G^{0\prime} = -RT \ln K_{eq} = -2.303 \, RT \log K_{eq}$$

$$-6000 \text{ cal} = -(2.303) \left(1.987 \frac{\text{cal}}{\text{K}}\right)$$

$$\times (311 \text{ K}) \log K_{eq}$$

$$\log K_{eq} = 4.216$$

$$K_{eq} = 1.64 \times 10^4$$

Biochemical oxidation–reduction and free energy changes: The ΔG for a biochemical reaction can also be estimated on the basis of differences in the oxidation–reduction potentials that exist between reactants. Those substances that contribute electrons to a reaction and consequently undergo oxidation (loss of electrons) are classified as reducing agents. In the same vein, those substances that accept electrons and become reduced are called oxidizing agents.

$$\text{(a)} \quad Fe^{3+} + 1e^- \xrightarrow{\text{reduced}} Fe^{2+}$$

$$\text{(b)} \quad Fe^{2+} \xrightarrow{\text{oxidized}} Fe^{3+} + 1e^-$$

Many other organic and biochemical compounds behave in an analogous fashion since they can serve as oxidizing agents and become reduced, or undergo oxidation, thereby acting as reducing agents:

Under standard electrochemical conditions as discussed in earlier sections, these oxidation–reduction reactions occur at a standard reduction potential. The standard reduction potentials are tabulated for specific half-cell reactions with reference to the standard hydrogen electrode, which is conventionally assigned a value of $E^0 = 0.000$ V. As in the previous cases, standard conditions are implied and the $E^0 = 0.000$ V occurs only at an idealized pH of 0.0.

Since standard electrochemical conditions are wholly unrealistic for studies of biochemical systems where pH values are in the range of 7.0, the biochemical standard reduction potential ($E^{0\prime}$) for $H^+ + 1 \, e \rightarrow \frac{1}{2} H_2$, defined at $[H^+] = 10^{-7} \, M$, becomes -0.420 V instead of 0.000 V. This value of -0.420 V acts as an important benchmark against which all other oxidation–reduction half-cell reactions of biochemical interest can be measured. Table 11.3 offers a list of oxidation–reduction potentials that have special biochemical importance. According to the tabular presentation of the half-cell reactions, it is clear that each reaction is presented as a reduction. Therefore, the coupling of any two half-cell reactions will proceed with the reaction having the more positive reduction potential going as indicated (a reduction), whereas the half-cell reaction with the least positive reduction potential will proceed in the reverse direction (an oxidation). A superficial inspection of the table also reveals that those compounds

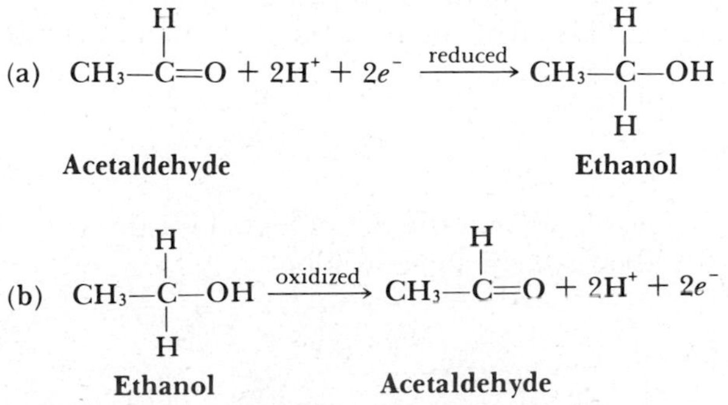

(a) $\quad CH_3{-}\overset{\overset{\displaystyle H}{|}}{C}{=}O + 2H^+ + 2e^- \xrightarrow{\text{reduced}} CH_3{-}\overset{\overset{\displaystyle H}{|}}{\underset{\underset{\displaystyle H}{|}}{C}}{-}OH$

Acetaldehyde   **Ethanol**

(b) $\quad CH_3{-}\overset{\overset{\displaystyle H}{|}}{\underset{\underset{\displaystyle H}{|}}{C}}{-}OH \xrightarrow{\text{oxidized}} CH_3{-}\overset{\overset{\displaystyle H}{|}}{C}{=}O + 2H^+ + 2e^-$

Ethanol   **Acetaldehyde**

Table 11.3. Oxidation–Reduction Potentials for Half-reactions that Have Biochemical Importance.

Half-reaction (written as a reduction)[a]	$E^{0\prime}$ (V) at pH 7.0
$\frac{1}{2} O_2 + 2H^+ + 2e^- \longrightarrow H_2O$	0.816
$Fe^{3+} + e^- \longrightarrow Fe^{2+}$	0.771
$NO_3^- + 2H^+ + 2e^- \longrightarrow NO_2^-$	0.42
Cytochrome a_3 $(Fe^{3+}) + e^- \longrightarrow$ cytochrome $a_3(Fe^{2+})$	0.55
$O_2 + 2H^+ + 2e^- \longrightarrow H_2O_2$	0.30
Cytochrome a $(Fe^{3+}) + e^- \longrightarrow$ cytochrome a (Fe^{2+})	0.29
Cytochrome c $(Fe^{3+}) + e^- \longrightarrow$ cytochrome c (Fe^{2+})	0.25
Cytochrome c_1 $(Fe^{3+}) + e^- \longrightarrow$ cytochrome c_1 (Fe^{2+})	0.22
Crotonyl—S-CoA $+ 2H^+ + 2e^- \longrightarrow$ butyryl—S-CoA	0.19
$Cu^{2+} + e^- \longrightarrow Cu^+$	0.15
Cytochrome b_2 $(Fe^{3+}) + e^- \longrightarrow$ cytochrome b_2 (Fe^{2+})	0.12
Ubiquinone $+ 2H^+ + 2e^- \longrightarrow$ ubiquinone—H_2 (coenzyme Q) (coenzyme QH_2)	0.10
Cytochrome b $(Fe^{3+}) + e^- \longrightarrow$ cytochrome b (Fe^{2+})	0.075
Dehydroascorbic acid $+ 2H^+ + 2e^- \longrightarrow$ ascorbic acid	0.058
Metmyoglobin—$Fe^{3+} + e^- \longrightarrow$ myoglobin—Fe^{2+}	0.046
Fumarate $+ 2H^+ + 2e^- \longrightarrow$ succinate	0.031
α-Ketoglutarate $+ NH_3 + 2H^+ + 2e^- \longrightarrow$ glutamate $+ H_2$	-0.140
Acetaldehyde $+ 2H^+ + 2e^- \longrightarrow$ ethanol	-0.163
Oxaloacetate $+ 2H^+ + 2e^- \longrightarrow$ malate	-0.166
$FAD + 2H^+ + 2e^- \longrightarrow FADH_2$	-0.180
Pyruvate $+ 2H^+ + 2e^- \longrightarrow$ lactate	-0.185
Riboflavin $+ 2H^+ + 2e^- \longrightarrow$ riboflavin—H_2	-0.200
Glutathione $+ 2H^+ + 2e^- \longrightarrow$ 2-reduced glutathione	-0.230
Acetoacetate $+ 2H^+ + 2e^- \longrightarrow \beta$-hydroxybutyrate	-0.290
Lipoic acid $+ 2H^+ + 2e^- \longrightarrow$ dihydrolipoic acid	-0.290
$NAD^+ + 2H^+ + 2e^- \longrightarrow NADH + H^+$	-0.320
$NADP^+ + 2H^+ + 2e^- \longrightarrow NADPH + H^+$	-0.324
$2H^+ + 2e^- \longrightarrow H_2$	-0.420
$CO_2 + 2H^+ + 2e^- \longrightarrow$ formate	-0.420
Ferredoxin—$Fe^{3+} + e^- \longrightarrow$ ferredoxin—Fe^{2+}	-0.432

[a]*Standard conditions*: All components are at unit activity (i.e., 1.0) except $[H^+]$, which is maintained at $10^{-7}M$; gases are at a pressure of 1.0 atm.

with the more negative reduction potentials act as effective reducing agents (e.g., H_2, NADH) and those having the most positive reduction potentials (e.g., O_2, Fe^{3+}) are excellent oxidizing agents.

As a classic example for determining the direction of product formation for two half-reactions, consider the flow of electrons throughout the mitochondrial electron transport chain, discussed in following sections, where reduced NADH + H$^+$ is involved in the reduction of molecular oxygen to form water. The half-reactions are

Equation 1:
$$NAD^+ + 2H^+ + 2e^- \rightleftharpoons NADH + H^+$$
$$E^{0\prime} = -0.320 \text{ V}$$

Equation 2:
$$\frac{1}{2}O_2 + 2H^+ + 2e^- \rightleftharpoons H_2O$$
$$E^{0\prime} = +0.816 \text{ V}$$

Since established electrochemical theories indicate that the net potential of an electrochemical cell is given by

$$E^{0\prime}_{cell} = E^{0\prime}_{oxidation} + E^{0\prime}_{reduction}$$

it follows that the net potential for the specified system above will be

$$E^{0\prime}_{cell} = E^{0\prime}_{NADH,NAD^+} + E^{0\prime}_{O_2,H_2O}$$

Using concepts also detailed in earlier sections, equation 1, which has the least positive potential, is reversed to indicate its oxidation, and equation 2 is written as given.

Equation 1:

$$NADH + H^+ \rightleftharpoons NAD^+ + 2H^+ + 2e^-$$
$$E^{0\prime} = +0.320 \text{ V}$$

Equation 2:

$$\frac{1}{2}O_2 + 2H^+ + 2e^- \rightleftharpoons H_2O$$
$$E^{0\prime} = +0.816 \text{ V}$$

Net:

$$NADH + H^+ + \frac{1}{2}O_2 \rightleftharpoons NAD^+ + H_2O$$
$$E^{0\prime} = +1.136 \text{ V}$$

Hence, the $E^{0\prime}$ gives an indication that the reaction proceeds spontaneously as written.

The transfer of electrons between half-reactions is inherently related to changes in free energy and $\Delta G^{0\prime}$ as indicated by the equation

$$\Delta G^{0\prime} = -n\mathfrak{F}E^{0\prime}$$

where

    n = the number of moles of electrons transferred in the overall reaction

    $\mathfrak{F}$ = Faraday's constant 2.306×10^4 cal/V mol

    $E^{0\prime}$ = net potential difference between the two half-reactions

When the net potential difference $(E^{0\prime})$ for mutually dependent and interacting half-cell reactions is positive $(+)$, $\Delta G^{0\prime}$ for the overall reaction will have a negative sign $(-)$. Application of the foregoing equation to the ultimate reduction of O_2 by NADH above indicates that an energy yield of approximately -5.239×10^4 cal will result; that is,

$$NADH + H^+ + \frac{1}{2}O_2 \rightleftharpoons NAD^+ + H_2O$$
$$E^{0\prime} = +1.136$$
$$n = 2 \text{ (since } 2e^- \text{ are transferred)}$$

and $\Delta G^{0\prime}$ is

$$\Delta G^{0\prime} = -n\mathfrak{F}E^{0\prime}$$
$$= -(2 \text{ mol}) (2.306 \times 10^4 \text{ cal/V mol})$$
$$\times (+1.136 \text{ V})$$
$$= -5.239 \times 10^4 \text{ cal}$$

The direction of electron transfer between two interacting half-reactions can be anticipated by comparing their relative standard reduction potentials. It must be recognized, however, that the $E^{0\prime}$ values used in such calculations naturally assume that the oxidant and the reductant in an oxidation–reduction reaction are both in their standard states and thus the ratio of oxidant/reductant = 1. This situation is identical to those restraints of unimolar standard state conditions that are required for calculating $\Delta G^{0\prime}$ values based on equilibrium constants for reactants and products. Accordingly, then, the E for an oxidation–reduction reaction where the oxidant (oxidized form) and the reductant (reduced form) are not present in their standard state can be evaluated through the use of the Nernst equation, where

$$E_{obs} = E^{0\prime} - \frac{2.303 \, RT}{n\mathfrak{F}} \log \frac{[\text{reductant}]}{[\text{oxidant}]}$$

$$E_{obs} = E^{0\prime} - \frac{0.0591 \text{ V}}{n} \log \frac{[\text{reductant}]}{[\text{oxidant}]}$$

Note: E_{obs} is the observed electrical potential resulting when biochemical half-cell reactions involving an oxidant and a reductant are not present in a 1:1 ratio; E_{obs} is the observed potential based on the oxidant–reductant concentrations.

The application and advantage of using the Nernst equation in calculation of parameters for an oxidation–reduction reaction becomes obvious after considering the apparent and actual behavior of the following reactions. Under standard conditions the $E^{0\prime}$

Equation 1:   Pyruvate + $2H^+$ + $2e^-$ $\rightleftharpoons$ lactate $E^{0\prime} = -0.185$ V
             10^{-1} M 10^{-2} M

                                                                       based on $E^{0\prime}$ values

Equation 2:   Acetaldehyde + $2H^+$ + $2e^-$ $\rightleftharpoons$ ethanol $E^{0\prime} = -0.163$ V
             10^{-4} M 10^{-1} M

values could be used to indicate that lactic acid is oxidized so as to reduce acetaldehyde and thereby yield ethanol as a final product (equation 1 is less positive than equation 2). One must, however, consider the influence of molar concentrations for oxidants and reductants in the individual half-reactions, so it is necessary to implement the Nernst equation discussed above. For example, consider the case above using reasonable arbitrary concentrations for the oxidants and reductants:

Nernst equation: $E_{obs} =$

$$E^{0\prime} - \frac{0.0591\ V}{n} \log \frac{[reductant]}{[oxidant]}$$

Equation 1: $E_{pyruvate,\ lactate}^{obs} =$

$$-0.185\ V - \frac{0.0591\ V}{2} \log \frac{[10^{-2}]}{[10^{-1}]} = -0.155\ V$$

Equation 2: $E_{acetaldehyde,\ ethanol}^{obs} =$

$$-0.163\ V - \frac{0.0591\ V}{2} \log \frac{[10^{-1}]}{[10^{-4}]} = -0.252\ V$$

Solving the respective equations reveals that equation 2 will proceed in an overall direction that is quite contrary to the expected direction based only on $E^{0\prime}$ comparison. Actually, since the $E_{acetaldehyde,\ ethanol}$ reaction is less positive (-0.252 V) than the calculated $E_{pyruvate,\ lactate}$ value (-0.155 V), ethanol will be oxidized to acetaldehyde and the electrons supplied by this step will reduce pyruvate to form lactate. This type of a dynamic system will be sustained until amounts of participating oxidants and reductants produce equivalent E_{obs} values based on the Nernst equation calculation.

Other thermodynamic calculations: According to thermodynamic definitions, the standard free energy of formation (ΔG_f^0) represents the free energy change associated with the production of one mole of a biochemical substance from its elements under standard state conditions. The standard free energy of formation can be calculated for

many biochemical reactions, but, as an example, consider the reaction

$C_6H_{12}O_6$ (s) + 6 O_2 (g) $\rightleftharpoons$
Glucose

$$6\ CO_2\ (g)\ +\ 6\ H_2O\ (l)$$

where (s), (g), and (l) refer to a solid, gaseous, or liquid state, respectively. Provided that the standard free energies of formation (ΔG_f^0) are known for every component of a reaction, it is possible to calculate the standard free energy change (ΔG^0) for a reaction. This calculation must account for the difference in free energy between reactants and products under standard state conditions. In mathematical terms, this operation is defined as

$$\Delta G_{rexn}^0 = \Sigma \Delta G_{f\ (products)}^0 - \Sigma \Delta G_{f\ (reactants)}^0$$
$$\text{(see Table 11.1)}$$

Based on the reaction cited above, the ΔG_f^0 values are, respectively,

ΔG_f^0 for glucose $= -219.42$ kcal/mol	ΔG_f^0 for CO_2 $= -94.35$ kcal/mol
	ΔG_f^0 for H_2O
ΔG_f^0 for $O_2 = 0.00$	$= -56.74$ kcal/mol
Reactants	**Products**

and the calculated ΔG^0 is

$$\Delta G_{rexn}^0 = \Sigma \Delta G_{f\ (products)}^0 - \Sigma \Delta G_{f\ (reactants)}^0$$
$$\Delta G^0 = 6\ mol(-94.35\ kcal/mol) + 6\ mol$$
$$\times\ (-56.74\ kcal/mol) - 1\ mol$$
$$\times\ (-219.42\ kcal/mol)$$
$$\Delta G^0 = -687.12\ kcal$$

Enthalpy changes (ΔH) can also be estimated for reactions involving biochemical compounds that are involved in a reaction. Calculations of this sort require the individual enthalpies of formation (ΔH_f^0) for each participant in a reaction. In the following reaction conducted under standard conditions (1 atm, 25°C) where

C_2H_5OH (l) + 3 O_2 (g) $\longrightarrow$

$$2\ CO_2\ (g)\ +\ 3\ H_2O\ (l)$$

the overall enthalpy change for the reaction can be calculated by the following steps:

ΔH_f^0 for ethanol $= -66.36$ kcal/mol ΔH_f^0 for oxygen $= 0.00$	ΔH_f^0 for CO_2 $= -94.05$ kcal/mol ΔH_f^0 for H_2O $= -68.32$ kcal/mol
Reactants	**Products**

$\Delta H_{rexn}^0 = \Sigma\Delta H_{f\text{ (products)}}^0 - \Sigma\Delta H_{f\text{ (reactants)}}$
$$(\text{see Table 11.1})$$

$\Delta H_{rexn}^0 = 2$ mol $(-94.05$ kcal/mol$) + 3$ mol $\times (-68.32$ kcal/mol$) - 1$ mol $\times (-66.36$ kcal/mol$)$

$\Delta H_{rexn}^0 = -326.70$ kcal

BIOCHEMICAL PRINCIPLES OF NUTRITIONAL ENERGETICS

Living organisms have the ability to grow and reproduce. Whether the organism is monocellular or multicellular, its growth depends on its ability to assimilate nutrients from the environment and to metabolize them through a series of enzyme-catalyzed reactions. These reactions provide both biochemical substances needed for cellular mass and the energy used for the organization and maintenance of all cell activities. *Nutritional energetics is concerned with the metabolic utilization of nutrients to provide energy that is then available for necessary physical and chemical processes.* As presented in this chapter, nutritional energetics deals with the sources and syntheses of high-energy compounds, such as ATP, and the caloric principles underlying human and animal nutrition.

High-energy Biochemical Compounds

Energy obtained and conserved by living systems usually takes the form of "high-energy" or "energy-rich" compounds such as (1) pyrophosphates (nucleotide di- and triphosphates); (2) acyl-phosphates; (3) enolic phosphates; (4) thio esters; and (5) guanidinium phosphates. These substances represent so-called high-energy compounds because a significant decrease in free energy occurs when they undergo hydrolysis. Note that the term high-energy compound indicates that a bond or compound has a large tendency to break apart and undergo reaction. For example, hydrolysis of a nonspecific energy-rich compound designated as X-O-Y-O-Z may produce a variety of hydrolytic products and energy. The marked decrease in free energy on hydrolysis of these compounds occurs because the products of hydrolysis have improved thermodynamic stability over the initial reactants. Table 11.4 contains the numerical values of the standard free energy of hydrolysis for some high-energy compounds.

$$\text{X—O—Y—O—Z} \xrightarrow[+2\ H_2O]{\text{hydrolysis}}$$

$$\text{X—OH + HO—Y—OH + HO—Z + energy}$$
Hydrolysis products

Many factors contribute to the improved stability of products over reactants. Improved reactant stabilities are often described in terms of decreased electrostatic repulsion, stabilization by ionization, isomerization, or resonance phenomena (charge delocalization).

One of the most important high-energy compounds is ATP, which acts as a store of energy derived from ingested food. When ATP undergoes hydrolysis, the stored energy becomes available for many cellular biochemical processes. Since ATP plays an important role in cellular energetics, an explanation will be presented regarding its stability. The structure of ATP has less stability than its hydrolytic products, ADP^{3-} and HPO_4^{2-} (P_i). The instability of ATP can be illustrated by considering the electrostatic

Table 11.4. Standard Free Energy of Hydrolysis of Some Compounds

Reaction	$\Delta G^{0\prime}$ at pH 7 (kcal/mol)
[a]ATP + $H_2O \longrightarrow$ ADP + P_i[b]	−7.3
ADP + $H_2O \longrightarrow$ AMP + PP_i[b]	−7.3
AMP + $H_2O \longrightarrow$ adenosine + P_i	−3.4
1,3-Diphosphoglycerate + $H_2O \longrightarrow$ 3-phosphoglycerate + P_i	−11.8
Phosphoenolpyruvate + $H_2O \longrightarrow$ pyruvate + P_i	−14.8
Acetyl-CoA + $H_2O \longrightarrow$ pyruvate + CoASH + H^+	−7.5
Creatine phosphate + $H_2O \longrightarrow$ creatine + P_i	−10.3

[a]Note that $\Delta G^{0\prime}$ for ATP hydrolysis is −7.3 kcal/mol under standard conditions; however, the value of ΔG in a living cell, where the [ATP], [ADP], and [P_i] are less than 1 M (usually in millimolar range), is estimated to be between −12 and −16 kcal/mol. The calculation of these values is based on the following equation: $\Delta G = \Delta G^{0\prime} + 2.303 RT \log Q$.
[b]P_i, Orthophosphate; PP_i, pyrophosphate.

repulsion inherent in the ATP structure. The protonated structure of ATP is

$$\text{Adenine-ribose}-O-\overset{\overset{O}{\|}}{\underset{\underset{OH}{|}}{P}}-O-\overset{\overset{O}{\|}}{\underset{\underset{OH}{|}}{P}}-O-\overset{\overset{O}{\|}}{\underset{\underset{OH}{|}}{P}}-OH$$

However, at the physiological pH of 7, the four —P—OH bonds of ATP undergo ionization and the structure can be written as ATP^{4-}. The electrons in the P=O bond of

$$\text{Adenine-ribose}-O-\overset{\overset{O}{\|}}{\underset{\underset{O^-}{|}}{P}}-O-\overset{\overset{O}{\|}}{\underset{\underset{O^-}{|}}{P}}-O-\overset{\overset{O}{\|}}{\underset{\underset{O^-}{|}}{P}}-O_-$$

the phosphates are drawn closer to oxygen (higher electronegativity) than phosphorus. This action produces a partial negative charge δ^- on oxygen and that induces a partial positive charge δ^+ on phosphorus as illustrated.

$$\text{Adenine-ribose}-O-\overset{\overset{O^{\delta-}}{\|\delta^+}}{\underset{\underset{O_-}{|}}{P}}-O-\overset{\overset{O^{\delta-}}{\|\delta^+}}{\underset{\underset{O_-}{|}}{P}}-O-\overset{\overset{O^{\delta-}}{\|\delta^+}}{\underset{\underset{O_-}{|}}{P}}-O^-$$

The four negatively charged oxygens produced in the process of ionization repel each other, and this induces a structural strain on the entire molecule, especially on the

—P—O—P— bonds. This electrostatic repulsion imposed on the ATP structure contributes significantly to the thermodynamic instability of this molecule. Upon hydrolysis of ATP, this strain is lessened by forming the product ADP^{3-}, which has less electrostatic repulsion than ATP^{4-}. Furthermore, the stability of the hydrolytic products is enhanced by the resonance of the dianion, HPO_4^{2-}. Hence, a large decrease in free energy is obtained when ATP is converted to its more stable hydrolytic products, ADP^{3-} and HPO_4^{2-}.

Interconversion of ATP and ADP with other high-energy nucleotides: The high-energy-yielding properties of ATP and ADP reactions are certainly important keystones in the energy flow mechanisms of cells, but there are other high-energy-yielding nucleotides that behave in a similar fashion. These include guanosine triphosphate (GTP), which acts as a source of energy for the movement of ribosomes and peptide synthesis, as well as cytidine triphosphate (CTP), uridine triphosphate (UTP), and deoxythymidine triphosphate (dTTP), which respectively supply energy in specialized synthesis reactions.

The cellular existence of these energy-rich pyrophosphates exists at the luxury of adequate cellular ATP stores. That is, the production of these pyrophosphate nucleosides depends on the transfer of phosphate(s) from ATP to an appropriate nucleoside monophosphate or nucleoside diphosphate. Figure 11.2 details some of the noteworthy nucleotide interconversions that exist between exergonic and endergonic pathways.

Coupled Biochemical Reactions

The operation of living biochemical systems requires the integration of exergonic (energy yielding) reactions with endergonic (energy consuming) reactions. The total complexity and delicate nature of these reaction interrelationships is overwhelming, yet superficial study of even the most complicated systems reveals that exergonic and endergonic reactions are universally linked to at least one common biochemical intermediate. Interrelated reactions such as these are generally called *coupled reactions*.

Law of mass action and coupled reactions: Each individual reaction in the overall scheme of a coupled reaction is subject to the principal laws of thermodynamics and the compensating mechanisms associated with the law of mass action. As a simple example of a coupled reaction, consider the following case in which A ultimately leads to the formation of C. So long as this coupled

mediate, namely, B. Therefore, the equilibrium constants for these individual reactions are

$$K_{eq_1} = \frac{[B]}{[A]}$$

$$K_{eq_2} = \frac{[C]}{[B]}$$

Solving both equilibrium expressions in terms of the common intermediate $[B]$ yields

$$[B] = [A]\, K_{eq_1} \quad \text{and} \quad [B] = [C]/K_{eq_2}$$

Therefore, a coupled reaction, as shown by this simple example, in which A is ultimately converted to C, will display a concentration ratio for the final products to the initial reactants, which is equal to the product of all intervening equilibrium constants; that is,

$$\frac{[C]}{[A]} = K_{eq_\omega} = K_{eq_1} \cdot K_{eq_2}$$

$$K_{eq_\omega} = K_{eq_1} \cdot K_{eq_2}$$

The implications of this relationship are wide ranging and fundamental to the understanding of the most complex biochemical reactions. Using the example cited above, the isolated equilibrium reaction of $A \rightleftarrows B$ (reaction I) can be encouraged to convert A to B by adding more A from an external source, or directly removing B from the sys-

---

A coupled reaction consists of two individual reactions linked by a common intermediate

$$A \underset{K_{eq_1}}{\overset{\text{Reaction I}}{\rightleftarrows}} B \underset{K_{eq_2}}{\overset{\text{Reaction II}}{\rightleftarrows}} C$$

Overall coupled reaction

---

reaction of $A \rightleftarrows C$ is at steady-state conditions, the concentration levels of the constituents in each reversible reaction will demonstrate concentration ratios identical to that particular reaction taken as an isolated case. The equilibrium constant is dictated by the nature of each particular reaction and the actual concentrations of the constituents are governed by the law of mass action for each independent reaction. Using this example, it is also clear that the net reaction of $A \rightleftarrows C$ consists of the individual reactions $A \rightleftarrows B$ and $B \rightleftarrows C$ that are linked by the common inter-

tem. Further, the conversion of A to B can be increased by coupling Reactions I and II as indicated,

$$A \overset{\text{Reaction I}}{\rightleftarrows} B \overset{\text{Reaction II}}{\rightleftarrows} C$$

provided that stores of A favor its high availability and that the $B \rightleftarrows C$ reaction encourages C formation.

Based on the previous considerations then, it is clear that (1) directional flow of reactants toward product formation in coupled reactions will be directly influenced by the math-

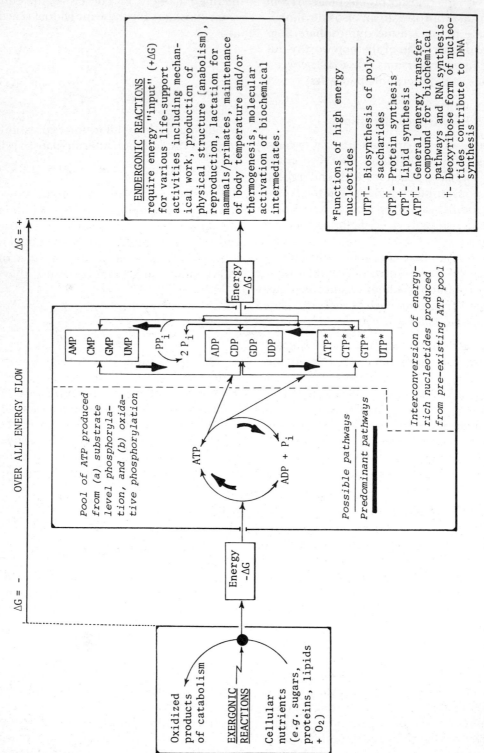

FIGURE 11.2. Noteworthy nucleotide conversions that exist between exergonic and endergonic pathways.

ematical product of the K_{eq} for each participating reaction in the series; (2) the overall product of the K_{eq} values for separate but coupled reactions is denoted as K_{eq_ω}; and (3) the available concentration levels of specified reactants and products determine the course of a coupled reaction at any given instant.

Free energy relationships and coupled reactions: The standard free energy values of coupled reactions are additive. For example,

Reaction I: $A \overset{K_{eq_1}}{\rightleftharpoons} B,$
$\Delta G_1^{0'} = -2.303 \, RT \log K_{eq_1} = -160.0 \text{ cal}$

Reaction II: $B \overset{K_{eq_2}}{\rightleftharpoons} C,$
$\Delta G_2^{0'} = -2.303 \, RT \log K_{eq_2} = +90.0 \text{ cal}$

Sum: $A \overset{K_{eq_\omega}}{\rightleftharpoons} C,$
$\Delta G_\omega^{0'} = -2.303 \, RT \log K_{eq_\omega} = -70.0 \text{ cal}$

Considering the above example, if Reaction I yields a $\Delta G^{0'}$ of -160.0 cal and Reaction II has a $\Delta G^{0'}$ equal to $+90.0$ cal, the calculated $\Delta G_\omega^{0'}$ value for the possible overall reaction of $A \rightleftharpoons C$ is -70.0 cal, since the standard free energy values are additive. This is not to imply that the reaction will definitely occur, but the reaction is thermodynamically feasible provided that the exergonic energy from Reaction I (exergonic) can be directed toward driving Reaction II (endergonic) to the eventual product formation of C.

The conservation of energy from an exergonic reaction followed by direct application of this energy toward promoting a dependent endergonic reaction is not a simple affair. It is a process bridled by (1) the characteristic K_{eq} for each contributing reaction in the coupled series, (2) the law of mass action, (3) the specific participation of enzymes (enhancing the rate of conversion of reactants into products in a reaction), and (4) the conservation of exergonic energy in the form of high-energy intermediates that are commonly linked to both the endergonic and exergonic reactions.

The process of coupling allows a spontaneous reaction to drive an endergonic reaction, and it serves as the basis of storage and use of chemical driving forces that control living systems.

Ultimate Sources of ATP

The *in vivo* production of ATP from ADP is generically described as a *phosphorylation reaction*. The formation of this high-energy nucleotide from ADP can be expressed merely as a transfer of a phosphate group from some compound to ADP. The forma-

$$\text{X}-\text{O}-\overset{\overset{\displaystyle O}{\|}}{\underset{\underset{\displaystyle O^-}{|}}{\text{P}}}-\text{O}^- + \text{ADP} \longrightarrow \text{ATP}$$

tion of ATP by this simple route does not reflect the complexity of a biochemical process that has perplexed biochemists for many years. Moreover, the nature of the process becomes further intriguing since the phosphorylation of ADP to form ATP can occur by one of the three ways:

1. Photophosphorylation.
2. Substrate-level phosphorylation.
3. Oxidative phosphorylation.

Photophosphorylation of ADP to yield ATP occurs in the chloroplasts of algae and higher vascular plants. Light energy (photons) acting on critical photoactive centers of plastids that contain chlorophylls and other photoactive accessory pigments promote photolysis. This eventually culminates in the "light-splitting" of water to yield oxygen and a source of high-energy electrons. The flow of these electrons through a series of mutually interacting electron donors and acceptors (oxidants and reductants) ultimately drive the phosphorylation of ADP to produce ATP and provide reduced NADH from oxidized NAD^+. The detailed mechanism for "light-driven" phosphorylation phenomena has been surveyed in Chapter 12.

Aside from photophosphorylation, *substrate-level* and *oxidative phosphorylations* act as primary mechanisms for the maintenance of requisite ATP pools in living heterotrophic cells.

Substrate-level phosphorylation: Substrate level phosphorylation is an important mechanism for ATP generation in many cells, and especially for anaerobic cells, where it provides the only route for ATP production.

The mechanisms for substrate level phosphorylation are not identical and, in fact, there are many variations on the theme. In glycolysis, for example, glyceraldehyde 3-phosphate is oxidized to 1,3-diphosphoglyceric acid and this product, in turn, is converted to 3-phosphoglyceric acid. The final formation of 3-phosphoglyceric acid is accompanied by the transfer of the 1-phosphoryl group of 1,3-diphosphoglyceric acid to ADP.

Inorganic phosphate is linked to GDP to yield GTP through an energy-conserving reaction mediated by succinyl-CoA synthetase in a three-step reaction. The transfer of P_i to GDP depends on the formation of a temporal phosphorylated enzyme prior to the actual phosphorylation of GDP to yield GTP.

Step 1: Succinyl-CoA + P_i + E $\rightleftharpoons$
 E—succinyl phosphate + CoASH
Step 2: E—Succinyl phosphate $\rightleftharpoons$
 E—P + succinate
Step 3: E—P + GDP $\rightleftharpoons$ E + GTP

Sum: Succinyl-CoA + P_i + GDP $\rightleftharpoons$
 succinate + GTP + CoASH

Once GTP has been formed, it donates its terminal phosphate group to ADP so as to produce ATP if adequate stores are depleted:

$$\text{GTP + ADP} \underset{\substack{\text{nucleoside}\\\text{diphosphate}\\\text{kinase}}}{\rightleftharpoons} \text{GDP + ATP}$$

1,3-Diphosphoglycerate
(or 3-phosphoglycerolyl phosphate)
3-Phosphoglycerate

Another approach to substrate level phosphorylation is coupled to the deacylation of succinyl-CoA in the tricarboxylic acid cycle.

Note that both of these high-energy nucleotide triphosphates, namely GTP and ATP, have similar standard free energies on hydrolysis.

Succinyl-CoA　　　　**Succinate**

In theory, high-energy phosphates or thioesters that provide more energy on hydrolysis than that required to phosphorylate ADP can conceivably promote ATP formation—provided that suitable enzymes are available to mediate the process. For example, ADP can be phosphorylated to form ATP by any of the high-energy compounds having $\Delta G^{0'}$ values on hydrolysis exceeding that of ATP's hydrolysis to ADP. Similarly, those metabolic substances having $\Delta G^{0'}$ values on hydrolysis of less than the $\Delta G^{0'}$ produced by ATP hydrolysis can be phosphorylated by ATP (Figure 11.3).

Oxidative phosphorylation: The principal method for ATP formation in aerobic cells involves oxidative phosphorylation. Normal processes of biochemical oxidation are mediated by specific enzymes that strip hydrogens and electrons from cellular nutrients and their related metabolic intermediates. The electrons generated as a by-product of these pooled oxidation reactions are funneled toward the ultimate reduction of molecular oxygen to form water. The actual formation of water, however, is preceded by a complicated series of electron transfer reactions that produce ATP.

The principal reactants in oxidative phosphorylation are characterized by their collective interacting ability to transfer electrons over an electrical reduction potential span from approximately -0.32 to $+0.82$ V. The site of this electron transport is along the inner and outer membrane of the mitochondrion. All of the mutually interacting electron carriers in the electron transport scheme are collectively referred to as the *electron transport chain* or *the respiratory chain*. Each electron carrier in the transport chain displays markedly different or sometimes only subtle structural differences from other members in the chain. Nevertheless, these disparities in biochemical structure result in different standard reduction potentials for each and every electron carrier in the chain. These differences in electrical reduction potentials foster the directional transport of electrons through the chain in a direction from the most negative electrical potential to a more positive potential.

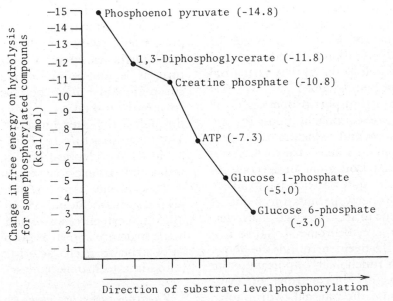

FIGURE 11.3. Relative abilities of some high-energy phosphorylated compounds to phosphorylate other compounds.

Electron Carriers in the Electron Transport Chain

The electron carriers in the respiratory chain are not biochemically identical. The only feature that they universally share is their ability to undergo repeated oxidations and reductions. The functional group of electron carriers in higher animals includes the nicotinamide adenine dinucleotides (NAD); the flavin nucleotide (FMN, FAD)-linked enzymes; coenzyme Q (CoQ), which is representative of the ubiquinones; a host of cytochromes individually specified as b, c_1, c, a, a_3; and iron–sulfur proteins (Fe–S). The oxidation–reduction mechanisms for many of these electron carriers in the respiratory chain have been discussed elsewhere—especially in Chapter 5, dealing with vitamins and coenzymes. Although many electron carriers, including the nicotinamide and flavin nucleotides as well as the ubiquinones, are definitely linked to vitamin and/or coenzyme activities, not all electron carriers fall into these classes, particularly the cytochromes and the iron–sulfur proteins.

The cytochromes: Cytochromes are heme-containing proteins (hemoproteins) that are localized in mitochondrial membranes. Rigorous scientific evidence suggests that the molecular structure of cytochromes may be the structural evolutionary predecessor of myoglobin as well as the tertiary structural predecessor of hemoglobin.

The electron transport properties of all cytochromes are contingent upon the reversible oxidation and reduction of the central iron atom in their heme structures. Here, the coordinated iron atom undergoes reduction to a Fe^{2+} state or oxidation to a Fe^{3+} form. This property permits the oxidized iron in the heme to act as an electron carrier. The rest of the cytochrome structure is a protein that effectively surrounds the heme. This protein is functionally important to the electron transport mechanisms of cytochromes because the internal hydrophobic properties of the protein minimize the oxidation of heme-iron by water. Without the protection of this hydrophobic *microclimate,* the important oxidation–reduction properties of heme would be ruined.

The cytochromes demonstrate amazing interspecies similarities in their fundamental structures, yet only minor structural differences account for the distinct differences in their standard reduction potentials that promote electron transfer.

Cytochrome nomenclature is largely credited to D. Keilin, who assigned the letters a, b, and c to these hemoproteins based on their characteristic absorption spectra maxima. The cytochrome absorbing the longest wavelength was called cytochrome-a, the cytochrome absorbing the next longest wavelength was called cytochrome-b and so on.

Absorption spectra for the cytochromes display obvious changes depending on whether the cytochrome is reduced or oxidized (Figure 11.4). These changes are best observed when the well-defined α, β, and γ absorption maxima (soret bands) for reduced cytochrome undergo peak reduction in the α and β spectral ranges, whereas the γ peak is displaced to the right of the γ absorption peak for oxidized species. These observed patterns of spectral behavior assume different expressions depending on the type of cytochrome that happens to be present in greatest abundance.

From the viewpoint of molecular structure, it is interesting to note that cytochromes-c and -c_1 have their heme structures covalently linked as a prosthetic group through a thioester linkage (Figure 11.5). The heme structure for both cytochromes-c and -c_1 are identical but -c_1 is associated with a higher-molecular-weight protein than -c. Cytochrome-b, on the other hand, has several forms including -b^{2+} 564 (α), -b^{2+} 530 (β), while cytochromes -a and -a_3 contain a heme structure known as porphyrin A. This latter structure is characteristic of the heme found in cytochrome oxidase.

The iron–sulfur proteins: Exclusive of the oxidation–reduction activities of the cytochromes, iron–sulfur proteins are strategi-

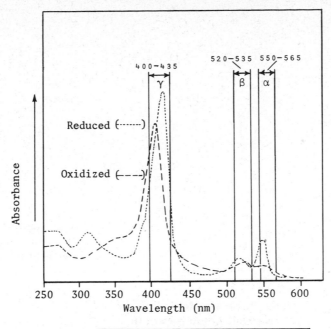

Cytochrome	Soret bands		
	α	β	γ
a	600	—	439
a + a₃ complex	604	—	443
b	563	532	429
c	550	521	415
c₁	554	524	418
b₅	556	526	423

FIGURE 11.4. Generalized comparison of oxidized and reduced absorption spectra for cytochromes. The absorption maxima indicated as α, β, and γ (Soret bands) undergo marked changes depending on the oxidation state of the cytochrome. Soret bands for various cytochromes appear in the table.

cally integrated into the electron transport chain. These proteins, also known as non-heme-iron proteins, contain an iron atom bound through sulfur atoms of cysteine residues on proteins or, alternatively, iron may be directly bonded to sulfur. Structures of iron–sulfur aggregates (Fe–S) may exist in reduced or oxidized forms as shown below, which account for the electron transport mechanism.

[Fe₂S₂Cys₄ complex]
ox

[Fe₂S₂Cys₄ complex]
red

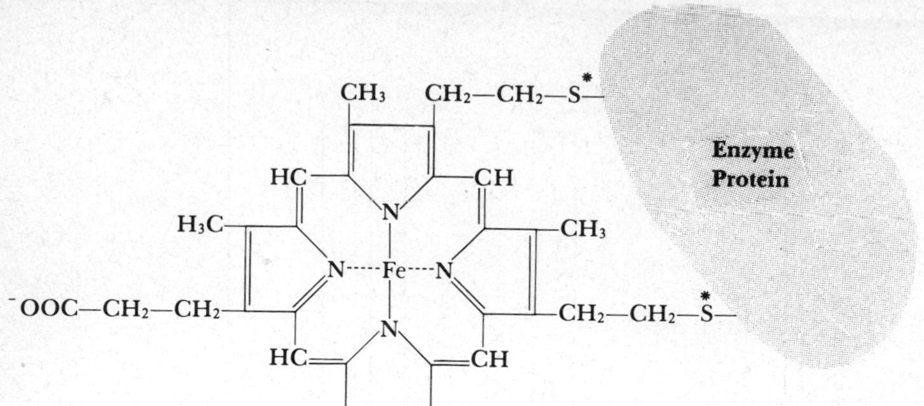

Porphyrin structure for heme
prosthetic group present in
cytochromes -*c* and -*c*₁

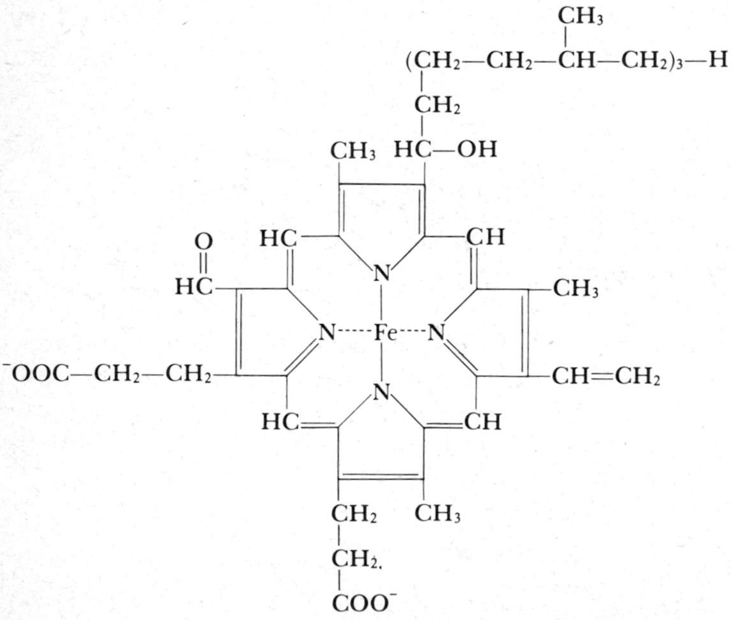

*Cystine residues link
heme structure to
protein structure

Porphyrin A present in cytochromes
-*a* and -*a*₃, and cytochrome oxidase
complex

FIGURE 11.5. Typical heme structures present in cytochromes.

More or less complex structural variations in iron–sulfur proteins other than that illustrated above are believed to occur, but the exact biochemical advantage of one iron–sulfur protein complex over another remains to be documented.

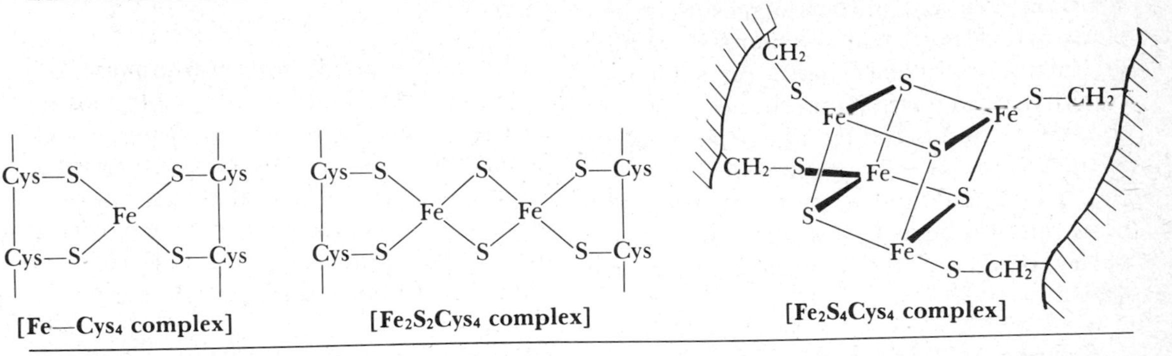

The remaining proton is released to solution and does not become specifically involved in the NADH structure. Formation of NADH (H^+) is significant in the scheme of electron transport because it is one of the earliest steps in the electrochemical pro-

[**Fe—Cys₄ complex**]   [**Fe₂S₂Cys₄ complex**]   [**Fe₂S₄Cys₄ complex**]

Roles of NAD⁺ and FAD in electron transport: The complete aerobic oxidation of glucose in the citric acid cycle (tricarboxylic acid cycle) yields carbon dioxide and hydrogen atoms. Mitochondrial activities direct the resulting hydrogen atoms through a series of electrochemically driven steps that ultimately contribute to reduction of molecular oxygen to form water. Hydrogen atoms may be liberated during metabolic processes as two hydrogen atoms containing their electrons, or as two electrons and two protons (H^+) in separate steps. The favored disposition for some hydrogens in the mitochondrion results in the formation of one proton and a hydride ion, $:H^-$ (hydrogen atom plus one extra electron). The hydride ion is then able to reduce the nicotinamide moiety of the nicotinamide adenine dinucleotide (NAD^+).

duction of ATP according to present theories.

The action of *NAD⁺-linked dehydrogenase* enzymes on mitochondrial substrates such as malate inevitably produces oxaloacetate and NADH(H^+). Other mitochondrial participants in the citric acid cycle other than malate can be specifically oxidized by flavin-linked enzymes (FAD-containing enzymes). This type of oxidation typically occurs during the oxidation of succinate to fumarate during the citric acid cycle where reduced $FADH_2$ is produced through the action of flavin-linked succinic acid dehydrogenase.

Both NADH(H^+) and $FADH_2$ formed by these routes serve as electron sources, which prime and perpetuate the activity of the electron transport chain and promote ATP production.

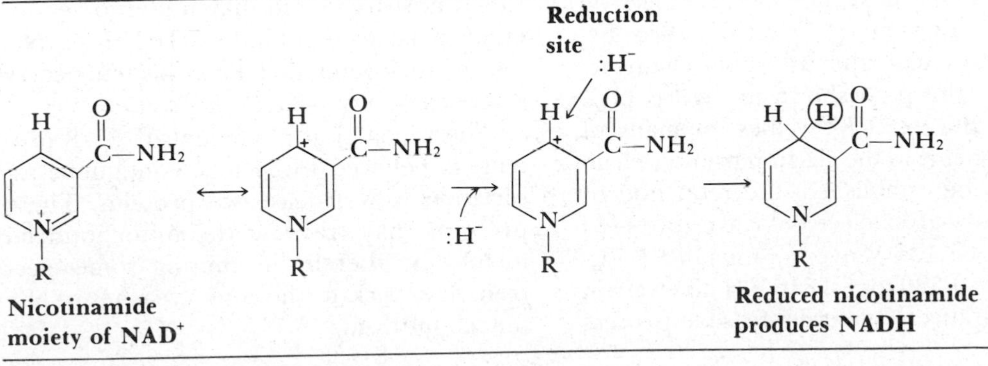

Nicotinamide moiety of NAD⁺

Reduced nicotinamide produces NADH

Organization of the Respiratory Chain

Once $NADH(H^+)$ and $FADH_2$ have been formed, their respective introductions of electrons into the respiratory chain assume slightly different routes owing to its structure, architecture, and electrochemical construction. Present concepts suggest that the electron transport chain is comprised of four interacting complexes based on submitochondrial fragmentation studies. These complexes, numbered I to IV, are positioned in the inner and outer regions of the mitochondrial membrane. Functional designations can also be assigned to the four complexes based on the oxidation–reduction enzyme activities:

Complex I. NADH-CoQ oxidoreductase complex. Reaction promoted: $NADH(H^+) + CoQ \xrightarrow{+H^+} NAD^+ + CoQH_2$

Complex II. Succinate CoQ oxidoreductase. Reaction mediated: Succinate + CoQ $\longrightarrow$ Fumarate + $CoQH_2$

Complex III. CoQH$_2$ cytochrome-c oxidoreductase. Reaction promoted: $CoQH_2 + 2$ Cyt-$c_{ox} \longrightarrow CoQ + 2$ Cyt-$c_{red} + 2H^+$

Complex IV. Cytochrome oxidase. Reaction conducted: 2 Cyt-$c_{red} + 2H^+ + \frac{1}{2}O_2 \longrightarrow 2$ Cyt-$c_{ox} + H_2O$.

These complexes apparently are integrated into an operational respiratory apparatus such as the one schematically indicated in Figure 11.6. The functional participants of these complexes are sequentially arranged according to their reduction potentials, and electron flow from each carrier progresses in the direction of the next most positive electron acceptor. High-energy electrons are believed to traverse the transport chain according to the path shown in Figure 11.7. Although the exact $E^{0'}$ values for many of the components in the electron transfer chain remain to be established, the reduction of water occurs after electrons have traversed a span of $+1.136$ V over a range of $E^{0'}$ values from -0.320 to $+0.816$ V. This electron transfer is an energetically feasible process and each electron transfer step has the potential to proceed spontaneously and exergonically. As shown in previous calculations, the transfer of only two electrons has a theoretical $\Delta G^{0'}$ value of -5.239×10^4 cal over the entire route from NADH to water formation.

The mechanics of electron transport: Energy-rich NADH and $FADH_2$ molecules respectively enter the respiratory chain through Complexes I and II. Based on close study of Figure 11.7, it is also clear that the flow of electrons follows the path shown below, but in fact, the mechanics are far more complex than the sequence of reactions indicates.

$$FADH_2 \text{ (Fe—S)}$$
$$\downarrow$$
$$NADH(H^+) \rightarrow FMN \rightarrow Fe—S \rightarrow CoQ \rightarrow Cyt\text{-}b \rightarrow$$
$$Fe—S \rightarrow Cyt\text{-}c_1 \rightarrow Cyt\text{-}c \rightarrow Cyt\text{-}a,a_3 \rightarrow O_2$$

Based on present scientific data, it seems plausible that the FMN structure is complexed to a protein that completely traverses both the inner and outer mitochondrial membranes. When NADH is oxidized by FMN to produce $FMNH_2$, the $FMNH_2$ is subsequently oxidized when the two hydrogen atoms on the internal surface of the mitochondrial membrane are translocated to its outer surface. Here, both of the hydrogen atoms undergo rapid ionization, followed by their discharge as protons into extramitochondrial regions. The electrons that remain are probably conveyed to Fe–S proteins.

The oxidized flavin-linked electron carrier is now freed and may undergo yet another reduction event with $NADH(H^+)$. Note, too, at this point that Fe–S proteins carry single electrons, *not electron pairs*.

The topological placement of Fe–S proteins is believed to be in a contiguous association with at least two proteins. These proteins may traverse the mitochondrial membrane thereby permitting some electron flow back to the inner reaches of the mitochondrion.

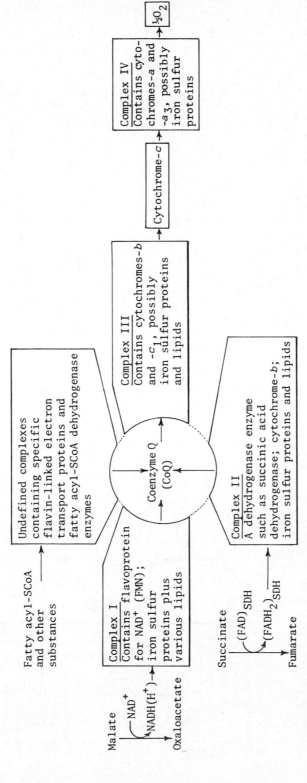

FIGURE 11.6. Complexes involved in electron transport.

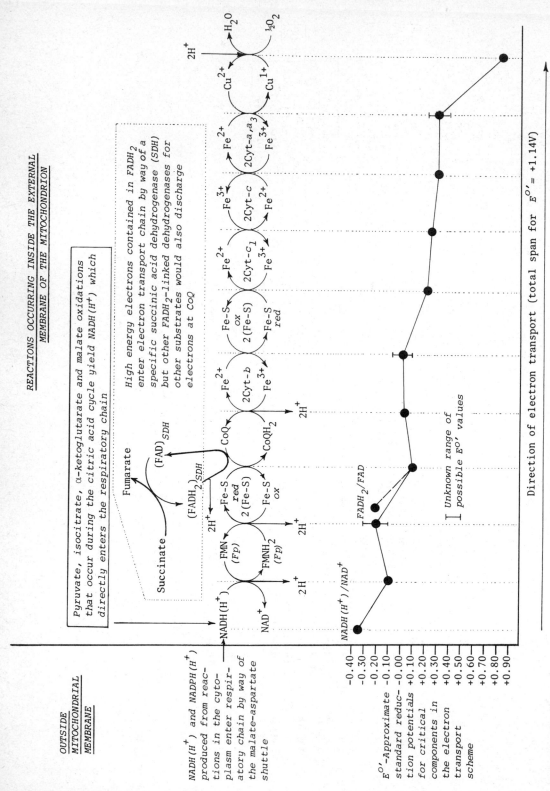

FIGURE 11.7. Key steps and directional transport of electrons through the electron transport chain.

Although specifics are still lacking, most evidence indicates that Fe–S$_{red}$ proteins release electrons to ubiquinone or coenzyme Q (Figure 11.8). Coenzyme Q exists in three distinct forms: (1) *a totally oxidized quinone form* denoted as Q; (2) *a semiquinone form*, QH· that results when one hydrogen atom reduces the oxidized quinone; and (3) *a fully reduced hydroquinone*, QH$_2$, where the quinone ring has been fully reduced by the up-

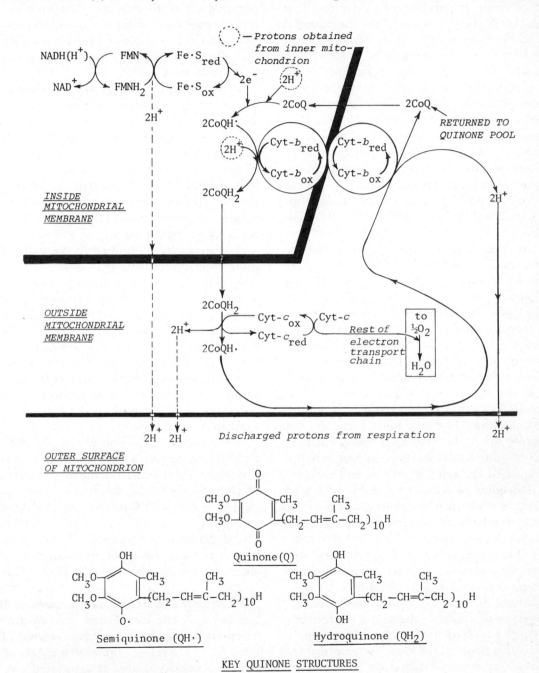

FIGURE 11.8. Interaction of coenzyme Q (CoQ) in the respiratory transport of electrons (key structures for the participating quinone structures are detailed).

take of hydrogen atoms. In the case of electron transfer processes, the initial interaction of Fe–S with coenzyme Q involves the release of an electron from the Fe–S proteins to two coenzyme Q molecules. This transfer of electrons is paralleled by a concomitant uptake of two protons from the internal reaches of the mitochondrial membrane, and two semiquinones are formed (2 QH·).

The first of the cytochrome species involved in electron transport, cytochrome-b, now offers two additional electrons to the two semiquinones—the source of these electrons will be detailed shortly. This overall event is then followed by the association of two more protons—again, obtained from the inner reaches of the mitochondrion. These oxidation and reduction events coupled with electron reshuffling produce two QH_2 molecules.

At this juncture the electron transfer scenario becomes more complicated as a result of further electron transfers. The two reduced quinone molecules (2 QH_2) are not steadfastly fixed between the inner and outer mitochondrial membranes. Instead the QH_2 seems to behave as a free-roving species and may be able to migrate from the inner side of the mitochrondrial membrane to the outside and back again. According to conventional theories, the two reduced quinones (QH_2) migrate from the inner surface of the membrane to the exterior surface. Each reduced quinone discharges one proton to the outside of the mitochondrion and each reduced quinone is also oxidized by virtue of a single electron transfer to cytochrome-c_1. This step produces two semiquinones (2 QH·). These two semiquinones then assume their original oxidized forms (2 Q) after two protons are lost to the external environment of the mitochondrion, and *after* their extra electrons are donated to cytochrome-b. The last step is responsible for the supply of electrons afforded by cytochrome-b in earlier steps.

At this point in the electron transport sequence, a proton and electron inventory indicates that six protons have discharged into the extramitochondrial environment, two protons from the early part of the respiratory chain where $FMNH_2$ and Fe–S protein interactions occurred, and now four additional protons have been lost as a result of quinone reactions. Consideration of the remaining electrons indicates that four electrons are retained within the transport chain. Two electrons reside with cytochrome-c and the other two electrons associate with Cyt-b_{ox} to produce Cyt-b_{red}.

Electrons in the overall flow mechanism that are retrieved by cytochrome-b are reintroduced into the cycle as previously discussed and illustrated in Figure 11.8. The two electrons held by cytochrome-c_1 are transferred to cytochromes-c, -a, and -a_3 in that order. Cytochrome-a_3 is dissimilar from the other cytochromes because it is located within the inner mitochondrial membrane contrary to cytochromes-c_1, -c, and -a, which are superficially positioned in the external mitochondrial membrane.

It should be recognized that some evidence based on mammalian mitochondrial studies indicates that cytochromes-a and -a_3 occur in a cytochrome oxidase complex. This complex also contains two atoms of copper (Cu^{1+}) per functional unit. Clearly, copper atoms act as oxidation–reduction intermediates in this step of the respiratory chain, but exact details regarding the mechanism need to be established.

The ultimate reduction of oxygen by cytochrome-a_3 requires transferred electrons, two protons obtained from the inner reaches of the mitochondrial membrane, and one oxygen molecule ($\frac{1}{2}O_2$), that is, in a balanced form, $4H^+ + 4e^- + O_2 \rightarrow 2H_2O$. A theoretical diagram of respiratory chain participants in both the inner and outer mitochondrial membranes is shown in Figure 11.9.

Electron transport and ATP formation: It has been clearly established that electron transport via the respiratory chain yields ATP from ADP. Therefore, although it is a cliché, the mitochondrion and its associated electron carriers actually represent the "power house of the cell." Contrary to this common

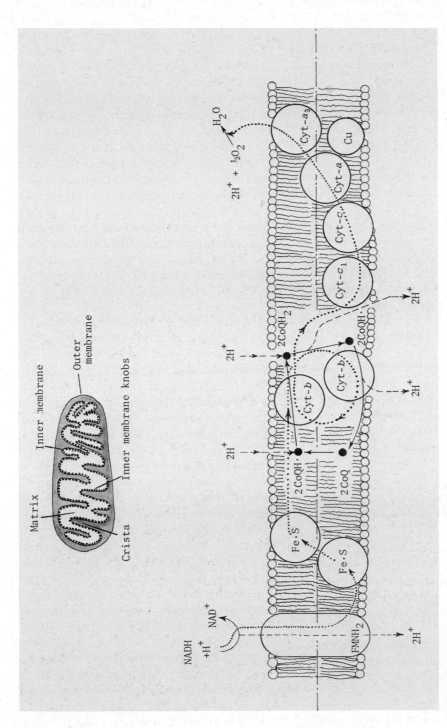

FIGURE 11.9. Planar schematic architecture of electron transport chain located within the inner and outer membranes of the mitochondrion as illustrated in the inset. Protons are pumped from the inner matrix of the mitochondrion (——) to the outer membrane surface. Electrons flow from left to right (···) and ultimately contribute to oxygen reduction and water formation. The interactions of coenzyme Q (CoQ) and its derivatives follow the direction of the solid arrows. Nonheme iron (Fe·S) may also be involved between cytochromes-c_1 and -c, whereas copper contributes to the process of electron flow between cytochromes-a and -a_3. The proton gradient established by this mechanism over the mitochondrial membranes drives the phosphorylation of ATP from ADP by a phosphorylation enzyme.

recognition, the actual mechanism of ATP formation by means of ADP phosphorylation has perplexed many investigators for many years. A host of theories have been propounded for phosphorylation events during electron transport including *chemical coupling, conformational coupling*, and *chemiosmotic coupling theories*. Of these three concepts, the first two are possible but not probable based on current data. The third theory of chemiosmotic coupling, proposed by the Nobel laureate Peter Mitchell (1978), however, has captured the practical imagination of most biochemists since it is compatible with contemporary scientific principles. In other words, Mitchell's chemiosmotic theory is quite probable since its stipulations are compatible with the recognized behavior of electron transport chains, both *in vivo* and *in vitro*, plus the fact that *it accounts for the significance of proton migration outside and into the mitochondrion.*

ATP formation by chemical coupling. According to this theory, high-energy phosphates such as ATP are produced from the transfer of electrons. Electron transport is then attributed with the temporal formation of a high-energy intermediate ($\sim$I) that acts as an ATP precursor.

Step 1: $A_{red} + B_{ox} + I \longrightarrow A_{ox} \sim I + B_{red}$
Step 2: $A_{ox} \sim I + ADP + P_i \longrightarrow A_{ox} + I + ATP$

Sum: $A_{red} + B_{ox} + ADP + P_i \longrightarrow$
$$A_{ox} + B_{red} + ATP$$

This concept of chemical coupling is paralleled in biochemical systems during reactions mediated by glyceraldehyde 3-phosphate dehydrogenase in which substrate level phosphorylation occurs in glycolysis. The experimental failure to isolate high-energy intermediates confounds reasonable acceptance of this hypothesis.

ATP formation by conformational coupling. Conformational coupling principles of ATP

formation assume that energy released during electron transport is directed toward the conformational modification of a protein. This modified protein is sometimes referred to as a *coupling factor* (CF). The energy conserved from electron transport enables the CF to release its energy in a way that promotes the phosphorylation of ADP to ATP. Unlike the previous theory, which relies on the formation of a high-energy covalently bonded intermediate to promote phosphorylation, this concept suggests that the critical energy for phosphorylation comes from weak noncovalent interactions that alter some molecular conformation. Energy unleashed during the reversion of the unstable molecular conformation is then directed toward ATP formation. This theory encounters the same pitfalls as the previous concept.

ATP formation by chemiosmotic coupling. The formation of ATP by chemiosmotic methods requires that the driving force for ADP phosphorylation to yield ATP emanates from the translocation of protons across the mitochondrial membranes. Since protons are ejected to the outside of the organelle during electron transport, a *proton concentration gradient is generated* in which the *outside of the organelle has a more positive electrical potential than the inside.* Aside from the presence of a proton concentration gradient, this effect also generates an electrochemical gradient. As a result of distinct differences in the electrochemical potential over the mitochondrial membrane, relatively high proton concentrations on the outside of the organelle membrane have a tendency to move back through the membrane and into the mitochondrial matrix. Of course, the influx of these protons can occur only at specific topological sites on the membrane that are associated with ATP production, and it is assumed that the free energy established by virtue of the proton gradient may be funneled into ATP formation. The pH gradient and the influx behavior of protons at these specific sites can be described in terms of the *proton motive force*.

$$\text{Proton motive force} = \Delta\Psi - \frac{RT}{F}\Delta\,\text{pH}$$

where

$\Delta\Psi$ = potential difference over membrane

ΔpH = pH gradient over membrane
Both terms expressed in volts and

(RT/F) = 0.0592 at 25°C

The influx of protons through the mitochondrial membrane is believed to occur by way of specialized protein complex that is embedded in the mitochondrial mem-

the membrane, *the overall direction of the phosphorylation event is determined by free energy balance*. The reaction specified above is expressed with a broken arrow ($—/\,/\rightarrow$) because some intervening steps occur in the mechanism, and they are not clearly established.

Three phosphorylation mechanisms for the formation of ATP have been outlined in the literature; these have varying degrees of compatability with the chemiosmotic theory (Figures 11.10A, B, and C).

Mitchell, for example, has proposed that a phosphorylation event occurs with the *influx* of two protons through the mitochondrial membrane. This activity culminates in the formation of a highly unstable and reactive phosphate species (Figure 11.10A).

Protons that traversed the mitochondrial membrane from outside to inside

Reactive inorganic phosphorous

brane. The protein complex contains an F_0 component that is intrinsically contained in the membrane itself, while another component recognized as F_1 is attached as a knob toward the inside of the mitochondrial membrane. The F_1 portion seems to have five distinct subunits, although some subunits are multiple copies of the others.

The vicinity of the F_1–F_0 protein complex is associated with the reversible enzymatic activity of the enzyme known as ATPase. This enzyme is responsible for the reaction

$$\text{ADP}^{3-} + P_i \xrightarrow[\text{ATPase}]{\quad—/\,/\quad} \text{ATP}^{4-} + H_2O$$

Based on these principles, it is recognized that every two protons that pass into the mitochondrial matrix by way of the F_1–F_0 complex will form *one ATP molecule* from ADP and *inorganic phosphate*. Although the F_1–F_0 complex may control proton transit across

The reactive phosphorous species, under the auspices of the ATPase enzyme, yields ATP from ADP.

An alternative reaction (Figure 11.10B) has been proposed that requires the ionization of water generated during the reaction.

$$\text{ADP} + P_i \longrightarrow \text{ATP} + H_2O$$
$$\Updownarrow$$
$$H^+ + OH^-$$

**ionization inside mitochondrion where
H^+ is naturally low**

According to this reaction, ATP formation is coupled with the controlled entry of protons into the mitochondrion. Since the proton concentration within the organelle is low, the influx of protons from outside is met by available concentrations of hydroxide anions inside the mitochondrion, to form *more* water molecules.

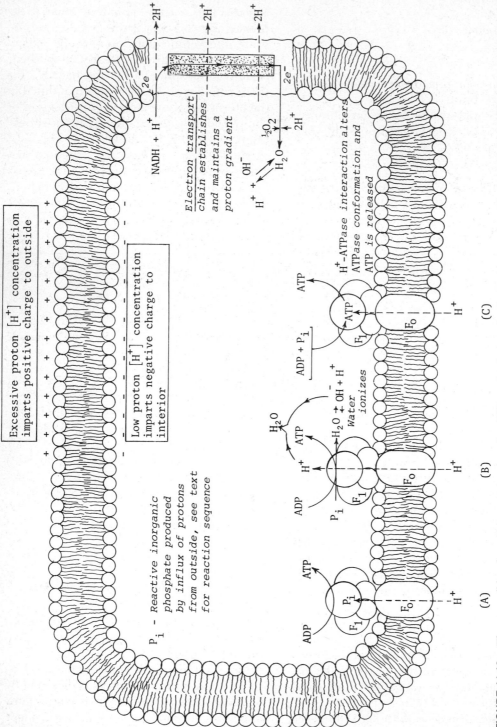

FIGURE 11.10. Three possible schematic representations for ATP production mechanisms where $P_i + ADP \longrightarrow$ ATP in the presence of an F_1–F_0 protein complex and ATPase. All three methods of phosphorylation rely ultimately on the chemiosmotic principles set forth by Mitchell. Consult text for details. Proton influx through the mitochondrial membrane prompts ATP formation by: (A) formation of a reactive inorganic phosphate species; (B) water formation, which consumes protons passing over the membrane with the concomitant formation of ATP; and (C) conformational stress applied to ATPase, caused by proton movement through F_1–F_0 proteins, encouraging formation and release of ATP.

Still another ADP phosphorylation concept holds that the proton influx through F_1–F_0 proteins creates a conformational stress on the ATPase enzyme. This unstable conformation somehow prompts the release of ATP from the ATPase active site (Figure 11.10C).

NADH produces three ATP molecules, FADH$_2$ produces two ATP molecules. Apart from the actual mechanism where ADP is phosphorylated to produce ATP, it is clear that mitochondrial oxidations of NADH and FADH$_2$ are linked to the simultaneous formation of ATP. Idealized *in vitro* studies of mitochondria using scrupulously controlled amounts of ATP, ADP, inorganic phosphate, NAD$^+$, FAD, oxygen, and highly purified reduced substrates have permitted further insight into phosphorylation events. Such studies have demonstrated that the amount of inorganic phosphate consumed in the oxidation of certain substrates parallels the oxygen consumption of the respiratory process. Accordingly, *phosphate/oxygen (P/O) ratios can be established for specific oxidizable substrates that enter the respiratory sequence.* Depending on the interpretation of these ratios, the values generally reflect (1) the number of moles of ATP produced for every gram atom of oxygen consumed during phosphorylation or (2) the number of moles of ATP produced per mole of substrate consumed in phosphorylation events. A substrate such as pyruvate, isocitrate, or malate, when oxidized by a specific NAD$^+$-dehydrogenase, will have a P/O ratio of $\simeq 3.0$, whereas reduced FADH$_2$, linked to succinic acid dehydrogenase, will normally indicate a P/O ratio of $\simeq 2.0$.

The intrinsic interdependence of ATP formation and electron transport processes is also exhibited by the actions of specific respiratory inhibitors (e.g., barbiturates, antimycin A, cyanide anions, azides, and carbon monoxide) as well as ionophores (e.g., valinomycin, gramicidin, 2,4-dinitrophenol), which concomitantly upset ATP formation at prescribed locations by blocking or uncoupling electron transport mechanisms.

Unimpaired electron transport from NADH to oxygen, which yields water, spans a standard reduction potential of $+1.136$ V and the process ideally yields a $\Delta G^{0'}$ value of -5.239×10^4 cal as previously calculated. This decline in free energy is captured at strategically located sites along the electron transport chain where large differences in the standard reduction potentials of consecutive electron carriers produce a free energy change in excess of the energy required to phosphorylate ADP. Inspection of the graph illustrated in Figure 11.11 indicates that electrons entering the respiratory chain sequence at NADH can yield three molecules of ATP, while electrons that enter the scheme as FADH$_2$ via flavin-linked oxidations produce only two molecules of ATP. Based on these observations, then, it is clear that the quantity of ATP molecules produced during electron transport depends entirely on the standard reduction potential of the electrochemical species that introduces electrons into the transport chain.

Moreover, the highly specific blocking actions of respiratory inhibitors permit ATP formation from reduced electron carriers up to the point where their blocking action is enforced. Some specific blocking actions for inhibitors of electron transport along with the limitations in ATP production have been outlined in Figure 11.12.

Total ATP yield from glucose oxidation. Energy in the form of ATP is the ultimate result of three integrated oxidative biochemical schemes: (1) glycolysis, (2) the tricarboxylic acid cycle, and (3) oxidative phosphorylation. *Glycolysis* is the anaerobic complement of ATP production since glucose and sugars, convertible into glucose, are oxidized into three-carbon fragments culminating in pyruvic acid. The aerobic complement of the scheme is the *tricarboxylic acid cycle,* in which pyruvate is decarboxylated and converted to acetyl-CoA. This intermediate combines with

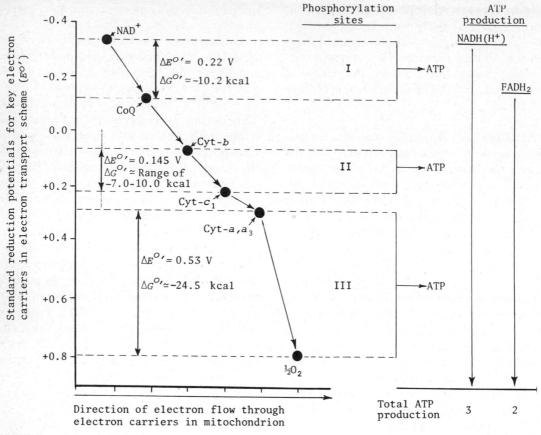

FIGURE 11.11. Relative locations of ATP formation with respect to changes in the standard reduction potentials throughout the electron transport chain. The changes in standard reduction potentials are indicated ($\Delta E^{0'}$) along with approximate changes in free energy at three phosphorylation sites (I, II, and III). Note that absolute uncertainties concerning *in vivo* potentials between cytochrome-b and -c_1 may give rise to a possible range of $\Delta G^{0'}$ values, ranging from -7.0 to -10.0 kcal, which is sufficient to effect the enzymatic phosphorylation of ADP to form ATP. Based on the entry of electrons at the potential of NADH(H$^+$) or FADH$_2$, three or two ATPs may, respectively, be produced by oxidative phosphorylation processes.

oxaloacetate to yield citrate, which is oxidized ultimately to carbon dioxide and water. Both the glycolytic and the tricarboxylic acid pathways also produce ATPs, either by substrate level phosphorylation (glycolysis only) or by means of generating reducing power as energy-rich pyridine and flavin nucleotides (i.e., NADH(H$^+$) or FADH$_2$). The detailed steps that give rise to these compounds are itemized in following sections dealing with glycolysis and the tricarboxylic acid cycle. The energy-rich nucleotides then enter the third step of the biochemical oxidation scheme,

namely, *oxidative phosphorylation,* where water and energy are formed.

With reference to following sections dealing with integrated metabolism, a prospectus on ATP production from substrate level phosphorylations plus oxidative phosphorylation for a single glucose molecule have been itemized in the energy balance sheet shown as Table 11.5. Note that ATP used for priming glucose oxidations is indicated with a negative ($-$) *loss sign* for specific reactions while positive ($+$) signs are used to denote *ATP production* from a given reac-

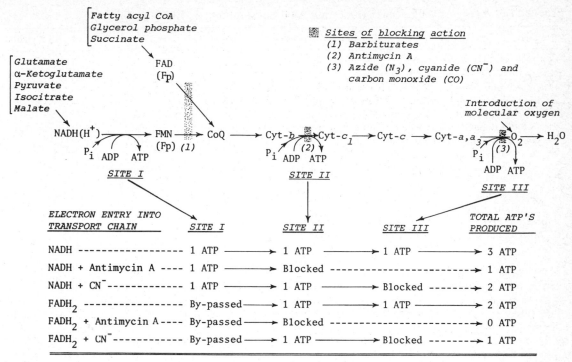

FIGURE 11.12. ATP formation originating with the introduction of electrons from NADH and FADH₂ into the respiratory transport chain, and effects of various respiratory chain blocking agents.

tion. The following section dealing with NADH(H⁺) shuttles *into* and *out of* the mitochondrion should also be consulted for a full perspective of total ATP yields from a single glucose molecule.

Enzymatic Shuttles for NADH

Until this point, the exposition of electron transport and oxidative phosphorylation processes have centered about mitochondrial activities. That is, NADH(H⁺) produced in the mitochondrion is oxidized to yield substantial stores of ATP. Contrary to this important activity, however, significant stores of NADH(H⁺) are formed in the cytosol of cells. These stores of reduced pyridine nucleotides do not have free access to the mitochondrial electron transport system. Instead the electrons carried by NADH are admitted to the respiratory chain by specialized mechanisms depending upon the type of cell found in tissues.

Glycerol–phosphate shuttle: The production of extramitochondrial NADH(H⁺) during the oxidation of glyceraldehyde 3-phosphate, which is a glucose metabolism intermediate, contributes electrons to mitochondrial ATP production by this route. Since the NADH(H⁺) cannot penetrate the mitochondrial membrane, its electrons are received by dihydroxyacetone phosphate, a jointly occurring metabolic intermediate of glucose. This reduction of dihydroxyacetone phosphate is mediated by a cytosolic *glycerol 3-phosphate dehydrogenase*. The product of the reaction is glycerol 3-phosphate, which is permeable to the mitochondrial membranes. Upon entrance into the mitochondrion, glycerol 3-phosphate is oxidized by the action of a flavin-linked glycerol 3-phosphate dehydrogenase. The net results of the oxidation are dihydroxyacetone phosphate and an FADH₂-linked enzyme. The dihydroxyacetone may exit the mitochondrion for another round of reduction while

Table 11.5. Energy Balance Sheet Indicating ATP Consumption and Production

Biochemical reactions	ATP yields
Cytosol (glycolytic pathway)	
Glucose + ATP $\longrightarrow$ glucose 6-phosphate	-1
Fructose 6-phosphate + ATP $\longrightarrow$ fructose 1,6-diphosphate	-1
2 Glyceraldehyde 3-phosphate $\longrightarrow$ 2 1,3-diphosphoglycerate + 2NADH + 2H$^+$	$+4$ or $+6$
Note: Since this pathway occurs in the cytoplasm the number of ATP molecules formed would depend on the type of shuttle system carrying the electron of NADH from cytoplasm into mitochondrion.	
For the glycerol–phosphate shuttle, there will be 2 ATPs formed for each NADH.	
For the malate–aspartate shuttle, there will be 3 ATPs formed for each NADH (e.g., liver and kidney cells)	
2 1,3-Diphosphoglycerate $\longrightarrow$ 2 3-phosphoglycerate + 2 ATP	$+2$
2 Phosphoenol pyruvate $\longrightarrow$ 2 pyruvate + 2 ATP	$+2$
	$+6$ or $+8$
Mitochondrion (tricarboxylic acid cycle)	
(From one molecule of glucose, 2 molecules of pyruvate are produced.)	
2 Pyruvate $\longrightarrow$ 2 acetyl-CoA + 2 NADH + 2H$^+$ + 2 CO$_2$	$+6$
2 Isocitrate $\longrightarrow$ 2 α-ketoglutarate + 2 CO$_2$ + 2 NADH + 2H$^+$	$+6$
2 α-Ketoglutarate $\longrightarrow$ 2 succinyl-CoA + 2CO$_2$ + 2NADH + 2H$^+$	$+6$
2 Succinyl-CoA $\longrightarrow$ 2 succinate + 2 GTP	$+2$
2 Succinate $\longrightarrow$ 2 fumerate + 2 FADH$_2$	$+4$
2 Malate $\longrightarrow$ 2 oxalacetate + 2 NADH + 2H$^+$	$+6$
	$+30$
	Grand total 36 or 38
Note: For each NADH, 3 ATPs are formed.	
For each FADH$_2$, 2 ATPs are formed.	
For each GTP, 1 ATP is formed.	
For each molecule of glucose 36 or 38 ATPs are formed in both cytoplasmic and mitochondrial reactions.	

the FADH$_2$-linked enzyme is oxidized at the level of coenzyme Q in the respiratory chain. The subsequent oxidative phosphorylation produces only two ATPs as opposed to the three ATPs normally expected from intra-mitochondrial stores of NADH(H$^+$). The main pathway of this so-called *glycerol–phosphate shuttle* is shown in Figure 11.13.

Malate–aspartate shuttle: Contrary to the glycerol–phosphate shuttle, mammalian heart and liver tissues display another type of shuttle mechanism for the conveyance of cytosolic reducing power into the mitochondrion. According to this mechanism, NADH(H$^+$) formed during glycolysis of

sugars in the cytosol is introduced to the mitochondrion by way of an oxaloacetate reduction to produce L-malate (Figure 11.14). The L-malate traverses the mitochondrial membranes where it is oxidized to form oxaloacetate. This oxidation is promoted by mitochondrial *malate dehydrogenase,* and the resulting NADH(H$^+$) is directed through the respiratory chain to yield three ATPs and *not* two ATPs as in the glycerol–phosphate shuttle. The mitochondrial oxaloacetate is converted to L-aspartate by way of a trans-amination reaction, which involves the α-amino group of L-glutamate. The L-aspartate exits the mitochondrion and enters the cytoplasm where it is deaminated through a

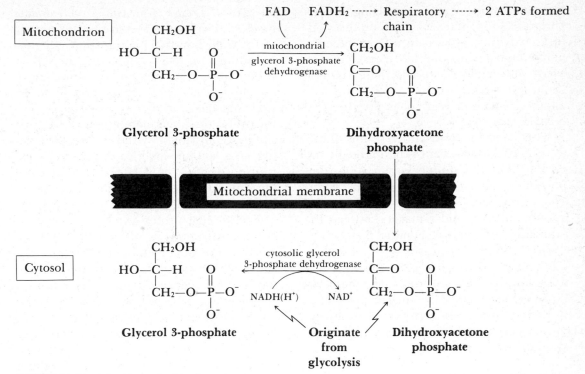

FIGURE 11.13.   The glycerol–phosphate shuttle.

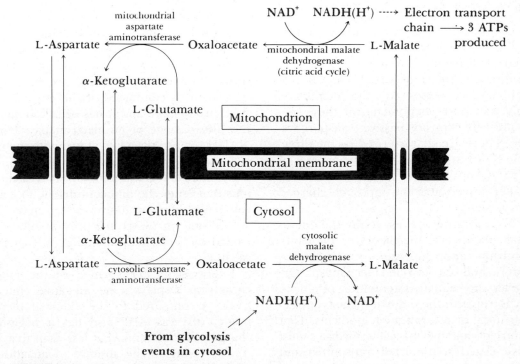

FIGURE 11.14.   The malate–aspartate shuttle system.

transamination reaction involving α-keto-glutarate, to form oxaloacetate. In this shuffling activity, L-glutamate may reenter the mitochondrion for another transamination event with oxaloacetate and the extramitochondrial oxaloacetate may undergo another reduction by glycolytically produced $NADH(H^+)$.

The energetic contribution of this shuttle to the respiratory chain is undoubtedly important, but the shuttle system is also important because it offers a reversible transport mechanism for L-aspartate, L-glutamate, α-ketoglutarate, and L-malate between the cytoplasm and the mitochondrial matrix. The shuttle system also proves to be of importance in gluconeogenic processes in which mitochondrial malate dehydrogenase reduces oxaloacetate. The L-malate product of this reaction is then translocated to the cytosol, where its oxidation is mediated by malate dehydrogenase.

Other Transport Mechanisms over Mitochondrial Membranes

The significant differences in concentration gradients and electrical potentials established over the mitochondrial membrane are also linked to the transport of many electrically charged species over the membrane. Many transport phenomena for these species are reflections of passive transport systems. These are characterized by the directional flow of charged species from areas of high concentration to a lower concentration or, alternatively, to some oppositely charged environment.

Living systems and especially the operational biochemical mechanics of the mitochondrion cannot operate by such passive, random, and uncontrolled mechanisms. Organelles and cellular systems must be able to function against the singular dictates of electrochemical or concentration gradients. That is, nutrients and metabolic intermediates must be assimilated, accumulated, transported, and discharged from organelles or cells *against*

the physical dictates of prevailing concentration gradients, if life processes are to succeed.

For mitochondrial ion transport, the flow of sodium ions (Na^+) is linked to the flux of protons over the membrane. Although an exchange protein probably participates in ion exchange mechanisms of this sort, each proton that crosses the membrane at a specific site is accompanied by the counterpositioning of a sodium ion (Na^+). The countertransport of Na^+ and H^+ reaches an equilibrium state when the proton and sodium ion gradient concentrations are equivalent.

Exclusive of proton gradient concentrations over the mitochondrial membrane, the maintenance of a membrane potential can also guide the direction of ion transport. Since the internal reaches of the mitochondrion are negatively charged with respect to the outside, divalent calcium cations (Ca^{2+}) are naturally attracted toward the inner surface of the mitochondrial membrane. Unlike the previous interaction of protons and sodium ions, calcium ion transport does not require proton displacement from the mitochondrion. Instead, it is theorized that the accumulation of calcium ions within the mitochondrial membrane must coincide with the joint influx of phosphate and an efflux of hydroxide anions from the mitochondrion. This countergradient accumulation of calcium is conducted at the expense of ATP production, and it has been estimated that the passage of an electron pair from $NADH (H^+)$ to oxygen will promote the accumulation of six calcium ions. Correspondingly, one ATP is lost for every two calcium ions that enter the mitochondrion. The accumulation of manganese (Mn^{2+}), iron (Fe^{2+}), and strontium (Sr^{2+}) are presumably controlled by a mechanism similar to that for calcium.

Energy-rich ATP formed in the mitochondrion displays a net negative charge (ATP^{4-}) compared to the external membrane, whereas ADP also has a negative charge (ADP^{3-}) somewhat less than that of ATP^{4-} owing to the absence of an added phosphate. The directed displacement of energy-rich ATP^{4-} through the mitochondrial membrane is probably met with a balanced movement of ADP^{3-} into the organelle, thereby ensuring adequate availability of substrate for phosphorylation events.

The energy-rich properties of $NADH(H^+)$ are critical to the operation of the respiratory chain, but the reducing power present in this compound is also important for reactions in the cytosol. Unfortunately, the large stores of reducing power held in $NADH(H^+)$ are *not able to migrate freely from the mitochondrion into the cytosol,* unless one of two events occurs. *First,* the reducing power of mitochondrial $NADH(H^+)$ may be conveyed to the cytosol by way of the malate–aspartate shuttle as presented earlier; or, *second,* the $NADH(H^+)$ reducing power inside the mitochondrion can be funneled to $NADP^+$. The $NADP^+$ species is suitable as an acceptor for "reducing power" only because it is linked to a *cytosolic isocitrate-dehydrogenase.* Based on this concept, mitochondrial isocitrate passes through the mitochondrial membrane under the auspices of a tricarboxylate carrier, and the isocitrate dehydrogenase mediates the reduction of $NADP^+$. The $NADPH(H^+)$ formed by this route maintains uniform cellular ratios of oxidized and reduced pyridine nucleotides ($NADH/NAD^+$ and $NADPH(H^+)/NADP^+$) at levels commensurate with existing energy states of the cell. Figure 11.15 illustrates some of the important transport schemes outlined here.

Aside from these specific transport mechanisms, the diverse biochemical activities of the mitochondrion require additional transport mechanisms for biochemical intermediates that must traverse the mitochondrial membrane. Many of these transport systems are cellularly specific and genetically determined since very specialized proteins and protein complexes are required. Examples of specific transport proteins exist throughout the mitochondrial membranes such as the tricarboxylate carrier mentioned above. This carrier mediates, not only mutual equimolar exchanges of isocitrate and citrate through the mitochondrial membrane, but

also equivalent molar ratios of tricarboxylates and dicarboxylates such as L-malate. The dicarboxylate counterpart of the tricarboxylate carriers behave in a similar fashion, but they afford mutual equimolar exchanges of malate, fumarate, succinate, or phosphate. Other types of transport systems do exist, and the enzyme systems that participate are generically referred to as *porters* or *translocases.*

Electron Transport without ADP Phosphorylation

Many extramitochondrial substrates can undergo hydroxylation reactions that necessarily require electron transport; however, these electron transport mechanisms do not entail any phosphorylation events connected to ATP formation. Electron transport mechanisms that abide by these requirements have been isolated from the microsome fraction of the hepatic endoplasmic reticulum. The biochemical participants involved here include a cytochrome called *P*-450 (Cyt-*P*-450); an NADPH-dependent flavin (FAD) containing cytochrome *P*-450 reductase; and a substrate destined to be hydroxylated. Some systems may also employ iron–sulfur proteins as electron transfer participants. The cytochrome *P*-450 represents a so-called "*b*-family" cytochrome because the reduced form of its carbon monoxide complex displays an absorption maximum at 450 nm. According to current theory, the cytochrome acts as a typical electron carrier in the process outlined in Figure 11.16. In this scheme, a substrate binds to ferricytochrome (Fe^{3+})-*P*-450. The reductase then catalyzes the transfer of reducing equivalents from $NADPH(H^+)$ by way of the FAD moiety to the cytochrome. Naturally, these steps are encouraged by the bonded substrate. Molecular oxygen is introduced into the complex and is reduced to yield a cytochrome–substrate complex containing a superoxide ($\cdot O_2^-$). The transient reactive nature of these intermediates results in the reduction of one oxygen atom

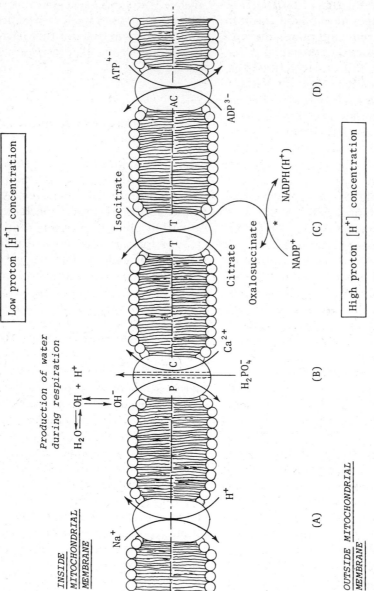

FIGURE 11.15. Some important transport mechanisms found in the mitochondrial membrane responsible for (A) sodium ion transport; (B) calcium and phosphate transport where "P" is a phosphate carrier and "C" represents a calcium carrier; (C) NADPH(H⁺) formation from intramitochondrial NADH(H⁺), where "T" corresponds to a tricarboxylate carrier; (D) ADP³⁻–ATP⁴⁻ transport is mediated by a carrier denoted as "AC."

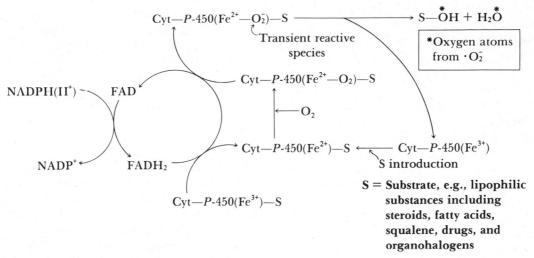

Hydroxylated substrate may be a more suitable biochemical intermediate for following reactions or less toxic and more water soluble

*Oxygen atoms from $\cdot O_2^-$

Transient reactive species

O_2

S introduction

S = Substrate, e.g., lipophilic substances including steroids, fatty acids, squalene, drugs, and organohalogens

FIGURE 11.16. Typical biochemical pathway for a nonphosphorylating electron transport pathway designed for hydroxylation of selected substrates.

to form water while the other oxygen atom hydroxylates the substrate.

Biochemical pathways such as the one illustrated are responsible for hydroxylations of lipophilic substances, for example, steroids, squalene derivatives, fatty acids; drugs, for example, morphines, codeines, amphetamines, and barbital derivatives; organohalogen derivatives (pesticides); and carcinogenic hydrocarbons. Hydroxylation is an effective biochemical mechanism for producing water-soluble derivatives of these substances and thereby acts as a preparatory route for their excretion by the kidneys. Moreover, it should be recognized that the flavin-linked NADPH-cytochrome reductase as well as cytochrome P-450 concentration levels will parallel increased demands for detoxification or excretory demands caused by the presence of these substances.

Other nonphosphorylating electron transport systems are recognized, especially for fatty acid desaturation outside the mitochondrion, where other cytochrome systems and reductases depend on cytochrome-b_5.

Oxygen Reactivity during Electron Transport

Electron transport schemes for *nonphosphorylation* events, such as hydroxylation or oxidative phosphorylation, where ATP is generated, are intrinsically dependent on the chemistry of oxygen. The schemes detailed above show neat, orderly inclusions of the oxygen atoms into hydroxides or a water molecule depending on the pathway. Evidence suggests, however, that these reactions are not always as orderly and direct as theorized, and many reactions involving oxygen may encourage the formation of *superoxide anions*. The superoxide anion is produced when molecular oxygen (diatomic oxygen, O_2) acquires an extra electron from some nonspecific source. These superoxide anions are probably produced in living cells

$$:\overset{..}{\underset{..}{O}}:\overset{..}{\underset{..}{O}}: \xrightarrow{1e^-} :\overset{..}{\underset{..}{O}}:\overset{..}{\underset{.}{O}}:$$

Molecular oxygen   **Superoxide**
(O_2)   **anion radical**
                       ($\cdot O_2^-$)

as temporal intermediates on the active sites of enzymes that are responsible for the reduction of oxygen to water or the inclusion of oxygen into hydroxyl groups. If the superoxide intermediate is not properly directed toward product formation, it holds a chemical potential for instituting a wide range of toxic intermediates including hydroxy free radicals ($\cdot$OH) and hydrogen peroxide (H_2O_2). Possible production of these species is indicated in Figure 11.17.

The hydroxy free radical is among the strongest oxidizing agents known, not to mention the intermediate formation of other potent oxidizing agents such as the superoxide anion or hydrogen peroxide. All of these species, by virtue of their high reactivities, offer a random threat to the biochemical operation of the living cell. The detrimental activities for many of these species are believed to be controlled, however, by antioxidants such as ascorbic acid (vitamin C), glutathione, and tocopherols (vitamin E) because they act as preferential electron donors. The superoxide anions, on the other hand, are progressively detoxified by a two-step enzymatic sequence that involves two distinct enzymes: *superoxide dismutase* (SOD) and catalase or peroxidase. It was established in 1969 by I. Fridovich and J. McCord that superoxide dismutase mediated this reaction:

$$2 \cdot O_2^- + 2H^+ \xrightarrow{\text{superoxide dismutase}} H_2O_2 + O_2$$

The hydrogen peroxide is certainly less toxic than the superoxide anion, but it, too, must be eliminated. This disposal of the hydrogen peroxide is conducted by either peroxidase or catalase enzymes:

(a) $H_2O_2 + SH_2 \xrightarrow{\text{peroxidase}} 2 H_2O + S$

                Reduced   Oxidized
                organic   organic
                substrate   substrate

or

(b) $2 H_2O_2 \xrightarrow{\text{catalase}} 2 H_2O + O_2$

Both intra- and extramitochondrial versions of superoxide dismutase exist. They are functionally similar but structurally different, and the cytosolic forms in particular require Zn^{2+} and Cu^{2+} cofactors. Aside from the recognized reactions detailed above, superoxide dismutase is believed to have clinical and therapeutic importance. That is, the presence of the enzyme at low levels of activity or its absence have been ascribed to general tissue inflammation, especially inflammations associated with the granulocytic production of superoxide anions during phagocyte development, as well as seriously inflamed arthritic conditions.

Dietary Energy: Methods of Measurement and Human Requirements

In their most fundamental terms, studies of animal and human nutrition attempt to describe and mathematically account for the

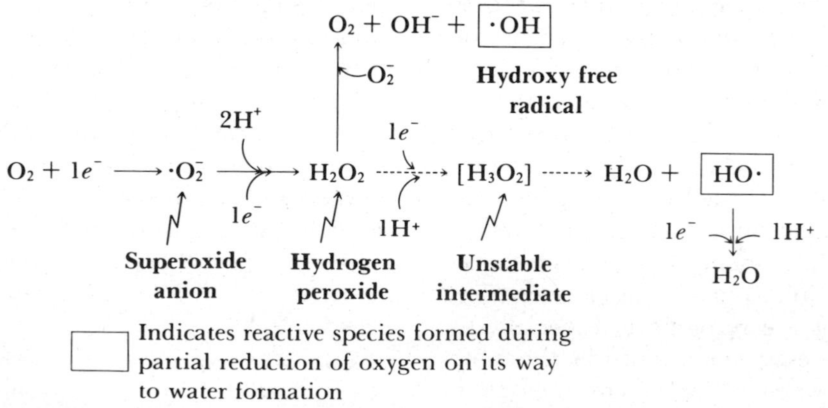

FIGURE 11.17. Superoxide anion reactions and formation of hydroxy free radicals.

requisite energy demands of the body. The energy flow through any animal body is governed by the basic laws of thermodynamics discussed earlier. The overall energy flow can be subdivided into the components illustrated in Figure 11.18. Since the first law of thermodynamics dictates the conservation of energy, it is clear that the potential energy (*PE*) contents of ingested dietary nutrients is equal to the sum of individual energy components (Consult Figure 11.18 for details.)

$$PE = HE + RE + WE + GE + FE$$

Because energy transitions within animals and humans are based on thermodynamic principles, dietary energy transformations can be studied on the basis of *heat transformations*. Energy studies of this type that pertain to the human diet in particular are expressed as *kilocalories* (kcal) or "large calories" (Calories (Cal)). In more recent years the unit of energy called the *joule* (J) has rivaled the calorie unit and, therefore, kilocalories (kcal or Cal) have been replaced by the *kilojoule* (kJ). The conversion factor for transforming a kcal (Cal) unit to a kilojoule (kJ) is 4.184; that is, 1 kcal (1 Cal) = 4.184 kJ. In spite of the transition toward the sole use of kilojoule terminology, both the calorie and the joule units are still used.

Energy content of food: The bomb calorimeter: The caloric value of all foodstuffs can be measured by using a bomb calorimeter (Figure 11.19). The device consists of a reaction or combustion chamber that is surrounded by an insulated water jacket. A specific food sample is placed in the combustion chamber along with gaseous oxygen (~30 psi) and the system is ignited with an electrical spark. Heat released during combustion of the food is transferred to the water and the water temperature rises. Since the quantity of water contained in the insulated water jacket is known along with its specific heat and change in temperature, the heat of food combustion, in terms of calories, can be estimated from

heat produced from food oxidation by combustion = heat gained by water plus calorimeter

Carefully conducted experiments show that the amount of heat produced in a bomb calorimeter for one gram of carbohydrate is 4.15 kcal, while 1 g of fat produces 9.45 kcal. Since the molecular constitution of different fats and carbohydrates may be somewhat variable, both of the values cited may be subject to variation. Similar studies conducted on proteins show that they liberate 5.65 kcal upon combustion. These ideal caloric yields from foods based on calorimetry are not realized in the human body. Caloric disparities between theoretical and actual metabolic energy yields stem from incomplete food digestion or absorption from the intestinal lumen. It is estimated, however, that carbohydrate absorption may be as high as 97%, fats may demonstrate up to 95% absorption, and 92% of proteins may be absorbed.

The calories produced by carbohydrates and fats in the bomb calorimeter are nearly the same as the amount produced in the body since they are completely converted to carbon dioxide and water, but this is not true for proteins. The oxidation of proteins in a bomb calorimeter is different from their actual roles in cellular metabolism since a certain amount of protein is excreted from the body as urea, uric acid, creatinine, and other compounds. More realistic and practical estimates of heat energy produced from carbohydrates, fats, and proteins are shown in Figure 11.20. Some typical caloric values for different foods based on calorimetry are outlined in Table 11.6.

Body heat measurement: Antoine Lavoisier proposed in the 1780s that heat generated by an animal is essentially the same in origin as heat derived from any combustion process. This hypothesis was tested by placing a guinea pig in one container surrounded by ice while another container holding a burning candle was similarly immersed in ice. The heat generated from the animal and the burning candle were respectively measured in terms of the quantity of melted ice. In addition, Lavoisier determined the amount of carbon dioxide produced in each chamber and calculated the

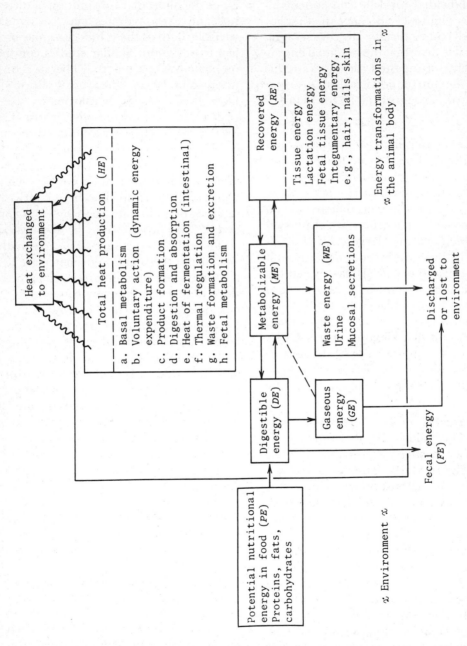

FIGURE 11.18. Energy flow throughout an animal. The potential energy of consumed foods can be translated into the sum of the individual unit processes that require energy plus fecal energy.

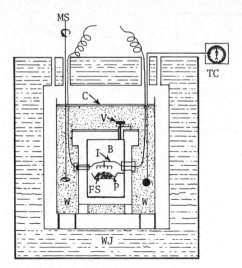

FIGURE 11.19. Cut-away view of a bomb calorimeter used for determining the caloric values of foods and other biochemical compounds. Notable features include: B, bomb chamber; C, pressure sensitive water cover; I, ignition wire; FS, food sample; MS, mechanical stirrer; P, platinum specimen dish; TC, thermocouple; V, bomb chamber access valve; W, weighed amount of water; WJ, water jacket insulator. (Adapted from Uber's *Biophysical Research Methods*, Wiley–Interscience, New York, 1950.)

ratio of the heat quantity generated in each system with respect to the amount of carbon oxidized to carbon dioxide. Based on these computations, it was concluded that the same oxidations probably transpired in both chambers albeit by distinctly different routes. As a corollary, it was also determined that the same total quantity of heat could be liberated from the combustion of a given amount of any food regardless of whether combustion occurred *in vivo* or *in vitro*. Following these animal calorimetry studies that established the quantitative role of oxygen in processes of nutrient oxidation, additional methods of direct and indirect calorimetry were perfected and more definitive nutritional laws were established.

Direct calorimetry. Aside from the contributions of Lavoisier and some others, no early investigations into direct animal calorimetry and respiratory exchanges in animals are more important than those of Rubner (1902).

Although small numbers of experimental animals by today's standards provided Rubner with the bulk of this data, direct animal calorimetry enabled him to establish some major conclusions concerning the laws of energy consumption in nutrition. It was clear from his work that proteins, fats, or carbohydrates could mutually replace each other in the heat production of the body according to the heat-producing value of each foodstuff. This is still recognized in many quarters as the *isodynamic law*. Furthermore, he is largely responsible for the following contributions:

1. The exact *caloric equivalents* per gram for proteins, fats, and carbohydrates.

2. The *surface law of cooling*, which states that the heat value of the resting metabolism is proportional to the surface area of the body.

3. The demonstration that the *first law of thermodynamics applies to the animal body.*

4. The *specific dynamic effect of foods* (discussed in later sections), which accounts for observed increases in resting heat production by the body in response to the ingestion of specific nutrients.

5. The discovery and description of *chemical (nonshivering) thermogenesis.*

6. The first definitive research on the *evaporative and convective-conductive cooling* of the animal body.

7. The *first quantitative descriptions of climatic stresses* on animals and men in cold environments.

Rubner's most fundamental claim dealing with the law of the conservation of energy, although postulated by Mayer in 1845 and independently by Helmholtz in 1847, is probably his most notable achievement. This struck the death knell for *vitalism* concepts and discounted independent sets of operational laws for animate and inanimate systems. Moreover, these findings acted as an important benchmark because they accounted for the existence of the animal body

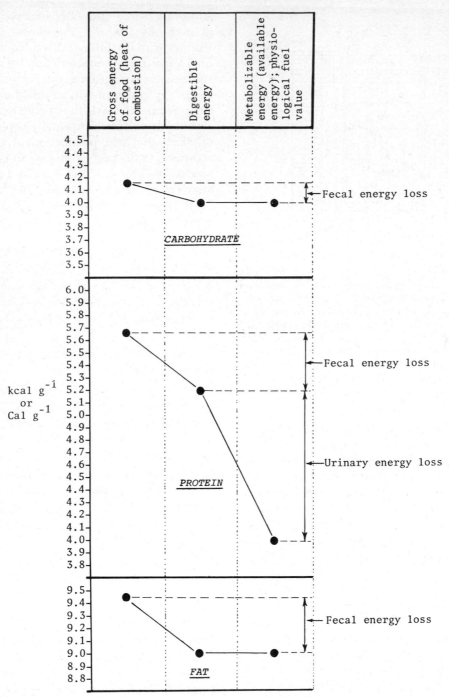

FIGURE 11.20. Relative energy values for carbohydrates, proteins, and fats indicating gross energy values as well as fecal and urinary losses of energy.

Table 11.6. Food Energy of Some Foods (Edible Portion) in Kilocalories per 100 g

Cereals		Fruits	
Corn flakes, Kellogg's	389	Apples, raw with skin	59
Corn flakes, Ralston Purina	390	(*Malus sylvestris*)	
Cream of Wheat, regular, dry	370	Avocado, raw, Florida	467
Crispy rice	396	(*Persea americana*)	
Farina, dry	369	Bananas, raw (*Musa X*	92
Raisin bran, Ralston Purina	318	*paradisiaca*)	
Rice, puffed	402	Blueberries, raw (*Vaccinium*	56
Wheat, shredded, small		spp.)	
biscuit	359	Cherries, sour, red, raw	50
		(*Prunus cerasus*)	
Dairy products and eggs		Cranberries, raw (*Vuccinium*	49
		macrocarpon)	
Butter, regular (salted)	717	Cranberry sauce, canned,	151
Cheese, bleu	353	sweetened	
Cheese, cheddar	403	Figs, raw	74
Cheese, cottage, lowfat, 2% fat	90	Grapes, American type, raw	63
Cheese, cottage, creamed	103	(*Vitis* spp.)	
Cheese, feta (from sheep's		Melons, cantaloupe, raw	35
milk)	264	(*Cucumis melo*)	
Cheese, mozzarella	281	Oranges, raw, Florida	46
Cheese, Parmesan, grated	456	Peaches, raw (*Prunus persica*)	43
Cheese, Roquefort	369	Pears, raw (*Pyrus communis*)	59
Cheese, Swiss	376	Plums, raw (*Prunus* spp.)	55
Milk, whole, 3.3% fat, fluid	61		
Milk, lowfat, 2% fat, fluid	50		
Milk, skim, fluid	35		
Eggs, raw		**Fruit juices**	
White	50	Apple juice, frozen or canned	50
Yolk	361	Orange juice, fresh	44
Whole	162	Orange juice, canned	44
		Prune juice, canned	71
Fats, oils, and shortenings		Tomato juice, canned	21
Beef, tallow	902		
Corn (*Zea mays*)	884		
Cottonseed (*Gossypium* spp.)	884	**Meat and meat products**	
Lard (*pork*)	902		
Mutton, tallow	902	Bologna, beef	313
Olive (*Olea europea*)	884	Bologna, pork	247
Peanut (*Arachis hypogaea*)	884	Chicken, broilers or fryers,	
Margarine	720	flesh, skin, giblets, raw	213
Mayonnaise	708	Duck, domesticated, flesh	
		and skin, raw	211
Fish		Frankfurter, beef	322
		Ham, sliced, extra lean	131
Clams, raw	81	(~5% fat)	
Cod, raw	74	Italian sausage, raw, pork	346
Flounder, raw	68	Turkey, all classes, flesh,	
Halibut, raw	126	skin, giblets, neck, raw	157
Lobster, raw	88	Goose, domesticated, flesh	
Oysters, raw	84	and skin, raw	371
Salmon, raw	223	Pheasant, flesh and skin, raw	181
Tuna, canned	198		

Table 11.6. (*Continued*)

Sugars and sweets		Brussels sprouts, raw	47
		Cabbage, raw	24
Chocolate with almonds	532	Celery, raw	18
Marshmallows	325	Lettuce, head	15
Honey	294	Onions, mature, raw	45
Caramels	415	Spinach, raw	29
Fudge, plain	411	Broccoli, raw	29
Jams and marmalades	278	Cauliflower, raw	25
Sugars, cane or beet	385	Corn, sweet, raw	92
		Eggplant, raw	24
Vegetables		Peas, green, raw	98
		Pumpkin, raw	31
Beets, red, raw	42	Squash, summer, raw	16
Carrots, raw	42	Squash, winter, raw	38
Potatoes, white, raw	83	Tomatoes, raw	20
Asparagus, raw	21		

SOURCE: *Composition of Foods, Agricultural Handbook,* Vols. 1–9, United States Department of Agriculture, Agricultural Research Service, Washington, D.C.

according to a mechanistic perspective whereby not only were cells, tissues, and organs functionally integrated but these mechanisms of life could be determined by applying methods and concepts rooted in the physical sciences. Other notable contributors to early concepts on nutritional energetics included Rubner's contemporaries and protégés such as Lusk, Pettenkofer, Voit, Büchner, and Atwater.

Atwater in particular, who was a student of Rubner, made considerable strides in assessing the energy requirements of humans. His studies were based on the construction of a human metabolic chamber that had successive layers of air in the walls (Figure 11.21). This effectively prohibited the escape of heat. A single test subject, placed in the chamber for a prescribed time, was fed a specific diet through a porthole. Body wastes were removed from the chamber by a similar route and chemically analyzed. The metabolic chamber was far more sophisticated than that of Lavoisier and his early counterparts. Atwater's chamber provided a circulating supply of water through pipes in order to dissipate heat liberated by the test subject. The chamber was also supplied with measured quantities of oxygen while expired carbon dioxide was quantitatively determined.

Complete "balance" studies were conducted on test subjects enclosed in the chamber; these dealt with the quantitative relationships among foods ingested, excreta eliminated, heat produced, oxygen inhaled, and carbon dioxide liberated. The use of the metabolic chamber permitted a direct measurement of heat transformations in a test subject as well as the relationship of gaseous oxygen consumption to the observed changes in heat. The technique ultimately demonstrated that the total energy expenditure (i.e., heat loss plus external work) was equal to the net energy retrieved from consumed foods (i.e., total chemical energy in food minus energy in urine and feces). Since the total energy expenditure was found to be related to oxygen consumption, oxygen consumption alone could be used as a criterion for assessing energy consumption by a test subject. Heat measurements based on this premise afforded studies based on *indirect calorimetry* that are still used today.

Indirect calorimetry and respiratory quotient. Methods for indirect calorimetry involve the precise measurement of oxygen consumption by a test subject. As discussed above, food is oxidized with a liberation of heat re-

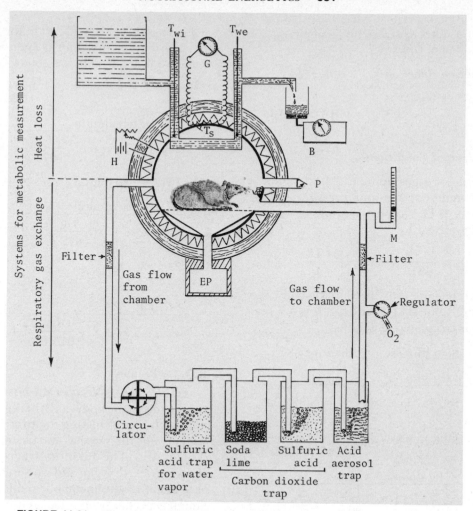

FIGURE 11.21. Illustrative diagram showing the functional structure of a modified At-water-Rosa respiration calorimeter. The test subject—a rodent in this case—is surrounded by two concentric walls. The outer wall is held at the same temperature (T_s) as the inner wall by means of a heat source (H) and a galvanometer control (G). The heat produced by the subject is carried away by water that enters the chamber at a temperature of T_{wi} and exit the chamber at temperature T_{we}. The flow rate is measured by the balance (B). The manometer (M) monitors pressure conditions within the system, the port (P) permits introduction of food, and the other port (EP) offers a removal site for excrement and body wastes (into an insulated trap) necessary for some BMR measurements. Similar chambers may be constructed for humans, but they are certainly far from routine pieces of scientific apparatus. (Adapted from M. Kleiber, *Calorimetric Measurements in Biophysical Research Methods*, F. M. Uber (Ed.), Wiley–Interscience, New York, 1950.)

gardless of whether the event occurs in a calorimeter or within the body. The rate of metabolism in the body is related to its oxygen consumption and carbon dioxide production. A device called a *respirometer* is used to measure these gases while a test subject performs specific tasks. The ratio of the volume of carbon dioxide expired to that of oxygen consumed (CO_2/O_2) is known as the *respiratory quotient (R.Q.)*, which reflects the type of food utilized by the body. The *R.Q.* values will vary from food to food.

R.Q. for Carbohydrates, for Example, Oxidation of Glucose

Oxidation Reaction

$$C_6H_{12}O_6 + 6 O_2 \rightleftharpoons$$
$$6 CO_2 + 6 H_2O + 687 \text{ kcal}$$

Recognized Conditions

A mole of a gas occupies 22.4 L at standard temperature and pressure (STP). Hence, 6 O_2 occupy 134.4 L or 6 CO_2 occupy 134.4 L.

$$R.Q. = \frac{6 CO_2}{6 O_2} \quad \text{or} \quad \frac{134.4 \text{ L}}{134.4 \text{ L}} = 1.0$$

When glucose is oxidized in the body, the caloric value of 1 L of oxygen is 5.1 kcal and is calculated as follows:

I. 180 g glucose   liberates 687

$$\frac{1 \text{ g glucose}}{x = 687 \text{ kcal} \times \frac{1g}{180g} = 3.82 \text{ kcal}}$$

II. 180 g glucose   reacts with 134.4 L O_2

$$\frac{1 \text{ g glucose} \qquad x}{x = 134.4 \text{ L} \times \frac{1 \text{ g}}{180 \text{ g}} = 0.75 \text{ L}}$$

III. 0.75 L O_2 is equivalent to 3.82 kcal

$$\frac{1 \text{ L } O_2 \qquad x}{x = 3.82 \text{ kcal} \times \frac{1 \text{ L } O_2}{0.75 \text{ L } O_2} = 5.1 \text{ kcal}}$$

R.Q. for Fats, for Example, Oxidation of Tripalmitin

Oxidation Reaction

$$2 C_{51}H_{98}O_6 + 145 O_2 \rightleftharpoons$$
$$102 CO_2 + 98 H_2O + 15,314 \text{ kcal}$$
$$1612 \text{ g} \quad 145 \times 22.4 \text{ L} \quad 102 \times 22.4 \text{ L}$$

$$R.Q. = \frac{102 CO_2}{145 O_2} = \frac{2284 \text{ L}}{3428 \text{ L}} = 0.703$$

When tripalmitin is oxidized in the body, the caloric value of 1 L of oxygen is 4.72 kcal, and it is calculated as follows:

I. 1612 g tripalmitin   gives 15,314 kcal

$$\frac{1 \text{ g tripalmitin} \qquad x}{x = 15.314 \text{ kcal} \times \frac{1 \text{ g}}{1612 \text{ g}} = 9.5 \text{ kcal}}$$

II. 1612 g tripalmitin   uses 3248 L O_2

$$\frac{1 \text{ g tripalmitin} \qquad x}{x = 3248 \text{ L} \times \frac{1 \text{ g}}{1612 \text{ g}} = 2.015 \text{ L}}$$

III. 2.015 L O_2 is equivalent to 9.5 kcal

$$\frac{1 \text{ L } O_2 \qquad x}{x = 9.5 \text{ kcal} \times \frac{1 \text{ L}}{2.015 \text{ L}} = 4.72 \text{ kcal}}$$

R.Q. for Proteins

Amino acid molecules do not oxidize completely in the body and, therefore, the approach for calculating the protein $R.Q.$ value is not the same as for fats and carbohydrates. The quantitative relationships between oxygen and carbon dioxide as a result of the metabolic oxidation of amino acids has been calculated from the data presented in Table 11.7.

Considering the equations where

$$C + O_2 \longrightarrow CO_2$$

and

$$2 H_2 + O_2 \longrightarrow 2 H_2O$$

it is apparent that 12 g of carbon will react with 32 g of oxygen to form 44 g of carbon dioxide, and 2 g of hydrogen will react with 16 g of oxygen to form 18 g of water. The tabulated value of 7.69 g of oxygen that is not excreted in urine and feces is directed to the oxidation of 0.96 g of hydrogen (2 × 7.69/16), leaving 3.44 g of hydrogen (4.40 − 0.96). The remaining 41.50 g of carbon and 3.44 g of hydrogen

Table 11.7. Quantitative Composition of Elements in 100 g of Meat Protein (Loewy)

Element	Quantity of elements (g)	Excreted in urine and feces (g)	Not excreted in urine and feces remaining for oxidation (g)	Remaining for oxidation (g)
C	52.38	10.877	41.50	41.50
H	7.27	2.87	4.40	3.44
O	22.68	14.99	7.69	—
N	16.65	16.65	—	—
		(Urine 16.28)		
S	1.02	1.02	—	—

need 138.19 g of oxygen in order to react. This amount of *oxygen* occupies a volume of *96.73 L* since 32 g of oxygen occupies 22.4 L at STP. The amount of *carbon dioxide* produced when 41.50 g of carbon is involved in the reaction is 152.17 g and its STP volume is *77.47 L*. Therefore, the *R.Q.* value can be calculated from

$$R.Q. = \frac{77.47 \text{ L } CO_2}{96.73 \text{ L } O_2} = 0.801$$

The respiratory quotient permits the estimation of the amount of carbohydrates, fats, and proteins consumed by the body. The percentage of carbohydrate, fat, and protein is reflected in the *R.Q.* value of the diet, which in turn reflects normal health or disease.

Normally, an *R.Q.* value of about 0.85 is typical for an individual who consumes a mixed diet. The *R.Q.* values for protein, fat, and carbohydrate are 0.801, 0.71, and 1.0, respectively. In the case of diabetes mellitus, the *R.Q.* is about 0.70, which indicates that the combustible foods are derived from fats. An *R.Q.* value between 0.63 and 0.701, for example, could indicate that amino acids are converted to glucose via gluconeogenesis and the glucose is excreted in the urine of a diabetic patient. Values of 1.0 and higher indicate that excessive amounts of carbohydrates are being oxidized with the exclusion of fats and proteins.

Human energy requirements

Basal metabolism. The rate of cellular metabolism or the basal metabolic rate (BMR), refers to the minimum level of energy requirement in a body for its normal maintenance activities such as respiration, glandular functions, gastrointestinal contractions, muscular tonicity, cardiac activity, liver and kidney functions, as well as the maintenance of normal body temperature. For accurate BMR measurements, the body must not be asleep, but rather in a state of physical and mental rest 12 h after the last meal and several hours after strenuous exercise. The test environment should be quiet and pleasant and offer a comfortable temperature.

Basal metabolic rate depends on a variety of factors such as age, sex, glandular secretions, sleep, nutritional status, health, and climate.

Age. In both sexes, the BMR illustrated in Figure 11.22 reaches its highest peak during the first and second years of life and these are followed by lesser peaks during puberty. After puberty, the BMR progressively decreases. With the onset of increased aging, biosynthetic growth is marginal or negligible.

Sex. Women have a BMR ~10% lower than that displayed by men. This fundamental difference may be attributable to differences in body composition between the

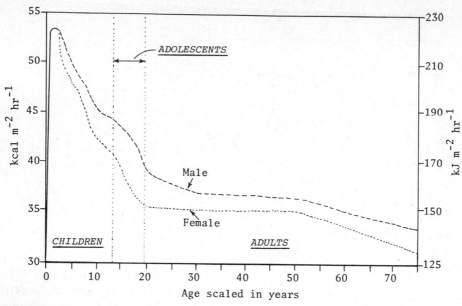

FIGURE 11.22. Approximate relationship between basal metabolism and age in children, adolescents, and adults.

sexes since average statistics indicate that females have more fat than males. This conclusion is supported by studies of athletes versus nonathletes, where the muscular mass of the athlete exhibits a higher BMR than the nonathlete, who has less muscular development.

The BMR is still higher among women who are pregnant. Some investigators attribute the increase to muscle and tissue development in the uterus, placenta, and fetus.

Glandular secretions. Basal metabolic rate is influenced by hormonal activity such as hypoactivity of the thyroid gland. This malady can result in BMR values that are lowered by 50% from normal. Hyperactivity of the thyroid gland, however, can contribute to increases in the BMR of ≥200%. Other hormones such as epinephrine (adrenalin) also contribute to increases in the BMR. Specifically, in cases of fear and flight responses of animals, epinephrine secretion increases. This increase of epinephrine parallels increased

glycogenolysis and mobilization of glucose to meet cellular energy demands. These basic animal responses to obvious danger extend to human recognition of a dangerous situation and may also be reflected during emotional excitement.

Sleep. Muscular relaxation occurs during sleep and this may reduce the BMR by as much as 10 to 15% from the normal rate. The sympathetic nervous system is less active during sleep and, consequently, the BMR decreases to some extent. The observed BMR decreases may also vary from individual to individual, partially as a result of temperament.

Nutritional status and health. The BMR is lowered by as much as 50% from normal rates during periods of fasting or chronic undernutrition. It is theorized that adaptive mechanisms of the body may direct energy-conserving modes of operation and display a lowered BMR since the mass of active tissue is decreased in concert with inadequate consumption of dietary nutrients. The BMR displays characteristically subnormal levels in conditions of hypothyroidism, hypoadrenalism, hypopituitarism, lipoid nephrosis, vitamin D deficiency, and shock. Higher than normal BMR values are displayed during fever, leukemia, anemia, essential hypertension, diabetes *insipidus*, and myocardial insufficiencies.

Climate. The BMR is widely and variably influenced by climatic factors, but since humans have developed reasonable control mechanisms for their environments, these factors generally do not upset the BMR of normal individuals.

Calculation of the basal metabolic rate (BMR): The BMR is directly related to the surface area of the body. Surface area for a body can be calculated from the Dubois and Dubois height–weight formula:

$$A = W^{0.425} \times H^{0.725} \times C$$

or, written as a logarithmic expression,

$$\log A = (\log W \times 0.425) + \\ (\log H \times 0.725) + 1.856$$

where

A = surface area in square meters (m^2)
W = weight in kilograms (kg)
H = height in centimeters
C = constant (71.84)

Since there is a linear relationship between metabolic rate and the three-fourth power of body weight, formulas have been derived by investigators to cover not only humans, as is the case of the Dubois' formula, but also all other animal species.

For men:
$$M = 71.2 \times W^{3/4} [1 + 0.004 (30 - a) \\ + 0.01 (S - 43.4)]$$

For women:
$$M = 65.8 \times W^{3/4} [1 + 0.004 (30 - a) \\ + 0.018 (S - 42.1)]$$

where

M = metabolic rate in kilocalories per day
W = body weight in kilograms
a = age in years
   Assumption: An $\simeq 0.4\%$ decrease of metabolic rate for each year above the age of 30
S = specific stature in centimeters per $W^{1/3}$
   Assumption: For each cm/kg$^{1/3}$
   For men, increase of 1% of metabolic rate
   For women, increase of 1.8% of metabolic rate

The BMR for an individual can also be estimated more expeditiously using the nomogram illustrated in Figure 11.23. For example, in order to determine the BMR energy for a 20-year-old woman whose height and weight are 165 cm and 55.0 kg, respectively, follow the steps detailed below:

1. Place a straight edge at the 55-kg point on scale A and place the other end of the straight edge at the 165-cm point on scale B.

2. Read the surface area of the subject at the point where the ruler intersects scale C, that is, 1.6 m^2.

3. Now place the straight edge at the 1.6-m^2 mark on scale C and the other end at the 20–29 year age on scale D.

4. Read the BMR value at the point where the ruler intersects scale E, which is about 1380 kcal/day. This value represents the BMR energy of the subject.

The amount and type of food constituents metabolized by the body can be estimated on the basis of BMR studies coupled with measurements of urinary nitrogen excretion, oxygen utilization in the oxidation of foods, and the subsequent production of carbon dioxide. As an example, consider the results of a BMR study where overnight fasting of a test subject led to a urinary nitrogen

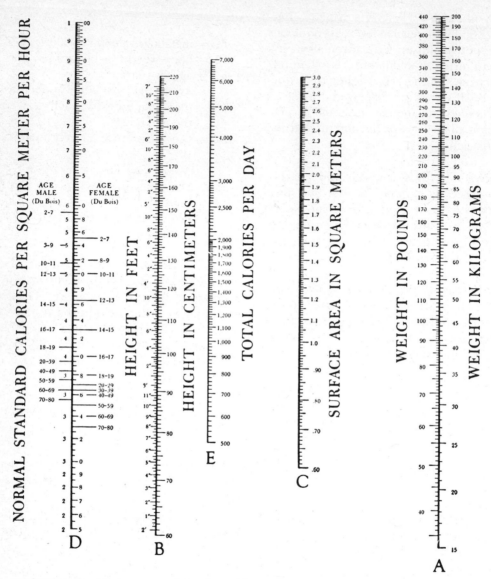

FIGURE 11.23. Nomogram used for the rapid approximation of BMR values. Locate the test subject's normal weight on scale A and his height on scale B. The line joining these two points intersects scale C at the subject's surface area. Specify the age and sex of the subject on Scale D. A line joining this point with the subject's surface area on scale C crosses scale E at the *basal* energy requirement. (Used with permission of the *New England Journal of Medicine* (**185**(12):349), and adapted from W. M. Boothby, and R. B. Sandiford, 1921, *Boston Med. Surgical J.* **185**:12).

value of 0.20 g/h, the oxygen consumption was 13.56 L/h, and the carbon dioxide production was 10.30 L/h. Based on this type of metabolic data, it is possible to determine

1. The number of kilocalories produced in a mixed diet of carbohydrates, fats, and proteins.

2. The percentage of total kcal derived from the oxidation of carbohydrates, fats, and proteins in a mixed diet.

For (1) above, the nonprotein *R.Q.* can be calculated by dividing the volume of nonprotein carbon dioxide by the volume of nonprotein oxygen. Using this calculated nonprotein $R.Q.$, the kilocalories for carbohydrates and fats are determined by consulting Table 11.8. The total kilocalories of the mixed diet are calculated by adding the kilocalories of carbohydrates and fats to those of protein.

Using the data presented in Table 11.7, it is clear that 100.0 g of meat protein can yield 16.28 g of urinary nitrogen. Thus, each gram of urinary nitrogen represents the metabolism of

100/16.28 or 6.15 g of meat protein

96.73/16.28 or 5.94 L of O_2

77.47/16.28 or 4.76 L of CO_2

However, for the average protein it is estimated that 1.0 g of urinary nitrogen represents the metabolism of 6.25 g of protein, a consumption of 5.91 L of O_2, production of 4.76 L of CO_2, and the liberation of 26.51 kcal. Since the urinary nitrogen in metabolic studies was determined to be 0.20 g, this value represents

$0.20 \times 6.25 = 1.25$ g of metabolized protein

$0.20 \times 5.91 = 1.18$ L of consumed O_2

$0.20 \times 4.76 = 0.95$ L of CO_2 produced

$0.20 \times 26.51 = 5.30$ kcal liberated

Table 11.8. Caloric Values for Oxygen and Carbon Dioxide for Nonprotein $R.Q.$

Nonprotein respiratory quotient	Caloric value of 1 L of O_2	Caloric value of 1 L of CO_2	Percentage of total O_2 consumed by fat	Percentage of total heat produced by fat
0.707	4.686	6.629	100.0	100.0
0.71	4.690	6.606	99.0	98.5
0.72	4.702	6.531	95.6	95.2
0.73	4.714	6.458	92.2	91.6
0.74	4.727	6.388	88.7	88.0
0.75	4.739	6.319	85.3	84.4
0.76	4.751	6.253	81.9	80.8
0.77	4.764	6.187	78.5	77.2
0.78	4.776	6.123	75.1	73.7
0.79	4.788	6.062	71.7	70.1
0.80	4.801	6.001	68.3	66.6
0.81	4.813	5.942	64.8	63.1
0.82	4.825	5.884	61.4	59.7
0.83	4.838	5.829	58.0	56.2
0.84	4.850	5.774	54.6	52.8
0.85	4.862	5.721	51.2	49.3
0.86	4.875	5.669	47.8	45.9
0.87	4.887	5.617	44.4	42.5
0.88	4.899	5.568	41.0	39.2
0.89	4.911	5.519	37.5	35.8
0.90	4.924	5.471	34.1	32.5
0.91	4.936	5.424	30.7	29.2
0.92	4.948	5.378	27.3	25.9
0.93	4.961	5.333	23.9	22.6
0.94	4.973	5.290	20.5	19.3
0.95	4.985	5.247	17.1	16.0
0.96	4.998	5.205	13.7	12.8
0.97	5.010	5.165	10.2	9.51
0.98	5.022	5.124	6.83	6.37
0.99	5.035	5.085	3.41	3.18
1.00	5.047	5.047	0	0

SOURCE: After N. Zuntz and H. Schumberg, with modifications by G. Lusk, E. P. Cathcart, and D. P. Cuthbertson, *J. Physiol. (London)* **73**:349, 1931.

The nonprotein *R.Q.* can be calculated according to the following steps:

$$\text{nonprotein } R.Q. = \frac{\text{volume of nonprotein } CO_2}{\text{volume of nonprotein } O_2}$$

$$\text{nonprotein } R.Q. = \frac{10.30 \text{ L} - 0.95 \text{ L}}{13.56 \text{ L} - 1.18 \text{ L}} = \frac{9.35 \text{ L}}{12.38 \text{ L}}$$

nonprotein *R.Q.* = 0.76

The *total calories* provided in a mixed diet corresponds to the *sum of* the *nonprotein calories* (fat + carbohydrate) *plus protein calories*.

The nonprotein *R.Q.* of 0.76 represents 4.751 kcal/L of O_2 (Table 11.7). This *R.Q.* in turn reflects an 80.8% caloric yield from fat oxidation and a 19.2% yield from the oxidation of carbohydrates.

On the basis of a total 12.38 L consumption of O_2, the *total* amount of nonprotein kilocalories can be calculated as

12.38 L O_2 × 4.751 kcal/L O_2

= 58.82 kcal (*total*)

From this 58.82-kcal total,

80.8% of 58.82 kcal *or* 47.53 kcal comes from 5.03 g of fat

19.2% of 58.82 kcal *or* 11.29 kcal comes from 2.72 g of carbohydrate

Recall that the respective amounts of fat and carbohydrate can be calculated on the basis that 1.0 g of general carbohydrate yields 4.15 kcal and 1.0 g of fat liberates 9.45 kcal.

Since the kilocalorie yield from proteins accounts for 5.30 kcal (or 0.20 × 26.51 kcal), the total kilocalories for the diet can be calculated as

$$\frac{58.82 \text{ kcal}}{\text{(nonprotein kcal)}} + \frac{5.30 \text{ kcal}}{\text{(protein kcal)}} = \frac{64.12 \text{ kcal}}{\text{(total kcal)}}$$

For part (2) of the problem posed in this section, the percent of kilocalories supplied by fats, carbohydrates, and proteins, respectively, can be calculated as outlined below:

5.03 g of *fat* yield:
(47.53 kcal/64.12 kcal) or 74.12%

2.72 g of *carbohydrate* provide:
(11.26 kcal/64.12 kcal) or 17.6%

0.20 g of *protein* account for:
(5.30 kcal/64.12 kcal) or 8.3%

For a more detailed clinical and nutritional biochemical approach dealing with practical determination of BMR values and respiratory exchange principles, consult Latner (1975).

Lean body mass as the predictor of basal metabolic rate: The relation of basal metabolic rate (BMR) to sex, age, height, and body mass was studied in the early part of the twentieth century. Benedict (1915) reported that body mass and body surface area were both inadequate indicators for BMR measurement and perceived the active protoplasmic tissue as a useful predictor. Many years later, Miller and Blyth (1952) proposed that *lean body mass* (LBM), also called *free-fat mass*, should be used as a metabolic standard since it demonstrated better predictability of oxygen consumption than surface area. Tzankoff and Norris (1977, 1978) subsequently showed that the inverse relationship between age and basal oxygen consumption was attributable solely to changes in LBM. The classic data of Harris and Benedict published in 1919 was then reanalyzed by Cunningham (1980) using regression analysis and a statistical relationship of BMR to sex, height, body mass, and the proposed metabolic standard based on LBM. The LBM in this study was calculated from the equations* of Moore *et al.* (1963) since the body composition of the subjects used in the 1919 studies were not reported. The statistical analysis performed by Cunningham revealed that LBM could be employed as the predictor of BMR and a simple equation was proposed for BMR calculation.

BMR (cal/day) = 500 + 22 (LBM)

Webb (1981) measured the energy expenditure of men and women by direct calorimetry and demonstrated that LBM contrasts favorably with the imprecise prediction of BMR according to age, sex, and surface area. In this study the LBM was calculated from

*Male LBM = (79.5 − 0.24M − 0.15A) × M ÷ 73.2; female LBM − (69.8 − 0.26M − 0.12A) × M ÷ 73.2, where M = body mass in kilograms and A = age in years.

body density determined from underwater weight.

Hence, all the aforementioned studies indicate that LBM can be used to calculate the BMR with less complexity since it avoids the use of the cumbersome tables for separate sexes and ages.

Recommended dietary allowances for energy: Energy balance is crucial for the maintenance of good health. The population of the United States can be described as light to sedentary with regard to their physical activities. The recommended daily caloric intake values based on these considerations and specified by the National Academy of Sciences have been detailed in Table 11.9.

Specific dynamic action (SDA): Specific dynamic action or the so-called *calorigenic effect* is a phenomenon in which extra body heat is produced following food ingestion. Some nutritionists claim that surges of heat production by the body reflect energy expended in the digestive and absorptive activities, but this opinion has been largely buried by others since intravenous doses of amino acids result in similar SDA effects. Carbohydrates, fats, and proteins vary in their respective specific dynamic actions, but proteins generally demonstrate the greatest effect.

Although the SDA values cannot be explained in a clear-cut fashion, the SDA must be considered for the maintenance of energy equilibrium. Usually, a 6.0% increase in kil-

Table 11.9. Mean Heights and Weights and Recommended Energy Intakes for the United States, 1979[a]

Category	Age (years)	Weight kg	Weight lb	Height cm	Height in.	Energy needs (with range) kcal	MJ
Infants	0.0–0.5	6	13	60	24	kg × 115(95–145)	kg × 0.48
	0.5–1.0	9	20	71	28	kg × 105(80–135)	kg × 0.44
Children	1– 3	13	29	90	35	1300 (900–1800)	5.5
	4– 6	20	44	112	44	1700 (1300–2300)	7.1
	7–10	28	62	132	52	2400 (1650–3300)	10.1
Males	11–14	45	99	157	62	2700 (2000–3700)	11.3
	15–18	66	145	176	69	2800 (2100–3900)	11.8
	19–22	70	154	177	70	2900 (2500–3300)	12.2
	23–50	70	154	178	70	2700 (2300–3100)	11.3
	51–75	70	154	178	70	2400 (2000–2800)	10.1
	76 +	70	154	178	70	2050 (1650–2450)	8.6
Females	11–14	46	101	157	62	2200 (1500–3000)	9.2
	15–18	55	120	163	64	2100 (1200–3000)	8.8
	19–22	55	120	163	64	2100 (1700–2500)	8.8
	23–50	55	120	163	64	2000 (1600–2400)	8.4
	51–75	55	120	163	64	1800 (1400–2200)	7.6
	76 +	55	120	163	64	1600 (1200–2000)	6.7
Pregnancy						+ 300	
Lactation						+ 500	

SOURCE: *Recommended Dietary Allowances*, 9th ed. National Academy of Sciences, Washington, D.C., 1980.

[a]The data in this table have been assembled from the observed median heights and weights of children together with desirable weights for adults for the mean heights of men (70 in.) and women (64 in.) between the ages of 18 and 34 years as surveyed in the U.S. population (HEW/NCHS data).

The energy allowances for the young adults are for men and women doing light work. The allowances for the two older age groups represent mean energy needs over these age spans, allowing for a 2% decrease in basal (resting) metabolic rate per decade and a reduction in activity of 200 kcal/day for men and women between 51 and 75 years, 500 kcal for men over 75 years, and 400 kcal for women over 75. The customary range of daily energy output is shown for adults in parentheses and is based on a variation in energy needs of ±400 kcal at any one age emphasizing the wide range of energy intakes appropriate for any group of people.

Energy allowances for children through age 18 are based on median energy intakes of children of these ages followed in longitudinal growth studies. The values in parentheses are tenth and ninetieth percentiles of energy intake to indicate the range of energy consumption among children of these ages.

ocalories is considered normal for an ordinary diet.

The specific dynamic action and the basal metabolism together account for the "resting metabolism" of the body.

Energy demands for various activities: The energy requirements for humans must take into account, not only basal metabolism and the SDA, but also the energy demands associated with different types of physical activities. In addition to the size of the body that is executing or performing physical or manual activities, the speed at which these activities are conducted holds paramount importance. Table 11.10 indicates the caloric expenditures for a variety of normal activities.

Total energy estimation: The total daily energy requirement for an individual is based on three distinct factors: (1) basal metabolism, (2) physical activity, and (3) the specific dynamic action of foods. The summation of these energies results in the daily energy requirement for an individual.

A cursory evaluation of energy requirements for an individual can be obtained by determining the energy expenditure for the subject over the course of a 24-h period using established BMR data (Tables 11.9 and 11.10). As an example, take the case where a 50-kg woman is employed as an agricultural worker and also maintains a home. The daily activities can be subdivided into (1) 8 h of sleep; (2) 6 h of very light activity; and (3) 10 h of moderate physical activity. Based on the total 24-h period for one day, the energies for (1) sleep, (2) very light activities, and (3) moderate activities can be calculated as follows (Recommended Dietary Allowances, 8th ed., 1974):

1. The energy required for sleep is calculated from information contained in Table 11.10. The BMR for the 50-kg woman (Table 11.11) is 1399 kcal/day, but considering that sleep is calculated at 90% of the BMR for 8 h (Table 11.11), this yields a value of (1399 kcal × 0.9)/3 or 420 kcal.

2. The energy required for very light activity is calculated using information appearing in Table 11.11. For very light activity, a value of 1.3 kcal/kg/h is obtained from the table and multiplied by 6 h of this activity and the kilogram weight of the subject: (1.3) (6.0) (50.0) = 390 kcal

3. The energy required for moderate activity is also calculated using Table 11.11.

Table 11.10. Energy Cost of Activities Exclusive of Basal Metabolism and Influence of Food

Activity	kcal/kg/hr	Activity	kcal/kg/hr
Bicycling (moderate speed)	2.5	Reading aloud	0.4
Boxing	11.4	Rowing in race	16.0
Carpentry (heavy)	2.3	Running	7.0
Dancing, waltz	3.0	Skating	3.5
Dishwashing	1.0	Stone masonry	4.7
Dressing and undressing	0.7	Sweeping with broom,	
Driving an automobile	0.9	bare floor	1.4
Eating	0.4	Swimming (2 mph)	7.9
Ironing (5-lb iron)	1.0	Typewriting rapidly	1.0
Knitting a sweater	0.7	Walking (3.0 mph)	2.0
Laundry, light	1.3	Walking rapidly	
Lying still, awake	0.1	(4 mph)	3.4
Playing ping-pong	4.4	Walking at high	
		speed (5.3 mph)	9.3
		Writing	0.4

SOURCE: Values based on C. M. Taylor and G. McLeod, *Rose's Laboratory Handbook for Dietetics*, 5th ed. Macmillan, New York. 1949, p. 18.

Table 11.11. Suggested Weights for Heights[a] and Basal Metabolic Rates (BMR)[b] of Adults[c]

| Height | | Men | | | Women | | |
| | | Median weight | | BMR | Median weight | | BMR |
in.	cm	lb	kg	(kcal/day)	lb	kg	(kcal/day)
60	152				109 ± 9	50 ± 4	1399
62	158				115 ± 9	52 ± 4	1429
64	163	133 ± 11	60 ± 5	1630	122 ± 10	56 ± 5	1487
66	168	142 ± 12	64 ± 5	1690	129 ± 10	59 ± 5	1530
68	173	151 ± 14	69 ± 6	1775	136 ± 10	62 ± 5	1572
70	178	159 ± 14	72 ± 6	1815	144 ± 11	66 ± 5	1626
72	183	167 ± 15	76 ± 7	1870	152 ± 12	69 ± 5	1666
74	188	175 ± 15	80 ± 7	1933			
76	193	182 ± 16	83 ± 7	1983			

[a]Modified from Table 80 of Hathaway and Foard (1960). "Heights and Weights of Adults in the U.S.," Home Economics Research Report No. 10, ARS, USDA. Weights were based on those of college men and women. Measurements were made without shoes or other clothing. The ± refers to the weight range between the 25th and 75th percentile of each height category.
[b]Adapted from Talbot (FAO/WHO, 1973).
[c]To determine the daily energy need of an individual, allow for hours of sleep at 90 percent of BMR and for time periods engaged in various activities as indicated in the table below. Data are expressed as kilocalories per kilogram per hour.

Activity	Men	Women
Very light—seated and standing activities, painting trades, auto and truck driving, laboratory work, typing, playing musical instruments, sewing, ironing	1.5	1.3
Light—walking on level, 2.5–3 mph, tailoring, pressing, garage work, electrical trades, carpentry, restaurant trades, cannery workers, washing clothes, shipping with light load, golf, sailing, table tennis, volleyball	2.9	2.6
Moderate—walking 3.5–4 mph, plastering, weeding and hoeing, loading and stacking bales, scrubbing floors, shopping with heavy load, cycling, skiing, tennis, dancing	4.3	4.1
Heavy—walking with load uphill, tree felling, work with pick and shovel, basketball, swimming, climbing, football	8.4	8.0

For moderate activity, 4.1 kcal/kg/hr over a timeframe of 10 h translates into (4.1) (10.0) (50.0) = 2050 kcal

Therefore, in summation:
(a) 8 h of sleep   =   420 kcal
(b) 6 h of light activity   =   390 kcal
(c) 10 h of moderate activity = 2050 kcal
          Answer   Total: 2860 kcal/day

Note: United States reference woman has energy requirement of ≃ 2000 kcal.

Total energy estimation offers a practical method for evaluating the caloric dietary requirements of individuals based on the amount of time spent in one or more specific activities. Any estimate of total energy must account for the energy demands dictated by normal maintenance activities of the body, plus those energy demands associated with specific voluntary activities. As long as the total energy requirement of the body is met with adequate energy supplies, an energy equilibrium will be maintained. Based on a plot of retained energy (RE) obtained from the diet versus food intake (I), this equilibrium point is evident when RE equals zero (Figure 11.24). The ratio of the difference between two levels of retained energy (ΔRE) with respect to two corresponding levels of food intake (ΔI) permit estimation of the net energy (NE) value of a food or diet. Net energy is expressed by

$$NE = \Delta RE / \Delta I$$

When the $\Delta RE / \Delta I$ value falls between $I = 0.0$ and $RE = 0.0$, this represents the *net energy for maintenance* (NE_m). Obviously as food intake approaches 0.0, energy loss develops from fasting conditions. Conversely, *ad li-*

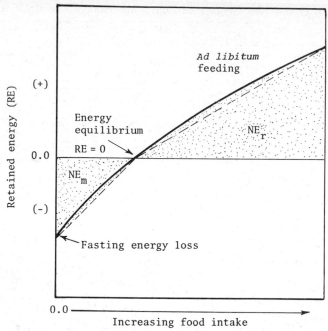

FIGURE 11.24. Plot of retained energy (*RE*) versus food intake (*I*). The rectilinear curve, drawn as a solid line, represents the actual trend observed for experimental data, but the straight line (dashed line) represents the conventional linear expression of the same data. Consult Lofgreen and Garrett (1968) for more details.

bitum feeding of a test subject above $RE = 0.0$ and beyond purely maintenance levels is reflected as a net energy retention value (NE_r).

In actuality, the plot of RE versus I is a rectilinear curve, but practical expressions of such data are shown as two straight lines having a common meeting point at $RE = 0.0$.

The RE value for any given point in the life cycle of a test subject is under the control of variable peripheral and metabolic factors. Variables that control the long-term "regulation" and acquisition of energy reserves by the body include (1) food intake, (2) muscular activity, (3) the state of energy stores (glycogen, fat), and (4) the depletion rate of energy through lactation, secretion, and gastrointestinal tract absorption and as heat. Clearly, a corresponding short-term "regulation" of food intake involves temporal variations in concentration levels of blood glucose, free fatty acids, amino acids, and

hormones, as well as osmotic pressure/gastrointestinal tract distention, neurohumoral factors, satiety factors, hunger factors, and body temperature along with sensory inputs of food texture, odor, and taste. The diverse interactions of these physicosensory and humoral factors have inevitably led to a *multiple-factor theory* regarding the control of energy balance in the body as opposed to any single control mechanism. Multiple factors in these theories of energy balance and food intake are also important because they not only act as a stimulus for "changes" in food intake but they may also be responsible for "resetting" the level where "regulation" takes place. These effects, in turn, may alter the relative importance of one factor with respect to another in the overall feedback mechanism responsible for moderating food consumption.

Assuming that the most favorable multiple-factor interaction can exist in order to promote digestion and assimilation of dietary nutrients, the actual *RE* value demonstrated for a single food or composite diet will reflect its overall digestibility. Two factors, one biochemical and the other physical, affect the energy retained from foods. It is hardly necessary to ponder the fact that calorimetric values for foods do not account for actual biochemical availability of food calories. For example, the calorimetric "yardstick" for evaluating foods may be based on the combustion calorimetry of benzoic acid. The combustion energy of benzoic acid measured as calories is real, but the actual biochemical availability of these calories is not realized. In a similar vein, cellulose has a high energy value based on calorimetry and ruminant digestive studies, but cellulose represents a useless source of energy for humans. As a result of the obvious disparities between calorimetric food values and actual amounts of food energy assimilated and retained by the body, some studies as early as those of Kellner (1909) proposed grading foods according to their required equivalents that replace 100 parts of starch. Starch was chosen as a grading standard because of its high natural digestibility.

Biochemical inabilities of the human body to obtain the full complement of dietary energy from certain foods is indicative of specific limitations in digestive enzymes as well as digestive limitations of intestinal flora. Therefore, many food substances have a high potential calorie yield, but resist normal digestive processes and transit the digestive system with no marked decrease in their calorie content. Ideally, the calorie values in food tables should be reported only after scrupulous attention has been directed to the specific indigestibility of foods (McCance and Widdowson, 1960; Widdowson, 1955). As an exemplary case, the dietary calorie yield of artichokes must be calculated at a level far below the combustion values in order to account for poor digestibility of their inulin content. Corn meal, on the other hand, offers a higher digestibility and dietary calorie yield than whole corn kernels of similar moisture content. The very nature of the corn kernel hull structure physically impedes the access of digestive enzymes to the endosperm and germ, and the dietary calorie yield is related to the degree of kernel mastication. Similar studies for whole peanuts versus peanut butter also support the concept that *RE* yields from foods will be determined by the (1) food matrix where calories are presented to the digestive system, (2) the surface-to-volume ratio of the foodstuff in the digestive system, and (3) the dietary fiber content coexisting with nutritionally accessible food calories. The influence of these factors has been demonstrated for many foods, and one typical example is cited below for calorie-rich peanuts, peanut butter, and peanut oil. The calorie absorption of peanut lipids reflects the fiber content as well as the available surface-to-volume ratios for different peanut product rations (Table 11.12).

Table 11.12. Percentage of Dietary Lipid Excreted per Day (±SEM)

| Fiber Content | Increasing surface-to-volume ratio of peanut matrix and lipids | | |
	Whole peanuts	Peanut butter	Peanut oil
Low	17.0 ± 11.7	4.0 ± 1.7	2.0 ± 0.4
High	18.0 ± 5.3	7.0 ± 1.4	4.8 ± 1.4

SOURCE: Adapted from A. S. Levine and S. E. Silvis (1980). Absorption of whole peanuts, peanut oil, and peanut butter. *N. Engl. J. Med.* **303**, 917.

REFERENCES

Adamson, A. W. 1979. *A Textbook of Physical Chemistry*, 2nd ed. Academic Press, New York.

Alberty, R. A. 1983. *Physical Chemistry*, 6th ed. Wiley, New York.

Atkins, P. W. 1982. *Physical Chemistry*, 2nd ed. Freeman, San Francisco.

Barrow, G. M. 1981. *Physical Chemistry for the Life Sciences*, 2nd ed. McGraw-Hill, New York.

Benedict, F. 1915. Factors affecting basal metabolism. *J. Biol. Chem.* **20**:263.

Castellan, G. W. 1983. *Physical Chemistry*, 3rd ed. Addison-Wesley, Reading, Mass.

Cunningham, J. J. 1980. A reanalysis of the factors influencing basal metabolic rate in normal adults. *Amer. J. Clin. Nutr.* **33**:2372.

Glasstone, S. 1940. *A Test-Book of Physical Chemistry*. Van Nostrand, New York.

Harris, J., and F. Benedict. 1919. A biomedic study of basal metabolism in man. *Carnegie Inst. Washington Publ.* **279**:40.

Hinkle, P. C., and R. E. McCarty. 1978. How cells make ATP. *Sci. Amer.* **238**(3):104.

Kellner, O. 1909. *The Scientific Feeding of Animals* (Trans. W. Goodwin). Duckworth, London.

Latimer, W. M. 1952. *The Oxidation States of the Elements and Their Potentials in Aqueous Solutions*. Prentice-Hall, Englewood Cliffs, N.J.

Latner, A. L. 1975. *Clinical Biochemistry*, 7th ed. Saunders, Philadelphia.

Lewis, G. N., and M. Randall. 1923. *Thermodynamics and the Free Energy of Chemical Substances*, 1st ed. McGraw-Hill, New York.

Lewis, G. N., and M. Randall. 1961. *Thermodynamics*, 2nd ed. (revised by K. S. Pitzer and L. Brewer). McGraw-Hill, New York.

Lofgreen, G. P., and W. N. Garrett. 1968. A system for expressing net energy requirements for beef cattle. *J. Anim. Sci.* **27**:793.

McCance, R. A., and E. M. Widdowson. 1960. The composition of foods. *Med. Res. Counc. Spec. Rep. Ser.*, No. 297. Her Majesty's Stationary Office, London.

Miller, A., and C. Blyth. 1952. Estimation of lean body mass and body fat from basal oxygen consumption and creatine excretion. *J. Appl. Physiol.* **5**:73.

Mitchell, P. 1979. Keilin's respiratory chain concept and its chemiosmotic consequences. *Science* **206**:1148.

Moore, T. 1977. The calorie as a unit of nutritional energy. *Wld. Rev. Nutr. Diet.* **26**:1.

Moore, W. J. 1972. *Physical Chemistry*, 4th ed. Prentice-Hall, Englewood Cliffs, N.J.

Moore, F., K. Olesen, J. McMurray, V. Parker, M. Ball, and C. Boyden. 1963. *The Body Cell Mass and Its Supporting Environment*. Saunders, Philadelphia.

Pippard, A. B. 1957. *The Elements of Classical Thermodynamics*. Cambridge Univ. Press, London.

Rubner, M. 1902. *Die Gesetze des Energieverbrauchs bei der Ernährung*. Deuticke, Leipzig/Wien.

Tinoco, Ignacio, Jr., Kenneth Sauer, and James C. Wang. 1978. *Physical Chemistry Principles and Applications in Biological Sciences*. Prentice-Hall, Englewood Cliffs, N.J.

Tzankoff, S., and A. Norris. 1977. Effect of muscle mass decrease on age-related BMR changes. *J. Appl. Physiol. Respir. Environ. Exercise Physiol.* **43**:1001.

Tzankoff, S., and A. Norris. 1978. Longitudinal changes in basal metabolism in man. *J. Appl. Physiol. Respir. Environ. Exercise Physiol.* **45**:53.

Weast, R. S. (Ed.). 1982–1983. *Handbook of Chemistry and Physics*, 63rd ed. Chemical Rubber Co., Cleveland.

Webb, P. 1981. Energy expenditure and fat-free mass in men and women. *Amer. J. Clin. Nut.* **34**:1816.

Widdowson, E. M. 1955. Assessment of the energy value of human foods. *Proc. Nutr. Soc.* **14**:142.

Photosynthesis, Plants, and Primary Productivity

INTRODUCTION

All foods having plant or animal origins are ultimately tied to the existence of terrestrial or marine plants. Through the action of photosynthesis, radiant energy from the sun is used to drive the biosynthesis of carbohydrates from CO_2 and water. This basic photosynthetic reaction that occurs in green plants and blue-green (*Cyanophycean*) algae can be superficially expressed as

$$6\ CO_2 + 6\ H_2O \longrightarrow C_6H_{12}O_6 + 6\ O_2$$
Glucose

In this reaction, water (an electron donor) undergoes oxidation to produce O_2 along with the collateral reduction of CO_2 (an electron acceptor).

Photosynthesis occurs in two distinct steps recognized as *primary* (*light reaction*) and *secondary* (*dark reaction* or *carbon dioxide fixation*) *phases*. The *primary phase* is also known to consist of two distinctly separate light reactions designated as photosystems I and II (PS I and II).

The primary process is responsible for trapping solar light energy by photoactive (photoreceptor) pigments. Some of the light energy is directed to the cleavage (photolysis) of water to produce O_2, protons (H^+), and electrons. Both the protons and the electrons contribute to the production of reducing power in the form of NADPH (from $NADP^+$) and chemical energy as ATP (from $ADP + P_i$). An overview for the primary photosynthetic phase can be written as

$$H_2O + NADP^+ + ADP + P_i \xrightarrow{\text{light energy}}$$
$$NADPH + H^+ + ATP$$

The secondary phase of photosynthesis consumes ATP and NADPH produced in the primary phase. Both ATP and NADPH serve as chemical energy sources for the fixation and reduction of CO_2 to yield carbohydrate. This reaction can be summarized as

$$CO_2 + NADPH + H^+ + ATP \longrightarrow$$
$$(CH_2O) + NADP^+ + ADP + P_i$$

Apart from carbohydrate production in the form of glucose, the foregoing reaction also produces the precursors for ATP and NADPH syntheses.

CHLOROPLASTS AND THE PHOTOSYNTHETIC APPARATUS

Photosynthetic reactions occur in the chloroplasts. These organelles are about 4.0 μm long and display a morphology similar to mitochondria. Twenty to fifty chloroplasts per cell are not atypical depending on the plant species and stage of cell maturity (Figures 12.1 and 12.2). The chloroplasts exhibit an *outer membrane* structurally similar to mitochondrial membrane architecture and an *inner membrane* that has developed into a complex series of vesicles called *thylakoids*. The stacked arrangement of thylakoids form *grana*.

Photosynthetic pigments such as chlorophylls, which are responsible for trapping light energy and directing it toward CO_2 fix-

ation, reside in the thylakoid membranes along with the enzymes necessary for the light reactions of photosynthesis. Enzymes necessary for the secondary (CO_2 fixation) photosynthetic phase occur in the fluid-filled compartment, known as *stroma,* that surrounds the thylakoid vesicles (Figure 12.2).

CHLOROPHYLLS AND OTHER PHOTOACTIVE PIGMENTS

Numerous forms of the light-absorbing, magnesium-containing tetrapyrrolic pig-

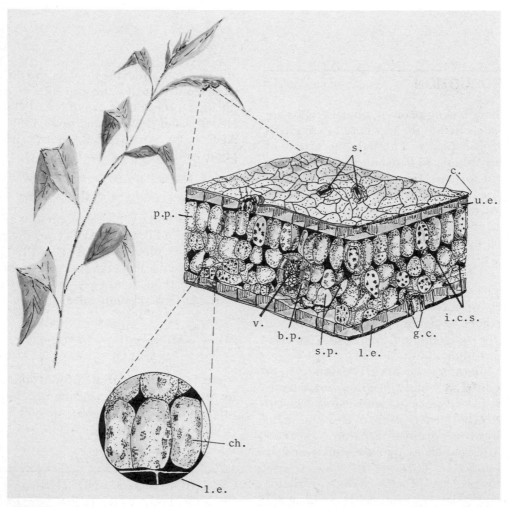

FIGURE 12.1. Structural relationships and anatomy of components within a typical leaf illustrating: b.p., bundle parenchyma; c., cuticle; ch., chloroplast; g.c., guard cell; i.c.s., intercellular spaces; l.e., lower epidermis; p.p., palisade parenchyma; s., stroma; s.p., spongy parenchyma; u.e., upper epidermis; and v., venule (vascular tissues within leaf).

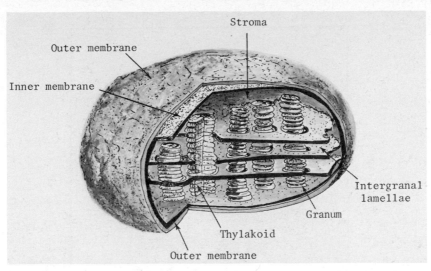

FIGURE 12.2. Chloroplasts contain a complex series of membranes that support column-shaped grana that are composed of many thylakoid disks. Cells of higher plants may contain up to 40 chloroplasts, whereas some unicellular algae contain only 1. Photosynthetic pigments reside within the membranes of thylakoids as well as intergranal lamellae of chloroplasts.

ment known as *chlorophyll* occur in higher plants. Chlorophylls-*a* and -*b* are most common and are structurally identical with the main exception that -*a* has a C-3 methyl group on ring II and -*b* has a C-3 formyl group. Both chlorophylls also display the presence of a fused *cyclopentanone ring* (V) and a hydrophobic *phytol* ($C_{20}H_{39}$) *ester* located at the propionic acid group of ring IV. The phytol ester is believed to anchor chlorophyll molecules to the lipid bilayer of functional photosynthetic membranes.

Visible light is effectively absorbed by chlorophylls over the range of 400 to 700 nm (Figure 12.3). This photoabsorptive behavior is augmented by *accessory pigments* such as yellow *carotenoids* that coexist with chlorophylls. Xanthophylls, red and purple carotenoids, as well as red or blue *open* tetrapyrrolic molecules called *phycobilins* may also exist in some plant cells. Many of these plant pigments are conjugated to structurally strategic proteins embedded in the photosynthetic membrane architecture (Figure 12.4).

Chlorophyll molecules actuate the photosynthetic process by absorbing light energy and assuming an *excited state* such that an electron is discharged from the complex molecular structure. This excitation of chlorophyll requires the absorption of a discrete quantum amount of light energy ($h\nu$). Electrons lost through this excitation are directed into subsequent steps of the light reaction of photosynthesis discussed below.

Although excitation of chlorophyll alone can furnish electrons to the light reaction of photosynthesis, the harvest of solar light energy is augmented further by accessory pigments. These pigments funnel their absorbed light energy to chlorophyll molecules at photosynthetic centers by electronic energy transfer mechanisms such as inductive resonance.

ELECTRON FLOW AND ATP FORMATION IN PHOTOSYNTHESIS

Conventional concepts of the primary photosynthetic phase rely on the enzyme-mediated *flow of electrons* from water oxidation to give ½ O_2 ($E^{o\prime} = +0.82$ V) and the subsequent reduction of $NADP^+$ to yield NADPH ($E^{o\prime} = -0.32$ V). This unorthodox flow of electrons from a positive electro-

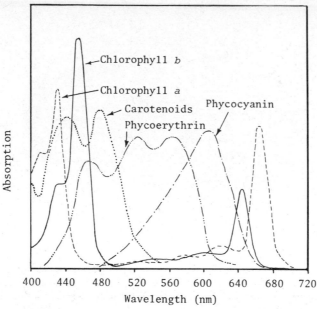

FIGURE 12.3. Comparative absorption spectra for some common plant pigments.

chemical potential to a *more negative value* is driven by the photoactivation of chlorophyll. The primary (light) phase of photosynthesis responsible for ATP formation, known as *photophosphorylation,* can assume two different routes. In one route described as *non-cyclic photophosphorylation* (NCP), electrons supplied from water oxidation and chlorophyll excitation yield ATP and NADPH plus molecular oxygen as a by-product. The second route is called *cyclic photophosphorylation* (CP) since electrons supplied from photoactivated chlorophyll are repetitively cycled in an ATP-yielding mechanism without a concomitant production of oxygen or NADPH.

TWO PHOTOSYSTEMS ARE INVOLVED IN PHOTOPHOSPHORYLATION

Two structurally distinct photosystems I and II (PS I and II) occur in chloroplasts interconnected by an *electron transport system,* not at all unlike that present in the mitochondrion. Both PS I and II contain photoactive

centers characterized by the presence of a key light-absorbing pigment, namely, P_{680} in PS II and P_{700} in PS I. The subscript indicates the absorbance maximum for each particular pigment.

The relationship of principal electron transport carriers to PS I and PS II are illustrated according to the conventional "zig-zag scheme" (Z scheme) outlined in Figure 12.5.

When PS I undergoes photoactivation, the P_{700} pigment is oxidized (electron loss) to produce P_{700^+}. P_{700^+} can be returned to its *ground state* (P_{700}) only by electrons supplied from PS II. The reaction center of PS II contains both chlorophylls-*a* and -*b* along with the photoactive pigment identified as P_{680}. Photoexcitation energy is channeled to P_{680} via chlorophyll-*a*. Oxidation of P_{680} to P_{680^+} then supplies electrons to PS I where P_{700^+} must be returned to its ground state (P_{700}).

Numerous electron transport carriers outlined in Figure 12.5 participate in the electrochemical transfer of electrons between PS II and PS I. These carriers include a quinone-based quenching substance (Q), a copper-containing plastocyanin (PC), se-

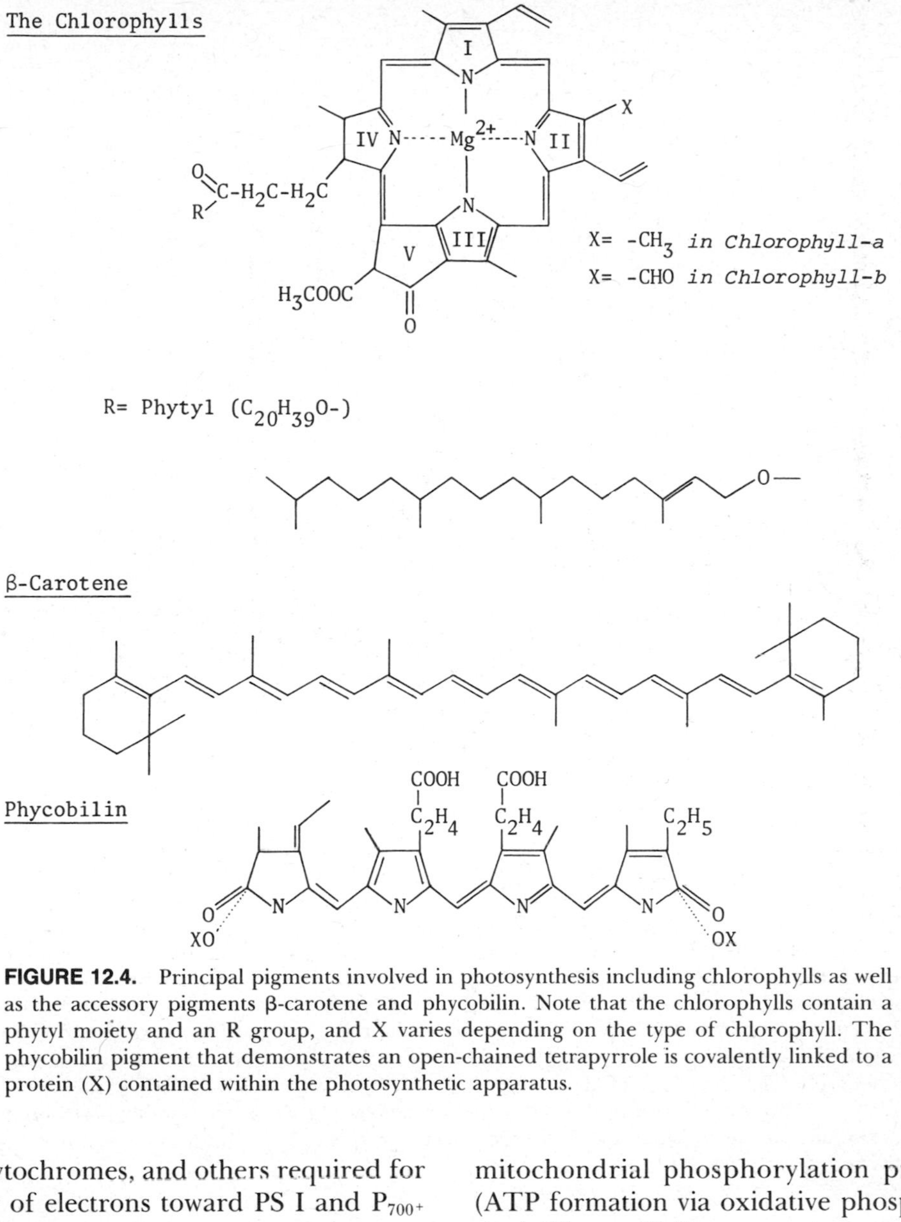

FIGURE 12.4. Principal pigments involved in photosynthesis including chlorophylls as well as the accessory pigments β-carotene and phycobilin. Note that the chlorophylls contain a phytyl moiety and an R group, and X varies depending on the type of chlorophyll. The phycobilin pigment that demonstrates an open-chained tetrapyrrole is covalently linked to a protein (X) contained within the photosynthetic apparatus.

lected cytochromes, and others required for the flow of electrons toward PS I and P_{700^+} reduction.

The passage of electrons through this electron transport system is coupled to the photophosphorylation of ADP to yield ATP. An electrochemical flux of high proton concentrations from the intrathylakoid space over the thylakoid membrane and into the stroma probably drives ATP formation. Proton flux over the membrane is assisted by a *coupling factor* (CF_1) similar to Mitchell's chemiosmotic-coupling hypothesis, which describes

mitochondrial phosphorylation processes (ATP formation via oxidative phosphorylation) (Figure 12.6).

Note, too, that the P_{680^+} species in PS II can be returned to its ground state (reduced) only by a complex and poorly understood photoreaction that oxidizes water to $2\ e^-$, $2\ H^+$, and $\frac{1}{2}\ O_2$ with the involvement of manganese cations.

The overall flow of electrons from water oxidation through PS II to PS I along with ATP and NADPH production is characteristic of NCP. ATP production by the CP

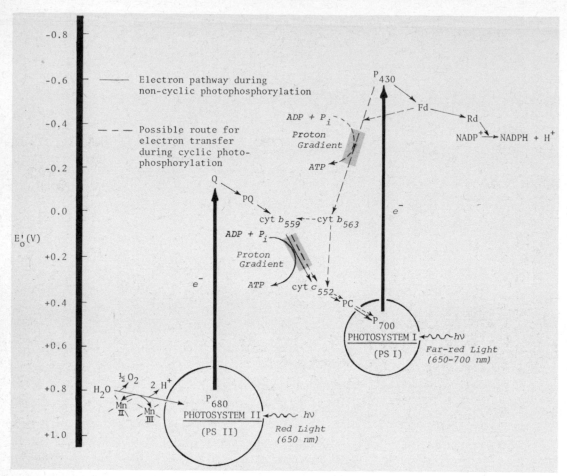

FIGURE 12.5. The so-called zigzag scheme (Z scheme), which depicts the flow of electrons during noncyclic and cyclic processes of photophosphorylation. P_{680} and P_{700} represent the key reaction centers that require the photointeraction of chlorophyll with red wavelengths of light. Photophosphorylation of ADP to ATP is driven by the production of a proton gradient as illustrated in Figure 11.10. Consult text for details regarding electron carriers. The exact electron transfer path for cyclic photophosphorylation is quite uncertain since two coupling sites for ATP formation may exist; or electrons may return directly to PS I by way of cytochrome-c_{552} resulting in only a single photophosphorylation event. Key abbreviations include: Q, quinone-based quenching substance; PQ, plastoquinone; cty, cytochrome; PC, plastocyanin; P_{430}, membrane-bound form of iron–sulfur protein (ferredoxin); Fd, ferredoxin; and Rd, ferredoxin oxidoreductase, which is responsible for the production of NADPH from $NADP^+$.

process involves only PS I whereby a photoactivated electron from P_{700} is donated to a primary electron acceptor (possibly P_{430}) and returned to reduce P_{700+} via cytochrome-b_{563} and plastocyanin (PC) (Figure 12.5). ATP is produced in this cyclic electron flow by mechanisms similar to NCP. CP processes may occur in plants, especially when NADPH stores necessary for CO_2 fixation are replete.

Light reactions for photosynthesis are far more complicated than this survey indicates and other detailed references cited at the end of the chapter should be consulted. Nonetheless it should be recognized that ATP and NADPH supplied by the light phase of photosynthesis are required for the formation of carbohydrates that occurs in the stromal environment. Special emphasis is directed toward carbohydrate formation and

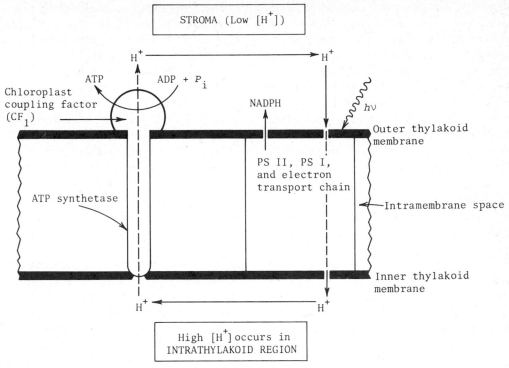

FIGURE 12.6. Proton concentration gradients generated over the thylakoid membrane, plus the action of ATP-synthetase, drive the phosphorylation of ADP to ATP in a manner similar to phosphorylation processes of mitochondria.

the *secondary phase* of photosynthesis since it serves as the biochemical source of many food and starch resources.

CARBON DIOXIDE FIXATION— "DARK REACTIONS" OF PHOTOSYNTHESIS PRODUCE SUGARS FROM CARBON DIOXIDE

Green plants characteristically convert carbon dioxide into organic carbon in the form of glucose. Glucose or derivatives of its carbon skeleton serve as the origin for hundreds of different compounds that are common to plant structure and physiology. Typical examples include dissacharides such as sucrose, starches, cellulose, proteins contained in leaf and seed tissues, fats, and oils. Amino acid and protein synthesis from carbohydrate precursors requires complex interactions with nitrogen and sulfur available to the plant as an organic or inorganic source. Reduction of carbohydrate precursors is at a maximum when fats and oils are produced. These compounds contain about 2.5 times the energy found in an equivalent amount of carbohydrate such as glucose. On this basis, sunflower, soybean, and other oil seeds offer photosynthetically based mechanisms for producing and storing reduced carbon that exceeds the energy content of proteins or carbohydrates.

Carbon dioxide is introduced into plants principally on the underside of leaf surfaces through stomates, although various plant species also display stomates on the stem and upper leaf surfaces.

For most agricultural crops and plants in general, air turbulence, diffusion, or a mixture of both factors expose the surfaces of leaves to carbon dioxide. Since carbon dioxide displays low concentrations within plant tissues, diffusion processes encourage its entry into the plant by stomate routes. Although oxygen is produced by green plants, the entry of carbon dioxide into plants is independent of oxygen. Therefore, physical absorption of carbon dioxide by plants is guided by equilibrium-based diffusion processes that depend on the environmental concentration of carbon dioxide.

Experimental studies indicate that the fully opened stomates having a surface area of 1.0×10^{-4} mm^2 can collectively account for huge volumes of carbon dioxide absorption. Based on a 300-ppm atmosphere of carbon dioxide (0.03%), crops such as corn will assimilate 9.5 tons of carbon dioxide per 100 bushels of corn at the end of the growing season.

Once introduced into the stomates, gaseous carbon dioxide diffuses through substomal spaces, through intercellular spaces, to mesophyll cells, and ultimately into chloroplasts (Figure 12.7). Dissolved gaseous forms of carbon dioxide are probably involved in carbon dioxide absorption, especially during the initial stages of its assimilation.

As long as photosynthetic processes favor photolysis of water and oxygen production accompanies carbon dioxide reduction during daylight, oxygen is diffused from the leaf tissues in the opposite direction of carbon dioxide movement. Water vapor is also lost through stomates in transpiration processes as a consequence of gas exchange.

Under conditions of low illumination or during night hours, carbon dioxide usually diffuses from leaves into the environment until an equilibrium is met with the environment. Furthermore, a lack of illumination requires plants to participate in respiratory activities that produce a high

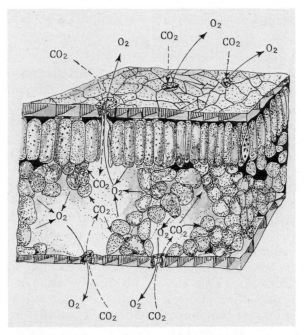

FIGURE 12.7. Gas exchange between the interior of the leaf structure and atmosphere occurs through stomates on leaf surfaces. It is understood in this illustration that both evaporative and condensed forms of water play a role in gas transport especially within the confines of leaf structure.

tension of carbon dioxide and energy necessary to maintain life processes.

$$C_6H_{12}O_6 + 6\,O_2 \longrightarrow 6\,CO_2 + 6\,H_2O + \text{energy}$$

Glucose   **Diffused gas**
stored in   **during plant**
plant   **respiration**

The carbon dioxide is readily diffused from most plants at night. Bromeliads and some other tropical and xeric plants close their stomates during the day to conserve moisture and then open them during nocturnal hours to absorb carbon dioxide. Carbon dioxide is partially assimilated by this mechanism until daylight and photo-induced reactions culminate in the reduction of carbon dioxide into sugars.

Carbon dioxide can be incorporated into different plant species by different routes. Carbon dioxide is incorporated into organic compounds by the Calvin cycle or C_3 pathway in many plants including green algae, spinach, and other common plants. The *first* compound produced from carbon dioxide assimilation is the three-carbon (C_3) compound known as *phosphoglyceric acid.*

In the cases of corn, sorghum, crabgrass, and tropical grasses such as sugar cane, radioactive-labeled carbon dioxide ($^{14}CO_2$) first appears in four-carbon (C_4) dicarboxylic acids (oxaloacetate, aspartate and malate).

The C_4 and C_3 pathways may occur jointly in some plant orders, but most plant geneticists theorize that these assimilative pathways for carbon evolved independently several times.

The C_4 plants predominate in tropical regions of the world where temperatures are usually high, light intensity is great, and water supplies are low.

C_3 Plants and the Calvin Cycle

A reductive pentose phosphate cycle known as the Calvin cycle is used by C_3 plants for carbon dioxide fixation. Experiments have revealed that these so-called dark reactions of photosynthesis occur in a cyclic sequence comprised of three distinct divisions denoted as (1) *carboxylative,* (2) *reductive,* and (3) *regenerative* phases (Figure 12.8).

Ribulose 5-phosphate is phosphorylated by ATP to produce ribulose 1,5-diphosphate. Carboxylation of this compound by atmospheric carbon dioxide, mediated by *ribulose 1,5-diphosphate carboxylase* results in a transient β-keto acid intermediate. This unstable intermediate is readily cleaved in the presence of water to produce two molecules of 3-phosphoglyceric acid. Pertinent structures for this carboxylative process, or Phase 1, of carbon dioxide fixation are outlined:

Phase 1—Carboxylation

Phosphoglyceric acid produced by this reaction is then phosphorylated by the action of 3-phosphoglyceric kinase at the expense of ATP to yield 1,3-diphosphoglyceric acid. Reduction of this compound by an $NADP^+$-specific glyceraldehyde 3-phosphate dehydrogenase completes the two-step reduction phase of the Calvin cycle.

molecule (or possibly cellulose). On the other hand, *if the cycle is to continue,* ribulose 1,5-diphosphate must be produced from some fructose 1,6-diphosphate formed during the cycle. Therefore, fructose 1,6-diphosphate yields fructose 6-phosphate through the action of a phosphatase enzyme (Figure 12.8). The fructose 6-phosphate is transformed into

---

Phase 2—A two-step reduction stage

Step 1

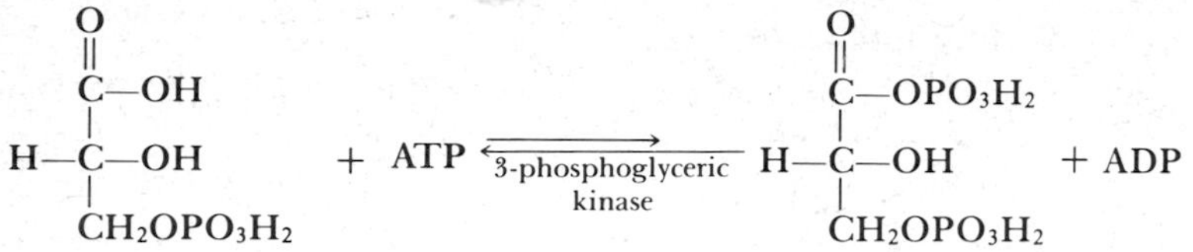

3-Phosphoglyceric acid + ATP →(3-phosphoglyceric kinase) 1,3-Diphosphoglyceric acid + ADP

Step 2

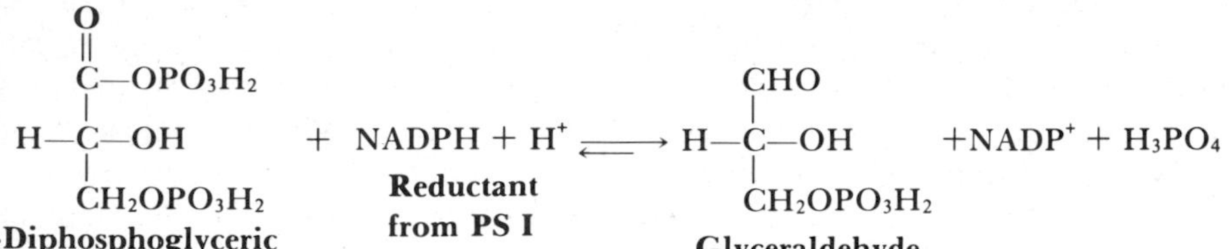

1,3-Diphosphoglyceric acid + NADPH + H^+ (Reductant from PS I) ⇌ Glyceraldehyde 3-phosphate + $NADP^+$ + H_3PO_4

Reduced nicotinamide nucleotide from PS I and about 50% of the ATP consumed in the Calvin cycle is used during the reduction phase. Glyceraldehyde resulting from the reductive process is representative of a triose-type sugar.

During the regenerative phase of the Calvin cycle, two glyceraldehyde molecules can condense into a single hexose molecule that may be incorporated into a growing starch

erythrose 4-phosphate and xylulose 5-phosphate by a *transketolase* in the presence of glyceraldehyde 3-phosphate. This transketolase action is followed by an *aldolase* and another *transketolase* reaction that culminate in the production of ribulose 1,5-diphosphate. Each of the three key steps involving these enzyme-mediated reactions during the regenerative steps of the Calvin cycle are illustrated below:

Phase 3—Regenerative stages of the Calvin cycle

Step 1

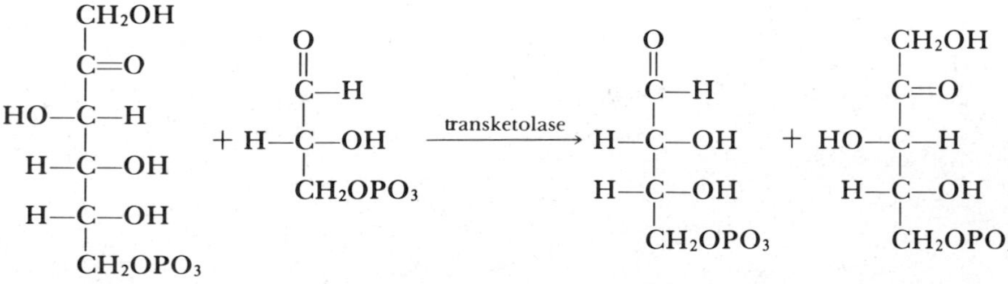

Fructose 6-phosphate + Glyceraldehyde 3-phosphate →(transketolase) Erythrose 4-phosphate + Xylulose 5-phosphate

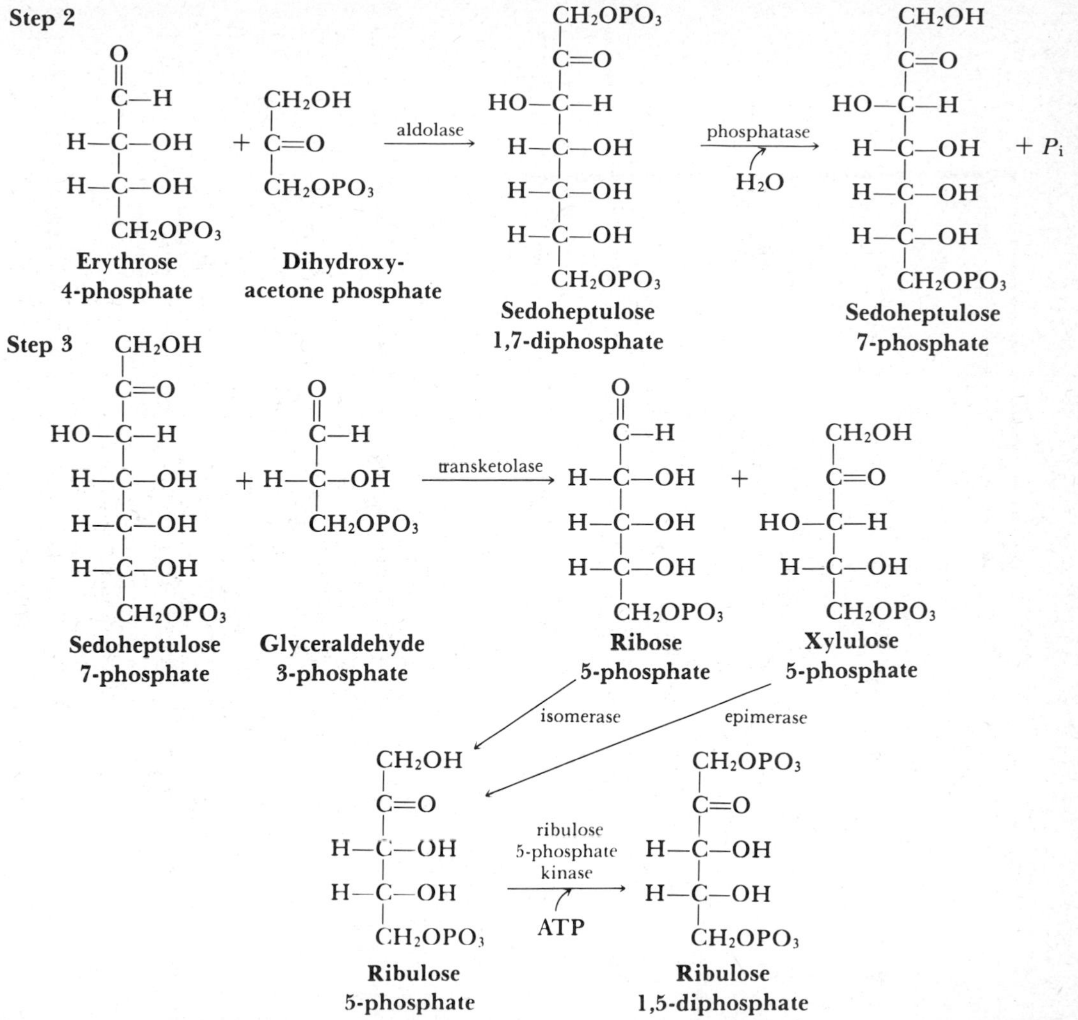

Step 2

Erythrose 4-phosphate + Dihydroxy-acetone phosphate →(aldolase) Sedoheptulose 1,7-diphosphate →(phosphatase, H_2O) Sedoheptulose 7-phosphate + P_i

Step 3

Sedoheptulose 7-phosphate + Glyceraldehyde 3-phosphate →(transketolase) Ribose 5-phosphate + Xylulose 5-phosphate

Ribose 5-phosphate →(isomerase) Ribulose 5-phosphate

Xylulose 5-phosphate →(epimerase) Ribulose 5-phosphate

Ribulose 5-phosphate →(ribulose 5-phosphate kinase, ATP) Ribulose 1,5-diphosphate

The total integrated reaction series leading to ribulose 1,5-diphosphate (pentose phosphate pathway) leads to a *net production* of 1 mole of fructose 6-phosphate:

6 ribulose 1,5-diphosphate + 6 CO_2 +
        18 ATP + 12 NADPH + 12 H^+
                    ↓
6 ribulose 1,5-diphosphate + 1 fructose
    6-phosphate + 17 P_i + 18 ADP + 12 $NADP^+$

According to the Calvin cycle, 36 carbon atoms (C_{36}) in 12 molecules of glyceraldehyde 3-phosphate ($C_{36}/12 = C_3$) are present at the end of the so-called reduction phase (Phase 2). During the regeneration phase (steps 1–3) 1 molecule of fructose 6-phosphate (C_6) and 6 molecules of ribulose 1,5-diphosphate (C_{30}) appear. In the very last

reaction of the regeneration phase (step 3), 6 molecules of ribulose 1,5-diphosphate are available for the initial carboxylative events of carbon dioxide fixation. Figure 12.8 illustrates the complete stoichiometry for the Calvin cycle as well as the overall photosynthetic carbon dioxide reduction cycle.

C_4–Dicarboxylic Acid Pathway—The Hatch–Slack Pathway

Studies conducted by Hatch and Slack, reported in 1966, indicated that important crop plants such as corn, sorghum, sugar cane, and many other plants of tropical origin assimilate carbon dioxide in a different fashion than the recognized pathways in C_3 plants. Unlike C_3 plants, these plants demonstrated

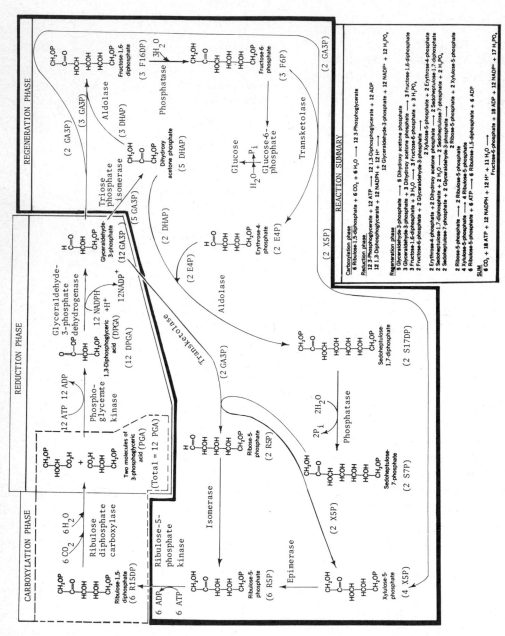

FIGURE 12.8. Principal reactions within the Calvin cycle and a summary of pertinent stoichiometry for reactions.

an ability to incorporate carbon dioxide into selected dicarboxylic acids, namely *aspartic acid* or *malate*, instead of 3-phosphoglyceric acid. The involvement of these four carbon skeletons served as a conceptual basis for the C_4 pathway or *Hatch–Slack pathway.*

Anatomical features of leaf vascular elements for C_4 plants are also different from the C_3 plants. Plants having the C_4 pathway exhibit a distinct layer of bundle sheath cells that surround the xylem and phloem of the vascular bundle. Furthermore, one or more layers of mesophyll cells surround the bundle sheath cells. This distinct *Kranz-type* anatomy has been recognized for many years as an evolutionary mechanism among tropical

Careful isolation of bundle-sheath cells from mesophyll cells has revealed that the C_3 pathway *occurs largely or entirely within the bundle-sheath cells.* Chloroplasts of mesophyll cells, however, contain the enzyme known as phosphoenol pyruvic acid carboxylase (PEPC). Given the circumstance of lower carbon dioxide concentration levels in C_4 plants than C_3 plants, PEPC exhibits a much higher affinity for incorporating carbon dioxide into an organic intermediate than the enzyme used to initiate the C_3 cycle: ribulose 1,5-diphosphate carboxylase. Furthermore, the initial entrapment of carbon dioxide relies on the ability of PEPC to catalyze the following reaction:

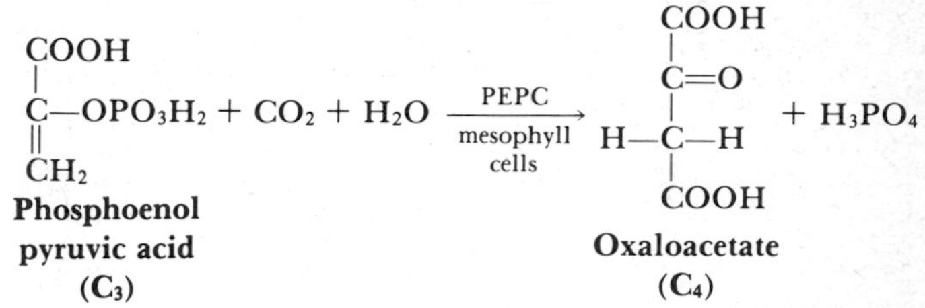

$$
\begin{array}{l}
\text{COOH} \\
| \\
\text{C—OPO}_3\text{H}_2 + \text{CO}_2 + \text{H}_2\text{O} \\
|| \\
\text{CH}_2
\end{array}
\xrightarrow[\substack{\text{mesophyll} \\ \text{cells}}]{\text{PEPC}}
\begin{array}{l}
\text{COOH} \\
| \\
\text{C=O} \\
| \\
\text{H—C—H} \\
| \\
\text{COOH}
\end{array}
+ \text{H}_3\text{PO}_4
$$

Phosphoenol pyruvic acid (C_3) **Oxaloacetate** (C_4)

plants that is linked to water conservation processes. Water conservation occurs since the vascular system is entirely separated from stomates that reside on leaf surfaces (Figure 12.9). Although Kranz-type anatomy conserves water within plant tissues, this modification of plant structure also restricts the availability of carbon dioxide necessary for reduction into carbohydrates.

Since the mesophyll cells of some C_4 plants have a high concentration of *malic dehydrogenase,* whereas other C_4 plants have *alanine–aspartic transaminase* (AAT), *two different groups of C_4 plants are recognized.* These differences permit the reduction of oxaloacetate to malate, or oxaloacetate may be transaminated by AAT to form aspartic acid. The two possible reactions can be expressed as follows:

Malate formation

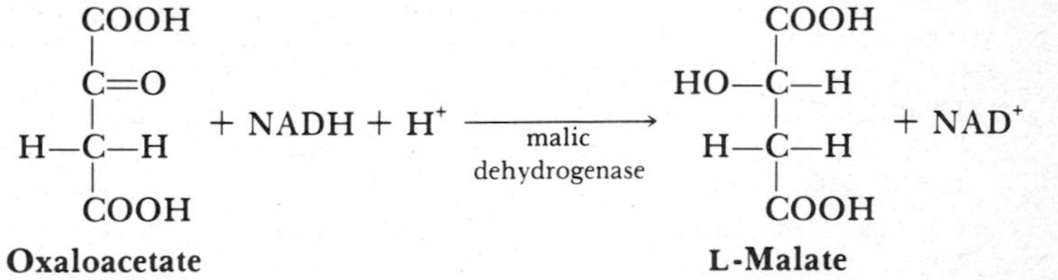

$$
\begin{array}{l}
\text{COOH} \\
| \\
\text{C=O} \\
| \\
\text{H—C—H} \\
| \\
\text{COOH}
\end{array}
+ \text{NADH} + \text{H}^+
\xrightarrow[\text{dehydrogenase}]{\text{malic}}
\begin{array}{l}
\text{COOH} \\
| \\
\text{HO—C—H} \\
| \\
\text{H—C—H} \\
| \\
\text{COOH}
\end{array}
+ \text{NAD}^+
$$

Oxaloacetate **L-Malate**

Aspartic acid formation

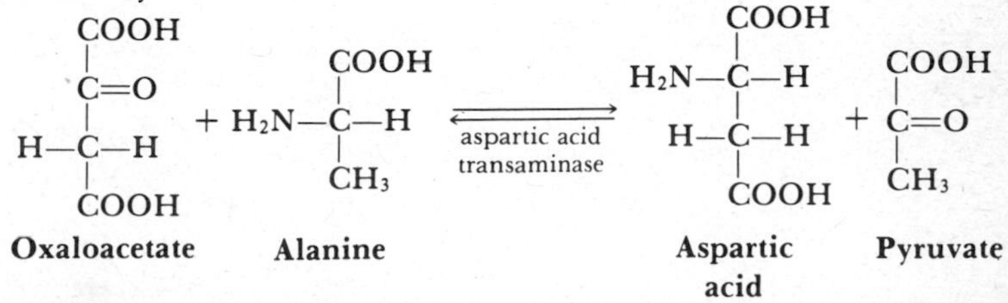

$$
\begin{array}{l}
\text{COOH} \\
| \\
\text{C=O} \\
| \\
\text{H—C—H} \\
| \\
\text{COOH}
\end{array}
+
\begin{array}{l}
\text{COOH} \\
| \\
\text{H}_2\text{N—C—H} \\
| \\
\text{CH}_3
\end{array}
\underset{\text{transaminase}}{\overset{\text{aspartic acid}}{\rightleftarrows}}
\begin{array}{l}
\text{COOH} \\
| \\
\text{H}_2\text{N—C—H} \\
| \\
\text{H—C—H} \\
| \\
\text{COOH}
\end{array}
+
\begin{array}{l}
\text{COOH} \\
| \\
\text{C=O} \\
| \\
\text{CH}_3
\end{array}
$$

Oxaloacetate **Alanine** **Aspartic acid** **Pyruvate**

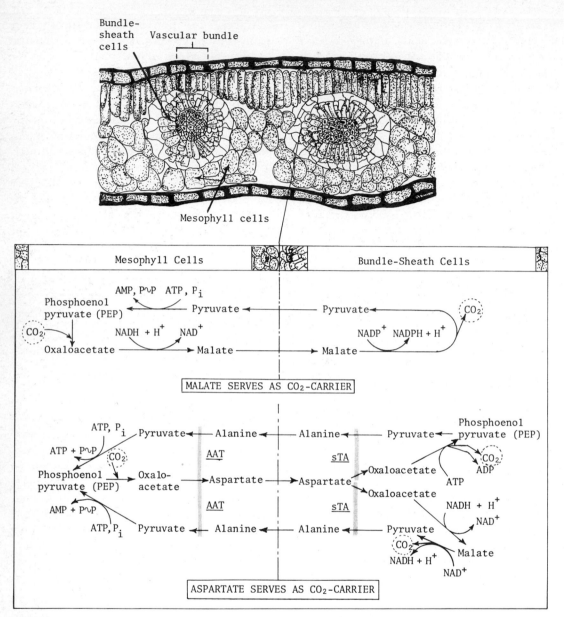

FIGURE 12.9. Anatomical cross section for a C_4 plant leaf illustrating the relationship of bundle-sheath cells to mesophyll cells and the respective biochemical pathways that occur in these differentiated tissues. Note that two reaction schemes are shown depending whether carbon dioxide is carried by malate or aspartate (both carboxylic acids). Key abbreviations for the aspartate scheme: AAT, aspartic acid transminase, is active over the gray area; whereas sTA, a specific transaminase, functions during aspartate → oxaloacetate and pyruvate → alanine transformations.

Aside from these notable reactions, it should be recognized that pyruvic acid from bundle-sheath cells may be converted to phosphoenol pyruvic acid, which serves as a carbon dioxide trapping agent. This reaction is accomplished by *pyruvate phosphate dikinase:*

$$\underset{\textbf{Pyruvate}}{\underset{\text{CH}_3}{\overset{\text{COOH}}{\overset{|}{\underset{|}{\text{C}=\text{O}}}}}} + \text{ATP} + \text{H}_3\text{PO}_4 \xrightarrow[\text{mesophyll cells}]{\overset{\text{pyruvate}}{\underset{\text{dikinase}}{\overset{\text{phosphate}}{}}}}$$

$$\underset{\underset{\textbf{pyruvate}}{\textbf{Phosphoenol}}}{\underset{\text{CH}_2}{\overset{\text{COOH}}{\overset{|}{\underset{||}{\text{C}-\text{OPO}_3\text{H}_2}}}}} + \text{AMP} + PP_i$$

The chloroplasts of bundle-sheath cells contain enzyme systems that may process malate or aspartate into pyruvate and CO_2. Carbon dioxide released by the preceding route is then assimilated by the C_3 cycle (Calvin cycle), which resides *only* in the bundle-sheath cells.

$$\underset{\textbf{L-Malate}}{\underset{\text{COOH}}{\overset{\text{COOH}}{\overset{|}{\underset{|}{\overset{\text{HO}-\text{C}-\text{H}}{\underset{\text{H}-\text{C}-\text{H}}{}}}}}}} + \text{NADP}^+ \xrightarrow{\overset{\text{NADP}^+\text{-specific}}{\text{malic enzyme}}}$$

$$\underset{\textbf{Pyruvate}}{\underset{\text{CH}_3}{\overset{\text{COOH}}{\overset{|}{\underset{|}{\text{C}=\text{O}}}}}} + \text{NADPH} + \text{H}^+ + \text{CO}_2$$

A cursory survey of the reactions presented for the bundle-sheath cells and the mesophyll cells indicates the presence of a unique mechanism for plant survival. The C_4 cycle of the bundle-sheath cells provides carbon dioxide necessary for the Calvin (C_3) cycle that resides there. The mesophyll cells, however, demonstrate the C_4 cycle and supply malate or aspartate (*depending on the plant species*) to the bundle-sheath cells where they

can be decarboxylated. The carbon dioxide produced during decarboxylation elevates intracellular carbon dioxide levels of the bundle-sheath cells by 15 to 60 times over the levels found in plants having a C_3 pathway. The combined reactions for the Hatch–Slack pathway and C_4 plants are outlined in Figure 12.9.

STARCH AND SUCROSE FORMATION FROM PHOTOSYNTHESIZED MONOSACCHARIDES

Glucose 6-phosphate produced in the Calvin (C_3) cycle is interconvertible with glucose 1-phosphate:

$$\text{Calvin (C}_3\text{) cycle} \rightarrow \rightarrow \rightarrow \rightarrow \underset{\underset{\text{glucose 1-phosphate}}{\uparrow \downarrow}}{\text{glucose 6-phosphate}}$$

and both phosphorylated glucose forms are fundamental factors in starch biosynthesis.

Starch formation must be initiated by establishing an energy-rich ADP derivative of glucose that facilitates the addition of glucose to an existing starch molecule.

$$\text{Glucose 1-phosphate} + \text{ATP} \xrightarrow{\overset{\text{ADP-glucose}}{\text{pyrophosphorylase}}} \text{ADP-glucose} + PP_i$$

Unlike the synthesis of glucose polymers (glycogen) in animals wherein UDP-glucose is required for glycosyl polymerization reactions, plants use the ADP-glucose for starch synthesis. The glycosyl group of this derivative is transferred to an existing linear starch molecule (having n glycosyl subunits) such as *amylose,* by a starch synthase enzyme. The elongated amylose molecule ($n + 1$ glycosyl subunits) shows consistent $\alpha(1\rightarrow4)$ linkages between glycosyl subunits unless glucose transfer from ADP-glucose to starch is mediated by a *branching enzyme.* In this case, the so-called branching enzyme establishes an $\alpha(1\rightarrow6)$ link between two glycosyl groups that can provide a future site for $\alpha(1\rightarrow4)$-glycosyl bonding. Steps such as this also initiate the structure of branched starches known as *amylopectins*. Glycosyl branching of this type may intermittently occur over 20 residue spans of $\alpha(1\rightarrow4)$-linked glucose molecules.

charidic components of the disaccharide may be cycled into the construction of new cells or metabolic intermediates.

The use of sucrose by consuming cells is begun by an enzyme-mediated hydrolysis that

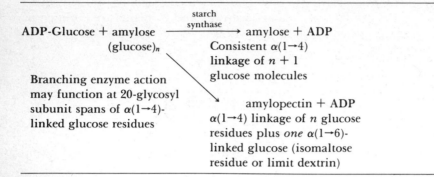

ADP-Glucose + amylose (glucose)$_n$ $\xrightarrow{\text{starch synthase}}$ amylose + ADP

Consistent $\alpha(1\rightarrow4)$ linkage of $n + 1$ glucose molecules

Branching enzyme action may function at 20-glycosyl subunit spans of $\alpha(1\rightarrow4)$-linked glucose residues

amylopectin + ADP $\alpha(1\rightarrow4)$ linkage of n glucose residues plus *one* $\alpha(1\rightarrow6)$-linked glucose (isomaltose residue or limit dextrin)

The previous reactions reflect starch synthesis events in chloroplasts, but starch is also formed in many other plant tissues and cells such as *leukoplasts* that do not conduct photosynthesis. These cells utilize photosynthate in the form of glucose 1-phosphate to elongate existing $\alpha(1\rightarrow4)$ glucose polymers under the auspices of a phosphorylase enzyme.

Glucose 1-phosphate + amylose (glucose)$_n$

$\updownarrow$ starch phosphorylase

amylose + P_i (glucose)$_{n+1}$

Exclusive of the role for ADP-glucose, many chloroplasts also reserve an ability to form UDP-glucose from glucose 1-phosphate. This glucose derivative is readily combined with fructose 6-phosphate to produce sucrose. Sucrose is a disaccharide common to nearly all higher plants that serves as a transport vehicle for the phloem translocation of photosynthate to those plant cells that do not participate in photosynthesis. Steps leading to sucrose formation are largely reversible until *sucrose phosphatase* (Figure 12.10) conducts an irreversible phosphate hydrolysis from sucrose phosphate. Nonphotosynthetic plant cells generally use sucrose formed by these steps for energy, or the monosac-

requires UDP. A UDP-glucose derivative and fructose are produced from the hydrolysis. Elevated pyrophosphate levels exist in sucrose-consuming cells due to deficits of pyrophosphatase, and this makes a previously irreversible formation of sucrose readily reversible. The enzymes known as UDP-glucose pyrophosphorylase and starch phosphorylase then cooperate in the polymerization of glucose and the elongation of starch structures.

PHOTORESPIRATION IN SELECTED PLANTS

Aside from photosynthetic events and the reduction of carbon dioxide to form simple sugars, plants also conduct respiratory processes that are similar to animals' and microorganisms'. Carbohydrates are oxidized by glycolysis and the tricarboxylic acid cycle, and other carbohydrates may be funneled into the synthesis of fats and proteins before being oxidized. The β-oxidation of fats and common reactions of protein and amino acid catabolism are often found in the organelles common to plants and animals (e.g., mitochondria, microsomes).

In addition to normal respiratory activities, many plant species exhibit a phenom-

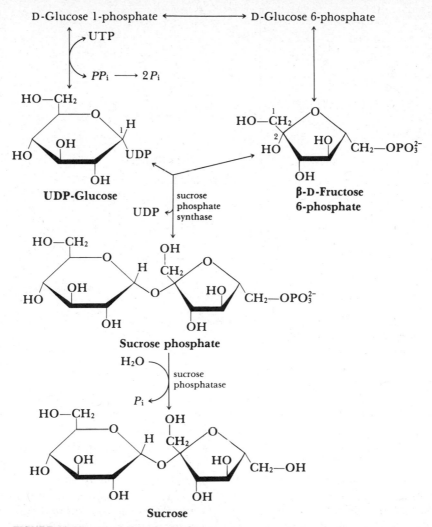

FIGURE 12.10. The biosynthesis of sucrose.

enon of *photorespiration*. Photorespiration is defined as a light-dependent uptake of oxygen along with a concomitant production of carbon dioxide. For food crop species that actively engage in photorespiration, this process generally *decreases* the apparent rate of photosynthesis; *carbon dioxide is released;* and both *crop growth and yields are markedly decreased.*

A key reaction in the photorespiratory process involves the conversion of ribulose 1,5-diphosphate into *phosphoglycolate* and 3-phosphoglycerate. This reaction is mediated by the enzyme known as *ribulose 1,5-diphos-* *phate carboxylase.* The enzyme requires oxygen, and the overall reaction reflects an oxygenase-type reaction (Figure 12.11).

The preceding reaction is light dependent because the enzyme responsible for the reaction must be activated by the pH of the stroma as well as specific magnesium ion concentrations released from the thylakoids into the stroma during illumination. Moreover, the enzyme critical to this overall reaction cannot be regenerated from phosphoglycerate unless ATP and NADPH, produced from light reactions, occur in sufficient amounts. A superficial empirical view

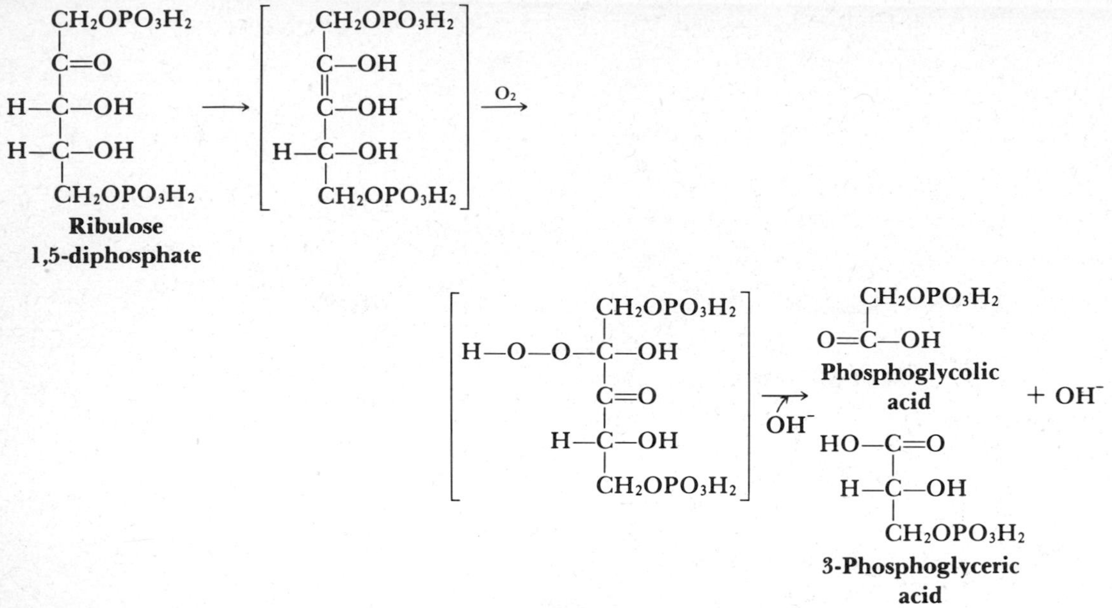

FIGURE 12.11. Glycolic acid formation is mediated by ribulose 1,5-diphosphate carboxylase in the presence high oxygen concentrations and low carbon dioxide.

of photorespiration suggests that it is the reverse of photosynthesis:

$$CO_2 + H_2O \xrightleftharpoons[\text{photorespiration}]{\text{photosynthesis}} (CH_2O) + O_2$$

Carbohydrate

but an inspection of Figure 12.12 indicates that photorespiration reactions are really quite different. In fact, the photorespiratory process is also largely dependent on the partial pressures of environmental oxygen and carbon dioxide according to the classical principles outlined by Le Chatelier.

A high concentration of O_2/CO_2 can encourage a net loss of carbon from the plant to the surroundings by photorespiration. When photosynthesis and photorespiration occur in a dynamic balance, this is called a *compensation point (CP)*. Using carbon dioxide as an index, 50 to 100 ppm represents a typical *CP* level for many species of temperate plants. Of course, the partial pressure of oxygen and light illumination can be critical to such measurements and under extremely high oxygen levels, photosynthesis may be inhibited.

Phosphoglycolate resulting from ribulose 1,5-diphosphate carboxylase reactions can liberate up to 25% of the carbon in glycolic acid as carbon dioxide along with the production of 3-phosphoglycerate. The latter compound readily enters the Calvin (C_3) cycle in order to maintain that reaction.

The *CP* levels for C_3 and C_4 plants are quite different, and these differences translate into significantly different productivity yields for different food crop plants. Most C_3 plants have a common *CP* value of 40 to 75 ppm at 25°C, whereas C_4 plants have a value of <10 ppm; atmospheric carbon dioxide levels are in the range of 0.03% or 300 ppm. Since the *CP* value for all plants can be influenced by a number of factors, increases in light intensity on the leaf canopies of crop plants result in an increase of environmental oxygen accompanied by a relative drop in carbon dioxide. Furthermore, daytime temperature increases lead to increases in the *CP* value for carbon dioxide. The net result of these factors is a decrease in the demonstrated photosynthetic activity of C_3 plants but not of the C_4 plants.

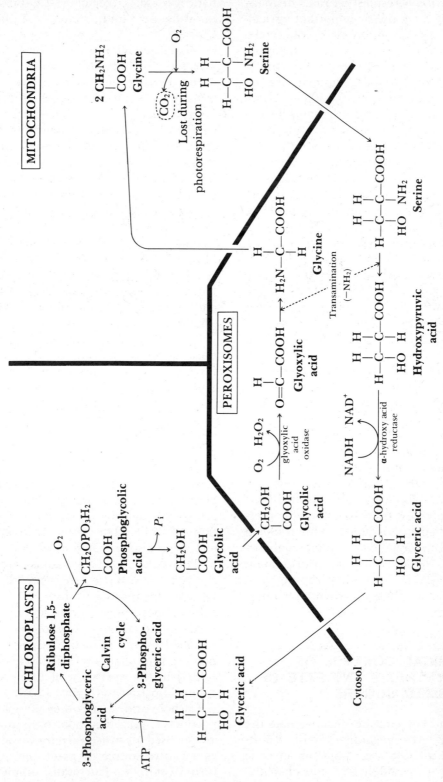

FIGURE 12.12. Photorespiration and the metabolism of glycolic acid involves the interaction of three organelles.

Because photorespiration rates of many plants limit their use as important agricultural crops, genetics and plant breeding could be used to improve the photosynthetic potential of many plant species by reducing or eliminating the process. Photorespiration alone is estimated to consume 50% of assimilated carbon dioxide in some plants including cereal plants, wheat, rice, leguminous plants, sugar beets, and many others.

Some food crops essential to American agriculture such as corn and other, less important crops like sugar cane and sorghum do not exhibit photorespiration. However, the presence of leaf peroxisomes, which are involved in photorespiratory pathways, indicate that at least some and perhaps all of the necessary individual reactions may be possible. Nevertheless, these plants principally depend on the C_4 pathway, which has a better ability to assimilate carbon dioxide and reassimilate carbon dioxide produced during photorespiration.

One practical way to reduce the effects of photorespiration in certain high-cash crops is to enrich the plant environment with carbon dioxide well over normal levels of oxygen. According to theory and practice, these conditions can greatly improve the growth of some plants. Unfortunately, however, these abnormally high carbon dioxide levels may be met with abnormal gaseous diffusion through stomates and throughout the intercellular air spaces of leaves. This problem may be accompanied by concomitant abnormalities in the operation of transpiration processes, which are entirely counterproductive for increasing photosynthesis rates.

BIOCHEMICAL CONTROL OF PHOTOSYNTHESIS AND FATE OF SYNTHESIZED SUGARS

Some of the key enzymes that regulate the Calvin (C_3) cycle include ribulose 1,5-diphosphate carboxylase; 3-phosphoglyceric acid kinase; glyceraldehyde 3-phosphate dehydrogenase; fructose 1,6-diphosphate phosphatase; and ribulose 5-phosphate kinase.

Most if not all of the previously listed enzymes are activated or deactivated by photo-induced changes in stroma pH and/or ionic balance (Cl^- and Mg^{2+} ions). In still other cases, photo-induced mechanisms may control the absolute concentrations of key enzymes. Whatever mechanisms may be involved, the principal enzymes necessary for the carboxylative and reductive phases of the C_3 cycle show increased concentrations within plants during illumination and decrease during darkness.

Activators and *inhibitors* also regulate the action of some enzymes such as ribulose 1,5-diphosphate carboxylase. Specifically, fructose 6-phosphate acts as an activator and fructose 1,6-diphosphate serves as an inhibitor for carboxylase activity.

Ferredoxin-linked mechanisms also function in plants. Photo-induced reduction of ferredoxin serves as an activator for fructose 1,6-diphosphatase, whose subsequent actions involve the conversion of fructose 1,6-phosphate into fructose 6-phosphate. Recall that the latter compound is responsible for activating the carboxylase enzyme.

Other enzymes including 3-phosphoglyceric kinase, ribulose 5-phosphate kinase, and glyceraldehyde 3-phosphate dehydrogenase are all activated by photo-induced production of ATP. The latter dehydrogenase enzyme is also activated by NADPH.

The flow of carbon atoms and sugars through the C_3 cycle is regulated by the interaction of those activators and inhibitors cited above plus other identified and unidentified factors. Carbon compounds supplied by the reductive reactions of the pentose phosphate cycle (C_3 cycle) are connected to the operation of numerous biochemical pathways (Figure 12.13).

Carbohydrates and other photosynthesis products (reduced carbon compounds) that are not used for structural purposes are stored as a future source of plant energy in the form of starches, fructosans, simple sugars,

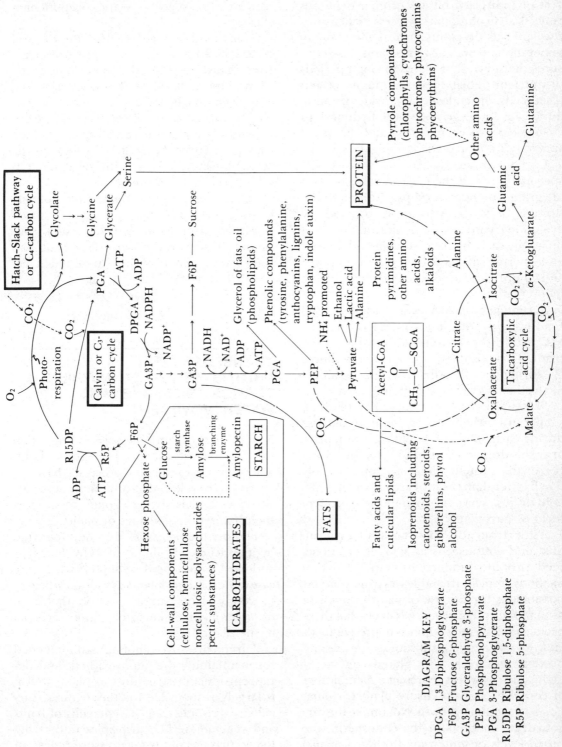

FIGURE 12.13. Key biochemical interconversions within plant cells and tissues that lead to protein, fat, and carbohydrate products.

DIAGRAM KEY
DPGA 1,3-Diphosphoglycerate
F6P Fructose 6-phosphate
GA3P Glyceraldehyde 3-phosphate
PEP Phosphoenolpyruvate
PGA 3-Phosphoglycerate
R15DP Ribulose 1,5-diphosphate
R5P Ribulose 5-phosphate

proteins, and fats or oils. The storage of direct or indirect products of photosynthesis reflects an evolutionary process that ensures the survival of plants from generation to generation, as well as a mechanism that ensures the survival of plants through periods of nutrient deficits. For both monocot and dicot seeds, the endosperm provides the most important reserve of carbohydrates to ensure embryo germination and the initial survival of a new plant seedling.

Carbohydrates stored in seed endosperms represent the most valuable portion of grain (monocot) crops. Although the storage of carbohydrates is dictated by evolutionary processes, significant improvements in carbohydrate storage within seed endosperms can be achieved by traditional and more advanced methods of genetic manipulation.

Perennial crop species such as those used for forage production store starch or fructosans in the overwintering portion of the plant. This action ensures an energy source for spring growth. Alfalfa, a classic example of this behavior, stores starch in a taproot.

Proteins may be warehoused in a fashion similar to carbohydrates for eventual use during early stages of seedling growth, plant growth, after grazing injury, or for spring regrowth. Protein concentrations, usually much lower than those of accompanying carbohydrates, serve as an important storage form of nitrogen.

Aside from the importance of soybean, flax, and cottonseeds as sources of animal feed protein, modern protein extraction methods from seeds and leaves can offer new possibilities for the use of plant proteins in human foods. However, it is likely that these proteins will serve as adjuncts to present foods and not as complete substitutes for existing protein resources. The gluten content of grains such as wheat represent a natural store of cereal protein to ensure plant germination and seedling growth. Not unlike the carbohydrate contents of the endosperm, the cereal levels of gluten (and other proteins) can be modified by genetic and hybridiza-tion methods to suit food crop demands instead of species-specific evolutionary dictates.

Many seeds contain lipids in the form of triacylglycerols and related glycerol structures that store reduced carbon. Some of the most important oil seed species are flax, sunflowers, soybeans, and peanuts. Seed oils also occur in cotton, safflower, mung beans, sesame, castor beans, and rape. In some cases, oil may be distributed throughout the seed; however, in corn, the oil content is largely relegated to the germ.

Large amounts of sugars and polysaccharides along with some proteins are stored in various plant organs, structures, and fruits (mature flower ovaries) that serve as human food sources. Some of these are listed in Table 12.1.

TRANSPORT OF PHOTOSYNTHESIZED NUTRIENTS, TURGOR PRESSURE, AND WATER MOVEMENT

The survival of land plants is contingent on the adequate exposure of leaves to sunlight. Therefore, evolutionary processes have favored the development of a physical superstructure that is strong enough to support the photosynthetic tissues of plants.

Photosynthetic tissues may be supported by interaction of billions of individual cells endowed with a rigid cellulose wall. In the case of herbaceous plants as well as the leaves of woody plants, support may be the result of *turgor pressure* generated within each cell of the plant.

Turgor pressure is not only fundamental for maintaining the photosynthetic and floral reproductive structures of plants, but it is largely responsible for the textures, succulence, crispness, and acceptability of fruits and vegetables. The principles underlying the occurrence of turgor pressure represent a complex interaction of water with cell sap,

Table 12.1. Some Plant Sources of Stored Sugars and Polysaccharides

Plant storage site	Representative plant types
Buds	Brussels sprouts
Floral structures	Broccoli and cauliflower
Leaves	Cabbage, kale, lettuce, parsley, spinach, and rape
Roots or rhizomes	Sweet potatoes and Irish potatoes
Ovary wall	Apricot, banana, cantaloupe, cherry, cucumber, eggplant, peach, plum, pumpkin, squash, tomato, watermelon
Root and hypocotyl	Carrots, parsnips, rutabagas, turnips, and sugar beets
Leaf petioles	Celery and rhubarb
Leaves and stems	Asparagus, kohlrabi, and onion
Flower receptacle	Apple, pear, strawberry, and blueberry
Wall of ovary and seeds	Beans (wax and green)

which is retained inside the semipermeable plasma membranes of plant cells.

The actual metabolizing portion of a plant protoplast usually represents only 5 to 15% of the whole cell depending on the cell type. The rest of the cell contains one or more large vacuoles holding cell sap. Cell sap is a mixture of amino acids, short peptides, organic acids, salts, and, most important, hydrophilic carbohydrates including simple sugars that are dissolved in water. These organic solutes represent direct or derived products of photosynthetic processes conducted by plant cells. The metabolizing protoplast, along with the cell vacuoles, is surrounded by a selective semipermeable membrane whose structure and shape is externally reinforced by a polysaccharide or cellulose wall.

The semipermeable membrane displays active transport properties for many inorganic and organic solutes. Therefore, the membrane is able to retain high concentrations of sugars and other solutes within the cell, contrary to the dictates of high solute concentration gradients produced by low extracellular solute concentrations.

Assuming the selective retention of a relatively high concentration of hydrophilic substances within a plant cell (as opposed to the aqueous surroundings of the cell), water molecules from the surroundings have a tendency to pass through the plasma membrane into the cell. This action serves as a physical means for equalizing water molecule concentrations on both sides of the membrane, since external and internal systems represent different thermodynamic states.

The tendency for water molecules to traverse the plasma membrane and enter a cell generates *osmotic pressure*. As water enters a plant cell, internal pressure called *turgor pressure* is generated (Figure 12.14). This pressure counterbalances the effect of osmotic pressure and inhibits the cellular entry of additional water molecules.

As long as the osmotic equilibrium is constant, vacuoles and protoplasts push against the slightly elastic walls of individual cells causing a slight degree of stretching and a firm cellular structure. This cellular state exists as long as plasma proteins responsible for selective permeability remain active. As soon as denaturation of these proteins occurs by heat, freezing, or dessicated storage, turgor pressure is lost and a permanent flaccid condition prevails throughout all injured cells.

Solute–water interactions are responsible for important phenomena other than turgor pressure. For example, solute–water con-

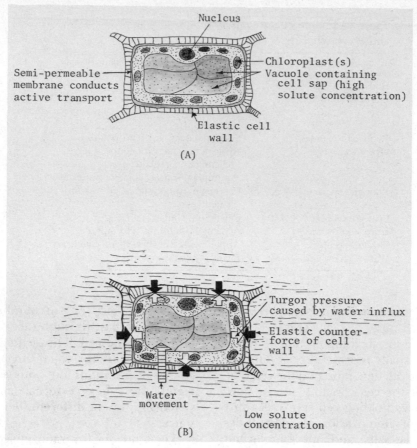

FIGURE 12.14. (A) Key components of a plant cell. (B) Direction of water movement from an area of high concentration into a cell having a relatively high concentration of solute. Note that the direction of the elastic force produced by the cell wall counteracts the expansive force of water entering the cell to produce turgor pressure within the cell.

centration gradients also contribute to the overall movement of water in plants as well as the maintenance of turgid leaf structures.

Water movement throughout plants results from *evapotranspiration* mechanisms caused by the evaporation and controlled loss of water from stomates located on leaves and stems. As long as the relative humidity of air surrounding a plant is less than 100%, and water is available from soil, a water gradient will exist over the periphery of plant structures. This gradient effectively drives water from the soil to the atmosphere via exposed leaf and plant surfaces covered by stomates. Water loss is minimal from other sites of plants owing to the presence of a waxy cuticle that

prohibits water loss. If the rate of evapotranspiration happens to exceed water replenishment within plant cells, wilting may occur throughout the plant (Figure 12.15 and Table 12.2).

Water transport and the translocation of plant sap is accomplished in vascular green plants (nonalgae plants) by xylem and phloem systems (Figure 12.16). The xylem vessels are chiefly involved in water and mineral transport from the roots through the stem and into the leaves. Phloem, however, translocates photosynthate (carbohydrates, amino acids, etc.) produced in the leaves, down the plant stem(s) for use by other tissues or storage (e.g., root or tuber storage).

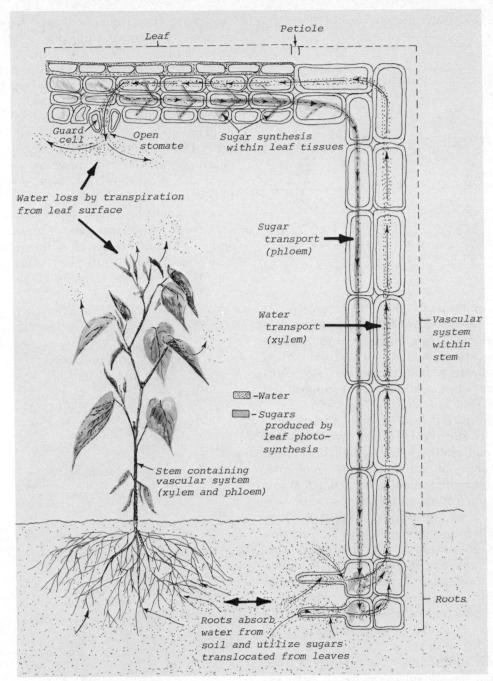

FIGURE 12.15. Sugars (photosynthate) produced by photosynthesis in leaves and other photosynthetic tissues are conducted to roots and other tissues for use by the phloem route, whereas water is translocated by the xylem.

Table 12.2. Water Requirements for Some Food Crops and Plants

Crop plant	Water consumption (lb)
Alfalfa	800
Barley	450
Corn	375
Millet	255
Oats	600
Potatoes	395
Soybeans	750
Sweet clover	750
Wheat	480

NOTE: Large amounts of water must be absorbed and transported by the vascular systems of plants. Based on the production requirements for some important food crop plants, average water requirements in pounds, for the production equivalent of *1 lb* of dry vegetative matter have been outlined.

The phloem of woody and perennial plants is also responsible for translocating nutrients up the stem. This is necessary especially during the spring season when stored carbon compounds in roots, tubers, and other storage sites must be mobilized to support bud growth and flowering.

The upward flow of water through the xylem is ensured by forces of *cohesion* and *adhesion*. Cohesive forces result from mutual associations of individual water molecules caused by their dipolar attractions. Water molecules lost from the leaves by evapotranspiration are replaced by water that enters root cells by osmosis and adhesion. Adhesion results from the mutual interactions between water and stationary components on the inner capillary walls of xylem vessels. Therefore, these combined attractive forces of adhesion and cohesion contribute to the upward movement of water in a plant (Figure 12.15).

Sugar and photosynthate transport from leaves to roots involves cell-to-cell diffusion processes, as well as intracellular cyclosis plus other poorly understood mechanisms. Since most translocations of sugars exceed calculated rates of diffusion, the circular streaming motion or *cyclosis* within phloem cells probably has a great influence on nutrient transport (Figure 12.17). By this process, sugars are effectively transferred from cell to cell at rates approaching 300 to 400 cm/h in some plants.

The liquid portion of vascular bundles (xylem and phloem) is called *plant sap*. Components of plant sap show seasonal, physiological, and species variabilities. Usually 98% of plant sap is water while the remaining 2% represents solutes including sugars, salts, enzymes, albuminoid proteins, organic acids such as citric and malic acids, and plant hormones (e.g., indoleacetic acid). Sugar concentrations for maple sap as an example vary from 1 to 6% sucrose (mean of 2.5%) and often contains traces of invert sugar. The acid constituents of plant saps account for pH values in the range of 4.5 to 7.0.

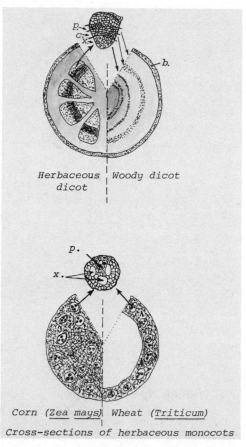

Herbaceous dicot | *Woody dicot*

Corn (Zea mays) | *Wheat (Triticum)*
Cross-sections of herbaceous monocots

FIGURE 12.16. Stem cross sections for typical herbaceous and woody dicot plants as well as two herbaceous monocots including corn and a grass (wheat). Key: x., xylem; p., phloem; c., cambium; b., bark.

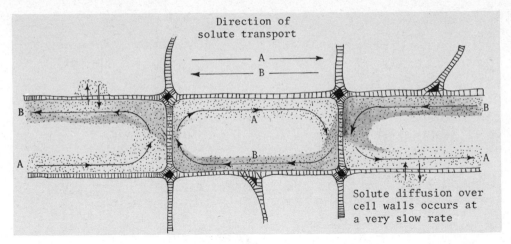

FIGURE 12.17. Solute transport from one phloem cell to another occurs by intracellular circular streaming motions or cyclosis. This mechanism permits the opposing flow of essential plant nutrients (solutes) within the same cell.

Apart from the most common plant saps, some plants that belong to the *Moraceae, Euphorbiaceae,* or *Apocyanaceae* have an additional latex system that assists circulatory processes in the plant. Latex is a milky juice contained in the specialized cells and vessels that permeate the bark, leaves, and soft parts of these plants or trees. Minute globules contained in latex represent a complex mixture of resins, oils, proteins, acids, salts, sugar, hydrocarbons, and caoutchouc mixed with water. The physiological roles for latex in these plants are poorly understood, but typical functions may include protective or nutritional mechanisms.

Latex has commercial value since caoutchouc can be used as a rubber source; opium is obtained from some latexes; and chicle obtained from the sapodilla or naseberry (*Achras zapota*) has served as an important chewing gum base.

HORMONES AFFECT PLANT PHYSIOLOGY AND STRUCTURE

The morphology of plants is geared to their species-specific survival as photosynthetic organisms. Their structures reflect time-tested success stories in evolution that ensure a photosynthetic mode of nutrition while jointly ensuring species survival.

Apart from genetic dictates, individual plant morphologies, flowering, fruiting, foliage, root, diurnal responses, and cellular differentiations are controlled by *plant hormones* (Figure 12.18).

Plant hormones represent a number of compounds that are transported upward from the roots by the xylem and downward from the leaves by the phloem. In other cases these hormones undergo intercellular transport over membranes by active transport mechanisms.

The biochemical processes responsible for plant hormone actions are largely unknown and confusing. Many hormones appear to overlap in function, and individual members of the same plant species may show different responses to the same hormone. These responses vary according to plant age, physiological growth stage, environmental conditions, and the nutritional status of plants.

Auxin

Auxin is considered one of the most important plant hormones because it is linked to normal patterns of cell growth, cell differ-

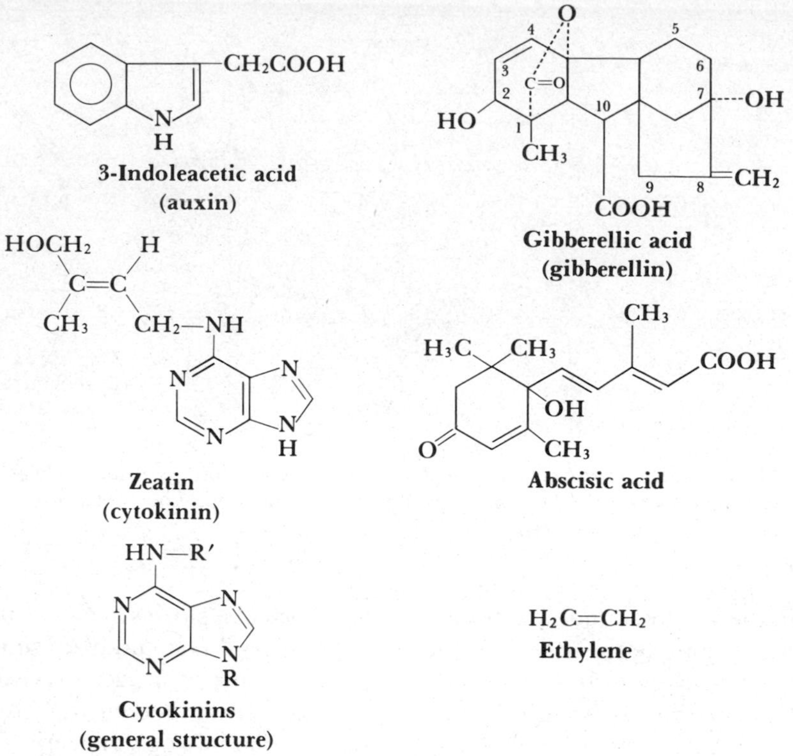

FIGURE 12.18. Some important chemical compounds that regulate plant growth and development.

entiation, and overall plant development. Biochemical mechanisms underlying these effects are not clear but increased auxin concentrations affect cell division, cell elongation, cell wall plasticity, and the water-holding capacity of cells. Auxins are produced by growing shoots, from which they eventually diffuse down stems at rates that approximate 0.5 to 1.5 mm/h. This action decreases lateral bud growth and ensures apical dominance.

Studies with actinomycin D, which is known to inhibit RNA synthesis, reveal that this antibiotic also prohibits auxin-induced cell elongation in plants. Therefore, it has been theorized that auxin(s) may enhance the transcriptional rates for RNA. Other biochemical evidence points toward auxin-induced permeability differences generated over cell membranes. This in turn may alter proton and ion migration over the membrane and eventually modify the actions of key enzymes.

Auxin concentrations as small as $10^{-12} M$ are also responsible for the *phototropic* and *geotropic* responses of plants. The light-induced bending of plants toward a light source is a typical phototropic effect. Unilateral perception of light by carotenoids, flavins, and other substances may be involved in the lateral transport of auxin from the illuminated part of a plant toward the dark side. Differential growth, water uptake, and other physiological events cause plant curvature toward light at auxin concentrations of as little as $4.0 \times 10^{-12} M$. Plant stem responses to gravity are also auxin induced, presumably due to auxin accumulation on the underside of stems. Here, auxin promotes cell elongation and division activities that cause an upright curvature of the stem and growth. For roots the situation is reversed and auxin concentrations on the underside of roots reduce elongation. This is true because concentrations of $\geq 10^{-10}$ M inhibit cell elongation, whereas lower auxin concentrations in roots stimulate growth.

Leaf shedding is another phenomenon partially controlled by auxin. Continued auxin production by leaves promotes petiole attachment of leaves to plants, but deficits of leaf-produced auxin spur the formation of an *abscission layer* at the petiole. Leaf shedding then follows.

Auxins have a variety of agronomic uses that encourage fruit and crop production. These uses include the stimulation of root growth from plant cuttings, enhanced rates of fruit ripening, fruit production without pollination and seed production (parthenocarpic fruits), and prevention of preharvest fruit dropping. Furthermore, since monocots (grasses) are less sensitive to the stimulative effects of auxins than dicot plants, the overstimulation of dicot plants by auxins can serve as a method for "weeding out" dicot plants in monocot-based food crops. The cell metabolism of dicots is easily overstimulated by artificial auxins such as 2,4-dichlorophenoxyacetic acid (2,4-D), which leads to their eradication. Once dicots are eliminated from monocot food crops such as corn, oats, wheat, rye, barley, and other grains, useless weed plant competition is minimized, and soil resources are directed toward maximum production of useful crops.

Gibberellins

The morphology of plants and their flowering and fruiting stages are all influenced by gibberellins. Over a dozen different gibberellins are recognized. Although gibberellins may elicit species-specific responses, they clearly serve as chemical messengers, promote flowering, and affect stem length.

As a messenger, classic studies of oats, wheat, and barley have shown that embryo-produced gibberellin stimulates the aleurone layer cells on the endosperm to produce α-amylase. This action presumably follows gibberellin-based activation of RNA synthesis mechanisms since actinomycin D, noted for its ability to displace DNA-dependent RNA polymerase, also inhibits α-amylase synthesis. Proteolytic enzymes are also mobilized by gibberellins. Protein hydrolysis is critical for at least the initial production of amino acids in germinating seeds so that the tryptophan precursors of auxins such as indoleacetic acid can be produced.

Mature leaves generally synthesize gibberellins prior to their downward translocation in the plant. This results in a gibberellin-based stimulus of RNA synthesis in many vegetables and lateral branching from the plant stem. Geotropic plant responses are also tied to gibberellins since horizontal roots apparently have high concentrations on their upper side. The exact mechanism responsible for gibberellin-induced geotropic responses is still not clear.

Food and agricultural uses for gibberellins are very important. These hormones can effectively set fruiting stages of plants, induce plant flowering and fruit maturation processes to meet market demands for produce, and generate full-size crop-bearing plants from dwarf plants. Dwarf corn, bush beans, and peas can be transformed into full-size specimens by gibberellin application to these crops.

Significant commercial crops of barley, sugar cane, and grapes use gibberellins. For barley, gibberellins may be used to induce the production of α-amylase during the malting of barley. Application levels of ~56 g of gibberellin per acre of sugar cane promotes increased yields approaching 4 to 5 tons per acre with a corresponding increase in refined sucrose yield of ~0.3 ton per acre. The succulence and size of grapes can be improved by similar gibberellin applications.

Far-ranging applications beyond the scope of this discussion favor the use of gibberellins for developing new crops and seeds. As opposed to auxins, which encourage the development of female flowers, gibberellins enhance male flower production. Nutrient absorption from soils by plants are not immune to the effects of gibberellins. In the case of wheat, these hormones increase absorption of potassium and other ions.

Cytokinins

Plant hormones classified as cytokinins probably exert their effect at the gene level on DNA, but their precise action is unknown. The cytokinins are derivatives of isopentenyladenosine, which may be substituted or hydroxylated in various ways at the 2 position (Figure 12.18). Methylthio- substitutions are not uncommon at this site. Unlike auxins, which are produced at the apical meristem and young leaves, cytokinins are produced in root tips and transported upward in the plant.

One key feature of cytokinins involves their apparent control of cell division (cytokinesis) and plant cell structures in culture media. Aggregates of loose cells occur at 10^{-9} M concentrations, 10^{-8} M concentrations of cytokinin encourage root development, and still higher concentrations seem to induce shoot growth. Further evidence indicates that a gibberellin/cytokinin balance may regulate early plant growth and development. This system of hormonal control eventually wanes in favor of auxins as plant maturation occurs. Whatever the biochemical effects may be, cytokinin effects probably do not involve tRNA functions.

Abscisic Acid

Unlike auxins, gibberellins, and cytokinins, which promote plant growth, abscisic acid exerts growth-inhibiting effects. Abscisic acid action may be related to a generalized gene repressor effect that initiates seasonal periods of plant dormancy. For this reason, it is also called a dormancy-inducing hormone. Application of this hormone to different plants does produce different effects. For growing woody twigs, internodal elongation is inhibited; some leaves develop an abscission layer and shed; foliage leaves become deformed; and terminal bud growth ceases.

In the cases of herbaceous plants and many crop plants, this acid hormone may contribute to plant dwarfism coupled with compact foliage production and/or enhanced fruit production.

Ethylene

Over the period of 1860 to 1917, ethylene gas was recognized to decrease stem elongation rates, increase plant stem diameters, and accelerate fruit ripening. Ethylene is a natural derivative of plant metabolism whose *in vivo* production is at least partially affected by exposure to red light (see discussion of phytochrome interactions in the following section) and auxin availability.

Evolution of ethylene may originate with the oxidative decarboxylation of α-keto acid analogs of methionine. Hydrogen peroxide has been linked to the development of ethylene from these analogs. Aside from ethylene production, two molecules of carbon dioxide are also produced plus a thiomethyl moiety that may ultimately be reincorporated into methionine again. The biochemical basis for ethylene action is not certain, but it is thought to retard mitosis and, therefore, decrease tissue expansion in plants.

Control of crop flowering and acceleration of fruit ripening have served as the most important agricultural applications for ethylene. Ethylene production is also related to contact and movement responses of plants known as *thigmomorphogenic effects.* Movement and pressure applied to plants (e.g., wind) stimulate ethylene production, which retards stem elongation and causes diametric growth in stems.

Photomorphogenic Hormones and Phytochrome

Other categories of hormones such as rooting hormones and flowering hormones are present in plants, but many still remain to be isolated for the first time. The biochemical actions for many of these hormones, especially the flowering hormones known as *florigens,* are coupled to the photochemistry of *phytochromes.*

Phytochromes are highly photosensitive polypeptides that contain a tetrapyrrolic molecule resembling the linear structure of phycobilin. The action of phytochromes is related to the control of plant flowering, branching, seed germination, and leaf formation.

Although phytochrome was thought until 1980 to have a monomeric structure, it is actually a dimeric structure with each subunit having its own photoactive site. Each subunit of the dimer can exist in two forms. One pigment form known as phytochrome 660 (PC_{660}) undergoes photo-induced structural changes at 660 nm, whereas the other structural form of phytochrome (PC_{735}) is structurally affected with 735-nm light. Both forms of phytochrome are repetitiously interconvertible by altering pigment exposure to light between 660 and 735 nm.

$$PC_{660} \underset{\text{infrared light (735 nm)}}{\overset{\text{red light (660 nm)}}{\rightleftharpoons}} PC_{735}$$

Because phytochrome is a dimer, one photo-induced form may exist independent of the other. That is, one subunit may be sensitive to 660-nm light, whereas the other is affected by 735-nm light. It turns out, then, that the dimer may exhibit one of three forms: $PC_{660}:PC_{660}$; $PC_{660}:PC_{735}$; or $PC_{735}:PC_{735}$. Based on lettuce seed germination studies, it has been speculated that light-induced $PC_{660}:PC_{735}$ and $PC_{735}:PC_{735}$ forms may complex with cell membranes in viable seeds. Since seeds chilled to 4.0°C and then elevated to 20°C accelerate red-light-induced seed germination by ~10,000 times, phytochrome interaction with temperature-altered membrane lipids may be involved.

It should be noted, too, that most evidence points to temperature-*independent* interconversion of PC_{660} and PC_{735} over the range of 0.0 to 37°C.

From perspectives of diurnal (day/night) occurrence, PC_{735} is produced at the onset of daylight, but night darkness leads to enzymatic formation of PC_{660} from PC_{735}. Photo- and enzyme-induced alterations of phytochrome structures such as these provide a biological benchmark for plants to detect light–dark cycles.

The photo-induced changes in phytochromes may involve changes in positional isomerization of double bonds in the tetrapyrrole, *cis–trans* isomerization of double bonds, or rotation of the tetrapyrrole about a key single bond.

The link between phytochrome and its ability to stimulate morphological changes in plants is far from certain, but PC_{735} or PC_{660} *may promote gibberellin release from plastids or alter membrane permeabilities*. In other cases, phytochrome-induced plant responses including flower and leaf closure may occur in only minutes. These effects are probably unrelated to gene transcription or translation events.

Phytochrome-induced hormone changes are responsible for the flowering behavior of *short-day, long-day*, and *day-neutral* plants.

Short-day plants exhibit flowering provided that light exposure does not exceed some critical period. Long-day plants flower only when light exposure exceeds a critical time period. Depending on the plant species, critical photoperiods may range anywhere from 9 to 16 h. Typical short-day plants include potatoes, chrysanthemums, dahlias; long-day plants include clover, beets, corn, lettuce, radish, mustard, and dill; whereas sunflowers, tomatoes, and dandelions represent day-neutral plants. Depending on the agricultural crop, phytochrome-induced changes in plant metabolism may be profound. As an example, photoperiod-induced flowering will determine fruit setting time, and for potatoes, starch deposition into tubers from leaves is hastened by short photoperiods.

NITROGEN FIXATION BY PLANT BACTEROIDS

The earth is shrouded in an atmosphere of ~80% nitrogen and yet very few organisms can use nitrogen in its molecular form (N_2). Most living organisms rely on other natural biological processes to reduce or "fix" nitrogen for its eventual incorporation into organic compounds such as amino acids, peptides, pyrimidines, and purines. The relatively few species of organisms that do fix nitrogen are quite efficient, however, and it is estimated that global nitrogen fixation by this route may approach 237 million metric tons.

The strong triple bond ($N \equiv N$) in molecular nitrogen, displaying a bond energy of ~225 kcal/mol, seems to offer an obstacle for the conversion of nitrogen into a form that green plants and most other forms of life can use.

A survey of simple nitrogen compounds indicates that many forms having different oxidation states exist, but in nature, the predominant forms are nitrates (NO_3^-) or reduced nitrogen in the form of ammonia:

food production, leguminous plants including alfalfa, clover, peas, vetch, beans, cow peas, soybeans, and lupines are among the most important host plants for nitrogen-fixing bacteria, but almost 200 other plant species serve as hosts for similar nitrogen-fixing bacteria.

Species of *Rhizobium* are naturally host specific, probably as a result of mutual cell surface recognition phenomena between bacterial cell surfaces and root hair tips on the plant. Theories suggest that plant proteins known as *lectins* serve as receptor sites on roots for polysaccharides on the surface of specific bacterial cells. The ensuing molecular interactions form a lectin bridge between the plant root and the bacterium. Such mechanisms may foster highly specific plant–bacterial associations. For clover, the lectin

NO_3^-	NO_2^-	$N_2O_2^{2-}$	N_2	NH_2OH	NH_3
Nitrate	Nitrite	Hyponitrite ion	Nitrogen gas	Hydroxyl-amine	Ammonia
+5	+3	+1	0	−1	−3

Oxidation numbers

Free-living bacteria such as aerobic *Azotobacter*, anaerobic *Clostridia* (sp.), the photosynthetic bacterium *Rhodospirillum rubrum*, as well as *Cyanophyceae* or blue-green algae represent nonsymbiotic organisms that fix nitrogen. Other nitrogen-fixing bacteria such as *Rhizobia* invade the roots of host plants where they participate in a symbiotic existence. These bacteria fix molecular nitrogen for plant use, whereas the host plant offers a permanent residence site for the rhizobial bacteria. Therefore, many plants that apparently fix nitrogen are actually unable to fix nitrogen without the rhizobial bacteria; and the rhizobial bacteria are unable to fix nitrogen outside their host plant.

These symbiotic associations of plants and bacteria are often evident as numerous root nodules owing to the development and residence of enlarged bacteroid cells within root tissues. From perspectives of agriculture and

called *trifoliin* selectively interacts with *Rhizobium trifolii*. Similar recognitive actions seem to operate in the case of soybeans and some strains of *R. japonicum* as well as other legumes.

Once rhizobial bacteria invade the roots of their host plant, they become enlarged, vacuolated, and branched cells called *bacteroids*. These bacteroids are contained in membranes that are surrounded by the cytoplasm of host plant cells.

BIOCHEMICAL REACTIONS FOR NITROGEN FIXATION

Far more is understood about the complex biochemistry of nitrogen fixation now than during the past decade, but many uncertainties remain. The first biochemically fixed form of nitrogen by microbial processes is ammonia, whose formation is mediated by a delicate but powerful nitrogenase enzyme system. Nitrogenase reduction of nitrogen can be summarized as shown below, but many intervening steps are involved. Although this

$$:N\equiv N: \xrightarrow[\substack{6 \text{ electrons} \\ 6 \text{ protons (6 H}^+) \\ \Delta G^{0\prime} = -8.0 \text{ kcal/mol}}]{\text{nitrogenase}} \;\;//\;\; :\overset{\displaystyle H}{\underset{\displaystyle H}{N}}-H$$

Nitrogen **Ammonia**

nitrogen reduction reaction is readily feasible from thermodynamic perspectives, it has a high activation energy, which impedes its occurrence in all but a few organisms.

The nitrogenase enzyme system is constructed of two proteins. One protein called *azoferredoxin* (Azofd) or component II represents an iron–sulfur protein constructed from two identical polypeptide dimers. Total molecular weight for the dimer is about 60,000. The second protein in the nitrogenase system is a tetramer composed of two different peptide (α_2 and β_2) dimers that are united to form a tetramer ($\alpha_2\beta_2$). Since each tetrameric structure statistically contains two molybdenum ions with iron–sulfur proteins, this second key element of nitrogenase is called *molybdoferredoxin* (Mofd) or component I. The subunits of the $\alpha_2\beta_2$ tetramer range from 50,000 to 60,000 molecular weight.

It is generally believed that electrons flow from a specific reductant such as NADPH to oxidized ferredoxin (Fd_{ox}) to produce reduced ferredoxin (Fd_{red}). From Fd_{red}, electrons participate in a two-step reduction of nitrogen (N_2) that produces ammonia (Figure 12.19).

In the first step, oxidized Azofd is reduced by Fd_{red} to yield $Azofd_{red}$. This reduced iron–sulfur protein then conveys its reducing power to oxidized Mofd, forming its reduced form denoted as $Mofd_{red}$. Reduction of the $Mofd_{ox}$ form requires the presence of a Mg^{2+} and ATP complex (Mg^{2+}–ATP), but the exact role for this com-

plex remains to be established. Nonetheless, $Mofd_{red}$ is speculated to bind to the nitrogenase complex, undergo hydrolysis, and then lower the energy of activation necessary for the chemically stubborn nitrogen reduction.

The lack of obvious phosphorylated intermediates, coupled with a redox potential change from -0.3 to -0.4 V for reductase in the overall system also suggests that ATP may induce conformational changes in key enzymes that foster the reaction.

Although the biochemical milieu of the *in vivo* nitrogenase system may not require the same calculated energy of activation for nitrogen reduction as *in vitro*, general observations suggest that about four molecules of ATP are consumed for each electron pair transferred. This is clearly a very significant amount of ATP and in the case of pea plants, 20% of all plant ATP may be consumed for fueling nitrogen-fixation processes of bacteroids.

By the final step of nitrogen reduction to ammonia, and its conversion to an ammonium radical, the stoichiometry for the nitrogenase-mediated reaction becomes

$$N_2 + 6\,e^- + 12\text{ ATP} + 12\text{ H}_2\text{O} \longrightarrow$$
$$2\text{ NH}_4^+ + 12\text{ ADP} + 12\,P_i + 4\text{ H}^+$$

Apart from the role of plant metabolism in supplying a reductant and ATP for nitrogen fixation, plant metabolism also supplies carbon skeletons that consume fixed nitrogen (NH_3). For example, Figure 12.19 illustrates the ultimate contribution of fixed nitrogen to amino acid synthesis. More details regarding this step appear in the next section.

The biochemistry of nitrogen fixation is complicated further by the high destructive sensitivity of Azofd to oxygen. Owing to this acute sensitivity, oxygen diffusion into bacteroids must occur at restricted yet consistent levels. This ensures the efficient aerobic production of ATP by an oxidase system without harming the sensitive nitrogenase components.

Careful studies have now shown that bacteroid inhabitants of legume nodules main-

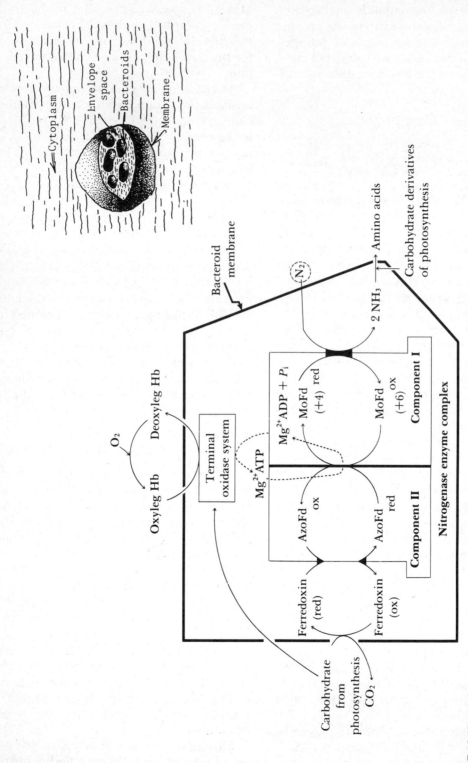

FIGURE 12.19. Only one nitrogenase complex is shown within a bacteroid membrane however numerous enzyme systems exist. Overall ammonia production requires 12 ATP's, 36 NADPH (H^+) for the reduction of ferredoxin, and 6 electrons for the production of 2 NH_3 molecules.

tain adequate oxygen levels by the action of *leghemoglobin.* Leghemoglobin is a very specialized form of an oxygen-binding hemoprotein under the structural genetic control of the plant. Most of the substance resides between the bacteroids jointly contained within a membrane (Figure 12.19), yet other studies suggest that the leghemoglobin is located outside bacteroid membranes within the plant cytoplasm.

The novelty of nitrogen fixation by living organisms plus the scrupulous experimental conditions required for studying the process has spurred some interesting analytical methods. Prior to the 1960s, nitrogen fixation and nitrogenase activity were studied by mass spectroscopy where ^{15}N-labeled nitrogen gas was reduced to ^{15}NH$_3$. Subsequent to the 1960s, an acetylene reduction test became popular for *in vivo* and *in vitro* studies of nitrogenase. This presumptive test was popularized after it was determined that nitrogen-fixing bacteria could reduce acetylene ($HC{\equiv}CH$) to ethylene ($H_2C{=}CH_2$). Since both acetylene and ethylene can be easily measured by gas chromatography, acetylene reduction tests have served as an important basis for evaluating the dynamic behavior of nitrogenase systems.

The wide potential application for the acetylene reduction test has also uncovered nitrogen-fixing activities in many other prokaryotes including Winogradsky's *Clostridium pasteuranium;* some strains of actinomycetes hosted by angiosperms; and selected strains of *Klebsiella* present as leaf nodules in some plants and in the gastrointestinal tract of some human populations. The ni-

trogenase activity of these organisms functions in a similar fashion to that already described for leguminous bacteria, but many variations do exist beyond the scope of this text. These variations largely center about biochemical sources of reducing power and electrons from the host cell; the occurrence of single-electron-accepting azotoflavins found with ferredoxin in *Azotobacters;* unique heterocyst formations in *Anabaena* and *Nostoc* that protect nitrogenase activity while absenting photosystem II; and other minor but very important variations.

ASSIMILATION OF FIXED NITROGEN INTO ORGANIC MOLECULES

One of the most fundamental entry sites for fixed nitrogen into organic molecules involves the overall conversion of the ammonium radical (NH_4^+) and glutamate into glutamine. This reaction requires the intervening formation of glutamate from NH_4^+ and α-ketoglutarate—a key intermediate in the citric acid cycle. Glutamate formation here is mediated by *glutamate dehydrogenase,* which requires NADPH.

$$NH_4^+ + \text{α-ketoglutarate} + NADPH(H^+)$$

$$\Updownarrow \begin{array}{l}\text{glutamate}\\\text{dehydrogenase}\end{array}$$

$$\text{L-glutamate} + NADP^+ + H_2O$$

Once the glutamate structure is established, a *glutamine synthetase* reaction proceeds at the expense of ATP hydrolysis:

L-Glutamate $+ NH_4^+ + ATP \xrightarrow[\text{synthetase}]{\text{glutamine}}$ **Glutamine** $+ ADP + P_i + H^+$

Both L-glutamate and glutamine serve critical roles in the biosynthesis of amino acids and other nitrogen-containing compounds. For example, the α-amino group of many amino acids is linked to a transamination reaction that necessitates glutamic acid and an α-keto acid. Glutamine, on the other hand, can contribute its amide nitrogen, by enzyme-mediated steps, to form asparagine from aspartate, glucosamine 6-phosphate from fructose 6-phosphate in addition to many other amino acid amides, purines, and so on.

Prokaryotic and eukaryotic organisms alike show glutamine synthetase activity; however, glutamate synthase, also present in most prokaryotes, negotiates the reductive amination of α-ketoglutarate. This reaction generates two molecules of glutamate when glutamine serves as a nitrogen donor.

$$\alpha\text{-Ketoglutarate} + \text{glutamine} + \text{NADPH(H}^+)$$

$$\downarrow \text{glutamine synthase}$$

$$2 \text{ glutamate} + \text{NADP}^+$$

Glutamate synthesis may proceed by an alternative route in cases where NH_4^+ concentrations are restricted. Under these conditions, the K_m value for a glutamine synthetase/NH_4^+ interaction is more favorable than the 10^{-3} K_m value for glutamate dehydrogenase. According to this kinetic dictate, the joint actions of glutamine synthetase, followed by glutamine synthase, produce glutamate:

$$NH_4^+ + \alpha\text{-ketoglutarate} + \text{NADPH(H}^+) + \text{ATP}$$

$$\downarrow$$

$$\text{glutamate} + \text{NADP}^+ + \text{ADP} + P_i$$

This overall reaction is favored in spite of its higher consumption of ATP and reducing power than the more economical glutamate dehydrogenase.

Reductive amination of glutamate is a key mechanism for installing nitrogen within amino groups, but plants also have other mechanisms at work. The most probable reactions involve the direct amination of α-keto acids and pyruvate by mechanisms reminiscent of glutamate dehydrogenase.

GLUTAMINE SYNTHETASE MAY CONTROL NITROGENASE AVAILABILITY

Aside from glutamine synthesis, the enzyme *glutamine synthetase regulates* the transcription of genes that control *nitrogenase synthesis*. For many years it has been observed that the *ammonium ion* will *restrict* the *biosynthesis* of *nitrogenase* by mechanisms that are still partially speculative. As discussed in the previous section, glutamine synthetase participates in amino acid biosynthesis of glutamine from glutamic acid.

Based on this recognition plus experimental evidence that nitrogenase production may be inhibited by mutations in genes responsible for glutamine synthetase production, it has become evident that this synthetase enzyme may control nitrogenase availability.

According to one popularized theory (Figure 12.20), glutamine synthetase probably binds to an operator portion of the microbial genome responsible for controlling nitrogenase biosynthesis. The corresponding structural genes for nitrogenase biosynthesis are known as the *NIFX region*. Glutamine synthetase binding is followed by gene transcription and mRNA translation into active nitrogenase. Normal activity of nitrogenase results in reduction of nitrogen into ammonia for amino acid biosynthesis, as previously discussed.

At this point, the presence of *ammonium ion*, formed from ammonia, *is believed to affect nitrogenase gene expression* by a series of steps that *modify* the DNA-binding properties of glutamine synthetase. This concept is consistent with the observation that mutant genes for glutamine synthetase probably affect its structure and thereby restrict its normal ability to initiate gene expression.

According to this model, glutamine synthetase activity is controlled by final events of nitrogen fixation (i.e., ammonia production). Conversely, low levels of fixed nitrogen in the form of ammonia induce favorable changes in glutamine synthetase,

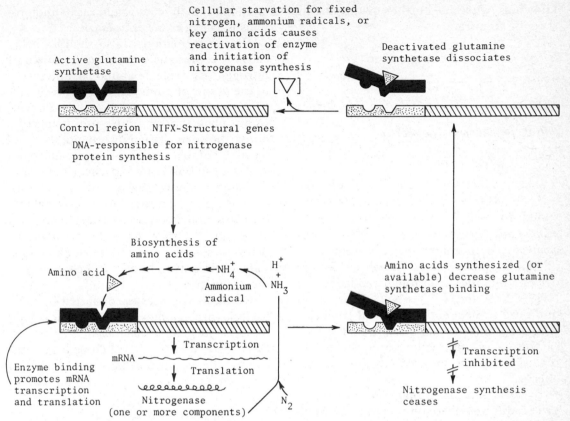

FIGURE 12.20. Nitrogen fixation in many microorganisms is controlled by concentration levels of nitrogenase synthesized within the cell. One speculative scheme suggests that glutamine synthetase regulates the biosynthesis of nitrogenase by affecting the transcription of its structural gene(s) (NIFX region) within the bacterial genome.

which promotes its binding to DNA and the expression of NIFX structural genes.

Glutamine synthetase activity is also regulated by cascade mechanisms and covalent modifications that involve its adenylation. For nitrogenase enzyme expression at the genetic level, it is doubtful that such control mechanisms are more important than the ability of glutamine synthetase to control NIFX structural genes, but other control mechanisms are undoubtedly involved.

The ability of many prokaryotes to control nitrogenase enzyme availability is overshadowed by the ability of nodular bacteria to continually produce nitrogenase. Whereas most organisms terminate the actions of nitrogenase structural genes when ammonia or fixed nitrogen levels are high, the nodular bacteria produce excess ammonia. Copious amounts of fixed nitrogen produced by this behavior either nourishes the host plant or may be released from the rhizobial organisms. Therefore, nodular bacteria exhibit the action of nitrogenase genes that seem to be de-repressed.

NITRIFICATION PROCESSES CONVERT AMMONIA TO NITRITE AND NITRATE

Very little fixed nitrogen persists in soil as ammonia. Instead, it is rapidly oxidized to nitrite (NO_2^-) and then nitrate (NO_3^-) by the action of nitrifying bacteria. The *Nitrosomonas* bacteria produce the nitrite ion from ammonia using molecular oxygen as an oxidizing agent. Another group of bacteria known as *Nitrobacter* further oxidize the nitrite ion to nitrate.

Nitrite production by Nitrosomonas

$$2\ NH_3 + 3\ O_2 \xrightarrow[\Delta G^{0\prime} = -66.5\ kcal/mol]{} 2\ NO_2^- + 2\ H_2O + 2\ H^+$$

Nitrate formation by Nitrobacter

$$2\ NO_2^- + O_2 \xrightarrow[\Delta G^{0\prime} = -17.5\ kcal/mol]{} 2\ NO_3$$

The oxidation state changes for $NH_3 \rightarrow NO_2^- \rightarrow NO_3^-$ respectively correspond to $-3 \rightarrow +3 \rightarrow +5$. Both *Nitrosomonas* and *Nitrobacter* are classified as *chemoautotrophs* since their respective reductions of carbon dioxide into carbohydrates are driven by the exergonic oxidation of $NH_3 \rightarrow NO_2^-$ and $NO_2^- \rightarrow NO_3^-$. Proteins and lipids are subsequently produced from carbohydrate intermediates. Note that similar carbon dioxide reductions are executed in plants, but photo-induced reactions supply energy.

Bacteria and plants that do not fix nitrogen readily assimilate nitrate, but it must be reduced to ammonia before it serves any useful biochemical function as a nitrogen source. This step is known as *denitrification* and is the result of enzyme action by *nitrate reductases*.

Nitrate reduction is probably a stepwise and sequential affair that requires the reducing power of pyridine nucleotides along with the reductase as outlined:

ammonia and then into amino acids and proteins is called *nitrate assimilation.*

Nitrate formation is also significant because it represents a more *stable* source of nitrogen than volatile ammonia, and owing to the *toxicity* of ammonia on many cells, nitrate can be warehoused to a limited extent within plant sap for future use. It must be noted, however, that NH_4^+ happens to be more favorable as an inorganic nitrogen source in soils because it participates in ion-exchange phenomena with anionic soil components. This improves its soil availability for plant absorption over nitrates, whose anionic radical either is rapidly absorbed by plants or simply percolates through anionic soil constituents to levels well below normal root residence.

Nitrification processes conducted by some bacteria are counteracted by denitrifying bacteria such as *Pseudomonas denitrificans, Denitrobacillus,* and others. These organisms conduct denitrification processes that yield nitrogen gas at the expense of ammonia.

NOVEL CULTIVATION METHODS, BIOCHEMISTRY, AND GENETICS WILL IMPROVE FOOD CROP PRODUCTIVITY

Plants are unique because they fix solar energy into carbohydrates, proteins, and some lipids. Only about 8% of solar energy reach-

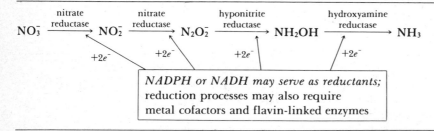

Once ammonia is formed, it is incorporated into organic compounds to form organic *nitrogen* compounds through the action of glutamic dehydrogenase. This process of nitrate utilization through reduction to

ing the surface of the earth interacts with green plants and only 2% of the 8% drives photosynthetic processes. The absorbed solar energy supports the life-sustaining mechanism of plants, and excess energy is stored as starches or lipids or used to increase plant biomass. Excess energy stored by plants beyond that energy required for life-maintenance processes reflects their *net primary productivity*.

In the case of sugarcane grown under optimal conditions, a 4000 kcal/m²/day exposure of solar energy can be converted to a net primary productivity of 190 kcal/m²/day. This corresponds to a net efficiency of only ~4.8% under the best growth conditions, and average efficiencies are often less than 2.0% because of less than optimal cultivation conditions.

Improvements in the net primary productivity of almost all food crops can be realized from modern agricultural machinery, cultivation techniques, and advanced chemical, biochemical, and genetic modifications of crop plants.

Primitive manual cultivation of wheat may produce harvest yields of 5 to 8 times above the energy input by the farmer. However, when wheat is cultivated using advanced mechanistic and agricultural practices, harvest yields can approximate 400 to 550 times the energy expenditure required for their cultivation.

Apparent increases in the net primary productivity of plant crops is *not always the answer for combating the nutritional maladies of mankind*. Most crops are carbohydrate-rich, but the human population as a whole generally suffers from a *dietary protein deficit*.

For these reasons, food production in advanced countries significantly depends on the conversion of carbohydrate-rich autotrophs (plants) into protein-rich heterotrophs that are then consumed as food. Production of protein-rich foods by this route is not an energy-efficient process. As an example, consider that 100% efficient digestion processes could yield 3600 kcal from 1.0 kg of corn. Consumption of this 1.0 kg of corn as feed grain by a cow producing protein-rich milk salvages only ~620 kcal or ~17.2% of the total 3600 kcal consumed. Meat produced from the consumption of 1.0 kg of corn by a cow reflects only ~360 kcal or

~10% of the caloric potential. In each consumption step for the corn, the caloric conversion yield from corn to milk to meat decreases as the food protein content increases. These marked energy losses have encouraged many novel experiments designed to enhance the use of autotrophs as a protein-rich dietary component.

Mass Cultivation of Algae

The mass culture of algae has been offered as one solution for enhancing plant-produced food protein resources for a variety of reasons. Algal photosynthesis is a highly efficient process that produces a protein-rich biomass when compared to other green plants. In contrast with land plants, the cellulose content of most algae is very low and 46 to 60% of algal dry weight is protein. Unlike sugar beets, whose effective net primary productivity from photosynthesis may approach 17 to 20 tons/acre/year, controlled growth of selected algae may yield *ten times* this biomass.

Although the Japanese have cultivated brown marine algae in shallow marine waters for centuries, more recent algal cultivation methods have proposed mass fermentation-type cultivation techniques. These methods support dense growths of algae in a fixed nutrient growth medium exposed to intense illumination.

The growth medium may be strictly inorganic salts or a combination of salts, vitamins, and other key organic growth factors. Under these conditions, individual algal cells such as *Chlorella*, *Scenedesmus*, and others conduct all the biochemical functions of the differentiated parts (leaves, roots, stem, etc.) of higher plants. Consumption of soybeans involves eating only the high-protein seeds of the plant, whereas the whole protein-rich autotrophic structure of algal cells may be consumed.

There can be no quarrel over the potential nutritional value of algae (Table 12.3), but the technological requirements and complexity of mass algal culture plus the pal-

Table 12.3. Relative Amounts of Amino Acids Present in Algae with Regard to Other Common Protein-Rich Foods (as g of amino acid/g of nitrogen)

Amino acid	Chlorella	Beef cuts	Soybean	Egg
Alanine	0.70	0.32	0.26	—
Arginine	0.40	0.40	0.45	0.40
Aspartic acid	0.64	0.60	0.76	0.44
Cystine	—	0.08	0.10	0.15
Glutamic acid	0.90	0.95	1.20	0.80
Glycine	0.45	0.40	0.26	0.20
Histidine	0.09	0.22	0.15	0.15
Isoleucine	0.30	0.34	0.34	0.40
Leucine	0.60	0.50	0.50	0.60
Lysine	0.40	0.55	0.40	0.40
Methionine	0.08	0.16	0.08	0.20
Phenylalanine	0.33	0.26	0.30	0.38
Proline	0.33	0.31	0.40	0.27
Serine	0.20	0.26	0.41	0.53
Threonine	0.20	0.29	0.25	0.30
Tryptophan	0.10	0.10	0.09	0.10
Tyrosine	0.18	0.20	0.20	0.28
Valine	0.44	0.35	0.33	0.47

atability and gastric tolerance of algae present *serious* problems. It is often necessary to eliminate bitterness from the cells as well as decolorize green algae by extraction processes using alcoholic solvents and 3% hydrogen peroxide. Depending on the processing, some components of algal biochemistry can produce diarrhea while other components are simply toxic. In still other cases, cultural contamination of algae with fungi can lead to a host of toxic substances.

The value of algae will probably rest on its use as a food protein adjunct rather than its use as a sole food source. Up to 50% (dry weight) of decolorized and debittered algae may be used in mashed potato–algal mixtures. Other foods that can be protein-enhanced by algae include date bars, green noodles, fruit bars, oatmeal and other cookies, and soups.

Exclusive of the strict growth requirements for high-density algal cultivation, another popular concept for growing algae involves the use of a sewage-based aqueous media held in vats or ponds exposed to sunlight. Algae produced by these cultivation methods are feasible, but the presence of pathogens, viruses, toxic substances, heavy metals, and the possibility that mixed algal cultures may yield an unacceptable product for human or animal food supplementation all counter the serious development of this potential food source.

Apart from the futuristic exploitation of algal cultivation as a food source, innovations in existing agricultural practices and genetic modifications of plant crops offer more immediate results for increasing plant food productivity. Other promising routes for enhancing crop plant productivities include (1) the regulation of fixed nitrogen availability, (2) the control of photorespiration, and (3) the cued biochemical control of plant life-cycle stages.

Control and Use of Fixed Nitrogen

The cost of fertilizer is tied to petrochemical energy. Although Chilean nitrate, guano, coprolitic nitrogen, and manure have served as nitrogen sources for fertilizer in the past, modern plant agriculture requires industrial nitrogen sources.

Since the introduction of the Haber process in the early part of this century, molecular nitrogen and hydrogen have been catalytically reacted under high pressure to produce ammonia:

$$N_2 + 3 H_2 \xrightarrow[\substack{200 \text{ atm} \\ 450°C}]{\text{osmium}} 2NH_3$$

Nitrogen from liquified air   **Reduced nitrogen**

Hydrogen is not readily available from liquified air as in the case of nitrogen, so hydrogen is obtained from the catalytic reformation of methane and other light petroleum fractions.

World fluctuations in petrochemical economics and uncertain supplies now serve as a major stimulus for enhancing biological nitrogen fixation processes. One of the most promising routes for achieving such goals includes the development of a universal rhizobial strain of nitrogen-fixing bacteria that will nodulate all leguminous and some non-leguminous food crops. This could ensure a relatively constant supply of fixed nitrogen to plants that would ordinarily require exogenous nitrogen sources for growth.

Another potential supply for biologically fixed nitrogen involves the *comminution* and *application* of high-nitrogen plants, which otherwise have no agricultural value, to plant crops that are used for food. One typical example of this practice would be the processing of leucaena, an underutilized, scrubby evergreen tree of the tropical regions that fixes nitrogen. Comminuted leaves of this plant can be used directly to fertilize crops with demonstrated crop yield increases only somewhat inferior to those with commercial fertilizer. Other plant species including blue-green algae have been similarly successful as crop fertilizers in Southeast Asia and India.

One of the most futuristic concepts for minimizing the use of petrochemical-based nitrogen fertilizers involves the transfer of regulatory and structural genes (NIFX) from nitrogen-fixing organisms into the genomes of plants that cannot fix nitrogen. Based on nitrogen-fixing yields of soybeans approaching 75 lb/acre/season, the introduction of NIFX genes into crop plants such as corn would certainly contribute to autofertilization and decrease the requirements for industrially reduced nitrogen.

Photorespiratory and Biochemical Control of Plants

In addition to adding fixed nitrogen to crops, the food yields from many crops could be boosted by *minimizing photorespiration* via chemical or genetic controls.

Photorespiration drains reduced carbohydrates produced during photosynthesis from the photosynthate stores of plants and releases it as carbon dioxide. An improved understanding of this process will be fundamental for minimizing carbohydrate losses in important food plants.

It is well recognized that controlled ratios of carbon dioxide to oxygen can be employed for minimizing photorespiratory processes, but abnormal ratios of these gases also upset the performance of nitrogen-fixing plants during critical stages of their life cycles. For soybeans and some other legumes, a high ratio of carbon dioxide to oxygen generally favors increased nodulation and nitrogen fixing activities, but this can be destructive to the survival of the plant over a long term.

Perhaps the most significant gains in food plant productivity and agronomic payoffs in the coming decades lie in the control of internal plant metabolism by artificial plant growth regulators. As shown in earlier sections, many naturally occurring plant hormones can have profound effects at tiny concentration levels on almost every aspect of plant development. Since the advent of the earliest commercial plant growth regulators in the 1940s and 1950s, such as naphthalene, acetic acid, and maleic hydrazide, a wide selection of plant growth regulators have been uncovered or developed. Some of the most important regulators are outlined in Table 12.4. Depending on the plant species, these substances may

1. Encourage plant rooting and propagation.

2. Promote or terminate seed dormancy, bud formation, and tuber growth.

3. Stimulate or postpone flowering, fruit setting, and/or development.

Table 12.4. Some Important Growth Regulators for Agricultural Plant and Crop Production

Name	Chemical name	Structure	Function
Abscisic acid	3-Methyl-5-(1'-hydroxy-4'-oxo 2',6',6'-trimethyl-2'-cyclohexen-1'-yl) cis,trans-2,-4-pentadienoic acid	(structure)	Acts as a growth inhibitor and defoliant
Benzyladenine	6-Benzylaminopurine	(structure)	Terminates dormancy
BNOA	2-Naphthoxyacetic acid	OCH_2COOH (naphthalene)	Improves fruit setting and growth
BOH	β-Hydroxyethylhydrazine	$HOCH_2CH_2NHNH_2$	Induces pineapple flowering
Chlormequat	2-Chloroethyltrimethyl ammonium chloride	$[CH_3-N(CH_3)(CH_3)-C_2H_4Cl]^+\ Cl^-$	Minimizes height of selected monocots (e.g., wheat); reduces lodging of cereal grain plants; and enhances sugarcane ripening
3-CPA	2-(3-Chlorophenoxy) propionic acid	(structure)	Fruit thinner for crops including peaches
4-CPA	4-Chlorophenoxy acetic acid	Cl—(ring)—OCH_2COOH	Encourages fruit development in tomatoes, squash, eggplant, and figs (generally not applicable to drupes); improves blooming and may be used to thin fruits
Cycloheximide	3-(2-[3,5-Dimethyl-2-oxocyclohexyl]-2-hydroxyethyl)-glutarimide	(structure)	0.1 lb/acre loosens most citrus fruits from tree (not used on Valencia oranges due to regulator-induced flower and immature fruit damage); general abscission agent

Name	Chemical name	Structure	Uses
2,4-D	2,4-Dichlorophenoxy acetic acid	(2,4-dichlorophenoxyacetic acid structure, OCH_2COOH)	Serves as a general dicot herbicide; delays abscission and preharvest fruit drop
SADH (Diaminozide)	Succinic acid-2,2-dimethyl hydrazine	$HOOCCH_2CH_2CONHN(CH_3)_2$	Apple, pear, and peach flower induction; 7 to 10-year delay of fruiting in trees is reduced to four years; increases firmness of apples and other fruits; decreases stem elongation on tomatoes
Dinoseb	2,4-Dinitro-6-sec-butylphenol	(2,4-dinitro-6-sec-butylphenol structure, $CHCH_2CH_3$, CH_3, OH, O_2N)	Serves as a herbicide; improves corn yield, as little as several grams per acre increases corn yield by 5–10%
Diquat	6,7-Dihydrodiprido-(1,2-a:2',1'-c) pyrazidinium dibromide	(dipyridinium structure, H_2C—CH_2, $\cdot 2Br^-$)	0.13 lb/acre prevents sugarcane flowering; may be used as a herbicide
Diuron	3-(3,4-Dichlorophenyl)-1,1-dimethylurea	(structure, $NHCN(CH_3)_2$, O, Cl)	Prevents sugarcane flowering and increases sucrose yield up to 1.3 tons/acre; herbicide
Endothall	7-Oxabicyclo(2.2.1)-heptane-2,3-dicarboxylic acid monoalkylamine salt	(bicyclic structure, C—N—$(CH_2)_{11}CH_3$, O_2CH_3, $COOH$, CH_3, O)	Enhances the ripening of sugarcane
Ethephon	2-Chloroethyl phosphonic acid	$Cl—CH_2CH_2PO_3H_2$	Fruit thinner for many fruits and grapes; increases latex production from rubber trees

Table 12.4. (*Continued*)

Gibberellic acid	2,4a,7-Trihydroxy-1-methyl-8-methylene gibb-3-ene-1,10-carboxylic acid-1,4-lactone		Accelerates amylase production during the malting of barley; enhances shoot growth; promotes flowering of plants; and serves as an enlarging agent for grapes (seedless types)
Glyphosine	N,N-Bis(phosphono-methyl)glycine	HOOCCH$_2$N(CH$_2$PO$_3$H$_2$)$_2$	Promotes the ripening of sugarcane
IAA	3-Indoleacetic acid		Increases the size of plant cells
IBA	3-Indolebutyric acid		Promotes rooting
Kinetin	6-Furfurylaminopurine		Terminates dormancy in many plants
Maleic hydrazide	1,2-Dihydro-3,6-pyridazine-dione		Prevents sugarcane flowering; prevents onion sprouting and potato sprouting during storage; general growth retardant

Name	Structure	Description
Mefluidide	N-2,4-Dimethyl-5-(trifluoromethyl)-sulfonylamino-phenylacetamide	Herbicide; sugarcane growth promoter
Monuron	3-(p-Chlorophenyl)-1,1-dimethylurea	See diuron; herbicide
NAA	1-Naphthalene-acetic acid	Prevents preharvest apple drop; thins fruits on trees; flowering enhancer for Bromeliads
TIBA	2,3,5-Triiodobenzoic acid	Application to soybean leaves produces stocky plants, increases branching and pod set (bean yield)
Mendock	2,3-Dichloro-2-methyl propionic acid, sodium salt	Gametocide used for selective hybridization of crop plants
Release	5-Chloro-3-methyl-4-nitro-1-H-pyrazole	Enhances abscission among many types of citrus fruits

4. Expedite or delay the abscission of leaves or fruits.

5. Synchronize crop and fruit maturity, chemical composition, and color with market demands.

6. Control nutrient absorption from soil.

7. Increase plant resistance to environmental stresses and pests.

Plant regulators achieve their respective actions by releasing the built-in genetic restrictions of plants, although exact mechanisms are not certain.

In addition to the synthesized growth regulators, many other natural substances capable of regulating plant activities remain to be discovered. Typical among these substances is the long-chain alcohol called triacontanol ($CH_3(CH_2)_{28}CH_2OH$). Until the late 1970s the ubiquitous occurrence of this alcohol in leaf waxes, beeswax, honey, apples, potatoes, and soils was not linked to its ability for enhancing plant growth at mg/acre application levels.

Advances in plant biochemistry also promise full-scale commercial crops grown by tissue culture; *hydroponics,* whereby plant growth is carried out in aqueous nutrient media without soil; *cloning of plant species* that avoids normally protracted periods of the life cycle; *ocean water cultivation of land plants* made possible by genetic modification or selection of salt-tolerant genes (e.g. tomatoes); *cultivation of plants* that have been *genetically recombined with* one or more *mammalian genes* responsible for the production of traditional animal protein(s) (e.g., blue-green algae with genes that can express the synthesis of casein—a milk protein); and *anther culture,* whereby a pollen germ cell serves as a basis for plant culture rather than somatic or tissue cells of a plant. The adaptability and scope of plant manipulations at the genetic, chemical, and biochemical levels are nearly unlimited.

During the World Food Congress in 1963, the late President Kennedy asserted that "The war against hunger is truly mankind's war of liberation . . . There is no battle on earth or in space more important, [for] peace and progress cannot be maintained in a world half-fed and half-hungry."

The battle ground for the war against hunger in the coming decades is the top 6 in. of precious topsoil that separates man from starvation, and the principal armaments against starvation are clearly the green plants.

REFERENCES

Arnon, D. I., H. Y. Tsujimoto, and G. M.-S. Tang. 1981. Proton transport in photoxidation of water: A new perspective on photosynthesis. *Proc. Natl. Acad. Sci. USA* **78**(5):2942–2946.

Arnon, D. I., H. Y. Tsujimoto, and G. M.-S. Tang. 1980. Contrasts between oxygenic and anoxygenic photoreduction of ferredoxin: Incompatibilities with prevailing concepts of photosynthetic electron transport. *Proc. Natl. Acad. Sci. USA* **77**(5):2676–2680.

Arnon, D. I., H. Y. Tsujimoto, and B. D. McSwain. 1967. Ferredoxin and photosynthetic phosphorylation. *Science* **214**:562–566.

Chollet, R. 1977. The biochemistry of photorespiration. *Trends Biochem. Sci.* **2**:155–159.

Clayton, R. K. 1981. *Photosynthesis: Physical Mechanisms and Chemical Patterns.* Cambridge University Press, New York.

Govindjee, and R. Govindjee. 1974. The primary events of photosynthesis. *Sci. Amer.* **231**:68.

Gregory, R. P. F. 1977. *Biochemistry of Photosynthesis.* Wiley, New York.

Halliwell, B. 1978. The chloroplast at work: a review of modern developments in our understanding of chloroplast metabolism. *Prog. Biophys. Mol. Biol.* **33**:1–54.

Hinkle, P. C., and R. E. McCarty. 1978. How cells make ATP. *Sci. Amer.* **238**(3):104.

Jensen, R. G., and J. T. Bahr. 1977. Ribulose 1,5-biphosphate carboxylase-oxygenase. *Annu. Rev. Plant Physiology* **28**:379–400.

Levine, R. P. 1969. The mechanism of photosynthesis. *Sci. Amer.* **221**:58–70.

Matthern, R. O., and R. B. Kock. 1964. Developing an unconventional food algae by continuous culture under high light intensity. *Food Technol.* **18**(5):58–62, 64–65.

Stumpf, P. K., and E. E. Conn (eds.). 1980–1981. *The Biochemistry of Plants,* 8 vols. Academic Press, New York.

Zelitch, I. 1975. Pathways of carbon fixation in green plants. *Annu. Rev. Biochem.* **44**:123–145.

C H A P T E R · 13

Food Fermentations

INTRODUCTION

The oxidative decomposition of organic compounds by enzymes, to produce simpler by-products, generally describes the process of *fermentation*. Fermentative actions of enzymes are classically linked with the growth and metabolism of microorganisms on carbohydrates and various other components of organic media. This action produces a wide range of potential fermentation products including alcohols, ketones, aldehydes, and organic acids.

According to the precepts of modern biochemical engineering and technology, the term *fermentation* also applies to the controlled *in vitro* action of cell-free enzyme extracts on specific substrates.

Long before the fermentation studies of Lavoisier in the 1780s, which established a crude stoichiometry between carbon dioxide, ethanol, and acetic acid production from sugar fermentations, prehistoric man accrued the benefits of fermentation. Since many foods are rich mixtures of proteins, carbohydrates, lipids, minerals, and vitamins, foods invariably lend themselves to heterotrophic and sapropnytic modes of microbial nutrition. Although unchecked microbial actions generally lead to the wholesale chemical destruction of foods, early man recognized that certain environmental conditions could foster *limited* growths of destructive microorganisms. The biochemistry of this process was unimportant compared to the fact that long-term food preservation was ultimately enhanced. Today it is realized that the preservative effects of food fermentations are linked to the marginal deterioration of key food constituents. This deterioration results from the preferred *selective growth* of certain yeasts, molds, and/or bacteria over more destructive organisms present in the same food system. The preferred organisms hold the growth of destructive organisms in check by ejecting metabolic by-products into the food system that hinder the progressive growth of the more unfavorable food spoilage organisms. These metabolic by-products may include acids, alcohols, and other substances that are harmless to humans.

Food systems and food products prepared by fermentative steps usually display very characteristic aromas, bouquets, flavors, and textures. These attributes are favorable in most cases so long as the fermentation products are principally derived from the decomposition of carbohydrates. However, obnoxious, foul-smelling, and sometimes toxic amine compounds may result from the fermentative destruction of amino acids, peptides, and proteins, which is known as *putrefaction.*

Many fermented food products still commonly occur in the dietary regimens of populations throughout the world. For practical purposes, these fermented foods can be generically classed according to acidic, alcoholic, mold, and enzymatic processes. One or more species of microorganisms are involved in most food fermentations. Distinctly different environmental factors, such as temperature, salt concentration, water, oxygen-to-carbon dioxide ratios, acidity, and fermentation time, all play critical roles in the preparation of these fermented food products. Some practical examples of these food fermentations include:

Type of food fermentation	Foods
Acidic	Fermented milks, yogurts, cheeses, sausages, vinegars, sauerkraut, pickles, and other vegetables
Alcoholic	Wines, spirits, beer, and yeast-raised breads and pastries
Mold	Cheeses, oriental sauces, and vegetable protein fermentation products
Enzymatic	Augments production of certain meats, fish and fish sauces, some teas, vanilla, cocoa, coffee, and other food products

FERMENTATION AND THE EMBDEN-MEYERHOFF-PARNAS (EMP) PATHWAY

The fermentative metabolism of carbohydrates serves as an energy-yielding mechanism for microbial survival. Almost every carbohydrate and its related derivatives can be fermented by at least one yeast, mold, or bacteria. Typical substrates include starch, cellulose, and chitin *polysaccharides; disaccharides* including maltose, lactose, and sucrose; *monosaccharides* such as glucose, galactose, fructose, arabinose, and xylose; sugar acids such as glucuronic and gluconic acids; as well as mannitol, glycerol, and other polyhydric alcohols. Although many isolated taxonomic varieties of microorganisms demonstrate characteristic fermentative properties, glucose is readily fermented by nearly all organisms, and this serves as a model for most fermentation systems.

The Embden–Meyerhoff–Parnas (EMP) pathway underlies the fermentation of glucose. Virtually any sugars convertible into glucose or an EMP intermediate (e.g., fructose) can participate in this fermentation scheme. As detailed in the chapter on nutritional energetics (Chapter 11), catabolic and anabolic pathways are present in all cells. The EMP scheme for fermentative processes represents an anaerobic, catabolic pathway (Figure 13.1). The major products of this pathway are far more oxidized and contain less potential energy than glucose.

Fermentation of glucose requires an initial ATP-dependent phosphorylation of the sugar. This event produces glucose 6-phosphate. A number of intervening steps mediated by specific enzymes culminate in the production of an α-keto acid, recognized as pyruvic acid, plus the liberation of ATP. The ATP produced by *substrate level phosphorylation* during the EMP sequence ideally provides a molar excess compared to that ATP required for initiating the oxidative sequence of sugars through the EMP pathway.

Since pyruvic acid is far more oxidized than the initial sugar substrate, pyridine nucleotides are necessarily reduced during the EMP reactions. In order for a dynamic balance to exist between oxidized and reduced forms of pyridine nucleotides over the course of the EMP pathway, some reduced pyridine nucleotides must be oxidized. Reduced nucleotides are often reoxidized by contributing their reducing power to pyruvic acid. This step converts pyruvic acid to lactic acid—a major end product of many fermentative pathways in many different microorganisms. Aside from lactic acid production, a variety of additional mechanisms, tied to the

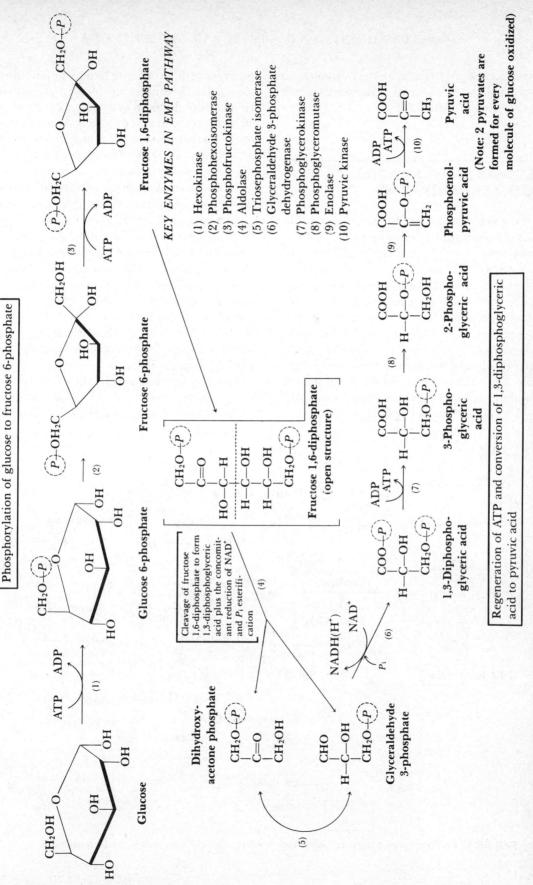

FIGURE 13.1. Conversion of glucose to pyruvic acid according to the Embden–Meyerhoff–Parnas (EMP) pathway.

reduction of pyruvic acid derivatives, also lead to the oxidation of reduced pyridine nucleotides (Figure 13.2).

ALCOHOLIC AND LACTIC ACID FERMENTATIONS

It has been recognized for many years that some microorganisms produce only one fermentative by-product, whereas others produce complex mixtures of fermentation products. Accordingly, those organisms yielding a single fermentative product are called *homofermentative,* whereas those producing mixed products are said to be *heterofermentative.* The homo- and heterofermentative designations have been commonly used since 1919 when Orla-Jensen attempted to classify lactic acid bacteria. Strict lactic acid–producing bacteria were called *homolactics* and

other bacteria producing lactic acid plus other aldehydes, ketones, and alcohols from pyruvate were denoted as *heterolactics*. Virtually all differences in end products shown by bacterial fermentations, including many molds and yeasts, reflect fundamental variations on the theme of pyruvate metabolism.

In the cases of alcoholic fermentations exhibited by yeasts as well as homolactic fermentations conducted by bacteria, the initial fermentative stages of the EMP pathway are identical. Two ATP molecules are consumed in initial stages of the pathway to produce fructose 1,6-diphosphate. The cleavage of this diphosphorylated hexose produces glyceraldehyde 3-phosphate and dihydroxyacetone phosphate. Both of these C_3 products are freely interconvertible by means of a *triose phosphate isomerase.*

Ensuing oxidation of the triose phosphate is tied to the reduction of NAD^+ and an esterification of inorganic phosphate to both

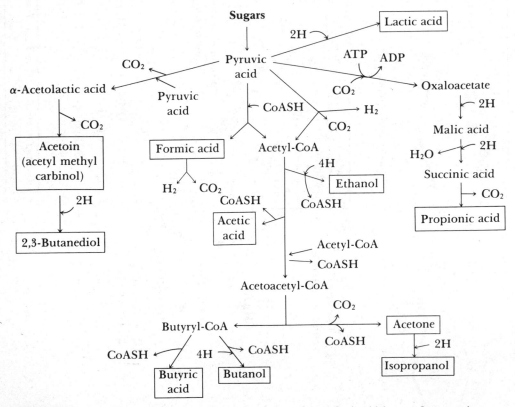

FIGURE 13.2. Descriptive origins for some major end products of microbial sugar fermentations.

C_3 structures. This action produces 1,3-diphosphoglycerate. The transformation of 1,3-diphosphoglycerate into pyruvate requires the transfer of both phosphate groups to ADP. Since 2 moles of phosphate are present for every triose (C_3) phosphate derivative of glucose, 4 moles of ATP are produced from 4 moles of ADP. It should be recalled, however, that 2 moles of ATP are required to initiate the oxidative process of glucose via the EMP pathway; therefore, only 2 moles of ATP are actually realized for every mole of fermented glucose.

In the case of homolactic fermentations, an *exact equivalence in reducing power must be maintained* within the fermentation scheme. This exact equivalence is maintained by the joint oxidation of NADH(H$^+$) to yield NAD$^+$ and the reduction of pyruvate to form lac-

tate. An alcoholic fermentation proceeds in a slightly different fashion because pyruvate is decarboxylated. The products of this reaction are acetaldehyde and carbon dioxide. This event is followed by the reduction of acetaldehyde to ethanol by reducing power from NADH(H$^+$) (Figure 13.3).

MIXED-ACID FERMENTATIONS

Those organisms classed as heterolactic fermenters, or showing mixed-acid fermentation, usually eliminate substantial amounts of pyruvate by converting it to lactic acid plus other products. Typical of these organisms are facultative anaerobes such as *Salmonella*, *Proteus*, *Vibrio*, *Yersinia*, and *Esche-*

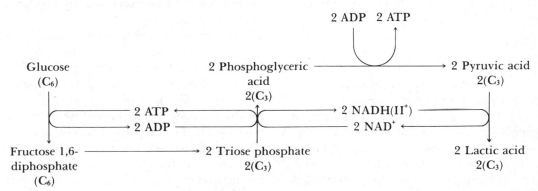

Homolactic fermentation

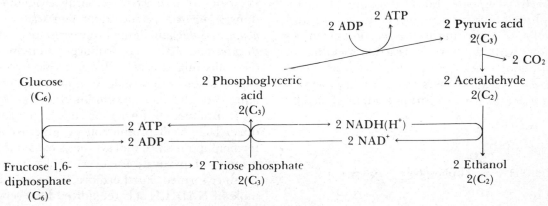

Homo-ethanolic fermentation

FIGURE 13.3. Homofermentative microbial reactions for lactic acid and ethanol rely on the production of pyruvic acid by the EMP pathway and the balanced production and consumption of high-energy phosphates (ATP ⇄ ADP) and reducing power (NADH(II$^+$) ⇄ NAD$^+$).

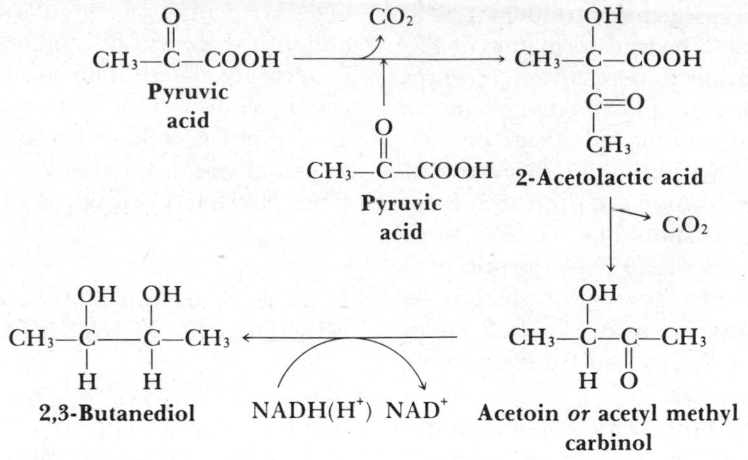

FIGURE 13.4. Butanediol formation from pyruvic acid.

richia. Escherichia coli are illustrative of this bacterial group in that they produce acetate and ethanol in addition to lactate from glucose.

2 Glucose + $H_2O \longrightarrow$
  2 lactate + acetate + ethanol + 2 CO_2 + 2 H_2

Some bacteria such as *Shigella* may refrain from carbon dioxide and hydrogen production by forming an equivalent amount of formic acid. It is interesting to note that this reaction depends on a coenzyme A-linked cleavage of pyruvic acid to produce acetyl-CoA and formate. The resulting acetyl-CoA is transformed into a store of high-energy *acetyl phosphate*. Equal portions of acetyl phosphate are converted to acetate and ethanol according to the reaction given below, with the former reaction yielding an additional ATP. Organisms such as *Escherichia coli* and others that contain *formic hydrogenlyase* convert formic acid, produced by

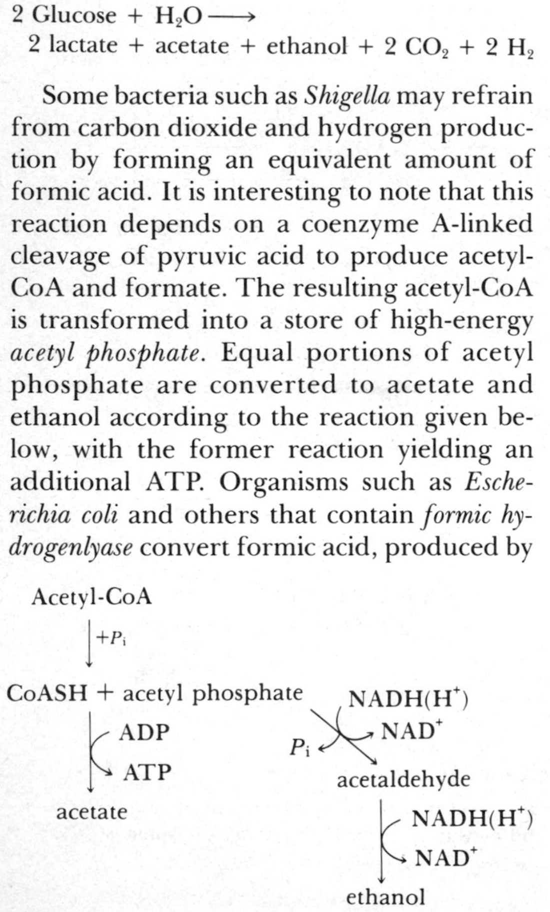

pyruvate cleavage, into carbon dioxide and hydrogen gas. Those organisms that lack this enzyme yield formic acid as a final product of metabolism.

The oxidation of glucose to ethanol, acetic acid, and formate (or carbon dioxide and hydrogen) can be ideally expressed as a balanced equation. In reality, however, most organisms that produce mixed acids also produce lactate from the reduction of pyruvic acid. The relative concentration of mixed acids to each other is widely variable depending on the microbial cultivation conditions and the availability of specific growth nutrients.

Aside from strict mixed-acid fermentations, some bacteria including *Aerobacter, Erwinia, Serratia,* and limited representatives of *Bacillus* and *Aeromonas* produce 2,3-butanediol plus mixed acids. Characteristic pyruvate reactions responsible for diol formation adhere to the pathway outlined in Figure 13.4. Butanediol formation markedly reduces lactic acid production in mixed-acid fermentations since diol synthesis consumes 2 moles of pyruvate for every molecule of product formed. Furthermore, since only 1 mole of NADH(H$^+$) is reoxidized for every two pyruvate molecules produced (a step necessary for sustaining cellular oxidation–reduction balance), butanediol-yielding organisms often demonstrate enhanced reduction of acetyl phosphate to ethanol. This

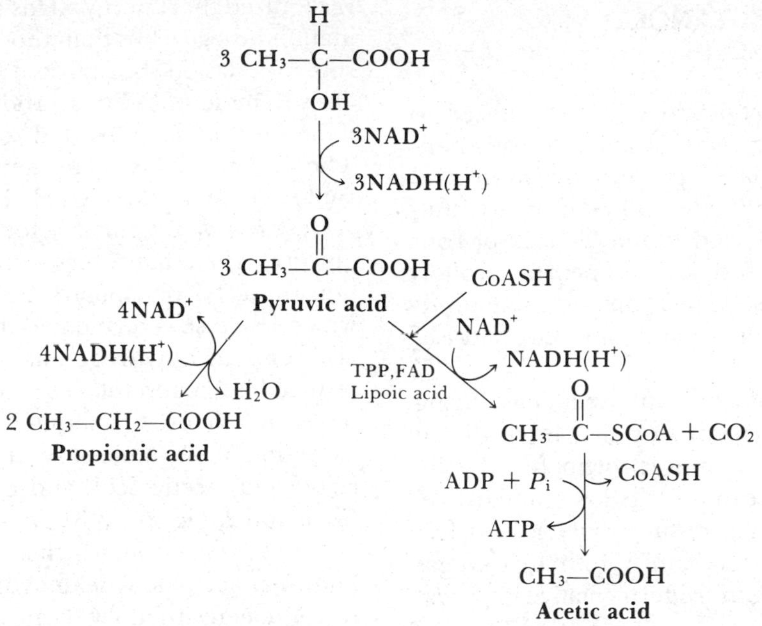

$$3\ CH_3-\overset{\displaystyle H}{\underset{\displaystyle OH}{\overset{\displaystyle |}{\underset{\displaystyle |}{C}}}}-COOH$$

3NAD$^+$

3NADH(H$^+$)

$$3\ CH_3-\overset{\displaystyle O}{\overset{\displaystyle ||}{C}}-COOH$$
Pyruvic acid

CoASH

NAD$^+$

4NAD$^+$

4NADH(H$^+$)

TPP,FAD
Lipoic acid

NADH(H$^+$)

H$_2$O

$$2\ CH_3-CH_2-COOH$$
Propionic acid

$$CH_3-\overset{\displaystyle O}{\overset{\displaystyle ||}{C}}-SCoA + CO_2$$

ADP + P_i

CoASH

ATP

$$CH_3-COOH$$
Acetic acid

FIGURE 13.5. Production of propionic acid from lactic acid.

step is coupled with a joint decrease in acetic acid formation.

Propionic acid is characteristically produced by the *Propionibacteria*. This genus of nonspore-forming bacteria converts glucose intermediates in the EMP pathway or lactate into acetic acid, carbon dioxide, and propionic acid. Lactic acid metabolism proceeds with its oxidation to pyruvic acid. Pyruvic acid is then partially oxidized to acetyl-CoA and carbon dioxide. Acetyl-CoA serves as an energy-rich intermediate for driving ATP synthesis in a manner similar to that dis-

cussed earlier for mixed-acid fermentations. Only 1 mole of ATP is generated for every 3 moles of lactate fermented. A balance between reduced and oxidized equivalents of pyridine nucleotides is sustained in the *Propionibacteria* by lactic acid reduction to propionic acid (Figure 13.5).

The generation of propionic acid from lactic acid is an indirect, complicated reaction that requires the participation of a biotin-linked enzyme. The total reaction leading to propionic acid plus a water by-product is expressed in Figure 13.6.

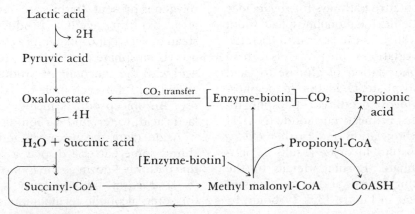

FIGURE 13.6. Detailed steps in the conversion of lactic acid to propionic acid.

ACETONE–BUTANOL FERMENTATIONS

Sugars and lactic acid may be oxidized to yield energy by *Clostridia* and some other anaerobic bacteria to produce pyruvate and crotonyl-CoA. These compounds are subsequently converted to butyric acid or contribute to the production of neutral alcohols and/or ketones. Typical among these products are butanol, ethanol, acetone, and isopropanol.

Reduction of acidic intermediates during butyric acid fermentation may be mediated by the iron-containing protein ferredoxin. Ferredoxin, because of its low standard reduction potential, reduces butyric acid fermentation intermediates with hydrogen gas liberated during initial fermentative steps that cleave pyruvate:

Pyruvic acid + CoASH ⟶
$$\text{acetyl-CoA} + CO_2 + H_2$$

All but the strictly anaerobic bacteria fail to demonstrate the reductive participation of ferredoxin. Figure 13.7 illustrates the possible routes involved in a butyric acid fermentation.

FERMENTATION PROCESSES MAY PROCEED BY ROUTES OTHER THAN THE EMP PATHWAY

Heterofermentative pathways for sugars may depend on the hexose monophosphate shunt in organisms such as the lactic acid bacteria. This sugar oxidation route proceeds with an initial phosphorylation of glucose to yield glucose 6-phosphate. Oxidation of the phosphorylated glucose produces 6-phosphogluconic acid plus reduced nucleotide (NADH (H^+)). Decarboxylation and further oxidation of 6-phosphogluconate results in ribulose 5-phosphate. Ensuing details of the pathway leading to lactic acid production and ethanol appear in Figure 13.8. It should be recognized that energy yields from sugar oxidation here are less than those produced by sugar oxidation through the EMP pathway. Only 1 mole of ATP is produced for each mole of glucose converted to pyruvic acid. The end products of fermentation by this route ideally lead to a 1:1:1 ratio of *lactic acid* : *carbon dioxide* : *ethanol* for each mole of glucose. The ethanol arises from the reduction of acetyl phosphate. In those cases in which fructose is fermented, this sugar may also serve as a hydrogen acceptor by being reduced to mannitol. The reduction of 2 moles of fructose to mannitol is linked to the oxidation of one mole of fructose to produce lactic acid, acetic acid, and carbon dioxide (Reaction 2, Figure 13.8). In organisms such as *Lactobacillus brevis*, which lacks the enzyme known as acetaldehyde dehydrogenase, glucose is not anaerobically fermented to ethanol. Instead, glucose can be fermented to acetic acid if a suitable hydrogen acceptor is available. Fructose serves as a hydrogen acceptor (oxidizing agent) under these conditions, ultimately producing mannitol. According to this route, the reduction of 2 moles of fructose to mannitol is linked to the oxidation of 1 mole of glucose to lactic acid, acetic acid, and carbon dioxide (Reaction 3, Figure 13.8). *L. mesenteroides*, on the other hand, displays fermentative behavior similar to *L. brevis;* however, it can reduce acetyl phosphate to ethanol as well as ferment fructose or lactose to lactic acid, ethanol, and carbon dioxide.

The heterofermentative lactic acid bacteria operate in a different fashion when oxygen is present. In this case oxygen serves as an oxidizing agent to produce water instead of acetyl phosphate. As a consequence, acetyl phosphate can be converted to acetic acid with the concomitant production of an added ATP for each mole of oxidized hexose. Aerobic products include acetic acid, lactic acid, water, and carbon dioxide.

Aside from the hexose monophosphate shunt, some bacteria display a variation of this pathway known as the *Entner–Doudoroff (E–D) pathway*. This pathway is important for some alcoholic fermentations exhibited

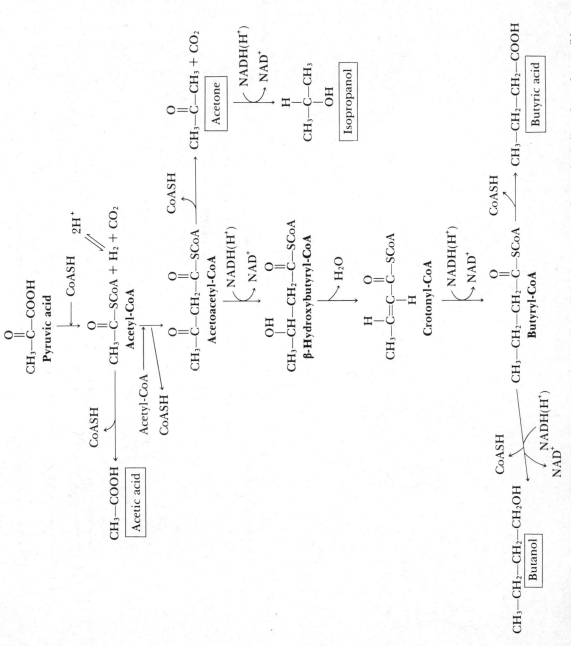

FIGURE 13.7. Transformation of pyruvic acid into acetone–butanol fermentaion products. Also indicated are the possible routes for acetic acid, isopropanol, and butyric acid products.

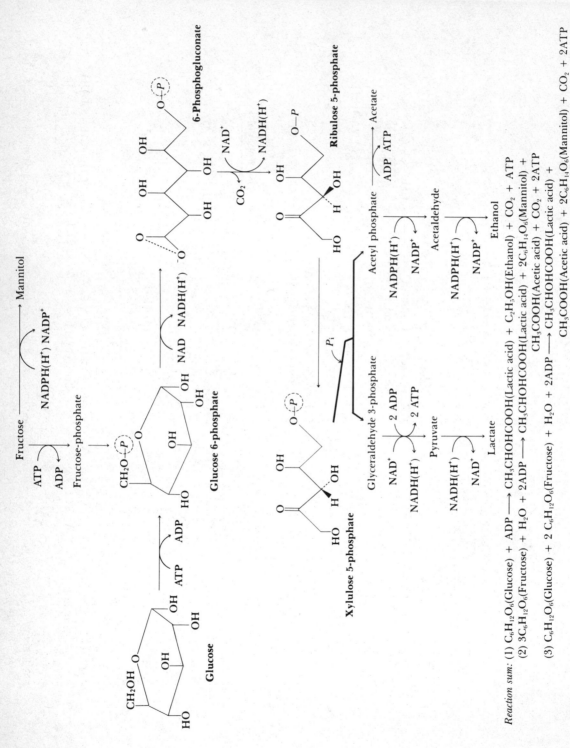

Reaction sum: (1) $C_6H_{12}O_6$(Glucose) + ADP ⟶ $CH_3CHOHCOOH$(Lactic acid) + C_2H_5OH(Ethanol) + CO_2 + ATP

(2) $3C_6H_{12}O_6$(Fructose) + H_2O + 2ADP ⟶ $CH_3CHOHCOOH$(Lactic acid) + $2C_6H_{14}O_6$(Mannitol) + CH_3COOH(Acetic acid) + CO_2 + 2ATP

(3) $C_6H_{12}O_6$(Glucose) + 2 $C_6H_{12}O_6$(Fructose) + H_2O + 2ADP ⟶ $CH_3CHOHCOOH$(Lactic acid) + CH_3COOH(Acetic acid) + $2C_6H_{14}O_6$(Mannitol) + CO_2 + 2ATP

FIGURE 13.8. Heterofermentative pathways for glucose and fructose in heterofermentative lactic acid bacteria.

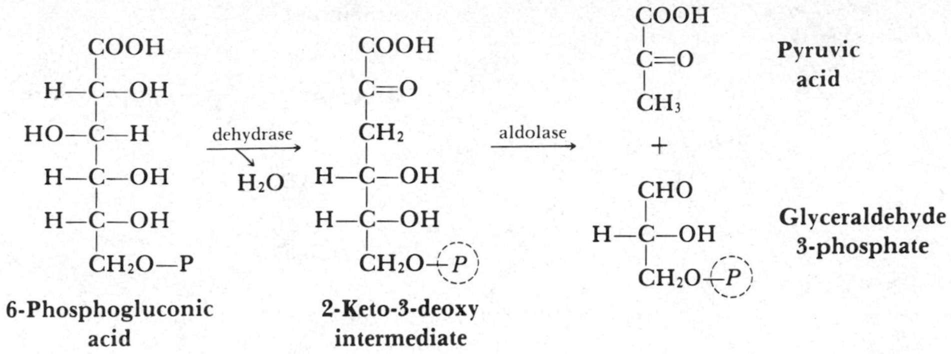

COOH
|
H—C—OH
|
HO—C—H   dehydrase
| ————————→
H—C—OH H_2O
|
H—C—OH
|
CH_2O—P

6-Phosphogluconic acid

COOH
|
C=O
|
CH_2
|
H—C—OH
|
H—C—OH
|
CH_2O—(P)

2-Keto-3-deoxy intermediate

aldolase
————————→

COOH
|
C=O
|
CH_3

Pyruvic acid

+

CHO
|
H—C—OH
|
CH_2O—(P)

Glyceraldehyde 3-phosphate

FIGURE 13.9. Bacteria that lack phosphofructokinase in an Embden–Meyerhoff–Parnas-type pathway cannot oxidize sugars according to the glycolytic sequence. The Entner–Doudoroff pathway is employed under these conditions to produce pyruvate and glyceraldehyde 3-phosphate through a 2-keto-3-deoxy intermediate.

by bacteria that lack *phosphofructokinase*—an enzyme critical for the operation of the EMP pathway. According to the E–D pathway, glucose catabolism proceeds by dehydration of 6-phosphogluconic acid (Figure 13.9). A *2-keto-3-deoxy intermediate* results from dehydrase activity, and an aldolase-type enzyme cleaves this 2-keto-3-deoxy intermediate into pyruvate and glyceraldehyde 3-phosphate. The E–D pathway can demonstrate various modifications that allow the oxidation of galactose and sugar acids such as D-glucuronic and D-galacturonic acids.

Once pyruvate and glyceraldehyde 3-phosphate have been formed, pyruvate may proceed toward ethanol formation by way of decarboxylation to acetaldehyde, or glyceraldehyde 3-phosphate can yield ethanol by enzyme actions common to the EMP pathway.

The complex biochemistry of all species and genera of microorganisms ensures that other mechanisms are also involved in hexose oxidations, but these are beyond the scope of the present text.

MISCELLANEOUS FERMENTATIONS OF PENTOSE SUGARS, ORGANIC ACIDS, AMINO ACIDS, AND AMIDES

Pentose sugars are fermented by many bacteria including the homo- and heterolactic bacteria. These pentoses typically yield a mixture of lactic and acetic acids by the pathway illustrated in Figure 13.10.

Common organic acids in vegetative tissues such as malate and citrate may also be fermented. Many of the lactic acid bacteria ferment these substrates to neutral end products such as acetoin, 2,3-butanediol, or ethanol in addition to salts of lactic, acetic, and formic acids (Figure 13.11).

Many protein-containing foods such as fermented milk products rely on the fermenting actions of lactic acid bacteria, but these bacteria are *generally* nonproteolytic. Although these bacteria do require exogenous amino acids for growth, fermentation of amino acids is somewhat restricted to serine and arginine. Deamination of serine and arginine serve as the first step in the respective conversion of these amino acids to ornithine and acetoin.

Arginine $\xrightarrow{\text{NH}_3}$ citrulline $\xrightarrow[\text{ATP} \quad \text{ADP}+P_i]{\text{NH}_3 \quad \text{CO}_2}$ ornithine

2 Serine $\xrightarrow{\text{NH}_3}$ 2 pyruvate $\xrightarrow{\text{CO}_2}$ acetoin

The limited fermentative actions of lactic acid bacteria on amino acids are not paralleled by their decarboxylative actions on amino acids to produce amines. Some streptococci readily decarboxylate tyrosine to tyramine, whereas other *Lactobacilli* decarboxylate histidine, lysine, and ornithine.

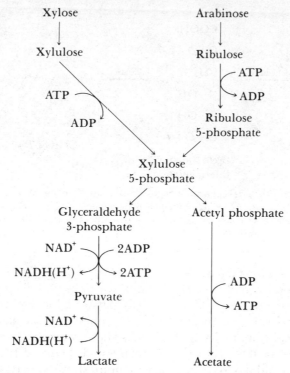

FIGURE 13.10. Generalized fermentative routes for pentoses in homo- and heterofermentative lactic acid bacteria.

The *Clostridia* are quite unlike the lactic acid bacteria as they are proteolytic, amino acid-fermenting bacteria. In fact, three distinct amino acid fermentations are recognized:

1. *Deamination of amino acids,* which liberates ammonia plus the organic acid skeleton of the original acid.

2. *Decarboxylation of amino acids* to produce an amine.

3. *Oxidation–reduction reactions* wherein a coupled Stickland-type reaction results in oxidation of one amino acid while another is reduced. Amino acids subject to oxidation include alanine, histidine, isoleucine, leucine, phenylalanine, serine, tryptophan, and tyrosine. Amino acids that undergo reduction include arginine, aspartic acid, cysteine,

glycine, methionine, ornithine, proline, tryptophan, and tyrosine. Based on this scheme, the oxidized amino acids are transformed into fatty acids having one less carbon than the original compound, and the reduced amino acid yields a fatty acid having the same number of carbons as the original amino acid.

Ammonia is produced in both reactions except those that deal with proline. The fermentation of alanine and glycine to acetic acid, ammonia, and carbon dioxide is exemplary of this last reaction (Figure 13.12). Table 13.1 also delineates some common amino acid-fermentation products produced by proteolytic *Clostridia*, but many of the reactions may occur to limited extents in other proteolytic bacteria.

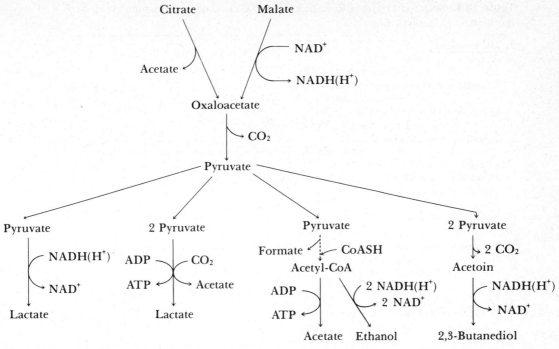

FIGURE 13.11. Possible routes for the oxidation of citrate and malate by homo- and heterofermentative lactic acid bacteria. (Adapted from P. McDonald, *The Biochemistry of Silage*, John Wiley & Sons, New York).

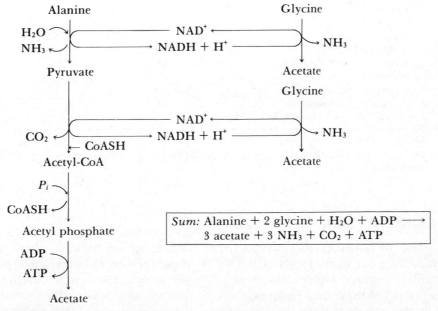

Sum: Alanine + 2 glycine + H_2O + ADP ⟶
3 acetate + 3 NH_3 + CO_2 + ATP

FIGURE 13.12. An oxidation–reduction (Stickland) reaction typical of amino acid catabolism by proteolytic *Clostridia*.

Table 13.1. Selected Products Resulting from the Catabolism of Amino Acids and Amides by Proteolytic *Clostridia*.

1. Deamination

Arginine → citrulline + NH_3
  ↳ ornithine + NH_3 + CO_2
Aspartic acid → fumaric acid + NH_3
  ↳ acetic acid + pyruvic acid
Glutamic acid → mesaconic acid + NH_3
  ↳ acetic acid + pyruvic acid
Histidine → urocanic acid + NH_3
  ↳ formiminoglutamic acid → formamide + glutamic acid
Lysine → acetic acid + butyric acid + $2NH_3$
Methionine → α-ketobutyric acid + methylmercaptan + NH_3
Phenylalanine → phenyl propionic acid + NH_3
Serine → pyruvic acid + NH_3
Threonine → α-ketobutyric acid + NH_3
Tryptophan → indolepropionic acid + NH_3
Tyrosine → *p*-hydroxyphenyl propionic acid + NH_3
Asparagine → aspartic acid + NH_3
Glutamine → glutamic acid + NH_3

2. Decarboxylation

Arginine → ornithine → putrescine + CO_2
Aspartic acid → alanine + CO_2
Glutamic acid → γ-aminobutyric acid + CO_2
Histidine → histamine + CO_2
Lysine → cadaverine + CO_2
Phenylalanine → β-phenylethylamine + CO_2
Serine → ethanolamine + CO_2
Tryptophan → tryptamine + CO_2
Tyrosine → tyramine + CO_2

3. Oxidation–reduction (Stickland)[a]

  (a) Oxidation
    Alanine + $2 H_2O^{-4H}$ → acetic acid + NH_3 + CO_2
    Leucine + $2 H_2O^{-4H}$ → isovaleric acid + NH_3 + CO_2
    Isoleucine + $2 H_2O^{-4H}$ → α-methyl butyric acid + NH_3 + CO_2
    Valine + $2 H_2O^{-4H}$ → isobutyric acid + NH_3 + CO_2
  (b) Reduction
    Glycine^{+2H} → acetic acid + NH_3
    Proline^{+2H} → δ-amino valeric acid
    Ornithine^{+2H} → δ-amino-valeric acid + NH_3

[a]Some examples.

SOURCE: From *The Biochemistry of Silage* by P. McDonald. Copyright © 1981, Wiley, New York, and reprinted with permission.

MILK FERMENTATION

Fermentation of lacteal fluids provides the production basis for many foods including soured milks, yogurts, buttermilks, butter, and cheeses. All of these milk fermentations depend, in one way or another, on lactic acid production by bacterial species of the genera *Streptococcus, Leuconostoc, Pediococcus,* and *Lactobacillus.* These bacteria produce lactic acid from lactose and other metabolizable sugars present in lacteal fluids plus some minor amounts of volatile aldehydes, ketones, and organic acids, which contribute characteristic flavors to these milk products. Of the four genera, the *Lactobacilli* are the most prodigious acid producers, followed by the three coccal genera. Moreover, since the lactic acid bacteria are generally microaerophilic, anaerobic conditions actually enhance their growth and production of lactic acid.

Normal fermentations of milk at room temperature proceed according to Figure 13.13. A population of *Streptococcus lactis* usually initiates the milk fermentation, followed by the *Lactobacilli,* and eventually by the yeasts and molds depending on the fermentation time. Progressive decreases in pH caused by lactic acid production cause casein to precipitate at its isoelectric point (~4.5 pH). It is this pH effect that suppresses the growth of many species of bacteria that otherwise exert a detrimental effect on the food system. Cessations of lactic acid production owing to sugar fermentations by the *Lactobacilli* may result in the growth of yeasts, molds, and other bacteria whose growth was formerly checked. During this period, pH may actually begin to *rise* as a result of microbial-linked lipolytic and peptonization reactions. Various yeasts, molds, and significant population numbers of *Proteus, Pseudomonas,* and *Achromobacter* may begin to materialize under these conditions. It should be recognized, however, that room temperature successions of microbial flora in milk are quite different from that sequence outlined above in cases of refrigerated fermentations.

Many milk fermentations can be largely controlled by inoculating or "seeding" the milk with specific "starter" bacterial cultures. These bacteria, ranked according to total acid production, include *Streptococcus lactis, Lactobacillus casei, L. bulgaricus, L. lactis, L. helviticus,* and their respective species variants.

The viscosity of fermented milk products is caused by pH-induced coagulation of milk proteins and the associated abilities of these proteins to interact with the fluid or serum fractions of milk. Moreover, heat treatments of raw and fermented milk products, which are not uncommon, also affect viscosity. Raw milks suitable for fermentative processes have a pH of ~6.6 where the protein component exists in *two forms,* (1) casein micelles and (2)

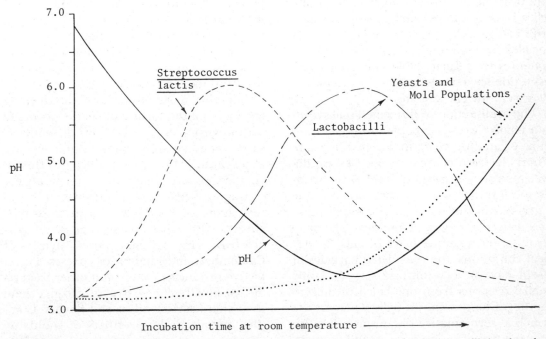

FIGURE 13.13. The relative succession patterns for metabolizing microbial populations in milk incubated at room temperature with respect to pH conditions.

whey or serum proteins. The casein micelles have an average size of ~100 nm and are constructed of three or more different proteins. Statistical calculations show collision frequencies of ~2500 times a second over intermicelle distances averaging ~50 nm. According to many theories, micelles fail to coalesce and form bigger particles because superficial negative charges on the micelles adequately repel the particles. The whey proteins, on the other hand, exist as a soluble protein fraction, but heating above 85°C (185°F) denatures these proteins and causes them to unfold. The β-lactoglobulin fraction in particular, which exists as a superficial component of micelles, assumes a "hairy-type" conformation.

The architecture of casein micelles favors superficial hydrophobic properties and, therefore, micelles tend to participate in mutual physical interactions instead of micelle–water interactions. This action causes linear portions of micelle structures to bind to each other in a gel network permeated with water. Micelles that exhibit a high surface component of "hairy" denatured β-lactoglobulin cannot come as close together as those micelles having a lower surface concentration of β-lactoglobulin. As a consequence, those micelles having a high β-lactoglobulin component form a fairly stable aqueous gel with less tendency to undergo syneresis than simple casein micelle systems.

The acidification of fermented milks leads to a charge decrease between micelles and an associated decrease in micelle repulsion. When electrical charges of micelles are depreciated to some critical level, overt coagulation of the milk proteins occurs. The characteristic viscosities of fermented milks also reflect a slow and uniform pH drop throughout the entire batch of milk. Under ideal conditions, the coagulated particles of casein have a very uniform size with only small variations from one batch of milk to another. Changes in raw milk quality or the principal fermentative organisms will influence the size of casein micelles and their aggregates along with the stability of the system. Buttermilk, yogurt, leben, and taettenjolk are typical examples of milks having homogeneously distributed, coagulated protein–water systems. These fluid food systems are quite different from cottage cheese, in which uncured curds of milk protein are separated from the aqueous serum fraction by simple drainage and a modicum of serum expression using physical pressure.

The lactic acid-induced production of a serum fraction may also serve as a preliminary step to the production of Ricotta cheese, which is classed as an albuminoid cheese, or Myosost, which is one of several sweet cheeses. These types of cheeses are produced by concentrating the protein portion of serum fractions.

The fermentative action of bacteria producing lactic acid and/or the milk curdling actions of fungal enzymes (e.g., *Mucor pusillus* and *M. miehei* proteases), porcine or bovine pepsins, rennin obtained from calves' stomachs (vells), or controlled addition of lactic acid to milk, serve as preludes to more complex versions of cheese manufacture. Exposure of milk to these agents is followed by careful control of water and salt in the milk curd as well as microbial ripening of the isolated curd. Only minor variations in the execution of these fundamental cheese-processing steps account for the existence of over 700 named cheeses.

Microbial metabolism within the ripening curd and on the surface of cheeses, coupled with environmental conditions of temperature, humidity, and sanitation practices are also critical factors for producing different cheeses. Elevated temperatures typically encourage gas production and propionic acid formation by *Propionibacteria* in Swiss-type cheeses. Other bacteria such as *Brevibacterium linens* grow on the exposed surfaces of Camembert and Limburger cheeses. Putrefactive products of these and other bacteria diffuse into the cheese structure to give them their characteristic odors and flavors. Complex metabolic by-products of molds *on* cheeses may also arise from growths of *Penicillium, Cladosporum, Alternaria, Monilia, Mu-*

cor, *Aspergillus,* and/or *Geotrichum.* For example, Camembert cheese requires a successive growth of *Oidium lactis* followed by *Penicillium camemberti* in order to develop the delicate mixture of metabolites that characterize the flavor of these cheeses. In some cases such as Roquefort, Gorgonzola, and bleu cheeses, lactic acid fermentation of a curd precedes inoculation of the ripening curd with molds such as *Penicillium roquefortii, P. camemberti,* or *P. glaucum.* Ripening may require 3 to 6 months for bleu cheeses or longer depending on the cheese.

Cream fermentation by lactic acid-producing bacteria also produces an isoelectric point-induced curd. This curd is partitioned from butterfats as a result of churning to produce a liquid fraction of buttermilk and a reverse emulsion of water in butterfat (~15.5% water in ~81% butterfat). *Streptococcus lactis* and *S. cremoris* are both responsible for lactic acid production, whereas the growth of *Leuconostoc citrovorum* and *L. dextranicum* in conjunction with *Streptococcus lactis* may produce biacetyl, acetoin (acetyl methyl carbinol), some alcohols, and aldehydes that are retained by the butterfat fraction. These compounds impart characteristic flavors to butterfat along with minor concentrations of free short-chained fatty acids.

The desired effects of lactic acid fermentations can be upset by many factors that culminate in the production of off-odors, off-flavors, and undesirable textures in the final fermented dairy product. In some cases, faulty fermentation is simply caused by an inadequate amount of "starter" culture to initiate fermentation and a corresponding lack of proper acid production. In other situations, faulty fermentation may be caused by improper incubation conditions and development of the "starter" culture. Aside from incubation conditions, development of the desired lactic acid bacterial culture may be defective as a result of milk that contains residual antibiotics, bacteriophage, ≥5.0 ppm of hypochlorite residues used for sanitation procedures, or trace amounts of poisonous quaternary amine compounds. Regardless of human health considerations, milk from cows having mastitis (often attributed to *Streptococcus agalactiae*) may also inhibit the growth of lactic acid-producing bacteria. Other cultural defects of fermentations originate with the growth of *Streptococcus lactis,* which produce nisin. This antibiotic compound inhibits *Clostridial* spoilage in processed cheeses, in which blowing is undesirable, but it may also inhibit the growth of *Streptococcus cremoris* and other streptococci necessary for normal milk fermentation.

The inadequate development of flavor in fermented milk products is complex. High concentrations of lactic acid may encourage excessive biacetyl formation, whereas low concentrations of citrate, metabolized by *Leuconostoc citrovorum* and *L. dextranicum* to produce acetoin and biacetyl, may inhibit proper flavor development. Long holding of fermented milks can also lead to decreased acetoin concentrations, thereby diminishing desired flavors. High populations of Gram-negative bacteria in fermented cultures also produce similar consequences.

Milk fermentations can be seriously affected by the initial presence of certain bacteria in the prefermented milk or starter culture. Ropiness in milk, which has been far more common in past years, originates from the growth of *Alcaligenes viscolactis.* Growth of this bacterium, spurred on by holding raw milk at a low temperature for several days, produces a slime layer or "rope," so-called because of its stringy consistency. Various varieties of *Streptococcus lactis* either produce undesirable organic compounds or adversely affect normal patterns of lactic acid production. The *hollandicus* variety produces a capsular material that is slimey and ropey; a *maltigenes* variety produces methyl butanol from leucine to give fermented milk a malty flavor; tardy lactic acid production characterizes the *tardus* variety; and high concentrations of citrate can be fermented to correspondingly high concentrations of acetoin and other strong aroma compounds typical of butter by the *diacetylactis* variety.

Fermentation-induced destruction of already fermented dairy products is a common event. Butters can undergo a putrid (surface taint) or rancid destruction. *Pseudomonas* bacteria are largely implicated with both forms of spoilage. The growth of *Ps. putrefaciens* for 7 to 11 days at 4 to 7°C results in the production of isovaleric acid to produce a putrid odor. Although a variety of lipases other than those of microbial origins can produce triacylglycerol hydrolysis, lipases from *Ps. fragi* and *Ps. fluorescens* are occasionally found. Malty flavors, skunk-like odors, and black discolorations of butter are also respectively linked to the *Streptococcus lactis* var. *maltigenes, Pseudomonas mephitica,* and *Ps. nigrificans.* Nearly all types of molds can ruin butter. In fact, the high lipid content and low water content of butter enhances its suitability as a growth medium for molds instead of bacteria. Almost all genera of molds can grow on butter to produce superficial colors that are typical of mold spore colors. Aside from *Aspergillus,* black yeasts of the genus *Torula* can attack butter.

Cottage cheese readily assumes stale, moldy, musty, and yeasty flavors caused by the metabolic by-products and fermentation actions of *Penicillium, Mucor, Alternaria,* and *Geotrichum.* Slimey curd may also result from capsular bacterial growths tied to the growth of *Alcaligenes, Pseudomonas, Proteus, Aerobacter,* and *Achromobacter* spp.

Finished aged cheeses exhibit minimal destruction by microorganisms so long as the water activity of the cheeses is low. If water activity is adequate, however, anaerobic *Clostridia* including *Cl. butyricum, Cl. pasteuranium,* and *Cl. sporogenes* may cause gassiness in cheeses. *Bacillus polymyxa* also causes gassiness, but it is an anaerobic sporeformer. In the cases of both the *Clostridia* spp. and the *Bacillus* spp., all of these organisms can ferment lactic acid to carbon dioxide.

VEGETATIVE AND FRUIT FERMENTATIONS

The biochemical constituents of fruits, vegetables, and their juices provide excellent growth media for yeasts, molds, and bacteria. These organisms superficially attack the vegetative tissues of intact fruits and vegetables, or they may invade damaged plant structures through wounds, cracks, or bruises. Most vegetables are susceptible to bacterial attack since water activity is high, pH ranges are compatible with many bacteria, and most vegetables display high oxidation–reduction potentials and lack a high poising capacity. This means that aerobic and facultative anaerobic bacteria are usually more important than anaerobic organisms in cases of vegetable spoilage, although a wide spectrum of fermentative products can be formed in any case. Bacterial soft rot mediated by species of the genus *Erwinia* are almost universally linked to vegetable spoilage. Invasions of vegetative tissues by *Erwinia carotovora* and other related organisms are closely tied to the production of pectinases (protopectinases) that destroy the cementing structure of plant pectins (protopectins). Moreover, the destruction of plant tissues by these bacteria are associated with malodors arising from volatile organic acids, ammonia, and other odorous substances. Complex cellulosic substances, aromatic-ring structures, and porphyrins are usually the last compounds to be destroyed. Sugars typically fermented include rhamnose, cellobiose, arabinose, and mannitol.

Other organisms demonstrating heterotrophic and/or saprophytic modes of nutrition on vegetables are summarized in Table 13.2.

Fruits, too, are subject to complex fermentative actions of yeasts, molds, and bacteria. Since fruits generally contain more hydrophilic constituents (e.g., sugars) than vegetables plus a higher acidity (lower pH), these factors seem to favor yeast and mold destruction of most fruits. Table 13.2 details the most common microbial agents of fruit destruction.

Many genera of yeasts ferment the sugars in fruits to produce carbon dioxide and alcohol. Compared to molds, yeast growth is quite rapid, and it is possible that yeasts may pave the way for later mold invasions of fruits.

Table 13.2. Representative Microorganisms That Can Exert Fermentative Actions on Fruits and Vegetables, Leading to Spoilage and Rotting

Apples 2, 4, 7, 12	Grapefruits 7, 10
Apricots 2, 4, 8, 9, 11, *13*[a]	Lemons 2, 7, 10
Artichoke *12*	Lettuce 1, 2, 3, *12*
Asparagus 1, 2, 3, 5	Limes 2, 7, 10
Avocados 9	Onions 1, 2, 3, 12
Beans (lima, green, wax) 1, 2, 3, 4, 6, 12, *13*	Oranges 2, 7, 10
Beets 1	Parsley 1, 3
Blackberries 7, 8	Parsnip 1, 2, 3
Broccoli 1, 2, 3, 4	Peaches 2, 4, 7, 8, 9, 11, *13*
Brussel sprouts 1, 2, 3, 4	Pears 2, 4, 7
Cabbage 1, 2, 3, 4	Peppers 2, 5, 6
Cantaloupes 1, 4, 5	Plums 2, 4, 7, 8, 9, 12
Carrots 1, 2, 3, 4, 13	Potatoes 1, 4
Cauliflower 1, 2, 3, 4	Prunes 2, 4, 7, 8, 9, 12
Celery 1, 2	Pumpkin 2, 4, 6
Cherries 2, 4, 7, 8, 9, 11, 12, *13*	Radishes 1, 2, 3, 4
Crucifers 12	Rhubarb 1, 2
Cucumbers 1, 2, 4, 6	Rudabagas 1, 2, 3, 4
Currents 2, 7	Spinach 1
Dewberries 7	Squash 2, 4, 6
Egg plant 5	Strawberries 2, *12, 13*
Endives 1, 2, 3	Sweet potatoes 2, 4, *13*
Garlic 1, 2, 3, 5	Tomatoes 1, 2, 3, 4, 5, 6 green—*12*, ripe—*13*
Globe artichoke 1, 2, 3	Turnips 1, 2, 3, 4
Grapes 2, 4, 7, 8, 9, *12*	Watermelons 1, 4, 5, 6

1. *Erwinia carotovora* (bacterial soft rot)
2. *Botrytis cinerea* (gray mold rot)
3. *Geotrichum candidum* (sour rot, oospora rot, watery soft rot)
4. *Rhizopus stolonifer* (rhizopus soft rot)
5. *Phytophora* spp. (phytophora rot)
6. *Colletotrichum coccodes* and other species (Anthracnose)
7. *Penicillium* spp. (blue rot mold)
8. *Aspergillus niger* (black mold rot smut)
9. *Cladosporium, Trichoderma* and other molds (green mold rot)
10. *Alternaria tenuis* (Alternaria rot)
11. *Cladosporum herbarum* (Cladosporum rot)
12. Generalized *Botrytis* rot
13. Generalized *Rhizopus* rot

[a]Italic number indicates that mold is the major cause of spoilage.

Whereas yeasts readily oxidize sugars, the molds metabolize higher-molecular-weight compounds in fruits as well as alcohols produced by primary microbial invaders.

Controlled food fermentations of vegetables principally rely on the heterofermentative actions of *Leuconostoc mesenteroides*. The rapid population growth of these bacteria ensures high lactic acid and carbon dioxide levels that exclude all but desirable fermentative organisms. Desirable fermenters include *Lactobacillus brevis, Pediococcus cerevis-* *iae,* and *Lactobacillus plantarum.* Depending on temperature and salt concentrations ranging from 1.5 to ≤10.0%, many fermented foods can be produced. These include cucumber pickles (dill and sour), sauerkraut, olives, Kimchi (mixed vegetables), and many other ethnic and culturally defined foods.

Other complex vegetative fermentations involve bacterial and mold cultivation. As an example, Tempeh, prepared from soybeans, requires initial acidification of the beans by bacteria. This step is followed by a dense mold growth of *Rhizopus oryzae* over the acidified beans. The desired type of *Rhizopus* is obtained by spore inoculation. Development of mycelial growth over the beans is associated with lipolytic and proteolytic enzyme activity that significantly modifies the protein and lipid components of the beans. Other mold species can be used on selected beans, de-oiled peanuts, or rice to yield products reminiscent of Tempeh.

Acetic acid production by species of the genus *Acetobacter* are fundamental to juice and vegetative fermentations that yield a high concentration of acetic acid. These aerobic bacteria characteristically ferment sugar derivatives from ethanol to acetic acid, which is the major vinegar acid.

The preservation of forage crops for animal feeds also depends on controlled microbial fermentations of vegetative matter. The process known as *ensiling* a crop implies that a forage crop has been preserved in the form of silage. Vegetative material including grasses and legumes, low-moisture grass, and oats, respectively produce grass silage, haylage, and oatlage. Other types of ensiled products are commonly described by similarly descriptive names.

The ensiling process is initiated by loading a closed silo with vegetative material. Oxygen is eventually exhausted in the vegetative mass by aerobic bacteria plus unchecked plant respiration. These combined events may normally raise the temperature of the vegetative mass from 70 to 155°F (21 to 68°C). Following a typical period of ~5 h, carbon dioxide levels plus consumed levels of oxygen ensure an anaerobic environment. Anaerobic acetic acid bacteria materialize for a short time along with a concomitant drop in pH. The resulting pH condition fosters the subsequent growth and reproduction of lactic acid-producing bacteria. The events are illustrated in Figure 13.14.

As in the case of nearly all fermentations, the production of good quality silage is both an art and a science. Improper oxygen levels at the outset of the process can ruin the whole preservation effort. This is true especially when initial levels of oxygen are too high and both putrifying and proteolytic bacteria may overrun the desired action of lactic acid bacteria.

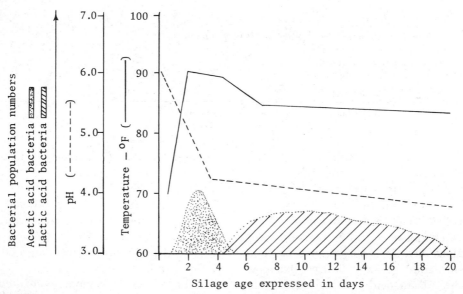

FIGURE 13.14. Relative occurrence of pH, temperature, and bacterial populations during the normal fermentation of silage. Consult text for details.

ETHANOLIC FERMENTATIONS AND RELATED PRODUCTS

The ethanolic fermentative actions of yeasts are fundamental to many foods. Yeast species of the genus *Saccharomyces* are responsible for the production of alcoholic beverages as well as the deterioration of many foods. Hybridized and cultured forms of true yeasts are principally used in the food industry where predictable fermentative reactions must occur. Common representatives of these yeasts include *Saccharomyces cerevisiae*, used for brewing and baking, and *S. cerevisiae* var. *ellipsoideus*, which is suited to high alcohol production for wine and distilled liquors. Many other yeasts including *S. lactis* and *S. fragilis* are lactose fermenters capable of preempting desired bacterial fermentations of lactose in foods. Unlike most bacteria, many yeasts including *S. rouxii* and *S. mellis* are halotolerant to levels of 18% salt. These organisms can exist as superficial growths on soy sauces and other high-salt foods. Many other yeasts including the *Schizosaccharomyces* are osmophilic destroyers of honey, fruits, syrups, and other high-sugar foods. Oxidative yeasts representing three genera (wild yeasts)—*Pichia*, *Hansenula*, and *Debaromyces*—are common film yeasts that oxidize organic acids. This action may prompt bacterial spoilage of foods since their growth results in an elevated pH condition.

Although yeasts of the genus *Saccharomyces* demonstrate aerobic oxidative metabolic pathways, involving both the tricarboxylic acid cycle and respiratory electron transport chain plus the anaerobic EMP pathway, alcohol production by yeasts requires only the anaerobic route. In fact, whenever concentration levels of fermentable sugars such as sucrose, glucose, fructose, and maltose *exceed* 0.1% concentration levels, yeast cells repress the synthesis of key enzymes within the tricarboxylic acid cycle plus some of those enzymes found in the respiratory chain. This phenomenon is widely known as the *Crabtree effect*.

The shut-down of aerobic oxidative pathways drives carbohydrate oxidation strictly through the EMP pathway. Accordingly, acetaldehyde serves as a hydrogen acceptor and both carbon dioxide and ethanol represent the major fermentation products.

Very basic differences in the production of fermented beverages result in the occurrence of many minor aldehydes, ketones, organic acids, trace alcohols, acetates, and other substances found in alcoholic beverages. These are generally considered to be primary products of yeast fermentations; however, aging, distilling, and other treatments of fermented liquors may modify these primary fermentation products to secondary products that characterize specific beverages (Tables 13.3 and 13.4).

The most popular alcoholic beverages fall into five categories: wines, spirits, beer, cider, and miscellaneous beverages. Yeast fermentations of hexose sugars present in fruits and grains are common to all these beverages but various beverages differ on the basis of

1. Fermentation adjuncts or fermentable material added to a primary fermentable source of carbohydrates.

2. Yeast fermentation methods.

3. Aging of the alcoholic ferment.

4. The distillation of alcohol from the fermented vegetative matter.

Wines

Yeast fermentations of hexose sugars present in fruits and their juices, notably grapes, produce wines. Three types of wine are common:

Group 1. Table wines in which alcohol fermentation proceeds to concentrations of 6 to 14% v/v by the direct fermentative action of surface organisms indigenous to grapes. These wines include Burgundy, claret, Hock, and Moselle.

Group 2. Wines produced from primary fermentations are subjected to a second

Table 13.3. Minor Primary Constituents in Alcoholic Beverages Produced by Yeast, Expressed as a Mean Value or as a Possible Range in Parts per Million

	Ale	Cider	Lager beer	Wine(s)
Acetaldehyde	—	—	13 ± 2	75 ± 20
n-Butanol	—	5–25	—	10–190
Diacetyl	0.05–0.25	—	0.05 ± 0.03	0.5–0.3
Ethyl acetate	17.0 ± 3.0	—	11.0 ± 2.0	10.0–200.0
Glycerol	3000 ± 400	—	—	9,000 ± 3000
Hydrogen sulfide	0.005 ± 0.002	—	0.005 ± 0.002	—
Isoamyl acetate	2.0 ± 1.5	—	1.5 ± 1.0	—
3-Methyl butanol	54.0 ± 9.0	50.0–350.0	45.0–9.0	35.0–900.0
2-Methyl propanol	22.0 ± 8.0	44.0 ± 20.0	8.0 ± 4.0	8.0–200.0
2-Phenylethanol	45.0 ± 8.0	15.0–200.0	30.0 ± 5.0	15.0–300.0
n-Propanol	40.0 ± 8.0	3.0–50.0	7.5 ± 4.0	—
Pyruvate	75.0 ± 30.0	75.0–450.0	50.0 ± 5.0	20.0–100.0
Succinic acid	250.0 ± 50.0	400.0–1500.0	—	400.0–1300.0

fermentation in the bottle to produce a high carbon dioxide tension. These products include champagne, Asti Spumante, and other sparkling wines having a visible excess of carbon dioxide.

Group 3. Direct addition of ethanol (spirits) to wines results in products called *fortified wines*. Alcohol content may reach 20% (v/v) in wines such as port, sherry, Madeira, and Marsala.

Specific types of wine fermentations are guided by the type of grape juice fermented, the sugar content, pH, total acidity, volatile acidity (measured as acetic acid), and total solids of the juice before, during, and after its fermentation.

Black grapes are used for producing red table wines, whereas white wines can be produced from black or white grapes. White wines made from black grapes, however, re-

Table 13.4. Selected Data Reflecting Notable Differences in Alcoholic Beverages

	Brandy	Cognac	Rum	Gin	Whisky
Specific gravity	0.9473	—	0.9403	0.9518	0.9428
% Total solids	0.845	—	0.366	0.054	0.108
% Ash	0.007	—	0.035	0.009	0.01
% Total acidity = mg acetic acid/ 100 mL alcohol	0.043 96.0	7.0–100.0	0.093 197.0	0.018 —	0.069 38.3 ± 12.0
Total esters as ethyl acetate	426 ± 100	—	235 ± 35	—	55 ± 22
Volatile esters[a]	89 ± 30	75–100	—	—	—
Furfural[a]	3.3 ± 2.0	~2.5	8.9 ± 2.0	—	~3.7 ± 2.0
Aldehyde[a]	51 ± 12	3–30	29 ± 6	—	87 ± 40
Higher alcohols[a]	483 ± 100	—	175 ± 15	—	380 ± 40
Methanol[a]	—	25–50	—	—	—
Diacetyl[a]	—	0.5–3.0	—	—	—
Acetoin[a]	—	1.5–7.5	—	—	—
2,3-Butanediol[a]	—	3.0–9.0	—	—	—
Isopentanol[a]	—	205 ± 80	—	—	4–18
iso-Butanol[a]	—	56 ± 10	—	—	50–120
n-Propanol[a]	—	31 ± 11	—	—	21–50

[a]Expressed in mg/100 mL of alcohol.

quire removal of skins before fermentation is started. The so-called pink wines or rosé wines require fermentation at slightly elevated temperatures using juice from only lightly pressed red grapes. The sugar content of wines may be highly variable. Dry wines typically display ≤0.2% sugar, up to 6% sugar occurs in sweet wines, whereas champagne exhibits up to 16%. The pH of wines is variable and very important for fermentative processes as well as storage. Tartaric acid is the principal fixed acid in most fruit juices, although the concentration may be variable during fruit ripening. Tartrate serves as an index of nonvolatile acidity in many wines where it is present as a mixture of free acid and cream of tartar with a small amount of calcium tartrate. Tartrate concentrations typically range from 0.3 to 0.55%. Volatile acidity, measured as acetic acid, may range in concentration from 0.03 to 0.35%; however, respective values of 0.12 and 0.14% for white and red table wines are indicative of poor-quality wines and vinegary tastes. The spoilage effect of acetic acid in wines may also be augmented by ethyl acetate and other minor associated acid compounds.

Alcoholic fermentation of grapes also produces variable amounts of glycerol as a fermentation by-product. Ratios of glycerol to alcohol depend on the sugar concentration levels, sulfite concentration in wine used for controlling wild yeasts, grape freshness, and the type of yeast used for fermentation. Concentrations of 0.3 to 1.5% glycerol are not uncommon, but levels higher or lower can serve as legal standards of identity in international wine commerce.

Many fruit wines including blueberry, cranberry, apple, and peach have similar production requirements as grape wines, but since hexose levels are inherently lower in these fruit juices than in grapes, some fermentable sugar is necessarily added to promote fruit juice fermentation.

In addition to yeasts, malolactic bacteria such as *Leuconostoc oenes* can play important roles in fruit wines. These bacteria reduce malic acid levels by converting it to lactic acid after the principal yeast fermentation. The reduced acid levels generally improve the acceptability of these wines.

Spirits

The condensation of distilled alcohol and volatile substances from fermented liquors leads to the production of potable spirits. Chief among the potable spirits are brandy, gin, rum, whiskies, and vodka. Alcohol content serves as a critical index for spirit classification and a variety of scales are used for reporting alcohol content. The *proof system* based on a 0 to 200 scale is used in the United States, but the *Gay–Lussac scale* is used throughout Europe. In the latter case, a value of 100 represents absolute ethanol and lesser values parallel corresponding percentages of alcohol (Table 13.5).

Brandy represents spirit distilled from grape wine. The wine is not aged before distillation, and if aging does occur, the fruity, grape fragrance of the brandy may be lost. The wine distillate is composed of low-boiling ("heads") and higher-boiling ("tails") volatile constituents. Both broad distillate fractions contain fusel oils, aldehydes, and other aromatic (odorous) compounds generally referred to as *congeners* (Table 13.6). Distillate fractions that vaporize between the "heads" and "tails" are called the "heart" or the "main run." The main run is desirable because of its high alcohol concentration and unobtrusive content of key congeners that give brandy its special character. A 150 to 170 proof distillate, which is most desirable for achieving fruitiness in finished brandy, is watered down to 100 proof before aging. Finally more water is added after aging but before bottling to produce 80 proof brandy. Three to eight years of oak cask aging produce "mature" brandies, whereas less than two years of aging qualifies brandy as "immature."

A 0.95 specific gravity is typical of many brandies having total solids of 0.1 to 0.7%. The acetic acid content is variable, usually increasing from ~4 to 10 mg/100 mL of

Table 13.5. Common Systems for Denoting the Alcohol Content of Spirits and Beverages

American proof system	British and Canadian proof system	Sykes scale	Percentage volume of ethanol[a]
200.0	75.25 overproof[b]	175.0	100 (absolute)
172.0	50.00 overproof	150.0	86.0
149.0	30.0 overproof	130.0	74.5
114.2	Proof[c]	100.0	57.1
100.0	12.5 underproof[d]	87.5	50.0
80.0	30.0 underproof	70.0	40.0
57.0	50.0 underproof	50.0	28.5
0.0	100.0 underproof	0.0	0.0

[a]Gay–Lussac scale—A value of 100 represents absolute ethanol and lesser values correspond to the percent concentration of alcohol.

[b]Overproof—More alcohol and less water than proof.

[c]Proof spirit—Defined by law as spirit at 51°F that weighs 12/13 of an equal measure of distilled water (sp. gr. 0.92308); e.g., 57% pure ethanol and 43% water.

[d]Underproof spirit—Less than 100% of the mixture designated as proof spirit; e.g., in 100 gal of 20 underproof whisky there are 80 gal at proof and 20 extra gal of water.

ethanol as aging commences, up to 100 mg/ 100 mL upon complete aging.

Although Cognac is a brandy, not all brandy is Cognac. Cognac specifically refers to grape brandy distilled in the Cognac region of France. New and turbid wine is distilled in two steps, however, since traditional stills in this geographic region are not equipped with fractionating columns. The first distillation produces a low wine or *brouillis*

(30% ethanol), which is redistilled. The heart of the second distillation (both "heads" and "tails" fractions being separated) is 68 to 70% alcohol (138–140 proof), which is then aged.

Many other brandies common throughout Europe depend on different fruit fermentations. *Calvados* is the traditional apple brandy of Normandy, while *Armagnac* is made in the south of France. In Switzerland, a cherry brandy known as Kirsch (or *Kirsch-*

Table 13.6. Some Common Organics Present in Potable Spirits (Congeners)

Acids	Alcohols	Esters	Miscellaneous: acetals, aldehydes, and ketones
Acetic[a]	Methanol	Ethyl formate	Diethyl acetal
Propionic	n-Propanol	Ethyl acetate[c]	Acetoin
iso-Butyric	2-Methyl-n-propanol[b]	Ethyl isobutyrate	Acetone
n-Butyric	n-Butanol	Ethyl isovalerate	Diacetyl
iso-Valeric	2-Butanol	Ethyl hexanoate	Acetaldehyde[d]
Hexanoic	2-Methyl-n-butanol[b]	Ethyl octanoate	Propionaldehyde
Heptanoic	3-Methyl-n-butanol[b]	Ethyl decanoate	iso-Butyraldehyde
Octanoic	n-Pentanol	Ethyl dodecanoate	iso-Pentanal
Nonanoic	2-Pentanol	Ethyl tetradecanoate	
Decanoic	n-Hexanol	Ethyl hexadecanoate	
Dodecanoic	2-Phenylethanol	iso-Butyl acetate	
Tetradecanoic		iso-Pentyl acetate	
Hexadecanoic		iso-Pentyl octanoate	
Octadecanoic		iso-Pentyl decanoate	
		2-Phenyl ethyl acetate	

[a]Principal acid accounting for 55–90% of total acidity.

[b]Predominant alcohols excluding ethanol.

[c]Principal acetate.

[d]Predominant aldehyde.

wässer) is produced; it has the distinct flavor of cherry. Plum brandy, on the other hand, is called *Mirabelle* in France, *Quetsch* in Alsace and Germany, and *Slivovitz* in Central Europe. Fruit-flavored brandies are not natural brandies since peach, apricot, blackberry, and other flavor extracts or concentrates are added to a grape brandy base.

Whisky is a spirit, aged in wood, which is ultimately obtained from the distillation of a fermented mash of grain. The manufacturing process for whisky requires several basic steps. The initial steps begin with the *milling* of choice cereal grains that offer a source of complex yet potentially fermentable polysaccharide. These grains include rye, corn, barley, or wheat. Polysaccharides are solubilized by *cooking* the mashed grain and then converted into fermentable sugars by the diastatic actions of amylases in malted barley. Yeast *fermentation* of the cooled mash yields ethanol and carbon dioxide. The fermented mash is then *distilled* to a level of ≤160 proof and *aged* in charred oak barrels in which it develops its characteristic taste, color, and aroma. This step relies on complex oxidations and reductions of congeners as well as the adsorption of undesirable organics to the internal char on barrel staves. According to most regulations, rye whisky must have 51 + % rye in the mash and bourbon must contain 51 + % corn. Fifty-one parts corn and forty-nine parts rye yield bourbon whisky. Swapping two parts corn for two parts rye results in rye whisky. In all cases, the percentage of any one major grain component usually exceeds 51%; however, aging may be quite variable for these spirits. Canadian whiskies are generally made from corn and lesser amounts of rye, wheat, and barley malt. The relative proportions of these components are proprietary formulations of distillers that produce characteristic flavor qualities of individual products. It should also be recognized that the distinctive flavors of Canadian whiskies are somewhat attributable to the grains that have been developed to withstand the rigors of the Canadian climate. The characteristic flavor of Scotch whisky does not originate entirely with the type of grain fermented or aging procedures. Instead, malted barley is dried in kilns fired by peat and acquires the peat aroma. This smokey flavor carries through to the final bottled product. For almost all types of whiskies (i.e. Scotch, Canadian, and American types), esters and volatile organics markedly increase in concentration during the first 6 months of aging with only minor increases thereafter.

The fermentation of grains necessary for production of whiskies should not be confused with those fermentations required for the production of Tequila or mezcal. Tequila is produced from the distilled, fermented juice of the mezcal variety of the agave plant. Authentic Mexican mezcal, however, is produced by processes similar to those for Tequila, but sap of the agave cactus, native to the Oaxaca region of Mexico's southern Pacific Coast, is used for its fermentable carbohydrates.

Gin is a highly rectified spirit, diluted to 80 + proof, which is obtained from a yeast fermented mash of grain. The characteristic flavor of gin is achieved from juniper berries and other herbs instead of congeners that characterize other spirits. In fact, the highly rectified preparation of gin, which requires at least two distillations, ensures the absence of almost all congenerics. Unlike whiskies that are simply distilled from fermented grains over a still head, juniper berries and proprietary botanicals are packed into the top of the gin still or "gin head." Alcoholic and steam vapors passing through the gin head steam distill the critical flavors into the condensate. The enormous complexity of essential oils, esters, alcohols, and other organics such as juniper berries, coriander, licorice, cassia, calamus, angelica, and other botanicals used in gin production make its chemical characterization almost impossible. Compounded gin should not be confused with "real" gin since it is gin produced by adding essential oils to gin spirit. So-called *sloe gin* is formulated by steeping sloes in gin and later adding sugar, whereas Old Tom gin merely contains added sugar. Gins generally have a typical specific gravity of ~0.955, and total solids of ≤0.1% in unsweetened varieties.

Distillation of alcoholic spirit through botanicals is responsible for gins, but the characteristic flavors of cordials or liqueurs are often achieved by more direct routes. Neutral grain spirits or brandies may acquire the flavor of botanicals (e.g., seeds, fruits, fruit peels, or herbs) by percolating these spirits through the botanicals held in a sieve-like container. The repetitive percolation of spirit through the macerated plant material removes critical essential oils, aldehydes, and other flavorful compounds. In other cases, macerated botanicals are simply steeped in spirits until flavors are extracted. For most liqueurs, the extraction step is only an initial step before the heavily flavored spirit extract is redistilled with great care to give a delicately flavored beverage.

Distilled spirits from fermented wheat, rye, barley, maize, or, more historically, potatoes, serve as a basis for vodka production. The preparation of vodka resembles that of grain whiskies but it is not aged and it is devoid of all congenerics and flavors. Exceptions to the flavorless vodkas are those neutral spirits flavored with Buffalo grass known as Zubrovka.

Rum is prepared by distilling fermented molasses of sugarcane syrups. Principal sites of production include Puerto Rico, Jamaica, and the Virgin Islands. Molasses, diluted to ~12% sugar, is admixed to "dunder" or the residue of the run still from a previous distillation. Spontaneous growth or inoculation by desired yeasts ferment the available sugars for a period of 10 to 14 days. The distillate of fermented cane sugars is aged in charred white-oak barrels in order to improve the flavor, odor, and smoothness of the rum. The final specific gravity is 0.87 to 0.94 and total solids range from 0.4 to 3.2%.

Beer

Yeast fermentations of cereal carbohydrates serve as the basis for brewing and the pro-

duction of beer and malt beverages. Unlike wine production, brewing processes are complex industrial fermentations that rely on many-unit operations.

Since yeasts are unable to ferment high-molecular-weight cereal polysaccharides, enzymatic degradation of polysaccharides to dextrins, maltose, and glucose must occur before fermentation begins. Conversion of complex cereal carbohydrates to fermentable forms relies on the action of α- and β-amylases provided by germinated barley. Fresh germinated barley may be used as an amylase source, but in most cases, germinated barley is dried to produce a dried barley malt.

The barley kernel contains about 55% by weight starch (amylose and amylopectin) and, therefore, serves as a potential source of fermentable carbohydrates due to amylase activities in the germinated kernel. Unfortunately, the fermentable carbohydrate produced by malting barley is inadequate for most brewing operations and additional cereal grains are often admixed to the malt. These *adjunct grains* are typically milled forms of corn and rice.

Cereal adjuncts must be processed to improve their susceptibility to malt enzyme attack. This is achieved by heating a milled cereal/water infusion in a cereal cooker. The starch component of the grains becomes gelatinized to produce a high-viscosity liquor. The addition of malt or malt extracts *before* heating over a range of 60 to 80°C partially saccharifies the complex starch to produce a fluid liquid. The quantitative hydrolysis of starches to dextrins and maltose adheres to the relationships below:

$$(C_6H_{10}O_5)_n \longrightarrow n/x\ (C_6H_{10}O_5)_x$$

162 g	162 g
Starch	Dextrin

$$(C_6H_{10}O_5)_n \longrightarrow n/2\ (C_{12}H_{22}O_{11})$$

162 g	171 g
Starch	Maltose

The amylases are resistant to fairly high temperatures in the malt–adjunct–water mixtures—described as a *mash*. α-Amylase is destroyed at 80°C while the β form is denatured at 75°C. β-Amylase activity is optimum at 60 to 65°C and a pH of 5.8. Therefore, high mash temperatures encourage high dextrin production, whereas long retention of the mash at 60 to 65°C produces a maltose-rich liquor. Because of these variable consequences, temperature can be used as a control mechanism for regulating fermentable maltose:dextrin ratios.

Barley malt also provides a source of proteases and peptidases that degrade barley proteins and other cereal proteins such as albumins, globulins, hordeins, and glutelins. Protein hydrolysis occurs throughout the mash up to 60°C, although it is highly pH dependent, with lower pH values (~5.4) providing increased hydrolysis.

After the malt–adjunct–water mixture is incubated for several hours, the liquid fraction is separated from the insoluble solids. The liquid fraction, known as *wort*, must have a specific gravity compatible with the dictates of the specific brewing process before it is acceptable. The typical carbohydrate profile of a wort is detailed in Table 13.7. As a result of protease activities, the total nitrogen of the wort is reflected in the form of 33% residual proteins and peptones, and 66% peptides and amino acids. The body of the beer and its head formation are respectively affected by the peptides and peptone concentration levels. Moreover, both components contribute to turbidity in finished beers.

Wort having a suitable carbohydrate, peptide, peptone, and amino acid profile is advanced to a boiling step conducted in a brew kettle. This step lasts 1.5 to 2.0 h depending on the brewing process. Here the wort is stabilized for yeast fermentation by

1. Inactivating all amylases derived from malt.

2. Sterilizing the wort.

3. Coagulating proteins that may contribute to excessive turbidity in the finished beer.

4. Concentrating wort to a highly specific soluble-solids content and specific gravity if necessary.

5. Slightly caramelizing some susceptible sugars.

Wort boiling is also critical for extracting and isomerizing resins from hops added just prior to wort boiling. The hop resins are contained in glands at the base of the blossoms on the female hop plants. These glands contain a viscous lemon yellow resin which, when extracted into the wort, compensates for the sweet, insipid taste of unhopped beer. The bitter tasting constituents of hop resins, such as humulone, lupulone, and their heated transition and isomerization products, are key flavor ingredients of conventional brews. Although a direct parallel between beer bitterness and extractable components has not been fully recognized, the transformation of α-acids to iso-α-acids is probably important. In many modern brewing operations, preisomerized hop extracts are added to finished beers, or processing steps beyond the brew kettle are taken, to ensure predictable levels of hop resins in the brew. Hop resins can be characterized as *insoluble hard* or *soluble soft resins* as detailed in Table 13.8. The level of hop loading into wort is quite variable depending on the brew, but 0.75 lb/31 gal of wort is not unusual. Hops also contribute their pectin (12–14%), proteins (13–24%), nitrogenous organics, and tannins (2–4%) to the wort. The tannin components of hops, as well as those derived from barley, are representative of leukoanthocyanins and derivatives of quercetin.

Yeast are admixed to the cooled, filtered wort after its aeration with sterile air. Exclu-

Table 13.7. Carbohydrate Content of a Typical Wort (in g/100 mL)

Monosaccharides (glucose)	1.00
Disaccharides (maltose)	5.75
Trisaccharides	1.30
Tetrasaccharides	0.28
Pentasaccharides	0.10
Hexasaccharides	0.18
Heptasaccharides	0.16
Octasaccharides	0.20
Nonsaccharides	0.12
Higher carbohydrates	1.10
Total carbohydrates	10.29
Fermentable carbohydrates	7.80

Table 13.8. Representative Compositions of Soft and Hard Resins Present in Hops

Hop			
Insoluble hard resins		Soluble soft resins	
α *soft resins*		β *soft resins*	
α-Acids	α-Resins	β-Acids	β-Resins
Humulone	uncharacterized	Lupulone	uncharacterized
Cohumulone		Colupulone	
Adhumulone		Adlupulone	
Prehumulone		Prelupulone	

α and β acids of hop resins

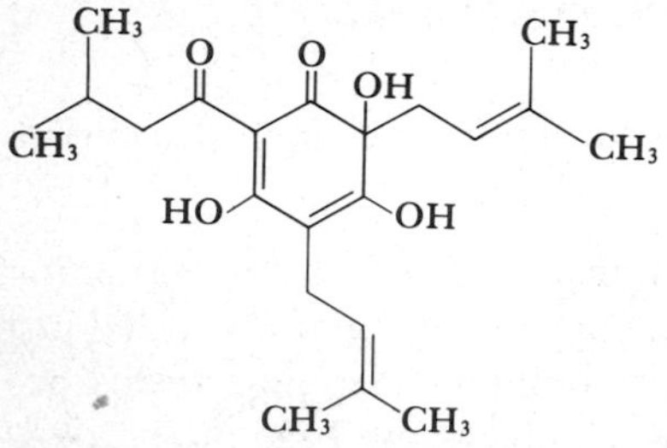

Humulone

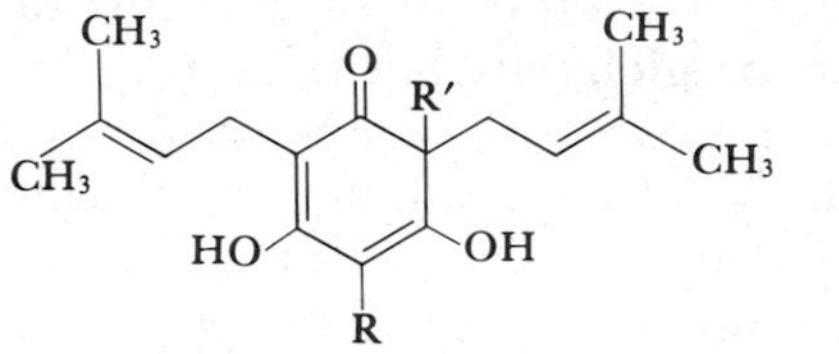

Cohumulone

	R	R′
Lupulone	COCH₂CH(CH₃)₂	CH₂CH=C(CH₃)₂
Colupulone	COCH(CH₃)₂	CH₂CH=C(CH₃)₂
Adhumulone	COCH(CH₃)CH₂CH₃	OH
Adlupulone	COCH(CH₃)CH₂CH₃	CH₂CH=C(CH₃)₂
Prehumulone	COCH₂CH₂CH(CH₃)₂	OH
Prelupulone	COCH₂CH₂CH(CH₃)₂	CH₂CH=C(CH₃)₂

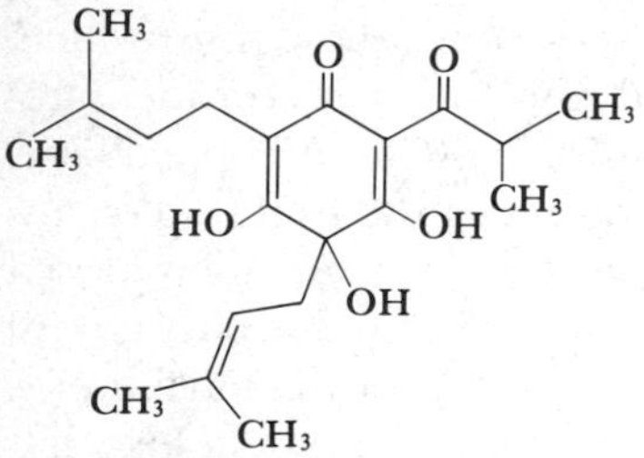

sive of fermentation processes peculiar to certain beers, fermentation often proceeds at 3 to 7°C for a 9- to 12-day period. The alcohol volume approaches 4.6% (9.2 proof), pH drops to about 4.0, and carbon dioxide levels account for 0.3% by weight in the so-called green beer.

Yeast fermentations may be classified as *top* or *bottom fermentations*. Both forms of fermentation are similar but the type of yeast, fermentation temperature, and harvesting methods for yeasts after wort fermentations are different. Pure strains of *Saccharomyces cerevisiae* are responsible for ale and stout production, whereas bottom-fermenting yeasts, including *S. carlsbergensis*, are used for lager and European beers. Bottom yeasts

simply reside at the bottom of a fermentation vessel or vat, as opposed to top yeasts, which similarly ferment sugars but mutually clump into multicellular aggregates and become buoyed-up by liberated carbon dioxide.

The stringent, bitter taste of green beer is mellowed and finished by a series of lagering and storage stages that involve beer volumes up to 899 barrels at a time. The primary lagering step initiates a maturing process in which yeasts and amorphous solids precipitate from the beer; carbon dioxide saturation is achieved; and aldehydes, alcohols, esters, and organic acids mutually interact or oxidize to produce less obnoxious taste qualities. Since lagering occurs at ~1°C, a *chill haze* commonly develops and coagulates during this holding period. After the primary lagering stage, the turbid beer undergoes *chill-proofing*. This process involves deproteinization of the beer by the addition of proteolytic enzymes such as papain and occasionally pepsin. The chill-proofed beer is further carbonated and stored another 3 to 8 weeks at ~1.5°C before its final fitration, carbonation, and packaging as draft or bottled beer.

Biological and physicochemical problems are common threats to normal brewing processes. Wild and undesirable yeast strains can foul the development of ideal fermentation products from worts. Beer turbidity, however, may develop from metal contaminants, chilling, or heating or as a result of oxalic acid present in barley. Chill-haze formation depends largely on the intermolecular oxidation of sulfhydryl groups (—SH) contained in beer-soluble proteins to yield large molecules linked by disulfide bonds (—S—S—). These molecules are insoluble and impart poor flavor and shelf-life to beers. The sulfur concentration level of chill haze usually falls in the range of 1.7 to 2.2%. Trace concentrations of heavy metals promote haze formation by promoting oxidation reactions. Off-flavors can be photochemically induced in beer by 410 to 510 nm light, which encourages mercaptan formation. Excessive foaming of finished and packaged beer is attributable to the use of weathered barley, the occurrence of *Fusaria* mycelia in the steeped brew, or high concentrations of microbubbles in packaged beers.

Aside from beers that rely solely on yeast fermentations, many brews require the combined actions of bacterial and yeast fermentation. Rice beer, or Saké, is prepared by the saccharification actions of molds and bacterial acidification prior to the all-important yeast fermentation.

Cider

Alcoholic fermentations of apples and/or pears, incited by yeasts that are indigenous to these fruit surfaces, result in cider production. Fermentation time may be variable depending on the product, but a 3-week fermentation at 15°C is not uncommon. If yeasts native to the fruit are undesirable, a more suitable yeast can be inoculated into the fruit juice. Storage time for the fermented juice may also be variable. Apple cider, or an equivalent pear cider known as *perry*, can be subjected to a secondary fermentation to produce a "champagne," or apple cider may be used in the preparation of an apple brandy known as Calvados. Specific gravities of most ciders range from 1.000 to 1.022 and exhibit 4 to 8% (v/v) alcohol with higher concentrations found in champagne ciders. Total solids may range from 2.0 to 8.0%, whereas tannins account for 0.18 to 0.27% of most ciders.

Other Ethanolic Fermentations

Other than classic alcoholic beverages, ethanolic fermentation of some milk beers, *kefir* prepared from cow, goat, and sheep milks, as well as *kumiss* produced from mare's milk, require both lactic acid and yeast fermentations. Since yeasts are unable to ferment lactose, they rely on lactic acid bacteria to hydrolyze lactose into glucose and galactose, which are fermentable by yeasts.

Most contemporary formulations for bread batters rely on the use of *Saccharomyces cerevisiae* and its related strains for leavening. These yeasts produce carbon dioxide as a fermentative by-product along with ethanol, aldehydes, and other organic compounds in unbaked doughs. Volumes of carbon dioxide, which undergo expansion within dough during baking, become entrapped by the elastic gluten component. Subsequent denaturation of the gluten leads to a low-density, leavened dough product. The chemical by-products of the yeast fermentation also experience complex thermal reactions during baking to give the final baked product its characteristic flavor. It should be recalled, however, that many breads require the fermentative actions of bacteria for their characteristic flavors. Pumpernickels and some steamed breads require the heterofermentative actions of lactic acid bacteria.

MISCELLANEOUS FOOD FERMENTATIONS

The historical development of many cultural and ethnic foods are linked to mold and bacterial fermentations as well as enzyme actions. Native enzymes present in fish organs lead to the proteolysis of fishes held in high-salt brines. Typical of these foods are *shott-suru*, *patis*, or *nuoc mam*.

Other cured meats may be processed into sausage mixes or emulsions and inoculated with desirable strains of *Pediococcus cerevisiae*. The acid produced by these bacteria and other species of the genera *Micrococcus*, *Lactobacillus*, and *Leuconostoc*, either singularly or jointly, impart characteristic taste properties to thuringer and summer sausages. However, many cheaply produced versions of these sausages have relied in recent years on artificial flavoring or acid additions for flavor instead of bacterial fermentations.

Some types of red meat products are also subjected to the tenderizing effects of proteolytic enzymes derived from molds. *Thamnidium elegans* is typical of such molds in that it produces exoproteases.

Combined fermentative effects of molds and bacteria are responsible for the preparation of Miso and soy sauces. These fermentations are complex and may require three or more types of organisms, such as soy sauce fermentations, which employ the mold *Aspergillus soyae*, the *soyae* variety of *Pediococcus cerevisiae* bacteria, and *Saccharomyces rouxii*-type yeasts.

The combined effects of lactic and acetic acid fermentations, plus the unchecked fermentative actions of native enzymes, are also important in the production of vanilla, teas, coffee, and cocoa. Enzyme action during the vanilla bean curing process transforms the naturally occurring vanillin glucoside into crystalline vanillin. The bitterness and astringency of Formosan, oolong, and black teas is reduced by varying degrees of fermentation after tea leaves are crushed. Fermentation is also important for the preparation of coffee cherries and cocoa beans where mucilages, parchments, and pectins surrounding the individual beans must be loosened before their removal. The chief enzymes responsible for these actions are plant pectinases and/or cellulases.

AN OVERVIEW OF INDUSTRIAL FERMENTATIONS

The origins and historical uses of fermentations for food preservation and spirit and beverage production are rooted in prehistory. Sophisticated application of these age-old food fermentation principles to modern industrial settings provide the basis for synthesizing bulk quantities of amino acids, organic acids, vitamins, antibiotics, and enzymes. These industrial fermentations rely on yeasts, molds, bacteria, and filamentous bacteria such as the *Streptomyces* to convert carefully formulated batches of nutrients into desirable biochemical compounds. Nutrient requirements for fermentative organisms are met by balanced mixtures of peptones, yeast products, agricultural materials such as cornsteep liquors (steepwater), soybean meal, cotton seed flour, acetates, ethanol, and/or buffer salts. Since fermentation must yield organic and biochemical compounds at a cost competitive with or better than wet chemical synthesis routes, a concerted effort must be made to keep financial expenditures for nutrients to a minimum. Aside from selecting and using the least expensive nutrients for fermentative syntheses, classical hybridizations and genetic engineering of microorganisms can provide new organisms that are biochemically efficient producers of desirable compounds. Among the genetic engineering methods, protoplast fusion can have distinct advantages for modifying yeast fermentations, wherease plasmid-mediated methods of gene transfer are suited for bacteria (refer to Chapter 23 for details).

Most industrial fermentations require a preliminary sterilization of the nutrient medium before its introduction into the fermentation vessel or fermentor. A temperature of ~150°C for several minutes or 126°C for a longer period is often adequate for sterilization. Heat can be introduced into the nutrient medium by live steam or, in the case of continuous culture systems, by a heat exchanger through which the nutrient medium is pumped at a high velocity (e.g., high-temperature, short-time pasteurizer) (Figure 13.15). For some short-time fermentations, only minor pasteurization and a large inoculum of fermentative organisms may adequately ensure the successful development of fermentative products. Heat-labile media, on the other hand, may be sterilized by membrane filtration techniques.

After a fermentor has been charged with growth medium, the system is inoculated with the desired microorganisms. The inoculum may range from 1 to 10% of the total nutrient medium volume, and organisms should be in their logarithmic growth phase. Inocula reflecting stationary growth or senescent phases often yield fermentative products having varying qualities.

Stringent controls over pH and temperature conditions of the fermentation medium are a prerequisite for success. Productive fermentations may occur only over a ±0.1 pH unit from the optimum. Optimum temperatures may also be limited to ±0.3°C deviations from a given optimum value. These small tolerances require automatic monitoring of pH and temperature parameters at all times as well as automatic modification of the fermentation liquor to ensure optimum conditions. Dissolved oxygen levels can also be critical, especially in cases of aerobic fermentations; therefore, oxygen sensors can direct the control of aeration or oxygenation in many types of fermenters. Air sterilized by filtration or heat, or sterile oxygen, may be introduced by perforated pipes that maximize the surface-to-volume ratio of the oxygenating gas to the fermentation liquor. Such perforated pipes are commonly known as spargers. This aeration process may be augmented further by a series of baffles that maximize oxygen tension throughout the fermenting liquor at a very rapid rate.

Fermentation vessels ranging in size from ≤380 m^3 (≤100 gal) to over 570 m^3 (150,000 gal) are not uncommon depending on the fermentation product.

The biochemical products of industrial fermentations may be operationally defined as *primary* or *secondary metabolites*. Primary metabolites are directly tied to the biochemical components of individual cells that permit their growth, metabolism, and reproduction. These products include purine nucleotides, vitamins, amino acids, and organic acids. Many primary metabolites, such as mycologically produced citric acid, can be enhanced by limiting certain nutrients within the growth medium such as manganese and iron. The combined effects of environmental and genetic modifications of microorganisms lead to more outstanding levels of primary metabolite production. For example, 18,000 + -fold increases in vitamin B$_2$ (riboflavin) production by *Ashbya gossypii* and 45,000 + -fold increases in vitamin B$_{12}$ (cyanocobalamin) production characterize *Propionibacterium shermanii* and *Pseudomonas denitrificans*. Secondary metabolites represent a structurally and functionally diverse group of compounds that are often linked to the survival of microorganisms in their natural habitat and somehow ensure species survival. When grown in a controlled fermentation, these compounds have little use for ensuring survival of microorganisms, but they are nevertheless synthesized to significant concentration levels. Antibiotics produced by *Streptomyces* are typical of these secondary metabolites.

Primary and secondary metabolites plus genetically enhanced productivities of metabolites have greatly increased the potential uses of fermentations for food uses. Bulk amounts of food and dictary supplements

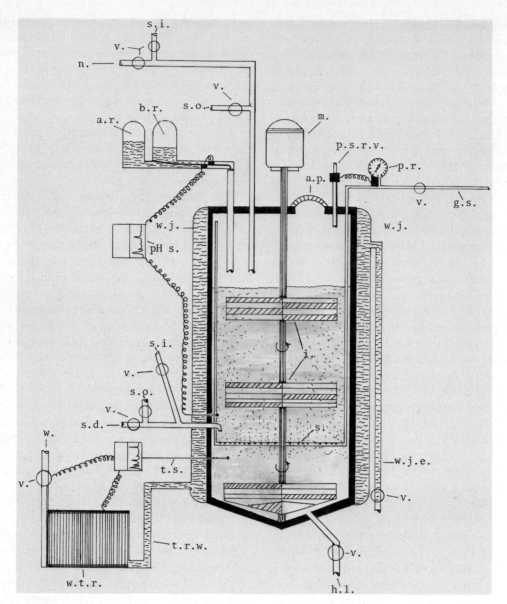

FIGURE 13.15. Batch reactor used for conducting enzymatically mediated reactions and especially industrial fermentations. The reactor system is carefully controlled with respect to nutrient medium pH, temperature, oxygen content, and sterility. Sterility is largely controlled by maintaining a positive pressure within the reactor and steam is used to purge all access lines into the tank. Microorganisms nurtured on the growth medium or compounds excreted into the growth medium by microorganisms are harvested from the bottom of the reactor vessel. Solid portions of the biomass generated during fermentation may be subsequently removed by filtration or other mechanical methods. Desirable biochemical compounds in the liquid fraction or the solid biomass may be extracted and purified for various uses. Other critical components of the reaction vessel include: a.p., access port; a.r., acid reservoir for pH control; b.r., base reservoir for pH control of fermentation medium; g.s., gas supply, sterile or filtered gas (O_2) or air supply; h.l., harvest line; i., impeller; n., nutrient supply to reactor vessel; m., motor for agitation; pH s., pH sensor and regulator; p.r., pressure regulator; p.s.r.v., pressure-sensitive release valve; s., sparger; s.d., sampling drain; s.i., steam inlet for sterilization; s.o., steam outlet for sterilization; t.r.w., temperature regulated water to water jacket surrounding the reactor; t.s., temperature sensor mechanism; w., water; w.j., water jacket; w.j.e., water jacket effluent; w.t.r., water temperature regulator; v., control valve: on/off and/or controlled flow.

including riboflavin, thiamine, ascorbic acid, lysine, and other compounds may be produced by fermentation. Antibiotics, flavor-enhancing purine nucleotides, malic acid, glutamic acid, polysaccharides (e.g., xanthan gum), fragrances, and numerous enzymes are produced by this route.

Fermentative methods for producing key industrial enzymes have great importance in food industries, especially since the advent of enzyme immobilization. Of particular importance is the application of fermentatively produced glucose isomerase to the manufacture of high-fructose corn syrups. These syrups have significantly replaced more expensive sucrose in many foods. Protein-coagulating proteases, used for preparing many cheeses, have their origins in mold fermentations. Molds and newer genetically modified bacteria, for example, have improved the limited industrial availability of rennet from calf stomachs in over half of the American cheese industry.

Other types of fermentations rely on the conversion of hydrocarbons, such as methane, into single-cell proteins. Although methane transformations are classically me diated by the bacterium *Methylophilus methylotrophus*, other selected organisms and hydrocarbon substrates may achieve similar objectives. Single-cell protein production by fermentative actions of bacteria or yeasts have been touted as a panacea for world hunger problems, but the real value of these protein sources rests on their ability to serve as protein adjuncts in existing foods.

CONCLUDING REMARKS

The historical benefits of food preservation by fermentative processes are well documented in the literature, and the principles of fermentation clearly underly many current methods of food preservation. The 1908 Nobel laureate Metchnikoff even claimed that fermented foods such as yogurt had distinct health benefits when consumed on a regular basis. It was postulated that "evil conse quences from constipation are due mainly to retention of fecal matter and undigested food residues in the blood . . . intestinal putrefaction if inhibited could prevent senility and premature death." The advice for avoiding these dreadful consequences rested on consuming sour milks having abundant acids and lactic acid-producing bacteria to check the putrefactive flora of the intestine. Although the impact of such practices on human health arc quitc questionable, there is no doubt that the intestinal fermentative activities of bacteria are important. Bacteria introduced with foods into the gastrointestinal tract, as well as the native intestinal flora, augment digestion by fermentative processes and provide some amino acids, vitamins, and other trace nutrients. Exact amounts of these nutrients obtained strictly by microbial actions are uncertain due to difficult analytical circumstances and the trace quantities of vitamins and other compounds being produced.

REFERENCES

Amerine, M. A. (ed.). 1980. *Technology of Winemaking*, 4th ed. Avi Publishing Co., Westport, Conn.

Food Technology. 1984. Fermentation technology features. **38**(6):41–50, 51–57, 59–63, 64–70. (Summary of current fermentation technologies for dairy starter cultures, yeasts, meats, and soybeans.)

Høyrup, H. E. 1978. Beer. In *Kirk-Othmer Encyclopedia of Chemical Technology*, Vol. 3, pp. 692–735. Wiley, New York.

McDonald, P. E. 1981. *The Biochemistry of Silage*. Wiley, New York.

Metzler, D. E. 1977. *Biochemistry*. Academic Press, New York.

Pederson, C. S. 1971. *Microbiology of Food Fermentations*. Avi Publishing Co., Westport, Conn.

Reed, G. (ed.). 1982. *Prescott & Dunns Industrial Microbiology*, 4th ed. Avi Publishing Co., Westport, Conn.

Roberts, T. A., and F. A. Skinner (eds.). 1983. *Food Microbiology, Advances and Prospects*. Academic Press, New York.

PART · 2

Integrated Metabolism
and Effects of
Dietary Constituents

C H A P T E R · 14

Digestion, Absorption, and Malabsorption of Food Nutrients

INTRODUCTION

Dietary constituents are digested throughout the course of the gastrointestinal tract, absorbed over the villular surfaces of the intestine, and then transported by blood and lymphatic circulations to various body tissues.

Digestion, assimilation, and transport of food constituents in the body are directed by hormonal activities in combination with voluntary and involuntary mechanisms. Hormonal activities are ultimately controlled by complex interactions within the sympathetic nervous system along with other neurological directives originating in the brain.

Brain stimulation controls a series of muscular movements that facilitate the ingestion and maceration of food. Food mixes with saliva in the mouth, where it assumes the form of a soft-rolled bolus that is then propelled into the esophagus by a swallowing reflex initiated in the pharyngeal region. The swallowing reflex requires that the vocal cords and the larynx close while the epiglottis swings backward to prevent the food from entering the trachea. The muscular sphincter that normally keeps the esophagus closed also

undergoes relaxation and the larynx moves upward, thereby opening the esophagus. As the food bolus enters the pharynx, pharyngeal muscles contract and force the food to descend into the gastric cavity. Food may be stored in the stomach for 3 to 4 h or longer depending on the nature of the foodstuff. Spinach, for example, may reside in the stomach for many hours.

Food contained in the gastric cavity is mixed with large quantities of digestive juices to produce a milky paste-like mixture called *chyme*. The physical and liquid properties of chyme offer ideal fluidity necessary for the discharge of food nutrients from the stomach.

Nerve impulses stimulate the contraction and relaxation of the gut muscles and push the food forward along the gastrointestinal tract. The entrance of gastric chyme into the intestine requires its passage through the pyloric sphincter at the bottom of the stomach (Figure 14.1). This action is controlled by an enterogastric reflex, which inhibits the antral peristalsis of the stomach and, accordingly, prevents the premature engorgement of the duodenum. This process is also controlled by humoral agents such as secretin, cholecystokinin-pancreozymin, and a gastric

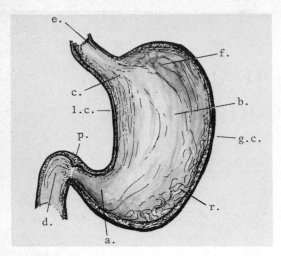

FIGURE 14.1. Some important features of stomach structure and anatomy. a., antrum; b., body or middle portion; c., cardia; d., duodenum; e., esophagus; f., fundus or upper round part; g.c., greater curvature; l.c., lesser curvature; p., pylorus or small end; r., rugae.

polypeptide. Other factors such as emotional stress, type of food, and inhibitory agents secreted into the duodenum all influence the emptying of the stomach. Before food is absorbed over the surface of the small intestine, it must be digested. Digestion involves the biochemical dismantling of large food molecules into smaller molecules capable of passing through the intestinal mucosa. Food digestion actually begins in the mouth, continues in the stomach, and terminates in the small intestine. Partially digested or undigestible food substances that resist normal absorptive processes travel to the large bowel where they are eventually eliminated. Some key digestive enzymes have been outlined in Table 14.1. Figure 14.2 also outlines the major digestive organs and structures responsible for food digestion and absorption.

MOUTH DIGESTION

Salivary glands introduce salivary amylase into the mouth, which hydrolyzes starch molecules into dextrins and maltose. The optimal pH for amylase activity in the mouth is approximately 6.8. When food enters the stomach, the hydrolytic action of amylase on starch persists so long as it is contained within the bolus and protected from gastric acidity. When the bolus disintegrates, salivary amylase is inhibited in response to the severe gastric acidity.

STOMACH DIGESTION

The surface of the gastric cavity is blanketed with an epithelial lining studded with pits that lead to gastric glands. These glands are composed of a central canal surrounded by different types of cells such as *parietal cells, chief cells*, and *mucosal cells* (Figure 14.2). These cells produce secretions that are released into the canal. Parietal cells produce hydrochloric acid, chief cells supply digestive enzymes, and mucosal cells secrete mucus discharges that protect the small intestine from gastric acid.

Hydrochloric acid secretions have critical importance in the conversion of *proenzymes* into *active enzymes* such as pepsinogen secretions, which are activated to yield pepsin. At pH values of 2.0 to 3.0, pepsin produces peptones and polypeptides (4–12 amino acid residues) from proteins. Other enzymes such as rennin coagulate milk, and tributyrase, which is most common in infants, digests butterfat.

Gastric gland secretions are stimulated by sight, smell, food taste factors, and food already present in the stomach as well as vagus nerve-stimulated gastrin production by ductless glands of the antral mucosa. Gastrin productivity is stimulated and/or enhanced by the presence of alcohol, caffeine, and other food extracts. The hormone is then carried through the bloodstream to parietal cells that initiate HCl production. As soon as gastric pH reaches a value of 2.0, gastrin produc-

Table 14.1. Digestion of food nutrients

Site of secretions	Name of secretion	Digestion products and activities
Mouth	Salivary amylase	Starches → dextrins + maltose
Stomach	Hydrochloric acid (HCl)	Minor hydrolysis of peptides, carbohydrates, lipids, glycosides, etc., and some proenzyme activation to enzymes
	Pepsinogen $\xrightarrow{\text{HCl}}$ pepsin	Proteins → proteoses + peptones + polypeptides
	Rennin	Casein $\xrightarrow{\text{Ca2+}}$ paracaseinate
	Tributyrase	Tributyrin → fatty acids + glycerol
Pancreas	α-Amylase	Starch → maltose
	Chymotrypsinogen ↓ trypsin	
	Chymotrypsin	Proteins + polypeptides → oligopeptides
	Cholesterol esterase	Free cholesterol → cholesterol esters (fatty acids)
	Collagenase	Hydrolysis of collagen
	Deoxyribonuclease	DNA → deoxyribonucleotides
	Proelastase ↓ trypsin	
	Elastase	Hydrolysis of fibrous proteins
	Lipase	Triacyglycerol(s) → free fatty acids + mono- or diacylglycerol derivatives
	Phospholipases A and B	Removal of fatty acids from lecithin
	Procarboxypeptidase A ↓ trypsin	Polypeptides → smaller peptides + aromatic amino acids
	Carboxypeptidase A	
	Procarboxypeptidase B ↓ trypsin	Polypeptides → smaller peptides + dibasic amino acids
	Carboxypeptidase	
	Ribonuclease	RNA → ribonucleotides
	Retinyl ester hydrolase	Hydrolysis of retinyl esters
	Trypsinogen ↓ trypsin	
	Trypsin	Proteins + polypeptides → oligopeptides
Small intestine	Alkaline phosphatase	Organic phosphates → free phosphates
	Aminopeptidases	Polypeptides → smaller peptides + free amino acids
	γ-Amylase	Amylose → glucose
	Dipeptidases	Dipeptides → amino acids
	β-Glucosidase	Glucosyl-ceramide → glucose + ceramide
	Lactase	Lactose → galactose + glucose
	Lecithinase	Lecithin → fatty acids + glycerol + H_3PO_4 + choline
	Isomaltase	Isomaltose and α-dextrins → glucose
	Maltase	Maltose → glucose
	Monoglyceride lipase	Monoglycerides → glycerol + fatty acids
	Nucleosidase	Nucleosides → purines + pyrimidines + pentose
	Nucleotidase	Nucleotides → nucleosides + H_3PO_4
	Sucrase	Sucrose → fructose + glucose
	α,α-Trehalase	Trehalose → glucose
Liver	Bile	Emulsifies fats, stabilizes emulsions, and accelerates action of pancreatic lipase

tion normally ceases. The gastric mucosa also secretes a glycoprotein known as *intrinsic factor* (IF), which binds vitamin B_{12} for absorptive purposes. The IF–B_{12} complex is carried to receptor sites on the brush border of ileal mucosal cells where B_{12} is ultimately assimilated into the circulation.

SMALL INTESTINE DIGESTION AND ABSORPTION

Chyme produced in the stomach enters the duodenum, where it is admixed with pancreatic, intestinal, and hepatic secretions. A

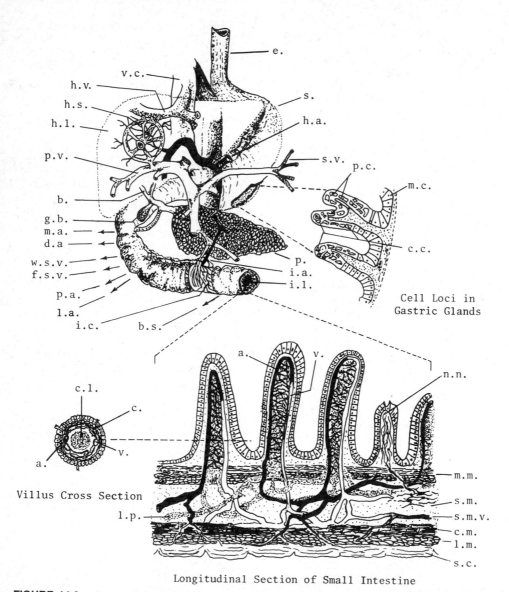

FIGURE 14.2. Anatomical relationship of major digestive organs and structures responsible for the absorption of food nutrients. Details of cell loci in gastric glands, a longitudinal section of small intestine, and a villus cross section are illustrated. The alphabetical structural key to the illustration is detailed at the bottom of page 781. Note that the sequence of nutrient absorption is only a relative sequence and not drawn to scale.

hormone known as secretin, specifically produced by the small intestinal mucosa, stimulates pancreatic juice production. This juice displays a high bicarbonate concentration, which effectively neutralizes gastric acid produced by the stomach and optimizes pancreatic enzyme activities.

Proteins

Considering that the total dietary protein load is about 70 to 100 g/day, and 35 to 200 g/day of endogenous proteins are produced as digestive enzymes and sloughed-off cells, protein losses from the body in feces are relatively minor. Crude calculations indicate that 6 to 12 g/day (4.0–5.7%) of these combined protein amounts exit the body as fecal proteins. Protein digestion and hydrolysis begin in the stomach and continue in the small intestine. Polypeptides resulting from digestion are converted to amino acids in three sequential phases: (1) the intestinal lumen phase, (2) the surface membrane phase of mucosal cells, and (3) the cytoplasmic phase of mucosal cells. Figure 14.3 outlines these three sequential phases. Within the intestinal lumen, proenzymes present in pancreatic juices are converted to enzymes that then digest polypeptides. Chymotryp-

sinogen, for example, is converted by trypsin into chymotrypsin, which catalyzes peptide cleavage on the carbonyl side of tyrosine, tryptophan, phenylalanine, methionine, and leucine residues. Trypsinogen is activated by enteropeptidase and preexisting trypsin *into trypsin,* which cleaves peptides on the carbonyl side of alanine, glycine, and serine residues. The proenzymes known as procarboxypeptidases A and B undergo trypsin-mediated activation and become transformed into carboxypeptidases A and B, respectively. Carboxypeptidase A cleaves the amino side of valine, leucine, isoleucine, and alanine residues, whereas the B-type enzyme hydrolyzes the corresponding side of arginine and lysine residues. Amino- and dipeptidases residing on the surfaces of digestive tract mucosal cells further convert oligopeptides into dipeptides and tripeptides. These polypeptide hydrolysis products then permeate cell membranes by a Na^+-dependent transport mechanism into the cytoplasm of cells. Intracellular hydrolysis of any remaining peptide bonds yields free amino acids that are then introduced into the capillary circulation.

According to Abidi (1980), protein digestion is more rapid and complete in the jejunal than ileal sections of the intestine. In

a. artery (arteriole)	l.a. lipid absorption
b. bile passage from hepatic tissues	l.m. longitudinal muscle
b.s. bile salt absorption	l.p. lymphatic plexus
c. capillaries	m.a. monosaccharide absorption
c.c. chief cells (enzyme production)	m.c. mucosal cell
c.l. central lacteal	m.m. mucosal muscles
c.m. circular muscle	n.n. nerve net of villus
d.a. disaccharide absorption	p. pancreas
e. esophagus	p.a. protein absorption (amino acids)
f.s.v. fat-soluble vitamin absorption	p.c. parietal cells
g.b. gall bladder	p.v. portal vein
h.a. hepatic artery to connective tissues	s. stomach
	s.c. serous coat
h.l. hepatic lobe	s.m. submucosa
h.s. hepatic sinuses	s.m.v. submucosal vein
h.v. hepatic vein	s.v. splenic vein
i.a. intestinal artery	v. vein (venule)
i.c. intestinal capillary bed(s)	v.c. vena cava
i.l. intestinal lumen	w.s.v. water-soluble vitamin absorption

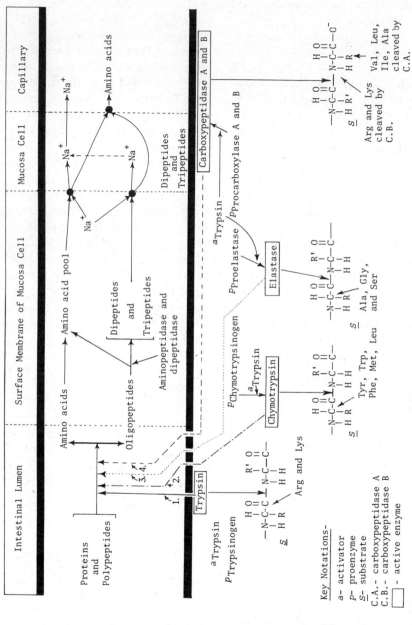

FIGURE 14.3. Key events of protein digestion and absorption.

some cases, nutritionally well-balanced test meals containing up to 50.0 g of protein failed to demonstrate complete digestion in the jejunal regions after 4 h. Chung *et al.* (1977) employed similar test meal models to demonstrate similar results and suggested that 60% of meal protein is digested in the upper small intestine, whereas the remainder of the dietary protein is digested in the distal small intestine. Based on this study and others, it is generally concluded that the ileum is especially important for the completion of the protein digestion process. For more details, consult Winick (1980).

Carbohydrates

Oligosaccharides are hydrolyzed by *surface enzymes* on small intestinal epithelial cells. Table 14.1 outlines many of the most important enzymes including lactase, isomaltase, maltase, sucrase, and α,α-trehalase along with their substrates and hydrolysis products.

Monosaccharides such as glucose, fructose, galactose, mannose, and pentoses (ribose, deoxyribose, xylose, etc.) are absorbed from the intestinal tract into mucosal cells of the villi by mechanisms that are not entirely clear. Some sugars such as pentoses transit the mucosal membrane by passive diffusion, whereas hexoses including glucose and galactose enter the cell by a Na^+-mediated flux down a concentration gradient. Although intracellular Na^+ concentrations are maintained by a Na^+/K^+-ATPase system, which ejects Na^+ from the cell, the resulting low Na^+ concentration in the cell encourages the entry of more Na^+ from the outside. However, Na^+ entry occurs in connection with certain sugars through a carrier protein. As shown in Figure 14.4, sugar and Na^+ influx occur as *cotransports* using the same carrier protein. Cotransport systems drive the cellular absorption of many sugars, amino acids, Cl^- and other substances. Fructose is absorbed by a mechanism that seems independent from Na^+ cotransport.

Comparative rates for sugar absorption are quite variable. Glucose is absorbed three times faster than fructose, whereas galactose is absorbed at a rate slightly greater than that of glucose. Mannose and pentoses are absorbed at a rate that may be up to ten times less than glucose. Sugars are released from the intestinal mucosa by diffusionary processes into the portal vein, which ultimately directs delivery of the sugars to the liver.

Depending on the temporal requirements of the body, absorbed carbohydrates may be

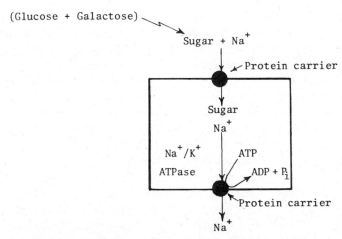

FIGURE 14.4. Absorption of glucose and galactose requires energy expenditure in the form of ATP. Transport mechanisms of this type, which operate against a concentration gradient (with energy expenditure), are defined as *active transport mechanisms*.

(1) converted to glycogen and stored for future use; (2) oxidized to produce carbon dioxide, water, and energy; (3) converted to fat and stored in adipose tissues; and/or (4) converted into amino acids and proteins.

Normal blood glucose (*normoglycemic*) levels are between 80.0 and 100.0 mg/dL. Postprandial blood glucose levels may transiently exceed this range following a starchy meal, but the body has built-in mechanisms that rapidly regulate and return blood glucose to normoglycemic levels. Excess glucose is stored largely as glycogen in hepatic tissues, but the presence of glycogen is not uncommon in cardiac, smooth, and skeletal muscles.

Lipids

A hormone known as cholecystokinin–pancreozymin is secreted by the duodenal mucosa and carried to the pancreas by the hematogenous route. This hormone not only elicits the production of pancreatic juices, but it spurs gall bladder contraction and the release of bile salts into the small intestine. Dietary fat particles mix with bile salts and then become reduced in size and emulsified. This action increases the surface-to-volume ratio of the fat so that hydrolysis activity of pancreatic lipase proceeds more rapidly. According to Borgstrom (1975), a small protein called *colipase*, secreted by the pancreas, binds to the bile salt–lipid surfaces and improves enzymatic hydrolysis efficiencies. The triacylglycerols undergo hydrolysis and yield fatty acids and β-monoacylglycerols (Figure 14.5). Free fatty acids and β-monoacylglycerols aggregate in the presence of bile acids forming micelles, which are absorbed into the intestinal mucosa cells. The bile salts are not absorbed in the same location as the micelles and travel farther down the ileum where they are absorbed into the blood. Bile salts are transported to the liver, discharged into the gall bladder, and eventually reintroduced back into the intestinal lumen for reuse. According to Glickman (1980), 20 to 30 g of bile salts enters the small intestine daily, and only 500 mg is lost in the feces. The recycling of bile salts is described as *enterohepatic circulation.*

The process of lipid digestion and absorption is very efficient, with only about 5% of the dietary lipids lost in fecal material. Lipids including fatty acids and monoacylglycerols migrate across microvillus membranes by a process that is largely passive. This is possible because the superficial lipid structures on microvillus membranes can readily interact with similarly nonpolar dietary constituents. According to Ockner *et al.* (1976), a fatty acid binding protein localized in the intestinal mucosal cells apparently binds fatty acids. This action facilitates the intracellular transport of long-chain fatty acids to the smooth endoplasmic reticulum where triacylglycerol synthesizing enzymes reside.

The synthesis of triacylglycerols is believed to reduce the overall concentration of the free fatty acids within the cell, thereby creating an effective concentration gradient for continued passive uptake of triacylglycerol constituents. It is also speculated that fatty acids may be converted into inert triacylglycerols that minimize potentially injurious fatty acid effects on the intestine while they await transport.

Triacylglycerols (~4%) and proteins (~1%) aggregate to form chylomicrons. The water solubilities of the lipids are greatly improved by this action, which significantly enhances lipid transport by hematogenous routes. The chylomicrons are coated with proteins that minimize their mutual aggregation as well as their tendencies to adhere on lymphatic or blood vessel walls. Completely assembled chylomicrons are secreted from the mucosal cells into the lacteals and lymphatics. From the lymphatics, the chylomicrons are introduced into the systemic blood circulation by the thoracic duct at the junction of the jugular and subclavian veins. The triacylglycerols containing long-chain fatty acids travel through the lymphatic route, whereas C_{12} or smaller fatty acids, which are water soluble, undergo transport via the portal blood circulation in association with albumin. In dis-

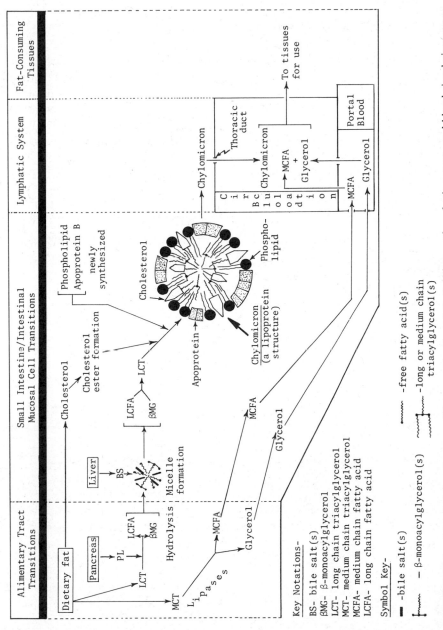

FIGURE 14.5. Schematic involvement of the pancreas, liver, intestinal lumen, lymphatic system, and blood circulation system in digestion, absorption, and ultimately transportation of fats to tissues where they are utilized.

eases associated with lymphatic obstructions such as lymphoma and Wipple's disease, the therapeutic use of medium-chain triacylglycerols (available commercially) has been widely employed since C_{12} and smaller fatty acids do not enter chylomicron formation. Instead, these fatty acids pass directly into the portal circulation and provide caloric supplementation without involving lymphatic transportation.

Free fatty acids enter the fat cells of adipose tissues where they are converted into triacylglycerols for storage. The lipid content of human fat cells may persist for up to 40 days in cases of starvation. Under well-fed conditions, however, all of the fat depots

in the body display a dynamic state of equilibrium since lipids are constantly being mobilized and deposited.

Fatty acid mobilization from triacylglycerols yields a variety of diacylglycerols, monoacylglycerols, and glycerol under conditions of strenuous exercise, starvation, or intense excitation. These stimuli for fatty acid mobilization actuate elevated levels of adrenalin (or epinephrine). Adrenalin attaches to specific receptor sites on individual fat cell membranes and spurs fatty acid mobilization from triacylglycerols (Figure 14.6). Free intracellular fatty acids produced via adrenalin effects eventually exit the cell and enter the blood circulation where they com-

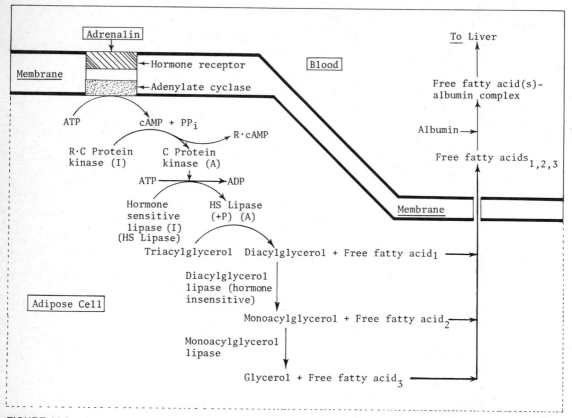

FIGURE 14.6. Fatty acid mobilization in adipose cells is initiated by adrenalin (epinephrine). Adrenalin binds to a specific receptor site, which ultimately activates a hormone-sensitive lipase responsible for hydrolytically producing free fatty acids from triacylglycerol. Serum albumin-type proteins comprising up to 50% of total plasma proteins binds six to eight fatty acid molecules per molecule of albumin protein. This binding mechanism converts otherwise insoluble, toxic fatty acids into transient species that are ultimately processed in the liver.

plex with the plasma albumin fraction. Albumin-complexed fatty acids are then translocated to hepatic tissues where they undergo catabolic reactions.

Nucleic Acids

Nucleic acids naturally occur in close association with proteins. Dietary assimilation of nucleic acid-rich protein complexes begins with their hydrochloric acid-mediated cleavage in the stomach. Nucleic acids liberated from protein complexes enter the upper part of the small intestine where pancreatic ribonuclease and deoxyribonuclease produce oligonucleotides from polynucleotides. Oligonucleotides are further degraded to mononucleotides by phosphodiesterase activities found in intestinal mucosal cells. Mononucleotide catabolism continues in the mucosal cell by nonspecific hydrolases and phosphatases. Products of these enzyme reactions include mixtures of free pentose sugars and nucleosides. Nucleotide catabolism according to this route underlies the general scheme of nucleic acid absorption.

Any proteins, fats, carbohydrates, nucleic acids, or other dietary constituents that have not been absorbed by the intestinal villi pass the ileal sphincter and enter the colon (large bowel). This segment of the digestive tract does not promote digestion per se, but it is directly involved in the absorption of fluids and electrolytes.

The microbial flora residing in the colon degrade some undigested substances and may lead to the production of many incomplete food oxidation and reduction products. In recent years many of these compounds have become suspect as potential carcinogens, but the record is far from clear. Evacuation of the colon is normally expedited by dietary bulk (fiber) such as cellulose, hemicelluloses, and lignin along with poorly digested materials including pectins, gums, and pentosans.

MALABSORPTION RELATED TO GASTROINTESTINAL DISEASES

Good nutritional practices contribute to the healthful maintenance of the gastrointestinal tract. Any unhealthful conditions along the tract can prompt a variety of malabsorption syndromes or complicate any preexisting diseases.

Inadequate absorption of food nutrients may also be incited by diarrhea caused by enterotoxins, laxatives, neoplasms, bile acids, inflammatory bowel diseases, sprue, or other factors. Diarrhea and steatorrhea are also hallmarks of malabsorption syndromes.

Apart from the factors outlined above, malabsorption problems must be considered with respect to many factors. For example, pancreatic enzymes are necessary for intraluminal digestion. Since pancreatic enzymes become operative at a pH of about 6.0, adequate bicarbonate concentrations associated with pancreatic secretions must ensure gastric acid neutralization for proper enzyme activity. In specific cases such as the Zollinger-Ellison syndrome, in which excessive gastrin is produced by a pancreatic tumor (gastrinoma), the resulting production of excessive gastric acid can lower intestinal pH to an extent that it impairs nutrient absorption. Still another factor that can affect nutrient absorption involves unusually low bile salt concentration levels. This produces a marked decrease in lipid micelle formation for lipids and retards their digestive hydrolysis.

A fundamental lack of active digestive enzymes can also lead to monumental malabsorption problems. As indicated by Isselbacher (1980), peptidase deficiencies may result in only a partial breakdown of gluten. Partially digested gluten may be highly significant because it serves as a precursor for at least one very toxid peptide that produces cell damage and immunological changes characteristic of sprue.

Lactase deficiency may induce diarrhea plus varying degrees of abdominal distension following lactose ingestion. Aside from physical evidence of this type associated with lactase deficiency, gaseous levels of breath hydrogen may reflect lactase abnormalities caused by bacterial actions on malabsorbed lactose (Bond and Levitt, 1977).

Inordinate increases in the bile acid content of fecal material may also be linked to high fluid losses and diarrhea. This effect reflects the stimulative effect of bile salts on colonic adenylate cyclase, which impairs electrolyte and water absorption.

Megaloblastic anemia evidenced as enlarged red blood cell precursors in bone marrow and abnormal-appearing nuclear chromatin both seem closely allied with alcohol-induced folate deficiencies. Severe alcoholism contributes to the dysfunction of enzymes that mobilize dietary folates from their polyglutamate forms into nutritionally active compounds (McGuire and Bertino, 1981). Folate deficiency is also associated with sprue conditions characterized by a sore mouth and gastrointestinal upset including steatorrhea and diarrhea.

Another form of anemia recognized as pernicious anemia can also result from vitamin B_{12} malabsorption. In cases of gastrectomy or a basic inability to produce intrinsic factor (IF), B_{12} absorption will be impaired since a critical IF–B_{12} complex must be formed to facilitate B_{12} transport over the brush border of the ileum.

Inflammatory bowel diseases such as Crohn's disease and ulcerative colitis also contribute to nutrient malabsorptions (Grand, 1980). Crohn's disease is indicated as a granulomatous lesion of the gastrointestinal tract. Aggressive treatment involving the excision of a diseased digestive tract region must be followed by careful postoperative nutritional care. Ulcerative colitis, on the other hand, may be characterized by severe colonic inflammation including bloody diarrhea. Causes for the malady are uncertain, but nervous tension or possibly immunological reactions may be involved. For information regarding techniques of malnutrition ther-

apy in inflammatory disease, consult Winick (1980).

Food nutrient malabsorption may also be complicated by diverticulosis. This problem occurs in the colon where distended sacs or pouches protrude outside the lumen. Fecal matter accumulates in these sacs, where it subsequently stagnates (stasis) and dries. Bacterial growth checked only by the fecal nutrients in these sacs can eventually cause sepsis, intermittent but gripping abdominal pains, abscesses, or, in acute cases, gangrene (Figure 14.7).

These direct and secondary effects associated with diverticulosis contribute to an improper nutritional regimen in patients. Although foods containing excessive amounts of raw fruits and vegetables should not be consumed, reasonable amounts of dietary fiber are recommended to maintain proper fecal evacuation in patients suffering from diverticulosis.

Additional surveys of other critical diges-

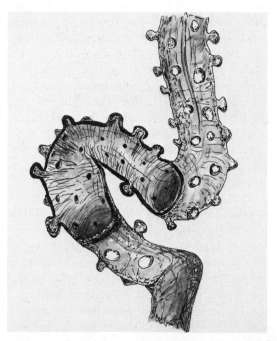

FIGURE 14.7. Multiple diverticula of the colon as well as lesser numbers of diverticula can contribute to diverticulitis.

tive disorders are outlined in the following chapters dealing with overall metabolism.

REFERENCES

Abidi, S. H. 1980. Role of small intestine in digestion of protein to amino acids and peptides for transport to portal circulation. In *Nutrition and Gastroenterology* (M. Winick, ed.), pp. 55–75. Wiley, New York.

Bond, J. H., and M. D. Levitt. 1977. Use of hydrogen (H_2) in the study of carbohydrate absorption. *Amer. J. Digest. Dis.* **22:**379.

Borgstrom, B. 1975. On the interactions between pancreatic lipase and colipase and the substrate, and the importance of bile salts. *J. Lipid Res.* **16:**411.

Chen, L. C., and N. S. Scrimshaw. 1983. *Diarrhea and Malnutrition: Interactions, Mechanisms, and Interventions.* Plenum Press, New York.

Chung, Y. C., Y. S. Kim, A. Shadcher, A. Garride, I. L. MacGregor, and M. H. Sleisenger. 1977. A study of digestion and absorption of protein meal and trypsin in human distal ileum. *Gastroenterology* **72,** A-15:1038.

Glickman, R. M. 1980. Intestinal fat absorption. In *Nutrition and Gastroenterology* (M. Winick, ed.), pp. 29–41. Wiley, New York.

Grand, J. 1980. Malnutrition and inflammatory bowel disease. In *Nutrition and Gastroenterology* (M. Winick, ed.). Wiley, New York.

Isselbacher, K. J. 1980. Malabsorption syndromes including disease of pancreatic and biliary origin. In *Nutrition and Gastroenterology* (M. Winick, ed.), pp. 93–104. Wiley, New York.

McGuire, J. J., and J. R. Bertino. 1981. Enzymatic synthesis and function of folylpolyglutamates. *Molec. Cell. Biochem.* **38:**19. (Excellent review of digestive folate assimilation and its biochemistry.)

Ockner, R. K., and J. A. Manning. 1971. Fatty acid binding protein: Role in esterification of absorbed long chain fatty acids in rat intestine. *J. Clin. Invest.* **58:**632.

Ockner, R. K., and J. A. Manning. 1976. Fatty acid binding protein: Role in esterification of long chain fat in rat intestine. *J. Clin. Invest.* **58:**632.

Winick, M. 1980. *Nutrition and Gastroenterology.* Wiley, New York.

C H A P T E R · 15

Hormonal Control of Metabolism and Nutrition

INTRODUCTION

A hormone is a biochemical substance originating from a group of cells in a tissue or organ that elicits specific biochemical actions in another part of the body. The hormone is usually transported by the blood to biochemically responsive cells, generally called *target cells*. Target cells in turn either (1) demonstrate modified (increased or decreased) functional activities or (2) secrete additional biochemical compounds that have very specific functions.

Hormonal actions represent a sophisticated evolutionary mechanism designed to maintain biochemical harmony, homeostasis, and responsiveness among the individual tissues and organs that comprise the animal body. These responses reflect the requirements of an animal to consume and digest food, grow and sexually differentiate, reproduce, control mineral balance, maintain vascular blood pressure, respond to psychological stimuli and life-threatening events as well as many other intra- and extracorporal factors (Table 15.1).

In man and other vertebrates the endocrine and central nervous systems have basic importance to all hormonal activities. The underlying connection between endocrine hormone production and the central nervous system (CNS) ensures prompt biochemical responses to nearly any stimulus. As indicated previously, *endocrine glands* secrete *endocrine hormones* into the blood where they are transported to specific target cells. This is in marked contrast to exocrine gland secretions and prostaglandins. Exocrine glands produce secretory products and then discharge them into a duct that leads to the outside of the body or to another internal organ. Prostglandins, on the other hand, seem to be produced by one cell type in a tissue and then affect the biochemical actions of other cells in the same tissue.

Endocrine hormone production is largely controlled and integrated with the CNS at the hypothalamus. In fact, the hypothalamus receives and integrates neurological directives from the CNS and then releases a variety of hypothalamic hormones. These hormones are directed to the pituitary gland, which is anatomically situated below the hypothalamus (Figure 15.1).

ORGANIZATION OF THE ENDOCRINE SYSTEM

Hypothalamic hormones are individually and specifically linked to the control of the an-

Table 15.1. A Survey of Important Hormones, Their Origins, and Simplified Functions

Gland of origin	Hormone name (symbol)	Primary actions
Hypothalamus	Thyrotropin releasing factor (TRF)	Release of pituitary thyrotropin (TSH)
	Gonadotropin releasing factor (LH/FSH-RF)	Release of both pituitary FSH and LH
	Somatostatin or somatropin release inhibitory factor (SRIF)	Inhibits the release of pituitary growth hormone
(Stored in posterior pituitary)	Vasopressin (antidiuretic hormone, ADH)	Contraction of blood vessels, kidney reabsorption of water
	Oxytocin	Stimulates uterine contraction, milk ejection
	Vasotocin	Maintains water balance (nonmammalian species)
(Median eminence)	Melanocyte stimulating hormone (MSH)	Dispersion of pigment granules
Anterior pituitary	Somatotropin or growth hormone (STH or GH)	Growth of body, organs, and bones
	Thyrotropin (TSH)	Size and function of thyroid
	Adrenocorticotropic hormone (ACTH)	Size and function of adrenal cortex
	Follicle stimulating hormone (FSH)	Growth of Graafian follicle, spermatogenesis (with LH)
	Luteinizing hormone (LH), interstitial-cell-stimulating hormone, (ICSH)	Causes ovulation *with* FSH, formation of testosterone and progesterone in interstitial cells
	Prolactin, mammotropin (luteotropin)	Growth of mammary gland, lactation, corpus luteum function
	Lipotropin (fat-mobilizing factor)	Release and oxidation of fats from adipose tissue
Parathyroid	Parathyroid hormone	Increases blood calcium Excretion of phosphate by kidney
Parathyroid and thyroid	Calcitonin	Lowers blood calcium
Thyroid	Thyroxine (T_4) Triiodothyronine (T_3)	Growth and maturation, and metabolic rate Metamorphosis
Pancreatic islets β cells	Insulin	Hypoglycemic factor Regulation of CHO, fats, proteins
α cells	Glucagon	Liver glycogenolysis
Adrenal medulla	Epinephrine Norepinephrine	Liver and muscle glycogenolysis
Adrenal cortex	Cortisol	Carbohydrate metabolism
	Aldosterone	Mineral metabolism
	Adrenal androgens	Androgenic activity (esp. females)
Pineal gland (epiphysis)	Indoles, serotonin, and melatonin	Effects on biological rhythms and brain function Counteracts MSH activity
Ovaries	Estrogens	Estrus cycle, female sex properties
	Progesterone	Secretory phase (with estrogens) of uterus and mammary glands
	Relaxin	Relaxes symphysis pubis for birth
Testis	Testosterone and androgens	Male sex properties and spermatogenesis
Placenta	Placental lactogenic hormone	Growth hormone–prolactin activity
	Chorionic gonadotropin, estrogen, progesterone	Adjunct to other endocrine glands in 2nd and 3rd stages of pregnancy

Table 15.1. (*Continued*)

Gland of origin	Hormone name (symbol)	Primary actions
Kidney	Renin	Hydrolysis of blood precursor protein to yield angiotensin
Prostate, gonads, many tissues	Prostaglandins (PG)	Many effects at membrane site of synthesis
Gastrointestinal (GI) tract	Gastrin	Stimulates parietal cell secretions
	Secretin	Stimulates pancreatic juice
	Cholecystokinin	Contraction of gallbladder
Brain opioids	Endorphins: β-endorphin, enkephalins	Endogenous peptides that bind to morphine receptor

SOURCE: From T. M. Devlin, 1982. *Textbook of Biochemistry with Clinical Correlations.* John Wiley & Sons, New York.

terior and posterior portions of the pituitary gland. These hypothalamic hormones serve as releasing factors or inhibitory factors that respectively release or terminate pituitary hormone production (Figure 15.1).

Pituitary hormones in turn affect the next level of endocrine system organization including the pancreas, thyroid gland, ovary, testis, and adrenal cortex. The heirarchical organization of the endocrine system, under the ultimate control of the CNS and the hypothalamus, is illustrated in Figure 15.2.

The range of biochemical control exerted by hormones is staggering. Thyroid-stimulating hormone (TSH), also known as *thyrotropin,* affects the size and vascularization of the thyroid gland. Other regulatory effects include the production of thyroid hormones and the lipolytic metabolism within adipocytes.

Adrenocorticotropic hormone (ACTH) has its regulatory effect on the adrenal cortex. Corticoid hormones produced by this gland are derived from the steroid nucleus of cholesterol, and they include both *mineral corticoids* and *glucocorticoids.* Mineral corticoids regulate sodium, potassium, and ion balance over the renal tubules, whereas glucocorticoids serve as an integrated control mechanism over protein and carbohydrate metabolism.

Growth hormone (GH) exerts physiological effects on many tissues. In particular it stimulates amino acid uptake by muscle cells, which coincides with increased protein synthesis. Anti-insulin effects can also result from GH actions that raise blood glucose concentrations. Excessive production of GH in children causes gigantism and acromegaly as adults, whereas underproduction in children leads to dwarfism.

A variety of anterior pituitary hormones are tied to complex mammalian reproductive processes. These hormones include follicle stimulating hormone (FSH), luteinizing hormone (LH), and prolactin (PRL).

Follicle stimulating hormone is an active hormone in both males and females. It promotes testicular spermatogenesis in conjunction with LH in males, whereas the same FSH promotes ovarian development of follicles in females. Luteinizing hormone stimulates formation of the corpus luteum and augments FSH actions during the female ovulation cycles. This hormone (LH) is also known as interstitial-cell-stimulating hormone (ICSH) owing to its role in stimulating development of interstitial testicular cells. Prolactin, on the other hand, promotes lactogenesis in the mammary gland and synthesis of lactose synthetase, which is responsible for lactose production.

Posterior pituitary hormones include both vasopressin and oxytocin. These peptide hormones have almost identical structures with the exception of amino acid residues 3 and 8 (Figure 15.3), yet their functions are quite dissimilar. Oxytocin promotes smooth

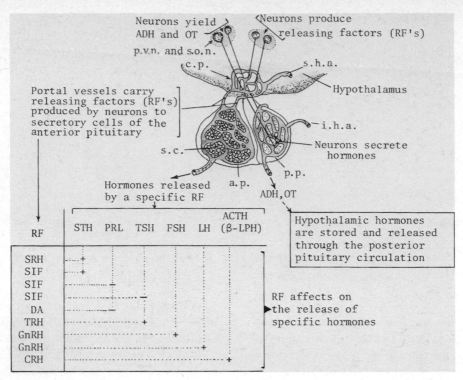

FIGURE 15.1. The anterior pituitary (a.p.) and posterior pituitary (p.p.) glands develop from embryologically distinct origins. Each portion displays its own capillary circulation and individually releases its own characteristic hormones. Hormones from this bean-sized gland control the hormonal and secretory actions of other glands throughout the body. The pituitary gland is intrinsically associated with the hypothalamus of the brain. Some neurons based in the hypothalmic regions have secretory functions that yield hormones. These hormones are introduced into blood circulation through the p.p. It should be recognized that hormonal transport prior to introduction into the hematogenous circulation involves transport over axons of the neurons. Other hypothalamic-based neurons instigate the formation of releasing factors (RFs). The hypothalamus consists of neuronal clusters called nuclei, which participate directly in hormonal control mechanisms. These include the supraoptic nucleus (s.o.n.) and the paraventricular nucleus (p.v.n.). The RFs regulate the synthesis and release of hormones from the a.p. Some RFs encourage (+) the synthesis, activation, or circulatory release of hormones, but others diminish (−) or suppress specific hormonal activities. Key abbreviations for the schematic illustration of the pituitary include c.p., capillary plexus of median eminence region; i.h.a., inferior hypophyseal artery; s.c., secretory cell; s.h.a., superior hypophyseal artery.

Releasing Factors Include

SRH	Somatotropin-releasing hormone
SIF	Somatostatin (somatotropin)-inhibiting factor
DA	Dopamine (prolactin-inhibiting factor)
TRH	Thyrotropin-releasing hormone (thyroliberin)
GnRH	Gonadotropin-releasing hormone (gonadoliberin)
CRH	Corticotropin-releasing hormone (corticoliberin)

Hormones Released by the Hypothalamus Include

STH	Somatotropin (or growth hormone, GH)
PRL	Prolactin
TSH	Thyroid-stimulating hormone (thyrotropin)
FSH	Follicle-stimulating hormone (follitropin)
LH	Luteinizing hormone (luteotropin)
ACTH	Adrenocorticotropic hormone (corticotropin)
β-LPH	β-Lipotropin
OT	Oxytocin
ADH	Antidiuretic hormone (vasopressin)

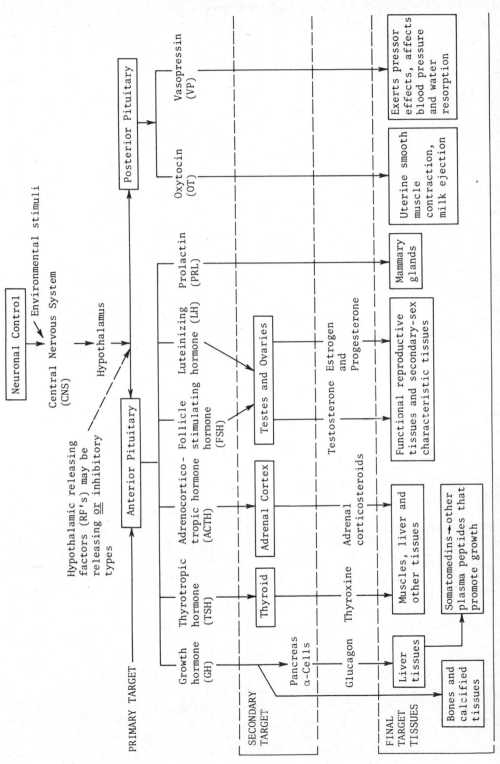

FIGURE 15.2. Neuronal stimuli involving the CNS stimulates the production of hypothalamic hormones. Some of these hormones act as releasing factors (RFs), whereas others inhibit releasing hormone actions. Hypothalmic hormone effects are exerted on the anterior pituitary gland (Figure 15.1), which produces a flourish of key hormones. Many of these hormones are directed at numerous secondary target cells. These secondary target cells of tissues also produce hormone-active substances aimed at controlling final target tissue activities.

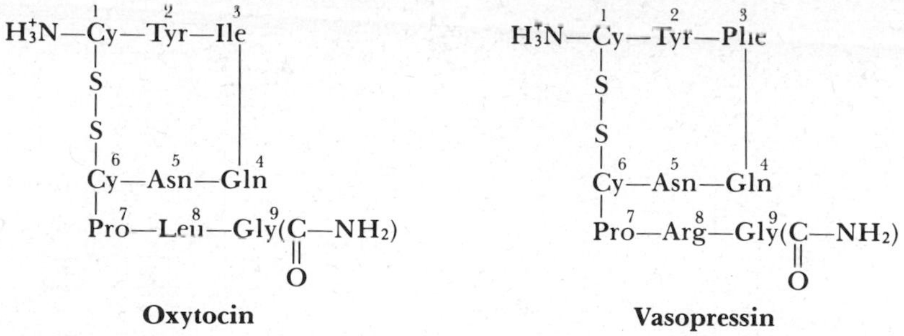

$$H_3^+N—Cy—Tyr—Ile$$

Oxytocin

$$H_3^+N—Cy—Tyr—Phe$$

Vasopressin

FIGURE 15.3. Structures for oxytocin and vasopressin.

muscle contraction (labor) and milk ejection from the mammary glands in lactating females. Vasopressin, however, contributes to the elevation of blood pressure and exhibits antidiuretic properties. These antidiuretic properties reflect effective absorption of water from the distal and collecting portions of the renal tubules.

The actions of the hormones outlined above are augmented by many other hormones that are mentioned throughout the following sections. Table 15.1 also details some general functions of hormones not treated in this chapter.

POSSIBLE MECHANISMS FOR HORMONE ACTION

All hormones exert profound effects on target cell biochemistry at tiny molar concentration levels ($10^{-10} \rightarrow 10^{-7} M$). Once a hormonal–target cell interaction has occurred, the cell undergoes a functional modification that culminates in *eliminating the stimulus that first initiated hormone release*. This version of a so-called negative feedback system is responsible for nearly all homeostatic mechanisms operating in an animal.

Endocrine hormone control systems can be subdivided into three distinct parts including (1) a secretory cell, (2) a transport mechanism for the hormone, and (3) the target cell for hormone interaction (Figure 15.4).

Although some hormones are freely transported in the blood as a solute, most hormones are bound to a serum protein. Hormone transit using proteins may require a highly discriminative type of protein–hormone binding interaction or merely nondiscriminative albumin–hormone associations.

The chemical diversity of hormones is also tied to their specific effects on target cells. Three types of hormones are commonly recognized:

1. Amines (thyroxine, epinephrine, norepinephrine, and melatonin).

2. Steroids (testosterone, aldosterone, cortisol, progesterone, and estradiol).

3. Polypeptides or small proteins (glucagon, insulin, vasopressin, oxytocin, prolactin, and adrenocorticotropic hormone).

When a hormone reaches its specific target cell, the cell metabolism is altered only after a specific hormone–cell binding interaction has occurred. Conceptually this interaction involves the hormone (H) and its cellular receptor site (RS). The RS may be a superficial feature of the cell membrane structure or it may exist as a free entity within the cytosol of the cell. The specific nature of H/RS interactions are speculated to involve (1) reversible H/RS interactions; (2) a specific presence of RS on key H-responsive cells and an absence of RS on H-insensitive cells; and (3) the RS on H-responsive cells must exhibit a topological binding site compatible with some stereochemical aspect of H.

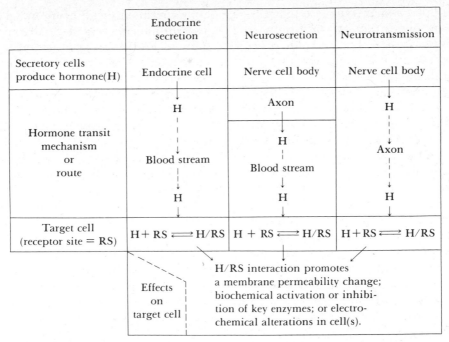

	Endocrine secretion	Neurosecretion	Neurotransmission
Secretory cells produce hormone(H)	Endocrine cell	Nerve cell body	Nerve cell body
Hormone transit mechanism or route	H ⋮ Blood stream ⋮ H	Axon ↓ H ⋮ Blood stream ↓ H	H ⋮ Axon ⋮ H
Target cell (receptor site = RS)	H + RS ⇌ H/RS	H + RS ⇌ H/RS	H + RS ⇌ H/RS

Effects on target cell: H/RS interaction promotes a membrane permeability change; biochemical activation or inhibition of key enzymes; or electrochemical alterations in cell(s).

FIGURE 15.4. Characteristic differences among endocrine secretion, neurosecretion, and neurotransmission of hormone-active compounds are schematically illustrated. Effects of a hormone (H) on a specific target cell are varied, but they usually involve an initial interaction between H and a receptor site (RS) located *on* or *in* the target cell. Once the H/RS complex has been formed, a characteristic hormone-induced response in the cell (targeted by the hormone) is initiated.

Intracellular responses to hormone receptor site interactions can result in:

1. Hormone-induced changes of *cell membrane permeability* for ions or specific solutes.

2. *Activation of a* key *"second messenger"* within a cell that "turns on" a certain biochemical reaction.

3. Hormone-*initiated* specific *gene actions.*

Membrane Permeability Changes

Alterations in target cell membrane permeability for key solutes serves as an initial effect of many hormone actions. Solute permeabilities commonly affected by hormones include glucose, amino acids, and ions. Insulin, as an example, is believed to control the entry of glucose into muscle cells and thereby alter sugar utilization at the cellular level. Some other hormones including growth hormone (GH), glucagon, glucocorticoids, and vasopressin are also believed to alter membrane permeabilities. It is still uncertain, however, whether observed changes in membrane permeability happen to reflect a primary or secondary effect of hormone–target cell interactions.

Messenger Activation

Hormones transported through the blood to a target cell can be operationally described as a "first messenger." Sometimes the "first messenger" elicits a specific biochemical reaction at the cellular level (e.g., membrane-induced permeability changes cited above). In other cases the first messenger initiates the activity of a "second messenger" within the cell. This second messenger is then responsible for sparking specific "targeted" biochemical reactions.

One popular scenario suggests that a messenger hormone is received by a cell membrane *receptor protein*. The receptor protein serves as a specific hormone-binding target. Once the hormone–receptor complex has been established, the enzyme activity of a key enzyme such as adenyl cyclase is either enhanced or activated. Adenyl cyclase then produces *cyclic AMP* (cAMP) (from ATP), which then functions as a second messenger. This second messenger subsequently instigates a specific hormone-directed biochemical response within the cell.

Formation of cAMP as a secondary messenger is one of the most important consequences of hormone-receptor site interactions at the cellular level. At least nine hormones are suspected to stimulate the production of cAMP and then use it as a second messenger. These hormones include antidiuretic hormone (ADH), ACTH, FSH, TSH, parathyroid hormone (PTH), glucagon, and calcitonin. Figure 15.5 outlines the role of epinephrine in mobilizing glucose from liver glycogen stores, which requires the participation of cAMP as a second messenger. Note, too, that the consequences of second messengers can be twofold. As the illustration shows, cAMP initiates the mobilization of glucose from glycogen and at the same time suppresses the action of glycogen synthetase, which is responsible for glycogenesis.

Effects of cAMP as a second messenger may involve more complexity than simply activating important enzymes. Experiments with cell-free *Escherichia coli* systems has revealed that cAMP can activate nonfunctional genes. Doubtless similar mechanisms operate in animals, but the scope of such mechanisms is still uncertain. The importance of cAMP in hormonal actions is unshaken but nevertheless uncertain in view of other evidence. For example, epinephrine-induced increases in the rate of liver glycogenolysis can be blocked by the action of phenoxybenzamine. However, it is interesting to note that this epinephrine antagonist accomplishes this effect without altering cAMP levels. Experiments with propranolol (Inderal) also prohibit increased cellular concentrations of cAMP, yet glycogenolysis remains unaffected. These results do not discount the necessity of a second messenger during hormonal interactions with their target cells, but it is clear that some hormone actions do not mirror intracellular cAMP concentrations.

Whereas liver cell glycogenolysis involves cAMP as a second messenger, other cells use cAMP for other regulatory mechanisms. In the case of adipocytes, epinephrine-induced cAMP actions affect the regulatory subunit of a protein kinase. This subunit becomes activated and then mediates the phosphorylation of a hormone-sensitive lipase. Phosphorylated lipase, in turn, catalyzes triacylglycerol hydrolysis to produce fatty acids and residual acylglycerols.

Steroid Hormones and DNA Activation

The small molecular size and lipophilic properties of some hormones such as steroids allows them to penetrate cell membranes. Hormone transit over the cell membrane is followed by an interaction between the steroid hormone and a *cytoplasmic binding protein* (CBP). The steroid–CBP complex is translocated to the nucleus. Here the complex associates with other proteins that are part of a *chromatin receptor* (CR) complex. This steroid–CBP–CR complex promotes activation of key structural genes, mRNA transcription, and the translation of mRNA into protein (Figure 15.6).

Apart from these actions, steroid hormones participate in "permissive effects" that influence the effects of other hormones. That is, fast-acting hormones such as epinephrine or glucagon rely on a static (normal) level of steroid hormones in order to be effective. Therefore, the effective concentration levels for all other hormones in the animal body will change according to the status of steroid hormone concentrations.

The interrelationships between steroid hormones, permissiveness, and the actions

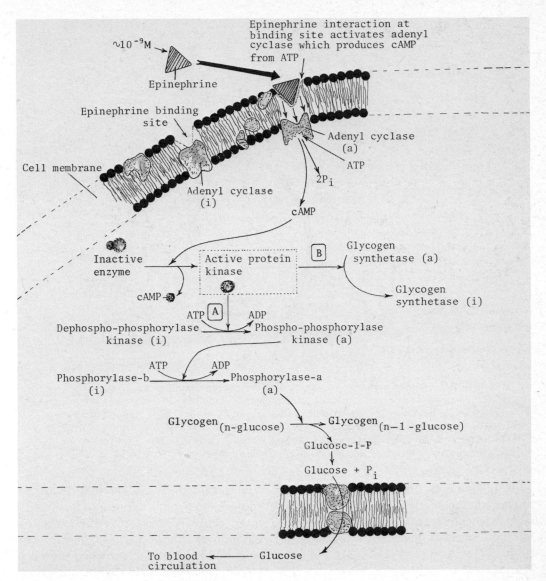

FIGURE 15.5. Epinephrine $\boxed{A}$ mobilizes blood sugar (glucose) from glycogen in the liver and at the same time suppresses $\boxed{B}$ glycogen synthesis by "turning-off" the action of glycogen synthetase. Small concentrations $(10^{-9}\,M)$ of epinephrine in the bloodstream bind at a specific site on a target cell. A complex cascade reaction results where adenyl cyclase *is activated* (a) from its inactive (i) form. The cyclic AMP (cAMP), produced from ATP and the activated enzyme, forms an active protein kinase. From step $\boxed{A}$, glycogenolysis is initiated. The active protein kinase forms active phosphophosphorylase (ATP consumed) and the active phosphophosphorylase mediates the formation of an active phosphorylase (with ATP consumption). The active phosphorylase cleaves and phosphorylates a glucose residue from glycogen leaving a glycogen molecule with one less glucose residue than the parent structure. Enzymatic *dephosphorylation* of glucose 1-phosphate followed by the membrane transit of the resultant *free glucose* ensures the maintenance of blood sugar.

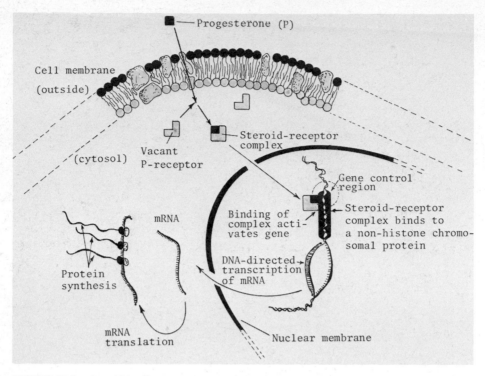

FIGURE 15.6. Steroid hormones such as estrogen and progesterone may act directly at the gene level. The progesterone (P) in this case traverses the cell membrane and complexes with a vacant progesterone receptor (P receptor). The steroid–receptor complex passes through the nuclear membrane, where it binds and activates a specific site on the DNA. DNA-directed transcription of mRNA, initiated by steroid complex binding, exits the nuclear region. The mRNA and ribosomal activity both participate in subsequent protein synthesis.

of rapidly acting hormones are confusing. However, there is little doubt that the permissive effects of steroids somehow rely on moderated levels of gene transcription.

THYROID HORMONES AND THEIR BIOSYNTHESIS

The thyroid gland is a bilobed structure that nearly surrounds the trachea just below the larynx (Figure 15.7). The gland contains numerous follicles, each of which has a single layer of epithelial cells surrounding a lumen containing colloid. *Thyroglobulin* (Tg), a glycoprotein having a molecular weight of 660,000, is the major constituent of the colloid. This substance acts as a storage site for thyroid hormones within the thyroid gland.

Thyroid hormone synthesis relies on a five-step production scheme that requires (1) trapping inorganic iodine and its conversion to an activated form ($I^- \rightarrow I^+$); (2) thyroglobulin synthesis; (3) *organification* or covalent linkage of the activated iodine to a tyrosine moiety of thyroglobulin; (4) covalent bonding of iodotyrosyl residues to form iodonated thyronines (T_3 and T_4); and (5) degradation of thyroglobulin to release T_3 and T_4 thyronines into the blood.

Apart from thyroid hormone activities, the thyroid gland is also associated with the parafollicular cells or "C" cells, which secrete calcitonin.

Iodine Is Essential for Thyroid Hormone Production

Thyrotropin-releasing factor (TRF) from the hypothalamus induces the anterior pituitary gland to form thyrotropic stimulating hor-

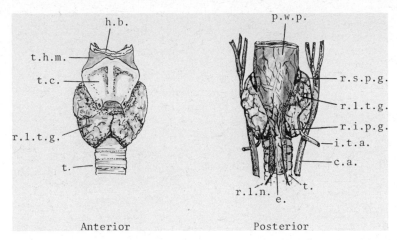

FIGURE 15.7. The bilobed structure of the thyroid gland surrounds the trachea (anterior view) and neighbors the pharyngeal wall (posterior view). The four parathyroid glands reside on the posterior side of the thyroid gland, next to the pharyngeal wall. The thyroid gland is controlled by the pituitary gland, which regulates thyroid hormone release by a negative feedback mechanism. Thyroxine, triiodothyronine, and other iodinated tyrosine derivatives having hormonal effects are released from the thyroid glands; the C cells of the thyroid produce calcitonin; and the parathyroid glands yield parathyroid hormone (PTH). Key structural features illustrated include c.a., carotid artery (right); e., esophagus; h.b., hyoid bone; i.t.a., inferior thyroid artery; p.w.p., posterior wall of pharynx; r.i.p.g., right inferior parathyroid gland; r.l.n., recurrent laryngeal nerve; r.l.t.g., right lobe of thyroid gland; r.s.p.g., right superior parathyroid gland; t., trachea; t.c., thyroid cartilage; t.h.m., thyrohyoid membrane.

mone (TSH) (Figure 15.1). This hormone subsequently activates thyroid follicle production of L-*thyroxine* (T_4) and L-*triiodothyronine* (T_3) as discussed below (Figure 15.8).

The secretion and function of thyroid hormones depends on the continuous availability of iodine. Based on a normal 150-μg daily intake of iodine, plasma levels of iodine hold at ~0.3 μg/100 mL. About 10 to 23% of plasma iodine is unavoidably lost, whereas the thyroid gland extracts almost 50% of the plasma iodine passing through the gland. Iodine may be obtained from exogenous sources such as foods and water (depending on locality) that pass through the gastrointestinal tract, or peripheral tissues may deiodinate thyroxine to free iodine for another round of hormone synthesis.

The entry of iodine as iodide into the thyroid follicle is mediated by an active transport mechanism. Iodide is readily trapped and oxidized by a specific thyroid peroxidase that requires O_2 and $NADPH(H^+)$. The I^+ product reacts with tyrosine already contained in a thyroglobulin molecule. Iodinated thyroglobulins may exist as *mono*- (MIT) and *diiodotyrosine* (DIT). Both of these compounds probably serve as precursors to thyroid hormone formation. That is, two DIT residues or one DIT and one MIT residue can be enzymatically coupled, by a specific thyroid peroxidase, to produce *thyroxine* (T_4) and *triiodothyronine* (T_3) (Figure 15.9). A series of other iodinated compounds also result during T_3 and T_4 syntheses, but they are largely inactive as hormones.

Iodinated thyronines (MIT, DIT, T_3, and T_4) are stored in high-molecular-weight proteins within the colloidal fraction of the thyroid. Endocytosis of the colloid to produce colloidal vesicles then serves as a prelude to T_4 and T_3 introduction into the blood. Moreover, mobilization of key thyroid hormones (T_3 and T_4) occurs only after (1) the

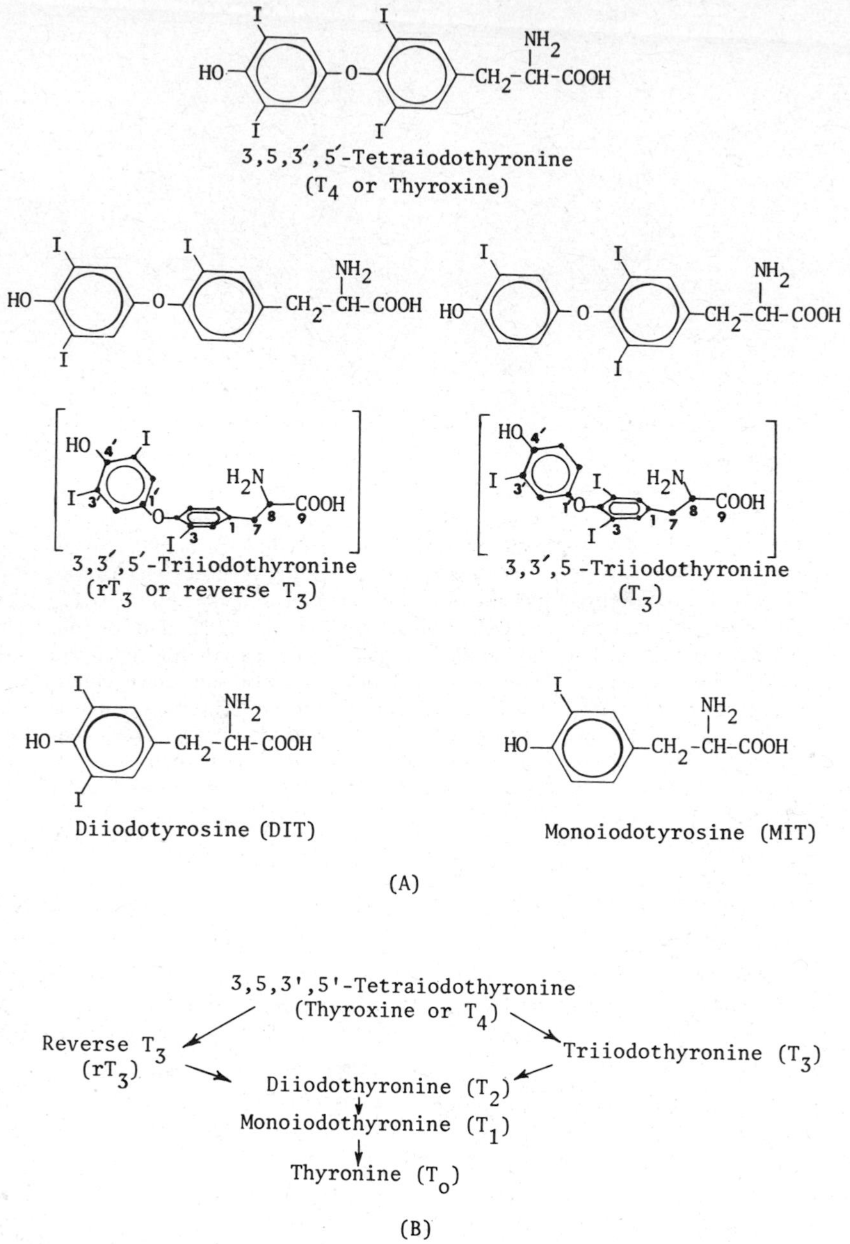

3,5,3′,5′-Tetraiodothyronine
(T_4 or Thyroxine)

3,3′,5′-Triiodothyronine
(rT_3 or reverse T_3)

3,3′,5-Triiodothyronine
(T_3)

Diiodotyrosine (DIT)

Monoiodotyrosine (MIT)

(A)

3,5,3',5'-Tetraiodothyronine
(Thyroxine or T_4)

Reverse T_3
(rT_3)

Triiodothyronine (T_3)

Diiodothyronine (T_2)

Monoiodothyronine (T_1)

Thyronine (T_0)

(B)

FIGURE 15.8. (A) Structures for some thyroid hormones and their derivatives. (B) Generalized metabolic pathways for thyroid hormones.

colloidal vesicle has been acted upon by proteolytic enzymes (having lysosomal origins), and (2) MIT and DIT have been deiodinated by a deiodinase with the liberated iodine being recycled (Figure 15.9). The T_3 and T_4 thyronines surviving these reactions are then bound to transient plasma proteins, namely, *thyroxine-binding globulin* (TBF), *thyroxine-binding prealbumin* (TBPA), and albumin.

The Fate of T_3 and T_4 Hormones

Plasma protein-bound T_3 and T_4 travel to target cells where the hormones readily dif-

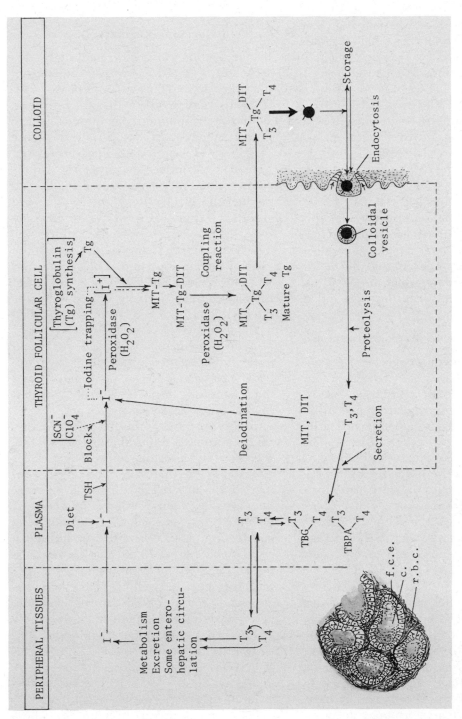

FIGURE 15.9. Major steps in the biosynthesis and secretion of thyroid hormones. Abbreviations: T_3, triiodothyronine; T_4, thyroxine; MIT, monoiodotyrosine; DIT, diiodotyrosine; TBG, thyroxine-binding globulin; TBPA, thyroxine-binding prealbumin. Microscopic view of thyroid cell structures (insert): f.c.e., follicular cell epithelium; c., colloid; r.b.c., red blood cells. Thyroid hormone synthesis initiated by the action of thyroid-stimulating hormone (TSH) promotes the trapping of inorganic iodine into an organic structure (organification). Thiocyanates and perchlorates upset this step by blocking iodine assimilation. See text for details concerning MIT, DIT, T_3, and T_4 formations and an explanation of the remaining event illustrated in the figure.

fuse *into* the cells. Possible target cell exceptions include the adult brain and testes. Nevertheless, T_3 and T_4 diffusion into receptive cells probably results in deiodination reactions. In particular, the distal ring of T_4 is deiodinated to yield T_3, whereas iodine loss from the proximal T_4 ring yields *reverse* T_3 (rT_3) (Figure 15.8). The peripheral deiodination of T_4 to T_3 occurs largely in the liver, kidneys, and skeletal muscles. Although T_4 does have intrinsic hormonal activity, many studies claim that T_3 has the most important (65–75%) metabolic role in humans.

Thyroid hormones are eliminated from the body by deiodination reactions and other structural modifications that yield hormone-inactive species. The most typical derivatives include tetraiodothyroacetic acid and $3,3'5'$-triiodothyronine (rT_3), which are either excreted in bile or urine or further dismantled.

Target Cell Effects of Thyroid Hormone

The physiological consequences of thyroid hormone activities are clearly recognized, but their biochemical effects on target cells remain confused. Most evidence suggests that thyroid hormone actions are instigated at the genetic level by derepression of DNA. This hormonal action assumes a preliminary interaction between the hormone and a cytoplasmic receptor, both of which participate in the genetic derepression event(s). Other models of hormone action claim that thyroid hormones spur on the synthesis of nucleic acids and proteins. This effect could stem from the interaction of the hormone with discreet receptors located in nuclear chromatin. Other studies indicate that thyroid hormones directly affect oxidative phosphorylation in the mitochondria. Still other investigators claim that cAMP may be involved in thyroid hormone effects at the cellular level. Although few contemporary theories are footed in cement, it is agreed by many authorities that these hormones do affect protein, lipid, and carbohydrate metabolism. Their main effect is reflected as an increase in the basal metabolic rate (BMR) and calorigenesis.

Thyroid Hormone Regulation

Blatant iodine deficiency and high consumption of *dietary thiocyanates* can contribute to low thyroid hormone productivity. Efforts to place iodide in selected food substances has minimized iodine deficiencies in the United States as well as the incidence of goiter. The impact of thiocyanates can have a cyclic effect on thyroid hormone suppression since thiocyanate intake can parallel the dietary intake of certain food crops. The *Cruciferae* in particular (i.e., cabbage, kale, brussels sprouts, cauliflower, broccoli, turnips) can yield thiocyanate (SCN^-) during digestion. Thiocyanates (as well as perchlorate (ClO_4^-) not obtained from foods) generally prohibit iodine uptake by the thyroid cells and create an iodine deficit in the iodine trap.

Apart from these factors, the thyroid gland exerts homeostatic control over the entire body. This control mechanism stimulates and moderates the calorigenesis, anabolism, and catabolism of nearly all tissues and organs. As indicated earlier, the formation of thyroid hormones is under the control of TSH. Schemes for regulatory control suggest that TSH binds to superficial receptor sites on thyroid follicular cell membranes. This event then activates adenyl cyclase. Production of cAMP by this enzyme may mediate TSH activity and play a role in thyroid hormonal regulation. Figure 15.10 outlines the suspected feedback mechanisms involving thyroid hormone production. When the influence of thyroid hormone drops below some necessary level, the hypothalamus responds by stimulating the pituitary gland with TRH. This TRH stimulation spurs the formation of TSH, which then induces follicular hormone synthesis. Very high productivity levels of thyroid hormone are countered by hypothalamic sensing mechanisms, which prohibit further hormone synthesis. This synthetic and inhibitory cycle is regulated in

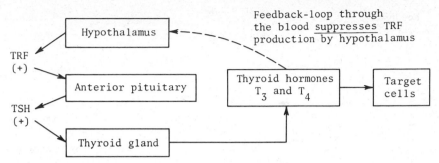

FIGURE 15.10. Regulation of thyroid hormone synthesis by feedback control mechanisms. A low thyroid hormone level in the blood spurs the hypothalamus to produce TRH. The TRH stimulates the pituitary production of TSH, which has a positive (+) effect on thyroid follicle production of thyroid hormones (e.g., T_3 and T_4). These thyroid hormones generally increase the BMR and other energy-producing cycles of the body. Once the action of thyroid hormones supersedes metabolic requirements, hypothalamic sensing mechanisms detect excessive hormone levels by a negative feedback (−) system. This effectively depresses TRF hormone levels, which correspondingly suppress TSH action on the thyroid follicles.

an overall fashion by the required metabolic rates of unit body operations.

CALCITONIN AND PARATHYROID HORMONE ACTIVITIES

Calcitonin and parathyroid hormone (PTH) are polypeptide hormones that regulate the disposition of calcium levels in the blood. Calcitonin (38 amino acid residues) is produced by the parafollicular or C cells of the thyroid gland, whereas PTH (84 amino acid residues) is secreted by four pea-sized parathyroid glands embedded within the thyroid gland region (Figure 15.7). These hormones exert antagonistic controls over calcium levels in the blood. Increases in calcium (Ca^{2+}) concentrations are spurred on by PTH while calcitonin effectively decreases blood Ca^{2+}.

Sensitive detection mechanisms ensure the release of PTH when Ca^{2+} levels are abnormally low (Figure 15.11). When Ca^{2+} levels are unusually high, calcitonin regulates Ca^{2+} disposition (Figure 15.11). The delicate regulatory interplay of both hormones over blood Ca^{2+} levels holds blood Ca^{2+} concentrations at ~9 mg/dL. This finely tuned and sensitive regulation of blood Ca^{2+} has critical survival

functions for cells. Blood fluctuations in Ca^{2+} can lead to the rapid cell death.

Normal Ca^{2+} levels in human sera range from 9.2 to 11.2 mg/dL during the first year of life and progressively decrease to 8.9 to 10.9 mg/dL by age 16. There is no significant clinical difference between males and females. The hematogenous circulation of Ca^{2+} occurs in three forms, namely, (1) free Ca^{2+} (46%), (2) Ca^{2+} bound to protein (40%), and (3) diffusible Ca^{2+} complexes (14%).

A superficial survey of hormonal functions shows that PTH inhibits Ca^{2+} excretion from the kidney but calcitonin increases cation excretion (Figure 15.11). These actions certainly maintain Ca^{2+} homeostasis in the body, but very complex underlying mechanisms are responsible for this process. It should also be recognized that these counteracting hormones principally involve three target tissues including the intestines, kidney, and bone cells.

Parathyroid hormone (Figure 15.11) exerts four effects on the kidney, one of which is tied to the mobilization of vitamin D activity. The four effects of PTH on the kidney result in

1. Inhibition of renal tubule phosphate resorption so as to produce hypophosphatemia.

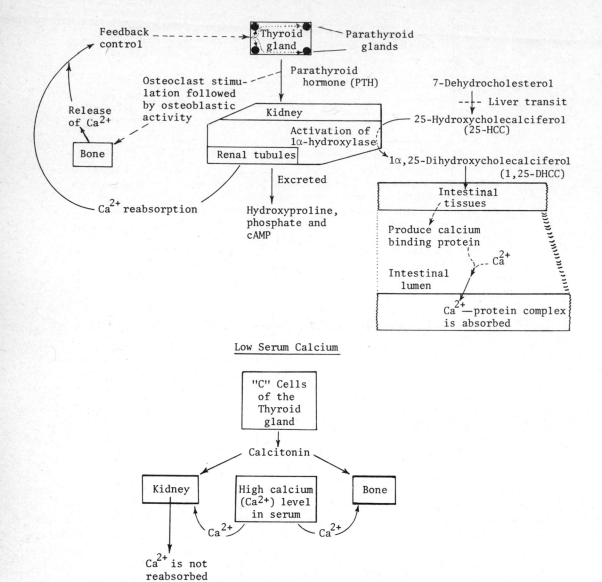

FIGURE 15.11. Effects of parathyroid hormone (PTH) during *low serum concentrations* of calcium and the actions of calcitonin when *serum calcium is high*.

2. An increase in resorption of free Ca^{2+} in the distal renal tubule.

3. A maintenance of acid–base balance in the blood by inhibiting both the renal tubular resorption of HCO_3^- and the exchange of tubular H^+ for luminal Na^+.

4. Mobilization of 1-α-hydroxylase, which mediates the biosynthesis of vitamin D-active steroids.

All of these factors play important roles in Ca^{2+} resorption, assimilation, and homeostasis, but the activation of 1-α-hydroxylase is clearly fundamental to the important hormonal effects of PTH.

As indicated in an earlier chapter, vitamin D_3 or cholecalciferol is ultimately produced from a photoreaction involving ultraviolet light. In the liver, cholecalciferol is hydroxylated to produce 25-hydroxycholecalciferol (25-HCC). This compound is the one that is hydroxylated in the kidney by the 1-α-hydroxylase to produce 1-α,25-dihydroxycholecalciferol (1,25-DHCC). Some 24,25-dihydroxycholecalciferol may be inadvertently formed, but this compound is quite inactive compared to 1,25-DHCC. Some studies suggest that 1-α-hydroxylation is favored over 24-hydroxylations of the steroid structure when serum Ca^{2+} and/or phosphorous concentrations are low. The exact significance of this control system is unknown. Regardless of control idiosyncrasies, 1,25-DHCC probably travels to the intestinal tract where it promotes the DNA-directed transcription of Ca^{2+}-binding proteins. Syntheses of these proteins, and possibly phosphate-binding proteins, are then followed by the transport of Ca^{2+} and PO_4^{3-} over the plasma membranes of digestive tract cells.

Aside from these actions, PTH also regulates Ca^{2+} mobilization from bones. Since Ca^{2+} homeostasis is critical to the life of normal cells, PTH essentially controls the exploitation of Ca^{2+} from calcium phosphate reserves in bones. Bones release their Ca^{2+} stores into the blood after PTH activates osteoclasts. These weird ameboid cells digest calcium phosphate and eventually liberate Ca^{2+} into the blood. Clinical evidence of these activities is urinary excretion of hydroxyproline, obtained from the decalcified bone matrix, along with phosphate and cAMP. Excessive decalcification is checked by osteoblasts that are controlled by calcitonin. Not only do osteoblasts check decalcification, but they ultimately contribute to bone formation—thereby reversing osteoclastic activity.

Parathyroid gland destruction or necrosis, such as that exhibited in autoimmune disease, can result in PTH deficiencies. These deficiency effects can cause muscular convulsions and eventual death. On the contrary, excessively high PTH activity can cause blatant decalcification of bones or osteoporosis. Osteoporosis is then followed by fibrous skeletal cysts, and in severe cases, death may result.

PANCREATIC HORMONES

The pancreas serves as both an exocrine and an endocrine gland. Its production of digestive enzymes that are ejected into the digestive tract give the gland an exocrine character, yet the gland contains groups of true endocrine cells. These cells are known as the *islets of Langerhans.* Islet cells consist of two distinct types denoted as *alpha* (α) and *beta* (β) *cells.* The α cells secrete *glucagon,* which promotes liver glycogenolysis, whereas the β cells secrete *insulin,* which is necessary for cellular assimilation or uptake of blood glucose. These actions of glucagon (29 amino acid residues) and insulin *complement* the action of each other during metabolism and maintain a finely tuned balance of blood sugar (Figure 15.12).

The formation of insulin begins with the synthesis of a single polypeptide chain called *proinsulin.* Proinsulin is cleaved by proteases at two sites to produce insulin and a so-called C peptide. The A chain of insulin contains

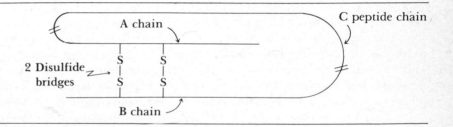

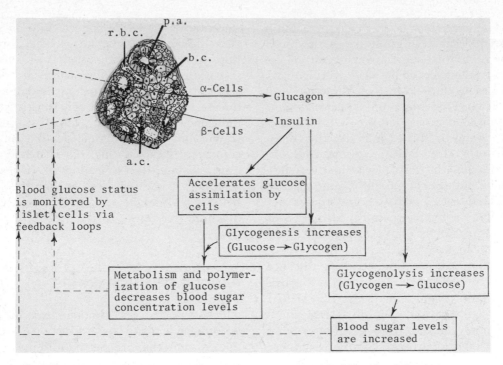

FIGURE 15.12. Interaction of insulin and glucagon produced by islet cells of the pancreas. Low glucose levels activate α-islet cells (a.c.), which promote glucagon. Glucagon promotes glycogenolysis in the liver and increases the mobilization of blood glucose. High blood glucose levels, on the other hand, encourage the β-islet cells (b.c.) to secrete insulin. Insulin promotes the uptake of glucose by metabolically active cells and also enhances liver glycogenesis. The actions of these hormones are not mutually exclusive, that is, one hormone does not inhibit the action of the other one. Moreover, the secretion products of these islet cells operate independently from the pituitary and endocrine glands notwithstanding the indirect affects of other hormones. The insert showing the islets of Langerhans also points out the locations of the pancreatic acini (p.a.) and sites in the tissue that permit the entry of red blood cells (r.b.c.).

21 amino acid residues, the B chain has 30 amino acids, and both chains are linked together by dual disulfide bridges. Beef and pork insulins used in treating human diabetes are immunologically distinct since they differ in only one and three amino acids, respectively.

Proinsulin probably exists because active insulin cannot be stored in the pancreas for prolonged periods without organ damage. Therefore, it is activated just prior to its release into the blood. As indicated earlier, the molecular structure and biochemical functions of insulin on target cells are difficult to assess, but the hormone does promote glucose uptake into metabolizing cells.

Insulin deficiency results in hyperglycemia and causes diabetes mellitus. In some instances the synthesis of insulin is impaired by defective or necrotic islet cells (insulin-dependent diabetes). This impairment may reflect genetic, viral, and/or life-style predispositions that promote the disorder. In other cases, insulin productivity may be normal but an enzyme such as *insulinase* destroys the active enzyme. Still other diabetic conditions exist whereby cell membranes targeted by insulin have developed a *deficiency* of *active receptor sites*. This is one of the contributing factors in adult diabetics (insulin-independent types), in which insulin levels are otherwise normal. Whereas increasing insulin levels of insulin-independent diabetics using drug therapy can improve the adult diabetic condition, this treatment does not work for insulin-dependent diabetics.

Here, exogenous insulin injections are the only route for coping with the pancreatic dysfunction.

Advanced and unchecked cases of diabetes reflect many direct and indirect clinical complications whose biochemical etiologies are very curious. The hyperglycemic condition of the blood is accompanied by inadequate glucose for normal cell metabolism. The physiological condition of the body then accelerates fat catabolism and the production of ketones, which serve as an energy source. Excess ketogenesis can result in ketosis, ketonuria, and kidney damage. Other consequences include weight loss, exhaustion, impaired circulation, high blood pressure, retinopathy, ulcerated skin, development of dangerous local skin infections, and a complex list of severe health consequences.

In cases where too much insulin is discharged into the blood, hypoglycemia or low blood sugar may result. This action causes cell starvation for glucose because available stores of glucose are readily processed into glycogen—a process expedited by insulin activity. Hypoglycemic conditions can cause dizziness, tremors, and behavioral dysfunctions owing to deficient levels of brain and nervous system glucose. All cases of actual hypoglycemia are not understood, but classic hypoglycemic conditions may result from the rampant growth of a pancreatic tumor, which causes production of far too much insulin.

The specialized pancreatic islet cells are also responsible for *somatostatin* (or growth hormone release-inhibiting hormone) production. This substance blocks TRH effects and suppresses the release of GH, PRL, insulin, glucagon, gastrin (which stimulates parietal cell secretions), and secretin (responsible for pancreatic juice production).

ADRENAL HORMONE PRODUCTION

The adrenal glands lie over the upper poles of the kidneys and produce a flourish of hormone-active and hormone-related but biologically inactive compounds. The outer part of the gland is called the adrenal cortex and the inner region is called the medulla (Figure 15.13).

Adrenocortical hormones are biochemically synthesized from cholesterol (Figure 15.14). At least 40 steroid structures related to these hormones are known to exist, but only a few of their functions are widely considered. The two most important corticoids for the present discussion include *cortisol* and *aldosterone*, both of which regulate certain mineral ion balances over the renal tubules.

Adrenocorticotropic hormone (ACTH) stimulates the synthesis and release of cortisol. This steroid in turn inhibits further ACTH stimulation in a feedback loop. After cortisol is released, 75% is bound to a glycoprotein called *transcortin*, 15% is bound to albumin, and 5 to 10% remains free. The half-life of cortisol is about 90 min, based on normal

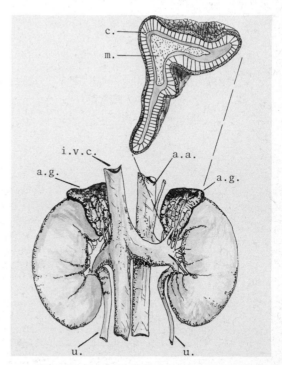

FIGURE 15.13. Anterior view of kidneys and their spatial relationship to the adrenal glands. (Note: Not all vasculature is shown.) An expanded portion of the adrenal gland is shown. Illustration key: a.a., abdominal aorta; a.g., adrenal gland (left and right); c., adrenal cortex; i.v.c., inferior vena cava; m., adrenal medulla; u., ureter(s) from kidneys.

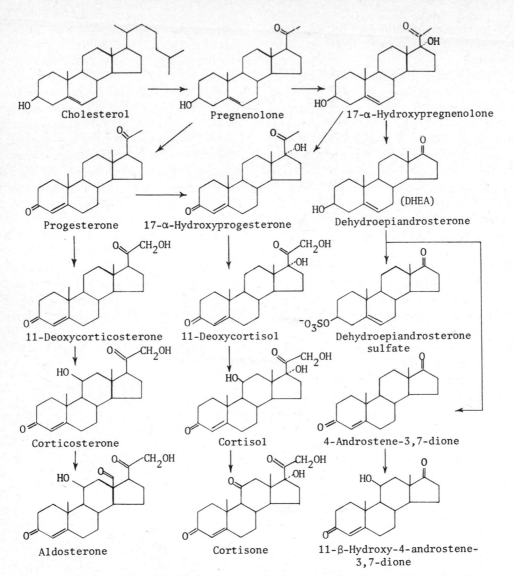

FIGURE 15.14. Three possible biosynthetic routes give rise to adrenal steroid hormones. Pregnenolone derived from cholesterol serves as a pivotal compound for all three pathways. By one synthesis route, progesterone (from pregnenolone) and 21-hydroxylation yields corticosterone and then aldosterone (occurs in glomerulosa zone of adrenal gland). In a second synthesis route 17α-hydroxyprogesterone is produced before 21- and 11β-hydroxylations to yield cortisol. Oxidation of the 11β-hydroxyl group of cortisol to an 11-ketone generates cortisone, although it is not the major secretory product of this reaction sequence. The third adrenal steroid synthesis pathway requires 17α-hydroxylation of pregnenolone followed by C_{17-20} desmolase activity, which cleaves the side chain. The 17-ketosteroid known as dehydroepiandrosterone (a C_{19} structure) is produced. This feeble androgenic steroid is further metabolized to 4-androstene-3,17-dione and 11β-hydroxy-4-androstene-3,17-dione. Dehydroepiandrosterone secretion, in the form of its conjugated sulfate, approaches 12–18 mg/day and 75–150 μg/dL is maintained as a plasma steroid. This action presumably ensures rapid availability of steroid precursors for estrogens and androgens in peripheral tissues.

degradative processes in the liver. The main cortisolic degradation products are biologically inert and conjugated with glucuronic acid prior to their urinary excretion.

Cortisol serves as a glucocorticoid because it counteracts insulin actions in maintaining normal glucose levels. Excess glucocorticoid effects can cause hyperglycemia, hyperlipidemia, nitrogen wasting caused by protein and amino acid catabolism, vitamin D suppression, muscular weakness, and psychiatric disturbances.

Aldosterone secretion is controlled by angiotensin II, which is produced from the renin–angiotensin system shown in Figure 15.15. The kidney produces the proteolytic enzyme called *renin* (not *rennin* used in milk

coagulation) in response to decreased blood pressure and other factors. Renin acts on the α_2-globular *angiotensin precursor* in the plasma to yield *angiotensin I*. As angiotensin I transits the pulmonary vascular system, this decapeptide is further cleaved to produce the octapeptide structure in *angiotensin II*. Angiotensin II functions as a powerful pressor agent, which mediates arterial constriction and the elevation of blood pressure. Angiotensin II also stimulates the secretion of aldosterone. This mineral corticoid promotes Na^+, Cl^-, and HCO_3^- absorption by the kidney. This action favors water retention and the maintenance or increase of blood volume. Aldosterone also controls the suppression of renin.

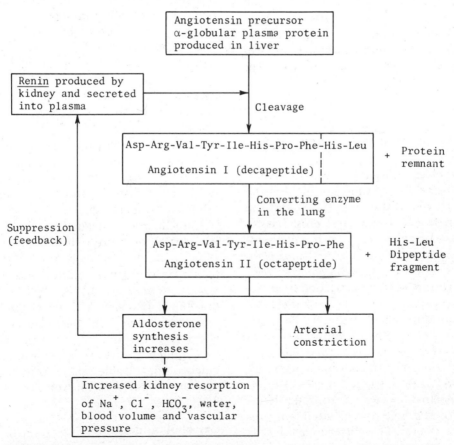

FIGURE 15.15. Regulation of blood volume and pressure by the renal–adrenal cycle of the renin–angiotensin–aldosterone system.

OTHER ADRENOCORTICAL HORMONES

The adrenal cortex also produces *hydrocortisone* (cortisol), which is primarily responsible for accelerating glucose metabolism but which also affects fat and protein metabolism. As a consequence of hydrocortisone activity, blood concentrations of amino acids, fats, and glucose may be increased. The clinical application of hydrocortisone serves as an anti-inflammatory agent for the treatment of rheumatoid arthritis and related diseases of articulated skeletal joints. Hydrocortisone causes some reduction of cartilage in joints and decreases the inflammatory activities of lymphocytes produced in the immune system.

Sex hormones including androgens and estrogens are also produced in small concentrations by adrenal glands. These hormones have important roles in sexual development, and both hormones are required in both sexes, albeit in different amounts (Figure 15.14).

ADRENAL MEDULLA HORMONES

The embryological development of the neural crest serves as the common origin for the sympathetic nervous system (SNS) and the adrenal medulla. The neural crest has an ectodermal origin and its primitive sympathogonia become transformed into *pheochromocytes* or into *sympathetic ganglion cells.* The neurotransmitter acetylcholine is secreted by preganglionic fibers of the SNS, whereas norepinephrine is secreted by postganglionic fibers. Pheochromocytes eventually differentiate into secretory cells after their migration into the adrenal cortex. By the age of 2.5 to 3.0 years, these differentiated pheochromocytes are thoroughly innervated by the SNS and its attendant preganglionic fibers. The differentiated cells contain a predominance of epinephrine (adrenalin), and about 12 to 16% contain norepinephrine (noradrenaline). According to current terminologies, both of these compounds are described as *catecholamines*.

Catecholamine synthesis in the CNS, SNS, adrenal medulla, and pheochromocytomas (catecholamine-producing tumors arising from SNS cells) occurs from tyrosine precursors (Figure 15.16). Catecholamine biosynthesis is rate-limited by the action of mitochondrial *tyrosine hydroxylase,* whereas the terminal *N-methyl transferase,* producing epinephrine, is induced by glucocorticoids. In fact there is significant evidence that glucocorticoids may control the whole epinephrine synthesis.

Dopamine, norepinephrine, and epinephrine serve as the *major* catecholamines of the CNS, SNS, and adrenal medulla (Figure 15.16), respectively. Some of the catecholamines exist as a free fraction, but the remainder are held as an inactive ATP complex. This equilibrium exerts a feedback control on tyrosine hydroxylase. Since catecholamines can have critical synaptic neurotransmitter functions, increases in synaptic nerve impulses result in accelerated tyrosine hydroxylase activity, enzyme synthesis, and increased releases of hormones.

Released catecholamines have two possible fates including (1) uptake and assimilation into sympathetic nerve endings and/or (2) catabolism by the kidney and liver followed by excretion.

Catecholamine metabolism is complex yet principally orchestrated by only two enzymes, namely, *catechol orthomethyltransferase* (COMT), found in the kidneys and liver, and *monoamine oxidases* (MAO), which are widely distributed among body tissues (Figure 15.16). A cursory survey of catecholamine metabolism shows that numerous intermediates can be conjugated and excreted as sulfates or glucuronides. Table 15.2 also indicates the primary origin sites for key catecholamine derivatives.

Catecholamine Effects

Catecholamines can serve as neurotransmitters: Stimuli originating from chemo-, thermo-, baro-, or thigmotactic receptors of

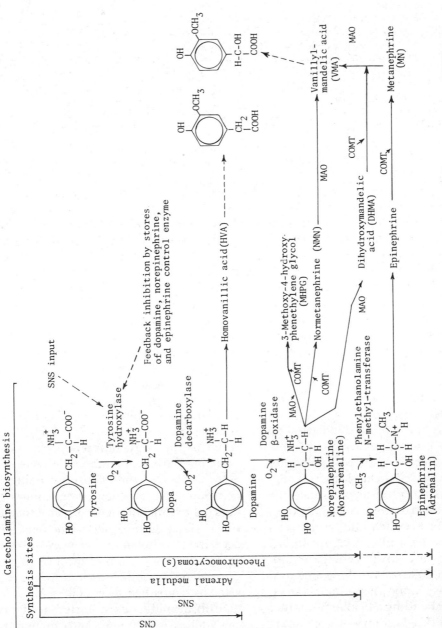

FIGURE 15.16. Catecholamine biosynthesis in the central nervous system (CNS), sympathetic nervous system (SNS), adrenal medulla, and a case of pheochromocytoma (see text for details). Abbreviations: Dopa, dihydroxyphenylalanine; MAO, monamine oxidase; COMT, catechol orthomethyltransferase. Note that catecholamines that serve as neurotransmitters are inactivated by COMT-assisted methylation of the 3-hydroxyl group of the catechol ring. S-Adenosyl methionine serves as the methyl donor in these reactions.

Table 15.2. Some Important Origin Sites for Excreted Catecholamines

Catecholamine metabolite	Possible catecholamine origin
Deaminated catecholamine metabolites	Nerve ending metabolism
Homovanillic acid (HVA)	Dopamine
3-Methoxy-hydroxymandelic acid (MHMA, vanillylmandelic acid, VMA) and 3-methoxy-4-hydroxyphenethylene glycol (MHPG)	All possible sources
Metanephrine (MN)	Epinephrine from adrenal medulla
Normetanephrine (NMN)	Neuronal release of epinephrine

the nervous system or other sites are transmitted by electrical impulses along neurons. Regardless of the stimulus, a sweeping depolarization event caused by an electrochemical flux of ions over a nerve membrane (Na^+/K^+ interchanges) is common to the operation of all nerve impulse transmissions. Nerve impulse transmission within a single neuron proceeds in this way until the axon of the neuron terminates in very fine axonal endings. Axonal endings are positioned close to but not in direct contact with (1) the dendrites or cell body of another neuron or (2) the superficial membrane of a muscle *or* gland. The space (0.02 µm) between an axonal ending and the next neuron or effector cell is called the *synaptic cleft*.

Since the electrochemical depolarization associated with a nerve impulse in one neuron cannot be directly imparted to another neuron or effector cell (in most cases), a neurotransmitter is required to propagate the electrical impulse from one cell to another. Neurotransmitters reside within tiny vesicles located in the axonal endings (Figure 15.17). According to many neurophysiological studies, a wave of depolarization along a neuron causes the presynaptic membrane, associated with axonal endings, to allow an influx of calcium ions (Ca^{2+}). Entry of Ca^{2+} spurs microtubules in the axonal endings to interact with the neurotransmitter-containing vesicles and direct them toward the presynaptic membrane (inside) where they then discharge their contents into the synaptic cleft.

The discharged neurotransmitter molecules diffuse across the cleft and interact with specific receptor sites on the postsynaptic membrane.

Assuming that some critical threshold interaction between the neurotransmitter molecules and receptor sites is achieved, a depolarization event will be initiated on this adjoining cell. For an abutting neuron this means that high concentrations of sodium ions (Na^+) outside the cell will diffuse into the cell, thereby initiating a sweep of electrochemical depolarization along the neuron. For an effector cell, the consequences of neurotransmitter actions are complex, with each cell or tissue responding in its own idiosyncratic but genetically dictated fashion.

Although many neurotransmitters exist, acetylcholine and norepinephrine (noradrenaline) are most commonly recognized. *Acetylcholine* is the neurotransmitter of neuromuscular junctions, at junctions of the internal organs, in the brain, and for other types of parasympathetic junctions. However, the catecholamine *norepinephrine* serves as the neurotransmitter of the SNS and those sites where smooth muscle cells are innervated by sympathetic nerve fibers. Many other neurotransmitters exist, including the catecholamines dopamine and epinephrine; indoleamine 5-hydroxytryptamine (serotonin); primary amines (histamine, octopamine, phenylethylamine, and phenethanolamine); polyamines (putrescine, spermine, and spermidine); as well as a number of amino

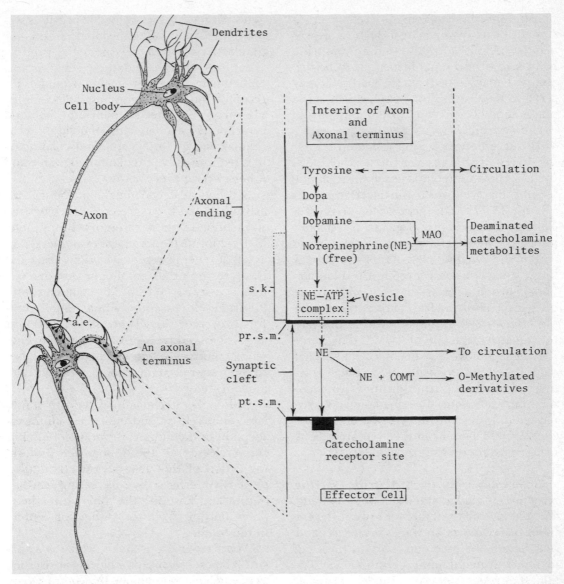

FIGURE 15.17. Axons of neurons synapse with an adjoining neuron by axonal endings that terminate in a number of axonal termini. Axonal endings contain vesicles that release neurotransmitters when stimulated by electrochemical depolarization phenomena. The schematic biochemical sequence shown to the right of the nerves details the biosynthesis, release, and simple metabolism of norepinephrine when it serves as a neurotransmitter. Key abbreviations for the diagram include a.e., axonal ending; pr.s.m., presynaptic membrane; pt.s.m., postsynaptic membrane. Although a neuron–neuron interaction is illustrated, similar synaptic mechanisms exist where neuromuscular junctions occur.

acids (glutamic and aspartic acids, glycine, β-alanine, γ-aminobutyric acid (GABA), taurine, and proline).

Since catecholamine activities are of concern here, it ought to be recognized that norepinephrine is produced from L-tyrosine *within* the sympathetic neuron (apart from adrenal gland syntheses of catecholamines). Tyrosine is acquired over the peripheral axonal membrane from the normal circulation. The key reactions illustrated in Figure 15.16 eventually yield norepinephrine. The norepinephrine can remain free for an instant after its synthesis or it can be stored as a more permanent ATP–norepinephrine complex. If norepinephrine fails to be discharged at any time from the neuron, it may serve as a substrate for MAO, which is the major catabolic enzyme for norepinephrine in the neuron.

If nerve depolarization results in norepinephrine discharge over the synaptic cleft, the catecholamine binds at the postsynaptic membrane receptor site (on the effector cell) (Figure 15.17). This action then triggers a characteristic biological response of a smooth muscle. The discharged norepinephrine within the synaptic cleft has *two* imminent fates. It may be recaptured by the axon terminals of the neuron, or if norepinephrine does successfully interact at receptor sites on the effector cell, it may serve as a substrate for the enzyme action of COMT. This latter possibility serves as a major extraneuronal degradative mechanism. Only a small fraction of norepinephrine diffuses from the synaptic cleft into the circulation as free plasma norepinephrine. When it does, however, COMT actions in the liver precede its urinary excretion as normetanephrine.

Catecholamines interact with specific receptors: Catecholamines of the SNS and adrenal medulla interact with two different types of receptors known as α-*receptors* and β-*receptors*. The α-receptors are intrinsically involved in maintaining peripheral circulation; they reside in mucosal, subdermal, renal, and splanchnic vascular beds. Here, catecholamine–receptor interactions participate in vasoconstriction. Interactions of β-receptors with catecholamine, on the other hand increase cardiac activity, promote bronchodilation and vasodilation, and retard gastrointestinal tract mobility. These receptors occur in skeletal, cardiac, and smooth muscles and adipose tissues.

General observations show that epinephrine stimulates both α- and β-receptors with β-receptors predominating. The consequences of epinephrine on hepatic glyco-genolysis have been outlined in previous sections, where it was explained that a β-receptor (β-adrenergic receptor) spurs the production of cAMP as a second messenger. Norepinephrine, however, mainly stimulates α receptors on cell membranes. The complex interplays of catecholamines on α- and β-receptors regulate the mobilization of cellular energy for essential muscular activities (e.g., vascular circulation, digestive and other visceral processes).

Synthetic catecholamine analogues can activate α- and β-type receptors or block the stimulation of these receptors. Propranolol hydrochloride (Inderal), for example, serves as a β-adrenergic receptor blocking drug and possesses no other autonomic nervous system activity. It specifically competes with β-adrenergic receptor stimulating agents for β-receptor sites. If access to β-receptor sites is blocked by propranolol, the chronotropic, inotropic, and vasodilator responses to β-adrenergic stimulation are proportionately decreased.

Although catecholamines tend to stimulate renin release and enhance sodium resorption in the kidney, propranolol inhibits renin release from the kidneys. The β-blocking action of this drug also exerts antihypertensive effects by decreasing cardiac output and lowering the tonic sympathetic nerve outflow from the vasomotor centers in the brain.

Apart from propranolol, which is a specific β blocker, α-receptor blockers exist such as phentolamine (Regitine). In a contrary vein, hormones of the SNS and adrenal medulla can have their specific receptor site effects enhanced by "agonistic" compounds. These include methoxamine for α-receptors and isoproterenol for β-receptors.

STEROID, SEX, AND GONADAL HORMONES

Androgens and estrogens constitute the major sex hormones of humans. The principal androgenic or male sex hormone is testosterone, whereas the primary estrogenic or female hormone is 17β-estradiol.

Testosterone is synthesized from *pregnenolone* by elimination of the C-17 side chain. The 17β-estradiol is formed by the oxidative removal of C-19 on testosterone and subsequent aromatization of the A ring. All of the estrogenically active steroid hormones exhibit this aromatic ring.

Since both sexes produce androgens and estrogens, the cellular differentiation of sexually distinct tissues as well as the maintenance of normal sexual characteristics in humans depend on a finely tuned interplay of sex hormones. For example, ~6.0 to 10.0 mg of testosterone is produced daily in males, whereas females produce ~0.45 mg. This distinct difference in androgen production encourages male over female secondary sex characteristics.

Interstitial cells (Leydig cells) located in the seminiferous tubules of the testes secrete androgenic steroids, particularly testosterone. Androgen biosynthesis in the Leydig cells begins with the biosynthesis of cholesterol from acetyl-CoA, which originates from the oxidation of fatty acids. This is quite contrary to minor amounts of androgens formed in the adrenal cortex, which employs blood cholesterol as a steroid precursor.

Synthesized testosterone exits the testes via the testicular veins and the lymphatic vesicles. Released testosterone is probably converted to *dihydrotestosterone* (DHT) before it can exert any biological activity. Some DHT is produced directly in the testes, but most DHT formation occurs in the cytoplasm of target cells or in the circulation (Figure 15.18)

Luteinizing hormone (LH) from the anterior pituitary stimulates Leydig cell activity. In spite of this action, LH produced in males is identical to that formed in females. The continued release of LH is in turn suppressed by testosterone.

Most testosterone is present in plasma as a bound complex with *sex hormone-binding globulin* (SHBG) and albumin, while ~2% remains free. Testosterone excretion occurs largely by hepatic reactions, where it is con-

jugated with sulfate or glucuronic acid (similar to the adrenal corticoids). The terminal excreted product in the urine is a 17-oxosteroid. Small amounts of *estradiol* are also produced from testosterone by the Serotoli cells of the seminiferous tubules, but its function in males remains to be firmly established.

The principal female sex hormones including *estrogens* and *progestins* are produced in the ovary. Estrogens are secreted by follicles of the ovary, whereas the corpus luteum, which develops within ruptured ovarian follicles, secrete progestins. Outside of pregnancy, the ovaries produce most of the estrogens, but during pregnancy the placenta synthesizes notable concentrations of these steroids.

Mammary glands, the uterus, and many other tissues throughout the body serve as target cells for estrogenic hormones. Estrogens are responsible for developing secondary female sexual characteristics and for cyclical changes in the vaginal epithelium and the endothelium of the uterus. Estrogens also act on the growing ends of the long bones to terminate their growth, enhance fat deposition on the female body, and produce smoother skin than that of the male. A number of natural estrogenic steroids are recognized, including *estradiol* (E_2), *estrone* (E_1), and the metabolic by-product of these steroids known as *estriol* (E_3). The ovary secretes E_1 and E_2, whereas the placenta secretes all three steroids. Ovarian production of androgens, particularly testosterone, occurs in the hilus of the ovary, but the productivity is only about 5% of that demonstrated in the testes.

Ovarian hormone secretion is controlled by the anterior pituitary gonadotropins, but significant fluctuations occur during the menstrual cycle.

Estradiol secreted into the plasma is largely carried by SHBG with minor amounts of albumin binding. Estradiol metabolism by liver cells produces estriol, a relatively inactive estrogen. Estriol is subsequently conjugated with glucuronic acid before its urinary ex-

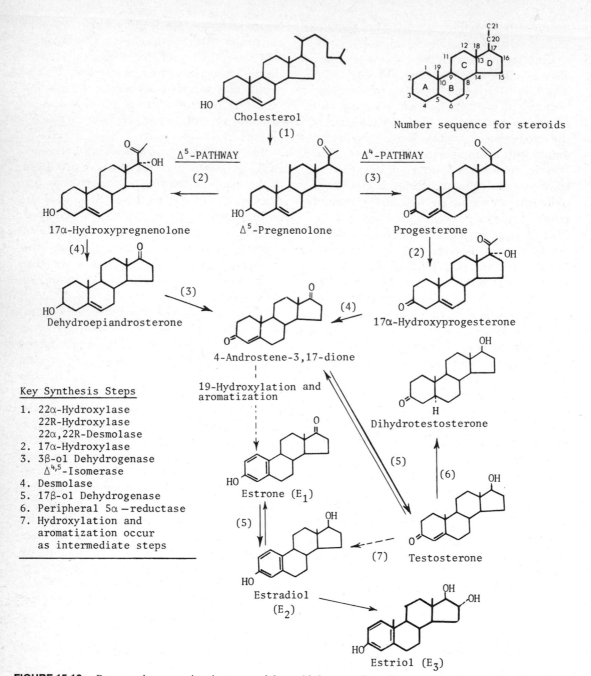

FIGURE 15.18. Pregnenolone can give rise to gonadal steroids by a number of important pathways. The Δ^5 pathway yields androstenedione via a dehydroepiandrosterone intermediate, whereas 17α-hydroxyprogesterone can also serve as an intermediate to androstenedione formation in the Δ^4 pathway.

cretion. Progesterone is converted to pregnanediol before it too is excreted in the urine as a glucuronide.

Gynecological Hormone Interactions

The menstrual cycle and lactation depend on complex interactions among female sex hormones. The menstrual cycle and its attendant hormonal stimuli can be subdivided into two distinct operative phases that involve the uterine endometrium. These two phases include a (1) follicular phase and a (2) luteal phase (Figure 15.19).

During the follicular phase, progesterone circulates in the plasma at a low concentration in the wake of corpus luteum regression in the *previous cycle*. This low progesterone level promotes the hypothalamic release of gonadotropin-releasing hormone (GnRH), which in turn spurs on the production of follicle-stimulating hormone (FSH). This latter hormone is critical for the growth of a mature graafian follicle in the ovary, which ultimately yields a mature ovum.

The midpoint of the menstrual cycle begins with a surge in LH, which mediates the rupture of the graafian follicle and liberates a free mature ovum. Follicular cells then undergo luteinization to form the corpus luteum. This small yellow body (i.e., *corpus*, body; *luteum*, yellow) develops within the ruptured ovarian follicle and secretes progesterone. Progesterone concentration in the plasma rises during the course of the luteal phase until the twentieth day of the menstrual cycle (Figure 15.19).

If fertilization of the ovum and pregnancy follow the release of the ovum (ovulation), plasma progesterone concentrations will continue to rise. At the twelfth week of pregnancy, placental progesterone production exceeds ovarian secretion. High concentrations of placental estrogens, including 250 mg of progesterone per day, exert a feedback effect on the hypothalamus. This action prohibits another LH surge followed by additional futile ovulations.

Based on the assumption that fertilization after ovulation does not occur, progesterone plasma levels decrease from the twentieth day toward the twenty-eighth day of the menstrual cycle. The corpus luteum atrophies during this period, only to be followed by another menstrual cycle after the luteal phase has been terminated.

Progesterone hormonal effects generally counteract estrogenic effects. That is, progesterone inhibits the proliferation of vaginal, cervical, and fallopian epithelial linings as well as the uterine endometrium. At the same time, progesterone also blocks smooth muscle contraction and augments the actions of estradiol on endometrial target cells. This estrogen is designed to convert these cells from a proliferative state to a secretory existence. A functional transformation of this type is necessary for the successful implantation of a fertilized ovum.

Steroidogenesis during pregnancy becomes a very complicated affair since the placenta, fetus, and mother all participate. The placenta and fetus exist as *one complete steroidogenic unit*. Both participants in this steroidogenic unit (fetus and placenta) have complementary enzyme interactions that lead to steroid synthesis, yet neither the fetus nor the placenta can singularly produce a full complement of steroids as can the mother. In particular, *only* the placenta has the necessary enzymes to convert pregnenolone to progesterone, and *only* the fetus can synthesize cholesterol from acetate.

As indicated in the preceding section, steroidogenesis occurs mainly in the ovaries of nonpregnant females as well as the adrenal glands. These steroids include progestins, androgens, and estrogens. The ovarian biosynthesis of estrogens may develop from pregnenolone (Δ^5-3β-ketone pathway) and/or progesterone (Δ^4-3-ketone pathway) (Figure 15.18). The corpus luteum and graafian follicle cells use the Δ^4 and Δ^5 pathways, respectively.

Steroids produced in females are secreted into the plasma; 37% of estradiol is bound to SHBG (as in the case of male testosterone) and 61% to albumin, whereas about 2 to 3%

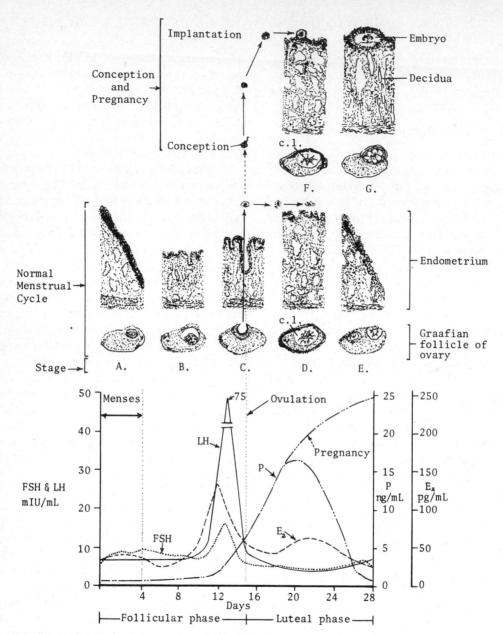

FIGURE 15.19. Dynamic interactions of female hormones and the menstrual cycle. Note that the plasma progesterone level remains high during the luteal phase and decreases if fertilization and implantation do not occur. Notable abbreviations include FSH, follicle-stimulating hormone (follitropin); LH, luteinizing hormone (luteotropin); E_2, estradiol; P, progesterone. In the normal menstrual cycle, the *menstrual phase* (A) is initiated when the uterine endometrium is shed and a graafian follicle begins to grow in the ovary. (B) Follicular phase—endometrium is reestablished and grows thicker while the follicle continues to enlarge. (C) Ovulation—endometrium grows thicker and graafian follicle ruptures thereby releasing ovum. (D) Luteal phase—endometrium develops further and provides a possible site for implantation of a fertilized ovum; however, (E) if fertilization does not occur, the uterine lining is shed (menses) at the end of the menstrual cycle and the whole sequence is repeated. A corpus luteum (c.l.) also begins to develop in the follicle (D) after ovulation and assuming that fertilization does not occur, it atrophies (E) before a new graafian follicle develops (in A again). If fertilization of the ovum occurs (F), the impregnated ovum becomes implanted in the uterine lining. The corpus luteum (c.l.) of pregnancy develops (F → G); and the endometrium (G) is transformed into decidua while the ova becomes increasingly embedded in the uterine lining.

of the estradiol remains free. The SHBG is produced in the hepatic tissues by both males and females, although its synthesis is inhibited by androgens and stimulated by estrogens. Therefore, adult females have nearly twice the SHBG plasma concentration level of males. Although SHBG concentrations of females are twice those of males, estrogen binding to SHBG is 30% of that observed for 5α-dihydrotestosterone. Since plasma clearance of sex steroids is inversely related to their relative binding on SHBG, women generally show a lower clearance rate than men.

Progesterone transit in plasma occurs as a protein complex involving albumin, orosumucoid, and *corticosteroid-binding globulin* (CBG, alternatively called *transcortin*). The CBG has the greatest affinity for progesterone, but most CBG binding sites suitable for progesterone are occupied by cortisol under physiological conditions. Transport of CBG-bound progesterone to the liver occurs as a first step toward its urinary excretion.

Effective liver catabolism of progesterone (66% of progesterone) ensures a short biochemical half-life for the hormone ($t_{1/2}$ = several minutes) and an expeditious formation of pregnanediol. This diol is also commonly conjugated with glucuronic acid or sulfates to produce water-soluble steroid derivatives. The major diol derivative is 5β-preganane-3α-diol, although at least eight other isomeric forms do exist.

As seen earlier in Figure 15.6, steroid hormones including progesterone affect target cells by forming a hormone–receptor complex that stimulates synthesis of DNA-directed mRNA. This mRNA can direct new sex-linked protein syntheses plus ensure continuous hormone synthesis and translocation. This stimulatory action is not shared by all steroid hormones. Indeed, some steroids may repress nucleic acid and protein syntheses by curious mechanisms. As an example, consider the ability of progesterone to inhibit the breast synthesis of the milk protein lactalbumin. This protein is otherwise synthetically induced by prolactin (PRL)

and/or hormones generated by endometrial carcinomas.

Lactogenic Hormone Interactions

Development of the mammary glands and the proliferation of their glandular components is prompted by estradiol. Developmental actions of this hormone are also augmented by growth hormone (GH) and insulin. Postpuberal development of these glands depends on the action of progesterone, and during pregnancy both estradiol and progesterone (from placenta) contribute to the progressive secretory development of the glands. During this period, the alveoli and ducts of the glands increase their *potential* for milk production, yet a high titer of placental progesterone inhibits milk production. Not until parturition, when progesterone levels markedly decrease, will milk production or lactation begin. Moreover, lactation is not possible until the placenta is discharged after parturition. At this point, placental gonadotropins, which formerly checked pituitary gonadotropic actions, cease to function. This is followed by pituitary secretion of prolactin (PRL) in large amounts, which stimulates milk production.

Neurochemical and sensory stimuli generated in the nipple during suckling direct the hypothalamus to produce *prolactin-releasing hormone* (PRH) or a *lactogenic releasing factor* (LTRF) (Figure 15.20). As the influence of this hormone is exerted on the pituitary gland, PRL production is increased. Thus, regular lactation of the gland coincides with continuous PRL secretion by the pituitary gland. Correspondingly, a lack of milk elimination from the gland leads to decreased PRL formation and a waning of milk productivity. The exact roles of ACTH, GH, and TSH in the growth of the mammary gland and milk secretion are not understood, but they are important factors in the overall metabolism of these glands.

Milk stores in the mammary gland are expelled by the contraction of myoepithelial cells, which squeeze the mammary alveoli.

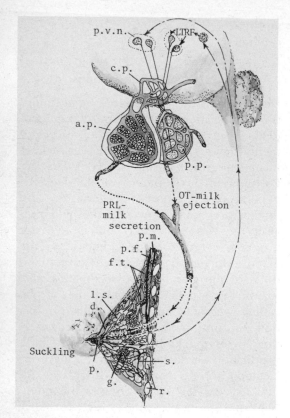

FIGURE 15.20. Direction of hormonal controls that stimulate lactation and milk ejection from the mammary gland. Hormonal and psychological factors complicate the control of the overall mechanism here but prolactin (PRL) promotes milk secretion while oxytocin (OT) stimulates milk ejection. A neural feedback response from suckling (1) promotes neuronal lactogenic releasing factor (LTRF) and (2) affects the paraventricular nucleus (p.v.n.) in the hypothalamus. The LTRF produced affects the capillary plexus (c.p.) of the median eminence of produce PRL. The p.v.n. response is geared to producing OT. Production of PRL occurs in the anterior pituitary (a.p.) while neuronal OT produced is transported from secretory neurons into the posterior pituitary (p.p.) Blood transport of PRL and OT to the mammary glands then elicits lactation. Key features of the mammary gland include d., duct(s); f.t., fatty tissues; g., glands; l.s., lactiferous sinuses; p., papilla terminating in nipple innervated by sensory nerves; p.f., pectoral fascia; p.m., pectoralis major; r., ribs; s., stroma.

This milk ejection reflex is largely controlled by a neurohormonal reflex, which is initiated after an infant is put to the breast. The hormone *oxytocin* also enters into the control of lactation and milk ejection. This hormone instigates myoepithelial cell contraction thereby increasing intramammary pressure along with eventual milk discharge.

SOME MISCELLANEOUS HORMONE ACTIONS

Aside from the hormonal actions presented in earlier sections, many other hormones are known. Their miniscule concentration levels in the body, poor assay methods, short biological half-lives, and many other practical factors account for the difficulties in understanding their biochemical roles as well as their physiological effects.

One of the anterior pituitary secretion products known as *proopiocortin* has deserved special attention for its precursor role in producing several other hormones. The name proopiocortin is used for this polypeptide (Figure 15.21) (~260 amino acid residues) because it yields opiate-like hormones (*endorphins*) and two *lipotropins*. The lipotropins control, or at least affect, the catabolism of lipids for their energy content, while endorphins exert analgesic effects and pain relief similar to the opiate drugs. In addition, the N-terminal pentapeptide of endorphin yields an *enkephalin* that displays distinct binding interactions with opiate receptors in the brain.

Many of the endorphins and enkephalins discovered in the past decade seem to act as neurotransmitters or chemical messengers in specific nerve pathways of the brain. These neuropathways often process information dealing with pain, emotional behavior, anxiety, and other sensations similarly affected by opiates. Several commonly recognized endorphins and enkephalins that aim at specific opiate receptors in the brain to produce effects appear in Figure 15.22.

Proopiocortin also produces adrenocorticotropic hormone (ACTH) and melanocyte-stimulating hormone (MSH) when acted upon by peptidase. The ACTH stimulates adrenal cortex growth and the conversion of

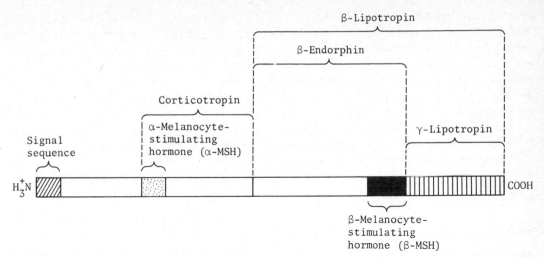

FIGURE 15.21. Conversion of proopiocortin (260 amino acid residues) yields a number of anterior pituitary hormones. Peptide cleavage at the sites shown produces two types of melanocyte-stimulating hormone (α- and β-MSH), β- and γ-lipotropins, corticotropin, and β-endorphin.

cholesterol to pregnenolone in steroidogenesis, whereas MSH stimulates melanin production in skin.

Many other hormones exist that have roles in reproductive biochemistry but whose exact functions are unknown. Some hormones such as *human chorionic gonadotropic hormones* (HCG) have functions similar to LH from the pituitary gland. *Human placental lactogen* (HPL) is a carbohydrate-deficient polypeptide (191 amino acid residues) that affects normal growth (similar to GH) and lactogenesis.

Melatonin or pineal hormone is produced by a gland that may have evolutionary origins as a third eye. A tryptophan derivative known as serotonin, which can also serve as an important neurotransmitter, undergoes acetylation and methylation reactions in this gland to form melatonin. The methylation reaction depends on the retinal photoreception and neuroactivation (over the inferior accessory optic tract) of hydroxy-*O*-methyl transferase. Melatonin produced by this photoactivated reaction has effects on ovarian functions by unknown mechanisms.

Prostaglandins are also important hormone-active compounds. Their biochemistry and origins have been presented elsewhere in the text.

SOME DISORDERS TIED TO HORMONE ACTIONS

Aberrations in the production of a hormone by a gland, defective interactions between a hormone and its receptor site on target cells, or the failure of a target cell to express a

H_3^+N-Tyr—Gly—Gly—Phe—Met—COO⁻   *Methionine enkephalin*
H_3^+N-Tyr—Gly—Gly—Phe—Leu—COO⁻   *Leucine enkephalin*
H_3^+N-Tyr—Gly—Gly—Phe—Leu—Met—Thr—Ser—Glu—Lys—Ser—Gln—Thr—Pro ⎤
 [His—Ala—Asn—Lys—Val—Ile—Ala—Asn—Lys—Phe—Leu—Thr—Val—Leu ⎦
  Lys—Lys—Gly—Gln—COO⁻   *β-Endorphin*

FIGURE 15.22. Comparative peptide structures for two types of enkephalins conventionally termed *met*-enkephalin and *leu*-enkephalin, plus the structure for β-endorphin.

hormone initiated function are at the root of many clinical ailments. Apart from the clinical aspects of hormonal dysfunctions, many endocrine based maladies involve a nutritional component, or must be surveyed with regard to nutritional factors. Some endocrine disorders that may involve a hormone–nutrition interaction are outlined in the following paragraphs. For a more detailed perspective, the reader is encouraged to consult one of the authoritative textbooks dealing with endocrinology.

Thyroid Disorders

Hyperthyroidism. An excess of thyroid hormone production by overactive thyroid glands results in hyperthyroidism. The onset of this problem occurs in early adult life with symptoms of nervousness, sweating, palpitation, weight loss, fatigue, tremor, diarrhea, dyspnea, and a general increase in the BMR. In some individuals an exophthalmic condition results from the development of increased tissue bulk in the retroorbital space.

Treatment methods are varied and almost any case of successful treatment is met with certain trade-offs and consequences. Four approaches to treatment can be used including (1) β-adrenergic drug treatment, (2) thiocarbamide administration, (3) radioiodine treatment, or (4) surgery.

The β-adrenergic drug treatment rapidly controls palpitations, sweating, and tremor, whereas the thiocarbamides (or earlier thiouracil agents) suppress several stages of thyroid hormone synthesis. Adverse reactions to thiocarbamides may be countered by the preferred use of potassium perchlorate (Figure 15.9), which blocks the iodine trap. Since the thyroid actively traps and retains iodine, 2 to 6 mCi ^{131}I can be administered to patients having a hyperthyroid condition. The accumulated radioactive iodine destroys the thyroid tissue in a dose-dependent reaction. This treatment is not risky from the point of surgical intervention, but the nonspecific control of radioactive iodine or its careless use inevitably causes hypothyroidism in many cases. Surgical intervention, as an alternative to radioactive iodine treatment, offers a more specific treatment route since defined amounts of the gland can be eliminated (approximately seven-eighths). Minimization of laryngeal nerve damage and preservation of the parathyroid glands are major considerations in this procedure. Because drug treatments fail to solve hyperthyroidism in a permanent way, other treatment methods are often recommended.

Hypothyroidism. Underactive thyroid glands, which produce insufficient amounts of thyroid hormones, result in a low BMR. This conditioin is more common in elderly subjects than younger ones. The classic hypothyroid symptoms often involve weakness, lethargy, graded degrees of obesity, constipation, coarseness of the skin, swollen eyelids, and progressive facial changes. Treatment often prescribed for this problem involves the administration of oral L-thyroxine (sodium) tablets.

Exclusive of hypothyroidism stemming from hyperthyroid treatment, a genetic abnormality is attributed to the *in utero* development of this malady. The failure of thyroid hormones to cross the placenta can cause cretinism characterized by physical and mental retardation.

Goiter. Iodine deficiency can lead to the obvious enlargement of the thyroid gland to produce a goiter condition. The increased size of the gland is designed to increase the extraction efficiency of available iodine from the circulation. Goiter formation can be prevented by consuming seafoods that are obviously rich in halogen salts including iodine, or 0.5 g of potassium iodide per kilogram of sodium chloride will allay the problem of dietary iodine insufficiency.

Adrenal Disorders

Cushing's syndrome. An excessive secretion of adrenocortical steroids, especially cortisol and possibly aldosterone, causes

Cushing's syndrome. This problem occurs more in women than men and is indicated by a variety of symptoms such as weight gain, overt obesity, amenorrhoea, infertility, muscular weakness, osteoporosis, curvature of the spine, hypertension, and diabetes mellitus. Treatment involves the removal of any tumors responsible for the problem or the use of metyrapone or aminoglutethimide, which suppress cortisol secretion. Fatalities from unchecked Cushing's disease occur in a few years from its outset owing to cardiovascular disease or sepsis.

Addison's disease. A hypofunction of the adrenal cortex leads to decreased production of cortisol and aldosterone and increased levels of ACTH. Some symptoms include muscular weakness, anorexia, loss of Na^+, retention of K^+, and hypotension, nausea, weight loss, depression, and altered pigmentation. Slate-gray pigmented spots on cheek mucosa, gums, and lips along with bronze-colored skin are not uncommonly observed. The etiology of Addison's disease may involve viral infections or autoimmune reactions of the adrenal glands. Since adrenal function is unlikely to return to normal in Addison's disease, lifelong replacement of adrenal corticoids is required. Oral cortisone or prednisone with small amounts of 9α-fluorocortisol are used for this purpose.

Conn's syndrome. The autonomous hypersecretion of aldosterone causes Conn's syndrome. Unprovoked hypokalaemia (depletion of circulating K^+ in blood), low renin content in the plasma, and hypertension are typical features of this disorder. Short of some surgical intervention, spironolactone or a related aldosterone antagonist can reverse all of the characteristics of aldosteronism.

Osteoporosis (Porosity of Bones)

The loss of bone mass can be caused by imbalances in the rate of bone formation and resorption. If resorption of the calcareous mineral elements in bone exceeds their deposition, osteoporosis begins to develop. When 30 to 50% of the bone minerals have disappeared, the void space once occupied by mineralized bone trabeculae become filled with fibrous and fatty marrow. This results in a bone structure having poor tensile strength. Moreover, evidence of bone demineralization, fatty tissue deposition, and incipient osteoporotic conditions readily appear from radiological studies of afflicted bones.

Osteoporotic disorders occur more in Anglo-Saxon women after menopause than women having a Southern European heredity. Older males (60+ years) can also be afflicted with this disease but not in numbers comparable to the female population.

Osteoporosis is a complex disorder triggered by a variety of causes and displaying many associated symptoms. In some cases osteoporosis may result from increased glucocorticoid activity or corticoids administered for therapeutic objectives. Classic symptoms arising from osteoporosis reflect backache due to vertebral body collapse, skeletal deformities such as kyphosis (e.g., humpback, hunchback, Pott's disease), height loss, and easily fractured wrist and hip bones. Prolonged nutritional disorders involving protein and caloric malnutrition, chronic alcoholism, intestinal absorption problems related to protein, calcium, and other minerals, immobilization, and homocystinuria can all promote the disorder. The multiple etiologies and pathophysiological effects of osteoporosis require different therapeutic approaches to this problem.

In postmenopausal or senile osteoporosis, in which circumstances the causes of these disorders are uncertain, therapies may be quite empirical. Physical therapy, bed rest, calcium intake of 1.5 to 2.0 g/day, exposure to sunlight, and vitamin D ingestion are all recommended. Fluoride intake may also counteract osteoporosis since it is believed to displace hydroxyl groups in crystals of calcium hydroxyapatite in bone structure. The fluoroapatite crystals that result presumably resist demineralization and bone resorption phenomena.

Osteomalacia and Rickets

Osteomalacia is characterized by a softening of the bones. In its advanced stages, bones can become so soft that they are brittle and flexible to the point that skeletal deformities occur. Osteomalacia occurs mainly in adults and chiefly in adult women. Rheumatic pains in limbs, spine, thorax, and pelvis are not uncommon complaints along with anemia and progressive muscular weakness.

Rickets reflects a juvenile bone disorder reminiscent of osteomalacia. In this case, however, deficient lime salt deposition in developing cartilage and new bones is at the root of the disorder. The inadequate deposition of minerals leads to skeletal deformities of individual bones as well as of the whole skeletal structure. Defects in vitamin D metabolism or its overt deficiency are linked to rickets since intestinal absorption of calcium and phosphorous plus phosphorous resorption from the renal tubules are inadequate for normal health. Enlarged liver and spleen, delayed dentition and the eruption of poorly formed teeth, *craniotabies* or paper-thin skull bones, and other bone deformities are associated with the disorder.

Ricket-induced softness and pliability of bones can also lead to contracted pelvic bones, bow legs, and enlargement of bone ends. A chemical analysis of bones reveals that their inorganic mineral content decreases while water and organic substances increase in their proportions. However, the ratio of calcium to phosphorous remains constant, indicating that bone salt deposition is normal but wholly insufficient. Serum studies also have some significance in diagnosing rickets. Levels of calcium in particular are often normal but phosphate is lower than normal. There may also be a demonstrated increase in *serum alkaline phosphatase* in children that can serve as a biochemical index of rickets; however, the validity of this parameter wanes during puberty, when levels of this enzyme become normally high.

Osteomalacia and rickets both reflect an abnormal disposition of calcium residence in bone structures. Vitamin D deficiency is often an important predisposing requirement for these disorders because of its role in the assimilation of dietary calcium and calcium deposition into bone structures. Many factors can lead to abnormal vitamin D activity such as insufficient dietary sources of the vitamin, inadequate dermal production of 7-dehydrocholesterol, malabsorption of the vitamin, impaired activation of 1,25-DHCC, or overactive catabolism of the vitamin in the liver.

Classical uncomplicated cases of rickets and osteomalacia can be cured by administration of at least 400 I.U. of vitamin D daily coupled with exposure to sunlight or the ultraviolet component of artificial light.

Vitamin D-resistant rickets, a familial disorder linked to renal tubular dysfunctions or an inability to convert CC to 25-HCC or 1,25-DHCC, requires more intensive treatment. Here large doses of vitamin D precursors are given to ensure a miniscule formation of active vitamin D.

In cases of chronic uremia and patients on prolonged hemodialysis, a condition known as *renal osteodystrophy* occurs. This disorder is reminiscent of osteomalacia, osteoporosis, osteosclerosis, and osteitis fibrosa resulting from secondary hypothyroidism. Treatment of this condition requires a high-calcium diet, reduction of phosphate, and an increase in vitamin D intake.

In all cases of vitamin D-regulated therapies, care must be taken to avoid hypervitaminosis D because excess calcium assimilation (hypercalcemia) and deposition of calcium in bones or arteries can cause serious consequences. Furthermore, since plasma levels of calcium may be tied to some cases of hypertension, this side effect of vitamin D therapy (as well as acceptable vitamin D additions to milk) must be carefully monitored.

Anorexia Nervosa

The hormonal disorders of anorexia nervosa are thought to rest within the hypothalamic regions of the brain. Anorexia nervosa is a self-imposed type of starvation that occurs largely among young women between 12 and 27+ years of age. Women of the upper and middle classes are often involved in this disorder, but recent surveys suggest that women of all ages and economic backgrounds may display anorexic tendencies. The actual number of women who suffer from eating disorders becomes remarkably higher from statistical perspectives with the inclusion of a related syndrome known as bulimia. Bulimia involves self-imposed gorging of food followed by a self-induced vomiting and/or the habitual use of laxatives.

An anorexic subject generally displays biochemical abnormalities such as higher than normal plasma corticoid concentrations; decreases in thyroxine, LH, and FSH levels; and an unpredictable menstrual cycle. Other symptoms include bradycardia, lowered blood pressure, hyperactivity, cold sensitivity, and edema in severe cases.

Many current theories on the subject of anorexia nervosa claim that patients have problems in establishing personal independence within their families. Also, anorexic females have a strong desire to emulate the appearance of "models" since they live in a society where the "thin is 'in'" cliché is practiced. The prompt recognition and treatment of this problem by psychiatric or medical professionals is the best route for avoiding hospitalization and other more severe long-term health consequences including death.

Obesity

Obesity is the result of an imbalance between food intake and energy actually expended by the body. In most cases obesity refers to body weights that are 20 to 30% over the average for a given sex, age, and height.

There are two classifications for obese conditions, namely, (1) *exogenous obesity*, caused by excessive food consumption, and (2) *endogenous obesity*, which results from some type of faulty body metabolism.

Apart from the clear-cut overindulgence of eating foods, endocrine-linked disorders are the most important causes of endogenous obesity. Endocrine disorders can reflect hypothyroid conditions in which a decreased metabolic rate (BMR) fails to provide an energy output to balance caloric intake. The more frequent cause of endocrine obesity is tied to adrenal hyperfunctions (Cushing's syndrome), testicular and ovarian hypofunction, hypothyroidism, and hypopituitarism. In spite of the common occurrence of obesity in these specific instances, it may not be an essential feature.

Polemic discussions regarding obesity and neurological influences have raged since the 1940s, when Hetherington and Ranson (1940) realized that bilateral lesions in the region of the ventromedial hypothalamus (VMH) caused rat obesity. Rats subjected to VMH lesions appeared to exhibit hyperphagia (excessive overeating) and obesity as opposed to obesity caused by alterations in neuroendocrine controls of metabolism. Similar observations have been made in other rodents, primates, and man.

Further work dealing with VMH lesions have also indicated that insulin responses to food increase after VMH lesions are induced. This may also be accompanied by a decline in GH. Since the counterresults of high insulin and low GH promote fat deposition, neuroendocrine effects of the VMH lesions may experimentally reflect normal physiological interactions among neuroendocrine stimuli, hyperphagia, and obesity.

Based on studies of normal males, liver and muscle glycogen stores account for ~400 and ~200 kcal, respectively, while fat makes up a potential ~100,000-kcal energy reserve. According to this compositional profile, where fat and carbohydrate account for the major sources of cellular metabolic energy, it is not unreasonable to assume that both fats and carbohydrates control normophagic (normal feeding) activities and energy balance. Two physiological hypotheses have been developed on this premise, which are known as the *glucostatic* and *lipostatic* hypotheses, respectively.

The *glucostatic theory* suggests that glucose metabolism is linked to feeding activity in three distinct ways:

1. Glucose utilization by the VMH and other brain tissues controls food intake and reflects the status of glucose metabolism.

2. Feeding operates under the auspices of glucostatic controls on a meal-to-meal cycle.

3. An inverse relationship exists between the rate of glucose use and food consumption (i.e., high glucose metabolism encourages feeding whereas low glucose use suppresses feeding).

Lipostatic theories on feeding suggest that the amount of depot fat relative to body weight is constant in adult mammals (but not geriatric subjects). Moreover, experimental tests and observations suggest to some investigators that a critical mass of fat inhibits adult feeding. Other investigators claim that the status of enlarged fat cells serves as a basis for lipostatic inhibition and control of feeding.

The glucostatic and lipostatic theories serve as useful explanations of feeding behavior and may shed some understanding on the problem of obesity. However, the crucial metabolic and humoral agents responsible for integrating body weight, food intake, and universal aspects of metabolism are still uncertain.

If gluco- and lipostatic mechanisms are engaged in the regulation of food intake, it is clear that humeroneural interactions must exist. Iontophoretic studies on hypothalamic neurons have shown that neuronal firing rates in the VMH region do show chemosensitivity. Firing rates are especially altered after iontophoretic exposure to fatty acids, glucose, glucose plus insulin, and insulin alone. Whether or not normal neurohormonal controls for feeding and chemosensitivity aberrations leading to obesity can be pegged to brain functions at this level is not known.

The problem of obesity is further complicated by recent studies suggesting that *hormone imbalances* contribute to obesity. These hormone imbalances are different in men and women, which lends credence to the belief that obesity may have different endocrinological origins in the two sexes.

Studies have shown that women retain intravenously administered estradiol in proportion to the degree of their obesity, but men fail to demonstrate the same effect. This supports some expectations that females have more estrogen receptors in their adipose tissues than males. If true, this could serve to amplify estrogen effects in females.

Related hormonal studies on obese females show the ratio of FSH to LH to be increased principally because of LH concentration decreases when compared with normal subjects. These ratio effects of FSH/LH are also observed in primates who are given exogenous pulses of gonadotropic releasing hormone (GnRH). Pulse discharges of GnRH less than the normal once per hour result in FSH/LH increases because LH concentrations decrease. In at least some cases, then, female obesity may be tied to GnRH availability.

Other studies suggest that increasing male obesity is inversely related to decreased testosterone levels without obvious effects on spermatogenesis, potency, or libido. This may be caused by the conversion of androgens to estrogens in the adrenal gland. Raised estrogen levels affect the pituitary gland by a feedback mechanism and suppress FSH levels. Lowered plasma concentrations of FSH then induce a lowering effect on testosterone levels. Some support for this theory stems from dexamethasone treatment of males in whom the drug reverses the testosterone abnormality.

Indeed, neurohumoral–obesity connections exist, but the biochemistry of these relationships remains shrouded by conflicting data and fundamental complexity. Moreover, the role of normal genetics versus genetic aberrations on demonstrably obese subjects is an uncertain but important component of many obesity situations.

POSSIBLE DIETARY FACTORS AND HORMONE-DEPENDENT CANCERS

Data accumulated in recent years suggests that steroid hormones collectively exert an umbrella of *permissive effects* over cells targeted by other hormones. This important influence of steroids over cell metabolism, coupled with their effects on nucleic acid transcription (Figure 15.6), stirs interest regarding their possible carcinogenic roles. Furthermore, assuming the veracity of present studies linking steroid hormones to carcinogenesis, questions arise about dietary factors and their contribution to hormone-induced cancers.

Interactions between dietary factors and hormone-induced cancers can be surveyed from many angles but only two cursory facets of this complex problem will be considered here:

1. *Dietary regimens and habits* that may augment hormone-induced cancer(s).

2. *Dietary acquisition of agents that* evoke nonessential, steroid-like responses in nonspecific target cells to *produce carcinoma(s)*.

Dietary Regimens and Habits

Laboratory and epidemiological studies have clearly implicated ovarian and hormone functions as one of the causative factors in breast cancers. Epidemiological studies also contend that these cancers occur more in western countries than in developing countries. These geographic distinctions indicate to many researchers that the high fat and calorie consumption inherent in the western diets, coupled with a low dietary fiber component, may be important underlying contributors to breast carcinomas.

The possible interplay between dietary fat consumption and estrogens has been examined in rodents with respect to mammary tumors induced by dimethylbenz(*a*)anthracene. Results of these studies have generally affirmed that (1) diet can alter estrogen metabolism and (2) animals fed high-fat diets exhibit higher incidences of mammary cancer formation than animals fed low-fat diets.

With this information as a backdrop, other research has shown that fecal estrogen excretion is greater in vegetarian women than omnivorous women. This type of information could be highly significant if enterohepatic circulation of estrogens coexcreted with bile happen to be implicated in breast carcinogenesis. The high fiber component of vegetarian diets probably favors biliary estrogen excretion in at least two ways: High fecal bulk and indigestible fiber may

1. Shield biliary estrogens from enterohepatic (intestinal) reabsorption phenomena, and/or

2. Minimize the β-glucuronidase activity of intestinal flora.

The second of these two possibilities may be very important since intestinal flora must deconjugate biliary estrogen conjugates as a prelude to estrogen reabsorption. Dietary factors responsible for decreasing this enzyme activity in vegetarians is unknown.

Dietary Acquisition of Carcinogenic Hormones

In 1945 the Food and Drug Administration (FDA) received an application to use diethylstilbestrol (DES) pellets in the necks of chickens. The use of this estrogenic compound was designed to produce a feminized male bird called a caponette. However, the ability of DES to promote the appearance of youth in old birds, with plumper and more attractive flesh, eventually encouraged the use of this hormone throughout the poultry industry. In later years the FDA sanctioned DES implants in not only the necks of poultry, but also the ears of cattle and sheep.

The livestock industry also obtained profits from hormone-medicated animals since they demonstrated faster weight gain and maturity, more efficient energy use of forage resources and a 20 to 30% drop in feeding costs, and many related cost competitive advantages that favored medicated animals over untreated livestock.

Unfortunately for the livestock producers, DES has been shown to cause cancer in selected test animals. This carcinogenicity problem plus the prospect of DES residues being served up at the dinner table is enough to make consumers and health professionals cringe. As a result of vigorous litigation by proponents and opponents of hormone use in livestock, the use of DES has been largely discontinued. Nevertheless, the threat of illicit DES use in livestock as well as at least nine other hormones is always a potential health threat.

The practical surveillance of such illicit practices in meat from domestic and imported origins is an overwhelming task. Even if surveillance were successful, threshold levels for cancers caused by estrogens (natural or synthetic) are not commonly recognized or established; maximum limits have not been established for chronic human exposures to livestock hormones that still ensure normal human health; and the possible synergistic effects of livestock hormones with the normal hormone balance of the human body are largely unexplored.

It should also be recognized at this point that *high concentrations* of *natural steroid hormones* (as well as synthetic types) residing in key animal organs could have a critical effect on the carcinogenic thresholds of *some* human subjects. These organs are not only high in natural steroid concentrations but serve as body filters that can accumulate high concentrations of growth hormones administered to animals. Common sense dictates that frequent consumption of such foods as chicken livers and ruminant livers by some ethnic groups could have undersirable health

Table 15.3. Animal Growth and Maintenance Hormones That *Have Been* Used to Improve Livestock Productivity Yields (Many of Which are Banned From Use)

Hormone-active compound	Administered by/in	Objective or effect
Chlormadinone acetate	Feed	Synchronization of estrus (heat) for beef cows and beef heifers
Dienestrol diacetate	Chicken feed	Promotes fat distribution, tenderness, and bloom on broilers, fryers, and roasting chickens
Estradiol benzoate	Subdermal injection	Used in conjunction with testosterone and progesterone
Estradiol monopalmitate	Subdermal injection at the base of the skull	Improves fat distribution and finish in roasting chickens
Medroxyprogesterone acetate	Feed	Used in breeding cattle and ewes for synchronizing estrus and ovulation
Melengestrol acetate	Feed	Suppresses estrus in heifers, improves feed use, and stimulates growth
Progesterone	Subcutaneous implantation	Promotes growth and feed efficiency in lambs and steers
Testosterone	Subcutaneous ear injection	Stimulates growth of beef cattle
Testosterone proprionate	Subcutaneous ear implantation	Promotes heifer growth and enhances feed efficiency

consequences over an unpredictably long period.

Growth enhancement of livestock and poultry by DES represents only one classic example of hormone use. Many other hormones can be employed for a variety of objectives (Table 15.3). Although the use of many of these hormones has been banned in some countries, literature reports still persist concerning the apparent consequences of possible hormone-contaminated foods in the human diet. Typical among these reports are those from Puerto Rico where foodborne acquisition of exogenous hormones may account for thousands of children affected by premature telarche, premature pubarche, prepubertal gynecomastia, and precocious pseudopuberty (Sáenz de Rodríguez, 1984).

REFERENCES

Aldercreutz, H. 1962. Studies on oestrogen excretion in human bile. *Acta Endocrinol. (Suppl.) (Copenhagen)* **72:**1.

Aldercreutz, H., and F. Martin. 1980. Biliary excretion and intestinal metabolism of progesterone and estrogens in man. *J. Steroid Biochem.* **13:**231.

Brobeck, J. R., J. Tepperman, and C. N. H. Long. 1943. Experimental hypothalamic hyperphagia in the albino rat. *Yale J. Biol. Med.* **15:**831.

Brody, P. K., and L. E. Rosenberg (eds.). 1949. *Duncan's Diseases of Metabolism*, 8th ed. Saunders, Philadelphia.

Brown, E. M. 1982. When you suspect hypercalcemia. *Patient Care* **16:**14.

Carrol, K. K. 1977. Dietary factors in hormone dependent cancers. In *Nutrition and Cancer* (M. Winick, ed.), pp. 25–40. Wiley, New York.

Carrol, K. K., and H. T. Khor. 1970. Effects of dietary fat and dose level of 7,12-dimethylbenz(a)anthracene on mammary tumor incidence in rats. *Cancer Res.* **30:**2260.

Carrol, K. K., and H. T. Khor. 1971. Effects of level and types of dietary fat on the incidence of mammary tumors induced in Sprague-Dawley rats by 7,12-dimethylbenz(a)-anthracene. *Lipids* **6:**415.

Felig, P., J. D. Baxter, A. E. Broadus, and L. A. Frohman. 1981. *Endocrinology and Metabolism*. McGraw-Hill, New York.

Felsenfeld, A., and F. Llach. 1982. Vitamin D and metabolic bone disease: a clinicopathologic overview. In *Pathology Annual: 1982* (S. C. Sommers and P. P. Rosen, eds.), Part 1, Vol. 17, pp. 383–410. Appleton-Century-Crofts, Norwalk, Conn.

Geschwind, N. 1965. Disconnexion syndromes in animals and man. *Brain* **88:**237–294, 585–644.

Goldin, B. R., H. Aldercreutz, S. L. Gorbach, J. H. Warram, J. T. Dwyer, L. Swenson, and M. N. Woods. 1982. Estrogen excretion patterns and plasma levels in vegetarian and omniverous women. *N. Engl. J. Med.* **307:**1542.

Grody, W. W., W. T. Schrader, and B. W. O'Malley. 1982. Activation, transformation and subunit structure of steroid hormone receptors. *Endocrine Rev.* **3**(2):141–163.

Guillemin, R. 1977. Endorphins: Brain peptides that act like opiates. *N. Engl. J. Med.* **296:**220.

Guillemin, R. 1978. Peptides in the brain: The new endocrinology of the neuron. *Science* **202:**390.

Hetherington, A. W., and S. W. Ranson. 1940. Hypothalamic lesions and adiposity in the rat. *Anat. Rec.* **78:**149.

Ip, C., and M. M. Ip. 1981. Serum estrogens and estrogen responsiveness in 7,12-dimethylbenz(a)anthracene-induced mammary tumors as influenced by dietary fat. *J. Nat. Cancer Inst.* **66:**291.

Jensen, E. V. 1977. Estrogen receptors in human cancers. *J.A.M.A.* **238:**59.

Jubiz, W. 1979. *Endocrinology—A Logical Approach for Clinicians*. McGraw-Hill, New York.

Kennedy, G. C. 1953. The role of depot fat in the hypothalamic control of food intake in the rat. *Proc. Roy. Soc. London Ser. B* **140:**578.

Martin, K. H., K. A. Hruska, J. J. Freitag, S. Klahr, and E. Slatopolsky. 1979. The peripheral metabolism of parathyroid hormone. *N. Engl. J. Med.* **301:**1092.

Mayer, J., and M. W. Bates. 1952. Blood glucose and food intake in normal and hypophysectomized, alloxan-treated rats. *Amer. J. Physiol.* **168:**812.

McMurray, W. C. 1982. *A Synopsis of Human Biochemistry with Medical Applications*. Harper & Row, New York.

Notkins, A. L. 1979. The causes of diabetes. *Sci. Amer.* **241**:62.

O'Malley, B. W., and W. T. Schrader. 1976. Receptors of steroid hormones. *Sci. Amer.* **234**:32.

Oomura, Y., M. Sugimori, T. Nakamura, and Y. Yamada. 1975. Contributions of electrophysiological techniques to the understanding of central control systems. In *Neural Integration of Physiological Mechanisms and Behavior* (G. J. Morgenson and F. R. Calaresu, eds.), pp. 375–395. University of Toronto Press, Toronto.

Sáenz de Rodríguez, C. A. 1984. Environmental hormone contamination in Puerto Rico. *N. Engl. J. Med.* **310**:1741.

Sterling, K. 1979. Thyroid hormone action at the cell level. *N. Engl. J. Med.* **300**:1117.

Williams, R. H. 1981. *Textbook of Endocrinology*. Saunders, Philadelphia.

Wyngaarden, J. B., and L. H. Smith. 1982. *Textbook of Medicine*, Vols. 1 and 2. Saunders, Philadelphia.

Yalow, R. S. 1978. Radioimmunoassay: A probe for the fine structure of biologic systems. *Science* **200**:1236.

C H A P T E R · 16

Carbohydrate Metabolism

INTRODUCTION

Carbohydrates provide 50 to 60% of the metabolizable energy in a typical American diet. Inhabitants of poor countries rely on carbohydrates for up to 90% of their energy requirements and show an increased risk to many health problems since essential dietary fats and proteins are absent.

Carbohydrates consumed by most American dietary regimens are derived from a variety of sources (Table 16.1), which can be classified as (1) *completely digestible,* (2) *partially digestible,* and (3) *indigestible.* The indigestible carbohydrates including cellulose, hemicelluloses, and pectins cannot be enzymatically hydrolyzed by human digestive enzymes. In the case of cellulose, this polyglucan displays β(1 → 4) glycosidic bonds that cannot be cleaved because the body lacks *cellulase.* This inability to digest cellulose may have distinct digestive benefits since the persistence of cellulose in the digestive tract promotes stool formation and waste transit through the intestines.

Digestible carbohydrates provide sources of dietary energy and support metabolic processes in several ways. Simple monosaccharides can be directly absorbed by the body and oxidized by biochemical pathways, such as the Embden–Meyerhof–Parnas pathway and other routes present in cells, to yield high-energy nucleotide triphosphates (e.g., ATP, GTP) and reduced pyridine nucleotides (e.g., NADH(H$^+$), NADPH(H$^+$)).

Monosaccharides in the intestinal tract are also susceptible to fermentative oxidations of intestinal flora, which produce various organic acids that can augment the supply of critical metabolic intermediates in the body. Many carbohydrate-rich diets composed of vegetables and fruits also contain citric, malic, and other acids that are metabolic intermediates of carbohydrate metabolism in plants. These acid compounds also have nutritional value in humans and animals and do not require digestion for absorption. Carbohydrates can also serve as sources of energy after their fermentative conversion to ethyl alcohol in wine, beer, and hard liquors. The ethyl alcohol derived from sugar fermentation is absorbed by the stomach and intestine in an unaltered form.

Other carbohydrates such as those broadly classed as *oligosaccharides* (di-, tri-, and tetrasaccharides) display variable digestibility for biochemical reasons as well as the compatibility of these substances in the digestive tracts of different individuals. Beans, peas, and

Table 16.1. Carbohydrates in the American Diet

Types	Main food sources
Monosaccharides	
Arabinose	Constituent of pentosans in fruits
Fructose	Fruits and honey
Galactose	Usually found in combination with other sugars (lactose)
Glucose	Fruits, honey, and corn syrup
Mannose	Derived from mannosans occurring in legumes
Ribose	Derived from nucleic acids of meat
Xylose	Constituent of pentosans in fruits
Disaccharides	
Lactose	Milk and milk products
Maltose	Malt products, infant formulas
Sucrose	Molasses, cane sugar, beet sugar, and maple syrup
Trehalose	Mushrooms and yeast
Polysaccharides	
Digestible	
Glycogen	Meat products and fish
Starch and dextrins	Grains and vegetables
Partially digestible	
Galactogens	Snails
Inulins	Onions, garlic, and Jerusalem artichokes
Mannosans	Legumes
Pentosans	Fruits and gums
Raffinose	Sugar beets, lentils, kidney beans, navy beans
Stachyose	Beans
Indigestible	
Celluloses and hemicelluloses	Vegetables
Pectins	Fruits
Derivatives from carbohydrates	
Citric acid	Fruits
Ethyl alcohol	Fermented liquors
Lactic acid	Milk and milk products
Malic acid	Fruits

soybeans all contain high concentrations of oligosaccharides that are digested in the small intestine, but utilization of these saccharides by bacteria in the large bowel can produce gas and lactic acid. These actions are responsible for cramps, diarrhea, and other intestinal irritations in many people. In those cases where consumption of these legumes does not present any problem, these foods can serve as a good source of carbohydrates and other nutrients.

Polysaccharides, including amylose, amylopectin, and glycogen, all serve as important sources of metabolizable carbohydrate once these polymers have been enzymatically hydrolyzed into their individual sugars.

The utilization of carbohydrates to supply energy in humans and other organisms is tied to anaerobic and aerobic biochemical pathways. The anaerobic oxidation of sugars supplies energy by breaking down glucose to smaller carbon fragments without a net reduction of molecular oxygen. The end products of anaerobic metabolism may include C_3 compounds such as lactic acid that cannot be further metabolized. The sequential steps leading to the anaerobic metabolism of glucose into smaller entities is operationally described as the *glycolytic* (fermentative) pathway or the *Embden–Meyerhof–Parnas* (EMP) pathway.

Further catabolism of glycolytic end products by aerobic mechanisms yields an additional harvest of energy and culminates in the overall production of carbon dioxide and water. The aerobic catabolism of glycolytic end products requires the operation of the tricarboxylic acid cycle as well as mitochondria that harbor the electron transport chain and mediate the reduction of molecular oxygen to water.

Nearly all organisms employ glycolytic mechanisms to produce pyruvic acid with the main exception being the *Cyanophycean* algae. The subsequent conversion of pyruvic acid to lactic acid by many cells and tissues is common, but the singular production of lactic acid is restricted to relatively few cells. Potato tubers and some other vegetative cells, lactic acid bacteria, and white skeletal muscles in animals produce notably high amounts of lactic acid. This is especially true for skeletal white muscle owing to its low numbers of mitochondria and a relatively poor supply of oxygen when compared with red skeletal or cardiac muscles. Low numbers of mitochondria restrict aerobic catabolism of glucose, and evolution has favored the presence of high concentrations of glycolytic enzymes that conduct anaerobic metabolism of sugars to yield energy. Cardiac muscle is just the opposite. Cells of this muscle are exposed to a high oxygen tension and, correspondingly, high mitochondrial populations; therefore, only limited amounts of glycolytically produced pyruvic acid are transformed into a terminal lactic acid product. Cardiac muscle and other tissues having adequate oxygen supplies catabolize pyruvic acid by forming acetyl-CoA in the aerobic phase of carbohydrate metabolism recognized as the tricarboxylic acid (TCA) cycle.

The present chapter offers a fundamental overview of the anaerobic and aerobic metabolism of sugars; the interconversions of sugars, carbohydrates, and polysaccharides; functions of important carbohydrate intermediates; minor oxidative pathways for sugars; abnormalities in sugar and polysaccharide metabolism; and physiological control mechanisms that regulate sugar availability, most notably in humans.

GLYCOLYSIS (EMP Pathway)

Glycolysis is a biochemical process that involves the degradation of glucose 6-phosphate to pyruvic and lactic acids along with the concomitant production of ATP. The term *glycolysis* has its etiology rooted in two Greek words, *glycos* and *lysis,* which mean "sweet" and "dissolution," respectively.

The glycolytic degradation of glucose occurs without oxygen and may be considered an anaerobic process. Moreover, the fundamental importance of the anaerobic scheme in microbiological fermentations yielding ethanol and carbon dioxide has also led to its designation as a *fermentative pathway.*

The glycolytic pathway operates in virtually all tissues of the human body and serves as an important source of ATP for driving endergonic reactions. Many of the intermediates in the glycolytic scheme, such as 1,3-diphosphoglycerate, also serve other unique biochemical functions. This compound serves as a precursor for 2,3-diphosphoglycerate (DPG), which stabilizes the physiological occurrence of deoxygenated hemoglobin.

Every glycolytic intermediate between glucose and pyruvate contains phosphate. The highly ionized state of the phosphate moieties largely prohibits the intermediates from leaving the cytosol of individual cells. Aside from limiting the migration of these intermediates, the phosphorylated compounds are also active participants in the substrate level phosphorylation of ADP to ATP. The glycolytic pathway, also referred to as the EMP pathway, depends on 11 different enzyme-mediated steps that convert glucose to lactic acid.

1. *Phosphorylation of glucose.*

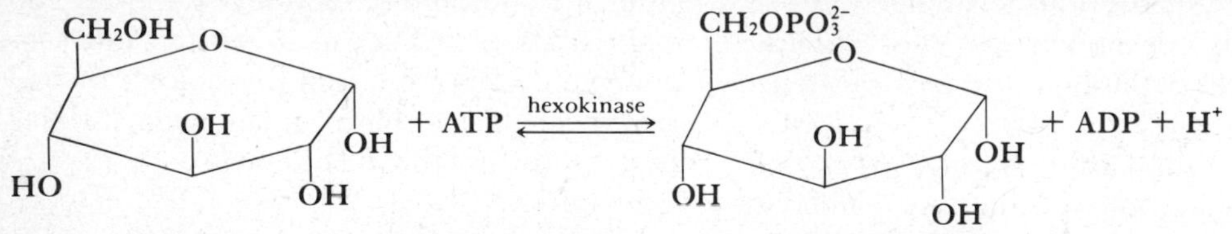

D-Glucose $\Delta G^{0'} = -4.0$ kcal/mol   **Glucose 6-phosphate**

Hexokinase catalyzes the phosphorylation of α-D-glucose, including D-fructose and D-mannose. Another enzyme that exhibits phosphorylating actions is *glucokinase,* but this enzyme has a very high K_m for sugars and demonstrates its activity in the liver following meals rich in carbohydrates.

2. α-D-*Glucose 6-phosphate undergoes isomerization to form* α-D-*fructose 6-phosphate.*

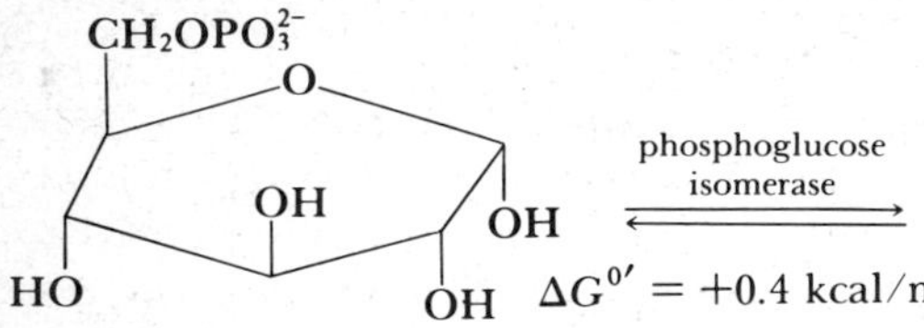

$\Delta G^{0'} = +0.4$ kcal/mol

α-D-Glucose 6-phosphate

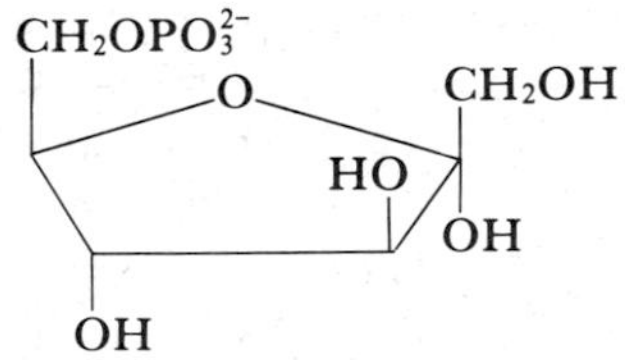

α-D-Fructose 6-phosphate

Phosphoglucose isomerase catalyzes the conversion of α-D-glucose 6-phosphate to α-D-fructose 6-phosphate.

3. α-D-*Fructose 6-phosphate is phosphorylated to produce* α-D-*fructose 1,6-diphosphate.*

This reaction is mediated by the allosteric enzyme *phosphofructokinase* (PFK), which catalyzes the phosphorylation of α-D-fructose 6-phosphate to form α-D-fructose 1,6-diphosphate. This reaction largely controls the glycolytic pathway. Citric acid, ATP, and long-chain fatty acids can exert a *negative* allosteric effect on PFK activity, whereas ADP and AMP have a *positive* allosteric effect. When ATP is plentiful in the cell, glycolysis decreases and, conversely, glycolysis is enhanced when ATP stores are low. Hence, the ATP/ADP or AMP ratio serves as a monitoring signal for the EMP pathway.

4. *An aldolase reaction cleaves* α-D-*fructose 1,6-diphosphate to form two* C_3 *fragments.*

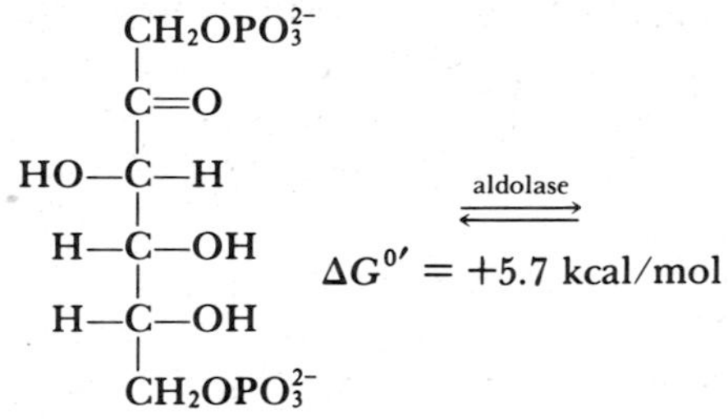

$\Delta G^{0'} = +5.7$ kcal/mol

α-D-Fructose 1,6-diphosphate

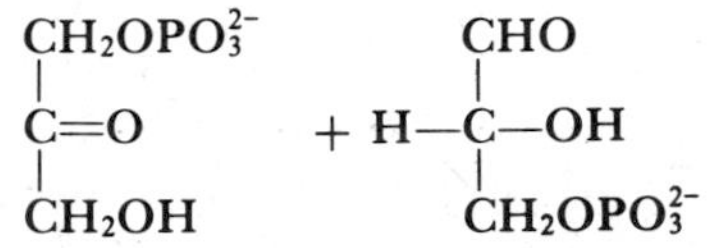

Dihydroxyacetone phosphate   **Glyceraldehyde 3-phosphate**

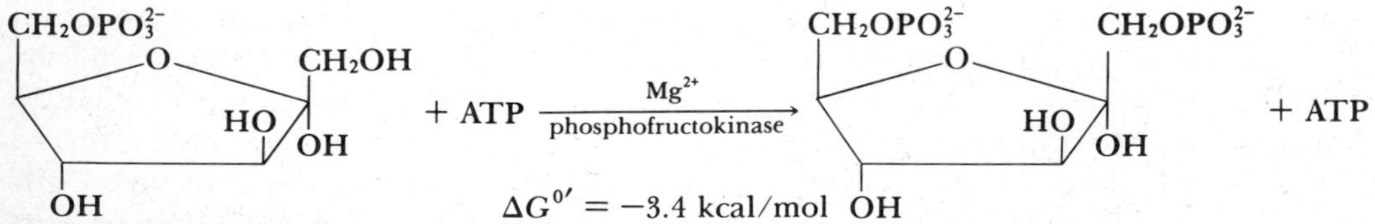

$\Delta G^{0'} = -3.4$ kcal/mol

α-D-Fructose 6-phosphate

α-D-Fructose 1,6-diphosphate

Aldolase is involved in the cleavage of α-D-fructose 1,6-diphosphate forming dihydroxyacetone phosphate and glyceraldehyde 3-phosphate. The pertinent mechanism for the reaction is explained later.

5. *Dihydroxyacetone phosphate and glyceraldehyde 3-phosphate readily undergo isomerization.*

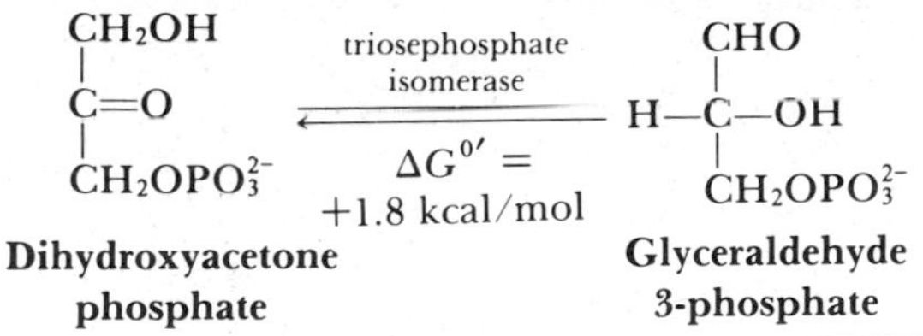

$$
\begin{array}{c}
CH_2OH \\
| \\
C=O \\
| \\
CH_2OPO_3^{2-}
\end{array}
\quad
\xrightleftharpoons[\Delta G^{0'} = +1.8 \text{ kcal/mol}]{\text{triosephosphate isomerase}}
\quad
\begin{array}{c}
CHO \\
| \\
H-C-OH \\
| \\
CH_2OPO_3^{2-}
\end{array}
$$

Dihydroxyacetone phosphate   **Glyceraldehyde 3-phosphate**

phate to glyceraldehyde 3-phosphate prevails owing to two factors: (1) intracellular concentrations of the substrates and products are *less than unit molarity*, which results in a favorable free-energy difference; and (2) glyceraldehyde 3-phosphate is readily *converted into* 1,3-diphosphoglycerate, which promotes the rapidity of the reaction.

6. D-*Glyceraldehyde 3-phosphate undergoes oxidation and phosphorylation to produce 1,3-diphosphoglyceric acid.*

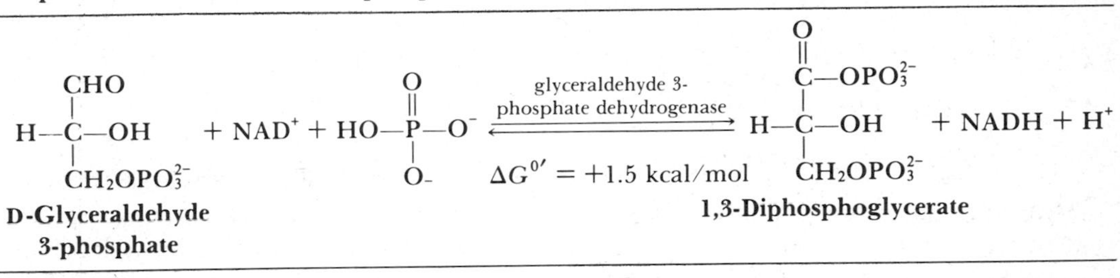

$$
\begin{array}{c}
CHO \\
| \\
H-C-OH \\
| \\
CH_2OPO_3^{2-}
\end{array}
+ NAD^+ +
\begin{array}{c}
O \\
\| \\
HO-P-O^- \\
| \\
O^-
\end{array}
\xrightleftharpoons[\Delta G^{0'} = +1.5 \text{ kcal/mol}]{\text{glyceraldehyde 3-phosphate dehydrogenase}}
\begin{array}{c}
O \\
\| \\
C-OPO_3^{2-} \\
| \\
H-C-OH \\
| \\
CH_2OPO_3^{2-}
\end{array}
+ NADH + H^+
$$

D-Glyceraldehyde 3-phosphate   **1,3-Diphosphoglycerate**

Triosephosphate isomerase catalyzes the isomerization reaction of the two C_3 fragments derived from α-D-fructose 1,6-diphosphate. Under standard conditions, aldolase and triosephosphate isomerase appear to catalyze thermodynamically unfavorable reactions. Nevertheless, the conversion of α-D-fructose 1,6-diphos-

The conversion of D-glyceraldehyde 3-phosphate into 1,3-diphosphoglycerate involves a two-step reaction sequence wherein an exergonic (oxidation) reaction is *coupled with* an endergonic (phosphorylation) reaction. The mechanism of the reaction involving D-*glyceraldehyde dehydrogenase* follows:

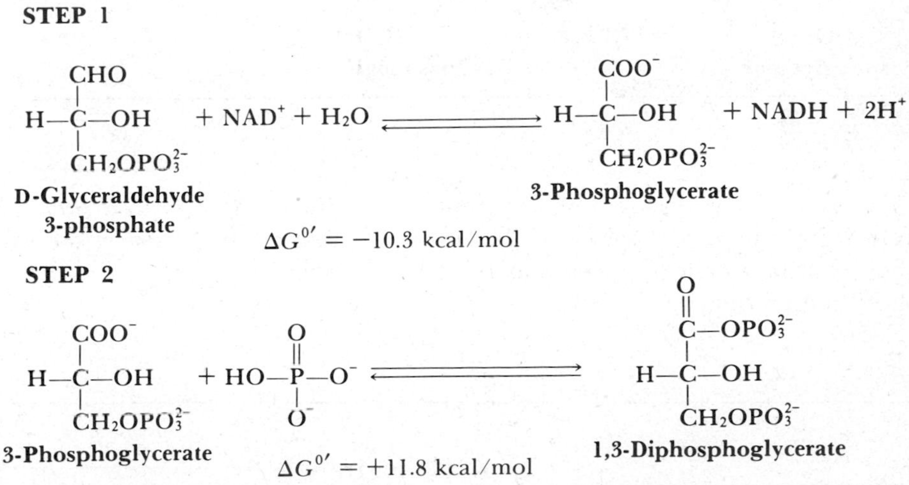

STEP 1

$$
\begin{array}{c}
CHO \\
| \\
H-C-OH \\
| \\
CH_2OPO_3^{2-}
\end{array}
+ NAD^+ + H_2O
\rightleftharpoons
\begin{array}{c}
COO^- \\
| \\
H-C-OH \\
| \\
CH_2OPO_3^{2-}
\end{array}
+ NADH + 2H^+
$$

D-Glyceraldehyde 3-phosphate   **3-Phosphoglycerate**

$$\Delta G^{0'} = -10.3 \text{ kcal/mol}$$

STEP 2

$$
\begin{array}{c}
COO^- \\
| \\
H-C-OH \\
| \\
CH_2OPO_3^{2-}
\end{array}
+
\begin{array}{c}
O \\
\| \\
HO-P-O^- \\
| \\
O^-
\end{array}
\rightleftharpoons
\begin{array}{c}
O \\
\| \\
C-OPO_3^{2-} \\
| \\
H-C-OH \\
| \\
CH_2OPO_3^{2-}
\end{array}
$$

3-Phosphoglycerate   **1,3-Diphosphoglycerate**

$$\Delta G^{0'} = +11.8 \text{ kcal/mol}$$

SUM: D-Glyceraldehyde 3-phosphate $+ NAD^+ + P_i \rightleftharpoons$ 1,3-diphosphoglycerate $+ NADH + 2H^+$

$$\Delta G^{0'} = -10.3 + 11.8 = +1.5 \text{ kcal/mol}$$

7. *Substrate level phosphorylation of ADP involves 1,3-diphosphoglyceric acid.*

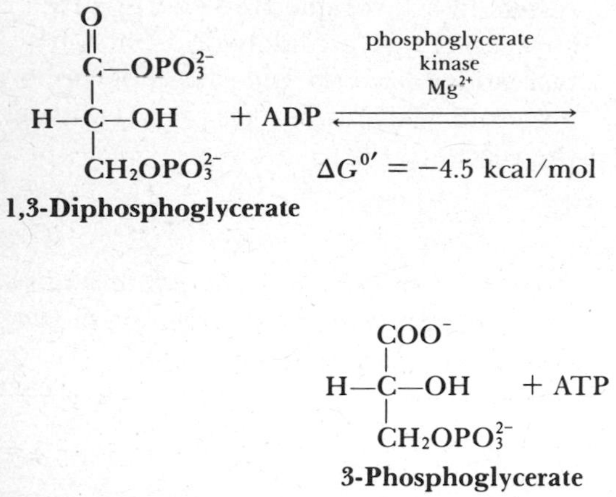

$$
\begin{array}{c}
\overset{\displaystyle O}{\underset{\displaystyle |}{\overset{\displaystyle \|}{C}}}-OPO_3^{2-} \\
H-\overset{|}{\underset{|}{C}}-OH \quad + ADP \\
CH_2OPO_3^{2-}
\end{array}
$$

phosphoglycerate kinase
Mg^{2+}

$\Delta G^{0'} = -4.5$ kcal/mol

1,3-Diphosphoglycerate

$$
\begin{array}{c}
COO^- \\
| \\
H-C-OH \quad + ATP \\
| \\
CH_2OPO_3^{2-}
\end{array}
$$

3-Phosphoglycerate

The ATP produced from ADP stores and 3-phosphoglycerate resulting from 1,3-diphosphoglycerate both reflect the enzyme actions of *phosphoglycerate kinase.*

8. *2-Phosphoglycerate is formed from 3-phosphoglycerate.*

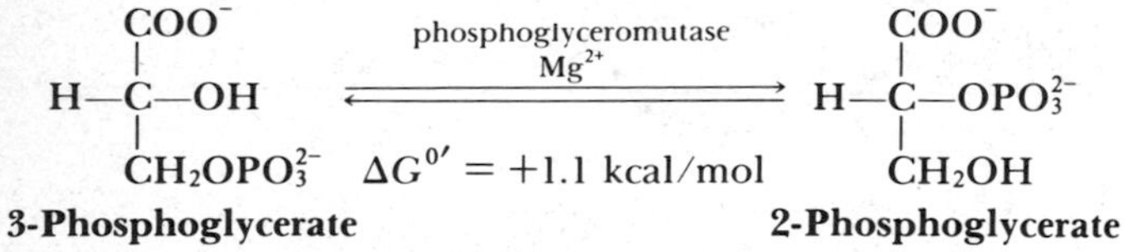

9. *Phosphoenolpyruvate (PEP) is produced from 2-phosphoglycerate.*

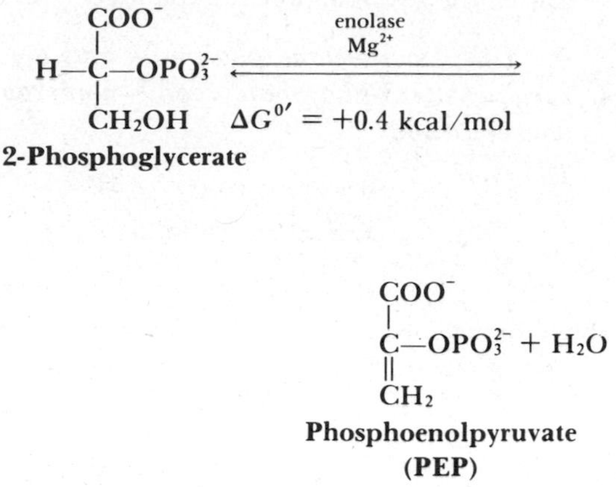

$$
\begin{array}{c}
COO^- \\
| \\
H-C-OPO_3^{2-} \\
| \\
CH_2OH
\end{array}
$$

enolase
Mg^{2+}

$\Delta G^{0'} = +0.4$ kcal/mol

2-Phosphoglycerate

$$
\begin{array}{c}
COO^- \\
| \\
C-OPO_3^{2-} + H_2O \\
\| \\
CH_2
\end{array}
$$

**Phosphoenolpyruvate
(PEP)**

Phosphoenolpyruvate formed by this reaction has a very large negative $\Delta G^{0'}$ on hydrolysis (-14.8 kcal/mol). This substantial change in free energy is adequate for driving ADP phosphorylation in the next step.

$$
\begin{array}{c}
COO^- \\
| \\
H-C-OH \\
| \\
CH_2OPO_3^{2-}
\end{array}
$$

phosphoglyceromutase
Mg^{2+}

$\Delta G^{0'} = +1.1$ kcal/mol

$$
\begin{array}{c}
COO^- \\
| \\
H-C-OPO_3^{2-} \\
| \\
CH_2OH
\end{array}
$$

3-Phosphoglycerate   **2-Phosphoglycerate**

This reaction relies on a phosphoryl group shift from the C-3 position of 3-phosphoglycerate to the C-2 position of 2-phosphoglycerate.

10. *ATP is formed from PEP hydrolysis and isomerization of the hydrolysis products.*

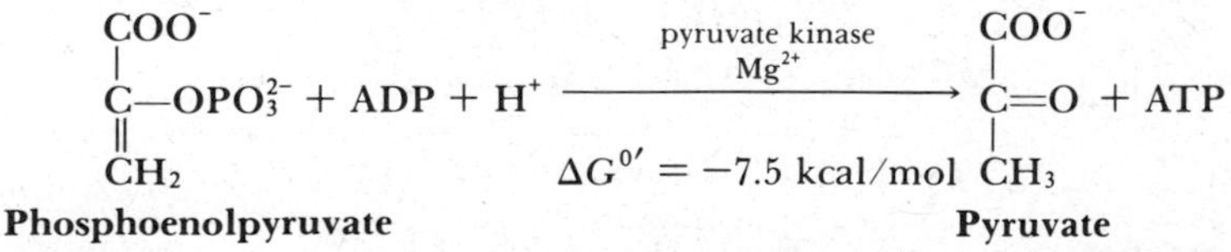

$$
\begin{array}{c}
COO^- \\
| \\
C-OPO_3^{2-} + ADP + H^+ \\
\| \\
CH_2
\end{array}
$$

pyruvate kinase
Mg^{2+}

$\Delta G^{0'} = -7.5$ kcal/mol

$$
\begin{array}{c}
COO^- \\
| \\
C=O + ATP \\
| \\
CH_3
\end{array}
$$

Phosphoenolpyruvate   **Pyruvate**

This reaction is an irreversible reaction and represents the second reaction of the EMP pathway responsible for ATP production. The *pyruvate kinase* enzyme is an allosteric enzyme that is subject to inhibition by ATP. Pyruvate, which is the product of the reaction, represents one major link in metabolism that *unites* the EMP pathway (or glycolysis), the tricarboxylic acid cycle, amino acid metabolism, and fatty acid oxidation into a mutually interacting scheme.

11. Lactate is produced by the reduction of pyruvate.

For anaerobic tissues including actively exercising muscle, or an inadequately perfused myocardium in instances of coronary artery disease, the energy-producing role of the tricarboxylic acid cycle and oxidative phosphorylation may become minimal. Accordingly, $NADH(H^+)$ generated from the glyceraldehyde 3-phosphate dehydrogenase reaction may become elevated, while stores of NAD^+ become strapped. Ordinarily, such circumstances should prohibit further glycolysis, but the lactate dehydrogenase (LDH) reaction consumes $NADH(H^+)$ during the reduction of pyruvate to lactate. This event

$$\begin{array}{c}
COO^- \\
| \\
C{=}O \\
| \\
CH_3 \\
\textbf{Pyruvate}
\end{array} + NADH + H^+ \underset{\Delta G^{0'} = -6.0 \text{ kcal/mol}}{\overset{\text{L-lactate dehydrogenase}}{\rightleftharpoons}} \begin{array}{c}
COO^- \\
| \\
HO{-}C{-}H \\
| \\
CH_3 \\
\textbf{Lactate}
\end{array} + NAD^+$$

This reduction reaction is catalyzed by any one of five possible forms of isoenzymes of L-*lactate dehydrogenase* (LDH) in humans. The NAD^+ formed from $NADH(H^+)$, which is a prerequisite for the reaction, is necessary for the conversion of D-glyceraldehyde 3-phosphate to 1,3-diphosphoglycerate. The balanced production and consumption of reduced and oxidized pyridine nucleotides ensures the schematic perpetuity of the glycolytic sequence.

In those instances where tissues have an adequate supply of oxygen, *pyruvate formation is the terminal step of glycolysis.* If this is the case, pyruvate is subsequently oxidized and decarboxylated to form an acetyl group. This acetyl group is combined with coenzyme A to form acetyl-CoA ($CH_3{-}\overset{\overset{\text{O}}{\|}}{C}{-}SCoA$), which enters the tricarboxylic acid cycle where aerobic oxidative events occur. These events result in the production of high-energy nucleotide triphosphates (ATP, GTP), reduced flavin compounds ($FADH_2$), carbon dioxide, and water.

supplies NAD^+. It should be clearly noted, then, that anaerobic glycolysis does not lead to an absolute increase or decrease of $NADH(H^+)$ stores.

ENZYME MECHANISM FOR GLYCERALDEHYDE 3-PHOSPHATE DEHYDROGENASE

The conversion of glyceraldehyde 3-phosphate into 1,3-diphosphoglycerate is catalyzed by glyceraldehyde 3-phosphate dehydrogenase. This tetrameric enzyme consists of four *identical* subunits. Each subunit exhibits a sulfhydryl group and requires a coenzyme in the form of NAD^+. The sulfhydryl group displays nucleophilic properties since it reacts at the carbonyl site on the aldehyde to produce an *adduct* (thiohemiacetal). The adduct is oxidized by the oxidant NAD^+ to yield a thioester (an *S*-acyl enzyme), while the reduced coenzyme form ($NADH(H^+)$) leaves the active site of the enzyme in exchange for a free NAD^+ molecule. Subsequent to this step, the acyl enzyme reacts with inorganic phosphate, P_i, to form an acyl phosphate such as 1,3-diphosphoglycerate as detailed in Figure 16.1.

MECHANISM FOR ALDOLASE ACTION

Fructose 1,6-diphosphate is cleaved by aldolase to form glyceraldehyde 3-phosphate and dihydroxyacetone phosphate. The reaction mechanism involves six fundamental steps (Figure 16.2).

1. The furanose ring structure of fructose 1,6-diphosphate opens when the C-1 phosphate of fructose attracts a proton from an

as yet unidentified source—possibly the lysine 107 residue of the aldolase.

2. Lysine 227 in the amino acid residue sequence of aldolase attacks the carbonyl group of fructose 1,6-diphosphate to form a Schiff base.

3. The Schiff base is protonated and serves as an electron sink, which assists in the carbon—carbon bond cleavage of fructose 1,6-diphosphate.

4. A sulfide group (—S⁻), possibly from cysteine residue 72 or 336 in aldolase, removes a proton, thereby initiating aldol cleavage to produce an eneamine and glyceraldehyde 3-phosphate.

5. The eneamine is protonated by the imidazole group of histidine 359. Tyrosine 361 in the aldolase structure permits histidine 359 to act as a proton transfer agent in ad-

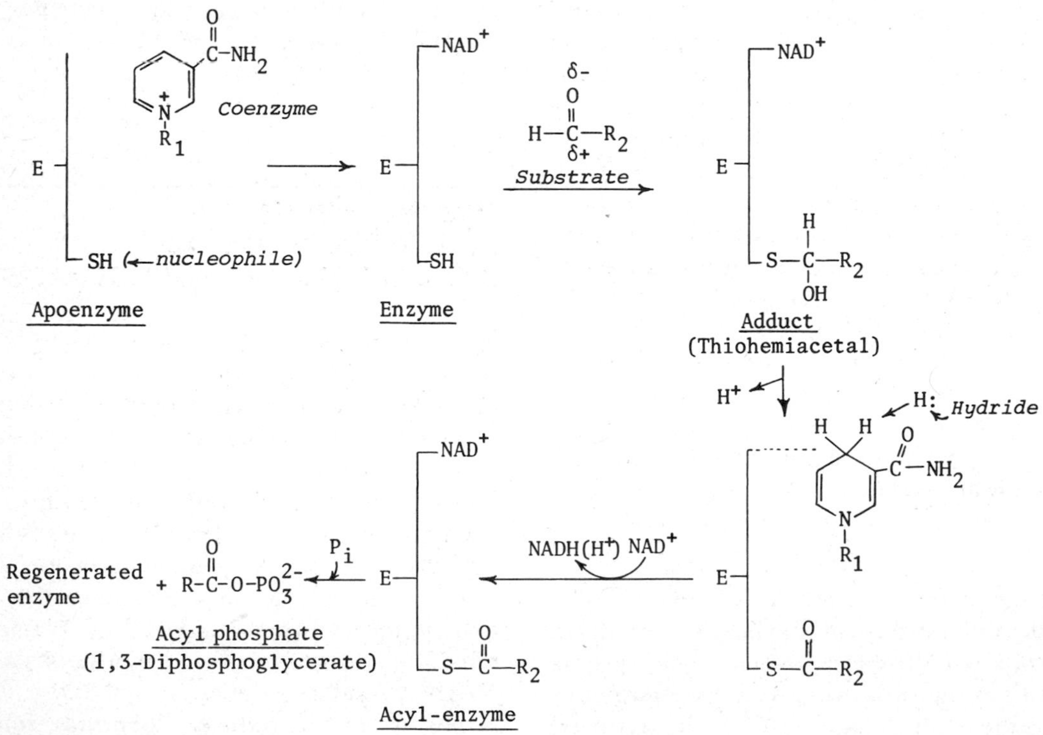

FIGURE 16.1. Glyceraldehyde 3-phosphate dehydrogenase enzyme mechanism.

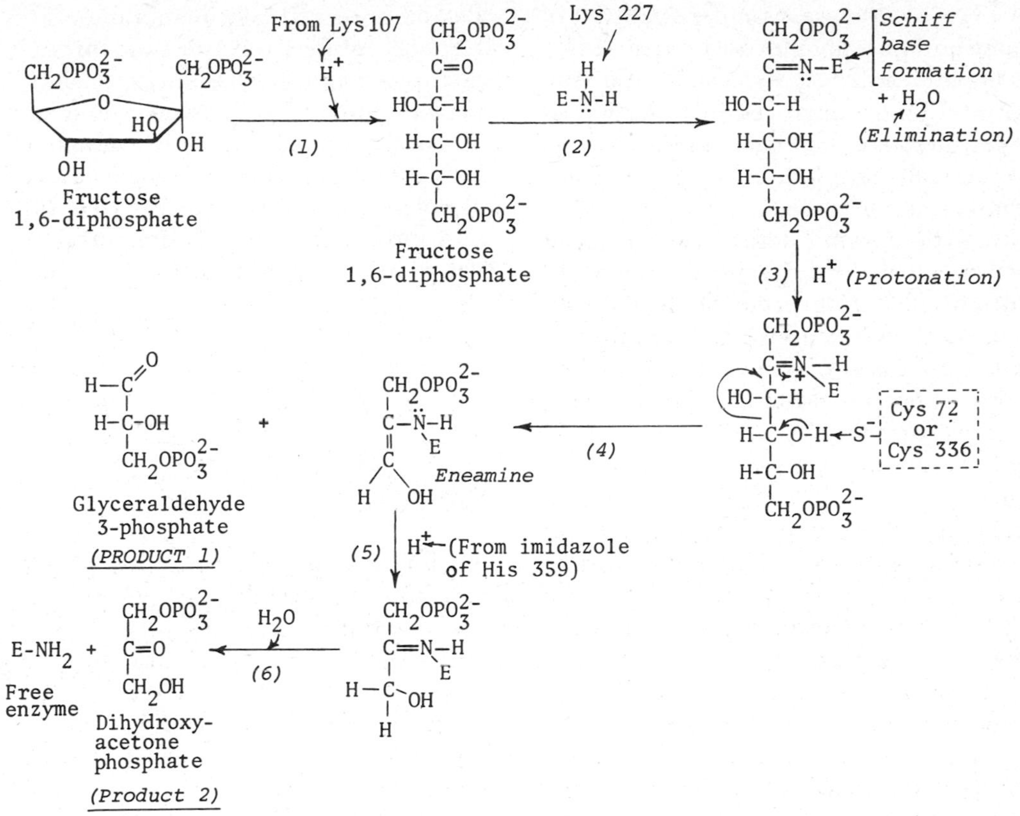

FIGURE 16.2. Aldolase enzyme mechanism.

dition to facilitating the interaction of lysine 107 with cysteine 72 (or cysteine 336).

6. The protonated eneamine is hydrolyzed to yield dihydroxyacetone phosphate, and the enzyme is regenerated.

LEVELS OF GLYCOLYTIC ENZYMES HAVE CLINICAL AND NUTRITIONAL SIGNIFICANCE

As indicated earlier, the glycolytic pathway and its enzymes are found in virtually all tissues and sera of humans and higher animals. Within individual cells, these enzymes are found in the cytosol as opposed to the mitochondrion. Owing to the essential role of the glycolytic reactions, any deficiency of one or more of the 11 glycolytic enzymes is incompatible with life processes.

Absolute serum levels of phosphoglucose isomerase, aldolase, and lactate dehydrogenase as well as the relative occurrence of these enzymes to each other, often reflects the presence of certain human diseases. Elevation of all three enzymes to a comparable degree in serum is not uncharacteristic of extensive carcinomatosis, granulocytic leukemia, megaloblastic anemia, infectious mononucleosis, myocardial infarction, and other hemolytic conditions. Serum levels of phosphoglucose isomerase most effectively indicate carcinomatosis, while lactate dehydrogenase levels are most indicative of megaloblastic anemia. Muscle conditions such as trichinosis, muscular dystrophy, and dermatomyositis are most notably indicated by high levels of aldolase. Comparable levels of aldolase and phosphoglucose isomerase suggest hepatic necrosis. Relative serum concentration levels, mirrored by other glycolytic enzymes, have *uncertain* significance, but *may* eventually have wide-ranging clinical and nutritional importance.

Glycolytic enzyme deficiencies obviously affect the normal energy metabolism of the mature human erythrocyte. Many glycolytic enzymes are conspicuously reduced in their activities in cases of congenital hemolytic conditions, but an absolute link between the homolysis mechanism and enzyme deficiency is not entirely clear.

Foremost among congenital hemolytic anemias are those resulting from *pyruvate kinase* (PK) *deficiencies.* Although PK deficiencies are relatively uncommon, 95% of the individuals suffering from this disorder exhibit an inherited recessive disposition. The consequences of PK deficiencies are multifaceted, but most evidence indicates that ATP production is sporadic and far from a constant state. Lack of constancy in ATP production results in a corresponding sporadic outward translocation of intracellular calcium ions from the erythrocyte. Such inconstant calcium ejection from cells is thought to spur the production of stiff, spiculated cells that typify this condition.

Many other congenital hemolytic anemias are linked to problems in red cell glycolytic enzymes. *Hexokinase deficiency,* for example, is second only to PK deficiencies as a glycolytically related cause of congenital anemias. Hexokinase deficiency causes a relatively low productivity of 2,3-diphosphoglyceric acid intermediates compared to PK deficiencies and a poorer exercise tolerance than in those individuals having PK deficiencies.

Red cells anaerobically oxidize about 90% of intracellular glucose by the EMP pathway to produce ATP and 2,3-diphosphoglyceric acid (DPG). Because erythrocyte stores of ATP are intrinsically linked to their membrane flexibility, low production rates of glycolytically produced ATP (caused by EMP pathway enzyme deficiencies) can lead to low flexibility of red cell membranes. This effectively inhibits red cell flow through the microvasculature of tissues, especially the reticuloendothelial organs (liver and spleen), and promotes the early destruction of erythrocytes. Enzyme-linked abnormalities of the glycolytic pathway also affect DPG production and the oxygen saturation properties of native hemoglobin molecules.

DPG AS A PHYSIOLOGICAL EFFECTOR

An intermediate of the glycolytic pathway, 1,3-diphosphoglycerate, serves as a substrate for the synthesis of *2,3-diphosphoglycerate* (DPG). DPG normally exists at concentration levels of ~4.5 mM in blood and ultimately influences the oxygen-binding properties of native hemoglobin (tetrameric $\alpha_2\beta_2$) molecules.

As illustrated by the curves in Figure 16.3, the partial pressure responsible for 0.5 saturation (P_{50}) of a fixed amount of adult hemoglobin (HbA) is about 1.0 Torr in the absence of DPG and nearly 26.0 Torr in the presence of DPG. The influence of DPG on hemoglobin oxygenation reflects its ability to stabilize the quaternary structure of deoxyhemoglobin. Molecular stabilization of deoxyhemoglobin is achieved by a binding interaction that involves DPG and three positively charged R groups on amino acid residues including histidine 143, lysine 82, and histidine 2 on each β-hemoglobin chain. The ramifications of this binding interaction are complex and far reaching.

Stabilization of deoxyhemoglobin by DPG occurs at the moment oxygen is released to oxygen-consuming cells and tissues. The deoxygenation of hemoglobin immediately yields a high-affinity (low P_{50}) form of hemoglobin. This species has a transient existence at best since its notable oxygen-binding properties are tempered by rapid DPG-binding interactions. The resulting hemo-

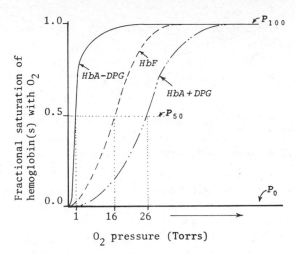

FIGURE 16.3. Effect of DPG on oxygen affinity of adult hemoglobin (HbA) in erythrocytes with (+) and without (−) the presence of DPG. The relative oxygen saturation curve for fetal hemoglobin (HbF) with respect to the HbA's has also been illustrated.

globin–DPG complex subsequently yields a hemoglobin form having a P_{50} of 26 Torr (high P_{50}). If the oxygen-binding affinity of hemoglobin were not stabilized at 26 Torr by DPG binding, the deoxygenated hemoglobin could conceivably cause rapid and counterproductive deoxygenation of respiring tissues (oxygen-consuming tissues).

Subsequent translocation of deoxygenated hemoglobin in erythrocytes through the lungs reestablishes the oxygenated form of hemoglobin since oxygen tension is in excess of the 26 Torr P_{50} value. Furthermore, under conditions of lung oxygenation, DPG is discharged from the β chains of the hemoglobin molecule due to allosterically induced spatial changes inside the molecule. The oxygenated hemoglobin is subsequently charged for another route through the circulatory system where it will distribute oxygen to any cells and tissues having a relatively deficient oxygen tension.

Since myoglobin and fetal hemoglobin (HbF) have respective P_{50} values of 1.0 and 16.0 Torr (far less than the 26 Torr P_{50} for adult hemoglobin), an effective oxygen transport mechanism to either muscle or fetal tissues is assured. The low P_{50} for myoglobin simply reflects its affinity for oxygen binding as a sole tertiary protein structure. This is quite unlike fetal hemoglobin, however, which is similar to the structure of adult, tetrameric hemoglobin with the minor exception of a γ-polypeptide chain in place of the β chain found in HbA. The F_2 and F_3 positions on the γ chain in HbF merely display alanine and glutamine in place of alanine and threonine in the β chain of HbA. The conformational consequences of this disparity minimize the strong binding interactions between DPG and HbF as opposed to HbA.

Contrary to the outward simplicity of oxygen transport mechanisms outlined here, the Bohr effect (increasing proton concentrations with decreasing oxygen binding affinities of hemoglobin) and the carbamylation of hemoglobin (a carbon dioxide transport mechanism) also augment the overall process of respiratory gas transport.

GLYCOGEN STORES GLUCOSE FOR EVENTUAL METABOLISM

Sugars obtained from dietary sources or sugars mobilized from polyglucans, such as gly-

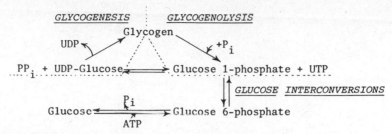

FIGURE 16.4. Glycogenesis and glycogenolysis.

cogen stored in muscles or liver tissue, can enter into the glycolytic reaction sequence. The direct utilization of dietary glucose is quite simple compared to glucose produced from the biochemical degradation of glycogen, which is referred to as *glycogenolysis*.

Glycogenolytic events are intrinsically tied to counterevents that lead to the polymerization of excess glucose and produce glycogen in the process called glycogenesis. The interplay between the biochemical pathways for glycogenolysis and glycogenesis are summarized in Figure 16.4.

GLYCOGENOLYSIS

The metabolic degradation of glycogen to its individual glucose residues requires enzymes such as *phosphorylase, transferase,* and an $\alpha(1 \to 6)$-*glucosidase* whose roles are schematically shown in Figure 16.5. Phospho-

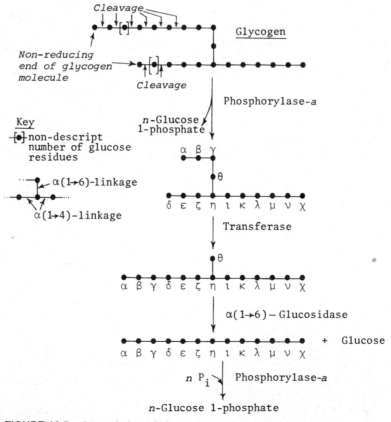

FIGURE 16.5. Degradation of glycogen.

rylase-*a* (active phosphorylase) catalyzes a repetitive reaction that cleaves the linear $\alpha(1 \rightarrow 4)$ linkages of glycogen. This enzyme action begins at the nonreducing termini of the glycogen molecule and continues until only three or four glucose residues remain glycosidically bonded to a residue denoted as theta (θ). This glucose residue prohibits further enzymatic mobilization of glucose molecules since it is held in an $\alpha(1 \rightarrow 6)$-glycosidic bond to form a *limit dextrin.*

Mobilization of glucose by phosphorylase results in the formation of glucose 1-phosphate, which can be directed into glycolysis.

The obstructive action of the $\alpha(1 \rightarrow 6)$-linked (θ) glucose toward phosphorylation of individual glucose residues is overcome by a transferase enzyme. The transferase mediates the "transfer" of the remaining α-, β-, or γ-polyglucose unit to the nonreducing δ residue of the main glycogen chain. This event exposes the θ-glucose from the main glycogen chain.

Again, phosphorylase-*a* resumes its attack on the $\alpha(1 \rightarrow 4)$ linkages of the glycogen molecule (beginning at the α-glucose residue), producing additional glucose 1-phosphate molecules.

Since glycogen phosphorylase actions are restricted to the nonreducing ends of glycogen molecules, it is speculated that evolution has favored this mechanism for the rapid mobilization of glucose owing to the highly branched glycogen structure.

Glucose 1-phosphate molecules *must be converted* to glucose 6-phosphate before their entry into the glycolysis pathway (Figure 16.6). This transformation involves a two-step reaction catalyzed by *phosphoglucomutase.* A phosphorylated serine residue in this enzyme ($E-CH_2OPO_3^{2-}$) is used for phosphorylating glucose 1-phosphate to produce an intermediate of glucose 1,6-diphosphate. The dephosphorylated enzyme ($E-CH_2OH$) is rephosphorylated by removal of the C-1 phosphate on glucose 1,6-diphosphate. This event subsequently leads to the formation of glucose 6-phosphate. Glucose 6-phosphatase can hydrolyze glucose 6-phosphate to form free glucose, which is released into the bloodstream. This hydrolytic step occurs during muscular activity or at extended intervals between meals. Free glucose liberated in this fashion is assimilated by individual cells and tissues, where it is metabolized. Since the liver can store and release glucose

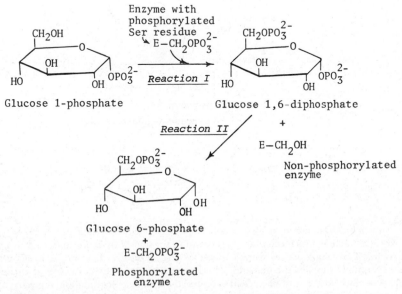

FIGURE 16.6. Schematic mechanism for the action of phosphoglucomutase on glucose 1-phosphate.

for eventual use by other tissues, it can be classified as an organ that truly performs an altruistic function.

PHOSPHORYLASE CONTROL MECHANISMS DEPEND ON EPINEPHRINE AND CALMODULIN

Phosphorylase activity initiates the degradation of glycogen by two discretely different control mechanisms including:

1. *Cyclic AMP* (cAMP)-*initiated cascade reactions* prompted by epinephrine (adrenalin) actions on hepatic cells.

2. *Calcium-* and *calmodulin-regulated systems* in muscle cells and tissues.

Epinephrine (adrenalin) secreted by the kidneys travels to the hepatic tissues via blood circulation. Epinephrine molecules contained in the blood at concentrations as low as 10^{-9} M interact at *specific receptor sites* on the surfaces of individual hepatic cell membranes. As illustrated in Figure 16.7, the epinphrine elicits specific conformational changes in cellular membrane components, which *activate adenylate cyclase*. Activated adenylate cyclase *mediates the formation of cAMP* at the expense of ATP. The cyclic nucleotide produced here subsequently binds to an inactive protein kinase and *activates the kinase*. Active protein kinase in turn converts *inactive* phosphorylase (phosphorylase-*b*) into *active* phosphorylase (phosphorylase-*a*). Activated phosphorylase-*a* initiates the cleavage of $\alpha(1 \rightarrow 4)$-glycosidic bonds in glycogen and

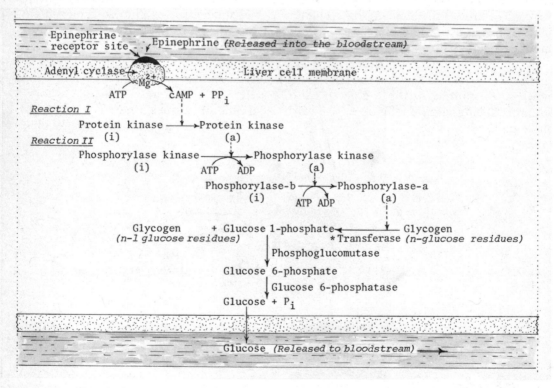

FIGURE 16.7. Hormonal stimulation of glycogenolysis by epinephrine. Cyclic AMP (cAMP) formed from ATP initiates a cascade reaction, which is ultimately responsible for phosphorylating and liberating free glucose into the bloodstream. This action occurs at the expense of glycogen stores held in the liver cell. Glucose in the bloodstream eventually supplies the demands of other body cells and tissues. (Note that the transferase enzyme*, an $\alpha(1 \rightarrow 6)$ glucosidase, must be present to ensure the complete degradation of glycogen molecules.) Abbreviations: (i), inactive enzyme form; (a), active enzyme form.

ultimately contributes to the elevation of blood glucose.

Phosphorylase activation and the mobilization of phosphorylated glucose by complex cascade reactions, as outlined above, can occur in a time span covering *minutes or seconds*. Phosphorylase actions can also be instituted over a *millisecond* time span *if* phosphorylase activation involves the actions of *calmodulin and calcium*.

Calmodulin is a single-chain protein composed of 148 amino acid residues. The primary amino acid sequence in the protein permits the binding of calcium ions to specific carboxylate groups. This process is closely linked to the activation of calmodulin from its inactive form. An obvious deficiency of cysteine residues in the calmodulin protein prohibits intra- and intermolecular bridge formation. This lack of structural bridges permits the molecule to retain a flexible conformational structure. Such flexibility at the molecular level allows calmodulin-binding interactions with enzymes, thereby affecting enzyme conformations and activities.

Calcium ions bind to calmodulin in association with the enzyme phosphorylase kinase (Figure 16.8). The kinase becomes activated and forms phosphorylase-*a* from phosphorylase-*b*. Production of phospho-rylase-*a* again permits the mobilization of glucose 1-phosphates from glycogen as in the case of the cAMP cascade.

Glycogenesis

In the interest of thermodynamic factors and energetic economy, *the synthesis of glycogen is not simply the reversal of glycogenolysis*. Instead, the synthesis of glycogen, or *glycogenesis*, is an entirely separate process from glycogenolysis. Glycogen synthesis is mediated by *glycogen synthetase*, an enzyme that is also under the hormonal influence of epinephrine.

The principal substrate in glycogenesis is uridine diphosphate glucose, which donates its glucose to a preexisting *glycogen primer*. Uridine diphosphate glucose is prepared for polymerization to the glycogen primer as shown in Figure 16.9. This polymerization reaction is mediated by *UDP-glucose pyrophosphorylase*. Although the reaction is reversible, the pyrophosphate readily undergoes hydrolysis into two inorganic phosphates and the reaction becomes *irreversible*.

Glucose 1-phosphate + UTP $\rightleftharpoons$
UDP-glucose + PP_i

$PP_i + H_2O \longrightarrow 2 P_i$

Net reaction:

Glucose 1-phosphate + UTP + $H_2O \longrightarrow$
UDP-glucose + $2P_i$

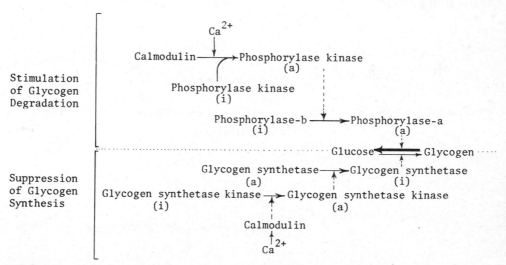

FIGURE 16.8. Glycogen metabolism in skeletal muscle. Calcium ions and calmodulin coordinate the suppression of glycogen synthesis with the stimulation of glycogen degradation.

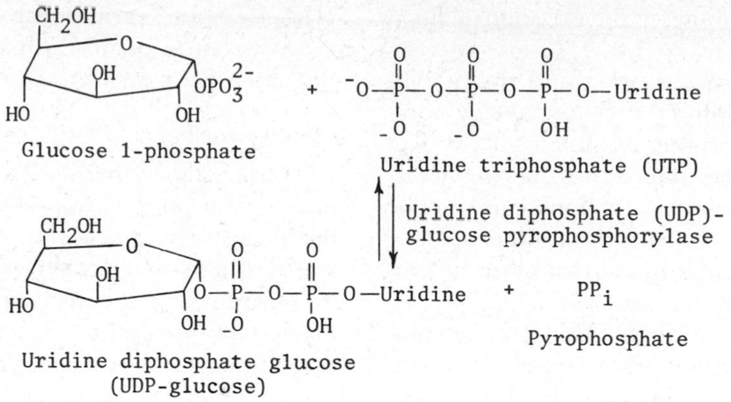

FIGURE 16.9. Biosynthesis of UDP-glucose.

After the formation of UDP-glucose, the glucose moiety is transferred to a preexisting glycogen primer (Figure 16.10). *Repetitive addition* of glucose molecules to the glycogen primer results in high-polymeric forms of glycogen.

Normal metabolic control mechanisms ensure that glycogenolysis and glycogenesis cannot competitively operate at the same time. Epinephrine-induced cascade mechanisms leading to activated protein kinase *activates* phosphorylase kinase, which in turn produces active phosphorylase-*a*. The protein kinase also *deactivates* glycogen synthetase responsible for the continued polymerization of glucose into glycogen.

Muscle glycogenesis, too, is controlled by the ultimate actions of calmodulin on *glycogen synthetase kinase*. Calmodulin activation of this enzyme leads to *deactivation of glycogen synthetase* and suppression of glycogen synthesis (Figure 16.8).

ABNORMALITIES IN GLYCOGEN STORAGE

Some of the most common diseases of carbohydrate metabolism reflect fundamental problems in glycogen storage caused by enzymatic abnormalities. Glycogen occurs in

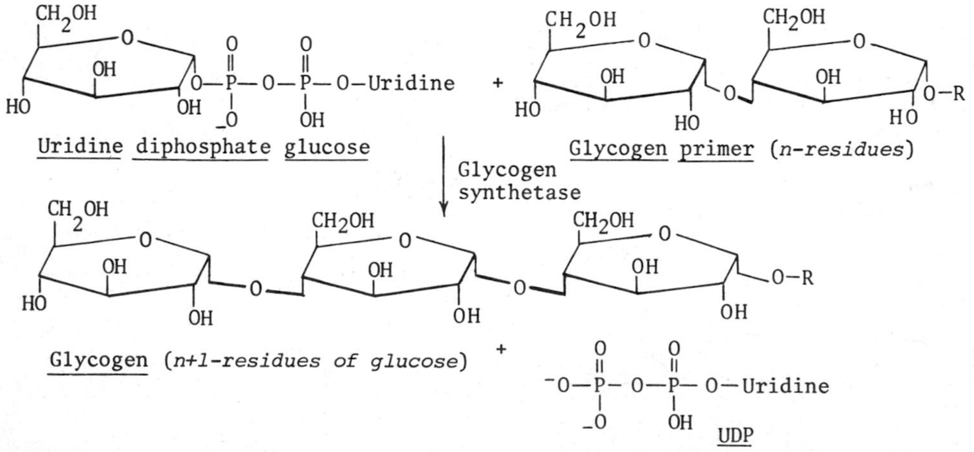

FIGURE 16.10. UDP-glucose serves as a source of glucose residues for polymerization into higher polymeric forms of glycogen.

virtually all cells of the body in variable concentrations and serves as the major storage form of reserve carbohydrate.

Many forms of glycogen storage disease are indicated as elevated levels of cellular glycogen concentrations, and still other forms are characterized by abnormal molecular structures for glycogen. Stores of liver glycogen can reflect the nutritional status of a human test subject; however, hyperalimentation is not recognized as a fundamental cause of elevated liver glycogen concentrations.

Glycogen storage diseases are designated according to a Roman numerical sequence introduced by Coris over 30 years ago. Recognition of specific enzyme abnormalities over many years has led to the identification of over *six different forms of glycogen storage diseases*. Aside from numerical categorizations, the glycogen storage diseases can be more simply classified as hepatic or muscular maladies of carbohydrate anabolism.

Hepatic Glycogen Storage Diseases

Glycogen storage diseases denoted as Types I, III, and IV are associated with *hepatic glycogen disorders,* although other organs may be involved.

Type I or *von Gierke's disease* stems from a *deficiency of glucose 6-phosphatase* in the liver, intestinal mucosa, and renal tissues. This prototypical condition of glycogen storage disease becomes apparent in early life as a disproportionately large liver (*hepatomegaly*). Continued enlargement of the liver reflects its progressive accumulation of glycogen and lipids. Furthermore, an obvious lack of glucose 6-phosphatase contributes to *hypoglycemia* since glucose cannot be obtained from glycogen. Advanced stages of the Type I disorder produce ketosis, convulsions, increased gluconeogenesis, and prominent hypoglycemia (starvation diabetes) upon fasting.

Other glycogen storage diseases, such as Types III, and IV, are similar to Type I but somewhat milder. Type III or *Cori's disease*

reflects a *deficiency of a glycogen debranching enzyme* (amylo-1,6-glucosidase). Abnormal molecular structures of glycogen accumulate because only the outer branches of the molecule can be removed by phosphorylase activity. In this disease, gluconeogenesis is unaltered, and hypoglycemia, along with its inherent consequences, is less severe. It should also be recognized that at least six subtypes of Type III storage disease are known in which relative differences are attributable to the tissue-specific occurrences of certain enzyme deficiencies.

Andersen's disease or Type IV storage disease is associated with a genetic *deficiency of the glycogen branching enzyme.* Inadequate glycogen branching may produce normal or abnormal hepatic glycogen levels, but the glycogen molecules present invariably display extraordinarily long outer branches.

Protracted occurrence of this problem causes the accumulation of glycogen and the eventual recognition of the hepatic tissues as a foreign body, followed by tissue reactions that culminate in cirrhosis. Andersen's disease notably contrasts with Type VI storage diseases, in which liver glycogen displays a more normal type of structure, although phosphorylase activity is marginally inadequate.

Muscular Glycogen Storage Diseases

The preeminent form of muscular glycogen storage disease is designated as Type II or *Pompe's disease*. This malady stems from an *absence of* $\alpha(1 \rightarrow 4)$-*glucosidase,* which normally resides in lysosomes, and ultimately causes an accumulation of glycogen in all tissues. The myocardium is notably affected in most individuals; therefore, cardiorespiratory difficulties and heart enlargement (cardiomegaly) are typical of advanced Type II disease.

Type V glycogen storage disease or *McArdle's disease* is the rarest form of the glycogen storage diseases. It manifests itself in the form of fast exhaustion in seemingly normal muscle and probably reflects the *ab-*

sence of adequate *phosphorylase activity*. Such inadequacies lead to energy deficits and may account for painful muscle cramps experienced after exercise.

Less common Type VII glycogen storage disease has characteristics similar to Type V forms, but muscular phosphofructokinase activity is absent instead of phosphorylase.

Although the genetic basis of inheritance is uncertain for all forms of glycogen storage diseases, Type V storage disease, phosphorylase-*b* kinase-type deficiencies, and others appear to be autosomally linked recessive traits.

THE TRICARBOXYLIC ACID (TCA) CYCLE AND THE AEROBIC PHASE OF CARBOHYDRATE METABOLISM

The end products of anaerobic glucose oxidation via the glycolytic pathway retain considerable stores of energy. Assuming that glycolysis products are limited to lactic acid production, *only about 7% of the free energy of glucose is liberated*. Evolutionary processes in all animals and many other organisms extract higher yields of free energy (stored as ATP) from glycolytic end products by the aerobic oxidation of glucose to carbon dioxide and water.

This process, recognized as respiration, requires the orchestration of three metabolically interrelated processes: the tricarboxylic acid (TCA) cycle, electron transport, and oxidative phosphorylation. All three of these coordinated processes occur in the mitochondria.

In cases of strictly anaerobic glycolysis, glucose is converted to lactic acid as a terminal end product (typical of fermentation). In other cases, in which glycolysis is only a prelude to the aerobic oxidation of glycolytic products by the TCA cycle, *pyruvate serves as a key link between anaerobic and aerobic phases of metabolism.*

The TCA cycle is a cyclic sequence of chemical reactions whereby acetic acid, linked to coenzyme A in the form of acetyl-CoA, is oxidized to carbon dioxide and water. In the case of sugar metabolism, pyruvic acid derived from glycolysis supplies one source of acetyl-CoA. Pyruvic acid is oxidized by a *multienzyme complex* in a complicated series of reactions. This *complex,* called *pyruvic acid dehydrogenase* (PAD), requires the combined participation of five vitamins and/or coenzymes:

$$\text{Pyruvate} + \text{CoASH} + \text{NAD}^+ \xrightarrow[\substack{\text{TPP,} \\ \text{lipoamide,} \\ \text{FAD}}]{\substack{\text{PAD} \\ \text{complex}}}$$

$$\text{acetyl-CoA} + \text{CO}_2 \ \text{NADH} + \text{H}^+$$

The PAD enzyme complex consists of three discrete enzymes each having a specific function: (1) *pyruvate dehydrogenase* coupled with thiamine pyrophosphate (TPP) decarboxylates pyruvate; (2) *dihydrolipoyl transacetylase,* linked to lipoamide as a prosthetic group, mediates oxidation of the C_2 unit (acetyl) and facilitates its transfer to coenzyme A; and (3) *dihydrolipoyl dehydrogenase,* plus a FAD coenzyme, assists in the regeneration of the oxidized form of lipoamide. The regenerated FAD is further tied to the reduction of NAD^+. Consult Chapter 4 (multienzyme systems) for a detailed prospectus of these reactions.

According to the TCA cycle, acetyl-CoA produced from PAD complex condenses with oxaloacetate (C_4) to form citrate (C_6). This reaction is catalyzed by *citrate synthase,* referred to in early literature as a "condensing enzyme." Citrate is converted to a *cis*-aconitate intermediate as an artifact of an *aconitase*-mediated reaction, but this intermediate is short-lived and ultimately yields isocitrate (Figure 16.11).

The oxidation of isocitrate to oxalosuccinate, as a result of *isocitrate dehydrogenase*, produces reduced nucleotide ($NADH(H^+)$) from NAD^+. Decarboxylation of oxalosuccinate forms α-ketoglutarate. This α-keto acid is subsequently decarboxylated by another enzyme complex known as α-*ketoglutarate dehydrogenase.* This complex uses many of the same coenzymes as the PAD complex, and the principal product of the reaction is a

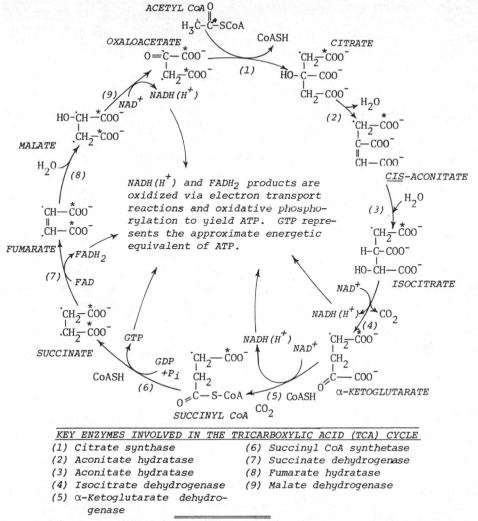

KEY ENZYMES INVOLVED IN THE TRICARBOXYLIC ACID (TCA) CYCLE

(1) Citrate synthase   *(6) Succinyl CoA synthetase*
(2) Aconitate hydratase   *(7) Succinate dehydrogenase*
(3) Aconitate hydratase   *(8) Fumarate hydratase*
(4) Isocitrate dehydrogenase *(9) Malate dehydrogenase*
(5) α-Ketoglutarate dehydro-
    genase

FIGURE 16.11. The tricarboxylic acid (TCA) cycle or "citric acid" cycle represents a series of important biochemical reactions that underly aerobic oxidation processes. Note that the introduction of carbon atoms into the cycle as "acetyl-CoA" have been specifically designated as (˙) and (*) in order to show their relative location in TCA intermediates throughout the cycle. Although these designations reflect results of radioactive tracer studies, the symmetrical appearance of succinate serves to disperse the "tagged carbon atoms" (i.e., from succinate to oxaloacetate).

thioester of succinic acid known as succinyl-CoA. Reduced pyridine nucleotide (NADH (H$^+$)) is also produced as a by-product of the reaction.

Hydrolysis of the succinyl-CoA thioester is accomplished by a substrate level phosphorylation of GDP to GTP plus succinate formation. Oxidation of succinate by a flavin-linked *succinate dehydrogenase* produces FADH$_2$ and fumarate.

The presence of water and a *fumarase* enzyme yield L-malate from fumarate, which is further oxidized to oxaloacetate with a concomitant production of NADH(H$^+$) from NAD$^+$. This last reaction serves as a regenerative step to ensure adequate amounts of oxaloacetate, which are critical for perpetuating the TCA cycle.

Aside from the role of oxaloacetate as a critical TCA intermediate, it also serves as a key intermediate in the synthesis of glucose from nonsugar compounds in the process known as *gluconeogenesis*. The consumption of oxaloacetate through gluconeogenesis plus its perpetuation of the TCA cycle requires the operation of "replenishment reactions" for oxaloacetate. These reactions are typically called *anaplerotic reactions*.

For oxaloacetate, an anaplerotic reaction is ensured by the carboxylation of pyruvate conducted by *pyruvate carboxylase* (650,000 M.W.):

$$\text{Pyruvate} + CO_2 + \text{ATP} + H_2O \xrightarrow[\Delta G^{0'} = -0.5 \text{ kcal/mol}]{Mn^{2+}}$$
$$\text{oxaloacetate} + \text{ADP} + P_i$$

This reaction requires two steps along with the enzyme-linked presence of biotin.

STEP 1:

$$E\text{—Biotin} + \text{ATP} + CO_2 + H_2O \rightleftarrows$$
$$E\text{—carboxybiotin} + \text{ADP} + P_i$$

STEP 2:

$$E\text{—Carboxybiotin} + \text{pyruvate} + H_2O \rightleftarrows$$
$$\text{oxaloacetate} + E\text{—biotin}$$

According to processes of the TCA cycle, all carbons entering into the scheme eventually produce carbon dioxide. The joint production of $NADH(H^+)$, $FADH_2$, and GTP all represent compounds rich in potential metabolic energy.

Although the discussion presented here suggests that activated acetates in the form of acetyl-CoA enter the TCA cycle from the glycolytic pathway, activated acetates arising from protein catabolism and β-oxidation of fatty acids may also enter into the cycle. The TCA cycle is also important because it supplies carbon skeletons for the biosynthesis of numerous amino acids including glutamate, aspartate, and others.

THE TCA CYCLE, ELECTRON TRANSPORT, AND OXIDATIVE PHOSPHORYLATION

Hydrogen atoms present in reduced pyridine ($NADH(H^+)$) and flavin ($FADH_2$) nucleotides represent electrons ultimately derived from the oxidation of glucose. These reduced nucleotides convey electrons to the electron transport chain where the reduced nucleotides are oxidized with the concomitant reduction of oxygen to water. This process is mediated by a series of mitochondrial enzymes that also produce ATP by the process of oxidative phosphorylation. ATP production here results from the descent of electrons from a negative electrochemical potential to a more positive potential characteristic of molecular oxygen ($-0.32 \rightarrow +0.82$ V). Details regarding oxidative phosphorylation mechanisms in the mitochondrial membrane are explained in Chapter 11.

THE GLYOXYLATE CYCLE IS A VARIATION OF THE TCA CYCLE

In some bacteria such as *Escherichia coli*, germinating plant seeds, and some other plant tissues, activated acetates are used for synthetic reactions *instead of* the catabolic reactions carried out in the TCA cycle. These reactions occur in the glyoxylate cycle. Unlike the TCA cycle, isocitrate degradation does not involve the action of isocitrate dehydrogenase. Instead, *isocitrate lyase* cleaves isocitrate to form succinate and glyoxylate. The glyoxylate subsequently condenses with acetyl-CoA under the auspices of *malate synthase* to produce malate.

Malate is then oxidized to oxaloacetate, which can condense with acetyl-CoA to yield citrate for another course about the cycle (Figure 16.12). Succinate produced within the glyoxylate cycle may be converted into oxaloacetate through fumarate and malate intermediates.

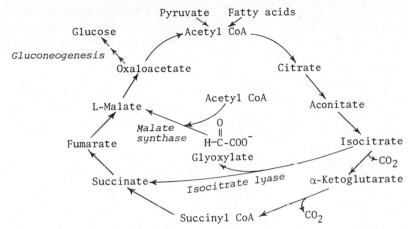

FIGURE 16.12. The glyoxylate cycle exists in certain bacteria and germinating plant seeds. The reactions that release CO_2 are bypassed by the glyoxylate pathway, and acetyl-CoA serves as a carbon source for the biosynthesis of compounds such as glucose. In the case of plants, the key glyoxylate enzymes including isocitrate lyase and malate synthase occur in cytoplasmic organelles known as glyoxysomes.

The conversion to oxaloacetate into phosphoenol pyruvate by the action of *phosphoenol pyruvate kinase* then serves as an initial step for the biosynthesis of glucose (gluconeogenesis).

For every turn through the glyoxylate cycle, two molecules of acetyl-CoA are required for each molecule of succinate produced.

Since isocitrate lyase and malate synthase *do not exist* in humans, the glyoxylate pathway does not occur and other mechanisms have evolved for carbohydrate production from simple compounds. As indicated above, the glyoxylate pathway has special importance in germinating plant seeds where lipids stored as triacylglycerols are oxidized to produce acetyl-CoA. This activated acetate is ultimately used for synthesizing glucose. The glyoxylate cycle clearly provides many organisms with a mechanism for subsisting on acetate as a major or sole source of carbon.

SOME ALTERNATIVE PATHWAYS FOR GLUCOSE CATABOLISM

Aside from the glycolytic metabolism of carbohydrates and the aerobic oxidation of glycolytic products by the TCA cycle, other routes for carbohydrate and sugar acid derivatives exist in many types of cells. These metabolic routes include the pentose phosphate pathway or hexose monophosphate pathway, and the glucuronic acid oxidation pathway.

The Hexose Monophosphate (HMP) Pathway

Glucose 6-phosphate and other suitable sugar derivatives participate in the HMP pathway, also called the *phosphogluconate pathway*. The HMP pathway has fundamental physiological importance in humans and many other animal cells and tissues because it

1. *Generates reducing power* in the form of $NADPH(H^+)$, which is used in reductive biosynthetic reactions (especially lipid synthesis).

2. *Produces pentoses*, including ribose 5-phosphate from hexoses, which are critical in the formation of ATP, NAD^+, FAD, RNA, and DNA nucleotides.

3. *Provides* a mechanism whereby four-, five-, six-, and seven-carbon sugars can be interconverted or synthesized to supply *glycolytic intermediates* (including D-fructose 6-phosphate and D-glyceraldehyde 3-phosphate).

The HMP pathway is prominent in the mammary glands, adipose tissues, testis, adrenal cortex, and hepatic tissues. The HMP route is also responsible for consuming nearly 10% of the glucose metabolized by the red blood cells.

The obvious occurrence of this pathway in red cells is tied to its production of NADPH(H$^+$). Adequate stores of this reduced pyridine nucleotide are necessary for maintaining the reduced form of glutathione, whose presence is necessary for the maintenance of erythrocyte architecture and hemoglobin function.

Initial stages of the HMP pathway (Figure 16.13) are instigated by the oxidation of glucose 6-phosphate to form glucono-δ-lactone-6-phosphate. This reaction is catalyzed by glucose 6-phosphate dehydrogenase, which requires NADP$^+$ and Mg^{2+} as a cofactor. The *second reaction* is prompted by 6-phosphogluconolactonase, which converts glucono-δ-lactone-6-phosphate into 6-phosphogluconic acid. Conversion of 6-phosphogluconic acid into ribulose 5-phosphate occurs in a *third reaction* mediated by 6-phosphogluconic dehydrogenase. This enzyme requires NADP$^+$ and Mg^{2+} as a cofactor.

These three reactions reflect oxidation events; however, the remaining reactions are *nonoxidative*. Nonoxidative reactions include the conversion of ribulose 5-phosphate into ribose 5-phosphate by phosphopentose isomerase, and the conversion of ribulose 5-phosphate to xylulose 5-phosphate under the auspices of a *phosphopentose epimerase*.

The three remaining reactions of the HMP pathway require a *transketolase, transaldolase,* and *another transketolase* reaction in that sequence. The transketolase transfers a two-carbon unit from the ketose sugar, xylulose 5-phosphate, to an aldose present in the form of ribose 5-phosphate. This action results in the formation of sedoheptulose 7-phosphate (a metabolically rare sugar) and glyceraldehyde 3-phosphate.

Transaldolase catalyzes the synthesis of fructose 6-phosphate and erythrose 4-phosphate from sedoheptulose 7-phosphate and glyceraldehyde 3-phosphate. Finally, the transketolase promotes the synthesis of fructose 6-phosphate and glyceraldehyde 3-phosphate from erythrose 4-phosphate and xylose 5-phosphate.

Common Enzyme Deficiencies Involving the HMP Pathway

The foremost congenital enzyme deficiency of the HMP pathway involves *glucose 6-phosphate dehydrogenase* (G-6-PD). Low activities or an absence of G-6-PD are linked to the occurrence of hemolytic anemia in those afflicted with this *X-linked recessive trait*. Evidence of hemolytic anemia is related to the inadequate maintenance of reduced glutathione, which normally relies on NADPH(H$^+$) production by the HMP pathway. Absence of glutathione, which behaves as a *thiol buffer,* may result in an undefended oxidation of erythrocyte constituents. In the event that *ferrohemoglobin* (Fe^{2+}) is oxidized to *ferrihemoglobin* (Fe^{3+}) (i.e., methomoglobin), normally maintained stores of reduced glutathione can easily become exhausted. Loss of this thiol-buffering capacity subsequently leads to an unchecked oxidation of globin sulfhydryl groups in hemoglobin. The oxidized hemoglobin then becomes denatured, assumes an insoluble form having a coccoid shape (*Heinz bodies*), and adheres to the inner membrane of the erythrocytes. Subsequent events leading to hemolytic anemia are unclear, but it is speculated that these membrane interactions prohibit the normal deformation capacity of erythrocytes and instigate hemolysis.

The G-6-PD deficiency is often observed in human populations where malaria is endemic. In these cases the deficiency may have a selective advantage since oxidative stresses engendered in red cells by the development of intraerythrocytic parasites cause hemolysis before parasite maturation and propagation.

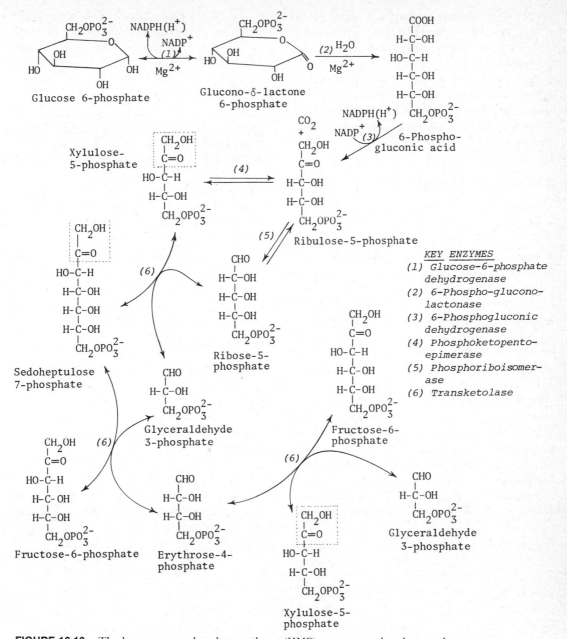

FIGURE 16.13. The hexose monophosphate pathway (HMP) or pentose phosphate pathway.

At least 100 million people suffer from G-6-PD deficiency, and about 10% of American blacks and 8 to 20% of West African blacks demonstrate the problem. *Explosive hemolytic episodes* can occur in those who are deficient and encounter oxidant drugs (antimalarial drugs including primaquine), water-soluble analogues of vitamin K, quinine (also used in popular carbonated beverages), and numerous drugs or environmental contaminants.

Enzyme deficiencies other than G-6-PD found in the HMP pathway can cause similar hemolytic results, but these situations are relatively uncommon compared with G-6-PD deficiency. Deficiencies of *6-phosphogluconate dehydrogenase* have been implicated with hemolysis but the mechanisms are not clear. Riboflavin deficiencies may also spur on *glutathione reductase* deficiencies that contribute to hemolysis (seemingly caused by G-6-PD), since an FAD coenzyme is required for glutathione reductase activity.

The Glucuronic Acid Oxidation Pathway

Glucose can be converted into D-glucuronic acid, which serves as an important compound in the renal excretion of phenolic compounds (Figure 16.14) including drugs, cellular detoxification of endogenous bilirubin, and the production of ascorbic acid and polysaccharides (e.g., hyaluronic acid), which use glucuronic acid as a precursor.

D-Glucuronic acid originates from the transformation of glucose to UDP-glucose (Figure 16.15). UDP-glucose is oxidized to UDP-glucoronic acid and eventually reduced by way of a D-glucoronic acid intermediate to L-gulonic acid. This C_6 sugar acid can undergo lactone ring formation by an *aldolactonase* to form L-gulonolactone. A flavoprotein-linked dehydrogenation by *gulonolactone oxidase* serves as an associated mechanism for the production of L-ascorbic acid (vitamin C). This synthesis route exists in many plants and animals with the main exceptions being guinea pigs, monkeys, humans, some tropical avian species, some fishes, and the Indian fruit bat—all of which require a dietary source of ascorbate.

GLUCOSE IS METABOLIZED BY VIRTUALLY ALL CELLS AND TISSUES OF THE HUMAN BODY

The schematic outline of glucose metabolism in Figure 16.16 shows that glucose oxidation by anaerobic and aerobic routes both provide an important source of metabolic energy as well as many different intermediates critical for normal metabolism. Specific fates of glucose in the major tissue cell divisions are outlined below.

I. Brain Tissue

Glucose enters brain tissue cells, where it is metabolized by the glycolytic pathway to form pyruvate. Pyruvate decarboxylation and the subsequent formation of activated acetates in the form of acetyl-CoA ensure a regular supply of C_2 intermediates for the perpetuation of the TCA cycle in individual cells.

Maintenance of the TCA cycle also provides a consistent supply of reduced nucleotides necessary for oxidative phospho-

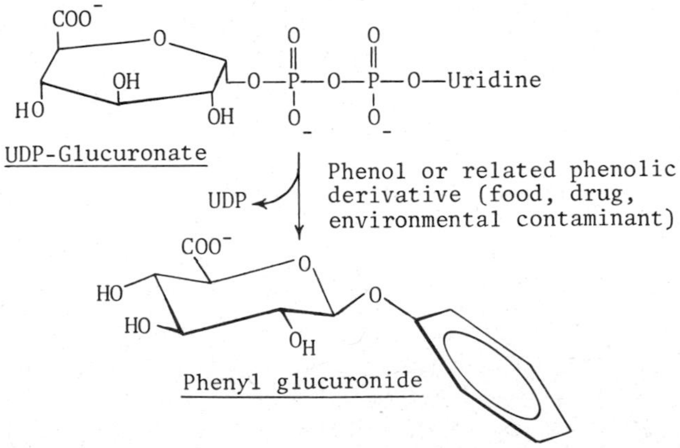

FIGURE 16.14. Phenyl glucosiduronide formation is an important step in the excretion of phenolic compounds.

rylation processes in the mitochondria as well as a supply of nucleotide triphosphate present as GTP. The HMP pathway also operates in brain tissue forming $NADPH(H^+)$, which is used as reducing power in many biosynthetic routes.

II. Erythrocytes (Red Blood Cells)

Glucose enters erythrocytes, where it is converted to lactate by the EMP pathway. The lactate is discharged into the plasma for eventual transfer and processing by hepatic tissues into glucose (see Cori cycle and gluconeogenesis that follow). Glucose produced from lactate by the liver can be returned to the circulatory system for eventual oxidation by erythrocytes (or other cells and tissues) or the glucose can be polymerized by the liver into glycogen.

Glucose metabolism also proceeds according to the HMP pathway to form $NADPH(H^+)$, which is used for the maintenance of reduced glutathione.

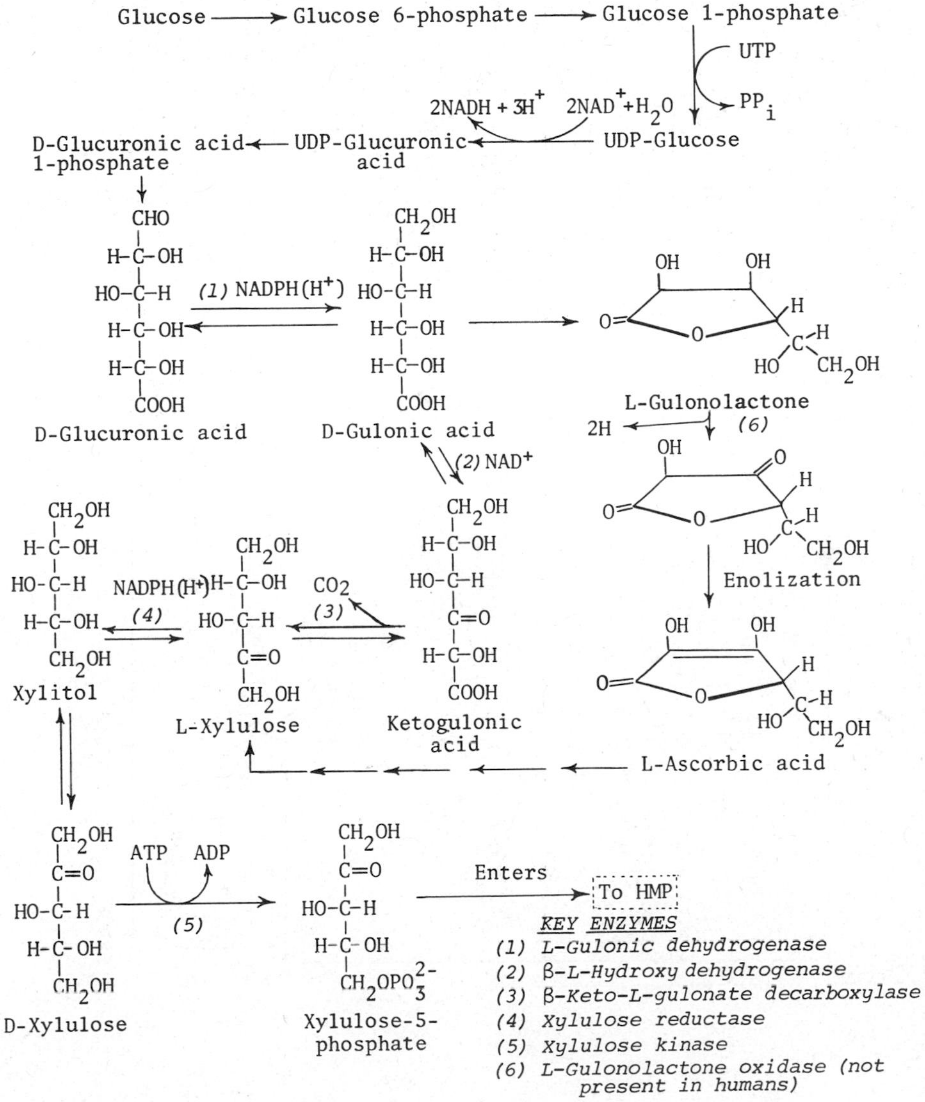

FIGURE 16.15. Glucose metabolism via the glucuronic acid oxidation pathway.

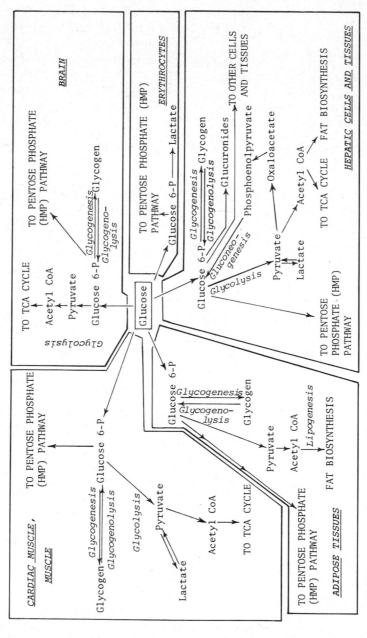

FIGURE 16.16. Potential fates for glucose in cells and tissues.

III. Hepatic Tissues and Cells

Glucose present in hepatic cells is oxidized by both the EMP and HMP pathways as well as the TCA cycle. Glucose may also be polymerized to yield glycogen during glycogenesis, or pyruvate produced from the EMP pathway can be used for lipogenesis and fat production. The liver also produces glucose from pyruvate and oxaloacetic acid in the process of gluconeogenesis.

IV. Adipose Tissue Cells

The glycolytic metabolism of glucose in adipose tissue cells leads to the *de novo* synthesis of fatty acids (lipogenesis). Reducing power (NADPH(H$^+$)) for fatty acid biosynthesis is provided by the HMP pathway occurring in individual adipose cells.

V. Skeletal and Cardiac Muscles

Glucose metabolism in skeletal and cardiac (heart) tissues proceeds by a variety of routes that include the EMP and HMP pathways in conjunction with the TCA cycle as well as glycogenesis. Aerobic oxidative pathways for the energy-efficient utilization of glucose predominate in cardiac muscle over skeletal muscle, but cardiac muscle can obtain temporary energy supplies of energy from anaerobic pathways under oxygen duress.

GLUCONEOGENESIS AND GLUCOSE HOMEOSTASIS ENSURE CONSTANT LEVELS OF BLOOD GLUCOSE

Minor fluctuations associated with the dietary absorption of carbohydrate can temporarily elevate blood glucose concentration levels to 120 to 130 mg/dL. Aside from this temporal condition, the liver is responsible for the maintenance of constant blood glucose levels through processes of glycogenolysis and gluconeogenesis.

Although both of these fundamental processes are responsible for mobilizing or producing glucose for metabolism, the broad scope of glucose regulation in the animal body is effected by *glucose homeostasis*. Homeostatic mechanisms of the body rely on complex hormonal interactions operating between organs and tissues that respond to the presence of specific circulating biochemical intermediates and substrates present in blood plasma.

In cases of dietary carbohydrate deprivation, hepatic glycogen stores can provide normal blood glucose concentrations for about 24 h by glycogenolysis. Continued deprivation beyond this period requires that blood glucose concentration levels be maintained by gluconeogenesis.

GLUCONEOGENESIS

Gluconeogenesis is the biochemical process responsible for the *de novo* synthesis of glucose from lactate, pyruvate, and other noncarbohydrate compounds including amino acids and glycerol. Superficial inspection suggests that gluconeogenesis is the reverse of glycolysis, but this is not true. Many gluconeogenic steps are the reverse of glycolysis, *provided* the energy differences between key glycolytic intermediates are only minor. In those steps where large changes in free energy exist between glycolytic intermediates, *bypass reactions have evolved* that avert otherwise insurmountable thermodynamic demands.

Figure 16.17 outlines the gluconeogenic pathway. Note that gluconeogenesis is an expensive process in terms of ATP consumption, and four distinct bypass reactions at steps 1, 2, 9, and 11 make gluconeogenesis possible.

STEP 1:

Pyruvate is used for the manufacture of oxaloacetate.

The production of oxaloacetate from pyruvate proceeds under the auspices of *pyruvate carbox-*

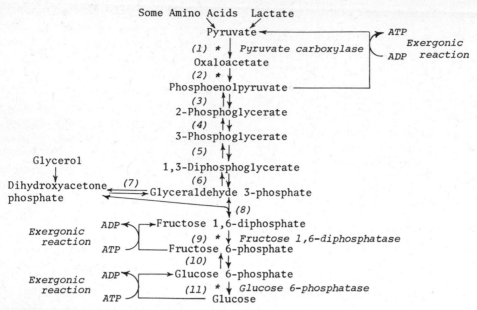

FIGURE 16.17. Gluconeogenesis. The cell has evolved a series of reactions to circumvent three exergonic reactions (*). The reactions with an arterisk are not shared by glycolysis, and the reactions without an asterisk are the reversible reactions of glycolysis.

ylase, whose coenzyme is biotin. Biotin serves as a carbon dioxide carrier during pyruvate carboxylation, and ATP is consumed.

STEP 2:
Conversion of oxaloacetate into phosphoenolpyruvate.

$$
\begin{array}{c}
\text{COO}^- \\
| \\
\text{C=O} \\
| \\
\text{CH}_3 \\
\textbf{Pyruvate}
\end{array}
+ \text{CO}_2 + \text{ATP} + \text{H}_2\text{O}
\xrightleftharpoons{\substack{\text{pyruvate} \\ \text{carboxylase}}}
\begin{array}{c}
\text{COO}^- \\
| \\
\text{C=O} \\
| \\
\text{CH}_2 \\
| \\
\text{COO}^- \\
\textbf{Oxaloacetate}
\end{array}
+ \text{ADP} + P_\text{i}
$$

$$\Delta G^{0'} = -0.5 \text{ kcal/mol}^{-1}$$

Pyruvate carboxylase is an *allosteric enzyme* that resides in the mitochondrion. This carboxylase is *activated by acetyl-CoA,* which induces the TCA cycle to synthesize more oxaloacetate. Since oxaloacetate is formed in the *mitochondrion* and gluconeogenesis requires the *cytosolic* presence of oxaloacetate, it must be translocated over the mitochondrial membrane. This process is accomplished by converting oxaloacetate into L-malate, which can diffuse into the cytosol where a *cytoplasmic malate dehydrogenase* reestablishes the malate structure by the reaction:

$$
\text{Malate} \xrightarrow[\substack{\text{NAD}^+ \quad \text{NADH(H}^+)}]{\substack{\text{cytoplasmic malate} \\ \text{dehydrogenase}}} \text{oxaloacetate}
$$

Oxaloacetate is converted into phosphoenolpyruvate through the action of *phosphoenolpyruvate carboxykinase:*

$$
\begin{array}{c}
\text{COO}^- \\
| \\
\text{C=O} \\
| \\
\text{CH}_2 \\
| \\
\text{COO}^- \\
\textbf{Oxaloacetate}
\end{array}
+ \text{GTP} \rightleftharpoons
$$

$$
\begin{array}{c}
\text{COO}^- \\
| \\
\text{C-OPO}_3^{2-} \\
\| \\
\text{CH}_2 \\
\textbf{Phosphoenolpyruvate}
\end{array}
+ \text{CO}_2 + \text{GDP}
$$

Note that the net reaction of Steps 1 and 2 above yield phosphoenolpyruvate:

Pyruvate + ATP + GTP + H_2O $\rightleftharpoons$
    phosphoenolpyruvate + ADP + GDP
                                    + P_i + 2H⁺

Steps 3 through 8 *in the gluconeogenesis pathway* are the reversible reactions of glycolysis.

STEP 9:

Conversion of fructose 1,6-diphosphate into fructose 6-phosphate.

Fructose 1,6-diphosphate yields fructose 6-phosphate in a reaction catalyzed by fructose 1,6-diphosphatase. This allosteric enzyme generally occurs in hepatic and kidney tissues but not adipose tissue.

Fructose 1,6-diphosphate $\xrightarrow[\text{fructose 1,6-diphosphatase}]{}$
                            fructose 6-phosphate

Step 10 of gluconeogenesis is a reversible reaction of glycolysis, but **Step 11** is a characteristic reaction of gluconeogenesis.

STEP 11:

Glucose 6-phosphate yields glucose.

Glucose 6-phosphatase promotes the conversion of glucose 6-phosphate into glucose and inorganic phosphate.

Glucose 6-phosphate $\xrightarrow[\text{glucose 6-phosphatase}]{}$
                        glucose + P_i

The key enzyme in this reaction is found in hepatic and kidney tissue but not in adipose tissue.

An *overview* of gluconeogenesis can be expressed in the net equation

action for the reversal of glycolysis exhibits a $\Delta G^{0\prime}$ of +20 kcal/mol:

2 Pyruvate + 2 ATP + 2 NADH(H⁺)
        + 2 H_2O $\longrightarrow$ glucose + 2 ADP
                        + 2 NAD⁺ + 2 P_i

The thermodynamically unfavorable (endergonic) energy requirement for the strict reversal of glycolysis, to yield glucose from pyruvate, is clearly circumvented by the occurrence of key exergonic reactions during gluconeogenesis.

Glucose Homeostasis

The overall biochemical, hormonal, and nutritional interactions that ensure constant glucose levels in the body are described by the process of *glucose homeostasis*. During conditions of adequate food consumption, glucose production by glycogenolysis and gluconeogenesis are negligible, and excess dietary glucose may be converted to fatty acids (lipogenesis) and triacylglycerols, or polymerized into glycogen. As outlined in Figure 16.18, insulin promotes a *decrease* of excess blood glucose concentration levels to maintain a normal blood sugar range of 80 to 100 mg/dL.

As insulin stimulates the storage of glucose, it also *inhibits* gluconeogenesis. This is quite contrary to the stimulative effects of epinephrine and glucagon on gluconeogenesis.

The obvious absence of blood glucose concentration levels below the normal range is met with a relative increase in the regulatory impact of the peptide hormone, glucagon, over that of insulin. In other words, energy consumption is favored over energy storage. The action of glucagon spurs the

$$2\ CH_3\overset{\overset{\displaystyle O}{\|}}{-}C-COOH + 4\ ATP + 2\ GTP + 2\ NADH + 2H^+ + 6H_2O \longrightarrow$$
$$C_6H_{12}O_6 + 2\ NAD^+ + 4\ ADP + 2\ GDP + 6\ P_i$$

The $\Delta G^{0\prime}$ of the net reaction for gluconeogenesis is −9.0 kcal/mol, while the net re-

mobilization of glucose from hepatic glycogen stores and gluconeogenesis from lactate,

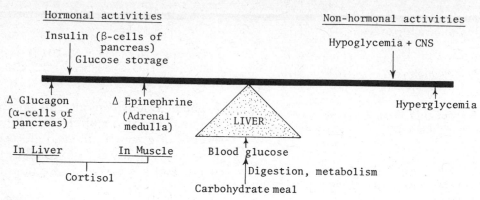

FIGURE 16.18. Hepatic glucose production contributes to the regulation of blood glucose. The process is under the influence of hormonal and nonhormonal factors including the central nervous system (CNS). Insulin and hyperglycemia inhibit hepatic glucose output, whereas glucagon and epinephrine levels sustained by cortisol and hypoglycemic conditions along with the CNS enhance hepatic glucose output. These counteracting regulatory mechanisms hold blood glucose levels within requisite demands of glucose-consuming cells and tissues of the body.

pyruvate, and amino acid precursors. Lipogenesis formerly instigated by the positive effect of insulin ceases, and glucose is preferably directed into the plasma. Depreciated rates of lipogenesis as triacylglycerols and/ or fatty acids are linked to a corresponding increase in fatty acid oxidation, and this serves as the major source of hepatic energy.

Apart from insulin, glucagon, and epinephrine, other participants in the homeostatic mechanisms of blood glucose involve hormones and other compounds produced by the adrenal glands, pituitary gland, gut, thyroid gland, adipose tissue, heart, and skeletal muscles (Figure 16.18). Figure 16.19 also details the participation of the liver in the metabolism of carbohydrates and other substances.

According to recognized time frames, the ingestion of ~100 g of glucose by a well-fed subject about to undergo fasting would reveal the utilization of the exogenous glucose within 4 h. The gradual absorption of glucose from the alimentary tract is met by increasing amounts of insulin in the peripheral circulation only *minutes* after consuming glucose. Continued absorption of glucose is mirrored by increased blood glucose concentrations and progressive elevation of insulin levels in blood. Blood glucose

and insulin levels jointly increase to a point when high blood sugar levels (hyperglycemia) transiently decrease before plasma insulin levels decrease to preeating normoglycemic levels. Since hepatic glycogenolysis and gluconeogenesis *cease during high insulin levels, falling levels* of insulin are ideally met with the reactivation of hepatic glucose production. In many instances, normal test subjects develop a *temporary hypoglycemic condition* between the time that plasma insulin levels approach normoglycemic levels and the time when hepatic glucose production must be resumed in a fasting state.

Regulatory abnormalities in reestablishing the normoglycemic state may produce a more prolonged period of hypoglycemia characterized as *reactive hypoglycemia*. This state is indicative of an overutilization of blood glucose with the net effect being an impaired ability for dealing with *recurrent fed–fasting states*.

Continued depletion of blood glucose during the next 8 to 12 h of the fed–fasting cycle should demonstrate the prevalence of hepatic glycogenolysis. Toward the end of this period, gluconeogenesis becomes the main source of glucose. After several days of continued fasting, metabolic energy stores are derived from the utilization of ketone

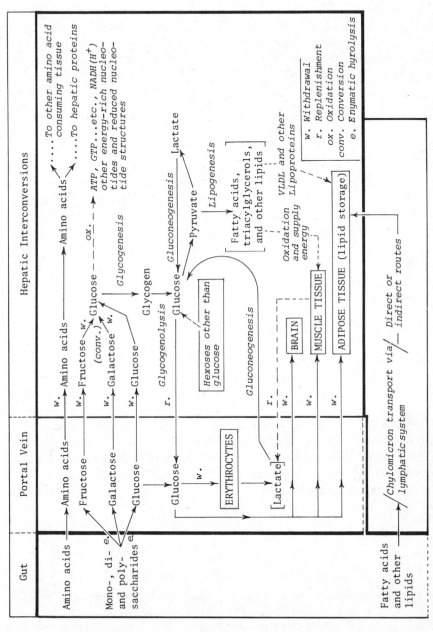

FIGURE 16.19. A simplified diagram illustrating the basic interconversions of glucose and some other sugars in the liver along with other food constituents.

bodies or fatty acid oxidation. So long as these potential energy sources are available as stores of metabolic energy, muscle proteins are spared. Progressive oxidation of fatty acids and ketone bodies ultimately results in muscle proteolysis and an overt physical-malnutritional degradation of the fasting subject. Clearly then, any diet designed for therapeutic purposes or mere weight loss must be planned with intelligence and recognition of potential metabolic consequences.

chemical route of lactate to the liver, followed by the return of glucose (produced by gluconeogenesis) to the muscle is known as the *Cori cycle* (Figure 16.20).

The Cori cycle is especially crucial to those tissues that periodically experience depressed oxygen levels. This is true for muscle tissues that produce lactate after frequent muscular contractions. Although muscular activities can significantly affect blood lactate levels, the glycogenolysis and any other glycogenic activities of muscle do not parallel those actions of the liver.

GLYCOLYSIS, GLUCONEOGENESIS, AND THE CORI CYCLE

As indicated above, insulin production by the pancreas has a greater regulatory impact than glucagon during a fed state. This enhances the rate of glycogenesis, whereas the rate of glycogenolysis decreases in hepatic and muscle tissues. The glycogen stores of muscles do not directly contribute to blood glucose concentration levels. Instead, lactate formed during glycolysis in the muscle travels by the bloodstream to the liver. Lactate conversion to glucose by gluconeogenesis may eventually lead to glucose storage as glycogen.

Glucose molecules tied into the structure of glycogen can be converted back into free glucose in the future and distributed to energy-starved metabolizing muscle cells for another round through glycolysis. This bio-

ABNORMAL GLUCOSE HOMEOSTATIC MECHANISMS CAN LEAD TO HYPOGLYCEMIA

Errors in normal homeostatic mechanisms of the body may result in *hypoglycemia* or *hyperglycemia*, depending on the particular metabolic problem.

Although hypoglycemia is widely touted as a common malady of American society, such claims are widely exaggerated. Furthermore, recognizable pathological hypoglycemia reflects depressed blood sugar concentrations to some point where definite symptoms of glucose deprivation are evident in the central nervous system. For adults, blood glucose levels of <45 to 50 mg/dL of serum or plasma are often indicative of a possible hypoglycemic disorder.

Habitual human eating habits involve re-

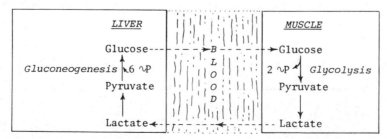

FIGURE 16.20. Schematic outline of the Cori cycle. Lactate produced within muscles is translocated to the liver, where it is converted into glucose via gluconeogenesis.

current feeding and fasting states. Consumed food (e.g., carbohydrates, glucose) is used for providing:

1. Oxidizable sources of food energy.

2. A source of reduced carbon for anabolic reactions.

3. Excess food substances to be stored for future use.

Abnormalities in the integrated and counteracting processes of glucose storage, as well as its future recall during periods of glucose deprivation, can lead to clear cases of hypoglycemia.

The occurrence of hypoglycemia may reflect any one of three possible types of *transient reactive hypoglycemia* (discussed above), including (1) *functional,* (2) *prediabetic* or *diabetic,* or (3) *alimentary* types. Indications of one of these three types of hypoglycemia usually becomes evident when a standard 100-g amount of glucose is given orally to a fasting test subject in a glucose tolerance test. A 5-h monitoring of plasma glucose levels at 30-min intervals shows that functional, mild diabetic and alimentary hypoglycemic conditions have distinctly different plasma glucose concentration profiles.

Hypoglycemia at 2 to 4 h after a meal, followed by epinephrine-induced metabolism, characterizes the functional type of hypoglycemia. When hypoglycemia occurs later than the functional type described, the positive glucose assimilation actions of insulin are delayed for about 1.5 h after the functional profile in cases of mild diabetes (Figure 16.21). The alimentary type of reactive hypoglycemia may occur in subjects who have had gastrointestinal surgery. Hypoglycemia of this type reflects an accelerated absorption rate for glucose.

At least 66% of hypoglycemia syndromes can be classified as functional hypoglycemia. Many cases are commonly linked to emotional conditions, but a substantial number of unrecognized causes may also contribute to the problem. Rebounds from the hypoglycemic state may become complicated by an asynchronous insulin release during otherwise normal glucose absorption.

Spastic or sluggish discharges of insulin under conditions of normal glucose absorption may be linked to mild or prediabetic hypoglycemia. Problems here typically begin when only a gradual reduction of hepatic glucose output into the plasma occurs, while the net removal of plasma glucose into hepatic glycogen stores is laggard. This reticent assimilation of plasma glucose into hepatic

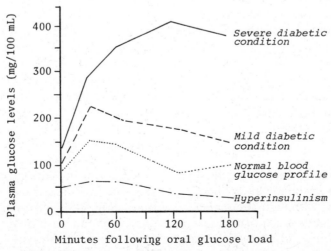

FIGURE 16.21. Some typical responses to a 100-g glucose oral intake.

glycogen encourages the development of increasing blood sugar concentration levels. The demonstrated presence of glucose in plasma after transhepatic circulation is not uncommon to mild prediabetes and early maturity-onset diabetes mellitus.

Alimentary hypoglycemia is not rooted in the abnormal biochemistry of metabolism. Instead, glucose absorbed from the gastrointestinal tract is very effectively utilized so that the low values of preeating (fasting) plasma glucose concentrations may be depressed to even lower values. Since alimentary hypoglycemia can originate from prior surgical procedures (e.g., vagotomy, pyloroplasty, gastrectomy), normally protracted periods of declining glucose absorption become preempted and fail to synchronously diminish with high insulin levels. The hyperinsulinism syndrome resulting from these events may produce reactive hypoglycemia the second or third hour after eating.

Reactive hypoglycemia can also be induced by certain dietary constituents that instigate an unusually flagrant production of insulin, such as leucine in the case of sensitive infants and children. Other reactive hypoglycemias may also stem from inherited abnormalities in enzyme kinetics and actions. This problem is typified by the failure of hepatic tissues to produce glucose when metabolic by-products accumulate (e.g., hereditary fructose intolerance, fructose 1-phosphate; galactosemia, galactose 1-phosphate).

DIABETES MELLITUS AND ABNORMAL GLUCOSE HOMEOSTASIS

Diabetes mellitus is a chronic metabolic disorder linked to biochemical anomalies in normal metabolism. Although the diabetes mellitus syndrome leads to abnormalities in carbohydrate, protein, and fat metabolism, glucose homeostasis is usually upset as a result of beta (β)-cell dysfunctions in the pancreas. Such pancreatic dysfunctions may result in production of little or no insulin whatsoever.

The onset of diabetes mellitus may result from a genetic diathesis (predisposition), obesity, or some other enigmatic cause including environmental exposures to chemicals, viruses, or other agents. There is still no conclusive cause for this malady, but the genetic component is probably one of the most important factors.

Two types of diabetes mellitus are recognized. These arbitrary divisions are based on the recognized disposition of fixed glucose levels of control or "normal" age-matched populations. One type, known as *juvenile* or *insulin-dependent diabetes,* appears in children, and the other type, known as *adult* (or maturity)-*onset* or *insulin-independent diabetes,* materializes in mature individuals.

Insulin-dependent types of diabetes display up to a 95% reduction of normal insulin levels or an absolute insulin deficiency. Insulin deficiency here results from *defective* or *absent* β-*cell function* in the pancreas. Uncontrolled forms of this diabetes are accompanied by distinct hyperglycemia, hyperlipoproteinemia, and a potential for fulminant ketoacidosis that can be fatal if not expeditiously treated.

For reasons that still remain to be firmly established, insulin interacts and binds to specific receptor sites on certain cell membranes. As in the case of many peptide hormones, this action stimulates the release or activity of at least one intracellular biochemical messenger that affects the concentration level or activities of certain enzymes in cells. Muscle and adipose cells as well as leukocytes depend on insulin for accelerating the cellular entrance of glucose. This is quite contrary to brain, nerve, retina, kidney, erythrocyte, and intestinal mucosa cells that are receptive to the entry of glucose at all times. Still other cells, including those of the liver, are not insulin dependent; but the hormone does affect the activation of glucokinase and thereby influence glucose metabolism.

While insulin-dependent diabetes reflects a deficit of insulin production, *insulin-independent diabetes* is probably related to *ineffectual insulin receptor sites* on hepatic, adipose, and muscle cell membranes. This predominantly adult form of diabetes exhibits a diabetic syndrome wherein hyperglycemia and hyperlipoproteinemia are common in the absence of ketoacidosis. Obesity may be a key predisposing factor for instigating the diabetic condition in genetically predisposed individuals, although nonobese adults may also be susceptible.

> *Whereas epinephrine and glucagon stimulate the mobilization of glucose from glycogen, insulin signals hyperglycemic conditions in the plasma and stimulates glycogenesis.*

As a consequence of insulin deficiency, hepatic glycogenesis is inhibited; glucokinase and phosphofructokinase activities are impaired; the glucose transit rate over cell membranes is suppressed; gluconeogenesis is enhanced; glucose 6-phosphatase activity increases in the liver; and both the hexose monophosphate pathway and lipogenesis in adipose tissues are suppressed.

These biochemical abnormalities reflected in glucose metabolism promote a *wide spectrum* of anatomic and metabolic problems that do not display distinct cause–effect relationships. Nevertheless, typical complications arising from diabetes include neuropathy, retinopathy (microvascular aneurysms and complications), kidney disease, and coronary artery disease. In fact, most of the deaths from advanced diabetes are linked in one way or another to vascular complications affecting the heart and kidneys.

Evaluation of Glucose in Diabetic Blood Samples

Glucose tolerance tests are useful for the routine evaluation of diabetic conditions. As previously discussed, suspected cases of hypoglycemia may require 4 to 5 h of monitoring of plasma glucose concentration levels; however, a 0.5-, 1.0-, 2.0-, and 3.0-h plasma glucose test is often adequate for most diabetic screening efforts.

Urine glucose levels should normally be nonexistent. If plasma glucose levels of 160 to 180 mg/dL exist, this range often exceeds the renal threshold and glucose appears in the urine. Lower renal thresholds are possible and up to 33% of low renal threshold subjects are liable to develop diabetes.

Typical glucose tolerance curves for normal subjects and a selected range of diabetic individuals are shown in Figure 16.21.

The determination of reducing sugars in urine is useful for a crude screening of potential diabetics and/or as a guide. It should be recognized, however, that a variety of reducing substances (Table 16.2) other than glucose can be found in urine and upset the validity of urine screening threshold limits. With the main exception of galactosuria, glucose is the only urinary sugar that has pathological significance.

Sporadic concentration levels of fructose, lactose, arabinose, xylose, and galactose can appear in urine specimens as a result of specific dietary constituents. Fructose is not uncommon in urine following the consumption of honey, fruits, and syrups; an infant's inability to metabolize galactose can result in galactosemia and obvious excretion of galactose; lactose can occur as a reducing sugar component in urine from women at the end of pregnancy or during lactation; and arabinose, ribose, and xylose are not uncommon urinary constituents in subjects who

Table 16.2. Reducing Sugars and Nonsugar Substances Found in Urine

Sugars	Nonsugars	
Arabinose	Ascorbic acid	Homogentisic acid
Fructose	Cinchophen	Isoniazid
Galactose	Creatinine	Ketone bodies
Glucose	Cysteine	Oxalic acid
Lactose	Formaldehyde	Salicylates
Maltose	Glucuronic acid	Salicyluric acid
Ribose	Hippuric acid	Sulfanilamide
Xylose		

have eaten prunes, plums, or cherries. Disaccharides such as maltose also reportedly occur with glucose in the urine of some diabetic individuals. Based on these recognized conditions, it is clear that a wide spectrum of reducing substances can be present in the urine, but these are not always linked to imbalances in glucose homeostasis or diabetes.

Since glucose tolerance curves generally reflect the status of glucose utilization over several hours, other analytical methods are required for long-term assessments of glucose status in blood. One typical long-term assessment method relies on the measurement of *glycosylated hemoglobin*.

Under normal conditions, erythrocytes contain hemoglobin A (HbA) as a major hemoglobin fraction and minor components designated as HbA_{1b} and HbA_{1c}. These minor fractions can account for 5 to 9% of the total hemoglobin, but in the case of diabetics, levels may become more than doubled. Of these minor components, the glycosylated hemoglobin present as HbA_{1c} represents 70% of the total glycosylated hemoglobin fraction and serves as *an index of past yet recent* glucose levels in the blood. HbA_{1c} is synthesized by a two-step process that is apparently independent of enzymatic activity (Figure 16.22). Glucose reacts with the β chain of HbA to form a labile aldimine (Schiff base), which readily dissociates to form free glucose and HbA. A small percentage of the labile intermediate is converted into a stable ketoamine (HbA_{1c}), which is slowly but consistently

formed over the 120-day life span of the erythrocyte(s). As a result of this reaction the concentration level of HbA_{1c} in the population of red cells mirrors the average blood glucose level during part of this period. Clinical evaluation of glycosylated hemoglobin is especially useful for evaluating the overall control of glucose in a diabetic subject. Unlike glucose tolerance tests and other simple blood tests for reducing sugars, glycosylated hemoglobins are not markedly affected by temporal fluctuations in blood glucose caused by exercise, insulin, or diet on the day of testing. Moreover, it is not imperative for the diabetic patient to fast before the test, and it is generally assumed that the level of glycosylated hemoglobin offers an index of glycemic blood levels in diabetics over a period of several weeks.

Under conditions of proper treatment for diabetes, glycosylated hemoglobin concentration levels drop to normal levels. For this reason many clinicians and studies of diabetic-related disorders favor the use of this test for monitoring the status of diabetic conditions as well as evaluating proper dosages of injected insulin.

Control of Diabetic Hyperglycemia

The treatment of diabetes mellitus must be guided by the type and severity of the syndrome. Insulin-dependent (juvenile) diabetes and its consequences are definitely more severe than adult-onset types (insulin-independent forms). Insulin-dependent forms

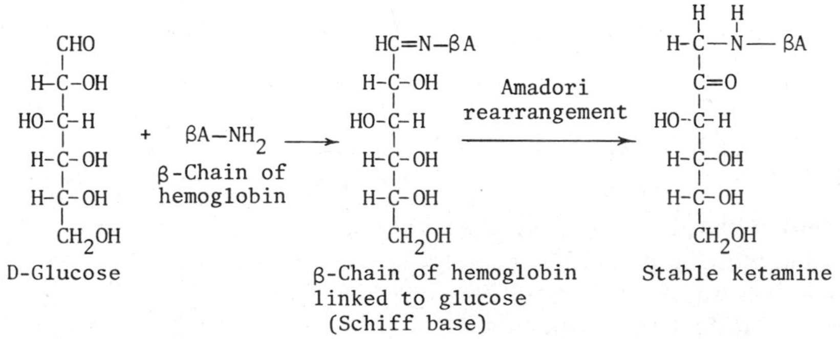

FIGURE 16.22. Formation of glycosylated hemoglobin.

of diabetes are necessarily controlled by injections of exogenous insulin isolated from beef pancreas or bacterial sources genetically programmed to produce insulin.

Diet, exercise, and other predisposing conditions exhibited by insulin-dependent diabetics can make this type of diabetes especially difficult to control. In fact, 50% of these diabetics demonstrate borderline responses to insulin that produce normoglycemia or hypoglycemia, and they are accordingly described as being unstable (labile) or brittle diabetics.

Insulin-independent diabetes, which usually appears in adults, may require insulin injections or the use of *oral hypoglycemic* agents (Figure 16.23) to control blood glucose levels. Typical oral hypoglycemic agents include sulfonylureas (sulfonamides) and the biquanide, phenformin hydrochloride. These substances are effective for stimulating the synthesis and release of endogenous insulin by mechanisms that are not entirely clear (Shen and Bressler, 1977).

Various oral hypoglycemic agents demonstrate different degrees of hypoglycemic action, biological half-lives, and metabolic routes. Moreover, demonstrated differences among these agents as hypoglycemic compounds may be linked to differential (1) absorption rates from the gastrointestinal tract, (2) renal excretion clearances, and (3) protein-binding interactions to blood components.

Most sulfonylureas are rapidly absorbed from the gastrointestinal tract. This absorp-

tion is followed by the formation of a protein-bound complex that transports the sulfonylureas through the blood. Some popular sulfonylureas such as chlorpropamide (Diabenese) appear in the blood 1 h after ingestion, when the agent begins to exert a hypoglycemic effect. The maximum effects of chlorpropamide and some other sulfonylureas are realized 2 to 4 h later, and their prolonged administration seems to enhance pancreatic β-cell functions, plus glucose tolerance for periods up to 24 h.

All oral hypoglycemics are eventually excreted or metabolized after their administration. Aside from uncertainties regarding the biochemical actions of hypoglycemics, their serum half-lives range from 3 to 35 h. The *minimal metabolism* of chlorpropamide ranks it among the most long-lived hypoglycemic agents. In fact, some of the popular sulfonylureas such as tolbutamide (Orinase) exhibit only 17% of chlorpropamide potency. Acetohexamide shows hypoglycemic activity, too, but its half-life is 6 to 8 h and its action requires preliminary reduction of the acetohexamide to 1-hydroxyhexamide in the liver. The reduced hexamide form exerts hypoglycemic effects ~250% over that of the acetohexamide for 12 to 24 h. This hypoglycemic effect is surpassed only by chlorpropamide, which has an effective duration of up to 60 h. Phenformin, on the other hand, has about the lowest biological half-life, 3.5 to 5 h, and its effective duration is 6 to 14 h depending on the method of oral administration.

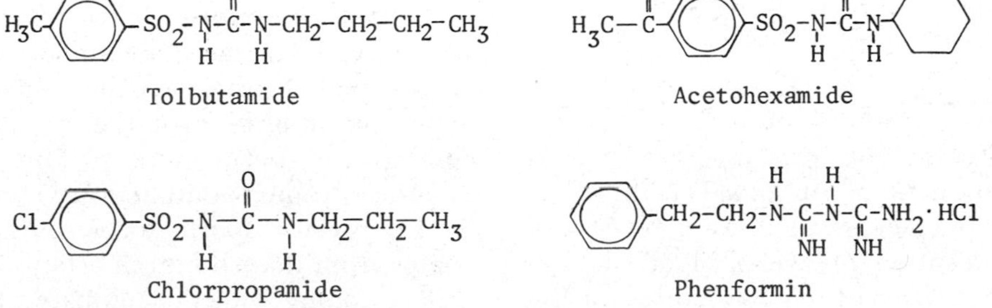

FIGURE 16.23. Some common oral hypoglycemic agents.

It should also be recognized in passing that many important hypoglycemic agents such as chlorpropamide and acetohexamide are excreted in the urine. Therefore, for those subjects who demonstrate *renal insufficiency, hypoglycemic actions may be prolonged.* Phenformin is a notable exception to this rule.

The oral hypoglycemic agents are generally suitable for moderately severe, non-ketotic insulin-independent (adult-onset) diabetes where dietary regulation alone is inadequate. They are not suitable, however, as an insulin-dependent diabetic treatment or for *brittle diabetics*, aside from some cases in which sulfonylureas may be used as an adjuvant to insulin injection. In the last case, the sulfonylureas can stabilize certain cases of labile diabetes.

For asymptomatic diabetics who demonstrate abnormal glucose tolerance, continuous use of chlorpropamide and related agents can produce a *normalization* of their glucose tolerance. In any case, treatment of diabetic conditions with sulfonylureas will not promote the return of normoglycemic levels when β-cell functions are *clearly absent.*

NORMAL AND ABNORMAL ROUTES FOR MONO- AND DISACCHARIDE METABOLISM

Monosaccharides such as fructose, galactose, and mannose are generally metabolized by the glycolytic pathway. Derivatives of these sugars that are immediately unacceptable for direct entry into the glycolytic route may eventually become acceptable if a suitable intermediate can be formed.

Fructose

Dietary fructose may exist as a free sugar or as a component of sucrose. Fructose does not elicit the release of insulin, yet it does maintain a sufficient level of blood sugar to prevent the stress of hypoglycemia. For this reason fructose has been labeled an ideal "crave-control food."

Humans generally seem to have a limited ability to handle fructose as a major or sole nutritive sugar. In some cases prolonged parenteral nutrition will result in liver damage. For this reason its chronic dietary use as a singular sugar is *questionable.*

Two principal pathways are responsible for fructose metabolism (Figure 16.24). In pathway 1, fructose is phosphorylated to fructose 1-phosphate by the action of a fructokinase. The phosphorylated fructose is cleaved to form dihydroxyacetone phosphate and glyceraldehyde by fructose 1-phosphate aldolase. The glyceraldehyde is then phosphorylated at the expense of ATP by a triose kinase to yield glyceraldehyde 3-phosphate. The two resulting C_3 molecules (dihydroxyacetone phosphate and glyceraldehyde 3-phosphate) condense to form fructose 1,6-diphosphate, which is suitable for entry into the glycolytic pathway.

Pathway 2, on the other hand, involves the phosphorylation of fructose in an ATP-consuming reaction mediated by hexokinase. The product is fructose 6-phosphate. Phosphofructokinase, at the expense of ATP, then produces fructose 1,6-diphosphate from fructose 6-phosphate. The fructose diphosphate subsequently enters the glycolytic pathway.

Hexokinase displays superior affinity for phosphorylating glucose over fructose. Since glucose occurs in the liver in higher concentrations than fructose, the amount of fructose phosphorylation (to fructose 6-phosphate) is small. On the contrary, fructose 6-phosphate formation prevails in adipose tissues, where fructose concentrations are relatively higher than glucose.

Notable amounts of fructose synthesis occur in the seminal vesicles of humans where it becomes incorporated into semen. Spermatozoic utilization of the sugar provides energy for sperm motility. The suggested synthesis route is outlined in Figure 16.25.

Energy for spermatozoic motility originates from mitochondrial actions that yield energy along with the ultimate production of carbon dioxide and water. The mitochondria in spermatozoa contain lactate de-

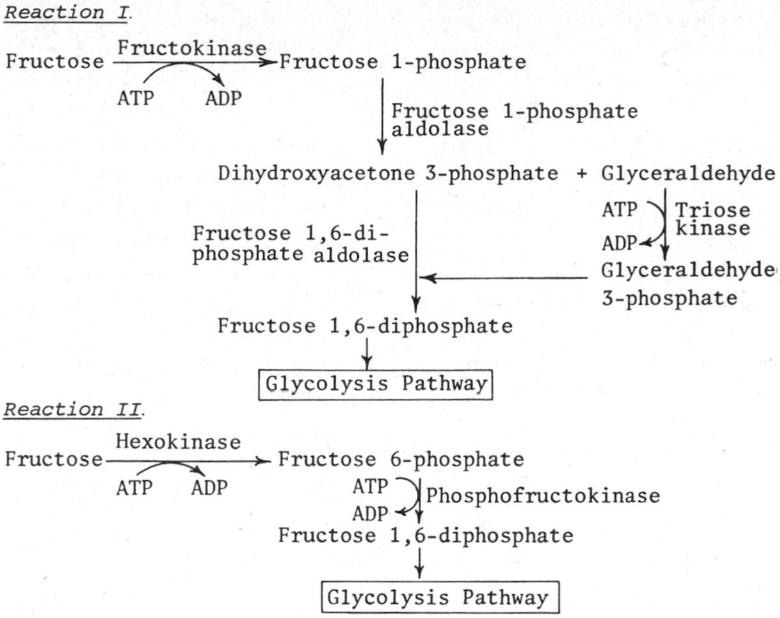

Reaction I.

Fructose $\xrightarrow[\text{ATP} \quad \text{ADP}]{\text{Fructokinase}}$ Fructose 1-phosphate

Fructose 1-phosphate aldolase

Dihydroxyacetone 3-phosphate + Glyceraldehyde

Fructose 1,6-di-phosphate aldolase

ATP ⟩ Triose
ADP ⟩ kinase

Glyceraldehyde 3-phosphate

Fructose 1,6-diphosphate

Glycolysis Pathway

Reaction II.

Fructose $\xrightarrow[\text{ATP} \quad \text{ADP}]{\text{Hexokinase}}$ Fructose 6-phosphate

ATP ⟩ Phosphofructokinase
ADP ⟩

Fructose 1,6-diphosphate

Glycolysis Pathway

FIGURE 16.24. Two principal pathways for fructose metabolism.

hydrogenase and, thereby, oxidize lactate obtained from fructolysis. By the use of this unique process, shuttle systems for the transport of reducing equivalents into the mitosol becomes unnecessary.

Abnormalities in fructose metabolism assume two different forms: (1) *essential fructosuria* and (2) *hereditary fructose intolerance*. These anomalies in fructose metabolism are generally inherited autosomal recessive disorders, although some exceptions may exist.

Fructosuria is a rather uncommon asymptomatic malady linked to a *fructokinase deficiency*. This enzyme occurs principally in liver intestinal mucosa and hepatic tissues where it mediates the earliest stages of fructose

metabolism in humans. Fructosuria is clinically evidenced during a *fructose-loading test*. When fructose administered to a subject accumulates in the blood and surpasses the renal threshold, fructose is discharged into the urine. Normal individuals respond to fructose loading by a peak in blood fructose concentration levels of ≤ 25 mg/dL, a peak value for blood fructose occurring about an hour after loading, and little or no urinary fructose excretion.

Hereditary fructose intolerance is not asymptomatic and is the result of low *fructose 1-phosphate aldolase* levels. In some cases fructoaldolase levels may show a 75% reduction from normal levels. Fructose loading in these

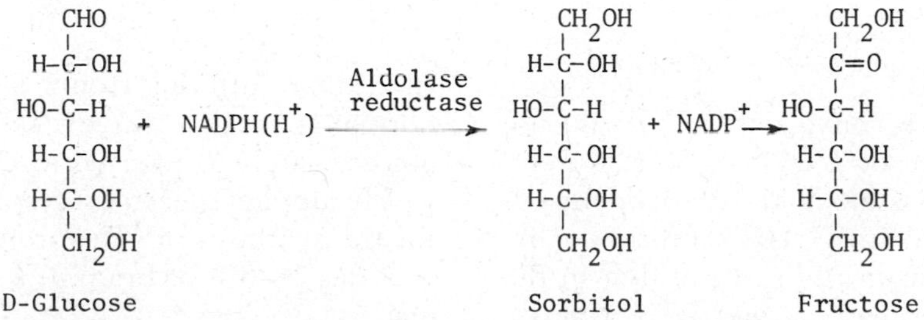

CHO
H–C–OH
HO–C–H
H–C–OH
H–C–OH
CH$_2$OH

+ NADPH(H$^+$) $\xrightarrow{\text{Aldolase reductase}}$

CH$_2$OH
H–C–OH
HO–C–H
H–C–OH
H–C–OH
CH$_2$OH

+ NADP$^+$ →

CH$_2$OH
C=O
HO–C–H
H–C–OH
H–C–OH
CH$_2$OH

D-Glucose   Sorbitol   Fructose

FIGURE 16.25. Synthesis of fructose in the male reproductive system.

subjects produces hypoglycemia and lowered concentration levels of inorganic plasma phosphate (P_i). Depressed plasma phosphate concentration levels probably reflect the rapid conversion of fructose to fructose 1-phosphate, whereas hypoglycemia probably reflects an *inhibition of phosphoglucomutase by fructose 1-phosphate*.

It has been noted for some time that fetal fructoaldolase preferentially splits fructose 1,6-diphosphate as opposed to fructose 1-phosphate. Thus, fetal enzyme persistence in adults may abet fructose 1-phosphate accumulation and fructosuria.

Galactose

Galactose can exist as a free sugar in some food systems, but it is more commonly a component of biopolymers, cerebrosides, and lactose. Biochemically complex forms of galactose can be liberated by enzymatic actions of the digestive tract, where it is readily absorbed and then converted into hepatic glycogen stores. One of the most common galactose sources is the hydrolysis of lactose into glucose and galactose.

The metabolism of galactose is initiated by *galactokinase*, which phosphorylates the monosaccharide to yield galactose 1-phosphate at the expense of ATP.

$$\text{Galactose} \xrightarrow[\text{ATP} \quad \text{ADP}]{\text{galactokinase}} \text{galactose 1-phosphate}$$

Galactose 1-phosphate reacts with UDP-glucose in the presence of *galactose 1-phosphate uridinyl transferase* to form UDP-galactose plus glucose 1-phosphate

Galactose 1-phosphate + UDP-glucose

$$\downarrow \text{galactose 1-phosphate uridyl transferase}$$

UDP-galactose + glucose 1-phosphate

UDP-galactose is converted to UDP-glucose in the presence of UDP-galactose 4-α-epimerase, which is an NAD^+-dependent enzyme (Figure 16.26). NAD^+ acquires the hydrogen atom from the C-4 position in the UDP-galactose forming $NADH(H^+)$ plus a 4-keto intermediate. Then, reducing power from $NADH(H^+)$ is directed to the reduction of the keto intermediate that yields UDP-glucose. Epimerization mechanisms at the C-4 position of the saccharidic moiety ensure glucose formation.

One classical inborn error of galactose metabolism is characterized by a decreased ability to convert galactose to glucose. This malady, called *galactosemia*, results in increased serum levels of galactose accompanied by galactosuria. Galactosemic disorders are generally attributed to a *hereditary galactokinase deficiency* or *deficiency of galactose 1-phosphate uridyl transferase*. Both enzyme deficiencies result in the reduction of galactose to *galactitol* (i.e., the corresponding sugar alcohol of the monosaccharide). Unchecked production of galactitol can affect the eye in two ways: (1) cataracts may result from galactitol deposition inside the lens and/or (2) osmotic pressure may become elevated inside the lens.

An overt absence of galactose 1-phosphate uridyl tranferase leads to hepatosplenomegaly (liver and spleen enlargement), impaired liver function, and eventual mental retardation, notwithstanding decreased dietary galactose consumption. The erythrocytes and liver cells alone respond to a uridyl transferase deficiency by trapping galactose 1-phosphate.

The sustained elimination of free galactose and lactose from the diet serves as a reasonable treatment for galactosemia. Eventually galactose deprivation results in the increased activity of an alternate enzyme for the formation of UDP-galactose.

$$\text{Galactose 1-phosphate} + \text{UTP} \underset{\text{phosphorylase}}{\overset{\text{UDP-galactose}}{\rightleftharpoons}} \text{UDP-galactose} + PP_i$$

Individuals suffering from galactosemia have normal epimerase activity; therefore, UDP-glucose can be converted to UDP-galactose, which supplies galactosyl residues necessary for the synthesis of glycoproteins and cerebrosides (4–6% of brain, 6% of white matter).

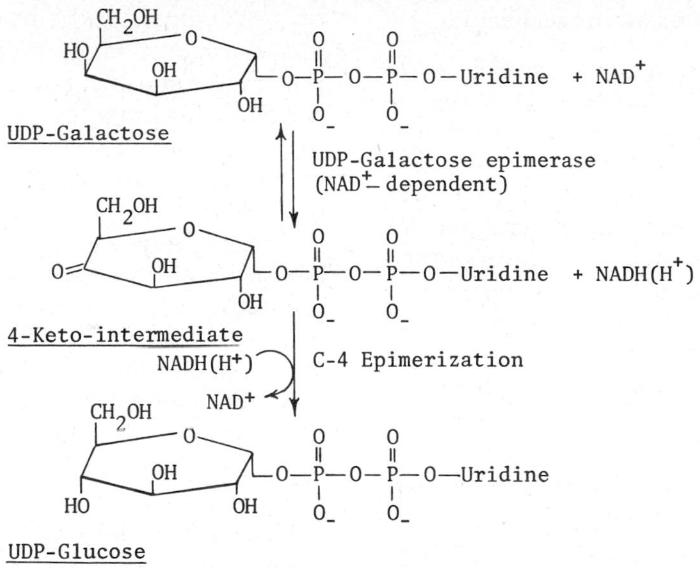

FIGURE 16.26. Transformation of UDP-galactose into UDP-glucose.

The clinical diagnosis of galactosemia largely relies on the recognition that galactose 1-phosphate uridyl transferase is absent in erythrocytes. Although past practices have relied on a UDP-glucose consumption test to detect this anomaly, current diagnostic methods rely on detecting UDP—[^{14}C]-galactose formed from [^{14}C]-galactose 1-phosphate in hemolyzed whole blood. Typical enzymatic levels for galactosemic individuals range from 0.7 to 1.4 mmol/L/h, whereas normal levels are nearly double these values.

Pentoses

Pentoses are ordinarily incorporated into a variety of metabolic routes, but the ingestion of large amounts of certain fruits can lead to pentosuria. Excretion rates of 1 to 4 mg/kg/day of pentose are not uncommon for adults adhering to a fruit-free diet. Normal excretion rates for arabinose and xylose become widely variable and higher, however, in cases of fever or drug therapy (cortisone, thyroid hormone, morphine).

Aside from xylose and arabinose excretion, L-xylulose (L-xyloketose) can reach high urinary excretion levels in cases of pento-suria. Excretion of this pentose is unrelated to the consumption of specific dietary pentose. Instead, this phenomenon represents an innocuous recessive autosomal trait that produces abnormal carbohydrate metabolism. Pentosurics commonly excrete 2.0 to 4.0 g of L-xylulose per day owing to an L-xylulose reductase deficiency. This form of abnormal carbohydrate metabolism is more common than fructosuria and principally affects Jews and Lebanese at population frequencies of 1:2000 to 1:5000.

Mannose

The monosaccharide mannose occurs in small quantities throughout many dietary constituents. Its metabolism is largely uncomplicated:

$$\text{D-Mannose} + \text{ATP} \xrightarrow{\text{kinase}}$$
$$\text{D-mannose 6-phosphate} + \text{ADP}$$

$$\text{D-Mannose-6-phosphate} \rightleftharpoons$$
$$\text{D-fructose 6-phosphate}$$

Mannose 6-phosphate equilibrates with fructose 6-phosphate and the phosphorylated fructose enters the glycolytic pathway.

Interconversions of Monosaccharides

Glucose, fructose, mannose, and galactose are interconvertible and may lead to the formation of various amino sugar derivatives (Figure 16.27). These derivatives can be used for syntheses of sialic acid, hyaluronic acid, heparin, chondroitin sulfate, dermatan sulfate, and keratan sulfate. Derivatives of these major monosaccharides can also be used for glycolipid, glycoprotein, and proteoglycan synthesis.

Glycolipids are sugar-containing lipids and glycoproteins are conjugated proteins that have one or more sugars as a prosthetic group. Proteoglycans display numerous protein cores covalently bonded to a long hetero-polysaccharide. One model of proteoglycan structure is illustrated in Figure 16.28. Note that keratan and chondroitan sulfates are also covalently bonded to the core protein structures.

Disaccharide Metabolism

Disaccharides such as lactose and sucrose are indigenous to foods in their most natural states. Lactose is the principal disaccharide of lacteal fluids and sucrose is the major disaccharide of many fruits and commercially important plant species from which it is refined for numerous food uses. Maltose, too, is a common disaccharide, but it often materializes as a result of enzyme starch hydrolysis.

The hydrolysis of dietary starches by amylase is efficient and nearly complete by the time starches reach the duodenum. Regardless of enzyme effectiveness, a mixture of maltotrioses, maltose, and limit dextrins commonly occur as a result of enzymatic actions. The di- and oligosaccharide derivatives of starch, sucrose, and lactose all require further hydrolysis to produce monosaccharide constituents before absorption from the intestinal lumen can occur. Specific disaccharide hydrolysis is achieved by *disaccharidases* (glycoproteins) whose presence is associated with intestinal mucosal cells.

Enzymes including α-*dextrinase* (*isomaltase*), *lactase*, and *sucrase* become positioned on the microvillar membranes in such a way that their hydrolytic active sites encourage substrate–enzyme interactions. The enzymes cited respectively produce glucose from

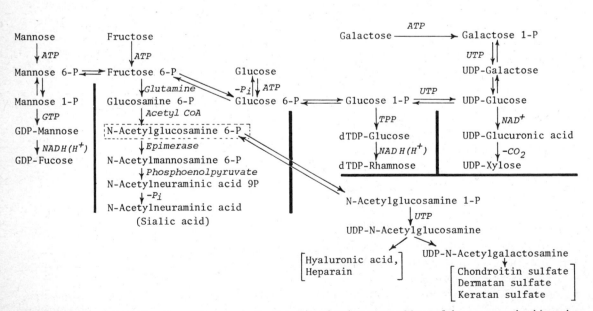

FIGURE 16.27. Interconversions of hexoses and metabolism of amino sugars. Many of the compounds ultimately formed including sialic acid, hyaluronic acid, chondroitin sulfate, dermatan sulfate, and keratan sulfate are especially important in the synthesis of glycolipids, glycoproteins, and proteoglycans.

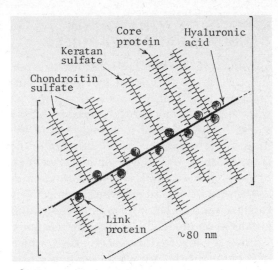

FIGURE 16.28. Studies of proteoglycan aggregates obtained from cartilage suggest that a hyaluronate backbone serves as a basis for the opposite but lateral attachment of proteoglycan monomers. Only a small portion of the schematic macromolecule has been detailed.

borders of the mucosal cells and are eventually transported through the portal vein to the hepatic tissues. Monosaccharide transport over brush border cells is stereospecific and requires the presence of Na^+ and specific transport proteins. Active absorption of glucose and galactose (plus xylose) rely on identical absorption mechanisms; however, the relatively high intraluminal fructose concentrations compared to intracellular levels of fructoses seems to drive the absorption of this ketose.

Disaccharidase Deficiencies

Inadequate disaccharidase concentrations or activities result in high concentrations of unabsorbed disaccharides—especially sucrose and lactose. As a consequence of this digestive aberration, these disaccharides serve as nutrients for intestinal bacteria that produce (1) copious amounts of short-chain fatty acids, (2) increased colonic acidity, and (3) methane, carbon dioxide, and hydrogen gas production. Clinical evidence of these actions can be indicated by the production of acid stools and elevated levels of hydrogen in the breath of enzyme-deficient subjects.

maltose and maltotriose (also isomaltose); glucose and galactose from lactose; and glucose and fructose from sucrose (Figure 16.29). The monosaccharides produced by these enzymatic actions are absorbed over the brush

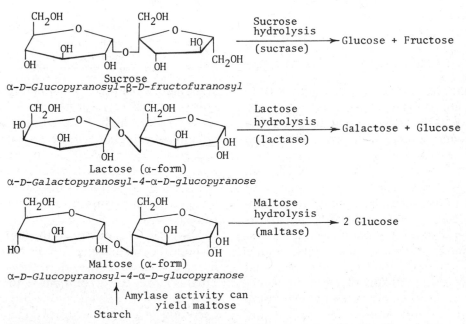

FIGURE 16.29. Actions of sucrase, lactase, and maltase on various disaccharide structures.

Anomalies in disaccharidase production can reflect a simple enzyme deficiency, the result of damage to the intestinal mucosa (sprue-type consequences), insufficient enzyme populations on microvillar membranes, excessive bacterial population growths on intestinal surfaces, as well as other unidentified factors that inhibit or eliminate enzyme actions.

Some apparent disaccharidase deficiencies, namely, lactase, prominently occur along racial lines. American Indians, Eskimos, Philippinos, Greek Cypriots, Asians, and American blacks display notable deficiencies of lactase on the brush border of the small intestine. These populations probably display lactase deficiencies as a result of historically

EFFECTS OF ETHANOL ON CARBOHYDRATE METABOLISM

Ethanol consumption has a distinct effect on the carbohydrate metabolism of humans; however, mechanisms involved are not always clear. Alcohol can be absorbed by the stomach, but most of it is absorbed in the intestine. The metabolism of ethanol in the liver involves two NAD^+-linked enzymes, namely, *alcohol dehydrogenase* and *acetaldehyde dehydrogenase*. These enzymes respectively catalyze the formation of acetaldehyde and acetate as outlined in the following reactions. Acetate is converted into acetyl-CoA, which may enter the TCA cycle or serve as a precursor to cholesterol and fatty acid synthesis.

$$CH_3CH_2OH + NAD^+ \xrightarrow[\text{[cytoplasmic reaction]}]{\text{alcohol dehydrogenase}} CH_3CHO + NADH + H^+$$

Ethanol → **Acetaldehyde**

$$CH_3CHO + NAD^+ \xrightarrow[\text{[mitochondrial reaction]}]{\text{acetaldehyde dehydrogenase}} CH_3COO^- + NADH + H^+$$

Acetaldehyde → **Acetate**

$$CH_3COO^- + ATP + CoASH \xrightarrow[\text{[mitochondrial reaction]}]{\text{condensing enzyme}} CH_3-\overset{\overset{\text{O}}{||}}{C}-SCoA + AMP + PP_i$$

Acetate → **Acetyl-CoA:**
to **fatty acid synthesis, TCA cycle, cholesterol synthesis**

low milk, consumption and/or a genetic component linked to an autosomal recessive trait.

Lactase is normally present in infants, but its activity wanes with age. For adult Caucasians, *alactasia* has a demonstrated occurrence of about 10%. In nearly all cases of lactase deficiencies, varying degrees of lactose intolerance are commonly demonstrated. Since lactose remains largely unhydrolyzed in the intestinal lumen, it can exert osmotic effects that cause water migration into the intestine. Increased fluid load, typically 150 mL of water for every 50.0 g of lactose, produces hyperperistalsis, diarrhea, and significant electrolyte and fluid losses.

Ethanol is oxidized by competing with lactate for stores of nicotinamide nucleotides (NAD^+). The higher the concentration of ethanol, the higher will be its competitive success for using NAD^+ as opposed to lactate. This means that lactate conversion to pyruvate may become minimal or nearly stopped if high concentrations of ethanol prevail.

Since lactate cannot pass through the mitochondrial membrane, it can accumulate in the cytoplasm and result in *lactate acidosis*. Furthermore, an accompanied depression of plasma pH can interfere with uric acid excretion in the kidney to produce gout in genetically susceptible individuals.

The oxidation of ethanol increases the $NADH(H^+):NAD^+$ ratio in the cytoplasm of hepatocytes to a point where the oxidation–reduction potential of the cytoplasm is altered. In an effort to reestablish a normal cellular $NADH(H^+):NAD^+$ ratio, part of the cytoplasmic $NADH(H^+)$ is oxidized in the malate–aspartate shuttle system, and part of the $NADH(H^+)$ is oxidized in coupled cytoplasmic reactions. Reactions mediated by lactic acid dehydrogenase and malate dehydrogenase are typical of these coupled reactions, as outlined below. In these reactions, the equilibrium is shifted toward formation of lactate and malate.

ably to the inhibition of hepatic gluconeogenesis caused by ethanol oxidation. Conflicting data suggest that ethanol may also affect the peripheral rate of glucose utilization. Still other evidence suggests that postprandial (after eating) hypoglycemia can be enhanced in normal subjects who consume a carbohydrate-rich meal along with high alcohol consumption. The actions of insulin on glucose (oral or intravenous) is potentiated under these conditions, but the pertinent mechanisms remain to be established.

Diabetics and obese subjects do not show potentiated insulin responses as a result of

---

Reaction 1

(a) $CH_3CH_2OH + NAD^+ \longrightarrow CH_3CHO + NADH + H^+$
  Ethanol   **Acetaldehyde**

(b) $CH_3-\overset{\overset{\displaystyle O}{\|}}{C}-COOH + NADH + H^+ \longrightarrow CH_3-\overset{\overset{\displaystyle OH}{|}}{\underset{\underset{\displaystyle H}{|}}{C}}-COO^- + NAD^+$
   Pyruvate   **Lactate**

Ethanol + pyruvate $\longrightarrow$ acetaldehyde + lactate

Reaction 2

(a) $CH_3CH_2OH + NAD^+ \longrightarrow CH_3CHO + NADH + H^+$
  Ethanol   **Acetaldehyde**

(b) Oxaloacetate + NADH + $H^+ \longrightarrow$ malate + NAD^+

Ethanol + oxaloacetate $\longrightarrow$ acetaldehyde + malate

---

Since reactions 1b and 2b do not favor the formation of pyruvate and oxaloacetate, respectively, phosphoenolpyruvate concentrations, formed from pyruvate and oxaloacetate precursors, become depressed. As a consequence of this action, gluconeogenesis, which requires adequate stores of phosphoenolpyruvate, can become inhibited.

Hypoglycemic conditions reportedly occur in fasting test subjects who consume ethanol. This hypoglycemic state is not linked to enhanced insulin production, but prob-

alcohol consumption. Furthermore, reactive hypoglycemia that follows glucose loading in normal individuals after alcohol consumption is not observed in obese, nondiabetic individuals. Speculation suggests that obesity may promote greater insensitivity to insulin and a corresponding inability to produce postglucose hypoglycemia observed in normal individuals. The low reserves of insulin in diabetes results in an ethanol-potentiated insulin response that is even less than that observed for obese nondiabetics.

Alcohol consumption and drugs including barbiturates and tranquilizers often employ similar enzyme systems for their me-

mers), pectins (polymerized galacturonic acids), and lignins, which are not carbohydrates but commonly occur with many poly-

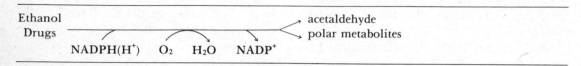

tabolism. The metabolism of ethanol and drugs requires $NADPH(H^+)$, O_2, the microsomal hemoprotein P-450, and other components. Complex alcohol–drug interactions resulting in abnormal metabolic circumstances can commonly result in severe health consequences.

ACTION MECHANISM OF DIETARY FIBER IN THE INTESTINE

Lung, breast, and colon cancers are among the most devastating and lethal forms of cancer. For colon cancer alone, at least 60,000 people die each year and another 100,000 cases are diagnosed. Since colon cancer is similar to lung cancer in that cells originating from the ectoderm are exposed to environmental factors outside the body, many researchers suspect that a dietary component is linked to the outset of colon cancers. Since the dietary regimens of industrialized countries are high in meat and fat but low in fiber, a consensus of most research also suggests that low dietary fiber and high fat levels may be closely tied to the etiology of colon cancers.

More than any other dietary substance, fiber influences human large bowel function. High dietary fiber levels increase stool production, promote dilution of colonic contents, spur on the expeditious motility of solid dietary wastes through the gut, and alter the metabolism of minerals, nitrogen, lipid derivatives, and bile acids in the colon.

Dietary fiber principally consists of undigestible carbohydrates such as cellulose (unbranched polymerized glucose), hemicelluloses (highly branched pentose poly-

meric forms of vegetative carbohydrates (especially cellulose).

Lowered probabilities for acquiring colon cancer in high-fiber diets may be linked to stimulated intestinal peristalsis and modified colonic motor functions. Fiber probably fosters colonic motility by increasing the water content of fecal matter and thus increases the bulk of colonic contents. Wheat and other cereal fibers endure the biophysical and biochemical rigors of gastrointestinal tract digestion and remain hydrated. This action largely reduces the transit time of fecal material and retards water absorption from the lumen of the large bowel.

Herbaceous fiber such as that provided by cabbage and other crucifers stimulate bacterial growth on the digestible vegetative matrix. Bacterial populations contain at least 80% water, and this property also accounts for an apparent increase in hydrated fecal bulk.

Although increased fecal bulk is attributable to at least two possible mechanisms (i.e., hydrated undigestible fiber or prodigious microbial growth), the natural hydration of colonic contents by bacterial growth is probably the most important factor in humans. This claim is based on the prevalence of vegetative fiber consumption in most diets over that of wheat and cereal fibers. Most foods rich in herbaceous fiber including apples, carrots, pectin-rich fruits, and foods containing guar gum are unable to completely survive bacterial digestive actions, but the lignified structures of cereal fibers resist most of these actions.

If improved transit of colonic contents is responsible for depreciating the incidence of colonic cancers, most research suggests that this observation probably parallels a decreased exposure to some type of a carcinogen. The speculated carcinogenic substance may be a natural component of foods, an artifact of conventional food preservation and preparation methods, or a by-product of human digestive agents. As a result of these possibilities, the potential gamut of carcinogens could range from aflatoxins to nitrosamines as well as the *in situ* colonic production of mutagens by bacterial metabolism. Bacterial production of mutagens is probably a leading factor in colonic carcinogenesis, but the experimental production, isolation, and structural analysis of critical substances is a major challenge. One compound typical of suspected mutagenic substances is (S)-3-(1,3,5,7,9-dodecapentaenyloxyl)-1,2 propanediol. This ether-linked conjugated lauryl glycerol demonstrates Ames test mutagenicity in the same league as benzo[a]pyrene. Bacterial productivity of this and other agents is unpredictable but probably requires (1) the availability of bile acids that enhance the solubility of one or more key mutagenic or carcinogenic precursors and (2) a specific strain of bacterial species (*Bacteroides*). Although one bacterial species may appear to be the principal agent responsible for carcinogen or mutagen production, the actual formation of these deleterious agents probably relies on the joint production of mutagenic or carcinogenic precursors by other key bacteria.

Aside from accelerating the transit and elimination of dangerous compounds from the body, the different chemical properties of carbohydrate food fibers exert different clinical and eliminative effects. Since pH values of 7.0 to 7.5 favor the existence of calcium phosphate, oxalate, iron phosphate, and similar compounds as insoluble species, a matrix of dietary fiber ensures the elimination of these substances. The elimination of other organic anions including butyrate, propionate, and acetate may also be encouraged.

Indigestible carbohydrate fibers can also provide a vehicle for eliminating proteins and fats. Nearly half of the fecal protein originates from bacteria, which account for up to 33% of the dry fecal weight. The remainder of the fecal proteins represent unabsorbed intestinal secretions, mucus, digestive fluids, small quantities of enzymes, and desquamated epithelial cells of the intestinal mucosa. If 75 to 175 g of feces is produced daily, less than 7.0 g of fat is usually eliminated.

Oligosaccharides including raffinose and stachyose generally resist human digestive enzyme activity, but intestinal flora can ferment these oligosaccharides. Fermentation products are variable and typically include small organic acids and prodigious amounts of gas. In cases where legumes account for >50% of the caloric requirement, up to 180 mL of gas per hour may be produced. The fiber matrix of the colon contents entrains some of this gas, whereas excess amounts contribute to intestinal cramping, borborygmus, and flatulation. Gas composition is variable (hydrogen, carbon dioxide, and methane in ~33% of adult humans) and seemingly dependent on both oxygen availability and strains of bacterial species inhabiting the intestinal tract.

The mechanism of action of dietary fiber in the human colon is complex, and many of its suspected beneficial functions in reducing colon cancer remain speculative yet supported by epidemiological observations (Burkitt, 1971; Pearce and Dayton, 1971; Hill, 1974; and Stephen and Cummings, 1980). The scientific record regarding the role of fiber in the human diet is fragmented at best and requires considerably more *in vitro* as well as clinical research.

REFERENCES

Atkins, G. L. 1981. A biochemist's view of modelling glucose homeostasis. In *Carbohydrate Metabolism*. (C. Cobelli, and R. N. Bergman, eds.), pp. 369–386. Wiley, New York.

Burkitt, D. 1971. Epidimiology of cancer of the colon and rectum. *Cancer* 28:3.

Devlin, T. M. (ed.). 1982. *Textbook of Biochemistry with Clinical Correlations.* Wiley, New York.

Furda, I. (ed.). 1983. *Unconventional Sources of Dietary Fiber.* ACS Symposium Series 214. ACS, Washington, D.C.

Hill, M. 1974. Colon cancer: a disease of fibre depletion or of dietary excess? *Digestion* **11**:289.

Hunt, S. M., J. L. Groff, and J. M. Holbrook. 1980. *Nutrition: Principles and Clinical Practice.* Wiley, New York.

Jeanes, A., and J. Hodge (eds.). 1975. *Physiological Effects of Food Carbohydrates.* ACS Symposium Series 15. American Chemical Society, Washington, D.C.

Lai, C. Y., N. Nakai, and D. Chang. 1974. Amino acid sequence of rabbit muscle aldolase and the structure of the active center. *Science* **183**:1204–1205.

Lieber, C. S. 1975. Alcohol and malnutrition in the pathogenesis of liver disease. *J.A.M.A.* **233**:1077–1082.

Pearce, M., and S. Dayton. 1971. Incidence of cancer in men on a diet high in polyunsaturated fat. *Lancet* **1**:464.

Senozan, N. M., and R. L. Hunt. 1982. Hemoglobin: its occurrence, structure and adaptation. *J. Chem. Ed.* **59**(3):173–178.

Shen, S. W., and R. Bressler. 1977. Clinical pharmacology of oral antidiabetic agents. *N. Engl. J. Med.* **296**(9):483–496.

Stephen, A. M., and J. H. Cummings. 1980. Mechanism of action of dietary fibre in the human colon. *Nature* **284**:283–284.

Symposium: Nutrition in the causation of cancer. 1974. *Cancer Res.* **35**(part 2):3237.

Vahouny, G. V., and D. Kritchevsky (eds.). 1982. *Dietary Fiber in Health and Disease.* Plenum, New York.

C H A P T E R · 17

Lipid Metabolism

INTRODUCTION

Lipids are not only essential structural components of living cells and tissues but also constitute an important source of *potential metabolic energy*. Weight for weight, lipids provide more than *twice* the metabolizable energy of carbohydrates or proteins. A typical American diet contains 40 to 45% of its calories in the form of lipids. Most current recommendations by nutritionists are directed toward prescribed amounts of dietary unsaturated fatty acids, but *no definite amount of total lipid or saturated-to-unsaturated fatty acid ratios have been dictated*.

Foodstuffs including carbohydrates, ingested beyond the caloric requirements of the body, are converted to lipids. These lipids are deposited in a variety of sites throughout the body such as the peritoneal cavity, kidneys, under the skin, and within muscular interspaces.

Body lipids are in a dynamic state of equilibrium. That is, stored lipids do not exist in a static state but, instead, are in a state of flux. The same type of lipids circulate in the blood plasma as those that are stored in the lipid depots of the body, but concentration levels may be quite variable, with the lowest

concentrations being in the blood. Typical circulated lipids include *cholesterol, phospholipids, fatty acid acylglycerols, free fatty acids, cerebrosides,* and *fat-soluble vitamins*. The blood lipids are also transported to the body tissues for eventual use as fatty acid–albumin complexes, ketone bodies, and lipoproteins.

Most stored lipids are readily metabolized whenever energy deficits develop within a cell or the entire animal body. Subcutaneous lipid reserves are most labile and, therefore, often the first to be used for energy. Accordingly, the acylglycerol structures of adipose tissues are among the first to be hydrolyzed by lipase activity. The fatty acid hydrolysis products plus glycerol resulting from lipase actions are released into the bloodstream and transported to body tissues. Specifically, glycerol may travel to the liver, where it is phosphorylated and metabolized by the glycolytic pathway. Fatty acids, on the other hand, are distributed to cells of various tissues where they can be completely oxidized to carbon dioxide and water. As an alternative route, acetyl-CoA formed as a product during intermediate steps of fatty acid oxidation can be used for the synthesis of essential compounds including steroids and their derivatives (e.g., cholesterol, bile acids, steroid hormones, adrenal corticoids).

Most of the fatty acids in the body have 16 (palmitic, palmitoleic), 18 (stearic, oleic, linoleic), or 20 (arachidonic) carbon skeletons. Aside from *in vivo* synthesis of these fatty acids, the fatty acid profile of the body can reflect dietary habits, intertissue fatty acid transfers, and the resident synthesis of fatty acids in cells from carbohydrates or proteins. The last route of fatty acid synthesis occurs principally in the liver and adipose tissue.

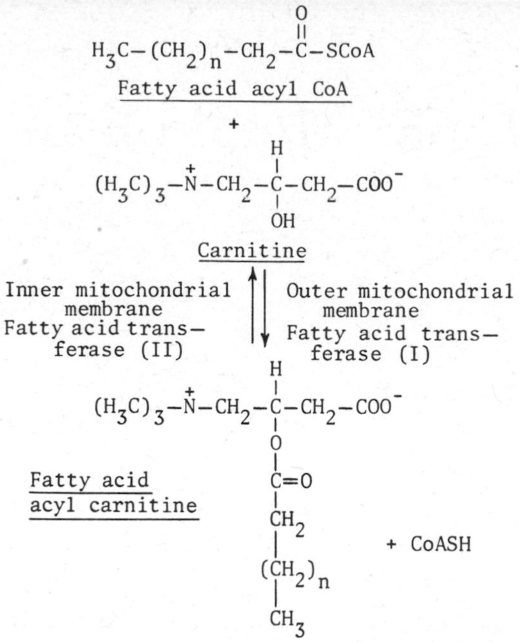

CATABOLISM OF FATTY ACIDS

The degradation of saturated and unsaturated fatty acids represents a significant phase of lipid metabolism for energy production as well as the formation of key metabolic intermediates.

Living cells catabolize fatty acids by one or more of three possible mechanisms described as α, β, or ω oxidation.

β Oxidation of Saturated Fatty Acids

Fatty acids are released from fatty acyl-glycerols in tissues by a hormone-sensitive tissue lipase and then translocated to various other tissues for oxidation. According to the β oxidation scheme for fatty acids, the β carbons of saturated fatty acids are cleaved and released as acetyl-CoA, two carbons (C_2 units) at a time. Since fatty acids have relatively inert chemical activities, they must be converted to thioesters in the cytosol to enhance their molecular reactivity. This action requires the presence of ATP, CoASH, and an *acyl-CoA synthetase*:

FIGURE 17.1. Transfer of an acyl fatty acid moiety from a CoA derivative to a carnitine derivative results in the production of fatty acid acyl carnitine, which is easily transported into the mitochondrial matrix.

The reaction outlined is irreversible since the pyrophosphate (PP_i) produced can readily undergo hydrolysis by *pyrophosphatase* to yield orthophosphate ($2\ P_i$). Because fatty acid acyl-CoA cannot pass through the mitochondrial membrane and into the mitochondrion where it is oxidized, a transport mechanism is required. This transport mechanism is facilitated by esterifying *carnitine* (γ-trimethylamino-β-hydroxybutyrate) to the fatty acid. Transfer of the acyl fatty acid moiety from a CoA derivative to a carnitine derivative results in the production of fatty acid acyl carnitine, which is easily transported into the mitochondrial matrix (Figure 17.1).

$$CH_3—(CH_2)_n—CH_2—CH_2—COOH + ATP + CoASH$$

Cytosolic side of outer mitochondrial membrane | acyl-CoA synthetase

$$CH_3—(CH_2)_n—CH_2—CH_2—\overset{\overset{\text{O}}{\|}}{C}—SCoA + AMP + PP_i$$
Acyl-CoA

Once the fatty acid acyl carnitine reaches the matrix, it reacts with CoASH to again form a *fatty acid acyl-CoA derivative* and free carnitine.

The fatty acid is then oxidized by β oxidation (Figure 17.2), while carnitine returns to its point of origin for recycling as a fatty acid carrier. It should be recognized in passing that the carnitine transport system can be *genetically defective*. If this occurs, clinical manifestations such as muscular cramps and ketosis can be common.

The β oxidation of fatty acid acyl-CoA derivatives occurs in four sequential steps (Figure 17.2).

1. *Oxidation* by FAD to form $FADH_2$ and *trans*-enoyl-CoA.

2. *Hydration* by the addition of water to the double bond of *trans*-enoyl-CoA to produce L-β-hydroxyacyl-CoA.

3. NAD^+-dependent *oxidation* to yield a ketoacyl-CoA derivative and $NADH(H^+)$.

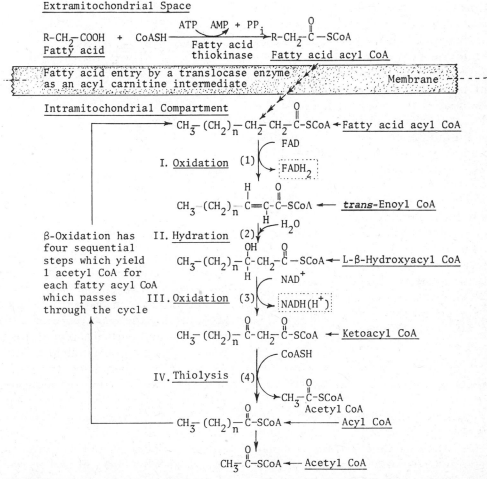

FIGURE 17.2. β oxidation of fatty acids yields acetyl-CoA's from fatty acyl-CoA intermediates. Production of acetyl-CoA's corresponds to $n/2$ acetyl-CoA units, where n is the number of carbon atoms in an even-numbered carbon-chain fatty acid. Complete β oxidation for even-numbered fatty acids requires $(n/2) - 1$ passes through steps I through IV before acetyl-CoA formation ceases. Note that one $FADH_2$ and one $NADH(H^+)$ are produced for every acetyl-CoA generated, and this translates into the respective production of two ATP and three ATP via oxidative phosphorylation. Key enzymes throughout the route include (1) *acyl-CoA dehydrogenase*, (2) *enoyl-CoA hydratase*, (3) *L-β-hydroxyacyl dehydrogenase*, and (4) *acetyl-CoA acyl transferase*.

4. *Thiolysis*, which requires CoASH to produce acetyl-CoA and a new acyl-CoA derivative from the remaining fatty acid shortened by two carbon atoms.

The newly formed fatty acid acyl-CoA is further shortened by two carbons involving steps 1 through 4 (Figure 17.2), until a final acetyl-CoA product is formed. Ideally, even-numbered fatty acids having n carbon atoms pass through the β oxidation route $[(n/2)\text{-}1]$ times (e.g., palmitic acid (C_{16}) passes through the scheme seven times $[(16/2)\text{-}1]$ to produce eight acetyl-CoA's).

Control of the β oxidation pathway rests on the availability of potential substrates, essential cofactors, and the rate of acetyl-CoA entry into the TCA cycle.

Those fatty acids having an *odd number of carbons* in the alkyl chain are oxidized in the same way as fatty acids having even-numbered carbon chains. Although both odd- and even-numbered fatty acids produce C_2 units as acetyl-CoA, the last round of acetyl-CoA production from odd-numbered fatty acids results in a *propionyl (C_3)-CoA derivative*. This propionyl-CoA can be carboxylated to form D-methylmalonyl-CoA, which then undergoes racemization to produce L-methylmalonyl-CoA. The enzyme *methylmalonyl-CoA mutase,* which requires B_{12} as a coenzyme, catalyzes the conversion of L-methylmalonyl-CoA into succinyl-CoA. The latter compound serves as an intermediate in the TCA cycle (Figures 17.3 and 16.11).

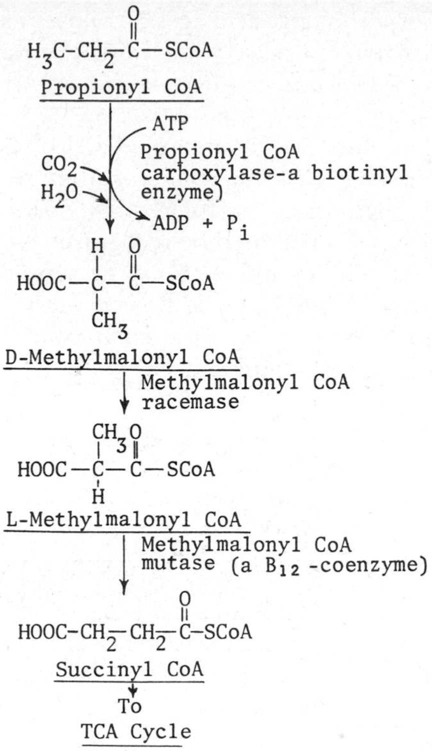

α Oxidation

Plants and some animal tissues such as the brain employ α oxidation for degrading fatty acids one carbon at a time. The reaction may proceed by the steps outlined in Figure 17.4. The D-α-hydroxy acid produced by this oxidative route can accumulate in the case of green leaves, but these acids can eventually be oxidized to an aldehyde. Aldehyde formation proceeds by decarboxylation and retention of the α hydrogen. Hydroxylation of the α carbon can alternately lead to an L-α-

FIGURE 17.3. Conversion of propionyl-CoA to succinyl-CoA.

hydroxy acid, which is dehydrogenated and decarboxylated. The product of these steps appears as a monocarboxylic acid having *one less carbon than the original fatty acid.*

The mechanism offered by α oxidation accounts for the occurrence of α-hydroxy fatty acids and partially accounts for the existence of some odd-numbered fatty acids. Furthermore, α oxidation plays an important role in the oxidation of methylated fatty acids such as *phytanic acid*, which is a product of chlorophyll degradation (Figure 17.5). Upon hydrolysis of chlorophyll, the phytol moiety is liberated and converted to phytanic acid. Phytanic acid is then decarboxylated via α oxidation to form *pristanic acid*. The latter substance is oxidized by β oxidation to yield acetyl-CoA, propionyl-CoA, and α-methylpropionyl-CoA.

One noteworthy hereditary disorder known as *Refsum's disease* is speculated to affect the ability of cells to produce an α-hy-

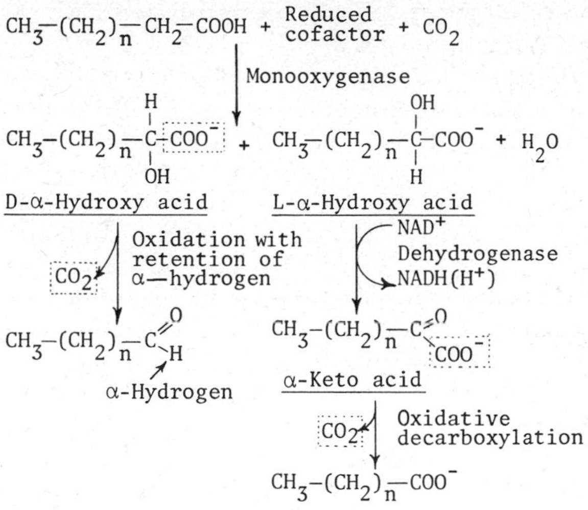

$$CH_3-(CH_2)_{\overline{n}}-CH_2-COOH + \text{Reduced cofactor} + CO_2$$

Monooxygenase

$$CH_3-(CH_2)_{\overline{n}}-\underset{\underset{OH}{|}}{\overset{\overset{H}{|}}{C}}-COO^- + CH_3-(CH_2)_{\overline{n}}-\underset{\underset{H}{|}}{\overset{\overset{OH}{|}}{C}}-COO^- + H_2O$$

D-α-Hydroxy acid · · · · · L-α-Hydroxy acid

CO_2 ↙ Oxidation with retention of α—hydrogen

NAD⁺
Dehydrogenase
NADH(H⁺)

$$CH_3-(CH_2)_{\overline{n}}-C\overset{O}{\underset{H}{\diagdown}}$$

α-Hydrogen

$$CH_3-(CH_2)_{\overline{n}}-C\overset{O}{\underset{COO^-}{\diagdown}}$$

α-Keto acid

CO_2 ↙ Oxidative decarboxylation

$$CH_3-(CH_2)_{\overline{n}}-COO^-$$

FIGURE 17.4. α oxidation routes.

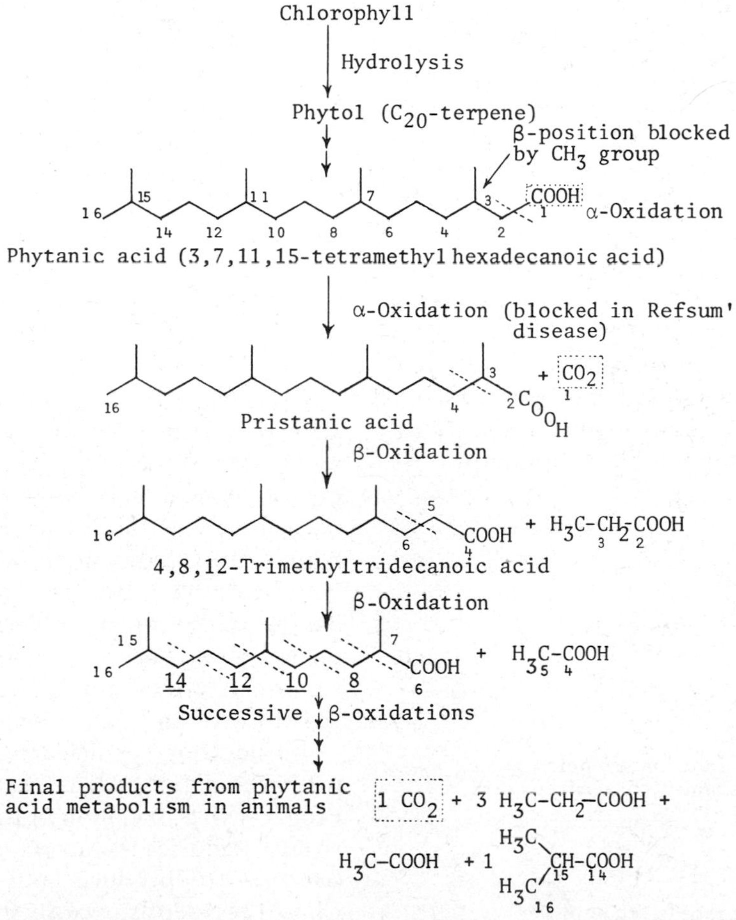

Chlorophyll

Hydrolysis

Phytol (C₂₀-terpene)

β-position blocked by CH_3 group

COOH α-Oxidation

Phytanic acid (3,7,11,15-tetramethyl hexadecanoic acid)

α-Oxidation (blocked in Refsum's disease)

Pristanic acid + CO_2

β-Oxidation

4,8,12-Trimethyltridecanoic acid + H_3C-CH_2-COOH

β-Oxidation

COOH + $H_3C-COOH$

Successive ↓↓ β-oxidations

Final products from phytanic acid metabolism in animals

$$1\ CO_2 + 3\ H_3C-CH_2-COOH +$$
$$H_3C-COOH + 1\ \overset{H_3C}{\underset{H_3C}{\diagup}}CH-COOH$$

FIGURE 17.5. Normal and abnormal aspects of phytanic acid metabolism. Dietary chlorophylls liberated from green vegetables supply available phytol for metabolism. Note that the combined effects of α and β oxidations are necessary for the complete oxidation process.

droxylating enzyme. Since α hydroxylation does not occur, phytanic acid accumulates in tissues and serum. Aside from plant materials, phytanic acid is a notable constituent of many animal fats and milk lipids.

ω Oxidation

Bacteria and microsomes present in the liver cells can oxidize long-chain fatty acids to their corresponding dicarboxylic acids by ω oxidation. Fatty acids such as hexanoic, octanoic, decanoic, and lauric acids typically undergo hydroxylation at the site of the methyl terminus (Figure 17.6).

A complex set of reactions are responsible for ω oxidation of fatty acids, and cytochrome P-450, NADPH(H$^+$), oxygen, and a monooxygenase are required. Bacterial hydroxylations also have a hydroxylase enzyme, which is crucial for most observed hydroxylation processes.

In animals, the roles for ω oxidation products are not entirely clear, but bacteria do use this route for degrading alkanes to fatty acids. Fatty acids produced in this way become susceptible to β oxidative processes and acetyl-CoA production.

Aside from metabolic fates of ω oxidation products, the overall trend of product formation proceeds from nonpolar species to more polar, water-soluble compounds.

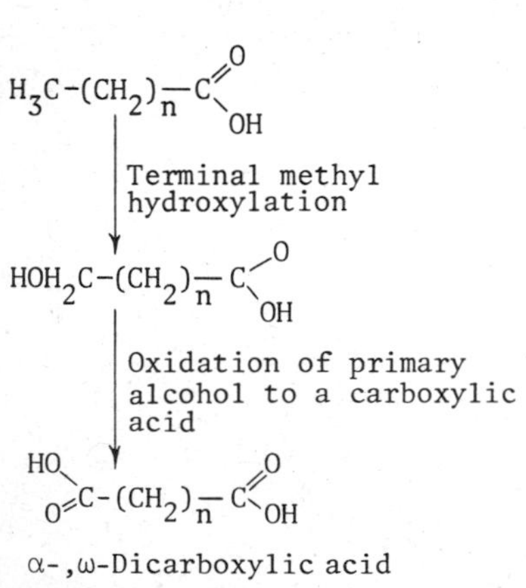

$$H_3C-(CH_2)\underset{n}{-}C\overset{O}{\underset{OH}{}}$$

Terminal methyl hydroxylation

$$HOH_2C-(CH_2)\underset{n}{-}C\overset{O}{\underset{OH}{}}$$

Oxidation of primary alcohol to a carboxylic acid

$$\underset{O}{\overset{HO}{}}C-(CH_2)\underset{n}{-}C\overset{O}{\underset{OH}{}}$$

α-,ω-Dicarboxylic acid

FIGURE 17.6. ω oxidation of a fatty acid yields a dicarboxylic acid.

Unsaturated Fatty Acid Oxidation

Unsaturated fatty acids are *also activated and transported by carnitine* across the mitochondrial membrane. Translocation of these fatty acids across the membrane can then be followed by β oxidation.

The β oxidation of fatty acids is a very routine process for saturated fatty acids, but β oxidation of unsaturated fatty acids can be stymied by the presence of one or more sites of unsaturation.

Direct or indirect influences of unsaturated sites in fatty acid structures make the actions of two additional enzymes imperative for their complete β oxidation. These enzymes include an *isomerase* and an *epimerase* (racemase). The important roles for these enzymes are illustrated in the total conversion of *linoleic acid to acetyl-CoA* (Figure 17.7).

According to recognized oxidative mechanisms, linoleic acid is activated by ATP and CoASH to form linoleoyl-CoA. This acyl-CoA derivative undergoes β oxidation three times to form *cis*-3-enoyl-CoA and acetyl-CoA. The *cis*-3-enoyl-CoA is then isomerized to form *trans*-2-enoyl-CoA, which is suitable for two rounds through β oxidation to form *cis*-2-enoyl-CoA and acetyl-CoA (Figures 17.7 and 17.8).

Isomerization to the *trans*-2-enoyl derivative is fundamental for the complete oxidation of unsaturated fatty acids, because a persistent *cis* double bond in the C-3,4 position prohibits the occurrence of a double bond in the C-2,3 position. The C-2,3 double bond is essential for continuation of β oxidation. In the case of linoleic acid, the *trans*-2-enoyl-CoA derivative goes through two rounds of β oxidation. This results in two acetyl-CoA's and *cis*-2-enoyl-CoA.

Introduction of water into *cis*-2-enoyl-CoA produces a D-3-hydroxyacyl-CoA product. Progressive β oxidation of this acyl-CoA derivative requires the action of an epimerase (*racemase*) to produce an L-3-hydroxyacyl-CoA. This step is necessary because L-*3-hydroxyacyl dehydrogenase*, which occurs in the β oxidation scheme, *cannot use* D-*3-hydroxy-*

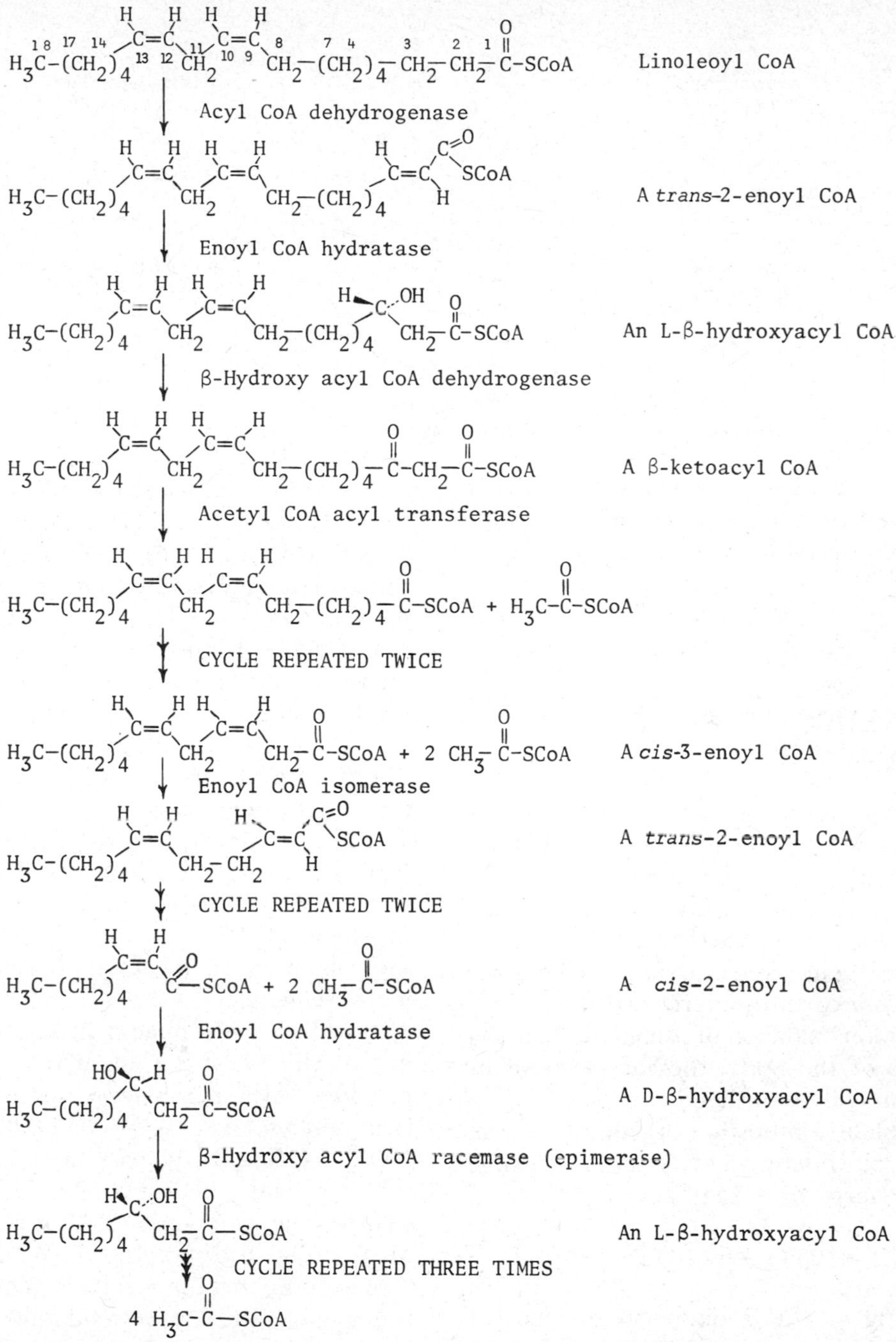

FIGURE 17.7. Oxidative degradation of linoleoyl-CoA, an unsaturated fatty acid indicating the role of key enzymes.

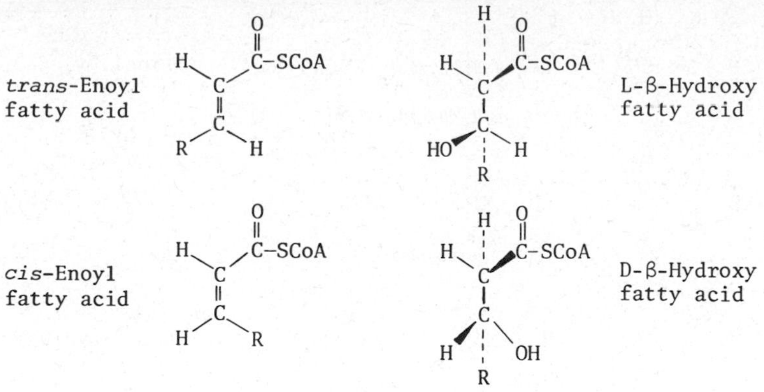

FIGURE 17.8. Reference structures for fatty acid intermediates that occur during their oxidation and biosynthesis.

acyl-CoA as a substrate. Once the L-3-hydroxy structure exists, three rounds of β oxidation yield four acetyl-CoA's.

ENERGETICS OF FATTY ACID OXIDATION

The highly reduced hydrocarbon chain of fatty acids constitutes a major source of metabolic energy when oxidized.

Palmitic acid is the *major* fatty acid transferred from the cytosolic side of the mitochondrial membrane into the mitochondrial matrix as a carnitine derivative. In view of the common oxidation of palmitic acid among the cells of the body, the energetics of its oxidation will be outlined.

Complete combustion of palmitic acid in a bomb calorimeter yields a standard change in free energy of −2340 kcal/mol.

$$C_{16}H_{32}O_2 + 23\ O_2 \longrightarrow 16\ CO_2 + 16\ H_2O$$

According to the β oxidation scheme for palmitic acid, a total of 129 ATP molecules are formed:

Acetyl-CoA can yield 12 ATPs during the course of the citric acid cycle, and therefore a total of 96 (12 × 8) ATPs are produced from palmitic acid by this route. The 7 $FADH_2$ molecules provide 14 (7 × 2) ATPs and 7 $NADH(H^+)$ produce 21 (7 × 3) ATPs during the course of β oxidation. The total ATP yield from all energy-rich molecules is 131 ATP (96 + 14 + 21). It must be recalled that 2 ATP equivalents are consumed in the activation of palmitic acid (ATP → AMP + PP_i) and the net production of ATP is actually 129 (Table 17.1). Biological efficiency for stripping energy from palmitate as ATP approximates 40.2% of the theoretical 2340 kcal/mol obtained on the basis of calorimetric measurements.

Based on mole:mole $\Delta G^{0\prime}$ ratios for palmitate:glucose, it is evident that palmitate provides 341% more energy upon oxidation than glucose (i.e., 2340 kcal/686 kcal × 100 = 341%). On a mass:mass basis, palmitate provides almost 2.41 times as much caloric yield as glucose (2340 kcal/mol/255 M. W.)/(686 kcal/mol/180 M. W.)).

Assuming that the metabolism of glycerol is negligible, the aerobic oxidation of tripalmitate could theoretically produce 2825 kcal/mol (3 palmitates × 941.7 kcal/mol).

$$C_{16}H_{31}COOH + 8\ CoASH + ATP + 7\ FAD + 7\ NAD^+ + 7\ H_2O \longrightarrow$$

$$8\ CH_3\!\!-\!\!\overset{O}{\overset{\|}{C}}\!\!-\!\!SCoA + 7\ FADH_2 + 7\ NADH(H^+) + AMP + PP_i$$

Table 17.1. Tabulated ATP Production for Palmitic Acid Oxidation by (1) β Oxidation Followed by (2) Oxidation of Acetyl-CoA in the TCA Cycle

1. β Oxidation
Palmitic acid $\longrightarrow$ 8 acetyl-CoA + 14 electron pairs

7 FADH$_2$ produce 2 ATPs/electron pair	14 ATP
7 NADH(H$^+$) produce 3 ATPs/electron pair	21 ATP
2 ATP equivalents consumed for palmitic acid activation (ATP + PP_i hydrolysis)	-2 ATP
Total (theoretical) ATP production	33 ATP

2. TCA Cycle Oxidation of Acetyl-CoA
8 Acetyl-CoA + 16 O$_2$ $\longrightarrow$ 16 CO$_2$ + 16 H$_2$O + 8 CoASH

8 (3 NADH(H$^+$)/acctyl-CoA = 9 ATP)	72 ATP
8 (1 FADH$_2$/acetyl-CoA = 2 ATP)	16 ATP
8 (1 GTP/acetyl-CoA = 1 ATP equivalent)	8 ATP
Total (theoretical) ATP production	96 ATP

3. Projected Efficiency for β Oxidation and TCA Cycle
33 ATP + 96 ATP = 129 ATP

$$\frac{(129 \text{ ATP})(7300 \text{ cal/mol})}{2340 \text{ kcal/mol}} \times 100 = 40.2\% \text{ energy conservation as ATP}$$

ANABOLISM OF LIPIDS AND THEIR METABOLIC DERIVATIVES

Many chemically diverse compounds exhibit molecular structures that are predominantly nonpolar and ordinarily classified as *lipids*. These nonpolar compounds are generally soluble in petroleum and diethyl ethers as their only common feature, and their structural diversity is matched by similarly diverse biosynthesis pathways. Included among the most important lipids and lipid-related substances are the fatty acids; some ketone bodies; mono-, di-, and triacylglycerols; phospholipids; sphingolipids; gangliosides; and steroids such as bile acids and cholesterol.

BIOSYNTHESIS OF FATTY ACIDS REQUIRES ACETYL-CoA AND NADPH(H$^+$)

Acetyl-CoA serves as the precursor for fatty acid synthesis. Using the *de novo* synthesis of palmitate from acetyl-CoA as an example,

the overall stoichiometry can be expressed as

8 Acetyl-CoA + 14 NADPH(H$^+$) + 7 ATP

$\downarrow$

C$_{16}$H$_{31}$COOH + 14 NADP$^+$ + 6 H$_2$O
Palmitic acid

+ 8 CoASH + 7 ADP + 7 P_i

Palmitic acid and other fatty acids can be synthesized in the cytosol from acetyl-CoA, which is derived from the decarboxylation of pyruvate in the mitochondrion (Figure 17.9).

Since acetyl-CoA produced in the mitochondrion cannot pass through its membrane into the cytosol, acetyl-CoA must be converted into citrate, which readily permeates the mitochondrial membrane. The subsequent decomposition of citrate in the cytosol eventually produces acetyl-CoA and oxaloacetate (Figure 17.10A).

Citrate + ATP + CoASH $\xrightarrow{\text{cleavage enzyme}}$

acetyl-CoA + oxaloacetate + ADP + P_i

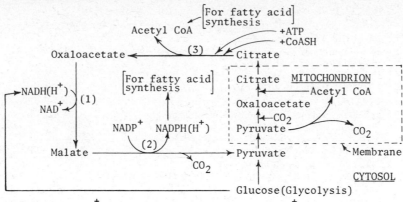

(1) Malate-NAD$^+$ dehydrogenase   (2) Malate-NADP$^+$ dehydrogenase

(3) Cleavage enzyme

FIGURE 17.9.   Source of acetyl-CoA and NADH(H$^+$) for fatty acid synthesis. Acetyl-CoA is derived from decarboxylation of pyruvate in the mitochondrion and NADPH(H$^+$) is formed in the cytosol as oxaloacetate is released from citrate and converted to pyruvate via malate.

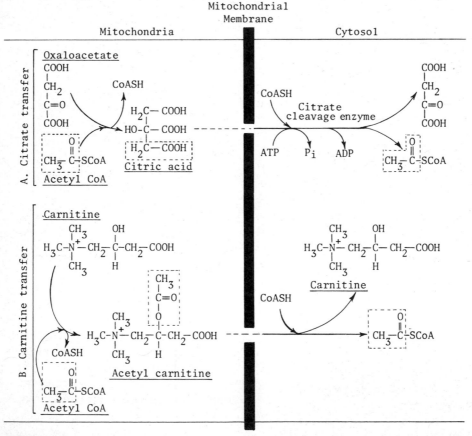

FIGURE 17.10.   Two independent mechanisms that translocate acetyl-CoA over the mitochondrial membrane to the cytosol where fatty acid biosynthesis occurs. *Without* (A) citrate and (B) carnitine transfer mechanisms, acetyl-CoA supplies for fatty acid synthesis could not traverse the mitochondrial membrane.

Carnitine, too, can supply acetyl-CoA to the cytosol for fatty acid synthesis by a mechanism similar to its role in fatty acid transfer into the mitochondrion from the cytosol (Figure 17.10A).

For the synthesis of palmitate, 8 acetyl-CoA's and 14 NADPH(H$^+$) are required. Some of the NADPH(H$^+$) requirement is provided by the cytosolic HMP pathway and the remainder originates from the reentry of cytosolic oxaloacetate into the mitochondrion by the reactions outlined below. Apart from NADPH(H$^+$) production, these reactions permit oxaloacetate passage over the inner mitochondrial membrane, which is otherwise impermeable to oxaloacetate.

1. Oxaloacetate + NADH(H$^+$) $\xrightarrow{\text{malate-NAD}^+ \text{ dehydrogenase}}$

$$\text{malate} + \text{NAD}^+$$

2. Malate + NADP$^+$ $\xrightarrow{\text{malate-NADP}^+ \text{ dehydrogenase}}$

$$\text{pyruvate} + CO_2 + \text{NADPH(H}^+)$$

3. Pyruvate + CO$_2$ + ATP + H$_2$O $\longrightarrow$

$$\text{oxaloacetate} + \text{ADP} + P_i + 2\text{H}^+$$

Reaction sum:

NADP$^+$ + NADH(H$^+$) + ATP + H$_2$O $\longrightarrow$

$$\text{NADPH(H}^+) + \text{NAD}^+ + \text{ADP} + P_i$$

For each acetyl-CoA transfer from the mitochondrion into the cytosol (Figure 17.10), one oxaloacetate is formed in the cytosol. Two reactions then lead to the conversion of oxaloacetate into pyruvate: (1) oxaloacetate is converted to malate by reduction involving NADH(H$^+$) from glycolysis; and (2) malate is oxidized to pyruvate by NADP$^+$. The pyruvate released in the last reaction *enters* the mitochondrion where *oxaloacetate is regenerated* for eventual condensation with acetyl-CoA *to produce citrate. Citrate formed by this route exists in the mitochondrion for another round of action in the TCA cycle.*

Using palmitate synthesis as a typical example of fatty acid production, 8 acetyl-CoA's must exist in the mitochondrion as part of citrate molecules, and it follows that 8

NADPH(H$^+$) molecules will be produced in the cytosol. A total of 14 NADPH(H$^+$) molecules are required for palmitate synthesis; therefore, 6 additional NADPH(H$^+$) must come from the HMP pathway (Figure 16.13).

Other saturated fatty acids are synthesized by *similar* pathways with acetyl-CoA and NADPH(H$^+$) consumption, which parallels the chain length of the new fatty acid.

FATTY ACID BIOSYNTHESIS REQUIRES THE MULTIENZYME SYSTEM—FATTY ACID SYNTHETASE

Fatty acid biosynthesis occurs as a stepwise process mediated by a multienzyme system called *fatty acid synthetase* (FAS). For primates the multienzymatic actions of FAS are embodied in a single polypeptide sequence. In the case of biosynthesis for palmitic acid (C$_{16}$), one of the most common fatty acids, the FAS multienzyme system exists as a dimer often specifically called *palmitic acid synthetase.* The molecular weight of the dimer approaches 550,000 and both subunits can separately produce a fatty acid.

The synthetic action of the FAS system also relies on the presence of an *acyl carrier protein* (ACP). This protein contains 77 amino acid residues with a 4'-phosphopantetheine prosthetic group bonded to the serine-36 hydroxyl group as shown in Figure 17.11. The 4'-phosphopantetheine moiety is important because it forms a thioester linkage with biosynthetic intermediates of growing fatty acid chains.

As a consequence of covalent bonding to the ACP, fatty acid (acyl) intermediates are permitted to circulate from one active site on FAS to another. Each complete circulation of an acyl intermediate around the FAS enzyme system results in a lengthening of an existing acyl-ACP-linked fatty acid precursor by two carbons (Figure 17.12).

Although early studies of fatty acid biosynthesis led to speculation that fatty acid

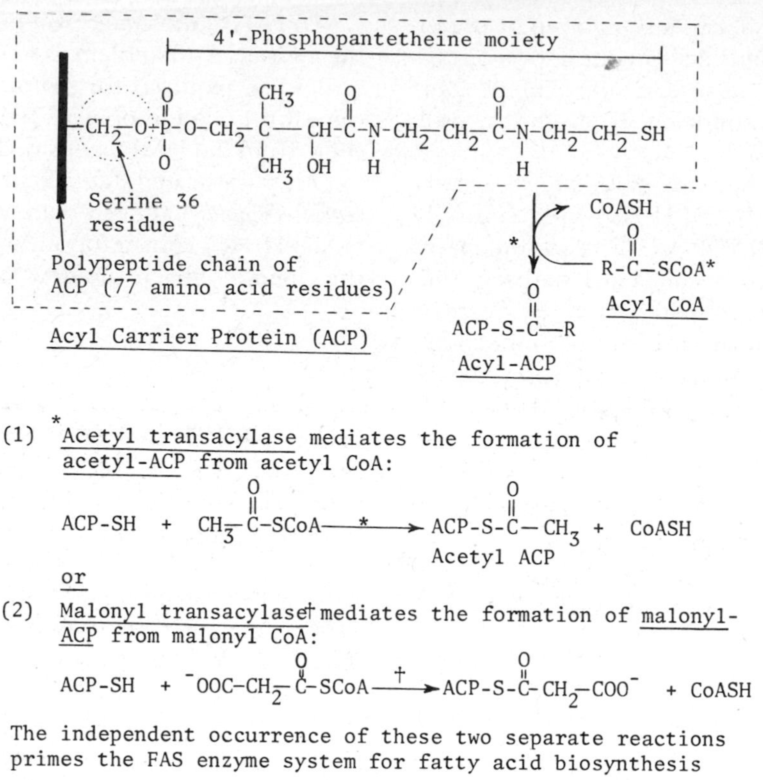

FIGURE 17.11. Formation of acetyl- and malonyl-ACP intermediates are crucial for initiating the early stages of fatty acid biosynthesis. Both acyl-ACP compounds rely on the formation of thioester linkages between the 4' phosphopantetheine moiety of FAS and acyl intermediates.

formation may represent a *simple reversal* of the β oxidative pathway, it is now clear that nothing could be further from the truth!

INDIVIDUAL REACTIONS OF FATTY ACID BIOSYNTHESIS

A stepwise process illustrating the key reactions of fatty acid biosynthesis by the FAS enzyme system has been detailed in the following paragraphs for palmitate. Other fatty acids having carbon chain lengths greater or shorter than C_{16} are produced in the same fashion, but acetyl-CoA and $NADPH(H^+)$ requirements correspondingly vary on the length of the carbon chain. In all cases, fatty acid synthesis begins with the formation of an ACP thioester of acetate and malonate in the forms of acetyl-ACP and malonyl-ACP. Fatty acid biosynthesis commences with these so-called initiation reactions below.

1. **Initiation step for fatty acid biosynthesis.**

    A. *Malonyl-CoA production (cytosolic reaction in liver cells).*

$$CH_3-\overset{O}{\overset{||}{C}}-SCoA + ATP + HCO_3^- \longrightarrow {}^-OOC-CH_2-\overset{O}{\overset{||}{C}}-SCoA + ADP + P_i + H^+$$

$\quad$ **Acetyl-CoA** $\qquad\qquad\qquad\qquad\qquad\qquad\qquad$ **Malonyl-CoA**

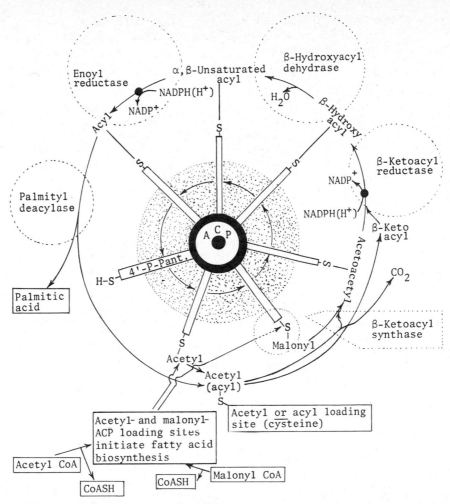

FIGURE 17.12. Schematic operation of mammalian fatty acid synthetase. This multienzyme system, found in liver cells, introduces acetyl-CoA and malonyl-CoA into a loading site. These C_2 and C_3 units combine under the auspices of β-ketoacyl synthase to produce acetoacetyl-ACP (step 3, Table 17.3). Transformation of the acetoacetyl-ACP into butyryl-ACP follows. The butyryl-ACP then reacts with another molecule of malonyl-ACP and β-ketoacyl synthase, and this repetitive sequence of events continues until a specific fatty acid carbon-chain length is produced. Using palmitic acid synthesis as an example, seven rounds through the synthesis scheme are necessary to produce the saturated C_{16} fatty acid. The equation for this reaction is

$$\text{acetyl-CoA} + 7 \text{ malonyl-CoA} + 14 \text{ NADPH(H}^+) + 14 \text{ H}^+ \longrightarrow$$
$$\text{palmitate} + 8 \text{ CoASH} + 7 \text{ CO}_2 + \text{NADP}^+ + 6 \text{ H}_2\text{O}$$

and the parent sources of palmitic acid C_2 units can be shown as

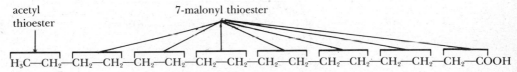

Table 17.2. Many Different Regulatory Agents Affect the Synthesis of Fatty Acids by Influencing Acetyl-CoA Carboxylase and Fatty Acid Synthetase

Enzyme	Regulatory Agent	Effect
Palmitate Biosynthesis		
Acetyl-CoA carboxylase		
Short term	Citrate	Allosteric activation
	C_{16}—C_{18} acyl-CoA's	Allosteric inhibition
	Insulin	Stimulation
	Glucagon	Inhibition
	cAMP-mediated phosphorylation	Inhibition
	Dephosphorylation	Stimulation
Long term	High-carbohydrate diet	Stimulation by increased enzyme synthesis
	Fat-free diet	Stimulation by increased enzyme synthesis
	High-fat diet	Inhibition by decreased enzyme synthesis
	Fasting	Inhibition by decreased enzyme synthesis
	Glucagon	Inhibition
Fatty acid synthetase	Phosphorylated sugars	Allosteric activation
	High carbohydrate diet	Stimulation by increased enzyme synthesis
	Fat-free diet	Stimulation by increased enzyme synthesis
	High-fat diet	Inhibition
	Fasting	Inhibition by decreased enzyme synthesis
	Glucagon	Inhibition by decreased enzyme synthesis
Biosynthesis of fatty acids other than palmitate		
Fatty acid synthetase	High ratio of $\dfrac{\text{methylmalonyl-CoA}}{\text{malonyl-CoA}}$	Increased synthesis of methylated fatty acids
	Thioesterase cofactor	Termination of synthesis with short-chain product
Stearoyl-CoA desaturase	Various hormones	Stimulation of unsaturated fatty acid synthesis by increased enzyme synthesis
	Dietary polyunsaturated fatty acids	Decreased activity

SOURCE: From F. N. LeBaron, 1982. Lipid metabolism I. In *Textbook of Biochemistry with Clinical Correlations* (T. M. Devlin, ed.). John Wiley & Sons, New York.

Acetyl-CoA is converted to malonyl-CoA by the action of a biotin-dependent *acetyl-CoA carboxylase*. The mechanism for carboxylation has been presented elsewhere.

Acetyl-CoA carboxylase directs acetyl-CoA into fatty acid synthesis and clearly exerts a significant controlling influence on fatty acid synthesis. The enzyme is inhibited by palmitoyl-CoA and also influenced by a variety of additional regulatory agents (Table 17.2). For example, a cAMP-regulated mechanism controls the phosphorylation of acetyl-CoA carboxylase. Since phosphorylated and dephosphorylated forms of the enzyme show differences in activity, *glucagon* spurs the phosphorylation of the enzyme to decrease its activity while *insulin* promotes the dephosphorylated active form of the enzyme.

B. *Formation of acetyl-ACP.*

$$CH_3-\overset{\overset{\displaystyle O}{\|}}{C}-SCoA + ACP\text{-}SH \rightleftharpoons$$
Acetyl-CoA

$$CH_3-\overset{\overset{\displaystyle O}{\|}}{C}-S-ACP + CoASH$$
Acetyl-ACP

An acetyl-CoA, in addition to the one in step A (above), reacts with ACP-SH to form acetyl-ACP in a reaction mediated by

acetyl transacylase. The ACP forms a thioester with the acetyl moiety to produce acetyl-ACP. Formation of this C_2-acyl intermediate is necessary before addition of acetyl (C_9) units into FAS can continue to

group, and the acyl group participates in a thioester linkage (Figure 17.11).

C. *Once acetyl-ACP is formed,* it is speculated that a certain cysteinyl residue on a key enzyme constituent of FAS, namely, β-ketoacyl ACP synthase, becomes linked to the acetyl moiety of the acetyl-ACP. The ACP-SH is now free to establish another thioester if necessary for future C_2 additions to fatty acyl-ACP intermediates:

$$H_3C-\overset{\overset{\displaystyle O}{\|}}{C}-SACP + synthase-SH \rightleftharpoons$$
Acetyl-ACP   **β-Ketoacyl synthase**

$$H_3C-\overset{\overset{\displaystyle O}{\|}}{C}-S-synthase + ACP-SH$$
Acetyl synthase   **Free ACP—SH**

2. Malonyl-ACP formation.

$$^-OOC-CH_2-\overset{\overset{\displaystyle O}{\|}}{C}-SCoA + ACP-SH \rightleftharpoons {}^-OOC-CH_2-\overset{\overset{\displaystyle O}{\|}}{C}-SACP + CoASH$$
Malonyl-CoA   **Free ACP—SH**   **Malonyl-ACP**

Malonyl-CoA reacts with ACP to form malonyl-ACP in a reaction mediated by malonyl transacylase (Figure 17.12).

3. Acetoacetyl-ACP formation (condensation).

$$CH_3-\overset{\overset{\displaystyle O}{\|}}{C}-S-synthase + {}^-OOC-CH_2-\overset{\overset{\displaystyle O}{\|}}{C}-SACP \xrightarrow{\text{condensation}}$$
Acetyl synthase   **Malonyl-ACP**

$$H_3C-\overset{\overset{\displaystyle O}{\|}}{C}-CH_2-\overset{\overset{\displaystyle O}{\|}}{C}-SACP + synthase-SH + CO_2$$
Acetoacetyl-ACP

produce a fatty acid. Note that the serine 36 residue of the ACP remains attached to the 4′-phosphopantetheine prosthetic

The acetyl and malonyl moieties condense to form acetoacetyl-ACP. This reaction is promoted by β-ketoacyl-ACP synthase.

4. Reduction of acetoacetyl-ACP (β-ketoacyl reduction).

3–6) (Table 17.3) is repeated until the C_{16} carbon chain for palmitate is completed.

$$H_3C-\overset{O}{\overset{||}{C}}-CH_2-\overset{O}{\overset{||}{C}}-SACP + NADPH(H^+) \xrightleftharpoons{\text{reduction}}$$
Acetoacetyl-ACP

$$H_3C-\overset{H}{\underset{|}{\underset{OH}{C}}}-CH_2-\overset{O}{\overset{||}{C}}-SACP + NADP^+$$
D-3-Hydroxybutyryl-ACP

Acetoacetyl-ACP is reduced by NADPH(H$^+$) to form D-3-hydroxybutyryl (C_4) ACP by a β-ketoacyl-ACP reduction.

5. Dehydration of the D-3-hydroxyacyl intermediate.

When the final C_{16} acyl-ACP structure is formed, a hydrolysis event occurs that yields ACP-SH and palmitate. Other fatty acids are synthesized by similar pathways, but little is known about the termination sequence that ultimately directs fatty acid carbon chain

$$H_3C-\overset{H}{\underset{|}{\underset{OH}{C}}}-CH_2-\overset{O}{\overset{||}{C}}-SACP \xrightleftharpoons{\text{dehydration}} H_3C-\overset{H}{\overset{|}{C}}=\overset{}{\underset{|}{\underset{H}{C}}}-\overset{O}{\overset{||}{C}}-SACP + H_2O$$

D-3-Hydroxybutyryl-ACP   *trans*-**2-Butenyl-ACP**
                                    (crotonyl-ACP)

D-3-Hydroxybutyryl-ACP is dehydrated to form *trans*-2-butenyl-ACP and the reaction is prompted by 3-hydroxyacyl-ACP hydratase.

6. Reduction of *trans*-2-butenyl-ACP.

lengths during their *de novo* synthesis.

Whereas *Escherichia coli* and other bacteria employ separate enzymes for fatty acid synthesis, FAS in primates (and other mammals) relies on a single multienzyme whose

$$H_3C-\overset{H}{\overset{|}{C}}=\overset{}{\underset{|}{\underset{H}{C}}}-CH_2-\overset{O}{\overset{||}{C}}-SACP + NADPH(H^+) \xrightarrow{\text{reduction}} H_3C-CH_2-CH_2-\overset{O}{\overset{||}{C}}-SACP + NADP^+$$

trans-**2-Butenyl-ACP**   **Butyryl-ACP**
(crotonyl-ACP)

Reduction of *trans*-2-butenyl-ACP by NADPH(H$^+$) to yield butyryl-ACP is conducted by enoyl reductase.

Butyryl-ACP can subsequently react with another molecule of malonyl-ACP and the sequence detailed above for condensation, reduction, dehydration, and reduction (Steps

topology is dictated by covalent bonding and hydrophobic forces. The schematic mechanism and functional structure for liver FAS is represented in Figure 17.12. Note that *palmityl deacylase* is responsible for liberating the free fatty acid as palmitate in this case, but *other fatty acids* could be produced and freed from FAS complexes by similarly *specific deacylases*.

Table 17.3. Summary of Major Reactions in Fatty Acid Biosynthesis

Step	Reactions	Enzymes
1. A.	Acetyl-CoA + ATP + HCO_3^- $\longrightarrow$ malonyl-CoA + ADP + P_i + H^+	Biotin dependent acetyl-CoA carboxylase
B.	Acetyl-CoA + ACP-SH $\rightleftharpoons$ acetyl*-ACP + CoASH	Acetyl transacylase
C.	Acetyl*-ACP + synthase-SH $\rightleftharpoons$ Acetyl*-synthase + ACP-SH	Acetyl transfer to synthase
2.	Malonyl-CoA + ACP-SH $\rightleftharpoons$ malonyl ACP + CoASH	Malonyl transacylase
3.	Acetyl*-synthase + malonyl-ACP $\rightleftharpoons$ acetoacetyl-ACP + synthase-SH + CO_2	β-Ketoacyl-ACP synthase
4.	Acetoacetyl ACP + NADPH(H^+) $\rightleftharpoons$ D-3-hydroxybutyryl-ACP + $NADP^+$	β-Ketoacyl-ACP reductase
5.	D-3-Hydroxybutyryl-ACP $\rightleftharpoons$ *trans*-2-butenyl-ACP + H_2O	3-Hydroxyacyl-ACP hydratase
6.	*trans*-2-Butenyl-ACP + NADPH(H^+) $\longrightarrow$ butyryl-ACP + $NADP^+$	Enoyl-ACP reductase

SYNTHESIS OF C₁₆₊ SATURATED, ODD-NUMBERED, AND UNSATURATED FATTY ACIDS

In the case of mammals, the FAS multienzyme system predominantly synthesizes fatty acids having even numbers of carbon atoms. Although mammalian systems are not noted for the synthesis of odd-numbered fatty acids, some organisms, especially of marine origin including porpoises, dolphins, and fishes can produce significant amounts of odd-numbered fatty acids.

Odd-numbered saturated and unsaturated straight-chained fatty acids may contain 11, 13, 15, 17, or 19 carbon atoms. Butterfat mirrors this structural variability since 0.5 to 1.0% of odd- and even-numbered fatty acids have 13, 14, 15, 16, or 17 carbon atoms, whereas up to 2.0% of saturated and unsaturated straight-chain acids have 11, 13, 15, 17, and 19 carbon atoms.

The production of odd-numbered fatty acids relies on fatty acid synthesis mechanisms that may be initiated with propionyl-CoA (C_3) in place of acetyl-CoA (C_2). Initiation of fatty acid biosynthesis with propionyl-CoA is possible since a general ACP-acyl transferase, exclusive of malonyl transacylase, can transfer initiating structures other than acetyl (C_2) units to ACP.

In the case cited above where an initial propionyl (C_3) group condenses with a malonyl moiety, a C_5 fatty acyl intermediate is typically produced along with CO_2. Clearly, the wide spectrum of possible initiating structures can yield many variations in fatty acid structures.

For higher animals and humans, palmitic acid is widely used for the synthesis of longer-chained fatty acids. These synthesis reactions can occur in the mitochondrion. If the endoplasmic reticulum serves as the site for biosynthesis, malonyl-CoA and NADPH(H^+) are used for the elongation of palmityl-CoA. However, fatty acid synthesis occurring in the mitochondrion is essentially the reverse of β oxidation, where NADPH(H^+) is used as a reductant for enoyl-CoA reductase.

Endoplasmic reticulum membrane

$$\text{Palmityl-CoA (C}_{16}\text{)} \xrightarrow[\substack{\nearrow \text{(FAS)} \\ \text{malonyl-CoA}}]{\text{NADPH(H}^+\text{)}} \to \to$$

$$\text{stearyl-CoA (C}_{18}\text{)}$$

Outer and inner mitochondrial membrane

$$\text{Palmityl-CoA (C}_{16}\text{)} \xrightarrow[\substack{\nearrow \text{(FAS)} \\ \text{acetyl-CoA}}]{\text{NADPH(H}^+\text{)}} \to \to$$

$$\text{stearyl-CoA (C}_{18}\text{)}$$

Unsaturated fatty acids including palmitoleic and oleic acids use palmitoyl-CoA and stearoyl-CoA, respectively, as precursors for

1. Palmityl CoA + NADPH(H$^+$) + O$_2$ $\longrightarrow$
   Palmitoleoyl CoA + NADP$^+$ + 2H$_2$O

2. Stearyl CoA + NADPH(H$^+$) + O$_2$ $\longrightarrow$
   Oleoyl CoA + NADP$^+$ + 2H$_2$O

FIGURE 17.13. General production of C_{16} and C_{18} unsaturated fatty acids (monounsaturated C_{16} and C_{18} fatty acids) through the action of desaturases.

their synthesis. These reactions leading to unsaturated fatty acid production occur largely in the microsomes (Figure 17.13).

Using stearyl-CoA as an example, and assuming that it has been synthesized within a liver cell, this *saturated* fatty acyl (C_{18}) structure can be *desaturated* to form oleoyl-CoA by the following reaction:

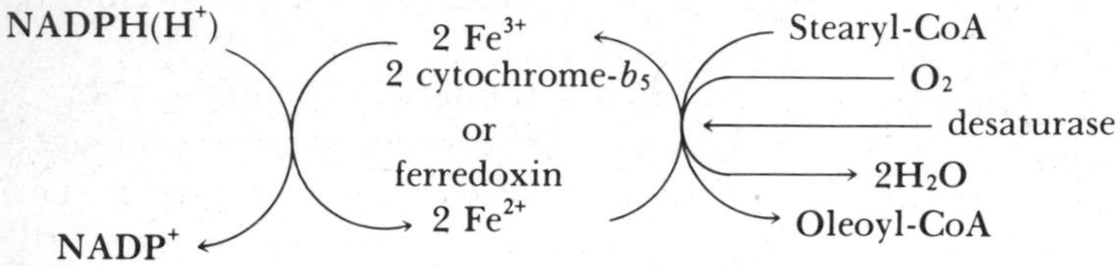

NADPH(H$^+$)   2 Fe^{3+} ← Stearyl-CoA
                 2 cytochrome-b_5 O$_2$
                      or   desaturase
                 ferredoxin   2H$_2$O
                 2 Fe^{2+}
NADP$^+$ ←   Oleoyl-CoA

A *desaturase* enzyme presumably mediates this reaction by causing a *cis* elimination at the C-9,10 position on stearyl-ACP. Such aerobic mechanisms account for the formation of oleic acid in animal cells, but plants employ ferredoxin instead of cytochrome-b_5 present in animals.

Many unsaturated fatty acids can be synthesized by human and animal tissues except those having one or more double bonds in the terminal seven carbons (methyl terminus end) of the acid. Although these fatty acids cannot be synthesized *de novo*, suitable stores necessary to ensure normal cell nutrition are obtained from exogenous or dietary sources. These unsaturated acids are normally classified as *essential fatty acids* (EFAs). Since linoleate ($C_{18}\Delta^9$, 12) can be desaturated by microsomal enzymes into γ-linoleate ($C_{18}\Delta^6$, 9, 12) and eventually transformed into arachidonate ($C_{20}\Delta^5$, 8, 11, 14), linoleate serves as the principal EFA for humans.

Ultimate formation of arachidonate requires the addition of a C_2 unit to γ-linoleate

($C_{18}\Delta^6$, 9, 12). This results in homo-γ-linoleate ($C_{20}\Delta^5$, 11, 14), which is further desaturated to arachidonate. In general, all desaturase enzyme actions proceed toward the carboxyl group.

Aside from EFA contributions to (1) cellular and subcellular membrane structures (e.g., phospholipids), (2) cholesterol metabolism and excretion, and (3) prostaglandin, thromboxane, and prostacyclin structures, EFA deficiencies are linked to dermatologic drying and flaking of the skin, upsets in cellular water balance, kidney damage, impaired fertility, and inefficient oxidative phosphorylation. Aside from enigmatic but important biochemical roles and mechanisms for unsaturated fatty acids, the recommended dietary consumption of linoleate should approach 3.0% of caloric energy (recommended for populations having low fat intake, i.e., less than 25% of calories).

TRIACYLGLYCEROL BIOSYNTHESIS

Adipose and liver tissues serve as the major sites for triacylglycerol biosynthesis, but they can be produced by other cells and tissues as well. Triacylglycerols are most notably produced, stored as cytoplasmic liquid droplets, and then hydrolyzed in adipose tissues since this specialized connective tissue is responsible for prolonged storage of energy-rich compounds. Cardiac and skeletal tissues can also store triacylglycerols to a limited extent in order to satisfy isolated energy demands of these tissues.

In spite of the apparent long-term storage of triacylglycerols in adipose tissues, the triacylglycerols do undergo dynamic biosynthesis and degradation. Typical half-lives range from 3 to 4 days and, therefore, triacylglycerol constituents are subject to *considerable turnover.*

Liver tissue biosynthesis of triacylglycerols is also important for reasons other than energy storage, namely, blood lipoprotein production.

Fatty acids used for triacylglycerol biosynthesis are acquired from dietary sources as well as cytoplasmic activities of the FAS system. Acetyl-CoA derived from glucose metabolism is used as the building block for fatty acids, whereas phosphorylated C_3 derivatives from glucose catabolism supply the glycerol moiety. Typical C_3 moieties include *dihydroxyacetone phosphate* (DHAP) or *glycerol 3-phosphate.* Reduction of DHAP originating from glycolytic events can yield glycerol phosphate, or glycerol phosphate can be produced from glycerol phosphorylation. Unlike many other tissues, however, adipose tissues rely extensively on glycerol phosphate derived from glycolytic mechanisms.

Fatty acid activation as a coenzyme A ester serves as a prelude to triacylglycerol synthesis. This reaction is irreversible and energetically feasible as a result of pyrophosphate hydrolysis to inorganic phosphate:

$$R—\overset{\overset{\displaystyle O}{\|}}{C}—OH + ATP + CoASH \xrightarrow[\text{acyl-CoA synthetase}]{}$$

$$R—\overset{\overset{\displaystyle O}{\|}}{C}—SCoA + AMP + PP_i + H_2O$$

Beginning with glycerol 3-phosphate, two successive acylations using activated fatty acids yield *phosphatidic acid.* This important lipid intermediate is subsequently converted to a diacylglycerol and eventually to a triacylglycerol (Figure 17.14).

If DHAP provides the C_3 backbone for triacylglycerol structures, *lysophosphatidic acid* is formed by a reductive reaction after a direct acylation of DHAP has occurred. Lysophosphatidic acid then provides a substrate for further acylation to yield triacylglycerols.

Aside from the synthesis of triacylglycerols from phosphatidic acid, *intestinal mucosa cells employ 2-monoacylglycerols as a substrate for acylation reactions.* Diacylglycerols are initially formed using acyl-CoA's and then triacylglycerols (Figure 17.14). The use of 2-monoacylglycerols is not unexpected since they are the major end products of lipid digestion and readily absorbed into mucosal cells.

The mobilization and biosynthesis of triacylglycerols is carefully controlled by complex hormonal interactions, and the availability of fatty acids residing on triacylglycerols depends on specific lipase actions. Total fatty acid hydrolysis yields three fatty acids plus free glycerol. Assuming that these products originate from adipose tissues they are translocated by the blood to energy-consuming cells (Figure 17.15).

Although fatty acids may exist as free forms in the blood, most are transported as serum albumin-bound species. Free glycerol derived from triacylglycerol hydrolysis is readily converted to DHAP in the liver and then directed into the EMP pathway for further catabolism.

PHOSPHOLIPID BIOSYNTHESIS

Phospholipids are produced when 1,2-diacylglycerol and a phosphodiester bridge at the C_3 position of glycerol (Figure 17.16) are linked to a nitrogenous compound such as serine, ethanolamine, or choline (Figure 17.17).

Although biosynthesis of phospholipids resides mainly in the endoplasmic reticulum of eukaryotic cells, prokaryotic cells demonstrate phospholipid synthesis in the area of the plasma membrane. In cases of all cells, however, the amphipathic nature of phospholipids is readily suited to their specific roles in membrane architecture.

A structural study of the phospholipid known as *phosphatidylcholine* (lecithin) (Fig-

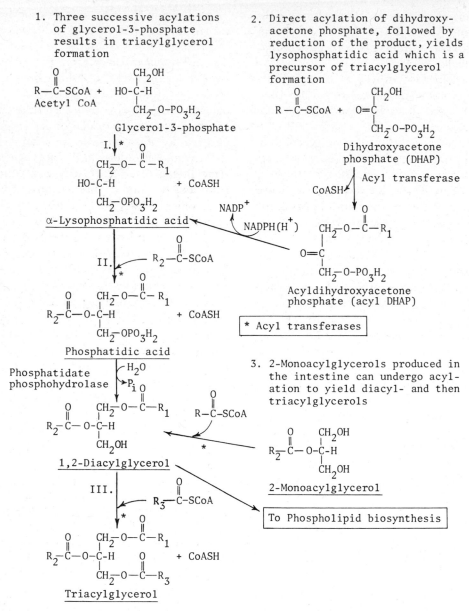

1. Three successive acylations of glycerol-3-phosphate results in triacylglycerol formation

2. Direct acylation of dihydroxyacetone phosphate, followed by reduction of the product, yields lysophosphatidic acid which is a precursor of triacylglycerol formation

3. 2-Monoacylglycerols produced in the intestine can undergo acylation to yield diacyl- and then triacylglycerols

FIGURE 17.14. Three routes lead to triacylglycerol formation beginning with (1) glycerol 3-phosphate, (2) dihydroxyacetone phosphate, or (3) 2-monoacylglycerol.

ure 17.18) reveals that long-chain fatty acids are esterified to the glycerol backbone at the C-1 and C-2 positions and the C-2-positioned fatty acid is usually polyunsaturated. Specific enzymes catalyze the characteristic assembly of key phospholipid components beginning with the phosphorylation of choline by *choline kinase* (Figure 17.19). Phos-

phocholine formed from this step is then used for the synthesis of *cytidine diphosphate choline* (CDP-choline). This reaction is mediated by *phosphocholine cytidyl transferase*. *Choline phosphotransferase* then mediates a reaction between CDP-choline and phosphatidate. This reaction results in lecithin production (i.e., *sn-3-phosphatidylcholine*).

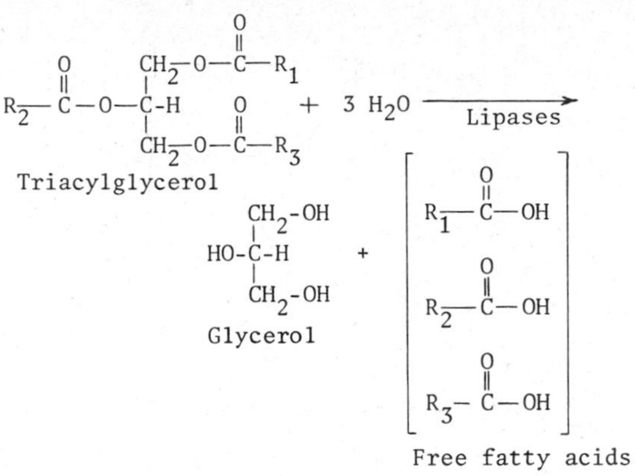

FIGURE 17.15. Generalized reaction for the complete hydrolysis of fatty acids from a triacylglycerol.

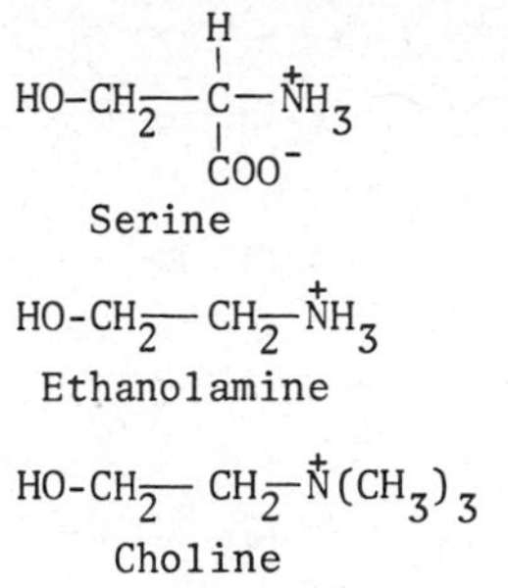

FIGURE 17.17. Structures for polar substituents commonly found in phospholipids (see Fig. 17.16).

An alternative route for lecithin formation has also been described. This pathway involves (1) the activation of phosphatidic acid (a diacylglycerol) by CTP through the enzyme action of CTP:phosphatidic acid cytidyltransferase (Figure 17.20) or (2) repetitive methylation of another phospholipid known as phosphatidylethanolamine (Figure 17.21A). In the first case (1), a nucleophilic reaction site for choline exists at the —OH group of the free choline, whereas the pyrophosphoryl structure of a CDP-diacylglyceride serves as the reaction center (Figure 17.20). According to the second alternative (2), phosphatidylethanolamine N-methyltransferase introduces methyl groups onto the phosphatidylethanolamine. A total of three methyl groups are added in a sequential fashion, and each methyl group originates from *S-adenosylmethionine* (SAM) (Figure 17.21A).

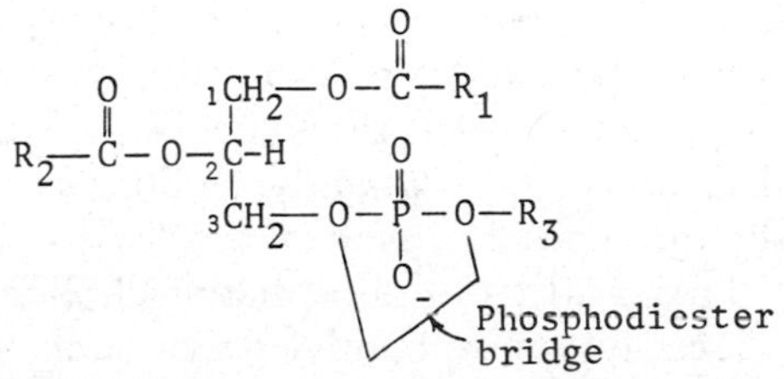

Some other phospholipids including phosphatidylethanolamine adhere to synthesis pathways that are similar to the formation of lecithin from CDP-choline (Figure 17.19) but the initial reaction steps require CDP-ethanolamine formation:

$$\text{Ethanolamine} + \text{ATP} \xrightarrow[\text{Mg}^{2+}]{\text{ethanolamine kinase}}$$
$$\text{phosphorylethanolamine} + \text{ADP}$$

and

$$\text{Phosphorylethanolamine} + \text{CTP} \xrightarrow[\text{Mg}^{2+}]{\substack{\text{phosphorylethanolamine}\\\text{cytidylytransferase}}}$$
$$\text{CDP-ethanolamine} + PP_i$$

CDP-ethanolamine is then linked to a 1,2-diacylglycerol by microsomal enzyme actions of *ethanolamine phosphotransferase* (Figure 17.21B). The foregoing reaction is notable in liver cells; however, phosphatidylethanolamine can also be produced in small amounts by the decarboxylation of phosphatidylserine (Figure 17.21C). Phosphatidylinositol, too, may be synthesized by a CDP-diacylglycerol plus inositol (Figure 17.21D).

FIGURE 17.16. A nonspecific phospholipid structure where R_1 and R_2 correspond to aliphatic fatty acid chains and R_3 represents a polar moiety such as serine, ethanolamine, or choline.

PHOSPHOLIPIDS DEMONSTRATE MANY DIFFERENT FUNCTIONS

Apart from contributions to membrane structure, phospholipids have many diverse biochemical functions. In the case of normal lung activity, at least 75% of extracellular alveolar phospholipid is comprised of di-

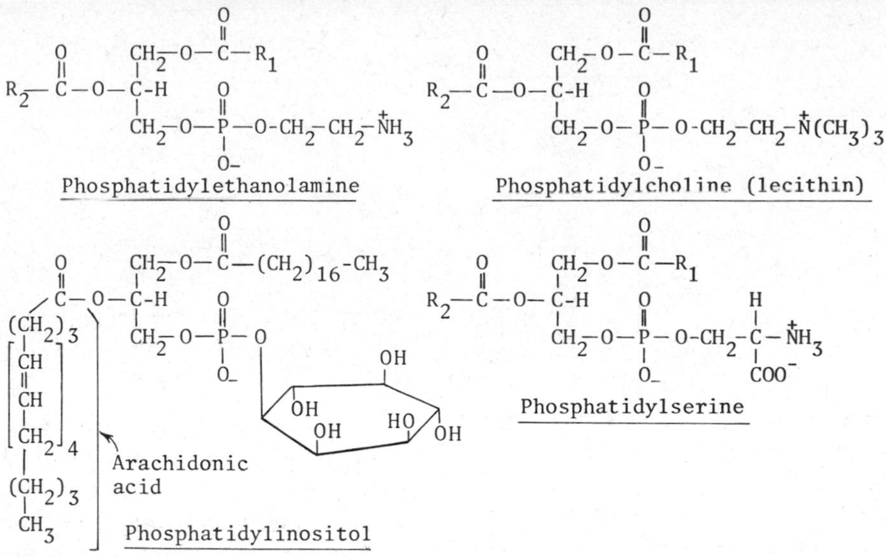

FIGURE 17.18. Selected structures for some phospholipids.

palmitoyllecithin. As indicated in the nomenclature, this molecule contains two residues of palmitic acid at the *sn*-1 and *sn*-2 positions. The amphipathic and associated surfactant properties of dipalmitoyllecithin are critical for holding the shape of alveoli because these structures cannot singularly maintain their shape against the surface tension of water. Dipalmitoyllecithin, however, reduces the surface tension of the fluid layer lining the alveoli and prevents their otherwise imminent collapse.

If premature infants have inadequate concentration levels of palmitoyllecithin, and the normal functioning of alveoli are prohibited, a *respiratory distress syndrome* (RDS) may develop. This problem is indicated as a rapid, shallow breathing, cyanosis, and ultimate death. Estimates suggest that up to 15 to 20% of neonatal deaths in the western world may be linked to this problem.

Fortunately, the potential for RDS in newborns can be estimated from amniotic screening tests that detect the presence of phosphatidylcholine. If type II pneumocytes located on the alveolar epithelium fail to produce adequate amounts of phosphatidylcholine, their "pulmonary surfactant" productivity can be improved by administering glucocorticoid drugs (e.g., betamethasone) to the mother. Glucocorticoids of this type accelerate fetal lung development and thereby promote the conversion of pneumocyte glycogen into dipalmitoyllecithin.

Phospholipids generally improve the solubilization and facilitate the translocation properties of steroids (cholesterol and cholesterol esters in bile and plasma), chylomicrons, triacylglycerols, as well as plasma lipids including LDL and VLDL.

Phospholipids can also function in transacylation reactions involving other lipids such as cholesterol. Such reactions depend on the action of *lecithin*:*cholesterol acyltransferase* (LCAT), which acylates the 3-hydroxyl group of free cholesterol at the expense of a fatty acyl group on phosphatidylcholine. Transacylation typically involves the loss of the *sn*-2 fatty acid from phosphatidylcholine to produce lysophosphatidylcholine and the cholesterol ester. Cholesterol esters produced in this way ensure the normal structure of plasma lipoproteins. It should also

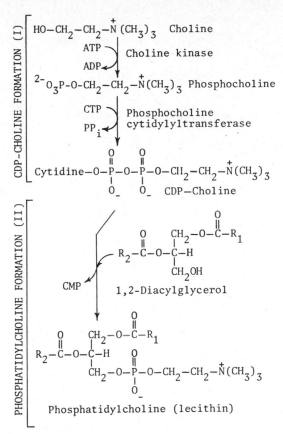

FIGURE 17.19. Choline is converted to CDP-choline as a preliminary step leading to phosphatidylcholine formation (step I). CDP-Choline is then covalently linked to a 1,2-diacylglycerol to produce phosphatidylcholine (step II).

be recognized in this vein that HDL serves as the substrate of LCAT.

Availability of the neurotransmitter acetylcholine may also be tied to the biochemistry of phosphatidylcholine. Since the hippocampal region of the brain is responsible for memory retention, it has been suggested that a reduction of cholinergic neurons in this brain region may contribute to geriatric memory loss and difficulties. Based on this premise, experiments were carried out with young people who were administered scopolamine—an agent that blocks cholinergic transmission; these demonstrated short-term memory impairment similar to that of ger-

iatric subjects. If speculation turns out to be true, any agent that elevates concentration levels of acetylcholine *may* lessen the severity of acetylcholine-linked memory disorders.

KETONE BODIES CAN BE PRODUCED FROM FATTY ACIDS

The metabolism of fatty acids can yield a variety of different products depending on the nutritional status of a test subject. Saturated and unsaturated fatty acids may be oxidized by β oxidation mechanisms to yield energy, or they can be esterified to glycerol, steroids such as cholesterol, phospholipids, and other plasma lipoproteins in the liver for vascular transport to adipose or muscle

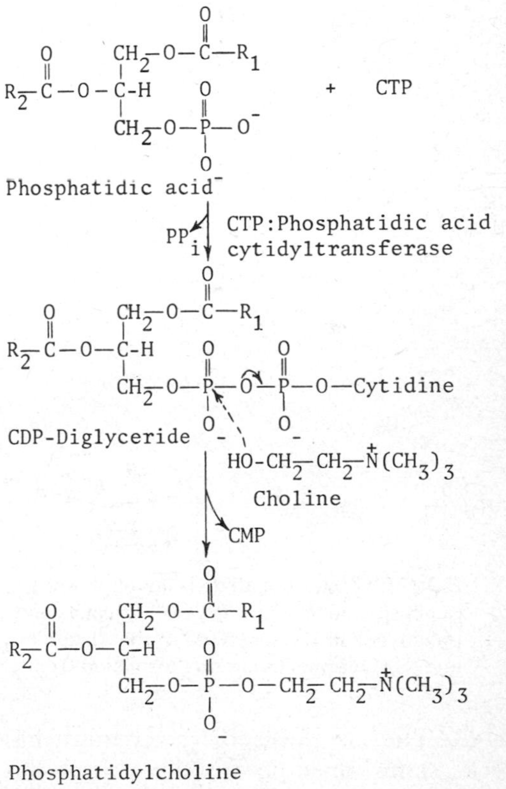

FIGURE 17.20. Biosynthesis of phosphatidylcholine from free choline and CDP-diacylglyceride formerly produced from phosphatidic acid.

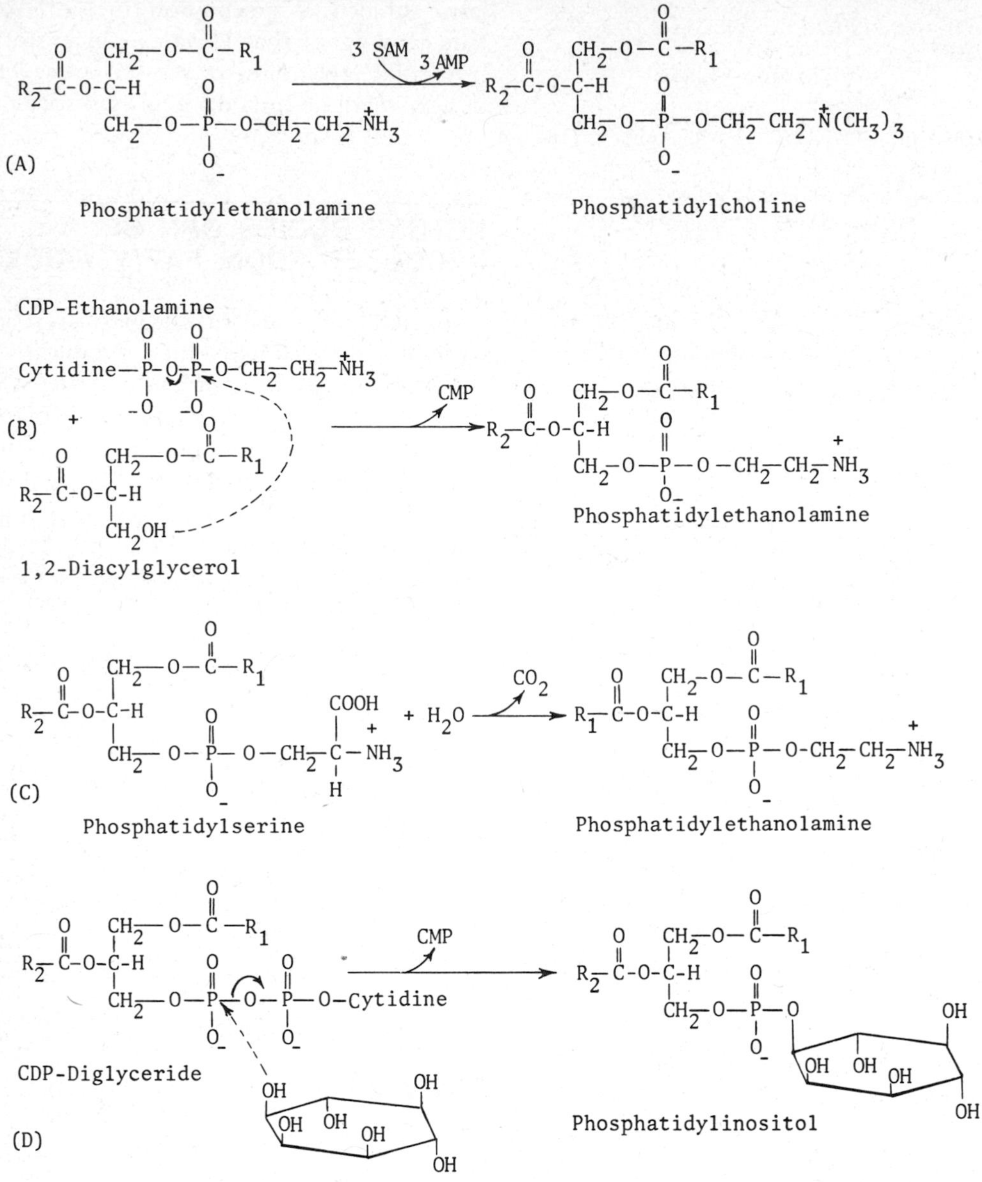

FIGURE 17.21. (A) Methylation of phosphatidylethanolamine by *S*-adenosylmethionine (SAM) leads to phosphatidylcholine. (B) CDP-Ethanolamine plus 1,2-diacylglycerol lead to the production of phosphatidylethanolamine. (C) Decarboxylation of phosphatidylserine results in phosphatidylethanolamine. (D) Phosphatidylinositol formation from a CDP-diglyceride and inositol.

tissues. The last transport mechanism has special significance because liver tissue has a finite ability to store fatty acids as triacylglycerols. Apart from these pathways, free fatty acids can be converted to ketone bodies. This transformation occurs primarily in the liver mitochondria, but some ketone formation also occurs in kidney tissues.

Ketone bodies hoid special importance because they represent a source of lipid-based potential metabolic energy. Unlike fatty acids, which are transported as an albumin complex in the plasma, ketone bodies are readily *soluble in water*. Furthermore they demonstrate lower toxicity than free fatty acids and rapidly diffuse across cell membranes.

Acetoacetate, 3-hydroxybutyrate (3-HB), and acetone are examples of ketone bodies commonly produced in humans. Ketone body production requires several sequential reactions beginning with the condensation of two acetyl-CoA molecules, resulting in *acetoacetyl-CoA* (Figure 17.22).

Acetoacetyl-CoA undergoes condensation with yet another acetyl-CoA to produce *3-hydroxy-3-methyl glutaryl-CoA* (HMG-CoA). The HMG-CoA undergoes decomposition

to yield acetoacetate and acetyl-CoA. Acetoacetate is subsequently reduced to 3-HB in the mitochondria, but the amount of 3-HB produced depends on the molar concentration ratio of mitochondrial $NAD^+/NADH(H^+)$.

Other than the obvious production of a D-hydroxy isomer of 3-HB, reduction of acetoacetate gives a product similar to the L-hydroxy isomer of 3-HB formed during the β oxidation of fatty acids. Some 3-HB is readily decarboxylated by nonenzymatic mechanisms to produce acetone and carbonic acid.

Acetoacetate and 3-HB can be converted

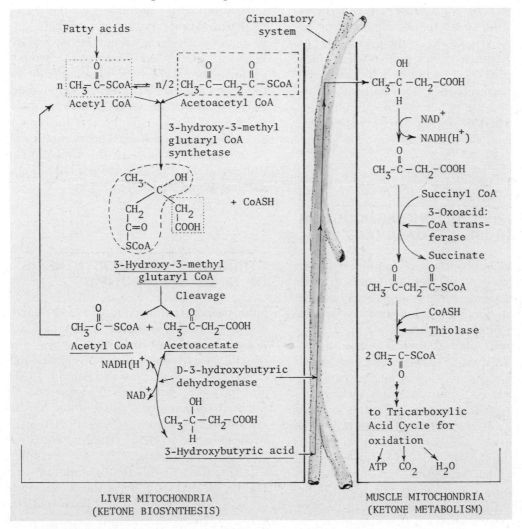

FIGURE 17.22. Fundamental steps in the biosynthesis and utilization of ketone bodies.

to acetyl-CoA as illustrated in Figure 17.22, after which it enters the TCA cycle for aerobic oxidation and energy production. Many key survival organs demonstrate ketone body utilization by this route, including skeletal and cardiac muscles and parts of the nervous system when low levels of glucose are apparent in the bloodstream. Hepatic tissues are fundamental to the productivity of ketone bodies as an energy source. It should be recognized that the utilization of ketone bodies is contingent upon the reactivation of acetoacetate to acetoacetyl-CoA. The critical reactivation step here involves succinyl-CoA as a source of coenzyme A and a *3-keto acid transferase*. This transferase resides in the mitochondrion of all tissues metabolizing ketones, but it is absent in the liver.

Aside from supplying metabolic energy, ketones can serve as building blocks for lipids in selected tissues. This requires the oxidation of 3-HB to butyryl-CoA. Butyryl-CoA is then used to initiate fatty acid synthesis reactions as shown earlier.

Under the duress of inadequate metabolizable glucose, extensive mobilization of fat in adipose cells supplies the liver and kidney with an overflow of oxidizable fatty acids. Since fatty acid oxidation of these tissues is checked by low amounts of oxaloacetate (used for gluconeogenesis), ketone body production is favored as an alternative to fatty acid oxidation. Unfortunately, concentration levels of ketone bodies in the body can build to a point that exceeds the threshold of the kidney. For example, 3-HB concentrations normally in the range of 3 mg/dL plasma can demonstrate 100 to 500-fold increases and cause *ketonuria*.

In addition to ketonuria, such high levels of plasma ketone bodies can severely *upset the normal acid–base balance* of the body. Many ketone bodies have functional carboxyl groups which display pK_a's substantially less than carbonic acid. Therefore, ketone bodies have a concentration-dependent ability to overpower the carbonic acid buffer system of normal plasma. The rampant and unchecked occurrence of this problem can cause *ketoacidosis* followed by overall physical debilitation and even death.

EXCESSIVE KETONE PRODUCTION CAN CAUSE KETOACIDOSIS

The metabolic utility of ketone bodies is overshadowed only by their ability to cause ketoacidosis in excessive amounts. This phenomenon is not a spontaneous biochemical problem but, rather, one that is spurred on by uncontrolled diabetes mellitus, prolonged fasting, a high-fat diet, post-ether anesthesia responses, and/or pregnancy.

In diabetic circumstances, high amounts of lipid in the form of fatty acids are mobilized from triacylglycerols. This occurs because insulin deficiency precludes normal cellular oxidation and use of carbohydrate, at a time when high glucagon concentrations decrease fatty acid introduction into fat depots. This action typically occurs in starvation situations when levels of plasma glucose concentration drop below 68% of nominal fasting glucose level (~88 mg/dL).

STRUCTURE AND BIOSYNTHESIS OF SOME SPHINGOLIPIDS

Sphingosine (D-sphingenine) is a long-chained aliphatic amine that serves as a fundamental constituent of sphingolipids. This compound arises from a series of enzyme-mediated reactions that require palmitoyl-CoA and L-serine. Intermediate formation of 3-dehydrosphinganine is followed by its reduction to D-sphinganine and then a dehydrogenation to yield D-sphingenine (or 4-sphingenine) (Figure 17.23). A corresponding formation of this *trans* compound may also be formed from desaturation of a fatty acyl-CoA before its condensation with L-serine. This desaturation step prior to D-sphingenine formation depends on a flavin-linked dehydrogenase and may serve as an

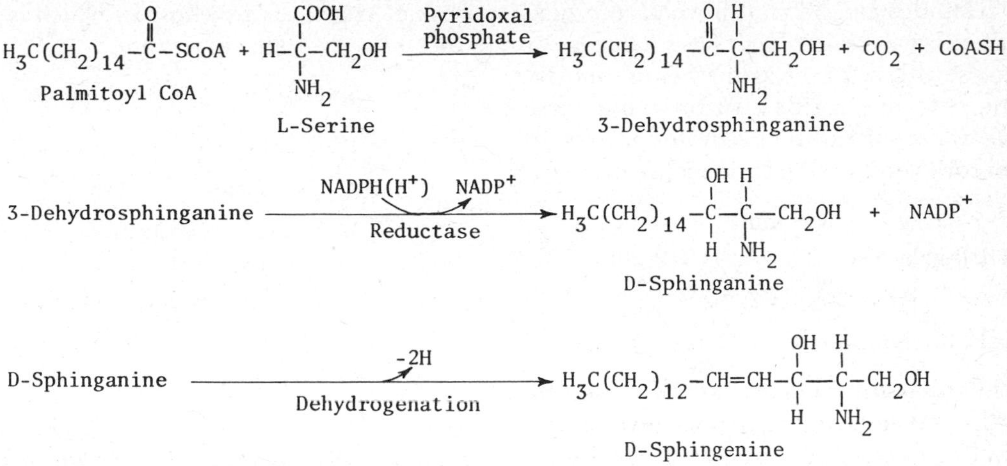

$$H_3C(CH_2)_{14}-\overset{O}{\underset{}{\overset{\|}{C}}}-SCoA + H-\overset{COOH}{\underset{NH_2}{\overset{|}{C}}}-CH_2OH \xrightarrow[\text{phosphate}]{\text{Pyridoxal}} H_3C(CH_2)_{14}-\overset{O}{\underset{}{\overset{\|}{C}}}-\overset{H}{\underset{NH_2}{\overset{|}{C}}}-CH_2OH + CO_2 + CoASH$$

Palmitoyl CoA   L-Serine   3-Dehydrosphinganine

3-Dehydrosphinganine $\xrightarrow[\text{Reductase}]{\text{NADPH}(H^+) \quad NADP^+}$ $H_3C(CH_2)_{14}-\overset{OH}{\underset{H}{\overset{|}{C}}}-\overset{H}{\underset{NH_2}{\overset{|}{C}}}-CH_2OH + NADP^+$

D-Sphinganine

D-Sphinganine $\xrightarrow[\text{Dehydrogenation}]{-2H}$ $H_3C(CH_2)_{12}-CH=CH-\overset{OH}{\underset{H}{\overset{|}{C}}}-\overset{H}{\underset{NH_2}{\overset{|}{C}}}-CH_2OH$

D-Sphingenine

FIGURE 17.23. D-Sphingenine formation results from dehydrogenation of D-sphinganine.

important general pathway for sphingenine synthesis.

Sphingenine now serves as a substrate for

nant synthesis route depends on the action of *CDP-choline:ceramide choline phosphotransferase* (CCPT).

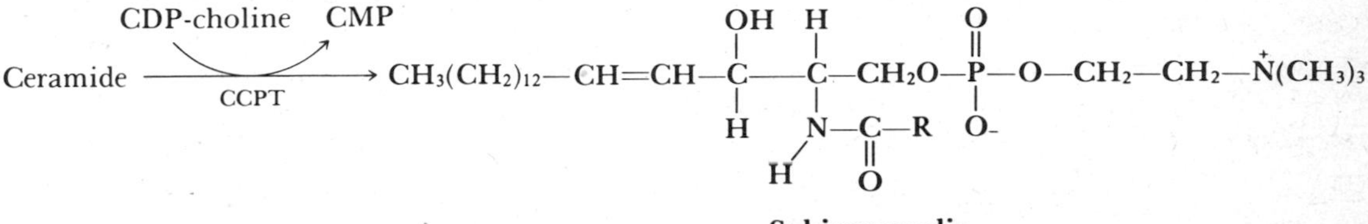

Ceramide $\xrightarrow[\text{CCPT}]{\text{CDP-choline} \quad \text{CMP}}$ $CH_3(CH_2)_{12}-CH=CH-\overset{OH}{\underset{H}{\overset{|}{C}}}-\overset{H}{\underset{N-C-R}{\overset{|}{C}}}-CH_2O-\overset{O}{\underset{O^-}{\overset{\|}{P}}}-O-CH_2-CH_2-\overset{+}{N}(CH_3)_3$

Sphingomyelin

acylation by a long-chain fatty acyl-CoA. The required enzyme in this step is *ceramide:N-acyl transferase* and the immediate end product is *ceramide*. This ceramide can also be described as a *N*-acyl sphingenine (or *N*-acyl sphingosine).

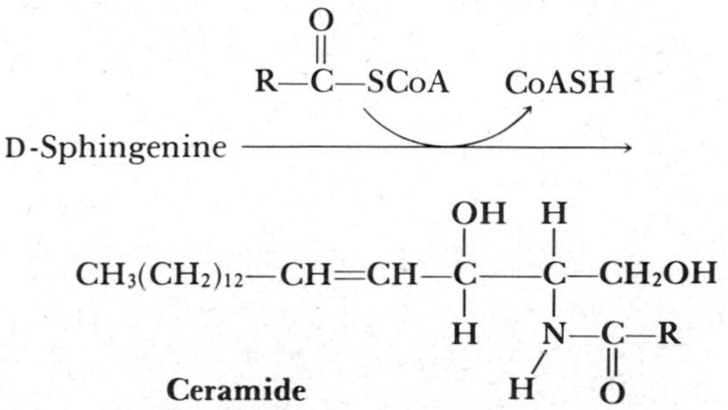

D-Sphingenine $\xrightarrow[]{\quad R-\overset{O}{\overset{\|}{C}}-SCoA \quad CoASH \quad}$

$$CH_3(CH_2)_{12}-CH=CH-\overset{OH}{\underset{H}{\overset{|}{C}}}-\overset{H}{\underset{N-C-R}{\overset{|}{C}}}-CH_2OH$$

Ceramide

Ceramide formation serves as a prelude to the biosynthesis of sphingomyelins. Although sphingomyelins can be produced by a variety of minor pathways, the predomi-

Activity of CCPT is common in liver mitochondria as well as those mitochondria in the spleen and kidney. Lesser amounts of sphingomyelins result from the following two reaction pathways:

Reaction 1

Ceramide + phosphatidylcholine $\rightleftharpoons$
    sphingomyelin + diacylglycerol

Reaction 2

CDP-Choline + sphingenine
    $\downarrow$
sphingenylphosphorylcholine + CMP
    fatty acyl-CoA
    CoASH
sphingomyelin

Ceramides can be transformed into hexose derivatives called *cerebrosides*. Glucose or galactose often serves as the hexose moiety within the cerebroside structure, but these sugars must exist as their activated UDP-sugars before cerebroside formation can occur.

Ceramide

UDP-galactose / \ UDP-glucose

UDP ← → UDP

galactocerebroside   glucocerebroside

Further structural elaboration of these cerebrosides results in numerous glycosphingolipid structures (Figure 17.24). It should be recognized that some cerebrosides are formed from sugar-substituted sphingosine followed by later acylation. One classic example is the galactose-substituted sphin-

gosine, known as psychosine, which is produced by the following route:

D-Sphingenine + UDP-galactose

⇅

psychosine + UDP

(galactose-substituted sphingenine)

│ fatty acyl-CoA

↓ CoASH

cerebroside

The ganglion cells of the central nervous system, and especially the nerve endings, contain a group of compounds called *gangliosides*. These glycosphingolipids (Figure 17.24) are also characterized by the presence of *sialic acid* (Sia). Sialic acids are essentially

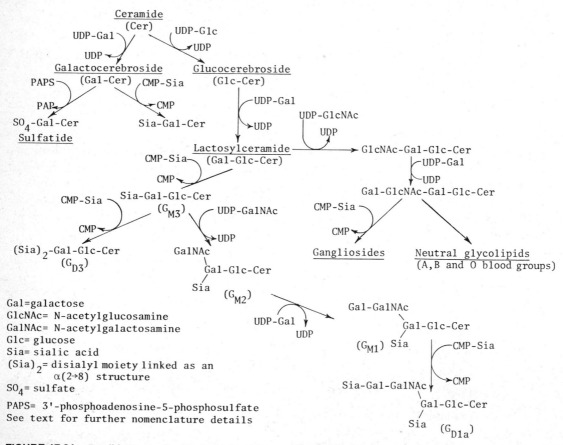

Gal=galactose
GlcNAc= N-acetylglucosamine
GalNAc= N-acetylgalactosamine
Glc= glucose
Sia= sialic acid
(Sia)$_2$= disialyl moiety linked as an α(2→8) structure
SO$_4$= sulfate

PAPS= 3'-phosphoadenosine-5-phosphosulfate
See text for further nomenclature details

FIGURE 17.24. Possible routes for biosynthesis of glycosphingolipids.

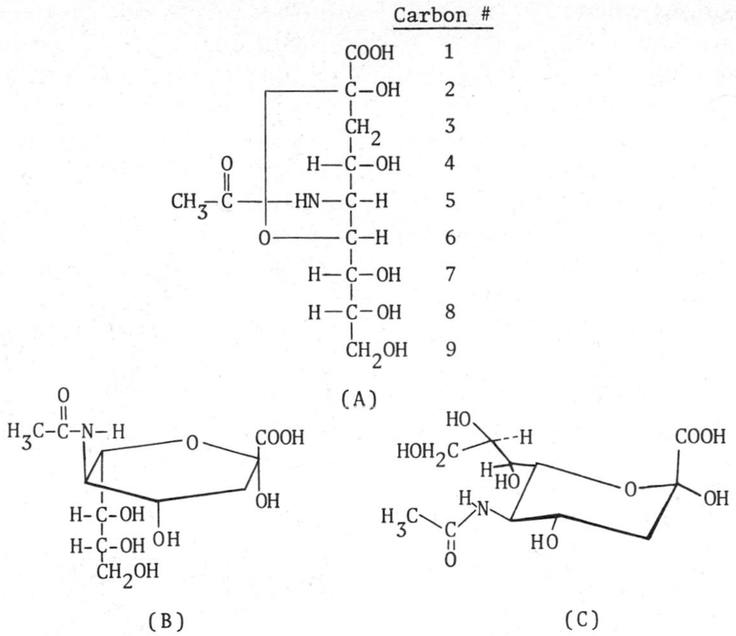

Carbon #

$$
\begin{array}{cc}
\text{COOH} & 1 \\
\text{C—OH} & 2 \\
\text{CH}_2 & 3 \\
\text{H—C—OH} & 4 \\
\text{HN—C—H} & 5 \\
\text{C—H} & 6 \\
\text{H—C—OH} & 7 \\
\text{H—C—OH} & 8 \\
\text{CH}_2\text{OH} & 9
\end{array}
$$

(A)

(B)   (C)

FIGURE 17.25. Common structural representations for a sialic acid, *N*-acetyl-D-neuramic acid. (A) Fischer, (B) Haworth, and (C) conformational representations are given.

ketoses that contain *nine* carbon atoms with an *N*-acetyl group as indicated in Figure 17.25. These sugars, also recognized as *ketononoses*, can also be described as acylated derivatives of neuraminic acid (3,5-dideoxy-5-aminononulosonic acid). At least half of the sialic acid in human tissues is present in lipid-bound ceramide structures as opposed to the oligosaccharide constituents of glycoproteins. Only about 10% of total sialic acid stores appear on the superficial membrane structures of nonneural cells.

Ganglioside biosynthesis relies on cerebroside formation followed by the addition of key sugars and sugar derivatives. These include galactose, glucose, and the neuraminic acids that are usually supplied to ganglioside biosyntheses as nucleotide (UDP)-activated forms (Figure 17.26).

Ganglioside nomenclature follows a series

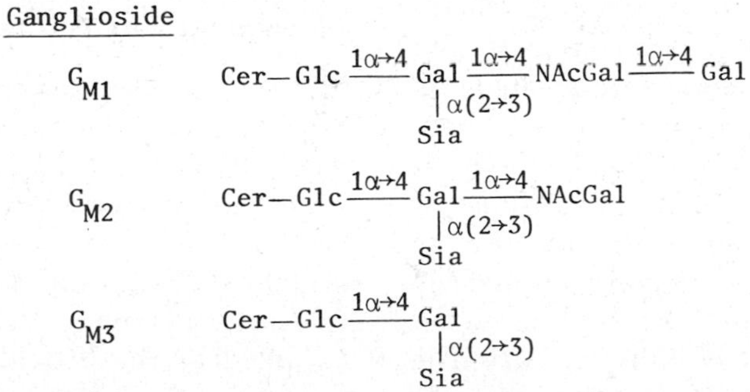

Ganglioside

$$G_{M1} \quad \text{Cer—Glc} \xrightarrow{1\alpha\to4} \text{Gal} \xrightarrow{1\alpha\to4} \text{NAcGal} \xrightarrow{1\alpha\to4} \text{Gal}$$
$$\qquad\qquad\qquad\qquad\quad |\,\alpha(2\to3)$$
$$\qquad\qquad\qquad\qquad\quad \text{Sia}$$

$$G_{M2} \quad \text{Cer—Glc} \xrightarrow{1\alpha\to4} \text{Gal} \xrightarrow{1\alpha\to4} \text{NAcGal}$$
$$\qquad\qquad\qquad\qquad\quad |\,\alpha(2\to3)$$
$$\qquad\qquad\qquad\qquad\quad \text{Sia}$$

$$G_{M3} \quad \text{Cer—Glc} \xrightarrow{1\alpha\to4} \text{Gal}$$
$$\qquad\qquad\qquad\qquad\quad |\,\alpha(2\to3)$$
$$\qquad\qquad\qquad\qquad\quad \text{Sia}$$

FIGURE 17.26. Structures for some important gangliosides. Abbreviations: Cer, ceramide; Gal, galactose; Glc, glucose; NAcGal, *N*-acetylgalactosamine residue; Sia, sialic acid residue reflects the structure of *N*-acetylneuramic acid.

of logical conventions where "G" (for ganglioside) is followed by subscripts M, D, T, and Q, indicating *mono-*, *di-*, *tri-*, and *quatra-* (or *tetra-*) sialic acid-containing gangliosides, respectively.

Because gangliosides have discrete sequences of sugar residucs attached to ceramide, numerical subscripts also define this structural property:

Subscript 1: Gal–NAcGal–Gal–Glc-ceramide

Subscript 2: NAcGal–Gal–Glc-ceramide

Subscript 3: Gal–Glc-ceramide

Some ganglioside structures have been indicated in Figure 17.26 according to this nomenclature scheme.

SPHINGOLIPID STORAGE DISEASES ARE LINKED TO GENETIC DISORDERS

At least nine hereditary disorders in lipid storage are recognized and generically called *lipidoses*. The signs and symptoms of the major lipidoses are detailed in Table 17.4.

The underlying cause for many of these diseases, with the exception of Fabry's disorder, are linked to *autosomally recessive* traits. That is, a homozygous child can result if both parents are heterozygous for the disease. In this case, each parent asymptomatically carries a gene for the same hereditary problem. The probable occurrence of lipid storage disease in a homozygous child will be 1:4. Fabry's disease, on the other hand, is transmitted as a *sex-linked feature* on the X chromosome. Here, it is necessary only for the mother to carry the genetic disorder so that an *affected* (heterozygous) male child will result. A 50% affliction probability occurs in these instances for each of the carrier's male progeny, whereas half of the carrier's female progeny will also be carriers of the causative gene for Fabry's disease.

Most evidence indicates that lipid storage

diseases mirror one or more defects in hydrolytic enzyme action necessary for sphingolipid degradation. Since these hydrolytic enzymes normally reside in the lysosomes, lipid storage defects can also be denoted as *lysosomal maladies*. In any event, these defective lysosomal enzymes cause neurological, cardiac, and kidney dysfunctions and/or mental retardation in infants. Analytical procedures for recognition of these genetic disorders have been described, but expeditious diagnosis and parental counseling based on genetic criteria offer the most effective means for management of these disorders (Glew and Peters, 1977).

CHOLESTEROL BIOSYNTHESIS

The *de novo* biosynthesis of cholesterol can occur in nearly all cells but the liver, small intestine, adrenal cortex, and reproductive tissues including testes, ovaries, and placenta display the most notable productivities.

Cholesterol biogenesis is ultimately a study of *activated acetate biochemistry*. That is, cholesterol biosynthesis is dependent on a supply of acetyl-CoA's that feed into an ongoing series of four sequential reactions: (1) *mevalonate biosynthesis*, (2) *isopentenyl pyrophosphate formation*, (3) *squalene formation* from isopentenyl pyrophosphate, and (4) *cholesterol production* from squalene, in that order.

Mevalonate Biosynthesis

Three molecules of acetyl-CoA (C_2 units) are enzymatically condensed to produce mevalonate (C_6) by way of an acetoacetyl-CoA (C_4) and a 3-hydroxy-3-methyl glutaryl-CoA (HMG-CoA) intermediate. Although two acetyl-CoA's can condense to form acetoacetyl-CoA, this compound can be alternatively supplied by the β oxidation of fatty acids. In any event, HMG-CoA undergoes an NADPH(H^+)-dependent reduction by an enzyme *3-hydroxy-3-methylglutaryl-CoA reductase* (HMG-CoA reductase).

Table 17.4. Glycolipid Storage Diseases (Lipidoses or Sphingolipidoses) and Their Characteristic Identification Properties

Sphingolipid structure	Defective enzyme (hereditary)	Symptoms and identification
Ceramide	Ceramidase	Skeletal deformation, mental retardation, hoarseness, dermatitis (Farber's disease)
R^{**}—O—$\overset{\beta}{}$Glu—$\overset{\beta}{}$Gal—$\overset{\alpha}{\vdots}$Gal **Ceramide trihexoside**	Ceramide trihexoside α-galactosidase	Lower extremity pain, renal failure, reddish-purple skin rash (Fabry's disease)
R—O—$\overset{\beta}{\vdots}$Gal **Galactocerebroside**	β-Galactosidase	White matter of the brain contains globoid bodies, myelin absence, mental retardation (Krabbe's disease or globoid leukodystrophy)
R—O—$\overset{\beta}{}$Glu—$\overset{\beta}{}$Gal—$\overset{\beta}{\vdots}$GalNAc $\quad\underset{\text{Sia}}{\mid \alpha}$ **Ganglioside G_{M2}**	Hexosaminidase A	Blindness, muscular weakness, red spot in retina, mental retardation (Tay–Sachs disease)
R—O—$\overset{\beta}{}$Glu—$\overset{\beta}{}$Gal—$\overset{\beta}{}$GalNAc—$\overset{\beta}{\vdots}$Gal $\quad\underset{\text{Sia}}{\mid \alpha}$ **Ganglioside G_{M1}**	Galactocerebroside β-galactosidase	Red spots in retina (50% incidence), hepatic enlargement, generalized gangliosidosis, skeletal deformities
R—O—$\overset{\beta}{\vdots}$Glu **Glucocerebroside**	Glucocerebroside β-glucosidase	Infantile versions cause mental retardation, pelvic and long-bone degeneration, hepatic and splenic enlargement (Gaucher's disease)
R—O—$\overset{\beta}{}$Glu—$\overset{\beta}{}$Gal—$\overset{\alpha}{}$Gal—$\overset{\beta}{\vdots}$GalNAc **Globoside (and ganglioside G_{M2})**	Hexosaminidase A and B	Rapidly progressing form of Tay–Sach's disease called Sandhoff's disease
R—O—$\vdots$PO$_3^-$—$\overset{\beta}{}$CH$_2$—CH$_2$—$\overset{+}{N}$(CH$_3$)$_3$ **Sphingomyelin**	Sphingomyelinase	Red spots in retina (30% incidence), hepatic and splenic enlargement, mental retardation (Nieman–Pick disease)
R—O—$\overset{\beta}{}$Gal—$\vdots$OSO$_3$ **Sulfatide**	Sulfatidase	Neuronal staining with cresyl violet results in yellow-brown appearance, psychological upset in adults, mental retardation (metachromatic leukodystrophy)

* Site of enzyme attack = $\vdots$.

** = R .

Since the activity of this *reductase* can be affected by dietary cholesterol levels, there is little doubt that this cytoplasmic enzyme serves as an important regulatory site in cholesterol biosynthesis. Cholesterol itself may have a negative feedback effect on the reductase, which effectively terminates its own synthesis; however, other mechanisms are speculated. Some theories contend that cholesterol containing lipoproteins, a protein component coexisting with bile, or bile acids themselves may be involved in reductase regulation. The overall scheme for mevalonate production is outlined in Figure 17.27.

Isopentenyl Pyrophosphate (IPP)

Mevalonate is phosphorylated to yield 3-phospho-5-pyrophosphomevalonate. Phosphorylation is achieved by the action of specific kinases and the consumption of three ATPs. Poor stability of a phosphorylated intermediate results in its decarboxylation to produce *isopentenyl pyrophosphate* (IPP) (Figure 17.28).

Squalene Formation from IPP

Progressive isomerization of IPP results in *dimethylallyl pyrophosphate* (DAP). According to a very complex condensation, four IPP units and two DAP units combine to form squalene (Figure 17.29). The apparent simplicity of squalene formation is not realized at the biochemical level. Instead, the mechanism requires the formation of two C_{10} units present as geranyl pyrophosphate followed by two C_{15} units of farnesyl pyrophosphate. Isomerization of one farnesyl unit produces nerolidol pyrophosphate (not illustrated in the figure), which subsequently condenses with the remaining farnesyl unit to form the C_{30} structure of squalene.

Cholesterol Synthesis from Squalene

Cholesterol formation from squalene proceeds by way of *lanosterol* formation (Figure 17.30). This reaction has been historically recognized as the most complicated reaction of cholesterol synthesis. Since C_{30}-squalene, having six trisubstituted double bonds, yields lanosterol, lanosterol formation actually represents the first synthesis step leading to *all other steroid compounds*. The characteristic structure of lanosterol displays four fused rings and seven asymmetric centers.

The production of lanosterol from squalene has garnered special attention as a result of key enzymatic reactions necessary for constructing this early steroid compound. Squalene, containing its six trisubstituted double bonds in an "almost" linear arrangement, predictably undergoes cyclization into a fused four-ringed system having seven sites of asymmetry. This structural transformation begins with the formation of 2,3-oxi-

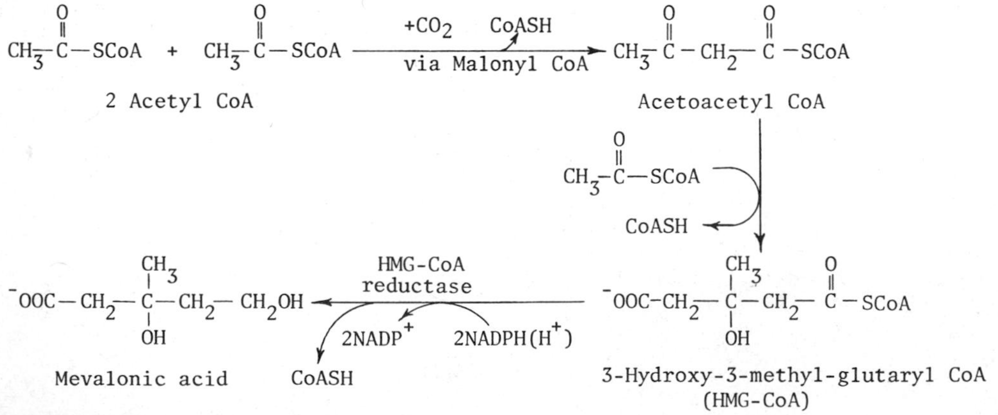

FIGURE 17.27. Mevalonate biosynthesis involves the consumption of acetyl-CoA's.

dosqualene. This 2,3-oxide results from the action of *2,3-squalene epoxidase* on squalene, but mechanistic disputes still exist regarding the actual introduction of molecular oxygen into the molecule. The linear structure of 2,3-oxidosqualene is converted to a characteristic steroid by *lanosterol cyclase*. According to the reaction outlined (Figure 17.30), fused steroidal rings are successively formed (according to arrows) in an order paralleling the A, B, C, and D rings of the lanosterol structure. This step is then followed by a hydrogen elimination as well as hydrogen and methyl migrations around the fused rings.

Cholesterol synthesis from lanosterol necessitates the elimination of three methyl groups (one a C-14 and two at C-4), translocation of the 8,9-double bond in lanosterol to the 5,6-position, plus elimination of the 24,25-unsaturated site in the side chain.

For plants and many microorganisms, farnesyl pyrophosphate and IPP condense to produce geranyl-geranyl pyrophosphate. This C_{20} structure can undergo further condensation with another similar unit to produce C_{40} unsaturated compounds that are descriptively divisible into *isoprene units.* Carotenoids and their hydroxylated derivatives including the xanthophylls can arise from this pathway. Steroid compounds present in yeasts, fungi, algae, and higher plants, including *ergosterol* (Figure 17.31), β-*sitosterol, campesterol,* and many other natural steroids follow similar synthesis tracts as cholesterol with only minor differences. One notable difference involves the conversion of squalene-2,3-epoxide into *cycloartenol,* instead of

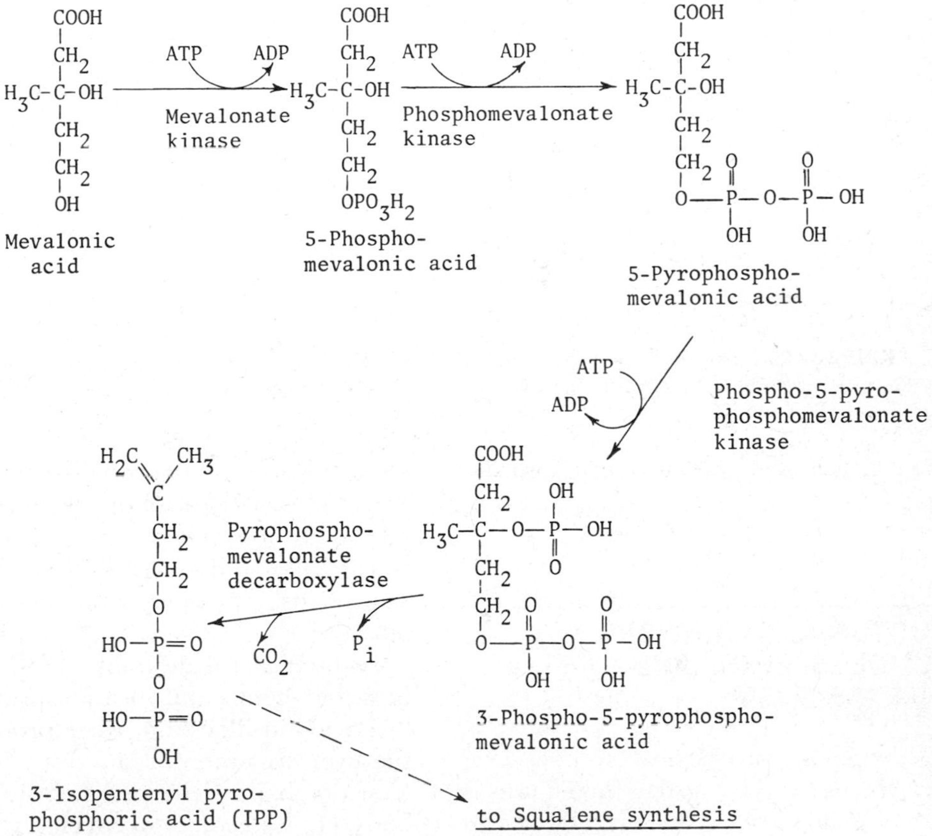

FIGURE 17.28. Isopentenyl pyrophosphate is formed from mevalonate.

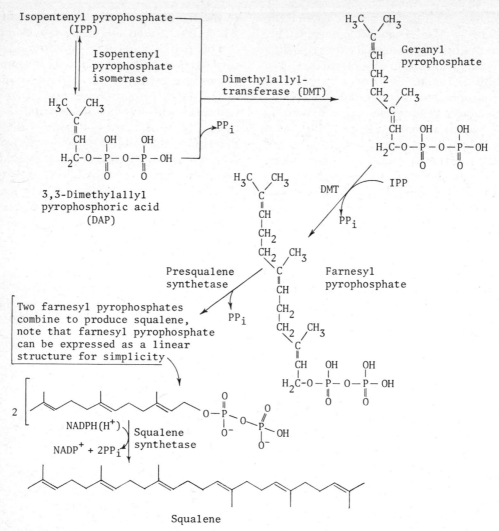

FIGURE 17.29. Isopentenyl pyrophosphate (IPP) is crucial for the synthesis of squalene. Squalene, in turn, serves as a precursor to many types of steroids including cholesterol.

lanosterol, as a plant steroid precursor (Figure 17.32).

STRUCTURAL VARIATIONS OF CHOLESTEROL AND ITS METABOLISM

The endogenous production of cholesterol by liver tissues is ~0.5 g/day (maximum is 5.0 g/day). Cholesterol synthesized by *de novo* hepatic mechanisms plus dietary cholesterol both contribute to the cholesterol pool of the body. This cholesterol may be directed into bile acid production, discharged in bile, translocated to nondescript tissue sites by lipoproteins, and/or used in cell membrane biosynthesis.

Aside from the desirable amphipathic nature and lipid emulsification properties of bile acids in digestion, their production in the liver accounts for the disposition of at least one-half of the per diem turnover in body cholesterol. Moreover, relative discharges of bile acids, cholesterol, and phos-

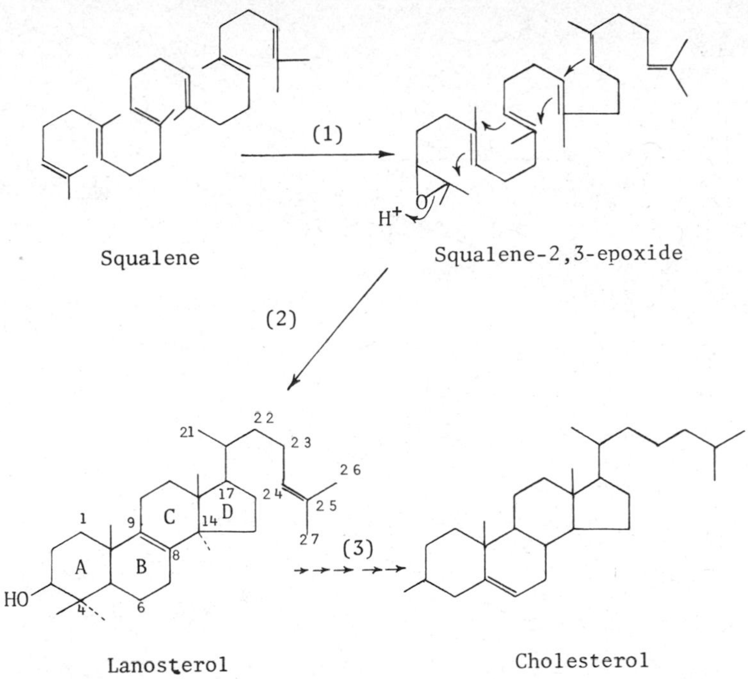

FIGURE 17.30. Key steps in the transformation of squalene into lanosterol and eventually cholesterol. Key enzymes include (1) 2,3-squalene epoxidase; (2) lanosterol cyclase, and (3) numerous steps.

phatidylcholine into bile affect the onset of *cholelithiasis*. Cholelithiasis typically results in pathological concretions that contain 50 to 100% cholesterol accompanied with minor amounts of calcium bilirubinate (plus other bilirubins), which becomes located in the biliary tract.

Bile acid synthesis from cholesterol occurs in the liver hepatocytes. Initial hydroxylation of the steroids is mediated by 7-α-hydroxylase to yield 7-α-hydroxycholesterol.

Progressive reactions ultimately form the two primary bile acids known as *cholate* and *chenodeoxycholate*. Although hydroxylation of the steroid occurs in the endoplasmic reticulum, oxidative degradation of the steroid takes place in the mitochondria (Figure 17.33).

Cholate and chenodeoxycholate are also conjugated in the liver with either taurine or glycine (*1:3 occurrence for each conjugate*), which enhances their solubility properties.

Bile salts are ejected under the influence

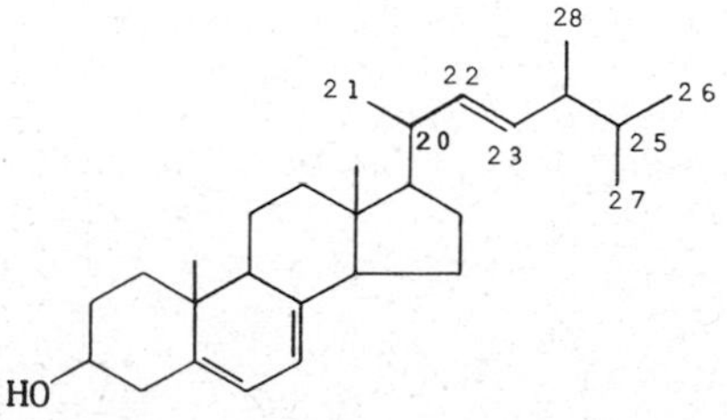

FIGURE 17.31. Ergosterol, a typical nonanimal steroid.

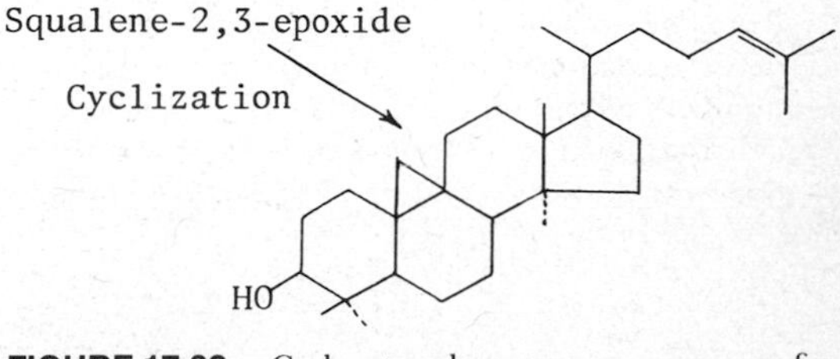

FIGURE 17.32. Cycloartenol serves as a precursor for numerous plant steroids that develop from the cyclization of squalene-2,3-epoxide.

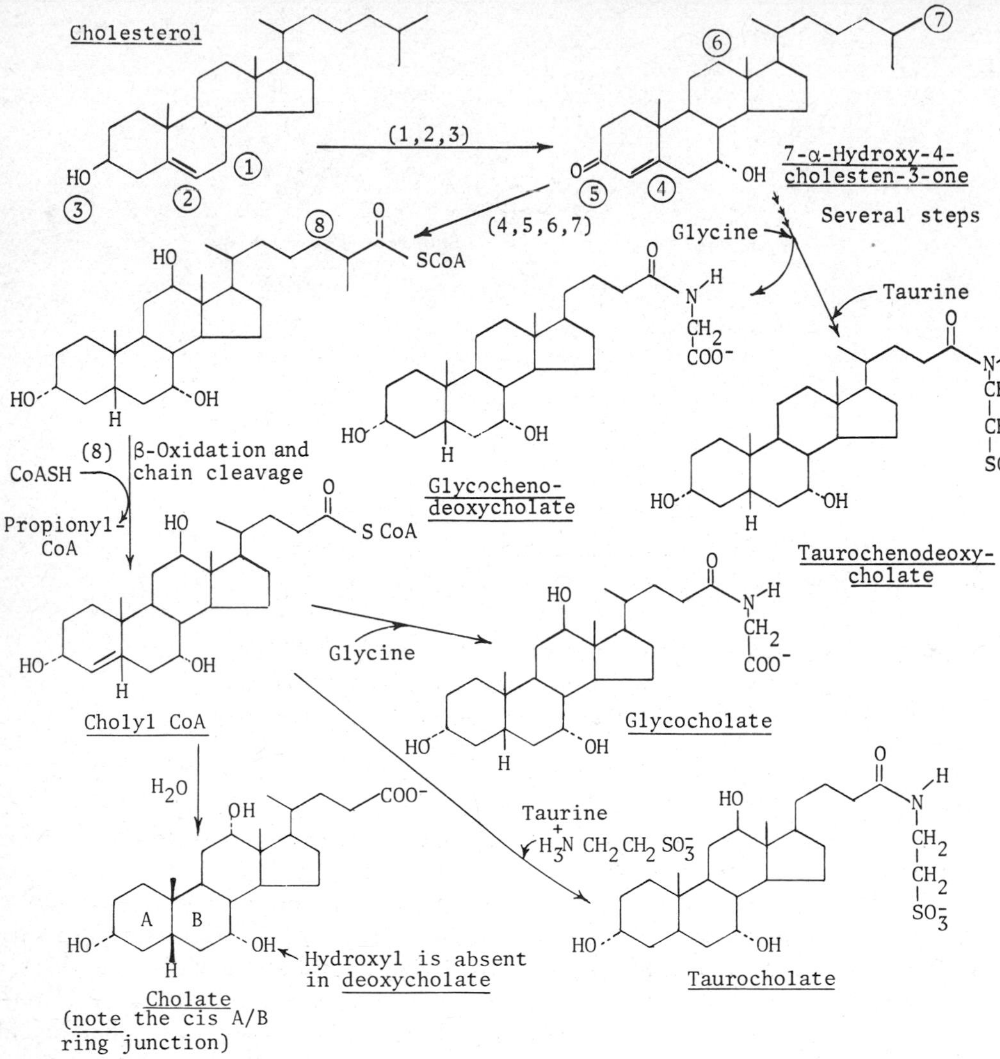

FIGURE 17.33. Conversion pathway for cholesterol into bile salts. These compounds are major metabolic products of cholesterol metabolism compared to steroid hormones. Bile acids are produced in the liver by α-hydroxylations and oxidative cleavage of the C-17 side chain. The initial conversion steps require $NADPH(H^+)$ for reductions as well as O_2 for hydroxylation reactions. Two key enzymes, namely, 7-α-hydroxylase (1) and 12-α-hydroxylase (6) are necessary for these hydroxylation reactions. Bile acid formation is then followed by their conversion to coenzyme A esters, which then undergo conversion to conjugates of either glycine or taurine to produce bile salts. The origin of glycine for conjugation is direct from the amino acid pools in the liver, but taurine is produced from cysteine by sulfur oxidation and decarboxylation. Circled numbers beside structures indicate reaction sites on steroid molecules and the numbers in parentheses indicate the sequence of reactions that specifically involve (1) hydroxylation; (2) dehydrogenation; (3) isomerization; (4) and (5) reduction; (6) hydroxylation; (7) hydroxylation and oxidation to acyl-CoA; and then (8) β-oxidation and chain cleavage.

of *cholecystokinin* via the bile duct into the gut. Here, they promote chylomicron formation of the dietary lipids and aid lipid absorption.

Passive diffusion of discharged bile salts occurs over the gut tissues with active transport predominating in the terminal 75 cm of the ileum. The combined effects of passive and active absorption ensure only minor losses of bile acids by the intestinal route. Once the bile acids are absorbed, they are bound to plasma albumins in the portal circulation and cleared from the plasma during passage through the hepatic circulation.

As might be expected, the biochemistry of unabsorbed bile salts does not cease in the intestine. Instead, it proceeds as a result of intestinal bacteria. These organisms go on to produce a host of *secondary bile salts* including *deoxycholate* and *lithocholate* in the intestine. If these compounds reach the absorptive surfaces of the colon, ~25 to 30% may be resorbed, conjugated in the liver, and ejected in another round of bile production (Figure 17.34).

Inventories of bile constituents suggest a normal mixture of ~38% cholate, ~38% chenodeoxycholate, ~18 to 20% deoxycholate, ~2 to 4% lithocholate, and some other related steroid derivatives.

The discharge of bile salts followed by their reabsorption touts an efficiency rate of 95 + %. Efficiency of this so-called *enterohepatic circulation* (Figure 17.34), however, depends on many different factors. In fact, bile salt reabsorption is closely linked to lecithin (phospholipids) codischarged with bile and cholesterol as well as intraluminal dietary constituents of the intestinal tract.

So long as lecithin concentrations (~3.0% of the total bile solids) permit an adequate interaction with bile salts (>2.0 mM concentrations) to ensure micelle solubility, cholesterol can be effectively carried in bile salt micelles. This permits transport of the hydrophobic compound in the aqueous environment of the bile. If the required quotas of lecithin and bile salts are inadequate to ensure cholesterol solubility, this can serve as a contributing factor in gallstone forma-

tion (cholelithiasis). Although bile, supersaturated with cholesterol, is important in inciting lithogenic bile and gallstone formation, most evidence suggests that a nucoprotein structure or some other *nidus* for cholesterol crystallization is probably necessary (Figure 17.35).

The remarkable reabsorption and enterohepatic circulation of bile salts and deoxycholate is not shared by the secondary bile salt, lithocholate, or by lecithin. Poor absorption for lithocholate is tied to reactions that produce its sulfated derivatives, whereas lecithin simply fails to survive gut biochemistry. In the last case, ester hydrolysis yields a free fatty acid and lysolecithin.

Poor enterohepatic absorption of all bile salts, bile-related steroids, and cholesterolic steroids can be encouraged by certain dietary constituents that bind to these compounds. Binding may involve temporal chemical bonding or electrostatic interactions with whole or partially digested foods, or it may simply involve physical factors. For example, steroids may become surrounded in a physical matrix of digested food that stymies their absorption, or steroids may become encased in a partially digested food matrix that has a high intestinal transit rate. All of these possibilities minimize effective enterohepatic absorption phenomena.

Apart from cholesterol excretion in bile, significant amounts can be translocated across the intestinal mucosa. Excretion of cholesterol by this route can result in the microbial reduction of steroids into *stanols* that have poor reabsorption properties. The stanols, also called fecal sterols, prominently include coprostanol and cholestanol, but numerous other related stanols and intermediates occur (Figure 17.36).

Added to the key role of cholesterol in bile acid metabolism is its contribution to the structure of membranes and many hormonal substances. *Pregnenolone* is among the most important of the steroid intermediates since it provides a nucleus for many other steroids. As indicated in Figure 17.37, pregnenolone serves as a precursor for (1) the

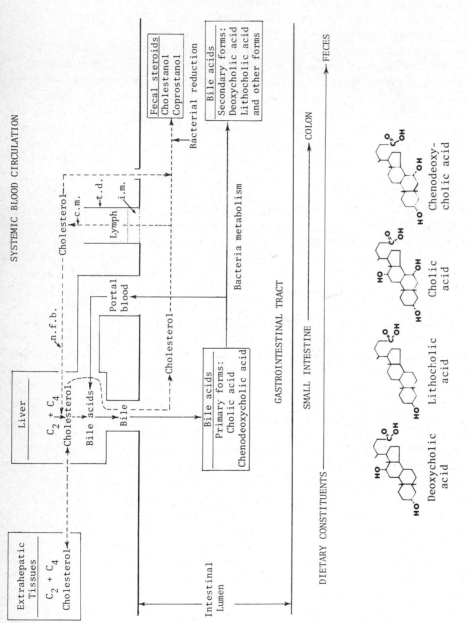

FIGURE 17.34. A general survey of cholesterol metabolism including the enterohepatic circulation of bile acids. Note that bile acids are excreted in the bile and conservatively returned to the liver via portal blood. Primary bile acids escaping reabsorption are converted to secondary forms. Cholesterol is excreted via bile and/or some amounts may be disposed of through the intestine. Key abbreviations and structures include $C_2 + C_4$, mevalonate biosynthesis followed by numerous steps yielding cholesterol, control site for biosynthesis assumed (here) to be HMG-CoA; c.m., chylomicrons (and some VLDL); i.m., intestinal mucosa; t.d., thoracic duct; n.f.b., negative feedback type response for cholesterol biosynthesis involving HMG-CoA reductase.

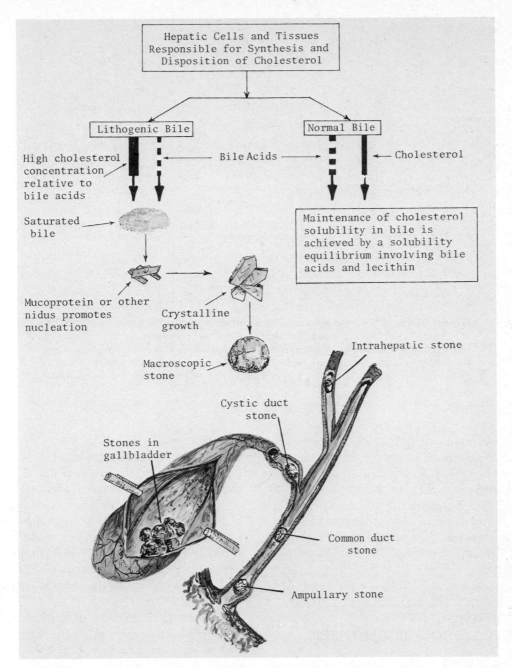

FIGURE 17.35. Rate-limiting production of cholesterol and bile acids depends respectively on HMG-CoA reductase and 7-α-hydroxylase. So long as proper proportions of bile acids and lecithin maintain cholesterol solubility in the bile, cholesterol is excreted from the biliary tract. Acquired physiological conditions, dietary factors, or congenital problems may enhance cholesterol formation or decrease bile acid production. If these events transpire, lithogenic bile (cholesterol saturated) can contribute to stone formation. Macroscopic stones can reside in a number of locations as shown. Gallbladder stones may be asymptomatic; however, stone migration into the cystic or common duct can cause biliary colic or cholecystitis.

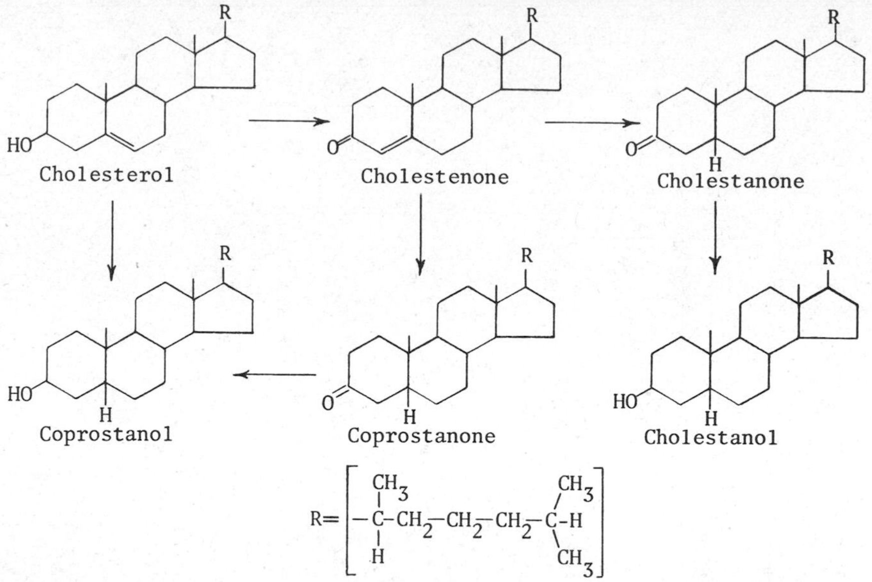

FIGURE 17.36. Cholesterol present in the intestine is subject to the reductive metabolism of intestinal flora. This action can produce fecal sterols such as cholestanol and coprostanol, which accound for over 50% of the fecal steroid derivatives in the intestinal tract. Once formed, both of these stanols are excluded from the enterohepatic circulation and reabsorption cycles. The R group for all stanols and cholesterol derivatives is indicated below the sterols.

progestational hormone called *progesterone,* which is produced by the corpus luteum and placenta, (2) male sex hormones (androgens) such as *testosterone* and *androsterone,* (3) female sex hormones (estrogens) including *estrone* and *estradiol,* and (4) nearly 40 adrenal corticosteroids, two of which include *corticosterone* and *aldosterone.*

REGULATION OF CHOLESTEROL BIOSYNTHESIS

As discussed in earlier sections, cholesterol biosynthesis occurs in nearly all tissues, but continuous synthesis does not occur in nervous tissues.

The hepatic tissues process both dietary and *de novo* synthesized cholesterols in a nondiscriminative fashion. Some cholesterol is subject to long-term hepatic storage as cholesterol esters, some enters the enterohepatic circulation as bile salts, and some is excreted as cholesterol in bile. Other portions of the remaining cholesterol are transported from the liver to peripheral tissues if body stores of cholesterol are low.

In the last case, transport of cholesterol to peripheral tissues may involve the liver as well as the intestine. Both of these tissues produce apolipoprotein B. This lipoprotein is the lipoprotein constituent of LDL and VLDL, which mediate the circulatory transport of cholesterol.

The initial appearance of apolipoprotein B in the plasma occurs as VLDL. This structure is then transformed into LDL by the elimination of a triacylglycerol and apoprotein C components. Both hepatic and peripheral tissues serve as sites for these structural transformations.

Aside from the cholesterol transport functions of LDL, *LDL-bound cholesterol* (LDL-C) is a key determinant in regulating the *de novo* synthesis of cholesterol in cells. Although speculative, most evidence does point to the interaction of LDL-C with *superficial*

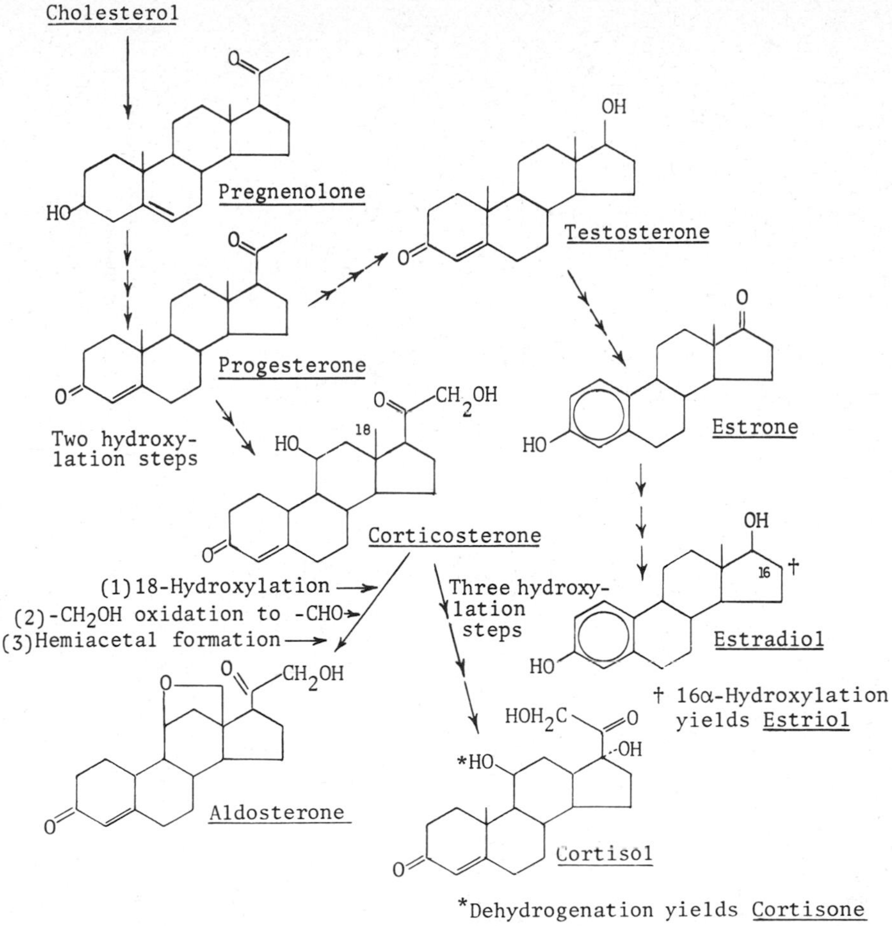

FIGURE 17.37. Schematic representation for the origins of steroidal hormones and adrenal corticoids from pregnenolone.

LDL receptor (LDL-R) sites located on cells (Figure 17.38).

Recognition of LDL-C by LDL-R probably depends on the presence of apolipoprotein B moieties of LDL and VLDL. In any event, a binding interaction between LDL-C and LDL-R incites an endocytic response by cellular vesicles called *endosomes*. This response causes the endosomes to engulf LDL-C, hold it in a vesiculated organelle, and then promote a *subsequent interaction* with another organelle, namely, a lysosome.

The hydrolytic enzymes of the lysosome include a cholesterol esterase that produces a free fatty acid and cholesterol from the cholesterol ester portion of LDL-C. The freed cholesterol diffuses into the cytoplasm where

it exerts a negative effect on cholesterol biosynthesis. The cytoplasmic appearance of cholesterol also suppresses HMG-CoA reductase activity.

This suppression response *does not* reflect an allosteric interaction between the enzyme and an effector, but instead, HMG-CoA reductase synthesis is controlled at the level of DNA. *Depressed* synthesis rates of the reductase are accompanied by a joint *activation* of the enzyme called *fatty acyl-CoA:cholesterol acyltransferase*. This enzyme stockpiles cholesterol oleate products within the cell and the esterified cholesterol eventually decreases LDL-R receptivity and cholesterol influx as LDL-C.

In cases of so-called *familial (genetic) hy-*

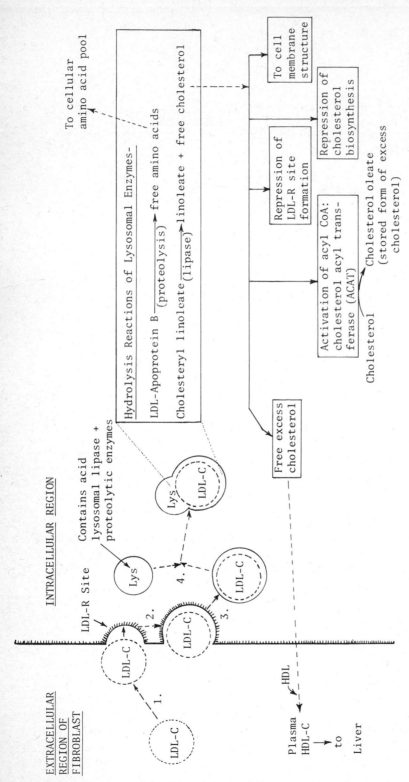

FIGURE 17.38. Cultured human fibroblasts have served as a model for understanding the cellular interaction and effects of cholesterol levels. (1) LDL-cholesterol (LDL-C) exists outside the fibroblast; (2) LDL-C binds to a LDL-receptor site (LDL-R), after which (3) endocytotic events internalize the LDL-C. (4) The vesicularized LDL-C interacts with a fibroblast lysosome (Lys), which initiates a series of degradative hydrolysis reactions (boxed reactions). Nearly all cells can manufacture cholesterol at one time or another, but the synthesis rate is regulated by cellular concentrations of cholesterol; see text for details. At cellular concentrations equivalent to only 2–4 mg/100 mL, LDL-C can shut down cholesterol biosynthesis while serum or plasma concentrations are 20–40 mg/100 mL on the other side of cells. This LDL-C level is far below the 130 mg/100 mL LDL-C concentration present in average sera samples. Therefore, even the complete elimination of cholesterol from the diet in most individuals would still provide four to five times the amount of LDL-C necessary to halt cholesterol synthesis. Defective LDL-C receptor sites (at LDL-R) permit rampant cholesterol production in cells of some individuals and serum cholesterol can rise to 400–800 mg/100 mL. According to the illustration, free cholesterol in the cell can contribute to cell membrane structures, repress further cholesterol formation, repress formation of more LDL-R sites, activate cholesterol storage mechanisms as an oleate ester, or form HDL-C, which is eventually translocated to the liver.

percholesterolemia, arteriosclerosis can be exacerbated by defects in LDL-R/LDL-C interactions. In the worst regulatory scenario, the *LDL-R sites are defective* for some reason that is unclear, *LDL-C does not interact with LDL-R,* and *cellular influx of cholesterol does not occur.* As a consequence, *de novo* cholesterol synthesis goes on unchecked at the cellular level.

Dietary factors also influence cholesterol biosynthesis at the level of hepatic HMG-CoA reductase. Fasting and cholesterol feeding both suppress HMG-CoA reductase activity, whereas refeeding after fasting, increased dietary carbohydrate, and/or triacylglycerol consumption spur on the conversion of acetyl-CoA to cholesterol. Opinions also prevail that dietary cholesterol can suppress cholesterol synthesis by influencing the synthesis step before lanosterol cyclase activity.

Aside from its biosynthesis in the body, the cholesterol pool is regulated by the efficiency of the enterohepatic circulation. If certain dietary constituents, such as fiber, impede bile acid and cholesterol absorption from the intestine, deficits of cholesterol can be met by increased *de novo* synthesis.

Notwithstanding isolated cases to the contrary, consistently poor absorption of cholesterol can lead to lowered cholesterol pools in the body in spite of its *de novo* biosynthesis. The influence of dietary saturated and unsaturated fatty acids on circulating concentrations of cholesterol cannot be overlooked either. A higher proportion of saturated to unsaturated acids is claimed to increase serum cholesterol concentrations.

SOME LIPID DISORDERS AND THEIR ASSOCIATION WITH DISEASE

Lipid transport in an aqueous system is challenging under the best of circumstances. Therefore, it is not surprising that effective vascular transport of plasma lipids, including cholesterol, in the blood can prove to be a complicated matter. Complexities in the physical–chemical interactions of lipids with the aqueous portions of the blood are surpassed in importance only by plasma lipid pathologies and cultural practices running counter to the normal disposition of body lipids.

Lipids are clearly involved in many pathological conditions, but few health problems have more epidemiological interest than *lipoproteinemias* and *accelerated instances of atherosclerosis.*

Incriminations of nearly every blood lipid fraction—ranging from triacylglycerols to phospholipids and free fatty acids—have produced many uncertainties regarding their *normal roles, mutual interactions,* and *critical concentration levels* in the body (e.g., maximum versus minimum) that encourage health or disease. Nonetheless, many of the common lipid disorders may have a *dietary component* and can be affected by nutritional factors or possibly controlled by dietary intervention.

LIPOPROTEINEMIAS

Lipoproteins are macromolecular complexes that serve as the transport vehicles for insoluble lipids contained in plasma. Lipoprotein disorders are indicated in plasma by their abnormal concentrations (Figure 17.39). When one or more types of lipoprotein are elevated, the condition is known as *hyper*lipoproteinemia. *Hypo*lipoproteinemia can also occur; in this case, normal lipoprotein concentrations are depressed.

Five different forms of hyperlipoproteinemias (*primary* hyperlipoproteinemias) are recognized. These develop from at least one heritable disorder, but some forms of hyperlipoproteinemias reflect the presence of multiple heritable disorders. Hyperlipoproteinemias can also be caused by other factors

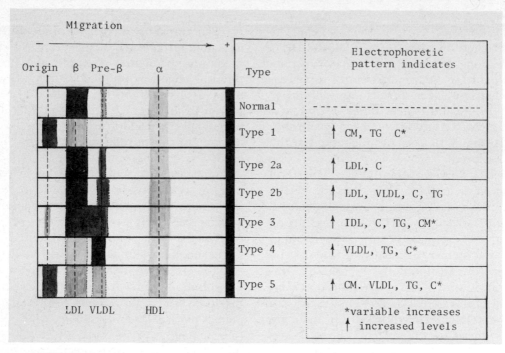

FIGURE 17.39. Schematic electrophoretic pattern for recognized lipoproteinemias. Separation medium (solid support) may be paper, cellulose, or an agarose gel. Fat-soluble, dye-stained lipoproteins are represented after their separation from seven distinctly different samples. Key abbreviations for components include CM, chylomicrons; TG, triacylglycerol(s); C, cholesterol; LDL, low-density lipoprotein; VLDL, very-low-density lipoprotein; IDL, intermediate-density lipoprotein.

such as predisposing diseases or poor dietary practices. These are recognized as *secondary* hyperlipoproteinemias.

Three types of *primary* inherited hypolipoproteinemias have been identified. Nearly all of these are rare, including *abeta-, hypobeta-,* and *analpha*-lipoproteinemias. Other hypolipoproteinemias can occur, but these are not heritable and tend to occur as a secondary feature of other diseases.

Clinical evidence for hyperlipoproteinemia relies on the evaluation of plasma cholesterol and triacylglycerol concentrations. This is followed by more sophisticated assays if either lipid category shows abnormal levels. Although reference values for upper plasma concentrations of these substances are not sacrosanct, any plasma sample that falls within 95% confidence limits of an arbitrarily assigned upper limit usually deserves further analytical attention.

This concern stems from recognized associations between certain plasma lipids and the onset of atherosclerosis. Based on most surveys, *coronary heart disease* (CHD) risks increase in tandem with the cube of plasma cholesterol levels. This rationale is affirmed for cholesterol concentrations in the range of 200 to 300 mg/dL, but similar estimates for triacylglycerols do not firmly withstand these generalizations.

Type 1 hyperlipoproteinemia is characterized by a congenital absence of a *lipoprotein lipase* (LPL). This enzyme, also called the "clearing factor," is responsible for clearing the chylomicrons from blood plasma. Therefore, deficiency effects of LPL result in increased chylomicrons and a condition known as *hyperchylomicronemia*.

The LPL disorder is apparently linked to an autosomal recessive trait, with parents of afflicted individuals demonstrating no definite lipid abnormalities. Clinical evidence for this rare disease includes papular swellings and eruptions caused by *xanthomas* or subdermal fatty acid deposits, as well as hepatosplenomegaly. The atherosclerotic rate of afflicted individuals is not notably enhanced by overt hyperchylomicronemia. A secondary hyperlipoproteinemia can also result from unchecked diabetes, which is believed to suppress normal LPL synthesis.

Type 2a hyperlipoproteinemias, also called hyper-β-lipoproteinemia (based on electrophoretic analysis), displays elevated levels of LDL. Another type 2 form, called 2b, is characterized by an increase in LDL and VLDL fractions. Type 2b is also known as hyperpre-β-lipoproteinemia.

Both forms of type 2 lipoproteinemia can be inherited. Moreover, essential familial hypercholesterolemia, which represents the classic form of the disorder, occurs in both hetero- and homozygous forms. Although the major plasma lipid pattern is type 2a, the type 2b form may materialize as a phenotypic variant in some families.

Still another hyperlipoproteinemia disorder called "familial combined hyperlipoproteinemia" has been uncovered. This disorder reflects types 2a, 2b, 4, and 5, which follow here.

Secondary forms of type 2a and 2b hyperlipoproteinemias can reflect obstructive disease of the bile tract, hypothyroidism, nephrotic syndrome, or other problems. Noneruptive xanthomas are clinically demonstrated in the skin, but cholesterol deposits also exist around the eyes, tendons, and blood vessels. Hypercholesterolemic consequences of these disorders favor an increased risk of atherosclerosis, especially CHD. Therefore, any dietary intervention involving decreased cholesterol intake, decreased consumption of saturated fat, and increased amounts of unsaturated fats would be desirable. Drug therapy may be suitable in some cases.

Type 3 hyperlipoproteinemia is a relatively rare heritable disorder of plasma lipoproteins, which is also called "floating β-hyperlipoproteinemia." This name refers to the characteristic densities of plasma lipoprotein constituents.

Type 3 disorders are clinically less severe than type 2 forms but an increased risk of occlusive vascular disease does occur. The so-called "floating β fraction" of lipoproteins indicates that this disorder may stem from problems of LDL formation from VLDL caused by defective IDL catabolism.

Apart from inherited predispositions for type 3 disorders, hypothyroidism can also serve as a preexisting complication of the disorder.

Type 4 hyperlipoproteinemia (hyperpre-β-lipoproteinemia) is characterized by high concentrations of VLDL without abnormal complications in other lipoprotein fractions. The primary form of the disorder is inherited as an autosomal dominant trait in some families that already have a type 4 problem. However, as indicated earlier, familial combined hyperlipoproteinemia can also display a type 4 component disorder if other family members show type 4, 2a, 2b, or 5 disorders.

Type 4 hyperlipoproteinemia has a number of secondary occurrences associated with excessive alcohol consumption, obesity, diabetes mellitus, oral contraceptive regimens, steroid therapies, hypothyroidism, and many other conditions.

The underlying biochemical problems in type 4 disorders are not entirely clear. Speculation does suggest, however, that poor rates of VLDL and hepatic triacylglycerol syntheses may be involved. Type 4 disorders can result in hyperchylomicronemia accompanied by eruptive xanthomas, pancreatitis, and increased risks of peripheral and coronary artery diseases.

Dietary intervention is an important factor in managing type 4 disorders since obesity, simple sugars, alcohol, and saturated fat consumption only exacerbate the condition. Rational treatment dictates careful caloric regulation to (1) normalize body weight and eliminate obesity, (2) use complex carbohydrates instead of simple sugars, and (3) use polyunsaturated fatty acids instead of saturated dietary fatty acids. Drug therapy using *clofibrate* or *nicotinic acid* may be feasible as a last resort if dietary controls break down in the control of the disorder.

Type 5 hyperlipoproteinemia is a rare disorder characterized by the occurrence of chylomicrons in the fasting state accompanied by an increase in VLDL fractions. This form of hyperlipoproteinemia, also called *mixed hyperlipoproteinemia,* has a primary form that is *inherited,* but it can also be expressed as a secondary feature of diabetes mellitus, chronic pancreatitis, and both hepato- and nephropathies. Dietary control of type 5 hyperlipoproteinemia is favorable to most existing drug therapies. Since chylomicron management is crucial, any effort to reduce triacylglycerol, alcohol, and dietary fat intakes are desirable, coupled with weight loss.

Some notable forms of hypolipoproteinemia are genetically tied to the absence of apolipoprotein B synthesis. This in turn creates an absence of LDL, VLDL, and chylomicrons.

In the case of Tangier's disease, HDL is 1 to 5% of normal values. Since plasma cholesterol and phospholipids occur in very low concentrations, cholesterol accumulates in the lymphoreticular system, where it ultimately causes hepato- and splenomegalies.

Lipoproteinemia diagnosis can also be analytically complicated in rare instances in which an unusual lipoprotein called lipoprotein X (density 1.040–1.045) appears with the β-lipoprotein fraction. This protein appears as a result of lecithin:cholesterol acyltransferase (LCAT) deficiencies.

ATHEROSCLEROSIS

Atherosclerosis is a natural phenomenon of the normal aging process in humans regardless of sex or race. Distinct differences in the onset and progress of atherosclerosis may occur, however, as a result of dietary and cultural habits.

Atherosclerosis is a type of arteriosclerosis in which plasma lipids aggregate on the superficial inner layer of an arterial wall. Lipid aggregates, known as *plaques,* principally consist of cholesterol and its esters, but other unidentified components are undoubtedly involved. The lipid matrix of the plaque eventually becomes more complex as it matures. Collagen and elastic fibrous tissues propagate, followed by calcification of the plaque to form a very complex arterial lesion (Figure 17.40).

The *initiation mechanism* for atherosclerosis is shrouded in mystery. Unfortunately, many plausible but unsubstantiated theories provide most explanations regarding plaque initiation. A history of plaque research suggests that the intimal lining of an artery is *somehow damaged.* Substances foreign to the lining, yet present in the blood, associate with this predisposed area and commence the atherosclerotic process. It has been suggested by some investigators that a plaque may originate from a transformation or mutation within a single smooth muscle cell on the artery wall. Tissue culture studies have provided evidence that blood platelets contain a factor that stimulates smooth muscle cell proliferation. It follows then that injured endothelial tissue may cause platelets to adhere on an arterial wall, a complex process of blood clotting is incited, and an immature plaque may be initiated.

Cholesterol and its esters aggregate in the vicinity of the plaque, presumably through the involvement of plasma lipoproteins such as VLDL, LDL, and HDL. Furthermore, plaque development depends on the deposition of collagen, elastin, glucosaminoglycans, and dead cells on the initial plaque site. Protrusion of the plaque into the arterial lumen ultimately restricts blood flow. If this action effectively depresses the requisite oxygen supply to vital organs, necrotic tissues may result. In situations where whole mural plaques or their dislodged fragments interrupt blood supplies to the heart or brain, a heart attack or cerebrovascular accident, respectively, may ensue.

Although plaque formation is the under-

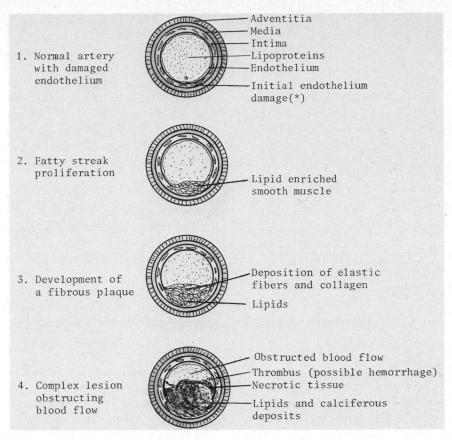

1. Normal artery with damaged endothelium
 — Adventitia
 — Media
 — Intima
 — Lipoproteins
 — Endothelium
 — Initial endothelium damage(*)

2. Fatty streak proliferation
 — Lipid enriched smooth muscle

3. Development of a fibrous plaque
 — Deposition of elastic fibers and collagen
 — Lipids

4. Complex lesion obstructing blood flow
 — Obstructed blood flow
 — Thrombus (possible hemorrhage)
 — Necrotic tissue
 — Lipids and calciferous deposits

FIGURE 17.40. Normal (1) and pathological consequences of plaque formation along an arterial wall (2–4). The intima is lined with a single layer of endothelial cells, which may be subjected to some type of damage from which plaque formation is initiated. The media layer is composed of smooth muscle cells that are surrounded by collagen, elastic fibers, and proteoglycans, whereas the adventitia consists of smooth muscle and fibroblasts mixed with collagen and proteoglycans.

lying physical reason for many atherosclerotic problems, this process is abetted by a multiplicity of factors including elevated plasma cholesterol, hypertension, smoking, obesity, hypertriacylglyceridemia, diabetes mellitus, genetic predisposition, and too little exercise.

Prediction of accelerated CHD is not a simple matter based solely on dietary history, habits, saturated versus unsaturated dietary fat intakes, or measurements of *total cholesterol* (TC) in serum specimens. Some researchers have offered an explanation of *why* HDL-cholesterol (HDL-C) levels are implicated with reduced risks of CHD, and why

other lower density cholesterols (VLDL-C and LDL-C) foster atherosclerosis.

The rationale suggests that HDL-C transports its steroid component from peripheral to hepatic tissues, where it is metabolized and excreted. The LDL and VLDL fractions of lipoproteins, however, transport cholesterol to a site where it is deposited. Since HDL represents the α-lipoprotein serum fraction and LDL plus VLDL are β-lipoprotein fractions, any increasing ratio of β/α may correspond to an increasing risk of CHD (e.g., β/α ratio: normal = 2.50–2.90 versus high risk of CHD = 5.00+).

Apart from early studies claiming the significance of HDL-C plasma levels (Miller and Miller, 1975), many other investigators have substantiated this principle (Glueck and Connor, 1978; Gordon *et al.*, 1977; Gotto and Jackson, 1978; Nikkila, 1978; Steinberg, 1978; Witzum and Schonfeld, 1979). Moreover, the renowned Framingham Heart Study, which has followed a controlled group of subjects since 1949, affirms that raised LDL-C levels parallel the onset of CHD, whereas HDL-C increases coincide with decreased CHD.

EPIDEMIOLOGICAL FACTORS IN ATHEROSCLEROSIS AND ITS TREATMENT

According to many clinical and diagnostic standards, plasma TC levels of <150 mg/dL will reflect low incidences of CHD (Gotto, 1979). In fact, optimal plasma cholesterol and lipoprotein levels are roundly debated depending on philosophical positions of investigators and the fundamental validities of analytical methods for evaluating key lipoproteins.

One panel convened by the American Health Foundation surveyed substantial amounts of clinical data. This group concluded that societal goals were attainable at average cholesterol levels of approximately 160 mg/dL (with a range over 130–200 mg/dL). Levels of 180 to 190 mg/dL are probably attainable, and this would translate into a LDL-C value of ≤90 to 100 mg/dL and HDL-C fraction of ≥50 mg/dL.

These projections for individuals can be upset by genetic (Glueck *et al.*, 1977) and hormonal variations in selected subjects. Premenopausal women have lower TC and higher HDL-C levels than menopausal women or men. Higher plasma cholesterol levels have also been recorded in obese individuals (Abrahams *et al.*, 1978) as well as those having type A personalities.

Dietary intervention can be useful in allaying the rapid progress of atherosclerosis.

The possible roles and advantages of unsaturated dietary fats versus saturated types have been discussed elsewhere in the text. Nutritionists generally recommend the dietary substitution of fat-rich meats, dairy products, and baked products with lower fat foods in most diets. Where possible, fish, poultry, and vegetable oils should be used. Grains, fruits, cereals, vegetables, and legumes should also serve as important portions of the daily caloric regimen.

Drug treatments can be useful in controlling some acute cases of hypercholesterolemia. Pharmaceuticals include clofibrate, D-thyroxine, cholestyramine, neomycin, nicotinic acid, and β-sitosterol (a natural steroid).

Clofibrate (Atromid-S) serves as an antilipidemic agent by reducing the triacylglycerol-rich VLDL plasma lipoprotein fraction. Serum cholesterol, especially the LDL-C fraction, is also decreased. Aside from a suspected interruption in cholesterol biosynthesis prior to mevalonate formation, the drug has been shown to cause increased excretion of neutral sterols. Desirable properties of clofibrate must be considered in perspective with its negative consequences, which range from upsets in the female reproductive steroids to increases in cholelithiasis. The latter problem probably stems from upsets in biliary lipoprotein:steroid:bile acid excretion ratios that would normally ensure steroid excretion.

D-Thyroxine accelerates cholesterol catabolism, cholestyramine prevents bile acid reabsorption by elevating fecal cholesterol losses, neomycin blocks intestinal absorption of cholesterol, nicotinic acid interferes with the mobilization of free fatty acids, and β-sitosterol can diminish cholesterol absorption. As in the case of clofibrate, the side effects for cholesterol drug treatments must be seriously weighed before their use.

If dietary intervention or drugs to control atherosclerotic problems are ineffectual and are followed by serious obstruction of key vascular system routes, surgical treatment may be called for. This is accomplished by removal of an atherosclerotic plaque area in an artery, followed by "patch grafting" or "revascularization" of the diseased artery using a vein or synthetic graft on the artery as a circulatory bypass route (Figure 17.41).

CHOLESTEROL OXIDES MAY CONTRIBUTE TO PLAQUE INITIATION

The storehouse of atherosclerotic data and the possible causes for plaque initiation are immense, but there is little doubt that dietary components serve as a major contributing factor to the disease. However, until reasons for the initiation of plaque formation are established, the etiology of atherosclerotic disease will remain obscure. For example, roles of cholesterol oxides as well as triol derivatives of cholesterol, which can result as a consequence of food processing, must be more seriously considered as initiators and cytotoxic instigators of atherosclerotic plaques.

Many foods considered to be atherogenic (red meats) have levels of cholesterol and

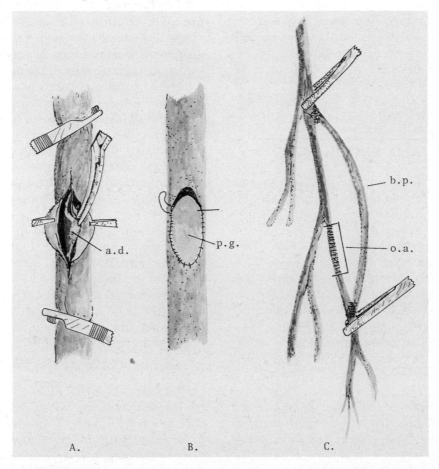

FIGURE 17.41. Surgical intervention to correct atherosclerotic deposits can take many approaches; two types are illustrated: (A) the atherosclerotic deposit (a.d.) is excised from the artery in an endarterectomy followed by a patch graft (p.g.); or a bypass (b.p.) graft may be performed around the occluded artery (o.a.) using a vein or synthetic (Dacron or Teflon) grafts. Procedure choice depends on health status of the patient, location of the occluded artery, and other medical considerations.

saturated fats that are correspondingly high relative to cholesterol and unsaturated fat levels in fish and poultry. Since customary food preparations for these different types of foods occur (e.g., broiling versus boiling), *different atherogenic tendencies among foods may actually reflect a food processing or preparation component.*

Therefore, in at least *some cases,* hyperatherogenic foods may be classified in this way as a result of

1. Historical processing methods that are consistently different from low saturated fat foods.

2. Concentration-dependent exposure of available cholesterol in foods to cooking and/or other methods of heat-induced processing that produce dangerous cholesterol oxides.

OBESITY AND BROWN FAT

Caloric consumption far in excess of that required for maintaining the animal body leads to a proliferation and enlargement of adipocytes. As populations and sizes of these cells change, the development of overt obesity can be observed and documented. Overconsumption of fats, carbohydrates and proteins can be linked to obesity, but the reasons for overconsumption of food calories is not always clear. Many believe, however, that behavioral or physiological factors play a role in obesity, and still other investigators claim that a genetic component is involved.

Animal studies and observations of some humans indicate that selected individuals *can consume more dietary calories than others without gaining weight.* This observation assumes that digestive efficiency of the subjects is unimpaired and that the physical activities for *both* types of test subjects are the same (i.e., two test groups: one gains weight *versus* one that fails to gain weight).

A cursory study of this situation indicates that apparently normal subjects who fail to gain weight may have an *in vivo* mechanism for dissipating excess caloric intake in the form of heat energy. Heat production (thermogenesis) seems to occur mainly in *brown fat* reserves of the animal body.

Brown fat is a mitochondria-rich fat whose color is attributed to high population numbers of red-brown cytochromes. Brown fat is common in newborn animals (including humans), some cold-acclimated animals, as well as those that ordinarily hibernate. For adult humans, brown fat is relegated to sites in the neck and back.

Mitochondria of brown fat cells are not mainly involved in ATP synthesis. Instead, their biochemical duty seems to dictate the dissipation of free energy, arising from electron transport, as heat. This mechanism is different from normal mitochondria and requires a distinctly different form of mitochondrial architecture. Brown fat mitochondria are unlike normal energy-producing mitochondria since they display specialized membrane pores that can be made permeable to protons. Proton fluxes over the porous mitochondrial membrane seem to be regulated by a 32,000-dalton polypeptide that is prominently fixed to the inner side of the membrane. Purine interactions are also believed to be involved in the regulatory activity of this peptide. That is, purine–polypeptide associations render the membrane impermeable to protons, whereas purine dissociation increases proton permeability.

Normal oxidative phosphorylation activities of mitochondria are coupled to the action of a proton pump according to the mechanism detailed by Mitchell. This process is questionable in brown fat mitochondria, however, where ATP production is diminished and proton circulation is altered. These mitochondria permit an influx of protons, an uncoupling of normal oxidative phosphorylation processes, and a bypass of ATP production. The lack of ATP production is met with an energy parity of heat.

Many nonobese animals that demonstrate hypercaloric intake effectively use this mechanism for heat production as opposed to using ATP. In cases of cold climatic duress, this action can be exhibited as *nonshivering thermogenesis*. The impact of brown fat metabolism in human biochemistry and its thermogenic consequences are not known for certain. It is speculated, however, that brown fat biochemistry *could* significantly contribute to weight regulation in at least some individuals, and its failure to operate as a thermogenic tissue in other subjects may spur obesity. In the case of humans, sex differences seem to occur in brown fat deposition, with women having less than men. This claim is based on relative differences between the sexes in demonstrating thermogenesis.

LIPID–CARBOHYDRATE INTERACTIONS

Carbohydrates and lipids participate in complex interactions during the course of normal and abnormal metabolism. In some cases, these interactions are very clear-cut. For example, galactose and small amounts of glucose directly contribute to cerebroside and ganglioside syntheses. The glycolytic reactions of carbohydrates, however, supply phosphorylated glycerol intermediates necessary for triacylglycerol syntheses in adipocytes. This source of phosphorylated glycerol (e.g., α-glycerophosphate) is critical for triacylglycerol formation since these cells cannot resynthesize their own phosphorylated glycerols from glycerol produced by hydrolytic (lipase) reactions. It should be recognized, too, that acetyl-CoA, produced from glycolytic routes, depends on the irreversible oxidative decarboxylation of pyruvate. This reaction is at the root of many claims that fatty acids cannot be directly transformed into carbohydrate.

Amino acids can contribute to lipid biosynthesis. For example, serine, sphingosine, choline, ethanolamine, key constituents of glycolipids, and phospholipids can originate from protein metabolism. Acetyl- and acetoacetyl-CoA's are readily produced from so-called ketogenic amino acids and funneled toward fatty acid and steroid syntheses. Other amino acids recognized as glucogenic types yield pyruvate, which can be directed to fatty acid or carbohydrate biosynthesis.

The normal role of insulin in the body complicates lipid–carbohydrate interactions. Excessive plasma concentrations of glucose are usually met with increased plasma insulin concentrations. This event not only encourages glucose assimilation of the carbohydrate at the cellular level but also (1) promotes fatty acid biosynthesis and (2) decreases fatty acid mobilization (lipase actions) in adipocytes. Increased glucose catabolism also promotes acylglycerol deposition in fat depots for reasons indicated above, since adipocytes cannot provide for their own necessary supplies of phosphorylated glycerols.

In cases of insulin deficiency (or extreme fasting), acetyl-CoA consumptions in the tricarboxylic acid cycle as well as fatty acid syntheses are both decreased. Both of these actions reflect an effort by the body to handle what it recognizes to be an energy deficit. Therefore, metabolic pathways become geared to normalizing blood glucose levels and providing energy-rich substances to cells that are starved for glucose. As a consequence, acetyl-CoA is directed into the formation of 3-hydroxy-3-methylglutaryl-CoA (HMG-CoA). This compound could lead to the production of steroids (e.g., cholesterol) or ketone bodies. Since an energy deficit linked to inadequate glucose metabolism is encountered during insulin deficiency, ketone body formation (acetoacetate, 3-hydroxybutyric acid) is favored.

Insulin insufficiency also promotes fatty acid mobilization from triacylglycerols followed by their β oxidation. Acceleration of fat metabolism depends on complex hormonal regulations involving both lipid and carbohydrate metabolism. The acetyl-CoA's produced by this route contribute to ketone body formation, and in some cases "ketosis"

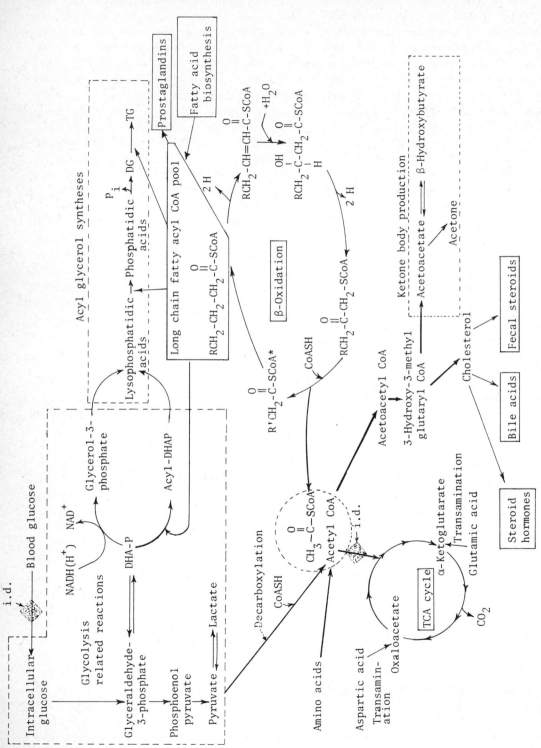

FIGURE 17.42. Schematic summary of lipid and carbohydrate interactions with some other key biochemical substances. Note that insulin deficiency (i.d.) blocks or at least minimizes blood glucose entry into metabolizing cells plus acetyl-CoA entry into the tricarboxylic acid (TCA) cycle. Assuming that these actions result in a cellular deficit of energy, fatty acid mobilization from triacylglycerols followed by their β oxidation provide a basis for ketone-body production during insulin deficiency. It should be recognized that fatty acid oxidation (β oxidation) yields R'—*, which contains two less carbons than the initial R group.

may occur if ketone body production is overbearing. Acetyl-CoA produced by β oxidation cannot be used for fatty acid biosynthesis because $NADPH(H^+)$ stores (from the HMP pathway) are lowered and lipogenic enzymes plus malonyl CoA formation are suppressed. Figure 17.42 outlines some important metabolic interactions of carbohydrates and lipids as surveyed above.

REFERENCES

Abrahams, J. P., D. T. Nash, G. G. Gensini, and T. Arno. 1978. Relationship of age and biomedical risk factors to progression of coronary artery disease. *J. Amer. Geriatrics Soc.* **26:**248.

American Health Foundation Committee on Food and Nutrition. 1972. Position statement on diet and coronary heart disease. *Prev. Med.* **1:**255.

Beg, Z. H., and H. B. Brewer. 1981. Regulation of liver 3-hydroxy-3-methyl glutaryl CoA reductase. *Curr. Top. Cell. Regul.* **20:**139.

Blackenhorn, D. H. 1978. Progression and regression of femoral atherosclerosis in man. *Atherosclerosis Rev.* **3:**169.

Blackenhorn, D. H. 1978. Reversibility of latent atherosclerosis: Studies by femoral angiography in humans. *Mod. Concepts Cardiovasc. Dis.* **47:**79.

Blackenhorn, D. H., and M. E. Sanmarco. 1979. Angiography for study of lipid-lowering therapy. *Circulation* **59:**212.

Bloch, K., and D. Vance. 1977. Control mechanisms in the synthesis of saturated fatty acids. *Annu. Rev. Biochem.* **46:**263.

Brady, R. O. 1978. Sphingolipidoses. *Annu. Rev. Biochem.* **47:**687.

Brown, M. S., and J. L. Goldstein. 1979. Familial hypercholesterolemia: Model for genetic receptor disease. *Harvey Lect.* **73:**163.

Castelli, W. P., J. T. Doyle, T. Gordon, G. C. Hmaes, M. C. Hjortland, S. B. Hulley, A. Kagan, and W. J. Zukel. 1977. HDL cholesterol and other lipids in coronary heart disease: The cooperative lipoprotein phenotyping study. *Circulation* **55:**767.

Crawford, D. W., and D. H. Blackenhorn. 1979.

Regression of atherosclerosis. *Annu. Rev. Med.* **30:**289.

Fleishman, A. I., P. B. Watson, A. Stier, H. Somol, and M. L. Bierenbaum. 1979. Effect of increased dietary linoleate upon blood pressure. Platelet function and serum lipids in hypertensive adult humans. *Prev. Med.* **8:**163.

Glew, R. H., and S. P. Peters. 1977. *Practical Enzymology of the Sphingolipidoses*. Liss, New York.

Glueck, C. J., and W. E. Connor. 1978. Diet–coronary heart disease relationships reconnoitered. *Amer. J. Clin. Nutr.* **31:**727.

Glueck, C. J., R. C. Tsang, R. W. Fallat, and M. J. Mellies. 1977. Diet in children heterozygous for familial hypercholesterolemia. *Amer. J. Dis. Child.* **131:**162.

Gordon, T., W. P. Castelli, M. C. Hjortland, and W. B. Kannel. 1977. The prediction of coronary heart disease by high-density and other lipoproteins: An historical perspective. In *Hyperlipidemia: Diagnosis and Therapy* (B. M. Rifkind and R. I. Levy, eds.), pp. 71–78. Grune & Stratton, New York.

Gotto, A. M., Jr. 1979. Is atherosclerosis reversible? *J. Amer. Dietet. Ass.* **74:**551.

Gotto, A. M., Jr., and R. L. Jackson. 1978. Plasma lipoproteins and atherosclerosis. *Atherosclerosis Rev.* **3:**231.

Hofmann, A. F. 1976. The enterohepatic circulation of bile salts in man. *Adv. Intern. Med.* **21:**501.

Kannel, W. B., W. P. Castelli, and T. Gordon. 1979. Cholesterol in the prediction of atherosclerotic disease. *Ann. Intern. Med.* **90:**85.

McGarry, J. D., and D. W. Foster. 1980. Regulation of hepatic fatty acid oxidation and ketone body production. *Annu. Rev. Biochem.* **49:**395.

Miller, N. E. 1979. Plasma lipoproteins, lipid transport and atherosclerosis: Recent developments. *J. Clin. Pathol.* **32:**639.

Miller, G. J., and N. E. Miller. 1975. Plasma-high-density-lipoprotein concentration and development of ischemic heart-disease. *Lancet* **1:**16.

Nestel, P. J., and A. Poyser. 1978. Cholesterol content of the human atrium is related to plasma lipoprotein levels. *Atherosclerosis* **30:**177.

Nikkila, E. A. 1978. Metabolic regulation of plasma high density lipoprotein concentrations. *Eur. J. Clin. Invest.* **8:**111.

Rifkind, B. M. 1977. *Hyperlipidemia: Diagnosis and Therapy*. Grune & Stratton, New York.

Rommel, K., and R. Bohmer (eds.). 1976. *Lipid Absorption: Biochemical and Clinical Aspects*. University Park Press, Baltimore, Md.

Smith, L. L. 1981. *Cholesterol Autoxidation*. Plenum, New York.

Stanbury, J. B., D. S. Wyngaarden, and D. S. Fredrickson (eds.). 1978. *The Metabolic Basis of Inherited Disease*, 4th ed., Part 4. McGraw-Hill, New York.

Steinberg, D. 1978. The rediscovery of high density lipoprotein: A negative risk factor in atherosclerosis. *Eur. J. Clin. Invest.* **8**:107.

Tamir, I., B. M. Rifkind, R. I. Levy, and R. Calhoun. 1979. Measurements of lipid disorders. In *Clinical Diagnosis and Management by Laboratory Methods* (J. B. Henry, ed.), pp. 189–227. Saunders, Philadelphia.

Torday, J., L. Carson, and E. E. Lawson. 1979. Saturated phosphatidylcholine in amniotic fluid and prediction of respiratory distress syndrome. *N. Engl. J. Med.* **300**:1013.

Volpe, J. J., and P. R. Vagelos. 1976. Mechanisms and regulation of biosynthesis of saturated fatty acids. *Physiol. Rev.* **56**:339.

Witztum, J., and G. Schonfeld. 1979. High density lipoproteins. *Diabetes* **28**:326.

C H A P T E R · 18

Metabolism of Nitrogenous Biomolecules

INTRODUCTION

Plants, animals, and microorganisms contain many nitrogenous molecules that display a variety of structures and functions. Although soluble inorganic nitrogen such as nitrate can supply the nitrogen requirement of plants, most organisms require organic nitrogen in the form of amino acids or proteins.

The nutritional well-being of humans and most animals rests on an *adequate supply* of dietary amino acids present in the free state or held in protein structures.

According to the National Academy of Sciences' (NAS) 1980 recommendations, the daily protein requirements for a *reference* adult man (70 kg) and woman (58 kg) translate into 56.0 and 46.0 g, respectively. Obvious variations in the body weights of individuals require some adjustment of these projected protein intakes. According to NAS guidelines, requisite protein intake can be estimated by multiplying body weight (kg) by a factor of 0.8. The 0.8 value is based on nitrogen balance studies that indicate a 0.47-g protein requirement/kg body weight/day. This value may be subject to a 30% increase based on individual variabilities as well as

further increases caused by a 75% use efficiency for dietary protein. Therefore, a calculated factor of 0.81 is derived from

$$\frac{(0.47)(0.30) + (0.47)(1.00)}{0.75}$$

Protein requirements for humans generally reflect the nutritional requirement for key amino acids present in their structures, as well as the nitrogen balance of the body. Any one of three nitrogen balance states can significantly affect protein requirements:

1. *Positive nitrogen balance* (nitrogen retention exceeds nitrogen loss).

2. *Negative nitrogen balance* (nitrogen retention is less than nitrogen loss).

3. *Nitrogen equilibrium* (dietary nitrogen intake is equivalent to metabolic nitrogen losses).

Positive nitrogen balance is observed during pregnancy, lactation, recuperation from illness, and especially during growth and development in children. In short, positive nitrogen balance parallels the occurrence of new growth or the replacement of wasted body protein reserves. In severe wasting diseases, long periods of *low-quality* protein intake, and/or general inadequacies of dietary protein, negative nitrogen balances are not unusual. Nitrogen equilibrium ordinarily indicates the *proper* nutritional availability and quality of dietary proteins.

Inadequacies in dietary protein quality and quantity underlie many current nutritional problems, especially in poor countries. Here nutritional inadequacies of proteins are very common among the aged, infants, and pregnant and lactating women. Similar protein inadequacies are also exhibited in poor population segments of developed countries to a smaller extent, but fundamental dietary ignorance also plays an important role in these countries.

When dietary protein deficiencies are coupled with reduced caloric intake, kwashiorkor, marasmus, and related body-wasting scenarios are initiated. Information concerning protein quality assessments and protein uses have been outlined in Chapter 3.

Nucleic acids contribute to the dietary supply of nitrogen compounds, but no basic dietary requirement has been established. Nucleic acid digestion provides the body with purines and pyrimidines, which are assimilated over the intestine and then transported to consuming cells. These cells either *degrade* the purines or pyrimidines or direct them into nucleotide syntheses. It should be recognized that *de novo* synthesis of nucleic acids involves purines and pyrimidines that are newly synthesized by the body into nucleotides.

AMINO ACID METABOLISM

Dietary proteins are primarily digested in the small intestine. The resulting amino acids and peptides are absorbed over the mucosal cells where the peptides are further cleaved into amino acids by peptidase activities. Amino acids are then transported by hematogenous routes to cells for biosynthetic and oxidative reactions.

Dietary amino acids can be classified as *essential* or *nonessential* types (Table 18.1). Since essential amino acids cannot be synthesized in adequate amounts to supply metabolic requirements of the body, they must be supplied in the diet. Nonessential amino acids are produced in the body from carbohydrate and lipid intermediates assuming that necessary amination reactions can be supported by existing nitrogen supplies.

Debates persist concerning the essentiality of some amino acids such as histidine, which is required by infants and possibly by adults. Arginine falls into the rank of an essential amino acid although it can be synthesized in the body. Unfortunately, the catabolic conversion of arginine into ornithine and urea supersedes its biosynthesis, and exogenous dietary sources are necessary for adequate nutrition. Essential amino acids other than lysine and threonine can be replaced by their α-keto analogues in the diet.

Amino acid metabolism involves the dynamic occurrence of catabolic and anabolic biochemical pathways. Because of the complexity and multiplicity of these pathways, only the major routes for amino acids and their related nitrogenous compounds will be considered.

SITES OF AMINO ACID METABOLISM

Virtually all cells of the animal body metabolize amino acids, but the liver, skeletal muscle, kidneys, and brain serve as major metabolic sites.

Liver

Many amino acids supplied to the liver by the portal circulation are converted into urea, whereas others are passed on to the systemic circulation as free amino acids. A small fraction of the hepatotransient amino acids are converted to structural proteins and tissues.

Serum albumin biosynthesis from amino acids is a critical function of liver activity.

Table 18.1. Essential and Nonessential Amino Acids

Essential	Nonessential
Arginine	Alanine
Histidine	Asparagine
Isoleucine	Aspartate
Leucine	Cysteine
Lysine	Cystine
Methionine	Glutamate
Phenylalanine	Glutamine
Threonine	Glycine
Tryptophan	Hydroxyproline
Valine	Proline
	Serine
	Tyrosine

Serum albumins contribute to the maintenance of osmotic pressure throughout the body and serve as translocation vehicles for metals, ions, hormones, bilirubins, drugs, and other metabolites. Since albumin status in the blood is greatly affected by the nutritional state of the body, malnutrition, surgical stress, trauma, and infections all *depress normal albumin levels*. Aside from clinical intervention, long-term restoration of albumin requires sustained dietary availability of high-quality proteins.

Most amino acids can be degraded in the liver with the exception of branched-chain amino acids (e.g., leucine, isoleucine, valine), which are degraded in the muscle cells. Keto acids produced by the deamination of amino acids can contribute to glucogenesis and glucose release into the systemic circulation or storage as glycogen. Other amino acids serve as feedstock for heme, glutathione, carnosine, creatine, purine, and pyrimidine biosyntheses.

Hepatic biosynthesis of nonessential amino acids may employ the ammonium ion present in the circulation as well as keto acids derived from carbohydrate metabolism. Other biogenic schemes for amino acids also rely on transamination reactions between amino acids and α-keto acids.

Some amino acids undergo catabolic reactions yielding keto acids that are structurally allied with fatty acids and ketone bodies.

These keto acids are readily assimilated into the TCA cycle in which they provide anaplerotic and energy-yielding functions.

Skeletal Muscle

The amino acid constituents and protein structure of muscles exist in a dynamic state of degradation and synthesis. Although the normal half-life for protein turnover is low (~180 days), dietary protein deficiencies can accelerate the rate.

Protein balance in the muscle not only controls muscle size but also affects energy homeostasis in the body as a whole. During fasting, muscular protein catabolism serves as an initial rate-limiting step in gluconeogenesis. Muscle proteins are also exploited during periods of severe infection or injury when immunological defenses, coagulation, and wound-healing mechanisms must be bolstered.

During starvation without injury, protein deficits are initiated in the liver as well as in labile visceral proteins. A substantial depletion of skeletal proteins begins only after 3 to 4 days of overt starvation.

Various investigators claim that severe protein losses should be combated by amino acid therapy. This treatment has been widely endorsed, but disputes are not uncommon concerning the suitable amino acid profile necessary to ensure the nutritional care of critically ill patients. Long (1983) and others claim that quantitative surveys of plasma, tissue amino acid profiles, and other metabolic parameters must direct the formulation of any amino acid therapy program to ensure its success.

Nonessential amino acids and branched-chain amino acids including leucine, isoleucine, and valine undergo a myriad of degradation reactions in the muscle. Unlike those in the liver and kidneys, the keto acids derived from deamination of amino acids in the muscle are not used for carbohydrate synthesis. Moreover, nonessential amino acids can be formed from keto acids that arise during carbohydrate metabolism.

The amino groups of amino acids located in the muscle are not involved in urea synthesis at this site. Instead, most of them form alanine, which enters the blood and travels to the liver. Here, alanine is converted to pyruvate and the amino group is directed into urea formation (see Figure 18.8). Pyruvate resulting from alanine eventually contributes to gluconeogenesis. Glucose produced by this route is subsequently carried by the blood to metabolizing muscle tissues, where it is oxidized. The *alanine–glucose* (A-G) cycle ensures body supplies of glucose when the exogenous source of glucose is low. The A–G cycle also assures the effective removal of nitrogen from muscle into the hepatic tissues (Figure 18.1).

Renal Tissues

Renal tissues facilitate the selective excretion of metabolic wastes from the blood. Gluconeogenesis as well as urea synthesis occurs in these tissues, but overall urea production is inferior to that of the liver.

Normal selective excretion functions are notably impaired in patients suffering from chronic renal failure. This produces an added excretory burden on the remaining renal tissues, which display a normal function. Therefore, in cases of renal failure, lowered dietary protein intake can be desirable for minimizing nitrogenous phosphate, sulfate, acid, and other excretory stresses that accompany protein metabolism. The logistics of minimized protein-intake therapies are often overshadowed by the appearance of a negative nitrogen balance and protein wasting. These complications can be marginally averted in some patients by dietary supplements of α-keto acids that correspond to essential amino acids. This dietary practice can promote a more neutral nitrogen balance and protein status during protracted renal therapy. It should be recognized in passing, however, that α-keto acids are readily degraded in the presence of a high-protein diet. Accordingly, α-keto acids should be administered only in cases when dietary protein is restricted.

Brain

The brain synthesizes nonessential amino acids using carbon atoms obtained from glucose metabolism and nitrogen atoms supplied from circulating ammonium ions. Normal brain activities and psychobehavioral actions are greatly influenced by amino acid concentration levels. For example, certain amino acids may trigger the release of hormones involved in feeding control mechanisms, or amino acids may actuate behavioral responses by contributing to neurotransmitter formation. Amino acids that are linked to neurotransmitter availability include tryptophan, a precursor of serotonin, and tyrosine, a molecular predecessor of catecholamines including dopamine, norepinephrine, and epinephrine.

The free amino acid concentration profile in the brain remains steadfast under most conditions. Contrary to rapid metabolic rates, rapid exchanges with plasma amino acids, and an essential amino acid half-life of a few minutes, brain amino acids are held in a dynamic equilibrium.

Postprandial rhythmic increases in plasma amino acids such as tryptophan are not always reflected in the brain since other amino acids compete for the same transport mechanism over the *blood–brain barrier* (BBB).

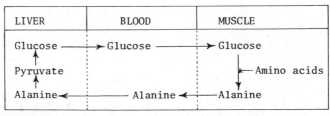

FIGURE 18.1. Alanine–glucose (A–G) cycle.

Apart from tryptophan, the other amino acids competing for BBB transport may include tyrosine, phenylalanine, leucine, isoleucine, and valine. The accentuated transport of tryptophan into the brain is favored by its high concentration ratio relative to all other amino acids competing for transport. In the same way that a high ratio of tryptophan favors its entry into brain cells, a low ratio to the other five amino acids favors an exit of tryptophan from the brain into the plasma.

Wurtman (1982) has reported that the ratio of tryptophan to the five competing amino acids can be modified by the proper diet. For example, a high ratio is attained when a carbohydrate-rich meal is consumed as opposed to a protein-rich meal. Although carbohydrates induce insulin secretion and concomitantly affect free plasma concentrations of tyrosine, phenylalanine, leucine, isoleucine, and valine, tryptophan levels remain generally unaffected and relatively high compared to the other amino acids. This relationship reflects the bound existence of tryptophan to plasma albumins. The resulting development of a high tryptophan ratio to other amino acids ensures preferential tryptophan translocation into the brain in these circumstances along with an *accentuated production* of *serotonin*. Because protein-rich meals provide high concentrations of tryptophan and other amino acids, tryptophan is not favored for transport into brain cells and serotonin levels are reduced.

In addition to these biochemical relationships, it has been shown that carbohydrate-rich and protein-poor meals can produce a neurochemical change that causes an animal subject to reduce its carbohydrate intake but not its protein intake. This apparent action raises the possibility that amino acids such as tyrosine or tryptophan, supplied alone or with insulin-releasing carbohydrates, may be used as drugs in the treatment of some clinically recognized conditions. These conditions include appetite disturbances, insomnia, neurologic disorders, depression, cardiac arrhythmias, and hypo- and hypertension.

In summary then it is clear that a selective increase of a single amino acid having precursor functions as a neurotransmitter can produce an increased cerebral level of that amino acid and its neurotransmitting product.

Contrary to cerebral changes in essential amino acids caused by extreme conditions, an influx of nonessential amino acids into brain is very slow, and no change in brain content occurs even during severe plasma concentrations of these compounds.

In cases of severe malnutrition, cerebral changes in the amino acid pool are quite specific. Some amino acids such as valine, serine, and aspartate decrease, and both histidine and homocarnosine show large increases. The endocrine effects of insulin are well established and most studies show that insulin hypoglycemia is linked to decreased levels of nonessential amino acids. This upset is probably tied to activity changes in the TCA cycle.

Glutamate concentrations along with glutamate's decarboxylation product γ-*aminobutyrate* (GABA) remain high in all brain regions relative to other body tissues under nearly all known conditions. Glutamate exhibits putative (possible) neurotransmitter activity, whereas GABA seems to exert inhibitory neurotransmitter actions in synapses of the CNS. The apparent decarboxylative origin of GABA from glutamate is complicated by its formation from glucose, pyruvate, and some other amino acids.

Complex biochemical interconversions surround the interplay of amino acids throughout the liver, muscle, kidney, and brain tissues. Thus, any biochemical aberrations in one tissue can exert a range of major or minor amino acid upsets throughout the entire body.

CATABOLISM OF AMINO ACIDS

Amino acids share a variety of general reactions including transamination, deamination, and decarboxylation.

Transamination

A typical transamination reaction involves an amino group transfer from one amino acid to a keto acid. This action results in the formation of a new amino acid and a new keto acid (Figure 18.2). Nearly all amino acids undergo transamination at some point in their metabolic life.

Transaminases or *aminotransferases* that mediate amino transfer reactions generally require pyridoxal phosphate and a divalent metal ion. The pyridoxal phosphate derivative of vitamin B_6 serves as a carrier of amino groups or amino acids. It also acts as a coenzyme for decarboxylases, dehydrases, racemases, and amino acid aldolases since it has the ability to enhance the lability of key bonds in different amino acids. Substantial evidence suggests that pyridoxal phosphate is loosely bound as a Schiff base to the ϵ-amino group of a lysyl residue in all enzymes that require the coenzyme. Appearance of a suitable amino acid substrate in the Schiff base region activates a transaldimation reaction; this displaces the lysyl amino group and forms a new Schiff base with the pyridoxal phosphate residue. These events pave the way for transamination, α-decarboxylation, racemization, and other enzyme-specific reactions. At least 22 specific pyridoxal phosphate-dependent amino acid reactions are known (Figure 18.3).

Transaminases for many different amino acids are located throughout the cytoplasm and mitochondria of hepatic, cardiac, cerebral, renal, and testicular cells. Of all the transaminases, alanine and glutamate transaminases are notably important and widely described in the literature. Glutamate transaminase is quite specific for glutamate and α-ketoglutarate as one of its two substrate pairs, but it will react with many other amino acids (Figure 18.4).

Alanine transaminase also reacts with alanine and almost any other amino acid although pyruvate–alanine reaction pairs are common. Highly specific glutamic–alanine transaminases are also recognized (Figure 18.5).

Transamination reactions have key importance in scavaging amino groups from many amino acids and holding them as glutamate. These cytoplasmic reactions ensure adequate supplies of glutamate that are specifically permeable over the inner mitochondrial membrane and into the matrix. The appearance of glutamate in the matrix facilitates mitochondrial transaminations involving aspartate transaminase or oxidative deamination by glutamic dehydrogenase.

Oxidative Deamination

Whereas transamination serves as a mainline pathway for the conversion of L-amino acids into α-keto acids, *oxidative deamination* (OD) offers a secondary route for α-keto acid formation. The OD route is characterized by a *flavin-dependent oxidase* or an *NAD$^+$-dependent dehydrogenase.*

Glutamate dehydrogenase is one of the most ubiquitous and important dehydrogenases. It is allosterically inhibited by GTP and ATP or activated by GDP and ADP. As expected, glutamate dehydrogenase action yields an α-keto acid from the amino acid undergoing deamination plus ammonia.

$$\text{L-Glutamate} + \underset{(\text{NADP}^+)}{\text{NAD}^+} \rightleftharpoons$$
$$\underset{(\text{NADPH})}{\alpha\text{-ketoglutarate}} + \text{NADH} + \text{NH}_3$$

$$R_1-\underset{\overset{|}{NH_3^+}}{\overset{\overset{H}{|}}{C}}-COO^- + R_2-\overset{\overset{O}{\|}}{C}-COO^- \rightleftharpoons R_1-\overset{\overset{O}{\|}}{C}-COO^- + R_2-\underset{\overset{|}{NH_3^+}}{\overset{\overset{H}{|}}{C}}-COO^-$$

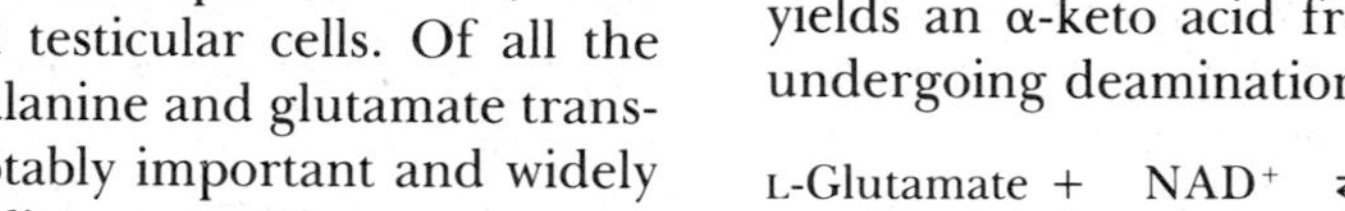

α-Amino acid$_1$ $\qquad$ α-Keto acid$_2$ $\qquad$ α-Keto acid$_1$ $\qquad$ α-Amino acid$_2$

FIGURE 18.2. Generalized reaction for a transamination reaction.

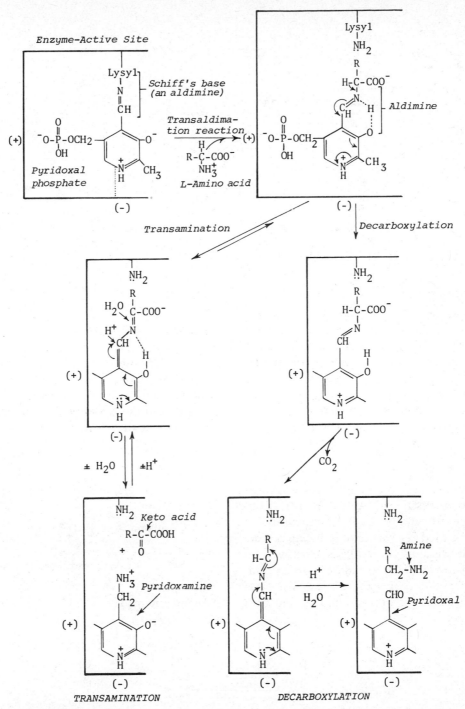

FIGURE 18.3. Schiff base formation and the roles of pyridoxal phosphate during transamination and dearboxylation reactions.

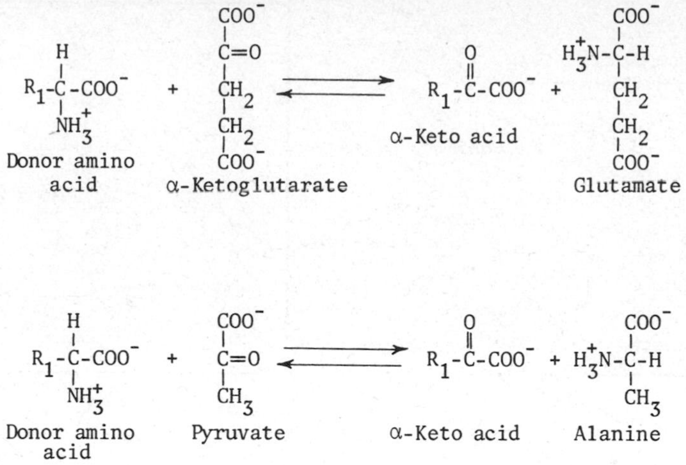

$$\text{Donor amino acid} + \alpha\text{-Ketoglutarate} \rightleftharpoons \alpha\text{-Keto acid} + \text{Glutamate}$$

$$
\begin{array}{c}
\text{H} \\
| \\
R_1\text{-C-COO}^- \\
| \\
\text{NH}_3^+
\end{array}
+
\begin{array}{c}
\text{COO}^- \\
| \\
\text{C=O} \\
| \\
\text{CH}_2 \\
| \\
\text{CH}_2 \\
| \\
\text{COO}^-
\end{array}
\rightleftharpoons
\begin{array}{c}
\text{O} \\
\| \\
R_1\text{-C-COO}^-
\end{array}
+
\begin{array}{c}
\text{COO}^- \\
| \\
\text{H}_3^+\text{N-C-H} \\
| \\
\text{CH}_2 \\
| \\
\text{CH}_2 \\
| \\
\text{COO}^-
\end{array}
$$

Donor amino acid | α-Ketoglutarate | α-Keto acid | Glutamate

$$
\begin{array}{c}
\text{H} \\
| \\
R_1\text{-C-COO}^- \\
| \\
\text{NH}_3^+
\end{array}
+
\begin{array}{c}
\text{COO}^- \\
| \\
\text{C=O} \\
| \\
\text{CH}_3
\end{array}
\rightleftharpoons
\begin{array}{c}
\text{O} \\
\| \\
R_1\text{-C-COO}^-
\end{array}
+
\begin{array}{c}
\text{COO}^- \\
| \\
\text{H}_3^+\text{N-C-H} \\
| \\
\text{CH}_3
\end{array}
$$

Donor amino acid | Pyruvate | α-Keto acid | Alanine

FIGURE 18.4. Exemplary glutamate and alanine transaminase activities.

The reversibility of the reaction is limited in favor of deamination. The ammonia could conceivably be directed toward the amidation of aspartate or glutamate to yield asparagine or glutamine, but physiological conditions favor ammonia formation and its existence as an ammonium ion (~98.5%). Most of this ammonia is eventually excreted through the urea cycle.

Apart from the deamination reaction outlined above, the concurrent production of reduced nucleotide (NADPH(H$^+$)) has significance for many anabolic reactions, in which it is used as a source of reducing power.

Flavin-dependent oxidases exist in two general forms. These oxidases, which catalyze deamination of D-amino acids, use FAD as an electron acceptor, but L-amino acid oxidases require FMN. The biochemical rationale for these differences is not clear; nevertheless the flavin-linked moiety yields hydrogen peroxide in a direct and irreversible reaction with oxygen. An overall summary of the flavin-dependent oxidase reactions is detailed below:

1. D-Amino acid + FMN-E* $\longrightarrow$
(oxidized flavin)
α-keto acid + FMNH$_2$-E + NH$_3$
(reduced flavin)

2. L-Amino acid + FAD-E* $\longrightarrow$
(oxidized flavin)
α-keto acid + FADH$_2$-E + NH$_3$
(reduced flavin)

*Amino acid oxidase.

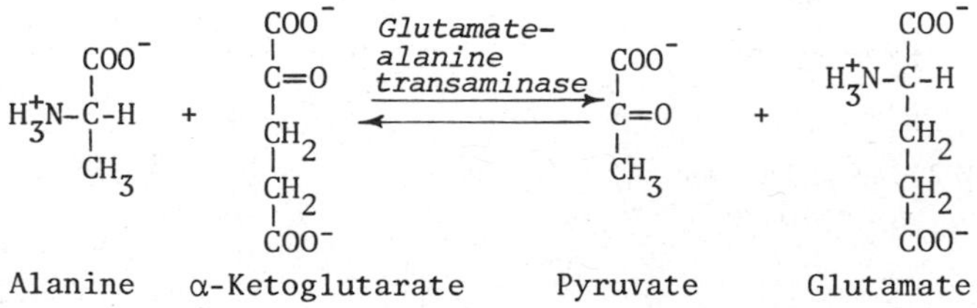

$$
\begin{array}{c}
\text{COO}^- \\
| \\
\text{H}_3^+\text{N-C-H} \\
| \\
\text{CH}_3
\end{array}
+
\begin{array}{c}
\text{COO}^- \\
| \\
\text{C=O} \\
| \\
\text{CH}_2 \\
| \\
\text{CH}_2 \\
| \\
\text{COO}^-
\end{array}
\underset{}{\overset{\textit{Glutamate-alanine transaminase}}{\rightleftharpoons}}
\begin{array}{c}
\text{COO}^- \\
| \\
\text{C=O} \\
| \\
\text{CH}_3
\end{array}
+
\begin{array}{c}
\text{COO}^- \\
| \\
\text{H}_3^+\text{N-C-H} \\
| \\
\text{CH}_2 \\
| \\
\text{CH}_2 \\
| \\
\text{COO}^-
\end{array}
$$

Alanine | α-Ketoglutarate | Pyruvate | Glutamate

FIGURE 18.5. Some glutamate–alanine transaminases are highly specific in their reaction requirements.

Catalase ensures the elimination of hydrogen peroxide in a final step:

$$FADH_2 \longrightarrow E + O_2 \longrightarrow$$

$$FAD\text{-}E + H_2O_2 \xrightarrow{\text{catalase}} H_2O + \tfrac{1}{2}O_2$$

$$FMNH_2 \longrightarrow E + O_2 \longrightarrow$$

$$FMN\text{-}E + H_2O_2 \xrightarrow{\text{catalase}} H_2O + \tfrac{1}{2}O_2$$

Oxidative deamination of primary amines is also executed by *monoamine oxidase* (MAO). This enzyme yields aldehydes from key primary amines in addition to ammonia and hydrogen peroxide.

$$R\overset{*}{\underset{}{-}}CH_2\text{—}NH_2 + O_2 \xrightarrow[H_2O]{MAO}$$

Primary amine

$$R\overset{*}{-}CHO + NH_3 + H_2O_2$$

Aldehyde

Reactions incited by MAO have crucial importance in brain metabolism as well as the overall disposition of catecholamines.

Decarboxylation

The α-decarboxylation of many amino acids is activated by pyridoxal phosphate-dependent decarboxylases. Aside from a relatively minor occurrence of decarboxylation in humans, the amine products of these reactions (Figure 18.6) elicit important physiological responses.

Foremost among cerebral decarboxylation enzymes is *glutamic acid decarboxylase* (GAD), which produces GABA from glutamate. Although the exact action mechanism of GABA remains enigmatic, experiments have clearly demonstrated its potential as a neurotransmission regulator. Whereas most evidence supports an inhibitory neurotransmission function for GABA, glutamate counteracts GABA action as an excitatory agent. Speculation also suggests that the dynamic concentration fluctuations in GABA, accompanied by the putative neurotransmitter actions of glutamate, may act as joint

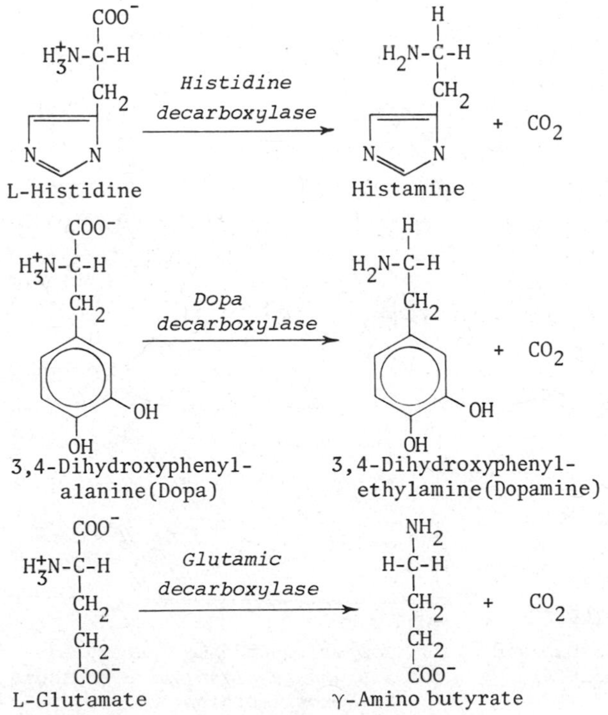

FIGURE 18.6. Three examples of important decarboxylation reactions that produce biologically important amines.

control signals in the neurochemical scheme of brain operations.

Decarboxylation of aromatic amino acids such as 5-hydroxytryptophan, 3,4-dihydroxyphenylalanine, in addition to histidine, have considerable biochemical importance.

The common *5-hydroxytryptophan* (5-HT) metabolite of tryptophan can be decarboxylated to produce *5-hydroxytryptamine,* also known as *serotonin* (Figure 18.7). Serotonin production is generally relegated to blood platelets and intestinal, brain, and renal cells. Substantial amounts of serotonin also occur in bananas, pineapples, plums, nuts, and certain mollusks. Serotonin exhibits vasoconstrictor activity, and it has been widely implicated as an important factor in neurochemical regulatory systems. Tryptophan decarboxylation yields *tryptamine,* which also incites neurochemical responses in the brain. Tryptamine occurrence is not restricted to the CNS. In fact, it serves as an indole acetic acid precursor in plant hormone biogenesis. Another notable decarboxylation reaction having neurological activity results from the action of *dopa carboxylase* on *3,4-dihydroxyphenylalanine* (dopa) to produce *3,4-dihydroxyphenylethylamine*. Dopa is a normal metabolic product of phenylalanine. Dopamine, however, goes on to serve as an intermediate in the formation of adrenalin, which exhibits important vasoconstrictor effects.

Although adrenalin release into the blood is under other control mechanisms, its amine precursor relies on decarboxylase activity. Histidine decarboxylation generates hista-

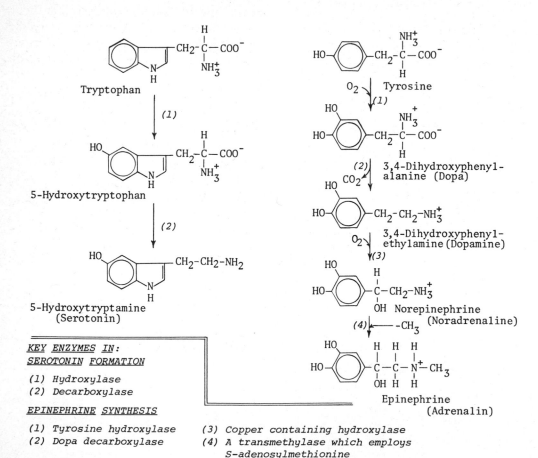

KEY ENZYMES IN:
SEROTONIN FORMATION

(1) *Hydroxylase*
(2) *Decarboxylase*

EPINEPHRINE SYNTHESIS

(1) *Tyrosine hydroxylase*   (3) *Copper containing hydroxylase*
(2) *Dopa decarboxylase*   (4) *A transmethylase which employs*
                                       S-adenosylmethionine

FIGURE 18.7. Synthesis of serotonin (left) as well as the production of catecholamines from tyrosine (right).

mine, which has notable vasodilator effects. Histamine serves as the principal agent secreted by mast cells during hypersensitive allergic reactions. Histamine is ultimately deactivated as a vasodilator by methylation and oxidation reactions in the liver.

THE UREA CYCLE ELIMINATES EXCESS AMMONIA

Humans and mammals in general rely on the biosynthesis of urea for the elimination of excess nitrogen (as ammonia) produced during *amino acid, purine,* and some pyrimidine metabolism (Figure 18.8).

Urea originates from a series of reactions where ammonia, as an ammonium ion, reacts with ATP, HCO_3^-, and water to produce carbamyl phosphate. This reaction operates under the auspices of *carbamyl phosphate synthetase,* and about 6 to 17 g/day of urea nitrogen pass through this reaction in adults. Carbamyl phosphate subsequently reacts with *ornithine* to produce *citrulline.* This reaction is facilitated by *ornithine transcarbamylase.* The enzyme *arginosuccinate synthetase* mediates a condensation reaction between citrulline and

genetic aberrations produce high blood ammonium ion concentration levels and *hyperammonemia.* Aside from hyperammonemia, a total deficiency of any one enzyme in the cycle will result in comatous neonatal death. Moreover, only partial enzyme deficiencies lead to mental retardation and other neurological dysfunctions. Clinical intervention in *very mild* urea cycle disorders often depends on a low-protein diet, which minimizes ammonium ion loads in the blood.

A cursory survey of the urea cycle shows that (1) urea carbon originates from carbon dioxide present as bicarbonate; (2) a 3:1 ratio exists for ATP requirements versus urea production; and (3) ammonia and the amino group of aspartate supply the immediate sources of urea nitrogen. In the last case, it must be noted that ammonia production is ultimately linked to *glutamate dehydrogenase activities.* Aspartate, on the other hand, can ultimately acquire its α-amino group by the transamination of oxaloacetate as a substrate. Note, too, that fumarate productivity in the cycle is conservative since it can be recycled through the TCA cycle and then undergo transamination to produce aspartate. A summary of the urea cycle can be expressed as

$$\overset{*}{C}O_2 + \overset{*}{N}H_3 + H_2O + aspartate \xrightarrow[\substack{5 \text{ enzymes}}]{\substack{(\overset{*}{N}) \quad 3ATP \quad AMP, PP_i \\ \qquad\qquad\qquad\qquad 2ADP, 2P_i}} H_2\overset{*}{N} - \underset{\underset{\textbf{Urea}}{}}{\overset{\overset{O}{\|}}{C}} - \overset{*}{N}H_2 + fumarate$$

aspartate to produce *arginosuccinate.* Cleavage of the latter compound by arginosuccinate lyase creates arginine and fumarate. *Arginase* activity subsequently cleaves arginine into ornithine and urea before ornithine makes a repeat trip through the urea cycle. Urea nitrogen cycled through this route corresponds to the disposition of ~35 to 100 g/day protein. Urea is formed in the liver, but minor amounts are routinely produced in the renal tissues.

Enzymatic disorders caused by genetic factors can upset any one of the five key steps in the urea cycle. All of the recognized

METABOLIC FATES OF AMINO ACIDS

Experimental nutritional studies have clearly demonstrated that carbohydrates and lipids are oxidized for energy purposes in deference to amino acids. In cases of caloric insufficiency and nutritional pathologies, protein wasting and amino acid oxidation can serve as a duress source of energy.

In normal situations amino acids are directed into (1) the biogenesis of peptides and proteins; (2) transamination reactions that convert α-keto acids into their correspond-

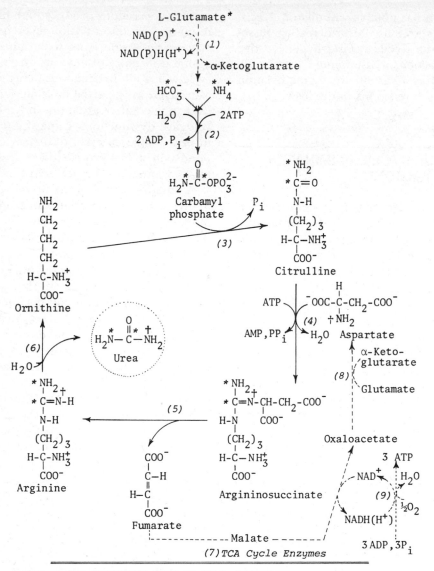

ENZYMES AND REACTIONS TIED TO UREA FORMATION

(1) Glutamate dehydrogenase
(2) Carbamyl phosphate synthetase
(3) Ornithine-carbamyl transferase
(4) Argininosuccinate synthetase
(5) Argininosuccinate lyase
(6) Arginase
(7) TCA enzymes yield aspartate
(8) Transaminase
(9) Oxidative phosphorylation

FIGURE 18.8. Schematic outline of the urea cycle. For each enzymatic step genetic defects have been reported including (1) *carbamyl phosphate synthetase* deficiency; (2) *ornithine transcarbamylase* deficiency; (3) citrullinemia; (4) arginosuccinic aciduria; and argininemia. With the exception of an X-linked dominant *ornithine transcarbamylase* deficiency, all other disorders are autosomal recessive traits.

ing amino acids; and (3) the biosynthesis of various nitrogenous and nonnitrogenous compounds. Supplies of amino acids in excess of those required for normal metabolism are promptly decimated and their carbon skeletons are metabolized.

Detailed studies have shown that amino acids can be classified according to the degradative fates of their carbon skeletons. Common classifications include those that are (1) *ketogenic,* (2) *glucogenic,* or (3) *ketoglycogenic.* Nearly all of the amino acid carbon skeletons are ultimately oxidized by the TCA cycle (Figure 18.9), or they produce pyruvate or acetyl-CoA. Although phenylalanine, tyrosine, leucine, and tryptophan are mainly converted into acetoacetate, a secondary degradation of this compound also yields acetyl-CoA.

Clearly, then, those amino acids that provide TCA intermediates suitable for glucogenesis are described as *glucogenic,* whereas those acids supplying ketone bodies are *ketogenic.* Tyrosine and phenylalanine are exemplary of ketoglycogenic amino acids because part of their carbon structure supplies *acetoacetate* and part is converted to *fumarate.*

Many disorders in amino acid metabolism designated as *amino acidopathies* are recognized. In general, these disorders are genetically determined, often transmitted as autosomal recessive traits, and epidemiologically rare. For example, *cystinuria,* demonstrated as a failure of the proximal tubule to reabsorb cystine and some other amino acids, is among the most common of the disorders. The prevalence of this problem is 1:7000. Abnormal phenylalanine levels including *phenylketonuria* and *hyperphenylalaninemia* occur in the population at 1:13,000 and 1:34,000, respectively. Still other maladies including *arginosuccinic aciduria, Fanconi syndrome, branched-chain ketoaciduria,* and *tyrosinemia* prevail at a rate of 1:250,000 to 1:350,000.

AMINO ACID CATABOLIC PATHWAYS

Essential and nonessential dietary amino acids including proteinaceous amino acids are catabolized into biochemical feedstock for many reactions or they are eliminated as fecal and

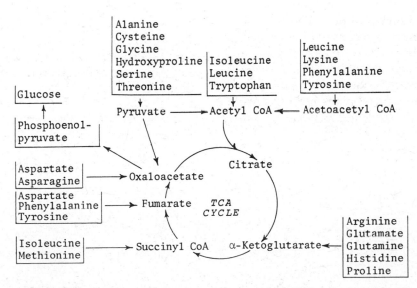

FIGURE 18.9. Origins of carbon atoms that enter the TCA (tricarboxylic acid) cycle.

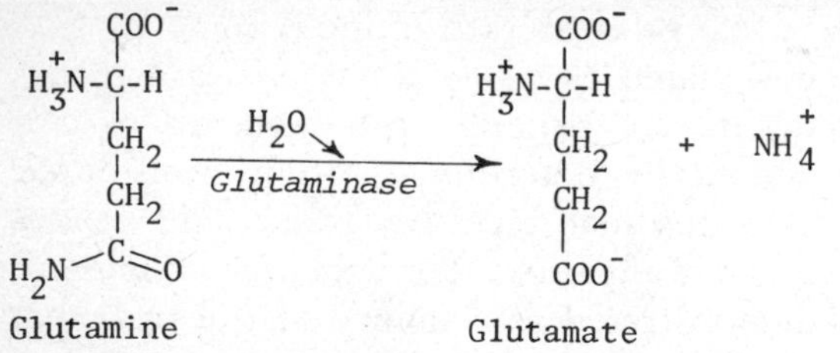

FIGURE 18.10. Glutamine conversion to glutamate.

urinary waste products. As one example, consider the branched-chain amino acids, which can undergo either transamination or deamination reactions leading to glutamate formation. Further transamination involving glutamate and pyruvate or oxaloacetate could provide alanine or aspartate, respectively. An essential amino acid such as methionine could serve as a precursor of homocysteine, homoserine, or taurine.

Enzymes for the degradation of amino acids occur in all organs and tissues, but most of the essential amino acids are degraded in the liver with the major exception of branched-chain acids that are catabolized in skeletal muscles. Nonessential amino acids are degraded in nearly all tissues.

Nitrogen derived from amino acid degradation at nonhepatic sites is transported to the liver for conversion into urea.

A detailed survey of proteinaceous amino acid metabolism is beyond the scope of this text; however, cursory outlines of amino acid metabolism are presented. The catabolism of α-amino acids is outlined according to the most common metabolites including α-ketoglutarate, oxaloacetate, pyruvate, succinyl-CoA, acetyl-CoA, and fumarate.

α-Ketoglutarate Group (Gln, Glu, Pro, Arg, His)

The amino acids in this group generally produce glutamate, which is readily converted by transamination into α-ketoglutarate, which serves as an important TCA intermediate.

Glutamine: Glutamine is converted to glutamate and ammonia by the action of glutaminase (Figure 18.10).

Glutamic acid: Glutamate transamination provides a major metabolic route for this amino acid. In this case, the amino group is donated to an α-keto acid by the action of transaminase to form α-ketoglutarate and a new amino acid (Figure 18.11).

Proline: Proline oxidase catalyzes the conversion of proline to Δ'-pyrroline-5-carboxylate, which spontaneously hydrolyzes to form glutamate-γ-semialdehyde. This product is then converted into glutamate by Δ'-pyrroline dehydrogenase (Figure 18.12).

Arginine: Arginase catalyzes the conversion of arginine to ornithine and urea. Ornithine then goes on to form glutamate-γ-semialdehyde by enzyme actions of ornithine transaminase. Glutamate-γ-semialdehyde is converted to glutamate by the involvement of Δ'-pyrroline dehydrogenase (Figure 18.13). Hereditary deficiencies in ornithine transaminase are known to occur; these cause retinal atrophy and eventual blindness.

Histidine: Histidine is one of the less abundant amino acids. It is deaminated by

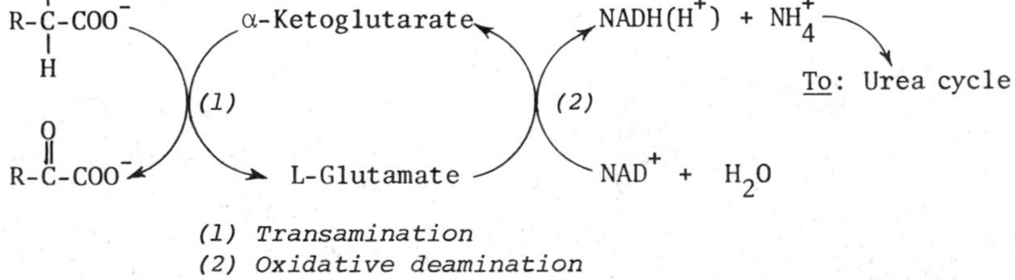

(1) Transamination
(2) Oxidative deamination

FIGURE 18.11. Glutamate transamination reaction.

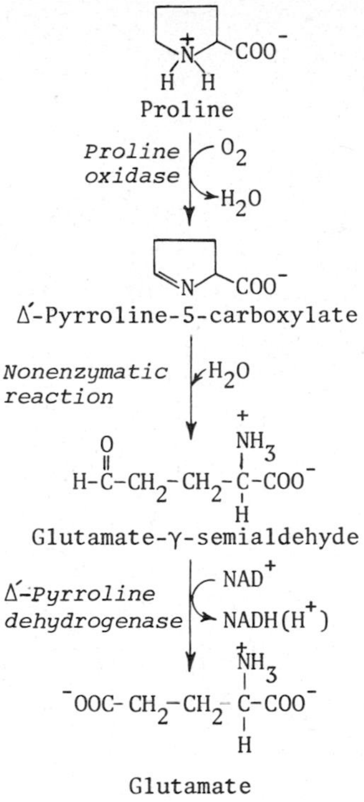

Proline

$$\text{Proline} \xrightarrow[\text{oxidase}]{\text{Proline}} \begin{array}{c} O_2 \\ H_2O \end{array}$$

Δ'-Pyrroline-5-carboxylate

$$\xrightarrow[\text{reaction}]{\text{Nonenzymatic}} H_2O$$

$$\overset{O}{\underset{||}{H-C}}-CH_2-CH_2-\overset{NH_3^+}{\underset{H}{C}}-COO^-$$

Glutamate-γ-semialdehyde

$$\xrightarrow[\text{dehydrogenase}]{\Delta'\text{-Pyrroline}} \begin{array}{c} NAD^+ \\ NADH(H^+) \end{array}$$

$$^-OOC-CH_2-CH_2-\overset{\overset{+}{N}H_3}{\underset{H}{C}}-COO^-$$

Glutamate

FIGURE 18.12. Catabolism of proline into glutamate.

histidase to an unsaturated intermediate known as *urocanate*. This intermediate undergoes hydration by *urocanase* to form 4-imidazolone-5-propionate, which is cleaved by imidazolone propionate hydrolase. The immediate cleavage product is *formiminoglutamate*. Donation of the formimino group from this cleavage product to tetrahydrofolate culminates in the production of *5-formiminotetrahydrofolate* plus glutamate. This reaction is catalyzed by *glutamate-tetrahydrofolate formiminotransferase*. Reactions for the histidine conversion into glutamate are outlined in Figure 18.14.

A genetic disorder linked to diminished *histidase* activity results in *histidinemia*. This problem blocks normal histidine conversion to urocanic acid. Histidine then accumulates in the blood only to be transaminated in increasing amounts of imidazolepyruvic acid. The mechanisms by which this enzyme defect leads to speech impairment and retardation in histidinemia are uncertain. Di-

etary histidine restriction may have minor beneficial effects on this disorder with a tradeoff being impaired growth and development.

Oxaloacetate Group (Asp, Asn)

Aspartic acid and asparagine: Aspartate is converted to oxaloacetate by transamination involving α-ketoglutarate (Figure 18.15). Catabolism of asparagine is only slightly more complicated than that of aspartate since the former requires hydrolysis by *asparaginase* to yield aspartate (Figure 18.16). Transamination of aspartate to oxaloacetate completes the catabolism. Asparagine is used in protein biosynthesis, but it can also participate in the formation of new amino acids by donating its amide to α-keto acids.

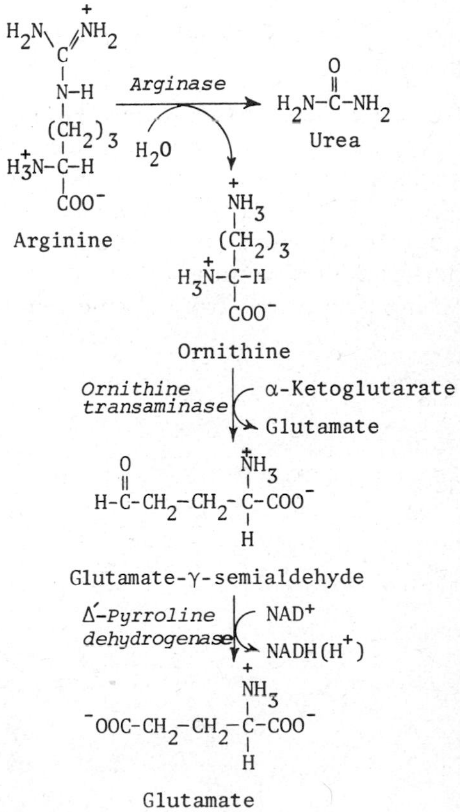

Arginine

$$\xrightarrow[H_2O]{\text{Arginase}} \overset{O}{\underset{||}{H_2N-C-NH_2}} \quad \text{Urea}$$

Ornithine

$$\xrightarrow[\text{transaminase}]{\text{Ornithine}} \begin{array}{c} \alpha\text{-Ketoglutarate} \\ \text{Glutamate} \end{array}$$

$$\overset{O}{\underset{||}{H-C}}-CH_2-CH_2-\overset{\overset{+}{N}H_3}{\underset{H}{C}}-COO^-$$

Glutamate-γ-semialdehyde

$$\xrightarrow[\text{dehydrogenase}]{\Delta'\text{-Pyrroline}} \begin{array}{c} NAD^+ \\ NADH(H^+) \end{array}$$

$$^-OOC-CH_2-CH_2-\overset{\overset{+}{N}H_3}{\underset{H}{C}}-COO^-$$

Glutamate

FIGURE 18.13. Conversion of arginine to glutamate.

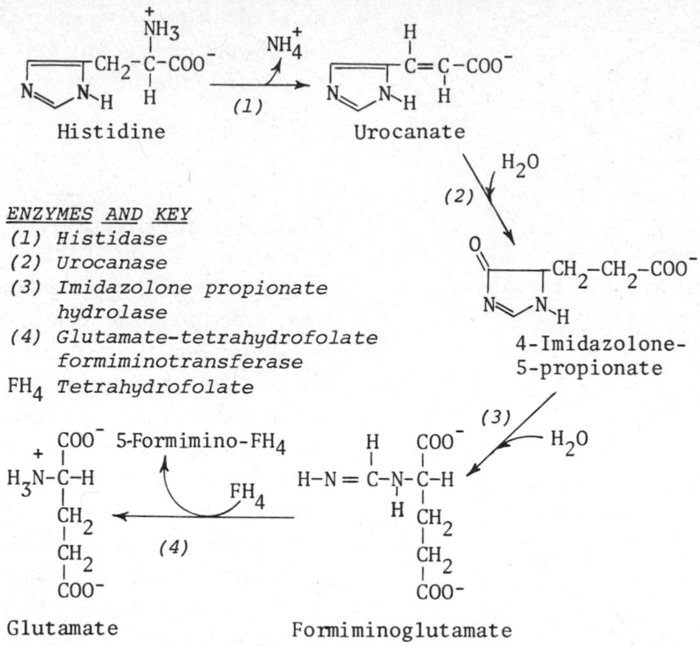

FIGURE 18.14. Conversion of histidine to glutamate.

Pyruvate Group (Gly, Ser, Thr, Ala, Trp, Cys)

Glycine, serine, and threonine: Glycine is the simplest of all the amino acids. Aside from its precursor activities in porphyrin synthesis, glycine participates in *some* detoxification mechanisms (via conjugation) and underlies the biosynthesis of serine.

Serine production from glycine requires the pyridoxal phosphate-dependent action of *transhydroxymethylase* as well as N^5, N^{10}-methylenetetrahydrofolate. As indicated in Figure 18.17, ketimine interaction with folate introduces a hydroxymethyl group onto the glycine moiety. Cleavage of this post-ketimine intermediate generates serine and pyridoxal phosphate.

Serine catabolism continues by a hydrogen atom loss at the α-carbon as well as a β-hydroxyl elimination conducted by *serine dehydrase*. The pyridoxal phosphate catalyzed β elimination provides an unstable amino acrylate. Amino acrylate then reacts with water to yield pyruvate (Figure 18.17).

Threonine can be metabolized in at least two ways. One pathway involves threonine cleavage into glycine and acetaldehyde. The

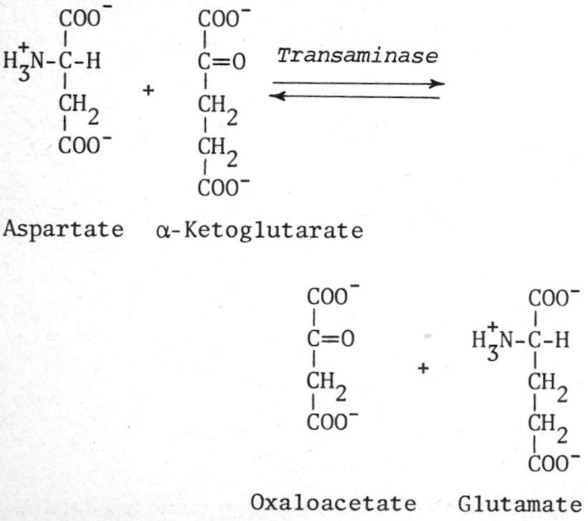

FIGURE 18.15. Aspartate conversion to oxaloacetate.

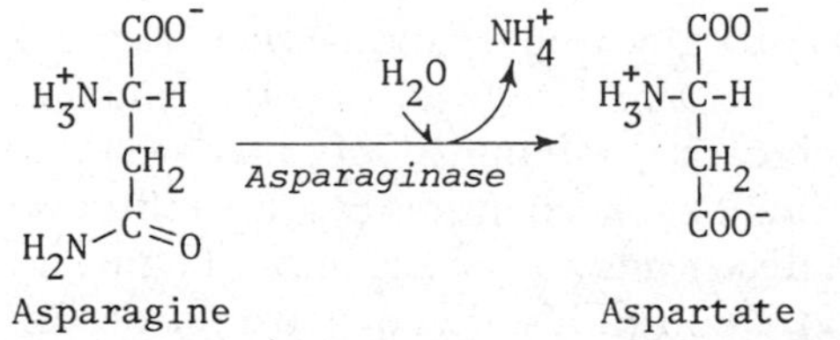

FIGURE 18.16. Asparagine conversion to aspartate.

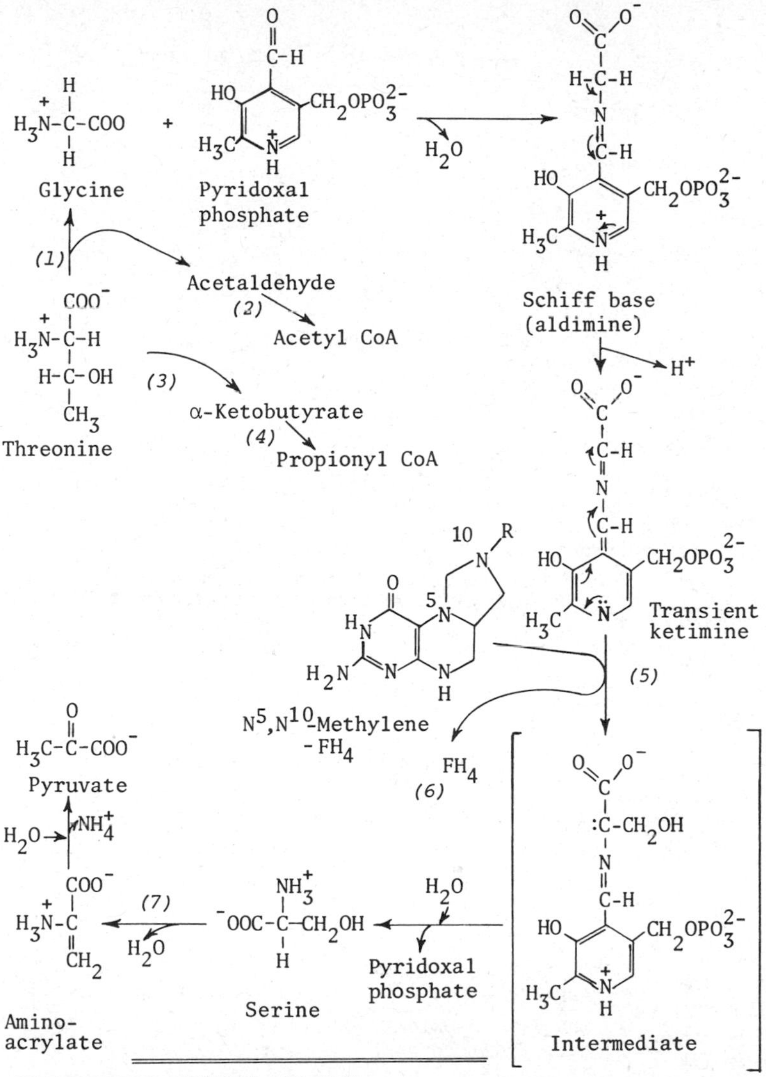

FIGURE 18.17. Catabolism of glycine, serine, and threonine.

SOME NOTABLE ENZYMES AND KEY

(1) Serine transhydroxymethylase
(2) Aldehyde dehydrogenase
(3) Threonine dehydratase
(4) Oxidative decarboxylation
(5) Serine transhydroxymethylase
(6) FH₄-Tetrahydrofolate
(7) Serine dehydratase

latter compound develops into acetyl-CoA under the auspices of an aldehyde dehydrogenase. In an alternate reaction, threonine can be deaminated to produce α-ketobutyrate, which is then oxidatively decarboxylated to give propionyl-CoA.

Alanine: The catabolism of alanine depends on the action of *glutamate-alanine transaminase,* which produces pyruvate from alanine (Figure 18.15). This enzyme represents another example in the glutamate family of transferases that produce a TCA intermediate from an amino acid. This reaction also ensures the availability of a carbon skeleton such as that provided by pyruvate which is suitable for glucose formation in gluconeogenesis. Since gluconeogenesis oc-

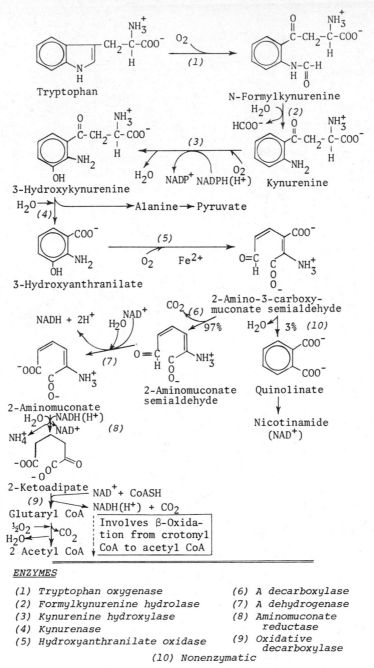

FIGURE 18.18. Conversion of tryptophan into pyruvate and 3-hydroxy-anthranilate, which is eventually converted into acetyl-CoA.

ENZYMES

(1) *Tryptophan oxygenase*
(2) *Formylkynurenine hydrolase*
(3) *Kynurenine hydroxylase*
(4) *Kynurenase*
(5) *Hydroxyanthranilate oxidase*
(6) *A decarboxylase*
(7) *A dehydrogenase*
(8) *Aminomuconate reductase*
(9) *Oxidative decarboxylase*
(10) *Nonenzymatic*

curs in the liver, it is not surprising that high levels of glutamate transaminase reside in the organ. Pyruvate produced by muscle tissues undergoes transamination to give alanine. Alanine is then translocated to the liver, where it is transaminated into pyruvate. Pyruvate subsequently serves as feedstock for glucose synthesis.

The contribution of alanine to gluconeogenesis is substantial since it supplies over 50% of the amino acid inventory suitable for gluconeogenic operations. Glucose produced from alanine then returns via the blood to glucose-consuming muscles, where pyruvate and alanine are again produced in another round of the alanine–glucose cycle.

Tryptophan: Tryptophan is not ranked among the most abundant amino acids, but its catabolism is among the most complex of all the amino acids (Figure 18.18). *Tryptophan oxidase,* a heme-containing enzyme, promotes the irreversible catalysis of tryptophan to give *N-formyl-kynurenine. N-Formyl-kyneurenine hydrolase* yields formate plus kynurenine. A *kynurenine hydroxylase* then catalyzes the conversion of kynurenine into *3-hydroxykynurenine.* The last compound serves as a substrate for *kynurinase,* which produces *3-hydroxyanthranilate* and *alanine.* During these catabolic events, the α-amino group of tryptophan appears as the α-amino group of alanine. Finally, 3-hydroxyanthranilate is converted into acetyl-CoA.

Tryptophan has notable importance as a potential niacin precursor. In cases of gross niacin deficiency and imminent pellagra, this precursor action is more important than levels of niacin obtained in a balanced diet. Nonetheless, it is estimated that 60 mg of tryptophan can yield 1 mg-equiv of niacin in normal individuals. The efficaceous conversion of niacin into NAD^+ relies on kynurenase activity and its requisite supply of B_6 coenzyme. Therefore, extreme nutritional duress involving B_6 or other factors can serve as an affront to tryptophan-based niacin supplies.

An infrequent (1:25,000) hereditary disorder of the renal tubule and gut mucosa known as *Hartnup disease* is characterized by impaired tryptophan transport as well as transport of alanine, serine, threonine, aspartate, glutamine, valine, leucine, isoleucine, phenylalanine, tyrosine, histidine, and citrulline. Massive aminoaciduria also characterizes this autosomal recessive malady. Since tryptophan fails to undergo normal gut absorption, *indole* and *indican* may appear in the urine due to intestinal flora conversions of tryptophan followed by absorption of the derivatives. Hartnup disease exhibits clinical syndrome features characterized by a pellagra-like rash, cerebellar ataxia, and psychological changes. The disorder can be marginally managed by a nicotinamide-supplemented diet (40–250 mg/day), which allays pellagra.

Cysteine: The major pathway for cysteine metabolism involves its oxidation to cysteine sulfinate mediated by an oxidase (Figure 19.19). Aspartate transaminase then catalyzes the conversion of cysteine sulfinated to

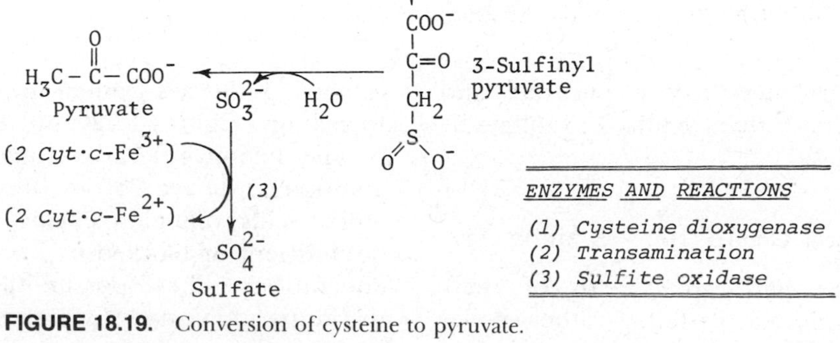

FIGURE 18.19. Conversion of cysteine to pyruvate.

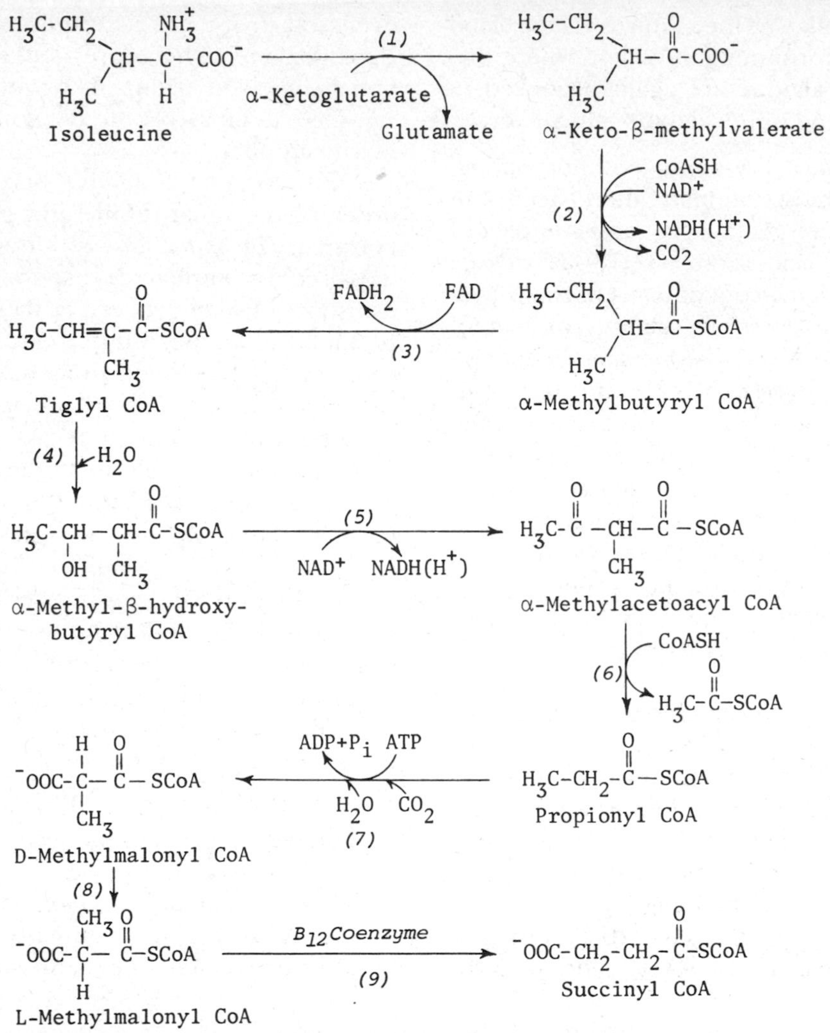

ENZYMES

(1) Transamination
(2) Oxidative decarboxylation, α-Ketoisocaproate dehydrogenase
(3) Acyl CoA dehydrogenase
(4) Enoyl CoA hydratase
(5) β-Hydroxyacyl dehydrogenase
(6) Acetyl CoA acyltransferase
(7) Propionyl CoA carboxylase
(8) Methylmalonyl CoA racemase
(9) Methylmalonyl CoA mutase produces succinyl CoA, enzyme requires coenzyme B_{12}

FIGURE 18.20. Conversion of isoleucine into succinyl-CoA.

3-sulfinylpyruvate to give pyruvate and sulfite. Sulfite is then oxidized to sulfate by sulfite oxidase.

Succinyl-CoA Group (Ile, Val, Met)

Isoleucine and valine: Both of these branched amino acids follow rather similar catabolic pathways leading to succinyl-CoA formation (Figures 18.20 and 18.21).

The oxidative decarboxylation of α-keto analogues, derived from branched-chain amino acids (valine, isoleucine, leucine), can be deficient or blocked by key biochemical aberrations. These possibilities incite ketoaciduria and urinary discharges of branched amino acid derivatives that produce a distinct maple syrup odor. For this reason, these genetically based disorders are classically described as *maple syrup urine* diseases. Infants afflicted with these disorders must be subject to strict dietary management and therapy throughout their lives. Special diets have been industrially formulated that contain synthetic amino acid mixtures free of branched-chain types, yet containing key carbohydrates, lipids, vitamins, and minerals. Strict

moderation of branched-chain plasma amino acids may permit added dietary supplementation using whole milk. This regimen at least encourages some semblance of normality in the dietary amino acid profile, growth, and development.

Inherited disorders in metabolism of methylmalonyl-CoA, produced from branched-chain amino acids, can also result in elevated *methylmalonic acid* (MMA) concentrations (increased blood acidity) along

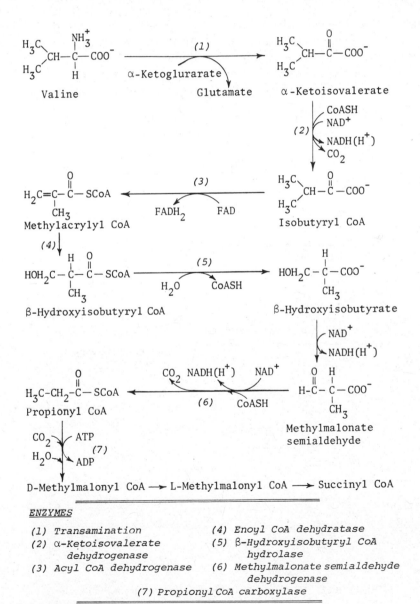

ENZYMES

(1) *Transamination*
(2) *α-Ketoisovalerate dehydrogenase*
(3) *Acyl CoA dehydrogenase*
(4) *Enoyl CoA dehydratase*
(5) *β-Hydroxyisobutyryl CoA hydrolase*
(6) *Methylmalonate semialdehyde dehydrogenase*
(7) *Propionyl CoA carboxylase*

FIGURE 18.21. Conversion of valine into succinyl-CoA.

with *ketoaciduria*. The MMA is ordinarily transformed into succinyl-CoA by the joint participation of *methylmalonyl-CoA mutase* and *adenosylcobalamin*.

Normalization of apparent acidosis by cyanocobalamin administration may indicate a transferase defect responsible for adenosylcobalamin synthesis. A failure to respond favorably to cobalamin, however, suggests a defective methylmalonyl-CoA mutase apoenzyme.

Methionine: There are two different catabolic pathways that involve methionine. The pathway that accounts for most methionine degradation relies on the incorporation of methionyl sulfur into cysteine, whereas the remaining carbon skeleton ultimately yields α-ketobutyrate and succinyl-CoA (Figure 18.22). Methionine also serves as an important donor of methyl groups used for the synthesis of choline and creatine.

Acetyl-CoA Group (Lys, Leu, Trp)

Lysine: Lysine occurs widely throughout many foods in a well balanced diet, and its mitochondrial metabolism produces high-energy phosphate. Several different routes direct the conversion of lysine to α-ketoadipate and eventually acetyl-CoA, and one key lysine catabolic route for mammals has been outlined in Figure 18.23. This metabolic scheme involves a condensation and reduc-

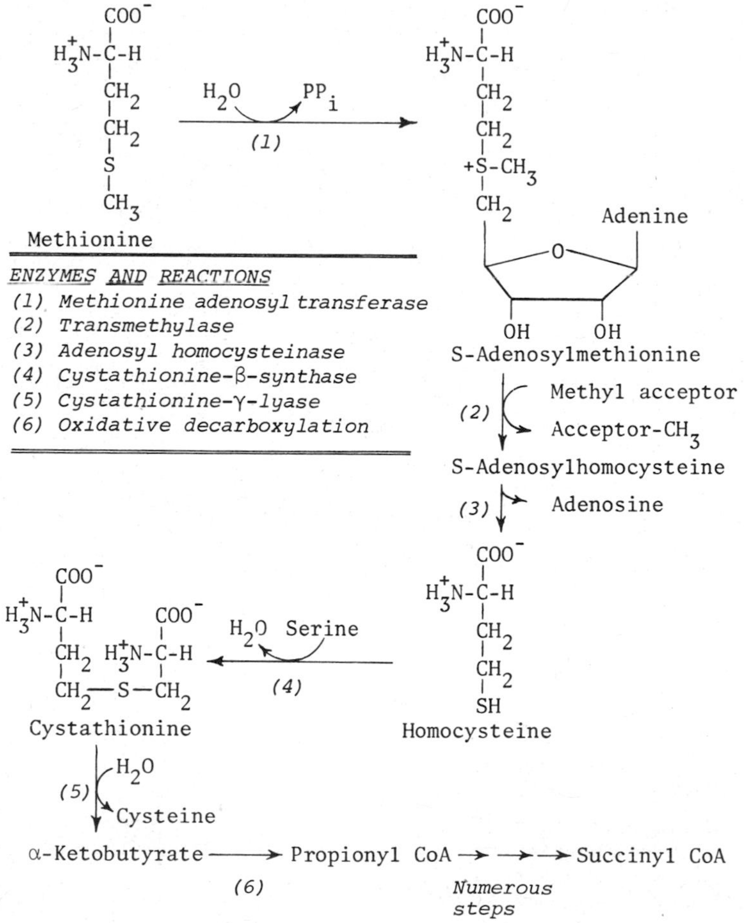

ENZYMES AND REACTIONS
(1) Methionine adenosyl transferase
(2) Transmethylase
(3) Adenosyl homocysteinase
(4) Cystathionine-β-synthase
(5) Cystathionine-γ-lyase
(6) Oxidative decarboxylation

FIGURE 18.22. Methionine catabolism is directed toward succinyl-CoA formation.

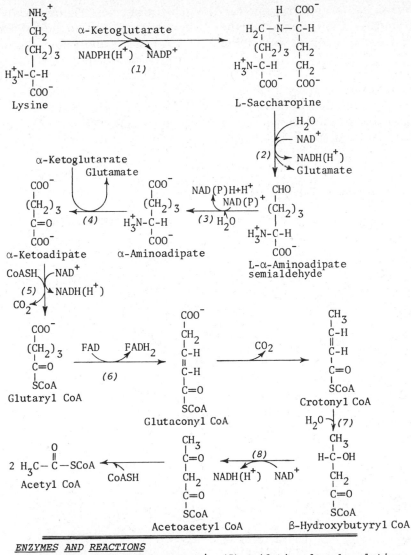

ENZYMES AND REACTIONS
(1) Saccharopine dehydrogenase†   (5) Oxidative decarboxylation
(2) Saccharopine dehydrogenase*   (6) Acyl CoA dehydrogenase
(3) α-Aminoadipate semialdehyde   (7) Enoyl CoA hydratase
    dehydrogenase   (8) β-Hydroxyacyl CoA
(4) α-Aminoadipate transaminase   dehydrogenase
              †Lysine-forming
              *Glutamate-forming

FIGURE 18.23. Conversion of lysine into acetyl-CoA. Note that catabolism beyond glutaryl-CoA involves β oxidation typical of fatty acid oxidation.

tion sequence involving lysine and α-keto-glutarate to give L-saccharopine. Glutamate cleavage from saccharopine produces α-aminoadipate semialdehyde, which oxidizes to α-aminoadipate. Transamination between α-aminoadipate and α-ketoglutarate then supplies glutamate and α-ketoadipate. Reactions reminiscent of β oxidation for fatty acids complete the oxidative sequence of α-ketoadipate into acetyl-CoA. A cursory survey of the overall reaction scheme reveals that both amino groups of lysine are lost as glutamate, two of the six lysine carbons are liberated as carbon dioxide, and the remaining four carbons appear as two acetyl-CoA's.

Leucine: The three initial catabolic steps for leucine involve transamination, oxidative decarboxylation tied to the formation of a fatty CoA derivative, and a flavin-dependent

dehydrogenation (Figure 18.24). These three steps are analogous to the early catabolic steps for isoleucine and valine. Biotin-dependent carboxylation of β-methylcrotonyl-CoA then gives β-methylglutaconyl-CoA. Hydration of the β-unsaturated site in the latter product is followed by an aldol cleavage of β-hydroxy-β-methylglutamyl-CoA to give acetyl-CoA and acetoacetate.

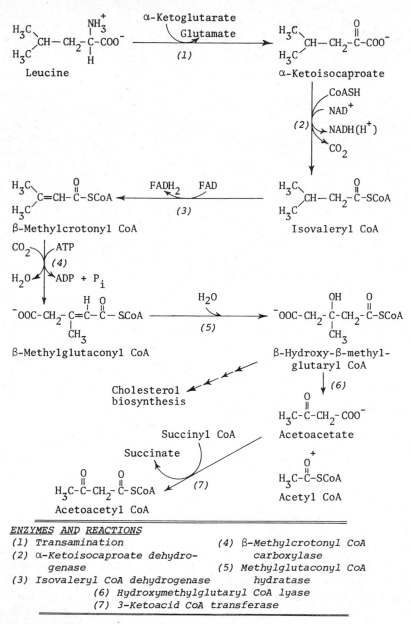

ENZYMES AND REACTIONS
(1) Transamination
(2) α-Ketoisocaproate dehydrogenase
(3) Isovaleryl CoA dehydrogenase
(4) β-Methylcrotonyl CoA carboxylase
(5) Methylglutaconyl CoA hydratase
(6) Hydroxymethylglutaryl CoA lyase
(7) 3-Ketoacid CoA transferase

FIGURE 18.24. Conversion of leucine into acetyl-CoA.

Tryptophan: Conversion of tryptophan into pyruvate and acetyl-CoA has been detailed in Figure 18.18.

Fumarate (and Acetoacetate) Group (Phe, Tyr)

Phenylalanine proceeds by its initial conversion to tyrosine and then to fumarate and acetoacetate (Figure 18.25). Tyrosine for-

mation depends on an irreversible hydroxylation of the aromatic ring (C-4) through the action of *phenylalanine 4-monooxygenase.* This reaction requires molecular oxygen and the cofactor activity of *5,6,7,8-tetrahydrobiopterin* (H_4-biopterin), which is oxidized during hydroxylation to *7,8-dihydrobiopterin* (H_2-biopterin) (Figure 18.25). Oxidation to H_2-biopterin requires the presence of oxygen (O_2) as an electron acceptor. One oxygen atom yields the β-hydroxy group of tyrosine

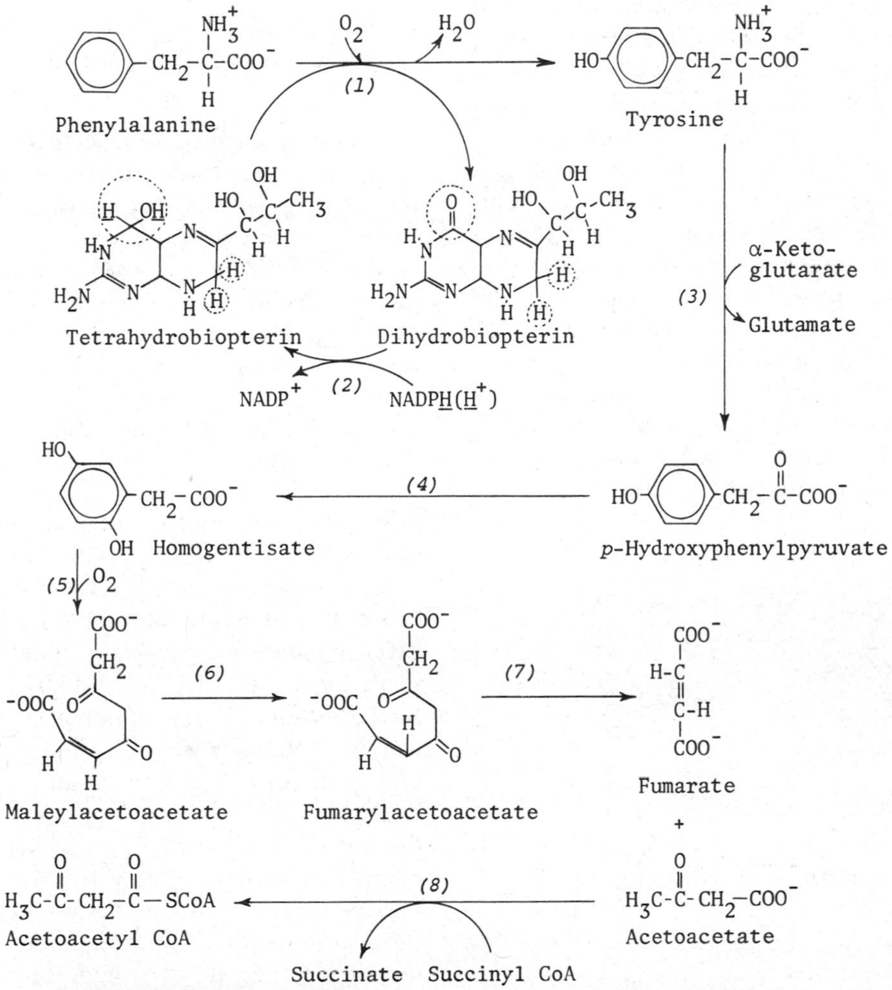

ENZYMES AND REACTIONS

(1) Phenylalanine hydroxylase
(2) Dihydropteridine reductase
(3) Tyrosine transaminase
(4) p-Hydroxyphenylpyruvate oxidase

(5) Homogentisate oxidase
(6) Maleylacetoacetate isomerase
(7) Fumarylacetoacetate hydrolase
(8) 3-Keto acid CoA transferase

FIGURE 18.25. Catabolism of phenylalanine and tyrosine.

while the other oxygen is reduced to water. Phenylalanine hydroxylation is also promoted by the intramolecular migration of a hydrogen atom from C-4 to C-3. This migration stems from an electrophilic attack on the aromatic ring by a hydroxycation (OH$^+$). The hydroxycation is produced from a Fe^{2+}–O$_2$ complex associated with phenylalanine 4-monooxygenase (Figure 18.26).

Tyrosine serves as a precursor of melanin, which is a polymer of indole nuclei produced in the melanocytes of the integumentary system, the eye, mucous membranes, and the nervous system. Graded degrees of skin pigmentations are caused by their melanin content, whereas inherited melanin deficiencies produce albinism (Figure 18.27).

Tyrosine undergoes transamination to give *p*-hydroxyphenylpyruvate while α-ketoglutarate is converted to glutamate after accepting the tyrosinic α-amino group. Decarboxylation of *p*-hydroxyphenylpyruvate by *p*-hydroxyphenylpyruvate oxidase, an added hydroxylation of the already hydroxylated aromatic ring, and an intramolecular rearrangement result in homogentisate. Finally, the ring cleavage of homogentisate produces maleylacetoacetate, which ultimately forms

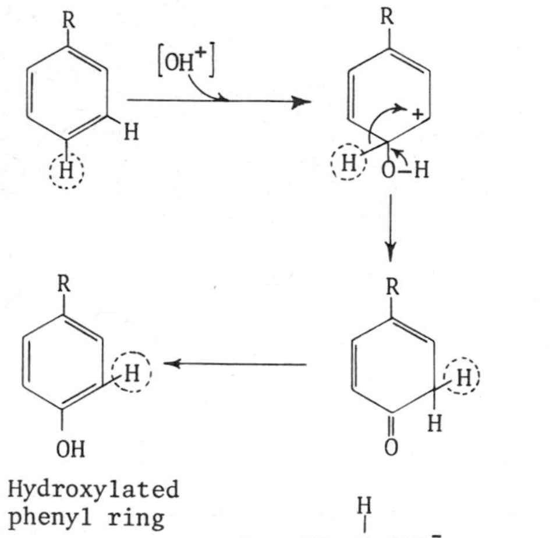

FIGURE 18.26. Model of enzymatic phenylalanine hydroxylation, which induces the migration of an intramolecular hydrogen prior to tyrosine formation.

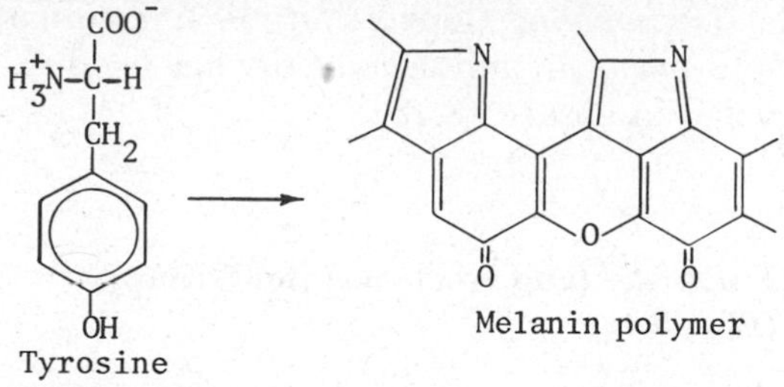

FIGURE 18.27. Tyrosine serves as precursor for the polymeric structure of melanin.

fumarate and acetoacetate (Figure 18.25). Acetoacetate is then converted to acetoacetyl-CoA.

Aberrant phenylalanine metabolism: Metabolic disorders can occur at several sites in the catabolic pathway for phenylalanine. One notable problem caused by an autosomal recessive deficiency of *phenylalanine hydroxylase* results in *phenylketonuria* (PKU). This congenital disorder was recognized by Folling in 1934 using ferric chloride to detect urinary phenylpyruvate. This compound appears in the urine since deficient phenylalanine oxidation to tyrosine results in the shunted formation of phenylpyruvate, phenyllactate, *o*-hydroxyphenylacetate, and phenylacetate. Black populations demonstrate minor evidence of PKU compared with Oriental and white populations, in whom PKU prevalence averages 1:12,000.

Patients with PKU are usually normal at birth, but mental retardation and other clinical indications such as tremors, seizures, electroencephalographic abnormalities, eczema, and hypopigmentation develop during the first year of life. Neurological complications associated with PKU have unknown origins, but they may be caused by the accumulation of phenylalanine or its metabolites. Other theories suggest that phenylalanine may inhibit the transport of other amino acids over CNS membranes, especially when phenylalanine shares a common carrier with other amino acids. This competitive inhibitory effect in amino acid transport may stifle brain growth in its very early stages. Hypopigmentation observed in PKU patients may also be caused by phenylalanine, which competitively inhibits tyrosinase thereby decreasing melanin synthesis.

Neonatal PKU screening is based on the detection of increased blood phenylalanine. The *Guthrie test,* which involves microbial inhibition, has served as a basis for PKU screening although more modern liquid chromatographic methods are possible for this purpose. Evidence of PKU is suggested when blood phenylalanine exceeds 4 mg/dL in the first week of life.

Dietary management of PKU requires minimizing phenylalanine intake to avoid its direct and indirect accumulation as metabolites. These restrictions also pertain to phenylalanine-based sweetening agents. Since phenylalanine accounts for ~5% of most proteins, special diets are often required. One infant formula known as Lofenalac (Mead & Johnson) supplies a 95% phenylalanine-free casein hydrolyzate as a protein source, yet fat, carbohydrate, essential vitamins, and minerals are included in appropriate amounts.

As a PKU infant grows, solid foods are introduced into the diet with the aid of *phenylalanine exchange food lists.* These exchange lists permit the calculated inclusion of cereals, vegetables, and fruits; however, blood phenylalanine levels must be carefully monitored with respect to dietary protein and calorie intakes. Inadequate dietary phenylalanine may be linked to anorexia and growth inhibition especially at blood levels consistently below 1 to 2 mg/dL. Rational nutritional precautions also dictate that prospective mothers with a history of PKU adhere to a low-phenylalanine diet. This practice ensures against any possible impairment of the developing CNS.

Apart from classical neonatal PKU, in which phenylalanine blood levels exceed 25 mg/dL, some apparent PKU cases can be caused by maturational delays in the phenylalanine hydroxylase system. These PKU conditions often disappear several weeks after birth.

Other hyperphenylalaninemias are recognized including *dihydropteridine reductase* defects, which upset the regeneration of the H_4-biopterin cofactor. Since H_4-biopterin is required for conversions of tryptophan to 5-hydroxytryptophan, and tyrosine to L-dopa, biopterin aberrations may incite defective levels of neurotransmitters and neurological problems in spite of phenylalanine restriction.

Another autosomal recessive disorder for phenylalanine catabolism known as *alkaptonuria* originates from *deficiencies in homogentisic acid oxidase.* This enzyme deficiency results in homogentisic acid accumulation in the blood and leads to its urinary excretion (4–8 g/day). The protracted accumulation and polymerization of homogentisic acid by *homogentisic acid polyphenol oxidase* can also produce cartilaginous deposits of the acid. This phenomenon, known as *ochronosis,* causes pigmentation and darkening of the sclera, cartilage, ligaments, and tendons. Although alkaptonuria is a relatively benign disorder, advanced stages of the disorder may spur on a degenerative form of arthritis. Since the reducing properties of homogentisic acid resemble those of urinary reducing sugars, *false positive tests for glycosuria and diabetes can occur in alkaptonuria.*

A SURVEY OF AMINO ACID ANABOLISM

The digestion of dietary proteins supplies the free amino acid feedstock necessary for the biogenesis of many new body proteins. Some of these amino acids are essential while others are nonessential, yet their structures all contribute to protein synthesis, glucogenesis, certain nitrogenous molecules, and fatty acid syntheses depending on the nutritional status of the body.

Since many dietary regimens supply amino acids that are not in proportions required by the body, some acids can undergo restructuring where possible to ensure the presence of certain other amino acids.

All the nonessential amino acids can be synthesized by the body from various metabolic intermediates or from essential amino acids. The nitrogen source for amino acid syntheses may originate from the amino group of degraded amino acids and amides, whereas glycolytic, TCA, and HMP pathway intermediates provide carbon skeletons.

Biosynthetic Pathways for Glutamate, Glutamine, and Proline Are Related

Nonessential amino acids are largely synthesized from their corresponding α-keto acid intermediates supplied by the TCA cycle. Studies of most mammals reveal that amino groups are commonly supplied by transamination reactions from glutamate. These reactions are facilitated by pyridoxal phosphate-dependent transaminases. Biogenic pathways for glutamic acid, glutamine, and proline share some common reaction steps such as the initial amidation of α-ketoglutarate. This reaction occurs in the mitochondrial matrix and requires ammonia, L-glutamate dehydrogenase, and reducing power supplied by $NADPH(H^+)$.

NH_4^+ + α-ketoglutarate + $NADPH(H^+)$

$$\Updownarrow$$

L-glutamate + $NADP^+$ + H_2O

Phosphorylation of glutamate by γ-*glutamyl kinase* and ATP produces a critical intermediate leading to glutamine synthesis. This intermediate, in conjunction with *glutamine synthetase* and an ammonium ion, react to produce *glutamine* (Figure 18.28).

Since ammonia exerts toxic effects on nearly all cells, glutamate formation using ammonia has critical importance in the hematogenous transport of free ammonia as a nontoxic form. Continued glutamate reduction to its γ-semialdehyde followed by cyclization and another reduction produce *proline* (Figure 18.28).

Arginine is produced on a continuous basis in the urea cycle, but most of this amino acid is cleaved into urea and ornithine. This effective catabolic route necessitates the dietary acquisition of arginine for normal nutritional supplies. For those systems in which arginine may be synthesized, activation and reduction of the γ-carboxylate on glutamate leads to ornithine and then arginine.

Biosynthesis of Alanine, Aspartate, and Asparagine

Oxaloacetate and pyruvate serve as key intermediates in transaminations involving glutamate. Oxaloacetate and pyruvate respectively lead to the formation of aspartate and alanine.

Glutamate + pyruvate $\rightleftharpoons$

α-ketoglutarate + alanine

Glutamate + oxaloacetate $\rightleftharpoons$

α-ketoglutarate + asparagine

Although *asparagine synthetase* reactions in many organisms result from the amidation of aspartate to asparagine using NH_4^+, this is not the case in mammals. Asparagine biogenesis in mammals relies on ATP-dependent asparagine synthetase activity in which an amide from glutamine is conveyed to the β-carboxyl site on aspartate.

Serine, Glycine, and Cysteine Arise from a Central Metabolite

An intermediate of the glycolytic pathway, 3-phosphoglycerate serves as a precursor for serine, glycine, and cysteine. This glycolytic intermediate is transformed into serine by the sequential actions of dehydrogenase, transaminase, and phosphatase (Figure 18.29). Serine produced from 3-phosphoglycerate then serves as (1) a source of C_1 units required for the conversion of tetrahydrofolate into its N^5, N^{10}-*methylene tetrahydrofolate* form and (2) a glycine precursor.

As indicated in Figure 18.29, glycine formation from serine relies on the cleavage of

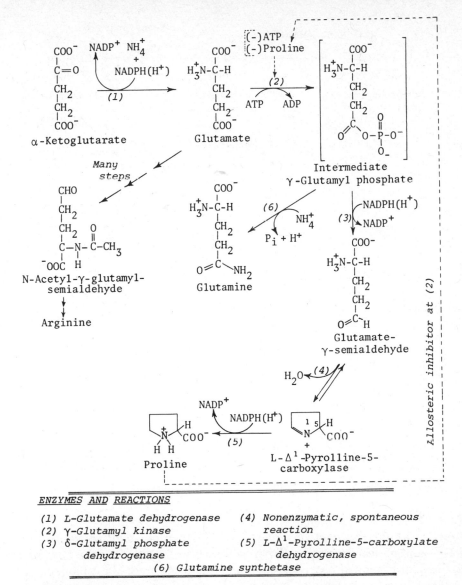

FIGURE 18.28. Synthesis of glutamate, glutamine, proline, and an indication of arginine origins in those organisms that are able to form the amino acid.

the β carbon from serine through the action of *serine hydroxymethyl transferase*.

Homocysteine, derived from methionine by a series of methyl donor reactions, along with serine, condenses to form cystathionine. This condensation is mediated by cystathionine-β-synthase. Deamination and cleavage of cystathionine by the action of pyridoxal phosphate-dependent cystathionine-γ-lyase yields cysteine and α-ketobutyrate.

Both cystathionine-β-synthase and cystathionine-γ-lyase are subject to the allosteric inhibitory effects of cysteine. Genetic disorders linked to cystathionine-β-synthase are recognized and may produce *homocysteinuria*, which contributes to mental retardation.

Tyrosine Is Formed from Phenylalanine

Although tyrosine is a nonessential amino acid, its phenylalanine precursor is an es-

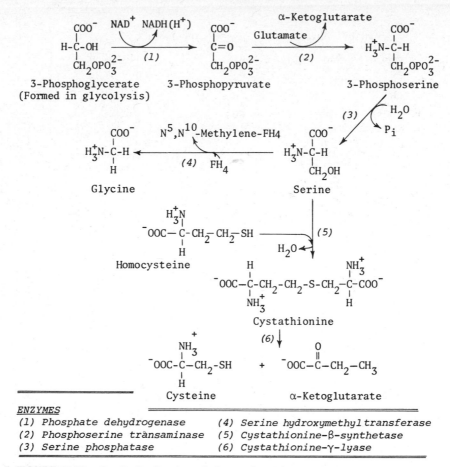

FIGURE 18.29. Synthesis of serine, glycine, and cysteine.

sential amino acid. *Phenylalanine oxygenase,* NADPH(H$^+$), and molecular oxygen all participate in the degradation of phenylalanine. The overall reaction for tyrosine formation is expressed as

Phenylalanine + NADPH(H$^+$) + O$_2$ $\longrightarrow$
tyrosine + NADP$^+$ + H$_2$O

Essential Amino Acid Biosyntheses

Apart from the amino acid biosyntheses already outlined, the remainder of amino acid biosyntheses are either insufficient or completely restricted in the most higher animals. This fact necessitates their acquisition from exogenous dietary sources.

Many of the crucial steps leading to es-

sential amino acid biogenesis in higher animals are stymied by the key absence of one or two enzymes. Many biogenic pathways for essential amino acids *would be* very long and complicated affairs *if they were possible,* especially in those cases where heterocyclic (histidine), aromatic (phenylalanine), or condensed ring (tryptophan) systems are displayed.

Although the biogenesis of essential amino acids will not be studied in detail here, most of their synthesis routes in plant and microbial sources have been uncovered. Furthermore, it is useful to recognize that lysine, methionine, and threonine originate from aspartate, whereas histidine and arginine develop from fumarate. Threonine serves as the basis for isoleucine biosynthesis, and both leucine and valine develop from a pyruvate precursor. Phenylalanine and tryptophan evolve from a complex reaction series initiated by erythrose 4-phosphate plus phosphoenolpyruvate to produce *chorismate.* The development of histidine, on the other hand, depends on the interaction of two atoms from the purine ring of ATP in the imidazole, five carbons from a phosphoribosylpyrophosphate, a glutamine amido group, and an amino group from glutamate to give the final amino acid.

SOME IMPORTANT BIOSYNTHETIC FUNCTIONS OF AMINO ACIDS

Certain amino acids serve as precursors for many other important molecules. Glycine is a critical requirement for *heme biosynthesis;* tryptophan is a precursor of the neurotransmitter *serotonin;* arginine and glycine are involved in the synthesis of *creatine phosphate;* and both histidine and β-alanine are required for *carnosine* and *anserine* syntheses.

Phenylalanine is the precursor for *thyroxine* and various *catecholamines;* methionine provides a source of methyl groups in *choline* synthesis; and both glutamate and aspartate along with their respective amides (glutamine and asparagine) supply amino groups. The role of glutamine as a biosynthetic precursor cannot be completely detailed here, but it must be noted that glutamine is a multifunctional precursor. No less than eight products of glutamine metabolism are recognized in *Escherichia coli* including tryptophan, histidine, glycine, alanine, glucosamine 6-phosphate, carbamyl phosphate (urea, arginine), ATP, and CTP. All these compounds exert independent negative feedback inhibition on the allosteric activity of *glutamine synthetase.* This enzyme supplies glutamine by the following reaction:

L-Glutamate + ATP + NH$_3$ $\longrightarrow$

L-glutamine + ADP + PO$_4$

Apart from model *E. coli* biosyntheses involving glutamine, glutamine biochemistry is of similar importance in mammals.

GLYCINE IS REQUIRED FOR TETRAPYRROLE BIOSYNTHESIS

The biosynthesis of tetrapyrroles and their cyclic derivatives, recognized as *porphyrins,* are derived from glycine and succinyl-CoA. Porphyrin structures serve as critical prosthetic moieties in myoglobin, hemoglobin, chlorophyll, all the cytochromes, and many enzymes such as the peroxidases. The corrinoid ring of vitamin B$_{12}$ is also *reminiscent* of the same porphyrin structure.

Glycine and succinyl-CoA react in three descriptive steps that eventually yield a porphyrin structure. These steps include (Figure 18.30)

1. Glycine and succinyl-CoA *condensation produce* δ-*amino levulinate* (ALA) under the auspices of ALA-synthetase.

2. Intermolecular *condensation of two ALA molecules* gives a porphobilinogen in the form of a substituted pyrrole.

3. *Condensation of four porphobilinogen molecules* forms a linear tetrapyrrole that cyclizes in a ring closure step to give a cyclic tetrapyrrole.

Cyclization of the tetrapyrrole is directed by highly specific enzymes including *uroporphyrinogen I* (UI) *synthetase* and *uroporphyrinogen III* (UIII) *cosynthetase.* The cosynthetase isomerizes one of the pyrrolic rings of UI to form UIII, which is necessary for heme synthesis. Decarboxylation of UIII to coproporphyrinogen III is subsequently prompted by uroporphyrinogen decarboxylase. In this step, the acetate side chains of UIII are decarboxylated to produce the characteristic methyl groups of coproporphyrinogen III. The enzyme action of coproporphyrinogen oxidase then converts the two propionate side chains of coproporphyrinogen III into vinyl

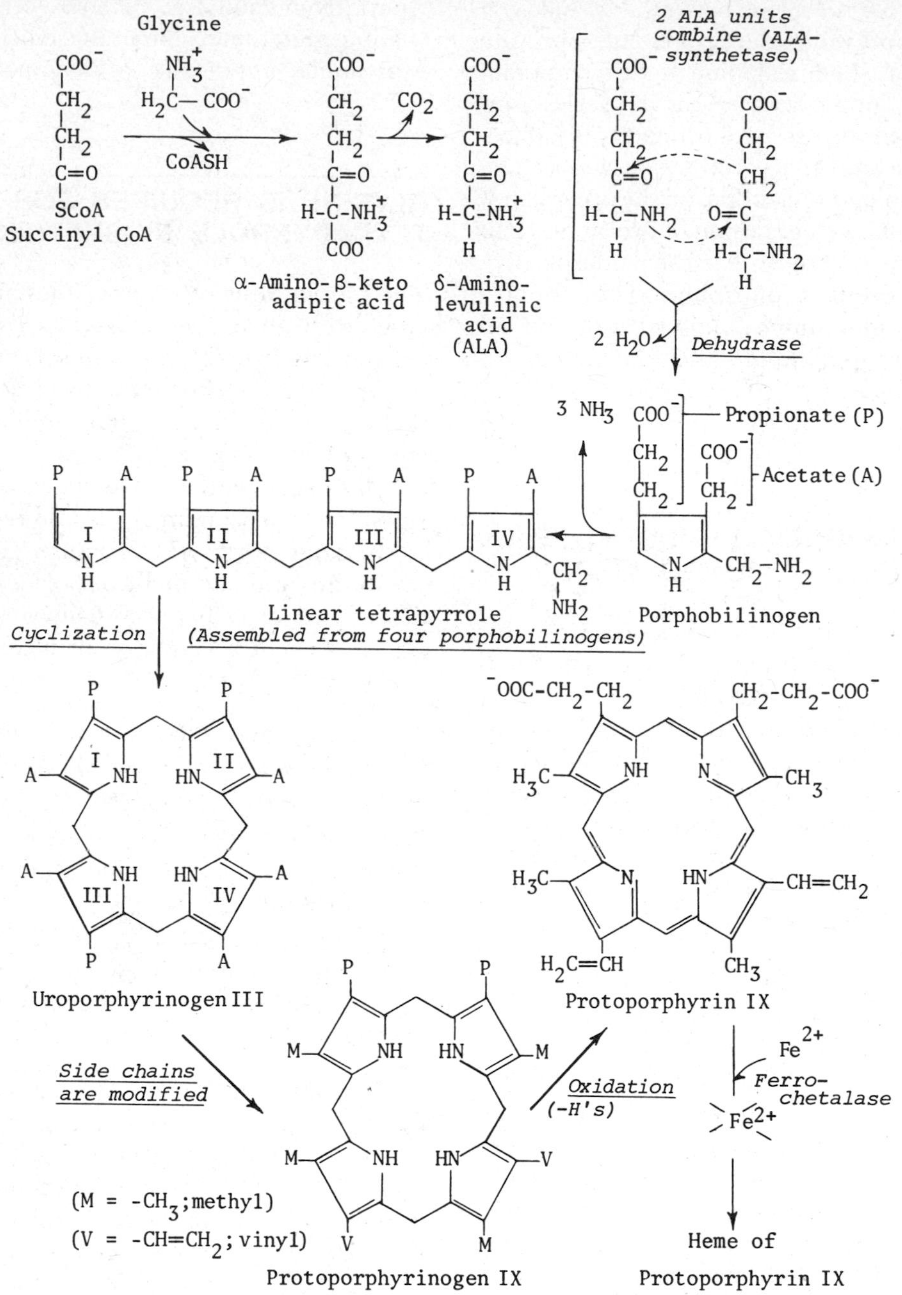

FIGURE 18.30. The biosynthesis of heme requires glycine in the earliest stages of the synthesis scheme.

groups. The resulting tetrapyrrolic structure is now recognized as *protoporphyrin IX*. Finally, protoporphyrin IX is transformed into the *heme* of protoporphyrin IX by *ferrochetalase*. This enzyme inserts iron into the tetrapyrrolic structure.

Biosynthesis of protoporphyrin IX is detailed here because it is the most common porphyrin encountered in biochemical systems. It should be noted in passing that the four methenyl linkages (—CH=), the four pyrrolic carbons, and the four nitrogens of the tetrapyrrole are all derived from glycine. The remaining carbons are obtained from succinyl-CoA. Furthermore, since succinyl-CoA is a key component of the TCA cycle, almost any sugar, fatty acid, or amino acid derivative capable of TCA oxidation could serve as a carbon source for succinyl-CoA.

Genetic disorders linked to heme biosyntheses are widely recognized. These disorders may lead to overproduction of porphyrins or porphyrin precursors, ALA, and porphobilinogens. Clinical evidence of these anomalies is displayed as neurologic and/or cutaneous problems.

Defects in the cosynthetase system can incite a *congenital erythropoietic porphyria* caused by excess production of UI instead of the normal UIII intermediate. This causes premature erythrocyte destruction and the presence of porphyrins in the urine, which give it a pink color. *Erythrodontia* (red teeth) may also be suggestive of accumulated porphyrins. Apart from genetic porphyrias, these disorders can also result from acute disturbances in hepatic metabolism following hepatitis, heavy-metal poisoning, certain anemias, and other conditions.

Ferrochetalase deficiencies can also produce cutaneous photosensitivity with symptoms of sunlight-induced burning and stinging sensations followed by erythema and/or edema. Repetitive sun exposure may cause thickening of the skin. This overall problem reflects an accumulation of ferrochetalase, but the problem may be minimized to some extent by administering β-carotene. This carotenoid apparently increases the dermal tolerance for sunlight by quenching photo- and porphyrin-induced active intermediates that cause cutaneous injury.

DISPOSITION OF EXPENDED PORPHYRINS INVOLVES BILIRUBIN FORMATION

Heme degradation proceeds by its conversion from a cyclic to an open tetrapyrrolic structure. Heme catabolism begins with its

removal from hemoglobin in the reticuloendothelial cells. A highly specific *microsomal heme oxygenase* then attacks the α-methane bridge of the porphyrin to *decyclize* the structure. Nearly 7.0 g/day hemoglobin is processed and replaced each day as a result of this degradative route. The globin and iron are returned to their respective metabolic pools while the open *biliverdin* tetrapyrrole serves as a substrate for *biliverdin reductase* (Figure 18.31). This enzyme produces almost 250 mg/day *bilirubin*. Smaller amounts of bilirubin also develop from excess heme synthesis, cytochrome decomposition, and defective red cells.

The fate of bilirubin as a nitrogenous waste relies on its establishment of an equilibrium

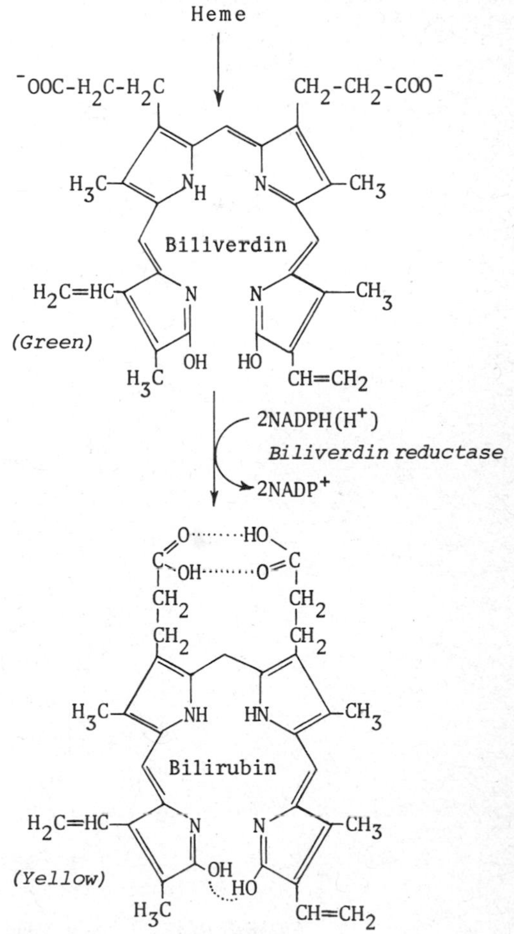

FIGURE 18.31. Action of bilirubin reductase on biliverdin.

with the albumin fraction of the blood. Bilirubin clearance through the kidney is normally insignificant owing to its firm hydrogen bonding with albumin. However, the kidney facilitates some monoglucuronide-conjugated bilirubin formation that binds less strongly to plasma proteins and does display limited renal clearance.

The principal site for metabolic clearance of bilirubin is the liver, where its discharge occurs in bile. Bilirubin excretion here also requires the preliminary formation of a diglucuronide derivative of bilirubin. This diglucuronide bilirubin derivative displays a weak albumin-binding interaction compared to free bilirubin. This effect not only enhances bilirubin solubility for bile excretion but also aids in the renal clearance of bilirubins that may fail to undergo biliary excretion.

Other derivatives of tetrapyrroles including *dipyrroles,* produced by capillary photoactivation of oxygen and its degradation of bilirubins, cause some unconjugated urinary or biliary pyrrole excretion. The actual transfer of bilirubins from plasma to the biliary canaliculus is a complex mechanism that remains to be firmly established. Nevertheless, numerous glycosyl-conjugating enzymes, transferases, and *cytosolic ligandin* proteins are necessary for *separating* bilirubins from their albumin vehicles and *forming* excretable bilirubin conjugates. Analyses of bilirubins in bile show that bilirubin species exist as di- ($\leq 90\%$), mono- ($\leq 10\%$), and miscellaneous ($\leq 2\%$) conjugates.

Accumulations of bilirubins result in *jaundice* with its characteristic yellowing of the skin and eyeballs. Impaired liver functions and bile excretion are typical causes of jaun-

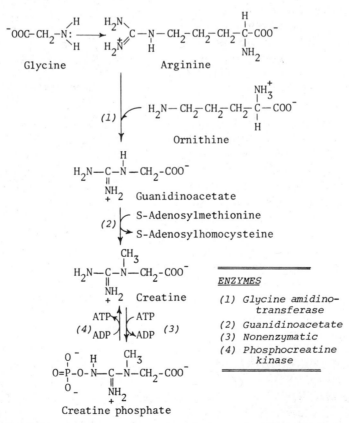

FIGURE 18.32. Creatine phosphate formation from glycine and arginine.

dice. Bilirubin accumulation is potentially toxic to neonates where its effective excretion is critical. Existing problems of neonatal jaundice can also be exacerbated by sulfonamides and salicylates that displace bilirubins from their albumin-binding agents. This action enhances the cytotoxic effects of bilirubin in the neonate.

The appearance of diglucuronide conjugates of bilirubins in the lower ileum marks the beginning of their reduction by intestinal flora. Bilirubin yields (+8H) *mesobilirubin,* which can be further reduced (+4H) to give *stercobilinogens.* Fecal products of bilirubin reduction finally persist as *urobilins.*

ARGININE, GLYCINE, AND CREATINE PHOSPHATE FORMATION

Interaction of an amino group from glycine with the imine carbon of arginine serves as a preliminary step to creatine phosphate formation. This early step, actuated by glycine amidinotransferase (Figure 18.32), generates ornithine plus guanidinoacetate. Guanidinoacetate receives a methyl group from *S*-adenosylmethionine in the presence of *guanidinoacetate methyltransferase* to give creatine.

This supply route for creatine is important in many higher animals since creatine can readily undergo phosphorylation by ATP to form creatine phosphate. Creatine phosphate supplies a crucial energy reserve for muscle contraction (especially white muscles) when *phosphocreatine kinase* promotes energy release as ATP from creatine phosphate.

HISTIDINE IS A PRECURSOR OF CARNOSINE AND ANSERINE

Carnosine synthetase mediates the peptide linkage of β-alanine with histidine to form carnosine (Figure 18.33). Carnosine then undergoes methylation by *S*-adenosylmethionine to give anserine. Academic disputes surround the biochemical functions of these dipeptides in muscle tissues, but they probably reinforce their intracellular buffering capacity in the range of pH 6.0 to 7.0. Carnosine is also found in the brain, where it may serve as a neurotransmitter (e.g., in olfactory pathways).

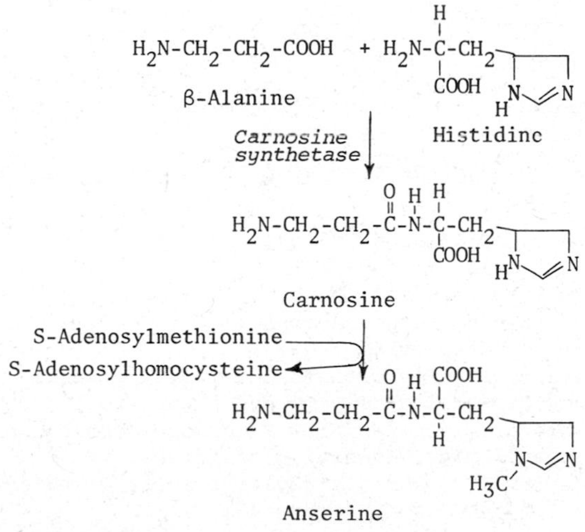

FIGURE 18.33. Synthesis of carnosine and anserine.

TYROSINE IS A PRECURSOR OF CATECHOLAMINES

Norepinephrine (noradrenaline), epinephrine (adrenalin), and dopamine represent *catecholamines* synthesized from tyrosine. Pathways responsible for catecholamine biogenesis occur in nervous tissue as well as in the adrenal medulla (Figure 18.34). Aside from their notable neurotransmitter func-

tions, catecholamines also stimulate lipid and glycogen degradations.

The earliest step in catecholamine biogenesis requires *tyrosine hydroxylase* activity and the production of L-3,4-dihydroxyphenylalanine (L-dopa). Molecular oxygen and 7,8-dihydrobiopterin are also required for this reaction, but the hydroxylase system is controlled by a negative feedback effect of norepinephrine—a later end-product in the

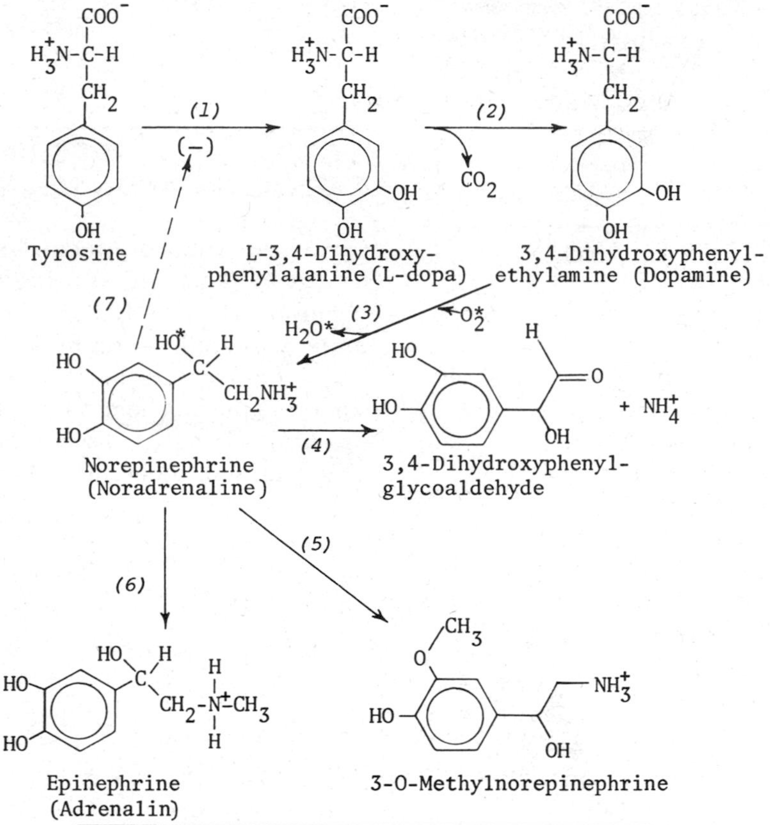

KEY ENZYMES AND REQUIREMENTS INVOLVED IN CATECHOLAMINE FORMATION AND METABOLISM

(1) Tyrosine hydroxylase,O_2,dihydrobiopterin
(2) Aromatic L-amino acid decarboxylase
(3) Dopamine-β-hydroxylase, ascorbate(Cu^{2+}) is oxidized to give dehydroascorbate
(4) Monoamine oxidase (MAO)
(5) Methylation of norepinephrine at its 3-hydroxyl group by catecholamine-O-methyl transferase (COMT), methyl group originates from S-adenosylmethionine
(6) Phenylethanolamine-N-methyl transferase
(7) Norepinephrine produces a negative feedback effect on (1) limiting L-dopa production

FIGURE 18.34. Catecholamine biosynthesis routes.

biogenic scheme. A pyridoxal phosphate-dependent L-amino acid decarboxylase subsequently converts L-dopa to dopamine.

A second hydroxylation of the tyrosine structure then occurs at the benzylic position of the molecule, which is facilitated by *dopamine-β-hydroxylase*. The resulting product is norepinephrine, which is known to (1) suppress the firing rate of sympathetic neurons and (2) actuate adenylcyclase activity.

These effects of norepinephrine are checked by two different catabolic routes. *Monoamine oxidase* (MAO) yields 3,4-dihydroxyphenylglycoaldehyde in one pathway, whereas *catecholamine-O-methyltransferase* (COMT) produces a 3-*O*-methylnorepinephrine derivative. Methyl groups are introduced into catecholamine metabolism in the last step by S-adenosylmethionine. A similar methylation of norepinephrine occurs in the adrenal medulla, where epinephrine (adrenalin) is produced.

Subpicomole ($<10^{-12}$ mol) levels of epinephrine, norepinephrine, and other catecholamines in the blood and urine show wide alterations in various diseases including Parkinson's disease, myocardial infarction, depression, hypertension, and neuroblastomas. Many monitoring methods for catecholamines have been reported including radioassays, but high-performance liquid chromatography employing electrochemical detectors (LCEC detection) offers superior analytical possibilities (Christensen and Blank, 1979; Cole and Snyder, 1981; Kissinger *et al.*, 1981; Robbins and Everitt, 1982).

PURINE NUCLEOTIDE ANABOLISM AND AMINO ACID REQUIREMENTS

Almost all the amino acids contribute in one way or another to the successful biosynthesis of nucleotides. *Adenosine 5'-monophosphate* (AMP) and *guanosine 5'-monophosphate* (GMP), respectively denoted as *adenylate* and *guanylate*, serve as the principal purine nucleotides in nucleic acids.

Adenylate contains the purine *adenine* whereas guanylate has *guanine*. Both of these purines have complex but common origins arising from glutamine, aspartate, and glycine (Figure 18.35).

Contrary to many early expectations, it has been revealed that purine nucleotides are constructed upon a ribose 5'-phosphate structure. The initial stages of AMP and GMP syntheses commence with the C-1 ribosyl linkage of an amino group donated by glutamine (Figure 18.36). Glycine is then attached to the C-1 ribosylamine. A variety of complex steps ensue that culminate in the establishment of a five-membered imidazole ring. Carboxylation of the resulting 5'-phosphoribosyl-5-aminoimidazole paves the way for an amination of the imidazole ring and a subsequent closure sequence producing a bicyclic ring system. The principal precursor to both AMP and GMP is *inosinic acid* (IA).

The IA is transformed into *adenylic acid* (AMP) through the addition of an amino group provided by aspartate (Figure 18.36). Production of *guanylic acid* (GMP) proceeds from IA in a series of similar reactions that also require ATP.

The *de novo* nucleotide structure of AMP can be converted to ADP by adenylate kinase

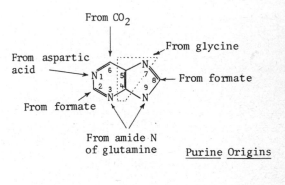

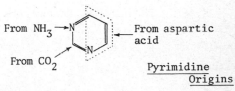

FIGURE 18.35. Origins of molecular regions in purine and pyrimidine molecules.

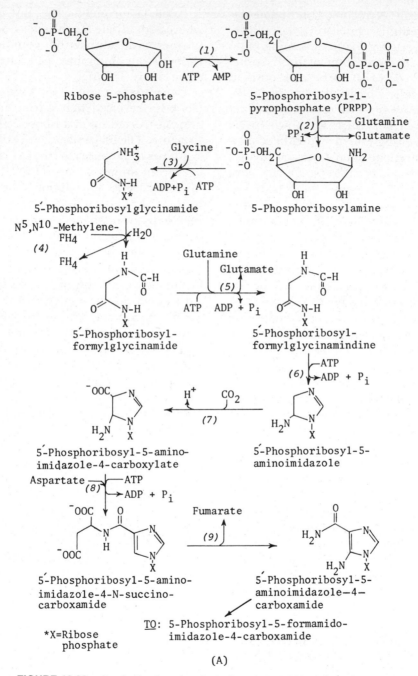

FIGURE 18.36. Synthesis of purines including AMP, XMP, and GMP.

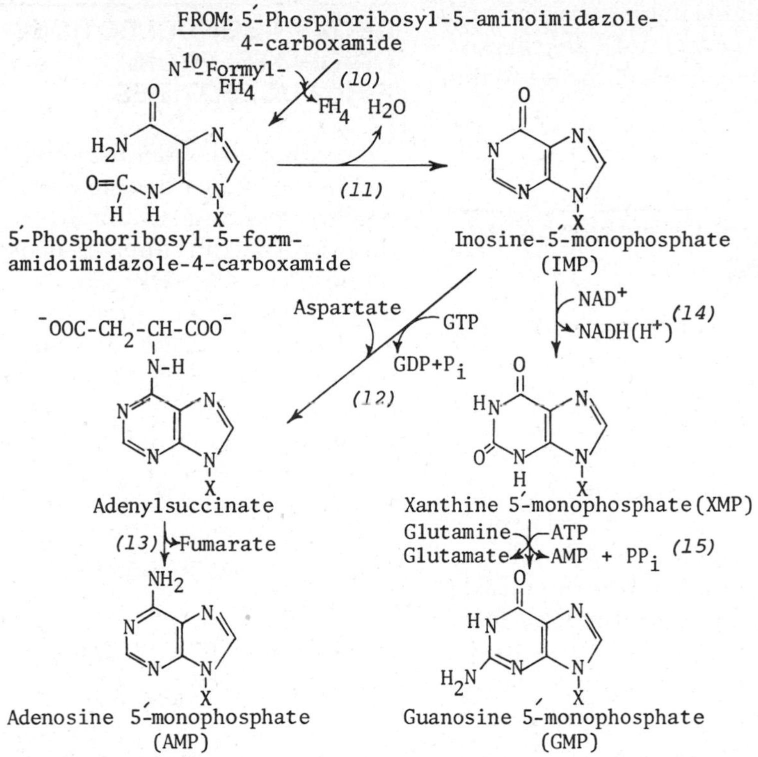

FROM: 5'-Phosphoribosyl-5-aminoimidazole-
4-carboxamide

5'-Phosphoribosyl-5-form-
amidoimidazole-4-carboxamide

Inosine-5'-monophosphate
(IMP)

Adenylsuccinate

Xanthine 5'-monophosphate (XMP)

Adenosine 5'-monophosphate
(AMP)

Guanosine 5'-monophosphate
(GMP)

ENZYMES INVOLVED IN PURINE BIOSYNTHESIS

*(1) Ribose phosphate pyrophospho-
kinase*
(2) Amidophosphoribosyl transaminase
*(3) Phosphoribosyl glycinamide
synthetase*
*(4) Phosphoribosyl glycinamide
formyl transferase*
*(5) Phosphoribosylformylglycinamide
synthetase*
*(6) Phosphoribosylaminoimidazole
synthetase*
*(7) Phosphoribosylaminoimidazole
carboxylase*

*(8) Phosphoribosylaminoimidazole
succinocarboxamide synthetase*
(9) Adenylsuccinase
*(10) Phosphoribosylaminoimidazole
carboxamide formyl transferase*
(11) Inosinicase
(12) Adenylosuccinate synthetase
(13) Adenylsuccinase
*(14) Inosine 5'-phosphate
dehydrogenase*
*(15) Guanosine 5'-phosphate
synthetase*

(B)

FIGURE 18.36. *(Continued)*

and high-energy phosphate provided by ATP. This reaction yields two ADPs for each 1:1 presence of ATP:ADP. Mitochondrial oxidative phosphorylations and substrate level phosphorylations of the glycolytic sequence eventually produce ATP from ADP. Experiments have revealed that the corresponding guanosine triphosphate form of ATP can also exist, but only at the expense of existing ATP

supplies. Conversions of GMP to GDP and GTP, respectively, require nucleoside monophosphate and diphosphate kinases expressed as

$$ATP + GMP \rightleftharpoons GDP + AMP$$
$$ATP + GDP \rightleftharpoons GTP + AMP$$

The overall production of AMP and GMP is under feedback control mechanisms such that 5′-phosphoribosylamine *formation* and IA *conversion* to GMP and AMP are checked by high concentration levels of these two nucleotides.

PYRIMIDINE NUCLEOTIDE ORIGINS AND GLUTAMINE

Cytidine 5′-monophosphate (CMP) and *uridine 5′-monophosphate* (UMP) represent the most ubiquitous pyrimidine nucleotides; they contain cytosine and uracil, respectively. Unlike AMP and GMP, *cytidylate* (CMP) and *uridylate* (UMP) originate from glutamine before being attached to the C-1 ribosyl position of the *de novo* nucleotide. The cytosolic action of carbamyl synthetase II is responsible for carbamyl phosphate formation from glutamine as an initial synthesis step for pyrimidines. Carbamyl phosphate then serves as a substrate for the allosteric enzyme aspartate transcarbamylase to give *N*-carbamylaspartate. This enzyme is subject to the *negative* allosteric modulator effects of high CTP levels in cells, whereas ATP exerts a more *positive* modulator effect (Figure 18.37).

Ring closure of *N*-carbamylaspartate through dehydration and NAD$^+$-dependent oxidation produces orotate. The ribosylphosphate of the growing nucleotide is introduced as *5′-phosphoribosyl 1-pyrophosphate* (PRPP). Orotate, PRPP, and a specific transferase then react to give *orotidine 5′-monophosphate* (orotidylate). Orotidylate decarboxylation followed by UTP-mediated phosphorylation and amination at the expense of glutamine supply CTP as the final pyrimidine nucleotide triphosphate.

It should be recognized that carbamyl phosphate formation, in the reaction outlined above, is also a prerequisite for urea formation in the urea cycle. The enzyme responsible for urea synthesis is a mitochondrial enzyme called carbamyl phosphate synthetase I and *not* the cytosolic II *form* involved in this discussion.

DEOXYRIBONUCLEOTIDES ORIGINATE FROM RIBONUCLEOTIDES

Deoxyribonucleotides arise from the reduction of the C-2′ carbon atom of ribose to give its corresponding C-2′ *deoxy* derivative. In this way ADP can supply 2′-deoxyadenosine diphosphate (dADP) and GDP yields 2′-deoxyguanosine diphosphate (dGDP). Formation of these deoxynucleotide diphosphates (dNDPs) from ribonucleotide diphosphates (NDPs) requires both the reducing power from NADPH(H$^+$) and a thioredoxin protein. This protein conveys hydrogen atoms from NADPH(H$^+$) to NDP by the reversible reductive and oxidative actions of paired intramolecular thiols as illustrated below:

1. NADPH(H$^+$) + thioredoxin $\longrightarrow$
$$\overset{|\quad|}{S-S}\text{ (ox)}$$
$$\text{NADP}^+ + \text{thioredoxin}$$
$$\overset{|\quad|}{HS\quad SH}\text{ (red)}$$

2. Thioredoxin + NDP $\longrightarrow$
$$\overset{|\quad|}{HS\quad SH}\text{ (red)}$$
$$\text{thioredoxin} + \text{dNTP} + H_2O$$
$$\overset{|\quad|}{S-S}\text{ (ox)}$$

Unlike RNA, which contains uridylate nucleotide residues (UMP), *DNA requires deoxythymidylate* (dTMP) *residues*. The dTMP structures are formed by a series of conversions where dUDP hydrolysis gives dUMP. This dUMP is then methylated by N^5,N^{10}-methylenetetrahydrofolate and thymidylate synthetase to form a deoxythymidylate (dTMP) structure that is otherwise similar to dUMP. Ensuing dTMP phosphorylation generates dTDP. This reaction can be summarized as

1. dUDP + H$_2$O $\longrightarrow$ dUMP + P_i
2. dUMP + N^5,N^{10}-methylenetetrahydrofolate
$$\downarrow$$
dTMP + dihydrofolate
3. dTMP + ATP $\longrightarrow$ dTDP + ADP

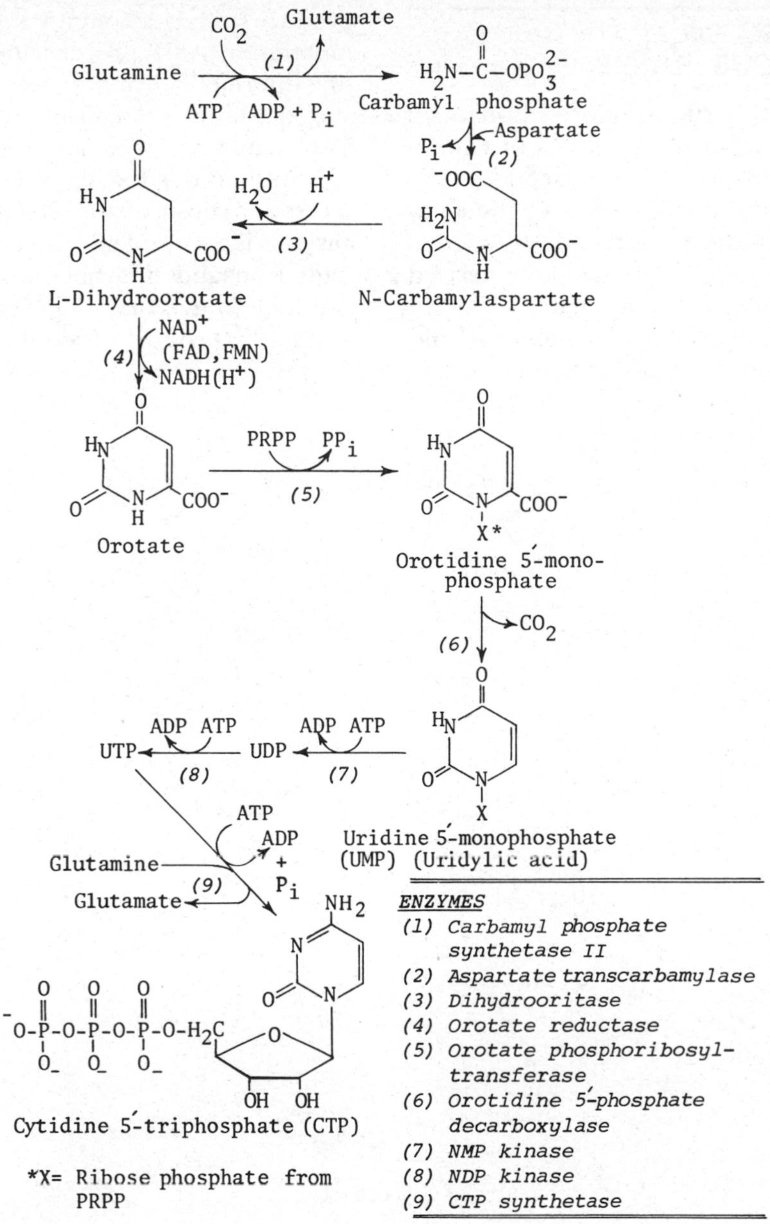

FIGURE 18.37. Synthesis of pyrimidines including UMP and CTP.

Dihydrofolate is converted back to the tetrahydrofolate structure by the action of tetrahydrofolate reductase and NADPH(II$^+$). The tetrahydrofolate structure is then reconstituted for another round of dUMP to dTMP conversions. The complete array of dNTPs required for DNA biosynthesis are produced according to the generalized reaction

$$ATP + dNDP \longrightarrow ATP + dNTP*$$

*dNTP = dADP, dCDP, dTDP, or dGDP.

CATABOLISM OF NITROGENOUS BASES

Polynucleotide structures are dismantled through the actions of endo- and exonucleases that produce free nucleotides. Depending on various circumstances, these nucleotides can undergo further destruction to give free nitrogenous bases, ribosyl-derived sugars, and inorganic phosphate.

Purine metabolism is well established and involves a variety of enzymes that ultimately produce uric acid from AMP and GMP. Degradation of AMP proceeds through an inosine intermediate and then to hypoxanthine and xanthine. Catabolism of GMP does not involve inosine formation prior to xanthine.

Xanthine derived from GMP and AMP serves as a substrate for *xanthine oxidase*. This enzyme is a complex protein that incorporates iron and molybdenum and converts xanthine to uric acid (Figure 18.38).

As a by-product of xanthine oxidase activity *superoxide radicals* (O_2^-) are produced

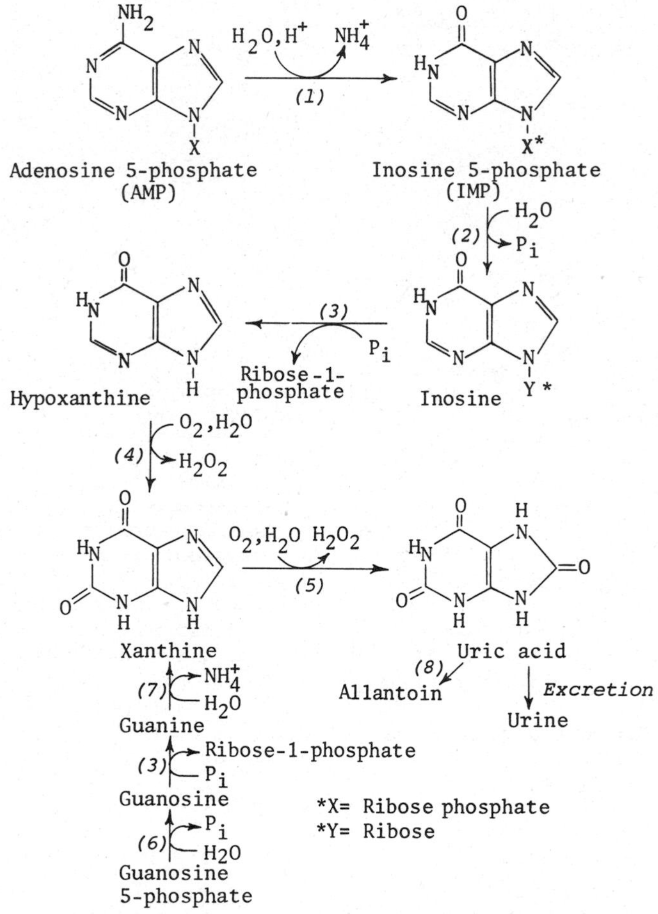

ENZYMES

(1) *Adenylate deaminase*
(2) *Nucleotidase*
(3) *Purine nucleoside phosphorylase*
(4) *Xanthine oxidase*
(5) *Xanthine oxidase*
(6) *Nucleotidase*
(7) *Guanase*
(8) *Urate oxidase*

FIGURE 18.38. Catabolism of purines and the production of uric acid.

and promptly converted to hydrogen peroxide (H_2O_2) by superoxide dismutase. *Catalase* in turn decomposes H_2O_2 to H_2O and $\frac{1}{2} O_2$. The normal actions of superoxide dismutase and catalase have critical importance in the disposition of O_2^- and H_2O_2. Both compounds are very reactive and demonstrate destructive effects on normal cells and their biochemical processes. In fact, the aberrant roles of superoxide dismutase have been widely suspected for inciting some types of arthritic conditions.

Uric acid is excreted as a key urinary end product of purine metabolism in primates; however, *urate oxidase* converts urate into *allantoin* in many other vertebrates.

Although a total of 0.5 to 0.7 g of uric acid is excreted over a 24-h period, many purine bases are recycled back into AMP, GMP, or IMP. Nucleotide formation paths involving *reclaimed purines* and PRPP are described as *salvage pathways*. For example, adenine or guanine can react with PRPP to yield AMP or GMP.

$$\text{Adenine} + \text{PRPP} \longrightarrow \text{AMP} + PP_i$$
$$\text{Guanine} + \text{PRPP} \longrightarrow \text{GMP} + PP_i$$

A correspondingly simple pathway involves the production of *inosinic acid* (IMP) from hypoxanthine:

$$\text{Hypoxanthine} + \text{PRPP} \longrightarrow \text{IMP} + PP_i$$

Phosphoribosyl transferases are generally responsible for these reactions.

Pyrimidine catabolism can be modeled by uracil, which is a substrate for *dihydrouracil dehydrogenase*. This $NADP^+$-dependent enzyme produces dihydrouracil from uracil. The dihydrouracil ring is broken by hydrase activity to give β-ureidopropionate. Ureidopropionase-mediated conversion of the last compound supplies β-alanine, NH_4^+, and CO_2. The β-alanine can then serve as a component of carnosine, anserine, or the pantothenic acid moiety of coenzyme A (Figure 18.39).

DISORDERS CAUSED BY NUTRITIONAL PROTEIN DEFICIENCIES AND ABERRANT PURINE METABOLISM

The quality and quantity of dietary proteins affects the amino acid balance necessary for maintaining protein turnover throughout the

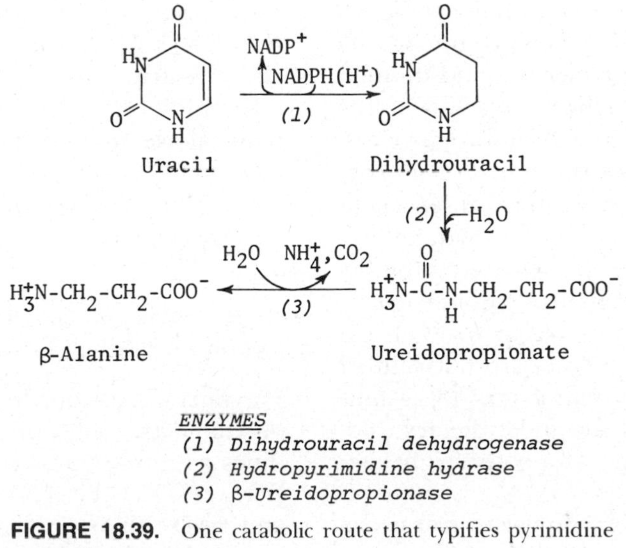

ENZYMES
(1) Dihydrouracil dehydrogenase
(2) Hydropyrimidine hydrase
(3) β-Ureidopropionase

FIGURE 18.39. One catabolic route that typifies pyrimidine metabolism.

body. Furthermore, the dietary supply of proteins, their constituent amino acids, as well as nucleotide status in the body have a direct impact on nitrogen metabolism and excretion. Nutritional disorders produced by gross protein and calorie deficiencies can be evidenced as marasmus or kwashiorkor. Both nutritional problems provide classic studies of *deficient* and *protracted* imbalances in protein and amino acid metabolism necessary for normal health. Other disorders in purine metabolism are evidenced as gout, xanthinuria, and other rare genetic factors. Although purine disorders are *not caused* by dietary purine intake, this factor can exacerbate the consequences of these disorders.

Marasmus and Kwashiorkor

Nutritional marasmus is common throughout most developing countries especially among children 1 year old or younger. Marasmus represents a form of starvation, and apart from a diversity of physiological causes, the etiology of marasmus is linked to food deficiencies. Although marasmus can exist secondary to celiac disease, cystic fibrosis, and challenging infections, gross nutritional deficiencies are the major cause of most marasmus cases.

Premature cessation of breast feeding often serves as a prelude to marasmic conditions in developing countries. This practice is often followed by the improper use and dilution of baby formulas, which only exacerbates the nutritional inadequacies of improper diets. Clinical indications of this nutritional problem include muscular wasting, an excellent appetite, no edema, anemia, growth failure, and diarrhea. Serum proteins are typically reduced; the ratio of hydroxyproline to creatinine is low compared to a normal well-fed subject, and hepatic tissues are not riddled with fat. Successful treatment of marasmus relies on a balanced supply of dietary calories and protein to allow for both weight gain and growth.

Kwashiorkor is very similar to marasmus but far more severe in its clinical indication and impending consequences. This problem occurs in poverty-stricken regions of the world where *gross* dietary protein deficiencies exist. Kwashiorkor-afflicted children 1 to 3 years old reflect the protracted effects of a predominant carbohydrate diet and little protein. The disease is often initiated during the weaning period when breast milk is eliminated and a starchy diet is supplied.

Mental changes, hepatomegaly, hair changes, dermatosis, subcutaneous fat, and *poor* appetite are common in kwashiorkor, and anemia is almost ubiquitous. Serum albumins and globulins are both depressed, but the albumin fraction is more severely depressed than the globulins. The essential-to-nonessential amino acid ratio is far more suppressed than similar ratios shown in marasmus; serum amylase and transferrin levels are lowered in proportion to the severity of the condition; fatty infiltration of the liver exists; and contrary to the appearance of edema, urinary albumin is absent. Anorexic conditions exemplified in kwashiorkor and impending development of irreversible mental problems make hospitalization a requirement for successful treatment. Once diarrhea and potassium losses have been checked, daily nutritional therapy should reflect ~120 cal and ~7 g protein/kg body wt. Dried skim milk, vegetable oil, and casein mixtures have been widely used to treat kwashiorkor, but other nutritional regimens are also effective.

In addition to developing countries, protein–calorie malnutrition may not be uncommon among hospital patients in the United States as well as in England (Bistrian and Blackburn, 1983).

Gout Conditions and Purine Excretion Disorders

Previous discussions in this text have demonstrated the roles of purines and pyrimidines as nitrogenous constituents of DNA, RNA, ATP, GTP, cAMP, NAD^+, $NADP^+$, and FAD. Inborn metabolic errors dictated by genetic disorders, regulatory disorders, various organic disorders of the body, and cancers all have impact on the *normal* status of purine metabolism. Unfortunately, the details surrounding upsets in purine metabolism are not always clear, with the main exceptions being gout and hyperuricemia.

Excessive accumulation of uric acid in the body can culminate in hyperuricemia, acute arthritis, chronic arthritis, tissue deposits of sodium urate called *tophi,* and/or neuropathy. These five conditions are generically categorized as *gout.* Although gout-strickened patients always display hyperuricemia, all hyperuricemic patients do not have gout.

When serum urate values exceed 6.0 mg/dL, urate crystallization may be initiated about some nidus within synovial fluids. In very severe cases, this action is exhibited as an acute inflammatory arthritis or uric acid lithiasis where renal uric acid calculi develop. Progressive growth of urate crystalline deposits in joints eventually produces bulbous tophi *in* and *around* joints, which can cause crippling effects that typify gout. These gout conditions demonstrate a 20:1 prevalence in men as opposed to women, but gout incidence accelerates in women after menopause. Although both sexes demonstrate a 50 to 80% daily turnover rate in the normal miscible pool of uric acid, men display a wider concentration range in the urate pool (85–1600 mg) than women (500–700 mg).

The appearance of uric acid as a culprit in gout is enigmatic since the biochemical reasons for its localized crystallizations are not certain. Therefore, symptomatic treatment of gout has relied on a variety of pharmacological agents including anti-inflammatory actions of *indomethacin* and *phenylbutazone.* Indomethacin seems to inhibit prostaglandin biosynthesis and phenylbutazone is uricosuric. Other uricosuric agents such as probenecid (Benemid) and sulfinpyrazone (Anturane) inhibit renal tubular absorption of urate and effectively depress the urate pool size of the body and plasma levels.

Allopurinol (AP) treatment is also widely used as a powerful xanthine oxidase inhib-itor. The inhibitory action of AP reflects its isomeric (7-C, 8-N) reminiscence of hypoxanthine. Aside from serving as a substrate for xanthine oxidase, AP is readily converted by the enzyme to the xanthine isomer *oxypurinol* (OP). This compound (OP) also exhibits notable enzyme inhibition effects, and its inhibitory half-life on xanthine oxidase supersedes that of AP. As a consequence of xanthine oxidase inhibition, hypoxanthine and xanthine pools accumulate in the body. The solubility of these compounds, however, is greater than that of urate and both substances are excreted in the urine.

Nearly all drug treatments for gout including the historical use of colchicine are linked to side effects that may counter their safe use. In addition to drug treatments for gout, dietary management of purine intake can be useful in controlling this biochemical problem. Foods suitable for dietary elimination include liver, heart, kidney, brain, sweetbreads, broths, bouillon, meat extracts, gravies, yeast, goose, partridge, fish roes, mackerel, scallops, sardines, herring, and muscular tissues. Other guidelines for gout management between attacks should involve

1. Low purine intake.

2. Only moderate protein intake since high levels enhance purine synthesis.

3. Low dietary fat consumption since high levels may prevent uric acid excretion.

4. High carbohydrate intake since carbohydrates may increase uric acid excretion.

Low-nucleoprotein-containing foods suitable for therapeutic diets include eggs, milk, and certain cheeses. It is also essential for gout patients to minimize alcohol beverage consumption, but fluid intake to produce a urinary output of ~2.0 L is desirable for minimizing renal uric acid precipitation.

Xanthinuria

A hereditary disorder of *xanthine oxidase* leads to the production of hypoxanthine and xanthine as the chief products of purine metabolism. This genetic disorder is apparently transmitted as an autosomal recessive trait with a prevalence of 1:40,000. Since deficiencies of xanthine oxidase fail to convert xanthine into uric acid, xanthine is excreted in the urine.

Although xanthine solubility exceeds the aqueous solubility of uric acid, xanthine solubility is markedly affected by pH conditions. At pH 5.0, xanthine solubility is low enough to permit urinary crystal or stone formation. In some cases, xanthine crystals may become interspersed in muscle tissues, where they produce muscle cramps upon exercise along with eventual myopathy.

Xanthinuria may be asymptomatic but suggested by *hypouricemia* of 1 mg/dL. Although asymptomatic subjects may not require any treatment, subjects who demonstrate xanthine stone formation can benefit from consumption of large volumes of water coupled with dietary purine restriction.

Xanthinuria can also reflect a combined deficiency of xanthine oxidase and sulfite oxidase owing to inadequate amounts of molybdenum. This metal serves as a coenzyme for both enzymes. In this type of xanthinuria, neurological abnormalities and mental retardation are not uncommon in those who are afflicted.

Lesch–Nyhan Syndrome

The Lesch–Nyhan syndrome is an X-lined genetic disorder present in the homozygous male and is characterized by hyperuricemia, excessive production of uric acid, renal stones, and neurological problems including spasticity, mental retardation, and self-mutilation.

A deficiency of hypoxanthine-guanine phosphoribosyltransferase, which catalyzes reactions involved in the conversion of hypoxanthine and guanine into their corresponding nucleotides, has been linked to the disorder. This enzyme specifically catalyzes the transfer of ribose 5-phosphate moiety on PRPP to hypoxanthine and guanine. The failure of these purines to be salvaged by the brain, which relies on a salvage pathway for synthesis of IMP and GMP, results in purine oxidation to form uric acid. Furthermore, the amount of PRPP increases considerably and seems to contribute to the accelerated rate of *de novo* purine biosynthesis observed in these patients. Treatment with allopurinol decreases uric acid production and alleviates stone formation to some extent, but pharmacological treatment does not avert neurological, mental retardation, and self-mutilation features of the syndrome.

Pyrimidine Metabolic Disorders

Pyrimidine nucleotides are no less important than purine nucleotides in the mechanism of molecular genetics. Aside from the rare occurrence of hereditary orotic aciduria, β-aminoisobutyriaciduria and pyrimidine 5'-nucleotidase deficiencies, most pyrimidine biochemical disorders have minor importance.

REFERENCES

General Amino Acid References

Anderson, G. H., N. T. Glanville, and E. T. Li. 1983. Amino acids and the regulation of quantitative and qualitative aspects of food intake. In *Amino Acids Metabolism and Medical Applications* (G. L. Blackburn, J. P. Grant, and V. R. Young, eds.), pp. 225–238. John Wright · PSG, Boston.

Bissell, M. D. 1982. Porphyria. In *Cecil Textbook of Medicine* (J. B. Wyngaarden and L. H. Smith, Jr., eds.), 16th ed., pp. 1121–1126. Saunders, Philadelphia.

Bistrian, B. R., and G. L. Blackburn. 1983. Assessment of protein-calorie malnutrition in the hospitalized patient. In *Nutritional Support and Medical Practice* (H. A. Schneider, C. E. Anderson, and D. B. Coursin, eds.), 2nd ed., pp. 128–139. Harper & Row, Philadelphia.

Blackburn, G. L., and L. L. Moldawer. 1983. An

evaluation of techniques for estimating amino acid requirements in hospitalized patients. In *Amino Acid Metabolism and Medical Applications* (G. L. Blackburn, J. P. Grant, and V. R. Young, eds.), pp. 265–290. John Wright · PSG, Boston.

Devlin, T. M. 1982. *Textbook of Biochemistry with Clinical Correlations.* Wiley, New York.

Epstein, C. M., R. K. Chawla, A. Wadsworth, and D. Rudman. 1980. Decarboxylation of α-ketoisovaleric acid after oral administration in man. *Amer. J. Clin. Nutr.* **33**:1968.

Goldberg, A. 1983. Factors affecting protein balance in skeletal muscle in normal and pathological states. In *Amino Acids Metabolism and Medical Applications* (G. L. Blackburn, J. P. Grant, and V. R. Young, eds.), pp. 201–211. John Wright · PSG, Boston.

Kinney, J. M. 1983. Amino acid support in the hypercatabolic patient. In *Amino Acids Metabolism and Medical Applications* (G. L. Blackburn, J. P. Grant, and V. R. Young, eds.), pp. 377–386. John Wright · PSG, Boston.

Long, C. L. 1983. Nutritional consideration of amino acid profiles in clinical therapy. In *Amino Acids Metabolism and Medical Applications* (G. L. Blackburn, J. P. Grant, and V. R. Young, eds.), pp. 291–307. John Wright · PSG, Boston.

Metzler, D. E. 1977. *Biochemistry. The Chemical Reactions of Living Cells.* Academic Press, New York. (Consult as source for essential amino acid biosynthesis in other organisms.)

Mitch, W. E. 1983. Amino acid analogues: metabolism and use in patients with chronic renal failure. In *Amino Acids Metabolism and Medical Applications* (G. L. Blackburn, J. P. Grant, and V. R. Young, eds.), pp. 439–450. John Wright · PSG, Boston.

Rothschild, M. A., M. Oratz, and S. Chreiber. 1972. Albumin synthesis. *N. Engl. J. Med.* **286**:748–757, 816.

Shepartz, B. 1973. *Regulation of Amino Acid Metabolism in Mammals.* Saunders, Philadelphia.

Waterlow, J. C., and J. M. Stephen. 1967. The measurement of total lysine turnover in the rat by intravenous infusion of L-(U-^{14}C) lysine. *Clin. Sci.* **33**:389.

Weiss, J. S., A. Gautam, J. J. Lauf, M. W. Sundberg, P. Jatlow, J. L. Boyer, and D. Seligson. 1983. The clinical importance of a protein-bound fraction of serum bilirubin in patients with hyperbilirubinemia. *N. Engl. J. Med.* **309**:147.

Wurtman, R. R. 1982. Nutrients that modify brain function. *Sci. Amer.* **246**(4):50.

Wurtman, R. R. 1983. Implications of parenteral and enteral amino acid mixtures in brain function. In *Amino Acids Metabolism and Medical Applications* (G. L. Blackburn, J. P. Grant, and V. R. Young, eds.), pp. 219–224. John Wright · PSG, Boston.

Wyngaarden, J. B. 1982. Disorders of purine and pyrimidine metabolism. In *Cecil Textbook of Medicine* (J. B. Wyngaarden and L. H. Smith, Jr., eds.), 16th ed., pp. 1107–1120. Saunders, Philadelphia.

Wyngaarden, J. B., and L. H. Smith, Jr. (eds.). 1982. *Cecil Textbook of Medicine,* 16th ed. Saunders, Philadelphia.

Biochemistry of Tyrosine Derivatives as Catecholamines

Axelrod, J. 1965. The metabolism, storage, and release of catecholamines. VI Neurohumors. *Recent Prog. Hormone Res.* **21**:597. (Useful overview of catecholamine biochemistry pre-1965.)

Christensen, H. D., and C. L. Blank. 1979. The determination of neurochemicals in tissue samples at sub-picomole levels. In *Biological/ Biomedical Applications of Liquid Chromatography* (G. L. Hawk, ed.), pp. 133–164. Dekker, New York.

Cole, J. T., and S. H. Snyder. 1981. Catecholamines. In *Basic Neurochemistry* (G. J. Siegel, R. W. Albers, B. W. Agranoff, and R. Katzman, eds.), pp. 205–217. Little, Brown, Boston. (Excellent reference for all aspects of neurotransmitting amines and amino acid derivatives.)

Kissinger, P. T., C. S. Bruntlett, and R. E. Shoup. 1981. Neurochemical applications of liquid chromatography with electrochemical detection—Minireview. *Life Sci.* **28**:455.

Robbins, J. W., and B. J. Everitt. 1982. Functional studies of the central catecholamines. In *International Review of Neurobiology,* Vol. 23, pp. 303–360. Academic Press, New York.

Ustin, E., I. J. Kopin, and J. Barchas. 1978. Catecholamines: Basic and clinical frontiers. In *Proceedings of the 4th International Catecholamine Symposium,* Vols. I and II. Pergamon Press, New York. (Excellent comprehensive reference.)

Macro- and Trace Elements in Nutrition

INTRODUCTION

The crude inorganic content of foods can be estimated from the weight of the *nonvolatile residue* produced after their analytical incineration at 500 to 600°C. Such intense heat oxidizes organic constituents to oxides of nitrogen and carbon dioxide. Sulfur and phosphorus are also converted to their oxides and volatilized unless alkali and alkaline earth elements are present. Nonvolatile constituents remain in the *ash residue* as oxides, sulfates, phosphates, silicates, and chlorides.

Inorganic food constituents can be classified as major (macro-) or minor (trace) elements depending on their nutritional requirements by plants and animals. For purposes of this discussion, calcium, magnesium, phosphorus, sodium, and potassium are considered *macro*elements. *Trace* elements, on the other hand, include iron, zinc, copper, manganese, molybdenum, chromium, selenium, vanadium, plus some others having uncertain biochemical functions.

The dietary acquisition of these inorganic nutritional factors stems from the combined intake of waterborne inorganic solutes plus those inorganics present in animal and plant foods. Apart from inorganic solutes in drinking water, plant foods offer one of the most direct dietary routes in obtaining both macro- and trace elements for proper nutrition.

PLANTS ASSIMILATE MINERAL ELEMENTS

Water, carbon dioxide, light, and key elements that constitute geophysical minerals are necessary for the successful growth of all food crop species. From perspectives of plant nutriture, a high crop yield per acre may demand 10^{3+} pounds of carbon, hydrogen, and oxygen; 10^1 to 10^2 pounds of nitrogen, potassium, calcium, phosphorus, magnesium, and sulfur; plus a diverse variety of additional mineral elements measured in pounds, ounces, or grams per acre.

As a prerequisite to the assimilation of mineral elements as well as nitrogen, these must exist in *phytoacceptable* forms as shown in Table 19.1. The crucial growth elements exist as cationic and anionic species that are often dissolved in the capillary component of soil water; however, the cationic species

Table 19.1. Phytoacceptable Forms of Plant Nutrients

Element	Phytoacceptable form of nutrient	General nutrient function
Calcium	Ca^{2+}	Cell wall structures and cell membrane activity
Magnesium	Mg^{2+}	Component of chlorophyll structure
Nitrogen	NO_3^-	Amino acids, pyrimidines, and purines
Phosphorus	$H_2PO_4^-$	ATP formation, phospholipids, and nucleic acids
Potassium	K^+	Cell membrane activity, osmotic pressure, and enzyme activation
Sulfur	SO_4^{2-}	Protein and amino acid structures
Boron	BO_2^{3-}, $B_4O_7^{2-}$	Translocation of carbohydrates, and cell elongation
Chloride	Cl^-	May be involved in chemiosmotic mechanism of photosynthesis
Copper	Cu^{2+}	
Iron	Fe^{2+}	Enzyme
Manganese	Mn^{2+}	activators
Molybdenum	Mo^{3+}	and (Nitrogen fixation, Mo^{3+})
Zinc	Zn^{2+}	cofactors (Auxin activity, Zn^{2+})

that have special significance for plant nutrition are bound to soil components. Binding interactions largely depend on the presence of *electrostatic interactions* between mineral ions and the countercharged structure of clay particles or organic matter (Figure 19.1).

Mineral–soil particle interactions are greatly dependent on soil pH conditions. Based on principles of ion exchange, a soil pH range of 5.0 to 7.0 ensures binding interactions between negatively charged clay particles and key cationic mineral species, whereas lower pH values mobilize these cations.

Low pH conditions can have disastrous effects on the retention of minerals in soil. Low pH's foster the expeditious leaching of minerals through topsoils. Aside from this possibility, low pH values may permit the temporary overabundance of some plant nutrients at any given time.

At low soil pH conditions, the sites on clay particles formerly responsible for binding mineral cations are replaced by hydrogen ions. This action is abetted by (1) the release of carbon dioxide into soil from plant respiration, which goes on to produce carbonic acid; (2) microbial decomposition of organic debris, which produces organic acids; and (3) acidic environmental stresses posed by acid rains.

Mineral nutrients are translocated from soil into plant tissues by mechanisms that remain to be firmly established. It is conceded, however, that water and mineral entries into root cells are largely independent processes governed by active transport mechanisms.

Postabsorptive mineral metabolism by plants depends on the functions of mineral elements in the plant at any given time. Plants use mineral elements in at least four ways, including (1) as direct or indirect *structural components* of plant biomass; (2) as *elements of biochemical structure,* for example, calcium is required for cell wall formation, iron is necessary for chlorophyll biosynthesis, magnesium is required for chlorophyll structure, orthophosphate contributes to the formation of high-energy phosphates, and so on; (3) as *enzyme activators or cofactors,* for instance, manganese and magnesium; and (4) for *maintenance of ionic strength and osmotic balance* in plant saps and fluids, for example, potassium.

Most food crop plants are nondiscrimi-

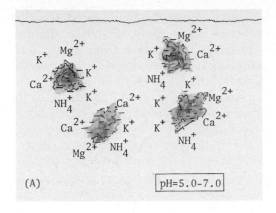

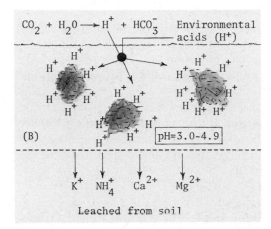

Leached from soil

FIGURE 19.1. Critical cations for plant growth and nutrition participate in electrostatic interactions with anionic soil constituents and soil particles. (A) So long as a pH of 5.0–7.0 is held, the cations *will* remain bound to soil particles and available for plant growth. (B) If pH values of the soil fall to less than 4.9, cationic interactions with anionic soil particles are minimized and eventually lost from surface soils. The cations subsequently leach into subsoils where they have limited availability for any or all plants depending on root depths. High proton concentrations responsible for displacing cations originate with environmental sources of acid (e.g., acid rains, saprophytic degradation products) and acid by-products of plant metabolism that are discharged about the root system.

native in their assimilation of available mineral nutrients. Thus, the inorganic element content of many plants will reflect the mineral element profile of the soils where they are grown. Gold may be present in plants growing on gold-bearing soil whereas selenaceous plants have a demonstrated ability to

accumulate selenium-bearing minerals. In both instances, the mineral element accumulation proceeds with plant growth, although neither gold nor selenium has recognized phytobiochemical actions.

MACROELEMENTS OF INORGANIC FOOD CONSTITUENTS

The macroelements of food structure derived from geophysical minerals are present largely as mono- and divalent cations and/or anions. Emphasis in the following discussion is directed toward calcium (Ca^{2+}), magnesium (Mg^{2+}), potassium (K^+), and sodium (Na^+) functions as well as their roles in nutritional processes. Although phosphates (PO_4^{3-}), sulfate (SO_4^{2-}), and bicarbonate (HCO_3^-) are not specifically detailed here, their importance and roles are outlined elsewhere throughout the text.

CALCIUM

The dietary requirement for calcium has been recognized since the cardiophysiology studies conducted by Ringer in the last century. These studies showed that sustained contraction of perfused amphibian hearts required ~1 μM of calcium ions. Apart from this early work, Ringer recognized little about the role of calcium in the myocardial activities of humans, muscle contractility, intracellular cementing matrices, blood coagulation, and other physiological processes.

In spite of many biochemical studies dealing with calcium metabolism, the empirical nutritional status of calcium still supersedes its recognized performance in the bioinorganic chemical mechanisms of nutrition.

On the basis of a 70-kg adult, the total body calcium content is ~1200 g. About 99% of the calcium resides in skeletal tissues including teeth, and the remaining 1% occurs in the extracellular fluids and soft tissues. Blood sera exhibit calcium concentrations of ~2.5 μM of which ~1 μM is chelated to proteins, carbohydrates, and other organic solutes, whereas the rest is free. The total calcium content of cells is somewhat less than serum levels. Cells actively exclude intracellular migration of calcium and maintain a total of ~3 μM calcium of which 33% is free.

The calcium structure of bones provides inorganic substance to skeletal structures and offers a reservoir for calcium in cases of its inadequate nutritional availability. Calcium salts existing in skeletal tissues undergo dynamic processes of skeletal residence followed by mobilization into physiological fluids. This type of constant exchange equilibrium occurs throughout life and usually involves carbonate and phosphate anions as well. Calcium normally exists in bone as a multiple apatite salt comprised principally of calcium carbonate and calcium phosphate held in a crystalline lattice network. The dentine enamel structure of the teeth also experiences calcium exchange reactions, but the turnover rate is lower than that of skeletal structures.

Based on U.S. Department of Agriculture estimates, 60% of dietary calcium is related to the intake of dairy products including milks and cheeses, which are among the richest sources. Of course, processed meats containing pulverized bone bulking agents can supply large calcium supplies to some frequent consumers of these products.

Since many mineral salts found in milk occur at concentrations superseding their solubility, calcium ions as well as magnesium, phosphate, and citrate often coexist with colloidal milk components. During digestive processes, and owing to rennin actions, many of these colloidal particles are readily precipitated upon milk coagulation.

Calcium Absorption and Excretion

Calcium absorption is mainly a duodenal event directed by active absorption and con-trolled by vitamin D. As indicated in earlier chapters, 1,25-dihydroxy-cholecalciferol (1,25-DHCC) formation from dietary vitamin precursors is responsive to parathyroid hormone (PTH) levels. The PTH levels at any time are concurrently determined by the existing calcium requirements of the body.

Activated vitamin D in the guise of 1,25-DHCC undergoes enterocytic localization at the nucleus. This in turn instigates the transcription of a *calcium-binding protein, alkaline phosphatase,* and *calcium-activated ATPase,* which jointly promote calcium absorption and transport.

Many types of calcium-binding proteins have been isolated from intestinal mucosa in animals and humans. These proteins have a commonly high content of aspartic and glutamic acids, which are probably responsible for their calcium-binding activities. Divalent calcium cation binding to these proteins is not always specific; instead, cation protein-binding affinities follow the order of $Ca^{2+} > Cd^{2+} > Sr^{2+} > Mn^{2+} > Fe^{2+} > Ba^{2+} > Mg^{2+} > Co^{2+}$.

Nearly 75% of available dietary calcium undergoes fecal excretion, but the absorptive efficiency for calcium is also determined in part by the nutritional requirement for calcium in the body, which may govern its absorption from the lumen. For example, pH conditions, high protein intake, and large concentrations of anions having low solubility (i.e., fatty acids) can all impede calcium absorption. Anions are presumed to bind luminal calcium and diminish its availability for absorption, but dietary protein effects on calcium absorption are less certain.

Rapid fluctuations (high or low) in the availability of dietary calcium lead to adaptive readjustments in the absorptive efficiency for the cation. Unfortunately, these adaptations can be protracted and incite temporary disturbances in normal calcium equilibria within the body.

Effective calcium absorption is also impeded by plant oxalates, which form insoluble and undigestible *calcium oxalate derivatives.* Plants including rhubarb and spinach have been implicated in this action, but nutritional effects among the population are probably insignificant.

The hexaphosphate of inositol, known as *phytic acid,* occurs widely in the bran layer of cereal grains. This agent readily precipitates calcium as an insoluble phytate in the intestine and effectively suppresses calcium absorption. Unlike oxalates, the habitual consumption of high phytate (Figure 19.2) foods such as brans, whole meal breads, and the like can promote calcium deficiencies especially in children.

The dietary availability of calcium must also be considered with regard to phosphorus availability. The so-called *calcium*: *phosphorus* (Ca:P) *ratio* clearly influences the dynamic equilibrium state of bone minerals and influences dietary calcium absorption. For humans, evidence shows that the Ca:P ratio may range from 2:1 to 1:2, but a Ca:P ratio of 1.0 has been generally recommended. For infants, however, the desirable Ca:P ratio is probably about 2.0, as demonstrated in human breast milk.

Urinary excretion of calcium and phosphorus can occur via the renal route with calcium elimination being regulated by PTH. Urinary excretion of calcium can be prompted by increased levels of dietary phosphorus, accompanied by a lowering of serum calcium. Thus, excessive phosphorus intake may accelerate bone resorption and fecal calcium losses spurred on by phosphorus-induced secondary hyperparathyroidism.

Glomerular filtration of calcium in adults averages ~10 g/24 h; however, 99% is resorbed over the renal tubules. This action results in a maximum per diem calcium excretion of only 0.1 g. The level of renal calcium excretion is markedly reduced to nearly undetectable amounts when serum calcium drops below 8 mg/dL.

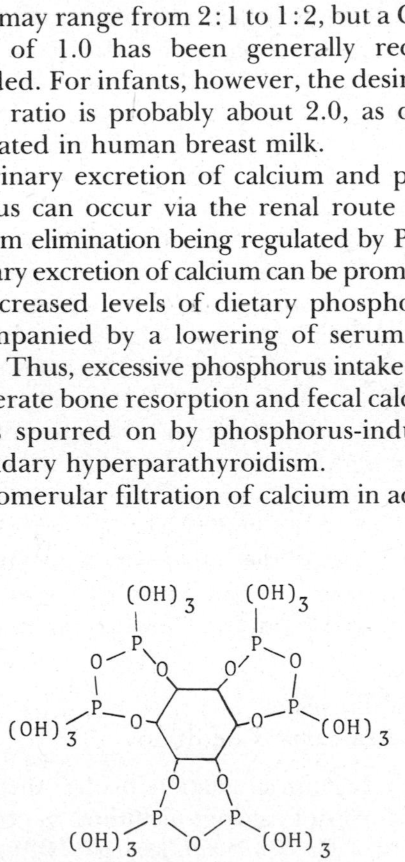

Biochemical Roles for Calcium

The bioinorganic chemistry for calcium is very complex and sometimes subtle. Nonetheless, the outstanding roles for calcium include its contribution to the (1) activation of metabolic processes, (2) neuronal impulse initiation in postsynaptic neurons, and (3) bone structure.

Cellular-specific processes can be activated by changes in membrane permeability to calcium and a concomitant influx of calcium into the cytosol of cells. For example, in the case of muscle contraction, 0.1 μM cellular concentrations increase to a 10 μM concentration when stored calcium is released from the endoplasmic reticulum. The liberated calcium binds to the troponin (C) and initiates a contraction event in the muscle, as detailed in earlier chapters.

The sweeping action of electrical depolarization over a neuron clearly depends on well-established principles of sodium and potassium ion fluxes over electrically excitable membranes. Perpetuation of this depolarization from neuron to neuron, or from neuron to an effector site, requires the presence of synaptic transmitters. These agents are typically recognized as acetylcholine, norepinephrine (noradrenaline), epinephrine (adrenalin), or dopamine.

The release of neurotransmitters at neuromuscular junctions or from synaptic reservoirs cannot occur unless calcium bathes the nerve endings. Most evidence suggests that increased calcium conductance occurs in nerve endings as a result of a depolarization event. Once this occurs, calcium can diffuse down the electrochemical gradient of the neuronal ending and into the presynaptic terminals. By mechanisms that are still

FIGURE 19.2. Structure of phytic acid, which is a hexaphosphate derivative of inositol.

unclear, synaptic vesicles are then triggered to release neurotransmitter. All of these events occur over a 1- to 2-ms span since neuron response must be able to be reactuated by subsequent depolarization responses in only several milliseconds following the initial incident.

Aside from the role of calcium in neurotransmitter release, calcium is one of the two basic factors in the common biochemical mechanisms whereby cells react to external stimuli—the other being the formation and liberation of cAMP.

Calcium cations assimilated from the diet are critical for the maintenance of bone structure. The collagen matrix of bone is riddled with a crystalline calcium phosphate whose empirical composition approaches that of hydroxyapatite $(Ca_{10}(PO_4)_6(OH)_2)$. Miniscule molar substitutions of magnesium for calcium and fluoride (F^-) for OH^- can occur within this compound. The fluoride substitution in particular is suspected of enhancing bone strength.

Osteocytes reside throughout the bone matrix. Among these are *osteoblasts,* which produce a fibrous substrate that facilitates calcium phosphate deposition and mineralization of osseous tissues. The established mineral phase of bone structure is held in equilibrium with the calcium and phosphate ions in blood plasma. Under the influence of equilibrium conditions, bone cells can actuate the mobilization or deposition of mineral components (e.g., Ca^{2+} and HPO_4^{2-}) by localized pH changes or chelating processes.

Apart from the crystalline hydroxyapatite structure ($\sim$10-nm-diameter crystals), bone also contains a 30 to 40% noncrystalline or amorphous fraction of calcium phosphate. The purpose of this fraction is uncertain, but its existence may precede crystalline apatite formation according to the following scenario.

Bone mineralization processes are speculated to involve a calcium-pumping mechanism whereby mitochondrial calcium (Ca^{2+}) and phosphate (HPO_4^{2-}) in bone cells are actively acquired until a *supersaturated condition exists.* The supersaturated ions eventually precipitate as an amorphous granular calcium phosphate only to be released to the exterior of the bone cell. Calcium phosphate translocation may be controlled by mechanisms reminiscent of hormonal secretion from vesicles. Once released from the cell, amorphous calcium phosphate undergoes recrystallization to heterogeneous hydroxyapatite spurred on by nucleation sites on bone collagen. Calcium is intrinsically linked to many other biochemical reactions in humans and food-related animal species.

Many enzymes including α-amylase are notably stabilized by tightly bound calcium. Calcium amylase binding in human saliva, for example, has 1 to 2 mol of calcium per mole of enzyme. Similar calcium-stabilizing effects are imparted to pancreatic deoxyribonuclease A, which probably protects the enzyme from proteolytic hydrolysis.

In the case of blood clotting, the soluble plasma protein called *fibrinogen* undergoes conversion into an insoluble fibrous matrix known as *fibrin.* This conversion is mediated by the enzyme called *thrombin.* The existence of thrombin is contingent upon the calcium-dependent conversion of a *prothrombin precursor* to thrombin in the presence of phospholipids, thromboplastin, and several additional factors. Prothrombin synthesis itself also requires *vitamin K.* The calcium-dependent transformation of prothrombin to thrombin probably involves the direct binding of calcium to the prothrombin at a ratio of about 10:1. Binding of Ca^{2+} presumably occurs along the nine γ-carboxyglutamyl residues occurring in the first 40 N-terminal amino acid residues of prothrombin.

Some Causes of Hypercalcemic Conditions

Serum calcium circulation involves three distinct forms of calcium including *free calcium* (Ca^{2+}), a Ca^{2+}–*albumin complex* (40%), and *diffusible* Ca^{2+} complexes involving anions of HCO_3^-, citrate, PO_4^{3-}, and SO_4^{2-}. Almost 80%

of the serum calcium is bound by the glutamic and aspartic acid-rich (20%) structure of serum albumin, whereas globulin binds the remaining 20% of serum calcium.

Upsets in the apparent nutritional status of calcium, especially toward *hypercalcemic conditions,* can be actuated by parathyroid abnormalities, malignancy, drug-induced conditions, and other situations.

Primary hyperparathyroidism, familial hypocalciuric hypercalcemia (FHH)—an uncommon autosomal dominant disease—and hypercalcemic secondary hyperparathyroidism underly the obvious parathyroid disorders. All three conditions show unusually high PTH concentrations.

Hypercalcemia is a common eventuality of malignant diseases including breast and lung cancers as well as epidermoid carcinoma of the lung, neck, cervix, and esophagus. Hypercalcemia in these conditions reflects the involvement of both skeletal and humoral factors. The origin of hypercalcemia is probably the osteolytic release of calcium and phosphate into the blood. This release supersedes the clearance rate of calcium by the gastrointestinal tract and kidneys combined. This inefficacy in excretion may result in *hypercalciuria, hyperphosphaturia,* and *hypercalcemia.* Tumors notably producing hypercalcemia via osteolytic bone metastases include those of thyroid gland, bronchus, kidney, and breast.

Humoral contributions to neoplastic hypercalcemic conditions may be linked to the tumorigenic release of biochemical agents. For example, urinary phosphate clearance, urinary cAMP, and fasting calcium excretion are all markedly increased, while 1,25-DHCC and plasma PTH remain low.

Drug-induced hypercalcemias are instigated by thiazide-based diuretics, vitamin D, calcium-containing antacids plus milk mixtures, steroids, parenteral administration of calcium, and a variety of other factors.

Thiazide-induced hypercalcemia stems from the reduced glomerular filtration rate of calcium augmented by increased calcium resorption over the renal tubules. Vitamin D fosters hypercalcemia at therapeutic levels of 500 to 50,000 USP units by enhancing gastrointestinal absorption of calcium and phosphate. High calcium levels resulting from this treatment in prolonged management of hypoparathyroidism, osteoporosis, and osteomalacia can cause overt calcinosis of the renal tissues, renal insufficiency, uremia, and calcium deposition in other tissues (e.g., cardiovascular system). Vitamin A/D cotherapies also exacerbate hypercalcemic possibilities.

Parenteral hyperalimentation using calcium for combating osteoporotic conditions must remain well below the recommended daily dietary intake of 800 mg. Since this 800-mg quantity accounts for the possibility of only 5 to 15% calcium assimilation in the gut, parenteral doses must remain much lower in order to avert hypercalcemia.

Hypercalcemia can be readily induced by the joint consumption of calcium carbonate antacids plus milk. Up to 33% of the calcium can be absorbed from these mixtures to produce a "milk–alkali syndrome." This syndrome is archetypical of patients having peptic ulcer pain and high milk consumption plus an absorbable antacid such as calcium carbonate.

Glucocorticoids, androgens, and estrogens can elevate serum calcium concentrations. The action mechanism is not certain, but some researchers suggest that key steroids may stimulate the growth of hormone-responsive tumors—most notably those in bone. For glucocorticoids, hypercalcemia effects may result from the depressed metabolism of vitamin D_3 thereby producing hypervitaminosis D.

Some Causes and Consequences of Hypocalcemia

Low circulating concentrations of calcium produce a host of neuromuscular consequences. Classical hypocalcemic signs include tetany characterized by hyperstimulatory responses originating from motor neurons, muscle twitching, muscular cramps, and overall changes in the electrical excitability of peripheral nerves. Infantile incidences of hypocalcemia not only serve as an affront to normal skeletal development, but sitting, walking, and standing behaviors may be delayed by insufficient muscle tone.

Although hypocalcemia can result from inadequate dietary supplies of calcium, vitamin D deficiencies, food faddism, and related nutritional scenarios, most severe hypocalcemic conditions are rooted in various pathological states, diseases, or hormonal disorders. As examples, hypocalcemia may occur secondary to the deficiency of an inherited end-organ refractoriness to PTH, uremia, acute pancreatitis, osteoblastic metastatic bone disease, hypoalbumenic states produced by cirrhosis and nephrosis, chronic renal failure, magnesium deficiency, and intestinal malabsorption syndromes.

From nutritional perspectives only, vitamin D deficiency can produce rickets in young children and osteomalacia in older children and adults. Prior to advanced vitamin deficiency effects of hypocalcemia (4–7.5 mg/dL), serum calcium levels may remain within normal bounds, although serum phosphate is characteristically decreased and serum alkaline phosphatase activity is increased. Normal calcium levels in serum are initially retained because impaired vitamin D-dependent assimilation of intestinal calcium actuates the parathyroid-directed mobilization of skeletal calcium with a concomitant lowering of serum phosphates. The increase in parathyroid activity is reflected by their glandular enlargement and/or hypersecretory responses. When the compromising action of PTH productivity on readily available stores of calcium has been exhausted, depressed levels of serum calcium become obvious. The lowered calcium adsorption may also produce a secondary effect of decreased phosphate absorption.

Vitamin D deficiency is largely relegated to poor portions of the American population, underdeveloped countries, and black-skinned populations in whom melanin pigments block dermal penetration of ultraviolet light.

Inadequate calcium absorption can also result from defective fatty acid absorption throughout the intestine. In these conditions of steatorrhea, insoluble calcium soaps arc produced in large quantities. This not only reduces the dietary availability of calcium, but the insoluble fat matrix may suppress vitamin D absorption. Celiac disease, sprue, protracted periods of obstructive jaundice, and pancreatic diseases all foster steatorrhea. It is generally believed that these maladies indirectly impair calcium absorption since defective bile secretion occurs with a sequel of physically oppressed villi. The villi become essentially flattened and have a lowered absorptive surface area.

Maternal and neonatal hypocalcemias can also occur, but these do not invariably reflect *severe* disease or pathological upsets in body biochemistry. The increased demand for calcium, phosphate, and vitamin D during pregnancy are mirrored by the existence of parathyroid hyperplasia in many individuals. Plasma concentrations of calcium may gradually decrease throughout pregnancy to only marginally acceptable or subnormal values. This is especially true during the late stages of pregnancy and lactation. The borderline insufficiency of calcium may produce temporary, mild onsets of tetany (maternal tetany). This may stem from a combined effect of calcium and vitamin D deficiency; however, parathyroid deficiencies may be involved.

Neonatal hypocalcemia accounts for the most prevalent cause of frank convulsions during the first 30 days of life. Two incidence peaks characterize neonatal hypocalcemia—*one occurring in the first 48 h of life* and *the second one during the fourth to tenth days*. The first case is usually associated with infants of (1) low birth weight for gestational age; (2) associated hypoglycemia; (3) high blood calcitonin concentrations; and (4) possible magnesium deficiency. The second instance of hypocalcemia is the more common neonatal problem and often involves full-term infants of average weight who have been formula fed. Since cow's milk contains more phosphate than human milk, the high-phosphate milk may exceed the excretory ability of the immature kidney to process the heavy phosphate load. Low-phosphate or human milk effectively circumvents most convulsive incidents.

Maternal hypoparathyroidism can also complicate neonatal hypocalcemia etiologies. Since maternal calcium levels may be high in this condition, the *in utero* depression of parathyroid activity may be insufficient for the requisite performance of parathyroid activity in the neonate.

Since osteoporosis appears to involve some degree of misconstrued calcium metabolism, it is tempting to categorize this problem in the ranks of an "iron-deficiency anemia." Some data regarding calcium supplementation in osteoporosis support this contention, but overwhelming evidence has resulted in contrary indications. For example, intakes of 400 to 1250 mg/day for calcium have little effect one way or the other in redressing osteoporotic calcium losses. The entire picture of calcium turnover in osteoporotic conditions will also continue to challenge biomedical and nutritional sciences until its hormonal and genetic components are uncovered.

Dietary Allowances for Calcium

According to the opinion of the National Research Council, "an allowance of 800 mg/day of calcium is recommended as a guide for planning food supplies and for the interpretation of food consumption for groups of adults." This amount is probably adequate for most American diets, which are rich in animal proteins and phosphorus with low calcium:phosphorus ratios. Nonetheless, some animal studies indicate that such diets may encourage bone resorption, hypercalciuria, fecal calcium losses, and

hyperparathyroidism. The scientific record is not well established.

About 30.0 g of calcium is accumulated during pregnancy. The calcium demand is greatest during the last trimester when calcification of the fetal skeleton occurs. Maternal and fetal demands are probably met by a 1200 mg/day calcium intake during gestation, that is, 400 mg/day more than normal adult requirements. This added dietary calcium also permits a daily calcium loss of 250 mg in milk during lactation (300 mg Ca^{2+}/L milk) without causing severe demineralization of the maternal skeleton. Calcium intake for infants should approach 60 mg/kg body weight, 800 mg/day for 1 to 10-year-old children, and about 1200 mg/day for adolescents.

Although recommended adult levels of calcium intake are generally uncontested, some studies suggest that protracted calcium intakes of ≤ 670 mg/day may correlate with hypertensive states in humans as opposed to nonhypertensive conditions with intakes of at least 900 mg/day.

The Calcium Structure of Teeth and Tooth Decay

Aside from the normal protracted turnover of calcium in teeth, their demineralization by decay processes deserves special attention. The preeminent calcified portions of tooth structure include *enamel* and *dentin*. At the molecular level, these contain ratios of 98:2 and 70:30 of prismatic hydroxyapatite to protein, respectively. The protein fraction known as *eukeratin* may provide a preliminary deposition matrix for hydroxyapatite but the process of tooth biosynthesis is not clear.

The native structure of hydroxyapatite-based tooth enamel is ordinarily resistant to most physical compression stresses, enzyme actions, and other common food constituents. Contrary to its remarkable structure, tooth enamel is under a constant siege of demineralization stresses fostered by microbial acids.

Microbial acids originate from the fermentative conversion of sugars to acids over the surface of the tooth. These acids promote demineralization reactions of the tooth by the following reaction:

$$Ca_{10}(PO_4)_6(OH)_2 + \quad H^+ \cdot$$

Hydroxyapatite   **Protons from organic acids**

$$\downarrow$$

$$10Ca^{2+} + 6HPO_4^{2-} + 2H_2O$$

The mobilization of hydroxyapatite constituents is insignificant at pH 6.0 to 7.0, is marginal at pH 5.5 to 6.0, and may be severe at pH values less than 5.5.

Lactobacilli have historically been indicted for their demineralization actions because they produce copious amounts of lactic acid from the action of lactic acid dehydrogenase on pyruvate. Pyruvate, of course, is produced during the anaerobic metabolism of sugars. Current evidence, however, supports the contention that *Streptococcus mutans* strains are the more likely culprits for destroying tooth enamel and dentin, especially in advanced decay processes. The *S. mutans* strain has been isolated from 90 to 95% of serious cavities in humans, and it is probably the most virulent contributor to tooth decay.

The virulent decay activities of *S. mutans* are caused by a variety of factors. Foremost among its actions is the ability to polymerize glucose fractions of sucrose thereby producing *glucans*. The adhesive properties of glucans results in the formation of a persistent polyglucan matrix over the tooth enamel. This biosynthesis is largely mediated under the auspices of glucosyltransferase.

The polyglucan structure formed then serves as the ground substance for bacterial *plaque* on tooth surfaces as well as the gum line, where it provides a microbiological habitat for many other bacterial species (Figure 19.3).

Depending on the dietary availability of sugars and carbohydrates, resident bacteria in the plaque produce organic acids as byproducts of metabolism. Unlike plaque-free enamel surfaces, where saliva action could remove high proton concentrations if present, the glucan matrix of plaque harbors fermentation acids. This action ensures the maintenance of a static acid environment and a demineralization stress over tooth enamel.

The destructive actions of *S. mutans* on enamel are complicated further by its ability to provide constancy in the productivity of acids, although dietary sugars may not be immediately present. This action is accomplished by the bacterial conversion of intracellular amylopectin-type carbohydrates to organic acids.

Lactobacilli are still recognized for their potential abilities to incite demineralization in the pits and fissues of molars, *Actinomycetes* can attack the roots of teeth, but *S. mutans* probably causes about 90% of most cavities. Successful inhabitation of the oral cavity by *S. mutans* requires some type of solid structure to which the bacteria can adhere, but not the mucous membranes of the mouth. Since membranes cannot support colonization of these bacteria, some investigators believe that the antidecay effects of fluorida-

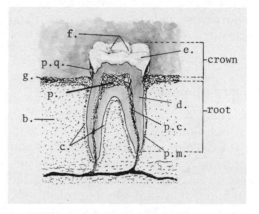

FIGURE 19.3. Major features of tooth structure for a molar including b., bone; c., cementum; d., dentin; e., enamel; f., fissure(s); g., gum; p., pulp; p.c., pulp canal; p.q., plaque; and p.m., periodontal membrane. *Streptococcus mutans* often initiates decay in the plaque region (p.q.), whereas *Lactobacilli* are not uncommonly associated with the demineralization of pits and fissues in molars.

tion may reflect the ability of fluorapatite derivatives of enamel structure to counteract plaque adhesion.

Fluorapatite can be formed from hydroxyapatite according to the reaction

$$Ca_{10}(PO_4)_6(OH)_2 + 2F^- \longrightarrow$$
$$Ca_{10}(PO_4)_6F_2 + 2OH^-$$

Other possible antidecay effects of fluoride suggest that it contributes to a more compact and stable crystal lattice than the hydroxyl ion of hydroxyapatite. This fluoridated structure exhibits distinctly lower acid solubility than hydroxyapatite and may even antagonize the growth and metabolism of decay-producing bacteria.

MAGNESIUM

Aside from potassium, magnesium (Mg^{2+}) is the principal intracellular cation. For humans, Mg^{2+} is absorbed mainly over the entire small bowel. Both Ca^{2+} and Mg^{2+} may share a common transport mechanism, but most studies have not been able to demonstrate intestinal Mg^{2+} transport against a concentration gradient.

About 70% of Mg^{2+} is dispersed throughout the skeletal tissues with the remaining 30% distributed in soft tissues and the central nervous system.

Dietary sources of Mg^{2+} are readily available from green plants, vegetables, whole grains, milk, meats, and nuts. The average adult ingests about 0.25 g of Mg^{2+} per day. About 33% of dietary Mg^{2+} is absorbed, but this is balanced by urinary excretion of a similar quantity from the body.

Dietary Mg^{2+} deficiencies are rare. Most deficiencies originate from (1) *pathological conditions* that necessitate protracted total parenteral nutrition or (2) *specific malabsorption syndromes*.

Serum concentrations of Mg^{2+} approach 0.85 mM with tissues maintaining 5 to 8 mM concentrations. Intracellular Mg^{2+} stores are largely bound to proteins, nucleotide tri-

phosphates, and the carboxylate and phosphate moieties of other cell constituents. Unlike Ca^{2+}, the intracellular access of Mg^{2+} is comparatively unrestricted, probably as a result of its smaller ionic radius.

Magnesium ions serve as critical *activators* for key enzymes involved in phosphoryl transfer reactions. These enzymes include numerous kinases, phosphatases, and mitochondrial processes of oxidative phosphorylation that ultimately yield Mg^{2+}-stabilized ATP. The glycolytic pathway is also highly dependent on Mg^{2+} as a cofactor, especially at the site of enolase enzyme activity. At the level of nucleic acid biochemistry, Mg^{2+} may also be important in stabilizing the native conformational structures in the vicinity of phosphate residues.

Total control mechanisms over blood Mg^{2+} levels and the relation of Mg^{2+} to other divalent cation concentrations in the body are not well defined. Contrary to an apparent free equilibrium for both $[Mg^{2+}]$ and $[H^+]$ over cell membranes and blood plasma, temporal metabolic duress can upset equilibrium conditions. For example, the release of diphosphoglycerate (DPG) upon the oxygenation of deoxymyoglobin can suppress plasma Mg^{2+} concentration. This action stems from an immediate coordination of plasma Mg^{2+} with DPG to form a Mg^{2+}–DPG complex. Furthermore, the expedient glycolysis of sugars to lactic acid in muscle cells can also produce a transient increase in Mg^{2+} concentrations. This is an indirect metabolic event caused by a lactic acid-induced cellular pH drop of ~1.0 pH unit to a pH of ~6.3. A pH drop of this magnitude effectively decreases Mg^{2+}–ATP binding and results in a temporary increase in Mg^{2+}.

Since renal absorption of Mg^{2+} varies inversely with Ca^{2+}, and fecal Mg^{2+} excretion decreases with lowered dietary Ca^{2+}, an interplay between these cations is suspected at the physiological level. Unfortunately, the control mechanism is not well understood for these cations, but whatever controls are enforced seem to favor calcium.

The loss of body Mg^{2+} can be monitored

using plasma or urine sample assays. Of these two sample assays, urine displays a superior index of Mg^{2+} losses from cells over plasma because plasma Mg^{2+} does not decrease below 1.0 mEq/L (normally 1.4–2.3 mEq/L) until >25% of cellular Mg^{2+} is lost.

Hypo- versus Hypermagnesemia and Dietary Intake of Calcium

Hypomagnesemia or low serum Mg^{2+} levels may result in neuromuscular dysfunctions indicated as hyperexcitability (convulsions, tremors, etc.). These symptoms may stem from malabsorption syndromes, chronic alcoholism, which promotes Mg^{2+} losses, delirium tremens, aldosteronism, chlorothiazide drug regimens, chronic glomerulonephritis, and excessive urinary Mg^{2+} losses.

Hypocalcemia is a closely allied indication of hypomagnesemia. This relationship between Ca^{2+} and Mg^{2+} status reflects mutual metabolic interactions between the cations. Absorption of both cations from the intestine is encouraged by vitamin D; Ca^{2+} and Mg^{2+} can both compete for reabsorption over the renal tubule(s); both cations exert counteractive influences on the central nervous system; normal PTH secretion incited by hypocalcemia requires the presence of Mg^{2+}; and hormonal effects of PTH at target cells requires Mg^{2+}. Based on these considerations and especially the last two, hypomagnesemia can produce apparent hypoparathyroidism and pseudohypoparathyroidism.

In contrast to hypomagnesemia, hypermagnesemia or high serum Mg^{2+} concentrations may occur from uremia, severe diabetic acidosis, dehydration, and any other glomerular maladies that retard Mg^{2+} filtration.

Dietary Mg^{2+} intake averages about 120 mg/1000 kcal with about 300 to 350 mg/day being suitable for holding normal Mg^{2+} levels in adults—although 150 mg/day excesses may be required during pregnancy and lactation. The extra Mg^{2+} required during lactation accounts for the 40 mg/L disposi-

tion of Mg^{2+} in human milk (bovine milk contains 120 mg/L).

SODIUM

Sodium (Na^+) serves as the major monovalent cation in the extracellular body fluids. The multifaceted physiological effects of Na^+ contribute to the maintenance of blood volume; the characteristic osmotic pressures of body fluids; pH balance and the electrolyte status of extracellular fluids; electrochemical potentials surrounding nerve and muscle tissues; plus the facilitation of active transport for amino acids and sugars over cell membranes.

Nearly 100% of Na^+ supplied to the gut is readily absorbed. Early infant requirements for Na^+ approximate 58 mg (2.5 mEq/day), but this progressively grows to 1400 mg (60 mEq/day) at 12 months. Voluntary adult intakes of sodium chloride usually provide 2.3 to 7.0 g/day (100–300 mEq/day), but even higher Na^+ loads may be consumed depending on the extent of Na^+ losses incurred during heavy work. Since the consequences of dietary sodium are under continuous scrutiny, the reader should consult the dictates of the National Research Council for Recommended Dietary Allowances of Na^+ on a periodic basis.

Of all the cations, Na^+ deficiencies are most unlikely in human nutrition since it is commonly present or admixed to foods as sodium chloride. Commercial processed and convenience foods, hydrated and emulsified meats, canned vegetables, soups, and many other foods (cheese, milk—483 mg (21.0 mEq/L), shellfish) all contribute to the total 5 to 10 g/day dietary salt loads of the populace.

Blood [Na^+] levels are stringently controlled by *renin–angiotensin–aldosterone mechanisms*. Low blood [Na^+] typically causes the kidney to release angiotensin. Angiotensin subsequently elicits aldosterone production by the adrenal cortex, which, in turn, accelerates Na^+ reabsorption from any existing glomerular filtrate. This control mechanism ensures restoration and/or maintenance of desirable [Na$^+$] blood levels. Depressed plasma [Na$^+$] levels may commonly result from pathological functions of renal or adrenocortical tissues, severe diarrhea, vomiting, or thermal dehydration.

POTASSIUM

Potassium (K$^+$) serves as the primary intracellular cation in the same way that Na$^+$ behaves in extracellular fluids. In fact, 90% of K$^+$ occurs within cells at concentrations of ~440 mg/100 g as opposed to lower concentrations in blood (200 mg/dL) and plasma (20 mg/dL). The translocation of extracellular K$^+$ into intracellular volumes is mediated by an energy-consuming transport mechanism. The intracellular role of K$^+$ is tied to enzymatic processes involving glycolysis and protein synthesis as well as the maintenance of acid–base and osmotic balances.

High dietary concentrations of K$^+$ are present in nearly all fresh fruits and vegetables (e.g., potatoes, bananas, oranges, tomatoes), which jointly exhibit low [Na$^+$] levels. Milk, eggs, and meat have high K$^+$ concentrations but high Na$^+$ concentrations in these foods may obviate their use as an effective tool for enhancing the [K$^+$]:[Na$^+$] ratio in certain individuals.

Dietary deficiencies of K$^+$ alone are rare, but overt K$^+$ deficiencies known as *hypokalemia* can result from excessive losses of K$^+$ spurred on by diarrhea, diabetic acidosis, diuretics, drugs, and purgatives. The resulting hypokalemic condition may be more or less severe depending on the extent of K$^+$ deficiency. When severe hypokalemia already exists, any rapid uptake of extracellular K$^+$ into intracellular volumes may incite a range of potential effects including slowed heart beat and impaired respiratory muscle activities. Apart from reversing minor hypokalemic conditions by oral K$^+$ supplements, severe K$^+$ deficits may require intravenous injections of the cation.

Although the [Na$^+$] levels in humans and other higher animals are variable, they are generally 0.1 to 0.2 times the [K$^+$] content. For blood plasma, however, this condition is reversed and [K$^+$] is typically 5.0 mM and [Na$^+$] is ~150 mM.

Bioinorganic principles of chemical evolution have clearly favored the intracellular residence of K$^+$ over Na$^+$ for reasons that are not entirely clear. It is speculated, however, that a smaller charge dispersion over small ions versus larger species may be a factor in small ion hydration and favor K$^+$ within cells. For example, the Na$^+$:K$^+$ ratio for possible hydration is estimated to 16:10.

REGULATION OF SODIUM AND POTASSIUM ION CONCENTRATIONS

The total [Na$^+$] and [K$^+$] content of the body is regulated by renal functions. In normal situations plasma [Na$^+$] is conserved by kidney excretion mechanisms and K$^+$ is discharged.

Detailed studies have shown that [Na$^+$] and [K$^+$] are differentially distributed throughout the intra- and extracellular compartments of the body. The apparent partitioning of these ionic species is dictated by:

1. *Gibbs–Donnan equilibrium conditions,* which are established and maintained between intra- and extracellular compartments through the collective effects of all cell membranes throughout the body.

2. *Energy-consuming cellular processes* regulated by hormonal factors plus other physicochemical effects of ongoing metabolism.

Energy Requiring Na$^+$/K$^+$ Transport

Both Na$^+$ and K$^+$ concentrations are notably controlled by active transport mechanisms presiding over cell membranes. Since distinct differences in intra- and extracellular $[Na^+]:[K^+]$ ratios must be maintained in animal systems, this action requires a *constant expenditure of ATP*. The ATP supplies energy to a *Na^+, K^+-ATPase transit system* located in cell membranes. Operation of the system is contingent on the hydrolysis of the terminal phosphate on ATP to yield a phosphorylated Na^+, K^+–ATPase molecule. Subsequent Na^+ binding to this phosphorylated species precedes an eventual translocation of Na^+ from the cell, whereas K^+ is assumed into the cell (Figure 19.4).

Not only are mechanisms of this type responsible for controlling intra- to extracellular ratios of $[Na^+]$ to $[K^+]$, but they underlie the renal tubular functions in Na^+ and K^+ excretion.

The renal disposition of Na^+ in particular is critical for holding the osmolality of body compartments at stable levels. Osmolality in this context refers to the *inexact measure* of ions (and/or molecules) held in solution by a solvent phase (water). Osmolality is related to the osmotic pressure that would be generated across a membrane totally impermeable to solutes but freely permeable to water.*

Sodium salts account for 90% of extracellular fluid solutes and, thus, extracellular fluid volume is closely tied to the renal excretion of Na^+. Excessive renal Na^+ losses reduce extracellular fluid volume, whereas an inordinate Na^+ retention dictates fluid retention and edema conditions.

The renal maintenance of blood Na^+ and K^+ concentrations versus their urinary excretions requires a large energy commitment. In fact, nearly 66% of high-energy nucleotides produced by kidney respiration are used for driving renal Na^+ and K^+ transport.

*For fluid dynamic relationships involving physiological chemistry, osmolality is usually preferred to osmolarity as an index of osmotic activity. This preference is based on the fact that practical colligative property assays (e.g., freezing-point depression) yield data in terms of solute per unit of solvent and *not* solution. Accordingly, results may be reported as milliosmols (mosm) per kilogram of water, *not* per liter of solution.

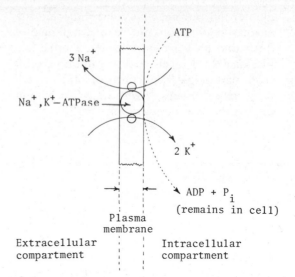

FIGURE 19.4. Ion transport for Na^+, K^+-ATPase systems occur in two steps: (1) Phosphorylation of ATPase by ATP, which permits Na^+ binding, and (2) K^+ binding to the enzyme system, which culminates in a $3Na^+:2K^+$ transit ratio over the membrane.

Renal Sodium Excretion and Plasma Osmolality

Apart from its numerous functions, the hypothalamus exhibits *osmosensor activity*, which is linked to its capillary blood pressures. Under conditions of low plasma osmolality (hypotonic), renal secretory processes discharge water and retain most Na^+. In cases in which plasma osmolality is elevated (a hypertonic condition exhibited as a water deficit), renal functions may respond by retaining water.

Conservation of renal filtered water is controlled by the hypothalamic secretion of *antidiuretic hormone* (ADH) into the posterior pituitary via nerve pathways. The posterior pituitary subsequently discharges the ADH into the blood. Once in the hematogenous circulation, *the hormone targets the epithelial cells of the distal convoluted tubule and collecting ducts of the nephron*. The effect of ADH on nephron functions is indicated in Figure 19.5. The ADH increases the permeability of the distal convoluted tubule until the water resorbed (and/or obtained from dietary routes) increases plasma–water content within the hypothalamic-dictated bounds. Once the requisite water content in the circulation is achieved, ADH release from the pituitary is slowed.

The role of ADH is only one contributing element in osmolal control of body fluids, and it operates in concert with the aldosterone regulation of sodium chloride. The ster-

oidal hormone aldosterone is secreted from the adrenocortical cells upon detection of lower than normal [Na$^+$] in the blood. This hormone targets the epithelial tissues of the distal convoluted tubule and increases their absorption of Na$^+$ and Cl$^-$ from the urine. As the [Na$^+$] level of the blood is increased, aldosterone secretion is suppressed. Apart from these actions it should be recognized

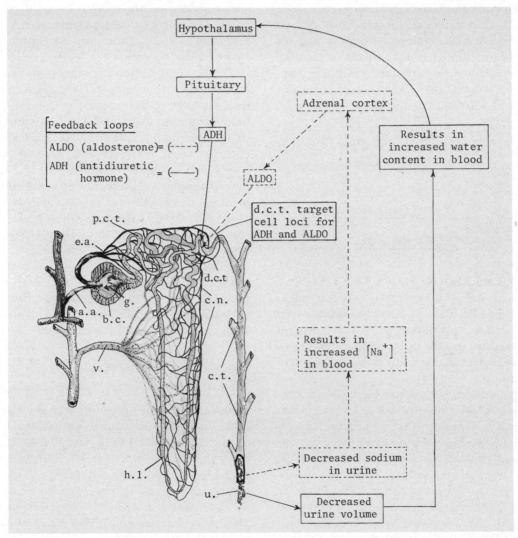

FIGURE 19.5. Interactions of aldosterone (ALDO) and antidiuretic hormone (ADH) on the sodium ion and water absorption over the nephron. See text for details concerning the operation of this system in regulating osmolality. Structural components of the nephron shown include a.a., afferent arteriole; b.c., Bowman's capsule; c.n., capillary net; c.t., collecting tubule; d.c.t., distal convoluted tubule; e.a., efferent arteriole; g., glomerulus; h.l., Henle's loop; u., urine; and v., venule.

that aldosterone also accelerates the urinary excretion of hydrogen ions, and ethanol suppresses ADH release thereby promoting diuretic consequences.

The regulation of $[Na^+]$ and its anions in plasma underlie the maintenance of normal osmolal concentrations. When abnormal plasma osmolality exists, the plasma contains an abnormal (total) amount of solute per volume of plasma water. Aside from urea and glucose, the principal contributor to plasma osmolality is Na^+ (along with its balance of attendant anions). The calculated osmolality of plasma can be estimated from the itemized concentration of each plasma constituent and substitution into the formula

$$\text{osmolality (mosm/kg)} = \\ 2 \times Na^+ \text{ (mmol/L)} + \text{urea (mmol/L)} \\ + \text{glucose (mmol/L)}$$

or

$$2 \times Na^+ \text{(mmol/L)} + \frac{\text{urea nitrogen (mg/dL)}}{2.8} \\ + \frac{\text{glucose (mg/dL)}}{18}$$

It should be recognized, however, that a disparity of ~10 mosm/kg between calculated and "actual" plasma values measured by colligative properties (e.g., freezing-point depression) usually indicates the presence of a polyhydric alcohol, ethanol, or some similar solute.

In general, Na^+ is the only plasma constituent commonly present to an extent that a significant concentration reduction can result in hypoosmolal conditions recognized as *hyponatremia*.

Using serum $[Na^+]$ as an index, hyponatremic conditions occur at <136 mmol/L. Such levels may develop from (1) depletion of body Na^+ levels or (2) excessive accumulation of water, which respectively produce *depletional* hyponatremia, sometimes caused by excessive fluid losses, or *dilutional* hyponatremia resulting from over hydration.

Dilutional hyponatremias are all spurred on by reduced renal perfusion from lowered cardiac output; impaired renal excretion of water; excessive fluid assimilation by intravenous or duodenal tube feedings; and hypoalbuminemia—all brought on by malnutrition, cirrhotic hepatic disease, or nephrotic diseases where albumin stores are lost in urine. These conditions can all direct a fluid shift from vascular fluid spaces into interstitial spaces of the body. This, in turn, promotes increased aldosterone and ADH secretion in response to an apparent hypovolemic condition.

Depletional hyponatremias may result from dietary deficiency of Na^+, gastrointestinal losses by vomiting, diarrhea, and so on; inadvertent but excessive diuretic actions of drugs designed to moderate Na^+ and water content of the body; as well as renal Na^+ losses incited by hypoaldosteronism, impaired tubular reabsorption of renal fluids (osmotic diuresis), renal failure, and so on.

Hypernatremia develops when Na^+ concentrations are >148 mmol/L. Although dehydration commonly produces this condition, excessive Na^+ intake abets hypernatremia. In the range of 155 to 165 mmol/L, Na^+ effectively promotes the dehydration of intracellular fluids owing to the high extracellular fluid osmolality.

Natremic Stresses in Pregnancy and Effects of High Dietary Sodium Intake

As indicated in the preceding sections, hypo- and hypernatremic disorders are closely tied to the volemic shifts of aqueous body fluids over membrane-partitioned cellular and noncellular body compartments. During pregnancy, normal natremic conditions must be maintained to meet (1) *increased rates of glomerular filtration* and (2) *the expanded fluid volume of maternal systems*.

During pregnancy, glomerular filtration may be 45 to 55% greater than nonpregnant conditions and Na^+ filtration is markedly increased. Moreover, elevated progesterone concentrations tend to *decrease* Na^+ reabsorption. In order to allay serious effects of imminent urinary Na^+ loss, levels of aldosterone production are progressively increased in concentration to meet these losses. Furthermore, as the extracellular fluid compartment during pregnancy increases, the renin–angiotensin–aldosterone control regime finally actuates even higher aldosterone levels. These combined effects ensure adequate tubular Na^+ reabsorption to meet osmoregulatory demands of the expanded fluid volume.

Since 750 mEq of total Na^+ are speculated to be adequate for the acquisition of a 11-kg weight gain (70% water), only 3 mEq/day (69 mg) above nonpregnant Na^+ requirements (up to 8 mEq/day) supplies adequate amounts of this cation.

It has been well established that *high dietary Na^+ levels can contribute to hypertensive conditions*. For humans, the hypertensive criterion depends on the sustained elevation of arterial blood pressure to >140 mm Hg systolic or >90 mm Hg diastolic. Mechanisms for dietary Na^+-induced hypertension are not entirely clear, but speculation suggests that its effects are twofold.

First, high $[Na^+]$ levels instigate a temporary *increase* in extracellular fluid volume within the body, right atrial pressure, mean systolic pressure, and cardiac output. A secondary response involves an autoregulated tissue response to the former cardiac events such that peripheral vascular resistance may be established.

During high cardiac output, the resulting high blood pressure fosters renal Na^+ and water excretion. In due time both $[Na^+]$ and fluid volume will wane, *apart from the possibility* that sustained high Na^+ and water loading may continue to foster peripheral resistance and hypertension.

High Na^+ concentrations are also suspected of increasing the intracellular water volume of atrial smooth muscle. Not only would this increase restrict the luminal diameter bounded by vascular walls, but it may result in excessive vasoconstrictive responses to otherwise normal vasoconstrictive effectors. This last consideration, in turn, actuates higher than normal degrees of peripheral circulatory resistance.

Diuretic therapies and/or Na^+ restriction probably exerts their proven antihypertensive effects by diminishing the extracellular fluid volumes of the body as well as excessive vascular ion content and hydration.

Sodium-induced hypertension may be far more complicated than the factors outlined above. In some individuals at least, Na^+-induced hypertension may be rooted in a genetically defective Na^+ transport system that impairs renal Na^+ excretion. In still other cases in which dietary calcium intake is low (<670 mg/day), hypertension may be an incidental consequence and subsequently misconstrued with the possible effects of high $[Na^+]$. The clinical and biochemical record for the role of Na^+ and other ionic species in hypertension remains to be finalized.

TRACE ELEMENTS—AN OVERVIEW

Historical studies (Table 19.2) have revealed that 14 elements are required by the body in quantities of less than a few milligrams per day. These include chromium, cobalt, copper, fluorine, iodine, iron, manganese, molybdenum, nickel, selenium, silicon, tin, vanadium, and zinc.

The ubiquitous but enigmatic roles for many of these elements have been complicated by *unreliable quantitative analytical chemistry* over the years and collateral ignorance concerning the biochemistry of one trace element with others. Moreover, the functions of trace elements are so diverse that their classification defies purely rational efforts.

Advances in chemical analysis using atomic absorption spectrophotometry, neutron activation analysis, X-ray fluorescence, proton-, electron-, and microwave-induced X-ray emission, and electrochemical

Table 19.2. Some Premier Reports Recognizing the Essentiality of Certain Trace Elements in Animal Nutrition

Iron	MacMunn, C. A. 1886. *Phil. Trans. Roy. Soc. London* **177**:267.
Iodine	Chatin, A. 1852. *C. R. Acad. Sci.* **34**:14.
Copper	Hart, E. B., *et al.* 1928. *J. Biol. Chem.* **77**:797.
Manganese	Kemmerer, A. R., *et al.* 1931. *J. Biol. Chem.* **92**:623.
Zinc	Todd, W. R., *et al.* 1934. *Amer. J. Physiol.* **107**:146.
Cobalt	Marston, H. R. 1935. *J. Counc. Sci. Ind. Res. (Aust.)* **8**:111.
	Filmer, J. F., *et al.* 1937. *Aust. Vet. J.* **13**:57.
	Lines, E. W., 1938. *J. Counc. Ind. Res. (Aust.)* **8**:117.
Selenium	Schwarz, K., *et al.* 1947. *J. Amer. Chem. Soc.* **79**:3292.
Molybdenum	Reichert, D. A., *et al.* 1953. *J. Biol. Chem.* **203**:915.
Chromium	Schwarz, K., *et al.* 1959. *Arch. Biochem. Biophys.* **85**:292.
Tin	Schwarz, K., *et al.* 1970. *Biochem. Biophys. Res. Commun.* **40**:22
Vanadium	Schwarz, K., *et al.* 1971. *Science* **174**:426.
	Hopkins, L. L., *et al.* 1971. *Fed. Proc.* **30**:462.
Fluorine	Schwarz, K., *et al.* 1972. *Bioinorganic Chem.* **1**:355.
Silicon	Schwarz, K., *et al.* 1972. *Nature* **239**:333.
	Carlysle, E. M., *et al.* 1972. *Science* **178**:619.
Nickel	Nielson, F. H. 1974. *Proc. Second International Symposium on Trace Element Metabolism in Animals (Madison, Wisc.).* University Park Press, Baltimore, Md.

techniques all hold promises for more accurate and sensitive trace analyses of elements in biological materials.

Aside from demonstrated deficiency effects of trace elements on crop plant biochemistry, microorganisms, and classic research animals, the effects of trace element deficiencies on humans were largely uncertain through the 1960s. From that time on, trace element deficiencies became more defined with the advent of more precise nutritional biochemical methods and the clinical development of *total parenteral nutrition* (TPN). A poor understanding of trace element actions resulted in the inadvertent but nonetheless significant development of overt *iatrogenic deficiencies* in patients subjected to TPN. Many of these TPN deficiencies were eventually recognized as previously observed but unexplained conditions in earlier medical literature.

The biochemical roles for many elements are intrinsically devoted to their cofactor actions with enzymes. Many of the most severe eventualities resulting from trace element deficiencies as cofactors develop only after long periods of deprivation. In most cases, short-term chronic trace element deficiencies have little value to nutritional studies. If short-term but severe nutritional deficiencies are instilled in test animals or human subjects, the ramifications of these trace element deficiencies may not materialize until much later in the life cycle. Furthermore, even when the health of a test subject seems compromised or affected by inadequate trace element content in the diet, the obvious effects of deficiency *may* actually *represent secondary* or *tertiary consequences* of the primary deficiency effect.

The roles for many trace elements are confused since many of the elements can extert *element–element* (EE) antagonisms. These EE antagonisms reflect *competitive absorption scenarios* over the gastrointestinal tract, but in other cases the EE antagonism may involve and *competition for limited amounts of plasma transport vehicles* or related phenomena.

Copper, iron, and manganese demonstrate mutual EE antagonisms. If one of these elements is present at higher relative concentrations, absorption of the others may be suppressed.

The basis of EE antagonisms and interactions has been theoretically addressed by Hill and Matrone (1970). Since the physical and chemical properties of elements depend on their electronic structures, those elements that are *most similar* will exhibit the *most notable antagonistic biochemical and biological actions.*

Experimental evidence also reveals that EE interactions can have important toxicological consequences. Rodent toxicity tests, for example, show that the toxicity of methyl mercury may be reduced by selenium supplied as selenite. Similar mechanisms are believed to operate in humans. Principles of EE antagonism may also counteract cadmium absorption when it coexists with dietary supplements of iron, copper, and manganese. These interactions may also exist in higher animals.

IRON

Iron is a critical constituent of heme-prosthetic groups found in myoglobin, hemoglobin, and the cytochromes. It also serves as a component of enzymes (catalase), *nonheme iron* (NHI) and sulfur proteins (succinate dehydrogenases) and as an inorganic cofactor for other enzymes (aconitase).

The adult human contains a total of about 5 g of iron with 70% in hemoglobin, 3 to 5% in myoglobin, and nearly 15% in iron storage proteins—*ferritin* and *hemosiderin.*

The oxidation states of iron dictate many of its bioorganic and biochemical reactions. The common oxidation states for iron include the ferrous (Fe^{2+}) or ferric (Fe^{3+}) forms, but the higher oxidation state of iron may

occur as a transient species in specific redox reactions.

Iron is noted for its ability to form complexes with organic molecules containing oxygen, nitrogen, and sulfur. The electronegative properties of these atoms coupled with their residence in large macromolecules permits interactions between electrons of oxygen, nitrogen, or sulfur with iron. This in turn leads to the establishment and maintenance of bonds between iron and organic macromolecules that are quite strong.

The nonbonding electrons residing in the incompletely filled $3d$ orbitals of iron can exist in *low-* or *high-spin states*. Weak iron-bonding interactions favor the distribution of nonbonding electrons throughout the $3d$ orbitals and their electron pairing does not occur. To the contrary, strong iron-bonding interactions prompt the pairing of outer nonbonding electrons and permits the existence of $3d$ orbitals having lower energy. Existence of these two electronic behaviors in both iron oxidation states is recognized as a "high-spin state" (electrons dispersed throughout all orbitals) or a "low-spin state" (electron pairing and $3d$ electron restriction to low-energy orbitals).

Based on electron spin resonance data, it is recognized that numerous iron–protein complexes experience changes in their *high–low spin states* without attendant oxidation changes in key reactions. This is especially true for the reversible oxygenation and deoxygenation of hemoglobin.

Since the oxidation state displayed by iron is fundamental to its biochemistry, it should be recognized that the Fe^{3+} state is favored at alkaline-to-neutral pH values, whereas acidic pH favors the Fe^{2+} state. For example, intestinal dietary absorption of iron requires the Fe^{2+} form and not Fe^{3+}. Furthermore, an overt predominance of Fe^{3+} serves as a precondition for joint interactions of hydroxide ions, water, and some other anions in tissues to produce precipitable pathological iron deposits.

Dietary sources of iron include liver, meat, fish, egg yolk, green vegetables, whole wheats, enriched breads, blackstrap molasses, and any other food source that can assimilate iron. Because Fe^{2+} is preferably absorbed over Fe^{3+}, it is worth noting that the dietary and gastric stability of iron is favored by sugar, sorbitol, cysteine, amino acids (histidine, lysine), organic acids (citric, gluconic, and succinic acids), mucoproteins, and other constituents of intestinal secretions. Effective dietary iron absorption is countered, however, by phytates, oxalates, tetracycline, tannates, carbonate, vegetable fibers, phosphates, pancreatic bicarbonate, and antacid preparations. Some foods, such as spinach, that are traditionally recognized as iron rich, often contain so much phytate and other chelators that their dietary iron content has little importance.

It is generally conceded that iron is absorbed more efficiently from foods of animal origin as compared to plant foods. This may reflect the fact that iron from hemoproteins can be absorbed as intact heme. Radioactive iron tracer studies reveal that iron absorption from cereal foods and vegetables is <5% compared to 15 to 20% for beef, liver, and fish. Apart from these observations, however, iron absorption is further complicated by the gastric and intestinal dietary milieu in which absorbable iron exists. For example, iron absorption from corn or certain legumes can be elevated as much as threefold when fed with fish or veal. The record is far from clear regarding the total dietary assimilation of iron, and the differential mechanism by which animal proteins seem to enhance iron absorption must be clarified.

Absorption and Transport

The assimilation of dietary iron across the intestinal mucosa begins with the absorption of ionic iron in the lumen of the small intestine. This absorption is initiated along the brush borders of the mucosal cells. The largest amounts of iron are absorbed in the duodenal region, but a gradient of lesser absorption coincides with more distal reaches of the intestine. In cases where heme iron is supplied, the metal ion is severed from the prophyrin in the cytoplasm of columnar mucosal cells. Iron is readily absorbed by mucosal cells, but the process is antagonized by pancreatic bicarbonates that favor Fe^{2+} conversion to Fe^{3+}, which is not assimilable (Figure 19.6).

Passive diffusion probably directs the entry of iron into brush border regions of mucosal cells, but active transport is still debated as a possible factor for its exit. It is also generally speculated that iron translocation over mucosal cells and into the bloodstream may require the participation of low-molecular-weight molecules. Persistent efforts to finally characterize mucosal iron carriers have produced querulous results and offered little support for present concepts on iron metabolism.

Assuming at least a marginal body demand for iron, some of the mucosal cell iron is conveyed to the iron-binding plasma protein *transferrin* (TF). The TF protein is a hepatosynthesized β-globular protein, also known as *siderophilin*. In this scenario, Fe^{2+} exits the mucosa into the bloodstream, where a copper-dependent enzyme known as *ferroxidase* (formerly called *ceruloplasmin*) mediates Fe^{3+} formation from Fe^{2+}. The Fe^{3+} then undergoes a carbonate-dependent binding reaction with TF:

$$TF + Fe^{3+} + CO_3^{2-} \longrightarrow TF—Fe^{3+}—CO_3^{2-}$$
$$TF—Fe^{3+}—CO_3^{2-} + Fe^{3+} + CO_3^{2-} \longrightarrow$$
$$TF—2(Fe^{3+}—CO_3^{2-})$$

Normal ferrokinetic binding of TF to iron shows a plasma profile of about 12% TF that is saturated with iron (*diferric TF*), 44% with at least one site occupied by iron (*monoferric TF*), and 44% that is iron deficient (*apoferric TF*). Furthermore, studies have revealed that the binding loci for iron on TF do not display equal binding affinities for all metals, and Fe^{2+} discharge from specific TF loci to iron-consuming tissues may be somewhat specific.

Regulation of iron transfer from mucosal cells *into the capillary circulation* of the intes-

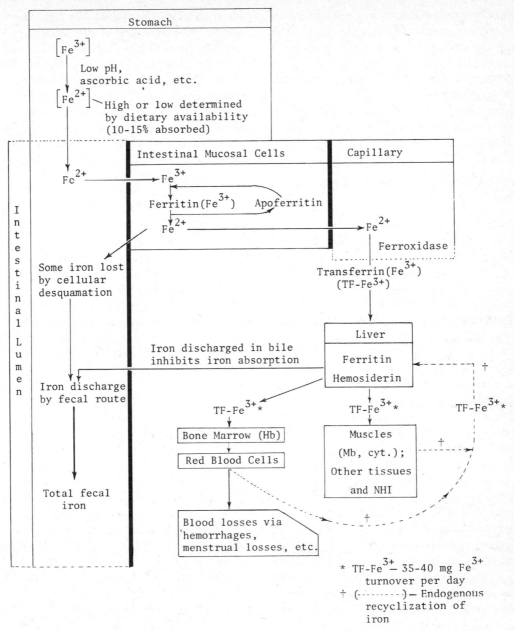

FIGURE 19.6. Simplified schematic route for iron absorption, storage, and transport. (Mb, Myoglobin; Hb, hemoglobin; NHI, nonheme iron.

tine depends on the iron status of an individual, his erythropoietic rate, and other iron-demanding reactions of the body that are ongoing. One popularized control mechanism suggests that the biosynthetic availability of a mucosal iron-binding protein, apoferritin, may control the release of iron to TF in the plasma. When iron deficits exist in the body, apoferritin is not synthesized by mucosal cells, and thus iron can be released to the capillary circulation. If adequate stores of iron exist, however, apoferritin biosynthesis in mucosal cells is actuated in order to bind iron as a ferritin complex. The iron plus apoferritin produces an exploitable iron store within mucosal cells. Assuming that these iron stores remain unused over the 9 to 11-day life span of mucosal cells, their desquamation eventually eliminates the iron stores from the body.

Other alternatives to apoferritin control of iron availability exist, but they are actively debated.

Occurrence and Circulatory Disposition of Iron

Iron-containing TF releases its iron to (1) *erythropoietic tissues* of bone marrow for hemoglobin biosynthesis; (2) *heme-containing enzyme biosyntheses* (cytochromes-*a–c* and *P*-450, cytochrome-*c* oxidase, catalase, peroxidases, tryptophan pyrrolase, lipoxidase, and homogentisic oxidase); (3) *iron flavoproteins* (cytochrome-*c* reductase, NADH dehydrogenase, succinate dehydrogenase, xanthine oxidase, and acyl-CoA dehydrogenase); and (4) *cofactor-requiring enzymes* (aconitase, ribonucleotide reductase, and succinic dehydrogenase). Excess TF-mobilized iron is directed to liver storage proteins, where it forms ferritin or hemosiderin.

Ferritin exists as a ferric–hydroxide–oxide phosphate micelle (up to 4300 iron atoms/molecule) surrounded by the 24-subunit apoprotein (M.W. = 440,000). Iron stored in this structure can be readily removed by reduction with reducing agents plus the presence of an appropriate chelating agent that can accept Fe^{2+}. Ferritin-bound iron also serves as the principal iron-excretion product on exfoliated mucosal cells of the intestine, as previously mentioned.

Iron accumulation and release from the micelle occurs through channels in the surrounding protein. When iron availability supersedes micelle storage capacity, iron deposition occurs outside the ferritin structure as amorphous iron deposits known as *hemosiderin*.

Most studies show that polyribosomic synthesis of ferritin apoproteins is checked only when the Fe^{2+} availability in cells does not warrant the presence of the protein.

Iron content in plasma and serum is equivalent with a typical range of values being 65 to 165 μg/dL. The mean for males is ~120 μg and ~110 μg for females. The plasma TF exhibits 33% saturation under normal circumstances with total iron-binding potentials in the range of 275 to 400 μg/dL. Diurnal variations are mirrored in plasma iron concentrations with evening values being ~30% below the morning.

Aside from the ferritin stores of tissues, small amounts of *low-iron ferritin* (mainly apoferritin) can be detected in plasma. Origins and functions of plasma ferritins are not clearly understood, but they generally coincide with the release of ferritin during normal hepatocyte turnover and erythrocyte senescence in reticuloendothelial cells.

Serum iron (SI), TF, and *serum ferritin* (SF) levels may be independently determined by routine clinical chemical procedures in order to assess iron status of individuals including the *iron-binding capacity* (IBC) of serum. The SF values hold special significance since these levels increase over nominal values during hepatocellular necrosis and other hemolytic disorders. Reference values are 120 ± 60 μg/dL for males and 60 ± 30 μg/dL for females. Any SF values <10 μg/dL support the possibility of iron deficiency, whereas 10 to 20 μg/dL values warrant further study. In addition, SF values hold significance for estimating stored iron. Each 8 mg of stored iron corresponds to 1 μg/L of SF:

$$\text{iron stores (mg)} = \text{SF}(\mu g/L) \times 8$$

Ferritin accounts for nearly 65% of stored iron with the balance being hemosiderin. Average iron stores are 750 mg for adult males and 300 mg for females.

Iron Excretion and Other Losses

Destruction of nonviable red blood cells provides 20 to 25 mg/day iron to the endogenous iron pool. This quantity is far greater than the daily 1 to 2 mg made available from dietary sources of iron. Endogenous iron is not excreted but, instead, recycled to replenish the iron-consuming demands of tissues.

Total daily iron excretion is ~0.5 to 1.0 mg/day via urine (0.1 mg), sweat, bile, and fecal routes (0.6 mg). Desquamation of the intestinal iron stores is a contributing iron source to fecal iron content. Menstrual blood losses in women account for 12 to 30 mg of iron lost each month (0.5 mg iron/mL blood), and an average of 0.4 to 1.0 mg/day loss. Iron losses up to 500 mg may not be uncommon following pregnancy.

Although iron is effectively recycled by very conservative biochemical mechanisms, a marginal iron balance in humans can easily lapse into a deficiency state.

Iron Deficiency versus Excess Iron

Excessive iron losses may occur with intermittent and occult blood losses occurring over the alimentary tract. These losses are generally linked to minor and major pathological situations. Chronic blood loss from menstrual events may be variable, especially depending on which contraceptive regimens are practiced. Combined estrogen–progesterone contraceptive practices can result in a 50 to 60% reduction of blood loss, whereas intrauterine devices increase blood losses.

Chronic illness impedes iron utilization and renal diseases interfere with hematopoiesis. These actions appear to produce iron deficiency and characteristic anemic conditions; however, for adult males and postmenopausal females, obvious iron-deficient anemia should be preemptively viewed as blood loss until contrary evidence is obtained.

In cases of excess iron, *hemosiderosis* or *hemochromatosis* may develop. The former condition results in raised tissue stores of iron without affecting tissue structure or function. Hemochromatosis, however, is indicated by tissue damage plus hepatic enlargement, bronze-pigmented skin, and diabetes (bronzed diabetes) in the most severe cases.

Recommended Dietary Allowances and Iron Sources

Uncomplicated iron deficiencies and their effects can be readily detected by decreased circulatory hemoglobin levels and packed red blood cell volume (hematocrit). In cases of iron-deficient microcytic anemias, subjects exhibit a pale washed-out appearance along with blood cells that are reduced from normal sizes. For children and adolescents, sporadic yet consistent growth can incite transient iron deficiencies. If these deficiencies are protracted, growth retardation, poor resistance to infections, and neurological lesions may ensue.

Daily allowances of 10 mg (infancy) to 18 mg (adolescence) have been recommended. For women of childbearing age, 18 mg/day has been suggested, but the iron stores of pregnant women and their diet cannot supply adequate iron without 30 to 60 mg/day supplementation.

Postparturition supplementation of iron may be useful for replenishing maternal iron stores; however, lactation iron expenses alone do not warrant increased iron intake above nonpregnant conditions.

Iron in milk is bound to the protein *lactoferrin* (LF). This protein is homologous to transferrin and offers two iron-binding sites that are rarely completely saturated. The role of LF in infant nutrition is not clear, especially regarding its role in augmenting iron availability to the gastrointestinal tract of infants. Many studies claim, however, that bacterial gastrointestinal infections are tied to iron availability. Thus, LF may bind iron and check iron-dependent infective microbial populations.

In addition to the notable iron content of most animal foods, legumes, cocoa, and cereal brans contain empirically significant amounts of iron. Iron availability, of course, is determined by the coexisting concentrations of phytates and other counterabsorptive chemical agents. The iron content of plant foodstuffs also depends on growth localities, seasonal variation, and crop fertilization programs. Therapeutic sources of iron include ferrous gluconate (also used in black olives), ferric ammonium citrate, and ferrous sulfate.

ZINC

About 2 g of zinc is distributed throughout all tissues (average 10–200 μg/g) of the adult human body, although higher than average concentrations are found in prostate and retina (600–800 μg/g), bones, teeth, pancreas, and integumentary proteins. At least 50% of zinc permanently resides in bone and almost 20% exists in integumentary tissues. At term, a fetus contains about 60 mg of zinc, which must be supplied from maternal nutrition during pregnancy.

Dietary Absorption and Assimilation

Absorption of dietary zinc occurs over the duodenal and jejunal regions of the gastrointestinal tract. *Zinc-binding ligands* (ZBL), presumably discharged by pancreatic secretions, participate in the intraluminal capture of zinc.

Evidence for low-molecular-weight ZBL species is largely circumstantial, but enduring candidates for this job include picolinic acid, citric acid(s), prostaglandins (E_2, F_2), and other amino acids or peptides. Active transport of the captive zinc into the portal blood is mediated by *metallothionein* productivity in intestinal mucosal cells. Since active absorption for zinc occurs, homeostatic regulation of zinc into portal blood may depend on the behavior of the sulfur-rich protein metallothionein. Other homeostatic regulation mechanisms are possible and may not be discounted.

Plasma zinc is complexed to organic ligands as opposed to existing as a free-roving metallic ion. Some is tenaciously bound to plasma proteins and some is loosely bound. Zinc–albumin complexes account for 50% of the zinc, and the metal is readily exchangeable throughout the peripheral circulation. About 7 to 8% of zinc is loosely bound to amino acid constituents in plasma. The remaining 40+% of plasma zinc is largely bound to macroglobulins (α_2-globulins). The tenacious metalloprotein interactions here counter the use of this zinc for nutritional purposes.

Serum and plasma zinc concentrations in adults range from 80 to 250 μg/dL, although circadian diurnal fluctuations in concentration occur. Similar fluctuations in the ceruloplasmic occurrence of copper suggest that this, or related metallo-binding proteins, may participate in regulating plasma availability of zinc.

Many nutrients display *enterohepatic* circulation, where they are coexcreted with bile into the intestinal lumen, then reabsorbed and repeatedly recycled through the liver. Zinc, in an analogous fashion to enterohepatic circulation, experiences *enteropancreatic* circulation.

Evidence for enteropancreatic circulation comes from dietary studies in which known zinc contents in meals have been surveyed as they transit the intestinal lumen. Perfusion of chyme from multiple intestinal points by coaxial intubation reveals an approximate 200% increase in zinc chyme content above the native zinc content of the meal after pancreatic secretions appear in the lumen. Regardless of these levels, chyme zinc is usually reduced below that of the food during distal absorptive processes unless metallo-binding agents are present in chyme.

Unabsorbed dietary zinc and that zinc expunged from the enteropancreatic cycle are excreted by the fecal route. This excretion mechanism maintains zinc balance along with the 0.5 mg/day obligatory renal excretion of zinc from zinc-replete subjects. Other losses of zinc include 1 to 5 mg/day via sweat and integumentary losses and lactation losses to the tune of 2 to 3 mg/L milk, which must be recompensed in maternal nutrition.

Biochemical Functions and Physiological Effects

Zinc deficiency can be rapidly demonstrated when animals are given a zinc-deficient diet. Plasma zinc levels decline by the first day and both static growth and anorexic states may be exhibited by the third day of a strict deficiency. Untoward metabolic disturbances materialize with high-protein diets as zinc deficits become more pronounced. Skin and orificial lesions develop only to be subjected to an unchallenged bacterial invasion. Contrary to apparent infectious conditions, such lesions fail to produce anything other than a marginal to nonexistent inflammatory response. Therefore, the zinc deficiency produces a patently obvious immunodeficiency in the cell-mediated immune system. Advanced deficiency consequences culminate in diarrhea, severe wasting, and ultimately death. This scenario is typical of at least 12 animal species including humans.

Zinc deficiency symptoms are nonspecific and may be coexhibited by many diseases. Symptomatic generalities of zinc deficiency in part reflect the fact that at least 85 zinc metalloenzymes exist and zinc status is intrinsically tied to nearly all levels of protein and amino acid metabolism. Apart from specific biochemical intervention as an enzyme cofactor, zinc has a notable affinity for sulfhydryl groups that are important determinants of protein structures (e.g., enzymes, membranes). Among the enzymes, zinc has special significance in the functions of carbonic anhydrase, carboxypeptidase A and B, alkaline phosphatase, alcohol dehydrogenase, retinene reductase, superoxide dismutase, glutamic-, lactic-, and D-glyceraldehyde 3-phosphate dehydrogenases.

Nucleic acid roles for zinc are varied and critical for both *protein synthesis* and *molecular genetics;* these roles include

1. Zinc-dependent *thymidine kinase phosphorylation of thymidine* following its formation from a uridine precursor (thymidine is uniquely present in DNA).

2. *Ribonuclease catabolism of RNA strands,* which is inversely related to zinc availability.

3. Zinc-mediated *maintenance of polysomic conformations* during protein syntheses.

Only more detailed research can uncover the mechanistic contributions of zinc to key metalloproteins including aminoacyl-tRNA synthetase, RNA and DNA polymerases, plus critical elongation factors.

Cell membrane integrity may depend on the presence of zinc, especially loosely bound ionic forms. Red blood cell integrity, white blood cell migrations, and immunity mechanisms somehow rely on adequate stores of zinc ion availability.

Neurological and chemoreceptor sensitivities are clearly affected by zinc deficits. Saliva contains the zinc metalloprotein *gustin,* which has uncertain developmental effects on tastebuds, while olfactory receptors of the nose are positively affected by mucus zinc levels through unknown mechanisms. As many as 33% of olfactory disorders have been traced to zinc deficiencies.

In many respects, the total picture of zinc deficiency is reminiscent of essential amino acid (EAA) deficits. As detailed by Gordon and Gordon (1981), features common to EAA and zinc deficiency involve at least eight collateral observations:

1. Anorexia and growth failure readily develop.

2. Zinc and EAA concentrations in their respective free pools markedly fall.

3. Hypersensitivity to dietary nitrogen intake develops.

4. No changes in bound reserves of zinc or EAA occur.

5. Dietary fat and carbohydrate show relatively uneventful sensitivity effects unlike dietary nitrogen.

6. Blood urea and ammonia concentrations increase.

7. Administered amino acids undergo rapid oxidation and marginal incorporation into *de novo* proteins.

8. RNA:DNA ratios are reduced.

Based on these substantial experimental observations, it is clear that the intrinsic deficiency effects of zinc unavoidably antagonize protein and nucleic acid syntheses.

Aside from the essential nutritional functions of zinc, it must be recognized that the anabolic or catabolic state of the body dictates the status of zinc requirement, the potential for zinc deficiency effects, and zinc plasma concentrations. For example, pediatric studies of kwashiorkor using soya-based (zinc-poor) diets and controlled use of zinc supplements show parallel but latent effects between zinc administration and later increases in body weight. A soya diet coupled with a lack of dietary zinc produces classical dietary deficiency effects with protein anabolism being stymied. Withholding food, on the other hand, eliminates obvious deficiency effects. In the last case, plasma zinc increases come at the expense of lean muscle catabolism, which releases zinc in order to supply short-term metabolic demands.

Dietary Zinc Sources and Nutrient Interactions

Many concentration levels cited for zinc content in foods are based on nonuniform methods of analysis. Thus, conflicting values for zinc concentrations in the same food using two different methods are not uncommon. In general, high-protein foods are much higher in zinc than those foods that are rich in carbohydrate. Notable dietary zinc stores are found in meat, fish, shellfish, poultry, eggs, and dairy products. The mixed adult diet supplies 6 to 15 mg/day, but questions exist regarding the percentage of the population that meets the higher end of this intake range. Guidelines usually suggest a zinc intake of 3 to 5 mg for infants, 10 mg for ages 1 to 10 years, 15 mg for 11 to 51+ age groups, and 20 mg for pregnant and lactating women. Colostrum offers zinc depletion equivalent to 10 to 20 mg/L while milk per se contains 3 mg/L declining with prolonged lactation.

Increased food processing often parallels increasing zinc losses from foods. This is especially true for sugar-refining and rice-polishing processes. Other dietary constituents including calcium, iron, phytate, and fiber counteract high-efficiency zinc absorption. Based on animal studies, zinc absorption may be reduced in the presence of soy products that have crept into commonly consumed foods. This problem is associated with the textured vegetable protein meat extenders, soy protein isolates, soy-based meat substitutes, and possibly soy-based milk formulas.

Congeners in wine, meat or meat extracts, and pyridoxine (vitamin B_6) all seem to enhance zinc absorption. Reasons for enhanced zinc absorption with the first two food substances are unclear, but pyridoxine is suspected of actuating pyridoxine dependent synthesis of picolinic acid which *may* mediate intestinal zinc absorption. Lactose mildly enhances zinc absorption, whereas ascorbate, oxalate, and conventional uses of ethylenediaminetetraacetic acid (EDTA) seem to have little effect. The reader should consult the authoritative work of Solomons cited in the references for more details regarding zinc biochemistry and dietary assimilation.

With regard to other nutrients, zinc deficiencies are implicated with decreased assimilation of conjugated folates as well as the impaired availability of hepatic retinol.

Although circulatory zinc concentrations have been widely used for the evaluation of zinc deficiency states, this approach is often deceptively affected by sampling protocols, steroid therapies, hemolysis, inflammatory or infectious conditions, and many other factors.

Severe zinc deficiencies are not uncommon among chronic alcoholics, especially those having cirrhosis, and other patients having chronic renal disease, severe malabsorption diseases, or protracted TPN. Early symptoms of TPN-based zinc deficiencies are first evidenced in humans as dermatitis.

COPPER

Copper is an essential nutritional element for many invertebrate and vertebrate animal species. For adult humans, the hepatic tissues contain most of the total copper (100–150 mg) as protein-bound copper and metalloenzymes.

Absorption, Transport, and Excretion

Radiotracer studies reveal that dietary copper is absorbed over the duodenal region. Absorption efficiency is about 30 to 50%, but depending on food or liquid volumes exposed to absorptive duodenal surfaces, copper absorption may be much higher. In some cases, up to 100% of copper in aqueous solution may be absorbed.

Copper transport over the mural regions of the intestinal tissues probably involves passive and/or active transport. In any event, copper absorption seems to be contingent upon a metallothionein protein that binds not only copper but also zinc and possibly some other metals. An antagonistic dietary absorption interaction between zinc and copper has been widely recognized. This antagonism may reflect (1) competitive binding interaction between the two metals for a limited number of metallo-binding sites on the primary binding protein; (2) hepatic synthesis of a copper-binding thionein protein that may be accelerated by zinc, which promotes both serum copper sequestration and eventual lysosomal-mediated excretion of copper via bile canaliculi; or (3) excessive intestinal mucosal synthesis of a thionein induced by

dietary zinc thereby inhibiting copper transfer to plasma. The last of these possibilities seems to be favored as a mechanism for the counterabsorptive effects of zinc on copper assimilation. Nickel–copper interactions may also exist, but the experimental record is still far from complete.

Since dietary forms of copper exist mainly as copper–protein complexes and metalloenzymes, effective absorption of the metal is favored by proteolytic digestion. Absorption of copper liberated by this process probably occurs along the jejunal regions, and yet another type of copper–amino acid-assisted absorption mechanism may be involved.

Postabsorption copper transport in plasma to the hepatocytes requires copper chelatins and copper thioneins. Both proteins are believed to be instrumental in binding copper to *hepatic apoceruloplasmin*. The copper–apoceruloplasmin interaction yields a copper-containing product known as *holoceruloplasmin*.

Nominal plasma copper concentrations range from 70 to 140 μg/dL, which is distributed in a 93:7 ratio of ceruloplasmins:albumin(s) and other key amino acids. The exact participation of these copper-binding species in copper transit and utilization are not clear.

Regulatory mechanisms for copper absorption and excretion by the body are still querulous. It is conceded, however, that 10 to 60 μg/kg/day is eliminated by the biliary route for a total of 0.7 to 4.2 mg from a 70-kg adult. The discharge of hepatic copper as biliary copper (BC) requires the participation of hepatic lysosomes and some type of ligand species. Copper-binding ligands may represent bile pigment derivatives or some other macromolecular structures. In any event, the copper component of BC is highly unlikely to undergo reabsorption. Some theories also suggest that excess biliary ligand productivity or unsaturated BC–ligand-binding sites may actually *preempt* the *dietary absorption of copper* in the distal absorptive regions of the intestine.

Biochemical Roles for Copper

Copper holds an undisputed rank among the most critical trace elements. Its main functions are linked to cofactor participation with metalloenzymes and proteins. Furthermore, a cursory glance shows that most copper-dependent enzymes, including oxidases, hydroxylases, or superoxide dismutase, use oxygen as a substrate for hydroxylation and oxidation reactions.

Many of the classical nutritional deficiency effects of dietary copper mirror the distant and protracted consequences of inadequate copper-dependent enzyme activities. Energy stores of the body can become depleted through inadequate cytochrome-*c* oxidase activity attributable to copper deficits. Bone demineralization and vascular fragility are also dependent on lysyl oxidase. This copper-dependent enzyme cross-links certain lysine residues in collagen and elastin with allysine to form a substantive structural protein matrix. The copper-containing enzyme ferroxidase is required for the conversion of Fe^{2+} (which can be absorbed) to Fe^{3+}, which undergoes binding to transferrin. Melanin productivity, which *contributes to* the color of skin and hair pigmentation, *and* catecholamine metabolism, demand the actions of two copper-dependent hydrolases.

Severe copper deficiencies among humans are rare. Such a deficiency, known as *hypocuprenemia,* is suspected if serum copper is 80 μg/dL. Since 93% of serum copper is bound to ceruloplasmin(s), hypoceruloplasminemia is usually synonymous with *hypocuprenemia.* Aside from the dietary deficiencies, hypocuprenemia can result from defective ceruloplasmin synthesis. Copper depletion and subsequent hypocuprenemia can also be abetted by protein–calorie malnutrition, nephrotic syndrome, sprue, and an inherited syndrome (Menke's "steely hair disease") that impairs intestinal absorption of copper.

Copper deficiency can also lead to bone lesions reminiscent of scorbutic deficiencies, neutropenia, and hypomyelination (in rodents) of the white matter. Detailed mechanisms remain to be worked out for all of these effects.

Copper inadequacies may produce anemic consequences that affect iron metabolism. The record is far from clear, but aberrant ceruloplasmin behavior may incite (1) defective iron discharges from reticuloendothelial cells or intestinal iron absorption; or (2) promote warehousing of parenchymal iron stores.

Since low *lecithin:cholesterol acyl transferase* (LCAT) activities have been related to ischemic heart disease, deficiencies in its copper-dependent cofactor requirements or biosynthetic enzymes may instigate atherosclerotic events. Copper deficiencies reportedly decrease LCAT activity by about 27% in rats, but unassailable human statistics remain to be gathered.

For additional perspectives on the enzymatic actions of copper, consult the authoritative references cited at the end of the chapter.

Glutathione Peroxidase, Superoxide Dismutase, and Copper Deficiency

The Fe^{2+} state of hemoglobin in erythrocytes can be readily oxidized to Fe^{3+} by hydrogen peroxide to produce methemoglobin, which cannot transport oxygen. Hydrogen peroxide can arise from a variety of *in vivo* reactions as well as the autoxidation of certain drugs that produce hydrogen peroxide from oxygen:

$$AH_2 + O_2 \longrightarrow H_2O_2 + A$$

The enzyme known as glutathione peroxidase (GSH-Px) functions with glutathione (GSH or L-glutamyl-L-cysteinylglycine) to protect hemoglobin (Fe^{2+}) from oxidation by the peroxide according to the reaction:

$$2GSH + H_2O_2 \xrightarrow{\text{GSH-Px}} GSSG + 2H_2O$$

Reduced glutathione   **Oxidized glutathione**

Adequate stores of GSH are ensured in the erythrocytes, which oxidize nearly 10% of glucose by way of glucose 6-phosphate (via glucose 6-phosphate dehydrogenase) to produce 6-phosphogluconate and NADPH(H$^+$). The reduced nucleotide offers reducing power to consistently convert oxidized glutathione (GSSG) to its reduced structure (GSH). The remaining 90% of erythrocytic glucose follows the glycolytic pathway and produces lactate.

Studies of GSH-Px have revealed that its activity is also dependent on a selenium-containing derivative of cysteine at its active site. This residue known as *selenocysteine* (H-Se-CH$_2$-CH(NH$_3^+$)COOH), seems to be critical for this enzyme mechanism as well as other selenium enzymes.

The fundamental importance of GSH-Px activity is somehow linked to copper-deficiency effects. According to Jenkinson and others, copper deficiencies yield decreased GSH-Px activity in the lungs and livers of test animals. Moreover, copper deficiencies also seem to minimize the activity of another critical enzyme, *superoxide dismutase* (SOD). This enzyme contains 2:1 ratio of zinc:copper per mole of protein. The copper atom participates in direct catalytic activities, whereas the zinc atoms have conformational functions.

Superoxide dismutase is essential for disposing superoxide radicals (:O$_2^-$), which develop from a single electron acquisition by O$_2$. The superoxide anion is extremely toxic to cells at the sites of unsaturated lipids located in membranes. Reactions of this type often produce fatty acid hydroperoxides. The SOD characteristically mediates elimination of the superoxide radical (a), and the resulting hydrogen peroxide product is then converted to water and oxygen by catalase (b):

(a) $\quad 2O_2^- + 2\ H^+ \xrightarrow{\text{SOD}} H_2O_2 + O_2$

(b) $\quad 2\ H_2O_2 \xrightarrow{\text{catalase}} 2\ H_2O + O_2$

Copper deficiencies in animals also seem to suppress the critical function of SOD in addition to GSH-Px. For the erythrocytes, reactions supplying superoxide anions are of more than academic interest since 2.5 to 3.0% of hemoglobin normally undergoes methemoglobin formation per day. This is prompted by the spontaneous reaction Hb(Fe^{2+}) + O$_2 \rightarrow$ MetHb(Fe^{3+}) + O$_2^-$. The phagocytic demise of pathogenic microorganisms by white blood cells is also believed to involve the formation of superoxide anions.

Food and Dietary Perspectives

Dietary availability of copper is hampered by sulfites and sulfates, which inhibit copper absorption by the formation of low-solubility complexes. Although ascorbic acid is beneficial for facilitating iron absorption, it interferes with copper absorption as a coexisting nutrient. Other studies emphasize, however, that *post*absorptive copper utilization by enzyme systems in tissues may actually be enhanced by ascorbic acid. Phytates, dietary fiber, and fiber constituents do not eliminate coper absorption; but undenatured copper-containing proteins inhibit assimilation of their copper content. So long as dietary rations for humans provide 1.24 to 1.35 mg/day, most biochemical requirements for copper will be met. For TPN, inadequate copper levels are more problematical (~20µg/kg/day for infants and 0.5–1.5 mg/day for adults).

Copper is notably present in foods in conjunction with polyphenoloxidases and tyrosinase, which are recognized for their discoloring (darkening) effects on cut fruit and vegetable tissues in particular. The prooxidant activity of copper on unsaturated fatty acids offers a constant threat to the integrity of native unsaturated fatty acids.

Plants and certain animals contain a variety of copper-containing compounds not found in mammals. For example, there is the copper-based oxygen carrier in the blood of *Gastropoda* (snails), *Cephalopoda* (octopus, etc.) and *Crustacea* exist as hemocyanin; ascorbic acid oxidase in plants that produces dehydroascorbate from ascorbate requires 8 to 12 mol of copper/mol of enzyme; and copper-containing *laccase* found in many fruits and vegetables (e.g., cabbage, beets, apples, asparagus, potatoes, fungi) has broad specificity for converting diphenols to quinones.

MANGANESE

Manganese deficiencies among the human populations are unknown since the normal diet provides adequate supplies for all unit biochemical operations. Foremost among manganese effects are its roles as a cofactor in enzyme reactions. Manganese and enzyme interactions can be classified in two ways: (1) *metalloenzymes* and (2) *metal–enzyme complexes*.

Classification by these two standards rests on mutual enzyme metal affinities, not functional interrelationships. Contrary to most other essential transition elements, examples of strict manganese-requiring enzymes are in short supply whereas *manganese-activated enzymes* (metal–enzyme complexes) are numerous. Typical among the last enzyme group are the kinases, hydrolases, decarboxylases, and transferases (e.g., glucosyl transferases for polysaccharide and glycoprotein syntheses).

Since enzyme activation for many enzymes may be marginally but not solely dependent on manganese (Mn^{2+}), obvious links between pathological manganese deficits and upsets in enzyme activities are not always clear. As evidence of multiple metallic ion suitabilities for certain enzymes, most kinases, over one-half of the recognized ligases (synthetases), and some dehydrogenases employ magnesium cations as cofactors (Mg^{2+}). However, Mg^{2+} can often be replaced by Mn^{2+} under *in vitro* conditions without a major loss in enzyme activity and/or zinc (Zn^{2+}) may perform similarly. The interchangeability among Zn^{2+}-, Mn^{2+}-, and Mg^{2+}-requiring ions attests to their similar ionic radii (e.g., Mg^{2+} = 0.065 nm, Mn^{2+} = 0.080 nm, Zn^{2+} = 0.069 nm), spherical electronic structures, or participation in octa- or tetra-hedral structures. Specific manganese containing metalloproteins include the lectins—*avimanganin* (avian liver), *manganin* (peanuts, *concanavalin A* (jackbean); and both *pyruvate carboxylase* (avian liver) and *superoxide dismutase* (*Escherichia coli*). Whether or not manganese is an absolute requirement in any way for human versions of the last two enzymes remains to be firmly established.

Intestinal absorption of manganese is inefficient at best, but dietary levels apparently ensure adequate manganese supplies. Speculation has swirled about the existence of a manganese transport protein in the blood known as *transmanganin*. This agent has proved to be either quite elusive, nonexistent, and/or possibly a β_1-globular protein. Postabsorptive manganese is nonetheless distributed to all tissues in a seemingly nonspecific way. Liver stores contain 1 to 3 ppm, whereas blood concentrations are about 2.5 to 100 μg/dL depending on the analytical method used. Mitochondrial structure is notably influenced by manganese deficiencies. This is evidenced by normal phosphate:oxygen consumption ratios during oxidative phosphorylation with reduced oxygen uptake as well as ultrastructural elongation plus cristae aberrations within the organelle.

Manganese homeostasis is controlled by the joint excretion routes of bile and intestinal secretions. Furthermore, the disposition of manganese may be related to genetic factors, at least in mice, where congenital ataxia—possibly related to defective otolith development within the inner ear—is tied to mutant coat-color genes (e.g., pallid types of mice). Some evidence also exists suggesting a connection between biogenic amines and manganese since chronic manganese poisoning and Parkinson's disease have symptomatic similarities. Furthermore, it is interesting to note that low levels of manganese in the blood can trigger serious convulsions in children that have been *temptingly* ascribed to lead poisoning.

Manganese deficiency can be experimentally induced in many animal species, but it may occur naturally in pigs and poultry. Infertility, ataxia, and skeletal disorders are typical. Poultry display "slipped tendon" perosis in which the hock joint is twisted and enlarged. Lameness and ataxia conditions are also demonstrated by pigs. Aside from similarities, however, manganese demands of poultry supersede those of mammals.

Rich sources of manganese in the human diet include nuts and unrefined grains, relatively less is found in fruits and vegetables, and most animal foods contain very little.

Balance studies using humans report manganese equilibrium or accretion with daily intakes ≥ 2.5 mg/day, whereas a negative balance occurs at 0.8 mg/day. Although mother's milk provides infants with about 15 µg/day, manganese is progressively lost during the first several weeks of infant life for unknown reasons.

MOLYBDENUM

Molybdenum (Mo) is the only heavy metal in the second transitional series that is essential for mammals. This element forms compounds that exhibit oxidation states in the range of $2+ \rightarrow 6+$. Molybdenum $(6+)$ is the most stable oxidation state, and it contains an electronically filled $4s$ shell although $4d$ orbitals remain for possible coordination with anionic ligands. Spectroscopic data indicate that coordination numbers of 4 and 6 are preferred, but eight ligands are possible in limited circumstances. The molybdenum oxycation MoO_2^{2+} is not uncommon to many observed complexes. It is also noteworthy that the coordination of MoO_2^{2+} with water (1 mol : 2 mol) yields such high proton acidity that a molybdate anion (MoO_4^{2-}) and protons are produced. Even in basic media, proton dissociation from coordination structures having lower oxidation states (e.g., $3+ \rightarrow 5+$) is far less than the $Mo(6+)$ state. The experimental record also shows that di- and polymeric bridge ions may be formed with molybdenum.

All recognized molybdenum-requiring enzymes display molecular weights of *over* 100,000. These enzymes notably include *xanthine oxidase, xanthine dehydrogenase, sulfite oxidase,* and *aldehyde oxidase.* The enzymes are mutually similar in that they also contain iron–sulfur proteins and FAD in addition to molybdenum. It is also theorized that iron components of molybdenum enzymes serve as electron carriers, while the molybdenum moiety provides a substrate-binding and redox site.

Emission spectrographic studies indicate that the liver, kidney, and adipose tissues display fairly consistent levels of molybdenum in most human subjects (e.g., approximately 2 mg liver; 0.3 mg blood and fat); however, the aortal, psoas muscle, tracheal, bone, ovarian, and uterine tissues contain little to undetectable amounts. Whole body stores are estimated to be less than 9 mg molybdenum for an idealized 70-kg person.

Studies using hexavalent molybdenum show that it is readily absorbed from the gastrointestinal tract and then deposited in the liver. Molybdenum contained in the hematogenous circulation is excreted mainly in the urine, but a molybdobiliary excretion route may contribute to a hepatointestinal cycle for this element. Since no rejection mechanism seems to exist for molybdenum absorption, hepatic control over molybdenum concentrations is probably critical. Mammalian requirements for molybdenum are based largely on rat studies, but for humans, the extrapolated daily consumption of molybdenum is probably about 120 µg for a 70-kg subject.

Overt molybdenum deficiency outside of TPN among the human population is unknown, but circumstantial evidence from animal and human studies attests to its nutritional importance.

As a key metal in *xanthine oxidase* (XO), molybdenum supports the conversion of xanthine to uric acid with the resulting production of hydrogen peroxide. This fact, coupled with geographic parallels involving high environmental molybdenum concentrations plus high incidences of gout-like syndromes (i.e., high molybdenum blood levels, XO, and uric acid), all spur on hypotheses that at least some disturbances in uric acid metabolism may be related to molybdenum. Since it is also recognized that dietary copper and sulfate are antagonistic to molybdenum biochemistry in sheep and cattle, and that excessive molybdenum incites copper deficiencies in important copper-requiring enzymes, these hypotheses have uncertain implications in the realm of human nutrition.

The physiological consequences of *aldehyde oxidase* activity in humans is not entirely clear. Nonetheless, since most aldehydes serve as vasodilators, aldehyde oxidase may be responsible for maintaining some degree of circulatory homeostasis. The generalized reaction mediated by this enzyme can be outlined as

$$RCHO \xrightarrow{O_2, H_2O} RCOOH + H_2O_2$$

The appearance of renal calculi (xanthine calculi) and dental caries may also mirror molybdenum deficiencies in some animal species, as indicated by Schroeder *et al.* (1970) and others.

The potential biochemical role for molybdenum is further complicated by its apparent requirement in the form of a possible pterinmolybdo cofactor for at least two enzymes, namely, *sulfite oxidase* (SO) and *xanthine dehydrogenase* (XD). Albeit rare, the lack of functional molybdenum cofactor seems to concur with severe mental retardation in documented cases of SO and XD deficiencies plus very low levels of hepatic molybdenum.

Unlike zinc, which is accumulated by the nuclei and DNA, or copper, which concentrates in mitochondria, molybdenum distribution throughout cells, even the liver, is seemingly uniform.

Richest dietary sources for molybdenum include meats, grains, and legumes, and among the poorest sources are vegetables, fruits, sugars, oils, and fats. Spinach (0.26 μg/g), molasses (0.20), mustard (0.60), wheat (5.15), sunflower seeds (1.03), corn syrup (Karo-type syrup) (0.85), and yams (0.60) are certainly among the highest molybdenum-containing foods, but the actual dietary availability of the metal in all foods requires further study.

For the major roles of molybdenum in plant and crop nutrition, the reader should review the role of molybdenum in nitrate reductase and nitrogenase activities as detailed in earlier sections.

CHROMIUM

Glycosuria, fasting hyperglycemia, corneal opacities, and aortic plaques have all been recorded in experimental animals as a result of severe chromium deficiencies. Aside from the outstanding parallels of these consequences with respect to human diabetes, the experimental record has not been able to affirm that human diabetes and chromium deficiencies are *absolutely* interdependent.

Although inorganic dietary chromium does not significantly improve human diabetic conditions, some evidence suggests that biochemical reactions for chromium are contingent on its incorporation into an aquodinicotinato compound. This compound is conventionally called the *glucose tolerance factor* (GTF). It is theorized that GTF interacts with *insulin* and somehow *potentiates* its characteristic effects on carbohydrate and fat metabolism.

Since GTF has been isolated from yeast, and yet its *in vivo* biosynthesis in animals is unconfirmed, many believe intestinal bacteria may convert chromium to GTF. Human evidence for this activity is marginal, but malnourished children having small bowel overgrowths of bacteria have responded to inorganic chromium with increased glucose tolerance.

Biologically active trivalent chromium (Cr^{3+}) supplied as chromic chloride is poorly absorbed over the digestive tract. In fact, less than 1 of 200-μg intakes are assimilated. Absorption mechanisms are unknown but hexavalent Cr^{6+} is absorbed better than Cr^{3+} forms. Scanty evidence shows that Cr^{3+} exists in plasma as a bound complex with transferrin, but Cr^{6+} may be the preferred form for absorption in erythrocytes, where it is reduced to Cr^{3+}. Redox enzymes tied to glutathione antioxidant effects may be responsible for Cr^{6+} to Cr^{3+} reductions.

Toxicological impacts of Cr^{6+} stem from its actions as a strong oxidizing agent. At high enough concentrations, Cr^{6+} effects can supersede the natural antioxidant effects of glutathione, ascorbic acid, α-tocopherol, and other protective mechanisms.

The extent of chromium deficiency among the American population is unknown, but based on (1) 0.75 to 0.85 μg/24-h urinary excretion rate, (2) an average 60 μg/day Cr^{3+} intake, and (3) less than 1% absorption, chromium status is marginally adequate at best for the population as a whole.

Chromium content of foods is variable and may represent the overall ability of plants and livestock to assimilate chromium from their respective nutrient and food supplies. Grasses may contain 0.1 to 0.5 ppm (dry weight), with some higher concentrations present in cereal grains and cereal germs. Until trace analysis for chromium in biological materials is absolutely reliable, the exact nutritional mechanism for chromium will remain clouded.

SELENIUM

Selenium was first recognized as a nutritional factor essential for preventing the death of liver cells in rats during 1957. Although selenium has been recognized for its poisonous properties, only 0.1 ppm of dietary selenium eliminates hepatic necrosis. Dietary selenium concentrations in the same range also allay the muscular dystrophic condition known as "white muscle disease" (WMD) in sheep and cattle that graze over selenium-deficient soils. The WMD is so-called because of white striations that develop in muscles.

The biochemical role of selenium is complicated by three factors. First, concentrations of ≤0.05 ppm are necessary to ensure the dietary health and welfare of many animal species. These minute concentrations of selenium impose an analytical hardship for elucidating its *in vivo* biochemical role plus detection of its deficiency effects. Second, although dietary requirements are miniscule for selenium, there is a relative closeness between its toxic and beneficial concentration levels. The third complicating factor is the fact that sharply contrasting deficiency effects for selenium are exhibited by different species. In addition, there is a biochemically unique interaction between vitamin E (α-tocopherol) and selenium. In some experimental animals, diseases can be partially prevented by vitamin E where selenium deficiencies are certain. Whether or not vitamin E can replace selenium in all cases is a matter for conjecture, but this possibility is unlikely. Apart from vitamin E–selenium interactions, dietary increases in cystcine or methionine may also minimize selenium-deficiency effects. Altogether, the *in vivo* variability of all these factors, coupled with the rapidity of chemical kinetics that selenium-dependent reactions seem to display, jointly frustrate many efforts designed to uncover the nutritional function of this element.

Selenium, a Group VI element, closely resembles sulfur in both its physical and chemical properties. In spite of its unknown functions, selenium has been shown to specifically replace sulfur atoms in many biomolecules.

Dietary studies of selenium must consider that it exists in multiple forms, each form displaying its own characteristic biochemical utility and bioavailability. *Inorganic forms* (H_2SeO_3 and H_2SeO_4) are obtained as strict soil mineral forms or by-products of selenium uptake by plants. Organic selenium compounds include *selenoamino acids* (Se-methionine, Se-cystine, Se-cysteine, Se-cystathione, and Se-methylselenocysteine), and *metabolic intermediates* of selenium include selenodiglutathione (GSSeSG), selenopersulfides (GSSeH), dimethyl selenide, and trimethyl selenonium. As indicated by Young, selenite or key selenoamino acids may be assimilated and converted by enzymatic and nonenzymatic systems to selenide. Selenide may then undergo (1) oxidation to elemental selenium, (2) interactive associations with plasma or cellular proteins, or (3) conversion to dimethylselenide or the trimethylselenonium (TMS) ion. The TMS ion serves as the physiological excretion product of selenium unless high or toxic levels are encountered, in which case selenide is exhaled.

Glutathione peroxidase (GSH-Px) is the most notable selenium-dependent enzyme in mammalian systems, although selected bacterial species also exhibit the selenoenzymes of glycine reductase, formate dehydrogenase, nicotinic acid hydroxylase, and thiolase.

As indicated in the previous section dealing with copper, GSH-Px requires copper as well as selenium for its activity. Some studies claim that the key Se-cysteine residue of GSH-Px is directed by a unique transcriptional codon, whereas others suggest that Se introduction into GSH-Px is a post-translational event. Whatever the case, the GSH-Px mediates the conversion of lipid hydroperoxides and hydrogen peroxide, respectively, to hydroxy acids and water. On this basis, most evidence supports the role of dietary selenium in a secondary antioxygenic capacity as a hydroperoxidase reducer, whereas vitamin E displays primary-chain antioxidant effects to counteract *in vivo* lipid peroxidation.

Firm pathological evidence supporting nutritional selenium deficiency in humans is rare *but* convincing. Keshan disease, an endemic cardiomyopathy occurring in the People's Republic of China, has been linked to low selenium levels in the body. Sodium selenite supplementation experiments have been shown to lower the morbidity of the disease in many patients, alleviate many of the clinical signs of the disease, and improve the prognosis for the disease. Many investigators of Keshan disease indicate, however, that selenium deficiency is not the only causative factor. Other evidence for human selenium requirements comes from Se-methionine responsive disappearances of muscle weakness in long-term TPN-sustained patients, as well as growth stimulatory effects of selenium on cultured human fibroblasts.

Selenium uptake from the diet is undoubtedly affected by coexisting foodstuffs, but balance studies reported by Levander *et al.* (1981) indicate that 70 μg/day selenium intakes are necessary to replace body losses and maintain whatever body stores exist.

Most vegetables and fruits contain 0.01 μg/g, and higher amounts are present in other foods such as seafoods (0.3–0.7 μg/g), meat (0.15–2.00 μg/g), and grain foods (0.03–0.06 μg/g). It should be recognized, however, that gross selenium may not be an accurate measure of bioavailable selenium in foods, and conventional food processing can produce notable losses of this trace element.

VANADIUM

Vanadium is a ubiquitous trace element of soils and ocean water. A general survey shows that animal tissues contain about 0.1 ppm. For humans, the adult body contains nearly 30 mg of vanadium. However, unlike cadmium and lead, which accumulate with age and exposure, vanadium seems to be controlled by some type of homeostatic mechanism.

Vanadium is noted for its catalytic actions and ability to establish chelates; however, vanadium metalloproteins are largely unknown. Vanadium displays oxidation states of $2+ \rightarrow 5+$, which suggests that it may be involved in redox reactions.

Deficiency effects for vanadium were first uncovered around 1971 with evidence of increased hematocrits and decreased bone developments observed in chicks and rats.

In vitro studies of Na^+, K^+-ATPase as well

as rodent studies have affirmed the inhibitory effect of vanadate (VO_4^{3-}) on ATPase (sodium pump) actions as well as its ability to exert powerful diuretic effects. In addition, studies have also revealed that vanadate synergistically interacts with potassium and magnesium to alter ATPase responses to extracellular potassium. These observations and others showing the rapid yet specific uptake of vanadium by the renal cortex support its role in kidney function. It is also theorized that vanadium deficiency may contribute to salt and water retention in cases such as nutritional edemas. Another physiological effect of vanadium is associated with its ability to increase the force of ventricular contractions and yet inhibit atrial contractions.

Biochemical studies of vanadium are complicated by the conversion of the vanadate (5+) ion to the vanadyl (4+) form in the body. Each species seems to produce different exhibitions of biochemical activity, and the reduced form results from an $NADH(H^+)$-dependent vanadate reductase located in cell membranes. From this point on, the biochemistry of vanadium is highly speculative.

Vanadium deficiencies have been described in chicks and rats in which typical indications of deficiency appear as elevated plasma cholesterol and hematocrit values, abnormal bone development, and decreased growth. Upsets in cholesterol metabolism and regulation have also been suspected to involve vanadium in humans. The essentiality of vanadium as a nutrient is clouded, however, since the conditions *and* unassailable deficiency signs for dietary inadequacies of vanadium are unknown.

Rat and chick estimates for vanadium range from 50 to 500 ng/g. Using these vanadium body levels as an index, and based on the apparent vanadium requirements for animals, Nielsen (1979) and Myron *et al.* (1978) have speculated that contemporary human diets may not provide adequate amounts of this element.

Concentrations of vanadium in human foods are highest in vegetable fats and oils, animal flesh contains about 10 ng/g, and both pulses and root crops contain 1.0 ng/g.

BIOCHEMICAL ROLES FOR MANY TRACE ELEMENTS ARE UNCERTAIN

Many other trace elements may be required in miniscule amounts to sustain the normal nutritional status of animals including humans. Until conventional methods of trace element analysis become more routinely available, the questionable nutritional contributions for many of these elements will remain confused and widely disputed.

Arsenic has been shown to be an essential nutrient for five animal species, and speculation concerning its essential nutrient functions in humans should continue to grow. Aside from its toxicity, evidence suggests that arsenic *may* catalyze the biosynthesis of glutathione. Furthermore, since the biochemical roles of selenium and arsenic are similar and closely attuned to the metabolism of sulfur compounds, the future unraveling of selenium's biochemical and anticarcinogenic roles may clear up some *in vivo* roles for arsenic.

Lithium has been recognized for about 35 years as a tool for controlling violent mood swings in manic-depressives and others exhibiting related mental disorders. Only recently, lithium has been shown to slow down choline (a precursor to the acetylcholine transmitter) transport both into and out of the brain, thereby affecting manic-depressive states.

Dietary aluminum has been claimed by some studies to induce neurological disorders through unknown etiological mechanisms, and yet other studies claim that dietary aluminum completely suppresses blood cholesterol levels. Aluminum–fiber complexes formed in the intestines may provide a transient matrix that binds dietary fat and promotes intestinal transport of cholesterol as a side effect.

Quantitative requirements for dietary silicon are not established, but it clearly serves as a cross-linking agent in mucopolysaccharides and other components of connective tissue and skin. Silicon is notably present in regions of active calcification.

Nickel is required by several animal species and probably humans in small amounts. For rats, nickel promotes optimal growth and reproduction, and in the case of chicks, nickel promotes optimal hepatic functions. Dietary absorption for nickel is low, 1 to 10% for ordinary diets, and postassimilative transport occurs as an unfilterable, albumin-bound compound and/or as a component of metalloproteins. Nickel distribution throughout body tissues is widespread, but its biochemical impact may be complicated by interactions with iron supplements.

The experimental nutritional record for tin is scanty, but it may function as a catalyst in redox, transesterification, and/or polymerization reactions.

Cobalt has no recognized function in human biochemistry other than its role in vitamin B_{12}, where it is coordinated into the structure of coenzyme B_{12}. The role of free cobalt as a cofactor in any other system is unknown.

Fluoride contributes to the structure of bone and teeth in which it is intrinsically involved in calciferous crystal formation. It is also recognized as an activator for (1) hepatic citrulline synthesis in some animals and (2) adenyl cyclase, but it may complex with magnesium cofactors of certain enzymes to inhibit their activities. Vegetable fluoride content is generally low, but spinach may contain up to 250 $\mu g/100$ g. Fish and marine foods are highest in fluoride (600–700 $\mu g/g$) followed by beef, tea (100–150 $\mu g/g$), and milk (15–20 $\mu g/100$ g).

Iodine biochemistry revolves about its previously described role in thyroid activities. Saltwater fish contain iodine concentrations of 75 to 175 $\mu g/100$ g, although shellfish may have up to 450 $\mu g/100$ g. Iodine concentrations are minimal at best in other foods and rarely meet the 60 $\mu g/day$ turnover of iodine body stores unless augmented by iodine-rich foods or supplementation.

SOME TRACE ELEMENTS CAN BE TOXIC

A fine line separates the toxic effects of some trace elements from their desirable dietary intake levels as well as other related elements of the periodic series. Many enzymes typically require *activators* that enhance the functional properties of their active sites. *Inhibitors* also exist that have the reverse effect of suppressing active-site functions.

Among the trace elements, heavy metals including mercury (Hg^{2+}), cadmium (Cd^{2+}), and lead (Pb^{2+}) are noted for their enzyme-inhibitory effects. Since —SCH_3 (in methionine) and —SH (in cysteine) groups are often present at enzyme-active sites, the innate tendency of these cations to react with sulfur can disturb the functional integrity of enzyme conformations. A sample reaction could be expressed as

$$\text{ENZYME}\left(\begin{array}{c}\text{SH}\\\text{SH}\end{array}\right) + M^{2+} \longrightarrow \text{ENZYME}\left(\begin{array}{c}\text{S}\\\text{S}\end{array}M^{2+}\right) + 2H^+$$

Normal active site   Distorted active site

The interchangeability of many Zn^{2+}-requiring metalloenzymes with Cd^{2+} can result in a Cd^{2+}-poisoned enzyme. Representative Cd^{2+}-poisoned enzymes include alcohol dehydrogenase, amylase, adenosine triphosphatase, carbonic anhydrase, carboxypeptidase, and glutamic-oxaloacetic transaminase. In the case of other elements, tri- and pentavalent arsenics inhibit pyruvate dehydrogenase; lead (Pb^{2+}) inhibits acetylcholinesterase, alkaline phosphatase, adenosine triphosphatase, δ-aminolevulinic acid dehydrase, carbonic anhydrase, and cytochrome oxidase; and mercury (Hg^{2+}) inhibits alkaline phosphatase, glucose 6-phosphatase, and lactic dehydrogenase.

The arsenite radical can also exert toxic effects on enzymes by reacting with key functional sulfhydryl groups:

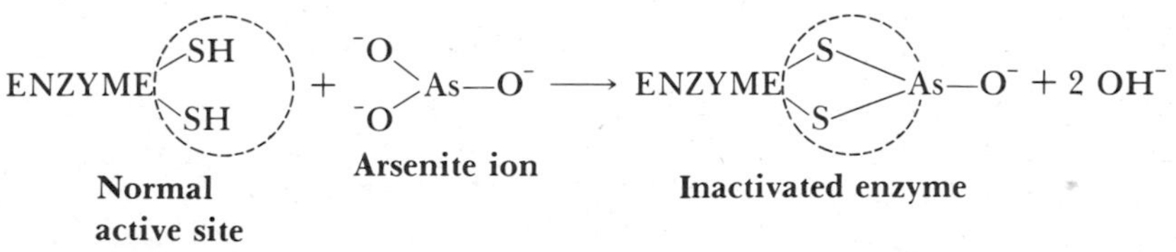

The biochemical effects of mercury are well known, although its exact toxic mechanisms are still open to study. Monomethyl (CH_3Hg^+) and dimethyl (($CH_3)_2Hg$) mercuries arise via the vitamin B_{12}-mediated methylation of elemental mercury. These organification reactions are especially carried out by anaerobic methane-producing bacteria. Studies have shown that acidic conditions prompt the formation of water-soluble CH_3Hg^+ from ($CH_3)_2Hg$. The methylmercury form is readily assimilated into the food chain (e.g., fishes, shellfish) by planktonic species. Covalent bond stability

$$\text{ENZYME}\begin{array}{c}\text{SH}\\\text{SH}\end{array} + \begin{array}{c}^-\text{O}\\^-\text{O}\end{array}\text{As}-\text{O}^- \longrightarrow \text{ENZYME}\begin{array}{c}\text{S}\\\text{S}\end{array}\text{As}-\text{O}^- + 2\text{ OH}^-$$

Normal active site **Arsenite ion** **Inactivated enzyme**

Reactions of this type notably occur on the lipoic acid structures of many enzyme systems and disturb their overall functions (e.g., pyruvate dehydrogenase, which is critical for oxidative degradation of pyruvate). Aside from the direct reaction of arsenite (AsO_3^{3-}) with sulfhydryl-rich molecular substituents, this species can also upset substrate level phosphorylations that yield ATP in glycolysis. Normal phosphorylation of glyceraldehyde 3-phosphate yields 1,3-diphosphoglycerate, but the collateral behavior of arsenite (ASO_3^{3-}) as phosphate (PO_4^{3-}) can alternatively produce 1-arseno-3-phosphoglycerate.

Calcium (Ca^{2+})–lead (Pb^{2+}) interactions readily occur since bone can serve as a reservoir for both substances. So long as bone structure is static or growing, any effects of lead accumulation may be quite innocuous. However, mobilization of lead and other components of bone structure can also free stored Pb^{2+}, which then exerts its enzyme-toxic effects. Since δ-*aminolevulinic acid dehydrase* is generally affected, heme synthesis is clearly disturbed by the liberation of Pb^{2+}

of alkyl mercury compounds permits their protracted existence in biological tissues, where they can exert severe biotoxic consequences. Although the alkyl mercury structure favors solubility in the lipid fractions of tissues and membranes, these compounds are transient to the extent that they can transgress placental membranes and enter fetal tissues.

Generic symptoms of mercury poisoning include cerebral palsy, mental retardation, convulsions and other neurological disorders, aberrant chromosomal segregations, and cleavage. Biochemical defects incited by mercury compounds probably reflect combined effects of mercurial binding at sulfur-rich loci on enzymes, serum albumins, and hemoglobin; inhibited membrane transport of sugars to neural cells; and hyperkalemic permeability over neuronal membranes.

CONCLUSION

Some striking epidemiological correlations exist between certain diseases in the human population and the mineral profiles of their local food crop-producing soils. For example, demographic studies show that there has been an inverse relationship between deaths from breast and colon cancer and soil deposits of selenium. In the United States, this demographic band runs through the central states of Texas to Iowa and Illinois. Although the active accumulation of soil selenium by selenaceous plants can lead to "blind staggers" and adverse selenium exposure consequences in livestock, these effects have not been generally observed in the human population. Using the implications of selenium as a backdrop, it is clear that the habitual consumption of foods produced from soils rich or poor in certain elemental species can affect health status of a human population.

Opposite correlate effects have been reported where certain soils seem to induce stomach cancers. This is true for certain soils in France, the Netherlands, Wales, and Scandanavian countries, which are typically acid, water-logged, and rich in humic constituents. Similar "cancer prone" soils are suspected in Japan and Chile, where stomach cancer has a notable occurrence. Fundamental differences exist in these soils (European vs. South American), but the chemistry of the allophanic clays common to the soils, their humus content, or trace element profiles may hold the key to their respective roles in health and disease. The record is far from clear.

Agricultural plant crops and livestock serve as incidental "ambassadors" of the mineral "profiles" in soils and geographic areas where they were produced. Fortunately, the present food distribution system of the United States tends to supply a homogeneous variety of foods that reflect the rich and poor mineral deficiencies of many geographic regions. So long as these counterbalances in key macro- and trace elements are realized, elemental nutrition among the contemporary population will remain generally adequate in a balanced dietary regimen. Unfortunately the macro- and trace elemental status of many food crops is becoming increasingly compromised and threatened by

1. Repetitive farming of a single crop on the same soil year after year, although good agricultural practices dictate crop rotation.

2. Miniscule amounts of attention paid to soil conservation while key minerals and elements are leached from soils by acid rains, runoff, and other environmental stresses.

3. Overgrazing by cattle and livestock that threatens to exacerbate elemental deficiencies in pastures and grazing lands.

Few examples of the consequences of grazing are more impressive with regard to elemental losses than winter grazing by livestock. Winter-grazed pastures demonstrate about 13 times more water runoff than summer-grazed pastures, nearly 24 times the nitrogen loss, 28 to 30 times the soluble salt loss, and 35 times the organic carbon loss.

Growing populations, environmental stresses, and high expectations of food productivity for every acre of arable land are steadily growing. Like any natural resource, the soil and its constituent elements must be managed and conserved for the future production of food resources. It is patently obvious that only a 6-in. blanket of topsoil, with its combined humus and inorganic elements, generally separates the entire human population from many uncertain nutritional deficiency effects and sheer starvation.

REFERENCES

General references and element–element interactions

Bunk, M. J., and G. F. Combs Jr. 1981. Relationship of selenium-dependent glutathione peroxidase activity and nutritional pancreatic atrophy in selenium deficient chicks. *J. Nutr.* **111:**1611.

Fischer, P. W., A. Giroux, and M. R. L'Abbe. 1981. The effect of dietary zinc on intestinal copper absorption. *Amer. J. Clin. Nutr.* **34:**1670.

Hill, C. H., and G. Matrone. 1970. Chemical parameters in the study of *in vivo* and *in vitro* interactions of trace elements. *Fed. Proc.* **29**:1474.

Irwin, M. I. (Ed.). 1980. *Nutritional Requirements of Man: A Conspectus of Research*. The Nutrition Foundation, New York/Washington. (Excellent review of calcium, zinc, iron, and copper.)

Jenkinson, S. G., R. A. Lawrence, R. F. Burk, and D. M. Williams. 1982. Effect of dietary deficiency on the activity of the selenoenzyme glutathione peroxidase and on excretion and tissue retention of $^{75}SeO_3^{2-}$. *J. Nutr.* **112**:197.

Johnson, J. L., H. P. Jones, and K. V. Rajagopalan. 1977. *In vitro* reconstitution of demolybdosulfite oxidase by a molybdenum cofactor from rat liver and other sources. *J. Biol. Chem.* **252**:4994.

McCarron, D. A., C. D. Morris, and C. Cole. 1982. Dietary calcium in human hypertension. *Science* **217**:267.

National Academy of Sciences. 1980. *Recommended Daily Allowances*, 9th ed. NAS, Washington, D.C.

Ochiai, Ei-ichiro. 1977. *Bioinorganic Chemistry: An Introduction*. Allyn & Bacon, Boston.

Riordan, J. F., and B. Vallee. 1976. Structure and function of zinc metalloenzymes. In *Trace Elements in Human Health and Disease*, Vol. I, *Zinc and Copper* (A. S. Prasad, ed.), pp. 227–256. New York, Academic Press.

Solomons, N. W., and R. A. Jacob. 1981. Studies on the bioavailability of zinc in humans: IV. Effects of heme and nonheme iron on the absorption of zinc. *Amer. J. Clin. Nutr.* **34**:475.

Sunde, R. A., and W. G. Hoekstra. 1980. Structure, synthesis and function of glutathione peroxidase. *Nutr. Rev.* **38**:265.

Valberg, L. S., P. R. Flanagan, J. Haist, J. V. Frei, and M. J. Chamberlain. 1981. Gastrointestinal metabolism of gallium and indium: Effect of iron deficiency. *Clin. Invest. Med.* **4**:103.

Williams, D. M., R. E. Lynch, G. R. Lee, and G. E. Cartwright. 1975. Superoxide dismutase activity in copper deficient swine. *Proc. Soc. Exp. Biol. Med.* **149**:534.

Selected references for copper

Disilvestro, R. A., and E. D. Harris. 1981. A postabsorption effect of L-ascorbic acid on copper metabolism in chicks. *J. Nutr.* **111**:1964.

Evans, G. W. 1973. Copper homeostasis in the mammalian system. *Physiol. Rev.* **53**:535.

Harvey, P. W., and K. G. D. Allen. 1981. Decreased plasma lecithin:cholesterol acyltransferase activity in copper-deficient rats. *J. Nutr.* **111**:1855.

Kuivaniemi, H., L. Peltonen, A. Palotie, I. Kaitila, and K. I. Kivirikko. 1982. Abnormal copper metabolism and deficient lysyl oxidase activity in a heritable connective tissue disorder. *J. Clin. Invest.* **69**:730.

Lau, B. W., and L. M. Klevay. 1981. Plasma lecithin:cholesterol acyltransferase in copper deficient rats. *J. Nutr.* **111**:1698.

Selected references for iron

Brittenham, G. M., E. H. Danish, and J. W. Harris. 1981. Assessment of bone marrow and body iron stores: Old techniques and new technologies. *Semin. Hematol.* **18**(3):194.

Cook, J. D., and C. A. Finch. 1979. Assessing iron status of a population. *Amer. J. Clin. Nutr.* **32**:2115.

Cook, J. D., and E. R. Monsen. 1976. Food iron absorption. III. Comparison of the effects of animal proteins on nonheme iron. *Amer. J. Clin. Nutr.* **29**:859.

Finch, C. A., and H. Huebers. 1982. Perspectives in iron metabolism. *N. Engl. J. Med.* **306**:1520.

Finch, C. A., H. Huebers, M. Eng, and L. Miller. 1982. Effect of transfused reticulocytes on iron exchange. *Blood* **59**:364.

Finch, C. A., L. R. Miller, A. R. Inamadar, R. Person, K. Seiler, and B. Mackler. 1976. Iron deficiency in the rat: Physiological and biochemical studies of muscle dysfunction. *J. Clin. Invest.* **58**:447.

Green, R., R. Charlton, H. Seftel, T. Bothwell, F. Mayet, B. Adams, C. Finch, and M. Layrisse. 1968. Body iron excretion in man. A collaborative study. *Amer. J. Med.* **45**:336.

Hallberg, L. 1981. Iron nutrition in women in industrialized countries. *Bibl. Nutr. Diet.* **30**:111.

Jacob, R. A., H. H. Sandstead, L. M. Klevay, and L. K. Johnson. 1980. Utility of serum ferritin as a measure of iron deficiency in normal males undergoing repetitive phlebotomy. *Blood* **56**:786.

Linder, M. C., and H. N. Minro. 1977. The mechanism of iron absorption and its regulation. *Fed. Proc.* **36**:2017.

Lozoff, B., G. M. Brittenham, F. E. Viteri, A. W.

Wolf, and J. J. Urrutia. 1982. The effects of short-term oral iron therapy on developmental defects in iron-deficient anemic infants. *J. Pediatr.* **100:**351.

Schricker, B. R., M. D. Gilbert, D. D. Miller, and D. Van Campen. 1982. Biological effects of substituting enriched [54Fe] for natural iron in rats. *J. Nutr.* **112:**151.

Selected references for manganese

Erway, L., L. S. Hurley, and A. S. Fraser. 1970. Congenital ataxia and otolith defects due to manganese deficiency in mice. *J. Nutr.* **100:**643.

Failla, M. L., and R. A. Kiser. 1981. Altered tissue content and cytosol distribution of trace metal in experimental diabetes. *J. Nutr.* **111:**1900.

Hurley, L. S. 1981. Teratogenic aspects of manganese, zinc and copper nutrition. *Physiol. Rev.* **61:**249.

Knox, D., C. B. Cowey, and J. W. Adron. 1981. The effect of low dietary manganese intake on rainbow trout (*Salmo gairdneri*). *Br. J. Nutr.* **46:**491.

Leach, R. M. 1976. Metabolism and function of manganese. In *Trace Elements in Human Health and Disease*, Vol. II, *Essential and Toxic Elements* (A. S. Prasad, ed.), pp. 235–247. Academic Press, New York.

Selected references for molybdenum

Abumrad, N. N., A. J. Schneider, D. Steel, and L. S. Rogers. 1981. Amino acid intolerance during prolonged total parenteral nutrition reversed by molybdate therapy. *Amer. J. Clin. Nutr.* **34:**2551.

Johnson, J. L., W. R. Waud, K. V. Rajagopalan, M. Duran, F. A. Beemer, and S. K. Wadman. 1980. Inborn errors of metabolism: Combined deficiencies of sulfite oxidase and xanthine dehydrogenase in a patient lacking the molybdenum factor. *Proc. Natl. Acad. Sci. (USA)* **77:**3715.

Nederbragt, H., and C. J. A. van den Hamer. 1981. Influence of dietary molybdenum on the metabolism of intravenously injested radioactive copper in the rat. *J. Inorg. Biochem.* **15:**281.

Nederbragt, H., and C. J. A. van den Hamer. 1981. Changes in the binding of copper in the plasma of molybdenum supplemented rats. *J. Inorg. Biochem.* **15:**293.

Schroeder, H. A., J. J. Balassa, and I. H. Tipton. 1970. Essential trace metals in man: Molybdenum. *J. Chron. Dis.* **23:**481.

Tsongas, T. A., R. R. Meglen, P. A. Walravens, and W. R. Chappell. 1980. Molybdenum in the diet: An estimate of average daily intake in the United States. *Amer. J. Clin. Nutr.* **33:**1103.

Selected references for nickel, vanadium, and arsenic

Horak, S. E., and F. W. Sunderman. 1973. Fecal nickel excretion in healthy adults. *Clin. Chem.* **19:**429.

Myron, D. R., T. J. Zimmerman, T. R. Schular, L. M. Klevay, D. E. Lee, and F. H. Nielsen. 1978. Intake of nickel and vanadium by humans: A survey of selected diets. *Amer. J. Clin. Nutr.* **31:**527.

Nielsen, F. H. 1979. Evidence for the essentiality of arsenic, nickel and vanadium and their possible nutritional significance. In *Advances in Nutritional Research* (H. Draper, ed.), Vol. II, pp. 157–173. Plenum, New York.

Rubányi, G., L. Ligeti, and Á. Koller. 1981. Nickel is released from the ischemic myocardium and contracts coronary vessels by a calcium-dependent mechanism. *J. Mol. Cell. Cardiol.* **13:**1023.

Solomons, N. W., F. Viteri, T. R. Shuler, and F. H. Nielsen. 1982. Bioavailability of nickel in Man: Effects of foods and chemically-defined dietary constituents on the absorption of inorganic nickel. *J. Nutr.* **112:**39.

Selected references for selenium

Diplock, A. T. 1981. Metabolic and functional defects in selenium deficiency. *Phil. Trans. R. Soc. London (Biol).* **294**(1071):105.

Dougharty, J. J., W. A. Craft, and W. G. Hoekstra. 1981. Effects of ferrous chloride and iron dextran on lipid peroxidation *in vivo* in vitamin E and selenium adequate and deficient rats. *J. Nutr.* **111:**1784.

Heinrich, H. C., E. E. Gabba, H. Bartels, K. H. Oppitz, O. Ch. Bender-Götze, and A. A. Pfau. 1977. Bioavailability of food iron-[59Fe], vitamin B12-[60Co] and protein bound selenomethionine-[75Se] in pancreatic exocrine insufficiency due to cystic fibrosis. *Klin. Wochenschr.* **55:**595.

Keshan Disease Research Group of the Chinese

Academy of Medical Sciences. 1979. Observation on effects of sodium selenite in prevention of Keshan disease. *Chinese Med. J.* **92**:471.

Keshan Disease Research Group of the Chinese Academy of Medical Sciences. 1979. Epidemiological studies on the etiologic relationship of selenium and Keshan disease. *Chinese Med. J.* **92**:477.

Levander, O. A., B. Sutherland, V. C. Morris, and J. C. King. 1981. Selenium balance in young men during selenium depletion and repletion. *Amer. J. Clin. Nutr.* **34**:2662.

Young, V. R., A. Nahapetian, and M. Janghorbani. 1982. Selenium bioavailability with reference to human nutrition. *Amer. J. Clin. Nutr.* **35**:1076.

Selected references for zinc

Bettger, W. J., M. S. Fernandez, and B. L. O'Dell. 1980. Effect of zinc deficiency on zinc content of rat red cell membranes. *Fed. Proc.* **59**:896. (abstr.).

Bettger, W. J., and B. L. O'Dell. 1981. A critical physiological role of zinc in the structure and function of biomembranes. *Life Sci.* **28**:1425.

Chandra, R. K., and B. Au. 1980. Single nutrient deficiency and cell-mediated immune responses. I. Zinc. *Amer. J. Clin. Nutr.* **33**:736.

Chvapil, M. 1973. New aspects in the biological role of zinc: A stabilizer of macromolecules and biological membranes. *Life Sci.* **13**:1041.

Cousins, R. J. 1979. Regulation of zinc absorption: Role of intracellular ligands. *Amer. J. Clin. Nutr.* **32**:339.

Golden, B. E., and M. H. N. Golden. 1981. Plasma zinc, rate of weight gain and the energy cost of tissue deposition in children recovering from severe malnutrition on a cow's milk or soya protein based diet. *Amer. J. Clin. Nutr.* **34**:892.

Golden, M. H., and B. E. Golden. 1981. Trace elements: Potential importance in human nutrition with particular reference to zinc and vanadium. *Br. Med. Bull.* **37**:31.

Gordon, E. F., R. C. Gordon, and D. B. Passal. 1981. Zinc metabolism: Basic, clinical and behavioral aspects. *J. Pediatr.* **99**:341.

Gordon, P. R., and B. L. O'Dell. 1980. Rat platelet aggregation impaired by short-term deficiency. *J. Nutr.* **110**:2123.

Horrobin, D. F., and S. C. Cunnane. 1980. Interaction between zinc, essential fatty acids and prostaglandins: relevance to acrodermatitis enteropathica, total parenteral nutrition, the glucagonoma syndrome, diabetes, anorexia nervosa, and sickle cell anemia. *Med. Hypotheses* **6**:277.

Klaiman, A. P., W. Victery, M. J. Kluger, and A. J. Vander. 1981. Urinary excretion of zinc and iron following acute injection of dead bacteria in dog. *Proc. Soc. Exp. Biol. Med.* **167**:165.

Leucke, R. W., and P. J. Fraker. 1979. The effect of varying dietary zinc levels on growth and antibody-mediated response in two strains of mice. *J. Nutr.* **109**:1373.

Neldner, K. H., and K. M. Hambidge. 1975. Zinc therapy of acrodermatitis enteropathica. *N. Engl. J. Med.* **292**:879.

Smith, K. T., and R. J. Cousins. 1980. Quantitative aspects of zinc absorption by vascularly perfused rat intestine. *J. Nutr.* **110**:316.

Solomons, N. W. 1979. On the assessment of zinc and copper nutriture in man. *Amer. J. Clin. Nutr.* **32**:856.

Solomons, N. W. 1981. Zinc and copper in human nutrition. In *Nutrition in the 1980's* (P. White and N. Selvey, eds.), pp. 97–127. Liss, New York.

Solomons, N. W. 1982. Biological availability of zinc in humans. *Amer. J. Clin. Nutr.* **35**:1048.

Weigand, E., and M. Kirchgessner. 1980. Total true efficiency of zinc utilization: Determination and homeostatic dependence upon the zinc supply status in young rats. *J. Nutr.* **110**:469.

C H A P T E R · 20

Toxicants and Undesirable Food Constituents

INTRODUCTION

Chemical analysis of uncooked foods reveals a bewildering array of different chemical molecules. Sugars, amino acids, lipids, vitamins, and minerals offer nutritional value, but many other components can create adverse effects on normal human health. Potentially dangerous organic molecules in foods include alkaloids, glycosides, sterols, amines, polypeptides, and many other miscellaneous categories of compounds. This spectrum of molecular types is also complicated by the presence of organic environmental contaminants that can be assimilated into plant and animal tissues.

Many of these molecules fail to exert any useful nutritional function and simply transit the gastrointestinal tract in an unaltered form. Other molecules present in foods at higher concentrations instigate a genuine health risk by virtue of their chronic daily consumption or their prolonged additive accumulation in the body brought about by multiple dietary exposures.

Apart from accumulating harmful dietary constituents, the hepatic tissues of animals and humans have limited abilities that permit the chemical dismantling of toxic substances. This action often yields water-soluble derivatives of toxicants, which are subsequently excreted in the urine or bound to fecal matrices after biliary excretion.

As a result of the historical evolution of mankind, humans have learned to omit certain harmful and poisonous foods from their diets. Unfortunately, the historical basis for selecting foods has been set into turmoil with the advent of fugitive and toxic industrial wastes in the environment. Food sources formerly regarded as safe have demonstrated abilities for accumulating these contaminants and expressing their toxic effects via dietary exposures on humans. These effects may materialize as allergic reactions, mutagenesis, cancer, or other health ramifications months or years after exposure to the toxicant.

Under the proper set of exposure conditions, virtually any food constituent can undermine human health. Thus, food toxicology deals with defining *degrees of hazard and safety for food constituents under prescribed sets of human-dose exposures*. Since different food constituents have different modes of adverse *in vivo* biochemical actions, the relative toxicities of individual food constituents are widely varied.

ACUTE TOXICITY AND SHORT-VERSUS LONG-TERM TOXICITY STUDIES

Almost any food constituent can be scored on the basis of its relative toxic effects compared to other substances. As expected, the biological effect of any potentially toxic compound depends on its dietary dosage and consumption frequency. Chemical compounds that exert adverse biological effects at *very low* concentrations and exposure frequencies are considered to be *more* toxic than other substances having similar effects at *much higher* dosages and exposure frequencies.

The ranking of toxic compounds according to their relative toxicities and potential health impacts is a difficult mission. Many low-dose exposures to toxic compounds ex-

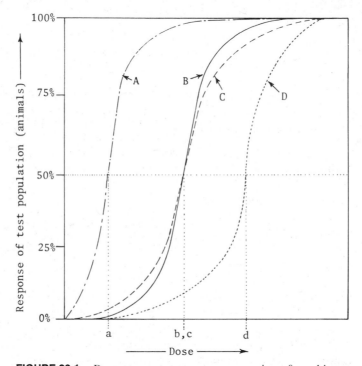

FIGURE 20.1. Dose–response curves assume a variety of graphic positions depending on testing conditions, toxicant exposure methods, etc. The response scale (ordinate) is *not specified here,* but it may measure *lethality* for a chemical agent or some other biological/biochemical parameter that can be accurately quantitated. For every lethal toxicant effect or specified sublethal effect, there will be a *respective* average lethal dose (LD_{50}) for the test population *or* some average toxic dose. In any event, 50% of the test subject population responds one way or the other to some prescribed toxicant dose range. Curve A illustrates a sharp dose–response relationship; curves B and C also represent sharp dose–response curves similar to A, but the average dose (b) of toxicant to achieve a 50% response is displaced to the right, indicating that it is bigger than the corresponding *a* value (less toxic). **Note:** Many relative displacements in dose–response curves can exist and yet reflect an identical average dose (see curves B and C where *b* and *c* values are identical). Curve D indicates an average toxic dose (*d*) at a higher concentration than *a, b,* or *c less* relative toxicity, and a larger response range.

hibit only subtle effects at best. For example, if only 1% of a test animal population shows a toxic response to a chemical agent, a minimum of several hundred to thousands of animal subjects may be required to establish (1) dose–response effects and (2) statistically precise toxic-exposure guidelines for humans.

The practical and statistical problem of assigning toxicity rankings to compounds dictates a more efficacious approach. For this reason lethal-dose concentrations of chemical agents are often administered to test animals in order to rank their relative acute toxicities (Figure 20.1). One common expression of acute toxicity is the LD_{50} value. This term refers to the *average lethal dose* of a specific compound that will kill 50% of a test animal population after its administration. The LD_{50} value is established using animals of the same species, strain, sex, and age. Moreover, the administration route for the toxicant may involve parenteral routes, but for nutritional investigations gavage (stomach tube) or normal dietary administrations can be employed.

Other commonly used parameters for acute toxicity based on lethal doses include

TDL_0, *toxic dose low.* The lowest known dose of a substance that has produced any toxic, carcinogenic, teratogenic, or neoplastic (tumorous) effects.

TCL_0, *toxic concentration low.* The lowest concentration of a substance that is reported to have any toxic effect.

TLV, *threshold limit value.* An estimate of the average safe toxicant concentration that can be tolerated on a repetitive basis, usually an 8-h period on a day-to-day basis.

TWA, *time weighted average.* A mathematical expression summing the products of toxicant concentrations and durations of exposure to those toxicants and dividing by the total exposure time. In simple terms, the concentrations of the various toxicants are multiplied by the duration of exposure to each individual toxicant,

and the results are added and then divided by the time of exposure.

LC_{50}, *average lethal concentration* (rarely used for foods except for certain volatiles). That concentration of an airborne toxicant that when administered to test animals for a defined period of time kills 50% of them. This value is usually expressed in parts per million for the specific time period.

Toxicological testing protocols can also be defined according to their duration and endpoints. That is, short-term tests may require 90 days to acquire useful data regarding exposure toxicity and animal responses to a specific substance. These tests may include simple 2-h LD_{50} range-finding tests that allow a rapid assessment of toxicity for a substance, or more complex tests such as 90-day continuous exposure or paired-feeding tests. Toxicity tests of 90+ days and up to 2 years are classified as long-term tests. These are among the most expensive and complicated tests to run.

Human toxicity of many intentional and accidental food constituents can be extrapolated from animal models and reported as a *minimum lethal dose* (MLD). Many MLD values assume the weight of a 150-lb test subject. Animal models must be carefully chosen for these tests since rodents and other animals have idiosyncratic metabolic pathways that may not reflect human tolerances for identical toxicants. Some MLD values are reported in Table 20.1.

All studies of toxicant effects on animal subjects can also be geared to a *predesigned endpoint.* According to this rationale, toxicity experiments are designed and executed on the basis of anticipated results. These results may involve the surveillance of functional effects (next section), teratogenicity (aberrant fetal development), and/or carcinogenicity. As demonstrated by the U.S. Food and Drug Administration, three-generation studies of animal populations may be necessary to assess a given endpoint for a single experimental period. Carcinogenic and ter-

Table 20.1. MLD Values for Some Toxic Compounds

Acetic acid	4 g	Ethyl alcohol (pure—300–400 mL)	
Alphanaphthylthiourea	20 g	Beer(s) 2–6% alcohol	
Camphor	2 g	Whiskey 40–50% alcohol	
Chlordane	8 g	Wine 10–20% alcohol	
DDT	15 g	Quinine	8 g
		Toxin, *Clostridium botulinum*	0.001 mg

atogenic studies are necessarily long-term investigations and their results are not easily expressed as simple acute toxicity indexes.

STRUCTURAL AND FUNCTIONAL ASSAYS FOR TOXICANT EFFECTS

Short of detailed biochemical profiles that indicate toxic effects in test animals, more simplified assays are often used as a preliminary estimate of toxicity. Among the most popular tests are *structural assays*. These include surveys of histological and morphological changes in organ structures and systems, and detection of gross and relative weight-change rates among experimental and control animals. Obvious weight differences in these two groups of animals can imply important toxicity effects involving aberrations in energy balance or nervous system functions.

Abnormal weights of specific organs in a test animal may also indicate functional changes of mutually interacting organs. For example, an increase in heart weight may be caused by decreased oxygenation of blood in the lungs, an increase in adrenal weight can mirror impaired steroid synthesis within the gland, and so on.

Overt indications of structural effects on organs caused by exposure to a toxic compound are usually a sequel to preexisting *functional effects* of a toxic substance. Toxic substances can modify organ functions by altering pulmonary efficiency, brain and neural activities, or cardiovascular efficiency. Organ-specific tests can be used on individual animals to detect biochemical anomalies that precede obvious structural damage to the organ. As an example, consider a toxicant effect on liver function in which both primary and secondary parameters can be monitored. The primary functional assay of the organ may involve the deposition and clearance of a specific dye, or a secondary effect indicated as some enzymatic function in serum can be surveyed.

Since toxic compounds notoriously upset hepatic, renal, and hematological conditions, many tests are available for elucidating evidence of functional difficulties in these systems. Liver functions may be tested for biliary obstruction (alkaline phosphatase, icterus index), liver damage (thymol turbidity, plasma protein ratios, cholesterol ratio, glucose concentrations, cholinesterase, and transaminase levels), excretory function (bilirubin tolerance, bromsulfalein clearance), and metabolic function (glucose clearance, glucose tolerance). Some of the most prominent toxicological tests monitor serum alkaline phosphatase and serum transaminases, as well as dye clearance (bromsulfalein) or glucose tolerance.

Renal tissues are inherently tied to urea excretion and the concentration or dilution of urine. Excretion tests may employ dyes such as phenolsulfonphthalein and measurements of creatinine and urea. Concentration and dilution tests involve the assay of urine specific gravities among test subjects over pre- and postprandial periods, or after fasting for definite periods. It should be recognized that these tests are run mainly when impaired kidney functions are suspected and not as routine toxicological protocols.

The discriminative significance of functional tests may proceed with or without modification of tissue metabolism. Thus, specific *promoters* and *inhibitors* of enzyme systems (e.g., SKF-525A for mixed function oxidase) may be used to enhance the susceptibility of an organ or system to toxicant damage.

The action of metabolic modifiers is crucial for the accelerated determination of toxic effects in animals that otherwise may require costly clinical investigation and long-term testing.

Hematological studies of animals for toxicant effects include surveys of bone marrow and circulatory cells throughout the hematogenous route. Qualitative and quantitative observation of all blood cell forms and their relative numbers are usually tabulated. Specific circulatory tests such as dye dilution for blood volume, specific gravity, osmotic fragility, hematocrits, clotting time, and sedimentation rate tests may be necessary depending on the dose–response and physiological consequences of the administered toxicant.

Additional discretionary biomedical tests for determining functional effects include analysis of cerebrospinal fluid density, sperm motility, calcium phosphate ratios in skeletal structures, tensile strength of bones, cardiac response to catecholamines, metabolic studies of extracorporally perfused organs, work and stress assays for specific muscles, and many other types of specialized assays. Behavioral alterations in test subjects may also be important criteria for toxicant actions, but, unfortunately, standardization and validation of behavioral responses are largely nonexistent.

IRRITANT AND SENSITIZING PROPERTIES OF CHEMICAL AGENTS

Many chemical agents can elicit an irritant-type response after contact with an animal test subject. Protocols designed for evaluating irritant responses often involve skin or eye applications of a chemical agent to an animal subject. In some cases, superficially abraded versus normal skin exposure to the agent may reveal its ability to penetrate through the cutaneous barrier. Irritant effects for skin and eye exposures to chemical agents can be semiquantitatively evaluated. For skin tests, erythema and edema effects may be ranked, whereas ocular exposures to irritants can be evaluated by opacity of the eye, iris reactions to light, hemorrhage, swelling, and discharges. Results of these tests are often indeterminate and have uncertain scientific significance despite their wide use in the past.

Contrary to the actions of so-called primary irritants outlined above, some chemical agents can elicit an irritant-type response after repeated contact with a test subject. These effects are often the result of *sensitization* phenomena.

Protocols for evaluating sensitization potentials deal with the exposure of a suitable test animal (e.g., guinea pig) to a suspected toxicant or food constituent at doses far below those required to achieve primary irritant responses. Following a prescribed time period (12–15 days), the animal is subsequently challenged with the same substance at another site. If the irritant or inflammatory response of the animal is greater than the initial response, the test substance can be classified as a *sensitizer*.

Sensitizing potentials have been routinely surveyed by protocols such as the guinea pig *maximization test*, where a suspected sensitizing agent is presented to the animal in an adjuvant. Freund's complete adjuvant has been widely used for these assays since it markedly enhances evidence for test animal responses.

Sensitization mechanisms rely on the ability of a specific chemical agent(s) or its metabolite(s) to serve as an antigen (A). This antigen elicits lymphocytic production of antibodies (ABs) that form an antigen–antibody (A–AB) complex. Formation of the A–AB complexes may involve free circulating AB or AB bound to lymphocytes. In any event A–AB complexation instigates the production of biologically active amines. These amines include histamines and others that cause edema and erythema at the locus of antigen exposure. In those cases where A reaches portions of the vascular system, overt anaphylaxis may ensue.

PERMISSIBLE DIETARY EXPOSURES TO TOXICANTS

On the basis of dose–response data obtained from animals and whatever data may exist for humans, specific dietary *tolerances* for toxic food substances may be established. These tolerance values weigh threshold against nonthreshold (Figure 20.2) and LD_{50} properties of food toxicants as well as possible lifelong consumption possibilities. Careful analysis of all animal and human data can be used to calculate a *no observed effect level* (NOEL) for any suspect substance. The NOEL value is expressed in milligrams per kilogram of body weight per day. Thus, a NOEL value reports that concentration of a substance that has *no observable effect* on test subjects.

The NOEL value is normally divided by a safety factor of 100 to yield an *acceptable daily intake* (ADI) quotient. It must be clearly recognized that the ADI values based on scrupulously attended animal experiments cannot ensure absolute safety for all people. The use of the safety factor (100) mirrors this uncertainty in translating animal data into a safety index suitable for human dietary guidelines.

The validity of animal toxicology models is further complicated by variability within all segments of the human population, lack of statistical toxicological data, and the failure of present experimental evidence to detect the instantaneous onset of genetic molecular damage and carcinogenic scenarios.

If a chronic 2-year toxicity study is conducted using animals, the typical safety factor is 100. When toxic threshold levels are based on human data, an acceptable safety factor of 10 may be invoked, but a factor of 1000 or more may be used in the absence of reliable toxicological data.

The *maximal permissible intake* (MPI) per day for a toxic substance is based on the ADI.

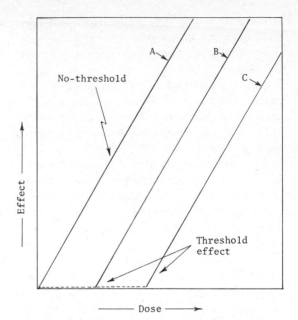

FIGURE 20.2. Curve A: Food toxicant effects on test animals may display parallel increases with administered doses of a chemical agent(s) or food constituent. This is contrary to curves B and C, which indicate that a critical concentration of toxicant must preexist before any correlation between toxic effects and doses are observed.

The MPI is calculated as the product of the ADI and the average weight of an adult (60–70 kg):

$$MPI = ADI \times (\text{body weight in kg})$$

This value can be compared to the maximum possible exposure of a toxicant likely to be consumed in a food. *Tolerance levels* for toxicants in foods are then calculated so that the toxicant content in the foodstuff is equal to or less than the ADI. This is mathematically carried out as

$$\text{tolerance} = \frac{ADI \times \text{average body weight of consumer}}{\text{food factor} \times 1.5 \text{ kg}}$$

60 kg = average body weight of consumer

food factor = percentage of the average daily diet made up by food containing the toxic substance

1.5 kg = average weight of food consumed per day

Tolerance can also be referred to in terms of the *maximum permissible level* (MPL).

Any toxicant identified as a natural biogenic constituent of a food can have pertinent ADI, MPI, and MPL values calculated on the basis of its occurrence in food. However, these terms have special reference to toxicants in foods arising from environmental contamination (e.g., biocides, industrial chemicals, heavy metals).

ACCUMULATION AND DEPOSITION OF TOXICANTS

Exposure to dietary toxicants may reflect an isolated event or a chronic pattern of toxicant consumption over a lifelong period. Therefore, a single-versus-chronic pattern of toxicant exposure may be a critical factor in assessing *in vivo* toxicant actions (Figure 20.3).

Dietary toxicants may be assimilated and directed into specific tissues and organs by mechanisms that involve active transport, passive diffusion, random binding interactions with other molecules, and distribution effects attributable to toxicant polarity.

Polar toxicant distributions are largely dependent on the *water pool* of the body, which includes interstitial, intercellular, and plasma water. Apart from water distribution, the albumin fraction of plasma proteins contains important binding agents for many toxicants. Albumin–toxicant associations often depend on their mutual participation in hydrogen-bonding interactions as well as van der Waals and electrostatic interactions. Although toxicant binding to plasma proteins has great importance, there are wide variations in the demonstrated binding affinities of polar toxicants to plasma protein fractions.

The scenario of toxicant distribution in the body is complicated by those toxic agents that undergo physical sequestration. These toxicants have an *in vivo* distribution behavior that is often directed by their lipophilic properties. This results in toxicant warehousing throughout tissues that may be far removed from the site(s) of potential toxic action.

Halogenated aromatic hydrocarbons and other metabolically persistent nonpolar organics behave in this fashion. The adipose cells and tissues of the body are ideal sites for their sequestration. Even though a toxicant resides in a fat depot, for example, it remains in an equilibrium with free toxicant in the peripheral circulation, organs, and other tissues. As expected, the toxicant distribution is dictated by *partitioning coefficients* of the lipid-soluble toxicant in aqueous body fluids, the vascular perfusion of toxicant-storage depots, blood lipid content, and other factors.

For both polar and nonpolar toxicants, the relationship between toxicant pool in the body and the plasma concentration can be defined as the *apparent volume of distribution* (V_d):

$$V_d = \frac{\text{amount of toxicant in body}}{\text{toxicant concentration in plasma}}$$

In practical terms, V_d represents the "apparent" volume required to hold a given amount of toxicant, assuming that it was uniformly distributed throughout an animal at the observed plasma concentration.

The V_d for toxicants can be experimentally determined by administering a parenteral (preferred) or dietary dose to a test subject. The administration route is discretionary depending on the study objectives. A resulting plot of plasma toxicant concentration versus time reveals an overall decrease in detectable toxicant concentration. As indicated in Figure 20.4, the hypothetical plot for the dichlorodiphenyltrichloroethane (DDT) insecticide consists of two phases: (1) a *toxicant distribution phase* and (2) an *elimination phase*. Extrapolation of the elimination phase data back to zero reveals the plasma concentration of toxicant at time zero (Cp_0). With the Cp_0 of 0.85 μg/L following a 10-mg dose (intravenous), the V_d can be calculated from

$$V_d = \frac{\text{toxicant dose}}{Cp_0}$$

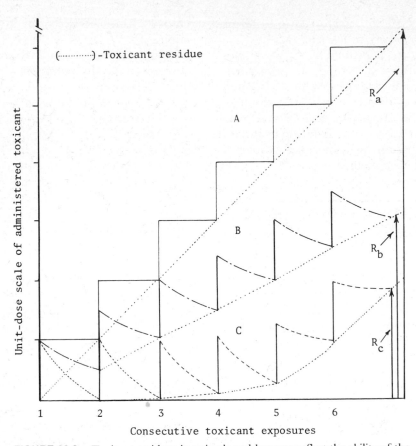

FIGURE 20.3. Toxicant residues in animals and humans reflect the ability of the body to metabolize or excrete the toxic compound. Hypothetical graphic plots in this figure illustrate six consecutive doses of equal amounts of toxicant that have been administered to a nonspecific test subject. Depending on toxicant metabolism, three distinctly different residue profiles can result. For Graph A, consecutive administrations of a nonmetabolizable, nonexcretable toxicant progressively increases in body tissues and produces incremental increases in the residue level (R_a = final residue level). Graph B illustrates toxicant dose applications of the same size and exposure frequency as in A, but partial metabolism of each toxicant dose is consecutively metabolized to produce a final residue (R_b = after six applications), which is half of R_a. In curve C, toxicant dose and exposure frequencies to an animal are identical to A and B *but* the first four doses are *effectively* metabolized by the body and residues remain small. After doses 5, 6, . . . , and so on, metabolism becomes marginal to nonexistent as a result of metabolic damage inflicted by the toxicant, and residue levels (R_c) markedly begin to increase in the body. This continues unless related effects of the toxicant become lethal. R_a and R_b as well as R_c respectively increase until a critical metabolic effect or lethal response is reached in the animal.

For the example then, a 10-mg dose of DDT having a Cp_0 value of 0.85 µg/L yields a V_d value of 11,764.

For toxicants that are relegated to the blood, V_d values range from 6 to 7 L; 12 to

15 L values are typical for toxicants held in extracellular fluids; and values of >50 L characterize the total aqueous phase of the body. The persistent chemical existence and lipophilic nature of many hydrocarbons (e.g.,

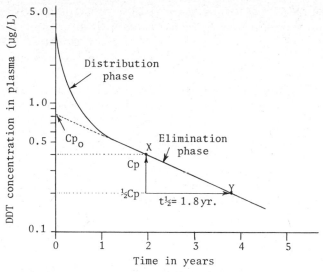

FIGURE 20.4. Plasma concentrations in a test subject following a 10-mg intravenous dose of DDT. Based on a semilog plot of toxicant plasma concentration versus time, the Cp_0 (plasma concentration at time 0) for DDT can be estimated by extrapolation of the "elimination phase" back toward zero time. The Cp_0 represents the theoretical toxicant distribution assuming that it was instantaneously distributed throughout the body at the time of injection. The Cp at point X is twice the DDT concentration at point Y. The intervening time between X and Y is equal to the *toxicant half-life* ($t^{1/2}$).

halogenated aromatic hydrocarbons) may display V_d values of >30,000. Here the V_d value reflects the reticent and protracted clearance of the lipophilic compound from its residence depots in fatty tissues. In some cases, many years may be required to achieve a notable clearance of high-V_d-value toxicants from animal tissues.

BODY CLEARANCE AND TOXICANT HALF-LIFE

The metabolic deposition of toxicants is controlled by (1) hepatic biochemical reactions and biliary excretion and (2) renal filtration of toxicants and/or their derivatives into the urine. Toxicant elimination and *clearance* (Cl) from the body, then, depends on the sum of all organ clearances including *renal* (Cl_r), *hepatic* (Cl_h), and *nonhepatorenal* (Cl_{nhr}) clearances. Moreover, the total toxicant clearance

is the quotient of the rate of toxicant elimination by *all routes* (R) divided by *plasma concentration of the toxicant* (Cp) (Figure 20.4).

$$Cl = R/Cp$$

Renal and hepatic clearance of toxicants is markedly influenced by the perfusion of clearing organs. A 5 L/min blood output by the heart ensures a high blood flow to most clearing organs in adult humans, and a very effective means for translocating toxicants via plasma or plasma protein fractions to clearing organs. Toxicant clearance is largely uneffected by toxicant distribution throughout the body, and clearance involves only those stores that are mobilized as extractable forms from the blood by clearing organs.

Clearance *and* distribution of toxicants in the body both dictate toxicant elimination. The dynamics of these processes are reflected by an *elimination rate constant* (k_e). This constant is a numerical expression of the portion of the *apparent volume of distribution* (V_d) cleared of toxicant per unit time:

$$k_e = Cl/V_d$$

Since k_e is linked to Cl and V_d (independent variables), the *toxicant half-life* ($t_{1/2}$) can be expressed as

$$t_{1/2} = 0.693 \, V_d/Cl$$

As expected, increases in V_d values necessarily increase the $t_{1/2}$ for toxicant clearance. Conversely, increases in Cl promote decreases in half-life for any specified V_d. Also note that k_e can be estimated from semilogarithmic plots (Figure 20.4) where it is expressed as the *slope of the linear elimination phase*.

Based on a first-order elimination rate constant, a toxicant that endures four half-lives ($4 \times t_{1/2}$) will result in a 93.75% elimination (6.25% residue). By five half-lives, 96.88% elimination (3.12% residue) of the toxicant will result, and only an infinite time frame can assure its asymptotic concentration decrease in tissues toward a zero residue.

As in the classical study of pharmacokinetics, the time frame for toxicant accumulation, as well as its elimination, depends on the clearance half-life. Unfortunately, *sequestration* and *chronic toxicant exposures* may lead to the accumulation of high toxic residues in animal and human tissues. Here, the ability of the body to clear toxicant may be successively challenged before it can ever eliminate 90% of an initial toxicant dose.

If toxicant exposures and toxicant accumulations are *steady* (no elimination, no metabolism), the toxic residue rapidly increases in a stepwise fashion (Figure 20.3.A). Minor to moderate residues also develop with chronic toxicant exposures and only *moderate clearance* (Figure 20.3B). Both cases, however, can ultimately lead to a cumulative residue that will display acute toxic and lethal effects.

Biochemically persistent lipophilic toxicants commonly exhibit a residue buildup in tissues owing to chronic exposures to toxicants with intervening periods of less than the $t_{1/2}$ for clearance. Moreover, frequent toxicant exposures to already damaged clearance organs and mechanisms may exacerbate toxicant accumulation and promote acute toxic effects at accelerated rates.

REVERSIBLE VERSUS IRREVERSIBLE TOXIC EFFECTS OF TOXIC AGENTS

The deposition of toxicants in the body at less than lethal concentrations generally adheres to the models discussed above. These models assume some type of toxicant interaction at an *in vivo* biochemical target site that is reversible or that causes biochemical damage that is recoupable. Other types of toxicants exist, however, and these display *delayed toxicity* effects on animals and humans following repeated toxicant exposures.

Repeated exposure to these toxicants is cumulative since the total dose is quantitated as a function of exposure time and individual dose concentrations. Toxicant actions here are considered as *irreversible* events since consequences of toxicant exposure are biochemically subtle, chemically complex, time-dalayed or *latent* and may be evidenced only as mutagenesis, carcinogenesis, or tumor production.

Toxicant concentrations responsible for initiating these events are difficult to define because any suspected toxicant *may be only a precursor to its short-lived biotoxic form produced in the body*. This biotoxification process is also complicated by predisposing coreactants, oxidation–reduction potentials, antioxidant concentrations, and many other factors including sequestration phenomena. Table 20.2 outlines a number of reversible and irreversible generic actions for toxicants that may occur in food systems.

Table 20.2. Toxic Agents that Exert Reversible and Irreversible Effects

Irreversible effects

Alkylation of biopolymers (e.g., DNA and other nucleic acids). Leads to defective tRNA transcription and/or its translation into aberrant cell proteins. *Consequences:* Cellular necroses, fibrosis, or generalized cellular degradation. *Note:* Nonspecific alkylations of other nongenetic molecules may occur at $-NH_2$ and $-SH$ sites, which upsets prescribed molecular functions.

Antimetabolites. Base analogues of normal purines and pyrimidines may be *introduced* into (1) DNA structure during replication or (2) RNAs. *Consequences:* Defective nucleic acid polymers or lethal protein biosyntheses.

Photoactivation reactions. A chemical agent assimilated by the body is photoactivated from an inactive precursor to yield an agent that promotes superficial dermal erythema (similar to sunburn).

Sensitization and allergic responses. Body responses to foreign chemical agent(s), which entail antibody stimulation and production.

Depression of ATP formation. Oxidative phosphorylation may be inhibited by specific chemical agents that uncouple the ATP biosynthesis route.

Reversible effects

Antivitamin and/or vitamin antagonists. Impede specific vitamin absorption or dietary acquisition by the body or suppress vitamin-dependent reactions (e.g., *avidin* suppresses biotin absorption, *thiaminase* destroys thiamine).

Competitive inhibition of enzyme activity. Stereochemical analogues or molecular substituents reminiscent of an enzyme's normal substrate may upset normal enzyme–substrate interactions, thereby suppressing biochemical product formation. *Note:* Irreversible enzyme inhibition can occur if substrate "analogue" becomes covalently bound to enzyme active site.

Metabolic interference. Temporal presence of toxicants in a biochemical system may cause an upheaval in normal hormonal regulation, ionic strength balance over renal tubules, acid–base and mineral balance, etc.

Neurotransmission blocks or interference. Synaptic cleft transmission of depolarization events from neuron to neuron can be upset by specific chemical agents (e.g., organic phosphate insecticides serve as acetylcholinesterase inhibitors, cyclodiene insecticides alter ion permeability of axonal membranes).

CYTOCHROME *P*-450 AND XENOBIOTIC OXIDATIONS

Body-foreign chemicals incorporated into the animal body are classified as *xenobiotics*. From dietary perspectives, these alien compounds may be assimilated from foods during digestive processes. Many xenobiotics have biological origins in food tissues being consumed, yet others such as halogenated hydrocarbon biocides, phthalate plasticizers that migrate from food packaging into foods, and other accidental food contaminants are abiotic constituents of foods systems.

Xenobiotics may be eliminated from the body unchanged via urine, feces, vomitus, milk, hair, and so on. In some cases excretion is possible only after the compound has been metabolized to yield a more polar, water-soluble derivative suitable for renal excretion. For those xenobiotics that are toxic, detoxification may precede excretion. In still other situations the alien compound may be defensively immobilized or physically entrapped by immune responses of the body.

The constant assault of xenobiotics on the human body requires an effective mechanism for the biochemical disposal of these compounds. One of the most important routes involves reactions mediated by *hydroxylases* and hemoproteins known as *P-450 cytochromes*. These proteins routinely promote the 11β-hydroxylation of steroids (Figure 20.5A) but seem to promote few other critical reactions on endogenous substrates.

In humans, however, the "simple" functions of the *P*-450 cytochromes and their attendant proteins are complicated by an ability to catalytically hydroxylate pesticides, steroid precursors, drugs, and polycyclic aromatic hydrocarbons. Most evidence sug-

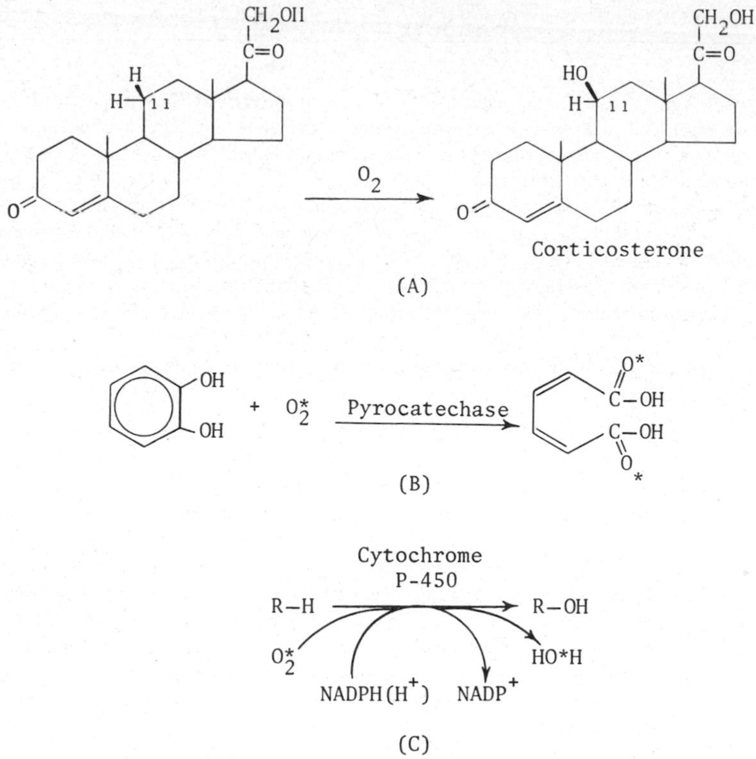

FIGURE 20.5. (A) *P*-450-dependent enzymes normally mediate the 11β-hydroxylation of steroids. (B) Dioxygenase activity is demonstrated by pyrocatechase actions on catechol. (C) *P*-450-dependent hydroxylation of a nonspecific substrate.

gests that *P*-450 cytochromes are important elements of the body's detoxification mechanisms since products of their reactions are often water soluble, unsuited for sequestration in fat, and excreted in the urine.

Notable concentrations of cytochrome *P*-450 occur in the microsomal fraction of the liver and adrenal cortex mitochondria, with smaller concentrations in other tissues. Cytochrome *P*-450 is a component of oxygenase enzyme systems, which introduce molecular oxygen into a substrate via oxygen activation. There are *two distinct types* of oxygenases—*dioxygenases* and *monooxygenases*.

Dioxygenases catalyze the insertion of both atoms of molecular oxygen into a substrate. One typical example of dioxygenase activity is demonstrated by the action of *pyrocatechase* on a catechol (Figure 20.5B).

Monooxygenases differ from dioxygenases since they perform two simultaneous reactions. These reactions involve (1), the insertion of an oxygen atom from molecular oxygen into a substrate and (2) the reduction of the remaining oxygen atom to water. Cytochrome *P*-450 is a critical component of numerous monooxygenation reactions found in the body. All facets of *P*-450-dependent hydroxylation mechanisms are not established, but it is conceded that these systems are dependent on an electron flow from NADPH(H$^+$) or NADH(H$^+$), possibly through nonheme iron proteins (Figure 20.5C).

Many different forms of microsomal P-450-mediated enzymes catalyze the hydroxylation of different xenobiotics. Furthermore, evidence indicates that *P*-450 microsomal enzymes may be specifically induced depending on the nature of organic substrates. For example, barbiturates and aromatic hydrocarbons (e.g., methylcholanthrene-like agents) induce different forms of P-450 hydroxylases.

The induction specificity for certain P-450 enzymes may be directed by one or a small number of genes. Operational definitions of this genetic system designate the pertinent genes as an *Ah* complex since particular responsiveness may be shown toward *aromatic hydrocarbons*.

The *Ah* complex is suspected to direct the construction of hundreds or possibly thousands of different P-450-dependent enzymes—with each enzyme being induced by mechanisms similar to steroid hormone actions on the nucleus. Fundamental similarities between steroid and certain xenobiotic hydrocarbons may permit unusual hydrocarbon interactions with steroid–hormone receptors. If this action occurs in the realistic operation of cells, then it follows that steroid–hormone receptor interactions with some xenobiotic hydrocarbons may initiate *Ah*-gene expression as specific P-450-mediated enzymes.

The induction of microsomal enzymes is a steady process. At birth and in early life, microsomal enzyme concentrations are low, but their substrate diversities and productivities steadily increase through adulthood. This observation probably reflects the body response to the bewildering assault of chemical threats from dietary, environmental, and microbiological sources.

Biochemical adaptation to xenobiotic stress is an important homeostatic mechanism. The maintenance of preexisting or potential microsomal enzymes, dictated by long-term adaptive exposures to chemical threats, is not without allocative costs to the body in spite of obvious benefits. These costs include unknown measures of energy, protein utilization, physiological stress, and other resource sapping factors.

MICROSOMAL AND NONMICROSOMAL ROUTES OF METABOLISM

The metabolism of toxic compounds is guided by kinetic and thermodynamic principles that underlie all biochemical processes. Many reactions of toxicants are geared to their elimination from the body. This may require simple excretory processes or preliminary detoxification and metabolism of the toxicant. Enzyme-mediated metabolic reactions largely depend on *oxidation, reduction, hydrolysis,* and *conjugation* reactions.

The metabolism of toxicants by the body is not a "cognizant" response or "biochemical strategy" of physiological biochemistry. It turns out that many toxicants can be dismantled, rendered harmless, and eliminated from the body. However, the opposite may also be true since some toxicants show enhanced toxicity after metabolism—or "biotoxification."

Some typical oxidation reactions mediated by microsomal enzymes of the liver are outlined in Table 20.3. From generic perspectives, microsomal oxidations involve some of the following reactions:

Reaction	Products
I. Hydroxylation of aromatic hydrocarbons	Phenols
II. Oxidation of aliphatic hydrocarbons	Acids, alcohols, ketones
III. N-, O-, and S-dealkylations	Alkyl group oxidation
IV. N- and S-oxidations	Hydroxylamines, and amine oxides, and sulfoxide and sulfones, respectively
V. Epoxide formation	Epoxides
VI. Thionosulfur and phosphorothionate oxidations	Inorganic sulfate

Table 20.3. Some Notable Microsomal and Nonmicrosomal Oxidation Routes for Organic Compounds

I. Hydroxylation of Aromatic Hydrocarbons

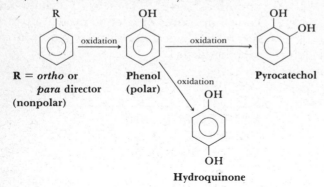

R = *ortho* or *para* director (nonpolar)

Phenol (polar)

Pyrocatechol

Hydroquinone

II. Oxidation of Aliphatic Hydrocarbons

$$R-CH_2CH_2CH_3 \xrightarrow{\text{oxidation}} R-CH_2CH_2CH_2OH \xrightarrow{\text{oxidation}} R-CH_2-CH_2COOH$$

Carboxylic acid

$\downarrow$ oxidation

$$R-CH_2CHCH_3$$
$$\overset{|}{\text{OH}}$$
Secondary alcohol

$\vdots$ Further oxidation

$$R-CH_2-CH_2COOH$$

$\vdots$ oxidation

$$R-COOH + 2CO_2$$

$\uparrow$

$$\left[\begin{array}{l}\text{Homologous carboxylic}\\\text{acid having 2 less carbons}\end{array}\right]$$

III. *N*-, *O*-, and *S*-Dealkylations Proceed by Formation of a Reactive Intermediate

$$R_1-NH\overset{*}{C}H_3 \xrightarrow{\text{oxidation}} \left[R-NH-CH_2OH\right] \longrightarrow R_1-NH_2 + CH_2O$$
Primary amine   **Aldehyde**

$$R_1-NH-\overset{*}{C}HR_2R_3 \xrightarrow{\text{oxidation}} \left[\begin{array}{c}R-NH-CH-R_2R_3\\\overset{|}{\text{OH}}\end{array}\right] \longrightarrow R_1-NH_2 + R_2-\overset{\overset{\displaystyle O}{\|}}{C}-R_3$$
Secondary amine   **Ketone**

$$R_1-O\overset{*}{C}H_2R_2 \xrightarrow{\text{oxidation}} \left[\begin{array}{c}R_1-O-CH-R_2\\\overset{|}{\text{OH}}\end{array}\right] \longrightarrow R_1-OH + \overset{\overset{\displaystyle O}{\|}}{C}H-R_2$$
                                    Hemiacetal   **Alcohol**   **Aldehyde**

$$R_1-S-\overset{*}{C}H_3 \xrightarrow{\text{oxidation}} \left[R-S-CH_2OH\right] \longrightarrow R-SH + CH_2O$$
                                                   Thiol   **Aldehyde**

*Site of oxidative attack is carbon atom.

IV. *N*- and *S*-Oxidations

Aromatic primary, secondary, and tertiary aromatic amines are susceptible to oxidation. Sulfides similarly undergo oxidations to yield sulfoxides and then sulfones. Both tertiary amines and sulfide oxidations may be promoted by the hydroxyl cation.

Table 20.3. *(Continued)*

a. Aromatic amines to hydroxylamine derivatives

(1) $\langle\!\langle\bigcirc\rangle\!\rangle$—$NH_2$ $\xrightarrow{\text{oxidation}}$ $\langle\!\langle\bigcirc\rangle\!\rangle$—$\overset{H}{\underset{..}{N}}$—OH

R_n R_n
Hydroxylamine derivative

(2) $\langle\!\langle\bigcirc\rangle\!\rangle$—$\overset{H}{\underset{}{N}}$—R $\xrightarrow{\text{oxidation}}$ $\langle\!\langle\bigcirc\rangle\!\rangle$—$\overset{..}{\underset{OH}{N}}$—R

R_n R_n
Hydroxylamine derivative

b. Amine oxide formation

(1) R_1—$N\overset{R_2}{\underset{R_3}{\raise2pt{:}}}$ $\xrightarrow{\text{oxidation}}$ $R_1R_2R_3N\underset{\delta+}{\rightarrow}O:$
Amine oxide

(2) $\langle\!\langle\bigcirc\rangle\!\rangle$—$NR_1R_2$ $\xrightarrow{\text{oxidation}}$ $\langle\!\langle\bigcirc\rangle\!\rangle$—$\overset{R_2}{\underset{\underset{\delta+}{R_1}}{N}}\rightarrow O$

R_n R_n
Amine oxide

c. *S*-Oxidations

(1) R_1—S—R_2 $\xrightarrow{\text{oxidation}}$ R_1—$\underset{O}{\overset{\downarrow}{S}}$—$R_2$ $\xrightarrow{\text{oxidation}}$ R_1—$\overset{\overset{O}{\uparrow}}{\underset{\downarrow}{\underset{O}{S}}}$—$R_2$

Aliphatic   **Sulfoxide**   **Sulfone**
or aromatic
sulfide

$\xrightarrow{\hspace{3cm}}$
Increasing polarity

d. Tertiary amine and sulfide oxidations by (HO^+)

(1) $R_1R_2R_3N$ $\xrightarrow{[\text{HO}^+]}$ $R_1R_2R_3$—$\overset{+}{N}$—OH $\longrightarrow$ $R_1R_2R_3$—$N\rightarrow O + H^+$
                                               Amine oxide

(2) R_1—S—R_2 $\xrightarrow{[\text{HO}^+]}$ R_1—$\underset{R_2}{\overset{+}{S}}$—OH $\longrightarrow$ R_1—$\underset{R_2}{S}\rightarrow O + H^+$
                                            Sulfoxide

V. Epoxidation of Aromatic and Aliphatic Double Bonds

This can be generically exhibited by naphthalene and the cyclodiene insecticide heptachlor. It should be noted that many compounds participate in these reactions.

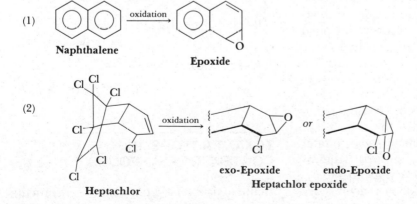

(1) **Naphthalene** $\xrightarrow{\text{oxidation}}$ **Epoxide**

(2) **Heptachlor** $\xrightarrow{\text{oxidation}}$ **exo-Epoxide** *or* **endo-Epoxide**
Heptachlor epoxide

Table 20.3. (*Continued*)

VI. Thionosulfur and Phosphorothionate Oxidations

(1) Thionosulfur oxidations are fundamental reactions of thioureas, thiosemicarbozones, and other biocides, whereas (2) "oxon" formation is a key activation step for mobilizing the toxic actions of phosphorothionate insecticides (e.g., parathion (toxic) $\xrightarrow{oxidation}$ *para*-oxon (more toxic)), which exert acetylcholinesterase inhibition effects.

$$(1) \quad \text{C=S} \xrightarrow{\text{oxidation}} \text{C=O} \;+\; SO_4^{2-}$$

$$(2) \quad -\text{P=S} \longrightarrow -\text{P=O} \;+\; SO_4^{2-}$$
Phosphorothionate 　　"**Oxon**"-**derivative**

VII. Aldehyde (1) and Alcohol (2) Oxidations

(1) $R-CHO + NAD^+ \xrightarrow{\text{aldehyde dehydrogenase}} R-COOH + NADH(H^+)$
Aldehyde 　　　　　　　　　　　　　　**Carboxylic acid**

(2) $R-CH_2CH_2CH_2OH + NAD^+ \xrightarrow{\text{alcohol dehydrogenase}} R-CH_2CHO + NADH(H^+)$
Primary alcohol 　　　　　　　　　　　　**Aldehyde**

VIII. Amine Oxidation

(1) $R-CH_2NH_2 \xrightarrow{\text{oxidation}} R-CH{=}NH \xrightarrow{+H_2O} R-CHO + NH_3$
Primary amine 　　　　　　　　　　　**Aldehyde** **Ammonia**

(2) $H_2N(CH_2)_n-CH_2-NH_2 \xrightarrow{\text{oxidation*}} H_2N(CH_2)_n-CHO + NH_3$

Diamine oxidases (DAO) mediate oxidations for one of two $-NH_2$ groups versus *monoamine oxidases* (MAO) that cause oxidative cleavage with deamination.

Nonmicrosomal oxidations (Table 20.3) are also important, but these reactions generally deal with aldehyde and alcohol oxidations (VII) as well as amine oxidations via dehydrogenation (VIII).

Oxidation routes are critical to the metabolism of many toxicants along with any other mechanism that enhances their polarity and water solubility. Conjugation reactions, for example, may effectively link polar sugar–acid derivatives to organic molecules. *Uridine diphosphate glucuronic acid* (UDPGA) commonly provides its sugar–acid moiety to organic substrates under the auspices of glucuronyl transferase. Similar conjugation reactions may also involve the transfer of ribosides, riboside phosphates, acid sulfates (e.g., $-O-SO_3^-$), methyl groups (from *S*-adenosyl methionine), or acyl groups to organic compounds. Reduction reactions are varied and may employ general or specific reductases plus $NADPH(H^+)$ or $NADH(H^+)$.

Hydrolysis reactions represent the most basic detoxification schemes commonly employed by many cells. Glycosidases, peptidases, lipases, carboxyesterases, and phosphoesterases readily mediate these reactions. Esters and amides are correspondingly hydrolyzed by esterases and amidases as shown below:

$$R_1-COOR_2 + H_2O \xrightarrow{\text{esterase}}$$
$$R_1COOH + R_2-OH$$
$$R_1-CONR_2R_3 + H_2O \xrightarrow{\text{amidase}}$$
$$R_1-COOH + R_2R_3NH$$

TOXIC FACTORS IN CONTEMPORARY FOODS

Acute toxic effects from dietary constituents are relatively minor occurrences throughout the United States when compared with less technological societies. Nonetheless, food toxicants are ever-present threats to health and welfare.

Unlike Neanderthal mistakes caused by selecting and eating the wrong type of ground nut, current food toxicology concerns are generally more subtle and complex. Contemporary problems may be exhibited in humans for various reasons including

1. *Overconsumption* of certain foods.

2. *Accidental consumption* of foods containing toxic agents.

3. *Immunological sensitivity* to a specific food constituent.

4. Outstanding *metabolic idiosyncracies* of individuals that conflict with the deposition of certain food constituents.

5. *Preexisting pathophysiological conditions* that adversely affect elimination of toxic agents (e.g., renal insufficiency).

Apart from these factors, food constituents can interact with each other or with therapeutic drugs to produce serious problems.

The dimensions of current food toxicology are also complicated by the delayed effects of certain food constituents and foodborne xenobiotics that cause molecular genetic damage. Genetic damage may occur at the level of specific cells in tissues or among germ cells, only to be evidenced decades later in adults or their progeny.

A variety of modern food toxicology problems also involve stored and preserved foods, which undergo chemical or microbial degradation, as well as the unintentional sequestration of environmental contaminants in foods.

Many foods have been overly incriminated for their potential toxic actions, such as spinach leaves and their liberation of oxalic acid, but there may be many other foods whose toxicities are underestimated.

The current status of food toxicants in American food commerce is complicated by food substances that are presently marketed as health foods. Herbs, herb teas, herbal remedies, imported foreign meats, cheeses, and similar products all offer potential health threats, largely unknown in this country. Many of these plant substances have not been tested, their physiological and biochemical effects on the body are unknown, and their content of environmental contaminants remains largely uncharted.

SOME TOXIC AND UNDESIRABLE FOOD CONSTITUENTS

The biogenesis of many natural food toxicants is a recognized property of numerous plants and microorganisms. Some researchers suspect that these toxicants represent (1) useless waste by-products of metabolism or (2) products of vestigial biochemical pathways or (3) serve as chemical deterrents to check the competitive coexistence of one plant species over another or to ward off insect and saprophytic invaders (*allelopathic responses*) (Table 20.4).

Whatever the biochemical and ecological rationales for toxic substances in foods, the descriptive toxicological effects (*toxicography*) for many chemical agents and foods have been widely recorded.

Table 20.4. Some Allelopathic Chemical Compounds Produced by Plants and Microorganisms that Have Been Linked to Toxic Effects of Foods

Alkaloid (caffeine)	Lactone(s) (patulin)
Aromatic acid(s) (cinnamic acid)	Monoterpene(s) (camphor)
Cyanogenic glycoside (dhurrin)	Quinone (juglone)
Flavonoid(s) (phlorizin)	Tannin (gallic acid)
Furanocoumarin (psoralin)	Thiocyanate (allylisothiocyanate)

Cyanogenic Glycosides

Nearly all drupe (multiple seeds) and pome (single pit)-type fruits biosynthesize cyanogenic glycosides such as amygdalin. This substance is harmless until crushed, moistened seeds containing the compound release *emulsin*. This enzyme in turn mobilizes cyanide from the glycoside as shown:

$$C_{20}H_{27}NO_{11} + 2H_2O \xrightarrow{\text{emulsin}}$$
Amygdalin

$$2\ C_6H_{12}O_6 + C_6H_5CHO + HCN$$
Glucose   Benzaldehyde   Cyanide

Concentrations of cyanogenic glycosides are variable from plant to plant. The concentration of cyanide for apricot, peach, wild apricot, and some butter beans reflects respective levels of 9, 88, 220, and 300 mg/100 g (wet seed/pit weight). Human toxicity is largely a matter of gross cyanide consumption via cyanide-rich plant tissue (MLD for HCN is 0.5–3.5 mg/kg).

Cyanide paralyzes oxidative phosphorylation at the cellular level of body structure (internal asphyxia) and prohibits medullary oxygenation with a sequel of respiratory failure. Subacute levels of cyanide in the body may be bound to plasma proteins, converted to cyanohemoglobin, or more importantly, converted into cyanocobalamin (vitamin B_{12}) in a fashion similar to cyanide deposition in smokers.

Acute pathologies from cyanide are realized in West African countries (Nigeria) where chronic consumption of poorly processed cassava root (tapioca stock) causes tropical atonic neuropathy and possibly goiter (thiocyanate etiology). Some of the highest exposures to cyanides in the American diet originate from fruit jams (e.g., quince, peach), which may include amygdalin-rich seed mucilages or pit extracts in the final jellied product.

Forage animals that consume decaying, trampled, or frost-struck grasses can develop cyanide toxicosis. Johnson grass, arrow grass, queens root, and sorghum are all suspected cyanogenic sources, especially when they exist in a state of environmental duress.

Oxalic Acid

Many leafy green plant leaves including rhubarb, spinach, and beet leaves contain 0.2 to 1.4% oxalic acid. Oxalate has been widely incriminated in juvenile poisonings, but these claims are widely overstated. The MLD for oxalates ranges from 15 to 30 g depending on coexisting digestive tract contents and body weight. Oxalic acid is noted for its corrosive action on mucous membranes, but its salts are less damaging. The acidic nature of this compound is not singularly responsible for its actions. Oxalate also has the ability to bind serum calcium thereby inducing hypocalcemia. Prolonged hypocalcemia produces eventual muscle stimulation, convulsion, and collapse. In cases of oxaluria, the renal excretion of oxalates can cause renal colic and hematuria. Pathological evidence for oxalates includes not only renal swelling and tubule sclerosis, but oral, pharyngeal, and gastric lesions.

Xanthines

The conversion of xanthine to its trimethylated derivative caffeine (1,3,7-trimethylxanthine) occurs in many plants. Caffeine and its related xanthines, namely, theophylline (1,3-dimethylxanthine) and theobromine (3,7-dimethylxanthine) are all natural constituents of coffee beans, tea leaves, kola nuts, cocoa beans, and maté (Figure 20.6).

As a consequence of consuming beverages produced from these plant sources, the xanthines exert important neurostimulatory actions on the CNS. Caffeine consumption is not a minor problem in the United States since coffee drinking alone accounts for >25 million lb of caffeine, whereas tea leaves and kola nuts account for ~7 million lb. Theophylline and theobromine contents of teas and cocoa provide another 15 million lb of xanthine intake.

The caffeine content of individual beverages is highly variable, but ground roasted coffee infusions may extract ~562 mg/100 g of coffee. Common dietary exposures ap-

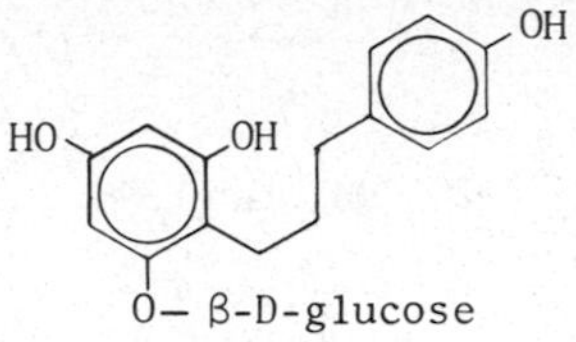

(1)

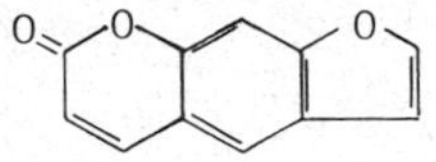

(2)

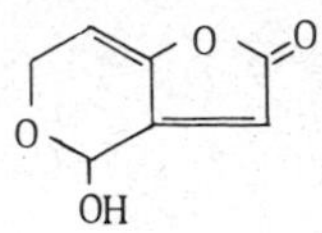

(3)

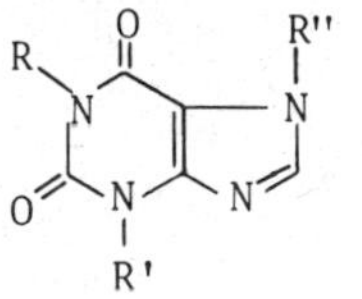

(4) R=R'=R''=CH_3

(5) R=H;R'=R''=CH_3

(6) R=R'=CH_3;R''=H

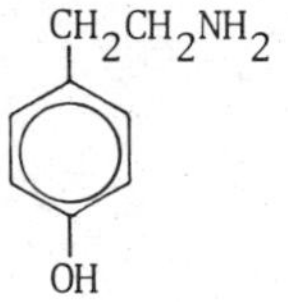

(7)

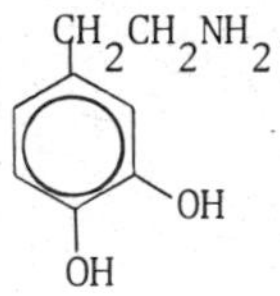

(8)

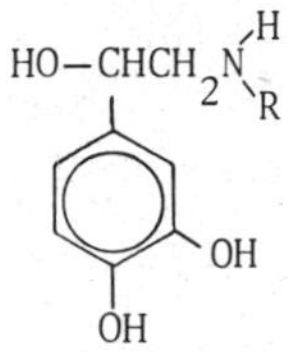

(9) R=CH_3

(10) R=H

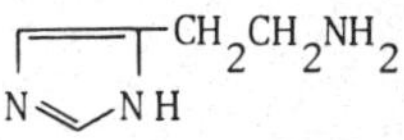

(11)

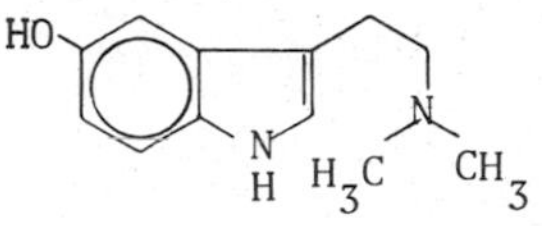

(12)

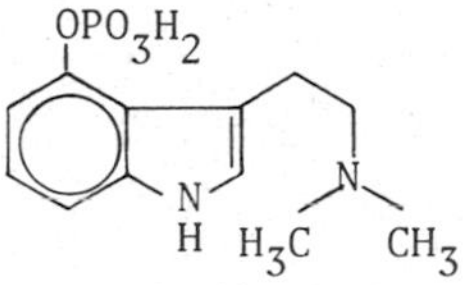

(13)

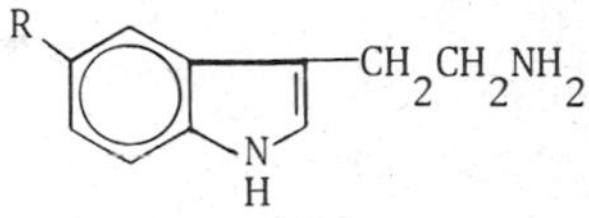

(14) R=H

(15) R=OH

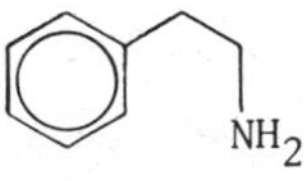

(16)

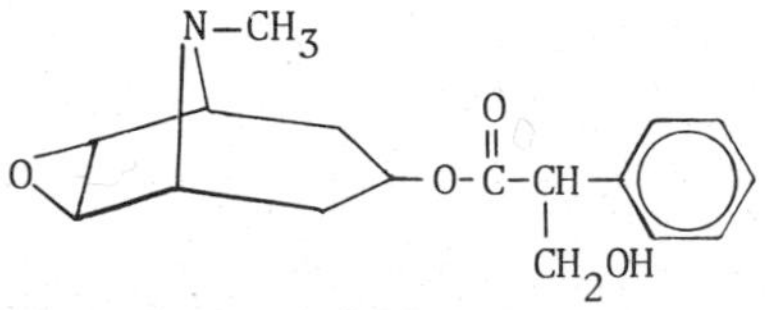

(17)

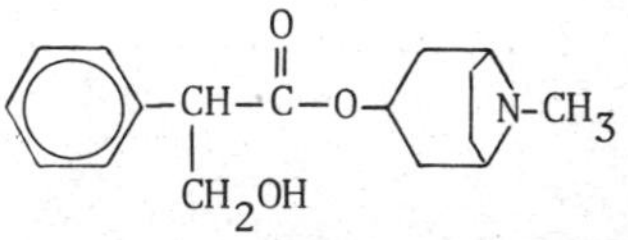

(18)

FIGURE 20.6. A selection of some toxic components that can be found in certain foods, many of which are discussed in the text. Molecular structures for toxic compounds continue on pages 1044–1045.

Acetylandromedol (24)
Andromedol (25)
Anhydroandromedol (26)
Atropine (tropane alkaloid) (18)
Bufotenine (12)
Cadaverine (23)
Caffeine (4)
Carotatoxin (falcarinol) (50)
Citrinin (20)
Coumarin (nucleus) (33)
Dioscorine (tropane alkaloid) (19)
Dopamine (8)
Epinephrine (9)
Glycyrrhizic acid (hypertensive agent in licorice plant) (48)
Histamine (11)
Ibotenic acid (46)
Ipomeamarone (28)
1-Ipomeanol (29)
4-Ipomeanol (27)
Muscarine (44)
Muscazone (47)
Muscimol (45)
Naringen (citrus polyphenol RhGl = rhamnoglucoside) (49)
Norepinephrine (10)
Ochratoxins (3 types) (35)
Patulin (3)
Phenethylamine (16)
Phlorizin (1)
Physostigmine (calabar bean) (31)
Psilocybin (13)
Psoralin (2)
Putrescine (22)
Pyrrolozidine alkaloid metabolism in liver tissues converts retrorsine (38) and heliotrine (39) into a very reactive metabolic pyrrole (40)
Quercetin (21)
Quinine (30)
Retronecine (37)
Saxitoxin (43)
Scopolamine (tropane alkaloid) (17)
Serotonin (15)
Solanine (R = trisaccharide) (36)
Sterigmatocystin (34)
Tannic acid (a polyphenol, "R" may also represent *m*-digallate) (41)
Tetrodotoxin (42)
Theobromine (5)
Theophylline (6)
Tricothecenes (4 types) (32)
Tryptamine (14)
Tyramine (7)

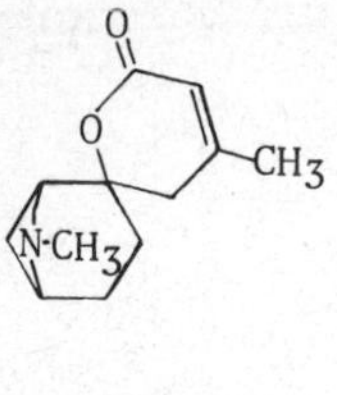

(19)

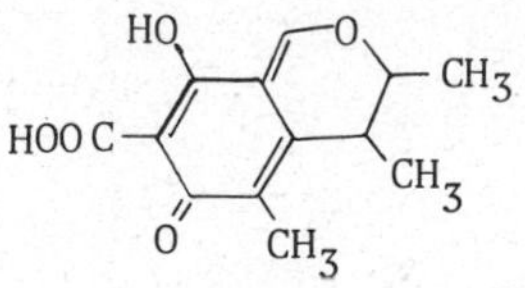

(20)

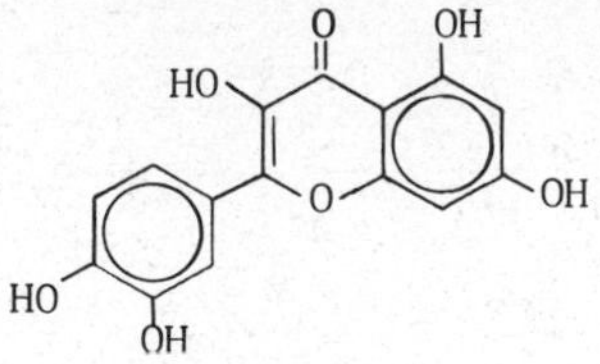

(21)

$NH_2(CH_2)_nNH_2$

(22) n = 4

(23) n = 5

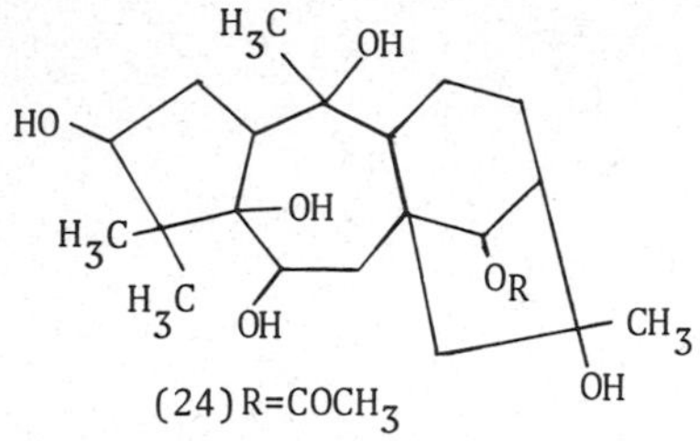

(24) R=COCH$_3$

(25) R=H

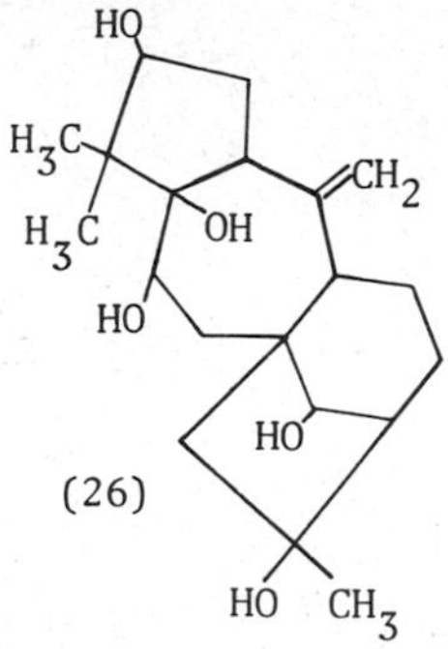

(26)

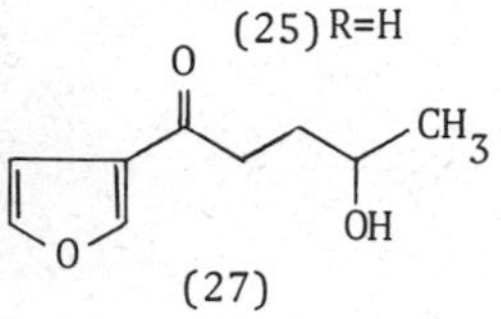

(27)

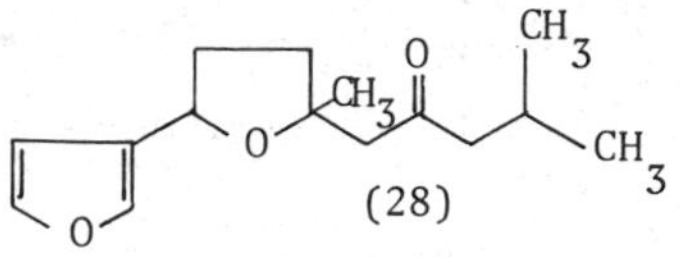

(28)

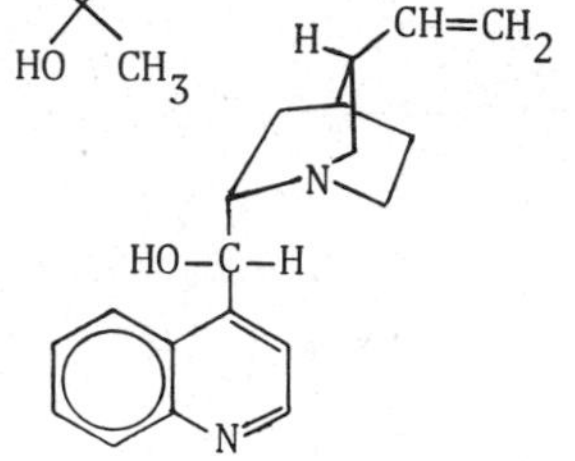

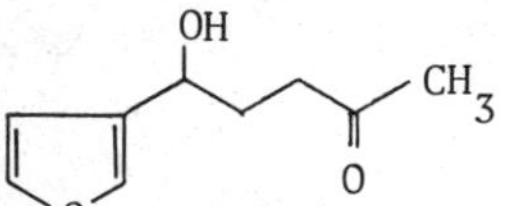

(29)

(30)

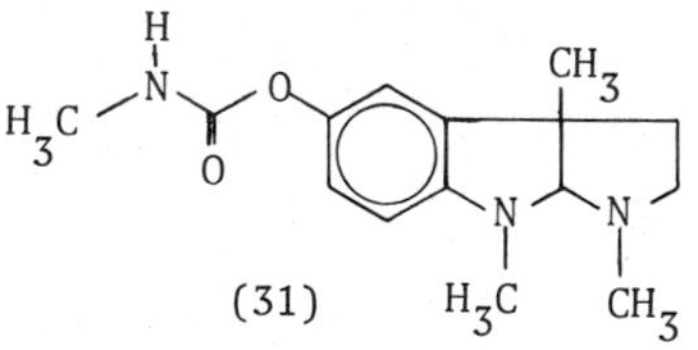

(31)

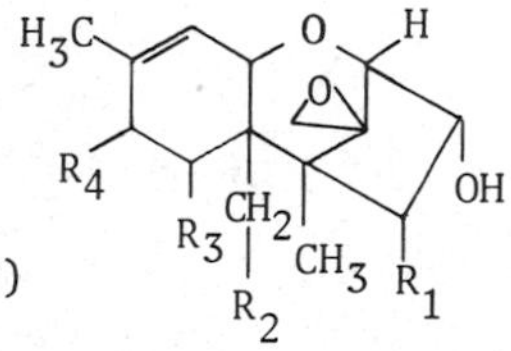

(32)

Type	R$_1$	R$_2$	R$_3$	R$_4$
T-2 Toxin	OCH$_2$CH$_3$	OCH$_2$CH$_3$	H	(CH$_3$)$_2$CHCH$_2$CO$_2$
HT-2 Toxin	OH	OCH$_2$CH$_3$	H	(CH$_3$)$_2$CHCH$_2$CO$_2$
DAS	OCH$_2$CH$_3$	OCH$_2$CH$_3$	H	H
DON	H	OH	OH	O

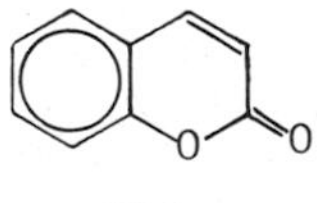

(33)

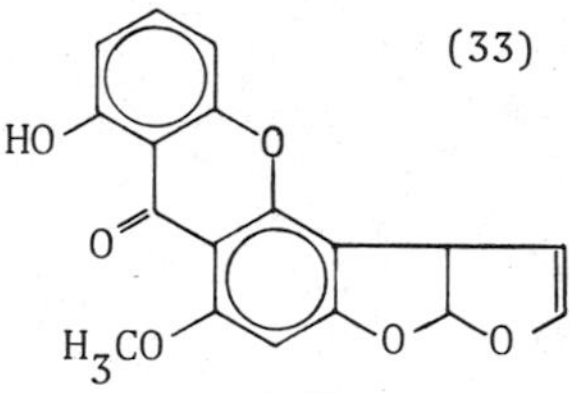

(34)

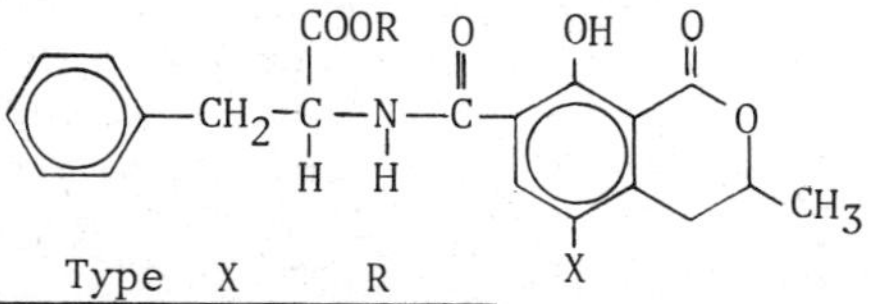

(35)

Type	X	R
A	Cl	H
B	H	H
C	Cl	C$_2$H$_5$

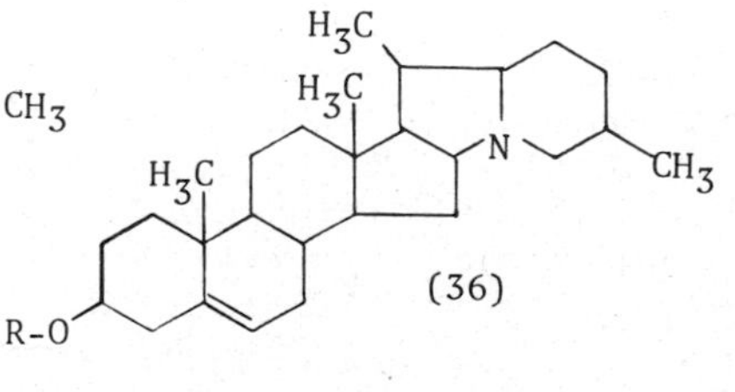

(36)

FIGURE 20.6. *(Continued)*

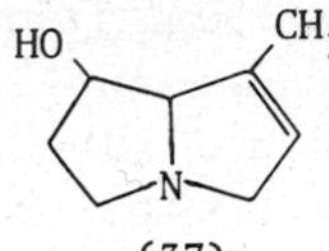

Hepatic metabolism

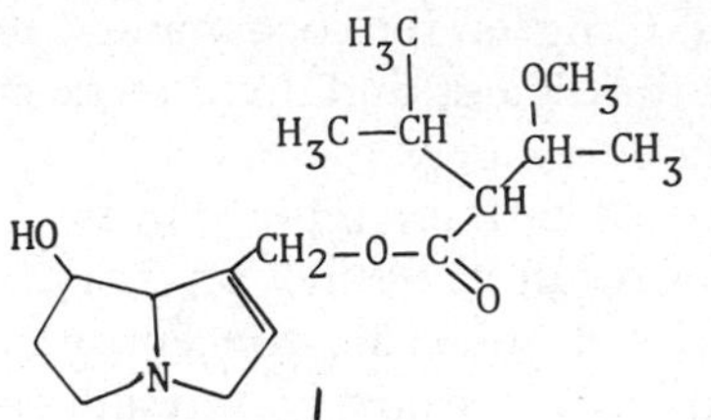

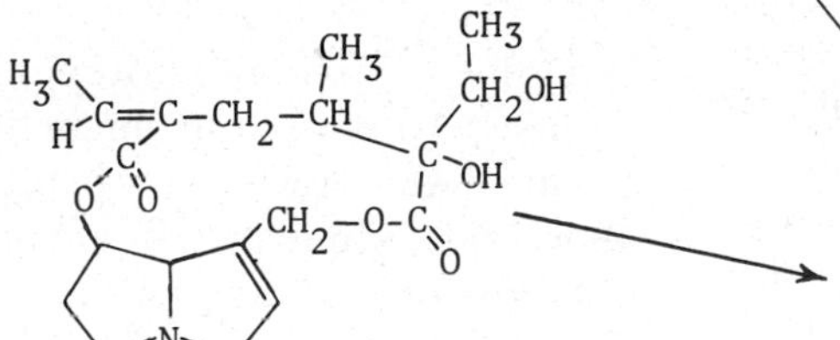

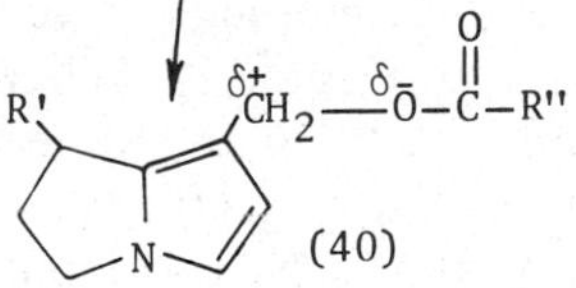

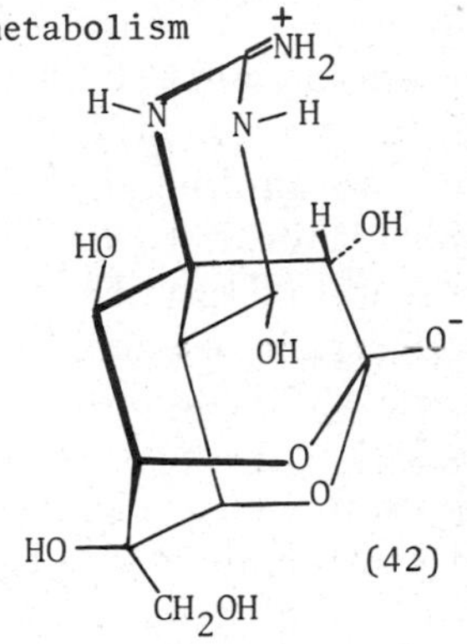

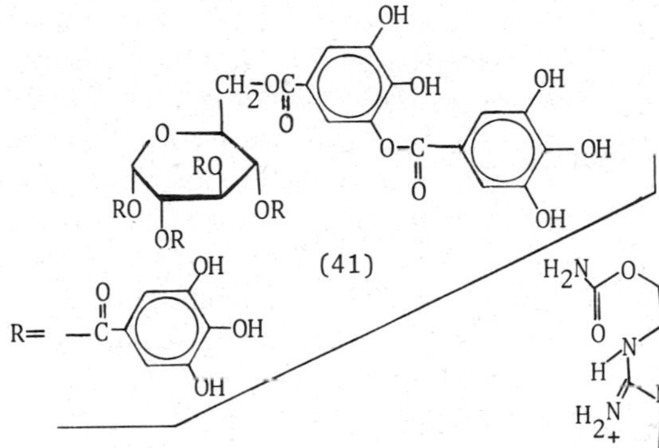

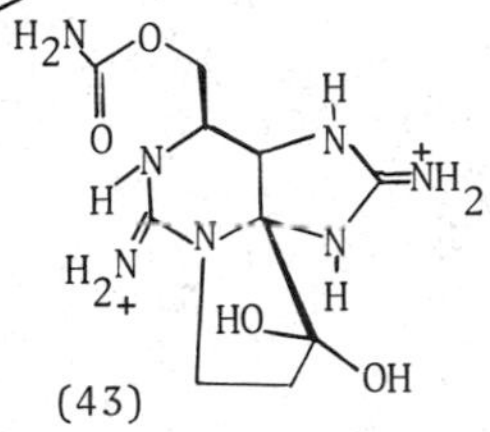

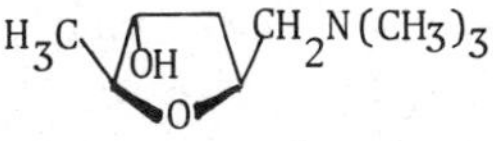

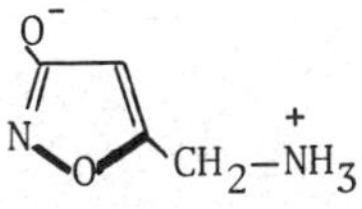

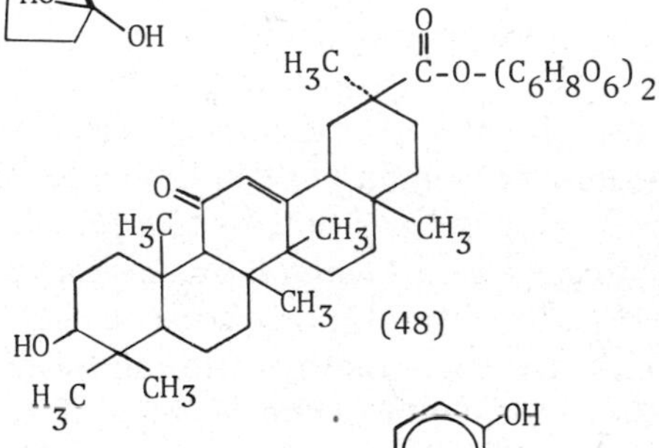

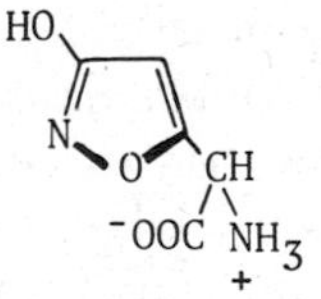

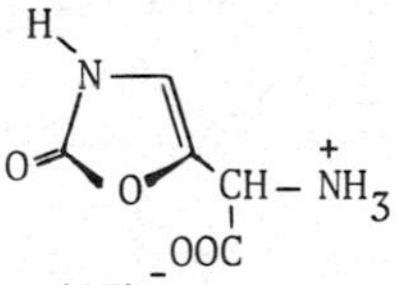

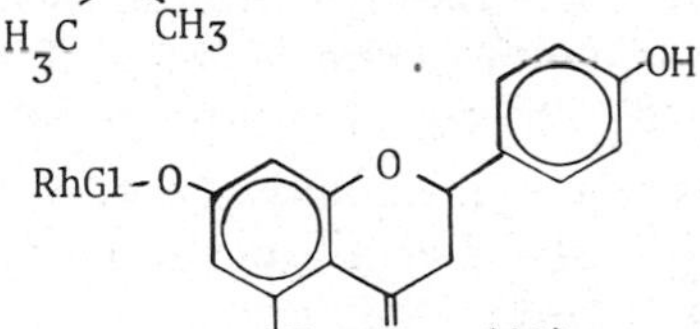

FIGURE 20.6. *(Continued)*

proach 18 mg/fluid oz for coffee and strong brewed teas, 6 mg/oz for cocoa, and 3 to 6 mg/fluid oz for dietetic and nondietetic carbonated cola beverages.

Fatal doses of caffeine require over 10 g, but <1 g can result in distinct physiological responses. Based on serum concentrations, >20 μg/mL can produce physiological effects in adults, which may include jitters and/or seizure, depending on individual sensitivities for test subjects. When serum concentrations reach 50 to 100+ μg/mL, a variety of more severe effects, described as *caffeinism*, become very common.

Xanthines are absorbed 30 to 60 min after oral ingestion and exhibit an average 3 to 6-h half-life in those individuals 6 months and older, into adulthood. A protracted 30 to 210-h half-life is typical for neonates and infants up to 6 months old.

In the realm of side-effects, xanthines can elicit sensory disturbances (e.g., tinnitus, light flashes), restlessness, insomnia, increased respiration, tachycardia, extra systoles, and other central and peripheral actions on the cardiovascular system. Apart from these actions, caffeine and/or other xanthines actuate or exacerbate glucosuric conditions in clinically controlled diabetics.

Methylxanthines are also reported to promote the growth of benign fibrocystic breast lumps in *some* women. The link between fibrocystic disease and methylxanthines in many women is *still debated*, but *some studies* claim occasional resolution of this malady by eliminating dietary methylxanthine sources including coffee, tea, chocolate, and cola beverages. The biochemical etiology of this problem *may* rest on the ability of methylxanthines to upset the "second messenger" effects of cAMP and cGMP.

Caffeine metabolism is largely dependent on the *P*-450 cytochrome system in all age groups. The low *P*-450 cytochrome population in neonates inhibits caffeine metabolism, which would ordinarily include various demethylation products, paraxanthine (1,7-dimethylxanthine) formation, and a C-8 oxidation. For adults, urinary excretion of caffeine metabolites occurs as *1-methylxan-thine* and *1-methyluric acid*. Transplacental effects of xanthines are probably marginal during fetal existence, although they may contribute to lower birth weights in neonates.

Pressor Amines

Tyramine, dopamine, norepinephrine, histamine, serotonin, and phenethylamine occur in a wide variety of plant tissues and other foods. These pressor amines, also called *vasoactive amines*, cause substantial increases in blood pressure when administered intravenously and have less potent effects when acquired from the diet.

One mainline route for amine metabolism in the body relies on the enzyme action of *monoamine oxidase* (MAO). Although vasoactive amines can exert a potential hypertensive effect on almost any individual who consumes amine-rich foods, amine effects are most notable for individuals medicated with *MAO inhibitors* (MAOI). Typical among MAOI drugs used for controlling depressive illnesses are phenelzine, nialamide, pheniprazine, tranylcypramine, iproniazid, and isocarboxazid. The MAOI prevents the expeditious oxidative deamination of the vasoactive dietary amines and may lead to a hypertensive crisis. This crisis may be a simple headache or an overt intracerebral hemorrhage depending on individual circumstances.

Vasoactive amines are also responsible for the hypertensive responses caused by some cheeses, where actions of bacterial L-*aromatic amino acid decarboxylases* on casein produce corresponding decarboxylated forms of original aromatic amino acids. Nearly all putrefied meats, pickled fishes (marinated herring), and fermented alcoholic beverages contain notable concentrations of vasoactive amines (Table 20.5). The phenethylamine content of chocolate is also implicated with antidepressive psychological responses.

Polyphenolic Compounds

Polyphenolic structures having molecular weights of >500 are generally classified as

Table 20.5. Tyramine Is One of the Most Potent Vasoactive Amines in Foods and Beverages

Beverages	Beer and sherry	2–5 μg/mL
	Chianti	20+ μg/mL
Cheeses	American processed	45 μg/mL
	Camembert	90–500+ μg/g
	Emmenthaler	200–700 μg/g
	Gruyere	475 μg/g
	Vermont cheddar	200–1500 μg/g
Fish	Marinated herring	2500+ μg/g
Meat	Beef liver (stored)	300 μg/g

tannins. The so-called *hydrolyzable tannins*, including tannic acid, can be spontaneously or enzymatically hydrolyzed to yield glucose and gallic acid. Other tannins classified as *polymeric flavonoids* or *condensed tannins* are stubborn survivors of physiological digestive conditions.

Acute oral LD_{50} values for single doses of tannic acid on rats are ~2300 mg/kg, which is far from toxic. However, direct or indirect effects of tannins and their constituents have been suspected to cause subacute problems. Apart from the unlikely assimilation and systemic toxicity of tannins, their gallic acid components are routinely detoxified in the body by conversion to 4-*O*-methylgallic acid. Since methylation here is at the luxury of choline and especially methionine, severe stress on crucial methylation potentials of the body can result unless methyl donor reserves and methionine (an essential amino acid) remain high. Any failure of tannins to complex with proteins during intestinal transit and produce insoluble tannates may also induce mucosal damage and expose these tissues to toxic intestinal wastes and/or xenobiotic chemical stresses.

Quercetin and other flavonols are common phenolic elements of many plant structures. Some bacterial evidence indicates that these substances possess mutagenic and possible carcinogenic actions, although their acute toxicities are quite low.

Quinine

The alkaloid of cinchona bark, known as quinine, serves as a basic protoplasmic poison, and its dietary effects can be described as mild *cinchonism*. In very sensitive individuals, quinine initiates sensory disturbances (tinnitus, blurry vision), gastric upsets, headache, and/or nausea. Cardiac upsets also seem to be caused by depression of myocardial contractability and conductivity.

Tonic waters are definite culprits for many hypersensitive subjects. These waters contain an average of 60 mg/L and some beverages such as Dubonnet may contain more. These quinine concentrations are widely recognized as sufficient to produce a generalized beverage purpura in some patients or "gin-and-tonic" purpura reactions.

Herbal Teas

The renaissance of herbal teas, herbal dietary remedies, and herbal "vitamin plants" has recently enhanced population exposures to natural toxicants that riddle many plant species. Global occurrences of hepatic occlusive diseases are often reported in India and Afghanistan where pyrrolizidine-containing plants are occasionally admixed to wheat crops and used for making breads. In the United States, sporadic occurrences of the *Budd–Chiari syndrome* have recently appeared. This malady, which has been endemic to Jamaica, results from drinking "bush teas." This herbal brew contains a variety of pyrrolizidine alkaloids that produce peritoneal fluid accumulation (ascites), general edema, and hepatic venoocclusive disease. Short of eliminating these teas from the diet, the pathological scenario can easily result in complex surgical intervention designed to rectify blood circulation through hepatic tissues. The pyrrolizidines may also be carcinogenic, but this fact is debated by many authorities.

The principal toxic pyrrolizidine alkaloids rely on an unsaturated carbinolic structure of *retronecine*. Retronecine is probably nontoxic, but the persistent unsaturated site and branched esters of retronecine seem to induce toxic properties. *Heliotrine* and *retrorsine* are typical of the toxic pyrrolizidines and they undergo hepatic conversion to yield numerous metabolically reactive pyrroles that exert alkylating agent potentials.

Budd–Chiari syndrome-like consequences also seem to be implicated with thread-leaf groundsel in the southwestern states.

Rejuvenated interest in comfrey tea (*Symphyta*) as an herbal vitamin B_{12} source may also offer dangerous threats to hepatic function, including hepatocarcinogenesis. Cataria (catnip) teas create less severe but disconcerting physiological reactions such as convulsions. Cataria-induced convulsions may be a composite effect of tannins, 1-sesquiterpenes, citral, limonene, geraniol, and short-chained fatty acid interactions plus unknown factors.

Essential Oils

The essential oil content of herbs is largely innocuous with the exception of trace organics that coexist with major oil constituents. Excessive consumption of apiol, nutmeg, absinthe, pennyroyal, rue, tansy, and menthol can produce untoward consequences including bronchiole irritation, skin flushing, vomiting, and circulatory collapse in sensitive individuals.

Some oils such as sassafras contain *safrene* and *safrole* constituents. This historical beverage-flavoring oil has been replaced by other flavors, since safrole is known to produce hepatocarcinogenesis in rats.

Aside from the recognized use of many essential oils for centuries as flavoring constituents in foods and alcoholic beverages or liqueurs (e.g., oils of calamus and wormwood in vermouths), relatively little is known about their low-level chronic exposure effects on body tissues. Moreover, since the essential oil constituents of many alcoholic beverages are proprietary, skepticism concerning the safety of some alcoholic beverages is not unjustified.

The natural oils of the star anise (Japanese), the mono- and polyhydric phenols of cashew nut oils, and eugenol (80%) content of clove oils can all exert undesirable effects. Cashew phenols in oil instigate contact dermatitis with edema and vesiculation, whereas the low solubility and anesthetic effects of clove oils can cause severe gastroenteritis in certain individuals.

Choleretic Effects of Some Herbs and Spices

Certain herbs, spices, and vegetables are noted for their *choleretic* actions that promote bile formation. Some spices also serve as *cholegogues* since they stimulate gall bladder and biliary duct discharges of bile.

Choleretics include gentian, peppermint, oregano, radish, and onion, but turmeric and wormwood both display choleretic *and* cholegogue activities.

Only 0.025% doses (0.1 dL, v/v%) of wormwood are reported to increase cholesterol excretion by >3000% over predose excretion levels. As a result of this effect, plasma and tissue cholesterols also show significant declines. Turmeric as well as curry powders (which contain turmeric) both produce similar cholesterol effects.

The onset of choleretic and cholegogic responses caused by these minor food constituents may be very beneficial to normal health. However, the clinical nutritionist should recognize the potential effects of these agents on individuals predisposed to biliary colic, cholecystitis, and other hepatobiliary diseases.

Lathyrogenic Agents

Some members of the *Cruciferae* and especially the *Leguminosae* exhibit lathyrogenic actions. The common sweet pea (*Lathyrus odoratus*), which serves as a substantial dietary constituent for some populations, contains β-cycloalanine and its decarboxylation product β-aminopropionitrile (βAPN) (N≡C—CH$_2$—CH$_2$—NH$_3^+$). This compound inhibits lysyloxidase activity necessary for the cross-linking of collagen and elastin. Thus, aortal ruptures and scoliosis are commonly recorded eventualities.

Since the notable action of βAPN is directed toward skeletal problems, it is often considered to be an *osteolathyrogenic agent*.

The presence of osteolathyrogens in some legumes is complicated by the joint presence of *neurolathyrogens*, which can cause tremors and other nervous system dysfunctions. At least two neurolathyrogens are recognized with L-α,γ-diaminobutyric acid being preeminent. This compound specifically inhibits hepatic ornithine transcarbamylase in mammals thereby disrupting the urea cycle and inducing ammonia toxicity.

Favism

Other leguminous plants, such as the broad bean *Vicia faba*, can provoke acute febrile hemolytic anemia accompanied by jaundice, hematuria, and hemoglobinuria depending on the sensitivity of certain individuals. Those having Mediterranean heritage are especially prone to this problem known as favism. Sensitization to this legume bean is hereditary and corresponds with glucose 6-phosphate dehydrogenase deficiencies in red blood cells. Symptoms may occur only sporadically in susceptible subjects, and the severity of individual responses is very unpredictable.

Hemagglutinins

The toxic actions of legumes are further complicated by their native concentrations of *phytohemagglutinins*. These chemical agents, commonly found in soybeans and black beans, show *specific abilities* to *agglutinate* red blood cells at high dilutions. These agglutination reactions are reminiscent of antigen–antibody responses, but nutritional and digestive consequences of these agents in uncooked vegetables is probably a matter of minor concern for most individuals.

Potato Saponins

Distinct neurological disorders result from the saponin-like glycoalkaloid structure of *solanine*. This steroid-based toxic substance produced in sprouting or green potatoes can produce subacute gastric and intestinal discomfort plus hemolytic and hemorrhagic damage to the gastrointestinal tract. The exact neurochemical effects of the steroid compound remain to be established.

Toxicants in Honey

The advent of exotic imported honeys requires awareness of cardioactive toxins including *acetylandromedol, andromedol,* and *anhydromedol* that may occur in these foods. Direct health threats are posed by these agents in rare circumstances or more commonly as interactions with therapeutic drug regimens. Toxic honeys are largely produced from nectars of the rhododendron, azalea, mountain laurel, and oleander. The effects of these toxic compounds should not be confused with occasional allergic responses caused by pollens that persist in native honeys.

Phytoalexins

Mechanical damage, fungal attack, and other stresses on some edible plants elicit the production of *phytoalexins* (PAs). These agents include phaseolin in green beans, lubminin and rishitin in white potatoes, and pisatin present in green peas. The PA contents of sweet potatoes are notable because at least three toxic forms exist, including ipomeamerone, 4-ipomeanol, and 1-ipomeanol. The first of these is a known hepatotoxic agent in rodents, whereas the last two produce edema. The impact of these compounds on selected portions of the consuming population are not clear.

Parsnip and Carrot Toxicants

Phototoxic, procarcinogenic, and mutagenic agents are recognized to occur in parsnips and other related vegetables such as parsley and celery. The principal toxic compounds are believed to be furocoumarins (e.g., psoralen, xanthotoxin, and bergapten). Furocoumarin content of 0.1 kg of parsnip root could produce an exposure potential of up to 5 mg. Furocoumarins found in parsnips should not be confused with the diacetylenic allyl alcohol known as carotatoxin found in celery and carrots.

Plant Toxicant Effects on Livestock

Forage animals generally prefer grazing on grasses and clovers. However, during conditions of range drought and unregulated territorial grazing, animals may consume toxic plants and weeds. In cases of well-managed grazing areas, toxic dicot weeds may be eradicated by mowing or the selective use of herbicides having effects typical of 2,4-dichlorophenoxyacetic acid (2,4-D).

Different types of toxic plants are distributed over native prairie ranges as opposed to cultivated fields or pastures. Lupine, loco weed, halogeton, false indigo, narrow-leaved milkweed, flowering spurge, and larkspur are only some of toxic prairie plants, whereas black night-shade, jimson weed, dogbane, cocklebur, white snakeroot, and other plants may occur as isolated patches in pastures or cultivated fields.

A wide spectrum of chemical agents in these plants are linked to livestock disorders including tremors, bloating, nausea, hemorrhaging (caused by anti-vitamin K substances), dermatitis (photosensitization), and abortion. Severe congenital malformation of fetal animals is also caused by maternal ingestion of toxic plants for only brief periods during early stages of pregnancy. Malformations typically include cleft palate, skeletal deformities, and cyclopian malformations.

Although human public health is not largely jeopardized by residues of plant toxicants in edible animal flesh, goat and cow milks may serve as a transmission vehicle for these toxicants. This route of toxicant exposure offers a direct threat to pregnant women who may consume these contaminated milks. Anagyrine from lupines and other toxic alkaloids can exert severe congenital malformation effects, especially if human embryonic exposure occurs during the first trimester of pregnancy. The scope of this problem may not be fully recognized in rural America or among health food zealots. Furthermore, the indictment of some herbicides for their apparent teratogenicity in some isolated cases may have failed to recognize the involvement of forage plant toxicants in the epidemiology of congenital malformations.

Historical surveys cite a massive variety of plant actions on the human body. For more details, consult the National Academy of Sciences publication cited in the references and the authoritative work by Kingsbury. Some of the toxic plant substances have relevance and practical importance in food commerce while others are rare and of incidental interest.

TOXIC AGENTS PRODUCED BY SAPROPHYTES

Achlorophyllous plants, including the parasitic and saprophytic fungi, produce a variety of toxic and psychoactive effects if consumed in sufficient amounts. Ergot-type alkaloids are produced during the parasitic attack of some grains by fungus. *Claviceps purpurea*, which grows on rye, is typical of such fungi. Ergot alkaloids are of notable but minor concern in American grain commerce at this time.

The principal ergot alkaloids resemble the structure and hallucinogenic effects of *d*-lysergic acid diethylamide (LSD). The biochemical origins of these alkaloids probably rests on the condensation of isopentenyl pyrophosphate with tryptamine as a precursor of these potent alkaloids. Apart from the historical abortifacient actions for these alkaloids, they serve as potent adrenergic blocking agents; stimulate smooth muscles of arterioles, intestines, and uterus; and produce conflicting excitation and depression of the CNS.

Some wild varieties of mushrooms (toadstools) can produce severe poisoning (mycetismus) and hallucinogenic effects after eating. *Agaricus campestrus*, the puffball (*Cavatia* family), the chicken mushroom, morels, and the "Shaggy-mane Coprinus" are among the notable edible varieties. Mistaken identities of other fungi for these edible types leads to severe circumstances, especially if members of the *Amanita* genus are involved such as *A. muscaria* and *A. phalloides*. The toxic factors in these fungi are generally aminitins (five different types). These toxic octapeptides (LD_{50} for mice = 0.3 mg/kg; 50 g of fresh *Amanita phalloides* can kill an adult human) rapidly block the transcriptional actions of eukaryotic RNA polymerases II and III. This ensures a major inhibition of all cell protein synthesis.

Amanitin actions are complicated by *phalloidins*, which are toxic cyclic heptapeptides. The fast-acting but unknown toxic actions of phalloidins are counteracted by a cyclic decapeptide called *antamanide*, which accompanies the phalloidins. This compound exhibits characteristic sodium-binding ionophoric actions, but it cannot counter the gross effects of mushroom poisoning. Phalloidic intoxication can be successfully countered, however, by thioctic acid (α-lipoic acid).

Other less known mushrooms are toxic including *Galerina venenata*, *Halvella esculenta*, *Lactarius vellereus*, and *Lepiota morgani*. These contain muscarine and phalloidine toxins, although other toxins are recognized including mycetoatropina (levohyoscyamine), bufotenin, helvellic acid, ibotenic acid, muscimol, muscazone, coprine, psilocybin, and disulfiram.

Tropane alkaloids are also characteristic of some Amanita (e.g., *A. muscaria*) but more recent evidence also indicates that burdock root teas and catnip can contain enough psychoactive alkaloids to cause weird behavior and hallucinations. Scopolamine, dioscorine, and atropine are representative of the tropanes. Acetylcholinesterase inhibition offers a successful route for controlling most tropane alkaloid actions.

Common edible "grocery" mushrooms can elicit sensitive responses in some individuals that may be linked to fungal nutrition, mushroom–drug interactions in the consumer, and alcohol–mushroom interactions. Even minor amounts of disulfiram in cultivated mushrooms can elicit adverse alcohol responses in the most sensitive individuals. Typical of mushroom–drug interactions are those instigated by hydroxychloroquinone. This drug, which exhibits antimalarial actions, also has beneficial effects in treating lupus erythematosus and acute chronic rheumatoid arthritis. Unfortunately, hydroxychloroquinone can toxify an ordinarily harmless mushroom by mechanisms that are not entirely clear.

Mold infected grains, nuts, ground-nuts, and some mold-fermented foods support the saprophytic growth and metabolism of certain molds. Typical among these molds are varieties of *Aspergillus* and *Penicillium*. Mold attack of foodstuffs is usually encouraged by poor climatic storage of bulk foods (e.g., temperature and humidity conditions), boring insects, and food-storage engineering problems.

Natural hepatotoxic substances produced as apparent by-products of mold metabolism may be important factors in the etiology of hepatomas and some chronic liver diseases.

Foremost among mold-produced hepatotoxic compounds are the aflatoxins produced by *Aspergillus flavus* and *A. parasiticus*. These attack agricultural crops such as wheat, rice, ground-nut, cottonseed, corn, coconut, dried peas, oat, sweet potato, millet, and cassava. Hepatotoxicity and apparent hepatomas caused by aflatoxins are widely recognized in animals that eat mold-infested grains and siled grains. It is not clear what health effects aflatoxins have on the American food consumer, but reason dictates minimizing aflatoxin presence in all animal and human food resources.

Gross effects of aflatoxins on humans have been eminently demonstrated in Southeast Asian countries. The principal effect seems to involve encephalopathy and visceral fatty degenerations. Deaths and hepatic damage to several hundred humans in India have been attributed to aflatoxin-contaminated corn. Aflatoxin concentrations approached 0.30 to 15 mg/kg corn in the last case.

Acute aflatoxicoses may be a minor concern of developed countries, but the chronic, low-level dietary intake of aflatoxins in foods coupled with their demonstrated abilities to produce animal carcinogenicity is a universal concern. Moreover, early studies have demonstrated that aflatoxin B_1 can bind covalently to DNA and form a single adduct at guanine residues. The adduct (2,3-dihydro-2-(N^7-guanyl)-3-hydroxyaflatoxin B_1) may somehow induce carcinogenic consequences by removing purines or distorting the normal stacking of DNA bases in the double helix. Assuming that this action at the genetic level exists, the mechanism of carcinogenic actions is complicated by results of other studies that claim that aflatoxin-induced rat hepatocarcinomas are averted by hypophysectomy.

At least six aflatoxins are commonly recognized as threats to human food resources, and many other types exist that may have minor toxicological significance (Figure 20.7). The detectable range of aflatoxin structures also seems to be enhanced during heat treatment, which produces a variety of thermal degradation products having properties somewhat similar to the parent aflatoxins.

Hepatic metabolism of aflatoxins may involve the temporal formation of a reactive electrophilic intermediate, which reacts with mutually reactive tissue constituents; or as an alternative, aflatoxins may be hydroxylated by liver microsomes and eliminated via urine, bile, or milk. Aflatoxin M_1 is a common product of aflatoxin B_1 metabolism, and it can be detected in human urine after ingestion of peanut butter contaminated with aflatoxin.

The inherently low microsomal activity of infants dictates that milk residues of aflatoxins should be minimized by ensuring that cows receive only good quality feedstock.

Other interesting toxic saprophytic by-products whose exact dietary impacts are unknown include

1. Forty different *tricothecenes* produced by eight different mold species, which reportedly evoke tissue irritant actions.

2. *Ochratoxins* and *citrinin*, which instigate nephrotoxic actions.

3. *Patulin* produced from *Penicillium* fruit rots.

4. *Sterigmatocystin* from *Aspergillus versicolor*, which elicits neoplastic skin and hepatic tumors.

An arbitrary survey of literature reveals virtually hundreds of fungal products and metabolites that may exert a public health threat. Based on animal studies, these byproducts of saprophytic organisms clearly have deleterious effects, but in the milieu of the entire contemporary American diet, their impact is uncertain and probably minimal.

MARINE FOOD INTOXICATIONS

Paralytic Shellfish Poisoning (PSP)

Sporadic seasonal blooms of marine dinoflagellates can inadvertently accumulate in the filter-feeding apparatus of edible shellfish. The indicted toxic dinoflagellates include *Gonyaulax tamarensis* and *G. catenella*, which produce *saxitoxin* (STX), and possibly 11 other minor toxins. The neurotoxic actions of STX are rivaled only by *tetrodotoxin* (TTX), which is concentrated in the organs

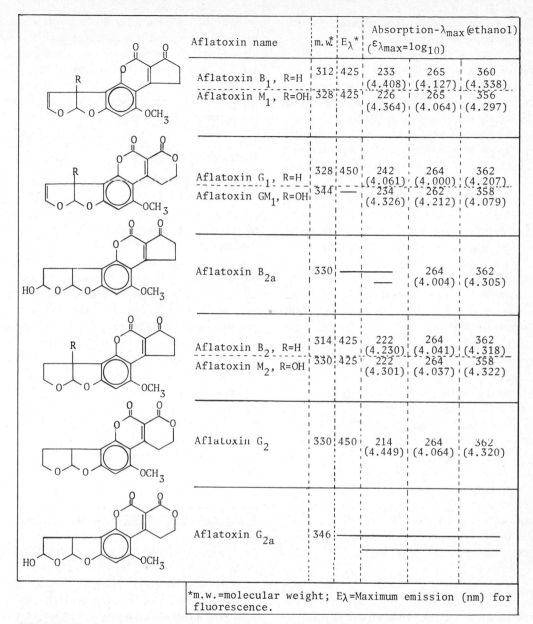

Aflatoxin name	m.w.*	Eλ*	Absorption-λmax (ethanol) ($\varepsilon\lambda_{max}=\log_{10}$)		
Aflatoxin B₁, R=H	312	425	233 (4.408)	265 (4.127)	360 (4.338)
Aflatoxin M₁, R=OH	328	425	226 (4.364)	265 (4.064)	356 (4.297)
Aflatoxin G₁, R=H	328	450	242 (4.061)	264 (4.000)	362 (4.207)
Aflatoxin GM₁, R=OH	344	—	234 (4.326)	262 (4.212)	358 (4.079)
Aflatoxin B₂ₐ	330	—	—	264 (4.004)	362 (4.305)
Aflatoxin B₂, R=H	314	425	222 (4.230)	264 (4.041)	362 (4.318)
Aflatoxin M₂, R=OH	330	425	222 (4.301)	264 (4.037)	358 (4.322)
Aflatoxin G₂	330	450	214 (4.449)	264 (4.064)	362 (4.320)
Aflatoxin G₂ₐ	346	—	—	—	—

*m.w.=molecular weight; Eλ=Maximum emission (nm) for fluorescence.

FIGURE 20.7. Some aflatoxin structures and characteristic spectral properties (approximate only). Routine thin-layer chromatographic separation using silica gel G (solvent-chloroform:methanol (96:4)) yields R_f values over the range of 0.6 to 0.1 with $B_1 > B_2 > G_1 > G_2 > M_1 > M_2 > B_{2a} > G_{2a} > GM_1$. The B and G designations stem from early recognition that some aflatoxins fluoresced blue and green when exposed to ultraviolet light after isolation by chromatographic methods.

of some marine fishes. The potent neurotoxic action of STX, as well as TTX, can kill all shellfish at high enough exposure levels, although certain mollusks are more neurotolerant to these toxins than others.

The edible oyster (*Crassotrea virginica*) is most susceptible and the edible mussel (*Mytilus edulis*) is among the most resistant shellfish based on blocking action-potential studies of nerves using these toxins. Quahogs (*Mercenaria mercenaria*), the little neck clam (*Prototbaca staminea*), and the sea scallop

(*Placopectem magellanicus*) show low to moderate neurotoxic tolerance.

For oysters 10^{-8} g/mL TTX and 10^{-7} g/mL STX can initiate a block of the action potential, whereas the mussel remains unaffected by 10^{-4} g/mL of both TTX and STX. Both toxins exert neurotoxic actions by blocking sodium pores of postsynaptic membranes to prevent nerve transmission.

Human consumption of neurotoxin-contaminated shellfish may cause transient numbness and tingling sensations in the lips and fingers at low concentrations, or complete respiratory failure at higher concentrations. Since oysters are readily killed by the neurotoxin, and "dead" oysters are undesirable as food, human intoxications usually involve the hardy shellfish survivors of neurotoxic dinoflagellate blooms. Although only 450 to 1000 μg/100 g shellfish may be fatal for humans, some New England shellfish contain 5000 to 10,000 μg/100 g.

Foremost among marine fish species, TTX is found in the Japanese "tora fugu" or tiger puffer fish (*Spheroides rubripes*). The paper-thin filleted flesh of this fish, which has gained notoriety as a historical delicacy that evokes slight facial tingling sensations (and occasional death), is an article of delectable Japanese cuisine.

Palytoxins and many other exotic marine toxins are known to exist, but very few of these have importance in American food commerce.

Scombroid Poisoning

Dark-meated pelagic fishes that belong to the *Scombroidea* family are linked to this type of poisoning. Scombroid poisoning in humans is probably caused by the proteolytic decomposition of fish flesh to produce a wide variety of pressor amines. Apart from fish enzymes that mediate the conversion of histidine to histamine, *Proteus morgani* and other Gram-negative bacteria produce pressor amines. Hypertension, numbness of fingers and lips, vomiting, and rapid pulse rates characterize this problem in humans.

Scombroid poisoning and marginal cases of paralytic shellfish poisoning both exhibit an ice-berg epidemiological occurrence in the United States. That is, because of the mild nature of many cases, they go unreported to medical authorities, especially in high fish-consuming coastal regions.

Ciguatera Poisoning

Marine varieties of *Lyngbia majuscula* produce a toxic substance called *ciguatoxin* or *ichthyosarcotoxin*. Since this toxin is immune to the digestive processes of fishes, it accumulates among predator fishes throughout the marine food chain. Although 400 + fishes can accumulate this toxin, many cases of suspected ciguatoxic poisoning involve other factors. Research indicates that some supposed ciguatoxic poisonings may result from the accumulation of (1) dinoflagellate toxins, (2) toxins produced by bacteria in the fish gut, or (3) benthic algal toxins.

Human symptoms of poisoning include influenza-type responses, gastroenteritis, blindness, paralysis, and/or mental depression. Red snapper is one notable species that occasionally demonstrates these toxic effects.

ESTROGENIC AGENTS IN FOODS

Many compounds in foods exhibit estrogenic activity. Some chemical agents such as diethylstilbestrol (DES) are synthetic estrogens, and these have been widely used as synthetic growth hormones to enhance livestock productivity. Other compounds with similar effects have natural origins as by-products of plants (phytoestrogens) or fungi (mycoestrogens). As indicated in Figure 20.8, most of these estrogenic agents contain a phenolic substituent(s).

Human effects of estrogenic substances are *uncertain*, but zearalenone, naturally produced by *Fusarium*-contaminated grain, has been used as a DES substitute to promote livestock weight increases.

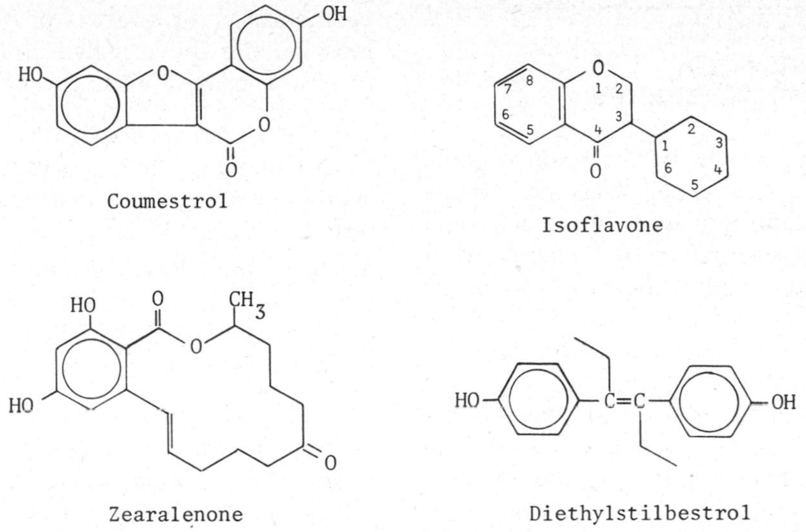

FIGURE 20.8. Structures for some notable phyto- and mycoestrogens occasionally present in foods as well as the synthetic estrogen diethylstilbestrol (DES). Coumestrol is a constituent of alfalfa, ladino clover, and many natural animal forage crops. The isoflavone structure is the basis for many soybean phytoestrogens, and zearalenone is representative of the estrogenic factor in *Fusarium* species.

Phytoestrogens are not restricted in their occurrence to any particular plant species, since they reportedly include lupulonic acid derivatives of hops, isoflavoncs from soybeans, and coumestrol in alfalfa and soybean sprouts.

Variable degrees of estrogenic activity are noted in rodents, sheep, swine, and cattle as a result of natural estrogens, and the effects may involve vulvar, ovarian, cervical, and mammary gland responses. Speculation suggests that the biochemical effects of estrogenic substances are probably similar to the natural activity of estradiol (F_2).

MINOR ALLERGIC AND SENSITIVITY RESPONSES TO FOODS

Although the basis for food allergic responses has been offered elsewhere in this text, allergic physiological responses short of anaphylaxis should not be overlooked.

Many foods evoke specific sensitivity responses in certain individuals each time they are consumed. Consequences typically include constipation or sporadic bouts of diarrhea, belching, nausea, cramps, and/or epigastric discomfort. With or without rectal spasms, mucous diarrhea may be accompanied by large population numbers of eosinophils. For many individuals pruritus ani can be a very common consequence of minor allergic responses.

The normal routes for minimizing these problems include outright elimination of suspect foods, desensitization, or administration of epinephrine or an antihistamine to control acute symptoms.

Certain derivatives of foods can also be assimilated into the body and cause light-provoked skin eruptions. These sensitivity responses are commonly encoutered as *fagopyrism* in animals, but comparable incidents seem largely unimportant in humans. The lack of obvious symptoms may stem from the degradation or distillation of key coumarin-photoreactive precursors from foods during cooking.

Some demonstrated links also seem to exist between food allergies and the chronic occurrence of rheumatoid arthritis.

THE PROBLEM OF CARCINOGENICITY VERSUS MUTAGENICITY

The potential reactivity of certain organic compounds in foods supersedes their effects as acute toxicants. For food consumers, some forms of agricultural biocides, agricultural chemicals, and fugitive industrial chemicals offer insidious mutagenic and carcinogenic threats at concentrations far below those required for producing acute toxicity.

Exposure limits to subacute toxic effects of reactive trace organic compounds in foods must be estimated largely on the basis of empirical carcinogenic and mutagenic bioassays.

Bioassays for carcinogens are often based on the chronic exposure of selected animal species (e.g., rodents) to different concentration levels of suspected carcinogenic agents until a *prescribed assay endpoint*. Assays typically run up to a 2-year endpoint when all animals are sacrificed and examined for evidence of cancers. Another endpoint approach allows animals to die naturally after which the average age at death is calculated and used as a carcinogenicity index for the chemical agent. The presumption in the last assay assumes that tumors kill the test animal, but accurate tests should discriminate benign tumorigenesis from malignant forms.

Regardless of endpoint rationales, bioassays attempt to estimate a *maximum tolerated dose* (MTD) for a suspected carcinogen. The MTD is the highest dose given during a chronic study that predictably will not alter the animals' longevity from affects other than cancer. In practice, the MTD is considered to be the highest dose that causes no more than a 10% loss in weight compared to control test animals—although this may be subject to many variations. Throughout carcinogenic bioassays, animals are regularly examined for signs of toxicity and sacrificed at prearranged times for histological and biochemical analyses.

Animal bioassays for carcinogenicity have been assailed for a variety of reasons. Foremost among these analytical criticisms are the two considerations:

1. The response of specific animal models to carcinogens may not parallel human responses or susceptibility to the chemical agent tested.

2. Test animals are usually administered inordinately larger doses of suspected carcinogens than consumed during the relative existence of humans.

Aside from these critiques of bioassay rationales, it is widely counterclaimed that man is not an exceptional species with regard to carcinogenesis, and the basic genetic and hormonal systems of animal bioassay subjects are similar to those of humans. Further, the observed fact that not all test animals display identical carcinogenic responses to the same chemical agents is not indicative of bioanalytical heresy. Instead this may only represent innate animal-specific variations to carcinogen susceptibility, which is also demonstrated among human beings.

Animal exposures to large doses of carcinogens may complicate the significance of carcinogenic bioassays, but this aspect of bioassays *does not* obviate the significance of their test results. It is claimed that high doses of suspect carcinogens are justifiably administered to test animals since few rodents live longer than 2 years. Thus, high exposure levels to potential carcinogens will be indicated within this limited time frame. Large doses of potential carcinogens administered to test animals are also justified on the basis that lifelong human exposures to carcinogenic agents is longer than that of test animals and is complicated by the impact of environmental threats. These outside threats coupled with the cumulative effects of carcinogens on the body all bolster the credibility of established "high-dose" bioassays.

Based on findings of the Office of Technology Assessments (OTA) for the U.S. Congress, animal tests still represent the most feasible route for evaluating potential carcinogens. All substances demonstrated to be carcinogenic in animals are regarded as potential human carcinogens and may not appear in foods.* No clear distinction has been annunciated for chemical agents that cause cancer in laboratory animals versus humans. Claims of the OTA for purposes of legislative information also support the rationale for feeding high doses of potential carcinogens to animals thereby producing possible carcinogenic conditions.

The rationale for feeding large doses of a substance in animal tests is as follows. As the dose of a substance that causes cancer is increased, the number of exposed animals that develop cancer also increases. To conduct a valid experiment at high dose levels, only a small number of animals (perhaps several hundred) is required. (The smallest incidence rate detectable with 10 animals is 10 percent or 1 animal. To detect a 1-percent incidence rate, at least 100 animals would be required.) Another important variable is the strength of the carcinogen. The stronger the carcinogen, the greater will be the number of animals getting cancer at a particular dose.

Conflicting testimonies regarding the credibility of carcinogenic bioassays nonetheless culminate in experimental efforts designed to legislate exposures to potential carcinogenic hazards. In a vein similar to LD_{50} tests for determining acute toxicities, E_{50} values can be reported for carcinogenic compounds whose effects on animals are subacute. This value is used to predict arbitrary doses per exposure for potential carcinogens that produce an "effect" in 50% of a specific test animal population. Unfortunately, E_{50} values fail to predict the maximum concentration level of a chemical agent that *will not produce* a future carcinogenic crisis. An accurate determination for such a critical threshold value (or no effect level) is largely impossible since there are no uniform methods for deter-

mining the threshold dose of a carcinogen. This perplexing problem reflects a poor understanding of carcinogenesis at its fundamental stages of biochemical initiation. Moreover, the delayed manifestations of cancer, months or even years after its initiation in animal systems, is enough to confound any statistical analysis of short-term dose–effect relationships for suspected carcinogens.

Short-Term Tests for Mutagenesis and Carcinogenic Potential

The practical complexity and fiscal expense involved in animal bioassays for carcinogens has spawned interest in *short-term bioassays.* Many of these tests rely on the ability of a chemical agent to inflict detectable genetic damage on test organisms.

Analytical rationales here are based on the tacit assumption that many mutagens in foods, such as aflatoxins, are known to be carcinogenic. Thus, chemical agents demonstrating mutagenesis may be indicted as possible carcinogens. This convenient parallel may not be true, however, since numerous food-borne mutagens are not carcinogenic in animal bioassays. Furthermore, many short-term bioassays have employed *in vitro* tissue cultures of mammalian cells, bacteria, and other simplified proxies for detecting molecular genetic insults caused by suspected mutagens and carcinogens. Critics charge that these practices cannot accurately predict the potential carcinogenic or mutagenic hazards imposed on humans by using identical chemical agents.

Carcinogenesis is a complicated pathological condition that involves an *initiation stage,* a phase of *tumorigenic promotion* and *progression,* and finally *metastasis.* These steps involve genetic and epigenetic factors in higher animals that are rarely considered in simplified bioassays. Reactive chemical agents including epoxides, alkylating agents, free radicals, reactive amines, and any other chemical agents that upset the normal

---

*Food Additive Amendment to Food, Drug and Cosmetic Act (1906, 1938, 1954) in 1958. Note extension of 1954 Miller Amendment and Delaney Clause inclusion, which legislates human and animal carcinogenic threats in foods.

molecular genetics of DNA are considered *genetic factors*. Beyond these factors, carcinogenesis is complicated by *epigenetic factors* such as permissive hormone influences (e.g., by steroids), nonreactive inorganic chemical effects on DNA, recurrent tissue damage at the site of parenteral injections for suspect chemical agents into animals, environmental stress effects, and other factors.

Foremost among *some* of the short-term tests for mutagenicity are those conducted on human populations, simplified animal studies, cultured mammalian cells, and microorganisms.

Although ethical considerations forbid mutagenicity tests on humans, portions of the human population already exposed to mutagens can serve as useful indicators of mutagenic agents. Circumspect studies of chromosome aberrations (cytogenetic studies) in white blood cells, blood and urine screening for mutagenic agents, and semen studies designed to detect abnormal sperm plus extra chromosomes are usually feasible.

Cursory animal studies can be employed on rodents (or insects including *Drosophila*) such as the dominant lethal test. Here male rodents, usually mice, are mated after being treated with a suspected mutagenic agent. When demonstrated fetal resorption or fetal death in pregnant females is detected, this is taken as evidence that germinal cell damage (sperm) may have been inflicted on the treated male animals.

Exposure of cultured mammalian cells to mutagenic agents serves as another simplistic approach to mutagenic testing. Although these cells largely maintain their native genetic properties and basic aspects of their mammalian metabolism, their response to a mutagenic agent is jaded by their isolation from the metabolic activity of an entire test animal. This is an important factor because enzymatic processes of the body may activate mutagens from nonmutagenic precursors as well as deactivate existing mutagens.

In an effort to improve the credibility of mammalian cell culture assays, suspected mutagenic agents may be incubated with a processed liver homogenate. This serves as a feeble but rather effective method for stimulating the activation or metabolism of mutagens before their introduction into a cell culture. A lack of liver homogenate uniformity from source liver-to-liver, homogenate storage problems, and an incomplete complement of enzymes having mutagenic activation potentials all detract from this method. In spite of these obvious problems, liver homogenate activation has been widely used.

Microbiological detection of mutagenic events may involve the use of fungi, bacteriophages, and specially bred strains of bacteria. Mutant *Salmonella typhimurium*, in particular, serves as the test organism in the "Ames spot test" for mutagenicity. This test employs a specially developed strain of *Salmonella* that is unable to biosynthesize histidine (His −). A "lawn" of these bacteria is applied to a growth-supporting medium (no histidine) that has a centered cellulose or Teflon disc saturated with a suspected mutagen. If mutagen activation is a prerequisite for testing purposes, processed liver homogenates may be provided to the system that supplies mammalian metabolic functions. A postinoculation and incubation period of about two days leads to the demise of most His(−) Salmonella bacteria, except those that have undergone mutagenesis by the chemical agent diffused from the disc. The mutagenic effect is one of genetic reversion such that His(−) bacteria become able to biosynthesize histidine (His +) and propagate into distinctly visible colonies on the growth medium.

The Ames test has been widely employed to detect *point mutations*. Thus, if carcinogenesis at the genetic level entails a singular point mutation, this test probably mimics the events that can instigate carcinogenesis in humans. Numerous variations of the Ames test have been detailed in the literature, and results from these tests generally tout a 70 to 90% accuracy for detecting carcinogens and noncarcinogens.

Ames test credibility depends significantly on the chemical class being studied as mutagens. Only 55 to 65% recognition of carcinogenic hydrazines, chloroethylene, lactone, and inorganic compounds is realized, whereas 25 ± 5% recognition of certain polyhalogenated aromatic, steroid, phenyl, azo, benzodioxole, carbamyl, and thiocarbamyl compounds are recognized as carcinogens. The higher possibility for recognition failure of carcinogens in the last two groups of compounds (*false negative Ames test*) probably reflects (1) the absence of hormonal interactions among animal tissues (i.e., permissive steroid hormone effects); (2) poor permeability of the mutagenic agent over bacterial cell membranes; (3) overt toxicity of the chemical agent to the bacteria; (4) lack of microorganism sensitivity to the agent; or one of many other complicating factors.

Many Mutagens Are Strong Electrophiles

The mutagenic and carcinogenic potential for chemical agents is intrinsically dependent upon their absorption, transport, distribution, metabolism, and partitioning coefficient within the animal body. Biochemical reactivities are also influenced by physicochemical, steric, and electronic properties of specific agents.

It is theorized that the most effective mutagens are those chemical agents that have limited reactivity in biological fluids and become translocated to target cells of the body. Some chemical agents show general interactions with *all cells* of the animal body, and yet other agents display *target-cell specificity*.

Cellular assimilation of chemical agents can result in a direct reaction with molecular genetic material, but scientific record shows that *cell-specific* biochemical reactions may account for the selective production of mutagenic agents in only certain cell types. This differential development of mutagenic and/or carcinogenic agents may be attributable to different forms of *P*-450-dependent monooxygenases. These enzymes catalyze hydroxylation of organic molecules including pesticides, steroids, many other incidental metabolites, and xenobiotics.

Microsomal enzymes also attack potentially carcinogenic compounds such as *polycyclic (polynuclear) aromatic hydrocarbons* (PAHs), nitrosamines, aromatic amines, and others.

Microsomal activity serves as a detoxification system since product formation generally favors the hydroxylation of chemical agents followed by their ultimate urinary excretion. Unfortunately, the limited reactivity for many of these chemical agents, followed by their hydroxylation, may yield chemical intermediates having potential carcinogenic activity *or* produce *ultimate carcinogens*. These ultimate carcinogens are usually strong electrophilic derivatives of parent compounds (e.g., diol epoxides of aromatic hydrocarbons) that readily form covalent adducts with nucleic acid bases. Thus, the relative activities of specific monooxygenases and precarcinogen activating enzymes in cells ultimately dictate which tissues may be seriously affected by specific chemical agents.

Strong electrophilic properties for many mutagenic and carcinogenic substances are suspected to involve formation of a cation species. Any one of many organic compounds is able to form this reactive species, including the alkylnitrosamine and the PAH illustrated in Figure 20.9A. Apart from the recognized reactivity of a $CH_3CH_2^+$ species derived from diethylnitrosamine, the PAHs have garnered special attention as potent carcinogens.

The enzymatic oxidation of PAH compounds can produce PAD epoxides, which are labile to ring opening by the nucleophilic nitrogen atom of nucleic acids (DNA). This reaction ultimately yields a covalent adduct involving the nucleic acid and the PAH (Figures 20.9B and C).

Many PAH representatives, having characteristically high carcinogenicity, exhibit a "bay region" methyl group plus an unsubstituted carbon positioned adjacent to, but opposite from, an unsubstituted aromatic ring. The unsubstituted ring may then serve

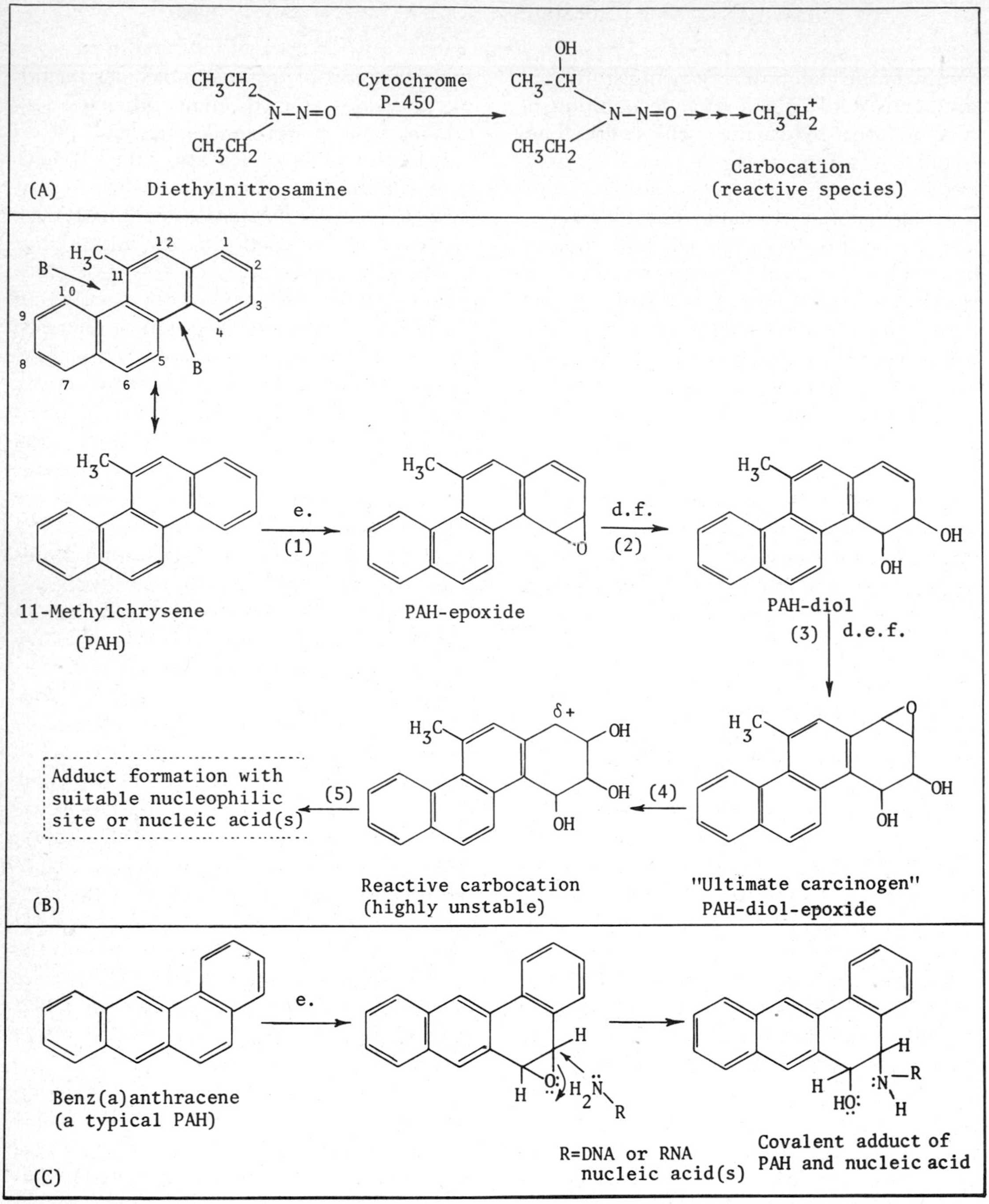

FIGURE 20.9. (A) Typical cytochrome *P*-450 enzyme mediated reaction for a nitrosamine yielding a reactive electrophilic species. (B) The "ultimate carcinogen" for 11-methylchrysene may involve dihydrodiol formation neighboring the "bay region" (B) of the molecule. The bay region refers to that part of the molecule bounded by four carbon atoms of three fused aromatic rings. The unsubstituted carbon atoms in the terminal ring ($C_{1\rightarrow4}$) offer a reactive site for enzymatic reactions. Epoxidation (e.) (1), *trans*-diol formation (d.f.) (2), and diol epoxidation (d.e.f.) (3) of the remaining double bond in the formerly unsubstituted ring ($C_{1\rightarrow2}$) serve as a prelude to "ultimate carcinogen" formation. Methyl groups can promote epoxide formation by increasing ring strain, or conversely, interfere with an enzymatically reactive locale on a PAH structure by directing enzymes to other positions. (C) Benz[*a*]anthracene behaves as a typical PAH that can undergo epoxidation (e.) to yield a reactive epoxide. The PAH epoxide resulting from this action can react with nucleic acids at nucleophilic nitrogen sites to produce PAH adducts and induce molecular genetic aberrations.

as a potential site for dihydrodiol epoxide formation (Figure 20.9B).

Abiotic Trace Organics in Foods with Mutagenic Potentials

Foods contain what seem to be an infinite number of potentially mutagenic and carcinogenic agents. *Some* of these are natural components of foods that undergo activation from a precarcinogenic form to a carcinogen in host cells and tissues. Initial steps in the reactive adduct chemistry of aflatoxin(s), a pyrrolizidine alkaloid, and safrole are quite typical of some naturally occurring precarcinogens (Figure 20.10).

Heat treatment of foods produces a wide range of mutagenic agents whose carcinogenic impacts on public health are largely unknown. Cooking food is one of the most important factors in daily life, and yet browning reactions during cooking can produce recognized mutagenic substances in foods. Typical among these mutagens are pyrazines, imidazoles, and other nondescript amino acid reaction products including polymers.

Severe heat treatment of foods spurs prolific pyrosynthesis reactions, fueled by the basic chemical components of foods. Of particular concern are those heated food treatments conducted at high temperatures. At temperatures exceeding 475°C, carbon—hydrogen and carbon—carbon bonds are broken to form free radicals. These radicals can undergo dehydrogenation and recombine to produce *pyrosynthetic aromatic rings* and, eventually, PAH compounds (Figure 20.11A).

Fat drippings onto hot gas or electric heating elements during cooking readily deposit PAH compounds, carried by smoke, onto meat surfaces. The pyrolyzed fat produces a spectrum of PAHs having notable mutagenicity (Figure 20.11B). Benzo[a]pyrene productivity alone, on the surface of a 2-lb "well-done" steak, may be equivalent to that amount of the PAH produced in the smoke of 600 cigarettes. Fortunately, the tissue exposure route for these PAH compounds in the digestive tract is not

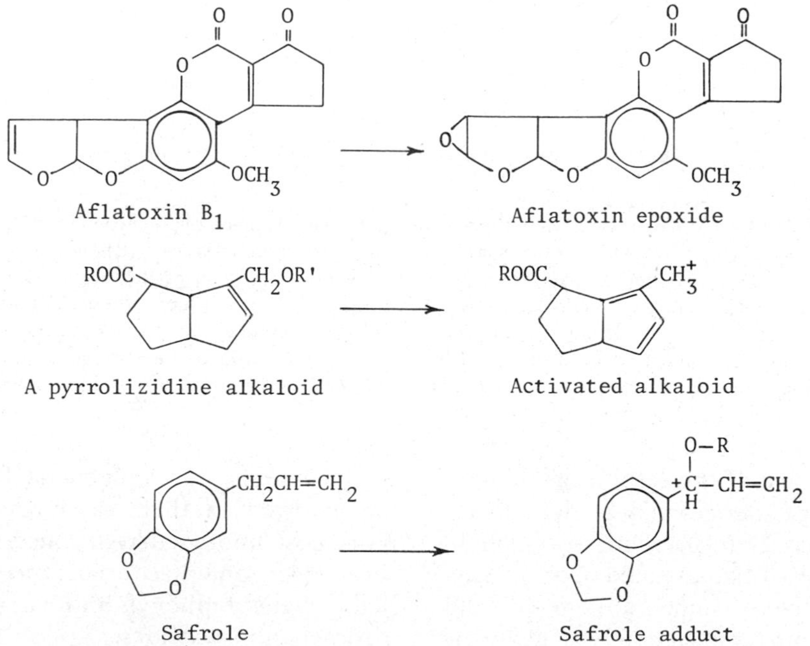

FIGURE 20.10. Some precarcinogens occurring naturally in foods undergo host-mediated activation steps before instigating their mutagenic and/or carcinogenic adduct actions at the molecular genetic level.

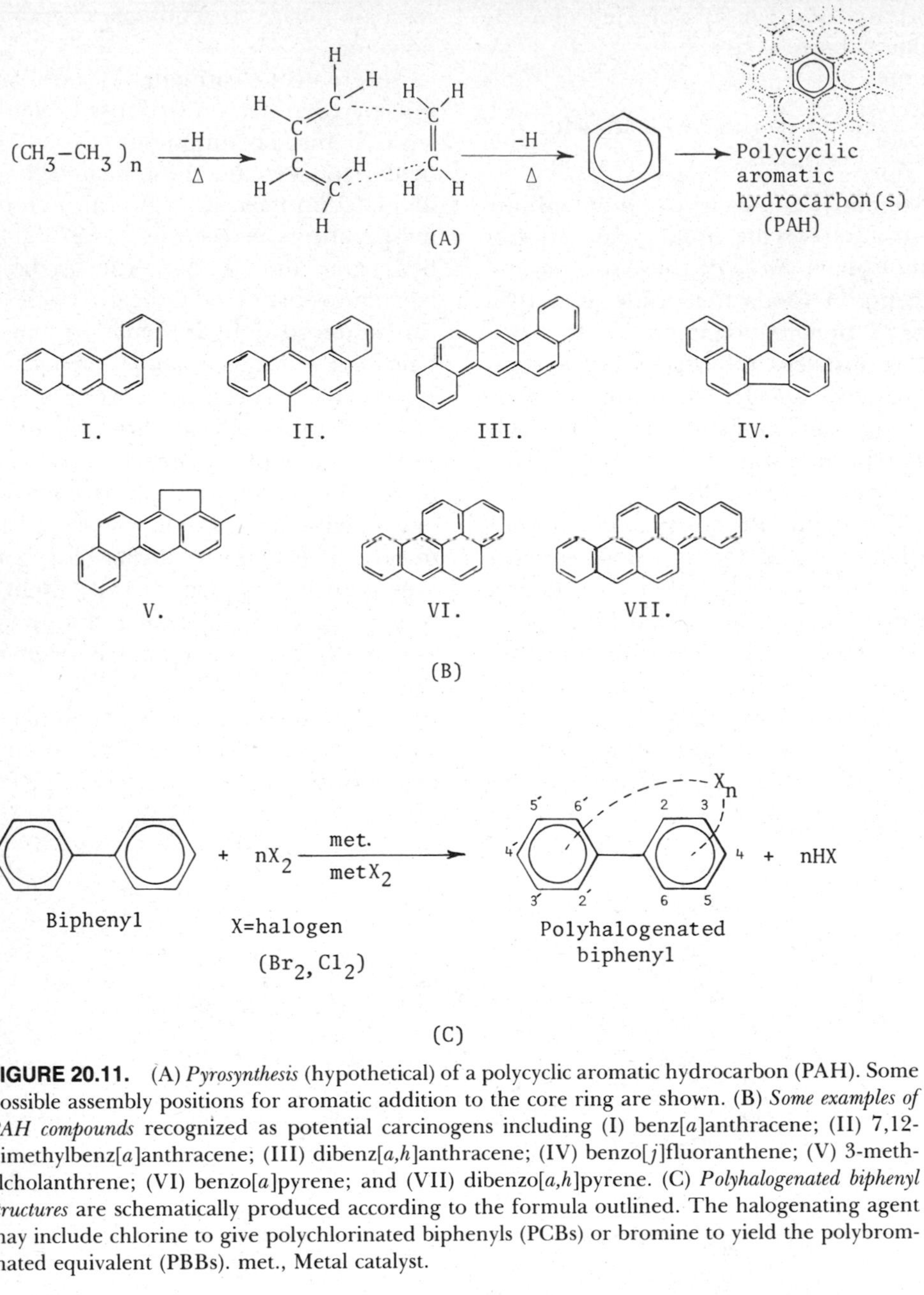

FIGURE 20.11. (A) *Pyrosynthesis* (hypothetical) of a polycyclic aromatic hydrocarbon (PAH). Some possible assembly positions for aromatic addition to the core ring are shown. (B) *Some examples of PAH compounds* recognized as potential carcinogens including (I) benz[*a*]anthracene; (II) 7,12-dimethylbenz[*a*]anthracene; (III) dibenz[*a,h*]anthracene; (IV) benzo[*j*]fluoranthene; (V) 3-methylcholanthrene; (VI) benzo[*a*]pyrene; and (VII) dibenzo[*a,h*]pyrene. (C) *Polyhalogenated biphenyl structures* are schematically produced according to the formula outlined. The halogenating agent may include chlorine to give polychlorinated biphenyls (PCBs) or bromine to yield the polybrominated equivalent (PBBs). met., Metal catalyst.

equivalent to that of the respiratory system; otherwise consequences may be demonstrably more severe than presently speculated.

Many types of halogenated aromatic hydrocarbons exert some mutagenic and possibly carcinogenic actions on cells. As previously noted, however, Ames test responsiveness to many of these agents is low

and suggests that epigenetic factors may be involved in their carcinogenic effects. Foremost among current concerns are fugitive environmental occurrences of polyhalogenated biphenyls and halogenated hydrocarbon insecticides.

The dielectric properties of polychlorinated biphenyls (tradename: *Arochlors*)

(PCBs), fire-retardant properties of poly-brominated biphenyls (PBBs), and the effective neurotoxic insecticide actions of the cyclodienes (dieldrin, aldrin, chlordane, heptachlor), γ-benzene hexachloride (lindane), mirex, dichlorodiphenyl trichloroacetic acid (DDT), and many other similar agents are well known. Apart from the commercial utility of the halogenated hydrocarbons, they display notable environmental persistence, sequestration in the lipid tissues of animals, and increase in concentration to-ward the apical segments of natural food chains.

The *in vivo* metabolism of many halogenated hydrocarbons proceeds by one or more initial dehydrohalogenation reactions. Such reactions, mediated by *dehydrohalogenase* enzymes, are clearly critical for the dechlorination of DDT, which *ultimately* yields a water-soluble dichlorodiphenylacetic acid (DDA) derivative. This assumes the protracted possibility of complete metabolism (Figure 20.12A).

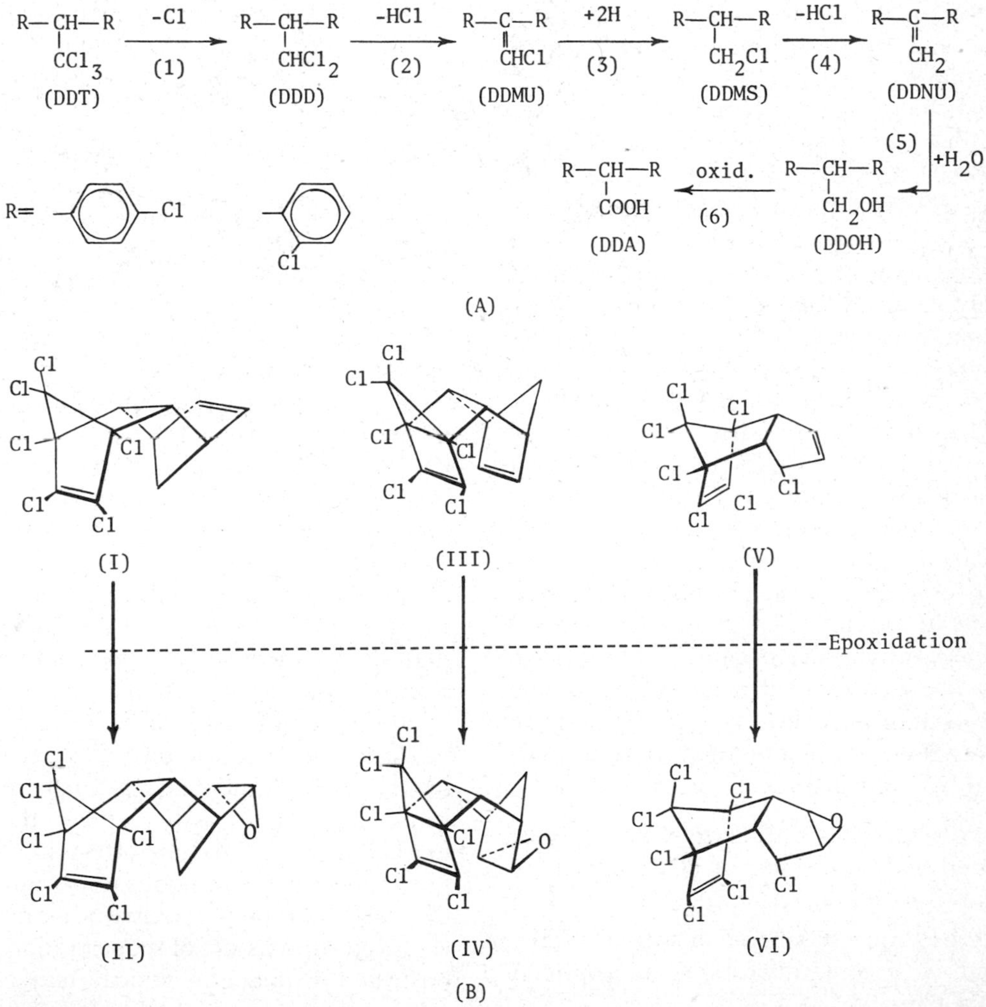

FIGURE 20.12. (A) Hypothetical course of degradation for a chlorinated hydrocarbon (DDT) involving metabolism and abiotic mechanisms: (1) dechlorination, (2) dehydrohalogenation mediated by dehydrochlorinase enzymes, (3) reduction, (4) dehydrohalogenation, (5) water addition, and (6) oxidation. (B) Epoxidation of selected cyclodienes: aldrin (I) to dieldrin (II), isodrin (III) to endrin (IV), and heptachlor (V) to heptachlor epoxide (VI).

Dehydrochlorinase enzymes responsible for removing chlorine from DDT also attack DDT-related hydrocarbons, halogenated cyclodienes, and possibly the polyhalogenated biphenyls. Recent research also speculates that *in vivo* dechlorination of certain hydrocarbons may also involve the participation of superoxide anions (O_2^-), which make a nucleophilic assault on these hydrocarbon compounds.

Since the industrial production for many of the halogenated hydrocarbons has ceased, their residues in foods are likely to diminish except for the impact of fugitive industrial stocks that may continue to escape into environmental systems. Unfortunately, persistent halogenated hydrocarbon residues continue to appear in imported meats, cheeses, and other foods that escape governmental residue-surveillance programs.

Conflicting testimonies concerning the mutagenicity of cyclodienes remain to be firmly resolved, but such debates have occasionally centered on their ability to undergo microsomal epoxidation. Typical among these reactions are the epoxidations of aldrin to dieldrin, isodrin to endrin, and heptachlor to heptachlor epoxide (Figure 20.12B). Aside from this possibility other authorities claim that epigenetic effects of halogenated hydrocarbons are most important, including their roles in upsetting permissive hormonal effects in selected cells and tissues.

Many other trace organic residues occur in foods. As expected, many of these have industrial origins and may cause mutagenic effects. Some of the most important contemporary threats include

1. Polychlorinated dibenzofurans (PCDF), formed as by-products of PCB contamination and incineration reactions.

2. Hexachlorodibenzo-*p*-dioxin (HCDD), which is a contaminant of chlorophenol manufacturing.

3. 2,3,7,8-Tetrachlorodibenzo-*p*-dioxin (TCDD) (LD_{50} = 1.0 μg/kg body weight), which is formed as a trace component of the herbicide 2,4,5-trichlorophenoxyacetic acid (2,4,5-T).

4. Pentachlorophenol (PCP), used as a recognized wood preservative and antirotting agent and often laced with trace quantities of hexa, hepta, and octa isomers of dioxins.

5. Chlorinated derivatives of organic compounds in organic-rich reclaimed municipal drinking waters that have received chlorination treatment.

For additional industrial toxicological threats to food resources, the reader should consult the environmental literature, GRAS listings promulgated by the government, *Priority Organic Pollutant* classifications dictated by the U.S. Environmental Protection Agency, and the Safe Drinking Water Act.

MOBILIZATION OF SEQUESTRATED TOXICANTS

The ability of many toxicants to undergo sequestration in the human body can cause severe consequences. This phenomenon often involves those toxic organic compounds that have a long biological half-life, limited detoxification potentials, and lipophilic partitioning properties (e.g., nonpolar halogenated hydrocarbons).

Chronic low-concentration-level exposures to such toxicants, at rates far greater than their biological half-lives, can result in selective tissue accumulation of toxicants (e.g., in adipose tissues and nonpolar organics).

Under these conditions, plasma concentrations of toxicants may gradually increase while the total toxicant load of the body markedly grows. Toxicant effects may be minor under these conditions, especially when the target site of a potential toxicant is far removed from its site of sequestration. Demonstrated evidence of toxicity may only be subtle so long as sequestration and warehousing of the toxicant occurs. However, in cases in which the sequestration capacity of the body is decreased, tissue stores of accumulated toxicants can be rapidly mobilized to dangerous levels in plasma and redirected toward toxicant-susceptible tissues.

For lipophilic organics, including halogenated hydrocarbons, mobilization from fat tissues may be initiated during periods of energy duress. Weight loss through dieting, pregnancy stress, lactation, and medical problems all promote negative energy balances within the body and rapid fat losses.

Assuming that a test subject does not realize *immediate* effects of acute toxicity from mobilized toxicants, this does not rule out a delayed pathological effect of the toxicant. Once mobilized fat stores of a toxicant reach a critical threshold point in the body, they may instigate a mutagenic and/or epigenetic prelude to carcinogenesis. Clearly then, evidence of the toxicant appears *only* when irreversible molecular damage or epigenetic aberrations are translated into a neoplastic transformation along with its daughter cells.

Mobilization of sequestrated toxicants during pregnancy or lactation offer special threats to the fetus or infant, respectively. High plasma levels of toxicant may unpredictably transgress the placental membrane and exert overt fetotoxic or teratogenic effects. Depending on the stage of fetal development, the time-exposure consequences to any toxicant may be more severe than others.

On the other hand, lactation may serve as a vehicle for discharging formerly sequestrated loads of maternal toxicants. Milk from dairy cattle as well as humans offers this potential problem—especially when prolonged exposures to persistent lipophilic industrial and agricultural chemicals may have preceded the onset of lactation.

Infants are particularly prone to toxicant actions provided in their milk or solid foods for a variety of reasons. They are plagued by (1) poorly developed microsomal enzyme systems, which are a prerequisite for detoxification of certain chemical agents; (2) poor digestive ability; (3) low gastric acidity; (4) minimal renal excretion abilities; (5) a marginally effective blood–brain barrier; and (6) an immature population of intestinal flora. The food consumption to body weight ratio for infants is also about three times greater than relative adult food intake, and this only enhances their exposure to toxic compounds.

CONCLUSION

This chapter offers only a superficial survey of the principles underlying some food toxicological problems. Toxic natural and synthetic organics have been stressed, but it should be recognized that heavy metals, organometallic complexes, and overconsumption of trace elements in foods can result in acute toxic effects.

Threats to food safety also develop with persistent antibiotic residues in meats and dairy products. Antibiotics are widely used for the control of livestock diseases as well as bacterial infection control (e.g., mastitis), and large quantities have been employed for promoting animal growth. Although allergic sensitivity to residual antibiotics in food is not a matter of rampant public health concern, this problem should be considered as a food safety factor. Furthermore, unscrupulous use of antibiotics can lead to the development of antibiotic resistant human pathogens that are ordinarily controlled by penicillins, tetracycline and other broadspectrum antibiotics.

The lack of detailed biochemical attention to the actions of bacterial food "poisonings" in this chapter is not meant to minimize their impact on food quality. The microbial biochemistry literature is rich in detailed discussions of *endotoxic* effects produced by *Salmonellae* and other enteric bacteria, *enterotoxic* effects of *Staphylococcal* bacteria, and the potent *neurotoxic* consequences of *Clostridium botulinum*. In addition, the possible effects of trace animal growth hormone residues in foods as well as vitamin effects have been detailed elsewhere in the text. Indeed, many other forms of food intoxications are possible, especially those incurred as a result of eccentric dietary habits or cultural and ethnic food practices.

Earlier sections have illustrated that toxicant actions routinely involve absorption, biotoxification, metabolic, and/or excretory factors. In addition to these factors, nutritional toxicology must also deal with the synergistic and antagonistic interactions of chemical agents in foods.

When the combined effects of two toxicants [X] and [Y] are greater than the singular effect of 2[X] or 2[Y], this is taken to represent a *synergistic* toxic interaction. In a contrary vein, the combined effects of toxicants may be less than 2[X] or 2[Y] and display the phenomenon of *antagonism*. The existence of these toxicant interactions should be recognized, but their biochemical principles are beyond the scope of the current text.

REFERENCES

Alexander, L. S., and H. M. Goff. 1982. Chemicals, cancer and cytochrome *P*-450. *J. Chem. Ed.* **59**(3):179.

Ames, B. N. 1972. A bacterial system for detecting mutagens and carcinogens. In *Mutagenic Effects of Environmental Contaminants* (H. E. Sutton and V. I. Harris, eds.). Academic Press, New York.

Ames, B. N., W. E. Durston, E. Yamasaki, and F. O. Lee. 1973. Carcinogens and mutagens: A simple test combining liver homogenates for activation and bacteria for detection. *Proc. Natl. Acad. Sci. USA* **70**:2281.

Ames, B. N., F. D. Lee, and W. E. Durston. 1973. An improved bacterial test system for the detection and classification of mutagens and carcinogens. *Proc. Natl. Acad. Sci. USA* **70**:782.

Belvedere, G., H. Miller, K. P. Vatsis, M. J. Coon, and H. V. Gelboin. 1980. Hydroxylation of benzo(a)pyrene and binding of (−)trans-7,8-dihydroxy-7,8-dihydrobenzo(a)pyrene metabolites to deoxyribonucleic acid catalyzed by purified forms of rabbit liver microsomal cytochrome *P*-450. Effect of 7,8-benzoflavone, butylated hydroxytoluene and ascorbic acid. *Biochem. Pharm.* **29**:1693.

Bender, A. E., and D. A. Bender. 1982. *Nutrition for Medical Students*. Wiley, New York.

Bloomer, A. W., S. I. Nash, H. A. Price, and R. L. Welch. 1977. Pesticides in people. *Pest. Monit. J.* **11**:111.

Brattsten, L. B. 1979. Ecological significance of mixed function oxidations. *Drug Metab. Rev.* **10**(1):35 (53 references).

Casciano, D. A. 1982. Mutagenesis assay methods. *Food Technol.* **36**(3):48.

Committee 17, Council of the Environmental Mutagen Society. 1975. Environmental mutagen hazards. *Science* **187**:503.

Commoner, B., A. J. Vitayathil, P. Dolara, P. Nair, P. Madyastha, and G. C. Cuca. 1978. Formation of mutagens in beef extracts during cooking. *Science* **201**:913.

Coon, M. J. (ed.). 1980. *Microsomes, Drug Oxidations and Chemical Carcinogenesis*, Vols. I and II. Academic Press, New York.

Creasey, W. A. 1979. *Drug Disposition in Humans—The Basis of Clinical Pharmacology*. Oxford University Press, New York.

Currie, R. A., V. W. Kadis, W. E. Breitkreitz, G. B. Cunningham, and G. W. Bruns. 1979. Pesticide residues in human milk, Alberta, Canada. *Pestic. Monit. J.* **13**(2):52.

Devoret, R. 1979. Bacterial tests for potential carcinogens. *Sci. Amer.* **241**(2):40.

Fry, D. L., and C. K. Toone. 1981. DDT-induced femization of gull embryos. *Science* **213**:922.

Gibaldi, M. 1977. *Biopharmaceutics and Clinical Pharmacokinetics*. Lea & Febiger, Philadelphia.

Hathcock, J. N. (ed.). 1982. *Nutritional Toxicology*, Vol I. Academic Press, New York.

Hilker, D. M. 1981. Carcinogens occurring naturally in food. *Nutr. Cancer.* **2**(4):217 (59 references).

Institute of Food Technologists. 1983. Caffeine—A scientific status summary by the Institute of Food Technologists' expert panel on food safety and nutrition. *Food Technol.* **37**(4):87.

Kilbey, B. J., M. Legator, W. Nichols, and C. Ramel. 1977. *Handbook of Mutagenicity Test·Procedures*. Elsevier, Amsterdam.

Kilian, D. J., and D. Picciano. 1976. Cytogenetic surveillance of industrial populations. In *Chemical Mutagens, Principles and Methods for their Detection* (A. Hollaender, ed.), Vol. IV. Plenum, New York.

Kingsbury, J. M. 1964. *Poisonous Plants of the United States and Canada*, 3rd ed. Prentice-Hall, Englewood Cliffs, N.J.

Legator, M. S. 1979. *Short Term Procedures for De-*

termining Mutagenic/Carcinogenic Activity. Office of Technology Assessment Working Paper, Washington, D.C.

Legator, M. S., L. Troung, and T. H. Connor. 1978. Analysis of body fluid including alkylation of macromolecules for detection of mutagenic agents. In *Chemical Mutagens, Principles and Methods for their Detection* (A. Hollaender, ed.), Vol. IV. Plenum Press, New York.

Lijinsky, W., and P. Shubik. 1964. Benzo(a)pyrene and other polynuclear hydrocarbons in charcoal-broiled meat. *Science* **145**(2):53.

Matsumura, F., and C. R. Krishna Murti. 1982. *Biodegradation of Pesticides*. Plenum Press, New York.

McMurray, W. C. 1982. *A Synopsis of Human Biochemistry*. Harper & Row, Philadelphia.

Menzie, C. M. 1969. *Metabolism of Pesticides*. Bureau of Sport Fisheries and Wildlife Special Scientific Report. Wildlife No. 127.

National Academy of Sciences. 1973. *Toxicants Occurring Naturally in Foods*, 2nd ed. NAS, Washington, D.C.

O'Brien, R. D. 1967. *Insecticides, Action and Metabolism*. Academic Press, New York.

Office of Technology Assessment. 1977. *Cancer Testing Technology and Saccharin*. U.S. Government Printing Office, Stock No. 052-003-00471-2. Washington, D.C.

Office of Technology Assessment. 1979. *Environmental Contaminants in Food*. U.S. Government Printing Office, Stock No. 052-003-00724-0. Washington, D.C.

Palmiter, R. D., and E. R. Mulvihill. 1978. Estrogenic activity of the insecticide kepone on the chicken oviduct. *Science* **201**:356.

Pariza, M. W., S. H. Ashoor, F. S. Chu, and D. B. Lund. 1979. Effects of temperature and time on mutagen formation in pan-fried hamburger. *Cancer Lett.* **7**:63.

Pariza, M. W. 1982. Mutagens in heated foods. *Food Technol.* **36**(3):53.

Parke, A. L., and G. R. V. Hughes. 1981. Rheumatoid arthritis and food: A case study. *Brit. J. Med.* **282**:2027.

Rappe, C., H. R. Buser, D. L. Stalling, L. M. Smith, and R. Dougherty. 1981. Identification of polychlorinated dibenzofurans in environmental samples. *Nature* **292**:524.

Report of the Secretary's Commission on Pesticides and Their Relationship to Environmental Health. 1969. Parts I & II. U.S. Department of Health Education & Welfare, Washington, D.C.

Roberts, H. R. 1981. *Food Safety*. Wiley, New York.

Roberts, J. L., and D. T. Sawyer. 1981. Facile degradation by superoxide anion of carbon tetrachloride, chloroform, methylene chloride and p,p'-DDT in aprotic media. *J. Amer. Chem. Soc.* **103**:712.

Shimbamoto, T. 1982. Occurrence of mutagenic products in browning model systems. *Food Technol.* **36**(3):59.

Sontag, J. M., N. P. Page, and U. Saffiotti. 1976. *Guidelines for Carcinogen Bioassay in Small Rodents*. DHEW Publication No. (NIH) 76-801. U.S. Government Printing Office, Washington, D.C.

Spingarn, N. E., L. A. Slocum, and J. Weisberger, 1980. Formation of mutagens in cooked foods. II. Foods with high starch content. *Cancer Lett.* **9**:7.

Stanbridge, E. J., and J. Wilkinson. 1978. Analysis of malignancy in human cells: Malignant and transformed phenotypes are under separate genetic control. *Proc. Natl. Acad. Sci. USA* **75**:1466.

Sulflita, J. M., A. Horowitz, D. R. Shelton, and J. M. Tiedje. 1982. Dehydrohalogenation. A novel pathway for the anaerobic biodegradation of haloaromatic compounds. *Science* **218**:1115.

Taylor, S. L. 1982. Mutagenesis versus carcinogenesis. *Food Technol.* **36**(3):65–68, 98–103, 127.

Ware, G. W. 1983. *Pesticides, Theory and Application*. Freeman, San Francisco.

Williams, R. 1981. Rheumatoid arthritis and food: A case study. *Brit. J. Med.* **283**:563.

C H A P T E R · 21

Food, Drugs, and Immune System Interactions

INTRODUCTION

The therapeutic treatment of many ailments affecting human health has traditionally been oriented toward achieving specific cures without much attention to the secondary impairment of nutritional status induced by drug therapy. The development of drugs for use on a chronic basis has produced an awareness of the serious consequences that may arise from long-term drug intake. Examples of such effects can be found in the case of alcohol-induced depletion of body stores of minerals such as magnesium and zinc as well as vitamins such as thiamine and pyridoxine. Other examples of drugs used chronically by a substantial proportion of the population in developed countries are seen in the use of oral contraceptives, tranquilizers, and antihypertensive agents. The potential malnutrition that may arise in individuals pursuing a regime of drug intake may be complicated by factors such as genetic predisposition (drug and food allergies), cultural background (ethnic foods and drug effectiveness), or Third World origins with chronic infections of a parasitic nature (roundworm or helminth infestations). The influx of immigrants to developed countries

from Third World origins has necessitated the use of multiple-drug intervention strategies to combat both the primary and chronic aspects of disease. This in turn has drawn attention to the potential problems of marginal nutritional status, which might be compromised further by multiple-drug therapy.

Prolonged hospitalization of patients exposed to intravenous feeding or restricted food intake after major surgery has produced the dilemma of optimizing nutrient intake in the presence of drugs that reduce the capacity and/or desire to consume food. The concept of iatrogenic malnutrition has arisen from these conditions of therapy-induced impairment of adequate nutritional status. It is essential to treat patients nutritionally prior to major surgery or prolonged drug therapy to optimize nutritional adequacy for prevention of iatrogenic effects. In particular, supplementation with vitamins and minerals before and after surgery is desirable in those cases where insufficiency of these nutrients is suspected (D'Arcy and Griffin, 1982).

The variety of drug and nutrient interactions that occur can be segregated into *two major classes*. The *first* is the class of effects produced by the direct and/or side reactions of drugs on nutrient intake. These are changes in appetite, digestion, absorption, metabolism, and excretion of nutrients. The *second* class of effects are those represented by the interactions of nutrients with drugs to diminish their effectiveness or aggravate side reactions. Such effects can be expressed by chelation of drugs by nutrients to diminish their absorption or to reduce drug uptake by nutrient competition and dilution in the gastrointestinal tract. Other mechanisms of food-induced reduction of drug activity are pH changes and drug stability, enhanced gastrointestinal transit rate, and increased urinary excretion.

THE INFLUENCE OF DRUGS ON NUTRIENT UTILIZATION

The therapeutic activity of many drugs when taken on either an acute or chronic basis can be compromised by a range of side reactions that alter nutritional competency. These side effects can range from mild to severe changes in one or more of the following processes: (1) desire for food (taste and appetite), (2) absorption of nutrients, (3) biological activity of nutrients (metabolism), and (4) elimination of nutrients (excretion).

A variety of therapeutic drugs and pharmaceuticals are commonly used that can upset normal nutrient disposition and interactions throughout the body. Some of these specific interactions have been outlined in Table 21.1.

APPETITE AND TASTE EFFECTS

A variety of drugs can reduce nutrient intake by either directly affecting appetite or changing taste perception to reduce the desire for food. The classical appetite suppressants represented by amphetamine act directly on hypothalamic centers in the brain involved in food intake. This effect is me-diated by increased release of dopamine to synapses of neuronal pathways in these brain centers. The relative effectiveness of amphetamines in appetite suppression is short lived and generally declines within 10 days of use. The availability of nonprescription appetite suppressants is widespread, but the use of amphetamines on a prescription basis has often been abused. This is not uncommon in professional sports where undesirable psychological effects have been observed as a consequence of chronic intake by athletes in attempts to improve physical performance. Amphetamines have also been applied to the treatment of hyperactive syndromes in young children. The behavioral effect of amphetamine in these cases is the opposite to that found in adults (sedation versus excitation) and can lead to undesirable growth suppression as a result of reduced nutrient intake.

The absence of appetite suppressant effects of drugs can nonetheless cause reduction in food consumption by actions on taste perception leading to reduced palatability. These taste changes can arise by either a change in taste sensitivity or distortion of taste quality. Decreases in sensitivity can occur by a direct effect of the drug or indirectly by causing losses of zinc, which is important for adequate taste function. The anesthetic lidocaine causes a reduction in salt and sweet taste, whereas phenytoin has a more general taste desensitizing action on all taste qualities (sweet, salty, sour, and bitter). Geriatric or drug-induced zinc deficiency can lead to loss of taste awareness (*hypogeusia*) or distorted perception of taste (*dysgeusia*). These changes can lead to a dramatic reduction of food intake and, consequently, malnutrition with increased susceptibility to infection. Treatment by zinc supplementation can restore gustatory appeal in many of these cases.

Drugs that affect taste or appetite to enhance palatability and food intake are not common. Cyproheptadine, a serotonin antagonist, is claimed to be a specific appetite stimulant that can produce weight gains by increased food consumption. The tranquil-

Table 21.1. Therapeutic and Nutritional Interactions of Some Drugs and Pharmaceutical Agents

Therapeutic drug class	Representative pharmaceutical	Nutritional effect	Physiological mechanism
Analgesic	Aspirin, salicylamide, acetamenophin	Elevated urinary loss of ascorbate	Depressed platelet uptake of ascorbate
Anorectics	Amphetamine, ritalin, phenmetrazine	Reduced appetite and food intake	Affects hypothalamic centers in CNS
Antacids	Aluminum hydroxide, magnesium trisilicate, magnesium hydroxide	Binds phosphate, reduces absorption	Neutralizes gastric acidity
Anticonvulsants	Dilantin (Phenytoin), primidone, carbamazepine	Elevated calciferol metabolism, osteomalacia	Activation of liver microsomal metabolism
Antidepressants	Pargyline, imipramine, tranylcypromine	Retards tyramine oxidation in gut	Monoamine oxidase inhibitor
Tumor inhibitors	Fluorouracil, methotrexate	Reduced nutrient intake	Alteration of taste perception
Antibiotics	Penicillin-G, chloramphenical, tetracycline	Hypokalemia	Increases urinary potassium loss
Antituberculars	Isoniazid, cycloserine, *para*-aminosalicylate	Pyridoxine and niacin depletion	Competes with niacin and pyridoxine in cellular metabolism
Laxatives	Phenolphthalein, bisacodyl	Elevated calcium and calciferol loss; steatorrhea	Increased intestinal peristalsis
Anti-inflammatory	Prednisone, indomethacin, phenylbutazone	Increased loss of potassium, magnesium, zinc	Increased protein catabolism and bone resorption
Diuretics	Hydrochlorothiazide, spironolactone, furosemide	Urinary loss of potassium, zinc, magnesium	Affects kidney regulation of ion excretion
Hypocholesterolemics	Cholestyramine, clofibrate	Anionic polystyrene depresses cholesterol absorption	Binds to bile acids thereby reducing fat availability
Hypotensive agents	Hydralazine, clonidine, guanethidine	Increased loss of B_6 and manganese	Forms complexes to enhance excretion
Contraceptives	Mestranol, chlormadinone	Depletes B_6 and alters tryptophan metabolism	Suppresses activity of B_6-dependent reactions
Tranquilizers	Chlorpromazine, diazepam	Elevated food intake	Reduces anxiety and promotes weight gain
Anesthetics	Nitrous oxide, halothane	B_{12} requirement is increased	Inactivates B_{12} activity

izer chlorpromazine has a similar effect on patients, although probably not mediated by the same mechanism. The anxiety- and tension-alleviating effects of this drug may be responsible for the increased food intake and hence weight gain of treated patients.

DIGESTIVE AND ABSORPTIVE EFFECTS

The presence of drugs and food in the gastrointestinal (GI) tract can create problems for both drug effectiveness and nutrient absorption. These interactive effects are most prevalent when food and drug ingestion are taken together but can be minimized by separating the time of intake of each. An example of some of these problems is found in penicillin, which is sensitive to gastric acidity and loses potency when consumed with food. On the other hand, the absorption of aspirin is enhanced in an acidic environment and its hemorrhagic effects on the gastric mucosa are reduced by the presence of food.

The action of drugs on the GI tract can be mediated by a direct physical action (dilution by stimulation of water entry or increased peristalsis and reduced transit time) or a metabolic effect (suppression of enzyme activity in mucosal cells). The first type of action is commonly encountered in laxatives such as magnesium sulfate and castor oil. The osmotic or irritating effects of these compounds increase the GI tract contents of fluid and hence peristaltic activity leading to more rapid transit of the contents. This in turn reduces the time for the digestive action of enzymes and also the absorption of their products. Losses of fat-soluble vitamins are increased when oil-based cathartics are used. Dependency on these agents can lead to osteomalacia due to reduced vitamin D absorption. The use of the bile-acid-binding drugs such as cholestyramine may also bind and reduce absorption of water-soluble vitamins by nonspecific ion-exchange effects. The lack of bile acid availability that is produced leads to steatorrhea and limited digestion of fats for the generation of linoleic acid (essential fatty acid) and other fatty acids.

The second class of drug actions (metabolic effects in the intestinal lining) are more widespread. Drugs in this class are the antitumor agents (methotrexate), anticonvulsants, and oral contraceptives. These drugs interfere with absorption and utilization of folic acid in cellular proliferation, which is required in organs with a rapid cell turnover like the intestinal mucosa, bone marrow, and skin. Secondary changes in nutrient absorption due to impaired replacement of luminal mucosal cells is a consequence of the use of these drugs for any extended period of time. Bone marrow changes resulting from impaired folate status led to the observation of megaloblastic anemia in one-quarter of the individuals on oral contraceptive maintenance. Two drugs with direct toxic effects on mucosal cells are neomycin and colchicine. Neomycin, a nonabsorbed antibiotic used in intestinal infections, causes generalized damage to the cells leading to both impairment of digestive enzyme synthesis and depression of absorption of all the major nutrient classes. Colchicine has been widely used for the treatment of gout and has an antimitotic action on cells by inhibiting the movement of microfilaments. Although its effect on mucosal cells may not be directly related to this process, it does cause depression of the transport of nutrients into these cells and hence into the blood.

METABOLISM AND EXCRETION EFFECTS

These effects are encountered when absorbed drugs act primarily at liver and kidney sites. The metabolic effects are produced by an action of the drug that alters the utilization of a normal cell component. Typical effects produced are competition with enzyme cofactors (vitamins and minerals) to displace the endogenous compound that is then lost by excretion. Drugs that act in this manner are methotrexate and the antimalarial pyrimethamine, which bind to dihydrofolate reductase to inhibit formation of biologically active tetrahydrofolate required for DNA synthesis. Under these conditions, body pools of folate are depleted more rapidly, leading to clinical manifestations of blood dycrasias (megaloblastic anemia).

An alternate mode of metabolic change produced by drugs is the activation of liver microsomal enzymes that oxidize the drugs to inactive metabolites. The anticonvulsants, such as phenobarbital and phenytoin, are effective inducers of hepatic cytochrome *P*-

450-dependent hydroxylases. The increase in activity of drug metabolizing enzymes may reduce the normal activity of calciferol hydroxylating pathways (*25-hydroxylase*) and increase activity of inactivating pathways (*24-hydroxylase*) and thus the control of calcium absorption. The clinical consequences of these changes are the production of osteomalacia in adults or rickets in children. Kidney hydroxylating enzymes are also affected in this pathway where the formation of the inactive form of calciferol (24,25-dihydroxy calciferol) may increase by elevation of 24-hydroxylase activity or reduction of 1-hydroxylase activity. The sequence of events in calciferol metabolism are shown in Figure 21.1.

A third possibility for metabolic inactivation of vitamins is the chemical coupling of drugs to reactive centers on a vitamin. The classical example of this effect is the coupling of isoniazid to pyridoxal during tuberculosis therapy. The complex so formed is excreted and depletes stores of pyridoxine resulting in dermatitis and polyneuritis in

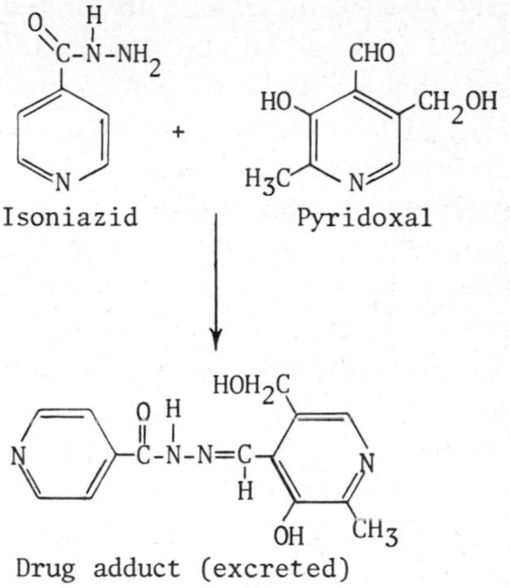

FIGURE 21.2. An example of vitamin deactivation involving a drug agent such as isoniazide. The drug adduct of the vitamin is subsequently excreted instead of exerting its normal coenzymatic effects.

uncorrected cases. The mechanism of this drug effect is shown in Figure 21.2.

A similar mechanism of pyridoxal depletion is encountered with the antihypertensive agent hydralazine. The hydrazine substituent on the phthalazine nucleus also couples to the aldehyde function of pyridoxal, forming an adduct that is removed from tissue sites and excreted.

The loss of plasma nutrients by renal excretion mechanisms can be significantly altered by a variety of drugs. The most common effect is loss of essential minerals such as potassium, calcium, magnesium, and zinc by either direct effects on the kidney (diuretics) or by secondary effects due to reduced tissue storage (alcohol). Competition effects of drugs acting to displace nutrients from albumin by mass action or chemical affinity can lead to elevated renal losses of nutrients. Such actions are seen with agents such as aspirin and penicillamine (used in heavy-metal toxicity), which respectively can displace albumin-bound folate and chelate zinc or copper. The treatment of mild hypertension with diuretics such as hydrochlorothiazide often can lead to potassium depletion unless compensated for by increased potassium intake. In this case car-

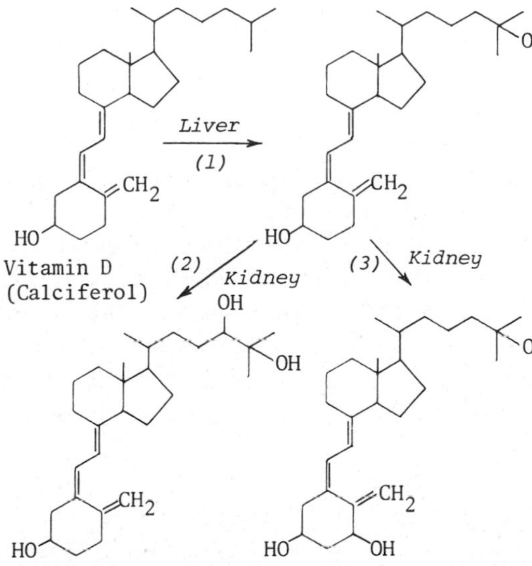

FIGURE 21.1. Hydroxylase enzyme activities on vitamin D (calciferol).

diac problems can occur if prolonged drug therapy is not adequately treated by increased dietary intake of potassium. Other diuretics such as furosemide can cause a significant loss of calcium by inhibition of renal reabsorption processes that can extend to losses of other divalent cations such as zinc and magnesium. Manganese depletion from body stores has been observed in hypotensive therapy with hydralazine. Net losses of iron and vitamin C are encountered in massive aspirin treatment (1–3 g/day) of arthritic conditions wherein vitamin C is depleted by increased renal excretion and iron by gastric hemorrhage and loss via the GI tract.

The use of alcohol by patients maintained on specific drug regimes may cause more severe depletion of nutrients than if alcohol were absent. The combined effects of altered tissue metabolism of nutrients by alcohol and enhanced losses of nutrients by specific drugs produces losses of both vitamins and minerals without adequate dietary replacement due to the appetite-suppressant actions of the alcohol.

EFFECTS OF NUTRIENTS ON DRUG UTILIZATION

The combination of drug therapy with simultaneous food consumption can lead to a variety of interactions that either suppress or enhance the effectiveness of the drug involved. The variable therapeutic response so obtained is mediated by the specific properties of drugs being physically altered by the presence of food. Drug absorption in the GI tract either in the stomach (acidic) or ileum (basic) regions is a function of its solubility in lipid and aqueous phases. This is represented by the cell membrane's permeability to the drug by passive diffusion processes and is a function of the *oil/water partition coefficient* (K_p) and acidic dissociation constant (pK_a) of the drug. Nonionic drugs with a high K_p will be readily absorbed, whereas ionizable drugs will be best absorbed when their solutions contain nonionic species in abundance. This condition prevails for low pK_a (fewer than six) drugs at acidic pH's and for high pK_a (more than seven) drugs at alkaline pH's. Acidic drugs (low pK_a) are thus preferentially absorbed in the stomach and basic drugs (high pK_a) in the ileum. This is illustrated in Table 21.2. Foods or food components such as caffeine that stimulate gastric acid secretion favor absorption of acidic drugs such as aspirin that are stable under these conditions. Acid-sensitive drugs such as the penicillin antibiotics are more quickly degraded under acidic conditions with consequent loss of therapeutic potency. The time when a drug can be most effectively absorbed relative to food intake is thus at least two hours before meals for acid-labile drugs (penicillin) or during meals for acid-stable drugs.

Foods that favor alkaline GI tract secretions such as most fruits and vegetables would favor the absorption of basic drugs (e.g., amphetamine and chlorpromazine). The best time for administering these drugs would be 2 to 3 h after the consumption of food.

Absorption of drugs taken with food can be impaired by the presence of crude fiber. These nondigestible bulk components aid in retaining water in the GI tract (hence dilution of the drug dose) and stimulate peris-

Table 21.2. Drug and pH Relationships

Drug type	pK_a	Ratio of nonionized to ionized forms	
		pH 2.0 (stomach)	pH 8.0 (ileum)
Acidic (aspirin)	3.0	10^1	10^{-5}
Basic (amphetamine)	9.0	10^{-7}	10^{-1}

talsis to speed up the transit time of food. The two effects combined serve to reduce the availability of the drug and thus its therapeutic effect.

Given the common practice in modern society of consuming alcoholic beverages together with food, it is important to consider the consequences of such habits for individuals being treated with a variety of drugs. The chronic use of alcohol leads to the elevation of liver microsomal oxidizing enzymes, which then influence the oxidation of a variety of drugs. The anticonvulsants phenobarbital and phenytoin are inactivated by these enzymes and have shorter effective half-lives than is found in nonalcohol-imbibing patients. Cases of severe alcohol abuse show degeneration of liver function (cirrhosis) with associated problems of retention of essential minerals and vitamins.

Dietary quality in terms of acid–base balance has effects on kidney function to either reduce or enhance the elimination of drugs from the vascular system. The same chemical properties of drugs affecting drug absorption in the GI tract operate in the kidney to modulate their rate of excretion. *Ionized* forms of drug are eliminated in the renal filtrate, whereas *nonionized* forms tend to be reabsorbed. Food components that generate acidic metabolites produce a low urinary pH, which favors the loss of basic drugs in their ionized form. Conversely, foods producing a high urinary pH favor excretion of acidic drugs. High-protein foods (meats, cheeses, nuts) tend to produce urine with a low pH, whereas vegetables and fruits produce a high-pH urine. High-pH urine is also encountered in those individuals who consume antacids on a frequent basis.

Many food components are physiologically innocuous in the average diet because of metabolic inactivation yet may become potentially dangerous or even fatal when taken in the presence of drugs that inhibit the action of metabolizing enzymes. The classic example of this is the presence of tyramine in many foods, which is usually oxidized by mucosal and vascular monoamine oxidases (MAO) to hydroxyphenylacetic acid and excreted. In the presence of MAO inhibitors (antidepressants) such as pargyline, tyramine levels in blood rise and exert an action on sympathetic neurons. The neurons controlling vascular smooth muscle produce vasoconstriction and elevate blood pressure, which may precipitate a heart attack or stroke in sensitized individuals. Foods that are rich in tyramine are those commonly found in the cocktail party setting such as beer, wines, cheeses, and herring. The effects of free tyramine can be aggravated by the presence of foods rich in caffeine when MAO inhibitors are used. This is due to inhibition or metabolic inactivation of cyclic adenylic acid (cAMP) in vascular smooth muscle cells. cAMP is the metabolic effector in these cells producing the cell response and is synthesized by neuronal activation with catecholamines (epinephrine and norepinephrine). The sequence of events producing the pressor response initiated by tyramine is seen in Figure 21.3.

The consumption of food having a high content of sodium on a chronic basis is conducive to the establishment of hypertension. Treatment of this problem often requires the use of diuretics and reduced sodium intake; however, the drugs also induce a loss of potassium. Optimal therapeutic effect is

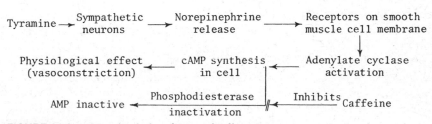

FIGURE 21.3. Tyramine-induced events leading to a pressor response.

obtained with these drugs if the potassium content of the food is considered and intake increased. Examples of foods with high potassium contents (0.2–0.3%, w/w) are orange juice, bananas, and potatoes.

DRUG AND FOOD INCOMPATIBILITIES—SOME SPECIFIC EXAMPLES

As indicated in previous sections, joint interactions of food substances with drugs can generate a variety of physiological and nutritionally related disorders. Certain foods can also counteract the desirable actions of some drugs and pharmaceuticals.

Many acidic foods in conjunction with the acid-rich gastric environment destroy the action of antibiotics including erythromycins and penicillins. Acid-rich foods plus aspirin, for example, can pose other problems such as the corrosion of the gastric lining followed by ulcerative scenarios or exacerbation of minor ulcerative conditions.

Contrary to minimizing the indigestion associated with tetracycline intake, calcium-rich foods can also reduce antibiotic effectiveness. For this reason, admixture of calcium-rich foods and tetracyclines should be avoided at the risk of developing a more severe infection than that for which the antibiotic was first prescribed.

Since vitamin K promotes blood clotting, foods rich in this vitamin tend to counteract anticoagulant agents designed to thin blood. The effects of these foods on anticoagulant actions may be minor or major depending on the amount of anticoagulant dosage as well as the amount of vitamin K-rich food intake. Extreme anticoagulant effects can also be countered to some extent by ascorbic acid consumption.

The tyramine content of many foods conflicts with the monoamine oxidase (MAO) inhibitory action of antidepressive drugs. These physiological conflicts can be evidenced as nosebleed, headache, elevated blood pressure, potentially lethal strokes, or heart attacks. Tyramine-rich foods also conflict with antihypertensive medication activity.

Although natural licorice flavorings have been largely supplanted by artificial flavorings in the United States, imported candies, confections, and certain liqueurs may contain natural licorice. The glycyrrhizic acid component of natural licorice can cause salt and water retention with the joint development of hypertension. This action obviously conflicts with the desired actions of antihypertensive agents.

Monosodium glutamate (MSG) and diuretics singularly eliminate water from body tissues. Their combined effects also act to remove water from tissues, but significant amounts of water-soluble vitamins (B vitamins and ascorbate), sodium, and potassium are lost from the tissues as well.

Foods rich in vitamin B_6 are also known to counteract the therapeutic doses of L-dopa used in treating Parkinson's disease. Furthermore, protein-rich diets can generally impede L-dopa assimilation into brain tissues.

Goitrogenic agents contained in *Cruciferous* plants also counteract thyroid preparations designed to enhance thyroid hormone production. Goiter may ultimately develop as a consequence of these foods and only exacerbate the metabolic upsets already present because of thyroid anomalies.

Many other interactions are recognized whereby foods inactivate specific drug and pharmaceutical actions, but Table 21.3 details some of the most notable incompatibilities between specific drugs and foods.

IMMUNE SYSTEM INTERACTIONS

The defense system of mammals and other vertebrates against damage from invading foreign organisms such as viruses, bacteria, parasites, and toxic proteins (or antigens) consists of both cellular and soluble secretion

Table 21.3. Some Notable Food–Drug Incompatibilities

Pharmaceutical agent	Specific food conflicts
Antibiotics	Acidic foods:
Erythromycin and penicillin	Caffeine, citrus fruits, cola drinks, fruit juices, pickles, tomatoes, and vinegar
Tetracycline	Calcium-rich foods: Almonds, buttermilk, cream, ice cream, milk, pizza, waffles, and yogurt (all cheeses are also included in this class)
Anticoagulants	Animal liver tissues, vegetable oils, and green leafy vegetables having high vitamin K content including brussels sprouts, cabbage, Chinese cabbage, kale, ruggola, spinach
Antidepressants (MAO inhibitors)	Tyramine-rich foods: Aged cheeses (Brie, Camembert, cheddar, Emmenthaler, Grùyere, processed American, and Stilton), aged meat, anchovies, avocados, bananas, beer, broad beans, caffeine, chicken liver, chocolate, cola drinks, canned figs, mushrooms, pickled herring, raisins, sausages (liverwurst, pepperoni, etc.), sour cream, soy sauce, wines, yeast extracts
Antihypertensives	Licorice (natural); tyramine-rich foods (above)
Aspirin	Acidic foods
Diuretics	Monosodium glutamate (MSG); licorice
Levodopa (L-dopa) (for Parkinson's disease)	Animal liver tissues (beef, pork), wheat germ, yeast, high-protein diet, B_6-rich foods
Thyroid agents	Brussels sprouts, cabbage, cauliflower, kale, mustard greens, rutabaga, soybeans, and turnips

factors produced in body organs such as the bone marrow, spleen, and liver. These host components comprised of circulating lymphocytes and plasma antibodies serve a variety of functions in integrating a series of events that optimize the body's response to trauma, infection, and allergies. In particular, the role of nutrient availability will affect the success of such responses where generalized malnutrition tends to impair them. Adequate nutritional status can also be compromised if food components produce allergies mediated by the immune sys-

tem, which may lead to secondary inhibition of resistance to infection. The use of drugs with potentially suppressive effects on the immune system is common and represented by the classes of anti-inflammatory drugs (aspirin, prednisone) and antimitotic drugs (methotrexate and fluorouracil). When combined with marginal nutritional status or even overt malnutrition, the effects of such drugs on host immune defenses can be debilitating. Individuals from Third World countries suffering from combined malnutrition and chronic infection need to be treated aggressively for the malnutrition and cautiously with drugs to combat the infection.

ANTIBODY CLASSES, ACTIONS, AND CELLULAR INTERACTIONS

Plasma antibodies are a class of soluble proteins of differing sizes and molecular weights that are responsible for the recognition and inactivation of foreign proteins or *antigens* (Ag) found free or associated with cell surfaces. These *antibodies* (Ab's) consist of four polypeptide chains combined in two pairs to form a molecule with a "Y" shape where each stem of the Y has a specific function. The lower stem of the Y is denoted as Fc and represents the attachment site of Ab to cell surfaces. The two upper Fab branches, however, offer recognition and *binding sites* of the Ab to an antigen, where the specificity of the Ab is determined by specific regions of the two pairs of polypeptides comprising the Ab (Figure 21.4).

The two chains designated H (*heavy*) and L (*light*) have regions in their amino acid sequence that can *vary among the different antibodies* (V region) or remain *constant* (C region) among Ab's of a given class. The synthesis of Ab's with the potential to recognize many millions of different Ag's is controlled by different groups of cellular genes that select combinations of V genes to combine with C genes and produce L and H chains of various combinations. A third class of genes known as J genes produce a sequence of amino acids responsible for linking specific antibody types via the H chain and add more variability to that already present in the V sequences. The ultimate result of the combination of different groups of L and H chains (with V and C sequences) is the formation of several classes of antibodies designated as immunoglobulins (Ig's) and found in different body fluid secretions. The composition of the *five basic types* (IgG, IgM, IgA, IgD, and IgE) and their functional roles are shown in Table 21.4. The different H chains are of

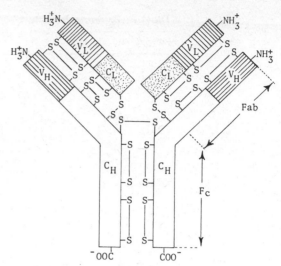

FIGURE 21.4. Structure of a typical IgG antibody. The IgG molecule consists of four polypeptide chains linked together by disulfide bonds. Each chain has two regions designated as variable amino acid sequences (V) and constant amino acid sequences (C). The locations of these regions are found in the heavy (H) and the light (L) chains to form domains of the V_L, V_H type and the C_L, C_H type. The stem of the Y-shaped molecule is denoted as the F_c region that binds to cell surfaces of lymphocytes, and the arms of the structure are the Fab regions that bind antigens specific for the V_L, V_H sequences.

five types designated by the Greek symbols of γ (G), μ (M), α (A), ϵ (E), δ (D) which combine with L chains of two types designated as λ or κ. A pair of H chains binds to a pair of either λ or κ chains to form a tetrameric complex such as $\gamma_2\lambda_2$ or $\gamma_2\kappa_2$. The λ or κ complexes so formed with each of the five types of H chain produce antibodies with specific functions in the immune response to foreign antigens and foreign cells. Variability in Ag specificity is conferred by the V region of a particular H and L chain combination. Most Ab's are distributed in the blood plasma, but the IgA class is also found in abundance in cells lining the mucous membranes of the eyes, mouth, lungs, gastrointestinal tract, and urinary tract. The role of the IgA Ab's thus is oriented to early contact with foreign Ag's or cells at their potential points of entry into the body prior to systemic distribution in the plasma. The IgG

Table 21.4. Characteristics of Immunoglobulins

Antibody class	H chain	L chain	Structure	Plasma concentration (mg/L)	Functions
IgG	γ	λ, κ	$\gamma_2\lambda_2$ or $\gamma_2\kappa_2$	6000–18,000	Main plasma Ab
IgM	μ	λ, κ	$(\mu_2\lambda_2)_5$ or $(\mu_2\kappa_2)_5$	500–1900	Early Ab response, fixes complement
IgA	α	λ, κ	$\alpha_2\lambda_2$ or $\alpha_2\kappa_2$	900–4200	Ab of respiratory and GI secretions
IgE	ϵ	λ, κ	$\epsilon_2\lambda_2$ or $\epsilon_2\kappa_2$	0.1–1.0	Ab for activating mast cells in allergic responses
IgD	δ	λ, κ	$\delta_2\lambda_2$ or $\delta_2\kappa_2$	30–400	Unknown function

and IgM class of Ab's in the plasma respond to the presence of foreign Ag's by insoluble complex formation leading to subsequent cellular disposal by phagocytosis. The IgM proteins are those that form the initial generalized or primary Ab response to foreign Ag while the IgG proteins comprise the specific or secondary Ab response to Ag, which persists for longer periods of time in the plasma than the IgM response. The IgG proteins are also those that can transfer maternal immunity across the placenta to the fetus. Allergic responses are mediated by the IgE proteins, which act by binding to specific cell surfaces causing the release of chemical mediators that generate an inflammatory or allergic response. The inflammatory or allergic response is a result of the permeation of the site of antigen entry by phagocytic cells and also release of histamine from local mast cells stimulated by Ag-bound IgE. Histamine release then may mediate changes in smooth muscle cells of blood vessels and lungs to cause vasodilation and bronchostriction. The role of the IgD proteins is essentially unknown in terms of cell and Ab function.

The synthesis of the different antibody types occurs in lymphocytes that originate in the bone marrow and consist of two kinds designated as B lymphocytes or T lymphocytes. B lymphocytes are cells that in development have not been modified by the thymus gland and have the capacity to synthesize IgG Ab's in response to an Ag. The T lymphocytes are cells that have been modified by the thymus gland to respond to Ag's in a manner that involves the whole lymphocyte cell reacting to a cell-bound Ag.

The primitive undifferentiated cell found in the bone marrow is the *stem cell*, which then can proliferate to form lymphocytes or macrophages. The lymphocytes are distributed via the thymus into T- and B-cell populations. T lymphocytes are further segregated into forms that *cooperate* (helper cells) with B cells or *inhibit* (suppressor cells) them in their Ab responses to Ag. These helper or suppressor functions play a role in distinguishing harmful Ag's from others and also from host proteins that may act in certain cases as Ag's (autoimmune disease). In the normal situation, a suppressor cell will destroy those lymphocytes early in development that form Ab's to host protein and are called *killer cells*. Helper T lymphocytes cooperate with either B cells or macrophages to amplify the Ab response to a given Ag. B cells that are stimulated by Ag, helper T cells, and macrophages are then converted to plasma cells that secrete Ab into the plasma. These interactions are illustrated in Figure 21.5.

Activation of phagocytosis by *lymphokines* (prostaglandins and related compounds) leads to ingestion of bacteria or other foreign particles by macrophages and neutrophilic leukocytes, which then destroy them by fusion with their lysosomes. Lysosomes release proteolytic and other enzymes into the vacuole so formed. Bacterial killing is accomplished

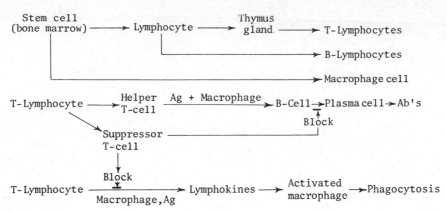

FIGURE 21.5.   Cellular interactions in the immune response. Relationships between lymphocytes and their responses to Ag. The stem cell differentiates into a lymphocyte that is produced by the thymus to become a T cell. Unprocessed lymphocytes remain as B cells. Stem cells can also become macrophage cells. T lymphocytes can aid B-cell responses to Ag in the presence of macrophage and are thus helper T cells. Suppressor T cells inhibit B-cell and T-cell responses to Ag. T cells also produce lymphokines, which further activate macrophages to stimulate phagocytosis of foreign material.

by the interaction of hydrogen peroxide, myeloperoxidase, and iodide in the vacuole containing the phagocytized bacterium with final dissolution by lysosomal enzymes.

INTEGRATION OF ANTIBODY AND CELLULAR COMPONENTS IN IMMUNE RESPONSES

Inflammation

The inflammatory response can be considered as a reaction of tissues to stimulation by cellular compounds produced by trauma, chemicals, or infection. These activating agents are derived either from plasma components (the complement system) or from cells (histamine, serotonin, prostaglandins, and leukotrienes). The physiological changes observed are erythema, edema, and vasoconstriction. When these changes occur rapidly on exposure to an antigen to which an individual may be sensitive (penicillin), then potentially lethal results may occur known as *anaphylaxis* or *anaphylactic shock*. Slower changes in the inflammatory reactions occur when antigen-activated cells such as mast cells

are triggered via IgE antibody to gradually release histamine and complement products. These agents cause vascular leakage and chemotactic attraction of neutrophilic leukocytes to the stimulated area.

The Complement System

The complement components consist of nine plasma proteins that are coordinated into a sequential series of binding reactions toward an Ab bound to an antigenic cell surface. The sequence of events leading to cell lysis by complement factors are shown in Figure 21.6.

Binding of plasma Ab to cell surface Ag forms a complex that binds the C1 component, which then successively binds components C2 to C9 with release of peptide fragments from C3 and C5. When all factors are bound, the complex leads finally to cell lysis. The production of C3 and C5 fragments released into plasma stimulates vascular leakage and neutrophil chemotaxis. The intermediate complex of C1, C2, C3, and C4 with cell-bound Ab facilitates phagocytosis of the cell by neutrophil leukocytes and macrophages.

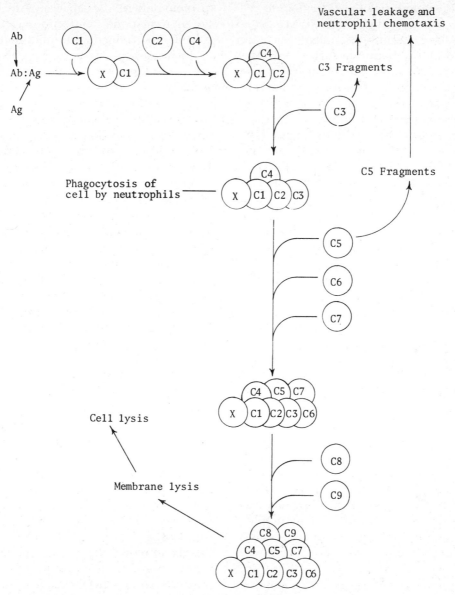

FIGURE 21.6. Complement system activation. Plasma antibody (IgM or IgG) forms a complex with the foreign cell surface antigen, which acts as a center for a cascade of binding events to complement factors C1 to C9. Proteolytic activity of some of the factors releases vasoactive fragments from factor C3 and C5 to cause vascular leakage and neutrophilic leukocyte infiltration. X = Ab:Ag complex.

Tissue-Activated Inflammation Responses and Allergies

When B lymphocytes are exposed to antigens, the Ab response is modulated by helper or suppressor T lymphocytes. Normally, innocuous antigens such as egg proteins in food do not stimulate Ab's because suppressor T cells inhibit B cell formation of Ab to them. In allergies this process is deregulated and B cells are transformed to plasma cells, which then secrete Ab to egg protein. The Ab class that is preferentially stimulated in allergies is the IgE proteins, which have a high affinity for attachment (via the Fc region) to cell surfaces on mast cells. Mast cells are granulated cells containing histamine, which are provoked by Ag (bound to surface IgE Ab's to release histamine and other vasoactive stimulants by degranulation). In addition to degranulation, several surface-active reactions are mediated by the IgE:Ag complex so that membrane-bound fatty acids of the arachidonic acid type are released and converted to *prostaglandins* and *leukotrienes* (Figure 21.7). Prostaglandins consist of several types designated as PGD_2, PGE_2, PGF_2, and PGI_2 that interact differentially with the smooth muscle cells of the lungs, the blood platelets, and mucus-secreting cells (nasal cavity). The leukotrienes (formerly known as *slow-reacting substance of anaphylaxis* or SRS-A) are also derivatives of arachidonic acid designated as leukotriene-A_4, -B_4, -C_4, and -D_4 and cause the constriction of smooth muscle cells in the bronchial tubes and branches (asthma). In terms of a decreasing concentration of released material on antigenic (allergenic) stimulation and increasing potency of inflammatory effect, the order of compounds released from mast cells is histamine to prostaglandins to leukotrienes.

The antigenicity of drugs such as penicillin can be attributed to the formation of protein-bound (covalent) complexes of the drug that effectively activate IgE to cause rapid release of histamines and an anaphylactic response. Treatment of such responses with epinephrine increases the intracellular concentration of cyclic AMP, which in turn suppresses histamine release. The slow release of histamine and inflammatory response in other types of allergies, as for example to cosmetics, can be reduced by a similar mechanism with corticosteroids such as prednisone and related derivatives.

MALNUTRITION AND THE IMMUNE RESPONSE

In severe malnutrition such as marasmus (protein–calorie deficits) and kwashiorkor (protein deficit), the capacity of the immune system to respond to infectious agents is impaired. The reduction in immune response to an invasive microorganism, virus, parasite, or antigen is a result of the reduced availability of amino acids for protein synthesis is lymphocytes to generate appropriate amounts of antibodies. A normal response to infection is the production of *pyrogens* leading to fever and a concomitant lowering of plasma iron. The net result of these two processes is to enhance the interactive effects of lymphocytes and neutrophils toward the foreign antigenic stimuli and to reduce the proliferative potential of foreign cells by restricting the availability of iron for DNA synthesis. In the malnourished person, however, the occurrence of fever can be further debilitating due to the demand on caloric reserves, whereas iron deficiency will suppress proliferation not only of foreign cells but of host cells as well. The reduced amounts of stored endogenous nutrients such as vitamins and minerals also contribute to the suboptimal response of the immune system to infection. In treating patients with infections who are also severely malnourished, it is important to replenish both caloric and protein reserves with simultaneous use of drugs to combat the infection. Care should be exercised so as not to employ those drugs that may compromise recovery (for example, isoniazid without supplemental pyridoxine) or to use iron therapy to reduce the infection-induced drop in plasma iron. Iron supplementation to treat the underlying deficiency state should be implemented only after treatment of the resident infection is successful. In marginal malnutrition states where negative nitrogen balance exists without excessive depletion of body muscle mass, then inflammatory responses to infections may still be normal. Treatment of the inflammatory symptoms with antiinflammatory drugs of the steroid class (prednisone) should be limited since the nutritional status can be aggravated and require additional dietary protein supplementation.

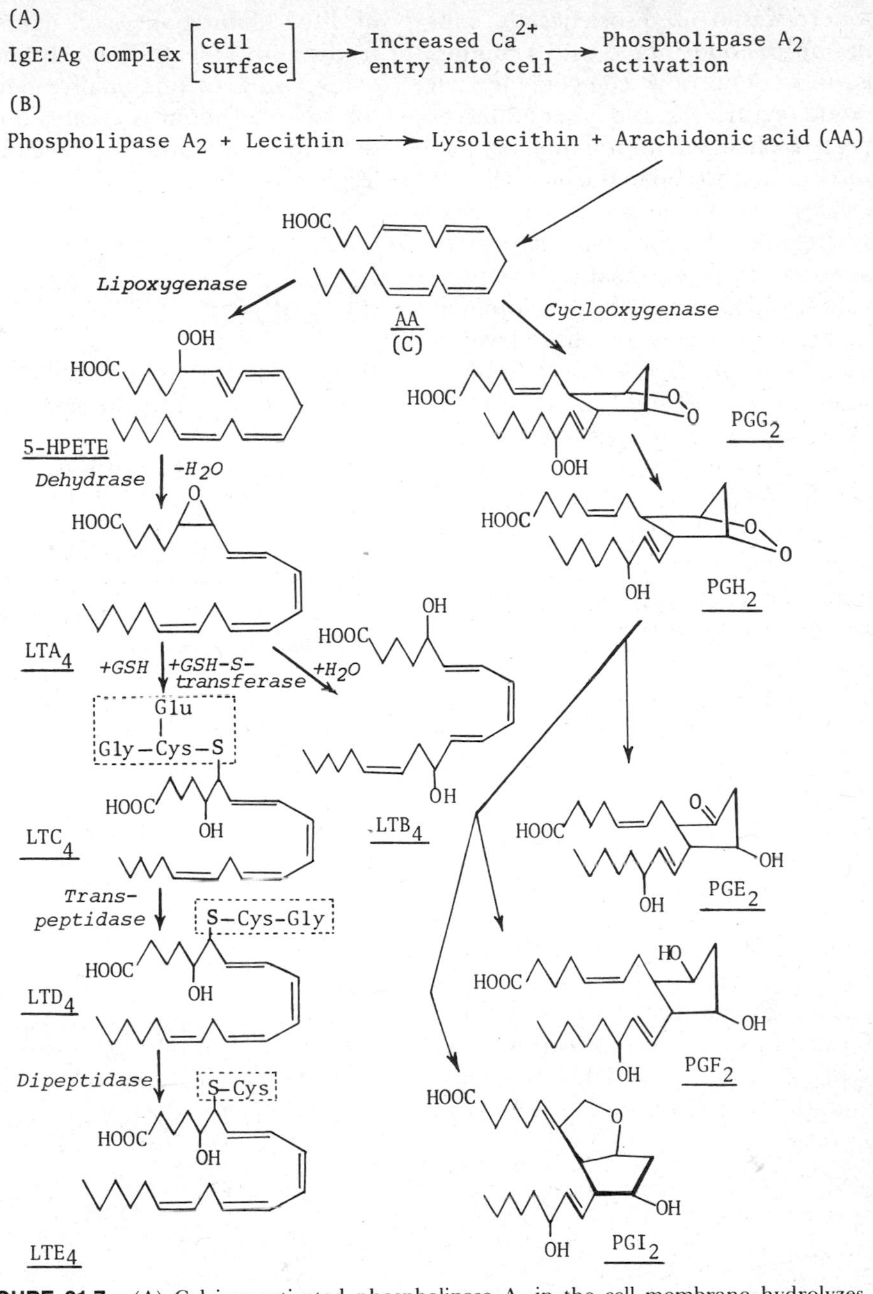

FIGURE 21.7. (A) Calcium-activated phospholipase A_2 in the cell membrane hydrolyzes lecithin (phosphotidylcholine) to give lysolecithin and arachidonic acid (B). Arachidonic acid (C) then serves as a substrate for lipoxygenase to form leukotrienes (LT) or cyclooxygenase may form prostaglandins (PG). The tripeptide glutathione (Glu–Cys(—SH)–Gly) binds to leukotriene-A_4 and sequentially loses glutamate (Glu) and glycine (Gly) to form leukotriene-E_4 with a single cysteine (Cys) residue attached. 5-HPETE, 5-Hydroperoxyeicosatetraenoic acid.

In modern westernized societies, the supplementation of food intake with vitamins, minerals, and nonnutritive components such as dyes, food texturizers, and other additives have created changes in immune system functions commonly seen in the incidence of food allergies. The symptoms of food allergy are characteristic of immediate hypersensitivity responses and appear as intestinal pain, hives, rashes, and asthma. The development of food allergies is usually established within the first 6 months of life when the transition from maternal milk to solid foods is introduced. Although a genetic sensitivity to acquisition of an allergy may be present in individuals, the genetic constitution for the inheritance of a specific allergy probably does not exist. Absorption of intact food proteins through the immature intestinal mucosa can stimulate IgE production and the IgE response on mast cells, resulting in allergic symptoms. The allergenic food components most frequently encountered are those from eggs, milk, wheat (gluten), fish, and fruits, with lesser contributions by components such as drugs (penicillin) and food additives.

Treatment for food allergies is most effective when the responsible component is removed from the diet; however, benefits can be obtained by prior denaturation of food proteins by cooking if elimination of foods is not practical. Awareness of those features of food consumption that can lead to food allergies provides the best defense against their generation in infants in their first year of life. Maintenance of maternal breastfeeding for at least 6 months and reduction of exposure to potentially allergenic foods for 12 to 24 months creates optimal conditions for inhibiting the development of allergic sensitivities.

REFERENCES

Darby, W. J. (ed.). 1981 and 1982. *Annual Review of Nutrition*. Annual Reviews Inc., Palo Alto, Calif.

D'Arcy, P. F., and J. P. Griffin (eds.). 1982. *Iatrogenic Diseases*. Oxford University Press, New York.

Devlin, T. M. 1982. *Textbook of Biochemistry with Clinical Correlations*. Wiley, New York.

Goldstein, A., L. Aronow, and S. M. Kalman. 1976. *Principles of Drug Action*. Wiley, New York.

Goth, A. 1976. *Medical Pharmacology*. Mosby, St. Louis, Mo.

Guttman, R. D. 1975. *Immunology*, Upjohn Co. Monograph. Upjohn Co., Kalamazoo, Mich.

Krause, M. V., and Mahan, L. K. 1979. *Food, Nutrition and Diet Therapy*. Saunders, Philadelphia.

McGeer, P. L., J. C. Eccles, and E. G. McGeer. 1978. *Molecular Neurobiology of the Mammalian Brain*. Plenum, New York.

Present Knowledge in Nutrition. 1976. The Nutrition Foundation, Washington, D.C.

Roe, D. A. 1976. *Drug Induced Nutritional Deficiencies*. Avi Publishing, Westport, Conn.

Ryan, G. B., and G. Majno. 1977. *Inflammation*, Upjohn Co. Monograph. Upjohn Co., Kalamazoo, Mich.

CHAPTER · 22

Nutritional Assessment and Recommended Dietary Allowances

INTRODUCTION

Hunger and malnutrition are among the most serious problems that confront people throughout the world. Both of these problems are endemic to poverty-stricken countries but trace incidences also occur in poor sectors of the most advanced countries. Regardless of the geographic distribution of hunger and malnutrition, children, pregnant and lactating women, as well as the elderly are most severely affected. The Citizen's Board of Inquiry into Hunger and Malnutrition in the United States of America has indicated that 33 to 50% of the poor people in this country were affected by hunger and malnutrition in 1968. Specific nutritional deficits and inadequacies of poor populations revolve about protein-energy malnutrition, vitamin A, C, and D deficiencies, and iron deficiencies. Other studies of contemporary populations still show similar tendencies in poverty-impacted geographical regions.

Apart from the scourges of hunger, poor nutrition of another type is common throughout the United States. This problem is evidenced as *overnutrition,* which leads to obesity. Obesity can be considered as a form of malnutrition since a poor nutritional spec-trum of foods and excess calories are habitually consumed. Van Italie (1979) has reported that 25 to 30% of adults and at least 10% of children are either obese or overweight.

The Greek physician Hippocrates realized centuries ago that there was an intrinsic link between diet and health. On this basis he resorted to using proper diets plus medicines for therapeutic purposes during an individual's illness and convalescence.

In the past decade, the American populace has become keenly aware of diet–health interrelationships. This concern has been translated into legislative directives by the U.S. Senate that established the Senate Select Committee on Nutrition and Human Needs. This committee ultimately established a compendium of U.S. Dietary Goals (1977) designed to direct the American populace toward good health and possibly lessen the incidence of cardiovascular diseases as well as cancer.

NUTRITIONAL STATUS EVALUATION

The term *nutritional status* refers to the health condition of the body as a consequence of prolonged dietary practices or regimens. Since the body largely reflects the effects of nutritional deficits or excesses, four techniques are commonly used for assessing nutritional status. These methods include (1) *clinical evaluation,* (2) *anthropomorphic measurements,* (3) *biochemical analyses* of blood and key body fluids, and (4) *surveys of dietary practices.* Information obtained from any one of these assessment methods can be useful for a preliminary indication of nutritional status, but a sound assessment relies on all four investigative methods.

Clinical evaluation usually involves a physical examination of the eyes, mouth (lips, tongue, gum, and teeth), skin, thyroid gland, nails, skeletal muscles, nervous system, and heart. A carefully conducted examination will ordinarily reveal most obvious nutritional diseases. For example, dryness of the cornea and conjunctiva reflects vitamin A deficiency; mottled tooth enamel may be caused by excess fluoride; a swollen thyroid gland may indicate iodine deficiency; a magenta-colored tongue suggests riboflavin deficiency; evidence of muscle wasting may be caused by protein–calorie deficiencies; and abnormally bent ribs may be indicative of vitamin D- and calcium-related deficiencies. Countless other nutritional deficiencies can be detected by similar observations.

Anthropomorphic measurements include height and weight determinations of an individual; circumference measurements of head, wrist, forearm, thigh, waist, and chest regions; and skinfold thickness on the thigh, triceps, and abdomen.

The height of an individual along with a head circumference measurement have served as good criteria for establishing past nutritional status of most individuals. Moreover, obesity is readily indicated by combined weight and skinfold measurements. Anthropomorphic measurements are notably useful for determining the nutritional status of infants, children, adolescents, and pregnant women in whom normal nutritional status is suspect.

Mathematical equations, derived from anthropometric measurements, are used for assessment of protein–calorie malnutrition.

1. ***The Creatinine-Height Index (CHI)***

$$\frac{\text{patient's urinary creatinine}}{\text{ideal urinary creatinine}} \times 100$$

Note: The ideal urinary creatinine clearance is obtained from standard tables (Blackburn *et al.,* 1977).

2. ***The Percentage Ideal Weight (PIW)***

$$\frac{\text{patient's weight} \times 100}{\text{ideal weight for patient's height}}$$

Note: The ideal weight is obtained from standard tables (Blackburn *et al.,* 1977).

3. ***Arm Muscle Circumference (AMC)***

AMC = arm circumference (cm) − 0.314 × triceps skinfold thickness (mm)

Biochemical analyses are very useful for evaluating nutritional status of the body. Biochemical studies can involve assays of plasma, erythrocytes, leukocytes, urine, liver tissues, and integumentary system constituents (hair, nails, skin, etc.) Some standards for assessing the significance of biochemical measurements have been tabulated. For details consult Christakis (1973) as well as the Ten-State Nutrition Survey (U.S. DHEW, 1968–1970).

In recent years, laboratory measurements of serum albumin, transferrin or total iron-binding capacity, thyroxine-binding prealbumin, and retinol-binding protein have been used as an index of protein–calorie malnutrition since they reflect the nutritional status of a patient (Bakerman and Stausbach, 1984). Of all the aforementioned proteins, thyroxine-binding prealbumin and retinol-binding protein with half-lives of 2 days and 12 h, respectively, serve as best indicators of protein–calorie malnutrition and can be used for monitoring the efficacy of nutritional therapy (Shetty *et al.,* 1979). The nutritional status of a patient can also be estimated on the basis of immunocompetence. Specifically, lymphocyte count and skin sensitivity reflect immunocompetence. Skin-sensitivity testing is carried on intradermally with "recall" antigens such as the mumps skin-test antigen and diphtheria toxoid. The diameter of skin induration is measured at 24 and 48 h, and a 5-mm skin induration signifies a positive test (Mullen *et al.,* 1979). For more information regarding the use of skin testing in nutritional assessment, refer to Twomey *et al.* (1982).

Dietary intake is useful for establishing a nutritional assessment of almost all individuals. This information can be compiled on the basis of 24-h recall or 3- or 7-day food records. Dietary histories conducted for longer periods of time can be very useful, or when circumstances permit, a weekly household food survey can be conducted.

SURVEYS FOR NUTRITIONAL STATUS EVALUATION

Two of the frequently cited surveys that purportedly define nutritional status in the United States include the Ten-State Nutrition Survey and the Health and Nutrition Examination Survey (HANES).

The Ten-State Nutrition Survey studied nutritional status of five low-income (Kentucky, Louisiana, South Carolina, Texas, and West Virginia) and five higher-income states (New York, Michigan, Massachusetts, Washington, and California). Results from this study suggested a number of fundamental conclusions, which are summarized below:

1. Malnutrition seemed to be related to income statistics with poor sectors of the population demonstrating the greatest incidence of malnutrition.

2. Nutritional status of children under the age of 17 was related to the education of the person who purchases and prepares family meals.

3. Malnutrition occurred among black populations more than Spanish-Americans and Caucasians in that order.

4. Undernutrition and malnutrition occurred widely in persons over 60 years of age.

5. Obesity was more predominant in women than men.

6. Tooth decay was found to be related to the consumption frequency of foods containing high amounts of added sugars.

7. Protein, thiamine, vitamin A, iron, ascorbate, riboflavin, and iodine levels were quite low in many test subjects.

For additional perspectives on the significance and implications of this study, consult references for Schaefer (1976) and Goldsmith (1973).

The HANES was conducted over the period of 1971 to 1972 on a large scale using a representative sample of noninstitutionalized Americans between the ages of 1 and 74. Results of this study concluded that

1. Children from families above poverty-level economic conditions were heavier and taller than their poor counterparts.

2. Obesity occurred more frequently in low-income black and white women.

3. White men demonstrated a higher incidence of obesity than black men.

4. Ten percent of the representative population studied also exhibited some degree of vitamin A, iodine, vitamin D, niacin, ascorbic acid, and thiamine deficiencies.

For further information consult *Preliminary Findings of the First Health and Nutrition Examination Survey, United States, 1971–1972: Anthropomorphic and Clinical Findings* published by the U.S. Department of Health and Human Services (1975).

RECOMMENDED DIETARY ALLOWANCES

The amounts of food nutrients required by the human body are fairly well established.

Table 22.1. Recommended Dietary Allowances[a]

Age (years)	Weight (kg)	Weight (lb)	Height (cm)	Height (in.)	Protein (g)	Fat-soluble vitamins Vitamin A (µg RE)[b]	Vitamin D (µg)[c]	Vitamin E (mg α-TE)[d]	Water-soluble vitamins Vitamin C (mg)	Thiamin (mg)	Riboflavin (mg)	Niacin (mg NE)[e]	Vitamin B6 (mg)	Folacin (µg)[f]	Vitamin B12 (µg)	Minerals Calcium (mg)	Phosphorus (mg)	Magnesium (mg)	Iron (mg)	Zinc (mg)	Iodine (µg)
Infants																					
0.0–0.5	6	13	60	24	kg × 2.2	420	10	3	35	0.3	0.4	6	0.3	30	0.5[g]	360	240	50	10	3	40
0.5–1.0	9	20	71	28	kg × 2.0	400	10	4	35	0.5	0.6	8	0.6	45	1.5	540	360	70	15	5	50
Children																					
1–3	13	29	90	35	23	400	10	5	45	0.7	0.8	9	0.9	100	2.0	800	800	150	15	10	70
4–6	20	44	112	44	30	500	10	6	45	0.9	1.0	11	1.3	200	2.5	800	800	200	10	10	90
7–10	28	62	132	52	34	700	10	7	45	1.2	1.4	16	1.6	300	3.0	800	800	250	10	10	120
Males																					
11–14	45	99	157	62	45	1000	10	8	50	1.4	1.6	18	1.8	400	3.0	1200	1200	350	18	15	150
15–18	66	145	176	69	56	1000	10	10	60	1.4	1.7	18	2.0	400	3.0	1200	1200	400	18	15	150
19–22	70	154	177	70	56	1000	7.5	10	60	1.5	1.7	19	2.2	400	3.0	800	800	350	10	15	150
23–50	70	154	178	70	56	1000	5	10	60	1.4	1.6	18	2.2	400	3.0	800	800	350	10	15	150
51+	70	154	178	70	56	1000	5	10	60	1.2	1.4	16	2.2	400	3.0	800	800	350	10	15	150
Females																					
11–14	46	101	157	62	46	800	10	8	50	1.1	1.3	15	1.8	400	3.0	1200	1200	300	18	15	150
15–18	55	120	163	64	46	800	10	8	60	1.1	1.3	14	2.0	400	3.0	1200	1200	300	18	15	150
19–22	55	120	163	64	44	800	7.5	8	60	1.1	1.3	14	2.0	400	3.0	800	800	300	18	15	150
23–50	55	120	163	64	44	800	5	8	60	1.0	1.2	13	2.0	400	3.0	800	800	300	18	15	150
51+	55	120	163	64	44	800	5	8	60	1.0	1.2	13	2.0	400	3.0	800	800	300	10	15	150
Pregnant					+30	+200	+5	+2	+20	+0.4	+0.3	+2	+0.6	+400	+1.0	+400	+400	+150	h	+5	+25
Lactating					+20	+400	+5	+3	+40	+0.5	+0.5	+5	+0.5	+100	+1.0	+400	+400	+150	h	+10	+50

[a]Food and Nutrition Board, National Academy of Sciences—National Research Council Recommended Daily Dietary Allowances, revised 1980. The allowances are intended to provide for individual variations among most normal persons as they live in the United States under usual environmental stresses. Diets should be based on a variety of common foods in order to provide other nutrients for which human requirements have been less well defined.

[b]Retinol equivalents. 1 retinol equivalent = 1 µg retinol or 6 µg β-carotene.

[c]As cholecalciferol. 10 µg cholecalciferol = 400 I.U. of vitamin D.

[d]α-Tocopherol equivalents. 1 mg d-α-tocopherol = 1 α-TE.

[e]1 NE (niacin equivalent) is equal to 1 mg of niacin or 60 mg of dietary tryptophan.

[f]The folacin allowances refer to dietary sources as determined by *Lactobacillus casei* assay after treatment with enzymes (conjugases) to make polyglutamyl forms of the vitamin available to the test organism.

[g]The recommended dietary allowance for vitamin B-12 in infants is based on average concentration of the vitamin in human milk. The allowances after weaning are based on energy intake (as recommended by the American Academy of Pediatrics) and consideration of other factors, such as intestinal absorption.

[h]The increased requirement during pregnancy cannot be met by the iron content of habitual American diets nor by the existing iron stores of many women; therefore the use of 30–60 mg of supplemental iron is recommended. Iron needs during lactation are not substantially different from those of nonpregnant women, but continued supplementation of the mother for 2–3 months after parturition is advisable in order to replenish stores depleted by pregnancy.

SOURCE: *Recommended Dietary Allowances*, 9th ed., Washington, D.C.: National Academy of Sciences—National Research Council, 1980.

In 1941, during World War II, the Food and Nutrition Board of the Nutritional Research Council (NRC) established standards designed to ensure that nutritional needs of people were met. These food standards eventually became recognized as the Recommended Daily Allowances (RDAs) (Tables 22.1–22.3).

The RDAs for key nutrients were expressed as a quantitated amount that would *probably* ensure the normal healthy status of *most* individuals, but not the nutritional requirements of ill people. The RDA levels established for all nutrients also failed to account for food nutrient losses resulting from the processing and preparation of foods. Based on the RDAs reported by the NRC in 1980, all of the following nutrient categories have been quantitatively addressed: vitamins

A, D, E, B_{12}, B_6, ascorbic acid, folacin, niacin, thiamine, riboflavin; inorganic nutrients such as calcium, phosphorous, magnesium, iron, zinc, and iodine; and perspectives on requisite caloric demands of the body. Other nutrient requirements outlined by the 1980 RDAs include estimated allowances for vitamin K, biotin, pantothenic acid, and trace elements such as copper, manganese, fluoride, chromium, selenium, and molybdenum. In addition to these food constituents, RDA ranges are projected for sodium, potassium, and chloride ions as the principal electrolytes of body fluids.

It should be recognized that the RDAs specified by the NRC are designed to ensure the normal healthy status of most individuals but not the specialized needs of ill people.

Although nutritional requirements for

Table 22.2. Mean Heights and Weights and Recommended Energy Intake[a]

Category	Age (years)	Weight kg	Weight lb	Height cm	Height in.	Energy needs (with range) kcal		Energy needs (with range) MJ
Infants	0.0–0.5	6	13	60	24	kg × 115	(95–145)	kg × 0.48
	0.5–1.0	9	20	71	28	kg × 105	(80–135)	kg × 0.44
Children	1–3	13	29	90	35	1300	(900–1800)	5.5
	4–6	20	44	112	44	1700	(1300–2300)	7.1
	7–10	28	62	132	52	2400	(1650–3300)	10.1
Males	11–14	45	99	157	62	2700	(2000–3700)	11.3
	15–18	66	145	176	69	2800	(2100–3900)	11.8
	19–22	70	154	177	70	2900	(2500–3300)	12.2
	23–50	70	154	178	70	2700	(2300–3100)	11.3
	51–75	70	154	178	70	2400	(2000–2800)	10.1
	76+	70	154	178	70	2050	(1650–2450)	8.6
Females	11–14	46	101	157	62	2200	(1500–3000)	9.2
	15–18	55	120	163	64	2100	(1200–3000)	8.8
	19–22	55	120	163	64	2100	(1700–2500)	8.8
	23–50	55	120	163	64	2000	(1600–2400)	8.4
	51–75	55	120	163	64	1800	(1400–2200)	7.6
	76+	55	120	163	64	1600	(1200–2000)	6.7
Pregnancy						+300		
Lactation						+500		

[a]The energy allowances for the young adults are for men and women doing light work. The allowances for the two older age groups represent mean energy needs over these age spans, allowing for a 2% decrease in basal (resting) metabolic rate per decade and a reduction in activity of 200 kcal/day for men and women between 51 and 75 years, 500 kcal for men over 75 years, and 400 kcal for women over 75 years. The customary range of daily energy output is shown in parentheses for adults and is based on a variation in energy needs of ±400 kcal at any one age, emphasizing the wide range of energy intakes appropriate for any group of people.

Energy allowances for children through age 18 are based on median energy intakes of children of these ages followed in longitudinal growth studies. The values in parentheses are 10th and 90th percentiles of energy intake, to indicate the range of energy consumption among children of these ages.

SOURCE: *Recommended Dietary Allowances*, 9th ed., Washington D.C.: National Academy of Sciences—National Research Council, 1980.

Table 22.3. Estimated Safe and Adequate Daily Dietary Intakes of Selected Vitamins and Minerals[a]

	Age (years)	Vitamins		
		Vitamin K (μg)	Biotin (μg)	Pantothenic Acid (mg)
Infants	0–0.5	12	35	2
	0.5–1	10–20	50	3
Children	1–3	15–30	65	3
and	4–6	20–40	85	3–4
adolescents	7–10	30–60	120	4–5
	11+	50–100	100–200	4–7
Adults		70–140	100–200	4–7

	Age (years)	Trace Elements[b]					
		Copper (mg)	Manganese (mg)	Fluoride (mg)	Chromium (mg)	Selenium (mg)	Molybdenum (mg)
Infants	0–0.5	0.5–0.7	0.5–0.7	0.1–0.5	0.01–0.04	0.01–0.04	0.03–0.06
	0.5–1	0.7–1.0	0.7–1.0	0.2–1.0	0.02–0.06	0.02–0.06	0.04–0.08
Children	1–3	1.0–1.5	1.0–1.5	0.5–1.5	0.02–0.08	0.02–0.08	0.05–0.1
and	4–6	1.5–2.0	1.5–2.0	1.0–2.5	0.03–0.12	0.03–0.12	0.06–0.15
adolescents	7–10	2.0–2.5	2.0–3.0	1.5–2.5	0.05–0.2	0.05–0.2	0.10–0.3
	11+	2.0–3.0	2.5–5.0	1.5–2.5	0.05–0.2	0.05–0.2	0.15–0.5
Adults		2.0–3.0	2.5–5.0	1.5–4.0	0.05–0.2	0.05–0.2	0.15–0.5

	Age (years)	Electrolytes		
		Sodium (mg)	Potassium (mg)	Chloride (mg)
Infants	0–0.5	115–350	350–925	275–700
	0.5–1	250–750	425–1275	400–1200
Children	1–3	325–975	550–1650	500–1500
and	4–6	450–1350	775–2325	700–2100
adolescents	7–10	600–1800	1000–3000	925–2775
	11+	900–2700	1525–4575	1400–4200
Adults		1100–3300	1875–5625	1700–5100

[a]Because there is less information on which to base allowances, these figures are not given in the main table of RDA and are provided here in the form of ranges of recommended intakes.

[b]Since the toxic levels for many trace elements may be only several times usual intakes, the upper levels for the trace elements given in this table should not be habitually exceeded.

SOURCE: Recommended Dietary Allowances, 9th ed., Washington D.C.: National Academy of Sciences—National Research Council, 1980.

many inorganic nutrients and trace metals appear in nationalized nutritional guidelines, debates still swirl about the credibility of the prescribed levels. Since the exact roles for numerous trace elements are specifically unknown, *exact* human nutritional requirements are difficult to establish outside of the information gleaned from animal studies.

Recommended vitamin concentrations in the RDAs are accurate in many cases as a result of classical animal and human studies, but inadequate analytical methods cloud the established biochemical requirements for some vitamins. This is especially true for vitamin B_{12}, ascorbic acid, vitamin A, folates, and vitamin E. Furthermore, even if the RDAs for these vitamins are accurate, the tabulated vitamin levels for many foods have querulous significance for use as guidelines in fulfilling vitamin RDAs. These problems stem from *antiquated analytical methodologies* for some of the most important vitamins, *nonuniform methods* of analytical vitamin extraction from foods prior to their assay, and *inconsistent implementation of standardized analytical methods* for vitamin assay. Unfortunately, these quantitative inefficacies riddle the tabulated levels for vitamin content of many foods and serve as an error-ridden database for assessing the actual fulfillment of dietary vitamin requirements.

U.S. RECOMMENDED DIETARY ALLOWANCE (USRDA)

Specific USRDA values have been adopted by the Food and Drug Administration (FDA) and used as a standard for nutrient requirements based on RDAs for an adult male. Therefore, the corresponding USRDAs of nutrients for children and women are generous. The USRDA-specified nutrients are listed on food labels as percentages of the daily RDA. Many food processors have begun to adopt this practice as a common labeling feature spurred on by consumer and governmental pressure.

DIETARY GUIDELINES FOR THE UNITED STATES

Both U.S. Departments of Agriculture and Health and Human Services (formerly Health, Education and Welfare) have promulgated new dietary guidelines that should ensure a healthful dietary status among most of the American populace. These recommendations suggest that

1. A variety of foods should be consumed.

2. Ideal body weight must be maintained.

3. Saturated fat and cholesterol in the diet must be minimized.

4. Starch- and fiber-rich foods are preferable to those foods rich in refined sugars.

5. Sodium intake should be reduced.

6. Alcohol consumption should be only moderate.

These guidelines are clearly useful in planning a menu for healthy people, but they do not apply to specialized diets.

The Basic Food Group Plan published by the U.S. Department of Agriculture (1957) has received wide use for planning menus. This plan suggests the inclusion of four main nutritional sectors in a balanced diet including (1) meat, (2) milk, (3) vegetable–fruit, and (4) bread–cereal groups.

In 1977 the Senate Select Committee on Nutrition and Human Needs also enumerated dietary goals for the United States. Many of these specific goals are reminiscent of the 1980 dietary guidelines outlined above and they suggest that

1. Overweight conditions should be avoided by consuming only as much energy (calories) as expended; in overweight conditions, energy intake should be decreased while energy expenditure is increased.

2. Complex carbohydrate and naturally occurring sugar intake should be increased from about 28% of energy intake to about 48% of energy intake.

3. Consumption of refined and processed sugars should be reduced by about 45% to account for only 10% of total energy intake.

4. Overall fat consumption should be reduced from about 40% to only 30% of energy intake.

5. Saturated fat consumption should be reduced to account for about 10% of total energy intake and balance that of mono- and polyunsaturated fats, which should each account for 10% of energy intake.

6. Cholesterol consumption ought to be reduced to levels below 300 mg/day.

7. Sodium intake should be reduced to levels less than 5 g/day.

A prophylactic approach to diseases implicated with certain dietary practices is wise, especially for those individuals who may be genetically predisposed to diseases. Hegsted (1978) considers the significance of the dietary goals from a progressive perspective, whereas Harper (1978) views the goals from a skeptical viewpoint.

MINIMIZATION OF SODIUM CONTENT IN FOODS

Food scientists can contribute to the U.S. Dietary Goals by formulating foods that contain low amounts of saturated fat, refined sugar, cholesterol, and salt. Saturated fat, cholesterol, and salt contents of foods are clearly important factors for increasing the risk of coronary heart disease (CHD) as indicated by historical records of the Framingham Heart Study (Castelli, 1982).

Since sodium content is largely an extrinsic additive to most processed foods, it is incumbent upon the food industry to regulate the amount of sodium introduced into foods. The association between dietary sodium intake and hypertension is well documented (Freis, 1976; Altschul and Grommet, 1980); moreover, Abernathy (1979) has indicated that the rate of death due to stroke, heart attack, or kidney disease may actually double as a result of sodium-induced hypertensive states. Aside from the fact that certain individuals may have a genetically predisposed tendency for hypertension, dietary sodium affects the physiology of all individuals to some extent (Tobian, 1979). These physiological upsets may be linked to (1) changes in water volume throughout various compartments in the body or (2) abnormal sodium–potassium concentration ratios, which are produced with the intake of excess salt. Although many investigators claim that potassium supplements can allay the hypertensive conditions produced by sodium, the mechanism for this action is not specifically understood (Meneely and Battarbee, 1976;

Abernathy, 1979; Lecos, 1983; Parfrey *et al.*, 1981). *Since the interplay of sodium–potassium concentration ratios seems to be important as a hypertensive factor, some researchers believe that the ratio of these cations is more important than the absolute concentration of dietary sodium intake.*

Sodium cannot be completely eliminated from many foods since preservational safety and consumer acceptance may be compromised, but salt has been successfully eliminated from infant foods, in which it is unnecessary. Although many food processors have made a distinct effort to reduce salt concentrations in foods, this has not been an intensive industry-wide effort for all food processors (*Food Chemistry News*, 1982). As a compromise for reducing sodium content in foods, some food processors have attempted to alter the relative sodium to potassium concentration ratio (Salovaara, 1982; Seman *et al.*, 1980; Seperich and Ashoor, 1983; Wyatt and Ronan, 1982).

The average dietary consumption of sodium is 10 to 20 times the amount required for physiological balance (NRC, 1980). Americans receive a daily food salt intake of up to 10 to 12 g in many cases. Of this amount, nearly 3 g occurs naturally in food, 3 g is added by the cook or at the table, and 4 to 6 g is added during commercial processing (Institute of Food Technologists, 1980).

Although high sodium levels contribute to hypertensive conditions, the exact relationship between sodium and other cations, excluding potassium cations, is not very clear. For example, calcium (Ca^{2+}) appears to decrease blood pressure, but its requisite dietary level for ensuring health status is still a matter of great interest. Belizan *et al.* (1983) has also suggested that the role of calcium in blood pressure regulation may involve its effect on parathyroid hormones and prostaglandin syntheses. Daily calcium supplements of only 1 g reportedly decrease blood pressure.

An intelligent reduction of dietary sodium is important since the elimination or avoidance of certain crucial food groups may result in nutritional deficiencies. For example, the omission of dairy products may produce calcium deficits and exacerbate hypertensive conditions.

CONCLUSION

Although medical sciences and nutrition have historically progressed along largely independent routes, these two areas have participated in a cooperative exchange in recent years. The pathomedical–nutritional interface involving dietary salt represents one of the more obvious links between medicine and nutrition, but similar relationships exist for nearly all food constituents. Countless other subtle dietary ties to common diseases also exist, but the specific interrelationships are obfuscated by complex biochemical pathways, prolonged periods of time during which nutrition-related diseases materialize, as well as protracted periods of poor nutritional practices.

REFERENCES

Abernathy, J. D. 1979. Sodium and potassium in high blood pressure. *Food Technol.* **33**(12):57.

Altschul, A. M., and J. K. Grommet. 1980. Sodium intake and sodium sensitivity. *Nutr. Rev.* **38**:393.

Bakerman, S., and P. H. Strausbach. 1984. Malnutrition: the laboratory and the clinic. *Lab. Management* **22**(4):13.

Belizan, J. M., J. Villar, O. Pineda, H. E. Gonzalez, E. Sainz, G. Garrera, and R. Sibrian. 1983. Reduction of blood pressure with calcium supplementation in young adults. *JAMA* **249**:1161.

Bender, A. E. 1984. Nutritional malnutrition. *Br. Med. J.* **288**:92.

Blackburn, G. L., B. R. Bistrian, B. S. Maini, H. Schlamm, and M. F. Smith. 1977. Nutritional and metabolic assessment of the hospitalized patient. *J. Parent. Ent. Nutr.* (*Baltimore*) **1**:11.

Castelli, W. P. 1982. Natural disease investigation, atherosclerosis, blood cholesterol and the environment. *Amer. J. Forensic Med. Pathol.* **3**:323.

Christakis, G. 1973. Nutritional assessment in health programs. *Amer. J. Pub. Health* **63**:80.

Citizen's Board of Inquiry into Hunger and Malnutrition in the United States. 1968. *Hunger U.S.A.* Beacon, Boston.

Food Chemistry News. 1982. Associations, firms report low-sodium, sodium labeling efforts. December 13, p. 31.

Freis, E. D. 1976. Salt, volume and the prevention of hypertension. *Circulation* **53**(4):589.

Gardenswartz, M. H., and R. W. Schrier. 1982. Renal regulation of sodium excretion. In *Sodium: Its Biological Significance* (S. Papper, ed.). CRC Press, Boca Raton, Fla.

Goldsmith, G. A. 1973. Nutrition and world health. *J. Amer. Diet Ass.* **63**:513.

Harper, A. E. 1978. Dietary goals—a skeptical view. *Amer. J. Clin. Nutr.* **31**:310.

Hegsted, D. M. 1978. Dietary goals—a progressive view. *J. Clin. Nutr.* **31**:1504.

Institute of Food Technologists. 1980. Dietary salt. A Scientific Status Summary. IFT Expert Panel on Food Safety and Nutrition. *Food Technol.* **34**(1):85.

Lecos, C. 1983. Potassium: keeping a delicate balance. *FDA Consumer* **17**(1):20.

Meneely, G. R., and H. D. Battarbee. 1976. Sodium and potassium. *Nutr. Rev.* **34**:225.

Mullen, J. L., M. H. Gertner, G. P. Buzby, G. L. Goodhart, and E. F. Rosato. 1979. Implications of malnutrition in the surgical patient. *Arch. Surg.* **114**(2):121.

National Research Council (NRC). 1980. *Recommended Dietary Allowances*, 9th ed. National Academy of Sciences, Washington, D.C.

Page, L., and E. F. Phipard. 1957. *Essentials of an Adequate Diet*, Home Economics Research Report No. 3. USDA, Washington, D.C.

Parfrey, P. S., M. J. Vandenburg, P. Wright, J. M. P. Holly, F. J. Goodwin, S. J. W. Evans, and J. M. Ledingham. 1981. Blood pressure and hormonal changes following changes in dietary sodium and potassium in mild essential hypertension. *Lancet* **1**:59.

Seperich, G. D., and S. H. Ashoor. 1983. *The Effect of Partial and Total Replacement of Sodium Chloride with Calcium Chloride in a Bologna Product*, Progress Report to the American Meat Institute from Arizona State University, Tempe.

Salovaara, H. 1982. Sensory limitations to

replacement of sodium with potassium and magnesium in bread. *Cereal Chem.* **59**:427.

Schaefer, A. E. 1976. Nutrition in the United States of America. In *Nutrition in the Community* (D. S. McLaren, ed.), pp. 369–383. Wiley, London.

Seman, D. L., D. G. Olson, and R. W. Mandigo, 1980. Effect of reduction and partial replacement of sodium on bologna characteristics and acceptability. *J. Food Sci.* **45**:1116.

Shetty, P. S., R. T. Jung, K. E. Watrasiewicz, and W. P. James. 1979. Rapid-turnover transport proteins: An index of subclinical protein-energy malnutrition. *Lancet* **2**:230.

Twomey, P., D. Ziegler, and J. Rombeau. 1982. Utility of skin testing in nutritional assessment: A critical review. *J. Parent. Ent. Nutr. (Baltimore)* **6**(1):50.

Tobian, L. 1979. The relationship of salt to hypertension. *Amer. J. Clin. Nutr.* **32**:2739.

U.S. Department of Agriculture, Department of Health and Human Services. 1980. *Nutrition and Your Health, Dietary Guidelines for Americans.* Superintendent of Documents, Washington, D.C.

U.S. Department of Health, Education and Welfare. 1968–1970. *Highlights of Ten-State Nutrition Survey.* DHEW Publication No. (HSM) 72-8134.

U.S. Senate. 1977. *Dietary Goals for the United States,* 2nd ed. Senate Select Committee on Nutrition and Human Needs, U.S. Senate, Washington, D.C.

Van Italie, T. B. 1979. Obesity: Adverse effects on health and longevity. *Amer. J. Clin. Nutr.* **32**:2723.

Wright, R. A., S. Heymsfield, and C. B. McManus (eds.). 1984. *Nutritional Assessment.* Blackwell Scientific, Oxford, England.

Wyatt, C. J., and K. Ronan. 1982. Evaluation of potassium chloride as a salt substitute in bread. *J. Food Sci.* **47**:672.

PART · 3

Informational Biomolecules and Applied Genetics

C H A P T E R · 23

Genetic Improvement of Agricultural Resources

INTRODUCTION

Plant geneticists and animal breeders have practiced *selective* propagation and growth of certain plant and animal species for many years. These efforts have been designed to produce agricultural crops and increase livestock populations that demonstrate (1) improved resistance to climatic and environmental stresses, (2) enhanced disease resistance, (3) superior reproductive vigor and fertility, or some other crucial factor that improves overall food quality or productivity. Microbiologists have also made a concerted effort to select bacteria and fungi that are used in food fermentation industries.

In recent years, molecular genetic techniques have allowed the tailored genetic modification of plants, animals, and microorganisms in order to amplify food production or develop new food resources.

An understanding of the principles underlying genetic engineering techniques and applied biotechnology for the food sciences requires some basic knowledge of empirical genetic principles as well as molecular genetics.

EVOLUTION OF MODERN APPLIED MOLECULAR GENETICS

Ancient Genetic Practices

Neolithic farmers recognized that seeds from successful crop plants produced in one growing season could be sowed to propagate another crop in the next season. These ancient farmers became very adept in detecting significant variations among plant populations and selecting seeds from only the best crop plants for future propagation.

During the past 10,000 years, which accounts for less than 1% of man's history on earth, gradual selection processes for food-producing species have advanced at a glacial speed. Nonetheless, plant species of the Neolithic period served as the breeding stock for many of our contemporary crops. These ancient crop species offered wide native variability and were randomly introduced into many different cultivation environments. Organisms demonstrating superior vigor, growth, food productivity, and reproductive potentials in a certain environment were consistently selected for cultivation.

Centuries of direct selection pressures by

farmers have produced thousands of different crop plant varieties. In the case of rice alone, at least 40,000 varieties have been developed from ancient rice varieties.

Each variety of any crop plan usually exhibits some distinct adaptation to a cultivation condition or requirement. Poorly organized Neolithic practices of plant breeding were gradually supplanted by more specific and systematic breeding methods. These practices directed the selection of seeds from only those plants in a mixed population that displayed superior crop assets.

Repetitive selections of seed stock from increasingly select crop plant specimens eventually led to *pure breeding lines*. Many of these selected breeding lines exhibited specific resistances to drought conditions, wind damage, or insect damage, or they excelled in food productivity over preexisting crop plants. Apart from the occurrence of one or more meritorious traits exhibited by any *pure breeding* line, the same line often displayed one or more coexisting *unfavorable* traits. Unfortunately, organisms isolated by pure breeding line selections classically failed to allow the transfer of a desirable trait rather than overbearing unfavorable traits.

Development of Mendelian Genetic Precepts

The historical studies of Gregor Mendel, reported in the late 1860s, ushered in the beginning of more advanced plant breeding practices. Mendel's studies, which were originally based on the analysis of many generations of controlled crosses between sweet pea plants, uncovered the empirical operational laws of inheritance. These laws offered explanations for (1) the orchestrated appearance of certain traits from one generation of a species to its progeny and (2) the *gene unit* of inheritance, which serves as the basic repository of transmissible hereditary characteristics between parents and progeny.

Taken as a group, the collective expressions of *all gene units* contained in an organism direct its development, overall appearance, reproduction, metabolism, and many other demonstrable properties. These observable features became commonly recognized as *phenotypic* characteristics.

Mathematical analyses of many sweet pea plant crosses conducted by Mendel revealed that specific plant genes displayed constancy from one plant generation to another. Aside from *generational constancy*, Mendel also recognized that distinct phenotypic traits and genes did not coincide on a one-to-one basis. Instead, studies indicated that *each trait* reflected the presence of *two genes*, with one gene originating from each parent. When both of the genes for a single trait were identical, the observed trait was recognized as being a *homozygous condition*. When distinctly different genes directed the appearance of some trait, this was alternatively characterized as a *heterozygous condition*.

In the case of an *ideal* Mendelian cross for homozygous organisms (plants or animals), the occurrence of distinct and predictable characteristics could be anticipated in the progeny. A Mendelian cross for heterozygous plants, however, failed to show a phenotypic blend for any given trait. Instead only one gene, contributed by one of the two parents, caused the predominant appearance of a trait over its jointly occurring gene. For example, classical Mendelian crosses using pea plants homozygous for yellow seed (YY) and those homozygous for green seed (gg) gave progeny (an F_1 or *first filial generation*) that were heterozygous for seed color (Figure 23.1A). The heterozygous F_1 peas clearly contained one "green gene" (g) from one parent and a "yellow gene" (Y) from the other parent in their characteristic genetic (Yg) constitutions.

Contrary to the uniform demonstration of phenotypic expression as yellow seeds, a successive cross of the F_1 generation plants showed that genetic differences *persisted as transmissible hereditary factors*. Crosses of the F_1 representatives produced 75% of the progeny with yellow seeds (YY, Yg, Yg) and 25% with green seeds (gg).

(A) Mendelian Inheritance of Pea Color

Y = Yellow gene
g = Green gene

YY = Homozygous condition directs yellow-seed formation
gg = Homozygous condition directs green-seed formation
Yg = Heterozygous condition directs yellow-seed formation

A parental homozygote cross of yellow (YY) and green-seed (gg) plants leads to 100% of progeny in the F_1 generation having heterozygotic conditions (Yg) and yellow-seeds.

Heterozygotic parent	Gametes	F_1 progeny
YY →	Y	→ Yg
gg →	g	→ Yg

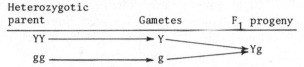

Mendelian crosses of the F_1 progeny involve combinations of "Y" and "g" gametes. Punnett Square analysis of all possible combinations reveals that 25% of the F_2 progeny are homozygous (YY) yellow-seed plants, 25% are homozygous (gg) green-seed plants, and 50% are heterozygous (Yg) yellow-seed plants.

F_1 Female		Y	g
F_2 Male	Y	YY	Yg
	g	Yg	gg

Possible F_2 genotypes

(B) Mendelian Inheritance of Plumage Color in a Fowl

B = Black gene
b = White gene

BB = Homozygous condition directs black plumage
bb = Homozygous condition directs white plumage
Bb = Heterozygous condition directs appearance of blue plumage

A parental homozygote cross of black (BB) and white (bb) birds produces 100% of F_1 progeny having heterozygotic conditions expressed as blue (Bb) plumage.

Heterozygotic parent	Gametes	F_1 progeny
BB →	B	→ Bb
bb →	b	→ Bb

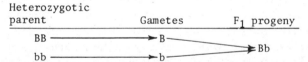

Crosses of two F_1 progeny can only involve combinations of "B" and "b" genes. Punnett Square analysis of all possible combinations shows that 25% of F_2 progeny are homozygous (BB) black birds, 50% (Bb) blue birds and 25% (bb) white birds.

F_1 egg		B	b
F_2 sperm	B	BB	Bb
	b	Bb	bb

Possible F_2 genotypes

FIGURE 23.1. Principles of Mendelian genetics underlie the appearance of many traits in plants and animals. The occurrences of certain phenotypes occur at certain mathematical frequencies in ideal circumstances as outlined in the illustration.

Repetitive observations of such phenotypic phenomena indicated that so long as the gene for a *dominant feature* such as yellow (*Y*) was present as a hetero- or homozygous condition, the pea seed color would be yellow. Only when the (*gg*) condition existed were the pea seeds green. Thus, the presence of at least one *Y* gene and yellow seeds indicated a clear-cut phenotypic *dominance* over green seeds (*gg*), which represented a *recessive trait.* The statistical occurrence of the recessive homozygous condition (*gg*) and green-colored seeds was eventually found to be predictable according to *random selections* of one of the two genes from each parent plant.

Many types of animal traits also demonstrated the same type of Mendelian inheritance as shown by plants, but absolute phenotypic dominance of a single gene was not always indicated. Figure 23.1B shows the results of mating Andalusian fowls, which are a special breed of chicken. Unlike the effects of a distinctly dominant gene (*Y*) in the case of pea plants, a cross between a white (hen or rooster) and a black mate (first parental generation, P_1) gave F_1 progeny having intermediate pigmentation (gray-blue). This cross required that both parents were the result of consistent breeding—white to white and black to black—such that no changes in feather color ever appeared for many years. White crosses must never have produced anything but white progeny and black crossings resulted in only blacks.

Inbreeding of the gray-blue F_1 generation further indicated a 1:2:1 occurrence ratio of white:blue:black birds. A further inbreeding of the F_2 subjects consistently resulted in similar observations such that white fowl crosses always produced whites, blacks produced only blacks, blacks crossed with whites gave all blues, and blues crossed with another blue allowed the appearance of whites:blues:blacks in a 1:2:1 ratio.

One-factor crosses, where mated individuals displayed a singular hereditary difference, always gave results similar to those outlined above. It should be recognized, however, that such generalizations required the F_1 hybrids to be characteristically different from both pure breeding line parents.

Genes Are Recognized as Units of Hereditary Information

According to classical breeding experiments, it became clear that the *discrete units* of hereditary information were transmitted from parents to progeny as genes. In highest forms of life, genes were found to be transmitted to progeny as genes borne on chromosomes of germ cells. All so-called *diploid organisms* were recognized to contain at least two genes (2*N*) that controlled the appearance of any one trait. While exerting its characteristic phenotypic effect, each gene (*N*) apparently retained its distinct integrity as a hereditary unit. The phenotypic expression of selected genes was also shown to be muted, blended, or indistinguishable with respect to coexisting *dominance properties of other genes*. Nevertheless, the functional genetic integrity of any gene was *not* permanently altered or modified by its existence in a diploid state (2*N*).

The process of meiosis was eventually regarded as the quintessential cellular mechanism underlying gametogenesis in diploid (2*N*) organisms. Distinctly similar meiotic processes directed gametogenesis in both sexes of parent organisms, and these processes ensured that genes for a trait would be segregated into compatible haploid (*N*) gametes. Thus, each gamete served as a vehicle for the transmission of only one of any given pair (2*N*) of genes from a parent.

It is now widely recognized that each gene in an organism occupies a specific *locus* on a specific chromosome. Those genes that reside at the same locus in a given pair of chromosomes are said to be *alleles* of each other. For the Andalusian fowl discussed above, the black gene (*B*) represents the allele of the white gene (*b*). These respective genes reside at similar loci on a chromosome pair. In the same way that diploid cells (2*N*) display homologous pairs of chromosomes, genes occur as allelic pairs, each one of the alleles of an allelic pair being contributed by one parental gamete.

Genes Can Exist in Chromosomal Linkage Groups

Research conducted in the twentieth century by T. H. Morgan demonstrated the experimental advantages of using *Drosophila melan-*

ogaster as a genetic test organism. These gnat-sized fruit flies displayed four readily apparent chromosome pairs per cell, high reproductive rates for successive generations, plus easily recognizable phenotypic traits.

Phenotypic data based on eye color, body color, "forked bristles," and curved wing traits of fruit flies were observed for many generations by numerous investigators. These studies revealed that certain of these traits seemed to be inherited together more frequently than other traits. Ruby eyes and yellow bodies were commonly observed. Furthermore, both of these traits were more commonly observed with the "forked bristle" trait than theorized on the basis of pure Mendelian occurrences of traits. In addition, the presence of curved wings occurred with all three of the preceding traits.

Repetitive observations of the foregoing inherited traits led to the conclusion that random assortment principles, outlined by Mendelian law, may not have been as universally applicable as once believed. In fact, it was later revealed that the joint inheritance of some traits actually reflected the presence of *linked genes* located on the *same chromosome*. Wherever the number of known genes exceeded the haploid number (N) of chromosomes, some genes for specific traits inevitably showed linkage; that is, they tended to be inherited as a group rather than individually.

Morgan's studies of *Drosophila* actually demonstrated that

1. The many genes of the fruit fly could be divided into four and only four linkage groups, which corresponded to four identifiable chromosomes.

2. Any evidence that linkage between two pairs of genes when incomplete was incomplete with an occurrence frequency that could be interpreted as a function of the *constant spatial relationship between genes*.

Further research by A. H. Sturtevant is credited with development of a *test cross* whereby the order of a third gene could be mapped with reference to two other genes located on the same chromosome. Such inventive gene-mapping techniques were based on the rationale that observed frequencies of gene occurrence mirrored their relative proximities on a chromosome at the time of gametogenesis. The initial meiotic steps of gametogenesis require the central crowding, coiling, and near fusion of chromosomes over their whole length. This temporal condition allows limited crossing over and natural recombination events between chromosomes. In this way chromosomal reshuffling is ensured before being introduced into succeeding generations.

Bacterial Transformation and DNA

The biochemical and molecular integrity of functional genes did not become specifically recognized until the 1940s. At this time O. T. Avery, C. M. MacLeod, and M. McCarty (1944) succeeded in elucidating the mechanism of bacterial transformation.

Bacterial transformation was a highly significant revelation since it marked the earliest evidence that factors other than reproductively conveyed genes on chromosomes could affect gene constitutions and phenotypic expressions. The studies of Avery *et al*. (1944) showed that two strains of *Diplococcus pneumoniae* could be easily distinguished according to the presence or absence of a visible capsule. Separate cultivation of the strains resulted in true breeding cultures. That is, encapsulated and unencapsulated bacterial cells reproduced, respectively, more of their own kind. A startling recognition was made, however, when unencapsulated cells were grown in a culture medium prepared from the debris of encapsulated cells. Under these cultivation conditions, previously unencapsulated cells began to develop capsules. Furthermore, although unencapsulated bacteria characteristically failed to show the pathogenicity associated with encapsulated bacteria, the newly encapsulated bacteria developed notable pathogenicity. After this *transformation* of the unencapsulated *Diplococci*, the newly encapsulated bacteria permanently retained the encapsulation from one generation to the next.

Although similar transformation events were conducted in the 1930s by the British physician F. Griffith, Avery and his associates theorized that heritable changes could be instilled within an organism by material transmitted through the cultivation medium. Prior to the recognition of bacterial transformation, the only recognized system for genetic transmission over generations involved the complex mechanisms of mitosis and meiosis followed by fusion of haploid germ cells.

Meticulous research by Avery and his colleagues further concluded that the agent responsible for bacterial transformation at the genetic level was *deoxyribonucleic acid* (DNA). The great German chemist Miescher had been credited with the discovery of DNA in 1868, only about 3 years after Mendel published his work, but the significance of DNA was not fully considered until the research of Avery and others.

The prosaic yet ubiquitous nature of DNA in all organisms did not garner respect as the key agent responsible for the universal molecular basis of genetics until 1953. At this time, Watson and Crick defined the characteristic structural features of DNA using X-ray diffraction and offered an explanation of how the polymer could direct heritable traits. Moreover, the work of Beadle and Tatum, which determined that a single gene is responsible for producing a single specific protein (the *one gene, one protein concept*), gave the work of Griffith, Avery, Watson, and Crick added importance.

Research from the 1950s to the present time has clearly revealed that *nucleic acids* such as DNA and various forms of *ribonucleic acid* (RNA) are critical in the overall control of amino acid sequences displayed by polypeptides. Nowhere is DNA control over amino acid sequence more important, however, than in enzyme syntheses. Amino acid sequence is quintessential to the special catalytic activities of these proteins, which direct all biochemical reactions at *in vivo* cellular conditions.

The amino acid sequence of any enzymatic protein is directed by some discrete *nucleotide* sequence buried in the total *polynucleotide* complexity of the nucleic acids.

AN OVERVIEW OF MOLECULAR GENETIC OPERATIONS

Nucleotides

The structural components of nucleic acids are nucleotides. Nucleotides are constructed from purine *or* pyrimidine bases, the pentose sugars ribose *or* deoxyribose, and phosphoric acid. *Pyrimidine bases* display a six-membered ring structure as shown in Figure 23.2. The three major pyrimidines including cytosine, thymine, and uracil contain oxygenated pyrimidine rings, whereas cytosine displays an NH_2 group at position 4. *Purine bases* are structural variations of a pyrimidine ring fused to an imidazole ring (Figure 23.2). The two major purines are adenine and guanine. Guanine notably contains a keto group at the 6 position whereas adenine contains an NH_2 group.

The oxygenated purines and pyrimidines can assume two different tautomeric forms that differ only in the respective location of a proton (Figure 23.3). At neutral pH for example, the lactam (*keto*) ($-NHC=O$) tautomeric form of a pyrimidine is more abundant than the so-called lactim (*enol*) ($-N=COH$) tautomer at the 2 position in the ring.

When purine or pyrimidine bases are bound to the C-1′ position of a pentose sugar, a *nucleoside* is formed. Nucleosides that incorporate adenine, guanine, cytosine, thymine, and uracil are called adenosine, guanosine, cytidine, thymidine, and uridine, respectively. Substitution of deoxyribose for ribose (Figure 23.4) is indicated in nucleoside nomenclature by the prefix *deoxy*, as in the case of *deoxy*thymidine. Note,

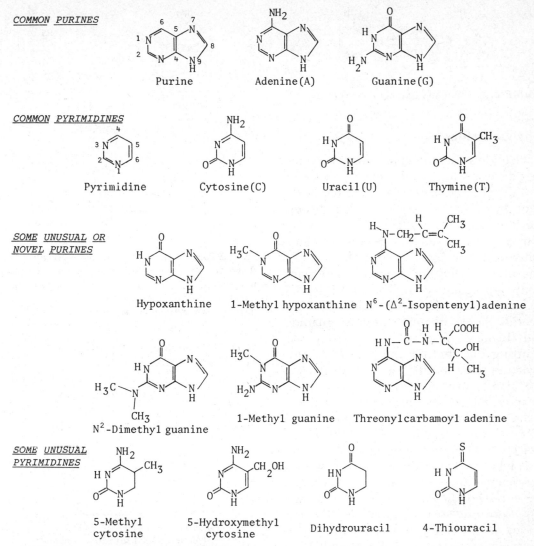

FIGURE 23.2. Typical purine and pyrimidine bases.

too, that nucleoside nomenclature requires the use of a prime number to indicate atom locations in the pentose (Figure 23.5).

Phosphorylation of the C-5′ position on a nucleoside yields a "nucleoside phosphate" or a *nucleotide*. The 5′-nucleoside monophosphate (NMP) of adenosine is known as adenosine monophosphate (AMP) or adenylic acid; guanosine monophosphate (GMP) is guanylic acid; cytidylic acid is CMP; thymidylic acid is TMP; and uridylic acid is UMP (Figure 23.6).

Structures for various NMPs can be further phosphorylated to give 5′-nucleoside diphosphates (NDPs) or triphosphates (NTPs) all of which can have varied bioenergetic and molecular genetic functions (Figure 23.7).

Some mononucleotides demonstrate a 3′,5′ linkage over the ribose ring through a phosphate. The structure of cyclic 3′,5′- AMP (cAMP) exhibits this type of structure; cAMP serves a unique "second messenger" function in response to many peptide hormone and catecholamine actions.

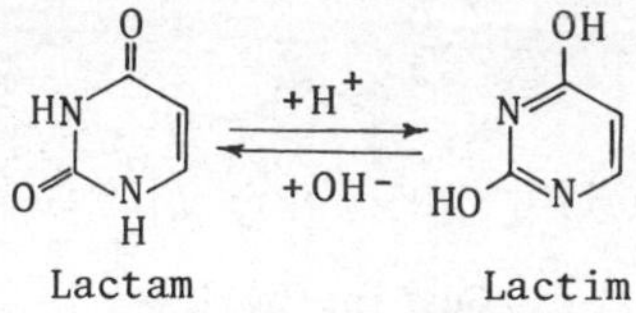

FIGURE 23.3. Tautomerization of a pyrimidine.

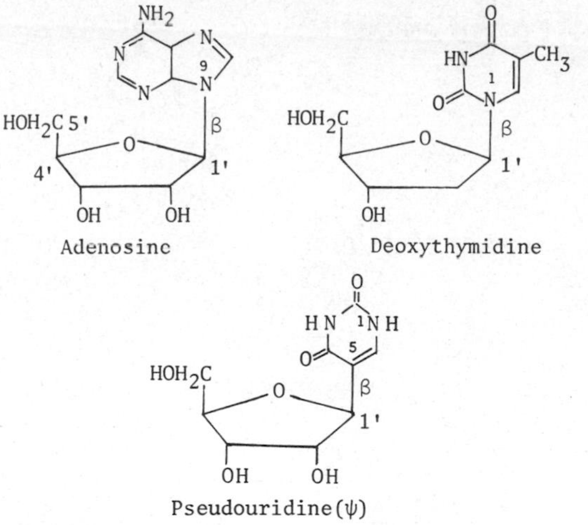

Polynucleotides

Nucleotide structures can become polymerized into very long chains by the formation of 3′,5′-phosphodiester bonds. That is, the phosphoric acid moiety of a nucleotide links the 5′-carbon of one pentose to the 3′-hydroxyl group of another pentose. As illustrated in Figure 23.8, "P" is indicative of a phosphoric acid link between shorthand "letter" designations A, C, G, and T (or U). These letters indicate the sequence of nucleotide bases: adenine, cytosine, guanine, and thymine (or uracil), respectively. Depending on whether the polynucleotide structure to be specified in DNA or RNA as shown in Figure 23.9, thymine or uracil may be specified. As seen in Figure 23.8, the 5′ end of a polynucleotide is written to the left of the 3′ end; pA designates a 5′-phosphate, and Up indicates a 3′-phosphate.

Polymerized nucleotides containing *ribose* characterize the structure of RNA whereas the corresponding polynucleotide containing *deoxyribose* is recognized as DNA. A survey of the base components and supermolecular structures of RNA and DNA show distinct differences. In DNA, thymine, is always present and uracil is absent. In RNA, uracil occurs in place of thymine. Moreover, the structure of RNA is a single-stranded polynucleotide, but DNA exists as a double-stranded helical polynucleotide

Adenosine   Deoxythymidine

Pseudouridine (ψ)

FIGURE 23.5. Some typical nucleoside structures. Note that the carbon atoms of sugars are designated as prime numbers (e.g., 1′, 2′, 3′, 4′, and 5′) as opposed to sequential numbered positions (e.g., 1, 2, 3, 4, 5, . . . etc.) in purines or pyrimidine structures.

with two polydeoxyribonucleotide strands entwined (Figure 23.10).

Studies of DNA using X-ray diffraction show that it exhibits a 20-Å diameter, 3.4-Å spacing between each of the 10 nucleotide residues in one turn of the molecule; plus

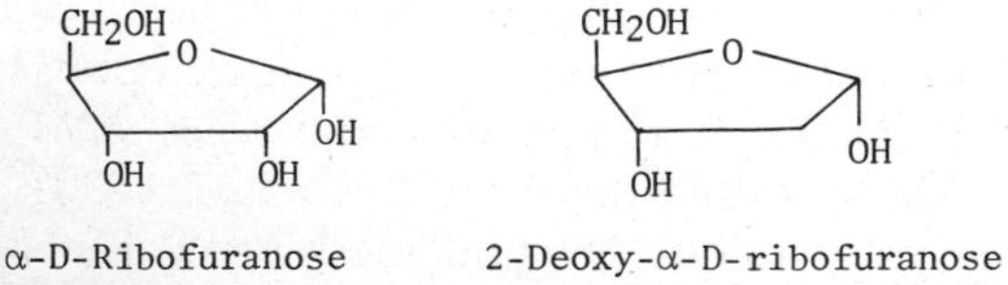

(A) Ribonucleoside 5′-monophosphate

(B) Deoxyribonucleoside 5′-monophosphate

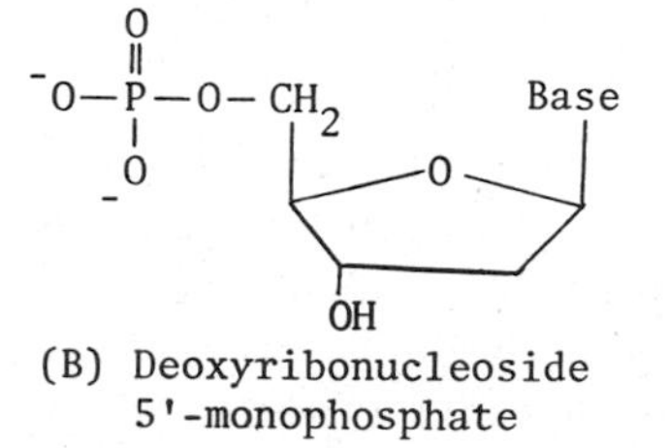

FIGURE 23.6. Generalized structures for a 5′-nucleoside monophosphate (NMP) containing *ribose* and another containing *deoxyribose*. An NMP containing ribose is referred to as a *ribonucleotide*, whereas an NMP containing deoxyribose is denoted as a *deoxyribonucleotide*. Any given nucleotide structure can undergo additional phosphorylation at the 5′ position on the sugar as shown in Figure 23.7.

α-D-Ribofuranose   2-Deoxy-α-D-ribofuranose

FIGURE 23.4. Structures for the two ribofuranose structures commonly found in nucleic acids.

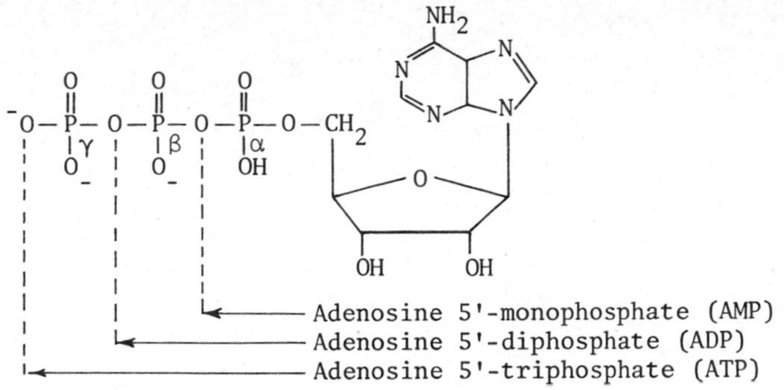

FIGURE 23.7. Mono-, di-, and triphosphorylated structure for the nucleoside known as adenosine, which forms the structure of a common nucleotide structure.

one major structural groove of 22 Å and a minor groove of 12 Å (Figure 23.10). This idealized structure of DNA often demonstrates additional supercoiling and macromolecular folding, which appears to be important during certain stages of molecular genetic mechanisms. In eukaryotic cells where genetic material is held within a defined nuclear membrane. DNA exists in close association with histone proteins, which contribute to the formation of chromatin fibers.

Analytical surveys of DNA also reveal that the numbers of certain nitrogen base pairs are *equivalent* within a double-helical DNA structure. Equivalent nucleotide base pairs are A = T, G = C, and A + G = T + C (purine and pyrimidine contents are equal). As seen in Figure 23.11, these pair equivalencies mirror the complementary hydrogen-bonding between base pairs of cytosine and guanine (C≡G) and thymine and adenine (T=A) within DNA. This predictable bonding of key base pairs accounts for the maintenance of the double-helical DNA structure. With reference to Figure 23.10, this equivalency of base pairing also contributes to the so-called *antiparallel structure* of DNA wherein each end of the helix contains the 5′ end of one DNA strand as well as the 3′ end of the other. Thus, the polynucleotide strands of DNA progress in distinctly *opposite* directions (antiparallelism).

The double-helical structure of DNA can undergo *denaturation* whereby the double-stranded DNA helix reverts back to a single-stranded polynucleotide. Denaturation is promoted by pH conditions and high temperatures that are generally uncharacteristic of *in vivo* cellular conditions.

The Central Dogma of Molecular Genetics

Nucleic acids serve as the principal biochemical agents involved in molecular genetic

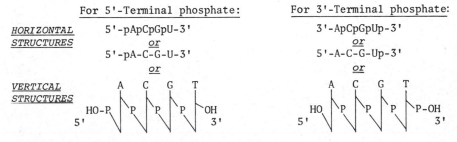

FIGURE 23.8. Shorthand notations for polynucleotide structures. Note that U is present only in RNA, whereas T occurs in DNA.

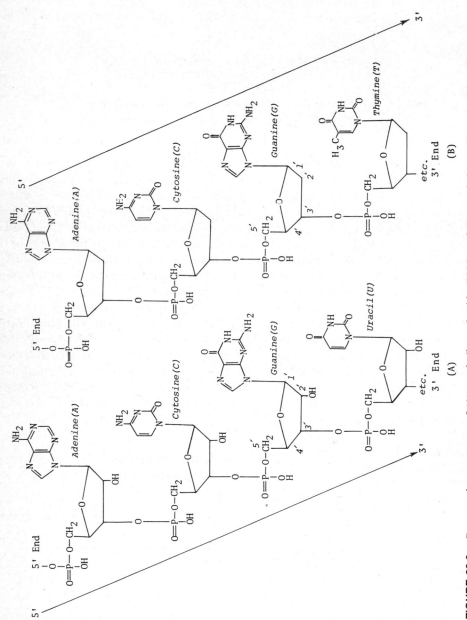

FIGURE 23.9. Comparative structures of (A) a polyribonucleotide (RNA) and (B) a polydeoxyribonucleotide (DNA). Note that the brief segments illustrated have a nucleoside skeleton linked by phosphodiester bonds. The nitrogen bases have variable sequences depending on the genetic information stored. The nitrogen bases of A, C, G, and U prevail in RNA, whereas T replaces U in the case of DNA.

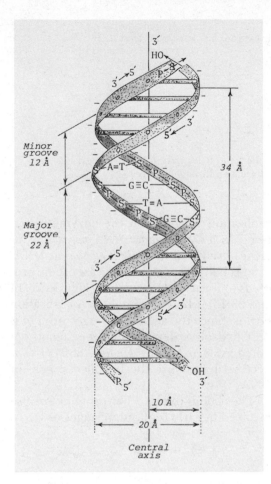

FIGURE 23.10. Double helical DNA (B-DNA) demonstrating hydrogen bonding between A═T and G≡C, which stabilizes the structure (P represents a phosphodiester intervening between deoxyribosyl residues indicated as S). Note that base pairs are stacked parallel to each other and perpendicular to the central axis. Distribution of negatively charged phosphate ($-PO_2^{2-}$) residues along DNA helices impart an overall polyanionic property to the molecule.

mechanisms. Detailed studies with viruses and bacteria have been responsible for modeling the known molecular genetic roles of nucleic acids. In turn, these models have served as useful tools for elucidating current concepts that explain the molecular genetic operations of eukaryotic cells. Many of the basic premises gleaned from viral and bacterial molecular genetic research are quite valid in the theater of plant and human genetic

mechanisms, but many other concepts do not apply.

The central dogma of modern genetics and biology dictates that DNA can reproduce itself by a process of *replication*, and DNA can also direct the protein synthesis of enzymes that regulate all metabolic reactions in cells. The latter process is comprised of two operational phases designated as (1) *transcription* and (2) *translation*. More recent evidence also suggests that viral agents may affect the central dogma concept by altering native DNA structures in a process described as *reverse transcription*. This process refers to the introduction of nucleic acids into a native nucleic acid structure of DNA, contained in a prokaryotic or eukaryotic cell, by a vector such as a virus. Figure 23.12 shows the relationships that exist between the processes outlined above.

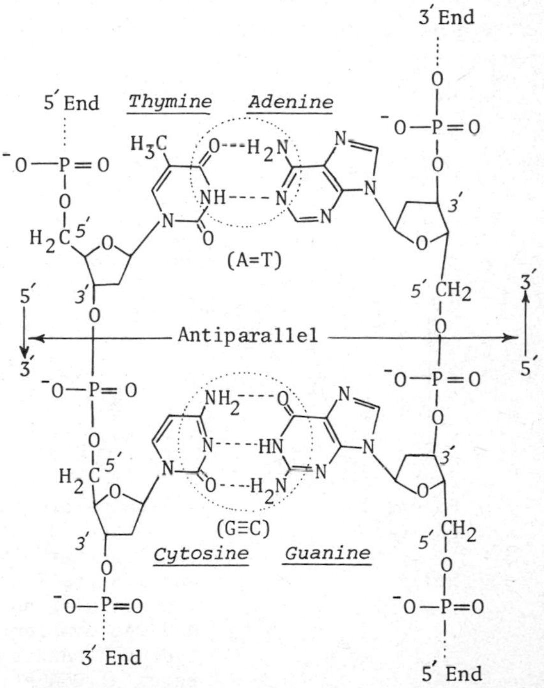

FIGURE 23.11. Hydrogen bonding interactions between complementary bases on two different antiparallel strands of DNA. Note that the shorthand designation for base pairing is expressed as A═T and G≡C to respectively indicate the presence of two or three hydrogen bonds.

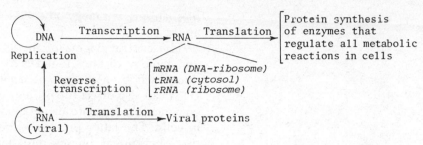

FIGURE 23.12. Schematic representation for the "central dogma of modern biology," which shows the directional flow of genetic information throughout a cell.

DNA Replication

The double-helical structure of DNA replicates itself in a *semiconservative* fashion. Each single strand of an original double strand of DNA (parental DNA) serves as a basis for the assembly of one newly synthesized strand of a DNA polynucleotide (Figure 23.13). Since histones notably shroud the DNA of animal cells, these basic proteins must be retracted or removed prior to the actual replication of parental DNA.

Similarities in bacterial and human DNA replication suggest that the enzyme recognized as a *DNA-directed RNA polymerase* attaches itself to a key *initiating site* on DNA. A so-called *rep* protein (helicase) then proceeds to form a *replication fork* (Figures 23.14 and 23.15), where parental DNA replication begins. Since the action of the *rep* protein is ATP hydrolysis dependent, other agents denoted as *single-stranded binding* (SSB) *proteins* necessarily stabilize the uncoiled region of DNA.

The actions of two other proteins are also involved in early replication stages of DNA.

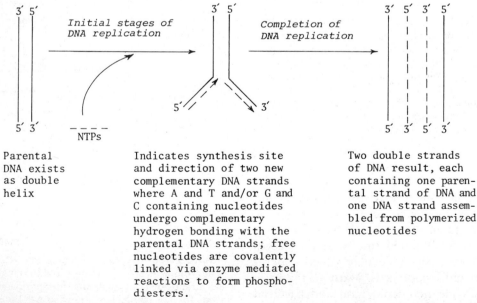

FIGURE 23.13. Schematic diagram indicating the semiconservative replication of DNA (NTP-nucleotide triphosphates).

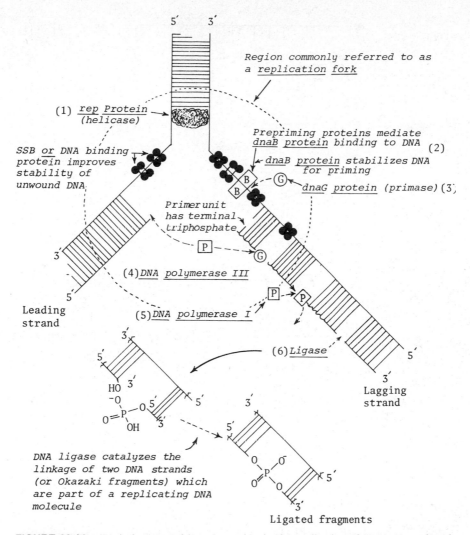

FIGURE 23.14.   Typical steps and important sites in the replication of DNA. Note that the Y-shaped structure where replication is initiated constitutes the so-called *replication fork*. (1) A *rep* protein (helicase) attacks double-stranded DNA at a specific site to form a replication fork. (2) Preprimary proteins complex *dnaB* protein to DNA strand; the *dnaB* protein stabilizes the replication fork. (3) The binding of preprimary proteins permits the formation of RNA primer units through the action of "primase." This special RNA polymerase promotes the linkage of nucleoside triphosphates (NTP's). (4) Deoxynucleoside triphosphates (dNTPs) are added at the 3'-OH end of an RNA primer as directed by DNA polymerase III forming Okazaki fragments provided that strand synthesis is discontinuous. For each Okazaki fragment, an RNA primer is involved. (5) RNA primer is removed and replaced with dNTPs. DNA polymerase I activity is involved in these steps and ultimately leads to the formation of new DNA fragments. (6) Okazaki fragments are joined to form continuous DNA strands by DNA ligase whose activity is promoted by NAD^+ (e.g., bacteria) or ATP (e.g., animal cells, bacteriophages).

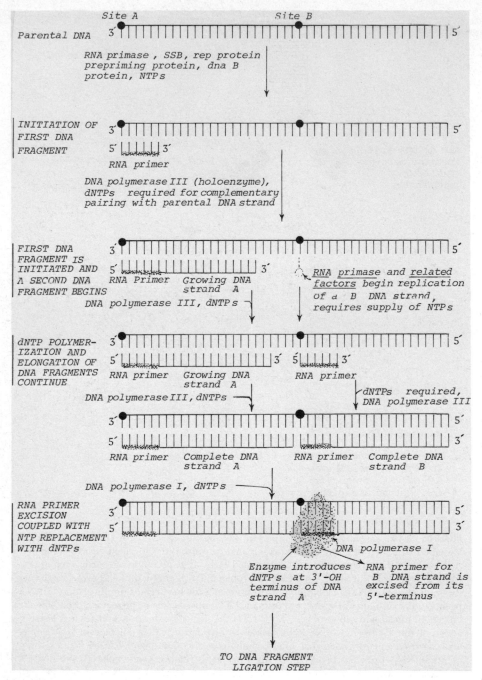

FIGURE 23.15. Overall replication scheme for a single parental strand of DNA, which details the participation and excision of RNA primers during the process as well as the ligation of new DNA (Okazaki) fragments.

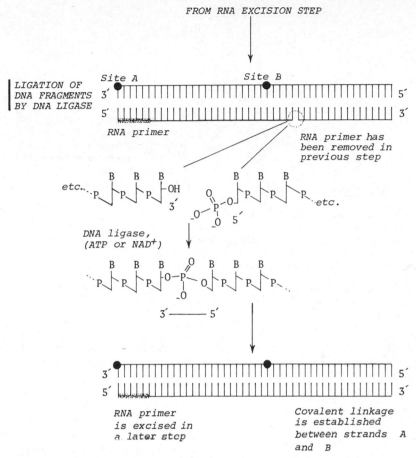

FIGURE 23.15. (*Continued*)

One protein called *dnaB* further stabilizes the open replicating structure of DNA, whereas a second group of *preprimary proteins* interact with DNA to promote the formation of an RNA primer (Figure 23.14).

Stabilization of the unwound parental DNA is followed by the actions of DNA-directed RNA polymerase, which assembles a *short RNA priming chain* on each of the two parental DNA strands being replicated. These RNA chains display complementary nucleotide base sequences with respect to each of the parental DNA strands. RNA polymerase assembles these priming RNA chains by the cleavage of pyrophosphate from ribonucleoside triphosphates (NTPs) as it sequentially adds ribonucleoside monophosphates (NMPs) to the growing RNA chain in a $5' \rightarrow 3'$ direction. Pyrophosphate hydrolysis then en-

sures that the overall polymerization will go to the right in the equation:

$$NTP + RNA \text{ polynucleotide}$$

$$\downarrow$$

$$NMP\text{-}RNA \text{ polynucleotide} + PP_i$$

$$PP_i + H_2O \longrightarrow 2 P_i$$

Establishment of an RNA primer, complementarily bound to a parental DNA strand, eventually reaches some critical length where the 3'-OH end of RNA allows polymerization of deoxyribonucleoside triphosphates (dNTPs) (Figure 23.16). Polymerization of dNTPs relies principally on DNA polymerase III, which assembles new DNA strands.

During semiconservative replication, the parental DNA molecule undergoes DNA polymerase-mediated replication in a $5' \rightarrow 3'$

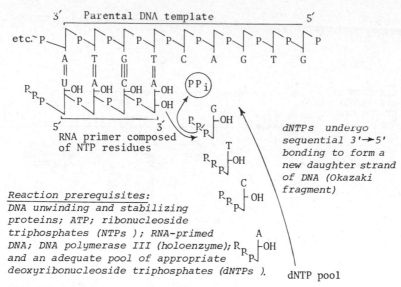

FIGURE 23.16. Covalent bonding of RNA primer to initial nucleotide residue in a growing DNA strand (Okazaki fragment).

direction along each strand to produce new DNA strands. As illustrated in Figure 23.14, replicated DNA in the region of the replication fork can be identified as *leading* and *lagging strands*. The leading strand is formed by the *continuous* polymerization of NMPs in a $5' \rightarrow 3'$ direction, whereas the lagging strand is produced in a piecemeal $5' \rightarrow 3'$ direction. This piecemeal synthesis of short replicated DNA strands is described as a *discontinuous process* and each small strand of new DNA is called an *Okazaki fragment*.

Individual Okazaki fragments are elongated along the lagging strand until the new DNA fragments reach any preexisting RNA primer strands ligated to their own respective DNA (Okazaki) fragments. Linkage of the Okazaki fragments into one continuous DNA strand along the lagging strand requires the excision of the RNA primer by *endonucleases*. This action is then followed by DNA polymerase I-mediated installation of dNTPs into the gap resulting from the excised polyribonucleotide primers. *DNA ligase* then proceeds to link the independently formed Okazaki fragments into a continuous DNA strand. Figure 23.15 details a schematic outline of some steps in DNA replication along a single strand of parental DNA.

Protein Synthesis Requires DNA Transcription and Translation

The expression of polypeptide structures based on the genetic information held in DNA base sequences requires both *transcriptional* and *translational* processes. *Transcription* involves the assembly of a *new* RNA chain, which has a complementary base sequence to those bases occurring in one strand of DNA. Transcription processes are critically dependent upon the action of a DNA-dependent RNA polymerase, which polymerizes the *new* RNA. Since this RNA base sequence will eventually be *translated* into an amino acid sequence at the ribosome, the RNA is called messenger RNA (mRNA). Thus, *translation* strictly pertains to the mechanism whereby genetic information in an mRNA molecule directs the polymerization sequence of specific amino acids during protein synthesis.

Transcription

The early stages of transcription involve the complete synthesis of an RNA molecule from specific nucleotide base sequences that represent a gene(s) contained in DNA. RNA synthesis here depends on the action of a DNA-directed RNA polymerase that not only seems to help open the DNA double helix but covalently links NTPs in a $5' \rightarrow 3'$ direction. This action produces RNA that is complementary to only *one* DNA strand at a time. Research has shown that RNA can arise from either one of the single strands held in a DNA structure, but only one chain at a time is faithfully selected by unknown mechanisms for the transcription of RNA.

RNA polymerase is composed of a multicomponent enzyme recognized as a *holoenzyme*. The holoenzyme contains protein molecule subunits indicated as α_2, β, β', and σ. The $\alpha_2\beta\beta'$ complex is called the *core* enzyme, whereas the $\alpha_2\beta\beta'\sigma$ structure is the *holoenzyme* (Figure 23.17). Note too in Figure 23.17 that the *sigma* factor (σ) dissociates from the holoenzyme after RNA synthesis begins.

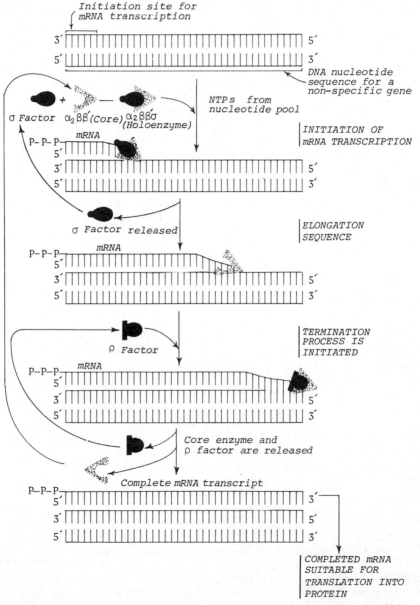

FIGURE 23.17. A DNA-directed RNA polymerase is required for the transcription of DNA into a strand of mRNA. Note that the 5'-triphosphate is a typical feature of the mRNA transcript.

The core enzyme then assumes responsibility for the continuation of RNA chain elongation along the DNA template. Upon completion of the RNA chain, a so-called *rho* factor (ρ) is usually necessary to terminate the final stages of RNA transcription.

A survey of RNA assembly along a DNA strand shows that the 5′ terminus of the RNA retains its triphosphate moiety. This group is retained through the time that the last NTP is polymerized onto the last position of the RNA structure as signaled by some key base sequence along the DNA strand. Once the last NTP is added to RNA, the favored stability of the double-helical DNA encourages the dissociation of the hydrogen bonded RNA.

Post-transcriptional Modification of RNA

Dissociated RNA polymers must be structurally transformed in order to ensure their

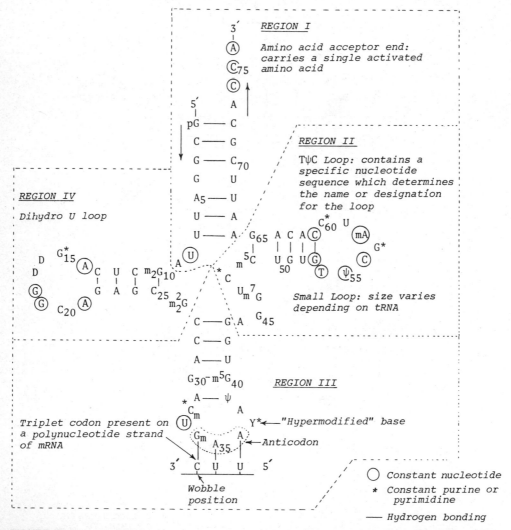

FIGURE 23.18. Cloverleaf structure of yeast phenylalanine tRNA. The circled ribonucleotide residues are constant in many tRNAs. Key abbreviations: A, adenosine; T, thymidine; G, guanosine; C, cytidine; U, uridine; D, dihydrouridine; Ψ, pseudo-uridine; Y, a purine nucleoside (hypermodified base); m, methyl; m₂, dimethyl. Solid lines indicate the presence of hydrogen bonding interactions.

respective activities as *messenger RNA* (mRNA), *ribosomal RNA* (rRNA), or *transfer RNA* (tRNA).

As an example, consider mRNA. The transcription of mRNA is followed by the addition of 185 + AMP residues according to some accounts. This so-called *poly (A) chain*, linked to the 3′ end of mRNA, probably promotes mRNA transit from the nucleus, where it is produced, to the cytoplasm. Here the poly(A) chain is removed.

The 100 to 2900 nucleotide residues in RNA that are destined to become rRNA also undergo post-transcriptional modification. The most notable modification involves rRNA binding to cytoplasmic proteins to yield ribosomes where proteins are eventually synthesized. The final ribosomal structure displays ~65% rRNA and ~35% protein. Based on sedimentation statistics, prokaryotic ribosomes contain 50 S and 30 S ribosomal subunits having a total 70 S value, whereas eukaryotes have 60 S and 40 S subunits with a total 80 S.

Crude transcribed tRNA must also be modified and thereby activated before it can actively participate in the translation of mo-

lecular genetic information. Activation typically entails the elimination of several nucleotides at opposite ends of the tRNA followed by addition of a trinucleotide (CCA) to the 3′ end. Other constituent nucleotides are further modified by sulfation, reduction, and methylation to produce inosinic acid, pseudouridylic acid, and methylguanylic acid. tRNAs display a classic linear cloverleaf structure (Figure 23.18), but their *in vivo* existence resembles an L-shaped conformation based on X-ray diffraction analysis (Figure 23.19).

mRNA, Anticodons, and Codons

The sequence of nucleotide bases held in the DNA polymer codes for the synthesis of many polypeptide and protein structures. When a given nucleotide base sequence codes for a single polypeptide chain, this nucleotide sequence is called a *gene*. The nucleotide sequence can be subdivided into successive units each containing three nitrogenous bases recognized as a *triplet*. *Each triplet base sequence in DNA involving A, G, C, or T bases is complementarily transcribed into mRNA us-

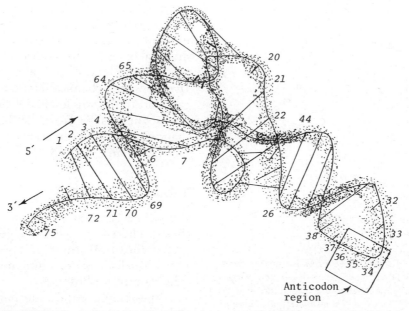

FIGURE 23.19. L-Shaped conformational structure for tRNA is maintained by the hydrogen bonding interactions of its constituent nitrogen bases.

ing A, G, C, or U. (**Note:** A in DNA undergoes complementary hydrogen bonding with U in RNA.)

The successive units of three bases in mRNA, reflecting the sequences of triplets appearing in DNA, are termed *codons*. Each codon in mRNA directs the addition of one amino acid residue during protein synthesis or *translation*. Since there are four possible bases that can undergo numerous combinations of three at a time, it is clear that 64 (or 4^3) trinucleotide sequences are possible for any single codon in mRNA. The structural concept of a triplet is illustrated in Figure 23.20, and the variety of codons that specify certain amino acids appear in Table 23.1. Although at least 61 codons direct the positioning of the 20 possible amino acids in a polypeptide structure, three codons (UAA, UAG, and UGA) serve as "flags" to terminate polypeptide chain elongation.

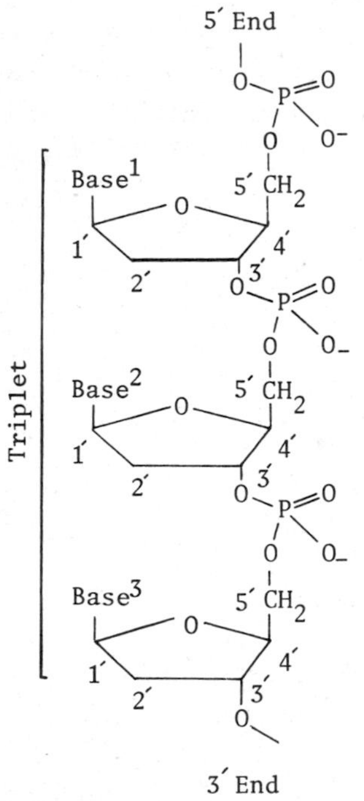

Since 61 codons direct the positioning of only 20 amino acids, each amino acid has several codons. This characteristic of the genetic code reflects the operation of a *degenerate coding system*. Evidence suggests that the third base of a codon usually accounts for the degeneracy, but the other two bases can also display isolated variations while coding for a specific amino acid. The codons AUA, AUC, and AUU all code for isoleucine and vary only in the third base, whereas a classic degeneracy in all three bases such as AGC, UCA, AGU, UCC, UCU, and UCG steadfastly code for serine. Other codons such as AUG strictly code for methionine as well as initiate protein translation in higher eukaryotic cells. Still other codon sequences direct the termination of polypeptide assembly.

If one or more triplet base sequences in DNA undergo a nonreplicative or random change, such errors in native DNA can induce distinct phenotypic changes. So-called *point mutations*, wherein any one base is somehow affected, are most common compared to double and/or triple base changes in a codon. Nonetheless, point mutations can cause the appearance of unusual termination codons and the premature cessation of polypeptide synthesis. Apart from the aberrant insertion of an additional nitrogen base into the triplet base sequence of DNA, individual bases can be modified by alkylating, methylating, and deaminating compounds. In other cases, chemical agents and radiation can lead to the wholesale deletion of triplet base sequences. Figure 23.21 shows the schematic changes that can occur in complementary base pairs of DNA.

FIGURE 23.20. Units of three successive nitrogen bases contained in the DNA structure are designated as *triplet* base sequences. These *triplets* are also referred to as *codons* since they code for the ultimate positioning of individual amino acids in the primary amino acid sequence of proteins.

Translation of mRNA

Ribosomes serve as the site of protein synthesis. These organelles are generally situated in the cytoplasm of prokaryotic and eukaryotic cells; however, humans also exhibit mitochondrial ribosomes.

Although the definite mechanism responsible for polypeptide synthesis in human ribosomes remains to be firmly established,

Table 23.1. The Genetic Code

First position (5′ end)	Second position				Third position (3′ end)
	U(T)	C	A	G	
U (T)	Phe	Ser	Tyr	Cys	U (T)
	Phe	Ser	Tyr	Cys	C
	Leu	Ser	Term[a]	Term	A
	Leu	Ser	Term	Trp	G
C	Leu	Pro	His	Arg	U
	Leu	Pro	His	Arg	C
	Leu	Pro	Gln	Arg	A
	Leu	Pro	Gln	Arg	G
A	Ileu	Thr	Asn	Ser	U
	Ileu	Thr	Asn	Ser	C
	Ileu	Thr	Lys	Arg	A
	Met[b]	Thr	Lys	Arg	G
G	Val	Ala	Asp	Gly	U
	Val	Ala	Asp	Gly	C
	Val	Ala	Glu	Gly	A
	Val[b]	Ala	Glu	Gly	G

[a]Term = chain terminating.

[b]Initiation codons; the AUG codon for methionine is a very common initiating point for translating a genetic message, although GUG may also initiate translation processes. In these cases, GUG codes for methionine, not valine.

the performance of *Escherichia coli* ribosomes in polypeptide synthesis serves as a useful model of translation mechanisms.

In general, polypeptide synthesis proceeds according to the amino acid codons specified in mRNA. Five operational steps in this translation are recognized, including (1) amino acid activation, (2) initiation of polypeptide synthesis, (3) polypeptide elongation, (4) termination of polypeptide synthesis, and (5) post-translational modifica-

tion of the new polypeptide chain to give an active protein product.

STEP 1: Amino Acid Activation: Formation of Aminoacyl-tRNA:

Amino acids used for polypeptide syntheses must exist in an activated form before they can undergo polymerization into a polypeptide structure. Activation requires that an amino acid must be bound to a tRNA molecule at the expense of ATP and pyrophosphate hydrolysis. An *amino acid specific*

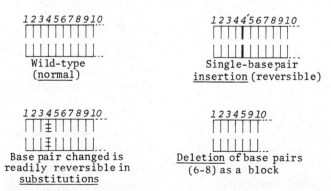

Wild-type (normal)

Single-base pair insertion (reversible)

Base pair changed is readily reversible in substitutions

Deletion of base pairs (6-8) as a block

FIGURE 23.21. Summary of changes in complementary base pairs that can lead to mutations and other hereditary molecular genetic anomalies.

aminoacyl-tRNA synthetase mediates condensation of an amino acid (carboxyl group) to the terminal 3'-OH group on an amino acid-specific tRNA.

STEP 2: Initiation of Polypeptide Synthesis:

Prokaryotic translation begins with the attachment of mRNA to the 30 S subunit of a dissociated ribosome. An *initiation factor 3* (IF3) associates with the 30 S subunit followed by another protein *initiation factor 1* (IF1). These actions are then succeeded by the binding of an initiating aminoacyl-tRNA to *initiation factor 2* (IF2). The initiating aminoacyl-tRNA is specifically known as *N*-formylmethionyl-tRNA (fMet-tRNA$_f$). In *E. coli*, the synthesis of almost every protein is begun with methionine. IF2, fMet-tRNA$_f$, and GTP then associate with the preexisting mRNA, 30 S subunit, and IF1-IF3 complex. This initial complex further unites with the dissociated 50 S ribosomal subunit to produce a functional 70 S ribosome. GTP hydrolysis to GDP and P_i drives the formation of the initiation complex, and IF1, IF2, and IF3 are eventually discharged from the ribosome. The fMet-tRNA$_f$ is formed according to Figure 23.22, and the fMet residue remains covalently bound to tRNA at the peptidyl site (P) on the ribosome. The tRNA ensures placement of the fMet residue at site P since the tRNA is bound by complementary hydrogen bonding to an initiation codon located on mRNA. Consult Figure 23.23 for details.

The fMet-tRNA$_f$ and every other specific aminoacyl-tRNA displays a "triplet" base sequence called an *anticodon*, which complements the mRNA codon sequence held in mRNA. Region III (Figure 23.18) of the tRNA structure holds the anticodon, whereas Region I offers a *specific amino acid acceptor end* responsible for conveying an activated amino acid to a growing polypeptide chain. Since 64 possible codons account for the linear program of 20 possible amino acids held in mRNA (Table 23.1), it follows that a comparable number of tRNA anticodon triplets must also exist.

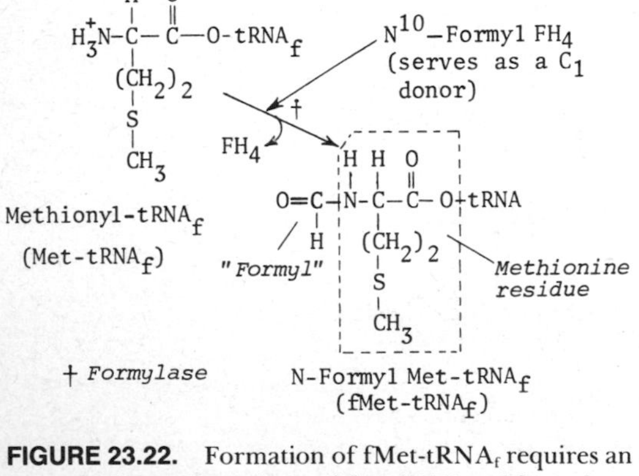

Numerous studies have revealed that mRNA can interact with ribosomes on a 1:1 basis, but it is not uncommon for a string of ribosomes to bind onto an mRNA strand and form a *polyribosome* or *polysome*. Each ribosomal subunit of a polysome individually produces its own polypeptide chain so that many polypeptides are being fabricated at any given time.

STEP 3: Polypeptide Elongation:

Following the occupation of site P by fMet-tRNA$_f$ (Figure 23.23), the aminoacyl binding site A remains open and available for the binding of a codon-specified aminoacyl-tRNA.

Peptidyltransferase then establishes a peptide bond between the fMet initiator and the specified aminoacyl-tRNA. The resultant peptide is then translocated from ribosomal site A to P with the concomitant ejection of tRNA$_f$ in addition to the ribosomal binding of the next codon specified aminoacyl-tRNA. The entire peptide translocation process is driven by GTP consumption in association with the operation of an *elongation factor* (EF) system. This overall process is repeated over and over for each peptide bond required to translate mRNA into a polypeptide.

STEP 4: Termination of Polypeptide Synthesis:

Polypeptide assembly from activated aminoacyl-tRNAs is terminated by one or more of the terminating codons present in mRNA. A number of proteins called *releasing factors* (RFs) are also required. RFs seem to interact with the terminating mRNA codon and thereby trigger bond hydrolysis between the complete polypeptide and tRNA at site P.

STEP 5: Post-translational Modification of Polypeptides:

Many polypeptides fail to exhibit proper biochemical activities unless they undergo structural modifications after their translation. For example, since the initial codon for many polypeptides may specify a formylmethionyl NH$_2$ terminus, re-

FIGURE 23.22. Formation of fMet-tRNA$_f$ requires an activated form of tetrahydrofolate (FH$_4$) and formylase.

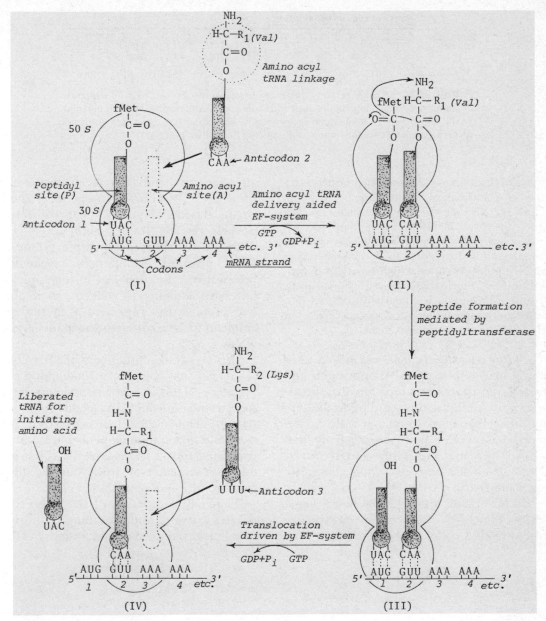

FIGURE 23.23. One cycle in the translation of mRNA into a primary polypeptide structure. For the example shown, formylmethionine serves as the initiating amino acid dictated by codon 1 and its corresponding anticodon on tRNA. Translocations of the growing peptide chains, tRNAs, and mRNAs through the ribosome require the action of an elongation factor system (EF system) driven by high-energy phosphates such as GTP. Specific steps illustrated in this figure include (I) fMet-tRNA$_f$ occupation of site P in the ribosome leaving site A open and available for binding the first amino acid in the new polypeptide chain (assuming that fMet will serve only as a preliminary residue in the amino acid sequence). (II) The first amino acid residue (Val) occupies site A as an aminoacyl-tRNA structure. (III) A peptide linkage is established to produce a formylmethionylvalyl structure at site A. (IV) Translocation of the peptide structure from sites A to P, ejection of tRNA$_f$, and ribosomal binding of the second amino acid (Lys) requires the EF system and consumes GTP. The process outlined is repeated for the entire length of the mRNA strand.

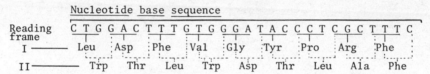

Nucleotide base sequence

Reading frame	C T G G A C T T T G T G G G A T A C C C T C G C T T T C
I ——	Leu : Asp : Phe : Val : Gly : Tyr : Pro : Arg : Phe
II ——	Trp : Thr : Leu : Trp : Asp : Thr : Leu : Ala : Phe

FIGURE 23.24. A single series of nucleotide bases can specify two different sequences of amino acids depending on the nucleotide base site where a reading frame is initiated. In the case shown, reading frame I may specify one functional protein while the other reading frame dictates the amino acid sequence in another protein.

moval of this amino acid residue may be necessary to ensure biochemical activity of the polypeptide. Thus, specific *deformylases* and an *amino peptidase* cleave the formylmethionyl group to produce active polypeptide. Furthermore, since hydroxyproline and hydroxylysine fail to have codons in mRNA, post-translational hydroxylations of proline and lysine yield these respective amino acids. Disulfide linkages are similarly introduced into translated polypeptides via enzymatic oxidation of —SH groups found in cysteine.

Coding sequences in genes and mRNA: For many prokaryotic genes, transcription and translation processes simply involve the assembly of amino acids according to the simple linear order of codons in mRNA. Research has revealed, however, that many viral, prokaryotic, and eukaryotic genes exist as *overlapping genes;* or in eukaryotic genes, *split genes* may exist.

Bacteriophage studies show that a single series of nucleotide bases can specify two different mRNA codon-directed sequences for amino acids depending on the nucleotide base site where a DNA reading/transcription frame begins. Based on Figure 23.24, two different possible reading frames could specify two distinctly different polypeptide syntheses. From practical evolutionary biochemical perspectives, the ability of a single nucleotide base sequence to code for more than one protein may represent a classical example of genetic conservatism at the level of DNA.

In cases of eukaryotic genes, the DNA base sequences that code for a final amino acid sequence (*exons*) in a polypeptide are interspaced with noncoding nucleotide sequences (*introns*) (Figure 23.25). It is apparent that exon and intron base sequences in DNA are translated into a *precursor* mRNA, but *only* the exon sequences of mRNA are finally ligated to give *mature* mRNA.

The coding structure of DNA also complicates current understanding of translational processes since certain sections of DNA

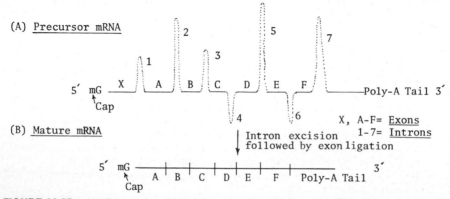

FIGURE 23.25. (A) Precursor mRNA conversion into (B) mature mRNA. Note that the 5'-methylated guanine and 3'-poly (A) tail protect the terminal ends of both mRNA forms.

can be inserted into other sections of DNA. Such translocatable DNA elements are called *transposons*.

Regulation of Gene Expression

Since the molecular genetic transcription and translation of DNA cannot be unleashed in an unremitting and chaotic fashion, the biochemical well-being of cells requires intricate biochemical regulation at the genetic level. Two distinct types of control mechanisms govern protein biosynthesis and enzymatic protein availability at the genetic level, namely, *induction* and *repression* mechanisms.

It has been shown that certain *structural genes* code for specific enzymes necessary to conduct biochemical processes, but these genes are under the control of an operator gene. Together, these operational DNA units constitute an *operon*.

The so-called *Lac* operon in *E. coli* has served as the classical concept for many conventional explanations of controlled gene expression. The *Lac* operon is responsible for producing key enzymes that transport lactose into the *E. coli* cell and then ultimately direct its metabolism.

The *Lac* operon represents a small region of the overall DNA molecule. Figure 23.26 shows that the *Lac* operon consists of three parts, which include the *regulatory gene* (I), an *operator gene* (O), and one or more *structural genes* (SG). The *promoter region* (P) of the operon is divided into two regions: the *CAP-binding site* and an *RNA polymerase binding site* (Figure 23.26). In the case of the *Lac* operon, at least three structural genes are involved, which respectively code for galactoside permease (z); β-galactosidase (y), which hydrolyzes lactose into galactose and glucose for use in energy production (Figure

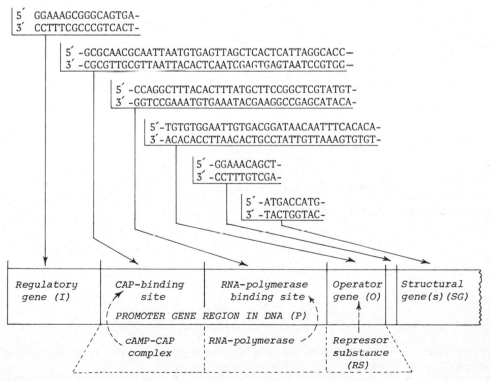

FIGURE 23.26. Conceptual diagram and schematic base sequence for the *Lac* operon regulatory region in *Escherichia coli*. Studies of this region suggest that it contains a bifunctional promoter gene region that binds with a cAMP–CAP complex and RNA polymerase.

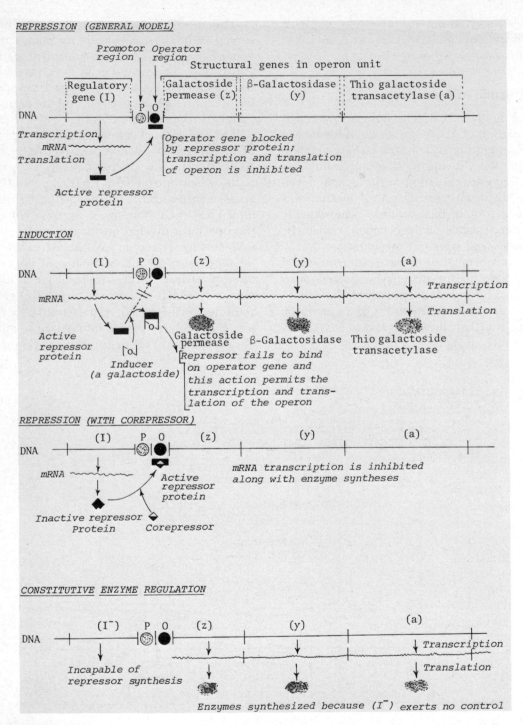

FIGURE 23.27. Gene transcriptional control mechanisms governing enzyme synthesis including repression, induction, and constitutive models. Consult text for details.

23.27); plus thio galactoside transacetylase (a), which has an unknown function.

The *regulatory gene* (I) codes for a *repressor protein substance* (RS) that binds to the operator gene. Based on Figure 23.26, RS probably binds to all the bases that intervene between the first and last two sets of triplet sequences.

Lac operon activity depends on the intracellular status of glucose, which is required for normal metabolism. The *Lac* operon is repressed when the cell is replete with glucose. That is, active RS synthesis directed by the regulatory gene blocks the operator gene thereby inhibiting structural gene expression(s). Glucose depletion in a cell coupled with the presence of lactose in the culture medium initiates structural gene expression. Gene expression is promoted since lactose serves as an *inducer*. As an inducer, lactose forms a *complex* with active repressor protein. The resulting complex cannot bind to the operator gene, and structural gene transcription is initiated. Figure 23.27 shows the fundamental relationships detailed above where the *Lac* operon undergoes *repression* or *induction*. Evidence suggests that active repressor protein forbids the binding of *DNA-directed RNA polymerase* and the resulting mRNA transcription of structural genes.

It should be recognized that the *Lac* operon performance is also contingent upon a *catabolic gene activator protein* (CAP). It has been observed that the *Lac* operon undergoes repression when glucose (a more efficient source of carbon than lactose) is present in bacterial culture media. Subsequent study has also revealed that glucose promotes decreased intracellular cAMP levels and that CAP is activated by cAMP, which augments the activity of RNA polymerase.

Gene expression can also reflect the actions of *corepressors* and modified regulatory gene functions that produce *constitutive mutants* from native genes. Corepressors are low-molecular-weight substances similar to inducers that interact with *inactive* repressor proteins to produce *active* repressor. The active repressor effectively binds to an operator gene thereby prohibiting RNA polymerase binding and *Lac* operon transcription.

In the event that the nucleotide base sequence at a regulatory gene undergoes some type of mutation or becomes aberrant compared to the native nucleotide sequence of the gene, nonfunctional repressor proteins may be produced. These defective proteins fail to bind to the operator gene. Thus, transcription and translation of structural genes goes on virtually unchecked by inducer–repressor interactions as illustrated in Figure 23.27.

The *Lac* operon concept remains useful as a model for the control of gene expression in many prokaryotes and eukaryotes, but many gene expression control systems are far more complicated. In view of gene overlapping, the existence of split genes, post-transcriptional modifications of mRNA, and post-translational alterations of protein structures, the controls governing mammalian gene expression remain largely uncharted.

THE ADVENT OF MODERN MOLECULAR GENETIC MANIPULATION

Since the 1970s, principles of DNA transcription and its translation into specific proteins have become so thoroughly understood that *in vitro* synthesis of DNA coding for proteins has been accomplished. Insertions of fabricated DNA into bacteria were also found to direct the synthesis of desired protein products.

The potential for directing the expression of certain genes, modifying the existing genetic makeup of organisms to produce useful biochemicals, and the potential for interspecies transfers to genes have always offered exciting possibilities for agriculture and health sciences. Unfortunately, pre-1970s genetic technology and the most sophisticated use of classical genetics did not permit the fruition of such noble goals. In fact, the tools for modern applied molecular genetics only began evolving in the 1960s. Recognition of viral transduction demonstrated that limited amounts of DNA (or genes) could also be transferred among bacterial hosts by viral agents known as bacteriophages.

Gene transfer was also recognized to occur from one bacterium to another by *conjugation processes*. Conjugation processes depend on the superficial contact of tiny pillar projections on one bacterium with a neighboring bacterium. DNA from the donor organism is transmitted to its neighbor by the aid of these projections in the conjugation process. The donor bacterial properties of a conjugating bacterium often represent genetically controlled traits that are largely controlled by *plasmids*. Plasmids are extrachromosomal features of the bacterial genome that exert their own unique forms of expression. Plasmids also undergo autonomous replication within the bacterial cell and exist as small molecules of double-stranded DNA.

Since the discovery of bacteriophages and plasmids, their potentials as vehicles are vectors for carrying genes among bacteria have been widely exploited. The use of one or a combination of these *vectors* have offered the first realistic tools for directing and manipulating the native genetic properties of cells.

GENE CLONING AND RECOMBINANT DNA

The transfer of a specific DNA fragment containing one or more genes into another functional DNA molecule describes the process of recombinant DNA formation. The transfer and incorporation of one DNA sequence into another may involve identical or different species, or viruses as outlined in later sections.

The genetic alteration of plants, animals, and microorganisms has been critical in the development of agricultural and fermentation industries for centuries. Classical methods of genetic hybridization, selection, and other Mendelian approaches for modifying the phenotypic expressions of species have rather limited potentials for producing rapid and desirable changes in any specific gene or genome.

Depending on the specific gene(s) transferred from one organism to another, recombinant DNA technology permits modification of a genome toward a directed goal. Genetic technologies based on recombinant DNA conceivably offer a direct approach for producing (1) new supplies and choices of food-producing species, (2) new sources of biochemicals, (3) improved genetic resistance to disease and environmental stresses in traditional species of agricultural plants and animals, (4) new products and increased productivity from fermentation technologies, and so on.

The application of recombinant DNA for any productive goal requires four crucial steps: *gene selection, gene transfer, gene recombination,* and *gene expression*.

Apart from the successful isolation and transfer of a desirable DNA sequence from one organism to another, tactical problems often surround the expression of the transferred DNA within its new recombinant milieu. Since each gene function of eukaryotic and prokaryotic cells varies widely from every other, the transcriptional control mechanisms governing gene expressions are nearly infinite and sometimes difficult to determine.

According to contemporary studies of recombinant DNA, *vectors* have been principally used for transferring DNA sequences (gene(s)) from one organism into the genome of another host.

SOME VECTORS USED FOR DNA RECOMBINATION

Plasmids are circular double-stranded DNA molecules that exist in most bacterial species as endosymbionts (Figure 23.28). That is, they

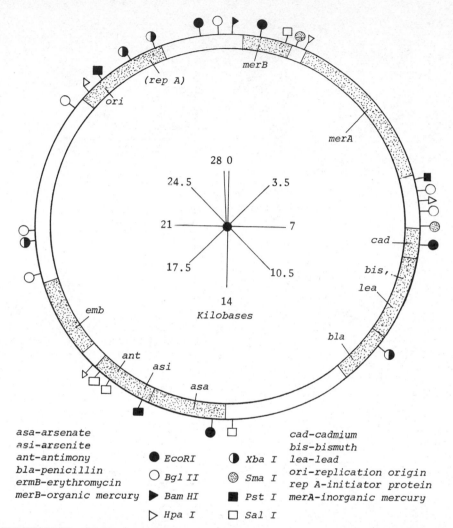

28 0
24.5   3.5
21   7
17.5   10.5
14
Kilobases

(rep A)
merB
ori
merA
cad
bis,
lea
bla
emb
ant
asi
asa

asa-arsenate
asi-arsenite
ant-antimony
bla-penicillin
ermB-erythromycin
merB-organic mercury

● EcoRI   ◐ Xba I
○ Bgl II   ◉ Sma I
▶ Bam HI   ■ Pst I
▷ Hpa I   □ Sal I

cad-cadmium
bis-bismuth
lea-lead
ori-replication origin
rep A-initiator protein
merA-inorganic mercury

FIGURE 23.28. Schematic genetic map for important recognized regions of the pI258 plasmid. Note that the apparent size of the plasmid genetic structure is indicated in kilobase dimensions. Symbols indicate the specific sites where restriction enzymes preferentially attack the structure, whereas the remaining denotations indicate specific plasmid genetic regions associated with penicillin resistance, arsenate, and so on. Structures for other plasmids are similarly reminiscent of this generalized structure, although the prescribed genetic features may be widely different.

exist as independent self-replicating entities within the host cell where they can account for up to 0.5 to 3% of the total genetic information.

Plasmids are also important to bacterial existence since they may impart specific adaptive genetic functions to the host cell. Phenotypic expression of these adaptive functions may appear as an unusual ability to use species-uncommon substances as growth nutrients, or the plasmid may confer unusual antibiotic or toxicant resistance to the host cell. The genetic information embodied in plasmids is probably transferred among bacteria in most natural situations through conjugation or mating events.

As in the case of viruses, plasmids manifest a borderline existence between a nonliving entity and a very primitive life form. Plasmids may occur as "small" or "large"

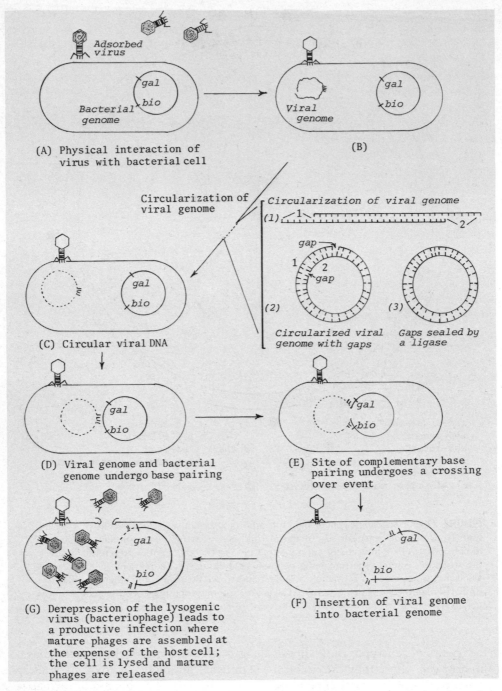

(A) Physical interaction of virus with bacterial cell

(B)

Circularization of viral genome

Circularization of viral genome

(1)

(2) Circularized viral genome with gaps

(3) Gaps sealed by a ligase

(C) Circular viral DNA

(D) Viral genome and bacterial genome undergo base pairing

(E) Site of complementary base pairing undergoes a crossing over event

(F) Insertion of viral genome into bacterial genome

(G) Derepression of the lysogenic virus (bacteriophage) leads to a productive infection where mature phages are assembled at the expense of the host cell; the cell is lysed and mature phages are released

FIGURE 23.29. Exemplary interactions of a bacteriophage with the genome of a host bacterium. Autonomous, dispensable genetic elements that are incapable of insertion into a bacterial host chromosome are called *plasmids*. This is contrary to the activity of bacteriophages, which actually insert their viral genome into the genome of a host bacterial cell. This is clearly exhibited in the case of certain DNA bacteriophages that display a *lysogenic cycle of infection* whereby intrabacterial penetration of the viral genome is followed by its incorporation within the bacterial genome. Here, it genetically reproduces in synchrony with genomic reproduction of the host, as long as total expression of the viral genes is *repressed*. In the event that an occasional cell undergoes *derepression* because of environmental or bio-forms. Small plasmids are capable of autonomous replication within a host cell, and although 10 to 40 plasmids normally occur within a host cell, these plasmids may demonstrate intracellular amplification by a factor of 100 or more. The so-called large plasmids also undergo autonomous replicative processes, but their native intracellular occurrence is substantially less than the "small" plasmids. Lower copy numbers in the range of 1 to 5 per cell are common, since higher amplification numbers approaching the thousands cannot be achieved. Moreover, the conjugation process between bacteria may result in the transfer of "large" plasmids. The ability to participate in this activity is often denoted as *Tra +*.

Unlike plasmids, viruses and bacteriophages can exist outside their host as inert protein-encapsulated particles that contain a strand of DNA or RNA. The random interaction of the virus with a suitable host cell eventually leads to viral genome entry into the host cell where viral nucleic acids direct the host cell to produce viral proteins. The consequences of this activity ensure the amplification of more identical viral particles at the expense of host cell metabolism and eventually culminates in the demise of the host cell (Figure 23.29). Unlike viruses, the existence of plasmids as endosymbionts is not synonomous with the death of the host cell; nevertheless, plasmids do share the ability of viruses to survive regardless of the fate of the host cell.

Plasmid viabilities outside a host cell coupled with their ability to interject polynucleotide directives from one bacterial host to

---

chemical factors, the lysogenic existence of the bacteriophage gives way to its lytic cycle. Under these conditions the viral genome directs the effective mass production of mature virions (bacteriophages) that eventually lyse and spew forth from the host cell. These mature virions can then randomly attack and genetically invade other suitable host cells. This overall virus–host relationship is identified as *lysogeny,* and infected cells that have a *latent capacity* to produce mature virus particles *subject to depression factors* are specifically defined as *lysogenic.* A bacteriophage that can exist in a lysogenic state is called *temperate,* and as long as it exists within cells as a lysogenic form it is called a *prophage.* Based on extensive study of the lambda (λ) DNA phage that exhibits a lysogenic relationship with *Escherichia coli,* the sequence of phage invasion, prophage existence, and lytic activity can be depicted according to the scheme in which

(A) The phage is absorbed to the bacterial cell wall.

(B) Viral DNA is injected into the cell.

(C) The injected viral genome undergoes circularization: (1) the linear viral genome has single-stranded DNA at each end that can complementarily bind to the other; (2) complementary base pair bonding of genome extremeties occurs by hydrogen bonding; and (3) the individual DNA strands are united by a ligase.

(D) Homologous regions of the viral and bacterial genomes undergo mutual pairing. (Note that for λ phages the site of pairing is quite specific, namely, between loci for galactose (gal) and biotin (bio)).

(E) A genetic crossing over event occurs in the homologous pairing region.

(F) Both the bacterial and viral genomes are incorporated into a single circular form of DNA. Limited transcription and translation of the viral genome at this point then forms so-called *early proteins* plus one or more varieties of *repressor substances* that suppress full-blown expression and replication of the viral genome.

(G) If one or more key repressor substance activities are eliminated, deactivated, or fail to be produced at a rate necessary to ensure lysogeny (prophage existence), the bacterial cell will show a "productive" infection and lyse. Substantial numbers of mature phage particles are produced at the expense of host cell biochemical resources. The lytic destruction of the cell may be prompted by the large physical numbers of mature virions in the cell in addition to viral genome-directed production of lysozyme that attacks host cell-wall construction.

another make these entities ideal vectors for recombinant DNA and genetic engineering technologies.

RESTRICTION ENZYMES AID RECOMBINANT DNA FORMATION

In order for a plasmid to serve as an effective vector for placing a foreign strand of DNA into another cell, it is necessary to (1) cleave the circular plasmid DNA, (2) introduce the desired foreign DNA into the intervening space between the terminal ends of the cleaved plasmid DNA, and then (3) covalently link the foreign DNA to the plasmid DNA.

The steps outlined above are expedited through the use of highly refined *restriction enzymes* that discriminatively excise specific DNA fragments according to preordained nucelotide base sequences (Table 23.2). Similarly, specific DNA ligases (e.g., T_4 DNA ligase) are used for the covalent linkage of DNA fragments to other polynucleotide sequences within the host cell.

The most useful restriction enzymes are those that offer reproducible recognitive abilities for cleaving *palindromic nucleotide sequences* such as CTTAAG and GAATTC where symmetric but opposite base sequences exist. One typical restriction endonuclease known as *Eco*RI displays hexanucleotide recognition and causes the staggered cleavage of DNA only when the six complementary base pairs exist, as shown in Figure 23.30. Other endonucleases also exist such as *Hae*III, which shows tet-

Table 23.2. Selected Base Sequences Recognized by Some Restriction Nucleases That Discriminatively Cleave Polynucleotide Chains

Restriction enzyme	Enzyme source	Base sequence recognized[a]
*Eco*RI	*Escherichia coli* KY13	5' - (T or A)—G—A—A—T—T—C— (A or T) - 3' 3' - (A or T)—C—T—T—A—A—G— (T or A) - 5'
*Eco*RII	*Escherichia coli*	5' - G—C—C—A—G—G—C—3' 3' - C—G—G—T—C—C—G—5'
*Hind*III	*Haemophilus influenzae* Rd	5' - A—A—G—C—T—T—3' 3' - T—T—C—G—A—A—5'
*Alu*I	*Arthrobacter luteus*	5' - A—G—C—T—3' 3' - T—C—G—A—5'
*Hae*III	*Haemophilus aegyptius*	5' - G—G—C—C—3' 3' - C—C—G—G—5'

[a]Vertical dotted lines between two polynucleotide strands show the axis of symmetry. Arrows indicate staggered and even (flush) cleavage. Nucleotide sequences that read the same in both the forward and reverse directions are called palindromes (from Greek, *palindromos*, meaning running back again).

ranucleotide recognition and yields a clean cut within the DNA nucleotide sequence.

Restriction endonucleases that offer hexanucleotide recognition are widely favored for use in recombinant DNA techniques. Their favorability stems from the staggered cleavage of palindromic base sequences that promote the efficiency of later reannealing steps (i.e., complementary base pairing) between foreign DNA and DNA provided by the plasmid vector.

DNA can also be modified by exonucleases that promote cohesion of the nucleotide during the terminal steps of recombinant DNA formation. DNA destined for transfer to another polynucleotide may be treated with an exonuclease that cleaves the phosphodiester bonds at the 5′ ends. Complementary bases are then added to the 3′ ends using a terminal transferase that produces an effect similar to endonuclease actions described above. This enables single-

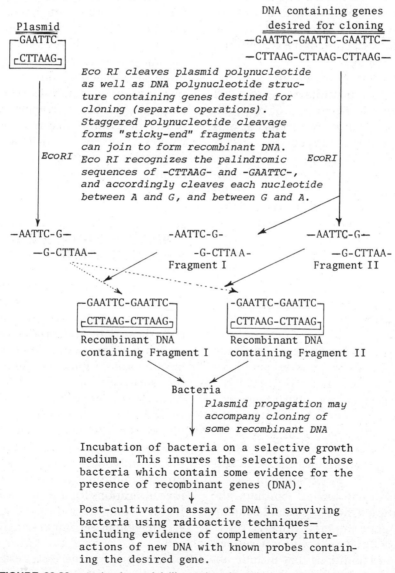

FIGURE 23.30. A simple model illustrating the principle of molecular cloning involving the use of restriction enzymes and plasmids.

stranded DNA fragments to readily "stick" together in the form of a combined molecule, assuming that a proper complementary base sequence is provided that permits reannealing.

Assuming that the polynucleotide sequence for one or more desirable genes can be sequenced and synthesized, it is possible in some cases to introduce the artificially produced DNA sequence into a plasmid. Clearly, this approach minimizes the role of endonucleases for preliminary excision of an *in vivo* source of DNA, but the use of restriction enzymes is still required for the cleavage and introduction of the synthesized DNA into a plasmid.

Restriction endonucleases are also favored since statistical studies have revealed that polynucleotide fragments of approximately 4000 base pairs are produced, plus the fact that endonuclease activity is largely limited to a single site in the most common plasmids. These statistical limitations prohibit highly random, multiple cleavage sites within polynucleotides and produce DNA fragments that are especially suited for genetic recombination.

PROPAGATION OF RECOMBINANT DNA PLASMIDS

Once a selected DNA fragment containing one or more specific genes is prepared, introduced, and ligated into a plasmid DNA strand, it is necessary to establish high copy numbers of the plasmid that contains the desired genetic material. Therefore, clones of recombinant DNA plasmid are produced by incubating the plasmid with selected bacterial cells grown in the presence of special nutrients and/or antibiotics. Consider the common example where a plasmid containing a desired gene imparts antibiotic resistance to the bacterial cells where it resides. Under nutrient conditions that include antibiotics, those bacteria without plasmid-directed antibiotic resistance will readily die, leaving only those bacteria with plasmid-

cloned resistance. In this example, antibiotic resistance conveyed to bacteria by plasmids is used as a *selective* mechanism to ensure the singular production of plasmids that incorporate recombinant DNA. As long as some type of selective pressure (e.g., antibiotics) is administered to cells containing a plasmid with the desired form of recombinant DNA, cloned versions of the plasmid will be directed toward genetic stability and uniformity.

It should be recognized at this point that the most earnest efforts for maintaining the genetic stability of the recombinant DNA plasmid can be defeated by one or more plasmid–host genetic transfers. This event is often encouraged by common homologies in polynucleotide sequences between plasmid and host genomes. Furthermore, the genetic purity of desired plasmids may be eliminated by transposon activity where nonhomologous interactions of foreign and host DNA occur.

EVIDENCE FOR GENETIC RECOMBINATION

Evidence for genetic recombination can be demonstrated in several ways. Clear evidence of functioning recombinant genes may be indicated by the production of specific proteins, enzymes, hormones, or some other biochemically significant compound tied to the successful transfer of one or more recombinant genes. In other cases, where recombinant genes have been successfully introduced into a target cell, their existence may be linked to other genes that impart overt antibiotic resistance or selective cultural or immunologic properties. Thus, positive indications for any one or more of these features may be indirect evidence of successful genetic recombination within a target cell.

Evidence for successful establishment of recombinant DNA in a target cell can also be revealed by DNA base sequence analysis. This requires that the base sequence of DNA for one or more genes must be known before its introduction into a target cell. Subsequent detection of this *recognized nucleotide base sequence* in target cells having undergone attempted genetic recombination affirms the success of the recombinant effort. The apparent success of genetic recombination, however, does not guarantee the *expression* of the foreign genes within the target cell.

Since the relationship between polypeptide synthesis and polynucleotide base sequences are fairly well established, several methods have been developed for the base sequence comparison of polynucleotide structures. Two of the most touted methods include the Maxam–Gilbert and Sanger methods.

Maxam–Gilbert Method for Polynucleotide Sequencing

A cleaved polynucleotide, such as double-stranded DNA, is treated with specific reagents to disrupt the hydrogen bonds between the two strands. The resulting polynucleotide strands are then separated and purified prior to any nucleotide base-sequencing effort.

According to the Maxam–Gilbert method, purified polynucleotide strands are subjected to four separate chemical treatments. These treatments are designed to selectively cleave the nucleotide sequence of a single- or double-stranded DNA whose 5′ or 3′* end is labeled with radioactive ^{32}P. The la-

beled polynucleotide fragments are electrophoretically separated on a polyacrylamide gel that can distinguish nucleotide fragments differing by only one nucleotide. Clearly then, only those fragments having nucleotide residues stretching from the ^{32}P-labeled end to the break are detected in the gel by their radioactivity. An autograph of the gel is prepared and the detectable autographic bands are identified as specific nucleotides by considering that the shortest polynucleotide fragments demonstrate the highest electrophoretic mobility (from all four reactions). Therefore, the 5′ end nucleotide residue of the polynucleotide fragment being studied begins at the bottom of the gel with each successively larger polynucleotide fragment appearing above the other.

The Maxam–Gilbert analysis for a theoretical DNA fragment can be completed according to the following procedure.

STEP 1:

The DNA polynucleotide is dephosphorylated at the 5′ end.

5′ P —C—G—A—C—T—A—C—T—G—A—3′

$$\text{alkaline phosphatase} \quad \begin{cases} \nearrow \text{H}_2\text{O} \\ \searrow P_i \end{cases}$$

5′ HO—C—G—A—C—T—A—C—T—G—A—3′

STEP 2:

The dephosphorylated DNA is labeled at the 5′ end with [^{32}P]:

5′ HO—C—G—A—C—T—A—C—T—G—A—3′

$$\text{adenosine—O—}\underset{\text{OH}}{\overset{\overset{O}{\|}}{P}}\text{—O—}\underset{O_{\delta-}}{\overset{\overset{O}{\|}}{P}}\text{—O—}\underset{O_{\delta-}}{\overset{\overset{O}{\|}}{P^{32}}}\text{—O}^{\delta-}$$

(ATP)

ADP

5′ $^{\delta-}$O—$\overset{\overset{O}{\|}}{\underset{O^{\delta-}}{P^{32}}}$—O—C—G—A—C—T—A—C—T—G—A—3′

*3′-^{32}P-end labeling requires the presence of ATP containing a terminal ^{32}P plus a terminal *transferase*.

STEP 3:

The *labeled DNA is divided into four distinct portions and each portion is subjected to a chemical treatment that selectively destroys one or two of the four bases: guanine, adenine, thymine, and cytosine.* This event cleaves the polynucleotide strand at the site of these bases to form a set of fragments, each displaying characteristically different sizes.

Evaluation of Fragmented Polynucleotide Strands

First portion: The chemical reagent known as dimethylsulfate characteristically methylates purines. The glycosidic bond of the methylated purine is unstable and is broken by applying heat at a neutral pH. This reaction leaves the sugar residue without its nitrogen base, thereby forming an *apurinic*

sugar. This event is followed by cleavage of the phosphodiester backbone (at the site of the removed purine) using 90°C alkaline conditions, and the sugar is released as a free entity. Nucleotide fragments of different sizes, labeled with ^{32}P, are produced (Figures 23.31 and 23.32). The separation of the fragments is then achieved through electrophoresis on a polyacrylamide gel. The positions of electrophoretically separated fragments containing the radioactive label (^{32}P) are easily detected within the gel by means of radioautography (Figure 23.33). Fragments containing the radioactive label produce a distinct image on the radioautograph as different tones of light and dark bands. Since the dimethylsulfate causes preferential methylation of guanine and severance of polynucleotides at that point, as opposed to adenine

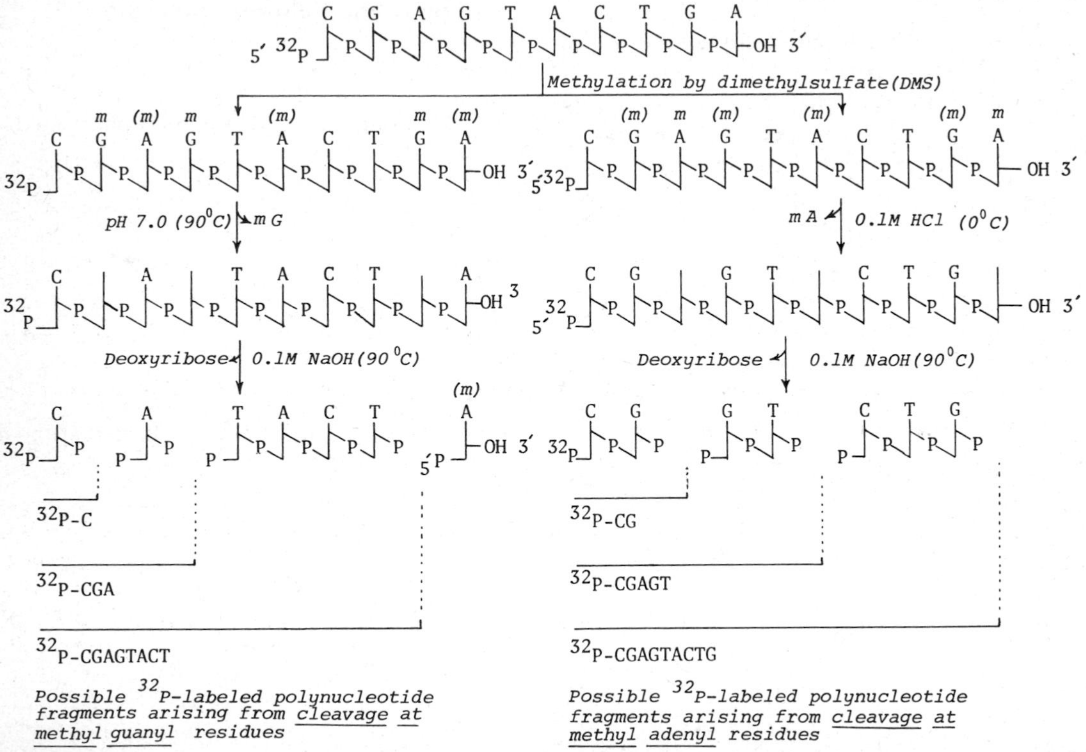

FIGURE 23.31. The methylation of adenyl and guanyl residues present in polynucleotides can be used as a preliminary step in the selective purine-specific cleavage of polynucleotides. Based on the hypothetical nucleotide base sequence of a DNA segment (a decanucleotide), the possible random yet purine-specific cleavage fragments originating from the first two treatments of the Maxam–Gilbert sequencing method have been outlined. Nucleotide fragments produced from the selective cleavage of the decanucleotide in the remaining two steps of the method are illustrated for comparative purposes in Figure 23.32.

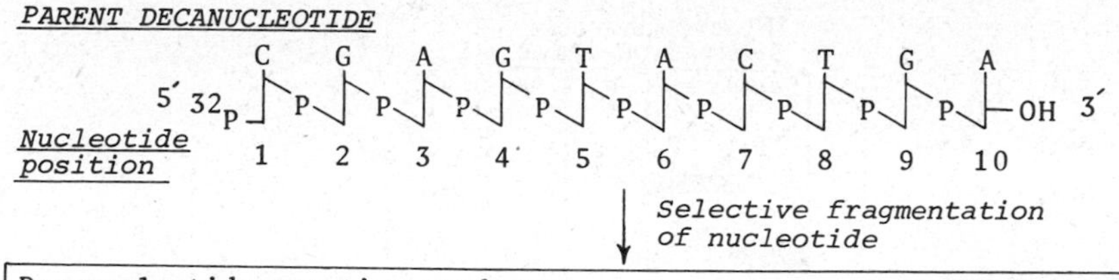

PARENT DECANUCLEOTIDE

FIGURE 23.32. Summary of analytical cleavage products for a hypothetical decanucleotide and their relationship to the parent polynucleotide structure.

(both purines), the darkest autoradiographic bands are indicative of guanine and the lighter bands are adenine.

Second portion: The polynucleotide sample is methylated with dimethylsulfate (Figures 23.31 and 23.32). Since the glycosidic bond of methylated adenine is less stable than that of guanine, a mild acidic treatment of the methylated sample at 0°C leads to the preferential *release* of the adenine. The resulting polynucleotide is then fragmented at this point using 90°C alkaline conditions as in the previous case. A typical autoradiograph of these electrophoresed polynucleotide fragments shows dark bands that indicate the presence of adenine in the base sequence while light-toned bands on the autoradiograph denote guanine (Figure 23.33).

Third portion: The sample is treated with hydrazine, which destroys both cytosine and thymine bases in the polynucleotide sequence. After this treatment, the strand is cleaved with piperidine in order to free the sugar residue that has lost its nitrogen base. Autoradiographs for separated fragments in this case show bands for cytosines and thymines that are equal in intensity (Figure 23.33).

Fourth portion: The polynucleotide is treated and fragmented with hydrazine and piperidine as before, but sodium chloride is introduced into the reaction. This added

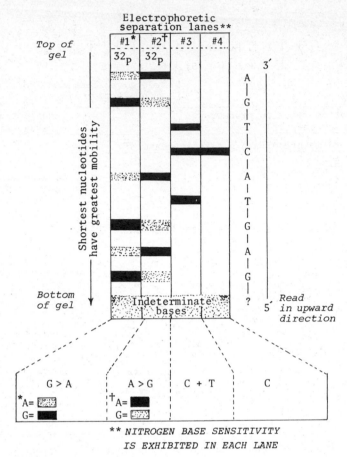

FIGURE 23.33. Electrophoretic bands for four separations of nucleotides cleaved from a polynucleotide. The example detailed reflects the hypothetical results for separation of polynucleotides produced by the Maxam–Gilbert method outlined in Figures 23.31 and 23.32. The four developed autoradiographs (each discriminative for a selected base) are shown, and with knowledge of what base(s) were probably destroyed to produce each of the fragments, the sequence of the bases is read from the bottom toward the top of the autoradiograph. Progressive detailed analysis of the autoradiographs for the four gels reveals all base positions in the DNA molecule with the exception of the 5′ end base. The 5′ end base can be determined by phosphorylating the 3′ end of the strand with ^{32}P-labeled ATP and the sequence is determined in the opposite direction (3′ → 5′).

reagent prevents hydrazine reaction with thymine; therefore, autoradiography of polynucleotides resulting from this fragmentation step indicates only the presence of cytosine (Figures 23.31–23.33).

The implementation of these four distinct sequencing steps is illustrated for the elucidation of nucleotide sequence in a decanucleotide fragment. Based on the four reactions illustrated, autoradiographic analysis of fragments originating in each step results in a schematic image such as the one shown in Figure 23.33. This visual observation of reaction products, plus knowledge as to which bases and/or base pairs were probably destroyed in order to produce each of the fragments, permits a nitrogen base sequence determination for the nucleotide when surveyed from the bottom of the radiograph toward the top. Assuming that the nucleotide base sequence of the one polynucleotide chain can be determined with certainty, it is clear that the base sequence of any complementary strand can also be determined.

Sanger "Dideoxy" Method for Polynucleotide Sequencing

The determination of DNA base sequence according to the Sanger "dideoxy" method relies on the principles of DNA replication. That is, the sequence of a single-stranded DNA is copied by DNA polymerase I, using a primer complementary to the DNA strand, and four dideoxynucleoside:deoxynucleoside triphosphate combinations (one of which is labeled ^{32}P) as indicated in Figure 23.34. The dideoxy analogue (ddNTP) of a specific deoxynucleoside triphosphate (dNTP) is used in a battery of reactions to terminate primer elongation at "specific distances" from the 5' end of the growing polynucleotide. Cessation of nucleotide polymerization occurs at specific sites since the 2',3'-dideoxy analogue of nucleotides does not have a hydroxyl group at the 3' position.

Completion of each of the four reactions

employing different dideoxynucleoside triphosphates results in a mixture of fragments that have the same ddNTP residue at the 3' ends. The strands of the DNA and the elongated primer are separated according to conventional polyacrylamide gel electrophoresis methods. Autoradiographs of these gels permit the resolution of polynucleotide sequences over a minimum–maximum range of 200 to 400 nucleotide residues. Once the autoradiograph is developed, nucleotide sequence is read according to the assumption that the shortest polynucleotide fragment corresponds to the fastest migrating band (from all four reactions). Therefore, the 5' end of the fragment starts at the bottom of the polyacrylamide gel while increasing fragment sizes migrate shorter distances from the top of the gel.

GENETIC RECOMBINATION IN MAMMALIAN CELLS

Unlike plasmid-vectored recombinant DNA methods suitable for bacteria, different methods must be available for introducing cloned DNA fragments or genes into mammalian cells. Furthermore, mechanisms still more challenging than those governing pro-

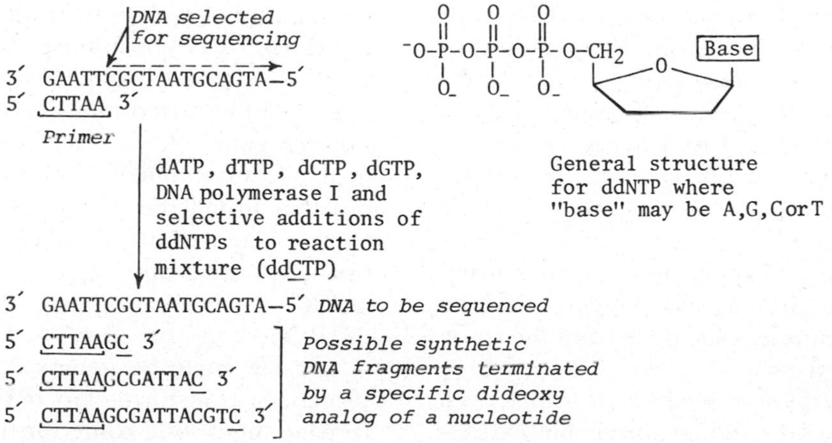

FIGURE 23.34. Survey of the Sanger "dideoxy" method for polynucleotide sequencing as well as the structure of a generalized 2',3'-dideoxynucleoside triphosphate.

karyotic gene expression must be established to ensure expression of recombinant DNA. This implies that DNA must be transcribed into mRNA and mRNA must then be cued to undergo translation into protein. Several variations in existing methods are available for inserting foreign DNA into mammalian cells, and other methods with improved reliability are being developed. Two popular recombinant methods are (1) somatic cell hybridization and (2) DNA-mediated gene transfer.

Somatic Cell Hybridization

Before the advent of current recombinant DNA techniques, *limited* gene transfers from one mammalian cell to another were possible using somatic cell hybridization. One of the earliest steps in this process requires the fusion of two somatic cells from different lines. This fusion is enhanced by a polyethylene glycol reagent and an inactivated *Sendai* virus that causes some of the cells to undergo fusion.

Fused heterokaryotic cells initially exhibit two distinctly different nuclei that represent the chromosomes of two different body cells. Ensuing cell division of the heterokaryote, however, is accompanied by disappearance of the separate nuclear membranes surrounding the two nuclei. This is followed by the formation of a single nucleus that contains the chromosomal material from both parent somatic cells. Since the resulting somatic cell hydrid may express some genes from each of the parent cells, the hybrid cell may be isolated and propagated by the use of a selective growth medium where only hybrid survival can occur. Genetic hybrids produced in this way have a temporal existence, and their main value has been realized as an aid for chromosomal mapping and gene expression studies that are beyond the scope of this discussion.

The practical problems with somatic cell hybridization, including unreliable expressions of individual genes and a lack of constant gene expression from one cell generation to the next, make other methods for inserting foreign DNA into cells an absolute necessity. One approach to this problem has been offered by uniting a DNA segment containing a desirable gene with the DNA of an animal virus. The virus can then serve as a vector for infecting a selected host cell. As only one limited example of this method, consider the linkage of the rabbit β-globulin gene to the DNA of the SV 40 monkey virus. Monkey cell infection by the viral recombinant DNA directs not only the coding for β-globulin mRNA but also the expression of the protein.

DNA-Mediated Gene Transfer

DNA recombination according to this method begins with the admixture of DNA fragments containing a desirable gene with a "carrier DNA." The DNA mixture is precipitated using calcium phosphate and then incubated with selected target cells. This incubation period leads to a limited transformation of some target cells such that they actually express the desired gene. An exemplary model for this type of genetic manipulation in mammalian cells is shown by the transformation of mouse L cells by the use of a purified DNA fragment obtained from a *Herpes simplex* virus. This virus embodies a gene responsible for thymidine kinase ($TK+$) activity.

A mixture of the highly purified TK-containing viral DNA is admixed to a salmon sperm DNA carrier. Precipitation of this DNA mixture onto isolated mouse L cells, devoid of the TK gene ($TK-$), results in a miniscule but significant number of cells (one in a million) that install the $TK+$ gene into their genome. These transformed cells then demonstrate positive phenotypic expression of the $TK+$ gene on an appropriate selective growth medium. Because $TK-$ mouse cells are considered to be genetically defective, the genetic transformation outlined above offers an interesting route for the "cure" of genetically based factors in a mammalian genome.

Exclusive of DNA-mediated gene transfer, injection of high gene copy numbers (>20,000) into the "male" pronucleus of fertilized mouse egg cells offers another possible method for gene transfer and recombinant DNA formation in mammalian cells. Based on existing studies, time-critical introduction of foreign DNA into merged egg and sperm cells (before nuclear fusion) may govern the long-term stability and phenotypic expression of the foreign DNA.

Regardless of the contemporary methods for establishing recombinant DNA, it is increasingly clear that

1. Functioning genes can be transferred among different species of plants, animals, and bacteria.

2. Mammalian genes, fused to the genes of a prokaryote, can be used as a "biochemical factory" to produce proteins and other biochemical products that are compatible with the biochemical systems and life processes of higher cells including those of mammals.

3. Historical challenges underlying the formation of recombinant DNA in prokaryotes *generally* reflect the problems encountered in establishing recombinant DNA in eukaryotic cells.

4. Insertion of foreign DNA into prokaryotic and eukaryotic cells may display long-term instability, and problems are often encountered with the cued phenotypic expression of a desirable gene. These problems stem from complex biochemical control mechanisms at work in the host cell, feedback mechanisms, and other idiosyncracies of individual transferred genes.

RECOMBINANT DNA TECHNOLOGY IN AGRICULTURE AND FOOD TECHNOLOGY

Many basic advances in agriculture and food sciences are tied directly to the development of new crop plants, livestock, poultry, and microorganisms. As seen in earlier sections, the historical route for improving food species has been largely restricted to classical principles of Mendelian inheritance, hybridization, and selection of certain organisms. For example, two or more varieties of plants or animals were strategically bred to give progeny having superior traits over the original parent organisms.

With the advent of recombinant DNA (rDNA) technology, modifications of genes for specific traits in organisms can be executed. The assurance of selected gene introductions into new organisms largely regularizes formerly unpredictable patterns of gene alterations and mutations that have directed evolution.

The universality of many molecular genetic operations also ensures that nearly any native genome of an organism can be modified by existing rDNA methods. Challenging technical and biochemical problems remain to be solved, however, for the reliable or cued expression of desirable genes once they are introduced into new organisms.

Modern Biotechnologies for Plant Breeding

Plant breeding serves as a focus for some of the most promising consequences of genetic engineering in food production. Crop plant varieties previously difficult or impossible to develop using conventional breeding techniques have become possible. Three genetic recombination methods have been successfully employed on plants including (1) *somaclonal variation*, (2) *protoplast technology* (or cell fusion), and (3) *plasmid transfer*.

Somaclonal variation relies on the enzyme-directed installation of a DNA fragment from a donor cell into the DNA of a host cell. The donated DNA is selected according to its ability to produce a desirable genetic trait. Genetically altered plant cells are then cultured as a clone on a selected medium designed to induce the regeneration of completely individual plants. Unlike conventional plant breeding methods for new plants that require 6 to 10 years, breeding time may be reduced to 3 years in the most successful cases using modern rDNA methods.

When incompatibilities between cells of two plant species forbid the execution of conventional breeding methods, protoplast technology (or cell fusion) may be used. According to this method, the cell walls of selected species of plant cells are dismantled by the actions of one or more enzymes. These enzymes include cellulases and/or pectinases. The naked protoplasts of the different cells are then *fused* together and a cell wall is then permitted to regenerate about the fused protoplasts. A viable hydridized protoplast will ideally display the genetic traits of *both* contributing protoplasts and mature into a plant with specifically improved characteristics.

Plasmid-mediated transfers of molecular genetic information also permit the development of new plant varieties. In these methods, genes for a desirable plant trait are isolated and introduced into an rDNA vector (e.g., plasmid or plant virus). Installation of the recombinant vector into a plant protoplast is then followed by a cellular or whole plant expression of the foreign gene. Continuous replication of the native plant genome along with the foreign genetic information can eventually yield a new regeneration-stable seed-borne trait (Figure 23.35).

Aside from the potential success of these genetic recombination methods for many varieties of plants, it must be recognized that conventional plant breeding methods will remain very important for producing plants that exhibit increased productivity yields and resistances to insects, diseases, and environmental threats, plus other features.

Micropropagation and tissue cultures of genetically modified root, stem, or leaf cells will also continue to serve as an important screening method for evidence of desirable plant traits. For example, positive evidence of a new genetic trait such as production of a key flavor constituent, or disease resistance in only a few cells out of millions, can be easily detected without the expenses associated with field cultivation, selection, and testing of mature plants. Micropropagation and its related plant selection techniques have already contributed to the commercial development of food plants such as asparagus, coffee, dates, pineapple, papaya, citrus fruits, and others.

Antherculture cannot be overlooked in any survey on the potential status of propagative methods for new genetic variants of plants. This culture technique permits the reversion of haploid pollen to its diploid form. The diploid form then serves as an important source of homozygous breeding stock used in the development of new plant species.

The prudential development of new varieties of plant species are tied to the impacts they may have on existing farming practices and general agronomic operations. Priority considerations are based on the abilities of new plant variants to

1. Maintain or increase crop yields by (a) resisting pests, diseases, frost, and drought effects or (b) demonstrating tolerance to unfavorable soil conditions and accelerated responses to fertilizers.

2. Increase the economic value of crop yields by selecting genetic traits that favor (a) existing milling, baking, or storing features for crops; (b) enhanced oil contents of key crops; or (c) improved nutritional status of a crop such as elevated protein content.

3. Reduce crop production costs by developing plant variants that (a) favorably survive mechanical harvesting methods; (b) require lower demands for pesticides and fertilizers; and (c) necessitate less tillage and cultivation for successful crop production, thereby conserving energy resources.

Genetic Improvement of Food-Producing Animals

Direct genetic manipulation and restructuring of plant genomes offer a formidable

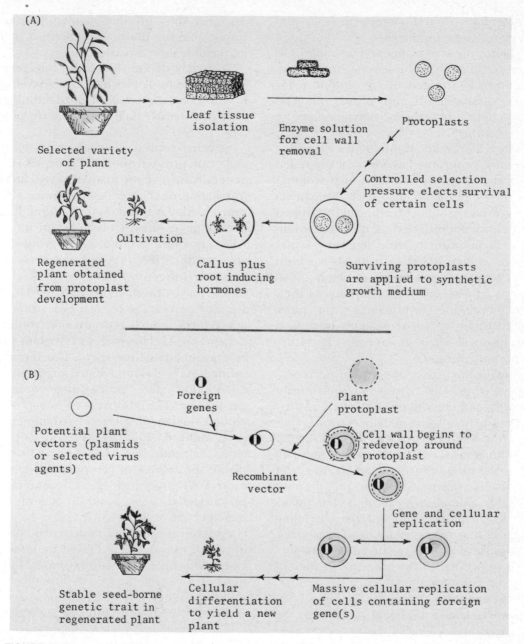

FIGURE 23.35. (A) Protoplast culture can be used as a basis for the regeneration of plants. Protoplasts that may have desired genetic or survival characteristics are selected and administered to a growth medium. Protoplasts proliferate into multicellular aggregates (*callus*) that eventually undergo cellular differentiation into roots, stem, and leaves and yield a regenerated plant. (B) Installation of recombinant DNA (rDNA) into a selected plant using a rDNA vector and a plant cell protoplast. Successful installation of the rDNA into the plant cell followed by cellular reproduction and retention of the new genetic information can result in a stable seed-borne genetic trait in the regenerated plant.

challenge for even the most genetically simple plant. Gene numbers of 10,000 or more commonly occur in plants, and desirable traits often result from multiple gene interactions. For animals, the genetic constitution is far more complicated than in plants, and gene interactions for specific traits are likewise more complex.

The widespread adoptions of new animal breeding programs have historically relied upon the obvious advantages of new methods over older methods. Since the past 25 years of farm animal breeding in the United States have not reflected overt economic failures, farm industry-wide interest in new routine animal breeding methods has been very conservative. There is no doubt, however, that the cautious adoption of new animal-breeding technologies and more sophisticated rDNA techniques in animal breeding will occur as warranted by future agronomic dictates.

Improvements in understanding reproductive physiology, direct manipulation of sex cells, and breeding controls have all augmented genetic gains in livestock and poultry up to the present time. Animal mating is no longer left to a chance event in a pen or on the range. Furthermore, *selective* and *designed* breeding practices using only the most desirable animals are executed. The favorable consequences of selective breeding practices may not be guaranteed in all cases, but statistical aspects of current genetic theory support the worthwhile consideration of these breeding methods.

Few examples demonstrate the advantages of selective breeding more than the increased milk yields of cows over the past 25 years. Although dairy herd populations have decreased nearly 50%, the average milk productivities of cows have nearly doubled. Surveys of the northeastern states alone suggest that at least 25% of the increased milk productivity is linked to permanent genetic improvements in Holstein cows. Poultry production, too, has benefited from selective breeding practices in conjunction with routinely applied artificial insemination (AI).

Together, these breeding methods have been responsible for the development of highly marketable large-breasted poultry.

A variety of individual and interrelated breeding technologies have been developed that foster the *selection* of outstanding animals and *amplify* their reproductive potentials.

Selected reproduction of nearly all animals can profit from AI. Principles of AI generally involve the manual *in vivo* uterine installation of selected sperm. Since sexual reproduction is based on the random fusion probability existing between sperm and ova, AI using sperm from a male having desirable germplasm can favor the occurrence of desirable progeny. Although 100% of domestic turkey production relies on AI, only a minor percentage of total beef cattle and other farm animal reproduction routinely depend on AI. However, as cryogenic preservation methods for sperm become more refined, AI will assume more importance in distributing and propagating desirable germplasm. The advantage of AI can be exploited further in selected herd animals by drug-induced *estrus synchronization*. Estrus or *heat* synchronization within a herd can expedite the success of (1) uniformly administered AI programs, (2) conventional reproductive programs, and (3) general herd management tactics.

Amplification of desirable female breeding stock can also be achieved by *superovulation*. This practice entails hormonal induction of ova release in abnormally high numbers. Whereas 1 or 2 ovulations are expected for goats and sheep, superovulation can increase the number to 12. Singular ovulations in cows can be similarly increased to 6 to 8.

Advanced reproductive and physiological technologies are especially valuable when used in conjunction with AI. Typical methods of advanced reproductive technology can entail the transfer of fertilized ova into surrogate mothers. Following this operation, the fertilized ova demonstrate progenic maturation rates and gestational successes similar to naturally ovulated and fertilized offspring. Since even the most genetically favorable female animals have limited procreation potentials, superovulation coupled with AI and embryo transfer into numerous genetically inferior surrogate mothers can enhance the reproductive rate of genetically superior animals.

There is no doubt that the reproductive efficiencies of farm animals and improvements in their genetic status can be strategically improved by the methods detailed above. Added to these methods, genetic enrichment of animal genomes can be implemented by *in vitro* fertilization techniques, improved oviduct or uterine recovery of fertilized ova followed by embryo transplantation and/or storage, predetermined sex selection of sperm for fertilization, and artificial induction of twinning. The routine use of rDNA technology, however, is not and will not be a common practice for maintaining major livestock and poultry populations for many years. The relatively minor implementation of rDNA technology in animal reproduction will persist until molecular genetic operations for specific farm animals have been worked out. Of particular developmental interest are the positions of gene loci for specific traits on chromosomes and identification of desirable phenotypic traits caused by multiple gene interactions, which are currently quite obscure.

Miscellaneous Applications for rDNA Biotechnology

Food processing and biochemical operations dependent upon fermentation technologies can benefit from rDNA in numerous ways. rDNA technologies can enhance existing microbial fermentative productivities for key *amino acids, vitamins,* and *enzymes* by either (1) increasing product yields of existing microorganisms or (2) developing new microbial sources having improved production efficiencies.

With the advent of effective enzyme immobilization methods, many unit industrial operations have begun to rely on constant bulk supplies of enzymes to yield food additives or fine chemicals. Immobilized enzyme applications are typically involved in (1) high-fructose syrup production, (2) selected amino acid biosynthesis, (3) lactose hydrolysis of cheese wheys, (4) saccharification of starch, (5) L-lysine formation from amino caprolactoran, (6) starch conversion to maltose, and many other reactions.

The bulk enzyme requirements of the cheese industries have already prompted the successful cloning of the rennin gene in *Esherichia coli*. Rennin, which is normally extracted from the fourth stomach of suckling calves slaughtered for veal, has come under increasing supply demands. These demands reflect increased cheese production plus static veal productivity. High microbial productivity of rennin could allay the demonstrated shortfalls in the availability of this enzyme on a permanent basis.

Alcoholic beverage and fermented baking product industries stand to gain from advances in yeast genetics. Whereas classical development of yeasts has depended upon hybridization practices, the possibilities of cell fusion now permit the development of *new* yeast strains. These new yeast strains may offer new and useful fermentation properties in spite of the fact that the original parent yeasts could never engage in *normal* mating.

Many practical considerations underlie the need for improving present yeast strains. For example, classical catabolic repression of yeasts by 0.1 to 0.3% concentrations of glucose may be eliminated by genetic modification of existing yeast strains. Catabolic repression is an unfavorable phenomenon since it not only represses respiratory enzyme activities, but it also suppresses aerobic growth.

Other yeasts, such as commercial ale brewing strains of *Saccharomyces cerevisiae*, can be designed to incorporate a "killer" property. This property is tied to a toxin production that exhibits a self-cleaning action for fermentation systems. In the case of baking industries, related varieties of yeast can be genetically adapted by rDNA to give accelerated carbon dioxide production rates and foster the leavening process in baked products.

Yeast mutants, developed from classical genetic research using rDNA, also offer improvements in the performance of wine yeasts. These new rDNA yeasts can display unusual properties including (1) improved sedimentation properties that augment their removal from wine and (2) increased tolerance to high alcohol content. This last feature permits the fermentation of grapes having high sugar concentrations. Some yeasts may also be modified to metabolize malic acid to ethanol and effectively decrease the notable acidities of many wines.

With respect to fermentative bacteria, the horizons for modified microbial biochemistry are nearly unlimited. Many of these potential modifications also hold great economic promise.

Although starter culture technology for fermented foods including dairy products, breads, pickled products (olives, pickles), and fermented meats (sausages) have been successful, there is room for improvement. Most goals for optimizing lactic acid fermentations have simply dealt with the characterization, isolation, and selection of improved varieties of *Pediococci*, *Lactobacilli*, and *Streptococci*.

Selection of improved fermentation bacteria combined with rDNA modifications can enhance lactose fermentation, casein digestion, and citrate fermentation to diacetyl (a buttery flavor). Some genes responsible for ideal bacterial fermentation properties are indigenous to bacterial genomes per se, but in other bacteria such as lactate-producing *Streptococci*, milk fermentation is tied to plasmid DNA.

Genetic manipulation of bacteria can also serve as a tool for inducing phage resistance in lactic *Streptococci*. Phage resistance of desirable *Streptococci* ensures their more reliable and cued performances in the dairy

industries, where nonresistant strains can undergo phage-mediated decimation.

Microbial fermentations directed by rDNA technology can also contribute to more economical supplies of important flavors, fragrances, and sweeteners. The sweetener aspartame is one typical sweetener whose production is expensive, and further biotechnological advances could jointly hold down costs and increase its production.

Principles of rDNA biotechnology have also served as a basis for producing a *Salmonella* species-specific assay that predicts the real or potential intoxication of foods by these

Table 23.3. Realistic Time Frame Appraisal for the Development of Genetically Engineered Products That Could Have a Major Economic Impact on Food Crop Plants, Livestock, and Food Processing

New product, improved biochemical process, new form of living organism, improved drug or vaccine processing, or other genetically improved feature	Time frame
Amino acids (p.i.)[a]; anabolic steroids (n.s.); vaccines for hoof and mouth disease, hog diarrhea, and other viral diseases (p.i.); methanol bacteria (m.o.); rennin production (n.s., p.i.); antibiotics (p.i.)	1988 ± 2 years
New pesticides (n.s., p.i.); improved yeasts (m.o.); improved control of salinization and osmoregulation in plants and unicellular organisms (m.o.); improved cellulosic metabolism by bacteria (m.o.)	1995 ± 5 years
Incorporation of nitrogen-fixing abilities into nonnitrogen-fixing plants, suppression of plant photorespiration, enhanced hydrogen assimilation (m.o.); livestock cloning (new rDNA and reproductive biological advances will be required)	2000 ± 5 years

[a]m.o., Modified organism using rDNA; n.s., new source; p.i., production improvement.

NOTE: Many other products are possible and the items listed represent only a rough survey.

bacteria. Such assays are based on the use of DNA probes that specifically hybridize with a section of the *Salmonella* genome assuming that the bacteria are present as food contaminants (Fitts *et al.*, 1983). Apart from a high specificity for detecting most of the serotypes of *Salmonella,* the principle of using specific DNA probes for hybridization provides a tactical basis for identifying a wide variety of Gram-negative enteric bacteria. Furthermore, these methods are far more rapid and reliable than historical diagnostic methods.

CONCLUSION

Only several years before this writing, many corporations took a jaundiced perspective of the long-term payoff that might result from genetically engineered food resources. Pharmaceutical, chemical, and some large food companies have departed radically from this posture. Realistic time projections for the availability of genetically engineered biotechnology objectives have been outlined in Table 23.3. Current objectives of rDNA research include the *production of new sources* of animal growth hormones for cattle and pigs, immunization agents, antibiotics, and many goals favoring safe, economical, and improved food resources. As indicated in earlier sections, plant and crop sciences stand to make major strides in the future. These advances include the installation of nitrogen-fixing capabilities in plants using rDNA; development of saline soil resistance in important crop plants; development of higher protein strains of oats, rice, and wheat; plus improvements in plant produce quality such as redder, tastier, and sturdier tomatoes.

Quite apart from the development of feasible rDNA technologies for improving foods, many food companies realize that their very economic survival and corporate welfare may rely on new molecular genetic advances. The past successes of agribusinesses and food producers have relied heavily on altering old food production processes and fine tuning and rejiggering time-honored food formulations or recipes. These old practices coupled with the progressive development of new plants, animal, and microbial variants, however, can provide the only realistic tack for agribusiness successes throughout the 1990s and into the next century.

REFERENCES

General references for molecular genetic principles

Adhya, S., and M. Gottesman. 1978. Control of transcription termination. *Annu. Rev. Biochem.* **47**:967.

Altman, S. 1978. Biosynthesis of tRNA. In *Transfer RNA* (S. Altman, ed.), pp. 48–77. MIT Press, Cambridge, Mass.

Aprion, D., and P. Gegenheimer. 1981. Processing of bacterial RNA. *FEBS Lett.* **125**:1. (Model of RNA processing in bacteria is detailed.)

Bloomfield, V. A., D. M. Crothers, and I. Tinoco, Jr. 1974. *Physical Chemistry of Nucleic Acids.* Harper & Row, New York.

Brenner, S., F. Jacob, and M. Meselson. 1961. An unstable intermediate carrying information from genes to ribosomes for protein synthesis. *Nature* **190**:576. (Classic paper outlining early recognition of mRNA.)

Britten, R., and D. Kohn. 1968. Repeated segments of DNA. *Sci. Amer.* **222**(4):24.

Brown, D. D. 1981. Gene expression in eukaryotes. *Science* **211**:667.

Cantor, C. R., and P. R. Schimmel. 1980. *Biophysical Chemistry,* Parts I, II, and III. Freeman, San Francisco.

Caskey, C. T. 1980. Peptide chain termination. *Trends Biochem. Sci.* **5**:234.

Chambon, P. 1981. Split genes. *Sci. Amer.* **244**:60.

Clark, R. E. C., and K. A. Marcker. 1968. How proteins start. *Sci. Amer.* **218**:36. (Role of formylmethionine is detailed.)

Crick, F. 1979. Split genes and RNA splicing. *Science* **204**:264.

Ebright, R. H., and J. R. Wong. 1981. Mechanism

for transcriptional action of cyclic-AMP in *E. coli:* Entry into DNA to disrupt DNA secondary structure. *Proc. Natl. Acad. Sci. USA* **78:**4011.

Fitts, R., M. Diamond, C. Hamilton, and M. Neri. 1983. DNA-DNA hybridization assay for detection of *Salmonella* spp. in foods. *Appl. Environ. Microbiol.* **46:**1146.

Gale, E. F., E. Cundliffe, P. E. Reynolds, M. H. Richmond, and M. H. Waring. 1981. *The Molecular Basis of Antibiotic Action,* 2nd ed. Wiley, New York. (Survey of toxin and antibiotic disruption of protein synthesis, pp. 402–549).

Gilbert, W., and B. Muller-Hill. 1966. Isolation of the *lac* repressor. *Proc. Nat. Acad. Sci. USA* **56:**1891.

Hunt, T. 1980. The initiation of protein biosynthesis. *Trends Biochem. Sci.* **5:**178. (Eukaryotic translation survey.)

Jacob, F., and Monod. 1961. Genetic regulatory mechanism in the synthesis of proteins. *J. Mol. Biol.* **3:**318.

Khorana, H. G. 1979. Total synthesis of a gene. *Science* **203:**614.

Klekowski, E. J. (ed.). 1982. *Environmental Mutagenesis, Carcinogenesis, and Plant Biology,* Vols. 1 and 2. Praeger, New York.

Kolata, G. B. 1980. Genes in pieces. *Science* **207:**392.

Kornberg, A. 1980. *DNA Replication.* Freeman, San Francisco.

Kornberg, R. D. 1977. Structure of chromatin. *Annu. Rev. Biochem.* **46:**931.

Kriegstein, H. J., and D. S. Hogness. 1974. The mechanism of DNA replication in *Drosophila* chromosomes: Structure of replication forks and evidence for bidirectionality. *Proc. Natl. Acad. Sci. USA* **71:**135.

Lake, J. A. 1981. The ribosome. *Sci. Amer.* **245:**84.

Lehman, I. R. 1974. DNA ligase: structure mechanism and function. *Science* **186:**790.

Lewin, B. 1980. Alternatives for splicing: Recognizing the ends of introns. *Cell* **22:**324.

Lewin, B. 1983. *Genes.* Wiley, New York.

Leighton, T. J., and W. F. Loomis (eds.). 1981. *The Molecular Genetics of Development: An Introduction to Recent Research on Experimental Systems.* Academic Press, New York.

Maxam, A. M., and W. Gilbert. 1977. A new method for sequencing DNA. *Proc. Natl. Acad. Sci. USA* **74:**560.

McHenry, C., and A. Kornberg. 1977. DNA polymerase III holoenzyme of *Escherichia coli:* Purification and resolution into subunits. *J. Biol. Chem.* **252:**6478.

Miller, O. L., Jr. 1973. The visualization of genes in action. *Sci. Amer.* **228**(3):34.

Monod, J., J.-P. Changeux, and F. Jacob. 1963. Allosteric proteins and cellular control systems. *J. Mol. Biol.* **6:**306.

Ogawa, T., and T. Okazaki. 1980. Discontinuous DNA replication. *Annu. Rev. Biochem.* **49:**421.

Pestka, S. 1983. The purification and manufacture of human interferons. *Sci. Amer.* **249**(2):36–43.

Quigley, G. J., A. H.-J. Wang, G. Ughetto, G. van der Marel, J. H. van Boom, and A. Rich. 1980. Molecular structure of an anticancer drug–DNA complex: Daunomycin plus d(CpGpTpApCpG). *Proc. Natl. Acad. Sci. USA* **77**(12):7204. (Excellent study of DNA intercalation by drugs.)

Revel, M., and Y. Groner. 1978. Post-transcriptional and translational controls of gene expression in eukaryotes. *Annu. Rev. Biochem.* **47:**1079.

Rich, A., and S. H. Kim. 1978. The three-dimensional structure of transfer RNA. *Sci. Amer.* **238**(1):52.

Rowen, L., and A. Kornberg. 1978. Primase, the *dna* G protein of *Escherichia coli:* An enzyme which starts DNA chains. *J. Biol. Chem.* **253:**758.

Sanger, F., S. Nicklen, and A. R. Coulson. 1977. DNA sequencing with chain terminating inhibitors. *Proc. Natl. Acad. Sci. USA* **74:**5463.

Schimmel, P. R., and D. Söll. 1979. Amino acyl-tRNA synthetases: general features and recognition of transfer RNAs. *Annu. Rev. Biochem.* **48:**601–648.

Schimmel, P. R., D. Söll, and J. N. Abelson. 1979. *Transfer RNA: Structure Properties and Recognition.* Cold Spring Harbor Press, Cold Spring Harbor, N.Y.

Taylor, J. H. (ed.). 1965. *Selected Papers on Molecular Genetics.* Academic Press, New York. (A survey of important and historical research papers related to the evolution of molecular genetics.)

Watson, J. B. 1976. *Molecular Biology of the Gene,* 3rd ed. Benjamin, New York.

Weissbach, H., and S. Ochoa. 1976. Soluble factors required for eukaryotic protein synthesis. *Annu. Rev. Biochem.* **45:**191.

Weissbach, H., and S. Pestka. 1977. *Molecular Mechanisms of Protein Biosynthesis.* Academic Press, New York.

Wintersberger, E. 1977. DNA-dependent DNA polymerases from eukaryotes. *Trends Biochem. Sci.* **2:**58.

Wold, F. 1981. *In vivo* chemical and modification of proteins. *Annu. Rev. Biochem.* **50:**783.

Wool, I. G. 1979. Structure and function of eukaryotic ribosomes. *Annu. Rev. Biochem.* **48:**719.

General references for recombinant DNA methods and biotechnology

Abelson, J., and E. Butz. 1980. Recombinant DNA. A review issue. *Science* **209:**1317.

Anderson, W. F., and E. G. Diacumakos. 1981. Genetic engineering in mammalian cells. *Sci. Amer.* **245**(1):106.

Avery, O. T., C. M. MacLeod, and M. McCarty. 1944. Studies on the chemical nature of the substance inducing transformation of pneumococcal types. Induction of transformation by a deoxyribonucleic acid fraction isolated from *Pneumococcus* Type III. *J. Exp. Med.* **79:**137.

Bull, A. T., D. C. Ellwood, and C. Ratledge. 1979. *Microbial Technology: Current State, Future Prospects.* 29th Symposium of the Society for General Microbiology at the University of Cambridge, April 1979. Cambridge University Press, Cambridge, England.

Cartwright, T. C. 1970. Selection criteria for beef cattle for the future. *J. Anim. Sci.* **30:**706.

Cohen, S. N., A. C. Y. Chang, H. W. Boyer, and R. B. Helling. 1973. Construction of biologically functional bacterial plasmids *in vitro. Proc. Natl. Acad. Sci. USA* **70:**3240.

Cundiff, L. V., and K. E. Gregory. 1977. *Beef Cattle Breeding,* USDA, Agricultural Information Bulletin 286 (Rev.). USDA, Washington, D.C.

Hafs, H. D. 1979. Potential impact of prostaglandins on prospects for food from dairy cattle. In *Proceedings Lutalyse of the Symposium* (J. W. Lauderdale and J. H. Sokolowski, eds.), pp. 9–14. Upjohn Co., Kalamazoo, Mich.

Helinski, D. R. 1979. Bacterial plasmids: Autonomous replication and vehicles for cloning. *Crit. Rev. Biochem.* **7:**83.

Holdgate, D. P. 1977. Propagation of ornamentals by tissue culture. In *Plant Cell, Tissue and Organ Culture* (J. Feinert and Y. P. S. Bajaj, eds.). Springer-Verlag, New York.

Hollaender, A., R. H. Burris, P. R. Day, R. W. F. Hardy, D. R. Helinski, M. R. Lamborg, L. Owens, and R. C. Valentine (eds.). 1977. Vol. 9, *Genetic Engineering for Nitrogen Fixation. Basic Life Sciences Series,* Plenum, New York.

Hollaender, A., R. D. DeMoss, S. Kaplan, J. Konisky, D. Savage, and R. S. Wolfe (eds.). 1982. Vol. 19, *Basic Life Sciences Series, Genetic Engineering of Microorganisms for Chemicals.* Plenum, New York.

Jensen, N. F. 1978. Limits to growth in world food production. *Science* **201:**317.

Lawrence, W. J. C. 1968. *Plant Breeding.* Arnold, London.

Lerner, M. I., and H. P. Donald. 1966. *Modern Developments in Animal Breeding.* Academic Press, New York.

Levy, S. B., R. C. Clowes, and E. Koenig (eds.). 1981. *Molecular Biology, Pathogenicity and Ecology of Bacterial Plasmids.* Plenum, New York. (Complete summary of plasmid research.)

Malik, V. S. 1980. Recombinant DNA technology. *Advan. Appl. Microbiol.* **27:**1.

Maniatis, T. 1980. Recombinant DNA procedures in the study of eukaryotic genes. In *Cell Biology* (L. G. Goldstein and D. M. Prescott, eds.), Vol. 3, pp. 563–608. Academic Press, New York.

Muhammed, A., R. Aksel, and R. C. von Borstel (eds.). 1977. *Basic Life Sciences Series,* Vol. 8, *Genetic Diversity in Plants.* Plenum, New York.

Murashige, T. 1977. Current status of plant cell and organ cultures. *Hort. Sci.* **12**(2):127.

National Academy of Sciences. 1978. *Conservation of Germplasm Resources: An Imperative.* National Academy of Sciences, Washington, D.C.

National Academy of Sciences. 1972. *Genetic Vulnerability of Major Crops.* National Academy of Sciences, Washington, D.C.

Novick, R. P. 1980. Plasmids. *Sci. Amer.* **243:**102.

Panopoulos, N. J. (ed.). 1981. *Genetic Engineering in the Plant Sciences.* Praeger, New York.

Perlman, D. 1973. The fermentation industries. *Amer. Soc. Microbiol. News* **39**(10):653, and 1977, **43**(2):82–89. (Practical surveys of enzyme and fermentation industries.)

Rains, D. W., R. C. Valentine, and A. Hollaender (eds.). 1980. *Basic Life Sciences Series,* Vol. 14, *Genetic Engineering of Osmoregulation.* Plenum, New York.

Russell, L. B. (ed.). 1978. *Basic Life Sciences Series,* Vol. 12, *Genetic Mosaics and Chimeras in Mammals.* Plenum, New York.

Schimke, R. T. 1980. Gene amplification and drug resistance. *Sci. Amer.* **243**(5):60.

Scrimshaw, N. S., and M. Béhar (eds.). 1976. *Basic Life Sciences Series,* Vol. 7, *Nutrition and Agricultural Development.* Plenum, New York.

Setlow, J. K., and A. Hollaender, 1979–1980. *Genetic Engineering*, Vols. 1–3. Plenum, New York.

Shepard, J. F., D. Bidney, and E. Shahin. 1980. Potato protoplasts in crop improvement. *Science* **208:**17.

Sinsheimer, R. L. 1977. Recombinant DNA. *Annu. Rev. Biochem.* **46:**415. (Status report through 1976.)

Smith, H. O. 1979. Nucleotide sequence specificity of restriction endonucleases. *Science* **205:**455.

Sprague, G. F., D. E. Alexander, and J. W. Dudley. 1980. Plant breeding and genetic engineering: A perspective. *Bioscience* **30**(1):17.

Torrey, J. G. 1977. Cytodifferentiation in cultured cells and tissues. *Hort. Sci.* **12**(2):138.

Wang, A. H.-J., G. J. Quigley, F. J. Kolpak, J. L. Crawford, J. H. van Boom, G. van der Marel, and A. Rich. 1979. Molecular structure of left-handed double helical DNA fragment at atomic resolution. *Nature* **282:**680.

Wang, A. H.-J., G. J. Quigley, F. J. Kolpak, G. van der Marel, J. H. van Boom, and A. Rich. 1981. Left-handed double helical DNA: variations in the backbone conformation. *Science* **211:**171.

Watson, J. D., and F. Crick. 1953. Genetical implications of structures of deoxyribose nucleic acid. *Nature* **171:**737.

U.S. Department of Agriculture, Agricultural Research Service. 1976. *Introduction, Classification, Maintenance, Evaluation and Documentation of Plant Germplasm,* (ARS) National Research Program. No. 20160. U.S. Gov. Printing Office, Washington, D.C.

APPENDIX · I

Unit Abbreviations, SI Prefixes, Conversion Factors, and Physical Constants

Unit abbreviations	
Acid number	A
Angstrom	Å
Atmosphere	atm
Calorie	cal
Centimeter	cm
Centipoise	cP
Coulomb	C
Deciliter	dL
Decimeter	dm
Gram	g
Hour	h
Joule	J
Kilocalorie	kcal
Kilojoule	kJ
Kelvin	K
Liter	L
Micrometer	μm
Micromole	μmol
Molar concentration	*M*
Molal concentration	*m*
Milliequivalent	mEq
Milligram	mg
Minute	min
Milliliter	mL
Molecular weight	M.W.
Millivolt	mV
Normal concentration	*N*
Nanometer	nm
Osmotic pressure	π
Saponification number	S
Second	s
Volt	V
Zeta potential	ζ

SI prefix	Multiple	Symbol
tera	10^{12}	T
giga	10^{9}	G
mega	10^{6}	M
kilo	10^{3}	k
deci	10^{-1}	d
centi	10^{-2}	c
milli	10^{-3}	m
micro	10^{-6}	μ
nano	10^{-9}	n
pico	10^{-12}	p
femto	10^{-15}	f
atto	10^{-18}	a

Conversion factors

Length

1 m = 100 cm

1 cm = 10 mm = 10^4 μm = 10^7 nm = 10^8 Å

Mass

1 kg = 10^3 g

1 g = 10^3 mg = 10^6 μg = 10^9 ng = 10^{12} pg = 10^{15} fg = 10^{18} ag

Volume

1 L = 10^3 mL = 10^6 μL (λ)

Temperature

K = °C + 273

$$°C = \frac{5}{9}(°F - 32)$$

$$°F = \frac{9}{5}(°C) + 32$$

Pressure

1 atm = 760 Torr (mm Hg)

Energy

1 J = 0.23901 cal

1 cal = 4.184 J

Physical Constants

Constant	Symbol	Values
Atomic mass unit	amu	1.66053×10^{-27} kg
Avogadro's number	*N* (or *L*)	6.0220×10^{23} mol^{-1}
Boltzmann constant	*k*	1.3807×10^{-23} JK^{-1}
Curie	Ci	3.70×10^{10} disintegrations/s
Electronic charge	*e*	1.6022×10^{-19} C
Faraday constant	$\mathcal{F}$	96,485 C mol^{-1}
Gas constant	*R*	8.3144 JK^{-1} 1.987 cal/mol/K
Planck's constant	*h*	6.6262×10^{-34} J s 1.584×10^{-34} cal s
Speed of light	*c*	2.997925×10^{8} m/s

Nutritive Values of the Edible Parts of Foods

EXPLANATION OF THE TABLE FOR NUTRITIVE VALUE OF FOODS

Table A.1 shows the food values in 730 foods commonly used.

Foods listed: Foods are grouped under the following main headings:

Dairy products
Eggs
Fats and oils
Fish, shellfish, meat, and poultry
Fruits and fruit products
Grain products
Legumes (dry), nuts, and seeds
Sugars and sweets
Vegetables and vegetable products
Miscellaneous items

Most of the foods listed are in ready-to-eat form. Some are basic products widely used in food preparation, such as flour, fat, and cornmeal.

The weight in grams for an approximate measure of each food is shown. A footnote indicates if inedible parts are included in the description and the weight. For example, item 246 is half a grapefruit with peel having a weight of 241 g. A footnote to this item explains that the 241 g include the weight of the peel.

The approximate measure shown for each food is in cups, ounces, pounds, some other well-known unit, or a piece of certain size. The cup measure refers to the standard measuring cup of 8 fluid oz or one-half liquid pint. The ounce refers to one-sixteenth of a pound avoirdupois, unless fluid ounce is indicated. The weight of a fluid ounce varies according to the food measured.

Food values: Table A.1 also shows values for protein, fat, total saturated fatty acids, two unsaturated fatty acids (oleic acid and linoleic acid), total carbohydrates, four minerals (calcium, iron, phosphorus, and potassium), and five vitamins (vitamin A, thiamin, riboflavin, niacin, and ascorbic acid or vitamin C). Food energy is in calories. The calorie is the unit of measurement for the energy furnished the body by protein, fat, and carbohydrate.

Those values can be used to compare kinds and amounts of nutrients in different foods. They sometimes can be used to compare different forms of the same food.

Water content is included because the

percentage of moisture present is needed for identification and comparison of many food items.

The values for food energy (calories) and nutrients shown in Table A.1 are the amounts present in the edible part of the item, that is, in only that portion customarily eaten—corn without cob, meat without bone, potatoes without skin, European-type grapes without seeds. If additional parts are eaten—the potato skin, for example—amounts of some nutrients obtained will be somewhat greater than those shown.

Values for thiamin, riboflavin, and niacin in white flours and white bread and rolls are based on the increased enrichment levels put into effect for those products by the Food and Drug Administration in 1974. Iron values for those products and the values for enriched cornmeals, pastas, farina, and rice (except riboflavin) represent the minimum levels of enrichment promulgated under the Federal Food, Drug, and Cosmetic Act of 1955. Riboflavin values of rice are for unenriched rice, as the levels for added riboflavin have not been approved. Thiamin, riboflavin, and niacin values for products prepared with white flours represent the use of flours enriched at the 1974 levels and iron at the 1955 levels. Enriched flour is predominantly used in home-prepared and commercially prepared baked goods.

Fatty acid values are given for dairy products, eggs, meats, some grain products, nuts, and soups. The values are based on comprehensive research by USDA to update and extend tables for fatty acid content for foods.

Niacin values are for preformed niacin occurring naturally in foods. The values do not include additional niacin that the body may form from tryptophan, an essential amino acid in the protein of most foods. Among the better sources of tryptophan are milk, meats, eggs, legumes, and nuts.

Values have been calculated from the ingredients in typical recipes for many of the prepared items such as biscuits, corn muffins, macaroni and cheese, custard, and many dessert-type items.

EQUIVALENTS BY VOLUME AND WEIGHT FOR TABLE A.1

Volume

Level measure	Equivalent
1 gallon (3.786 liters; 3,786 milliliters)	4 quarts
1 quart (0.946 liter; 946 milliliters)	4 cups
1 cup (237 milliliters)	8 fluid ounces ½ pint 16 tablespoons
2 tablespoons (30 milliliters)	1 fluid ounce
1 tablespoon (15 milliliters)	3 teaspoons
1 pound regular butter or margarine	4 sticks 2 cups
1 pound whipped butter or margarine	6 sticks Two 8-ounce containers 3 cups

Weight

Avoirdupois weight	Equivalent
1 pound (16 ounces)	453.6 grams
1 ounce	28.35 grams
3½ ounces	100 grams

Values for toast and cooked vegetables are without fat added, either during preparation or at the table. Some destruction of vitamins, especially ascorbic acid, may occur when vegetables are cut or shredded. Since such losses are variable, no deduction has been made.

For meat, values are for meat cooked and drained of the drippings. For many cuts, two sets of values are shown: meat including fat and meat from which the fat has been removed either in the kitchen or on the plate.

A variety of manufactured items—some

of the milk products, ready-to-eat breakfast cereals, imitation cream products, fruit drinks, and various mixes—are included in Table A.1. Frequently those foods are fortified with one or more nutrients. If nutrients are added, this information is on the label. Values shown here for those foods are usually based on

products from several manufacturers and may differ somewhat from the values provided by any one source.

SOURCE: Nutritive Values of Foods, Home and Garden Bulletin No. 72, Table 2, pp. 4–29. U.S. Department of Agriculture, Washington, D.C., revised 1981.

Table A.1. NUTRITIVE VALUES OF THE EDIBLE PART OF FOODS (Dashes denote lack of reliable data for a constituent believed to be present in measurable amounts.)

DAIRY PRODUCTS (CHEESE, CREAM, IMITATION CREAM, MILK; RELATED PRODUCTS)

Butter. See Fats, oils; related products, items 103-108.

Item No. (A)	Foods, approximate measures, units, and weight (edible part unless footnotes indicate otherwise) (B)	Grams	Water (C) Percent	Food energy (D) Calories	Protein (E) Grams	Fat (F) Grams	Saturated (total) (G) Grams	Oleic (H) Grams	Linoleic (I) Grams	Carbohydrate (I) Grams	Calcium (K) Milligrams	Phosphorus (L) Milligrams	Iron (M) Milligrams	Potassium (N) Milligrams	Vitamin A value (O) International units	Thiamin (P) Milligrams	Riboflavin (Q) Milligrams	Niacin (R) Milligrams	Ascorbic acid (S) Milligrams
	Cheese:																		
	Natural:																		
1	Blue---------- 1 oz-----------	28	42	100	6	8	5.3	1.9	0.2	1	150	110	0.1	73	200	0.01	0.11	0.3	0
2	Camembert (3 wedges per 4-oz container). 1 wedge----------	38	52	115	8	9	5.8	2.2	.2	Trace	147	132	.1	71	350	.01	.19	.2	0
	Cheddar:																		
3	Cut pieces------------ 1 oz-----------	28	37	115	7	9	6.1	2.1	.2	Trace	204	145	.2	28	300	.01	.11	Trace	0
4	1 cu in---------	17.2	37	70	4	6	3.7	1.3	.1	Trace	124	88	.2	17	180	Trace	.06	Trace	0
5	Shredded------------- 1 cup-----------	113	37	455	28	37	24.2	8.5	.7	1	815	579	.8	111	1,200	.03	.42	.1	0
	Cottage (curd not pressed down):																		
	Creamed (cottage cheese, 4% fat):																		
6	Large curd------------ 1 cup-----------	225	79	235	28	10	6.4	2.4	.2	6	135	297	.3	190	370	.05	.37	.3	Trace
7	Small curd------------ 1 cup-----------	210	79	220	26	9	6.0	2.2	.2	6	126	277	.3	177	340	.04	.34	.3	Trace
8	Low fat (2%)--------- 1 cup-----------	226	79	205	31	4	2.8	1.0	.1	8	155	340	.4	217	160	.05	.42	.3	Trace
9	Low fat (1%)--------- 1 cup-----------	226	82	165	28	2	1.5	.5	.1	6	138	302	.3	193	80	.05	.37	.3	Trace
10	Uncreamed (cottage cheese dry curd, less than 1/2% fat). 1 cup-----------	145	80	125	25	1	.4	.1	Trace	3	46	151	.3	47	40	.04	.21	.2	0
11	Cream----------- 1 oz-----------	28	54	100	2	10	6.2	2.4	.2	1	23	30	.3	34	400	Trace	.06	Trace	0
	Mozzarella, made with—																		
12	Whole milk----------- 1 oz-----------	28	48	90	6	7	4.4	1.7	.2	1	163	117	.1	21	260	Trace	.08	Trace	0
13	Part skim milk------- 1 oz-----------	28	49	80	8	5	3.1	1.2	.1	1	207	149	.1	27	180	.01	.10	Trace	0
	Parmesan, grated:																		
14	Cup, not pressed down-- 1 cup-----------	100	18	455	42	30	19.1	7.7	.3	4	1,376	807	1.0	107	700	.05	.39	.3	0
15	Tablespoon----------- 1 tbsp----------	5	18	25	2	2	1.0	.4	Trace	Trace	69	40	Trace	5	40	Trace	.02	Trace	0
16	Ounce---------------- 1 oz-----------	28	18	130	12	9	5.4	2.2	.3	1	390	229	.3	30	200	.01	.11	.1	0
17	Provolone------------ 1 oz-----------	28	41	100	7	8	4.8	2.1	.1	1	214	141	.1	39	230	.01	.09	Trace	0
	Ricotta, made with—																		
18	Whole milk----------- 1 cup-----------	246	72	430	28	32	20.4	7.1	.7	7	509	389	.9	257	1,210	.03	.48	.3	0
19	Part skim milk------- 1 cup-----------	246	74	340	28	19	12.1	4.7	.5	13	669	449	1.1	308	1,060	.05	.46	.2	0
20	Romano--------------- 1 oz-----------	28	31	110	9	8	—	—	—	1	302	215	—	—	160	—	.11	Trace	0
21	Swiss---------------- 1 oz-----------	28	37	105	8	8	5.0	1.7	.2	1	272	171	Trace	31	240	.01	.10	Trace	0
	Pasteurized process cheese:																		
22	American------------- 1 oz-----------	28	39	105	6	9	5.6	2.1	.2	Trace	174	211	.1	46	340	.01	.10	Trace	0
23	Swiss---------------- 1 oz-----------	28	42	95	7	7	4.5	1.7	.1	1	219	216	.2	61	230	Trace	.08	Trace	0
24	Pasteurized process cheese food, American. 1 oz-----------	28	43	95	6	7	4.4	1.7	.1	2	163	130	.2	79	260	.01	.13	Trace	0
25	Pasteurized process cheese spread, American. 1 oz-----------	28	48	80	5	6	3.8	1.5	.1	2	159	202	.1	69	220	.01	.12	Trace	0
	Cream, sweet:																		
26	Half-and-half (cream and milk)- 1 cup-----------	242	81	315	7	28	17.3	7.0	.6	10	254	230	.2	314	260	.08	.36	.2	2
27	1 tbsp----------	15	81	20	Trace	2	1.1	.4	Trace	1	16	14	Trace	19	20	.01	.02	Trace	Trace
28	Light, coffee, or table---- 1 cup-----------	240	74	470	6	46	28.8	11.7	1.0	9	231	192	.1	292	1,730	.08	.36	.1	2
29	1 tbsp----------	15	74	30	Trace	3	1.8	.7	.1	1	14	12	Trace	18	110	Trace	.02	Trace	Trace

(A)	(B)	(C)	(D)	(E)	(F)	(G)	(H)	(I)	(J)	(K)	(L)	(M)	(N)	(O)	(P)	(Q)	(R)	(S)
	Whipping, unwhipped (volume about double when whipped):																	
30	Light------ 1 cup	239	700	5	74	46.2	18.3	1.5	7	166	146	0.1	231	2,690	0.06	0.30	0.1	1
31	1 tbsp	15	45	Trace	5	2.9	1.1	.1	Trace	10	9	Trace	15	170	Trace	.02	Trace	Trace
32	Heavy------ 1 cup	238	820	Trace	88	54.8	22.2	2.0	7	154	149	.1	179	3,500	.05	.26	.1	1
33	1 tbsp	15	80	Trace	6	3.5	1.4	.1	Trace	10	9	Trace	11	220	Trace	.02	.1	Trace
34	Whipped topping, (pressurized)- 1 cup	60	155	2	13	8.3	3.4	.3	7	61	54	Trace	88	550	.02	.04	Trace	0
35	1 tbsp	3	10	Trace	1	.4	.2	Trace	Trace	3	3	Trace	4	30	Trace	Trace	Trace	0
36	Cream, sour------ 1 cup	230	495	7	48	30.0	12.1	1.1	10	268	195	.1	331	1,820	.08	.34	.2	2
37	1 tbsp	12	25	Trace	3	1.6	.6	.1	1	14	10	Trace	17	90	Trace	.02	Trace	Trace
	Cream products, imitation (made with vegetable fat):																	
	Sweet:																	
	Creamers:																	
38	Liquid (frozen)------ 1 cup	245	335	2	24	22.8	.3	Trace	28	23	157	.1	467	[1]220	0	0	0	0
39	1 tbsp	15	20	Trace	1	1.4	Trace	0	2	1	10	Trace	29	[1]10	0	0	0	0
40	Powdered------ 1 cup	94	515	5	33	30.6	.9	Trace	52	21	397	.1	763	[1]190	0	.16	0	0
41	1 tsp	2	10	Trace	1	.7	Trace	0	1	Trace	8	Trace	16	[1]Trace	0	[1]Trace	0	0
	Whipped topping:																	
42	Frozen------ 1 cup	75	240	1	19	16.3	1.0	.2	17	5	6	.1	14	[1]650	0	0	0	0
43	1 tbsp	4	15	Trace	1	.9	.1	Trace	1	Trace	Trace	Trace	1	[1]30	0	0	0	0
44	Powdered, made with whole milk. 1 cup	80	150	3	10	8.5	.6	.1	13	72	69	Trace	121	[1]290	.02	.09	Trace	1
45	1 tbsp	4	10	Trace	Trace	.4	Trace	Trace	1	4	3	Trace	6	[1]110	Trace	0	Trace	Trace
46	Pressurized------ 1 cup	70	185	1	16	13.2	1.4	.2	11	4	13	Trace	13	[1]330	0	0	0	0
47	1 tbsp	4	10	Trace	Trace	.4	.1	Trace	1	Trace	1	Trace	1	[1]120	0	0	0	0
48	Sour dressing (imitation sour cream) made with nonfat dry milk. 1 cup	235	415	8	39	31.2	4.4	1.1	11	266	205	.1	380	[1]120	.09	.38	.2	2
49	1 tbsp	12	20	Trace	2	1.6	.2	.1	1	14	10	Trace	19	[1]Trace	.01	.02	Trace	Trace
	Ice cream. See Milk desserts, frozen (items 75-80).																		
	Ice milk. See Milk desserts, frozen (items 81-83).																		
	Milk:																		
	Fluid:																		
50	Whole (3.3% fat)------ 1 cup	244	150	8	8	5.1	2.1	.2	11	291	228	.1	370	[2]310	.09	.40	.2	2
	Lowfat (2%):																		
51	No milk solids added- 1 cup	244	120	8	5	2.9	1.2	.1	12	297	232	.1	377	500	.10	.40	.2	2
	Milk solids added:																		
52	Label claim less than 10 g of protein per cup. 1 cup	245	125	9	5	2.9	1.2	.1	12	313	245	.1	397	500	.10	.42	.2	2
53	Label claim 10 or more grams of protein per cup (protein fortified). 1 cup	246	135	10	5	3.0	1.2	.1	14	352	276	.1	447	500	.11	.48	.2	3
	Lowfat (1%):																		
54	No milk solids added- 1 cup	244	100	8	3	1.6	.7	.1	12	300	235	.1	381	500	.10	.41	.2	2
	Milk solids added:																		
55	Label claim less than 10 g of protein per cup. 1 cup	245	105	9	2	1.5	.6	.1	12	313	245	.1	397	500	.10	.42	.2	2
56	Label claim 10 or more grams of protein per cup (protein fortified). 1 cup	246	120	10	3	1.8	.7	.1	14	349	273	.1	444	500	.11	.47	.2	3
	Nonfat (skim):																		
57	No milk solids added- 1 cup	245	85	8	Trace	.3	.1	Trace	12	302	247	.1	406	500	.09	.34	.2	2

[1]Vitamin A value is largely from beta-carotene used for coloring. Riboflavin value for items 40-41 apply to products with added riboflavin.

[2]Applies to product without added vitamin A. With added vitamin A, value is 500 International Units (I.U.).

DAIRY PRODUCTS (CHEESE, CREAM, IMITATION CREAM, MILK; RELATED PRODUCTS)—Con.

Milk—Continued
Fluid—Continued
Nonfat (skim)—Continued

Item No. (A)	Foods, approximate measures, units, and weight (edible part unless footnotes indicate otherwise) (B)	Grams	Water (C) Percent	Food energy (D) Calories	Protein (E) Grams	Fat (F) Grams	Saturated (total) (G) Grams	Oleic (H) Grams	Linoleic (I) Grams	Carbohydrate (I) Grams	Calcium (K) mg	Phosphorus (L) mg	Iron (M) mg	Potassium (N) mg	Vitamin A value (O) IU	Thiamin (P) mg	Riboflavin (Q) mg	Niacin (R) mg	Ascorbic acid (S) mg
58	Milk solids added: Label claim less than 10 g of protein per cup — 1 cup	245	90	90	9	1	0.4	0.1	Trace	12	316	255	0.1	416	500	0.10	0.43	0.2	2
59	Label claim 10 or more grams of protein per cup (protein fortified) — 1 cup	246	89	100	10	1	.4	.1	Trace	14	352	275	.1	446	500	.11	.48	.2	3
60	Buttermilk — 1 cup	245	90	100	8	2	1.3	.5	Trace	12	285	219	.1	371	[3]380	.08	.38	.1	2
	Canned: Evaporated, unsweetened:																		
61	Whole milk — 1 cup	252	74	340	17	19	11.6	5.3	0.4	25	657	510	.5	764	[3]610	.12	.80	.5	5
62	Skim milk — 1 cup	255	79	200	19	1	.3	.1	Trace	29	738	497	.7	845	[4]1,000	.11	.79	.4	3
63	Sweetened, condensed — 1 cup	306	27	980	24	27	16.8	6.7	.7	166	868	775	.6	1,136	[3]1,000	.28	1.27	.6	8
	Dried:																		
64	Buttermilk — 1 cup	120	3	465	41	7	4.3	1.7	.2	59	1,421	1,119	.4	1,910	[3]260	.47	1.90	1.1	7
	Nonfat instant:																		
65	Envelope, net wt. 3.2 oz[5] — 1 envelope	91	4	325	32	Trace	.4	.1	Trace	47	1,120	896	.3	1,552	[6]2,160	.38	1.59	.8	5
66	Cup — 1 cup	68	4	245	24	Trace	.3	.1	Trace	35	837	670	.2	1,160	[6]1,610	.28	1.19	.6	4
	Milk beverages: Chocolate milk (commercial):																		
67	Regular — 1 cup	250	82	210	8	8	5.3	2.2	.2	26	280	251	.6	417	[3]300	.09	.41	.3	2
68	Lowfat (2%) — 1 cup	250	84	180	8	5	3.1	1.3	.1	26	284	254	.6	422	500	.10	.42	.3	2
69	Lowfat (1%) — 1 cup	250	85	160	8	3	1.5	.7	.1	26	287	257	.6	426	500	.10	.40	.2	2
70	Eggnog (commercial) — 1 cup	254	74	340	10	19	11.3	5.0	.6	34	330	278	.5	420	890	.09	.48	.3	4
	Malted milk, home-prepared with 1 cup of whole milk and 2 to 3 heaping tsp of malted milk powder (about 3/4 oz):																		
71	Chocolate — 1 cup of milk plus 3/4 oz of powder	265	81	235	9	9	5.5	—	—	29	304	265	.5	500	330	.14	.43	.7	2
72	Natural — 1 cup of milk plus 3/4 oz of powder	265	81	235	11	10	6.0	—	—	27	347	307	.3	529	380	.20	.54	1.3	2
	Shakes, thick:[8]																		
73	Chocolate, container, net wt. 10.6 oz — 1 container	300	72	355	9	8	5.0	2.0	.2	63	396	378	.9	672	260	.14	.67	.4	0
74	Vanilla, container, net wt. 11 oz — 1 container	313	74	350	12	9	5.9	2.4	.2	56	457	361	.3	572	360	.09	.61	.5	0
	Milk desserts, frozen: Ice cream: Regular (about 11% fat):																		
75	Hardened — 1/2 gal	1,064	61	2,155	38	115	71.3	28.8	2.6	254	1,406	1,075	1.0	2,052	4,340	.42	2.63	1.1	6
76	1 cup	133	61	270	5	14	8.9	3.6	.3	32	176	134	.1	257	540	.05	.33	.1	1
77	3-fl oz container	50	61	100	2	7	3.4	1.4	.1	12	66	51	Trace	96	200	.02	.12	Trace	Trace
78	Soft serve (frozen custard) — 1 cup	173	60	375	7	23	13.5	5.9	.6	38	236	199	.4	338	790	.08	.45	.2	5
79	Rich (about 16% fat), hardened — 1/2 gal	1,188	59	2,805	33	190	118.3	47.8	4.3	256	1,213	927	.8	1,771	7,200	.36	2.27	.9	5
80	1 cup	148	59	350	4	24	14.7	6.0	.5	32	151	115	.1	221	900	.04	.28	.1	1
	Ice milk:																		
81	Hardened (about 4.3% fat) — 1/2 gal	1,048	69	1,470	41	45	28.1	11.3	1.0	232	1,409	1,035	1.5	2,117	1,710	.61	2.78	.9	6
82	1 cup	131	69	185	5	6	3.5	1.4	.1	29	176	129	.1	265	210	.08	.35	.1	1

(A)	(B)	(g)	(C)	(D)	(E)	(F)	(G)	(H)	(I)	(J)	(K)	(L)	(M)	(N)	(O)	(P)	(Q)	(R)	(S)
83	Soft serve (about 2.6% fat)—1 cup	175	70	225	8	5	2.9	1.2	0.1	38	274	202	0.3	412	180	0.12	0.54	0.2	1
84	Sherbet (about 2% fat)—1/2 gal	1,542	66	2,160	17	31	19.0	7.7	.7	469	827	594	2.5	1,585	1,480	.26	.71	1.0	31
85	—1 cup	193	66	270	2	4	2.4	1.0	.1	59	103	74	.3	198	190	.03	.09	.1	4
	Milk desserts, other:																		
86	Custard, baked—1 cup	265	77	305	14	15	6.8	5.4	.7	29	297	310	1.1	387	930	.11	.50	.3	1
	Puddings:																		
	From home recipe:																		
	Starch base:																		
87	Chocolate—1 cup	260	66	385	8	12	7.6	3.3	.3	67	250	255	1.3	445	390	.05	.36	.3	1
88	Vanilla (blancmange)—1 cup	255	76	285	9	10	6.2	2.5	.2	41	298	232	Trace	352	410	.08	.41	.3	2
89	Tapioca cream—1 cup	165	72	220	8	8	4.1	2.5	.5	28	173	180	.7	223	480	.07	.30	.2	2
	From mix (chocolate) and milk:																		
90	Regular (cooked)—1 cup	260	70	320	9	8	4.3	2.6	.2	59	265	247	.8	354	340	.05	.39	.3	2
91	Instant—1 cup	260	69	325	8	7	3.6	2.2	.3	63	374	237	1.3	335	340	.08	.39	.3	2
	Yogurt:																		
	With added milk solids:																		
	Made with lowfat milk:																		
92	Fruit-flavored[9]—1 container, net wt., 8 oz	227	75	230	10	3	1.8	.6	.1	42	343	269	.2	439	[10]120	.08	.40	.2	1
93	Plain—1 container, net wt., 8 oz	227	85	145	12	4	2.3	.8	.1	16	415	326	.2	531	[10]150	.10	.49	.3	2
94	Made with nonfat milk—1 container, net wt., 8 oz	227	85	125	13	Trace	.3	.1	Trace	17	452	355	.2	579	[10]20	.11	.53	.3	2
	Without added milk solids:																		
95	Made with whole milk—1 container, net wt., 8 oz	227	88	140	8	7	4.8	1.7	.1	11	274	215	.1	351	280	.07	.32	.2	1

EGGS

(A)	(B)	(g)	(C)	(D)	(E)	(F)	(G)	(H)	(I)	(J)	(K)	(L)	(M)	(N)	(O)	(P)	(Q)	(R)	(S)
	Eggs, large (24 oz per dozen):																		
	Raw:																		
96	Whole, without shell—1 egg	50	75	80	6	6	1.7	2.0	.6	1	28	90	1.0	65	260	.04	.15	Trace	0
97	White—1 white	33	88	15	3	Trace	0	0	0	Trace	4	4	Trace	45	0	Trace	.09	Trace	0
98	Yolk—1 yolk	17	49	65	3	6	1.7	2.1	.6	Trace	26	86	.9	15	310	.04	.07	Trace	0
	Cooked:																		
99	Fried in butter—1 egg	46	72	85	5	6	2.4	2.2	.6	1	26	80	.9	58	290	.03	.13	Trace	0
100	Hard-cooked, shell removed—1 egg	50	75	80	6	6	1.7	2.0	.6	1	28	90	1.0	65	260	.04	.14	Trace	0
101	Poached—1 egg	50	74	80	6	6	1.7	2.0	.6	1	28	90	1.0	65	260	.04	.13	Trace	0
102	Scrambled (milk added) in butter. Also omelet.—1 egg	64	76	95	6	7	2.8	2.3	.6	1	47	97	.9	85	310	.04	.16	Trace	0

FATS, OILS; RELATED PRODUCTS

(A)	(B)	(g)	(C)	(D)	(E)	(F)	(G)	(H)	(I)	(J)	(K)	(L)	(M)	(N)	(O)	(P)	(Q)	(R)	(S)
	Butter:																		
	Regular (1 brick or 4 sticks per lb):																		
103	Stick (1/2 cup)—1 stick	113	16	815	1	92	57.3	23.1	2.1	Trace	27	26	.2	29	[11]3,470	.01	.04	Trace	0
104	Tablespoon (about 1/8 stick)—1 tbsp	14	16	100	Trace	12	7.2	2.9	.3	Trace	3	3	Trace	4	[11]430	Trace	Trace	Trace	0
105	Pat (1 in square, 1/3 in high; 90 per lb)—1 pat	5	16	35	Trace	4	2.5	1.0	.1	Trace	1	1	Trace	1	[11]150	Trace	Trace	Trace	0
	Whipped (6 sticks or two 8-oz containers per lb):																		
106	Stick (1/2 cup)—1 stick	76	16	540	1	61	38.2	15.4	1.4	Trace	18	17	.1	20	[11]2,310	Trace	.03	Trace	0
107	Tablespoon (about 1/8 stick)—1 tbsp	9	16	65	Trace	8	4.7	1.9	.2	Trace	2	2	Trace	2	[11]290	Trace	Trace	Trace	0
108	Pat (1 1/4 in square, 1/3 in high; 120 per lb)—1 pat	4	16	25	Trace	3	1.9	.8	.1	Trace	1	1	Trace	1	[11]120	0	Trace	Trace	0

[3] Applies to product without vitamin A added.
[4] Applies to product with added vitamin A. Without added vitamin A, value is 20 International Units (I.U.).
[5] Yields 1 qt of fluid milk when reconstituted according to package directions.
[6] Applies to product with added vitamin A.
[7] Weight applies to product with label claim of 1 1/3 cups equal 3.2 oz.
[8] Applies to products made from thick shake mixes and that do not contain added ice cream. Products made from milk shake mixes are higher in fat and usually contain added ice cream.
[9] Content of fat, vitamin A, and carbohydrate varies. Consult the label when precise values are needed for special diets.
[10] Applies to product made with milk containing no added vitamin A.
[11] Based on year-round average.

NUTRIENTS IN INDICATED QUANTITY

Item No. (A)	Foods, approximate measures, units, and weight (edible part unless footnotes indicate otherwise) (B)	Grams	Water (C) Percent	Food energy (D) Calories	Protein (E) Grams	Fat (F) Grams	Fatty Acids Saturated (total) (G) Grams	Unsaturated Oleic (H) Grams	Linoleic (I) Grams	Carbohydrate (J) Grams	Calcium (K) Milligrams	Phosphorus (L) Milligrams	Iron (M) Milligrams	Potassium (N) Milligrams	Vitamin A value (O) International units	Thiamin (P) Milligrams	Riboflavin (Q) Milligrams	Niacin (R) Milligrams	Ascorbic acid (S) Milligrams
	FATS, OILS; RELATED PRODUCTS—Con.																		
109	Fats, cooking (vegetable shortenings). 1 cup---	200	0	1,770	0	200	48.8	88.2	48.4	0	0	0	0	0	—	0	0	0	0
110	1 tbsp---	13	0	110	0	13	3.2	5.7	3.1	0	0	0	0	0		0	0	0	0
111	Lard------ 1 cup---	205	0	1,850	0	205	81.0	83.8	20.5	0	0	0	0	0	0	0	0	0	0
112	1 tbsp---	13	0	115	0	13	5.1	5.3	1.3	0	0	0	0	0	0	0	0	0	0
	Margarine: Regular (1 brick or 4 sticks per lb):																		
113	Stick (1/2 cup)---	113	16	815	1	92	16.7	42.9	24.9	Trace	27	26	.2	29	[12]3,750	.01	.04	Trace	0
114	Tablespoon (about 1/8 stick)---	14	16	100	Trace	12	2.1	5.3	3.1	Trace	3	3	Trace	4	[12]470	Trace	Trace	Trace	0
115	Pat (1 in square, 1/3 in high; 90 per lb).	5	16	35	Trace	4	.7	1.9	1.1	Trace	1	1	Trace	1	[12]170	Trace	Trace	Trace	0
116	Soft, two 8-oz containers per lb. 1 container---	227	16	1,635	1	184	32.5	71.5	65.4	Trace	53	52	.4	59	[12]7,500	.01	.08	.1	0
117	1 tbsp---	14	16	100	Trace	12	2.0	4.5	4.1	Trace	3	3	Trace	4	[12]470	Trace	Trace	Trace	0
	Whipped (6 sticks per lb):																		
118	Stick (1/2 cup)---	76	16	545	Trace	61	11.2	28.7	16.7	Trace	18	17	.1	20	[12]2,500	Trace	Trace	Trace	0
119	Tablespoon (about 1/8 stick)---	9	16	70	Trace	8	1.4	3.6	2.1	Trace	2	2	Trace	2	[12]310	Trace	Trace	Trace	0
	Oils, salad or cooking:																		
120	Corn------ 1 cup---	218	0	1,925	0	218	27.7	53.6	125.1	0	0	0	0	0	—	0	0	0	0
121	1 tbsp---	14	0	120	0	14	1.7	3.3	7.8	0	0	0	0	0	—	0	0	0	0
122	Olive------ 1 cup---	216	0	1,910	0	216	30.7	154.4	17.7	0	0	0	0	0	—	0	0	0	0
123	1 tbsp---	14	0	120	0	14	1.9	9.7	1.1	0	0	0	0	0	—	0	0	0	0
124	Peanut------ 1 cup---	216	0	1,910	0	216	37.4	98.5	67.0	0	0	0	0	0	—	0	0	0	0
125	1 tbsp---	14	0	120	0	14	2.3	6.2	4.2	0	0	0	0	0	—	0	0	0	0
126	Safflower--- 1 cup---	218	0	1,925	0	218	20.5	25.9	159.8	0	0	0	0	0	—	0	0	0	0
127	1 tbsp---	14	0	120	0	14	1.3	1.6	10.0	0	0	0	0	0	—	0	0	0	0
128	Soybean oil, hydrogenated (partially hardened). 1 cup---	218	0	1,925	0	218	31.8	93.1	75.6	0	0	0	0	0	—	0	0	0	0
129	1 tbsp---	14	0	120	0	14	2.0	5.8	4.7	0	0	0	0	0	—	0	0	0	0
130	Soybean-cottonseed oil blend, hydrogenated. 1 cup---	218	0	1,925	0	218	38.2	63.0	99.6	0	0	0	0	0	—	0	0	0	0
131	1 tbsp---	14	0	120	0	14	2.4	3.9	6.2	0	0	0	0	0	—	0	0	0	0
	Salad dressings: Commercial: Blue cheese:																		
132	Regular--- 1 tbsp---	15	32	75	1	8	1.6	1.7	3.8	1	12	11	Trace	6	30	Trace	.02	Trace	Trace
133	Low calorie (5 Cal per tsp) 1 tbsp---	16	84	10	Trace	1	.5	.3	Trace	1	10	8	Trace	5	30	Trace	.01	Trace	Trace
	French:																		
134	Regular--- 1 tbsp---	16	39	65	Trace	6	1.1	1.3	3.2	3	2	2	.1	13	—	—	—	—	—
135	Low calorie (5 Cal per tsp) 1 tbsp---	16	77	15	Trace	1	.1	.1	.4	2	2	2	.1	13	—	—	—	—	—
	Italian:																		
136	Regular--- 1 tbsp---	15	28	85	Trace	9	1.6	1.9	4.7	1	2	1	Trace	2	Trace	Trace	Trace	Trace	—
137	Low calorie (2 Cal per tsp) 1 tbsp---	15	90	10	Trace	1	.1	.1	.4	Trace	2	1	Trace	2	Trace	Trace	Trace	Trace	—
138	Mayonnaise--- 1 tbsp---	14	15	100	Trace	11	2.0	2.4	5.6	Trace	3	4	.1	5	40	Trace	.01	Trace	—
	Mayonnaise type:																		
139	Regular--- 1 tbsp---	15	41	65	Trace	6	1.1	1.4	3.2	2	2	4	Trace	1	30	Trace	Trace	Trace	—
140	Low calorie (8 Cal per tsp) 1 tbsp---	16	81	20	Trace	2	.4	.4	1.0	2	3	4	Trace	1	40	Trace	Trace	Trace	—
141	Tartar sauce, regular--- 1 tbsp---	14	34	75	Trace	8	1.5	1.8	4.1	1	3	4	.1	11	30	Trace	Trace	Trace	Trace
	Thousand Island:																		
142	Regular--- 1 tbsp---	16	32	80	Trace	8	1.4	1.7	4.0	2	2	3	.1	18	50	Trace	Trace	Trace	Trace
143	Low calorie (10 Cal per tsp) 1 tbsp---	15	68	25	Trace	2	.4	.4	1.0	2	2	3	.1	17	50	Trace	Trace	Trace	Trace
	From home recipe:																		
144	Cooked type[13]--- 1 tbsp---	16	68	25	1	2	.5	.6	.3	2	14	15	.1	19	80	.01	.03	Trace	Trace

FISH, SHELLFISH, MEAT, POULTRY; RELATED PRODUCTS

(A)	(B)	(C)	(D)	(E)	(F)	(G)	(H)	(I)	(J)	(K)	(L)	(M)	(N)	(O)	(P)	(Q)	(R)	(S)
	Fish and shellfish:																	
145	Bluefish, baked with butter or margarine. 3 oz	85	68	135	22	4	—	—	—	0	25	244	0.5	40	0.09	0.08	1.6	—
	Clams:																	
146	Raw, meat only. 3 oz	85	82	65	11	1	—	—	—	2	59	138	5.2	90	.08	.15	1.1	8
147	Canned, solids and liquid. 3 oz	85	86	45	7	1	—	—	—	2	47	116	3.5	—	.01	.09	.9	—
148	Crabmeat (white or king), canned, not pressed down. 1 cup	135	77	135	24	3	0.2	Trace	Trace	1	61	246	1.1	—	.11	.11	2.6	—
149	Fish sticks, breaded, cooked, frozen (stick, 4 by 1 by 1/2 in). 1 fish stick or 1 oz	28	66	50	5	3	.6	0.4	0.1	2	3	47	.1	0	.01	.02	.5	—
150	Haddock, breaded, fried[14]. 3 oz	85	66	140	17	5	1.4	2.2	1.2	5	34	210	1.0	—	.03	.06	2.7	2
151	Ocean perch, breaded, fried[14]. 1 fillet	85	59	195	16	11	2.7	4.4	2.3	6	28	192	1.1	—	.10	.10	1.6	—
152	Oysters, raw, meat only (13-19 medium Selects). 1 cup	240	85	160	20	4	1.3	.2	.1	8	226	343	13.2	740	.34	.43	6.0	—
153	Salmon, pink, canned, solids and liquid. 3 oz	85	71	120	17	5	.9	.8	.1	0	[15]167	243	.7	60	.03	.16	6.8	—
154	Sardines, Atlantic, canned in oil, drained solids. 3 oz	85	62	175	20	9	3.0	2.5	.5	0	372	424	2.5	190	.02	.17	4.6	—
155	Scallops, frozen, breaded, fried, reheated. 6 scallops	90	60	175	16	8	—	—	—	9	—	—	—	—	—	—	—	—
156	Shad, baked with butter or margarine, bacon. 3 oz	85	64	170	20	10	—	—	—	0	20	266	.5	30	.11	.22	7.3	—
	Shrimp:																	
157	Canned meat. 3 oz	85	70	100	21	1	.1	.1	Trace	1	98	224	2.6	50	.01	.03	1.5	—
158	French fried[16]. 3 oz	85	57	190	17	9	2.3	3.7	2.0	9	61	162	1.7	—	.03	.07	2.3	—
159	Tuna, canned in oil, drained solids. 3 oz	85	61	170	24	7	1.7	1.7	.7	0	7	199	1.6	70	.04	.10	10.1	2
160	Tuna salad[17]. 1 cup	205	70	350	30	22	4.3	6.3	6.7	7	41	291	2.7	590	.08	.23	10.3	—
	Meat and meat products:																	
161	Bacon, (20 slices per lb, raw), broiled or fried, crisp. 2 slices	15	8	85	4	8	2.5	3.7	.7	Trace	2	34	.5	0	.08	.05	.8	—
	Beef, cooked: Cuts braised, simmered or pot roasted:																	
162	Lean and fat (piece, 2 1/2 by 2 1/2 by 3/4 in). 3 oz	85	53	245	23	16	6.8	6.5	.4	0	10	114	2.9	30	.04	.18	3.6	—
163	Lean only from item 162. 2.5 oz	72	62	140	22	5	2.1	1.8	.2	0	10	108	2.7	10	.04	.17	3.3	—
	Ground beef, broiled:																	
164	Lean with 10% fat. 3 oz or patty 3 by 5/8 in	85	60	185	23	10	4.0	3.9	.3	0	10	196	3.0	20	.08	.20	5.1	—
165	Lean with 21% fat. 2.9 oz or patty 3 by 5/8 in	82	54	235	20	17	7.0	6.7	.4	0	9	159	2.6	30	.07	.17	4.4	—
	Roast, oven cooked, no liquid added: Relatively fat, such as rib:																	
166	Lean and fat (2 pieces, 4 1/8 by 2 1/4 by 1/4 in). 3 oz	85	40	375	17	33	14.0	13.6	.8	0	8	158	2.2	70	.05	.13	3.1	—
167	Lean only from item 166[18]. 1.8 oz	51	57	125	14	7	3.0	2.5	.3	0	6	131	1.8	10	.04	.11	2.6	—
	Relatively lean, such as heel of round:																	
168	Lean and fat (2 pieces, 4 1/8 by 2 1/4 by 1/4 in). 3 oz	85	62	165	25	7	2.8	2.7	.2	0	11	208	3.2	10	.06	.19	4.5	—

[12] Based on average vitamin A content of fortified margarine. Federal specifications for fortified margarine require a minimum of 15,000 International Units (I.U.) of vitamin A per pound.

[13] Fatty acid values apply to product made with regular-type margarine.

[14] Dipped in egg, milk or water, and breadcrumbs; fried in vegetable shortening.

[15] If bones are discarded, value for calcium will be greatly reduced.

[16] Dipped in egg, breadcrumbs, and flour or batter.

[17] Prepared with tuna, celery, salad dressing (mayonnaise type), pickle, onion, and egg.

[18] Outer layer of fat on the cut was removed to within approximately 1/2 in of the lean. Deposits of fat within the cut were not removed.

FISH, SHELLFISH, MEAT, POULTRY; RELATED PRODUCTS—Con.

NUTRIENTS IN INDICATED QUANTITY

Item No. (A)	Foods, approximate measures, units, and weight (edible part unless footnotes indicate otherwise) (B)	Grams	Water (C) Per cent	Food energy (D) Calories	Protein (E) Grams	Fat (F) Grams	Fatty Acids Saturated (total) (G) Grams	Unsaturated Oleic (H) Grams	Linoleic (I) Grams	Carbohydrate (J) Grams	Calcium (K) Milligrams	Phosphorus (L) Milligrams	Iron (M) Milligrams	Potassium (N) Milligrams	Vitamin A value (O) International units	Thiamin (P) Milligrams	Riboflavin (Q) Milligrams	Niacin (R) Milligrams	Ascorbic acid (S) Milligrams
	Meat and meat products—Continued																		
	Beef,[18] cooked—Continued																		
	Roast, oven cooked, no liquid added—Continued																		
	Relatively lean such as heel of round—Continued																		
169	Lean only from item 168--- 2.8 oz---	78	65	125	24	3	1.2	1.0	0.1	0	10	199	3.0	268	Trace	0.06	0.18	4.3	---
	Steak:																		
	Relatively fat—sirloin, broiled:																		
170	Lean and fat (piece, 2 1/2 by 2 1/2 by 3/4 in) 3 oz---	85	44	330	20	27	11.3	11.1	.6	0	9	162	2.5	220	50	.05	.15	4.0	---
171	Lean only from item 170--- 2.0 oz---	56	59	115	18	4	1.8	1.6	.2	0	7	146	2.2	202	10	.05	.14	3.6	---
	Relatively lean—round, braised:																		
172	Lean and fat (piece, 4 1/8 by 2 1/4 by 1/2 in) 3 oz---	85	55	220	24	13	5.5	5.2	.4	0	10	213	3.0	272	20	.07	.19	4.8	---
173	Lean only from item 172--- 2.4 oz---	68	61	130	21	4	1.7	1.5	.2	0	9	182	2.5	238	10	.05	.16	4.1	---
	Beef, canned:																		
174	Corned beef--- 3 oz---	85	59	185	22	10	4.9	4.5	.2	0	17	90	3.7	---	---	.01	.20	2.9	---
175	Corned beef hash--- 1 cup---	220	67	400	19	25	11.9	10.9	.5	24	29	147	4.4	440	---	.02	.20	4.5	---
176	Beef, dried, chipped--- 2 1/2-oz jar---	71	48	145	24	4	2.0	2.0	.1	0	14	287	3.6	142	---	.05	.23	2.7	0
177	Beef and vegetable stew--- 1 cup---	245	82	220	16	11	4.9	4.5	.2	15	29	184	2.9	613	2,400	.15	.17	4.7	17
178	Beef potpie (home recipe), baked[19] (piece, 1/3 of 9-in diam. pie). 1 piece---	210	55	515	21	30	7.9	12.8	6.7	39	29	149	3.8	334	1,720	.30	.30	5.5	6
179	Chili con carne with beans, canned. 1 cup---	255	72	340	19	16	7.5	6.8	.3	31	82	321	4.3	594	150	.08	.18	3.3	---
180	Chop suey with beef and pork (home recipe). 1 cup---	250	75	300	26	17	8.5	6.2	.7	13	60	248	4.8	425	600	.28	.38	5.0	33
181	Heart, beef, lean, braised--- 3 oz---	85	61	160	27	5	1.5	1.1	.6	1	5	154	5.0	197	20	.21	1.04	6.5	1
	Lamb, cooked:																		
	Chop, rib (cut 3 per lb with bone), broiled:																		
182	Lean and fat--- 3.1 oz---	89	43	360	18	32	14.8	12.1	1.2	0	8	139	1.0	200	---	.11	.19	4.1	---
183	Lean only from item 182--- 2 oz---	57	60	120	16	6	2.5	2.1	.2	0	6	121	1.1	174	---	.09	.15	3.4	---
	Leg, roasted:																		
184	Lean and fat (2 pieces, 4 1/8 by 2 1/4 by 1/4 in). 3 oz---	85	54	235	22	16	7.3	6.0	.6	0	9	177	1.4	241	---	.13	.23	4.7	---
185	Lean only from item 184--- 2.5 oz---	71	62	130	20	5	2.1	1.8	.2	0	9	169	1.4	227	---	.12	.21	4.4	---
	Shoulder, roasted:																		
186	Lean and fat (3 pieces, 2 1/2 by 2 1/2 by 1/4 in). 3 oz---	85	50	285	18	23	10.8	8.8	.9	0	9	146	1.0	206	---	.11	.20	4.0	---
187	Lean only from item 186--- 2.3 oz---	64	61	130	17	6	3.6	2.3	.2	0	8	140	1.0	193	---	.10	.18	3.7	---
188	Liver, beef, fried[20] (slice, 6 1/2 by 2 3/8 by 3/8 in). 3 oz---	85	56	195	22	9	2.5	3.5	.9	5	9	405	7.5	323	[21]45,390	.22	3.56	14.0	23
	Pork, cured, cooked:																		
189	Ham, light cure, lean and fat, roasted (2 pieces, 4 1/8 by 2 1/4 by 1/4 in).[22] 3 oz---	85	54	245	18	19	6.8	7.9	1.7	0	8	146	2.2	199	0	.40	.15	3.1	---
	Luncheon meat:																		
190	Boiled ham, slice (8 per 8-oz pkg.). 1 oz---	28	59	65	5	5	1.7	2.0	.4	0	3	47	.8	---	0	.12	.04	.7	---
191	Canned, spiced or unspiced: Slice, approx. 3 by 2 by 1/2 in. 1 slice---	60	55	175	9	15	5.4	6.7	1.0	1	5	65	1.3	133	0	.19	.13	1.8	---

(A)	(B)	(C)	(D)	(E)	(F)	(G)	(H)	(I)	(J)	(K)	(L)	(M)	(N)	(O)	(P)	(Q)	(R)	(S)
	Pork, fresh,[18] cooked:																	
	Chop, loin (cut 3 per lb with bone), broiled:																	
192	Lean and fat --- 2.7 oz	78	305	19	25	8.9	10.4	2.2	0	9	209	2.7	216	0	0.75	0.22	4.5	---
193	Lean only from item 192 --- 2 oz	56	150	17	9	3.1	3.6	.8	0	7	181	2.2	192	0	.63	.18	3.8	---
	Roast, oven cooked, no liquid added:																	
194	Lean and fat (piece, 2 1/2 by 2 1/2 by 3/4 in) --- 3 oz	85	310	21	24	8.7	10.2	2.2	0	9	218	2.7	233	0	.78	.22	4.8	---
195	Lean only from item 194 --- 2.4 oz	68	175	20	10	3.5	4.1	.8	0	9	211	2.6	224	0	.73	.21	4.4	---
	Shoulder cut, simmered:																	
196	Lean and fat (3 pieces, 2 1/2 by 2 1/2 by 1/4 in) --- 3 oz	85	320	20	26	9.3	10.9	2.3	0	9	118	2.6	158	0	.46	.21	4.1	---
197	Lean only from item 196 --- 2.2 oz	63	135	18	6	2.2	2.6	.6	0	8	111	2.3	146	0	.42	.19	3.7	---
	Sausages (see also Luncheon meat (items 190–191)):																	
198	Bologna, slice (8 per 8-oz pkg) --- 1 slice	28	85	3	8	3.0	3.4	.5	Trace	2	36	.5	65	---	.05	.06	.7	---
199	Braunschweiger, slice (6 per 6-oz pkg) --- 1 slice	28	90	4	8	2.6	3.4	.8	1	3	69	1.7	---	1,850	.05	.41	2.3	---
200	Brown and serve (10-11 per 8-oz pkg), browned --- 1 link	17	70	3	6	2.3	2.8	.7	Trace	---	---	---	---	0	---	---	---	---
201	Deviled ham, canned --- 1 tbsp	13	45	2	4	1.5	1.8	.4	0	1	12	.3	---	---	.02	.01	.2	---
202	Frankfurter (8 per 1-lb pkg), cooked (reheated) --- 1 frankfurter	56	170	7	15	5.6	6.5	1.2	1	3	57	.8	---	0	.08	.11	1.4	---
203	Meat, potted (beef, chicken, turkey), canned --- 1 tbsp	13	30	2	2	---	---	0	0	---	---	---	---	---	Trace	.03	.2	---
204	Pork link (16 per 1-lb pkg), cooked --- 1 link	13	60	2	6	2.1	2.4	.5	Trace	1	21	.3	35	0	.10	.04	.5	---
	Salami:																	
205	Dry type, slice (12 per 4-oz pkg) --- 1 slice	10	45	2	4	1.6	1.6	.1	Trace	1	28	.4	---	---	.04	.03	.5	---
206	Cooked type, slice (8 per 8-oz pkg) --- 1 slice	28	90	5	7	3.1	3.0	.2	Trace	3	57	.7	---	---	.07	.07	1.2	---
207	Vienna sausage (7 per 4-oz can) --- 1 sausage	16	40	2	3	1.2	1.4	.2	Trace	1	24	.3	---	---	.01	.02	.4	---
	Veal, medium fat, cooked, bone removed:																	
208	Cutlet (4 1/8 by 2 1/4 by 1/2 in), braised or broiled --- 3 oz	85	185	23	9	4.0	3.4	.4	0	9	196	2.7	258	---	.06	.21	4.6	---
209	Rib (2 pieces, 4 1/8 by 2 1/4 by 1/4 in), roasted --- 3 oz	85	230	23	14	6.1	5.1	.6	0	10	211	2.9	259	---	.11	.26	6.6	---
	Poultry and poultry products:																	
	Chicken, cooked:																	
210	Breast, fried,[23] bones removed, 1/2 breast (3.3 oz with bones) --- 2.8 oz	79	160	26	5	1.4	1.8	1.1	1	9	218	1.3	---	70	.04	.17	11.6	---
211	Drumstick, fried,[23] bones removed (2 oz with bones) --- 1.3 oz	38	90	12	4	1.1	1.3	.9	Trace	6	89	.9	---	50	.03	.15	2.7	---
212	Half broiler, broiled, bones removed (10.4 oz with bones) --- 6.2 oz	176	240	42	7	2.2	2.5	1.3	0	16	355	3.0	483	160	.09	.34	15.5	---
213	Chicken, canned, boneless --- 3 oz	85	170	18	10	3.2	3.8	2.0	0	18	210	1.3	117	200	.03	.11	3.7	3
214	Chicken a la king, cooked (home recipe) --- 1 cup	245	470	27	34	2.7	14.3	3.3	12	127	358	2.5	404	1,130	.10	.42	5.4	12
215	Chicken and noodles, cooked (home recipe) --- 1 cup	240	365	22	18	5.9	7.1	3.5	26	26	247	2.2	149	430	.05	.17	4.3	Trace

[18] Outer layer of fat on the cut was removed to within approximately 1/2 in of the lean. Deposits of fat within the cut were not removed.
[19] Crust made with vegetable shortening and enriched flour.
[20] Regular-type margarine used.
[21] Value varies widely.
[22] About one-fourth of the outer layer of fat on the cut was removed. Deposits of fat within the cut were not removed.
[23] Vegetable shortening used.

Item No. (A)	Foods, approximate measures, units, and weight (cubic part unless footnotes indicate otherwise) (B)	Grams	Water (C) Per cent	Food energy (D) Calories	Protein (E) Grams	Fat (F) Grams	Fatty Acids Saturated (total) (G) Grams	Fatty Acids Unsaturated Oleic (H) Grams	Fatty Acids Unsaturated Linoleic (I) Grams	Carbo-hydrate (J) Grams	Calcium (K) Milligrams	Phos-phorus (L) Milligrams	Iron (M) Milligrams	Potas-sium (N) Milligrams	Vitamin A value International units	Thiamin (P) Milligrams	Ribo-flavin (Q) Milligrams	Niacin (R) Milligrams	Ascorbic acid (S) Milligrams
	FISH, SHELLFISH, MEAT, POULTRY; RELATED PRODUCTS—Con.																		
	Poultry and poultry products—Continued																		
	Chicken chow mein:																		
216	Canned—— 1 cup——	250	89	95	7	Trace	——	——	——	18	45	85	1.3	418	150	0.05	0.10	1.0	13
217	From home recipe—— 1 cup——	250	78	255	31	10	2.4	3.4	3.1	10	58	293	2.5	473	280	.08	.23	4.3	10
218	Chicken potpie (home recipe), baked, [19] piece (1/3 or 9-in diam. pie). 1 piece——	232	57	545	23	31	11.3	10.9	5.6	42	70	232	3.0	343	3,090	.34	.31	5.5	5
	Turkey, roasted, flesh without skin:																		
219	Dark meat, piece, 2 1/2 by 1 5/8 by 1/4 in. 4 pieces——	85	61	175	26	7	2.1	1.5	1.5	0	——	——	2.0	338	——	.03	.20	3.6	——
220	Light meat, piece, 4 by 2 by 1/4 in. 2 pieces——	85	62	150	28	3	.9	.6	.7	0	——	——	1.0	349	——	.04	.12	9.4	——
	Light and dark meat:																		
221	Chopped or diced—— 1 cup——	140	61	265	44	9	2.5	1.7	1.8	0	11	351	2.5	514	——	.07	.25	10.8	——
222	Pieces (1 slice white meat, 4 by 2 by 1/4 in with 2 slices dark meat, 2 1/2 by 1 5/8 by 1/4 in). 3 pieces——	85	61	160	27	5	1.5	1.0	1.1	0	7	213	1.5	312	——	.04	.15	6.5	——
	FRUITS AND FRUIT PRODUCTS																		
	Apples, raw, unpeeled, without cores:																		
223	2 3/4-in diam. (about 3 per lb with cores). 1 apple——	138	84	80	Trace	1	——	——	——	20	10	14	.4	152	120	.04	.03	.1	6
224	3 1/4 in diam. (about 2 per lb with cores). 1 apple——	212	84	125	Trace	1	——	——	——	31	15	21	.6	233	190	.06	.04	.2	8
225	Applejuice, bottled or canned[24]—— 1 cup——	248	88	120	Trace	Trace	——	——	——	30	15	22	1.5	250	——	.02	.05	.2	[2]2
	Applesauce, canned:																		
226	Sweetened—— 1 cup——	255	76	230	1	Trace	——	——	——	61	10	13	1.3	166	100	.05	.03	.1	[2]3
227	Unsweetened—— 1 cup——	244	89	100	Trace	Trace	——	——	——	26	10	12	1.2	190	100	.05	.02	.1	[2]2
	Apricots:																		
228	Raw, without pits (about 12 per lb with pits). 3 apricots——	107	85	55	1	Trace	——	——	——	14	18	25	.5	301	2,890	.03	.04	.6	11
229	Canned in heavy sirup (halves and sirup). 1 cup——	258	77	220	2	Trace	——	——	——	57	28	39	.8	604	4,490	.05	.05	1.0	10
	Dried:																		
230	Uncooked (28 large or 37 medium halves per cup). 1 cup——	130	25	340	7	1	——	——	——	86	87	140	7.2	1,273	14,170	.01	.21	4.3	16
231	Cooked, unsweetened, fruit and liquid. 1 cup——	250	76	215	4	1	——	——	——	54	55	88	4.5	795	7,500	.01	.13	2.5	8
232	Apricot nectar, canned—— 1 cup——	251	85	145	1	Trace	——	——	——	37	23	30	.5	379	2,380	.03	.03	.5	[2]36
	Avocados, raw, whole, without skins and seeds:																		
233	California, mid- and late-winter (with skin and seed, 3 1/8-in diam.; wt. 10 oz). 1 avocado——	216	74	370	5	37	5.5	22.0	3.7	13	22	91	1.3	1,303	630	.24	.43	3.5	30
234	Florida, late summer and fall (with skin and seed, 3 5/8-in diam.; wt., 1 lb). 1 avocado——	304	78	390	4	33	6.7	15.7	5.3	27	30	128	1.8	1,836	880	.33	.61	4.9	43
235	Banana without peel (about 2.6 per lb with peel). 1 banana——	119	76	100	1	Trace	——	——	——	26	10	31	.8	440	230	.06	.07	.8	12
236	Banana flakes—— 1 tbsp——	6	3	20	Trace	Trace	——	——	——	5	2	6	.2	92	50	.01	.01	.2	Trace

(A)	(B)	(grams)	(C)	(D)	(E)	(F)	(G)	(H)	(I)	(J)	(K)	(L)	(M)	(N)	(O)	(P)	(Q)	(R)	(S)
237	Blackberries, raw — 1 cup	144	85	85	2	1	—	—	—	19	46	27	1.3	245	290	0.04	0.06	0.6	30
238	Blueberries, raw — 1 cup	145	83	90	1	1	—	—	—	22	22	19	1.5	117	150	.04	.09	.7	20
	Cantaloup. See Muskmelons (item 271).																		
	Cherries:																		
239	Sour (tart), red, pitted, canned, water pack. — 1 cup	244	88	105	2	Trace	—	—	—	26	37	32	.7	317	1,660	.07	.05	.5	12
240	Sweet, raw, without pits and stems. — 10 cherries	68	80	45	1	Trace	—	—	—	12	15	13	.3	129	70	.03	.04	.3	7
241	Cranberry juice cocktail, bottled, sweetened. — 1 cup	253	83	165	Trace	Trace	—	—	—	42	13	8	.8	25	Trace	.03	.03	.1	[27]81
242	Cranberry sauce, sweetened, canned, strained. — 1 cup	277	62	405	Trace	1	—	—	—	104	17	11	.6	83	60	.03	.03	.1	6
	Dates:																		
243	Whole, without pits — 10 dates	80	23	220	2	Trace	—	—	—	58	47	50	2.4	518	40	.07	.08	1.8	0
244	Chopped — 1 cup	178	23	490	4	1	—	—	—	130	105	112	5.3	1,153	90	.16	.18	3.9	0
245	Fruit cocktail, canned, in heavy sirup. — 1 cup	255	80	195	1	Trace	—	—	—	50	23	31	1.0	411	360	.05	.03	1.0	5
	Grapefruit:																		
	Raw, medium, 3 3/4-in diam. (about 1 lb 1 oz):																		
246	Pink or red[28] — 1/2 grapefruit with peel	241	89	50	1	Trace	—	—	—	13	20	20	.5	166	540	.05	.02	.2	44
247	White[28] — 1/2 grapefruit with peel	241	89	45	1	Trace	—	—	—	12	19	19	.5	159	10	.05	.02	.2	44
248	Canned, sections with sirup — 1 cup	254	81	180	2	Trace	—	—	—	45	33	36	.8	343	30	.08	.05	.5	76
	Grapefruit juice:																		
249	Raw, pink, red, or white — 1 cup	246	90	95	1	Trace	—	—	—	23	22	37	.5	399	([29])	.10	.05	.5	93
	Canned, white:																		
250	Unsweetened — 1 cup	247	89	100	1	Trace	—	—	—	24	20	35	1.0	400	20	.07	.05	.5	84
251	Sweetened — 1 cup	250	86	135	1	Trace	—	—	—	32	20	35	1.0	405	30	.08	.05	.5	78
	Frozen, concentrate, unsweetened:																		
252	Undiluted, 6-fl oz can — 1 can	207	62	300	4	1	—	—	—	72	70	124	.8	1,250	60	.29	.12	1.4	286
253	Diluted with 3 parts water by volume. — 1 cup	247	89	100	1	Trace	—	—	—	24	25	42	.2	420	20	.10	.04	.5	96
254	Dehydrated crystals, prepared with water (1 lb yields about 1 gal). — 1 cup	247	90	100	1	Trace	—	—	—	24	22	40	.2	412	20	.10	.05	.5	91
	Grapes, European type (adherent skin), raw:																		
255	Thompson Seedless — 10 grapes	50	81	35	Trace	Trace	—	—	—	9	6	10	.2	87	50	.03	.02	.2	2
256	Tokay and Emperor, seeded types — 10 grapes[30]	60	81	40	Trace	Trace	—	—	—	10	7	11	.2	99	60	.03	.02	.2	2
	Grapejuice:																		
257	Canned or bottled — 1 cup	253	83	165	1	Trace	—	—	—	42	28	30	.8	293	—	.10	.05	.5	[25]Trace
	Frozen concentrate, sweetened:																		
258	Undiluted, 6-fl oz can — 1 can	216	53	395	1	Trace	—	—	—	100	22	32	.9	255	40	.13	.22	1.5	[31]132
259	Diluted with 3 parts water by volume. — 1 cup	250	86	135	1	Trace	—	—	—	33	8	10	.3	85	10	.05	.08	.5	[31]10
260	Grape drink, canned — 1 cup	250	86	135	Trace	Trace	—	—	—	35	8	10	.3	88	10	[32].03	[32].03	.3	([32])
261	Lemon, raw, size 165, without peel and seeds (about 4 per lb with peels and seeds). — 1 lemon	74	90	20	1	Trace	—	—	—	6	19	12	.4	102	10	.03	.01	.1	39
	Lemon juice:																		
262	Raw — 1 cup	244	91	60	1	Trace	—	—	—	20	17	24	.5	344	50	.07	.02	.2	112
263	Canned, or bottled, unsweetened — 1 cup	244	92	55	1	Trace	—	—	—	19	17	24	.5	344	50	.07	.02	.2	102
264	Frozen, single strength, unsweetened, 6-fl oz can. — 1 can	183	92	40	1	Trace	—	—	—	13	13	16	.5	258	40	.05	.02	.2	81
	Lemonade concentrate, frozen:																		
265	Undiluted, 6-fl oz can — 1 can	219	49	425	Trace	Trace	—	—	—	112	9	13	.4	153	40	.05	.06	.7	66
266	Diluted with 4 1/3 parts water by volume. — 1 cup	248	89	105	Trace	Trace	—	—	—	28	2	3	.1	40	10	.01	.02	.2	17

[19] Crust made with vegetable shortening and enriched flour.
[24] Also applies to pasteurized apple cider.
[25] Applies to product without added ascorbic acid. For value of product with added ascorbic acid, refer to label.
[26] Based on product with label claim of 45% of U.S. RDA in 6 fl oz.
[27] Based on product with label claim of 100% of U.S. RDA in 6 fl oz.
[28] Weight includes peel and membranes between sections. Without these parts, the weight of the edible portion is 123 g for item 246 and 118 g for item 247.
[29] For white-fleshed varieties, value is about 20 International Units (I.U.) per cup; for red-fleshed varieties, 1,080 I.U.
[30] Weight includes seeds. Without seeds, weight of the edible portion is 57 g.
[31] Applies to product without added ascorbic acid. With added ascorbic acid, based on claim that 6 fl oz of reconstituted juice contain 45% or 50% of the U.S. RDA, value in milligrams is 108 or 120 for a 6-fl oz can (item 258), 36 or 40 for 1 cup of diluted juice (item 259).
[32] For products with added thiamin and riboflavin but without added ascorbic acid, values in milligrams would be 0.60 for thiamin, 0.80 for riboflavin, and trace for ascorbic acid. For products with only ascorbic acid added, value varies with the brand. Consult the label.

NUTRIENTS IN INDICATED QUANTITY

Item No. (A)	Foods, approximate measures, units, and weight (edible part unless footnotes indicate otherwise) (B)		Water (C) Percent	Food energy (D) Calories	Protein (E) Grams	Fat (F) Grams	Fatty Acids Saturated (total) (G) Grams	Unsaturated Oleic (H) Grams	Linoleic (I) Grams	Carbohydrate (J) Grams	Calcium (K) Milligrams	Phosphorus (L) Milligrams	Iron (M) Milligrams	Potassium (N) Milligrams	Vitamin A value (O) International units	Thiamin (P) Milligrams	Riboflavin (Q) Milligrams	Niacin (R) Milligrams	Ascorbic acid (S) Milligrams
		Grams																	
	FRUITS AND FRUIT PRODUCTS—Con.																		
	Limeade concentrate, frozen:																		
267	Undiluted, 6-fl oz can --- 1 can	218	50	410	Trace	Trace	---	---	---	108	11	13	0.2	129	Trace	0.02	0.02	0.2	26
268	Diluted with 4 1/3 parts water by volume. 1 cup	247	89	100	Trace	Trace	---	---	---	27	3	3	Trace	32	Trace	Trace	Trace	Trace	6
	Limejuice:																		
269	Raw --- 1 cup	246	90	65	1	Trace	---	---	---	22	22	27	.5	256	20	.05	.02	.2	79
270	Canned, unsweetened --- 1 cup	246	90	65	1	Trace	---	---	---	22	22	27	.5	256	20	.05	.02	.2	52
	Muskmelons, raw, with rind, without seed cavity:																		
271	Cantaloup, orange-fleshed (with rind and seed cavity, 5-in diam., 2 1/3 lb). 1/2 melon with rind[33]	477	91	80	2	Trace	---	---	---	20	38	44	1.1	682	9,240	.11	.08	1.6	90
272	Honeydew (with rind and seed cavity, 6 1/2-in diam., 5 1/4 lb). 1/10 melon with rind[33]	226	91	50	1	Trace	---	---	---	11	21	24	.6	374	60	.06	.04	.9	34
	Oranges, all commercial varieties, raw:																		
273	Whole, 2 5/8-in diam., without peel and seeds (about 2 1/2 per lb with peel and seeds). 1 orange	131	86	65	1	Trace	---	---	---	16	54	26	.5	263	260	.13	.05	.5	66
274	Sections without membranes --- 1 cup	180	86	90	2	Trace	---	---	---	22	74	36	.7	360	360	.18	.07	.7	90
	Orange juice:																		
275	Raw, all varieties --- 1 cup	248	88	110	2	Trace	---	---	---	26	27	42	.5	496	500	.22	.07	1.0	124
276	Canned, unsweetened --- 1 cup	249	87	120	2	Trace	---	---	---	28	25	45	1.0	496	500	.17	.05	.7	100
	Frozen concentrate:																		
277	Undiluted, 6-fl oz can --- 1 can	213	55	360	5	Trace	---	---	---	87	75	126	.9	1,500	1,620	.68	.11	2.8	360
278	Diluted with 3 parts water by volume. 1 cup	249	87	120	2	Trace	---	---	---	29	25	42	.2	503	540	.23	.03	.9	120
279	Dehydrated crystals, prepared with water (1 lb yields about 1 gal). 1 cup	248	88	115	1	Trace	---	---	---	27	25	40	.5	518	500	.20	.07	1.0	109
	Orange and grapefruit juice: Frozen concentrate:																		
280	Undiluted, 6-fl oz can --- 1 can	210	59	330	4	1	---	---	---	78	61	99	.8	1,308	800	.48	.06	2.3	302
281	Diluted with 3 parts water by volume. 1 cup	248	88	110	1	Trace	---	---	---	26	20	32	.2	439	270	.15	.02	.7	102
	Papayas:																		
282	Papayas, raw, 1/2-in cubes --- 1 cup	140	89	55	1	Trace	---	---	---	14	28	22	.4	328	2,450	.06	.06	.4	78
	Peaches: Raw:																		
283	Whole, 2 1/2-in diam., peeled, pitted (about 4 per lb with peels and pits). 1 peach	100	89	40	1	Trace	---	---	---	10	9	19	.5	202	[34]1,330	.02	.05	1.0	7
284	Sliced --- 1 cup	170	89	65	1	Trace	---	---	---	16	15	32	.9	343	[34]2,260	.03	.09	1.7	12
	Canned, yellow-fleshed, solids and liquid (halves or slices):																		
285	Sirup pack --- 1 cup	256	79	200	1	Trace	---	---	---	51	10	31	.8	333	1,100	.03	.05	1.5	8
286	Water pack --- 1 cup	244	91	75	1	Trace	---	---	---	20	10	32	.7	334	1,100	.02	.07	1.5	7
	Dried:																		
287	Uncooked --- 1 cup	160	25	420	5	1	---	---	---	109	77	187	9.6	1,520	6,240	.02	.30	8.5	29
288	Cooked, unsweetened, halves and juice. 1 cup	250	77	205	3	1	---	---	---	54	38	93	4.8	743	3,050	.01	.15	3.8	5

(A)	(B)	(C)	(D)	(E)	(F)	(G)	(H)	(I)	(J)	(K)	(L)	(M)	(N)	(O)	(P)	(Q)	(R)	(S)
	Frozen, sliced, sweetened:																	
289	10-oz container	1 container	284	77	250	1	Trace	—	—	—	64	11	37	1.4	352	1,850	0.03	0.11	2.0	[35]116
290	Cup	1 cup	250	77	220	1	Trace	—	—	—	57	10	33	1.3	310	1,630	.03	.10	1.8	[35]103
	Pears:																		
	Raw, with skin, cored:																		
291	Bartlett, 2 1/2-in diam. (about 2 1/2 per lb with stems).	1 pear	164	83	100	1	1	—	—	—	25	13	18	.5	213	30	.03	.07	.2	7
292	Bosc, 2 1/2-in diam. (about 3 per lb with stems).	1 pear	141	83	85	1	1	—	—	—	22	11	16	.4	83	30	.03	.06	.1	6
293	D'Anjou, 3-in diam. (about 2 per lb with stems).	1 pear	200	83	120	1	1	—	—	—	31	16	22	.6	260	40	.04	.08	.2	8
294	Canned, solids and liquid, sirup pack, heavy (halves or slices).	1 cup	255	80	195	1	1	—	—	—	50	13	18	.5	214	10	.03	.05	.3	3
	Pineapple:																		
295	Raw, diced	1 cup	155	85	80	1	Trace	—	—	—	21	26	12	.8	226	110	.14	.05	.3	26
	Canned, heavy sirup pack, solids and liquid:																		
296	Crushed, chunks, tidbits	1 cup	255	80	190	1	Trace	—	—	—	49	28	13	.8	245	130	.20	.05	.5	18
	Slices and liquid:																		
297	Large	1 slice; 2 1/4 tbsp liquid.	105	80	80	Trace	Trace	—	—	—	20	12	5	.3	101	50	.08	.02	.2	7
298	Medium	1 slice; 1 1/4 tbsp liquid.	58	80	45	Trace	Trace	—	—	—	11	6	3	.2	56	30	.05	.01	.1	4
299	Pineapple juice, unsweetened, canned.	1 cup	250	86	140	1	Trace	—	—	—	34	38	23	.8	373	130	.13	.05	.5	[27]80
	Plums:																		
	Raw, without pits:																		
300	Japanese and hybrid (2 1/8-in diam., about 6 1/2 per lb with pits).	1 plum	66	87	30	Trace	Trace	—	—	—	8	8	12	.3	112	160	.02	.02	.3	4
301	Prune-type (1 1/2-in diam., about 15 per lb with pits).	1 plum	28	79	20	Trace	Trace	—	—	—	6	3	5	.1	48	80	.01	.01	.1	1
	Canned, heavy sirup pack (Italian prunes), with pits and liquid:																		
302	Cup[36]	1 cup	272	77	215	1	Trace	—	—	—	56	23	26	2.3	367	3,130	.05	.05	1.0	5
303	Portion	3 plums; 2 3/4 tbsp liquid.	140	77	110	1	Trace	—	—	—	29	12	13	1.2	189	1,610	.03	.03	.5	3
	Prunes, dried, "softenized," with pits:																		
304	Uncooked	4 extra large or 5 large prunes.[36]	49	28	110	1	Trace	—	—	—	29	22	34	1.7	298	690	.04	.07	.7	1
305	Cooked, unsweetened, all sizes, fruit and liquid.	1 cup[36]	250	66	255	2	1	—	—	—	67	51	79	3.8	695	1,590	.07	.15	1.5	2
306	Prune juice, canned or bottled	1 cup	256	80	195	1	Trace	—	—	—	49	36	51	1.8	602	—	.03	.03	1.0	5
	Raisins, seedless:																		
307	Cup, not pressed down	1 cup	145	18	420	4	Trace	—	—	—	112	90	146	5.1	1,106	30	.16	.12	.7	Trace
308	Packet, 1/2 oz (1 1/2 tbsp)	1 packet	14	18	40	Trace	Trace	—	—	—	11	9	14	.5	107	Trace	.02	.01	.1	Trace
	Raspberries, red:																		
309	Raw, capped, whole	1 cup	123	84	70	1	1	—	—	—	17	27	27	1.1	207	160	.04	.11	1.1	31
310	Frozen, sweetened, 10-oz container	1 container	284	74	280	2	1	—	—	—	70	37	48	1.7	284	200	.06	.17	1.7	60
	Rhubarb, cooked, added sugar:																		
311	From raw	1 cup	270	63	380	1	Trace	—	—	—	97	211	41	1.6	548	220	.05	.14	.8	16
312	From frozen, sweetened	1 cup	270	63	385	1	1	—	—	—	98	211	32	1.9	475	190	.05	.11	.5	16

[27] Based on product with label claim of 100% of U.S. RDA of ascorbic acid.
[28] Weight includes rind. Without rind, the weight of the edible portion is 272 g for item 271 and 149 g for item 272.
[29] Weight includes rind. Without rind, the weight of the edible portion is 272 g for item 272.
[34] Represents yellow-fleshed varieties. For white-fleshed varieties, value is 50 International Units (I.U.) for 1 peach, 90 I.U. for 1 cup of slices.
[35] Value represents products without added ascorbic acid, value in milligrams is 116 for a 10-oz container, 103 for 1 cup.
[36] Weight includes pits. After removal of the pits, the weight of the edible portion is 258 g for item 302, 133 g for item 303, 43 g for item 304, and 213 g for item 305.

Item No. (A)	Foods, approximate measures, units, and weight (edible part unless footnotes indicate otherwise) (B)	Grams	Water (C) Per-cent	Food energy (D) Cal-ories	Pro-tein (E) Grams	Fat (F) Grams	Satu-rated (total) (G) Grams	Unsaturated Oleic (H) Grams	Lino-leic (I) Grams	Carbo-hydrate (J) Grams	Calcium (K) Milli-grams	Phos-phorus (L) Milli-grams	Iron (M) Milli-grams	Potas-sium (N) Milli-grams	Vitamin A value (O) Inter-national units	Thiamin (P) Milli-grams	Ribo-flavin (Q) Milli-grams	Niacin (R) Milli-grams	Ascorbic acid (S) Milli-grams
	FRUITS AND FRUIT PRODUCTS—Con.																		
	Strawberries:																		
313	Raw, whole berries, capped------ 1 cup	149	90	55	1	1	---	---	---	13	31	31	1.5	244	90	0.04	0.10	0.9	88
	Frozen, sweetened:																		
314	Sliced, 10-oz container------- 1 container	284	71	310	1	1	---	---	---	79	40	48	2.0	318	90	.06	.17	1.4	151
315	Whole, 1-lb container (about 1 3/4 cups)- 1 container	454	76	415	2	1	---	---	---	107	59	73	2.7	472	140	.09	.27	2.3	249
316	Tangerine, raw, 2 3/8-in diam., size 176, without peel (about 4 per lb with peels and seeds)- 1 tangerine	86	87	40	1	Trace	---	---	---	10	34	15	.3	108	360	.05	.02	.1	27
317	Tangerine juice, canned, sweetened- 1 cup	249	87	125	1	Trace	---	---	---	30	44	35	.5	440	1,040	.15	.05	.2	54
318	Watermelon, raw, 4 by 8 in wedge with rind and seeds[37] (1/16 of 32 2/3-lb melon, 10 by 16 in)- 1 wedge with rind and seeds[37]	926	93	110	2	1	---	---	---	27	30	43	2.1	426	2,510	.13	.13	.9	30
	GRAIN PRODUCTS																		
	Bagel, 3-in diam.:																		
319	Egg----------- 1 bagel	55	32	165	6	2	0.5	0.9	0.8	28	9	43	1.2	41	30	.14	.10	1.2	0
320	Water--------- 1 bagel	55	29	165	6	1	.2	.4	.6	30	8	41	1.2	42	0	.15	.11	1.4	0
321	Barley, pearled, light, uncooked- 1 cup	200	11	700	16	2	.3	.2	.8	158	32	378	4.0	320	0	.24	.10	6.2	0
	Biscuits, baking powder, 2-in diam. (enriched flour, vegetable shortening):																		
322	From home recipe----- 1 biscuit	28	27	105	2	5	1.2	2.0	1.2	13	34	49	.4	33	Trace	.08	.08	.7	Trace
323	From mix[38]--------- 1 biscuit	28	29	90	2	3	.6	1.1	.7	15	19	65	.6	32	Trace	.09	.08	.8	Trace
324	Breadcrumbs (enriched)[38] Dry, grated--------- 1 cup	100	7	390	13	5	1.0	1.6	1.4	73	122	141	3.6	152	Trace	.35	.35	4.8	Trace
	Soft. See white bread (items 349-350).																		
	Breads:																		
325	Boston brown bread, canned, slice, 3 1/4 by 1/2 in.[38]- 1 slice	45	45	95	2	1	.1	.2	.2	21	41	72	.9	131	[39]0	.06	.04	.7	0
	Cracked-wheat bread (3/4 enriched wheat flour, 1/4 cracked wheat):[38]																		
326	Loaf, 1 lb---------- 1 loaf	454	35	1,195	39	10	2.2	3.0	3.9	236	399	581	9.5	608	Trace	1.52	1.13	14.4	Trace
327	Slice (18 per loaf)- 1 slice	25	35	65	2	1	.1	.2	.2	13	22	32	.5	34	Trace	.08	.06	.8	Trace
	French or vienna bread, enriched:[38]																		
328	Loaf, 1 lb---------- 1 loaf	454	31	1,315	41	14	3.2	4.7	4.6	251	195	386	10.0	408	Trace	1.80	1.10	15.0	Trace
	Slice:																		
329	French (5 by 2 1/2 by 1 in)- 1 slice	35	31	100	3	1	.2	.4	.4	19	15	30	.8	32	Trace	.14	.08	1.2	Trace
330	Vienna (4 3/4 by 4 by 1/2 in)- 1 slice	25	31	75	2	1	.2	.3	.3	14	11	21	.6	23	Trace	.10	.06	.8	Trace
	Italian bread, enriched:																		
331	Loaf, 1 lb---------- 1 loaf	454	32	1,250	41	4	.6	.3	1.5	256	77	349	10.0	336	0	1.80	1.10	15.0	0
332	Slice, 4 1/2 by 3 1/4 by 3/4 in- 1 slice	30	32	85	3	Trace	Trace	Trace	.1	17	5	23	.7	22	0	.12	.07	1.0	0
	Raisin bread, enriched:[38]																		
333	Loaf, 1 lb---------- 1 loaf	454	35	1,190	30	13	3.0	4.7	3.9	243	322	395	10.0	1,057	Trace	1.70	1.07	10.7	Trace
334	Slice (18 per loaf)- 1 slice	25	35	65	2	1	.2	.3	.2	13	18	22	.6	58	Trace	.09	.06	.6	Trace

(A)	(B)	(C)	(D)	(E)	(F)	(G)	(H)	(I)	(J)	(K)	(L)	(M)	(N)	(O)	(P)	(Q)	(R)	(S)
	Rye Bread:																	
	American, light (2/3 enriched wheat flour, 1/3 rye flour):																	
335	Loaf, 1 lb — 1 loaf — 454	36	1,100	41	5	0.7	0.5	2.2	236	340	667	9.1	658	0	1.35	0.98	12.9	0
336	Slice (4 3/4 by 3 3/4 by 7/16 in) — 1 slice — 25	36	60	2	Trace	Trace	Trace	.1	13	19	37	.5	36	0	.07	.05	.7	0
	Pumpernickel (2/3 rye flour, 1/3 enriched wheat flour):																	
337	Loaf, 1 lb — 1 loaf — 454	34	1,115	41	5	.7	.5	2.4	241	381	1,039	11.8	2,059	0	1.30	.93	8.5	0
338	Slice (5 by 4 by 3/8 in) — 1 slice — 32	34	80	3	Trace	.1	Trace	.2	17	27	73	.8	145	0	.09	.07	.6	0
	White bread, enriched:[38] Soft-crumb type:																	
339	Loaf, 1 lb — 1 loaf — 454	36	1,225	39	15	3.4	5.3	4.6	229	381	440	11.3	476	Trace	1.80	1.10	15.0	Trace
340	Slice (18 per loaf) — 1 slice — 25	36	70	2	1	.2	.3	.3	13	21	24	.6	26	Trace	.10	.06	.8	Trace
341	Slice, toasted — 1 slice — 22	25	70	2	1	.2	.3	.3	13	21	24	.6	26	Trace	.08	.06	.8	Trace
342	Slice (22 per loaf) — 1 slice — 20	36	55	2	1	.2	.3	.2	10	17	19	.5	21	Trace	.08	.05	.7	Trace
343	Slice, toasted — 1 slice — 17	25	55	2	1	.2	.2	.2	10	17	19	.5	21	Trace	.06	.05	.7	Trace
344	Loaf, 1 1/2 lb — 1 loaf — 680	36	1,835	59	22	5.2	7.9	6.9	343	571	660	17.0	714	Trace	2.70	1.65	22.5	Trace
345	Slice (24 per loaf) — 1 slice — 28	36	75	2	1	.2	.3	.3	14	24	27	.7	29	Trace	.11	.07	.9	Trace
346	Slice, toasted — 1 slice — 24	25	75	2	1	.2	.3	.3	14	24	27	.7	29	Trace	.09	.07	.9	Trace
347	Slice (28 per loaf) — 1 slice — 24	36	65	2	1	.2	.2	.2	12	20	23	.6	25	Trace	.10	.06	.8	Trace
348	Slice, toasted — 1 slice — 21	25	65	2	1	.2	.3	.2	12	20	23	.6	25	Trace	.08	.06	.8	Trace
349	Cubes — 1 cup — 30	36	80	2	1	.2	.3	.2	15	25	29	.8	32	Trace	.12	.07	1.0	Trace
350	Crumbs — 1 cup — 45	36	120	4	1	.3	.5	.3	23	38	44	1.1	47	Trace	.18	.11	1.5	Trace
	Firm-crumb type:																	
351	Loaf, 1 lb — 1 loaf — 454	35	1,245	41	17	3.9	5.9	5.2	228	435	463	11.3	549	Trace	1.80	1.10	15.0	Trace
352	Slice (20 per loaf) — 1 slice — 23	35	65	2	1	.2	.3	.3	12	22	23	.6	28	Trace	.09	.06	.8	Trace
353	Slice, toasted — 1 slice — 20	24	65	2	1	.2	.3	.3	12	22	23	.6	28	Trace	.07	.06	.8	Trace
354	Loaf, 2 lb — 1 loaf — 907	35	2,495	82	34	7.7	11.8	10.4	455	871	925	22.7	1,097	Trace	3.60	2.20	30.0	Trace
355	Slice (34 per loaf) — 1 slice — 27	35	75	2	1	.2	.3	.3	14	26	28	.7	33	Trace	.11	.06	.9	Trace
356	Slice, toasted — 1 slice — 23	24	75	2	1	.2	.3	.3	14	26	28	.7	33	Trace	.09	.06	.9	Trace
	Whole-wheat bread:[38] Soft-crumb type:[38]																	
357	Loaf, 1 lb — 1 loaf — 454	36	1,095	41	12	2.2	2.9	4.2	224	381	1,152	13.6	1,161	Trace	1.37	.45	12.7	Trace
358	Slice (16 per loaf) — 1 slice — 28	36	65	3	1	.1	.2	.2	14	24	71	.8	72	Trace	.09	.03	.8	Trace
359	Slice, toasted — 1 slice — 24	24	65	3	1	.1	.2	.2	14	24	71	.8	72	Trace	.07	.03	.8	Trace
	Firm-crumb type:[38]																	
360	Loaf, 1 lb — 1 loaf — 454	36	1,100	48	14	2.5	3.3	4.9	216	449	1,034	13.6	1,238	Trace	1.17	.54	12.7	Trace
361	Slice (18 per loaf) — 1 slice — 25	36	60	3	1	.1	.2	.3	12	24	57	.8	68	Trace	.06	.03	.7	Trace
362	Slice, toasted — 1 slice — 21	24	60	3	1	.1	.2	.3	12	24	57	.8	68	Trace	.05	.03	.7	Trace
	Breakfast cereals: Hot type, cooked: Corn (hominy) grits, degermed:																	
363	Enriched — 1 cup — 245	87	125	3	Trace	Trace	Trace	.1	27	2	25	.7	27	[40]Trace	.10	.07	1.0	0
364	Unenriched — 1 cup — 245	87	125	3	Trace	Trace	Trace	.1	27	2	25	.2	27	[40]Trace	.05	.02	.5	0
365	Farina, quick-cooking, enriched — 1 cup — 245	89	105	3	Trace	Trace	Trace	.1	22	147	[41]113	(42)	25	0	.12	.07	1.0	0
366	Oatmeal or rolled oats — 1 cup — 240	87	130	5	2	.4	.8	.9	23	22	137	1.4	146	0	.19	.05	.2	0
367	Wheat, rolled — 1 cup — 240	80	180	5	1	—	—	—	41	19	182	1.7	202	0	.17	.07	2.2	0
368	Wheat, whole-meal — 1 cup — 245	88	110	4	1	—	—	—	23	17	127	1.2	118	0	.15	.05	1.5	0
	Ready-to-eat:																	
369	Bran flakes (40% bran), added sugar, salt, iron, vitamins — 1 cup — 35	3	105	4	1	—	—	—	28	19	125	5.6	137	1,540	.46	.52	6.2	0
370	Bran flakes with raisins, added sugar, salt, iron, vitamins — 1 cup — 50	7	145	4	1	—	—	—	40	28	146	7.9	154	[43]2,200	(44)	(44)	(44)	0

[37] Weight includes rind and seeds. Without rind and seeds, weight of the edible portion is 426 g.
[38] Made with vegetable shortening.
[39] Applies to product made with white cornmeal. With yellow cornmeal, value is 30 International Units (I.U.).
[40] Applies to white varieties. For yellow varieties, value is 150 International Units (I.U.).
[41] Applies to products that do not contain di-sodium phosphate. If di-sodium phosphate is an ingredient, value is 162 mg.
[42] Value may range from less than 1 mg to about 8 mg depending on the brand. Consult the label.
[43] Applies to product with added nutrient. Without added nutrient, value is trace.
[44] Value varies with the brand. Consult the label.

Item No. (A)	Foods, approximate measure, units, and weight (edible part unless footnotes indicate otherwise) (B)		Grams	Water (C) Percent	Food energy (D) Calories	Protein (E) Grams	Fat (F) Grams	Fatty Acids Saturated (total) (G) Grams	Oleic (H) Grams	Linoleic (I) Grams	Carbohydrate (J) Grams	Calcium (K) Milligrams	Phosphorus (L) Milligrams	Iron (M) Milligrams	Potassium (N) Milligrams	Vitamin A value (O) International units	Thiamin (P) Milligrams	Riboflavin (Q) Milligrams	Niacin (R) Milligrams	Ascorbic acid (S) Milligrams
	GRAIN PRODUCTS—Con.																			
	Breakfast cereals—Continued																			
	Ready-to-eat—Continued																			
	Corn flakes:																			
371	Plain, added sugar, salt, iron, vitamins.	1 cup	25	4	95	2	Trace	—	—	—	21	[44]	9	[44]	30	[44]	[44]	[44]	[44]	[44]13
372	Sugar-coated, added salt, iron, vitamins.	1 cup	40	2	155	2	Trace	—	—	—	37	1	10	[44]	27	1,760	.53	.50	7.1	[44]21
373	Corn, oat flour, puffed, added sugar, salt, iron, vitamins.	1 cup	20	4	80	2	1	—	—	—	16	4	18	5.7	—	880	.26	.30	3.5	11
374	Corn, shredded, added sugar, salt, iron, thiamin, niacin.	1 cup	25	3	95	2	Trace	—	—	—	22	1	10	.6		0	.33	.05	4.4	13
375	Oats, puffed, added sugar, salt, minerals, vitamins.	1 cup	25	3	100	3	1	—	—	—	19	44	102	4.0		1,100	.33	.38	4.4	13
	Rice, puffed:																			
376	Plain, added iron, thiamin, niacin.	1 cup	15	4	60	1	Trace	—	—	—	13	3	14	.3	15	0	.07	.01	.7	0
377	Presweetened, added salt, iron, vitamins.	1 cup	28	3	115	1	0	—	—	—	26	3	14	[44]	43	[44]1,240	[44]	[44]	[44]	[44]15
378	Wheat flakes, added sugar, salt, iron, vitamins.	1 cup	30	4	105	3	Trace	—	—	—	24	12	83	4.8	81	1,320	.40	.45	5.3	15
	Wheat, puffed:																			
379	Plain, added iron, thiamin, niacin.	1 cup	15	3	55	2	Trace	—	—	—	12	4	48	.6	51	0	.08	.03	1.2	0
380	Presweetened, added salt, iron, vitamins.	1 cup	38	3	140	3	Trace	—	—	—	33	7	52	[44]	63	1,680	.50	.57	6.7	[44]20
381	Wheat, shredded, plain.	1 oblong biscuit or 1/2 cup spoon-size biscuits.	25	7	90	2	1	—	—	—	20	11	97	.9	87	0	.06	.03	1.1	0
382	Wheat germ, without salt and sugar, toasted.	1 tbsp	6	4	25	2	1	—	—	—	3	3	70	.5	57	10	.11	.05	.3	1
383	Buckwheat flour, light, sifted	1 cup	98	12	340	6	1	0.2	0.4	0.4	78	11	86	1.0	314	0	.08	.04	.4	0
384	Bulgur, canned, seasoned	1 cup	135	56	245	8	4				44	27	263	1.9	151	0	.08	.05	4.1	0
	Cake icings. See Sugars and Sweets (items 532–536).																			
	Cakes made from cake mixes with enriched flour:[46]																			
	Angelfood:																			
385	Whole cake (9 3/4-in diam. tube cake).	1 cake	635	34	1,645	36	1	—	—	—	377	603	756	2.5	381	0	.37	.95	3.6	0
386	Piece, 1/12 of cake.	1 piece	53	34	135	3	Trace	—	—	—	32	50	63	.2	32	0	.03	.08	.3	0
	Coffeecake:																			
387	Whole cake (7 3/4 by 5 5/8 by 1 1/4 in).	1 cake	430	30	1,385	27	41	11.7	16.3	8.8	225	262	748	6.9	469	690	.82	.91	7.7	1
388	Piece, 1/6 of cake.	1 piece	72	30	230	5	7	2.0	2.7	1.5	38	44	125	1.2	78	120	.14	.15	1.3	Trace
	Cupcakes, made with egg, milk, 2 1/2-in diam.:																			
389	Without icing	1 cupcake	25	26	90	1	3	.8	1.2	.7	14	40	59	.3	21	40	.05	.05	.4	Trace
390	With chocolate icing	1 cupcake	36	22	130	2	5	2.0	1.6	.6	21	47	71	.4	42	60	.05	.06	.4	Trace
	Devil's food with chocolate icing:																			
391	Whole, 2 layer cake (8- or 9-in diam.).	1 cake	1,107	24	3,755	49	136	50.0	44.9	17.0	645	653	1,162	16.6	1,439	1,660	1.06	1.65	10.1	1
392	Piece, 1/16 of cake	1 piece	69	24	235	3	8	3.1	2.8	1.1	40	41	72	1.0	90	100	.07	.10	.6	Trace
393	Cupcake, 2 1/2-in diam.	1 cupcake	35	24	120	2	4	1.6	1.4	.5	20	21	37	.5	46	50	.03	.05	.3	Trace

(A)	(B)	(C)	(D)	(E)	(F)	(G)	(H)	(I)	(J)	(K)	(L)	(M)	(N)	(O)	(P)	(Q)	(R)	(S)
	Gingerbread:																	
394	Whole cake (8-in square)------- 1 cake	570	1,575	18	39	9.7	16.6	10.0	291	513	570	8.6	1,562	Trace	0.84	1.00	7.4	Trace
395	Piece, 1/9 of cake------------- 1 piece	63	175	2	4	1.1	1.8	1.1	32	57	63	.9	173	Trace	.09	.11	.8	Trace
	White, 2 layer with chocolate icing:																	
396	Whole cake (8- or 9-in diam.)-- 1 cake	1,140	4,000	44	122	48.2	46.4	20.0	716	1,129	2,041	11.4	1,322	680	1.50	1.77	12.5	2
397	Piece, 1/15 of cake------------ 1 piece	71	250	3	8	3.0	2.9	1.2	45	70	127	.7	82	40	.09	.11	.8	Trace
	Yellow, 2 layer with chocolate icing:																	
398	Whole cake (8- or 9-in diam.)-- 1 cake	1,108	3,735	45	125	47.8	47.8	20.3	638	1,008	2,017	12.2	1,208	1,550	1.24	1.67	10.6	2
399	Piece, 1/16 of cake------------ 1 piece	69	235	3	8	3.0	3.0	1.3	40	63	126	.8	75	100	.08	.10	.7	Trace
	Cakes made from home recipes using enriched flour:[7]																	
	Boston cream pie with custard filling:																	
400	Whole cake (8-in diam.)-------- 1 cake	825	2,490	41	78	23.0	30.1	15.2	412	553	833	8.2	[48]734	1,730	1.04	1.27	9.6	2
401	Piece, 1/12 of cake------------ 1 piece	69	210	3	6	1.9	2.5	1.3	34	46	70	.7	[48]61	140	.09	.11	.8	Trace
	Fruitcake, dark:																	
402	Loaf, 1-lb (7 1/2 by 2 by 1 1/2 in).------ 1 loaf	454	1,720	22	69	14.4	33.5	14.8	271	327	513	11.8	2,250	540	.72	.73	4.9	2
403	Slice, 1/30 of loaf.----------- 1 slice	15	55	1	2	.5	1.1	.5	9	11	17	.4	74	20	.02	.02	.2	Trace
	Plain, sheet cake:																	
	Without icing:																	
404	Whole cake (9-in square)------- 1 cake	777	2,830	35	108	29.5	44.4	23.9	434	497	793	8.5	[48]614	1,320	1.21	1.40	10.2	2
405	Piece, 1/9 of cake------------- 1 piece	86	315	4	12	3.3	4.9	2.6	48	55	88	.9	[48]68	150	.13	.15	1.1	2
	With uncooked white icing:																	
406	Whole cake (9-in square)------- 1 cake	1,096	4,020	37	129	42.2	49.5	24.4	694	548	822	8.2	[48]669	2,190	1.22	1.47	10.2	2
407	Piece, 1/9 of cake------------- 1 piece	121	445	4	14	4.7	5.5	2.7	77	61	91	.8	[48]74	240	.14	.16	1.1	Trace
	Pound:[49]																	
408	Loaf, 8 1/2 by 3 1/2 by 3 1/4 in.------ 1 loaf	565	2,725	31	170	42.9	73.1	39.6	273	107	418	7.9	345	1,410	.90	.99	7.3	0
409	Slice, 1/17 of loaf----------- 1 slice	33	160	2	10	2.5	4.3	2.3	16	6	24	.5	20	80	.05	.06	.4	0
	Spongecake:																	
410	Whole cake (9 3/4-in diam. tube cake).------ 1 cake	790	2,345	60	45	13.1	15.8	5.7	427	237	885	13.4	687	3,560	1.10	1.64	7.4	Trace
411	Piece, 1/12 of cake------------ 1 piece	66	195	5	4	1.1	1.3	.5	36	20	74	1.1	57	300	.09	.14	.6	Trace
	Cookies made with enriched flour:[50][51]																	
	Brownies with nuts:																	
	Home-prepared, 1 3/4 by 1 3/4 by 7/8 in:																	
412	From home recipe----------- 1 brownie	20	95	1	6	1.5	3.0	1.2	10	8	30	.4	38	40	.04	.03	.2	Trace
413	From commercial recipe----- 1 brownie	20	85	1	4	.9	1.4	1.3	13	9	27	.4	34	20	.03	.02	.2	Trace
414	Frozen, with chocolate icing,[52] 1 1/2 by 1 3/4 by 7/8 in.------ 1 brownie	25	105	1	5	2.0	2.2	.7	15	10	31	.4	44	50	.03	.03	.2	Trace
	Chocolate chip:																	
415	Commercial, 2 1/4-in diam., 3/8 in thick.------ 4 cookies	42	200	2	9	2.8	2.9	2.2	29	16	48	1.0	56	50	.10	.17	.9	Trace
416	From home recipe, 2 1/3-in diam.------ 4 cookies	40	205	2	12	3.5	4.5	2.9	24	14	40	.8	47	40	.06	.06	.5	Trace
417	Fig bars, square (1 5/8 by 1 5/8 by 3/8 in) or rectangular (1 1/2 by 1 3/4 by 1/2 in).------ 4 cookies	56	200	2	3	.8	1.2	.7	42	44	34	1.0	111	60	.04	.14	.9	Trace
418	Gingersnaps, 2-in diam., 1/4 in thick.------ 4 cookies	28	90	2	2	.7	1.0	.6	22	20	13	.7	129	20	.08	.06	.7	0
419	Macaroons, 2 3/4-in diam., 1/4 in thick.------ 2 cookies	38	180	2	9	—	—	—	25	10	32	.3	176	0	.02	.06	.2	0
420	Oatmeal with raisins, 2 5/8-in diam., 1/4 in thick.------ 4 cookies	52	235	3	8	2.0	3.3	2.0	38	11	53	1.4	192	30	.15	.10	1.0	Trace

[44]Value varies with the brand. Consult the label.
[45]Applies to product with added nutrient. Without added nutrient, value is trace.
[46]Excepting angelfood cake, cakes were made from mixes containing vegetable shortening; icings, with butter.
[47]Excepting spongecake, vegetable shortening used for cake portion; butter, for icing. If butter or margarine used for cake portion, vitamin A values would be higher.
[48]Applies to product made with a sodium aluminum-sulfate type baking powder. With a low-sodium type baking powder containing potassium, value would be about twice the amount shown.
[49]Equal weights of flour, sugar, eggs, and vegetable shortening.
[50]Products are commercial unless otherwise specified.
[51]Made with enriched flour and vegetable shortening except for macaroons which do not contain flour or shortening.
[52]Icing made with butter.

Item No. (A)	Foods, approximate measures, units, and weight (edible part unless footnotes indicate otherwise) (B)		Grams	Water (C) Percent	Food energy (D) Calories	Protein (E) Grams	Fat (F) Grams	Fatty Acids Saturated (total) (G) Grams	Unsaturated Oleic (H) Grams	Linoleic (I) Grams	Carbohydrate (J) Grams	Calcium (K) Milligrams	Phosphorus (L) Milligrams	Iron (M) Milligrams	Potassium (N) Milligrams	Vitamin A value (O) International units	Thiamin (P) Milligrams	Riboflavin (Q) Milligrams	Niacin (R) Milligrams	Ascorbic acid (S) Milligrams
	GRAIN PRODUCTS—Con.																			
	Cookies made with enriched flour[50][51]—Continued																			
421	Plain, prepared from commercial chilled dough, 2 1/2-in diam., 1/4 in thick.	4 cookies	48	5	240	2	12	3.0	5.2	2.9	31	17	35	0.6	23	30	0.10	0.08	0.9	0
422	Sandwich type (chocolate or vanilla), 1 3/4-in diam., 3/8 in thick.	4 cookies	40	2	200	2	9	2.2	3.9	2.2	28	10	96	.7	15	0	.06	.10	.7	0
423	Vanilla wafers, 1 3/4-in diam., 1/4 in thick.	10 cookies	40	3	185	2	6	—	—	—	30	16	25	.6	29	50	.10	.09	.8	0
	Cornmeal:																			
424	Whole-ground, unbolted, dry form.	1 cup	122	12	435	11	5	.5	1.0	2.5	90	24	312	2.9	346	[53]620	.46	.13	2.4	0
425	Bolted (nearly whole-grain), dry form.	1 cup	122	12	440	11	4	.5	.9	2.1	91	21	272	2.2	303	[53]590	.37	.10	2.3	0
	Degermed, enriched:																			
426	Dry form	1 cup	138	12	500	11	2	.2	.4	.9	108	8	137	4.0	166	[53]610	.61	.36	4.8	0
427	Cooked	1 cup	240	88	120	3	Trace	Trace	.1	.2	26	2	34	1.0	38	[53]140	.14	.10	1.2	0
	Degermed, unenriched:																			
428	Dry form	1 cup	138	12	500	11	2	.2	.4	.9	108	8	137	1.5	166	[53]610	.19	.07	1.4	0
429	Cooked	1 cup	240	88	120	3	Trace	Trace	.1	.2	26	2	34	.5	38	[53]140	.05	.02	.2	0
	Crackers:[38]																			
430	Graham, plain, 2 1/2-in square	2 crackers	14	6	55	1	1	.3	.5	.3	10	6	21	.5	55	0	.02	.08	.5	0
431	Rye wafers, whole-grain, 1 7/8 by 3 1/2 in.	2 wafers	13	6	45	2	Trace	—	—	—	10	7	50	.5	78	0	.04	.03	.2	0
432	Saltines, made with enriched flour.	4 crackers or 1 packet	11	4	50	1	1	.3	.5	.4	8	2	10	.5	13	0	.05	.05	.4	0
	Danish pastry (enriched flour), plain without fruit or nuts:[54]																			
433	Packaged ring, 12 oz	1 ring	340	22	1,435	25	80	24.3	31.7	16.5	155	170	371	6.1	381	1,050	.97	1.01	8.6	Trace
434	Round piece, about 4 1/4-in diam. by 1 in.	1 pastry	65	22	275	5	15	4.7	6.1	3.2	30	33	71	1.2	73	200	.18	.19	1.7	Trace
435	Ounce	1 oz	28	22	120	2	7	2.0	2.7	1.4	13	14	31	.5	32	90	.08	.08	.7	Trace
	Doughnuts, made with enriched flour:[38]																			
436	Cake type, plain, 2 1/2-in diam., 1 in high.	1 doughnut	25	24	100	1	5	1.2	2.0	1.1	13	10	48	.4	23	20	.05	.05	.4	Trace
437	Yeast-leavened, glazed, 3 3/4-in diam., 1 1/4 in high.	1 doughnut	50	26	205	3	11	3.3	5.8	3.3	22	16	33	.6	34	25	.10	.10	.8	0
	Macaroni, enriched, cooked (cut lengths, elbows, shells):																			
438	Firm stage (hot)	1 cup	130	64	190	7	1	—	—	—	39	14	85	1.4	103	0	.23	.13	1.8	0
	Tender stage:																			
439	Cold macaroni	1 cup	105	73	115	4	Trace	—	—	—	24	8	53	.9	64	0	.15	.08	1.2	0
440	Hot macaroni	1 cup	140	73	155	5	1	—	—	—	32	11	70	1.3	85	0	.20	.11	1.5	0
	Macaroni (enriched) and cheese:[55]																			
441	Canned	1 cup	240	80	230	9	10	4.2	3.1	1.4	26	199	182	1.0	139	260	.12	.24	1.0	Trace
442	From home recipe (served hot)[56]	1 cup	200	58	430	17	22	8.9	8.8	2.9	40	362	322	1.8	240	860	.20	.40	1.8	Trace
	Muffins made with enriched flour:[38]																			
	From home recipe:																			
443	Blueberry, 2 3/8-in diam., 1 1/2 in high.	1 muffin	40	39	110	3	4	1.1	1.4	.7	17	34	53	.6	46	90	.09	.10	.7	Trace
444	Bran, 1 1/2 in high.	1 muffin	40	35	105	3	4	1.2	1.4	.8	17	57	162	1.5	172	90	.07	.10	1.7	Trace
445	Corn (enriched degermed cornmeal and flour), 2 3/8-in diam., 1 1/2 in high.	1 muffin	40	33	125	3	4	1.2	1.6	.9	19	42	68	.7	54	[57]120	.10	.10	.7	Trace

(A)	(B)	(C)	(D)	(E)	(F)	(G)	(H)	(I)	(J)	(K)	(L)	(M)	(N)	(O)	(P)	(Q)	(R)	(S)
446	Plain, 3-in diam., 1 1/2 in high. — 1 muffin	40	38	120	3	4	1.0	1.7	1.0	17	42	60	0.6	50	40	0.09	0.12	0.9	Trace
	From mix, egg, milk:																		
447	Corn, 2 3/8-in diam., 1 1/2 in high.[58] — 1 muffin	40	30	130	3	4	1.2	1.7	.7	20	96	152	.6	44	[57]100	.08	.09	.7	Trace
448	Noodles (egg noodles), enriched, cooked. — 1 cup	160	71	200	7	2	—	—	—	37	16	94	1.4	70	110	.22	.13	1.9	0
449	Noodles, chow mein, canned. — 1 cup	45	1	220	6	11	—	—	—	26	—	—	—	—	—	—	—	—	—
450	Pancakes, (4-in diam.):[38] Buckwheat, made from mix (with buckwheat and enriched flours), egg and milk added. — 1 cake	27	58	55	2	2	.8	.9	.4	6	59	91	.4	66	60	.04	.05	.2	Trace
	Plain:																		
451	Made from home recipe using enriched flour. — 1 cake	27	50	60	2	2	.5	.8	.5	9	27	38	.4	33	30	.06	.07	.5	Trace
452	Made from mix with enriched flour, egg and milk added. — 1 cake	27	51	60	2	2	.7	.7	.3	9	58	70	.3	42	70	.04	.06	.2	Trace
	Pies, piecrust made with enriched flour, vegetable shortening (9-in diam.):																		
	Apple:																		
453	Whole — 1 pie	945	48	2,420	21	105	27.0	44.5	25.2	360	76	208	6.6	756	280	1.06	.79	9.3	9
454	Sector, 1/7 of pie — 1 sector	135	48	345	3	15	3.9	6.4	3.6	51	11	30	.9	108	40	.15	.11	1.3	2
	Banana cream:																		
455	Whole — 1 pie	910	54	2,010	41	85	26.7	33.2	16.2	279	601	746	7.3	1,847	2,280	.77	1.51	7.0	9
456	Sector, 1/7 of pie — 1 sector	130	54	285	6	12	3.8	4.7	2.3	40	86	107	1.0	264	330	.11	.22	1.0	1
	Blueberry:																		
457	Whole — 1 pie	945	51	2,285	23	102	24.8	43.7	25.1	330	104	217	9.5	614	280	1.03	.80	10.0	28
458	Sector, 1/7 of pie — 1 sector	135	51	325	3	15	3.5	6.2	3.6	47	15	31	1.4	88	40	.15	.11	1.4	4
	Cherry:																		
459	Whole — 1 pie	945	47	2,465	25	107	28.2	45.0	25.3	363	132	236	6.6	992	4,160	1.09	.84	9.8	Trace
460	Sector, 1/7 of pie — 1 sector	135	47	350	4	15	4.0	6.4	3.6	52	19	34	.9	142	590	.16	.12	1.4	Trace
	Custard:																		
461	Whole — 1 pie	910	58	1,985	56	101	33.9	38.5	17.5	213	874	1,028	8.2	1,247	2,090	.79	1.92	5.6	0
462	Sector, 1/7 of pie — 1 sector	130	58	285	8	14	4.8	5.5	2.5	30	125	147	1.2	178	300	.11	.27	.8	0
	Lemon meringue:																		
463	Whole — 1 pie	840	47	2,140	31	86	26.1	33.8	16.4	317	118	412	6.7	420	1,430	.61	.84	5.2	25
464	Sector, 1/7 of pie — 1 sector	120	47	305	4	12	3.7	4.8	2.3	45	17	59	1.0	60	200	.09	.12	.7	4
	Mince:																		
465	Whole — 1 pie	945	43	2,560	24	109	28.0	45.9	25.2	389	265	359	13.3	1,682	20	.96	.86	9.8	9
466	Sector, 1/7 of pie — 1 sector	135	43	365	3	16	4.0	6.6	3.6	56	38	51	1.9	240	Trace	.14	.12	1.4	1
	Peach:																		
467	Whole — 1 pie	945	48	2,410	24	101	24.8	43.7	25.1	361	95	274	8.5	1,408	6,900	1.04	.97	14.0	28
468	Sector, 1/7 of pie — 1 sector	135	48	345	3	14	3.5	6.2	3.6	52	14	39	1.2	201	990	.15	.14	2.0	4
	Pecan:																		
469	Whole — 1 pie	825	20	3,450	42	189	27.8	101.0	44.2	423	388	850	25.6	1,015	1,320	1.80	.95	6.9	Trace
470	Sector, 1/7 of pie — 1 sector	118	20	495	6	27	4.0	14.4	6.3	61	55	122	3.7	145	190	.26	.14	1.0	Trace
	Pumpkin:																		
471	Whole — 1 pie	910	59	1,920	36	102	37.4	37.5	16.6	223	464	628	7.3	1,456	22,480	.78	1.27	7.0	Trace
472	Sector, 1/7 of pie — 1 sector	130	59	275	5	15	5.4	5.4	2.4	32	66	90	1.0	208	3,210	.11	.18	1.0	Trace
473	Piecrust (home recipe) made with enriched flour and vegetable shortening, baked. — 1 pie shell, 9-in diam.	180	15	900	11	60	14.8	26.1	14.9	79	25	90	3.1	89	0	.47	.40	5.0	0
474	Piecrust mix with enriched flour and vegetable shortening, 10-oz pkg. prepared and baked. — Piecrust for 2-crust pie, 9-in diam.	320	19	1,485	20	93	22.7	39.7	23.4	141	131	272	6.1	179	0	1.07	.79	9.9	0

[38] Made with vegetable shortening.
[50] Products are commercial unless otherwise specified.
[51] Made with enriched flour and vegetable shortening except for macaroons which do not contain flour or shortening.
[53] Applies to yellow varieties; white varieties contain only a trace.
[54] Contains vegetable shortening and butter.
[55] Made with corn oil.
[56] Made with regular margarine.
[57] Applies to product made with yellow cornmeal.
[58] Made with enriched degermed cornmeal and enriched flour.

Item No. (A)	Foods, approximate measures, units, and weight (edible part unless footnotes indicate otherwise) (B)	Grams	Water (C) Per cent	Food energy (D) Cal- ories	Pro- tein (E) Grams	Fat (F) Grams	Satu- rated (total) (G) Grams	Oleic (H) Grams	Lino- leic Grams	Carbo- hydrate (I) Grams	Calcium (K) Milli- grams	Phos- phorus (L) Milli- grams	Iron (M) Milli- grams	Potas- sium (N) Milli- grams	Vitamin A value (O) Inter- national units	Thiamin (P) Milli- grams	Ribo- flavin (Q) Milli- grams	Niacin (R) Milli- grams	Ascorbic acid (S) Milli- grams
	GRAIN PRODUCTS—Con.																		
475	Pizza (cheese) baked, 4 3/4-in sector; 1/8 of 12-in diam. pie.[19] 1 sector	60	45	145	6	4	1.7	1.5	0.6	22	86	89	1.1	67	230	0.16	0.18	1.6	4
	Popcorn, popped:																		
476	Plain, large kernel 1 cup	6	4	25	1	Trace	Trace	.1	.2	5	1	17	.2	—	—	—	.01	.1	0
477	With oil (coconut) and salt added, large kernel. 1 cup	9	3	40	1	2	1.5	.2	.2	5	1	19	.2	—	—	—	.01	.2	0
478	Sugar coated 1 cup	35	4	135	2	1	.5	.2	.4	30	2	47	.5	—	—	—	.02	.4	0
	Pretzels, made with enriched flour:																		
479	Dutch, twisted, 2 3/4 by 2 5/8 in. 1 pretzel	16	5	60	2	1	—	—	—	12	4	21	.2	21	0	.05	.04	.7	0
480	Thin, twisted, 3 1/4 by 2 1/4 by 1/4 in. 10 pretzels	60	5	235	6	3	—	—	—	46	13	79	.9	78	0	.20	.15	2.5	0
481	Stick, 2 1/4 in long. 10 pretzels	3	5	10	Trace	Trace	—	—	—	2	1	4	Trace	4	0	.01	.01	.1	0
	Rice, white, enriched:																		
	Instant, ready-to-serve, hot:																		
482	1 cup	165	73	180	4	Trace	Trace	Trace	Trace	40	5	31	1.3	—	0	.21	(59)	1.7	0
	Long grain:																		
483	Raw 1 cup	185	12	670	12	1	.2	.2	.2	149	44	174	5.4	170	0	.81	.06	6.5	0
484	Cooked, served hot 1 cup	205	73	225	4	Trace	.1	.1	.1	50	21	57	1.8	57	0	.23	.02	2.1	0
	Parboiled:																		
485	Raw 1 cup	185	10	685	14	1	.2	.1	.2	150	111	370	5.4	278	0	.81	.07	6.5	0
486	Cooked, served hot 1 cup	175	73	185	4	Trace	.1	.1	.1	41	33	100	1.4	75	0	.19	.02	2.1	0
	Rolls, enriched:[38]																		
	Commercial:																		
487	Brown-and-serve (12 per 12-oz pkg.), browned. 1 roll	26	27	85	2	2	.4	.7	.5	14	20	23	.5	25	Trace	.10	.06	.9	Trace
488	Cloverleaf or pan, 2 1/2-in diam., 2 in high. 1 roll	28	31	85	2	2	.4	.6	.4	15	21	24	.5	27	Trace	.11	.07	.9	Trace
489	Frankfurter and hamburger (8 per 11 1/2-oz pkg.). 1 roll	40	31	120	3	2	.5	.8	.6	21	30	34	.8	38	Trace	.16	.10	1.3	Trace
490	Hard, 3 3/4-in diam., 2 in high. 1 roll	50	25	155	5	2	.4	.6	.5	30	24	46	1.2	49	Trace	.20	.12	1.7	Trace
491	Hoagie or submarine, 11 1/2 by 3 by 2 1/2 in. 1 roll	135	31	390	12	4	.9	1.4	1.4	75	58	115	3.0	122	Trace	.54	.32	4.5	Trace
	From home recipe:																		
492	Cloverleaf, 2 1/2-in diam., 2 in high. 1 roll	35	26	120	3	3	.8	1.1	.7	20	16	36	.7	41	30	.12	.12	1.2	Trace
	Spaghetti, enriched, cooked:																		
493	Firm stage, "al dente," served hot. 1 cup	130	64	190	7	1	—	—	—	39	14	85	1.4	103	0	.23	.13	1.8	0
494	Tender stage, served hot. 1 cup	140	73	155	5	1	—	—	—	32	11	70	1.3	85	0	.20	.11	1.5	0
	Spaghetti (enriched) in tomato sauce with cheese:																		
495	From home recipe 1 cup	250	77	260	9	9	2.0	5.4	.7	37	80	135	2.3	408	1,080	.25	.18	2.3	13
496	Canned 1 cup	250	80	190	6	2	.5	.3	.4	39	40	88	2.8	303	930	.35	.28	4.5	10
	Spaghetti (enriched) with meat balls and tomato sauce:																		
497	From home recipe 1 cup	248	70	330	19	12	3.3	6.3	.9	39	124	236	3.7	665	1,590	.25	.30	4.0	22
498	Canned 1 cup	250	78	260	12	10	2.2	3.3	3.9	29	53	113	3.3	245	1,000	.15	.18	2.3	5
499	Toaster pastries 1 pastry	50	12	200	3	6	—	—	—	36	54[60]	67[60]	1.9	74[60]	500	.16	.17	2.1	(60)
	Waffles, made with enriched flour, 7-in diam.:[38]																		
500	From home recipe 1 waffle	75	41	210	7	7	2.3	2.8	1.4	28	85	130	1.3	109	250	.17	.23	1.4	Trace
501	From mix, egg and milk added 1 waffle	75	42	205	7	8	2.8	2.9	1.2	27	179	257	1.0	146	170	.14	.22	.9	Trace

Wheat flours:
All-purpose or family flour, enriched:

(A)	(B)	(wt, g)	(C)	(D)	(E)	(F)	(G)	(H)	(I)	(J)	(K)	(L)	(M)	(N)	(O)	(P)	(Q)	(R)	(S)
502	Sifted, spooned — 1 cup	115	12	420	12	1	0.2	.1	.5	88	18	100	3.3	109	0	0.74	0.46	6.1	0
503	Unsifted, spooned — 1 cup	125	12	455	13	1	.1	.1	.3	95	20	109	3.6	119	0	.80	.50	6.6	0
504	Cake or pastry flour, enriched, sifted, spooned — 1 cup	96	12	350	7	1	.1	.1	.3	76	16	70	2.8	91	0	.61	.38	5.1	0
505	Self-rising, enriched, unsifted, spooned — 1 cup	125	12	440	12	1	—	—	—	93	331	583	3.6	—	0	.80	.50	6.6	0
506	Whole-wheat, from hard wheats, stirred — 1 cup	120	12	400	16	2	.4	.2	1.0	85	49	446	4.0	444	0	.66	.14	5.2	0

LEGUMES (DRY), NUTS, SEEDS; RELATED PRODUCTS

(A)	(B)	(wt, g)	(C)	(D)	(E)	(F)	(G)	(H)	(I)	(J)	(K)	(L)	(M)	(N)	(O)	(P)	(Q)	(R)	(S)
507	Almonds, shelled: Chopped (about 130 almonds) — 1 cup	130	5	775	24	70	5.6	47.7	12.8	25	304	655	6.1	1,005	0	.31	1.20	4.6	Trace
508	Slivered, not pressed down (about 115 almonds) — 1 cup	115	5	690	21	62	5.0	42.2	11.3	22	269	580	5.4	889	0	.28	1.06	4.0	Trace
	Beans, dry: Common varieties as Great Northern, navy, and others: Cooked, drained:																		
509	Great Northern — 1 cup	180	69	210	14	1	—	—	—	38	90	266	4.9	749	0	.25	.13	1.3	0
510	Pea (navy) — 1 cup	190	69	225	15	1	—	—	—	40	95	281	5.1	790	0	.27	.13	1.3	0
	Canned, solids and liquid: White with—																		
511	Frankfurters (sliced) — 1 cup	255	71	365	19	18	—	—	—	32	94	303	4.8	668	330	.19	.15	3.3	—
512	Pork and tomato sauce — 1 cup	255	71	310	16	7	2.4	2.8	.6	48	138	235	4.6	536	330	.20	.08	1.5	5
513	Pork and sweet sauce — 1 cup	255	66	385	16	12	4.3	5.0	1.1	54	161	291	5.9	—	—	.15	.10	1.3	—
514	Red kidney — 1 cup	255	76	230	15	1	—	—	—	42	74	278	4.6	673	10	.13	.10	1.5	—
515	Lima, cooked, drained — 1 cup	190	64	260	16	1	—	—	—	49	55	293	5.9	1,163	—	.25	.11	1.3	—
516	Blackeye peas, dry, cooked (with residual cooking liquid) — 1 cup	250	80	190	13	1	—	—	—	35	43	238	3.3	573	30	.40	.10	1.0	—
517	Brazil nuts, shelled (6–8 large kernels) — 1 oz	28	5	185	4	19	4.8	6.2	7.1	3	53	196	1.0	203	Trace	.27	.03	.5	—
518	Cashew nuts, roasted in oil — 1 cup	140	5	785	24	64	12.9	36.8	10.2	41	53	522	5.3	650	140	.60	.35	2.5	—
	Coconut meat, fresh:																		
519	Piece, about 2 by 2 by 1/2 in — 1 piece	45	51	155	2	16	14.0	.9	.3	4	6	43	.8	115	0	.02	.01	.2	1
520	Shredded or grated, not pressed down — 1 cup	80	51	275	3	28	24.8	1.6	.5	8	10	76	1.4	205	0	.04	.02	.4	2
521	Filberts (hazelnuts), chopped (about 60 kernels) — 1 cup	115	6	730	14	72	5.1	55.2	7.3	19	240	388	3.9	810	—	.53	—	1.0	Trace
522	Lentils, whole, cooked — 1 cup	200	72	210	16	Trace	—	—	—	39	50	238	4.2	498	40	.14	.12	1.2	0
523	Peanuts, roasted in oil, salted (whole, halves, chopped) — 1 cup	144	2	840	37	72	13.7	33.0	20.7	27	107	577	3.0	971	—	.46	.19	24.8	0
524	Peanut butter — 1 tbsp	16	2	95	4	8	1.5	3.7	2.3	3	9	61	.3	100	—	.02	.02	2.4	0
525	Peas, split, dry, cooked — 1 cup	200	70	230	16	1	—	—	—	42	22	178	3.4	592	80	.30	.18	1.8	—
526	Pecans, chopped or pieces (about 120 large halves) — 1 cup	118	3	810	11	84	7.2	50.5	20.0	17	86	341	2.8	712	150	1.01	.15	1.1	2
527	Pumpkin and squash kernels, dry, hulled — 1 cup	140	4	775	41	65	11.8	23.5	27.5	21	71	1,602	15.7	1,386	100	.34	.27	3.4	—
528	Sunflower seeds, dry, hulled — 1 cup	145	5	810	35	69	8.2	13.7	43.2	29	174	1,214	10.3	1,334	70	2.84	.33	7.8	—
	Walnuts: Black:																		
529	Chopped or broken kernels — 1 cup	125	3	785	26	74	6.3	13.3	45.7	19	Trace	713	7.5	575	380	.28	.14	.9	—
530	Ground (finely) — 1 cup	80	3	500	16	47	4.0	8.5	29.2	12	Trace	456	4.8	368	240	.18	.09	.6	—
531	Persian or English, chopped (about 60 halves) — 1 cup	120	4	780	18	77	8.4	11.8	42.2	19	119	456	3.7	540	40	.40	.16	1.1	2

[19] Crust made with vegetable shortening and enriched flour.
[38] Made with vegetable shortening.
[59] Product may or may not be enriched with riboflavin. Consult the label.
[60] Value varies with the brand. Consult the label.

SUGARS AND SWEETS

Item No. (A)	Foods, approximate measures, units, and weight (edible part unless footnotes indicate otherwise) (B)	Grams	Water (C) Percent	Food energy (D) Calories	Protein (E) Grams	Fat (F) Grams	Fatty Acids Saturated (total) (G) Grams	Unsaturated Oleic (H) Grams	Linoleic (I) Grams	Carbohydrate (J) Grams	Calcium (K) Milligrams	Phosphorus (L) Milligrams	Iron (M) Milligrams	Potassium (N) Milligrams	Vitamin A value (O) International units	Thiamin (P) Milligrams	Riboflavin (Q) Milligrams	Niacin (R) Milligrams	Ascorbic acid (S) Milligrams
	Cake icings:																		
	Boiled, white:																		
532	Plain------------- 1 cup-------	94	18	295	1	0	0	0	0	75	2	2	Trace	17	0	Trace	0.03	Trace	0
533	With coconut------ 1 cup-------	166	15	605	3	13	11.0	.9	Trace	124	10	50	0.8	277	0	0.02	.07	0.3	0
	Uncooked:																		
534	Chocolate made with milk and butter. 1 cup-------	275	14	1,035	9	38	23.4	11.7	1.0	185	165	305	3.3	536	580	.06	.28	.6	1
535	Creamy fudge from mix and water. 1 cup-------	245	15	830	7	16	5.1	6.7	3.1	183	96	218	2.7	238	Trace	.05	.20	.7	Trace
536	White------------- 1 cup-------	319	11	1,200	2	21	12.7	5.1	.5	260	48	38	Trace	57	860	Trace	.06	Trace	Trace
	Candy:																		
537	Caramels, plain or chocolate---- 1 oz-------	28	8	115	1	3	1.6	1.1	.1	22	42	35	.4	54	Trace	.01	.05	.1	Trace
	Chocolate:																		
538	Milk, plain------- 1 oz-------	28	1	145	2	9	5.5	3.0	.3	16	65	65	.3	109	80	.02	.10	.1	Trace
539	Semisweet, small pieces (60 per oz). 1 cup or 6-oz pkg--	170	1	860	7	61	36.2	19.8	1.7	97	51	255	4.4	553	30	.02	.14	.9	0
540	Chocolate-coated peanuts------- 1 oz-------	28	1	160	5	12	4.0	4.7	2.1	11	33	84	.4	143	Trace	.10	.05	2.1	Trace
541	Fondant, uncoated (mints, candy corn, other). 1 oz-------	28	8	105	Trace	1	.1	.3	.1	25	4	2	.3	1	0	Trace	Trace	Trace	0
542	Fudge, chocolate, plain---- 1 oz-------	28	8	115	1	3	1.3	1.4	.6	21	22	24	.3	42	Trace	.01	.03	.1	Trace
543	Gum drops--------- 1 oz-------	28	12	100	Trace	Trace	---	---	---	25	2	Trace	.1	1	0	0	Trace	Trace	0
544	Hard-------------- 1 oz-------	28	1	110	0	Trace	---	---	---	28	6	2	.5	1	0	0	0	0	0
545	Marshmallows------ 1 oz-------	28	17	90	1	Trace	---	---	---	23	5	2	.5	2	0	0	Trace	Trace	0
	Chocolate-flavored beverage powders (about 4 heaping tsp per oz):																		
546	With nonfat dry milk------- 1 oz-------	28	2	100	5	1	.5	.3	Trace	20	167	155	.5	227	10	.04	.21	.2	1
547	Without milk------ 1 oz-------	28	1	100	1	1	.4	.2	Trace	25	9	48	.6	142	0	.01	.03	.1	0
548	Honey, strained or extracted-- 1 tbsp------	21	17	65	Trace	0	0	0	0	17	1	1	.1	11	0	Trace	.01	.1	Trace
549	Jams and preserves- 1 tbsp------	20	29	55	Trace	Trace	---	---	---	14	4	2	.2	18	Trace	Trace	.01	Trace	Trace
550	1 packet----	14	29	40	Trace	Trace	---	---	---	10	3	1	.1	12	Trace	Trace	Trace	Trace	Trace
551	Jellies----------- 1 tbsp------	18	29	50	Trace	Trace	---	---	---	13	4	1	.3	14	Trace	Trace	.01	Trace	1
552	1 packet----	14	29	40	Trace	Trace	---	---	---	10	3	1	.2	11	Trace	Trace	Trace	Trace	1
	Sirups:																		
	Chocolate-flavored sirup or topping:																		
553	Thin type-------- 1 fl oz or 2 tbsp--	38	32	90	1	1	.5	.3	Trace	24	6	35	.6	106	Trace	.01	.03	.2	0
554	Fudge type------- 1 fl oz or 2 tbsp--	38	25	125	2	5	3.1	1.6	.1	20	48	60	.5	107	60	.02	.08	.2	Trace
	Molasses, cane:																		
555	Light (first extraction)----- 1 tbsp------	20	24	50	---	---	---	---	---	13	33	9	.9	183	---	.01	.01	Trace	---
556	Blackstrap (third extraction)- 1 tbsp------	20	24	45	---	---	---	---	---	11	137	17	3.2	585	---	.02	.04	.4	---
557	Sorghum---------- 1 tbsp------	21	23	55	---	---	---	---	---	14	35	5	2.6	---	---	---	.02	Trace	---
558	Table blends, chiefly corn, light and dark. 1 tbsp------	21	24	60	0	0	0	0	0	15	9	3	.8	1	0	0	0	0	0
	Sugars:																		
559	Brown, pressed down--------- 1 cup-------	220	2	820	0	0	0	0	0	212	187	42	7.5	757	0	.02	.07	.4	0
	White:																		
560	Granulated------- 1 cup-------	200	1	770	0	0	0	0	0	199	0	0	.2	Trace	0	0	0	0	0
561	1 tbsp------	12	1	45	0	0	0	0	0	12	0	0	Trace	Trace	0	0	0	0	0
562	1 packet----	6	1	23	0	0	0	0	0	6	0	0	Trace	Trace	0	0	0	0	0
563	Powdered, sifted, spooned into cup. 1 cup-------	100	1	385	0	0	0	0	0	100	0	0	.1	3	0	0	0	0	0

VEGETABLE AND VEGETABLE PRODUCTS

(A)	(B)	(C)	(D)	(E)	(F)	(G)	(H)	(I)	(J)	(K)	(L)	(M)	(N)	(O)	(P)	(Q)	(R)	(S)
	Asparagus, green:																	
	Cooked, drained:																	
	Cuts and tips, 1 1/2- to 2-in lengths:																	
564	From raw — 1 cup — 145	94	30	3	Trace	---	---	---	5	30	73	0.9	265	1,310	0.23	0.26	2.0	38
565	From frozen — 1 cup — 180	93	40	6	Trace	---	---	---	6	40	115	2.2	396	1,530	.25	.23	1.8	41
	Spears, 1/2-in diam. at base:																	
566	From raw — 4 spears — 60	94	10	2	Trace	---	---	---	2	13	30	.4	110	540	.10	.11	.8	16
567	From frozen — 4 spears — 60	92	15	2	Trace	---	---	---	2	13	40	.7	143	470	.10	.08	.7	16
568	Canned, spears, 1/2-in diam. at base. — 4 spears — 80	93	15	2	Trace	---	---	---	3	15	42	1.5	133	640	.05	.08	.6	12
	Beans:																	
	Lima, immature seeds, frozen, cooked, drained:																	
569	Thick-seeded types (Fordhooks) — 1 cup — 170	74	170	10	Trace	---	---	---	32	34	153	2.9	724	390	.12	.09	1.7	29
570	Thin-seeded types (baby limas) — 1 cup — 180	69	210	13	Trace	---	---	---	40	63	227	4.7	709	400	.16	.09	2.2	22
	Snap:																	
	Green:																	
	Cooked, drained:																	
571	From raw (cuts and French style). — 1 cup — 125	92	30	2	Trace	---	---	---	7	63	46	.8	189	680	.09	.11	.6	15
	From frozen:																	
572	Cuts — 1 cup — 135	92	35	2	Trace	---	---	---	8	54	43	.9	205	780	.09	.12	.5	7
573	French style — 1 cup — 130	92	35	2	Trace	---	---	---	8	49	39	1.2	177	690	.08	.10	.4	9
574	Canned, drained solids (cuts). — 1 cup — 135	92	30	2	Trace	---	---	---	7	61	34	2.0	128	630	.04	.07	.4	5
	Yellow or wax:																	
	Cooked, drained:																	
575	From raw (cuts and French style). — 1 cup — 125	93	30	2	Trace	---	---	---	6	63	46	.8	189	290	.09	.11	.6	16
576	From frozen (cuts) — 1 cup — 135	92	35	2	Trace	---	---	---	8	47	42	.9	221	140	.09	.11	.5	8
577	Canned, drained solids (cuts). — 1 cup — 135	92	30	2	Trace	---	---	---	7	61	34	2.0	128	140	.04	.07	.4	7
	Beans, mature. See Beans, dry (items 509-515) and Blackeye peas, dry (item 516).																	
	Bean sprouts (mung):																	
578	Raw — 1 cup — 105	89	35	4	Trace	---	---	---	7	20	67	1.4	234	20	.14	.14	.8	20
579	Cooked, drained — 1 cup — 125	91	35	4	Trace	---	---	---	7	21	60	1.1	195	30	.11	.13	.9	8
	Beets:																	
	Cooked, drained, peeled:																	
580	Whole beets, 2-in diam. — 2 beets — 100	91	30	1	Trace	---	---	---	7	14	23	.5	208	20	.03	.04	.3	6
581	Diced or sliced — 1 cup — 170	91	55	2	Trace	---	---	---	12	24	39	.9	354	30	.05	.07	.5	10
	Canned, drained solids:																	
582	Whole beets, small — 1 cup — 160	89	60	2	Trace	---	---	---	14	30	29	1.1	267	30	.02	.05	.2	5
583	Diced or sliced — 1 cup — 170	89	65	2	Trace	---	---	---	15	32	31	1.2	284	30	.02	.05	.2	5
584	Beet greens, leaves and stems, cooked, drained. — 1 cup — 145	94	25	2	Trace	---	---	---	5	144	36	2.8	481	7,400	.10	.22	.4	22
	Blackeye peas, immature seeds, cooked and drained:																	
585	From raw — 1 cup — 165	72	180	13	1	---	---	---	30	40	241	3.5	625	580	.50	.18	2.3	28
586	From frozen — 1 cup — 170	66	220	15	1	---	---	---	40	43	286	4.8	573	290	.68	.19	2.4	15
	Broccoli, cooked, drained:																	
	From raw:																	
587	Stalk, medium size — 1 stalk — 180	91	45	6	1	---	---	---	8	158	112	1.4	431	4,500	.16	.36	1.4	162
588	Stalks cut into 1/2-in pieces — 1 cup — 155	91	40	5	Trace	---	---	---	7	136	96	1.2	414	3,880	.14	.31	1.2	140
	From frozen:																	
589	Stalk, 4 1/2 to 5 in long — 1 stalk — 30	91	10	1	Trace	---	---	---	1	12	17	.2	66	570	.02	.03	.2	22
590	Chopped — 1 cup — 185	92	50	5	1	---	---	---	9	100	104	1.3	392	4,810	.11	.22	.9	105
	Brussels sprouts, cooked, drained:																	
591	From raw, 7-8 sprouts (1 1/4- to 1 1/2-in diam.). — 1 cup — 155	88	55	7	1	---	---	---	10	50	112	1.7	423	810	.12	.22	1.2	135
592	From frozen — 1 cup — 155	89	50	5	Trace	---	---	---	10	33	95	1.2	457	880	.12	.16	.9	126

NUTRIENTS IN INDICATED QUANTITY

Item No. (A)	Foods, approximate measure, units, and weight (edible part unless footnotes indicate otherwise) (B)	(Grams)	Water (C) Per cent	Food energy (D) Cal-ories	Pro-tein (E) Grams	Fat (F) Grams	Fatty Acids — Satu-rated (total) (G) Grams	Unsaturated Oleic (H) Grams	Lino-leic (I) Grams	Carbo-hydrate (J) Grams	Calcium (K) Milli-grams	Phos-phorus (L) Milli-grams	Iron (M) Milli-grams	Potas-sium (N) Milli-grams	Vitamin A value (O) Inter-national units	Thiamin (P) Milli-grams	Ribo-flavin (Q) Milli-grams	Niacin (R) Milli-grams	Ascorbic acid (S) Milli-grams
	VEGETABLE AND VEGETABLE PRODUCTS—Con.																		
	Cabbage:																		
	Common varieties:																		
	Raw:																		
593	Coarsely shredded or sliced- 1 cup	70	92	15	1	Trace	—	—	—	4	34	20	0.3	163	90	0.04	0.04	0.2	33
594	Finely shredded or chopped- 1 cup	90	92	20	1	Trace	—	—	—	5	44	26	.4	210	120	.05	.05	.3	42
595	Cooked, drained- 1 cup	145	94	30	2	Trace	—	—	—	6	64	29	.4	236	190	.06	.06	.4	48
596	Red, raw, coarsely shredded or sliced- 1 cup	70	90	20	1	Trace	—	—	—	5	29	25	.6	188	30	.06	.04	.3	43
597	Savoy, raw, coarsely shredded or sliced- 1 cup	70	92	15	2	Trace	—	—	—	3	47	38	.6	188	140	.04	.06	.2	39
598	Cabbage, celery (also called pe-tsai or wongbok), raw, 1-in pieces- 1 cup	75	95	10	1	Trace	—	—	—	2	32	30	.5	190	110	.04	.03	.5	19
599	Cabbage, white mustard (also called bokchoy or pakchoy), cooked, drained- 1 cup	170	95	25	2	Trace	—	—	—	4	252	56	1.0	364	5,270	.07	.14	1.2	26
	Carrots:																		
	Raw, without crowns and tips, scraped:																		
600	Whole, 7 1/2 by 1 1/8 in, or strips, 2 1/2 to 3 in. long- 1 carrot or 18 strips	72	88	30	1	Trace	—	—	—	7	27	26	.5	246	7,930	.04	.04	.4	6
601	Grated- 1 cup	110	88	45	1	Trace	—	—	—	11	41	40	.8	375	12,100	.07	.06	.7	9
602	Cooked (crosswise cuts), drained- 1 cup	155	91	50	1	Trace	—	—	—	11	51	48	.9	344	16,280	.08	.08	.8	9
	Canned:																		
603	Sliced, drained solids- 1 cup	155	91	45	1	Trace	—	—	—	10	47	34	1.1	186	23,250	.03	.05	.6	3
604	Strained or junior (baby food)- 1 oz (1 3/4 to 2 tbsp)	28	92	10	Trace	Trace	—	—	—	2	7	6	.1	51	3,690	.01	.01	.1	1
	Cauliflower:																		
605	Raw, chopped- 1 cup	115	91	31	3	Trace	—	—	—	6	29	64	1.3	339	70	.13	.12	.8	90
	Cooked, drained:																		
606	From raw (flower buds)- 1 cup	125	93	30	3	Trace	—	—	—	5	26	53	.9	258	80	.11	.10	.8	69
607	From frozen (flowerets)- 1 cup	180	94	30	3	Trace	—	—	—	6	31	68	.9	373	50	.07	.09	.7	74
	Celery, Pascal type, raw:																		
608	Stalk, large outer, 8 by 1 1/2 in, at root end- 1 stalk	40	94	5	Trace	Trace	—	—	—	2	16	11	.1	136	110	.01	.01	.1	4
609	Pieces, diced- 1 cup	120	94	20	1	Trace	—	—	—	5	47	34	.4	409	320	.04	.04	.4	11
	Collards, cooked, drained:																		
610	From raw (leaves without stems)- 1 cup	190	90	65	7	1	—	—	—	10	357	99	1.5	498	14,820	.21	.38	2.3	144
611	From frozen (chopped)- 1 cup	170	90	50	5	1	—	—	—	10	299	87	1.7	401	11,560	.10	.24	1.0	56
	Corn, sweet:																		
	Cooked, drained:																		
612	From raw, ear 5 by 1 3/4 in- 1 ear[61]	140	74	70	2	1	—	—	—	16	2	69	.5	151	[62]310	.09	.08	1.1	7
	From frozen:																		
613	Ear, 5 in long- 1 ear[61]	229	73	120	4	1	—	—	—	27	4	121	1.0	291	[62]440	.18	.10	2.1	9
614	Kernels- 1 cup	165	77	130	5	1	—	—	—	31	5	120	1.3	304	[62]580	.15	.10	2.5	8
	Canned:																		
615	Cream style- 1 cup	256	76	210	5	2	—	—	—	51	8	143	1.5	248	[62]840	.08	.13	2.6	13
	Whole kernel:																		
616	Vacuum pack- 1 cup	210	76	175	5	1	—	—	—	43	6	153	1.1	204	[62]740	.06	.13	2.3	11
617	Wet pack, drained solids- 1 cup	165	76	140	4	1	—	—	—	33	8	81	.8	160	[62]580	.05	.08	1.5	7
	Cowpeas. See Blackeye peas. (Items 585-586).																		
	Cucumber slices, 1/8 in thick (large, 2 1/8-in diam.; small, 1 3/4-in diam.):																		
618	With peel- 6 large or 8 small slices	28	95	5	Trace	Trace	—	—	—	1	7	8	.3	45	70	.01	.01	.1	3

(A)	(B)		(C)	(D)	(E)	(F)	(G)	(H)	(I)	(J)	(K)	(L)	(M)	(N)	(O)	(P)	(Q)	(R)	(S)
619	Without peel — 6 1/2 large or 9 small pieces	28	96	5	Trace	Trace	—	—	—	1	5	5	0.1	45	Trace	0.01	0.01	0.1	3
620	Dandelion greens, cooked, drained — 1 cup	105	90	35	2	1	—	—	—	7	147	44	1.9	244	12,290	.14	.17	—	19
621	Endive, curly (including escarole), raw, small pieces — 1 cup	50	93	10	1	Trace	—	—	—	2	41	27	.9	147	1,650	.04	.07	.3	5
	Kale, cooked, drained:																		
622	From raw (leaves without stems and midribs) — 1 cup	110	88	45	5	1	—	—	—	7	206	64	1.8	243	9,130	.11	.20	1.8	102
623	From frozen (leaf style) — 1 cup	130	91	40	4	1	—	—	—	7	157	62	1.3	251	10,660	.08	.20	.9	49
	Lettuce, raw:																		
	Butterhead, as Boston types:																		
624	Head, 5-in diam — 1 head[63]	220	95	25	2	Trace	—	—	—	4	57	42	3.3	430	1,580	.10	.10	.5	13
625	Leaves — 1 outer or 2 inner or 3 heart leaves.	15	95	Trace	Trace	Trace	—	—	—	Trace	5	4	.3	40	150	.01	.01	Trace	1
	Crisphead, as Iceberg:																		
626	Head, 6-in diam — 1 head[64]	567	96	70	5	1	—	—	—	16	108	118	2.7	943	1,780	.32	.32	1.6	32
627	Wedge, 1/4 of head — 1 wedge	135	96	20	1	Trace	—	—	—	4	27	30	.7	236	450	.08	.08	.4	8
628	Pieces, chopped or shredded — 1 cup	55	96	5	Trace	Trace	—	—	—	2	11	12	.3	96	180	.03	.03	.2	8
629	Looseleaf (bunching varieties including romaine or cos), chopped or shredded pieces. — 1 cup	55	94	10	1	Trace	—	—	—	2	37	14	.8	145	1,050	.03	.04	.2	10
630	Mushrooms, raw, sliced or chopped — 1 cup	70	90	20	2	Trace	—	—	—	3	4	81	.6	290	Trace	.07	.32	2.9	2
631	Mustard greens, without stems and midribs, cooked, drained. — 1 cup	140	93	30	3	1	—	—	—	6	193	45	2.5	308	8,120	.11	.20	.8	67
632	Okra pods, 3 by 5/8 in, cooked — 10 pods	106	91	30	2	Trace	—	—	—	6	98	43	.5	184	520	.14	.19	1.0	21
	Onions: Mature: Raw:																		
633	Chopped — 1 cup	170	89	65	3	Trace	—	—	—	15	46	61	.9	267	[65]Trace	.05	.07	.3	17
634	Sliced — 1 cup	115	89	45	2	Trace	—	—	—	10	31	41	.6	181	[65]Trace	.03	.05	.2	12
635	Cooked (whole or sliced), drained. — 1 cup	210	92	60	3	Trace	—	—	—	14	50	61	.8	231	[65]Trace	.06	.06	.4	15
636	Young green, bulb (3/8 in diam.) and white portion of top. — 6 onions	30	88	15	Trace	Trace	—	—	—	3	12	12	.2	69	Trace	.02	.01	.1	8
637	Parsley, raw, chopped — 1 tbsp	4	85	Trace	Trace	Trace	—	—	—	Trace	7	2	.2	25	300	Trace	.01	Trace	6
638	Parsnips, cooked (diced or 2-in lengths). — 1 cup	155	82	100	2	1	—	—	—	23	70	96	.9	587	50	.11	.12	.2	16
	Peas, green: Canned:																		
639	Whole, drained solids — 1 cup	170	77	150	8	1	—	—	—	29	44	129	3.2	163	1,170	.15	.10	1.4	14
640	Strained (baby food) — 1 oz (1 3/4 to 2 tbsp)	28	86	15	1	Trace	—	—	—	3	3	18	.5	28	140	.02	.03	.3	3
641	Frozen, cooked, drained — 1 cup	160	82	110	8	1	—	—	—	19	30	138	3.0	216	960	.43	.14	2.7	21
642	Peppers, hot, red, without seeds, dried (ground chili powder, added seasonings). — 1 tsp	2	9	5	Trace	Trace	—	—	—	1	5	4	.3	20	1,300	Trace	.02	.2	Trace
	Peppers, sweet, about 5 per lb, whole), stem and seeds removed:																		
643	Raw — 1 pod	74	93	15	1	Trace	—	—	—	4	7	16	.5	157	310	.06	.06	.4	94
644	Cooked, boiled, drained — 1 pod	73	95	15	1	Trace	—	—	—	3	7	12	.4	109	310	.05	.05	.4	70
	Potatoes, cooked:																		
645	Baked, peeled after baking (about 2 per lb, raw). — 1 potato	156	75	145	4	Trace	—	—	—	33	14	101	1.1	782	Trace	.15	.07	2.7	31
	Boiled (about 3 per lb, raw):																		
646	Peeled after boiling — 1 potato	137	80	105	3	Trace	—	—	—	23	10	72	.8	556	Trace	.12	.05	2.0	22
647	Peeled before boiling — 1 potato	135	83	90	3	Trace	—	—	—	20	8	57	.7	385	Trace	.12	.05	1.6	22
	French-fried, strip, 2 to 3 1/2 in long:																		
648	Prepared from raw — 10 strips	50	45	135	2	7	1.7	1.2	3.3	18	8	56	.7	427	Trace	.07	.04	1.6	11
649	Frozen, oven heated — 10 strips	50	53	110	2	4	1.1	.8	2.1	17	5	43	.9	326	Trace	.07	.01	1.3	11
650	Hashed brown, prepared from frozen. — 1 cup	155	56	345	3	18	4.6	3.2	9.0	45	28	78	1.9	439	Trace	.11	.03	1.6	12
	Mashed, prepared from— Raw:																		
651	Milk added — 1 cup	210	83	135	4	2	.7	.4	Trace	27	50	103	.8	548	40	.17	.11	2.1	21

[61] Weight includes cob. Without cob, weight is 77 g for item 612, 126 g for item 613.
[62] Based on yellow varieties. For white varieties, value is trace.
[63] Weight includes refuse of outer leaves and core. Without these parts, weight is 163 g.
[64] Weight includes core. Without core, weight is 539 g.
[65] Value based on white-fleshed varieties. For yellow-fleshed varieties, value in International Units (I.U.) is 70 for item 633, 50 for item 634, and 80 for item 635.

VEGETABLE AND VEGETABLE PRODUCTS—Con.

Item No. (A)	Foods, approximate measures, units, and weight (edible part unless footnotes indicate otherwise) (B)		Grams	Water (C) Percent	Food energy (D) Calories	Protein (E) Grams	Fat (F) Grams	Saturated (total) (G) Grams	Oleic (H) Grams	Linoleic (I) Grams	Carbohydrate (J) Grams	Calcium (K) Milligrams	Phosphorus (L) Milligrams	Iron (M) Milligrams	Potassium (N) Milligrams	Vitamin A value (O) International units	Thiamin (P) Milligrams	Riboflavin (Q) Milligrams	Niacin (R) Milligrams	Ascorbic acid (S) Milligrams
	Potatoes, cooked—Continued																			
	Mashed, prepared from—Continued																			
	Raw—Continued																			
652	Milk and butter added	1 cup	210	80	195	4	9	5.6	2.3	0.2	26	50	101	0.8	525	360	0.17	0.11	2.1	19
653	Dehydrated flakes (without milk), water, milk, butter, and salt added.	1 cup	210	79	195	4	7	3.6	2.1	.2	30	65	99	.6	601	270	.08	.08	1.9	11
654	Potato chips, 1 3/4 by 2 1/2 in oval cross section.	10 chips	20	2	115	1	8	2.1	1.4	4.0	10	8	28	.4	226	Trace	.04	.01	1.0	3
655	Potato salad, made with cooked salad dressing.	1 cup	250	76	250	7	7	2.0	2.7	1.3	41	80	160	1.5	798	350	.20	.18	2.8	28
656	Pumpkin, canned	1 cup	245	90	80	2	1	—	—	—	19	61	64	1.0	588	15,680	.07	.12	1.5	12
657	Radishes, raw (prepackaged) stem ends, rootlets cut off.	4 radishes	18	95	5	Trace	Trace	—	—	—	1	5	6	.2	58	Trace	.01	.01	.1	5
658	Sauerkraut, canned, solids and liquid.	1 cup	235	93	40	2	Trace	—	—	—	9	85	42	1.2	329	120	.07	.09	.5	33
	Southern peas. See Blackeye peas (items 585-586).																			
	Spinach:																			
659	Raw, chopped	1 cup	55	91	15	2	Trace	—	—	—	2	51	28	1.7	259	4,460	.06	.11	.3	28
660	Cooked, drained: From raw	1 cup	180	92	40	5	1	—	—	—	6	167	68	4.0	583	14,580	.13	.25	.9	50
	From frozen:																			
661	Chopped	1 cup	205	92	45	6	1	—	—	—	8	232	90	4.3	683	16,200	.14	.31	.8	39
662	Leaf	1 cup	190	92	45	6	1	—	—	—	7	200	84	4.8	688	15,390	.15	.27	1.0	53
663	Canned, drained solids	1 cup	205	91	50	6	1	—	—	—	7	242	53	5.3	513	16,400	.04	.25	.6	29
	Squash, cooked:																			
664	Summer (all varieties), diced, drained.	1 cup	210	96	30	2	Trace	—	—	—	7	53	53	.8	296	820	.11	.17	1.7	21
665	Winter (all varieties), baked, mashed.	1 cup	205	81	130	4	1	—	—	—	32	57	98	1.6	945	8,610	.10	.27	1.4	27
	Sweetpotatoes:																			
	Cooked (raw, 5 by 2 in; about 2 1/2 per lb):																			
666	Baked in skin, peeled	1 potato	114	64	160	2	1	—	—	—	37	46	66	1.0	342	9,230	.10	.08	.8	25
667	Boiled in skin, peeled	1 potato	151	71	170	3	1	—	—	—	40	48	71	1.1	367	11,940	.14	.09	.9	26
668	Candied, 2 1/2 by 2-in piece	1 piece	105	60	175	1	3	2.0	.8	.1	36	39	45	.9	200	6,620	.06	.04	.4	11
	Canned:																			
669	Solid pack (mashed)	1 cup	255	72	275	5	1	—	—	—	63	64	105	2.0	510	19,890	.13	.10	1.5	36
670	Vacuum pack, piece 2 3/4 by 1 in.	1 piece	40	72	45	1	Trace	—	—	—	10	10	16	.3	80	3,120	.02	.02	.2	6
	Tomatoes:																			
671	Raw, 2 3/5-in diam. (3 per 12 oz pkg.).	1 tomato[66]	135	94	25	1	Trace	—	—	—	6	16	33	.6	300	1,110	.07	.05	.9	[6]28
672	Canned, solids and liquid	1 cup	241	94	50	2	Trace	—	—	—	10	[6]14	46	1.2	523	2,170	.12	.07	1.7	41
673	Tomato catsup	1 cup	273	69	290	5	1	—	—	—	69	60	137	2.2	991	3,820	.25	.19	4.4	41
674		1 tbsp	15	69	15	Trace	Trace	—	—	—	4	3	8	.1	54	210	.01	.01	.2	2
	Tomato juice, canned:																			
675	Cup	1 cup	243	94	45	2	Trace	—	—	—	10	17	44	2.2	552	1,940	.12	.07	1.9	39
676	Glass (6 fl oz)	1 glass	182	94	35	2	Trace	—	—	—	8	13	33	1.6	413	1,460	.09	.05	1.5	29
677	Turnips, cooked, diced	1 cup	155	94	35	1	Trace	—	—	—	8	54	37	1.6	291	Trace	.06	.08	.5	34
	Turnip greens, cooked, drained:																			
678	From raw (leaves and stems)	1 cup	145	94	30	3	Trace	—	—	—	5	252	49	1.5	—	8,270	.15	.33	.7	68
679	From frozen (chopped)	1 cup	165	93	40	4	Trace	—	—	—	6	195	64	2.6	246	11,390	.08	.15	.7	31
680	Vegetables, mixed, frozen, cooked	1 cup	182	83	115	6	1	—	—	—	24	46	115	2.4	348	9,010	.22	.13	2.0	15

MISCELLANEOUS ITEMS

(A)	(B)	(C)	(D)	(E)	(F)	(G)	(H)	(I)	(J)	(K)	(L)	(M)	(N)	(O)	(P)	(Q)	(R)	(S)
	Baking powders for home use: Sodium aluminum sulfate:																	
681	With monocalcium phosphate monohydrate.	1 tsp---- 3.0	2	5	Trace	Trace	0	0	0	1	58	87	---	5	0	0	0	0	0
682	With monocalcium phosphate monohydrate, calcium sulfate.	1 tsp---- 2.9	1	5	Trace	Trace	0	0	0	1	183	45	---	---	0	0	0	0	0
683	Straight phosphate----	1 tsp---- 3.8	2	5	Trace	Trace	0	0	0	1	239	359	---	6	0	0	0	0	0
684	Low sodium----	1 tsp---- 4.3	2	5	Trace	Trace	0	0	0	2	207	314	---	471	0	0	0	0	0
685	Barbecue sauce----	1 cup---- 250	81	230	4	17	2.2	4.3	10.0	20	53	50	2.0	435	900	.03	.03	.8	13
686	Beverages, alcoholic: Beer----	12 fl oz---- 360	92	150	1	0	0	0	0	14	18	108	Trace	90	---	.01	.11	2.2	---
	Gin, rum, vodka, whisky:																		
687	80-proof----	1 1/2-fl oz jigger---- 42	67	95	---	---	---	---	0	Trace	---	---	---	1	---	---	---	---	---
688	86-proof----	1 1/2-fl oz jigger---- 42	64	105	---	---	---	---	0	Trace	---	---	---	1	---	---	---	---	---
689	90-proof----	1 1/2-fl oz jigger---- 42	62	110	---	---	---	---	0	Trace	---	---	---	1	---	---	---	---	---
	Wines:																		
690	Dessert----	3 1/2-fl oz glass---- 103	77	140	Trace	0	0	0	0	8	8	---	---	77	---	.01	.02	.2	---
691	Table----	3 1/2-fl oz glass---- 102	86	85	Trace	0	0	0	0	4	9	10	.4	94	---	Trace	.01	.1	---
	Beverages, carbonated, sweetened, nonalcoholic:																		
692	Carbonated water----	12 fl oz---- 366	92	115	0	0	0	0	0	29	---	---	---	---	0	0	0	0	0
693	Cola type----	12 fl oz---- 369	90	145	0	0	0	0	0	37	---	---	---	---	0	0	0	0	0
694	Fruit-flavored sodas and Tom Collins mixer.	12 fl oz---- 372	88	170	0	0	0	0	0	45	---	---	---	---	0	0	0	0	0
695	Ginger ale----	12 fl oz---- 366	92	115	0	0	0	0	0	29	---	---	---	0	0	0	0	0	0
696	Root beer----	12 fl oz---- 370	90	150	0	0	0	0	0	39	---	---	---	0	0	0	0	0	0
	Chili powder. See Peppers, hot, red (item 642).																		
	Chocolate:																		
697	Bitter or baking----	1 oz---- 28	2	145	3	15	8.9	4.9	.4	8	22	109	1.9	235	20	.01	.07	.4	0
	Semisweet, see Candy, chocolate (item 539).																		
698	Gelatin, dry----	1 7-g envelope---- 7	13	25	6	Trace	0	0	0	0	---	---	---	---	---	---	---	---	---
699	Gelatin dessert prepared with gelatin dessert powder and water.	1 cup---- 240	84	140	4	0	0	0	0	34	---	---	---	---	---	---	---	---	---
700	Mustard, prepared, yellow----	1 tsp or individual serving pouch or cup. 5	80	5	Trace	Trace	---	---	---	Trace	4	4	.1	7	---	---	---	---	---
	Olives, pickled, canned:																		
701	Green----	4 medium or 3 extra large or 2 giant.[69] 16	78	15	Trace	2	.2	1.2	.1	Trace	8	2	.2	7	40	---	---	---	---
702	Ripe, Mission----	3 small or 2 large[69]---- 10	73	15	Trace	2	.2	1.2	.1	Trace	9	1	.1	2	10	Trace	Trace	---	---
	Pickles, cucumber:																		
703	Dill, medium, whole, 3 3/4 in long, 1 1/4-in diam.	1 pickle---- 65	93	5	Trace	Trace	---	---	---	1	17	14	.7	130	70	Trace	.01	Trace	4
704	Fresh-pack, slices 1 1/2-in diam. 1/4 in thick.	2 slices---- 15	79	10	Trace	Trace	---	---	---	3	5	4	.3	---	20	Trace	Trace	Trace	1
705	Sweet, gherkin, small, whole, about 2 1/2 in long, 3/4-in diam.	1 pickle---- 15	61	20	Trace	Trace	---	---	---	5	2	2	.2	---	10	Trace	Trace	Trace	1
706	Relish, finely chopped, sweet---	1 tbsp---- 15	63	20	Trace	Trace	---	---	---	5	3	2	.1	---	---	---	---	---	---
	Popcorn. See items 475-478.																		
707	Popsicle, 3-fl oz size---	1 popsicle---- 95	80	70	0	0	0	0	0	18	0	---	Trace	---	0	0	0	0	0

[66] Weight includes cores and stem ends. Without these parts, weight is 123 g.
[67] Based on year-round average. For tomatoes marketed from November through May, value is about 12 mg; from June through October, value is 32 mg.
[68] Applies to product without calcium salts added. Value for products with calcium salts added may be as much as 63 mg for whole tomatoes, 241 mg for cut forms.
[69] Weight includes pits. Without pits, weight is 13 g for item 701, 9 g for item 702.

NUTRIENTS IN INDICATED QUANTITY

Item No. (A)	Foods, approximate measures, units, and weight (edible part unless footnotes indicate otherwise) (B)		Water (C)	Food energy (D)	Protein (E)	Fat (F)	Fatty Acids Saturated (total) (G)	Unsaturated Oleic (H)	Linoleic (I)	Carbo-hydrate (J)	Calcium (K)	Phos-phorus (L)	Iron (M)	Potas-sium (N)	Vitamin A value (O)	Thiamin (P)	Ribo-flavin (Q)	Niacin (R)	Ascorbic acid (S)
		Grams	Per-cent	Cal-ories	Grams	Grams	Grams	Grams	Grams	Grams	Milli-grams	Milli-grams	Milli-grams	Milli-grams	Inter-national units	Milli-grams	Milli-grams	Milli-grams	Milli-grams
	MISCELLANEOUS ITEMS—Con.																		
	Soups:																		
	Canned, condensed:																		
	Prepared with equal volume of milk:																		
708	Cream of chicken——————— 1 cup————	245	85	180	7	10	4.2	3.6	1.3	15	172	152	0.5	260	610	0.05	0.27	0.7	2
709	Cream of mushroom——— 1 cup————	245	83	215	7	14	5.4	2.9	4.6	16	191	169	.5	279	250	.05	.34	.7	1
710	Tomato——————————— 1 cup————	250	84	175	7	7	3.4	1.7	1.0	23	168	155	.8	418	1,200	.10	.25	1.3	15
	Prepared with equal volume of water:																		
711	Bean with pork——————— 1 cup————	250	84	170	8	6	1.2	1.8	2.4	22	63	128	2.3	395	650	.13	.08	1.0	3
712	Beef broth, bouillon, consomme. 1 cup————	240	96	30	5	0	0	0	0	3	Trace	31	.5	130	Trace	Trace	.02	1.2	—
713	Beef noodle——————————— 1 cup————	240	93	65	4	3	.6	.7	.8	7	7	48	1.0	77	50	.05	.07	1.0	Trace
714	Clam chowder, Manhattan type (with tomatoes, without milk). 1 cup————	245	92	80	2	3	.5	.4	1.3	12	34	47	1.0	184	880	.02	.02	1.0	Trace
715	Cream of chicken——————— 1 cup————	240	92	95	3	6	1.6	2.3	1.1	8	24	34	.5	79	410	.02	.05	.5	Trace
716	Cream of mushroom——— 1 cup————	240	90	135	5	10	2.6	1.7	4.5	10	41	50	.5	98	70	.02	.12	.7	Trace
717	Minestrone—————————— 1 cup————	245	90	105	5	3	.7	.9	.4	14	37	59	1.0	314	2,350	.07	.05	1.0	—
718	Split pea———————————— 1 cup————	245	85	145	9	3	1.1	1.2	.4	21	29	149	1.5	270	440	.25	.15	1.5	1
719	Tomato———————————— 1 cup————	245	91	90	2	2	.5	.5	1.0	16	15	34	.7	230	1,000	.05	.05	1.2	12
720	Vegetable beef————————— 1 cup————	245	92	80	5	2				10	12	49	.7	162	2,700	.05	.05	1.0	—
721	Vegetarian————————— 1 cup————	245	92	80	2	2				13	20	39	1.0	172	2,940	.05	.05	1.0	—
	Dehydrated:																		
722	Bouillon cube, 1/2 in———— 1 cube————	4	4	5	1	Trace				Trace	—	—	—	4	—	—	—	—	—
	Mixes:																		
	Unprepared:																		
723	Onion——————————————— 1 1/2-oz pkg	43	3	150	6	5	1.1	2.3	1.0	23	42	49	.6	238	30	.05	.03	.3	6
	Prepared with water:																		
724	Chicken noodle————————— 1 cup————	240	95	55	2	1				8	7	19	.2	19	50	.07	.05	.5	Trace
725	Onion————————————— 1 cup————	240	96	35	1	1				6	10	12	.2	58	Trace	Trace	Trace	.5	2
726	Tomato vegetable with noodles. 1 cup————	240	93	65	1	1				12	7	19	.2	29	480	.05	.02		5
727	Vinegar, cider———————— 1 tbsp————	15	94	Trace	Trace	0	0	0	0	1	1	1	.1	15	Trace	—	—	—	—
728	White sauce, medium, with enriched flour. 1 cup————	250	73	405	10	31	19.3	7.8	.8	22	288	233	.5	348	1,150	.12	.43	.7	2
	Yeast:																		
729	Baker's, dry, active—————— 1 pkg————	7	5	20	3	Trace				3	3	90	1.1	140	Trace	.16	.38	2.6	Trace
730	Brewer's, dry———————— 1 tbsp————	8	5	25	3	Trace				3	[7]17	140	1.4	152	Trace	1.25	.34	3.0	Trace

[7]Value may vary from 6 to 60 mg.

Glossary of Abbreviations

ACP	Acyl carrier protein	ddNDP	Dideoxynucleoside diphosphate
ACTH	Adrenocorticotropic hormone	ddNTP	Dideoxynucleoside tirphosphate
ADH	Antidiuretic hormone	DE	Dextrose equivalents
ADI	Acceptable daily intake	DHA	Dihydroxyacetone phosphate
ADP	Adenosine diphosphate	1,25-DHCC	1,25-Dihydroxycholecalciferol
AMC	Arm muscle circumference	25-DHCC	25-Hydroxycholecalciferol
ATCC	American Type Culture Collection	DHF	Dihydrofolic acid
ATP	Adenosine triphosphate	DIFP	Diisopropylfluorophosphate
ATPase	Adenosine triphosphatase	DIT	Diiodothyronine
ATT	Alanine-aspartate transaminase	*DM*	Degree of methylation
Azofd	Azoferredoxin	DMB	Dimethyl benzimidazole
BEPT	Birefringent endpoint temperature	DNA	Deoxyribonucleic acid
BMR	Basal metabolic rate	dNDP	Deoxynucleoside diphosphate
BV	Biological value	DNFB	Dinitrofluorobenzene
BU	Brabender units	dNTP	Deoxynucleoside triphosphate
Cal	Calorie (kcal)	*DP*	Degree of polymerization
cAMP	Cyclic 3′,5′-adenosine monophosphate	DPG	2,3-Diphosphoglyceric acid
		DS	Degree of hydroxyl substitution
CBP	Cytoplasmic binding protein	E–D pathway	Entner–Doudoroff pathway
CC	Cholecalciferol	EFA	Essential fatty acid
CCP	Carboxyl carrier protein	emf	Electromotive force
CDP	Cytidine triphosphate	EMP pathway	Embden–Meyerhof–Parnas pathway
CF	Coupling factor	FAD	Flavin adenine dinucleotide (oxidized)
CHD	Coronary heart disease		
CHI	Creatine-height index	FADH$_2$	Flavin adenine dinucleotide (reduced)
CI	Competitive inhibitor		
CICs	Characteristic impact compounds	FAS	Fatty acid synthetase
CMC	Carboxymethyl cellulose	FBP	Folate-binding protein
CoASH	Coenzyme A	Fd	Ferredoxin
CoE-B$_{12}$	Coenzyme B$_{12}$	FH$_4$	Tetrahydrofolate
CoE-M	Coenzyme M	FMN	Flavin mononucleotide (oxidized)
COMT	Catechol orthomethyl transferase	FMNH$_2$	Flavin mononucleotide (reduced)
cP	Centipoise unit	FPC	Fish protein concentrate
CP	Compensation point	FSH	Follicle-stimulating hormone
CP	Cyclic photophosphorylation	GH	Growth hormone
C-PER	Computed protein efficiency ratio	GLC	Gas–liquid chromatography
cps	Centipoise seconds	GM$_x$	Ganglioside(s)
CR	Chromatin receptor	G-6-PD	Glucose 6-phosphate dehydrogenase
cSt	Centistoke unit	GRAS	Generally recognized as safe
CTP	Cytosine triphosphate	GTP	Guanosine triphosphate
dATP	Deoxyadenosine triphosphate	HbA	Hemoglobin (adult form)

HbF	Hemoglobin (fetal form)	PGA	Pteroyl glutamic acid
HDL	High-density lipoprotein	P_i	Inorganic orthophosphate
HFS	High fructose syrup	*PIW*	Percentage ideal weight
HLB	Hydrophile–lipophile balance	PK	Pyruvate kinase
HMG-CoA	3-Hydroxy-3-methyl glutaryl-CoA	PKU	Phenylketonuria
HMP pathway	Hexose monophosphate pathway	POC	Primary odor characteristic
HPLC	High-pressure liquid chromatography	PP_i	Inorganic pyrophosphate
		PRL	Prolactin
HVA	Homovanillic acid	PRPP	Phosphoribosylpyrophosphate
I	Inosine (hypoxanthine)	PS	Photosynthesis
ICSH	Interstitial cell-stimulating hormone	PSP	Paralytic shellfish poisoning
IDL	Intermediate-density lipoprotein	PTH	Parathyroid hormone
IF	Intrinsic factor	PTH	Phenylthiohydantoin (for amino acids)
II	Inhibition index		
IMP	Inosine monophosphate	PUFA	Polyunsaturated fatty acids
IPP	Isopentenyl pyrophosphate	Q	Quenching substance
I.R.	Initial rise	RDA	Recommended dietary allowance
I.U.	International unit	rDNA	Recombinant DNA
JND	Just noticeable difference	RDS	Respiratory distress syndrome
kcal	Kilocalorie (Cal)	RNA	Ribonucleic acid
LCAT	Lecithin:cholesterol acyl transferase	*RNV*	Relative nutritional value
LD_{50}	Lethal dose, 50% mortality	*RPV*	Relative protein value
LDH	Lactic acid dehydrogenase	*R.Q.*	Respiratory quotient
LDL	Low-density lipoprotein	RS	Receptor site (hormone)
LH	Luteinizing hormone	rT_3	Reverse T_3
LLD	*Lactobacillus lactis Dorner* (factor)	R-TBP	Retinol–thyroxine-binding protein
LPL	Lipoprotein lipase	SAD	Succinic acid dehydrogenase
MAO	Monoamine oxidase	SAM	*S*-Adenoxyl methionine
MAOI	Monoamine oxidase inhibitors	*SCI*	Solid content index
MDD	Minimal detectable difference	SCP	Single-cell protein
MDR	Minimum daily requirement	SDA	Specific dynamic action
Met	Methionine (methionyl)	SHBG	Sex hormone-binding globulin
MIT	Monoiodothyronine	Sia	Sialic acid
MLD	Minimum lethal dose	SO	Sulfite oxidase
Mofd	Molybdoferredoxin	SOD	Superoxide dismutase
mRNA	Messenger RNA	SPC	Soy protein concentrate
MS	Molar substitution value	SPI	Soy protein isolate
MTD	Minimum tolerated dose	STX	Saxitoxin
NAD^+	Nicotinamide adenine dinucleotide (oxidized)	T_3	Triiodothyronine
		T_4	Thyroxine
$NADH(H^+)$	Nicotinamide adenine dinucleotide (reduced)	TBA	Thiobarbituric acid
		TBG	Thyroxine-binding globulin
$NADP^+$	Nicotinamide adenine dinucleotide phosphate (oxidized)	TBPA	Thyroxine-binding prealbumin
		TC	Transcobalamin
$NADPH(H^+)$	Nicotinamide adenine dinucleotide phosphate (reduced)	TCA	Tricarboxylic acid cycle
		TDL_0	Toxic dose low
NCI	Noncompetitive inhibitor	TG	Thyroglobulin (hormone)
NCP	Noncyclic photophosphorylation	THF	Tetrahydrofolic acid
NE	Niacin equivalent	*TLV*	Threshold limit value
NE_m	Net energy for maintenance	TMA	Trimethylamine
NPR	Net protein ratio	TPN	Total parenteral nutrition
NPU	Net protein utilization	TPP	Thiamine pyrophosphate
NOEL	No observed effect level	tRNA	Transfer RNA
PABA	*p*-Amino benzoic acid	TRF	Thyrotropin-releasing factor
PAD	Pyruvic acid dehydrogenase	TSH	Thyroid-stimulating hormone
PAH	Polyaromatic hydrocarbons	*TTV*	Taste threshold value
PBB	Polybrominated biphenyl	TTX	Tetrodotoxin
PC	Plastocyanin	TWA	Time weighted average
PCB	Polychlorinated biphenyl	UCI	Uncompetitive inhibitor
PE	Pectin esterase	UDP	Uridine diphosphate
PEP	Phosphoenol pyruvate	U.S.P.	United States Pharmacopeia
PEPC	Phosphoenol pyruvate carboxylase	UTP	Uridine triphosphate
PER	Protein efficiency ratio	V_d	Apparent volume of distribution
PF	Protection factor	VLDL	Very low-density lipoprotein
PFK	Phosphofructokinase	XD	Xanthine dehydrogenase
PG	Polygalacturonase	XO	Xanthine oxidase

N A M E   I N D E X

S U B J E C T

I N D E X